The
ENCYCLOPEDIA
of
**SOLID EARTH
GEOPHYSICS**

ENCYCLOPEDIA OF EARTH SCIENCES SERIES
Series Editor: Rhodes W. Fairbridge

THE ENCYCLOPEDIA OF OCEANOGRAPHY / *Rhodes W. Fairbridge*
THE ENCYCLOPEDIA OF ATMOSPHERIC SCIENCES AND ASTROGEOLOGY
  / *Rhodes W. Fairbridge*
THE ENCYCLOPEDIA OF GEOMORPHOLOGY / *Rhodes W. Fairbridge*
THE ENCYCLOPEDIA OF GEOCHEMISTRY AND ENVIRONMENTAL
  SCIENCES / *Rhodes W. Fairbridge*
THE ENCYCLOPEDIA OF MINERALOGY / *Keith Frye*
THE ENCYCLOPEDIA OF SEDIMENTOLOGY / *Rhodes W. Fairbridge and
  Joanne Bourgeois*
THE ENCYCLOPEDIA OF PALEONTOLOGY / *Rhodes W. Fairbridge and David
  Jablonski*
THE ENCYCLOPEDIA OF WORLD REGIONAL GEOLOGY, PART 1: Western
  Hemisphere (Including Antarctica and Australia) /
  Rhodes W. Fairbridge
THE ENCYCLOPEDIA OF STRUCTURAL GEOLOGY AND PLATE
  TECTONICS / *Carl K. Seyfert*
THE ENCYCLOPEDIA OF CLIMATOLOGY / *John E. Oliver
  and Rhodes W. Fairbridge*
THE ENCYCLOPEDIA OF SOIL SCIENCE, PART 1: Physics, Chemistry, Biology,
  Fertility, and Technology / *Rhodes W. Fairbridge and Charles W. Finkl, Jnr.*
THE ENCYCLOPEDIA OF APPLIED GEOLOGY / *Charles W. Finkl, Jnr.*
THE ENCYCLOPEDIA OF FIELD AND GENERAL GEOLOGY /
  Charles W. Finkl, Jnr.
THE ENCYCLOPEDIA OF BEACHES AND COASTAL ENVIRONMENTS /
  Maurice L. Schwartz
THE ENCYCLOPEDIA OF SOLID EARTH GEOPHYSICS /
  David E. James

ENCYCLOPEDIA OF EARTH SCIENCES

The ENCYCLOPEDIA of SOLID EARTH GEOPHYSICS

EDITED BY

David E. James
Carnegie Institution of Washington

VAN NOSTRAND REINHOLD COMPANY
New York

Copyright © 1989 by Van Nostrand Reinhold
Library of Congress Catalog Card Number 88-26751
ISBN 0-442-24366-9

All rights reserved. No part of this work covered by the copyright hereon
may be reproduced or used in any form or by any means—graphic, electronic,
or mechanical, including photocopying, recording, taping, or information
storage and retrieval systems—without written permission of the publisher.

Printed in the United States of America

Van Nostrand Reinhold
115 Fifth Avenue
New York, New York 10003

Van Nostrand Reinhold International Company Limited
11 New Fetter Lane
London EC4P 4EE, England

Van Nostrand Reinhold
480 La Trobe Street
Melbourne, Victoria 3000, Australia

Nelson Canada
1120 Birchmount Road
Scarborough, Ontario M1K 5G4, Canada

16 15 14 13 12 11 10 9 8 7 6 5 4 3 2 1

Library of Congress Cataloging-in-Publication Data
The Encyclopedia of solid earth geophysics / edited by David E. James.
    p.   cm. — (Encyclopedia of earth sciences; v. 16)
  Includes bibliographies and index.
  ISBN 0-442-24366-9
  1. Geodynamics—Dictionaries. I. James, David E., 1939-   . II. Series.
QE501.E58 1989
551'.03'21—dc19   88-26751

PREFACE

Unprecedented advances in geophysics during the 1970s and 1980s have revolutionized our concepts about the Earth. This scientific revolution is largely a consequence of the discovery of seafloor spreading and the subsequent formulation of the theory of plate tectonics. In addition, however, the exploration of the Moon and planets has had a profound impact upon geophysics, as has the development of innovative space technologies, where orbiting satellites have yielded a wealth of new and truly global geophysical data. A vital catalyst for these great advancements in knowledge has been the astonishing growth of computer technology, with the attendant development of powerful new digital techniques for collecting, processing, and analyzing massive quantities of digital data. The overall effect has been to alter forever not only our understanding of the Earth but also the fundamental way in which we do geophysical research.

The transformation of geophysics has resulted in an explosive increase in the volume of published literature. Scientific advances have been so rapid that even experts have had difficulty keeping abreast of developments in their own fields. The discoveries that paced the scientific revolution, moreover, have underscored the interdisciplinary nature of geophysics and have blurred the distinctions between what were once neatly compartmentalized geophysical disciplines. Thus some of the most exciting new work is to be found at the interfaces between, and in the integration of results from, quite disparate branches of geophysics. One aim of *The Encyclopedia of Solid Earth Geophysics* is to make the interconnections in geophysics and between geophysics and geology more apparent and understandable by drawing together articles that embrace a wide spectrum of solid earth geophysics. Our overall objectives in publishing this work are to summarize the diverse and far-reaching results of the past decades of research in solid earth geophysics, to integrate those results with classical geophysics, and to organize the whole into a comprehensive and authoritative *modern* encyclopedia of geophysics.

Solid earth geophysics is here taken to comprise chiefly seismology, geodesy, gravimetry, geo- and paleomagnetism, magnetotellurics, tectonophysics, and a few closely related areas in physics, geology, and space sciences. Although topics include a few of the more important research areas of exploration geophysics, our focus is principally on basic research in geophysics. Additional articles in the key areas of plate tectonics may be found in *The Encyclopedia of Structural Geology and Plate Tectonics* (Seyfert, 1987, Van Nostrand Reinhold) published as part of this encyclopedia series.

The Encyclopedia of Solid Earth Geophysics is intended to serve as an authoritative reference work for scientists, both specialists and nonspecialists, as well as for students and knowledgeable laymen. While the articles have been written by experts actively engaged in research in the areas discussed, we have endeavored to keep the presentation of most of the articles at a level suitable for a general scientific audience. Emphasis has been on explaining *what* a subject is all about, *why* it is important, and *how* it contributes to our broader understanding of the Earth. Mathematics has been kept to the minimum consistent with scholarly treatment of the subject. The entries are organized alphabetically by subject. A comprehensive index and various cross-references are included to aid the reader in finding a particular subject. In addition, each article contains a list of references that provide a point of entry to the scientific literature on the subject. An appendix at the end of this volume gives a summary of the S.I. system of units used in most of the articles here.

The preparation of this work involved the generous support of many people, first and foremost the international cast of prominent contributors who have given so willingly of their time and energy. Special thanks are due Bernie Chovitz (geodesy/gravity), Mike Fuller (geomagnetism/paleomagnetism), Ben Page (tectonophysics), and Paul Silver (seismology) for substantial help in developing the initial list of topics and authors for the *Encyclopedia*. For the day-to-day tasks of putting this volume together, I am deeply indebted to Janice Dunlap, a paragon of the modern computerized office and without whose energy and vast productivity this project would not have been possible. Bernice Pettinato was extremely helpful in the final stages of preparation and did a most efficient job of keeping the volume on schedule. Finally, the acknowledgments would certainly not be complete without heartfelt thanks to my wife Jeri Thomson, and to our young daughters, Kaitlin and Kristen, who endured with understanding and good humor the many evenings I spent sequestered alone with mountains of manuscripts.

DAVID E. JAMES

MAIN ENTRIES

Absolute Age Determination: Radiometric
Absolute Gravity Measurements
Accretionary Tectonics: Examples from the North American Cordillera
Adjustment Methods
Aeromagnetic Surveying
Archaeomagnetism

Brittle Phenomena

Chemical Remanent Magnetization (CRM)
Continental Collision Zones: Seismotectonics and Crustal Structure
Continental Crustal Structure
Continental Drilling: Ultradeep
Continental Lithosphere
Continental Rifting: Types and Rates
Continental Rifts: Structural Traits
Controlled Source Electromagnetic Mapping of the Crust
Controlled Source Seismology
Core-Mantle Coupling
Crustal Movements and Tectonic Deformation
Curie Temperature

Deep Earthquakes
Deep Seismic-Reflection Profiling
Deformation of Rocks and Minerals
Demagnetization
Density Distribution in the Earth
Detrital Remanent Magnetization (DRM)
Doppler Positioning: Satellite

Earth and Moon: Origins
Earth Orientation
Earthquake Mechanisms
Earthquake Mechanisms and Plate Tectonics
Earthquakes: Hazards and Predictions
Earthquakes: Location Techniques
Earthquakes: Magnitude, Energy, and Intensity
Earthquakes: Volcanogenic
Earthquakes and Crustal Deformation

Earthquake Seismology
Earthquakes and Seismicity
Earth's Core
Earth's Core: Structure
Earth's External Gravity Field
Earth Structure, Global
Earth Tides
Elasticity and Wave Propagation: Principles
Elastic Waves in Homogeneous and Inhomogeneous Media
Energy Budget of the Earth
Equilibrium Figure of the Earth
Exploration Seismology

Far-Traveled Terranes
Figure of the Earth
Finite-Difference Forward Modeling in Seismology
Free Oscillations of the Earth

Geodesy: Geometric
Geodesy: Historical Introduction
Geodesy: Physical
Geodesy: Satellite
Geodetic Ground Positioning
Geodetic Leveling
Geodetic Networks and Control Surveys
Geodetic Reference Systems
Geodetic Refraction
Geomagnetic Field: Asymmetries
Geomagnetic Field: Elements
Geomagnetic Field: Measurement
Geomagnetic Field: Westward Drift
Geomagnetic Field, Main: Theory
Geomagnetic Field Analysis
Geomagnetic Measurement Techniques and Surveys
Geomagnetic Polarity Reversals: Observations
Geomagnetic Polarity Reversals: Theory and Models
Geomagnetic Reversal Sequence: Statistical Structure

Geomagnetic Secular Variation: Direction and Intensity
Geomagnetic Secular Variation: Theory
Geomagnetism: Historical Introduction
Geothermal Energy
Global Positioning System (GPS)
Gravity Anomalies: Interpretation
Gravity Anomalies: Statistical Analysis
Gravity Fields: Implications for Planetary Interiors
Gravity and Isostasy

Heat Flow in the Earth

Inverse Theory and Methods: Seismology

Lithosphere: Mechanical Properties
Lithosphere, Oceanic: Formation and Evolution
Lunar Paleomagnetism
Lunar Seismology

Magma Physics
Magmatic Processes
Magnetic Analysis of Rock Fabric
Magnetic Domains
Magnetic Properties of Minerals
Magnetic Self-Reversal
Magnetohydrodynamic Waves within the Earth
Magnetostratigraphy
Magnetotelluric Method: Fundamental Concepts
Mantle, Lower: Structure
Mantle, Upper: Structure
Mantle Convection and Plumes
Mantle Discontinuities
Mantle Dynamics
Mantle Viscosity
Map Projections in Geodesy
Meteorites: Nature and Origin
Mineralogic Phase Transformations in the Earth

Natural Remanent Magnetization (NRM)

Ocean–Continent Transition: Structure

Paleomagnetic Field: Intensity
Paleomagnetic Secular Variation
Paleomagnetism: Deep-Sea Sediments
Paleomagnetism: Meteorites
Paleomagnetism: Orogenic Belts
Paleomagnetism and Continental Drift: Historical Introduction
Paleomagnetism and Plate Tectonics
Photogrammetry
Planetary Geodesy
Planetary Magnetic Fields

Rock Magnetism
Rock Magnetism: Measuring Techniques and Apparatus

Satellite Altimetry
Satellite Laser Positioning
Satellite Magnetic Measurements
Seafloor Spreading: Magnetic Evidence
Seismic Anisotropy in the Earth
Seismic Attenuation: Observation and Measurement
Seismic Diffraction
Seismic Imaging
Seismic Instrumentation
Seismic Instrumentation: History
Seismicity: Intraplate
Seismicity: Mid-Ocean Ridge
Seismicity: Subduction Zone
Seismicity and Plate Tectonics
Seismic Monitoring of Nuclear Explosions
Seismic Noise
Seismic Properties of Rocks
Seismic Ray Theory
Seismic Signal Processing
Seismic Source: Observations
Seismic Source: Theory
Seismic Tomography
Seismic Wavefield Migration
Seismic Wave Scattering
Seismograms: Interpretation
Seismograms: Synthetic
Seismology: History
Seismology: Physical Model Studies
Solar System: Origins
Spherical Harmonic Analysis
Stress in the Earth's Lithosphere
Strong Motion Seismology
Subduction Zones
Surface Waves

MAIN ENTRIES

Thermoremanence
Thin-Skin Tectonics
Tsunamis

Very-Long-Baseline Interferometry (VLBI)
Viscous Remanent Magnetization (VRM) and
  Viscous Remagnetization

CONTRIBUTORS

DUNCAN CARR AGNEW, IGPP A-025, University of California, La Jolla, California 92093. *Seismic Instrumentation; Seismology: History.*

KEIITI AKI, Department of Geological Sciences, University of Southern California, Los Angeles, California 90089-0740. *Elasticity and Wave Propagation: Principles.*

LEROY R. ALLDREDGE, U.S. Geological Survey, MS 964, Box 25046, Denver Federal Center, Denver, Colorado 80225-0046. *Geomagnetic Field: Elements.*

RICHARD J. ANDERLE, Federal and Electronic Systems Division, General Electric Company, P.O. Box 8048-38041, Philadelphia, Pennsylvania 19101. *Global Positioning System (GPS).*

JOHN G. ANDERSON, Seismological Laboratory, Mackay School of Mines, University of Nevada-Reno, Reno, Nevada 89557-0047. *Strong Motion Seismology.*

P. V. ANGUS-LEPPAN, World Bank/ADAB, Land Titling Project, Department of Lands, Phra Phiphit Road, Bangkok, 10200 Thailand. *Geodetic Refraction.*

K. ARNOLD, Akademie der Wissenschaften der DDR, Zentralinstitut für Physik der Erde, DDR-1561 Potsdam, German Democratic Republic. *Density Distribution in the Earth.*

SUBIR K. BANERJEE, Department of Geology and Geophysics, University of Minnesota, 108 Pillsbury Hall, 310 Pillsbury Drive, S.E., Minneapolis, Minnesota 55455. *Seafloor Spreading: Magnetic Evidence.*

MUAWIA BARAZANGI, Institute for the Study of the Continents, Snee Hall, Cornell University, Ithaca, New York 14853-1504. *Continental Collision Zones: Seismotectonics and Crustal Structure.*

DAVID R. BARRACLOUGH, Geomagnetism Research Group, British Geological Survey, Murchison House, West Mains Road, Edinburgh EH9 3LA, U.K. *Geomagnetism: Historical Introduction.*

C. E. BARTON, Division of Geophysics, Bureau of Mineral Resources, Geology and Geophysics, G.P.O. Box 378, Canberra A.C.T. 2601, Australia. *Geomagnetic Secular Variation: Direction and Intensity.*

BRUCE G. BILLS, Lunar and Planetary Institute, 3303 NASA Road 1, Houston, Texas 77058. *Planetary Geodesy.*

BRUCE A. BOLT, Seismographic Station, 473 Earth Sciences, University of California, Berkeley, California 94720. *Earthquake Seismology.*

ALAN P. BOSS, Department of Terrestrial Magnetism, Carnegie Institution of Washington, 5241 Broad Branch Road, N.W., Washington, D.C. 20015. *Earth and Moon: Origins; Solar System: Origins.*

JOHN D. BOSSLER, Director, Center for Mapping, Ohio State University, 1958 Neil Avenue, Columbus, Ohio 43210-1247. *Geodetic Networks and Control Surveys.*

S. I. BRAGINSKY, Institute of Geophysics and Planetary Physics, University of California, Los Angeles, California 90024. *Magnetohydrodynamic Waves within the Earth.*

RAY BULAND, U.S. Geological Survey, MS 967, Box 25046, Denver Federal Center, Denver, Colorado 80225-0046. *Earthquakes: Location Techniques.*

F. H. BUSSE, Physikalisches Institut Universität Beyreuth, Postfach 101251, D-8580 Bayreuth, Federal Republic of Germany. *Geomagnetic Field, Main: Theory.*

CONTRIBUTORS

JOSEPH C. CAIN, Department of Geology, The Florida State University, Tallahassee, Florida 32306-3026. *Geomagnetic Field Analysis.*

RICHARD CARLSON, Department of Terrestrial Magnetism, Carnegie Institution of Washington, 5241 Broad Branch Road, N.W., Washington, D.C. 20015. *Absolute Age Determination: Radiometric.*

WILLIAM E. CARTER, National Geodetic Survey, U.S. Department of Commerce, C124 Office of Charting and Geodetic Services, NOAA, National Ocean Survey, Rockville, Maryland 20852. *Earth Orientation.*

V. ČERVENÝ, Institute of Geophysics, Charles University, Ke Karlovu 3, 121 16 Praha 2, Czechoslovakia. *Seismic Ray Theory.*

J. E. T. CHANNELL, Department of Geology, 1112 Turlington Hall, University of Florida, Gainesville, Florida 32611. *Paleomagnetism: Deep-Sea Sediments.*

BERNARD H. CHOVITZ, 8813 Clifford Avenue, Chevy Chase, Maryland 20815. *Figure of the Earth.*

NIKOLAS I. CHRISTENSEN, Department of Earth and Atmospheric Sciences, Civil Engineering Building, Purdue University, West Lafayette, Indiana 47907. *Seismic Properties of Rocks.*

ULRICH CHRISTENSEN, Max-Planck-Institut für Chemie, Abteilung Geochemie, Saarstrasse 23, Postfach 3060, D-6500 Mainz, Federal Republic of Germany. *Energy Budget of the Earth.*

STANLEY M. CISOWSKI, Department of Geological Sciences, University of California, Santa Barbara, California 93106. *Lunar Paleomagnetism* (with M. Fuller); *Paleomagnetism: Meteorites.*

TIMOTHY J. CLARKE, Department of Terrestrial Magnetism, Carnegie Institution of Washington, 5241 Broad Branch Road, N.W., Washington, D.C. 20015. *Seismograms: Synthetic.*

MARK CLOOS, Department of Geological Sciences, University of Texas, Austin, Texas 78712. *Subduction Zones.*

VERNON F. CORMIER, Department of Geology and Geophysics, U-45, Room 207, The University of Connecticut, 345 Mansfield Road, Storrs, Connecticut 06268. *Seismic Attenuation: Observation and Measurement.*

R. S. CROSSON, Graduate Program in Geophysics, University of Washington, Seattle, Washington 98195. *Earthquakes: Volcanogenic.*

GEOFFREY F. DAVIES, Res. School of Earth Sciences, Australian National University, G.P.O. Box 4, Canberra A.C.T. 2601, Australia. *Mantle Dynamics.*

PETER DAVIS, Seismologisches Zentralobservatorium, Krankenhausstrasse 1–3, D-8520 Erlangen, Federal Republic of Germany. *Free Oscillations of the Earth.*

B. de VOOGD, Départment de Géologie, École Normale Supérieure, 24, rue Lhomond, 75231 Paris Cedex 05, France. *Deep Seismic-Reflection Profiling* (with C. Keen).

DURK J. DOORNBOS, Institute of Geophysics, University of Oslo, Post Box 1022, Blindern, 0315 Oslo 3, Norway. *Seismic Diffraction.*

SEWERYN J. DUDA, University of Hamburg, Institute of Geophysics, Bundesstrasse 55, D-2000 Hamburg, Federal Republic of Germany 13. *Earthquakes: Magnitude, Energy, and Intensity.*

DAVID J. DUNLOP, Department of Physics, University of Toronto, Mississauga, Ontario, Canada L5L 1C6. *Viscous Remanent Magnetization (VRM) and Viscous Remagnetization.*

ADAM M. DZIEWONSKI, Department of Earth and Planetary Sciences, Hoffman Laboratory, 20 Oxford Street, Harvard University, Cambridge, Massachusetts 02138. *Earth Structure, Global.*

R. N. EDWARDS, Department of Physics, Geophysics Division, University of Toronto,

Toronto, Ontario Canada M5S 1A7. *Controlled Source Electromagnetic Mapping of the Crust.*

IRENE K. FISCHER, 301 Philadelphia Avenue, Takoma Park, Maryland 20912. *Geodesy: Historical Introduction.*

A. J. FORBES, 23 Birch Grove, Boat-of-Garten, Inverness-shire, Scotland. *Geomagnetic Field: Measurement.*

DONALD W. FORSYTH, Department of Geological Sciences, University of California, Santa Barbara, California 93106. *Lithosphere: Mechanical Properties.*

M. FULLER, Department of Geological Science, University of California, Santa Barbara, California 93106. *Lunar Paleomagnetism* (with S. M. Cisowski)

RICHARD G. GORDON, Department of Geological Sciences, Northwestern University, Evanston, Illinois 60208. *Paleomagnetism and Plate Tectonics* (with G. Acton).

G. GRAU, Institut Francaise du Petrole, 1-4, Avenue de Bois Preau, B.P. 311, 92506 Rueil-Malmaison Cedex, France. *Seismic Wavefield Migration* (with P. Lailly).

M. GUFFANTI, U.S. Geological Survey, MS 922, Reston, Virginia 22092. *Geothermal Energy.*

S. L. HALGEDAHL, Lamont-Doherty Geologial Observatory of Columbia University, Palisades, New York 10964. *Magnetic Domains.*

AKIRA HASEGAWA, Observation Center for Earthquake Prediction, Faculty of Science, Tohoku University, Sendai 980, Japan. *Seismicity: Subduction Zone.*

RONALD R. HATCH, Magnavox Advanced Products and Systems Company, 2829 Maricopa Street, Torrance, California 90503. *Geodesy: Satellite.*

ROBERT D. HATCHER, Jr., Department of Geological Sciences, University of Tennessee, Knoxville, Tennessee 37916-1410. *Continental Drilling: Ultradeep.*

SOREN HENRIKSEN, 2863 Beachwood Circle, Arlington, Virginia 22207. *Photogrammetry.*

JOHN F. HERMANCE, Department of Geological Sciences, Brown University, Providence, Rhode Island 02912. *Magnetotelluric Method: Fundamental Concepts.*

KENNETH A. HOFFMAN, Physics Department, California Polytechnic State University, San Luis Obispo, California 93407. *Geomagnetic Polarity Reversals: Theory and Models.*

L. GARY HOLCOMB, U.S. Geological Survey, Albuquerque Seismological Laboratory, Building 10002, Kirtland AFB-East, Albuquerque, New Mexico 87115-5000. *Seismic Noise.*

PETER HOOD, Head, Regional Geophysics Subdivision, Resource Geophysics and Geochemistry Division, Energy, Mines and Resources CANADA, Earth Sciences, Geological Survey of Canada, 601 Booth Street, Ottawa, Ontario, Canada K1A OE8. *Aeromagnetic Surveying.*

B. F. HOWELL, Jr., Department of Geosciences, 403 Deike Building, The Pennsylvania State University, University Park, Pennsylvania 16802. *Seismic Instrumentation: History.*

BRYAN L. ISACKS, Institute for the Study of the Continents, Department of Geological Sciences, Snee Hall, Cornell University, Ithaca, New York, 14853-1504. *Seismicity and Plate Tectonics.*

H. M. IYER, U.S. Geological Survey, Branch of Seismology, MS 977, 345 Middlefield Road, Menlo Park, California 94025. *Seismic Tomography.*

DAVID E. JAMES, Department of Terrestrial Magnetism, Carnegie Institution of Washington, 5241 Broad Branch Road, N.W., Washington, D.C. 20015. *Continental Lithosphere; Controlled Source Seismology; Exploration Seismology; Surface Waves.*

CHRISTOPHER JEKELI, AFGL, Geodesy and Gravity Branch, Earth Sciences Division,

CONTRIBUTORS

Hanscom Air Force Base, Massachusetts 01731-5000. *Earth's External Gravity Field.*

WILLIAM M. KAULA, Department of Earth and Space Sciences, 3806 Geology Building, University of California, 405 Hilgard Avenue, Los Angeles, California 90024-1567. *Gravity Fields: Implications for Planetary Interiors.*

ICHIRO KAWASAKI, Department of Earth Sciences, Toyama University, Gofuku, Toyama, 930 Japan. *Seismic Anisotropy in the Earth.*

B. L. N. KENNETT, Research School of Earth Sciences, The Australian National University, Institute of Advanced Studies, G.P.O. Box 4, Canberra A.C.T. 2601 Australia. *Elastic Waves in Homogeneous and Inhomogeneous Media.*

M. A. KHAN, Hawaii Institute of Geophysics, 2525 Correa Road, University of Hawaii, Honolulu, Hawaii 96822 (former Minister of Petroleum and Natural Resources, Government of Pakistan, Islamabad, Pakistan). *Equilibrium Figure of the Earth.*

CHESTER J. KOBLINSKY, Geodynamics Branch, Laboratory for Terrestrial Physics, NASA, Goddard Space Flight Center, Greenbelt, Maryland 20771. *Satellite Altimetry* (with J. G. Marsh).

J. KOUBA, Geophysics Division, Energy, Mines and Resources Canada, 1 Observatory Crescent, Ottawa, Ontario, Canada K1A 0Y3. *Doppler Positioning: Satellite.*

ROBERT L. KOVACH, Department of Geophysics, Mitchell Building, Stanford University, Stanford, California 94305-2215. *Lunar Seismology.*

CARLO LAJ, Centre des Faibles Radioactivites, Laboratoire Mixte CNRS-CEA, Avenue de la Terrasse, 91198, Gif-sur-Yvette, CEDEX, France. *Geomagnetic Polarity Reversals: Observations.*

ROBERT A. LANGEL, Geophysics Branch Code 622, Goddard Space Flight Center, Greenbelt, Maryland 20771. *Satellite Magnetic Measurements.*

C. G. LANGEREIS, Paleomagnetisch Laboratorium, 'Fort Hoofddijk,' Rijksuniversiteit Utrecht, Postbus 80.020, 3508 TA Utrecht, Netherlands. *Demagnetization* (with J. H. Linssen, T. A. T. Mullender, and J. D. A. Zijderveld).

CHARLES A. LAWSON, U.S. Geological Survey, MS-959, Reston, Virginia 22092. *Magnetic Self-Reversal.*

THORNE LAY, Department of Geological Sciences, 1006 C.C. Little Building, The University of Michigan, Ann Arbor, Michigan 48109-1063. *Mantle, Lower: Structure.*

TYPHOON LEE, Institute of Earth Sciences, Academia Sinica, P.O. Box 23-59, Taipei, Taiwan 10769. *Meteorites: Nature and Origin.*

H. E. Le GRAND, H.P.S., The University of Melbourne, Parkville, Victoria, Australia, 3052. *Paleomagnetism and Continental Drift: Historical Introduction.*

ALAN R. LEVANDER, Department of Geology and Geophysics, Geology Building #106, Rice University, 6100 South Main Street, Houston, Texas 77005. *Finite-Difference Forward Modeling in Seismology.*

SHAUL LEVI, Geophysics Department, College of Oceanography, Oregon State University, Oregon Street, Corvallis, Oregon 97331-5503. *Chemical Remanent Magnetization (CRM).*

DAVID E. LOPER, Geophysical Fluid Dynamics Institute, The Florida State University, Tallahassee, Florida 32306-3017. *Earth's Core.*

WILLIAM LOWRIE, Institut fuer Geophysik, ETH-Hönggerberg, CH-8093, Zurich, Switzerland. *Magnetic Analysis of Rock Fabric.*

STEVE P. LUND, Department of Geological Sciences, University of Southern California, Los Angeles, California 90089-0740. *Paleomagnetic Secular Variation.*

DENNIS D. McCARTHY, International Earth Rotation Service, Rapid Service Sub-Bureau, U.S. Naval Observatory, 34th and Massachusetts Ave., N.W., Washington, D.C. 20392-5100. *Geodetic Reference Systems.*

JAMES S. McCLAIN, Department of Geology, University of California-Davis, Davis, California 95616. *Lithosphere, Oceanic: Formation and Evolution* (with J. A. Orcutt).

P. L. McFADDEN, Division of Geophysics, Bureau of Mineral Resources, Geology, and Geophysics, G.P.O. Box 378, Canberra A.C.T. 2601, Australia. *Geomagnetic Field: Asymmetries* (with R. T. Merrill and M. W. McElhinny); *Geomagnetic Reversal Sequence: Statistical Structure.*

S. J. MACKWELL, Department of Geosciences, 503 Deike Building, The Pennsylvania State University, University Park, Pennsylvania 16802. *Deformation of Rocks and Minerals.*

KAREN C. McNALLY, Earth Science Board of Studies, Applied Sciences Building, University of California, Santa Cruz, California 95064. *Earthquakes and Seismicity.*

MARCIA McNUTT, Department of Earth, Atmospheric, and Planetary Sciences, Massachusetts Institute of Technology, Cambridge, Massachusetts 02139. *Gravity and Isostasy.*

RAUL MADARIAGA, Laboratoire de Sismologie, Institut de Physique du Globe de Paris, 4, Place Jussieu, Tour 14, F-75230 Paris, Cedex 05, France. *Seismic Source: Theory.*

BRUCE D. MARSH, Department of Earth and Planetary Sciences, The Johns Hopkins University, Baltimore, Maryland 21218. *Magma Physics.*

GUY MASTERS, IGPP, University of California, San Diego, La Jolla, California 92093. *Earth's Core: Structure.*

R. MEISSNER, Christian-Albrechts-Universität Zu Kiel, Institut für Geophysik, Olshausenstrasse 40, D-2300 Kiel 1, Federal Republic of Germany. *Continental Crustal Structure* (with Th. Wever).

WILLIAM MENKE, Lamont-Doherty Geological Observatory of Columbia University, Palisades, New York 10964. *Seismology: Physical Model Studies* (with B. Duberdorff).

RONALD T. MERRILL, Chairman, Geophysics, Department of Oceanography, University of Washington, Seattle, Washington 98195. *Rock Magnetism.*

M. MEGHAN MILLER, Jet Propulsion Laboratory 183-501, California Institute of Technology, 4800 Oak Grove Drive, Pasadena, California 91109. *Accretionary Tectonics: Examples from the North American Cordillera* (with J. B. Saleeby).

PETER MORA, Laboratoire de Sismologie, Institut de Physique du Globe, 4 Place Jussieu, Paris, 75252, Cedex 05, France. *Seismic Imaging* (with A. Tarantola).

PAUL MORGAN, Department of Geology, Box 6030, Northern Arizona University, Flagstaff, Arizona 86011. *Heat Flow in the Earth.*

T. NAGATA, 66 Akagishitamachi, Shinjuku-Ku, Tokyo 162, Japan. *Natural Remanent Magnetization (NRM).*

S. P. NISHENKO, U.S. Geological Survey, MS 967, P.O. Box 25046, Denver Federal Center, Denver, Colorado 80225-0046. *Earthquakes: Hazards and Predictions.*

G. NOLET, Department of Theoretical Geophysics, Budapestlaan 4, P.O. Box 80.021, 3508TA Utrecht, The Netherlands. *Mantle, Upper: Structure* (with M. J. R. Wortel).

AMOS NUR, Department of Geophysics, Stanford University, Stanford, California 94305. *Far-Traveled Terranes* (with Z. Ben-Avraham).

PETER OLSON, Department of Earth and Planetary Sciences, The Johns Hopkins University, Baltimore, Maryland 21218. *Mantle Convection and Plumes.*

W. O'REILLY, University of Newcastle-upon-Tyne, School of Physics, Newcastle, NE1 7RU, England. *Magnetic Properties of Minerals.*

CONTRIBUTORS

GARY L. PAVLIS, Department of Geology, Indiana University, Bloomington, Indiana 47405. *Inverse Theory and Methods: Seismology.*

NORMAN W. PEDDIE, U.S. Geological Survey, Branch of Global Seismology and Geomagnetism, MS 968, Box 25046, Denver Federal Center, Denver, Colorado 80225. *Geomagnetic Measurement Techniques and Surveys.*

W. R. PELTIER, Department of Physics, University of Toronto, Toronto, Ontario, Canada M5S 1A7. *Mantle Viscosity.*

N. PETERSEN, Institut für Geophysik, Theresienstrasse 41, D-8000 Munich 2, Federal Republic of Germany. *Curie Temperature.*

RICHARD H. RAPP, Department of Geodetic Sciences and Surveying, Ohio State University, 1958 Neil Avenue, Columbus, Ohio 43210-1247. *Gravity Anomalies: Statistical Analysis.*

R. E. REILINGER, Department of Earth, Atmospheric, and Planetary Sciences, Earth Resource Laboratory, E34-342, Massachusetts Institute of Technology, Cambridge, Massachusetts 02142. *Crustal Movements and Tectonic Deformation.*

PAUL G. RICHARDS, Lamont-Doherty Geological Observatory of Columbia University, Palisades, New York 10964. *Seismic Monitoring of Nuclear Explosions.*

PAUL ROBERTS, Institute of Geophysics and Planetary Physics, University of California, Los Angeles, California 90024. *Core-Mantle Coupling.*

D. S. ROBERTSON, National Geodetics Survey, C124, Rockville, Maryland 20852. *Very-Long-Baseline Interferometry (VLBI).*

BRUCE R. ROSENDAHL, Project PROBE, Geology Department, Room 310 Old Chemistry Building, Duke University, Durham, North Carolina 27706. *Continental Rifting: Types and Rates; Continental Rifts: Structural Traits.*

LARRY J. RUFF, Department of Geological Science, 1006 C.C. Little Building, University of Michigan, Ann Arbor, Michigan 48109. *Seismic Source: Observations.*

C. T. RUSSELL, Institute of Geophysics and Planetary Physics, University of California, Los Angeles, California 90024. *Planetary Magnetic Fields.*

T. SASATANI, Department of Geophysics, Faculty of Science, Hokkaido University, Sapporo, 060 Japan. *Deep Earthquakes.*

V. A. SCHMIDT, Department of Geology and Planetary Sciences, 321 Old Engineering Hall, University of Pittsburgh, Pittsburgh, Pennsylvania 15260. *Thermoremanence.*

D. SCHMITZ, Box 1277, Golden, Colorado 80402. *Spherical Harmonic Analysis.*

LEONARDO SEEBER, Lamont-Doherty Geological Observatory of Columbia University, Palisades, New York 10964. *Thin-Skin Tectonics* (with D. Byrne).

JOHN SHAW, Geomagnetism Laboratory, Department of Earth Sciences, Oliver Lodge Laboratories, The University of Liverpool, Oxford Street, P.O. Box 147, Liverpool, L69 3BX, U.K. *Paleomagnetic Field: Intensity.*

RUTH B. SIMON, 2723 Esmond Ave., Richmond, California 94804. *Seismograms: Interpretation.*

DAVID E. SMITH, Head, Geodynamics Branch, Laboratory for Terrestrial Physics, NASA, Goddard Space Flight Center, Greenbelt, Maryland 20771. *Satellite Laser Positioning.*

RICHARD A. SNAY, Geodetic Research and Development Laboratory, National Geodetic Survey, U.S. Department of Commerce, NOAA, National Ocean Service, Rockville, Maryland 20852. *Earthquakes and Crustal Deformation.*

J. A. SNOKE, Department of Geological Sciences, Virginia Polytechnic Institute and State University, Blacksburg, Virginia 24061-0420. *Earthquake Mechanisms.*

CONTRIBUTORS

JOHN P. SNYDER, 2370 Old Trail Drive, Reston, Virginia 22091. *Map Projections in Geodesy.*

SETH STEIN, Department of Geological Science, Northwestern University, Evanston, Illinois 60201. *Earthquake Mechanisms and Plate Tectonics* (with D. F. Woods); *Seismicity: Mid-Ocean Ridge* (with D. F. Woods).

HERBERT W. STOUGHTON, 2821 Carey Avenue, Cheyenne, Wyoming 82001. *Geodetic Ground Positioning.*

HANS SÜNKEL, Institute of Mathematical Geodesy, Technical University at Graz, Rechbauerstrasse 12, A-8010 Graz, Austria. *Geodesy: Physical.*

MANIK TALWANI, Department of Geology and Geophysics, Room 206, Rice University, Houston, Texas 77251. *Ocean–Continent Transition: Structure.*

D. H. TARLING, Faculty of Science, Department of Geological Sciences, Plymouth Polytechnic, Drake Circus, Plymouth, Devon, PL4 8AA, U.K. *Archaeomagnetism.*

LISA TAUXE, Scripps Institution of Oceanography, A-020, University of California-San Diego, La Jolla, California 92093. *Magnetostratigraphy.*

R. THOMPSON, Department of Geophysics, James Clark Maxwell Building, University of Edinburgh, Mayfield Road, Edinburgh EH9 3JZ, U.K. *Geomagnetic Field: Westward Drift.*

ROB Van der VOO, Department of Geological Sciences, 1006 C.C. Little Building, The University of Michigan, Ann Arbor, Michigan 48109-1063. *Paleomagnetism: Orogenic Belts.*

PETR VANÍČEK, Department of Surveying Engineering, University of New Brunswick, P.O. Box 4400, Fredericton, N.B., Canada E3B 5A3. *Adjustment Methods.*

L. P. VINNIK, Institute of Physics of the Earth, USSR Academy of Science, Moscow D-242, B. Gruzinskaya 10, U.S.S.R. *Mantle Discontinuities.*

KENNETH L. VEROSUB, Department of Geology, 175 Geology/Physics (0995), University of California, Davis, California 95616. *Detrital Remanent Magnetization (DRM).*

JOHN M. WAHR, Department of Physics, University of Colorado, Boulder, Colorado 80309-0390. *Earth Tides.*

DAVID WALKER, Lamont-Doherty Geological Observatory of Columbia University, Palisades, New York 10964. *Magmatic Processes.*

STEVEN N. WARD, C.F. Richter Seismological Laboratory, The Institute of Tectonics, Earth Sciences, University of California, Santa Cruz, California 95064. *Tsunamis.*

DONALD J. WEIDNER, Department of Earth and Space Sciences, State University of New York (SUNY), Stony Brook, New York 11794. *Mineralogic Phase Transformations in the Mantle* (with G. Gwanmesia, V. Haniford, and A. Remsberg).

DOUGLAS A. WIENS, Department of Earth and Planetary Science, Washington University in St. Louis, Campus Box 1169, St. Louis, Missouri 63130. *Seismicity: Intraplate.*

LUMAN E. WILCOX, 211B Braeshire Drive, Manchester, Missouri 63021. *Gravity Anomalies: Interpretation.*

TENG-FONG WONG, Department of Earth, Atmospheric and Planetary Sciences, Massachusetts Institute of Technology, Cambridge, Massachusetts, 02139. *Brittle Phenomena.*

M. H. WORTHINGTON, Department of Geology, Imperial College of Science and Technology, Royal School of Mines, Prince Consort Road, London, England SW7 2BP. *Seismic Signal Processing.*

RU-SHAN WU, Applied Science Building, Room 191, University of California, Santa Cruz, California 95064. *Seismic Wave Scattering.*

G. YOUNG, Vertical Network Branch, National Geodetic Survey, U.S. Department of

Commerce, NOAA, National Ocean Service, Rockville, Maryland 20852. *Geodetic Leveling.*

TAKESI YUKUTAKE, Earthquake Research Institute, The University of Tokyo, No. 1-1, Yayoi 1-chome, Bunkyo-Ku, Tokyo, 113, Japan. *Geomagnetic Secular Variation: Theory.*

T. ZELINKA, Geophysical Institute, Czechoslovak Academy of Sciences, Boční II čp. 1401, 141 31 Praha 4, Czechoslovakia. *Rock Magnetism: Measuring Techniques and Apparatus* (with V. Kropáček and J. Tauer).

SHENG-YUAN ZHU, Institut für Physikalishe Geodäsie, Technische Hochschule Darmstadt, Petersenstrasse 13, D-6100 Darmstadt, Federal Republic of Germany. *Geodesy: Geometric.*

MARK D. ZOBACK and MARY LOU ZOBACK, Department of Geophysics, Stanford University, Stanford, California 94305. *Stress in the Earth's Lithosphere.*

MARK A. ZUMBERGE, IGPP A-025, University of California, La Jolla, California 92093. *Absolute Gravity Measurements.*

A

ABSOLUTE AGE DETERMINATION: RADIOMETRIC

Breaking from the biblical estimates of an Earth no more than 6000 years old, Hutton's theory of uniformitarianism, developed in the late eighteenth century, predicted that geologic processes would require hundreds of million years to create the landforms present on the Earth. The idea of an "ancient" Earth prospered until the late nineteenth century when William Thomson (Lord Kelvin) produced a series of straightforward calculations showing that the Earth could be no older than 40 million years, based on the time required to cool the Earth to its present state. The discovery of radioactivity by Henri Becquerel in 1896 settled this conflict by providing both a heat source not accounted for by Lord Kelvin and a tool to precisely determine the absolute age of a rock. This tool was used almost immediately by the renowned physicist Ernest Rutherford, who showed in 1905 that the amount of helium present in certain uranium ores would require the ores to be 500 million years old if the helium were produced by the radioactive decay of uranium. Not until the discovery that elements are composed of mixtures of atoms of differing weight (isotopes) by J. J. Thomson in 1913 was it clear that radioactive elements decay only to specific isotopes of the new, or daughter, element. Coincidentally, in this same year, Arthur Holmes published his book *The Age of the Earth*, which used radioactive dating to push the age of the Earth back to 1600 million years. It was, however, the discovery of isotopes and the improvement of the instruments capable of measuring their abundance (mass spectrometers) that allowed radiometric dating to be applied to rocks and minerals more common than uranium ores. This, in turn, launched radiometric geochronology as a tool of widespread use in the geosciences in the years following World War II.

Radioactive decay occurs when the nucleus of an isotope of one element transforms into an isotope of another element. An example of this is the mass 87 isotope of the element rubidium, which loses an electron (β-particle) from its nucleus, transforming one of its neutrons to a proton to become the mass 87 isotope of the element strontium. Another means of decay, called α-*decay*, involves the loss of two protons and two neutrons (one α-particle or ^{4}He nucleus) from the nucleus. Examples of elements that decay in this manner include uranium, thorium, and samarium. A final common means of nuclear transformation is electron capture, wherein the nucleus captures an electron from its electron shell to transform a proton to a neutron, conserving mass but losing one nuclear charge. The element potassium contains an isotope (^{40}K) that decays both by electron capture to ^{40}Ar and by β-decay to ^{40}Ca.

In all radioactive decay, the rate of decay is proportional to the number of atoms of the element present. This can be described by the equation

$$-\frac{dN}{dt} = \lambda N \qquad (1)$$

where N is the number of atoms, t is the time, and λ is the rate constant of decay. Decay constants are assumed to be independent of time, pressure, and temperature effects. Except for a slight sensitivity of the electron capture decay mechanism to pressure variations, the available data suggest that the assumption of constancy in the decay rate is valid. Particularly noteworthy is that many rocks whose age has been determined give identical ages with several radiometric systems, based on a variety of decay mechanisms (Fig. 1). Coincidence of ages from different systems would not be expected if the decay constants were functions of external parameters such as time, temperature, or pressure. Integrating Equation 1 gives

$$N = N_0 e^{-\lambda t}$$

where N_0 is the number of atoms present when $t = 0$. In practice, it is often more convenient to determine the abundance of the decay product, or daughter, isotope than that of the parent. If no atoms are lost, then the abundance of the daughter isotope increases with time by

$$D = N_0 - N = N(e^{\lambda t} - 1)$$

or if some atoms of the daughter isotope are present initially (D_0), then

$$D = D_0 + N(e^{\lambda t} - 1) \qquad (2)$$

Because mass spectrometers can measure isotopic ratios of some elements to precisions of 0.001%, whereas elemental abundances can only be deter-

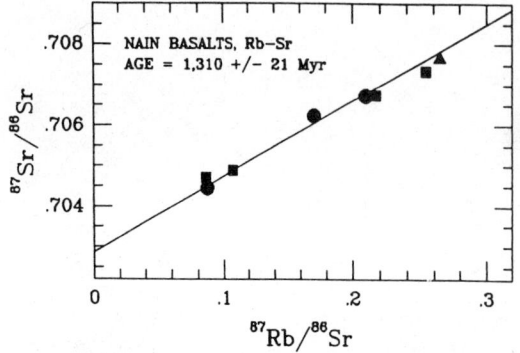

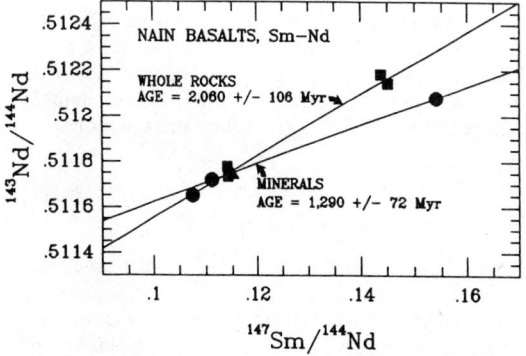

FIGURE 1. Rb-Sr and Sm-Nd isochron diagrams for whole rocks (squares and triangle) and mineral separates (dots) from basaltic dikes in the Nain Province of Labrador. In the Rb-Sr system, both minerals and whole rocks define a line with slope corresponding to an age of 1310 ± 21 million years. In the Sm-Nd system, the mineral separates from one whole-rock sample define an eruption age that agrees with that determined by Rb-Sr, but enough initial isotopic heterogeneity was present in the whole rocks to define a second line of age near 2 billion years. The age significance of the Sm-Nd whole rock line is not clear, but may represent a time of incompatible element enrichment in the sources of these basalts that preceeded their eruption by over 700 million years.

mined to about 0.1% to 1%, Equation 2, using the Rb-Sr decay system as an example, is often expressed as ratios:

$$\frac{^{87}Sr}{^{86}Sr} = \left(\frac{^{87}Sr}{^{86}Sr}\right)_0 + \left(\frac{^{87}Rb}{^{86}Sr}\right)(e^{\lambda t} - 1) \quad (3)$$

where ^{86}Sr is a stable isotope of strontium and ^{87}Sr is produced by the decay of ^{87}Rb. Equation 3 can be seen to be that of a straight line if the two measurable quantities $^{87}Sr/^{86}Sr$ and $^{87}Rb/^{86}Sr$ are plotted against one another as in the example in Fig. 1. Such lines, called *isochrons*, provide both the age (slope $= e^{\lambda t} - 1$) and the initial ratio of radioactive isotope to stable isotope ($^{87}Sr/^{86}Sr)_0$. Isochrons can be determined for individual rocks whose constituent minerals have a range in the Rb/Sr ratio—for instance, the minerals biotite (Rb/Sr > 10) and plagioclase (Rb/Sr < 0.01), which are common constituents of granitic composition rocks. Isochrons also can be determined for related whole rock samples that show a sufficient range in parent-daughter ratio. For example, one volcano may erupt lavas varying in composition from basalt (Rb/Sr < 0.1) to rhyolite (Rb/Sr > 1).

The advantage of "whole-rock" isochrons over "mineral" isochrons is that, because of their small size, the constituent minerals of a rock are more easily modified during alteration or high-temperature metamorphic processes. Equation 3 will provide an accurate age only if the minerals, or rocks, used to form the isochron have remained a "closed system" since the formation of the rock—that is, no transport of either parent or daughter isotope into or out of the rock or mineral takes place. At the high temperatures involved in some metamorphic processes, diffusion of elements is rapid enough to continually "reset" a radiometric system by allowing the radioactive parent and daughter elements to migrate between minerals. Because of the very small diffusion coefficients of most elements in minerals, diffusion only occurs over length scales sufficient to cause intermineral equilibria (millimeters) but not over lengths (several centimeters) that would disturb the radiometric systems at a whole-rock scale. Thus many examples have been found of rock suites whose minerals may give one age, the age of metamorphism, but whose whole-rock isotopic relationships record an older age, possibly the original age of igneous activity that formed the rocks.

A major disadvantage of the whole-rock isochron technique, however, is that it is more difficult to verify the requirement of Equation 3 that all samples used to construct an isochron initially had identical ratios of daughter isotope to a stable isotope of the daughter element. The initial isotopic composition of a daughter element in a rock depends on the average parent-daughter ratio in the source materials of the rock. For example, if a rhyolitic composition magma is produced by the melting of a very old granite that has a high Rb/Sr ratio, that magma will have a much higher $^{87}Sr/^{86}Sr$ ratio than will a rhyolite produced by the differentiation of a mantle-derived basalt magma, since the mantle has a very low Rb/Sr ratio. This difference in initial isotopic composition is the basis of a different field, *isotope geology*, which is concerned with the origin and provenance of igneous rocks. Variable initial isotopic composition is much more prevalent on a whole-rock scale than on a mineral scale because minerals of one rock crystallize from one magma that is well-enough mixed to ensure isotopic homogeneity. Different rocks, however, can crystallize from totally unrelated magmas that may have different initial isotopic compositions. The effect of variable initial isotopic composition is to introduce scatter about a line on an isochron diagram, scatter that could reflect the age of the rock's source materials (a so-called pseudoisochron), but that also could produce a meaningless age (Fig. 1).

Table 1 lists all the naturally occurring radionuclides that have been used for radiometric age determination. Also listed is the decay constant for each system expressed as the time required for half of the amount present to decay. This factor, called the *half-life* is

$$t_{1/2} = \frac{\ln 2}{\lambda}$$

Some of the isotopes with half-lives significantly shorter than the age of the Earth (e.g., ^{10}Be, ^{14}C), which otherwise would not still exist, are created continually by nuclear interactions between other, naturally occurring isotopes and cosmic rays. While these reactions occur predominantly in the upper atmosphere, they can also occur by direct cosmic ray irradiation of surface rocks at high altitudes, where shielding by the atmosphere is not so thorough. Other short-lived isotopes are created as decay products of other isotopes, such as in the decay of ^{238}U, where successive losses of α-particles produce useful isotopes like ^{234}U, ^{230}Th, and ^{210}Pb, with the final decay product being the stable ^{206}Pb. The remainder of the radioactive species date back to the early history of the universe, where they were produced in the violent final stages of a star's evolution, the supernovae. Some of these isotopes (e.g., ^{26}Al, ^{129}I, ^{244}Pu) have long since become extinct but have left the signature of their decay in the varying abundance of their daughter isotopes. The others, including the most heavily used systems K-Ar, Rb-Sr, Sm-Nd and U, Th-Pb, continue to decay and are used to determine the age of a wide variety of geological processes.

The oldest events whose age can be determined are those associated with the formation of the elements in our solar system. Because the production rate of the two long-lived isotopes of uranium, ^{235}U and ^{238}U, should be nearly the same during a supernovae, the very high ratio of ^{238}U to ^{235}U found today (^{238}U/^{235}U = 138) reflects the more rapid decay of ^{235}U compared with ^{238}U. Assuming that ^{235}U/^{238}U was equal to 1 when most of the U in our solar system formed, the present abundance of the two U isotopes indicates that the average age of the elements is about 6 billion years, about half that of the inferred age of the universe.

The oldest solid objects studied are the meteorites, most of which provide radiometric ages near 4.56 billion years. A particularly interesting feature of meteorites is that many contain measurable excesses of daughter isotopes from the decay of the now extinct nuclides ^{244}Pu, ^{146}Sm, ^{129}I, and ^{26}Al. The general interpretation of these data is that the gas that formed the early solar system mixed with newly created elements injected from a nearby supernovae only a few million years before solid objects condensed out of the high-temperature gas. Some

TABLE 1. Radioactive Systems Used for Geochronological Studies

Radioactive Isotope	Production Mechanism	Half-Life (Years)	Decay Product(s)
^{222}Rn	U-decay	0.01	^{206}Pb, ^{4}He
^{210}Po	U-decay	0.38	^{206}Pb, ^{4}He
^{3}H	Cosmic Rays, Atomic Bombs	12.3	^{3}He
^{210}Pb	U-decay	22	^{206}Pb, ^{4}He
^{14}C	Cosmic Rays	5730	^{14}N
^{230}Th	U-decay	7.52×10^4	^{206}Pb, ^{4}He
^{234}U	U-decay	2.48×10^5	^{206}Pb, ^{4}He
^{36}Cl	Cosmic Rays	3.01×10^5	^{36}Ar
^{26}Al	Natural-Extinct, Cosmic Rays	7.3×10^5	^{26}Mg
^{10}Be	Cosmic Rays	1.6×10^6	^{10}B
^{53}Mn	Cosmic Rays	3.8×10^6	^{53}Cr
^{107}Pd	Natural-Extinct, Cosmic Rays	6.5×10^6	^{107}Ag
^{129}I	Natural-Extinct	1.59×10^7	^{129}Xe
^{244}Pu	Natural-Extinct	8.3×10^7	Fission Products
^{146}Sm	Natural-Extinct	1.03×10^8	^{142}Nd
^{235}U	Natural	7.04×10^8	^{207}Pb, ^{4}He
^{40}K	Natural	1.28×10^9	^{40}Ca, ^{40}Ar
^{238}U	Natural	4.47×10^9	^{206}Pb, ^{4}He
^{232}Th	Natural	1.4×10^{10}	^{208}Pb, ^{4}He
^{176}Lu	Natural	2.9×10^{10}	^{176}Hf
^{187}Re	Natural	4.6×10^{10}	^{187}Os
^{87}Rb	Natural	4.9×10^{10}	^{87}Sr
^{138}La	Natural	1.05×10^{11}	^{138}Ba, ^{138}Ce
^{147}Sm	Natural	1.06×10^{11}	^{143}Nd

believe that the interaction between the gas in our solar system and the shock wave emitted by this supernovae led to the collapse of the solar system and the eventual formation of the sun and planets.

Ages approaching 4.56 billion years have been determined for some rocks of the lunar crust. Like some meteorites, however, certain lunar rocks have been modified by meteorite impacts. The age of these collisions can be determined by radiometric dating because, if the impact is severe enough, daughter isotopes can either be lost completely, which occurs when the noble gas ^{40}Ar is volatilized, or homogenized throughout the effected rock, thereby "resetting" the radiometric clock. Although all memory of the original igneous age of a rock can be lost in such an event, the timing of impacts can be used to understand the rates of surface bombardment of the planets. Information from the moon shows that much of its crust was created prior to 4.4 billion years ago, with some remaining igneous activity continuing until about 4.1 billion years ago. A series of major impacts roughly 3.9 billion years ago created many of the large basins seen on the Earth-facing side of the moon. These basins were then filled in by basaltic lavas from 3.7 to 3.1 billion years ago. No record has been found in the lunar sample collection to suggest that igneous activity on the moon continued much past 3.1 billion years ago.

The similarity in age of the meteorites and of the oldest lunar rocks suggests that the Earth also is 4.56 billion years old, though no rocks this old have been found on Earth. Besides the meteorite-lunar analog, however, the average Pb isotopic composition of rocks from the Earth's crust and mantle show that the Earth has had a U-Pb ratio much higher than either the average solar value or that found for most meteorites for at least 4.4 to 4.5 billion years. Thus the U-Pb isotopic system provides supporting evidence to conclude that the Earth existed as a planet roughly 4.5 billion years ago.

The lack of terrestrial rocks dating back to the earliest history of the Earth reflects the continuing dynamic activity of the planet. The oldest rocks found on the Earth, the 3.77-billion-year-old Isua metasediments from western Greenland have seen several cycles of alteration and metamorphism. Problems with partial resetting or disturbance of the radiometric systems due to alteration and metamorphism in the oldest rocks are common. In order to avoid these problems, radiometric systems based on elements that remain relatively immobile during these processes, such as the rare earth element ^{147}Sm, which decays to ^{143}Nd, are finding more common use. Another method employed depends on the mineral zircon, which strongly concentrates U over Pb and is very resistant to both alteration and metamorphism. Zircon is so resistant to disturbance of its U-Pb system that examples have been found of granites containing zircons whose overgrown rims give the age of granite information, but whose cores, inherited from the source materials melted to produce the granite, still record the age of the source materials. Detailed analysis of individual zircon grains with an ion microprobe capable of determining the Pb isotopic composition in 10- to 20-micron-diameter spots on individual zircons recently has allowed the discovery of a few zircons, 4.2 to 4.3 billion years old, in quartzites from western Australia. These very old zircons are but a minor component of the zircons in these rocks, most of which are much younger, but they provide evidence for terrestrial crust older than any previously known.

Although metamorphic resetting is a problem when one is trying to determine the original igneous age of a rock, it is a boon to those interested in the time-temperature history of metamorphic processes. The various radiometric systems "close" to diffusive resetting at different temperatures in different minerals, some, such as U-Pb in zircon, at temperatures above 700–800°C, others like K-Ar in K-feldspar as low as 150°C. When determining the "age" of a rock, one actually determines the time at which the rock cooled below the closure temperature of the radiometric system being used. Using different systems with different closure temperatures thus can give information on the cooling and uplift history of a rock.

The precision of a radiometric age determination is a function of the amount that the daughter element's isotopic composition has been changed by radioactive decay. This, in turn, is a function of the amount of enrichment of parent over daughter element and the age of the rock relative to the half-life of the radiometric system being used. For example, the Sm-Nd system, based on the decay of ^{147}Sm with a half-life of 106 billion years, is useful for dating rocks a few hundred million to billions of years old. The increase in ^{143}Nd caused by the decay of ^{147}Sm in rocks less than about 100 million years old is generally too small to be measured precisely. In contrast, the K-Ar system, with a half-life of 12 billion years, has been used to determine the age of rocks as young as 10,000 years old due to the fact that the daughter element, the noble gas Ar, is almost completely degassed during eruption of a volcanic rock, leaving extremely high ratios of K to Ar in the cooled lava. This same feature, however, makes the K-Ar system more easily disturbed, since the gaseous Ar can leave the rock with only mild heating or alteration.

Systems based on elements with much shorter half-lives, such as ^{14}C, can be used to determine precise ages of geologic events only a few hundred to a few thousand years old. Because these short-lived isotopes are produced by nuclear reactions in the atmosphere, they are not present in the materials of the Earth's interior. Consequently, these systems most commonly are used to determine the time at which a material became isolated from exchange with the atmosphere, for example, to determine the age of groundwater, a sediment, deep ocean water, or the death of a living organism. Besides its extensive use in archeology, this aspect of the ^{14}C system has been used to determine the age of a variety of young geologic phenomena, including the age of recent volcanic activity by examination of plant material burned by a lava flow, and the timing of prehistoric seismic activity by determining the age of ash or soot layers laid down by forest or grass fires and later displaced by fault activity.

Radiometric age determination thus provides a tool to determine precise, absolute ages for a wide variety of geologic and geophysical phenomena.

RICHARD CARLSON

References

Becquerel, H., 1896, Sur les radiations invisibles emises par phosphorescence; sur les radiations invisibles emisses par les corps phosphorescents; Sur les radiations invisibles emisses par les sels d'uranium, *Compte Rendu* **122**, 420, 501, 689.

Faure, G., 1986, *Principles of Isotope Geology*, 2nd Ed. New York: Wiley, 589 p.

Friedlander G., J. W. Kennedy, E. S. Macias, and J. M. Miller, 1981, *Nuclear and Radiochemistry*, 3rd Ed. New York: Wiley, 684 p.

Holmes, A., 1913 *The Age of the Earth*. London: Harper and Brothers, 194 p.

Rutherford, E. 1906, *Radioactive Transformations*. New York: Scribner's, 287 p.

Thomson, J. J., 1914, Rays of positive electricity, *Roy. Soc. [London] Proc.* ser. S, **89**, 1-20.

Cross-references: *Earth and Moon: Origins; Meteorites: Nature and Origin.*

ABSOLUTE GRAVITY MEASUREMENTS

Modern geodesists view g, the local acceleration due to gravity, as an important parameter in the physics of the Earth. It is defined as the acceleration undergone by a freely falling body under the influence of both the gravitational attraction of the Earth (and other astronomical bodies) and the Earth's rotation. The value of g, nominally 9.8 m/sec^2, varies by 0.5% over the Earth's surface. Variations in the eighth decimal place are of great interest to geodesists and other scientists, making absolute determinations of g an important activity (though pursued by few).

History

Historically, g is one of the most measured parameters of the physical world. In the early 1600s, Galileo noted that the period of a pendulum was proportional to the square root of the length. Huygens worked out the period of the compound pendulum in the 1670s, paving the way for centuries of work using pendulums for gravity measurements. In a compound or reversible pendulum, a rigid asymmetric bar was alternately swung from two points on either side of its center of mass. By either adjusting the positions of these axes (with, for example, movable knife edges) or the bar's moment of inertia (by means of movable weights along the pendulum), a configuration could be found where the period of the pendulum was the same for oscillations around both axes. In such a situation, the period can be shown to be exactly equal to that of a simple pendulum of length l, except that in the case of the reversible pendulum, l is the distance between the axes. The period of the pendulum, τ, is given by

$$\tau = 2\pi\sqrt{l/g}.$$

This equation conveniently reduced the absolute determination of gravity to a measurement of time (the pendulum's period) and length (the axis separation), rather than requiring the metrologically more complex problem of finding the pendulum's moment of inertia to mass ratio.

Many researchers, most notably Capt. Henry Kater of England, took advantage of the simplicity of the reversible pendulum to make absolute surveys of the Earth's gravity field. Still others constructed *invariable* pendulums, whose moments of inertia were not adjustable. These were used to make relative gravity measurements by noting the variations in period from location to location. Early controversies surrounding the Earth's shape were partially resolved by early pendulum gravity measurements (see Lenzen and Multhauf, 1965, for an interesting review of pendulum gravity measurements).

Modern Methods

Ultimately, the fractional uncertainty in pendulum measurements could not be reduced below approximately 10^{-6} owing to the insurmountable difficulties from imperfections in knife edges, which prohibited the true rotation axes from being defined exactly. Electronic timing methods have improved to the point where the acceleration of a freely falling body can be determined to an accuracy approaching one part in 10^9 or about 1 μGal. The Gal is the unit for gravity (after Galileo) and is equal to 1 cm s^{-2}.

The first free fall gravity instruments varied widely in design. Cook (1967) recorded the times at which a freely falling glass sphere focused light onto slits separated by quartz blocks whose thicknesses could be determined optically. Tate (1968) used a falling bar on which accurately determined length markers had been machined. Faller (1965) employed a white light interferometer to record the position of a falling optical element.

The advent of lasers has brought about a convergence of all modern absolute gravity meters into two nearly identical designs. A Michelson interferometer is arranged such that one of the reflectors falls freely in a vacuum (see Fig. 1). For each half wavelength of descent, a photodetector senses a variation in intensity of light (a "fringe"). The times of occurrence of the interference fringes are recorded by a computer to produce a table of the falling mass's position, z, as a function of time, t. The position, being derived from the fringe spacing, depends on the wavelength of the light from the laser. When a quadratic of the form

$$z = z_0 + v_0 t + \tfrac{1}{2} g t^2$$

is fit to the position versus time record, the quadratic coefficient contains the desired parameter, g, and does not depend on the initial position, z_0, or velocity, v_0 (the effect of the local gradient in gravity is significant but can easily be corrected for). The major variation on the theme presented in Fig. 1 is that some researchers (Sakuma, 1984; Alasia et al., 1982) launch the test mass vertically and determine its acceleration during both its rise and its fall, rather than during the fall alone.

If one hopes to achieve the highest accuracy possible, the simple description of the experiment—the measurement of the acceleration of a freely falling mass—is not an adequate understanding of the process. The actual measurement is the second time derivative of the optical path difference defined by two optical elements, one attached to the Earth while the other is accelerated by the vector sum of all forces on it. The transfer of this measurement into the local value of gravity is a process requiring a detailed understanding of the components involved. Governing the procedure are four underlying considerations: (1) nongravitational forces on the test mass must be minimized; (2) the optical path difference must be translated into a physical length; (3) the acceleration of the reference frame from which the measurement is made must average to zero; and (4) the electronic system that counts and times the occurrence of interference fringes must be capable of doing so accurately and without introducing systematic errors.

The most obvious nongravitational force on a falling body is air drag; obviously the measurements must be made in a vacuum. The quality of the vacuum is important, however, since even at a pressure of 10^{-9} atm (10^{-4} Pa) the drag can be 10^{-8} g after a typical falling distance of 0.5 m (Hammond et al., 1984). This level of vacuum can be obtained with only modest difficulty, but further reduction of the air drag is desirable. Immunity to air drag is the primary advantage of the free rise and fall method over the simple free fall. As the test mass rises, the air drag force acts downward; the sign of this force

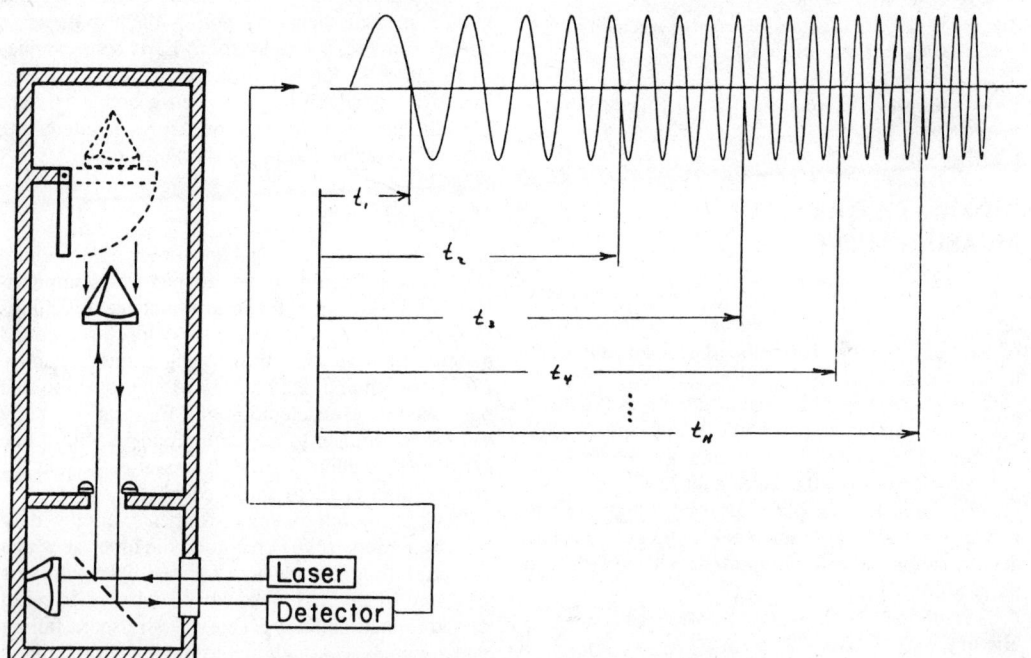

FIGURE 1. The basic method for making modern absolute gravity measurements. A mechanism allows an optical element—usually a cornercube retroreflector—to fall in a vacuum. A laser interferometer produces an increasing frequency sinusoidal signal; each cycle corresponds to the falling object having descended an additional wavelength of light. Nearly a million fringes are generated in a typical interval of 0.3 m; usually only a small subset (of order 100) are actually timed.

then changes when the test mass begins to fall back downward, canceling to first order.

In the simple free fall configuration, the air drag has been reduced in some instruments by surrounding the mass as it falls with a chamber that also falls (Tate, 1968; Niebauer et al., 1986; Zumberge et al., 1986). In such a state the residual air molecules in the vacuum chamber around the falling mass are falling at the same rate. This reduction in relative velocity between test mass and air molecules reduces air drag to a negligible level in a very modest vacuum.

Magnetic and electric forces can create problems if not attended to. Usually, the falling mass is constructed of conducting, nonmagnetic material (aluminum being the most common) while the necessary optical parts of the mass (mirrors or a corner-cube retroreflector) are coated with transparent, conducting material to minimize the accumulation of electric charge. Magnetic fields in the vicinity of the test mass's trajectory must be reduced to moderate levels (of the order of the Earth's field) so that field gradients do not generate eddy current forces.

The optical systems in modern absolute gravity meters are invariably designed around a wavelength-stabilized He-Ne laser. Since g has units of cm s^{-2}, the apparatus must have both a length and a time standard. The laser provides the length standard, and, in fact the actual measurement of g is in terms of laser wavelengths per second[2]. The laser wavelength must be known absolutely to an accuracy of one part in 10^9 to avoid an error larger than one μGal in g. Attention must be paid to a number of effects that can produce errors in the measurement by disturbing the optical distance to the falling mass. These include mirror and window vibrations, inadvertent steering of the beam, and optical component imperfections. If a corner cube is used as the element to reflect the light from the falling mass, some rotation of the falling mass is tolerable. The upper limit depends on how closely the optical center and center of mass have been made coincident.

The reference corner cube would ideally be mounted to a nonaccelerating reference frame. The Earth's surface, of course, does not qualify. Most instruments make use of some sort of seismometer to decrease the noise caused by ground motion.

Also displayed schematically in Fig. 1 is the function performed by a typical absolute gravity meter's electronic system. The increasing frequency sinusoidal signal from the photodetector that senses interference fringes contains all of the information on g. The times of occurrence of a subset of the

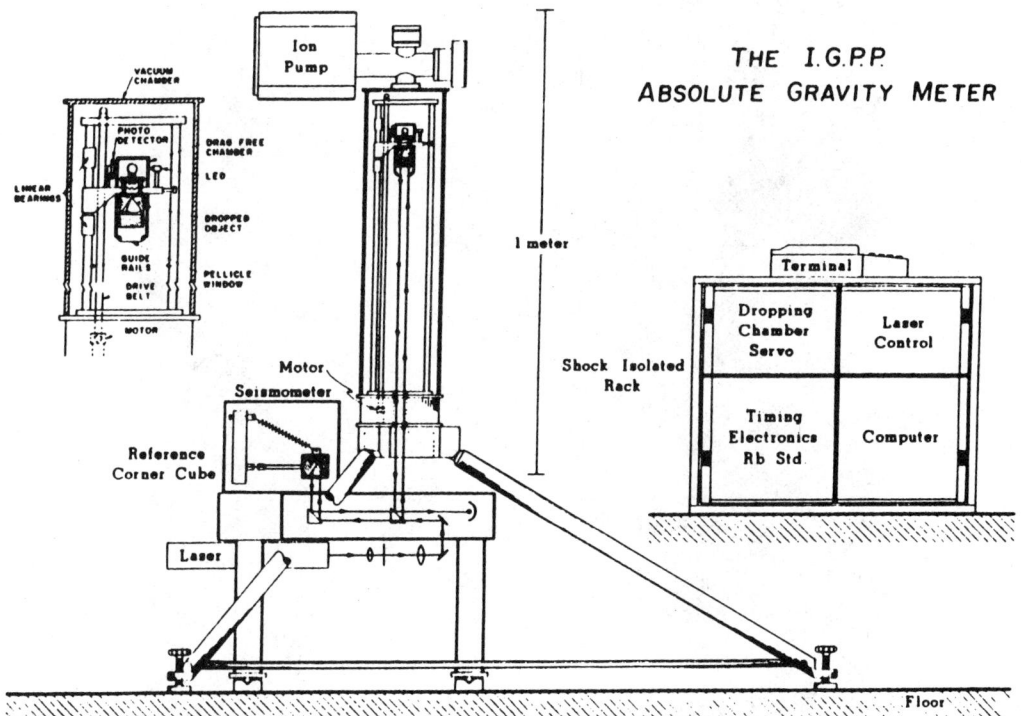

FIGURE 2. Schematic drawing of the absolute gravity meter built at the Institute of Geophysics and Planetary Physics, University of California, San Diego. The falling mass is shielded from nongravitational forces by a servo-driven co-falling chamber.

fringe zero crossings must be measured with an accuracy of ≈ 0.1 nsec in order to obtain an accuracy approaching 1 μGal. The electronic system that determines the times must be referenced to an absolute frequency standard. The frequency response of the photodetector and its amplifier are important since a frequency dependent phase shift can cause a systematic timing error. The varying delay time caused by the finite speed of light in the interferometer must also be taken into account.

Although many of the practical details vary from instrument to instrument, most fit the same basic description (a nearly complete list can be found in Marson and Faller, 1986). Usually an absolute gravity meter is portable in the sense that it can be disassembled and reassembled in a matter of hours, can be transported in a small truck (400 kg is a typical size), and make several hundred single-drop measurements in a day. So far none is a true "field" instrument since all require several hundred watts of power, a fairly stable temperature environment, and a solid floor on which to operate. Figure 2, a schematic drawing of the Institute of Geophysics and Planetary Physics (California) absolute gravity meter, displays a typical example of a portable, absolute gravity meter; Fig. 3 is a photograph of it. Other examples not already referenced above include Arnautov et al., 1983; Feng et al., 1982; and Hanada and Tsubokawa, 1982.

The accuracy of results from absolute measurements of the Earth's gravity is not an easily judged quantity. Probably the best evidence for accuracy lies simply in the few occasions upon which several instruments have been operated in the same location at the same time. Although absolute accuracies of 1 to 5 μGal have been claimed, the comparisons

FIGURE 3. Photograph of the absolute gravity meter shown in Figure 2 at a "field" site.

indicate that perhaps a more general estimate for the absolute accuracy of the measurements is 10 μGal (see, for example, Boulanger et al., 1983).

Applications

Outside of standards laboratories where determinations of certain fundamental quantities require accurate knowledge of local gravity, the most common application of absolute gravity measurements is for calibration of relative meters. Relative gravity meters determine the difference in gravity between locations in terms of the elongation of an extremely delicate spring suspending a pivoted mass. The transfer of such a measurement into knowledge of a true gravity difference requires that the relative instrument be calibrated along a series of sites that have been surveyed with an absolute instrument. Both a scale factor and a baseline datum are required. The relative instruments can resolve a gravity difference to a precision of less than 10 μGal over a range of 100 mGal when properly calibrated.

Some attempts are being made to apply absolute gravity measurements directly to geodynamical studies. An observer moving away from the Earth's center experiences a change in gravity of about 3 μGal per cm. Thus absolute gravity measurements can be used in principle to detect vertical deformations of the Earth's crust due to postglacial rebound or tectonic activity. Because these proceed at fairly slow rates—no more than a few centimeters per year—a number of years will be required for a program of absolute gravity measurements to detect them.

MARK A. ZUMBERGE

References

Alasia, F., L. Cannizzo, G. Cerutti, and I. Marson, 1982, Absolute gravity acceleration measurements: experiences with a transportable gravimeter, *Metrologia* **18,** 221–229.

Arnautov, G. P., Yu. D. Boulanger, E. N. Kalish, V. P. Koronkevitch, Yu. F. Stus, and V. G. Tarasyuk, 1983, "Gabl", an absolute free-fall laser gravimeter, *Metrologia* **19,** 49–55.

Boulanger (or Bulanzhe), Yu. D., G. P. Arnautov, Ye. N. Kalish, V. P. Koronkevich, Yu. F. Stus, V. G. Tarasyuk, and S. N. Shcheglov, 1983, The results of the first international comparison of absolute gravity meters, Sèvres, 1981, *Izvestiya, Earth Physics* **19,** 196–202.

Cook, A. H., 1967, A new absolute determination of the acceleration due to gravity at the National Physical Laboratory, England, *Royal Soc. London Philos. Trans. ser. A* **261,** 211–252.

Faller, J. E., 1965, Results of an absolute determination of the acceleration of gravity, *Jour. Geophys. Research* **70,** 4035–4038.

Feng, Y. Y., G. Y. Zhang, D. X. Li, X. M. Qiu, J. H. Zhou, J. L. Gao, D. L. Huang, C. Q. Huang, and Y. G. Guo, 1982, A transportable absolute gravimeter for determining the acceleration due to the earth's gravity, *Metrologia* **18,** 139–144.

Hammond, J. A., R. L. Iliff, and R. W. Sands, 1984, New techniques for absolute gravity measurement, in B. N. Taylor and W. D. Phillips, eds., *Precision Measurement and Fundamental Constants II.* U.S. Natl. Bur. Standards Spec. Pub. 617, 423–426.

Hanada, H., and T. Tsubokawa, 1982, Absolute determination of gravity with transportable apparatus, *General Meeting of the International Association of Geodesy, Tokyo, May 7–15, Proc.* 358–365.

Lenzen, V. F., and R. P. Multhauf, 1965, Development of gravity pendulums in the 19th century, *U.S. Natl. Mus. Bull.* **240,** 307–347.

Marson, I., and J. E. Faller, 1986, g—the acceleration of gravity: its measurement and its importance, *Jour. Physics, E: Sci. Instrum.* **19,** 22–32.

Niebauer, T. M., J. K. Hoskins, and J. E. Faller, 1986, Absolute gravity: a reconnaissance tool for studying vertical crustal motions, *Jour. Geophys. Research* **91,** 9145–9149.

Sakuma, A., 1984, Present status of the absolute measurement of gravitational acceleration, in B. N. Taylor and W. D. Phillips, eds., *Precision Measurement and Fundamental Constants II.* U.S. Natl. Bur. Standards Spec. Pub. 617, 397–404.

Tate, D. R., 1968, Acceleration Due to Gravity at the National Bureau of Standards, *U.S. Natl. Bur. Standards Jour. Research—C: Engineering and Instrumentation* **72C,** 1–20.

Zumberge, M. A., G. Sasagawa, and M. Kappus, 1986, Absolute gravity measurements in California, *Jour. Geophys. Research* **91,** 9135–9144.

Cross-references: *Gravity Anomalies: Interpretation; Gravity and Isostasy; Gravity Fields: Implications for Planetary Interiors.*

ACCRETIONARY TECTONICS: EXAMPLES FROM THE NORTH AMERICAN CORDILLERA

Accretion of crustal fragments to continents along active plate margins, together with continental arc magmatism, form the primary mechanisms for the development of continental crust. Accretionary processes are complex and varied. Three primary processes control accretionary tectonics: (1) generation of new crustal material, largely within an intraoceanic setting; (2) addition or accretion of these juvenile crustal fragments to the continental margin; and (3) dispersal and attrition of accreted materials along the continental margin. Because accretionary processes obscure original relations between displaced crustal fragments, early paleogeographic ties are often difficult to establish. As a result, fault-bounded slivers of rocks that have differing geologic histories are suspected of being far-travelled, and have thus come to be called "suspect terranes" (Coney et al., 1980).

Several lines of geologic and geophysical evidence are used to recognize accreted terranes and to detail the history of evolution, accretion, and subsequent deformation within terranes. Stratigraphy, igneous and metamorphic history, and biogeography are used to delineate the petrotectonic and paleogeographic origins of individual terranes. Structural relations, paleogeographic "links" between terranes, and paleomagnetics are used to decipher the history of accretion to the continent and subsequent deformation.

The North American Cordillera provides a complex record of long-lived accretionary tectonism, which began as early as mid-Paleozoic time. A growing understanding of Paleozoic and Mesozoic rocks within the western Cordillera is revealing Cordilleran accretionary history in detail. The complex systems of convergent and transform margins within the southwest Pacific provide modern analogs for processes that have been inferred to have acted within the Cordillera. This review focuses on the late Paleozoic and early Mesozoic evolution of the North American Cordilleran accretionary orogen, which represents an advanced stage in the current style of tectonism seen in the southwest Pacific today.

The next section reviews the processes involved in accretionary tectonics. Then, drawing on our understanding of the North American Cordillera, some of the better examples of accretionary tectonism within the western Cordillera are presented to illustrate accretionary processes. For recent comprehensive reviews of Cordilleran evolution, to Saleeby (1983), Monger (1984) and Miller et al. (in press). Examples of accretionary tectonism in other regions are discussed by Nur (this volume).

Accretionary Tectonics

Three primary processes are involved in peripheral accretion of continental crust: (1) the evolution of new lithosphere, either on oceanic crust, within island arcs, or in marginal basins; (2) the addition (accretion) of these new materials to the continental margin by collision, offscraping, or back-arc and intraarc thrusting; and (3) removal or rearrangement (attrition) of accreted fragments by tectonic erosion along the inner trench wall, rifting, or transcurrent faulting within the continental margin.

New crustal materials that ultimately form terranes within accretionary orogens include seamounts and oceanic plateaus and their flanking carbonates, basinal sediments that are incorporated into imbricate sedimentary prisms, subduction-related metamorphic terranes, oceanic volcanic island arcs, and fragmented or coherent ophiolites that formed within marginal or larger, open ocean basins. Recognition of the original tectonic affinity of various accreted terranes is based on detailed characterization of stratigraphy, geochemistry of igneous rocks, and structural and metamorphic history.

A number of geologic processes are active during the addition of these crustal fragments to the continental margin. In some instances, terranes are juxtaposed to each other before they are added to the continental margin. This juxtaposition is referred to as amalgamation and includes the formation of accretionary prisms adjacent to volcanic arcs, offscraping of bathymetric irregularities (such as seamounts) within subduction-related accretionary complexes, intra-arc thrusting, and arc-arc collision. A group of terranes that have been amalgamated prior to addition to the continental margin is referred to as a super-terrane. Final juxtaposition of terranes to the continental margin, called accretion, results from arc-continent or continent-continent collision, emplacement from along the continental margin on strike-slip faults, or telescoping and thickening of marginal arc/basin systems to form continental crust.

Finally, the removal of material from the edge of an accretionary orogen, known as terrane attrition, occurs through several tectonic processes. These include tectonic erosion by underthrusting of materials that form the inner trench wall, rifting of the continental margin, or removal of material by strike-slip faulting and displacement along strike.

The processes of accretion and attrition obscure the original paleogeographic relations between terranes. As a result, traditionally accepted paleogeographic ties between terranes have come into question, and careful reevaluation of inferred ties characterizes our growing understanding of accretionary provinces. Data that have been used to suggest large amounts of relative transport between terranes include differences in stratigraphic and structural history, preservation of contrasting biogeographic provinces, and paleomagnetic data that demonstrate large amounts of differential poleward migration or differential rotation between terranes, or between a terrane and the continental margin. In contrast, data that are used to suggest ties between terranes include deposition of strata that overlap more than one terrane (overlap sequence), deposition of detritus derived from one terrane within another (provenance ties), shared stratigraphic, igneous, metamorphic, or magnetic histories, and biogeographic ties.

In the following sections, we emphasize some well-documented examples of accretionary tectonism within the North American Cordillera to illustrate accretionary processes.

Paleozoic and Mesozoic Accretionary Tectonics Within the North American Cordillera

Paleozoic and Mesozoic evolution of the North American Cordillera provides a record of long-lived accretionary tectonism. Within the western United States, the relations between Paleozoic and early

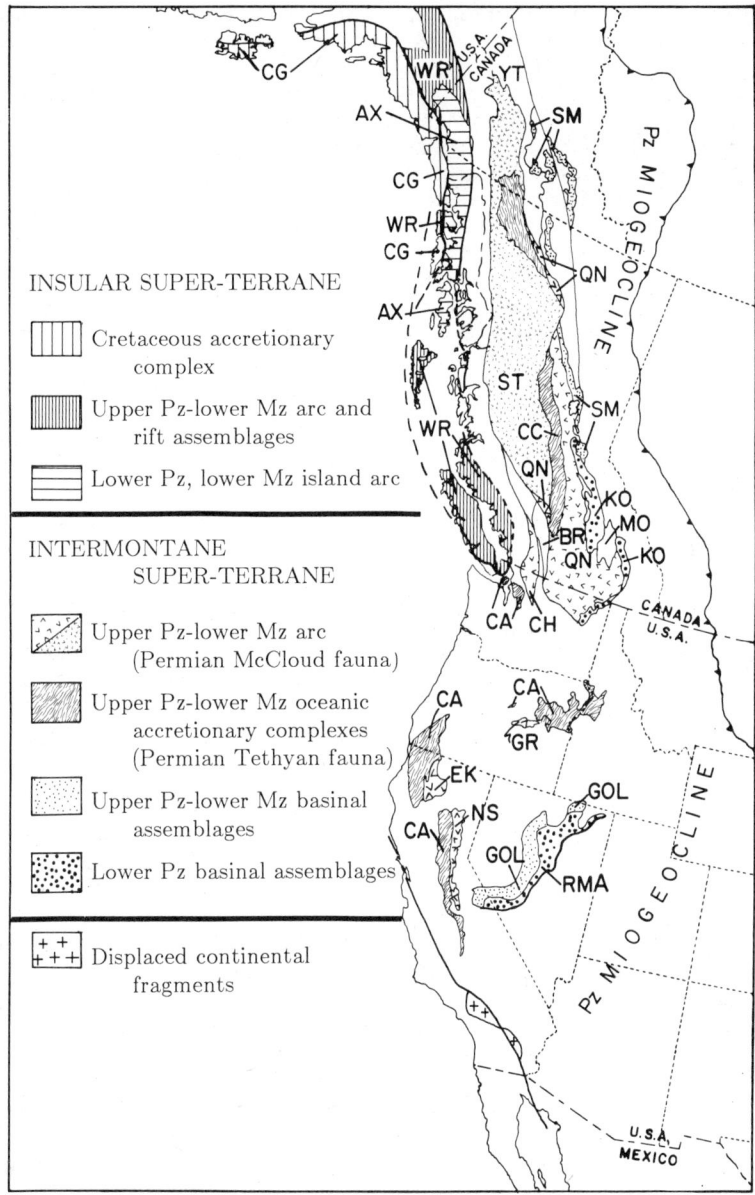

FIGURE 1. Generalized terrane map of the Cordillera. AX = Alexander terrane, BR = Bridge River terrane, CA = Cache Creek affinity terranes, CC = Cache Creek terrane, CG = Chugach terrane, CH = Chilliwack terrane, EK = eastern Klamath terrane, GOL = Golconda allochthon, GR = Grindstone terrane, KO = Kootenay terrane, MO = Monashee terrane, NS = northern Sierra terrane, QN = Quesnellia, RMA = Roberts Mountains allochthon, SM = Slide Mountain terrane, WR = Wrangellia, YT = Yukon-Tanana terrane. (After Saleeby, 1983; Monger, 1984; and Miller et al., in press).

Mesozoic terranes are in many places obscured by younger deformation and extensive cover (Fig. 1). Within British Columbia, a complex history of strike-slip faulting (e.g., Gabrielse, 1985) has masked some of the original relations between terranes. However, a growing body of geological and geophysical data is revealing Cordilleran accretionary history. The late Precambrian rifted edge of western North America formed basement for a westward-thickening, upper Precambrian and Paleozoic miogeoclinal wedge, largely comprised of intercalated carbonates and shales. This Paleozoic continental margin of North America formed the buttress against which the westward allochthonous terranes were subsequently accreted (Fig. 1).

Terranes containing Paleozoic and Mesozoic rocks can be grouped based on their age, original tectonic setting, structural evolution, and, in places, by biogeographic affinity. Using these criteria, three Cordilleran scale belts are defined in Fig. 1. Regionally, these include: (1) Paleozoic basinal sequences that have been emplaced over the continental margin sequentially during middle and latest Paleozoic time, (2) upper Paleozoic and lower Mesozoic oceanic accretionary assemblages and volcanic island arc successions of the Intermontane, or Stikine super-terrane in Canada, and related rocks within the western United States, and (3) lower Paleozoic to Cretaceous island arc and related rift assemblages of the Insular super-terrane in western British Columbia and southeastern Alaska. Displaced continent-related fragments occur more sporadically (Fig. 1). All four belts are present within the Canadian Cordillera, however, only the belt of Paleozoic basinal rocks and Stikine-related terranes are widespread in the western United States (Fig. 1). This section focuses on the late Paleozoic and early Mesozoic evolution of terranes within these two more extensive belts. Constituent petrotectonic assemblages, biogeographic relations, and, finally, the structural and paleomagnetic relations that bear on their amalgamation and accretion histories are discussed.

Allochthonous Paleozoic Basinal Sequences. Within the belt of Paleozoic basinal rocks, two groups of terranes that are differentiated by age and structural evolution are present. The easterly and structurally lower is early Paleozoic in age and is represented by rocks of the Roberts Mountains allochthon in Nevada and the Kootenay terrane in southeastern British Columbia. The second group is largely upper Paleozoic and represented by rocks of the Golconda allochthon in Nevada and coeval rocks of the Slide Mountain terrane in British Columbia (Fig. 1). The depositional and deformational history recorded within these Paleozoic rocks in the Canadian Cordillera is somewhat different from that recorded within Nevada (Monger and Price, 1979).

The Roberts Mountains allochthon is comprised of chert, argillite, sandstone, and greenstone, ranging in age from Cambrian to latest Devonian and locally earliest Mississippian (Fig. 2). Several lines of stratigraphic evidence suggest a persistent paleogeographic relation to a continent throughout deposition (Miller et al., in press). Inferred continental provenance for quartzose sandstone, early Proterozoic U/Pb ages for detrital zircon, reworked fauna from the North American shelf and the presence of phosphatic chert are all thought to imply proximity to a continent throughout the early Paleozoic.

The Roberts Mountains allochthon was folded and imbricately faulted in an east-vergent deformation during mid-Paleozoic time, as summarized recently by Nilsen and Stewart. Age constraints for the timing of the Antler orogeny come from the development of a clastic foredeep that accompanied emplacement of the Roberts Mountains allochthon over the continental margin, and suggest that deformation occurred during Early Mississippian time.

The Roberts Mountains allochthon may have been emplaced as an accretionary wedge in front of an encroaching volcanic arc above a west-dipping subduction zone, according to the model of Speed and Sleep. However, if the encroaching arc is represented by Devonian volcanic successions within the Sierra Nevada and eastern Klamath Mountains, subduction probably occurred along a predominantly east-dipping zone for reasons discussed below.

Within southeastern British Columbia, the Kootenay terrane (Fig. 1) contains lower Paleozoic rocks (Lardeau Group) including phyllite, quartzite, metabasalt, and limestone (Fig. 2). These strata are lithologically similar to strata within the Roberts Mountains allochthon, and may represent distal facies equivalents to lower Paleozoic strata of the North American shelf, although this correlation has recently come into question (Monger, 1984). The Kootenay basinal rocks were deformed, metamorphosed, and intruded prior to deposition of Lower Mississippian rocks within the overlying Milford Group, and the age of deformation is thus bracketed between Ordovician(?) and Early Mississippian (Monger and Price, 1979).

Initial correlations between the Antler orogeny in Nevada and mid-Paleozoic orogenesis within the Kootenay terrane have been questioned more recently based on the lack of demonstrable overthrusting of the Lardeau Group within this time interval (Monger and Price, 1979). Primary differences between the two areas include conspicuous metamorphism and plutonism within the Kootenay terrane, the lack of demonstrable thrusting of Lardeau rocks over the miogeocline, and the broader, but permissive, age constraints for deformation within the Kootenay terrane. Metamorphism, plutonism, and shear zone development within the Kootenay terrane may simply reflect orogenesis at a deeper structural level than that preserved within the Antler belt, or the two events may be unrelated. Coeval extension along the North American continental margin has been ascribed to a transtensional tectonic setting (e.g., Gordey et al., 1987). Both the extensional and compressional events may have been related to the same transcurrent or oblique North American plate margin.

To the west of exposures of lower Paleozoic basinal strata lies a belt of upper Paleozoic rocks comprised of the Golconda allochthon in Nevada and the Slide Mountain terrane in British Columbia (Fig. 1). The Golconda allochthon includes the Schoonover and Havallah sequences that range in age from latest Devonian to Late Permian. Together these

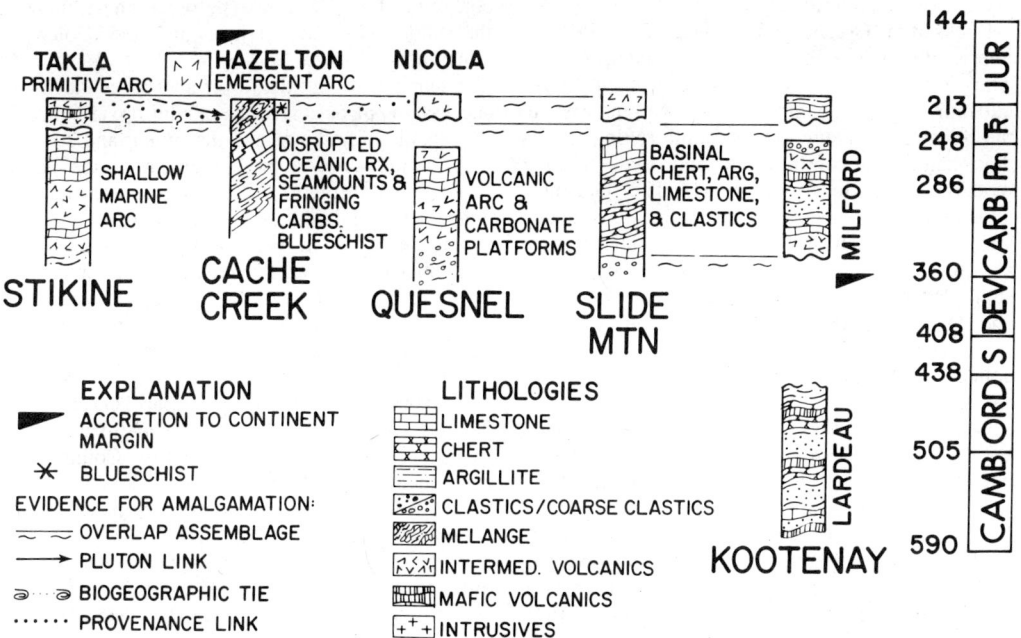

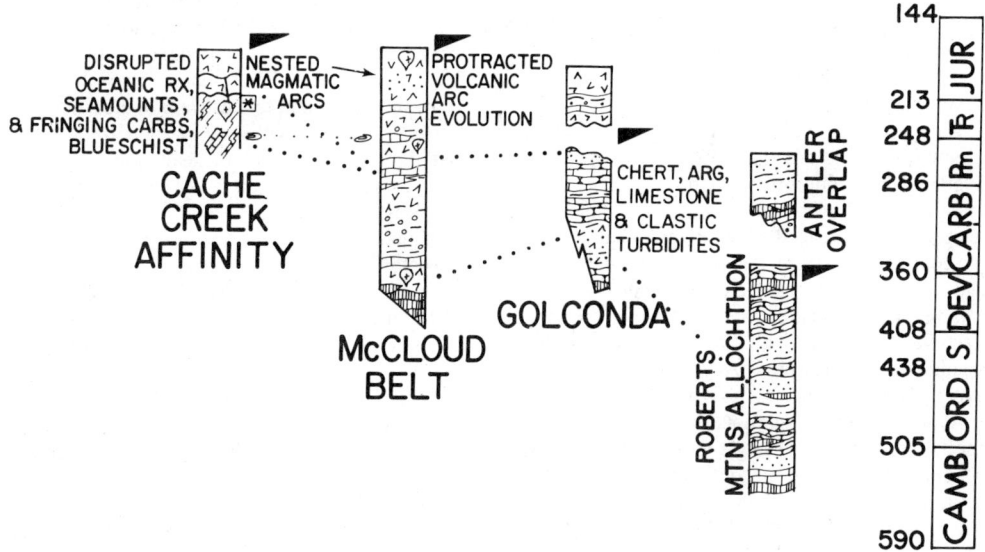

FIGURE 2. Diagram showing the petrotectonic, amalgamation, and accretion histories of Paleozoic basinal rocks, McCloud belt, and Cache Creek terranes near latitudes 51° (British Columbia) and 41° (northern Nevada-California). This figure integrates data acquired by the following workers: Behrman and Parkison, 1978; Gehrels, 1986; Harms, 1985, 1986; Irwin, 1972; Klepacki and Wheeler, 1985; Little, 1987; Miller et al., 1984, in press; Miller and Wright, 1987; Monger, 1977, 1984; Monger and MacMillan, 1984; Mortimer, 1986; Saleeby, 1982, 1983; Sanborn, 1960; Silberling, 1973; Speed, 1977, 1979; Struik and Orchard, 1985; Tomlinson and Wright, 1986; Watkins et al., 1986; and Wright, 1982. [Full references available from the authors upon request.]

units contain basalt, chert and argillite, volcaniclastic turbidites, and limestone turbidites (Fig. 2; see review in Miller et al., in press).

Rocks of the Golconda allochthon were folded and imbricately faulted during latest Permian or earliest Triassic time during eastward emplacement of the allochthon over autochthonous rocks to the east, in the Sonoman orogeny. The tectonic setting of this deformation is controversial and two differing models are variously accepted. The Golconda allochthon may represent an accretionary wedge that was formed trenchward of an encroaching volcanic arc above a west-dipping subduction zone, as first suggested by Speed. Alternatively, the Schoonover and Havallah sequences were deposited in a marginal basin and were deformed and emplaced during collapse of the back-arc basin (Burchfiel and Davis, 1981).

Coeval basinal rocks of the Slide Mountain terrane within the Canadian Cordillera (Eastern assemblage of Monger, 1977) are geographically widespread and regionally share a similar upper Paleozoic stratigraphic succession. The Slide Mountain terrane contains chert, argillite, sandstone, conglomerate, minor limestone, ultramafic rocks, and basaltic greenstone (Monger, 1977), and ranges in age from latest Devonian(?)-Early Mississippian to Early Permian (Fig. 2). Some workers suggest a correlation of allochthonous strata within the Slide Mountain terrane to the Milford Group in southern British Columbia (e.g., Monger, 1977). Because the Milford Group depositionally overlies lower Paleozoic rocks of the Kootenay terrane (Fig. 2), this correlation implies a paleogeographic tie between rocks of the Slide Mountain terrane and the continental margin of North America.

The youngest rocks deformed within the Slide Mountain terrane are Early Permian in age and are overlain by undeformed Upper Triassic strata (Monger, 1977, 1984). Rocks of the Slide Mountain terrane were emplaced over Lower Permian strata of the continental margin. These relations constrain the age of deformation and emplacement of the Slide Mountain assemblages as Late Permian to Middle Triassic. In places, final emplacement is much younger, or overprinted by younger deformational events (Monger, 1984).

As for the Golconda allochthon, two models exist for the depositional setting and deformation of Slide Mountain lithologies. In southern British Columbia, stratigraphic ties to both the North American margin and to volcanic arc rocks to the west, recognized by D. W. Klepacki and others, have been used to suggest evolution within a marginal, back-arc basin. Alternatively, to the north, work by T. A. Harms suggests telescoping of tracts of oceanic rock assemblages, presumably of originally great lateral extent, and is thought to suggest subduction of a vast expanse of proto-Pacific oceanic crust beneath an east-facing volcanic arc. Relations within coeval rocks to the west, in McCloud and Cache Creek affinity terranes discussed below, provide new constraints for the setting of deformation within both the Slide Mountain terrane and the Golconda allochthon.

In summary, both the lower and upper Paleozoic successions of basinal strata exhibit broad similarities in lithology, ages, and the timing and style of deformation along a large segment of the North American Cordillera. Although direct correspondence between the lithologic packages and orogenic events recorded in Nevada to those in British Columbia is not well documented, similar basinal evolutionary history and broadly coeval deformation occurred in both regions and marks the onset of Phanerozoic accretionary tectonism in western North America.

Intermontane (Stikinian) Super-Terrane and Related Rocks of the Southwest Cordillera. The Intermontane super-terrane of British Columbia is made up of several smaller terranes with contrasting Permian biogeographic affinity and includes the Paleozoic basinal strata discussed above. During Permian time, the paleo-Pacific contained three different shallow water equatorial provinces (Ross and Ross, 1983; Stevens, 1985), thus, biogeographic differentiation was well established. The Pangean landmass and cold polar seas inhibited migration of shallow water, tropical fauna, such as corals and fusulinids, resulting in the development of distinct faunal provinces distinguished either on the family or generic level (Stevens, 1985). Tethyan fauna, characterized by verbeekinid fusulinids and waagenophyllid corals, populated the westerly Permian province. These fauna lived both in the Tethyan seaway and far eastward in the paleo-Pacific Ocean, preserved within reefs and carbonate banks that flanked a network of seamounts and oceanic plateaus during Permian time (Saleeby, 1981; Ross and Ross, 1983; Miller and Wright, 1987). The McCloud province, best represented by the McCloud Limestone, California, characterized the eastern paleo-Pacific and is largely known from volcanic island arc successions (Miller, 1987). McCloud fusulinids resemble those of western North America on a family level and are even more closely related to those from miogeoclinal rocks within southeastern California. The third province, the North American province, is known from authochthonous Permian shelf rocks of the North American miogeocline.

Within the Intermontane super-terrane, fauna of both the McCloud and Tethyan provinces are preserved. Distribution of Permian rocks with differing faunal affinity corresponds on a regional scale to the distribution of lithologies and the tectonic affinity of individual terranes. Upper Paleozoic volcanic arc successions contain Lower Permian limestones characterized by McCloud fauna (Ross and Ross, 1983). The Cache Creek and lithologically similar terranes (Cache Creek affinity terranes, Fig. 1) are comprised of upper Paleozoic to Upper Triassic (or locally younger) disrupted chert and argillite sequences containing olistostromal and tectonic blocks of Permian Tethyan limestone and are associated with Upper Triassic blueschist blocks or terranes (Fig. 2; Monger, 1977, 1984; Davis et al., 1978; Saleeby, 1983).

The biogeographic significance of Permian fauna was first recognized by Monger and Ross (1971), who noted the anomalous geographic position of the Stikine terrane. In most places, the original east-to-west distribution of biogeographic provinces is preserved, and McCloud affinity terranes lie inboard of Cache Creek affinity terranes (Fig. 3). The Stikine terrane, however, represents an important exception to this pattern and lies outboard of the Cache Creek terrane.

The belt of upper Paleozoic arc terranes with Permian McCloud affinity fauna include the Quesnel, Chilliwack, Grindstone, Bilk Creek, eastern Klamath and northern Sierra terranes (Fig. 1). The somewhat different history of the Stikine terrane is discussed separately. Detailed comparison of the upper Paleozoic stratigraphy within McCloud affinity terranes suggests a common evolution within a volcanic island arc setting (Miller, 1987; Fig. 4).

Magmatism within the arc began during Early to Middle Devonian time, and was succeeded by subsidence and basinal deposition (Figs. 2, 4; Miller, 1987). Uplift of the volcanic arc basement occurred during the Late Devonian and Mississippian. During the Late Carboniferous, shallow marine and deltaic volcaniclastic apron deposition records uplift and accompanying volcanism. During the Early Permian, volcanism waned and shallow water carbonate platforms were developed (Fig. 4). Large volumes of Upper Permian volcanic rocks record another pulse in volcanism throughout the arc, this time accompanied by rapid subsidence. A break in stratigraphy occurred in most places during the latest Permian and Early Triassic, and volcanism intermittently persisted until the Middle Jurassic.

Deformation of rocks within the McCloud belt, and inferred accretion to North America occurred during Middle or Late Jurassic time (Monger et al., 1982; Saleeby, 1983). However, several lines of evidence suggest proximity between this island arc and the North American margin throughout the late Paleozoic. Within the northern Sierra terrane, Devonian arc rocks unconformably overlie deformed strata of the lower Paleozoic Shoo Fly Complex. The Shoo Fly and similar strata within the Klamath Mountains contain quartzose sandstone of continental provenance, and detrital zircon that yields early Proterozoic ages, suggesting an ultimate cratonal source. An influx of volcaniclastic turbidites, recorded within Mississippian rocks of the Golconda allochthon (Schoonover sequence), has been taken to indicate proximity of a westerly, volcanic arc source during this time period (Fig. 2). This influx was contemporaneous with uplift of the Devonian McCloud arc and its substrate and suggests a possible provenance tie (Fig. 2). During Permian time, a similar influx of volcaniclastic detritus is recorded within rocks of the Golconda allochthon. Within British Columbia, similar Upper Triassic strata overlie the Harper Ranch Group (Quesnel terrane) and the Slide Mountain terrane (Monger, 1984), confirming an early Mesozoic tie.

Thus, although final accretion of McCloud arc terranes to North America did not occur until Jurassic time, a paleogeographic relation to rocks to the east is suggested as early as the middle or early Paleozoic and throughout arc evolution. This provides a paleogeographic constraint on the tectonic setting of both Antler-age and Permo-Triassic orogenesis recorded to the east. If the McCloud arc was adjacent to North America throughout the mid- and late Paleozoic, then basinal successions to the east were deposited within a marginal, back-arc basin. Deformation of these successions occurred during back-arc compression (see Burchfiel and Davis, 1981), or a short-lived, Permo-Triassic interval of arc polarity reversal. Consistent with this interpretation, evidence for impingement of the arc during the Permo-Triassic Sonoman orogeny is recorded by the stratigraphic and structural evolution of the Golconda allochthon.

Cache Creek and related terranes form an extensive belt (Fig. 1) which is spatially and, in places, paleotectonically associated with McCloud arc terranes (Davis et al., 1978; Miller and Wright, 1987). Cache Creek affinity terranes are comprised of disrupted assemblages of oceanic rocks. Within the Canadian Cordillera, the Cache Creek terrane has been subdivided into numerous subterranes. In southern British Columbia, it is comprised of three belts based on the work of Monger and McMillan (Fig. 2): (1) The westerly belt contains Triassic chert, argillite, and sandstone that enclose Upper Triassic volcanic rocks; (2) The central belt is made up of the massive Marble Canyon Formation, an extensive Permian limestone of Tethyan faunal affinity; (3) The eastern belt is made up of Upper Permian basalt, ultramafic rocks, and melange with Upper Permian to Upper Triassic matrix. Volcanic rocks within the melange resemble the Upper Triassic Nicola volcanics within Quesnellia to the east. Another conspicuous element of the Cache Creek terrane is Upper Triassic blueschist.

Within the western United States, similar disrupted oceanic assemblages are referred to as Cache Creek affinity terranes (Davis et al., 1978; Saleeby, 1983) and are associated with McCloud belt island arc terranes (Fig. 3). The Cache Creek affinity assemblages contain upper Paleozoic to lower Mesozoic chert-argillite matrix melange, disrupted mafic and ultramafic igneous rocks, Permian limestone blocks of Tethyan faunal affinity, and Upper Triassic

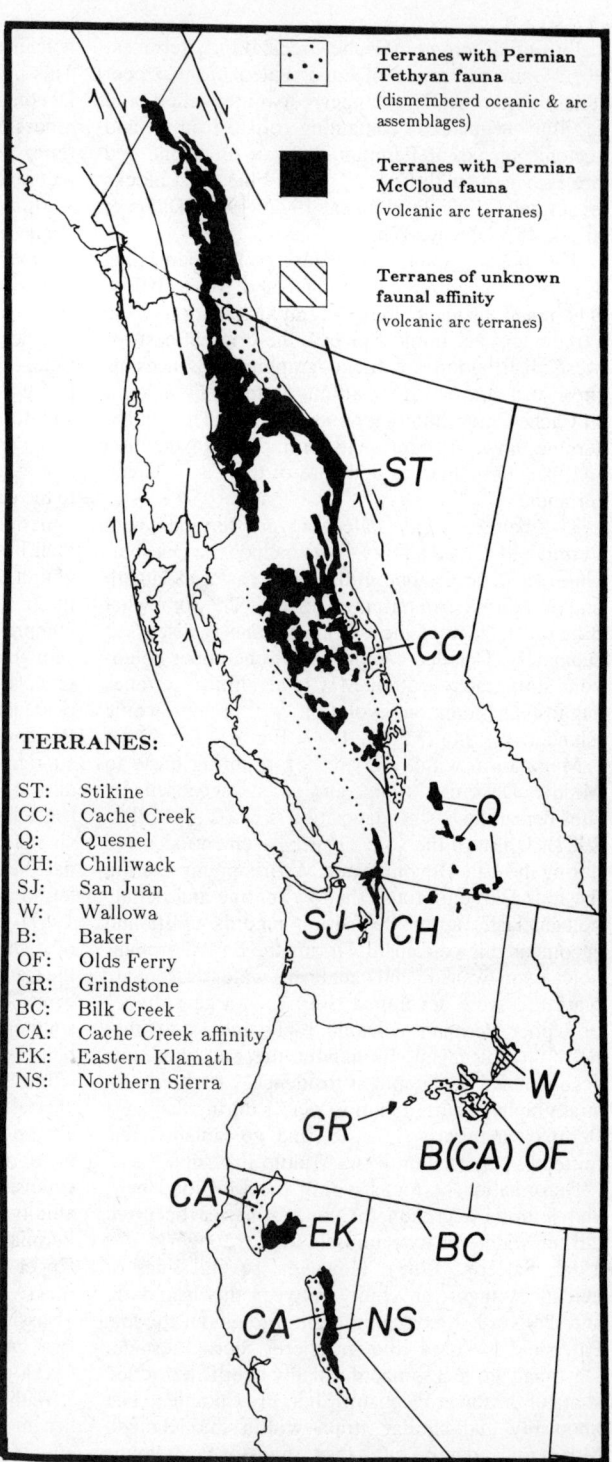

FIGURE 3. Distribution of terranes containing Permian Tethyan affinity and McCloud affinity fauna. (From Miller, 1987)

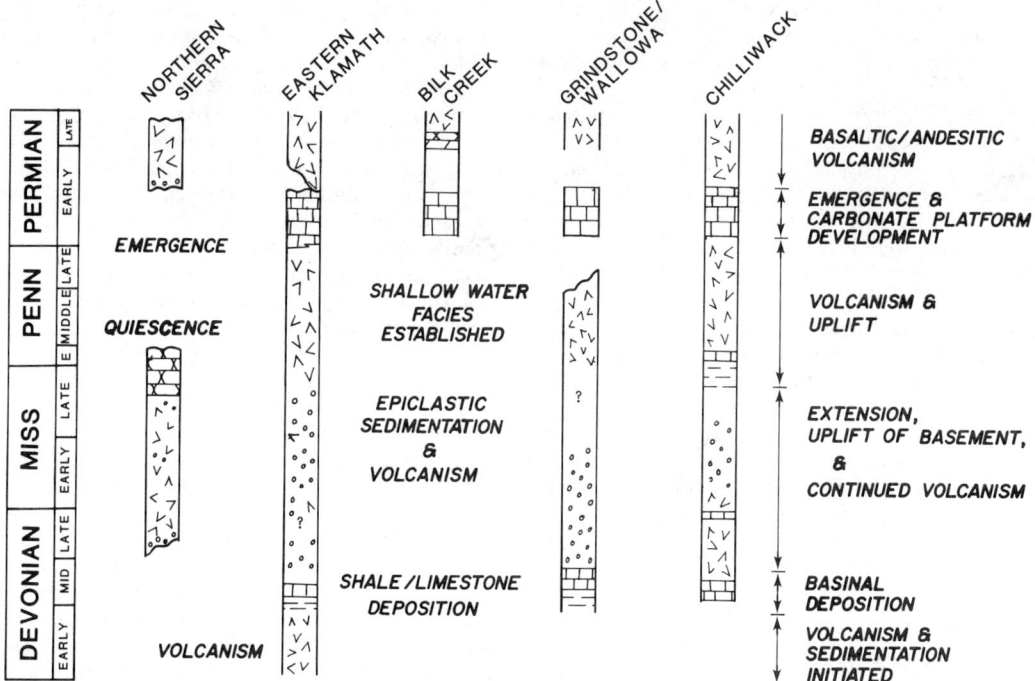

FIGURE 4. Schematic summary of the petrotectonic evolution of McCloud belt terranes within the western United States (From Miller, 1987)

blueschist as blocks and as coherent terranes (Fig. 2; Saleeby, 1983). The Tethyan limestones are thought to have been deposited as shallow water reefs and carbonate platforms that fringed seamounts or blanketed oceanic plateaus. Shallow water limestone blocks generally occur within a matrix of younger, basinal chert and argillite, suggesting that they originated as slide blocks, which were incorporated into deeper water sediments as olistoliths.

The wide variety of petrotectonic elements within Cache Creek and related terranes indicates that these assemblages have polygenetic origins, unrelated to their final assembly. The juxtaposition of shallow water oceanic rocks, pelagic sediments, mafic-ultramafic substrate, subduction zone metamorphic rocks, and fragments of arc volcanics, is inferred to have taken place within a subduction-related accretionary assemblage (Davis et al., 1978). Thus, although Cache Creek affinity terranes contain elements that are clearly exotic to North America, several lines of evidence suggest paleogeographic and paleotectonic ties between the accretionary complex and the McCloud belt island arc (Fig. 1), thus, the actual amalgamation of the McCloud arc/Cache Creek accretionary complex occurred at a single evolving convergent plate margin during and prior to Late Triassic time. These relations further support a general pattern of east-dipping subduction polarity for Permo-Triassic time, providing an independent constraint on the paleogeographic and paleotectonic setting of the North American plate margin. This further implies that rocks within the southerly reaches of the Slide Mountain terrane and within the Golconda allochthon were deposited within a late Paleozoic marginal basin. The original width of this basin is not constrained and was likely on the order of hundreds of kilometers. It may have widened substantially to the north where continental influence is less evident; alternatively, that may be an artifact of the sparser data base.

The youngest rocks within the Cache Creek belt are Upper Triassic, and Jura-Trias volcanic arc and basinal successions were constructed across and deformed with the accretionary belt (Monger, 1984; Saleeby, 1983). In places the younger rocks were deformed together with Cache Creek affinity terranes during Middle Jurassic or younger deformation. Evolution of the Cache Creek accretionary complex must have persisted until Late Triassic time, but no maximum age constraint exists for its inception. It may well have been active during much of the late Paleozoic.

Outboard of the Cache Creek terrane in British Columbia and the southern Yukon territories lies the Stikine terrane (Figs. 1, 2). Like other terranes of Permian McCloud biogeographic affinity, it contains an upper Paleozoic volcanic arc succession, however, this terrane is areally dominated by Triassic volcanic and younger clastic rocks. Stikine experienced a very different Mesozoic history than other terranes within the McCloud belt. The details of Stikine separation from the McCloud arc and its northward travel to an anomalous position outboard of the Tethyan affinity Cache Creek terrane are unclear. The Upper Triassic Takla arc volcanics that overlie the McCloud-type arc section of Stikine (Fig. 2) have been considered to form an overlap assemblage, equivalent to both the Nicola arc rocks overlying Quesnellia and to Upper Triassic volcanic blocks that were incorporated into the Cache Creek terrane (Monger, 1977). This requires that Stikine was adjacent to the Cache Creek terrane in the final stages of its Triassic deformation, posing fundamental problems concerning the paleogeographic and tectonic setting of the plate margin(s) at which these terranes were juxtaposed. If Stikine and Quesnellia converged across oppositely dipping subduction zones, then McCloud affinity Permian rocks lay west of the Tethyan seamount chain during Triassic time. Alternatively, if a single plate margin is envisaged, then the accretionary complex occupies an unusual axial position with respect to the two belts of arc volcanics (Saleeby, 1983). The actual thrust boundary between the Stikine terrane and adjacent Cache Creek rocks is Middle Jurassic in age (Monger et al., 1982), allowing for alternative interpretations.

If the Upper Triassic Takla and Nicola volcanics are related to geographically separated subduction zones and do not form an overlap assemblage, this problem is somewhat alleviated. The Takla perhaps lay along tectonic strike from Nicola arc rocks or within another arc system during the Late Triassic. In this scenario, Stikine's final emplacement outboard of the Cache Creek terrane occurred between the latest Triassic and Middle Jurassic time. Significant oblique convergence (transpression) or translation along the edge of the North American plate is required to account for Stikine's outboard position with respect to the Tethyan (originally westerly) rocks of the Cache Creek terrane. This is consistent with the paleomagnetic data of Monger and Irving which suggest northward displacement of 1300 km for the Stikine terrane with respect to North America since Late Triassic time. Subsequent revision of the North American polar wander path by Gordon, Cox, and O'Hare has reduced this poleward transport to 600 km, and the relative transport was accomplished by Middle Jurassic time. (See *Paleomagnetism and Plate Tectonics.*)

Insular Super-Terrane. To the west, and separated from the Stikinian super-terrane by the intervening Coast Plutonic Complex, lies the Insular, or Wrangellian super-terrane (Figs. 1, 5). The Insular super-terrane extends from Vancouver Island to southern Alaska and contains three principal basement components: Wrangellia, the Alexander terrane, and the Peninsular terrane (Fig. 5). These are overlapped by Jura-Cretaceous strata and crosscut by Cretaceous plutons.

The Alexander terrane contains a long-ranging succession of volcanic arc, stable oceanic, and rift rocks. The history of the Alexander terrane has been divided into three phases (Gehrels and Saleeby, 1987). The early Paleozoic history is characterized by arc volcanism that was punctuated by two compressional deformations, of Cambro-Ordovician and Late Silurian-Early Devonian age. The Devonian and late Paleozoic were characterized by relative tectonic stability, recorded by deposition of marine clastics and carbonates. Triassic volcanic and sedimentary successions are interpreted to represent a rift environment.

Lower Paleozoic rocks within the Alexander terrane are paleogeographically unrelated to coeval rocks within other western North American terranes. They may have been derived from as far away as the Gondwanan paleo-Pacific margin, based on lithologic and structural similarities to the Lachlan trans-Antarctic fold belt (Gehrels and Saleeby, 1987).

Wrangellia is another principal component of the Insular super-terrane. The oldest basement rocks consist of a fragmentary upper Paleozoic volcanic arc sequence. A distinctive section of 8 to 10 km of subaerial and shallow marine tholeiitic basalts and capping Upper Triassic carbonates, which have been interpreted as an oceanic plateau, was constructed across this basement. The section contains increasingly deeper water facies upwards, recording submergence, rifting, and open ocean drifting (see review in Saleeby, 1983).

The Upper Triassic Wrangellian carbonates contain biogeographically distinct fauna, which suggest a southern hemisphere, eastern Pacific origin according to C. R. Newton. The paleomagnetic data of Hillhouse also imply significant poleward transport with respect to North America. Wrangellia was amalgamated to the Alexander terrane by Pennsylvanian time (Fig. 5). Thus, the Alexander terrane represents, at least in part, the basement across which the Triassic oceanic plateau was constructed.

The third Insular basement terrane, the Peninsular terrane, consists of an early Mesozoic (Triassic and Jurassic) andesitic volcanic arc succession (Fig. 5). Sparsely preserved fragments of the ocean basin that lay between the Insular super-terrane and easterly Mesozoic terranes of the North American Cordillera may be represented by the Bridge River terrane in British Columbia and by the Talkeetna flysch basin. All three of the terranes are overlapped by Jura-Cretaceous marine strata, which record widespread arc volcanism. The polarity of the convergent margin that is implied by the presence of these arc volcanics is constrained by the amalgamation of parts of the Chugach terrane accretionary prism (Fig. 1) to the Insular super-terrane during Jurassic and Cretaceous time, based on the work of Plafker and others. These relations suggest that a second, east-dipping subduction zone was developed beneath the encroaching super-terrane prior to its final accretion. Subsequent

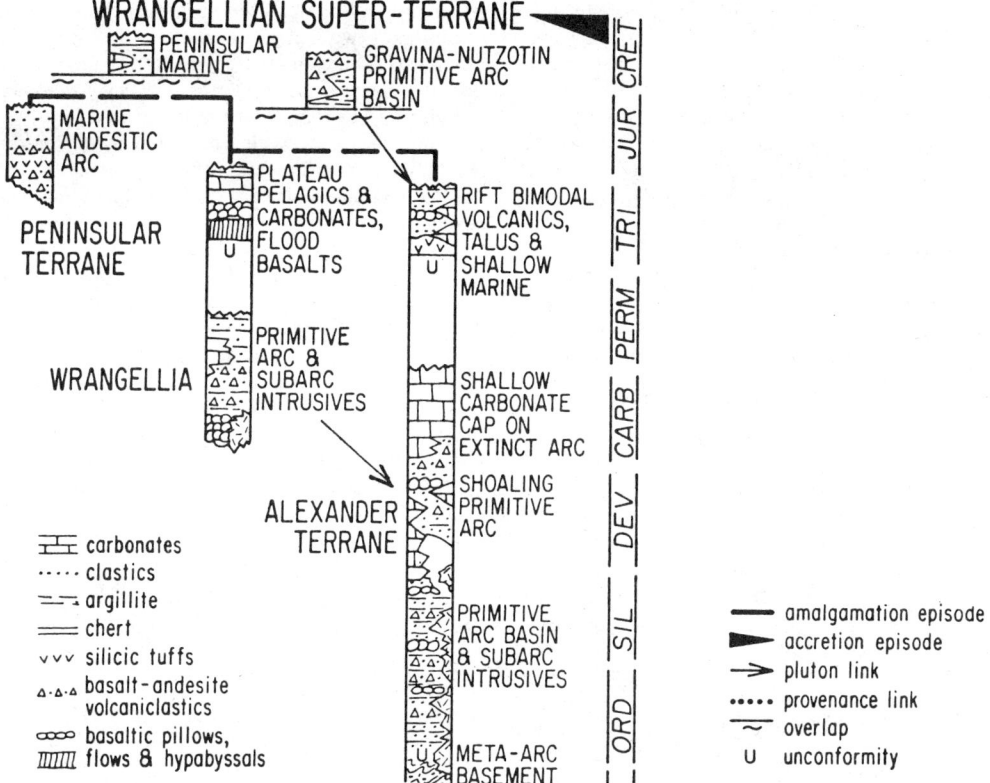

FIGURE 5. Insular super-terrane—petrotectonic, amalgamation, and accretion histories of the Alexander, Wrangel, and Peninsular terranes. (Modified from Saleeby, 1983; with new data from MacKevett et al., 1986)

to accretion of the Insular super-terrane, east-dipping subduction persisted along a single plate margin, and younger (Cretaceous and Tertiary) components of the Chugach terrane and Pacific Rim Complex of Vancouver Island were constructed oceanward of the Paleozoic to mid-Mesozoic accretionary orogen.

Together, the Alexander, Wrangellia, and Peninsular terranes constitute a super-terrane that extends over 2000 km along tectonic strike. The three basement terranes form a large tract of the Cordillera, which apparently contains no evidence for a paleogeographic relation to North America until Cretaceous time, when the super-terrane was accreted by the processes of convergence and large-scale dextral translation along the continental margin (Saleeby, 1983). A west-vergent thrust belt, which was recognized by Rubin and Saleeby, was active prior to 95 Ma, and characterizes the inner boundary of the Insular super-terrane. In most places, however, the boundary is obscured by plutonism and metamorphism associated with the Coast Plutonic Complex or major Cenozoic dextral strike slip faults.

A simple contrast of the volume of terranes represented within the Canadian and Alaskan Cordillera, with those of the southwestern Cordillera (Fig. 1), emphasizes the importance of northward megadrift in modifying internal distribution and overall width of the accretionary orogen. This reflects largely oblique Mesozoic convergence of the paleo-Pacific plates that interacted with western North America, as well as subsequent offset by strike-slip faulting.

Summary

Phanerozoic accretionary tectonism within the North American Cordillera began during mid-Paleozoic time with the accretion of the Roberts Mountains allochthon in Nevada and deformation and emplacement of the Kootenay arc in southern British Columbia. The accretion of these two terranes occurred within a largely compressional regime, however, coeval extension was widespread within western North America (Gordey et al., 1987), and both may have been related to a single transcurrent plate margin. The presence of craton-derived detritus and Proterozoic detrital zircon within basement complexes of associated volcanic arc rocks suggest that this deformation occurred above an eastward-dipping (oblique) subduction zone developed beneath the North American continental rise and slope, however, alternative hypotheses exist.

Broadly coeval with Antler-age deformation within

the Cordillera, volcanism initiated in the McCloud arc system and persisted throughout the late Paleozoic and locally into Middle Jurassic time. The arc was constructed on transitional and, along strike, oceanic crust, and shows evidence for a protracted geographic relation to western North America during the late Paleozoic. During the mid-Permian, the history of the Stikine terrane diverged from that of other McCloud belt terranes. Predominantly east-dipping subduction beneath the McCloud arc is implied by paleogeographic relations, however, short-lived arc polarity reversals may have occurred. To the east, upper Paleozoic successions within the Golconda allochthon and Slide Mountain terrane were deposited within a marginal basin eastward of the arc and have stratigraphic ties to both the continent and a volcanic arc source. During the Late Permian and Early Triassic, the basinal strata were deformed and emplaced over the continental margin in a back-arc setting, with the impingement of the McCloud arc against western North America.

Accumulation and deformation of the Cache Creek accretionary complex west of the McCloud belt persisted until the Late Triassic, and in places younger volcanic arc and oceanic sequences were deformed with Cache Creek rocks at a subsequent time. Triassic and Jurassic volcanic arc sequences were widespread both within the Intermontane super-terrane and within western continental North America. During the Middle Jurassic, the Stikine terrane was finally emplaced outboard of the Cache Creek terrane. This collision was broadly coeval with final accretion of the McCloud-type arc terranes to western North America. McCloud arc accretion occurred within the westward convergent margin of the North American plate by structural telescoping of the fringing subduction zone complex, volcanic arc, and marginal basin systems. By means of such crustal shortening these terranes acquired continental crustal thickness and character. Subsequent Mesozoic arc magmatism in western North America occurred within a convergent continental margin, welding the accretionary mass to the continent.

Accretion of the Insular super-terrane occurred between the late Jurassic and mid-Cretaceous, again along a convergent margin, with probable major dextral transform component. The Paleozoic and Mesozoic history recorded within the constituent Alexander terrane, Wrangellia, and the Peninsular terrane suggests a protracted history of arc volcanism punctuated by periods of tectonic stability when the terrane formed an oceanic plateau. The super-terrane shows no apparent paleogeographic relation to western North America prior to Cretaceous accretion, and represents a far-travelled, exotic crustal fragment. Northward dispersion of the Insular super-terrane occurred prior to development of the Cretaceous batholithic belt across the sutured crustal boundary.

In summary, the Mesozoic history of the North American Cordillera is dominated by the Mesozoic accretion of two major volcanic arc successions and previously amalgamated subduction-related accretionary complexes (Monger et al., 1982), although accretionary tectonism was established as early as the mid-Paleozoic with emplacement of basinal assemblages over the continental margin. Rocks of the Intermontane super-terrane, including the basinal eastern assemblage, the McCloud arc and the Cache Creek related terranes, were amalgamated by Late Triassic time with the possible exception of the Stikine terrane. Compressional deformation within a west facing convergent margin during Middle Jurassic time transformed the fringing arc system to continental crust that subsequently formed basement to the late Mesozoic magmatic arc. The second (Insular) super-terrane was accreted during the mid-Cretaceous, by collision of a truly exotic piece of crust with the North American continent and subsequent large-scale northward displacement.

M. MEGHAN MILLER
JASON B. SALEEBY

References

Burchfiel, B. C., and G. A. Davis, 1981, Triassic and Jurassic tectonic evolution of the Klamath Mountains-Sierra Nevada geologic terrane, in W. G. Ernst, ed., *The Geotectonic Development of California*, Rubey Volume No. 1. Englewood Cliffs, N.J.: Prentice-Hall, 50–70.

Coney, P. J., D. L. Jones, and J. W. H. Monger, 1980, Cordilleran suspect terranes, *Nature* **288**, 329–333.

Davis, G. A., J. W. H. Monger, and B. C. Burchfiel, 1978, Mesozoic construction of the Cordilleran 'collage,' central British Columbia to central California, in *Pacific Coast Paleogeographic Symposium 2: Mesozoic Paleogeography of the Western United Section*. Tulsa, Okla.: Society of Economic Paleontologists and Mineralogists, 1–32.

Gabrielse, H., 1985, Major dextral transcurrent displacements along the Northern Rocky Mountain Trench and related lineaments in north-central British Columbia, *Geol. Soc. America Bull.* **96**, 1–14.

Gehrels, G. E., and J. B. Saleeby, 1987, Geologic framework, tectonic evolution, and displacement history of the Alexander terrane, *Tectonics* **6**, 151–173.

Gordey, S. P., J. G. Abbott, D. J. Tempelman-Kluit, and H. Gabrielse, 1987, "Antler" clastics in the Canadian Cordillera, *Geology* **15**, 103–107.

Miller, M. M., 1987, Dispersed remnants of a northeast Pacific fringing arc: Upper Paleozoic terranes of Permian McCloud faunal affinity, western U.S., *Tectonics* **6**, 807–830.

Miller, M. M., and J. E. Wright, 1987, Implications of a new Permian Tethyan coral from the Klamath Mountains, California, *Geology* **15**, 266–269.

Miller, E. L., M. M. Miller, C. F. Stevens, J. E. Wright, and R. Madrid, in press, Late Paleozoic tectonic evolution of the western U.S. Cordillera, in *The Cordilleran Orogen: Conterminous United States*, B. C. Burchfiel, P. W. Lipman, and M. L. Zoback, eds.

Monger, J. W. H., 1977, Upper Paleozoic rocks of the western Canadian Cordillera and their bearing on Cordilleran evolution, *Canadian Jour. Earth Sci.* **14**, 1832–1859.
Monger, J. W. H., 1984, Cordilleran tectonics: a Canadian perspective, *Soc. Géol. France Bull.* **7**, 255–278.
Monger, J. W. H., and R. A. Price, 1979, Geodynamic evolution of the Canadian Cordillera—progress and problems, *Canadian Jour. Earth Sci.* **16**, 770–791.
Monger, J. W. H., and C. A. Ross, 1971, Distribution of fusulinaceans in the western Canadian Cordillera, *Canadian Jour. Earth Sci.* **8**, 259–278.
Monger, J. W. H., R. A. Price, and D. J. Tempelman-Kluit, 1982, Tectonic accretion and the origin of the two major metamorphic and plutonic welts in the Canadian Cordillera, *Geology* **11**, 427–429.
Ross, C. A., and J. R. P. Ross, 1983, Late Paleozoic accreted terranes of western North America, in C. H. Stevens, ed., *Pre-Jurassic Rocks in Western North American Suspect Terranes.* Los Angeles, Ca.: The Pacific Section, Society of Economic Paleontologists and Mineralogists, 7–22.
Saleeby, J. B., 1983, Accretionary tectonics of the North American Cordillera, *Ann. Rev. Earth and Planetary Sci.* **15**, 45–73.
Stevens, C. H., 1985, Reconstruction of Permian paleogeography based upon distribution of Tethyan faunal elements, in J. T. Dutro, Jr., and H. W. Pfefferkorn, eds., *Neuvieme Congres international de stratigraphie et de geologie du carbonifere, May 17-26, 1979, Compte Rendu,* vol. 5: *Paleontology, Paleoecology, Paleogeography,* Carbondale Il.: Southern Illinois University Press, 383–393.

Cross-references: *Far-Travelled Terranes; Continental Collision Zones: Seismotectonics and Crustal Structure.*

ADJUSTMENT METHODS

In experimental sciences we deal with three kinds of variables: observable quantities, physical or geometrical constants, and unknown parameters. Observable quantities, or *observables*, $\mathbf{l} \equiv \{l_1, l_2, \ldots, l_n\}$, as the name suggests, are quantities that are observable, i.e., capable of being observed, such as temperature or length. Once they are observed, meaning that real numbers get associated with individual l_i, they become known as *observations*, or measurements, and $\mathbf{l} \in \mathbb{R}^n$ where $\mathbb{R}^n$ is n-dimensional real space. *Constants* are those values that are either known theoretically (e.g., number of degrees in the sum of three internal angles of a triangle) or have been adopted by convention (number of millimeters in one meter) or have been determined from an earlier, usually more accurate, experiment (speed of light in a vacuum). Unknown *parameters* $\mathbf{x} \equiv \{x_1, x_2, \ldots, x_u\}$ are those quantities for which we wish to determine from experiment the real values $\mathbf{x} \in \mathbb{R}^u$ (e.g., coordinates of a point, Lamé's coefficients of the material at a given location).

These three kinds of variables are related through a postulated *mathematical model*, which the constants are normally taken as being part of. When we are lucky, the mathematical model can be formulated in such a way that the unknown parameters are expressed explicitly as functions of observables:

$$\mathbf{x} = \mathbf{g}(\mathbf{l})$$

If this is the case, then all we have to do is to evaluate the parameters from observations using the postulated mathematical model. The situation becomes more complicated when the mathematical model can only be formulated such that the observables are functions of parameters:

$$\mathbf{l} = \mathbf{f}(\mathbf{x}) \qquad (1)$$

or, even worse, when both the observables and parameters are wrapped together in a system of implicit functions:

$$\mathbf{F}(\mathbf{x}, \mathbf{l}) = \mathbf{0} \qquad (2)$$

The latter is the most general formulation and, as such, is dealt with extensively in the adjustment literature.

The challenge is in extracting the parameters from the implicit (Eq. 2) or the explicit-in-l (Eq. 1) models when the values of observables have been measured. If the implicit case can be readily transformed into the explicit-in-l case, then it may be advisable to do so. We shall show later what can be done if the decision is to leave the formulation implicit. Investigating now the different situations we may encounter in the explicit-in-l case, we discover that the solution $\mathbf{x}$ may be

1. *Unique*, when just enough observations have been made for the desired number of parameters and the model has been properly formulated
2. *Undetermined*, when there are not enough observations available, or the model is improperly formulated so that even though seemingly enough observations have been collected, the inversion of Eq. 1 needed for $\mathbf{x}$ determination is defective
3. *Overdetermined*, when more than enough observations have been obtained so that there is a redundancy in the formulation

Clearly, case 1 may be challenging from the inversion point of view, but once that is solved we get one and only one solution $\mathbf{x}$. Case 2 gives infinitely many solutions; additional information, criteria, or conditions have to be supplied to select from these the one solution deemed appropriate. Case 3, when looked at from the purely mathematical point of view, does not generally have any solution. It constitutes the proper *problem of adjustment,* and we focus on this case in this article.

The problem of adjustment is encountered in geodesy as well as in other experimental sciences much more often than any of the other two problems.

It arises from a desire to make sure that the observations have been carried out properly and that the parameters can be evaluated to the required accuracy. As we shall see, the least squares approach to the solution of an adjustment problem allows a statistical judgment on the two points.

Let us begin by assuming an overdetermined case with mathematical model explicit in $\mathbf{l}$ and dim $\mathbf{l} = n <$ dim $\mathbf{x} = u$. This case may be depicted diagrammatically as in Fig. 1, where by $\mathbb{X}$ we denote the *parameter space*, and by $\mathbb{L}$ the *observation space*. Both $\mathbb{L} \equiv \mathbb{R}^n$ and $\mathbb{X} \equiv \mathbb{R}^u$ will be considered real linear spaces (of finite dimensions); complex or other linear spaces may be used as well. To obtain any solution $\mathbf{x} \in \mathbb{X}$, we have to relax the original formulation by allowing the measured l_i to change. This relaxation is justified by acknowledging that the observations have, by necessity, only finite accuracy—i.e., that the measurements l_i depart from the correct values $\tilde{l}_i$ of the observables by residual errors (residual after having done everything to make them as small as possible), called, in short, *residuals* r_i. Since Eqs. 1 would be satisfied for the correct values $\tilde{l}_i$, we can rewrite them as the system

$$\mathbf{f}(\mathbf{x}) = \mathbf{l} + \mathbf{r} \qquad (3)$$

which is solvable. (Note that Eq. 1 should have been written with the correct $\tilde{\mathbf{l}}$ on the left-hand side.) These equations are known as *observation equations*.

Having relaxed the original overdetermined formulation, we have, however, created another obstacle: Since we are now to solve for u unknown parameters and n unknown residuals, the formulation becomes underdetermined and, as such, has infinitely many solutions. To choose one of these infinitely many solutions is a task of *optimization*. Many different optimization schemes exist (Kreyszig, 1978, Ch. 6; Davis, 1975, Ch. VII). The one normally used in adjustments is the *least squares optimization*. It seeks a solution $\mathbf{x}$ that minimizes the sum of weighted squares of residuals $\mathbf{r}$. By the way, now we can see where the name "adjustment" has come from: It really means the adjustment of observations to fit the mathematical model.

The least squares optimization is more readily understood when the modern concepts of functional analysis are employed. This is the approach we shall use here. Let us, to begin with, metricize the observation space $\mathbb{L}$ by the *weight matrix* $\mathbf{P}_l$ of observations. This matrix, which should take care of not only the weights of observations but also the correlations between them, is merely the inverse of the *covariance matrix* $\mathbf{C}_l$ of observations (Mikhail and Gracie, 1981, p. 140), a positive-definite, square (regular) matrix. It can be then shown that a distance

$$d(\mathbf{a}, \mathbf{b}) = \sqrt{(\mathbf{a} - \mathbf{b})^T \mathbf{P}_l (\mathbf{a} - \mathbf{b})}$$

between any two elements $\mathbf{a}, \mathbf{b}$ of $\mathbb{L}$ satisfies the axioms for a metric, and $\mathbb{L}$ becomes a metric space (Kreyszig, 1978, p. 3). (Loosely speaking, a metric space is a space populated by elements the distance between any two of which can be measured, hence the name.) If we let this metric induce a norm on $\mathbb{L}$ by the equation

$$\forall \mathbf{a} \in \mathbb{L}: \qquad \|\mathbf{a}\| = d(\mathbf{a}, \theta) = \sqrt{\mathbf{a}^T \mathbf{P}_l \mathbf{a}}$$

where θ is the null element of $\mathbb{L}$, and where it can be shown that $\|\cdot\|$ satisfies the axioms for a norm, then $\mathbb{L}$ becomes a normed space (Kreyszig, 1978, p. 59). This norm is known as the *quadratic norm*. The structure is then easily completed by defining an inner product as

$$\forall \mathbf{a}, \mathbf{b} \in \mathbb{L}: \qquad \langle \mathbf{a}, \mathbf{b} \rangle = \mathbf{a}^T \mathbf{P}_l \mathbf{b}$$

and $\mathbb{L}$ becomes an inner product space (Kreyszig, 1978, p. 128). Since $\mathbb{L}$ has finite dimension n, $\mathbb{L}$ is complete (Kreyszig, 1978, p. 33) and, therefore, a Hilbert space.

Now, the required minimization of the sum of weighted squared residuals can also be understood as the minimization of the norm of residuals under the P_x metric: Because $\mathbf{r} \in \mathbb{L}$, we have

$$\min_{x \in \mathbb{X}} (\text{sum of weighted squared residuals})$$
$$= \min_{x \in \mathbb{X}} \|\mathbf{r}\| = \min_{x \in \mathbb{X}} \mathbf{r}^T \mathbf{P}_l \mathbf{r} \qquad (4)$$

The reason why least squares optimization is

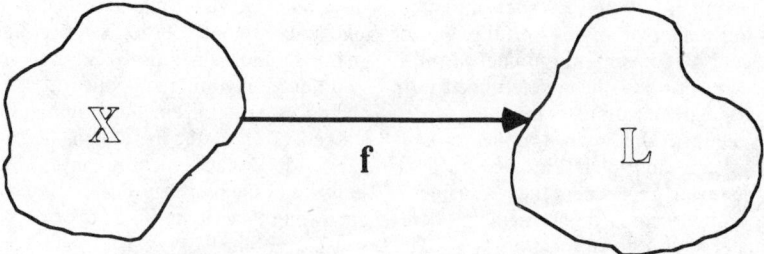

FIGURE 1. Mathematical model $\mathbf{f}$ explicit in $\mathbf{l}$.

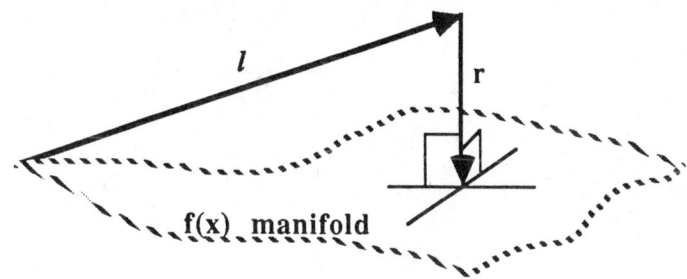

FIGURE 2. The projection theorem.

usually preferred to other kinds of optimization (e.g., uniform, Chebyshev) can now be explained. By metricizing $\mathbb{L}$ by $\mathbf{P}_l$, $\mathbb{L}$ is also converted into a probability space (with probability measure $\mathbf{P}_l$) (Wilks, 1962, p. 11) populated by n-dimensional cross-correlated random variables $\mathbf{l}$, $\mathbf{r}$, etc. Thus all kinds of statistically meaningful questions about the goodness of observations, blunders, goodness of estimated parameters, etc., can be asked and answered in $\mathbb{L}$. Various pertinent statistical tests that deal with $\mathbf{l}$ or $\mathbf{r}$ vectors can be performed with easy statistical interpretation. The interested reader should refer to Vaníček and Krakiwsky (1986, Ch. 13).

How can the *least squares estimate* $\hat{\mathbf{x}}$ that minimizes the norm of residuals be estimated (computed)? If $\mathbf{f}(\mathbf{x})$ is a u-dimensional manifold in $\mathbb{L}$ (see Fig. 2), the application of Eq. 4 leads to the well-known *projection theorem* (Kreyszig, 1978, p. 146), which ascertains that it is the head of the "$\mathbf{r}$ arrow," normal to the manifold $\mathbf{f}(\mathbf{x})$ that gives the least squares estimate $\hat{\mathbf{x}} \in \mathbb{X}$ that figures in the $\mathbf{f}(\hat{\mathbf{x}})$. Algebraically, the solution $\hat{\mathbf{x}}$ is normally sought in the linear mode.

Let us assume that $\mathbf{f}$ is a nonlinear function of $\mathbf{x}$. Then

$$\mathbf{l} \doteq \mathbf{f}(\mathbf{x}^{(0)}) + \left.\frac{\partial \mathbf{f}}{\partial \mathbf{x}}\right|_{\mathbf{x}=\mathbf{x}^{(0)}} (\mathbf{x} - \mathbf{x}^{(0)}) \quad (5)$$

is the linear approximation of Eq. 1 for the *initial estimate* $\mathbf{x}^{(0)}$ of $\mathbf{x}$. The Jacobian matrix $\partial \mathbf{f}/\partial \mathbf{x}|_{\mathbf{x}=\mathbf{x}^{(0)}} = \mathbf{A}$, called a *design matrix* in adjustment calculus, is the matrix of a linear operator from $\mathbb{X}$ to $\mathbb{L}$. Denoting the *correction* $\mathbf{x} - \mathbf{x}^{(0)}$ *to the initial estimate* by $\delta \in \mathbb{X}$ and the *misclosure* $\mathbf{l} - \mathbf{f}(\mathbf{x}^{(0)})$ by $\mathbf{w} \in \mathbb{L}$, we can write the observation Eq. 3 in linearized form as

$$\mathbf{A}\delta = \mathbf{w} + \mathbf{r} \quad (6)$$

This equation is valid only in the differential neighborhood of $\mathbf{x}^{(0)}$, as shown in Fig. 3, and the operator $\mathbf{A}$ transforms it onto a u-dimensional manifold in $\mathbb{L}$. We note that if the model $\mathbf{f}$ is linear to begin with—i.e., if Eq. 1 is of the form

$$\mathbf{A}\mathbf{x} = \mathbf{l} \quad (7)$$

—then we can write Eq. 6 directly, denoting $\mathbf{l} - \mathbf{A}\mathbf{x}^{(0)}$ by $\mathbf{w}$. Once the observation equations are linearized, or linear, we seek the least squares solution δ for which

$$\hat{\mathbf{x}} = \mathbf{x}^{(0)} + \hat{\delta} \quad (8)$$

The least squares solution $\hat{\delta}$ can be derived by seeking the minimum of $\|\mathbf{r}\|$ through differentiation with respect to $\mathbf{x}$. Alternatively, we can use a more elegant approach employing a commutative diagram. To do so, let us first define a Hilbert space $\mathbb{L}^*$ dual

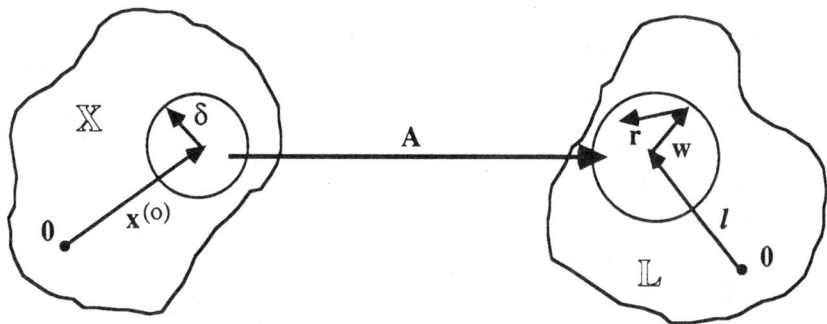

FIGURE 3. Linearized version of a model explicit-in-l.

to $\mathbb{L}$ by the following prescription (Aubin, 1979, p. 124):

$$P_l \mathbf{1} = \mathbf{1}^* \in \mathbb{L}^*$$

Let us further assume that $\mathbb{X}$ is metricized, similar to $\mathbb{L}$, by a weight matrix $\mathbf{P}_x = \mathbf{C}_x^{-1}$, as yet unknown. We can then upgrade $\mathbb{X}$ to a Hilbert space by defining the quadratic norm and inner product analogous to that for $\mathbb{L}$. A dual Hilbert space to $\mathbb{X}$, $\mathbb{X}^*$, is then defined as

$$\mathbb{X}^* \equiv \{\forall \mathbf{x}^* = \mathbf{P}_x \mathbf{x} \; (\forall \mathbf{x} \in \mathbb{X})\}$$

The four Hilbert spaces defined above can be chained together into a commutative diagram as in Fig. 4 (Vaníček, 1979). That the transformations from $\mathbb{X}^*$ to $\mathbb{X}$ and from $\mathbb{L}^*$ to $\mathbb{L}$ are mediated by $\mathbf{P}_x^{-1}$ and $\mathbf{P}_l^{-1}$, respectively, is the consequence of $\mathbb{X}^*$, $\mathbb{L}^*$ being dual to $\mathbb{X}$, $\mathbb{L}$ (Aubin, 1979, p. 52). The transformation from $\mathbb{L}^*$ to $\mathbb{X}^*$ is through $\mathbf{A}^T$, a well-known fact from functional analysis (Aubin, 1979, p. 57).

Now we are ready to use the commutative diagram to derive the formula for the least squares solution $\hat{\mathbf{x}}$. Going from $\mathbb{L}$ to $\mathbb{X}$ (counterclockwise and clockwise), we get

$$\forall \mathbf{w} \in \mathbb{L}: \quad \boldsymbol{\delta} = \mathbf{Q}\mathbf{w} \in \mathbb{X}$$

where

$$\mathbf{Q} = \mathbf{P}_x^{-1} \mathbf{A}^T \mathbf{P}_l$$

But going from $\mathbb{X}$ to $\mathbb{X}^*$ yields

$$\mathbf{P}_x = \mathbf{A}^T \mathbf{P}_l \mathbf{A}$$

so that, requiring that $\mathbf{A}^T \mathbf{P}_l \mathbf{A}$ be regular, we get

$$\hat{\boldsymbol{\delta}} = (\mathbf{A}^T \mathbf{P}_l \mathbf{A})^{-1} \mathbf{A}^T \mathbf{P}_l \mathbf{w} \qquad (9)$$

The operator $\mathbf{Q}$ is known as the *orthogonal left inverse of* $\mathbf{A}$ (Aubin, 1979, p. 87), and the solution $\hat{\boldsymbol{\delta}}$ is the least squares solution identical to the standard solution obtained by means of *normal equations* (Mikhail and Gracie, 1981, p. 244). In fact, $\mathbf{P}_x$ is the matrix of normal equations besides being the weight matrix of the unknown parameters and the desired metric of $\mathbb{X}$. Equation 8 then gives the final result $\hat{\mathbf{x}}$.

As a by-product of getting the least squares solution, we have also obtained the covariance matrix of $\hat{\mathbf{x}}$, which is equal to

$$\mathbf{C}_{\hat{\mathbf{x}}} = (\mathbf{A}^T \mathbf{P}_l \mathbf{A})^{-1}$$

Formally, it is possible to use the commutative diagram in Fig. 4 to derive the *law of propagation of errors* (Mikhail, 1976, p. 77) for the linear model in Eq. 7. Assuming $\mathbf{x}$, $\mathbf{C}_x$ known and $\mathbf{l}$, $\mathbf{C}_l$ unknown, we obtain, going from $\mathbb{L}^*$ to $\mathbb{L}$, the correct expression

$$\mathbf{C}_l = \mathbf{A}\mathbf{C}_x \mathbf{A}^T \qquad (10)$$

It is easy to see that once the least squares solution $\hat{\boldsymbol{\delta}}$ is found, the least squares estimate $\hat{\mathbf{r}}$ of $\mathbf{r}$ is obtained from Eq. 6 as

$$\hat{\mathbf{r}} = \mathbf{A}\hat{\boldsymbol{\delta}} - \mathbf{w}$$

The adjusted observations are then

$$\hat{\mathbf{l}} = \mathbf{l} + \hat{\mathbf{r}} \qquad (11)$$

and their covariance matrix $\mathbf{C}_{\hat{\mathbf{l}}}$ is given by Eq. 10, with $\mathbf{C}_x$ replaced by $\mathbf{C}_{\hat{\mathbf{x}}}$. It is of interest sometimes

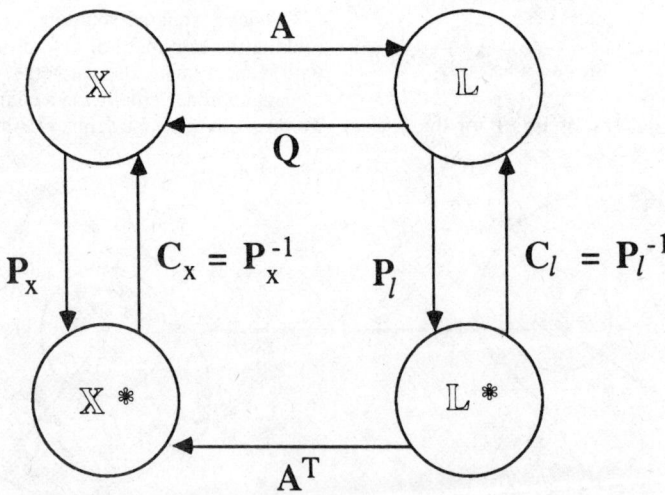

FIGURE 4. Commutative diagram for a linear (linearized) model explicit-in-l.

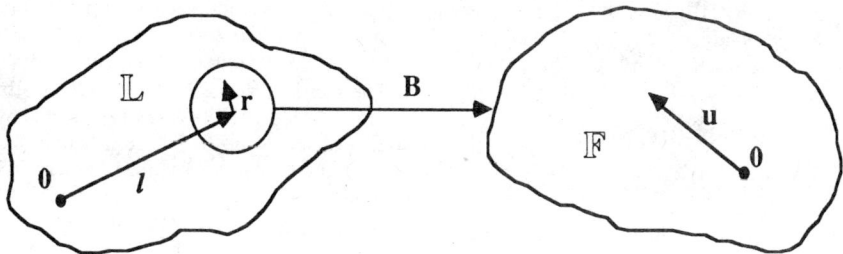

FIGURE 5. Linearized version of a condition model.

to evaluate the covariance matrix of estimated residuals $\hat{\mathbf{r}}$ given by

$$\mathbf{C}_{\hat{r}} = \mathbf{C}_l - \mathbf{A}\mathbf{C}_{\hat{x}}\mathbf{A}^T$$

If the original model is nonlinear and linearization is used to estimate the solution, it is important to make sure that $\mathbf{x}^{(0)}$ is sufficiently close to $\hat{\mathbf{x}}$; in other words, $\hat{\boldsymbol{\delta}}$ must be "sufficiently small" in terms of its norm. For this reason, the solution is usually iterated by replacing $\mathbf{x}^{(0)}$ by $\hat{\mathbf{x}}$, $\mathbf{l}$ by $\hat{\mathbf{l}}$, and $\mathbf{C}_l$ by $\mathbf{C}_{\hat{l}}$ in all the pertinent expressions and $\hat{\boldsymbol{\delta}}^{(1)}$, the new solution, is obtained. When the norms of two successive solutions $\hat{\boldsymbol{\delta}}^{(i)}$, $\hat{\boldsymbol{\delta}}^{(i+1)}$ are sufficiently close to zero, the iterations are stopped.

In adjustment calculus, the above-described problem (Eq. 1, its linearized version, Eq. 6, and its solution, Eq. 9) is known as the *parametric adjustment*. Under certain circumstances, we may wish to formulate the problem as a *condition adjustment*, where the model represents merely the conditions that all the observations $\mathbf{l}$ have to meet. Such a model can be written generally as

$$\mathbf{F}(\mathbf{l}) = \mathbf{0}$$

clearly a special case of Eq. 2. Its linearized and reformulated version reads

$$\mathbf{u} = -\left(\frac{\partial \mathbf{F}}{\partial \mathbf{l}}\right)\mathbf{r} = -\mathbf{B}\mathbf{r} \qquad (12)$$

where $\mathbf{u} = \mathbf{F}(\mathbf{l})$ is the misclosure of the actual observations $\mathbf{l}$, and $\mathbf{B}$ is the Jacobian matrix of the (linear) operator from $\mathbb{L}$ to $\mathbb{F}$ (Fig. 5). Here $\mathbb{F}$ is the Hilbert space of all functional values $\mathbf{F}(\mathbf{l})$, and $\mathbf{u} \in \mathbb{F}$.

An application of the same steps as those in the parametric adjustment leads to a commutative diagram displayed in Fig. 6. The following equations can be derived from the diagram:

$$\hat{\mathbf{r}} = -\mathbf{T}\mathbf{u} = -\mathbf{C}_l\mathbf{B}^T\mathbf{P}_F\mathbf{u}$$

where

$$\mathbf{P}_F = \mathbf{C}_F^{-1} = (\mathbf{B}\mathbf{C}_l\mathbf{B}^T)^{-1}$$

and $\mathbf{B}\mathbf{C}_l\mathbf{B}^T$ is required to be regular. The operator $\mathbf{T} = \mathbf{C}_l\mathbf{B}^T(\mathbf{B}\mathbf{C}_l\mathbf{B}^T)^{-1}$ is known as the *orthogonal right*

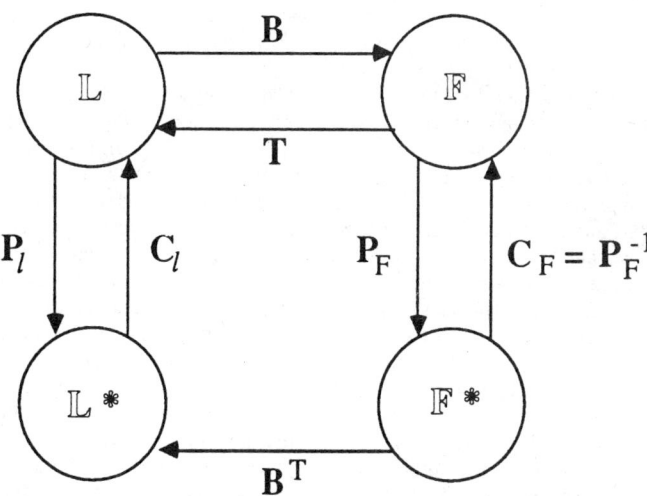

FIGURE 6. Commutative diagram for a linear (linearized) condition model.

inverse of **B** (Aubin, 1979, p. 89). The adjusted observations $\hat{l}$ are obtained from Eq. 11, and the two pertinent covariance matrices are (Mikhail, 1976, p. 217)

$$C_{\hat{r}} = C_l B^T P_F B C_l$$

$$C_{\hat{l}} = C_l - C_{\hat{r}}$$

The third fundamental method of adjustment used in practice is the *combined adjustment*. It uses the most general mathematical model, the implicit model given by Eq. 2. Its linearized form is

$$A\delta + Br = u$$

where **A** is given immediately following Eq. 5 and **B** by Eq. 12. Its commutative diagram (not shown here) uses all three spaces ($\mathbb{L}$, $\mathbb{X}$, and $\mathbb{F}$) and their duals. The final equations are (Vaníček and Krakiwsky, 1986, Ch. 12)

$$\hat{x} = x^{(0)} + \hat{\delta} = x^{(0)} + (A^T P_F A)^{-1} A^T P_F u$$

$$\hat{l} = l + \hat{r} = l + C_l B^T P_F (A\hat{\delta} - u)$$

$$C_{\hat{x}} = (A^T P_F A)^{-1}$$

$$C_{\hat{l}} = C_l[I - B^T P_F(I - AC_{\hat{x}} A^T P_F)BC_l]$$

It is always possible, even though not necessarily desirable, to reduce the combined case to the parametric case by defining

$$\tilde{r} = -Br$$

We also note that the condition adjustment can be viewed as a special case of the combined adjustment, such that $A = 0$.

In practice, many variations on the three fundamental methods exist. It would not be appropriate, however, to show them here, and the interested reader may consult Mikhail (1976), Meissl (1982), and Vaníček and Krakiwsky (1986). Two main variants should nevertheless be mentioned here. The first arises from the desire to process either the observations or the parameters in installments, thus updating the least squares estimates $\hat{x}$, $\hat{l}$ successively. The former approach has become known as *sequential adjustment*, and the latter is called *phase adjustment*. These two approaches lead directly to expressions, for the same least squares estimates, different from the ones derived here. These expressions can be interpreted as *filters* of different kinds (Vaníček and Krakiwsky, 1986, Ch. 14), which are used in dynamical and other applications.

The other main variation arises from the desire to hold some of the parameters δ_i as close to zero as dictated by statistical confidence in the initial values $x_i^{(0)}$. This desire leads to an interpretation of $x^{(0)}$ as a statistical quantity with its own covariance matrix $C_x^{(0)}$. The various approaches based on this idea are known as *weighted constraint methods* (Mikhail, 1976, p. 217); in practice they come in many guises.

The metricization of $\mathbb{X}$ by $P_x = C_x^{-1}$ allows us to regard $\mathbb{X}$ as a probability space populated by random multivariates **x** and **δ**. Hence a variety of different statistical tests can be performed on either $\hat{x}$ or $\hat{\delta}$. In practice, many of these tests, in both $\mathbb{X}$ and $\mathbb{L}$, are routinely performed to make sure that, for example, all the observations behave in a proper statistical fashion and fit the chosen mathematical model, that selected P_l is properly scaled, or that standardized estimated residuals make a sample from normally distributed population.

PETR VANÍČEK

References

Aubin, J.-P., 1979, *Applied Functional Analysis*. New York: John Wiley & Sons, Inc.

Davis, P. J., 1975, *Interpolation and Approximation*. New York: Dover.

Kreyszig, E., 1978, *Introductory Functional Analysis with Applications*. New York: John Wiley & Sons, Inc.

Meissl, P., 1982, *Least-Squares Adjustment: A Modern Approach*. Graz, Austria: Mitteilungen der geodätischen Institute der Technischen Universität Graz, Folge 43.

Mikhail, E. M., 1976, *Observations and Least Squares*. New York: IEP—A Dun-Donnelley Publisher.

Mikhail, E. M., and G. Gracie, 1981, *Analysis and Adjustment of Survey Measurements*. New York: Van Nostrand Reinhold Co.

Vaníček, P., 1979, Tensor structure and the least squares, *Bull. Géod.* **53,** 221-225.

Vaníček, P. and E. J. Krakiwsky, 1986, *Geodesy: The Concepts*. 2nd ed. Amsterdam: North Holland.

Wilks, S. S., 1962, *Mathematical Statistics*. New York: John Wiley & Sons, Inc.

Cross-references: *Geodetic Leveling; Geodetic Networks and Control Surveys; Geodetic Ground Positioning.*

AEROMAGNETIC SURVEYING

The purpose of aeromagnetic surveys is to map the small variations of the Earth's magnetic field that are produced by the differing magnetizations of the underlying igneous rocks essentially in a horizontal plane above the surface. The magnetic properties of most igneous rocks are caused by the presence of titanomagnetite alone, but in some cases the presence of the magnetic iron sulphide, pyrrhotite, may also contribute. Sedimentary rocks usually have negligible magnetic properties because any magnetite present when they are laid down will be oxidized to the nonmagnetic form, hematite. Thus if the average magnetite content of one igneous rock formation is slightly different from the magnetite content of an adjacent rock formation, then the two rock formations will have slightly different magnetic properties and this will be reflected by a variation in the magnetic field above them, which can be measured by a magnetometer. In the case of Precambrian Shields where the igneous rocks are close to the surface but are covered by a relatively thin veneer of overburden a few meters thick, aeromagnetic surveys are carried out to delineate the underlying geology (Boyd, 1970).

The magnetic field produced by the igneous rocks of the Earth's crust is about 2000 nT in amplitude. This is superimposed on the main Earth's field that derives from the Earth's core and the combination varies from about 25,000 nT to 70,000 nT at the Earth's surface. In addition, a small time-varying field caused by polarized charges emanates from the sun and impinges on the Earth's atmosphere. Over the course of a day, this diurnal variation usually amounts to less than 100 nT in amplitude except in the so-called auroral zone, which extends in a circle around the geomagnetic pole in northern Greenland passing through southern Greenland, Labrador Sea, Hudson Bay, Great Slave Lake, and across northern Alaska.

Figure 1 shows the various elements comprising a basic aeromagnetic survey system. Several types of magnetometer sensors are available but all measure the total intensity of the Earth's magnetic field. It is necessary to nullify the effects of the magnetic fields of the aircraft itself on the magnetometer sensor by a compensation system.

The sensor output is processed by the electronic unit in a suitable fashion to produce a value that in state-of-the-art equipment is normally a digital number for recording on digital magnetic tape and a voltage that activates the pen on an analog chart recorder. A control unit is required to trigger the electronic unit to obtain a given parameter reading at set intervals of time regulated by a clock or less frequently at constant distance intervals along the flight line using a positioning system such as a Doppler navigation system. In compiling a geophysical map it is necessary that the geographic coordinates of the survey parameter values be known, using some form of positioning system. Usually the terrain clearance of the survey aircraft is also recorded using a radar altimeter to assist the subsequent interpretation of the magnetic data.

Gradient measurements may be made by utilizing two magnetometers and recording the difference between them (Hood, 1981). The gradient of the magnetic field is then obtained by dividing the resultant difference value by the sensor separation. Figure 2 shows an aeromagnetic survey aircraft in which magnetometers are mounted in twin booms separated by 2 m to measure the vertical gradient of the total intensity of the Earth's magnetic field, although single sensor systems (enclosed in a single boom) are much more common.

Airborne Magnetometers

Three varieties of airborne magnetometers are utilized in aeromagnetic surveying, namely (1) fluxgate, (2) proton precession—free and spin, and (3) optical absorption. All measure the total field

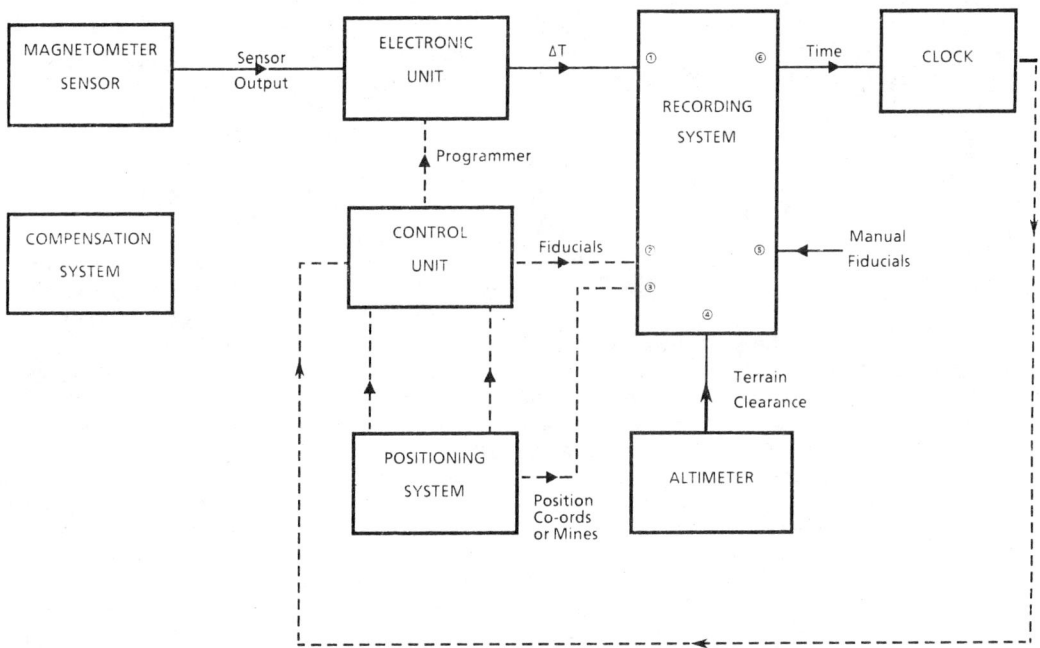

FIGURE 1. Block diagram of the basic aeromagnetic survey system.

FIGURE 2. Beechcraft B80 Queenair aircraft of the Geological Survey of Canada equipped with twin boom system for aeromagnetic gradiometer surveys.

value of the Earth's magnetic field. The theory of operation of the foregoing magnetometers are described in *Geomagnetic Field: Measurement*.

Airborne Positioning Systems

Positioning systems used for airborne geophysical surveys can be divided into two main categories: (1) those that are self-contained within the aircraft (these include tracking cameras, Doppler, and inertial navigation) and (2) those that require external references. These methods are all electronic positioning systems that normally require fixed radio transmitters on the ground. Decca, Loran C, and Shoran navigation systems have been most used up to the present time; Raydist and Toran have seen limited use; and for reconnaissance surveys Loran A, Omega, and VLF navigation systems have been employed. For the aeromagnetic surveys of extensive bodies of water, the use of an electronic positioning system is mandatory. The satellite-based Global Positioning System has seen limited use to date in aeromagnetic surveys but once the full constellation of 18 satellites is launched it will be possible to carry out surveys using the system in any part of the world.

The most widely used method of recovering the aircraft track up to the present time has been by the use of vertically mounted 35 mm cameras, either continuous strip or the more usual single frame variety, or by the use of video cameras. Doppler navigation systems (see *Doppler Positioning: Satellite*) are quite commonly used in airborne geophysical survey systems primarily to assist the pilot to accurately fly straight parallel flight lines but also to assist the subsequent flight path recovery process. The cumulative errors on most Doppler systems are usually between 1% and 2% of distance traversed. Inertial navigation systems that employ accelerometers whose output is doubly integrated with respect to time are also extensively used to assist in recovering the aircraft but like Doppler systems, the errors are cumulative.

Aeromagnetic Survey Operations

Aircraft Installation. Starting in 1960 there was a gradual switch from the use of larger aircraft of World War II vintage to light twin-engine airplanes such as the Aero Commander, Beechcraft Queenair, and Piper Aztec in aeromagnetic surveys because of their low operational cost. The magnetometer head is usually mounted in a boom extending from the tail of the aircraft (see Fig. 2), although some survey companies still offer towed bird systems to avoid magnetic compensation difficulties.

Inboard installations have been made possible by the improvement in the magnetic compensation of aircraft. The aircraft itself produces three different sources of interference. The first is the permanent magnetism of the various components made of steel, such as the engines, whose direction remains fixed with respect to the aircraft. The second interference source is the induced field caused by the magnetic susceptibility of these same components and the Earth's field. Its polarity and magnitude depend on the orientation of the aircraft with respect to the Earth's magnetic field. The third source of interference is that caused by the magnetic effect of eddy currents generated in the skin and other conducting parts of the aircraft by their motion in the Earth's magnetic field.

In passive compensation systems, the permanent magnetism is eliminated by the use of a set of three-orthogonal compensating coils mounted near the magnetometer head and on the roll axis of the aircraft through which the appropriate DC currents may be passed. The induced components are eliminated by the use of permalloy strips, and the eddy currents are compensated for by the use of coils of wire mounted in close proximity to the sensitive element. It is necessary to carry out a set of rolls, pitches, and yaws of the aircraft in low gradient areas in order to separate the effects of the various components. The excellence of magnetic compensation of a given survey aircraft is measured by its figure of merit (FOM). This index is obtained by summing, without regard to sign, the amplitudes in nanoteslas of the 12 magnetic signatures recorded when the aircraft carries out 20° rolls, 10° pitches, and 10° yaws peak-to-peak on north, east, south, and west headings.

In recent years, active 9-term aircraft magnetic compensation systems have been manufactured by Canadian Aviation Electronics Industries Ltd. of Montreal. The use of these active compensation systems improves the FOM of a given aircraft considerably and drastically reduces the time required for aircraft compensation.

Aeromagnetic Survey Specifications. Airborne magnetometer surveys in mining exploration are usually carried out at a mean terrain clearance of 150 m and a line spacing of 400 m or less. In Canada, where over 9,000,000 line km of aeromagnetic survey have been flown by government agencies since 1947, a line spacing of 800 m and a survey altitude of 300 m has been used to obtain regional aeromagnetic coverage. The choice of survey altitude and line spacing are interrelated because, in contouring an aeromagnetic map, features have to be followed across the flight lines. Lower flight elevations will produce more detailed profiles, i.e., greater resolution of the anomalies, and it is therefore necessary to reduce the distance between the flight lines in order to maintain the accuracy of the contoured data. For optimum results, the flight line spacing should be about twice the distance of the aircraft above the magnetic basement. For adequate sampling along the flight line, the sample interval should be about one-quarter of the vertical distance from the survey aircraft to the magnetic basement, i.e., 75 m for aeromagnetic surveys flown at 300 m over Precambrian Shield terrane.

The flight line direction should normally be oriented to cross the geological strike of the basement rocks approximately at right angles. However at low magnetic latitudes (inclination less than 20°) because the magnetic anomalies are produced by susceptibility changes across east-west striking rock contacts (and not by susceptibility changes across north-south striking rock contacts), it is mandatory to orient the flight lines approximately in a north-south direction regardless of the geological strike.

The contour interval used for aeromagnetic maps is normally between 5 and 10 times the basic sensitivity of the airborne magnetometer that acquired the data. Aeromagnetic surveys carried out in the two decades following World War II were mostly flown with one-nanotesla sensitivity magnetometers and the contour interval of the resultant maps was commonly 10 nT. Such surveys are now referred to as standard sensitivity aeromagnetic surveys, whereas those that use more recently developed airborne magnetometers having a sensitivity of 0.1 nT or better are designated high sensitivity (or high resolution) aeromagnetic surveys.

Ground Stations to Monitor the Diurnal Variation of the Earth's Magnetic Field. Magnetic diurnal stations are usually set up to monitor the changes in the Earth's field during the survey operations, and the resultant records are used to decide on any reflights necessary because of magnetic disturbances. The ground monitor is often located at the base of survey operations because of logistical convenience. In recent years digital recording of the diurnal variation (and time) has been used to permit in certain instances (e.g., where the distance between the ground station and survey area is small) direct subtraction of the diurnal variation from the airborne record (Reford, 1980). Accurate digital clocks must be used in both the survey aircraft and the ground station if such a procedure is employed.

Calibration of Aeromagnetic Survey Instrumentation

The calibration of aeromagnetic survey systems should preferably be carried out immediately before and at the close of survey operations. The first step is to compensate the survey aircraft as described earlier to obtain the lowest FOM possible. A check should then be made that the background noise of the geophysical system has also been reduced to an acceptable level. The background noise is controlled to a considerable extent by the compensation FOM of the survey aircraft and the turbulence conditions encountered by the survey aircraft. The background noise of the aeromagnetic survey system is most effectively monitored by plotting one-sixteenth of the fourth difference of the measured field values (Hood et al., 1979) on the analog recorder as the survey lines are flown.

It is also necessary to establish that the measured total field values recorded by an aeromagnetic survey system correspond accurately to those actually existing at the magnetometer sensor. In Canada, for government aeromagnetic surveys utilizing proton precession or optical absorption magnetometers, an aeromagnetic calibration range has been set up in a low gradient area at a crossroad near Bourget, Ontario, which is approximately 45 km east of Ottawa and easily recognizable from the air. The values at the ground level at the crossroad have been tied to the Blackburn Magnetic Observatory using calibrated proton precession magnetometers. The values at 150 m and 300 m elevation above the crossroad have been measured by flying a survey aircraft at various heights across the crossroad to ascertain the vertical change, which is about 12 nT in 300 m positively upward. Thus the total field values at the two levels above the Bourget crossroad have been tied to within a few nanoteslas to the continuously recording magnetometer at the Blackburn Magnetic Observatory. Because the diurnal variation at the Bourget crossroad can be expected to follow closely the diurnal variation at Blackburn, the value at Bourget can be calculated at any instant of time by subtracting a constant difference value from the Blackburn Observatory value.

Aeromagnetic survey aircraft are normally flown along the four cardinal headings across the Bourget crossroad with their flight path cameras operating and the field values for each cardinal heading are ascertained at the crossroad. The difference value is subtracted from the Blackburn reading at the exact time that the survey aircraft crossed the Bourget crossroad to get the true reading. The heading errors for the survey aircraft are also calculated as part of the same calibration procedure.

In general the calibration errors should not exceed 10 nT and the heading errors should be within 5 nT in an acceptable aeromagnetic survey system. For fluxgate magnetometers, which measure the total field above an arbitrary datum, Helmholtz coils can be utilized to check the sensitivity of the aeromagnetic survey system in a similar way that ground systems are calibrated.

Compilation and Interpretation of Aeromagnetic Survey Data

Compilation of Aeromagnetic Survey Data. Five end products are available at the completion of an aeromagnetic survey operation: digital magnetic tapes containing the total field, terrain clearance, time, and in some cases position coordinates; positioning film; the aircraft operators log; analog monitoring records of the recorded digital data; and the diurnal ground station data.

Figure 3 summarizes the various processes involved in compiling aeromagnetic data into contour map form. The compilation process proceeds in two parallel activities. The flight paths of the survey aircraft are first plotted on air photographs or good air photo mosaics by ascertaining points on the flight path film such as roads, railways, rivers, lake shorelines or even clearings every 3 km or so that the aircraft has passed over. The position of these fiducial points are transferred onto large scale (e.g., 1:50,000) topographic maps on stable base material and the points are joined to provide a reasonably accurate flight path map. The flight path is then digitized for use in the subsequent digital compilation process. It is usual for a speed check to be made at this stage to identify along-track errors in the flight path recovery process. The speed check consists essentially of computing the average speed of the aircraft between the picked fiducial points along the profile. Since the average speed along a given flight line (for a fixed-wing aircraft) remains constant within a few percent, a wrongly positioned fiducial point will result in adjacent high and low calculated speeds with respect to the average.

The second parallel activity consists of editing the total field data and removing spikes, etc., which can be done automatically utilizing the fourth difference technique. Then the two sets of data are merged so that the total field data on the digital tape has its corresponding geographic coordinates added.

The next step is the levelling process, which uses the data from a set of control lines flown at right angles to the main series of traverse lines and thus provide a series of individual data loops. In the levelling process, the errors in the data going around the individual loops must be minimized and since a considerable amount of subjective decisions are often made, it is usually best to utilize a combination of digital and manual iteration techniques in levelling the data.

The data is gridded using a suitable interpolation process. A common size of grid is 2.5 mm at the publication scale. In general, the gridding method involves fitting smooth, continuous interpolation functions along parallel lines normal to the flight line direction. Each function has values equal to the magnetic measurements on the flight lines at the points where the two sets of lines intersect. The interpolation function lines are spaced apart at an interval equal to the ultimately desired fine grid interval and the functions are evaluated along these lines at points separated by the same interval, thus producing the fine grid directly. By this method the contours when traced pass within 0.01 cm of their true flight line intercept position and all fine detail is preserved.

The resultant contoured data is then inspected for noticeable aberrations since the compilation process itself is a rather stringent check on the quality of the data used in making the contour map.

Typical compilation scales for the compilation of standard sensitivity aeromagnetic maps are 1:50,000 but it is sensible to publish at the same scale as the national geological map series and/or the national topographic map series. The resultant printed aeromagnetic maps should also have the flight line information shown, preferably in a color different from the aeromagnetic contours so that users can assess how well a given anomaly has been defined. It is also usual to have the planimetry and drainage from the concomitant topographic maps appearing as a subdued background on the aeromagnetic maps so that the anomalies can be located with respect to geographic features. The topographic contours can also be included but a subdued presentation format is used; otherwise such geophysical maps become too cluttered.

It is also usual to produce aeromagnetic composite maps at a smaller scale of one-quarter to one-fifth that of the main end product prepared. Thus to accompany 1:50,000 aeromagnetic maps, 1:250,000

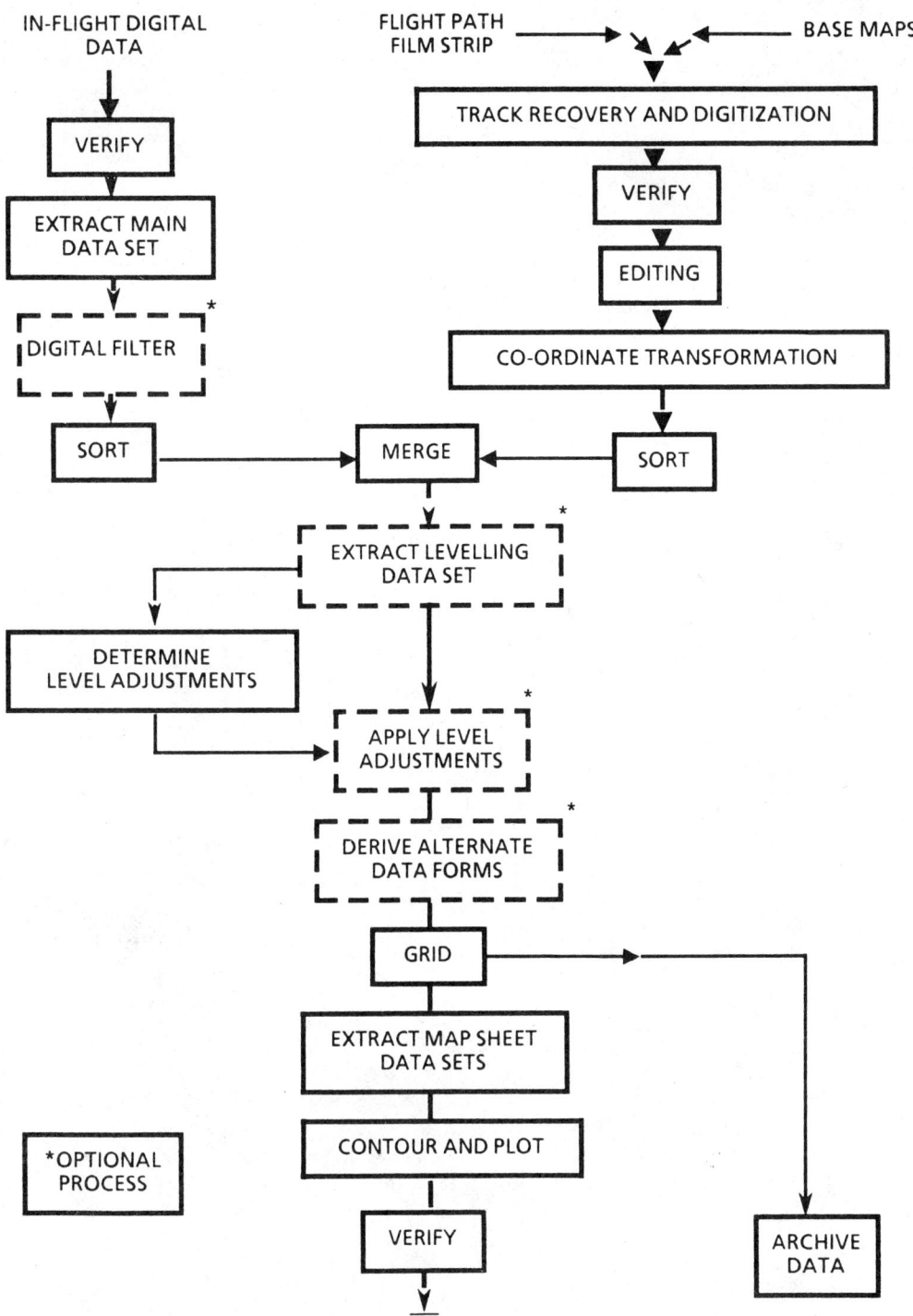

FIGURE 3. Flow chart for aeromagnetic data reduction. (From Hood et al., 1979)

maps are often prepared by photographic reduction of the 1:50,000 maps. The flight lines are usually omitted from these composite maps because their resultant closer spacing usually distracts from the ready identification of regional geological trends.

Regional Magnetic Anomaly Maps. The regional magnetic anomaly patterns of a given area of interest may be brought out by compiling the data at an even smaller scale. A typical scale is 1:1,000,000, which is usually suitable for the national magnetic anomaly map for many countries in the world. It is normally desirable in compiling regional magnetic anomaly maps to first remove the dominating effect of the magnetic field produced by the Earth's core, which is done by removing the International Geomagnetic Reference Field (IGRF) for the appropriate year of the aeromagnetic survey (Regan and Cain, 1975; Hood et al., 1979). Now that it is standard practice to record aeromagnetic data in digital form, the recently developed color plotters can be used to produce gradational color tones on the resultant pixel maps with intervals ranging from 25 nT to 100 nT. A typical color contour scheme utilizes essentially the spectrum of white light. Color separations for each of the primary colors may be prepared and these are combined at

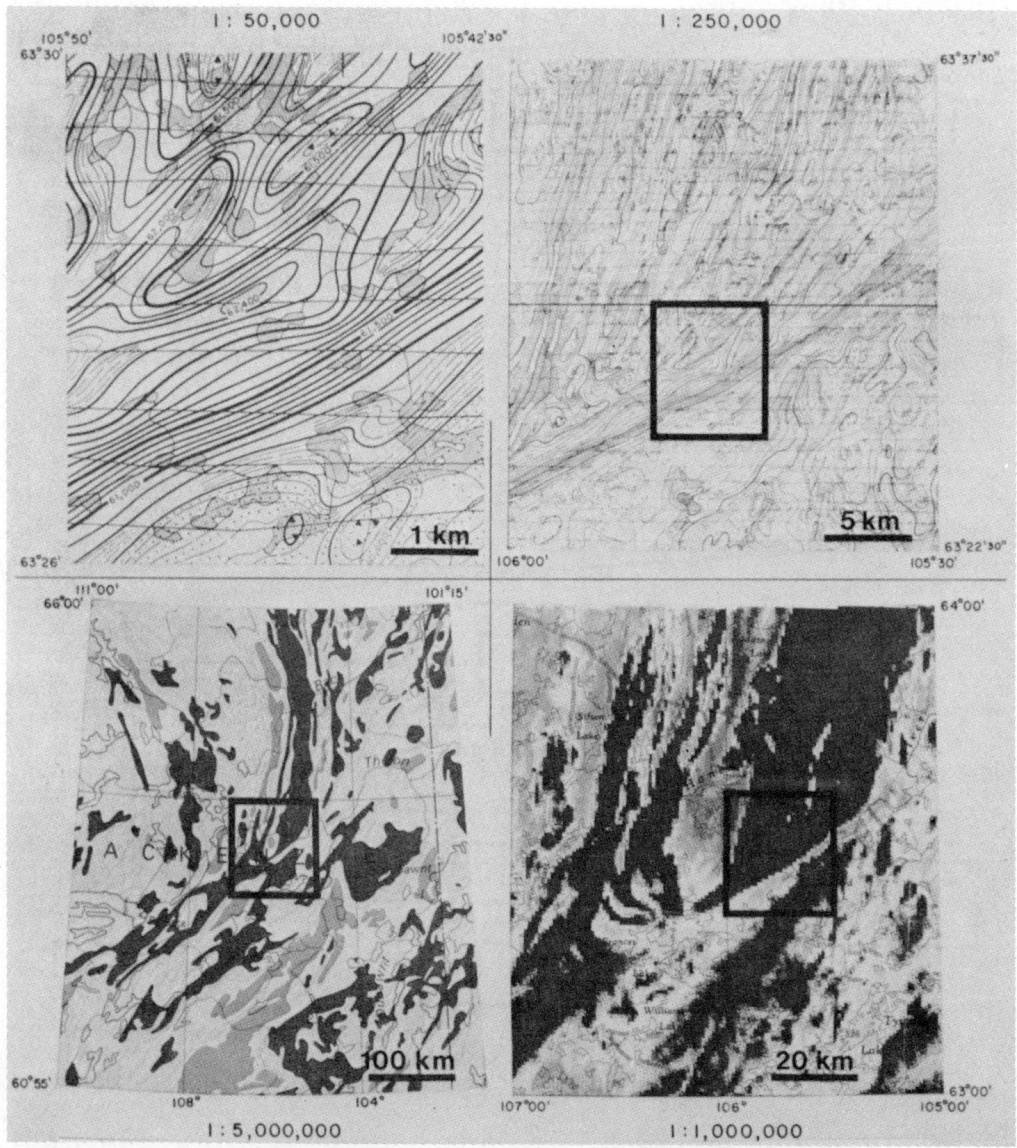

FIGURE 4. Example of the various scales used in Canadian government aeromagnetic maps. The square on each map is the same area for the preceeding large scale map. The 1:1M and 1:5M maps are colored for the actual published map.

the printing stage to produce the final end product. Figure 4 presents an example of the various scales used for aeromagnetic maps published by the Canadian government.

Interpretation of Aeromagnetic Survey Data. It is axiomatic that the ultimate value of a given aeromagnetic survey lies in the geological information that can be derived from an interpretation of the resultant data. First the aeromagnetic data may be treated in some fashion to improve the resolving power of the technique (Grant, 1972; Paterson and Reeves, 1985). Such treatment will result in some form of filtered map being produced, for instance, a second vertical derivative or downward continuation map, which emphasizes the magnetic effects of the near-surface geology and removes the gradient effects of the main Earth's field. Then the interpretation proper of the aeromagnetic data is carried out. This important step may be divided into two phases. The first phase is a *qualitative* interpretation in which areas underlain by a common rock type are delineated and structural features such as faults are recognized. Then a *quantitative* interpretation is carried out in which the geometrical shape and position of each causative body producing an individual anomaly is calculated together with its intensity of magnetization; the latter being, of course, the diagnostic physical parameter for the magnetic survey technique. The value obtained for the intensity of magnetization will be a guide to the lithology of the causative body. However, as a control on the quantitative interpretation, and to set permissible limits for the upper and lower values acceptable for the intensity of magnetization, it is a recommended procedure to obtain representative remanent magnetization and susceptibility measurements of the main formations in the survey area. The results emanating from the qualitative and quantitative interpretation should, of course, be compatible. At the synthesis phase, all types of geoscience information including air photo and Landsat interpretation should be combined with the aeromagnetic interpretation results and the available geological information in order to deduce the most probable geological edifice for the area. However additional information can always be gleaned from the aeromagnetic data and the subsequent acquisition of other types of geoscience information may indicate what additional facets of information can be derived from the aeromagnetic data.

Conclusions

Many countries have carried out a systematic aeromagnetic survey program to obtain coverage of both their landmass and offshore areas to assist mineral and petroleum exploration programs and the geological mapping programs of the national geological survey. Starting in 1946, the aeromagnetic survey technique was the first airborne geophysical survey technique to be introduced and after more than 40 years it is still being improved in many new and novel ways. It remains the most utilized airborne survey technique in terms of the line kilometrage flown each year throughout the world. This continued popularity is a result of a variety of reasons. Of all the airborne geophysical survey techniques, the aeromagnetic method has by far the greatest depth penetration being able to detect features down to the Curie point geotherm some 20 km or so beneath the Earth's surface. Moreover in contrast to other airborne techniques, the aeromagnetic method is unaffected by the presence of surficial material such as overburden and tropical weathering or by the presence of lakes and swamps. Aeromagnetic surveys also provide a continuity of information at low cost that is impossible to achieve in ground geophysical or geological surveys. One of the outstanding advantages of aeromagnetic surveys becomes apparent when large areas are surveyed because large regional geological features are often discovered. These may not be recognizable on the ground because they are so large or are perhaps obscured by sedimentary formations.

PETER HOOD

References

Boyd, D., 1970, The contribution of airborne magnetic surveys to geological mapping, *Canada Geol. Survey Econ. Geology Rept.* **26,** 213-227.

Grant, F. S., 1972, Review of data processing and interpretation methods in gravity and magnetics 1964-71, *Geophysics,* **37**(4), 647-661.

Hood, P. J., 1981, Aeromagnetic gradiometer: a superior geological mapping tool for mineral exploration programs, in *Squid Applications to Geophysics Proc.* Los Alamos: Society of Exploration Geophysicists, 72-77.

Hood, P. J., M. T. Holroyd, and P. H. McGrath, 1979, Magnetic methods applied to base metal exploration, *Canada Geol. Survey Econ. Geology Rept.* **31,** 77-104.

Paterson, N. R. and C. V. Reeves, 1985, Applications of gravity and magnetic surveys: the state of the art in 1985, *Geophysics,* **50**(12), 2558-2594.

Reford, M. S., 1980, Magnetic method, *Geophysics* **45**(11), 1640-1658.

Regan, R. D. and J. C. Cain, 1975, The use of geomagnetic field models in magnetic surveys, *Geophysics* **40**(4), 621-629.

Cross-references: *Geomagnetic Field: Measurement; Geomagnetic Measurement Techniques and Surveys.*

ARCHAEOMAGNETISM

Archaeomagnetism is the study of the magnetic properties of materials found within archaeological contexts (Aitken, 1974; Tarling, 1975, 1983a, 1985). It is thus similar to *paleomagnetism* (q.v.), but is concerned with archaeological rather than geological time scales, although these necessarily overlap. It provides records of the geomagnetic field parameters thus giving a direct measure of geomagnetic secular variations and forming a method of archaeological dating. Most archaeological materials contain iron oxide impurities that are capable of being magnetized and some artifacts can be manufactured entirely of magnetic materials, such as iron and steel, which means that most materials within an archaeological site are capable of carrying a magnetization. Thus their magnetic properties can be used in a wide variety of ways. The most well known archaeological application is for magnetic dating, but the magnetic properties can be used to assist in prospecting, reconstructions, provenancing, past temperature evaluation, detection of forgeries, etc.

At an atomic level, all materials are magnetic (diamagnetic or paramagnetic), but these magnetizations are weak compared with those associated with ferromagnetic materials such as iron oxides, iron, nickel, cobalt, etc., and only *ferromagnetism* will be considered here (see *Magnetic Properties of Minerals; Magnetic Domains*). When placed in a magnetic field (< 10mT), a magnetization is induced in such materials that is proportional to the strength of the field applied, the constant of proportionality being termed the *susceptibility* (x). However, such ferromagnetic materials are also capable of retaining a memory of the field after its removal, i.e., they can possess a remanent magnetization, which, if acquired naturally, is termed a *natural remanent magnetization* (q.v.). Such remanence will normally be acquired by archaeological materials in one of four main ways. (1) Thermoremanence is acquired when a material containing ferromagnetic particles is heated and then cooled within a magnetic field. As cooling takes place, the alignment of the electron spins by the field becomes blocked within the crystal lattice and, if the grain sizes are between about 0.5 and 1 μm, this alignment can persist for thousands or even billions of years (see *Thermoremanence*). (2) As minute crystallites of ferromagnetic grains increase in size within a magnetic field, the electron spin alignments by the external magnetic field similarly become locked indefinitely within the crystal structure as its size becomes of the order of 0.5 to 1.0 μm (see *Chemical Remanent Magnetization*). (3) As all ferromagnetic substances lie within a magnetic field, such as that of the Earth, their previous remanences can gradually decay, although the rate of decay is negligible for magnetic particles in the region of 0.5 to 1 μm. At the same time, the electron spin alignments of smaller or larger ferromagnetic particles gradually relax into alignment with the ambient field and can retain this *viscous* remanence for time scales of similar orders as those in which they were acquired (see *Viscous Remanent Magnetization (VRM) and Viscous Magnetization*). (4) Already magnetized particles can, if separately deposited in air or water, become aligned with the ambient geomagnetic field in the same way as compass needles become aligned. Sediments can thus acquire a *detrital remanence* (see *Detrital Remanent Magnetization*), although the particle alignment can be affected by water current activity or the mechanical processes of deposition. In many instances, the particles carrying the more stable magnetization (0.5–1.0 μm) in such sediments are sufficiently fine, that postdepositional realignment with the ambient field can occur. Such grains will then be capable of retaining their magnetization for archaeologically infinite lengths of time. In consolidated sediments, chemical changes in both the detrital minerals and cement can give rise to a dominant chemical remanence acquired during diagenesis (see *Paleomagnetism*).

Archaeomagnetic dating depends on the presence of a remanent magnetization in the archaeological materials, acquired by one of the four processes outlined above. The dating can be either absolute, if compared with known directions or intensities of the geomagnetic field, or relative, if two or more different sites or levels are being compared. In both cases, the dating refers to the time at which the magnetic remanence was acquired. In the case of fired materials, this is the time of firing, but viscous remanences are acquired over long, less definite periods and this form of remanence must generally be removed for dating purposes, although it has clear relevance in terms of distinguishing forgeries and the length of time that materials have lain in situ (see below). As viscous remanences generally have a lower stability than either thermal or chemical remanences, the presence of viscous remanence can usually be detected and removed by incremental thermal and alternating magnetic field demagnetization (see *Demagnetization*), thereby isolating a single component of remanence with a specific direction and intensity.

Relative dating depends on the fact that the intensity and direction of the geomagnetic field is gradually changing (see *Geomagnetic Secular Variation*) so that it is only occasionally that exactly the same direction occurs at the same location, and even rarer that both the direction and intensity are repeated identically. Thus archaeological materials magnetized at the same location at different times will normally have recorded different intensities and directions of the geomagnetic field at the times that they acquired their remanence. Comparison of either the directions of remanence or of the paleo-field intensity can therefore be used to determine whether the materials were magnetized at the same time. As the direction of the past geomagnetic field can often be determined within 1–2° (see *Rock Magnetism: Measuring Techniques and Apparatus*), and its intensity within some 5% (see Paleointensity of the geomagnetic field), and the present day geomagnetic field is uniform to this order over regions of at least 1,000 km^2 (see *Geomagnetic Field*), such relative dating can generally be made between different archaeological sites up to at least 100 km apart, and can be extended to distances of c. 1,000 km although with increasing uncertainty. A typical example of such dating is that of comparing both the directions and paleointensities of samples from Late Minoan IB destruction levels on Crete with each other and with the properties of samples of "Minoan" volcanic eruptions at Thira, some 120 km north of Crete (Downey and Tarling, 1984).

Absolute dating requires comparison between the directions and paleointensities recorded in archaeological materials with some known time scale for such changes. Unfortunately, the causes of geomagnetic secular variations (q.v.) are too imprecise to allow prediction of the intensity or direction of the field at any particular time or place. The time scale of such secular variations must therefore depend on either direct measurements at known times—such as geomagnetic observatory records—or by the construction of curves of declination, inclination, and paleointensity for given localities, as a function of time, using the remanence of independently dated archaeological materials. As observatory records are sparsely distributed and only extend back as far as 1600 A.D. in London and Paris, most archaeomagnetic dating depends on the construction of regional master curves (Fig. 1). Such curves are, of course, subject to errors in both the original magnetic determinations and in the previous method of their dating. They thus require constant updating as further data becomes available. The southwestern United States master curve (Dubois, 1975), for example, is largely based on ^{14}C dating and hence subject to corrections based on dendrochronological determinations, etc. The fact that the geomagnetic field appears to change moderately uniformly over regions of some 10^6 km^2, means that such master curves can be constructed from observations made on a subcontinental scale, the regional variation being corrected by assuming a geocentric dipole model for the regional field (Tarling, 1983a). Such dating has now been applied in many regions of the world—the United States, western Europe, the Middle East, Japan, etc.

Archaeomagnetic dating has a high potential accuracy, particularly using directional parameters. The normal rate of change for Britain during the last 2,000 years (Figure 1a) is 0.05°/year, so that a ±1° directional error would correspond to age error of some ±20 years. In contrast, records of the last 100 years suggest that the intensity of the field varies by some 1% during such an interval, so that a 1% error in intensity corresponds to ±100 years. Naturally, both faster and slower rates occur at different times, so that both higher and lower accuracy can be expected for different times. At the moment, the largest error in directional dating is that of the initial dating of the archaeomagnetic master curves. Where these have been established, the realistic errors are mostly of the order of ±25–50 years. However, the precision of most intensity determinations are still rather poor and cannot yet be considered to be accurate to less than ±100 years except in unusual circumstances. Intensity master curves are also dependent on the precision of the original archaeological dating method. However, one major advantage of the archaeomagnetic dating is that the

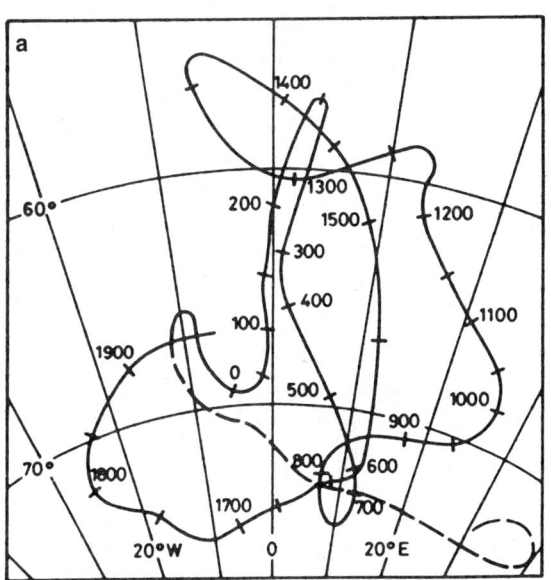
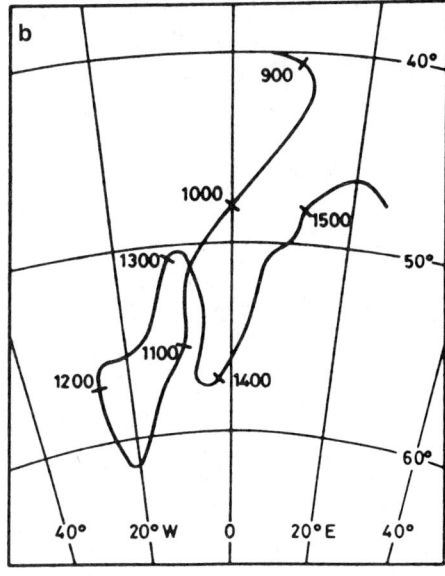

FIGURE 1. Archaeomagnetic master curves of directional change. Examples are shown for (a) Britain (after A. J. Clark, pers. comm. and Tarling, 1983a, 1983b, and (b) the SW United States. (With kind permission of R. L. Dubois, University of Oklahoma at Norman)

increase in the data base results in further improvement in the precision of the master curves and hence their reliability will improve. Ultimately, both methods are still limited by the affects of local geomagnetic deviations, in particular, large, strongly magnetic structures that cool slowly can locally distort the geomagnetic field, and hence studies of these structures may include errors greater than those involved in small, more rapidly cooled structures.

Magnetic prospecting mostly makes use of the distortion of the local geomagnetic field by the presence of magnetic materials that are inhomogeneously distributed either laterally or vertically within a site (see *Aeromagnetic Surveying*). Such inhomogeneous distributions are commonly caused by the presence of buried artifacts of different magnetic susceptibility to the surrounding soil—a contrast that will often be enhanced where the structures have been fired so that remanent magnetization also contributes to the magnetic field anomalies observed at the surface. Local firing, deliberate or accidental, enhances the susceptibility because of thermally induced chemical changes (Tite and Mullins, 1971; Mullins, 1977). Such minerals then become associated with the enhanced runoff following such firing, and therefore tend to become concentrated in natural or depressions of human origin, such as post-holes, ditches, plough lines, etc., thus increasing the possibility of detection of such features.

Reconstructions of artifacts or structures, e.g., pottery, kilns, crucibles, etc., can be assisted by comparison of the magnetic directions of individual fragments. When last fired, each structure and its contents will have acquired a magnetization in the uniform direction of the geomagnetic field at that time. Repositioning of the now dispersed fragments into such positions that their magnetic remanent directions are again in uniform alignment (Fig. 2) can be of particular significance when normal "jigsaw" fitting cannot be undertaken (Burnham and Tarling, 1975).

Provenancing can be undertaken in certain circumstances. The magnetic properties of obsidian flows, for example, vary according to the precise composition and grain sizes of microcrystallites within the glasses. They also acquire a remanence, as they cool after eruption, which will depend on the magnetic properties and also the strength of the ambient field. This means that artifacts manufactured from such materials can be attributed to specific source localities and hence trade routes distinguished and potentially quantified. Preliminary studies (McDougall et al., 1983) indicate that the intensity of natural remanence, the saturation magnetization (in 3.5 T) and the low field susceptibility can frequently "fingerprint" a specific source with accuracies comparable with that of vastly more expensive neutron activation methods.

Past temperatures can be determined if the heating of interest did not exceed the Curie point (q.v.) of the contained minerals, normally 575° or 675°C. On cooling after heating to lower temperatures, all reheated grains with a blocking temperature at or below the maximum heating temperature will have become aligned with the geomagnetic field direction at that time, while the magnetization associated with higher blocking temperature grains will be unaffected. Analysis of the different magnetic vectors (see *Demagnetization*) will thus distinguish the temperature at which the observed remanent direction changes from one to the other.

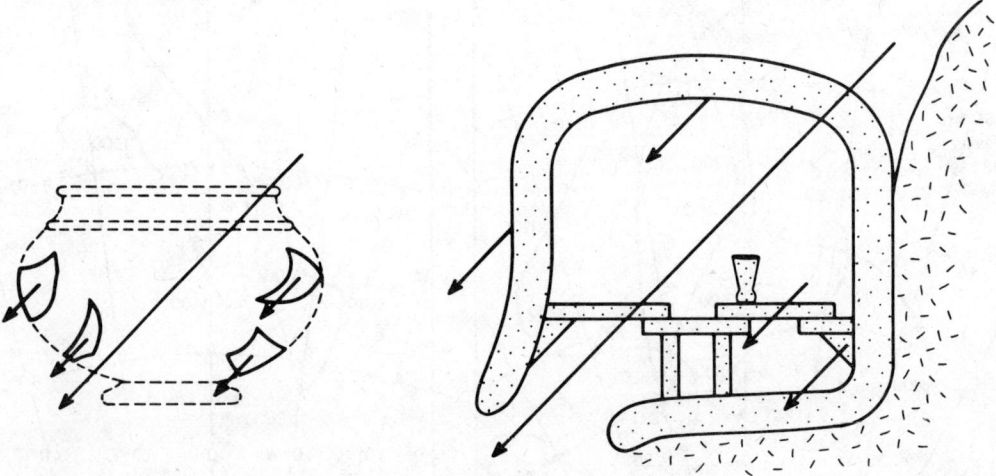

FIGURE 2. The reconstruction of objects and structures. Fragments from objects can be repositioned so that they are in the position in which they originally acquired their remanence. This allows reconstruction of either specific objects, such as pots (left) or parts of structures, such as kilns (right).

Forgeries, technology, etc. are all susceptible to analysis or closer study by magnetic methods. Archaeological items that have been buried for a long time will, for example, have acquired a viscous remanence dependent on the duration of their burial, while relatively new articles will not have acquired a significant viscous remanence. Technologies can be studied by determining the position of different items at the time that they were heated. Cast statues, for example, acquire a remanence as they cool, and this can be used to determine their orientation when cast. Similarly coins can be cooled in different ways following their manufacture, with characteristic remanence acquisition properties—thus allowing the distinction of centrifugal systems, the orientation while being silvered, etc. (Tarling, 1983*b*). The main evidence for long-term *geomagnetic secular variations* is necessarily confined to archaeomagnetic determinations because observatory records of the geomagnetic field extend back only to 1600 A.D. in London and Paris, and to even more recent times elsewhere. On a geological time scale, such records extend to include reversals of polarity, but such drastic changes are not included within the archaeological time scale. However, the understanding of the geomagnetic field ideally requires continuous global coverage throughout the last 100,000 years or so, which may include geomagnetic events, possibly failed polarity reversals. Studies of lake sediments (e.g., Thompson, 1982) provide a rapid, partially continuous record of the geomagnetic changes in one locality, but may be affected by depositional and postdepositional processes that render absolute directional properties of low reliability. Generally, palaeointensity determinations are impracticable with such materials. More reliable directional and intensity parameters are becoming established using archaeological materials, particularly in Japan, western and southeastern Europe and the area of Mesoamerica and the southwestern United States. There are still numerous discontinuities in such records and the age correlation between different areas is still only poorly established for periods prior to 1000 A.D., where dating frequently depends on methods, such as ^{14}C, that are themselves subject to errors or 100 years or more. However, such data are accumulating so that the rates of change of different geomagnetic parameters can become established. Over the last 2,000 years, for example, the average angular change of the geomagnetic field in Western Europe has been 0.05°/year, with minimum and maximum rates of 0.008 and 0.2°/year. Rapid fluctuations in intensity have also been observed, but it remains unclear if these are real or related to instrumental difficulties.

D. H. TARLING

References

Aitken, M. J., 1974, *Physics and Archaeology*. Oxford: Clarendon Press, 291p.

Burnham, R. J. P., and D. H. Tarling, 1975, Magnetization of shards as an assistance to the reconstruction of pottery vessels, *Studies in Conservation* **20,** 152-158.

Downey, W. S., and D. H. Tarling, 1984, Archaeomagnetic dating of Santorini volcanic eruptions and fired destruction levels of Late Minoan civilization, *Nature* **309,** 519-523.

Dubois, R. L., 1975, Secular variation in southwestern United States as suggested by archaeomagnetic studies, in R. M. Fisher et al., eds., *Magnetic Fields: Past and Present*. New York: Goddard Space Flight Center, 133-144.

McDougall, J., D. H. Tarling, and S. E. Warren, 1983, The magnetic sourcing of obsidian samples from Mediterranean and Near Eastern sources, *Jour. Archaeological Science* **10,** 441-452.

Mullins, C. E., 1977, Magnetic susceptibility of the soil and its significance in soil science—a review, *Jour. Soil Sci.* **28,** 223-246.

Tarling, D. H., 1975, Archaeomagnetism: The dating of archaeological materials by their magnetic properties, *World Archaeology* **7,** 185-197.

Tarling, D. H., 1983a, *Palaeomagnetism*. London: Chapman & Hall, 379p.

Tarling, D. H., 1983b, The possible utilisation of the magnetisation of archaeological metallic artifacts, *Jour. Archaeo. Sci.* **10,** 41-42.

Tarling, D. H., 1985, Archaeomagnetism, in G. Rapp and J. A. Gifford, eds., *Archaeological Geology*. New Haven: Yale University Press, 237-263.

Thompson, R., 1971, A comparison of geomagnetic field behaviour as recorded by historical, archaeomagnetic and palaeolimnological data, *Roy. Soc. [London] Philos. Trans.* **A303,** 103-112.

Tite, M. S., and C. E. Mullins, 1971, Enhancement of magnetic susceptibility of soils on archaeological sites, *Archaeometry* **13,** 208-219.

Cross-references: *Aeromagnetic Surveying; Chemical Remanent Magnetism (CRM); Demagnetization; Detrital Remanent Magnetism; Geomagnetic Field: Westward Drift; Geomagnetic Secular Variation: Direction and Intensity; Magnetic Domains; Magnetic Properties of Minerals; Magnetostratigraphy; Paleomagnetic Secular Variation; Paleomagnetism and Plate Tectonics; Rock Magnetism: Measuring Techniques and Apparatus; Viscous Remanent Magnetization (VRM) and Viscous Magnetization.*

B

BRITTLE PHENOMENA

In treating the inelastic deformation and failure behavior of rocks, geologists have conventionally classified the mechanical behavior into the brittle and the ductile fields. This convenient classification is commonly adopted by the geophysicist also. It is principally based on the macroscopic mechanical responses and failure modes. Before discussing various geophysical phenomena associated with brittle behavior, we should first outline some of the major differences between brittle and ductile behavior.

A rock in the brittle field fails either by tensile fracture or by macroscopic shear fracture along a marked discontinuity after it has reached a strain of several percent. On the other hand, a rock in the ductile field can sustain significant permanent strain that is homogeneously distributed without macroscopic fracture.

The ultimate deviatoric stress for brittle failure is strongly dependent on the mean pressure and pore pressure. The inelastic deformation before macroscopic fracture occurs in a brittle rock is usually characterized by appreciable dilatancy. In contrast, the deformational behavior in the ductile field is strongly dependent on temperature and strain rate but relatively insensitive to pressure. Inelastic volume change is generally negligible. The micromechanical processes are thermally activated and involve various lattice defects—vacancies, dislocations, and possibly grain boundaries (see *Deformation of Rocks and Minerals*).

Cracks of dimension spanning a wide range of scales exist in the upper crust. These include microcracks on the grain scale, joints on the mesoscopic scale and faults on the crustal scale. The growth and healing of individual cracks as well as crack interaction and coalescence all play a dominant role in controlling the inelasticity and brittle failure process.

Historically, rock mechanics research on brittle behavior has focused on the empirical study of the fracture strength, i.e., the ultimate deviatoric stress sustained before the onset of macroscopic fracture. The brittle failure process is taken as a discrete event without significant prior deformation and without warning. The fracture strength is a physical quantity that is of interest as an upper bound on solutions of boundary value problems.

Such an approach has been used extensively in the qualitative analysis of geologic faulting and engineering design (Jaeger and Cook, 1979). In the past two decades, important advances have been made through a materials science approach to research that uses nondestructive testing (e.g., acoustic emission, ultrasonic tomography, holographic interferometry) and electron microscopy techniques. This approach focuses on a fundamental understanding of the microscopic mechanism (Paterson, 1978). The evolution of microstructure is treated as a continuous process culminating in the coalescence of microfissures to form a thoroughgoing macroscopic fracture.

An appreciation of the dominant role of cracks in controlling brittle processes in rocks has also led researchers to follow the advances in engineering fracture mechanics and to adopt a similar approach in the analysis of geophysical processes. The most systematic application of such a fracture mechanics approach has been in earthquake mechanics (Rice, 1980; see *Seismic Source: Theory*). Important results have also been obtained for other geological problems, reviewed by Atkinson (1987).

Empirical Criteria of Brittle Failure

Depending on the signs of the principal stresses, a sample in the laboratory can fail in tension or in shear. Tensile fracture occurs in uniaxial tension tests. It may also occur by *axial splitting* in an uniaxial compression test. Under all compressive loading, the dominant mode of failure is by formation of a thoroughgoing shear fracture at an angle between 20° and 30° to the direction of the maximum (compressive) principal stress (Fig. 1).

Because of its fissured state, the fracture strength of a rock in the brittle field shows a strong pressure dependence. It is not uncommon to achieve a tenfold increase in strength by a small increment in mean stress (Fig. 2). Experiments at elevated temperatures show that the fracture strength decreases somewhat with an increase in temperature (Fig. 3). The temperature effect is relatively small in comparison with the pressure effect. The strain rate effect is also small. Systematic decrease in the fracture strength with a decrease in strain rate is usually attributed to the chemical effect of pore fluid.

Most of the experimental data have been obtained in conventional triaxial tests (with the compressive principal stress $\sigma_1 > \sigma_2 = \sigma_3$). Such an experimental setup is discussed in detail under *Deformation of Rocks and Minerals*. It is customary to characterize

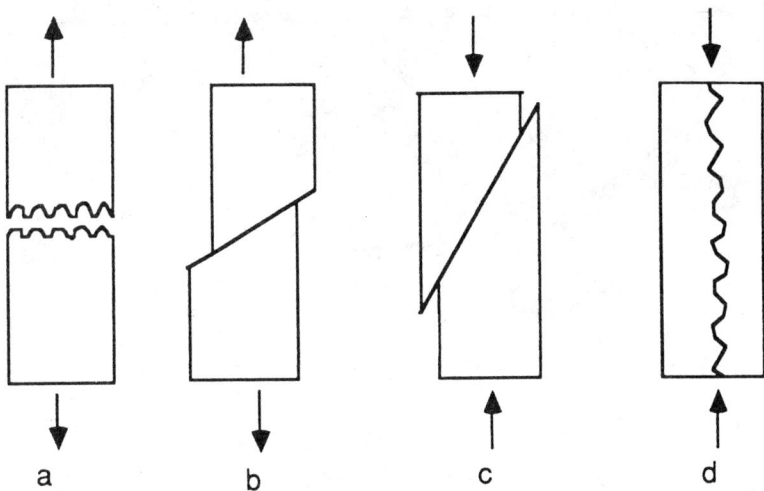

FIGURE 1. Types of macroscopic fracture: (a) tensile fracture in extension test; (b) shear fracture in triaxial extension test (least principal stress vertical, compression positive); (c) shear fracture under all compressive loading (maximum principal stress vertical); (d) "axial splitting" fracture in uniaxial compression test.

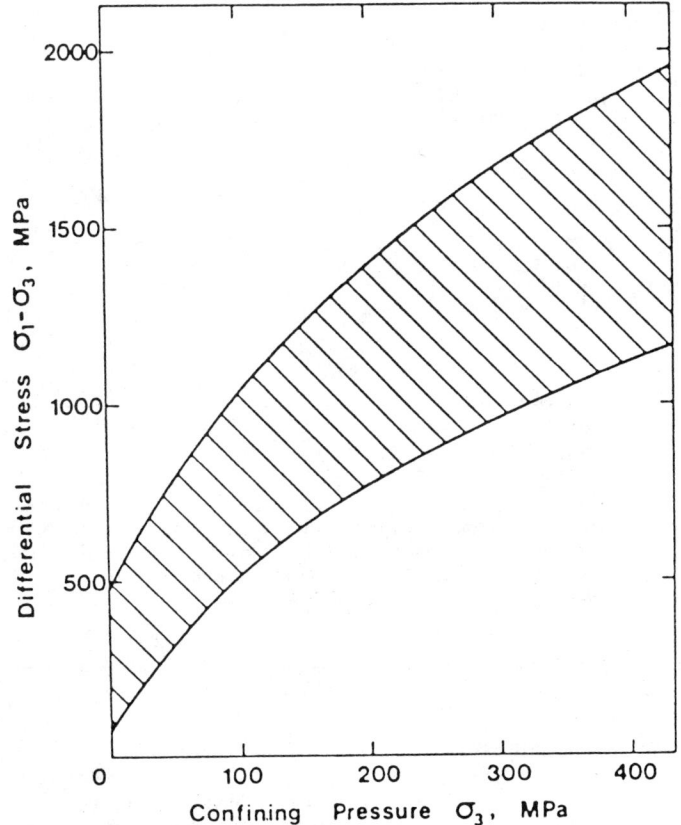

FIGURE 2. Dependence of differential stress at shear fracture in compression on confining pressure for a wide range of igneous rocks: limits taken from Ohnaka's (1973) compilation. (After Paterson, 1978)

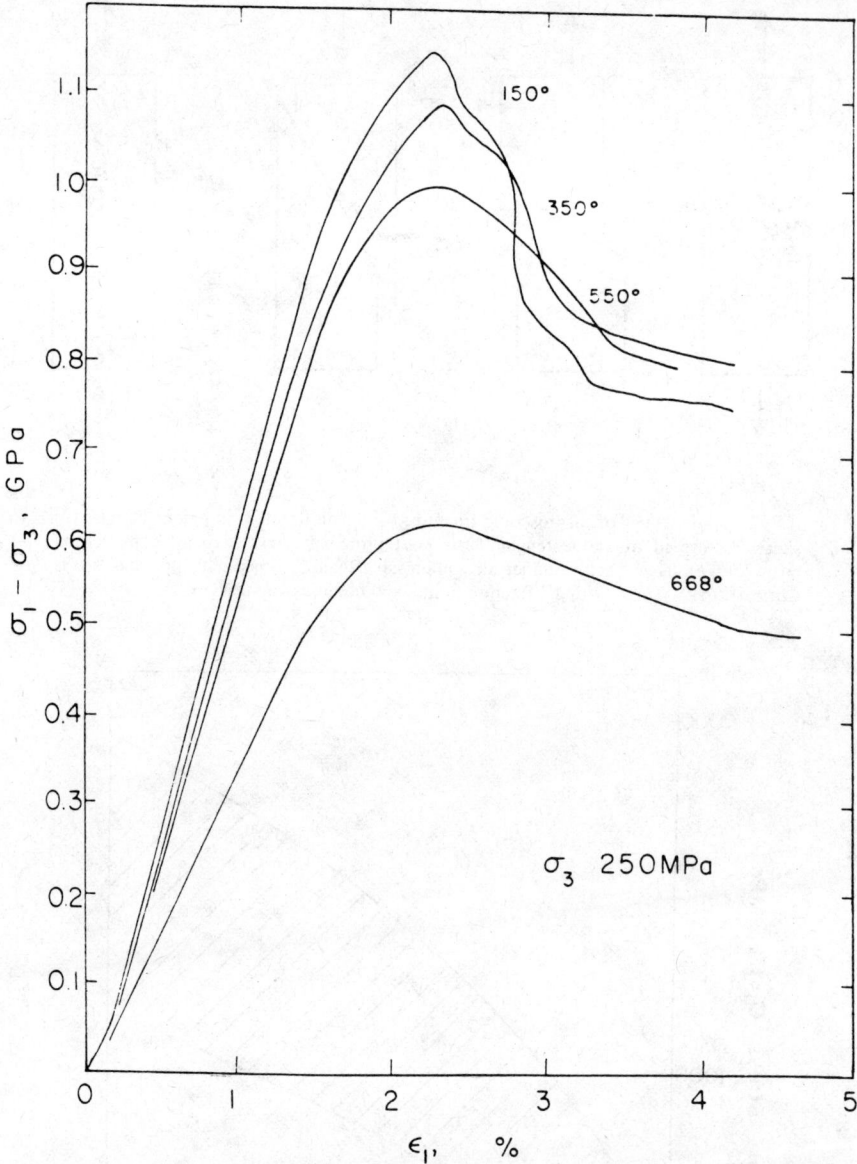

FIGURE 3. Complete load-displacement curves of Westerly granite in triaxial compression at 250 MPa confining pressure and at temperatures shown. Note the slight temperature dependence of the peak stress on temperature for brittle fracture observed at or below 550°C. The overall postfailure slope also becomes gentler as temperature increases.

the brittle strength data for a given rock type by a criterion of failure in principal stress (σ_1, σ_3) space. Over a moderate range of pressure, the Coulomb criterion of the following linear form seems to fit the data adequately:

$$\sigma_1 - \sigma_3 = \sigma_0 + \kappa P$$

where σ_0 is a constant, equal to the uniaxial compressive strength if the criterion applies to uniaxial compression, and κ is constant with a value usually between 1 and 10; $P = \sigma_3$ in a compression test, and $P = \sigma_1$ in a tensile test (Paterson, 1978).

Shear fracture data over the whole range within the brittle field usually show a nonlinear trend, being concave toward the σ_3-axis. The situation is more complicated when data of both failure modes (extension and shear) are considered. It is not uncommon in engineering applications to adopt a nonlinear failure criterion (Hoek and Brown, 1980). Use of the

Mohr criterion assumes implicitly that the effect of intermediate principal stress is unimportant. Anisotropy in the failure behavior is also not considered.

Effects of Pore Fluid on Brittle Failure

Most rocks contain an interconnected network of microcracks and pores. Under natural conditions, mesoscopic and macroscopic fractures (including joints, faults, and shear zones) provide additional conduits to enhance fluid flow. Results from deep drilling, electrical conductivity, and petrology studies suggest that an interconnected pore space can possibly extend down to tens of kilometers in the crust.

Rock deformation is profoundly influenced by the presence of pore fluid, which can exert both mechanical and chemical effects. The mechanical effect is due to fluid pressure, which is independent of fluid-rock chemical interaction. It should be described by the same physical law whether the fluid is an aqueous fluid, an inert gas, an organic fluid, or even a partial melt.

Experiments on a variety of rocks (ranging from porous sedimentary rocks to low-porosity crystalline rocks) show that the pore pressure effect on brittle failure follows Terzaghi's principle of effective stress. If a stress state σ_{ij} is on the failure envelope for samples tested under "dry" conditions, then the stress state $\sigma_{ij}' = \sigma_{ij} - p\delta_{ij}$ would also be on the failure envelope for saturated samples tested under "drained" conditions at a pore pressure p (δ_{ij} is Kronecker's delta).

If pore fluid diffusion is relatively slow so that the pore pressure is no longer uniform, then the deformation is "undrained," and pore space deformation and permeability must be considered. Tectonic deformation under undrained conditions can potentially lead to the generation and maintenance of anomalous pore pressure. Hydraulic permeability is the key physical quantity in advective processes in geophysical problems, including fault mechanics and hydrothermal circulation at mid-ocean ridges, as well as in geological problems such as magma crystallization and mineral deposit genesis. It is also one of the few quantities the magnitude of which can vary by as much as ten orders of magnitude. A comprehensive review of the data base was given by Brace (1980).

Theories have been developed for deformational behavior of a saturated porous rock. In particular, Biot's theory for poroelastic material has been extensively used for analysis of geophysical phenomena, including earthquake mechanics, subsidence, and hydraulic fracturing (Rice and Cleary, 1976). Such a theory would also imply that physical quantities such as elastic moduli and permeability satisfy an effective stress law with the effective stress given by $\sigma_{ij} - \alpha p \delta_{ij}$, where α is a constant not necessarily equal to unity. Some experimental data indicate that α can be as high as 4 for permeability of certain rock types.

If the interstitial fluid is reactive with respect to the mineral constituents of the rock, the pore fluid can exert a chemical effect in addition to the purely

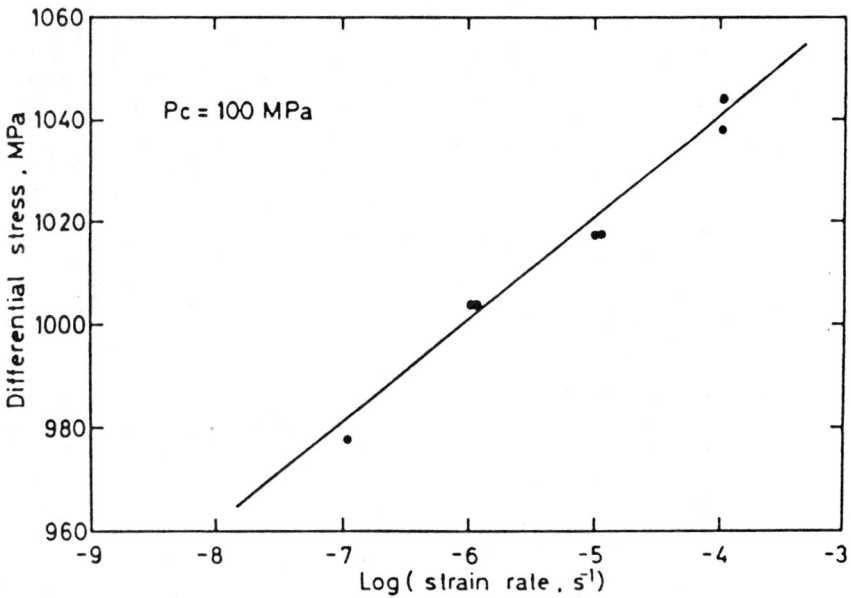

FIGURE 4. Dependence of differential stress at shear fracture in compression on strain rate. The confining pressure was 100 MPa, and the rock was Inada granite. (After Masuda, Mizutani, and Yamada, 1987)

mechanical one already discussed. The ultimate stress for rock specimens tested under triaxial compression decreases systematically as a function of strain rate (Fig. 4). If the stress is held constant at a level somewhat less than the peak value for fast loading, appreciable time-dependent strain can be observed. Such a creep process can potentially lead to macroscopic fracture. This phenomenon is referred to as *static fatigue*. The time to failure increases as the stress level decreases.

Most of the studies have been on quartz- and calcite-bearing rocks. There are some data indicating that static fatigue is enhanced at elevated temperatures and is reduced under confining pressure. The chemical effect is usually attributed to environmentally assisted subcritical crack growth. The mechanism that has attracted most attention is stress corrosion, which will be discussed later.

Pre- and Postfailure Deformation: Micromechanics of Dilatancy and Shear Localization

Prefailure behavior refers to the inelastic processes that occur before the peak deviatoric stress is achieved. In the brittle field, the ability of a rock sample to support a load usually decreases with further deformation once the peak stress has been reached. *Postfailure behavior* can be observed in the laboratory by the use of a relatively stiff or an appropriately servocontrolled test machine (Fig. 3). If a soft test machine is used, catastrophic stress drop occurs.

Under compressive loading, an inelastic increase in volume is generally observed when the deviatoric stress is increased beyond a threshold level (Fig. 5). This phenomenon of *dilatancy* is the most important attribute of prefailure behavior in the brittle field. The onset of dilatancy is accompanied by evident stress-induced anisotropy in the inelastic deformation. Consequently, the sound velocities decrease and become path dependent. Shear wave birefringence can also be observed. Two types of seismological phenomena have been suggested to be due to rock dilatancy: velocity anomaly as an earthquake precursor, and shear wave splitting observed by three-component seismic arrays.

The key micromechanical process in the prefailure stage seems to be stress-induced microcracking. Scanning electron microscopy observations on stressed samples show that microcracks propagate subparallel to the maximum compression direction. The stress-induced cracking initiates at the stress level for onset of dilatancy, and the mode is predominantly tensile. The preferred orientation for crack growth results in the observed macroscopic anisotropy. The microcracking activity induces emission of acoustic signals that are readily picked up by a sensitive transducer. The onset of dilatancy is generally marked by an evident increase in the acoustic emission count. Recent location studies show that the spatial distribution of acoustic emission sources is relatively homogeneous in the prefailure stage.

In the past decade, several rock mechanics laboratories pioneered the use of new nondestructive testing

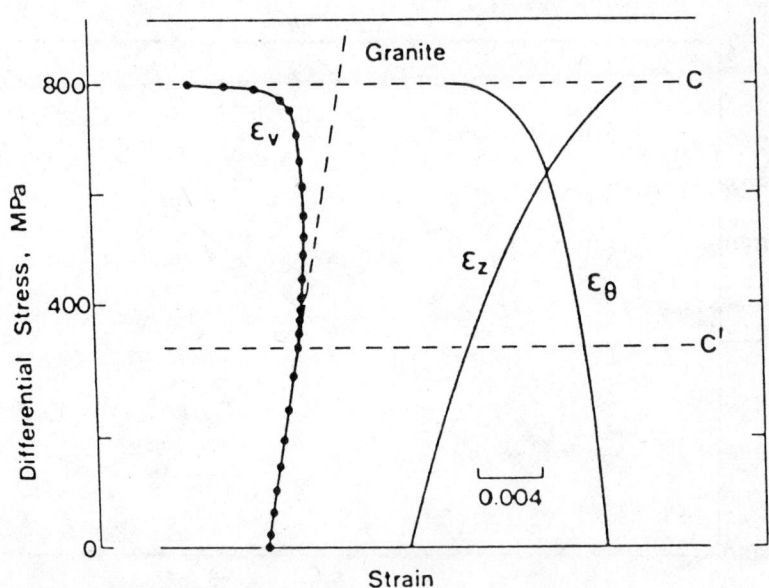

FIGURE 5. Dilatancy in Westerly granite in triaxial test at 100 MPa confining pressure. ϵ_v represents the volumetric strain, ϵ_z the axial strain, and ϵ_θ the circumferential strain. (Relative dimensional decrease is taken to be positive.) C is the peak stress, and C' the stress at the onset of dilatancy. (After Brace, Paulding, and Scholz, 1966)

techniques to investigate the evolution of brittle failure under compressive loading. These include acoustic emission source location, holographic interferometry, and ultrasonic tomography. A general conclusion reached by such studies is that the spatial distribution of both inelastic deformation and acoustic emission activity undergo a transition from homogeneous to highly localized as the peak stress is approached. The zone of localization coincides with the macroscopic shear fracture that ultimately develops.

Microscopy observations of the samples deformed into the postfailure stage point out the complexity of the shear localization process. Various micromechanical processes dependent on mineralogy and grain orientation are involved, including geometric instability mechanisms, such as microbuckling and kinking, as well as frictional slip along cracks at a high angle to the maximum compression direction (Fig. 6). The coalescence of a multiplicity of microfissures leads to the ultimate development of a thoroughgoing shear fracture.

From a continuum mechanics point of view, the initiation of shear localization can be analyzed as the onset of bifurcation of deformation (Rice, 1980). Such a localization analysis is very sensitive to the details of the constitutive equation. It is necessary to formulate a physically based constitutive model that characterizes the inelastic behavior of a pressure-sensitive dilatant material with a reasonable number of parameters. Attempts have also been made to consider the micromechanics and to adopt a damage mechanics approach to model the inelastic deformation and shear localization process.

Fracture Mechanics Applied to Rock Deformation

The foregoing discussion highlights the important role played by cracks of all scales in the inelastic deformation and brittle failure process. It is therefore quite natural for fracture mechanics to become an important tool in the treatment of brittle phenomena. As an engineering discipline, fracture mechanics analyzes quantitatively the conditions under which a load-supporting body can fail due to the growth of a dominant crack embedded in that body. Its development goes back to A. A. Griffith, who first established quantitatively the relation between strength and flaw dimension. It has evolved as a mainstream engineering field as a result of Irwin's fundamental contributions on linear elastic fracture mechanics following World War II (Kanninen and Popelar, 1985).

Crack propagation is categorized into three fundamental modes on the basis of the crack surface displacement. As shown in Fig. 7, mode I (tensile) corresponds to a displacement discontinuity normal to the crack surface, whereas mode II (in-plane shear) and mode III (antiplane shear) correspond to shear slip perpendicular to and parallel to the crack front, respectively. Mode I cracking is important in crustal processes, including igneous dike intrusion, joint formation, oceanic ridge propagation, and hydraulic fracturing. Modes II and III are important in all earthquake-related shear rupture processes.

The stress field in the vicinity of the tip of a sharp crack in a homogeneous, linearly elastic medium is proportional to $1/\sqrt{r}$, r being the radial distance from the crack tip. Such a stress singularity does not exist in any realistic material. Beyond a threshold stress, nonlinear irreversible processes dominate the crack tip deformation. As long as this breakdown

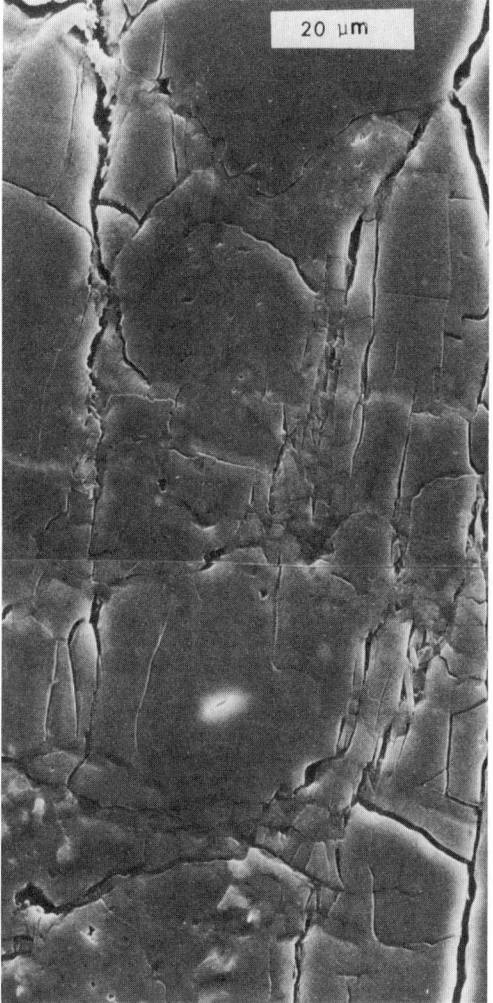

FIGURE 6. Scanning electron micrograph showing coalescence of microcracks and pores in a plagioclase grain in a Westerly granite specimen deformed in triaxial compression to the postfailure region. Maximum compression was vertical. Note the preferred (subvertical) orientation of the stress-induced cracking and the development of an incipient shear localization zone at about 30°.

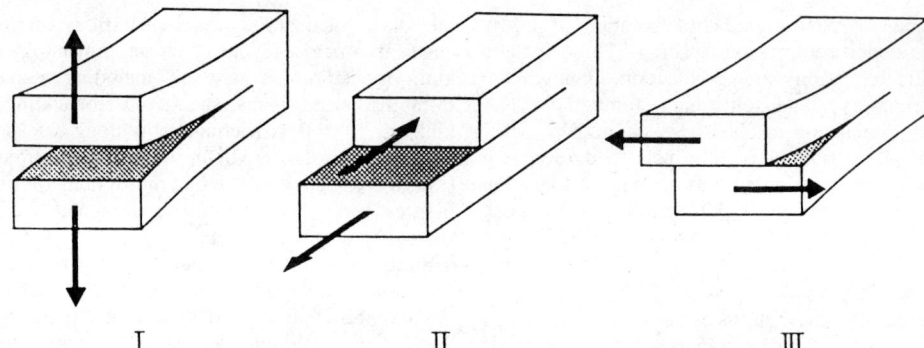

FIGURE 7. Three fundamental modes of crack propagation: mode I (tensile), mode II (in-plane shear), and mode III (antiplane shear).

region is limited in spatial extent, then linear elastic fracture mechanics applies. The stress field for a given fracture mode (α = I, II, or III) is then given by

$$\sigma_{ij} = K_\alpha (2\pi r)^{-1/2} f_{ij}(\theta)$$

where $f_{ij}(\theta)$ characterizes the angular distribution of stress and depends on the fracture mode. K_α is called the *stress intensity factor*. It is a function of the external loading configuration and the crack geometry; e.g., the stress intensity factor for a two-dimensional crack (of length $2a$) embedded in an infinite body subjected to a remote tensile stress σ is given by $K_I = \sigma\sqrt{\pi a}$.

An alternative approach to analyze crack propagation is based on energetics. The crack propagation requires energy to be released from the system made up of the cracked body and the external loading mechanism. The derivative of this energy loss with respect to crack area is G, which is usually referred to as the *crack extension force* or the *strain energy release rate*. Physically it represents the energy flux for breakdown processes at the crack tip. A simple relation between G and the stress intensity factors (of a mixed-mode crack) for a linearly elastic material was established by Irwin:

$$G = \frac{1-\nu^2}{E}(K_I^2 + K_{II}^2) + \frac{1+\nu}{E}K_{III}^2$$

(plane strain)

where E and ν are the Young's modulus and Poisson's ratio, respectively. A related concept was introduced by Rice in the late sixties to treat cases in which the breakdown zone is relatively large. His formulation of the path-independent J-integral represents an important breakthrough in elastic-plastic fracture mechanics.

A key concept in fracture mechanics is that a crack propagates when the stress intensity factor reaches a critical value. This critical stress intensity factor or fracture toughness (K_c) is a material constant independent of flaw size or loading configuration. An equivalent condition in a linearly elastic material is $G = G_c$, the critical strain energy release rate or fracture energy. An abundance of data have been obtained in the past decade for rocks and minerals in a tensile mode using standard techniques of engineering fracture mechanics. The mode I fracture energy for common minerals is about 1–10 J · m^{-2}, whereas that for silicate and calcerous rocks is typically 10–100 J · m^{-2}. Comprehensive compilations of fracture mechanics data can be found in Atkinson (1987).

Under long-term loading, a crack can propagate even if the stress intensity factor is somewhat less than K_c. This phenomenon of subcritical crack growth can be due to several mechanisms. In the brittle field, the most important mechanism is possibly stress corrosion, by which the crack tip is weakened by chemical reaction between the mineral constituents and environmental species.

The stress corrosion process is usually characterized by the stress intensity factor (K_I) as a function of the crack velocity (v) during subcritical crack propagation. The behavior is separated into three different regimes (Fig. 8): region I (where crack growth is controlled by the reaction rate at the crack tip), region II (where v is limited by the diffusion rate of the reactive species to the tip), and region III (where stress-induced cracking occurs independent of velocity at a value of K_I close to K_{Ic}). Such a scenario is well documented in glasses. However, region II is seldom observed in rocks and minerals. The data in region I can usually be fitted reasonably well with an equation of the form

$$v = v_0 \exp\left(\frac{-H}{RT}\right)K_I^n$$

where v_0 and n are constants (for given rock type and chemical environment), R is the universal gas

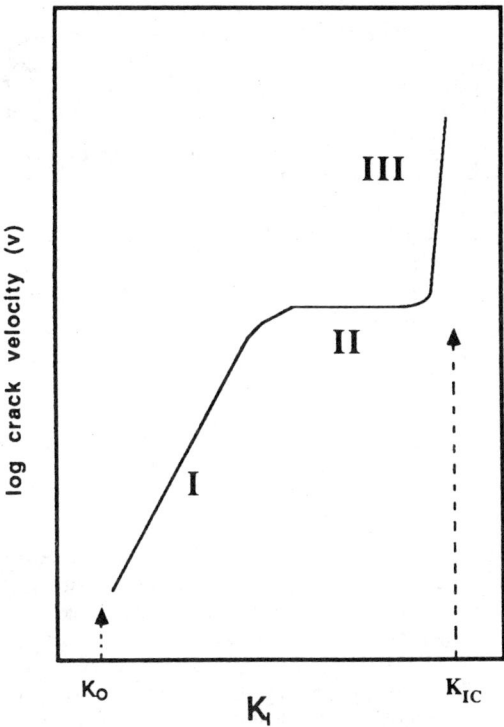

FIGURE 8. Schematic diagram for crack velocity (v) versus stress intensity factor (K_I) behavior observed for subcritical crack growth by stress corrosion in glass. K_{Ic} is the critical stress intensity factor under fast loading. K_o is the stress corrosion limit below which subcritical crack growth does not occur. The three regimes marked are discussed in the text.

constant, T is the absolute temperature, and H is an activation enthalpy. The stress corrosion index n ranges from 10 to 100, and H ranges from 30 to 100 kJ · mol^{-1} for common rocks and minerals. Several extensive reviews on the chemical effects of water on brittle deformation are given in Kirby and Scholz (1984).

Data on the pressure dependence of fracture toughness and subcritical crack growth are limited. A very useful result can be obtained, however, by applying fracture mechanics principles to the compressive loading situations of importance in geophysical applications. This result predicts that a crack, even if initially in a mixed mode, will "kink" and gradually take up a predominantly tensile mode along a path orthogonal to the minimum compression direction. Laboratory-deformed samples showing stress-induced microcracking confirm this tendency for a crack to take a path corresponding to maximum strain energy release rate (and minimum shear loading). The en echelon geometry of overlapping and curving oceanic ridge segments observed at some spreading centers has also been interpreted with an analogous model.

The nonlinear processes operative in the vicinity of a crack in mode I is an important problem still under active research. It has been postulated that a "process zone" made up of a cloud of grain scale microcracks surrounds a macrocrack in a brittle material (such as rocks and ceramics). Such a crack shielding process enhances the effective fracture toughness. Pollard and Segall (1987) suggested that a process zone of mesoscopic scale joints can play a similar role in shielding an igneous dike, which would imply that the fracture energy associated with dike propagation can be as high as 10^2–10^4 J · m^{-2}.

Earthquake rupture is usually modeled as the dynamic propagation of a shear crack. Most earthquakes occur on a preexisting fault representing a plane of weakness. Tensile cracking probably occurs on a local scale, but the extent of propagation is limited by high macroscopic compressive loading. Consequently, the overall macroscopic rupture is a combination of modes II and III.

Recent laboratory studies of in-plane shear localization and antiplane crack propagation show that the processes are very complex and involve an effective shear fracture energy of about 10^3–10^4 J · m^{-2}. A variety of approaches have been used to infer G_c from seismological data. As summarized by Wong (1982), these values are typically 10^5–10^8 J · m^{-2}, indicating that an even more complex process is involved on a crustal scale.

Frictional Sliding and Instability Behavior

Direct observation in drill holes and electrical conductivity measurements indicate that an interconnected pore space possibly extends down to several tens of kilometers. In this region of fractured rock, a lower bound on the strength of the rock mass is given by the peak stress for frictional sliding on discontinuity surfaces, such as faults, joints, and bedding foliations (Brace and Kohlstedt, 1980).

A principal focus of rock friction studies has been on the determination of the peak frictional strength as a function of rock type, temperature, normal stress, pore pressure, and fluid chemistry. The experimental configurations include direct shear, double shear, and torsional shear as well as sliding on sawcut and fractured surfaces in a triaxial setup (Fig. 9). As summarized by Byerlee (1978), data accumulated through more than a decade of research show that the frictional strength can be fitted reasonably well by a bilinear function of the effective normal stress:

$$\tau = 0.85\sigma_n \quad \text{for } 3 < \sigma_n < 200 \text{ MPa}$$
$$= 60 (\pm 10) + 0.6\sigma_n \text{ for } \sigma_n > 200 \text{ MPa}$$

where τ is the shear stress and σ_n is the normal stress (minus the pore pressure) at which frictional resistance is overcome on a sliding surface. This

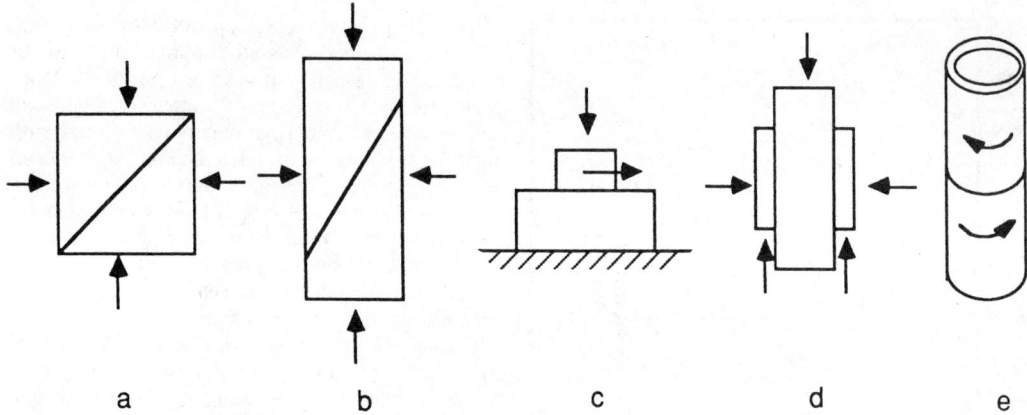

FIGURE 9. Common types of friction test configuration used in rock mechanics laboratories: (a) sliding on sawcut in biaxial test; (b) sliding on sawcut or previously induced shear fracture in triaxial test; (c) conventional direct shear test; (d) double shear test; (e) torsional shear on thin-walled specimens.

relationship, now usually referred to as Byerlee's law, is valid over a wide range of pressure and temperature in the brittle field. It is also independent of rock type or fault gouge mineralogy (except for a few clay gouges).

Together with laboratory data on the flow laws for steady-state creep of common crustal and upper mantle phases, Byerlee's law has been used extensively in recent analyses of geophysical problems, including the state of stress in the lithosphere, the depth of the seismogenic layer, and the bending moment and bathymetry associated with the lithospheric flexure process.

Two aspects of frictional sliding behavior are relevant to earthquake mechanics. First, energy dissipation due to frictional sliding represents a principal energetic sink during an earthquake cycle. The energy dissipated that way probably becomes heat, hence the heat flow measurement on a fault zone can be used to place constraints on the crustal stress level (see *Stress in the Earth's Lithosphere*). Second, a fundamental understanding of high-pressure frictional instability behavior is important in earthquake source dynamics. Since Brace and Byerlee suggested that stick-slip instability observed in the laboratory can be considered as a physical analogue of the earthquake rupture process, extensive experimental research has been carried out in the past two decades on the effects of normal stress, temperature, strain rate, and physical characteristics of the sliding surface on frictional instability behavior.

It is now generally recognized that stick-slip instability results from the interaction of an elastic loading system with a friction surface. The mechanical response of the loading system is characterized by its elastic stiffness, whereas that for the sliding surface is characterized by its friction constitutive equation. The classical approach is to hypothesize that the frictional sliding behavior can be separated into two separate regimes characterized by the static and kinetic coefficients of friction, respectively. In the transition from one regime to another, a drop in frictional strength occurs instantaneously in time before any slip has initiated. Careful observations of the evolution of dynamic instability (e.g., Okubo and Dieterich, 1984) show that the strength degradation process actually occurs over a finite slip displacement, which is a function of the physical characteristics of the sliding surface (and maybe other parameters). This is usually referred to as the *slip-weakening* process.

The slip-weakening model was originally developed by Ida, Palmer, and Rice as a generalization of the cohesive zone model (for a mode I crack in engineering fracture mechanics) to the shear crack propagation problem. It provides a straightforward interpretation of the shear fracture energy G_c: If the breakdown zone dimension is small relative to the fault length, then the shear fracture energy is simply given by an integral under the slip-weakening curve (Fig. 10). Since it is based on a physical description of the breakdown process in the evolution of frictional instability and at the same time has such a simple connection with linear elastic fracture mechanics, the slip-weakening model has been used extensively to tackle a broad range of fault mechanics problems (Rice, 1980).

Since time does not enter as a variable into the friction constitutive equation for the slip-weakening model, the cyclic behavior of earthquakes has to be attributed to the time-dependent (e.g., viscoelastic) response of the loading system. Intuitively, one expects some form of healing process operative on a fault zone during the sticking phase before the onset of instability. In this sense, the slip-weakening model seems to be inadequate to capture all the important elements of the physical process. A rate- and state-

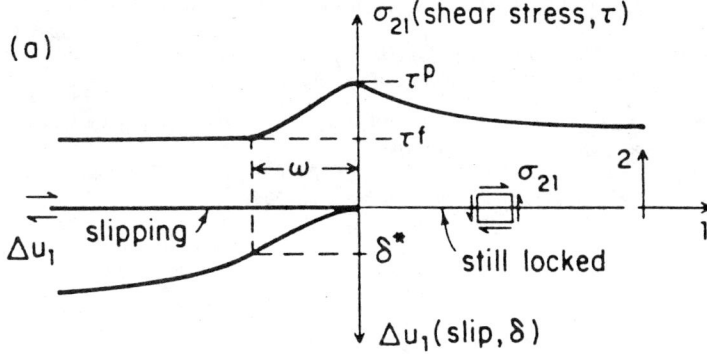

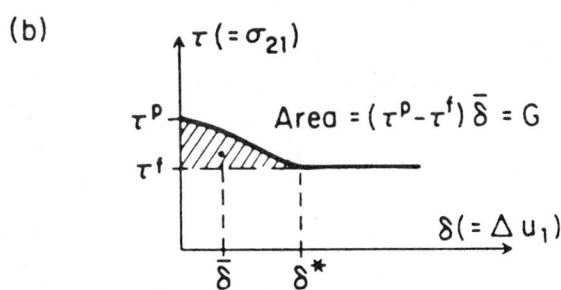

FIGURE 10. The slip-weakening model: (a) nonuniform stress along a fault as a slipping region advances into a locked section; (b) slip-weakening relation followed between shear stress (τ) and slip (δ). When breakdown zone of dimension ω is small compared relative to fault dimensions, the shear fracture energy G_c is given by an integral under the slip-weakening curve as shown. (After Rice, 1980)

dependent friction law was formulated by Dieterich and Ruina with these processes in mind. Using laboratory observations, they postulated that a nonlinear constitutive equation relating the friction stress to the velocity (slip rate) and the surface microstructure (characterized by a set of one or more state variables evolving with time) is necessary to model the complexities of rock friction.

In the past decade, significant advances have been made in understanding the nonlinear dynamics of mechanical, electrical, and biological systems. Numerical simulation of the dynamics of a system following the Dieterich-Ruina type of friction law shows features similar to those documented for other systems (ranging from a simple pendulum to Rayleigh-Benard convection) in their routes to chaos. Related research efforts are summarized in Tullis (ed., 1986).

From an experimental point of view, the dependence of frictional strength on velocity is a rather subtle effect of which the accurate measurement in the laboratory poses significant challenge to researchers. At this point, one major shortcoming of this type of rate- and state-dependent constitutive equation is that the micromechanical basis for the state variables has not been clearly specified.

One common feature in this type of model and the slip-weakening model is that both introduce a characteristic slip displacement in the constitutive equation. Furthermore, comparison of experimental data, seismological observations, and numerical simulations suggests that the characteristic slip in laboratory samples (which is anywhere from several to several hundred microns) is less than that involved in crustal processes by at least two orders of magnitude. The discrepancy is possibly due to the complex physical and geometric characteristics of a geologic fault that must be considered before one can extrapolate laboratory results to the field.

TENG-FONG WONG

References

Atkinson, B. K. ed., 1987, *Fracture Mechanics of Rock*. London: Academic Press, 534p.

Brace, W. F., 1980, Permeability of crystalline and argillaceous rocks, *Internat. Jour. Rock Mechanics and Mining Sci.* **17**, 241–251.

Brace, W. F., and D. L. Kohlstedt, 1980, Limits on lithospheric stress imposed by laboratory experiments, *Jour. Geophys. Research* **85**, 6248–6252.

Brace, W. F., B. W. Paulding, and C. Scholz, 1966, Dilatancy in the fracture of crystalline rocks, *Jour. Geophys. Research* **71**, 3939–3953.

Byerlee, J. D., 1978, Friction of rocks, *PAGEOGH* **116**, 615-626.

Hoek, E., and E. T. Brown, 1980, *Underground Excavation Engineering*. London: Institute of Mining and Metallurgy, 527p.

Jaeger, J. C., and N. G. W. Cook, 1979, *Fundamentals of Rock Mechanics*, 3rd ed. London: Chapman and Hall, 593p.

Kanninen, M. F., and C. H. Popelar, 1985, *Advanced Fracture Mechanics*. New York: Oxford University Press, 563p.

Kirby, S. H., and C. H. Scholz, ed., 1984, Chemical Effects of Water on the Strength and Deformation of Crustal Rocks, special issue, *Jour. Geophys. Research* **89**, 3991-4358.

Masuda, K., H. Mizutani, and I. Yamada, 1987, Experimental study of strain-rate dependence of failure properties of granite, *Jour. Physics Earth* **35**, 37-66.

Okubo, P. G., and J. H. Dieterich, 1984, Effects of physical fault properties on frictional instabilities produced on simulated faults, *Jour. Geophys. Research* **89**, 5817-5827.

Paterson, M. S., 1978, *Experimental Rock Deformation—The Brittle Field*, New York: Springer-Verlag, 254p.

Pollard, D. D. and P. Segall, 1987, Theoretical displacements and stresses near fracture in rock: with applications to faults, joints, veins, dikes, and solution surfaces, in B. K. Atkinson, ed., *Fracture Mechanics of Rock*, London: Academic Press, 277-350.

Rice, J. R., 1980, The mechanics of earthquake rupture, in A. M. Dziewonski and E. Boschi, eds., *Physics of the Earth's Interior*. Bologna: Italian Physical Society, pp. 555-649.

Rice, J. R., and M. P. Cleary, 1976, Some basic stress diffusion solutions for fluid-saturated elastic porous media with compressible constituents, *Rev. Geophys. Space Phys.* **14**, 227-241.

Tullis, T. E. ed., 1986, Friction and Faulting, special issue, *PAGEOGH* **124**, 375-608.

Wong, T.-F., 1982, Shear fracture energy of Westerly granite from post-failure behavior, *Jour. Geophys. Research* **87**, 990-1000.

Cross-references: *Deformation of Rocks and Minerals; Lithosphere: Mechanical Properties; Seismic Source: Theory; Stress in the Earth's Lithosphere.*

CHEMICAL REMANENT MAGNETIZATION (CRM)

Chemical remanent magnetization (CRM) refers to remanent magnetism imparted isothermally to magnetic particles produced by chemical processes below their Curie temperature in an ambient magnetic field. In rocks, CRM is usually a secondary remanence superimposed on a preexisting remanence, such as thermal remanent magnetization (TRM) in igneous rocks or depositional/post depositional remanent magnetization (DRM/PDRM) in sediments. It is generally a nuisance remanence that has to be removed to get at the primary remanence, which hopefully was recorded at the time of initial rock formation. Occasionally, CRM is the remaining characteristic remanence, the primary remanence having been obliterated, and it can be useful for paleomagnetism provided that the time of CRM production is known, or that the chemical processes have not changed the primary remanence direction, as is the case for some submarine basalts. In nature CRM is usually acquired in the presence of the Earth's magnetic field, but it can also be produced in the laboratory and in zero external field, where CRM alignment might be achieved by the sample's own internal magnetic field.

In terrestrial rocks the most common magnetic minerals are the iron-titanium (Fe-Ti) oxides of the ternary system $FeO-TiO_2-Fe_2O_3$ (Fig. 1), primarily the cubic titanomagnetite series along the line joining magnetite (Fe_3O_4) and ulvöspinel (Fe_2TiO_4) and the rhombohedral titanohematite series with compositions between hematite (αFe_2O_3) and ilmenite ($FeTiO_3$) (see *Magnetic Properties of Minerals*). Although other magnetic species, such as certain iron sulfides, sometimes contribute to the paleomagnetic signal, our examples of CRM in rocks will be confined to the Fe-Ti oxides. However, the physical and chemical mechanisms responsible for CRM production, and which determine its magnitude and stability, are general and apply equally to other magnetic minerals.

The earliest CRM experiments were done over a century ago in association with electrolytic deposition of iron and other magnetic metals. Beetz (1860) produced CRM in electrolytically deposited iron supporting Weber's hypothesis that some atoms possess intrinsic magnetic moments. Similar experiments producing CRM by electrolytic deposition of iron and nickel in the presence of external fields were also reported by Maurain (1901, 1902). Koenigsberger (1938, part 1, p. 122) envisioned a *crystallization remanence* "impressed by the Earth's field at temperatures between about 100°C and 500°C during the time of lattice changes in magnetite which result very probably from unmixing of Fe_2O_3," and he advanced this as a possible mechanism for the strong, coherent remanence of some sediments (Koenigsberger, 1938, part 2, p. 319). With the growth of research in paleomagnetism after World War II it became apparent to many rock magnetists in Europe, North America, and Japan that the remanence of many sediments, especially red beds, was at least partially controlled by magnetic minerals chemically precipitated subsequent to deposition (e.g., Blackett, 1956).

CRM research has been relatively neglected in comparison with that of other naturally occurring remanences, such as TRM, viscous RM (VRM), and DRM/PDRM, because it is often difficult to distinguish CRM from these remanences, because of overlapping stabilities and intensities and because of difficulties in designing well controlled laboratory experiments.

Grain Growth CRM

In this section we discuss grain growth CRM, because, conceptually, it is the simplest form of CRM, and because the evolution of its intensity and stability can be understood as a particular example of a more general treatment of remanent magnetism. In addition, grain growth CRM has been used to explain the often stable remanence of many chemically altered sedimentary and igneous rocks.

Néel (1949, 1955) developed a theory for the remanence of noninteracting single domain (SD) particles. The theory has been highly successful in explaining many features of TRM, particularly its relatively high intensity and stability as compared with IRM (isothermal remanent magnetization) produced in comparable fields at room temperature. (See *Thermoremanence* and *Viscous Remanent Magnetization (VRM) and Viscous Remagnetization*.) Following Néel's treatment, the remanence, M_R, of a magnetized body will decay in zero field, seeking the new state of thermal equilibrium (zero remanence), at a rate determined by the relaxation time, τ, such that

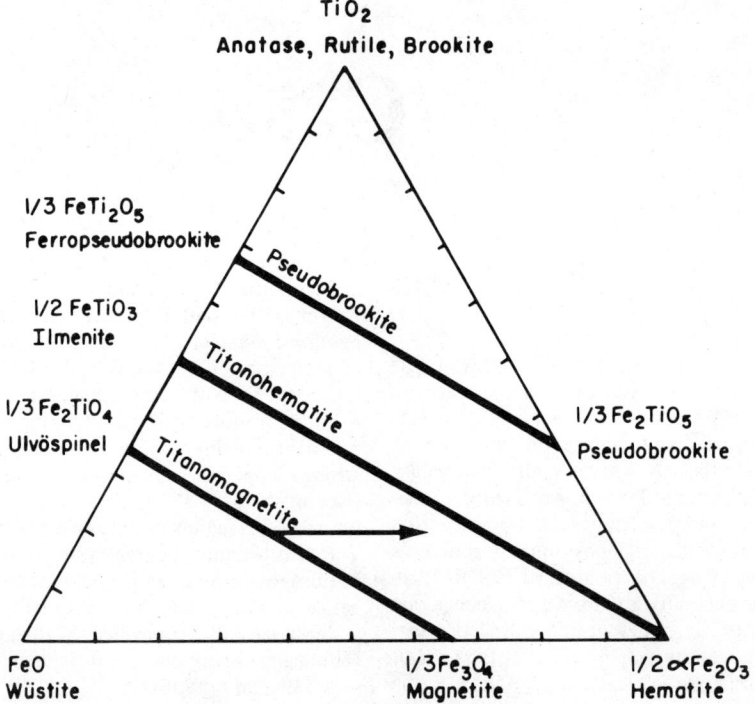

FIGURE 1. Compositional diagram for the ternary system FeO-TiO_2-Fe_2O_3, normalized with respect to cations, such that horizontal lines have constant Fe:Ti ratios, indicating pure oxidation/reduction reactions. The three principal Fe-Ti oxide solid solution series are shown by heavy lines. All members of the pseudobrookite series are paramagnetic at room temperature.

$$M_R = M_0 \exp[-t/\tau] \qquad (1)$$

where M_0 represents the initial remanence, and t is the time. Néel derived an expression for τ, which for mathematical simplicity considers an assemblage of identical noninteracting spherical SD particles with uniaxial anisotropy and aligned easy axes of magnetization. Then in zero external magnetic field

$$\tau = C^{-1} \exp[Kv/kT] \qquad (2)$$

where K = the uniaxial anisotropy energy coefficient; v = the particle volume; k = Boltzmann's constant; T = the absolute temperature (°K). C is the frequency factor whose value is on the order of 10^9 sec^{-1}. Although C depends on the temperature and temperature-dependent material properties, it is usually considered constant in comparison with the exponential factor. In contrast, τ, which is a measure of the remanence stability, varies orders of magnitude as a result of modest changes of the argument (Kv/kT). Thus, the Néel theory provides an explanation for the observed blocking/unblocking process and the additivity of partial TRMs (PTRMs) of noninteracting SD particles.

CRM production is an isothermal process, where τ increases exponentially with volume for homogeneously magnetized grains. The precipitating ferro/ferrimagnetic particles grow from atomic/molecular paramagnetic nuclei to larger superparamagnetic (SPM) grains having spontaneous magnetization but no remanence, $\tau \ll \tau_L$, where τ_L is a characteristic laboratory time usually on the order of minutes. As the volume increases, the blocking volume, v_B, may be exceeded, $\tau \gg \tau_L$, the magnetic moments are stabilized, and the magnetization is said to be blocked. This is exactly analogous to TRM, where v is constant and τ increases as the temperature decreases. ($T_B \equiv$ blocking temperature; $T_C \equiv$ Curie point or temperature). Above the blocking temperature ($T_B < T < T_C$), $\tau \ll \tau_L$, thermal equilibrium is quickly established, and the ferri/ferromagnetic grains are in the SPM state with no remanence. For $T < T_B$, $\tau \gg \tau_L$, and the remanence is blocked. Thus, as the temperature decreases from above T_C the particle assemblage progresses from paramagnetic to SPM to the stably magnetized state.

As long as the magnetic ensemble consists of homogeneously magnetized SD grains, τ will increase exponentially with particle volume (recall that Eq. 2 for τ was specifically derived for noninteracting SD particles). For multidomain (MD)

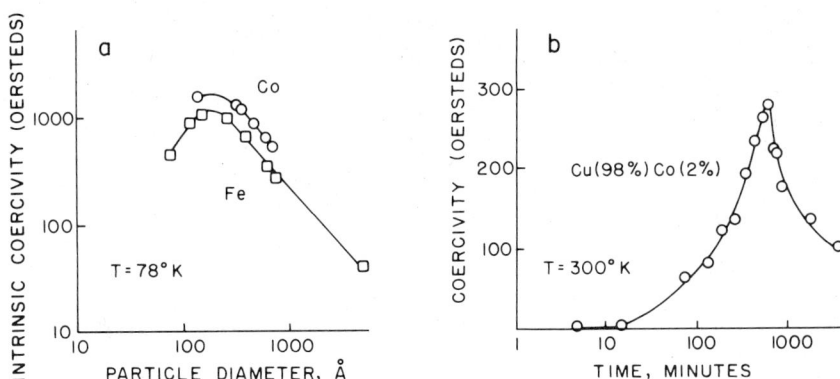

FIGURE 2. (a) Intrinsic coercive force of iron and cobalt as a function of particle sizes at liquid nitrogen temperature (from Meiklejohn, 1953). (b) Change in coercive force of the alloy Cu (98%) Co (2%) as a function of annealing time (particle size) at 700°C, measured at 300°K. (From Becker, 1957)

particles τ will decrease with increasing volume, principally because it becomes progressively easier to alter the remanence by domain wall movements in contrast to rotating the magnetic moments of SD particles. (For further discussion consult *Rock Magnetism* and *Thermoremanence*.) These conceptual considerations of the particle size dependence of the remanence stability (relaxation time) were demonstrated experimentally by coercivity measurements of dispersed fine particles of magnetic metals (Fe, Ni, Co), as the grain sizes were increased by progressive heat treatments (e.g., Meiklejohn, 1953; Becker, 1957). Zero initial coercivity in the SPM region is followed by increasing coercivity with particle sizes, presumably in the SD size range, which then decreases for larger presumably MD particles (Fig. 2). Similar results have also been obtained for magnetite particles.

Haigh (1958) was first to apply Néel's theory and the grain size dependence of the coercivity to rock magnetism, to explain the CRM properties observed during laboratory reduction of hematite to magnetite, where the CRM resided in growing particles of magnetite. Kobayashi (1959) also examined CRM in magnetite obtained from hematite reduction and showed that CRM stability with respect to both alternating fields and thermal demagnetization was much greater than for IRM and very similar to the stability of TRM (Fig. 3). Kobayashi (1961) produced CRM in cobalt grains precipitated from

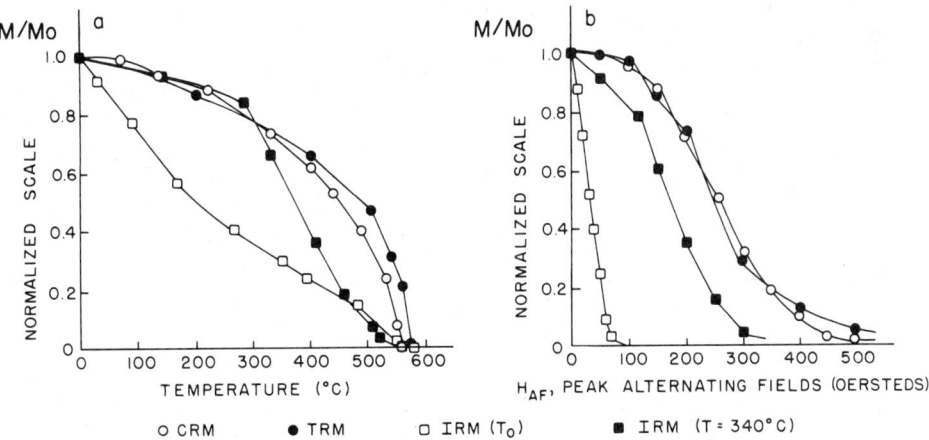

FIGURE 3. (a) Thermal, (b) alternating fields demagnetization of remanence in magnetite. All remanence measurements were made at room temperature. CRM was produced at 340°C with external field of 3 Oe (a), 10 Oe (b); total TRM was produced with 3 Oe (a), 0.5 Oe (b); IRM (T_0) at room temperature was produced with 200 Oe (a), 30 Oe (b); IRM(T) was produced at 340°C with fields of 20 Oe (a), 10 Oe (b). CRM (T = 340°C) and IRM (T = 340°C) were cooled to T_0 in zero field prior to demagnetization. (From Kobayashi, 1959)

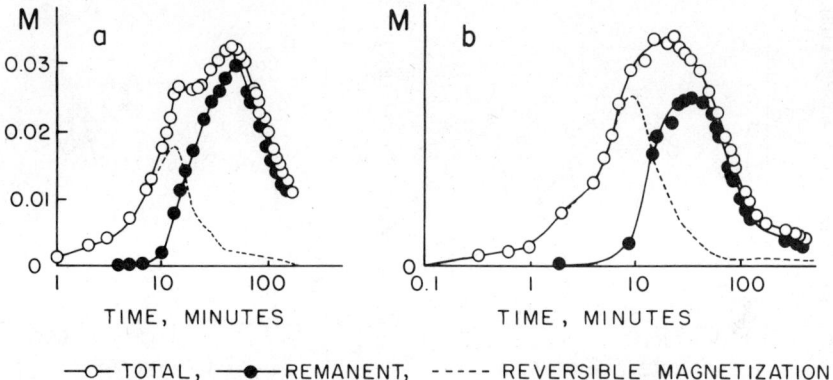

FIGURE 4. Total, remanent, and reversible magnetization of Cu-Co alloy as a function of annealing time (particle size) at 750°C, (measured at 750°C). Experiments were done in external fields of (a) 1 Oe and (b) 20 Oe. The reversible magnetization is the difference between the total magnetization (in presence of an applied field) and the remanence. (From Kobayashi, 1961)

Cu-Co alloy and showed that the specific CRM intensity had a similar bell-shaped grain size dependence as the coercivity, increasing from zero for SPM particles, attaining a maximum value and then decreasing for inhomogeneously magnetized MD particles (Fig. 4).

Evidence has been accumulating during the last few years that nearly pure SD magnetite produced by magnetotactic bacteria may be a significant source of magnetic particles for some marine and lacustrine sediments. The bacterial magnetite is thought to attain its stable SD configuration by biologically mediated grain growth in the bacteria.

Although grain growth is conceptually the simplest form of CRM, it can be used to explain many examples in paleomagnetism, in which the NRM is predominantly a secondary remanence in chemically altered rock, and where new particles of a chemically nucleated magnetic phase have grown beyond the SPM size range.

CRM In Sedimentary Rocks

Oxidized Red Sediments. Red beds comprise a broad category of loosely defined sediments with colors that range from brown to purple, usually resulting from secondary fine particles of hematite or ferric oxyhydroxide. Red sediments have been used extensively for paleomagnetism since the late 1940s, because (1) they are extensively distributed geographically and with respect to geologic time; (2) their bedding orientation can often be inferred from field observations, and (3) their remanence is often stable and sufficiently intense for paleomagnetic measurements, even prior to the development of the super sensitive cryogenic magnetometer. As we noted earlier, already in the 1950s it was deduced that low temperature oxidation was responsible for removing the original magnetite and the precipitation of fine hematite particles, which give rise to the color and CRM of red beds (Blackett, 1956). Larson and Walker (1975) conducted a detailed study of CRM development during early stages of red bed formation in late Cenozoic sediments; they showed that in their samples CRM resided in several authigenic phases including hematite and goethite. Moreover, the CRM, which obscured the original depositional remanence, had formed over a protracted period spanning more than a single polarity interval, as indicated by the different polarities of the different authigenic phases. Complex patterns of remanent magnetization, associated with several generations of CRM, including multiple polarities within single specimens, have been observed in Paleozoic and Mesozoic red beds. The degree to which the secondary CRM affects the remanence depends on the stability of the precipitated iron oxides and on the relative proportions of authigenic CRM carriers to the primary DRM recorded in detrital particles of specularite hematite. Examples have been reported where the primary DRM in specularite hematite remained the characteristic remanence with respect to CRM (e.g., Collinson, 1974). In some cases distinct CRM components can be isolated by thermal demagnetization, selective leaching in acid (e.g., Collinson, 1967) and selective removal of altered phases of sediment by selective destructive demagnetization, SDD (Larson, 1981).

Nonred Sediments. The discussion in the following paragraphs deals with CRM in a subset of nonred carbonate sediments, whose remanence is usually much weaker than that of red beds. The low intensity remanence of these sediments was one of the main reasons that they were often bypassed for paleomagnetic studies until the introduction in the early 1970s of cryogenic magnetometers with the ability to measure minute remanence signals, down to the 10^{-9}–10^{-10} Gauss range. Since then, there has been a flourishing of paleomagnetic investigations of weakly magnetized nonred sediments. For example, several studies of early Paleozoic carbonates have shown that the characteristic remanence was acquired in the late Paleozoic, sometimes hundreds of millions of years later than the biostratigraphic ages of the formations (e.g., McCabe et al., 1983). Magnetic extracts from these diagenetically altered sediments contained essentially pure magnetite particles, whose botryoidal and spheroidal forms (Fig. 5) have been used to infer their secondary origin, and they are thought to be responsible for the secondary characteristic remanence of these formations. In the absence of evidence of significant heating of these sediments, their remanence has been interpreted as being predominantly a low temperature CRM in secondary magnetite, which might have been produced by "diagenetic alteration of preexisting iron sulfides (e.g., framboidal pyrites)" (McCabe et al., 1983). For the analogous Miocene dolomites and limestones of the Monterey Formation, Hornafius (1984) concluded that the secondary remanence, presumably a low temperature CRM, resides in diagenetic magnetite produced by the partial oxidation of pyrite upon the introduction of oxygenated meteoric groundwaters to the formation.

The presence of magnetite (and siderite) in oil-impregnated sediments was discovered by Bagin and Malumyan (1976), and by Donovan et al. (1979) reported correspondence of near surface magnetic anomalies over an oil field with a higher concentration of magnetic minerals in the sediments. More recently, paleomagnetic studies of remagnetized hydrocarbon-impregnated Paleozoic sediments (McCabe et al., 1987; Benthien and Elmore, 1987) indicate a relationship between hydrocarbon migration and the precipitation of authigenic magnetite particles, wherein the secondary CRM resides. This scenario is supported by extracted magnetite spherules up to several tens of microns in diameter (Fig. 5) and the presence of other authigenic textures in the sediments. All these studies suggest a net reduction of ferric ions in oxides, hydroxides, and silicates, caused by the biodegradation (oxidation) of the hydrocarbons with a net production of more reduced phases such as magnetite (Fe_3O_4), siderite ($FeCO_3$), and wustite (FeO). Of these, only magnetite is ferrimagnetic, and therefore it is responsible for the measured secondary CRM and also for being preferentially extracted during magnetic separations.

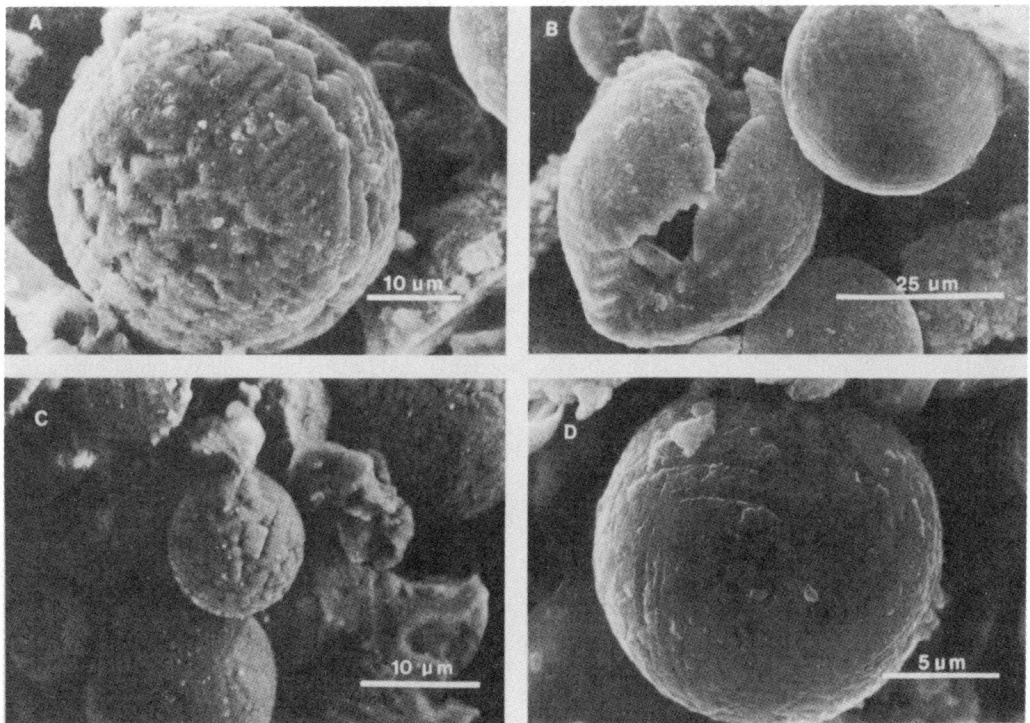

FIGURE 5. Scanning electron micrographs of magnetite crystal aggregates isolated from Thornton Quarry bitumen (A and B), Bonneterre Dolomite (C), and Cynthia Quarry bitumen (D). Many of the spheroidal aggregates have hollow cores. (From McCabe et al., 1987).

Comparisons of the paleomagnetic directions of some of these Paleozoic sediments with available apparent polar wandering paths, suggest that the CRM may have been imparted on the order of 10^8 years after deposition. In contrast, paleomagnetic and rock magnetic studies of Recent hemipelagic sediments, accumulating in suboxic environments (Karlin et al., 1987) indicate the authigenic production of magnetite downcore, as the sediment becomes progressively more reducing, accompanying the oxidative decomposition of organic matter. The time resolution of paleomagnetic results could be significantly reduced in sediments where CRM occurs in authigenic magnetic minerals, formed below the initial fixing depths of the remanence.

CRM in Igneous Rocks

General Considerations. CRM in igneous rocks often results from grain growth of magnetic particles or the transformation of one magnetic phase to another at lower temperatures than the final blocking temperatures of the new magnetic species. Merrill (1975) presented an excellent review of "magnetic effects associated with chemical changes in igneous rocks" in a paper by that title. Grain growth CRM in igneous rocks is often complicated by probable effects associated with temperature changes, such as reheating during metamorphism and laboratory experiments or the slow cooling of an intrusive unit resulting in the exolution of a new magnetic phase.

The direction and intensity of the secondary CRM depends among other factors on the resultant magnetic field and the dominant magnetic interactions at the site of the newly forming magnetic phase. The external magnetic field exerts an important influence; however, if the new phase is created in the presence of an existing magnetic substance with higher blocking temperatures, the CRM will also depend on the magnetic interactions between the various magnetic species. Intimate contact between two magnetic phases such as might occur during exolution of the titanohematite series might cause the CRM to be controlled by local fields produced by the higher blocking temperature phase through magnetostatic or exchange interactions (see *Rock Magnetism*). Occasionally, such interactions might produce CRM of opposite polarity to that of the existing magnetic phase (negative magnetic interactions), and in rare cases a self-reversal may ensue (Néel, 1955) (see *Magnetic Self-Reversal*).

· It is usually difficult to distinguish CRM in igneous rocks from TRM and VRM, because of overlapping stabilities, and also because of possible associations between CRM, partial TRM (PTRM) and high temperature VRM. Consider, for example, grain growth CRM produced at an elevated temperature $T_1 > T_0$, where T_0 denotes room temperature. The CRM carriers have blocking temperatures $T_B > T_1$. The chemically precipitated particles with $T_B < T_1$ contribute no remanence at T_1. However, on cooling from T_1 to T_0 a PTRM will be blocked in the particles with $T_1 > T_B > T_0$, in addition to the growth of spontaneous magnetization of most magnetic particles. Moreover, ubiquitous time effects might result in significant VRM being superimposed on the CRM. Such commonplace complexities make it difficult to uniquely isolate CRM from other remanences in rocks. Hence, it is probable that CRM occurrences in the paleomagnetic record are considerably more numerous than is usually recognized.

CRM Origin of Marine Magnetic Anomalies. Submarine basalts are an example of igneous rocks that undergo chemical change whereby the initial magnetic phase (titanomagnetite, $x\text{Fe}_2\text{TiO}_4$ $(1-x)\text{Fe}_3\text{O}_4$) is transformed by topotactic low temperature oxidation at the sea floor to cation deficient titanomaghemites. This reaction is described in Fig. 1 by the arrow to the right, emanating from near the middle of the titanomagnetite series. Horizontal lines in Fig. 1 denote pure oxidation-reduction processes at constant Fe/Ti ratios. Freshly extruded submarine basalts at spreading centers comprise stoichiometric titanomagnetite with $x \cong 0.6 \pm 0.1$, that is, about 60% molar ulvöspinel, with Curie points between 100°–150°C. Oxidation is thought to proceed by a net cation migration out of the crystal lattice, to accommodate the decrease in ferrous to ferric ratio, with possible changes of the Fe/Ti proportions. The resulting cation deficient phase has higher Curie points, up to nearly 500°C, lower saturation magnetization, and diminished lattice size. These time-dependent chemical modifications of the magnetic properties are responsible for the rapid diminution of the observed marine magnetic anomaly amplitudes away from spreading centers. Therefore marine magnetic anomalies over the world's oceans can be considered to be preserved predominantly as CRM in oxidized Fe-Ti oxides. However, the natural remanent magnetization (NRM) of some coarse textured and magnetically less stable submarine basalts is occasionally dominated by VRM, which can frequently be removed by low to moderate alternating field demagnetization.

The consistency of various marine magnetic anomaly time scales and their agreement with polarity time scales from continental lavas and marine sediments, as well as agreement with paleomagnetic measurements of many oriented dredged and drilled submarine basalts indicate that the CRM is usually of the same polarity as the primary TRM recorded upon extrusion. This result is also supported by several laboratory CRM experiments, which investigated the low temperature oxidation of titanomagnetites (e.g., Marshall and Cox, 1971; Johnson and Merrill, 1974; Özdemir and Dunlop, 1985).

The preservation of marine magnetic anomalies despite ubiquitous low temperature oxidation of submarine basalts suggests that CRM production is not primarily influenced by the external magnetic field, whose effect during protracted oxidation would be expected to shift anomaly boundaries to older ages—greater distances from respective spreading centers. This might lead to systematically older Brunhes/Matuyama boundary ages and a general tendency of obscuring short polarity events or subchrons in marine magnetic anomaly profiles. However, (1) there is a remarkable correspondence between geomagnetic polarity time scales from marine magnetic anomalies, independent sedimentary sections, and subaerial volcanic sequences, and (2) there is no evidence for a systematically older Brunhes/Matuyama boundary age in marine magnetic anomaly records. Consequently, CRM modification of anomaly boundaries by submarine basalt oxidation may not be a significant effect in areas of coherent, "high quality" anomalies.

Alternatively, prevailing positive superexchange interactions during submarine basalt oxidation might successfully explain the above observations. The low temperature oxidation leads to cation depletion and lattice shrinkage, but the titanomaghemites retain the cubic crystal structure of the original titanomagnetite minerals. The superexchange interactions will vary with changes in cation distribution and lattice dimensions, as indicated by the reduced saturation magnetization and higher Curie points. However, it is possible that the orientations of the sublattice magnetic moments of the titanomagnetite minerals remain intact during low temperature oxidation. Therefore the ensuing CRM will retain or inherit the original TRM direction, depending on one's point of view. Hence, positive superexchange interactions between the original stoichiometric titanomagnetite phase and the oxidized titanomaghemites might be an important mechanism for preserving the marine magnetic anomaly patterns, even though the remanence, to a large degree, is CRM.

CRM Control of Paleointensity Determinations. At present, only TRM can be used to determine the absolute paleointensity of the Earth's magnetic field. (See *Paleomagnetic Field: Intensity*.) All the paleointensity methods require that the NRM be essentially a pure TRM, and they all call for laboratory heating of the specimens to produce a new TRM in the known laboratory field. Chemical and mineralogical alterations of the specimens during the laboratory heatings often preclude reliable paleointensity determinations. Therefore, the different paleointensity procedures apply various tests of the magnetization before and after heating to assess the extent of chemical changes on the remanence and paleointensity experiments. Several paleointensity methods require one heating of the specimens to above their highest Curie temperatures to produce a total laboratory TRM. This procedure tends to maximize chemical alterations, because the reaction rates increase at higher temperatures.

In contrast, the Thelliers' double heating method (Thellier and Thellier, 1959) was developed to remedy this problem, by increasing the temperature gradually in steps from room temperature. At each temperature (T_M) two heatings are conducted to determine the thermally demagnetized PNRM and the other for the acquired PTRM between T_M and room temperature. The Thellier procedure depends on the following observed properties of remanence, which have been explained theoretically by the Néel theory for noninteracting SD particles: first, PNRMs and PTRMs produced in different temperature intervals are independent and additive; second, the PNRM and PTRM blocked in any temperature interval are linearly proportional to the external fields in which they were blocked; that is, PNRM (T_0, T_M) = $A_M h$ and PTRM (T_0, T_M) = $A_M h_L$, where A_M is constant for a particular temperature interval (T_0, T_M). h_L is the known laboratory field and h is the unknown paleointensity, which is assumed constant during the original TRM production for rapidly cooled extrusives and archeomagnetic artifacts. Therefore, in the absence of chemical changes on heating, and the satisfaction of the above assumptions the ratio PNRM (T_0, T_M)/PTRM (T_0, T_M) = h/h_L should be the same for each temperature interval; hence, at each temperature step an independent paleointensity value is calculated. Thus, a primary advantage of the Thellier technique is the consistency checks provided by the independent paleointensity estimates at the different temperatures. The results can be shown on a PNRM versus PTRM diagram, with data corresponding to the different temperature steps (Fig. 6). Ideal behavior in the Thellier sense implies linear data with slope equal to h/h_L (line A, Fig. 6). Chemical modifications, which are common at relatively higher temperatures, can often be detected as deviations from the ideal straight line, and often a paleointensity can still be determined by using the linear lower temperature points (e.g., Fig. 6, curve B, $T_1 - T_3$). Another feature of the Thelliers' procedure is that PTRM checks can be conducted. A PTRM check is a repeat PTRM at a lower temperature $T_P \leq T_M$, which provides information about changes in the PTRM capacity due to chemical alterations and CRM production in grains with $T_B \leq T_P$. Therefore, the Thellier method is usually considered to be more reliable than other paleointensity procedures.

High temperature CRM during the Thelliers' paleointensity procedure is one of the more common causes for unsuccessful or abbreviated paleointensity experiments. In these cases a CRM is produced at an elevated temperature $T_M > T_0$ in new magnetic particles with blocking temperatures $T_B > T_M$. In general, the CRM produced at T_M may not be

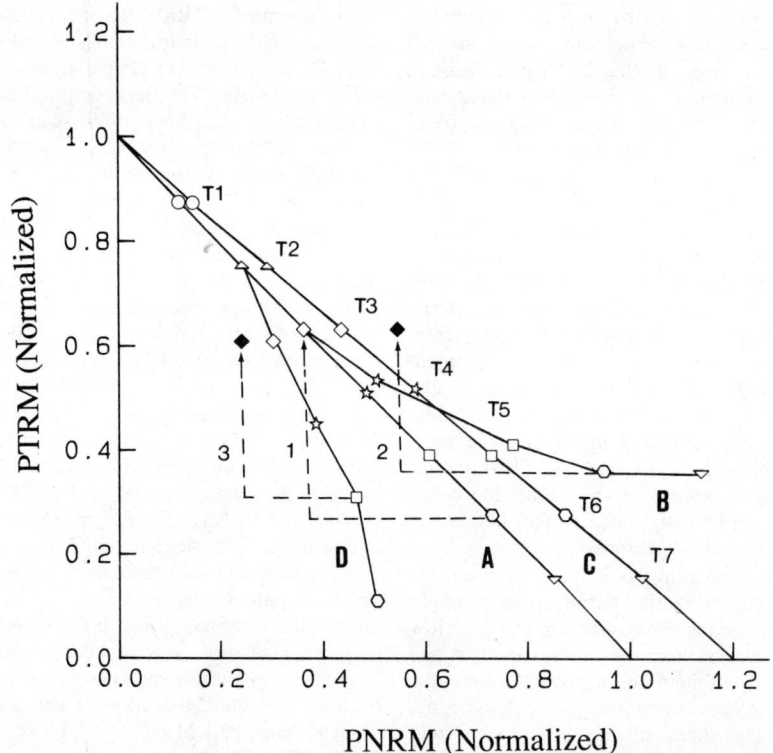

FIGURE 6. PNRM-PTRM diagram for four hypothetical Thellier paleointensity experiments A, B, C, and D, discussed in the text. Like symbols are given to identical temperatures. Dashed lines 1, 2, and 3 indicate PTRM checks between the designated temperatures, corresponding to experiments A, B, and D, respectively. Closed symbols refer to unsuccessful PTRM checks.

completely demagnetized at T_{M+1}, the next higher thermal demagnetization step. The resulting nonideal concave-up PNRM-PTRM diagram is shown in Fig. 6, curve B. A PTRM check at a lower temperature T_P will only detect CRM produced in particles with blocking temperatures $T_B \leq T_P$ (PTRM check 2, Fig. 6). In the special case where CRM is completely demagnetized at the next thermal demagnetization step at $T_{M+1} > T_M$, a linear PNRM-PTRM behavior might result (line C, Fig. 6), provided that proportionately more CRM is produced at progressively higher temperatures. However, these linear data plot above the ideal line of unit slope (line A, Fig. 6), and the calculated paleointensity would be lower than its actual value. This emphasizes that linear PNRM-PTRM data are a necessary but not sufficient condition for obtaining reliable paleointensities.

In the above discussion we considered CRM production that increases the PTRM capacity of the specimen. Such alteration may result from precipitation of new magnetic particles or from unmixing of titanomagnetite grains to a mixture comprising a more Fe-rich phase with higher saturation magnetization. Chemical modifications, which lead to higher PTRM, usually cause the PNRM-PTRM points to lie above the ideal line, and the calculated paleointensity will be lower than the actual paleo-field. Alternatively, if the chemical modifications decrease the PTRM potential by destroying magnetic particles or by transforming them to have lower intrinsic magnetic moments, then the PNRM-PTRM points will plot below the ideal line, with higher apparent paleointensities than the actual values (e.g., curve D and PTRM check 3, Fig. 6).

Concluding Thoughts

The foregoing CRM article is not meant to be exhaustive, and it reflects some of the interests, biases, and limitations of the author. In the future as paleomagnetists tackle more difficult tectonic and geomagnetic problems, which require information from structurally more complex, metamorphosed, and older formations, it will be increasingly likely that the NRM will be partly CRM. Indeed, paleomagnetists have become more clever at isolating different remanence components by using detailed and varied demagnetization procedures. It is usually assumed that the most resistant remanence component, whether with respect to increasing alternating fields, elevated temperatures, or a particular leaching agent is also the oldest. However, the current knowledge of CRM is so rudimentary that it is usually not known whether this assumption is correct. A better understanding of CRM is needed to assist paleomagnetists in interpreting complex, often multicomponent, remanence from rocks with probable CRM overprinting. This understanding might be accomplished through controlled field and laboratory CRM experiments for different magnetic minerals: (1) to identify the varied geochemical environments, including pressures and temperatures, necessary to form different magnetic minerals and the associated CRMs; (2) to determine the ranges of magnetic and mineralogical stabilities with respect to thermal demagnetization or with respect to a chosen solvent; (3) to discover the best methods of demagnetization to isolate a particular CRM; and (4) to develop the needed experimental procedures for ascertaining the temporal sequence of multicomponent CRMs. In short, a more thorough knowledge of CRM is required to interpret paleomagnetic records with complex geologic histories, to further the contributions of paleomagnetism to Earth magnetism and tectonic reconstructions.

Acknowledgements

Preparation of this article benefited from discussions with P. Arason, H. Audunsson, and P. Roperch and financial support of the National Science Foundation.

SHAUL LEVI

References

Bagin, V. I., and L. M. Malumyan, 1976, Iron containing minerals in oil-impregnated sedimentary rocks from a producing rock mass of Azerbaidzhan, *Izv. Acad. Sci., USSR, Phys, Solid Earth* (English translation) **12**, 273-277.

Becker, J. J., 1957, Magnetic method for the measurement of precipitate particle sizes in a Cu-Co alloy, *Jour. Metals* **9**, 59-63.

Beetz, W. von, 1860, Ueber die inneren Vorgänge, welche die Magnetisirung bedingen, *Annalen der Physik und Chemie* **111**, 107-121.

Benthien, R. H., and R. D. Elmore, 1987, Origin of magnetization in the Phosphoria Formation at Sheep Mountain, Wyoming: A possible relationship with hydrocarbons, *Geophys. Research Letters* **14**, 323-326.

Blackett, P. M. S., 1956, *Lectures on Rock Magnetism*, The Weizmann Science Press of Israel, 131p.

Collinson, D. W., 1967, Chemical demagnetization, in K. M. Creer and S. K. Runcorn, eds., *Methods in Paleomagnetism*. Amsterdam:Elsevier, 306-310.

Collinson, D. W., 1974, The role of pigment and specularite in the remanent magnetism of red sandstones, *Royal Astron. Soc. Geophys. Jour.* **38**, 253-264.

Donovan, T. J., R. L. Forgey, and A. A. Roberts, 1979, Aeromagnetic detection of diagenetic magnetite over oil fields, *Am. Assoc. Petroleum Geologists Bull.* **63**, 245-248.

Haigh, G., 1958, The process of magnetization by chemical change, *Philos. Mag.* **3**, 267-286.

Hornafius, J. S., 1984, Origin of remanent magnetization in dolomite from the Monterey Formation, in R. E. Garrison, M. Kastner, and D. H. Zenger, eds., Dolomites of the Monterey Formation and Other Organic-Rich Units, *Soc. Econ. Paleontologists and Mineralogists Pacific Sec. Pub. 41*, 195-212.

Johnson, H. P., and R. T. Merrill, 1973, Low-temperature oxidation of a titanomagnetite and the implications for paleomagnetism, *Jour. Geophys. Research* **78**, 4938-4949.

Karlin, R., M. Lyle, and G. R. Heath, 1987, Authigenic magnetite formation in suboxic marine sediments, *Nature* **326**, 490-493.

Kobayashi, K., 1959, Chemical remanent magnetization of ferromagnetic minerals and its application to rock magnetism, *Jour. Geomagnetism and Geoelectricity* **10**, 99-117.

Kobayashi, K., 1961, An experimental demonstration of the production of chemical remanent magnetization with Cu-Co alloy, *Jour. Geomagnetism and Geoelectricity* **12**, 148-164.

Koenigsberger, J. G., 1938, Natural residual magnetism of eruptive rocks, *Terr. Magn. Atm. Electr.* **43**, part 1: 119-130, part 2: 299-320.

Larson, E. E., 1981, Selective destructive demagnetization, another microanalytic technique in rock magnetism, *Geology* **9**, 350-355.

Larson, E. E., and T. R. Walker, 1975, Development of CRM during early stages of red bed formation in late Cenozoic sediments, Baja, California, *Geol. Soc. America Bull.* **6**, 639-650.

McCabe, C., R. Van der Voo, D. R. Peacor, R. Scotese, and R. Freeman, 1983, Diagenetic magnetite carries ancient yet secondary remanence in some Paleozoic sedimentary carbonates, *Geology* **11**, 221-223.

McCabe, C., R. Sassen, and B. Saffer, 1987, Occurrence of secondary magnetite within biodegraded oil, *Geology* **15**, 7-10.

Marshall, M., and A. Cox, 1971, Effect of oxidation on the natural remanent magnetization of titanomagnetite in suboceanic basalt, *Nature* **230**, 28-31.

Maurain, Ch., 1901, Propriétés des dépots électrolytiques de fer obtenus dans un champ magnétique, *Jour. Physique*, ser. 3, **10**, 123-135.

Maurain, Ch., 1902, Sur les propriétés magnétiques de lames trés minces de fer et de nickel, *Jour. Physique*, ser. 4, **1**, 90, 151-156.

Meiklejohn, W. H., 1953, Experimental study of the coercive force of fine particles, *Revs. Mod. Physics* **25**, 302-306.

Merrill, R. T., 1975, Magnetic effects associated with chemical changes in igneous rocks, *Geophys. Surv.* **2**, 277-311.

Néel, L., 1949, Théorie du trainage magnétique des ferromagnétiques en grains fins avec applications aux terres cuites, *Ann. Geophys.* **5**, 99-136.

Néel, L., 1955, Some theoretical aspects of rock-magnetism, *Advances in Physics* **4**, 191-243.

Özdemir, Ö., and D. J. Dunlop, 1985, An experimental study of chemical remanent magnetizations of synthetic monodomain titanomaghemites with initial thermoremanent magnetizations, *Jour. Geophys. Research* **90**, 11513–11523.

Thellier, E., and O. Thellier, 1959, Sur l'intensité du champ magnétique terrestre dans le passé historique et géologique, *Ann. Géophys.* **15**, 285–376. 1959.

Cross-references: *Demagnetization; Detrital Remanent Magnetization (DRM); Magnetic Analysis of Rock Fabric; Magnetic Domains; Magnetic Properties of Minerals; Magnetic Self-Reversal; Natural Remanent Magnetization; Paleomagnetic Field: Intensity; Paleomagnetism: Deep-Sea Sediments; Rock Magnetism; Thermoremanence; Viscous Remanent Magnetization (VRM) and Viscous Magnetization.*

CONTINENTAL COLLISION ZONES: SEISMOTECTONICS AND CRUSTAL STRUCTURE

Continental collision zones usually develop along convergent plate boundaries when a considerable mass of continental lithosphere that is connected to a subducting oceanic lithosphere collides with an overriding continental plate. This process often results in the development of dramatic surface and subsurface geological processes including major mountain building, continental plateau uplift, and foreland basins. Continental collision episodes ultimately end in changes in the configuration of the nearby plate boundaries. The collision zones then evolve through erosion and isostatic adjustment into mature, relatively inactive orogenic belts. The whole cycle may last tens of millions of years and possibly longer.

The Himalayan and Zagros continental collision zones are often cited in the geological literature as type examples of such processes, and along with the Alpine belt (from Turkey to Spain) probably represent the most active collisional belts during the Cenozoic and Quaternary times. Of course, not all mountain belts are the result of collision between two major continental plates along convergent plate boundaries. Some of the currently most active mountain-building processes are associated with noncollisional (i.e., ocean-continent) convergent plate boundaries, such as the Andes of western South America, or are located in an intraplate setting (i.e., within a continent), such as the Rockies of western North America and the Atlas system of northwest Africa, or are even associated with an extensional tectonic setting, such as the East African rift system. This paper, however, focuses on active collisional mountain belts and specifically on the Himalayan and Zagros collisional zones. It should be emphasized that collision zones, in general, are preceded by episodes of ocean-continent subduction, producing accretionary prisms, volcanic arcs, back-arc mobilization, and tectonism.

Any discussion of the deep structure and evolution of the Himalayas and Zagros cannot be complete without addressing the nature and evolution of the Tibetan and Iranian plateaus located behind these two collision zones, respectively. These uplifted regions appear to be intimately related to the evolution of the nearby collisional zones, and hence understanding their deep structure is an essential part of deciphering the deep structure of the collisional zones. The following includes a discussion of the seismotectonics and deep structure of the Himalayas and Zagros and a hypothesis on how these two collisional belts may represent different stages in an evolutionary cycle.

The Himalayan Collision Zone

The Himalayan system is the result of collision between the Indian subcontinent and what is now Tibet, about 50 million years ago. The initial contact is represented by the Indus-Tsangpo suture, which was the site of the plate boundary along which the Tethyan oceanic lithosphere that separated India and Tibet was subducted beneath Tibet (see Fig. 1). The Trans-Himalayan (Kangdese) plutons of southernmost Tibet, located just north of the suture, represent the core of the volcanic arc of that annihilated subduction system. Convergence between India and Tibet continued after their initial contact, up to the present (about 5 cm/yr) (Molnar and Tapponnier, 1975) and has resulted in the development of two northerly dipping new intraplate convergent zones: the Main Central Thrust (MCT) and the Main Boundary Thrust (MBT) (Valdiya, 1980). The MCT appears to have developed since about Miocene time with minor, if any, recent movements. It separates the High Himalaya from the Lesser Himalaya. The MBT has developed since about Pliocene time and is still active. It is made up of a series of thrusts that separate the Lesser Himalaya from the Tertiary Siwalik sedimentary belt (the Sub-Himalaya belt) and the adjacent Gangetic foredeep. Thus the impressive Himalayan mountains have developed entirely within the advancing Indian subcontinent during about the last 20 million years. That is, these mountains represent an intraplate deformation located to the south of the initial contact between India and Tibet.

Though most of these surface geological structures are reasonably well investigated at the surface, the behavior of these structures at depth is to a large extent unknown. Knowledge of the architecture of these structures at depth is essential if we want to understand why the High Himalaya has attained its impressive height, and why, for example, the Lesser Himalaya has not developed into a High Himalayan type. Moreover, the estimated convergence between

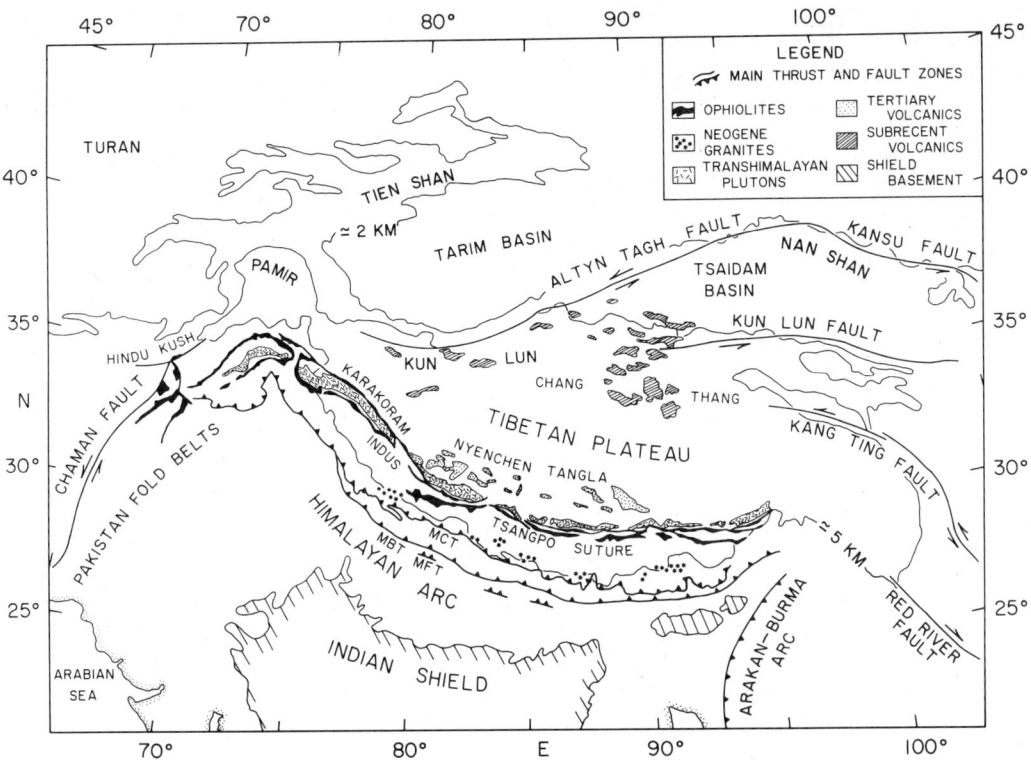

FIGURE 1. Map of Himalayan arc, Tibetan plateau, and surrounding regions showing major tectono-structural features. The Tibetan plateau is approximately defined by the 5-km contour.

India and Asia since the initial continental contact varies between 2000–3000 km. A major unresolved question is how this convergence is being accommodated. Different hypotheses have been advanced to explain it. The hypotheses fall into two major categories: (1) The Tibetan terranes are shortened and thickened in an accordion fashion in front of the advancing rigid indentor of India. Subsequently, or possibly concurrently, some parts of the Tibetan terranes are extruded and "escaped" away from the advancing Indian plate along major strike-slip faults that emanate out of Tibet (Tapponnier et al., 1986). (2) The Greater Indian subcontinent is underthrust beneath Tibet. These models are not exclusive, and a possible combination of different aspects of both models may ultimately provide the required explanation of two major observations of the Tibetan terranes—the high elevation (average of about 5 km) and the double crustal thickness of about 70 km.

The relatively simple arcuate geometry of most of the Himalayan mountain belt is yet another remarkable observation. In particular, the Himalayan topographic front, i.e., the zone where the averaged elevations relatively abruptly change from about 2 km to about 5 km, defines a small circle on a sphere and closely coincides with the MCT (Seeber et al., 1981). This occurrence suggests a genetic relationship between the bending and underthrusting of the Indian subcontinent beneath the Himalayan blocks and the formation of the High Himalaya just north of the MCT.

The Tibetan Plateau. It is reasonably well documented that the Tibetan crust was formed by the progressive accretion of a few (three?) continental or island-arc-type fragments to the Asian lithosphere during Mesozoic times, before the initial contact between the Indian subcontinent and southern Tibet. It is not surprising, therefore, that the structural trends in Tibet, including faulting and folding, are predominantly east-west oriented. Mesozoic and some Tertiary sediments are locally strongly deformed, but Tibet lacks major fold and thrust belts of the magnitude, say, of the Andean or the Rockies systems. Moreover, available information suggests that intense local deformation on the surface of Tibet is documented along or near the Mesozoic sutures that separate the different fragments of the Tibetan crust (Changfa et al., 1986). Most of this deformation is not accurately dated, and it is possible that most of it is Paleogene in age. Tertiary volcanism occurred in many parts of Tibet. In particular, the Nyenchen Tangla volcanic belt in southern Tibet is characterized by predominantly silicic volcanic rocks of Paleogene and possibly younger age (see Fig. 1), while the Chang Thang volcanic belt of northern Tibet is of Neogene and Quaternary age and includes large basaltic lava flows. In summary, Tibet appears to have experienced crustal mobilization, including volcanism, folding, and faulting (both thrust and strike-slip types), during most of the Paleogene time and possibly at a lesser intensity during some parts of the Neogene and Quaternary times.

Results based on geophysical studies, utilizing surface waves, refraction, and gravity, show that the crustal thickness of Tibet is about 70 km and that the uppermost mantle compressional and shear velocities are about 8.2 and 4.7 km/s, respectively. These uppermost mantle velocities are typically obtained in stable and/or shield continental regions, such as those found beneath the Indian shield farther to the south. Some researchers (Chen and Molnar, 1981; Lyon-Caen, 1986) argue that the relatively high uppermost mantle velocities beneath Tibet are restricted to a mantle lid that ranges in thickness only between about 10–20 km. The argument is based on analysis of shear wave travel times that indicate a slower shear velocity structure in the upper mantle beneath Tibet relative to that beneath the Himalayas and/or India. However, the shear velocity results may be explained by more pronounced lateral velocity variations in the asthenosphere that exists beneath India and Tibet. Thus a thicker mantle lid, say 60–80 km, could be present beneath Tibet. The observation that Sn (i.e., high-frequency shear waves that predominantly propagate in the mantle lid part of the lithosphere) efficiently propagates beneath most of Tibet (see Fig. 2) (Barazangi and Ni, 1982) lends support to the interpretation that a thicker, shield-like mantle lid exists beneath most of Tibet. This is based on some recent modelling of Sn and on the empirical observation that for a relatively thin mantle lid, about 40–60 km, there appears to exist a shadowing (cutoff) effect of the low-Q asthenosphere on Sn propagation. That is, available information suggests that to observe Sn at considerable epicentral distances the mantle lid of the lithosphere in which Sn propagates has to be thicker than about 40–60 km. It is clear, however, that new data with more resolution are required before the question of the thickness of the Tibetan mantle lid can be satisfactorily resolved.

The recent tectonics of Tibet appear to be dominated by the uplift of the plateau, some volcanism, and by strike-slip faulting and north-trending normal faults and graben-type structures. Shallow earthquakes, with focal depths between

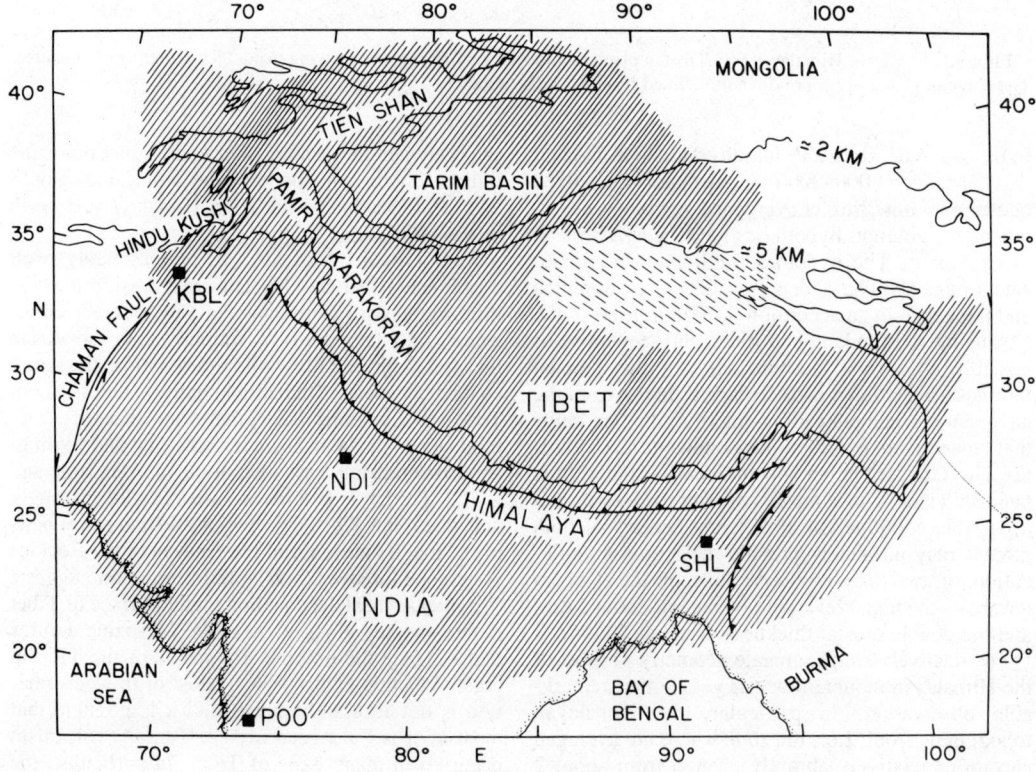

FIGURE 2. Map showing zones of efficient (solid lines) and inefficient (dashed lines) Sn propagation. Seismic stations used are also shown. Results shown on this map are based on data produced by all of these stations. (From Barazangi and Ni, 1982.)

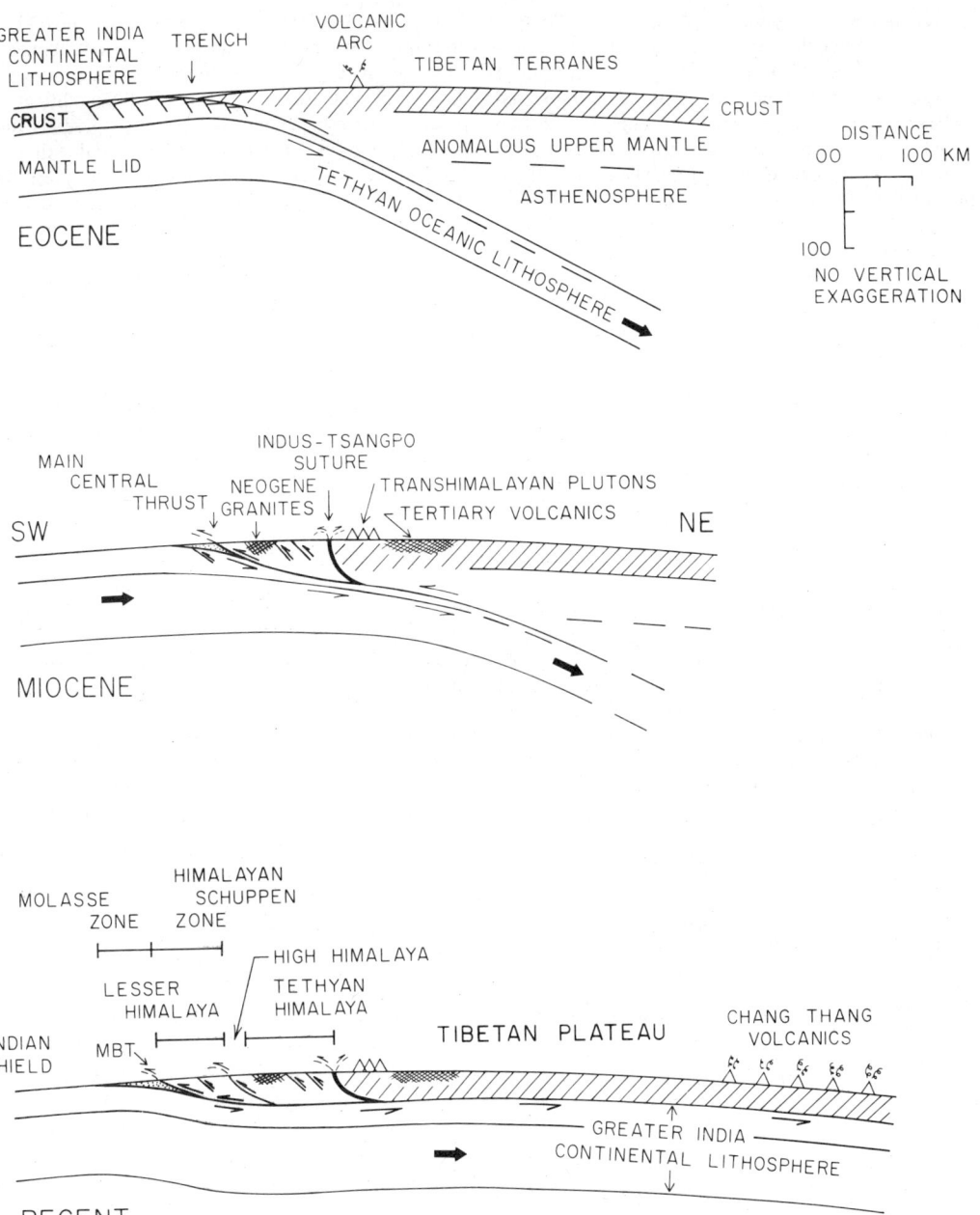

FIGURE 3. Schematic cross sections perpendicular to Himalayan arc showing an evolutionary sequence of how convergence between Indian and Asian plates may result in shallow-angle underthrusting of Greater India continental lithosphere beneath present-day Tibetan plateau. (From Barazangi and Ni, 1982.)

about 5–15 km occur throughout Tibet. The focal mechanisms of the larger events show combinations of normal and strike-slip faulting with T axes approximately oriented east-west. The mechanisms are interpreted to indicate that Tibet is currently under an east-trending extension caused by gravity acting on the thickened Tibetan crust, itself caused by the continued convergence of India and Asia. Limited evidence suggests that the major episode of uplift of Tibet is post-Miocene. The available data, however, appear not to further resolve the details of the history of the uplift.

A conceptual model for the deep structure and evolution of Tibet is shown in Fig. 3. The schematic representation of the model emphasizes the following points: (1) After the initiation of continental collision and up to Miocene time, the convergence between India and Asia was mainly accommodated by shortening along the leading edges of the two colliding plates, partial subduction of the Indian continental lithosphere, and primarily by mobilization of the Tibetan crust by volcanism, shortening (especially along Mesozoic sutures), and lateral extrusion of some Tibetan crustal fragments. (2) The continued convergence between India and Tibet resulted in the fracturing of the leading edge of India along an intracontinental megathrust, the MCT, in Miocene time, and the Greater India continental lithosphere began to underthrust the Tibetan terranes at a relatively shallow angle. The Tethyan Himalaya block represents the detached leading edge of Greater India, while the Lesser Himalaya block represents a schuppen zone that consists of material scraped off the underthrusting Indian plate. The high elevation of the Tibetan plateau and the double thickness of the Tibetan crust are the result of underthrusting of India beneath Tibet.

The conceptual model of underthrusting as presented above must be considered an oversimplification of the actual geological processes involved. Recent refraction results near the Indus-Tsangpo suture show complex lower crustal boundaries that may provide a clue to the detailed mechanical aspects of underthrusting. Recently Zhao and Morgan (1985) proposed a modification of the underthrust model where the leading edge of the underthrusting Indian plate is injected into the very weak lowermost crust and uppermost mantle material that exists beneath Tibet and hydraulically uplifts the total width of the Tibetan plateau at the same time. If this modification of the underthrust model is correct, it predicts that high-attenuation, asthenospheric material exists at the base of the crust in northern Tibet, i.e., the underthrusting Indian plate has not yet reached northern Tibet. A mapped zone of high-attenuation (low Q) Sn propagation in northern Tibet (see Figs. 1 and 2), which closely coincides with Quaternary basaltic volcanics of the Chang Thang region, is consistent with the above prediction. It should also be emphasized here that the underthrust model, or any other variation on it, does not necessarily eliminate the continued mobilization and extrusion of the Tibetan crust, though it reduces the magnitude of this effect.

The underthrust model has dramatic implications for the deep structure and evolution of the continental lithosphere. If India can underthrust Tibet in a relatively very short geological time, say 10-20 my, then such a process may have repeatedly occurred throughout geologic history. The geological consequences of such a model (see Fig. 4), especially for fluids migration, metamorphism, and magma genesis, among other effects, are yet to be explored.

Seismotectonics and Deep Structure of the Himalayas. The Himalayas initially developed in response to intraplate deformation (Miocene time) after the initial collision (Eocene time) between India

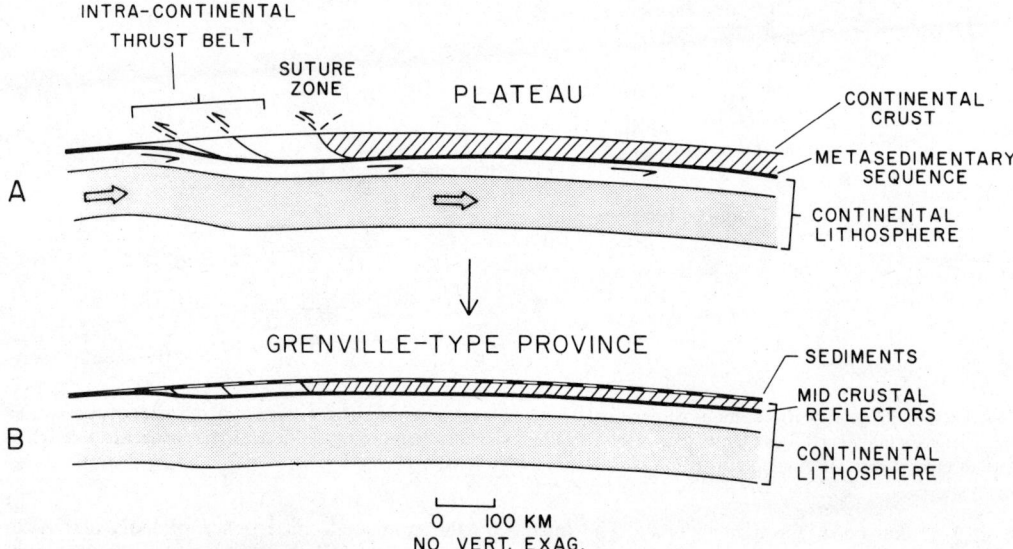

FIGURE 4. Upper cross section schematically shows possible present geometries of converging plates in Himalayan arc-Tibetan plateau region. Metasedimentary sequence may exist on top of Greater India plate as it underthrusts Tibetan plateau. Lower cross section schematically shows possible evolution of Tibetan-type terrane into Grenville-type province after considerable erosion of upper crust and possibly deposition of a new sedimentary sequence. (From Barazangi and Ni, 1982.)

CONTINENTAL COLLISION ZONES: SEISMOTECTONICS AND CRUSTAL STRUCTURE

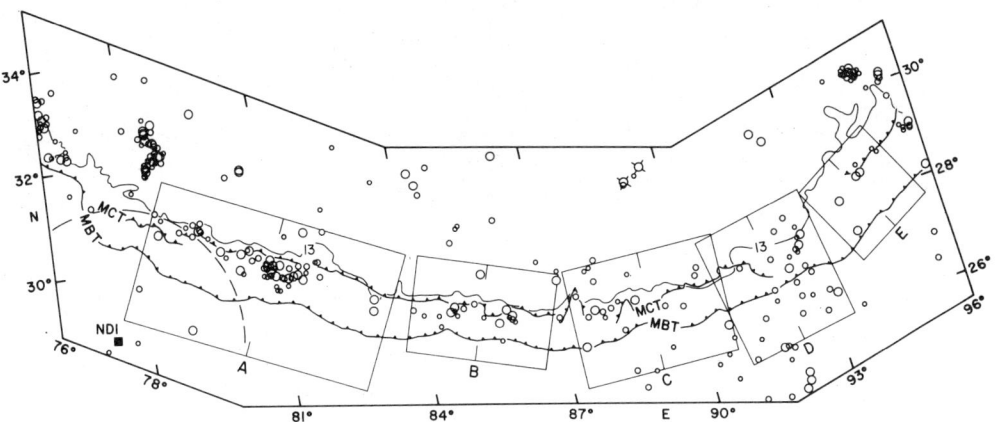

FIGURE 5. A selected epicentral map of the Himalayan region showing International Seismological Centre locations for earthquakes occurring between 1961 and 1981. Large, medium, and small open circles represent very good, good, and fair quality of epicenters on the basis of station distribution and travel time residuals. The two open circles with crosses represent intermediate-depth events. The MCT and many of the Himalayan earthquakes lie south of the 13,000-ft (≈ 4 km) contour. Most of the epicenters lie along a relatively narrow belt, about 50 km wide, between the MBT and the MCT, and the majority of these events are located just south of the MCT. Infrequent seismic activity occurs in the Ganga basin to the south of the MBT, except in the Shillong plateau region. Seismicity within the five rectangles (A, B, C, D, and E) along the Himalayan arc is projected linearly on the center line of each rectangle. (From Ni and Barazangi, 1984.)

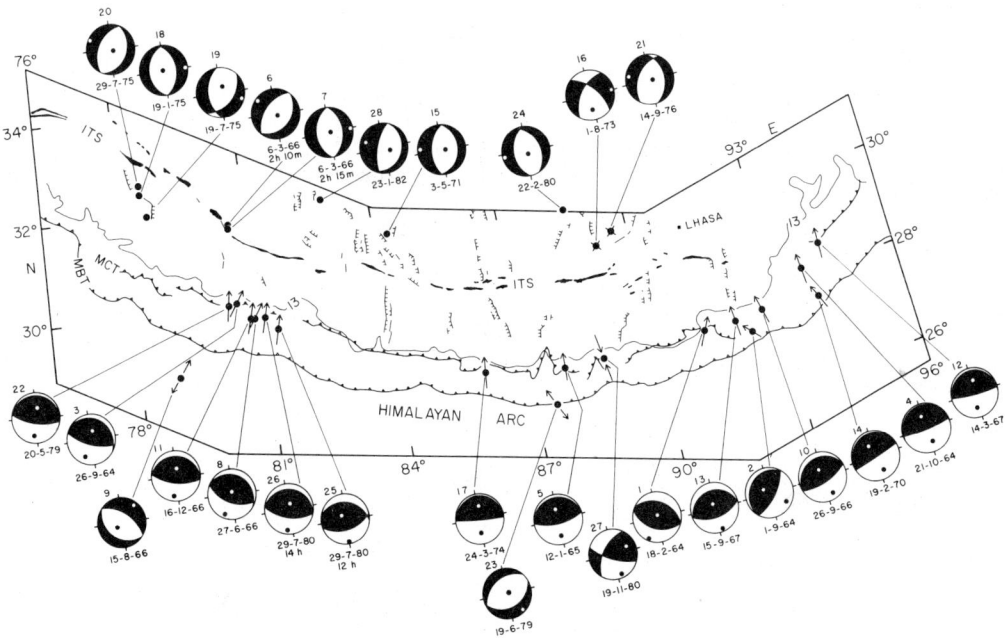

FIGURE 6. Seismotectonic map of the Himalayan region. Darkened region in the fault plane solution indicates compression. T and P axes are shown by open and solid circles, respectively. Arrows indicate direction of slip if one assumes the shallow-angle nodal planes as fault planes for the thrust-type events. Two double-pointed arrows show the directions of extension for two normal faulting events that occurred beneath the Ganga basin and Sub-Himalaya. Two intermediate-depth events located to the north of ITS are represented by solid circles with crosses. Lines with hatches on one side are normal faults. Positions of ITS, MCT, MBT, and spatially averaged 13,000-ft (≈ 4 km) contour are also shown. (From Ni and Barazangi, 1984.)

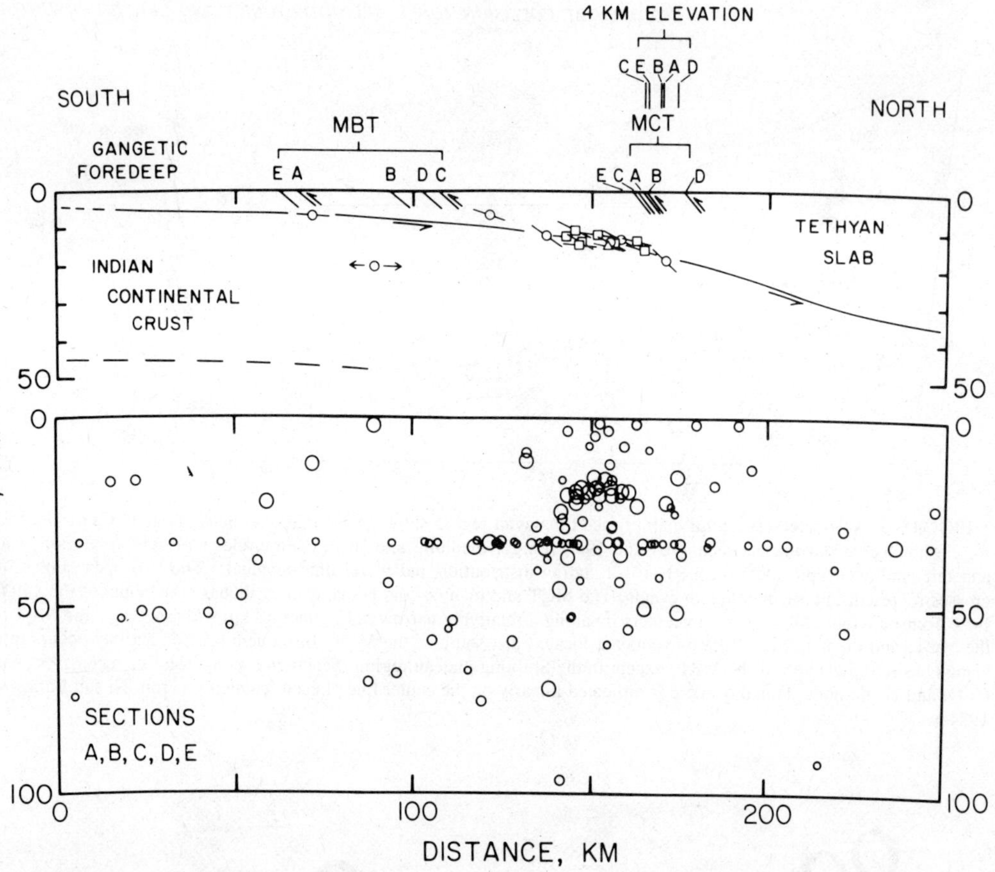

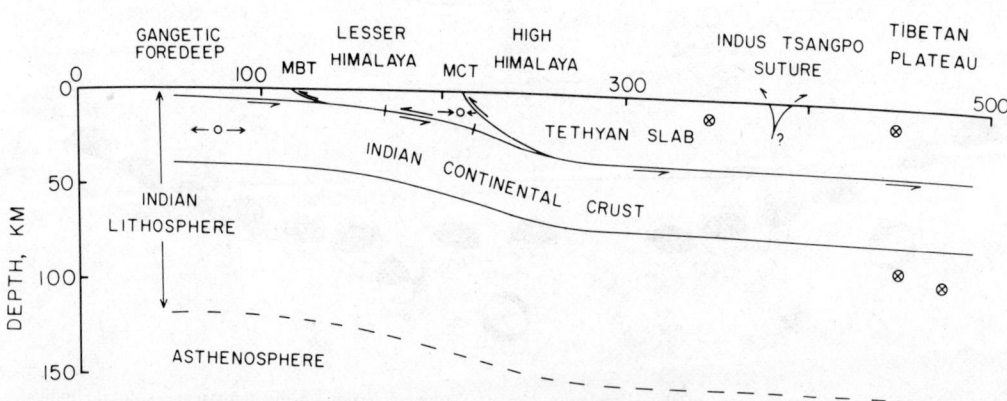

FIGURE 7. A composite cross section of all five individual sections using the MCT and the 4-km elevation contour as a reference for projection. Bottom section shows ISC hypocenters. The larger the open circle the better the epicenter location. Notice that the majority of earthquakes are located south of the MCT. In the top section the projected low-angle nodal planes for thrust-type focal mechanisms are shown as line segments drawn across the accurately located depths of hypocenters. (From Ni and Barazangi, 1984.)

FIGURE 8. A schematic cross section of the present geometries of converging plates in the Himalaya-Tibetan region. Events with normal-type focal mechanisms located in the shallow crust and uppermost mantle of Tibet are shown as open circles with crosses. The most important conclusion of this study is that all Himalayan thrust-style events can be interpreted most simply to define a part of the detachment between the underthrusting Indian plate and the upper Himalayan blocks. (From Ni and Barazangi, 1984.)

and Tibet along the Indus-Tsangpo suture. There are controversies regarding the behavior of the different Himalayan thrusts at depth. However, recent studies of gravity (Lyon-Caen and Molnar, 1985) and the spatial distribution and focal mechanisms of earthquakes have provided some new insight into the geometry of these structures (Molnar, 1984; Ni and Barazangi, 1984).

The Himalayas are seismically very active. During the past century, four great earthquakes (M > 8) have occurred along the Himalayas, with a large number of small and medium size events. Seeber et al. (1981) proposed that a master detachment underlies the Himalayas. The detachment separates the underthrusting Indian continental crust from the Himalayan crustal blocks. The great Himalayan earthquakes occur along it. Local seismic networks exist only in the northwestern part of the Himalayas in Pakistan and temporarily along some small areas in other parts of the Himalayas. Hence available earthquake hypocenters for most of the Himalayas are based on teleseismic locations. Usually, the epicenters of the better located events are accurate to within about 10–20 km. However, the depths of such events could be in error by as much as 50 km. Careful classification of how well the events are located, based on qualitative selection criteria, shows that the better located epicenters are concentrated in a relatively narrow zone, about 50 km wide, lying just south of the northerly dipping MCT thrust zone

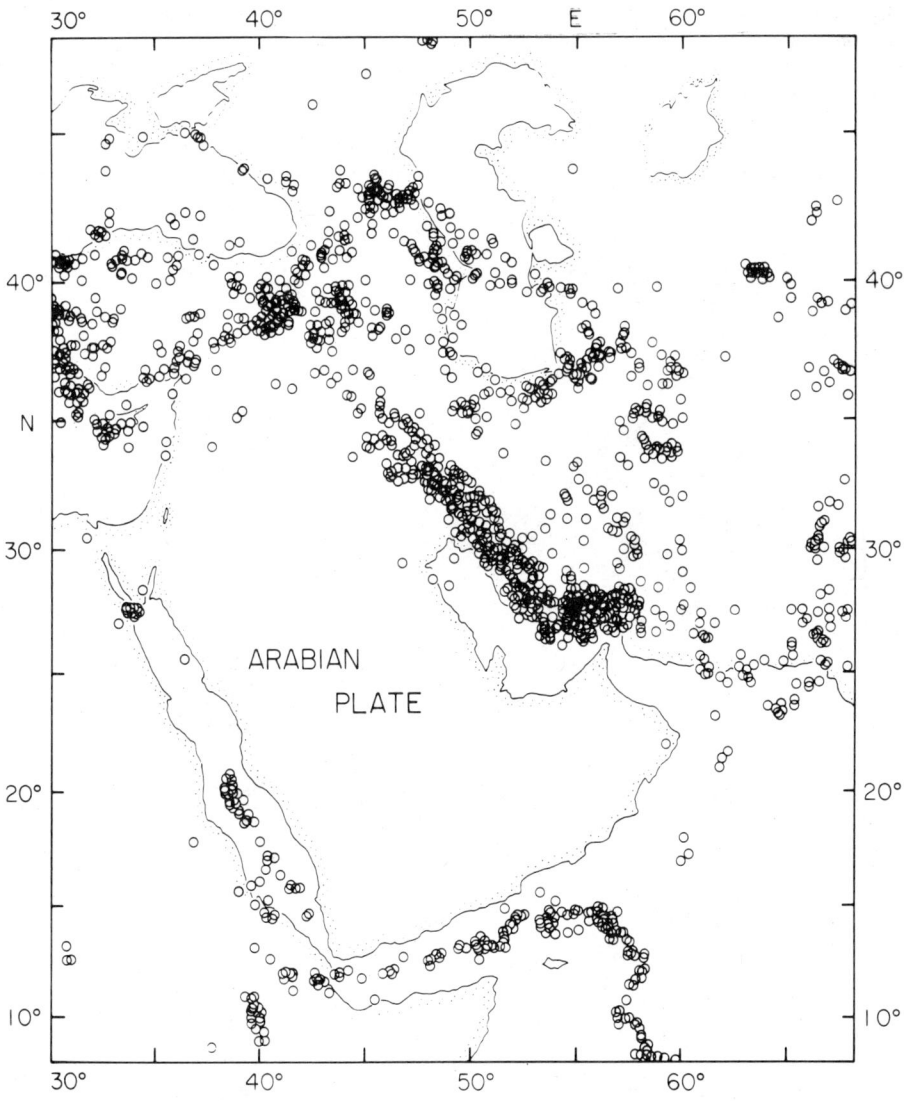

FIGURE 9. Seismicity map of the Middle East region, including the Zagros and central Iran (from U.S. Geological Survey, Preliminary Determination of Epicenters 1962–1977.)

(see Fig. 5). The spatial distribution of the events clearly indicates that they do not occur along the subsurface extension of the MCT (Ni and Barazangi, 1984). Medium size events ($5.5 < M < 6.9$) produce large enough P waves to be teleseismically recorded at both short- and long-period seismograms. Hence the focal mechanisms of such events can be obtained (see Fig. 6) and accurate focal depths (within about 3 km) can be determined by comparison of synthetic seismograms, calculated from theoretical far-field response of a shear dislocation source in layered elastic medium, and vertical long-period P waveforms. Medium size events with accurately determined focal depths, which are located along about 1700 km length of the Himalayan arc, define a relatively simple, planar zone ranging from about 10 km–20 km depth and having an apparent northward dip of about 15° (see Fig. 7). Proper care is essential in projecting these events onto a single, composite section; otherwise the events will not define a simple geometry.

Except for one earthquake, all available focal mechanisms of events within this zone indicate shallow (<30°), north dipping thrusts that are approximately coincident with the planar seismic zone. This shallow, north dipping zone can be interpreted to delineate a part of the master detachment of the Himalayas that separates the underthrusting Indian plate from the overriding Lesser Himalayan crustal block (see Fig. 8). The spatial extent and the geometry of this "interplate" thrust zone clearly indicate that the MBT, rather than the MCT, is currently the active seismogenic structure of the Himalayan arc. It is probable that the great Himalayan events occur along the same detachment surface as defined by the thrust-type, medium size events.

Limited borehole data beneath the Ganga basin and the Sub-Himalaya indicate that the upper surface of the underthrusting Precambrian Indian plate dips northward at an angle of about 2°–3°. In Fig. 8 we extrapolate this dip to join the geometry of the seismogenic zone, as defined by the spatial distribution and focal mechanisms of medium size events, in order to infer the geometry of the upper surface of the underthrusting Indian plate beneath the Himalayas. Detailed analysis of available gravity data by Lyon-Caen and Molnar (1985), assuming

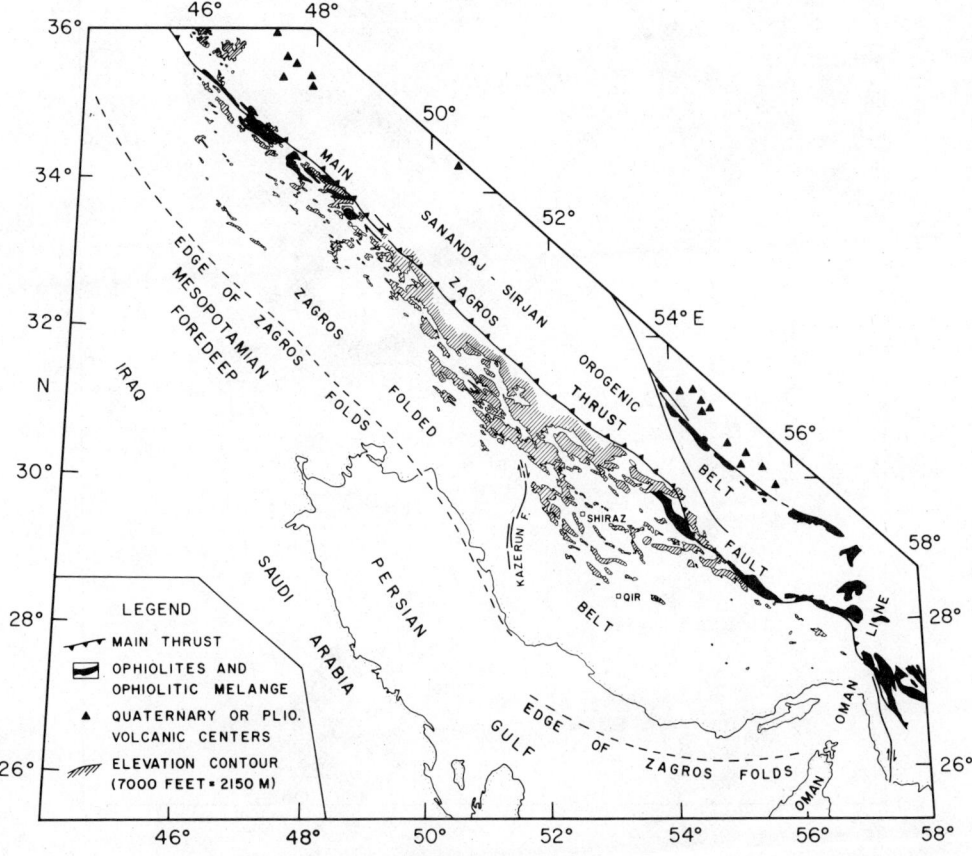

FIGURE 10. Map of the Zagros region showing major tectonic structural features. Areas located to the southwest of the Main Zagros Thrust (MZT) and higher than 7000 feet (2150 m) are shaded. (From Ni and Barazangi, 1986.)

that the Indian plate was flexed down as an elastic plate under the load of the overlying Himalayas, produces a geometry of the upper surface of the Indian plate that is remarkably similar to that based on earthquake data, including about 15° dip of the Indian plate beneath the High Himalaya. The change in dip of the underthrusting Indian plate produces a crustal-scale, ramp-like structure beneath the High Himalaya and, with the continued convergence of the Indian plate, such a structure requires the material above it to rise in comparison to the material (farther to the south) that exists above the gently (2°-3°) dipping detachment surface. This may explain the dramatic uplift of the Himalaya along a topographic front that is closely associated with the change of dip of the underthrusting Indian plate at depth.

The available data and observations appear to support a model in which the underthrusting Indian plate beneath the Himalayas continues to underthrust most of Tibet. The occurrence of two medium size, intermediate-depth events near Lhasa in southern Tibet (see Figs. 5 and 8) can be interpreted by the underthrust model to have occurred in the uppermost mantle of the underthrust Indian plate. This underthrust model as discussed above is schematic in nature and presented to emphasize a concept. In reality, the underthrusting process is probably a very complex one, with varied interaction between the underthrusting and overriding plates, both temporally and spatially.

Zagros Collision Zone

The Zagros mountain belt, which extends for a distance of about 1500 km along the southwestern part of Iran and northeastern Iraq, is the result of the collision in Miocene time between the Arabian plate and the central Iranian blocks (see Figures 9 and 10). Before this relatively young collision, geological evidence clearly indicates that oceanic lithosphere,

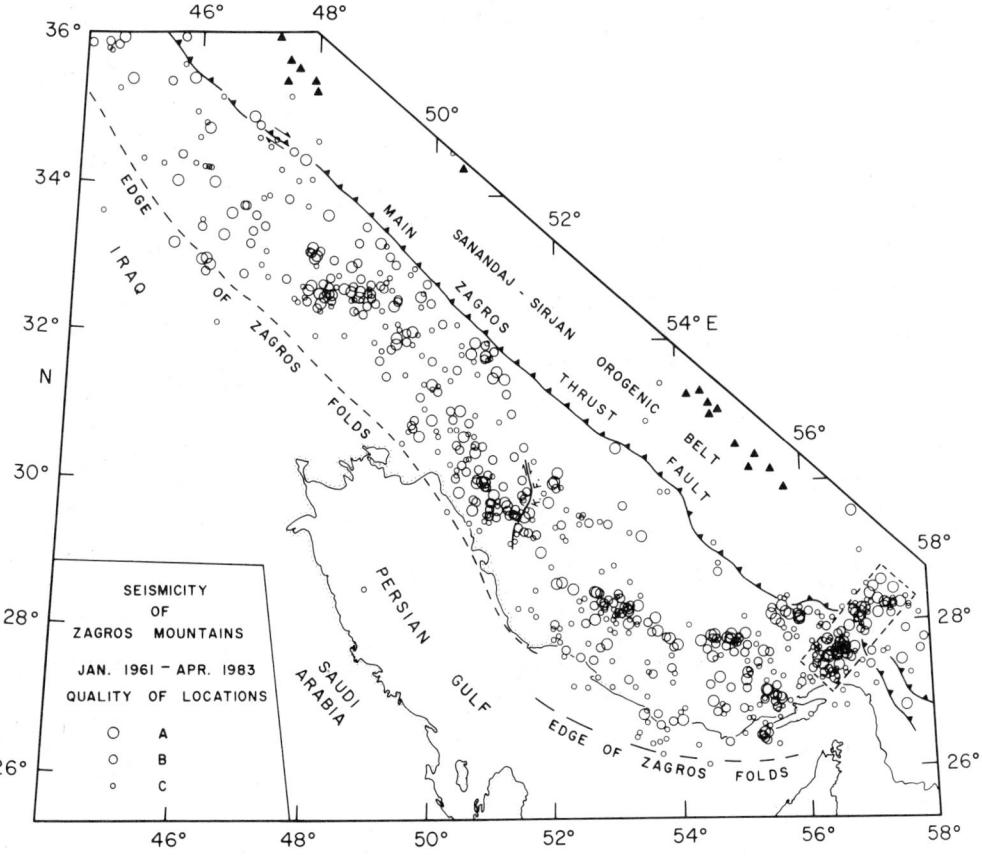

FIGURE 11. Seismicity map of the Zagros region showing ISC locations for earthquakes occurring between 1961 and April 1983. Large, medium and small open circles represent very good (A), good (B), and fair (C) quality of epicenters on the basis of azimuthal station distribution and travel time residuals. Very poor locations, generally with less than 20 stations reporting, are not included in this map. Epicenters are distributed in a relatively broad belt about 200 km in width and show great variation in the spatial distribution of events relative to the MZT. (From Ni and Barazangi, 1986.)

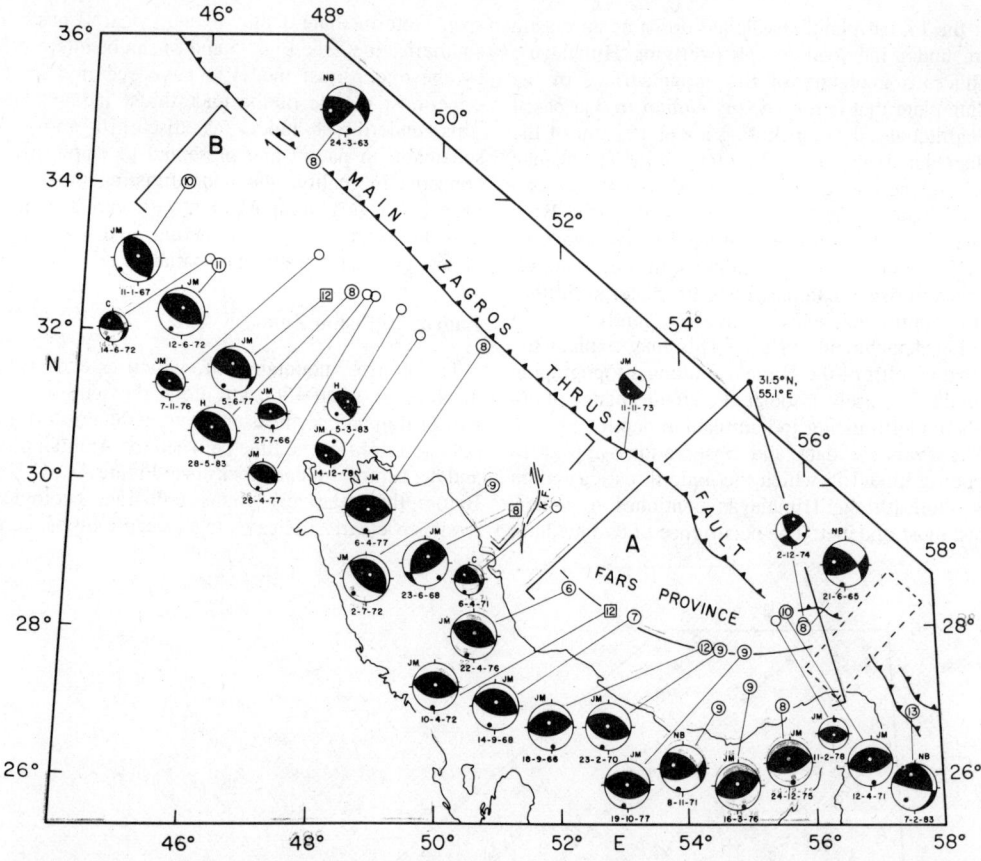

FIGURE 12. Seismotectonic map of the Zagros showing epicenters of larger events and their fault plane solutions. Darkened regions in fault plane solutions indicate compression. T and P axes are shown by open and solid circles, respectively. Small focal spheres are for those solutions that have poorly constrained nodal planes. The numbers inside large open squares and circles are focal depths in km determined by comparing observed long-period, vertical component seismograms with synthetic seismograms. Hypocenters of large events with constrained focal depths in the blocked regions (A and B) are projected onto cross sections. (From Ni and Barazangi, 1986.)

part of the Arabian plate, was subducting beneath central Iran. The continued magmatic activity in Neogene and Quaternary times along the well-defined Zagros volcanic arc, as well as the occurrence of at least one well-determined, intermediate-depth earthquake beneath the volcanic arc suggest that the pre-Miocene subduction process has not yet terminated at depth, and it is possible that some oceanic lithosphere is still attached to the colliding Arabian plate.

Geological and paleomagnetic evidence indicates that most of the central Iranian blocks, such as the Lut and Tabas blocks, were accreted to the southern margin of Asia, probably in Paleozoic time. Since that time these blocks have been subjected to numerous orogenic activities that are closely associated with nearby convergent plate boundaries. The Main Zagros Thrust (MZT) marks the suture zone between the two colliding plates of Arabia and Iran. Two major collisional episodes, during late Mesozoic and Miocene times, are recognized (Sengor and Kidd, 1979). The continued convergence of the Arabian plate during Neogene time (about 5 cm/yr) has resulted in more intense deformation of the suture and nearby regions.

In Pliocene time folding of the Zagros belt was initiated as the latest episode in the continued convergence of Arabia and Iran. The thick sedimentary cover of Zagros (about 5–10 km) are predominantly platform carbonates that were deposited on the Precambrian basement of the Arabian continental margin from late Precambrian to Miocene time. The Zagros sedimentary section is separated from the Precambrian basement by a detachment along the relatively thick (about 1 km) infra-Cambrian Hormuz salt deposits. Salt deposits are also present in the Mesozoic and Tertiary portions of the sedimentary section. A remarkable observation about the Zagros folded belt is the lack of mapped faults and the absence of fault breaks during

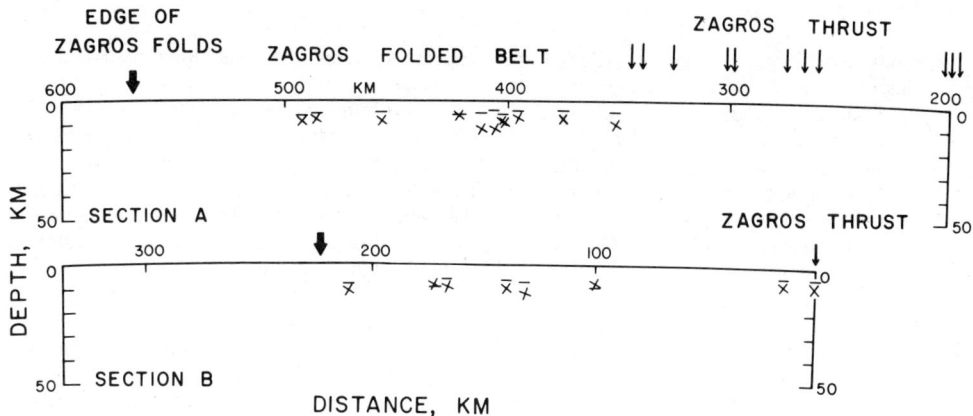

FIGURE 13. Cross sections (no vertical exaggeration) of accurately located hypocenters showing both nodal planes of fault plane solutions and position of MZT relative to each event. Focal depths range between 6 and 12 km. Cross section A is for the southern Zagros (Fars region) where seismicity follows approximately a semicircular band. In this section we project hypocenters along an arc with its center at 31.5°N and 55.1°E (see Fig. 12). The arrows indicate the projected positions of the Zagros Thrust. Cross section B is for those large events in northwestern Zagros where seismicity extends in a zone about 200 km wide parallel to the MZT. In cross section B (azimuth = N41°E) we use the MZT as a reference because its trend is parallel to both regional fold axes and the distribution of better located earthquakes. Solid bars located above or beneath the projected hypocenters indicate depth to upper surface of infra-Cambrian Hormuz salt beds. (From Ni and Barazangi, 1986.)

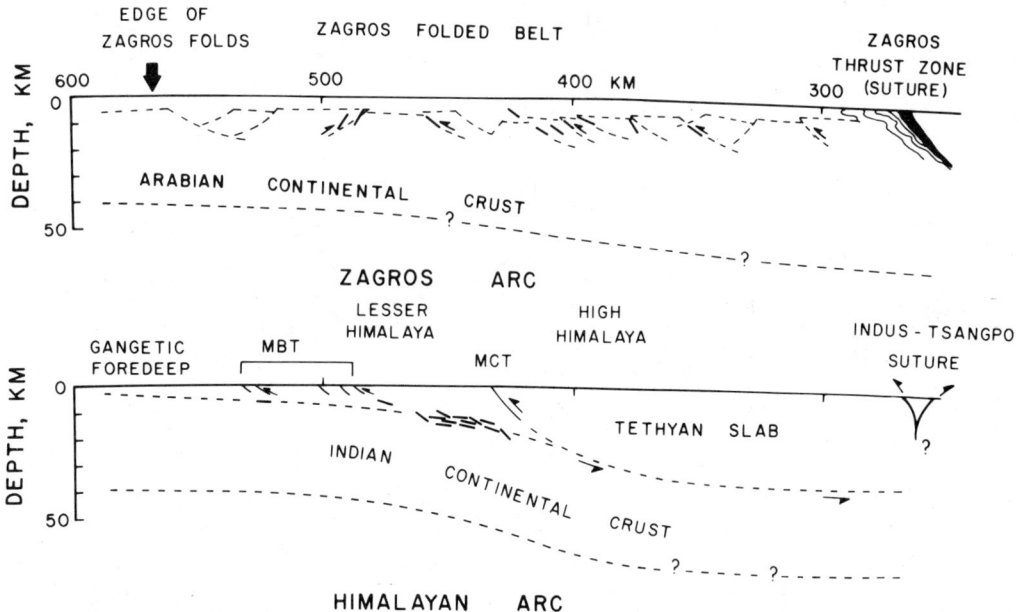

FIGURE 14. Schematic cross sections perpendicular to the Zagros (top) and Himalayas (bottom) collisional zones showing the relationship of accurately located hypocenters to crustal structures. The size and dip of fault-bounded blocks on the Zagros cross section are based on reflection profiles from the eastern continental margin of North America (Bally, 1981) and across the Bay of Biscay (Montadert et al., 1979). Here we emphasize the geometry of the upper surface of the basement beneath Zagros, and only one of the steeply dipping nodal planes (thick bar) of each event is shown. The seismogenic portion of the Zagros is about 150-200 km wide and is spatially comparable to an average Atlantic-type continental margin. This cartoon-type cross section is drawn to emphasize the consistency of the available seismological data with the hypothesis (Jackson, 1980) that Zagros is underlain by the Arabian continental margin that is being reactivated under compression. In the Himalayan cross section, thick bars denote the dip of shallow-angle nodal planes of thrust-type focal mechanisms. In contrast to the intraplate nature of the larger Zagros events, the Himalayan events represent an interplate deformation along a detachment surface that marks the upper surface of the underthrusting Indian plate. (From Ni and Barazangi, 1986.)

the occurrence of large earthquakes (Berberian, 1981). This observation and the relative simplicity of folding indicate that the Zagros sedimentary section acts as a plastic layer and is decoupled from the underlying basement along the Hormuz salt beds. The intensity of folding in the Zagros is at a maximum near the suture and decreases toward the Mesopotamian foredeep and the Persian Gulf, which indicates that the deformation front is migrating toward the foredeep, i.e., the amount of relative displacement along the detachment is decreasing southwestward toward the foredeep.

The Iranian Plateau. With a few depressions surrounded by mountainous terranes having an overall average elevation of about 1.5 km, the Iranian plateau is more of a basin-and-range-type province than a single, relatively uniform plateau. The idea that central Iran is made up of a few, relatively stable crustal blocks surrounded by deformation belts appears to be an oversimplification of recent geological processes. For example, pre- and post-Neogene and Quaternary deformation has been clearly documented within the Lut block, though it appears to be less than that along the boundaries of the block.

Central Iran is a very active tectonic environment (Jackson and McKenzie, 1984). Numerous active faults of both the strike-slip and thrust type, in addition to young folding, are mapped throughout the region (Berberian, 1981). Some of these structures are also associated with Quaternary volcanoes. Central Iran is seismically active, with well-documented instrumental and historic large earthquakes. Some of these large events have magnitudes that approach about 8. Most such events, moreover, are associated with surface faults. The surface faulting is usually of the reverse and/or strike-slip type. The mobilization of the crustal blocks of central Iran appears to be a consequence of the continued convergence between Arabia and Asia. In a way, all of central Iran represents a broad deformation belt between the relatively rigid plates of Arabia and Asia. The crustal thickness of central Iran is about 40 km, and uppermost mantle compressional and shear velocities are lower than those of shieldlike material (Asudeh, 1982). Sn waves do not

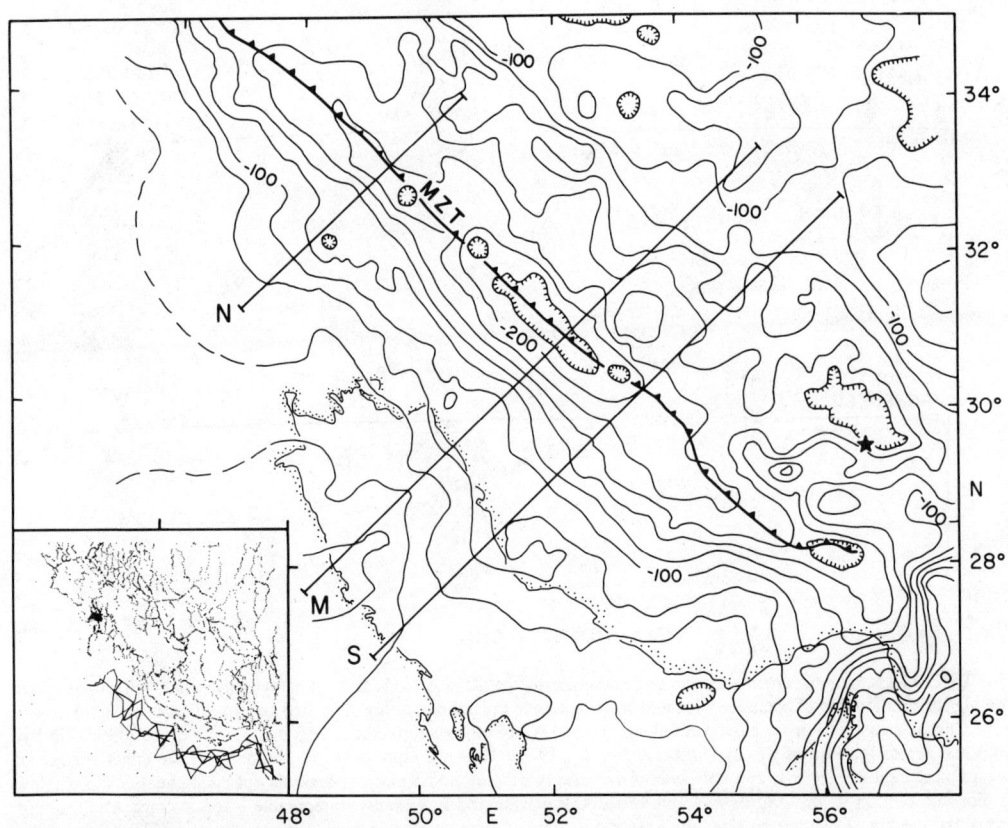

FIGURE 15. Simple Bouguer anomaly map of the Zagros region. Contour interval is 20 mGal. Solid lines, labelled N, M, and S, are gravity profiles. Star marks the location of an intermediate depth earthquake. Bouguer reduction density is 2.67 g/cm^3. Inset in figure shows the distribution of gravity observations within the region. (From Snyder and Barazangi, 1986.)

propagate across the northern part of the Iranian plateau (Kadinsky-Cade et al., 1981), where most Quaternary volcanism is located. Sn, however, efficiently propagates beneath the rest of the plateau, though with smaller amplitudes, relative to Pn, than is typical of stable and shield areas. The apparent lack of significant crustal thickening in central Iran, in spite of the continued convergence of Arabia and Iran after the Miocene collisional episode, may be the result of the ability of the central Iran blocks to "escape" away from the advancing Arabian plate along the numerous strike-slip faults in the region.

Seismotectonics and Deep Structure of Zagros. Figure 11 shows the seismicity of the Zagros region over more than 20 years. Only the teleseismically better located events are shown. As in the central Himalayas, no permanent seismic network has operated in the Zagros region; hence the carefully selected data of Fig. 11 probably represent the best available seismicity data for the region. The epicenters are distributed along about a 200-km wide belt that closely coincides with the Zagros folded belt. In the northwestern Zagros the seismicity extends to the Main Zagros Thrust (MZT), while in the southeastern Zagros (the Fars province) the seismicity edge is separated from the MZT by a notable gap.

Medium to large size earthquakes frequently occur in the Zagros region, but the size of the events rarely exceeds about magnitude 7; that is, no great earthquakes (M > 8) have occurred in Zagros, based on both well-documented instrumental and historical records. An important observation is that no surface fault break is known to be associated with any large Zagros event. Another remarkable observation is that the spatial distribution of Zagros events for a period of about 20 years is very similar to that based on careful study of destructive historic earthquakes (from about 700 A.D. to the 1970s) (Ambraseys and Melville, 1982). In this respect, Zagros is an exception to most active continental regions where the record of 20 years or more of seismicity hardly defines the "stationary" seismicity pattern as obtained from well-documented historical record.

Focal mechanisms of medium to large size events (see Fig. 12) show relatively steeply dipping (30°–60°) nodal planes. Most mechanisms show simple thrusting with nodal planes that are closely parallel to the local trend of the Zagros folded belt (Jackson and McKenzie, 1984). Because of the lack of permanent local seismic networks in the Zagros region, reliable focal depths of medium to large size Zagros events are determined using a synthetic seismogram algorithm. Focal depths vary between 6

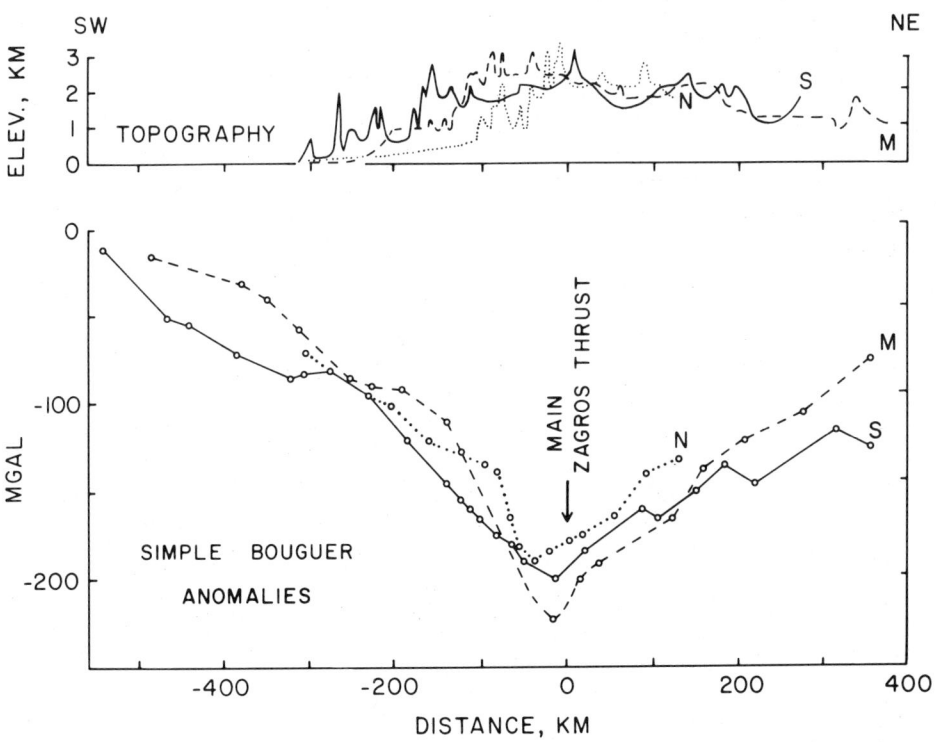

FIGURE 16. Coincident simple Bouguer and topographic profiles along the north (N), middle (M), and south (S) gravity traverses. The onset of the highest elevations is generally offset 50–150 km southwest of the gravity minima. (From Snyder and Barazangi, 1986.)

and 12 km and show no tendency to deepen toward the MZT (Ni and Barazangi, 1986). Fig. 13 shows cross sections of accurately determined hypocenters taken perpendicular to the Zagros trend. The sections show both the nodal planes of focal mechanisms and the estimated depth to the upper surface of the infra-Cambrian Hormuz salt beds, which separate the Zagros sedimentary cover from the underlying basement. Because the Hormuz beds are about 1 km thick Fig. 13 indicates that most medium to large size Zagros events occur in the uppermost part of the basement; however, because of uncertainties in the determined focal depths it is possible that some events may have occurred in the sedimentary cover. The relatively high-angle, reverse faulting of the Zagros events along with the spatial distribution of the events indicate that medium to large size Zagros earthquakes do not occur on a single, shallow-dipping detachment surface. Rather, the events appear to occur on a large number of relatively steep, subparallel basement faults distributed over a zone about 150–200 km wide. These faults accommodate horizontal shortening of the Zagros basement. Since the Zagros region is underlain by the passive continental margin of the converging Arabian plate, the reverse faulting may represent the reactivation of preexisting, high-angle normal faults of the Arabian margin, as proposed by Jackson (1980) (see Fig. 14). Moreover, this interpretation may explain the lack of great earthquakes (M > 8) in Zagros since passive continental margins have faults that are usually less than about 100 km in length, and therefore such faults are probably not capable of generating great earthquakes.

Detailed analysis of gravity data along the Zagros, constrained by other geological and geophysical data, such as the relatively high uppermost mantle Pn velocity of 8.3 km/s beneath Zagros (Asudeh, 1982), provides new insight into the deep crustal structure of the region (Snyder and Barazangi, 1986). Bouguer minima (see Figs. 15 and 16) closely coincide with the MZT (i.e., the suture zone) but are clearly offset from the beginning of Zagros higher topography, suggesting that simple isostatic considerations are insufficient to explain the gravity field and that other forces must be acting on the converging Arabian plate. Modeling of the Bouguer anomalies indicates that the Moho (at 40 km depth) beneath the Persian Gulf dips about 1–2° to the northeast beneath the Zagros folded belt and

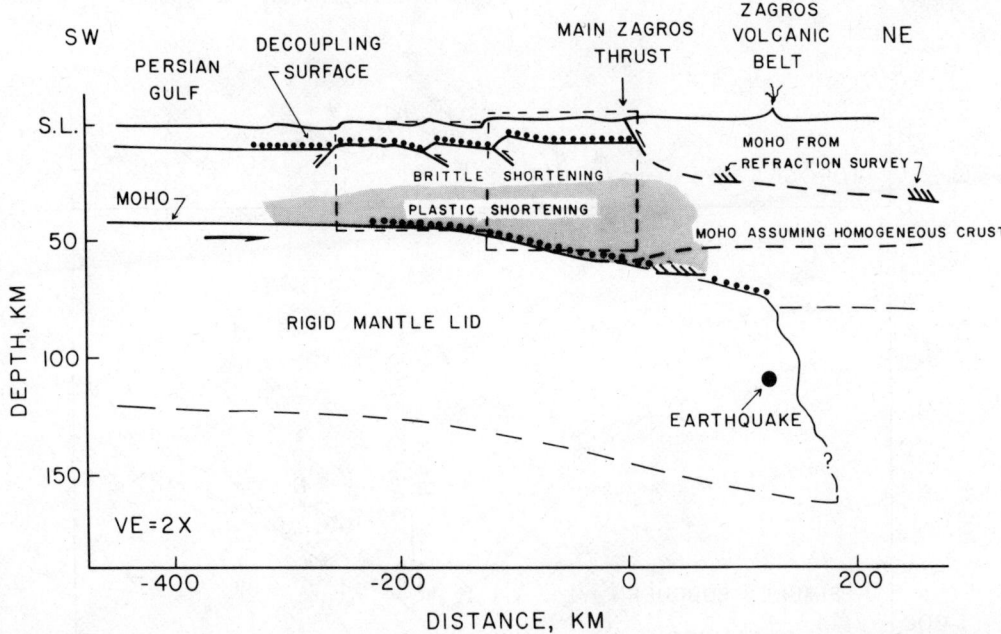

FIGURE 17. Schematic interpretative cross section transverse to Zagros and parallel to modeled gravity profiles, incorporating both gravity model and seismic results. Dotted lines represent detachment surfaces formed by the plastic behavior of salt horizons (upper surface) or lower crustal material. The crust may thicken by differing mechanisms at different levels: by folding above the 10-km-deep salt layers, by faulting along high-angle reverse faults to about 25-km depths, and by aseismic creep or plastic flow to Moho depths. The rigid uppermost mantle of Arabia may thus slide horizontally beneath both shortened Arabian crust and the lithosphere of the central Iranian blocks. The region of plastic shortening is here hypothesized to act much like a hydraulic fluid between two semirigid plates and, thus, gradually elevates the surface and depresses the Moho by isostatically thickening the crust in increments southwestward from the Main Zagros Thrust (dashed boxes). (From Snyder and Barazangi, 1986.)

increases in dip to about 5–6° near the MZT, reaching a depth of about 60–65 km. This thickening of the Zagros crust has more than restored the original thickness of the passive continental margin of the converging Arabian plate. This thickening, moreover, may in effect provide the necessary "subsurface load" that along with the Zagros surface topographic load could explain the observed flexing of the Arabian plate and the formation of the Persian Gulf and the Mesopotamian foredeep.

We propose a model for the deep structure and active tectonics of the Zagros folded belt that is probably unique to the Zagros region. Principal features of the model are the presence of two detachment surfaces and a ductile lower crust (see Fig. 17). In response to the continued convergence between Arabia and Iran, the Zagros sedimentary cover is shortened by folding, and is decoupled from the underlying basement by a detachment surface along the infra-Cambrian Hormuz salt beds. The basement is shortened by brittle deformation in the upper part of the crust, while the lower crust is plastically shortened and thickened. The lowermost crust is probably decoupled from the rigid uppermost mantle along a second detachment surface that may coincide with the Moho (see Fig. 17). This rigid mantle lid of the converging Arabian plate continues to underthrust and slides under the suture zone and the Iranian blocks, while the crustal part of the converging Arabian plate is being shortened and deformed in front of the suture zone. Since the plastic lower crust is confined by the brittle upper crust and the rigid mantle lid, it behaves like a compressed hydraulic fluid. The thickened lower crust depresses the Moho and raises the surface to form the uplifted Zagros ranges.

A Comparative Study and a Possible Evolutionary Model

The reader may have concluded by now that there are similarities and differences in the evolution and present tectonics of the Himalayan and Zagros continental collision zones. We speculate that the deep crustal structures of the Himalayas and Zagros may represent collisional processes at different stages of maturity. For clarity, we may divide the collision cycle into four major phases: (1) In the early phase the initial contact and the development of a suture zone between the two colliding plates take place. A

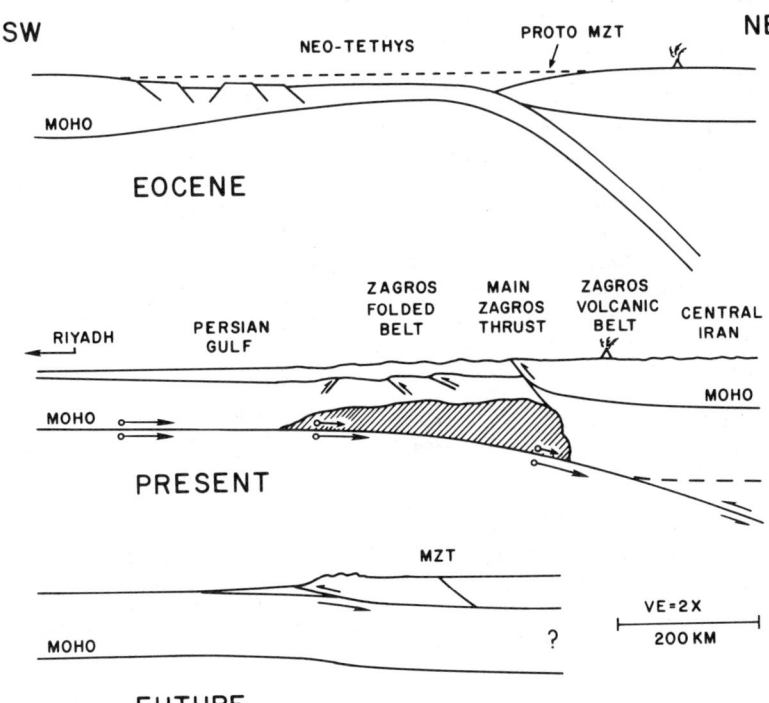

FIGURE 18. Schematic, three-stage evolutionary model of the Zagros. The hypothesized Eocene crustal structure of the Zagros is modeled after passive continental margins of the present-day North Atlantic. The geometry of the passive margin makes possible shearing along the top of the mantle lid as well as continual isostatic equilibrium as ocean is replaced by thickened crust. The hypothetical future crustal structure of the Zagros is approximately that of the present-day Himalaya. (From Snyder and Barazangi, 1986.)

modern example of this phase may be the collision between the northwestern Australian continental margin and the Banda arc in the Timor island region. (2) In the mobilization phase the continued convergence is accommodated along the suture zone, at the leading edge of the advancing plate and in the back-arc region. Zagros represents a good example of this phase. (3) In the relatively advanced detachment phase intraplate megathrusts develop within the leading edge of the advancing plate. This phase is characterized by more intense mountain building processes and by the development of high plateaus in the back-arc region, as is the case in the Himalayas. (4) In the final phase the convergence between the colliding plates ceases, and processes of erosion and isostatic adjustment prevail. The Appalachian system may represent such a phase.

Some of these four phases may overlap, but the second and third phases are most representative of the Zagros and the Himalayan zones, respectively. The Zagros collision zone is at a stage where the passive continental margin of the colliding Arabian plate is being shortened and thickened. The style of shortening appears to be strongly influenced by paleogeographic setting (i.e., the presence of the decoupling Hormuz salt beds) and inherited zones of weakness (i.e., reactivation of preexisting normal faults into reverse faults). The Zagros phase is also characterized by mobilization of the back-arc region of central Iran, including volcanism and shortening, rotation, and lateral extrusion of the different crustal blocks. Though these deformations are concentrated along the boundaries of the blocks, the interiors of some of the blocks are also being deformed. The present tectonic setting of the Zagros and central Iran is probably comparable, in gross features, to that which prevailed along the Himalayan arc and the Tibetan terranes about the Oligocene time and before the development of the intraplate MCT megathrust in Miocene time. It is possible that the similar precollision accretionary history of both the Tibetan terranes and the Iranian blocks, as well as the tectonic settings of the nearby plate boundaries that allow the "tectonic escape" of some of the Tibetan terranes (toward southeast Asia) and central Iranian blocks (toward the Makran subduction system in southern Iran), may have significantly contributed to the similarity in the evolution of central Iran and the Tibetan terranes.

One may speculate that as the convergence between Arabia and Iran continues and the collision becomes more mature, a crustal block of the leading edge of the Arabian plate in Zagros may detach along a newly formed intracrustal thrust and the Arabian plate start to underthrust itself and the crustal blocks of central Iran. According to this speculation the future development of the Zagros would be similar to the development of the MCT in the Himalayas in the Miocene time (see Fig. 18).

Acknowledgments

I thank Bryan Isacks, Dan Karig, Peter Molnar, Douglas Nelson, James Ni, Jack Oliver, and David Snyder for useful discussions, and Theresa Alt for editorial and typing assistance. This research was supported by National Science Foundation grants EAR 85-07814, EAR 87-07035 and EAR 88-45389. Institute for the Study of the Continents of Cornell University contribution 83.

MUAWIA BARAZANGI

References

Ambraseys, N. N., and C. Melville, 1982, *A History of Persian Earthquakes,* New York: Cambridge University Press, 219p.

Asudeh, I., 1982, Seismic structure of Iran from surface and body wave data, *Royal Astron. Soc. Geophys. Jour.* **71,** 715-730.

Bally, A., 1981, Geology of passive continental margins: History, structure and sedimentologic record (with special emphasis on the Atlantic margin), *American Association of Petroleum Geol. Course Note Series 19,* 1-48.

Barazangi, M., and J. Ni, 1982, Velocities and propagation characteristics of Pn and Sn beneath the Himalayan arc and Tibetan plateau: Possible evidence for underthrusting of Indian continental lithosphere beneath Tibet, *Geology* **10,** 179-185.

Berberian, M., 1981, Active faulting and tectonics of Iran, in *Zagros, Hindu Kush, Himalaya, Geodynamic Evolution,* H. Gupta and F. Delaney, ed. Washington, D.C.: American Geophysical Union, (Geodynamics Series, vol. 3), 33-69.

Chen, W. P., and P. Molnar, 1981, Constraints on the seismic wave velocity structure beneath the Tibetan plateau and their tectonic implications, *Jour. Geophys. Research* **86,** 5937-5962.

Chengfa, Chang, et al., 1986, Preliminary conclusions of the Royal Society and Academic Sinica 1985 geotraverse of Tibet, *Nature* **323,** 501-507.

Jackson, J., 1980, Reactivation of basement faults and crustal shortening in orogenic belts, *Nature* **283,** 343-346.

Jackson, J. A., and D. P. McKenzie, 1984, Active tectonics of the Alpine-Himalayan belt between western Turkey and Pakistan, *Royal Astron. Soc. Geophys. Jour.* **77,** 185-264.

Kadinsky-Cade, Katharine, Muawia Barazangi, Jack Oliver, and Bryan Isacks, 1981, Lateral variations of high-frequency seismic wave propagation at regional distances across the Turkish and Iranian Plateaus, *Jour. Geophys. Research* **86,** 9377-9396.

Lyon-Caen, H., 1986, Comparison of the upper mantle shear wave velocity structure of the Indian Shield and the Tibetan Plateau and tectonic implications, *Royal Astron. Soc. Geophys. Jour.* **86,** 727-749.

Lyon-Caen, H., and P. Molnar, 1985, Gravity anomalies, flexure of the Indian plate, and the structure, support and evolution of the Himalaya and Ganga basin, *Tectonics* **4,** 513-538.

Molnar, P., 1984, Structure and tectonics of the Himalaya: Constraints and implications of geophysical data, *Ann. Rev. Earth Planet. Sci.* **12,** 489-518.

Molnar, P., and P. Tapponnier, 1975, Cenozoic tectonics of Asia: Effects of a continental collision, *Science* **189**, 419-426.

Montadert, L., O. de Charpal, D. Roberts, P. Guennoc, and J.-C. Sibuet, 1979, Northeast Atlantic passive continental margins: Rifting and subsidence processes, in M. Talwani, W. Hays, and W. Ryan, eds., *Deep Drilling Results in the Atlantic Ocean: Continental Margins and Paleoenvironment*, Maurice Ewing Series, vol. 3. Washington, D.C.: American Geophysical Union, 154-186.

Ni, J., and M. Barazangi, 1984, Seismotectonics of the Himalayan collision zone: Geometry of the underthrusting Indian plate beneath the Himalaya, *Jour. Geophys. Research* **89**, 1147-1163.

Ni, J., and M. Barazangi, 1986, Seismotectonics of the Zagros continental collision zone and a comparison with the Himalayas, *Jour. Geophys. Research* **91**, 8205-8218.

Seeber, L., J. Armbruster, and R. Quittmeyer, 1981, Seismicity and continental subduction in the Himalayan arc, in H. K. Gupta and F. M. Delany, eds., *Zagros, Hindu Kush, Himalaya, Geodynamic Evolution*, Washington, D.C.: American Geophysical Union, (Geodynamics Series, vol. 3), 215-242.

Sengor, A. M. C., and W. Kidd, 1979, Post-collisional tectonics of the Turkish-Iranian Plateau and a comparison with Tibet, *Tectonophysics* **55**, 361-376.

Snyder, D. B., and M. Barazangi, 1986, Deep crustal structure and flexure of the Arabian plate beneath the Zagros collisional mountain belt as inferred from gravity observations, *Tectonics* **5**, 361-373.

Tapponnier, P., G. Peltzer, and R. Armijo, 1986, On the mechanics of the collision between India and Asia, in M. P. Coward and A. C. Ries, eds., *Collision Tectonics*. Geological Society Special Publication No. 19, 115-157.

Valdiya, K. S., 1980, The two intracrustal boundary thrusts of the Himalaya, *Tectonophysics* **66**, 323-348.

Zhao, W.-L., and W. J. Morgan, 1985, Uplift of the Tibetan Plateau, *Tectonics* **4**, 359-369.

Cross-references: *Accretionary Tectonics: Examples from the North American Cordillera; Brittle Phenomena; Far-Travelled Terranes; Lithosphere: Mechanical Properties; Stress in the Earth's Lithosphere; Subduction Zones.*

CONTINENTAL CRUSTAL STRUCTURE

Although the continental crust covers only slightly more than one-third of the Earth's surface and is very accessible for geoscientific studies, its structure is still less well known than that of the oceanic crust. One reason is its higher degree of heterogeneity in contrast to the regularity of oceanic areas. Moreover, the continental areas cover a time period of more than 3.8×10^9 years while oceanic crust has a maximum age of 0.2×10^9 years (ca. 5% of continental maximum age). Both types of crust differ in several other aspects. The age of the oceanic crust increases with the distance from its center. For the continental crust we find a reverse relation: The oldest "cores" or "nuclei" in the center of the continents are surrounded by increasingly younger units, as illustrated in Fig. 1. The average height of the continents above sea level is 0.87 km; the average depth of the oceans below sea level is 3.7 km.

Both oceanic and continental crust are well defined and distinguished from the underlying mantle. The crust-mantle transition is named *Mohorovičić* discontinuity (or *Moho* for short), after the Yugoslavian seismologist. The crust differs from the mantle in seismic velocity and rock density. Seismically, the crust is defined by velocities of compressional waves (V_p) smaller than 7.8 km/s or by shear wave velocities (V_s) smaller than 4.3 km/s. Because density is closely related to seismic velocity, one may also identify the crust by density values below 3.1 g/cm^3. With an average thickness of about 35 km, the continental crust forms only a thin shell of the Earth. To illustrate the dimensions, the Earth's crust in relation to the Earth's diameter corresponds to a stamp on a football!

The rock types of the continental crust are sediments, gneisses, granites, and granodiorite with $V_p < 6.3$-6.5 km/s in its upper part, and gabbro, amphibolites, and granulites with 6.3 km/s ($V_p < 7.8$ km/s) in its lower part. In these rocks the proportion of feldspars to other minerals is more than 50%. From the chemical point of view, SiO_2 reaches about 60 wt %.

In our definition the sediments are part of the crust, which seems logical because thick sedimentary sequences tend to change in their deeper parts into diagenetically modified, often metamorphosed, rock units and cannot be distinguished from the "real" crustal rocks on the basis of their velocities. Hence, the upper part of the "real," i.e., the crystalline, crust is not always well defined, at least not below thick layers of sediments.

In some areas an intracrustal boundary, the Conrad discontinuity, is found. We may define it as the boundary between the upper and the lower continental crust where V_p starts to assume values of > 6.5 km/s. The search for a Conrad-like boundary seems justified because of its importance for an internal (second ?) differentiation of the crust into light sialic (sial) granites and gneisses with $V_p < 6.3$-6.5 km/s and heavier cumulates of gabbroic, mafic (sima) composition in the lower crust with $V_p > 6.5$ km/s.

Methods of Crustal Studies

A short summary of methods for investigating the crustal structure follows. For a detailed discussion of the methods, refer to standard textbooks such as Telford et al. (1980).

Seismology. The analysis of earthquakes gave the first information on the continental crust. In 1909

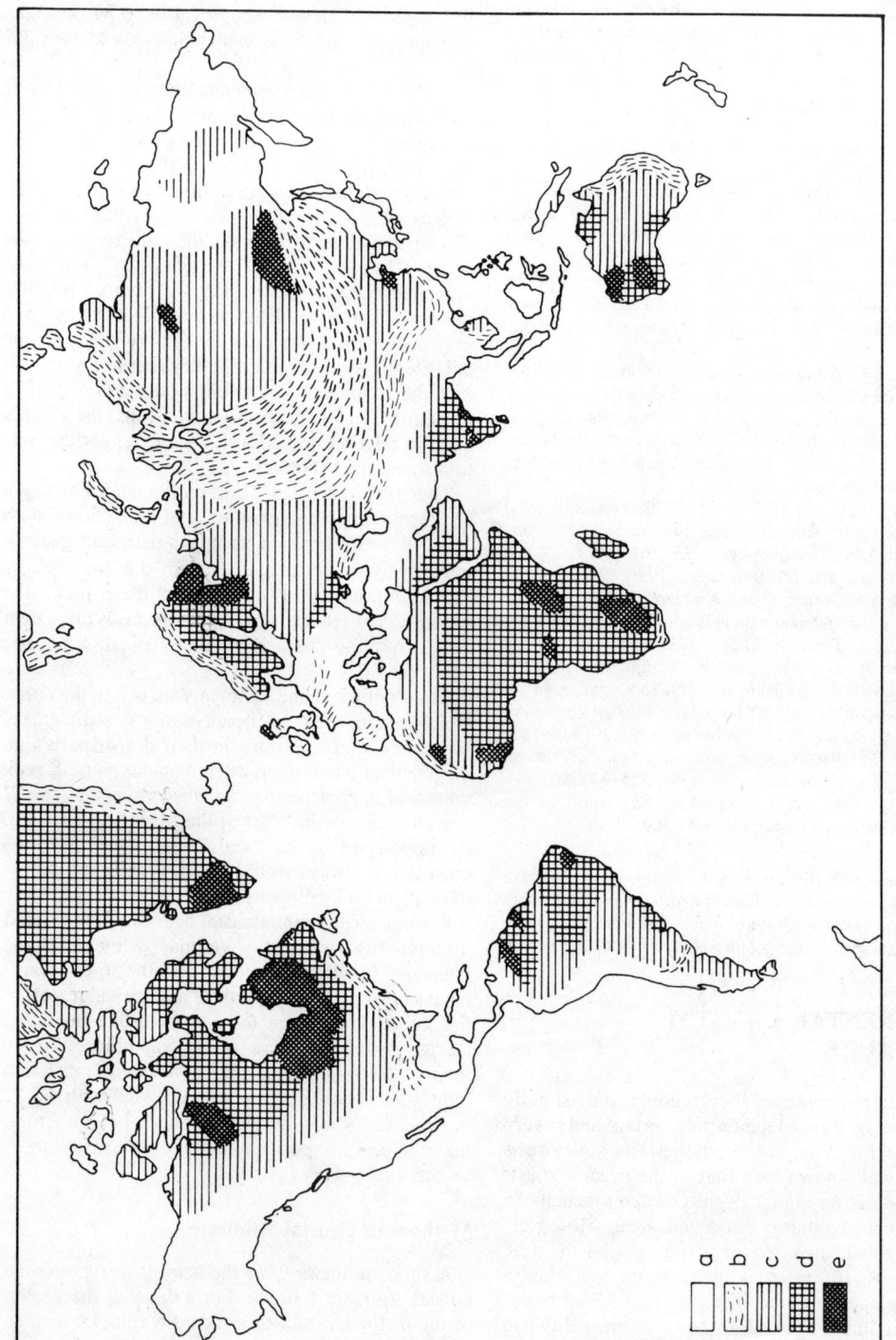

FIGURE 1. Age provinces of the continental crust. (a) Mesozoic and Cenozoic orogenic belts, (b) Paleozoic orogenic belts, (c) Proterozoic platforms, (d) Proterozoic shields, and (e) Archean shields. (After Miyashiro et al., 1982)

Mohorovičić identified the head wave running in the upper mantle immediately below the crust with a high velocity of about 8 km/s. This wave (the P_n) was used to estimate the thickness of the Earth's crust. In 1925 Conrad identified an intracrustal discontinuity (below which the velocity assumes values of more than 6.5 km/s). But this Conrad discontinuity does not seem to have the same significance as the Moho; it is sometimes not found and only poorly constrained in shield areas. Today, controlled source seismology (see below) is used to investigate the velocity structure of the crust in more detail.

Seismology also contributed important constraints (1) to stress estimates by means of the earthquake distribution and the focal mechanisms, (2) to crustal rheology by means of the hypocenter-depth distribution, and (3) to the recognition of the overall structure of the continents and oceans by the strong body and surface waves of earthquakes.

Refraction Seismic Measurements. The power of this method lies in its resolution of crustal velocities. Beginning in the late 1940s, refraction profiles several hundred kilometers in length were shot to resolve crustal structure. They confirmed the first seismological results by the observation of the P_n waves. In addition, the interpretation of reflected waves in the wide-angle range (P_mP waves) became more and more important for the studies of the lower crust. This method is discussed in more detail in the section on the power of the seismic methods: refraction vs. reflection.

Reflection Seismic Measurements. Because the power of this method lies in the resolution of structures, it became very popular for deep crustal studies in the mid-1970s. The observations in the near-vertical range provided a wealth of data, mainly on the structure of the continents and differences of its various units. It will also be discussed in more detail later.

Gravity. Gravity observations are still very valuable for investigating isostatic and nonisostatic features, for selecting between competing seismic models, and for revealing crustal inhomogeneities not seen in seismic sections.

Electromagnetic and Magnetotelluric Methods. These methods make use of natural variations of the Earth's magnetic field, inducing electric currents in zones of high conductivity. The result of model calculations are conductivity-depth estimates. Often, the lower crust seems to be a layer of increased electrical conductivity.

Magnetic Methods. Magnetic studies, like gravity, provide information on crustal inhomogeneities. But the depth of the magnetic basement and its undulations below sedimentary basins also can be estimated, and data on the Curie depth (where temperature causes the magnetic properties of ferrimagnetic rocks to disappear) can be collected.

Mapping this depth can be used for a (rough) cross-check of temperature estimates.

Geothermal Observations. Geothermal studies are the basis for the temperature-depth estimates (T-z curves), which again are the key for rheological calculations. These studies could demonstrate a concentration of radiogenic heat sources in the upper crust but also discriminate between regional heat flow provinces. Worldwide comparisons show an age-dependent heat flow.

Stress Measurements. Knowledge of stress distribution contributes substantially to our understanding of tectonism and crustal dynamics. Stress analyses are typically achieved by overcoring in competent rock units (e.g., in tunnels or quarries), by making direct observations in boreholes, by analyzing earthquake focal plane solutions, and by observing geological folds and faults.

Geological Mapping. The mapping of outcrops of rocks that once were at lower or mid-crustal depths where they were modified in high-pressure/high-temperature regimes (such as amphibolites or granulites) is a very important tool for understanding the deeper parts of the continental crust and for offering candidates for the reflecting lamellae by means of their inhomogeneous, often mylonitic, zones within them.

Laboratory Ultrasonic Measurements. The determination of V_p and V_s at high-pressure and high-temperature conditions forms the basis for correlations of seismic velocities to petrology. Seismic anisotropy, V_p/V_s-, the Poisson ratio, and absorption are also very useful parameters.

Creep Experiments. Laboratory creep data are essential for estimates on crustal viscosities and the ductile behavior of the lower crust. They can be combined and checked with the hypocenter-depth curves mentioned before.

Age Determinations. Finally, the age dating of rock samples reveals major constraints for metamorphic events and sometimes helps to provide pressure-temperature cycles of crustal rocks. These investigations also are the basis for all models of crustal evolution and recycling.

The Power of the Seismic Methods: Refraction vs. Reflection

Controlled source observations by means of portable seismometers along lines of several hundred kilometers using special quarry blasts started in the 1940s; this *refraction shooting method* dominated crustal studies for the next 30 years. In the 1960s the importance of wide-angle events was noted. The wide-angle event P_mP, partly a true reflection from the Moho, partly (toward the end of the travel-time curve) a diving wave in the V_p-gradient zone above the Moho turned out to be the strongest event in crustal refraction shooting. Sometimes, similarly strong wide-angle events are found for intracrustal boundaries, e.g., from the Conrad discontinuity (generally called P_1P). The value of such wide-angle events lies in its power to determine the velocity structure of the crust, especially for its deeper parts, a task which only very seldom can be obtained by (near-vertical) reflection work.

Crustal low-velocity zones are generally derived from a travel-time difference between a wide-angle reflection and the refracted arrival from the overlying refractor in the hanging wall. At best, these Δt values provide some estimates on the average interval velocity between the zone of deepest penetration of the refraction or diving wave in the hanging wall and the next underlying reflector. It is a good approximation if no three-dimensional velocity structure is present, which then may cause a lateral detour of rays, resulting in erroneous interval velocities. There is no doubt, on the other hand, that zones of velocity inversions do exist in the continental crust, especially in warm areas and in the thick orogenic belts.

The uppermost 10 km of the crystalline crust is still a domain of a "true" refracted event: the P_g wave, which generally dives smoothly into the basement, revealing a small gradient zone in the upper 5–10 km that matches laboratory data from ultrasonic tests on sialic samples nicely. Figure 2 shows an example of a reduced, unreversed seismogram section of a crustal refraction profile. In general, reversed profiles are to be preferred in the field because of the possibility of obtaining more reliable, i.e., "true," velocity-depth functions.

The interpretation of the profiles is generally done in three steps: (1) Picking of arrivals and establishing certain correlations along travel time branches $t(x)$, (2) assessing the velocity structure by direct, simple methods, (3) raytracing in one or two dimensions, and (4) comparing the data (travel time and amplitude) with synthetic seismogram sections. The better the fit between observed and calculated sections, the better the V-z model. Models with 3–5 different layers, sometimes dipping or showing velocity gradients or low velocity zones in the middle crust, are the outcome of today's refraction studies with dense station spacings.

Systematic *reflection studies* for the deeper parts of the continental crust started as early as the 1960s. The shorter, near-vertical ray paths, the shorter wavelengths, and the higher frequencies determine the high structural resolution of the reflection method. It is unique in its representation of deeper reflectors unequivocally at increasing travel times. Explosive and vibroseis patterns on land and air gun arrays at sea are used together with arrays of geophones (hydrophones at sea) connected to recording units of up to 1000 traces. The continuous profiling technique is generally applied, and data are collected by the common midpoint (CMP) technique by means of which a reflection element is covered by a number of rays with different geometry. Many digital techniques enhance the signal-to-noise ratio, much more than in refraction work: static and dynamic corrections, deconvolution, filtering, all kinds of inverse filtering and migration, to name only a few. (See *Seismic Signal Processing*.)

Often the final display is an x-t section. For migration processes, good velocity information is essential, but many well-established working groups still use spread length of only up to 10 km, too short to gather velocity data on the lower crust from the hyperbolalike reflection curves, i.e., from the move-out times.

The necessary additional velocity information is generally collected from (1) longer spreads, (2) wide CMP gathers, or (3) reversed refraction observations. Longer spreads are useful in areas of long, correlatable reflectors, e.g., at the Urach surveys (SW Germany), where routine spreads of 23 km in length were used and reflectors were found to be continuous up to 20 km. CMP gathers from 0 to about 100 km generally provide the best V-z information, although only in a limited area and at high cost. They were used in the surrounding of the German Deep Drillhole Project in the Oberpfalz area. Reversed refraction profiles may also be applied to obtain V-z information along a reflection profile, as carried out along the French profile ECORS 1 and the German line DEKORP 2-S. Which method provides the best V-z information depends on the continuity of reflectors and the particular jumps, gradients, and inhomogeneities of an area.

By comparison between reflection and refraction studies it was found that crustal reflections nearly always seem to end at the Moho, at least in tectonically young areas. This observation offers another possibility to map the Moho depth. In general, crustal reflectivity varies strongly from area to area, but a certain correlation between reflectivity patterns and age has been found, as will be discussed later.

For an overview the record sections are often shown in the form of line drawings. These space-saving plots may give the general structural trend along the profile but depend heavily on the skill of the interpreter. Nevertheless, it is the only way to present the large volume of reflection data in a concise way. (This presentation will be dominant throughout this paper.) Figure 3 shows an example from the shelf area of the Irish Sea, collected by BIRPS.

Old Shields and Platforms

Shields and platforms are treated in the same context because of their similar structure in composition and seismic velocity. Platforms differ from shields in having a thin veneer (1–3 km) of sediments indicating a temporary subsidence and sedimentation below sea level. Both shields and platforms show only little relief and have not been affected by orogenic events for at least 1000 my.

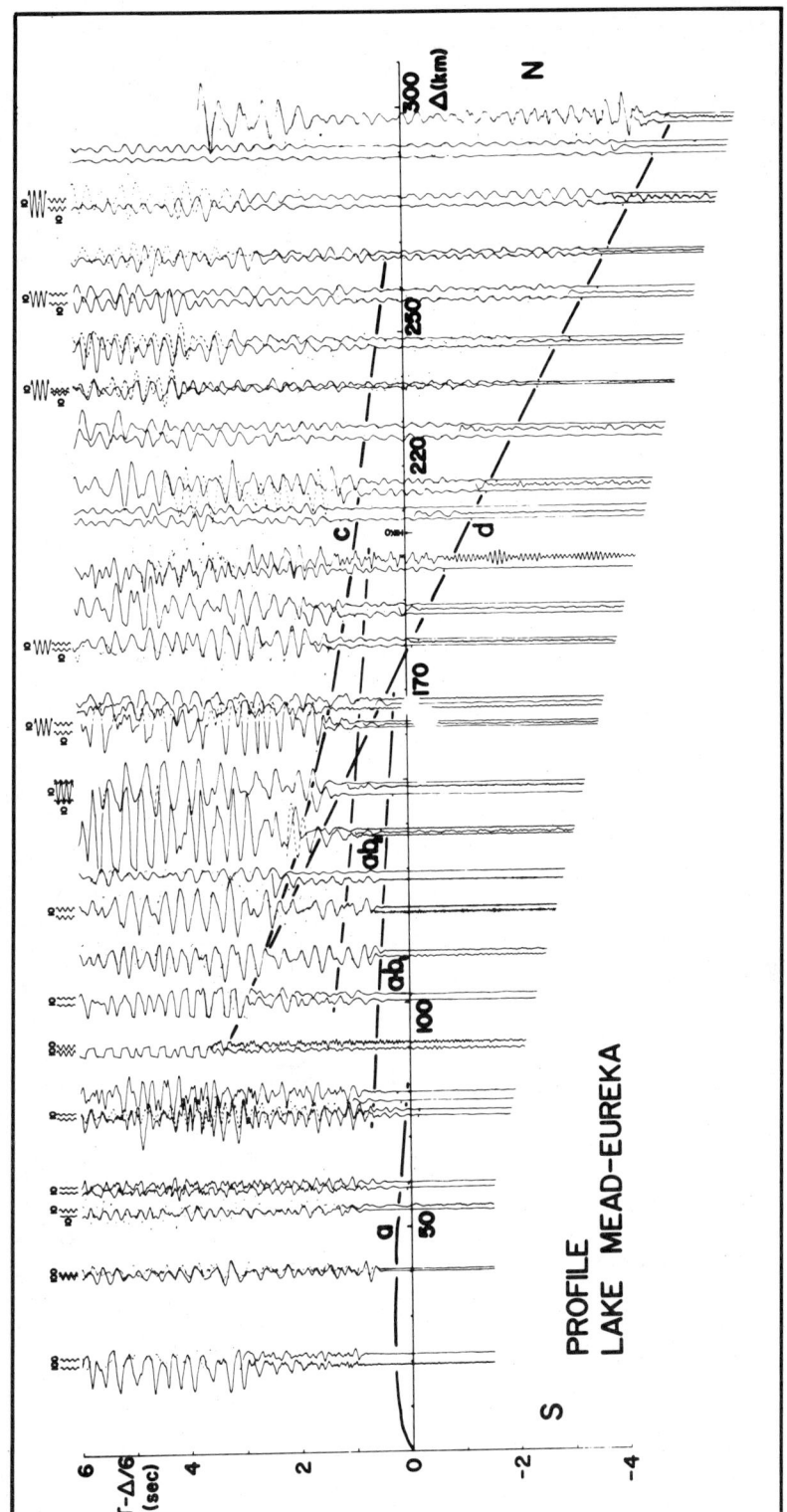

FIGURE 2. Example of a reduced record section with correlation from Prodehl (pers. comm.). Here a represents the P_g wave, ab_1 and ab_2 are diving waves from intermediate boundaries or gradient zones, c represents the P_mP wave, and d represents the P_n wave.

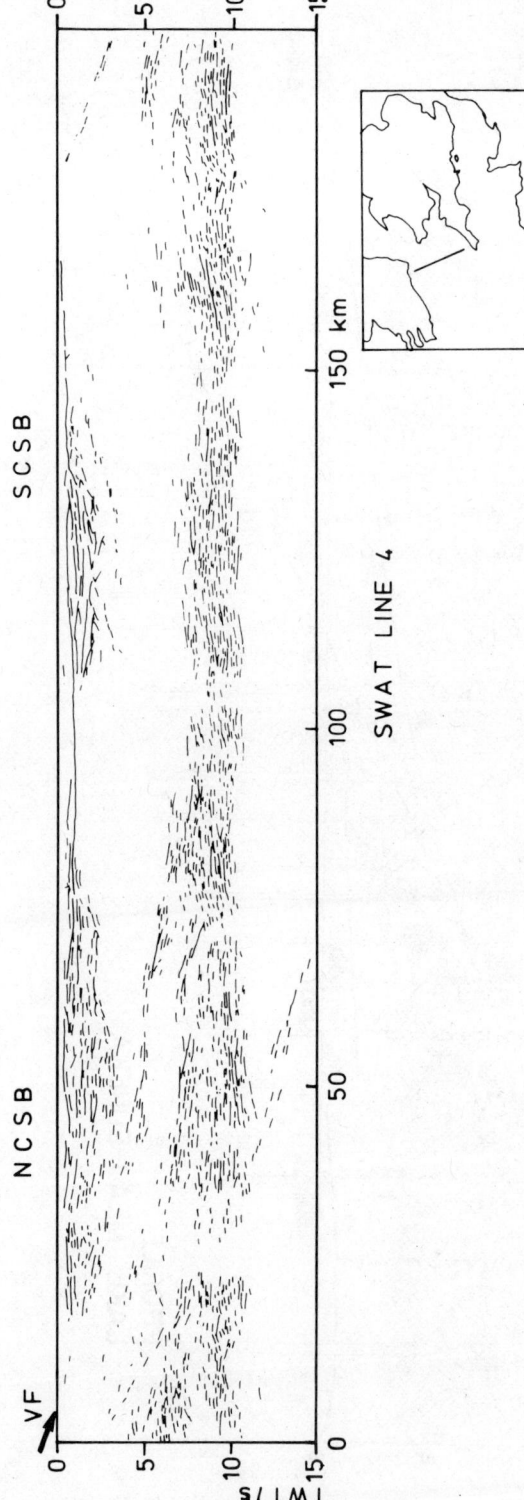

FIGURE 3. Example of large-scale lower crustal layering on BIRPS line SWAT 4. VF: Variscan Front, NCSB: North Celtic Sea Basin, SCSB: South Celtic Sea Basin.

Good coverage of modern reversed refraction profiles and even lithospheric profiles more than 1000 km in length have been obtained in the Baltic Shield (Sweden and Finland) and in the East European Platform (USSR). Only very few (near-vertical) reflection lines are available for crustal studies in these areas in Europe. For the American shields and platforms the situation is reversed: A considerable number of reflection lines exist, most of them carried out by COCORP, but only a few refraction profiles have been observed. Other continents, such as Australia and India, also contribute to the seismic information, mostly by refraction. An increasing number of reflection data are available from Australia.

Meissner (1986) summarizes and discusses V-z data and crustal depths of shield and platform areas from refraction work. An average V-z profile has been constructed from about 50 reversed crustal refraction profiles (Fig. 4a). The velocities are standardized and contain five equally spaced velocity ranges:

$V_p < 5.7$ km/s   Sediments or near-surface crystalline rocks
$5.7 < V_p < 6.4$ km/s   Upper crust, average (gneisses, granites)
$6.4 < V_p < 7.1$ km/s   Lower crust, (diorites, gabbros, amphibolites, granulites)
$7.1 < V_p < 7.8$ km/s   Lowermost crust in shield and platform areas
$V_p > 7.8$ km/s   Uppermost mantle (peridotite).

The upper crust, with its more sialic composition made up of granites and gneisses, is mostly characterized by a small positive velocity gradient down to 6 to 10 km as derived from the smooth P_g wave. The old shields and platforms have a thinner upper crust than have the later accumulated areas, but their total crustal thickness is greater. Pronounced low velocity zones seem to be absent. The lower crust with its velocity range from 6.4 to 7.8 km/s seems less clear, and many rock types, both metamorphic and igneous, fulfill the seismic requirements in laboratory investigations. From areas of outcrops and regions of strong erosion (through continuous uplift, as in the Baltic shield) the metamorphic granulites are the best candidates for the rocks of the lower crust. Laboratory measurements reveal that below about 200 MPa (= 2 kbar), equivalent to about 6 km depth, seismic velocities increase with lithostatic pressure linearly, meaning that our average V-z model would show equally spaced velocity intervals, if the rock type in the crust would stay the same. This is definitely not the case, indicating different rock types, i.e., a certain "layering," perhaps differentiation of the old crusts. In Fig. 5 three V-z profiles are shown for the Baltic shield (similar results have been reported from Precambrian areas of Australia, North America, and the Voronesh (USSR) and Arabian shields). They are typical for Precambrian areas showing a continuous increase in velocity throughout the crust without prominent low velocity zones. Only minor steps occur in the V-z functions, i.e., first-order boundaries like a Conrad are absent.

The high resolution and the long range of travel

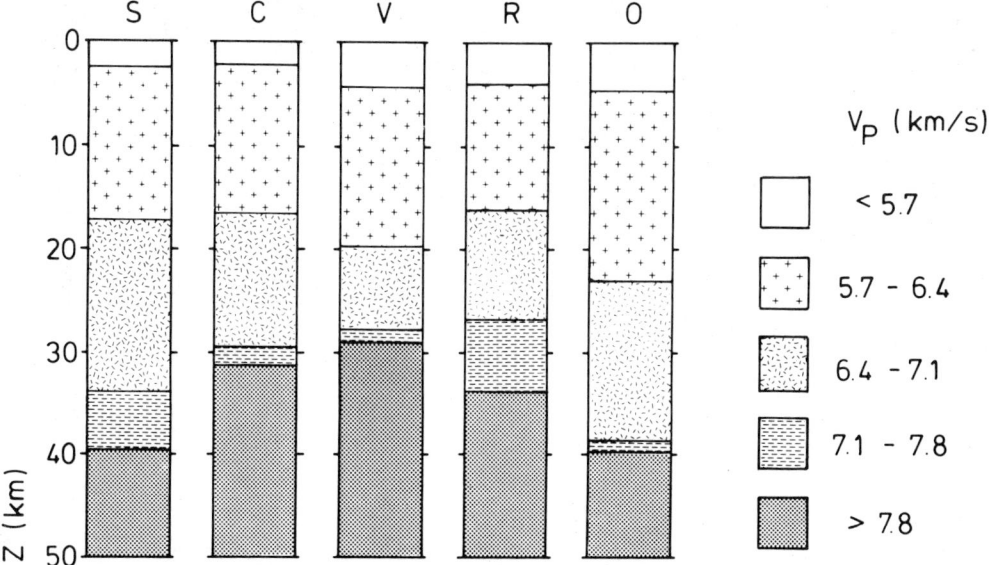

FIGURE 4. Standardized velocity-depth distribution for different crustal provinces. S: shield areas, C: Caledonian provinces, V: Variscan provinces, R: rifts, O: orogens.

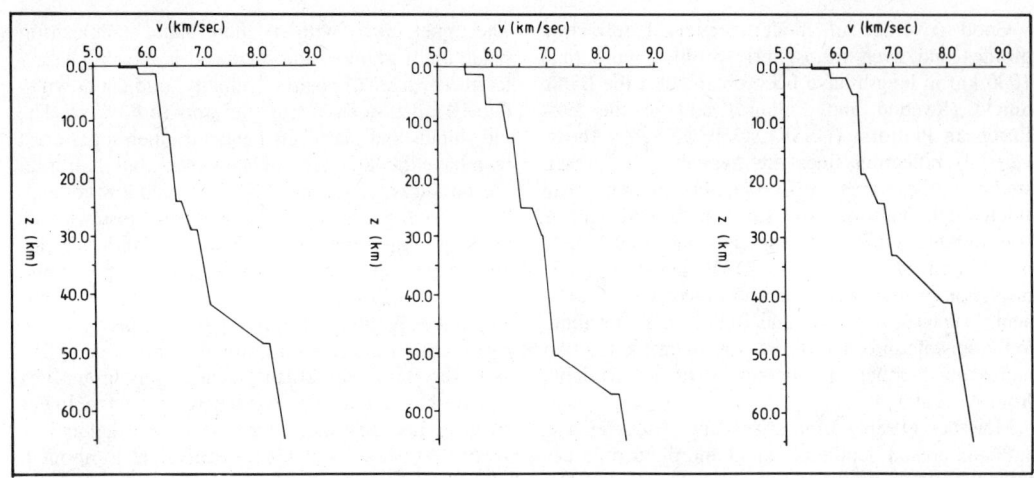

FIGURE 5. Three velocity-depth functions along profile Fennolora in Scandinavia. (R. Bittner, pers. comm.)

time branches with good energy is a special feature of old crusts, which generally show a low noise level, i.e., a high signal-to-noise ratio caused by a small absorption of seismic waves. Generally, in addition to the P_mP, P_n and other P phases, also prominent S phases (S_mS or S_n) can be observed and provide additional information on crustal structure and allow the determination of Poisson's ratio. In such areas explosions in the range of one ton have yielded signals down to the base of the lithosphere and even to the first transition zone in the mantle at about 400 km depth as on the Fennolora profile through Scandinavia.

Recently many COCORP reflection profiles have

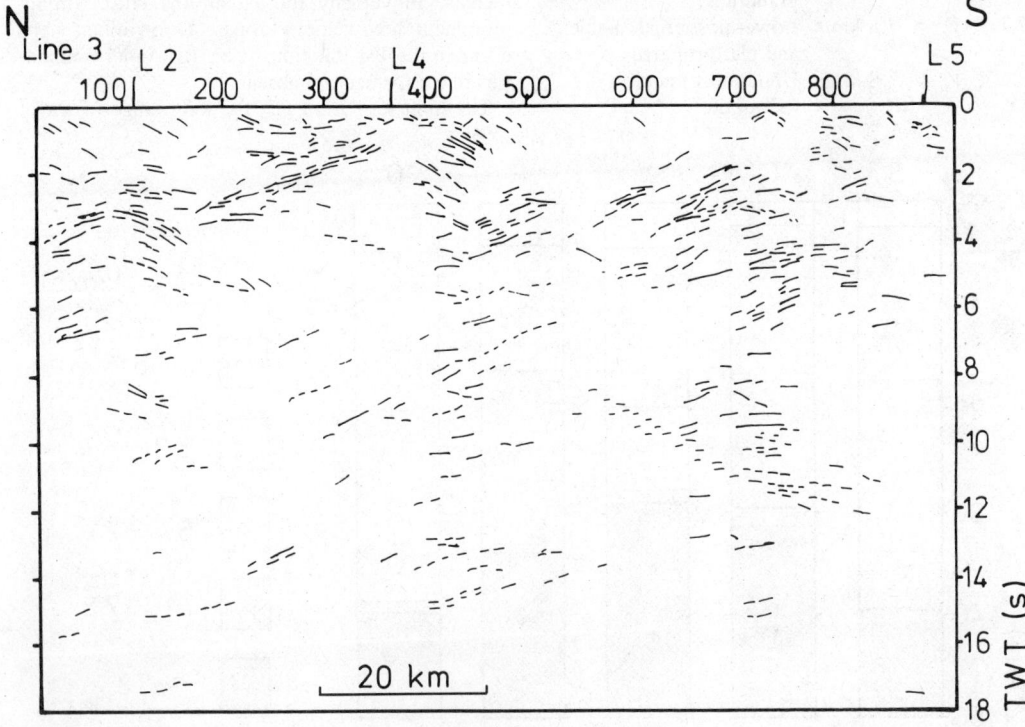

FIGURE 6. Line drawing of Minnesota line 3 from COCORP. In contrast to lower crustal layering (Fig. 3), reflectivity decreases with depth. Note the strong dip of reflecting elements.

been observed in the United States through Proterozoic (and some Archean) terrains. In general, reflectivity in the near-vertical range seems scattered and rather poor, mostly petering out toward greater depth, not even showing the Moho, which always appears in refraction data as the P_n and/or the P_mP waves. Examples of COCORP lines with scattered reflectivity, decreasing with depth, are the Minnesota survey (Fig. 6) and the Wind River lines. Figure 7 shows reflectivity-depth diagrams of the Minnesota and Wind River profiles. The decreasing reflectivity with depth is clearly demonstrated. The near-vertical reflection seismograms in old areas are characterized by only short (generally less than 2 km) reflections, which often show strong dips. On the other hand, those profiles that are observed in areas where the old platforms were reactivated by younger tectonic events do show prominent crustal reflections, e.g., those through the Colorado uplift and all the lines across the southwestern and southern rim of the Bohemian massif, which is discussed in the section on Phanerozoic areas.

Observations of few near-vertical reflection lines in the Baltic shield support the observation that the old shields and platforms in general are only poorly reflective, and especially their lower crust. The Moho cannot be defined by an abrupt termination of reflections.

By long-time creep and diffusion, a certain homogenization, increasing with time, seems to have taken place, smoothing out first-order boundaries, i.e., no jumps in impedance, essential for near-vertical reflection observations, are within a seismic wavelength.

The assumption that smooth and gradientlike boundaries dominate in the older units is supported by the fact that the P_mP and P_lP waves are seldom observed in the subcritical-angle range, i.e., they do not show apparent velocities higher than about 8.5 km/s, meaning that these waves are "overcritical," i.e., they consist of diving waves from a strong-gradient zone gaining their high energy by a focusing effect.

These and other characteristic differences between old and younger crusts are discussed again when dealing with the greater crustal thickness in the old units.

Phanerozoic Areas

Numerous late Proterozoic and Phanerozoic belts have accumulated around the old shields and platforms during the last 1000 my. They are several thousand kilometers long and several hundred kilometers wide. As seen in Fig. 1, the Phanerozoic areas make up a considerable part of today's continents. Central and western Europe as a whole, large parts of Asia, and the southeastern, southern, and southwestern rim of North America consist of Phanerozoic areas.

Phanerozoic belts can be subdivided into Paleozoic (older than 250 my and Mesozoic/Cenozoic (younger than 250 my) belts. Because of their higher age and long-time erosion, Paleozoic belts show only low mountains and exhibit a certain tectonic stability. The younger and still active orogenic belts in recent continent-continent or continent-ocean collisions show a high surface topography and also deep mountain roots. They are also characterized by tectonic activity. They are discussed in the next section, Rifts and Orogenic Belts.

Except for the Variscides, long belts of ophiolites demonstrate the existence of modern plate tectonic processes with subduction of oceanic plates, continent-continent collisions and accretion of terranes, at least from the Pre-Phanerozoic Grenvillean, i.e., from about 1000 my.

Especially in densely populated Europe the area of the Variscan and Caledonian orogenies have been widely investigated by all kinds of seismic methods. Giese et al. (1976) summarized many of the early refraction/wide-angle studies in central Europe; Meissner (1986) summarized the reflection studies. Based on joint interpretation of both seismic methods along the same lines, the coincidence of near-vertical reflectors with wide-angle boundaries could be demonstrated, and the high reflectivity of the Variscan lower crust was attributed to lamellae, a sequence of reflecting lenses with alternating impedance.

Average V-z profiles, based again on the standard-

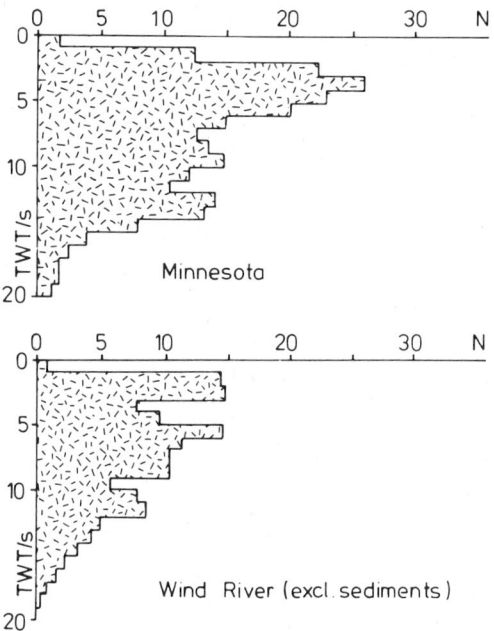

FIGURE 7. Histograms of reflectivity distribution in the crust in shield areas (Minnesota and Wind River surveys of COCORP).

ized, equal-interval velocities, were also established for the Caledonian and Variscan provinces using about 40 reliable refraction profiles. Figures 4b and 4c show this result. Total crustal thickness is reduced in the younger provinces, average velocities are smaller and the upper crust with $V_p < 6.4$ km/s is thicker. Often the thickness of the high velocity lower crust ($6.4 < V_p < 7.1$ km/s) is less than ⅓ of the total crustal thickness. Only few profiles, most of them in or around the Rhinegraben, show a velocity higher than 7.1 km/s. Wide-angle events (P_mP or P_1P) are generally strong.

In the younger crusts some low velocity zones are found at the base of the upper crust. The Conrad discontinuity often becomes more pronounced. The Moho appears as a first order boundary with undercritical reflections.

Although some tectonically significant but short crustal reflection profiles have been observed in Europe since 1964, it was the success of the COCORP group in the United States that induced European groups (BIRPS in Great Britain, ECORS in France, and DEKORP in Germany) to collect data along longer profiles. The older reflection work had revealed some interesting tectonic features: a normal fault cutting the whole crust at the transition from the Devonian Hunsrück mountains to the Permian Saar-Nahe trough; thin-skinned tectonics at the North Variscan Deformation Front along the Aachen-Faille du Midi fault, very similar to that in the Appalachians (COCORP); and a prominent low velocity body in the geothermal region of Urach by using very long spread lengths. In the United States, COCORP extended their studies to Phanerozoic areas and found record sections in tectonically young areas very similar to those observed in Europe. The Basin and Range survey for instance, looks very similar to European lines. They show reflecting fault zones in the otherwise transparent upper crust, huge bands of subhorizontal laminated reflections in the lower crust, and a shallow Moho at about 30 km or less. Figure 8 shows this profile together with a European and an Australian line.

Figure 9 illustrates the reflectivity of typical Phanerozoic profiles. Other observations in reflection seismograms in Phanerozoic areas are that (1) phase correlatable reflections have a typical length of 5–7 km, seldom exceeding 20 km; (2) reflections are nearly horizontal, showing dips less than 5°; (3) reflections are longer along the tectonic strike direction than perpendicular to it; and (4) from refraction as well as reflection studies the Moho is extremely flat. No crustal roots of former orogenies have survived. The Moho must be a young feature. Only the rigid upper crust keeps some memory of former thrusts and other fault zones.

The observation of a somewhat transparent upper crust fits well into rheological calculations and distributions of hypocenters of crustal intraplate earthquakes: The upper crust is rigid and shows earthquakes. It is reflective only if a large fault pattern, i.e., thrusts, backslip, or even normal faults occurred during the last tectonic event. The lower crust in the Variscan-Caledonian areas still reacts ductilely, allowing no earthquakes.

Rifts and Orogenic Belts

Special tectonic features of continents are the rifts and the orogenic belts. Continental rifts are narrow (normally < 60 km) but sometimes very long features (up to several thousand kilometers). They develop in a tensional regime (as can be seen by fault plane solutions) and are fault bounded.

In rifts, grabens, and along passive margins, special mixed velocities of 7.1 to 7.8 km/s are found in the form of cushion (Rhinegraben) in the crust-mantle area. Such cushions appear as high-velocity layers as shown in Fig. 4d (here it definitely is not part of the crust!). Often it is doubtful which of the boundaries, e.g., a jump to 7.7 km/s or a deeper one to 8.3 km/s can be identified as the Moho. Most rifts show a crustal antiroot, i.e., crustal thickness is shallower below the rifts than in the surrounding area. Rifts show a strong tendency towards isostasy.

Near-vertical reflection data show a highly reflective lower crust with a flat Moho. This observation indicates the importance of temperature for the creation of reflectors and also shows that the Moho is a young phenomenon.

Continental seismicity is concentrated along rifts. High heat flow values of more than 100 mW/m^2 can also be used to outline rifts. The high temperatures may cause partial melting, thus explaining the high conductivity values often observed within rifts.

The young orogenic belts, on the other hand, all have crustal roots, many of which are made up of low-velocity/low-density material. These relatively narrow crustal roots are a manifestation of a dynamic compression with crustal shortening including nappe displacement near the surface and a suture with a root below. Also in orogenic belts a tendency towards isostasy is observed although the root mostly is laterally displaced slightly with regard to the highest elevation, i.e., the maximum load. Seismicity often concentrates on prominent zones of detachment. Preliminary reflection data in the Alps and the Pyrenees indicate a high reflectivity of the lower crust.

Figure 4e shows the standardized V-z distribution of orogenic provinces.

Comparison and Evolution of Crusts

Some comparisons between the old shields and platforms and the Phanerozoic areas around them have already been discussed. We did not show details of the most recent compressional tectonics with their mountain roots or extensional features with their antiroots. Figure 10 shows these rather narrow

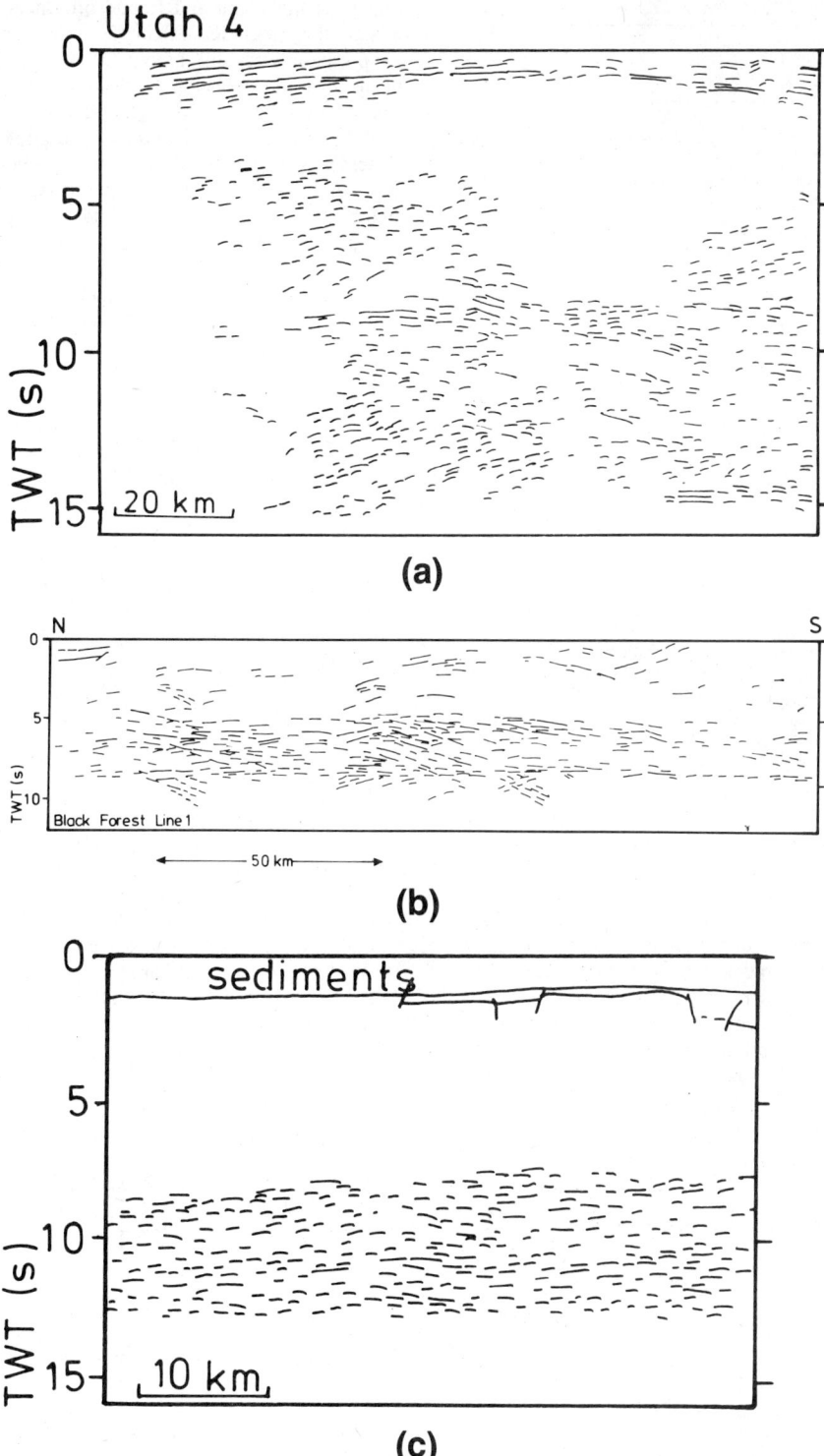

FIGURE 8. Line drawing of (a) COCORP line Utah 4, (b) DEKORP/KTB line Black Forest 1, and (c) a part of line 1 across the Eromanga basin in Australia. All lines show a more or less transparent upper crust and a highly reflective lower crust.

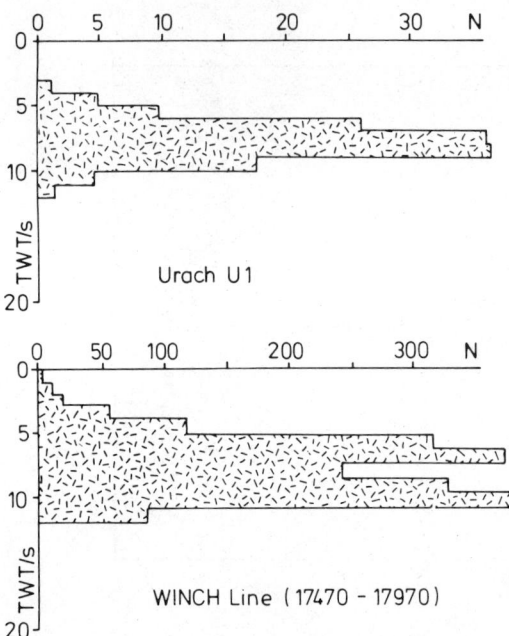

FIGURE 9. Histograms of reflectivity distribution in Phanerozoic areas (line Urach 1 in SW Germany and BIRPS line WINCH west of Britain). Compare with Fig. 7.

features on the left-hand side—the roots below the Pyrhenees and the antiroots below the Rhone- and the Rhinegraben. This figure also demonstrates the flat and shallow Moho in the Variscan area where today no signs of roots are observed. This void seems surprising at first because collisional phenomena must have been similar to today's tectonics, and petrologically determined p-T paths of some surface rocks of the Moldanubian (the southern unit of the Variscides) show metamorphic pattern that definitely indicate a Variscan root zone of at least 60 km.

These observations and others prove that crustal roots and antiroots must be transient and rather young. In the Basin and Range province, with its crustal thickness less than 30 km, apparently compression was still active 20 Ma ago, which means that the Moho and the whole lower crust cannot be older than 20 Ma. The shields, on the other hand, have large crustal thicknesses despite erosional processes. Figure 10 depicts the Baltic Shield on the right-hand side. As mentioned in the section on old shields and platforms, the shields and platforms generally have a higher average velocity, i.e., they also must have a higher density than the later accumulated areas. These younger areas are more or less balanced with respect to the old shields because of their lower density, as shown in Fig. 11.

From a compilation of crustal depths (Z_M) and tectono-thermal age (TTA), i.e., the age of the last great metamorphic-tectonic event (excluding the most recent—and transient—roots and antiroots), a correlation between Z_M and TTA has been established as shown in Fig. 12. Meissner et al. (1987) give an extensive discussion. Similar, but not identical, curves of Z_M versus TTA, may be derived by correlations between Z_M and heatflow q_0, and by correlations between q_0 and TTA. The general dependence of crustal depth on age is beyond doubt when excluding the most recent roots and antiroots.

The explanation of this age dependence of Mohos is not straightforward, and at least four possibilities exist.

1. The younger crusts with their lower velocity and density and their thicker and more sialic upper crust are more strongly differentiated. A multiple differentiation has transferred light material from the lower crust upward. Ultramafic cumulates from this melting process are incorporated into the (now higher) uppermost mantle.
2. A general extensional process that is similar to McKenzie's stretching model has reduced lithospheric and crustal thickness. Postorogenic periods of extension are very common.
3. The mild accretion of terranes that have gener-

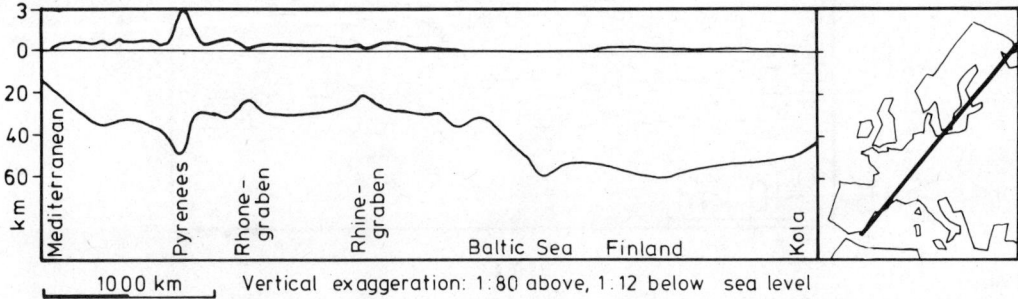

FIGURE 10. Crustal cross-section through Europe showing a thick crust in the Baltic shield, antiroots below Rhine- and Rhonegraben, and the deep root below the Pyrenees.

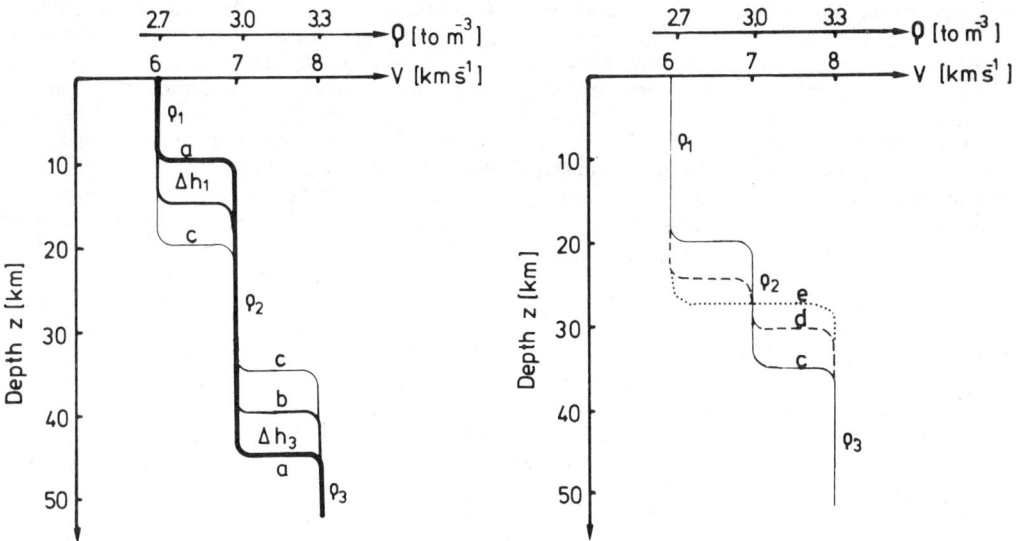

FIGURE 11. Isostatic model calculations of differentiation of a mafic ancient lower crust into a low-density (low-velocity) sialic upper part and a high-density (high-velocity) ultramafic residue, to be added to the mantle. A complete melting of the lower crust may transform the shield curve a to curves d or e.

ally a small thickness might also account for some areas with lower crustal thickness.

4. A general difference in composition between old and young areas might originate from the steady depletion and modification of the original Earths's mantle by a quasi-permanent extraction of material to create crusts. The hotter and rather undepleted mantle of Archean times may have produced a greater volume of a more primitive, i.e., mafic, crustal composition.

Any combination of these four processes are possible although we favor explanations 1 and 4.

Crustal Reflectivity

Figures 7 and 9 have already demonstrated the contrasting reflectivities of different crustal age provinces. It seems that the general reflective behavior of a crust is related—as is crustal thickness—to its tectono-thermal age, which again correlates to heatflow (q_0) and temperature (T) as well as to its viscosity (η).

Small intraplate earthquakes cause small zones of rupture, i.e., small-scale inhomogeneities, in the brittle *upper* crust. Only occasionally, in areas of clear and strong tectonic stress systems, mostly in the neighborhood of plate boundaries, more continuous rupture zones originate in the upper crust. They might develop into large zones of detachment, reflect seismic energy, and provide a key to tectonic developments. In the old and cold shield areas, the brittle regime reaches deeper; the old zones of detachment might be obscured by small-scale ruptures and/or long-time diffusion processes.

Warm, Phanerozoic areas generally show a highly reflective lower crust. Reflection coefficients of the

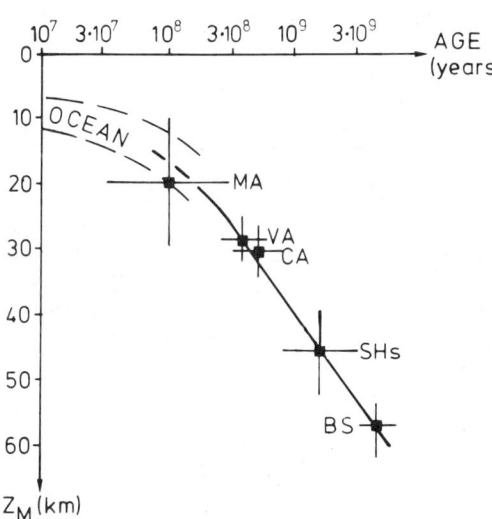

FIGURE 12. Crustal depth (Z_M) as a function of tectono-thermal age as based on more than 100 seismic profiles. BS, Baltic Shield; CA, Caledonian areas; MA, stable margins; SHs, Shield areas; and VA, Variscan areas.

lenticular-shaped lamellae may reach 0.1 and even 0.2 i.e., they indicate a considerable impedance contrast. Older crustal units, the London-Brabant Massif and the Bohemian Massif, incorporated into and affected by the Variscan orogeny, show a completely different, atypical reflectivity: While the former shows a fish-bone-like reflectivity with short elements in the whole crust, the latter shows a strong reflectivity with definite dips on its western flank.

The reflecting lamellae of Phanerozoic crusts may be explained in various ways.

1. *Igneous layering.* Intrusions from the mantle and melting of parts of the lower crust may lead to a mafic-ultramafic layering. The Muscox intrusion in Canada is a good example of such a layering. COCORP results in the Basin and Range province also support this idea; the best reflections supposedly originate from magma chambers in areas of active volcanism. Massive intrusions from the mantle into a strongly stretched crust have been postulated by several authors, e.g., McKenzie (1984).
2. *Metamorphic layering.* Internal metamorphic processes may modify any material in a high T and/or high p process, leading to metamorphism, partial melts, or even anatexis. Early melting products segregate into bands, causing subhorizontal layering. Observations of Archean terrains, which are supposedly former lower crust, support this idea: Granulites and amphibolites show plane heterogeneities at all scales. Also sediments, guided down to lower crustal levels in suture zones and metamorphosed, are candidates for metamorphic layering.
3. *Dynamic layering.* Shear movements along planes of weakness creating mylonite bands, i.e., changes of rocks within the shear zone and even creating partial melts along them, are also possible candidates for crustal reflectors. The sporadic reflections in the upper crust associated with definite fault zones often turn horizontal when reaching the lamellae of the lower crust. While concentrating on single elements in the upper crust they might spread out and diversify in the lower crust.
4. *Pore pressure layering.* Fluid or gaseous phases may enter the crust by a number of processes associated with suture zones or degassing of the mantle. The ductile, i.e., crackless, lower crust may act as a kind of barrier, and areas with high pore pressure may alternate with low-pressure zones, which would produce alternating V_p velocities but an uneffected V_S structure. S-wave tests are planned to prove or disprove this hypothesis.

Certain combinations of these explanations are highly probable. It seems important that the subhorizontal or at least plane layering of the lower crust must have come into being by a large-scale ordering process in a high T/low p surrounding regardless of the specific process. The history of many young crustal units involves one or several p-T cycles by which at least one of the processes 1 to 3 becomes intense. A cooling along the final p-T paths associated with slow uplifts might transform the ductile material into the brittle field again, partly destroying any long (and reflective) banding by small rupture processes in earthquakes. In the brittle field of shields, or generally in the upper crusts, three-dimensional (nonreflective) plutons also might survive a very long time, contributing to the low reflectivity. In general, information on the last tectonic process is stored in the rigid upper crust while the ductile lower crust and Moho of Phanerozoic areas tend to transform into flat, subhorizontal structures by creep.

Conclusions

The structure of the continental crust from the points of velocity, reflectivity, and depth can generally be related to age, i.e., to heat flow, temperature, and viscosity. While the Moho in young areas reflects seismic waves both in the near-vertical and the wide-angle range, intracrustal boundaries like the mysterious Conrad mostly create only wide-angle events and might often consist of strong-gradient zones rather than being first order boundaries. In the Phanerozoic crusts the reflective lamellae generally start slightly above the Conrad and abruptly end at the Moho.

The old shield and platform areas are less differentiated and appear more primitive. Smooth boundaries inside the crust dominate. Their velocity seems to be higher, and sometimes even the Moho must be a gradient zone, only producing wide-angle events in P and S waves.

The lack of deep continuous reflections in the old areas is attributed to smooth velocity transitions, to a more primitive, mafic chemistry, where no second or repeated differentiation has taken place, as revealed, by definition of their tectono-thermal age. The chemical differentiation of older crustal (or mantle-derived) material into light sialic upper crusts and mafic residues in the lower crust apparently is the precondition for a high reflectivity of the younger crusts. A high-T/high-p ordering process acting in the weak and ductile lower crust has to be considered a young feature, creating both reflecting lamellae and crustal thinning. Crustal mountain roots, except for recent and late Tertiary collisions, have disappeared. The rigid upper crust, on the other hand, may preserve characteristic fault patterns of the latest tectonic event and only here the memory of the latest tectonic event is preserved. The continental lithosphere does not resemble the rigid plate as its oceanic counterpart does. Especially the lower crust is ductile, weak, and vulnerable. Thin-skinned tectonics with nappe displacement above, sutures below, and reflections dipping in opposite directions near zones of compression, provide pictures of "crocodile tectonics" so different from the rigid-plate concept developed mainly from the oceanic scenarios.

R. MEISSNER
TH. WEVER

References

Barazangi, M., and L. D. Brown, eds., 1986. *Reflection Seismology: A Global Perspective*, Geodynamics Series vol. 13, 313p, and *Reflection Seismology: The Continental Crust*, Geodynamics Series vol. 14, 340p. Washington, D.C.: American Geophysical Union.

Giese, P., C. Prodehl, and A. Stein, eds., 1976. *Explosion Seismology in Central Europe*. Berlin: Springer, 429p.

Matthews, D. H., and C. Smith, eds., 1987. Deep Seismic Reflection Profiling of the Continental Lithosphere," 495p. Special Issue of *Geophys. Jour. Royal Astron. Soc.* **89.**

McKenzie, D., 1984. A possible mechanism for epeirogenic uplift, *Nature* **307,** 616-618.

Meissner, R. 1986. *The Continental Crust: A Geophysical Approach,* London: Academic Press, 426p.

Meissner, R., Th. Wever, and E. Flüh, 1987. The Moho in Europe—Implications for crustal development, *Annales Geophysical* **5 B,** 357-364.

Miyashiro, A., K. Aki, and A. M. Sengör, eds., 1982. *Orogeny.* Chichester: Wiley, 242p.

Telford, W., L. Geldart, R. Sheriff, and D. Keys, 1980. *Applied Geophysics.* Cambridge: Cambridge University Press, 860p.

Cross-references: *Accretionary Tectonics; Continental Collision Zones; Continental Drilling; Continental Lithosphere; Continental Rifting; Controlled Source Seismology; Deep Seismic Reflection Profiling; Earth Structure; Exploration Seismology; Far-Travelled Terranes; Seismic Imaging; Seismic Properties of Rocks; Seismotectonics and Crustal Structure; Thin-Skin Tectonics.*

CONTINENTAL DRIFT—See PALEOMAGNETISM AND CONTINENTAL DRIFT: HISTORICAL INTRODUCTION; PALEOMAGNETISM AND PLATE TECTONICS; SEAFLOOR SPREADING: MAGNETIC EVIDENCE.

CONTINENTAL DRILLING: ULTRADEEP

Ultradeep drill holes are those whose depths exceed 6 km. Numerous industry holes have been drilled to these depths worldwide during the past several decades in search of hydrocarbons and economic mineral deposits. Our knowledge of continental structure has also been increased in recent years by the use of deep seismic reflection profiling, high quality potential field data, tomographic, and other techniques to image the crust. These methods, however, provide indirect knowledge of the crust and are subject to wide variations in interpretations. The successes of scientific drilling programs in the oceans have underscored the need for similar drilling programs to provide direct knowledge of the less accessible continental crust. A summary of the state of activity and knowledge, scientific drilling activity, and drilling technology in the early 1980s was presented in a compendium of papers by Raleigh (1985).

Ultradeep drilling is being used as a primary scientific tool for the first time. Scientific drilling to great depths will be carried out following careful planning, which will include detailed site and engineering studies to determine the optimum places and conditions to assure the greatest chance of success. The projects to be undertaken must address one or more fundamental geological problems considered soluble only with drilling, have a target geometry capable of being intersected at great depths with the drill, and a site providing conditions that will permit completion of the ultradeep hole using technology that is either available or can be readily designed and implemented for use in the project.

The identification of one, or occasionally more than one, unique problem per site commonly enables many other problems to be addressed. For example, an ultradeep hole proposed to address a fundamental problem in tectonics will likely permit important research to be conducted in geophysical logging technology, paleomagnetism, in situ stress, thermal properties, and other disciplines. Consequently, each of these projects will add immensely to our knowledge of the crust, and, as several are completed with experiments designed to be conducted in any ultradeep hole, will permit important comparative studies to be conducted.

The principal engineering difficulties that will be encountered in ultradeep drilling, even in benign environments, will be those of preventing the hole from deviating from the vertical, bit life, coring, and core retrieval at depths greater than 5-6 km, and trip time at great depths. The Soviet hole discussed below experienced some deviation, but they managed to control the problem to a degree that it was completed to almost 14 km depth. Bit life is a major problem because the entire drill string must be pulled to replace the bit, requiring considerable time (and cost) at great depths. Several alternative designs are currently being considered in the western world, since a breakthrough in bit design would be very beneficial in keeping costs down. A major workshop, "Core Drilling at Ultradeep Depths: An Engineering Challenge," was convened in 1986 in northeastern

Georgia under the auspices of the U.S. Engineering Foundation and National Science Foundation to consider problems related to ultradeep drilling (Rowley et al., 1986). This workshop brought together drilling engineers from both the petroleum and mining industries.

Coring presents an even greater challenge at ultradeep depths. The problems of bit life and keeping the hole straight are still present, but the problems of core barrel design and core retrieval must be added. Some holes, like the proposed Appalachian ultradeep core hole, require core recovery in a large percentage of the length of the hole. Recovery of a large percentage of core has not been accomplished at depths greater than 5 to 6 km, even by the Soviets using a specially designed downhole turbine and core sample retrieval system.

Soviet Program in Ultradeep Continental Drilling

The most successful ultradeep drilling program to date is that of the Soviet Union. In the mid-1980s an ultradeep scientific hole was cored to a depth of greater than 12 km into the Baltic shield on the Kola Peninsula, concluding a drilling project that had begun almost two decades before (Kozlovsky, 1984). The primary goal of the project was to explore a segment of very old crust (3 to 1.6 Ga) that at the present level of erosion had experienced both the P-T conditions of high metamorphic grade and intrusive activity of the lower crust. Specific goals include study of the deep structure of the nickel-bearing rocks and Baltic shield of the Kola Peninsula; investigating seismic boundaries and the thermal regime in the deep crust, the material composition and physical state of the rocks in the deep crust; and testing and perfecting new technology for drilling and geophysical investigation of rocks and mineral deposits at great depths (Kozlovsky, 1987).

A spectacular result of this project is the discovery of small quantities of free water at depths as great as 9 km, a factor directly related to the presence of open microspace, the macroscale shear strength of the crust, and processes generating seismic activity in different parts of the crust. Occurrences of water in narrow confined zones were also noted at about 10 km (Borevsky et al., 1987). Seismic velocities abruptly increase at depths greater than 9 km in the vicinity of the Kola hole. Abrupt changes in seismic velocity traditionally have been attributed to compositional or phase differences in the crust locating crustal "discontinuities." In light of the Soviet results, we must now reevaluate the causes of discontinuities to include boundaries produced by the presence of microcracks and fluids as a cause of seismic layering in the crust. This, of itself, increases the difficulty of interpreting deep seismic reflection and refraction data worldwide. A direct correlation between dipping seismic layers and geologic boundaries was established in the upper part of the hole, but horizontal layering in the lower part of the Kola hole could not be correlated with geology, and, consequently, is attributed to changes in physical properties of the crust (Kozlovsky, 1987). The occurrence of the frequency and size of well-bore breakouts with depth is also directly related to the presence of microcracks and fluids.

A zone of low seismic velocity discovered at a depth of 4,500 m is attributed to low temperature (retrograde) metamorphism of the basaltic rocks at that depth. Densities and P- and S-wave velocities all show a systematic decrease at this depth, along with a systematic increase in porosity, permeability, and anisotropy. Other changes in elastic constants and electromagnetic characteristics throughout the hole are correlative with changes in composition, metamorphic overprints, and the occurrence of ductile shear zones encountered during drilling (Kozlovsky, 1987).

A major factor contributing to the success of the Soviet ultradeep drilling program has been the use of downhole turbine drilling motors. Motors of this type do not require turning the entire drill string to turn the core bit. Instead, pressure is increased in the drilling fluid transmitting energy to the down hole motor, which turns the bit. This permits the use of light alloy tubing, resulting in a huge weight savings at depths of several km.

The Soviet ultradeep scientific drilling program is without a doubt the most experienced in the world at present, with the success of one ultradeep hole, another completed to 8,500 m at Saatly in the Baku petroleum field near the Caspian Sea, and several others planned to exceed depths of more than 7,000 m (Kozlovsky, 1984). Each has different goals and should produce a variety of interesting results.

Other European Drilling Programs

In 1986, the Federal Republic of Germany ultradeep drilling program (*Kontinentalen Tiefbohrprogramms der Bundesrepublik Deutschland,* KTB) selected a site for drilling a 15 km hole in the Oberpfalz region. Its purpose is to investigate the properties of the crust in central Europe and to address several problems in the evolution of the Paleozoic Variscan orogen.

The ultradeep Siljan Project in Sweden is planned to reach 7–8 km. Its principal goal is to test the possibility of accumulation of abiogenic (mantle-derived) methane in the Siljan impact crater. The site in Precambrian granitic rocks of the Baltic shield was chosen because of a presumed lesser chance for contamination from biologically-generated hydrocarbons.

U.S. Continental Scientific Drilling Program

The U.S. Continental Scientific Drilling Program has been operating for several years under support by a variety of agencies. More scientific drilling has been completed to moderate depths (<5 km) by the Department of Energy than by any other group, principally investigating geothermal areas in the West and the Salton Sea Project, a study of an active hydrothermal system.

The U.S. Continental Scientific Drilling Program (National Academy of Sciences, 1979) has evolved from its beginning in the late 1970s, when the National Academy of Sciences Continental Scientific Drilling Committee was formed, to the interagency group DOSECC (Deep Observation and Sampling of the Earth's Continental Crust) formed in 1984 to manage the U.S. Continental Scientific Drilling Program. The U.S. program has remained a fledgling organization for little more than a modest drilling program because of lack of major funding. Even so, several small shallow hole projects have been initiated. The first of these at Cajon Pass, east of San Bernardino, California, was a 4-km hole located near the San Andreas fault to investigate the in situ stress-heat flow paradox. Technology developed in drilling this hole will be used in ultradeep drilling projects in crystalline rocks, which should begin in the 1990s. A second shallow hole program is to explore the hydrothermal geochemistry of the mineralized volcanic-plutonic complex near Creede, Colorado.

Several ultradeep holes have been proposed in the U.S. DOSECC program of continental scientific drilling to address major problems in crustal evolution. The Appalachian Ultradeep Core Hole (ADCOH) Project intends to penetrate a large intact crystalline thrust sheet, the platform sedimentary succession, and autochthonous basement beneath. This project was endorsed by the U.S. National Academy of Sciences (1984) as the first project that should be undertaken in a U.S. national program of continental scientific drilling. The principal goals of this project include: (1) to investigate the processes of crustal response to large scale overthrusting of a composite crystalline sheet and its overridden platform; (2) to investigate fundamental structural, geochronological, and geophysical properties related to the emplacement mechanics of large thrusts over a wide variety of compressional fault types (ductile vs. brittle; crystalline vs. platform; dip-slip vs. strike-slip); (3) to calibrate a multitude of seismic reflectors; (4) to study the structural and stratigraphic character of the platform and the geochemical nature of the paleofluid regime preserved beneath the thrust sheet; and (5) to study the present-day thermal and in situ stress regimes in this segment of the crust. The ADCOH Project was endorsed by the DOSECC Science Advisory Committee early in 1987 as the first ultradeep hole that should be undertaken once funding becomes available. Drilling a pilot core hole to 5-6 km should begin in the early 1990s with the goals of engineering study of rock properties and scientific investigation of the upper section of the ultradeep hole site.

The Gulf Investigations and Deep Exploration (GUIDE) Project proposes to locate a site for ultradeep drilling on the San Marcos Arch in southeastern Texas, one of the few places in the western Gulf Coast where it is possible to penetrate the Mesozoic and Tertiary sedimentary section to the Paleozoic and Precambrian (?) crust beneath using existing technology. It is thought that a site in this area may penetrate a zone of transitional crust beneath the Coastal Plain sedimentary section. This project intends to investigate the processes related to opening the Gulf of Mexico, crustal extension, the evolution of basin margins, and the crustal processes attendant to formation of trailing margin successions.

To investigate the processes of initiation and development of intracratonic basins, the Illinois Basin Ultradeep (IBUD) Project proposed a 10-km deep core hole in the southern part of the Illinois Basin. The site chosen will permit penetration of the entire Paleozoic section that was deposited in the basin, an arm of the Late Proterozoic-early Paleozoic Reelfoot Rift system that may have influenced, or even initiated, formation of the Illinois Basin and a substantial section of the underlying Proterozoic crust. This project would also test the thermal contraction/isostatic model for crustal subsidence and investigate both the paleo- and present-day fluid regimes in the Illinois Basin.

Deep to ultradeep drill holes have also been proposed for Hawaii and the Katmai region in the Aleutian Islands. These projects propose to investigate the interrelationships between the processes of basaltic magma generation in the intraplate and plate boundary regimes. The Hawaiian project is to investigate the possible mantle hot spot that exists beneath the Hawaiian islands.

All of the ultradeep hole projects proposed to date in the United States focus on one or more fundamental unknowns that would be best addressed by drilling. All propose targets that may be accessible using present-day technology, although the ADCOH Project would probably be drilled in the most benign environment and therefore have the greatest probability of reaching its target depth. The IBUD and GUIDE targets may be in thermally safe areas but both may encounter geopressured and caustic fluids. The Hawaiian hole would be drilled into a thermally active environment.

Drilling and coring of ultradeep holes is a long-term undertaking that is already subject to some controversy among U.S. earth scientists. In contrast to the worldwide seismic network program and most of the large projects undertaken in physics and astronomy that are said to address fundamental problems that pervade all of science, ultradeep drilling must address more specific fundamental questions related to crust formation, faulting, basin evolution, magma generation, and other processes. The earth sciences community should soon arrive at an agreement that this is a worthwhile undertaking that will benefit the entire science community over the several decades that are required for a program of this magnitude to be completed.

ROBERT D. HATCHER, JR.

References

Borevsky, L. V., G. S. Vartanyan, and T. B. Kulikov, 1987, Hydrogeological essay, in Ye. A. Kozlovsky, ed., *The Superdeep Well of the Kola Peninsula.* New York: Springer-Verlag, 271-287.

Kozlovsky, Ye. A., 1984, The world's deepest well, *Sci. American* **251**(6), 98-105.

Kozlovsky, Ye. A., 1987, *The Superdeep Well of the Kola Peninsula.* New York: Springer-Verlag, 558p.

National Academy of Sciences, 1979, *Continental Scientific Drilling Program,* U.S. Geodynamics Committee, Geophysics Research Board. Washington, D.C.: National Research Council, 192p.

National Academy of Sciences, 1984, *Priorities for a National Program of Continental Scientific Drilling,* Continental Scientific Drilling Committee, Geophysics Research Board. Washington, D.C.: National Research Council, 13p.

Raleigh, C. B., ed., 1985, *Observation of the Continental Crust through Drilling I.* Heidelberg, West Germany: Springer-Verlag, 364p.

Rowley, J. C., F. Schuh, and M. Walton, eds., 1986, *Ultradeep Coring and Drilling,* summary of an Engineering Foundation Conference and Symposium. New York: Engineering Foundation.

Cross-references: *Continental Crustal Structure; Continental Lithosphere; Controlled Source Seismology; Crustal Movements and Tectonic Deformation; Deep Seismic-Reflection Profiling; Deformation of Rocks and Minerals; Exploration Seismology; Stress in the Earth's Lithosphere; Thin-Skin Tectonics.*

CONTINENTAL LITHOSPHERE

The continental lithosphere consists of the continental crust and, typically, some nonconvecting part of the underlying upper mantle. In plate tectonics terms, the continental lithosphere is part of the rigid outer rind of the Earth, which is segmented into several major plates. The lithosphere "floats" atop a more mobile (low strength) asthenosphere. In this context, oceanic lithosphere is relatively straightforward to define, as thickness for the most part obeys simple thermal models (although even here we must be cautious; see *Lithosphere, Oceanic: Formation and Evolution*). By contrast, the continental lithosphere is enormously complex. This contrast reflects the fact that oceanic lithosphere is formed in much the same way worldwide and that it is continually recycled into the interior of the Earth via the plate tectonic "conveyer belt." Continental lithosphere, on the other hand, is too buoyant to be subducted. Thus, while no known oceanic lithosphere is older than about 200 my, the continental lithosphere has evolved over billions of years, one consequence of which is that it is seen today as the resulting collage of many superimposed tectonic events, including rifting, subduction, continental collision, accretion, and hot spot magmatism, that recur repeatedly over geologic time.

The definition of continental lithosphere is necessarily vague, for as we shall see, no single definition of continental lithosphere can be agreed upon by the scientific community. Particularly ill-defined is the base of the lithosphere, for despite many important advances, the very deep structure and composition of the continents remains a frustrating enigma. The dearth of information contributes to the proliferation of definitions of continental lithosphere that have been proposed in the literature—seismological, mechanical, rheological, thermal, and compositional. Depending on one's view, the continental lithosphere may be a *mechanical boundary layer,* a *thermal boundary layer,* or a *chemical boundary layer.* In the remainder of this article, we shall explore these various concepts in some detail. Our goal here is to describe the continental lithosphere in terms of global Earth structure, plate tectonics, and the long-term evolution of the continents. By this view, the continental lithosphere is envisaged to be a long-lived plate tectonic unit, consisting not only of continental crust but also a very significant mantle "root" that is attached to and translates with the continent. The lithospheric mantle, therefore, is a stable component of the continent and does not participate in the convective processes that go on in the rest of the mantle.

Mechanical Models

Perhaps the simplest expression of the lithosphere is *elastic thickness.* This thickness is a function of the mechanical response of the lithosphere, which can be modeled by an elastic plate overlying a weak (asthenospheric) substrate. As discussed in detail in *Lithosphere: Mechanical Properties,* the elastic thickness of the lithosphere can be moderately well determined through a number of different measurements. Chief among these are topographic response to surface loads (such as volcanoes), the relationship between topography and Bouguer gravity anomalies, and the shape of depositional surfaces in basins. As determined from these measurements, the elastic lithosphere of the continents is rarely greater than 100 km thick and in areas of extension or hotspot activity may be only a few km or less. The elastic thickness of the continents is clearly controlled by the thermal state of the lithosphere (see *Lithosphere: Mechanical Properties,* Fig. 4). As Forsyth (*Lithosphere: Mechanical Properties*) points out, however, the apparent elastic thickness is also affected by the state of stress and the rheological stratification in the crust and mantle, so the mechanical thickness may not bear a simple relationship to one particular isotherm.

One important aspect of the mechanical structure of the continental lithosphere as it relates to lithospheric evolution is the large decrease in strength of crustal rocks in the deep crust (see Fig. 3 in *Deformation of Rocks and Minerals*). The steep geothermal gradient in the crust produces increasing ductility of crustal rocks with depth. The brittle zone in tectonic areas typically extends only to about 15 km, below which the rocks deform plastically. The topmost part of the underlying mantle, however, is relatively much stronger as temperatures there are still far from the melting point of mantle rocks. Thus there is a zone of weak lower crust sandwiched between strong upper crust and strong upper mantle, creating a possible zone of crust-mantle decoupling. P. Bird, in a series of recent papers (e.g. Bird, 1984), has argued that during the period of horizontal subduction of the Farallon plate beneath the western United States, the continental lithospheric mantle (and perhaps part of the lower crust) have been detached from the crust and translated northeastward in response to basal shear stress from the oceanic plate. If this model is correct (and it remains highly controversial), then whole regions of the continent can be stripped of their lithospheric mantle during periods of near-horizontal subduction. At least some of that lithosphere may subsequently be accreted or underplated to adjacent lithospheric blocks.

While the flexural lithosphere is an extremely useful *mechanical* concept, it fails as a plate tectonic description of lithosphere. From the perspective of plate tectonics, it may be argued that the full definition of the lithosphere should account for the long-term structural coherence and evolution of continental plates and the unique seismic, petrologic, and geochemical identity of the nonconvecting upper mantle keel beneath the continents. In many tectonic regions the flexural lithosphere is not even as thick as the continental crust. Clearly, however, the low density sialic crust cannot be resorbed into the mantle and must therefore be a permanent part of the lithosphere.

Thermal Models

A convenient definition for *thermal lithosphere*, or the thermal boundary layer, is the outer layer of the Earth in which heat transfer is dominated by conduction (see *Heat Flow in the Earth*). Thus the stronger lithosphere acts as a barrier to thermally induced buoyancy forces that drive convective heat transfer in the underlying asthenosphere (Morgan, 1984). In the case of the oceanic lithosphere, this conductively cooled layer of the Earth is thought to have finite thickness (i.e. the lithosphere ceases to undergo thermal contraction as it ages beyond about 80 my, implying that heat is supplied to the lithosphere from the underlying asthenosphere (see *Lithosphere, Oceanic: Formation and Evolution*). For the continents, the situation is far more complex. The continental lithosphere does not undergo a simple monotonic cooling cycle, but may be subjected to repeated episodes of thermal or tectonic disturbances. Thus superimposed upon the geological age of the lithosphere is its thermal or tectonic age. Because of its great age and complex history, the continental lithosphere is also compositionally heterogeneous, both in major elements and in abundances of radiogenic heat-producing elements.

We consider first the thickness of the thermal lithosphere beneath the stable cratonic cores of the continents. Here the lithosphere is both at its thickest and its strongest. If the effective viscosity of the asthenosphere beneath the cratonic lithosphere is taken to be about 10^{21} Pa·s, based on glacial unloading and gravity/topography correlations (see *Mantle Viscosity: Constraints from Glacial Isostasy, Plate Tectonics, and Seismic Tomography*), it implies a temperature of transition from conductive to convective heat transfer of about 1300° to 1400°C for typical mantle compositions (Morgan, 1984). If crustal heat production is taken into account, the thickness of the thermal lithosphere of the stable continental cratons as calculated from surface heat flow may range from about 90 km to 220 km.

The lithospheric thicknesses obtained by this simple thermal model clearly depend on other factors as well, the most important of which is the composition, including volatile content, of the subcrustal mantle. If the continental lithospheric mantle consists of low density peridotite depleted of its low-melting basaltic fraction and devolatilized over time by metamorphic and magmatic events (see discussion below), the continental mantle will be less dense and substantially more refractory than the rest of the mantle. The chemical boundary layer represented by this depleted subcontinental mantle peridotite may stabilize the continental mantle root against thermal disruption to depths below the thermal boundary layer.

The concept of a stabilizing chemical boundary layer is at the heart of the *temperature-depletion compensation* hypothesis (see below) laid out by Jordan in an important series of papers on the continental *tectosphere* (see Jordan, 1981 for references) and discussed in some detail below. The term tectosphere was introduced by Jordan in an attempt to circumvent some of the difficulties associated with the conflicting definitions of lithosphere alluded to above. It is a particularly convenient concept for understanding those seismological and petrological/geochemical aspects of deep continental structure that appear to indicate continent-ocean heterogeneity to depths of several hundred km (see following sections).

Thus far, we have considered only the lithosphere

of the stable continental cratons. The effects on lithosphere thickness of tectonothermal events are described in *Heat Flow in the Earth*. Thermal disturbances will result in lithospheric thinning. Regions where prior thermal disturbances have occurred, therefore, may be more prone to reactivation, whereas stable cratons should be comparatively less vulnerable to thermal perturbations from the underlying asthenosphere (Pollack, 1982). This seems to be particularly true of continental rifting, where repeated episodes apparently recur along the same zones of weakness. While hotspot traces may be found in even the oldest continental interiors, these occurrences (such as the Snake River Plain/Yellowstone terrane) are relatively rare, suggesting that only very large thermal disturbances can penetrate the stable platforms (Gass et al., 1978). A wide range of epeirogenic movements within continental interiors has been interpreted as caused by thermal disturbances followed by conductive relaxation and asymptotic thickening of the lithospheric thermal boundary layer.

Seismological/Compositional Models

Tectosphere. It has long been known that average seismic velocities, especially shear velocity, are much greater under cratons than under oceanic or tectonically active areas. This fact was first shown by surface wave dispersion measurements and later by measurements of vertically travelling *ScS* phases. Sipkin and Jordan (1980) showed that *ScS* one-way travel-time anomalies associated with the upper mantle beneath stable continental interiors could be as large as 3.5 s (fast) relative to the upper mantle beneath the western Pacific. Moreover, the high velocity continental paths are also associated with high Q_s, suggesting that the anomalies are caused by thermal and/or compositional variations in the upper mantle (Jordan, 1981). In addition to the studies of Sipkin and Jordan, a number of other studies based on three-dimensional inversion of travel-time anomalies for velocity structure have been applied regionally in Europe and the United States. The results nearly all show that regional velocity contrasts, even between provinces entirely within the continents, extend to at least 250 km.

If only thermal variations are considered, the thermal boundary layer beneath continents must be at a depth of 300 km or more to account for the S-wave travel-time differences. Jordan argues that thickening of the tectosphere by growth of the thermal boundary layer to 300 km, while possible over time scales of ca. 1000 Ma if there are no small-scale convective instabilities in the deeper tectosphere, is ruled out by the gravitational requirement that the crust thicken with time to compensate isostatically for volume contraction (increasing density) of the thermal boundary layer. Such crustal thickening is not observed. If this view is correct, the only remaining stabilizing mechanism for thick tectosphere beneath the cratons is compositional.

There is a considerable body of knowledge about the composition of deep cratonic mantle from the study of xenoliths, mantle fragments that have been carried to the surface in explosive volcanic pipes. A comprehensive summary of recent work done on mantle xenoliths will be found in volumes devoted to the subject by Boyd and Meyer (1979), Dawson (1980), or Nixon (1987). The development of accurate mineralogical geothermometers and geobarometers, which allow the equilibrium depth and temperature of xenolith source regions to be determined, has made it possible to "map" xenolith compositions as functions of position in the mantle. Among the major discoveries is the identification of two distinct suites of ultramafic nodules associated with several cratonic areas: basalt-depleted *granular* nodules and more "fertile" *sheared* nodules (e.g. Boyd and Nixon, 1975). The granular nodules, which are by far the most common, appear generally to be derived from depths less than about 200 km. While they are depleted in a low-melting temperature basaltic fraction, they are commonly enriched relative to oceanic peridotite in certain incompatible trace elements, including heat-producing elements such as K and Rb, presumably as the result of metasomatism in the mantle. Sheared nodules, at least some of which are enriched in Fe, Ti, and Na, seem to be associated with greater depths and significantly higher temperatures. The boundary between granular and sheared nodules is marked by a deflection or "kink" in the geotherm, with the gradient of the deeper portion of the geotherm sharply steeper. There has been and continues to be considerable controversy surrounding the tectonic and compositional significance of the sheared nodules, particularly whether they represent shear-heated samples from the base of the lithosphere (see Dawson, 1980, pp. 202–207) and whether they are in fact "enriched." Petrologists generally agree, however, that the common "depleted" granular peridotites represent a significant percentage of the subcontinental mantle to depths of at least 150–200 km. This fact led Jordan (1979) to postulate a thick *chemical boundary layer* for the continental tectosphere.

The effect of removing a basaltic partial melt from a fertile mantle peridotite (such as the hypothetical pyrolite) is to reduce the density of the residue and to increase seismic velocities. Jordan compares densities of depleted peridotite nodules with densities of a hypothetical pyrolite composition and shows that average continental garnet lherzolite (depleted peridotite) has a calculated normative density about 1.3% less than that of pyrolite. This reduction in density will approximately compensate for the density contrast in a homogeneous mantle that would result from the temperature contrast of about -400 °C between a typical shield geotherm and the average oceanic adiabat (Jordan, 1979). The model then, is based on the supposition that the lower density of the subcontinental mantle caused by basalt depletion is gravitationally balanced by the density decrease in the oceanic mantle caused by a much higher geotherm, hence the expression *temperature-depletion compensation*. The thickness of the chemical boundary layer is generally taken to be about 150 to 200 km.

Thickness of the tectosphere by this model is still very much open to question. The presence of substantial quantities of metasomatically introduced heat-producing elements in the basalt-depleted upper mantle will result in significant heat generation in the tectosphere. If correct, the contribution to surface heat flow from the deep mantle must be even smaller than is generally assumed (see *Heat Flow in the Earth*), and the calculated geotherm even lower beneath continents. It has been proposed on this basis that the thermal boundary layer may be as much as 400 km thick, consistent with the very large shear-wave travel-time anomalies mentioned above (Jordan, 1978).

Considerable controversy surrounds virtually all aspects of the tectosphere model. The mechanisms by which a chemical boundary layer can be built up are poorly understood, although there are many modes by which magma can be extracted from continental mantle over time. More importantly, questions of the dynamical stability of a chemical boundary layer whose thickness varies laterally by 150 km or more is the subject of present research. Preliminary models show that a thick zone of depleted mantle beneath the continents may be difficult to maintain over billions of years. Convective instability should act to disperse and thin the zone of depleted mantle, although the calculations that lead to this conclusion do not include effects such as compositional dependence of viscosity or the order of magnitude difference between vertical and horizontal compositional gradients.

Seismic Evidence from Surface Waves and Converted Phases. In the preceding discussion we have ignored many important seismological results that relate to the structure and thickness of the continental lithosphere. The travel-time anomalies of *ScS* through the upper mantle provide information about average velocities over long ray paths; however, integrated travel times cannot provide evidence on velocity distribution as a function of depth in the upper mantle, nor, perhaps more importantly, whether thermal boundary layers or chemical boundary layers correspond to seismic discontinuities. A recent review of this subject is found in Sacks and Snoke (1984), who also include tables of depth to the lithosphere/asthenosphere boundary based on results from a number of different seismic methods.

Among seismologists, the definitive test for lithospheric thickness has long been the presence of a high velocity "lid" overlying a distinct seismic zone (asthenosphere) characterized by low shear-wave velocities and high seismic attenuation. For many, this continues to be a comfortable definition. The earliest studies that demonstrated the presence of a low velocity zone in the upper mantle beneath ocean basins and tectonic regions of the continents were done using surface waves (see *Surface Waves*). Despite the relatively small number of measurements of surface wave dispersion compared to body wave studies, much of what we know about the upper mantle, particularly shear-wave velocities, low velocity zones, and density stratification, comes from the analysis of surface waves.

Surface waves are dispersive, the nature of the dispersion depending on the velocity and density structure of the Earth along the propagation path. Nonunique models of velocity and density as functions of depth along the propagation path can be obtained from analysis of dispersion. Shear velocity is by far the best determined parameter. Surface wave studies over the years have shown that low velocity zones appear to exist to a greater or lesser extent beneath all continental regions save, perhaps, the oldest and most stable continental regions. There is a caveat here, however, for in most studies the velocity resolution of surface wave inversion is simply not adequate to resolve a low velocity zone. Hence the presence of low velocity zones in the derived models simply reflects the assumptions of the starting model, typically designed to include an upper mantle low velocity zone. For the stable continental platforms, keeping the above caveat in mind, surface wave studies indicate possible low velocity zones centered at depths of 150 to 250 km (Sacks and Snoke, 1984). Similar results are obtained from studies of high shear-wave attenuation, which is nearly always related to low shear-wave velocities.

A more direct approach to determining the base of the lithosphere is through the use of *converted phases*. Converted phases are created by conversion of shear waves to compressional waves, or vice versa, at a velocity or density discontinuity (Fig. 1). In principle, the study of converted phases can provide detailed information about the structure (sharpness) and the velocity contrast, including the sign of the velocity contrast, across a discontinuity. The method cannot be used, however, to determine absolute velocities, so additional evidence is usually needed to confirm the identity of a given discontinuity. A thorough summary of results from converted phase studies of the base of the lithosphere is given in Sacks and Snoke (1984). Among the most interesting of these studies is that by Sacks and Snoke (1977) based on converted phases from deep South American earthquakes observed at seismic stations in western South America. The results demonstrate that there is a sharp velocity reversal at about 400 km depth beneath the Brazilian shield (the well known 400 km discontinuity should exhibit a velocity increase). While this finding remains something of an enigma, the simplest interpretation

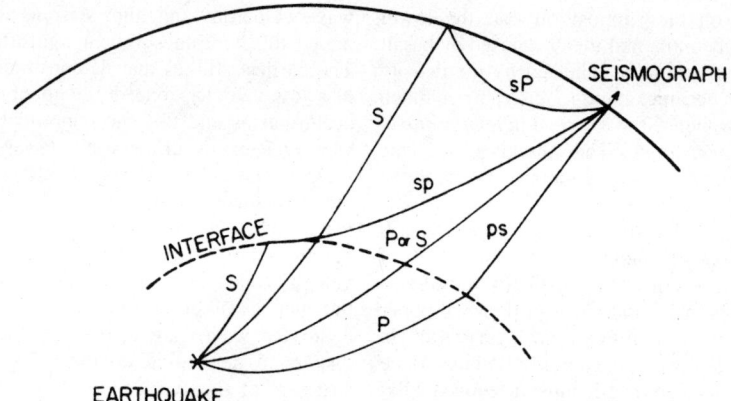

FIGURE 1. Predicted ray paths for converted phases. (From Sacks and Snoke, 1977)

is that the velocity reversal is associated with the base of the lithosphere (tectosphere?) beneath the South American craton. If so, it is the only known direct observation of this very deep and mostly hypothetical lower plate boundary. A similar study by the same authors using data from the fixed NORSAR seismic array in Norway indicated a depth to the base of the lithosphere beneath the Baltic shield of about 250 km.

Other Evidence

We have discussed the continental lithosphere mostly from the point of view of structure and evolution of cratonic lithosphere. One aspect that has not been addressed is the role of underplating of continental lithosphere by oceanic lithosphere. There is evidence, for example, that some of the eclogites brought up in kimberlite pipes in Africa may be relicts of an ancient oceanic crust emplaced beneath the continental lithosphere (MacGregor and Manton, 1986). The data on which this conclusion is based come primarily from measurements of oxygen isotopic ratios combined with trace element and radiogenic isotopic ratios. The oxygen isotopic ratios in particular are difficult to explain as originating anywhere other than in oceanic crust. Similar "dead" oceanic lithospheric slabs have been postulated beneath parts of western United States, but concrete evidence is lacking.

An active field of study concerns the growth of new lithosphere beneath subduction zones. Here, lithosphere appears to be created by freezing out of asthenosphere in the wedge between the downgoing and overriding plates (Fig. 2). At present, most of the evidence for growth of lithosphere comes from

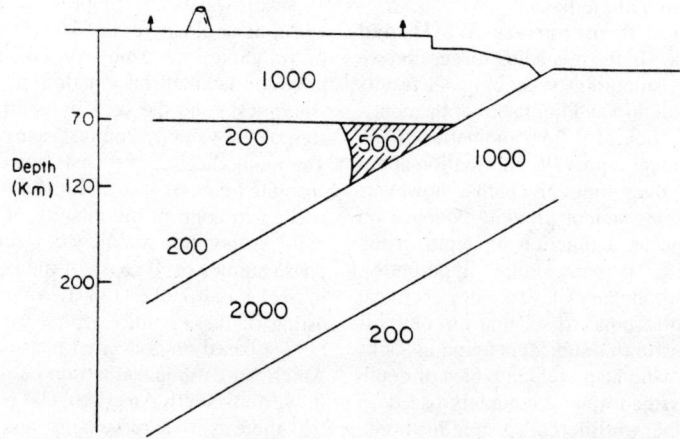

FIGURE 2. Hypothetical growth of continental lithosphere by freezing out asthenosphere in the mantle wedge between downgoing and overriding plates. (From Sacks and Snoke, 1984)

studies of attenuation (Q). Higher Q values are associated with lithosphere and lower values with asthenosphere. In typical island arcs, the contrast in Q between lithosphere and underlying asthenosphere is very large, commonly a factor of five or more. Beneath parts of Japan, however, a zone of anomalous intermediate Q has been mapped in what should have been asthenosphere (Sacks and Okada, 1974). This zone has been interpreted to mean that lithosphere, or incipient lithosphere, is being frozen out of the asthenosphere at the apex of the mantle wedge between converging plates (Fig. 2). If this interpretation is correct, cooling in the subduction wedge could be an efficient mechanism for creating new lithosphere beneath continents and island arcs.

It should be apparent from the preceding discussion that the study of the continental lithosphere, particularly the lithospheric mantle, is in its infancy. Massive (by earth science standards) national programs have been initiated in the United States and elsewhere to study the continental lithosphere in great detail. Included among these programs are plans for seismic imaging of the deep lithosphere using portable arrays employing hundreds or even thousands of microprocessor-based seismographs systems. Such array studies will go far toward advancing our knowledge and understanding of the continental lithosphere.

DAVID E. JAMES

References

Bird, P., 1984, Laramide crustal thickening event in the Rocky Mountain foreland and Great Plains, *Tectonics* **3**, 741-758.

Boyd, F. R., and H. O. A. Meyer, eds., 1979, *The Mantle Sample: Inclusions in Kimberlites and other Volcanics*. Washington, D.C.: American Geophysical Union.

Boyd, F. R., and P. H. Nixon, 1975, Origins of the ultramafic nodules from the kimberlites of northern Lesotho and the Monastery Mine, South Africa, *Physics Chem. Earth* **9**, 431-453.

Dawson, J. B., 1980, *Kimberlites and Their Xenoliths*. Berlin: Springer-Verlag, 252p.

Gass, I. G., D. S. Chapman, and H. N. Pollack, 1978, Geological and geophysical parameters of mid-plate volcanism, *Royal Soc. London Philos. Trans.* **288A**, 581-597.

Jordan, T. H., 1978, Composition and development of the continental tectosphere, *Nature* **274**, 544-548.

Jordan, T. H., 1979, Mineralogies, densities and seismic velocities of garnet lherzolites and their geophysical implications, in F. R. Boyd and H. O. A. Meyer, eds., *The Mantle Sample: Inclusions in Kimberlites and Other Volcanics*. Washington, D.C.: American Geophysical Union, 1-14.

Jordan, T. H., 1981, Continents as a chemical boundary layer, *Royal Soc. London Philos. Trans.* **301A**, 359-373.

MacGregor, I. D., and W. I. Manton, 1986, Roberts Victor eclogites: Ancient oceanic crust, *Jour. Geophys. Research* **91**, 14063-14079.

Morgan, P., 1984, The thermal structure and thermal evolution of the continental lithosphere, *Physics Chem. Earth* **16**, 107-193.

Nixon, P. H., ed., 1987, *Mantle Xenoliths*. New York: Wiley.

Pollack, H. N., 1982, The heat flow from the continents, *Ann. Rev. Earth Planetary Sci.* **10**, 459-481.

Sacks, I. S., and H. Okada, 1974, A comparison of the anelasticity structure beneath western South America and Japan, *Physics Earth and Planetary Interiors* **9**, 211-219.

Sacks, I. S., and J. A. Snoke, 1977, The use of converted phases to infer the depth of the lithosphere-asthenosphere boundary beneath South America, *Jour. Geophys. Research* **82**, 2011-2017.

Sacks, I. S., and J. A. Snoke, 1984, Seismological determinations of the subcrustal continental lithosphere, *Physics Chem. Earth* **15**, 3-37.

Sipkin, S. A., and T. H. Jordan, 1980, Multiple ScS travel times in the western Pacific: Implications for mantle heterogeneity, *Jour. Geophys. Research* **85**, 853-861.

Cross-references: Continental Collision Zones: Seismotectonics and Crustal Structure; Continental Crustal Structure; Controlled Source Seismology; Deep Seismic-Reflection Profiling; Exploration Seismology; Heat Flow in the Earth; Lithosphere: Mechanical Properties; Lithosphere, Oceanic: Formation and Evolution; Mantle, Upper: Structure; Mantle Discontinuities; Mantle Dynamics; Mantle Viscosity; Ocean Continent Transition: Structure; Seismic Tomography.

CONTINENTAL RIFTING: TYPES AND RATES

It is now widely accepted that ocean basins evolve from oceanic rifts, which are the offspring of juvenile spreading systems such as those in the Red Sea and Gulf of California. The latter have evolved from continental rifts. The overall progression seems to be as incontrovertible as hominid evolution, but the details of each step are equally controversial. These details concern such basic issues as whether there are certain kinds of rifts that evolve to make oceans, how long the process takes, what the various expressions are along the way, and what drives the process. A myriad of related issues pertain to such topics as the role and timing of doming and volcanism in rifting, the importance of cycles or periodicities in rift evolution, and the effect of preexisting structures on determining rifting geometries. These details are not only of academic interest; some are crucial in developing hydrocarbon exploration strategies for rifts and passive margins.

Because the geology and geography of rifts are so diverse, limited treatments such as this one must pick and choose among many topical, methodological, and geographical possibilities. This study focuses on the types and rates of continental rifting, using the East African rift as a comparative base. *Continental Rifts: Structural Traits* addresses the topics of continental rifts structure, morphology, and form, again from the African perspective. The goal of these two studies is to bring the reader to a few key research precipices, within the context of a broader exposure to the subject.

Types of Continental Rifts

Like the study of other geological features, detailed examinations of individual continental rifts tend to reveal more differences than similarities. At some level, every rift zone is unique, even those as closely related as the Tanganyika and Malawi rift zones of East Africa. (Locations of all rifted areas mentioned in the text are provided in Fig. 1.) Hence, all classifications ultimately break down under increasing scrutiny. It is sensible to open this discussion with an examination of several of the more popular classification schemes and the conditions or situations under which they come unraveled.

Active Vs. Passive Rifting. It has become commonplace to divide continental rifts into active and passive categories (Fig. 2). Active rifts are those in which the horizontal tensile stresses are derived mainly by vertical uplift (e.g. Sengor and Burke, 1978). Such uplift is usually attributed to the buoyant forces of a rising plume of partially molten asthenospheric material, which stopes its way upward. The expressions predicted by the active rifting model should include doming concurrent with fracturing and the creation of regional unconformities produced by erosion of the rising dome. Active rifting also is often associated with extensive rift volcanism. Most known rifts and virtually all those that are situated atop long-wavelength bulges (e.g. the African rifts) have been classified as active rifts at one time or another.

We must digress at this point to clarify the issue of uplift scale. The smallest scale involves rift flanks, which occupy ribbons of ground on the order of tens of kilometers wide. All active rifts have flanks that are elevated relative to their centers of subsidence, but in some cases there is no actual uplift of the flanks. In other cases the flanking uplift can be extreme. For example, the rift mountains around the Tanganyika Rift Zone can reach heights of 4 km, the Ruwenzori Mountains between Lakes Albert and Edward more than 5 km. Modest flanking elevations can be explained as elastic byproducts of stretching (i.e. passive rifting). In other words, special conditions are not necessary to explain rift mountains up to 1 km high. However, this is not the case for Ruwenzori-type flanks, which are clearly elevated well above the prerift surface and which must be held up by some ongoing, dynamic process.

The second scale of uplift is the dome. Examples are the Ethiopian Dome, Rhine Upwarp, and Baikal Arch, all with widths measured in hundreds of kilometers. The third scale of uplift is measured in terms of a thousand kilometers and the best present-day example is the African swell, upon which the entire East African Rift is situated. It is these latter two scales of uplift that are usually associated with the active rifting theorem.

Passive rifts are those in which the horizontal tensile stresses are derived purely by differential plate motions. In the context of this proposition, rifts are a response to interactions between plates, so the mechanism of rifting relates directly to whatever drives the plates. (We believe this driving force to be the vector summation of forces that act on the bases and edges of plates.) Passive rifts can be modeled by various types of stretching experiments and the initial surficial expressions are ordinarily topographic lows (McKenzie, 1978). Stretching attenuates the lithosphere, and asthenospheric material rises to passively infill the area of necking. If regional doming occurs, it happens relatively late in the development of rifts, according to the passive rifting model. Most advocates of this proposition also associate volcanism with the later stages of rifting. The scheme presented in Fig. 19 in *Continental Rifts: Structural Traits* is essentially passive.

The great attraction of the active/passive categorization is that it is rooted in rifting dynamics, which also may be its undoing. Lately, the classification has come under considerable attack. The problem is that it is almost impossible to prove that a rift began in the active mode, whereas it is quite clear that at least some rifts originated by passive stretching. Examples of the latter include the Galicia Bank off Spain, the Basin and Range Province of the western United States, Gulf of Suez, and many of the onshore Chinese rift basins, as well as the South China Sea. It is certainly true that some rifts are associated with regional doming, such as the East African rifts, but it can be argued that this is a later manifestation of passive rifting. In truth, there are few hard data that require active doming at the onset of rifting, whereas considerable data demonstrate the need for passive rifting.

It is worthwhile pointing out that although the proposition of active rifting may be under attack, the occurrences of vigorously upwelling mantle plumes (i.e. hot spots) cannot be denied. The Hawaiian Islands and Iceland are proof that such plumes exist. I have adopted the stance that most rifts are fundamentally passive, in the sense that they originate by plate stretching. The development of bulges happens after stretching commences and depends upon whether or not a hot spot becomes trapped beneath the rift, or develops in the region of lithospheric necking. We will add some frills to this idea as we evaluate the remaining categories.

Degree of Magmatism. The above discussion leads to another category of continental rifts—their propensity toward magmatism. By definition, all successful rifts ultimately reach a stage of 100% magmatism, so this comparison must be restricted to rifts that did not or have not yet reached the spreading stage. For this reason and others we also should be

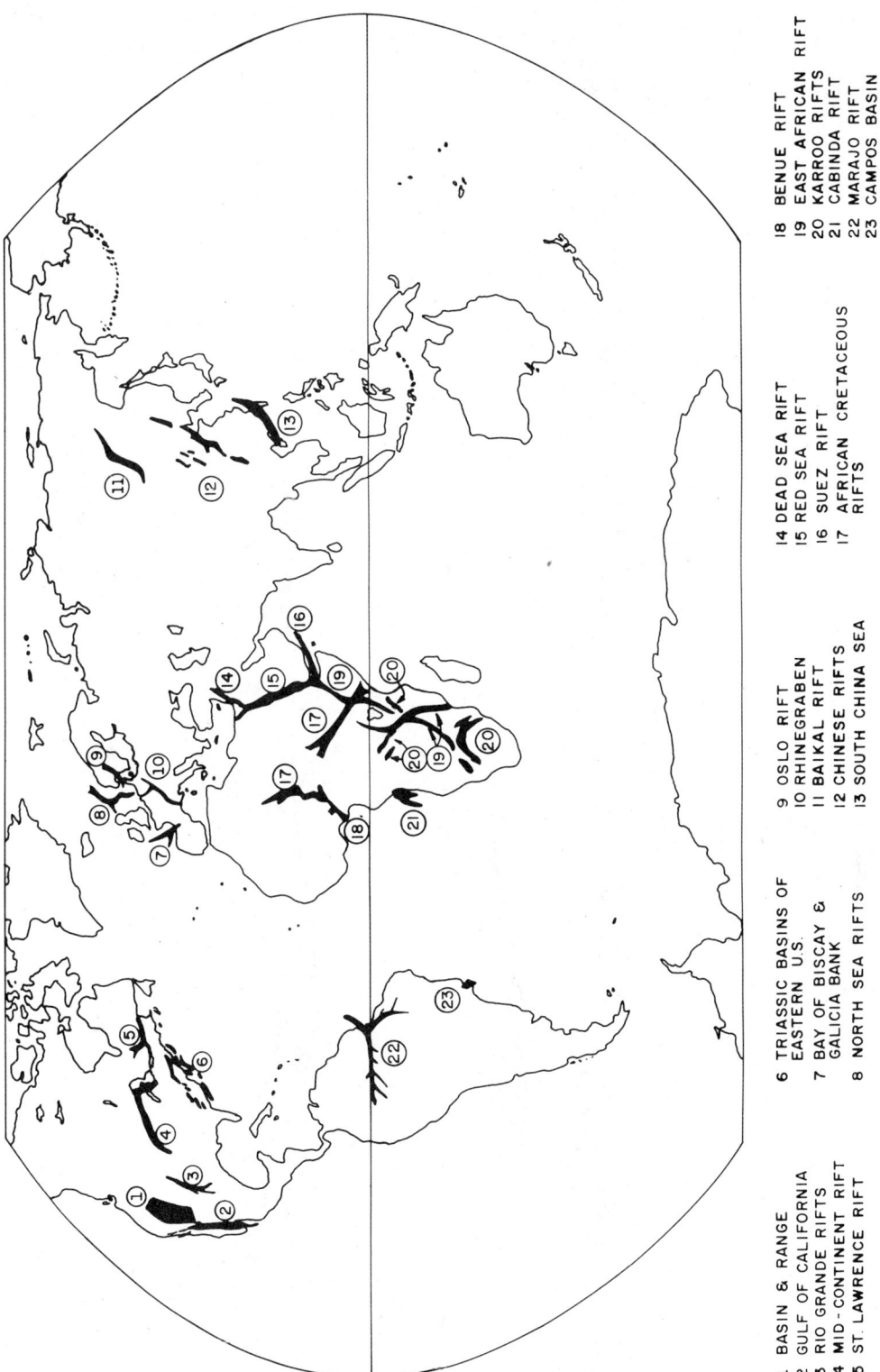

FIGURE 1. Location of rifts and rifted terrains mentioned in text. Note that there are many more rifted margins than shown.

1 BASIN & RANGE
2 GULF OF CALIFORNIA
3 RIO GRANDE RIFTS
4 MID-CONTINENT RIFT
5 ST. LAWRENCE RIFT
6 TRIASSIC BASINS OF EASTERN U.S.
7 BAY OF BISCAY & GALICIA BANK
8 NORTH SEA RIFTS
9 OSLO RIFT
10 RHINEGRABEN
11 BAIKAL RIFT
12 CHINESE RIFTS
13 SOUTH CHINA SEA
14 DEAD SEA RIFT
15 RED SEA RIFT
16 SUEZ RIFT
17 AFRICAN CRETACEOUS RIFTS
18 BENUE RIFT
19 EAST AFRICAN RIFT
20 KARROO RIFTS
21 CABINDA RIFT
22 MARAJO RIFT
23 CAMPOS BASIN

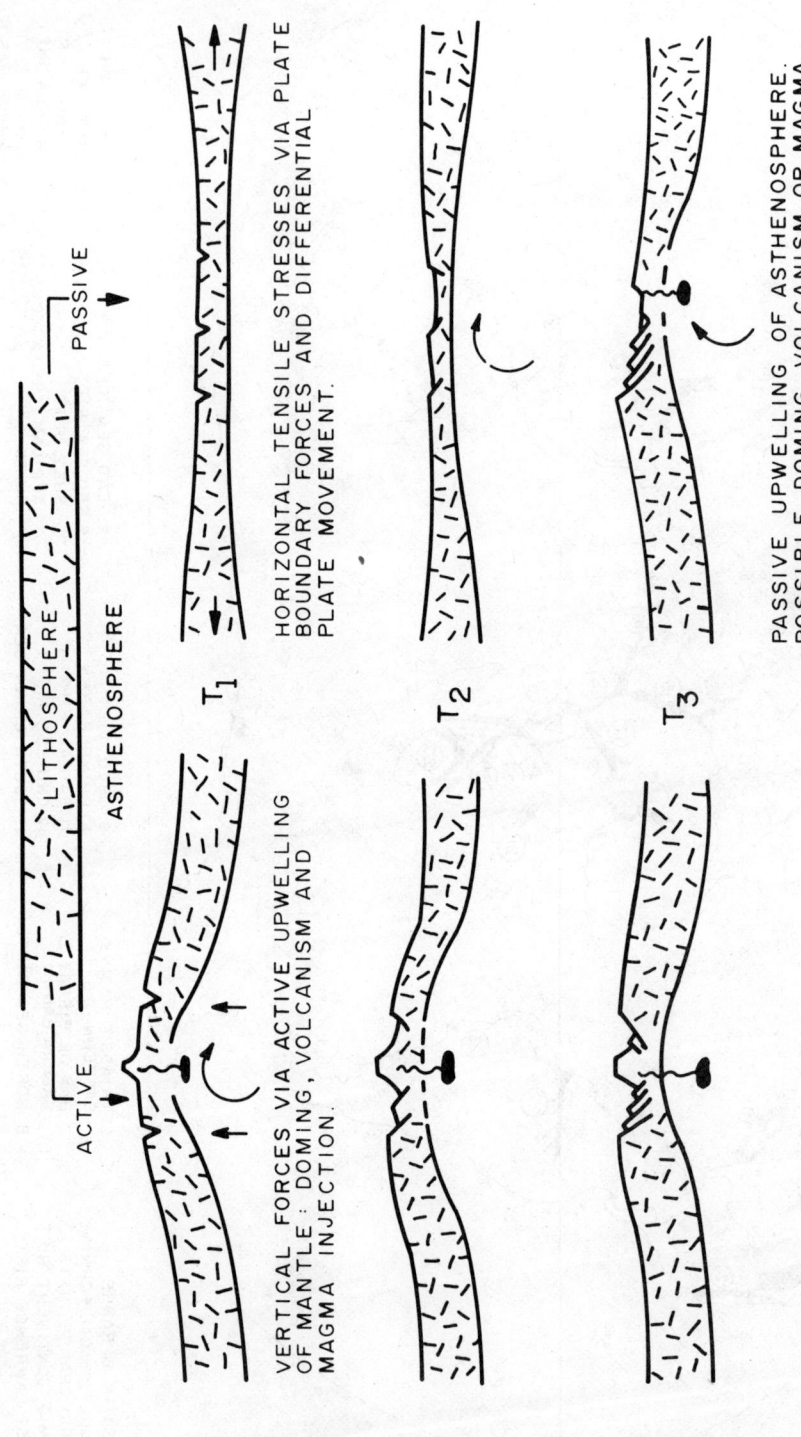

FIGURE 2. Idealized models of active and passive rifting modes.

careful in using examples associated with passive margins.

The magmatic classification of rifts is essentially into the "wet" versus "dry" categories; that is, rifts that are rich in volcanic rocks vs. those that are not. The differences are undeniable and remarkable, as pointed out by Mohr (1982). The Eastern Branch of the East African Rift (Fig. 1) is literally buried in volcanic effluents, and intrusive equivalents probably abound. In fact, the Gregory (Kenya) Rift Zone is probably the wettest continental rift known. The other category, the dry rift, also is illustrated best in Africa, this time by the Western Branch of the East African Rift and by some of the Karroo basins. Except for small pods of volcanic rocks that occur mainly at their connections or at the major kinks, the Western Branch rift zones of Malawi, Tanganyika, and Kivu/Edward/Albert/West Nile are devoid of rift-related volcanics. Other continental rifts fall between these two extremes, with the Baikal Rift and Rhinegraben tending toward the dry side and the Rio Grande, Mid-Continent, and Oslo Rift tending toward the wet. Suez is also a volcanically deficient rift, which is surprising in view of associations made later.

There is little apparent rhyme or reason regarding which rifts are wet or dry. Some workers have at least implicitly correlated wet and dry to active vs. passive rifts, but the differences between the two East African Rift branches argue against any clear-cut correlation, prompting me to treat volcanism as the "wildcard" of rifting and to discount the usefulness of the wet-dry classification.

A measure of sensibility can be reached if volcanism is viewed as a byproduct of hot spots trapped beneath an originally passive rift. Hence, the difference between the Eastern and Western branches of the East African Rift may be that a plume, or plumes, underlie the former but not the latter. It must be cautioned that the petrologic data show that this story is very complicated, both in space and time. Also, the fact that both the Eastern and Western branches are situated on the same swell argues that part of the difference in volcanic "richness" may relate to where the plume or plumes are allowed egress.

Morphologic Form. Another rift classification is based on the overall morphologic form of rift zones. There are a number of possible schemes, but most are rooted in meaningless divisions. A case in point is the categorization based upon apparent cross-sectional shapes of rifts, which range from asymmetric half-graben to symmetric full-graben. For reasons that should be clear from the discussions of rift architecture in *Continental Rifts: Structural Traits*, unqualified classifications based on morphologic criteria are not only meaningless, but possibly misleading.

A more useful morphologic distinction is that between relatively narrow, well-defined rift zones such as the East African rifts and broad, diffuse extensional terrains such as the Basin and Range Province. Other examples of the former type include the Baikal, Oslo, Rhinegraben, Mid-Continent, and Rio Grande rifts. Examples of the diffuse type include the Mesozoic Basins of the eastern seaboard of the United States (i.e. the swath of extended terrain that includes the classic Triassic Basins), Bay of Biscay, South China Sea, and possibly the Benue Trough. The difference between these types is analogous to that between mid-ocean spreading ridges and back-arc basins. In fact, we have used a possible back-arc basin setting—the South China Sea—as an example of a diffuse rift.

There have been suggestions that narrow, well-defined rifts evolve into diffuse extensional terrains as the amount of extension increases. There may be a degree of truth to this in regards to a rift such as Suez, which seems to "open" toward the south, or to the Kenya-Ethiopian-Afar rift situation. However, it is relatively certain that the Basin and Range did not evolve from an East African-type rift.

Some authors have cited the reverse progression, arguing for centralization of rifting along some axis as extension progresses. Although it is likely that African rift zones began as somewhat broader and more diffuse subsidence troughs (see Stage I, Fig. 18 in *Continental Rifts: Structural Traits*), perhaps comparable to what Cochran (1983) has proposed for the Red Sea, it is clear that they were not born in Basin and Range-type terrains.

Another factor that argues against a complete progression between these two types of rifts is the lack of intermediate examples: Most rifts fall rather neatly into one or the other categories. Those that do not are usually explicable as special cases.

The interrelationships of these two morphologic types to the active/passive and wet/dry categorizations is not clear-cut. It might be argued that narrow, well-defined rifts have these characteristics because they are not obscured by blankets of lavas, implying an association with the magmatically drier rifts. This implication is incorrect, as evidenced by the two branches of the East African Rift. It is more valid to note the correlation between narrow, well-defined rift expressions and what have traditionally been classed as active rifts. Conversely, the terrains of broad, diffuse extension are those which are usually cited as examples of passive rifts.

Are There Fundamentally Different Types of Rifts? Although the individual classifications of rifts may be somewhat artificial, certain traits tend to group together. For example, most rifts that are now the innards of passive margins usually are best interpreted in terms of little uplift of a regional nature prior to rift trough subsidence. They also tend to show extreme block rotations, which imply large amounts of extension and stretching (often more than 100%). The areas affected often cover swaths of ground that are hundreds of kilometers wide, or more. Some examples include the Galicia Bank; the Triassic Basins of the Eastern United States, including those beneath the coastal plain; the marginal basins of the east coast of South America and west coast of Africa (e.g. Campos Basin of Brazil and Cabinda area of Angola); and the northwest Australian margins. The best modern analog may be the Basin and Range Province of the western United States. The South China Sea also may fit this billing.

On the other side of the coin are the paleorifts that are not directly associated with passive margins and successful oceans. These include the Rhinegraben, Oslo Rift, Mid-Continent Rift, many of the Cretaceous and Karroo rifts of Africa, the St. Lawrence Rift, and many others. Most of these rifts are defined by a single, narrow zone of fracture (typically 30–60 km wide) and the presence of inferred flanking uplifts. There is usually some evidence of regional uplift also, albeit the timing is always questionable. Fractures tend to be linear and high angle and block rotations are relatively small, implying small amounts of extension and stretching (e.g. usually less than 10%). Modern or recent analogues would be the East African rifts, Baikal Rift, Rhinegraben, and Rio Grande Rift.

Almost by definition, Group 1 rifts are the scars of successful rifting, whereas Group 2 rifts are either outright failures in regard to reaching the spreading stage or they have not reached it yet. The distinction is useful in the sense that most oceans may very well evolve from Basin and Range type rifts, whereas most Group 2 types may be destined to fail. However, the correlations are not fool-proof and they do not preclude the possibility of a progression between the two groups.

The key here may be the misfits or anomalies—the rifts that do not fall neatly into one of the above groupings and those oceans which may not have evolved from Group 1 rifts. The Gulf of Suez, Red Sea, Gulf of California, Dead Sea, Benue Trough, and North Sea all fit the misfit category in one sense or another. Suez has the dimensional attributes of Group 2, but the block rotations of Group 1. It is reasonably certain that it did not begin with regional doming, although it has been associated with some degree of uplift during its rifting history. Suez is about what the Tanganyika Rift would be if the latter were subjected to two or three times as much extension. The bulk of the data from the Red Sea suggests that it may have evolved from a Suez-type situation, which is implied above to have evolved from a Group 2 rift. One of the Red Sea's prespreading predecessors probably was comparable to the Ethiopian Rift. The implication here is clear—apparently some ocean basins have evolved from Group 2 rifts.

The Gulf of California also is evolving along a narrow zone of rifting, but the rifting here originates more from transform faulting than pure extension. This observation raises the question of where pull-apart basins fit into the picture and whether the orientation of the stress field has any bearing on these rift groupings. The issue of pull-apart tectonics can be addressed by examining the Dead Sea Rift. The differences between this feature and the typical Group 2 rift are that the Dead Sea is somewhat narrower, has a higher length-to-width ratio (e.g. about 4), and lacks any evidence of significant regional uplift any time during its history. This kind of rift also is volcanically dry, as are continental transform faults in general, but this is not diagnostic in terms of the above groupings. At face value none of these differences seem to carry the significance of the differences between Group 1 and 2 rifts. The question of stress orientation is particularly intriguing: Is it possible that Group 1 rifts are the expressions of nearly orthogonal plate divergence, whereas Group 2 rifts express more oblique divergence? Although we cannot disprove such a notion with existing data, it is not a likely explanation for the differences between Groups 1 and 2. However, there must be a continuum between orthogonal plate divergence and transform slippage and this may explain some of the variations within the groups. It certainly contributes to the difference between, say, the Tanganyika and Dead Sea rifts. For these reasons, we shall classify transform rifts with Group 2.

The Benue Trough is an example of the reverse situation. It is an aulacogen, or failed arm of a triple junction, associated with the break-up of the South Atlantic. There are many such aulacogens around the world (e.g. see Burke, 1977). What is curious about Benue is that it is a relatively wide trough with no evidence of regional uplift and which seems to have experienced considerable extension. In other words, it may be a Group 1 rift that failed. The Amazon aulacogen and the Marajo Basin, in particular, may be another example.

The North Sea is a broad zone of rifting that appears to have failed just prior to successful spreading. It has most of the characteristics of a Group 1 rift and in many ways it is reminiscent of the Basin and Range Province. The important question is whether the North Sea is generically comparable to the Basin and Range, or did it evolve to this form from an East African-type rift. Both cases have been implied by various workers. Either way, it is anomalous because either it is an example of a failed Group 1 rift, or it shows that there can be a progression between the two groups.

The issues raised above are fundamental to tectonics because we are asking how ocean basins are born. Is it by a process that heats and softens large patches of a continent, resulting in its being pulled apart similar to a piece of taffy, or is it by a process in which a continent is ripped apart along a narrow tear, such as a run in a nylon stocking? We are forced to conclude that both mechanisms apparently happen. Further research is likely to show that there are still other modes of continental fragmentation.

Comments on the Rates of Continental Rifting

How fast does rifting occur and how long does it take to make a new ocean basin? It depends. Parts of the East African Rift have been active since at least the Lower Miocene, and perhaps discontinuously into Karroo times in certain localities. Yet, the measured amount of surface extension is on the order of 5–10%, based on balancing seismic sections. The average seems to be about 7–8%. This equates to a pseudospreading rate on the order of 0.1mm/yr or less. The overall rate for the Baikal Rift may be even slower. This sort of snail's pace also appears applicable to the Rhinegraben, Oslo Rift, Rio Grande Rift, and even the Suez Rift. In most of these cases, the extension histories appear to be episodic, periodic, or markedly discontinuous. Careful examination of the seismic profiles shown in *Continental Rifts: Structural Traits* suggests that the Malawi and Tanganyika rift zones are characterized by long periods of either tectonic quiescence or very uniform subsidence, interspersed with pulses of deformation. The evolutionary term "punctuated equilibrium" may be apropos here.

Such reasoning leads to an obvious correlation with Group 2 rifts and the conclusion that ripping continents apart along narrow sutures is a slow and tortuous process, at least in the prespreading stages. The Red Sea may be particularly instructive in this regard because it seems to show a two-stage opening history. The first began in early Tertiary and continued to the Pliocene. This stage is comparable in every regard to a Group 2 rift, including the snail's pace of apparent surface extension. The second stage began in the Pliocene and is marked by a jump in the apparent rate of extension. This second stage presumably correlates to the onset of spreading and creation of oceanic, or quasi-oceanic crust, possibly in the fashion described by Bonatti (1985).

From the preceding statements it is tempting to visualize Group 2 rifting as a process in which extension of the brittle crust sputters along in a slow, episodic fashion until complete plate separation is achieved. At this point, the process rapidly approaches steady state and extension rates increase by several orders of magnitude. If the stresses causing plate separation are insufficient to achieve complete separation, or if they become reoriented, then failure and fossilization may result.

Not all rifts classed as Group 2 behave this way, however. Transform rifts apparently can rip continents apart very quickly, as evidenced by the Gulf of California. The separation there has taken only a few million years, at typical plate motion rates. In this regard, there seems to be a fundamental difference between transform rifts and the other Group 2 rifts cited herein.

Another complication to this conceptualization is that apparent surface extension is taken to be a faithful measure of whole-plate extension. It almost certainly is not. Ductile stretching below the brittle zone could be very much greater than what is expressed in near-surface faulting. Although this does not change the conclusion that Group 2 rifting is a long process, it means our visualization of it as discontinuous and non-steady-state may be applicable only in regards to the zone of brittle deformation.

Do Group 1 rifts behave any differently? It all depends, at least based on what is known about the Basin and Range Province. On the one hand, the Basin and Range has been active since at least the middle Miocene and some of the normal faulting dates from the earliest Oligocene. Although much less precisely known, there is no reason to believe that the east coast Triassic Basins did not have an equally long life. In this regard (i.e. longevity), Group 1 rifts appear to be comparable to Group 2. On the other hand, it seems undeniable that the Basin and Range has experienced much more extension of the brittle crust, and almost certainly of the ductile part of the lithosphere, as well. Values of up to 300% have been proposed for the total amount of stretching.

In this sense it might be surmised that Group 1 rifting is more efficient than Group 2, especially if the transform rifts are excluded from the Group 2 category. Also, descriptions of Basin and Range tectonics often allude to episodes, but a reader is still left with the impression that stretching in the Basin and Range is markedly more uniform than in places such as East Africa.

Conclusions

As pointed out at the beginning of this paper, all groupings, classifications, or categorizations of continental rifts seem to lose meaning at some point. It is particularly disconcerting to those of us who attempt to understand the systematics of rifting that the levels at which the classifications break down have become progressively finer with increasing knowledge of individual rifts. It is conceivable that beyond the few basic criteria that allow rifts to be identified as such, there are no true systematics. It might be that the conditions or situations giving rise to continental fragmentation are sufficiently different with each occurrence that meaningful generalizations are impossible, especially where the media being rifted are so geologically variable. Now the challenge is to tackle the issue of process uniformity, or lack thereof as the case may be. When different stacks of plate glass sheets are smashed with a hammer, they may display considerable deformational variety (although we may still be able to recognize that the overall fracture pattern was caused by the same agent). Is the variety due to differing types, quality, or thickness of glass or is the hammer the true variable? Therein lies one of the chief frontiers in rift research for the next decade.

Acknowledgements

This entry is dedicated to the students and staff of Project PROBE, without whom none of this would be possible. I also am deeply indebted to the many companies and organizations who have sponsored PROBE's research over the last 6 years. They include Agip, Amoco, Arco, Conoco, Eason, Elf Aquitaine, Esso, Marathon, Mobil, Pecten, Pennzoil, Petrofina, Placid, Shell, Texaco, and World Bank. The countries of Burundi, Kenya, Malawi, Mozambique, Tanzania, Zaire, and Zambia are thanked for their cooperation and support of our research in their territories.

B. R. ROSENDAHL

References

Bonatti, E., 1985, Punctiform initiation of seafloor spreading in the Red Sea during transition from a continental to an oceanic rift, *Nature* **316**, 33–37.
Burke, K., 1977, Aulacogens and continental breakup, *Ann. Rev. Earth Planetary Sci.* **5**, 371–396.
Cochran, J. R., 1983, A model for the development of Red Sea, *Am. Assoc. Petroleum Geologists Bull.* **67**, 41–69.
McKenzie, D., 1978, Some remarks on the development of sedimentary basins, *Earth and Planetary Sci. Letters* **40**, 25–32.
Mohr, P. A., 1982, Musing on continental rifts, in G. Pálmason, ed., *Continental and Oceanic Rifts*. Washington D.C.: American Geophysical Union (Geodynamics Series, vol. 8), 293–309.
Sengor, A. M. C., and K. Burke, 1978, Relative timing and rifting and volcanism on Earth and its tectonic implications, *Geophys. Research Letters* **5**, 419–421.

Cross-references: *Continental Crustal Structure; Continental Lithosphere; Continental Rifts: Structural Traits; Crustal Movements and Tectonic Deformation; Deep Seismic Reflection Profiling; Ocean Continent Transition: Structure; Stress in the Earth's Lithosphere; Thin-Skin Tectonics.*

CONTINENTAL RIFTS: STRUCTURAL TRAITS

It is well known that continental rifts are places where tensile stresses have combined with various vertical forces to produce block-faulted terrains. Where the rifts are young and active, these terrains can be among the most spectacular on the face of the Earth—witness the great rips in the African continent bordered by such features as the Mountains of the Moon (i.e. Ruwenzori Mountains). Conversely, where rifts have been fossilized for many millions of years, their surface expressions can be markedly unimpressive. In some cases, they are identifiable only with subsurface geophysics. Some of the Triassic rifts of the eastern United States are examples. (Fig. 1 shows the location of all rifted terrains mentioned in this paper.)

Generalized models of rifts ordinarily depict them as symmetric about a central axis of maximum subsidence (e.g. Bott and Mithen, 1983). They generally are not. Standard models also treat rifts as linear, two dimensional features. Again, they are not, at least not to the extents that are usually implied. Many treatments also fail to recognize the importance of different morphologic elements and scales of rifts. In actuality, continental rifts are composed of individual fault blocks, usually tilted, that group together in complicated patterns to form quasi-discrete and markedly asymmetric subsidence units. These units, also termed "half-graben" (Bally, 1982), link together in different ways to form rift zones. An arrangement of zones sausaged together end-to-end creates a rift branch (e.g. the Eastern and Western branches of the East African Rift System). Groups of zones also can be arranged side-by-side, or even partly on top of each other, creating broad swaths of extended terrain such as that in the Basin and Range Province of the United States (Eaton, 1982).

All of these component scales are interrelated so a discussion of one must draw on knowledge of the others. Because the key scale is believed to be the subsidence unit, or half-graben, it is the logical level upon which to focus this discussion. We shall begin with a description of the basic structural ingredients of a subsidence unit, including how they mix together to create the unit, and conclude with a discussion of how units link together to create rift zones and branches.

Master or Controlling Fault Systems

Most rift units are half-graben in cross-section, with border fault systems (BFS) mainly developed on one side (Fig. 2). Most BFS are zones of seismic nondefinition or even seismic wipe-out (Fig. 3A, 3B), caused by various combinations of the following factors: local breakdowns in the geometric assumptions of the common depth point method (because CDPs become spatially smeared on steep, complicated slopes); the nature and intensity of deformation in these zones; and the masking or distorting effects of chaotic sediment deposits (e.g. border fault fanglomerates, debris flows, and slumps).

Where BFS have been examined in the field or where conditions are favorable for seismic imaging, they are often composed of numerous closely-spaced faults sandwiching slivers of cataclastized rock or discete blocks (Fig. 3C). BFS also can appear as single, coherent fractures, at least at the resolution scale of seismics. In African rifts this seems to happen mainly on newly developing BFS (Fig. 3D),

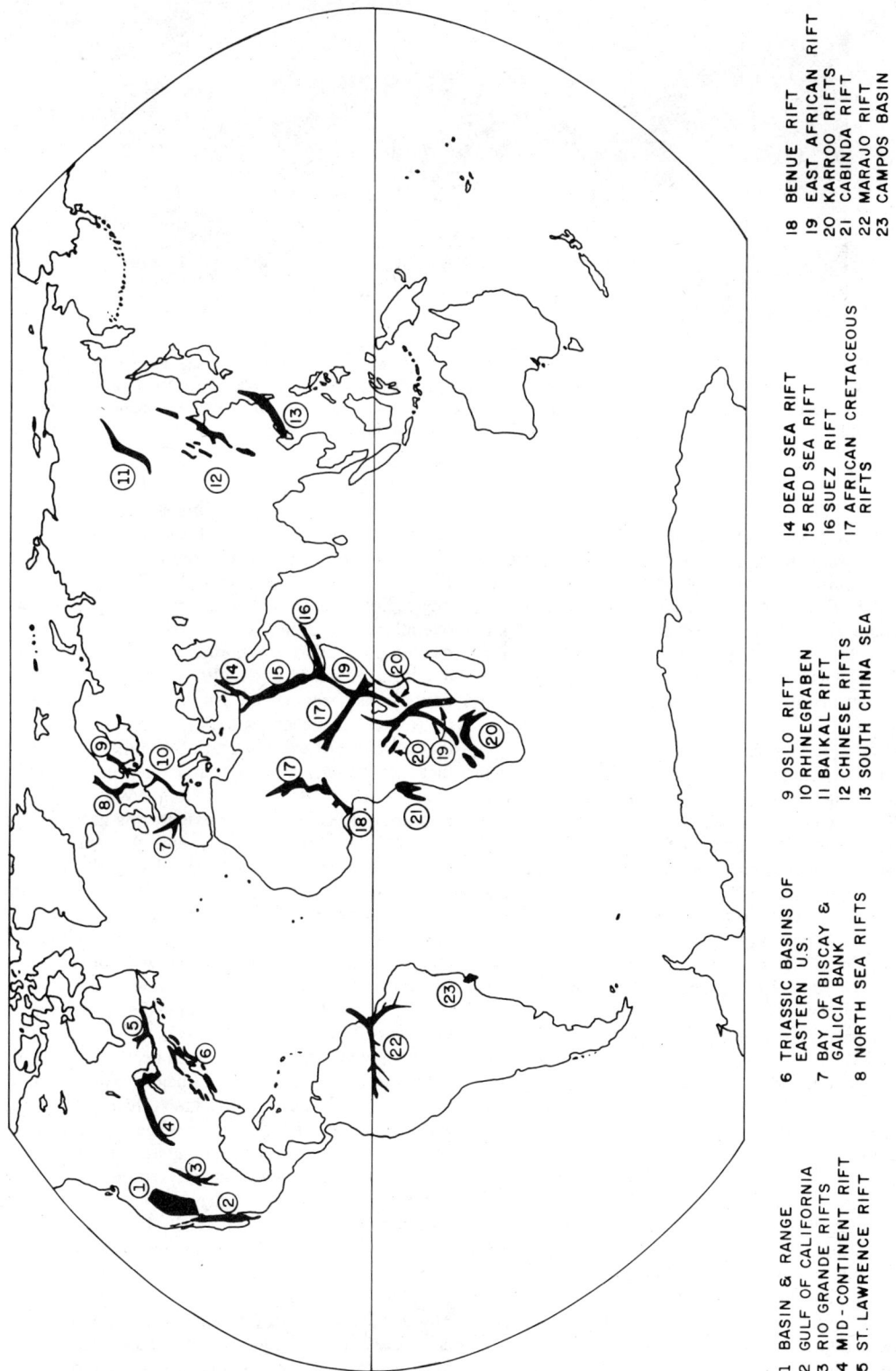

FIGURE 1. Location of rifts and rifted terrains mentioned in text. Note that there are many more rifted areas than shown here.

1 BASIN & RANGE
2 GULF OF CALIFORNIA
3 RIO GRANDE RIFTS
4 MID-CONTINENT RIFT
5 ST. LAWRENCE RIFT
6 TRIASSIC BASINS OF EASTERN U.S.
7 BAY OF BISCAY & GALICIA BANK
8 NORTH SEA RIFTS
9 OSLO RIFT
10 RHINEGRABEN
11 BAIKAL RIFT
12 CHINESE RIFTS
13 SOUTH CHINA SEA
14 DEAD SEA RIFT
15 RED SEA RIFT
16 SUEZ RIFT
17 AFRICAN CRETACEOUS RIFTS
18 BENUE RIFT
19 EAST AFRICAN RIFT
20 KARROO RIFTS
21 CABINDA RIFT
22 MARAJO RIFT
23 CAMPOS BASIN

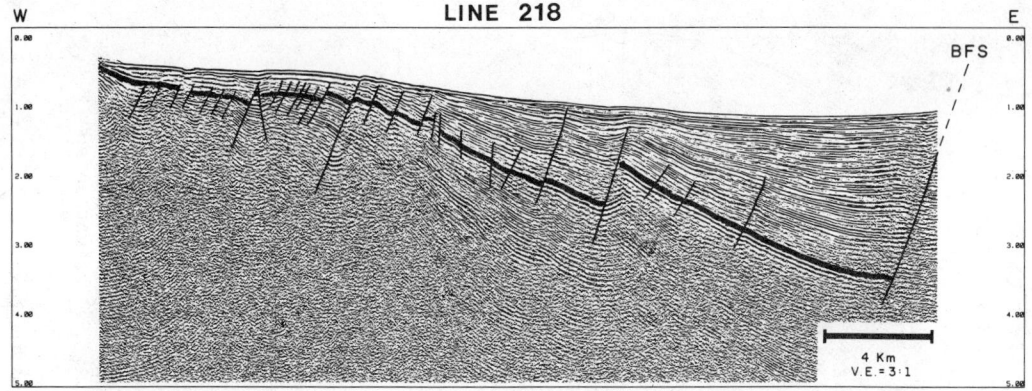

FIGURE 2. Interpreted seismic reflection profile (Line 218) from the Lake Tanganyika Rift Zone. Note the asymmetric, half-graben form created by concentration of subsidence along a single border fault system (BFS). Heavy, solid line generally represents acoustic basement. Position of profile relative to idealized rift architecture is shown in Fig. 11, actual position in Fig. 18. Profile is migrated, 24-fold stack displayed in AGC format.

but seismic data from rifted terrains such as the Basin and Range and the Triassic Basins of the eastern seaboard seem to show simple BFS with large throws and presumably long subsidence histories. Even where BFS are complex zones, they tend to be surprisingly narrow, often on the order of only 5 km wide or less (Fig. 3). In such cases, the throw on a BFS accounts for much of a unit's total subsidence, as shown in Fig. 2. In some African rift units the amount of total subsidence accounted for by individual BFS exceeds 90%

A few African rift units show situations in which total subsidence is apparently divided between a few faults, which are interspaced across the full width of the unit (Fig. 4). In such instances there are no BFS per se. This situation is rare also in the rifts that are usually compared to East Africa, including the Rhinegraben, Oslo Rift, Baikal Rift, and the Rio Grande Rift; however, it is a prevalent characteristic of the Triassic Basins of the eastern United States. Dispersed BFS also may be common in the Basin and Range Province.

It is usually presumed that motions along BFS are essentially dip-slip. Based on observational data and mechanical considerations of rift unit geometries (see discussion below), this is not always so. Indeed, it is likely that some segments of BFS are characterized by considerable strike-slip dislocations.

The form of BFS in plan view is generally arcuate, sometimes because the component faults are truly curvi-linear and sometimes because linear segments of the faults comprising a BFS are offset by transverse faults. In this fashion the component faults can be linear, but the overall pattern crudely arcuate or at least zigzag. Mechanical and volumetric considerations argue that where BFS are crescentic-shaped, some portion of the system probably experiences oblique slip subsidence.

The dips of BFS range from relatively low-angle in areas such as the Basin and Range to values of up to 60° or higher in African rifts. However, because BFS can have listric forms, the dip might be a misleading number for comparative purposes. The overall cross-sectional form of BFS should be a more meaningful yardstick of comparison. Unfortunately, the form of major rift faults, especially bounding faults, has become the subject of considerable controversy in recent years. Some workers argue that these faults are essentially linear; others postulate that they flatten with depth and sole out into some type of basal decollement (e.g. Wernicke and Burchfiel, 1982). To some extent, this argument is between the field geologist and the exploration seismologist and may, in part, reflect their different scales of study. In outcrop, most major rift faults appear linear over the vertical interval of observation, which is usually on the order of hundreds of meters and seldom exceeds a few kilometers. In seismic reflection data some major rift faults seem to have listric forms, at least where the depth interval of observation is significant (i.e. 5–15 km). The problem with the seismic-derived results is that they are biased toward what can be imaged, which is more likely to be low-angle listric faults than high-angle, linear faults. A related problem is that some workers have not always been careful about differentiating between fault behaviors in rift sedimentary cover versus basement rocks; behaviors that can be very different. It does not automatically follow that listric forms in sediments will be continued in the underlying basement.

Another way of attacking this problem is to examine the degree of rift fault-block rotation, which is a necessary consequence of bounding faults that flatten with depth. In this context it is clear that some rifted terrains show rather extreme rotations (e.g.

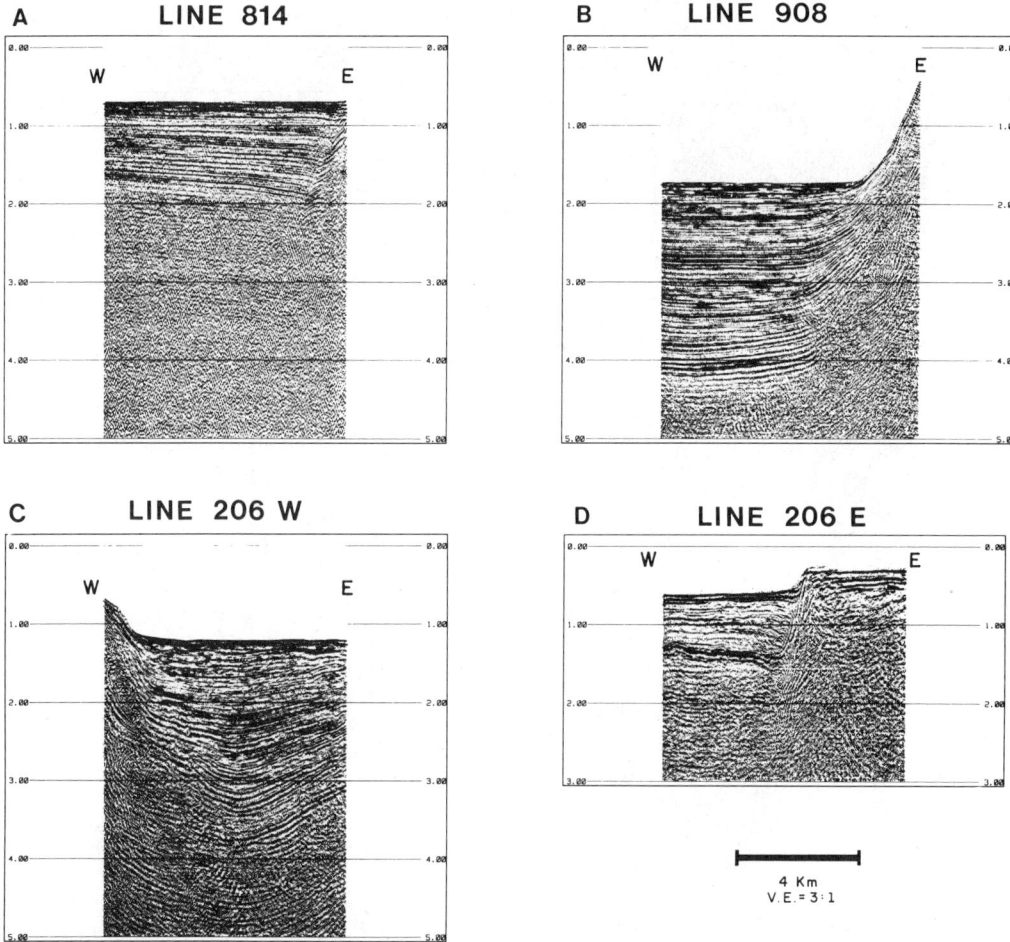

FIGURE 3. Uninterpreted seismic reflection profiles of border fault systems. Figure 3A (Line 814) shows the Livingstone BFS from the northern end of the Lake Malawi Rift Zone; Fig. 3B (Line 908) the East Kigoma BFS from the Lake Tanganyika Rift Zone. Figure 3C (Line 206) shows the North Kigoma BFS of Lake Tanganyika. Note that this BFS is composed of small fault blocks. Figure 3D is from the eastern side of Line 206 and shows the newly developing end of the East Kigoma BFS. Note the simplicity of the fault expression here compared to its seismic "nondefinition" 30 km further south (i.e. Line 908, Fig. 3B). All panels show 24-fold stacks displayed in AGC format. Data in Panel 3A are unmigrated; all other data have been migrated. Positions of seismic panels are shown in Fig. 18.

Basin and Range, Galicia Bank, and Bay of Biscay), whereas others show very little. We have been particularly impressed by the degree of fault linearity and lack of block rotations observed in our African rift lakes seismics compared to Bay of Biscay or Galicia Bank data (e.g. Chenet and Montadert, 1981). In some locales, we can observe up to 4 km of relief on border fault scarps above lake levels, another 1.5 km below the water surface, and up to 4.5 km of throw below the lake floor. Hence, the total vertical observational interval may approach 10 km in certain locales, and regularly exceeds 5 km. Yet, the amount of fault flattening with depth required by the data is usually extremely small. The amount permitted by the data is another matter, but it is not as large as that observed in, say, the Basin and Range Province.

Even in African rifts, there are good mechanical and geometrical reasons for postulating that major rift faults eventually undergo modest flattening with depth, especially where BFS are truly arcuate in plan view. Nonetheless, it must be concluded that there is less flattening of faults over comparable depth intervals in African rifts than in the Basin and Range. Curiously, most rifts seem to follow one or the other of these cases.

Most rift BFS are broken and often offset by transverse faults, trending anywhere from 30°–90° to the

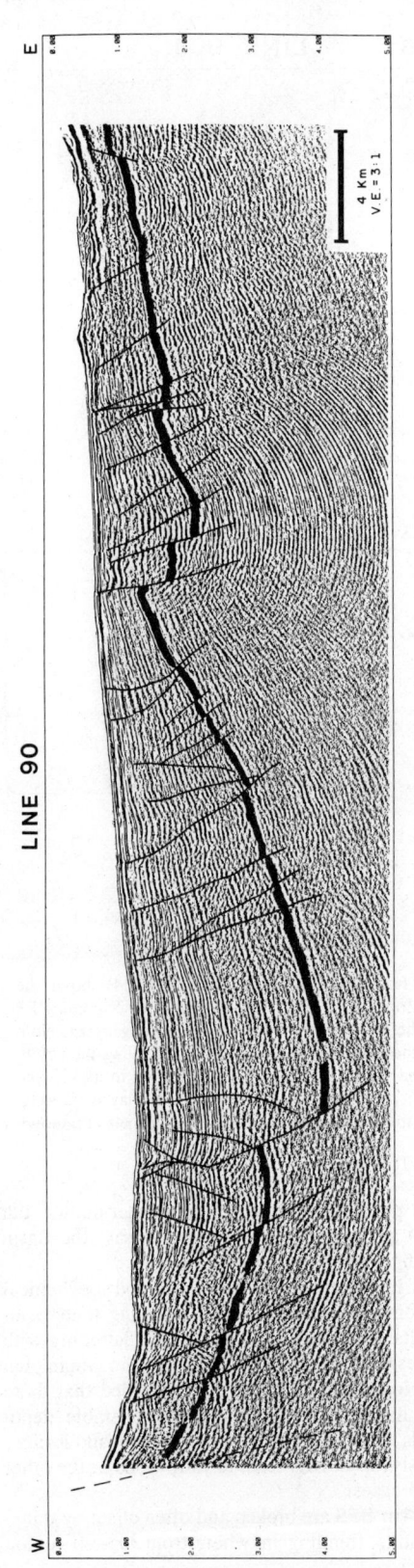

FIGURE 4

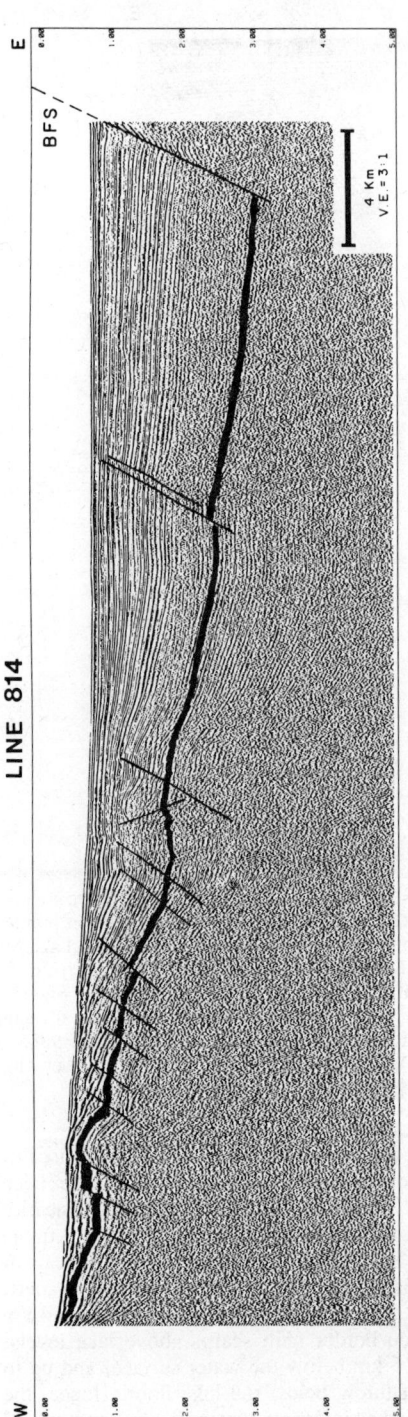

FIGURE 5

overall strike of the BFS. Some workers have adopted the term "transfer fault" (Gibbs, 1984) for such offsets. In some cases, these faults seem to efficiently transfer the subsidence of one segment of BFS to another, in which case they are mechanically analogous to ridge-ridge transform faults. In other cases transverse faults may originate as strike-slip faults, offsetting an originally continuous BFS and continuing beyond the BFS, possibly linking with strike-slip networks described below. In still other cases, transverse faults may alternate between transfer and strike-slip faults, perhaps depending on the orientation of the stress field at any given time and the geometry of adjacent rift units.

Infrastructure of Subsidence Units

The infrastructure of rift units consists of (1) the second-order "normal" faults that create an internal fault-block mosaic and that contribute to the overall subsidence of the unit, albeit in a minor way compared to most BFS; and (2) the wrinkles, crinkles, and other spatial adjustments that accompany imperfect subsidence, including the networks of strike-slip faults that crisscross the rift floor and interlink the second-order normal faults.

All scales of internal fault blocks exist, but there often seems to be a preferred scale and pattern of fracture within individual subsidence units. In terms of block widths, this preferred scale ranges from a few kilometers to a few tens of kilometers. There is good evidence that block widths (i.e. fault spacings) can change systematically across the widths of subsidence units, at least across those units that have the form of simple half-graben. The pattern in many rifts, including those in East Africa, is a decrease in fracture spacing away from the master border fault systems (or toward the ramping sides of the half-graben). The first significant internal fault typically occurs about 10–20 km from the border fault, with a progressive decrease to values of about 5 km or less on the far side of the unit (Fig. 5). Not surprisingly, fault throws may be related to this progression, in the sense that the larger the spacing between internal faults, the larger the throws.

The strike dimensions of individual blocks are poorly constrained in most rifts, but are probably as variable as block widths. There is some evidence that length-to-width ratios of blocks may be a more stable datum than either parameter, with ratios usually falling between 2–4 (c.f., Aydin and Nur, 1982).

In most half-graben the internal normal faults seem to be parallel or subparallel to the controlling border fault, creating cross-sectional geometries like those shown in Fig. 2–4. As will become clear from subsequent discussions, the more isolated a subsidence unit is from other units, the more prone it is to show a simple half-graben form with this synthetic arrangement of internal normal faults. Antithetic faults occur, but they usually are very subordinate (Fig. 6). Significant occurrences of antithetic faults probably are a manifestation of the linking together of rift units, rather than an intrinsic property of the internal structure of units.

On some seismic profiles across some half-graben, the throws on the internal faults can be exceedingly small and the faults are identifiable only from subtle disruptions of the sediment column. Some of these "zero-offset" faults are probably more closely related to internal strike-slip networks (see below) than to the fault block mosaics under discussion here. Nonetheless, we cannot overlook the fact that significant portions of subsidence units can be almost devoid of any internal fracture, at least at the level of seismic resolution (Fig. 7). For some mechanical reason, these regions seem to subside in toto, rather than shattering into the usual mosaics of internal fault blocks. Geologic common sense suggests that this condition should be an early-stage expression of subsidence, when the mechanical strength of the region is relatively high compared to the rifting stresses. However, seismic data from the African rifts and interpretations of the Rhinegraben show that this condition (i.e., lack of internal fracturing) can persist for rather long intervals of subsidence history.

The internal fault blocks of rift units almost always display a marked plunge in one of the strike directions. Values are highly variable but plunges of 3–4° are typical over distances of 60 km (Fig. 8). There is little rhyme or reason to plunge geometry of individual blocks that we have been able to ascertain, except that the block immediately riftward of the BFS often shows the greatest plunge. Linking of adjacent half-graben units creates complicated arrangements of plunges and usually overprints any original strike-asymmetries of the prelinked units.

Although it is naive to believe that all deformation in the upper 10 km of rifts occurs via brittle failure along normal faults, there has been very little discussion in the modern continental rift literature

FIGURE 4. Interpreted seismic reflection profile (Line 90) from the Lake Tanganyika Rift Zone showing a dispersed BFS, which is rare in African-type rifts but may be prevalent in rifted areas characterized by broader swaths of extended terrain. Profile also represents an example across a half-graben linking arrangement most comparable to Case 9 in Fig. 11. Heavy, solid lines represent acoustic basement. Position of profile relative to idealized rift architecture is shown in Fig. 11, actual position in Fig. 18. Profile is migrated, 24-fold stack displayed in AGC format.

FIGURE 5. Interpreted seismic reflection profile (Line 814) across the Livingstone half-graben of the Lake Malawi Rift Zone. Note the decrease in spacings of internal faults toward the ramping side of the half-graben, to the west. Heavy, solid lines represent inferred acoustic basement. Position of profile relative to idealized rift architecture is shown in Fig. 11, actual position in Fig. 18. Profile is unmigrated, 24-fold stack displayed in AGC format.

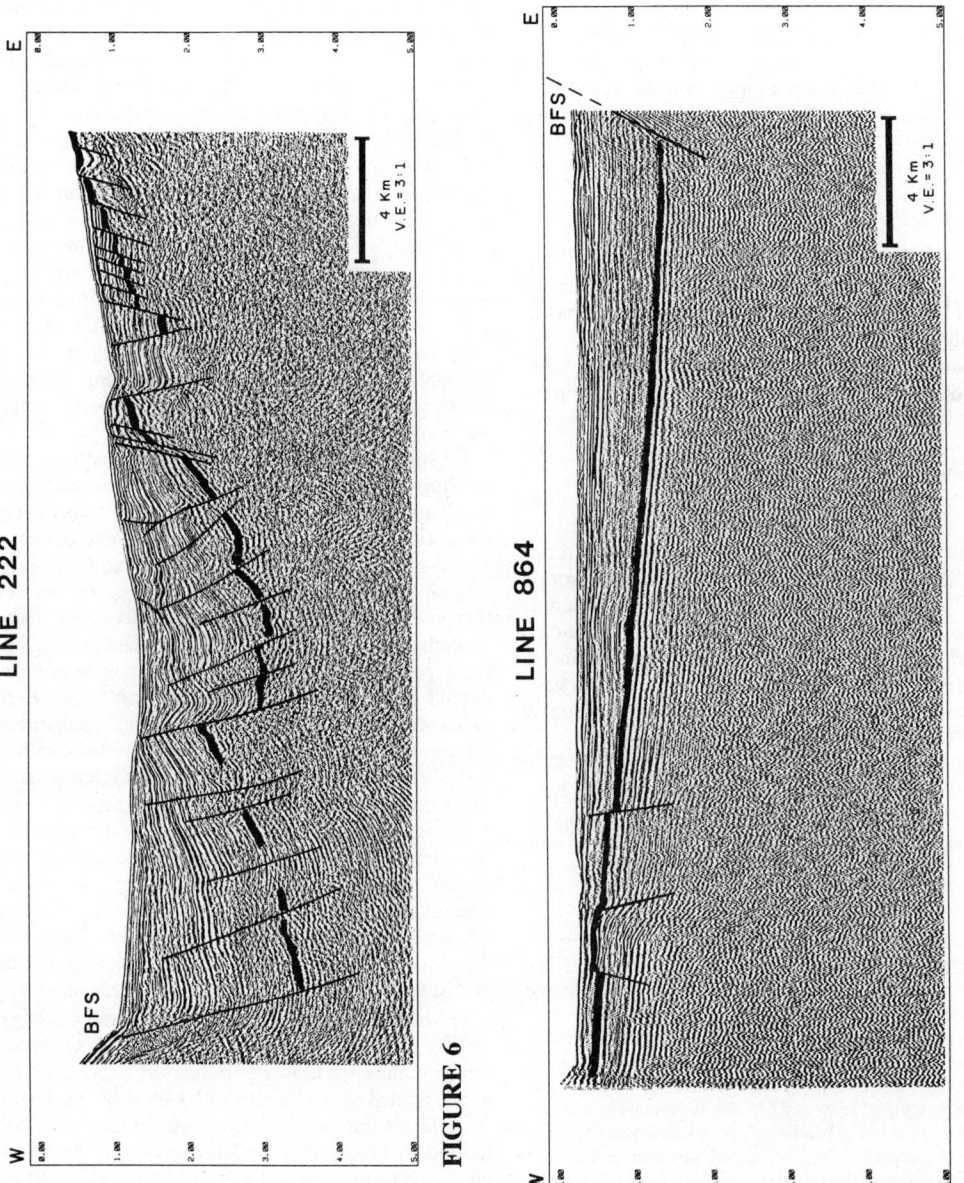

FIGURE 6

FIGURE 7

about such things as flexures, folds, and reverse and strike-slip faults—in short, all those things that might result from complicated applications of stress to imperfect media. Some of these expressions are associated with the mechanics of linking half-graben units together and are best discussed in that context. However, to a certain degree, these traits also are part of the typical infrastructure of many rift units, and to that extent they need airing here.

How much flexing occurs in association with rift subsidence at basement and shallower levels? Our African seismic data suggest that flexure probably occurs in both the dip and strike senses, that it occurs on the block to the basinal scales, and that it originates from a host of factors. Of special concern here are the simple basement warps that occur between adjacent fractures or along the plunge of a particular block. Figure 9 shows a typical half-graben cross section in which there seems to be considerable undulation or flexing of the basement surface. It must be understood that it is impossible to prove that the curvilinear form of the basement reflector on this profile is partly a flexural response, but the alternative explanations are weak. Also, the undulatory form of the basement reflector is a three-dimensional attribute. The appropriate image might be fault blocks whose surfaces resemble gently twisted ribbons. As the seismic profiles in this paper depict, the measurable amounts of apparent flexing are small but the phenomenon is a ubiquitous element of most blocks.

Figure 9 shows what appear to be flexures in both vertical directions. The concave downward forms are easily attributable to bending moments associated with differential motions on paired faults. The concave upward and compound forms, however, seem to imply local compression and shortening. This is consistent with the dips of some of the internal faults in this basin, which tend to be very high angle with respect to the basement offsets. Several of the major internal faults in Fig. 9 actually

---

FIGURE 6. Interpreted seismic reflection profile (Line 222) from the Lake Tanganyika Rift Zone showing the occurrence of antithetic faults on the ramping side of a half-graben that is dominated by fractures that parallel the BFS. Heavy, solid lines represent acoustic basement. Position of profile relative to idealized rift architecture is shown in Fig. 11, actual position in Fig. 18. Profile is migrated, 24-fold stack displayed in AGC format.

FIGURE 7. Interpreted seismic reflection profile (Line 864) from the Lake Malawi Rift Zone showing the almost complete absence of significant internal normal faults. Compare to Figs. 2–6, which show the more usual "shattering" that characterizes rift units. Heavy, solid lines represent acoustic basement. Position of profile is shown in Fig. 18. Profile is unmigrated, 24-fold stack displayed in AGC format.

are reverse faults with respect to their dip angles relative to the basement. These conditions may signify a type of ongoing imbrication resulting from asymmetric subsidence into a restricted space. We use the analogy to a deck of splayed cards that are pushed down along one edge and pinned at the other. Some cards are likely to slip toward each other, in effect turning normal faults into reverse faults and causing local crinkles. Sediment loading may play a role in forcing subsidence to proceed even though the space accommodation is not entirely adequate. It is also possible that some rift units experience compressional episodes associated with readjustments of rifting stresses.

The related issue of strike-slip faulting within rift units is a controversial subject, not because workers doubt that such faults exist but because their expressions in subsurface data are usually ambiguous. Earlier, we addressed this topic from the perspective of transverse offsets of BFS. Later, we will show the large strike-slip or oblique-slip fault systems that are expressed seismically as flower structures and which are a manifestation of various modes of half-graben linking. Now we must tackle the strike-slip faults that are part of the rift unit infrastructure. This subject is elusive because identifying individual strike-slip faults on two-dimensional subsurface data, mainly seismics, can be very difficult and mapping them in three dimensions requires unusually tight observational grids. Figure 10 shows a number of examples of what are interpreted to be faults with significant or predominant strike-slip senses of motions. In these cases the expressions are relatively subtle, involving minor disruptions of the sediment column rather than large and systematic dislocations of beds. The lack of expression belies the significance of these structures; they are probably the primary way in which spatial problems associated with subsidence are accommodated internally. We know that these faults form networks that help define the length dimension of blocks, but generally we cannot map individual faults. Hence, precise orientations are uncertain in the African rifts we have studied. However, petroleum exploration work in other rifts (e.g., Rhinegraben) shows predictable 30° and 60° relations to the normal faults. Almost all previous reports have understated the mechanical importance of strike-slip faulting in rifts. Future work is likely to concentrate heavily on this topic, particularly in regard to structural closure associated with rift hydrocarbon plays.

Linking Arrangements of Rift Units

The subject of half-graben linking arrangements has been discussed by Rosendahl, 1987. A condensed and updated review is presented here. Continental rifts of the East African type can be modeled as a series of half-graben units linked together in various ways (Fig. 11). The key to inter-

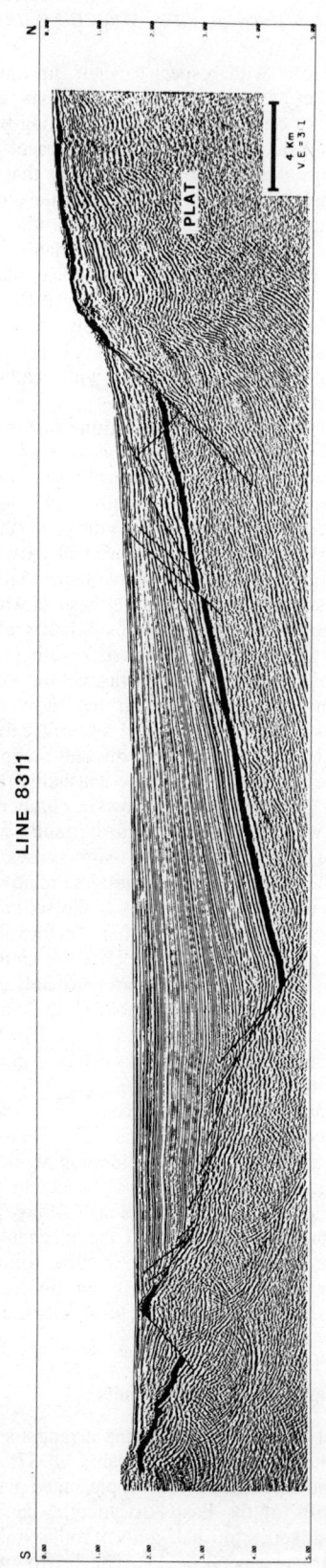

FIGURE 8

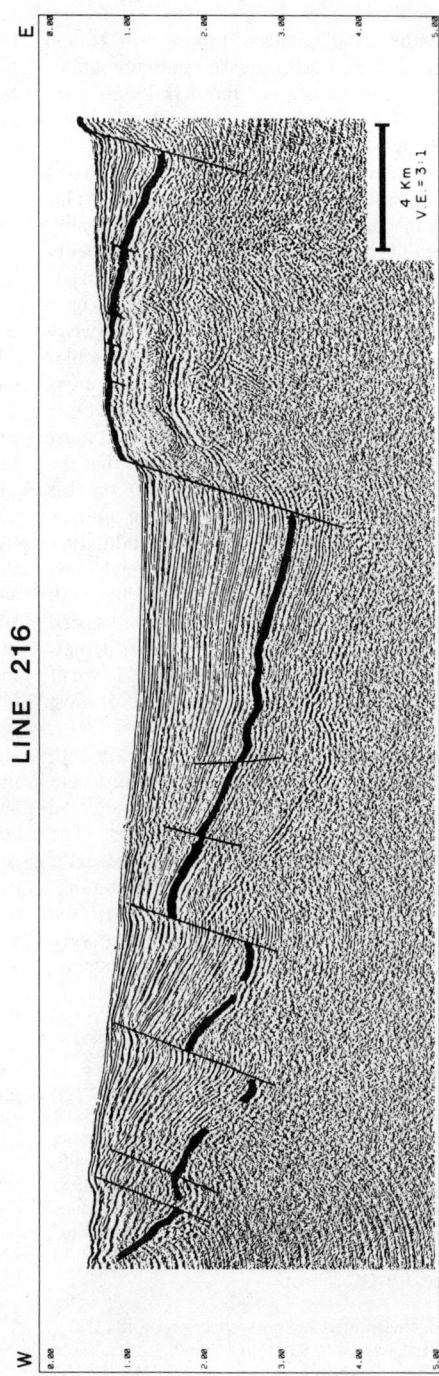

FIGURE 9

preting two-dimensional data from such areas (e.g. seismics) is recognizing that linking geometries have a direct bearing on rift basin morphology. For example, dip profiles across overlapping, opposing half-graben (Fig. 11, Cases 1-3) show two half-graben that face each other, creating an apparent full-graben with some sort of complex axial structure, usually antiformal in character. Figures 12-14 show a range of seismic expressions observed with facing geometries.

The axial features associated with the arrangement shown in Fig. 11 have been termed "hinged highs," "low-relief accommodation zones," and "interference accommodation zones" (INAZ) in various publications. The latter term is adopted here because it best describes the mechanical purpose of such zones—they are interference structures that accommodate the subsidence of the adjacent, facing half-graben. In some cases, particularly in skewed versions of Cases 2 or 3 (Fig. 11), the accommodation involves considerable oblique slip to strike-slip faulting, giving rise to what are termed "flower structures" (Harding, 1985) in seismic jargon. Where the flowers are positive (i.e., antiformal), which is the case for most of the examples shown in Figs. 12 (pages 116-117) and 13 (pages 118-119), some compressional pinning presumably occurs. However, the overall and long-term stress regimes are still dominantly extensional because INAZ eventually subside. INAZ also can be expressed as negative flower structures, particularly where the linking arrangements are comparable to Case 4 (Fig. 11). Several examples are given in Fig. 15 (page 121).

FIGURE 8. Interpreted seismic reflection profile (Line 8311) from the Lake Tanganyika Rift Zone. Profile is a strike line paralleling the East Kigoma BFS. The tilt of the basement reflector toward the south shows the approximate plunge of the first fault block riftward of the BFS. Heavy, solid lines represent acoustic basement. Position of profile relative to idealized rift architecture is shown in Fig. 11, actual position in Fig. 18. Profile is migrated, 24-fold stack displayed in AGC format.

FIGURE 9. Interpreted seismic reflection profile (Line 216) from the Lake Tanganyika Rift Zone. The undulatory character of the basement reflector is believed to represent a combination of fracture and flexural deformation. The large, elevated block on the eastern side shows a concave downward basement surface, which may be a result of differential subsidence between the western and eastern edges of the block. Other blocks show concave upward or compound basement undulations. Note that the main faults bounding these blocks are extremely high-angle relative to the basement reflector and several could be described as reverse faults (with respect to the basement). Many other seismic examples used herein also show evidence of flexure. Heavy, solid lines represent acoustic basement. Position of profile relative to idealized rift architecture is shown in Fig. 11, actual position in Fig. 18. Profile is migrated, 24-fold stack displayed in AGC format.

Depending on the linking geometry and the nature of the pre-rift fabric, INAZ can range from short, squat features to relatively large and long zones (compare Cases 1-3 in Fig. 11). They usually terminate in structural platforms (Fig. 14, page 120), which often are associated with some of the more significant drainage systems entering rift basins. Apparently these platforms substitute for flanking uplifts, which are usually paired with BFS, permitting rivers to enter the rift at platform locations.

The areas where border fault systems, accommodation zones, and platforms intersect seem to be subject to complex deformational forces. At least several such areas are characterized by the kind of structural styles shown in Fig. 16 (page 122). The seismics across these zones tend to be riddled with acquisitional and processing problems. For this and other reasons pertaining to uncertainties about trackline orientation versus structural trends, the western half of Fig. 16 may not be as compressive as visual impressions suggest. However, it is clear that the area has been severely deformed by stresses other than simple tension. It is likely that these intersection areas are subject to considerable horizontal shear stress and some of the deformation is probably caused by strike-slip transpression.

Dip profiles across nonoverlapping, opposing half-graben (Family 2, Fig. 11) show two half-graben facing away from each other. They are usually separated by backbones of relatively unsubsided country rock (Fig. 17, page 123). These features have been termed "interbasinal ridges," "high-relief accommodation zones," and "isolation accommodation zones" (ISAZ). The latter term is used here because these zones physically isolate the component half-graben. They can be sites of considerable strike-slip faulting. Unlike INAZ, which usually are not modern-day barriers to fluvial sediment dispersal, ISAZ often act as nearly complete dams throughout much of the history of a rift. In effect, they subdivide rifts into depositional basins. In the Lake Tanganyika Rift Zone, two ISAZ divide the lake into three basins. Although there are numerous INAZ in Lake Malawi, there are no ISAZ per se. Hence, there are no widespread, structural barriers to modern sediment dispersal within Lake Malawi. Fluvial clastic input to the half-graben shown in Cases 5-6 is mainly from the shoaling or ramping sides, which generally lack flanking rift mountains.

Cases 7-9 in Fig. 11 show a third family of linking modes. A seismic reflection profile corresponding to Case 9 is shown in Fig. 4. Although it is common for one sense of polarity to dominate in African rift zones, there is never an exclusive direction of asymmetry and usually no more than two adjacent half-graben display the same polarities. This occurrence contrasts with the Triassic rift basins of the eastern United States, where all but a few half-graben

(Text continues on page 126.)

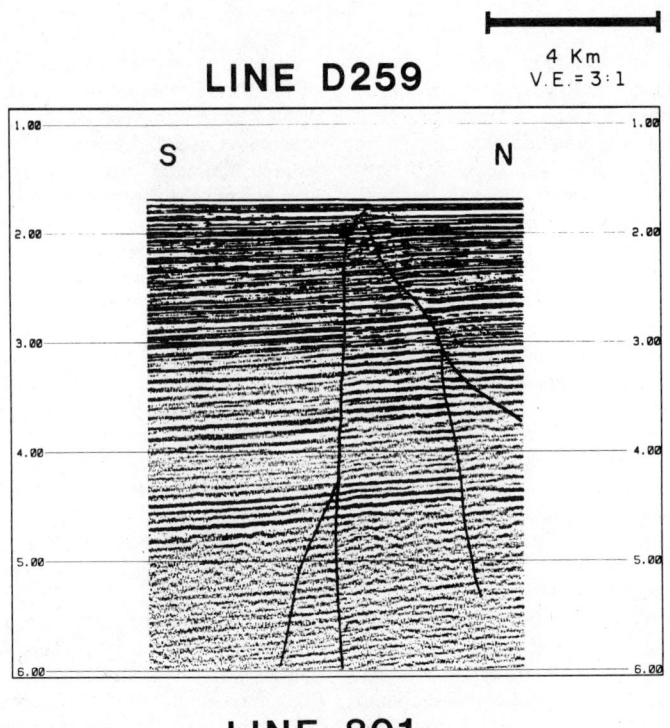

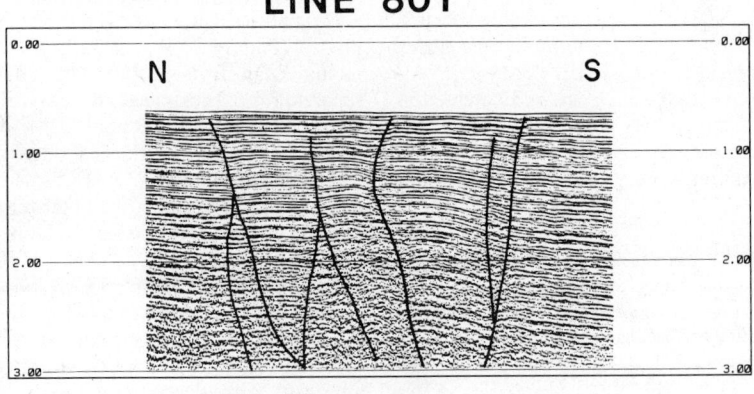

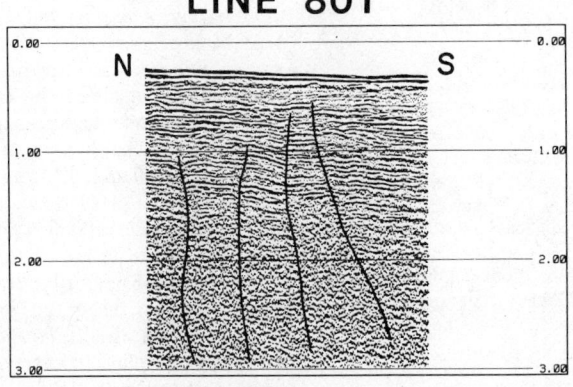

FIGURE 10. Seismic reflection expressions of faults believed to show significant strike-slip displacements. Profiles are strike lines from the Lakes Tanganyika and Malawi Rift Zones. Compare to large flower structures shown in Fig. 13. Positions of seismic panels are shown in Fig. 18. Panels are unmigrated, 24-fold stacks displayed in AGC format.

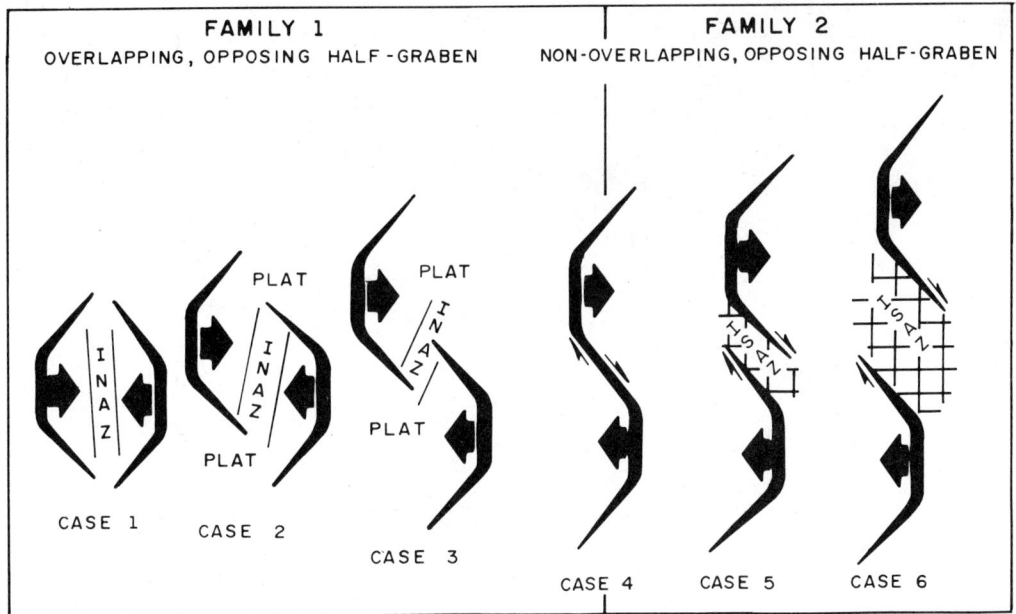
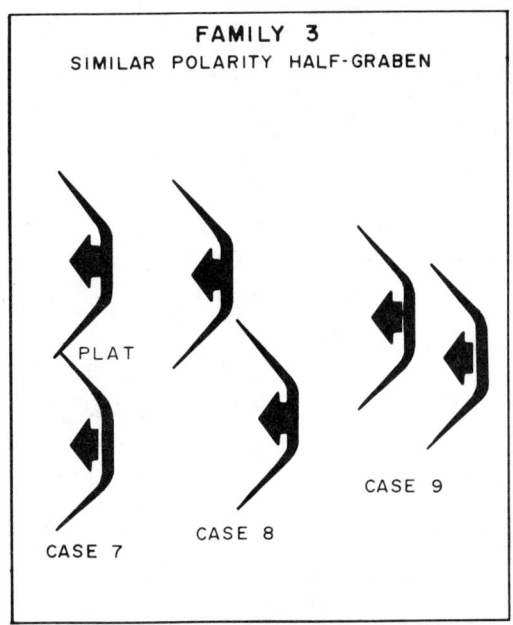

FIGURE 11. Examples of some half-graben linking modes observed in continental rifts (idealized plan views). Cases 1–3 show sets of facing half-graben resembling pairs of offset parentheses. Interference accommodation zones (INAZ) occur in Cases 1–3 in the areas of overlap, platforms (PLAT) in the nonoverlapping corners. Cases 5 and 6 show sets of opposed half-graben that do not overlap. The "backbones" between the units are termed isolation accommodation zones (ISAZ) because they usually isolate the component half-graben from each other in both the structural and depositional senses. Case 4 is transitional between the overlapping and nonoverlapping cases. Cases 7–9 show arrangements of similar polarity half-graben. Some of the regions between adjacent BFS can act like platforms in regard to fluvial processes, particularly in Case 7. Heavy, solid arrows point toward ramping sides of half-graben and are not necessarily equivalent to plate motion directions.

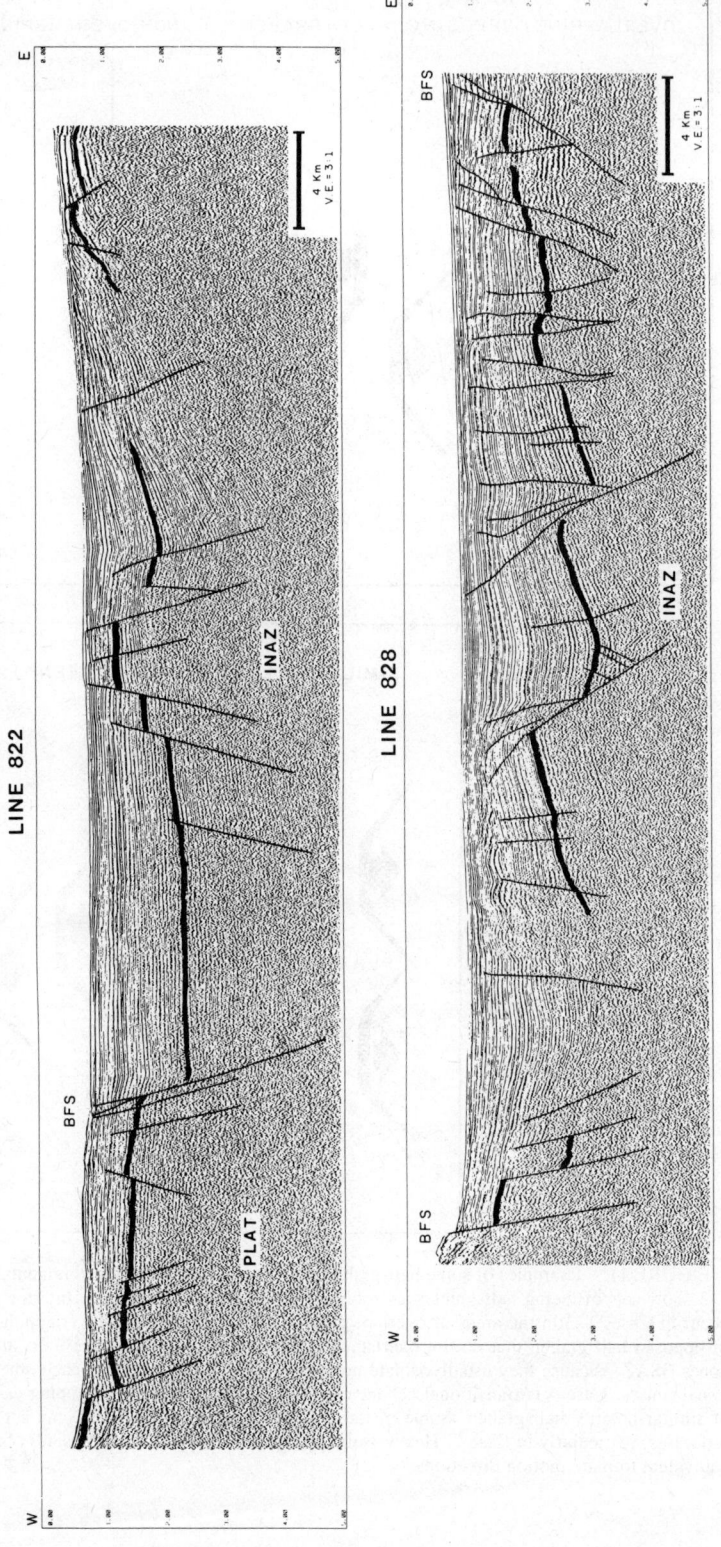

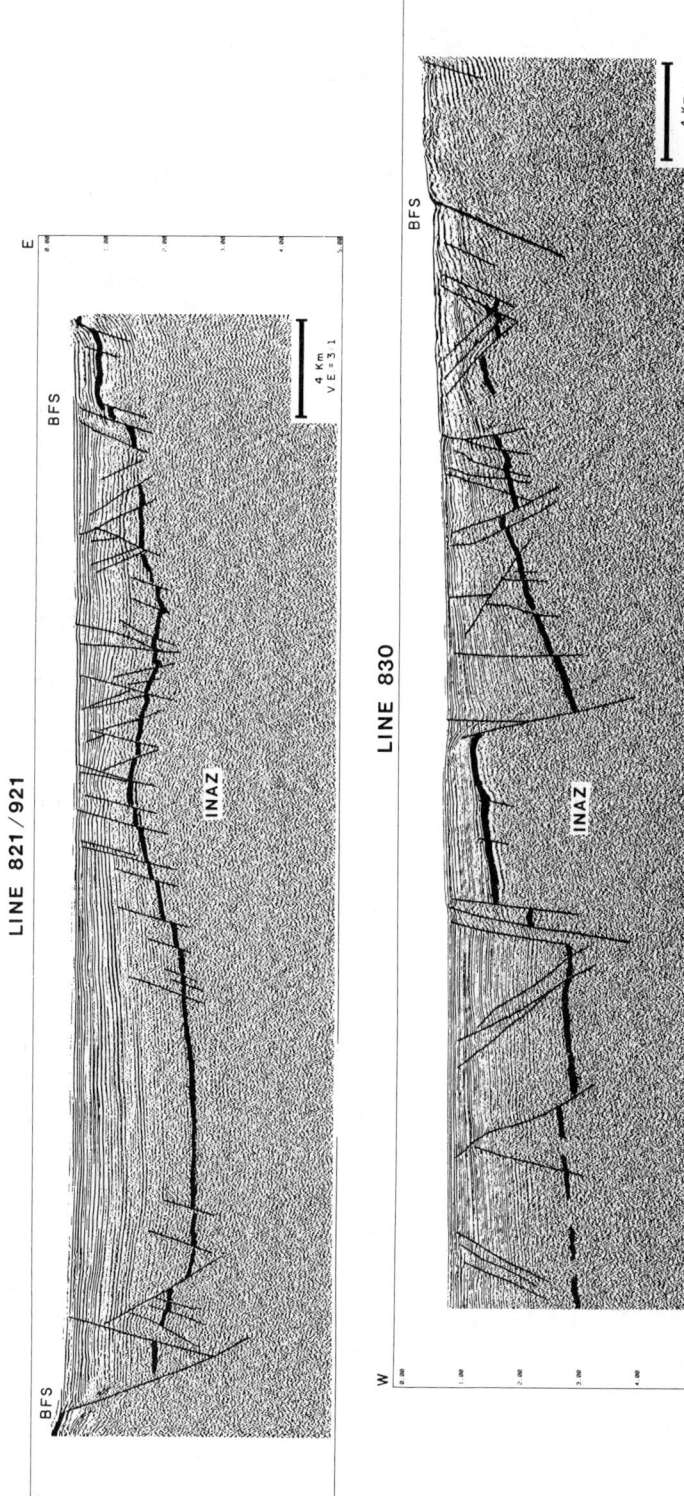

FIGURE 12. Examples of some interpreted seismic reflection profiles from Lake Malawi across half-graben linking arrangements most comparable to Case 1 in Fig. 11 (Lines 822, 828, 821/921, and 830). BFS, INAZ, and PLAT refer to border fault systems, interference accommodation zones, and platforms, respectively. Heavy, solid lines represent acoustic basement. Positions of profiles relative to actual rift architecture are shown in Fig. 18. Profiles are migrated, 24-fold stacks displayed in AGC format.

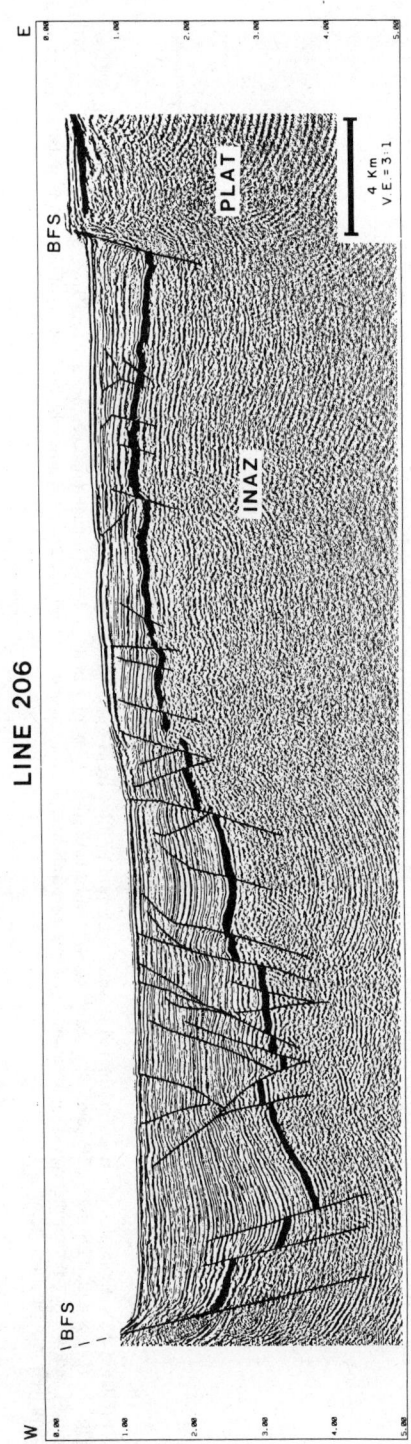
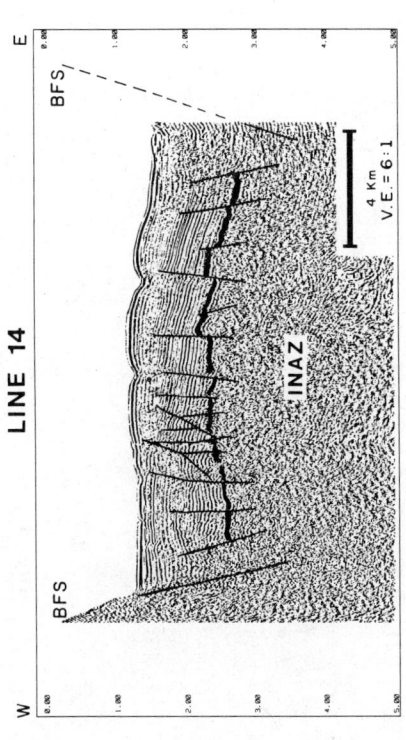

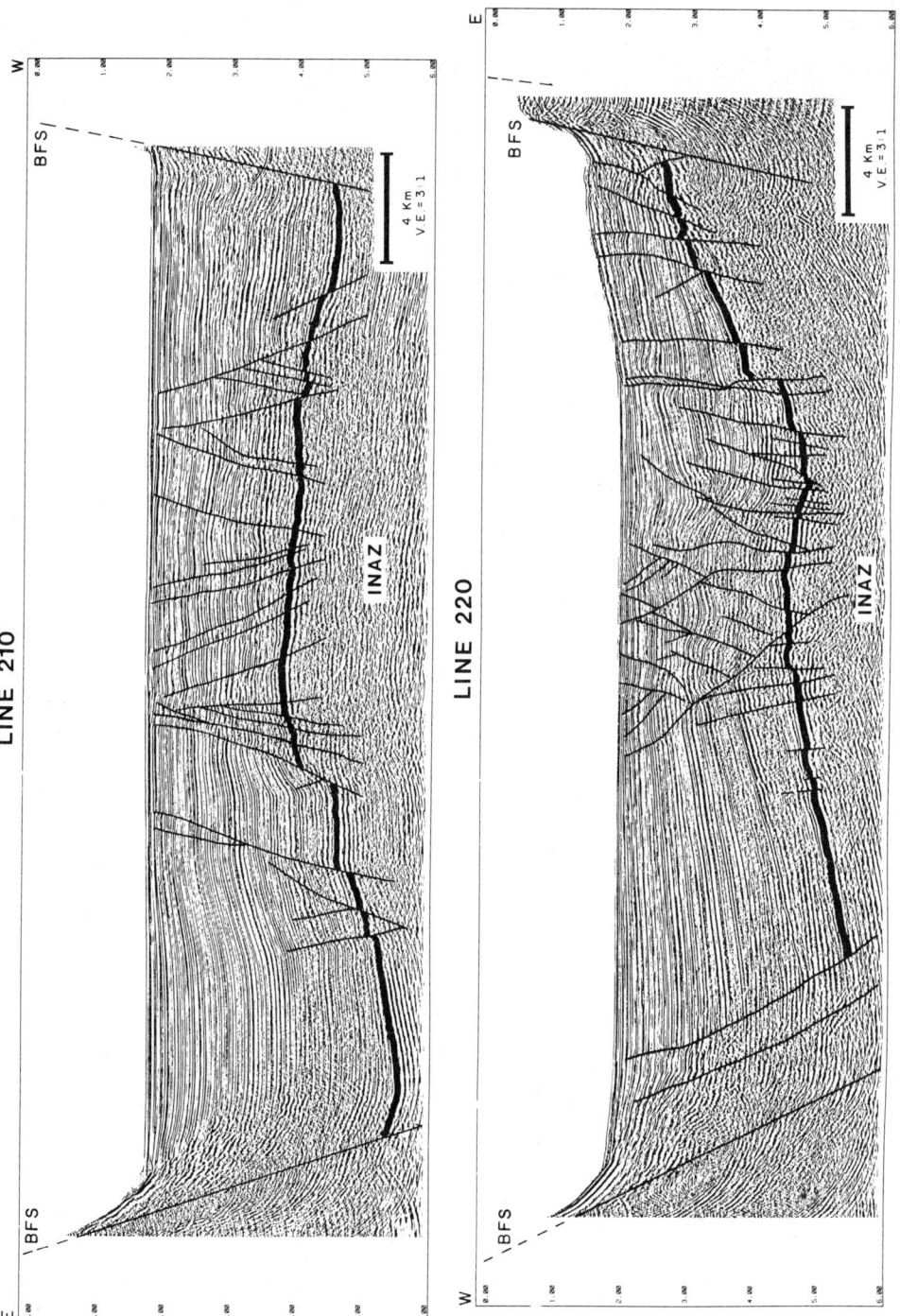

FIGURE 13. Examples of some interpreted seismic reflection profiles from Lake Tanganyika across half-graben linking arrangements most comparable to the overlapping areas of Case 2 in Fig. 11 (Lines 206, 14, 210, and 220). BFS, INAZ, and PLAT refer to border fault systems, interference accommodation zones, and platforms, respectively. Heavy, solid lines represent acoustic basement. Positions of profiles relative to actual rift architecture are shown in Fig. 18. Profiles are migrated, 24-fold stacks displayed in AGC format.

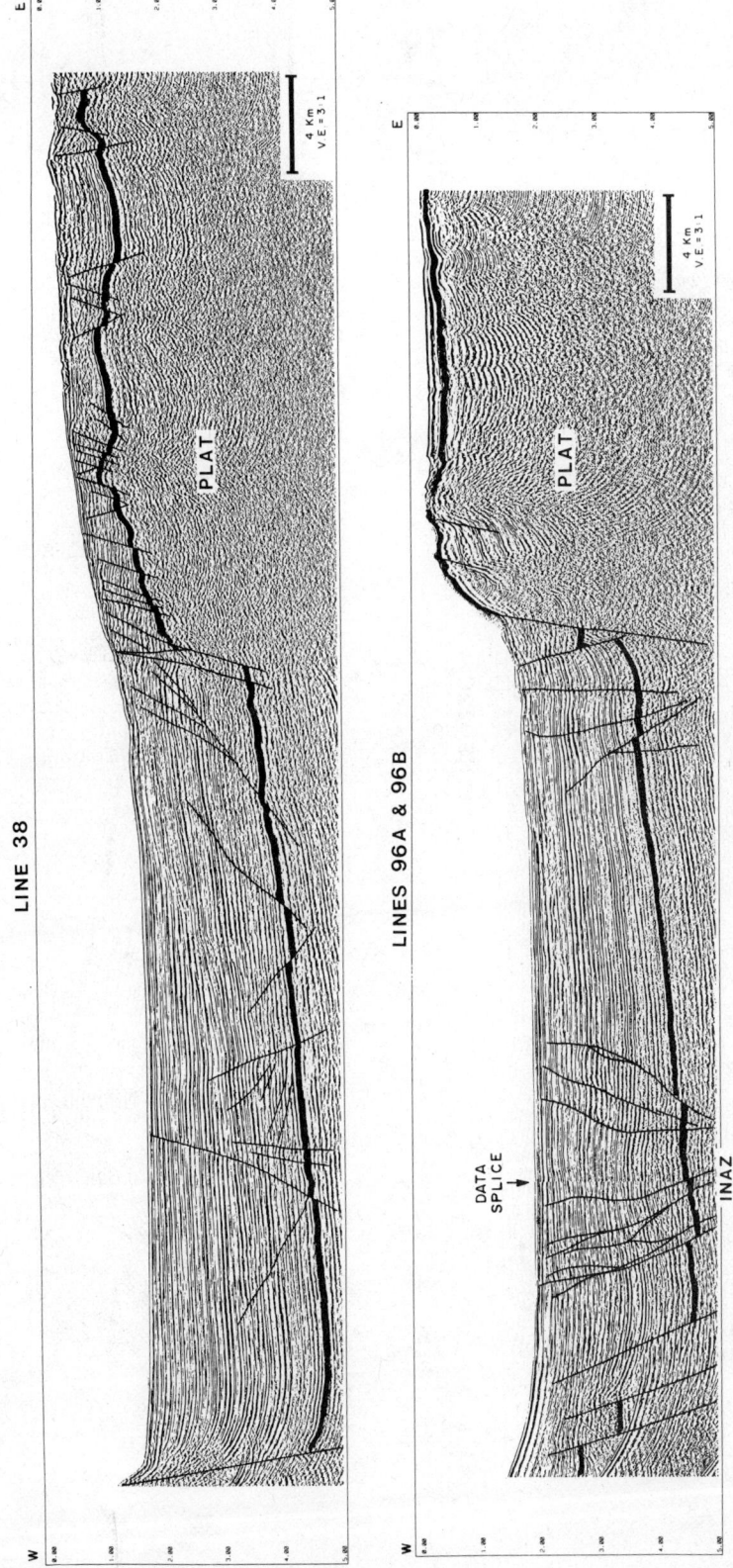

FIGURE 14. Examples of some interpreted seismic reflection profiles from Lake Tanganyika across half-graben linking arrangements most comparable to the nonoverlapping areas of Cases 2 and 3 in Fig. 11 (Lines 38 and 96A/B). BFS, INAZ, and PLAT refer to border fault systems, interference accommodation zones, and platforms, respectively. Heavy, solid lines represent acoustic basement. Positions of profiles relative to actual rift architecture are shown in Fig. 18. Profiles are migrated, 24-fold stacks displayed in AGC format.

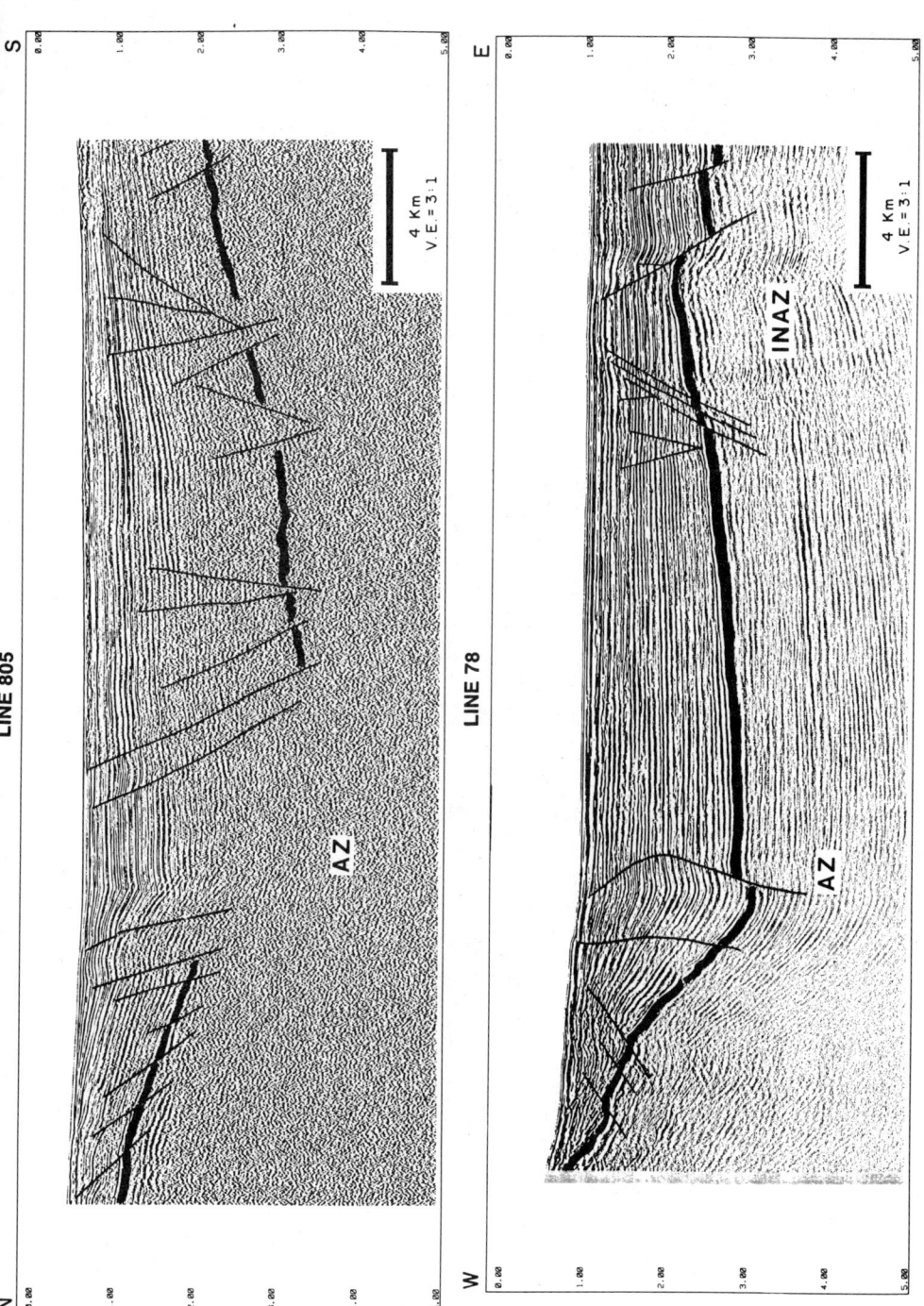

FIGURE 15. Examples of some interpreted seismic reflection profiles across half-graben linking arrangements most comparable to Case 4 in Fig. 11. Line 805 is from Lake Malawi, 78 from Lake Tanganyika. Note that the accommodation zone (AZ) associated with this type of linking arrangement is topographically low in comparison with typical interference and isolation accommodation zones (INAZ and ISAZ, respectively). Heavy, solid lines represent acoustic basement. Positions of profiles relative to actual rift architecture are shown in Fig. 18. Both profiles are 24-fold stacks displayed in AGC format; Line 78 is migrated, 805 is not.

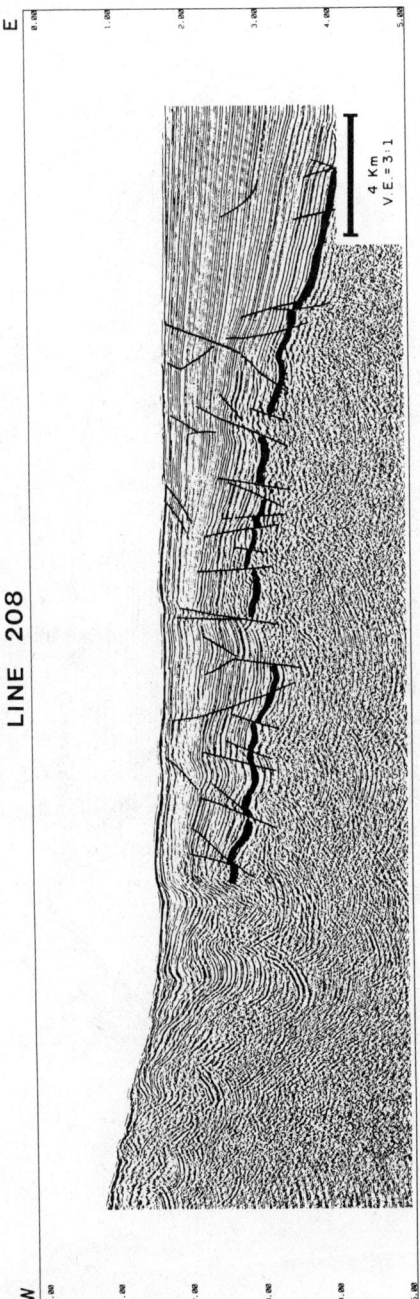

FIGURE 16. Partially interpreted seismic reflection profile (Line 208) from the Lake Tanganyika Rift Zone. The ramping side of this half-graben (i.e. the uninterpreted region) demarks a zone of intersection of an INAZ with a BFS that is best developed further to the south. The western third of the profile shows the deformational style associated with such intersection zones; the eastern two-thirds of the profile shows the more typical expression of asymmetric half-graben. It is likely that the intersection zone is an area of oblique-slip strike-slip faulting and transpression. Heavy, solid lines represent acoustic basement. Profile position relative to actual rift architecture is shown in Fig. 18. Profile is migrated, 24-fold stack displayed in AGC format.

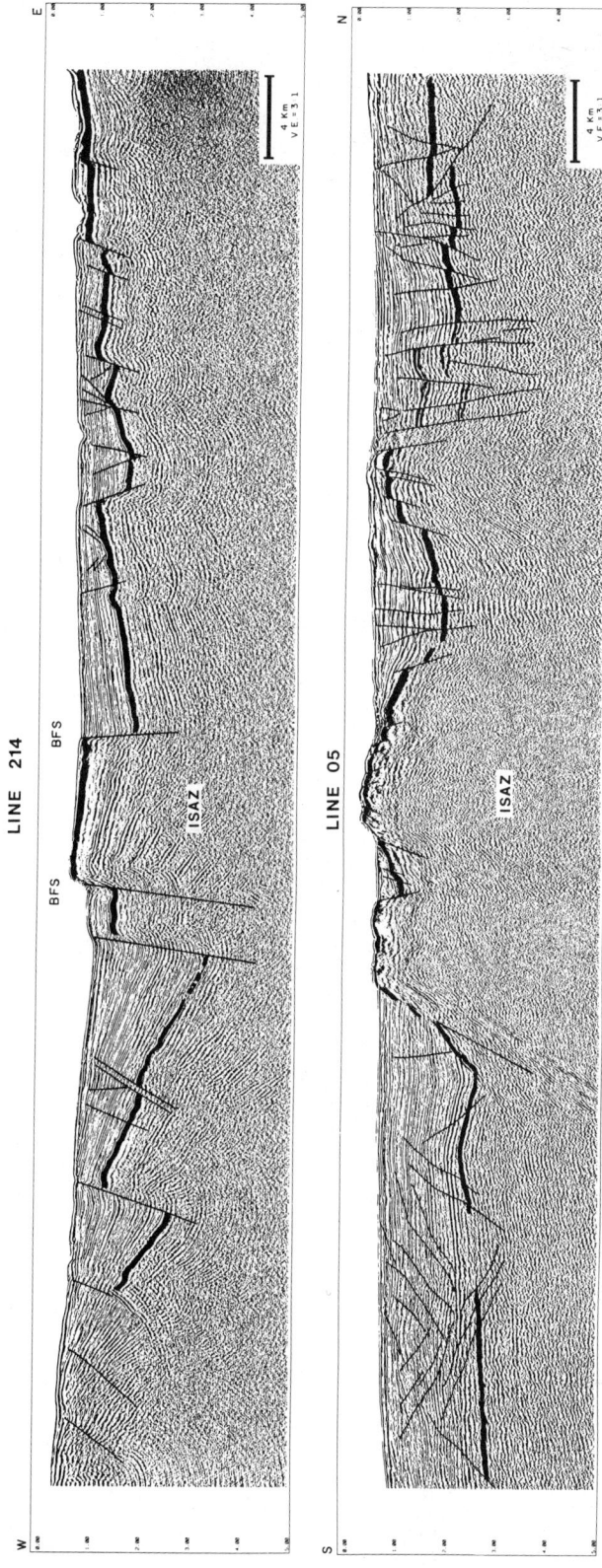

FIGURE 17. Examples of some interpreted seismic reflection profiles from Lake Tanganyika across half-graben linking arrangements most comparable to Case 5 in Fig. 11 (Lines 214 and 05). Heavy, solid lines represent acoustic basement. Positions of profiles relative to actual rift architecture are shown in Fig. 18. Both profiles are 24-fold stacks displayed in AGC format; Profile 214 is migrated, 05 is not.

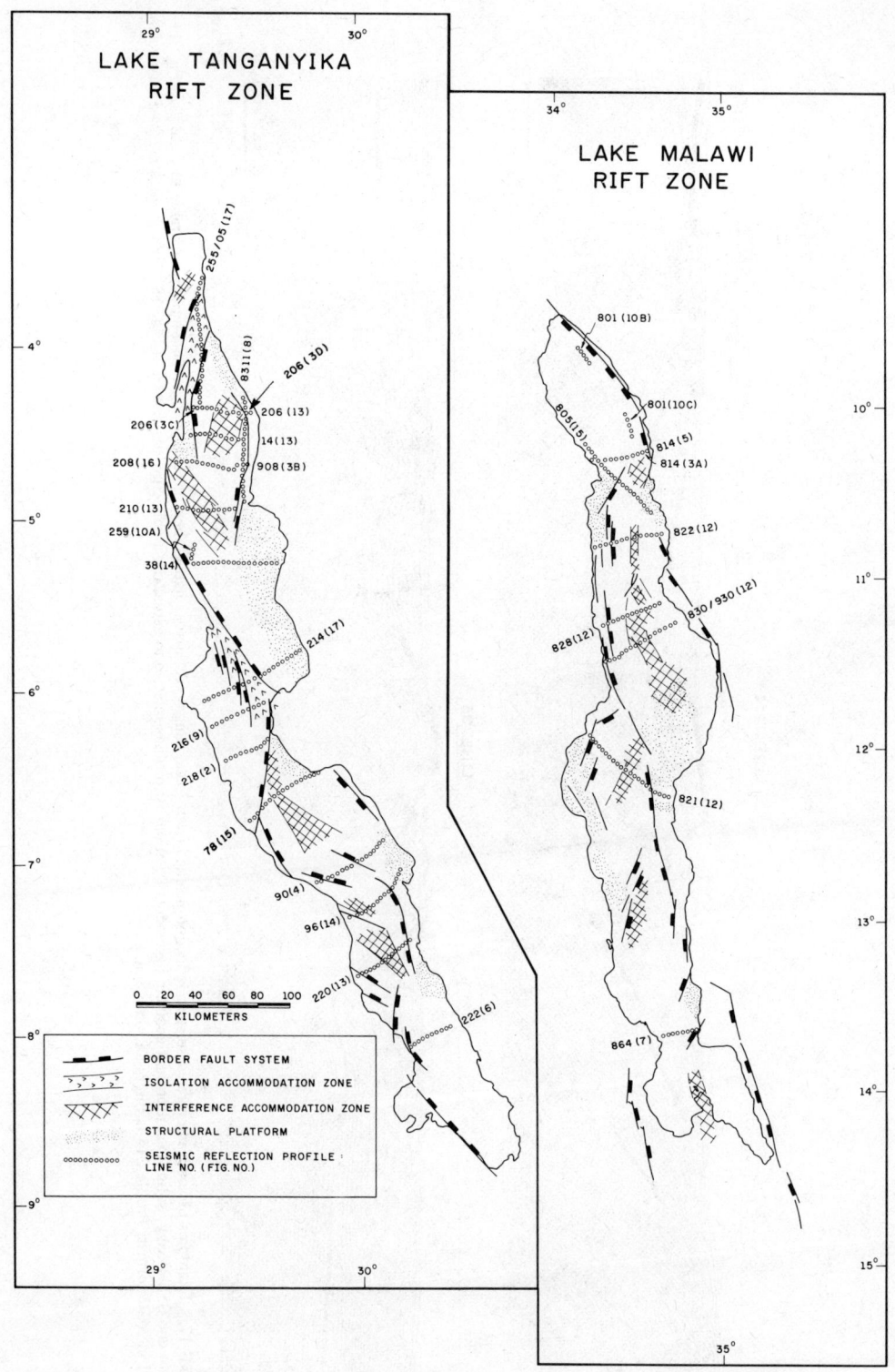

FIGURE 18. Generalized structure maps of the Lakes Tanganyika and Malawi rift zones. Successful rifting of such zones probably would result in asymmetric conjugate margins.

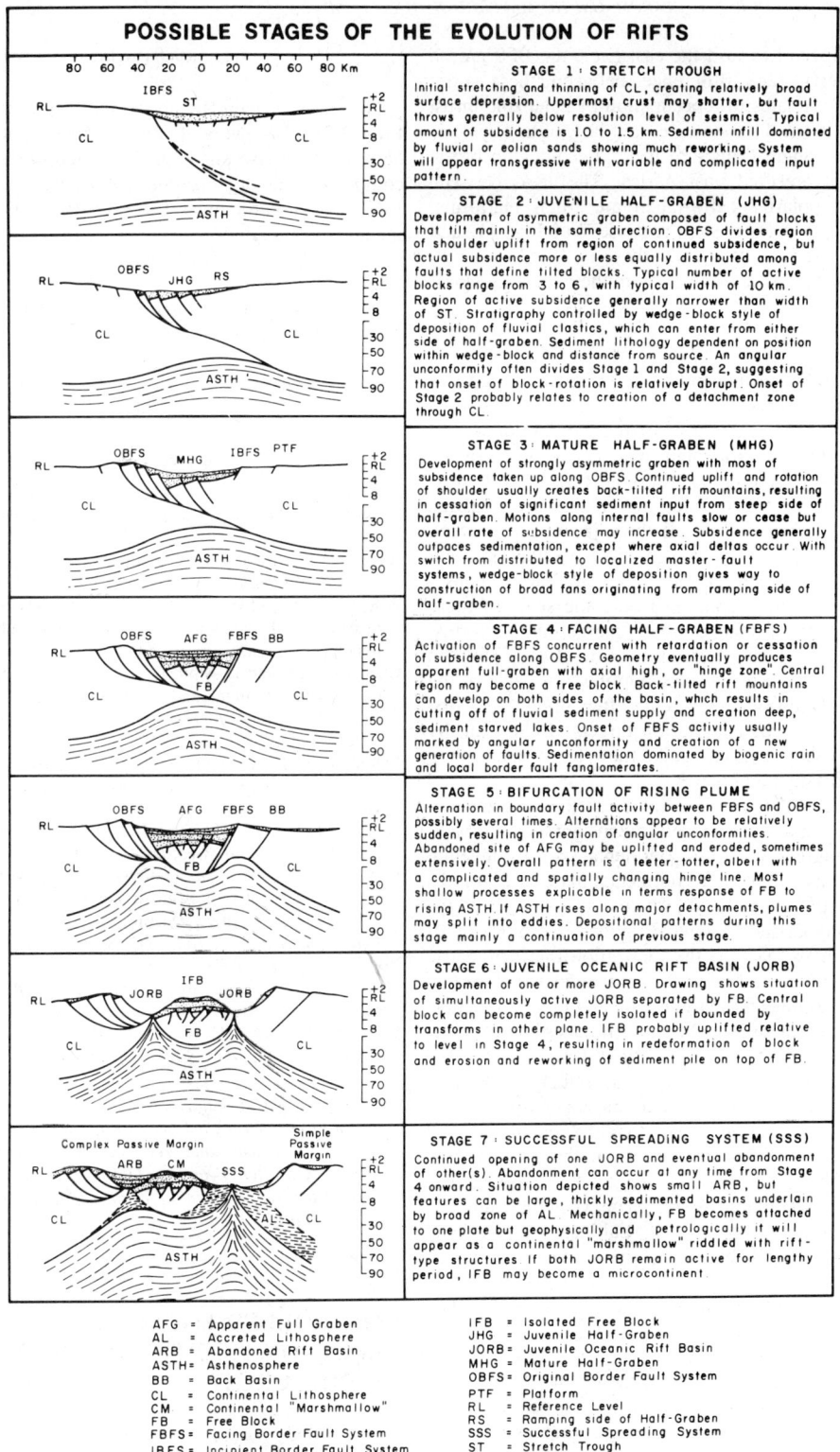

FIGURE 19. A possible structural chronology for the evolution of the African rift system into a successful oceanic rift of the Red Sea type. (After Rosendahl, 1987)

ramp upward toward the east (i.e., the BFS are on the west). The difference between these areas probably relates to reoccupation of Paleozoic thrust planes in the Triassic case, and lack of any systematic low-angle thrusts in the old, cold, and brittle crust of prerifted East Africa. The areas between similar polarity half-graben are often platforms, especially in cases such as Case 7 of Fig. 11, and fluvial clastics often enter these half-graben across such platforms, as well as from their shoaling sides.

Conclusions

The linking arrangements described above have an important impact on a host of subjects related to rifting. Clearly, the style of linking determines the loci of fluvial clastic input, which bears on the distribution of such things as reservoir rocks in rift basins. We might also consider that INAZ are dynamic features—they evolve with the adjacent depocenters, usually remaining high relative to the adjacent basins. The stratigraphic and hydraulic consequences of this, combined with the scale of these features and the fact that they merge into platforms, make them one of the most attractive petroleum targets in continental rifts. On the other side of the coin are the ISAZ, which are often among the worst possible synrift exploration targets. Blanketing by postrift sediments can make these features excellent postrift targets, however.

Perhaps the greatest value of the architectural concepts presented above is the insight they offer into the progression from continental rift to juvenile ocean basin to passive continental margin. Figure 18 shows the generalized structure maps of the Lake Tanganyika and Malawi rift zones. Examination of these maps leads to the realization that symmetric conjugate rift margins (i.e. mirrored structures on facing continental margins) are an unlikely outcome of ripping these East African rift zones apart, unless the primary tears faithfully follow interference accommodation zones. This is unlikely, especially where isolation accommodation zones are present. Such a deduction leads to the conclusion that asymmetry should rule; most of the rifted terrain probably is left on only one of the margins, in a pattern that crudely alternates between the conjugate margins along strike. The Cabinda, Angola, and Campos, Brazil, margins (Fig. 1) may be a case in point.

A related insight is that facing half-graben arrangements (Figs. 11-13) and hence INAZ must be an intermediate step toward successful rifting, which apparently begins with juvenile half-graben expressions and ends with the success of a spreading center (and abandonment of one or more others). We conclude this discussion with the presentation of a model that uses these ideas to explain the possible evolution of the East African-Ethiopian-Afar-Red Sea System (Fig. 19).

Acknowledgements

This entry is dedicated to the students and staff of Project PROBE, without whom none of this would be possible. I also am deeply indebted to the many companies and organizations who have sponsored PROBE's research over the last six years. They include Agip, Amoco, Arco, Conoco, Eason, Elf Aquitaine, Esso, Marathon, Mobil, Pecten, Pennzoil, Petrofina, Placid, Shell, Texaco, and World Bank. The countries of Burundi, Kenya, Malawi, Mozambique, Tanzania, Zaire, and Zambia are thanked for their cooperation and support of our research in their territories.

B. R. ROSENDAHL

References

Aydin, A., and A. Nur, 1982, Evolution of pull-apart basins and their scale independence, *Tectonics* **1**, 91-105.

Bally, A. W., Musings over sedimentary basin evolution, *Royal Soc. London Philos. Trans.* **305**, 325-338.

Bott, M. H. P, and D. P. Mithen, 1983, Models of rifting: mechanism of graben formation—the wedge subsidence hypothesis, *Tectonophysics* **94**, 11-22.

Chenet, P. Y., and L. Montadert, 1981, Rifting tectonics on Galicia, Portugal and N. Biscay margin, *Papers Presented to the Conference on the Processes of Planetary Rifting*, 47-50.

Eaton, G. P., 1982, The Basin and Range Province: origin and tectonic significance, *Ann. Rev. Earth Planetary Sci.* **10**, 409-440.

Gibbs, A. D., 1984, Structural evolution of extensional basin margins, *Jour. Geology* **141**, 609-620.

Harding, T. P., 1985, Seismic characteristics and identification of negative flower structures, positive flower structures, and positive structural inversion, *Am. Assoc. Petroleum Geologists Bull.* **69**, 582-600.

Rosendahl, B. R., 1987, Architecture of continental rifts with special reference to East Africa, *Ann. Rev. Earth Planetary Sci.* **10**, 445-503.

Wernicke, B., and B. C. Burchfiel, 1982, Modes of extensional tectonics, *Jour. Struct. Geology* **4**, 105-115.

Cross-references: *Continental Crustal Structure; Continental Lithosphere; Continental Rifts: Types and Rates; Crustal Movements and Tectonic Deformation; Deep Seismic-Reflection Profiling; Ocean Continent Transition: Structure; Stress in the Earth's Lithosphere; Thin-Skin Tectonics.*

CONTROLLED SOURCE ELECTROMAGNETIC MAPPING OF THE CRUST

Controlled source electromagnetic (CSEM) mapping of the electrical conductivity of the Earth's interior relies on a simple, basic concept of physics. If a time-varying *primary* EM field is generated in the vicinity of a conductor, then eddy currents are induced in the conductor in accordance with Faraday's law, and the progress of the currents with time depends on the scale or size of the conductor and on the magnitude of its electrical conductivity. Measurements of the *secondary* EM fields generated by the eddy currents may be inverted for the physical characteristics of the conductor.

The tools available for CSEM are many and have been developed from systems designed principally for prospecting for base metal mineral deposits. They are known by acronyms such as TURAM, UTEM, (C)SIROTEM, and DEEPEM, reflecting system configuration, origin, or even anticipated performance! The methods differ in the way the transmitter excites the ground and in the way the receiver measures the secondary field. *Frequency domain* systems rely on the transmission and reception of a single frequency or, perhaps, several frequencies sequentially. In the common, alternative *transient* method, a static primary EM field previously established over a substantial period of time is rapidly removed. The subsequent secondary field of the induced eddy currents in the Earth is then measured in the absence of any primary field. A measurement of the secondary field alone may be advantageous at a given location if the primary field is sufficiently large compared with the secondary field for irregularities in it to degrade a combined measurement. In a short article, one cannot cover the response of many Earth models to many different CSEM systems. The core of the discussion is limited to one method in which the transient secondary fields due to the interruption of a primary current in a horizontal loop of wire resting on the ground surface are detected. The models are the one-dimensional layered Earth and a plate conductor in a background half-space. A study of the responses of these two models reveals nearly all the basic physics of controlled source induction in the Earth.

The Layered Earth Problem

The conductivity model most often fitted to CSEM observations is the depth-dependent, one-dimensional, plane-layered profile. On a global scale, the electrical conductivity of the surface layers of the Earth certainly does not vary only with depth. Lateral conductors such as the oceans are dominant features. Their conductivity of about 3 S · m^{-1} is 10,000-fold greater than the conductivity of typical continental crystalline material. However, on the scale of a few tens of kilometers, several continental sedimentary areas are approximately plane layered. The shales and limestones that form the sedimentary sequence have conductivities ranging from 0.005 to 0.5 S · m^{-1}, depending on the porosity of the rock and the salinity of the contained pore water, which carries most of the electric current. In other regions, a series of plane-layered models fitted piecewise to the data is a useful first approximation of a more complex three-dimensional structure.

How does the EM field induced in a plane-layered model behave? The differential equation satisfied by the field in a uniform layer can be determined from basic physics. The Maxwell equations governing the interrelationships between the electric field **E** and the magnetic field **B** in a homogeneous material may be combined as the wave equation

$$-\nabla \times \nabla \times \mathbf{E} = \mu\sigma \frac{\partial \mathbf{E}}{\partial t} + \mu\epsilon \frac{\partial^2 \mathbf{E}}{\partial t^2} \quad (1)$$

where σ, μ, and ϵ are the conductivity, permeability, and permittivity of the material, respectively. A similar equation may be written for the magnetic field vector **B**. Equation 1 may be rationalized by scaling length and time in terms of a characteristic length L and a characteristic time $\tau = \mu\sigma L^2/2$ (the factor 2 is preemptive) to give

$$-\nabla \times \nabla \times \mathbf{E} = 2 \frac{\partial \mathbf{E}}{\partial t} + \frac{4\epsilon}{\mu\sigma^2 L^2} \frac{\partial^2 \mathbf{E}}{\partial t^2} \quad (2)$$

The second term on the left-hand sides of Eqs. 1 or 2 may be neglected in comparison with the first term and the physics simplified to a diffusion process, provided the scale of the CSEM experiment is large compared with $2(\epsilon/\mu\sigma^2)^{1/2}$ or $2/377\sigma$. The omission is equivalent to the neglect of the magnetic effects of displacement current in comparison with conduction current. The critical scale is largest for very resistive crystalline rock, having a value about 70 m. A feel for the time taken for an EM disturbance to diffuse through a uniform medium may be gained by evaluating the characteristic time τ for a few typical cases. If the scale L is set to 1 km and the parameter μ takes its freespace value, then τ has values of 62 ms, 6.2 ms, and 620 μs for conductivities of 0.1, 0.01, and 0.001 S · m^{-1}, respectively.

The transient fields inside a layered Earth excited by the sudden removal of a current in a circular loop cannot be written directly as a closed-form solution of the diffusion equation. The transient secondary fields are the Fourier synthesis of the response to individual harmonic source currents of time dependence $e^{i\omega t}$ weighted by the Fourier transform of the source current step-off function. If the loop has a radius a and carries a primary current of amplitude I, then the only nonvanishing, azimuthal component of the electric field $E_\phi(r)$ is given at the surface of the Earth by the inverse Hankel transform

$$E_\phi(r) = \frac{i\omega\mu I a}{2} \int_0^\infty \frac{2\lambda Q}{1 + \lambda Q} J_1(\lambda a) J_1(\lambda r) \, d\lambda \quad (3)$$

The expression for E_ϕ is just the free-space form modified by the term $2\lambda Q/(1 + \lambda Q)$. The parameter

$Q(\lambda)$ must contain all the information about the Earth. It is defined in terms of the ratio of the Hankel transform of the azimuthal component of the electric field at the surface of the Earth to the Hankel transform of the orthogonal radial component of the magnetic field B_r there, through the relationship

$$Q(\lambda) = \frac{-E_\phi(\lambda)}{i\omega B_r(\lambda)} \quad (4)$$

Clearly, Q is proportional to the *magnetotelluric* impedance, provided a source field wave number λ is included in the formulation of that function. The value of Q can be determined recursively for any layered Earth model, given the thicknesses and conductivities of the layers, by the rule

$$Q_i = \frac{1}{\theta_i} \left[\frac{\theta_i Q_{i+1} + \tanh(\theta_i d_i)}{1 + \theta_i Q_{i+1} \tanh(\theta_i d_i)} \right]$$

where Q_i and Q_{i+1} are the values of Q at the top and bottom of the ith layer, with thickness and conductivity d_i and σ_i, respectively. The end member of the recursion Q_1 is the value of Q inserted in Eq. 3. The wave number θ_i is given by $\theta_i^2 = \lambda^2 + i\omega\mu\sigma_i$. The electric field inside any layer can be written as an inverse Hankel transform similar to Eq. 3.

The azimuthal electric field induced in a layered Earth by a loop source is axisymmetric about a vertical axis through the center of the loop. Consequently, the change in the form of the field with time, following interruption of a primary current of 1 A in a loop of radius 250 m, may be displayed as a contour map on a vertical plane through this axis of symmetry. Three families of such contour maps, arranged in sequence, for three layered models, 1, 2, and 3, respectively, are shown in Figs. 1, 2, and 3. Model 1 is a simple half-space of resistivity 100 $\Omega \cdot$ m. The second and third models are of the layer over a half-space type. The layer has thickness 500 m and resistivity 100 $\Omega \cdot$ m. The lower half-space has resistivities 10 $\Omega \cdot$ m and 1000 $\Omega \cdot$ m for models 2 and 3, respectively. The units of the contour annotation are 10^{-8} V·m^{-1}. At a very early time, an intense current is induced immediately under the loop. Its size is just what is needed to retain the EM field everywhere inside the conductive Earth at the same value as that which existed before the transient. As time progresses, the intense current *ring* diffuses downward and outward from the source, energizing ever deeper zones. The relative effect of a conductive zone at depth is to inhibit downward and outward progression of the induced current ring. The resistive zone has the opposite effect.

One possible measure of the surface expression of the secondary field, a measure employed by the coincident loop SIROTEM system, is the emf induced in a receiver loop located almost in the same position as the transmitter loop. The emf is a line integral of the electric field and, for harmonic excitation, is the product of Eq. 3, with r set equal to a, and $2\pi a$. The transient secondary emf, plotted as a function of time following an interruption of current in the transmitter, is shown in Fig. 4 for the three models defined earlier and for additional 10- and 1000-$\Omega \cdot$ m half-space models, labeled 4 and 5, respectively. The decrease in the emf with increasing time for the half-space models clearly follows an inverse 5/2 power law at late time. The analytic, asymptotic form is

$$\text{emf} = \frac{\pi}{5\sigma a} \left[\frac{\tau}{t} \right]^{5/2}$$

where σ is the conductivity of the half-space and τ is the characteristic time $\mu\sigma a^2/2$. The response for both models of a layer over a half-space initially follows the response of a half-space of resistivity equal to the resistivity of the upper layer. Then, at late time, the response merges with the response of a half-space of resistivity equal to that of the lower half-space. Measurement of emf as a function of time is clearly influenced by the variation of resistivity with depth. Field data can be converted rapidly to an *apparent resistivity* or processed to obtain an *image* of the layered Earth; such techniques ease the task of interpretation.

The temporal variation of the magnetic field or its time derivative, the field measured by an induction coil, at some point remote from the transmitter can also be used to infer Earth structure. The vertical magnetic field component may be obtained from the azimuthal electric component through the differential form of Ampere's law; thus

$$B_z(r) = \frac{\mu I a}{2} \int_0^\infty \frac{2\lambda^2 Q}{1 + \lambda Q} J_1(\lambda a) J_0(\lambda r) \, d\lambda$$

The radial magnetic component is related to E_ϕ through Eq. 4, the definition of Q, so that

$$B_r(r) = \frac{\mu I a}{2} \int_0^\infty \frac{2\lambda}{1 + \lambda Q} J_1(\lambda a) J_1(\lambda r) \, d\lambda$$

The rise in the vertical magnetic component with time, expressed as a fraction of the static late-time field, following a sudden *step-on* of current in the loop, is shown in Fig. 5. The five models are the set introduced before, and the loop radius and horizontal range are 250 m and 1500 m, respectively. At this range the late time field has a value of about 20 mV·A^{-1} of current in the loop. The characteristics of the sounding curves can be deduced from our previous description of the outward diffusion of the induced currents in the ground for the layered models. At early time, the magnetic field of the induced current ring in the ground opposes the field of the current in the transmitter wire, and the total

(*Text continues on page 133.*)

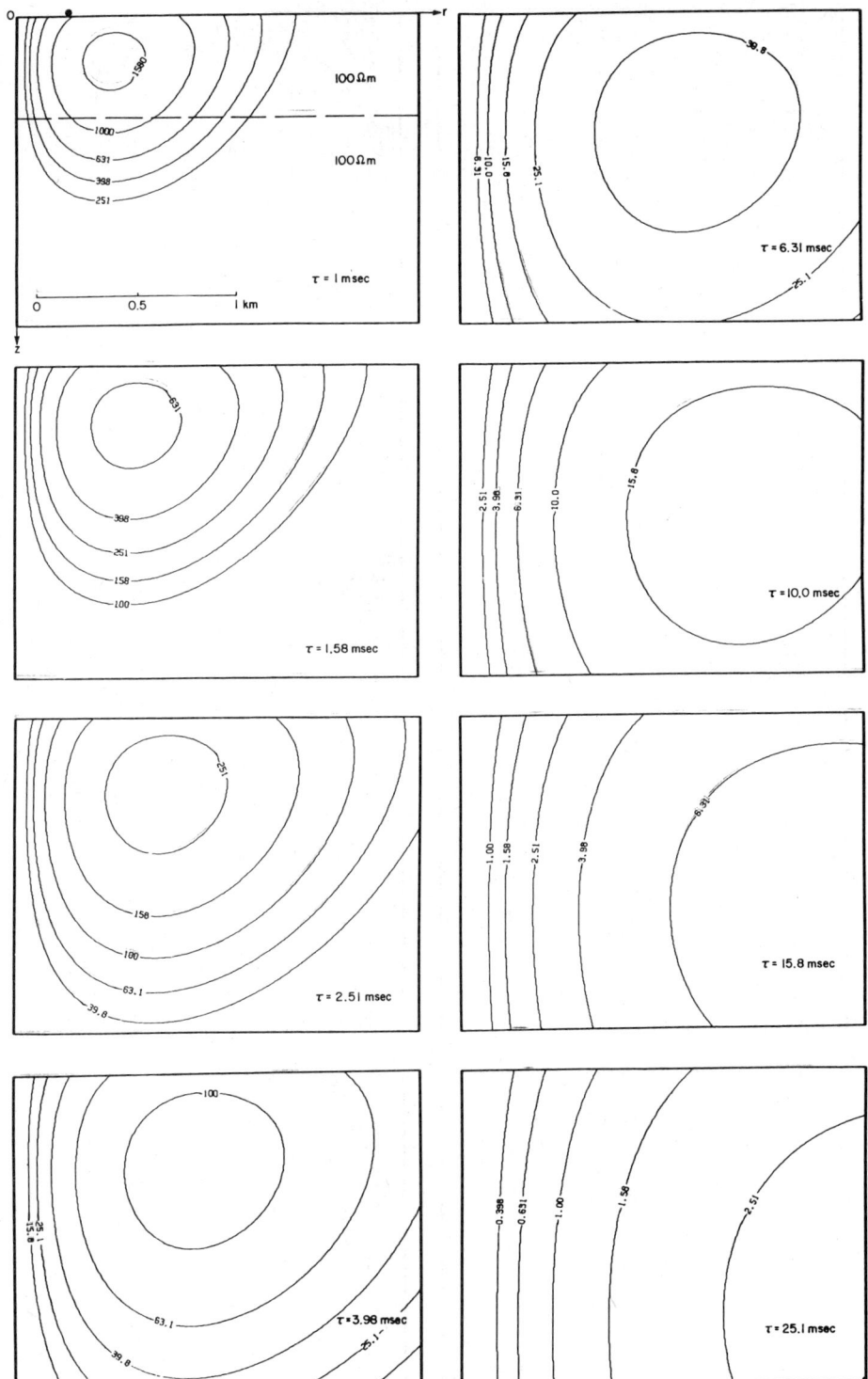

FIGURE 1. The diffusion into a conductive half-space of the induced azimuthal component of the electric field following a rapid turn-off of a unit current in a horizontal circular loop at the surface of the Earth. The radius of the loop and the resistivity of the half-space are 250 m and 100 $\Omega \cdot$ m, respectively. The logarithmically spaced contours are in units of 10^{-8} V/m.

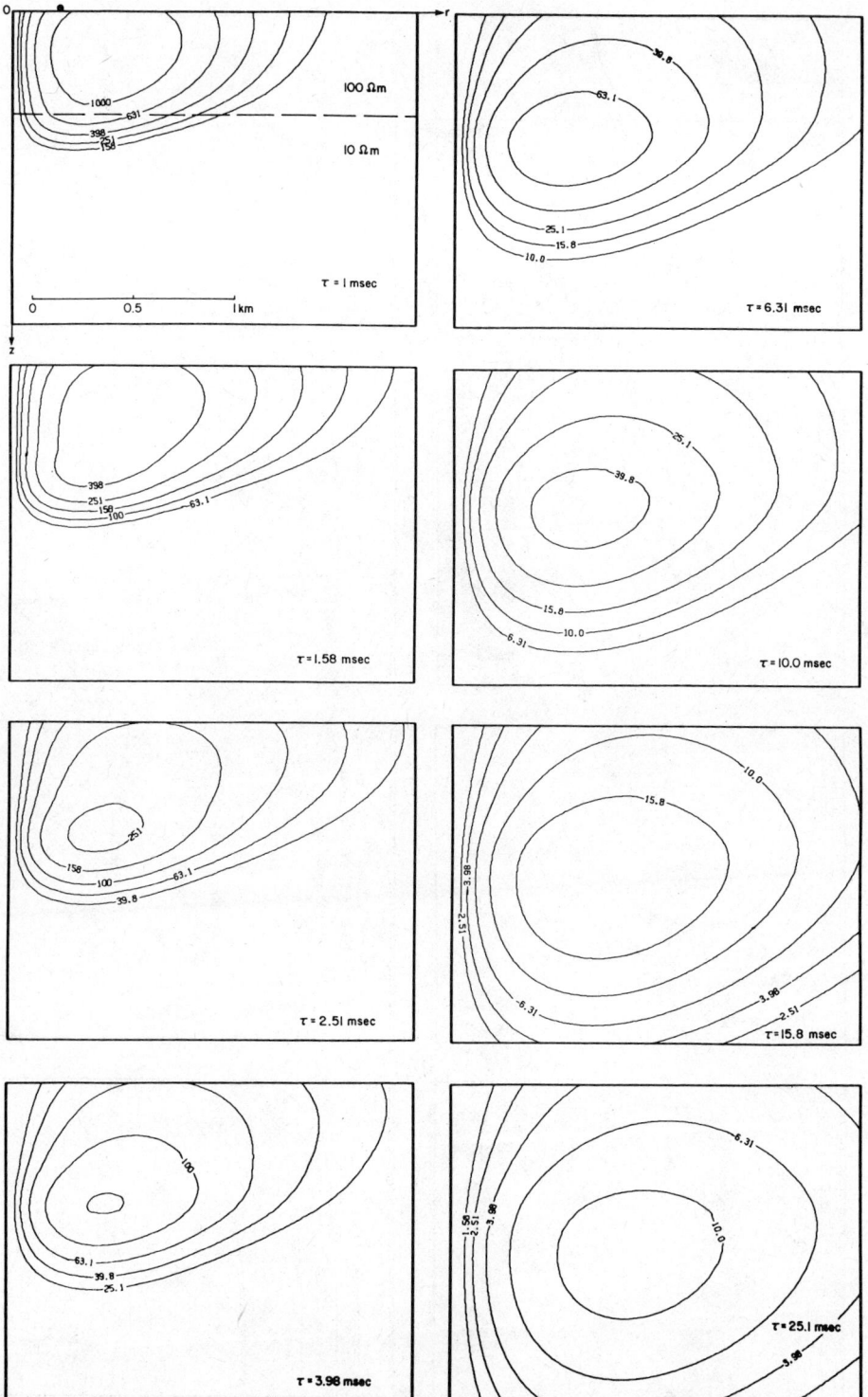

FIGURE 2. The diffusion into a layer over a conductive half-space of the induced azimuthal component of the electric field following a rapid turn-off of a unit current in a horizontal circular loop at the surface of the Earth. The radius of the loop, the thickness of the layer, and the resistivities of the layer and the half-space are 250 m, 500 m, 100 $\Omega \cdot$ m, and 10 $\Omega \cdot$ m, respectively. The logarithmically spaced contours are in units of 10^{-8} V/m.

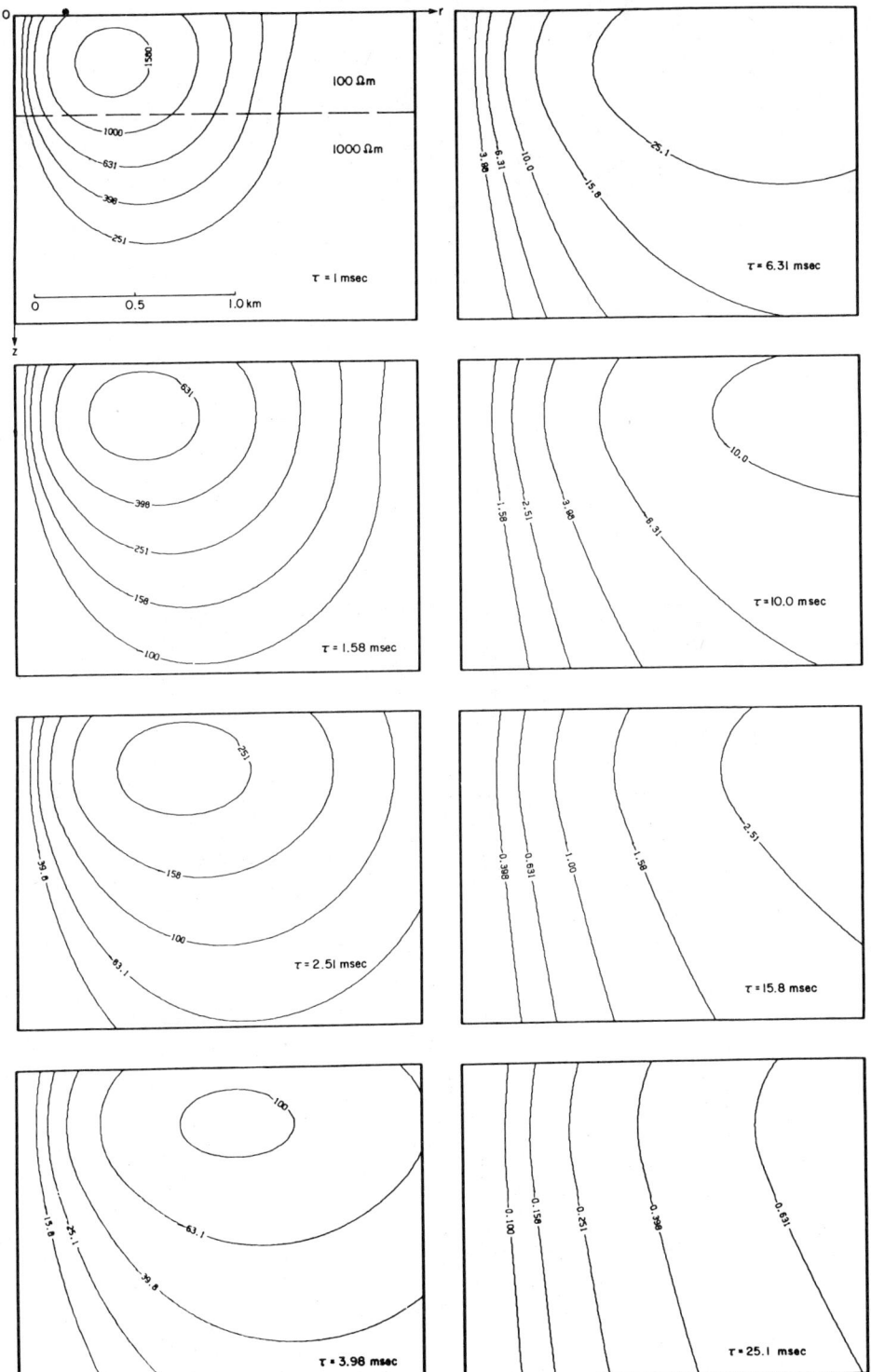

FIGURE 3. The diffusion into a layer over a resistive half-space of the induced azimuthal component of the electric field following a rapid turn-off of a unit current in a horizontal circular loop at the surface of the Earth. The radius of the loop, the thickness of the layer, and the resistivities of the layer and the half-space are 250 m, 500 m, 100 Ω · m, and 1000 Ω · m, respectively. The logarithmically spaced contours are in units of 10^{-8} V/m.

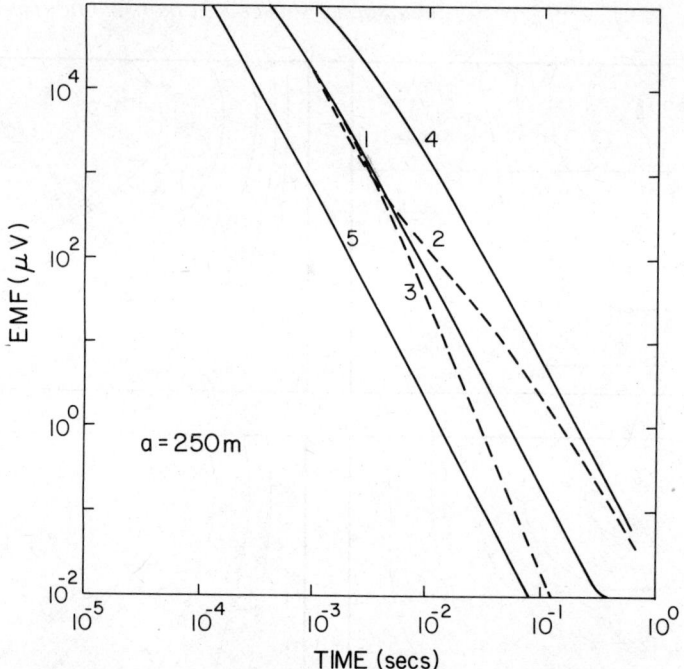

FIGURE 4. The emf induced as a function of time in a horizontal circular loop, radius 250 m, coincident with a second loop following a rapid turn-off of unit current in the latter. The emf depends on the conductivity of the layered Earth. The curves labeled 1, 4, and 5 are for half-space models of resistivity 100, 10, and 1000 $\Omega \cdot$ m, respectively. The curves labeled 2 and 3 are for a 100-$\Omega \cdot$ m layer, thickness 500 m, over 10- and 1000-$\Omega \cdot$ m half-spaces, respectively.

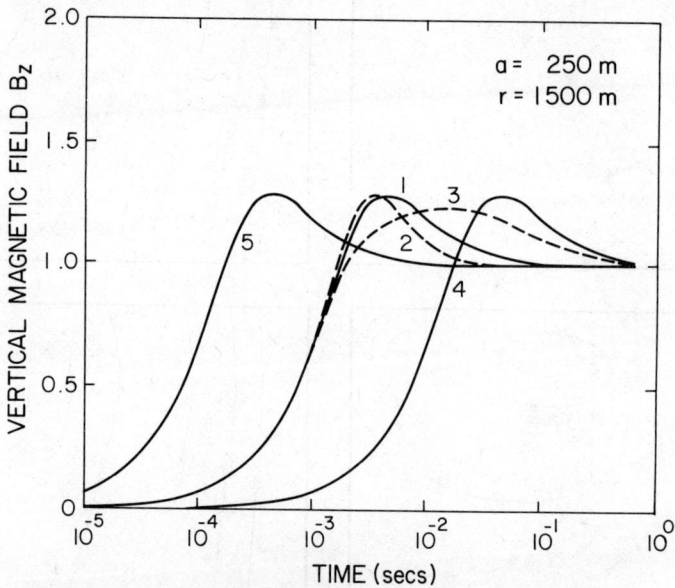

FIGURE 5. The vertical magnetic field B_z as a function of time at a point 1500 m from the center of a loop, radius 250 m, on the surface of the earth following a rapid *turn-on* of unit current in the latter. The field depends on the conductivity of the layered Earth. The curves labeled 1, 4, and 5 are for half-space models of resistivity 100, 10, and 1000 $\Omega \cdot$ m, respectively. The curves labeled 2 and 3 are for a 100-$\Omega \cdot$ m layer, thickness 500 m, over 10- and 1000-$\Omega \cdot$ m half-spaces, respectively.

measured field is small. At a later time, which is model dependent, the ring has passed under the receiver and its field enhances the field of the transmitter, producing the overshoot in the sounding curve.

The behavior of the time derivative of the radial magnetic field at a distance of 400 m from the transmitter, following the more common step-off of the current in the loop, is shown in Fig. 6. An initial negative portion is followed by a positive peak that decays as an inverse cube law of time. The measurement of the horizontal component complements the measurement of the vertical component, and their ratio is a diagnostic of three-dimensional effects.

Before leaving the layered Earth problem, we address the question of uniqueness. Perfect magnetotelluric data uniquely determine the variation in electrical conductivity with depth (see *Magnetotelluric Method: Fundamental Concepts*). The transient technique described is capable in theory of determining exactly the same information. Consequently, the uniqueness theorem holds for this method too. In practical terms, the data are inaccurate and imprecise, and a range of different profiles can be fitted to them within the experimental errors. In particular, the conductivity of intermediate resistive zones is relatively poorly confined. For geological problems that require accurate physical properties of resistive zones to be found, an alternative configuration of the transient method can be employed. If the electric field of a grounded long-wire transmitter is measured in addition to the magnetic field, then an additional function of the thicknesses and conductivities of the layered Earth can be determined. The electric field of the grounded source in (ω, λ) space depends not only on the parameter $Q(\lambda)$ but also on a similar parameter $P(\lambda)$. The parameters P and Q are associated with two independent modes of current flow in the layered Earth, characterized by the absence of a vertical electric field and a vertical magnetic field, respectively. The electric field as a function of frequency is a combination of inverse Hankel transforms involving Q and P.

The function P is defined by the recursion rule

$$P_i = \frac{1}{\rho_i \theta_i} \left[\frac{\rho_i \theta_i P_{i+1} + \tanh(\theta_i d_i)}{1 + \rho_i \theta_i P_{i+1} \tanh(\theta_i d_i)} \right]$$

The recursion relation for P has the same form as the recursion for Q, except the wave number θ is replaced in places by the product $\rho\theta$. The inclusion

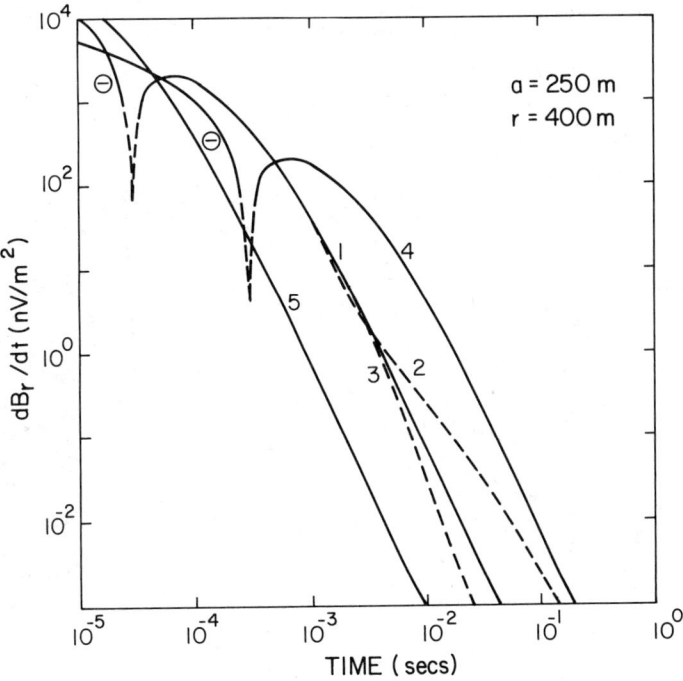

FIGURE 6. The time derivative of the radial magnetic field dB_r/dt as a function of time at a point 400 m from the center of a loop, radius 250 m, on the surface of the Earth following a rapid turn-off of unit current in the latter. The field depends on the conductivity of the layered Earth. The curves labeled 1, 4, and 5 are for half-space models of resistivity 100, 10, and 1000 $\Omega \cdot$ m, respectively. The curves labeled 2 and 3 are for a 100-$\Omega \cdot$ m layer, thickness 500 m, over 10- and 1000-$\Omega \cdot$ m half-spaces, respectively.

of the additional factor involving the resistivity causes the effect on the surface value of P of a resistive zone embedded in a conductive background to be more pronounced than the effect of the same zone on Q.

Three-Dimensional Effects

My choice of a plate for modeling the EM effects of a three-dimensional conductor in a conductive background medium stems from the extensive use of this model as a simple representation of a base metal mineral deposit in a shieldlike environment. The application of the model is certainly not restricted to mining exploration. Very realistic models of large-scale tectonic features can be made up of one or more bent, lamellar substructures.

It is possible to compute accurately, but not inexpensively, the CSEM response of a plate in a conductive environment. The plate and the electric field induced in the background medium interact in a complicated manner. However, the physics of the problem can be split approximately into three parts, and reasonable estimates of the response can be

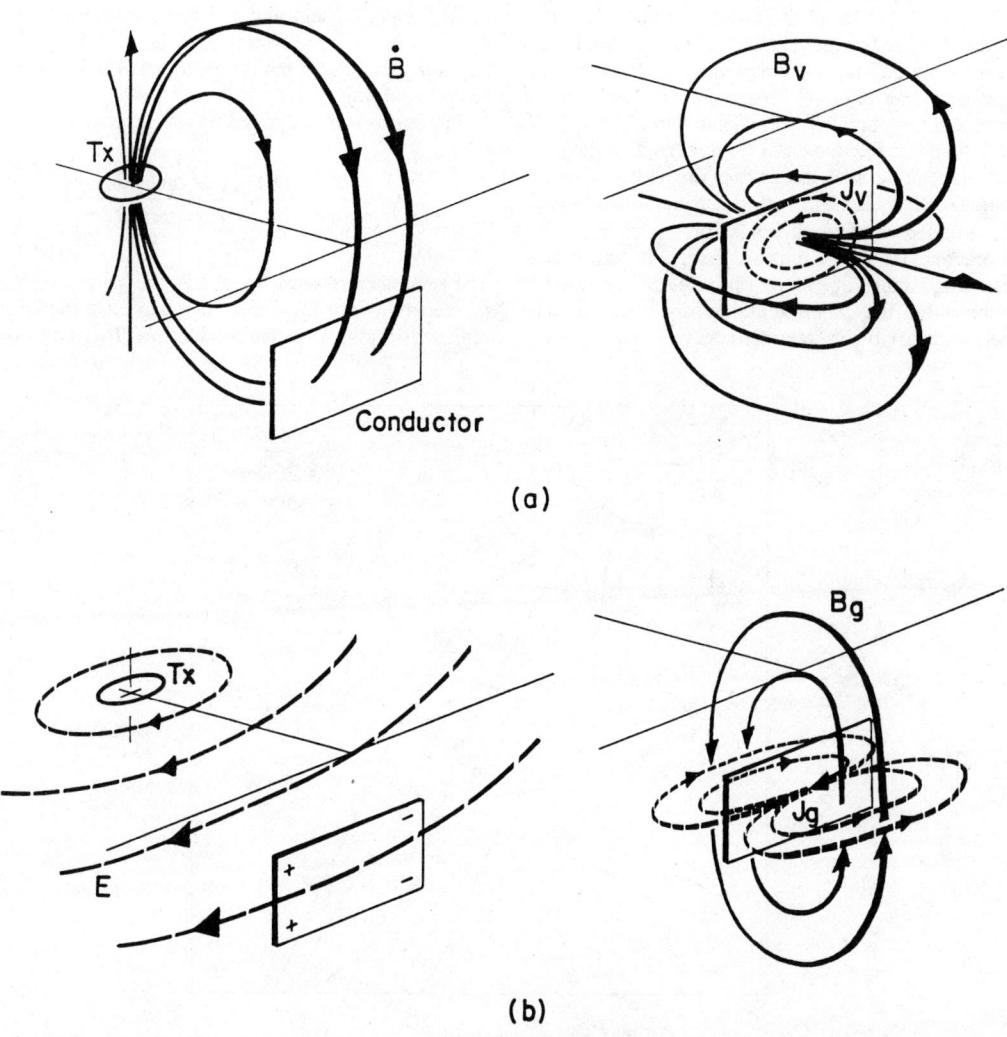

FIGURE 7. *a.* The electromagnetic fields in the vicinity of a conductive plate in a conductive half-space following the rapid turn-off of current in the horizontal loop transmitter (Tx). The broadened impulsive time derivative of the magnetic field dB/dt induces a vortex of current j_v in the plate that decays exponentially with time. The secondary magnetic field B_v produced by the vortex is similar to that of a buried horizontal magnetic dipole. *b.* The electromagnetic fields in the vicinity of a conductive plate in a conductive half-space following the rapid turn-off of current in the horizontal loop transmitter (Tx). The expanding ring of intense electric field is reduced over the plane of the plate by impressed free charge on the ends of the plate. The associated anomalous galvanic current flow j_g produces a magnetic field B_g at the surface of the Earth similar to that of a buried horizontal current dipole.

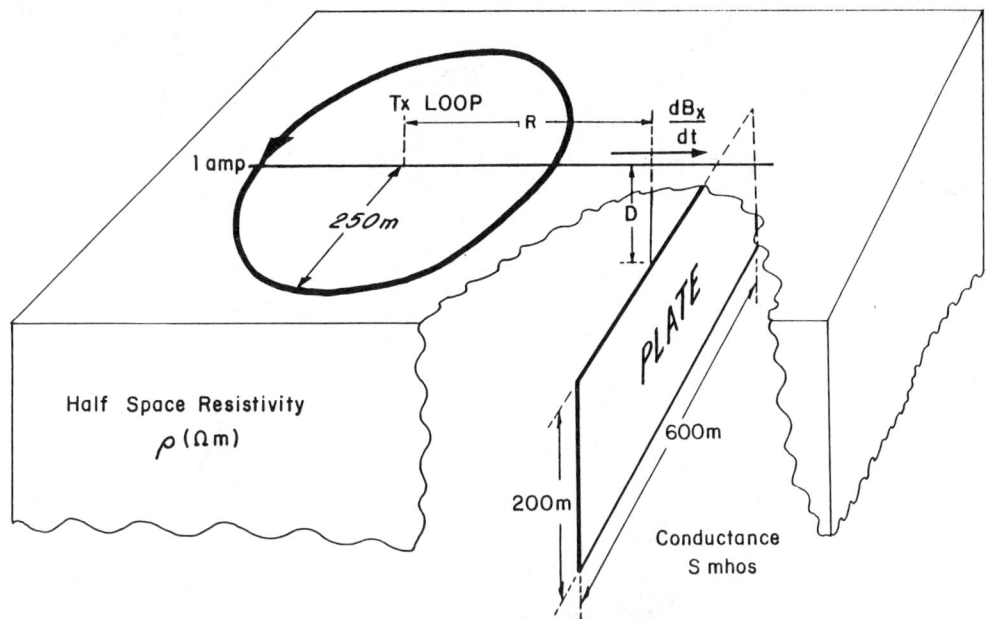

FIGURE 8. The plate in a half-space: definition of parameters.

obtained by adding together the three separate estimates. The three parts to the response are the layered background medium response, which has already been derived, and the plate *vortex* and *channeling* responses, respectively.

The physics of the induction of vortex currents in the plate and the channeling of regional current flow by the plate are best illustrated with a diagram. Following the removal of constant current in the transmitter, the electric field in the background medium in the vicinity of the plate will have both a nonvanishing component of its curl normal to the plate, or equivalently a rate of change of the normal magnetic component, and a nonvanishing component parallel to the plate. The plate will react to minimize the values of both these components in some average sense for all time. To first order, the minimization of the curl is achieved through the induction of a vortex j_v of currents in the plate, as shown in Fig. 7a, and the reduction of the electric field parallel to the plate is achieved by setting up a free-charge distribution near the ends of the plate, as shown in Fig. 7b. The associated perturbation, galvanic, or channeled current flow is denoted by j_g.

The time taken for the electric field induced in the background medium to pass by the plate depends on the conductivity of the background medium. If the medium is relatively resistive, then diffusion time is short and the temporal variation of the electric field is impulsive. The emf driving the induced vortex current flow must be impulsive too. The induced current j_v in the conductive plate can be approximated by the current in a loop of wire of inductance L and resistance R. The loop response of an impulsive source is exponential, so j_v obeys a rule of the type

$$j_v = J_v \exp\left(\frac{-t}{\tau_p}\right) \qquad (5)$$

The parameter τ_p is the principal time constant of the plate and is approximately $\mu aS/4$, where a is the length of the shorter side of the plate and S is the product of the conductivity and the thickness of the plate. The amplitude J_v is a function only of the geometry of the source and the plate for the resistive background medium. If the background medium is not resistive, then the vortex current flow can be obtained approximately by convolving the temporal behavior of the curl of the electric component in the host medium normal to the plate with the exponential form. Then J_v also depends on the conductivity of the host medium.

The currents channeled by the plate at any time are approximately determined by the electric field in the background medium tangential to the plate at that time only. Consequently, if the background medium is very resistive, a channeling response is observed only at very early times when the electric field in the host medium is not vanishingly small. The channeling response is extended to relatively later and later times as the conductivity of the host increases. The size of the channeled current j_g can be found by a static, dc calculation. This current varies with the strength of the applied electric field,

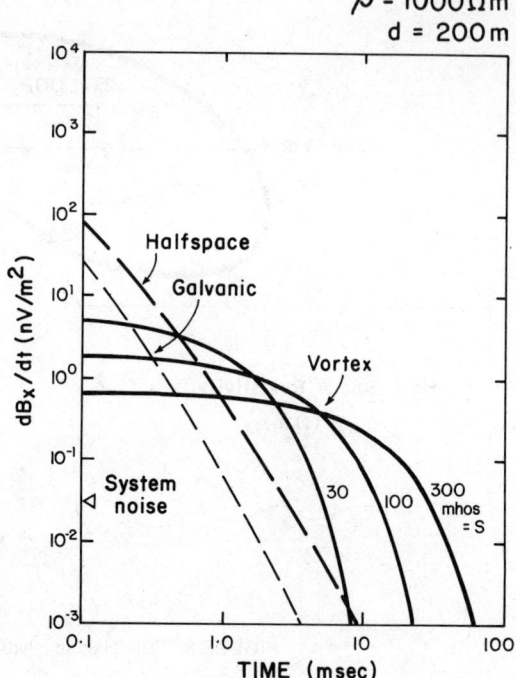

FIGURE 9. The contributions to the time derivative of the horizontal magnetic component dB_x/dt from the half-space, vortex, and galvanic currents plotted as a function of time. The half-space has a resistivity of 1000 $\Omega \cdot$ m. The depth to the top of the plate and the offset of the plate from the center of the transmitter are 200 and 400 m, respectively. The characteristic inductive time constants of the plate are 1, 3, and 10 ms.

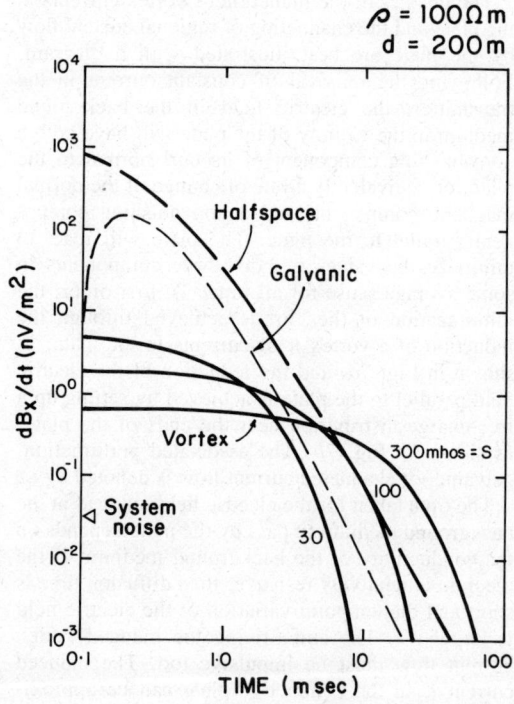

FIGURE 10. The contributions to the time derivative of the horizontal magnetic component dB_x/dt from the half-space, vortex, and galvanic currents plotted as a function of time. The half-space has a resistivity of 100 $\Omega \cdot$ m. The depth to the top of the plate and the offset of the plate from the center of the transmitter are 200 and 400 m, respectively. The characteristic inductive time constants of the plate are 1, 3, and 10 ms.

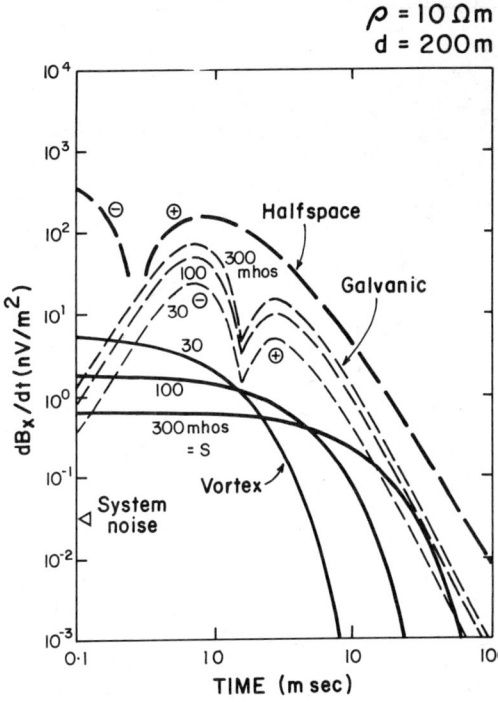

FIGURE 11. The contributions to the time derivative of the horizontal magnetic component dB_x/dt from the half-space, vortex, and galvanic currents plotted as a function of time. The half-space has a resistivity of 10 Ω · m. The depth to the top of the plate and the offset of the plate from the center of the transmitter are 200 and 400 m, respectively. The characteristic inductive time constants of the plate are 1, 3, and 10 ms. The vortex response is valid only at times greater than about 4 ms.

the geometry of the plate, and the ratio of the conductivity of the plate to that of the host medium σ_h. The current j_g is approximately of the form

$$j_g = J_g \left[\frac{\alpha}{1 + \alpha} \right] \quad (6)$$

where α is the dimensionless number $S/\sigma_h a$. The amplitude J_g is a function of time and depends on the geometries of the plate and transmitter and the conductivity of the host medium.

Equations 5 and 6, for the vortex and channeled currents in the plate, are made up of time-dependent amplitude factors and response functions that essentially describe the two physical processes involved in the total plate reaction to an impressed electric field. The electrical conductivity of the plate appears only in the response functions. Consequently, these representations of the current flow have a practical benefit. The response of several plates obtained in a known host medium as a function of time can be fitted to a sum of terms like Eqs. 5 and 6, the response for other plate conductivities then being obtained by interpolation, with a considerable saving of computing effort.

The secondary fields produced by the two types of current flow can also be described approximately in terms of two fundamental secondary sources, a horizontal current dipole parallel to the plate and a magnetic dipole whose axis is normal to the plate. The relative importance to a given survey of the channeled and vortex current flows will depend on the relative size of the EM field generated at the surface of the Earth by the two flows. We know, for example, that the EM field of a current dipole decays more slowly with distance than does the field of a magnetic dipole.

The relative magnitudes of the layered Earth, vortex, and channeling responses are best illustrated by an example from McNeill et al. (1984). The geometry is illustrated in Fig. 8. The transmitter loop has a radius of 250 m. The vertical conductive plate has a strike length of 600 m and a depth extent of 200 m. The depth to the top d is 200 m. The radial distance r of the plate from the center of the transmitter is 400 m. I assume an initial current of 1A is removed at time $t = 0$.

The time derivative of the horizontal magnetic field dB_x/dt immediately over the center of the plate is shown in Figs. 9, 10, and 11 as a function of time for three different half-space host medium resistivities of 1000, 100, and 10 Ω·m, respectively. The half-space response is transcribed from Fig. 6 and was described earlier in this entry. The galvanic or channeling response behaves predictably, its persistence to late time and its relative importance increasing with increasing half-space conductivity. An initial negative portion is followed by a positive peak that decays as an inverse 7/2 power law of time, which is slightly faster than the inverse cube law of the half-space. The crossover in the galvanic response is delayed relative to the crossover in the

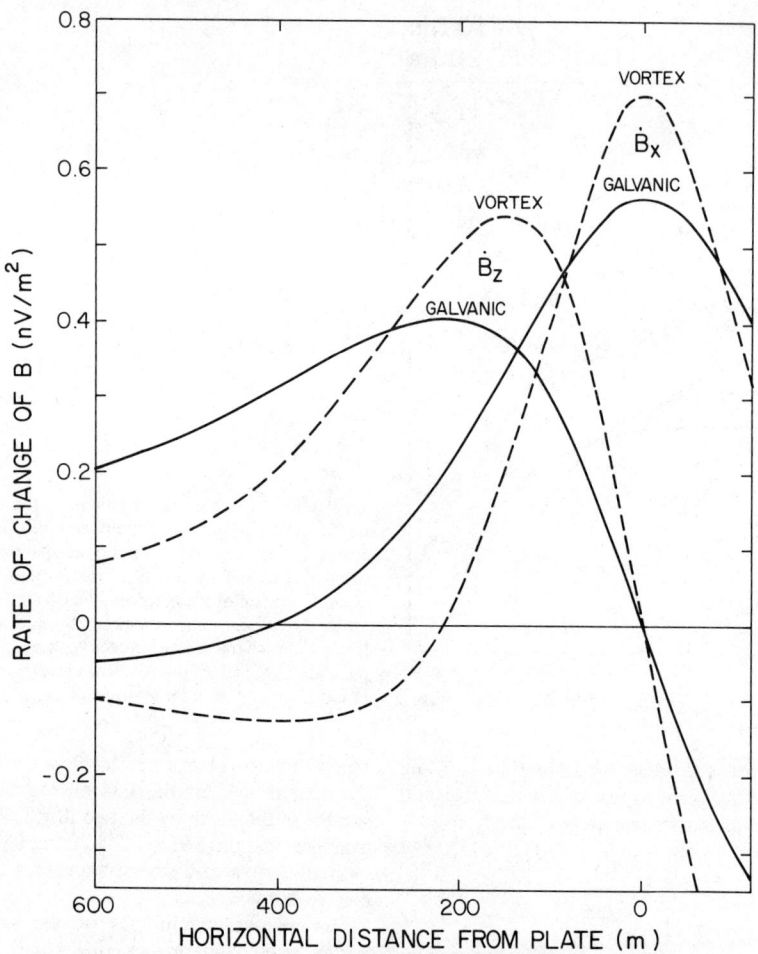

FIGURE 12. The time derivatives of the anomalous magnetic fields dB_v/dt and dB_g/dt caused by the plate along a profile parallel to the x-axis of Fig. 8, which crosses over the center of the plate. The vortex and galvanic contributions resemble the fields produced by buried horizontal magnetic and current dipoles, respectively. The plate conductance, the inductive characteristic time constant, and the half-space resistivity are 100 S, 3 ms, and 100 $\Omega \cdot$ m, respectively. The time of measurement is 3 ms following turn-off of current in the transmitter.

half-space response. The maximum in the y-component of the electric field at the plate occurs at a later time than does the maximum in the horizontal magnetic field at the surface. In the two resistive half-spaces the galvanic component is independent of the conductance S of the plate because the channeling number α is significantly greater than unity for all three values of S considered: 30, 100, and 300 S.

The characteristic exponential behavior of the vortex component, controlled by the conductance of the plate, with time constants 1, 3, and 10 ms, respectively, is clearly very different from the power law behavior of the other two components. The vortex component is dominant if the half-space is a poor conductor. The vortex responses have not been convolved with the temporal variation in the x-component of $\nabla \times \mathbf{E}$, so they are invalid at times earlier than 40 μs, 400 μs, and 4 ms in Figs. 9, 10, and 11, respectively. However, at these times, the relative importance of the vortex component is small.

Figure 12 illustrates the geometrical behavior of the time derivatives of the vertical and horizontal magnetic field components along a profile across the center of the plate. The parameters are listed in the figure caption. The magnetic components are integrals over the distributed vortex and galvanic current flow, but they differ little from the fields generated by buried horizontal magnetic and current point dipoles, respectively. Intuitively, one would not expect negative side lobes in the horizontal field of the current dipole, but they do occur and are due to influence of the surface of the Earth on the dipolar return current flow in the host medium.

R. NIGEL EDWARDS

References

Buselli, G., 1977, SIROTEM: A new portable instrument for multi-channel transient electromagnetic measurements, *Exploration Geophysics* **8**, 82–87.

Chave, A. D., 1983, Numerical integration of related Hankel transforms by quadrature and continued fraction expansion, *Geophysics* **8**, 1671–1686.

Cheesman, S. J., R. N. Edwards, and A. D. Chave, 1987, On the theory of sea floor conductivity mapping using transient EM systems, *Geophysics* **52**, 204–217.

Fullagar, P. K., and D. W. Oldenberg, 1984, Inversion of horizontal loop electromagnetic frequency soundings, *Geophysics* **49**, 150–164.

Gomez-Trevino, E., and R. N. Edwards, 1983, Electromagnetic sounding in the sedimentary basin of southern Ontario—A case history, *Geophysics* **48**, 311–330.

Kaufman, A. A., and G. V. Keller, 1983, *Frequency and Transient Soundings, Methods in Geochemistry and Geophysics*, vol. 16, Amsterdam: Elsevier.

Knight, J. H., and A.P. Raiche, 1982, Transient electromagnetic calculations using the Gaver-Stehfest algorithm, *Geophysics* **47**, 47–50.

Lajoie, J. J., and G. F. West, 1976, The electromagnetic response of a conductive inhomogeneity in a layered earth, *Geophysics* **41**, 1133–1156.

Lee, T., and R. Lewis, 1974, Transient electromagnetic response of a large loop on a layered ground, *Geophysical Prospecting* **22**, 430–444.

Macnae, J. C., and Y. Lamontagne, 1987, Imaging quasi-layered conductive structures by simple processing of transient electromagnetic data, *Geophysics* **52**, 545–556.

McNeill, J. D., R. N. Edwards, and G. M. Levy, 1984, Approximate calculations of the transient electromagnetic response from buried conductors in a conductive half-space, *Geophysics* **49**, 918–924.

Nabighian, M. N., 1979, Quasi-static transient response of a conducting half-space—An approximate representation, *Geophysics* **44**, 1700–1705.

Newman, G. A., G. W. Hohmann, and W. L. Anderson, 1986, Transient electromagnetic response of a three-dimensional body in a layered earth, *Geophysics* **51**, 1608–1627.

Raiche, A. P., D. L. B. Jupp, H. Rutter, and K. Vozoff, 1985, The joint use of coincident loop transient electromagnetic and Schlumberger sounding to resolve layered structures, *Geophysics* **50**, 1618–1627.

Wait, J. R., 1982, *Geoelectromagnetism*. New York: Academic Press.

West, G. F., and R. N. Edwards, 1985, A simple parametric model for the electromagnetic response of an anomalous body in a host medium, *Geophysics* **50**, 2542–2557.

West, G. F., J. C. Macnae, and Y. Lamontagne, 1984, A simple time-domain electromagnetic system measuring the step response of the ground, *Geophysics* **49**, 1010–1026.

Cross-references: *Continental Crustal Structure; Continental Lithosphere; Magnetotelluric Method: Fundamental Concepts; Thin-Skin Tectonics.*

CONTROLLED SOURCE SEISMOLOGY

Controlled source seismology is the study of the Earth's interior by means of artificially generated seismic waves. The most common types of artificial (controlled) sources include explosions, ground vibrators, airguns, and weight drops. While the term *controlled source seismology* encompasses all seismology in which artificial sources of seismic energy are used, in practice it is generally used to connote academic (as opposed to exploration or engineering) research. We adopt that meaning for this article, although of course the methodologies of controlled source seismology, even those employed in studies of deep structure, substantially overlap the methods of exploration seismology (see *Exploration Seismology*).

Artificial sources produce dominantly compressional wave energy, and until quite recently controlled source studies were limited almost exclusively to compressional velocity structures in the earth. The relatively small size of artificial sources compared to earthquakes has also meant that controlled source studies have been confined to the shallower parts of the earth. It is not surprising, therefore, that the major contributions of controlled source seismology have been to the understanding of the Earth's crust and uppermost mantle; in fact, controlled source seismology is virtually synonymous with seismic crustal studies. Because controlled sources generate high frequencies (observations are typically above 1 Hz), controlled source techniques are usually considered to have substantially higher resolving power than do other seismic methods used to study the crust and uppermost mantle (see Nolet, 1984, however, for a cautionary view of that statement).

Traditional methods for controlled source studies involve analysis of travel-time information from refractions and wide-angle reflections recorded along profiles hundreds of kilometers long. Signals are generated by large explosions whose location at the Earth's surface, time of detonation, and size are precisely known. In recent years, near-vertical deep reflection profiling, based in large part upon seismic exploration techniques, has come into wide use. Seismic sources for reflection profiling include large truck-mounted ground vibrators (Vibroseis) that are typically configured in groups of several trucks, and very large volume airguns or airgun arrays (in water), as well as explosions.

Refraction/wide-angle reflection and vertical-incidence reflection methods provide complementary information. The former yield information about velocity structure, both vertically and horizontally along the line of observations. The latter give detailed information about discontinuities beneath the line of sources and receivers. To capitalize on both kinds of information, many experiments today are designed to cover a wide range of apertures, from near-vertical to wide-angle, over the same reflection points. (Wide-angle reflections differ in important ways from near-vertical reflections. Wide-angle reflections exhibit large amplitudes at ranges near those expected for total reflection from an abrupt discontinuity; however, the test is not a good one for determining fine structure because many models with gradational velocity changes will produce focused refractions with times, ranges, and amplitudes that differ only in subtle ways from the simple reflection case.)

The evolution of controlled source seismology is closely linked to advances in digital instrumentation, computer processing, analytical techniques, and experiment design. Various aspects of these advances are outlined in the sections below, although the discussion is necessarily limited. Further review of pertinent seismic processing and analysis is given in *Seismic Imaging; Seismic Tomography; Seismic Signal Processing; Seismograms: Synthetic; Finite-Difference Forward Modeling in Seismology;* and *Exploration Seismology*. Additional results from controlled source studies are contained in other sections in this volume, including *Deep Seismic-Reflection Profiling; Continental Crustal Structure; Lithosphere, Oceanic: Formation and Evolution; Ocean-Continent Transition: Structure;* and *Mantle Discontinuities*.

Refraction Seismology

History. Most of what is known about the velocity structure of the crust and uppermost mantle has been gained from seismic refraction studies. A review of early work is given in James and Steinhart (1966), and only a brief summary is included here. Modern controlled source seismology in the United States was initiated in 1947 by Merle Tuve and his co-workers at the Carnegie Institution of Washington. Similar efforts were begun in Japan, Europe, and Russia at about the same time.

An important early result of explosion studies was the recognition of large-amplitude events identified as critical or postcritical reflections from the M discontinuity. For a simple two layer case in which the lower layer has the higher velocity, the critical angle is defined by

$$\theta_{cr} = \sin^{-1}(\alpha_1/\alpha_2)$$

where $\alpha_1 < \alpha_2$ and θ is measured from the vertical.

Refractions from the lower (high velocity) layer will first appear on the seismic record just at the critical distance (see Fig. 1). Reflection amplitudes are typically very large at critical distances. Postcritical reflections used jointly with refraction arrivals form a double criterion for the identification and depth determination of discontinuities. It was on the basis of disagreement between the results of critical reflections and those for refraction measurements that it was first demonstrated that velocities must increase in the lower crust.

Until the mid-1960s, most controlled source seismology consisted of long-range refraction profiling over observing lines hundreds of kilometers in length. Typical energy sources were large explosions, up to 5 or 10 tons or more, detonated in bodies of water or in boreholes spaced widely along the observing profiles and at their ends. Where possible, shots were laid out to achieve "reversed" profiles, such that travel-time observations were made in both directions along the line, a procedure that allowed dip angles of refracting interfaces to be determined. In Europe and the United States, receiver spacings tended to be large, 10 km or more, and source spacings were even greater, commonly 100 km or more. Typical experiments involved a relatively small number of recording instruments. Several shots would be fired at a fixed shot point and the instruments deployed and redeployed along the observing line after each shot until coverage was complete. Analysis was generally based on travel-times of the first arrivals and selected large-amplitude secondary arrivals that could be interpreted in terms of homogeneous layers with uniformly dipping interfaces. The resulting highly simplified two or three layer models provided a measure of mean velocities in the crust and upper mantle, but were too crude to resolve horizontal crustal structures at scales of less than several tens of kilometers. Even major crustal features remained difficult to resolve. Only rarely, for example, were low velocity zones in the crust identified with any degree of reliability, and then only when shadow zones were observed in first arrival data. As lateral discontinuities in structure along the observing profile can produce offsets in travel-time curves that mimic shadow zones, the question of low velocity regions in the crust remained highly controversial.

In startling contrast to the very sparse arrays of refraction profiling in the United States were the high density observing arrays of the deep seismic sounding method (DSS), first used extensively in the Soviet Union in the 1950s. The DSS method is based on a phase correlation technique developed by G. A. Gamburtsev and his co-workers around 1940. The technique specifies seismometer spacings of 100–200 m along spreads up to about 100 km in length, with gaps in spreads of 10–30 km. Explosions that were detonated at several points in-line with the receivers yielded overlapping and reversed profiles on which individual phases could be traced with great reliability (see James and Steinhart, 1966, for further details). The data were further refined by separating waves on the basis of differing apparent velocities (velocity filtering). (Apparent velocity is simply the maximum velocity with which a seismic wave *appears* to travel through an array of stations.) By velocity filtering, near-vertical reflections could be

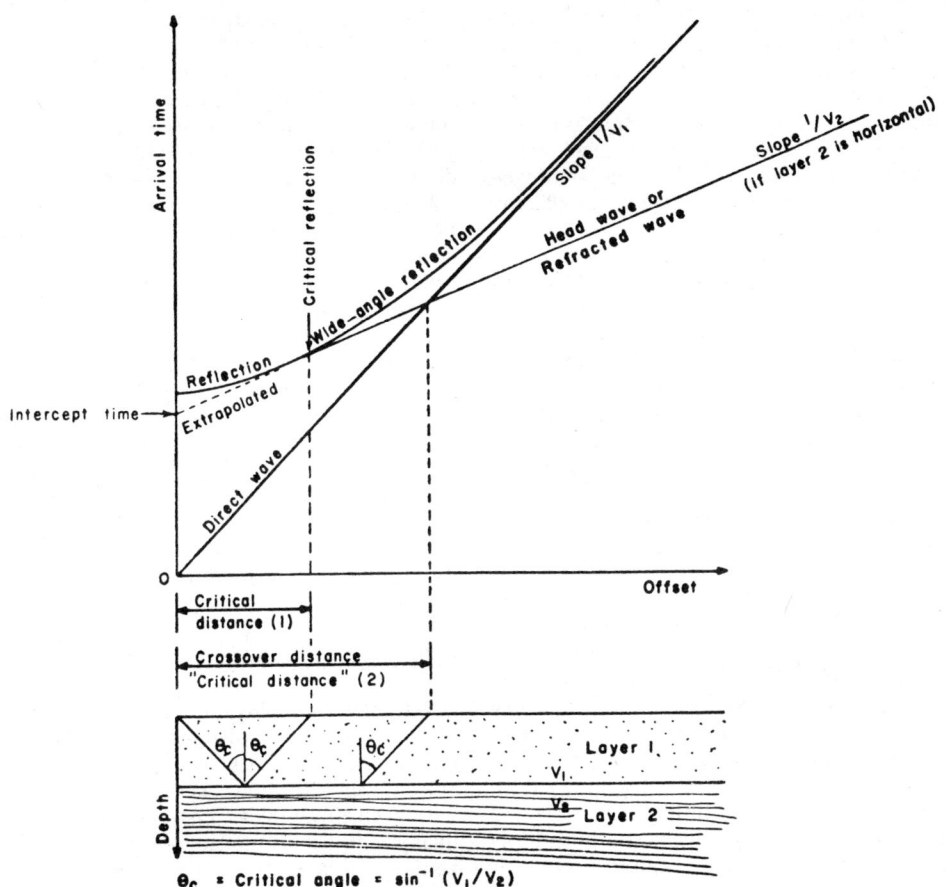

FIGURE 1. Two layer medium showing geometry of critical reflection. (From R. E. Sheriff, 1973, *Encyclopedic Dictionary of Exploration Geophysics*, p. 41, with permission.)

identified and associated with specific discontinuities. The crustal models produced from the DSS data in the USSR were the most detailed of their time, and the DSS method remains one of the most important techniques of controlled source seismology.

Controlled source refraction studies in the United States reached their apogee in the 1960s when the Vela Uniform program, funded by the Department of Defense, resulted in many thousands of kilometers of refraction surveys in the United States. Field methods had not changed much from the earliest surveys, however, so that crustal models, while greatly increased in number, remained highly simplified. Station spacings even through the 1970s remained nearly constant at about 5 km and shot point spacings at 100 km. Field techniques on land typically involved a few shot points and a moving station array of relatively few stations. In some instances a fixed array-moving shot point technique was employed in which explosions were set off from a ship at regular intervals along the continuation of a line of fixed seismic stations, thus allowing apparent velocities to be determined along both the line of stations and the line of shots. The method was limited, however, by the requirement of a large body of water suitably located.

The revitalization of controlled source seismology in recent years can be traced directly to the advent of microprocessor-based instrumentation, advanced digital data processing, and powerful new methods of wavefield analysis. Until quite recently, however, most of these advances had remained in the domain of reflection seismology, where modern instrumentation and analytical techniques allow imaging even of comparatively complex three-dimensional structures.

Modern Refraction Seismology. Refraction profiling, except at sea, still involves almost exclusively the use of explosive sources. Large volume airgun arrays have been used with some success in onshore-offshore experiments to ranges of hundreds of kilometers, and several attempts have been made to use Vibroseis over comparable distances, but to date these nonexplosive sources appear to be of only limited use for long-range profiling. That Vibroseis and large volume airgun arrays can be considered at all for long range profiling is a tribute to modern digital data processing. Repetitive Vibroseis or airgun signals can be recorded repeatedly for the same source/receiver geometry and stacked (summed) to increase signal/noise ratios to levels approaching those obtainable from small explosions. Despite advances in these areas, however, explosions remain the most common and effective source for refraction profiling.

Because the number of shot sites is nearly always limited by logistics, local politics, or budgetary restrictions, the greatest efforts toward improving methods of data acquisition and enlarging the quantity of data have been devoted to increasing the density of the station array. Digital seismographs (or analog systems that produce records suitable for routine digitization) may be deployed in large numbers (hundreds or more) in dense arrays of relatively closely spaced instruments. In most routine refraction profiling today, station spacings are less than about 1 km. Seismographs are equipped with recording systems and are synchronized through radio telemetry or programmed turn-on to record automatically within a prescribed time window.

The data collected in modern refraction surveys typically are analyzed via a three-step forward modeling process (e.g., Meissner, 1986). The first step, which may simply draw on results of previous work, is to obtain an approximate one-dimensional velocity-depth structure to serve as a starting model. The next two steps—two-dimensional ray tracing and computation of synthetic seismograms—are typically done in tandem and iterated until the synthetic seismograms give a satisfactory match to the observed waveform data. The purpose of the analysis is to exploit the entire seismic record, so that arrival times of later phases (particularly wide-angle reflections) *and their waveforms* become an integral part of the interpretation. An example of a typical refraction result in the United States is shown in Fig. 2.

A somewhat different approach is taken to the interpretation of data in the Soviet Union. There, long range DSS profiling routinely involves nonlinear velocity filtering to identify wide-angle reflections. By that technique, seismograms from closely spaced receivers deployed over relatively short segments of the observing profile are appropriately time-shifted and linearly stacked for discrete apparent velocities (stacking velocities) that cover the range of values expected for wide-angle reflections. Next, the signal-to-noise ratio is measured for each stacking velocity, the best-fit apparent velocities obtained, and the corresponding stacked seismograms calculated. Those data are then frequency filtered to produce a final series of seismic traces on which wide-angle reflections can be traced. An example of the resulting seismic records is given in Fig. 3, which shows a velocity-filtered P-wave seismic section from an experiment in Siberia (Egorkin et al., 1987).

The studies summarized by Egorkin et al. and similar studies in the Soviet Union represent a unique class of controlled source experiments that are duplicated nowhere else in the world. The explosive sources, at least until the early 1980s, were nuclear, with yields sufficient to transmit strong signals to distances of 3000 km or more (V. Ryaboy, pers. comm.). The profiles were reversed and overlapping. The experiments have provided a remarkably detailed description of the upper mantle, particularly beneath the Siberian platform and adjacent regions. Of great importance has been the discovery that the upper mantle may consist of several low velocity zones (see Fig. 4) and that lateral velocity contrasts beneath continents extend to depths of at least 300 km in the mantle. The significance of these results in terms of lithospheric studies is very great indeed, for they imply that large contrasts in composition or temperature or both must exist between regions of the Earth to depths of 300 km or more, far greater than commonly accepted depths for the base of the continental lithosphere (see *Continental Lithosphere*).

Reflection Seismology

Reflection seismology is the study of discontinuities. When one asks what is the nature of a discontinuity, one is really asking what the behavior of the velocity-depth function is in the vicinity of a discontinuity. One feature of a first-order velocity discontinuity is that it will reflect seismic waves at near-vertical incidence. The resolution that can be obtained depends on the wavelengths of the seismic waves observed, and most observations today emphasize frequencies of 8–25 Hz, although deep profiling experiments may utilize lower frequencies to improve depth penetration.

Deep reflection profiling has been adapted from seismic exploration methods, so it is not surprising that the first convincing evidence of deep crustal reflections came from the application of standard reflection techniques to the study of deep structure. Among the best of the early results were those of Dix (1965), who applied rigorous reflection profiling procedures in a study of deep structure in the Mojave Desert of California. He found a group of fairly strong and consistent reflections, which he interpreted to be from the M discontinuity. Subsequent deep reflection work was confined mostly to Europe until about the mid-1970s when large-scale programs such as COCORP (Consortium for Continental Reflection Profiling) were launched in the United States and elsewhere to capitalize on advanced exploration methods (e.g. Dohr and Meissner, 1975).

Deep reflection profiling has contributed enormously to bridging the gap between seismology and geology by producing deep crustal images of sufficient detail and at high enough resolution that

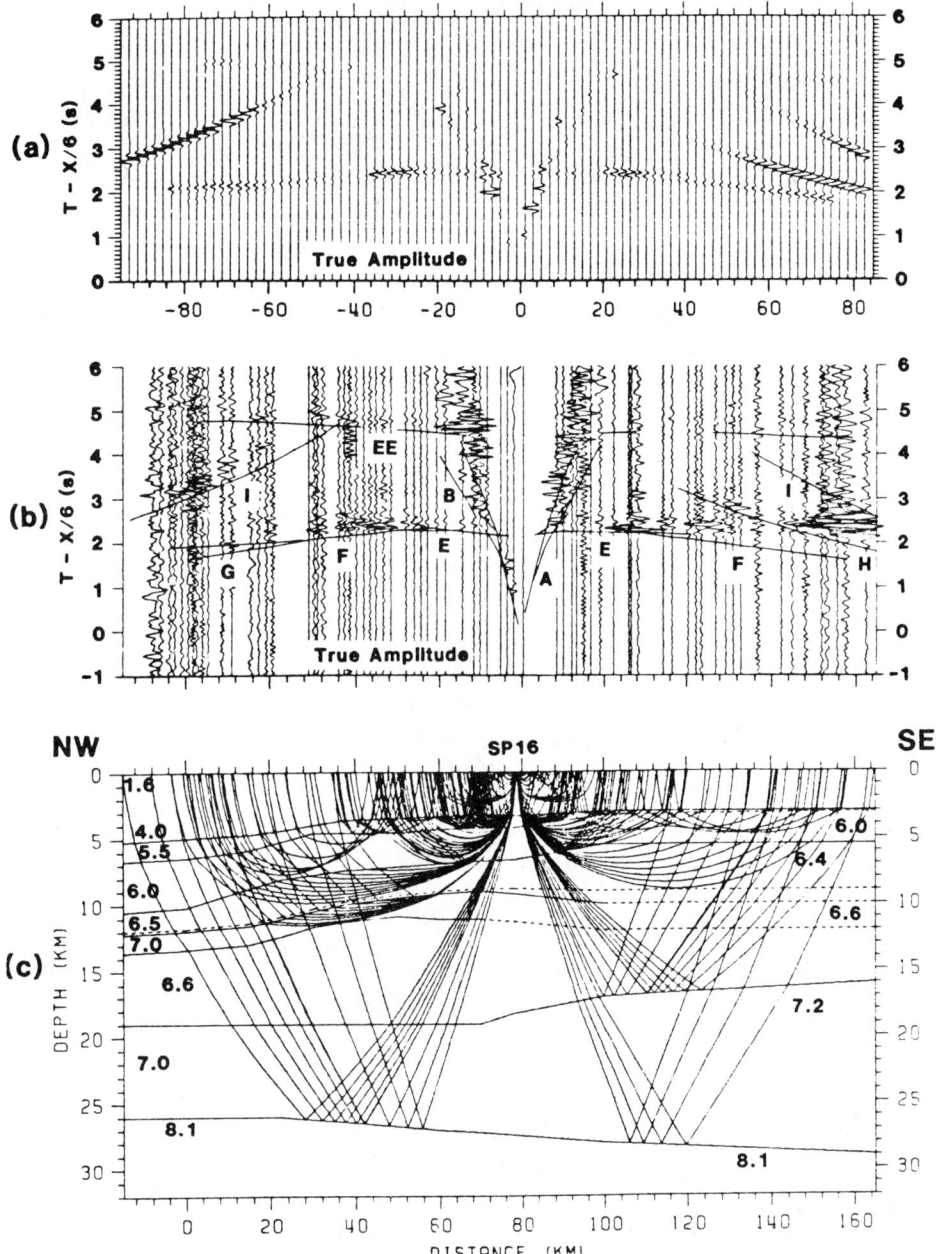

FIGURE 2. Example of modern refraction data, with synthetic seismogram and ray tracing analysis. The results are for the Great Valley of California. (From Holbrook and Mooney, 1987)

they can be related to geologic structure. Deep reflection profiling has played a major role in resolving a number of long-standing problems in geotectonics (see Barazangi and Brown, eds., 1986a, 1986b). In recent years, deep reflection profiling has become a more important tool for marine studies of continental structure, where advanced array technology involving extremely large volume airguns has made feasible the study of deep crustal and upper mantle structures (see *Deep Seismic-Reflection Profiling*). Seismic reflection and refraction profiling of the deep ocean basins is discussed in *Lithosphere, Oceanic: Formation and Evolution*.

Methods employed in deep reflection profiling on land are typified by COCORP, where station and source spacings are commonly about 100 m. The

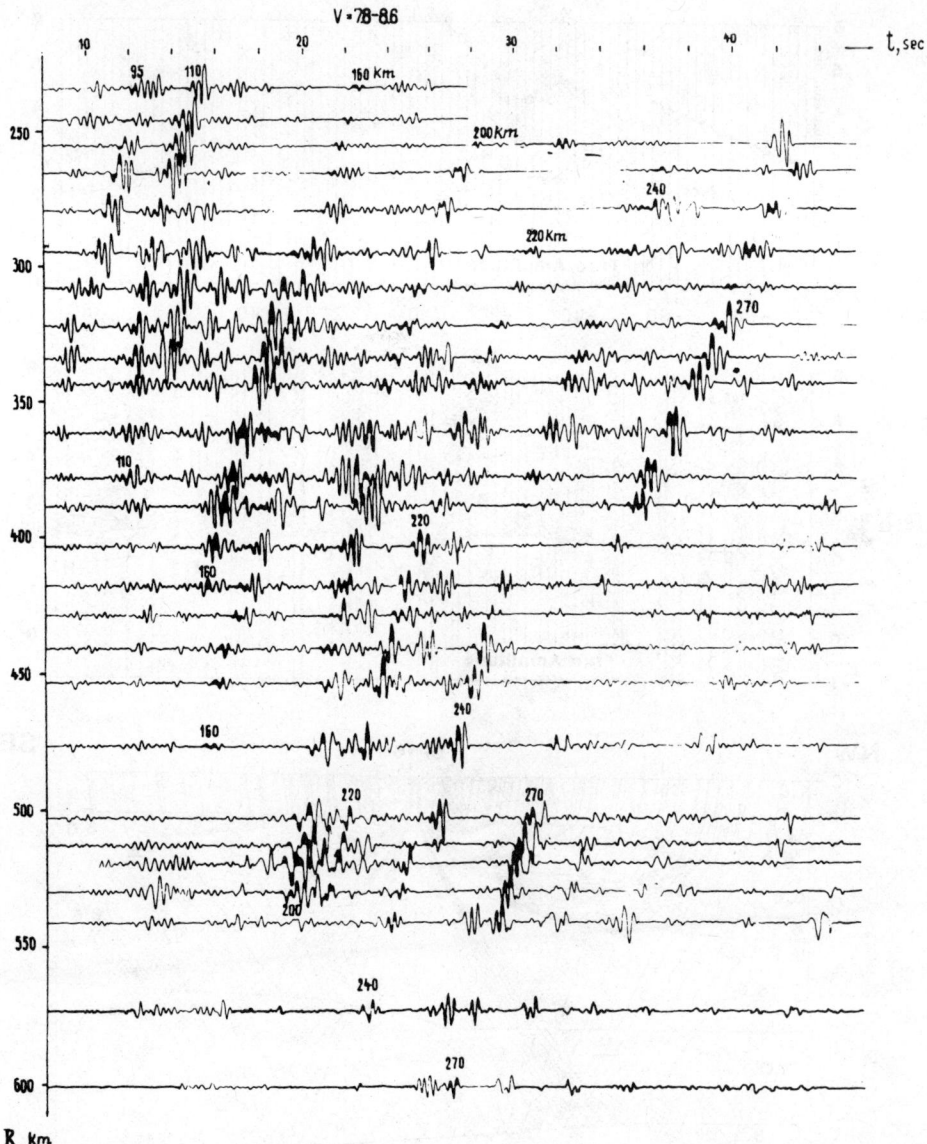

FIGURE 3. Processed wide-angle reflection seismograms for very long refraction profiles across the Siberian platform in the USSR. The records have been frequency and nonlinear velocity filtered for the band of apparent velocities 7.8–8.6 km/sec. (From Egorkin et al., 1987, with permission.)

source consists of five truck-mounted ground vibrators, configured in an array that can be fine-tuned to specific site conditions (Brown, 1986). The recording system (as of the mid-1980s) consists of 96 channels. Sources and receivers are normally deployed according to the requirements of the common midpoint (CMP) method that is used throughout the exploration industry (see *Exploration Seismology*). When the spacing between source and receiver sites is the same, the 96 channel system yields a 48-fold CMP stack. By this we mean that 48 individual source-receiver pairs will share the same (common) midpoint, geographically. These so-called common midpoint gathers can, after appropriate corrections for near-surface geology and other effects, be stacked (summed) to form a single *zero-offset* trace, where a zero-offset trace is the hypothetical record that would be obtained for a coincident source/receiver pair located at the common midpoint. Reflection time sections are typically made up of zero-offset traces. COCORP data processing is shown diagramatically in Fig. 5. COCORP data are

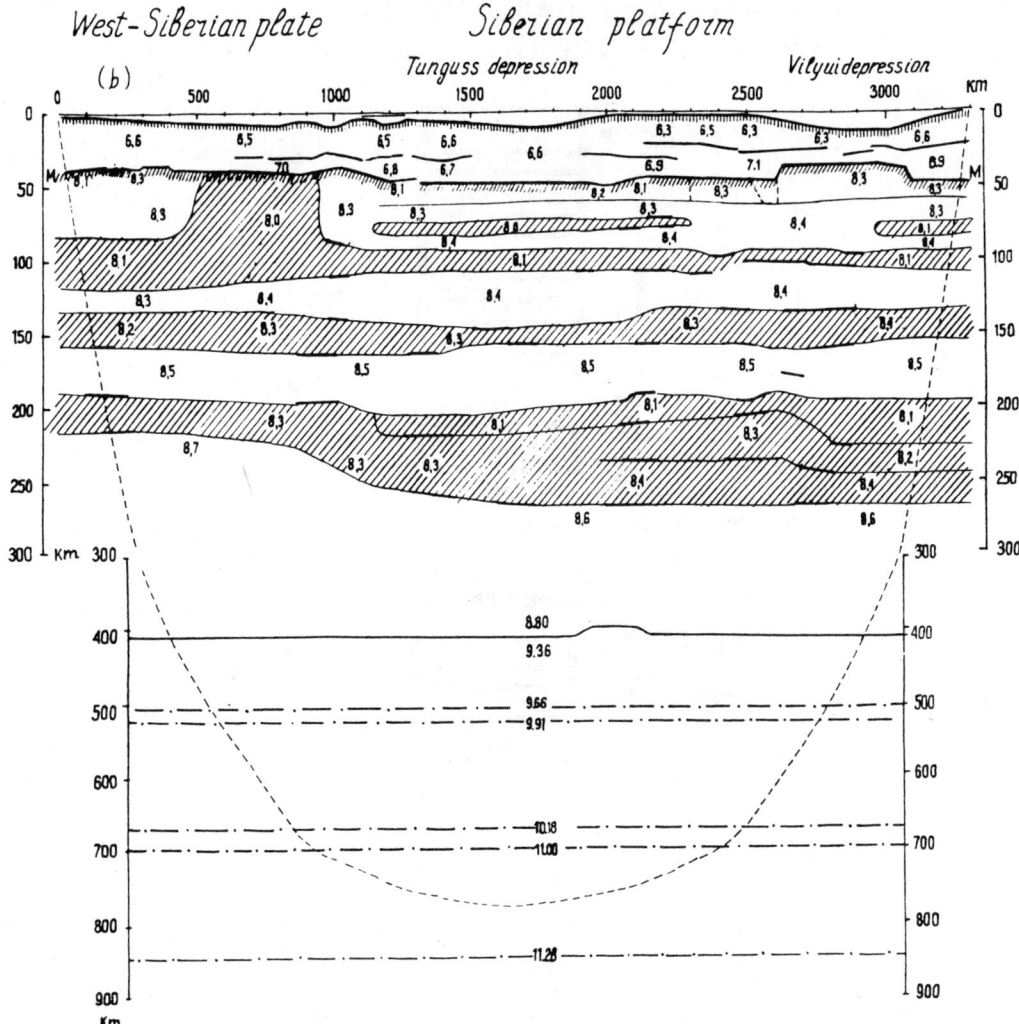

FIGURE 4. Deep structure beneath the Siberian platform from very long refraction profiles in the USSR. Structure based on seismograms in Fig. 3 and other data. Reliable structural interpretations are possible to depths of about 300 km, and maximum ray penetration is nearly 800 km. Note that lateral velocity contrasts are apparent to depths of nearly 400 km. (From Egorkin et al., 1987, with permission.)

generally displayed as unmigrated zero-offset time sections. It is particularly difficult in deep reflection work to migrate *time sections* to *depth sections* (see *Seismic Wavefield Migration*) because of a lack of velocity information at depth; the length of the CMP gathers is not sufficient to allow accurate determination of deep crustal velocities necessary for reliable depth migration. Unmigrated time sections, while giving the appearance of structural images, are distorted to varying degrees by dips and diffractions. For a more complete discussion of data acquisition and processing in reflection profiling refer to *Exploration Seismology; Deep Seismic-Reflection Profiling;* and *Seismic Signal Processing*.

Among the innovations of recent years are the *sign bit* recording systems consisting of 1000 channels or more. The large number of channels is made possible by virtue of the fact that only the sign of ground motion (positive or negative) is recorded for each data sample. The very large number of channels makes possible a substantially increased aperture of the recording spread while retaining a small receiver interval. On the other hand, the usefulness of the data is limited by the absence of amplitude information.

One of the most remarkable general results to emerge from deep reflection profiling is that the upper half of the crust appears to be "transparent" (few reflectors), while the lower part of the crust commonly appears to be highly laminated (Fig. 6,

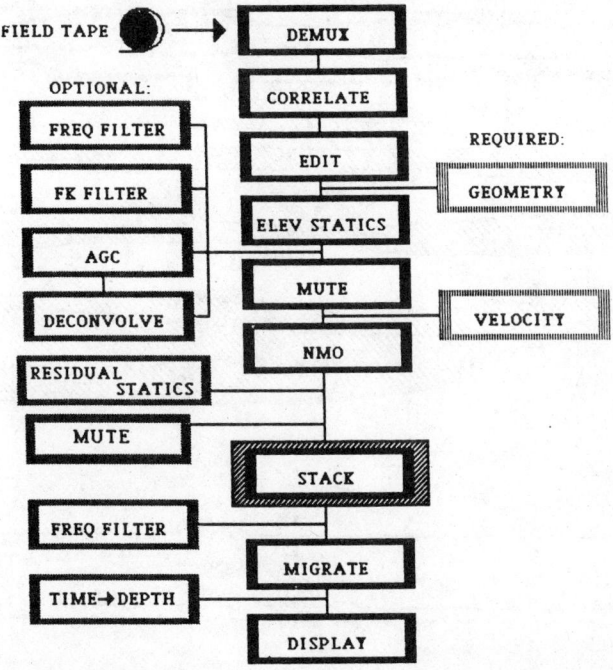

FIGURE 5. Schematic flow diagram for COCORP data processing. (From Brown, 1986)

also review by Mooney and Brocker, 1987). The laminated nature of lower crust includes the M discontinuity itself which typically appears as a cluster of closely-spaced reflecting horizons, nearly always subhorizontal and below which reflectors virtually disappear. A comprehensive summary of results from deep reflection research is given in Barazangi and Brown, 1986a, 1986b.

The enormous success of deep reflection profiling sometimes tends to mask the shortcomings and fundamental limitations of the method. The exploration techniques from which deep profiling has been adapted were developed to explore layered strata in sedimentary basins. In such environments, the CMP methods now in almost universal use provide a powerful means of enhancing the signal-to-noise ratio, while damping out unwanted effects such as multiples reflections. In complex geologic terranes, however, where the attitude of reflectors may be at any angle to the vertical, and layering of a sedimentary type virtually nonexistent, it is not clear that CMP stacking, particularly over short spreads, will result in reflection sections that are in any way an accurate "image" of the crustal structure. The problem is one of spatial aliasing (see section on Miscellaneous Aspects), which can be overcome only by densifying the observing array. It is not surprising, therefore, that the greatest successes of deep reflection profiling have been in mapping low- to medium-angle faults (the surfaces of which are typically good reflectors) and in mapping flat-lying lower crustal laminations, including the M discontinuity. Reflecting bodies in the crust with high-angle boundaries may be missed entirely in the reflection section, particularly for short recording spreads where reflected energy may return to the surface outside the bounds of the receiver array. A further problem in complex terranes is the presence of side reflections, which originate out of the plane of the profile, and which can cause false images on the seismic record section or interfere with in-plane reflections.

Miscellaneous Aspects

Among the important recent developments in experiment design is the simultaneous recording of near-vertical to wide-angle reflections in the same experiment. By this method, reflection points in the earth are sampled over a wide range of angles of incidence, providing detailed information about the fine structure of the underlying discontinuities and the velocity contrasts (including shear velocity contrasts) across them. A recent study in the Ouachita Mountains of the United States by G.

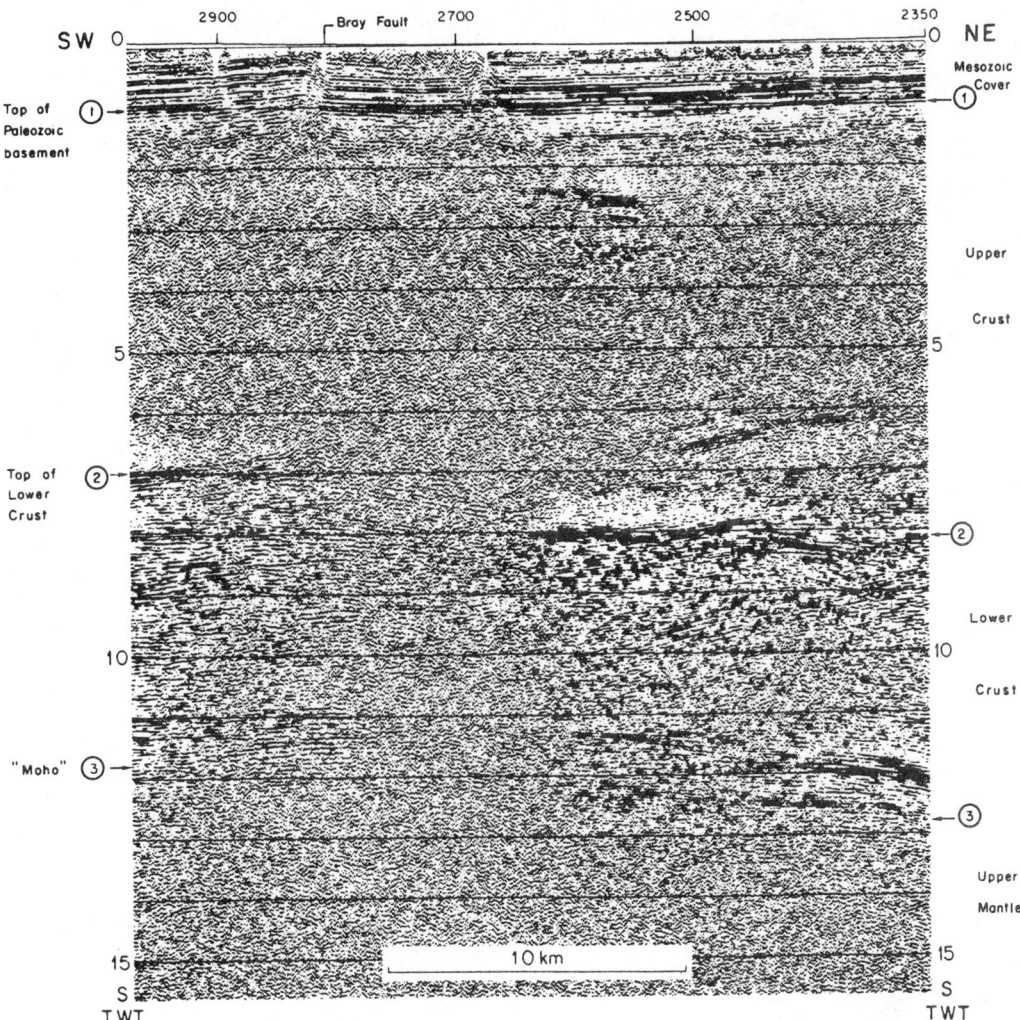

FIGURE 6. Deep seismic reflection cross section showing evidence for a transparent upper crust (excluding sedimentary layers) and layered lower crust. (From an ECORS profile for northern France, Bois et al., 1986)

McMechan and colleagues involved 400 stand-alone seismic group recorders deployed in 100-km-long arrays along which near-vertical and wide-angle reflections could be observed from the same deep interfaces. A method similar in intent to the wide aperture technique is *expanding spread profiling,* where receiver arrays are stepped out from the source point to distances several times the length of the receiver array. Both the wide aperture and expanding spread methods are aimed at accurate velocity measurements of rocks in the deep crust, as well as high resolution reflection imaging. The results provide a more accurate picture of the crust than would be possible with either refraction or reflection surveying alone.

Seismic tomography has been used in novel ways for controlled source studies (see also *Seismic Tomography*). One study, applied to a volcanic structure in France, involved firing shots in a ring around the region of interest and measuring P-wave delay times of upgoing reflections from the M discontinuity that passed through the structure of interest. The method is strictly limited, however, to areas where a good planar reflector exists beneath the region of study. Another innovative tomographic method has been suggested by Pavlis (1986) for imaging lateral variations in seismic velocity. The method involves sources and receivers deployed in a circle around the region of interest and redeployed a sufficient number of times that a high density of ray path cross fire is obtained through the volume. Refraction and reflection lines may be run along diameters of the circle to provide control on the tomographic results.

A great deal of current research is devoted to detailed seismic interpretation of laterally heterogeneous media. Such studies have been made possible by significant improvements both in two- and three-dimensional ray tracing (where the scale of heterogeneities may be as small as a wavelength) and in the generation of more accurate synthetic seismograms (see *Seismic Ray Theory* and *Seismograms: Synthetic*).

Some important developments have been made in the area of wavefield imaging of refraction profiles, where the wavefield itself is directly inverted to obtain subsurface velocity structure. These wavefield inversion methods remain of limited practical value, however, because they require spatially unaliased wavefield data. *Spatial aliasing* is conceptually identical to frequency aliasing of digital time series, where energy at frequencies above the Nyquist frequency (one wavelength equals exactly two sample intervals) will "fold" into spurious low frequency energy. Thus, signals sampled at a rate of less than two samples per wavelength will appear on the replayed digital record as spurious low frequencies. By the same token, spatial aliasing will occur whenever the spacing between receivers is greater than one-half wavelength at the seismic wavelengths of interest, typically 500 m or less. Such close spacings over the long distances required for refraction profiles is a very tough experimental requirement. In spite of this, the prospect of true imaging of earth structures by wavefield inversion has inspired a number of national efforts aimed at producing digital instruments in sufficient number to carry out large-scale experiments to image those highly elusive deep continental structures.

DAVID E. JAMES

References

Barazangi, M., and L. D. Brown, 1986a, *Reflection Seismology: A Global Perspective*. Washington, D.C.; American Geophysical Union, (Geodynamics Series, vol. 13), 311p.

Barazangi, M., and L. D. Brown, 1986b, *Reflection Seismology: The Continental Crust*. Washington, D.C.; American Geophysical Union, (Geodynamics Series, vol. 14), 339p.

Bois, C., M. Cazes, B. Damotte, A. Galdéano, A. Hirn, A. Mascle, P. Matte, J. F. Raoult, and G. Torreilles, 1986, Deep seismic profiling of the crust in northern France: The ECORS project, in M. Barazangi and L. D. Brown, eds., *Reflection Seismology: A Global Perspective*. Washington, D.C.: American Geophysical Union, (Geodynamics Series, vol. 13) 21-29.

Brown, L. D., 1986, Aspects of COCORP deep seismic profiling, in M. Barazangi and L. D. Brown, eds., *Reflection Seismology: A Global Perspective*. Washington, D.C.: American Geophysical Union (Geodynamics Series, vol. 13), 209-222.

Dix, C. H., 1965, Reflection seismic crustal studies, *Geophysics* **30**, 1068-1084.

Dohr, G. P., and R. Meissner, 1975, Deep crustal reflections in Europe, *Geophysics* **40**, 25-39.

Egorkin, A. V., S. K. Zuganov, N. A. Pavlenkova, and N. M. Chernyshev, 1987, Results of lithospheric studies from long-range profiles in Siberia, *Tectonophysics* **140**, 29-47.

Holbrook, W. S., and W. D. Mooney, 1987, The crustal structure of the axis of the Great Valley, California, from seismic refraction measurements, *Tectonophysics* **140**, 49-63.

James, D. E., and J. S. Steinhart, 1966, Structure beneath continents: A critical review of explosion studies 1960-1965, in J. S. Steinhart and T. J. Smith, eds., *The Earth Beneath the Continents*, Am. Geophys. Union Geophys. Mon. 10, 293-333.

Meissner, R., 1986, *The Continental Crust, A Geophysical Approach*. New York: Academic Press, 426p.

Mooney, W. D., and T. M. Brocker, 1987, Coincident seismic reflection/refraction studies of the continental lithosphere: A global review, *Rev. Geophysics* **25**, 723-742.

Nolet, G., 1984, Global seismology and the investigation of deep continental structure, in R. Cassinis, ed., *Problems and Methods for Lithospheric Exploration*. New York: Plenum Press, 5-22.

Pavlis, G. L., 1986, Geotomography using refraction fan shots, *Jour. Geophys. Research* **91**, 6522-6534.

Cross-references: Continental Collision Zones: Seismotectonics and Crustal Structure; Continental Crustal Structure; Continental Lithosphere; Deep Seismic-Reflection Profiling; Exploration Seismology; Lithosphere, Oceanic: Formation and Evolution; Mantle, Upper: Structure; Ocean Continent Transition: Structure; Seismic Signal Processing; Seismic Tomography.

CORE-MANTLE COUPLING

Some Questions

It has long been known that the Earth is not a perfect timekeeper. The spectrum of the variations in the mantle's angular velocity Ω spans a wide range of frequencies. We shall be concerned with the comparatively large fluctuations on time scales of the order of ten years, the *decade variations*, in which changes of up to 5 ms occur in the length of the day (lod). That the atmosphere and oceans cannot be responsible becomes clear when we consider an extreme case.

Suppose that all motions in the atmosphere and oceans relative to the solid Earth ceased, their excess angular momentum $\delta M (= \delta M_{atm} + \delta M_{ocn})$ being shared by the entire Earth, so leading to a change in the lod $P\ (= 2\pi/\Omega)$ of $\delta P = -P(\delta M/I_{tot}\Omega)$, where $I_{tot}\ (\approx 8.04 \times 10^{37}\ \text{kg} \cdot \text{m}^2)$ is the principal moment of inertia of the entire Earth. Although the moment of inertia $I_{ocn}\ (\approx 300 I_{atm})$ of the hydrosphere is large compared with that of the atmosphere, ocean currents are so slow compared with atmospheric motions that $|\delta M_{ocn}| \ll |\delta M_{atm}|$. According to Barnes et al. (1983), $\delta M_{atm} \approx 1.5 \times 10^{26}\ \text{kg} \cdot \text{m}^2 \cdot \text{s}^{-1}$, so $|\delta P|$

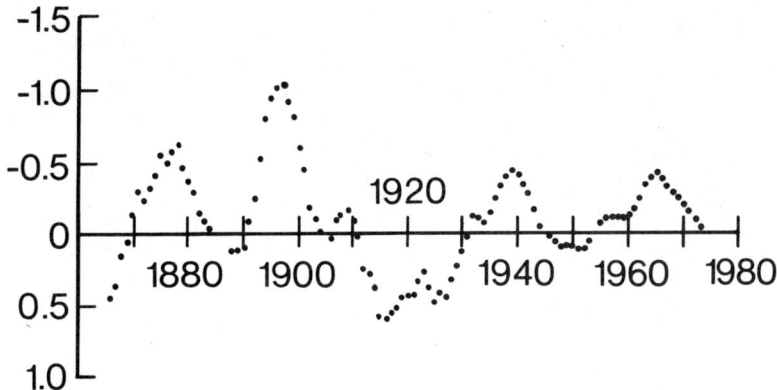

FIGURE 1. The unbalanced torque on the mantle required to explain changes in the length of the day. The units of Γ_z are 10^{18} N·m. (After Morrison, 1979)

is only 2.2 ms. To account for the largest decade variations in lod, the global wind system might have to reverse!

The origin of the decade variations must be sought in the Earth's interior, and Fig. 1 suggests that the task will not be a light one. This figure shows a reconstruction of Γ_z as a function of time t based on observations of Ω during the recent past; here Γ_z is the component of the torque Γ on the mantle that is parallel to the axis of rotation Oz. Figure 1 also sets a target for theorists, "How can torques as large as 10^{18} N · m be generated in the interior of the Earth?"

The time scale of the decade variations in lod shown in Fig. 1 is so strongly reminiscent of that of the secular variation of the geomagnetic field **B** that it is natural to seek a common cause for each in magnetohydrodynamic processes in the core.

Magnetohydrodynamics (MHD) is the marriage of electrodynamics and fluid mechanics. The motion **v** of an electrically conducting fluid induces from a prevailing magnetic field **B** an electromotive force perpendicular to both (i.e., parallel to **v** × **B**); the resulting electric currents (density **J**) create a magnetic field. (Probably the entire field **B** is created in this way—the dynamo origin of geomagnetism. See *Geomagnetic Field, Main: Theory*. The flow of electric current across a magnetic field creates a force (the *Lorentz force*, volume density **J** × **B**, which is perpendicular to **J** and to **B**) that affects the fluid flow **v**. In this way field and motion are intimately linked and, in particular, share the same time scales. The observed secular variation time scales of **B** must therefore also be time scales characteristic not only of the unobserved and unobservable **v**, but also of the stresses acting across the core-mantle boundary (CMB), of the resulting torque Γ on the mantle, and of changes in the mantle's angular velocity Ω.

The existence of the decade variations in lod not only betrays the existence of stresses on the base of the mantle but also establishes that these stresses are large enough to be detectable. Challenging questions may be asked, such as, "What does the changing lod teach us about the deep interior of the Earth?") "By what processes do fields and motions in the core affect the mantle?" and "In what way does the presence of a mantle affect the MHD state of the core?"

Although our discussion will center on variations in the lod (i.e., on changes in the magnitude Ω of Ω), the direction of Ω also fluctuates on similar time scales. Again, this may in some way be related to MHD stresses acting across the CMB. It would certainly be remarkable if Γ were parallel to Ω; in general, one must expect that Γ has components both parallel and perpendicular to Ω and that both the lod and pole positions will be affected.

Restoring Torques and Their Time Scales

To any stress produced by the core on the base of the mantle, there is an equal but opposite stress exerted by the mantle on the surface of the core. Since the outer core is fluid, it will (unlike the mantle) not respond by moving rigidly. To simplify the argument, we shall, however, suppose in this section that the core is a rigid body having an angular velocity Ω_c. A change $\delta\Omega$ in Ω will then be accompanied by a simultaneous change $\delta\Omega_c = - (I_M/I_c)\delta\Omega$ in the angular velocity of the core. Since the principal moment of inertia of the mantle, I_M ($\approx 7.2 \times 10^{37}$ kg · m^2), is almost an order of magnitude larger than that of a solid core, I_c ($\approx 8.5 \times 10^{36}$ kg · m^2), $|\delta\Omega_c|$ exceeds $|\delta\Omega|$ by a like factor; more precisely, $\delta\Omega_c \approx -8.47\delta\Omega$, the minus sign showing that the angular velocities of core and mantle fluctuate in opposite senses.

The relative motion of mantle and core is associated with an important process of self-regulation and smoothing. Suppose for simplicity that initially $\Omega_c = \Omega$ and that, due to some unspecified cause, constant *excitation torques*, Γ_{ex} and $-\Gamma_{ex}$, suddenly start acting on the mantle and core at $t = 0$. The relative angular velocity

$$\Omega_{rel} = \Omega - \Omega_c = \frac{I_{tot}}{I_c}\Omega \approx 9.5\Omega$$

between mantle and core will start to increase. Not surprisingly, stresses will arise at the CMB to eliminate this differential motion. *Restoring couples*, Γ_{rst} and $-\Gamma_{rst}$, start to grow and, in the simplest models, are proportional to Ω_{rel}:

$$\Gamma_{rst} = -\Lambda\,\Omega_{rel}, \quad \Lambda > 0 \qquad (1)$$

Provided that there is no further excitation, Ω_{rel} will asymptote to the constant value

$$\Omega_{rel}^{\infty} = \frac{\Gamma_{ex}}{\Lambda}$$

over an e-folding time of

$$\tau_{rst} = \frac{I}{\Lambda} \qquad (2)$$

where $I = I_M I_c / I_{tot} \approx 7.6 \times 10^{36}$ kg · m^2 is the "reduced" moment of inertia of the core-mantle system.

To establish this, combine the angular momentum equations

$$I_M \frac{d\Omega}{dt} = \Gamma_{ex} + \Gamma_{rst} = -I_c \frac{d\Omega_c}{dt}$$

for the mantle and core to obtain

$$I\frac{d\Omega_{rel}}{dt} = \Gamma_{ex} + \Gamma_{rst} = -\Lambda(\Omega_{rel} - \Omega_{rel}^{\infty}) \qquad (3)$$

The appropriate solution of Eq. 3 is

$$\Omega_{rel} = \Omega_{rel}^{\infty}(1 - e^{-t/\tau_{rst}}) \qquad (4)$$

When the excitation torques are continuously varying for all time, we might replace Eq. 4 by

$$\delta\Omega(t) = \frac{1}{I_M}\int_{-\infty}^{t}\Gamma_{ex}(t')e^{-(t-t')/\tau_{rst}}\,dt'$$

which also satisfies Eq. 3. This expression illustrates how the inertia of the core-mantle system filters out the high frequencies of the excitation $\Gamma_{ex}(t)$ so that they do not appear in the signal $\Omega(t)$. The smoothing time τ_{rst} depends on Λ, which in turn depends on the nature of the coupling that links core and mantle. It seems from Fig. 1 that this time does not greatly exceed a decade or that $\Lambda_{ex}(t)$ has no significant time scales less than a decade.

We later recognize that the excitation torques and the restoring torques rely on the same physical processes of coupling between core and mantle, namely those associated with core MHD. The division of Γ_z into Γ_{ex} and Γ_{rest} is therefore a little artificial. It nonetheless correctly portrays a continuous process in which Γ_z is perturbed from zero and relaxes back toward zero. It is no accident that, according to Fig. 1, there is no torque on average.

We should caution against taking our present model too literally. In reality, the processes creating Γ_{rst} would take a finite time to act; the assumption 1, that Γ_{rst} depends only on the contemporaneous Ω_{rel}, is simplistic. Also, a fluid core would not respond to stresses on its surface by moving as a solid body, a point to which we return below. (See the discussion of geostrophic motions.)

Three Coupling Mechanisms

Stress is exerted by the core on the base of the mantle in three ways: through the viscosity of the core, through the geomagnetic field (and the associated flow of electric currents in the mantle), and through the pressure of the core on the mantle. Correspondingly, the torque Γ experienced by the mantle has three parts: viscous (V), magnetic (B), and topographic (T):

$$\Gamma = \Gamma^V + \Gamma^B + \Gamma^T \qquad (5)$$

To express these torques formally, we consider an element of area $d\mathbf{A}_{CMB} = \mathbf{n}\, dA_{CMB}$ situated at the base of the mantle at vector distance $\mathbf{r}$ from the geocenter ($\mathbf{r} = \mathbf{0}$). If $\mathbf{T}$ is the surface traction (force per unit area) exerted by the core across that element, we may write

$$\Gamma = \oint_{CMB}(\mathbf{r} \times \mathbf{T})\,dA_{CMB} \qquad (6)$$

where, quite generally,

$$dT_i = -S_{im}n_m\,dA_{CMB} \qquad (7)$$

Here S_{im} is the total stress tensor at the CMB; the minus sign appears because (by our convention) the unit normal $\mathbf{n}$ on the CMB is directed from core into mantle rather than the reverse. By Eqs. 6 and 7 we have

$$\Gamma_i = -\epsilon_{ijk}\oint_{CMB}r_j S_{km} n_m\,dA_{CMB} \qquad (8)$$

In the next three sections, respectively, explicit expressions will be given for the three parts of S_{ij} corresponding to Eq. 5:

$$S_{ij} = S_{ij}^V + S_{ij}^B + S_{ij}^T$$

but these stresses and the resulting couples will only be roughly evaluated, since numerous geophysical uncertainties make sophisticated modeling of dubious value. For example, except when we try to assess the size of the topographic couple Γ_z^T, which would be zero if the CMB were spherical (since then the surface traction $\mathbf{T} = p\mathbf{n}$ would be parallel to $\mathbf{r}$), we shall replace the CMB by a sphere S_0 of radius c ($= 3.485 \times 10^6$ m) and express Eq. 8 as a double integral over all colatitudes θ and longitudes ϕ. For example,

$$\Gamma_z^V = -\int\int S_{r\phi}^V c^3 \sin^2\theta \, d\theta \, d\phi \quad (9)$$

Were $S_{r\phi}^V$ constant over S_0, Eq. 9 would give

$$\Gamma_z^V = \pi^2 c^3 S_{r\phi}^V$$

but $S_{r\phi}^V$ must vanish at the poles, so if $S_{r\phi}^V$ is typical of mid-latitudes then $\pi^2 c^3 S_{r\phi}^V$ will overestimate Γ_z^V. We shall therefore, in Eq. 9 and elsewhere, omit the π^2 factor; e.g., our estimate of Eq. 9 would be

$$\Gamma_z^V = O(c^3 S_{r\phi}^V)$$

The inaccuracies of these geometrical simplifications are less serious than those arising from uncertainties in various geophysical quantities (e.g., core viscosity, mantle conductivity, bottom topography) and from overlooking possible cancellations that may occur when the integrals in Eqs. 8 or 9 are evaluated. We give two examples. First, as we shall see, $S_{r\phi}^B = O(B_r B_\phi/\mu)$ where μ is the magnetic permeability (assumed equal to its value in free space of $4\pi \times 10^{-7}$ H · m^{-1}; we use SI units), but it may be quite wrong to conclude that $\Gamma_z^B = O(c^3\mathcal{B}^2/\mu)$, where $\mathcal{B}$ is a typical field strength on the CMB. Were the mantle nonconducting, the integral (9) of S_{km}^B would be identically zero! Second, we shall find that apparently $S_{r\phi}^T = O(2\rho\Omega\mathcal{U}\mathcal{H})$, where ρ is the core density, $\mathcal{U}$ is a typical core flow velocity, and $\mathcal{H}$ is a typical height of CMB topography, i.e., the deviation of the CMB from the sphere S_0. It would be unwise to conclude too hastily that $\Gamma_z^T = O(2\rho\Omega c^3 \mathcal{U}\mathcal{H})$; in the simplest model of topographic coupling, global cancellation of $S_{r\phi}^T$ in Eq. 9 makes $\Gamma_z^T \equiv 0$! (See Eq. 41.)

Viscous Coupling

Viscous coupling is not unlike the frictional drag experienced by a body as a fluid moves past it. Even though the fluid attaches itself to the body so that $\mathbf{v} = 0$ on its surface (the no-slip condition), the normal gradient $\mathbf{n} \cdot \nabla \mathbf{v}$ is in general nonzero, and the surface traction, which is about $\rho\nu\mathbf{n}\cdot\nabla\mathbf{v}$, does not vanish when integrated over the body; this is the viscous drag of the fluid on the body.

Assuming that the core is a Newtonian fluid moving with velocity $\mathbf{v}$, in the frame rotating with the mantle, the viscous stress tensor is

$$S_{ij}^V = [\rho(\zeta - \tfrac{2}{3}\nu)\nabla\cdot\mathbf{v}]\delta_{ij} + \rho\nu(\nabla_i v_j + \nabla_j v_i) \quad (10)$$

When we model the CMB by the sphere S_0, the term involving the bulk viscosity $\rho\zeta$ disappears on substituting Eq. 10 into Eq. 8 for Γ^V. Only the shear viscosity $\rho\nu$ contributes, and from Eq. 9 we find that the z-component of the viscous torque is

$$\Gamma_z^V = -\int\int (\rho\nu \, \nabla_r v_\phi)_{S_0} c^3 \sin^2\theta \, d\theta \, d\phi \quad (11)$$

Evidently, to estimate Γ_z^V, we require knowledge of $\nabla_r v_\phi$ and the kinematic viscosity ν on the CMB. The former requires some understanding of core hydrodynamics.

The rotation of the Earth has a profound effect on core flow (the reason why the magnetic compass needle points approximately north-south!), and we must draw on the theory of rotating fluids.

To appreciate this, we first estimate the magnitude $\mathcal{U}$ of the relative velocities $\mathbf{v}$ in the core. If the observed westward drift of the geomagnetic field ($\approx 0.2°$ longitude per year) is a reliable guide to flow speeds near the core equator, we have $\mathcal{U} \approx 4 \times 10^{-4}$ m · s^{-1}. This is not very different from the extreme values ($\mathcal{U} \approx 5 \times 10^{-4}$ m · s^{-1}) of the velocity field "at the top of the core," obtained by analyzing secular variation data using the *frozen flux* approximation (see below). Since $\mathbf{v} = 0$ on the CMB, the velocity obtained in this way is really not the flow at the top of the core at all; it is $\mathbf{v}$ *beneath* the boundary layer at the top of the core. In what follows we shall use $\mathcal{U} = 3 \times 10^{-4}$ m · s^{-1} as an estimate of the core velocity at the lower edge of that boundary layer. This velocity is minute compared with the underlying solid body rotation of the core. For example, the core equator moves at over 250 m · s^{-1} relative to the geocenter.

Core MHD is most easily discussed in the frame of reference moving with the mantle (although, since the mantle is obviously at rest in that frame, variations in the lod are most readily described in the inertial frame). We should note some implications of choosing the rotating frame. First, the frame is unsteady, but, because the fluctuations $\delta\Omega$ in Ω are so small (about 10^{-8} Ω), this is inconsequential. Second, although the magnetic field in (instantaneously coinciding) rotating and inertial frames should be different according to relativity theory, even the large rotational velocities are small compared with the speed of light, so $\mathbf{B}$ is effectively frame independent.

In the rotating frame, the fluid velocity $\mathbf{v}$ is small, and rotational effects are contained in three fictitious force densities: the centrifugal force $-\rho\mathbf{\Omega} \times (\mathbf{\Omega} \times \mathbf{r})$, the Coriolis force $-2\rho\mathbf{\Omega} \times \mathbf{v}$, and the Poincaré force $-\rho\dot{\mathbf{\Omega}} \times \mathbf{v}$. The changing $\mathbf{\Omega}$ associated with the lunisolar precession provides a persistent and significant Poincaré force, as will be mentioned later, but the fluctuations $\delta\Omega$ in Ω make negligible contributions; we may assume that Ω is constant when evaluating the three fictitious force densities. The centrifugal force, though apparently the largest of these forces, can be absorbed into the gravitational force density $\rho \mathbf{g} = \rho \nabla V$; the two terms together, $\rho \nabla [V + \frac{1}{2}(\mathbf{\Omega} \times \mathbf{r})^2]$, are completely offset by the pressure gradient in the dominant hydrostatic balance of the Earth. Only the Coriolis force $-2\rho \mathbf{\Omega} \times \mathbf{v} \approx 2\rho \Omega \mathcal{U} \approx 4 \times 10^{-4}$ N · m^{-3} remains.

The Coriolis force density $-2\rho \mathbf{\Omega} \times \mathbf{v}$ is large compared with both the "inertial" force density $-\rho \mathbf{v} \cdot \nabla \mathbf{v}$ and the viscous force density $\rho \nu \nabla^2 \mathbf{v}$ in the bulk of the core. We have placed quotes around "inertial," because in the nonrotating frame the corresponding term (the true inertial force) contains the centrifugal, Coriolis, and Poincaré forces and is very far from being negligible. The dominance of rotation over inertia and friction may be confirmed from the magnitudes of the Rossby number Ro (the size of $-\rho \mathbf{v} \cdot \nabla \mathbf{v}$ compared with $-2\rho \mathbf{\Omega} \times \mathbf{v}$) and the Ekman number E (the size of $\rho \nu \nabla^2 \mathbf{v}$ compared with $-2\rho \mathbf{\Omega} \times \mathbf{v}$):

$$\text{Ro} = \frac{\mathcal{U}}{\Omega c}, \quad E = \frac{\nu}{\Omega c^2}$$

Our previous estimate of $\mathcal{U}$ gives Ro $\approx 10^{-6}$; i.e., inertial forces are small compared with the Coriolis forces in the main body of the core. To estimate E, we require knowledge of ν, and this is uncertain by a large factor; values between 10^{-6} m^2 · s^{-1} and 10^{-2} m^2 · s^{-1} are perhaps most often quoted and give $10^{-15} < E < 10^{-11}$.

Viscous forces cannot be neglected in a boundary layer on the CMB (and presumably also on the inner core surface) because they bring the flow in the main body of the core to rest on the CMB. The boundary layer is most probably of Ekman layer type, in which the Coriolis force $-2\rho \mathbf{\Omega} \times \mathbf{v} \approx 2\rho \Omega \mathcal{U}$ is of the same order as the viscous force $\rho \nu \nabla^2 \mathbf{v} \approx \rho \nu (\partial^2 \mathbf{v}/\partial r^2) \approx \rho \nu \mathbf{v}/\delta_E^2$. The thickness of the layer is therefore

$$\delta_E \approx \left(\frac{\nu}{\Omega}\right)^{1/2} = E^{1/2} c \quad (12)$$

This is barely 12 m even if we assume that $\nu = 10^{-2}$ m^2 · s^{-1}.

Though $\mathbf{v}$ is zero on the CMB, it takes its full core magnitude at depth δ_E beneath the CMB. This is indeed the *core surface velocity* $\mathbf{v}_s$ deduced from the secular variation by use of the frozen flux approximation (see below) on which our estimate (3×10^{-4} m · s^{-1}) of $\mathcal{U}$ was based. Evidently, on the CMB

$$\nabla_r v_\phi = O\left(\frac{\mathcal{U}}{\delta_E}\right)$$

so, by Eqs. 11 and 12,

$$\Gamma_z^V = O(c^3 \rho (\Omega \nu)^{1/2} \mathcal{U})$$

This does not depend sensitively on ν and is only 10^{17} N · m, even for $\nu = 10^{-2}$ m^2 · s^{-1}.

Some reasons why this estimate of Γ_z^V may be too large or too small may be noted. First, if the westward drift of the geomagnetic field is a reliable indicator, $\mathbf{v}_s$ has a predominantly negative ϕ-component, which, by Eq. 11, creates a steady torque that tends to slow down the mantle, which is impossible (see the second section, "Restoring Torques and Their Time Scales") unless some compensating magnetic or topographic torque exists that maintains the total Γ_z at its zero average. Even if we have underestimated viscous coupling, the magnetic and/or topographic torque cannot be less important. The possibility that a major (westward) part of $\mathbf{v}_s$ generates a steady but negated torque suggests that $\mathcal{U} \approx 3 \times 10^{-4}$ m · s^{-1} overestimates the remaining part of $\mathbf{v}_s$, which produces the fluctuating torque.

The possibility that angular momentum transport by turbulence enhances Γ_z^V deserves some attention. If our estimate of $\mathcal{U}$ is reliable, the Ekman flow is likely to be laminar. Its Reynolds number $R_E = \mathcal{U} \delta_E / \nu = \mathcal{U}/(\Omega \nu)^{1/2}$ is scarcely 10, even if $\nu = 10^{-6}$ m^2 · s^{-1} is assumed, and instabilities arise only when $R_E > 56.3$. (See Greenspan, 1968, §6.3.) It was therefore correct to use the molecular viscosity, not some "turbulent viscosity," when estimating δ_E, and the magnitude of the surface tractions

$$\rho \nu \mathbf{v}_S / \delta_E \approx \rho (\Omega \nu)^{1/2} \mathcal{U}$$

Nevertheless, it is likely (Braginsky, 1964) that the compositional (and/or thermal) buoyancy that stirs the fluid core creates anisotropic turbulence beneath the Ekman layer. Because of its small scale and remoteness, this turbulence cannot yet be inferred through analysis of geomagnetic field data. Increasing $\mathcal{U}$ might add significantly to the viscous stresses on the CMB. Whether, because of the randomness of the turbulence, these stresses would tend to cancel in the integrals in Eqs. (8) or (9) is not known. According to our estimates of ν, the turbulent part of $\mathbf{v}_S$ would have to be large to make the Ekman layer unstable.

Magnetic Coupling

Magnetic coupling arises from electric currents flowing in the deep mantle through the MHD activity of the core. The associated Lorentz force density $\mathbf{J}_M \times \mathbf{B}_M$ integrated over the mantle V_M creates a *magnetic torque* $\mathbf{\Gamma}^B$ about $\mathbf{r} = \mathbf{0}$ of

$$\mathbf{\Gamma}^B = \int_{V_M} \mathbf{r} \times (\mathbf{J}_M \times \mathbf{B}_M) \, dV_M \quad (13)$$

This torque can also be expressed as an integral (Eq. 8) of magnetic stresses over the CMB. In fact, by using $\nabla \cdot \mathbf{B} = 0$ and Ampère's law

$$\mathbf{J} = \nabla \times \mathbf{H} \tag{14}$$

where $\mathbf{H} = \mathbf{B}/\mu$ is the *magnetizing force*, we may express the Lorentz force as the divergence

$$(\mathbf{J}_M \times \mathbf{B}_M)_i = \nabla_j S^B_{ij} \tag{15}$$

of the Faraday-Maxwell *magnetic stress tensor*

$$S^B_{ij} = \frac{1}{\mu}\left[B_i B_j - \frac{1}{2}B^2 \delta_{ij}\right] \tag{16}$$

After substituting Eq. 15 into Eq. 13, we extend the domain of integration to the entire infinite exterior of the core. This extension is permitted since the currents flowing above the Earth's surface make a negligible contribution to $\mathbf{\Gamma}^B$. Applying the divergence theorem, we obtain the form of Eq. 8 for $\mathbf{\Gamma}^B$:

$$\mathbf{\Gamma}^B = -\frac{1}{\mu}\oint_{S_0} (\mathbf{r} \times \mathbf{B})\mathbf{B} \cdot d\mathbf{A}_0 \tag{17}$$

Here the CMB is again the sphere S_0. Since **B** is continuous across the CMB, no subscript M is required for **B** in Eq. 17.

Equation 17 shows clearly that nonvanishing $\mathbf{\Gamma}^B$ requires $\mathbf{B} \cdot d\mathbf{A}_0$ ($=B_r\, dA_0$) to be nonzero; i.e., lines of force must thread the CMB. This *poloidal* part $\mathbf{B}_P$ of the total magnetic field within the Earth is also the part observed at the Earth's surface. Additionally, **B** possesses a *toroidal* part $\mathbf{B}_T$ that has no radial component. It is therefore magnetically unobservable at the Earth's surface, and does not affect $\mathbf{B} \cdot d\mathbf{A}_0$. The current density **J** may similarly be divided into $\mathbf{J}_P$ and $\mathbf{J}_T$. According to Eq. 14, $\mathbf{J}_T$ creates $\mathbf{B}_P$, and $\mathbf{J}_P$ produces $\mathbf{B}_T$. Since $J_{Tr} \equiv 0$, its presence in the (source-free) mantle cannot depend on leakage of current from the core; $\mathbf{J}_T$ is created by electromagnetic induction through a time-varying $\mathbf{B}_P$. Unlike $\mathbf{J}_P$, it would exist and contribute to $\mathbf{\Gamma}^B$ even if a nonconducting layer existed at the base of the mantle! From the observed secular variation, we can in principle calculate $\mathbf{J}_{TM}$, or could do so were the conductivity σ_M of the mantle known. [For a discussion of the poloidal-toroidal decomposition of **B**, see Backus (1986).]

Both $\mathbf{B}_P$ and $\mathbf{B}_T$ contribute to the $\mathbf{r} \times \mathbf{B}$ factor in Eq. 17, and we may, if we like, divide $\mathbf{\Gamma}^B$ into poloidal and toroidal parts:

$$\mathbf{\Gamma}^B_P = -\frac{1}{\mu}\oint_{S_0} (\mathbf{r} \times \mathbf{B}_P) B_r\, dA_0$$

$$\mathbf{\Gamma}^B_T = -\frac{1}{\mu}\oint_{S_0} (\mathbf{r} \times \mathbf{B}_T) B_r\, dA_0$$

We should emphasize that the main part of $\mathbf{B}_M$ is the observed potential field of the Earth, which originates from currents flowing within the core. This field, as Eq. 13 makes plain, does not by itself contribute to $\mathbf{\Gamma}^B_P$. The small remaining part of $\mathbf{B}_P$, the part created by currents flowing in the mantle, is necessary for a nonzero $\mathbf{\Gamma}^B_P$.

By Ohm's law,

$$\mathbf{J}_M = \sigma_M \mathbf{E}_M \tag{18}$$

Here and below, **E** is the electric field in the frame rotating with the mantle; $\mathbf{E}_M$ owes its existence to the core, in which Ohm's law is

$$\mathbf{J}_c = \sigma_c(\mathbf{E}_c + \mathbf{v} \times \mathbf{B}_c) \tag{19}$$

On the CMB where **v** is zero, Eq. 19 has the same form as Eq. 18, but σ_M and σ_c are probably very different in magnitude. Since $\mathbf{r} \times \mathbf{E}$ is continuous across the CMB, it follows from Eqs. 18 and 19 that $\mathbf{J}_M$ and $\mathbf{J}_c$ are discontinuous:

$$\mathbf{r} \times \mathbf{J}_M = \left(\frac{\sigma_M}{\sigma_c}\right)\mathbf{r} \times \mathbf{J}_c \tag{20}$$

Instead of Eq. 20 we shall employ

$$\mathbf{r} \times \mathbf{J}_M = \sigma_M \mathbf{r} \times \left(\frac{\mathbf{J}_S}{\sigma_c}\right) - \sigma_M(\mathbf{r} \cdot \mathbf{B})\mathbf{v}_S \tag{21}$$

where $\mathbf{J}_S$ is the current density at the lower edge of the Ekman layer, where the flow is the core "surface" velocity $\mathbf{v}_S$. To derive Eq. 21, recall that the Ekman layer is thin and that, therefore, $\mathbf{r} \times \mathbf{E}$ cannot change appreciably within it: $\mathbf{r} \times \mathbf{E}_S = \mathbf{r} \times \mathbf{E}_M$. We obtain Eq. 21 from Eqs. 18 and 19 and $\mathbf{r} \cdot \mathbf{v}_S = 0$. [Strictly, it is not v_{Sr} but v_r that vanishes on the CMB; the *Ekman suction* v_{Sr} is about $\mathcal{U}\delta_E/c$, which we assume is negligible (but see Braginsky, 1978). No appreciable net current flows within the Ekman layer, so $\mathbf{B}_S \approx \mathbf{B}_M$.]

It is clearest to assess the two sources on the right-hand side of Eq. 21 separately. In the *frozen flux theory* of secular variation, it is supposed that $\sigma_c = \infty$. Magnetic field lines move with a perfect conductor as though frozen to it (Alfvén's theorem), and it is argued that the secular variation observed at the Earth's surface is predominantly the result of the *rearrangement* by $\mathbf{v}_S$ of preexisting field lines emerging from the core. On assuming that the secular variation is of this character, the poloidal part of $B_r \mathbf{v}_S$ can be uniquely deduced from measurements of **B** at the Earth's surface, but the toroidal part cannot without additional assumptions. For example, if it is supposed that the surface flow is steady, it can be uniquely determined by the frozen flux hypothesis; see Voorhies and Backus (1985) and Voorhies (1986). A prime geomagnetic ambition is to extract, from the observed $\mathbf{B}_P$, the best possible $\mathbf{v}_S$ with the fewest assumptions.

A procedure that parallels the frozen flux hypothesis is to use $\sigma_c = \infty$ to justify setting $\mathbf{J}_S/\sigma_c = 0$ in Eq. 21, thus obtaining

$$(\mathbf{r} \times \mathbf{J}_M)_{\text{CMB}} = -\sigma_M c(B_r \mathbf{v}_S) \qquad (22)$$

If $\mathcal{B}$ is a typical field strength, we see from Eq. 14 that

$$\mathbf{J}_M = O(\sigma_M \mathcal{U} \mathcal{B}_r) \qquad (23)$$

To order of magnitude, Eq. 13 now gives

$$\Gamma_z^{B,\text{fr}} = O(c^3 \mathcal{U}[\sigma_M d] \mathcal{B}_r^2) \qquad (24)$$

where d is the thickness of the layer at the base of the mantle in which σ_M is not negligible. We have here added fr to Γ_z^B to emphasize that this *flux rearrangement* part of Γ^B is the contribution of only the last term in Eq. 21. The superscript fr may also be read as "friction," a role that Eq. 26 will make clear

The $\mathcal{B}_r$, derived by extrapolating downward the field observed at the Earth's surface, is typically 3×10^{-4} T (= 3 G), but this is almost certainly an underestimate. Fields of short wavelength are invisible at the Earth's surface, but must also contribute to the magnetic coupling of core to mantle; we shall therefore set $\mathcal{B}_r = 5 \times 10^{-4}$ T. Taking $\mathcal{U} = 3 \times 10^{-4}$ m · s^{-1} as before, we obtain

$$\Gamma_z^{B,\text{fr}} \approx 3 \times 10^9 [\sigma_M d] \text{ N} \cdot \text{m}$$

To achieve a torque of 10^{18} N · m requires the conductance $\sigma_M d$ to be about 3×10^8 S; e.g., we might take $d = 10^6$ m and $\sigma_M = 3 \times 10^2$ S · m^{-1}, which is much smaller than the core conductivity σ_c that we shall assume, namely 10^5 S · m^{-1}.

Currents induced in the mantle by fields originating outside the Earth show that σ_M increases rapidly with depth. In principle, information about σ_M in the deep mantle can be obtained from the secular variation data. The basic idea is familiar to every electrical engineer: An oscillating field induces currents in a solid conductor that tend to shield that field from observation. If the *electromagnetic skin depth* $\delta_\sigma = (\mu\sigma\omega)^{-1/2}$, where ω is the frequency of the source, is small compared with the length scale d, of the conductor, the induced currents are confined to a "skin" of thickness δ_σ on the surface of the conductor, beyond which the source field cannot penetrate. In other words, a conducting sheet (thickness d) placed between a magnetic source and an observer will shield the latter from the former, provided the *electromagnetic time constant* $\tau_\sigma = \mu\sigma d^2$ of the sheet is significantly larger than the period $P = 2\pi/\omega$ of the source. In a similar way, the mantle filters out the short-period ($P \ll \tau_\sigma$) core signal, and τ_σ can in principle be obtained by analyzing that signal (i.e., the secular variation). In practice, results are prone to contamination by fields of external origin and higher frequency. Some analyses have suggested that τ_σ may be only a few months, but there appears to be no conclusive evidence as yet that τ_σ is less than a year or a few years. Our earlier estimate ($\sigma_M d \approx 3 \times 10^8$ S, for $d = 10^6$ m) gives $\tau_\sigma \approx 10$ yr.

Were τ_σ really as large as 10 yr, the secular variation would be smoothed more thoroughly in crossing the mantle than the observations allow. This highlights a difficulty: Can the target torque be produced magnetically with a $\sigma_M(r)$ that does not smooth the secular variation too much?

This dilemma is eased by taking the semiconducting character of the mantle into account; the observed increase of σ_M with depth should continue right down to the CMB. The implications may be assessed by rewriting Eq. 24 as

$$\Gamma_z^{B,\text{fr}} = O\left(\frac{c^3 \mathcal{U} \tau_\sigma \mathcal{B}_r^2}{\mu d}\right) \qquad (25)$$

The more compact is the conductivity distribution giving τ_σ (i.e., the smaller d for the same τ_σ), the larger is $\Gamma^{B,\text{fr}}$. For example, $d = 10^5$ m and $\tau_\sigma = 1$ yr gives $\Gamma_z^{B,\text{fr}} \approx 10^{18}$ N · m as surely as did $d = 10^6$ m, $\tau_\sigma = 10$ yr previously. This more acceptable value for τ_σ arises because electric current, Lorentz force, and magnetic torque in the mantle are determined by its integrated conductivity, i.e., essentially by the conductance

$$C = \int_c^{c+d} \sigma_M(r) \, dr$$

whereas the attenuation of the secular variation is a diffusion process determined (at high frequencies) by

$$\tau_\sigma = \left[\int_c^{c+d} [\mu\sigma_M(r)]^{1/2} \, dr\right]^2$$

To the extent that d is small, Eq. 22 implies that

$$\mathbf{J}_M \times \mathbf{B}_M = \sigma_M B_r^2 \mathbf{v}_S \qquad (26)$$

This allows a simple interpretation of the torque $\Gamma^{B,\text{fr}}$: The drag of the magnetic field lines through the conducting layer in the mantle acts as a frictional stress, $CB_r^2 \mathbf{v}_S$, parallel to the viscous stress $\rho\nu\mathbf{v}_S/\delta_E$ derived earlier, but plausibly larger. Taking $C = 3 \times 10^8$ S as before and $\mathcal{B}_r = 5$ G, we find $C\mathcal{B}_r^2 \approx 75$ kg · m^{-2} · s^{-1}, which may be compared with the most and least optimistic values of $\rho\nu/\delta_E$ derived in the preceding section, namely 8.5 kg · m^{-2} · s^{-1} and 8.5×10^{-5} kg · m^{-2} · s^{-1}.

As before, if the predominantly westward drift of the nondipole field implies that the same is true of the surface motion, then the average of $\Gamma_z^{B,\text{fr}}$ will be negative. A compensating torque must exist that cancels out the average $\Gamma_z^{B,\text{fr}}$ and leaves only the observed fluctuating part. One possibility arises from the term in Eq. 21 that we have so far neglected. Zonal shears, and especially the geostrophic component $\mathbf{v}_G$ of $\mathbf{v}$, are easily excited in a rapidly rotating fluid and are therefore probably large. [*Geostrophic flow* is the name given to a zonal motion, $\mathbf{v}_G = \omega(s, t) \times \mathbf{r}$, that is independent of z and is therefore unaffected by the Coriolis force. Here (s, ϕ, z) are cylindrical coordinates with Oz parallel to $\mathbf{\Omega}$.]

The geostrophic part of $\mathbf{v}$ has a special significance for a rotating fluid (of uniform density in an axisymmetric container): it, and it alone, carries the angular momentum. (See Greenspan, 1968, ch. 2.) When the angular momentum of the mantle changes, the equal and opposite change in the angular momentum of the core is contained in an altered $\mathbf{v}_G$. This provides a more precise picture of the way the fluid core responds to an excitation torque than the crude "solid core" model described in the second section of this article; a change in Ω is accompanied by a "torsional wave" in the core, in which the cylinders of constant s move relative to one another. (See *Magnetohydrodynamic Waves Within the Earth.*) The change in $\mathbf{v}_G$ can be trivially determined from the associated change in $\overline{v_{S\phi}}$, the longitudinal average of the zonal surface velocity. And, on certain plausible assumptions (see below), $\overline{v_{S\phi}}$ can be extracted from the geomagnetic data. Jault et al. (1988) have, in fact, monitored $\overline{v_{S\phi}}$ from 1975 to 1985, and have shown that the corresponding change in angular momentum associated with $\mathbf{v}_G$ is approximately equal and opposite to the change in angular momentum of the mantle over the same period, thus providing a comforting confirmation of the dynamical picture.

Consider now the creation of magnetic field by a zonal shear such as $\mathbf{v}_G$. It is the magnitude of the zonal flow that $\mathcal{U}$ measures, and its magnetic Reynolds number

$$R_m = \mu \sigma_c c \mathcal{U}$$

is large ($R_m \approx 120$). The frozen flux idea therefore works well and shows that the lines of force of $\mathbf{B}_p$ should be stretched into the ϕ-direction by a sheared zonal flow to create a $\mathbf{B}_T$ of order $R_m \mathbf{B}_P$. This process is called the ω-effect. By Eq. 14 the current density associated with this $\mathbf{B}_T$ is $O(\sigma_c \mathcal{U} \mathbf{B}_P)$; i.e., it is proportional to σ_c. Far from being negligible, $\sigma_M \mathbf{r} \times \mathbf{J}_S/\sigma_c = O(\sigma_M c \mathcal{U} \mathbf{B}_P)$; i.e., it is of the same order as the term retained in Eq. 22. Thus Eq. 23 will again apply, and Eq. 24 is a valid estimate of the corresponding torque $\Gamma^{\mathrm{B},l}$. Unlike $\Gamma^{\mathrm{B,fr}}$, this torque cannot be directly linked to the surface of the core: it is created by potential differences set up on the CMB by motions deep beneath the core surface.

An early and celebrated illustration of $\Gamma^{\mathrm{B},l}$ was provided by Bullard et al. (1950). In Fig. 2 we look down on the North pole N of a core, modeled simply

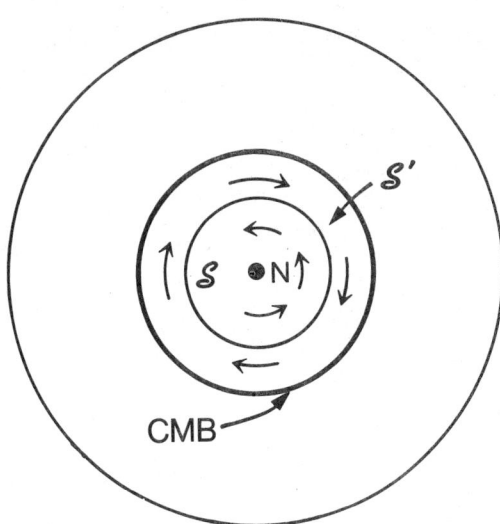

FIGURE 2. The model of the westward drift proposed by Bullard et al. (1950).

as a rigid sphere S surrounded by a shell S' in perfect electrical contact across their interface $\mathcal{I}$. A torque Γ'_z is applied to S, and an equal and opposite torque is applied to S', causing them at first to accelerate in the directions shown. The shear across $\mathcal{I}$ generates from $\mathbf{B}_p$ a toroidal field and associated restoring couples proportional to the jump in angular velocity across $\mathcal{I}$; see Eq. 1. In the ultimate steady state this jump has exactly that value for which the exciting and restoring couples balance. The toroidal field generated by the relative motion at $\mathcal{I}$ diffuses across S' to the CMB, where it creates a torque $\Gamma^{\mathrm{B},l}$ that (by Lenz's law) accelerates the mantle to motion in the same sense as S. A jump in angular velocity develops at the CMB (the westward drift), and in the steady state the associated torque $\Gamma^{\mathrm{B,fr}}$ exactly cancels the torque $\Gamma^{\mathrm{B},l}$ created by the zonal shear at $\mathcal{I}$.

Although this model may appear unsophisticated to modern eyes, much of the underlying physics is as valid today as it was in 1950. We would tend today only to use different language: The zonal shears beneath the CMB set up potential differences on the CMB. The resulting *leakage current* $\mathbf{J}_{MP}$ creates the *leakage torque* $\Gamma^{\mathrm{B},l}$ that accelerates the mantle to the east relative to the surface of the core. The relative motion $\mathbf{v}_S$ equilibrates when its compensating, flux-rearranging torque balances the leakage torque. This picture suggests that one should regard leakage as providing the excitation torque of the second section of this article, and flux rearrangement as providing the restoring torque. The dynamical reason why zonal shears in the core should create a relative westward, rather than an eastward, drift of the core surface is an objective of geodynamo theory and will not be considered here.

By writing $\mathcal{U} = \Omega_{rel}c$, we may cast Eq. 25 in the form of Eq. 1:

$$\Gamma^{B,fr} \approx -\Lambda^B \Omega_{rel}$$

where

$$\Lambda^B = \frac{c^4 \tau_\sigma \mathcal{B}_r^2}{\mu d}$$

The smoothing time (2) associated with magnetic coupling is therefore

$$\tau_{rst}^B = \frac{I\mu d}{c^4 \tau_\sigma \mathcal{B}_r^2}$$

For the case $d/\tau_\sigma = 10^5$ m/1 yr considered earlier, τ_{rst}^B is 26 yr, but an allowance for the fluidity of the core reduces this. Which of τ_σ or τ_{rst}^B is the larger is still not known for certain.

The question arises, "Would not a correlation, or a lack of correlation, between secular change and variations in the lod demonstrate unequivocally whether the magnetic torque dominates other forms of coupling?" Unfortunately, changes in Γ_T^B (e.g., from leakage) produce variations in lod that lack any detectable magnetic signal at the Earth's surface. It should, however, be possible in principle to relate some of the observed changes in $\mathbf{B}_P$ (and in the inferred $\mathbf{v}_S$) to variations in lod caused by the associated changes in Γ_P^B (flux rearrangement). A second question might be asked: "If a correlation exists, will the change in lod precede or follow the secular change of $\mathbf{B}$?" LeMouël et al. (1981) present evidence that magnetic variations lead changes in lod by about a decade, implying that $\tau_\sigma < \tau_{rst}^B$ by a similar amount. Langel et al. (1986) present conflicting evidence that changes in lod precede magnetic variations, also by about a decade, thus suggesting that $\tau_{rst}^B < \tau_\sigma$. The future doubtless holds the resolution.

Topographic Coupling

Topographic coupling depends on deviations from sphericity in the shape of the CMB, i.e., the presence of inverted mountains and valleys, often collectively termed *bumps* (see Mantle, Lower: Structure and Earth's Core: Structure). Since core pressure p will exert a force on a bump that will not in general be purely radial, the mantle will be subjected to a *topographic torque* Γ^T given by

$$\Gamma^T = \oint_{CMB} p\mathbf{r} \times \mathbf{dA}_{CMB} \quad (27)$$

Since Hide's pioneering paper (1969) suggested this mechanism, new geophysical objectives have been set: Efforts have been made to detect and map bumps seismically; correlation has been attempted between the Earth's gravitational and magnetic fields (since both should be influenced by bumps); connections have been sought between bumps and terrestrial heat flow (since bumps might be associated with patterns of mantle convection, e.g., plumes); and, recognizing that, in analogy with the way that atmospheric flow is altered by mountains on the Earth's surface, bumps on the CMB will influence core field and flow, experimentalists and theorists have attempted to evaluate their effects. *Geophrenology*, as the new subject is sometimes jokingly called, has brought together diverse branches of the earth sciences and has itself grown into a significant geophysical field. We shall describe some of the problems encountered in trying to estimate Γ^T.

The CMB possesses one giant topographic feature that creates a torque dwarfing all others—the torque associated with the lunisolar precession. The precession of the mantle is due to the rotationally created flattening of the Earth. But, by the same token, the core is flattened also and would, were it disconnected from the mantle, precess at a different rate. The fact that it does not do so is due to a torque of order 4.5×10^{20} N · m acting across the CMB. Theory suggests that, if there were no other source of motion, this torque would create a large-scale circulation in the core of order 0.01 m · s^{-1}. In the inertial frame, this pattern would rotate slowly over the precessional period (25,800 yr); its frequency, as seen in the frame rotating with the mantle, is diurnal, making it of little interest in core MHD. We consider it no further, despite its magnitude.

Obviously the torque of Eq. 27 would vanish if, as in the previous sections, we modeled the CMB by a sphere; it is important to distinguish between $\mathbf{dA}_{CMB}$ and the radial $\mathbf{dA}_0$ of S_0. Denoting by $\mathbf{1}_r$ the radial unit vector, we write the equation of the CMB as

$$\mathbf{r} = [c - h(\theta, \phi)]\mathbf{1}_r$$

where $h(\theta, \phi)$, the inverted height of the bumps, is characteristically $\mathcal{H}$ (say). Were there bumps on the CMB several kilometers in height, seismology would readily discern them; clearly $\mathcal{H} \ll c$. This allows us easily to refer values at a point on the CMB to corresponding values at the underlying ($h < 0$) or overlying ($h > 0$) point on the reference sphere S_0, and to rewrite Eq. 27 as

$$\Gamma^T = -c \int_{S_0} p\, \nabla h \times \mathbf{dA}_0 \quad (28)$$

We shall again be most interested in the z-component of this torque:

$$\Gamma_z^T = \int_{S_0} p \frac{\partial h}{\partial \phi} dA_0 \quad (29)$$

This fixes attention on zonal variations in topography; an axisymmetric core could not topographically create any changes in lod. More convenient forms are obtained by integrating Eqs. 28 and 29 by parts:

$$\Gamma^T = c \oint_{S_0} h \, \nabla p \times d\mathbf{A}_0$$

$$\Gamma_z^T = -\oint_{S_0} h \frac{\partial p}{\partial \phi} dA_0 \quad (30)$$

By Eq. 30 only the asymmetric part of p contributes to Γ_z^T.

It may be conceptually helpful to divide the pressure field on the CMB into two parts: the *topographically independent* p, and the *topographically created* p. The former originates from MHD processes in the core, and would exist even if h were zero. The observed geomagnetic field is not axisymmetric, a reflection of the fact that the geodynamo cannot be axisymmetric (Cowling's theorem; see *Geomagnetic Field, Main: Theory*); this essential lack of symmetry would ensure that $\partial p/\partial \phi \neq 0$ on the CMB even if h were zero. The presence of bumps does, however, disturb the core's MHD state, thus bringing the topographically created p into being.

Consider first the contribution to Γ_z^T made by the topographically independent p. The two factors h and $\partial p/\partial \phi$ in Eq. 30 are uncorrelated. The latter, since it is independent of h, may be derived on the assumption that the CMB is spherical ($h = 0$). There is the exciting possibility (Speith et al., 1986) of computing Γ_z^T from available geophysical data by using seismically determined h and geomagnetically inferred $\partial p/\partial \phi$, the latter being obtained from the following chain of reasoning.

We recall that, since $Ro \ll 1$ and $E \ll 1$, the equation of motion

$$\rho \left(\frac{\partial \mathbf{v}}{\partial t} + \mathbf{v} \cdot \nabla \mathbf{v} + 2\mathbf{\Omega} \times \mathbf{v} \right)$$
$$= -\nabla p + \mathbf{J} \times \mathbf{B} + \rho \nu \nabla^2 \mathbf{v} + \rho \mathbf{g} \quad (31)$$

in the main body of the core reduces to

$$2\rho \mathbf{\Omega} \times \mathbf{v} = -\nabla p + \mathbf{J} \times \mathbf{B} + \rho \mathbf{g} \quad (32)$$

[Because of variations in ρ, the viscous term in Eq. 31 is incomplete, but it is small anyway. In neglecting $\partial \mathbf{v}/\partial t$, we have supposed that the time scale of $\mathbf{v}$ is $c/\mathcal{U}$ so that $|\partial \mathbf{v}/\partial t|/|2\mathbf{\Omega} \times \mathbf{v}| = Ro \ll 1$. Later we shall consider time scales of order $\lambda/\mathcal{U}$, where λ is the horizontal scale of the bump. It will then be necessary to restore $\partial \mathbf{v}/\partial t$.]

The largest terms in Eq. 32 are ∇p and $\rho \mathbf{g}$, and they are almost equal because of the predominantly hydrostatic state of the Earth. After they have been removed, the equation of *magnetogeostrophic balance* follows:

$$2\rho_0 \mathbf{\Omega} \times \mathbf{v} = -\nabla p' + \mathbf{J} \times \mathbf{B} + \rho' \mathbf{g}_0 \quad (33)$$

where $\rho' = \rho - \rho_0$ and $p' = p - p_0$ are the deviations in density and pressure from those of the hydrostatic state (i.e., ρ_0 and p_0). We have here ignored the change $\mathbf{g}'$ in gravitational force that ρ' creates; probably even $\rho' \mathbf{g}_0$ is small in comparison with the Coriolis and Lorentz forces.

In using Eq. 33 to evaluate Γ_z^T from Eq. 30, we note first that, since p_0 is axisymmetric, we may replace p by p' in Eq. 30. In order of magnitude, Eq. 33 gives

$$\frac{\partial p'}{\partial \phi} = O(2\rho_0 \Omega \mathcal{U}) \quad (34)$$

A preliminary estimate of Γ_z^T follows from substituting Eq. 34 into Eq. 30:

$$\Gamma_z^T = O(2\rho_0 \Omega c^3 \mathcal{U} \mathcal{H}) \quad (35)$$

This gives $\Gamma_z^T = 10^{18}$ N · m for $\mathcal{H} = 50$ m, an apparently modest requirement.

This estimate is too simplistic: p' is single-valued in ϕ, and $\partial p'/\partial \phi$ is "as often positive as negative." Even if Eq. 34 is true locally on the CMB, contributions to the integral in Eq. 30 from different areas of the CMB will cancel each other out partially or totally (see the previous section, "Three Coupling Mechanisms"). Nevertheless, although Eq. 35 is certainly an upper bound on the size of Γ_z^T, there is apparently a considerable latitude for cancellation. If, as seismic interpretations allow, $\mathcal{H}$ is 500 m (instead of 50 m), 90% of the integrand of Eq. 30 could cancel, still leaving a torque of 10^{18} N · m. The likelihood of significant cancellation in Eq. 30 does, however, make the computation of Γ_z^T from geophysically determined h and $\partial p/\partial \phi$ a delicate and daunting undertaking.

Returning to Eq. 33, we recall that, although the toroidal field may be so large in the main body of the core that Lorentz and Coriolis forces are comparable, $\mathbf{B}_T$ must be small on the CMB. (It would vanish if σ_M were zero; for $\sigma_M \neq 0$, it should be smaller than its value deep in the core by a factor of about σ_M/σ_c.) The Lorentz force $\mathbf{J} \times \mathbf{B}$ is therefore negligible compared with the Coriolis force $2\rho_0 \mathbf{\Omega} \times \mathbf{v}$ on the CMB. On discarding it from Eq. 33 we obtain

$$2\rho_0 \mathbf{\Omega} \times \mathbf{v} = -\nabla p' + \rho' \mathbf{g}_0 \quad (36)$$

Since $g_{0\phi} = 0$, Eq. 30 and the ϕ-component of Eq. 36 give (Speith et al., unpublished)

$$\Gamma_z^T = 2\Omega \rho_0 c \oint_{S_0} h v_\theta \sin \theta \cos \theta \, dA_0 \quad (37)$$

The problem of finding $\partial p/\partial \phi$ for use in Eq. 30 has been transformed into that of determining v_θ for Eq. 37. This may be accomplished with the help of the frozen flux hypothesis, but, as noted below Eq. 21, this approximation must be supplemented by one or more additional assumptions—e.g., that $\mathbf{v}$ is steady. An alternative suggestion is based on Eq. 36. Again recognizing that $\mathbf{g}_0$ is radial (and ignoring Ekman suction v_{Sr}, as before), we may operate on Eq. 36 by $\mathbf{r} \times$ and deduce *tangential geostrophy* on the CMB:

$$\mathbf{v}_S = \frac{\mathbf{r} \times \nabla p'}{2\rho_0 \Omega c \cos\theta} \qquad (38)$$

Backus and LeMouël (1986) combined Eq. 38 with the frozen flux hypothesis to determine $\mathbf{v}_S$ uniquely over patches covering some 75% of the CMB. This so-called "leaky flux" or "Swiss-cheese" model is less complete than Voorhies' procedure, but it does not require $\mathbf{v}_S$ to be steady; see the preceding section. After the leaky patches have been plugged in some way, the first step in the calculation of Γ_z^T is complete, namely that of obtaining v_θ from Eq. 38 for use in Eq. 37.

Despite the increasing precision of mantle tomography, the second step, that of obtaining h reliably from the seismic data, is extremely demanding; in some approaches, information about the physical state of the lower mantle must also be injected. It is perhaps not surprising that a satisfactory consensus on $h(\theta, \phi)$ has yet to be reached, though a preliminary calculation has been reported (Speith et al., 1986) that gives a Γ_z^T of the targeted magnitude.

Supposing that a change in the core's MHD state can supply a topographic excitation torque of the required magnitude, we now ask whether topography can supply the restoring torque also. In the spirit of the second section, we may suppose that $\mathbf{v}_S$ is initially known and that the associated topographically independent p creates from h a nonzero excitation torque $\Gamma_{\text{ex},z}^T$ under the action of which a new and growing relative motion $\Omega_{\text{rel}} \times \mathbf{r}$ between core and mantle is added to $\mathbf{v}_S$. This new motion interacts with the bumps on the CMB, and a topographically created p results, together with a restoring torque $\Gamma_{\text{rst},z}^T$. To estimate $\Gamma_{\text{rst},z}^T$, we, for simplicity, ignore $\mathbf{v}_S$ and seek the disturbance created by the topography on an otherwise zonal flow $\Omega_{\text{rel}} \times \mathbf{r}$. Regarding the magnetic linkage between core and mantle as being adequately treated in the preceding section, we may assume that the mantle is an insulator.

When an object lies in a slow (Ro $\ll$ 1), steady ($\partial/\partial t = 0$) flow in a rotating, but almost inviscid ($E \ll 1$), fluid, there is almost no motion within the cylinder whose generators are parallel to Ω and tangent to the body; the fluid outside the cylinder flows round it almost as though the cylinder too were a solid. This *Taylor column* is a remarkable demonstration of the *Proudman-Taylor theorem*, which follows from the equation of motion

$$2\rho_0 \Omega \times \mathbf{v} = \nabla p' \qquad (39)$$

appropriate to the conditions stated. Assuming ρ_0 is constant and operating on Eq. 39 by $\nabla \times$, we obtain $\partial \mathbf{v}/\partial z = 0$. It follows that $\mathbf{v}$ and p' depend only on x, y, and t.

Hide (1969) observed that, in such circumstances, bumps on the CMB would (provided $\mathcal{H} \gg \delta_E$) similarly create Taylor columns from one hemisphere of the core to the other, a remarkable consequence of the long-range influence of rotation. Although one might have expected that a bump of height $\mathcal{H}$ would create a disturbance in $\mathbf{v}$ and p' proportional to $\mathcal{H}$, this is not the case: they are roughly independent of $\mathcal{H}$. From Eq. 39 one rederives Eq. 34 with $\mathcal{U}$ replaced by $\Omega_{\text{rel}} c$, and at first sight Eq. 35 again holds, so

$$\Lambda^T = O(2\rho_0 \Omega c^4 \mathcal{H}) \qquad (40)$$

If Eq. 40 is correct, topography can provide the restoring couple. There are, however, serious doubts.

A geostrophic contour on the CMB is defined as a curve through every point of which the line parallel to Ω from one hemisphere to the other is of the same length L, say. For a spherical bumpless CMB, the geostrophic contours are latitude circles, $\theta = \cos^{-1}(L/2c)$ and $\pi - \cos^{-1}(L/2c)$. Bumps distort these contours, but, because $\mathcal{H} \ll c$, they remain closed. Whenever all geostrophic contours are closed, Eq. 39 implies

$$\Gamma_z^T \equiv 0, \qquad \Lambda^T \equiv 0 \qquad (41)$$

The cancellation in Eq. 30 is then complete. (See, for example, Anufriev and Braginsky, 1977b.)

Hide (1970) reported that in joint, but still unpublished, experiments with C. G. Collier, "boundary layer separation" was observed in flows over topographic features in a rotating fluid. He conjectured that this separation would, in analogy with the resolution of the classic d'Alembert paradox, reinstate Eq. 35. This paradox concerns a sphere moving uniformly with speed U in a nonrotating incompressible fluid. Were the fluid truly inviscid in the sense of superfluidity, the flow round the sphere would be symmetric with respect to front and rear; the Bernoulli pressure deficit $\frac{1}{2}\rho U^2$ exerted by the fluid on a unit area of the forward hemisphere would be exactly canceled by that on the rear hemisphere, and the net force on the sphere would be precisely zero. If the fluid has *any* viscosity whatever, boundary layer separation occurs on the rear hemisphere, the sphere sheds eddies, front/rear symmetry is destroyed, and the pressure deficits on the two hemispheres, though still of order ρU^2 are not in balance. The sphere therefore experiences a net drag force of about $\rho U^2 R^2$. By analogy, it is argued that, if boundary layer separation occurs at a bump, the cancellation of $O(2\rho_0 \Omega \mathcal{U})$ stresses that led to Eq. 41 would be incomplete and that then Eq. 35 and Eq. 40 might follow. James (1980) showed numerically that a topographic feature could shed eddies at a rate comparable with $\mathcal{U}/\lambda$. (In the geophysical context, $\lambda = 10^6$ m would be linked to 100-yr time scales.) For such rapidly varying flows, the $\partial \mathbf{v}/\partial t$ omitted from Eq. 31 must be restored to Eq. 39; thus Eq. 41 no longer follows. James found that Eq. 35 gives the order of magnitude of the fluctuating torques.

This does not completely resolve matters. Although Eq. 36 may be valid at the CMB, its use becomes increasingly suspect when applied to long-range effects, like Taylor columns, that penetrate deep into the core. As $\mathbf{B}_T$ increases with depth so does the Lorentz force $\mathbf{J} \times \mathbf{B}$, until (in the currently favored strong field geodynamo picture) it becomes comparable with the Coriolis force $2\rho_0 \mathbf{\Omega} \times \mathbf{v}$, thus effectively destroying the Taylor-Proudman two-dimensionality and the Taylor column. No longer will $\partial p'/\partial \phi$ be independent of $\mathcal{H}$ as in Eq. 34, and $\partial p'/\partial \phi$ may be expected to be proportional to $\mathcal{H}$ as in Eq. 42.

Unfortunately, numerical integration of the MHD equations is difficult, and laboratory experiments are nearly impossible. The effects of the magnetic field on topographic coupling have so far been studied analytically; see particularly Anufriev and Braginsky (1975, 1977a,b). These authors recall that the Faraday-Maxwell stresses (Eq. 16) of a strong, subsurface zonal field $\mathbf{B}_T$ would give the core an "elasticity" and as a result the flow disturbance created by a bump would be spread out in longitude over a distance $\mathcal{L}$ that may be large compared with λ. They concluded that the estimates of Eq. 34, and therefore of Eq. 40 also, are too optimistic by a factor of about $\mathcal{H}/\mathcal{L}$ and therefore should be replaced by

$$\frac{\partial p'}{\partial \phi} = O\left(\frac{2\rho_0 \Omega \Omega_{\mathrm{rel}} c \mathcal{H}}{\mathcal{L}}\right) \quad (42)$$

$$\Lambda^T = O\left(\frac{2\rho_0 \Omega c^4 \mathcal{H}^2}{\mathcal{L}}\right) \quad (43)$$

They also argued that, except for small bumps (i.e., those having a small magnetic Reynolds number $\mu \sigma_c c \lambda \Omega_{\mathrm{rel}}$),

$$\mathcal{L} \sim \frac{\mathcal{B}_T^2}{2\mu \rho_0 \Omega \Omega_{\mathrm{rel}} \lambda}$$

For $\mathcal{B}_T = 250$ G, $\Omega_{\mathrm{rel}} = \mathcal{U}/c = 10^{-10}$ s^{-1}, and $\lambda = 10^6$ m, this gives $\mathcal{L} \sim 4 \times 10^6$/m; for $\mathcal{H} \sim 1$ km, the estimate of Λ^T (based on Eq. 40) is reduced by a factor $\mathcal{H}/\mathcal{L}$ of about 2×10^{-4}.

Anufriev and Braginsky (1977a) also argued that when the magnetic Reynolds number $\mu \sigma_c c \lambda \Omega_{\mathrm{rel}}$ of a bump is small, $\mathcal{L}$ is of order λ and the reduction factor $\mathcal{H}/\mathcal{L}$ is less serious. If, for instance, $\mathcal{H} = 0.6$ km and $\lambda = 3$ km, then $\mu \sigma_c c \lambda \Omega_{\mathrm{rel}} \approx 0.1$ and $\mathcal{H}/\mathcal{L} \approx \mathcal{H}/\lambda \approx 0.2$. Thus Γ_z^T is 10^{19} N · m according to Eq. 35, and even 2×10^{18} N · m by Eq. 43. Surface roughness on this scale would pass undetected tomographically but would contribute to seismic wave scattering and apparently to the topographic torque also.

Summarizing this discussion, although the interaction of CMB topography with the pressure field associated with the geodynamo mechanism may be capable of creating an excitation torque of the target magnitude, it is far less certain that it can produce the restoring torque needed to regulate the angular velocity of the mantle. Perhaps for this reason, magnetic stresses are necessary.

Conclusions

This article has described current thinking about core-mantle coupling and has highlighted several significant open questions. The topic deserves further study, not only because it has the potential to provide greater insight into the nature of the core-mantle interface but also because core-mantle coupling may be significant to the structure and behavior of the main geomagnetic field. It is centrally important to "model Z," a geodynamo model of current interest. In that model variations in core-mantle coupling might trigger geomagnetic field reversals (Braginsky, 1978). Core-mantle coupling would then indeed play an important role in the study of the Earth's deep interior.

PAUL ROBERTS

Acknowledgments

I am grateful to Drs. S. I. Braginsky, R. Hide, H. K. Moffatt, and C. V. Voorhies for helpful criticisms.

References

Anufriev, A. P., and S. I. Braginsky, 1975, Influence of irregularities of the boundary of the earth's core on fluid velocity and magnetic field, Part I, *Geomagnetism and Aeronomy*, **15**, 754-757.

Anufriev, A. P., and S. I. Braginsky, 1977a, Influence of irregularities of the boundary of the earth's core on fluid velocity and magnetic field, Part II, *Geomagnetism and Aeronomy* **17**, 78-82.

Anufriev, A. P., and S. I. Braginsky, 1977b, Influence of irregularities of the boundary of the earth's core on fluid velocity and magnetic field, Part III, *Geomagnetism and Aeronomy* **17**, 492-496.

Backus, G. E., 1986, Poloidal and toroidal fields in geomagnetic field modeling, *Rev. Geophysics* **24**, 75-109.

Backus, G. E., and J.-L. LeMouël, 1986, The region on the core-mantle boundary where a geostrophic velocity field can be determined from the frozen-flux data, *Geophys. Jour. Royal Astron. Soc.* **85**, 617-628; addendum **88**, 1987, 321-322.

Barnes, R. T. H., R. Hide, A. A. White, and C. A. Wilson, 1983, Atmospheric angular momentum fluctuations, length-of-day changes and polar motion, *Royal Soc. [London] Proc.* **A387**, 31-73.

Braginsky, S. I., 1964, Magnetohydrodynamics of the earth's core, *Geomagnetism and Aeronomy* **4**, 698-712.

Braginsky, S. I., 1978, Nearly axisymmetric model of the hydromagnetic dynamo of the earth, *Geomagnetism and Aeronomy* **18**, 225-231.

Brosche, P., and J. Sündermann, eds., 1982, *Tidal Friction and the Earth's Rotation*, II. Berlin: Springer, 359 p.

Bullard, E. C., C. Freedman, H. Gellman, and J. Nixon, 1950, The westward drift of the earth's magnetic field, *Royal Soc. [London] Philos. Trans.* **A243**, 67-92.

Cazenave, A., ed., 1986, *Earth Rotation: Solved and Unsolved Problems*, Dordrecht: Reidel, 330 p.

Greenspan, H. P., 1968, *The Theory of Rotating Fluids*, Cambridge: University Press, 339 p.

Hide, R., 1969, Interaction between the earth's liquid core and solid mantle, *Nature* **222**, 1055-1056.

Hide, R., 1970, On the earth's core-mantle interface, *Quart. Journal Royal Meterol. Soc.* **96**, 579-590.

Hide, R., ed., 1984, Rotation in the solar system, *Royal Soc. [London] Philos. Trans.* **A313**, 1-186.

James, I. N., 1980, The forces due to topographic flow over shallow topography. *Geophys. Astrophys. Fluid Dynamics* **14**, 225-250.

Jault, D., C. Gire, and J.-L. Le Mouël, 1988, Westward drift core motions and exchanges of angular momentum between core and mantle, *Nature* **333**, 353-356.

Langel, R. A., D. J. Kerridge, D. R. Barraclough, and S. R. C. Malin, 1986, Geomagnetic temporal change: 1903-1982, a spline representation, *Jour. Geomagnetism and Geoelectricity* **38**, 573-597.

LeMouël, J.-L., T. R. Madden, J. Ducruix, and V. Courtillot, 1981, Decade fluctuations in the geomagnetic westward drift and the earth's rotation, *Nature* **290**, 763-765.

Morrison, L. V., 1979, Re-determination of the decade fluctuations in the rotation of the earth in the period 1861-1978, *Geophys. Jour. Royal Astron. Soc.* **58**, 349-360.

Speith, M. A., R. Hide, R. W. Clayton, B. H. Hager, and C. V. Voorhies, 1986, Topographic coupling of core and mantle, and changes in the length of the day, *EOS (Am. Geophys. Union Trans.)* **67**(44), 908.

Voorhies, C. V., 1986, Steady superficial core motions: an alternative method, *Geophys. Research Letters* **13**, 1537-1540.

Voorhies, C. V., and G. E. Backus, 1985, Steady flows at the top of the core from geomagnetic models: the steady motions theorem, *Geophys. Astrophys. Fluid Dynamics* **32**, 163-173.

Cross-reference: *Earth's Core; Earth's Core: Structure; Earth Structure: Global; Geomagnetic Field, Main: Theory; Magnetohydrodynamic Waves within the Earth; Mantle, Lower: Structure; Mantle Convection and Plumes; Mantle Dynamics; Seismic Tomography.*

CRUSTAL MOVEMENTS AND TECTONIC DEFORMATION

The Earth's surface is in constant motion. This motion consists of rapid oscillations (seismic waves) and slow, continuous movements (plate tectonics). For the most part, seismic or oscillatory movements are of such small amplitude and continuous movements are sufficiently slow (mm/yr to cm/yr) that they are beyond the range of human perception. Periodically, however, major earthquakes occur causing both sudden and large continuous movements (~ 10m) and large amplitude oscillatory movements, which can cause extensive damage and loss of life over areas exceeding several thousand square kilometers. It is estimated that, as a result of earthquakes, 24,000 lives have been lost on average each year since the beginning of the twentieth century.

Earthquakes result in catastrophic movements of the Earth's surface, but they are by no means the only cause of such movements. In fact, earthquakes themselves, at least the most destructives ones near the surface, are more properly thought of as the result rather than the cause of surface movements. That is, earthquakes represent the sudden release of accumulated strain caused by differential movements of various parts of the Earth's surface. The theory of plate tectonics, which has gained wide acceptance among earth scientists since the late 1960s, provides a framework for understanding these continuous movements of the Earth's surface. The plate tectonic theory, together with the tendency for the Earth to respond to changes in the distribution of mass at and near the surface (isostasy) provide the basis for understanding many of the processes that result in surface movements.

This entry considers the more or less continuous movements of the Earth's surface as monitored by geodetic measurements. For a discussion of seismic or oscillatory surface movements, refer to the entries on seismology and earth tides.

Plate Tectonics and Crustal Movements

According to the plate tectonics hypothesis, the Earth's surface is composed of a relatively small number of rigid plates that move relative to one another. These plates comprise the lithosphere, the strong outer layer of the Earth composed of the crust and upper mantle. The lithospheric plates overlay and "ride on" a weaker layer called the asthenosphere. Individual plates can contain both ocean and continental crust and consequently plate boundaries do not necessarily coincide with the boundaries between oceans and continents. The boundaries between lithospheric plates are the location of the vast majority of orogenic activity (earthquakes, volcanoes, active mountain belts) and hence are the locus of most tectonic deformation at the Earth's surface (Fig. 1). The motion of the plates is driven

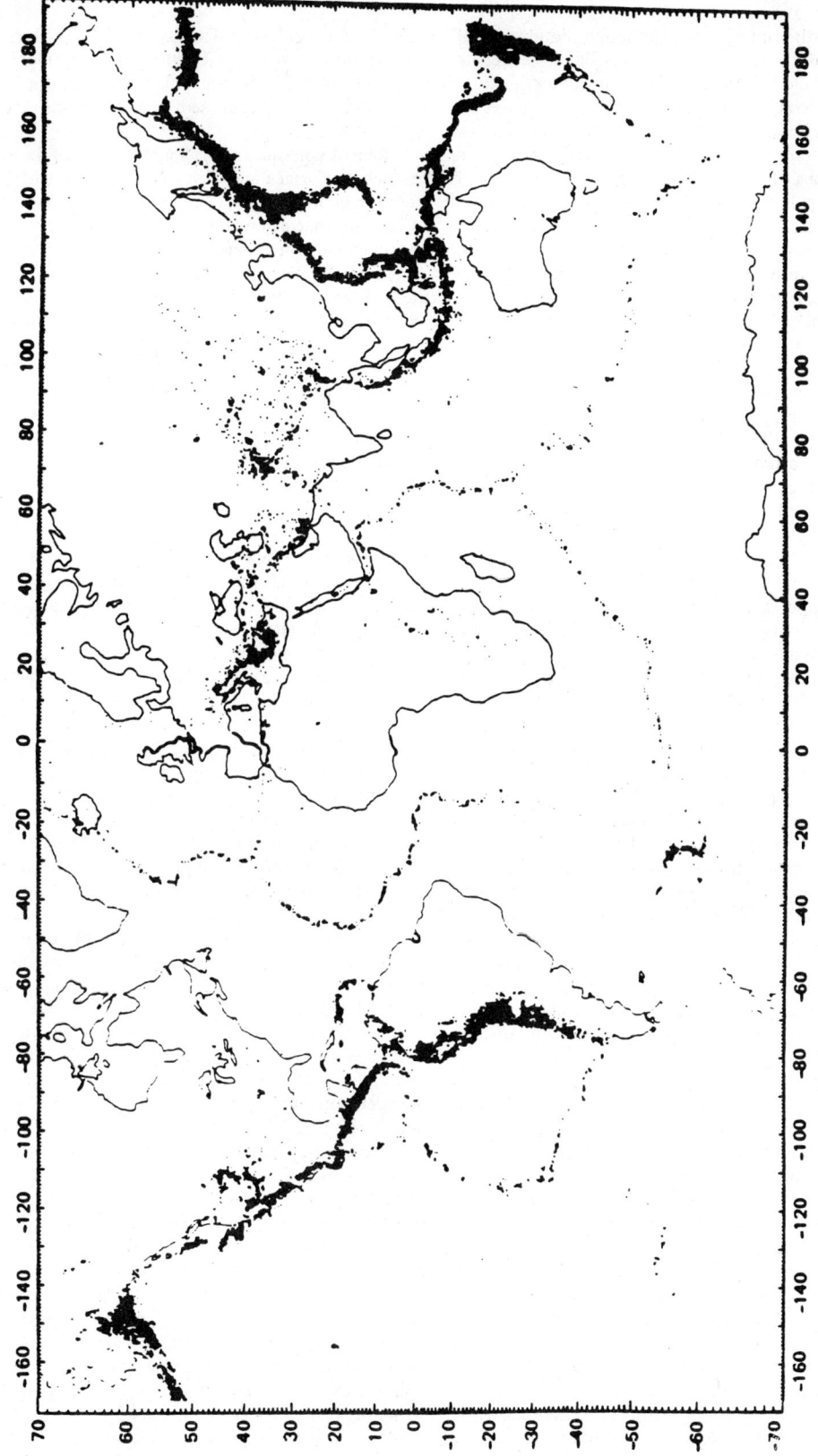

FIGURE 1. Epicenters of 29,000 earthquakes, 1961–1967, depth 0–700 km, which define the boundaries of the lithospheric plates covering the Earth's surface. (From Barazangi and Dorman, 1969)

by the Earth's internal heat although the physics of this process is the subject of considerable debate (refer to Plate Tectonics and Geodynamics).

The various lithospheric plates move relative to one another over the Earth's surface at rates of a few cm/yr. These rates are primarily determined from observations of magnetic anomalies on the sea floor and from paleomagnetism. Such measurements can only provide average rates over about 1 my. Since the early 1970s, the Crustal Dynamics Project of the National Aeronautics and Space Administration (NASA) has been involved in the measurement of relative plate movements using space geodetic techniques (Geodynamics Program Office, 1979). This project is providing the first direct measurements of contemporary plate motions. Although still very preliminary, early results suggest that at points far from plate boundaries, the plates move in a continuous fashion (versus jerky motion) at rates that are comparable to the long-term average rates (Christodoulidis et al., 1985). In contrast, conventional geodetic measurements indicate that at points near plate boundaries, motions are much more complex over short time periods.

The nature of crustal movements and associated tectonic activity along any plate boundary depends on the type of plate boundary. In the simplest sense, there are three types of plate boundaries: (1) divergent boundaries where plates move apart creating new crust; (2) convergent boundaries where plates move toward each other destroying old crust; and (3) transform boundaries where plates slip past each other horizontally.

When a divergent plate boundary begins to form, the lithospheric plate is stretched and thinned. As the cool, dense lithosphere thins, hot mantle rock comes closer to the surface and heats up the overlying plate. As the plate continues to stretch, the brittle lithosphere breaks to form a rift valley or graben. This breaking of the thinned lithosphere is accompanied by earthquakes and large vertical movements as the rift valley subsides and the regions on either side of the rift rise to maintain isostatic balance (for example, the African rift valleys). Direct measurements of crustal spreading associated with the rifting process have been made in a number of areas of active rifting. A notable example is the Afar rift in northern Africa where repeated geodetic measurements documented over 1.5 m of crustal extension and uplift of the rift margins during a major swarm of earthquake and volcanic activity in 1978 (Abdallah et al., 1979).

If the lithosphere stops spreading at this stage of its development, the hot mantle below the thinned lithosphere slowly cools and becomes more dense. The now heavy mantle material acts like a sinker pulling down on the overlying crust causing regional subsidence. The bordering mountain ranges erode and deposit sediments in the subsiding area. Isostatic adjustment to the weight of the sediments causes additional subsidence. This type of "failed rift" process (that is, the rift failed to continue widening) is one of a number of mechanisms proposed for the formation of continental sedimentary basins (Bally et al., 1980).

If the rift continues to widen, the hot mantle rock reaches the surface and forms new ocean crust. The Red Sea is an example of a rift in the initial stages of forming ocean crust. The process can continue and eventually generate ocean basins such as the Atlantic, which was formed over the past 200 my by the rifting of North and South America from Eurasia and Africa. The young ocean crust is hot and buoyant and forms a mid-ocean ridge, which can be as much as 3 km above the adjacent ocean floor. Iceland represents a segment of the North Atlantic Mid-Ocean ridge that is above sea level. Direct geodetic measurements of crustal spreading and associated vertical movements that accompanied volcanic activity along the rift zones traversing Iceland have been made during a number of rifting events (Björnsson et al., 1979).

The oceanic lithosphere cools and subsides as it becomes older and moves progressively away from the mid-ocean ridge. This subsidence together with erosion eventually destroys the block faulted mountain ranges along the edge of the rifted continent and leaves a gently sloping continental margin and shelf such as that along the east coast of North America.

Along a convergent plate boundary one plate overrides the other or conversely, one plate sinks below the other. The downgoing or subducted plate descends into the mantle along deep ocean trenches. The subducted lithosphere can reach depths of at least 700 km before it is apparently assimilated into the surrounding mantle. The zones of deep and intermediate earthquakes (100–700 km), called Benioff-Wadati zones, mark the locations where brittle lithosphere has sunk into the asthenosphere.

At points far from the interface, the converging plates apparently move toward each other at a more or less constant rate. This hypothesis is being tested by the NASA Crustal Dynamics Project, which is making measurements of the relative movement between converging plates across a number of plate boundaries [e.g., Nazca-S. American plates, Pacific-N. American plates (Alaska), Pacific-Eurasian plates]. However, for the most part the plates are locked together at their interface, although there are some places where the plates are thought to slide past each other continuously, such as along the Mariana arc. The continuous motion of the plates slowly builds up stress along the locked plate interface. This stress is periodically released by sudden slip during earthquakes. The largest earthquakes ever recorded have occurred along convergent plate boundaries (e.g., 1960, $M = 8.5$ Chile earthquake; 1964, $M = 8.4$, Alaska earthquake).

When an oceanic plate is overridden by a continental plate (for example, along the west coast of South America, southern Alaska or eastern Japan) the large earthquakes along the plate boundary cause uplift of the continental margin. Repeated leveling, tide gauge observations, and field observations of exhumed marine features have documented large uplifts associated with subduction earthquakes (observed uplift exceeded 5 m during the 1960 Chile earthquake). This type of plate boundary is also characterized by volcanoes, which form along the edge of the continent, presumably due to partial melting of the subducted oceanic plate as it descends into the Earth. In addition, earthquakes with associated faulting and deformation can extend far into the overriding continental lithosphere. The combined effect of episodic uplift accumulated by many plate boundary earthquakes over millions of years, faulting and folding due to compression within the overriding plate, and the building of volcanic edifices produce the high mountains along such continental margins (for example, the Andes of South America).

If a subducting plate carries a continent, the buoyancy of the continental lithosphere prevents subduction. The most intense deformation and greatest uplift, and consequently the highest mountains, occur where two continental plates collide, such as along the Himalayas. These regions of continental collision also produce very broad deformation. The uplifted and faulted Tibetan Plateau, which extends up to 1000 km into the continent of Asia, is believed to have been caused by the collision of India into the southern part of the Asian continent (Molnar and Tapponier, 1978).

A classic example of a transform plate boundary or transform fault is the San Andreas fault in California, where the Pacific plate is moving toward the northwest relative to the North American plate at a rate of about 5 cm/yr. Repeated triangulation surveys since the early 1900s and repeated trilateration measurements since about 1970 made along the fault have provided much information on the nature of horizontal movements and crustal deformation along this plate boundary (Savage, 1983). In fact, measurements along the San Andreas fault, together with the long history of geodetic observations in Japan, represent the primary observational basis for our present understanding of the cyclic deformation associated with earthquakes (see below).

At points distant from a transform plate boundary, the two plates are believed to move at a more or less constant rate relative to each other (this is being tested by intensive observations by the NASA Crustal Dynamics Project in the vicinity of the San Andreas fault). Similarly, along the boundary itself, the two adjacent plates move past each other continuously below a certain depth (i.e., below the depth of the deepest earthquakes, which is about 15 km for the San Andreas fault). However, at shallower levels the fault can be locked (although some sections of the San Andreas are thought to slip more or less continuously up to the surface). The relative motion of the plates increases the stress on the locked portion of the fault until the fault strength is exceeded and seismic slip occurs. Very large earthquakes can occur along transform faults such as the 1906 $M = 8.3$ San Francisco earthquake, which involved 4 m of slip along a 200 km segment of the fault.

If a transform fault is not aligned exactly along the direction of the relative plate motion, then movement of the plates will involve either extension or compression across the fault. Crustal thickening and uplift occur where the fault is under compression, and crustal thinning and subsidence where the fault is extended. The San Gabriel mountains in southern California and the Southern Alps of New Zealand are thought to be caused by compression and conversely, the Imperial Valley of southern California has been attributed to extension across predominantly transform faults.

Geodetic measurements in the vicinity of large earthquakes along transform and convergent plate boundaries are beginning to reveal a cyclic pattern of deformation (Fig. 2), which can be understood within the plate tectonic framework (e.g., Thatcher, 1984). According to our present understanding of the earthquake deformation cycle, the steady movement of adjacent lithospheric plates cause stresses (and hence, strain or deformation) to increase along the locked sections of the plate boundaries. This portion of the cycle corresponds to the interseismic period (Fig. 2). When this stress exceeds the strength of the fault, sudden slip occurs, resulting in the large, essentially instantaneous, coseismic deformation. The rapid coseismic slip increases stresses in the more ductile region below the fault. As these stresses are relieved after the earthquake (either through aseismic fault slip below the coseismic fault or through more broadly distributed viscous flow at depth), the fault is rapidly restressed (strained). This portion of the cycle corresponds to the postseismic period of rapid deformation (Fig. 2). This period of rapid deformation eventually grades into the steady interseismic stress (or strain) accumulation, which progresses to the next earthquake. Permanent deformation results when the postseismic and interseismic strains do not exactly balance the coseismic strain.

Understanding this cyclic deformation is important as it provides a framework for forecasting future earthquakes and thereby mitigating their effects. In theory, one can estimate the time to the next earthquake on a particular fault from the deformation that accompanied the previous earthquake, the rate of permanent deformation (determined from geologic observations), and the rate of interseismic strain accumulation (determined from geodetic observations). This estimation has actually proven quite difficult in practice because of the generally long time period between earthquakes and the relatively short history of geodetic measurements. In addition, the

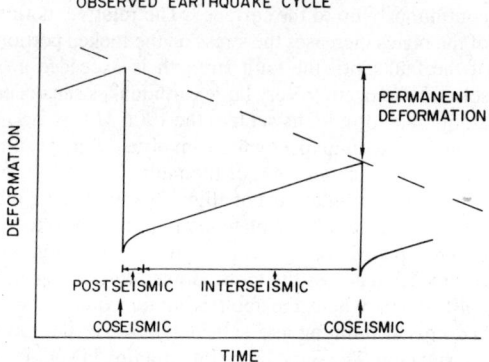

FIGURE 2. Schematic illustration of the idealized earthquake deformation cycle. (From Thatcher, 1984)

cyclic deformation illustrated in Fig. 2 is highly idealized. Because of the paucity of appropriate observations, it is still not known whether interseismic strain accumulation or permanent deformation can be characterized by uniform rates on the time scale of individual earthquakes. In any case, extending the history of geodetic observations to monitor strain accumulation and release in seismically active areas is recognized as an important element in earthquake prediction research.

Intraplate Deformation

While the vast majority of tectonic deformation occurs along plate boundaries, they are by no means the whole story as far as active crustal motions are concerned (Bally et al, 1980). The large and destructive earthquakes that occurred in the eastern United States (e.g., 1811–1812 New Madrid, Missouri, earthquakes; 1896 Charleston, South Carolina, earthquake) and the broadly distributed seismic activity in the western United States and Southeast Asia (Fig. 1) attest to this fact. Whether these phenomena are truly intraplate or whether they represent some kind of distributed plate boundary effects, their place in the plate tectonic framework remains unresolved.

Even more clearly divorced from simple plate tectonics concepts are the deformations associated with intraplate volcanism. Hawaii is a dramatic example. The volcanoes of the Hawaiian Islands-Emperor Seamount chain are generally believed to result from upwelling of magmas from a source beneath the Pacific lithospheric plate. Although the geometry of the volcanic chain appears to be caused by the motion of this plate over the underlying mantle "hot spot," the hot spot itself is arguably a phenomena independent from the overlying global plate framework. Movements associated with such volcanoes (as well as volcanoes along subduction zones) have been closely monitored by geodetic measurements for some time and are generally thought to reflect the episodic inflation and deflation (eruption) of subsurface magma chambers. Observations of ground motion near volcanoes, together with attendant seismic activity, have proven to be effective predictors for future eruptions.

Somewhat less well known, perhaps, are other geodetic results that suggest subsurface magma injection in areas of not-so-recent volcanism. Examples of surface deformation attributed to magma at depth come from the Rio Grande rift of central New Mexico, the Yellowstone Park region, and the Mammoth Lakes region of eastern California. The inflation in the Mammoth Lakes region may have been responsible for a series of four magnitude 6 earthquakes in 1980. Because there have been several explosive eruptions and extrusion of rhyolite domes in this area during the past 400 years, and because the Mammoth Lakes region is about 300 km from major population centers such as Sacramento and San Francisco, deformation and seismic activity are being monitored in order to predict possible future activity.

Contemporary deformation of the stable interior is virtually synonymous in many minds with *postglacial isostatic rebound*. The broad doming of recently (12,000 years ago) deglaciated parts of North America and Scandinavia has been documented by geologic studies of warped beach terraces and geodetic measurements of continued uplift (Fig. 3). This motion is perhaps the best understood, geomechanically speaking, of any type of intraplate deformation (see isostasy). However, controversy still revolves around distinguishing those possible deep earth rheologies that are most consistent with the observed rebound effects (Mörner, 1980).

An important aspect of recent studies of postglacial rebound is that this is a global phenomena, not restricted to the immediate area of glacial retreat. Melting of the ice sheets not only unloads the underlying crust, but also affects the distribution of water throughout the world's oceans. The response of the Earth to this redistribution of mass results in a complex pattern of relative motion of land and sea the world over.

Ice is not the only crustal load capable of driving surface deformation. Geologic (deformed shorelines) and releveling evidence indicate continued local uplift in the area of former Lake Bonneville, Utah. This rebound is inferred to follow the removal of the waterload associated with climatic changes beginning about 10,000 years ago.

Redistribution of crustal loads by sedimentation, erosion, and faulting is fundamental in geology. Sedimentary infilling clearly augments thermalmechanical subsidence during the formation of sedimentary basins and during the evolution of passive continental margins. The importance of crustal loading during the mountain building process is recognized as a major factor in the development of the sedimentary basins that form along the fronts

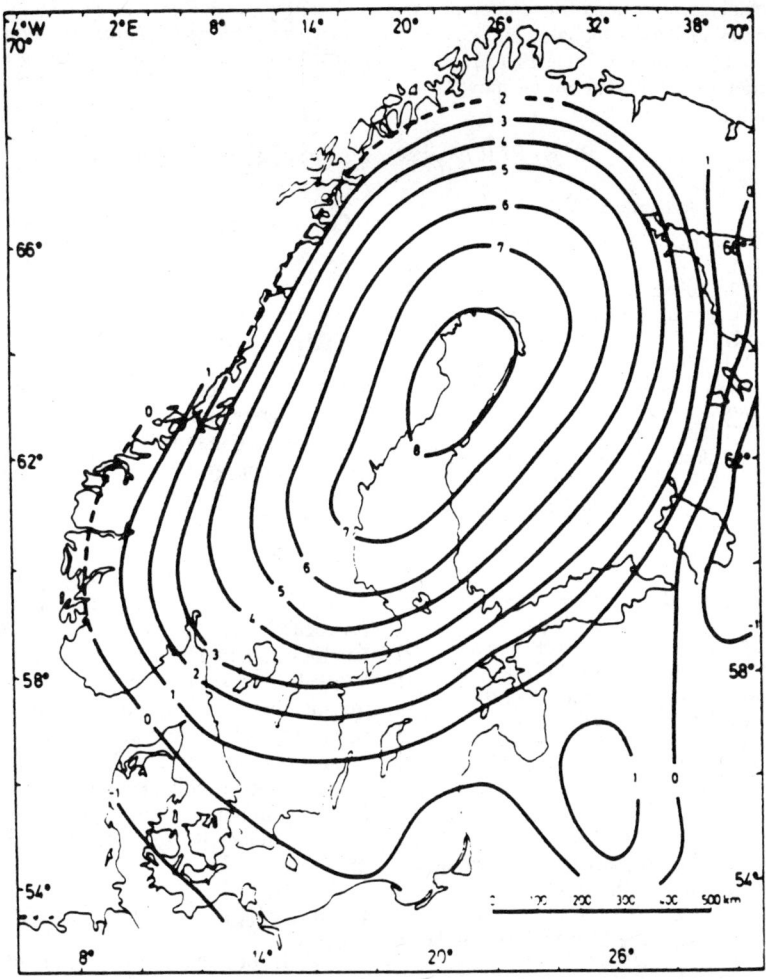

FIGURE 3. Contours of present-day rate of crustal uplift (mm/yr) in Fennoscandia due to postglacial isostatic rebound. (From Balling, 1980)

of mountain belts (foreland basins). Understanding the nature of basin subsidence is important as basins contain much of the world's supply of hydrocarbons.

Conclusion

The geophysical phenomena we are attempting to understand with geodetic measurements are characterized by long repeat times (earthquake recurrence times on the order of hundreds to thousands of years) and by relatively slow rates of deformation (millimeters to centimeters per year). The 100-year history of classical geodetic measurements, although rather sparse, has already proven extremely useful in many respects. However, continued geodetic monitoring, particularly in areas of active tectonic deformation, and the development of new and more accurate measurement techniques are clearly necessary to establish the observational basis required to further our understanding of these geological phenomena. Particularly exciting in this regard are the space geodetic surveying methods such as *Very-Long-Baseline Interferometry (VLBI)* and *Satellite Laser Positioning* (q.v., this volume). These systems are capable of measuring baselines between widely separated stations (1000–10,000 km) with precisions of a few centimeters and they are already providing important observations on the relative movements of the Earth's lithospheric plates. The Global Positioning System (GPS) (q.v., this volume) presently under development, is expected to provide low cost, relative horizontal and vertical positions of a few centimeters over distances of tens to hundreds of kilometers. With these new systems, classical geodetic techniques can be concentrated in specific areas to monitor small scale deformation (one to a few tens of kilometers), particularly in regions of potential seismic and volcanic activity.

Given the social ramifications of contemporary geologic processes such as earthquakes and volcanic eruptions, it is in the world's best interest to continue an active program of monitoring crustal movements and tectonic deformation.

R. E. REILINGER

References

Abdallah, A., V. Courtillot, M. Kasser, A. Le Dain, J. Lépine, B. Robineau, J. Ruegg, P. Tapponier, and A. Tarantola, 1979, Relevance of Afar seismicity and volcanism to the mechanics of accreting plate boundaries, *Nature* **282,** 17-23.

Balling, N., 1980, The land uplift in Fennoscandia, gravity field anomalies and isostasy, in N. Mörner, ed., *Earth Rheology, Isostasy, and Eustasy.* New York: John Wiley & Sons, 295-321.

Bally, A. W., P. L. Bender, T. R. McGetchen, and R. I. Walcott, eds., 1980, *Dynamics of Plate Interiors.* Washington, D. C.: American Geophysical Union (Geodynamics Series, vol. 1), 162p.

Barazangi, M., and J. Dorman, 1969, World seismicity map of ESSA Coast and Geodetic Survey epicenter data for 1961-1967, *Seismol. Soc. America Bull.* **59,** 369-380.

Bjornsson, A., G. Johnsen, S. Sigurdsson, G. Thorbergsson, and E. Tryggvason, 1979, Rifting of the plate boundary in North Iceland 1975-1978, *Jour. Geophys. Research* **84,** 3029-3038.

Christodoulidis, D. C., D. E. Smith, R. Kolenkiewicz, S. M. Klosko, M. H. Torrence, and P. J. Dunn, 1985, Observing tectonic plate motions and deformations from Satellite Laser Ranging, *Jour. Geophys. Research* **90,** 9249-9263.

Geodynamics Program Office, National Aeronautics and Space Administration, 1979, *Applications of Space Technology to Crustal Dynamics and Earthquake Research,* Washington, D. C.: NASA Technical Paper 1464, 257p.

Geophysics Study Committee, National Research Council, 1986, *Active Tectonics.* Washington, D.C.: National Academy Press.

Molnar, P., and P. Tapponnier, 1978, Active tectonics of Tibet, *Jour. Geophys. Research* **83,** 5361-5375.

Mörner, N., ed., 1980, *Earth Rheology, Isostasy, and Eustasy.* New York: John Wiley & Sons, 599p.

Panel on Crustal Movement Measurements, National Research Council, 1981, *Geodetic Monitoring of Tectonic Deformation—Toward a Strategy.* Washington, D.C.: National Academy Press, 109p.

Savage, J. C., 1983, Strain accumulation in the Western United States, *Ann. Rev. Earth Planetary Sci.* **11,** 11-43.

Thatcher, W., 1984, The earthquake deformation cycle, recurrence, and the time predictable model, *Jour. Geophys. Research* **89,** 5674-5680.

Cross-references: *Continental Collision Zones; Continental Crustal Structure; Continental Drilling: Ultradeep; Continental Lithosphere; Continental Rifting; Deformation of Rocks and Minerals; Earthquakes and Crustal Deformation; Earthquakes and Seismicity; Earth Tides; Lithosphere: Mechanical Properties; Lithosphere, Oceanic: Formation and Evolution; Ocean Continent Transition: Structure; Satellite Laser Positioning; Seismicity: Intraplate; Seismicity: Subduction; Stress in the Earth's Lithosphere; Thin-Skin Tectonics; Very-Long-Baseline Interferometry (VLBI).*

CRUSTAL STRUCTURE—See CONTINENTAL CRUSTAL STRUCTURE; LITHOSPHERE, OCEANIC.

CURIE TEMPERATURE

The Curie temperature of a rock is the temperature (upon heating) at which ferromagnetic mineral components of the rock lose their magnetization and exhibit paramagnetic properties. Rocks frequently have no single, well-defined Curie temperature. Instead they possess a range of temperatures that depend on and reflect the chemical composition of the ferromagnetic mineral components. Because the Curie temperature is an intrinsic material property, it is better suited than most other magnetic parameters for estimating the chemical makeup of magnetic minerals in rocks. In common rocks, minerals of the iron-titanium oxide group and, to a much smaller percentage, of the iron sulfide group exhibit ferromagnetic properties and hence control the Curie temperature. In sedimentary rocks, iron hydroxides also play an important role. For the Curie temperature of minerals see *Magnetic Properties of Minerals*.

Igneous Rocks

Igneous rocks have best been studied in terms of the variation of Curie temperature with rock type and the composition of the ferromagnetic components. The following summarizes our understanding of these dependencies.

Primary and Secondary Curie Temperatures. The primary (original) ferromagnetic mineral in igneous rocks is titanomagnetite ($Fe_{3-x}Ti_xO_4$, $0 \leq x \leq 1$), which crystallizes together with members of the hematite-ilmenite series ($Fe_{2-y}Ti_yO_3$, $0 \leq y \leq 1$) from silicate melts at temperatures of 1100°-700°C. The contribution of primary pyrrhotite to the magnetic properties of almost all igneous rocks is negligible.

The composition of the primary Fe-Ti oxides varies in general with the SiO_2 content of the rock (Table 1) but is roughly constant for a given rock type. Hence, the Curie temperature varies in a characteristic way according to rock type (Fig. 1). It is important to distinguish primary and secondary Curie temperatures: The *primary* Curie temperature expresses the original composition of the primary magnetic minerals before any subsequent alteration. *Secondary* Curie temperatures arise from alteration

TABLE 1. Composition and Curie Temperatures of the Primary Fe-Ti Oxides in Igneous Rocks

Bulk Chemistry	acid	intermediate	basic
Rock Type	Granite rhyolite	Dacite trachite	Gabbro basalt
Titanomagnetite $Fe_{3-x}Ti_xO_4$	$0 \leq x \leq 0.2$	$0.2 \leq x \leq 0.6$	$0.5 \leq x \leq 0.85$
Curie Temperature (°C)	580–463	463–165	244--−41
Ilmenite–Hematite Series $Fe_{2-y}Ti_yO_3$	$0.5 \leq y \leq 0.9$	$0.5 \leq y \leq 0.9$	$0.8 \leq y \leq 1.0$
Curie Temperature (°C)	220–130	220--−130	−40--−220

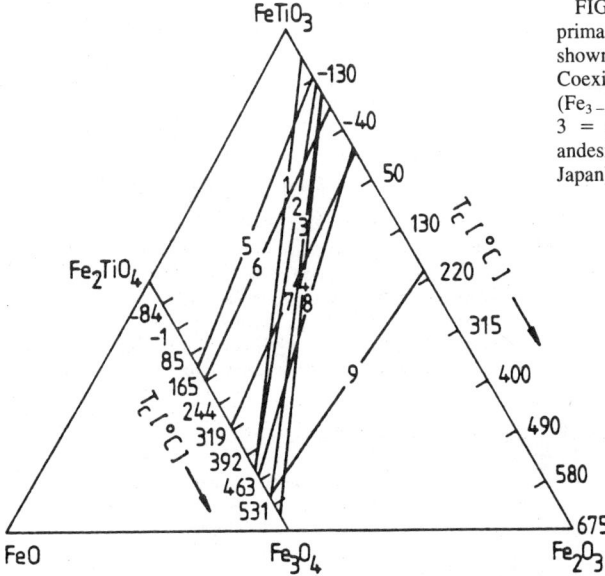

FIGURE 1. Composition and Curie temperatures of primary Fe-Ti oxides from different igneous rocks, shown in the ternary system FeO-Fe_2O_3-TiO_2. Coexisting members of the series $(Fe_{2-y}Ti_yO_3)$ and $(Fe_{3-x}Ti_xO_4)$ are connected by tielines. 1, 2 = syenite, 3 = granite, 4 = pegmatite, 5, 6 = basalt, 7 = andesite, 8 = rhyolte, 9 = dacite (Haruna Volcano, Japan). (After Buddington and Lindsley, 1964)

of the original magnetic components by oxidation, reduction, or phase inversion. A comparison of the actual Curie temperature of a rock with its theoretical primary Curie temperature (read from Table 1 or Fig. 1) gives a simple method of assessing the degree of alteration (see Fig. 2).

Changes in Magnetic Properties by Oxidation and Reduction of the Primary Fe-Ti Oxides. By far the most common type of alteration affecting the magnetic minerals in igneous rocks is oxidation at the Earth's surface. Generally speaking, oxidation of the primary Fe-Ti oxides causes a rise in the Curie temperature. Conversely, reduction generally causes the Curie temperature to fall (except in the extreme case where native iron forms, with a Curie temperature of $\approx 700°C$).

The change in Curie temperature and other magnetic properties depends strongly on the temperature at which oxidation or reduction takes place.

In basaltic rocks (sensu lato) the following types of oxidation of the magnetic minerals are distinguished:

1. High-Temperature Oxidation (T > 350°C)

Progressive oxidation ⟶

Titanomagnetite $T_C = 100°$–$200°C$

Magnetite $T_C = 550°$–$580°C$

Hematite $T_C = 650°$–$680°C$

Ilmenite $T_C = -180°C$

Rutile (pseudobrookite if $T > 600°C$)

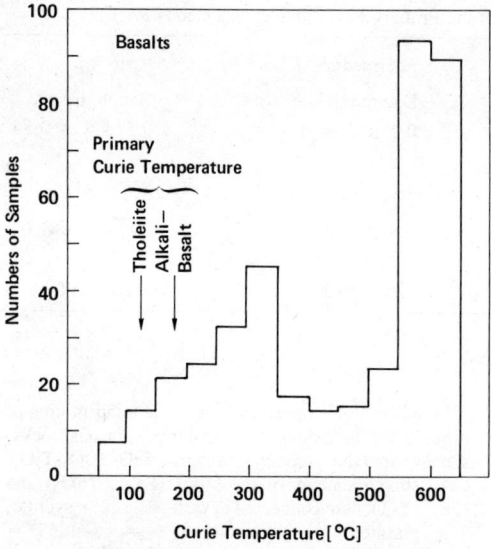

FIGURE 2. Histogramm of Curie temperatures from Tertiary and Cretaceous continental basalts. Also shown for comparison are the primary Curie temperatures of tholeiitic and alkali basalts. (After Petersen and Bleil 1982)

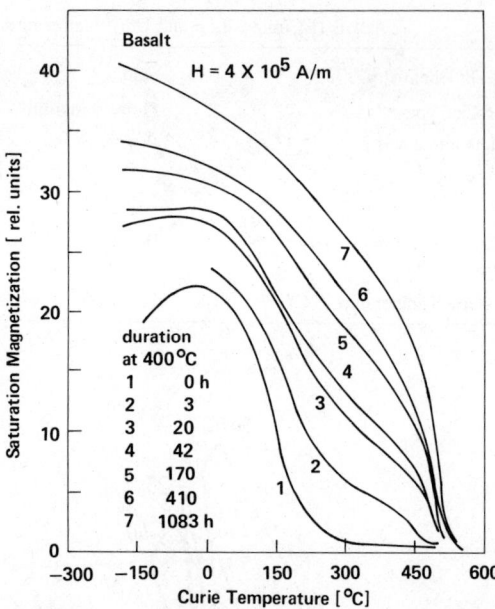

FIGURE 3. The effect of high-temperature oxidation on the primary Fe-Ti oxides in basalt. Prior to measurement the samples were heated in air for various periods of time. The magnetization curves show a characteristic pattern with increasing preheating: Starting at the primary Curie temperature of about 190°C, the Curie temperatures rise steadily, along with the magnetization intensity.

TABLE 2. Oxidation Scale Proposed by Wilson and Watkins (1976)

Class	Observation	Curie Temperature (°C)
I	Homogeneous titanomagnetite	100–200
II	Titanomagnetite with a few ilmenite exsolution lamellae	100–500 generally poorly defined
III	Titanomagnetite with abundant ilmenite exsolution lamellae	500–580
IV	The ilmenite lamellae lose their sharp outlines, are brighter in color and speckled—"meta-ilmentite"	550–580 also indications of 600–680
V	The crystallites of the oxidation products increase in size; rutile appears intergrown with titanhematite; relicts of titanomagnetite may still be present	550–580 and 600–680
VI	Complete oxidation. No relict titanomagnetite. The typical mineral is pseudobrookite	650–680

The formation of magnetite during high-temperature oxidation increases both the Curie temperature and the magnetization intensity (Fig. 3). Hematite forms with further oxidation, with another increase in Curie temperature but a decrease in intensity of magnetization.

The degree of high-temperature oxidation can be established easily with reflected-light microscopy. Wilson and Watkins (1967) proposed a widely used six-stage oxidation scale that is based on microscopic observations (Table 2). High-temperature oxidation of primary Fe-Ti oxides takes place primarily in subaerial basalt flows, with complicated effects on their magnetic behavior.

A particular variety of high-temperature oxidation takes place in gabbros and within thick-basalt flows without contact with the atmosphere. Here, under closed-system, internally buffered conditions, the primary titanomagnetites react on cooling to form intergrowths of magnetite and ilmenite.

This "internal oxidation" leads to a rise in Curie temperature and magnetization intensity (Hargraves

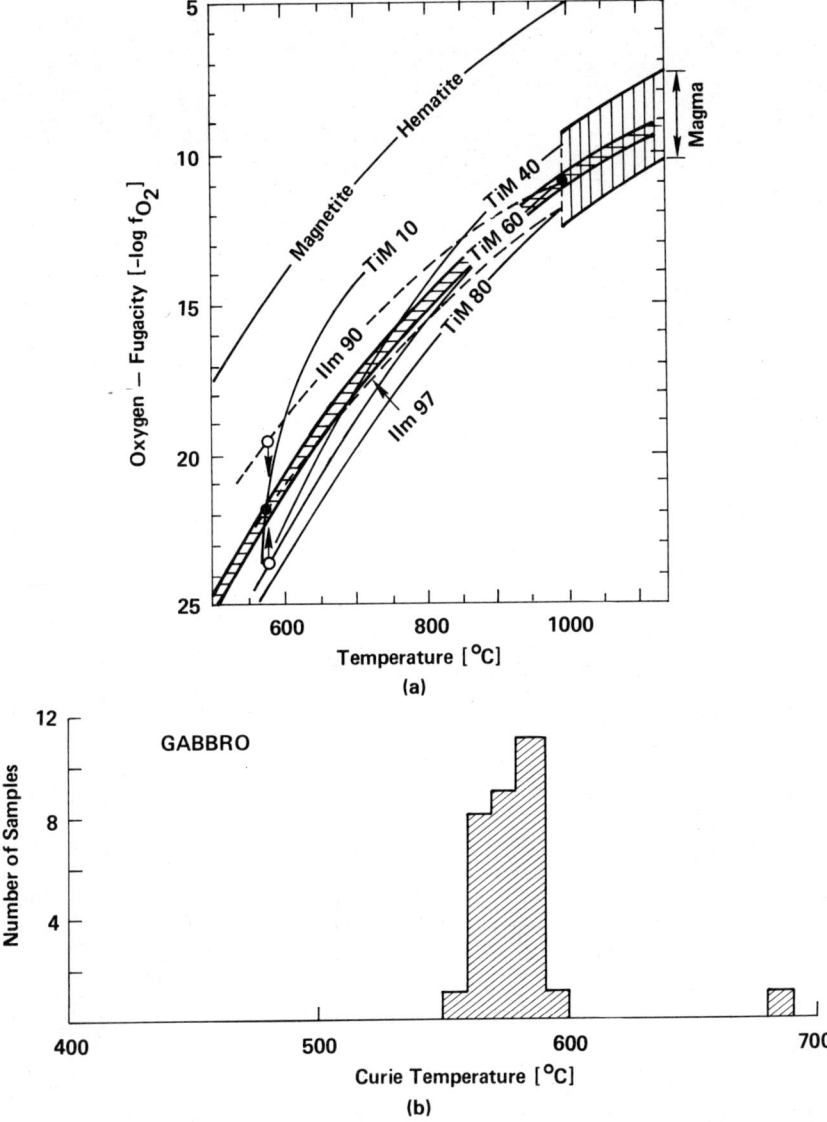

FIGURE 4. Internal oxidation/reduction of primary Fe-Ti oxides in a slowly cooling basaltic magma under closed-system conditions. Such conditions are closely approached by gabbro intrusions and thick basalt flows. (a) Oxygen fugacity (f_{O2})-temperature dependence. The range of f_{O2} in basaltic magmas is vertically ruled (after Carmichael and Nicholls, 1967). The QFM (quartz-fayalite-magnetite) buffer curve is shown in the horizontally ruled pattern. This curve closely approximates f_{O2}-T conditions in a cooling basaltic magma. The TiM and Ilm curves show equilibrium f_{O2}-T relations for the titanomagnetite and ilmenite-hematite series, respectively (after Buddington and Lindsley, 1964). The primary Fe-Ti oxides crystallized from basaltic magma at about 1050°C (solid circle) in equilibrium with the local oxygen fugacity are TiM 60 ($Fe_{3-x}Ti_xO_4$ with $x = 0.60$) and Ilm 90 ($Fe_{2-y}Ti_yO_3$ with $y = 0.90$). After cooling to 600°C, however, the compositions of oxides in equilibrium with the local oxygen fugacity (QFM buffer) have changed to TiM 10 and Ilm 97 (solid circle at 600°C). Thus the primary titanomagnetite TiM 60 has changed in composition, by oxidation and exsolution of ilmenite during cooling, to a Ti-poor magnetite TiM 10 (open circle). The primary ferro-ilmenite Ilm 90 similarly changes to Ilm 97 during cooling by reduction and exsolution of magnetite. In the first case the result is the formation of ilmenite lamellae in magnetite, and in the second case fine needles of magnetite in ilmenite. The entire process leads to an increase in the Curie temperature to 550°–600°C (Hargraves and Petersen 1971). (b) Histogram of Curie temperatures of gabbro samples. The samples plotted are from the Skaergaard intrusion (Chevalier et al., 1954) and the North Atlantic (Kent et al., 1978).

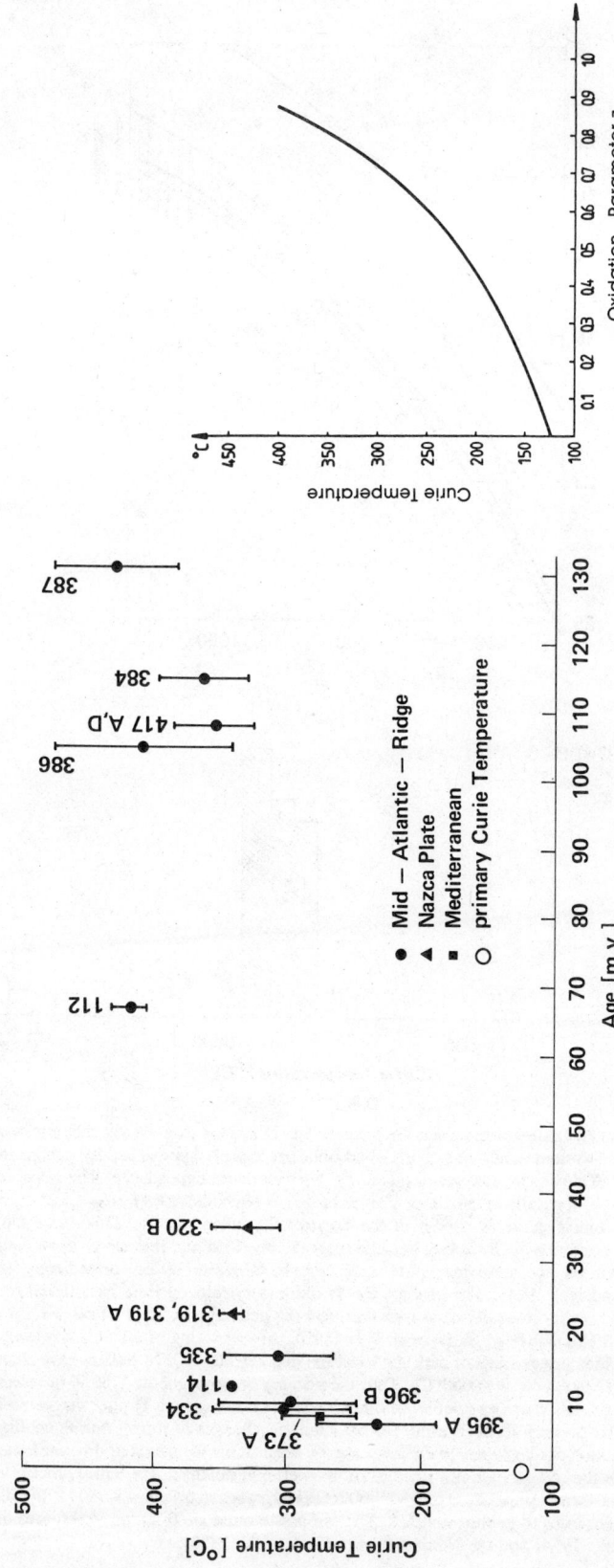

FIGURE 5. The effect of low-temperature oxidation of titanomagnetites in oceanic basalts (after Petersen and Bleil 1982): (a) The age dependence of the Curie temperature in oceanic basalts. The basalt samples are from the deep sea drilling project; DSDP drill hole numbers are shown. Error bars give the range of measured Curie temperatures. The primary Curie temperature of oceanic basalts is shown for comparison. (b) The Curie temperature of titanomagnetites separated from oceanic basalts plotted against the titanomagnetite oxidation parameter z ($0 \leq z \leq 1$). z is defined as the fraction of the initial Fe^{2+} ions in titanomagnetite converted to Fe^{3+} ions (O'Reilly and Banerjee, 1967).

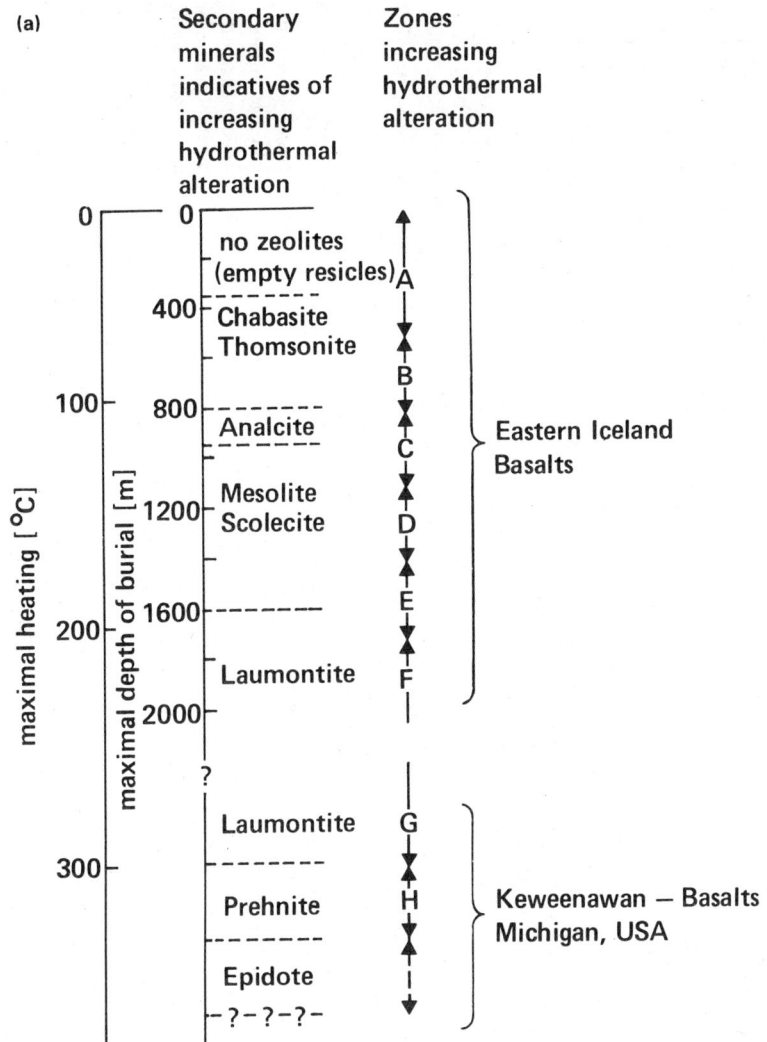

FIGURE 6. The influence of hydrothermal alteration on Curie temperature (after Ade-Hall et al., 1971). (a) Standard sequence of zones of increasing hydrothermal alteration, based on studies of East Iceland lavas and Keweenawan lavas, Michigan, United States. The temperature and pressure increase from zones A to I.

and Petersen, 1971). Thus Curie temperatures of 580° are typical for gabbros and dolerites (Figs. 4a, b).

2. *Low-Temperature Alteration ($T \leq 300°C$)*

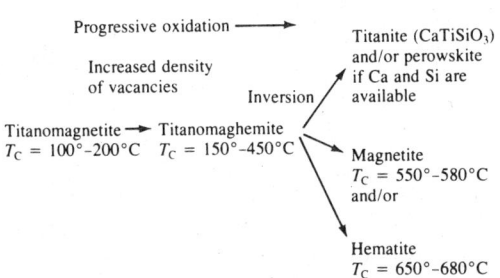

Low-temperature oxidation processes also lead to higher Curie temperatures. However, the intensity of magnetization decreases with increasing oxidation until the density of vacancies is too great and phase inversion of the titanomaghemite occurs. If magnetite forms, the magnetization intensity increases; if hematite forms, it decreases further.

Low-temperature oxidation is the typical form of alteration in oceanic basalts. A striking phenomenon observed here is the steady increase in the Curie temperature of oceanic basalts with increasing age (Figs. 5a, b). Low-temperature oxidation also occurs in subaerial basalts, often transitional to hydrothermal alteration.

3. *Hydrothermal Alteration ($50 \leq T \leq 400°C$).* This type of oxidation of primary Fe-Ti

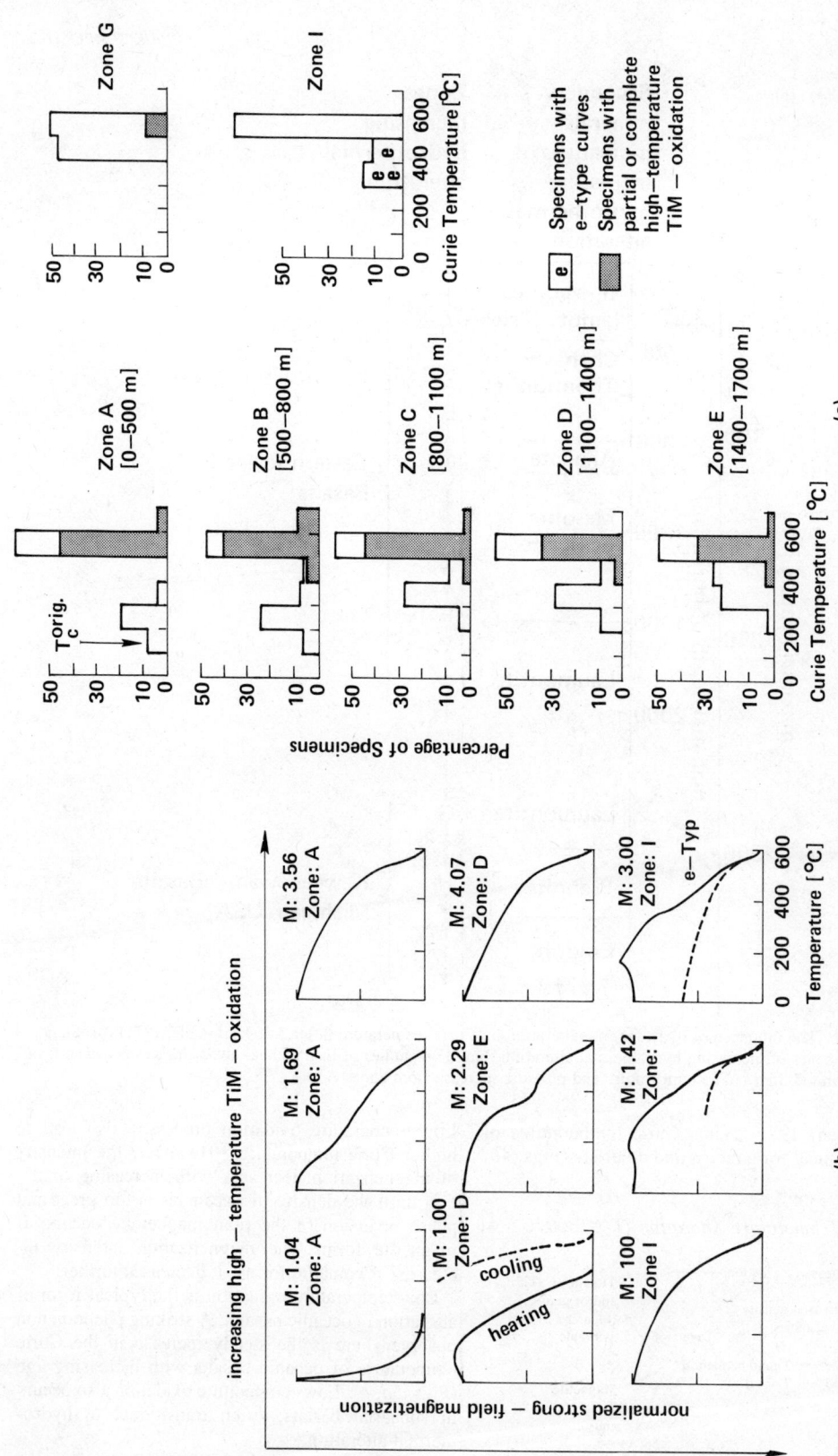

FIGURE 6 (Continued). (b) Influence of hydrothermal alteration on the Curie temperature. The development of thermomagnetic curves (measured in a magnetic field of 80×10^5 $A \cdot m^{-1}$) for basalt samples from various zones of hydrothermal alteration. M is the degree of high-temperature oxidation of titanomagnetite according to the scale of Wilson and Watkins (1.0 = no oxidation, 6.0 = maximal oxidation). (c) Influence of hydrothermal alteration on the Curie temperature in basalts. Histograms of the Curie temperatures of basalt samples from various zones of hydrothermal alteration. $T_C^{orig.}$ indicates the primary Curie temperature of tholeiitic basalts.

oxides is intermediate between high- and low-temperature oxidation. It is important in intrusive volcanic sequences and ophiolite complexes (Figs. 6a, b, c). It is often difficult to distinguish between these three types of oxidation in natural samples, because they are transitional with one another. The dividing temperatures given here are to be understood as rough guides. Taken together, the three oxidation types cause a rise in Curie temperatures from the primary 100°–200° to a maximum of 680°C.

NIKOLAI PETERSEN

References

Ade-Hall, J. M., H. C. Palmer, and T. P. Hubbard, 1971, The magnetic and opaque petrological response of basalts to regional hydrothermal alteration, *Royal Astron. Soc. Geophys. Jour.* **24**, 137–174.

Buddington, A. F., and D. H. Lindsley, 1964, Iron-titanium oxide minerals and synthetic equivalents, *Jour. Petrology* **5**, 310–357.

Carmichael, I. S. E., and J. Nicholls, 1967, Iron-titanium oxides and oxygen fugacities in volcanic rocks, *Jour. Geophys. Research* **72**, 4665–4687.

Chevallier, R., S. Mathieu, and E. A. Vincent, 1954, Iron-titanium oxide minerals in layered gabbros of Skaergaard intrusion, East Greenland, *Geochim. Cosmochim. Acta* **6**, 27–34.

Hargraves, R. B., and N. Petersen, 1971, Notes on the correlation between petrology and magnetic properties of basaltic rocks, *Zeitschr. Geophysik* **37**, 367–382.

Kent, D.V., B. M. Honnorez, N. D. Opdyke, and P. J. Fox, 1978, Magnetic properties of dredged oceanic gabbros and the source of marine magnetic anomalies, *Royal Astron. Soc. Geophys. Jour.* **55**, 513–537.

O'Reilly, W., and S. K. Banerjee, 1967, The mechanism of oxidation in titanomagnetites: a magnetic study, *Mineralog. Mag.* **36**, 29–37.

Petersen, N., and U. Bleil, 1982, Magnetic properties of rocks, in G. Angenheister, ed., *Landolt-Börnstein: Numerical Data and Functional Relationships in Science and Technology.* Berlin: Springer-Verlag, 366–432.

Wilson, R. L., and N. D. Watkins, 1967, Correlation of petrology and natural magnetic polarity in Columbia Plateau basalts, *Royal Astron. Soc. Geophys. Jour.* **12**, 405–424.

Cross-references: *Chemical Remanent Magnetization; Demagnetization; Magnetic Domains; Magnetic Properties of Minerals; Magnetic Self-Reversal; Rock Magnetism; Rock Magnetism: Measuring Techniques and Apparatus; Thermoremanence; Viscous Remanent Magnetization (VRM) and Viscous Remagnetization.*

D

DEEP EARTHQUAKES

Earthquakes occurring at depths below about 60 km are called deep earthquakes. They are frequently classified into two groups, intermediate-depth and deep (in the restricted sense) earthquakes, depending on their focal depths. The boundary between the intermediate and deep foci is usually set at 300 km. The deep seismic activity extends to about 700 km in the mantle. The deeper parts of the Earth are apparently completely aseismic.

Wadati in 1928 first showed clear evidence for the existence of deep earthquakes in and near Japan. Subsequently, it became clear that deep earthquakes occurred in several different areas. The worldwide distribution of deep earthquakes is apparent in Fig. 1, which reveals that the deep seismicity is restricted to island-arcs and island-arc-like structure around the circum-Pacific belt and in the Alpide belt. Since their discovery, deep earthquakes have furnished important information about the interior of the Earth and mantle dynamics.

Deep Seismic Zone

Some features of the spatial distribution of deep earthquakes can be obtained by plotting their epicenters on a map. Figure 2 demonstrates the epicentral distribution in the typical island-arc region (the Kurile, northern Honshu, Izu-Bonin, Mariana, and Ryukyu arcs). As indicated by contour lines of equal focal depth, deep earthquake foci in each region are distributed about a plane (a seismic zone) dipping from a region around the trench to depths of several hundred kilometers in the upper mantle. The vertical section of hypocenters nearly perpendicular to the trench axis shows that the thickness of the inclined seismic zone is a few tens of kilometers.

The depth contours shown in Fig. 2 run almost parallel to the trench axis. Most active volcanoes lie along the 150 km depth contour. The nearly equal intervals between contours indicate that the dip of the seismic zone does not change appreciably with depth, except in the region of Izu-Bonin. The dip value, however, varies regionally: about 45° in the Kurile arc; about 30° in the northern Honshu arc; and about 45° in the Ryukyu arc. It is evident that the distribution of earthquake foci is uneven within these same zones.

Beneath the Izu-Bonin arc, the inclined seismic zone varies as a function of depth and location along the strike. A similar contortion of the seismic zone can be seen at the northern end of the Tonga arc and

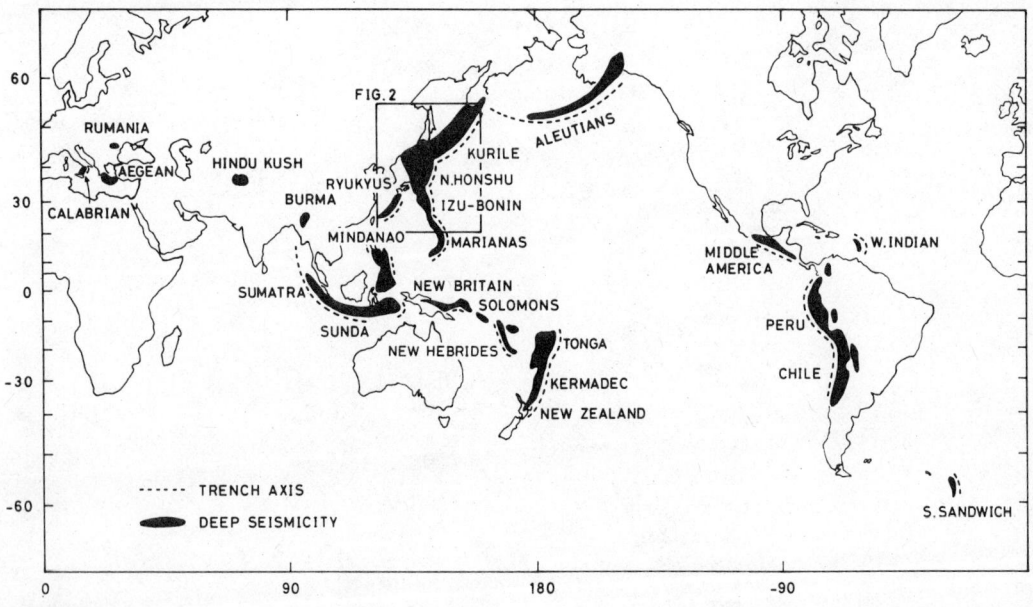

FIGURE 1. Schematic illustration of the world deep seismicity.

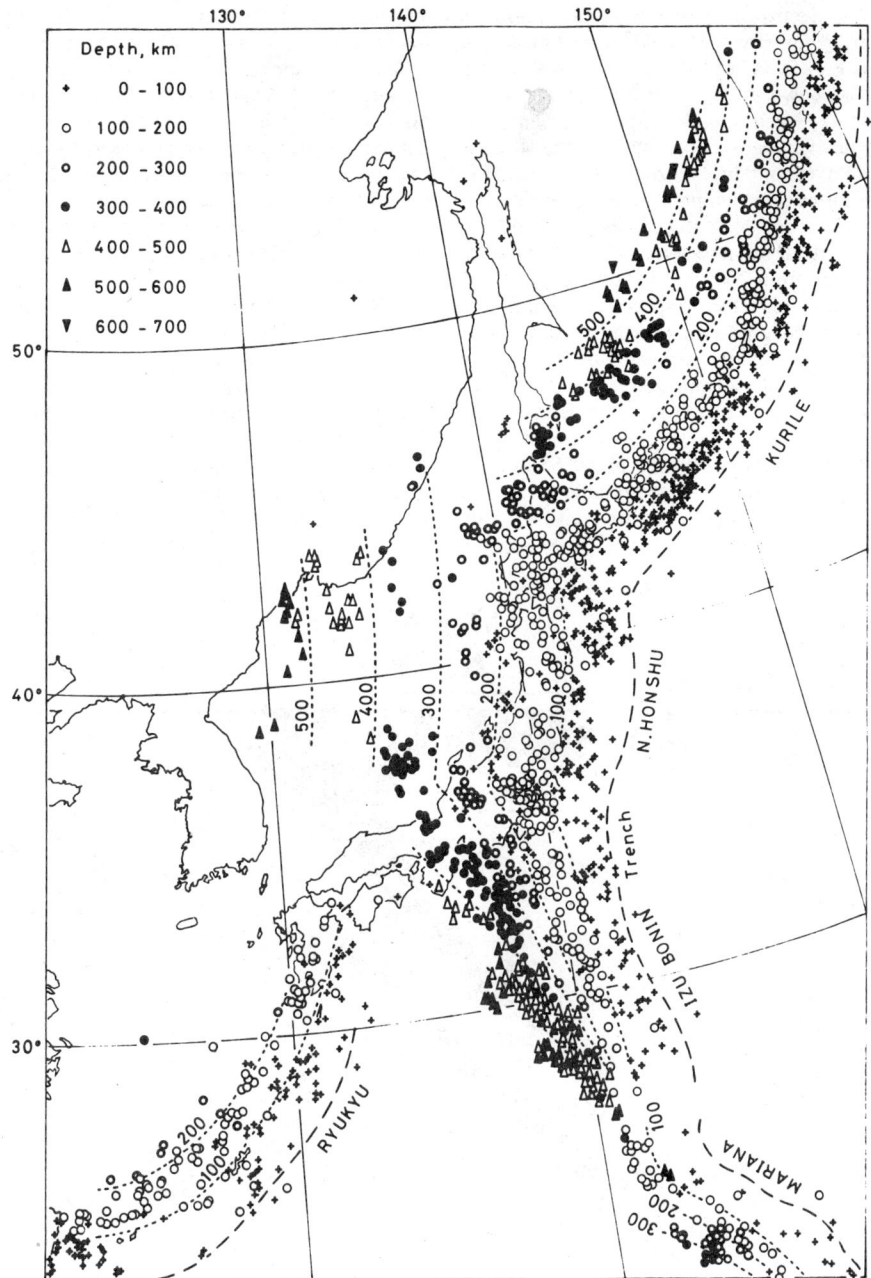

FIGURE 2. Distribution of epicenters of earthquakes (1964–1973) in the Kurile, northern Honshu, Izu-Bonin, Mariana and Ryukyu arcs. Dotted lines indicate contour lines of equal focal depth (redrawn from Utsu, 1977).

beneath the Banda Sea at the eastern end of the Sunda arc. In these two regions, the lateral edges of the seismic zones are bent upward.

In several island-arc regions, the seismic activity roughly shows an exponential decrease down to a depth of 150 km to 300 km. There is no evidence for seismic activity below that depth in the Aleutians, Middle America, Ryukyus or Hindu Kush. In some regions such as Tonga-Kermadec, Izu-Marianas, Kuriles, Sunda, and South America, a second maximum of activity occurs near the bottom of the seismic zone at about 400 km to 600 km. These seismic zones are characterized by a minimum or complete lack of activity at some intermediate depth ranges. The depths of maximum and minimum activity vary regionally even in the same zone as can be seen in Fig. 2. A global summary of the deep seismicity is shown in Fig. 2 of *Subduction Zones*. In that figure, each cross-section represents the seismicity in a vertical section aligned perpendicular to the strike of the subduction zone. Not included in the figure are certain complex regions near prominent junctions or ends of zones. A distinctive feature is a nearly vertical seismic zone in the Marianas in the depth range from 200 km to about 700 km.

Until recently, the deep hypocenters within any given arc were thought to form a single plane. In the mid-1970s, a two-layered structure of the seismic zone, i.e., the double seismic zone, was discovered at intermediate depths in the northern Honshu and the Kurile arcs; the two layers are separated by 30 km to 40 km for depths between about 65 km and 185 km (Fig. 3). In the observed double zones, the level of seismic activity in the two layers varies greatly; in northern Honshu the lower zone is considerably less active, while in the Kuriles the lower zone appears to be more active. Precise hypocenter determinations made possible by the installation of local seismic networks led to the discovery of double seismic zones in these arcs. The double seismic zone is not considered to be a feature of all island arcs, but the accuracy of hypocenter determination varies from arc to arc.

A number of seismological studies have indicated the existence of the large-scale anomalous structure beneath island-arc regions, i.e., the existence of a high Q, high velocity (high V) zone about 100 km thick that extends downward from the vicinity of the trench to a depth of several hundred kilometers (Fig. 4). Deep earthquakes, indicated by dots in Fig. 4, occur at the uppermost portion of the high Q, high V zone. In the plate tectonics hypothesis, the high Q, high V zone is taken to be the oceanic lithospheric slab that descends into the mantle. Therefore, deep and intermediate-depth earthquakes mark the location of the descending lithosphere.

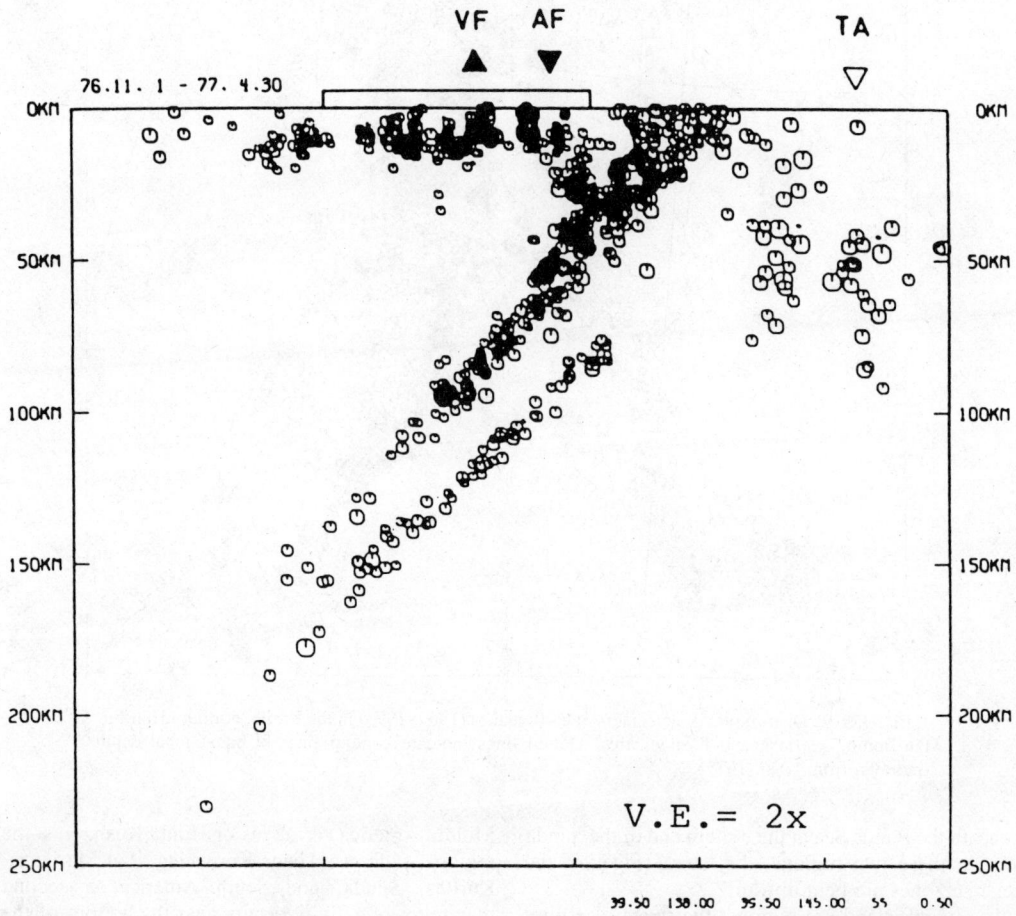

FIGURE 3. Double seismic zone in the northern Honshu arc. Earthquakes located in the region between 39°N and 40°N are projected on the vertical section in E-W direction (from Hasegawa, et al., 1979).

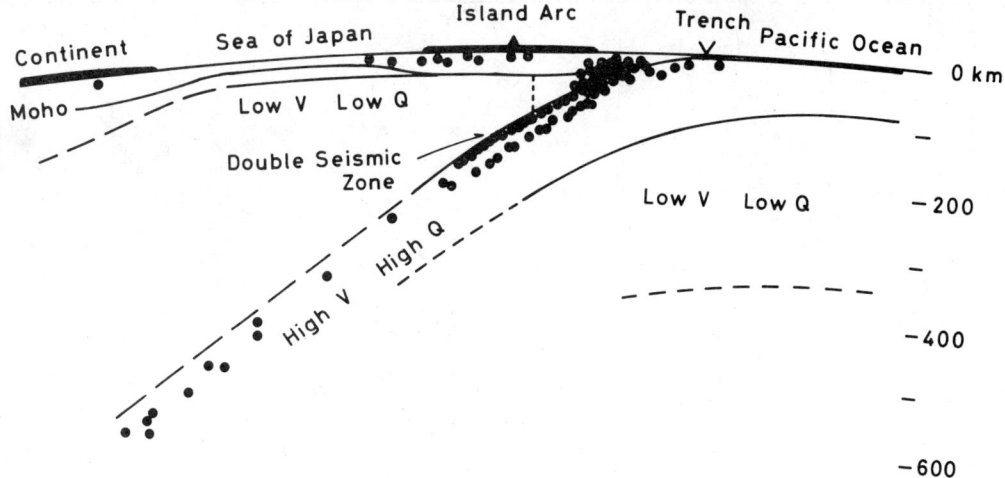

FIGURE 4. A model for the deep structure of the northern Honshu arc. Dots represent earthquake foci. The double seismic zone is schematically shown from Fig. 4 (redrawn from Utsu, 1977).

Earthquake Mechanism

The distribution of P wave first motions and S wave polarization angles from deep earthquakes can be adequately explained by the double-couple model, as is the case for shallow earthquakes. The parameters of the double-couple solution (the earthquake mechanism solution) are a pair of orthogonal planes at the focus (the nodal planes) and three stress axes (the P-, B-, and T-axes). In the dislocation model or the shear faulting model equivalent to the double couple, one of the nodal planes is the fault plane, and the pole of the other nodal plane is parallel to the slip vector. It is simply assumed that the P-, B-, and T-axes correspond to the principal axes of maximum, intermediate, and minimum compressive stress. The mechanism solution provides information about a possible fault motion and the earthquake-generating stresses.

A great number of mechanism solutions for deep earthquakes have been obtained in various island arcs. From these solutions, it has been found that an intimate relationship exists between orientations of the stress axes and the inclined seismic zones; either the P- or T-axis is most nearly parallel to the planar geometry of the seismic zone. Figure 5 shows examples of the relationship for the Kurile and northern Honshu arcs. As usual we refer to orientations of the stress axes in which the P or T axis is parallel to the local dip of the seismic zone as "down-dip compression" or "down-dip extension," respectively.

A global summary of the variation of down-dip stress type as a function of depth is shown in Fig. 2 of *Seismicity: Subduction Zones*. The figure shows that the deep solutions below about 300 km are very strongly dominated by down-dip compression but that intermediate-depth solutions are more variable and include both down-dip compression and extension. Even here, however, there is an indication that down-dip extension is predominant in regions characterized either by gaps in the seismicity as a function of depth or by an absence of deep earthquakes (e.g., northern Chile and Middle America). The fault plane solutions in complex regions such as the junction between the northern Honshu and Izu-Bonin segments and the northern end of the Tonga arc exhibit rather localized effects caused by bending in the seismic zone.

The double seismic zones observed in the northern Honshu and the Kurile arcs are interesting not only in their distribution of hypocenters, but also in their stress types; the upper zone is in down-dip compression and the lower zone is in down-dip extension.

Source Process

The source process has been studied by assuming a moving dislocation model or a propagating fault model. The parameters that characterize the source process are seismic moment, source area, rupture velocity, average dislocation, and stress drop. For shallow earthquakes, these parameters can be estimated from aftershock area, geodetic data, tsunami source area, and surface wave analyses, as well as body wave analyses. For deep earthquakes, however, the first four of the above methods cannot be applied in most cases; hence, body waves provide the fundamental information for the source process.

The source parameters are estimated generally by matching synthetic seismograms to observed seismograms recorded by long-period seismometers. The duration of the far-field time function of body waves and its azimuthal variation caused by the rupture propagation are used to estimate the source size and rupture velocity. The seismic moment is

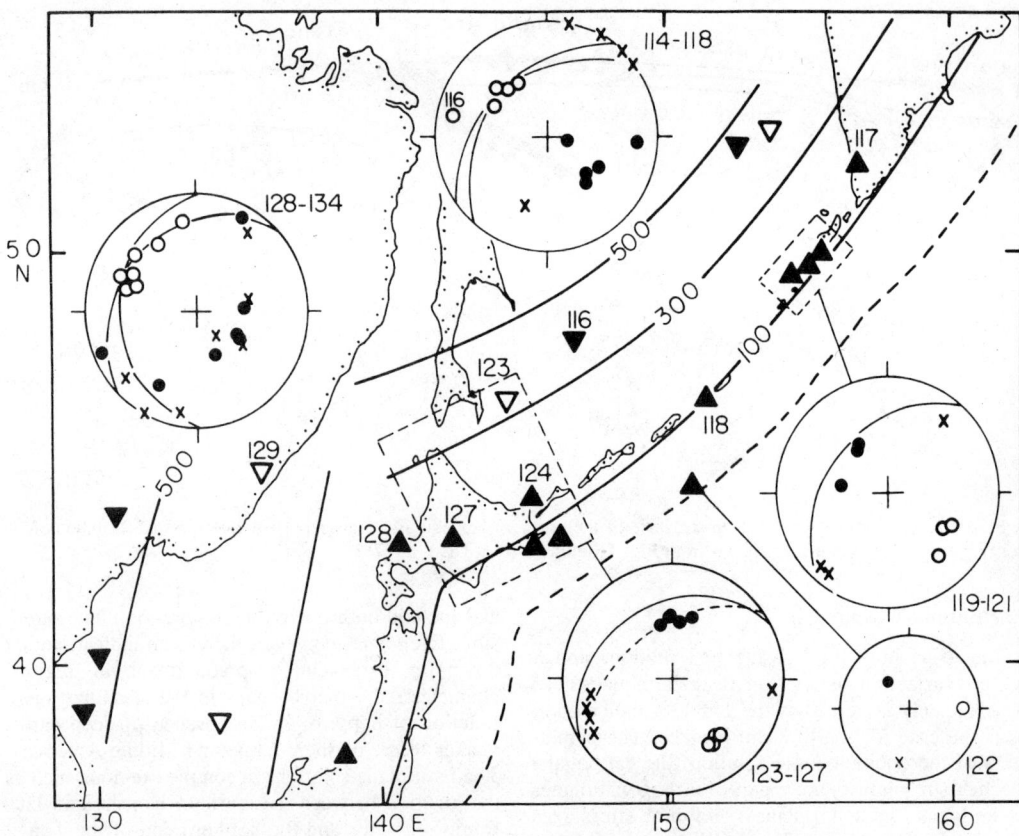

FIGURE 5. Equal area projection of the P axes (unfilled circles), T axes (filled circles) and B axes (crosses) for deep earthquakes in the Kurile and the northern Honshu arcs. The solid line in the projection is the trace of the plane that best fits the orientation of the seismic zone, but in some cases two lines are present due to the contortion of the seismic zone (from Isacks and Molnar, 1971).

determined by comparing the theoretical and observed amplitudes of body waves. The average dislocation and the stress drop are calculated from estimated values of the seismic moment and the source area.

Studies of the source process for many deep earthquakes have been done, but reliable source parameters are few. Figure 6 shows the relation between source areas, seismic moments, and stress drops, estimated for 18 deep earthquakes. Usually the seismic moment (M_0) can be well determined. The largest event at intermediate depths is the Banda Sea earthquake of November 4, 1963 ($M_0 = 3.1 \times 10^{28}$ dyne cm). The Colombia earthquake of July 31, 1970 ($M_0 = 2.2 \times 10^{28}$ dyne cm) is the largest among the deep focus events. These large seismic moments were determined from long-period surface wave analyses. The source areas and stress drops are more dependent on the rupture model assumed and on the method used, hence have the large uncertainties. However, the stress drops, which are the most important parameters for investigating the state of stress and material properties within the deep seismic zone, range from 100 to 1000 bars, much larger than those observed for shallow earthquakes (10–100 bars). Furthermore, the events plotted in Fig. 6 show the following characteristics: the average dislocations are 1 m to 7 m; the rupture velocities are less than S-wave velocity around the focus; the rupture times range from several seconds to several tens of seconds. These source characteristics represent the general features of deep earthquakes. Far more studies, and more detailed studies, of the source process will be required to investigate the variation of these parameters as a function of seismicity, depth, and region.

Seismograms from some deep earthquakes exhibit a very simple waveform for body waves (Fig. 7a). Complicated waveforms, however, are frequently recorded where seismograms may show abrupt changes in phase or sudden substantial increases in amplitude, or sometimes even consist of a sequence of simple pulses (Fig. 7b). The waveform complexity is apparently caused by time variation

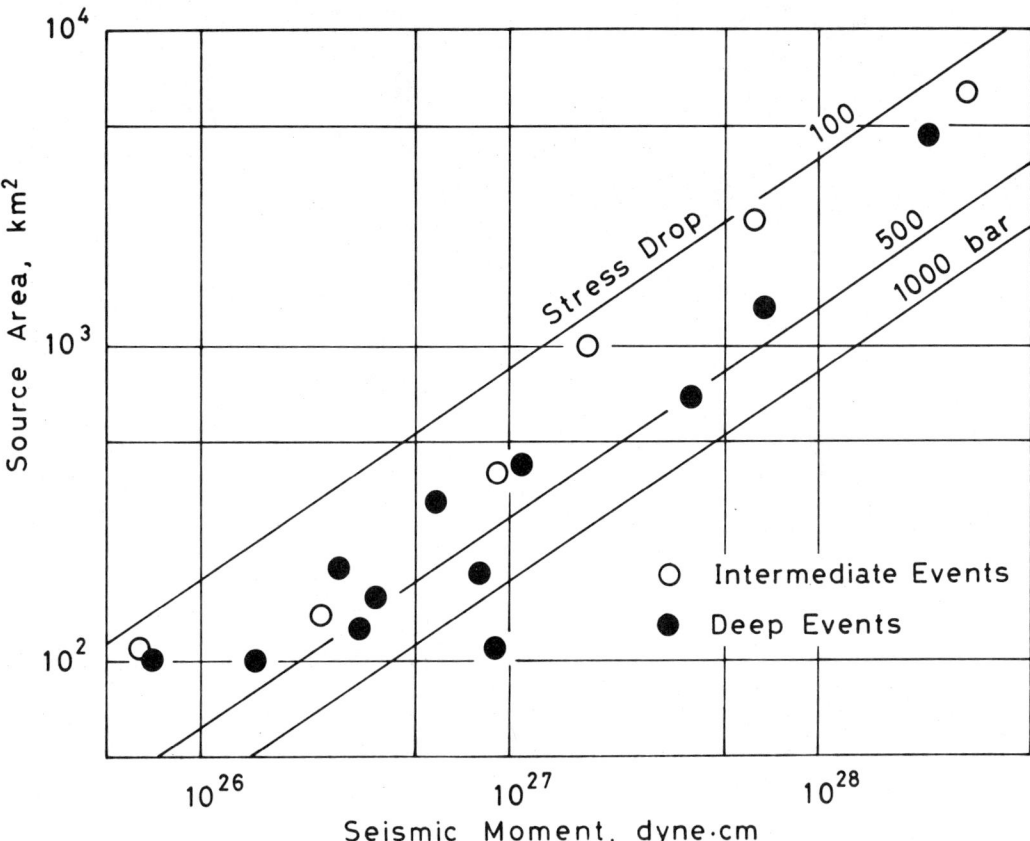

FIGURE 6. Relation between source areas and seismic moments. The straight lines give the relations for circular cracks with constant stress drop.

earthquake rupture process, such as a multiple shock consisting of several identifiable events in the process of strain release. For example, the deep Colombia event of July 31, 1970 (Fig. 7b) consists of several events during 50 sec, which are separated from one another by a few tens of kilometers. Strain release in the form of a multiple shock may indicate that the stress field or material strength in a region around the focus of deep earthquakes is quite inhomogeneous.

If a multiple shock occurs in the form of shear faulting, the events forming the multiplet are likely to share the same fault plane; the events subsequent to the first event are therefore expected to concentrate on one of two nodal planes. Many multiple shock analyses show such a tendency, supporting the shear faulting model for deep earthquakes.

Physical Process of Deep Earthquakes

As far as the radiation pattern of P and S waves are concerned, the earthquake mechanism of deep events is superficially similar to that of shallow ones; that is, the dislocation model or the shear faulting model. Laboratory experiments of rocks, however, indicate that it is difficult to attribute deep earthquakes to frictional instability, widely accepted as the faulting mechanism of shallow earthquakes. Under the physical conditions at great depth, frictional resistance to slip is extremely high and the rocks yield plastically. In order that fracture may occur, the resistance to slip after loss of cohesion must be less than the flow strength. To overcome these difficulties, a number of different hypotheses have been proposed for the instability mechanism of a deep seismic source, including shear melting instability in which a thermally activated creep promotes further acceleration of creeping in rocks, or embrittlement or weakening due to dehydration of hydrous minerals. Both the theoretical and the experimental bases for these hypotheses with respect to deep earthquakes are uncertain, and hence the physical mechanism of deep earthquakes still remains an open question.

In recent years, the seismic source has been represented in general form using the moment tensor (see *Earthquake Mechanisms; Seismic Source: Theory*), not constraining the mechanism to be the double

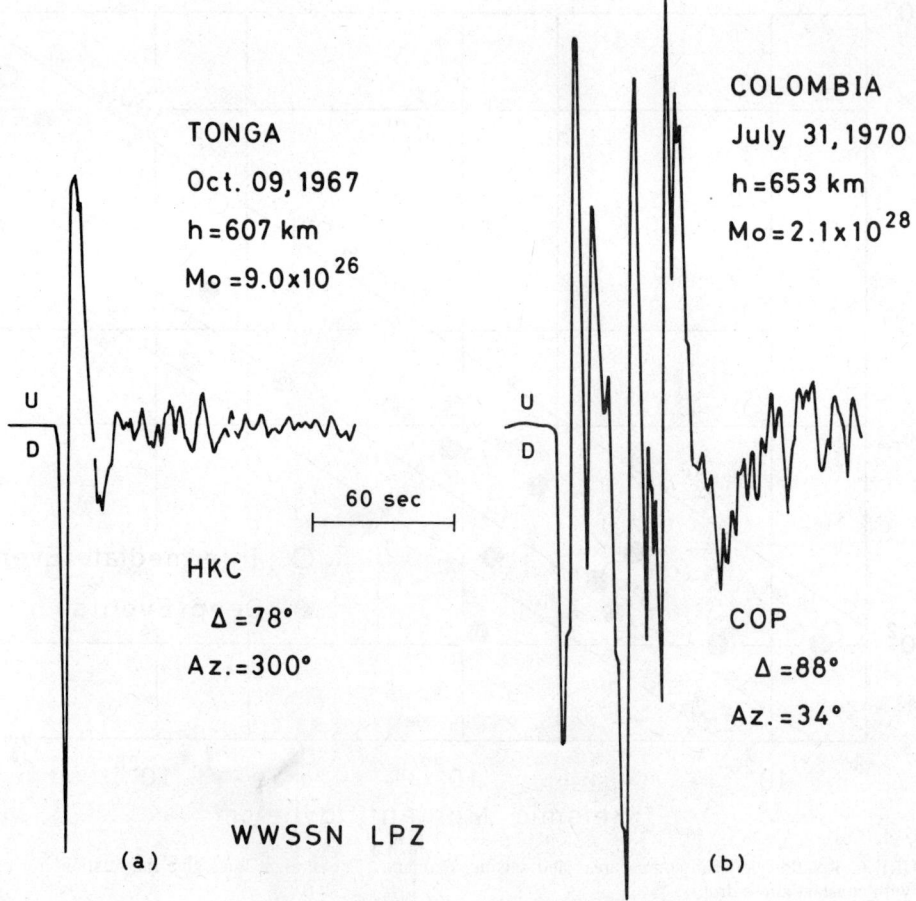

FIGURE 7. WWSSN long-period seismograms (vertical component) of P waves for a simple event (a) and a multiple shock (b).

couple, which is part of the moment tensor. The moment tensor inversion using high quality digital data can provide important information about properties of the seismic source, which are not obtainable by assuming a double couple, e.g., volume change of the source and the deviation from the double-couple model. In fact, the inversion result for the deep Colombia earthquake of July 31, 1970, obtained from data derived from free oscillations, shows evidence for large precursive compression in the focal region, indicating a transition of the matter in the mantle into a denser phase. At present the interpretation of the deviation from the double-couple model, however, is still very tentative and controversial.

Cause of Deep Earthquakes

Since their discovery, the cause of deep earthquakes as well as their physical mechanism has been a problem of utmost importance in geophysics. The classical (pre-plate tectonic) interpretation is that a deep seismic zone marks the location of a large-scale thrust fault between the continental and oceanic blocks; thus one of the nodal planes in the mechanism solution should be parallel to the dip of the deep seismic zone. This hypothesis has been directly contradicted by the fact that the P or T axis, rather than one of the nodal planes, is parallel to the dip of the seismic zone (see Fig. 5).

In the late 1960s, the plate tectonics hypothesis provided a good explanation for deep earthquakes. By this hypothesis, deep and intermediate-depth earthquakes mark the location of a descending tongue of lithosphere (a slab) and occur inside the descending lithosphere in response to stresses aligned to the slablike geometry of the lithosphere. Gravitational body forces caused by thermal density differences between the colder slab and surrounding mantle are important forces in determining the stress within the lithosphere. A correlation between the depth variations of down-dip stress type and

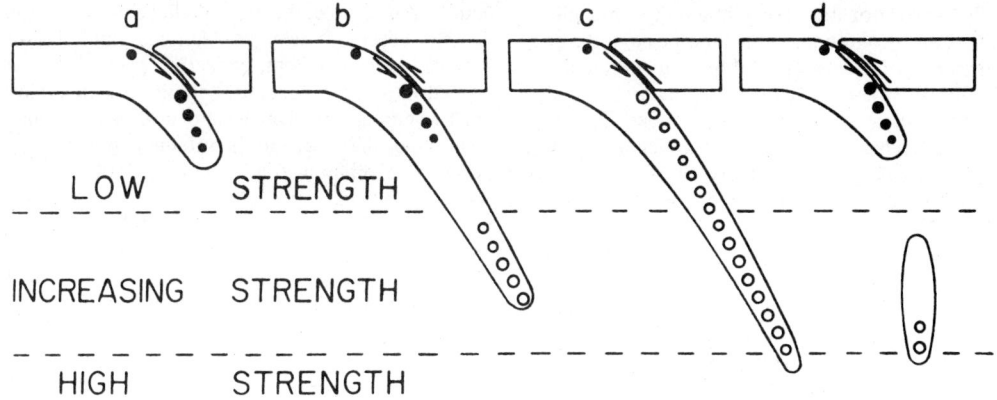

FIGURE 8. A model showing plausible distribution of stresses within slabs where gravitational forces act on excess mass within the slabs. A filled circle represents down-dip extension, an unfilled circle represents down-dip compression, and the size of the circle qualitatively indicates the relative amount of seismic activity. The horizontal dashed lines in the figures indicate possible phase changes in the upper mantle near 350-400 km and 650-700 km (from Isacks and Molnar, 1971).

seismicity is tentatively explained by the models shown in Fig. 8. In Fig. 8a, the slab, more dense than the surrounding mantle, sinks into the asthenosphere under its own weight, the load of excess mass is mainly supported by forces applied to the slab above the sinking portion, and the slab is thereby under extension. As the slab penetrates stronger material, part of the load is supported from below, part from above and the stress changes from extension to compression as a function of depth (Fig. 8b). When the entire load is supported from below, the slab is under compression throughout (Fig. 8c). Figure 8d shows the model for a detached slab that is sinking independently. The gaps in seismic activity as a function of depth may correspond to the models in Figs. 8b and d.

T. SASATANI

References

Hasegawa, A., N. Umino, A. Takagi, and Z. Suzuki, 1979, Double-planed deep seismic zone and anomalous structure in the upper mantle beneath northeastern Honshu (Japan), *Tectonophysics* **57**, 1-6.

Isacks, B. L., and P. Molnar, 1971, Distribution of stresses in the descending lithosphere from a global survey of focal-mechanism solutions of mantle earthquakes, *Rev. Geophys. Space Phys.* **9**, 103-174.

Utsu, T., 1977, *Seismology*. Tokyo: Kyoritu Shuppan (in Japanese), 286p.

Wadati, K., 1928, Shallow and deep earthquakes, *Geophys. Mag.* **1**, 161-202.

Cross-references: *Earthquake Mechanisms; Earthquake Mechanisms and Plate Tectonics; Earthquake Seismology; Earthquakes and Seismicity; Earth Structure, Global; Mantle, Upper: Structure; Mantle Convection and Plumes; Mantle Dynamics; Seismicity and Plate Tectonics; Seismic Source: Observations; Seismicity: Subduction Zone; Seismic Source: Theory; Subduction Zones*.

DEEP SEISMIC-REFLECTION PROFILING

Deep seismic reflection profiling is the application of exploration seismic reflection techniques to studies of the crust and upper mantle down to depths of 15 km to 100 km.

Seismic reflection has been the preferred method of obtaining subsurface geologic information in sedimentary basins since 1930 and is widely used today in petroleum exploration. The rapid advances in computer technology in the last two decades have led to a burgeoning of techniques for digital signal enhancement (see *Exploration Seismology*). This development has allowed extension of the seismic reflection method into hard rock terrains and to greater depths. In the late 1960s a number of investigators in Europe and North America reported the observation of deep reflections from the crust. In 1975 a systematic survey of major geological structures was begun by COCORP (Consortium for Continental Reflection Profiling) in the United States. Since that time national and regional groups have emerged in Australia (ACORP, Australian Continental Reflection Profiling), Canada (LITHO-PROBE), France (ECORS, Etude Continentale et Oceanique par Reflexion et refraction Sismiques), W. Germany (DEKORP, Deutches Kontinentales Reflexions-seismiches Program), United States (COCORP and CALCRUST, California Consortium for Crustal Studies), United Kingdom (BIRPS, British Institutions Reflection Profiling Syndicate), and in eastern European countries. This list is by no means exhaustive and is probably inadequate as new groups continue to be formed and because the various universities or national geological surveys involved are too numerous to be individually listed here.

The reason for this research activity on a global scale is the recognition of the importance of the deep structure of the crust and lithosphere in studies of the dynamics and structural framework of major geological features, such as orogenic belts or sedimentary basins. The seismic reflection method is able to image the geometry of reflectors deep within the crust and to relate this deep structure to near surface geology.

Method

Data acquisition and processing techniques are in principle identical to those used in exploration seismology (see *Exploration Seismology*). Differences are noted below and pertain to the choice of acquisition parameters, different emphasis in the type of information sought, and additional difficulties in data processing. A careful choice of acquisition and processing parameters is critical in deep seismic reflection experiments because the signal-to-noise ratio in the data may be poor due to weak source signals at depth and low reflectivities.

Data have been collected both on land and at sea and methods differ in the two regions. Work is cheaper at sea and the deep data are often of better quality than on land because of lower noise levels, straighter lines, and better source coupling. On the other hand, it is possible to employ longer receiver arrays on land.

Data Acquisition. The seismic source may be explosives or a series of vibrators (as in the VIBROSEIS method, TM Conoco) on land, or an array of air guns at sea. High energy and low frequency content, by comparison with exploration seismic methods, characterize sources used for deep reflection work. As a general rule, the larger the energy output of the source, the greater will be the maximum depth of observed seismic reflections. Similarly, the lower frequency content, typically 7 Hz to 25 Hz, is needed in order to penetrate through the uppermost crust where high frequencies are absorbed. On land, coupling of the source to the ground can be a major problem in attaining optimum signal penetration. Also, where an array of sources is used, proper tuning of the array to obtain the desired signal shape and frequency content is essential. Finally, the shot spacing should be sufficiently small to give data redundancy, which can be used during processing to enhance signal-to-noise.

The receiving device consists of an array of geophones (or hydrophones at sea). Long arrays are best, as they give more moveout on the deeper reflections and provide velocity information at greater depths. A typical array length on land is 10 km (COCORP) while at sea shorter arrays, 3 km long, are used. There is, however, a trade-off between array length and receiver spacing. Unless the number of receivers along the array is increased, which may be very costly or impractical, longer arrays imply greater station spacing. This spacing is about 50 m to 100 m on land and 25 m or less at sea.

After the shot is "fired," data are recorded for about 20 sec although greater recording times have been employed; 30–60 sec in recent BIRPS experiments (Fig. 1). The use of a lower frequency range allows a lower sampling rate (4 or even 8 ms versus 1–2 ms for most exploration work). Therefore, the number of samples per seismic trace is decreased by a factor of two or three, which compensates for the greater record length. Thus the total volume of data to be processed remains manageable.

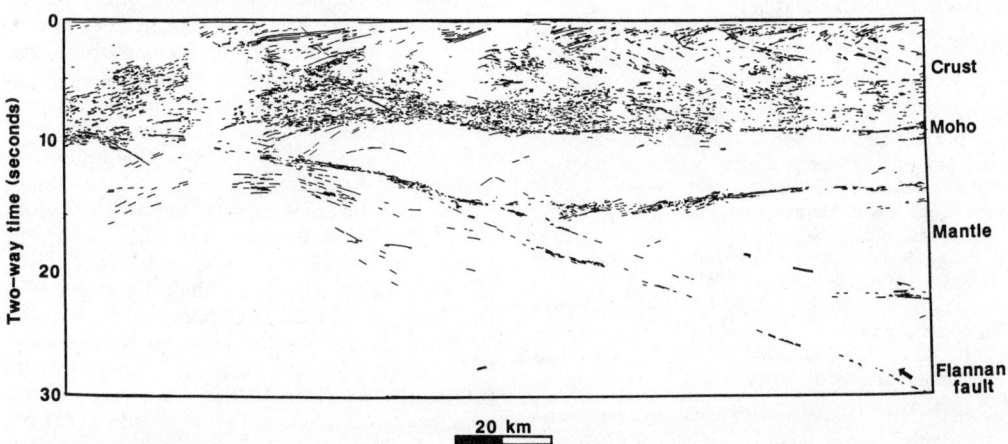

FIGURE 1. Unmigrated line-drawing of a portion of the DRUM (Deep Reflections from the Upper Mantle) line that imaged a mantle reflector known as the Flannan fault. Three-dimensional seismic control on the position of this event precludes the possibility that it is an out-of-plane reflection or other artifact. The Flannan fault reflection can be traced to the bottom of this record (to 80 km depth after migration). In 1986 BIRPS collected data to 60 s 2-way travel-time to trace the Flannan fault deeper into the mantle. (Courtesy of BIRPS)

Data Processing. Data processing involves methods similar to those employed in exploration seismology (see *Exploration Seismology*). Typical major steps in a processing sequence would include wavelet deconvolution, common midpoint gathers, velocity analysis, normal moveout corrections, common midpoint stacking, frequency filtering, and amplitude scaling. In addition migration may be performed on the stacked data. While methods are essentially the same as in exploration seismology, a few difficulties arise when deep data are processed, as described below.

The length of the array controls the moveout and so limits the maximum depths for which velocities can be obtained from the data. The seismic ray paths from deep reflectors become increasingly vertical as reflector depth increases. A simple resulting rule is that velocities cannot be obtained below a depth comparable with the length of the array. It is therefore highly desirable to obtain velocity information independently in experiments where wide angle reflection or refracted energy is recorded (see *Controlled Source Seismology*).

Current methods of migrating seismic data perform poorly on deep reflectors, because near surface features distort the seismic wave field and consequently distort the deeper reflectors. These distortions are erroneously assumed to be real by conventional migration techniques and they result in a migrated section that exhibits a number of nonexistent reflections. This undesirable effect is also present in exploration data but is almost unnoticeable because the problem worsens with increasing depth and velocity.

The photographic reduction of large seismic sections to a size suitable for publication often leads to loss of resolution and of information that is critical to the readers' understanding. Some effort has been directed toward alleviating this problem and special processing steps such as coherency filtering for enhancing the coherent energy are useful in this respect. Because of this problem, data are often shown as a line drawing in which lines have been drawn through the strongest and most coherent events on the original seismic section (Fig. 1).

Interpretation. A few examples of published interpretations and the associated seismic sections are presented later. In most publications the seismic data are shown as a time section in which the vertical scale is two-way travel time (sec). A rough conversion to depth can be made by taking 1 sec as 2 km in sediments and as 3 km in basement.

Proper interpretation of a seismic section requires knowledge of the processing steps used in producing it; for example it is usually prudent to examine both migrated and unmigrated sections to ensure that migration has not introduced any prominent but artificial reflectors, which might otherwise be interpreted as real events.

Usually deep seismic lines are widely spaced, and one has to assume that reflectors occur directly below the line. In fact, reflections that may come from either side of the line (side swipes) are of great concern in deep reflection interpretation. These reflectors often cannot be distinguished from those beneath the line, unless a cross line is available showing the true position of the reflector. As depth increases, side reflections are increasingly out of position and may come from a reflector 20 to 50 km away from the vertical plane of section. A related problem is that the observed dips of reflectors are apparent dips only, as the structures may not strike perpendicular to the line. Without a cross line the true dip is unknown.

Finally, we must consider the resolution of the seismic reflection method. The vertical resolution is limited by the low frequency content of signals that can penetrate to the required depths. For maximum frequencies around 25 Hz, the thinnest layer one can hope to resolve is about 50 m thick. Lateral resolution decreases with depth as the wavefronts expand. For example, the lateral resolution at the Moho is about 2 km. These values show that seismic reflection profiling gives us greater resolution than any other geophysical observation.

Major Results

Major contributions of deep reflection profiling are briefly reviewed below, using selected examples of published data. Although plate tectonic models have been highly successful in explaining observations from oceanic and plate margin regions, intraplate deformations are poorly understood. Therefore, many efforts have focused on continental areas where deep structures were virtually unexplored.

Continental Lithosphere. Many fundamental features of deep seismic data across continental crust in a variety of tectonic settings are shown by the examples selected. They illustrate some typical (but not universal) characteristics of the crust observed on a global scale. Outside sedimentary basins, the upper crust is generally seismically transparent. Dipping reflections in this region are often attributed to faults. Conversely, the lower crust of many areas exhibits a multitude of strong reflections and diffractions. Where such a deep crustal layer exists, the Moho appears to lie just below this reflective zone. Below the Moho, the uppermost mantle is usually seismically transparent. The above are highly generalized statements, not supported by all observations. Perhaps the most important result of deep reflection profiling is the observation of lateral and vertical heterogeneity of the lithosphere.

The seismic section shown in Fig. 1 exhibits some of the characteristics mentioned above. It is unique in that it was recorded to 30 sec (an approximate depth of 100 km). These data were recorded at sea, north of Scotland, across the foreland to the early Paleozoic Caledonian Orogen of western Europe. It shows a relatively transparent upper crust and a reflective lower crust. However, one can see that the upper crust is not devoid of reflections, as evidenced by many dipping events which may be normal faults that bound small sedimentary basins. Also reflections from the flat-lying surface sediments are present.

More importantly, spectacular reflections are observed in the upper mantle. The major dipping event, the Flannan Fault, extends down to at least 80 km. It appears to flatten near the Moho. Other events whose nature are unclear are also present: in particular a subhorizontal event at about 14 sec appears to be truncated by the Flannan Fault. These striking data show that it would be unwise to think of the upper mantle as always "transparent."

Oceanic Crust. Oceanic crust has often been visualized as a simple laterally homogeneous, vertically layered pile of mafic rocks that formed at the mid-ocean ridges. In contrast, recent deep reflection profiles indicate that oceanic crust is compositionally complex and exhibits variations both laterally and vertically. The example shown in Fig. 2 comes from the western Atlantic Ocean Basin, and traverses 115–125 my old crust. In this section, flat-lying sediments cover the top of the basement, which occurs at a depth of about 8 sec. Major dipping reflectors can be seen throughout the section, some of which begin near the top of the crust (triangles in Fig. 2), while most are confined to the lower crust, below 10 sec. In addition there are some subhorizontal events. There is one in the upper crust at about 9 seconds and one at the anticipated depth of oceanic Moho, at 10.5 seconds. Little in the way of reflection energy is seen below this Moho reflector. These reflectors may be related to compositional variations observed within ophiolite suites. Alternatively, some may be normal faults.

Mountain Belts. One of the most important results of deep reflection profiling was observed on the first COCORP transect recorded across the Appalachian orogen. This profile showed that 300 or 400 my ago, as proto-North America and Africa collided, thrust sheets were being transported westward at least 260 km along a low-angle detachment within the crust. The mechanics of such thrusting are still the object of numerous studies, but more recent deep reflection studies have confirmed this observation.

A different structural style was documented by COCORP in Wyoming across the Wind River Mountains (Fig. 3). The Wind River thrust is traced to a depth of at least 24 km as a moderately dipping fault that does not flatten with depth. There is no evidence for a low-angle detachment in this example. The lens-shaped structure imaged near the deeper part of the thrust (5–8 sec. on Fig. 3) is interpreted as the result of ductile shear within a brittle-ductile transition zone. This result indicates that horizontal compressional forces were the principal cause of Rocky Mountain foreland deformation in the Mesozoic.

Active Convergent Margin. The southwestern coast of Canada is an active convergent margin. There the Juan de Fuca oceanic plate is being subducted eastward beneath the North American plate. The profile in Fig. 4 crosses the ocean plate and the subduction zone to the west. The top of the subducting oceanic plate can be traced inland beneath the continent (JdF in Fig. 4). Oceanic Moho can also

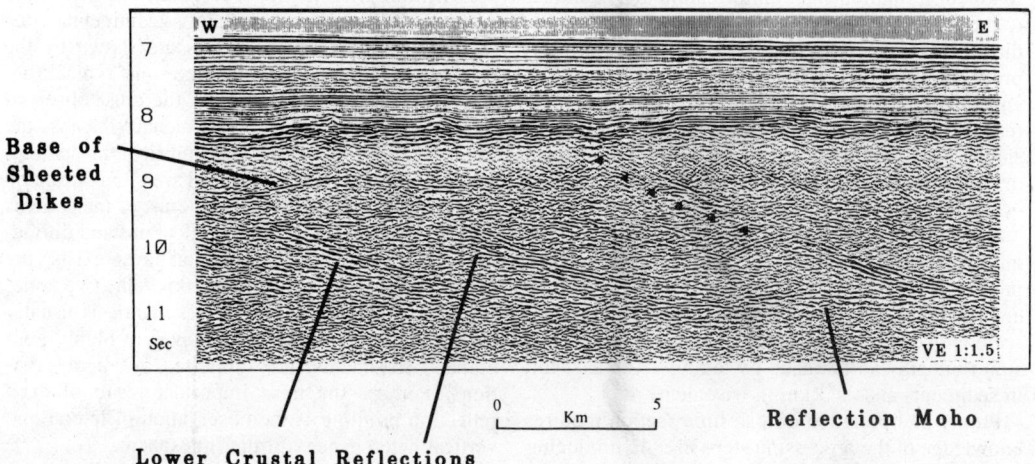

FIGURE 2. Migrated portion of the North Atlantic Transect (NAT). The section reproduced here samples oceanic crust about 120 my old, located halfway between Florida and Bermuda. Prominent dipping reflections are observed between an undulating upper crustal reflection at about 9 sec and a Moho reflection at 10.9 s. A basement-cutting normal fault is also present and is marked with triangles. No vertical exaggeration for rock velocities of 6 km · s^{-1}. (Courtesy of John Mutter, Lamont-Doherty Geological Observatory)

FIGURE 3. Unmigrated section of the COCORP line across the Wind River Mountains, reprocessed by John Sharry at Gulf Oil Corporation. A sedimentary basin is imaged on both sides of the Wind River uplift. The Wind River thrust is clearly imaged, at least in the upper half of the section, and has an average dip of 30°–35°. (Courtesy of J. Sharry)

be traced for a limited distance from the ocean basin (M in Fig. 4).

There is a deep reflective zone beneath the eastern part of the line whose limits are indicated by C and E in Fig. 4. The significance of this region is not yet clear. The interpretation shown here implies that the reflective zone consists of sediments and oceanic crust that has been scraped off the top of the subducting plate and accreted to the overriding continent.

Extensional Terrains and Sedimentary Basins. Reflection profiling provides key evidence on deep fault geometry which is necessary to discriminate between various speculative models for the mode of lithospheric extension. The 1000 km long COCORP transect across the Basin and Range Province of the western United States, as well as several other surveys worldwide, have imaged the following elements: (1) asymmetric half-grabens, often formed along moderate or low-angle normal faults that may either sole into a midcrustal zone of detachment or in some places are traceable to Moho depth (Figs. 5 and 6); (2) a pervasive subhorizontal seismic fabric in the lower crust of many regions; and (3) unusually bright reflections attributed to deep crustal intrusions emplaced during extension. Pulling the crust apart along low-angle normal faults had been thought mechanically impossible. Yet, Fig. 5 shows reflections from a low-angle normal fault, the Sevier Desert detachment, that can be traced as deep as 15–20 km. These reflections are not disrupted by the steep normal faults known at the surface, which are also evident on the seismic section of Fig. 5 where they displace a strong reflection above the detachment. Therefore, motion along the Sevier Desert detachment must accompany crustal extension in this area.

While sedimentary basins have been extensively explored for petroleum resources, their deep structure is largely unknown. The example in Fig. 6 comes from the Grand Banks off eastern Canada, a region occupied by several extensional sedimentary basins of Mesozoic age. The basin is an assymetric half graben, with a major bounding fault on its west side. It is filled with syn-rift sediments that are truncated by a prominent unconformity at about 0.7 sec, which separates syn and postrift sediments. The major feature of interest here is the basin bounding fault, which extends deep within the crust, to at least 9 sec. This example illustrates the existence of faults or shear zones deep in the crust; an observation that constrains rheological models of the crust.

Passive Rifted Continental Margin. Figure 7 shows a crossing of the Mesozoic rifted margin east of the Grand Banks off Eastern Canada. To the west on Fig. 7 we identify a region underlain by continental crust and characterized by a reflective lower crust, with Moho at about 10 seconds. A thin veneer (1 sec) of sediment blankets continental basement. A confused region occurs below the continental slope and upper continental rise with strong water bottom multiples obscuring deep reflections. At the eastern end of the profile, oceanic crust is inferred overlain by about 3 seconds of sediments.

Two important observations can be made from these data. First, there is a strong landward dipping reflector marking the landward end of oceanic crust. This reflector is observed on a number of profiles in this and other regions and is interpreted to mark the continent-ocean boundary. The second observation is that the reflective lower crustal zone continues to the edge of the continent. While the significance of this is not yet clear, it has been suggested that the reflective zone may result from extension of the lithosphere. Its presence at the continental edge of a rifted margin may therefore be an important observation.

Fossil Structures and the Cratonic Interior. Parts of the continental interior have a tectonic history that goes back more than 3 billion years and

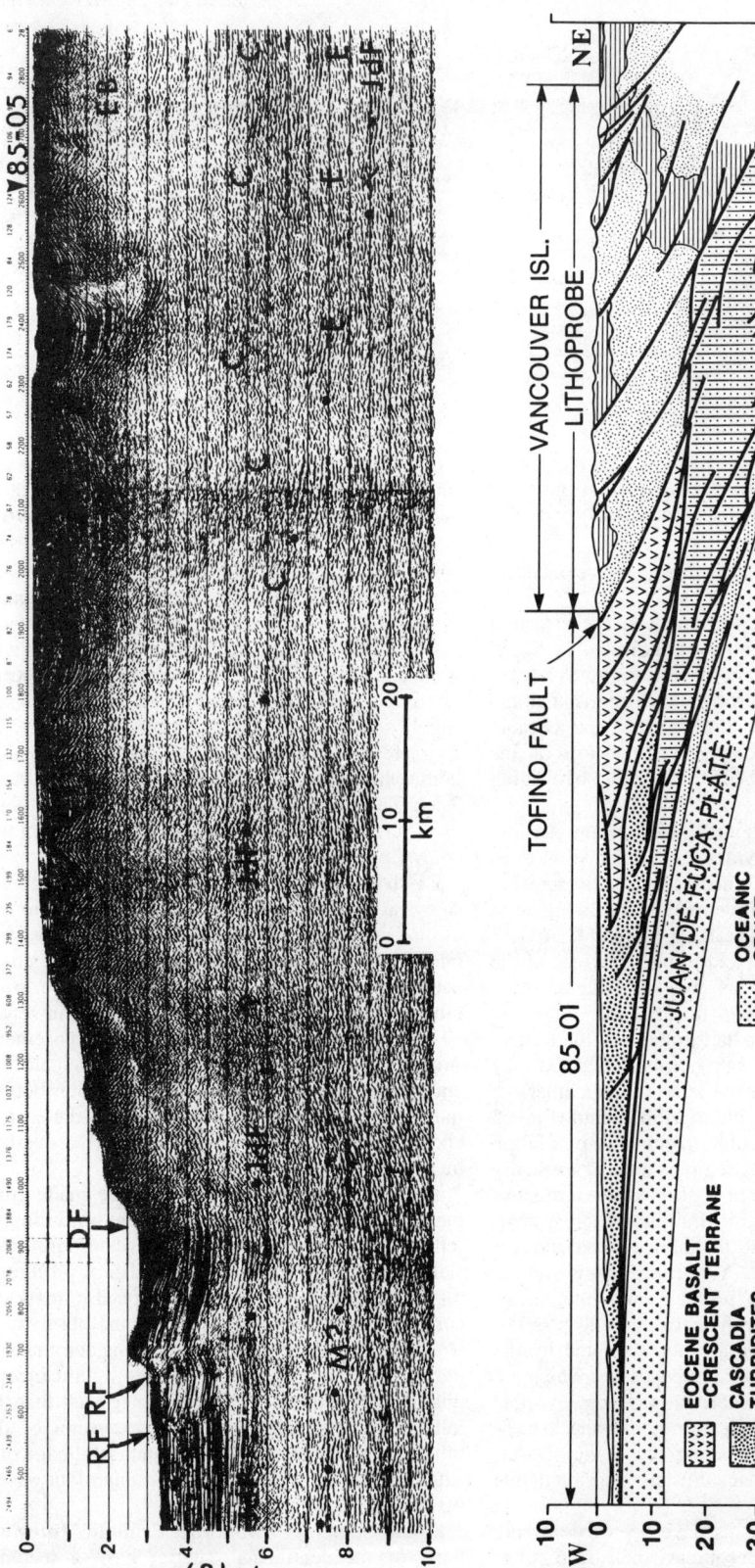

FIGURE 4. *Top*: Migrated seismic section across the active subduction zone beneath Vancouver Island, W. Canada. Ramp faults (RF) occur in the oceanic sediments near the deformation front (DF). JdF with dots denotes the top of the oceanic Juan de Fuca plate being subducted. M? with dots is possibly oceanic Moho. C and E ar the upper and lower bounds of a high reflectivity zone. *Bottom*: Interpretative cross-section across the margin. (Courtesy of R. M. Clowes, University of British Columbia)

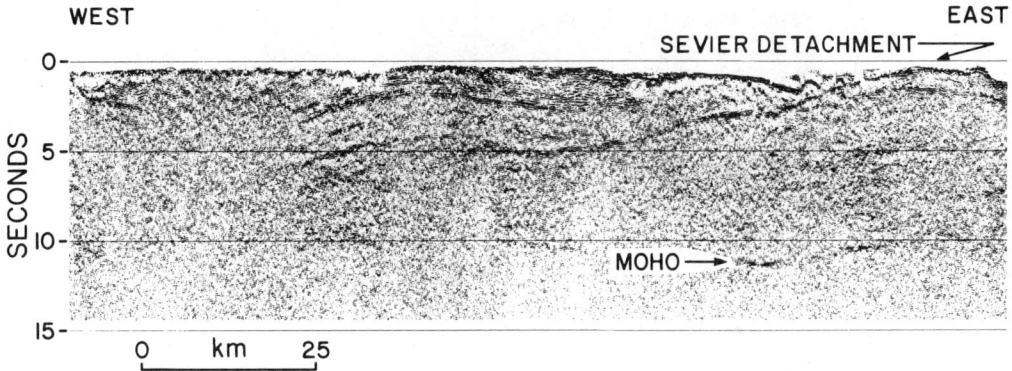

FIGURE 5. Portion of COCORP's seismic section (unmigrated) from central Utah in the Basin and Range extensional province. The strong westward dipping reflection that dominates the section is a low-angle Cenozoic normal fault, the Sevier Desert detachment. Minor apparent disruptions of this detachment are explained as velocity pull-down effects resulting from overlying sediments. (Courtesy of COCORP)

continents are a complex collage of terranes separated by collision zones, with later compressional or extensional events overprinting earlier structures. Deep reflection profiling has proven successful in recognizing fossil structures in the continental interior.

Ancient sutures marking the site of Paleozoic collisions have been interpreted on profiles from several regions. These often appear as dipping and imbricate reflection zones that penetrate the whole crust. Figure 8 illustrates one such example. The Iapetus suture separates continental blocks formerly on opposite sides of the Iapetus (or Paleozoic proto-Atlantic) ocean. At the surface, the exact location of the suture is masked by younger sediments. The NEC profile (Fig. 8) shows a prominent zone of north-dipping reflections, interpreted as representing the Iapetus suture. The lower-crustal seismic character is markedly different across the suture but the reflection Moho is surprisingly flat for a region where continental collision occurred. These and other results indicate that the Moho may be a young, mobile feature.

Origin of Deep Reflections and Outstanding Problems

Mapping the lithosphere by reflection profiling has shed light on many fundamental geologic problems. But it has also raised a host of new questions about the structure, composition, and evolution of the earth. Perhaps the most tantalizing question is "Why do deep reflections exist?" Even where reflections can be related to a specific feature such as an outcropping fault, the exact cause of the reflectivity is unknown. Possible causes include: lithological or compositional boundary, phase change, zone of localized strain, change in anisotropy, and rheological boundary.

Faults. Dipping crustal reflections are commonly attributed to fault zones. This differs from exploration seismic data where faults are usually identified in the shallow section on the basis of offsets or truncations of reflectors, diffractions, misties, or fault plane reflections generated by the juxtaposition of rocks of different acoustic impedance (e.g., sediments against basement). Why deep fault zones (e.g., Fig. 3) produce reflections remains a controversial issue. Recent profiling over exposed mylonites and known shallow detachment faults confirms that weak to fair reflections can be attributed to ductile shear zones. In contrast, brittle faults are usually nonreflective, unless they juxtapose rocks of different impedance.

To what depth crustal faults can be traced in the Earth is a fascinating debate. For the first time, deep reflection profiling has provided direct evidence that some faults penetrate the entire crust. In other places, it appears that faults end in the middle or lower crust where deformation is accommodated instead by ductile flow.

Lower Crustal Reflections. The lower continental crust is commonly observed to be highly reflective (e.g., Figs. 1 and 7). First thought to be the result of magmatism or enhanced ductility associated with extension, recent results suggest that a highly reflective lower crust may form under a variety of tectonic regimes. Possible origins for the layered appearance and high reflectivity of the lower continental crust include: (1) pervasive shear, resulting in mylonite formation; (2) igneous layering; (3) free fluids; and (4) some combination of the above.

Compositional and rheological heterogeneity in the lower continental crust is consistent with the limited information provided by xenoliths, exposed high grade metamorphic terrains, electromagnetic sounding experiments, and laboratory studies of mineral

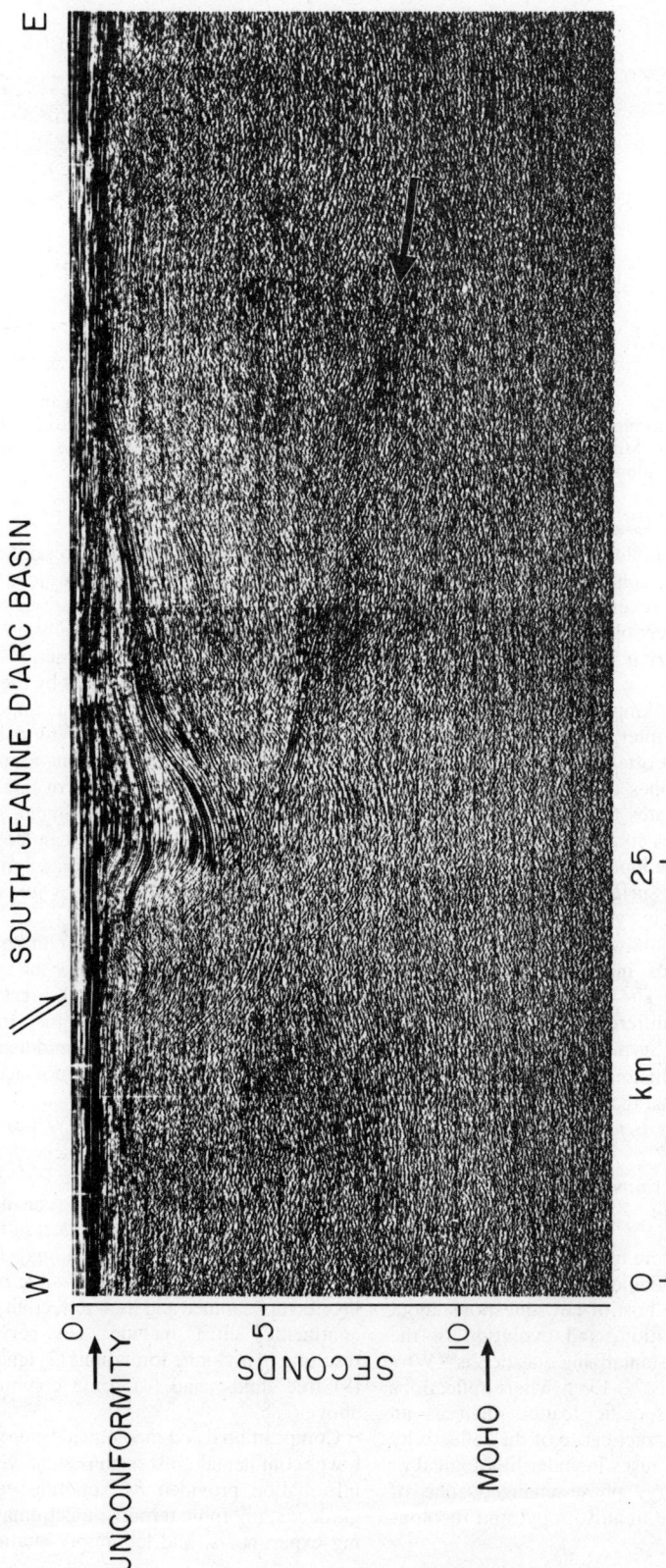

FIGURE 6. Mesozoic extensional sedimentary basin on the Grand Banks off eastern Canada. This section is migrated; no vertical exaggeration at 5 km · s^{-1}. Note a major fault indicated by arrows that bounds this half-graben and extends deep into the crust. (Data from the Geological Survey of Canada)

FIGURE 7. Transect across the rifted margin of eastern Canada. Migrated section; no vertical exaggeration at 5 km · s^{-1}. Note the reflective lower continental crustal layer. Oceanic crust on the left terminates in a landward dipping reflector, indicated by the arrow. (Data from the Geological Survey of Canada)

stability fields. Because the observed high reflection amplitudes in the deep crust and upper mantle are difficult to explain, and because the reflective lower crust often corresponds to zones of high electric conductivity, fluids are also thought to have a role to play.

Fluids in the Deep Crust. This topic is of great interest because fluids greatly affect the physical properties of rocks at depth, and therefore crustal rheology. Evidence for fluids in the deep crust comes from high conductivity zones identified by magnetotelluric surveys, studies of high grade metamor-

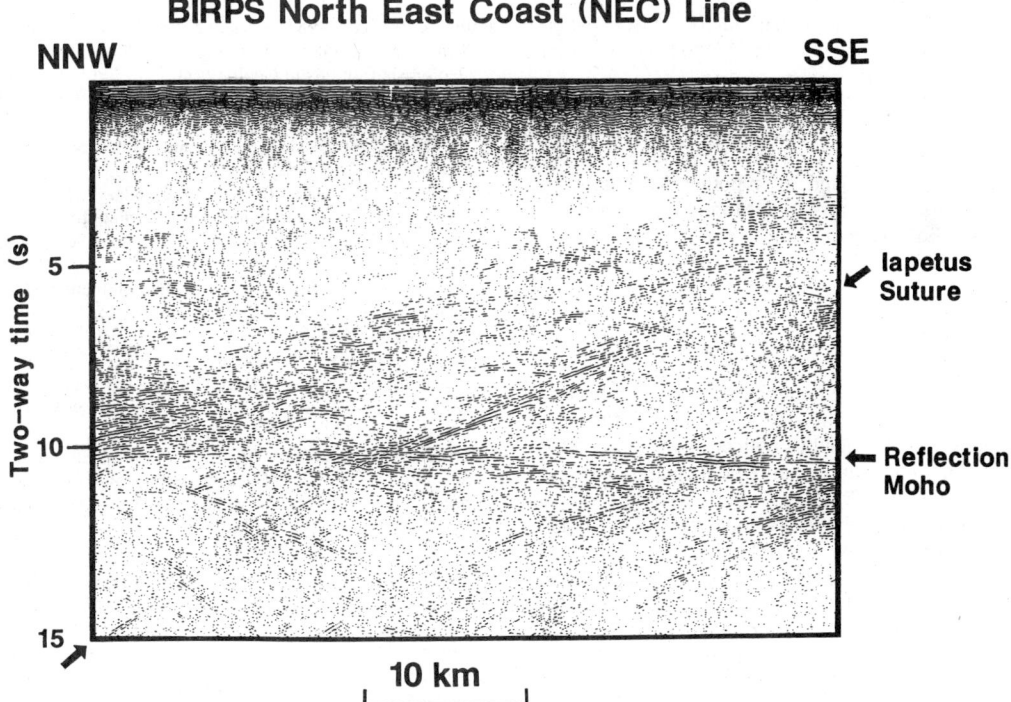

FIGURE 8. Detail of the North East Coast (NEC) line (unmigrated) showing the north-dipping series of very bright events interpreted as the Iapetus suture. The dipping events shown here move into the crust after migration. The Moho is continuous beneath the suture zone, indicating that reequilibration of the crust-mantle transition may have occurred after the continental collision in the Paleozoic. No vertical exaggeration at 6 km · s^{-1}. (Courtesy of BIRPS)

phic rocks, and deep seismic reflection profiling. Likely candidates are metamorphically derived volatiles, formational water or brine, hydrocarbons, and magma. Recent deep continental drilling results show that free fluids exist to depths of at least 12 km. The large reflectivity provided by fluids can best explain the unusual reflection amplitudes observed in several data sets.

Moho. Where data are available, the deepest, bright subhorizontal reflections, often observed at the base of lower crustal reflective zones approximately coincide in time with the crust-mantle boundary defined by seismic refraction experiments (see *Controlled Source Seismology*) and therefore define the reflection Moho. However, recent detailed comparisons of refraction and reflection data indicate some discrepancies.

Reflection Moho has the same travel time (therefore most likely a similar depth) across zones that separate crustal blocks with very different tectonic histories (e.g., the suture zone shown in Fig. 8). Therefore, many have suggested that Moho is an evolving boundary, though a model that satisfies all observations has yet to be found. It is also possible that the reflector labeled "Moho" has a different nature in different areas. A new picture then emerges, where the continental lithosphere is evolving not only through episodes of compressional and extensional tectonics, but also through the addition and removal of material to and from the base of the crust, thereby changing Moho position through time.

Moho that results from magmatic underplating or ponding at the base of the crust is consistent with the sharp reflector observed on many reflection data. Many favor this kind of hypothesis, though other models for the origin of the Moho, such as a compositional boundary or a shear zone, cannot be ruled out.

Conclusions

While a good beginning has been made, this discipline is relatively new in crustal and lithospheric geophysics. We need to collect much more data so that consistent patterns on regional and global scales will begin to emerge, and at the same time we need to pursue our attempts to understand why parts of the crust and mantle are reflective.

To date some fundamental concepts in crustal geophysics have been challenged by the new deep reflection data. The notion of a simple layered, laterally homogeneous crust in both continental and oceanic regions is invalid. The nature of the Moho is being seriously questioned, a very significant development given the Moho's role as the most important physical boundary within the lithosphere. Finally, deep seismic data have allowed us to determine much that previously was only conjecture concerning the rheological zonation of the lithosphere.

In general, it has been demonstrated quite clearly that near surface geological features are reflected in structure at depth, often in ways that we can understand and that advance our understanding of their origin and evolution. Exploration of the deep crust has resulted in a revolutionary change in our approach to geological problems. Hopefully, the petroleum and mining industries will greatly benefit from this better knowledge of the broad tectonic framework within which targeted exploration is being conducted.

Acknowledgements

Illustrations shown were provided by BIRPS, Ron Clowes (U.B.C.), COCORP, John Mutter (Lamont-Doherty Geological Observatory), John Sharry (Gulf Oil Corp.). We thank all who contributed information and/or illustrations; we regret that not all prints received during preparation of this article could be used with figures in the final version. W. Kay, M. Keen, and K. Dickie read the manuscript and made helpful suggestions. Geological Survey of Canada contribution No. 13287.

B. DE VOOGD
C. KEEN

References

Bally, A. W., ed., 1983, *Seismic Expression of Structural Styles*. Tulsa, OK: American Association Petroleum Geologists (Studies in Geology, no. 15, 3 vols.).

Barazangi, M., and Brown, L. eds., 1986, *Reflection Seismology: A Global Perspective*. Washington, D.C.: American Geophysical Union (Geodynamics Series vol. 13).

Barazangi, M., and Brown, L., eds., *Reflection Seismology: The Continental Crust*. Washington, D.C.: American Geophysical Union (Geodynamics Series vol. 14).

Deep seismic reflection profiling of the continental lithosphere, special issue of the *Royal Astron. Soc. Geophys. Jour.* **89**(1), 1987.

Sheriff, R. E., and Geldart, L. P., 1982, Exploration Seismology, vol. 1: "History, Theory and Data Acquisition." Cambridge, Great Britain: Cambridge University Press, 253p.

Sheriff, R. E., and Geldart, L. P., 1983, Exploration Seismology, vol. 2: "Data-processing and Interpretation." Cambridge, Great Britain: Cambridge University Press, 221p.

Cross-references: *Continental Collision Zones; Continental Crustal Structure; Continental Lithosphere; Controlled Source Seismology; Elastic Waves in Homogeneous and Inhomogeneous Media; Exploration Seismology; Finite-Difference Forward Modeling in Seismology; Inverse Theory and Methods: Seismology; Lithosphere: Mechanical Properties; Lithosphere: Oceanic; Mantle, Upper: Structure; Seismic Imaging; Seismic Properties of Rocks; Seismic Signal Processing; Seismic Tomography; Seismic Wavefield Migration; Thin-Skin Tectonics.*

DEFORMATION OF ROCKS AND MINERALS

The nonbrittle deformation behavior of materials can be subdivided into three classes (Nowick and Berry, 1972), all of which are of interest to the geophysicist:

1. In *elastic deformation,* the strain response to an applied stress is time independent and reversible. The strain response is time independent in that the strain reaches its equilibrium level instantaneously upon application or removal of the stress, and the response is reversible in that, if the stress is removed, the strain returns to zero. Figure 1, curve I illustrates the strain-time path for one cycle of applied stress in the elastic regime.
2. In *anelastic deformation,* the strain response to an applied stress is time dependent and reversible. The strain response is time dependent in that it takes a finite time for the strain to reach the equilibrium level for the given stress, and the response is reversible in that, if the stress is removed, the strain returns to its initial value. Figure 1, curve II illustrates the strain-time path for one cycle of applied stress in the anelastic regime.
3. In *plastic deformation,* the strain response to an applied stress is time dependent and irreversible. The strain response is time dependent in that the strain changes continuously with time and does not reach an equilibrium level for the given stress, and the response is irreversible in that, if the stress is removed, the strain does not return to its initial value, even after long times. Figure 1, curve III illustrates a possible strain-time path for one cycle of applied stress in the plastic regime.

Elastic Deformation

Elastic deformation of rocks and minerals is discussed in detail in *Seismic Properties of Rocks*.

Anelastic Deformation

Mechanisms of Anelastic Deformation. An understanding of anelastic deformation effects in rocks is necessary to interpret the seismic behavior of the Earth, as the rocks that make up the Earth do not behave as perfectly elastic solids. For instance, determination of the phases present in the Earth by comparing seismic velocities measured in the Earth with the velocities measured for minerals or rocks in the laboratory is likely to give misleading conclusions. Anelasticity measurements may also give indications of the variation of mantle viscosity with depth in the Earth (Orowan, 1967). Deviations from perfectly elastic behavior can result from a number of macroscopic and microscopic characteristics of the rock (Nowick and Berry, 1972; Gueguen et al., 1981):

1. *Grain boundaries.* Reversible, time-dependent slipping along grain boundaries will absorb energy from sound waves moving through a rock. The magnitude of this *attenuation* (*internal friction*) will depend on environmental parameters, such as temperature, pressure, and the frequency of the propagating wave. In particular, the attenuation at ultrasonic frequencies may be very different from the attenuation at seismic frequencies.
2. *Grain boundary fluid phase.* Anelastic motion of a fluid phase, either as a melt or a saturated hydrous fluid, will give rise to energy losses from sound waves propagating through a rock and to a parallel decrease in the velocity of the waves through the rock. This phenomenon has been widely postulated as the source of the upper mantle low velocity zone.
3. *Dislocations within crystal grains.* Anelastic interactions between sound waves propagating through a crystal grain and dislocations within the grain may result in reversible, time-dependent bowing of the dislocations. After each wave has passed, the dislocations return to their initial position but energy is dissipated in the stretching process. The attenuation of the wave energy may be paralleled by a decrease in the velocity of the waves through the rock. This attenuation mechanism has also been proposed to account for the low velocity zone in the upper mantle.
4. *Point defects within crystal grains.* Anelastic interactions between sound waves propagating through a crystal grain and point defects within the grain may result in reversible, time-depen-

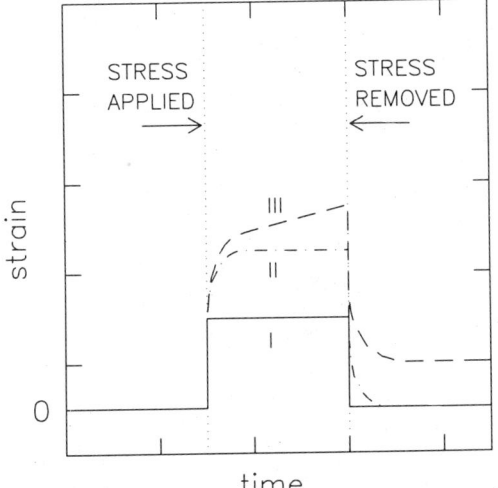

FIGURE 1. Plot of strain versus time for one cycle of applied stress, showing curves for I. elastic deformation, II. anelastic deformation, and III. plastic deformation.

dent motion of the defects. After each wave has passed, the defects return to their initial position but energy is dissipated in the process, giving rise to attenuation of the sound waves.

Experimental Research. Although there have been studies of the anelastic behavior of metals for the previous 50 years and of ceramics for the last 10 to 20 years, the study of the anelastic behavior of geological materials is in its infancy. This slow start is partly because of the complexity of the geological systems of interest and the consequent difficulty in uniquely identifying the source of any attenuation. In addition, accurate measurement of attenuation effects at the appropriate geological conditions (temperature, pressure, and frequency) requires a complex, yet sensitive, apparatus.

Most experimental studies are performed at room pressure to eliminate the difficulties with running experiments in a pressure vessel with the consequent sealing and friction problems. A schematic diagram of a 1 atm attenuation apparatus is shown in Fig. 2 (Weiner et al., 1987). In this apparatus, a periodic torque is applied to the specimen, which has been heated to the experimental temperature, using the electromagnet near the top of the column. The attenuation of the sample at the applied frequency and temperature is then determined from the strain response of the specimen, as measured by the rotation of the mirror attached to the top of the column. Unfortunately, in a 1 atm apparatus (Gueguen et al., 1981), there is no confining pressure to prevent or inhibit brittle deformation processes, such as grain boundary cracking, which limit the usefulness of this apparatus in studying polycrystal materials. At the higher confining pressures of the gas-confining medium apparatus, inelastic processes at grain boundaries are suppressed and anelastic processes, rather than brittle failure, can be measured. A recent review of anelastic techniques and processes in geological materials at high pressures may be found in Jackson (1986).

Plastic Deformation

Rheological Model for the Lithosphere. (See also *Lithosphere: Mechanical Properties*) The *lithosphere* is made up of the crust and uppermost mantle and is considered to be relatively stiff and capable of elastically supporting large loads. By comparison, the bulk of the earth is taken to be viscous or plastic. The lithosphere is the only region in the earth where the rock can support stresses greater than a few megapascals without deforming

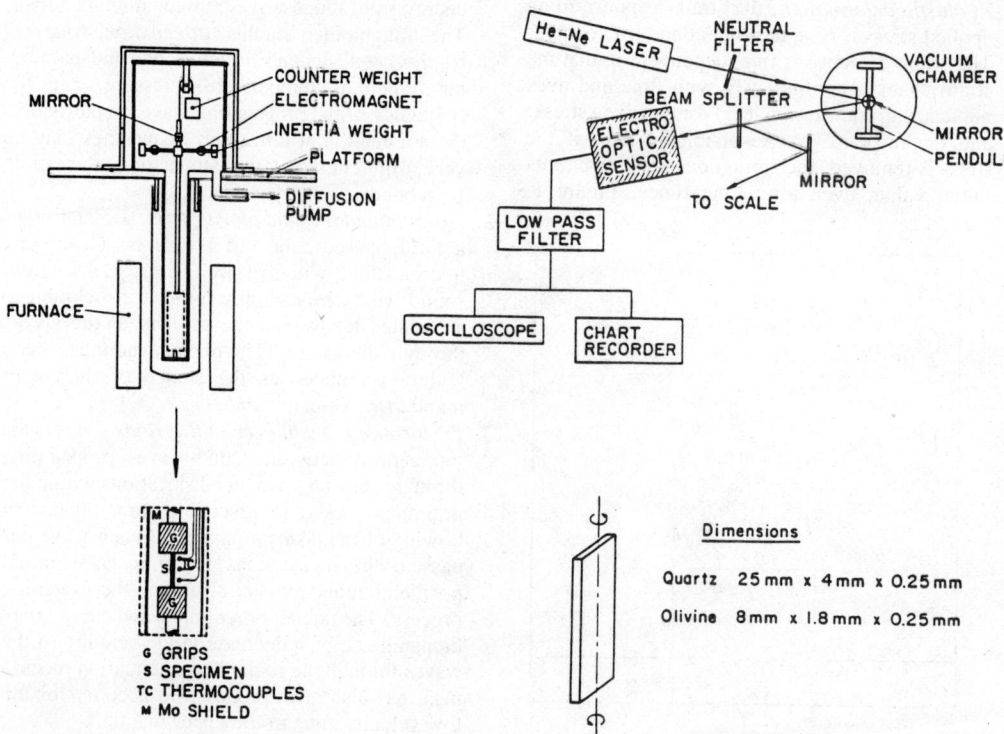

FIGURE 2. Schematic diagram of an inverted torsional pendulum for measurement of attenuation at 1 atm and high temperatures, showing the electro-optic system for stress application and strain amplitude measurement, and the dimensions for specimens. (After Weiner et al., 1987)

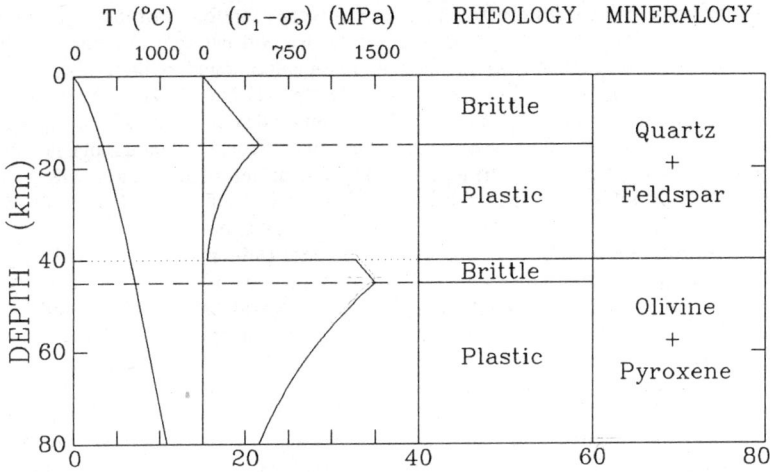

FIGURE 3. Plot showing distribution of temperature in a two-component lithosphere versus depth; distribution of the maximum deviatoric stress ($\sigma_1 - \sigma_2$) that can be supported by the rocks before deformation occurs at geological deformation rates versus depth; nature of the rheology as a function of depth; and mineralogy as a function of depth.

plastically at geological deformation rates. At greater depth, the temperatures are high enough that plastic flow processes will prevent larger stresses from building up. A typical geotherm for the continental crust and upper mantle is plotted in Fig. 3. Temperature is the most important parameter in determining the behavior of a material; as temperature increases toward the melting point, the stress required to plastically deform a material at a given set of conditions decreases rapidly.

The rheology of a region within the Earth can be conveniently defined in terms of the maximum deviatoric stress that the rock within that region can support without deforming at geological deformation rates or faster (Brace and Kohlstedt, 1980). Within the context of this discussion, deviatoric stress is defined as the difference between the maximum and minimum principal stresses ($\sigma_1 - \sigma_3$) acting on the rock in three dimensions; in the Earth, the minimum stress σ_3 is usually equal to the overburden pressure. A simplified plot of the maximum deviatoric stress that can be supported within the continental lithosphere and upper mantle as a function of depth in the lithosphere is given in Fig. 3.

The overburden pressure increases with depth at a rate of approximately 30 MPa/km. In general, when the deviatoric stress is significantly greater than the overburden pressure, the rock will fail in a brittle fashion because of the inability of the overburden pressure to prevent the opening of cracks on microscopic and macroscopic scales. Consequently, rocks will fail in a brittle manner at relatively low deviatoric stresses near the Earth's surface. At greater depth, the overburden pressure suppresses the opening of cracks and higher deviatoric stresses are required for the rock to fail in a brittle manner. The maximum deviatoric stress that rocks within the upper crust can support can be considered, to a first approximation, to be independent of deformation rate and temperature, and to increase only with the overburden pressure, as indicated by the linear brittle regime in Fig. 3. A more detailed discussion of the study of brittle behavior in the Earth is given in *Brittle Phenomena*.

The progressive increase in temperature with depth has little effect on the brittle failure of crustal rock until the temperature reaches a point, near 10 to 15 km depth in the continental crust, where dislocations within the crystalline phases become sufficiently mobile that the deviatoric stress can be relaxed predominantly in a plastic, rather than a brittle, manner. Plastic deformation mechanisms, other than those that involve dislocation motion, may become the dominant deformation mechanisms at other temperature, pressure, stress and fluid conditions; these will be discussed in more detail later. Near the temperature at which plastic mechanisms begin to control the rheology of the rock, both brittle and plastic processes will be operating. As the depth of the brittle-plastic transition corresponds to the zone in the crust that can support the highest deviatoric stresses (usually at about 15 km in the continental crust), brittle failure here will result in the maximum energy release, giving rise to a higher concentration of earthquakes with foci near 15 km than at other crustal depths.

In the simple model depicted in Fig. 3, the maximum deviatoric stresses are shown using the rheology of a single rock type throughout the crust.

With increase in depth below the transition to plastic deformation mechanisms, the maximum deviatoric stress that can be supported by the rock will decrease because of the reduction in the plastic flow strength with increasing temperature. In general, the lower crust is aseismic as applied stresses result in progressive plastic deformation, such that the stress does not increase to a level that will lead to sudden brittle failure.

The model plotted in Fig. 3 assumes that there is a well-defined boundary between the dominantly felsic minerals of the crust and the dominantly mafic minerals of the mantle. In the continental crust, this boundary is generally regarded as the source of the Mohorovičić discontinuity (Moho). For temperatures corresponding to this boundary, it has been established experimentally that the mafic minerals (olivine and pyroxene) are significantly stronger than the felsic minerals (quartz and feldspar). Consequently, there is a sharp increase in the deviatoric stress that can be supported by the rock across this boundary; this strength contrast may result in detachment between the felsic and mafic layers. In addition, the temperature may be locally low enough that, even though the overlying felsic rocks deform plastically at low flow stresses, the mafic rocks may fail in a brittle manner, resulting in earthquakes with relatively deep foci. For instance, Chen and Molnar (1983) report earthquakes with foci just below the crust-mantle boundary under the Himalayas and the Atlas mountains. With increasing depth, the maximum deviatoric stress that can be supported by the mafic rocks without plastic deformation at geological deformation rates will decrease because of the reduction in the plastic flow stress with increasing temperature.

Plastic Deformation Mechanisms. The mechanism that controls the deformation at a particular depth interval in the Earth will depend on the environmental parameters, such as temperature, pressure, stress, deformation rate, grain size, and mineralogy of the rock, as well as the presence of fluid phases. Other environmental factors, such as the partial pressures of oxygen and water and the thermodynamic activity of the mineral constituents may also have a significant effect in determining the operative deformation mechanism.

Figure 4 shows a deformation mechanism map for olivine of 1 mm grain size. The figure is plotted in temperature-stress space with superimposed contours for deformation rate. The figure has been compiled from experimental data and theoretical flow laws for the various deformation mechanism fields, indicated as regions in the figure. Similar figures (Frost and

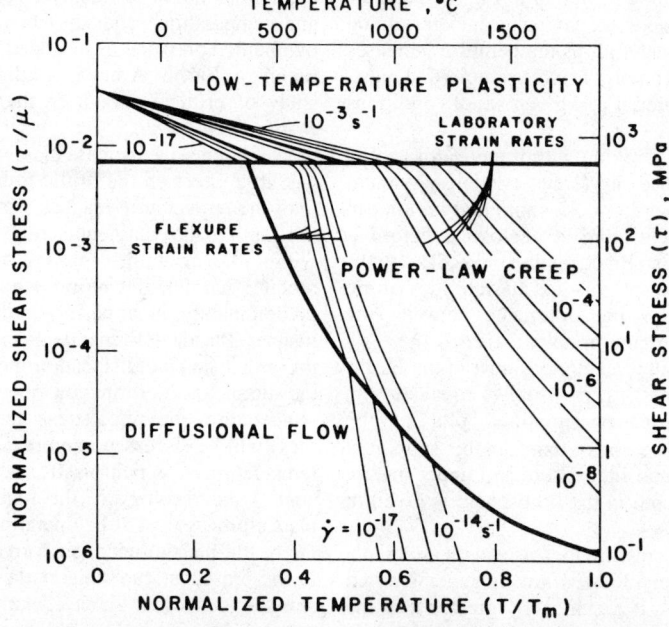

FIGURE 4. Deformation mechanism map for olivine with a 1 mm grain size. Shear deformation rates $\dot{\gamma}$ (in s^{-1}) are contoured over shear stress τ normalized by shear modulus μ and absolute temperature T normalized by melting temperature T_m. Low-temperature plasticity corresponds to dislocation glide, power-law creep to dislocation creep, and diffusional flow to grain boundary diffusional (Coble) creep, as described in the text. (From Kirby, 1983, after Ashby and Verall, 1978).

Ashby, 1982) can be drawn for other minerals and with axes corresponding to the other environmental parameters, where there are sufficient experimental measurements to constrain the models.

In general, the plastic deformation of geological materials under conditions appropriate to natural deformation will follow a flow law of the form

$$\dot{\epsilon} = A \left(\frac{\sigma}{\mu}\right)^n d^{-m} \exp-\left(\frac{Q}{RT}\right) \quad (1)$$

called the *Power-law Creep Equation*, where $\dot{\epsilon}$ is the deformation rate, σ is the deviatoric stress, μ is the shear modulus for the mineral, d is the grain size of the rock, Q is the activation energy for the deformation mechanism, T is the temperature in Kelvins, R is the gas constant and A is a constant (Nicolas and Poirier, 1976; Poirier, 1985). Under high stress conditions where $\sigma > 10^{-3}\mu$, the power-law creep equation may break down to give an exponential dependence of deformation rate on the stress

$$\dot{\epsilon} = A' \exp\left(K\frac{\sigma}{\mu}\right) \exp-\left(\frac{Q'}{RT}\right) \quad (2)$$

where Q' is the activation energy for the high-stress deformation mechanism and A' and K are constants. The deformation mechanisms in which the exponential law holds generally involve the motion of dislocations for which there is usually no dependence of deformation rate on the grain size.

Consider each of the deformation mechanisms illustrated in Fig. 4 in detail:

Dislocation glide. This mechanism is the dominant plastic deformation mechanism at low temperatures and high stresses. Deformation arises from the glide of dislocations along slip planes within the individual grains; as deformation continues, the dislocations interact and form tangles within the grains, resulting in work-hardening. As work-hardening continues, the stress increases until the failure stress is reached and brittle failure occurs. This effect is illustrated in Fig. 5a, a plot of stress versus time for an experiment in which a specimen is deformed at constant temperature and deformation rate in the dislocation glide regime. The deformation in this regime is generally described by an exponential law equation such as (2).

Dislocation creep. This mechanism is the dominant plastic deformation mechanism at relatively high temperatures (greater than one half of the melting temperature) and lower stresses. Deformation arises by the coupled glide of dislocations in slip planes and the climb of dislocations normal to the slip planes to circumvent microscopic obstacles such as other dislocations or precipitates. The dislocation climb component results from increased rates of diffusion at higher temperatures and prevents the dislocation tangles that result in the

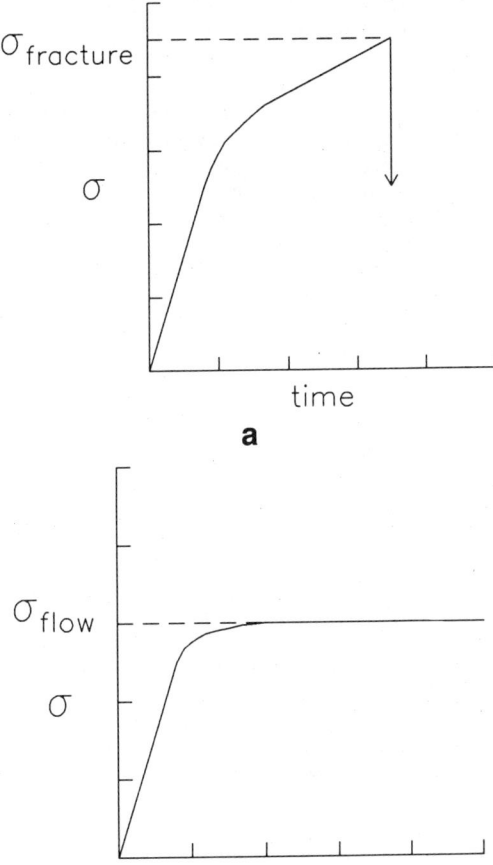

FIGURE 5. Plots of stress versus time at constant deformation rate and temperature for *a*. dislocation glide, showing the initial elastic loading and subsequent plasticity, with work hardening and failure at the fracture stress, and *b*. dislocation creep, showing the initial elastic loading and subsequent steady state deformation at the flow stress.

work hardening that occurs in the dislocation glide regime. The ability of dislocations to climb also permits mutual annihilation of dislocations, allowing a steady-state dislocation density, and, consequently, a steady-state deformation rate, to be attained for a given set of environmental conditions. This behavior is illustrated in Fig. 5b, a plot of stress versus time for an experiment in which a specimen is deformed at constant temperature and deformation rate. Deformation in this regime is generally described by a power-law equation such as (1), with $n \simeq 3$ and $m = 0$. Deformation within the mafic lithosphere and lower crust is generally considered to occur in this fashion, as indicated for olivine polycrystals in Fig. 4 (flexure strain rates). Because of the time limitations for experimental deformation studies, experimental research is performed at rather higher deformation rates and temperatures (laboratory strain rates).

Grain boundary diffusional creep (Coble creep). This mechanism controls polycrystal deformation at lower temperatures and stresses. Deformation arises from diffusion of ions along the grain-boundaries from grain faces under high compressive stress (perpendicular to the maximum stress direction) to grain faces under lower compressive stress, resulting in a net shortening of the rock parallel to the maximum compressive stress direction. The deformation in this regime is generally described by a power-law equation such as (1), with $n \simeq 1$ and $m = 3$. This mechanism is expected to control the deformation in the upper mantle where the elevated temperatures and low stresses favor diffusional, rather than dislocation mechanisms. In crustal shear zones, where grain sizes are small because of dynamic recrystallization, the distances for diffusion may become small enough that Coble creep, rather than dislocation creep, will control the deformation.

Other deformation mechanisms may be operative under different deformation conditions than those considered in Fig. 4.

Fluid-enhanced grain boundary diffusional creep (pressure-solution creep). This mechanism is expected to control deformation in low-grade metamorphic terrains. As with diffusional grain boundary creep, this mechanism involves grain boundary diffusion of the mineral species; however, in pressure-solution creep, the transport is primarily through an intercrystalline fluid phase. Consequently, as ionic transport through a fluid phase is orders of magnitude faster than grain-boundary diffusion, the deformation rate is considerably more rapid than for diffusional creep. Thus, pressure-solution creep is likely to be the dominant deformation mechanism under hydrous conditions at lower temperatures in the crust where dislocation creep or even brittle failure would dominate under anhydrous conditions.

Lattice diffusional creep (Nabarro-Herring creep). This mechanism is the dominant plastic deformation mechanism at very high temperatures, near the melting temperature of the mineral, and at low stresses. Deformation arises from the intracrystalline diffusion of mineral ions from grain faces under high compressive stress to grain faces under lower compressive stress, resulting in a net shortening of the rock parallel to the maximum compressive stress direction. The deformation in this regime is generally described by a power-law equation such as (1), with $n \simeq 1$ and $m = 2$. As intracrystalline diffusion only dominates over grain-boundary diffusion at temperatures near the melting point of the mineral, this deformation mechanism is not expected to operate in the lithosphere.

Variations in the grain size of the rock and the partial pressures of fluid phases will result in differences in the absolute deformation rates at given temperature and stress conditions and will also move the boundaries between the deformation mechanism fields. For instance, the presence of trace amounts of water have been shown to reduce the strength of olivine or quartz polycrystals by a factor of two or more.

Experimental Deformation of Rocks and Minerals. Laboratory studies of the deformation behavior of rocks and minerals are performed in a wide array of apparatuses, each with its advantages and limitations. Each apparatus provides an essentially hydrostatic environment around the sample and, using opposing pistons, applies a deviatoric stress to the sample. For simplicity, only the case of compressive loading of the samples is discussed in this article; currently, there is little tensile or shear testing of geological materials under conditions favoring plastic deformation. A brief description of the main types of deformation apparatus and their uses follows:

One atmosphere deformation apparatus. In this apparatus, the sample is deformed between opposing pistons at 1 atm pressure within a furnace. Depending on the nature of the sample and the application, temperatures from significantly below ambient to above 2500°C can be attained in such an apparatus. In general, temperatures are known to $\pm 5°C$ and thermal gradients within the specimen are normally less than 2°C. The deformation rate is monitored by measuring the piston displacement with time, and the stress on the sample is provided by placing calibrated weights in a load pan attached to one of the pistons. By incorporating a furnace tube, gas-tight seals and a gas-mixing system, the thermodynamic activity of gaseous components, such as oxygen or sulphur, can be very accurately controlled. The oxygen activity around the sample can be continuously monitored using in situ zirconia oxygen fugacity sensors. Although this apparatus is ideal for deformation studies on single crystals of minerals that do not undergo phase transitions at 1 atm between room temperature and run conditions, its usefulness for polycrystal specimens is limited as grain boundary cracking during heating and subsequent deformation may give rise to cavitation of the sample and brittle failure at relatively low strains.

Gas-medium deformation apparatus. This apparatus utilizes a gas confining medium and a furnace within the pressure vessel, as indicated schematically in Fig. 6. In this apparatus, confining pressures to 1000 MPa can be attained during deformation at temperatures to 1400°C. The temperature at the sample can generally be controlled to $\pm 5°C$ to $\pm 10°C$ with thermal gradients within the specimen kept to less than 5°C. As in the 1 atm apparatus, the sample strain is monitored by measuring the displacement of the advancing piston and making a simple correction for the apparatus distortion caused by the elastic deformation of the loading pistons. In

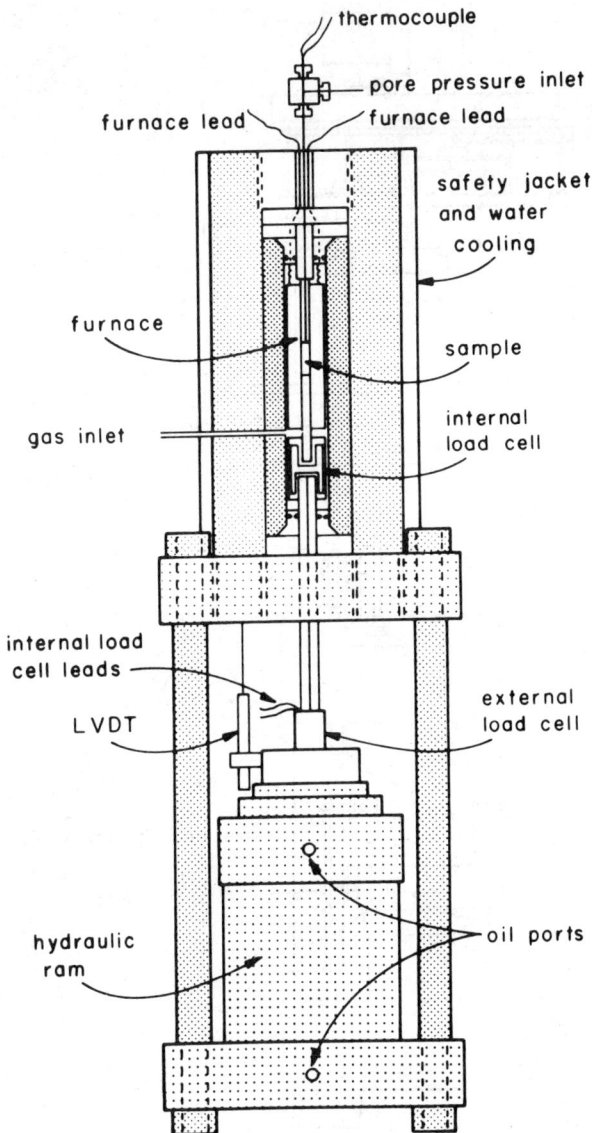

FIGURE 6. Schematic diagram of an internally heated gas apparatus, with the top part showing a cut away view of the vessel and deformation assembly, and the lower part showing an external view of the hydraulic ram for the moving piston. The confining pressure is provided by a series of high-pressure gas pumps (not shown). For scale, a 0.36 m diameter hydraulic ram is typical. (After Tullis and Tullis, 1986).

general, a load cell is located within the pressure vessel to monitor the stress on the sample directly; in this configuration, no subsequent corrections need to be made for the friction caused by the gas seals. Samples are generally sealed from the gas confining medium by enclosing them inside a metal jacket so that cracking and brittle failure during the experiment are suppressed by the confining pressure. However, because of the jacketing of the sample, it is more difficult to control or monitor the environmental conditions around the sample during the experiment. More detailed discussion of the design and limitations of the gas-medium apparatus can be found in Paterson (1970) and Tullis and Tullis (1986).

Solid-medium deformation apparatus (Griggs' apparatus). In this apparatus, confining pressures to 4000 MPa are provided by compressing a solid assembly containing the sample, a graphite or metal furnace element, and a salt, pyrophyllite, or talc-confining medium within a carbide pressure vessel. A schematic diagram of one possible assembly is shown in Fig. 7. The temperature at the sample can generally be controlled to $\pm 10°C$ to $\pm 20°C$ with thermal gradients within the specimen from $10°$ to $30°C$. As in the other designs, the deviatoric stress is applied through pistons that move within the confining medium. The sample strain is monitored by measuring the displacement of the advancing piston after correction for the elastic deformation of the

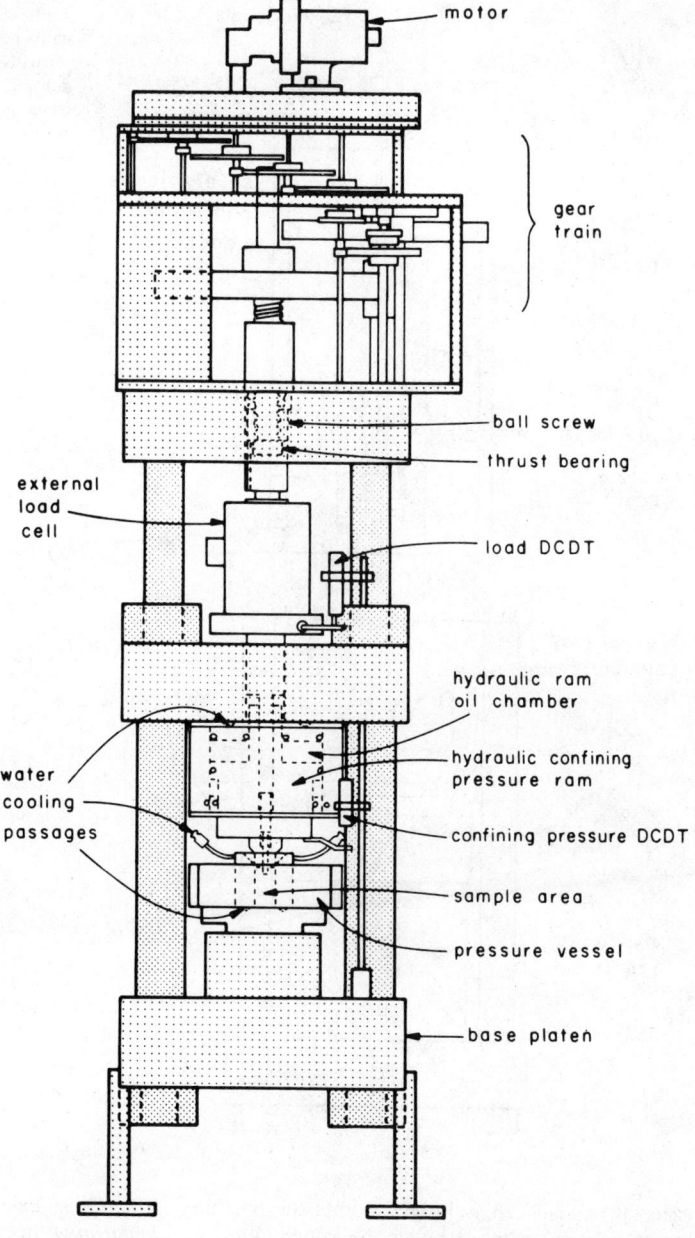

FIGURE 7. Solid medium apparatus of the Griggs' design, slightly modified. The hydraulic ram is driven by a hand pumping system (not shown). The confining pressure is applied using the hydraulic ram, and the deformation piston is advanced using the gearing system at the top of the apparatus. Stress is measured using the external load cell, and strain using the load DCDT. The specimen assembly, including the heating element and insulation ceramic is loaded into the sample area. For scale, the sample area is usually about 15 mm in diameter. (After Tullis and Tullis, 1986).

pistons. The stress is monitored by measuring the force applied to the piston using an external load cell. As the stress is measured externally, corrections must be made for the friction between the moving piston and the seals and confining medium, severely limiting the resolution of the apparatus; stresses less than 100 MPa are usually not measurable in this apparatus. However, recent innovations, such as the use of a partially liquid confining medium, have markedly improved the resolution of this apparatus. Samples are generally sealed within capsules to prevent or limit interactions between the surrounding assembly and the specimen. As in the gas apparatus, it is difficult to control and monitor the chemical environment around the specimens during deformation. More detailed discussion of the design and limitations of the solid-medium apparatus can be found in Tullis and Tullis (1986).

Each of the apparatuses discussed briefly above has its uses and limitations. Clearly, resolution in temperature, stress, strain, and chemical environment decreases rapidly as the confining pressure of the apparatus increases. However, many problems can only be studied at the high pressures available in the gas- and solid-medium apparatuses. In particular, the effects of volatiles on the deformation behavior of quartz and olivine can only be studied at pressures above 100 MPa, where the solubility of carbon and hydrogen in these materials becomes measurable.

Rheology of the Oceanic Lithosphere. The relatively simple thermal history and mineralogy of the oceanic lithosphere permits easy application of the results of experimental deformation studies to the oceanic plates. Coupled with the geophysical observations of plate bending caused by subduction and seamount loading, much information can be derived about the structure and physical behavior of the oceanic lithosphere. In terms of the model presented in Fig. 3, the oceanic lithosphere can be considered as a single compositional layer with a region dominated by brittle deformation overlying a region in which plastic deformation mechanisms will dominate; this plastic region extends from the oceanic lithosphere into the upper mantle.

The oceanic lithosphere is composed primarily of three phases: olivine, orthopyroxene, and clinopyroxene, and it is expected that the rheology of peridotite is controlled by the rheology of olivine, the weakest and most abundant of the three minerals. At the conditions appropriate for plastic deformation in the oceanic lithosphere, olivine is expected to deform in the dislocation creep and dislocation glide regimes in Fig. 4.

Deformation studies of olivine single crystals and polycrystals have investigated a wide range of intergranular and intragranular deformation processes, including the role of fluid environment and the effect of melts. Although the details of all of these experimental studies are beyond the scope of this article, a summary of the major contributions of experimental deformation to our understanding of the behavior of the oceanic lithosphere follows:

1. The chemical environment is extremely important in determining the high-temperature rheology of olivine. In particular, the thermodynamic activities of water and oxygen around the olivine single crystals and polycrystals have marked effects on the deformation rate.
2. At experimental deformation rates, olivine deforms at high temperatures ($>1100\,°C$) predominantly by the motion of dislocations with Burgers vector [100] along the (010) glide plane. At lower temperatures ($<1000\,°C$), dislocations with a [001] Burgers vector become more dominant, moving in the {110} glide planes.
3. At fine grain sizes and/or low stresses, a transition is predicted from dislocation creep to diffusional creep. In the diffusional creep regime, the presence of a fluid phase enhances the grain boundary diffusion rates. Diffusional creep, rather than dislocation creep, is expected to control the deformation at the higher temperatures and lower stresses present in the sublithospheric mantle.
4. Although the presence of a small amount of pyroxenes is not expected to markedly affect the rheology of peridotites, the presence of a melt phase has been observed to reduce the strength of the rock at conditions favoring deformation in the diffusional creep regime. This weakening results from an effective shortening of the grain boundary path for the mineral ions to diffuse, as diffusion of ions in a melt phase is much more rapid than diffusion along a dry grain boundary.

From this brief summary, it is clear that detailed information on the temperatures, pressures, creep rates, stresses, fluid environments, and oxygen activities present in the oceanic lithosphere is necessary to extrapolate the results of laboratory deformation experiments to the earth. In particular, the extrapolation of 7 orders of magnitude or so in deformation rate from the laboratory to the earth is always a concern. The major question that must be addressed in this extrapolation is whether the same deformation mechanism is operative in both situations. If the answer to this question is "yes," on the basis of a comparison of microstructures from the field and laboratory, then the extrapolation can be made with reasonable confidence.

From a comparison of the observed behavior of the oceanic lithosphere and the experimental measurements, a greater understanding of the structure, chemical environment, mineralogy, and physical behavior of the oceanic lithosphere can be derived. In particular, a clearer picture of the physical processes involved in the bending of the oceanic plates as a result of subduction or isostasy can be obtained from application of the results of experimental studies of the deformation of olivine-rich rocks. More detailed reviews of recent experimental research on the oceanic lithosphere can be found in Kirby (1980, 1983).

Rheology of the Continental Lithosphere. Mineralogical and compositional heterogeneities of the continental crust, the lack of reliable rheologies for representative crustal rocks, and the complex thermal history of the continental crust make it difficult to use geological and geophysical observations to constrain the rheology of the continental lithosphere. Although the oceanic crust can sensibly be described in terms of a single mineralogical layer with a brittle deformation regime grading into a plastic regime with increasing depth, the continental crust can only be described as a laterally-varying structure of multiple mineralogical layers, each having its own brittle and plastic regimes. In this sense, the simple crustal model presented in Fig. 3 is a gross oversimplification. However, it does provide a conceptual starting point for a study of the rheology of the continental lithosphere.

The study of the deformation behavior of continental crustal rocks in the plastic regime has involved research on over a dozen rock-forming minerals plus numerous natural polyphase rocks. The studies have investigated a wide range of intergranular and intragranular deformation processes, the role of fluid environment and the effect of melts on the physical behavior of crustal rocks and minerals. Although the details of all of these experimental studies are beyond the scope of this report, the major contributions of experimental deformation to our understanding of the behavior of the continental lithosphere are summarized below.

One of the primary concerns in experimental rock deformation is to provide ways in which the field geologist can evaluate the microscopic textures observed in naturally deformed rocks. In particular, from experimental deformation studies of various minerals, paleopiezometers, or calibrated stress indicators, can be established that allow the geologist to determine the tectonic stress to which a particular rock was subjected. In addition to being clear indicators of stress, piezometers must be relatively insensitive to other environmental conditions. Piezometers that have been developed by rock deformation studies include calibrations of tectonic stress against dislocation density, subgrain size, recrystallized grain size, and the spacing of deformation twin lamellae and slip bands.

At temperatures to 500°C, most silicate minerals are strong and brittle when tested under dry deformation conditions in the laboratory. By contrast, there is abundant microstructural evidence, from observations of the same minerals in the field, that silicates deform plastically and by intragranular mechanisms at temperatures far below 500°C and at lower stresses than in the laboratory. Consequently, much experimental research has focused on finding the reason for this disparity, concentrating specifically on the effects of fluid environment. The accumulated evidence from deformation studies indicates that the presence of a hydrous fluid results in a marked intragranular weakening of the major rock-forming minerals in the crust. Activation energies for dry deformation are generally about 200 kJ/mol, whereas under wet deformation conditions, the activation energies are rather lower, near 150 kJ/mol. Although the effect of water on mineral strength has been the focus of many deformation studies, the exact nature of the weakening mechanism has yet to be established. A number of recent papers on this problem have been compiled in a special volume (Kirby and Scholz, 1984).

The presence of hydrous fluids or melts along grain boundaries in crustal rocks has been shown, experimentally, to greatly enhance the plasticity of the rock by promoting processes such as pressure-solution and recrystallization.

The rheology of the continental lithosphere below the Moho is dominated by the high-temperature rheology of olivine, which has been discussed, in detail, in the previous section on "Rheology of the Oceanic Lithosphere." On the scale of whole lithospheric plates, the deformation behavior of the continental lithosphere may be controlled by the rheology of the mafic lithosphere below the Moho.

Although the use of experimental rheologies to model the behavior of the continental lithosphere is difficult because of the complex physical, chemical, and thermal structure of the continental crust, experimental deformation studies have provided key tools for geologists to characterize the deformation history of crustal rocks and, coupled with geophysical observations of isostatic rebound and isostasy, have greatly increased our understanding of crustal processes and conditions. More detailed reviews of recent experimental research on the continental lithosphere can be found in Kirby (1980, 1983).

Rheology of the Mantle. The sublithospheric mantle is usually regarded by geophysicists as a slowly-convecting solid mass of mafic material. With increasing depth, pressure, and temperature, the mineral phases present in the upper mantle undergo a sequence of transformations to more dense phases. The experimental study of the deformation of mantle materials is, strictly speaking, limited to those materials that are stable at pressures below 4000 MPa because of apparatus capabilities. This limitation means that olivine, clinopyroxene, and orthopyroxene can be studied directly, but not spinel, perovskite and other phases that are stable at greater pressures.

The experimental rheology of olivine-rich rocks has been discussed in the section on the oceanic lithosphere. The general results of those studies indicate that deformation in the upper mantle will be dominated by the deformation of olivine in the dislocation creep or diffusional creep regimes. The detailed behavior of the upper mantle is, as yet, difficult to model until the chemical environment of the rocks, particularly the partial pressures of water, oxygen, and carbon, are better constrained, and experimental studies have been completed of the effects of these volatiles on the deformation behavior of olivine-rich rocks under conditions that can be reliably extrapolated to the upper mantle.

As experimental studies of the major high-pressure phases that exist in the mantle, spinel $(Mg,Fe)_2SiO_4$ and perovskite $(Mg,Fe)SiO_3$, are not possible with the current apparatuses, a number of studies have been performed on *analog* materials. Analog materials, in this context, are materials that have the same chemical bonding structure as the mantle phase under consideration but have different ions in some, or all, of the chemical sites in the structure; for instance, Mg_2GeO_4 is often used as an analog for the spinel phase as it undergoes the transition from the olivine structure to the spinel structure at room pressure and 820°C, allowing the spinel phase to be studied in low pressure apparatuses. In the study of the higher pressure phases more exotic analogs have to be used; compounds such as $KZnF_3$ are used as analogs for perovskite.

Obviously, the use of materials that are so chemically different from the rocks that are really of interest introduces some question as to the validity of studying such materials. However, they are the only windows that we have to the behavior of the lower mantle and the results from these studies, although not highly quantitative, give indications as to viscosity in the deep earth. From these studies, the concepts of whole-mantle convection, diapirism, and deep-mantle plumes can be tested.

S. J. MACKWELL

References

Ashby, M. F., and R. A. Verall, 1978, Micromechanisms of flow and fracture, and their relevance to the rheology of the upper mantle, *Royal Soc. London Philos. Trans.* **A288**, 59-95.

Brace, W. F., and D. L. Kohlstedt, 1980, Limits on lithospheric stress imposed by laboratory experiments, *Jour. Geophys. Research* **85**, 6248-6252.

Chen, W-P., and P. Molnar, 1983, Focal mechanisms and intraplate earthquakes and their implications for the thermal and mechanical properties of the lithosphere, *Jour. Geophys. Research* **88**, 4183-4214.

Frost, H. J., and M. F. Ashby, 1982, *Deformation Mechanism Maps*. Oxford: Pergamon Press, 165p.

Gueguen, Y., J. Woirgard, and M. Darot, 1981, Attenuation mechanisms and attenuation in the upper mantle, in F. D. Stacey, M. S. Paterson, and A. Nicholas, eds., *Anelasticity in the Earth*. Washington, D.C.: American Geophysical Union (Geodynamics Series, vol. 4) 86-94.

Jackson, I. N. S., 1986, The laboratory study of seismic wave attenuation, in B. E. Hobbs and H. C. Heard, eds., *Mineral and Rock Deformation: Laboratory Studies*. Washington, D.C.: American Geophysical Union (Geophys. Monogr. Ser., vol. 36) 11-23.

Kirby, S. H., 1980, Tectonic stresses in the lithosphere: constraints provided by the experimental deformation of rocks, *Jour. Geophys. Research* **85**, 6353-6363.

Kirby, S. H., 1983, Rheology of the lithosphere, *Rev. Geophysics Space Physics* **21**, 1458-1487.

Kirby, S. H., and C. H. Scholz, 1984, Chemical effects of water on the strength and deformation of crustal rocks, *Jour. Geophys. Research* **89**, 3991-4358.

Nicolas, A., and J-P. Poirier, 1976, *Crystalline Plasticity and Solid State Flow in Metamorphic Rocks*. London: Wiley, 444p.

Nowick, A. S., and B. S. Berry, 1972, *Anelastic Relaxation in Crystalline Solids*. New York: Academic Press, 677p.

Orowan, E., 1976, Seismic damping and creep in the mantle, *Royal Astron. Soc. Geophys. Jour.* **14**, 191-218.

Paterson, M. S., 1970, A high-temperature, high-pressure apparatus for rock deformation, *Jour. Rock Mech. Min. Sci.* **7**, 517-526.

Poirier, J-P., 1985, *Creep of Crystals*. Cambridge: Cambridge University Press, 260p.

Tullis, T. E., and J. Tullis, 1986, Experimental rock deformation techniques, in B. E. Hobbs and H. C. Heard, eds. *Mineral and Rock Deformation: Laboratory Studies*. Washington, D.C.: American Geophysical Union (Geophys. Monogr. Ser., vol. 36) 297-324.

Weiner, A. T., M. H. Mangnani, and R. Raj, 1987, Internal friction in tholeiitic basalts, *Jour. Geophys. Research* **92**, 11,635-11,643.

Cross-references: *Brittle Phenomena; Continental Drilling: Ultradeep; Continental Crustal Structure; Continental Lithosphere; Lithosphere: Mechanical Properties; Mantle Convection and Plumes; Mantle Dynamics; Mantle Viscosity; Seismic Properties of Rocks; Stress in the Earth's Lithosphere; Subduction Zones; Thin-Skin Tectonics.*

DEMAGNETIZATION

The total natural remanent magnetization (NRM) measured in paleomagnetic studies generally consists of different remanent magnetization components acquired during different events in the geologic history of the rock. A *primary remanent magnetization* (primary component) is acquired in the ambient geomagnetic field at the time of formation of the rock and can reside in more than one type of the magnetic minerals present. Several generations of *secondary remanent magnetizations* (secondary components) may have been acquired in the period between the formation of the rock and the present time, e.g., because of growth of new minerals or as an overprint of an older natural remanent magnetization. In addition to weathering as a common origin of recent secondary remanence, secondary components may have been acquired during regional burial and subsequent uplift, tectonic deformation, and various diagenetic and metamorphic alterations. Moreover, most rocks contain magnetic grains with low magnetic stability producing a generally superimposed *viscous remanent magnetization* (VRM), which is acquired in relatively short time, ranging from laboratory timescale to recent geological timescale. Viscous components are always secondary, but the reverse need not be true: Secondary components may originate from chemical alteration producing a stable chemical remanent magnetization (CRM). Finally, rock samples may acquire secondary and/or viscous remanence components during sampling, transport, storage, preparation, and further laboratory treatments.

A major part of paleomagnetic work deals with investigations based on the primary remanence components, although the interest in (the significance of) secondary remanence components is growing.

Since the different NRM components in a specific rock have been induced by the ambient local geomagnetic field at widely different times, each individual remanence component usually has its distinctive direction. This difference in directions is important evidence for the recognition and identification of the different components.

The magnetic mineral suite in natural rocks usually is a composite of various magnetic minerals of which the most important for paleomagnetic purposes are (titano)magnetite, (titano)hematite, maghemite, goethite, and pyrrhotite. Each mineral is represented by a distribution of single domain (SD), pseudo single domain (PSD), and multidomain (MD) particles, with a considerable variation in grain size and shape. Mineral type, grain size, and shape determine the magnetic properties of individual grains such as coercive force (H_c) and blocking temperature (T_b). Since each NRM component may reside in different subassemblies of these magnetic grains, the various components may reveal a different coercive force spectrum, a different blocking temperature spectrum and/or a different resistance to chemical leaching. Essentially, these differences in properties and hence in demagnetization characteristics afford the possibility of separating and determining the various individual components of the multicomponent NRM present in a rock sample. This can be done by progressive demagnetization of the total NRM either by means of alternating magnetic fields (a.f. demagnetization), by heating (thermal demagnetization), or by chemical leaching (chemical demagnetization) of the rock sample. Alternating magnetic field and thermal demagnetization are by far the most frequently used methods.

A frequently observed example of a two-component NRM system is a fresh rock with a primary natural remanent magnetization and a superimposed secondary component of viscous origin, usually acquired in the present-day geomagnetic field. Viscous remanence components can be removed by low alternating magnetic fields or by moderate heating. Continuation of the demagnetization treatment then merely removes the primary component, so that the direction of the remaining remanence does not change anymore; only its intensity decreases. The primary component then is said to be obtained as "stable endpoint" direction in the course of demagnetization. The removal of the viscous remanence acquired in the present day geomagnetic field (the "noise") from the primary remanence (the "signal") often is paraphrased as "cleaning," which expression has become synonymous with demagnetization.

No simple relationship exists between the stability of a natural remanent magnetization component and its age. An example is the natural magnetic mineral goethite, a common weathering product of magnetic minerals and Fe-bearing silicates. A very young secondary component of the NRM of a rock may thus reside in goethite grains. This goethite component shows an extreme hardness with respect to a.f. demagnetization and often cannot be removed by alternating field cleaning. On the other hand, such a goethite component is readily eliminated by very moderate heating (100°–150°C). The natural remanence of hematite grains—another common magnetic mineral that may carry a primary or secondary NRM depending on its mode of occurrence—is hardly or not at all affected by the highest alternating fields commonly in use (0.1–0.3 Tesla). Hematite also displays the highest blocking temperature spectrum of the natural magnetic minerals. The blocking temperatures, however, do not exceed 675°C, a temperature easily obtained with laboratory furnaces. Conversely, components residing in (titano)magnetite—which is often assumed to carry a primary remanence—can be successfully demagnetized by alternating fields, as well as by heating to a maximum temperature of 580°C. So, whereas a.f. demagnetization may fail to eliminate several of the common natural (secondary) remanences, thermal demagnetization can remove the remanences of all existing natural magnetic minerals.

Alternating magnetic field demagnetization has the advantage of being nondestructive, whereas thermal demagnetization may cause alterations in the rock samples studied. These alterations may affect the original magnetic minerals present and/or may create a new generation of magnetic minerals derived from the nonmagnetic matrix. Therefore, removal of NRM by thermal demagnetization may, in some cases, be a consequence of progressive alteration of the magnetic minerals rather than a result of passing through their blocking temperature spectrum. Finally, alterations caused during thermal demagnetization treatment of rock samples prevent subsequent rockmagnetic studies of the original magnetic mineral suite.

Progressive demagnetization of rock samples not only yield the directions of the various NRM components. Coercive force spectra and blocking temperature spectra can be obtained for each individual remanence component from the analysis of the progressive demagnetization data set. These spectra give information about the magnetic mineral grains responsible for the various components observed. Consequently, this information may give indications concerning the origin of the different magnetic mineral phases and therefore the origin of the various natural magnetization components. Such an approach is rather complicated, however, and yet far from being a routine procedure in paleomagnetic research. Essentially, however, this method gives more and much better information on the carriers of the NRM than the generally applied method of determining the isothermal remanent magnetization (IRM) acquisition curve of a rock sample. The latter method concerns all magnetic minerals, plus those that do not contribute to the NRM. In practice, this method is only used to reveal the presence of magnetite and is based on the commonly accepted but challengeable idea that natural remanences residing in magnetite must be of primary origin, i.e., date from the formation of the rock.

The different directions of the NRM are not only used to discriminate between these different components but also to determine or interpret their relative ages. A component with a direction (and polarity) identical with or very close to the present local geomagnetic field direction and observed in older rocks, obviously is of recent (and therefore secondary) origin. Those components with a direction deviating from the present day local geomagnetic field (and proved not to be related to the sampling procedure or to preparation and storage of the samples) must have been acquired at an earlier episode in the history of the rock studied. Since at this stage the age of this (consistent) magnetization direction is still uncertain and a primary origin cannot be proven satisfactorily, most conveniently the component is neutrally referred to as the *characteristic remanent magnetization* (ChRM).

To assess the primary character of this ChRM, a number of classical field tests for stability can be applied. Some of these tests yield information about the significance and stability of the remanence component in general and others give a more detailed confirmation about its age. Commonly, five such field tests are traditionally recognized (see, e.g., McElhinney, 1973, and references therein).

Consistency test. A natural remanent magnetization component gains paleomagnetic significance if its direction is observed to be consistently present within a geological formation, preferably in a variety of rock types, sampled over a wide area and through a thickness of strata assumed to represent a considerable time span. The consistency test may be enhanced if it can be shown that paleomagnetic directions of coeval geological formations sampled at remote distances on a single continent correspond to a geocentric dipole field. The characteristic remanence certainly is of secondary origin, however, if the corresponding paleomagnetic pole fits the continent's apparent polar wander path at a younger age than the origin of the rock studied.

Reversal test. The observation of characteristic directions with different polarity and, in particular, antiparallel directions is taken as a strong indication for the primary origin of the remanence component. This test is greatly enhanced if a polarity zonation can be established, and if this zonation is independent of possible changes in the composition of the rock sequence. Very convincing is the observation of polarity transitions (e.g., decrease of intensity, intermediate directions), especially if they are situated at identical levels with respect to other stratigraphical datums in a number of sections at different localities.

Baked contact test. Another direct evidence for the primary origin of a natural remanence component can be obtained from the study of the baked contact zone of an igneous rock. Consistency of the directions of the NRM in the igneous rock and its contact zone, with usually a different composition, indicates that this remanence component typically is related to the cooling of the intrusion and thus dates from the period of the igneous activity.

Fold test. If the characteristic remanence directions from differently tilted beds converge after correction for the dip of the strata, this remanence obviously was acquired prior to the tilting and has remained stable since that time. Strictly this fold test does not directly prove a primary origin of a component, but only that it dates from before tilting. The high stability inferred for the remanence component often is taken as a strong indication that it concerns a primary remanence.

Conglomerate test. Conglomerate pebbles are randomly deposited and therefore random magnetization directions in these pebbles suggest that this magnetization component has been stable since the formation of the conglomerate. This observation is then extended to the same remanence component in the corresponding source rock. In practice, this test is of limited value: it does not provide a definite answer on the age of the remanence component and, moreover, has the disadvantage that the information about the stability is indirect and not based on observation of the rocks studied for paleomagnetic purposes itself.

Principles of Demagnetization

For an assemblage of aligned identical single domain magnetic grains there is a certain time-constant called the relaxation time t_r:

$$t_r = 1/C \exp \left(\frac{v \, H_c J_s \, (1 - H_e/H_c)^2}{2 \, k \, T} \right) \quad (1a)$$

where C is a frequency factor equal to about $10^{+10} s^{-1}$, v is the volume of the grain, H_c the coercive force, J_s the saturation magnetization, H_e the applied external field along the easy axis, k the Boltzmann's constant, and T the absolute temperature. Frequently $H_e << H_c$, and without the need for alignment (1a) changes into:

$$t_r = 1/C \exp \left(\frac{v \, H_c J_s}{2 \, k \, T} \right) \quad (1b)$$

This relaxation time is a measure of the time needed for the magnetization of the assemblage to reach equilibrium. This equilibrium magnetization depends on the external field; and it will be zero for zero external field. If the assemblage has an initial moment M_0, on removal of the applied field, it will decay at a rate determined by t_r according to:

$$M_r = M_0 \exp(-t/t_r) \quad (2)$$

where M_r is the moment remaining after time t.

We can see that by lowering the relaxation time of grains of a specific type and size to a laboratory timescale (e.g., 100 to 1000 seconds) in a zero ambient field, the magnetization will rapidly decrease according to (2). The contribution of such a particular assemblage of grains to the total remanent magnetization is then nullified. By subsequently nullifying the contribution of assemblages of grains with increasing stability (i.e., longer relaxation times), the total remanent magnetization of a rock sample can be demagnetized. From (1) it can be seen that the relaxation time can be lowered by either lowering the volume of the grains (chemical demagnetization), by lowering the coercive force of the grains (or, alternatively, by applying a field approaching the coercive force of the grains, a.f. demagnetization), or by increasing the temperature (thermal demagnetization). The different methods of demagnetization will be discussed in more detail in the next paragraphs.

Alternating Field Demagnetization. If a rock sample is subjected to an alternating magnetic field with peak value H_p, all its magnetic grains with an effective coercive force $H_c < H_p$ will change their magnetic state at each halfcycle of the alternating field. These changes involve domain wall movements in multidomain grains and rotation of the magnetic moment in single domain grains. As the alternating field is gradually decreased to zero, the magnetic domain moments and walls in grains with decreasing effective coercive force will stagnate in the last obtained magnetic state. Although multidomain grains may be entirely demagnetized, mostly they are left with randomly directed moments which, together with the randomized moments of the cycled single domain grains, will constitute a selfcancelling assembly effectively removed from the NRM of a rock sample. By treatment of the rock samples with successively stronger alternating fields, the magnetic moments of grains with successively higher coercive forces are demagnetized or randomized and thus removed from the NRM. By measuring the remaining NRM after each treatment, remanences of increasing hardness can be distinguished.

The application of alternating field demagnetization theory to natural samples is difficult, since it depends on a variety of factors such as demagnetizing (or internal) field and movement of domain walls. For a more detailed discussion, see, e.g., Collinson (1983) and references therein. Some theoretical considerations are in good agreement with practical experience. The coercive force associated with multidomain magnetite has a maximum of 30 mT-60 mT (Collinson, 1983). In single domain magnetite the coercive force is controlled by anisotropy due to grain shape and has a theoretical maximum of 300 mT (Evans and McElhinney, 1969). Therefore, all remanences carried by magnetite can in principle be removed by means of alternating field demagnetization. In hematite high to very high coercive forces exceeding 500 mT can be observed due to a strong crystalline anisotropy. Thermal demagnetization is therefore required for totally removing hematite remanences.

Alternating field demagnetizing equipment consists basically of a coil (and stationary or tumbling sample holder), a capacity bank for tuning the coil, a power supply and current reduction mechanism, and a coil system or magnetic shield to provide a zero (or low) ambient field. Full symmetry of the alternating field is obligatory to avoid a steady (direct) field or even harmonics of the alternating field. The latter are easily eliminated by tuning the demagnetization coil and the former by cancelling the geomagnetic field. The presence of a steady (direct) field produces an anhysteretic remanent magnetization (ARM) and must be avoided. The use of a tumbler assists in removing this effect and is frequently used in a.f. demagnetizing equipment.

The most effective orientation of the field is parallel to the NRM direction. Since this is not practical, essentially two methods are in use. The use of *tumbling sample holders* assists in allowing the demagnetizing field to act along as many directions in the rock as possible. A *stationary sample holder* is used for demagnetization along three mutually perpendicular axes in the sample, at each value of the peak field. In this way any direction in the rock is less than 55° from a field direction. Experiments by Zijderveld (1975) show that this method produces virtually the same results as when the field is parallel to the NRM direction. Also, the acquisition of a rotational remanent magnetization (RRM; Wilson and Lomax, 1972) as a result of tumbling can be avoided.

Thermal Demagnetization. From equation (1) it follows that the logarithm of the relaxation time of single domain grains is inversely proportional to temperature. The blocking temperature T_b for an assemblage of identical grains is defined as that temperature at which the relaxation time becomes small, i.e., laboratory timescale. A rock sample usually contains a distribution of grain assemblages of different size and composition with a corresponding distribution of blocking temperatures. Furthermore, the relaxation time t_{r1} at temperature T_1 is simply related to the relaxation time t_{r2} at temperature T_2 as:

$$T_1 \ln (Ct_{r1}) = T_2 \ln (Ct_{r2}) \tag{3}$$

provided J_s is constant between T_1 and T_2. Thus the same effect at temperature T_1 for a time t_{r1} is obtained at a higher temperature T_2 for a shorter time t_{r2}. For instance, maintaining the temperature at 150°C for one million years is equivalent to maintaining at 450°C for ca. 1000 seconds. Thermal demagnetization consists of heating a rock sample to a certain temperature T_p. Grains or grain assemblages with a blocking temperature $T_b < T_p$ are then unstable. In a zero ambient field the magnetic moments of the grain assemblage will orient at random and not contribute any longer to the NRM. On cooling in a zero ambient field the relaxation time becomes large again and the random orientation will be fixed, thus being effectively removed from the NRM.

The process of stepwise thermal demagnetization involves increasing T_p in steps and measuring the remaining NRM at each step after cooling down. The last step usually involves heating to within a few degrees of the Curie point of the mineral carrying the NRM. The main requirement for thermal demagnetization is to cool the sample in a zero ambient field to avoid acquisition of a partial TRM (PTRM). If during cooling of the rock samples the field is not sufficiently "zero," an important PTRM may be acquired either by existing or newly formed magnetic minerals, as a result of chemical alteration by heating to elevated temperatures. This may be the case especially in sediments, in which magnetite may be formed through the breakdown of clay minerals. Such a PTRM acquired in even a low residual field in the order of several nT may easily dominate the NRM of the sediment. Another effect may be the acquisition of a rather large VRM component while transporting the sample from the demagnetization furnace to the magnetometer. This VRM may quite quickly disappear in the (near) zero field of a shielded magnetometer, but can slow down the measuring procedure considerably.

Equipment (furnaces) for thermal demagnetization purposes involves creating a uniform temperature over enough room to hold a reasonable amount of samples; the winding used in electric furnaces should be noninductive to prevent or reduce generation of magnetic fields. A (near) zero ambient field can be created either by a coil system or by mumetal magnetic shielding. Many designs for furnaces exist in the literature (cf. Collinson, 1983); a commercially available furnace consists of two chambers, one for heating a batch of (8-12) samples and one for rapid cooling of these samples, allowing a continuing procedure of demagnetizing sets of samples with the necessity to cool down the furnace after each heating step. Mumetal shields are used to create a low ambient field in the chambers.

Continuous thermal demagnetization involves measuring the NRM continuously (or at short time intervals) while the sample is being heated. The sample—only one at a time—takes a short time, typically 1-2 hours, to be totally demagnetized. The advantages are that no PTRM is acquired while measuring and there is less chance of measuring disturbing magnetizations caused by chemical changes provided that the new mineral is formed above its blocking temperature. A disadvantage is the much lower sensitivity due to the distance between sample and detector because of the furnace surrounding the sample and the space needed for different orientations of sample plus sampleholder. Moreover, the NRM at higher temperatures will be increasingly less because of the decay of spontaneous magnetization with temperature.

Chemical Demagnetization. Chemical demagnetization involves the leaching of minerals carrying the different remanences in a rock sample. Hydrochloric acid is forced through the sample provided the rock is porous enough; if not, holes or cuts may be made to increase the surface area for reaction with the acid (Henry, 1979). If the mineral most resistant to destruction by acid is also carrying the primary NRM, chemical demagnetization may be successful in determining different NRM components. In some cases it may even reveal components not distinguished by a.f. or thermal demagnetization (Roy and Park, 1974). The disadvantages, long solution times, and complications in handling hydrochloric acid make the application of this method rather limited.

Presentation and Analysis of Demagnetization Results

The multicomponent NRM is a vectorial sum of linear magnetic components. These linear components in the multicomponent NRM have a unique combination of direction and spectrum. A spectrum can be either a blocking temperature spectrum, a coercive force spectrum or a resistance to chemical leaching. This unique combination allows the separation and identification of these components with progressive demagnetization. The identification of the magnetic carriers from the spectra—which can be derived from the decay curves of the separate NRM components—as well as the sorting of the components in order of age (which is not necessarily the same as the order of relative stability) are geologic and rockmagnetic problems and cannot be solved from the demagnetization data set alone.

The progressive (stepwise) demagnetization of the NRM of a single specimen yields a data set composed of n demagnetization steps, $3n$ coordinates of the magnetization vectors, and a set of at least n either scalar or vectorial measurement errors. The NRM components are extracted from this data set with geometric techniques.

When NRM components have *no spectral overlap* the endpoints of the measured magnetization vectors line up to linear segments in three-dimensional space with directions and intensities corresponding to the NRM components. This ideal behavior rarely occurs, unfortunately. In the more common situation of *partial spectral overlap* between the components such a linear segment is present only when there is a coercive force or temperature range in which only one magnetic component is demagnetized (mutual nonoverlap or *window* in the blocking spectrum). In the rare situation in which two components have different directions but *total spectral overlap* it is not possible to determine the original directions: the two components cannot be separated. In the unlikely

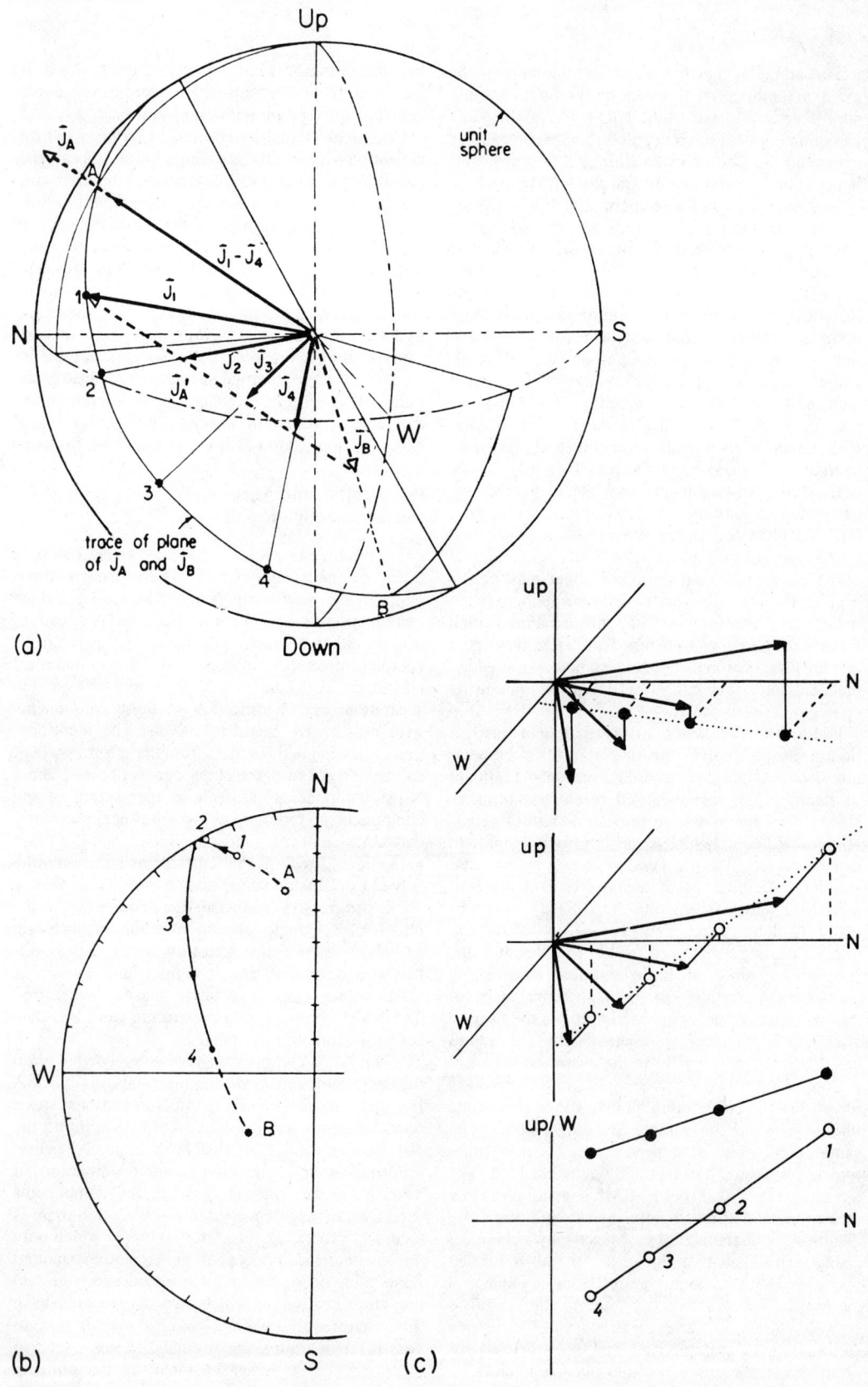

situation of identical total spectral overlap the components will show up as a single linear segment (one component) with an anomalous direction.

The direction and intensity of a component *without* a window in its spectrum can only be found if a softer as well as a harder component is present, both *with* a window in their spectrum: the intersection line of the two planes that can be constructed from three superimposed components corresponds to the direction of the second component (Hoffman and Day, 1978). Consequently, if there is no window in the spectrum of the most stable component its direction cannot be determined from one demagnetization data set without presuppositions about its spectrum.

To visually recognize linear segments and/or planar chunks through the endpoints of the magnetization vectors in three-dimensional space, it is practical to project the data set on to a plane. The three projection methods currently in use are the orthogonal, stereographic (or Wulff), and equal area (or Lambert) projections.

In *orthogonal projection plots* (Zijderveld, 1967, 1975) the endpoints of the measured magnetization vectors are projected on the horizontal plane and on a vertical plane in the same diagram (Fig. 1a,c). For the vertical plane conveniently the one containing either the N-S or E-W axis is used. The intersection axis of the two planes, either N-S or E-W, is taken as the *x*-axis, the *y*-axis then corresponds to the up/W-down/E axes (*up/W projection*) or to the up/N-down/S axes (*up/N projection*), respectively (Fig. 2a,b). By convention the solid and open symbols

---

FIGURE 1. The magnetization vectors J_A and J_B constitute a two-component NRM which is partially demagnetized in four steps. The NRM vectors J_A and J_B and the magnetization vectors J_1, J_2, J_3, J_4 measured after demagnetization are plotted (a) in a spherical coordinate system, (b) as unit vectors on an equal-area projection and (c) as vector components on a Zijderveld diagram. In (a) A, B, 1, 2, 3, 4 are the unit magnetization vectors, dot-dash lines outline horizontal and vertical planes. The measured magnetization vectors and the subtracted vector $J_1 - J_4$ derived from them are shown as solid lines and vectors not known a priori as dashed lines. Note that the subtracted vector J_{1-4} parallels J_A and J'_A (i.e. J_A translated to the endpoint of J_B). The circle connecting A, 1, 2, 3, 4, B is the remagnetization circle, also plotted in (b). In (b) the unknown vectors A and B are denoted as stars and extensions of the remagnetization circle beyond the range of measured vectors are shown as dashed lines. In (c) the vectors J_1, J_2, J_3, J_4 are projected on the horizontal plane (upper diagram) and on the vertical plane (middle diagram) and on both in the same diagram (lower diagram). The intersection axis of the two planes is the N-S axis. The pair of linear segments are the projections of the vector J_1–J_4 which parallels magnetization vector J_A and J'_A. Open (closed) symbols denote projections of upward (downward) vectors in an equal-area projection, and projections on a vertical (horizontal) plane in orthogonal projections. (After Dunlop, 1979).

respectively denote projections on the horizontal and vertical plane. A linear segment made up by the endpoints of the magnetization vectors will be projected as a pair of lines in the diagram.

The declination (D) of a magnetization vector or linear segment is easily measured as the clockwise angle between North and the projection of this magnetization vector or linear segment on the horizontal plane. The angle between the horizontal axis and the projection of the magnetization vector or linear segment on the vertical plane is the apparent inclination (I_{app}). The inclination (I) can be calculated with

$\tan I = \tan I_{app} |\cos D|$ for an Up/N projection
$\tan I = \tan I_{app} |\sin D|$ for an Up/W projection.

Decay curves of the separate components can be constructed from the diagram by extrapolation of the linear segments as shown in Fig. 2c. The main advantage of the orthogonal projection method is that all information concerning the data set is used and visualized. For example, linear segments are shown as a pair of lines on the diagram and hence easily recognized (Fig. 3). The advantages and restrictions of the method are discussed in detail by Dunlop (1979).

In Lambert's *equal area projection* or in Wulff's *stereographic projection* (Collinson, 1983) the magnetization vectors are set equal to unit length (Fig. 1a,b). As a result of this operation the endpoints of the unit magnetization vectors are distributed on the surface of a unit sphere from which they are downward projected as points in the horizontal plane (the projection circle). The declination of the magnetization vector equals the clockwise measured angle between north and the projection of the magnetization vector on to the horizontal plane. The inclination of the magnetization vector corresponds to a radius in the projection circle equal to

$$R = R_0 \, \text{sqr}(1 - |\sin I|)$$

for Lambert's projection and

$$R = R_0 \, 2\sin I/(1 + \cos I)$$

for Wulff's projection.

R_0 is the radius of the projected circle. By convention open and closed symbols respectively denote negative (upward) and positive (downward) inclinations. The Lambert projection plots equal areas on a sphere as equal areas on a plane, whereas Wulff's stereographic projection plots the angles on a sphere as true angles on a plane. A linear segment through the origin of coordinates and defined by the endpoints of the magnetization vectors in three-dimensional space will be projected as a point.

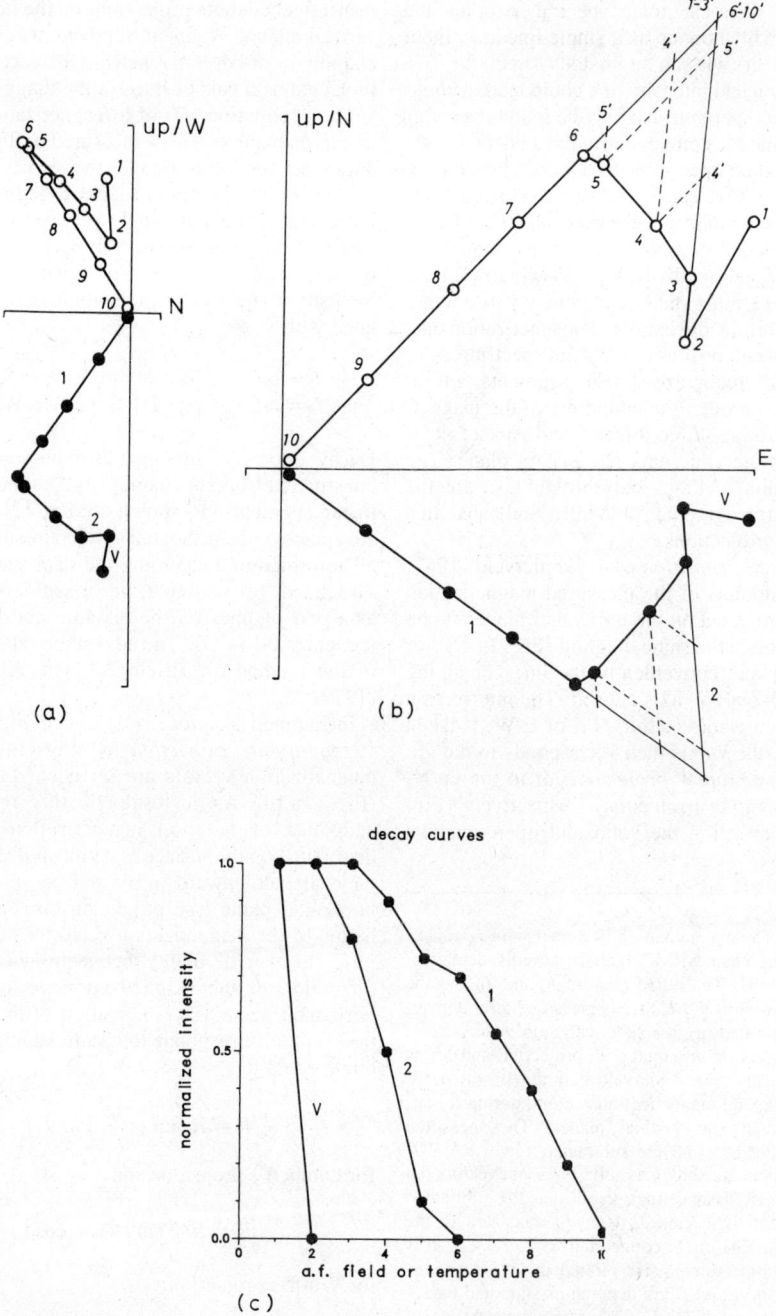

FIGURE 2. Orthogonal projection of an artificial three-component NRM as an (a) up/W-projection (common N-S axis) and (b) up/N-projection (common E-W axis). Separate decay curves can be constructed from the projection by reconstructing or extrapolating the total original vectors of a viscous (V), a secondary (2) and primary (1) magnetization: dashed lines passing the projected endpoints of the magnetization vectors are drawn parallel to the linear segments. The viscous component is removed after step 2, the secondary component decays from step 2 to 6 (according to 2, 3, 4', 5', 6'), the primary component decays from step 3 to 10 (according to 3', 4', 5', 6, 7, 8, 9, 10). In (c) the separate and normalized decay curves from the demagnetization set are plotted, showing no overlap (V vs. 2 or V vs. 1) or partial overlap (2 vs. 1) of the blocking spectra.

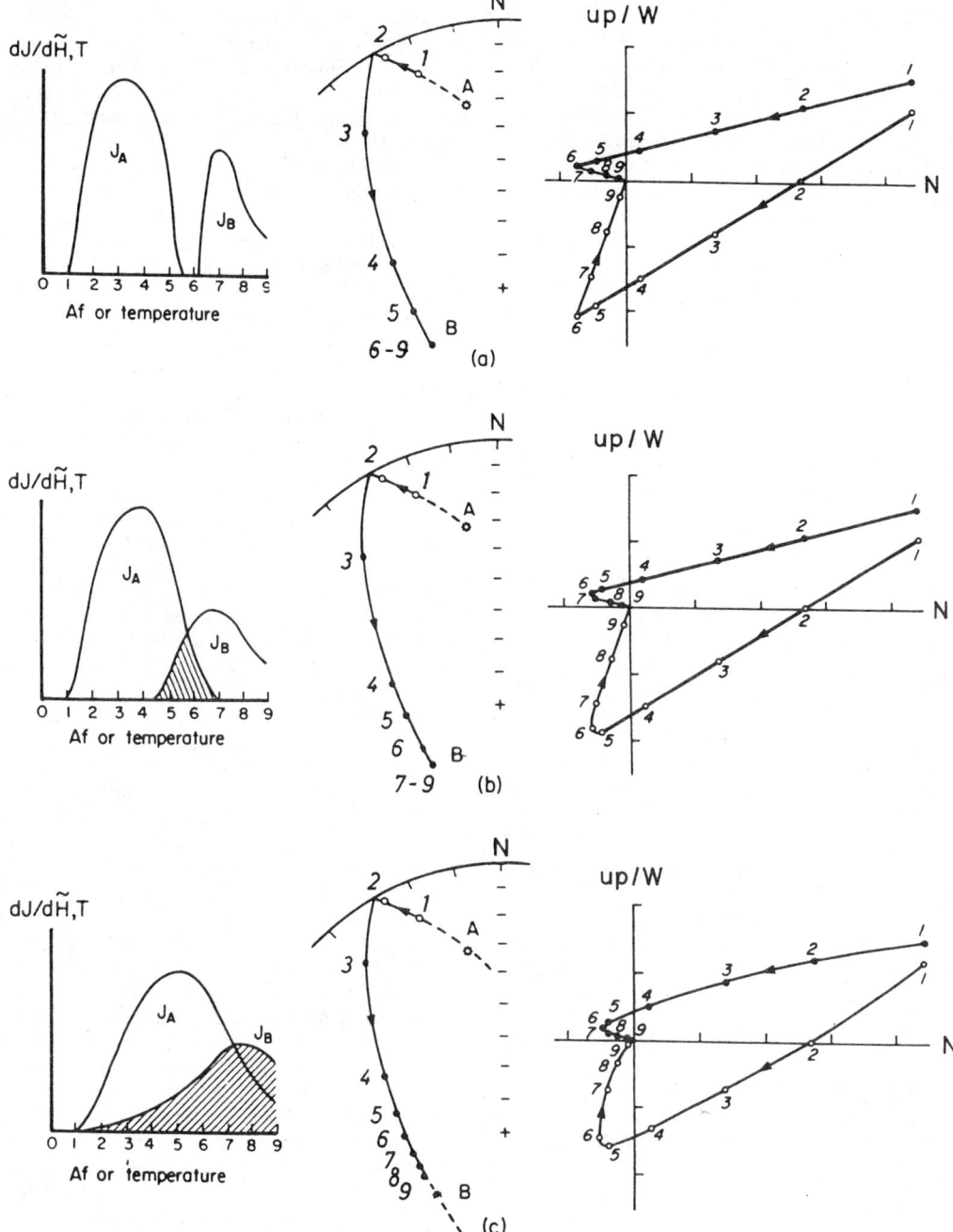

FIGURE 3. Blocking spectra, equal-area and orthogonal projection plots of a demagnetized two-component NRM are plotted with (a) nonoverlapping spectra of the two NRM components A and B, (b) partial overlap but wide windows in the blocking spectra and (c) total overlap (no windows) of the blocking spectra. In (a) and (b) the component B can be determined from the stable end-point in the equal-area projection. The component A is only known to lie on the extension of the remagnetization circle. The orthogonal projection diagrams of (a) and (b) show linear segments which are parallel to the projections of the vectors A and B. In (c), neither of the projections resolve the directions of A or B. (After Dunlop, 1979).

All NRM components define planes through each subsequent pair of components. As a result the data points of the two most stable components will lie in a plane through the origin of coordinates and will be projected along a great circle, a so-called remagnetization circle. When the data set is projected as normalized difference vectors (Hoffman and Day, 1978) all planes will be projected as remagnetization circles. A considerable disadvantage of the Lambert and Wulff projections is that the relative intensity information of the demagnetization data set is lost, which makes it impossible to construct decay curves for the separate components. The decay curve of the total NRM intensity is often used: Its usefulness is very restricted, since this curve depends on the directional difference of the components and hence the information on the separate components is lost. A more practical disadvantage of those projection methods is that the fitting of remagnetization circles through the data points in a stereoplot is not as easy as fitting a line passing the data points in an orthogonal projection.

Once a linear segment or planar chunk is recognized, its orientation, intensity, and confidence (or error angle) can be determined. Although many paleomagnetists use the "eye-ball" statistics, the less subjective method of least-squares fitting of the data set by lines and planes is coming more in use.

Measurement errors (reproducibility) can be used as a weighting factor in least-squares fitting of lines or planes. However, their physical meaning has to be considered prior to their application. The reproducibility of the present-day spinner and cryogenic magnetometers is high, even at low intensities. The demagnetization procedure introduces its own bias or systematic errors as a result of nonperfect demagnetization equipment, alterations in a magnetic field, viscous magnetizations, etc. Either the data set has to be corrected for this introduced bias, or this bias has to be small compared to the measurement error. Only when these conditions are met, a weighting of the data set is meaningful.

Principal component analysis (Kirschvink, 1980; Schmidt, 1982; Kent et al., 1983; McFadden and Schmidt, 1986) of a possibly weighted data set gives the line and/or plane of best least-squares fit. In this analysis the original orthogonal coordinate system is translated to a position in which the new origin of coordinates corresponds to the weighted mean of the data set. In that position the coordinate system is rotated to least-squares fit the coordinate axes to the data. In the case of a fitted line, one principal axis will be oriented in a direction of maximum variance of the data set which has its maximum value when all points lie precisely on one line, another principal axis is oriented in the direction of minimum variance; the orientation of the third axis is then known by definition. The fitting of a plane through the data points follows the same reasoning.

The directions of the principal axes will equal the directions of the eigenvectors of the matrix of sums of squares and products of the data set. The three eigenvalues of the eigenvectors are related to the variances of the data and an angular confidence region of the line or plane can be calculated, or estimated when the total measurement error is unknown.

Statistically based computer algorithms to find lines and planes of best fit from the demagnetization data set are in common use (Kirschvink, 1980; Schmidt, 1982; Kent et al., 1983). Ideally, with such a search algorithm there is no risk that components that are nonideally projected are overlooked. The algorithms, however, are still far from perfect and there is a certain risk to believe the computer while overlooking obvious incidental or systematic errors or unlikely results.

Analytical Methods Using Demagnetizations from Suites of Specimens. When a component has no window in its spectrum and a second superimposed component is present, the technique of *converging remagnetization circles* (Halls, 1976) can be applied to recover the component's direction. The method requires that the searched component has the same direction in all specimens and the second component has not. This may be the case when the variation in the directions of the latter component is produced by tectonics or secular variation, for example. Planes through the endpoints of the magnetization vectors of different specimens will then intersect according to an axis having the direction of the searched component.

With the *exponential least squares modeling technique* (Stupavsky and Symons, 1978) one can in principle resolve NRM components without windows in their blocking spectra. This method presumes a log-normal grain size coercivity distribution and requires a.f. demagnetization data. It is not applicable to material with a bad response to alternating field demagnetization and deviating grain size distributions. These rockmagnetic properties of the carriers of the NRM components must be known, which is only possible when decay curves of the NRM components are determined from other similar specimens with well separated components. However, if these other similar specimens do show well separated components, it may be questioned whether they have the same rockmagnetic properties. Therefore, this procedure is very susceptible to circular reasoning and cannot give unique results.

Stability indices define the paleomagnetic stability of NRM components and are derived from the demagnetization data set. They may be applied in the determination of optimum demagnetization procedures. Most stability indices are related to the alternating demagnetization field or temperature required to resolve linear segments of a primary component in a multicomponent NRM. Hence, they are related to the spectrum and to the grain size of the carrier of this component. Examples of stability indices can be found in Collinson (1983); in general they are not widely used. Most stability indices can be derived from decay curves of separate components. As decay curves do not necessitate presuppositions about the magnetic carriers of the NRM they are considered to be superior to the stability indices.

C. G. LANGEREIS
J. H. LINSSEN
T. A. T. MULLENDER
J. D. A. ZIJDERVELD

References

Collinson, D. W., 1983, *Methods in Rock Magnetism and Palaeomagnetism.* London/New York: Chapman and Hall, 503p.
Dunlop, D. J., 1979, On the use of Zijderveld vector diagrams in multicomponent paleomagnetic studies, *Physics Earth and Planetary Interiors* **20**, 12-24.
Evans, M. E., and M. W. McElhinney, 1969, An investigation of the origin of stable remanence in magnetite-bearing igneous rocks, *Jour. Geomagnetism and Geoelectricity* **21**, 757-773.
Halls, H. C., 1976, A least-squares method to find a remanence direction from converging remagnetization circles, *Geophys. Jour. Royal Astron. Soc.* **45**, 297-304.
Henry, S. G., 1979, Chemical demagnetization: methods, procedures and applications through vector analysis, *Canadian Jour. Earth Sci.* **16**, 1832-1841.
Hoffman, K. A., and R. Day, 1978, Separation of multicomponent NRM: a general method, *Earth and Planetary Sci. Letters* **40**, 433-438.
Kent, J. T., J. C. Briden, and K. V. Mardia, 1983, Linear and planar structure in ordered multivariate data as applied to progressive demagnetization of palaeomagnetic remanence, *Geophys. Jour. Royal Astron. Soc.* **75**, 593-621.
Kirschvink, J. L., 1980, The least-squares line and plane and the analysis of palaeomagnetic data, *Geophys. Jour. Royal Astron. Soc.* **62**, 699-718.
McElhinney, M. W., 1973, *Palaeomagnetism and Plate Tectonics.* Cambridge: Cambridge University Press, 358p.
McFadden, P. L., and P. W. Schmidt, 1986, The accumulation of palaeomagnetic results from multicomponent analyses, *Geophys. Jour. Royal Astron. Soc.* **86**, 965-979.
Roy, J. L., and J. K. Park, 1974, The magnetization process of certain redbeds: vector analysis of chemical and thermal results. *Canadian Jour. Earth Sci.* **11**, 437-471.
Schmidt, P. W., 1982, Linearity spectrum analysis of multi-component magnetizations and its application to some igneous rocks from southeastern Australia, *Geophys. Jour. Royal Astron. Soc.* **70**, 647-665.
Stupavsky, M., and D. T. A. Symons, 1978, Separation of magnetic components from AF step demagnetization data by least-squares computer methods, *Jour. Geophys. Research* **83**, 4925-4931.
Wilson, R. L., and R. Lomax, 1972, Magnetic remanence related to slow rotation of ferromagnetic material in alternating magnetic fields, *Geophys. Jour. Royal Astron. Soc.* **12**, 405-424.
Zijderveld, J. D. A., 1967, A.C. demagnetization of rocks. In D. W. Collinson et al., eds., *Methods in Palaeomagnetism.* Amsterdam: Elsevier, 254-286.
Zijderveld, J. D. A., 1975, *Paleomagnetism of the Esterel Rocks,* Ph.D. dissertation, State University of Utrecht, Utrecht, 199p.

Cross-references: *Archaeomagnetism; Chemical Remanent Magnetization; Curie Temperature; Detrital Remanent Magnetization; Magnetic Domains; Magnetic Properties of Minerals; Magnetic Self-Reversal; Rock Magnetism; Rock Magnetism: Measuring Techniques and Apparatus; Thermoremanence; Viscous Remanent Magnetization (VRM) and Viscous Magnetization.*

DENSITY DISTRIBUTION IN THE EARTH

In this article we model density anomalies in the Earth in terms of low-order spherical harmonics. The density anomalies are determined in relation to global variation of gravity potential and to lateral variation of seismic velocity in the upper layers of the mantle. We first discuss the basic observational evidence that bears upon density distribution in the Earth. We next present mathematical models for computing density from measurements of gravity potential and seismic wave velocity. Finally, we discuss the result of a computation of a model for the distribution of density anomalies in the mantle.

Basic Data Material

Four different types of data material are introduced.

Gravity Potential. The spherical harmonics series development for the gravity potential W of the Earth, valid in the exterior space of the Earth, has the form

$$W = \frac{GM}{r} \left[1 + \sum_{n=2}^{N} \left(\frac{1}{r}\right)^n w_n S_n(\varphi, \lambda) \right] + Z \quad (1)$$

where G is the gravitational constant, M is the mass of the Earth, (r, φ, λ) is the geocentric polar coordinate system, S_n are the spherical harmonics of degree n and order m ($m = 0, 1, 2, \ldots, n$), w_n are the Stokes constants obtained from measurements, (φ, λ) is the latitude and longitude, and Z is the centrifugal potential.

Isostasy. The Airy-Heiskanen isostatic potential W_I is generated by the gravitational effect of both the mountain masses, situated above the ocean level, and the compensating mountain roots, situated below a depth of 30 km. The crustal masses have standard density $\rho_0 = 2650 \text{ kg} \cdot \text{m}^{-3}$. The extension in depth of the mountain roots depends on the heights h by a linear function, in sufficient approximation. In the exterior of the Earth's body, the potential W_I has the function (Arnold and Schoeps, 1984; see *Gravity*

Fields: Implications for Planetary Interiors and Spherical Harmonic Analysis

$$W_I = 4\pi f \rho_0 \sum_n \left(\frac{1}{r}\right)^{n+1} \alpha(n) h_n S_n(\varphi, \lambda) \qquad (2)$$

The topographical heights have the expression

$$h = R \sum_n h_n S_n(\varphi, \lambda)$$

where R = the radius of the Earth, $\alpha(n)$ is a simple function of n, and h_n are empirical constants.

Station Anomalies of the Velocities of the Seismic Waves. Seismic station anomalies are found by measurements in the following way. At a seismological station, the observed travel times of all the P waves are tabulated for all earthquakes, irrespective of the geographical region where the quakes happened (see Fig. 1). These observed travel times are compared to travel times predicted by standard Earth models in order to obtain travel time residuals for each event at every station. For each station, the average value of these residuals is defined to be the station anomaly u. The u values here introduced are obtained at many continental stations, but only at some island stations in the midst of the oceans. Especially, the addition of more seismological stations situated on islands in the midst of the oceans will improve the precision of the finally computed model for the density anomalies in the mantle. The station anomalies at present do not have the details and precision of the gravity and isostasy data. The station anomalies can be interpolated between the seismological stations and expressed in terms of spherical harmonics (Table 1):

TABLE 1. Spherical Harmonics Development for the Station Anomalies u

n	m	u (s)
1	0	0.159
1	1c	−0.014
1	1s	0.086
2	0	−0.149
2	1c	0.002
2	1s	−0.159
2	2c	−0.062
2	2s	0.100
3	0	−0.040
3	1c	−0.089
3	1s	0.080
3	2c	0.113
3	2s	−0.053
3	3c	−0.015
3	3s	−0.013

$$u = \sum_n u_n S_n(\varphi, \lambda)$$

It is generally assumed that certain density anomalies situated below the seismological station are the cause of the station anomaly, since they give rise to velocity anomalies. Velocity anomalies caused by density anomalies in the crust and upper mantle down to a depth of about $t = 400$ km will contribute to the station anomalies, because these layers are crossed by all the P waves recorded at a given station. For depths greater than about 400 km, however, the P-wave paths diverge in a fan-shaped fashion, according to the geographical positions of the different foci. Thus, velocity anomalies below about 400 km will affect any given station anomaly as random variances that are averaged out in the mean (Toksöz et al., 1969).

In the upper layers of the Earth to a depth of about 400 km, P-wave velocity and density can be related by Birch's formula (Arnold and Schoeps, 1984):

$$V_P = -0.665 + 0.00264\rho \qquad (3)$$

where V_P is the P-wave velocity (km/s) and ρ is the density (kg · m^{-3}). Obviously, by knowing the u value, which has the meaning of a time delay, one can compute the anomaly of the P-wave velocity for depths less than 400 km, δV_P. Further, if we assume for depths less than 400 km that Eq. 3 is valid, then δV_P, as calculated from station residuals u, can be used to calculate the corresponding density anomaly $\delta \rho_B$. Thus, Eq. 3 relates $\delta \rho_B$ and u; see Eq. 10.

Anomalies of the Velocity of the Seismic P Waves in the Deep Mantle. Velocity anomalies of the seismic P waves in the deep mantle are well determined for low-order spherical harmonics (Dziewonski et al., 1977); see Table 2. Later, in this

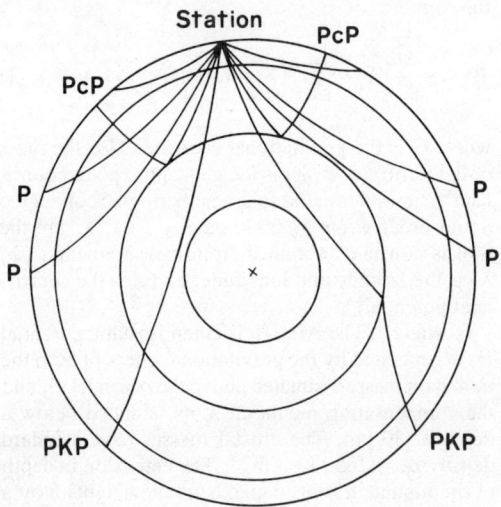

FIGURE 1. The fan-shaped paths of the P waves reaching one seismological station.

TABLE 2. Spherical Harmonics Development for the Anomalies
of the Velocity of the Seismic P Waves in the Deep Mantle

		q_n		
n	m	Region IIB Depth 1100–1500 km, 10^{-3} km·s^{-1}	Region III Depth 1500–2200 km, 10^{-3} km·s^{-1}	Region IV Depth 2200–2886 km 10^{-3} km·s^{-1}
1	0	3.2	0.5	4.1
1	1c	3.4	−0.4	−12.3
1	1s	2.1	1.0	−3.2
2	0	0.8	5.6	0.6
2	1c	−0.8	−0.2	5.4
2	1s	−5.1	0.8	3.6
2	2c	−10.5	2.0	−2.0
2	2s	−3.9	−3.4	0.3
3	0	−1.1	−4.7	−7.2
3	1c	4.6	−1.0	−3.7
3	1s	1.4	−0.7	1.8
3	2c	3.2	−4.0	−2.9
3	2s	5.2	3.7	−1.3
3	3c	—	—	—
3	3s	−3.5	−0.3	−7.2

context, we will compare them with the density anomalies in the deep mantle obtained in this article.

As given by Table 2, the seismic velocity anomalies within the three layers of the deep mantle (IIB, III, IV) have empirically determined expressions in spherical harmonics:

$$Q = \sum_n q_n S_n(\varphi, \lambda) \quad (4)$$

The rms value of Q is

$$(Q)_a = \left(\sum_n q_n^2\right)^{1/2} \quad (5)$$

The three layers, described by Table 2, give

$(Q_{IIB})_a = \pm 0.0159$ km · s^{-1}

$(Q_{III})_a = \pm 0.0101$ km · s^{-1}

$(Q_{IV})_a = \pm 0.0189$ km · s^{-1}

Mathematical Model

The model of the gravity potential W, Eq. 1, has the following expressions in terms of the different gravitating sources

$$W = U + W_I + W_B + W_A \quad (6)$$

$$W_I = G \iiint_{V_I} \frac{1}{e} \, dm_I \quad (7)$$

$$W_B = G \iiint_{V_B} \frac{1}{e} \, dm_B \quad (8)$$

$$W_A = G \iiint_{V_A} \frac{1}{e} \, dm_A \quad (9)$$

where e is the straight distance between the test point and the moving integration point, U is the standard potential, m_I are isostatic masses within the volume V_I (mountains and their roots), and m_B is the mass anomaly in the volume V_B to a depth of 400 km.

The integral for W_I, Eq. 7, is replaced by Eq. 2.

In the integral for W_B, Eq. 8, the mass anomalies are expressed by the density anomalies

$$dm_B = \delta\rho_B \, dV$$

where dV is the volume element, and the density anomalies $\delta\rho_B$ are obtained from Eq. 10, which derives from Birch's law, Eq. 3,

$$\delta\rho_B = -\tfrac{1}{15} \times 10^3 u \quad (10)$$

where $\delta\rho_B$ has units of kilograms per cubic meter and u is given in seconds.

As to Eq. 9, the mass anomalies and the density anomalies at depths greater than 400 km are connected by

$$dm_A = \delta\rho_A \, dV$$

The mass anomalies m_A are contained in the volume V_A, within the depth range 400 km $< t <$ 2900 km. Thus, $\delta\rho_A$ are the density anomalies in the deep mantle, which are the unknown values to be determined here. W, U, W_I, W_B are functions known beforehand (Eqs. 1, 7, and 8). Thus, the unknown

TABLE 3. Final Spherical Harmonics Development for the Density Anomalies in the Earth's Mantle

n	m	$\delta\rho_B$ Depth 0–400 km, kg·m^{-3}	$\delta\rho_{A.1}$ Depth 400–1233 km, kg·m^{-3}	$\delta\rho_{A.2}$ Depth 1233–2067 km, kg·m^{-3}	$\delta\rho_{A.3}$ Depth 2067–2900 km, kg·m^{-3}
1	0	−7.7	3.0	2.6	2.1
1	1c	3.5	−1.4	−1.2	−1.0
1	1s	−3.9	1.6	1.3	1.1
2	0	11.9	−6.2	−4.5	−3.0
2	1c	1.2	−0.6	−0.5	−0.3
2	1s	12.1	−6.3	−4.6	−3.1
2	2c	2.3	−1.0	−0.7	−0.5
2	2s	−6.9	3.5	2.5	1.7
3	0	2.0	−1.1	−0.7	−0.4
3	1c	5.3	−3.1	−1.9	−1.1
3	1s	−4.8	3.1	1.9	1.1
3	2c	−9.6	6.2	3.8	2.1
3	2s	5.5	−3.6	−2.2	−1.2
3	3c	1.5	−0.8	−0.5	−0.3
3	3s	3.2	−1.8	−1.1	−0.6

function W_A has the following constraint, Eq. 11, which is also a constraint for $\delta\rho_A$ (Eqs. 6 and 9):

$$W_A = G \iiint_{V_A} \left(\frac{1}{e}\right) \delta\rho_A \, dV$$

$$= W - U - W_I - W_B \quad (11)$$

We shall assume that the mantle below 400 km can be represented by three layers. The density anomalies in these layers are unknown and depend only upon φ and λ. The mathematical model is described by the following terms:

$\delta\rho_{A.1}$ for depths 400 km < t < 1233 km

(12a)

$\delta\rho_{A.2}$ for depths 1233 km < t < 2067 km

(12b)

$\delta\rho_{A.3}$ for depths 2067 km < t < 2900 km

(12c)

The harmonics development for $\delta\rho_{A.i}$ is (independent of radius r)

$$\delta\rho_{A.i} = \sum_n \delta\rho_{i.n} S_n(\varphi, \lambda) \quad (i = 1, 2, 3) \quad (13)$$

The constants $\delta\rho_{i.n}$ may be determined by the method of least squares, so

$$\sum_i \iiint_{V_{A.i}} (\delta\rho_{A.i})^2 P_i \, dV \longrightarrow \text{minimum}$$

$(i = 1, 2, 3)$ (14)

where $V_{A.i}$ are the volumes of the three layers (Eqs. 12a–c) and P_i are convenient weighting factors. The minimum of Eq. 14 has to be found, with the constraint 11 observed (Arnold and Schoeps, 1984).

Inferred Density Anomalies

Table 3 shows the coefficients $\delta\rho_{i.n}$ of the finally computed spherical harmonics development for $\delta\rho_{A.i}$ ($i = 1, 2, 3$) (Eq. 13). This table also gives values for $\delta\rho_B$ as computed from Eq. 10. The rms values $(\delta\rho_{A.i})_a$ of the density anomalies $\delta\rho_{A.i}$, defined in Eqs. 12a–12c, are computed for the three layers. These amounts reach about 10 kg · m^{-3}, as shown in Fig. 2.

The masses of material with density $\delta\rho_B$ are rather well compensated by $\delta\rho_{A.1} + \delta\rho_{A.2}$, as shown in Table 3; hence,

$$\delta\rho_B \simeq -\delta\rho_{A.1} - \delta\rho_{A.2}$$

A comparison of the rms values of Q (Eqs. 4, 5) and of $\delta\rho_{A.i}$, Eq. 13, is of interest. The whole region

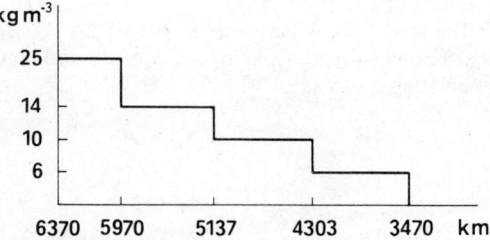

FIGURE 2. The rms values of the lateral inhomogeneities of the density as a function of the distance from the center of the Earth.

of the values Q_{IIB} and Q_{III} covers depths between 1100 km and 2200 km (Table 2). This is about the depth area of $\delta\rho_{A.2}$ (Eq. 12b). Further, the function Q_{IV} covers about the same depth range as $\delta\rho_{A.3}$. The computations lead to

$$\tfrac{1}{2}[(Q_{IIB})_a + (Q_{III})_a] = 0.0014(\delta\rho_{A.2})_a$$

$$(Q_{IV})_a = 0.0031(\delta\rho_{A.3})_a$$

The mean of the coefficients 0.0014 and 0.0031 is 0.0022. The corresponding Birch coefficient for the upper 400 km is 0.00264, a value similar to 0.0022 (Eq. 3). We do not imply by this, however, that Eq. 3 is valid for the lower mantle, too.

K. ARNOLD

References

Arnold, K., and D. Schoeps, 1984, Lateral inhomogeneities of density in the interior of the Earth, *Gerlands Beitr. Geophysik* **93**, 185-201.

Dziewonski, A. M., B. H. Hager, and R. J. O'Connell, 1977, Large-scale heterogeneities in the lower mantle, *Jour. Geophys. Research* **82**, 239-255.

Toksöz, M. N., J. Arkani-Hamed, and C. A. Knight, 1969, Geophysical data and long-wave heterogeneities of the Earth's mantle, *Jour. Geophys. Research* **74**, 3751-3770.

Cross-references: *Continental Lithosphere; Earthquake Seismology; Earth's External Gravity Field; Earth Structure, Global; Geodesy, Physical; Gravity Fields: Implications for Planetary Interiors; Gravity and Isostasy; Mantle, Lower: Structure; Mantle, Upper: Structure; Mantle Discontinuities.*

DETRITAL REMANENT MAGNETISM (DRM)

The paleomagnetism of sediments is known as detrital remanent magnetism (DRM), and it arises from the collective alignment with the Earth's magnetic field of small magnetic carriers in the sediment. In general, the magnetic carriers that contribute most to the paleomagnetic signal are smaller than 10 μ across. Except where the sediment itself is derived from older material containing hematite grains, the primary magnetic mineral is fine-grained magnetite (or titanomagnetite). In a typical situation, the magnetite weathers out of parent rock material and is carried by water into a relatively quiet sedimentary environment where it is deposited with the sediment. As a magnetic carrier settles through the water column, the magnetic moment of the carrier tends to rotate to become parallel to the Earth's magnetic field. Thus when a carrier reaches the sediment/water interface, its orientation has been influenced by the direction of the existing magnetic field. Although not every carrier is actually aligned with the magnetic field, the net magnetization of all of the carriers is parallel to the external field. If the magnetic carriers were simply incorporated into the sediment with the orientations that they had acquired in the water column, detrital remanent magnetism would provide an accurate record of the Earth's magnetic field.

Unfortunately magnetic carriers may interact with the substrate when they first reach the sediment/water interface, and they may be subjected to other influences once they have been incorporated in the sediment. A magnetization acquired as a magnetic carrier reaches the sediment/water interface is called a depositional DRM or dDRM, whereas a magnetization acquired at a later time is called a postdepositional DRM or pDRM.

Based on early experimental studies (King, 1955), it is generally believed that sediments that possess a dDRM consistently record the magnetic field with small but measurable errors. These alignment errors are associated with the presence of a clearly defined boundary at the sediment/water interface. As spherical magnetic carriers encounter the substrate at the boundary, they roll in random directions (Griffiths et al., 1960). If the magnetic carriers are elongate, they rotate as their long axes become horizontal. At first glance, it is not obvious how these effects can produce an alignment error because for every carrier that rotates in one direction, there is another carrier that rotates in the opposite direction. However, two carriers that rotate in opposite directions with respect to the vertical plane of the Earth's magnetic field each experience a shallowing of their inclination in addition to opposite and cancelling changes in their declination. Thus, when the effects of rotation are averaged statistically, there is a net shallowing of the inclination that can amount to as much as 10°. This shallowing is known as the *inclination error*. Other factors can also affect the magnetization but are generally less important. For example, deposition on a sloping substrate introduces an asymmetry in the rotation and can lead to errors in both inclination and declination. If the magnetic carrier is deposited in the presence of a water current, the laminar flow layer of the current near the substrate produces a shear on the magnetic carrier that can affect its magnetic alignment.

In most cases, however, the sediment/water interface does not represent a clearly defined substrate but is rather a transition zone, which at the top is a very dilute dispersion of sediment with a very high water content and at the bottom is a fully consolidated and dewatered sediment. Recent studies have shown that if a predominantly sandy sediment has a high water content, remagnetization will occur in response to changes in the external magnetic field (Payne and Verosub, 1982). However, as the water content decreases, the degree of remagnetization decreases. Eventually the water content reaches a critical value below which no remagnetization occurs. This result is consistent with a theoretical model that assumes that at high water contents the magnetic carriers are mobile within fluid-filled voids (Irving and Major, 1964). As the sediment is dewatered, first the larger grains and then the smaller ones become immobilized within the voids. If the sediment has a high clay content, the magnetic carriers are apparently immobilized by their interaction with the clay floccules. Thus at least for sandy sediments, the water content plays an important role in the development of a pDRM.

It is also clear that certain types of disturbances, such as those produced by bioturbation, can remobilize the magnetic carriers and produce a realignment with the Earth's magnetic field. In the deep sea, many sediments are thoroughly bioturbated, and yet they possess a satisfactory record of the Earth's magnetic field, suggesting that bioturbation can produce a new magnetization at the same time that it destroys an old one. Such a phenomenon has been observed in field studies in estuarine environments. In laboratory studies, vigorous stirring of a sediment in a uniform magnetic field erases any effects of a prior magnetization and creates a magnetization parallel to the applied field. The stirring effectively mimics certain types of bioturbation, and small-scale localized liquefaction, which briefly raises the water content of the sediment above the thixotropic limit, may be responsible for the reorientation of the magnetic carriers (Payne and Verosub, 1982).

The remagnetization of a sediment through bioturbation is, however, a very complex process. The biological and ecological literature concerning bioturbation shows that, in general, bioturbation is related to feeding behavior, burrow construction, and movement through the sediment, and that for each of these functions, several broad categories can be used to classify the interaction of the organisms with the sediment. Each of these categories has different effects on the physical (and chemical) properties of the sediment. For example, some types of burrowing behavior produce large increases in the water content of the surrounding sediment, while others involve compaction and a reduction in water content. Thus the first type of behavior might be quite effective in producing a remagnetization while the second would have only a minimal effect.

About a decade ago, the available evidence seemed to indicate that sediments that possessed a dDRM had probably recorded the field with a classic inclination error while sediments that possessed a pDRM had probably recorded the field accurately (Verosub, 1977). This simple dichotomy led to the belief that the inclinations of bioturbated marine sediments were reliable while those of lacustrine sediments and unbioturbated marine sediments were probably not. Recent work, however, has shown that in some cases a sediment can acquire a dDRM without an associated inclination error and that in other cases the presence or absence of the inclination error depends on whether the deposition occurs from a thin dispersion or a thick slurry. In addition it has been shown that a sediment with a pDRM may develop an inclination error as the sediment becomes compacted and conversely that the collapse of clay floccules during compaction can lead to the erasure of an inclination error acquired during deposition.

These latter studies indicate that compaction is another process that can be important in determining the final magnetization of a sediment. Furthermore, the presence or absence of a compaction-related inclination error is a particularly relevant question when the latitudinal translation of plates and microplates is inferred from the measured inclinations of sedimentary rocks. In recent years, the conclusions drawn from such studies have tended to be based more and more on relatively small differences between measured and expected directions. However, the progressive downcore shallowing of inclination that might result from compaction is not easily distinguished from the shallowing that would result from poleward motion of a plate, and at mid-latitudes, 10° of inclination error would be equivalent to 1000 km of translation.

The existence of a compaction-induced inclination error actually represents something of a paradox. Presumably the change in inclination arises from the fact that the compaction leads to physical rotation of the magnetic carriers. If the axes of rotation are all parallel, the magnetization of each grain would be rotated by the same amount, and the net magnetization would show an inclination error directly proportional to the amount of compaction. Such behavior might be exhibited by a collection of elongated magnetic grains whose axes of magnetization were parallel to their long axes. On the other hand, if the axes of rotation are randomly oriented, the rotations must be averaged over all directions. This problem is mathematically analogous to the theoretical treatment of the depositional inclination error, which assumes that the magnetic carriers roll when they reach the sediment/water interface (Griffiths et al., 1960). When the rotations are averaged over all directions, there is still a net shallowing of inclination; however, the effect is considerably smaller than that of the first mechanism and is not paleomagnetically significant. The paradox is that the second mechanism appears to be more realistic physically, but the rather limited available data suggests that compaction does produce a measurable change in inclination.

As noted above, the recent research that has been done on pDRM only serves to underscore the complexity of the subject and the need to conduct additional studies of it. Unfortunately, many paleomagnetic studies, particularly those done on marine sediments and sedimentary rocks, do not address the question of the acquisition of the magnetization, and there has been a dramatic decrease in the number of publications concerned with this topic. Much of the work that has been done has involved synthetic sediments, which typically consist of a magnetic carrier of uniform shape and size dispersed in an inert matrix. The advantage of using such materials is that the nature of the magnetic carriers can be easily characterized, which allows relatively easy comparison with theoretical models. The disadvantage of using such materials is that the interactions that occur in natural sediments are far more complex than those that occur in synthetic sediments. For example, any effects resulting from the interaction of magnetic carriers with clay minerals and the organic residue in a natural sediment are unlikely to be detected in experiments with synthetic sediments. Thus even if we could understand fully the processes in synthetic sediments, we would still not understand the processes in natural sediments. Experiments with synthetic sediments must be complemented with studies of natural sediments.

The importance of understanding the magnetization process can be illustrated by considering the transition zone between dilute dispersion and fully dewatered sediment, where the interplay between water content and bioturbation depends on both the type of organisms present and the nature of the sediment. If the magnetization is primarily reset by bioturbation, the magnetization will not become locked until enough new sediment has accumulated that the magnetic carriers are beyond the range of the bioturbation. If there is no bioturbation or if the water content is still high below the zone of bioturbation, the rate at which the sediment dewaters will determine the point at which the magnetization becomes fixed. Thus the magnetization is acquired not at the sediment/water interface but at some deeper level in the sediment.

The existence of an offset between the sediment/water interface and the point at which the magnetization becomes locked was recently demonstrated in a study that compared variations in the concentration of ^{10}Be with variations in the paleomagnetic intensity of sediments (Raisbeck et al., 1985). Both parameters are believed to be directly related to variations in the intensity of the Earth's magnetic field. In the study the same pattern of variation was found, but the paleomagnetic changes were displaced several centimeters with respect to the changes in beryllium-10 concentration.

Where the dewatering is the most important process, the magnetization may actually be acquired over a finite interval because different magnetic carriers may have different critical water contents. The effect of these various processes on the information content of the sediments has recently been analyzed from the point of view of linear systems theory. In this approach the paleomagnetic signal is viewed as the convolution of the original geomagnetic signal with a filter function that depends on the acquisition mechanism. If the acquisition process is understood well enough to allow a mathematical formulation of the filter function, the paleomagnetic signal can be inverted to find the original geomagnetic signal. Although this approach is promising, the primary problem is the uncertainty in the magnetization process.

Many years ago Watkins (1968) pointed out that the magnetization process could erase short polarity intervals from the magnetic polarity record of marine sediments and that this factor should be considered when making correlations to the magnetic polarity time scale. His call for caution is still valid and applies equally well to the increasing use of rapidly-deposited marine sediments to provide detailed information about secular variation and polarity transitions. With the improved chronologic framework provided by high-resolution biostratigraphy, by the presence of correlated tephra or tektite layers, or by the existence of a suite of radiocarbon dates (for younger sediments), it is tempting to assume that the same time-stratigraphic horizon at two sites can be used to determine the simultaneous behavior of the geomagnetic field at the two sites. Unfortunately the validity of such an inference is directly related by the extent to which the acquisition of the magnetization was controlled by the same processes. For example, if the magnetization at one site was acquired at the sediment/water interface while that at the other site was acquired at a lock-in zone located well below the sediment/water interface, the correlative paleomagnetic directions will not be located on the apparent time stratigraphic horizon. In addition, the paleomagnetic signal may be smoothed by the acquisition process and it may contain a compaction-related inclination error. While none of these factors is serious enough to preclude the use of high-resolution paleomagnetic studies of sedimentary sequences, they do underscore two important principles. The first is that it is important to understand the processes by which sediments have acquired their magnetization; the second is that there are inherent limitations on the agreement in both space and time that we can expect from two paleomagnetic records.

KENNETH L. VEROSUB

References

Griffiths, D. H., R. F. King, A. I. Rees, and A. E. Wright, 1960, Remanent magnetism of some recent varved sediments, *Royal Soc. [London] Proc. Ser. A,* **256,** 359-383.

Irving, E., and A. Major, 1964, Post-depositional detrital remanent magnetization in a synthetic sediment, *Sedimentology* **3,** 135-143.

King, R. F., 1955, Remanent magnetism of artificially deposited sediments, *Royal Astron. Soc. Monthly Notices, Geophys. Suppl.,* **7,** 115-134.

Payne, M. A., and K. L. Verosub, 1982, The acquisition of post-depositional detrital remanent magnetization in a variety of natural sediments, *Geophys. Jour.* **68,** 625-642.

Raisbeck, G. M., F. Yiou, D. Bourles, and D. V. Kent, 1985, Evidence for an increase in cosmogenic Be-10 during a geomagnetic reversal, *Nature* **315,** 315–317.

Verosub, K. L., 1977, Depositional and postdepositional processes in the magnetization of sediments, *Rev. Geophys. Space Phys.* **15,** 129–143.

Watkins, N. D., 1968, Short period geomagnetic polarity events in deep-sea sedimentary cores, *Earth Planetary Sci. Letters* **4,** 341–349.

Cross-references: *Chemical Remanent Magnetization; Curie Temperature; Demagnetization; Magnetic Domains; Magnetic Properties of Minerals; Natural Remanent Magnetization; Paleomagnetic Field: Intensity; Paleomagnetism: Deep-Sea Sediments; Paleomagnetism and Continental Drift: Historical Introduction; Rock Magnetism; Rock Magnetism: Measuring Techniques and Apparatus; Thermoremanence; Viscous Remanent Magnetization (VRM) and Viscous Magnetization.*

DOPPLER POSITIONING: SATELLITE

Doppler satellite positioning has its origin in early satellite orbit determinations that used measurements of the Doppler shifts on radio signals received from the first artificial Earth satellites. It was soon realized that Doppler shift measurements could be used to determine ground positions if the satellite orbits were known (McClure, 1958). This discovery gave the impetus for the development of the *Navy Navigation Satellite System (NNSS)*, also known as TRANSIT. Geodetic Doppler positioning methods almost exclusively utilize NNSS satellites, though Doppler shift measurements have been used for tracking and orbit determination of geophysical and oceanographic satellites (see *Satellite Altimetry*).

The NNSS was introduced for military applications in 1964 and released for nonmilitary use in 1967. It is scheduled to be phased out by 1994 when the GPS [see *Global Positioning System (GPS)*] will be fully operational. The NNSS currently (August, 1988) consists of six satellites in polar, near-circular orbits at an altitude of about 1100 km. Each satellite transmits highly stable radio signals on two coherent frequencies (149.988 and 399.968 MHz) modulated with a satellite message that contains predicted orbital information pertaining to the particular satellite. The *broadcast ephemerides* are computed by the U.S. Navy Astronautics Group from Doppler data observed at four OPNET tracking stations located in California, Minnesota, Maine, and Hawaii. These broadcast ephemerides are calculated from 30-hr Doppler data spans every 12 hr (24 hr during periods of low solar activity) by extrapolating the satellite orbit ahead for up to 30 hr before uploading it into the satellite memory. The broadcast orbital parameters are rounded off to the nearest 10 m or 10^{-4} degree, which can result in positioning errors of 6 m. Typically, at least one of the NNSS satellites is visible every 2 hr for up to 20 min, which facilitates determination of a position in two dimensions (latitude, longitude) at the 50-m level. Such position fixes from satellite Doppler measurements are analogous to marine hyperbolic navigation; here the known satellite positions during a single pass provide reference similar to known shore transmitters.

The first satellite Doppler receivers were rather bulky, required portable power generators, and used paper tape to record Doppler observations. Often a minicomputer was used for preprocessing and data reductions. Modern receivers, on the other hand, are compact, operate from rechargeable batteries, and incorporate microcomputers for real-time data processing and evaluation. Typically the satellite Doppler observations are recorded on magnetic tape cassettes or in nonvolatile computer memory. The navigation capabilities of the system have been described and analyzed extensively (e.g., Kershner and Newton, 1962; Stansell, 1978).

For stationary users it is possible to observe and combine Doppler data from many satellite passes. Such multipass, often multistation, data sets are selected and assembled during postprocessing. Methods have been developed to facilitate high-precision positioning in three-dimensional terrestrial reference frames satisfying global as well as regional geodetic and geodynamic requirements. These meticulous approaches have resulted in significant improvements in our knowledge of the size and shape of the Earth and its gravitational field and have led to the establishment of a truly global, geocentric, reference frame.

Satellite Doppler Measurements

Geodetic Doppler measurement techniques usually count cycles of a beat frequency $f_0 - f_r$, which results from mixing the ground reference frequency f_0 with the Doppler shifted satellite frequency f_r as received at the ground station. These *Doppler counts* D, accumulated between reception times T_1 and T_2, form the basic measurements:

$$D = \int_{T_1}^{T_2} (f_0 - f_r)\, dT \qquad (1)$$

In practice the integration interval $T_2 - T_1$ can vary between 4.6 s and 120 s, but intervals between 20 s and 30 s are commonly used to optimize system performance and economy.

Since the duration of a satellite pass is less than 20 min, the ground frequency f_0 can be considered constant, and the integration of Eq. 1 yields

$$D = f_0(T_2 - T_1) - \int_{T_1}^{T_2} f_r\, dT \qquad (2)$$

The integral in Eq. 2 can also be expressed in number

of cycles of the satellite frequency f_s emitted between the corresponding satellite time marks $t_1 = T_1 - S_1/c$ and $t_2 = T_2 - S_2/c$, where S_1 and S_2 are distances between the receiver and the satellite at times t_1 and t_2, and c is the velocity of light in a vacuum. Thus

$$\int_{T_1}^{T_2} f_r \, dT = \int_{t_1}^{t_2} f_s \, dt \qquad (3)$$

For simplicity, propagation, relativistic, and instrumental effects are not introduced here, though they have to be taken into account in precise Doppler measurement reduction. Since f_s can also be considered constant during a satellite pass, the expression for the integrated Doppler counts becomes

$$D = (f_0 - f_s)(T_2 - T_1) + \frac{(S_2 - S_1)f_s}{c} \qquad (4)$$

An equivalent expression in terms of signal emission times t_i is obtained from Eq. 4 through substitution of $T_i = t_i + S_i/c$:

$$D = (f_0 - f_s)(t_2 - t_1) + \frac{(S_2 - S_1)f_0}{c} \qquad (5)$$

Here

$$S_i = |\mathbf{r}_s(t_i) - \mathbf{r}_g(T_i)|$$

are distances between the ground station position vector $\mathbf{r}_g$ at T_i and the satellite position vector $\mathbf{r}_s$ at t_i. Equations 4 and 5 express the relationship between the integrated Doppler count and the satellite and receiver coordinates. They also form the basis of *mathematical models* employed in Doppler positioning.

Error Sources

Instrumentation errors, propagation delays caused by the Earth's atmosphere and uncertainties in satellite coordinates contribute to the random and systematic errors in Doppler positioning.

Instrumentation Errors. The most significant errors are due to the instability of reference frequency standards during a satellite pass since both f_0 and f_s were assumed to be constant in Eqs. 2 and 3. Crystal oscillators, due to their low power consumption and small size, are often used in satellite Doppler receivers, but they introduce slowly varying systematic measurement errors that bias solutions and limit the Doppler positioning accuracy at the 0.1-m to 0.5-m level. The effect of frequency instabilities can be eliminated by using highly stable atomic frequency standards in ground stations and by suitable observing and data reduction strategies requiring at least two simultaneously observing stations (e.g., by differencing Doppler counts or by introducing a bias unknown for each Doppler interval). Other instrumentation error contributions are due to antenna phase center changes (multipass) and receiver phase and time delay variations. These error sources can introduce positioning errors at the 0.1-m level and, in general, cannot be eliminated by any observing strategy.

Atmospheric Effects. Propagation delays due to the ionosphere, the ionized portion of the atmosphere above 50 km, are larger than those due to the troposphere, the neutral portion below 50 km. The *ionospheric refraction* strongly depends on solar activity. It shows diurnal and annual variations with minima at night and in winter, and maxima at midafternoon and in summer. The ionospheric refraction index is a function of signal frequency f and can be approximated from Doppler observation at two different frequencies by a series expansion. When Doppler counts D_H^0 and D_L^0 are measured on the frequencies f_H and f_L, the first-order Doppler count ionospheric refraction correction d_H is

$$d_H = \frac{f_L^2 D_H^0 - f_L f_H D_L^0}{f_H^2 - f_L^2}$$

The neglected second-order refraction term does not exceed several percent of the first-order term, which represents a few meters in position. Most of the residual ionospheric refraction errors can be eliminated in relative positioning when data from at least two simultaneously observing stations are available.

Tropospheric refraction effects are frequency independent in this frequency range and are smaller than ionospheric delays. The effect of the tropospheric refraction lengthens the observed ranges (see *Geodetic Refraction*). Neglecting the tropospheric refraction will cause significant position errors that can easily exceed 10 m, particularly when satellite passes are not balanced in east-west and north-south directions. Tropospheric refraction models make use of surface meteorological measurements (pressure, temperature, and humidity) to calculate approximate refraction corrections while making some simplifying assumptions about the troposphere. The model developed by Hopfield (1969) is most often used in Doppler positioning. It is based on the assumption of a nominal temperature gradient and the perfect gas law, which leads to a convenient quartic expression for the dry refractivity as a function of the height above the Earth's surface. The refractivity $N = (n - 1) \times 10^6$, where n is the index of refraction. An analogous expression, though without any theoretical foundation, is also used by Hopfield (1969) for the wet refractivity. The expressions for dry and wet refractivities can be integrated analytically to give propagation delay, but several fast and precise approximations have been developed that are within 0.1% or a few centimeters in range for elevations above 5°; all these approximations are well below model errors. The wet refraction correction is difficult to evaluate due to uncertainties in measurements of relative humidity. From extensive *very long baseline interferometry* (VLBI) experience it is apparent that the mismodeling of wet tropospheric refraction can contribute up to 10 cm to the errors in relative position over distances in excess of 50 km.

Satellite Orbits and Coordinate Systems. Three types of satellite orbits are used in Doppler positioning: broadcast ephemeris, which is available in real time; historical precise ephemeris, which can be made available only for some of the NNSS satellites after a delay of several weeks; and short arc orbits, which are computed within Doppler reduction software.

The *broadcast ephemeris* is based on comprehensive orbit modeling that includes the gravitational model WGS72 (Sepplin, 1974) and extrapolated Bureau International de l'Heure (BIH) pole positions. Still the errors in broadcast orbits can reach 100 m (Jenkins and Leroy, 1979) due to orbit extrapolation. These systematic orbit errors will not average out, and broadcast point positioning will show position- and time-dependent biases of up to 10 m.

The U.S. Defense Mapping Agency (DMA) independently computes a postfitted ephemeris for one or more NNSS satellites. This *precise ephemeris* is computed from 48-hr data sets observed by the worldwide network of TRANET stations using software and program constants (Anderle, 1974) different from those used for the broadcast ephemeris. Since no extrapolation is involved and data from about 20 globally distributed TRANET stations are used, the precise ephemerides are consistent at the 2-m level.

Short arc orbit computation involves the calculation of a portion of the orbit for which observations are available; this typically represents a 20–30-min arc for the NNSS satellites. The short arc approach takes advantage of both geometrical methods (see *Geodesy: Satellite*) and dynamical methods—i.e., the simplicity of the geometrical and the better solution strength of the dynamical methods where many unknown satellite positions are replaced with six unknown orbital parameters. For short arc computations the gravitational potential expansions can be truncated, and other forces can be considered constant during the portion of the orbit, thus being effectively absorbed into the six initial orbital elements. In the case of the NNSS satellites a gravitational expansion to at least degree and order 10 should be used (Kouba, 1983).

Precise and broadcast ephemerides also imply two distinct *coordinate systems*. Position, orientation, and scale are mainly established by the set of station coordinates adopted for the TRANET and OPNET stations, respectively. The coordinate systems, implied by the precise ephemeris, NWSC9Z-2 and its predecessor NWL9D were extensively compared to VLBI and other independent satellite and astronomical systems (e.g., Boucher and Feissel, 1984). The NWSC9Z-2 and NWL9D systems require corrections of about -0.6×10^{-6} in scale, $+0.8''$ in longitude east, and $+4$ m in Z-coordinate to agree with geocentric and international conventions for scale given by the speed of light, mean pole, and longitude orientation (see *Geodetic Reference Systems*). The broadcast ephemeris is not in the WGS72 coordinate system as is often assumed. Instead, the coordinate system implied by the broadcast ephemeris is nominally compatible with NWSC9Z-2 or NWL9D except for a small-scale difference of 0.2×10^{-6} and a -2-m Z-coordinate shift (see Jenkins and Leroy, 1979). Since January 1, 1987, the precise and broadcast ephemerides refer to the World Geodetic System 1984 (WGS 84) (Decker, 1986), thus conforming to the international conventions.

Satellite Doppler Applications

Doppler positioning for a stationary receiver, or *point positioning*, is based on Doppler data from many satellite passes and on the assumption that orbital, propagation, and instrumental errors are reduced by averaging, thus yielding a repeatability below 1 or 2 m. This precision may not be achievable where the broadcast ephemeris is used or where unfavorable atmospheric conditions are encountered. Most of the orbital and part of the propagation errors can be eliminated when relative positioning over distances less than 250 km are considered from at least two simultaneously observing stations. In the simplest case, called *translocation*, two simultaneous point positioning solutions are subtracted without any regard to commonality of passes and Doppler data. Relative positioning can also be carried out by *short arc* and *semi–short arc positioning methods* that utilize locally computed or external orbits, respectively, in a rigorous least squares adjustment involving many passes from several stations. In relative positioning, errors in scale and orientation still persist at the 1-ppm level when broadcast, rather than precise, orbits are used.

The *mathematical models* used in geodetic Doppler positioning include, in addition to station coordinates and satellite orbital parameters, bias parameters to model instrumental effects such as receiver delay, frequency offset and drift, tropospheric refraction scaling bias, and corrections for relativistic and aberration effects. In addition, some Doppler programs also use higher-order ionospheric refraction and modeling, which reflect spatial and temporal correlation of tropospheric refraction errors (e.g., Kouba, 1983). A rigorous least squares adjustment with proper variance covariance matrix is often carried out pass by pass (so-called adjustment in phases) to increase efficiency and facilitate improved statistical testings (see *Adjustment Methods*).

Geodetic Applications. The most significant contributions of the satellite Doppler technique are to continental size adjustments of geodetic networks, such as the North American Datum 1983 (NAD 83) Adjustment (Fig. 1). Doppler positioning based on the precise ephemeris, scaled and oriented to conform with the international conventions (WGS 84), was used to position and orient the NAD 83. Doppler positioning at a spacing of 200–500 km also provided the control of error growth in such geodetic networks (see *Geodetic Networks and Control Surveys*). Similar work has been carried out in Europe and Africa.

Present-day hardware and precise short arc and semi-short arc reduction methods facilitate relative positioning with accuracy below 0.5 m by using small data sets of 15 to 30 passes collected in one to two days of observation (Fig. 2). At this level of accuracy and efficiency, satellite Doppler positioning is used to strengthen classical first-order geodetic networks with station spacing of about 100 km and can also provide the control required for inertial surveying.

Because Doppler positioning provides three-dimensional coordinates, it determines the ellipsoidal height h of stations above the reference ellipsoid.

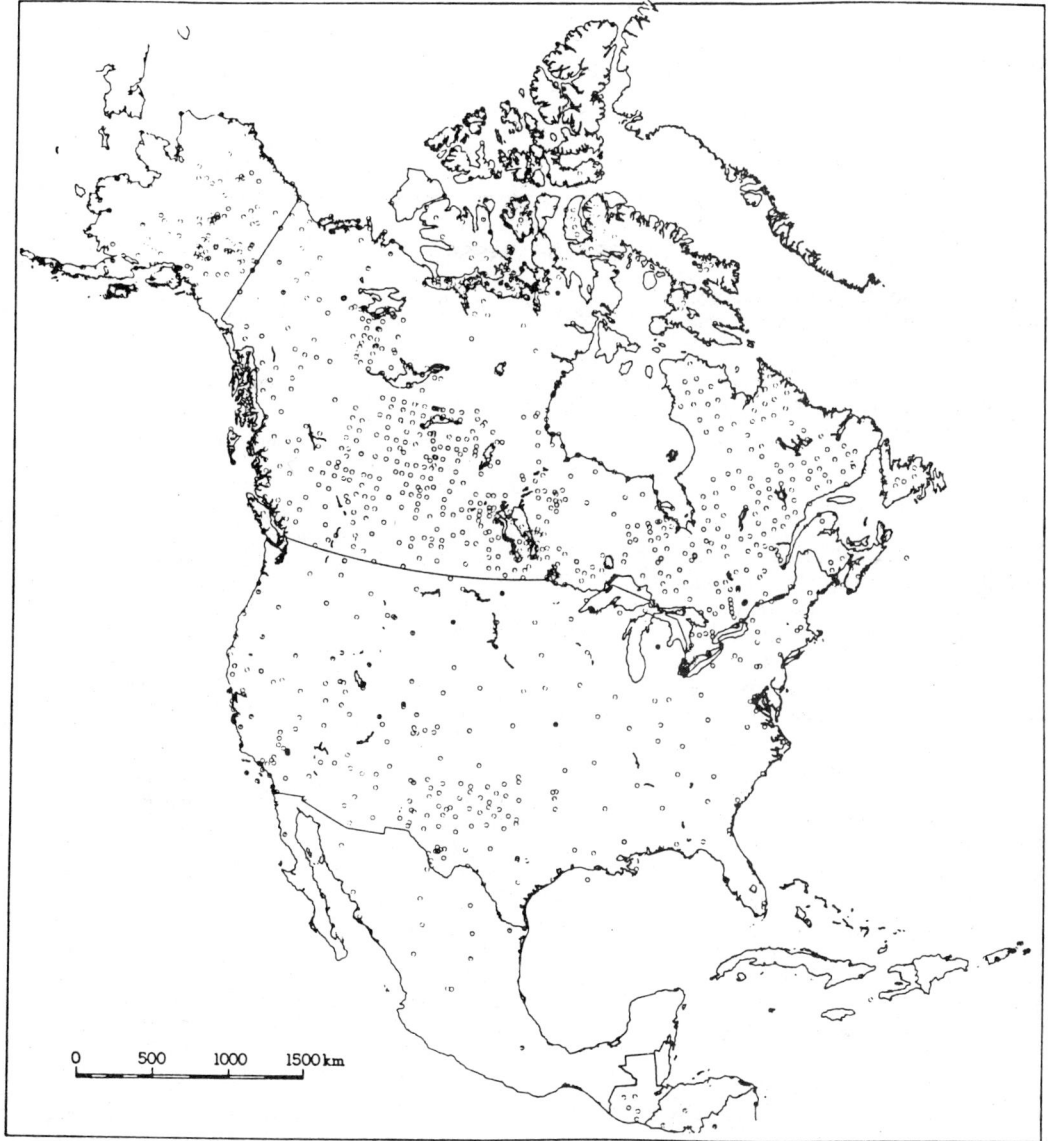

FIGURE 1. Doppler horizontal control networks included in the North American Datum 1983 (NAD 83) Adjustment. (Courtesy of The Geodetic Survey of Canada)

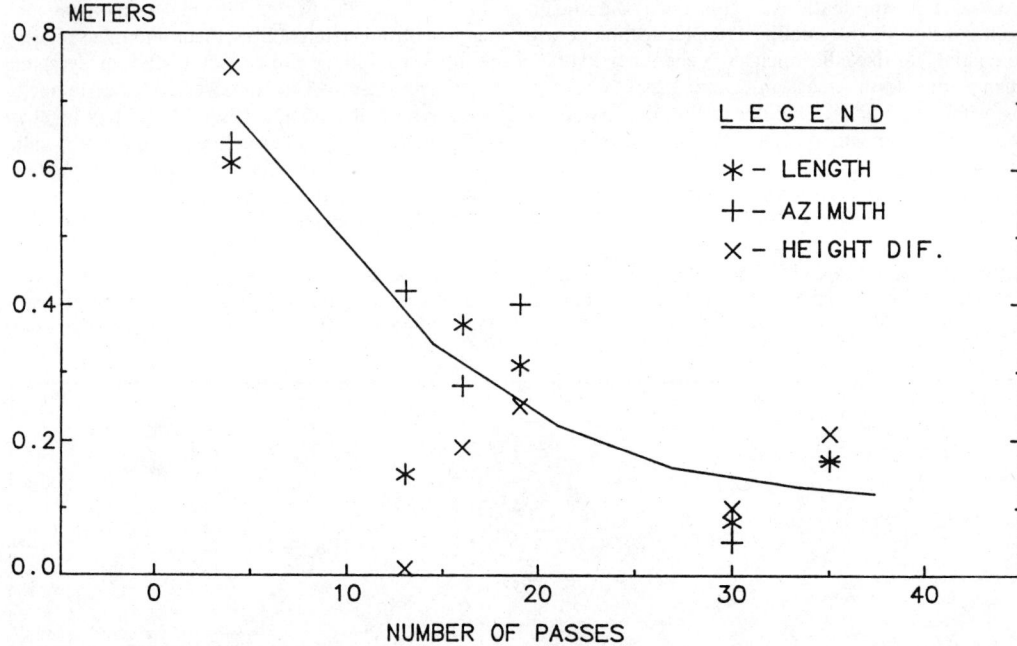

FIGURE 2. Typical relative position accuracy (rms) for short station separations (< 250 km) based on short arc and semi-short arc methods. (Compiled from Kouba and Hothem, 1986)

When orthometric heights H are known at some stations, usually by geodetic leveling, it is possible to determine the geoidal heights

$$N^0 = h - H$$

Such Doppler-derived geoidal heights are used to control error propagation in continental geoid surveys, and for determination of zero-order geoid undulation, the mean Earth ellipsoid, and the displacement of the Doppler coordinate system.

Geodynamics Applications. During precise satellite ephemeris computation, DMA has solved for the pole coordinates for every 48-hr data span since 1972. The *Doppler pole positions* were aligned with the BIH pole position at 1966.7. These two-day pole positions are estimated to be consistent to 0.5m and have been regularly included in the calculation of the BIH pole coordinates. Scientifically important by-products of the precise ephemeris generation are systematic changes in station coordinates, which may relate to crustal dynamics. This requires analyses of long series of Doppler positions, such as those of the TRANET tracking stations. Orbital errors are the main source of uncertainty. Doppler positioning is also used for determination of much larger movements, such as those found in polar ice dynamics (e.g., Popelar and Kouba, 1983).

J. KOUBA

References

Anderle, R. J., 1974, Transformation of terrestrial survey data to Doppler satellite datum, *Jour. Geophys. Research* **79**(35), 5319-5331.

Boucher, C., and M. Feissel, 1984, Realization of BIH terrestrial System, J. Somogyi, ed., *Inter. Symp. Space Techniques for Geodynamics Proc.* Research Inst. of Hungarian Academy of Sciences, Sopron, 33-43.

Decker, L. B., 1986, World Geodetic System 1984, *4th Inter. Geod. Symp. Satell. Positioning Proc.*, Applied Research Laboratory, University of Texas, Austin, TX, 69-92.

Hopfield, H. S., 1969, Two-quadratic tropospheric refractivity profile for correcting satellite data, *Jour. Geophys. Research* **74**, 4487-4499.

Jenkins, R. E., and C. F. Leroy, 1979, "Broadcast" versus "precise" ephemeris—apples and oranges, *2nd Inter. Geod. Symp. Satell. Doppler Positioning Proc.*, Applied Research Laboratory, University of Texas, Austin, TX, 39-62.

Kershner, R. B., and R. R. Newton, 1962, The TRANSIT System, *Jour. Inst. Navig.* **15**, 129-144.

Kouba, J., 1983, Review of geodetic and geodynamic satellite Doppler positioning, *Rev. Geophysics and Space Physics* **21**(1), 27-40.

Kouba, J., and L. D. Hothem, 1986, Doppler results on the Federal Geodetic Committee (FGCC) Test Net, *4th Inter. Geod. Symp. Satell. Positioning Proc.*, Applied Research Laboratory, University of Texas, Austin, TX, 889-902.

McClure, T. F., 1958, Method of Navigation, U.S. Patent, no. 3,172,108, filed May 12, 1958. (Issued March 2, 1965).

Popelar, J., and J. Kouba, 1983, Satellite Doppler determination of differential sea ice motion in the vicinity of the north pole, *Marine Geodesy* **7**(1-4), 171-197.

Sepplin, T. O., 1974, The Department of Defense World Geodetic Datum 1972, *Canadian Surveyor* **28**(5), 496-506.

Stansell, T. A., 1978, The many faces of TRANSIT, *Navigation Washington* **25**, 55-70.

Cross-references: *Adjustment Methods; Geodesy: Satellite; Geodetic Networks and Control Surveys; Geodetic Reference Systems; Geodetic Refraction; Global Positioning System (GPS); Satellite Altimetry; Satellite Laser Positioning; Very-Long-Baseline Interferometry (VLBI).*

E

EARTH AND MOON: ORIGINS

Because of the great similarities in the orbits and physical composition of Mercury, Venus, Earth and Moon, and Mars, the origin of the Earth and Moon is most naturally studied within the context of the formation of the terrestrial planets as a whole, which requires in turn an understanding of the events that led to the formation of the entire solar system. While we cannot claim to have a fully developed theory of solar system formation at this moment, we do know enough to outline at least one chain of events through which the terrestrial planets may have formed, namely, through the accumulation of a population of smaller sized bodies, termed *planetesimals*.

Alternatively, the terrestrial planets could have formed as rocky cores within giant gaseous protoplanets, which may have formed themselves as a result of gravitational instability of the gaseous portion of the solar nebula. Because this second mechanism has certain drawbacks for terrestrial planet formation (e.g., the need to strip the gaseous envelopes to reveal the rocky cores, and the question of how to form the rocky core in the first place), and because it has not been as fully developed as the accumulation hypothesis, we will not describe it further here (see *Solar System: Origins* for more details).

Solar System Formation

The solar system formed from the collapse of an interstellar cloud of gas and dust, a cloud that was initially rotating slow enough to allow a single protosun to form at the center following the collapse, but fast enough to produce a flattened, rotating disk of gas and dust surrounding the protosun, termed the *solar nebula*. This collapse process took on the order of 100,000 years. The planets formed in the solar nebula, out of residual matter associated with the star formation process. Formation in the solar nebula ensured that the resulting planets acquired their present orbital characteristics, namely confinement to a nearly planar region and revolution in the same sense as the rotation of the sun.

Elements destined to form the terrestrial planets (e.g., Si, O, and Fe) were condensed into dust grains at the temperatures and pressures characteristic of interstellar clouds. The interstellar dust grains, with characteristic sizes of about 10^{-5} cm, began to undergo significant growth once they entered the solar nebula, where their greatly enhanced spatial density led to more frequent collisions and growth caused by sticking due to intermolecular forces. Within about 1000 years, the dust grains sedimented to the midplane of the nebula, growing to sizes of perhaps 1 cm in the process. The dust grains formed a thin disk, which is thought to have become gravitationally unstable and to have broken up into a large number of self-gravitating planetesimals, after the dust disk became sufficiently dense. Further growth must then have occurred through collisions between the planetesimals in this swarm of perhaps 10^{12} bodies, revolving about the sun in nearly circular orbits, in the midplane of the solar nebula. According to the accumulation hypothesis, these planetesimals are the basic building blocks for forming the protoplanets that grew into the final terrestrial planets. Further details about the earliest phases of planetary formation may be found in *Solar System: Origins*.

Closely Packed Phase of Terrestrial Planet Accumulation

Regardless of the details of these early processes, it is inevitable that terrestrial planet accretion in the context of the accumulation hypothesis involved an intermediate population of planetesimals. Thus one can fruitfully study terrestrial planet formation by calculating the time evolution of different ensembles of planetesimals and learning whether or not they can lead to the observed planets.

The planetesimal accumulation process can be divided into two different phases. In the earliest phase (closely packed phase), the planetesimals are relatively small (perhaps 1–10 km in size) and numerous, and collisions can occur between planetesimals that are confined within a narrow ring of orbits about the sun ($\approx 0.0001 - 0.01$ AU wide, where 1AU = 1.5×10^{13} cm is the astronomical unit, defined to be the distance between Earth and the sun). In this phase, the planetesimal masses are too small to be able to gravitationally perturb the planetesimals out of their initially circular orbits or to high relative velocities. Collisions then lead to growth, because kilometer-sized bodies are massive enough ($\approx 10^{16}$ g) to gravitationally trap the debris of gentle collisions. In the later phase (loosely packed phase), the planetesimals are considerably larger in mass, but smaller in number, so that in order for collisions to occur, planetesimals from different rings must collide, which requires eccentric orbits.

One important reason for distinguishing between these two phases is that the study of each phase requires a distinctly different theoretical approach; in the closely packed phase, the planetesimals can be treated in a statistical sense, as a collection of particles in a box (similar to the kinetic theory of gases). The loosely packed phase, however, requires that the three dimensional orbits of the planetesimals about the sun be accurately computed, and hence is best treated by the methods of celestial mechanics.

In either phase, perhaps the most fundamental question is whether runaway accretion can occur. Runaway accretion occurs when a single body in a swarm of planetesimals begins growing at a faster rate than the other planetesimals and becomes much more massive than the rest. This is possible because growth is dependent on the cross section for collisions, and the self-gravity of planetesimals greatly increases their effective cross sections (in the two body scattering approximation, this cross section is proportional to the fourth power of the planetesimal radius, rather than the second power appropriate for a purely geometric cross section). Thus the more massive a body becomes, the greater its cross section for accumulating other planetesimals, compared to smaller planetesimals, and hence the more rapid its growth. In the absence of outside forces, runaway growth of the most massive body will only stop when the reservoir of accretable bodies is depleted.

Current work implies that whether runaway accretion occurs in the closely packed phase is strongly dependent on the initial mass spectrum of planetesimals. If the planetesimals within a given ring start out with a distribution where the most massive planetesimal is about twice as massive as the next largest body, then that body may very well runaway and sweep up all the planetesimals within that ring. If the initial mass distribution is more nearly uniform, however, the runaway will be much more subdued or suppressed altogether. Thus whether or not a runaway occurs in the closely packed phase seems to depend critically on the initial mass spectrum of planetesimals, which unfortunately is largely unknown at present.

Even if runaway accretion does occur in the closely packed phase, the outcome is somewhat uncertain. Because of the need for substantial orbital changes in order to accumulate all the planetesimals in the terrestrial region, it is unlikely that runaway accretion could occur all the way from $\approx$ kilometer-sized bodies to the terrestrial planets. Rather, an early runaway might lead to $\approx$ 100 protoplanets of mass $\approx 10^{26}$ g (lunar sized), on widely spaced, stable orbits, a configuration that might not evolve into the present terrestrial planet system. The stability of this configuration is dependent in part on whether or not Jupiter has already grown to a size comparable to its present mass; Jupiter can be a powerful source of orbital perturbations for both the planetesimals and the growing protoplanets in the terrestrial zone. Alternatively, an early runaway could lead to a collection of perhaps 10^4 bodies of size $\approx 10^{24}$ g, distributed throughout the terrestrial planet region. The latter situation has been investigated in the most detail, and its implications for the loosely packed phase will be described in the remainder of this section.

Loosely Packed Phase of Terrestrial Planet Accumulation

The closely packed phase may well have occurred in the presence of the gaseous portion of the solar nebula, because evolution in the closely packed phase is thought to have required no more than about 10^5–10^6 years, about the same as the time during which young stars remove their gaseous placentas through strong stellar winds. Astronomical observations imply that solar-type stars experience sustained or episodic stellar winds with mass loss rates up to $\approx 10^{-5} M_\odot$ per year, where $M_\odot = 2 \times 10^{33}$ g is the mass of the sun. Such a stellar wind loses mass at a rate about 10^9 times greater than the current solar wind and is thought to have cleared the remaining gas from the solar nebula. Thus the loosely packed phase for the terrestrial planets may very well have occurred in the absence of significant gas; while gas drag is negligible for very large bodies, the effects of gravitational interactions between protoplanets and a significant gaseous nebula (i.e., the generation of spiral density waves), are largely unknown and could lead to substantial orbital evolution. In the remainder of this section, we will assume that the loosely packed phase occurred in the absence of such gas.

The orbital evolution of a loosely packed swarm of about 10^4 bodies of mass 10^{24} g is thought to occur as follows. The initially small orbital eccentricities are gradually increased following close encounters (near misses) between planetesimals. If the eccentricities become too large, subsequent collisions will occur at such relative velocities that collisional fragmentation could dominate over collisional accumulation, thereby halting the growth process. At the same time that close encounters are pumping up the eccentricity, collisions damp the eccentricity of the accumulating planetesimals. If the eccentricity is damped too much, no further collisions will occur, again halting the growth process. It turns out that the combined effect of these two processes is to regulate eccentricities toward values that allow continued growth of the protoplanets.

Runaway accretion does not appear to be significant for the loosely packed phase, in part because the remaining mass range (e.g., from 10^{24} g to 10^{28} g $\approx$ Earth's mass) is considerably smaller than the range of masses encountered in the closely packed phase ($\approx 10^{16}$ g to 10^{24} g). Instead, a swarm of planetesimals evolves in a more uniform fashion, with collisions often occurring between nearly equal mass planetesimals and protoplanets. In order for accumulation to proceed in the loosely packed phase, fairly large eccentricities must be maintained, leading to relative velocities upon impact on the order of 10 km/sec. These parameters mean, however, that the accumulation process involves potentially catastrophic collisions. In particular, the protoearth is quite likely to have been struck at about 10 km/sec by a planetesimal as large as ⅓ the mass of the Earth. Other collisions that built up the Earth (say of lunar-sized bodies) would have been somewhat less spectacular, but qualitatively similarly violent events. Considering that these giant impacts can be expected to result in considerable melting, vaporization, and fragmentation, depending on the details of the impact, the planetary accumulation process may well involve occasional episodes of major mass loss as well as mass gain. We shall see that such a collision also presents intriguing possibilities for making the Moon. Accumulation of the Earth (and the other terrestrial planets) in this fashion requires about 10^7 years to accumulate 90% of the Earth's mass, and another 10^8 years to gather the remaining 10% of mass. Thus compared to the earlier phases of protostellar collapse, planetesimal formation, and the closely packed phase, the loosely packed phase requires by far the largest interval of time.

Collisions at these velocities are expected to be even more catastrophic for bodies smaller than the Earth, because their reduced gravitational force will be less able to resist the disruptive forces of the collision. Thus Mercury may have formed as a primarily iron fragment from a high velocity, head-on collision between two comparable sized protoplanets; its silicate mantle may have been removed by the vaporization and shattering produced by the collision. Mars may have ended up with a mass only about ⅒ that of Earth and Venus because of its closer proximity to Jupiter during the final stages of accumulation and the ability of Jupiter to gravitationally interfere with the growth of Mars.

Because of the stochastic nature of the final phase of accumulation of the terrestrial planets, one can never hope to produce a theoretical model that is exactly equivalent to our solar system. Nevertheless, simulations of the accumulation of about 1000 bodies with masses on the order of 10^{25} g (and with the angular momentum and total energy of the present terrestrial planets) have shown that it is quite probable for such a system to evolve into four planets with masses and orbits similar to those of the terrestrial planets.

The planetesimals that formed the Earth are thought to have been roughly uniform in chemical and isotopic composition, having an averaged abundance similar to that of chondritic meteorites. Variations in the composition of any given planetesimal could result from condensation of different elements or minerals at different temperatures (i.e., different distances from the sun), providing that the primordial dust grains encountered temperatures high enough to produce vaporization. Such thermal processing might have occurred in the accretion shock surrounding the nebula or in the nebula itself. Condensation sequences in a solar nebula with temperature decreasing outward from the sun have long been proposed as an explanation for the compositional gradient with heliocentric distance of the terrestrial planets (i.e., Mercury is largely iron, while Earth is largely silicates). Some solar nebula models that include strong turbulent viscous dissipation produce midplane temperatures high enough to support the condensation sequence theory. However, orbital evolution during the loosely packed phase can be expected to stochastically redistribute planetesimals throughout the terrestrial planet zone, which erodes this explanation of the apparent compositional gradient. Thus well-defined feeding zones may not exist for the terrestrial planets, except in the general sense that the feeding zone for each terrestrial planet was the entire terrestrial zone.

Earth's Earliest Thermal History

The concentration of radioactive isotopes can be used to date when a rock was last crystallized, and samples of the Earth, Moon, and meteorites have been studied in order to find the oldest possible rocks. The age of the oldest rock gives a lower bound on the age of the solar system. Many primitive meteorites give ages clustering around 4.5 billion years old, strongly implying that this value is the approximate age of the solar system. Silicate inclusions in the Allende meteorite currently give the oldest accepted dates of crystallization, about 4.56 billion years ago (see *Meteorites: Nature and Origin*).

Contrary to the ideas of a few decades ago, it is no longer necessarily thought that Earth's accumulation proceeded primarily through the impact of bodies very much smaller in size. This removes one fundamental problem: forming the Earth's molten outer core quickly enough to be able to provide a terrestrial dynamo capable of explaining the remnant magnetism found in the earliest terrestrial rocks. If accumulation by the impact of very small bodies had been the case, the impacts would bury little energy deep within the protoplanet, instead depositing the dissipated kinetic energy at the surface, where it could be readily radiated to space and lost from the planet's thermal budget. In such a case, the Earth would have formed relatively cold, i.e., with temperatures everywhere below the solidus, unless the formation time was very short (e.g., 10^6 years). The fact that such short formation times are hard to obtain in the accumulation hypothesis meant that it was difficult to explain how the iron core formed early in Earth's evolution. It now appears that the late phases of Earth formation involved energetic collisions between rather massive planetesimals and formation of the initial iron core during the accumulation phase.

Formation of the Earth through collisional impacts involving very large planetesimals ($> 10^{24}$ g) must have been a violent process; the gravitational energy liberated is potentially sufficient to melt the entire planet. The degree of melting obtained in the protoearth is a fundamental question, because melting can result in geochemical differentiation, greatly enhanced heat transport, and qualitatively different outcomes with respect to dynamical questions such as the possibility of rotational fission. Rigorous models of impacts on this gigantic scale are just now being seriously investigated, and the details of these solutions are certain to have profound importance for the early thermal history of the Earth in the context of the accumulation hypothesis of origin.

The most important differentiation event in Earth's evolution was the separation between iron and silicates and the migration of the iron to form the core. One manner in which the core may have formed involves a catastrophic event. Once the protoearth grew to about the size of Mars, the energy released by the impacting planetesimals would have been sufficient to lead to pools of liquid iron, which then sink because of their high density compared to the partially molten, silicate outer layer. However, the liquid iron will cool as it sinks into the cold, undifferentiated lower layers and it will eventually solidify again. The spherically symmetric iron layer overlying the cold undifferentiated core is thought to be unstable to a catastrophic failure, where the cold core shatters and the iron displaces the core and forms the first iron core. The energy released by this process, coupled with the ongoing accretional energy, is transported downward by descending molten iron, and leads to a partially molten, largely silicate mantle. The iron produced by subsequent impacts can more easily sink to the core because of the partially molten state of the overlying silicate mantle. The remnants of the cold, undifferentiated core may persist until thermal diffusion brings them into equilibrium with the rest of the mantle.

Formation of the iron core thus occurred at the same time that the Earth was accreting, and consequently the two processes must be considered together. However, it is not so certain whether differentiation of the silicate mantle occurred to a significant extent during the formational period. The answer to this question depends primarily on the question of whether the silicate mantle, once it became partially molten, was able to resolidify (perhaps only locally) before becoming molten again as a consequence of the next large impact. If not, then the mantle may have remained largely molten throughout the formational period, and differentiation would only have been possible once accretion ended and the planet began to cool and resolidify.

Lunar Formation: Classical Hypotheses

George H. Darwin (son of Charles Darwin) analyzed the orbit of the Moon over 100 years ago and found that at some time in the past the Moon must have been considerably closer to the Earth (the net effect of the tides raised on the solid and fluid portions of the Earth by the Moon is to transfer angular momentum from the Earth's spin to the Moon's orbit, thereby moving the Moon further away from the Earth as time goes on.) Darwin therefore hypothesed that the Moon arose from the mantle of the Earth, and proposed a mechanism for removal of the Moon: rotational fission. Rotational fission is the first of three hypotheses that have come to be known as the classical hypotheses of lunar origin.

Rotational fission involves the splitting into two of a body that is rotating too fast to be stable. The concept was based upon studies of the possible equilibrium configurations of self-gravitating, incompressible (uniform density), fluid bodies in solid-body rotation. Clearly when such a body is spun fast enough, the centrifugal acceleration needed to hold the equatorial regions onto the body can no longer be provided by the finite gravitational acceleration inward. What happens thereafter could not be discerned on the basis of the equilibrium models available 100 years ago, but because of the existence of equilibrium models for tidally distorted bodies in orbit about one another, it was hypothesized that the natural outcome of rotational instability would be separation into a binary system. Recently numerical calculations have been able to follow the growth of the fission instability in rapidly rotating, inviscid bodies with compressibilities similar to that of the Earth. Contrary to the classical fission hypothesis, however, the outcome of at least one fission instability is not simple bifurcation into two bodies; instead, the embryo central binary loses its orbital angular momentum to growing spiral arms located further outward. The binary effectively decays into a single body before it can become well defined, while the spiral arms remove the excess angular momentum from the inner body, allowing it to reach a stable equilibrium. The spiral arms degenerate into a ring with a small fraction of the total mass, but a large fraction of the total angular momentum. While this unforeseen outcome appears to eliminate the fission hypothesis as a means of producing nearly equal mass binary systems (e.g., close binary stars), it is still an attractive means for forming the Moon (providing the Moon can be assembled out of the ring), because the mass of the Moon is only about $1/80$ that of the Earth.

Rotational fission has several strong arguments in its favor. The first is that forming the Moon out of the Earth's mantle would produce a Moon with a low mean density (about 3 g cm^{-3}), similar to that of the Earth's mantle, as is observed. Furthermore, analysis of the lunar samples returned by the Apollo missions has shown that the lunar surface rocks are strongly depleted in iron, as is the Earth's mantle, consistent with the low mean density of both. Siderophilic elements have abundances in the Apollo samples that are very close to those of the Earth's mantle. In several ways then, the Moon's chemistry appears to be very similar to the Earth's mantle, which argues in favor of a fission origin. However, the Apollo samples also showed that the Moon is greatly depleted in volatiles such as water, compared to Earth's mantle, but it has been argued that devolatilization could have been caused by the energetic processes associated with the fission instability and ring formation.

While fission appears to be basically consistent with the history of the Moon's orbit, there are several dynamical problems that in sum appear to rule out the possibility of forming the Moon in this manner. One early objection to the fission hypothesis was that fission requires about four times more angular momentum in the initial protoearth than is present in the Earth-Moon system. It is hard to explain the amount of angular momentum in the Earth-Moon system, much less four times as much. Second, assuming that a mechanism could be found for providing the requisite angular momentum, the excess angular momentum must be disposed of eventually. In the numerical calculations of fission, it is conceivable that inefficient formation of the Moon out of the ring of matter orbiting the Earth could account for the loss of the excess angular momentum (the ring is likely to be considerably more massive than the Moon), but this has not been investigated.

The third objection concerns the limitation of the previous work on the fission instability to inviscid, fluid bodies. If the protoearth was totally molten, this approximation might be adequate. However, in spite of the undoubted ability of the major impacts associated with the late stages of terrestrial planet formation to produce widespread melting, it does not seem likely that the protoearth would have been completely molten, for several reasons. A large fraction of the impact energy is thought to be lost by ejecta and radiation during the explosive collision, and partially molten bodies are quite effective in convecting heat upward to be radiated to space. Estimates of the minimum viscosity needed to transport the heat of formation out of the planet place the effective viscosity of the protoearth well out of the range of being totally molten. A totally molten Earth may also have undergone a mantle-wide differentiation inconsistent with the inferred structure of the Earth's mantle. The viscosity associated with a partially molten protoearth is sufficient to stifle the dynamic fission instability originally hypothesized by Darwin, or as modified by the numerical calculations of spiral arm ejection. Instead, if a viscous, rigidly rotating protoearth is spun fast enough, it will begin to lose mass from its equator, a process called mass shedding. This process leads to the formation of a ring in Earth orbit from which the Moon could later form, and so retains some similarity to the modified fission hypothesis. As in the original objection to fission, however, the angular momentum necessary to initiate mass shedding is prohibitively large. For example, if the angular momentum is derived from the tangential impact at 10 km/sec of Mars-sized bodies with the protoearth, then four successive impacts, each in the same spot on the limb of the protoearth, would be required to initiate mass shedding; this prospect is quite unlikely compared to the single giant impact hypothesis (to be discussed shortly).

The second classical hypothesis of lunar origin is capture. This hypothesis proposes that the Moon formed elsewhere in the solar system, and then was captured in orbit about the Earth. The former assumption can be challenged on the basis of the strong chemical similarities between the Moon and the Earth's mantle; if the Moon was formed out of a different sample of planetesimals than the Earth, it is improbable that it would have acquired certain elemental and isotopic abundances so similar to those of Earth. The latter assumption (i.e., a mechanism allowing capture) is an even more crucial failing of this hypothesis.

Intact capture of a protomoon about the Earth requires the dissipation of the relative energy of motion between the protomoon and Earth. The most promising (and most often invoked) mechanism is tidal dissipation, which is certain to occur in a close encounter between partially molten or solid protoplanets. However, even the most optimistic estimates of the efficiency of tidal dissipation imply that in order to be captured, the protomoon must have approached the Earth from an orbit with a relative velocity of no more than about 0.1 km/sec. Considering that the orbital velocity of the Earth is about 30 km/sec, this means that the protomoon must have started from an orbit very nearly the same as that of the Earth. But then it is hard to see how the protomoon could have accumulated to its present size without having been subjected to a close encounter with protoearth earlier in their evolution, an encounter that would have kicked the Moon to a much higher relative velocity. We have seen that encounters between planetesimals and protoplanets in the loosely packed phase occur at higher relative velocities (up to 10 km/sec), where capture by tidal dissipation can be ruled out.

The intact capture hypothesis was hence modified by the added hypothesis of tidal disruption of the protomoon into a shower of debris, some of which might remain in orbit about the Earth and later reaccumulate into the Moon. This hypothesis is based on the absence of equilibrium configurations for a fluid satellite on a circular orbit within a critical distance from the primary body. The critical distance, named the Roche limit after the scientist who first derived it, is about 3 times the radius of the Earth for the Earth-Moon system. It was thus hypothesized that if a planetesimal passed within the Roche limit of a growing protoplanet, it would be torn asunder by strong tidal forces. One appeal of disruptive capture is that the probability of passing within the Roche limit is quite large, and is in fact considerably more likely than a direct collision, because of the larger cross section. However, the Roche criterion applies to an inviscid satellite on a circular orbit, where there is an infinite amount of time to tidally disrupt the satellite. In terrestrial planet accumulation, a planetesimal passing close by the Earth will only spend on the order of an hour within the Roche limit, and furthermore is unlikely to be totally molten and hence an inviscid fluid. When reasonable effective viscosities are included in detailed models of the close encounter of a planetesimal with the Earth, even for grazing incidence, the amount of mass lost from the planetesimal is small; tidal disruption of a dissipative planetesimal does not occur during the brief period available during a flyby. Hence disruptive capture cannot occur unless a means is found to totally melt lunar-sized bodies during the accumulation process.

The third classical hypothesis of lunar origin is binary accretion, where the Moon forms in orbit about the protoearth as both grow to their final size. This hypothesis has the advantage of explaining the similar chemistries of the Moon and Earth's mantle as being caused by their accumulation from essentially the same sample of planetesimals. In order to account for the lack of iron in the Moon, however, significant compositional filtering must occur. It has been argued that a swarm of planetesimals orbiting the protoearth could preferentially trap silicate, rather than iron, planetesimals, because of the former's lower mean density. This mechanism assumes that a prior generation of protoplanets have already differentiated into mantles and cores and later had been disrupted into much smaller pieces, and that a swarm remains in place about the protoearth to accomplish the processing. The former requirement appears to be consistent with a scenario where runaway accretion to protoplanetary sizes occurs, with accumulation proceeding from a distribution dominated by many very small bodies. While such an eventuality cannot be disproven at present, it is at odds with the scenario developed so far in this section, where the later phases of accumulation do not involve runaway accretion. The latter requirement is even harder to achieve dynamically, because accumulation in Earth orbit is likely to proceed on a much shorter time scale than that for the introduction of new planetesimals from heliocentric orbit, because of the much shorter orbital period in geocentric versus heliocentric orbit.

A third objection, apparently fatal, is that it is impossible to achieve the angular momentum necessary for maintaining the orbiting swarm, unless a special population of orbits provides the bulk of the planetesimals interacting with the binary planet system. This failure occurs because incoming planetesimals will have roughly equal chances of colliding with the swarm and ending up in orbit in either possible sense of revolution about the protoearth; thus, on average, the angular momentum is zero. Obtaining appreciable angular momentum in this way requires accreting planetesimals from very nearly circular orbits or from a population depleted at the Earth's heliocentric distance. Both of these possibilities appear to be unlikely, because of the relatively highly eccentric orbits of planetesimals late in planetary accumulation, and because of the tendency for stochastic evolution to fill in any depleted regions with more planetesimals. Thus all of the classical hypotheses for forming the Moon appear to be severely flawed.

Lunar Formation: Giant Impact Hypothesis

The frequent occurrence of giant impacts on the growing terrestrial planets provides a natural means for forming a major satellite like the Moon, and for providing the large amount of angular momentum necessary for its orbit. If a giant impact of a Mars-sized body at 10 km/sec occurs close to the limb of the Earth, the angular momentum of the incoming body with respect to the center of the Earth is equal to that of the Earth-Moon system. Providing that the colossal explosive impact that follows ejects substantial debris into stable Earth orbit, it may be possible for the Moon to later reaccumulate in geocentric orbit. The implications of this hypothesis for explaining the chemistry of the Moon are extremely attractive: the impact would presumably result in the injection of portions of the silicate mantles of the Mars-sized interloper and the Earth into Earth orbit, accounting for the low mean density and similarity to terrestrial elemental abundances of the Apollo samples. The thermal processing implied by a giant impact makes devolatilization of the prelunar matter quite understandable. Provided that the impact places a fair amount of protoearth mantle in orbit, it appears possible that the giant impact hypothesis can explain all of the chemistry of the Moon.

Dynamically speaking, the fact that the giant impact hypothesis involves collisions that are predicted to occur with high probability and that can account for the angular momentum of the Earth-Moon system, means that this hypothesis has already overcome two of the most formidable obstacles to a consistent theory of lunar formation. It is also consistent with the observation that of the four terrestrial planets, only the Earth has a major satellite. Venus may not have a Moon-like satellite for

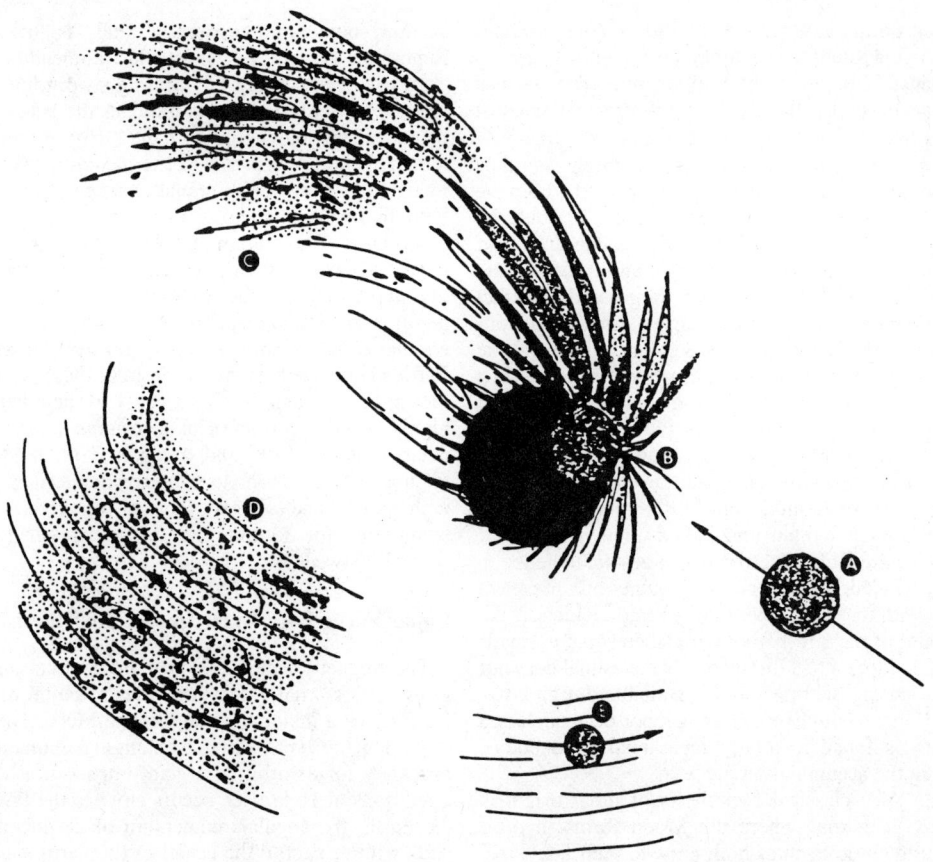

FIGURE 1. Giant impact model for lunar formation. Following the catastrophic collision of a Mars-sized protoplanet (A) with the growing Earth (B), debris ejected by the impact (C) is lost from Earth orbit or reimpacts the Earth, or is left in Earth orbit (D), where it later reaccumulates into the Moon (E). (From Boss, 1986; copyright © 1986 by AAAS)

several good reasons; its largest impact simply may not have occurred at the right orientation to the Venusian limb to create a satellite system, or even if it did once have a large satellite, Venus may have lost the satellite through tidal decay of the satellite's orbit onto the slowly rotating Venusian surface.

Because the giant impact hypothesis is little more than a decade old, and because serious interest in it is even more recent, we are just beginning to understand the processes associated with impacts on a scale much larger than anything previously contemplated. In particular, the details of the ejection of matter from the impacting body and the protoearth are currently a subject of debate. Little is known about whether the ejected matter is capable of accumulating into a single body that will have the mass and other physical characteristics of our Moon. The next decade may very well see sufficient effort put into solving the remaining problems associated with the giant impact hypothesis to enable a reasonable assessment of whether we have finally found the origin of the Moon.

ALAN P. BOSS

References

Boss, A. P., 1986, The origin of the Moon, *Science* **231**, 341-345.
Hartmann, W. K., R. J. Phillips, and G. J. Taylor, eds., 1986, *Origin of the Moon*. Houston, Tex.: Lunar and Planetary Institute, 781p.
Ringwood, A. E., 1979, *Origin of the Earth and Moon*. New York: Springer-Verlag, 295p.
Wetherill, G. W., 1980, Formation of the terrestrial planets, *Ann. Rev. Astron. Astrophys.* **18**, 77-113.

Cross-references: *Absolute Age Determinations: Radiometric; Earth's Core; Energy Budget of the Earth; Meteorites: Nature and Origin; Solar System: Origins.*

EARTH ORIENTATION

The Earth continuously changes orientation in space in response to external torques generated by gravitational coupling with the Sun, Moon, and other planets of the solar system, and to internal torques that result from dynamic geophysical processes involving the redistribution of masses and angular momentum. Changes in Earth orientation are conveniently divided into three categories: changes in the direction of the Earth's axis of rotation in space (precession and nutation), wobble of the "solid Earth" about the axis of rotation (polar motion), and changes in the rate of rotation (rotational phase angle (UT1) and length-of-day (LOD)). See Fig. 1. By observing these responses of the Earth to applied forces it is possible to estimate, or at least to put bounds on, certain parameters that describe the structure of the Earth.

Extracting the geophysical information embedded in Earth orientation time series has proven difficult for two reasons. The first is a fundamental problem: The observed motions represent the combined effects of a variety of phenomena. For example, the period of the Chandler component of polar motion depends strongly on the ellipticity and elasticity of the mantle and crust, but the effects of the oceans and fluid core enter at about the 10% level, and even such effects as the anelastic response of the mantle change the expected period at the level of a few days. Deconvolving the observed motion to estimate specific parameters at best involves complex analytical

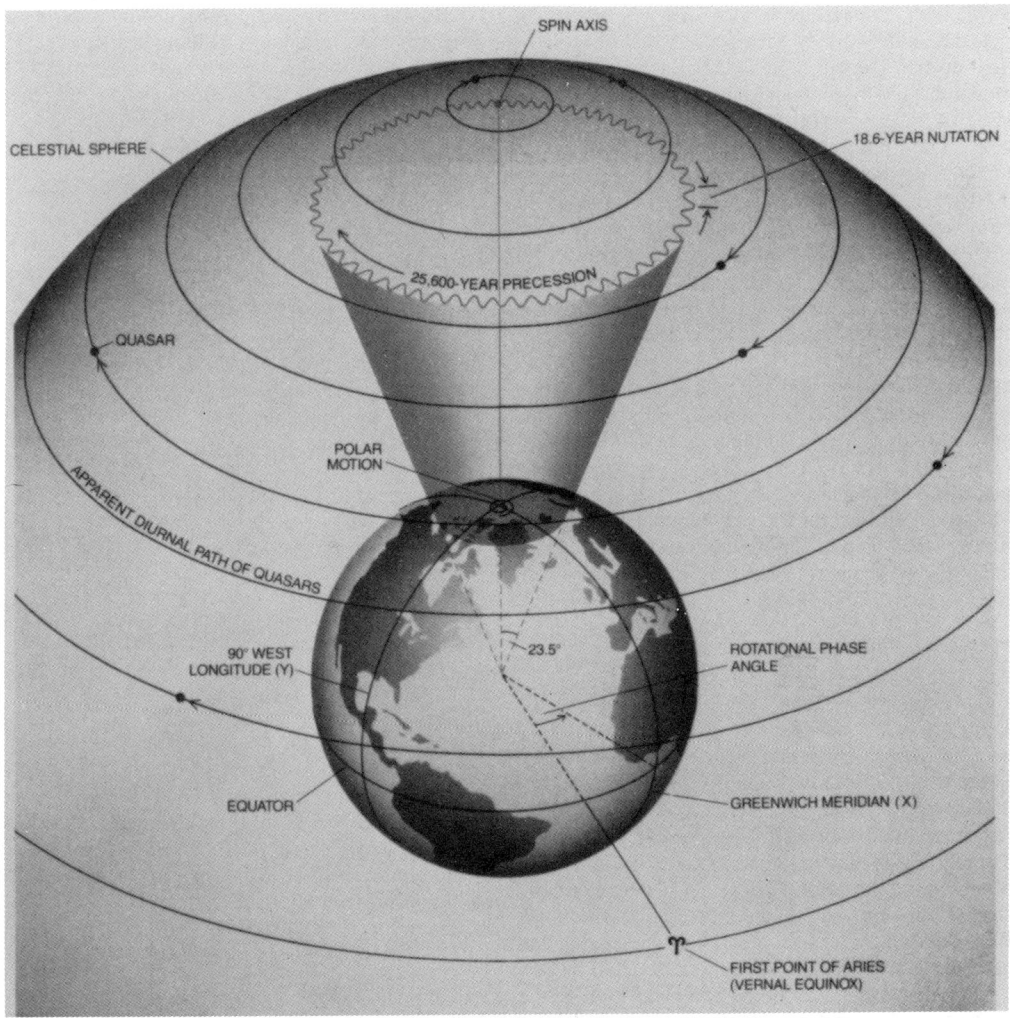

FIGURE 1. This sketch shows how the Earth's rotation axis moves in space (precession and nutation) and how the solid-earth wobbles about the axis (polar motion). The quasars define the most nearly fixed reference frame currently known. (From Carter and Robertson, *Scientific American*, November, 1986)

procedures, and at worst may not be possible without the use of information from other types of measurements. The second difficulty has been to achieve adequate observational accuracies and temporal resolutions. To be useful in constraining modern Earth models, orientation values must be accurate to about 1 msec of arc over an averaging period of 24 hours, and for some applications the requirements are even more stringent. The historical methods of measurement, i.e., visual, photographic, and photoelectric observations of nearby stars simply could not reach these standards. Only with the introduction of advanced technologies during the past decade, including lunar laser ranging (LLR), satellite laser ranging (SLR), and most importantly, very-long-baseline interferometry (VLBI) have the Earth orientation time series approached the requirements for modern geophysical applications.

Today, a network of only a few VLBI observatories in the United States and Europe routinely measures polar motion and nutation accurate to 1–2 msec of arc, and UT1 accurate to 0.05 msec–0.10 msec, in 24-hour observing periods, every 5 days. In addition, daily measurements of UT1 accurate to 0.10 msec are obtained from only 45 minutes of observations. And the Earth orientation parameters (EOP) obtained from VLBI are referred to a celestial reference frame based on extragalactic radio sources (quasars) that are so distant from Earth that their apparent positions remain fixed at the sub-msec of arc level over time scales of decades. For a more detailed discussion, see *Very-Long-Baseline Interferometry (VLBI)*. LLR and SLR observations provide limited checks on the accuracy of the VLBI EOP time series and contribute unique information on such phenomena as the rate of recession of the Moon and changes in the gravitational field of the Earth, which help in sorting out the complexities of the observed changes in Earth orientation.

Precession and Nutation

Precession is the slow "top-like" rotation of the Earth's axis in space that sweeps out a circle on the celestial sphere with a radius equal to the inclination of the ecliptic, approximately 23.5°, in a period of 25,600 years. Nutations are relatively small higher frequency motions superimposed on the precession. The largest nutation has an amplitude of about 9 sec of arc and a period of 18.6 years. There are many smaller nutations, in fact the model currently in use (IAU 1980) includes over 100 terms. Precession and nutation are caused by torques produced by the gravitational attractions of the Sun and Moon on the equatorial bulge of the Earth. The torques that drive these motions are well understood, but the precise response of the Earth depends on such parameters as the elasticity and anelasticity of the mantle; the size and shape of the fluid core and its coupling to the mantle; and the complex responses of the nonuniformly distributed oceans. Spectral analysis of the differences between the computed and observed nutations, for the 7-year time series of VLBI observations currently available, yield 3 significant peaks at periods of 12 months, 6 months, and 14 days. The 12-month term has an amplitude of approximately 2 msec of arc, and is thought to be caused by imperfections in the model of the core. The core may be slightly flatter (by about 500 m) than the hydrostatic equilibrium value assumed in the model. Topographic features at the core-mantle boundary or convective cells within the fluid core may also be involved, and researchers are working on models that fit both the nutational and rapidly improving seismic tomography data. The 6-month term is about 1 msec of arc in amplitude, and its cause is still unknown. The 14-day term is smaller still, and considering the relatively high frequency, is thought most likely to be related to the mantle, perhaps an error in the elasticity or anelasticity parameters. The differenced nutation spectrum also displays significant power in the low frequency region, but the detailed structure will be revealed only as the time series extends over a sufficient time span.

Polar Motion

Polar motion is the wobble of the "solid earth," i.e., the mantle and crust, about the axis of rotation. The theoretical foundations for understanding the phenomenon can be traced back to the eighteenth century Swiss mathematician, Leonhardt Euler. He developed the now famous equations of motion for a rotating body, which when referred to axes (x, y, z) that are fixed in a rigid body that is rotationally symmetric about z (a reasonable zeroth-order approximation for the Earth) simplify to the form:

$$d\omega_x/dt + [(C - A)/A]\omega_y\omega_z = L_x/A,$$

$$d\omega_y/dt - [(C - A)/A]\omega_x\omega_z = L_y/A,$$

$$d\omega_z/dt = L_z/C$$

ω_x, ω_y, ω_z specify the angular velocities with which the axes turn about themselves. C is the moment of inertia about the maximum principal axis, z, and A is the moment of inertia about the two equatorial principal axes, x and y. L_x, L_y, and L_z are torques about the respective axes. For zero torques, the solution of these equations is:

$$\omega_x = a_0 \cos \sigma_r t + b_0 \sin \sigma_r t$$

$$\omega_y = a_0 \sin \sigma_r t - b_0 \cos \sigma_r t$$

$$\omega_z = \text{constant} = \Omega$$

where a_0, b_0 and Ω are constants of integration and the frequency of the motion σ_r is defined by:

$$\sigma_r = [(C - A)/A]\Omega$$

Thus, for a small displacement of the principal axis z from the rotation axis $\bar{\omega}$, the latter revolves in a circular path about the z axis in a period of about 10 months. Or, in a reference frame fixed in space, the axis of figure of the Earth "wobbles" about the rotation axis. The wobble causes the latitudes of points on the surface of the Earth to change with time, and these changes can be detected by observations of extraterrestrial objects, historically stars, but today also artificial satellites, the Moon, and extragalactic radio sources.

A number of astronomers attempted, without success, to detect polar motion over the better part of two centuries until the late nineteenth century. In 1891 the American geodesist, astronomer, and actuary, Seth Carlo Chandler, Jr., startled the scientific community when he announced that observations from several observatories displayed polar motion, but that the period of the motion was more nearly 14 months than the predicted 10 months. The well known American astronomer Simon Newcomb qualitatively explained the discrepancy in the period as being caused by the elasticity of the Earth and the fluidity of the oceans. In 1892 Chandler reported that further analysis showed that polar motion was not a simple periodic motion, but rather a compound motion consisting of a 14-month component, about 0.15 seconds of arc in amplitude, and a somewhat smaller annual component, of about 0.10 seconds of arc. The annual term was qualitatively ascribed to the redistribution of masses (e.g., atmosphere, ground water, snow cover) related to the changing of the seasons.

The 14-month and 12-month components of polar motion "beat" together, causing the path of the pole to alternately spiral outward and then inward, completing a full cycle approximately every 6–7 years (Fig. 2). But the new high accuracy polar motion time series clearly show that the pole does not follow perfectly the smooth path predicted by the two-component model. Rather, the actual path wanders back and forth across the computed path, sometimes leading and other times lagging behind the expected progress. Figure 3 shows the along track, across track and total deviations of the observed motion of the pole relative to a least squares best fitting model, consisting of 14 and 12-month circular motions and a linear drift, over roughly one beat period. The residuals display systematic variations on time scales of a few weeks to several months that reach peaks as large as a few tens of milliseconds of arc. Spectral analyses of these residuals indicate that there may be a significant component with a period of 6 months (researchers have suggested that the annual term may have some degree of ellipticity, which may even vary in degree and orientation with time) but otherwise the power drops off roughly as the inverse of the frequency, as for a Brownian process.

The accurate EOP time series currently span only slightly more than 1 full beat period of polar motion, and spectral analysis barely separates the 14 and 12-month terms, providing essentially no usable information about the widths of the spectral peaks. Analyses of the historical time series have consistently yielded a significantly broader peak at the 14-month period than at the 12-month period. Several researchers, beginning even with Chandler, have concluded that the 14-month peak may contain two or more components with periods ranging from 13 to 15 months. Others have maintained that this is an artifact either of changes in phase of the 14-month term, caused by random excitations (e.g., by earthquakes) or observational noise. The width of the spectral peak is of interest because it provides an estimate of the Q of the mantle at that period. The new high accuracy time series eliminate observational noise as a serious problem in interpreting the spectrum, but it will be some time before the series will span enough time to yield a reliable estimate of the width of the 14-month peak.

It has also been suggested that changes in the path of the pole have been correlated with the occurrences of great earthquakes. There have been no truly great earthquakes since the new techniques have begun producing high accuracy polar motion time series. The largest has been the magnitude 8.1 (Richter scale) earthquake that occurred September 19, 1985, in Mexico. There was no discernible change in the path of the pole at that time. Currently the most widely suspected sources for variations in polar motion on time scales of weeks to months, of the character displayed in Fig. 3, are interactions between the crust and the atmosphere, specifically asymmetrical barometric loading of the crust and transfer of angular momentum between the north-south components of the global wind field and the crust. However, attempts to correlate polar motion and atmospheric time series have met with only limited success, at times seeming to show significant correlation, but over periods of a few years yielding correlation coefficients of only 0.4–0.6.

The historical polar motion observations have also indicated that there is a secular drift of the pole of about 3 msec of arc per year, roughly toward Hudson Bay. It is thought that this motion is related to glacial rebound. Mass in the mantle that was squeezed southward by the heavy loading of ice in the Canadian and Scandinavian regions of the northern hemisphere during the last ice age, is now slowly flowing back. According to the best available models, this change in the distribution of mass should cause the pole to move approximately in the direction observed at a rate that depends on the viscosity of the mantle. A rate of 3 msec of arc per year corresponds to a viscosity in the range of 10^{22} and 10^{24} poise. The relocation of mass in the mantle should also cause a change in the J2 component in the spherical harmonic representation of the Earth's

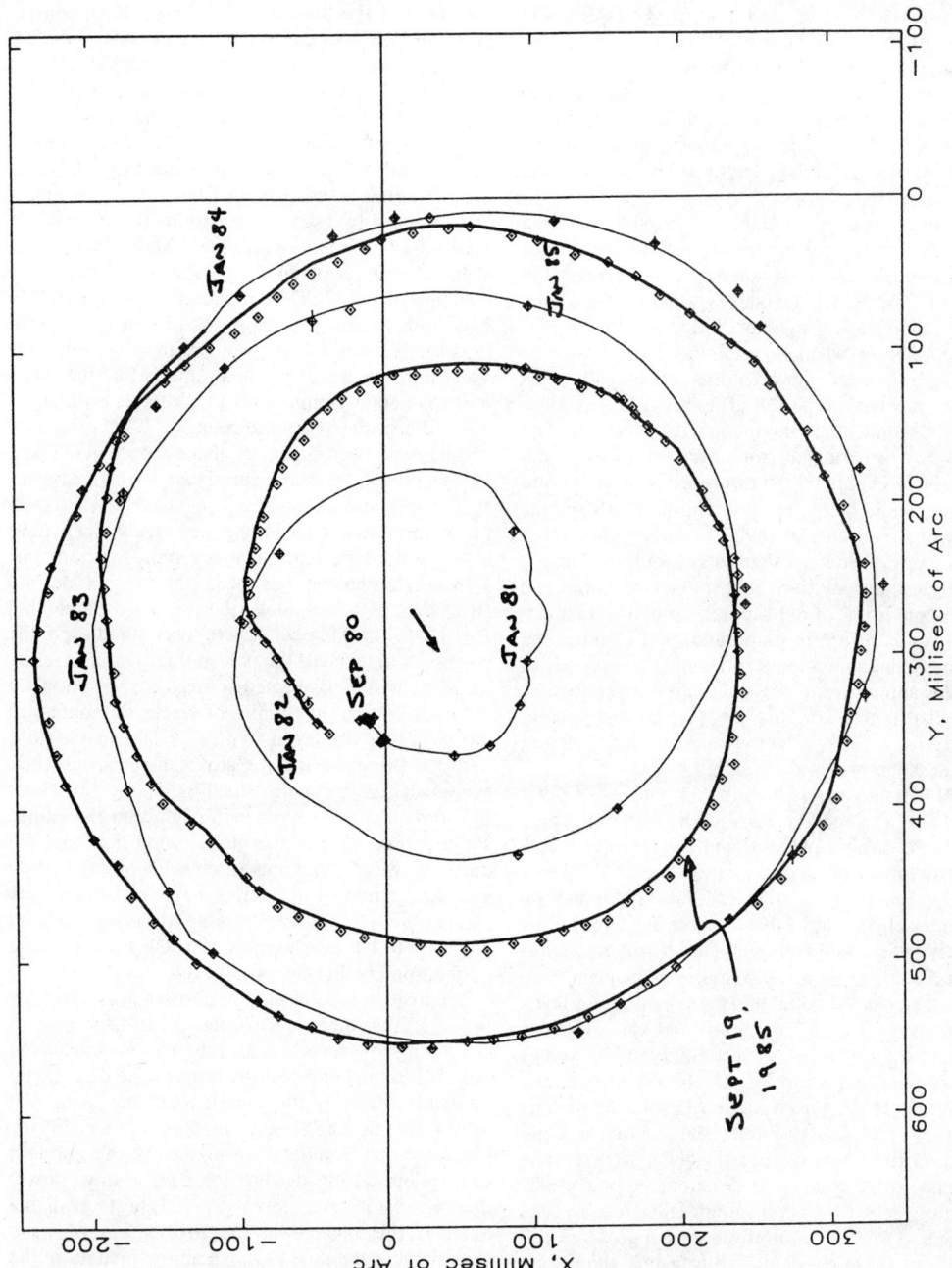

FIGURE 2. This plot shows the path of the pole over approximately one "beat" period.

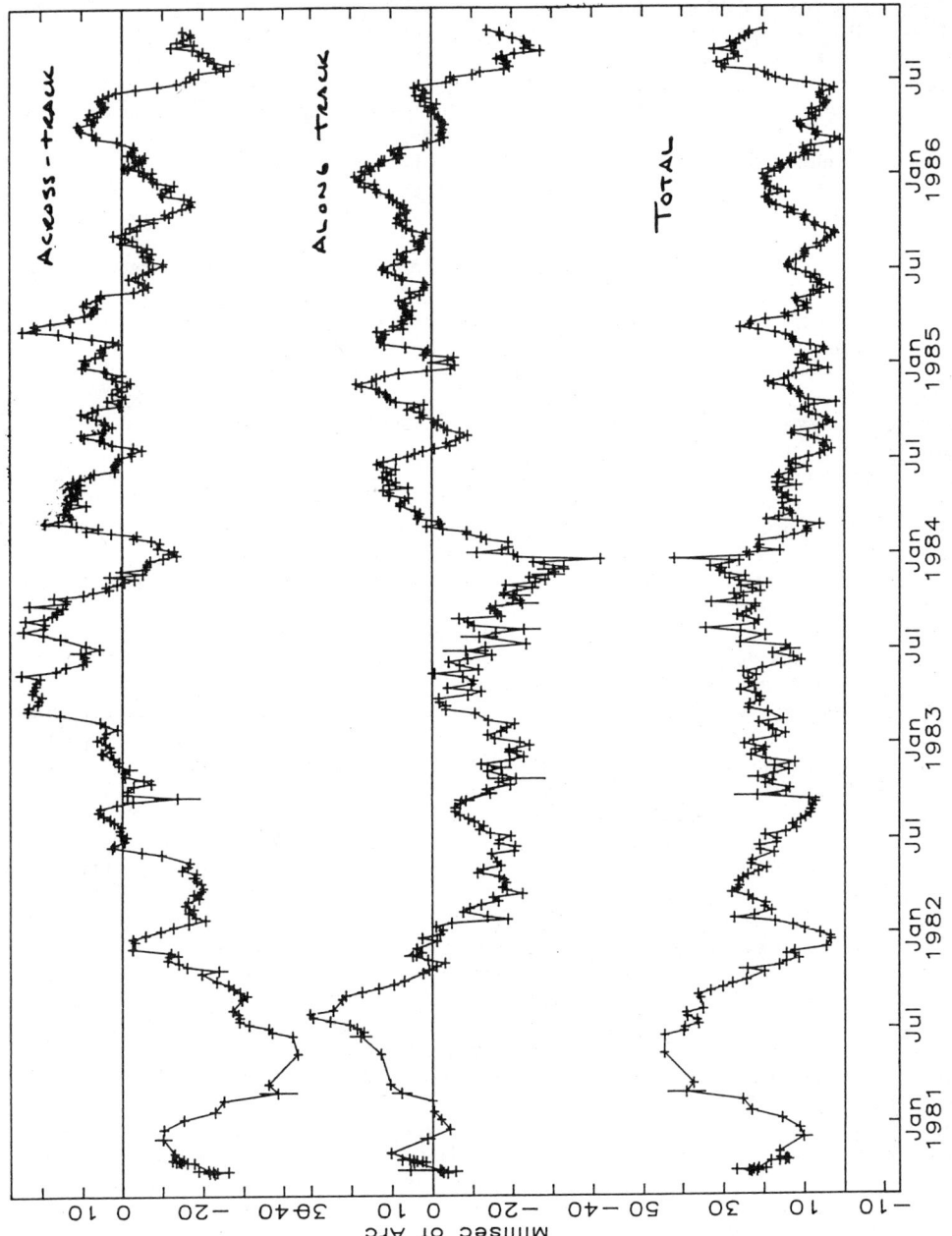

FIGURE 3. These plots show the across-track, along-track, and total residuals of the observed motion of the pole relative to a model consisting of 14-month and 12-month circular components, and a linear drift.

gravitational field. This effect has recently been detected using a combination of SLR and LLR observations. Another possible source of secular motion of the pole is contemporary melting or building of the Greenland and Antarctic ice caps. Changes of only about 10 cm per year in the thickness of these ice masses would cause millisecond of arc motions of the pole. The small vector in the center of Fig. 3 shows the direction and magnitude of the secular motion computed from 7 years of SLR and VLBI observations. Considering the limited length of the new series, the motion is not significantly different from the historical values.

UT1 and LOD

Historically, measuring variations in the spin rate of the Earth involved all the same observational difficulties as measuring polar motion (instrumental errors, atmospheric refraction, errors in star positions and proper motions, etc.) but had the additional difficult technological problem of keeping time sufficiently accurately. The best mechanical clocks ever built were only slightly more stable in frequency than the rotation of the Earth. Such clocks permitted researchers to detect the large seasonal variations in LOD, but lacked the stability to capture the true complexities of the variations. Progress in understanding the variations in LOD was severely limited until a breakthrough was made in timekeeping with the introduction of electronic (quartz oscillator) clocks in the 1940s. It then became obvious that in addition to the annual term, there were semiannual, lunar monthly, and fortnightly variations as well. The development of atomic frequency standards during the 1950s improved the long term stability of clocks by another 2–3 orders of magnitude, and the limiting constraint on improving the determination of LOD became the accuracy of the astronomic observations. It is interesting to note that new timekeeping technology also played a key role in the recent breakthrough in observational capabilities. Determining the differences in arrival times of radio signals at VLBI observatories located thousands of kilometers apart requires clocks that run at constant rates, or at least are well enough behaved that their variations can be easily modeled, to about 1 part in 10^{15}, over a period of 24 hours. Only with the invention of the Hydrogen-MASER frequency standard, and its refinement within the past decade, has it been possible to meet that requirement.

In the relatively short period that the new methods of observation have been operational, they have contributed most dramatically to studies of the exchange of angular momentum between the atmosphere and the solid earth. As early as 1960 researchers argued that the nontidal variations in LOD on seasonal scales were associated with changes in the east-west wind patterns, but it was not until the 1970s that adequate global meteorological data became available to show an undeniable connection. Apparent higher frequency nontidal variations were also noted, but the uncertainties in the LOD determinations were sufficiently large that they were generally discounted. By the mid-1970s meteorologists at the U.S. National Meteorological Center (NMC) were routinely twice daily producing global atmospheric parameters, including east-west wind velocities at 2.5° grid points, for 12 pressure levels ranging from 1000 to 100 millibars. Numerically integrating these winds with respect to latitude, longitude, and pressure yields the atmospheric angular momentum (M) about the polar axis, in an earth-fixed reference frame. Assuming that angular momentum is conserved in the atmosphere-solid-earth system for time scales of days to perhaps a few years (i.e., changes in the angular momentum of the atmosphere are accompanied by equal but opposite changes in the angular momentum of the crust and mantle) the expected change in the length of day (ΔLOD) can be computed for a given change in atmospheric angular momentum (ΔM) from the simple relationship:

$$\Delta \text{LOD} = (1.68 \times 10^{-26}) \Delta M$$

where ΔLOD is in units of milliseconds and ΔM is in units of kg m^2 s^{-1}.

Figure 4 shows a plot of the changes in LOD inferred from measurements of atmospheric angular momentum data and as measured by VLBI for a 5-year period. Strong annual variations are evident in both series. Closer inspection reveals 40–50 day quasi-periodic variations common to the two series, and an unusually strong peak in January-February, 1983. Both of these phenomena have been traced to global scale atmospheric-oceanic disturbances, the latter being related to an unusually strong El Niño. During an El Niño the prevailing westerly winds in the Pacific equatorial region collapse, and the warm surface water usually pushed into the western Pacific migrates eastward, resulting in strong climatological changes over much of the globe. El Niños happen at intervals of 3–6 years. Since the discovery of the El Niño signal in the LOD time series, researchers have able to piece together a history extending back several decades, for which both the LOD and meteorological data indicate the occurrences of El Niños.

Numerical analyses of the LOD and atmospheric angular momentum time series indicate that approximately 90% of the nontidal changes in LOD at periods shorter than a few years are directly related to exchanges of angular momentum between the atmosphere and solid earth. The residuals display a 6-month periodic component, which appears to be linked to changes in the equatorial stratospheric jet stream, above the pressure levels routinely monitored. In limited tests, inclusion of this component of the atmospheric angular momentum reduced the

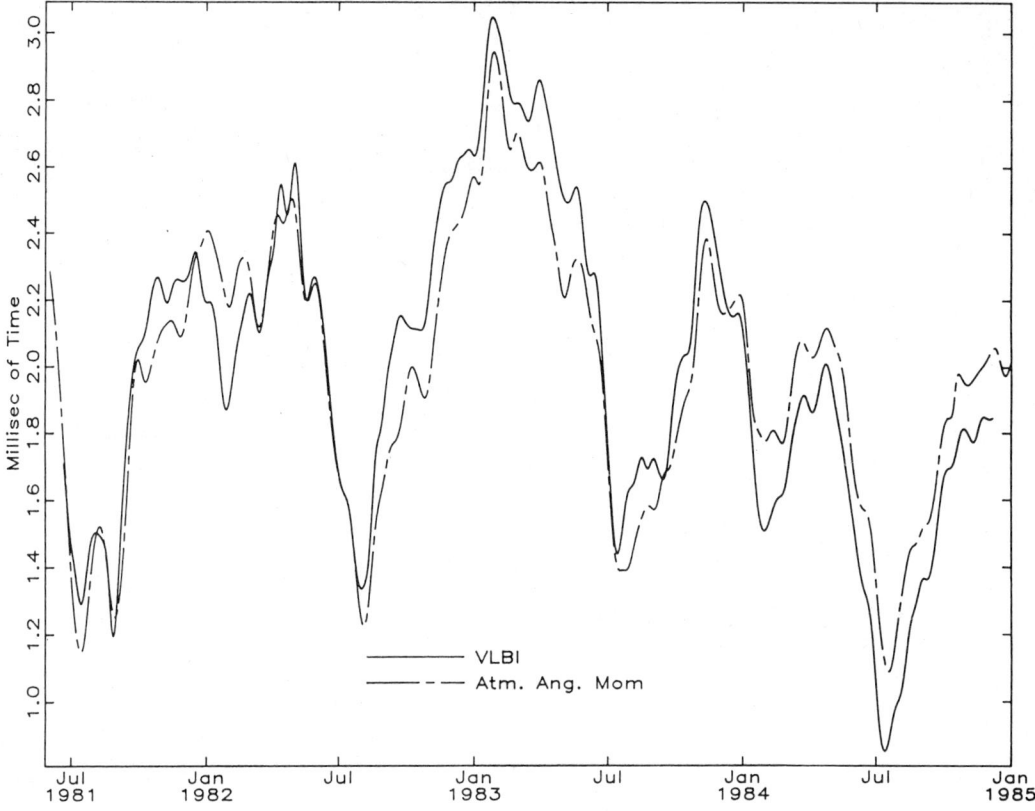

FIGURE 4. This plot shows the changes in the length of day observed by VLBI and as inferred from change in the atmospheric angular momentum.

differences between inferred and observed LOD to about 5%. Currently, in addition to the NMC, meteorological centers in Europe and Japan independently produce global wind and barometric pressure data sets, and the treatment of these data to infer changes in LOD is being refined.

Having found a high degree of correlation between observed changes in LOD and changes in atmospheric angular momentum, we can remove those effects and examine the "cleansed" time series for geophysical information. As was noted before, there are variations in the Earth's rotation at the luni-solar tidal frequencies. A simplified explanation of this effect is that the gravitational attractions of the Moon (and to a lesser extent the Sun) deform the Earth, raising solid Earth tides about a half a meter in height that move continually to stay aligned along the lines connecting the centers of the bodies. As the Earth tides oscillate back and forth across the equator, the effect is to change the Earth's ellipticity slightly, and to conserve angular momentum, the rate of rotation must change. The expected changes have been worked out for an elastic Earth with liquid oceans and core, and the phases and amplitudes of 80 components with periods ranging from a few days to 18.6 years computed. Forty-one components have periods shorter than 35 days, and are clustered in 3 groups near 9, 14, and 27 days. Figure 5 is a plot of the expected tidal variations in UT1 (the time integral of LOD) and the values observed by VLBI, after removing lower frequency components. While the observed and calculated variations clearly agree at the fraction of a millisecond level, there remain systematic discrepancies. The periods and phases of the Earth tide components are fixed by geometrical relationships that are quite well known. The amplitudes depend primarily on the elastic response of the mantle, expressed by the parameter k/C, but may be increased by a few percent by the anelastic response. In addition, ocean tides may cause out-of-phase variations at the same frequencies and ocean currents may contribute complex variations at seasonal and longer periods. Preliminary analysis of 2 years of daily UT1 values measured by VLBI indicate that the observational precision is adequate to detect the expected anelastic response of the mantle, but process noise degrades the determination by an order of magnitude or more. This may well be unmodeled oceanic effects, suggesting that better ocean models will have to be used, or perhaps even increased mon-

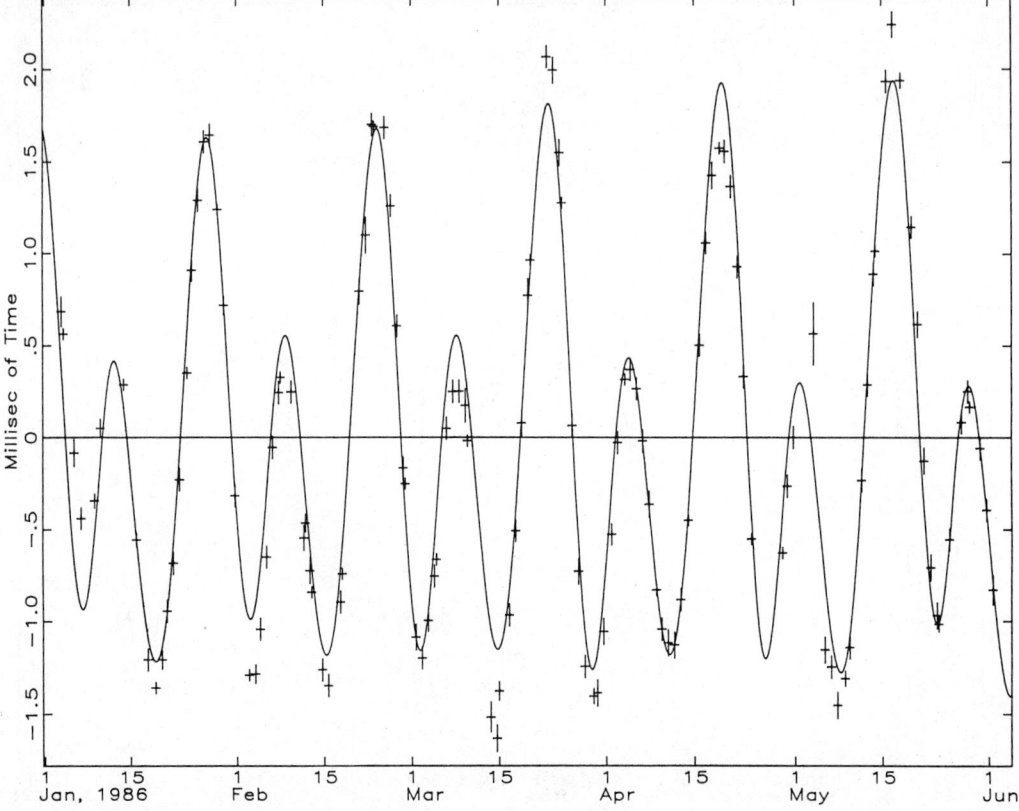

FIGURE 5. This plot shows the high frequency changes in the LOD as observed with VLBI, after removing the variations inferred from the changes in atmospheric angular momentum, along with the theoretical variations expected as a result of the solid-earth tides.

itoring of the oceans on a global scale (similar to that currently practiced with the atmosphere) may be required before the anelastic response of the mantle can be measured. Very little is known about the anelasticity of the mantle on time scales of weeks, and the promise that EOP data will eventually yield an accurate value is good news for geophysicists.

On interdecadal time scales and longer, core-mantle interactions are thought to cause changes in the LOD of several milliseconds, i.e., a few times as large as the seasonal variations. Comparisons of the observed LOD and atmospheric angular momentum time series show that the two track each other well for periods of years, and then rather suddenly veer apart for some period of time, suggesting some sort of episodic coupling mechanism. There have been attempts to correlate these changes in LOD with changes in the magnetic field of the Earth, which would indicate that they were caused by electromagnetic coupling of the core and mantle. It has also been suggested that the core-mantle boundary might have undulations, similar to the long relatively smooth undulations of the geoid. And very recently, researchers working with seismic tomography data, have even suggested that there may be more rugged terrainelike features (mountains) at the boundary. Such complexities at the core-mantle boundary could result in temporal changes in the coupling, producing changes in the rotation rate of the Earth.

At much longer periods, of the order of 10^5 years, the formation and melting of continental ice sheets cause changes in the Earth's rate of rotation. As water is withdrawn from the oceans and redeposited in the polar regions, the levels of the oceans drop by hundreds of meters. The relocation of mass from the equatorial to the polar regions results in an increase in the rate of rotation. At the same time, the weight of the ice pressing on the surface of the Earth causes mass in the mantle to be pushed toward the equator, partially compensating for the change in sea level. When the ice masses melt, which appears to take place in about one fifth of time for their formation, the effects are interchanged. Currently, of course, the ice sheets have been destroyed, and the mantle is still adjusting to the removal of the load. The northern regions are rebounding at rates as high as about 1 centimeter per year in some locations, while zones just south of the southernmost extent of the glaciers, including much of the United States, are subsiding by millimeters per year. This ongoing change in shape of the entire northern hemisphere, and indeed at some level the entire globe, has direct effects on the orientation the Earth, and has the secondary effect of changing the geometry of the network that defines the terrestrial reference frame.

At an even longer time scale is a variation in LOD caused by the transfer of energy and angular momentum between the spin of the Earth and the orbital motion of the Moon. This is a highly nonlinear phenomenon, but at the current configuration of the Earth-Moon system the interaction results in the Moon receding from the Earth, with the mean distance increasing by about 3.7 centimeters per year, while the Earth's spin rate is decreasing by approximately 6.3×10^{-22} radians per second^{-2}. The ability to directly measure the change in the distance to the Moon is an important contribution of LLR to separating long period changes in Earth rotation.

Concluding Remarks

The introduction of advanced technology has completely redefined the monitoring of Earth orientation during the past decade, and the immediate future promises still further progress. By the close of this century a network of 20 to 30 globally distributed VLBI observatories will define a terrestrial reference frame that will continually be maintained at the 1 cm level, properly accounting for the effects of plate tectonics, glacial rebound, earth tides, and ocean loading. For the first time a complete suite of EOP time series accurate to a fraction of a millisecond of arc, at intervals of a few hours, will be available for geophysical applications. This breakthrough in observational capabilities has already begun a revitalization of theoretical and analytical efforts, offering new hope of unraveling some of the mysteries about the internal structure and dynamics of Earth.

WILLIAM E. CARTER

References

Carter, W. E., D. S. Robertson, J. E. Pettey, B. D. Tapley, B. E. Schutz, R. J. Eanes, and Miao Lufeng, 1984, Variations in the rotation of the Earth, *Science* **224,** 957-961.

Carter, W. E., and D. S. Robertson, 1986, Studying the Earth by Very-Long-Baseline Interferometry, *Sci. American* **254,** 46-54.

Lambeck, K., 1980, *The Earth's Variable Rotation: Geophysical Causes and Consequences,* Cambridge, England: Cambridge University Press.

Moritz, H., and I. I. Mueller, 1987, *Earth Rotation: Theory and Observation.* New York: Unger Publishing Co.

Rosen, R. D., D. A. Salstein, T. M. Eubanks, L. O. Dickey, and J. A. Steppe, 1984, An El Nino signal in atmospheric angular momentum and earth rotation, *Science* **225,** 411-414.

Reid, M. J., and J. M. Moran, eds., 1988, *The Impact of VLBI on Astrophysics and Geophysics,* Proceedings of the 129th IAU Symposium 129, Cambridge, Mass., May 1987. Dordrecht, The Netherlands: Kluwer, 599 p.

Cross-references: *Core-Mantle Coupling; Earth Structure: Global; Equilibrium Figure of the Earth; Figure of the Earth; Mantle Convection and Plumes; Mantle Dynamics; Mantle Viscosity; Planetary Geodesy; Very-Long-Baseline Interferometry (VLBI).*

EARTHQUAKE MECHANISMS

Introduction

The most common model for the origin of crustal earthquakes is based on the "elastic rebound" theory as put forth in 1911 by Reid in his study of the 1906 San Francisco earthquake: Earthquakes occur in regions subject to deformation due to external causes, generally regional tectonic stresses. When the accumulated strain at some point exceeds the strength of the rock there is brittle fracture of competent rock and/or slip on preexisting zones of weakness. The region over which the slip or displacement occurred is called the earthquake *fault*. Before and after the earthquake the region is in static equilibrium. In accordance with this model, the estimate of the plane of faulting and the direction of slip within that plane inferred from seismic data is referred to as the *focal mechanism solution*. Focal mechanism solutions determined for several earthquakes in a geological region allow both the possible identification of large-scale faults and also some insight into the spatial geometry of the in situ stresses that cause the deformation.

This model is also applied to deep-focus earthquakes in subducting lithospheric slabs where "faults" in the conventional sense may not exist. However, a single point-force model for the focal mechanism, the *double-couple* model, appears applicable to all earthquakes [with the possible exception of some earthquakes related to volcanoes (see *Earthquakes: Volcanogenic*)].

In this chapter we develop the link between the earthquake source moment tensor introduced by Madariaga (see *Seismic Source: Theory*) and the applications of earthquake mechanisms to constrain tectonic plate motions as discussed by Stein and Woods (see *Earthquake Mechanisms and Plate Tectonics*). The emphasis here is on the methods used to obtain estimates of focal mechanisms from body wave polarity data, from body wave amplitude ratios, and from waveform fitting.

A Brief History

As early as the 1920s, some seismologists noticed that the first motions for P waves from earthquakes—up or down on vertical seismograms—showed systematic patterns with regard to the source-

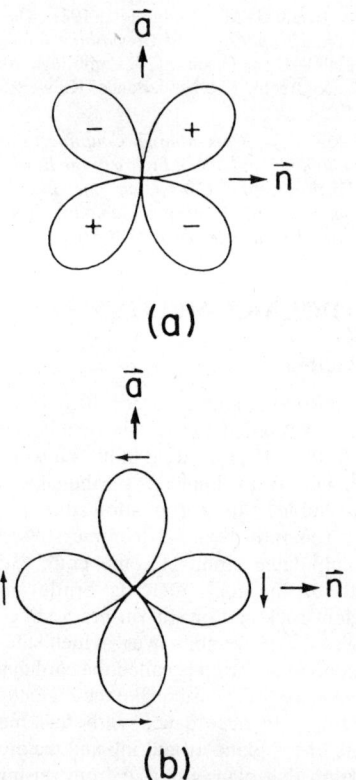

FIGURE 1. Radiation patterns for (a) P waves and (b) SH waves for a vertical strike-slip fault oriented as in Fig. 3. In (a) the + and − indicate the direction of first motion for the P arrivals with respect to the source, and the arrows in (b) indicate the first motion direction for SH.

station geometry. In particular, the first motions could be grouped into quadrants about the focus alternating between up motion (compressions) and down motion (dilatations) as shown in Fig. 1a. Many point-force equivalent models for the earthquake source process were compared with observation, and the *single-couple* model (often called a *type I source*) seemed consistent with the data and physically reasonable, based on the elastic rebound theory described earlier.

In addition to fitting P-wave first motion data, a successful source model must also be consistent with the observed S-wave particle motion. Because S waves are not first arrivals, S-wave first-motion data are generally inferior in quality to P-wave data. In addition, three-component seismographs are required to determine estimates of the S-wave particle motion. Hence it was not until the 1960s that there was an adequate data set of S-wave particle motion of sufficient quality to test source models. Once available, these data were consistent with the quadrupole pattern of the *double couple* (or *type II source*) as shown in Fig. 1b rather than the dipole pattern predicted by the single-couple model, ruling out the single couple as an appropriate source model. When the mathematical theory now used for source studies was developed, it was realized that a single couple was not consistent with static equilibrium before and after the earthquake—the same logic that leads to the stress tensor being symmetric. Benioff (1964) showed that the double-couple model, rather than the single-couple model, was implied by a *correct* interpretation of the classic elastic rebound theory.

Good reviews of the earliest days of earthquake mechanisms studies are given by Stauder (1962) and Honda (1962).

The use of P-arrival polarities remains today the most common way to constrain the focal mechanism for a given event. S-wave particle motion, in the form of polarization angle estimates, has been supplanted by more robust methods. Among the methods currently used are polarity comparisons for teleseismic observations of P with pP and/or sP, amplitude ratios from among pP/P, sP/P, SV/P, or SH/P, and inversion schemes based on waveform fitting. After a review of the underlying theory, these methods are described.

Theory

The ground displacement vector from an earthquake can be written as the convolution of a moment tensor and a gradient of a Green's function (see *Seismic Source: Theory*). The Green's function includes propagation effects, and the moment tensor contains information about the source process. The moment tensor can be interpreted as the volume integral of the stress drop associated with the earthquake. The trace of the moment tensor represents an isotropic source, such as an explosion. Since no volume change is observed with normal earthquakes, one generally restricts consideration to traceless moment tensors. (Removing the trace is a linear constraint and is therefore easily incorporated in an inversion procedure.) The resulting point-source moment tensor includes more general force systems than a pure double couple, but for now we consider only a pure double-couple source model. A double-couple model corresponds to a point-source moment tensor of the form

$$M(t)_{jk} = M_0(t) [n_j a_k + n_k a_j] \qquad (1)$$

where $M_0(t)$ is the total seismic moment (see *Seismic Source: Theory*), and a_j and n_k are respectively the jth and kth components of $\hat{a}$, a unit vector in the direction of the fault slip, and $\hat{n}$, a unit vector normal to the fault plane (Fig. 2).

To see explicitly the connection between the fault geometry and the equivalent force picture for an earthquake, we evaluate Eq. 1 in two different coordinate frames. For the coordinate frame in which **a** lies along **x** and **n** along **y**, the right-hand side of Eq. 1 becomes $M_0(t) [\delta_{j1}\delta_{k2} + \delta_{j2}\delta_{k1}]$, where δ_{jk} is

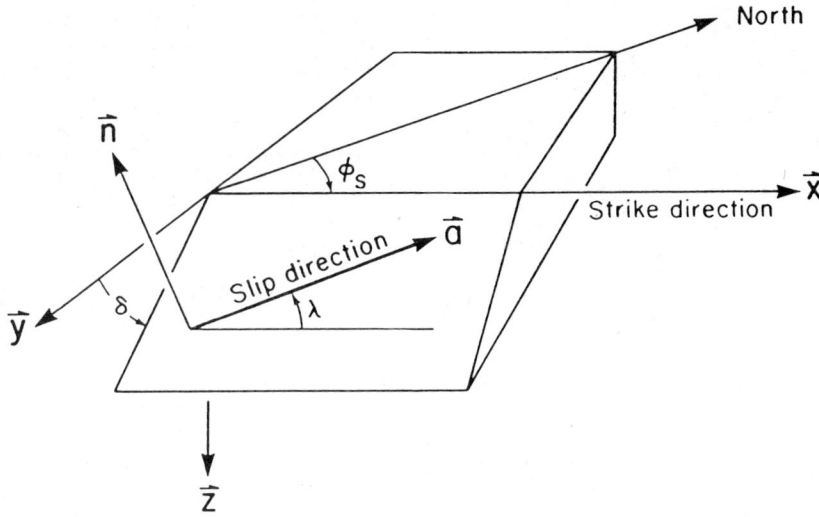

FIGURE 2. Schematic of the foot wall side of a fault showing the conventions for the fault parameters: strike (ϕ_S), dip (δ), rake (λ), and slip direction **a**. The normal to the fault place is **n**. (Adapted from Aki and Richards, 1980, Fig. 4.13)

unity for $j = k$ and 0 otherwise. As shown in Fig. 3a, in this coordinate frame the stresses implied by the moment tensor take the form of a double couple with one couple representing the slip along **a**. If now we represent the moment tensor in a coordinate system rotated clockwise by 45° about **z**, the right-hand side of Eq. 1 is of the form $M_0(t) [\delta_{j1}\delta_{k1} - \delta_{j2}\delta_{k2}]$, a purely diagonal traceless tensor with an equivalent force representation as shown in Fig. 3b. P waves are longitudinally polarized, so the direction of first motion for P arrivals will be away from the source (compressions) in the NE and SW quadrants and toward the source (dilatations) in the NW and SE quadrants. The outward-pointing arrows in the compressional quadrants are along **T**, the *tension axis*, and the inward-pointing arrows in the dilatational quadrants are along **P**, the *pressure axis*. The direction orthogonal to both **P** and **T** is **B**, the *null axis*, which is along the direction within the fault plane that is perpendicular to the slip direction. **B** is also orthogonal to **a** and **n**.

There are three equivalent conventions for defining the three parameters that specify a unique fault plane solution. Two conventions involve specifying the directions of vectors on the *focal sphere*, a conceptual sphere of unit radius centered at the point source. These two parameterizations are to specify the trend and plunge either of **â** and **n̂** or of **P** and **T**. (Orthogonality provides one constraint condition, so there are indeed only three independent parameters in each case.) The **P**, **T** parameterization is the one most used in the analysis of deep-focus earthquakes for which the local stresses are modeled as tensional or compressional (see *Earthquake Mechanisms and Plate Tectonics*). The third way is to specify the dip (δ) and strike (ϕ_s) of the fault plane and the rake (λ) associated with the direction of motion in that place (Fig. 2). There are several conventions for these angles, but the current standard is the one used by Herrmann (1975) and Aki and Richards (1980, p. 106). A fault has two surfaces: the lower surface, shown in Fig. 2, is the *foot wall*, and the upper surface, not shown, is the *hanging wall*. The strike is the azimuth of the fault with the convention that if one faces downdip, the strike direction is to the left. The dip is measured down from the horizontal and is bounded by 0° and 90°. The rake angle is measured within the fault plane and is bounded by $-180°$ and $+180°$ measured from the strike direction. If $0° < \lambda < +180°$, the fault is called a *reverse fault* or *thrust fault;* if $-180° < \lambda < 0°$, the fault is referred to as a *normal fault*. If $\lambda = +90°$ or $\lambda = -90°$, the fault is called *dip-slip*. A *right-lateral* fault is one for which an observer standing on one side of the fault sees the block on the other side move to the right. If $\lambda = 0°$, the fault is *left-lateral strike-slip* (as in Fig. 3); if $\lambda = +180°$, the fault is designated *right-lateral strike-slip*. Herrmann (1975) includes a complete discussion about relationships among these different parameterizations.

As a consequence of the symmetry of **â** and **n̂** in Eq. 1, there is no way to distinguish on the basis of far-field data between the above interpretation and one in which the roles of **â** and **n̂** are reversed. The plane normal to **â** is referred to as the *auxiliary fault plane*. The decision as to which is the "real" fault plane requires other kinds of data, such as correlation with other events in the region, location on a known fault, and/or assumptions about the regional stress.

If the far-field displacement is evaluated for the

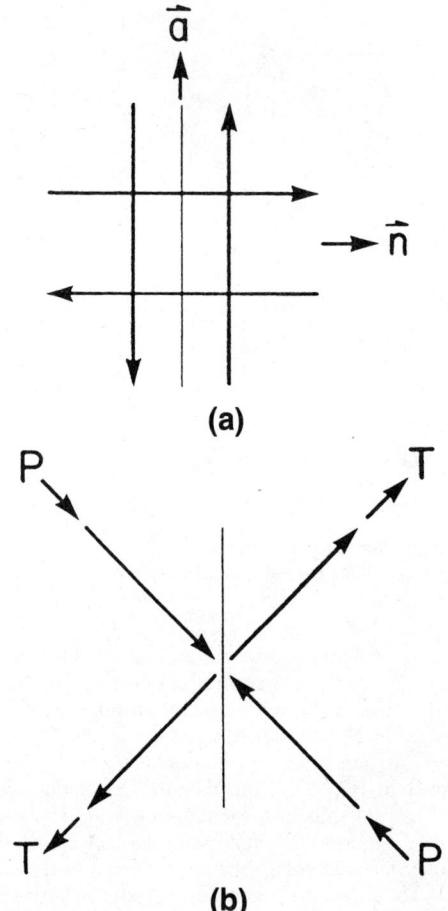

FIGURE 3. Two representations of the double-couple point source model for a vertical strike-slip fault (**z** is into the paper): (a) double couple corresponding to the moment tensor evaluated in a coordinate frame such that **x** lies along **a**, the slip direction, and **y** lies along **n**, the normal to the fault plane; (b) normal forces corresponding to the same moment tensor in a frame rotated by 45° about **z**. **T** is the tension axis, and **P** is the pressure axis.

moment tensor in Eq. 1 [as in Aki and Richards (1980, Chapter 4)], the far-field displacement, $\mathbf{u}_{ff}$, evaluated in an Earth-based coordinate system on the focal sphere can be written in the form $\hat{\mathbf{u}}_{ff} = u_P \hat{\mathbf{r}} + u_{SV} \hat{\boldsymbol{\theta}}, + u_{SH} \hat{\boldsymbol{\phi}}$, where r, θ, ϕ are spherical coordinates in a reference frame such that **z** points down and **x** is along the strike direction (so that $\phi = \phi_A - \phi_s$, where ϕ_A is the azimuth). The unit vectors $\hat{\mathbf{r}}$, $\hat{\boldsymbol{\theta}}$, and $\hat{\boldsymbol{\phi}}$ are an orthonormal set evaluated at the point (r, θ, ϕ) with directions along increasing values of the variables. There is inconsistency in the literature regarding the convention for "positive" SH and SV (just as there is inconsistency as to whether vertical "up" or "down" is positive). According to the convention used here, for an observer with his or her back to the source and facing the station, SV is positive up and backwards, and SH is positive to the right. Expressions for u_P, u_{SV}, and u_{SH} are then of the following form: $u_P = (v_S/v_P)^3 F R_P$, $u_{SV} = F R_{SV}$, and $u_{SH} = F R_{SH}$, where v_P and v_S are, respectively, the P and S velocities at the source, F is a common factor independent of angle, and the R's are normalized radiation factors of the form (e.g., Aki and Richards, 1980, pp. 113–114)

$$R_P = r_j M_{jk} r_k = 2(\hat{\mathbf{a}} \cdot \hat{\mathbf{r}})(\hat{\mathbf{n}} \cdot \hat{\mathbf{r}})$$

$$R_{SV} = r_j M_{jk} \theta_k = (\hat{\mathbf{a}} \cdot \hat{\boldsymbol{\theta}})(\hat{\mathbf{n}} \cdot \hat{\mathbf{r}}) + (\hat{\mathbf{n}} \cdot \hat{\boldsymbol{\theta}})(\hat{\mathbf{a}} \cdot \hat{\mathbf{r}})$$

$$R_{SH} = r_j M_{jk} \phi_k = (\hat{\mathbf{a}} \cdot \hat{\boldsymbol{\phi}})(\hat{\mathbf{n}} \cdot \hat{\mathbf{r}}) + (\hat{\mathbf{n}} \cdot \hat{\boldsymbol{\phi}})(\hat{\mathbf{a}} \cdot \hat{\mathbf{r}})$$

The predicted P-wave and SH-wave radiation patterns for the fault plane orientation shown in Fig. 3 are given in Fig. 1. In this case the projection of the SV radiation pattern is identical in shape to the P radiation pattern, but it has a different dependence on θ. The surface upon which the radiation for a given wave type is zero is called a *nodal surface*. For a vertical strike-flip fault the nodal surfaces are planes; for P and SV the nodal planes are the fault and auxiliary planes; for SH the nodal planes are those containing the **B**-axis (**z** in this case) and the **P**- and **T**-axes. For P radiation, the nodal surfaces are always planes; for SV and SH the nodal surfaces are generally not planes. As can be seen from the example discussed later and shown in Fig. 4, the stereographic representation of the SV or SH nodal surfaces is much more complex than the projection of the P nodal planes.

Solutions from Polarities

The method to obtain estimates of the focal mechanism from observed polarities is as follows:

1. For each observed polarity one determines the azimuth ϕ_A and the takeoff angle with the downward vertical θ. Before about 1970, θ estimates were gotten from tables; now θ is usually calculated on the computer by a ray-tracing program. (Many earthquake location programs include θ and ϕ_A in the output.)
2. Symbols representing compression or dilatation are plotted for each station on a planar projection of the focal sphere. The first studies of focal mechanisms used teleseismic data, so the rays leading to the stations started downward. Hence the convention is to use a lower hemisphere projection. Geologists generally use a Wulf projection because it preserves angles, but for the teleseismic data such a projection complicates interpretation because the plot becomes too cluttered near the center. Accordingly, the equal-area Lambert-Schmidt projection is used. For this projection, if the takeoff angle θ is less than 90°, the (r, ϕ) position for a polarity is $\phi = \phi_A$ and $r = \sqrt{2} r_0 \sin(\theta/2)$, where r_0 is the radius of the circle. If $\theta > 90°$, $\phi = 180° + \phi_A$ and $r =$

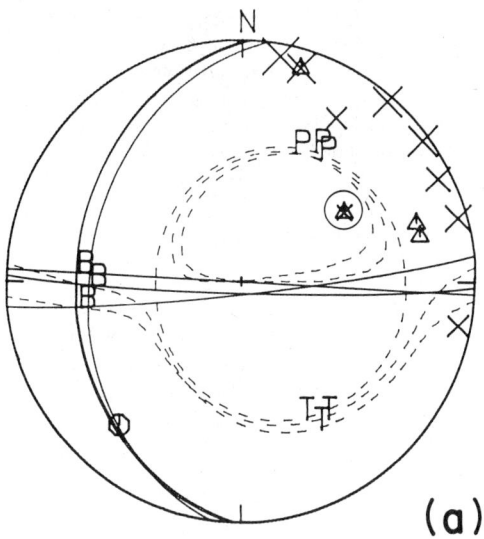 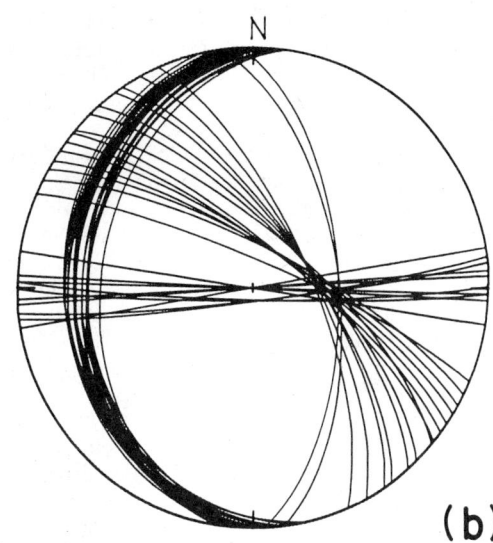

FIGURE 4. Data and focal mechanism solutions for an $m = 2.2$ earthquake recorded on vertical seismometers: (a) triangles represent dilatations, circles represent compressions, and the X sizes are proportional to the log of the $(SV/P)_z$ ratios. Solid lines are P nodal plane projections, and dashed lines are SV nodal surface projections for solutions for which no polarity errors are allowed and for which the ratio misfit is less than 70%; (b) P nodal plane projections for solutions for which no polarity errors are allowed, but one ratio misfit is allowed to be greater than 70%. The station polarity ratio circled in (a) is in error for the solutions in (b) that have one nodal plane poorly constrained.

$\sqrt{2}\, r_0 \sin(90 - \theta/2)$. (For a point source, rays reflected through the origin have identical properties.) Figure 6 in *Earthquake Mechanisms and Plate Tectonics* shows examples of focal sphere projections for strike-slip and thrust fault mechanisms. (The compressive quadrants are shaded.)

3. Determine the best-fit focal mechanisms. Previously this was done by picking visually the perpendicular nodal planes that separate the dilatational and compressional quadrants. Now possible solutions are generally calculated by computer programs that systematically search the focal sphere for all possible solutions consistent with the data. The input typically includes the polarity data and the search criteria: the number of "acceptable" errors, the region of the focal sphere to be searched, and the fineness of the grid. The output includes all solutions that fit the criteria.

An extension of the above procedure is used when polarity data are available from several events in a given region. In this procedure the data are combined and a *composite solution* is obtained. Such solutions are particularly useful for attempts to infer regional stresses from focal mechanisms.

Solutions from Ratios and from Waveform Fitting

If the station coverage were dense and uniform in both distance and azimuth for a given event, observed direct-arrival first motions would tightly constrain the event's focal mechanism. Generally this is not the case, but there is considerably more information derivable from seismograms which can potentially be used. For example, it follows from the above discussion that comparisons between predicted and observed amplitude ratios from among P, SV, and SH arrivals provide possible constraints. For many events data are only from vertical seismometers, which still allows observations of $(u_{SV}/u_P)_z$. A disadvantage of this ratio is that SV is often not well recorded on the vertical and it is the most affected by earth structure due to S-to-P converted energy. However, for small earthquakes with observations at local or regional distances, these ratios have proved to be a useful supplement to polarity data in many studies. For discussions of $(SV/P)_z$, see Kisslinger (1980, 1982). Other researchers have used synthetics to aid in the identification of depth phases such as pP and sP at teleseismic distances and then used amplitude ratios and relative polarities between them or the direct SH and the direct P to constrain the mechanisms (e.g., Langston et al., 1982).

Waveform-fitting procedures to find focal mechanisms include implicit polarity and ratio comparison techniques, and they are the only methods available for the analysis of complex events. The development and implementation of these procedures have grown considerably in the last decade because of the advent of broadband three-component digital data, refined estimates of local and regional velocity structures, and advances in synthetic seismograms (see *Seismograms: Synthetic*). Waveform-fitting procedures to invert for focal mechanisms have become almost routine for large ($m_B \geq 5.5$) events and are increasingly used for smaller events as well.

For large earthquakes recorded at teleseismic distances the best-quality data have been long-period seismograms. Such data are easier to model than are short-period data because long-period waves are less sensitive to source and receiver site effects, attenuation, and details of the rupture process. Green's functions for synthetics have been constructed using normal mode theory (e.g., Dziewonski et al., 1981), WKBJ theory (e.g., Sipkin, 1982), and generalized ray theory (e.g., Langston et al., 1982).

As discussed in detail by Sipkin (1982), there are two different inversion procedures commonly used: the construction method and the appraisal method. Both methods start with the same set of equations: Each far-field displacement component at each station is written as linear superpositions from among five terms (assuming a purely deviatoric moment tensor), where the terms are proportional to linear combinations of elements of the *moment rate tensor*—the time derivative of the moment tensor (see *Seismograms: Synthetic* and Aki and Richards, 1980, p. 79). For the construction method, one varies the relative weighting of the moment rate tensor terms to optimize the waveform fit at all stations and for all components simultaneously. Solutions have been evaluated by a systematic grid search, by a least squares inversion procedure, or by some combination of the two. In the appraisal method, which utilizes linear inverse theory, filters are constructed that, when convolved with the time series data, yield unique time averaged estimates of the individual elements of the moment rate tensor. A disadvantage of this method is that all three components cannot be used simultaneously in the formal inversion, because SH is decoupled from P and SV in the far field. The appraisal method, as developed by Dziewonski et al. (1981), is applied routinely by the National Earthquake Information Center in preparing its preliminary determinations of epicenters.

The outputs from both the construction and appraisal methods are estimates of elements of the moment rate tensor. The necessary condition that a traceless moment rate tensor correspond to a single double-couple point source is that one of the principle values be zero. Such a condition cannot be included as a linear constraint in the inversion process, so the moment rate tensor found by inversion will in general not correspond to a single double-couple. As Dziewonski et al. (1981, p. 2837) point out, there is no unique way to project out a double-couple solution from a purely deviatoric moment tensor; the method currently used by the National Earthquake Information Center is to identify **T** with the maximum eigenvalue and **P** with the minimum eigenvalue.

As noted by Sipkin (1982), focal mechanisms found from the inversion of entire waveforms correspond to averages over the spatiotemporal dimensions of the source, so the results may not be entirely consistent with solutions implied by first motions that include information about only the initiation of rupture.

Reliability of Focal Mechanisms

The reliability of focal mechanisms inferred from polarity and amplitude ratio data depends on the quality of the data. Aspects to be considered include the reliability of the individual polarities and/or amplitudes themselves, the accuracy of the location of the event, the correctness of the assumed velocity structure, and the distribution and density of the data on the focal sphere. Arrivals may be impulsive or emergent, and for emergent arrivals the pick of a direction for the initial motion may be in error. A low signal-to-noise ratio, particularly on short-period records, may cause errors in polarity picks, as may small amplitudes caused by the proximity to a nodal surface. If the station coverage of the focal sphere is not uniform, the nodal surfaces may be poorly constrained. In addition to the problems inherent in polarity data, ratio data may be in error because of interference among arrivals and/or inaccurate anelasticity or free surface corrections. Errors in hypocenter estimates will shift the positions of the stations on the focal sphere and will accordingly distort the patterns in the projected data. This can be quite important for stations located near nodal surfaces. Changing the velocity structure at the source has a usually small but systematic effect on the takeoff angles. For crustal events observed at regional distances, using a velocity structure composed of constant-velocity layers (a practice commonly used in locating earthquakes), the takeoff angles will appear unnaturally "quantized" at the critical refraction takeoff angles for each boundary below the hypocenter. This feature can be removed by using gradient velocity models.

In some currently available focal mechanism computer programs the user may specify an allowed number of polarity errors and an allowed range in amplitude ratio error for a solution to be acceptable. Such programs may also allow for weighted errors, depending upon the proximity to the predicted nodal surface, the logic being that errors far from the nodal surface should be more significant. One generally plots all acceptable solutions on a single plot: P nodal planes and/or projections of the **P**-, **T**-, and **B**-axes. The range in acceptable solutions will then show—qualitatively at least—how well constrained the solutions are.

The effects of sparse and nonuniform coverage can be seen from the following example. Figure 4 shows polarity and ratio data along with inferred focal mechanisms for a small ($m_b = 2.2$) Virginia earthquake recorded on vertical seismometers in the Virginia Tech Seismic Network. There are five reliable P polarities and nine $(SV/P)_z$ ratios, and the coverage of the focal sphere is far from uniform. Because four of the polarities are dilatations, the focal mechanism would be very poorly constrained on the basis of polarities alone. Shown in Fig. 4a are the P nodal planes, the SV nodal surfaces, and the projections of the **P**-, **T**-, and **B**-axes for a grid search of the focal sphere with no allowed polarity or ratio errors (where a ratio error is here defined as a ratio misfit greater than 70%). The three solutions shown are all essentially the same. Allowing one polarity error but no ratio errors produces no significantly new solutions. Figure 4b shows the P nodal planes for no polarity errors but one ratio error. Although one nodal plane remains well constrained, the other is now almost unconstrained. For this station event geometry, the fact that one observed ratio (circled in Fig. 4a) is very small requires that it lie near an SV nodal surface. Allowing that ratio to be in error removes that constraint and allows for many different solutions that are consistent with the data. A (qualitative) interpretation is that one nodal plane is well constrained, but the other one is only marginally constrained.

At present there is no standard method to assess quantitatively the reliability of a solution. The discrete nature of polarity data precludes the use of least squares inversion procedures [as in earthquake locations (see *Earthquakes: Location Techniques*)]. Kisslinger (1980) employs a nonlinear least squares inversion procedure for the dip, strike, and rake when only amplitude ratio data (which are continuously varying) are used, which allows him to specify confidence regions for these angles. Julian (1986) shows how linear programming procedures can be applied to solve for focal mechanisms; such procedures can include both continuous and discrete data. Other researchers have formulated probability models in which a likelihood function can be constructed based on polarity and/or S-wave polarization data. Such a formulation allows the calculation of confidence regions for focal mechanisms, but the implicit assumptions in such a procedure are most likely to be valid when there are many observations.

The reliability of focal mechanisms inferred from waveform fitting is affected by all the factors described here, except that errors in focal depth and in the assumed near-station velocity structure are more important than in the other inversion schemes. Only through waveform fitting is it possible to deal with events that are too complex to be approximated adequately by a point source (see *Seismic Source: Observations*). For solutions found by the appraisal method, confidence regions for the solutions can be obtained directly. For focal mechanisms found from waveform fitting by the construction method, a 5 × 5 covariance matrix can be obtained for the elements of the deviatoric moment rate tensor, and then estimates of the variances in the focal mechanisms can be calculated.

J. A. SNOKE

References

Aki, K., and P. G. Richards, 1980, *Quantitative Seismology: Theory and Methods,* San Francisco: W. H. Freeman, 932 p.

Benioff, H., 1964, Earthquake source mechanisms, *Science* **143,** 1399-1406.

Dziewonski, A. M., T. A. Chou, and J. H. Woodhouse, 1981, Determination of earthquake source parameters from waveform data for studies of regional and global seismicity, *Jour. Geophys. Research* **86,** 2825-2852.

Herrmann, R. B., 1975, A student's guide to the use of P and S wave data for focal mechanism determination, *Earthquake Notes* **46,** 29-39.

Honda, H., 1962, Earthquake mechanism and seismic waves, *Jour. Physics Earth* **10,** 1-97.

Julian, B. R., 1986, Analyzing seismic-source mechanisms by linear programming methods, *Royal Astron. Soc. Geophys. Jour.* **84,** 431-443.

Kisslinger, C., 1980, Evaluation of S to P amplitude ratios for determining focal mechanisms from regional network observations, *Seismol. Soc. America Bull.* **70,** 999-1014.

Kisslinger, C., 1982, Errata, *Seismol. Soc. America Bull.* **72,** 344.

Langston, C. A., J. S. Barker, and G. B. Pavlin, 1982, Point-source inversion techniques, *Physics Earth Planetary Interiors* **30,** 228-241.

Sipkin, S. A., 1982, Estimation of earthquake source parameters by the inversion of waveform data: synthetic waveforms, *Physics Earth Planetary Interiors* **30,** 242-259.

Stauder, S.J., W., 1962, The focal mechanism of earthquakes, *Advances in Geophysics* **9,** 1-72.

Cross-references: *Earthquake Mechanisms and Plate Tectonics; Earthquakes: Hazards and Prediction; Earthquake Seismology; Seismic Monitoring of Nuclear Explosions; Seismic Source: Observations; Seismic Source: Theory.*

EARTHQUAKE MECHANISMS AND PLATE TECTONICS

Earthquake seismology has played a major role in the development of our current understanding of global plate tectonics. Because earthquakes occur primarily at the boundaries between lithospheric plates, their distribution is used to map plate boundaries and their focal mechanisms provide information about the motion at individual boundaries.

Plate boundaries are divided into three types (Fig. 1). Oceanic lithosphere is formed at *spreading centers,* or mid-ocean ridges, and is destroyed at *subduction zones,* or trenches. Thus, at spreading centers plates move away from the boundary, whereas at subduction zones the subducting plate moves toward the boundary. At the third boundary type, *transform faults,* plate motion is parallel to the

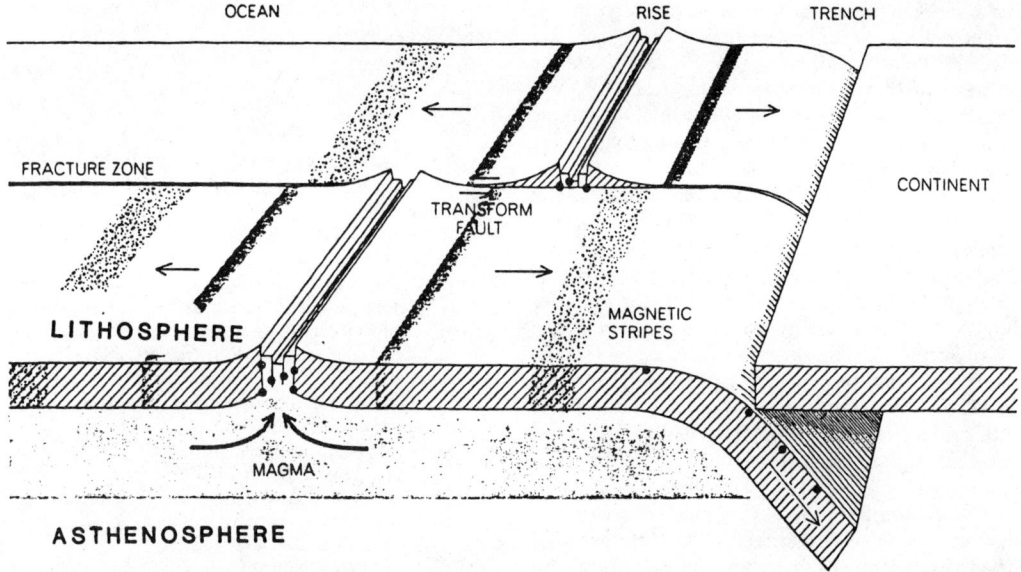

FIGURE 1. Plate tectonics at its simplest. Plates are formed at ridges and subducted at trenches. At transform faults, plate motion is parallel to the boundaries. Each boundary type has characteristic earthquakes.

boundary. The *slip vectors* of the earthquakes on plate boundaries, which show the motion on the fault plane, reflect the direction of relative motion between the two plates.

The basic principle of plate kinematics is that the relative motion between any two plates can be described as a rotation on a sphere about an *Euler pole* (Fig. 2). Specifically, at any point along the boundary between plates i and j, with latitude λ and longitude μ, the linear velocity of plate j with respect to plate i is

$$\mathbf{v}_{ji} = \boldsymbol{\omega}_{ji} \times \mathbf{r}$$

the usual formulation for rigid body rotations in mechanics. The vector $\mathbf{r}$ is the position vector to the point on the boundary, and $\boldsymbol{\omega}_{ji}$ is the *Euler vector*. Both are defined from an origin at the center of the Earth.

The direction of relative motion at any point on a plate boundary is a small circle, a parallel of latitude *about the Euler pole* (not a geographic parallel about the North Pole!). For example, in Fig. 3a the pole shown is for the motion of plate 2 with respect to plate 1. The first-named plate ($j = 2$) moves counterclockwise about the pole with respect to the second plate ($i = 1$). The segments of the boundary where relative motion is parallel to the boundary are transform faults. Thus, transforms are small circles about the pole, and earthquakes occurring on them should have pure strike-slip mechanisms. Other segments have relative motion away from the boundary and are thus spreading centers. Figure 3b shows an alternative case. The pole here is for plate 1 ($j = 1$) with respect to plate 2 ($i = 2$), so plate 1 moves toward some segments of the boundary, which are subduction zones. Note that the ridge and subduction zone boundary segments are not small circles.

The magnitude, or rate, of relative motion increases with distance from the pole, since

$$|\mathbf{v}_{ji}| = |\boldsymbol{\omega}_{ji}| \, |\mathbf{r}| \sin \gamma$$

where γ is the angle between the Euler pole and the site (corresponding to a colatitude about the pole). Thus, although all points on a plate boundary have the same angular velocity, the linear velocity varies.

If we know the Euler vector for any plate pair, we can write the linear velocity at any point on the boundary between the plates in terms of the local EW and NS components by a coordinate transformation. With this, the rate and azimuth of plate motion become

$$\text{Rate} = |\mathbf{v}_{ji}| = \sqrt{(\mathbf{v}_{ji}^{NS})^2 + (\mathbf{v}_{ji}^{EW})^2}$$

$$\text{Azimuth} = \tan^{-1}\left(\frac{(\mathbf{v}_{ji}^{NS})}{\mathbf{v}_{ji}^{EW}}\right)$$

where azimuth is measured in degrees counterclockwise from east (right-handed rotation).

Given a set of Euler vectors with respect to one plate, those with respect to others are found by vector arithmetic. For example, the Euler vector for the reverse plate pair is the negative of the Euler vector ($\boldsymbol{\omega}_{ij} = -\boldsymbol{\omega}_{ji}$). Euler vectors for other plate pairs are found by addition:

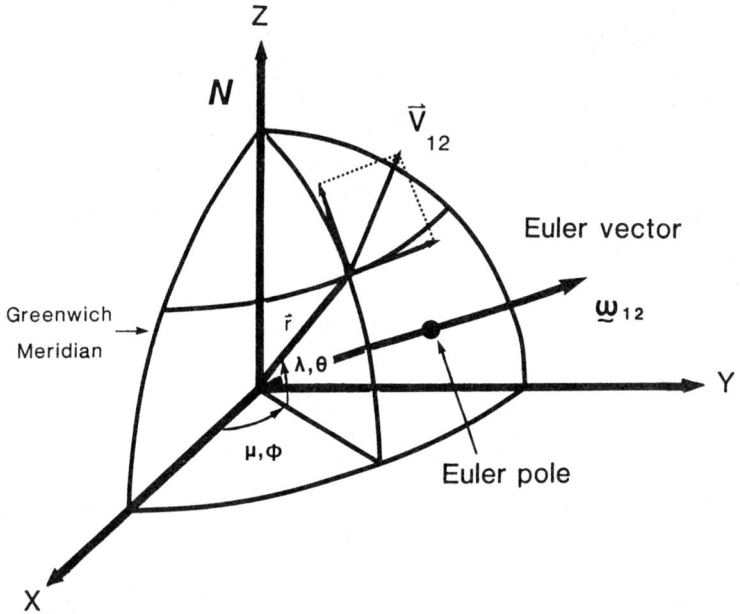

FIGURE 2. Geometry of plate motions. Linear velocity at point **r** is given by $\mathbf{v}_{ji} = \boldsymbol{\omega}_{ji} \times \mathbf{r}$. The Euler pole is the intersection of the Euler vector with the Earth's surface. (After Minster et al., 1974)

$$\omega_{jk} = \omega_{ji} + \omega_{ik}$$

So, given a set of vectors all with respect to plate i, any Euler vector needed is found from

$$\omega_{jk} = \omega_{ji} - \omega_{ki}$$

For further information on plate kinematics see an introductory text like Cox and Hart (1986) or Cox's (1973) collection of early papers on the subject.

Motions between plates are determined by combining different types of data from different boundaries. The rate of spreading at ridges is given by seafloor magnetic anomalies, and the directions of motion are found from the orientations of transform faults and the slip vectors of earthquakes on transforms and at subduction zones. As is evident, earthquake slip vectors are only one of three types of plate motion data available. Euler vectors are determined from the relative motion data by using geometrical conditions. Since slip vectors and transform faults lie on small circles about the pole, the pole must lie on a line at right angles to them (Fig. 3). Similarly, the rates of plate motion increase with the sine of the distance from the pole. These constraints make it possible to locate the poles.

We only have certain types of data for individual boundaries. Although spreading centers provide rates from the magnetic anomalies and azimuths from both transform faults and slip vectors, only the direction of motion is directly known at subduction zones. As a result, the rate of convergence at subduction zones is estimated by combining data from different plate pairs. For example, the rate of Cocos Plate subduction beneath North America, which gives rise to the large earthquakes in Mexico, is partially determined from the rates of Cocos–Pacific motion on the East Pacific Rise and Pacific–North America spreading in the Gulf of California. Determination of Euler vectors for all the plates can thus be treated as an overdetermined least squares problem, and the best solution is found by using the generalized inverse (Chase, 1972; Minster et al., 1974). Figure 4 shows schematically which boundaries contribute which type of data and how uncertainties in knowledge of the motion on one boundary contribute to uncertainties on the others.

Oceanic Spreading Center Focal Mechanisms

Earthquake mechanisms from the mid-ocean ridge system reflect the spreading process. Figure 5 schematically shows a portion of a spreading ridge offset by transform faults. Since new lithosphere forms at the ridges and then moves away, the relative motion of lithosphere on either side of a transform is in opposing directions. The direction of transform offset, and not the spreading direction, determines whether there is right- or left-lateral motion on the fault. This relative motion, defined as *transform faulting*, is not what produced the offset of the ridge crest. In fact, if the spreading at the ridge is symmetric (equal rates on either side), the length of the transform will not change with time. This is a very different geometry from a *transcurrent fault*,

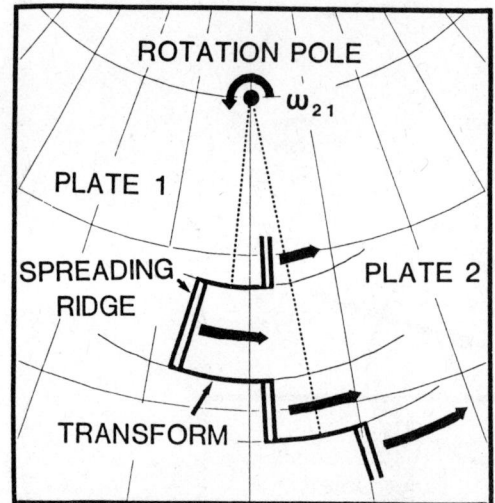

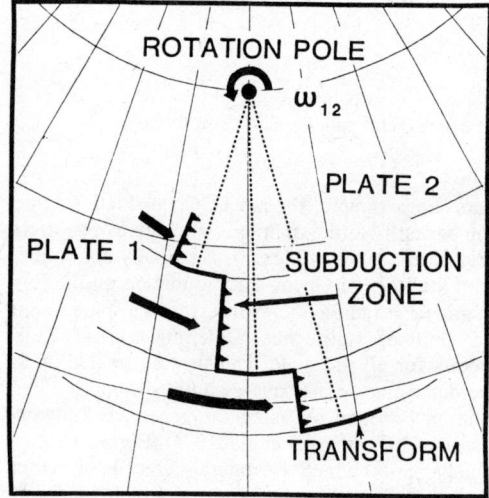

FIGURE 3. Relationship of motion on plate boundaries to the Euler pole. Relative motion occurs on small circles; the rate increases with distance from the pole. Note the difference the sense of rotation makes: ω_{ji} is the Euler vector corresponding to the rotation of plate j counterclockwise with respect to i.

where the offset is produced by motion on the fault, and the length of the offset between ridge segments would increase with time.

The model is illustrated by focal mechanisms. Figure 6 (top) shows a portion of the Mid-Atlantic Ridge composed of north-south-trending ridge segments, offset by transform faults, such as the Vema Transform, which trend approximately east-west. Both the ridge crest and the transforms are seismically active. The mechanisms show that the relative motion along the transform is right-lateral. Seafloor spreading must be occurring on the ridge segments to produce the observed relative motion. For this reason, earthquakes occur almost exclusively on the active segment of the transform fault between the two ridge segments, rather than on the inactive extension, known as a *fracture zone*. Although no relative plate motion occurs on the fracture zone, it is frequently marked by a distinct topographic feature. Unfortunately, some transform faults that were named before this distinction became clear are known as fracture zones along their entire length. Earthquakes also occur on the spreading segments. Their focal mechanisms show normal faulting, with nodal planes trending along the ridge axis.

The seismicity is different on fast spreading ridges. Figure 6 (bottom) shows a portion of the Pacific-Antarctic boundary along the East Pacific Rise. Here, strike-slip earthquakes occur on the transforms, but we do not observe the ridge crest normal faulting events. These observations can be explained by the thermal structure of the lithosphere.

The mechanisms are consistent with the predictions of plate kinematics. The area in Fig. 6 is a portion of the boundary between the South American and Nubian (West African) Plates. An Euler vector for Nubia with respect to South America with a pole at 62°N, 37.8°W and a magnitude of 0.328° per million years predicts that at 0°N, 20°W Africa is moving N81°E, or almost due east, at 3.3 cm·yr^{-1} with respect to South America. The Vema is a boundary segment parallel to this direction, and so is a transform fault characterized by strike-slip earthquakes with directions of motion along the trace of the transform. The short segments at essentially right angles to the direction of relative motion are then spreading ridge segments. The spreading rate determined from magnetic anomalies, and thus the slip rate across the transform, is that predicted by the Euler vector.

Additionally, oceanic ridge seismicity places constraints on the thermal structure of the lithosphere.

Subduction Zone Focal Mechanisms

Both the largest earthquakes and the majority of large earthquakes occur at subduction zones. Their focal mechanisms reflect various aspects of the subduction process. Figure 7 is a composite cartoon showing some of the features observed in different subduction zones.

Most of the large, shallow, subduction zone earthquakes indicate thrusting of the overriding plate over the subducting lithosphere. The best such examples are the two largest earthquakes ever recorded: the 1960 Chilean ($M_0 = 2.7 \times 10^{30}$, $M_s = 8.3$) and the 1964 Alaskan ($M_0 = 7.5 \times 10^{29}$, $M_s = 8.4$) earthquakes. These were impressive events; in the Chilean earthquake 24 m of slip occurred on a fault 800 km long along strike and 200 km long downdip.

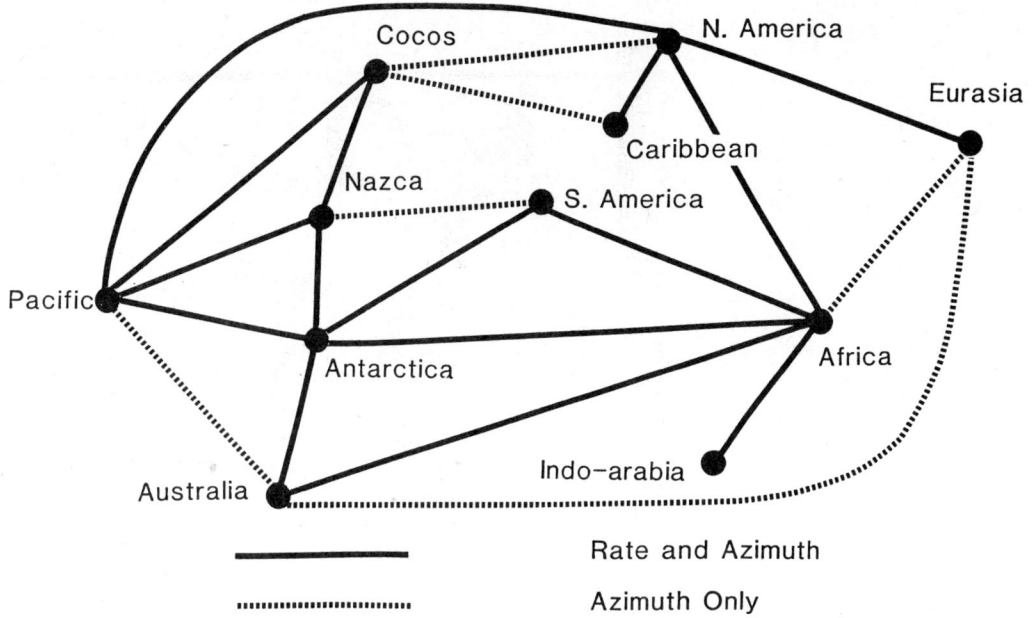

FIGURE 4. Global plate circuit geometry. Relative motion data exists only on the boundaries indicated.

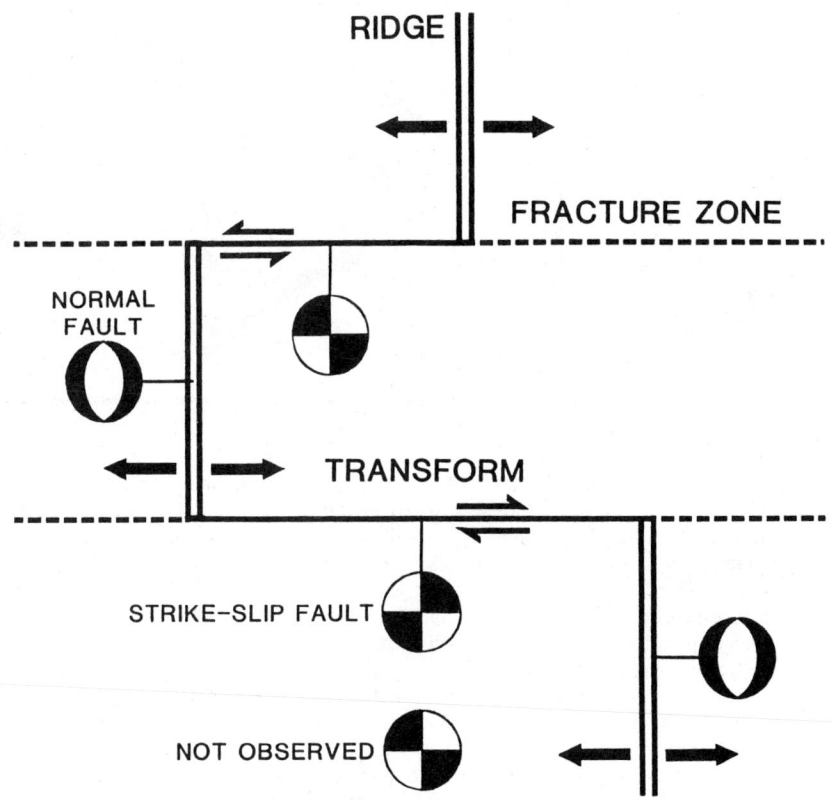

FIGURE 5. Possible tectonic settings of earthquakes at an oceanic spreading center. Most events occur on the active segment of the transform and have strike-slip mechanisms consistent with transform faulting. On a slow spreading ridge, like the Mid-Atlantic, normal fault earthquakes occur. Very few events occur on the inactive fracture zone.

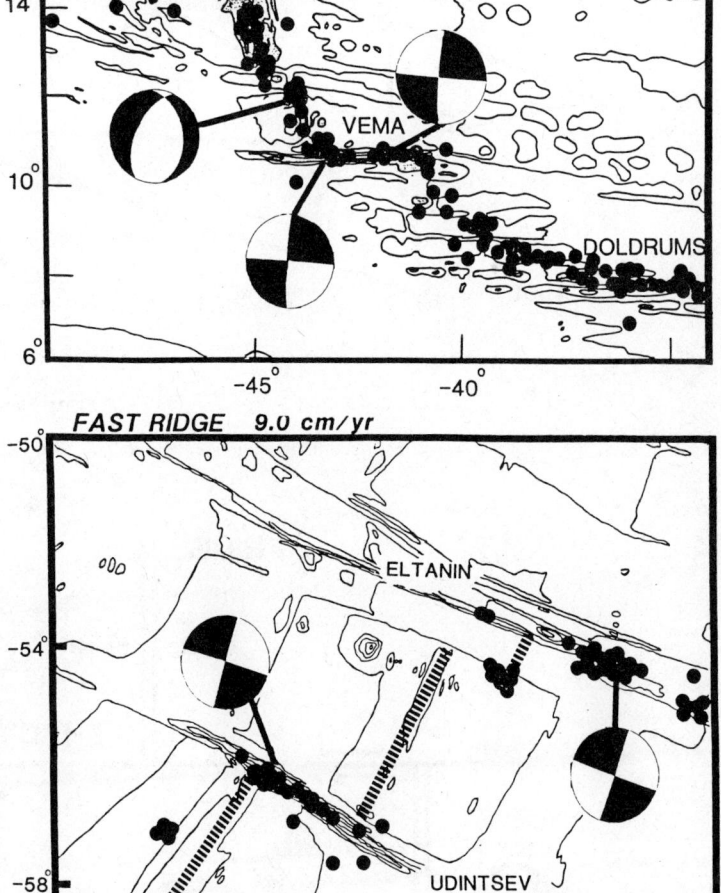

FIGURE 6. Maps contrasting on slow and fast spreading centers. Top: The Mid-Atlantic ridge has earthquakes both on the active transform and ridge segment. Strike-slip faulting on a plane parallel to the transform azimuth is characteristic. On the ridge segments, normal faulting with nodal planes parallel to the ridge trend is seen. Bottom: The East Pacific Rise has only strike-slip earthquakes on the transform segments. Mechanisms from Engeln et al. (1986), Huang et al. (1986), and Stewart and Okal (1983).

Smaller, but large, thrust events are characteristic. For example, Fig. 8 shows the focal mechanisms of large shallow earthquakes along a portion of the Middle America Trench, where the Cocos Plate is subducting beneath the North American Plate. The mechanisms show thrust faulting on fault planes with a consistent geometry, parallel to the coast, which corresponds to the trench axis, with shallow dips to the northeast.

These thrust events directly reflect the plate motion. At a point on the trench (17°N, 100°W) the motion of the Cocos Plate with respect to North America given by an Euler vector with pole at 28.56°N, 119.6°W, with magnitude 1.52° per million years is a rate of 6.2 cm·yr^{-1} and an azimuth of N36°E. The direction of motion is into the trench, as expected at a subduction zone. The major thrust earthquakes at the interface between subducting and overriding plates thus provide the best direct indication of the nature of subduction. Slip vectors from their focal mechanisms can be used to determine the direction of plate motion. The rate of subduction is harder to assess. Although the rate can be computed from relative motion models, which rely on the global closures (Fig. 4), all of the plate motion is not always released seismically in earthquakes. In this case, the seismic slip rate estimated from seismic moments can be only a fraction of the real plate motion. Nonetheless, it is useful to determine the seismic slip rate to assess the fraction of seismic slip, since it reflects the mechanics of the subduction process. It is also interesting to know how this slip varies as a function of time and position along a subduction zone.

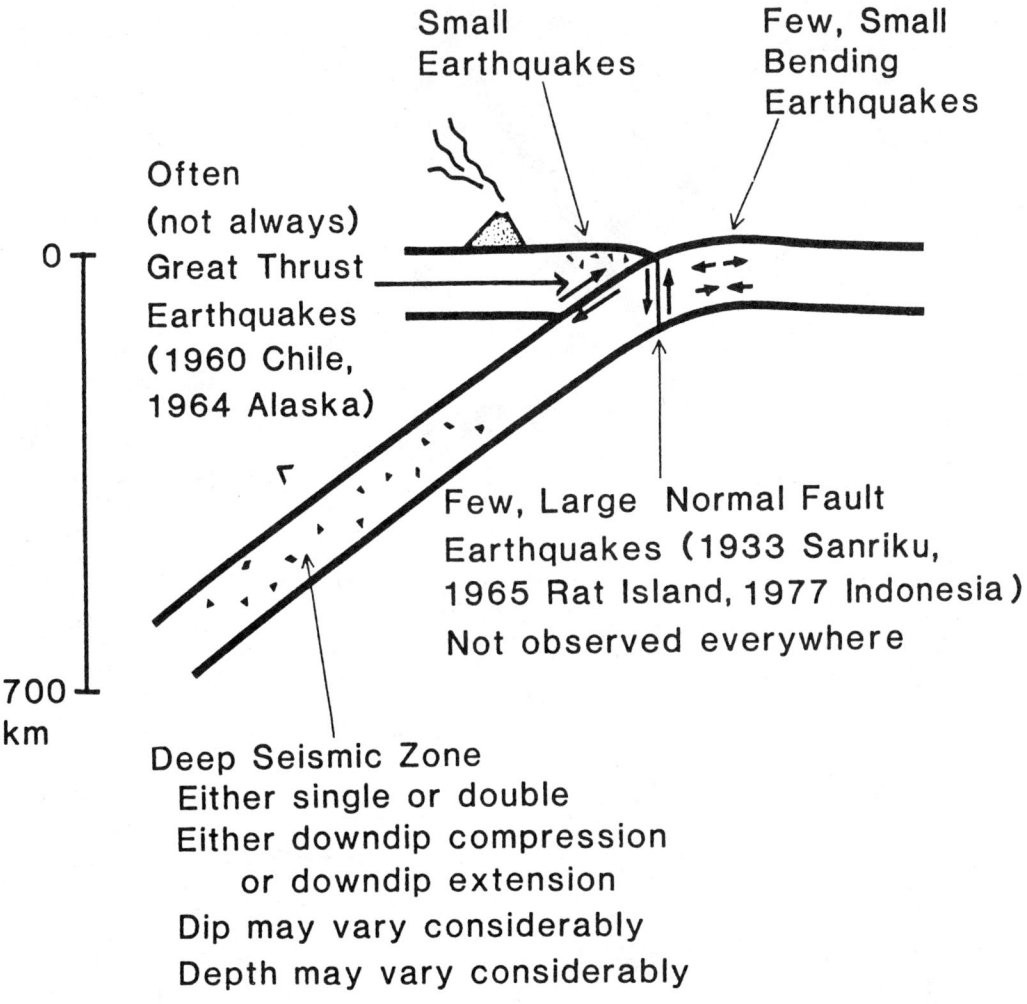

FIGURE 7. Schematic of some of the features observed at subduction zones. Not all features are seen at all subduction zones.

Figure 7 also shows other types of shallow subduction zone earthquakes. An interesting class of subduction zone earthquakes results from the flexural bending of the downgoing plate as it enters the trench (Fig. 9). Precise focal depth studies show a pattern of normal faulting in the upper part of the plate to a depth of 25 km and of thrusting in its lower part, between 40 km and 50 km. These observations constrain the position of the neutral surface that separates the upper extensional zone from the lower flexural zone and thus provide information on the mechanical state of the lithosphere.

Occasionally, trenches are the sites of large normal fault earthquakes (e.g., Sanriku 1933, Peru 1970, and Indonesia 1977). There has been some controversy as to whether to interpret these earthquakes as bending events in the upper flexural sheet or as "decoupling" events showing rupture of the entire downgoing plate due to "slab pull." Aftershock distributions and studies of the rupture process indicate that faulting extended through a major portion, and perhaps all, of the lithosphere (Fig. 10). Such data may be able to determine whether these events are flexural. Rupture extending through the entire lithosphere favors the decoupling model. If only a portion of the lithosphere breaks, the interpretation is more complicated. The rupture may have been restricted to one side of the neutral surface (in the flexural model), or it may reflect only the fact that material is too hot and weak to support seismic rupture. In the latter case, the entire lithosphere could have failed, with the deeper rupture being aseismic. This issue remains unresolved, due to the difficulty in determining actual fault area and interpreting the results mechanically.

The deeper earthquakes, which form the Wadati-Benioff zone, go down to depths of 700 km within the downgoing slab. Their mechanisms provide

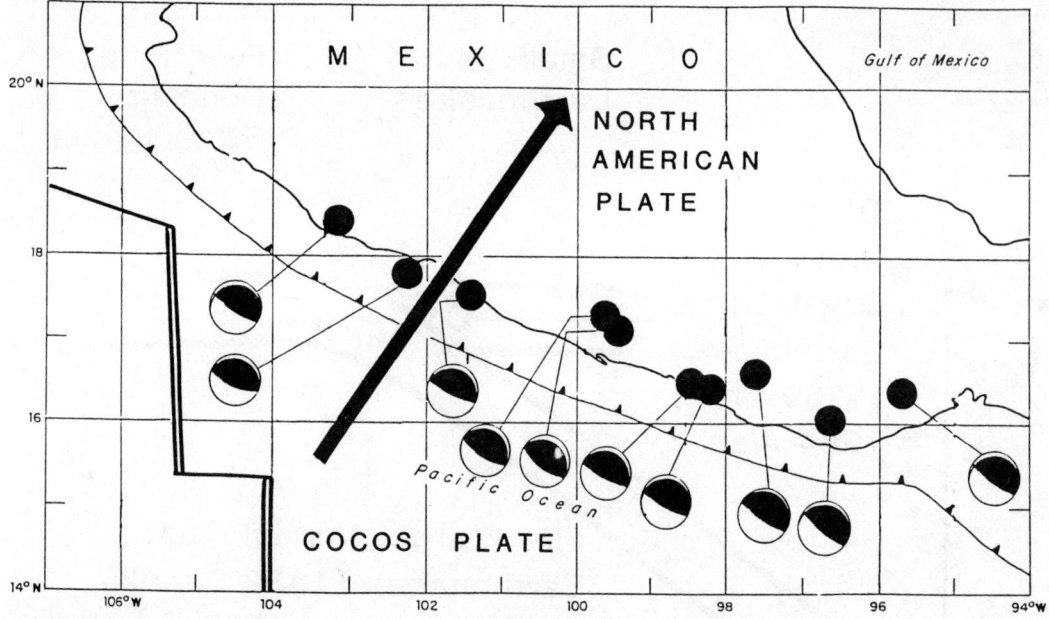

FIGURE 8. Map of the Middle America trench where the Cocos Plate subducts under the North American Plate at 6.2 cm·yr^{-1}. Large shallow-thrust earthquakes on planes parallel to the coast are seen, as is predicted by the geometry. (From Singh et al., 1984)

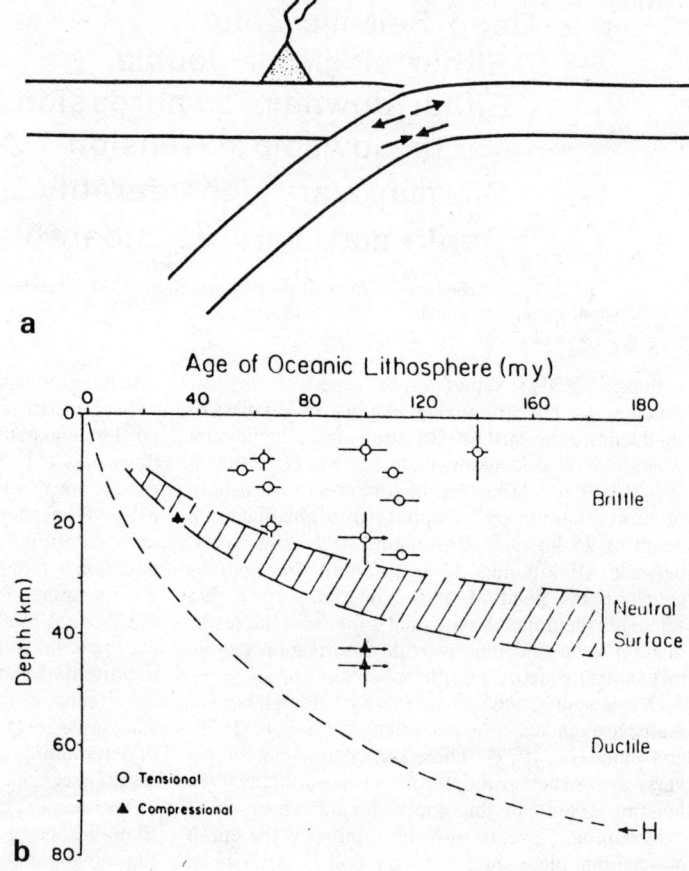

FIGURE 9. Focal depths of flexural earthquakes (a) due to the bending of the subducting plate as it enters the trench (Chapple and Forsyth, 1979) and (b) predictions of a mechanical model (Bodine et al., 1981). Tensional events occur above the neutral surface, and compressional events occur below. The plate mechanical thickness H increases with age, as expected from thermal models.

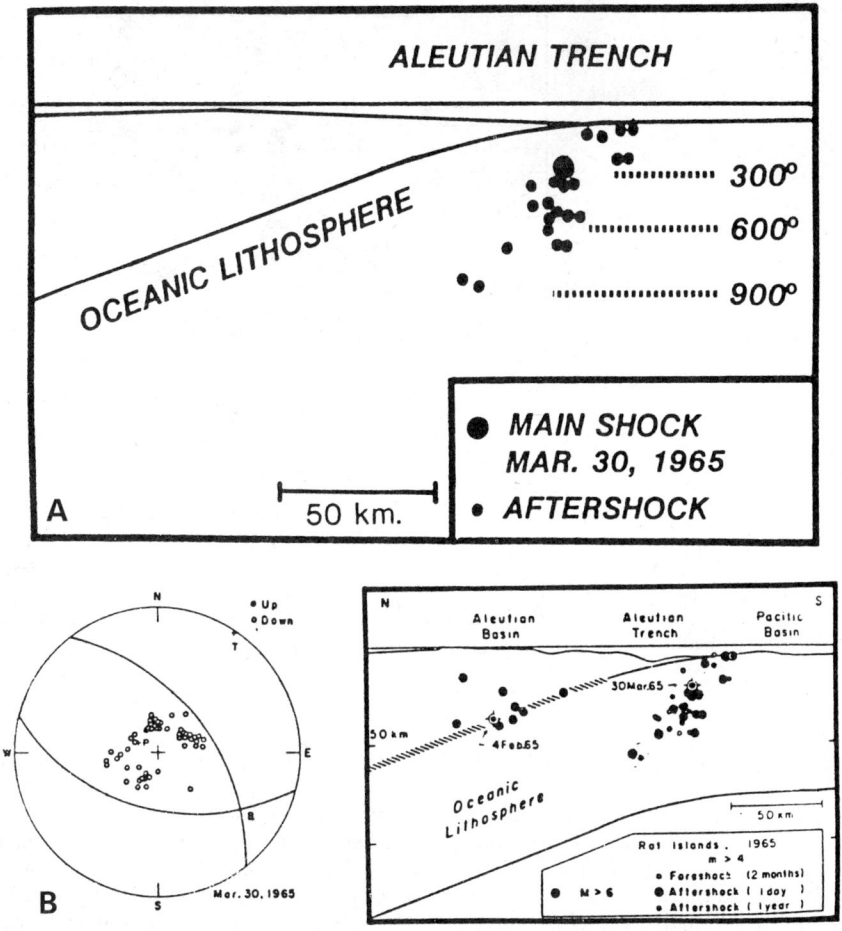

FIGURE 10. Large normal faulting earthquakes at trenches, such as the 1965 Rat Island earthquake (A: Abe, 1972), may be due to flexure or failure of the lithosphere under its own weight. The extent of aftershocks, which appear not to cut the entire lithosphere, may reflect the extent of rupture or be a temperature effect (B: Wiens and Stein, 1985).

important information about the physics of the subduction process. The essence of the process is the penetration and slow heating of a cold slab of lithosphere in the warmer mantle. This temperature contrast has important consequences. The subducting plate is identified by the locations of earthquakes in the Wadati-Benioff zone below the zone of thrust faulting at the interface between the two plates. Earthquakes occur to greater depths than elsewhere because the slab is colder than the surrounding mantle. The mechanisms of earthquakes within the slab similarly reflect this phenomenon. The thermal evolution of the downgoing plate and its surroundings is controlled by the relation between the rate at which cold slab material is subducted to that at which it heats up, primarily by conduction as it equilibrates with the surrounding mantle. In addition, adiabatic heating due to the increasing pressure with depth and phase changes contribute.

Numerical temperature calculations show that the downgoing plate remains much colder than the surrounding mantle until considerable depths, where the downgoing slab heats up to the ambient temperature. Comparison of calculated temperatures and the observed locations of seismicity for the Aleutian subduction zone (Fig. 11) shows that the earthquakes occur in the cold regions of the slab. The thermal model also helps explain their focal mechanisms. The force driving the subduction is the integral over the slab of the force due to the density contrast between the denser subducting material and the density of "normal" mantle material outside. This force, known as *slab pull*, is the plate driving force due to subduction. Its significance for stresses in the downgoing plate and for driving plate motions depends on its size relative to the resisting forces at the subduction zone. There are several such forces. As the slab sinks into the viscous mantle, material

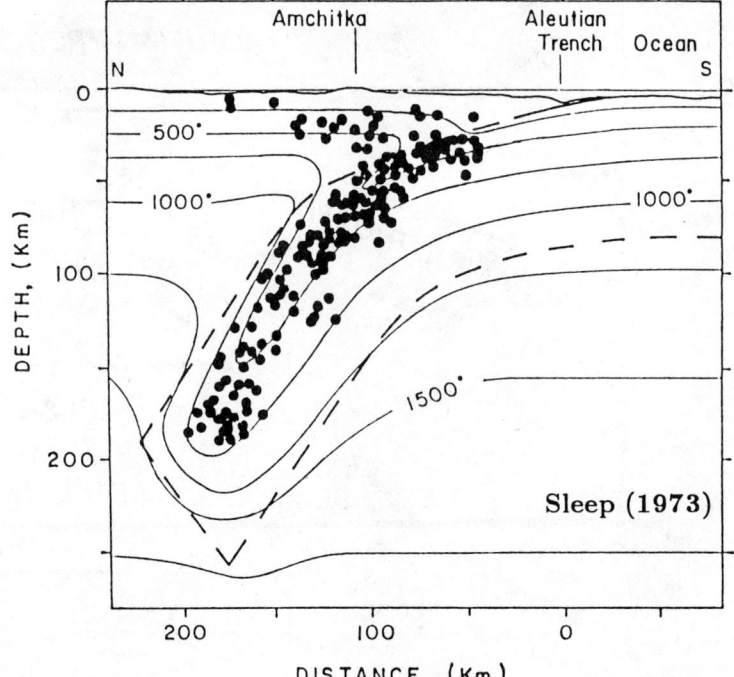

FIGURE 11. Seismological tests of a numerical thermal model for the Aleutian subduction zone. Locations of Wadati-Benioff zone seismicity and predicted temperatures. Earthquakes occur within the cold interior of the slab (Sleep, 1973).

must be displaced. The resulting force depends on the viscosity of the mantle and the subduction rate. The slab is also subject to drag forces on its sides and resistance at the interface between the overriding and downgoing plates. The latter is often manifested as shallow thrust earthquakes.

One way to study the relative size of the negative buoyancy and resistive forces is to use focal mechanisms to examine the state of stress in the downgoing slab. The general results are shown in Fig. 12: There are both downdip compressional and downdip extensional events. Those above 300 km show generally downdip tension, whereas those below 300 km show generally downdip compression. A proposed explanation is that there are *two* basic processes operating: Near the surface the slab is being extended by its own weight; at depth the slab begins to "run into" stronger material, and downdip compression occurs. Numerical models of stress in downgoing slabs are consistent with this interpretation since they can reproduce the shallow downdip tension and deep downdip compression (Fig. 13).

Not all features shown in Fig. 7 have been observed at all places. For example, the dips and shapes of subduction zones vary substantially. Some show double planes of deep seismicity; some do not. Even the very large thrust earthquakes, considered characteristic of subduction zone events, are not observed in all subduction zones. In recent years, considerable effect has been made to understand such variations.

Continental Earthquake Focal Mechanisms

Although the basic relationships between plate boundaries and earthquakes apply to continental as well as oceanic lithosphere, the continents are more complicated. The continental crust is much thicker, less dense, and has very different mechanical properties from the oceanic crust. Since continental crust and lithosphere are not subducted, the continental lithosphere records a long, involved tectonic history. In contrast, the oceans record only the last 200 million years. One major result of these factors is that plate boundaries are much broader zones than in the oceanic lithosphere. It is thus common to use the oceanic lithosphere as a simple model for the continents. This approach gives some useful insights but by no means resolves all questions.

One obvious example of these ideas is the East African rift, a spreading center between the Nubian (West Africa) and Somalia (East Africa) Plates. The extension rate is quite slow, less than 0.5 cm $\cdot$ yr^{-1}. Such regions of continental extension are much more complicated than oceanic spreading centers. The seismicity shows clearly that the boundary zone is much broader and more diffuse than at a mid-ocean ridge (Fig. 14). For example, the seismicity simply ends in southern Africa and has no clear connection to the Southwest Indian Ridge, where the plate boundary must go. Unlike a mid-ocean ridge, some of the additional complexity of continental extensional zones results from the fact that the lithosphere

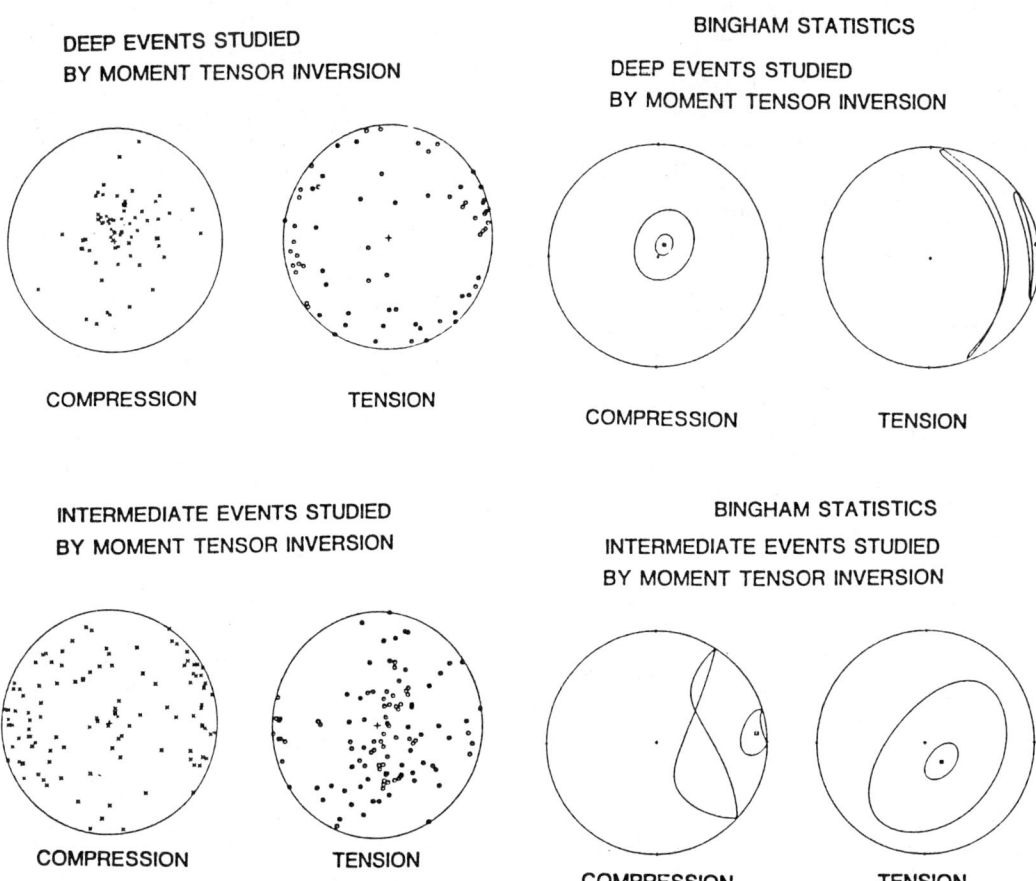

FIGURE 12. Data for the stress orientations inferred from focal mechanisms of subduction zone earthquakes. The *P*- and *T*-axes have been rotated so that the downdip direction is at the center of each plot, and their distribution has been contoured. *Top:* Events below 300 km are dominated by downdip compression. *Bottom:* Events from 70–300 km are dominated by downdip tension (Vassiliou et al., 1984).

has reasonable thickness and thus is stretched and thinned in the extending zone. The rifting process can eventually progress far enough that new oceanic crust is formed and true seafloor spreading occurs. This is not, however, the inevitable fate of continental rifts. The geologic record shows many rifts that, although active for some time, failed to develop into oceanic spreading centers and simply died.

The same situation occurs for continental transforms, like the Pacific–North America plate boundary in the western United States (Figs. 15 and 16). The basic plate motion is transform motion along the San Andreas Fault, but the picture is muddied by the existence of a broad region of faulting with diverse styles. The active seismic zone is hundreds of kilometers wide, in contrast to less than 10 km for oceanic transforms, suggesting complexity beyond the simple transform model. Although the focal mechanisms show strike-slip along the San Andreas itself, they demonstrate the further complexities, including thrust faulting for events like the 1971 San Fernando earthquake.

Of the three boundary types, continental convergence zones may be the most complicated compared to their oceanic counterparts. The primary difference is that, since continental lithosphere is much less dense than the upper mantle, it is not subducted and a Wadati-Benioff zone is not formed. As a result, continental convergence zones in general do not have intermediate and deep-focus earthquakes. In addition, the boundary tectonics occur over a broader and more complex region than in an oceanic case.

There are also interesting differences between continental convergence zones. For example, Fig. 16 contrasts the convergent zones between the Eurasian and Indian Plates at the Himalayas and between Arabia and Eurasia at the Zagros. In both cases a zone of thrust fault events occurs. In the Himalayas these events define a thrust surface that appears to indicate the top of the underthrusting Indian conti-

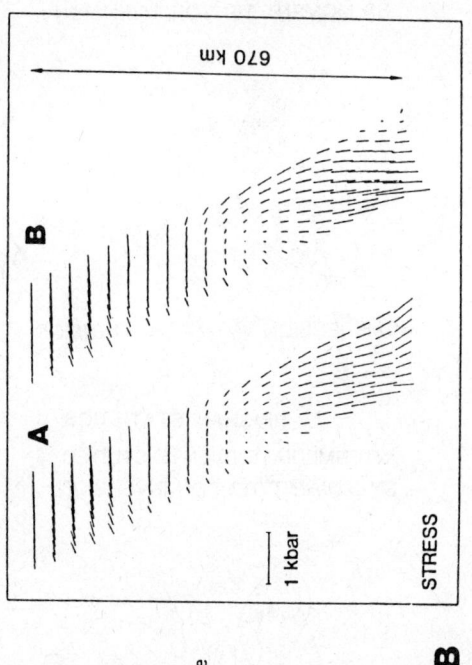
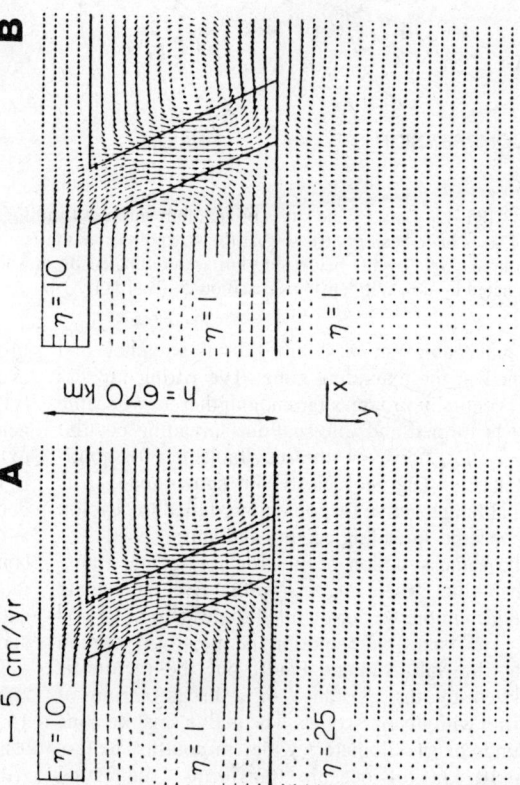

FIGURE 13. Numerical model of stresses within a downgoing slab for the cases of a slab that (A) encounters higher-viscosity material below 670 km and (B) cannot penetrate below this depth. Both predict downdip tension in the upper portion of the slab and downdip compression in the lower portion. The calculated stresses are highest near the bottom of the slab (Vassiliou et al., 1984).

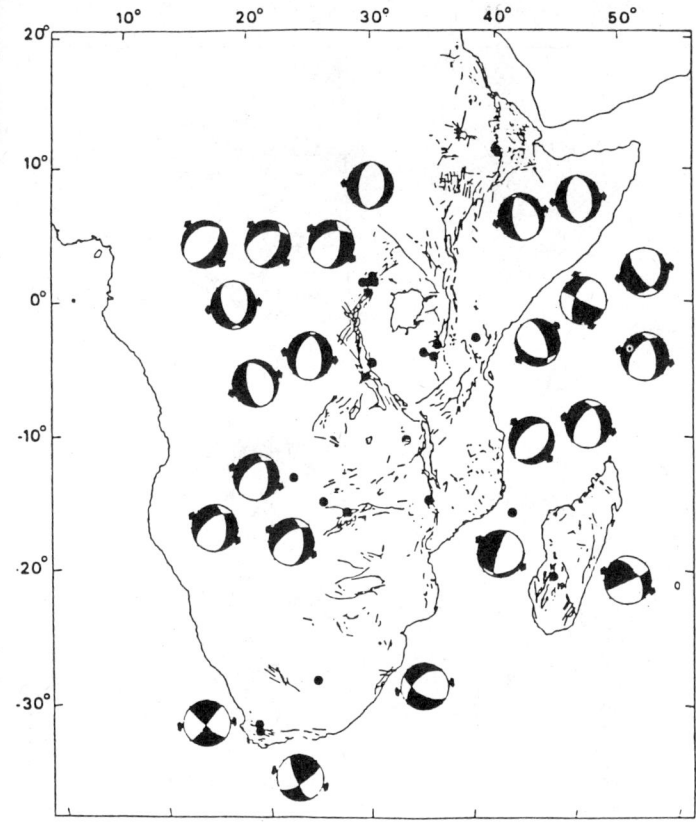

FIGURE 14. Seismicity and focal mechanisms for the East African Rift system (Shudofsky, 1985). The diffuse seismicity defines a broad plate boundary zone. Complexity is added by the apparent end of the boundary in southern Africa, a condition not allowed by rigid plate tectonics.

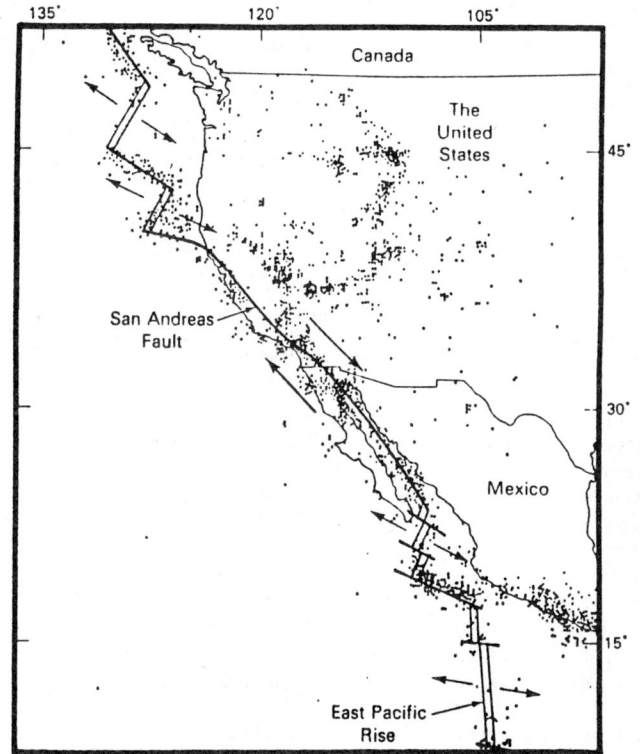

FIGURE 15. The diffuse North America–Pacific Plate boundary zone in the western U.S. (Uyeda, 1978).

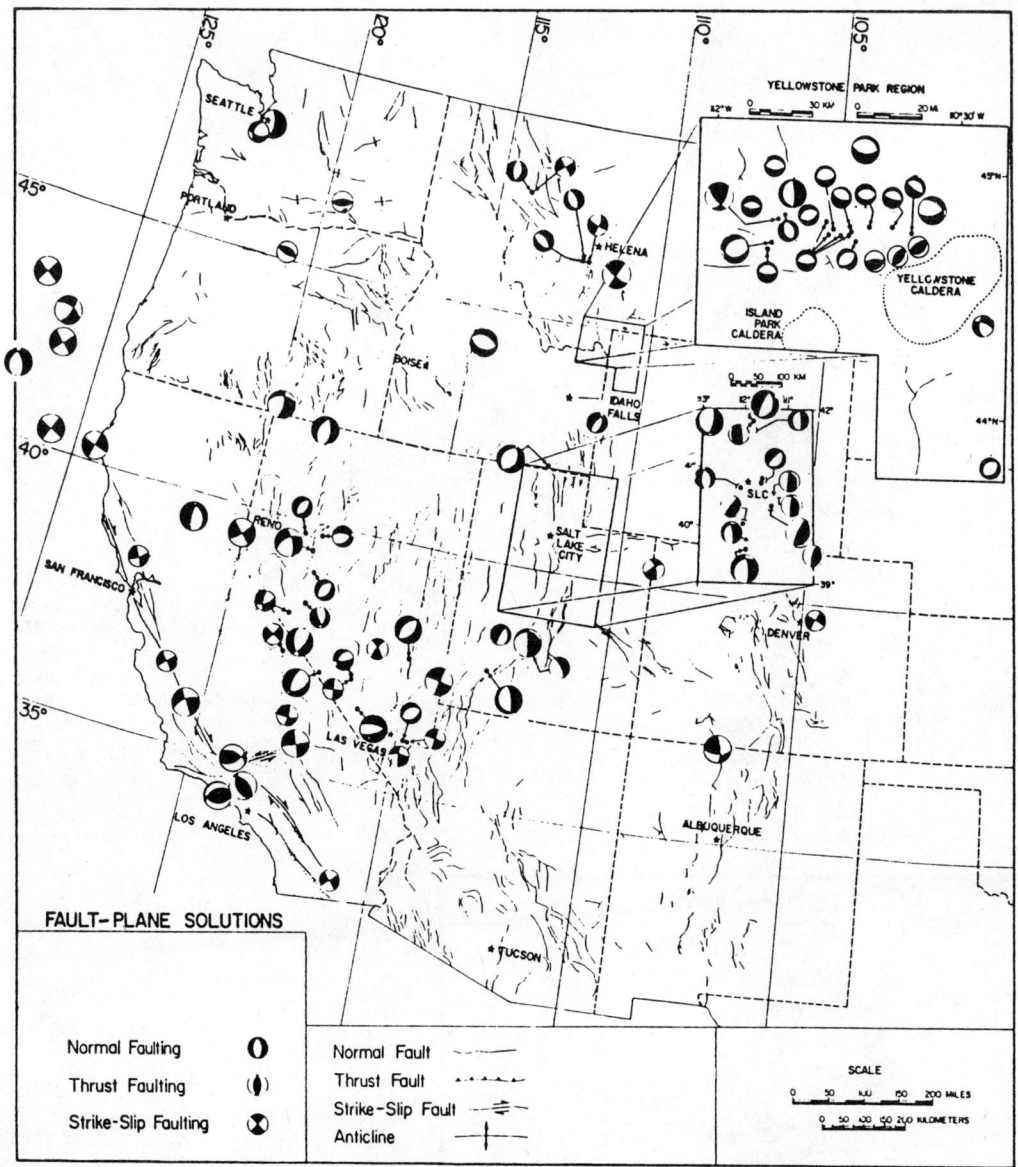

FIGURE 16. Focal mechanisms for the western United States (Smith and Lindh, 1987).

nental crust. In contrast, in the Zagros the thrusting is distributed over a broad zone of compression and shortening within the Arabian continental crust. The difference may be partially due to an evolutionary sequence; the Zagros are a young collisional zone that may evolve into a Himalayan geometry. An interesting feature of the Himalayan collision is that it is associated with a considerable amount of deformation within the Eurasian Plate, as shown by the normal faulting behind the convergent zone, in the Tibetan Plateau. It has been suggested that this is part of a very broad region of deformation within the Eurasian Plate shown by seismicity and faulting.

Back-arc basinlike phenomena also occur in continental lithosphere. The Basin-and-Range seismicity of the western United States (Fig. 15), well inland from the San Andreas system, is often interpreted as the result of an extensional process similar to back-arc spreading resulting from the complex history of subduction and strike-slip faulting along the continental margin. Continental extension is also observed in areas including Greece or Turkey.

The continents also contain a variety of other seismic regions reflecting internal deformation. These effects are not as directly relatable to plate tectonic processes. One such example is the New Madrid area

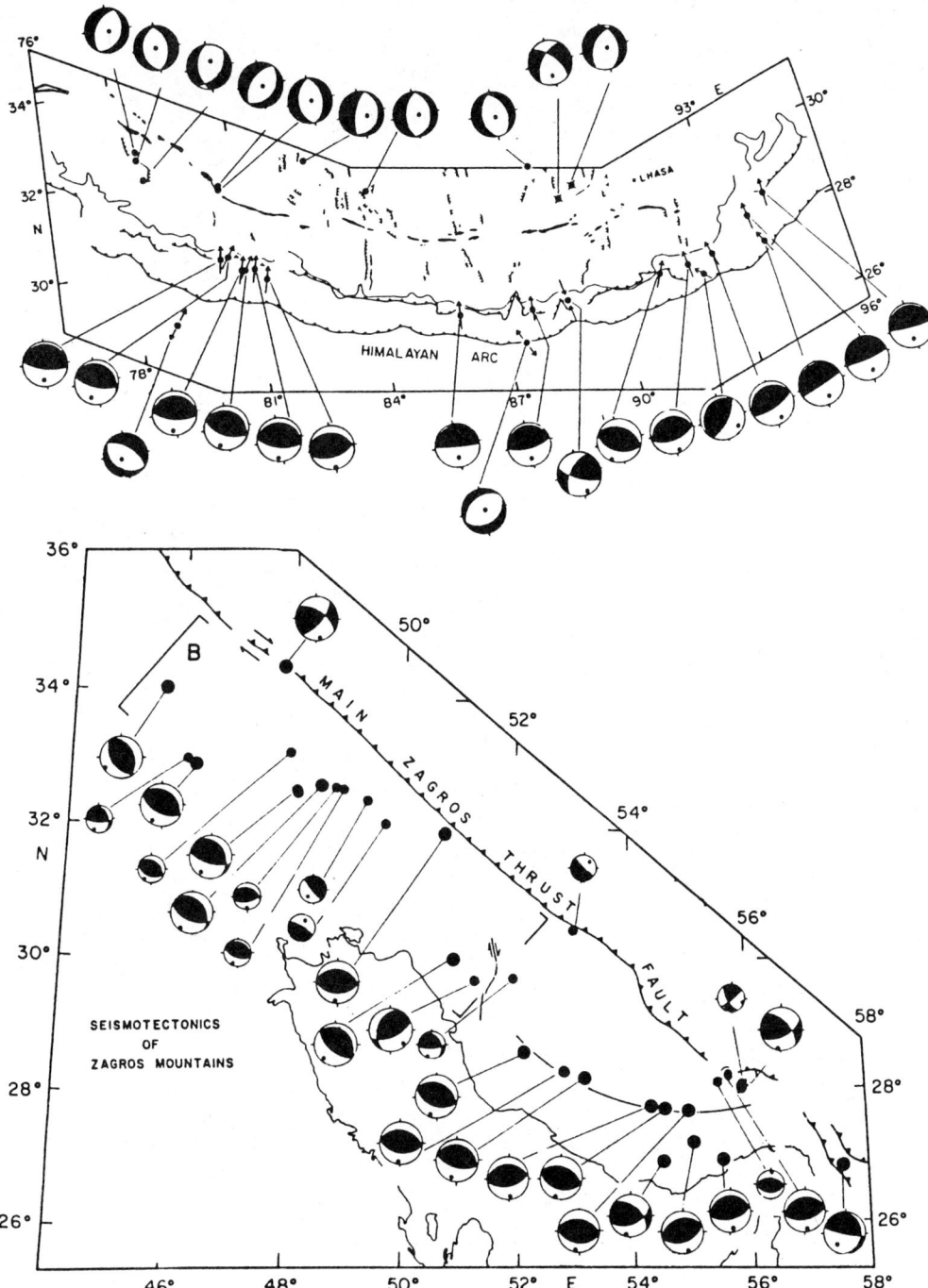

FIGURE 17. Focal mechanisms for the Himalayan and Zagros continental convergence zones (Ni and Barazangi, 1984, 1986). The Himalayan focal mechanisms define a thrust surface between two portions of continental crust. The Zagros collision zone has no such surface.

in Missouri, which had very large earthquakes in 1811-1812. The seismicity of such regions is generally thought to be due to the reactivation of preexisting faults or weak zones in response to intraplate stresses. As a result, such major intracontinental earthquakes should occur substantially less frequently than plate boundary events. The recurrence estimates for 1811-1812-type earthquakes average 600-700 years. A special case of this phenomenon occurs at passive margins, where the continental and oceanic lithosphere join. Although these areas are in general tectonically inactive, magnitude 7 earthquakes can occur, as on the eastern coast of North America. Such earthquakes are thought to be associated with stresses at the continental margin, including those due to the removal of glacial loads, which reactivate the faults remaining along the continental margin from the original rifting.

SETH STEIN
DALE F. WOODS

References

Abe, K., 1972, Lithospheric normal faulting beneath the Aleutian Trench, *Physics Earth Planetary Interiors* **5**, 190-198.

Bodine, J. H., M. S. Steckler, and A. B. Watts, 1981, Observations of flexure and the rheology of the oceanic lithosphere, *Jour. Geophys. Research* **86**, 3695-3707.

Chapple, W. M., and D. W. Forsyth, 1979, Earthquakes and bending of plates at trenches, *Jour. Geophys. Research* **84**, 6729-6749.

Chase, C. G., 1972, The N plate problem of plate tectonics, *Royal Astron. Soc. Geophys. Jour.* **29**, 117-122.

Cox, A., 1973, *Plate Tectonics and Geomagnetic Reversals*. San Francisco: W. H. Freeman, 702 p.

Cox, A., and R. B. Hart, 1986, *Plate Tectonics*. Palo Alto, Ca.: Blackwell Scientific Publications.

Engeln, J. F., D. A. Wiens, and S. Stein, 1986. Mechanisms and depths of Atlantic transform earthquakes, *Jour. Geophys. Research* 91, 548-577.

Huang, P. Y., S. C. Solomon, E. A. Bergman, and J. L. Nabelek, 1986, Focal depths and mechanisms of Mid-Atlantic Ridge earthquakes from body waveform inversion, *Jour. Geophys. Research* **91**, 579-598.

Minster, J. B., T. H. Jordan, P. Molnar, and E. Haines, 1974, Numerical modeling of instantaneous plate tectonics, *Royal Astron. Soc. Geophys. Jour.* **36**, 541-576.

Ni, J., and M. Barazangi, 1984, Seismotectonics of the Himalayan continental collision zone: geometry of the underthrusting Indian plate beneath the Himalayas, *Jour. Geophys. Research* **89**, 1147-1164.

Ni, J., and M. Barazangi, 1986, Seismotectonics of the Zagros continental collision zone and a comparison with the Himalayas, *Jour. Geophys. Research* **91**, 8205-8218.

Shudofsky, G. N., 1985, Source mechanisms and focal depths of east African earthquakes using Rayleigh-wave inversion and body wave modelling. *Royal Astron. Soc. Geophys. Jour.* **83**, 563-614.

Singh, S. K., M. Rodriguez, and L. Esteva, 1983, Statistics of small earthquakes and frequency of occurrence of large earthquakes along the Mexican subduction zone, *Seismol. Soc. America Bull.* **73**, 1779-1796.

Sleep, N. H., 1973, Teleseismic P-wave transmission through slabs, *Seismol. Soc. America Bull.* **63**, 1349-1373.

Smith, R. B., and A. G. Lindh, 1978, Fault plane solutions of the western United States: a compilation, in R. B. Smith and G. P. Eaton, eds., *Cenozoic Tectonics and Regional Geophysics of the Western Cordillera*. Geol. Soc. America Mem. 152, 107-111.

Stewart, L. M., and E. A. Okal, 1983, Seismicity and aseismic slip along the Eltanin Fracture Zone, *Jour. Geophys. Research* **88**, 10495-10507.

Uyeda, S., 1978, *The New View of the Earth*. San Francisco: W. H. Freeman.

Vassiliou, M. S., B. H. Hager, and A. Raefsky, 1984, The distribution of earthquakes with depth and stress in subducting slabs, *Geodynamics* **1**, 11-28.

Wiens, D. A., and S. Stein, 1985, Implications of oceanic intraplate seismicity for plate stresses, driving forces and rheology, *Tectonophysics* **116**, 143-162.

Cross-references: *Earthquake Mechanisms; Earthquake Seismology; Earthquakes and Seismicity; Lithosphere: Mechanical Properties; Lithosphere, Oceanic: Formation and Evolution; Paleomagnetism and Plate Tectonics; Seismicity: Intraplate; Seismicity: Mid-Ocean Ridge; Seismicity: Subduction Zone; Seismicity and Plate Tectonics; Seismic Source: Observations; Seismic Source: Theory; Stress in the Earth's Lithosphere.*

EARTHQUAKE PREDICTION—See EARTHQUAKES: HAZARDS AND PREDICTIONS.

EARTHQUAKES: HAZARDS AND PREDICTIONS

Following the great 1746 Lima, Peru, earthquake, a Jesuit priest, Lozano, described how Lima prior to the earthquake

had arrived at the acme of perfection of which a city of this New World was capable, for the sumptuousity of its buildings, the dwellings which adorned its well laid out streets, its beautiful fountains, the dignity of its churches, and the construction of the monasteries, which could well compete with the grandest works of its kind in the world. But all this. . .*was reduced to dust in an instant.* [Silgado, 1973, p. 191]

The above scene has been repeated numerous times in various locations, and underscores the basic human desire to understand, foretell, and possibly control earthquakes in an effort to minimize the resultant loss of life and property.

The hazards associated with earthquakes are those natural and cultural phenomena that produce adverse effects on human activities. Natural or geologic phenomena include landsliding, liquefaction of soils, ground shaking and subsidence, active faulting, and tsunamis. Cultural hazards involve possible failure of buildings, fires, floods, the disruption of lifelines (communications, power, transportation, medical services, water supplies, etc.), and radioactive leaks from nuclear facilities. Seismic risk refers to the chance that the social or economic consequences of earthquakes will equal or exceed some specified threshold value for a particular time and location.

Earthquake prediction refers to the specification of the expected magnitude, geographic location, and time of occurrence of a future event with sufficient precision that the ultimate success or failure of a prediction can be evaluated. While all three parameters are important, a premium is placed on correctly identifying when the next damaging event will occur. Accordingly earthquake predictions are usually categorized in terms of the time interval for which the estimates are made. These estimates range from long-term predictions or forecasts, which cover several years to tens of years; to intermediate-term predictions, which cover months to years; to short-term predictions, which span days to months. The purpose of this article is to highlight some of the basic concepts and advances that have led to the present efforts in predicting and forecasting earthquakes. At this writing, research efforts in long-term earthquake forecasting and hazards evaluation have been the most productive. In addition to providing valuable scientific information, long-term hazards assessments are socially invaluable for long-term planning and the development of engineering design and building codes. In contrast, progress toward routine intermediate- and short-term earthquake prediction has not been as rapid. More detailed reviews, summarizing progress toward these goals can be found in Rikitake (1976) and Simpson and Richards (1981).

Historical Perspective

Early legends attributed earthquakes to the movements of animals that resided underground and supported the earth. A medieval Japanese legend, for example, told how earthquakes were caused by the sudden motions of a big catfish that lived in the Earth. In other parts of Asia, frogs, moles, and oxen were blamed. In North America, the Algonquin Indians attributed earthquakes to a giant tortoise who supported the earth.

The Greeks, Aristotle and Strabo, offered a more mechanical explanation, suggesting that subterranean winds ignited combustible materials causing earthquakes and volcanic eruptions. Greek and Roman mythology, however, also interpreted earthquakes as punishments meted out by the angry gods Poseidon and Neptune. The concept of divine intervention as the cause of earthquakes is mentioned in the Bible (the destruction of Sodom and Gomorrah and the collapse of the walls of Jericho, for example) and was a popular explanation in eighteenth century Europe. For example, the 1755 Lisbon, Portugal, earthquake was "proof" of God's retribution for sinful behavior. During the seventeenth, eighteenth, and nineteenth centuries, the seeds for the serious study of earthquakes were being sown. A detailed account of the great 1835 Chile earthquake was published in Charles Darwin's *Voyage of the Beagle*, and Charles Lyell included a number of descriptions of earthquakes in *Principles of Geology*. A detailed account of the history of seismology can be found in Richter (1958).

In general, formal efforts in seismology and earthquake prediction during the late nineteenth and twentieth centuries developed in response to the occurrence of specific earthquakes. The Seismological Society of Japan was founded following the 1880 (M 5.4) Yokohama earthquake. Following the great (M 7.9) Nobi, Japan, earthquake of 1891, the Imperial Earthquake Investigation Committee was established and charged with investigating seismic and volcanic phenomena and finding a way to lessen earthquake hazards. After the great (M_w 7.9) 1923 Kanto (Tokyo) earthquake, the Earthquake Research Institute was founded and began investigation of earthquakes on a more physical basis. In the United States, the Seismological Society of America was founded following the great (M_w 7.7) 1906 San Francisco earthquake. (See Table 1 for definitions of specific magnitudes.)

Systematic research directed toward actual earthquake prediction, however, has only occurred since the 1960s and again was prompted by a series of disasters in China, Japan, and the United States. Motivated by the losses experienced during the great (M_w 9.2) 1964 Alaskan earthquake, earth scientists in the United States began to focus on the problems of earthquake prediction and hazards reduction and proposed long-term research programs. The goals of these proposals were emphasized by the 1971 (M_s 6.5) San Fernando, California, earthquake, which demonstrated the vulnerability of rapidly growing areas such as southern California to earthquake hazards. Efforts in the United States during this period culminated in the passage of the Earthquake Hazards Reduction Act of 1977, which established into law the goals of earthquake prediction and hazards reduction. Formal efforts in Japan also began in the early 1960s, as an extension of the Disasters and Countermeasures Basic Law of 1961, which was enacted to mitigate natural catastrophes such as floods, typhoons, landslides, volcanic eruptions as well as earthquakes and tsunamis. Government-supported efforts for earthquake prediction in Japan were intensified following the 1964 (M_s 7.5) Niigata earthquake and 1965–1967 Matsushiro earthquake swarm. The Chinese intensified their efforts following the 1966 (M_s 6.8) Xingtai shock. At the same time, Soviet Union prediction efforts were concentrated in central Asia and in the Kurile-Kamchatka regions. Since the 1960s, several international meetings have been organized to discuss recent research and to promote cooperation among the various scientific organizations involved in earthquake prediction.

At this writing, all of the above countries have active prediction research and hazard mitigation programs. A review of two active earthquake prediction efforts, one in the Tokai region of central Japan and the other in the Parkfield area of central California, is presented at the end of this article.

Classification

The theory of plate tectonics provides a basic framework for understanding the global distribution of earthquakes and the mechanism by which the majority of these events occur. Most of the shallow (<40 km) seismic energy release along major plate boundaries occurs in large and great earthquakes ($M_s \geq 7.0$ and $M_s \geq 7.75$, respectively). The elastic strain energy that is relieved in these shallow shocks comes from the relative motion of the plates, at rates of 2 cm to 12 cm/yr, and is built up slowly over periods of tens to hundreds of years. Friction along plate boundaries prevents many of these zones from moving more frequently than once in tens or hundreds of years. Once the strain along a fault segment reaches a critical level, the plate interface moves with 1 m to 20 m of motion in a large or great earthquake. Earthquakes that occur along the edges or boundaries of plates (interplate shocks) reflect the motion of the plates either away from one another (divergence or spreading), past one another (transform or strike-slip motion), or one beneath the other (convergence or subduction). Earthquakes that occur along spreading centers are usually not larger than M_s 7. The majority of the world's large and great shocks originate along convergent or transform plate boundaries, and most of the successful earthquake predictions and forecasts to date have been associated with these types of tectonic features. Earthquakes that occur within the interior of plates are termed intraplate earthquakes. To a first approximation, plate tectonics treats the plates as rigid bodies. The processes that lead to strain accumulation within plate interiors are not as well understood as those processes occurring along the edges. Additionally, the time scales for intraplate processes are substantially longer (i.e., the recurrence periods for large intraplate shocks may be on the order of hundreds to thousands of years compared to tens to hundreds of years for interplate events). A number of successful short- and intermediate-term predictions, however, have also been made for intraplate earthquakes.

Interplate Earthquakes

The concept of seismic gaps forms the cornerstone of earthquake prediction and forecasting research. According to the seismic gap hypothesis, segments of active transform or convergent plate boundaries that have not experienced a large or great earthquake for a period of time (usually a few decades) are the expected sites for future events. Plate tectonic theory indicates that individual plate boundaries are continuous and that they do not end abruptly unless motion is transferred to some other type of tectonic feature such as a spreading ridge, subduction zone, or transform fault. In the long-term, which is taken to be thousands of years, plate motion is thought to be fairly uniform along the entire length of a simple plate boundary. Hence, segments of a plate boundary that have not experienced large earthquakes for tens to hundreds of years are likely to be either the sites of future large shocks or regions where plate motion is accommodated either aseismically or by the occurrence of only small to moderate sized shocks.

The basic principles describing how strains are accumulated and released in earthquakes were developed prior to the advent of plate tectonics and can be traced to Reid's (1910) elastic rebound hypothesis. Gilbert's (1909) discussion about forecasting earthquakes developed a number of basic concepts (i.e., periodicity or recurrence intervals, alternation or seismic gaps, and earthquake precursors) that form the core of present-day prediction and forecasting efforts. One of the earliest applications of the seismic gap concept was by A. Imamura, who, prior to the 1923 Kanto earthquake, noted that the Kanto area in Japan had no historical activity, while the surrounding region was one of high activity (see Aki, 1980). In 1928 Imamura again noted that the potential was high for the repeat of an earthquake in the Nankaido-Tokaido region of southwest Japan. This prediction was based on the observation that previous great earthquakes had occurred in that area every 100 to 150 years, and that it had already been more than 70 years since the last earthquake. Imamura's forecast was subsequently verified by the occurrence of the 1944 Tonankai (M_w 8.1) and 1946 Nankaido (M_w 8.1) earthquakes. During the 1960s, S. A. Fedotov (1965) investigated the occurrence of great shallow earthquakes along the convergent margins of northeastern Japan, the Kuriles, and Kamchatka. Noticing that earthquakes tended to occur in regions that were not the sites of great earthquakes for at least several decades, and that earthquakes tended to rupture discrete segments (as defined by their aftershock zones), Fedotov suggested several areas as candidates for future large events. Since 1965, several of these seismic gaps have been "filled" in by large and great earthquakes (see Table 1). Work on seismic gaps increased during the 1970s with numerous studies conducted throughout the circum-Pacific region. McCann et al. (1979) present an exhaustive summary of research done on seismic gaps up until the late 1970s.

Understanding the factors that contribute to the global variations in large and great earthquake

TABLE 1. Successful Earthquake Forecasts and Predictions

Location	Latitude	Longitude	Date	Magnitude	Type	Reference
Nankai Trough, Japan	33–34 N	133–137 E	1944,46	8.1, 8.1 M_w	SG	1
Tokachi-Oki, Japan	40.9 N	143.4 E	1968	8.2 M_w	SG	2
Southern Kuriles	43.4 N	147.8 E	1969	8.2 M_w	SG	2
Central Kamchatka	56.0 N	163.2 E	1971	7.8 M_s	SG	2
Sitka, Alaska	56.8 N	135.7 W	1972	7.6 M_s	SG	3
Nemuro-Oki, Japan	43.0 N	145.7 E	1973	7.7 M_s	SG	2,4
*Blue Mt. Lake, New York	43.8 N	74.5 W	1973	2.6 m_b	V	5
Colima, Mexico	18.4 N	103.2 W	1973	7.5 M_s	SG	6
Lima, Peru	12.4 S	77.6 W	1974	8.1 M_w	SG	7
*Haicheng, China	40.6 N	122.8 E	1975	7.3 M_s	Various	8
*Central Asia, USSR	37–44 N	55–80 E	1960–70	4–7 m_b	Various	9
Vrancea, Romania	45.8 N	26.7 E	1977	7.1 M_s	R	10
*Oaxaca, Mexico	15.7 N	96.8 W	1978	7.8 M_s	SG, Q	6,11
Petatlan, Mexico	17.3 N	101.3 W	1979	7.6 M_s	SG	6
Tumaco, Colombia	1.6 N	79.4 W	1979	7.9 M_s	SG	7
Southern Tonga	24.1 S	175.9 W	1982	7.5 M_w	SG	12
Valparaiso, Chile	33.1 S	71.8 W	1985	8.0 M_s	SG, R	7,13
Michoacan, Mexico	18.3 N	102.3 W	1985	8.1 M_s	SG	6
*Stone Canyon, California	36.6 N	122.2 W	1986	4.6 m_b	Q	14
*Andreanof Is., Aleutian Is.	51.5 N	174.8 W	1986	7.7 M_s	Q	15
Northern Kermedec	28.1 S	176.4 W	1986	8.1 M_s	SG	12

Notes: *denotes prediction, all others forecasts. Type: SG = Seismic Gaps; Q = Seismic Quiescence anomaly; V = V_p/V_s anomaly; R = Recurrence times.

M_w = seismic moment magnitude
M_s = surface wave magnitude
m_b = body wave magnitude
M_L = local magnitude
M = Gutenberg and Richter unified magnitude

References: 1. Imamura (1928) 2. Fedotov (1965) 3. Sykes (1971) 4. Mogi (1968) 5. Aggarwal et al. (1975) 6. Kelleher et al. (1973) 7. Kelleher (1972) 8. Raleigh et al. (1977) 9. Sadovsky and Nersesov (1983) 10. Purcaru (1979) 11. Ohtake et al. (1977) 12. McCann et al. (1979) 13. Nishenko (1985) 14. Wyss and Burford (1985) 15. Kisslinger (1988)

occurrence has become crucial for improving our ability to forecast and predict earthquakes. Observations correlating the maximum sizes and recurrence times of earthquakes along plate boundaries with the width of the plate interface zone, the degree of heterogeneity of the fault surface, the age of the plates, and relative rates of plate motion have increased our understanding of both the earthquake mechanism and the earthquake recurrence cycle.

Overall, as seen in Table 1 (and Fig. 1), the seismic gap concept has been successful in forecasting the location and size of a number of large and great earthquakes that have occurred along the simple plate boundaries of the circum-Pacific during the last few decades. It is the expected time of occurrence, however, that has been the most elusive parameter, and it is the one that has the most scientific and social impact. Compilation of earthquake catalogs and geologic investigations of faulting have substantiated the idea of regular recurrence behavior for large and great earthquakes along segments of many simple plate boundaries. Incorporating the concept of recurrence intervals into the seismic gap model has enabled the development of time-dependent earthquake forecasts, and probabilistic statements of earthquake hazards. The probability for the occurrence of a future event along a specific fault segment during some time interval is a function of both the amount of time elapsed since the previous large or great earthquake and the average repeat time. Examples of this approach for earthquake forecasts in Chile and California can be found in Nishenko (1985) and Sykes and Nishenko (1984), and the Working Group on California Earthquake Probabilities (1988). At present, the observed periodicity of moderate-sized earthquakes occurring along the Parkfield section of the San Andreas fault in central California is being used as the basis for an intensive earthquake prediction experiment.

Because of the observed variability of earthquake recurrence intervals at a single location or fault segment, the temporal resolution available using historic and geologic data ranges from a few decades to a few years, at best. Hence, recurrence time estimates based on these types of data usually fall into the category of earthquake forecasts. More specific estimates for the time of a future event require the observation of precursory phenomena.

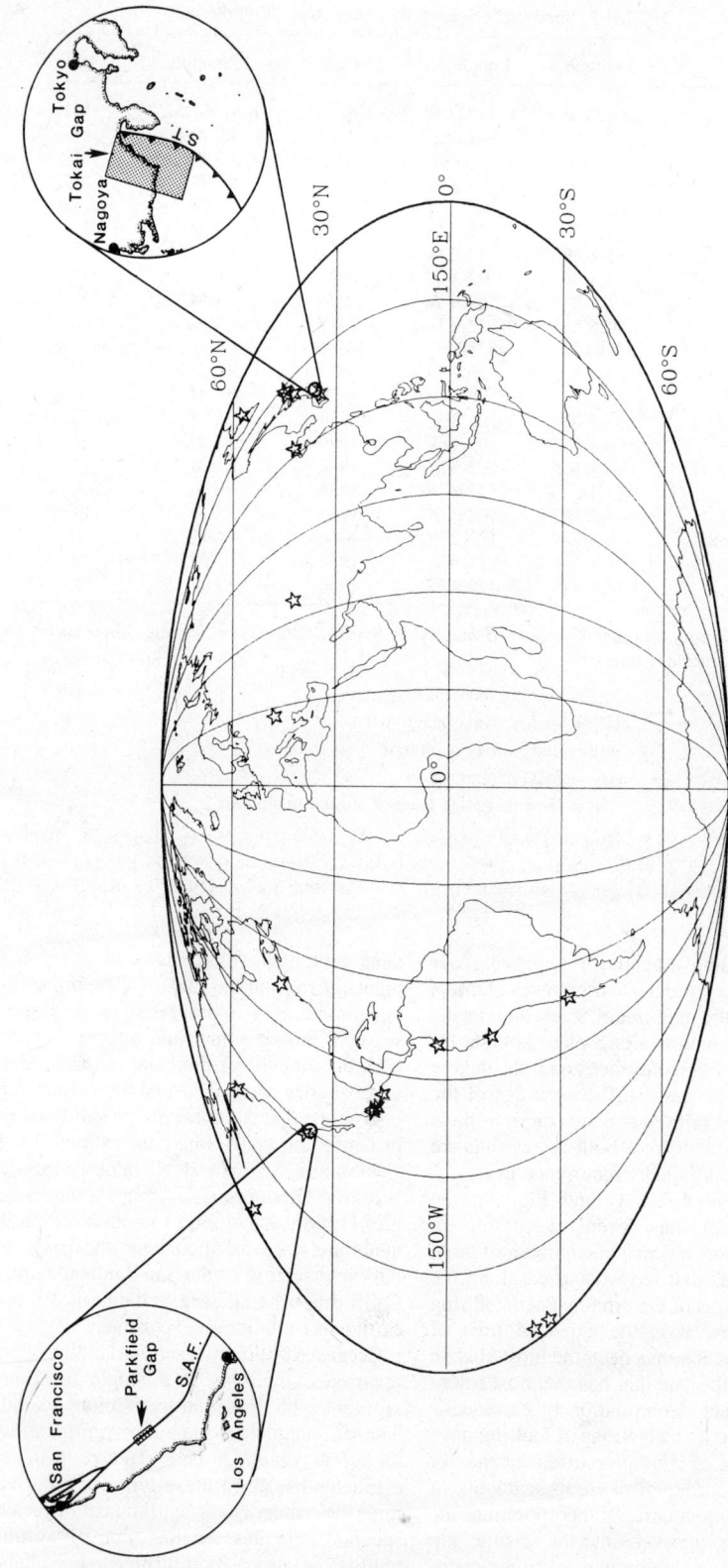

FIGURE 1. Successful earthquake forecasts and predictions and the location of two current prediction programs (1987). Stars show the locations of the successful forecasts and predictions listed in Table 1. Inset maps show the locations of current prediction programs along the San Andreas fault (S.A.F.) in the Parkfield, central California, area (left) and along the Suruga Trough (S.T.) in the Tokai, central Japan, region (right).

Short and Intermediate Term Estimates—Precursors

The observation of a number of anomalous physical phenomena prior to the occurrence of earthquakes (termed precursors) has prompted research into identifying and verifying them as possible indicators for routine use in earthquake prediction. The types of precursors that have been identified fall into the categories of crustal deformation, seismic activity, geochemical and hydrological anomalies, and other types of geophysical anomalies. While some of these phenomena have been observed prior to many earthquakes, no one phenomenon has yet been observed to precede every earthquake. Finding anomalies that systematically occur prior to specific types of earthquakes and that can be reliably used for intermediate- and short-term prediction represents one of the main challenges in seismology today. Recent reviews of progress in this area can be found in Aki and Stuart (1987).

Crustal Deformation. Observations of crustal deformation measure the amount of strain accumulating in the Earth's crust. Over a time interval of tens of years, leveling surveys and tide gauge observations provide information about vertical deformations, while information about horizontal deformations is provided by triangulation and trilateration networks. Shorter term data are obtained by continuous monitoring of tide gauges, strainmeters, tiltmeters and gravity surveys. Anomalous ground deformations preceding the 1944 Tonankai, Japan, and 1964 Niigata, Japan, earthquakes are often cited as examples of these types of phenomena. Anomalous crustal activity, however, has not been noted for all earthquakes. For example, catalogs of earthquakes occurring in Japan since 1923 only list about 26% of all destructive shocks as being preceded by anomalous deformation. The reliability of these types of data are further complicated by the lack of a systematic relationship between the time of appearance, distance, and magnitude of a forthcoming event. Additionally, a number of instances of nonseismic crustal movements have also been reported (cases of uplift without associated earthquakes).

Seismic Activity. By far the broadest category of observed premonitory phenomena, seismic activity encompasses changes in spatial and temporal patterns of seismicity and the occurrence of foreshocks, as well as variations in other seismologically measured quantities such as seismic wave velocities, focal mechanisms, b values (from the Gutenberg-Richter frequency-magnitude relation, $\log N = A - bM$), and the spectral characteristics of earthquakes.

Spatio-Temporal Changes in Seismicity. Changes in patterns of small earthquakes prior to larger events have been noted in a number of instances. In 1969, K. Mogi recognized that, prior to several large Japanese earthquakes, the region surrounding the future rupture zones were active, while the rupture zones themselves were seismically quiet. This encircling pattern of seismicity was named the Mogi doughnut and is thought to be indicative of the long-term preparation process before large events. As seismicity patterns are complex in many instances, statistical treatment has become essential for recognizing these types of anomalies. Seismic quiescence, a decrease in the background seismicity or rate of occurrence of smaller shocks prior to a larger event, has been noted for a number of earthquakes and was used as a basis for the prediction of the 1978 (M_s 7.8) Oaxaca, Mexico, and 1986 (M_L 4.6) Stone Canyon, California, earthquakes (see Table 1). The statistical identification of these types of anomalies depends on the completeness of the earthquake catalogs and rigorous definitions of seismicity patterns. On a global basis, complete earthquake catalogs (i.e., catalogs containing all events above a certain threshold magnitude, e.g., $M \geq 5$) exist only since the early 1960s as a result of the development of the World Wide Standardized Seismograph Network, WWSSN). This time interval is shorter than the observed recurrence times for many large and great earthquakes, and it has not been established if fluctuations in seismicity rates and patterns occur periodically or if they are primarily related to the preparation process preceding larger events.

Foreshocks. Foreshocks or the occurrence of smaller earthquakes immediately before a larger event have been reported prior to a large number of events. Globally, approximately 44% of large shallow earthquakes are preceded by foreshocks. Patterns of foreshock activity appear to be related to rock type. In laboratory experiments, foreshock activity that increases continuously up until the time of the main fracture is characteristic of heterogeneous rock types, while discontinuous activity that decreases immediately before the main fracture is characteristic of more homogeneous rock types. Discriminating foreshocks from other nonprecursory small earthquakes and swarms is necessary if foreshocks are to be used in actual real time prediction efforts. Based on prior earthquake histories, it is possible to identify regions that may be expected to produce foreshocks prior to the main event. Discrimination on the basis of changes in b value has also been reported in a number of cases. Foreshocks have been reported to have smaller b values (or a lower number of smaller magnitude events) than aftershocks or other types of earthquakes. Additional investigations into discrimination based on changes in stress drop and spectral characteristics (i.e., ratios of high-frequency to low-frequency waves) also appear promising at this date.)

Variations in Seismic Wave Velocities, (V_p/V_s) Ratios. In the early 1960s, variations in the velocities of seismic waves (principally P waves) passing through the preparation region of a forthcoming earthquake were noted in the Garm region of Soviet Central Asia. Similar observations were made in the United States in the early 1970s, at Blue Mountain Lake, New York, where they were used to predict a small earthquake. Experiments designed to find similar anomalies elsewhere along active fault zones have met with mixed results. Physically, changes in seismic wave velocities are accounted for by the dilatancy model of rock fracture, in which changes in tectonic stress cause microcracks to open and close prior to a catastrophic failure.

Systematic Changes in Earthquake Source Mechanisms and Source Parameters. Changes in focal mechanisms of small shocks sometimes reflect a rotation of the principal compressive stress prior to large events. The occurrence of small earthquakes with higher than average stress drops are also thought to reflect the presence of high stress levels in the region prior to failure. Both are presently being investigated as tools for monitoring seismic gaps.

Geoelectric and Geomagnetic Precursors. Changes in resistivity of the Earth's crust are related to changes in the volume of microcracks and water. Typically, resistivity changes are many times larger than actual volumetric strain changes. Changes prior to local shocks may be related to the opening and closing of microcracks as described by the dilatancy model. Problems with noise, including rainfall contamination of signals, and sensitivity to the type of underlying rock affect the reliability of these type of data. Geomagnetic anomalies have been reported before some earthquakes, however, background noise and few reported cases also make these data hard to interpret at present.

Geochemical and Hydrological Precursors. The history of research for geochemical precursors prior to major earthquakes is shorter than others previously mentioned. Two classes of anomalies have been discussed in the literature, (1) changes in level of water wells and rates of flow of springs (whether these anomalies reflect coseismic or preseismic changes, i.e., changes in strain at the time of or prior to an earthquake, respectively, is not clear in all cases) and (2) changes in the concentration of elements in groundwater (Rn, Cl and ratios of isotopes, $^2H/^1H$, and other elements He/Ar, N^2/Ar) have been claimed as precursors. For example, the 1966 m_b 5.5 Tashkent earthquake in Soviet Central Asia was preceded by a 100%-200% increase in radon concentration.

A Successful Prediction: The 1975 Haicheng, China, Earthquake

The successful prediction of the 1975 M_s 7.3 Haicheng earthquake illustrates the basic strategy and development of a specific prediction program from a long-term forecast through the actual short-term prediction, within an intraplate environment.

Long-Term Forecast Phase—1966–1973: Following the 1966 Xingtai earthquake, seismic activity in northern China increased notably. Analysis of historical and recent seismic activity, in addition to the tectonic setting of the region, suggested that the area may be due for a large damaging earthquake in the future. This judgment was substantiated by increases in seismic activity since 1966 and observable crustal deformation. Intermediate-Term Prediction Phase—1973–1974: Intensified research and observation revealed increases in crustal movement, seismic activity, geomagnetic changes, and rises in sea level (from tide gauge stations). These changes prompted the institution of the third phase, Short-Term and Imminent Prediction—1974–1975: Continuous increases in seismic activity and leveling anomalies. In addition, anomalies in water well levels and animal behavior began to be reported. Radon anomalies and swarms of earthquakes (foreshocks) were reported in the final stages, which led to the issuing of a short-term prediction nine hours before the actual occurrence. A more detailed review of the prediction can be found in Raleigh et al. (1977).

Current Prediction Programs

Currently, earthquake prediction programs in Japan and the United States are focused on two interplate fault segments that have high long-term probabilities for recurrence. Present efforts are concentrated on observation of intermediate- and short-term precursors.

Japan-Tokai gap: The Tokai region of central Japan is the primary target of the Japanese earthquake prediction program (see inset in Fig. 1). The Tokai region is located along the Suruga Trough, which is the eastern extension of the Nankai Trough, where the Philippine plate is subducted beneath central Japan. The Japanese prediction program is based on a two-stage strategy. In the first stage, the target area is specified based on historical, seismological, and geologic criteria. In the second stage, the goal is to predict the time of occurrence by detecting intermediate and short-term precursors by means of concentrated observations in the target region. The primary long-term criteria for selecting the Tokai region are three-fold. First, the Tokai region ruptured in the great 1707 and 1854 Nankai Trough earthquakes that extended into the Suruga Trough region, but was unruptured in the 1944 and 1946 series of Nankai Trough events. Historically, great earthquakes tend to occur along the Nankai Trough every 100 to 150 years, and it has been 132 years since the last event. Hence, the Tokai region is considered a seismic gap. The magnitude of the expected gap-filling Tokai earthquake is about $M_w \approx 8.0$. Second, measurements of crustal deformation indicate steady subsidence along the western margin of Suruga Bay since 1900, as well as horizontal shortening across the bay of about 1 m in 100 yr. Finally, seismicity patterns in and around the Tokai district indicate the presence of a Mogi doughnut since about 1973. The actual prediction of the expected Tokai earthquake will depend on the observation of short-term precursors. Few data exist for the 1854 earthquake to identify characteristic precursors, however, comparison with anomalous activity immediately before the 1944 and 1946 shocks (crustal uplift and anomalous hot spring and water well behavior) may provide the crucial signs to allow a successful prediction of a future Tokai event. Further background information can be found in Ishibashi (1981) and Mogi (1986).

United States-Parkfield, California: The Parkfield earthquake prediction experiment is being undertaken on a section of the San Andreas fault between the towns of Parkfield and Cholame (see inset in Fig. 1). The last earthquake that ruptured this fault segment occurred in 1966 (M_L 5.6). The average recurrence interval for earthquakes of this size is about 22 yr, based on the occurrence of events along this segment of the San Andreas fault since 1857. The expected date of the next Parkfield earthquake is approximately 1988 and estimates of the conditional probability for the recurrence of this event during the next 5 yr (1988-1993) are at the 68% level, and for the next 10 yr (1988-1998) are at the 92% level. Historically, Parkfield earthquakes have been preceded by one or a series of foreshocks originating at the epicenter, near the northern end of the rupture zone. In this case, past patterns of activity allow the identification of a possible precursor that may provide a basis for the short-term prediction of the next Parkfield earthquake. A review of the Parkfield earthquake prediction experiment is presented in Bakun and Lindh (1985).

S. P. NISHENKO

References

Aggarwal, Y. P., L. R. Sykes, D. W. Simpson, and P. G. Richards, 1975, Spatial and temporal variations in t_s/t_p and in P wave residuals at Blue Mountain Lake, New York: Application to earthquake prediction, *Jour. Geophys. Research* **80**, 718-732.

Aki, K., 1980, Presidential address: Possibilities of seismology in the 1980s, *Bull. Seismol. Soc. America* **70**, 1969-1976.

Aki, K., and W. D. Stuart, eds., 1987, *Physical and Observational Basis for Intermediate-Term Earthquake Prediction*, Proceedings of Workshop XXXVII, 2 vols., U.S. Geol. Survey Open File Rpt. 87-591, 989p.

Bakun, W. H., and A. G. Lindh, 1985, The Parkfield, California earthquake prediction experiment, *Science* **229**, 619-624.

Fedotov, S. A., 1965, Regularities of the distribution of strong earthquakes in Kamchatka, the Kuril Islands, and northeast Japan, *Akad. Nauk. SSSR Inst. Fiziki Zemli Trudy* **36**, 66-93.

Gilbert, G. K., 1909, Earthquake forecasts, *Science* **29**, 121-138.

Imamura, A., 1928, On the seismic activity of central Japan, *Japanese Jour. Astron. Geophys.* **6**, 119-137.

Ishibashi, K., 1981, Specification of a soon-to-occur seismic faulting in the Tokai district, central Japan, based upon seismotectonics, in D. W. Simpson and P. G. Richards, eds. *Earthquake Prediction, An International Review*, Washington, D.C.: American Geophysical Union (Maurice Ewing Series, vol. 4), 297-332.

Kelleher, J. A., 1972, Rupture zones of large South American earthquakes and some predictions, *Jour. Geophys. Research* **77**, 2087-2103.

Kelleher, J. A., L. R. Sykes, and J. Oliver, 1973, Possible criteria for predicting earthquake locations and their applications to major plate boundaries of the Pacific and Caribbean, *Jour. Geophys. Research* **78**, 2547-2585.

Kisslinger, C., 1988, An experiment in earthquake prediction and the May 7, 1986 Andreanof Islands earthquake, submitted *Bull. Seismol. Soc. America* **78**, 218-229.

Mogi, K., 1969, Some features of recent seismic activity in and near Japan (2), *Tokyo Univ. Earthquake Research Inst. Bull.* **47**, 395-417.

Mogi, K., 1986, Recent earthquake prediction research in Japan, *Science* **233**, 324-330.

McCann, W. R., S. P. Nishenko, L. R. Sykes, and J. Krause, 1979, Seismic gaps and plate tectonics: Seismic potential for major boundaries, *Pageoph* **117**, 1082-1147.

Nishenko, S. P., 1985, Seismic potential for large and great interplate earthquakes along the Chilean and southern Peruvian margins of South America: A quantitative reappraisal, *Jour. Geophys. Research* **90**, 2589-3615.

Ohtake, M., T. Matumoto, and G. V. Latham, 1977, Seismicity gap near Oaxaca, southern Mexico as a probable precursor to a large earthquake, *Pageoph* **115**, 375-385.

Purcaru, G., 1979, The Vrancea, Romania earthquake of March 4, 1977—A quite successful prediction, *Physics Earth and Planet. Interiors* **18**, 274-287.

Raleigh, B., G. Bennett, H. Craig, T. Hanks, P. Molnar, A. Nur, J. Savage, C. Scholtz, R. Turner, and F. Wu, 1977, Prediction of the Haicheng earthquake, *EOS (AGU Trans.)* **58**, 236-272.

Reid, H. F., 1910, The California earthquake of April 18, 1906, in *The Mechanics of the Earthquake*, Washington, D.C.: Carnegie Institution, Report of the State Earthquake Comm., vol. 2, 16-19.

Richter, C. F., 1958, *Elementary Seismology*, San Francisco: W. H. Freeman, 768p.

Rikitake, T., 1976, Earthquake Prediction. New York: Elsevier, 357p.

Sadovsky, M. A., and I. L. Nersesov, 1983, Earthquake prediction problems in the USSR, in *Proceedings of the Seminar on Earthquake Prediction Case Histories*. Geneva, Switzerland: UNDRO, 35-49.

Silgado F., E., 1973, Historia de los sismos mas notables ocurridos en el Peru (1513-1970), *Geofisica Panamericana* **2**, 179-243.

Simpson, D. W., and P. G. Richards, eds., 1981, *Earthquake Prediction, An International Review*, Washington, D.C.: American Geophysical Union (Maurice Ewing Series, vol. 4), 680p.

Sykes, L. R., 1971, Aftershock zones of great earthquakes, seismicity gaps and earthquake prediction for Alaska and the Aleutians, *Jour. Geophys. Research* **76**, 8021-8041.

Sykes, L. R., and S. P. Nishenko, 1984, Probabilities of

occurrence of large plate rupturing earthquakes for the San Andreas, San Jacinto and Imperial faults, California, 1983-2003, *Jour. Geophys. Research* **89,** 5905-5928.

Working Group on California Earthquake Probabilities, 1988, Probabilities of large earthquakes occurring in California on the San Andreas fault, *U.S. Geol. Survey Open File Rpt.* 88-398, 62p.

Wyss, M., and R. O. Burford, 1985, Current episodes of seismic quiescence along the San Andreas fault between San Juan Bautista and Stone Canyon, California: Possible precursors to local moderate mainshocks? in C. F. Shearer, ed., *Minutes of the National Earthquake Prediction Evaluation Council*, U.S. Geol. Survey Open File Rpt. 85-754, 368-426.

Cross-references: *Earthquake Mechanisms; Earthquakes: Location Techniques; Earthquakes: Magnitude, Energy, and Intensity; Earthquakes: Volcanogenic; Earthquakes and Crustal Deformation; Earthquake Seismology; Geodetic Leveling; Geodetic Networks and Control Surveys; Lithosphere: Mechanical Properties; Seismicity: Intraplate; Seismicity: Subduction Zone; Seismic Source: Observations; Seismology: History; Stress in the Earth's Lithosphere; Strong Motion Seismology; Tsunamis.*

EARTHQUAKES: LOCATION TECHNIQUES

Locating an earthquake implies that the point where the rupture process first began (nucleated) is estimated. That is, either the point within the Earth (the hypocenter) or the point on the surface of the Earth immediately above the hypocenter (the epicenter) is determined. Usually the time of nucleation, the origin time, and estimates of size, called magnitudes, are found as a by-product of the location process. The distinction between estimating the nucleation point and the rupture zone extent can be important as rupture zones may exceed 1000 km in length while the dimensions of the nucleation zone are probably on the order of tens to hundreds of meters.

Until the early twentieth century, earthquakes were located by examining surface faulting, damage, or felt reports. However, essentially all present-day earthquake locations are determined by mathematical modeling of instrumental recordings. Seismic instruments, located at a seismic observatory or station, typically record a signal proportional to ground motion in one rectilinear direction over a limited frequency band. For the purposes of earthquake location, the arrival times of relatively high frequency (typically .5-10 Hz) wave groups, called body waves, are the most useful. The most useful of these by virtue of their prominance at the beginning of event recordings and the ease of measuring their arrival times are the compressional or P waves, which are best observed on vertical component instruments. Even for events with complex rupture histories the beginning of the body wave train can be considered to be emitted by the nucleation process.

Earthquake location holds a unique and somewhat paradoxical position in the hearts and minds of seismologists. Virtually all seismic research, all seismic hazards work, and much of earthquake engineering begins with a knowledge of the location of one or more earthquakes. However because the task of preliminary event location is performed routinely by service organizations such as the (U.S.) National Earthquake Information Service and the International Seismological Centre, it is considered to be "seismic practice" rather than research. Subsequent work, including single event relocation, joint hypocentral determination, and simultaneous earthquake location and inversion for earth structure are considered to be research. Because of the long history of earthquake location, a similar dichotomy exists with respect to research on the location algorithm itself. Researchers generally consider the problem to be solved. On the other hand, seismic analysts, who routinely locate earthquakes, find existing algorithms inadequate in numerous ways, which can only be resolved by further research.

The nature of this article necessitates the presentation of only the barest bones of the topic. References given are generally either classic papers or good, but rather technical, summaries of an important aspect of the subject. More comprehensive summaries of earthquake location have been given by Dewey (1979) and Lee and Stewart (1981).

History

In the late nineteenth century the instrumental recording of large earthquakes at stations around the world was becoming a reality. The mathematical underpinnings of least squares regression had been laid by Gauss and others many years before. In 1910 Dr. L. Geiger, working at the Geophysical Observatory of Saint Louis University, developed a mathematical algorithm for locating earthquakes from instrumental data (see Geiger [1912] for an English translation). This algorithm, which has come to be called Geiger's method, is the basis of all commonly used earthquake location procedures to this day. Jeffreys (1932) pointed out that the statistical errors in measured arrival times were distributed in a manner that greatly reduced the effectiveness of the least squares procedure. He proposed a simple modification that has come to be called uniform reduction. This method is a special case of modern, statistically robust M-estimator regression that it predated by 50 years. Bolt (1960) applied the uniform reduction algorithm to the single event location problem with considerable success.

In the late 1950s and early 1960s electronic, digital computers first became available for use in geophysical research. Naturally earthquake location was among the first of the seismological algorithms to be programmed. After a brief flurry of activity, research interest in the location problem once again waned to its present low level. However, many computer programs still in use today for routine earthquake location were originally written in this era. This fact is indicative of the success of Geiger's basic method and is a result of the need for uniformity in routinely prepared, cumulative earthquake catalogs.

The past quarter century has been a period of monumental changes in programming languages, computer hardware, and operating systems concurrent with equally substantial developments in numerical analysis, statistics, and seismology itself. Because earthquake location has largely become the province of seismic practice, seismologists have been slow to incorporate these advances into implementations of Geiger's method. In the following section, an updated version of Geiger's algorithm, which approaches the state-of-the-art, is presented.

Theory

Given the N body wave arrival times, t_i, $i = 1, 2, \ldots, N$, observed at stations with positions, $\mathbf{x}_i$, $i = 1, 2, \ldots, N$, we wish to estimate the true hypocentral position, $\hat{\mathbf{x}}_0$, and origin time, $\hat{t}_0$, of the causative event. The position vectors, $\mathbf{x}_i$, represent the three spatial coordinates designated as $x_{ji}, j = 1, 2, 3$. This defines the single event earthquake location problem in which the data from one event only is used to determine the location of that event. Define travel time residual to be the difference between the observed and computed arrival times

$$r_i^k = t_i - (t_0^k - \Im(\mathbf{x}_i, \mathbf{x}_0^k)) \qquad (1)$$

where $\Im$ is a model of travel time and t_0^k and $\mathbf{x}_0^k$ are estimates of the hypocentral parameters after k successive approximations. In general, travel time is a nonlinear function of the distance between the source and the receiver. Therefore the $k + 1$ estimate is found by using a linear approximation to the residual

$$\tilde{r}_i^k(\delta t^k, \delta \mathbf{x}^k) = r_i^k - \left(\delta t^k + \sum_{j=1}^{3} \delta x_j^k \frac{\partial \Im(\mathbf{x}_i, \mathbf{x}_0^k)}{\partial x_{j0}^k}\right) \qquad (2)$$

where the $\delta x_j^k, j = 1, 2, 3$ are the components of vector $\delta \mathbf{x}^k$ and the derivatives are evaluated at $\mathbf{x}_0^k$. The corrections δt^k and $\delta \mathbf{x}^k$ to t_0^k and $\mathbf{x}_0^k$ are found by minimizing

$$\sum_{i=1}^{N} \rho(\tilde{r}_i^k) \qquad (3)$$

where ρ is a penalty function which, in general, is nonlinear. The choice of ρ is determined by the statistical model chosen. Minimizing Eq. 3 is equivalent to solving the set of simultaneous equations

$$\sum_{i=1}^{N} \frac{\partial \rho(\tilde{r}_i^k)}{\partial \delta x_j^k} = 0, j = 1, 2, 3 \text{ and } \sum_{i=1}^{N} \frac{\partial \rho(\tilde{r}_i^k)}{\partial \delta t^k} = 0$$

(4)

The $k + 1$ approximation is then given by

$$t_0^{k+1} = t_0^k + \delta t^k \text{ and } \mathbf{x}_0^{k+1} = \mathbf{x}_0^k + \delta \mathbf{x}^k \qquad (5)$$

Assuming convergence, this iteration continues until $|\delta \mathbf{x}^k|$ becomes small. The final values of the hypocentral parameters, t_0^{k+1} and $\mathbf{x}_0^{k+1}$ after the last iteration, are taken to be estimates of the true values $\hat{t}_0$ and $\hat{\mathbf{x}}_0$. There are variants of the algorithm in which origin time is determined separately, either during or after the iteration, so as to make the mean or median of the residuals zero. Note that data from at least three different stations is required to uniquely define the hypocentral position. Relocation of earthquakes as part of a research study is done in exactly the same manner as single event location. Also the location of explosions may be done using exactly the same algorithm.

Seismological Considerations

In practice, Geiger's method is interpreted differently in different seismological regimes. In particular, different computer programs are often used for local, regional, and teleseismic data. The regime is defined by the (angular) distance, Δ_{max}, between the epicenter and the most distant station. For our purposes, we define local distances to be $\Delta_{max} \leq 5°$, regional distances to be $5° < \Delta_{max} < 30°$, and teleseismic distances to be $30° \leq \Delta_{max}$. At local distances the curvature and ellipticity of the earth may be neglected, but lateral structural variations cannot. Local location programs generally assume highly simplified earth models that permit the rapid calculation of travel time for each datum during each iteration. A different laterally homogeneous Earth model may then be used for each local network of stations. At teleseismic distances the curvature and ellipticity of the Earth cannot be neglected, but lateral variations often can. In addition travel times become very complicated as a result of the multiplicity of body waves generated by the deep structure of the Earth. In this case one set of average travel times (usually in a table format) is generally used. Although the regional location problem has not been as well studied, it is clear that it encompasses all of the difficulties of both local and teleseismic earthquake location. There is no compelling reason why all three regimes could not be accommodated by a single computer program in which the travel times were computed directly from an arbitrary, spherical Earth model (see Buland and Chapman, 1983).

To a fair approximation, the structure of the Earth varies only with depth. The treatment of general lateral structural variations is very complicated and currently an area of active seismological research. However, a few simple, but very important, aspects of lateral variations are routinely approximated including station elevation, ellipticity of the Earth, and anomalous structure beneath each station. The first two approximations account for perturbations in the distance that each body wave must travel due to the laterally varying shape of the Earth. In the third approximation, near station lateral structure is accounted for by either a constant or some simple function of azimuth. Also, nonphysical solutions are always excluded by constraining hypocenters to lie underground.

Statistical Considerations

Regression analysis is a statistical procedure. Thus far we have seen how a model (the hypocentral parameters) can be constructed (relative to the assumed Earth model) by fitting a set of observed body wave arrival times. Assessing the validity of this model requires relating the combined measurement and modeling errors to the reliability of the derived hypocentral parameters. This assessment is done by assuming that (1) all of the observations can be scaled (variance weighted) to have approximately the same error characteristics, (2) the distribution of observed, scaled residuals is representative of the distribution of errors, and (3) the observations are unbiased and uncorrelated. Given these assumptions least squares regression ($\rho(r) \equiv r^2$) is the optimal, unbiased, linear estimator by the Gauss-Markoff theorem and corresponds to the classical Geiger's method. If information about the residual distribution is derived from each earthquake separately, then parameter reliability is proportional to the Fisher distribution. This approach to assessment is the most commonly used, most correct, but least informative (particularly when the number of observations is small). If information about the residual distribution is known before the event location then parameter reliability is proportional to the χ^2 distribution. This approach is the most informative but requires accurate knowledge of the residual distribution. It is also possible to combine *a priori* information with information from each event. This Baysian approach is a good compromise in many cases (Jordan and Sverdrup, 1981).

Actually no existing earthquake location program relies on pure least squares because observed travel time residual distributions are so non-Gaussian that least squares is simply not adequate. To circumvent this problem, location algorithms all incorporate some means of reducing the effect of outliers (data with large residuals) on the solution. Popular methods include discarding data with large residuals (truncation) and weighting data according to their current residuals and sometimes other criteria (iterative reweighting). The proper way to solve this problem is to use some form of robust regression. Both maximum likelihood (M-estimator) and rank-sum (R-estimator) regression can be applied quite effectively. Both techniques yield unbiased estimates at the expense of nonlinear penalty functions, and both can be tailored to yield optimal, unique solutions (which are asymptotically equivalent, as $N \rightarrow \infty$). Even better M-estimators (including uniform reduction) can be derived, but they are not unique (Buland, 1986). Estimates of the reliability of the hypocentral parameters determined by robust methods are found in nearly the same way as for least squares.

Numerical Considerations

The poor performance of earthquake location programs is generally attributed to numerical rather than statistical difficulties. Numerical problems arise from two sources, (1) nonlinearity and (2) poorly conditioned simultaneous equations. The nonlinearity arises from the nonlinearity of the travel time function and possibly the penalty function. It results in a finite radius of convergence (volume around the best estimate in which the iterative algorithm will converge). Convergence can be improved by combining the steepest descent (optimal far from the solution) and least squares (optimal near the solution) directions (Marquardt's algorithm) and then shortening (damping) the step length as needed. Poor conditioning is a result of intrinsic geometrical limitations in the ability of a given set of observations to resolve a given set of hypocentral parameters. The most common example is attempting to locate an earthquake that is much farther from the nearest station than the stations are from each other. Poor conditioning can degrade convergence or preclude a solution altogether. Conditioning can be improved by various matrix techniques (notably the QR algorithm). Conditioning can also be dramatically improved in many cases by applying a constraint such as fixing hypocentral depth. If all else fails, techniques such as the singular valued decomposition and ridge regression can guarantee a solution (which may be nearly useless). In practical terms there is no substitute for gathering more and better data.

Later Arrivals

Later arriving body waves can be used for earthquake location in exactly the same way as first arrivals. However, in practice, only two types of later arrivals are commonly used. At local and regional distances, shear (S) waves are often well recorded. Because S waves travel at a different speed than P waves, the S minus P time difference is directly proportional to epicentral distance. At teleseismic distances, body waves that are reflected from the surface near the source and then travel to the receiver (e.g., pP) are often observed. These arrivals turn out to provide a powerful constraint on hypocentral depth.

Station Corrections

The utility of station corrections has been discussed earlier. Station corrections are estimated by averaging the residuals found by locating a suite of well constrained earthquakes. This estimate can be improved by iteratively relocating the events using the station corrections and then determining new station corrections until convergence is obtained. Station corrections have seen little use in routine teleseismic applications because only about one third of the reporting stations contribute sufficient data from which to estimate a reliable correction. Stations with no corrections may be useful for only a few events each year, but they are often critical for these events.

Joint Hypocentral Determination

Unfortunately the distribution of earthquakes and seismic stations is such that it is not uncommon for a number of events to be much closer to each other than to the nearest seismic station. In such cases the body waves traveling from all events to each station sample lateral structural variations in approximately the same way. By simultaneously locating all earthquakes and determining a constant correction for each station, the relative accuracy of the event locations with respect to one another can be greatly improved. Accurate relative locations are useful for understanding the shape of tectonic structures that generate earthquakes. Once a joint hypocentral determination has been done for a region, relative locations for subsequent events in the same region can be estimated by applying the same station corrections. Alternatively, by orthogonalizing the problem (decoupling relative location from the determination of station corrections) it is possible not only to locate each subsequent event relative to the others, but to improve the station corrections and the relative locations of the other events as well with a minimum of computation (Jordan and Sverdrup, 1981). Additionally it is possible to extended joint hypocentral determination to estimate some features of earth structure in the vicinity of the events being located. A related method, called the master event technique, involves locating one very well recorded earthquake by the single event technique and then treating its residuals as station corrections for subsequent events in the same region (i.e., locate other earthquakes relative to the master event). This method is particularly effective if the master event can be localized by other, independent information. Joint hypocenter and master event techniques have seen little use in routine applications because the station corrections change significantly over relatively short distances (≈ 50–100 km for teleseismic data).

Earthquake Location and Inversion for Structure

Although the simultaneous inversion for earth structure and the location of earthquakes within the structure is significantly more complicated than inverting for structure alone there are several applications where it is warranted. A well-known example is inversion for laterally homogeneous crustal structure using locally recorded earthquakes (e.g., Pavlis and Booker, 1980). Another important application is in the study of the three-dimensional structure underlying island arcs. Note that in principle, earthquake location in a three-dimensional structure is no more complicated than a one-dimensional structure provided that travel times can be estimated. In practice significant new difficulties can arise. Intriguingly the result of these procedures is often to focus seismic activity from apparently diffuse patterns into clusters that seem to delineate tectonic structures. As for joint hypocentral determination, simultaneous earthquake location and structural inversion can be orthogonalized to decouple the two parts of the problem allowing events that occur subsequent to the inversion to be located while refining both the structure and all previously determined earthquake locations.

RAY BULAND

References

Bolt, B. A., 1960, The revision of earthquake epicenters, focal depths and origin-times using a high-speed computer, *Royal Astron. Soc. Geophys. Jour.* **3**, 433–440.

Buland, R., 1986, Uniform reduction error analysis, *Seismol. Soc. America Bull.* **76**, 217–230.

Buland, R., and C. H. Chapman, 1983, The computation of seismic travel times, *Seismol. Soc. America Bull.* **73**, 1271–1302.

Dewey, J. W., 1979, A consumer's guide to instrumental methods for determination of hypocenters, *Geol. Soc. America Eng. Geol. Rev.* **4**, 109–117.

Jeffreys, H., 1932, An alternative to the rejection of observations, *Roy Soc. [London] Proc.* **187**, 78–87.

Jordan, T. H., and K. A. Sverdrup, 1981, Teleseismic location techniques and their application to earthquake clusters in the south-central Pacific, *Seismol. Soc. America Bull.* **71**, 1105–1130.

Geiger, L., 1912, Probability method for the determination of earthquake epicenters from the arrival time only, *St. Louis Univ. Bull.* **8**, 60–71.

Lee, W. H. K., and S. W. Stewart, 1981, *Principles and Applications of Microearthquake Networks*, San Francisco: Academic Press, 293p.

Pavlis, G. L., and J. R. Booker, 1980, The mixed discrete continuous inverse problem: application to the simultaneous determination of earthquake hypocenters and velocity structure, *Jour. Geophys. Research* **85**, 4801–4810.

Cross-references: *Earthquakes: Hazards and Prediction; Earthquake Seismology; Earthquakes and Seismicity; Elasticity and Wave Propagation: Principles; Inverse Theory and Methods: Seismology; Seismic Instrumentation; Seismic Monitoring of Nuclear Explosions; Seismic Noise; Seismic Signal Processing; Seismograms: Interpretation; Seismology: History.*

EARTHQUAKES: MAGNITUDE, ENERGY, AND INTENSITY

Development of the Concepts

Preliminaries. The focus of an earthquake radiates mechanical energy in the form of seismic waves. The radiation of seismic energy during the event continues from a fraction of a second to several tens of seconds. The duration of seismic wave radiation is identical with that of the instantaneous faulting.

The time between two radiation events from a given focus is substantially larger than the radiation duration of a given event. Thus, earthquakes are transient radiation phenomena, and the methods for the determination of the strength of sources radiating transient signals are correspondingly applicable. On this basis the earthquake magnitude scale was introduced. A steady development of the magnitude concept is necessitated by ongoing improvements in seismological instrumentation and in the theories of the focal process and of seismic wave propagation.

A comprehensive review of the magnitude problem was given by Båth (1981), together with an extensive bibliography on the topic. Reference is made to this paper, as far as pre-1981 findings are concerned. In the following, only essential information is repeated; otherwise the development since 1981 stressed.

Basic Principles of Seismic Rating. The seismic energy radiated from the focus F is sensed at the earth surface. Figure 1 shows the basic geometry of the problem. At short epicentral distance Δ the ground motion can be strong enough to cause specific sensations to human beings and/or to inflict damage to them and to objects. Sensation and damage are usable to rate the macroscopic strength of ground motion at a given place. In order to quantify the strength of shaking, the noninstrumental seismic intensity scale is available, first introduced more than 100 years ago (i.e., prior to seismographs), and thus prior to the definition of the earthquake magnitude. Usually, a suit of strengths will correspond to a given earthquake, the strengths generally decreasing with

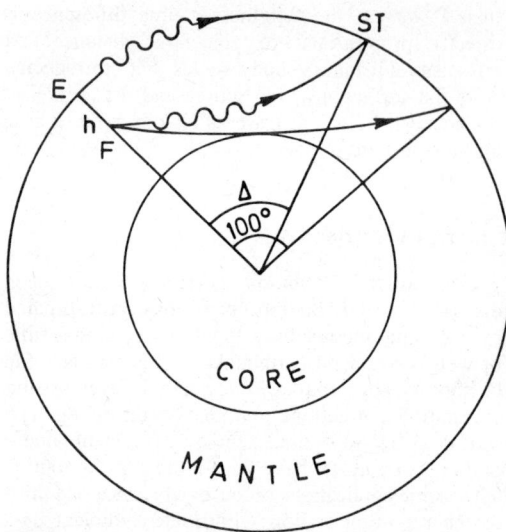

FIGURE 1. Propagation of body waves and surface waves radiated from focus F with focal depth h and arriving at station ST with epicentral distance Δ.

the distance from the epicenter of the earthquake. The seismic intensity reflects the total effect of the ground motion generated by the earthquake at the point of observation (see the last section, "Seismic Intensity").

The seismic energy arrives at the point of observation ST in the form of seismic phases, notably of body waves (P waves and S waves) and of surface waves (Love and Rayleigh waves). The seismographic recordings of the waves constitute the basis of an instrumental rating of the strength of the radiation from the seismic source F.

The foci F can have depths h below the earth surface ranging from 0 km (surface events, explosions) to 700–800 km (deepest earthquakes so far observed). For a given focal depth the amplitude of the seismic phase varies with epicentral distance. In order to arrive at a quantity independent of the epicentral distance and characterizing the radiation strength of the seismic source only, the amplitude measured at a particular epicentral distance Δ must be compensated for this variation. Thus, the amplitude variation curve is essential for the definition of any magnitude scale. Examples of amplitude variation curves are shown in Fig. 2. Here and in the following, it is assumed that with respect to the amplitude and the frequency content of the signal, the radiation from the focus is the same in all directions.

Because the shadow of the Earth core prevents the observation of body waves at greater distances, the amplitude variation curves for P and S waves are obtained only up to epicentral distances of about 100°. The application of reflected and diffracted P

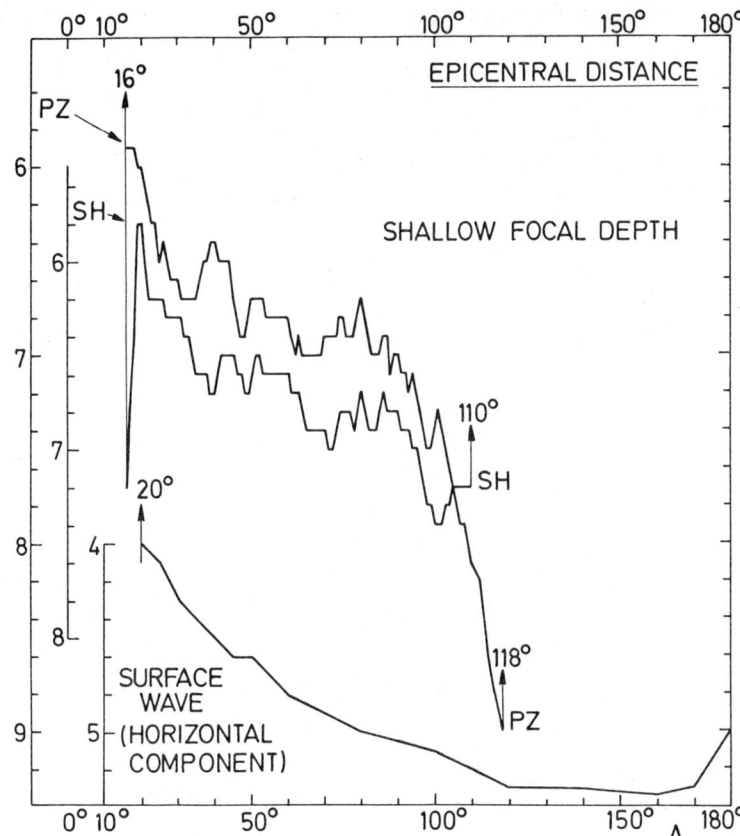

FIGURE 2. Amplitude variation of body waves and surface waves as function of epicentral distance. The ordinates are of σ_b (6), as given in Fig. 3 for the PZ and SH waves, and σ_s (3), as given in Table 2 for the surface waves.

and S waves, as well as that of body waves that are refracted through the core and arriving at greater distances, is inopportune for magnitude determinations, notwithstanding some attempts to utilize such waves correspondingly. Surface wave amplitude variation curves, on the other hand, are obtained out to epicentral distances of 180°.

With the knowledge of the amplitude variation curve, the magnitude scale for the corresponding seismic phase can be defined, if two further conditions are specified: (1) when to consider two earthquakes with the same focal depth to have magnitudes 0 and 1, respectively, and (2) when to consider two earthquakes with different focal depths to have identical magnitudes. As in any analogous definition of a physical quantity, the answer can only be arbitrary.

All magnitude scales employ decadic logarithms. Thus, the increase by one magnitude unit entails a tenfold increase of the measured quantity (seismogram trace amplitude, ground displacement amplitude, ground velocity amplitude, etc.). The zero-magnitude event is taken small enough to assure that most magnitudes are numerically positive.

According to the intention of Gutenberg and Richter (1956), two events are assigned the same body wave magnitude if the seismic energies released in the events are identical. The seismic energy is thus intended to link with each other events having different focal depths. This implies that a unique relation exists between the body wave magnitude and the seismic energy released. It also implies that it is possible to determine the seismic energy independently and with sufficient accuracy so as to make it applicable as a standard quantity entering the definition of the magnitude.

Magnitude Calibrating Functions.

The Local Magnitude M_L. The first magnitude scale was defined by Richter (1935) for earthquakes in California:

$$M_L = \log A(\Delta) + \sigma_L(\Delta)$$

$A(\Delta)$ is the vectorial sum of the maximum trace amplitudes on the horizontal seismograms of a given earthquake, in millimeters, as obtained on a particular seismometer located at the epicentral distance Δ (Wood-Anderson seismometer, $T_0 = 0.8$ s, $V_{\max} = 2800$, $\eta = 0.8$, where $T\sigma$, $V_{\max}$ and η are the free period of oscillation, the maximum magnification, and the damping ratio, respectively).

The calibrating function is

TABLE 1. Calibrating the Function Equation 1 for Local Magnitude M_L

Δ (km)	$\sigma_L(\Delta)$	Δ (km)	$\sigma_L(\Delta)$	Δ (km)	$\sigma_L(\Delta)$	Δ (km)	$\sigma_L(\Delta)$
0	1.4	90	3.0	260	3.8	440	4.6
10	1.5	100	3.0	280	3.9	460	4.6
20	1.7	120	3.1	300	4.0	480	4.7
30	2.1	140	3.2	320	4.1	500	4.7
40	2.4	160	3.3	340	4.2	520	4.8
50	2.6	180	3.4	360	4.3	540	4.8
60	2.8	200	3.5	380	4.4	560	4.9
70	2.8	220	3.65	400	4.5	580	4.9
80	2.9	240	3.7	420	4.5	600	4.9

Source: From Richter, 1958.

$$\sigma_L(\Delta) = -\log A(\Delta; M_L = 0) \quad (1)$$

where $A(\Delta; M_L = 0)$ is the maximum trace amplitude, in millimeters, for a magnitude-zero event at the distance Δ.

The calibrating function $\sigma_L(\Delta)$ does not depend on the focal depth, as the earthquakes in California were assumed to occur at common depths.

Table 1 gives the calibrating function (Eq. 1), defined up to an epicentral distance of 600 km. As seen from the table, a magnitude-zero event is one that would record at a distance of 100 km with a maximum trace amplitude of 1 μm.

In the strict sense the local magnitude, as defined here, is applicable in a region outside of California only if in both regions the maximum trace amplitudes on the Wood-Anderson seismograph vary in the same way and if the focal depths of earthquakes are the same.

The difficulties in verifying the conditions have led to the development of numerous independent local magnitude scales (Båth, 1981).

The Surface Wave Magnitude M_s. In order to extend the magnitude scale to larger epicentral distances, the surface wave magnitude, applicable to shallow events, was introduced. Here use is being made of the observation that surface waves with a period of about 20 s are persistently recorded for a large portion of earthquakes (Airy phase in the dispersion curve of the surface wave train).

The magnitude M_s is defined as

$$M_s = \log A(\Delta) + \sigma_s(\Delta) \quad (2)$$

where $A(\Delta)$ denotes the ground displacement amplitude, in micrometer, of the horizontal components of the surface waves combined vectorially, with a period of 20 s $\pm$ 3 s. Table 2 gives the calibrating function

$$\sigma_s(\Delta) = -\log A(\Delta; M_s = 0) \quad (3)$$

for epicentral distances Δ ranging from 20° to 180° (Richter, 1958).

Accordingly, a magnitude-zero event is defined as one producing at an epicentral distance of 20° a horizontal ground displacement with an amplitude of 10^{-4} μm at a period of about 20 s. The particular amplitude value was chosen in an attempt to obtain surface wave magnitudes M_s numerically identical to local magnitudes M_L. The problem of relating the surface wave magnitude to the local magnitude, however, was never solved satisfactorily, and both scales have to be considered as largely independent of each other.

In order to utilize surface waves with periods diverging from 20 s $\pm$ 3 s, the surface wave magnitude was redefined as

$$M_s' = \log \frac{A(\Delta)}{T} + \sigma_s'(\Delta) \quad (4)$$

TABLE 2. Calibrating the Functions in Equations 3 and 5 for Surface Wave Magnitude

Δ (degs)	$\sigma_s(\Delta)$	$\sigma_s'(\Delta)$	$\sigma_s'(\Delta)$ for $T = 20$ s	Δ (degs)	$\sigma_s(\Delta)$	$\sigma_s'(\Delta)$	$\sigma_s'(\Delta)$ for $T = 20$ s
20	4.0	5.46	4.16	90	5.05	6.54	5.24
25	4.1	5.62	4.32	100	5.1	6.62	5.32
30	4.3	5.75	4.45	110	5.2	6.69	5.39
40	4.5	5.96	4.66	120	5.3	6.75	5.45
50	4.6	6.12	4.82	140	5.3	6.86	5.56
60	4.8	6.25	4.95	160	5.35	6.96	5.66
70	4.9	6.36	5.06	170	5.3		
80	5.0	6.46	5.16	180	5.0		

where

$$\sigma'_s(\Delta) = -\log \frac{A(\Delta; M'_s = 0)}{T}$$

$$= 1.66 \log \Delta° + 3.3 \quad (5)$$

(Vanek et al., 1962). $A(\Delta)$ is the amplitude of the horizontal component of the ground displacement of the Rayleigh wave, in micrometers, and the period T ranges from 10 to 30 s.

To be consistent with Eq. 2, the ground displacement amplitude $A(\Delta)$ for any Δ is required for all earthquakes to increase linearly with period from 10 to 30 s. Equation 4, known as the Moscow-Prague formula, applies for earthquakes with focal depths smaller than 50 km and observed in the distance range 20°–160°. It was adopted and recommended for general use during the 1967 Zürich General Assembly of the International Union of Geodesy and Geophysics (IUGG) (Båth, 1973). The calibrating function $\sigma'_s(\Delta)$ is given in Table 2. It can be seen that $\sigma'_s(\Delta)$ for 20 s exceeds, on the average, $\sigma_s(\Delta)$ by 0.2; i.e., Eq. 5 yields surface wave magnitudes numerically larger by 0.2 than the original calibrating function (Eq. 3).

The Body Wave Magnitude m_b. An extension of the magnitude scale to events with arbitrary focal depths became possible by employing body waves:

$$m_b = \log \frac{A(\Delta)}{T} + \sigma_b(\Delta, h) \quad (6)$$

This scale is based on the maximum ground velocity $(A(\Delta)/T)$ observed at the epicentral distance Δ, whereby $A(\Delta)$ denotes the maximum ground displacement amplitude in the respective phase (P or S), in micrometers, and T is the corresponding period in seconds.

Figure 3 gives the calibrating functions σ_b for the vertical and horizontal components of the P wave and the horizontal component of the S wave (Gutenberg and Richter, 1956). A magnitude-zero event occurring at the Earth surface is seen from the figure to produce at 90° epicentral distance maximum ground velocities of 10^{-7}, $10^{-7.3}$, and $10^{-6.85}$ $\mu m \cdot s^{-1}$ associated with the PZ, PH, and SH phase, respectively.

The definition of the body wave magnitude implies that one and the same magnitude figure can be obtained for a given earthquake regardless of whether the P phase or the S phase is employed. It further implies that the magnitude figure will be the same regardless of the period of the phase utilized. The general use of the calibrating functions in Fig. 3 for the determination of body wave magnitudes was recommended by the 1967 Zürich General Assembly of IUGG (Båth, 1973).

Based on observations of shallow focus earthquakes, several conversion formulas were developed for the body wave and surface wave magnitudes.

The use of the following formula was recommended by the 1967 Zürich General Assembly:

$$m_b = 0.56 M_s + 2.9 \quad (7)$$

equivalent to

$$M_s = 1.79 m_b - 5.18$$

Relation between Magnitude, Energy and Intensity. If the earthquakes would occur at only one, say shallow, focal depth, the maximum seismic intensity at the surface could serve as a rating of the strength of the given earthquake, provided the ground conditions were similar at all observational sites. Because focal depths vary and ground conditions are largely different at various points of observations, the maximum intensity shows generally only a weak correlation with the strength of the earthquake as expressed by the magnitude.

Numerous efforts have been undertaken to find whether seismic intensities do correlate with instrumentally measurable quantities, such as, e.g., the maximum ground acceleration or the maximum ground velocity occurring in the course of the earthquake. Table 3 shows the relation, as published by Bolt (1978). However, the acceleration and velocity values for the wave motion are for firm ground only. They vary greatly depending on surface conditions and on the type of earthquake source.

Richter (1958) gives a relation between the ground acceleration a (cm·s^{-2}) and the intensity I expressed in the M.M. scale (see the last section):

$$\log a = \frac{I}{3} - \frac{1}{2}$$

which is valid for California. Båth (1973) cites a relation between the maximum intensity I_0 and the corresponding acceleration a_0 (cm·s^{-2}):

$$I_0 = 3 \log a_0 + 1.5$$

which is valid for Southern California.

The same author gives the interrelation between the seismic energy released E (in ergs), the surface and body wave magnitudes M_s and m_b, the maximum intensity I_0, and the maximum ground acceleration a_0, as shown in Table 4. Thereby, he utilizes the empirical formulas

$$\log E = 12.24 + 1.44 M_s$$

and Eq. 7 to obtain

$$\log E = 4.78 + 2.57 m_b$$

$$M_s = 1 + \frac{2I_0}{3}$$

$$I_0 = 3 \log a_0 + 1.5$$

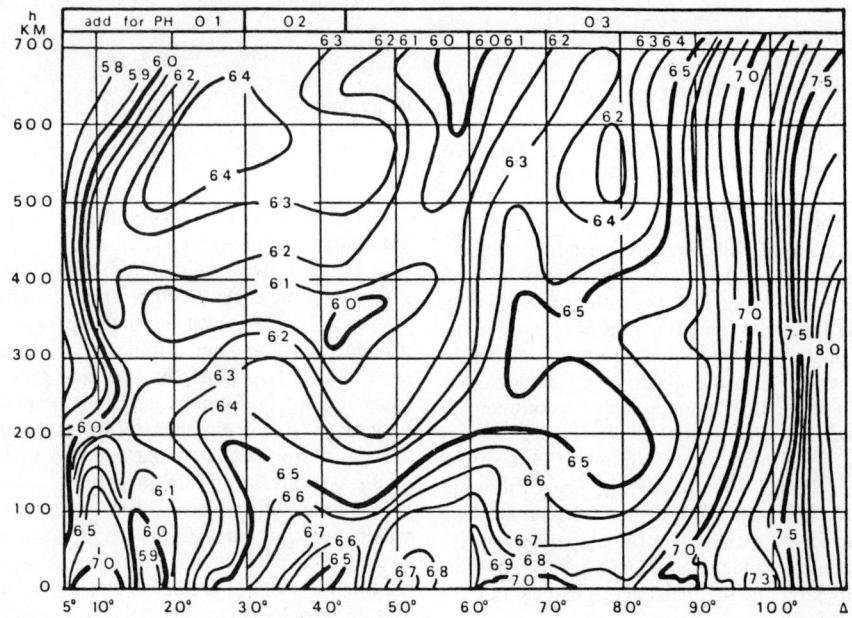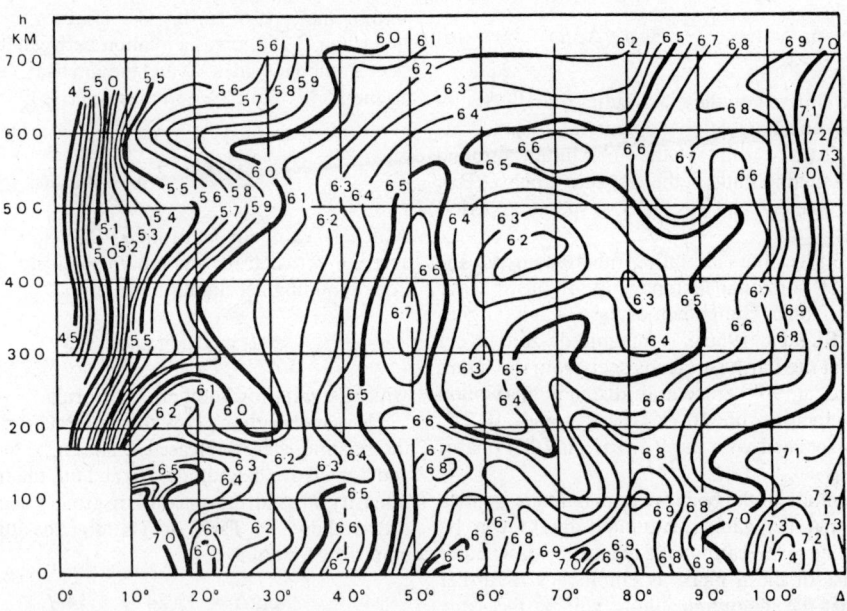

FIGURE 3. Magnitude calibrating functions for PZ, PH, and SH waves (Richter, 1958). The functions are generally used for the determination of m_b.

TABLE 3. Relation between Seismic Intensity, Maximum Ground Velocity and Maximum Ground Acceleration

Maximum Velocity (average; cm·s^{-1})	Seismic Intensity (according to the abridged modified Mercalli intensity scale)	Maximum Acceleration (average; $g = 980$ cm·s^{-2})
1–2	IV	$0.015g$–$0.02g$
2–5	V	$0.03g$–$0.04g$
5–8	VI	$0.06g$–$0.07g$
8–12	VII	$0.10g$–$0.15g$
20–30	VIII	$0.25g$–$0.30g$
45–55	IX	$0.50g$–$0.55g$
>60	X	>$0.6g$

Source: From Bolt, 1978.

The figures in Table 4 have to be considered with care. They are gaining significance, however, if only incomplete information on the earthquake is available, as, e.g., historical earthquakes. Here, the description of damage inflicted, as eventually found in historical records at large, may lead to an estimate of the maximum intensity of shaking and subsequently to at least a rough estimate of the magnitude of the earthquake.

Earthquake Magnitude and Energy

Present Practice of Determination and Publication. Magnitudes are being determined from observations at individual stations or as averages from observations at a network of stations. The present practice of magnitude determination at the agencies publishing annually the largest number of magnitudes is given below.

National Earthquake Information Service (NEIS), U.S.A. The surface wave magnitude is computed from Eq. 4, whereby the vertical component of the surface wave must have a period ranging from 18 to 22 s, and the epicentral distances range from 20° to 160°. No depth correction is applied. The magnitude is determined for earthquakes with focal depth not greater than 50 km, making allowance for uncertainties in the depth determination. The M_s value published is the average of individual magnitudes computed from reported amplitude and period data.

The body-wave (P-wave) magnitudes are computed from Eq. 6. The period though is restricted to the range 0.1–3.0 s, and the amplitude is not necessarily the maximum in the P group. The epicentral distance must be larger than 5°. All NEIS magnitudes obtained as averages are computed by a 25% trimmed mean (Preliminary Determination of Epicenters, Monthly Listing, U.S. Department of the Interior/Geological Survey, National Earthquake Information Center).

United Services for Earthquake Research (ESSN), USSR. In the routine work of ESSN (Edinaya slushba seismicheskich nabludenii) one surface wave and two body wave magnitudes are being determined. Each of the magnitudes is the arithmetic average of a possibly large number of individual determinations. According to Vanek et al. (1980), the surface wave magnitude is found from the maximum ground velocity associated with the Rayleigh wave having periods in the range 10–20 s. Thereby the recordings made on medium-period seismographs located at epicentral distances larger than 25° and the calibrating function (Eq. 5) are utilized.

The two body wave magnitudes are based on recordings in the distance range 20°–100° of the vertical component of the P wave obtained on short-period (SKM-3) and medium-period (SK, SKD) seismographs, respectively. In some cases also the horizontal component of P, PP, and S waves is used. The body wave magnitude of a deep-focus earthquake is based exclusively on the vertical component of the P wave. In all cases the calibrating functions of Gutenberg and Richter (1956), shown in Fig. 3 for P and S waves, are used.

Since 1976, for a selection of major earthquakes, body wave magnitudes are determined based on an experimental setup at the Obninsk station. Here, a broadband seismograph (SKD) with a flat velocity response in the frequency range 2–20 s is connected to eight galvanometers. The transfer functions of the seismometer-galvanometer systems are given in Fig.

TABLE 4. Relation between Seismic Energy E, Surface Wave Magnitude M_s, Body Wave Magnitude m_b, Maximum Intensity I_0, and Maximum Ground Acceleration

E (erg)	M_s	m_b	I_0	a_0^a (cm·s^{-2})
10^{20}	5.4	5.9	VI–VII	40
10^{21}	6.1	6.3	VII–VIII	100
10^{22}	6.8	6.7	VIII–IX	200
10^{23}	7.5	7.1	IX–X	400
10^{24}	8.2	7.5	X–XI	1,000
10^{25}	8.9	7.9	XII	2,000

Source: From Båth, 1973.

[a] Refers to periods of ground motion ranging from 0.1 to 0.5 s.

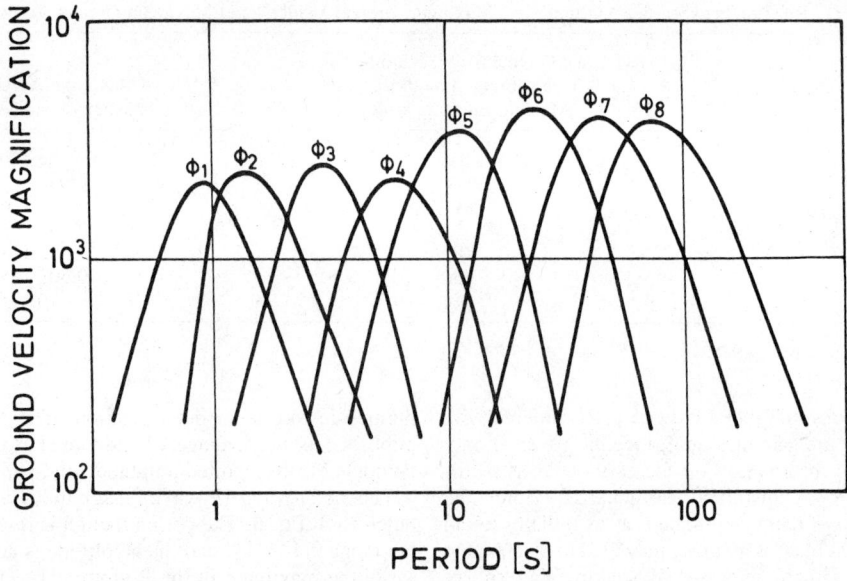

FIGURE 4. Ground velocity magnification for the system of one broadband seismograph and eight galvanometers attached. The system is operational at Obninsk (USSR).

4. Eight analog recordings are obtained at one time. Amplitude and period measurements are made for the P wave, and up to eight P-wave magnitudes are obtained. The maximum P-wave magnitude, together with its period, is published in addition to the surface wave and the two conventional body wave magnitudes of the selected earthquakes.

International Seismological Centre (ISC), U.K. The surface wave magnitude is determined from Eq. 4 by using the vertical or the resultant horizontal amplitudes with periods ranging from 10 to 60 s in the distance range 5° to 160°.

The body wave magnitude is based on Eq. 6, assuming the amplitude to be observed on a short-period vertical instrument. No station corrections are applied, and the epicentral distances must be more than 20° and less than 101°. The two magnitudes are published as averages (Bulletin of the International Seismological Centre, vol. 21, July 1984). It is estimated that generally the magnitudes published by ISC are based on a larger number of individual observations than any other magnitude figures.

Physical Significance of the Earthquake Magnitude.

Amplitude Spectral Density and Bandwidth of the Wavelet. The ground motion at the point of observation ST (Fig. 1) is a time function with finite duration. Its Fourier transform will thus have an infinite duration. The significant frequency range can be defined as the one within which the amplitude density lies not more than 3 dB below its maximum value (ANS, 1966). The bandwidth B of the signal is then obtained as the difference of the band-edge frequencies.

On the other hand, each seismograph system constitutes a bandpass filter for the ground motion. Usually the bandwidth of the seismograph system is narrower than that of the arriving signal. Table 5

TABLE 5. Bandwidth Characteristics, and Bandwidth Correction b for Determination of Spectral Magnitude (11)

Seismograph Type	Midband Period (s)	Passband (s)	Passband (Hz)	Bandwidth B(Hz)	Bandwidth Octaves	$b = \log B$
WWSSN-SP	0.66	0.37–1.19	0.84–2.70	1.86	1.69	−0.27
WWSSN-LP	14.2	6.03–33.6	0.03–0.17	0.14	2.48	+0.87
SRO-SP	0.71	0.34–1.48	0.68–2.94	2.26	2.12	−0.35
SRO-LP	28.2	18.5–43.0	0.02–0.05	0.03	1.22	+1.51
Wielandt/Streckeisen (broadband velocity output)	2.00	0.20–19.9	0.05–5.00	4.95	6.64	−0.69

shows the bandwidths for a selection of seismographic instruments currently in use. Whereas a broadband wavelet generally has a complex form, narrowband wavelets tend to have an approximately sinusoidal form.

The maximum amplitude a_{max} of the wavelet is, approximately, the product of the spectral density F within the bandwidth Δf of the wavelet and the bandwidth:

$$a_{max} = 2 F \Delta f \qquad (8)$$

Here, $\Delta f = f_2 - f_1$, where f_1, f_2 are the band-edge frequencies of the wavelet, a_{max} corresponds to the displacement, the velocity, or the acceleration of the ground at a frequency within the range $f_1 - f_2$, and F is assumed to be constant in the range (Aki and Richards, 1980).

The ground amplitude measured for the purpose of the surface wave magnitude determination corresponds to a nearly monochromatic wave with a period of 20 s $\pm$ 3 s, and the bandwith is constant. Thus, the amplitude is an estimate of the amplitude spectral density of the ground displacement $A(\Delta)$ (in Eq. 2) or the ground velocity $A(\Delta)/T$ (in Eq. 4). For a constant focal depth, the surface wave magnitude is, consequently, a rating of the spectral density of the radiation from the seismic source at a period of about 20 s.

The situation is less favorable with the body wave magnitude. From Eqs. 6 and 8 it is seen that this magnitude is an estimate of the amplitude spectral density of the ground velocity multiplied by the bandwidth of the seismograph system. Only body wave magnitudes determined from seismograph systems with the same bandwidth are eventually free from the bias due to variable bandwidths.

For example, the 1-s body wave magnitudes of a given earthquake will differ from each other if they are obtained on the basis of records from instruments with different bandwidths. This was clearly demonstrated by Båth (1977): Operating on one and the same pier two seismographs with a period of oscillation of 1 s (Benioff and Grenet-Coulomb seismographs), he found a systematic difference of body wave magnitudes amounting, on the average, to as much as 0.43, the larger magnitudes being obtained from the Grenet-Coulomb system, featuring the larger bandwidth.

In general, the body wave magnitude, as usually published, is thus a rating of the radiation at variable periods and is based on observations from seismographs systems having unknown bandwidths. The deficiencies in the definition of the body wave magnitude certainly contribute to the numerical instability of this magnitude, if determined for a given earthquake at several stations. It appears that the potential of the body-wave magnitude is not yet fully exploited (see the subsection "Spectral Magnitudes and Magnitude Spectra").

Far-Field Spectrum of Body Waves. A minimum of assumptions about the focal process is required when rating the source strength by way of the magnitude or the seismic energy. If, on the other hand, a specific source model is assumed, additional parameters, independent of each other, for the source strength can be defined and eventually determined. For example, consider the model of unidirectional faulting along a rectangular fault plane. Assuming a homogeneous, isotropic, unbounded elastic medium, the displacement waveform of P or S waves at the point x in the far field is

$$\Omega(\mathbf{x}, t) = \iint_{\Sigma} \Delta\dot{\mu} \left(\xi, t - \frac{r}{c} \right) d\Sigma$$

Δu is the source function, which describes the time history of the motion at the point ξ located in the fault plane Σ, and c is either α, the P-wave, or β, the S-wave velocity. The amplitudes of the waves decrease inversely proportional to r, the distance from $d\Sigma$ to the point of observation x (Aki and Richards, 1980).

For a source model having the form of a ramp with rise time τ and final slip D, and for a rupture along a fault plane with length L and width W, the far-field spectrum of the displacement waveform is

$$\left| \Omega(\mathbf{x}, \omega) \right| = WLD \, \frac{\sin X}{X} \left| \frac{1 - e^{i\omega\tau}}{\omega\tau} \right| \qquad (9)$$

Here

$$X = \frac{\omega L}{2} \left[\frac{1}{v} - \frac{\cos \Psi}{c} \right]$$

with v the rupture velocity and Ψ the angle between the rupture direction and the direction to the point of observation. For $\omega \rightarrow 0$ the spectrum becomes flat and proportional to WLD, which in turn is related to the moment of an equivalent double-couple point source:

$$M_0 = \mu WLD$$

with μ the shear modulus of the medium.

Figure 5a shows the envelopes of Eq. 9 for the choice of fault parameters indicated. The corner frequency is generally seen to decrease with the increase of the strength of the earthquake. For an earthquake with a given seismic moment the corner frequency changes with the stress drop of the earthquake (Sarkar and Duda, 1985).

In Fig. 5b the velocity density spectra are given. Here the significance of the body wave magnitude can be clearly seen: The body wave magnitude based on the ground velocity $A(\Delta)/T$ (in Eq. 6) samples the spectrum in Fig 5b at the frequency corresponding to the period T. The broader the bandwidth

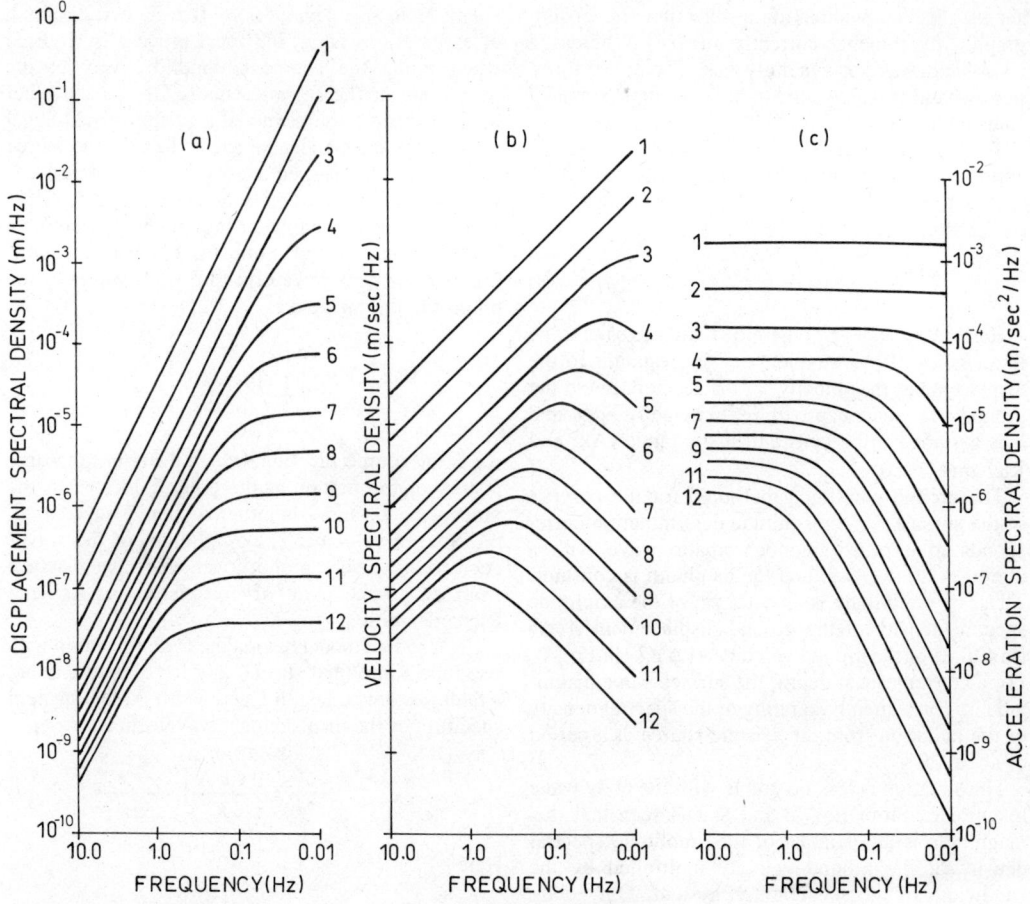

FIGURE 5. Spectral densities at a distance of 1 km from the point source complying with the ω-square model. For a stress drop of p bars the ordinate has to be multiplied by p.

of the seismograph is, the more blurred is the sampling of the velocity density, as the assumption about the constancy of F in Eq. 8 is progressively violated. The narrower the bandwidth is, the better the assumption in Eq. 8 is satisfied, and the sampling of the velocity density is more precise. Thereby the body wave magnitude is, as a rule, a function of the period T. The maximum body wave magnitude is thereby attained for the corner period (frequency).

Figure 5c presents the ground acceleration density spectra for the source model considered.

Quantification. The spectra in Fig. 5 range in frequency over ten octaves. A conventional seismogram, however, covers only part of the spectrum radiated from the focus. For example, from Fig. 6 it is seen that the short-period WWSSN seismograph passes only the spectrum around 1 s, and the body wave magnitude based on the record is thus a rating of the velocity spectral density at about this period. The period of this seismograph, however, was chosen because of a relatively low ground noise, not because of the particular relevance of the period for source processes. Thus, the rating of the source strength at this period is largely incidental.

On this background it is evident that a single magnitude figure cannot suffice to fully describe the strength of the source. Rather, a suit of figures sampling the radiated spectrum over a range of frequencies will yield an adequate rating of the source strength.

The practice to independently determine and publish more than one magnitude figure for a given earthquake is a step in this direction (compare with the previous section).

New Developments of the Magnitude Concept.

Synthetic Magnitude Calibrating Functions. The calibrating function projects the far-field surface motion at the epicentral distance Δ to a common distance near the source. Changes of the signal along the Earth's surface are due primarily to the radial heterogeneity of the Earth structure and, to a lesser degree, to lateral heterogeneities.

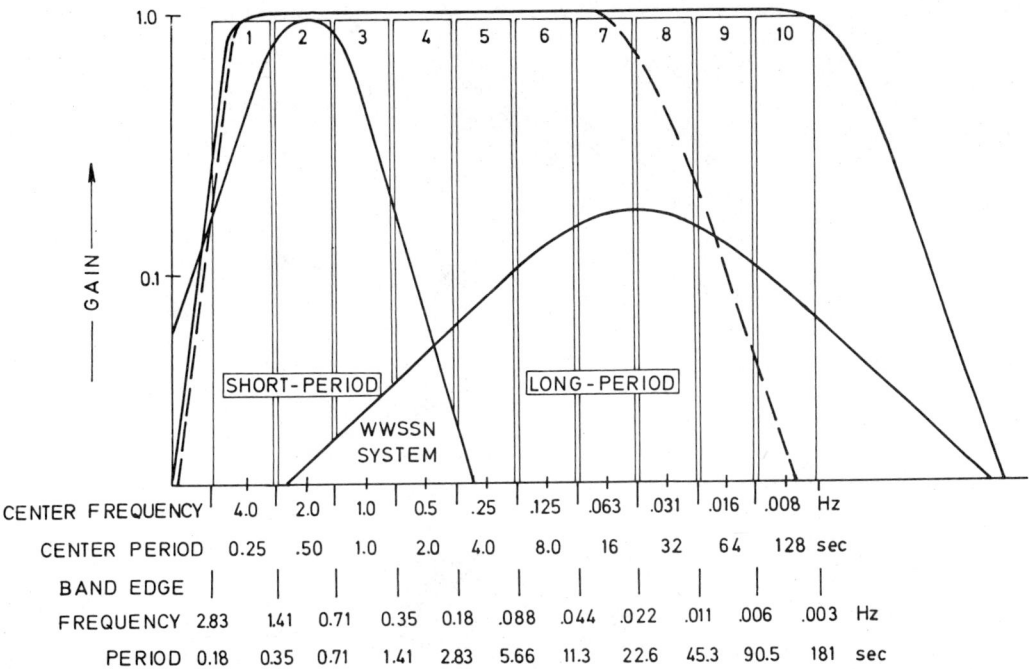

FIGURE 6. Magnification of short-period and long-period seismograph systems of the World Wide Standardized Seismograph Network. Magnification of Wielandt-Streckeisen seismograph (dashed curve). Applying inverse filtering, we obtain a seismogram with a flat magnification over nearly ten octaves. Superimposed is a system of ten one-octave filters (compare Fig. 8a).

The dependence on lateral heterogeneities is summarized in the "regional" and "station" corrections that are added in some cases to magnitude determinations at a given station. The calibrating functions in Eqs. 1, 3, and 5 and the ones in Fig. 3 were obtained prior to standardization of seismographs. They are based on amplitude observations on older instruments and, thus, possibly degrade magnitude measurements made on modern seismographs. For this reason, attempts are undertaken to arrive at new calibrating functions. Reliable functions have been obtained on the basis of amplitude variation curves for large explosions. They apply though to surface events only. Much effort has been expended on homogenizing the magnitudes determined from a selection of Eurasian stations (Vanek et al., 1980). In an involved iterative procedure, calibrating functions for P wave were calculated together with station corrections.

In a global scale, calibrating functions may be derived from global velocity and anelasticity models. The amplitude variation of body waves along the Earth's surface for a radially heterogeneous Earth is expressed by the so-called geometric spreading function:

$$G(\Delta, h) = \frac{1}{r_0} \left| \frac{\rho_h}{\rho_0} \frac{v_h}{v_0} \frac{\sin i_h}{\sin \Delta \cos i_0} \frac{di_h}{d\Delta} \right|^{1/2}$$

where ρ and v are the density and velocity, and the subscripts h and 0 refer to the focal depth and to the Earth surface, respectively. The remaining symbols are defined in Fig. 1. In case of an anelastic medium the factor $\exp(-\pi t^*/T)$ has to be superimposed with

$$t^* = \int_S \frac{ds}{Q(T, s)v(s)}$$

where T is the period of the wave, $Q(T, s)$ and $v(s)$ are the intrinsic quality factor and the seismic velocity of the medium, respectively, at the distance s along the ray path with a total length S (Båth, 1974).

In addition to the epicentral distance and focal depth, the amplitude variation is seen to depend on the period of the wave.

Numerical computations have been performed by Nortmann and Duda (1982). Figure 7a,b shows $g(\Delta, h)$ and $a(\Delta, h)$ for the two kinds of body waves and corresponding to the radial velocity heterogeneity and to the anelasticity, respectively. The two functions are utilized in the definition of the spectral magnitudes (see the section after the next).

Broadband, Digital Recording of Seismic Body Waves. The dynamic range of conventional seismographic observations amounts to 60–80 dB. The range can be increased to more than 130 dB by digital

(Text continues on page 286.)

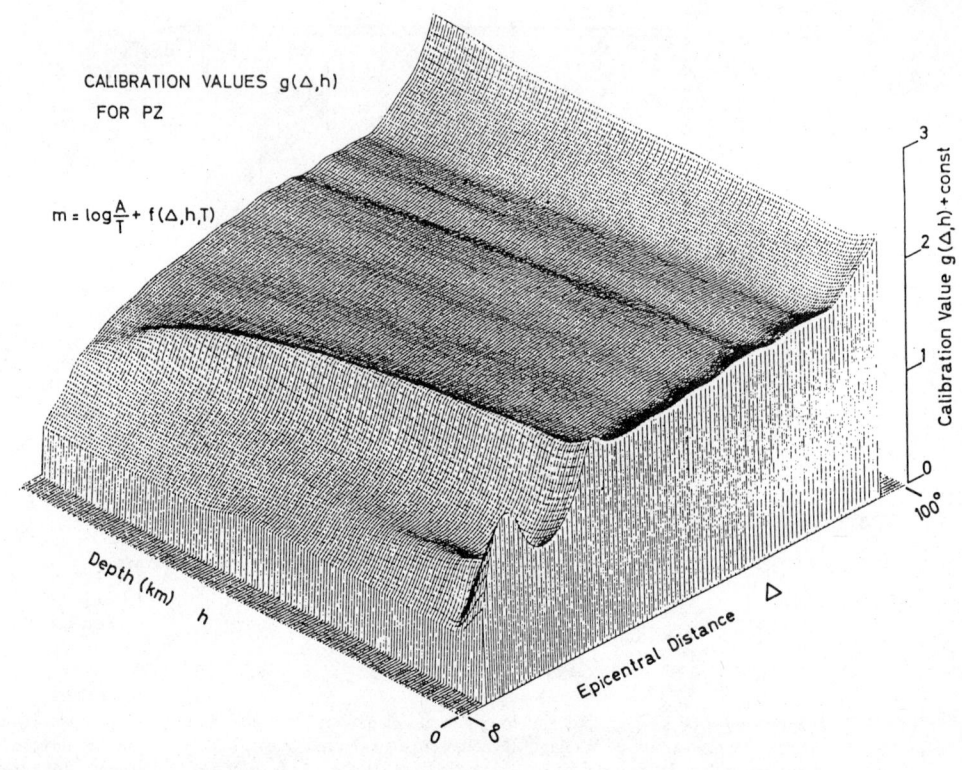
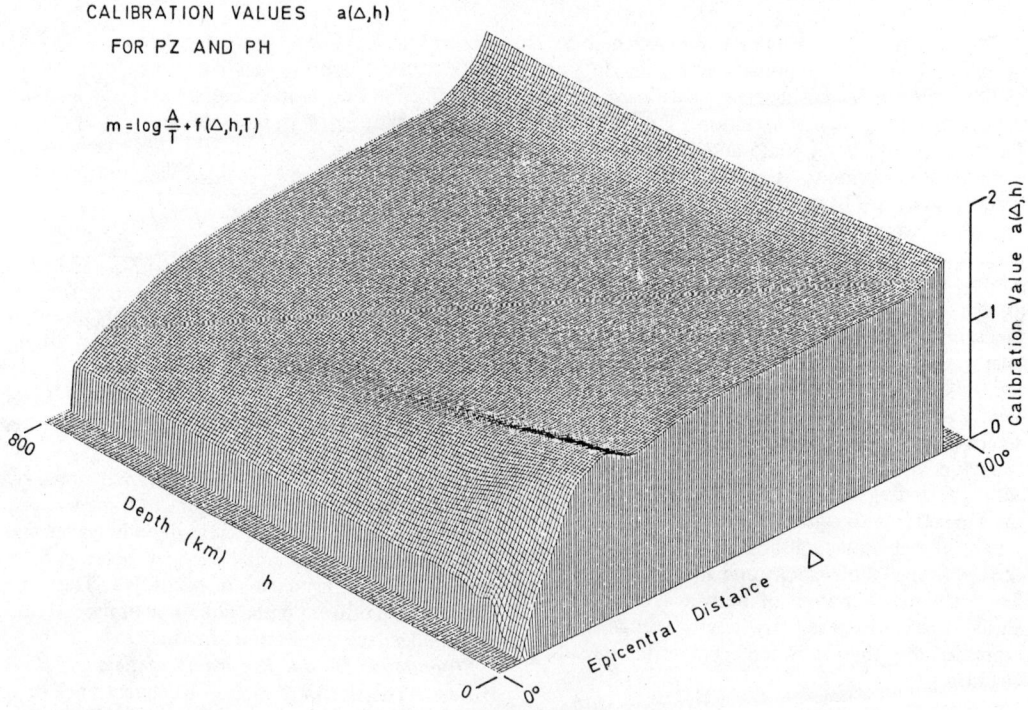

FIGURE 7. Synthetic magnitude calibrating functions. The functions $g(h, \Delta)$ and $a(h, \Delta)$ enter Eq. 11: (*left*) PZ and PH waves; (*right*) SH wave.

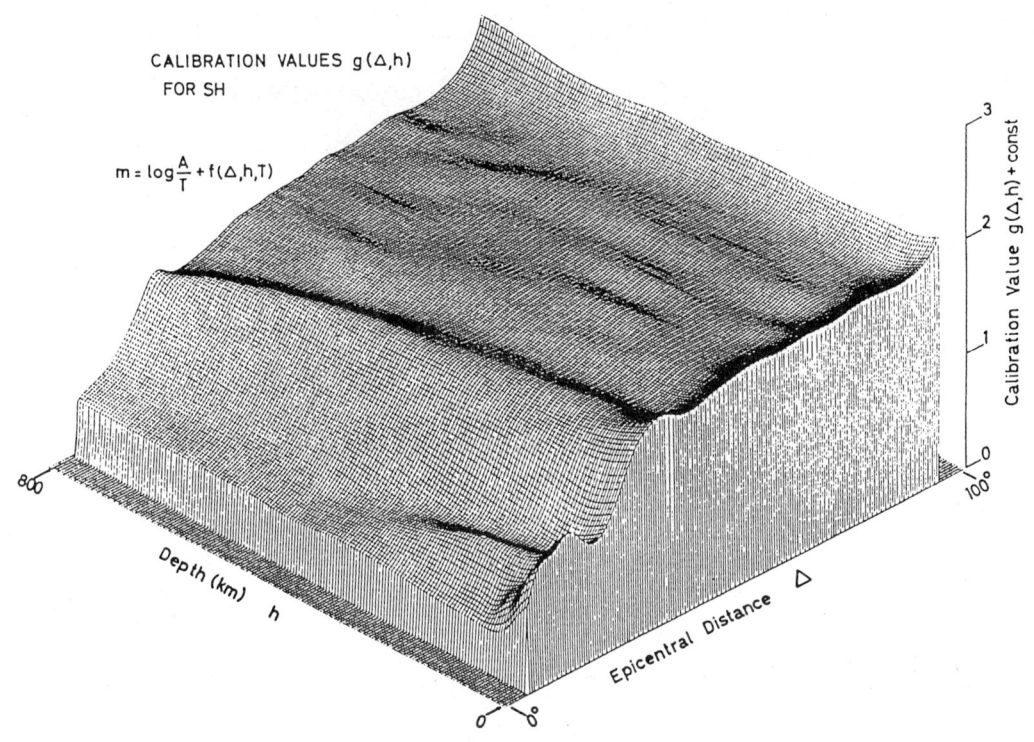

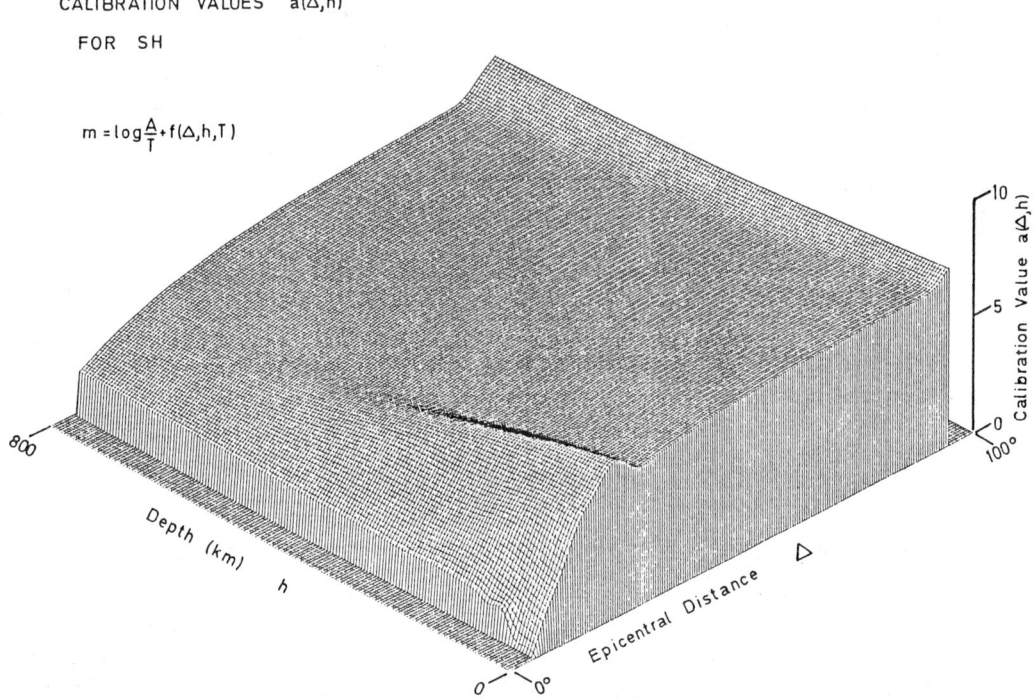

FIGURE 7. (*Continued*)

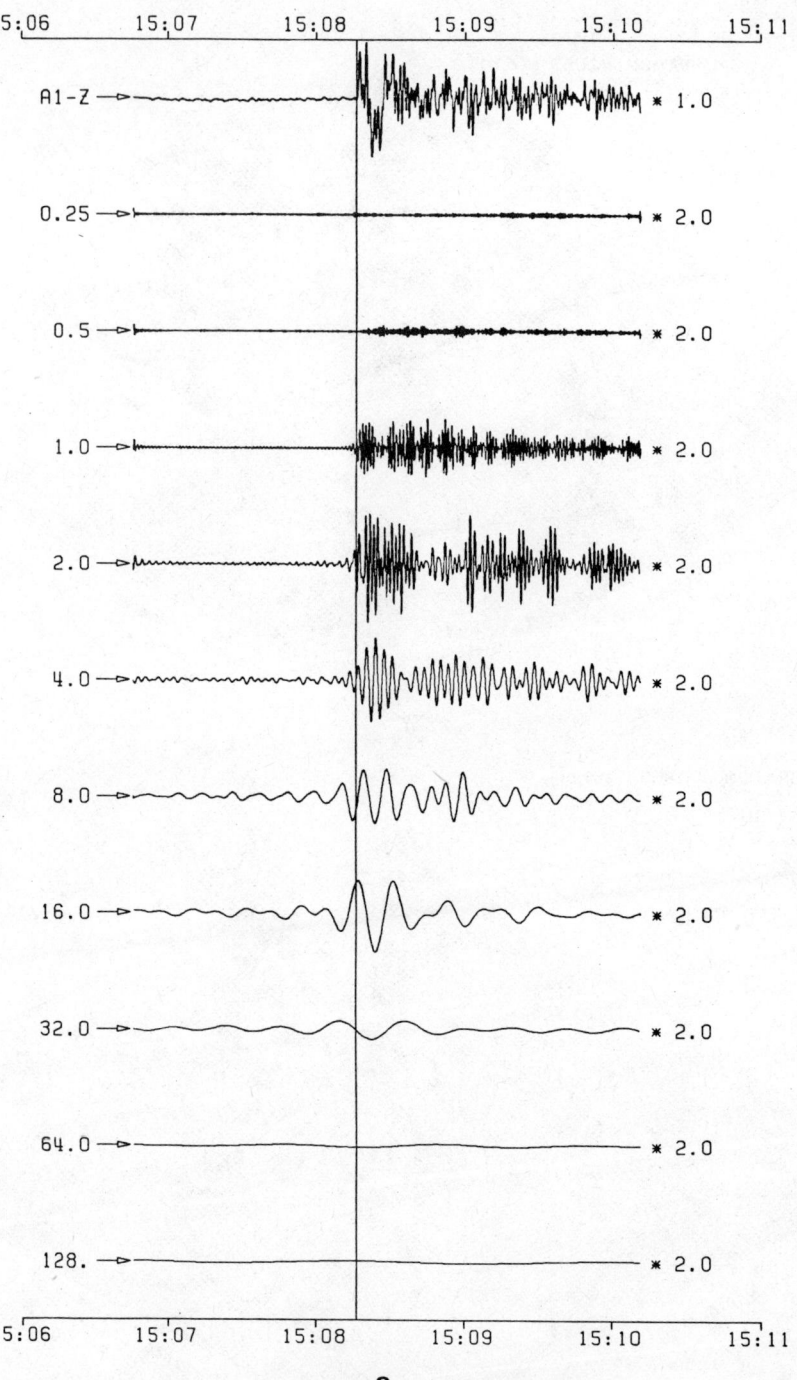

FIGURE 8. Record of Nepal earthquake (29.07.1980; 14:58:40.8; 29.60N, 81.09E; 18 km according to NEIS) as obtained at the Central Seismological Observatory of the Federal Republic of Germany at Erlangen (GRF): (*a*) Broadband (top line) and ten bandpass seismograms of vertical component of P wave. The bandpass characteristics are shown in Fig. 6. (*b*) P-wave signal and P-wave magnitude spectrum for four time windows as indicated. Only the longest window yields a stable estimate of the low-frequency component of the signal.

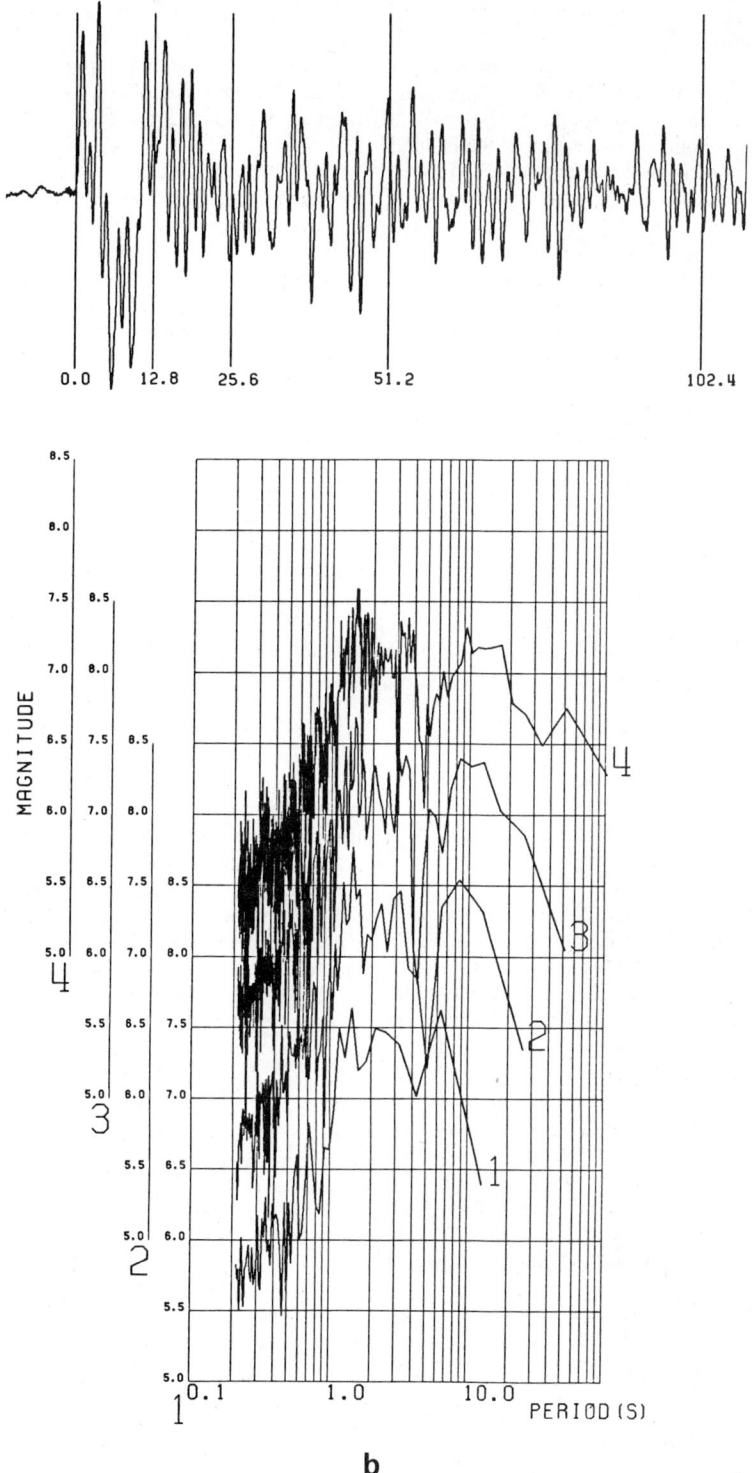

FIGURE 8. (*Continued*)

recording and by utilization of new seismographs (Harjes and Seidl, 1978).

Figure 6 shows the velocity magnification of the broadband, wide dynamic range seismograph (Wielandt-Strecheisen seismograph) operational at the Central Seismological Observatory of the Federal Republic of Germany at Erlangen (GRF) (dashed curve). Superimposed is the magnification of a virtual seismograph, simulated by inverse filtering of the actual record. An example of the corresponding broadband recording of the P wave (vertical component) is shown in Fig. 8a (top line). Applying to the recording ten one-octave bandpass filters, as defined in Fig. 6, ten bandpass seismograms are obtained as shown in Fig. 8a. From the bandpass seismograms body wave magnitudes can be determined as functions of the center period of the respective filter.

Spectral Magnitudes and Magnitude Spectra. On the bases of bandpass seismograms and of the new calibrating functions, the body wave magnitude is redefined:

$$m_{P,S}(T) = \log \frac{A_{P,S}(\Delta)}{T} + f_{P,S}(\Delta, h, T) + b \quad (10)$$

with the calibrating function

$$f_{P,S}(\Delta, h, T) = g_{P,S}(\Delta, h) + \frac{a_{P,S}(\Delta, h)}{T^*} + b \quad (11)$$

Equation 10 defines spectral magnitudes separately for P and S waves. T^* in Eq. 11 is a function of T, the period measured, and takes into account the period dependence of Q; b is the bandwidth correction, as given, e.g., in Table 5. For further details see Nortmann and Duda (1983).

The spectral magnitude yields estimates of the radiation strength of the seismic source at the given period. The bandwidth correction ensures that spectral magnitudes are not biased by instrumental characteristics. Thus, a spectral magnitude can be determined from conventional recordings if the bandwidth of the seismometer-galvanometer system is known.

Digital recordings permit the bandwidth of the bandpass seismograms to be varied. The narrower the filter will be, the more details of the radiated spectrum are reflected. Ultimately, the bandwidth may be reduced to the Nyquist frequency. The set of spectral magnitudes yields in this case a sampling of the velocity density spectrum radiated at the frequencies within the given range. The set is the velocity density spectrum rated in magnitude units. It is called the P-wave or S-wave magnitude spectrum.

As an example, Fig. 8b shows the P-wave magnitude spectrum of the same earthquake as in Fig. 8a. Four time windows are applied to the broadband record of the P wave. The windows range in length from 12.8 sec (spectrum 1) to 102.4 sec (spectrum 4).

All four windows yield the maximum of the spectrum at about 1.5 sec, the corner period of the P wave. The location of a secondary maximum changes from about 6 sec (spectrum 1) to about 9 sec (spectrum 4). It is concluded that only the longest window (102.4 sec) can yield a stable estimate of the longperiodic radiation intensity. However, as it is usually the case in the magnitude problem the likely contamination of the signal by later arrivals is not taken into account.

Earthquake Magnitude and Other Focal Parameters. Undoubtedly, the best single quantity characterizing the strength of the seismic source is the total seismic energy released. To determine the energy, we must integrate the velocity density spectrum of the given earthquake (see Fig. 5b). Thus, no unique relation can exist between the magnitude, sampling the spectrum at some period, and the energy. Though for some class of earthquakes the semilogarithmic relation

$$\log E = a + bM$$

(E is the seismic energy, M is the magnitude in general, and a,b are constants) was found to hold, the relation cannot apply in all cases.

An analogous conclusion holds for parameters such as the seismic moment M_0, fault length L, final slip D on the fault plane, stress drop $\Delta\sigma$, each parameter being an a priori independent characteristic of the strength of the source.

According to Aki (1967), the fault length is related to the corner frequency f_0 by

$$L \text{ (km)} = \frac{C_1}{f_0}$$

The constant C_1 ranges from 0.65 to 4.47 km·s^{-1} (Hanks and Wyss, 1972). The maximum spectral magnitude $m(f)_{max}$ is related to the product of the seismic moment M_0 and the corner frequency by

$$m(f)_{max} = C_2 + \log(M_0 f_0) \quad (12)$$

A good control is given of the constant C_2 (Sarkar and Duda, 1985), and, based on Eq. 12, the seismic moment can be conveniently found. For a circular fault with radius r_0, the seismic moment is also expressed as

$$M_0 = \tfrac{16}{7} \Delta\sigma \, r_0^3$$

with $\Delta\sigma$ the stress drop along the fault plane (Keilis-

Borok, 1959). Thus, from the seismic moment and the fault dimension the stress drop can be calculated.

Finally, with the energy spectral density $E(f)$ related to the spectral magnitude of the given body wave by

$$E(f) = 10^{2m(f)-1.4} \ \mathrm{J \cdot Hz^{-1}}$$

the seismic energy E in the frequency band f_1-f_2 is

$$E = \int_{f_1}^{f_2} E(f) \, df$$

The total energy is composed of that radiated in the form of P waves and S waves. In each case the principal part of the energy is reflected by the maximum spectral magnitude.

Seismic Intensity

The effects of the earthquake at near epicentral distances need to be quantified, primarily for engineering purposes. Although there is an increasing number of instruments in earthquake-prone areas capable of recording strong motions during the event, an adequately dense distribution of instruments proves to be too costly. For this and other reasons, noninstrumental observations continue to play an important role in assessing the intensity of shaking during an event as well as in predicting the intensities to be expected in future earthquakes. The noninstrumental intensity scale eventually establishes a connection between places in which the ground acceleration, velocity displacement, and the duration of shaking are instrumentally determined. Thereby it must be borne in mind that the frequency content of the signal recorded changes with epicentral distance, and the relation between the intensity and the respective element of ground motion needs to take into account the period of the signal (Hurtig and Stiller, 1984).

The seismic intensity reflects the integral effect of all elements of ground motion. The intensity scales presently in use have 12°. The scales differ from each other slightly, due to the variation of building standards in different parts of the world. Table 6 gives the abridged modified Mercalli intensity scale, as it was developed for California (and most of the United States) from the Mercalli scale introduced in Italy around the turn of the century (Bolt, 1978).

Concentric stripes at the Earth's surface experiencing the same seismic intensity are separated from

TABLE 6. Abridged Modified Mercalli Intensity Scale

Value	Description of Sensation to Human Beings and Damage to Objects
I.	Not felt except by a very few under especially favorable circumstances.
II.	Felt only by a few persons at rest, especially on upper floors of buildings. Delicately suspended objects may swing.
III.	Felt quite noticeably indoors, especially on upper floors of buildings, but many people do not recognize it as an earthquake. Standing motorcars may rock slightly. Vibration like passing of truck. Duration estimated.
IV.	During the day felt indoors by many, outdoors by few. At night some awakened. Dishes, windows, doors disturbed; walls make creaking sound. Sensation like heavy truck striking building. Standing motorcars rocked noticeably.
V.	Felt by nearly everyone, many awakened. Some dishes, windows, and so on broken; cracked plaster in a few places; unstable objects overturned. Disturbances of trees, poles, and other tall objects sometimes noticed. Pendulum clocks may stop.
VI.	Felt by all, many frightened and run outdoors. Some heavy furniture moved; a few instances of fallen plaster and damaged chimneys. Damage slight.
VII.	Everybody runs outdoors. Damage negligible in buildings of good design and construction, slight to moderate in well-built ordinary structures, considerable in poorly built or badly designed structures; some chimneys broken. Noticed by persons driving cars.
VIII.	Damage slight in specially designed structures, considerable in ordinary substantial buildings with partial collapse, great in poorly built structures. Panel walls thrown out of frame structures. Fall of chimneys, factory stacks, columns, monuments, walls. Heavy furniture overturned. Sand and mud ejected in small amounts. Changes in well water. Persons driving cars disturbed.
IX.	Damage considerable in specially designed structures; well-designed frame structures thrown out of plumb; great in substantial buildings, with partial collapse. Buildings shifted off foundations. Ground cracked conspicuously. Underground pipes broken.
X.	Some well-built wooden structures destroyed; most masonry and frame structures destroyed with foundations; ground badly cracked. Rails bent. Landslides considerable from river banks and steep slopes. Shifted sand and mud. Water splashed, slopped over banks.
XI.	Few, if any, (masonry) structures remain standing. Bridges destroyed. Broad fissures in ground. Underground pipelines completely out of service. Earth slumps and land slips in soft ground. Rails bent greatly.
XII.	Damage total. Waves seen on ground surface. Lines of sight and level distorted. Objects thrown into the air.

each other by lines called *isoseismals*. They are usually elongated in the direction of the fault that is activated during the earthquake. The distance between isoseismals increases with the focal depth of the earthquake. Respective empirical relations show a pronounced regional variability.

The distribution of isoseismals for a multitude of earthquakes in a region is of basic importance for estimating the distribution of shakability in the region. The latter, in turn, enters the problem of seismic zoning, together with parameters like the maximum magnitude of the earthquakes to be expected, their frequency distribution, the frequency content of the radiation from the source, and others.

SEWERYN J. DUDA

References

Aki, K., 1967, Scaling law of seismic spectrum, *Jour. Geophys. Research* **72,** 1217–1231.
Aki, K., and P. G. Richards, 1980, *Quantitative Seismology—Theory and Methods*, vols. I and II. San Francisco: W. H. Freeman, 932 p.
ANS, 1966, *American National Standard, specification for octave, half-octave, and third-octave band filter sets*. New York: American National Standards Institute, Inc., 26 p.
Båth, M., 1973, *Introduction to Seismology*. Basel and Stuttgart: Birkhäuser Verlag, 395 p.
Båth, M., 1974, *Spectral Analysis in Geophysics*. Amsterdam-Oxford-New York: Elsevier, 563 p.
Båth, M., 1977, Teleseismic magnitude relations, *Annali di Geofis.* **30,** 299–327.
Båth, M., 1981, Earthquake Magnitude—Recent Research and Current Trends, *Earth-Sci. Rev.* **17,** 315–398.
Bolt, B. A., 1978, *Earthquakes—A Primer*. San Francisco: W. H. Freeman, 241 p.
Gutenberg, B., and C. F. Richter, 1956, Magnitude and Energy of Earthquakes, *Annali Geofisica* **9,** 1–15.
Hanks, T. C., and M. Wyss, 1972, The use of body wave spectra in the determination of seismic source parameters, *Seismol. Soc. America Bull.* **62,** 561–589.
Harjes, H.-P., and D. Seidl, 1978, Digital recording and analysis of broad-band seismic data at the Graefenberg (GRF)-array, *Jour. Geophys.* **44,** 511–523.
Hurtig, E., and H. Stiller, 1984, *Erdbeben und Erdbebengefährdung*. Berlin: Akademie-Verlag, 328 p.
Keilis-Borok, V., 1959, On estimation of the displacement in an earthquake source and of source dimensions, *Annali di Geofis.* **12,** 205–214.
Nortmann, R., and S. J. Duda, 1982, The amplitude spectra of P- and S-waves and the body-wave magnitude of earthquakes, *Tectonophysics* **84,** 251–275.
Nortmann, R., and S. J. Duda, 1983, Determination of spectral properties of earthquakes from their magnitudes, *Tectonophysics* **93,** 251–275.
Richter, C. F., 1935, An instrumental earthquake scale, *Seismol. Soc. America Bull.* **25,** 1–32.
Richter, C. F., *Elementary Seismology*. San Francisco and London: W. H. Freeman, 768 p.
Sarkar, D., and S. J. Duda, 1985, Spectral P-wave magnitudes, Aki's w-square model and source parameters of earthquakes, *Tectonophysics* **118,** 175–193.
Vanek, J., A. Zatopek, V. Karnik, N. V. Kondorskaya, Yu. V. Riznichenko, E. F. Savarensky, S. L. Solov'yov, and N. V. Shebalin, 1962, Standardization of magnitude scales (in Russian), *Akad. Nauk CCCP, Izv. Ser. Geofiz.* **2,** 153–158.
Vanek, J., N. V. Kondorskaya, and L. V. Christoskow, 1980, *The Earthquake Magnitude in Seismological Practice - PV- and PV_s- Waves* (in Russian). Sofia: Bulgarian Academy of Sciences, 263 p.

Cross-references: Earthquake Mechanisms; Earthquakes: Hazards and Prediction; Earthquakes and Crustal Deformation; Earthquake Seismology; Earthquakes and Seismicity; Seismic Instrumentation; Seismic Instrumentation: History; Seismicity and Plate Tectonics; Strong Motion Seismology.

EARTHQUAKES: VOLCANOGENIC

Volcanoes are among the most seismically active features on earth, producing up to thousands of small earthquakes per day during eruptive episodes. For this reason, the studies of both volcanoes and earthquakes share a long history; seismographs were installed near the beginning of operation at some of the first volcano observatories, such as those at Kilauea Volcano in Hawaii and Mount Vesuvius in Italy. Volcanoes give rise to a remarkable diversity of seismic signals, and the origin of some of the more unusual types is still debated within the scientific community. Certain types of earthquakes such as volcanic tremor seem to be uniquely associated with volcanoes. Seismology plays a crucial role in the monitoring and short-term prediction of volcanic eruptions and seismological observations have improved our understanding of the internal workings and structure of volcanic systems, providing opportunities to gain new insights on earthquake sources.

Observations and Monitoring

Modern short-period seismograph networks (Lee and Stewart, 1981) are now extensively employed for volcano observations. Large numbers of mainly vertical component seismographs distributed around the volcanic centers are preferred to fewer observatory type instruments because they provide the ability to accurately locate the sources of volcanic earthquakes, knowledge which is important in understanding volcanic processes. As an illustration, the network around Kilauea Volcano has grown from 13 stations in 1963 to 47 stations scattered throughout the island of Hawaii in 1983 (Klein et al., 1987). In Washington, there are approximately 15 stations located within 20 km of the cone of Mount St. Helens primarily for volcanological monitoring and research. These instruments are typically narrow in their frequency response (1 Hz to 20 Hz) with peak magnification of ground motion of 10^2 to 10^6 at about 10 Hz. However, noise levels around volcanoes are frequently high due to local site conditions (often on pyroclastic material) in combination with noise sources such as running water, glaciers, and geothermally produced vibration. Thus, the highest magnifications possible are often not useful.

Networks with large numbers of stations are possible because of the development of efficient methods of telemetering data to a central site for recording and analysis. At Kilauea, for example, large numbers of stations and a high rate of earthquake recording (possibly thousands per day during eruption episodes) are handled efficiently by a computerized data analysis and recording system. The advantage of networks in volcano seismology is that they provide the ability, in nearly real time, to accurately locate, categorize, and further analyze earthquakes related to the volcanic activity. Accurate observations at large numbers of stations have led to advances in understanding the structure of volcanoes through formal travel time inversion and modeling of magma plumbing systems. For more quantitative study of the sources of volcanic earthquakes, broadband three-component digital instrumentation is beginning to be used. The largest volcanic related earthquakes are observed at stations worldwide, and the development of ultrahigh quality global stations will also increase our ability to study major explosive eruptions.

Types of Volcanogenic Earthquakes

Classification of volcanogenic earthquakes is based on an understanding of their sources and on their different signal characteristics. Minakami's (1961) classification is still widely referenced. He recognized type A earthquakes occurring in or near the volcanic magma system as being indistinguishable from normal shallow tectonic earthquakes. Type A earthquakes are high frequency and show clear evidence of direct P and S phases. Minakami also recognized type B earthquakes, which have a lower frequency coda, no distinct S phase, and often emergent onsets. These earthquakes are unique to volcanic environments. In addition, Minakami classified volcanic tremor (a continuous oscillation with rather narrow frequency spectrum), and explosion quakes resulting directly from volcanic blasts. Although Minakami's classification is useful, continuous variation between these types is possible. Malone (1983) produced a series of examples from Mount St. Helens, shown in Fig. 1, which he breaks into categories "t," "h," "m," and "l" for "tectonic," "high," "medium," and "low" frequency volcanic. An example of volcanic tremor is also shown in Fig. 1. Using a source classification, we can identify the following major categories of volcanogenic earthquakes.

Direct Eruption Blast. The direct eruption blast or events leading up to a blast can give rise to earthquakes of considerable size where both atmospheric and seismic signals are generated. With modern wideband seismographs, these signals can be recorded and studied at teleseismic distances (Eissler, 1986) with modern wideband seismographs. The far field seismic radiation of these sources may be distinctively different from faulting sources found for tectonic earthquakes, identifying them as uniquely volcanic in nature.

Tectonic Earthquakes. Tectonic earthquakes of small size are normally the most common types of earthquakes in volcanic regions. These events have double-couple source mechanisms, typically less than magnitude 5 in size, and are indistinguishable from tectonic earthquakes found elsewhere. They often occur in shallow swarms, at depths less than 5–10 km, appearing to arise from near or even within the magma plumbing system. They represent the release of tectonic shear stress, which is either of regional origin or the result of alteration of the stress near the volcano resulting from magma emplacement.

Low-Frequency Earthquakes. Low-frequency (LF) transient earthquakes are a distinctive feature of volcanic regions, although similar earthquakes have been observed to originate from glaciers. These earthquakes are identifiable by their distinctive coda which appears to be narrow band, typically in the range of 1–5 Hz, with emergent onset and lack of a clear S phase. In volcanoes, their exact source is not well understood, but they frequently occur in connection with eruption, intrusion, or degassing episodes. Measuring the magnitude of LF earthquakes presents particular problems because of their narrow band nature and unusual codas, although quite often, the range of magnitudes observed is limited. Each volcanic system may produce LF earthquakes with characteristic frequencies, although much overlap in the signal features exist from one locality to another.

Volcanic Tremor. Tremor, frequently called harmonic tremor, is perhaps the most mysterious, yet characteristic seismic activity of volcanoes. Like LF earthquakes, it typically has a narrow frequency spectrum in the range of 1–5 Hz, and a quasi-steady state nature that can persist for hours or even days. Episodes commonly last from minutes to hours and are associated with eruptive or intrusive episodes. Although usually detected only in the immediate vicinity (within 10 km) of the volcanic center, much stronger tremor has been observed. For example, prior to the 18 May 1980 St. Helens blast, tremor episodes that could be detected on stations over 100 km from the volcano were observed. The frequency and amplitude of tremor can change during a single episode, possibly reflecting changes in the magma supply system or rates of magma flow.

Magma Fracturing or Direct Magma Injection. Although not yet well documented, there appears to be a class of earthquakes, identified primarily from their source mechanisms, that may

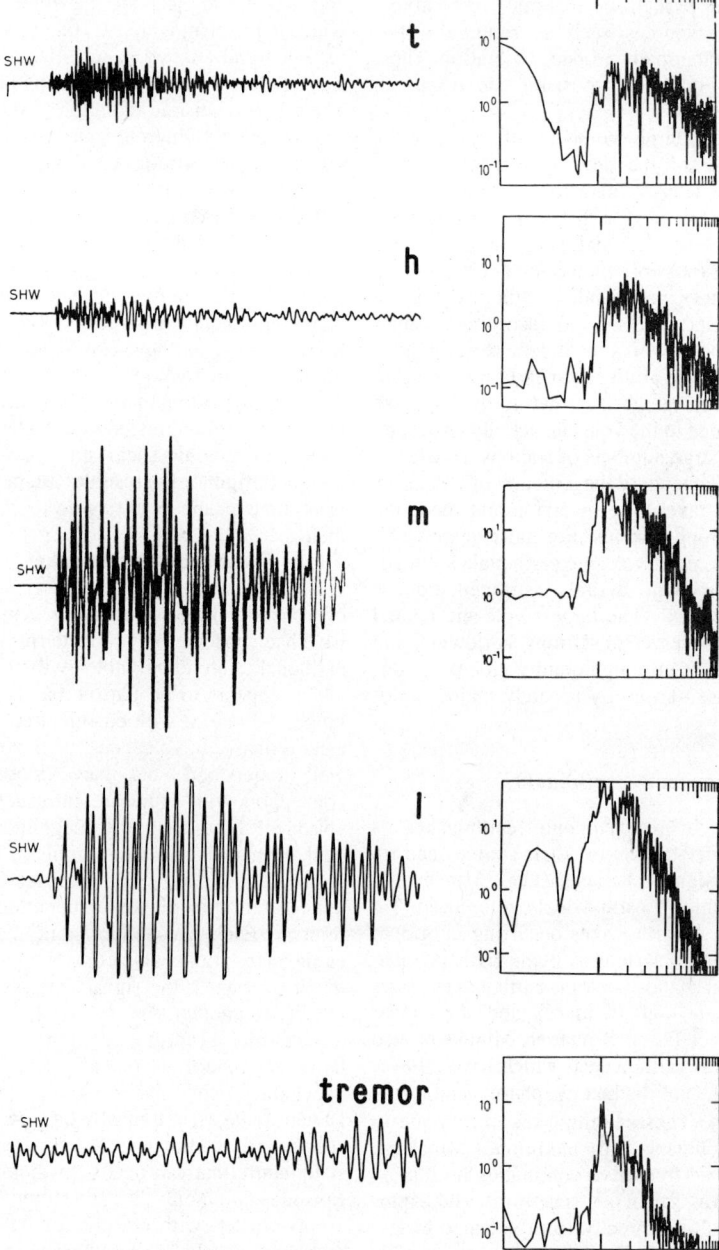

FIGURE 1. Seismograms for five different classes of earthquakes at Mount St. Helens (after Malone, 1983). The types are: tectonic (t), high frequency volcanic (h), medium frequency volcanic (m), low frequency volcanic (l), and tremor. Frequency spectra on the same (but arbitrary) vertical scale are shown to the right of each seismogram. All seismograms are from the same station (SHW) located on the west flank of the volcano. See Malone (1983) for details of data acquisition.

reflect the direct forceful injection of gas or magma. Such an earthquake was recently reported near Tori Shima, Japan by Kanamori et al. (1986). This earthquake had a mechanism obtained from inversion of the long period body waves of a compensated linear vector dipole, which can be interpreted as injection of fluid into a fracture with the two sides of the fracture moving apart. Confirmation of such mechanisms would categorize these earthquakes as uniquely volcanic in nature. Although mechanisms other than double couple could be common in volcanic systems, their existence is difficult to verify due to the difficulty of getting reliable source mechanisms and the expected contamination of the mechanisms by double-couple moment resulting from the presence of shear strain energy.

Indirect Tectonic Sources. Near volcanoes, local but indirect tectonic effects can give rise to large earthquakes. A notable example of this type of activity is found in Hawaii. The 1975 Kalapana earthquake resulted from loading of the seaward side of Kilauea Volcano, by repeated dike injection along rift zones extending from the central vent (Tilling et al., 1976). After an interval of strain buildup (approximately 100 years), the entire block of crust moved laterally in a magnitude 7.5 earthquake causing damage and a major local tsunami. This earthquake was found by Eissler and Kanamori (1987) to have a source mechanism that resembled a lateral force, similar to a laterally directed volcanic explosion. Their interpretation is that the block was momentarily decoupled from the earth's surface during rupture.

Low Frequency Earthquakes and Tremor

Low-frequency (LF) earthquakes and volcanic tremor appear to be closely related. They are often found in close spatial and temporal association, and spectral analysis reveals a similar frequency content. The narrow band spectrum of the seismic signals indicates a characteristic source dimension. Tremor is quasi-continuous, which implies a long duration forcing function that could be provided by fluid flow or repeated fracturing. While most LF earthquakes and tremor episodes appear to originate at depths of a few kilometers within or near the magma reservoirs of volcanoes, depths of 30 to 60 km are found in Hawaii (Koyanagi et al., 1987).

Precise location of the source of tremor is difficult to determine since normal seismic "phases" cannot be timed. Source locations are usually inferred by relative amplitudes at different network stations or from the association of tremor with LF earthquakes that are located using conventional methods. In the seismological literature on sources of LF earthquakes and tremor, the division between these two types of events is often indistinct. There has been past uncertainty over the influence of propagation path effects in producing the distinctive coda of LF earthquakes, and there is little doubt that path attenuation does play some role. However, convincing evidence of the dominant role of source effects has been found from the analysis of broadband seismograph measurements from Mount St. Helens. The dominant frequency in the range of 1–5 Hz, found for both tremor and LF earthquakes at a wide variety of volcanoes, is a remarkable feature in view of the vast differences in physical structure that exist. This points to a source mechanism that is consistent in its scale properties over the wide range of physical structures, magma viscosities, and rock properties that exist in volcanoes.

The notion that tremor is basically a superposition of LF volcanic earthquakes has been formalized by Chouet (1985) who proposed that LF earthquakes are the "Green's function" for harmonic tremor. A number of models have been suggested to explain the source of LF earthquakes and tremor. Kubotera (1974) used the fundamental mode of a spherical liquid-filled magma chamber to explain long period tremor at Aso Volcano in Japan. Riuscetti et al. (1977) suggested that free oscillations in the cylindrical magma chamber explained the two-peaked spectra that they observed at Mount Etna. Aki and co-workers proposed that volcanic tremor originates by the repeated opening of tensile cracks by fluid injection and Chouet later used this model in numerical simulations to predict the details of ground motion near a tremor source. Recently, Chouet studied the resonant oscillations of a more complex model consisting of a gaseous chamber over a fluid-filled cylinder, both within an elastic half space. (These and other articles on source models are referenced in Crosson and Bame, 1985.) Chouet et al. (1987), using numerical simulation, study several source models that include the interaction of fluid magma with cracks within the magma storage system. These models explain many features of the seismic radiation of tremor, but the extended duration of the source of tremor is an admitted difficulty.

Julian (1986) has formalized a model of fluid flowing through a restricted channel, such as a dike, as the source of tremor. In this model, an increase in fluid flow speed produces a local pressure decrease (Bernoulli effect), which causes the channel walls to elastically restrict thereby reducing the flow volume. When the flow is reduced sufficiently, the walls relax and the cycle is repeated. This cyclic oscillation is the source of tremor, with low frequency earthquakes arising from subcritical flow velocities in Julian's model. (This effect may be reproduced in a simple backyard experiment by pinching a garden hose and adjusting the flow velocity until oscillations are produced.) The fluid flow model, although difficult to analyze mathematically, predicts that under some conditions the characteristic frequency decreases as the amplitude increases—behavior that could be compared with observations to test the model.

Another recent finding that may have significance for LF earthquakes and tremor is the discovery by Ferrazzini et al. (1986) of very slow speed dispersive waves in a fluid layer such as a magma filled dike. These waves are analogous to tube waves found in well bores, and if preferentially excited, may explain the long period nature of LF earthquakes and possibly tremor. Crosson and Bame (1985) found that a fluid cavity containing a gas bubble also exhibited a long-period resonance, at about the right frequency for LF earthquakes that was surprisingly insensitive to the cavity dimension. In this active research field, there are now a number of possible alternative source models for LF earthquakes and tremor. Further progress must await careful comparison of theory and observations.

Seismicity and Magma Systems

The study of seismicity has provided remarkable glimpses into the inner workings of volcanoes. In no place is this more true than at Kilauea Volcano in Hawaii where extensive monitoring capability and a long record has combined to provide recent investigators with a wealth of data. The basic model is that the magma storage system is a honeycomb of fluid pockets and cracks in a solid rock matrix. Much of the small, high frequency earthquake activity that is recorded reflects the intrusion and redistribution of the fluid magma, resulting in opening, closing, and stressing the solid rock matrix. With the use of a high quality seismograph network, hypocenters of these earthquakes can be accurately located in space and time, allowing the mapping of the magma system and its kinematic behavior. Where large pockets of magma exist, no earthquakes, or at least no high frequency tectonic type earthquakes, are observed. Using the extensive Hawaii catalog, Ryan (1987) produced a detailed three-dimensional picture of the magma distribution system at Kilauea based on earlier work by Ryan et al. (1981). He was careful to point out that interpretation must be applied to the hypocenter pattern and not all earthquakes near the volcano reflected magma locations. Only swarms of earthquakes or earthquakes closely related in time and space were associated with magma movement. Three-dimensional plexiglass models afforded a better understanding of the spatial configuration of the magma zone. This model, as viewed from the south toward the summit caldera of Kilauea Volcano, is reproduced as Fig. 2. Space and time analysis of earthquakes has been extended by Klein et al. (1987) to track not only the static dimensions of the magma supply system, but also the kinematics of magma intrusion into the two major rift zones of Kilauea.

Earthquakes from Volcanic Eruptions

Although small in number, earthquakes caused directly by volcanic eruptions may be quite energetic, equivalent to tectonic earthquakes larger than magnitude 5. The triggering earthquake and subsequent explosion on 18 May 1980 at Mount St. Helens was widely recorded at teleseismic distances. An unusual aspect of volcanic explosion earthquakes is that their source mechanisms may depart from the usual double-couple mechanism characteristic of tectonic earthquakes. For example, Kanamori et al. (1986) found from modeling teleseismic records that the St. Helens explosion was equivalent to a single vertical force impulse resulting from uncapping the top of the magma chamber. One of Minakami's categories of volcanic earthquakes is explosion earthquakes, which are often numerous during eruption episodes. A particular category of explosion earthquakes is those related to Strombolian volcanism. Strombolian eruptions are the result of repetitive gas eruptions from an open and unconfined but fluid-filled volcanic vent. Such activity is observed on an almost continuous basis at the type locality of Stromboli Volcano in the Tyrrhenian Sea. At Stromboli, observations of LF volcanic earthquakes preceding vent explosions suggest that gas bubble expansion at some level within the throat of the volcano may be at least one mechanism for earthquake generation. When the gas bubble reaches the surface, a visible eruption occurs.

Tides and Volcanic Earthquakes

Stress changes in the earth caused by solid earth tides and ocean loading are over three orders of magnitude less than stresses typically released in earthquakes (10^{-3} MPa). Nevertheless, the periodic nature of tides and the possibility that they may push a metastable system to the point of instability has made tidal forcing a subject of investigation as a possible mechanism for triggering both eruptions and volcanic earthquakes. In a fluid-solid system, tidal strains may be magnified considerably in fluid conduits that are connected to larger reservoirs. This effect is often observed in water wells when small tidal dilatations over a large reservoir are converted to relatively large changes in water level. Volcanic systems require a delicate balance between gravitational body forces, fluid buoyancy, and tectonic stress to maintain equilibrium. Even small changes in stress, such as those caused by tides, could superimpose a quasi-regularity to volcanic systems that are already primed to generate eruptions or earthquakes.

Klein (1976) found significant correlations between peaks of activity in swarms of earthquakes in several volcanic regions with solid earth tides. He suggested a mechanism of tidal stress concentration

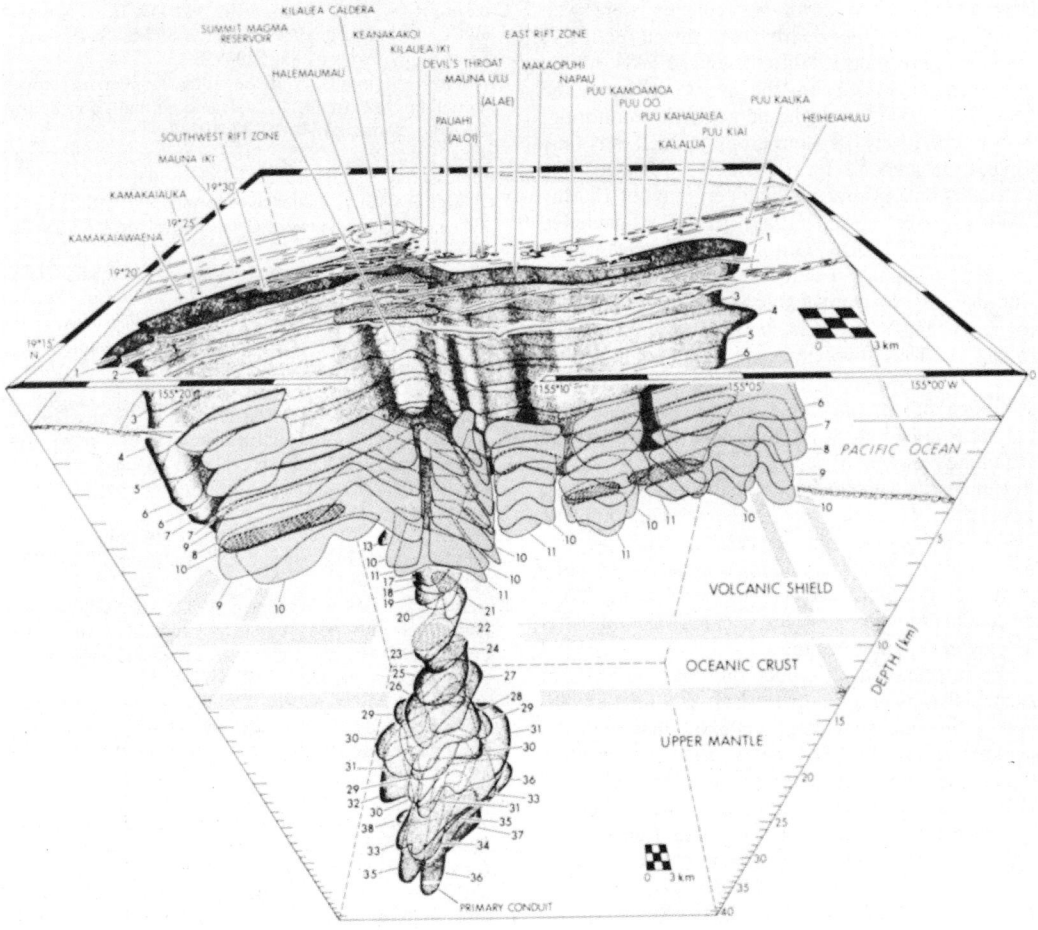

FIGURE 2. Diagram of three-dimensional model of Kilauea Volcano, Hawaii, constructed from analysis of earthquake locations and theory of magma emplacement (by Ryan, 1987). This view is looking approximately northward and details of construction and further references are given in Ryan, 1987.

through plates acting as stress guides. At Pavlof Volcano in the Aleutians, McNutt and Beavan (1984) found significant statistical correlation of solid earth tides and volcanic earthquake swarms at Pavlof Volcano, Alaska, for periods before and just after explosive eruptions. They reported a change in sign of the correlation of earthquake frequency with stress rate that they interpreted as caused by the change in the volcanic cycle from inflation before the eruption to deflation after an eruption. However at Mount St. Helens they found only a weak correlation between the rates of occurrence of low frequency earthquakes and tides for a period prior to the 18 May 1980 eruption. Tidal correlations appear to be an ineffective predictor of eruptions at St. Helens. By contrast, Dzurisin (1980) concludes that fortnightly tides have a 90% likelihood of being correlated with eruptions at Kilauea Volcano. We conclude that the influence of tides on volcanic earthquakes may be observable in some instances, but is highly dependent on specific local conditions.

Earthquakes and Eruption Prediction

Eruption prediction or forecasting remains a major objective of applied volcanology (Decker, 1973; Swanson et al., 1985), and no discussion of volcanic earthquakes would be complete without including the role of seismology in this effort. For the period following the 18 May 1980 eruption at Mount St. Helens, Swanson et al., (1985, p. 408), stated that "Seismicity has provided the single best means of predicting within a few hours to a day or so the time of an impending extrusion or explosion." At St. Helens, eruption prediction has been very successful in this period once observation capability and the characteristic behavior of the volcano were well established. Of the nearly two dozen extrusive eruptions from June 1980 to the end of 1984, all but two were predicted on the basis of integrated geophysical, geochemical, and geologic monitoring. However the May 18 main eruption itself was only forecast in general terms. Having reviewed the seismicity data prior to the main eruption of 18 May 1980, Crosson et al. (1980, p. 531) concluded shortly thereafter that "Our preliminary conclusions are that seismicity provides an intermediate term warning of explosive volcanic hazard (scale of weeks to months) but no apparent short term warning (hours to days)." These two views are not in conflict since they apply to different stages of the volcanic cycle. Although subsequent dome building and extrusive eruptions at Mount St. Helens have been predicted with a high degree of success, it is generally agreed that the initial cataclysmic explosive phase could not have been precisely predicted with any existing means. In spite of the success of seismological measurements, there is no single means of eruption prediction that guarantees success, particularly in the critical phase when an volcano is reawakening after an extended period of quiet.

We can learn much about the characteristics of earthquakes as eruption precursors from the experience at St. Helens. There, both harmonic tremor and swarms of earthquakes have acted as eruption precursors. For example, on 12 June 1980 an eruption warning was issued on the basis of a sudden increase then decrease in harmonic tremor. The eruption occurred within two hours of this warning. On 22 July 1980 an eruptive episode was preceded by several hours of increasing low frequency earthquake activity, whereas an eruption on 7 August 1980 was preceded by harmonic tremor and sporadic low frequency earthquake activity. Malone et al. (1981) speculate that shallow earthquakes occur as eruption precursors when the volcanic vent is plugged and pressure is high. Alternatively when the volcanic vent is unplugged, leading to low pressure conditions, harmonic tremor is observed prior to eruption. Although these suggestions hint at the mechanisms underlying the seismic precursors to eruptions, caution must be exercised since we have only a poor understanding of the physical processes and conditions leading up to eruptions.

R. S. CROSSON

References

Chouet, B., 1985, Excitation of a buried magmatic pipe: A seismic source model for volcanic tremor, *Jour. Geophys. Research* **90**, 1881-1893.

Chouet, B., R. Y. Koyanagi, and K. Aki, 1987, Origin of volcanic tremor in Hawaii, Part II. Theory and discussion, in R. W. Decker, T. L. Wright, and P. H. Stauffer, eds., *Volcanism in Hawaii, Vol. 2, U.S. Geol. Survey Prof. Paper 1350*, 1259-1280.

Crosson, R. S., E. T. Endo, S. M. Malone, L. L. Noson, and C. S. Weaver, 1980, Eruption of Mt. St. Helens: Seismology, *Nature* **285**, 529-536.

Crosson, R. S., and D. A. Bame, 1985, A spherical source model for low frequency volcanic earthquakes, *Jour. Geophys. Research* **90**, 10237-10247.

Decker, R. W., 1973, State-of-the-art in volcano forecasting, *Bull. Volcanol.* **37**, 372-393.

Dzurisin, D., 1980, Influence of fortnightly Earth tides at Kilauea Volcano, Hawaii, *Geophys. Research Letters* **7**, 925-928.

Eissler, H. K., 1986, Investigations of earthquakes and other seismic sources in regions of volcanism, Ph.D. diss., California Institute of Technology, Pasadena, Cal.

Eissler, H. K., and H. Kanamori, 1987, A single-force model for the 1975 Kalapana, Hawaii, earthquake, *Jour. Geophys. Research* **92**, 4827-4836.

Ferrazzini, V., K. Aki, B. Chouet, and M. Fehler, 1986, Very slow waves trapped in a fluid layer sandwiched between two solid half spaces with application to the long-period events (abstract), *EOS (Am. Geophys. Union Trans.)* **44**, 1203.

Julian, B. R., 1986, Volcanic tremor: Flow induced vibration (abstract), *EOS (Am. Geophys. Union Trans.)* **67**, 1264.

Kanamori, H., and J. W. Given, 1982, Analysis of long-period seismic waves excited by the May 18, 1980 eruption of Mt. St. Helens—A terrestrial monopole? *Jour. Geophys. Research* **87**, 5422-5432.

Kanamori, H., J. W. Given, and T. Lay, 1984, Analysis of seismic body waves excited by the Mount St. Helens eruption of May 18, 1980, *Jour. Geophys. Research* **89**, 1856-1860.

Kanamori, H., G. Ekstrom, A. Dziewonski, and J. S. Barker, 1986, An anomalous seismic event near Tori Shima, Japan—A possible magma injection event, *EOS (Am. Geophys. Union Trans.)* **67**, 1117 (abstract).

Klein, F. W., 1976, Earthquake swarms and the semidiurnal solid earth tide, *Royal Astron. Soc. Geophys. Jour.* **45**, 245-295.

Klein, F. W., R. Y. Koyanagi, J. S. Nakata, and W. R. Tanigawa, 1987, The seismicity of Kilauea's magma system, in R. W. Decker, T. L. Wright, and P. H. Stauffer, eds., *Volcanism in Hawaii, Vol. 2, U.S. Geol. Survey Prof. Paper 1350*, 1019-1185.

Koyanagi, R. Y., B. Chouet, and K. Aki, 1987, Origin of volcanic tremor in Hawaii, Part I. Data from the Hawaiian Volcano Observatory 1969-1985, in R. W. Decker, T. L. Wright, and P. H. Stauffer, eds., *Volcanism in Hawaii, Vol. 2, U.S. Geol. Survey Prof. Paper 1350*, 1221-1257.

Kubotera, A., 1974, Volcanic tremors at Aso volcano, in L. Civetta, P. Gasparini, G. Luongo, and A. Rapolla, eds., *Developments in Solid Earth Geophysics, Physical Volcanology.* New York: Elsevier, 29-48.

Lee, W. H. K. and S. W. Stewart, 1981, in *Principles and Applications of Microearthquake Networks*, New York: Academic Press, 293p.

McNutt, S. R., and R. J. Beavan, 1984, Patterns of earthquakes and the effect of solid earth and ocean load tides at Mount St. Helens prior to the May 18, 1980, eruption, *Jour. Geophys. Research* **89**, 3075-3086.

Malone, S. D., 1983, Volcanic earthquakes: Examples from Mount St. Helens, in *Earthquakes: Observations, Theory and Interpretation*, H. Kanamori and E. Boschi, eds., Amsterdam: Elsevier/North Holland, 436-455.

Malone, S. D., E. T. Endo, C. S. Weaver, and J. W. Ramey, 1981, Seismic monitoring for eruption prediction, in *U.S. Geol. Survey Prof. Paper 1250*, 803–813.

Minakami, T., 1961, Study of eruptions and earthquakes originating from volcanos, I, *Internat. Geology Rev.* **3**, 712–719.

Riuscetti, M., R. Schick, and D. Seidl, 1977, Spectral parameters of volcanic tremors at Etna, *Jour. Volcanology and Geothermal Research* **2**, 289–298.

Ryan, M. P., 1987, Elasticity and contractancy of Hawaiian olivine tholeiite and its role in the stability and structural evolution of subcaldera magma reservoirs and rift systems, in R. W. Decker, T. L. Wright, and P. H. Stauffer, eds., *Volcanism in Hawaii, Vol. 2, U.S. Geol. Survey Prof. Paper 1350*, 1395–1447.

Ryan, M. P., R. Y. Koyanagi, and R. S. Fiske, 1981, Modeling the three-dimensional structure of magma transport systems: Application to Kilauea Volcano, Hawaii, *Jour. Geophys. Research* **86**, 7111–7129.

Swanson, D. A., T. J. Casadevall, D. Dzurisin, R. T. Holcomb, C. G. Newhall, S. D. Malone, and C. S. Weaver, 1985, Forecasts and predictions of eruptive activity at Mount St. Helens, USA: 1975–1984, *Jour. Geodynamics* **3**, 397–423.

Tilling, R. I., P. W. Lipman, J. P. Lockwood, J. G. Moore, and D. A. Swanson, 1976, Earthquake and related catastrophic events, Island of Hawaii, November 29, 1975: A preliminary report, *U.S. Geological Survey Circular 710*, 33p.

Cross-references: *Earthquakes: Hazards and Prediction; Earthquakes and Crustal Deformation; Earthquake Seismology; Seismic Source: Observations; Seismic Source: Theory; Seismic Wave Scattering.*

EARTHQUAKES AND CRUSTAL DEFORMATION

The San Francisco earthquake of 1906 caused land offsets over a length of more than 300 km. Points on one side of the San Andreas fault were displaced as much as 5 m relative to points on the opposite side. This seismic event commands particular importance because much of the current theory on the nature of earthquake-related crustal deformation has roots in the geodetically observed displacements associated with it. In particular, from his analysis of the associated displacements, Reid (1910) hypothesized that, prior to the earthquake, the Earth's crust had been gradually deformed over time by applied stresses and that the earthquake resulted when this slowly accumulating deformation generated sufficient elastic strain energy to cause slip on an existing surface of weakness, the San Andreas fault. Reid's rebound hypothesis differentiates between *coseismic* and *aseismic* deformation, that is, between deformation occurring during an earthquake and that occurring in the absence of simultaneous earthquakes. Moreover, Reid's hypothesis implies that, to produce an earthquake, aseismic crustal deformation, at least as large as the ensuing coseismic deformation but complementary in sense, must occur beforehand. Our discussion of earthquake-related deformation, consequently, includes appropriate consideration of the movements that precede and follow an earthquake, as well as those that occur coseismically. At first, however, we address only coseismic deformation because experimental evidence corroborates the corresponding theory rather successfully and because contemporary theory for aseismic deformation is based, in part, on the adopted coseismic theory.

Coseismic Deformation

Hayford and Baldwin (1908) were first to analyze the geodetic measurements made before and after the 1906 San Francisco earthquake. They found (1) that points on opposite sides of the disrupted San Andreas fault moved in opposite directions, (2) that displacements were less the greater the distance from the fault, and (3) that the directions of the displacements correlated well with fault orientation. These discoveries helped establish the then controversial theory that slip on faults causes the vibrations in the Earth that we know as earthquakes. The association between fault slip and earthquakes extends even further. Numerous studies have subsequently demonstrated that coseismic crustal deformation relates quantitatively to fault slip. Steketee (1958a, 1958b) constructed the mathematical foundation for this relationship by translating relevant concepts from the theory that deals with *dislocations* or offsets in the lattice structure of crystals. Even the name, dislocation theory, has been carried from crystallography to the realm of geophysics. There is a basic difference, however. Whereas crystallography deals with dislocations in a discrete structure, geophysics is concerned with dislocations in a continuum.

A dislocation surface in a continuous three-dimensional body is, briefly, a surface of discontinuity in displacement. It may be understood conceptually by visualizing its formation in the following process:

1. make a cut inside the body to form an arbitrary new surface S,
2. apply relative displacement of slip-type to the two faces of the cut,
3. rejoin the faces in their new positions.

In this manner, the body regains its stress continuity but it undergoes deformation to accommodate the displacement discontinuity across S.

In geophysical applications the body encompassing the dislocation surface is customarily a homogeneous, isotropic, elastic half-space whose free surface represents the Earth's surface. For the special case of uniform slip over a rectangular dislocation surface whose upper edge parallels the free surface (Fig. 1), closed analytical formulas for computing the associated displacement field have been published by Okada (1985).

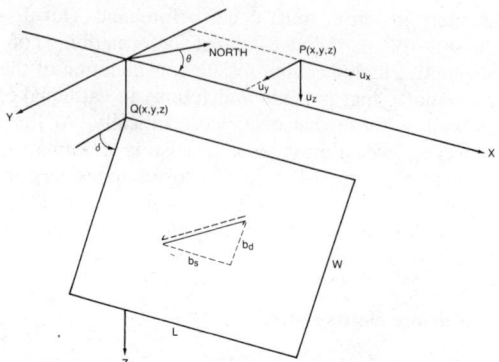

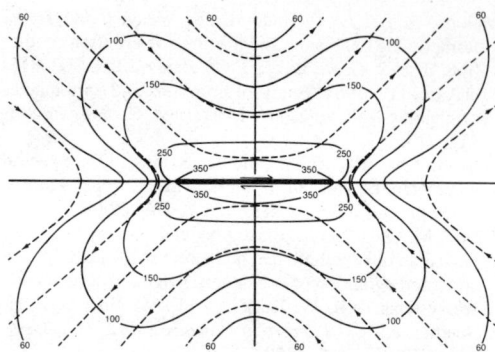

FIGURE 1. With uniform slip over a rectangular dislocation surface in an elastic half-space, the displacement vector (u_x, u_y, u_z) at a point $P(x, y, z)$ depends on the location of a designated reference point $Q(x, y, z)$ on the rectangle; the rectangle's size (L = length, W = width); the rectangle's orientation (θ = strike, δ = dip); and the components of the slip vector (b_s = strike slip, b_d = dip slip).

For the case of pure strike slip of uniform magnitude b on an infinitely long, vertical fault intersecting the Earth's surface and extending to depth D, the dislocation equations become

$$u_x(x,y,z=0) = b/\pi \arctan(D/y)$$
$$u_y(x,y,z=0) = 0$$
$$u_z(x,y,z=0) = 0$$

where the x-axis parallels the surface trace of the fault, y denotes horizontal distance from the fault, and z quantifies depth below the Earth's surface. For uniform slip on a finite-length fault, the corresponding dislocation equations produce a more complicated displacement field, especially near the two fault ends. Contours of resultant horizontal displacement form a butterfly-like pattern about the fault trace (Fig. 2). Also with pure strike slip on a finite-length fault, significant vertical displacements occur near the fault ends (Fig. 3). With a vertical fault, both horizontal and vertical displacement fields are antisymmetric about the fault plane. With an inclined fault, relatively greater displacements occur on the down-dip side of the fault trace.

For the case of pure dip slip on an infinitely long fault, the vertical component of the displacement field varies solely as a function of distance from the fault trace. On a finite-length fault, contours of vertical displacement mostly close on the fault trace. Again, relatively larger displacements occur on the down-dip side of an inclined fault (Fig. 4).

For the case of oblique slip, the associated displacement field equals the vector sum of two separate displacement fields, namely, those fields obtained individually from the strike-slip and the dip-slip components of the slip vector. This relationship reflects the fact that the displacement field is a linear function of the slip vector. As another consequence of this linearity, an increase or decrease of the slip vector by some scalar factor results in the equivalent proportional increase or decrease of the displacement vectors.

Dislocation equations for uniform slip on a rectangular surface constitute a special case of the integral equation

$$\vec{u}(P) = \iint_S [G(P, Q; \lambda, \mu)] \vec{b}(Q) \, dS$$

where $\vec{u}(P)$ denotes the displacement vector at the half-space point $P(x,y,z)$, $\vec{b}(Q)$ denotes the slip vector at the point $Q(x,y,z)$ on the dislocation surface S, and kernel $[G(P,Q;\lambda,\mu)]$ corresponds to a matrix of mathematical expressions that embody the behav-

FIGURE 2. Contours of resultant horizontal displacements for uniform strike slip on a vertical fault plane intersecting the Earth's surface. Fault width equals fault length. Broken lines show direction of movement. Contour values in units of 10^{-3} times the slip (after Chinnery, 1961).

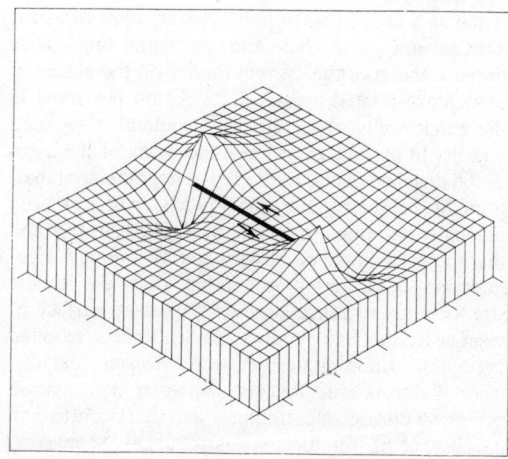

FIGURE 3. Vertical deformation pattern caused by strike slip on a vertical fault plane.

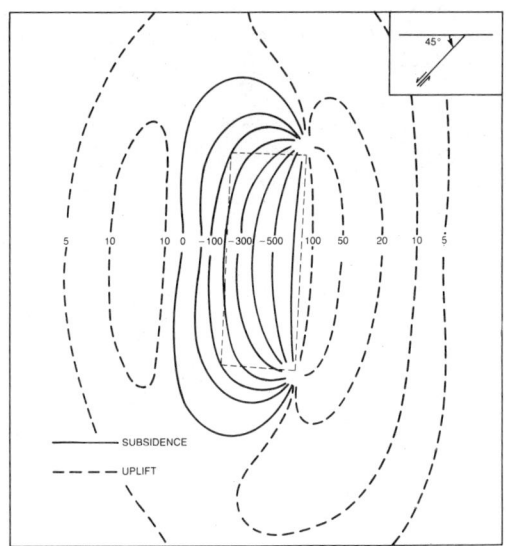

FIGURE 4. Contours of vertical displacement for uniform dip slip on a 45° inclined fault plane intersecting the Earth's surface. Fault width equals half the fault length. Contour values in units of 10^{-3} times the slip.

ioral properties of the body whose elasticity is characterized by the Lamé parameters λ and μ. This integral may be numerically evaluated to obtain the displacement field produced by a spatially varying slip vector over an arbitrarily shaped fault surface, or alternatively, closed analytical formulas, such as those given by Okada, may be used to approximate this same displacement field as the sum of a finite number of displacement fields, each resulting from uniform slip over a rectangular surface. In practice, however, the quality of geodetic data has rarely supported the resolution of a dislocation model containing more than three rectangles, each with uniform slip.

Aseismic Deformation

Tectonic plate interactions, volcanic processes, and certain human activities like the creation of artificial lakes represent some familiar mechanisms that generate aseismic deformations that may lead to earthquakes. We focus here, however, on only the tectonically generated aseismic deformations, as this mechanism is responsible for most major earthquakes.

According to the theory of plate tectonics, a mosaic of twenty or more plates form the outermost layer of the Earth, called the *lithosphere*. These plates float on the weak, plastic layer known as the *asthenosphere*. Relative movement among the plates is thought to be caused by thermal convection below the lithosphere. To a first approximation the plates move as rigid bodies except in the vicinity of their boundaries. At boundaries where two adjacent plates diverge, new material, injected from the Earth's interior, adds to the plates. At convergent boundaries, one plate subducts or dives beneath the other. Along transform boundaries, two plates slip past one another horizontally. Most major earthquakes occur along subduction and transform boundaries where interplate friction inhibits relative plate motion. The interacting plates deform aseismically until they have accumulated sufficient elastic strain energy to overcome the existing friction along some segment of their common boundary. An earthquake is the rapid release of this elastic strain energy. After the earthquake, aseismic deformation resumes along the disrupted segment in preparation for the next earthquake. Consequently, along a plate boundary, earthquakes recur at rather regular intervals and the associated crustal deformation exhibits a cyclic nature. Repeated geodetic survey measurements reveal that this deformation cycle consists of three principal phases: coseismic strain release, postseismic transient movement, and relatively uniform interseismic strain accumulation.

Whereas the elastic dislocation model enjoys wide acceptance for characterizing coseismic deformation, proposed models for the postseismic and interseismic phases are somewhat speculative. Model uncertainty results from a lack of observational constraints. Aseismic strain rates are small; they seldom exceed a few parts in 10^7 per year. Consequently, even with modern geodetic technology, almost a decade of periodically repeated measurements is required to quantify these rates. Also the recurrence interval between great interplate earthquakes is long, generally a century or more, and only for a few areas around the world does a comparably long geodetic record exist.

Many observed postseismic and interseismic features may be explained in terms of a two-dimensional model, that is, a model whose properties do not vary along the plate boundary. We consider a model (Fig. 5) that consists of two coupled bodies: an elastic slab of thickness H represents the lithosphere and overlies a viscoelastic half-space representing the asthenosphere. The infinitely long fault representing the plate boundary is contained completely within the lithosphere and is partitioned into three infinitely long patches, each of which is characterized by a particular mode of slip. The uppermost patch ($0 \leq z < D$, where z measures depth from the free surface) does not slip except during the earthquake. The intermediate patch ($D \leq z < D_o$) slips continuously throughout the deformation cycle in such a way that its slip rate is greatest immediately after an earthquake and this rate decreases monotonically with time until the time of the next earthquake. The lowermost patch ($D_o \leq z < H$) also slips continuously, but at a constant rate.

In this model, the two deepest fault patches, with their continuous aseismic slip, serve to represent the fact that, along a given segment of a plate boundary,

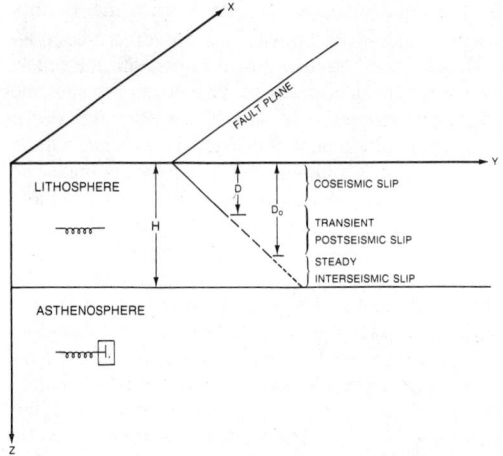

FIGURE 5. Model for aseismic deformation near plate boundary. Elastic lithosphere of thickness H is coupled to viscoelastic asthenosphere (Maxwell body). The plate-bounding fault is partitioned by depth into three patches, each characterized by a specific mode of slip.

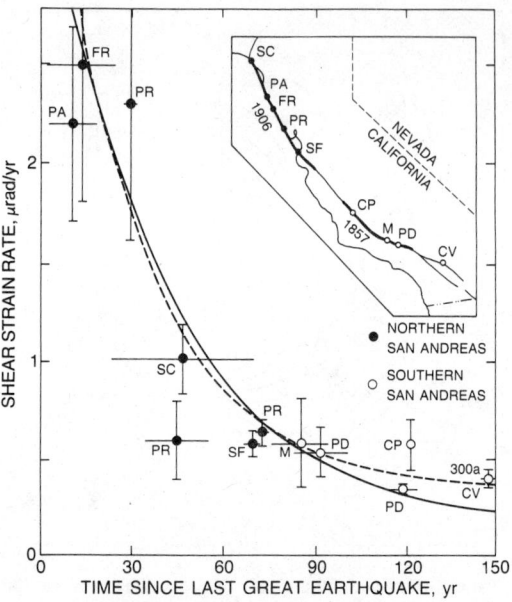

FIGURE 6. Shear strain rate versus time since last great earthquake on San Andreas fault. Vertical bar is one standard deviation; horizontal bar gives time interval between geodetic surveys. Dashed curve and solid curve show expected temporal decay in strain rate for two variations of the model. The variation corresponding to the dashed curve features a thick lithosphere ($H = \infty$), so its temporal decay results solely from accelerated postseismic slip at depth. The variation corresponding to the solid curve contains no variably slipping patch ($D = D_o$), so its temporal decay results from accelerated postseismic flow in the asthenosphere (after Thatcher, 1983).

earthquakes occur only above some specific depth. In California, for example, earthquakes occur only in the upper 20 km of the lithosphere. In California, however, as well as other regions, it is possible that the locked portion of the fault extends down to the lithosphere-asthenosphere boundary (that is, $D=H$), in which case a mechanism of continuous aseismic slip at depth becomes unnecessary.

Also in this model, the intermediate patch is introduced to represent the transient increase in the deformation rate that has been observed on the Earth's surface immediately following some of the larger earthquakes. This transient surface motion may also be explained in terms of the asthenosphere's viscoelastic response to the stress relaxation produced by the earthquake. Thatcher (1983) demonstrates that either of the two mechanisms, with the appropriate choice of values for model parameters, can independently explain existing geodetic measurements of aseismic deformation along California's San Andreas fault, a transform plate boundary (Fig. 6). Thatcher and Rundle (1984) show, however, that, in theory, geodetic measurements along a subduction boundary can better discriminate between the two mechanisms. When they applied their theory to geodetic measurements from Japan, these authors found that relaxed asthenospheric flow below a relatively thin lithosphere ($H \cong 30$ km) better accounts for certain observed aseismic features while accelerated postseismic slip within a relatively thick lithosphere ($H \cong 60$ km) better matches other features. Consequently, even the superposition of the two mechanisms does not adequately explain the Japanese data unless the elastic thickness H is allowed to somehow vary. This latter result suggests that a model featur-

ing a more gradual transition in mechanical properties with depth may be more appropriate than our two-layer model. Also, our model's inability to explain the observations suggests that this model may oversimplify the geometry of the plate boundary.

For convergent boundaries, our model may be refined to include some representation for the underthrust extent of the subducting plate. This is one of several possible enhancements that may be introduced without abandoning the realm of two-dimensional models. Other geometric characteristics of the plates, however, signify the need for three-dimensional models for both convergent and transform plate boundaries. First, plate boundaries come in various, generally nonplanar, shapes. Second, a plate boundary usually consists of a fault system as opposed to a single fault, and relative deformation between two plates may be distributed over many faults—some essentially parallel to the plate boundary, others oblique to the boundary. Third, a given fault usually consists of several distinct, finite-length segments. During a specific earthquake, one or more of these segments might slip while neighboring segments remain locked by friction.

One three-dimensional model consists of an elastic slab of thickness H, which overlies a viscoelastic half-space. Within the slab a collection of dislocation surfaces or patches, each having finite dimensions, represents the fault system in the vicinity of the plate boundary. Some of these patches are locked and slip only during specific earthquakes. Other patches slip continuously; some at steady rates, others at rates that vary with time following specific earthquakes. Moreover, some continuously slipping patches are allowed to intersect the Earth's surface to represent particular boundary segments, such as that in central California, that do not lock and that, consequently, do not suffer significant earthquakes. Rundle (1986) employed such a three-dimensional model to characterize California deformation as resolved from precise geodetic measurements spanning the past two decades (Fig. 7).

The models that we have considered so far are based on the premise that, within a plate, strain accumulation and release are perfectly balanced. Actually, recent studies indicate that significant amounts of permanent deformation occur within the plates, that is, aseismic deformation that is not recovered during earthquakes. For example, uplifted terraces corresponding to former shorelines of late Quaternary and Holocene age record how stretches of the Japanese coast have gradually risen over the course of many seismic cycles in association with offshore plate convergence. Moreover, geodetic leveling documents this uncompensated uplift over the most recent seismic cycle. Thatcher (1984) notes that the proportion of permanent deformation to coseismic deformation within a cycle can vary from less than 10% to more than half. Consequently, modeling the contribution of permanent, nonrecoverable deformation constitutes a significant problem for future research.

Perspective

Geodetic measurements document coseismic deformation for many earthquakes in addition to the 1906 event. In each case, dislocation theory explains observed motion sufficiently well that geophysicists feel relatively confident in their knowledge of how the crust moves during an earthquake. Aseismic crustal deformation, on the other hand, is poorly understood. Several geodetic monitoring programs have been initiated, especially during the past two decades, to overcome this lack of understanding. Some of these programs involve the deployment of electro-optical distance measuring instruments that can achieve a precision of a few millimeters over intersite lines up to 30 km in length. For longer intersite lines, distinct geodetic techniques involving either earth-orbiting satellites or extragalactic radio sources, such as quasars, have been employed to measure all three components of intersite vectors to a precision of a few centimeters [see *Global Positioning System (GPS)*, *Very-Long-Baseline Interferometry (VLBI)*]. The monitoring programs have been designed to determine current plate velocities and to track regional strain accumulation. Other geophysical measurements—for example, observations of seismicity, gravity, stress, and heat flow—complement geodetic data in providing clues as to the nature of plate-driving forces and the Earth's mechanical response to these forces. Consequently, these geophysical measurements also contribute significantly toward the understanding of earthquake-related aseismic deformation.

RICHARD A. SNAY

References

Chinnery, M. A., 1961, Deformation of the ground around surface faults, *Seismol. Soc. America Bull.* **51**, 355–372.

Hayford, J. F., and A. L. Baldwin, 1908, Geodetic measurements of earth movements, in A. C. Lawson, ed., *The California Earthquake of April 19, 1906: Vol. I*, Washington, D.C.: Carnegie Institution of Washington, 114–145.

Okada, Y., 1985, Surface deformation due to shear and tensile faults in a half-space, *Seismol. Soc. America Bull.* **75**, 1135–1154.

Reid, H. F., 1910, Mechanics of the earthquake, in A. C. Lawson, ed., *The California Earthquake of 1906: Vol. II*, Washington, D.C.: Carnegie Institution of Washington.

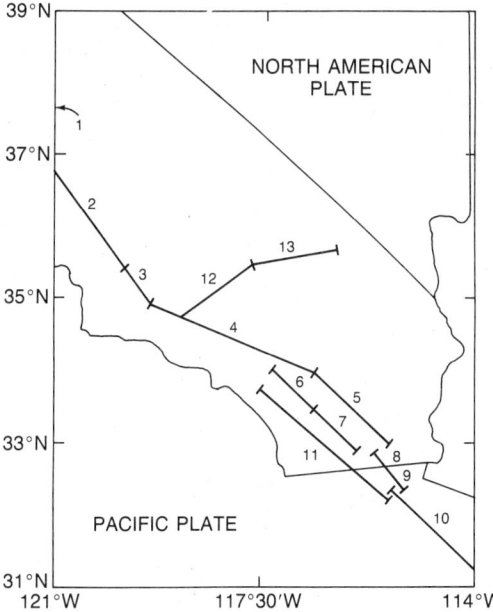

FIGURE 7. Three-dimensional model for crustal deformation in southern California. Numbers identify surface traces for the various dislocation surfaces that represent elements of the fault system along this part of the North American-Pacific plate boundary. Each surface trace may correspond to one or more dislocation surfaces as distinguished by depth and by mode of slip (after Rundle, 1986).

Rundle, J. B., 1986, An approach to modeling present-day deformation in southern California, *Jour. Geophys. Research* **91**, 1947–1959.

Steketee, J. A., 1958a, On Volterra's dislocations in a semi-infinite elastic medium, *Canadian Jour. Phys.* **36**, 192–205.

Steketee, J. A., 1958b, Some geophysical applications of the elasticity theory of dislocations, *Canadian Jour. Phys.* **36**, 1168–1198.

Thatcher, W., 1983, Nonlinear strain buildup and the earthquake cycle on the San Andreas fault, *Jour. Geophys. Research* **88**, 5893–5902.

Thatcher, W., 1984, The earthquake deformation cycle, recurrence, and the time-predictable model, *Jour. Geophys. Research* **89**, 5674–5680.

Thatcher, W., and J. B. Rundle, 1984, A viscoelastic coupling model for the cyclic deformation due to periodically repeated earthquakes at subduction zones, *Jour. Geophys. Research* **89**, 7631–7640.

Cross-references: *Brittle Phenomena; Continental Collision Zones: Seismotectonics and Crustal Structure; Continental Rifting; Crustal Movements and Tectonic Deformation; Earthquake Mechanisms; Earthquake Mechanisms and Plate Tectonics; Earthquakes: Hazards and Prediction; Earthquakes: Volcanogenic; Earthquake Seismology; Earthquakes and Seismicity; Seismicity: Intraplate; Thin-Skin Tectonics.*

EARTHQUAKE SEISMOLOGY

Modern seismology encompasses the scientific study of any significant shaking of the ground surface, whether produced by natural or artificial causes. Originally, however, the term meant observations and analyses of natural earthquakes, especially geological and engineering aspects. In this section of the treatment of seismology, the emphasis is on the latter meaning. (Since 1977, when the first Apollo seismographic instruments operated on the Moon, it is necessary to include natural lunar earthquakes, called *moonquakes*.)

Earthquakes can be disastrous over a large area in a short time, with casualties in a single earthquake sometimes amounting to hundreds of thousands. For this reason the first serious earthquake studies were primarily concerned with the *macroseismic* (noninstrumental) effects of earthquakes and the immediate reduction of hazards. Notable early contributions were from engineers, such as that by Robert Mallett after the great earthquake of 1857 near Naples in Italy (Mallett, 1862). Quantitative descriptions of damaged buildings were made and comprehensive earthquake catalogs prepared.

Gradually, as mechanical properties of rocks, such as elasticity and failure criteria, became physically understood it became clear that brittle fracture of rocks caused the ground shaking. The basic model emerged after studies of the 1906 San Francisco earthquake by the California State Earthquake Investigation Commission (Reid, 1910): Strain in the crustal rocks slowly accumulates until a sudden slip occurs on a (usually preexisting) zone of weakness or geological fault planes. This sudden slip produces waves by the relative movement of rock surfaces in contact along the rupturing fault. Most earthquakes are produced in this way and are called *tectonic earthquakes*.

A difficulty with the word "earthquake" is that it is commonly used to mean both the shaking of the ground and the cause of the shaking. In the following description, the word earthquake is used in the sense of shaking while the cause of the earthquake is called the *seismic source*.

When an earthquake occurs, seismic waves are transmitted through the rocks and emerge at the Earth's surface at places both near and far from the seismic source. Special instruments, called *seismographs*, have been designed to record the ground motion as a function of time (see Fig. 1). The early mechanical instruments have gradually been replaced by a wide range of sophisticated electromagnetic devices that enable both very weak and very strong wave motion to be recorded. (Ground displacements in the earthquake magnitude range of -2 to $+8$ range from 10^{-10} to 10^{-1} meters or about 180 db.) The ground shaking is a result of three main factors: the properties of the seismic source, the effect of Earth structure on the transmitted waves, and the effect of the local conditions at the receiving point. As a consequence, an earthquake at any place usually consists of a complicated train of waves (Aki and Richards, 1980; Bullen and Bolt, 1985). A major challenge in seismology is to explain every wiggle on a seismogram or every seismic vibration input

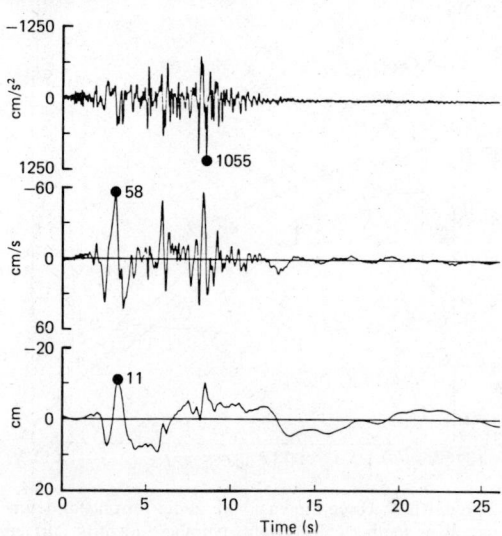

FIGURE 1. Ground motions in the 1971 San Fernando, California, earthquake, recorded on the S74°W component accelerometer on the abutment of Pacoima Dam. Traces are, from top: acceleration, velocity, and displacement.

into an engineered structure. This interpretation stage is of special interest to statisticians and mathematicians. Although significant progress has been made, geological complexities at the source and along the wave path make complete explanations difficult.

After the interpretation stage, recorded seismic motion is used to reconstruct the geological properties along the transmission path and the physical properties of the seismic source.

Distribution of Earthquakes

By the 1960s, the location of earthquake sources based on recordings from earthquake observatories around the world had provided a reasonably uniform mapping of the major tectonic seismic activity (Gutenberg and Richter, 1954). The mapped parameters are *focus* (point where rupture begins) and *epicenter* (point on the surface above the focus). The global maps define belts of earthquake foci along the mid-oceanic ridges, island arcs, and the deep trenches of the oceans (Fig. 2). These seismically-active belts are connected by transcontinental active zones such as that across the Himalaya Mountains into the Mediterranean region. This pattern was a key ingredient in the development of the geological theory of plate tectonics in the 1960s.

Most earthquakes originate within 60 km of the Earth's outer surface; these are called *shallow* earthquakes. Earthquakes at focal depths from 60–300 km are called *intermediate; deep focus* earthquakes occur down to 680 km. About 12% of the energy released in earthquakes comes from intermediate earthquakes and 3% comes from deeper earthquakes. Deep focus earthquakes occur along zones, called *Benioff* zones, that dip into the Earth. The dip angles are about 45°, with both shallower and nearly vertical dips. These zones, or subduction slabs, occur under tectonically-active island arcs such as Japan and Alaska and are also normally, but not always (e.g., Romania, Hindu Kush), associated with deep ocean trenches such as parallel the South American Andes.

After most moderate-to-large shallow earthquakes many lesser earthquakes, called *aftershocks,* occur within the source region. The original fault rupture alters the stress within the crustal rocks, producing slip on relatively small fault sections until stress equilibrium is reached. Aftershocks sometimes occur with frequencies of hundreds or even thousands per day. Sometimes major earthquakes are preceded by less intense earthquakes called *foreshocks* but many major earthquakes have no detectable warning of this kind. In some regions, a series of small-to-moderate earthquakes occurs over days or months without a major earthquake. Such series are called earthquake *swarms.* Often swarms are associated with volcanic activity but may also occur in nonvolcanic regions.

Most earthquakes occur near boundaries of the

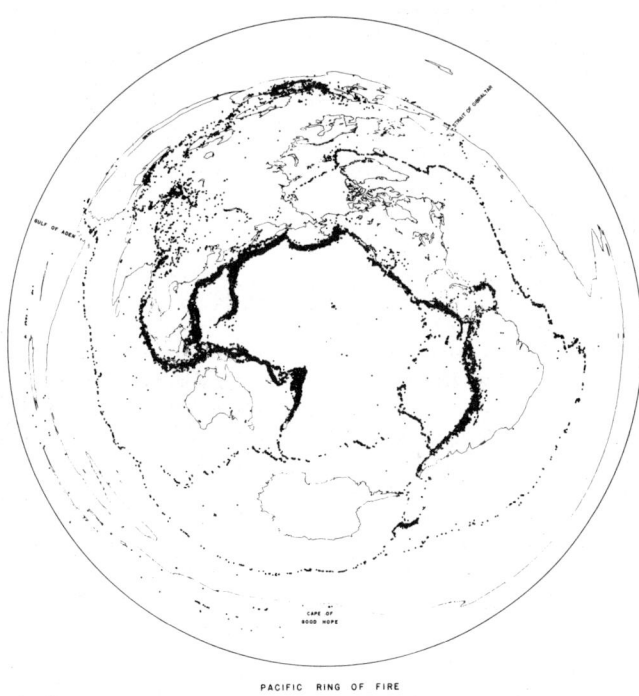

FIGURE 2. Seismicity of the Earth, 1963–1973, based on computer plot of epicenters located by the National Oceanic and Atmospheric Agency (1974).

tectonic plates such as the mid-oceanic regions and subduction zones. These earthquakes are called *interplate earthquakes*. The interior of a tectonic plate is, however, not completely seismically quiescent and large earthquakes, called *intraplate earthquakes*, can occur within a plate. A recent example of the former was the Great Alaska Earthquake of 1964 (magnitude 8.6), which was caused by the thrusting of the Pacific Plate along the subduction zone under the Alaskan continental mass, producing a large *tsunami* (seismic sea wave). The 1985 subduction zone earthquake (magnitude 7.8) along the Pacific Coast of Mexico heavily damaged a portion of Mexico City with over 500 deaths. Accelerometers there demonstrated the importance of local soil conditions on seismic intensity. An example of an intraplate earthquake is the Tanshan earthquake in Northeast China in 1976. This earthquake (magnitude 7.6) was not predicted; because of the high population in housing that was not of seismically-resistant design there were over 250,000 deaths.

Interpretation of Seismic Waves

Nowadays, an experienced seismologist can interpret most of the wave pattern from a distant earthquake. This ability depends on assumptions of linear elasticity and ray theory (Bullen and Bolt, 1985; Kennett, 1983). A distant source can be approximated by a point or small sphere and, because wave fronts are effectively planar at large distances, elastic wave motions can be separated into longitudinal and transverse components.

Like sound and electromagnetic waves, when seismic waves encounter a boundary that separates rock of different elastic properties they undergo reflection, refraction, and diffraction. Within a homogeneous isotropic elastic medium two distinct types of body waves propagate. The faster is a *dilatational wave*, called a P or primary wave, and the slower is a *shear wave*, called an S or secondary wave. When such body waves encounter a boundary, conversion between these types occurs: Incident P or S waves yield reflected P and S waves as well as refracted P and S waves. In addition, variations in rock structure produce wave scattering, thus introducing seismic energy into regions of shadow.

The free surface of the Earth permits the existence of two additional types of seismic waves, called *surface waves*. *Rayleigh waves* have particle motions near the ground surface that are elliptical in a vertical plane; when surficial soil or rock layers are present, or there is a gradient in elastic properties, horizontally polarized surface waves called *Love waves* also propagate. At considerable distances from the source, the P, S, Rayleigh, and Love waves are separated on seismograms according to their respective velocities. Seismograms also often show sets of waves that have traveled along multiple paths called, for example PP, SS, which correspond to single reflections of P and S waves, respectively, at the outer surface of the Earth.

In addition to the usual phases mentioned above, other types of seismic waves (often pulselike) are observed near the seismic source. These include *stopping phases*, which are caused by the intermittent stopping of the fault rupture and the final "sudden" cessation of the rupture. A special form of the stopping phase is called a *breakout phase*, which arises from the generation of a wave pulse when the rupture reaches the free surface of the Earth.

Causes of Earthquakes

The most realistic model for the source of a tectonic earthquake is the representation of fault displacement by a coherent slip of a segment of the fault with incoherent slips over subsections (Aki and Richards, 1980). Physically, the rupture begins suddenly and then spreads with periods of acceleration and retardation along the previously welded fault zone. In this model, statistical randomness of fault slip or "chattering" in irregular steps along the fault plane is an essential property. Rupture can be represented as progressing along a fault plane with *barriers* distributed along it. When rupture begins near a barrier it propagates over the fault plane until it is brought to rest or slowed at the next barrier. Sometimes the barriers are broken by the moving dislocation; sometimes they remain unbroken, but the dislocation reinitiates at the far side and continues; sometimes the barrier is not broken ititially but because of local repartitioning of the stresses and nonlinear elastic effects, it eventually breaks, perhaps with the advent of aftershocks.

This elastic rebound model, involving a moving dislocation along the fault plane over which roughnesses of various types are distributed stochastically, is the starting point for the interpretation of near-field ground shaking. It cnables the computation of synthetic seismograms for points near to the source that are comparable with observations. Such synthetic seismograms provide the input motion for engineered designs of critical structures.

Geologically, there are different modes of fault rupture; some involve horizontal slip (strike slip); some involve vertical slip (dip slip) (see Fig. 3). Wave patterns generated by fault mechanisms of different kinds differ owing to the different radiation patterns in the generated waves. The movement of the fault dislocation also has consequences that depend on the speed of the fault rupture and the fault geometry. The physical problem is analogous (but more difficult) to the problem of sound emission from moving sources. In the far field, the seismic intensity will be the same as when the source is at rest. However, in the near field, the time dependence of both frequency and wave amplitude will be a

FIGURE 3. Fence offset by strike-slip on the San Andreas fault, Marin County, California after great 1906 earthquake.

function of the azimuth of the site relative to the moving source.

The development of *synthetic seismograms* proceeds along three main lines (Bolt, 1987b). The first is a kinematic approach in which the time history of the slip on a generating fault is specified a priori. The assumed defining parameters are shape, duration, and amplitude of a source, the velocity of the slip over the fault surface, and the final area of the region over which the slip occurred. A Green's function representation is usually used to calculate displacements of the medium. Green's functions (system response to unit impulse) for the various fault types have been constructed for this purpose. A variation is to use aftershocks along the fault plane as small prototypes of the local slip in the calculations. A second approach is to use the differential equations that involve the forces producing the rupture. The basic dynamic model is a shear crack that is initiated in the preexisting stress field by a stress concentration at the tip of the crack. These concentrations in turn cause the crack to grow. The modeling of the physical processes of a realistic crack involves specification of the criteria of fracture and the stress variation along the fault. All methods involve numerical work, using finite elements or finite differences, so that realistic geological structure and elastic conditions can be incorporated.

From a mathematical point of view, the source can be represented by a moment tensor and the seismic source can be expanded in terms of spatial moments. In modeling small earthquakes or solving far-field problems it is sufficient to represent the source in terms of a single first-degree moment of the equivalent force. In the near field, however, higher terms must be included, which correspond to additional tensor components.

As well as the tectonic earthquakes caused by strain accumulation in brittle rocks of the Earth, earthquakes have other causes such as large rock avalanches and the sudden movement of magma and gases under volcanoes. Most earthquakes associated with volcanic activity, however, are also caused by elastic rebound along strained rock zones. Some human activities, such as the injection of fluids into deep wells, excavation in mines, filling of large reservoirs, and detonation of underground nuclear explosions produce earthquakes. In all cases, the trigger mechanism for fault slip is thought to be small changes in the local conditions of already strained rock. Over 20 cases of moderate earthquakes produced after the impounding of large reservoirs behind high dams are now documented. Clear-cut examples of reservoir induced earthquakes with magnitudes in excess of 6 are associated with Koyna Dam in India, Aswan High Dam in Egypt, and the Hsingfeng Kiang Dam in China.

Earthquake Parameters

Two common measures of the size of an earthquake are *intensity*, which is a qualitative measure of shaking based on felt reports and damaged structures, and *magnitude*, which is an instrumental measure based on the measured maximum amplitude recorded on a standard seismograph. Recently these basic parameters have been elaborated (Bolt, 1987a).

For example, the original earthquake magnitude introduced by C. F. Richter in 1930, now has been fissioned into at least four alternative definitions of magnitude, each measuring an aspect of earthquake size. Two quantities that are more dynamical than magnitude are the seismic energy E_s released and the moment M_o of the seismic source. Energy is difficult to estimate because of the nonconservative nature of the propagation path and the fault rupture processes. (This difficulty is obviated by using records of strong motion near the source.) As in mechanics, M_o is a measure of the force couples that strain the fault region. Seismic moment can be calculated both theoretically and estimated from seismograms and it is used for both local and large regional earthquakes.

Other parameters of an earthquake used in engineering contexts are peak ground acceleration, velocity, and displacement and the duration of the shaking (usually defined as the bracketed duration, i.e., the time interval between the first and last peaks of given wave amplitude).

In the 1930s a powerful inverse method was developed by P. Byerly, which enabled seismologists to infer the strike and dip of a slipped fault from simple readings on groups of seismograms.

This algorithm, called the *first motion method,* for defining fault-planes (Kasahara, 1981) uses waves recorded at the surface to estimate the fault orientation and principal stresses.

When the extended plane of the earthquake source (i.e., fault rupture) is taken into account an important physical point location is not the point of initiation of the rupture (the focus) but the *centroid* of the release of energy. For long fault rupture, such as in the 1964 Alaskan earthquake, the centroid of energy release may be 100 km away from the focus.

Prediction of Time and Place

It was pointed out earlier that tectonic earthquakes are generated by the elastic rebound of strained rocks along a geological fault. Not only does the theory give a general explanation for the generation of earthquakes, it also contains elements that predict the time interval before rupture (i.e., the slow buildup of strain in the rocks) and the amount of seismic energy released. The elastic rebound theory, however, does not predict exactly the time and place of a rupture (Rikitake, 1976). At present, a completely general predictive theory is not available; even if it were, geological complexities and variations in tectonic histories may make impractical specific predictions in many seismically hazardous regions.

Periodicities of earthquake occurrence in space and time have been analyzed for over a century. Only in recent decades, however, has the seismicity data base been homogeneous and complete enough to permit worthwhile statistical tests for hidden periodicities. The longest historical record is found in China beginning about 700 B.C. This 2,700 year history has been modeled as cyclical time series consisting of over 1,000 destructive earthquakes.

Another approach to the probabilistic occurrence of earthquakes is the search for trigger forces, which initiate the rupture along faults already strained elastically. Tidal forces, volcanic activity, and weather conditions have been considered but no trigger mechanisms for the most significant earthquakes have been found that satisfy strict significance tests necessary to establish a causative link.

Perhaps the most widely discussed prediction model that suggests feasible field observations is one involving the *dilatancy* of rock prior to rupture. The term is used to denote an increase in volume of rocks as they become stressed to the fracture point. For shallow sources, as the crustal rocks are stressed to failure, microcracking occurs with changes of pore pressure due to intrinsic changes in rock properties and the migration of water trapped in the cracks. The effect is to reduce the velocity of seismic waves, produce an uplift of ground level, increase electrical resistivity, and alter the outflow of dissolved gases such as radon.

Work on earthquake prediction has been vigorous in many earthquake countries of the world, particularly Japan, the United States, and China. At present there are two classes of credible earthquake prediction. The first involves the calculation of nonspecific probabilities of earthquake occurrence using observations of activity over a considerable time span. Such probability assessments in California are based on earthquake catalogs for the 190-year historical record, extended in some places by dating geomorphological and liquifaction features. For example in Northern California, the percentage probability of an earthquake above magnitude 6.5 occurring in one year is 8% and above magnitude 7.0 in a year is about 3%.

The second method yields more definitive predictions based on the prediction of the elastic rebound theory that there will be regular repetition of earthquakes of about the same size along the same fault. In central California there is such a pattern about every 22 years since 1857 along the San Andreas fault near Parkfield. Correlation of recorded earthquakes there gives a time window between 1987 and 1993 for the next earthquake of magnitude 5.5–6.0. Various instruments have been put in place to measure the effect of dilatation of the ground including creep along the fault, changes in horizontal strain, magnetic field, and background seismicity (Bolt, 1987a). In 1934 and 1966, when previous large Parkfield earthquakes occurred, there was a foreshock cluster in what has been called the *preparation zone* where the fault rupture initiated. Sensitive seismographs are now operating to try to detect such foreshock activity.

Earthquakes and a Comprehensive Nuclear Test Ban Treaty

Since the early 1950s, seismology has played a central role in the problem of detecting and verifying the explosion of underground nuclear devices (Bolt, 1976). The first nuclear explosion for which source data became generally available was that on 24 July 1946 below the ocean surface near Bikini Atoll. It was recorded at 8 seismographic stations at distances between 69° and 78.6°. In 1958 the field of *forensic seismology* was initiated by a Conference of Experts convened in Geneva to discuss the technical basis for a verifiable test-ban treaty.

Such a treaty for underground nuclear explosions requires a way to distinguish between explosions and natural earthquakes of comparable magnitudes. The problem initially proved difficult and required significant upgrading of seismographic stations around the world. In 1958, only about 700 such stations were in operation with seismographs of various types, imperfect calibration, and poor timing. A major improvement occurred in the late 1950s with the installation of the Worldwide Standardized Seismographic Network (WWSSN). Foreign observatories cooperated with the U.S. Coast and Geodetic Survey in installing and operating the new seismographs. Timing and accuracy were maintained by crystal clocks and calibration pulses were placed daily on each record. Each observatory sent originals of the seismograms to the U.S. Coast and Geodetic Survey (and later to the U.S. Geological Survey) to be copied and seismologists in any country could request copies. By 1967 this standard network was composed of 126 stations in 60 countries. (In other countries, such as Canada, independent upgrading took place.) The operation of the WWSSN stimulated significant new research on earthquake mechanisms, global tectonics, and the structure of the Earth's interior.

In the 1980s the world network began to be upgraded again using digital technology. The new aim is to equip global observatories with seismographs able to record, with high fidelity, seismic waves having periods from about 0.1 to 100 sec. The resulting digital signals on magnetic tapes are able to be input rapidly to high speed computers to estimate earthquake parameters and to report underground explosions.

At the present time, except under most unusual circumstances, seismic methods that use remote stations enable discrimination between natural and artificial earthquake sources down to magnitudes of about 3. Remaining problems involve better calibration of the yield of explosions and its comparison with earthquake size, better maps of attenuation in the Earth, and of the variability in the frequency spectra of small, shallow events.

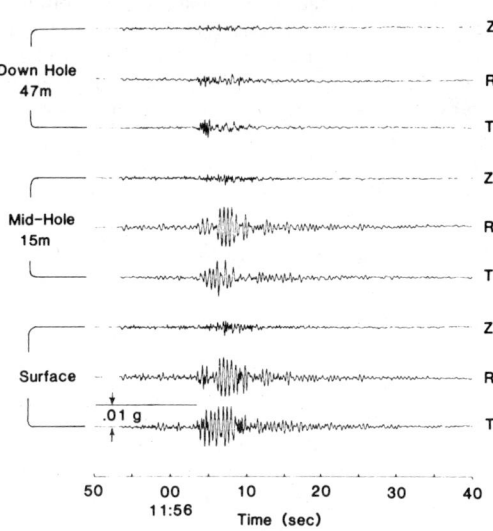

FIGURE 4. Seismograms obtained at three depths in a vertical borehole 70 km from the Mt. Lewis, California, earthquake ($M_L = 5.7$), March 31, 1986. Symbols Z, R, T denote vertical, radial, and transverse ground motions, respectively. The deepest recordings are on basement rock; the shallower ones in a surficial mud layer.

Earthquakes and Geology

Study of the properties of earthquakes has been most valuable in understanding the structure and deformational history of the outer part of the Earth. In particular, phases on seismograms have established that the Earth has an outer homogeneous shell, called the *crust*, having a lower boundary (the Mohorovičić discontinuity) at the depth of 30–50 km under the continents and 5 km under the oceans. Recent work with seismic waves has mapped heterogeneities in the crust and soil down to small-scale stratigraphy (see Fig. 4). Other matters of crucial geological interest have been the structured global distribution of earthquakes including deep focus earthquakes in subduction zones, and measurements of the strength of rocks and their physical properties deep in the crust.

Recent studies of seismogenic zones have explored the effect of temperature and pressure on earthquake sources. The depth of this zone is closely correlated to the transition between brittle and ductile behavior in quartz-bearing rocks. Mechanisms of the earthquakes give an indication of the slip vectors along the zones of weakness and deep seated faults, and the calculation of the principal stresses associated with fault slips provides a three-dimensional mapping of regional crustal strain (Cox and Hart, 1986).

Seismological studies have often been crucial in understanding Quaternary geology and tectonics. The definition of modes of Earth movements and mountain building has been achieved by the application of focal mechanism algorithms. Determination of the stress field along plate margins has complemented geologic inferences on the deformation of the whole interplate zone.

Major landforms, such as the uplifted features of the Basin and Range province and the ancient landslides in the Columbia River basin in northwestern North America, have been the subjects of imaginative geomorphological studies related to earthquake occurrence. Ancient fault movements have been dated by radiocarbon methods applied to sand and peat layers exposed in trenches dug across active faults and by measurement of the erosional geometry of fault scarps.

In addition, volcanic history and associated orogenies, usually studied using geological and geochemical methods, have in recent decades been assisted by the examination of local earthquake patterns. Structures determined from seismological studies have produced more adequate physical models of the eruptive mechanisms.

Finally, seismological work has led to fundamental results on seismic risk theory and the precise mapping of earthquake epicenters has delineated presently active faults. The main responsibility of many geological organizations is now to create special geological maps for earthquake hazard study zones. Correlations between earthquake magnitudes and geological studies of fault dimensions, such as fault offset and fault length, have allowed the estimation of strong ground motion parameters needed for design of critical engineered structures.

Use of Earthquake Waves in Geophysical Imaging

As earthquake waves travel through the deep interior, they are changed by the Earth's structure that they encounter (Bolt, 1982). The first major use of earthquakes to X-ray the Earth was in 1906 when R. D. Oldham inferred from the measured travel times of waves that passed through the central part of the Earth that there was a large *central core* below an outer *mantle*. He plotted the travel times of the longitudinal and transverse waves against the angular distance traveled. The S waves could be plotted as arrivals along a linear trend only out to about 110° distance. This change in behavior occurs because the outer core of the Earth is liquid.

The exploration technique is analogous to that used in modern medicine for obtaining images of anomalous growth inside the human body by CAT scanning (Cathode Applied Tomography). In medicine, sensors outside the human body measure variations in the physical properties of human tissues by detecting changes in intensity of X-rays or atomic particles introduced at the other side. In geophysical tomography, the Earth's structure is observed by comparing earthquake waves at seismographic stations remote from the seismic sources.

One useful tomographic method is to use only the earthquake waves in which motion is limited to the surface region of the Earth. The reason is that the measured speeds and waveforms of the Love and Rayleigh waves carry much information on the structure and elasticity of the upper mantle. Careful decoding of such tomographic signals from earthquakes has begun to yield a quantitative map of tectonically complex regions and physical variations within the mantle. The variations shed light on geodynamical processes such as convection of the rocks within the mantle and the geological history of the Earth's constitution.

Earthquake Engineering Applications

In its modern form, earthquake engineering has become increasingly dependent on seismological studies. This interaction was emphasized by John Milne, Professor of Mining Geology at the Imperial College of Engineering in Tokyo in 1882 (Herbert-Gustar and Mott, 1980). His interest in strong motion seismology was stimulated by his field studies of the 28 October 1891 Mino-Owari earthquake in Japan, one of the largest and most devastating in the modern era.

Milne was concerned with the interaction between wave motion and structural effects and drew up a circular containing queries on intensity and damage. The Imperial Earthquake Investigation Committee on which Milne served consisted of engineers and architects as well as professors of seismology, geology, and physics. It was charged with finding ways to predict earthquakes and determining how damage could be minimized by the choice of suitable methods of construction, building materials, and building sites. Milne did experiments with model structures, investigating the fracturing of mortar by horizontally applied forces; he built a metal-frame trolley on which high columns of bricks were built and which was vibrated by a system of cranks. He tested the concept of base-isolation of structures from earthquakes (now vigorously being developed) by designing buildings standing on cast-iron balls held between metal plates that separated the building from its supports.

These seminal contributions were extended by engineers in the following decades, with more and more emphasis on quantitative dynamical aspects. Major advances came with the recording of strong ground motions in damaging earthquakes, notably in the 1933 Long Beach and 1940 El Centro (Imperial Valley) earthquakes in California.

By the 1960s, in earthquake-prone regions, structural design of large engineered structures, such as high-rise buildings, large dams, and bridges, incorporated considerable dynamical theory and the results of large-scale testing. Seismic building codes dictated that quantitative analyses of critical structures be performed (Wiegel, 1970). The 1940 El Centro accelerogram was broadly accepted as characteristic of strong shaking; it showed peak horizontal ground accelerations of up to 0.3 g in the frequency range from 10 hz to 1 hz. Insights into strong motion seismology subsequently grew with the wide deployment of strong motion instruments in the United States, Japan, Europe, and other seismic regions. These accelerographs supplied a variety of recordings of strong ground motions at various distances from the source of moderate magnitude earthquakes. Among the first key results was a more reliable measure of the attenuation with distance of ground acceleration and velocity in earthquakes.

One specially significant advance followed the 1971 San Fernando earthquake (magnitude 6.5) in southern California, which produced considerable damage (see Fig. 5) and hundreds of strong motion records. The most striking record (see Fig. 1) showed a value of 1.1g horizontal acceleration near the abutment of the Pacoima Dam. Interpretation of

FIGURE 5. Collapsed modern reinforced concrete structure at Olive View Hospital after the 1971 San Fernando, California, earthquake.

this record required the consideration of topographic amplification of seismic waves and the role of high near-field energy input, as well as high peak accelerations, in structural failure.

Large capacity digital computers now enable earthquake engineers to undertake extensive theoretical modelling, particularly through the use of response spectral computations (Bolt, 1987b). In addition, large shaking tables with computer-aided input provide large-scale testing of structural models. Of special note in the United States and in many other earthquake-prone countries, strict criteria for seismological input into design analysis is set by nuclear regulatory agencies for the licensing of nuclear power stations. The result has been major financial support for earthquake research.

At the present time, the introduction of reliable *digital* strong-motion seismographs is providing wider dynamic range and simple access to high-speed computers. Most instruments incorporate pre-event memories and radio receivers that allow direct time-correlations between sites. Special arrays of such seismographs, on the surface and down-hole, have become operational in a few highly seismic areas. They have immediate advantages: P waves are not missed and complete strong ground motions can be correlated spatially and temporally across the whole array. The correlations provide quantitative measures of the crucial effects of soil layers and geological variations on earthquakes and give estimates of the incoherence between wave motions at the multisupports of large structures.

BRUCE A. BOLT

References

Aki, K., and P. G. Richards, 1980, *Quantitative Seismology—Theory and Methods*, 2 vols. San Francisco: W. H. Freeman, 932p.

Bolt, B. A., 1976, *Nuclear Explosions and Earthquakes: The Parted Veil*. New York: W. H. Freeman.

Bolt, B. A., 1982, *Inside the Earth*. New York: W. H. Freeman.

Bolt, B. A., 1987a, *Earthquakes*. New York: W. H. Freeman.

Bolt, B. A., 1987b, *Strong Motion Synthetics*. Orlando, Fla.: Academic Press.

Bullen, K. E., and B. A. Bolt, 1985, *Introduction to the Theory of Seismology*. New York: Cambridge University Press.

Cox, A., and R. B. Hart, 1986, *Plate Tectonics*. Palo Alto, Ca.: Blackwell.

Gutenberg, B., and C. F. Richter, 1954, *Seismicity of the Earth and Associated Phenomena*. Princeton, N.J.: Princeton University Press.

Herbert-Gustar, L. K., and P. A. Mott, 1980, *John Milne, Father of Modern Seismology*. Tenterden: Norbury.

Kasahara, K., 1981, *Earthquake Mechanics*. New York: Cambridge University Press.

Kennett, B. L. N., 1983, *Seismic Wave Propagation in Stratified Media*. New York: Cambridge University Press.

Mallett, R., 1862, *The Great Neopolitan Earthquake of 1857: The First Principles of Observational Seismology*. London: Chapman and Hall.

Reid, H. F., 1910, *The California Earthquake of April 18, 1906. The Mechanics of the Earthquake*. Washington, D.C.: Carnegie Institution of Washington.

Rikitake, T., 1976, *Earthquake Prediction*. Amsterdam: Elsevier.

Wiegel, R. L., ed., 1970, *Earthquake Engineering*. Englewood Cliffs, N.J.: Prentice-Hall.

Cross-references: *Continental Collision Zones: Seismotectonics and Crustal Structure; Deep Earthquakes; Earthquake Mechanisms; Earthquake Mechanisms and Plate Tectonics; Earthquakes: Hazards and Prediction; Earthquakes: Location Techniques; Earthquakes: Magnitude, Energy, and Intensity; Earthquakes: Volcanogenic; Earthquakes and Seismicity; Earth Structure: Global; Elasticity and Wave Propagation: Principles; Seismic Instrumentation; Seismic Instrumentation: History; Seismicity: Intraplate; Seismicity: Mid-Ocean Ridge; Seismicity: Subduction Zone; Seismicity and Plate Tectonics; Seismic Monitoring of Nuclear Explosions; Seismic Source: Observations; Seismic Source: Theory; Seismic Tomography; Seismograms: Synthetic; Seismology: History.*

EARTHQUAKES AND SEISMICITY

The thin surface of the Earth, constantly in motion, emits a fluctuating field of earthquake fracturing activity. This activity was labeled *seismicity* by Gutenberg and Richter in their classic work of 1954. Seismicity is described by space and time distributions of energy release, numbers of events, stress orientations, and other parameters. Studies of seismicity seek answers to fundamental geophysical questions pertaining to the nature and forces of Earth dynamics; the processes of straining, failure, and stress redistribution in earthquakes; and the nature and history of deformation in the fragmented outer shell of the Earth's crust and lithosphere. Studies of seismicity frequently proceed by observation, classification and measurement, and finally, postulation of causative mechanisms. The distributions of seismicity can be used to test the validity of various hypotheses, which are based on other geophysical or geological data. In particular regions, such as China, Italy, and the Middle East, lengthy catalogs of seismicity, compiled from reports of earthquake damage, span up to 3,000 years of history. Instrumental records of global seismicity, however, span less than a century, a relatively short time compared with the scales of global geologic processes. The largest earthquakes worldwide ($M_s \geq 8.0$) from 1897 through 1986 are shown in Figures 1 and 2 and are listed in Table 1.

Spatial Distributions

Seismology is a young science; the first modern seismographs began operating less than a century ago. Today we take for granted that earthquakes occur on faults, however, it was only in 1910 that Reid, using measurements of crustal displacements associated with the great northern California earthquake of 1906, conclusively demonstrated that major earthquakes occur on these large, semipermanent crustal fractures. The existence of deep earthquakes was not discovered until 1922, by Turner; further work by Wadati (1928), Scrase (1931), and Stechschulte (1932) proved that the seismicity field was not simply a surface phenomenon but rather that it was distributed in depth throughout the Earth's outer shell. Current information indicates that earthquakes can occur at depths down to at least 680 km, or about 10% of the Earth's thickness from its surface to the center (see *Deep Earthquakes*). Such deep earthquakes are the exception; over most of the globe earthquakes seldom occur at depths greater than 30 km. Gutenberg and Richter (1954) studied the spatial distribution of worldwide seismicity and found that 80% of earthquakes occur around the circum-Pacific belt. In the 1960s, with the confirmation of seafloor spreading and global plate tectonics, it became clear that the circum-Pacific belt of seismicity was caused by preferential motion along a major fault system separating relatively competent "plates" of the Earth's lithosphere (see *Seismicity and Plate Tectonics*). Earthquakes reflect the relative motions of these plates, which cause straining, slipping, and breaking at their boundaries. Deep seismicity is found only near the boundaries of plates. Shallow earthquakes are found both along plate boundaries (interplate seismicity) and within the interiors of plates (intraplate seismicity). An average of 4.7 earthquakes each year (22%) are deeper than 70 km and have $m_b \geq 7.0$ (Abe and Kanamori, 1979), compared with an average of 17 per year (78%) with shallower depths and $M_s \geq 7.0$ (Kanamori, 1977a). Figure 3 shows the number and energy release of large earthquakes worldwide ($m_b \geq 7$, from 1904–1974) as a function of depth.

Temporal Distributions

The Search for Periodicities. Gutenberg and Richter (1954) also studied the temporal fluctuations in global seismicity and determined that, although the rate of seismic energy release is extremely irregular, variations (p.21) "apparently are within the limits of normal statistical fluctuation . . . [with] no evident relation to minor periodicities." They used harmonic analyses to test for correlations between global seismicity patterns ($M \geq 7$) from 1904 to 1952 and tidal fluctuations; seasonal influences; annual, daily, and hourly regularities; and sunspot activity. In 1979 Kanamori and Abe studied global seismicity data from 1897–1977 ($M_s \geq 8$) using new and improved magnitude calculations and determined that an apparent peak in seismic activity around the turn of the century was only of marginal significance. Interim work had suggested that periodic increases in worldwide seismicity might be

FIGURE 1. Locations of the largest worldwide earthquakes ($M_s \geq 8.0$) from 1897 through 1986 (see Table 1).

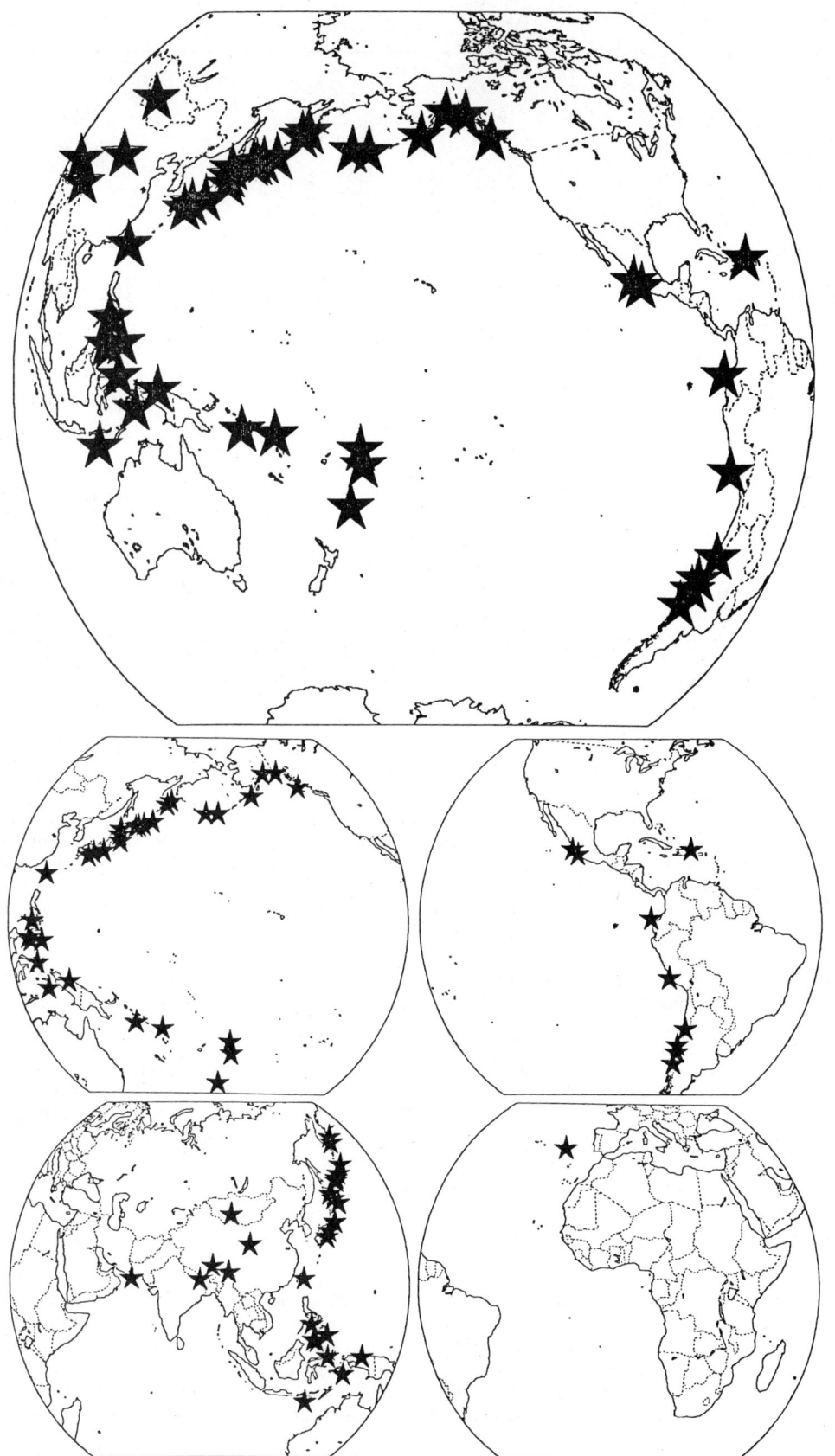

309

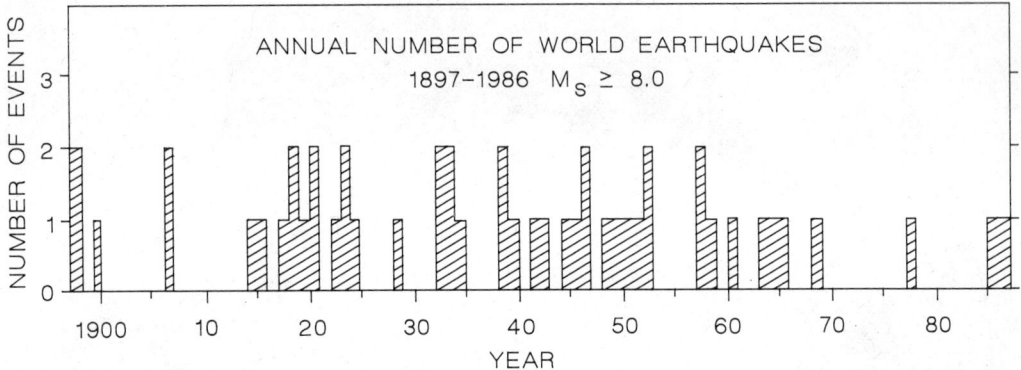

FIGURE 2. Temporal distribution of the largest worldwide earthquakes ($M_s \geq 8.0$) from 1897 through 1986.

correlated with changes in the rotation rate of the earth and peaks in the amplitude of the Chandler wobble.

Statistical analyses of long-term temporal distributions of regional seismicity $M \geq 6$ in Central China from A.D. 1000 to A.D. 1976 by Lee and Brillinger (1979) suggest that there are pronounced episodes of numerous earthquakes at intervals of about 300 years; conversely, earthquake activity is low at intervals of about 175 years and 375 years. This episodicity has not yet been correlated with any specific physical process, but such results provide opportunities for future research.

Deviations from Randomness. The seismicity field has also been tested to determine whether fluctuations in earthquake density and energy release are purely random or independent in time. Observers noted long ago that large earthquakes (mainshocks) are often followed by many large, moderate, and small shocks (aftershocks), and that a number of moderate or small nearby events (foreshocks) may precede the mainshock. One of the earliest quantitative descriptions of an earthquake distribution was the law describing aftershock behavior. In 1894 a Japanese scientist, Omori, proposed an empirical relation for the decay in earthquake activity during an aftershock sequence that, with small modifications by Utsu in the 1960s, still stands today:

$$n(t) = n_1(t + c)^{-p}$$

The number of earthquakes per unit time at time t after the mainshock is $n(t)$; n_1 is the rate of activity at unit time after the mainshock; c is approximately 0; and p is typically between 1.0 and 1.4. An example of aftershock behavior is shown in Fig. 4. It is remarkable that the Omori relation, despite many advances in seismographic instrumentation, has been reconfirmed many times and still stands as one of the fundamental laws of earthquake distribution. In fact, the modified Omori relation is so well established that it is commonly used to test the validity of models proposed for the earthquake process, such as the nature of microfracturing in rock fracture experiments and the self-similarity of the seismic process.

Aftershock sequences contribute a substantial fraction of the earthquakes listed in seismicity catalogs. Aftershocks usually continue for a number of years following a large mainshock. Ogata and Shimazaki (1984) found that aftershocks ($m_b \geq 4$) continued for about six years after the large [magnitude $M_w = 8.7$ (Kanamori, 1977a)] Rat Islands mainshock of 1965. Small aftershocks ($M_L \geq 2$) are still continuing some 35 years after the large 1952 earthquake in Kern County ($M_s = 7.7$), southern California. Aftershock sequences often contain secondary aftershock sequences in which some large aftershocks produce aftershocks of their own. The secondary aftershock sequences also obey the Omori relation. Although nearly one hundred years have passed since Omori's discovery, the exact physical process of aftershock sequences is still a subject of active research.

Aftershocks occur as a burst of activity in the background distribution of earthquakes. Another type of burst of activity is called a *swarm*. Swarms are distinguished from aftershocks by the absences of a large mainshock. Swarms can be short or prolonged, although usually they are much shorter than aftershock sequences. Swarms of earthquakes with small magnitudes are relatively common, but swarms of large earthquakes also occur. Several causes of swarms have been suggested, but as with aftershocks, the exact physical mechanism is not known. Swarms of small earthquakes are frequently found in volcanic environments or in regions of extensional tectonics; at locations of and following fluid injection into the Earth's crust or rapid filling of a reservoir; and in the region of and during a decade or so prior to large earthquakes.

While foreshock sequences are usually similar to swarms, in some cases they may resemble a small mainshock-aftershock sequence. Foreshocks differ

TABLE 1. World's Largest Earthquakes, 1897-1986, $M_s \geq 8.0$

Year	Mo/Da	Hr:Mn:Sec	Lat.	Long.	M_s	M_w
1897	9/20	19: 6: 0.0	6.000	122.000	8.1	
1897	9/21	5:12: 0.0	6.000	122.000	8.2	
1899	9/10	21:41: 0.0	60.000	−140.000	8.0	
1906	1/31	15:36: 0.0	1.000	−81.500	8.2	8.8
1906	8/17	0:40: 0.0	−33.000	−72.000	8.1	8.2
1914	5/26	14:22:42.0	−2.000	137.000	8.0	
1915	5/1	5: 0: 0.0	47.000	155.000	8.0	
1917	6/26	5:49:42.0	−15.500	−173.000	8.4	
1918	8/15	12:18:12.0	5.500	123.000	8.0	
1918	9/7	17:16:13.0	45.500	151.500	8.2	
1919	4/30	7:17: 5.0	−19.000	−172.500	8.2	
1920	6/5	4:21:28.0	23.500	122.000	8.0	
1920	12/16	12: 5:48.0	36.000	105.000	8.6	
1922	11/11	4:32:36.0	−28.500	−70.000	8.3	8.5
1923	2/3	16: 1:41.0	54.000	161.000	8.3	8.5
1923	9/1	2:58:36.0	35.250	139.500	8.2	7.9
1924	4/14	16:20:23.0	6.500	126.500	8.3	
1928	12/1	4: 6:10.0	−35.000	−72.000	8.0	
1932	5/14	13:11: 0.0	0.500	126.000	8.0	
1932	6/3	10:36:50.0	19.500	−104.250	8.0	8.1
1933	3/2	17:30:54.0	39.250	144.500	8.5	8.4
1934	1/15	8:43:18.0	26.500	86.500	8.3	
1934	7/18	19:40:15.0	−11.750	166.500	8.1	
1938	2/1	19: 4:18.0	−5.250	130.500	8.2	8.5
1938	11/10	20:18:43.0	55.500	−158.000	8.3	8.2
1939	4/30	2:55:30.0	−10.500	158.500	8.0	
1941	11/25	18: 3:55.0	37.500	−18.500	8.2	
1942	8/24	22:50:27.0	−15.000	−76.000	8.2	
1944	12/7	4:35:42.0	33.750	136.000	8.0	8.1
1945	11/27	21:56:50.0	24.500	63.000	8.0	
1946	8/4	17:51: 5.0	19.250	−69.000	8.0	
1946	12/20	19:19: 5.0	32.500	134.500	8.2	8.1
1948	1/24	17:46:40.0	10.500	122.000	8.2	
1949	8/22	4: 1:11.0	53.750	−133.250	8.1	8.1
1950	8/15	14: 9:30.0	28.500	96.500	8.6	8.6
1951	11/18	9:35:47.0	30.500	91.000	8.0	7.5
1952	3/4	1:22:43.0	42.500	143.000	8.3	8.1
1952	11/4	16:58:26.0	52.750	159.500	8.2	9.0
1957	3/9	14:22:28.0	51.300	−175.800	8.1	9.1
1957	12/4	3:37:48.0	45.200	99.200	8.0	8.1
1958	11/6	22:58: 6.0	44.400	148.600	8.1	8.3
1960	5/22	19:11:14.0	−38.200	−72.600	8.5	9.5
1963	10/13	5:17:51.0	44.900	149.600	8.1	8.5
1964	3/28	3:36:14.0	61.100	−147.500	8.4	9.2
1965	2/4	5: 1:22.0	51.300	178.600	8.2	8.7
1968	5/16	0:48:57.0	40.900	143.400	8.1	8.2
1977	8/19	6: 8:55.0	−11.200	118.400	8.1	8.3
1985	9/19	13:17:47.0	18.190	−102.533	8.1	
1986	10/20	6:46: 9.0	−28.117	−176.367	8.1	

M_s, 1897-1912: Abe, K., and S. Noguchi, 1983, Revision of magnitudes of large shallow earthquakes, 1897-1912, *Physics Earth and Planetary Interiors* **33**, 1-11.

M_s, 1932 (6/3): Wang, S., K. McNally, and R. Geller, 1982, Seismic strain release along the Middle America Trench, Mexico, *Geophys. Research Letters* **9**, 182-185.

M_s, 1904-1980: Abe, K., 1981, Magnitudes of large shallow earthquakes from 1904 to 1980, *Physics Earth and Planetary Interiors* **27**, 72-92.

M_s, 1981-1986: NEIS, U.S. Geological Survey.

M_w: Kanamori, H., 1986, Global seismicity, in H. Kanamori and E. Boschi, eds., *Earthquakes: Observation, Theory and Interpretation*. Amsterdam: North Holland, 596-608.

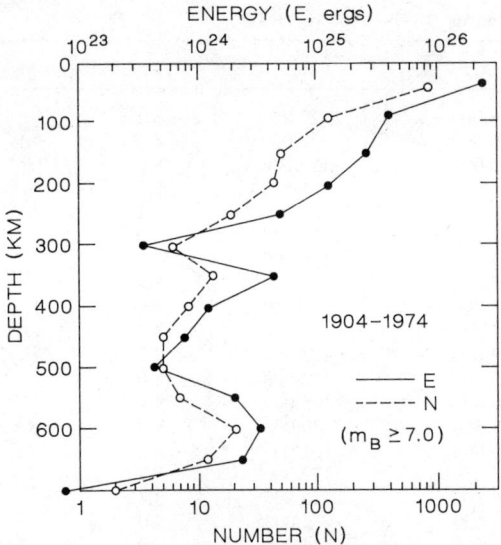

FIGURE 3. The variation in numbers (N) and energy release (E) of large worldwide earthquakes [$m_b \geq 7$, from 1904 through 1974] as a function of depth. Each point includes earthquakes within 25 km above and below. The trends in number and energy are similar: (a) Energy released (per 50 km of depth) above 70 km is nearly 100 times greater than that released below 300 km [the total energy released in shallow earthquakes (above 70 km depth) is 3.2×10^{26} ergs, which is similar to the total energy released below 70 km, which is 1.0×10^{26} ergs, however]; (b) two peaks are found at 350 km and 600 km depth. (After Abe and Kanamori, 1979.)

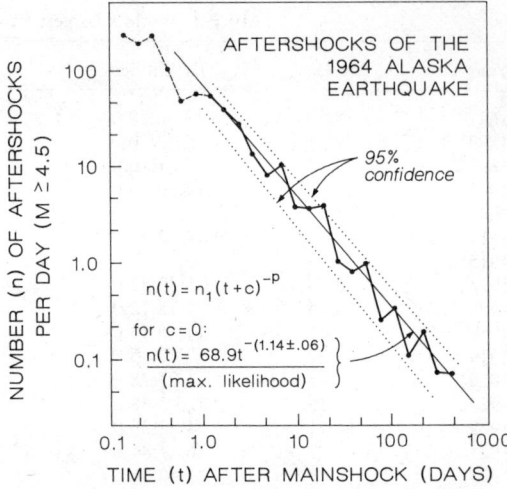

FIGURE 4. Frequency of occurrence of aftershocks as a function of time following the great Alaska earthquake (after Page, 1968).

from swarms or aftershocks because, by definition, they occur some short time prior to a larger mainshock. Studies to distinguish foreshocks from swarms are important for earthquake prediction. Unfortunately, it is difficult to distinguish foreshocks from swarms with statistical significance because data are sparse. A recent extension of the foreshock concept has been suggested by Ishida and Kanamori (1978, 1980): A tight spatial clustering of preshocks may occur for several years, concentrating in the region of an eventual mainshock. Another observation is that a ring of increased activity likened to a doughnut, often encircles a future rupture zone for years or decades before a large earthquake (Mogi, 1969; Kelleher and Savino, 1975). Earthquake swarms often constitute portions of the activity defining such rings.

Magnitude Distributions

The spatial distribution of the global seismic field is highly nonuniform. The strongest earthquake activity, both in number and size, is concentrated along plate boundaries such as the circum-Pacific belt. In contrast, intraplate activity within great plate masses is smaller and less frequent. Slower processes of intraplate strain accumulation and stress release are not yet well understood and constitute an important area of research. Certainly intraplate earthquakes in the United States and Canada such as New Madrid, Mo., 1811; Charleston, S.C., 1886; New Brunswick, Canada, 1982; and Borah Peak, Idaho, 1983 pose significant hazards to populations.

Plate tectonic motions are fairly well understood in terms of their average (geologic) rates of relative motion and straining, and one might expect that the seismicity along plate boundaries might be highly regular and somehow proportional to these rates. Unfortunately, the sizes and rates of earthquakes are neither uniform around interplate boundary zones, nor even at any particular location. The sizes of earthquakes are physically related to the size of the fault area that ruptures, as well as to the amount of fault slippage. A bimodal distribution in the sizes of the largest earthquakes around plate boundaries would not be surprising. Where the plates move laterally with respect to one another, the length along the brittle contact zone can be a few hundred kilometers long but only ~30 km wide (deep). Conversely, in places where one plate slides under another in subduction zones, the brittle contact zone can be 200 km wide. For example, the largest historic earthquake in California ruptured laterally along a 360–400 km section of the San Andreas Fault in 1857 and possessed a moment magnitude of $M_w = 7.9$. The largest known earthquake along a subduction zone ruptured an area 1,000 km long and 210 km wide beneath Chile in 1960, and possessed a moment magnitude of $M_w = 9.4$ (see *Earthquakes: Magnitude, Energy, and Intensity* for a definition of moment magnitude).

A strict bimodality of the distribution in earthquake size, along strike slip and subduction plate margins, is modified by at least two factors. In subduction zones, the width of the contact zone will be greater if the downgoing plate dips shallowly beneath the overriding plate and lesser if it dips steeply. The area of the brittle contact zone and the sizes of the largest earthquakes are thus modified by dip angle of the downgoing plate. The second factor, suggested by Kanamori (1977b) and now fairly well confirmed, is that the strength of coupling between plates may vary from one fault zone to another. Ruff and Kanamori (1980) compared the sizes of earthquakes along subduction zone plate boundaries and found that the largest earthquakes occur where young oceanic plates are rapidly subducting. In fact, surface measurements along the main trace of the plate boundary in northern California, the San Andreas Fault, indicate that there can also be variations in coupling along the same fault. Some sections of the San Andreas slide continuously and produce only moderate (M = 5–6) and small (M < 5) sized earthquakes (albeit at a very high rate). Other sections appear to be locked for long periods, occasionally producing great (M = 8) earthquakes with virtually no interevent seismicity or slippage.

In general, fault sections that are both currently locked and have produced large earthquakes in the past can be expected to yield large size events in the future. In the 1920s Imamura noted that large earthquakes tend to repeat in the same location and warned of a future large earthquake in Japan. In 1923 the devastating Tokyo earthquake left 120,000 dead, largely due to ensuing fires. Subsequent work by Sykes (1971) and Kelleher et al. (1973) reintroduced the concept of large earthquakes repeating along fault segments into the context of plate tectonic boundaries, where the strain mechanisms and rates are relatively well known. Aki (1978, p. 4) provides a clear and simple definition of the associated phenomenon known as the *seismic gap*:

> ... the aftershock areas of large earthquakes in the past tens to hundreds of years tend to fill up a seismic zone without significant overlap. In the context of plate tectonics, this implies that the relative motion of plates is accomplished by a series of ruptures along the plate margin, each associated with a major earthquake. Then, it is natural to anticipate the next earthquake in a gap of aftershock zones unfilled for more than the expected recurrence time of major earthquakes for the seismic zone. Thus, the gap is important for a long range prediction of major earthquakes in the plate margin. ...

Recent work indicates that a particular stretch of fault, or a seismic gap, does not always break in characteristic events of the same size. Kanamori and McNally (1982) found that several fault segments occasionally break simultaneously, producing a particularly large earthquake. At other times the same segments break separately in smaller events. They labeled this behavior *variable rupture mode*.

Earthquakes are not uniformly distributed in time, space, and magnitude, although seismicity can be examined in terms of global averages; for example, there are approximately 2 earthquakes per year with $M \geq 8$, 20 per year with $M \geq 7$, and so forth. One of the early laws of seismicity, that of the relation between numbers of earthquakes and their magnitudes, has not only stood the test of time but, like Omori's law of aftershocks, has proven extremely useful in basic and applied research. Three works by Ishimoto and Iida (1939) and Gutenberg and Richter (1942, 1944) documented that:

$$\text{LOG } N = a - bM$$

where N is the number of earthquakes of magnitude M or greater, a is a constant, and b typically ranges from 0.8 to 1.2 (see also *Earthquakes: Magnitude, Energy, and Intensity*). Figure 5 illustrates this relation for Southern California earthquakes, $M \geq 3.5$, from 1932 to 1971. This relationship holds true, on average, at both regional and global scales and

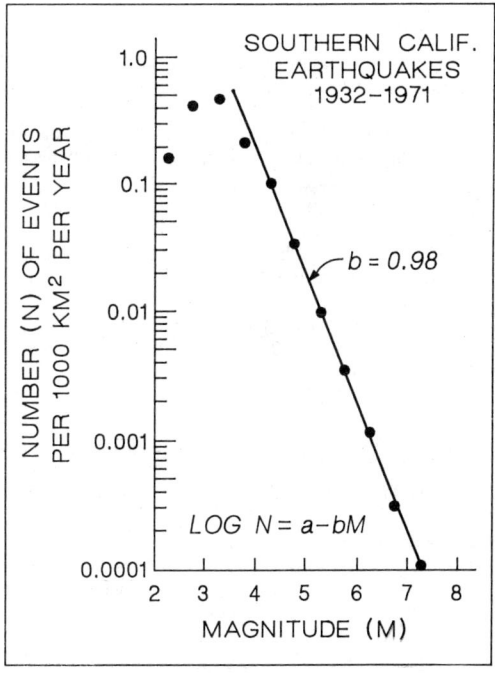

FIGURE 5. Frequency of occurrence of earthquakes in southern California as a function of magnitude. The number of earthquakes with magnitude greater than 3.5 from 1932 through 1971 is 3,404; the area is 238,600 km^2 (after Hileman et al., 1973).

was made possible by the earliest quantification of earthquake sizes with the development of the magnitude scale for local earthquakes by Richter (1935, 1958). Additional work by Gutenberg and Richter (1942) and Gutenberg (1956) related earthquake magnitude to seismic energy release (E):

$$\text{LOG } E = 11.8 + 1.5M$$

and was extended to include present knowledge of the largest earthquakes measured by modern instruments by Kanamori (1977a). These classic distributions led to the understanding that

> ... smaller shocks almost never are sufficiently frequent to approximate the energy released in larger shocks. This means that great shocks are essentially independent events, uninfluenced by the occurrence of smaller earthquakes, which are at most symptomatic of the regional strains released in major shocks ...

(Gutenberg and Richter, 1954, p. 19), as well as ongoing research and discoveries about the earthquake process.

KAREN C. McNALLY

References

Abe, K., 1979, Size of great earthquakes of 1837-1974 inferred from tsunami data, *Jour. Geophys. Research* **84**, 1561-1568.

Abe, K., and H. Kanamori, 1979, Temporal variation of the activity of intermediate and deep focus earthquakes, *Jour. Geophys. Research* **8**, 3589-3595.

Aki, K., 1978, Origin of the seismic gap: what initiates and stops a rupture propagation along a plate boundary? Conf. VI Proc., Methodology for Identifying Seismic Gaps and Soon-to-Break Gaps, U. S. Geol. Survey Open-File Rept., 3-46.

Aki, K., 1980, Presidential address: possibilities of seismology in the 1980s, *Seismol. Soc. America Bull.* **70**, 1969-1976.

Allen, C. 1968, The tectonic environments of seismically active and inactive areas along the San Andreas Fault system, in W. R. Dickinson and A. Grantz, eds. *Conference on Geologic Problems of the San Andreas Fault System. Proceedings.* Stanford, Calif.: Stanford Univ., 70-82.

Gutenberg, B., 1956, Great earthquakes 1896-1903, *EOS (Am. Geophys. Union Trans.)* **37**, 608-614.

Gutenberg, B., and C. F. Richter, 1942, Seismicity of the earth, *Geol. Soc. America Bull.* **32**, 163-191.

Gutenberg, B., and C. F. Richter, 1944, Frequency of earthquakes in California, *Seismol. Soc. America Bull.* **34**, 185-188.

Gutenberg, B., and C. F. Richter, 1954, *Seismicity of the Earth and Associated Phenomena*, 2nd ed. Princeton, N.J.: Princeton University Press.

Hileman, J., C. Allen, and J. Nordquist, 1973, *Seismicity of the Southern California Region, 1 Jan. 1932-31 Dec. 1972,* Pasadena, California: California Institute of Technology.

Hirata, T., 1987, Omori's power law aftershock sequences of microfracturing in rock fracture experiment, *Jour. Geophys. Research* **92**, 6215-6221.

Ishida, M., and H. Kanamori, 1978, The foreshock activity of the 1971 San Fernando earthquake, California, *Seismol. Soc. America Bull.* **68**, 1265-1279.

Ishida, M., and H. Kanamori, 1980, Temporal variation of seismicity and spectrum of small earthquakes preceding the 1952 Kern County, California earthquake, *Seismol. Soc. America Bull.* **70**, 509-527.

Ishimoto, M., and K. Iida, 1939, Observations sur les seismes enregistres par le microseismographe construit dernierement, *Tokyo Univ. Earthquake Research Inst. Bull.* **17**, 443-478.

Isacks, B., J. Oliver, and L. R. Sykes, 1968, Seismology and the new global tectonics, *Jour. Geophys. Research* **73**, 5855-5899.

Kagan, Y. Y., and L. Knopoff, 1981, Stochastic synthesis of earthquake catalogs, *Jour. Geophys. Research* **86**, 2853-2862.

Kanamori, H., and K. Abe, 1979, Reevaluation of the turn of the century seismicity peak, *Jour. Geophys. Research* **84**, 6131-6139.

Kanamori, H., and K. C. McNally, 1982, Variable rupture mode of the subduction zone along the Ecuador-Colombia coast, *Seismol. Soc. America Bull.* **72**, 1241-1253.

Kanamori, H., 1977a, The energy release in great earthquakes, *Jour. Geophys. Research* **82**, 2981-2987.

Kanamori, H., 1977b, Seismic and aseismic slip along subduction zones and their tectonic implications, in W. C. Pitman II and M. Talwani, eds., *Island Arcs, Deep Sea Trenches and Back-Arc Basins*. Washington, D.C.: American Geophysical Union (Maurice Ewing Ser. 1), 163-174.

Kelleher, J., and J. Savino, 1975, Distribution of seismicity before strike slip and thrust-type earthquakes, *Jour. Geophys. Research* **80**, 260-271.

Kelleher, J. A., L. R. Sykes, and J. Oliver, 1973, Possible criteria for predicting earthquake locations and their applications to major plate boundaries of the Pacific and Caribbean, *Jour. Geophys. Research* **78**, 2547-2585.

Kelleher, J., J. Savino, H. Rowlett, and W. R. McCann, 1974, Why and where great thrust earthquakes occur along island arcs, *Jour. Geophys. Research* **79**, 4889-4899.

Lee, W. H. K., and D. R. Brillinger, On Chinese earthquake history—an attempt to model an incomplete data set by point process analysis, *PAGEOPH* **117**, 1229-1257.

Mogi, K., 1969, Some features of recent seismic activity in and near Japan (2), Activity before and after great earthquakes, *Tokyo Univ. Earthquake Research Inst. Bull.* **47**, 395-417.

Nishenko, S. P., 1985, Seismic potential for large and great interplate earthquakes along the Chilean and Southern Peruvian margins of South America: a quantitative reappraisal, *Jour. Geophys. Res.* **90**, 3589-3616.

Ogata, Y., and K. Shimazaki, 1984, Transition from aftershock to normal activity: the 1965 Rat Islands earthquake aftershock sequence, *Seismol. Soc. America Bull.* **74**, 1757-1765.

Omori, F., 1894, On the aftershocks of earthquakes, *Tokyo Univ. Coll. Sci. Imp. Jour.* **7**, 111-200.

Page, R., 1968, Aftershocks and microaftershocks of the great Alaskan earthquake of 1964, *Seismol. Soc. America Bull.* **58**, 1131-1168.

Page, R., 1986, Comments on earthquake frequency and prediction by Liu Z. R., *Seismol. Soc. America Bull.* **7**, 1491-1496.

Reid, H. F., 1910, *The Mechanics of the Earthquake, The California Earthquake of April 18, 1906, Report of the State Investigation Commission*, 2. Washington, D.C.: Carnegie Institution of Washington, 16-28.

Richter, C. F., 1935, An instrumental earthquake scale, *Seismol. Soc. America Bull.* **25**, 1-32.

Richter, C. F., 1958, *Elementary Seismology*. San Francisco: W. H. Freeman.

Ruff, L., and H. Kanamori, 1980, Seismicity and the subduction process, *Physics Earth and Planetary Interiors* **23**, 240-252.

Scholz, C. H., 1968, Microfracture, aftershocks, and seismicity, *Seismol. Soc. America Bull.* **58**, 1117-1130.

Scrase, F. J., 1931, The reflected waves from deep-focus earthquakes, *Royal Soc. [London] Proc.* **132**, 213-235.

Sieh, K. E., 1978, Slip along the San Andreas fault associated with the great 1857 earthquake, *Seismol. Soc. America Bull.* **68**, 1421-1448.

Stechschulte, V. C., 1932, The Japanese earthquake of March 29, 1928, and the problem of depth of focus, *Seismol. Soc. America Bull.* **22**, 81-137.

Sykes, L. R., 1971, Aftershock zones of great earthquakes, seismicity gaps, and earthquake prediction for Alaska and the Aleutians. *Jour. Geophys. Research* **79**, 8021.

Sykes, L. R., and R. C. Quittmeyer, 1981, Repeat times of great earthquakes along simple plate boundaries, in D. W. Simpson and P. G. Richards, eds., *Earthquake Prediction: An International Review*. Washington, D.C.: American Geophysical Union (Maurice Ewing Ser. 4), 217-247.

Tocher, D., 1960, Creep on the San Andreas fault California—creep rate and related measurements at Vineyard, California, *Seismol. Soc. America Bull.* **50**, 396-404.

Turner, H. H., 1922, On the arrival of earthquake waves at the antipodes, and the measurement of the focal depth of an earthquake, *Royal Astron. Soc., Monthly Not., Geophys. Suppl.* **1**, 1-13.

Utsu, T., 1969, Aftershocks and earthquake statistics: some parameters which characterize an aftershock sequence and their interrelations, *Hokkaido Univ. Fac. Sci. Jour. Ser. VII (Geophysics)* **3**, 129-195.

Utsu, T., 1970, Aftershocks and earthquake statistics (II). Further investigation of aftershocks and other earthquake sequences based on a new classification of earthquake sequences, *Hokkaido Univ. Fac. Sci. Jour. Ser. VII (Geophysics)* **3**, 197-266.

Wadati, K., 1928, Shallow and deep earthquakes, *Geophys. Mag.* **1**, 162-202.

Wesnousky, S. G., 1986, Earthquakes, quaternary faults, and seismic hazard in California, *Jour. Geophys. Research* **91**, 12587-12632.

Cross-references: *Earthquake Mechanisms and Plate Tectonics; Earthquakes: Hazards and Prediction; Earthquakes: Magnitude, Energy, and Intensity; Earthquakes and Crustal Deformation; Earthquake Seismology; Seismicity and Plate Tectonics; Seismology: History; Stress in the Earth's Lithosphere; Strong Motion Seismology.*

EARTH'S ABSOLUTE GRAVITY—See ABSOLUTE GRAVITY MEASUREMENTS.

EARTH ROTATION—See EARTH ORIENTATION.

EARTH'S CORE

The Earth consists of a mantle some 2885 km thick surrounding a core of radius 3485 km. The existence of the core was inferred from seismic observation as early as 1897; refraction of seismic P waves by the core creates a "shadow zone" at distances between 105° and 140° from the epicenter. Further analysis of seismic waves has shown that the core is divided into a liquid outer core about 2260 km thick surrounding a solid inner core of radius 1225 km. During the past decade there has been a shift from descriptive studies of the Earth's deep interior to serious attempts to relate the structures deduced within the Earth to the processes that may cause or result from them.

The Earth acts as a great engine that converts thermal and potential energy into mechanical and magnetic energy. This engine operates as two separate convective systems in the mantle and core. Motions in the fluid outer core produce the magnetic field of the Earth by dynamo action. The energy supply for these motions is intimately coupled to the thermal evolution of the Earth, and according to one theory, the waste heat from this process rises through the mantle via thermal plumes to cause *hot-spot* volcanism at such places as Hawaii and Iceland. Furthermore, dynamic coupling between the core and mantle may be the cause of irregular changes in the rotation rate of the Earth, particularly changes in the length of day with periods on the order of decades. Thus the processes within the core have a measurable effect on the features and processes observed at the Earth's surface.

Much is still unknown about the early history of the Earth, but it probably was at one time very hot, most likely as a result of core formation. This process, in which the denser core and lighter mantle separate from an initially homogeneous protoearth, would have released sufficient gravitational potential energy to heat the entire Earth several thousand degrees. Because heating mobilizes material, core formation is a "run-away" process that accelerates rapidly once begun and goes to completion. Since then, the Earth has been gradually cooling, and the structure and dynamics within the core are a direct consequence of this cooling process.

The most plausible explanation of the existence of the Earth's inner core is that it has solidified from the outer core as the Earth has gradually cooled. The core, although coolest at the top, will freeze first at the bottom provided the liquidus (i.e., freezing) gradient with pressure, dT_L/dP, is steeper than the adiabat, dT_A/dP (assuming that the outer core is convecting). If this were not true, either the cooling core would have a solid outer core, or else the inner core would not be the solid phase of the liquid within the outer core. In 1971 Higgins and Kennedy caused a great stir by asserting that $dT_L/dP < dT_A/dP$ in the core, giving rise to the core paradox. Current opinion is that the liquidus gradient is steeper than the adiabat, and there is no core paradox.

Core Structure

The core consists of a liquid outer core and a solid inner core. An anomalous region between these two, called the F-layer by seismologists, is no longer believed to exist; the seismic signals that led to this interpretation are now believed to be caused by small-scale irregularities at the base of the mantle. Lateral variations in properties, particularly density, cannot be dynamically sustained in the outer core; only radial variations are possible. The simplest picture is of a well-stirred outer core that is very nearly homogeneous and adiabatic throughout. This picture is supported by seismic evidence, particularly the low value of seismic attenuation observed in the outer core. However, there may be layers within the outer core, especially at its top, that are stably stratified. This possibility is indicated by an apparent decrease in the fraction of nonmetals with depth in the outer core inferred from high-pressure studies. Also studies of the dynamical cause of the secular variation of the Earth's magnetic field have also led to this conclusion, although this interpretation is controversial.

Useful insight into the possible structure of the inner core comes from metallurgical studies of the solidification of metallic alloys. When an alloy solidifies, the solid phase typically is not monolithic, but rather forms a rigid interconnected framework of dendritic crystals, called a mushy zone. This structure is the result of an instability of the hypothetical smooth interface, which could exist between the liquid and solid phases. Mathematically this is called a *Stefan problem* and the physical process leading to this *morphological* instability is called *constitutional supercooling*. It has been estimated that the conditions for this instability are easily satisfied within the core, suggesting that the inner core is not completely solid, but is a mixture of solid and liquid very nearly in melting/freezing equilibrium.

Two seismic observations constrain the structure of the inner core. First, the inner core has anomalous seismic absorption, i.e., low Q, in the outermost 200–300 km, and second, short-period P waves can be reflected off the inner-core boundary. The first of these is interpreted in terms of loss mechanisms associated with the detailed structure of the mushy inner core, most likely by a mechanism called melt squirt. The decrease of anelasticity with depth in the inner core is probably caused by the closing of channels by compaction of the crystalline solid. It has been argued on thermodynamic grounds that the inner core is likely to be partially molten to its center. The second seismic constraint is less easy to reconcile with the concept of a mushy inner core; it requires that the fraction of solid is large a short distance below the boundary, so that it appears to be a sharp boundary to the P waves. However, this constraint can be satisfied if convective motions in the core penetrate into the mushy inner core.

Seismic inversion techniques have recently developed to the point that the nonspherically symmetric structure of the Earth's deep interior can be imaged. In particular, the nonspherical shape of the core-mantle boundary is beginning to be modeled, although there is considerable disagreement between the preliminary models. The shape of the core-mantle boundary is an important issue in determining the amount and type of dynamic coupling that exists between the core and mantle. This coupling is believed to be responsible for the irregular changes in the length of day on timescales of a decade or longer, and possibly the Chandler wobble. If the core-mantle boundary is not spherical, the dynamic coupling may be due to the topography, as core fluid flows past the bumps and hollows on the boundary. The other possible type of coupling is electromagnetic, caused by the leakage of electric currents from the core into an electrically conducting lower mantle.

Core Composition

During core formation nearly all the denser metallic constituents collected at the center of the Earth to form the core and the lighter silicate constituents rose to form the mantle. Some siderophile (iron-loving) nonmetals went into the core, and some normally dense elements, such as the transuranium elements, which bind easily to lighter elements, segregated into the mantle. High-pressure studies, combined with cosmochemistry, show that the core is composed principally of iron with an admixture of less dense nonmetals. The core may contain other metals than iron; studies of iron meteorites suggest that there may be as much as 10% nickel in the core. The most likely nonmetals present in the core in significant amounts are silicon, sulfur, and oxygen. Amounts of these elements that satisfy the pressure-density relations for the Earth are approximately 21%Si, 11%S, and 7%O.

One unresolved question is whether sufficient

amounts of radioactive elements are present in the core to produce significant ($> 10^{12}$ W) heating. This question has an important bearing on the nature of the energy supply for the geomagnetic dynamo and on the cooling history of the core. The most likely candidates are ^{40}K, ^{232}Th, ^{235}U, and ^{238}U.

Whereas significant amounts of light elements occur in the liquid outer core, the seismic and high-pressure data for the inner core can be fit with pure (or more nearly pure than the outer core) iron, implying that the core has an iron-rich composition; that is, it is on the iron-rich side of the eutectic point. The growth of an iron-rich inner core has significant implications for the dynamics of the core and the energetics of the dynamo. It is difficult to form an iron-rich inner core from an iron-poor outer core, and even more difficult to extract sufficient energy during this process to drive the dynamo.

Core Energetics

The geomagnetic dynamo theory is the only viable theory of the Earth's magnetic field capable of explaining its important features such as reversals, secular variation, and the near alignment of the dipole and rotation axes. In this theory the field is maintained against ohmic losses by the dynamo action of convective motions within the liquid outer core, which requires some source of energy to sustain the fluid motions. The geomagnetic record appears to require a steady supply of energy for at least the past 2.7×10^9 years. It is difficult to estimate the power needed to sustain the dynamo, particularly as the magnitude of the unseen toroidal component of the field is not known. If the field has a large toroidal component, as is the most popular assumption at present, the power needed is roughly $1-5 \times 10^{11}$ W. Three possible energy sources have been proposed: compositional convection, thermal convection, and precessional stirring.

The Earth's axis of rotation precesses with a period of 25,800 years because of the gravitational pull of the Moon. The core and mantle each have a different oblateness, due to their differing densities, which causes them to tend to precess at differing rates. This difference in dynamic response has the potential to stir the fluid outer core and provide energy to the dynamo. The power supplied to the dynamo by this mechanism has been estimated to be rather low (about 10^9 W), and there are serious questions about the dynamical efficiency of this mechanism; consequently it is not currently in favor.

Thermal convection in the outer core, driven by a combination of internal heating by radioactive decay, bottom heating by latent heat release, and cooling by the mantle, was for many years the most popular energy supply for the dynamo. One serious problem with this mechanism is its inherent inefficiency. The core is metallic and hence is a good conductor of heat. A significant amount of heat can be conducted radially outward along the adiabatic thermal gradient in the core and thence to the mantle. Only the heat flux in excess of this is available to stir thermal convective fluid motions in the outer core. Even then, the conversion of heat energy to fluid motions is limited by the Carnot efficiency. It is difficult to supply energy to the dynamo by this mechanism at a rate necessary to support a vigorous magnetic field.

As previously mentioned, it is likely that the core is gradually cooling and the inner core is freezing from the outer core. Since the inner core is more iron rich, the freezing process forces the light constituents back into the outer core. This buoyant fluid rising from the inner-core boundary causes compositional convection within the outer core. These motions, occurring in a rotating environment, are very likely to possess helicity and hence be capable of dynamo action. Assuming that the inner core has grown to its present size over the past 4×10^9 years, this mechanism is capable of supplying approximately $1-3 \times 10^{11}$ W of useful power to the dynamo, sufficient to sustain a magnetic field with a strong toroidal component. Because of the small value of the material diffusivity compared with the value of the thermal diffusivity, this mechanism has an inherent efficiency (not limited by the Carnot efficiency) that makes it a much more viable power supply than thermal convection.

Core Dynamics

The core must be dynamically able to sustain a magnetic field by dynamo action, which requires that a significant fraction, if not all, of the outer core be convecting. There have been several models of dynamo action proposed in a stratified outer core, but none are viable. Convection implies a well-mixed, adiabatic, and homogeneous outer core. However, there may be layers that are stably stratified, particularly at the top of the outer core.

The convective instability of a layer of fluid is governed by the vertical variation of density; a static layer of fluid is convectively unstable if a parcel of fluid becomes lighter than its surroundings when it is adiabatically displaced upward. If the fluid is modeled as a binary alloy, the density and entropy are functions of pressure, temperature, and composition. Since the core is, to a very good approximation, in hydrostatic balance, the stability depends on the variation of temperature and composition with pressure. In the static outer core the ambient gradients of temperature and composition result from diffusive fluxes of heat and material associated with the cooling of the core, freezing of the inner core, and rearrangement of light material released by the freezing process. These fluxes tend to be destabilizing. There are two counterbalancing, stabilizing effects: the conduction of heat down the adiabat and pressure diffusion of material. This latter effect results from the tendency for the light component to rise and the heavy to sink diffusively in a gravitational field. It may be shown that for parameter values appropriate for the core, thermal buoyancy and thermal conduction are negligibly small when compared with compositional buoyancy and pressure diffusion, which means that the dynamic state of the outer core is completely determined by compositional effects. Also it may be shown that the condition for convective instability is easily satisfied in the outer core, except possibly at the very top, implying that the dynamo is driven by compositional convective motions.

DAVID E. LOPER

References

Higgins, G., and G. C. Kennedy, 1971, The adiabatic gradient and the melting point gradient in the core of the Earth, *Jour. Geophys. Research* **76,** 1870–1878.

Jacobs, J. A., 1975, *The Earth's Core.* London: Academic Press.

Lay, T., ed., 1986, Special section: Geophysics of the core and core-mantle boundary, *Geophys. Research Letters* **13,** 1488-1560.

Loper, D. E., 1984, Structure of the core and lower mantle, *Adv. Geophys.* **26,** 1–34.

Melchior, P., 1986, *The Physics of the Earth's Core.* Oxford: Pergamon Press.

Runcorn, S. K., K. M. Creer, and J. A. Jacobs, eds., 1982, *The Earth's Core: Its Structure, Evolution and Magnetic Field.* London: The Royal Society.

Stacey, F. D., 1977, *Physics of the Earth*, second ed. New York: Wiley & Sons.

Verhoogen, J., 1980, *Energetics of the Earth.* Washington, D.C.: National Academy of Sciences.

Cross-references: *Core-Mantle Coupling; Earth's Core: Structure; Earth Structure: Global; Energy Budget of the Earth; Mantle, Lower: Structure; Mantle Convection and Plumes; Mantle Dynamics; Seismology: Physical Model Studies.*

EARTH'S CORE: STRUCTURE

The principal source of information about the structure of the core comes from seismology and the spherically averaged structure is now quite well known. Both long-period (free oscillation) and short-period (body wave) seismology have contributed to our understanding of core structure. A detailed review of work through the early 1970s can be found in Jacobs (1975) so we concentrate on later developments in this article.

The basic structure of the core was known by the late 1930s. Using travel time information, Lehmann had postulated a high velocity inner core in 1936 and the distribution of compressional velocity throughout the core was quite accurately determined by both H. Jeffreys and B. Gutenberg in the late 1930s. Modern travel time analyses use many thousands of measurements. For example, travel time measurements of core phases collected by the I.S.C. are plotted in Fig. 1. The branches of PKP are clearly defined as are SKS and, to a lesser extent, SKKS (which are essential to constrain the compressional velocity in the top few hundred kilometers of the core). Many techniques exist for converting such data into velocity models. All of these techniques require that the effect of the mantle be removed from the data. It is usually assumed that this can be done perfectly so the resulting models may be contaminated by correction of the data with an imperfect mantle model. Figure 2 shows the result of reparameterizing the travel time data in terms of delay time, $\tau(p)$, and then constructing bounds on compressional velocity as a function of depth using linear programming. Also shown in Fig. 2 is a typical compressional velocity model, PREM (see *Earth Structure, Global*). Although it lies within the bounds of possible models, it does not actually fit the data.

It is clear from Fig. 2 that the compressional velocity is quite well constrained by the travel time data. The bounds have been constructed using the reasonable assumption that there are no low velocity zones inside the core and lead to fairly strong constraints on the radius of the inner core assuming that this is a first-order discontinuity. The result of such studies is that the inner core has a radius of between 1210 km and 1240 km.

Additional constraints on the radii of the inner and outer cores come from the study of waves reflected off these interfaces at nearly vertical incidence. This work results in values of 3485 ± 3 km for the radius of the outer core and $1225 (\pm 15$ km) for the inner core. The high frequency content (~ 1 sec) of PKiKP also indicates that the inner core boundary is very sharp with a transition width of, at most, a few kilometers.

The outer core is usually assumed to be a fluid with a viscosity close to that of water. Seismological observations are consistent with an extremely low rigidity, though experiments indicate that a shear velocity of a few hundred meters per second in the outer core would have almost no observable effect on the seismic observations including accurately measured free oscillation frequencies and the decay of diffracted waves. The lack of rigidity is therefore most strongly indicated by the short time scale of geomagnetic secular variation. The inner core, on the other hand, is generally accepted to be solid, though direct evidence for its solidity is lacking. Indirect evidence comes mainly from low frequency seismology which we discuss later. No unambiguous observation of a body wave, which has propagated as a shear wave through the inner core, has yet been made.

The travel time data provide no direct constraints on the density structure of the core or on the rigidity of the inner core. Early density models were constructed using the assumption of adiabaticity and homogeneity while ad hoc arguments about the

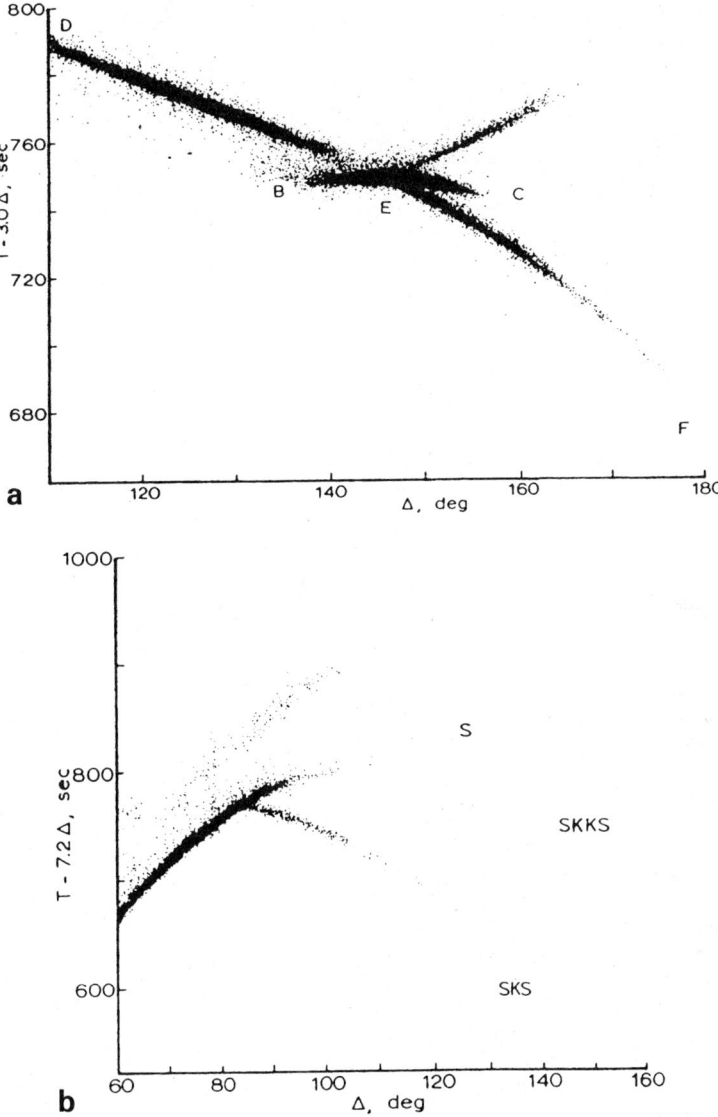

FIGURE 1. *A*. Reduced travel times of approximately 60,000 P waves in the core. The letters denote the end points of the various branches of the travel-time curve according to standard notation. *B*. Reduced travel times of approximately 30,000 S waves in the lower mantle. The branch of the travel-time curve corresponding to the direct S wave is labeled in addition to SKS and SKKS, the two waves that have penetrated the core as P waves.

behavior of bulk modulus were used to estimate inner core rigidity. The assembly of a large dataset of free oscillation frequencies by F. Gilbert and A. M. Dziewonski in the early 1970s (see *Earth Structure, Global*) resulted in an improved density distribution and the demonstration that a solid inner core provides a much better fit to the data. An apparent observation of $_{11}S_2$, a "core mode" (i.e., a mode of oscillation of the Earth with most of its energy as shear in the inner core and which would not exist in the inner core were it not solid) was later shown to be a result of anomalous splitting of a nearby mode, $_{10}S_2$, which has most of its energy in compression in the outer core (Masters and Gilbert, 1981). Spectral peaks showing weak attenuation have been identified on vertical component recordings close to the theoretical frequencies of the core modes, $_6S_2$ and $_7S_3$. The identification of these peaks as the core modes seemed certain because no other spheroidal modes are near the observed peaks but later work

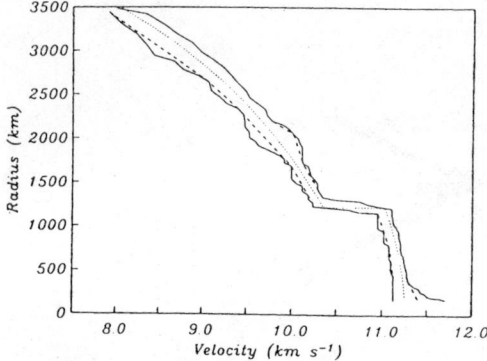

FIGURE 2. Extremal bounds on compressional velocity in the Earth's core. The dashed lines represent the geometrical effect of imposing the constraint that the second derivative of velocity is negative, which can be shown to be thermodynamically reasonable. The dotted line is model PREM.

has shown that Coriolis coupling between spheroidal and toroidal modes can lead to significant spectral lines on vertical component recordings close to the frequencies of toroidal modes. Both $_6S_2$ and $_7S_3$ lie close to toroidal modes so their identification can no longer be regarded as unambiguous.

The strongly anomalous splitting of $_{10}S_2$ is demonstrated in Fig. 3 where spectra of various recordings indicate that singlets of this mode are scattered over $\sim 14\mu$Hz, in contrast to the $\sim 6\mu$Hz predicted for a rotating Earth in hydrostatic equilibrium. It has been found that nearly all modes with a large fraction of their energy in the core that have been observed to date are anomalously split. The observations suggest that there is large amplitude, large-scale aspherical structure in or on the boundaries of the outer core, which we discuss later. The data analysis techniques used to estimate the degenerate frequencies of multiplets can give biased results when multiplets are broadly split. It is therefore likely that many of the degenerate frequencies of free oscillations that sample the core are biased measurements because of the prevalence of anomalous splitting. Careful modeling of the spectrum of split multiplets yields accurate estimates of degenerate frequencies and modeling experiments with new measurements result in structures similar to models based on the old measurements (Fig. 4). The models show a density structure corresponding closely to a state of neutral convective stability and a density jump at the inner core boundary of about 0.5 g/cc. Because formal resolution analyses of these models have not yet been completed, we must rely on earlier work to assess the uniqueness of these results. The modal data mainly constrain the travel time of shear waves across the inner core, thus there are tradeoffs between inner core radius and mean inner core shear velocity. A rather conservative range for the mean shear

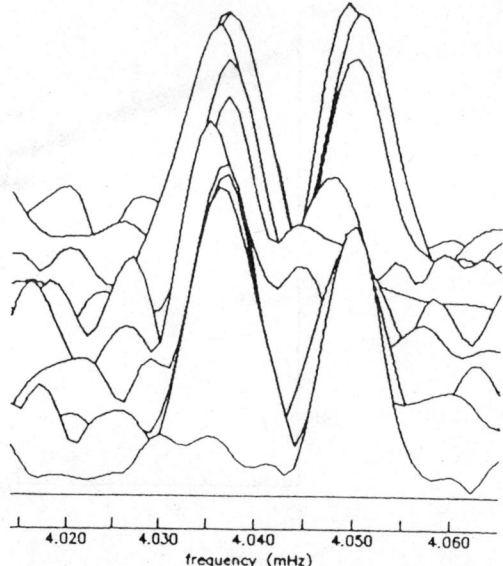

FIGURE 3. Amplitude spectra of 13 recordings in the 50-μHz band around $_{10}S_2$ following two large, deep earthquakes (the Tonga event of 1977 and the Banda Sea event of 1982, both on June 22). The spectra are plotted as a function of absolute value of latitude so that an equatorial station would be in back and a polar station directly in front. The only spheroidal mode in this band is $_{10}S_2$ and can be seen to be anomalously wide, spanning more than 14 μHz though predicted to be less than 6 μHz wide for a rotating, hydrostatic Earth model. The spectrum in front is for the south polar recording following the Tonga event; the single peak is consistent with the hypothesis that axisymmetric structure is dominantly responsible for the splitting. The peak at 4.050 mHz had previously been identified as $_{11}S_2$.

velocity in the inner core is 3.5–3.6 km/sec. The mode data do not constrain details of the shear velocity and it is possible that the shear velocity is quite low near the surface of the inner core, then climbs to its mean value (see below). The density jump at the inner core boundary was studied with the old mode dataset, which constrained the density jump to be about 0.7 ± 0.35 g/cc while acoustic impedance measurements indicate only that the density jump is less than 1.8 g/cc. The new mode data are more precise than the old data but less numerous so a value of 0.5 ± 0.3 g/cc is probably not unreasonable in the absence of a formal analysis.

The velocity structure within the vicinity of the inner core boundary has been the subject of several detailed analyses that use the waveforms or amplitude of PKP. Amplitude data constrain the velocity jump at the inner core boundary (ICB) to be 0.64 ± 0.05 km/sec (Häge, 1983). These data also suggest a fairly low value for the shear velocity immediately below the ICB, and for reasonable values of the density jump at the ICB, the shear velocity at the top of the inner core is apparently between 2.6 and 3

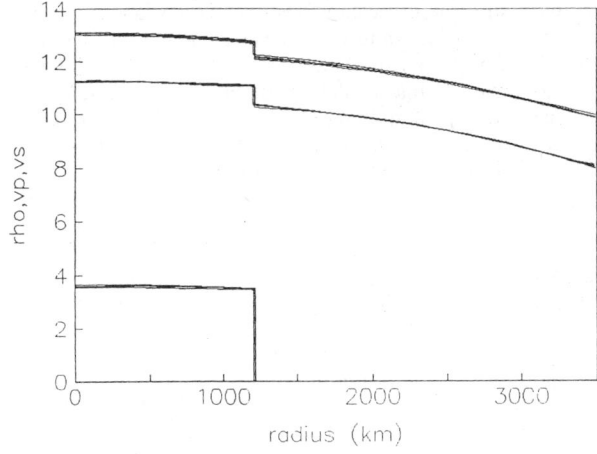

FIGURE 4. A suite of smooth core models designed to fit mode data. Note that possible details of structure such as steep velocity gradients near the top of the core are not reflected in these models. From top to bottom the curves are density (in g/cc), compressional velocity (in km/sec) and shear velocity (in km/sec). Demanding smoothness results in narrow corridors of possible models.

km/sec. Modeling of broad band PKP waveforms by Choy and Cormier (1983) resulted in a lower estimate of the compressional velocity jump at the ICB (0.52 ± .07 km/sec). The disagreement of this result with the global travel time and amplitude data may be due to poor sampling of the ICB, and a value of 0.6 → 0.7 km/sec is now preferred. Second-order features of the PKP waveforms support the suggestion of a low value of the shear velocity at the top of the inner core. This reduction in velocity may also be accompanied by very strong attenuation of seismic energy in the top of the inner core. The analysis of spectral ratio data and waveform modeling give high compressional wave attenuation throughout most of the inner core with an average Q_α of about 300 at high frequency (0.2 → 2 Hz) (see *Seismic Attenuation: Observation and Measurement*). This Q_α for the inner core should be contrasted with the almost negligible attenuation in the fluid outer core ($Q_\alpha >$ 10^4 for 1 Hz body waves) determined from spectral ratios of PmKP. Attenuation within the core can also be constrained using measurements of the decay of free oscillations that sample this region. All free oscillations that strongly sample the core, including those with significant fractions of their energy in the inner core are weakly attenuated. Even if suspect observations of core modes are excluded, these data still imply a shear Q (i.e., Q_μ) of greater than 1000 for the inner core at a frequency of about 4 mHz. If all attenuation is in shear processes, the body wave measurements imply that Q_μ is about 40. Clearly, if both body wave and free oscillation measurements are unbiased we must postulate that attenuation in the inner core is strongly frequency dependent and/or that bulk attenuation (as well as shear attenuation) is important in this region. Possible ways of reconciling the measurements are partial melting or a seismically induced phase transition in the inner core (see *Earth's Core*). Such mechanisms may not be physically unreasonable if the inner core is close to its melting temperature and the suggestion of low shear velocity near the top of the inner core could be taken as corroboration that partial melting is occurring. [The average shear velocity of the inner core (~3.55 km/sec) leads to an apparently anomalous Poisson ratio of 0.44. This feature is to be expected for a normal crystalline material under extreme compression and does not, by itself, indicate partial melting.]

The velocity structure near both boundaries of the fluid outer core has also been the subject of discussion over the last few years. The *P* velocity structure just below the mantle-core boundary (MCB) is constrained by observations of SmKS. The times of such phases can be biased if pulse distortion is ignored and can lead to low values of compressional velocity in the upper outer core. It should also be realized that such phases have portions of their ray paths near to grazing incidence in the region just above the MCB and so are sensitive to the well-documented lateral heterogeneity at the base of the mantle. It now appears that the velocity gradient in the outermost core is not anomalously steep and is consistent with that of a convecting region. The velocity structure near the base of the outer core has also undergone major revisions as features in the data have been reinterpreted in terms of scattering from aspherical structure at the base of the mantle and by diffraction. It now seems that the velocity structure throughout the whole outer core is consistent with that of a homogeneous, convecting region. Direct evidence on the convective stability structure of the outer core would be given by observations of gravity oscillations in the outer core or of the translational oscillation of the inner core (known as the Slichter mode). The Slichter mode is expected to have a period of a few hours and is extremely sensitive to the density jump at the inner core boundary. Despite several careful searches, this mode has escaped detection. Oscillations in the outer core where the restoring force is the Coriolis force will have frequencies of about half a day while convectively stable regions of the outer core can support internal gravity waves with frequency related to the local Brunt-Väisälä frequency. The problem with observing such modes (apart from severe signal-to-noise constraints) is a plethora of spectral peaks due to tides. The tidal signal must therefore be very accurately known so that the tidal lines can be removed. A recent analysis of a superconducting gravimeter record by Melchior and Ducarme (1986) shows a residual spectral peak with a period of around 13.9 hours, which may be related to a core oscillation. Such observations need to be verified at different sites before any inferences about core structure are warranted.

In summary, the spherically averaged velocity and density distribution within the whole outer core appear to be consistent with that of an adiabatic and homogeneous region. In the inner core, there is some evidence for anomalously steep gradients in both shear and compressional velocity near the top of this region, which may be indicative of partial melting. Such a hypothesis could also explain the high apparent attenuation of body waves that sample this region. Some words of caution are necessary as it is always possible to confuse the effects of aspherical structure with anomalous spherically averaged structure. Thus the effects of lateral heterogeneity near the base of the mantle can map into anomalous structure near both boundaries of the outer core. It may also be possible that aspherical structure near the top of the inner core may lead to erroneous estimates of the spherically averaged structure of this region. The evidence for aspherical structure somewhere in the core is now very strong though there is little agreement as to the nature or location of the anomalous structure. There are theoretical reasons why large lateral variations in density are not likely to be observed in the outer core (see e.g., Stevenson, 1987). It is always possible that the outer core chemistry is such that lateral variations in velocity exist without lateral variations in density but there is no experimental evidence to support this so most workers have chosen to exclude aspherical structure from the fluid outer core. The two datasets that seem to point to the existence of large-scale dominantly axisymmetric heterogeneity somewhere within the core are the anomalous travel times of PKP phases and the anomalous splitting of free oscillations (e.g., Fig. 3). Already, a large number of hypotheses have been entertained to explain these signals including large amplitude structure on the MCB and/or ICB, a chemical boundary layer above or below the MCB, large isotropic volumetric perturbations in the inner core, and anisotropy in the inner core (Lay, 1986; Giardini et al., 1987; and Ritzwoller et al., 1986). It remains unclear which, if any, of these features are physically plausible or even if any are capable of explaining all of the data. Both free oscillation and travel time datasets are deficient in the sense that neither at present is capable of resolving aspherical structure in the core. The travel time dataset is noisy and suffers from very uneven geographical coverage, which can lead to near-source structure being mapped into models of the deep earth. The observations of splitting of free oscillation are currently too few to unambiguously determine structure and, in any case, are only sensitive to certain large-scale components of aspherical structure. Clearly, much work remains to be done before a clear picture of the three-dimensional structure of the core emerges.

GUY MASTERS

References

Choy, G. L., and V. F. Cormier, 1983, The structure of the inner core inferred from short-period and broadband GDSN data, *Royal Astron. Soc. Geophys. Jour.* **72**, 1–21.

Häge, H., 1983, Velocity constraints for the inner core inferred from long-period PKP amplitudes, *Physics Earth Planetary Interiors* **31**, 171–185.

Jacobs, J. A., 1975, *The Earth's Core.* New York: Academic Press.

Giardini, D., X.-D. Li, and J. H. Woodhouse, 1987, Three-dimensional structure of the earth from splitting in free oscillation spectra, *Nature* **325**, 405–411.

Lay, T., ed., 1986, Special issue *Geophys. Research Letters*, December.

Masters, G., and F. Gilbert, 1981, Structure of the inner core inferred from observations of its spheroidal shear modes, *Geophys. Research Letters* **8**, 569–571.

Melchior, P., and B. Ducarme, 1986, Detection of inertial gravity oscillations in the Earth's core with a superconducting gravimeter at Brussels, *Physics Earth Planetary Interiors* **42**, 129–134.

Ritzwoller, M., G. Masters, and F. Gilbert, 1986, Observations of anomalous splitting and their interpretation in terms of aspherical structure, *Jour. Geophys. Research* **91**, 10203–10228.

Stevenson, D., 1987, Limits on lateral density and velocity variations in the earth's core, *Royal Astron. Soc. Geophys. Jour.* **88**, 311–319.

Cross-references: *Core-Mantle Coupling; Earth's Core; Earth Structure: Global; Free Oscillations of the Earth; Mantle, Lower: Structure; Mantle Convection and Plumes; Mantle Dynamics.*

EARTH'S EXTERNAL GRAVITY FIELD

An important component of geodesy is the determination of the Earth's external gravity field. Most geodetic measurements and the objectives they serve are tied inextricably to an accurate knowledge of this field. For example, determining the size and shape of the Earth traditionally requires the determination of the *geoid*, the equipotential surface coinciding with mean sea level. Increasingly, there are requirements to know gravity or its potential accurately not only at or near the Earth's surface but also at aircraft and satellite altitudes. All autonomous inertial navigation and guidance systems achieve their function only if their sensitive axes can be related to the local vertical; that is, the in situ direction of gravity must be known. As another example, the precise description of a satellite's orbit, its position and velocity, is based on an accurate model of the gravitational potential in space. One object of geodesy, then, is to determine the gravity field on and outside the Earth's surface. The most accurate measurements that support this purpose are still made predominantly with earthbound instrumentation. Even the satellite altimeter, which measures the height of sea level, yields data that are inherently coordinated on the Earth's surface. Given this situation, potential theory, in the context of mathematical physics, provides the tools, in the form of the boundary value problem (BVP) to build an accurate and useful description of the Earth's external gravity field from measurements on the Earth's surface. The problem and its solution increase in complexity as measurements are obtained also on airplanes and satellites. Once these added "boundary values" are properly incorporated, they naturally enhance the accuracy of the solution, which, from the start, is never exact.

The following exposition treats the determination of the Earth's external gravity field in the context of classical potential theory, but, because of obviously necessary limitations of space, it is descriptive and heuristic rather than analytical and mathematically rigorous. Thorough mathematical treatments are listed in an abbreviated bibliography whose entries contain numerous additional references. Considering the accuracy of current geodetic measurements, the Newtonian theory of mass attraction is entirely adequate for present geodetic application and therefore forms the basis of this article. Several definitions and concepts central to understanding the geodetic connection are explained in Fig. 1 and the associated Table 1. These explanations preface the remaining text, as they define most of the notations appearing therein.

Historical Notes

The force of gravity was contemplated already in antiquity, for example, by Aristotle, who ascribed the fall of a heavy body to an inherent property of the body to gravitate toward the center of the Earth until it finds sufficient resistance and comes to rest. However, he and his successors through the early Renaissance sought mostly to explain how, rather than why, nature operated; specifically they sought laws that describe the static and dynamic mechanics of bodies. In fact, in their endeavor to explain the world, the early philosophers relied mostly on their intuition and daily experience rather than on systematic experimentation. It was not until the time of Galileo, in the late sixteenth century, that the force of gravity was described as a uniform acceleration displacing a body under its influence in direct proportion to the square of time. Although Galileo was not the first to propose this relationship, he was the first to deduce it from thorough and methodical experiments, accurately timing balls as they rolled down carefully calibrated inclined planes. Galileo's law of free fall, however, still only explained how. The great mathematician Isaac Newton in his *Philosophiae Naturalis Principia Mathematica* of 1687 finally answered the question why—namely, that the force of attraction is directly proportional to the masses involved. Not only was this to hold near the Earth, but Newton's genius was able to extend this law to all space, and in one bold stroke the dynamics of the entire universe was virtually explained. By mathematically relating mass attraction to an inverse relationship with the square of distance, he could immediately explain the physical causes of Kepler's carefully observed planetary orbits, as well as the force that governs earthly objects in free fall.

On the basis of Galileo's law of falling bodies and

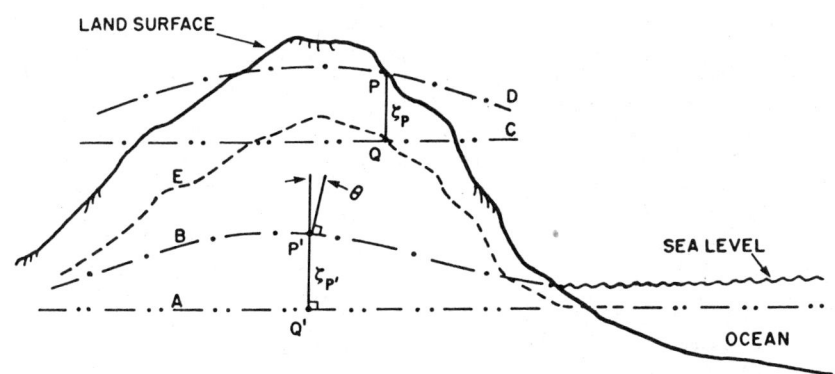

FIGURE 1. Surfaces for the geodetic boundary value problem; see Table 1 for a full explanation.

TABLE 1. Definitions for the Geodetic Boundary Value Problem

Surfaces

A = reference (normal) ellipsoid

B = geoid

C = equipotential surface of normal gravity field

D = equipotential surface of actual gravity field

E = telluroid, where at each point the normal potential equals the actual potential of the corresponding point on the Earth's surface

The ellipsoid (A) as an equipotential surface yields the normal gravity field, which has potential U and gravity γ.

The geoid (B) is an equipotential surface of the actual gravity field, which has potential W and gravity g.

Surfaces A and B have the same potential value, by definition.
Surfaces C and D have the same potential value, by definition.

Disturbances and Anomalies

$T_P = W_P - U_P$ = disturbing potential at point P

$\delta g_P = g_P - \gamma_P$ = gravity disturbance at point P

$\Delta g_P = g_P - \gamma_Q$ = gravity anomaly at point P

$\Delta g_{P'} = g_{P'} - \gamma_{Q'}$ = gravity anomaly at point P'

$\theta_{P'}$ = deflection of the vertical at P'

ζ_P = height anomaly at point P

$\zeta_{P'}$ = geoid undulation at point P'

Linear Approximations

$\zeta_P = \dfrac{T_P}{\gamma_Q}$ (Bruns' formula)

$\Delta g_P = -\left.\dfrac{\partial T}{\partial n}\right|_P + \left.\dfrac{1}{\gamma}\dfrac{\partial \gamma}{\partial n}\right|_Q T_P$ (fundamental equation of geodesy; n = vertical direction)

$\xi_P = -\dfrac{1}{\gamma_Q}\left.\dfrac{\partial T}{\partial s_N}\right|_P$ (north component of the deflection of the vertical; s_N = north horizontal direction)

$\eta_P = -\dfrac{1}{\gamma_Q}\left.\dfrac{\partial T}{\partial s_E}\right|_P$ (east component of the deflection of the vertical; s_E = east horizontal direction)

Note: Symbols refer to Fig. 1.

on Huygens' studies of pendulum motion and centrifugal force, Newton concluded that the Earth, considered as a homogeneous rotating body in equilibrium, must resemble an oblate ellipsoid (i.e., one with flattened poles). This model explained, in part, the observed increase in gravity as one travels from the equator to either pole, and thus closer to the Earth's center. However, the full explanation of this gravity variation is due to Huygens, who correctly attributed it primarily to the centrifugal force of the rotating Earth. Clairaut, in the eighteenth century, showed that the geometric flattening of this ellipsoid can be calculated from two measurements of gravity, one at the equator and one at either pole. The link thus formed between the Earth's geometric shape and its physical gravity field suggests the first primitive notion of a geodetic BVP, a concept that was not fully developed until 1849, with Stokes' famous integral formula.

The origin of the BVP, in general, and its application to mathematical physics can be traced to Fourier's work in the mathematically flourishing times of the late 1700s and early 1800s. With the concept of potential fully established by Lagrange and Laplace's demonstration that it must satisfy his famous partial differential equation came the well-known BVPs of Dirichlet and Neumann. Solutions (including uniqueness and existence theorems) were also studied by Green, using the identities bearing his name, and later by Fredholm, who transformed the study of differential equations with boundary conditions into one of integral equations.

Stokes' formulation and solution of the BVP required that boundary values be given on an equipotential surface, the geoid. The inaccessibility of this surface on land areas posed a difficulty to the practical application of his solution that was eliminated in 1960 by Molodensky, who derived a solution for which boundary values are given on the (accessible) Earth's surface. Stokes' integral provides the succinct relationship between gravity anomalies observed everywhere on the geoid and the disturbing potential or, what is essentially equivalent, the shape of the geoid. Therefore, Stokes' and, particularly, Molodensky's formulas are concerned primarily with determining the bounding surface itself. This distinguishes the geodetic BVP from the customary BVPs of mathematical physics where the boundary is a known geometric entity.

The Boundary Value Problem in Geodesy

The problem of expressing the Earth's external gravitational potential field in functional form is a classical application of potential theory. With a complete knowledge of the Earth's mass distribution that generates the gravitational potential, the solution is straightforward, being simply a direct application of Newton's law of gravitation. In geodesy, no hypothesis is generally made regarding the Earth's masses for the simple reason that the three-dimensional density distribution of the lithosphere and deeper regions cannot be inferred with sufficient accuracy. Instead, the starting point for all geodetic solutions for the exterior potential V is Laplace's differential equation, which is the special case of zero density ($\rho = 0$) in Poisson's differential equation (the latter being valid in all space):

$$\nabla^2 V = -4\pi G\rho \quad \text{(Poisson)}$$
$$\nabla^2 V = 0 \quad \text{(Laplace)}$$

A solution to Laplace's equation is a *harmonic* function; it represents the Earth's gravitational potential in free space, provided it also vanishes at infinity. With Newton's gravitational constant G assumed really constant, one can prove that any solution to this differential equation must be a Newtonian potential. The converse is also true. Indeed, one method of experimentally testing the validity of Newton's law of gravitation is a so-called null experiment that checks the sum of the second-order gradients obtained from gravity gradiometer measurements in a vacuum. Therefore, postulating Poisson's differential equation implicitly embraces Newton's law of universal gravitation (for constant G), and it is sufficient to start with Laplace's differential equation when solving for the Earth's exterior potential. In fact, Green's third identity (Green's three identities are a direct consequence of Gauss's divergence theorem) yields immediately the representation of any solution to Laplace's equation as a sum of (Newtonian) potentials of simple and double-layer density distributions. Clearly, such a density representation is strictly fictitious, but the formulation is useful because the potential thus obtained is unique, and therefore it is Earth's potential, given that the boundary conditions pertain to the Earth.

The problem of determining the Earth's exterior gravitational field may therefore be formulated generally as a BVP: Solve Laplace's differential equation for the potential in the Earth's exterior space given field-related values everywhere on a surface enclosing all generating masses. Solving this problem would mean determining the potential field everywhere in free space using only observations of this field on a bounding surface and no hypotheses regarding the densities of the generating masses. Obviously, at least in conventional geodesy, this bounding surface is the Earth's surface, or one close to it, where observations are most easily carried out. Proper implementation, however, requires no masses external to this surface. When warranted by the observational accuracy, atmospheric masses and extraterrestrial bodies such as the Moon, Sun, and possibly Venus (others can be neglected) are mathematically removed from the problem and treated as subsequent corrections. Also, the Earth's gravitational field rotates with the Earth, and Laplace's (or Poisson's) equation is vitiated in an Earth-fixed coordinate frame because the centrifugal potential is not harmonic. An appropriate remedy is to subtract the centrifugal force from all gravimetric observations.

The most familiar BVPs associated with Laplace's equation are Dirichlet's problem (first BVP), where the surface enclosing all masses is known and the potential is a function given on this surface; Neumann's problem (second BVP), where, instead of the potential, its directional derivatives normal to the surface constitute the boundary values; and the third BVP, where the boundary value is a linear combination of the potential and its normal derivative. Each problem is divided into two cases according as the empty region bounded by the given surface is finite or infinite. The latter is called the *exterior* BVP, and the present geodetic application is obviously restricted to this case.

The most straightforward and accurate measurements of the Earth's gravity field are measurements of the magnitude of gravity obtained by a gravimeter. The gravimeter's sensitive axis is aligned with the vertical, or the perpendicular to an equipotential surface of the gravity field. If measurements existed everywhere on a nonrotating geoid with no external masses, then the problem of determining the exterior potential would be a Neumann problem. The object of geodesy, however, includes the determination of the Earth's size and shape, that is, the geoid, which is the classical geodetic reference surface for land elevations. Being an equipotential surface of the unknown potential, the geoid and, hence, the Earth's surface are also unknown quantities in any coordinate system in ordinary space.

The geodetic BVP is formulated as the determination of the Earth's surface from gravity-related observations on this surface. It is termed a *free* BVP since the boundary values are given on a surface that itself is an unknown quantity to be solved for. A solution exists, at least with some approximation, because the surface is not arbitrary but is tied to the potential field. Once the boundary is known, the problem of determining the exterior potential field reverts to the more conventional BVP. The geodetic BVP is *nonlinear* in that the functional relationship between the unknown surface and the boundary values is nonlinear and rather complicated. Further difficulties arise when the observed gravity (derivative of the potential) is not normal to the bounding surface, as on the Earth's surface. The problem is then known as an *oblique-derivative* BVP. Another adaptation to the real world involves the use of different types of boundary values over different parts of the bounding surface, resulting in the *mixed* BVP. For example, gravity observations are most prevalent on land areas, whereas geoid heights from satellite altimetry are the dominant data on the oceans. In general, then, the geodetic BVP most accurately corresponds to a nonlinear, free, oblique-derivative, and, possibly, mixed BVP.

All practical solutions to the geodetic BVP are obtained first of all through a *linearization* of the problem. Mathematically, the potential (and its derivatives, e.g., gravity) and the bounding surface are expanded in a Taylor series, and second- and higher-order terms are neglected. The "point" about which the expansion occurs is, respectively, the reference or *normal* potential (and the normal gravity, etc.) and the *ellipsoid* (if S is the geoid) or the *telluroid* (if S is the Earth's surface). The difference between the actual and normal quantities is called a disturbance (or anomaly, if the geometric points to which the two quantities refer belong to the actual and normal surfaces, respectively). Hence, the disturbances, or anomalies, entering the solution to the linearized geodetic BVP embody only the linear term in a Taylor series expansion. All terms of second or higher order are neglected; for example, the free BVP is thus transformed into a fixed BVP by neglecting the effect of the height anomaly on the value of the observed gravity anomaly. By definition the normal potential includes the centrifugal potential so that the disturbing potential is harmonic in the Earth's exterior space tied to an Earth-fixed coordinate frame, and may be considered a solution to Laplace's equation.

Even with the linearization, the problem remains mathematically formidable and is, therefore, for practical purposes, subjected to a subsequent *spherical approximation*. With this approximation coordinates on the linearized surface are simply identified (formally, through a mapping) as coordinates on a sphere; thus, vertical derivatives become radial derivatives, for example, and the ellipsoid's flattening is neglected. If justified by the accuracy of the linearization, the *ellipsoidal approximation* provides an improvement by omitting all terms on the order of the square of the flattening. In some applications where nonlocal effects can be safely ignored, the *planar approximation* may suffice. It results from the formal limit of the spherical approximation as the radius of the sphere approaches infinity.

Methods of Solving the Geodetic BVP

Spherical Harmonic Series. There are several approaches to solving BVPs. When situations in the physical sciences allow, partial differential equations, such as Laplace's equation, are often solved by the method of separation of variables. This reduces the problem to one of solving three univariate ordinary differential equations, which can be a simpler task with an appropriate choice of coordinates. The final solution is an infinite series of all possible solutions where the coefficients of the series are determined from the boundary conditions. The commonest coordinates for the Earth's exterior potential are the spherical polar coordinates whose origin is at the Earth's center. For a given radial coordinate value, the solution to Laplace's equation is an infinite set of orthogonal functions called spherical harmonic functions, which are products of sinusoidal and associated Legendre functions. Naturally, for Laplace's equation to hold, this radial distance must not describe a sphere (centered at the origin) that intersects the Earth's surface at any point. Including the dependence on radial distance from the origin, the solution is given by

$$T(r, \theta, \lambda) = \sum_{n=0}^{\infty} \sum_{m=0}^{n} \left(\frac{r_0}{r}\right)^{n+1}$$

$$\cdot (c_{nm} \cos m\lambda + s_{nm} \sin m\lambda)$$

$$\overline{P}_{nm}(\cos \theta)$$

The associated Legendre functions are usually normalized to yield orthonormality in the spherical harmonics. With the condition $r \geq r_0$, the spherical harmonic series converges uniformly. The coefficients of the series are obtained by taking advantage of the orthogonality of the spherical harmonics and using boundary values given on an origin-centered sphere with radius r_1:

$$\left(\frac{r_0}{r_1}\right)^{n+1} \begin{Bmatrix} c_{nm} \\ s_{nm} \end{Bmatrix} = \frac{1}{4\pi} \iint_\sigma T(r_1, \theta, \lambda)$$

$$\cdot \begin{Bmatrix} \cos m\lambda \\ \sin m\lambda \end{Bmatrix} \overline{P}_{nm}(\cos \theta) \, d\sigma \quad (1)$$

The integration extends over the unit sphere. In case

a gravimetric quantity, such as the gravity anomaly, is related to the disturbing potential by a linear operator that, when applied to the series above, returns a series of orthogonal functions, then a similar integral exists for the coefficients, with the gravimetric quantity replacing the disturbing potential as boundary function.

The set of coefficients thus obtained determines the potential and all its derivatives everywhere in the region of series convergence. Spherical harmonic series are the preferred representation of the Earth's gravity field on a global scale. With this type of expression, any linear functional or operator of the potential is immediately represented as a series incorporating the same coefficients. In principle, then, the computation, at least with the approximation of linearization, of quantities like the gravity anomaly, the geoid, and the deflection of the vertical anywhere above the bounding sphere requires no additional data once the coefficients are determined. The true globality of this representation is ideally suited for computing satellite orbits that, depending on their inclination, trace through most or all of the Earth's potential field within a particular range of altitudes. Conversely, in fact, many of the harmonic coefficients, especially with low degrees, are derived most accurately from the observed perturbations of a satellite's orbit as it deviates from a true Keplerian orbit (the latter being valid only in a central force field). The orbit often resonates with certain field harmonics as it repeatedly passes over the same features of the generating masses, and the corresponding resonant coefficients are naturally determined with greater accuracy. The decomposition of the potential (or gravity) into constituent spherical harmonics also admits of easy interpretation and comparison with other geophysical phenomena described in this way, such as the magnetic field, density, and iostatic compensation models, and even ocean currents caused, in part, by deviations of the instantaneous sea level from the geoid.

A spherical harmonic representation of the gravity field is, however, only approximate, since for practical utilization the series must be terminated at some finite degree n. Typical degrees of truncation are 20, 36, and 180, which imply series of 441, 1369, and 32,761 coefficients, respectively. In addition, the boundary values defining the coefficients should reside on a sphere, but often the Earth's surface, or the geoid, is used, since the data on these surfaces enter directly into the integrals for the coefficients. Sometimes correction formulas are applied that account for the Earth's ellipticity. On another, but not very popular level, solutions to Laplace's equation can also be expressed as a series of ellipsoidal harmonic functions. The data, on the other hand, are distributed discretely and unevenly over the globe, being largely deficient or very inaccurate in parts of some continents and in the polar regions. All of these shortcomings and also errors in the data influence the accuracy of the calculated coefficients, which usually becomes rapidly worse as the degree n increases.

Although convergence is guaranteed above the bounding sphere, divergence everywhere below this sphere has yet to be proven. Nevertheless, a theorem due to Runge and modified by Krarup ensures the existence of a potential whose spherical harmonic series representation converges everywhere above the Earth's surface and that uniformly approximates the actual potential with arbitrary accuracy. How to determine the coefficients of this series is not known. This theoretical difficulty has not hindered the practical evaluation and utility of spherical harmonic series near the Earth's surface, since the series is, in any case, terminated at some finite degree, and numerical effects of divergence have not yet been discerned.

The spherical harmonic representation decomposes the disturbing potential (or any of its derivatives) into components characterized by wavelength. The larger the wavelength, the deeper within the Earth is the density contrast that generates the disturbance. The magnitude of the series coefficient at a particular degree n and order m indicates the strength of the represented signal at that composite wavelength. Formally, the spherical harmonic series for the potential field (at some fixed radius r) is the inverse Legendre (or generalized inverse Fourier) transform of the set of coefficients, or the Legendre transform of the field is the set of coefficients. This set, called the *spectrum*, is defined with respect to a particular sphere and thus includes the $(n + 1)$st power of the ratio of radii; for example, Eq. 1 defines the spectrum of the disturbing potential on the sphere of radius r_1. As such, the attenuation of the spectrum with altitude is evident, being essentially exponential and depending on the degree n. Short wavelengths (large n) attenuate more rapidly than long wavelengths (small n). The Earth's disturbing gravity field is often considered to be a stochastic process on the sphere. One may then speak of a *power spectrum*, which, by definition, is the spectrum of the autocorrelation function of the field. Being also related to the spectrum of the field itself, the power spectrum may be computed directly, at least for the longer wavelengths, from a given set of harmonic coefficients. This statistical interpretation of the gravity field and its connection to spherical harmonic series figure significantly in another method to solve the BVP, least squares collocation, discussed later.

Green's Function. A second approach to solving the BVP makes direct use of Green's third identity applied to the disturbing potential:

$$T_P = -\frac{1}{4\pi} \iint_S \left[\frac{1}{l}\frac{\partial T}{\partial n} - T\frac{\partial}{\partial n}\left(\frac{1}{l}\right)\right] dS$$

P outside S

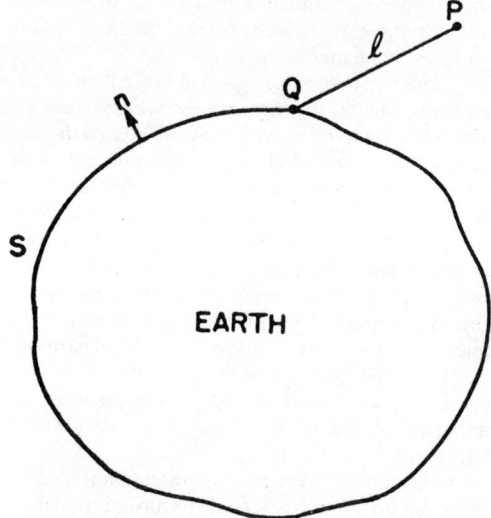

FIGURE 2. The Earth's surface as boundary in Molodensky's solution to the geodetic boundary value problem.

where the evaluation point P is located in the space exterior to the surface S, n is the normal to S directed outward, and l is the distance between P and a point Q on the surface (see Fig. 2). Given the boundary values f of one of the three classical types (Dirichlet, Neumann, third), one seeks a function G that yields an expression of the disturbing potential directly and explicitly as the integral of the boundary values:

$$T_P = \iint_S H_{P,Q} f_Q \, dS$$

For the first BVP, H is the normal derivative of G, whereas for the other BVPs it is G itself. G is called Green's function.

Green's function exists in analytic form when S is a sphere. In this case the solution to Dirichlet's exterior BVP problem is Poisson's integral, which is valid for any harmonic function whose values are known on the boundary (sphere of radius r_Q):

$$T_P = \frac{r_Q(r_P^2 - r_Q^2)}{4\pi} \iint_\sigma \frac{1}{l^3} T_Q \, d\sigma$$

In the linear and spherical approximations, the product of the radius and the gravity anomaly, $r \Delta g$, is a harmonic function, and in the planar approximation, the first derivatives of the disturbing potential are harmonic functions. Therefore, Poisson's integral, with the appropriate substitutions, serves as an analytic upward continuation of the potential and quantities such as the gravity anomaly and the deflection of the vertical. The solution to Neumann's exterior BVP, when S is a sphere, is given by

$$T_P = \frac{r_Q}{4\pi} \iint_\sigma G(r_P, \psi) \left. \frac{\partial T}{\partial r} \right|_Q d\sigma$$

$$G(r_P, \psi) = \ln\left(\frac{l + r_Q - r_P \cos \psi}{r_P - r_P \cos \psi}\right) - \frac{2r_Q}{l}$$

where ψ is the spherical angle between r_P and r_Q. Finally, the most famous geodetic integral, Stokes' integral, is the solution to a third BVP:

$$T_P = \frac{r_Q}{4\pi} \iint_\sigma S(r_P, \psi) \Delta g_Q \, d_\sigma$$

where the boundary values are expressed as

$$\Delta g = -\frac{\partial T}{\partial r} - \frac{2}{r} T \qquad (2)$$

The kernel S is known as Stokes' function:

$$S(r_P, \psi) = \frac{2r_Q}{l} + \frac{r_Q}{r_P} - 3\frac{r_Q l}{r_P^2} - \frac{r_Q^2}{r_P^2}$$
$$\cdot \cos \psi \left(5 + 3 \ln \frac{r_P - r_Q \cos \psi + l}{2r_P}\right)$$

The boundary in the classic geodetic BVP is a well-defined equipotential surface, the geoid. However, by the necessary linearization of the problem, the observed boundary values are gravity anomalies rather than gravity disturbances, because the lack of knowing the geoid height means that normal gravity must be computed on a known surface, the ellipsoid. The boundary values, in the linear approximation, are found to be

$$\Delta g = -\frac{\partial T}{\partial n} + \frac{1}{\gamma} \frac{\partial \gamma}{\partial n} T \qquad (3)$$

where n is the normal to the geoid and γ is normal gravity; the spherical approximation simplifies this to Eq. 2. The classic geodetic BVP under these approximations is therefore of the third type, and its solution is Stokes' integral.

One can prove that each of these three solutions also holds when the evaluation point P is located on the surface of the sphere. (Dirichlet's case, the Poisson integral, requires the appropriate definition of a spherical Dirac delta function.) Hence, with Stokes' integral and observations of gravity one obtains the size and shape of the geoid relative to the known ellipsoid, since the geoid height, in linear approximation, is directly proportional to the disturbing potential on the geoid, according to Bruns' formula. The slope of the geoid or, equivalently, the deflection of the vertical is obtained by differentiating Stokes' formula with respect to the horizontal displacement at P. The resulting integrals for the two orthogonal components of the deflection are known as the Vening-Meinesz formulas:

$$\xi_P = \frac{1}{4\pi\gamma_P} \iint_\sigma \frac{dS}{d\psi} \cos \alpha \, \Delta g_Q \, d\sigma$$

$$\eta_P = \frac{1}{4\pi\gamma_P} \iint_\sigma \frac{dS}{d\psi} \sin \alpha \, \Delta g_Q \, d\sigma$$

where α is the azimuth.

The BVP solutions based on Green's third identity, by definition, require a complete and global set of boundary observations. But, whereas the coefficients in the spherical harmonic series are sensitive to all boundary values over the entire surface, the integral solutions may often be approximated by integrals over smaller neighborhoods of the computation point P or of its radial projection onto the bounding sphere. Green's function describes the precise physical correlation between boundary value and disturbing potential at the computation point and shows that, generally, boundary values at great distances from P have less influence than those near P. The P-neighborhood of integration can be reduced even further by linearizing the potential field with respect to a reference field that, in addition to the ellipsoidal components, includes the higher-degree wavelength components of the actual field. The boundary values then need only provide the short-wavelength variations of the potential and therefore need be available only in a small area, typically within a few degrees around the computation point.

Because BVPs require a completely empty space outside the boundary, the solution to Stokes' geodetic BVP presupposes that terrestrial masses above the geoid have been removed by computation. The resulting effect on the observed gravity in conjunction with the analytic continuation of observations from the topographic surface to the geoid is known as a *gravity reduction*. It involves hypotheses with respect to mass densities and estimates of gravity gradients, both of which are problematic aspects of Stokes' solution to the geodetic BVP.

Integral Equations. To overcome these theoretical imperfections of the practical application of Stokes' solution, Molodensky devised a solution for the physically most accessible surface, the Earth's surface, as the boundary. Although solutions to this more complicated problem exist with Green's third identity as a starting point, a mathematically simpler approach postulates that a surface density coating φ generates the observed disturbing potential:

$$T_P = \iint_S \frac{\varphi}{l} \, dS$$

The linearization fixes the surface S as the telluroid, which is a known surface in the normal gravity field. Expressed as the potential of a density layer, T is harmonic in empty space, but its first derivative in the vertical direction is discontinuous at the boundary. Therefore, the substitution of T into the spherical approximation of the boundary condition (Eq. 3) yields

$$2\pi\varphi_P \cos \beta - \iint_S \left[\frac{3}{2r_Pl} + \frac{r_Q^2 - r_P^2}{2r_Pl^3} \right] \varphi \, dS = \Delta g_P$$

where β is the inclination of the surface with respect to the horizon. The problem thus transforms into an integral equation for an unknown and fictitious density function with the gravity anomaly being the perturbing function. Because of the strong singularity of the integrand, however, this integral equation is not of the Fredholm type, and the classical methods of solution do not necessarily apply. Formally, the solution for the density layer is iterative where the first approximation uses the geoid as boundary. Even with the spherical approximation, the final result is sufficiently complicated to be delegated to more mathematically complete texts, such as the one by Moritz (1980). Nevertheless, the elegance of this method, originated by Molodensky, led to other practical solutions, notably by Brovar. Brovar sought to simplify the integral equation for the density layer by choosing an appropriate harmonic kernel alternative to the reciprocal distance used by Molodensky. The ensuing integral equation is

$$\varphi_P \cos \beta - \frac{1}{4\pi} \iint_S \frac{r_Q^2 - r_P^2}{r_Pl^2} \varphi \, dS = \Delta g_P$$

whose zero-order solution (spherical surface) is simply

$$\varphi = \Delta g$$

The final solution follows along lines similar to Molodensky's method on the basis of the spherical approximation. Both solutions require heights of the telluroid above the ellipsoid, which may be substituted with terrain elevations above the geoid. The convergence of the iterations in each case is not guaranteed, but in practice the first few iterations seem to yield a stable solution, at least for relatively smooth topography.

Least Squares Collocation. Although the solutions to the BVP provide a rigorous mathematical and physical framework for geodetic application, they are never free from several additional very important sources of error. The boundary values, being derived from measurements, are discrete samples of the continuous boundary function and invariably are corrupted by some level of observation and instrumentation errors. Furthermore, the collection of measurements is often an incomplete, that is, nonglobal, set.

These shortcomings, a lack of technique to deal with them in the general context of integral equations, and the diversity and disjointedness of geodetic data sets, which further confound the application of BVP solutions, have precipitated an operational approach to physical geodesy, an approach predicated on the discreteness of observational data that treats interpolation, extrapolation, and estimation with the formality of modern mathematics. The method of this approach is called least squares collocation (LSC) and is very closely related to Wiener-Kolmogorov estimation. The following synopsis intends to give a flavor of the theory behind LSC, which is found with more substance in modern geodetic textbooks, such as Moritz (1980).

The disturbing potential is supposed to belong to a Hilbert space of functions harmonic in the exterior space, with a reproducing kernel. Given a finite, discrete set of data, each element a measurement of a linear functional of the disturbing potential (such as, the gravity anomaly in linear approximation), the technique of LSC gives a best estimate of the disturbing potential in the exterior space on the basis of this set. This estimate belongs to a subspace of the Hilbert space that contains the infinity of solutions all compatible with the finite set of data. That is, the finite collection of measurements alone will not determine the disturbing potential completely (incomplete, or ill-posed BVP). The estimate is unique only with an additional constraint; for example, the minimum norm constraint, where the square of the norm is the inner product associated with the Hilbert space. The spectrum of the reproducing kernel defines the inner product of two functions such that the kernel reproduces

$$f_P = \langle K_{P,Q}, f_Q \rangle$$

By assumption the region of harmonicity is exterior to a sphere of radius r_0, so functions belonging to the Hilbert space are expandable in terms of spherical harmonics. The possibility, then, of defining the inner product in this way exists when the kernel depends only on the angle ψ between the two radii through the points P and Q:

$$K_{P,Q} = \sum_{n=0}^{\infty} \left(\frac{r_0^2}{r_P r_Q}\right)^{n+1} k_n P_n (\cos \psi)$$

One can prove that the LSC estimate also has minimum error variance when the reproducing kernel is the autocovariance function of the disturbing potential. Therefore, the LSC estimate, because it has minimum norm, represents the smoothest disturbing potential function that fits the observed data, and because the error of estimation has minimum variance, it is the best in terms of estimation accuracy. In addition to the estimation error resulting from an insufficient amount of observations, the LSC estimator also minimizes the effect of the error due to observational inaccuracy when the (statistical) observational error covariance is incorporated into the (physical) covariance of the data. The estimation equation is

$$s = C_{sx}(C_{xx} + D)^{-1} x$$

C is the matrix of covariances between the quantity to be estimated, s, and the data vector x or between the data vector and itself; D is the covariance matrix of data errors (which are assumed to be uncorrelated with the data and have zero mean). The covariance matrix of the error of estimation is given by

$$E_{ss} = C_{ss} - C_{sx}(C_{xx} + D)^{-1} C_{xs}$$

The quantities s and x are assumed to have global means of zero value, but otherwise may stand for not only values of the disturbing potential T but any linear functional of T. The theory of LSC also has been extended to include the estimation of systematic model parameters.

The law of propagation of covariances provides the means to compute the covariances of linear functionals of the disturbing potential. It states that the covariance between two quantities is the iterated functional of the covariance function of T. For example,

$$\text{cov}(s, x) = L_s(L_x(K_{P,Q}))$$

where the functionals are

$$s = L_s(T), \quad x = L_x(T)$$

The covariance function of T is the average of products of T under all rotations of the coordinate system. It acquires a statistical interpretation if the disturbing potential is viewed as a stochastic process on the sphere, with the phase variable being space rather than time. The power spectrum of the disturbing potential, or a suitable model, defines the spectrum of the covariance function.

Least squares collocation has found widespread application because the law of propagation of covariances permits the construction of an estimator that operates on any set of linear functionals of the disturbing potential and provides an estimate of any (different) set. For example, the set of observational data may consist of measurements of the gravity anomaly, the deflection of the vertical, and the eastward gradient of the gravity anomaly at several points on the Earth's surface; and the estimated quantity may be the deflection of the vertical at aircraft altitude. On the basis of the linear and spherical approximations and the assumption of a correct covariance model, the LSC estimate is optimal since it minimizes the effects of errors due to the discreteness and finite extent of the available data, as well as observational errors. The practical difficulties with LSC are mostly of a numerical nature, stemming from the requirement to solve a large system of linear equations. The size of the autocovariance matrix, and therefore the number of independent linear equations to be solved, equals the number of observational data. Ill-conditioned solutions result if the data refer to a set of spatially dense points, and an excessive amount of data could make the problem numerically intractable. Frequently cited theoretical problems concern the incomplete knowledge of the disturbing potential covariance function and the assumption usually applied to models that the potential is a stationary and isotropic stochastic process. Yet with these limitations and approximations LSC is a successful method for determining the disturbing potential on and outside the Earth's surface. Advanced applications attempt to alleviate the spherical approximation by separately treating the terrain relative to a reference surface in the completely deterministic manner of a gravity reduction. This again requires the adoption of a density hypothesis.

Notes on Existence and Uniqueness

The usefulness of a BVP rests with the demonstration that a solution exists or, more strongly, with the actual construction of a solution. Moreover, the solution must be unique so that, once constructed, it does describe the Earth's exterior potential and not some other potential satisfying the given boundary values. Theorems on existence and uniqueness depend on the type of boundary. Generally, one requires that the mass-enclosing surface be regular; i.e., there exists a small sphere of predefined radius such that the portion of the surface it intersects, when it is centered at any point on the surface, may be expressed as a well-defined continuous function with continuous derivatives. These conditions, sometimes known as Lyapunov conditions, apply only approximately to the Earth's surface.

With Green's identities one can show that the solutions to the first, second, and third exterior BVPs are unique for continuous boundary values. To prove the existence of solutions is generally more difficult and devolves upon the study of solutions to integral equations. Dirichlet's and Neumann's exterior BVPs do have solutions, as does the third exterior BVP provided its boundary values satisfy certain conditions. For Stokes' and Molodensky's solutions, the gravity anomalies must not contain harmonics of the first degree. Centering the reference ellipsoid at the Earth's center of mass easily meets this stipulation.

Advanced studies on the existence of solutions to the general oblique-derivative nonlinear geodetic BVP may be found in the works by Sanso (1981), Hörmander (1976), and Moritz (1980).

CHRISTOPHER JEKELI

References

Heiskanen, W. A., and H. Moritz, 1967, *Physical Geodesy*. San Francisco: W. H. Freeman and Co.

Hörmander, L., 1976, The Boundary Problems of Physical Geodesy, *Archive Rational Mechanics and Analysis* **62**, 1-52.

Kellogg, O. D., 1953, *Foundations of Potential Theory*. New York: Dover Publications.

Krarup, T., 1969, *A Contribution to the Mathematical Foundation of Physical Geodesy*, report no. 44. Copenhagen, Denmark: Geodaetisk Institut.

Molodensky, M. S., V. F. Eremeev, and M. I. Yurkin, 1962, *Methods for the Study of the External Gravitational Field and Figure of the Earth*, translation from Russian. Jerusalem: Israel Program for Scientific Translation.

Moritz, H., 1980, *Advanced Physical Geodesy*. Turnbridge Wells, Kent, England: Abacus Press.

Sansó, F., 1981, Recent Advances in the Theory of the Geodetic Boundary Value Problem, *Rev. Geophys. Space Physics* **19**(3), 437-449.

Cross-references: *Equilibrium Figure of the Earth; Figure of the Earth; Geodesy: Physical; Gravity Fields: Implications for Planetary Interiors; Gravity and Isostasy; Mantle Convection and Plumes; Mantle Dynamics; Mantle Viscosity; Spherical Harmonic Analysis.*

EARTH STRUCTURE, GLOBAL

Introduction

If this article were written some six years ago, it would be likely to contain only a description of the Earth as a spherically symmetric body (a body whose properties vary only with radius, or depth, and do not depend on geographical coordinates, the effects due to the Earth's ellipticity and rotation being implicitly accounted for). It is known, however, that there are lateral differences in physical properties at all levels within the Earth; indeed its dynamic behavior—mantle convection, for example—tells us that this must be so. Now, because of the availability of new sources of data, very large wavelength lateral variations in the deep interior of the Earth have been identified, and there is strong evidence that further improvement in the knowledge of the spherically symmetric properties is just part of the much larger and complex problem of determining the Earth's structure in three dimensions.

One way to look at the problem is as follows:

Real Earth = Spherically symmetric model
+ 3-dimensional perturbation

A spherically symmetric Earth model (SSEM)

approximates the real Earth quite well; the relative size of the three-dimensional (3-D) part with respect to SSEM varies from several percent in the upper mantle to a fraction of a percent in the lower mantle.

A concept of an SSEM, often referred to as an "average" Earth model, is a necessary tool in seismology. These models are used to compute functionals of the Earth structure and their differential kernels, which are needed to locate earthquakes and to determine focal mechanisms. Knowledge of the internal properties of the Earth is needed in geodesy and astronomy. Important inferences with respect to the chemical composition and physical conditions within the deep interior of the Earth are made by using information on radial variations of the elastic and anelastic parameters and density.

Most of the forward and inverse problems in seismology are nonlinear. However, if the "real Earth" structure departs only slightly—say, less than several percent—from a reference SSEM, the effect of such differences on the Earth functionals can be adequately represented by a linear approximation. All progress thus far in modeling the Earth in three dimensions has been achieved with this method.

Unraveling the internal structure of the Earth by seismology began about 1900. The first decade, the 1930s, the 1960s, and now the 1980s have been periods of rapid progress, which generally coincided with the availability of a new class of data. The one-dimensional (1-D) image of the Earth's interior, except for the fine structure of the upper mantle and anelasticity, had been essentially completed by the late 1930s as a result of work of Harold Jeffreys, Beno Gutenberg, Inge Lehmann, and Keith Bullen.

An SSEM is a mathematical fiction—a useful and, often, a necessary one, but fiction nevertheless. This fact is most obvious at the Earth's surface, where one must face the dilemma of how to reconcile the occurrence at the same depth, or elevation, of water and rocks; the systems of equations governing the wave propagation in liquid and in solid are different. The commonly adopted solution is to introduce a layer of water whose thickness is such that the total volume of water in all the oceans and that calculated for the SSEM are equal. It is a reasonable decision, but it will be necessary to introduce corrective measures even when constructing the model, since practically all seismographs that record ground motion are located on land.

Another problem related to the design of an SSEM is the question of discontinuities. For example, the average depth to the Mohorovičić ("Moho") discontinuity is 35 km on land but only about 10 km in the oceans. The weighted average, considering the area of continents and oceans, is 18 km, a very atypical figure, not representative of any commonly encountered real structure. An alternative would be to provide a smooth transition between 10 and 35 km from the crustal to upper-mantle elastic parameters. But such a model would be useless in explaining, for example, observations of the waves reflected from or refracted at the Moho. The general philosophy in defining an SSEM is that it should represent the radial variation of the composition with depth, and where an abrupt change in the physical properties is indicated, the model should give an average radius, or depth, of such a discontinuity.

This article uses the preliminary reference Earth model (PREM), published in 1981 by Dziewonski and Anderson, and reviews the current state of mapping of the deep interior of the Earth in three dimensions.

A Spherically Symmetric Earth Model

Our knowledge of the properties of the Earth's interior depends on the availability of measurements of functionals of its elastic and anelastic parameters. Such data have to be properly identified before a model can be constructed that is consistent with those functionals within their observational errors. Calculation of a functional for a given model is called a *forward problem*; derivation of a model from a set of functionals is called an *inverse problem*. The theory of inverse problems is a branch of mathematics; geophysics provides some of the most challenging examples, since all parameters of the Earth structure below the depths accessible through drilling (~ 10 km) must be inferred through solving an inverse problem. A comprehensive review of inverse theory is given by Menke (1984) (see also *Inverse Theory and Methods: Seismology*).

Gross Earth Data: The Constraints.

Travel Times. The earliest, and still the most commonly reported, functionals of the Earth's structure are the travel times of body waves: compressional (P) or shear (S) waves that propagate through the interior of the Earth. Such data led Oldham to discover, in 1906, the liquid core of the Earth and Mohorovičić, in 1909, to detect the presence of a sharp discontinuity in P velocity at a depth of about 30 km.

An introductory level account of the forward and inverse problems related to the seismological applications of geometrical optics can be found in Bullen and Bolt (1985, chap. 7). Some formulas are introduced here to explain how the inverse problem is posed.

The distribution of seismic velocity as a function of radius is given as $v(r)$ and, in general, will be piecewise continuous. A *ray parameter p* is defined as

$$p = \frac{r \sin i}{v}$$

where i is the incidence angle, which, because of Snell's law, remains constant along the ray path. A ray *bottoms* or reaches the *turning radius* $r = r_p$

when sin $i = 1$. A useful designation is $\eta = r/v$. Generally, η is a monotonically increasing function of radius: velocities tend to decrease toward the surface, although there are exceptions known as *low-velocity zones* (*LVZ*). Rays originating above such zones do not bottom in it, and one practical consequence of this is that the velocity profile within an LVZ cannot be resolved. A ray originating at the Earth's surface with radius a and ray parameter p will emerge at the surface again after a time T, where

$$T = 2 \int_{r_p}^{a} \eta^2 r^{-1} (\eta^2 - p^2)^{-1/2} \, dr$$

and at an angular distance Δ, where

$$\Delta = 2p \int_{r_p}^{a} r^{-1} (\eta^2 - p^2)^{-1/2} \, dr$$

These relationships can be used to calculate tables of travel times, where time is given as a function of epicentral distance. Complete tables of travel times were computed in the 1930s, and those prepared by Jeffreys and Bullen in 1939 are still in use.

At discontinuities including the Earth's surface, part of the wave energy may be reflected and part may be converted, either from shear to compressional or vice versa. These multiple reflections and conversions are responsible for the complex appearance of the seismograms at teleseismic ($>25°$) distances. Figure 1 shows examples of ray paths for P waves confined to the mantle (top), for P-wave rays entering the core or reflected from the core-mantle boundary (middle), and for S-wave rays (bottom), including waves converted to P waves at the core-mantle boundary. Figure 2 is a plot of travel times for an assembly of phases calculated by Jeffreys and Bullen (1939).

Derivation of a velocity profile from a set of observed travel times can be accomplished in at least two ways. The traditional method is due to Herglotz and Wiechert. The observed travel times $T(\Delta)$ must be smoothed such that $p(\Delta) = dT/d\Delta$ is obtained. Then evaluation of the integral

$$\int_0^{\Delta_1} \cosh^{-1}\left(\frac{p}{\eta_1}\right) d\Delta = \pi \ln \frac{a}{r_1}$$

yields radius r_1 at which the ray, with ray parameter $p_1 = \eta_1 = r_1/v_1$ bottoms, and, hence, velocity v_1 at r_1. The method fails when η is not a monotonically increasing function of radius. Practical difficulties arise for triplications associated with rapid or discontinuous changes in velocity, because observations of secondary arrivals are difficult and often uncertain.

An alternative is the perturbation approach. Given a departure $\delta v(r)$ from a reference model $v_0(r)$, the

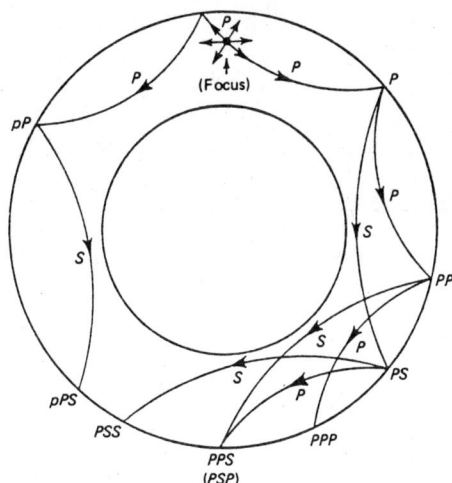

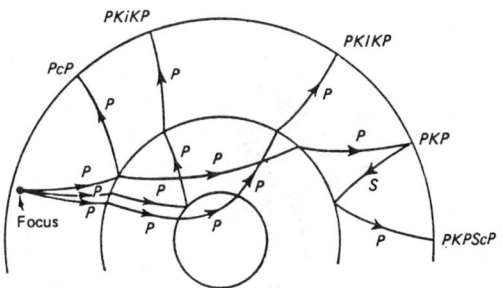

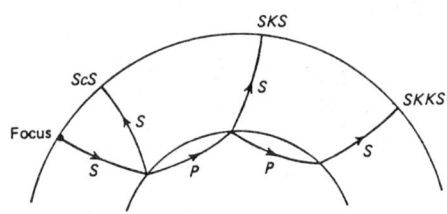

FIGURE 1. Seismic ray paths in the Earth and the corresponding names of seismic phases. (From Bullen and Bolt, 1985)

change in the travel time at a fixed epicentral distance is

$$\delta T = \int_{r_p}^{a} G_0 \delta v(r) \, dr \qquad (1)$$

where the differential kernel, G_0, is given by

$$G_0 = 2\eta v^{-2} (\eta^2 - p^2)^{-1/2}$$

Given a set of differences between the observed travel times and predictions of the reference model,

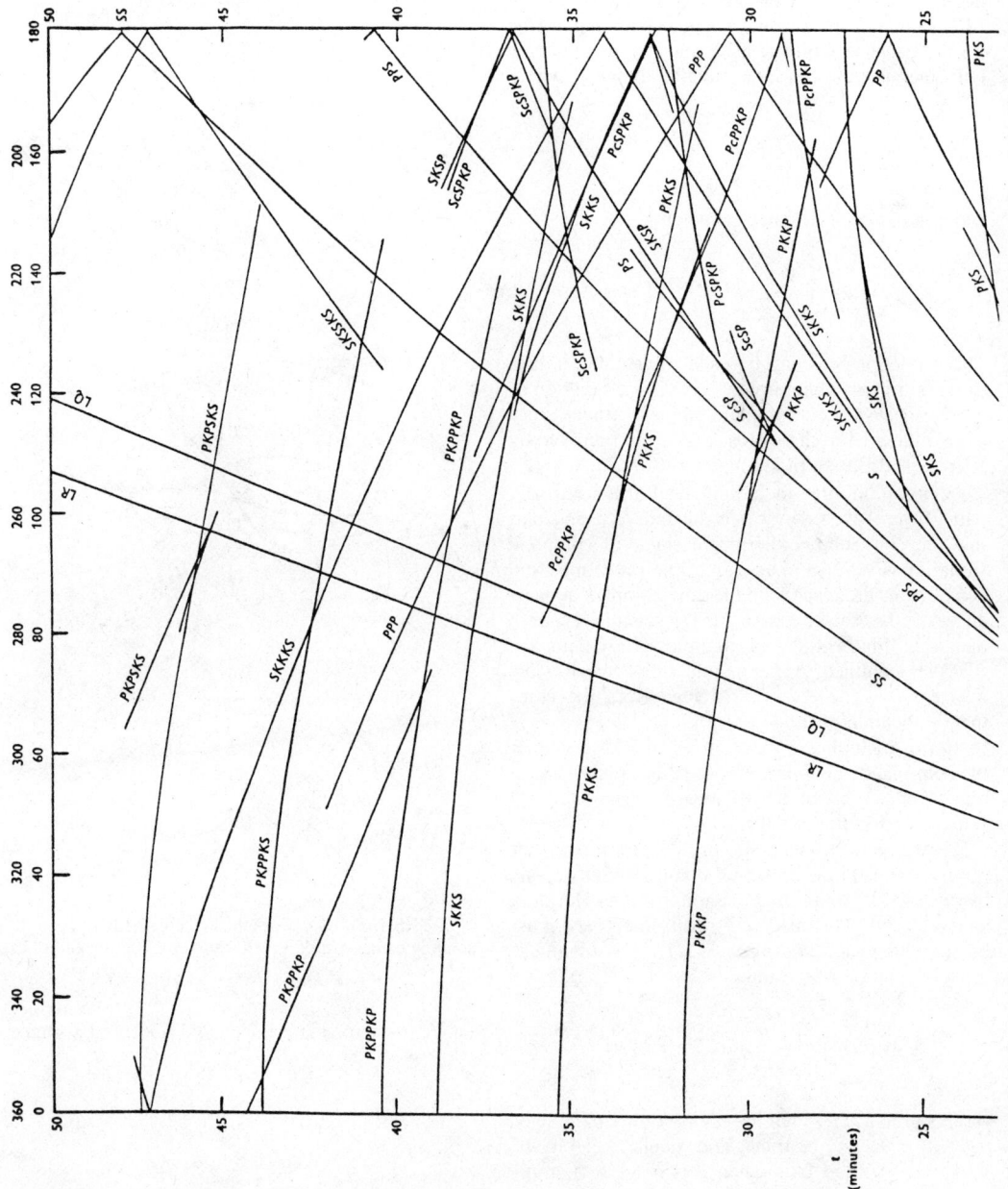

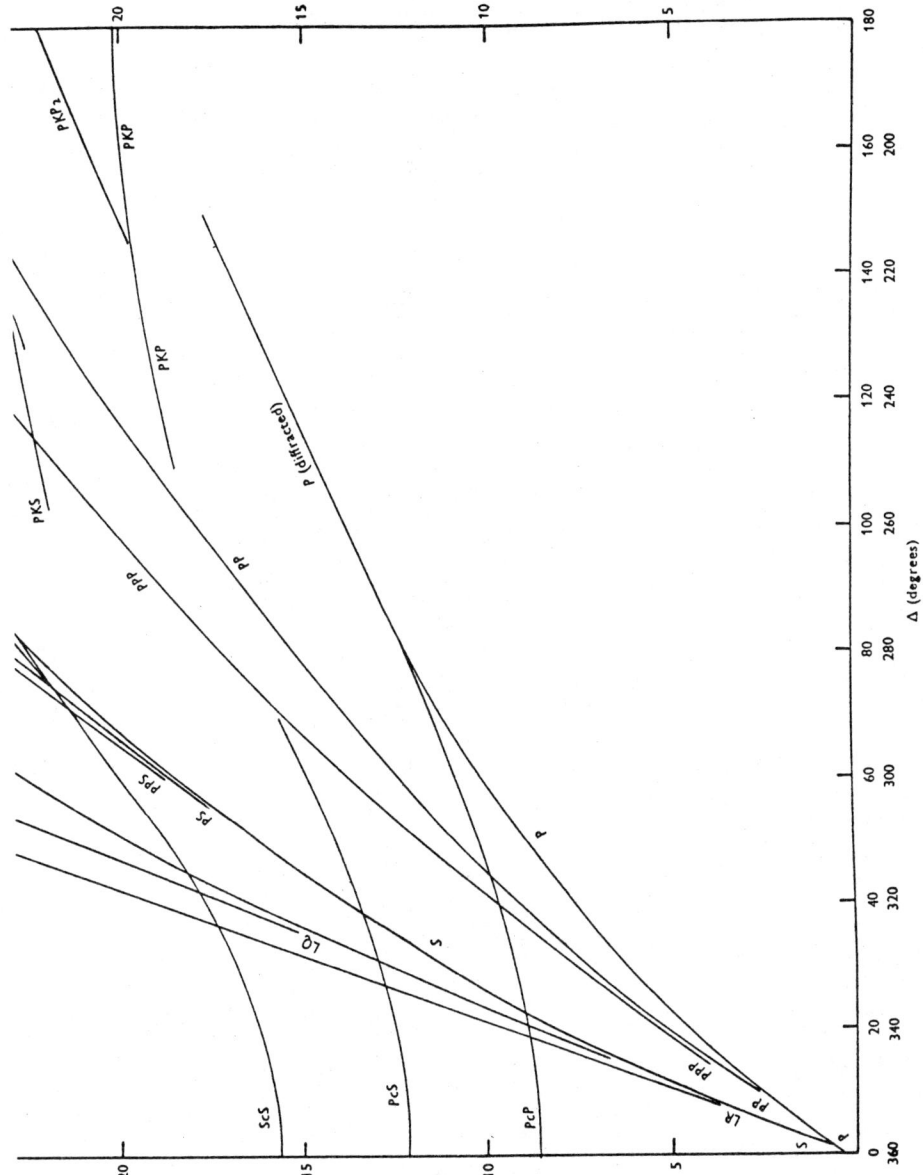

FIGURE 2. Travel times of seismic phases. (From Jeffreys and Bullen, 1939)

it is possible to formulate the inverse problem for $\delta v(r)$. The principal disadvantage of the linearized approach is that it will not yield the correct answer if the starting model is not *linearly close* to the real Earth. The advantage is that it allows us to use simultaneously data from different branches of travel times as well as an entirely different data set, the eigenfrequencies of free oscillations of the Earth.

Free Oscillations of the Earth. The free oscillation (or normal mode) approach to representing propagation of seismic disturbances within the Earth is most satisfactory and complete (see *Free Oscillations of the Earth*). Numerically, however, the superposition of all normal modes above a certain cutoff frequency can be formidable. In order to synthesize the ground motion at 1 Hz, one would have to use roughly 10 million different modes. Figures 3a and 3b, from Gilbert and Dziewonski (1975), show all the spheroidal and toroidal modes with periods longer than 100 s. At low frequencies—say, less than 3 mHz—the number of existing normal modes is relatively small; they are, in general, well separated in frequency, and it is possible to resolve individual spectral peaks in the Fourier transform of a record of a major earthquake.

In the late 1950s, long-period seismographic instrumentation was developed to provide the long-period data necessary for solving the forward problem of free oscillations of the Earth. About the same time the methods for analyzing digital data on relatively fast computers were developed. Thus when the Chilean earthquake of May 22, 1960—the largest ever documented—struck, several research groups reported nearly simultaneously observations of free oscillations of the Earth and related them to the existing models of the Earth interior. The differences between the model predictions and observations were sufficiently large that it became clear that free oscillation data provide additional constraints on the Earth structure. An account of the basic elements of the theory of free oscillations can be found in Lapwood and Usami (1981) and in *Free Oscillations of the Earth* (q.v.).

Attenuation and Anisotropy. The periods of free oscillations are two to three orders of magnitude greater than those of body waves (typically 1 Hz). It is important, therefore, to consider the effects of velocity dispersion due to attenuation. The Earth is not perfectly elastic. The body waves and free oscillations are, therefore, attenuated in addition to decreasing in amplitude as a result of geometrical spreading. The quality factor Q has been measured for many normal modes and some travel time branches; there is evidence of a significant increase in Q at periods shorter than 1 s. (See *Seismic Attenuation: Observation and Measurement*.)

For an isotropic region of the Earth, perturbation in a period of free oscillation or travel time of a body wave can be expressed by

$$\frac{\delta T}{T} = \int_0^a dr \, [\tilde{P}\delta v_P + \tilde{S}\delta v_S + \tilde{R}\delta\rho$$

$$+ (\tilde{M}\delta q_\mu + \tilde{K}\delta q_K)\ln \tau]$$

+ term related to changes in radii of discontinuities

where v_P is compressional velocity, v_S is shear velocity, ρ is density, q_μ and q_K are the reciprocals of Q_μ and Q_K, the quality factors for the shear and compressional energy, respectively, $\tilde{P}, \tilde{S}, \tilde{R}, \tilde{M}$, and $\tilde{K}$ are the appropriate differential kernels, and τ represents either the period of a particular normal mode ($\tau = T$ in that case) or the dominant period of the body wave. Perturbation in an observed attenuation factor q is

$$\delta q = \int_0^a dr \, (\mu\tilde{M}\delta q_\mu + K\tilde{K}\delta q_K)$$

where μ and K are the shear and bulk moduli, respectively.

To derive the PREM model, we extend this approach to transverse anisotropy. A transversely anisotropic medium with the axis of symmetry coinciding with the radial direction is characterized by elastic constants A, C, F, N, and L, which can be expressed in terms of velocities (see also *Seismic Attenuation*)

$$v_{PH} = \left(\frac{A}{\rho}\right)^{1/2}, \quad v_{PV} = \left(\frac{C}{\rho}\right)^{1/2},$$

$$v_{SH} = \left(\frac{N}{\rho}\right)^{1/2}, \quad v_{SV} = \left(\frac{L}{\rho}\right)^{1/2}$$

and the anisotropic parameter $\eta = F/(A - 2L)$. The letters V and H correspond to the radial and horizontal direction of propagation.

PREM: An Example. The design of the starting model is critical: once a discontinuity is included in the starting model, it will remain there. The same is true with regard to anisotropy or, in general, any complexity of the model. The strategy, therefore, is to begin with simple models and to introduce complexities only when it is demonstrated that the data demand it.

It was first suggested in the early 1970s that the seismic velocities and density in a reference Earth model be described by a set of piecewise continuous functions, such as low-order polynomials in radius. The advantage of this approach is that the model parameters are uniquely defined, without recourse to interpolation, at any depth. Also, the model parameters can be differentiated or integrated analytically. The inverse problem may then be formulated in terms of polynomial coefficients, allowing us to test various hypotheses. For example, perturbation to the compressional velocity in a region spanned by radii r_1 and r_2 can be sought in the form

$$\delta v_P = a_0 + a_1 r + a_2 r^2 + a_3 r^3, \quad \text{for } r_1 \leq r \leq r_2$$

or, more generally,

$$\delta v(r) = \sum_{n=0}^{N} a_n r^n \qquad (2)$$

The following principal regions have been defined in the design of the starting model for PREM. This division, except for the fine details of the upper-mantle structure, is quite similar to that proposed by Bullen in 1942 (Bullen and Bolt, 1985, p. 317). For comparison, Bullen's designations from **A** through **G** are given in parentheses.

1. Ocean layer: 0–3 km
2. Upper crust: 3–15 km (**A**)
3. Lower crust: 15–24.4 km (**A**)
4. Region above *LVZ*, considered to be the main part of the lithosphere: 24.4–80 km (**B**)
5. LVZ: 80–220 km (**B**)
6. Region between *LVZ* and 400-km discontinuity: 220–400 km (**B**)
7. Transition zone spanning the region between the discontinuities at 400 and 670 km (**C**)
8. Lower mantle [the region of the Earth between 670 km depth and the core-mantle boundary (CMB)] subdivided into three parts
   - transitional, high-gradient region from 670 to 771 km (**C**)
   - principal part of the lower mantle: 771–2741 km (**D'**)
   - region above the CMB: 2741–2891 km (**D''**)
9. Outer core: 2891–5150 km (**E, F**)
10. Inner core: 5150–6371 km (**G**)

Dziewonski and Anderson adopted the polynomial parameterization of the radial variations of seismic velocities, density, and shear and bulk attenuation. To refine the starting Earth model, they considered the following types of data:

1. *Astronomic-geodetic data.* Radius of the Earth of equal volume: $a = 6371$ km; mass: $M = 5.974 \times 10^{24}$ kg; moment of inertia ratio: $I/Ma^2 = 0.3308$.
2. *Body waves.* Travel times for 11 different branches, including differential travel times, compiled from ten sources. This includes the analysis of 2,000,000 P-wave and 250,000 S-wave travel times derived from the *Bulletins of the International Seismological Centre* for the years 1964–1979.
3. *Free oscillation and surface wave data.* A set of about 900 normal mode periods compiled from six sources. In addition, long-period surface wave data from five sources. Data on attenuation of normal modes and surface waves compiled from eight sources.

Model PREM is listed in its parametric form in Table 1. This form is valid for a period of 1 s. Because of velocity dispersion, velocities at other periods must be modified:

$$v_S(T) = v_S(1)(1 - \frac{\ln T}{\pi} q_\mu) \qquad (3)$$

$$v_P(T) = v_P(1)\{1 - \frac{\ln T}{\pi}[(1 - E)q_K + E q_\mu]\} \qquad (4)$$

where $E = \frac{4}{3}(v_S/v_P)^2$.

Model PREM is isotropic everywhere but in a depth range from Moho to 220 km. Given the five elastic constants for a transversely anisotropic solid, it is possible to calculate the shear and bulk moduli for an equivalent isotropic solid:

$$K = \frac{1}{9}(4A + C + 4F - 4N)$$

$$\mu = \frac{1}{15}(A + C - 2F + 5N + 6L)$$

These are Voigt moduli, and they represent upper bounds on their effective values.

Table 2 is an expansion of "isotropic" PREM at a period of 1 s. The values for the anisotropic part of PREM are listed in Table 3. Model PREM is shown in Fig. 4 for a full range of radii; upper mantle is shown in an expanded scale in Fig. 5. Several derived parameters are listed in Table 2:

attenuation of compressional waves

$$Q_\alpha^{-1} = (1 - E)Q_K^{-1} + E Q_\mu^{-1}$$

seismic parameter Φ defined as

$$\Phi = v_P^2 - \frac{4}{3}v_S^2$$

Poisson ratio

$$\sigma = \frac{1}{2}\frac{K - \frac{2}{3}\mu}{K + \frac{2}{3}\mu}$$

gravity

$$g(r) = \frac{4\pi G}{r^2} \int_0^r \rho x^2 \, dx$$

(Text continues on page 346.)

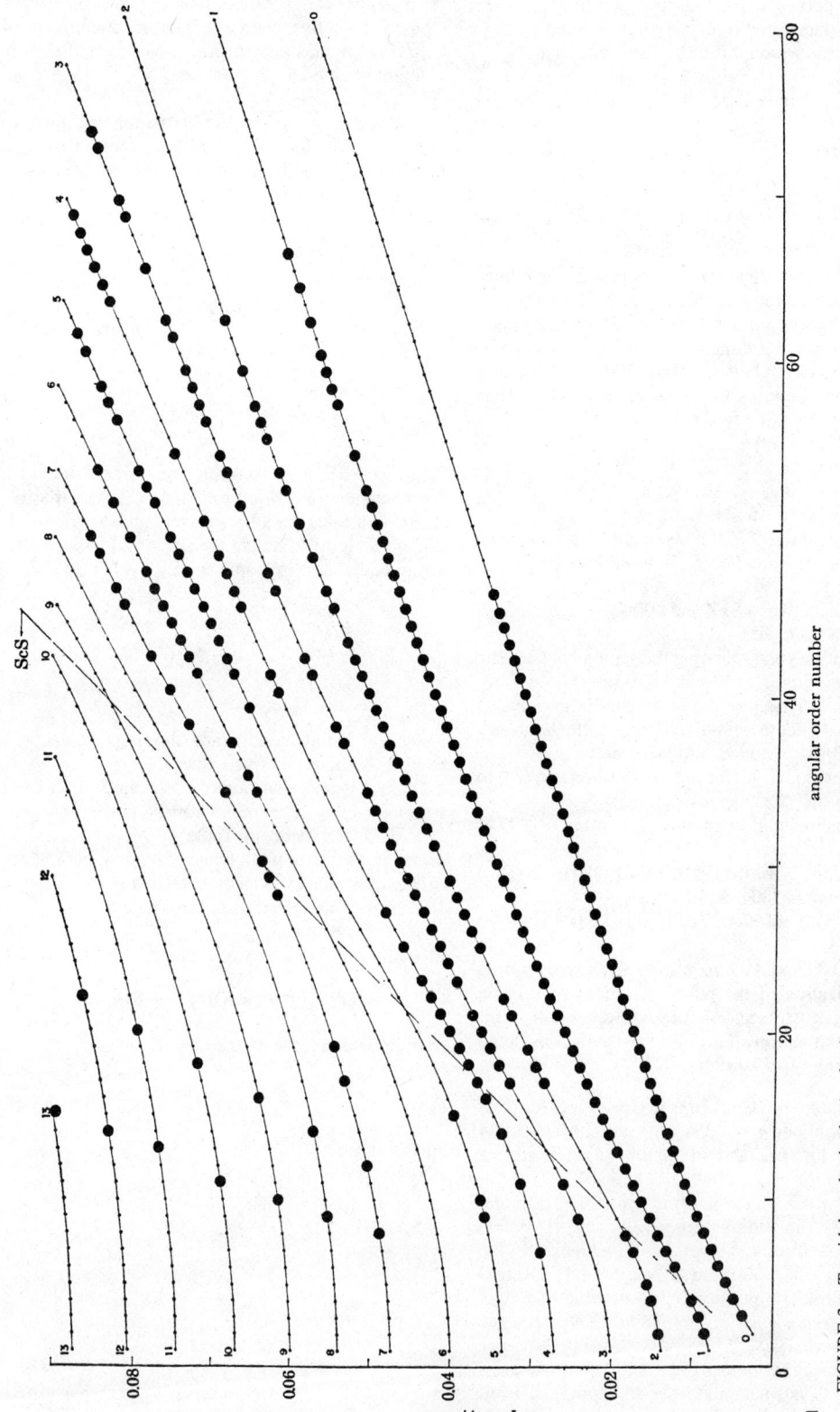

FIGURE 3a. Toroidal modes in the (ω, l) plane. The large dots indicate modes for which eigenfrequencies have been measured. The dashed line designated ScS divides the modes into two groups according to the normal mode–body wave analogy: modes to the left of this line correspond to ScS_H reflections; those to the right correspond to mantle S_H waves. (After Gilbert and Dziewonski, 1975)

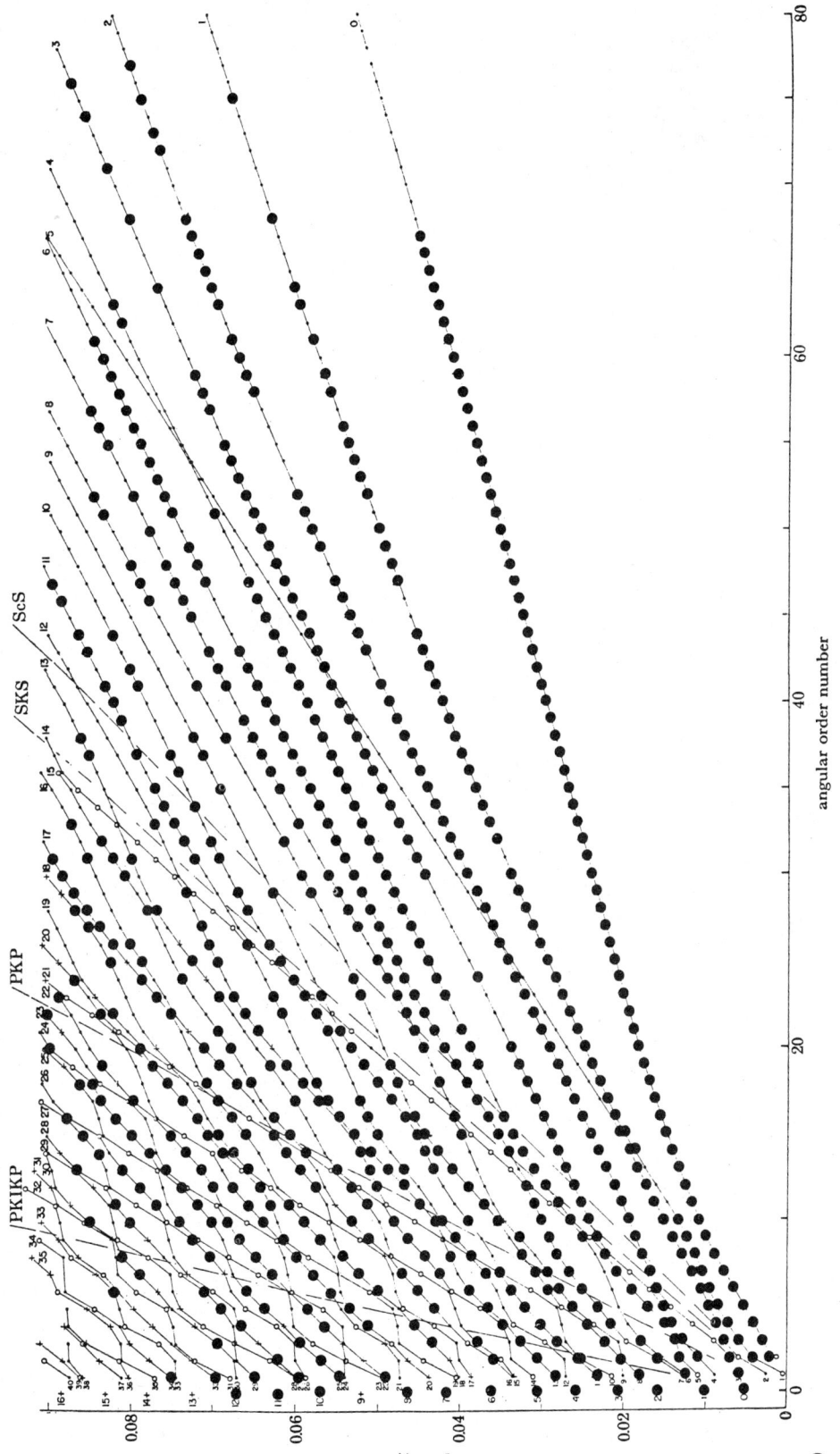

FIGURE 3b. Spheroidal normal modes in the (ω, l) plane. This image, much more complex from that of Fig. 3a, reflects the behavior of P-SV waves. Well to the left of the ScS line, the wave types tend to decouple, and we see the S_V pattern (similar to S_H) with the modes designated by small filled circles; PKIKP-equivalent modes have more than 50 energy and are shown with a +-sign. The inner core modes are shown with an open circle; this class of modes corresponds asymptotically to S_V waves trapped in the solid inner core. (After Gilbert and Dziewonski, 1975)

TABLE 1. Coefficients of Polynomials Describing the Preliminary Reference Earth Model

Region	Radius (km)	Density (g·cm^{-3})	V_p (km·s^{-1})	V_S (km·s^{-1})	Q_μ	Q_K
Inner Core	0–1221.5	13.0885 −8.8381x^2	11.2622 −6.3640x^2	3.6678 −4.4475x^2	84.6	1327.7
Outer Core	1221.5–3480.0	12.5815 −1.2638x −3.6426x^2 −5.5281x^3	11.0487 −4.0362x +4.8023x^2 −13.5732x^3	0	∞	57823
Lower Mantle	3480.0–3630.0	7.9565 −6.4761x +5.5283x^2 −3.0807x^3	15.3891 −5.3181x +5.5242x^2 −2.5514x^3	6.9254 +1.4672x −2.0834x^2 +0.9783x^3	312	57823
	3630.0–5600.0	7.9565 −6.4761x +5.5283x^2 −3.0807x^3	24.9520 −40.4673x +51.4832x^2 −26.6419x^3	11.1671 −13.7818x +17.4575x^2 −9.2777x^3	312	57823
	5600.0–5701.0	7.9565 −6.4761x +5.5283x^2 −3.0807x^3	29.2766 −23.6027x +5.5242x^2 −2.5514x^2	22.3459 −17.2473x −2.0834x^2 +0.9783x^3	312	57823
Transition Zone	5701.0–5771.0	5.3197 −1.4836x	19.0957 −9.8672x	9.9839 −4.9324x	143	57823
	5771.0–5971.0	11.2494 −8.0298x	39.7027 −32.6166x	22.3512 −18.5856	143	57823
	5971.0–6151.0	7.1089 −3.8045x	20.3926 −12.2569x	8.9496 −4.4597x	143	57823
Low Velocity Zone[a]	6151.0–6291.0	2.6910 +0.6924x	V_{PV} 0.8317 +7.2180x	V_{SV} 5.8582 −1.4678	80	57823
			V_{PH} 3.5908 +4.6172x	V_{SH} −1.0839 +5.7176x	η 3.3687 −2.4778x	
Lid[a]	6291.0–6346.6	2.6910 +0.6924x	V_{PV} 0.8317 +7.2180	V_{SV} 5.8582 −1.4678x	600	57823
			V_{PH} 3.5908 +4.6172x	V_{SH} −1.0839 +5.7176x	η 3.3687 −2.4778	
Crust	6346.6–6356.0	2.900	6.800	3.900	600	57823
	6356.0–6368.0	2.600	5.800	3.200	600	57823
Ocean	6368.0–6371.0	1.020	1.450	0	∞	57823

Note: The variable x is the normalized radius: $x = r/a$, where $a = 6371$ km is the equal volume Earth radius. The parameters listed are valid at a reference period of 1 s.

[a] The region between 24.4 km and 220 km depth is transversely isotropic with the symmetry axis vertical. The effective isotropic velocities over this interval can be approximated by $V_P = 4.1875 \pm 3.9382x$, $V_S = 2.1519 + 2.3481x$.

TABLE 2. Earth Model PREM and Its Functionals Evaluated at a Reference Period of 1 s

Radius km	Depth km	Density g·cm^{-3}	V_P km·s^{-1}	V_S km·s^{-1}	Q_μ	Q_K	Q_α	Φ km^2·s^{-2}	K kb	μ kb	σ	Pressure kb	$\frac{dK}{dP}$	η_B	Gravity cm·s
0.	6371.0	13.08848	11.26220	3.66780	85	1328	431	108.90	14253	1761	0.4407	3638.524	2.3360	0.99	0.
100.0	6271.0	13.08630	11.26064	3.66670	85	1328	431	108.88	14248	1759	0.4407	3636.131	2.3363	0.99	36.56
200.0	6171.0	13.07977	11.25593	3.66342	85	1328	431	108.80	14231	1755	0.4408	3628.956	2.3365	0.99	73.11
300.0	6071.0	13.06888	11.24809	3.65794	85	1328	432	108.68	14203	1749	0.4409	3617.011	2.3369	0.99	109.61
400.0	5971.0	13.05364	11.23712	3.65027	85	1328	432	108.51	14164	1739	0.4410	3600.315	2.3375	0.99	146.04
500.0	5871.0	13.03404	11.22301	3.64041	85	1328	433	108.29	14114	1727	0.4412	3578.894	2.3382	0.99	182.39
600.0	5771.0	13.01009	11.20576	3.62835	85	1328	434	108.02	14053	1713	0.4414	3552.783	2.3391	0.99	218.62
700.0	5671.0	12.98178	11.18538	3.61411	85	1328	436	107.70	13981	1696	0.4417	3522.024	2.3402	0.99	254.73
800.0	5571.0	12.94912	11.16186	3.59767	85	1328	437	107.33	13898	1676	0.4420	3486.665	2.3414	0.99	290.68
900.0	5471.0	12.91211	11.13521	3.57905	85	1328	439	106.91	13805	1654	0.4424	3446.764	2.3428	0.99	326.45
1000.0	5371.0	12.87073	11.10542	3.55823	85	1328	440	106.45	13701	1630	0.4428	3402.383	2.3443	0.99	362.03
1100.0	5271.0	12.82501	11.07249	3.53522	85	1328	443	105.94	13586	1603	0.4432	3353.596	2.3460	1.00	397.39
1200.0	5171.0	12.77493	11.03643	3.51002	85	1328	445	105.38	13462	1574	0.4437	3300.480	2.3480	1.00	432.51
1221.5	5149.5	12.76360	11.02827	3.50432	85	1328	445	105.25	13434	1567	0.4438	3288.513	2.3486	1.00	440.02
1221.5	5149.5	12.16634	10.35568			57822	57822	107.24	13047		0.5000	3288.502	3.7545	1.03	440.03
1300.0	5071.0	12.12500	10.30971			57822	57822	106.29	12888		0.5000	3245.423	3.6539	1.02	463.68
1400.0	4971.0	12.06924	10.24959			57822	57822	105.05	12679		0.5000	3187.493	3.5478	1.01	494.13
1500.0	4871.0	12.00989	10.18743			57822	57822	103.78	12464		0.5000	3126.159	3.4649	1.01	524.77
1600.0	4771.0	11.94682	10.12291			57822	57822	102.47	12242		0.5000	3061.461	3.4017	1.00	555.48
1700.0	4671.0	11.87990	10.05572			57822	57822	101.12	12013		0.5000	2993.457	3.3552	1.00	586.14
1800.0	4571.0	11.80900	9.98554			57822	57822	99.71	11775		0.5000	2922.221	3.3230	1.00	616.69
1900.0	4471.0	11.73401	9.91206			57822	57822	98.25	11529		0.5000	2847.839	3.3028	1.00	647.04
2000.0	4371.0	11.65478	9.83496			57822	57822	96.73	11273		0.5000	2770.407	3.2927	1.00	677.15
2100.0	4271.0	11.57119	9.75393			57822	57822	95.14	11009		0.5000	2690.035	3.2911	1.00	706.97
2200.0	4171.0	11.48311	9.66865			57822	57822	93.48	10735		0.5000	2606.838	3.2966	1.00	736.45
2300.0	4071.0	11.39042	9.57881			57822	57822	91.75	10451		0.5000	2520.942	3.3080	1.00	765.56
2400.0	3971.0	11.29298	9.48409			57822	57822	89.95	10158		0.5000	2432.484	3.3242	1.00	794.25
2500.0	3871.0	11.19067	9.38418			57822	57822	88.06	9855		0.5000	2341.603	3.3441	1.00	822.48
2600.0	3771.0	11.08335	9.27876			57822	57822	86.10	9542		0.5000	2248.453	3.3670	1.00	850.23
2700.0	3671.0	10.97091	9.16752			57822	57822	84.04	9220		0.5000	2153.189	3.3919	1.00	877.46
2800.0	3571.0	10.85321	9.05015			57822	57822	81.91	8889		0.5000	2055.978	3.4180	1.00	904.14
2900.0	3471.0	10.73012	8.92632			57822	57822	79.68	8550		0.5000	1956.991	3.4448	1.00	930.23
3000.0	3371.0	10.60152	8.79573			57822	57822	77.36	8202		0.5000	1856.409	3.4714	1.00	955.70
3100.0	3271.0	10.46727	8.65805			57822	57822	74.96	7846		0.5000	1754.418	3.4972	1.00	980.51
3200.0	3171.0	10.32726	8.51298			57822	57822	72.47	7484		0.5000	1651.209	3.5215	1.00	1004.64
3300.0	3071.0	10.18134	8.36019			57822	57822	69.89	7116		0.5000	1546.982	3.5437	0.99	1028.04
3400.0	2971.0	10.02940	8.19939			57822	57822	67.23	6743		0.5000	1441.941	3.5629	0.99	1050.69
3480.0	2891.0	9.90349	8.06482			57822	57822	65.04	6441		0.5000	1357.510	3.5769	0.98	1068.23

(*Continued*)

TABLE 2 (*Continued*)

Radius km	Depth km	Density g·cm^{-3}	V_P km·s^{-1}	V_S km·s^{-1}	Q_μ	Q_K	Q_α	Φ km^2·s^{-2}	K kb	μ kb	σ	Pressure kb	$\frac{dK}{dP}$	η_B	Gravity cm·s^{-2}
3480.0	2891.0	5.56645	13.71660	7.26466	312	57822	826	117.78	6556	2938	0.3051	1357.509	1.6435	0.99	1068.23
3500.0	2871.0	5.55641	13.71168	7.26486	312	57822	826	117.64	6537	2933	0.3049	1345.619	1.6434	1.00	1065.32
3600.0	2771.0	5.50642	13.68753	7.26575	312	57822	823	116.96	6440	2907	0.3038	1287.067	1.6424	1.01	1052.04
3630.0	2741.0	5.49145	13.68041	7.26597	312	57822	822	116.76	6412	2899	0.3035	1269.742	1.6420	1.01	1048.44
3630.0	2741.0	5.49145	13.68041	7.26597	312	57822	822	116.76	6412	2899	0.3035	1269.741	3.3344	1.01	1048.44
3700.0	2671.0	5.45657	13.59597	7.23403	312	57822	819	115.08	6279	2855	0.3026	1229.719	3.2957	1.01	1040.66
3800.0	2571.0	5.40681	13.47742	7.18892	312	57822	815	112.73	6095	2794	0.3012	1173.465	3.2443	1.01	1030.95
3900.0	2471.0	5.35706	13.36074	7.14423	312	57822	811	110.46	5917	2734	0.2998	1118.207	3.2029	1.00	1022.72
4000.0	2371.0	5.30724	13.24532	7.09974	312	57822	807	108.23	5744	2675	0.2984	1063.864	3.1716	1.00	1015.80
4100.0	2271.0	5.25729	13.13055	7.05525	312	57822	803	106.04	5575	2617	0.2971	1010.363	3.1503	1.00	1010.06
4200.0	2171.0	5.20713	13.01579	7.01053	312	57822	799	103.88	5409	2559	0.2957	957.641	3.1393	1.00	1005.35
4300.0	2071.0	5.15669	12.90045	6.96538	312	57822	795	101.73	5246	2502	0.2943	905.646	3.1383	1.00	1001.56
4400.0	1971.0	5.10590	12.78389	6.91957	312	57822	792	99.59	5085	2445	0.2928	854.332	3.1472	1.00	998.59
4500.0	1871.0	5.05469	12.66550	6.87289	312	57822	788	97.43	4925	2388	0.2913	803.660	3.1657	1.00	996.35
4600.0	1771.0	5.00299	12.54466	6.82512	312	57822	784	95.26	4766	2331	0.2898	753.598	3.1935	0.99	994.74
4700.0	1671.0	4.95073	12.42075	6.77606	312	57822	779	93.06	4607	2273	0.2881	704.119	3.2302	0.99	993.69
4800.0	1571.0	4.89783	12.29316	6.72548	312	57822	775	90.81	4448	2215	0.2864	655.202	3.2750	0.99	993.14
4900.0	1471.0	4.84422	12.16126	6.67317	312	57822	770	88.52	4288	2157	0.2846	606.830	3.3276	0.99	993.01
5000.0	1371.0	4.78983	12.02445	6.61891	312	57822	766	86.17	4128	2098	0.2826	558.991	3.3871	0.99	993.26
5100.0	1271.0	4.73460	11.88209	6.56250	312	57822	761	83.76	3966	2039	0.2805	511.676	3.4527	0.99	993.83
5200.0	1171.0	4.67844	11.73357	6.50370	312	57822	755	81.28	3803	1979	0.2783	464.882	3.5236	0.99	994.67
5300.0	1071.0	4.62129	11.57828	6.44232	312	57822	750	78.72	3638	1918	0.2758	418.606	3.5989	0.99	995.73
5400.0	971.0	4.56307	11.41560	6.37813	312	57822	743	76.08	3471	1856	0.2731	372.852	3.6775	0.98	996.98
5500.0	871.0	4.50372	11.24490	6.31091	312	57822	737	73.34	3303	1794	0.2701	327.623	3.7582	0.98	998.36
5600.0	771.0	4.44317	11.06557	6.24046	312	57822	730	70.52	3133	1730	0.2668	282.928	3.8403	0.97	999.85
5600.0	771.0	4.44316	11.06556	6.24046	312	57822	730	70.52	3133	1730	0.2668	282.927	2.9819	0.97	999.85
5650.0	721.0	4.41241	10.91005	6.09418	312	57822	744	69.51	3067	1639	0.2732	260.783	3.0086	0.97	1000.63
5701.0	670.0	4.38071	10.75131	5.94508	312	57822	759	68.47	2999	1548	0.2798	238.342	3.0358	0.98	1001.43
5701.0	670.0	3.99214	10.26622	5.57020	143	57822	362	64.03	2556	1239	0.2914	238.334	2.4000	0.37	1001.43
5736.0	635.0	3.98399	10.21203	5.54311	143	57822	362	63.32	2523	1224	0.2911	224.364	2.3868	0.37	1000.88
5771.0	600.0	3.97584	10.15782	5.51602	143	57822	362	62.61	2489	1210	0.2909	210.426	2.3734	0.37	1000.38
5771.0	600.0	3.97584	10.15782	5.51600	143	57822	362	62.61	2489	1210	0.2909	210.425	8.0910	1.98	1000.38
5821.0	550.0	3.91282	9.90185	5.37014	143	57822	363	59.60	2332	1128	0.2917	190.703	7.8833	1.92	999.65
5871.0	500.0	3.84980	9.64588	5.22428	143	57822	364	56.65	2181	1051	0.2924	171.311	7.6761	1.86	998.83
5921.0	450.0	3.78678	9.38990	5.07842	143	57822	365	53.78	2037	977	0.2933	152.251	7.4695	1.79	997.90
5971.0	400.0	3.72378	9.13397	4.93259	143	57822	366	50.99	1899	906	0.2942	133.527	7.2633	1.73	996.86
5971.0	400.0	3.54325	8.90522	4.76989	143	57822	372	48.97	1735	806	0.2988	133.520	3.3718	0.83	996.86
6016.0	355.0	3.51639	8.81867	4.73840	143	57822	370	47.83	1682	790	0.2971	117.702	3.3369	0.82	995.22
6061.0	310.0	3.48951	8.73209	4.70690	143	57822	367	46.71	1630	773	0.2952	102.027	3.3017	0.80	993.61

6106.0	265.0	3.46264	8.64552	4.67540	143	57822	365	45.60	1579	757	0.2933	86.497	3.2662	0.79	992.03
6151.0	220.0	3.43578	8.55896	4.64391	143	57822	362	44.50	1529	741	0.2914	71.115	3.2305	0.78	990.48
6151.0	220.0	3.35950	7.98970	4.41885	80	57822	195	37.80	1270	656	0.2797	71.108	−0.7364	−0.12	990.48
6186.0	185.0	3.36330	8.01180	4.43108	80	57822	195	38.01	1278	660	0.2797	59.466	−0.7200	−0.12	989.11
6221.0	150.0	3.36710	8.03370	4.44361	80	57822	195	38.21	1287	665	0.2796	47.824	−0.7035	−0.12	987.83
6256.0	115.0	3.37091	8.05540	4.45643	80	57822	195	38.41	1295	669	0.2795	36.183	−0.6868	−0.13	986.64
6291.0	80.0	3.37471	8.07688	4.46953	80	57822	195	38.60	1303	674	0.2793	24.546	−0.6700	−0.13	985.53
6291.0	80.0	3.37471	8.07689	4.46954	600	57822	1447	38.60	1303	674	0.2793	24.539	−0.6700	−0.13	985.53
6311.0	60.0	3.37688	8.08907	4.47715	600	57822	1447	38.71	1307	677	0.2792	17.891	−0.6603	−0.13	984.93
6331.0	40.0	3.37906	8.10119	4.48486	600	57822	1446	38.81	1311	680	0.2790	11.239	−0.6505	−0.13	984.37
6346.6	24.4	3.38076	8.11061	4.49094	600	57822	1446	38.89	1315	682	0.2789	6.043	−0.6428	−0.13	983.94
6346.6	24.4	2.90000	6.80000	3.90000	600	57822	1350	25.96	753	441	0.2549	6.040			983.94
6356.0	15.0	2.90000	6.80000	3.90000	600	57822	1350	25.96	753	441	0.2549	3.370			983.32
6356.0	15.0	2.60000	5.80000	3.20000	600	57822	1456	19.99	520	266	0.2812	3.364			983.31
6368.0	3.0	2.60000	5.80000	3.20000	600	57822	1456	19.99	520	266	0.2812	0.303			982.22
6368.0	3.0	1.02000	1.45000			57822	57822	2.10	21		0.5000	0.299			982.22
6371.0	0.	1.02000	1.45000			57822	57822	2.10	21		0.5000				981.56

Note: Above 220 km the mantle is transversely anisotropic; the parameters given here are "equivalent" isotropic moduli and velocities. See Table 3 for complete elastic constants in this region.

TABLE 3. Crust and Upper Mantle of PREM Including Directional Velocities, Anisotropic Elastic Constants, and "Equivalent" Isotropic Velocities (evaluated at a reference period of 1 s)

Radius m	Depth km	Density g·cm^{-3}	V_{PV} km·s^{-1}	V_{PH} km·s^{-1}	V_{SV} km·s^{-1}	V_{SH} km·s^{-1}	η	Q_μ	Q_k	A k bar	C kb	L kb	N kb	F kb	V_P km·s^{-1}	V_S km·s^{-1}
6151.0	220.0	3.35950	7.80050	8.04862	4.44110	4.43629	0.97654	80	57822	2176	2044	663	661	831	7.98970	4.41885
6171.0	200.0	3.36167	7.82315	8.06310	4.43649	4.45423	0.96877	80	57822	2186	2057	662	667	835	8.00235	4.42580
6191.0	180.0	3.36384	7.84581	8.07760	4.43189	4.47218	0.96099	80	57822	2195	2071	661	673	839	8.01494	4.43285
6211.0	160.0	3.36602	7.86847	8.09209	4.42728	4.49013	0.95321	80	57822	2204	2084	660	679	843	8.02747	4.44000
6231.0	140.0	3.36819	7.89113	8.10659	4.42267	4.50807	0.94543	80	57822	2213	2097	659	685	847	8.03992	4.44724
6251.0	120.0	3.37036	7.91378	8.12108	4.41806	4.52602	0.93765	80	57822	2223	2111	658	690	851	8.05231	4.45458
6271.0	100.0	3.37254	7.93644	8.13558	4.41345	4.54397	0.92987	80	57822	2232	2124	657	696	854	8.06463	4.46201
6291.0	80.0	3.37471	7.95909	8.15006	4.40885	4.56191	0.92210	80	57822	2242	2138	656	702	857	8.07688	4.46953
6291.0	80.0	3.37471	7.95911	8.15008	4.40884	4.56193	0.92209	600	57822	2242	2138	656	702	857	8.07689	4.46954
6311.0	60.0	3.37688	7.98176	8.16457	4.40424	4.57987	0.91432	600	57822	2251	2151	655	708	860	8.08907	4.47715
6331.0	40.0	3.37906	8.00442	8.17906	4.39963	4.59782	0.90654	600	57822	2260	2165	654	714	863	8.10119	4.48486
6346.6	24.4	3.38076	8.02212	8.19038	4.39603	4.61184	0.90047	600	57822	2268	2176	653	719	866	8.11061	4.49094
6346.6	24.4	2.90000	6.80000	6.80000	3.90000	3.90000	1.00000	600	57822	1341	1341	441	441	459	6.80000	3.90000
6356.0	15.0	2.90000	6.80000	6.80000	3.90000	3.90000	1.00000	600	57822	1341	1341	441	441	459	6.80000	3.90000
6356.0	15.0	2.60000	5.80000	5.80000	3.20000	3.20000	1.00000	600	57822	875	875	266	266	342	5.80000	3.20000
6368.0	3.0	2.60000	5.80000	5.80000	3.20000	3.20000	1.00000	600	57822	875	875	266	266	342	5.80000	3.20000
6368.0	3.0	1.02000	1.45000	1.45000	0.	0.	1.00000	0	57822	21	21	0	0	21	1.45000	0.
6371.0	0.	1.02000	1.45000	1.45000	0.	0.	1.00000	0	57822	21	21	0	0	21	1.45000	0.

Note: To obtain values at other periods use Eqs. 3 and 4.

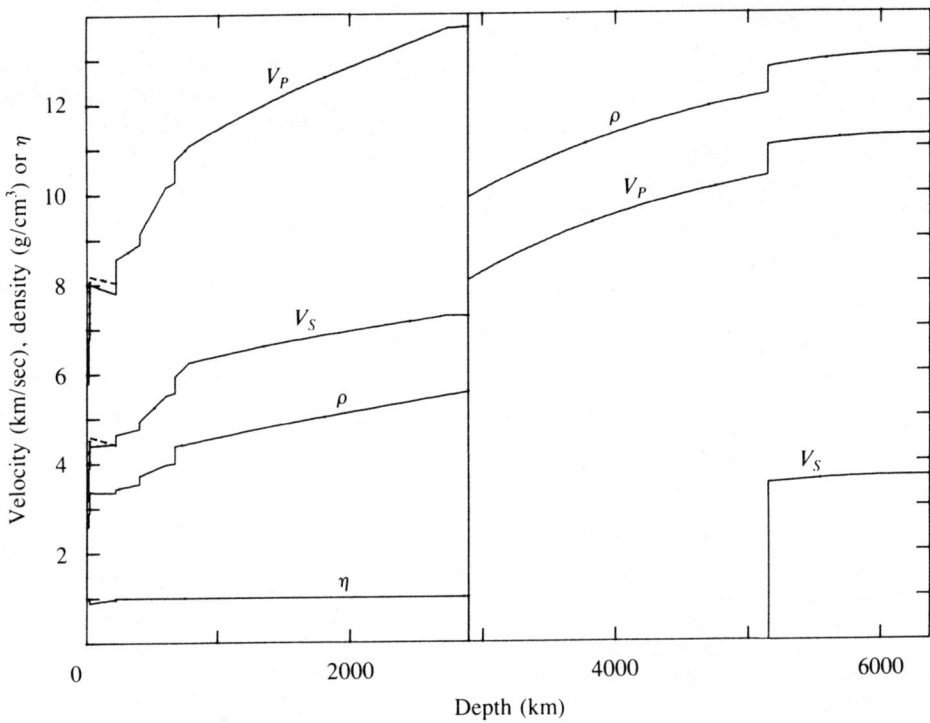

FIGURE 4. The PREM model. Dashed lines are the horizontal components of velocity (V_{PH} and V_{SH}). When η is 1, the model is isotropic. The core is isotropic. (From Dziewonski and Anderson, 1981)

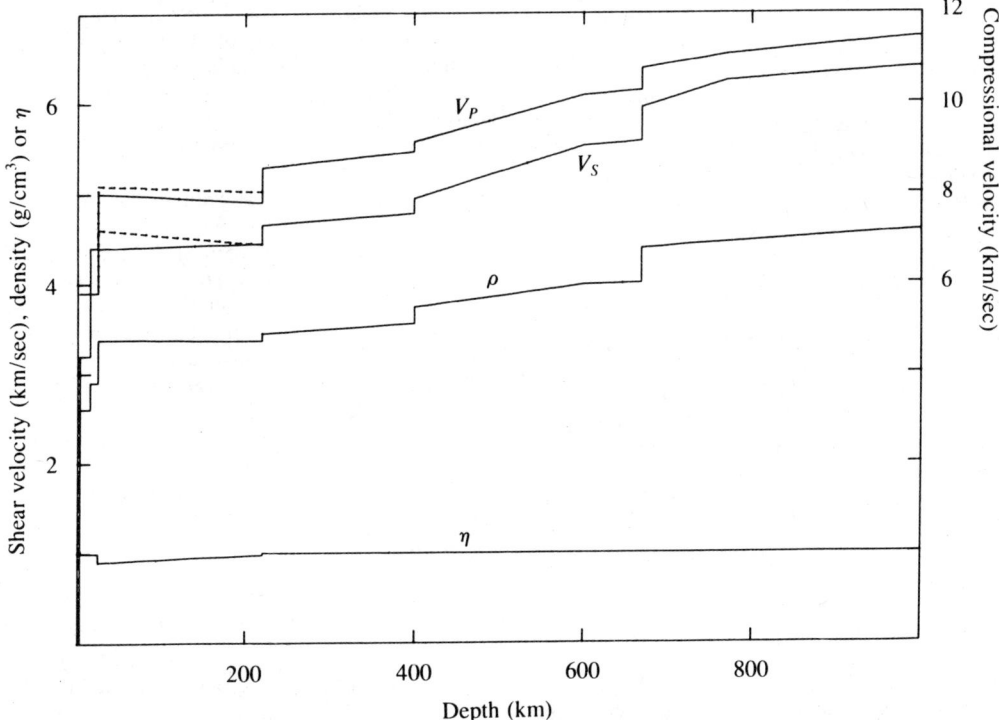

FIGURE 5. Upper-mantle velocities, density, and anisotropic parameter η in PREM. The dashed lines are the horizontal components of velocity. The solid curves are η, ρ and the vertical, or radial, components of velocity. (After Dziewonski and Anderson, 1984)

pressure

$$P(r) = \int_r^a g\rho \, dx$$

Bullen parameter η_B, which represents a measure of deviation of a model from the conditions of adiabatic compression of a homogeneous material

$$\eta_B = \frac{dK}{dP} + \frac{1}{g}\frac{d\Phi}{dr}$$

Earth in Three Dimensions

If the internal properties of the Earth were spherically symmetric, our planet would be tectonically stagnant. Both short time scale (earthquakes, volcanoes) and long time scale (mountain building, seafloor spreading) observations indicate that this is not the case. Furthermore, the existence of a geodynamo, with its implications for vigorous convection in the liquid core and significant flux of heat into the lower mantle, indicates that convection currents are present at all depths within the Earth. This convective system must be driven by lateral differences in temperature and density. Unfortunately, the distribution of these parameters cannot be unambiguously inferred from observations at the surface.

Seismic velocities depend on temperature and composition and, therefore, density. Variations in the wave speed can be resolved with seismic methods. Regional studies, addressing relatively shallow structures whose tectonic nature is understood, demonstrate that the hypothesis of linking high seismic velocities with low temperatures, and vice versa, is justified. High seismic velocities have been found under continental shields, older than 1 billion years with very low heat flow, while the material in the vicinity of mid-oceanic ridges has very low velocities at the same depths. The extension of this hypothesis to depths greater than several hundred kilometers is confirmed by, among others, a significant correlation between the velocity anomalies and the very large wavelength features of the gravity field. There are, therefore, very good reasons to believe that three-dimensional maps of seismological properties of the Earth provide data that can answer some of the fundamental questions of geodynamics.

No attempt to present a complete review of the field can be made here. The subject is rapidly developing, and some important references are still available only as abstracts of oral presentations at meetings. A comprehensive review of the progress in modeling the Earth in three dimensions can be found in Dziewonski and Woodhouse (1987); other reviews are Anderson and Dziewonski (1984) and Dziewonski and Anderson (1984).

Inverse Problem in Three Dimensions. Equation 1 can be modified such that the integral is with respect to the running arc distance ξ with $r = r(\xi)$:

$$\delta T = \int_0^\Delta G_0(\xi)\delta v[r(\xi)] \, d\xi$$

where Δ is the epicentral distance and the differential kernel $G_0(\xi)$ is

$$G_0(\xi) = \eta(\xi)v^{-2}(\xi)[\eta^2(\xi) - p^2]^{-1/2}$$

If δv is a function of three dimensions and not only of the radius, then the travel time anomaly depends on the angular distance between the epicenter (e) and seismograph (s) and on their vector coordinates $\mathbf{x}_e = (r_e, \theta_e, \phi_e)$ and $\mathbf{x}_s = (r_s, \theta_s, \phi_s)$. The appropriate expression for the travel time anomaly is

$$\delta T(\mathbf{x}_e, \mathbf{x}_s) = \int_{\mathbf{x}_e}^{\mathbf{x}_s} \delta v(\mathbf{x})G_0(\xi) \, d\xi$$

where $\mathbf{x} = \mathbf{x}(\mathbf{x}_e, \mathbf{x}_s, \xi)$. This equation, with certain implicit approximations, is linear in $\delta v(\mathbf{x})$. By analogy with the procedure described in the previous section, the inverse problem can be solved given an appropriate parameterization of δv in three dimensions and an adequate distribution of the ray paths for which δT is measured. Such approach to three-dimensional modeling is sometimes called *seismic tomography*, because of similarities with the procedures used in medicine (see *Seismic Tomography*). Unfortunately, when working on the global scale, seismologists have little control over the distribution of sources and sensors.

There are two conceptually different approaches to discretization. In the first, the model is discretized in the space domain (division of the Earth into an array of cells). The second approach involves discretization in the wave-number domain (equivalent to determination of the harmonic expansion coefficients). Each has advantages and disadvantages. For a review of different aspects of seismic tomography, see Nolet (1987) and *Seismic Tomography*. The wave-number domain expansion is consistent with the approach used to derive PREM, and further examples will be given using this representation.

The three-dimensional equivalent of Eq. 2 is

$$\delta v(r, \theta, \phi) = \sum_{i=0}^N a_i g_i(r, \theta, \phi)$$

where a_i are coefficients to be determined and g_i are known basis functions. For spherical geometry, natural choices of basis functions are spherical harmonics and orthogonal polynomials in radius:

$$\delta v(r, \theta, \phi) = \sum_{k=0}^{K} \sum_{l=0}^{L} \sum_{m=0}^{l} f_k(r)(A_{klm} \cos m\phi$$
$$+ B_{klm} \sin m\phi)p_{lm}(\cos \theta) \quad (5)$$

where $f_k(r)$ are Legendre polynomials, for example, and $p_{lm}(\cos \theta)$ are associated Legendre polynomials. The inverse problem requires determining the coefficients A_{klm} and B_{klm}, with the cutoff numbers K and L determined by the distribution of data and their resolving properties.

Data and Models. A data set used as early as the mid-1970s to obtain one of the first three-dimensional models of the Earth's interior consisted of travel time anomalies derived from the *Bulletins of the International Seismological Centre* (BISC). The BISC contain arrival time data reported since 1964 by several thousand globally distributed stations, although as much as 90% of the readings come from only about 10% of the stations, most of which are in Europe. This data set may be confusing, because of the low signal-to-noise ratio. Even when gross mistakes are rejected, the contribution to the variance of a single observation of reading errors and small-scale lateral heterogeneities is ten times as great as that due to heterogeneities with wavelength greater than several hundred kilometers. Averaging is one way to demonstrate that the effect of large-scale lateral heterogeneities is discernible. Figure 6 is a map of summary residuals of P-wave travel times. Travel time anomalies associated with 70 earthquakes that occurred within ±2.5° of 57.5°S and 27.5°W (Sandwich Islands region), representing a *source region*, were averaged in *receiver regions* (3° in epicentral distance, 7.2° in azimuth). Only points with at least five observations are plotted. Significant areal consistency is present in the pattern

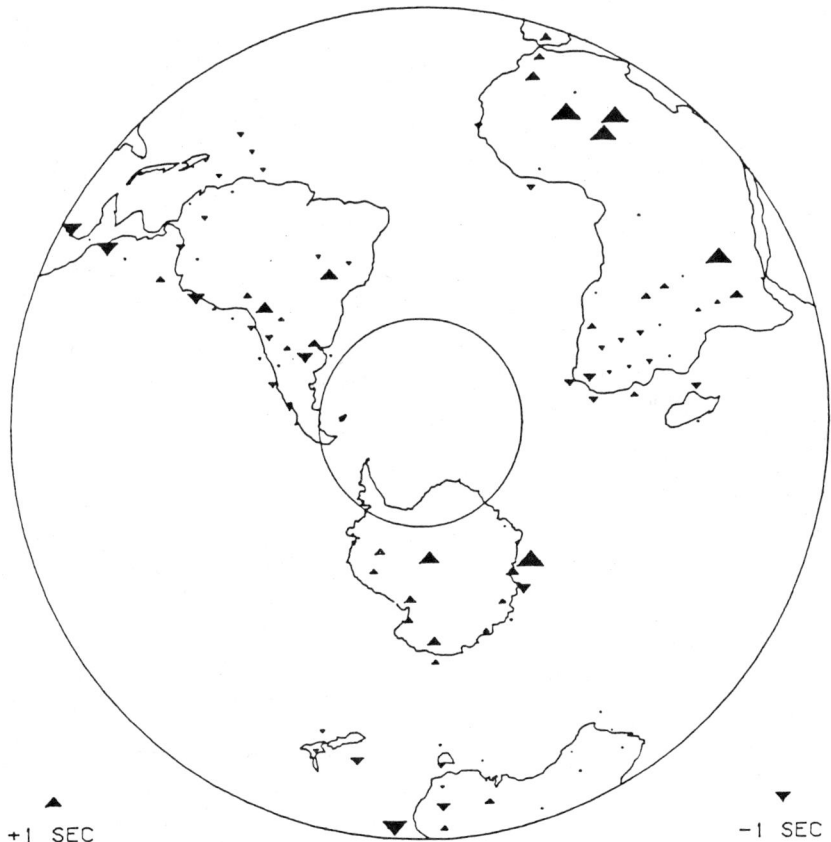

FIGURE 6. Summary residuals for a source region centered on 57.5°S and 27.5°W. Upward-pointing triangles represent positive residuals, indicating slower than average propagation; downward-pointing triangles correspond to paths faster than normal. The size of the symbol is proportional to the size of the anomaly. Note very slow arrivals in northern and western Africa; these arrivals—as it is found following an inversion using data from many other source regions—are caused by a very low velocity anomaly near core-mantle boundary under southern Africa. (From Dziewonski, 1984)

observed in Africa and Antarctica. The introduction of summary residuals also allows us to reduce the potential bias due to the effect of the uneven distribution of earthquakes and seismographic stations.

Dziewonski (1984) used BISC data for 1964–1979 for selected earthquakes. He was able to obtain summary data such as shown in Fig. 6 for about 600 source regions. He derived a model of the lower mantle, designated L02.56; the limits on the radial function expansion, as in Eq. 5, were $K = 4$ and, for the highest order of spherical harmonic used, $L = 6$. This model is listed in Table 4, and Fig. 7 shows the pattern of the P-velocity anomalies at a depth of 2500 km.

Legendre polynomials in reduced, normalized radius were chosen to represent the functions $f_k(r)$. For the lower mantle a function $u(r)$ is defined so that it spans the interval $[-1, 1]$:

$$u(r) = \frac{2r - (r_{670} + r_{CMB})}{r_{670} - r_{CMB}} \quad (6)$$

where CMB is the core-mantle boundary.

Functions $f_k(r)$ are normalized such that

$$\int_{-1}^{1} f_k^2(u)\, du = 1.$$

The specific representations of the first five functions f_k are

$$f_0 = (\tfrac{1}{2})^{1/2}$$

$$f_1 = (\tfrac{3}{2})^{1/2} u$$

$$f_2 = (\tfrac{5}{8})^{1/2}(3u^2 - 1)$$

$$f_3 = (\tfrac{7}{8})^{1/2}(5u^3 - 3u)$$

$$f_4 = (\tfrac{9}{128})^{1/2}(35u^4 - 30u^2 + 3) \quad (7)$$

The normalized associated Legendre polynomial is defined as

$$p_{lm} = \left[(2 - \delta_{m,0})(2l + 1)\frac{(l - m)!}{(l + m)!}\right]^{1/2} P_{lm} \quad (8)$$

where, with $\mu = \cos\theta$,

$$P_{lm} = \frac{\sin^m \theta}{2^l l!} \frac{d^{l+m}}{d\mu^{l+m}}(\mu^2 - 1)^l \quad (9)$$

The other kinds of data that led to early progress in seismic tomography are dispersed long-period surface waves (mantle waves). When excited by a major earthquake, these surface waves, either Rayleigh or Love type, travel on multiple paths around the globe. Figure 8 compares examples of synthetic seismograms against the corresponding observed ones. The synthetic traces are obtained by summing the contribution of all (about 600) normal modes with periods longer than 135 s (the observed traces have been appropriately low-pass filtered). Such calculation requires knowledge of the earthquake mechanism, also obtained by the analysis of digital data (see *Earthquake Mechanisms*). The upper two pairs of traces in Fig. 8 compare the observed seismograms (the top trace in each pair) with the synthetic ones calculated theoretically for model PREM; there are substantial discrepancies between the observed and synthetic traces. These are especially clear for the wave groups dominated by the fundamental modes R, for Rayleigh waves, and G, for Love waves. The discrepancy is particularly acute for the Love waves: there is a 180° difference in phase between the observed and synthetic arrivals identified as G_4.

The synthetic seismogram is sensitive to the parameters of the Earth's structure expressed, for example, by Eq. 3. Woodhouse and Dziewonski (1984) have shown that it is possible to construct differential seismograms, each corresponding to perturbation of one of the parameters describing an average structure along the minor or major arc. Then, by using a large set of waveforms, for many earthquakes and many stations, the data are inverted for a laterally heterogeneous model. Several iterations of the inversion are necessary, because of the nonlinearity of the problem.

Table 5 lists the parameters of model M84C of Woodhouse and Dziewonski (1984). There are several differences with respect to the representation of the lower-mantle model in Table 4:

1. The parameter perturbed with respect to PREM is δv_S^2.
2. The reduced, normalized radius is defined between the Moho (depth of 24.4 km) and 670 km discontinuity: $u(r) = [2r - (r_{Moho} + r_{670})]/(r_{Moho} - r_{670})$.
3. The limits on expansion are $K = 3$ and $L = 8$.
4. The model is derived after the waveform data are corrected for the effect of varying crustal thickness. The crustal thickness, in kilometers, is evaluated from the expression

$$h(\theta, \phi) = \sum_{l=0}^{8} \sum_{m=0}^{l} (A_{lm} \cos m\phi + B_{lm} \sin m\phi) P_{lm}(\cos\theta) \quad (10)$$

where the coefficients A_{lm} and B_{lm} are listed in the first pair of columns of Table 6.

Figures 9 and 10 show relative perturbations in shear velocity at a depth of 150 km and 550 km. The

TABLE 4. Model L02.56 of Dziewonski (1984)

		k=0 A VAL	k=0 A SD	k=0 B VAL	k=0 B SD	k=1 A VAL	k=1 A SD	k=1 B VAL	k=1 B SD	k=2 A VAL	k=2 A SD	k=2 B VAL	k=2 B SD	k=3 A VAL	k=3 A SD	k=3 B VAL	k=3 B SD	k=4 A VAL	k=4 A SD	k=4 B VAL	k=4 B SD
$l=0$	$m=0$	−4.70	0.61							−2.54	1.42			8.31	1.63			−5.15	1.69		
$l=1$	$m=0$	4.15	0.58			5.25	1.06			−0.15	1.22			−5.85	1.38			0.70	1.47		
	$m=1$	−0.74	0.73	0.25	0.59	−5.78	0.93	−0.95	0.97	2.51	1.61	2.68	1.34	−2.04	1.87	1.18	1.53	−0.21	1.90	0.32	1.65
$l=2$	$m=0$	4.35	0.63			−2.87	0.91			3.56	1.24			1.24	1.44			−1.12	1.53		
	$m=1$	−2.06	0.76	1.09	0.65	1.27	1.10	−3.21	0.93	−1.69	1.42	0.52	1.24	0.52	1.58	−0.46	1.44	1.49	1.65	−1.58	1.56
	$m=2$	−7.52	0.73	−2.70	0.74	8.82	1.12	3.80	1.10	−0.81	1.53	−3.51	1.49	4.44	1.73	2.02	1.74	−0.11	1.80	1.13	1.82
$l=3$	$m=0$	−5.04	0.76			7.51	0.96			−1.18	1.24			2.95	1.43			2.32	1.53		
	$m=1$	−4.10	0.76	−0.41	0.72	−0.27	0.93	2.13	0.88	0.57	1.27	−0.99	1.20	2.18	1.48	1.39	1.37	0.47	1.62	−0.15	1.48
	$m=2$	−2.06	0.83	5.11	0.82	−2.09	1.07	−4.02	1.04	0.15	1.38	−2.35	1.34	0.48	1.53	−2.62	1.55	1.27	1.63	−2.37	1.69
	$m=3$	0.08	0.79	0.23	0.78	0.62	1.12	3.44	1.02	1.22	1.55	2.27	1.38	−0.01	1.72	1.15	1.67	1.77	1.83	−4.40	1.77
$l=4$	$m=0$	−3.35	0.89			−0.30	1.02			−0.37	1.25			−0.21	1.40			−1.87	1.51		
	$m=1$	0.23	0.90	−0.11	0.84	1.17	1.00	−1.73	0.91	1.51	1.24	0.21	1.10	−0.66	1.46	1.59	1.27	0.51	1.56	1.46	1.40
	$m=2$	2.67	0.88	−2.06	0.91	0.19	0.98	4.57	1.03	−2.64	1.24	−1.87	1.27	−0.56	1.39	−0.32	1.45	−3.66	1.54	0.28	1.58
	$m=3$	0.69	0.93	−3.79	0.95	−0.42	1.10	−1.52	1.12	2.54	1.36	0.20	1.36	4.48	1.54	−1.10	1.54	2.82	1.65	0.19	1.68
	$m=4$	−0.47	0.86	1.41	0.83	1.06	1.10	−1.36	1.05	−0.14	1.48	2.93	1.38	−4.67	1.66	2.02	1.62	1.29	1.79	−1.33	1.75
$l=5$	$m=0$	0.03	0.94			−0.54	1.05			0.20	1.19			1.01	1.32			0.27	1.47		
	$m=1$	2.80	0.97	0.30	0.96	−4.05	1.09	1.08	1.05	1.60	1.22	−0.95	1.15	−2.73	1.39	1.47	1.25	−0.92	1.49	−3.56	1.39
	$m=2$	−1.80	0.95	−1.57	0.96	3.56	1.03	−1.26	1.02	−1.04	1.17	0.22	1.15	−0.16	1.33	−1.51	1.34	−1.17	1.47	−2.64	1.48
	$m=3$	0.78	0.96	4.02	1.02	−1.36	1.07	−2.85	1.15	3.29	1.24	2.35	1.29	−0.94	1.39	−0.54	1.41	−0.87	1.53	2.92	1.57
	$m=4$	−0.62	1.01	0.54	1.00	0.23	1.20	−0.28	1.19	−0.43	1.36	−2.58	1.32	3.14	1.51	−3.03	1.48	−1.80	1.63	1.07	1.61
	$m=5$	3.12	0.87	1.43	0.87	−1.82	1.07	−1.03	1.07	−0.59	1.32	−2.94	1.31	0.91	1.52	−0.58	1.52	1.63	1.66	−1.26	1.68
$l=6$	$m=0$	3.53	0.91			−0.90	1.00			0.50	0.99			−1.41	1.07			−0.97	1.24		
	$m=1$	−1.77	0.91	1.49	0.95	2.21	1.03	−1.14	1.05	0.63	1.07	0.21	1.00	2.03	1.17	−2.21	1.05	0.90	1.27	2.61	1.14
	$m=2$	−3.39	0.95	−0.52	0.92	−2.28	1.05	−1.77	1.03	2.84	1.07	−0.69	1.06	2.25	1.15	2.91	1.16	2.16	1.26	0.62	1.27
	$m=3$	2.64	0.92	−1.65	0.95	−1.22	1.03	4.43	1.07	−1.87	1.06	−0.02	1.09	−1.71	1.18	1.51	1.18	−1.94	1.29	0.59	1.29
	$m=4$	2.60	0.98	3.30	0.97	−1.17	1.15	−1.47	1.14	1.48	1.20	1.35	1.17	−2.35	1.27	0.76	1.26	0.62	1.37	−0.26	1.36
	$m=5$	1.83	0.98	2.30	0.97	2.50	1.17	−3.68	1.15	−0.74	1.23	5.08	1.20	0.54	1.35	0.03	1.33	0.13	1.46	2.40	1.43
	$m=6$	0.65	0.90	−0.58	0.89	0.05	1.09	2.07	1.08	−2.62	1.22	1.14	1.26	−0.91	1.39	−0.66	1.39	−0.73	1.53	0.01	1.53

Note: Coefficients of the spherical harmonic expansion of the lateral heterogeneities in P velocity in the lower mantle for $L = 6$ and $K = 4$. Units are in meters per second. In order to synthesize the value of velocity δv_P at any (r, θ, ϕ) between 670 km and CMB use Eqs. 5 to 9.

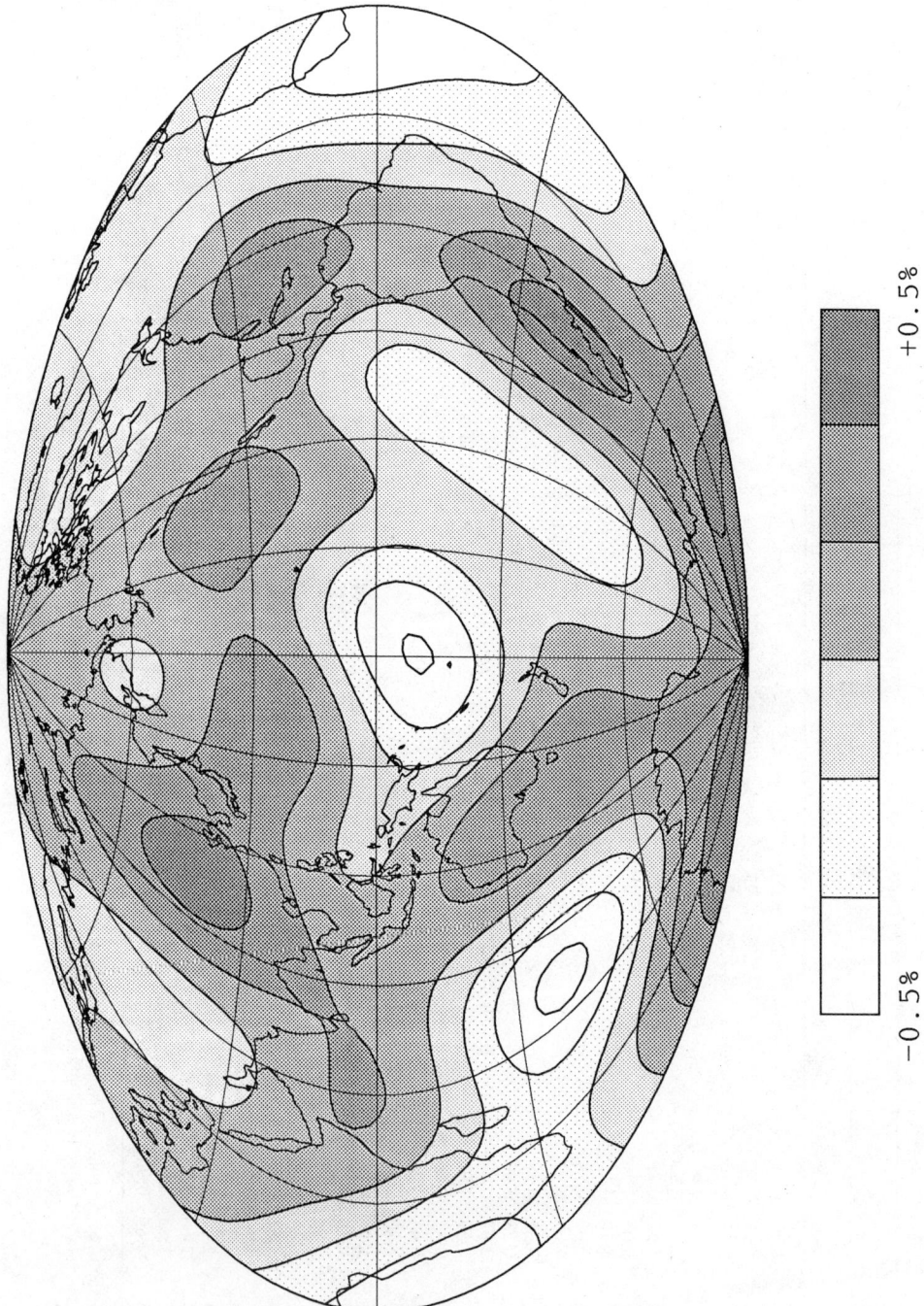

FIGURE 7. Map of P-velocity anomalies at a depth of 2500 km predicted by model *L02.56* listed in Table 4. Notice the ring of high velocities circumscribing the Pacific Basin and the slow region in its middle. This pattern dominates the velocity anomaly maps from a depth of about 1000 km to 2750 km: nearly entire lower mantle.

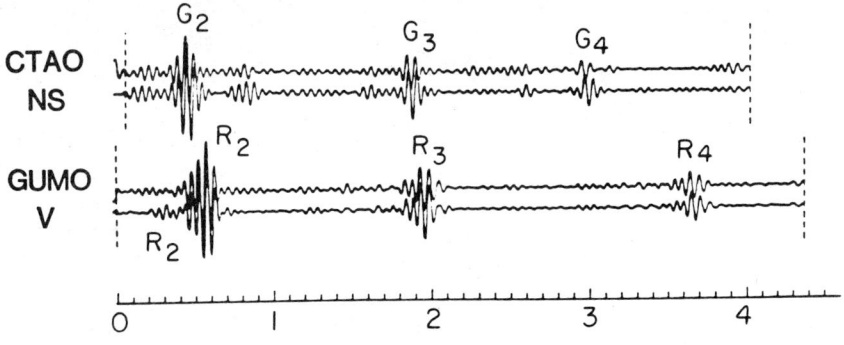

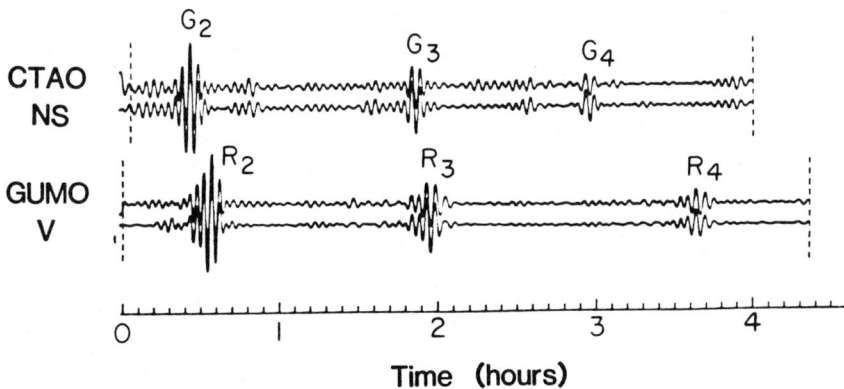

FIGURE 8. In each of these four pairs of seismograms the top trace is the observed seismogram of a large earthquake near Eureka, California, in 1980, and the bottom trace is a synthetic seismogram obtained by summing the contribution of about 600 individual normal modes. The recording of the transverse component of ground motion (Love waves) was made at Charters Towers, Australia (*CTAO-NS*); the recording of the vertical component (predominantly Rayleigh waves) was made on the island of Guam (*GUMO-V*). In the top two pairs of traces the synthetic seismograms computed for PREM show significant deviations from the actual recording, both for the Love waves (*G*) and Rayleigh waves (*R*); note the 180° difference in phase for G_4. These differences are greatly diminished by "correcting" the structure along the minor and major arcs of a particular great circle path, as shown by the bottom two traces. Derivation of a three-dimensional model of the Earth is accomplished by the analysis of such deviations for many criss-crossing paths. (After Woodhouse and Dziewonski, 1984)

former map remains in excellent correlation with the surface tectonic features: the regions with the slowest velocities correspond to mid-ocean ridges and back-arc basins, and the highest velocities are inferred for continental shields. The largest signal in the map for a depth of 550 km belongs to degree 2, as shown in earlier studies. There is little correspondence with the surface tectonics: One of the patches with high velocities is located under the South Atlantic.

Recent studies include extension of the waveform inversion to long-period body waves, such that the shear velocity in the lower mantle can now be also mapped. The existing models of δ_{SH} show great similarity in the pattern to those for P velocities, except that the ratio $(\delta v_S/v_S)/(\delta v_P/v_P)$ is about 2, which is significantly higher than 1.25, predicted by relatively low pressure laboratory measurements.

A similarly high ratio is also obtained for models derived from observations of splitting of normal modes. These models, which are derived only for the even degrees l in the spherical harmonic expansion, show high correlation with the appropriately filtered models obtained from the analysis of travel time residuals and waveform inversion. These three kinds of data span roughly three orders of magnitude in frequency and wavelength. The way in which they sample the Earth structure is entirely different. One must conclude, therefore, that the fundamental

TABLE 5. Model *M84C* (with Crustal Correction) of Woodhouse and Dziewonski (1984).

	Crust			$k = 0$				$k = 1$				$k = 2$				$k = 3$		
	A	B	A	RM	B	RM	A	RM	B	RM	A	RM	B	RM	A	RM	B	RM
$l = 0$ $m = 0$	0.000		0.383	0.99			−0.641	0.97			1.142	0.95			−1.804	0.92		
$l = 1$ $m = 0$	3.918		0.998	0.88			1.408	0.70			0.130	0.62			−0.611	0.46		
$m = 1$	3.040	1.833	1.903	0.89	0.661	0.90	0.882	0.71	0.888	0.73	−0.063	0.63	0.419	0.63	0.011	0.47	−0.022	0.49
$l = 2$ $m = 0$	3.164		0.045	0.95			0.907	0.78			−0.474	0.70			−0.960	0.50		
$m = 1$	1.347	1.699	−0.408	0.96	0.062	0.96	0.326	0.80	−0.189	0.80	0.016	0.71	−1.088	0.71	−0.253	0.52	−0.828	0.52
$m = 2$	−1.661	−1.003	0.567	0.95	−2.231	0.95	0.474	0.79	1.225	0.79	0.703	0.70	0.257	0.70	0.479	0.51	−0.984	0.52
$l = 3$ $m = 0$	−1.020		0.225	0.75			−0.291	0.54			−0.614	0.56			−0.289	0.13		
$m = 1$	−1.076	0.184	0.340	0.75	0.262	0.75	0.469	0.49	0.055	0.53	0.402	0.50	−0.187	0.56	0.141	0.09	−0.126	0.10
$m = 2$	−2.005	2.789	−0.377	0.75	0.700	0.75	−0.570	0.52	0.640	0.54	−0.295	0.53	0.216	0.57	0.024	0.10	0.008	0.09
$m = 3$	0.541	2.707	0.219	0.75	0.410	0.76	−0.981	0.51	0.552	0.53	−1.381	0.53	0.245	0.55	−0.465	0.10	−0.041	0.11
$l = 4$ $m = 0$	2.122		0.694	0.92			0.728	0.76			0.320	0.68			0.512	0.45		
$m = 1$	−1.028	−0.436	0.025	0.95	0.051	0.94	0.041	0.79	0.481	0.77	0.817	0.70	0.534	0.69	0.687	0.49	0.221	0.48
$m = 2$	−2.129	0.527	−0.783	0.95	0.515	0.95	−0.449	0.79	0.496	0.78	0.338	0.71	0.440	0.70	0.094	0.50	0.004	0.49
$m = 3$	1.553	−0.073	0.025	0.96	0.105	0.94	−0.751	0.80	0.071	0.78	−0.484	0.71	0.425	0.69	−0.086	0.51	0.300	0.48
$m = 4$	−0.614	3.357	−0.189	0.91	1.456	0.94	0.248	0.74	0.820	0.78	−0.128	0.66	0.480	0.69	−0.320	0.42	0.175	0.47
$l = 5$ $m = 0$	−2.731		−1.137	0.75			−1.425	0.51			−0.680	0.54			0.035	0.09		
$m = 1$	−0.100	−0.421	1.165	0.75	0.228	0.75	0.748	0.50	0.180	0.42	−0.241	0.48	−0.015	0.43	−0.358	0.11	−0.054	0.10
$m = 2$	−0.585	−0.864	−1.024	0.75	0.397	0.75	−1.105	0.51	−1.014	0.51	−0.480	0.52	−1.528	0.54	−0.009	0.09	−0.494	0.10
$m = 3$	0.352	0.722	−0.146	0.75	0.826	0.75	−0.208	0.47	0.883	0.45	−0.066	0.49	0.356	0.46	0.049	0.09	−0.007	0.09
$m = 4$	3.208	−0.896	1.481	0.75	−1.287	0.75	1.258	0.47	−0.936	0.45	0.238	0.48	0.012	0.46	−0.156	0.08	0.221	0.09
$m = 5$	0.074	1.112	0.231	0.74	0.484	0.74	−0.136	0.43	0.274	0.42	−0.281	0.40	−0.058	0.41	−0.074	0.07	−0.062	0.07

$l=6$	$m=0$	1.243		0.019	0.90	−0.535	0.93	0.465	0.74	−0.186	0.77	0.484	0.65	−0.072	0.67	0.088	0.37	−0.064	0.44
	$m=1$	−0.076	−0.317	−0.040	0.94	0.382	0.94	−0.082	0.78	0.730	0.78	0.081	0.69	1.075	0.67	0.123	0.47	0.506	0.46
	$m=2$	−0.376	−0.113	−0.346	0.93	0.952	0.94	−0.475	0.77	0.441	0.78	−0.089	0.68	0.185	0.70	0.032	0.45	0.116	0.45
	$m=3$	0.196	1.162	−0.001	0.95	0.207	0.94	0.256	0.80	0.012	0.78	0.258	0.69	−0.621	0.69	0.165	0.51	−0.287	0.47
	$m=4$	0.722	−0.541	−0.043	0.93	0.368	0.92	0.577	0.77	0.453	0.77	0.499	0.67	0.120	0.65	0.075	0.46	−0.036	0.42
	$m=5$	−0.288	−0.791	0.702	0.93	0.224	0.89	0.232	0.77	0.347	0.73	0.228	0.68	−0.018	0.65	0.223	0.44	−0.229	0.37
	$m=6$	0.233	0.426	0.044	0.91			−0.877	0.74			−1.041	0.67			−0.190	0.40		
$l=7$	$m=0$	−1.153		−0.428	0.74	0.670	0.74	−0.524	0.33	0.825	0.47	−0.240	0.27	0.379	0.49	0.013	0.06	−0.038	0.09
	$m=1$	−0.109	0.895	0.068	0.74	0.231	0.74	−0.241	0.36	0.890	0.34	−0.339	0.33	0.884	0.31	−0.112	0.07	0.254	0.07
	$m=2$	0.622	0.358	−0.106	0.74	0.964	0.74	0.229	0.30	0.239	0.36	0.333	0.26	−0.485	0.31	0.094	0.07	−0.279	0.07
	$m=3$	−0.518	0.425	0.174	0.74	−0.378	0.74	0.071	0.39	0.203	0.37	−0.024	0.36	0.554	0.35	−0.018	0.07	0.236	0.08
	$m=4$	−1.086	0.107	0.007	0.74	−0.284	0.74	0.050	0.39	−0.434	0.38	0.080	0.37	−0.302	0.36	0.042	0.07	−0.072	0.07
	$m=5$	0.434	−0.352	0.184	0.74	−0.400	0.74	−0.224	0.42	−0.906	0.40	−0.399	0.40	−0.743	0.39	−0.125	0.07	−0.159	0.07
	$m=6$	−0.207	−1.105	0.433	0.74	0.148	0.74	0.430	0.41	0.180	0.37	0.101	0.39	0.124	0.29	−0.062	0.07	0.042	0.05
	$m=7$	−0.306	−0.738	−0.171	0.74			−0.206	0.35			−0.105	0.28			0.018	0.06		
$l=8$	$m=0$	−0.723		−0.518	0.88	−0.203	0.88	−0.217	0.70	−0.353	0.72	−0.128	0.65	−0.260	0.63	−0.240	0.31	−0.069	0.32
	$m=1$	−0.075	−0.297	−0.034	0.93	0.201	0.92	0.062	0.77	0.262	0.76	0.229	0.67	0.306	0.66	0.072	0.43	0.190	0.42
	$m=2$	0.269	0.105	−0.066	0.92	0.203	0.93	−0.467	0.76	−0.165	0.77	−0.554	0.69	−0.221	0.68	−0.108	0.43	−0.020	0.44
	$m=3$	0.169	0.324	0.137	0.93	0.429	0.93	0.045	0.77	−0.102	0.78	0.362	0.68	−0.727	0.68	0.291	0.44	−0.391	0.46
	$m=4$	0.048	0.838	−0.106	0.94	−0.040	0.93	0.301	0.78	0.233	0.78	0.159	0.68	−0.020	0.68	−0.079	0.47	−0.259	0.43
	$m=5$	−0.409	−0.137	−0.224	0.95	0.521	0.93	−0.343	0.79	−0.499	0.77	−0.220	0.70	−0.073	0.67	−0.163	0.50	0.310	0.45
	$m=6$	0.566	0.566	−0.361	0.94	0.283	0.92	0.252	0.78	−0.326	0.76	0.319	0.67	−0.346	0.65	−0.093	0.46	0.047	0.40
	$m=7$	0.994	0.376	0.294	0.92	0.138	0.89	0.108	0.76	0.267	0.75	−0.282	0.67	0.356	0.63	−0.315	0.44	0.090	0.34
	$m=8$	−0.220	−0.389	0.074	0.87			0.269	0.70			−0.094	0.60			−0.221	0.30		

Note: Spherical harmonic coefficients of the perturbations in crustal thickness and the perturbation in squared shear velocity. Crustal thickness coefficients are in kilometers, see Eq. 10. The units of the coefficients of expansion of $\delta(v_S)^2$ are m$^2 \cdot$s$^{-2} \times 10^5$. In order to synthesize the value of $\delta(v_S)^2$ at any (r, θ, ϕ) between the Moho and 670-km depth, use Eqs. 5 to 9. The columns RM contain the diagonal elements of the resolution matrix corresponding to each parameter.

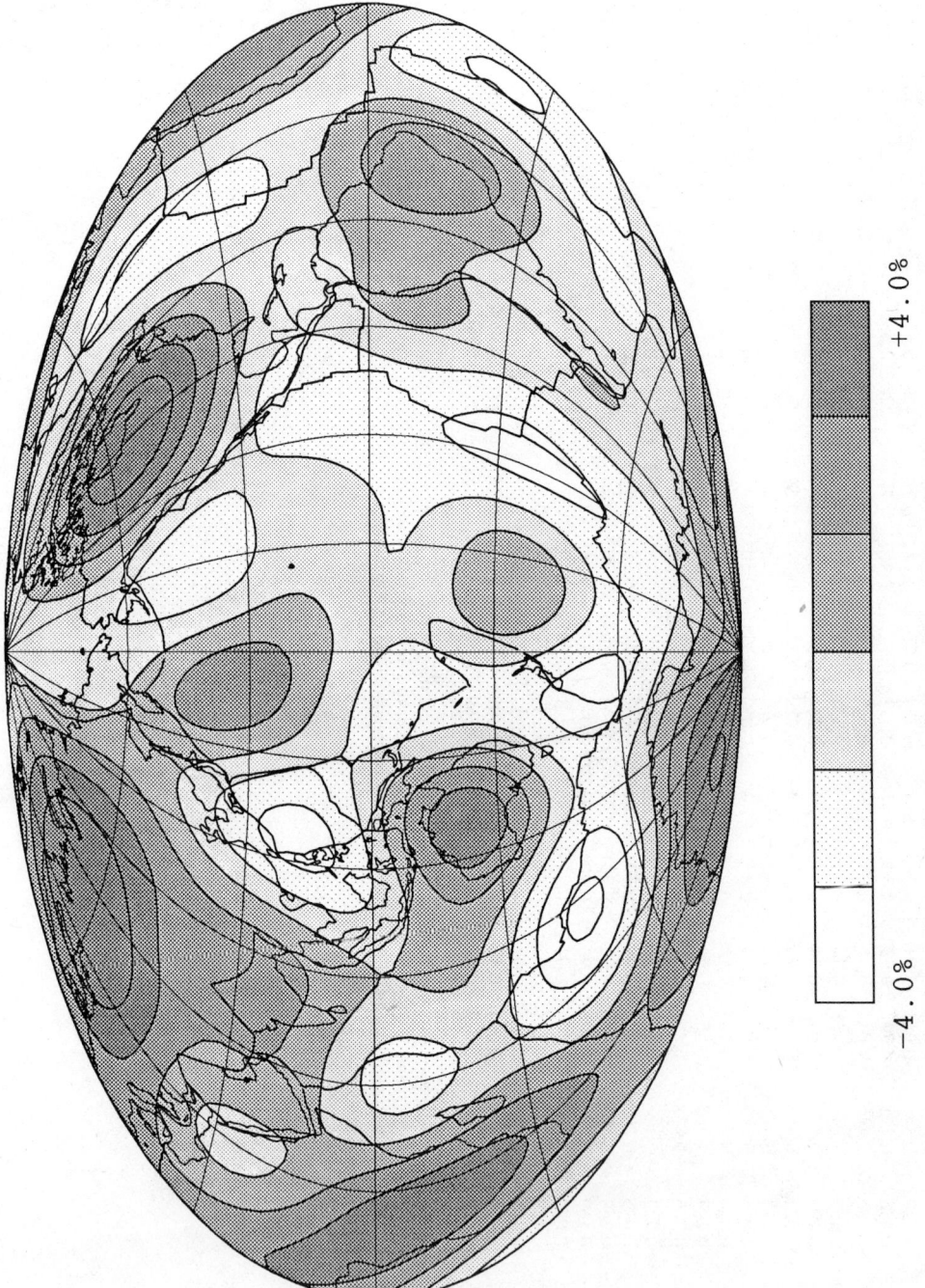

FIGURE 9. Relative perturbations in the shear velocity at a depth of 150 km predicted by model *M84C* listed in Table 5. Notice that high velocity occurs under the stable continental interiors and very old oceanic floor (western and southern Pacific), while the lowest velocities are under the mid-ocean ridges and back-arc basins. In the figure, plate boundaries are drawn with a heavy line to aid in identification of tectonic features, but the model was derived without *a priori* information on the age of the continents or oceanic lithosphere.

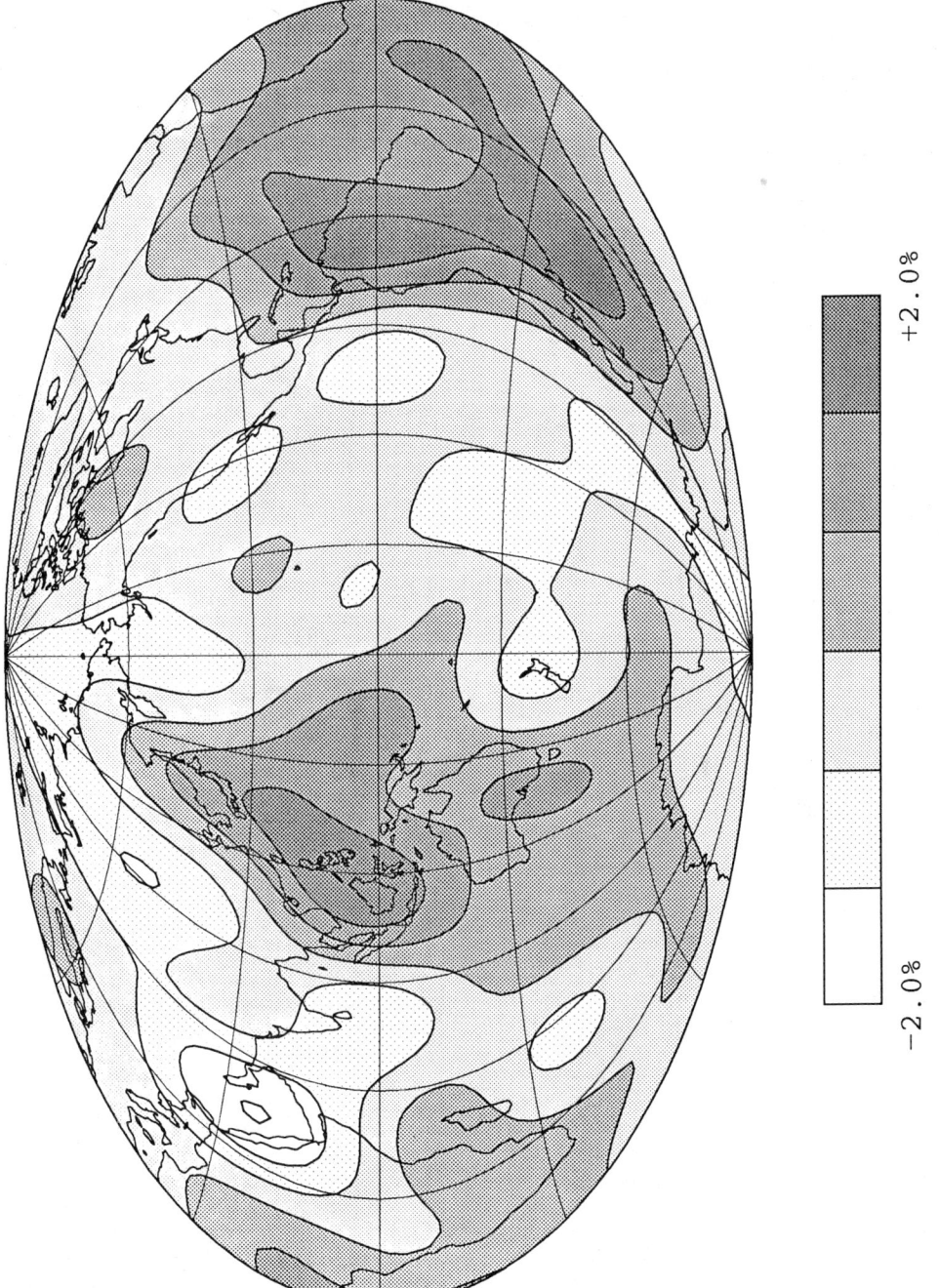

FIGURE 10. Same as Fig. 9 but for a depth of 550 km. At this depth there is little correlation with the surface tectonic features.

elements of the three-dimensional structure of the Earth have been resolved.

The shape of the core-mantle boundary is currently considered to be a controversial issue. Morelli and Dziewonski (1987) derived separate maps of the CMB topography by using the waves reflected from the CMB (*PcP*) and transmitted through it (*PKP*). They found significant correlation between the two sets of coefficients.

Because the sensitivity of travel times of *PcP* and *PKP* to perturbations in the core radius are of opposite sign, the inference is made that the resulting topography is not an artifact of systematic errors in source location or imperfect knowledge of 3-D mantle structure. Figure 11 shows CMB topography obtained by Morelli and Dziewonski (1987) from a combined data set of *PcP* and *PKP$_{BC}$* travel time residuals. The patterns obtained by Creager and Jordan (1986) are different.

The search for lateral heterogeneity extends now into the inner core. Rather surprisingly, the largest signal detected is consistent with the existence of transverse anisotropy. Unlike in PREM, where the plane of symmetry is always parallel to the Earth's surface, the axis of symmetry in the inner core is aligned with the axis of rotation and the plane of symmetry is parallel to the equatorial plane. For the waves traveling through the center of the Earth, the transit time for rays traveling along the rotation axis is about 1% faster than for the rays traveling in the equatorial plane. The evidence comes from the travel time anomalies of *PKIKP*—the P waves traveling through the inner core—and from splitting of the spectral peaks of free oscillations. The smoothed travel time anomalies of *PKIKP* observed in the range of epicentral distances between 170° and 180° are shown at the top of Fig. 12; the largest signal is the zonal harmonic of degree 2: P_2^0. The map of the splitting function for the mode $_{13}S_2$, one of the *PKIKP*-equivalent modes, determined by Giardini et al. (1987), is shown at the bottom of Fig. 12. The pattern is very similar.

In order to satisfy the travel time data and normal mode data, the bulk of the effect needs to be concentrated near the top of the inner core. The anisotropic perturbation varies as r^2 in the model of Morelli et al. (1986); at the surface of the inner core the velocity in the direction parallel to the rotation axis is 3.2% higher than in the equatorial plane. Shearer et al. (1988) use constant anisotropy throughout the inner core, and the size of the perturbation is only 1%. Woodhouse et al. (1986) need about 5% anisotropy near the top of the inner core to match the free oscillation data. It would seem that the radial dependence of anisotropy varying with radius as r^4 could satisfy both travel time and free oscillation data. An attempt has been made to explain the anisotropy in the inner core by degree-1 convection (Jeanloz and Wenk, 1988).

Significance of Three-Dimensional Earth Structure Studies. The study of lateral heterogeneity is of great significance to seismology. For example, in investigation of earthquakes the effect of lateral heterogeneity, like an imperfect lens, can distort the image of an event. The estimates of location, fault length, and the pattern of released stresses can be false if the medium is not adequately known. Even introduction of corrections for the low-order lateral heterogeneity can result in shifts of epicenters as much as 20 km and changes of the origin time in excess of 1 s.

But there is also an already felt impact of the recent discoveries in seismic tomography on other fields of Earth sciences. Some examples of such influence are

Mantle convection. The 3-D images of the seismic velocity anomalies, under the assumption that they are proportional to temperature perturbations, provide powerful constraints on the modeling of mantle convection. A degree-2 pattern dominates the lower mantle. A ring of high velocities circumscribes the Pacific basin, while the material under its center is slow (hot) from the core-mantle boundary to within 100 km of the surface.

Petrology and geochemistry. Mapping of thermal anomalies in the upper mantle helps to determine the depth of magma sources under the mid-ocean ridges and seems to confirm the hypothesis of "continental roots." About 80% of the hot spots occur over the area that corresponds to negative velocity anomalies near the core-mantle boundary. There also appears to be correlation between a band from 10°S to 30°S of low-velocity anomalies near the CMB and the occurrence of an isotopic anomaly (Dupal anomaly).

Geomagnetism. The regions at the CMB where the magnetic field changes with time most rapidly coincide with the low-velocity anomalies and, presumably, elevated temperatures in the lowermost mantle. The inference is made that the thermal state of the lowermost mantle, which can be mapped tomographically, determines the stability of convection patterns, and hence geomagnetic field, in the liquid core.

Gravity. The origin of the large-wavelength geoid anomalies cannot be determined uniquely from the gravity field alone. Their statistically significant correlation with the velocity anomalies identifies the lower mantle as the source region. Even though this correlation was first noted in the mid-1970s its sign was negative, because the calculation was made for a rigid Earth. Allowing for the flow and deformation of the boundaries, with the free surface and CMB being the most important because of the density contrast, leads to a change of sign of the correlation coefficient.

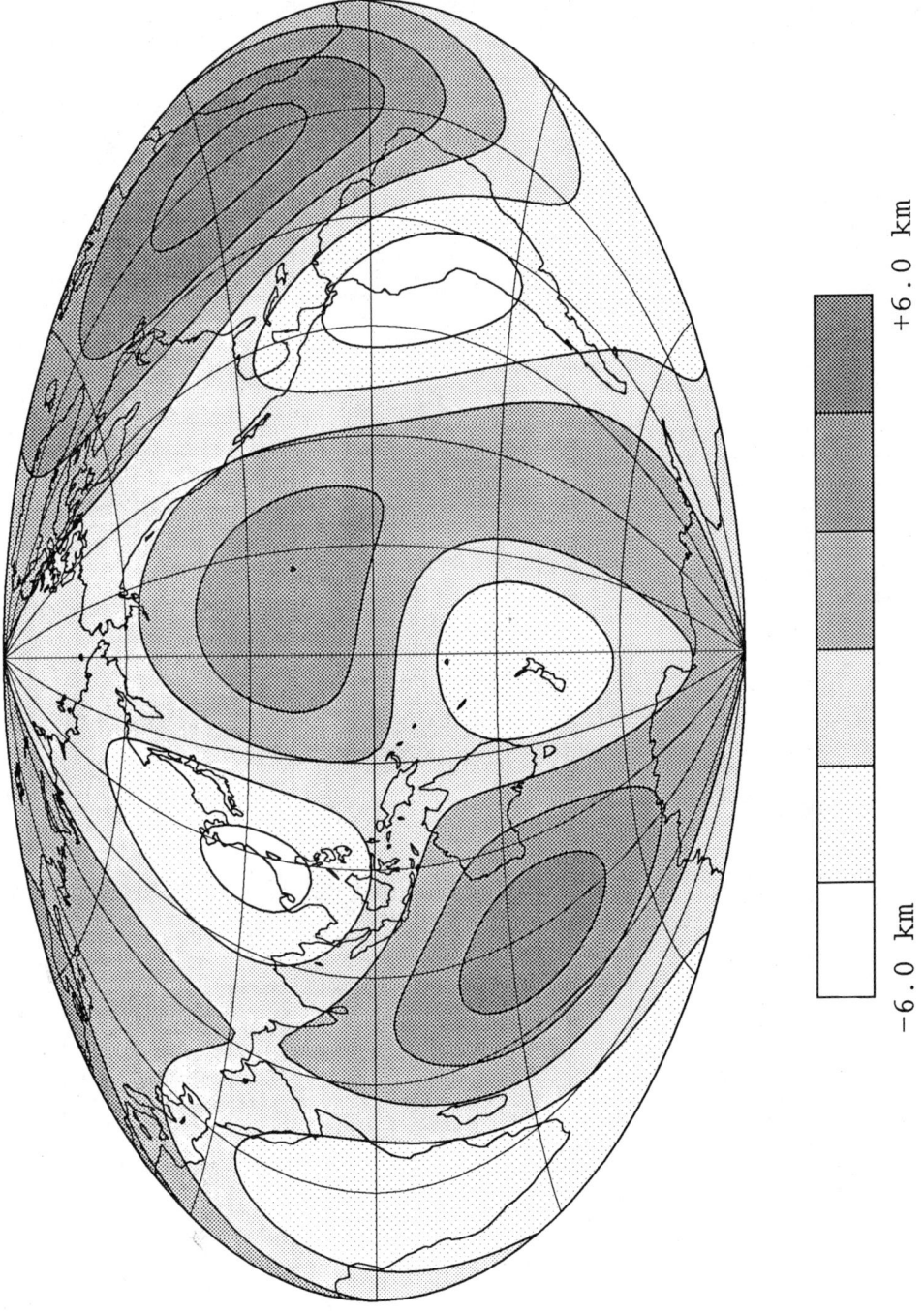

FIGURE 11. Topography of the core-mantle boundary obtained by Morelli and Dziewonski (1987) from the analysis of PcP and PKP_{BC} travel time anomalies. Notice that the CMB is depressed under most of the Pacific rim.

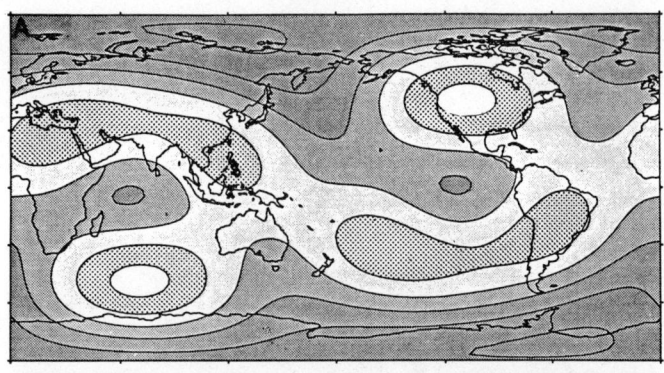

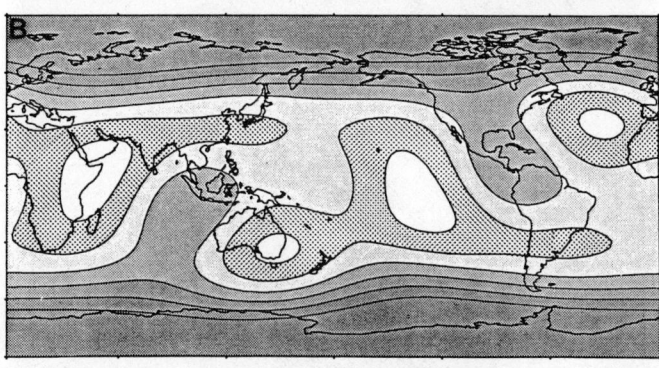

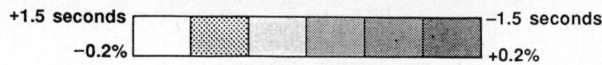

FIGURE 12. (*Top*) Travel time anomalies, corrected for lateral heterogeneity above the inner core, of *PKIKP* arrivals at nearly vertical incidence: distance range from 170° to 180°. (After Morelli *et al.*, 1986). (*Bottom*) The splitting function for the *PKIKP*-equivalent mode $_{13}S_2$ from Giardini et al. (1987). The splitting function is dominated by the axisymmetric terms P_2^0 and P_4^0, and the overall pattern is similar to the top part of the figure. (From Dziewonski and Woodhouse, 1987)

Mineral physics. An example of an in situ measurement made possible through seismic tomography is the ratio of perturbations in the shear and compressional velocities: $(\delta v_S/v_S)/(\delta v_P/v_P)$. The tomographic results yield a ratio much higher than determined, at relatively low pressures, in the laboratory. It appears that under the temperature and pressure conditions appropriate for the lower mantle, the shear modulus is much more sensitive to changes in temperature than the bulk modulus. Similar observations for the upper mantle were interpreted as being associated with partial melting.

Geodesy and astronomy. From the analysis of the data on Earth rotation obtained by the very-long baseline interferometry (VLBI) techniques, it was determined that the flattening of the shape of the Earth's core, the P_2^0 term, departs from its equilibrium value by (400 ± 100) m. With the reconstruction of the CMB shape using tomographic methods, these two results can be compared. While the CMB topography shown in Fig. 11 has a range of ±6 km, its P_2^0 term is very small.

The resolving length of seismic tomography is insufficient at the present time to draw many critical conclusions. The world seismological community is undertaking to remedy this by deployment of permanent as well as temporary networks equipped with state-of-the-art equipment.

ADAM M. DZIEWONSKI

References

Anderson, D. L., and A. M. Dziewonski, 1984, Seismic tomography, *Sci. Am.* **251**(4), 60–68.

Bullen, K. E., and B. A. Bolt, 1985, *An Introduction to the Theory of Seismology.* Cambridge: Cambridge University Press, 499 p.

Creager, K. C., and T. H. Jordan, 1986, Aspherical structure of the core-mantle boundary from *PKP* travel times, *Geophys. Research Letters* **13**, 1497–1500.

Dziewonski, A. M., 1984, Mapping the lower mantle: Determination of lateral heterogeneity in *P*-velocity up to degree and order 6, *Jour. Geophys. Research* **89**, 5929–5952.

Dziewonski, A. M., and D. L. Anderson, 1981, Preliminary Reference Earth Model (PREM), *Physics Earth Planetary Interiors* **25**, 297–356.

Dziewonski, A. M., and D. L. Anderson, 1984, Seismic tomography of the Earth's interior, *Am. Sci.* **72**, 483–494.

Dziewonski, A. M., and J. H. Woodhouse, 1987, Global images of the Earth's interior, *Science* **236**, 37–48.

Giardini, D., X.-D. Li, and J. H. Woodhouse, 1987, Three-dimensional structure of the Earth from splitting in free oscillation spectra, *Nature* **325**, 405–411.

Gilbert, F., and A. M. Dziewonski, 1975, An application of normal mode theory to the retrieval of structural parameters and source mechanisms from seismic spectra, *Royal Soc. London Philos. Trans.* **278 A,** 187-269.

Jeanloz, R., and H.-R. Wenk, 1988, Convection and anisotropy of the inner core, *Geophys. Research Letters* **15,** 72-75.

Jeffreys, H., and K. E. Bullen, 1939 *Seismological Tables.* London: British Association for Advancement of Science.

Lapwood, E. R., and T. Usami, 1981, *Free Oscillations of the Earth.* Cambridge: Cambridge University Press, 243 p.

Menke, W., 1984, *Geophysical Data Analysis: Discrete Inverse Theory.* New York: Academic Press.

Morelli, A., and A. M. Dziewonski, 1987, Topography of the core-mantle boundary and lateral homogeneity of the liquid core, *Nature* **325,** 678-683.

Morelli, A., A. M. Dziewonski, and J. H. Woodhouse, 1986, Anisotropy of the inner core inferred from *PKIKP* travel times, *Geophys. Research Letters* **13,** 1545-1548.

Nolet, G., ed., 1987, *Seismic Tomography with Applications in Global Seismology and Exploration Geophysics.* Dordrecht: D. Reidel, 386 p.

Shearer, P. M., K. M. Toy, and J. A. Orcutt, 1988, Axisymmetric Earth models and inner core anisotropy, *Nature* **333,** 228-232.

Woodhouse, J. H., and A. M. Dziewonski, 1984, Mapping the upper mantle: Three dimensional modelling of Earth structure by inversion of seismic waveforms, *Jour. Geophys. Research* **89,** 5953-5986.

Woodhouse, J. H., D. Giardini, and X.-D. Li, 1986, Evidence for inner core anisotropy from free oscillations, *Geophys. Research Letters* **13,** 1549-1552.

Cross-references: *Earth's Core: Structure; Free Oscillations of the Earth; Inverse Theory and Methods: Seismology; Mantle, Lower: Structure; Mantle, Upper: Structure; Mantle Convection and Plumes; Mantle Discontinuities; Mantle Dynamics; Mantle Viscosity; Seismic Anisotropy in the Earth; Seismic Tomography; Spherical Harmonic Analysis; Surface Waves.*

EARTH TIDES

The Earth tide is the deformation of the solid Earth caused by the gravitational attraction of the Sun and Moon. The most striking gravitational effects of the Sun and Moon are the orbital motion of the Earth's center of mass. The Moon causes the Earth to orbit the Earth-Moon center of mass, and the Sun causes that center of mass to orbit the Sun.

The Tidal Force

The Sun and Moon also cause points within the Earth to be displaced relative to each other. In other words, they cause the Earth to deform. For example, the side of the Earth nearest the Moon is attracted toward the Moon more than is the center of the Earth, and the side farthest from the Moon is attracted less than the center of the Earth. Consequently, both the far and near sides of the Earth are pulled radially outward away from the center. Similarly, those regions of the Earth at right angles to the Earth-Moon vector are pulled radially inward.

This deformation pattern is illustrated in Fig. 1. The total gravitational acceleration vectors due to the Moon are shown in Fig. 1a. The vectors all point toward the Moon, but are of unequal length and direction, although the differences are greatly exaggerated in the figure. The orbital acceleration is, to a high degree of accuracy, equal to the acceleration vector at the Earth's center of mass. Subtracting that vector from the other vectors in Fig. 1a results in the pattern of vectors shown in Fig. 1b. These residual vectors represent that portion of the lunar gravitational force tending to deform the Earth. The residual acceleration field multiplied by the local material density is defined as the *lunar tidal force,* and the deformation it induces is called the *lunar tide.*

Note from Fig. 1b that because of the Earth's diurnal rotation the tidal force at a fixed point in the Earth varies through two complete cycles in one day. This semidiurnal time dependence, through, is split into many periodic terms with frequencies closely spaced about two cycles per day, due to the time variability of the oribital motion of the Moon. Furthermore, because the Moon is not always in the plane of the Earth's equator, there is also significant variability at frequencies closely spaced about one cycle per day (imagine, for example, that the Earth-Moon vector is inclined at 45° to the Earth's rotation axis) and closely spaced about zero cycles per day (imagine that the moon is directly above the North Pole).

The solar tidal force and the solar tide are defined in a similar manner and can also be decomposed into semidiurnal, diurnal, and long-period terms.

For a quantitative description, it is useful to work with the tidal potential, denoted here as V_T and defined so that its gradient is the tidal force per unit mass. Define a nonrotating coordinate system with origin at the center of the Earth. The total lunar gravitational potential at the point **r** inside the Earth is (assuming the Moon to be a point mass)

$$V = \frac{GM}{|\mathbf{r} - \mathbf{R}|}$$

where M is the mass of the Moon and **R** is the position vector of the Moon. Let r, θ, λ be the radial distance, colatitude, and eastward longitude of the Earth-fixed point **r** so that $\lambda + \Omega t$ (where $\Omega = 1$ cycle/day is the Earth's angular velocity of rotation) is the azimuthal angle of **r** in nonrotating, inertial space. Then the factor $1/|\mathbf{r} - \mathbf{R}|$ can be expanded as a sum of complex spherical harmonics $Y_n^m(\theta, \lambda)$ so that

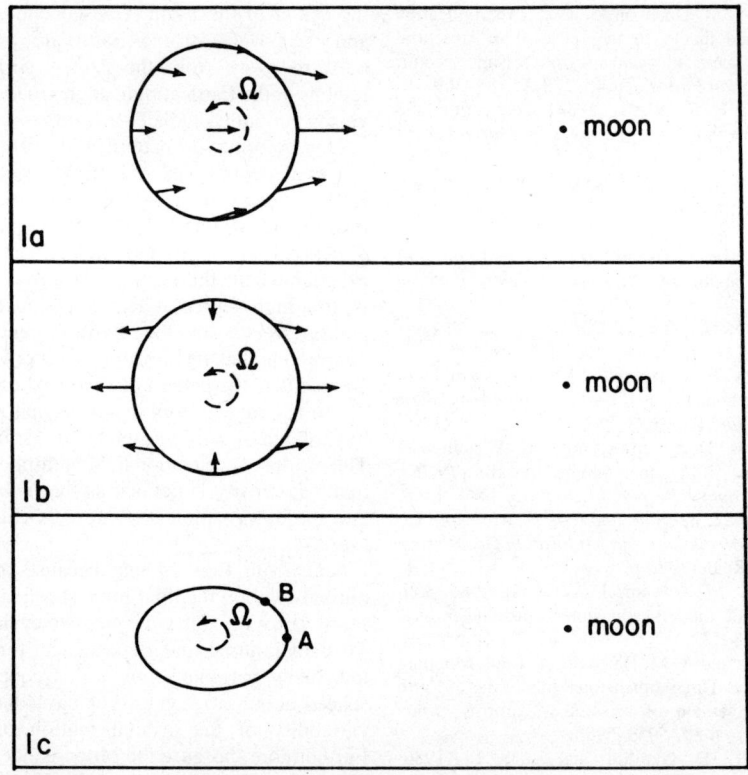

FIGURE 1. All three panels show the Earth as seen from above the North Pole. The Earth's diurnal rotation is represented by Ω. The arrows in (a) illustrate the magnitude and direction of the gravitational acceleration toward the moon. The differences in the lengths and directions of the arrows are greatly exaggerated. The lunar tidal acceleration is defined by subtracting the acceleration vector at the Earth's center of mass from the other acceleration vectors. The resulting vectors are shown in (b). The tidal force (the local density times the acceleration) deforms the Earth into the elliptical shape shown greatly exaggerated in (c). Points A and B are used in the text to illustrate tidal strain and tilt.

$$V = \mathrm{Re}\left[\frac{GM}{R}\sum_{n=0}^{\infty}\sum_{m=0}^{n}\left(\frac{r}{R}\right)^n d_n^m Y_n^m(\theta,\lambda) e^{im\Omega t}\right] \quad (1)$$

where Re denotes the real part (included in Eq. 1 so that the sum over m includes only $m \geq 0$), R is the radial coordinate of the moon, and the d_n^m are coefficients that depend on the angular position of the Moon. The coordinate R and the Moon's angular position depend on time, due to the Moon's orbital motion.

The $n = 0$ term in Eq. 1 is a spatial constant and so is physically meaningless. It can be ignored. The $n = 1$ term represents a spatially constant force. That force is equal to the lunar force at the Earth's center of mass. Consequently, the $n = 1$ term is removed from Eq. 1 when the tidal potential V_T is defined.

Since $r \leq a$, where a is the Earth's radius, and $a/R \cong 1/60$ for the Moon (and is much smaller for the Sun), the factor $(r/R)^n$ in Eq. 2 implies that the contributions to Eq. 1 fall off rapidly with increasing n. For most purposes, it is sufficient to keep only the $n = 2$ terms in V_T; thus V_T has the approximate form

$$V_T = \mathrm{Re}\left[\left(\frac{r}{a}\right)^2 \sum_{m=0}^{2} c_2^m Y_2^m(\theta,\lambda) e^{im\Omega t}\right] \quad (2)$$

where the $c_2^m\ [=GM/R(a/r)^2 d_2^m]$ are complex, time-varying coefficients that depend on the orbital coordinates of the Moon.

The dominant time dependence in Eq. 2 is provided by the $e^{im\Omega t}$ term. This term results in semidiurnal, diurnal, or long-period tides, depending on whether m is 2, 1, or 0, respectively. The time-dependent c_2^m in Eq. 2 split the $e^{im\Omega t}$ time dependence into terms with frequencies closely spaced about $m\Omega$.

The Earth's Response

The tidal force causes ocean tides as well as Earth tides. The observed ocean tide is actually the difference between the response of the ocean and the response of the solid Earth. Ocean tides are complicated and difficult to model for a number of reasons that need not concern us here. Open ocean tides are typically less than a meter in height, whereas ocean tides near coasts can be several meters peak to peak.

Earth tides are much easier to understand and model. The tidal force shown in Fig. 1b tends to deform the solid Earth into the elliptical shape shown greatly exaggerated in Fig. 1c. Tidal displacements of the solid Earth are typically several tens of centimeters.

But, unlike ocean tides, Earth tides cannot be observed without sensitive instruments. The reason is that both the ground and the observer are displaced by the same amount. One way to detect Earth tides is to use a gravimeter. There are three contributions to the observed tidal variation in the gravitational acceleration: (1) the direct attraction of the Sun and Moon; (2) the change in the Earth's gravity field due to tidal deformation within the Earth; (3) the change in the gravitational acceleration at the gravimeter caused by the radial tidal displacement of the Earth's surface beneath the gravimeter (if the surface moves outward, gravity decreases, and vice versa). All three contributions are roughly the same order.

The total tidal variation of the observed gravitational acceleration is about 100 μGal or more (a Gal is 1 cm/s^{-2}). The unperturbed gravitational acceleration at the Earth's surface is about 10^3 Gal. So, tides affect gravity at about the 10^{-7} level. This is roughly the relative importance of tides of any physical observable, and can be traced to the fact that tidal displacements (a few tens of centimeters) are about 10^{-7} of the Earth's radius (about 6×10^8 cm).

A simple gravimeter might be a mass hanging from a spring. The amount of spring is stretched at equilibrium is proportional to the local gravitational acceleration. Thus, by continually monitoring the length of the spring, we could determine tidal variations in gravity. (Although real gravimeters are more complicated than this, the basic idea is the same. Most instruments, in fact, do use springs.)

To illustrate the effects of tides on gravity, Fig. 2 shows one month of hourly gravitational acceleration data as observed with a spring-type instrument at Boulder, Colorado. The tides are clearly the dominant signal in the data. The semidiurnal and diurnal tides are especially evident. The longer-period (about two-week) oscillation evident in the figure is caused by beating between two diurnal frequencies, separated by one cycle per 13.7 days. Long-period gravity tides have small amplitudes at the latitude of Boulder and are hard to distinguish in Fig. 2.

One problem common to all relative gravimeters is calibration. For example, for spring-type instruments, the proportionality constant between the length of the spring and gravity depends on the spring constant, which can never be precisely known. This, for example, could introduce errors into the vertical scale in Fig. 2.

Another method of detecting Earth tides is to use a strainmeter, which is an instrument that continually monitors the distance between two points on the Earth's surface, separated by anywhere from a few meters to a kilometer. Today, the most accurate strainmeters are laser interferometers.

The fact that Earth tides perturb the distance between points can be seen by considering points A

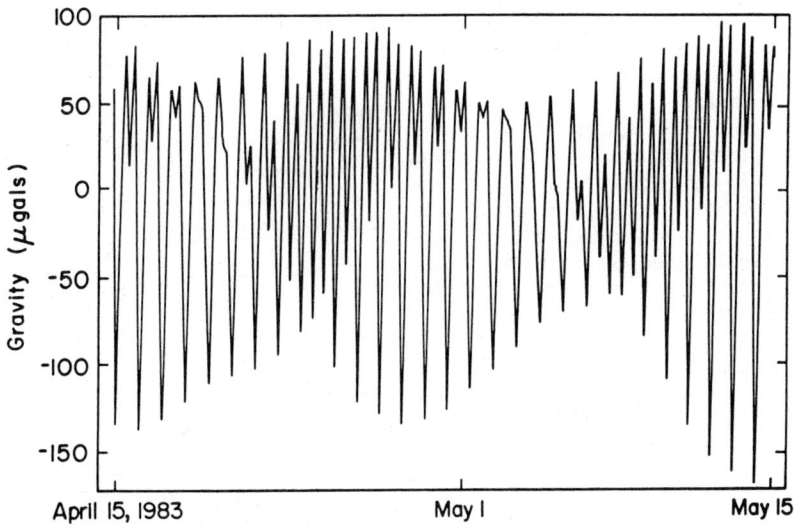

FIGURE 2. One month of hourly gravity data, taken with a spring-type gravimeter at Boulder, Colorado, U.S.A. The tides are the dominant signal and are described further in the text. (Data courtesy of Judah Levine)

and B in Fig. 1c. The distance between those points has changed during the tidal deformation, because A has moved radially outward more than B. Also, not evident in the figure, A and B are displaced horizontally by different amounts during the deformation. Typical tidal variations in the distance between two points are about 10^{-7} times the unperturbed distance.

A third way to detect tidal deformation is with tiltmeters. Tiltmeters measure the tilt of the Earth's surface with respect to the direction of gravity. There are various types of tiltmeters, ranging from pendulums to fluid-filled pipes with sensors at each end to measure fluid levels. Tidal tilt of the surface can be seen from Fig. 1c by drawing an imaginary line between A and B. The direction of gravity also tips during the deformation, but by a different amount. It is this difference that is observed. Typically, observed tidal tilts are a few tenths of a microradian.

Tidal deformation can now also be observed by space geodetic techniques, such as satellite laser ranging (SLR) and very long baseline interferometry (VLBI). These techniques are sensitive to displacements of the ground observing stations (lasers for SLR and radio antennas for VLBI). The primary geophysical objective of these techniques is to detect tectonic motion of the stations. The effects of Earth tides are evident in the data, but they cannot presently be determined with as much relative accuracy as can the tidal effects on surface gravity, strain, or tilt.

SLR can also detect tidal variations in the Earth's gravity field. Those variations affect the orbit of a satellite, which, in turn, affects the ranging data. In fact, the ranging data is more sensitive to these tidal gravity variations than to tidal displacements of the ground stations.

These various observable tidal effects can be conveniently described, mathematically, by using dimensionless parameters known as Love numbers, defined as follows (see any of the references listed at the end of the article).

The deformation of a spherically symmetric, nonrotating Earth in response to an applied external potential $Y_n^m(\theta, \lambda)$ can be described with that same $Y_n^m(\theta, \lambda)$ angular dependence. Mathematically, this is true because the Y_n^m separate spherically symmetric differential equations. The change in the Earth's gravitational potential at $r = a$ (the unperturbed surface of the Earth) and the radial displacement at $r = a$ are both proportional to Y_n^m, and the horizontal displacement vector at $r = a$ is proportional to ∇Y_n^m. The proportionality constants depend on n but are independent of m. Thus, the same constants are pertinent for all three Y_2^m terms in Eq. 2.

Let $\Phi(\theta, \lambda)$, $U_r(\theta, \lambda)$, $U_\theta(\theta, \lambda)$, and $U_\lambda(\theta, \lambda)$ denote tidal effects at the Earth's surface ($r = a$) on, respectively, the Earth's gravitational potential, the radial displacement of the point (θ, λ), the southward displacement, and the eastward displacement.

Then, for our assumed spherical Earth, Φ, U_r, U_θ, and U_λ have the forms

$$\Phi(\theta, \lambda) = k V_T^a(\theta, \lambda)$$

$$U_r(\theta, \lambda) = \frac{h}{g} V_T^a(\theta, \lambda)$$

$$U_\theta(\theta, \lambda) = \frac{l}{g} \partial_\theta V_T^a(\theta, \lambda)$$

$$U_\lambda(\theta, \lambda) = \frac{l}{g \sin \theta} \partial_\lambda V_T^a(\theta, \lambda) \qquad (3)$$

where k, h, and l are the dimensionless Love numbers, g is the unperturbed gravitational acceleration at $r = a$, and

$$V_T^a(\theta, \lambda) = \text{Re} \left[\sum_{m=0}^{\infty} c_2^m Y_2^m(\theta, \lambda) e^{im\Omega t} \right]$$

is the tidal potential (Eq. 2) evaluated at $r = a$.

Tidal variations in gravity, strain, and tilt can be parameterized in terms of k, h, and l. As one example, consider the gravitational acceleration at a fixed point on the Earth's surface. There are three contributions to this acceleration, as was discussed earlier. First, the direct gravitational acceleration from the Moon (or Sun) in the inward radial direction is $-\partial_r V_T(r = a) = -(2/a)V_T^a$ (using the r^2 radial dependence of V_T shown in Eq. 2). Second, Φ (the tidal change in the Earth's gravitational potential at $r = a$) has angular dependence Y_2^m and so, outside of the Earth, has radial dependence r^{-3}. Thus, the effect of Φ on the radially inward gravitational acceleration is

$$-\partial_r \Phi(r = a) = \frac{3}{a} \Phi = \frac{3}{a} k V_T^a$$

Third, in the absence of tides, the surface gravitational acceleration g varies with radius as r^{-2}. So, if the surface point is displaced radially by U_r, then the resulting perturbation in the gravitational acceleration at the displaced surface point is

$$U_r \partial_r g = -\frac{2}{a} g U_r = -\frac{2}{a} h V_T^a$$

Adding these three contributions together gives the total tidal effect on the observed acceleration as

$$\Delta g(\theta, \lambda) = -\delta \frac{2}{a} V_T^a(\theta, \lambda)$$

where $\delta = 1 - \frac{3}{2} k + h$ is called the gravimetric factor.

Similar exercises show that tidal tilt is described

by the diminishing factor $\gamma = 1 + k - h$, and that tidal strain, although more complicated than either tilt or gravity, depends on the Love numbers h and l.

The numerical values of the Love numbers are determined by the Earth's internal properties. There have been attempts to constrain those properties by comparing observed gravity, strain, and tilt with predictions based on theoretical results for the Love numbers. Those efforts have not been particularly successful, for several reasons.

One problem is instrument calibration, as was described earlier for gravimeters. Any error in the calibration maps directly into a frequency-independent, multiplicative error in the estimated Love numbers.

However, there are other effects, most of which are related to the fact that the Earth is not really spherically symmetric or nonrotating. Some of these effects are potentially useful. For example, the Earth does rotate, and because of that rotation the Earth's internal properties are closer to being elliptically, rather than spherically, symmetric. In this case, the results of Eq. 3 are still approximately valid, but the Love numbers k, h, and l are notably dependent on frequency near one cycle per day. This diurnal frequency dependence is particularly sensitive to the shape of the core-mantle boundary; hence, diurnal tidal results can be used to help constrain that shape.

Another omission in the theory described so far is the ocean. Tides in the ocean cause time-varying pressure loads on the surface of the solid Earth with the same frequencies as the Earth tides. These loads cause the Earth to deform, and this *load tide* (the tidal deformation that would occur in the absence of oceans is called the *body tide*) affects all Earth tide observations to some extent. Tidal gravity observations, for example, can be perturbed by up to 10% near coasts and, typically, by a few percent in the interior of continents. Tidal tilt and strain can be perturbed by several hundred percent near coasts.

It is often difficult to model and remove the load tide well enough to use the remaining body tide to learn about the Earth's deep interior. For example, there are uncertainties in ocean tide models. Furthermore, the Earth's response to the ocean tides, particularly the contributions to tilt and strain near coasts, can be sensitive to the local material properties, which may or may not be adequately known. On the other hand, because of these possible uncertainties, people have sometimes been able to use tidal observations to help constrain the nearby ocean tide or the underlying material properties.

For tilt and strain, local effects can be important on both the load tide and the body tide. By comparing observed tidal amplitudes from instruments in an array of tiltmeters or strainmeters, geophysicists can learn about the local geology and the underlying structure. There are also ongoing attempts to look for time-dependent variations in tidal amplitudes near active earthquake faults, which are caused by sudden changes in local material properties preceding an earthquake. The hope is that any such variability might be useful in predicting earthquakes.

A related problem is that tilt and strain amplitudes are also affected by the local topography and by the shape and size of any cavity the instrument is placed in. (Tiltmeters and strainmeters are often placed in boreholes or tunnels to minimize the effects of the surface environment.) These effects are rarely interesting, and they cannot be observationally separated from the effects of local geology. Instead, they must be modeled separately and removed from the data.

Summary

The effects of Earth tides can be detected in several different types of geophysical measurements. These tidal observations can be inverted, in principle, to learn about large-scale material properties of the Earth's deep interior and about local geological structure. In practice, a useful inversion is often difficult, due to the many possible geophysical interpretations and to uncertainties associated with instrument calibration, topographic and cavity effects, and ocean tidal loading of the solid Earth.

JOHN M. WAHR

References

Baker, T. F., 1984, Tidal deformations of the earth, *Sci. Prog. Oxf.* **69**, 197-233.
Harrison, J. C., 1984, *Earth Tides* (Benchmark Papers in Geology). Stroudsburg, PA: Hutchinson Ross.
Melchior, P., 1983, *The Tides of the Planet Earth*. Oxford: Pergamon Press.

Cross-references: *Absolute Gravity Measurements; Earth Orientation; Figure of the Earth; Geodesy: Physical; Mantle Viscosity; Spherical Harmonic Analysis.*

ELASTICITY AND WAVE PROPAGATION: PRINCIPLES

For a Cartesian coordinate system x_i ($i = 1, 2, 3$), we define a *displacement* component u_i ($i = 1, 2, 3$) of a particle initially located at $\mathbf{x}(x_1, x_2, x_3)$ in the undeformed state. Particle location at time t is $x_i + u_i(\mathbf{x}, t)$. For infinitesimally small u_i, we have the basic equations

$$\rho \ddot{u}_i = f_i + \tau_{ij,j} \qquad (1)$$

$$\tau_{ij} = C_{ijpq} e_{pq} \qquad (2)$$

$$W = \tfrac{1}{2} C_{ijpq} e_{ij} e_{pq} \qquad (3)$$

where $\cdot$ represents $\partial/\partial t$, j represents $\partial/\partial x_j$, ρ is the density, f_i is the ith component of the body force, C_{ijpq} is the elastic constant relating the ijth stress component τ_{ij} to the pqth strain component, W is the strain energy function, and the repeated index implies summation with respect to the index over 1, 2, and 3, e.g.,

$$\tau_{ij,j} = \frac{\partial \tau_{i1}}{\partial x_1} + \frac{\partial \tau_{i2}}{\partial x_2} + \frac{\partial \tau_{i3}}{\partial x_3}$$

It is remarkable that almost all results in theoretical seismology have been deduced from these three equations.

Equations 1 and 2 originate from Newton and Hooke, respectively. These two great men were contemporaries in seventeenth-century England, but they were not friends. Hooke believed that Newton stole his idea about the inverse square law on gravitational force. In any case, we owe them the origin of dynamic elasticity.

The concepts of stress components τ_{ij} and strain components e_{ij} were unknown in the seventeenth century. They were developed by Cauchy in the 1820s. The 1820s in France was an extraordinary period for elasticity theory. Fresnel explained various polarization effects of light by considering light to be a transverse elastic wave propagating through ether. Earlier, only a longitudinal wave was considered to propagate through the interior of a body. Poisson then showed that two types of waves, which we now call P and S waves, can propagate in the Newtonian atomic structure and that the ratio of the P-wave velocity to the S-wave velocity is $\sqrt{3}$. Thus, elastic waves played a central part in the fundamental physics of this period.

With the introduction of stress and strain components, Cauchy founded elasticity theory almost in its present-day form, except for the concept of the strain energy function W, shown in Eq. 3. In the nineteenth century, when Green laid the foundation of elasticity on the existence of the strain energy function, the basis for development of theoretical seismology was well prepared.

Stress and Strain

Stress and strain were introduced by Cauchy to describe the force acting in the interior of a continuum and its corresponding deformation. To define stress at a point, one must imagine a plane and define the magnitude and direction of the force exerted by the body at one side of the plane upon the other side. He found that six stress components are needed to completely define the state of stress at a point.

Cauchy defined strain by using displacement components u_i:

$$e_{ij} = \tfrac{1}{2}(u_{i,j} + u_{j,i}) \qquad (4)$$

With this definition, the local rigid body rotation is excluded from strain (see, e.g., Aki and Richards, §10.1.4), thus making the unique relation between stress and strain given in Eq. 2 possible.

The existence of the strain energy function W postulated by Green leads to the microscopic reciprocal relation for the elastic constants

$$C_{ijkl} = C_{klij} \qquad (5)$$

Equation 5 and the symmetry of the stress and strain tensors give 21 independent elastic constants for an arbitrarily anisotropic elastic body (Love, 1934). (See also *Seismic Anisotropy in the Earth*.)

The Assumption of Linearity

In most practical cases in seismology, the assumption of small displacement and, consequently, the linear equations 1, 2, and 4 are justified because seismic strain is generally small (less than 10^{-4}) and the region with strain exceeding 10^{-4} is confined in a small volume or a narrow zone in comparison with the observable wavelength, and may be excluded from the elastic system without a significant effect on the resultant motion. For example, the North Izu, Japan, earthquake ($M = 7.1$) of 1930 was accompanied by a fault movement that cut across the Tanna Tunnel, which was under construction at the time of the earthquake. The tunnel was at a depth of 150 m, and displacement at the tunnel was 2.7 m. The exposed fault surface, the "slickenside," was visible in the tunnel. Fault breccia were well developed and soaked with underground water. After examining the fault surface, Ishimoto (1935) concluded that the movement of such a weak material could not have sent the violent waves recorded at stations around the world. We know now that his conclusion was wrong. Waves recorded at teleseismic stations have wavelengths longer than a few kilometers. For these long waves, the nonelastic weak fault gouge zone can be assumed as infinitesimally thin. If we apply a displacement discontinuity between the two faces of the fault, composed of normal crustal material, and calculate seismic motions generated by such faulting for a realistic Earth model, we find that indeed the fault slip and length measured in the epicentral area give the amplitude of seismic waves consistent with those observed at distant stations, as shown by Kanamori and Anderson (1975). Such a quantitative test of the fault model was impossible in the 1930s. It became possible in the mid-1960s when the seismograms from a global network of calibrated stations, a large-scale digital computer, and mathematical formulae for computing seismograms for realistic Earth models became available.

Basic Theorems and Green's Function

The fundamental theorem of elastodynamics is the uniqueness theorem, which guarantees the existence of a solution for a given set of initial and boundary conditions.

Consider an elastic body occupying a region V enclosed by a surface S. Equations 1, 2, and 4 apply to the displacement u_i in V. The uniqueness theorem states that if displacement and particle velocity are known in V at $t = 0$, and if displacement *or* traction is known on S (not both on the same part of S), then a unique solution exists for the problem. The proof is by imagining two possible solutions and considering the strain energy function W for the differential field. The positiveness of W at any strained state guarantees that the differential field vanishes everywhere.

The next most important theorem is the reciprocal theorem. Here we consider two sets of solutions with different body force distributions and boundary conditions. The theorem can be stated as the work done by the body force and surface traction of Set 1 over the displacement of Set 2 is equal to the work done by the body force and surface traction of Set 2 over the displacement of Set 1.

The microscopic reciprocal relation given in Eq. 5 plays a key role in the derivation of this theorem.

The reciprocal theorem leads to the most useful theorem in elastodynamics, the representation theorem, where Green's function is one of the two sets of solutions.

Green's function is a displacement field often written as $G_{ij}(\mathbf{x}, t; \boldsymbol{\xi}, \tau)$, where $\mathbf{x}(x_1, x_2, x_3)$ represents the observation point, $\boldsymbol{\xi}(\xi_1, \xi_2, \xi_3)$ represents the source point, the index i refers to the ith component displacement, and the index j refers to the x_j-direction in which a unit impulsive force is applied at $\boldsymbol{\xi}$. In other words, $G_{ij}(\mathbf{x}, t, \boldsymbol{\xi}, \tau)$ is the ith component displacement observed at time t at $\mathbf{x}$ due to an impulsive point force applied at time τ at $\boldsymbol{\xi}$. One may call it a tensorial-spatial-temporal impulse response. Unlike the response of an electronic circuit to a voltage impulse applied at an input terminal, Green's function must be specified with regard to the tensorial and spatial characters in addition to the temporal response. Applying the reciprocal theorem to the solution pair $\mathbf{u}(\mathbf{x}, t)$ and using Green's function, we obtain the following representation theorem:

$$u_n(\mathbf{x}, t) = \int_{-\infty}^{\infty} d\tau \iiint_V f_i(\boldsymbol{\xi}, \tau) G_{in}(\boldsymbol{\xi}, t - \tau; \mathbf{x}, 0)\, dV(\boldsymbol{\xi})$$

$$+ \int_{-\infty}^{\infty} d\tau \iint_S \{G_{in}(\boldsymbol{\xi}, t - \tau; \mathbf{x}, 0) C_{ijkl}(\boldsymbol{\xi}) n_j u_{k,l}$$

$$- u_i(\boldsymbol{\xi}, \tau) C_{ijkl}(\boldsymbol{\xi}) n_j G_{kn,l}(\boldsymbol{\xi}, t - \tau; \mathbf{x}, 0)\}\, dS(\boldsymbol{\xi}) \quad (6)$$

Equation 6 appears very complex, involving up to fourfold integrals and sums up to four indices. This equation is basically a simple formula equivalent to the convolution theorem for a linear system for which the output (in our case, the displacement in V) is expressed as a convolution of the input (in our case, body force in V and traction and displacement on S) with the impulse response of the system (in our case, Green's function). The formula is expressing the basic simplicity of elastodynamics that is due to the linearity and time invariance of the system.

If we apply the representation theorem to Green's function itself and assume a homogeneous boundary condition (displacement, traction, or their linear combination vanishes on S), we obtain the remarkable relation due to Knopff and Gangi (1959)

$$G_{ij}(\mathbf{x}, t - \tau, \boldsymbol{\xi}, 0) = G_{ji}(\boldsymbol{\xi}, t - \tau, \mathbf{x}, 0) \quad (7)$$

When the observation point and source point are switched, the direction of force and displacement must also be switched. Equation 7 is practically useful because if we know the solution of one problem, we can immediately obtain the solution of its reciprocal problem. Sometimes, the reciprocal problem is easier to solve than the original problem.

Isotropic, Homogeneous, and Unbounded Medium

The simplest elastic body is the isotropic, homogeneous, and unbounded (IHU) medium. For this body, we can write Green's function in a compact form. Consider an impulsive point force applied at the origin and $t = 0$ in the direction x_j. Then the displacement field at $\mathbf{x}$ and t can be written as

$$u_i(\mathbf{x}, t) = \frac{1}{4\pi\rho} (3\gamma_i\gamma_j - \delta_{ij}) \frac{1}{r^3} \int_{r/\alpha}^{r/\beta} \tau \delta(t - \tau)\, d\tau$$

$$+ \frac{1}{4\pi\rho\alpha^2} \gamma_i\gamma_j \frac{1}{r} \delta\left(t - \frac{r}{\alpha}\right)$$

$$+ \frac{1}{4\pi\rho\beta^2} (\gamma_i\gamma_j - \delta_{ij}) \frac{1}{r} \delta\left(t - \frac{r}{\beta}\right) \quad (8)$$

where $\delta(t)$ is Dirac's delta function, $\gamma_i = x_i/r$, $r = |\mathbf{x}|$, $\alpha\, (= \sqrt{(\lambda + 2\mu)/\rho})$ is the velocity of P waves, and $\beta\, (= \sqrt{\mu/\rho})$ is the velocity of S waves. The first term on the right-hand side is the *near-field* term, which attenuates faster with distance than do the second term (far-field P waves) and third term (far-field S waves). Both far-field terms have the impulse form (same as the applied force) in time, and the near-field term vanishes after the arrival of S waves. It is this simple nature of Green's function that makes seismic waves a high-fidelity distortionless carrier of information when the medium is not too far from the idealized IHU body. The foregoing solution was first given by Stokes in 1849.

Plane Wave Decomposition of Green's Function

The real Earth, however, is not an IHU body. It is clearly bounded by the free surface and some hidden internal discontinuities. It is also heterogeneous and anisotropic. A first step to the realistic Earth model is to consider a stack of homogeneous layers with parallel interfaces overlying a homogeneous half-space.

The propagation of plane waves in such a layered medium is well understood and can easily be calculated. Then the natural way to obtain Green's function for the layered medium is to decompose Green's function for the IHU medium into plane waves, calculate the wave field generated in the layered medium due to each component plane wave, and then superpose them. The superposition is valid because of the linearity of our system.

In order to find the plane wave decomposition of Eq. 8, we first change the time dependence from $\delta(t)$ to $e^{-i\omega t}$, by convolving Eq. 8 with $e^{-i\omega t}$. We then have

$$u_i = \frac{1}{4\pi\rho}\left[(3\gamma_i\gamma_j - \delta_{ij})\left(\frac{r}{i\omega\beta} + \frac{1}{\omega^2}\right)r^{-3}\right.$$
$$\left. - (\gamma_i\gamma_j - \delta_{ij})(\beta^2 r)^{-1}\right]\exp\left(i\omega\frac{r}{\beta}\right)$$
$$-\frac{1}{4\pi\rho}\left[(3\gamma_i\gamma_j - \delta_{ij})\left(\frac{r}{i\omega\alpha} + \frac{1}{\omega^2}\right)r^{-3}\right.$$
$$\left. - \gamma_i\gamma_j(\alpha^2 r)^{-1}\right]\exp\left(i\omega\frac{r}{\alpha}\right) \quad (9)$$

Consider the case $j = 3$, in which the point force is directed in the x_3-direction. To express this displacement field as an integral superposition of plane waves, we use the Weyl integral:

$$\frac{1}{r}\exp\left(i\omega\frac{r}{c}\right) = \frac{1}{2\pi}\int\int_{-\infty}^{\infty}\frac{\exp(ik_1x_1 + ik_2x_2 + ik_cx_3)}{-ik_c}$$
$$\cdot dk_1\, dk_2 \quad \text{for } x_3 > 0$$

where

$$k_c = \left[\left(\frac{\omega}{c}\right)^2 - k_1^2 - k_2^2\right]^{1/2} \quad \text{and} \quad \text{Im } k_c > 0 \quad (10)$$

Recognizing that terms in Eq. 9 can be constructed by taking second derivatives of Eq. 10 with respect to x_1, x_2, and x_3, we find that u_i in Eq. 9 can be written as

$$u_i = \frac{i}{8\pi^2\rho\omega^2}\int\int_{-\infty}^{\infty} k_i \exp[i(k_1x_1 + k_2x_2 + k_\alpha x_3)]$$
$$\cdot dk_1\, dk_2$$

$$-\frac{i}{8\pi^2\rho\omega^2}\int\int_{-\infty}^{\infty}\left[k_i - \frac{\delta_{i3}\omega^2}{k_\beta\beta^2}\right]\exp[i(k_1x_1$$
$$+ k_2x_2 + k_\beta x_3)]\, dk_1\, dk_2 \quad \text{for } x_3 > 0 \quad (11)$$

where

$$k_\alpha = \left[\frac{\omega^2}{\alpha^2} - k_1^2 - k_2^2\right]^{1/2}$$

and

$$k_\beta = \left[\frac{\omega^2}{\beta^2} - k_1^2 - k_2^2\right]^{1/2}$$

with the choice of sign according to Im $k_\alpha > 0$ and Im $k_\beta > 0$. Equation 11 shows Green's function for the IHU body as a superposition of plane P and S waves.

Since the integration range with respect to k_1 and k_2 is from $-\infty$ to ∞, it includes "inhomogeneous plane waves," for which k_α and k_β become pure imaginary. These waves will propagate along the plane normal to the x_3-axis with velocity lower than the medium P velocity or S velocity, and their amplitudes decay exponentially in the positive and negative x_3-directions away from the source.

Inhomogeneous plane waves contained near the source play important roles when the source is near the free surface or interfaces. For example, the dominant seismic motion observed on the free surface from a shallow source is due to Rayleigh waves, which are coupled P and S inhomogeneous waves. Rayleigh waves have velocity lower than either P or S waves and can never be generated by the incidence of regular plane P or S waves upon the free surface.

Green's Function for Realistic Earth Models

So far the most widely used Earth model for interpreting short-period (say, less than 20 s) seismograms has been the multilayered half-space. Since the exact response of the multilayered half-space to a plane wave source can be obtained in a closed form, the formal solution for Green's function for the multilayered half-space can be written in a straightforward manner as an integral superposition of plane waves similar to Eq. 10 but with weight functions determined by the parameters of the layered medium.

There are basically three methods to evaluate this integral. The most direct approach is to replace the integral by a discrete sum (reflectivity method and discrete wave-number method). Using the second method, the formal solution is decomposed into a series of terms, each of which can be associated with a ray path. These terms, called *ray integrals*, can be efficiently evaluated by Cagniard's method. The third method decomposes the formal solution into a series of terms, each of which represents the residue contribution from poles in the complex plane to which the originally real integral variables k_1 and k_2 are extended. Surface waves such as Love and Rayleigh waves and their higher modes are normal modes associated with poles on the real axis. In addition to normal modes, leaky modes are required to represent the complete wave field. This requirement is in contrast to the seismic motion of a layered sphere, which can be completely described as the infinite sum of normal modes. The leaky mode exists for the layered half-space because of its infiniteness. Green's function for a layered sphere differs from that for a layered half-space in its dependence on lateral coordinates, namely, spherical harmonics instead of sinusoidal functions. The multilayered sphere has been extensively used for interpreting long-period seismograms, such as surface waves and free oscillations.

In addition to the layered Earth models, a great variety of models have been used for interpreting seismograms in particular situations. For example, vertically heterogeneous Earth models in which both P and S velocities vary continuously and discontinuously have been used for analyzing the travel time data by ray methods since the beginning of modern seismology. Ray methods have also been used for studying the lateral variation in the velocity structure (see *Seismic Ray Theory*). Green's function can be calculated for three-dimensional heterogeneous media with irregular interfaces by ray methods or their extension, such as the Gaussian beam method (Cerveny and Psencik, 1983), if the medium property varies smoothly within a wavelength and the radius of curvature of the interface is much greater than the wavelength.

For Earth models with heterogeneity scale length comparable to or less than the wavelength, the finite difference and finite element methods are applicable for relatively short travel distances. However, as travel distance increases, deterministic modeling becomes difficult, and one has to resort to statistical approach by characterizing the Earth as a random medium. Recommended approaches for different regions of the kL-ka diagram, where L is the travel distance, a is the scale length of heterogeneity, and k is the wave number, have been discussed in §13.3.5 of Aki and Richards (1980).

Summary

The representation theorem (Eq. 6) is central to the theory of seismic wave propagation based on principles of elasticity. The theorem is equivalent to the convolution theorem for a linear system for which the output is expressed as a temporal convolution of the input with the impulse response of the system. Although the theorem involves up to fourfold integrals and sums up to four indices, it is expressing the basic simplicity of elastodynamics due to the linearity and time invariance of the system. The more difficult part of the problem is the actual calculation of the impulse response, namely, Green's function. We have outlined various approaches for obtaining Green's function for realistic Earth models.

KEIITI AKI

References

Aki, K., and P. G. Richards, 1980, *Quantitative Seismology: Theory and Methods*, Vols. 1 and 2. San Francisco: W. H. Freeman.

Cerveny, V., and I. Psencik, 1983, Gaussian beams and paraxial ray approximations in three-dimensional elastic inhomogeneous media, *Jour. Geophys.* **53**, 1-15.

Ishimoto, M., 1935, *Study of Earthquakes*, (in Japanese). Tokyo: Kokin-Shoin.

Kanamori, H., and D. L. Anderson, 1975, Theoretical basis of some empirical relations in seismology, *Seismol. Soc. America Bull.* **65**, 1073-1096.

Knopoff, L., and A. F. Gangi, 1959, Seismic reciprocity, *Geophysics* **24**, 681-691.

Love, A. E. H., 1934, *A Treatise on the Mathematical Theory of Elasticity*, Cambridge, England: Cambridge at the University Press.

Cross-references: *Earthquake Seismology; Elastic Waves in Homogeneous and Inhomogeneous Media; Exploration Seismology; Finite-Difference Forward Modeling in Seismology; Free Oscillations of the Earth; Inverse Theory and Methods: Seismology; Seismic Anisotropy in the Earth; Seismic Monitoring of Nuclear Explosions; Seismic Ray Theory; Seismic Wavefield Migration; Seismic Wave Scattering; Seismograms: Interpretation; Seismograms: Synthetic; Seismology: Physical Model Studies.*

ELASTIC WAVES IN HOMOGENEOUS AND INHOMOGENEOUS MEDIA

By the time that the first seismographs were operational (1889) the mathematical theory of elastic waves was well developed and available for use in interpretation. The complexity of seismic records then encouraged further research into the propagation of elastic waves in stratified media, which has enabled the computation of theoretical seismograms for complex stratified media.

Much of the early work on elastic waves was inspired by trying to understand the propagation of light by using the model of a crystalline medium. Thus Green in 1838 made a detailed study of two-dimensional wave propagation. He introduced the concept of a strain energy function and showed that 21 elastic constants were needed for an anisotropic medium, which reduced to 2 for isotropy. Green isolated P, SV, and SH plane wave components and established the equations that hold at a planar interface between two dissimilar media. He did not attempt to solve these equations. Knott, in 1899, appears to have been the first to derive the reflection and transmission coefficients for incident plane waves.

The excitation of elastic waves by concentrated sources was investigated in 1849 by Stokes, who was able to determine the displacements generated by an arbitrarily oriented force. Two distinct disturbances corresponding to compressional (P) and transverse (S) waves move out through the medium, linked by motion of intermediate character. The extension to more realistic couple sources was made by Love in 1904, and his solution forms the basis of modern treatments of seismic source radiation.

Rayleigh (1885) showed that it was possible for a free surface wave to propagate along the surface of an isotropic elastic half-space without dispersion. He pointed out that such a wave would at large distances dominate any P or S waves traveling through the bulk of the solid. The forced motion of an isotropic elastic half-space was studied in some detail by Lamb (1904); this work showed that the surface Rayleigh wave could be significantly excited. However, the displacements Lamb computed did not show the oscillatory character nor the large transverse component of motion seen on contemporary seismograms.

Lamb suggested that the oscillatory nature arose from the recording system. This would have been partly true for the lightly damped systems of the time but, in general, would have been smaller than the effects of dispersion arising from the increase of seismic velocities with depth in the Earth. To explain the transverse surface wave component, Love (1911) introduced stratification into the model. He showed that with a superposed layer on a uniform half-space there could be a free transverse surface wave (a Love wave) with amplitude that decayed exponentially into the half-space. This Love wave shows frequency dispersion with horizontal wave number.

The excitation of elastic waves due to a source in half-space continued to receive considerable attention as the details of the wave strains produced were elucidated by Nakano, Lapwood, and others. The excitation of Love waves was treated in some detail by Sezawa, who investigated the effect of source depth. For such purely analytical developments of elastic wave theory the level of manageable algebraic complexity is reached in a model with one or two uniform layers overlying a uniform half-space.

Such problems are considered in detail by Ewing, Jardetsky, and Press (1957); their book also contains a comprehensive bibliography. Exact solutions are only available for restricted forms for the velocity distribution with depth, such as uniform layers. Some progress with general stratification problems, however, was made, notably by Jeans, in 1923, who considered elastic wave propagation in a sphere, and by Jeffreys, in 1928, who examined Love wave propagation in a model having a uniform layer overlaying a nonuniform half-space.

When the elastic parameter distribution depends only on one coordinate, Alterman et al. (1959) showed how the stress-strain relations and the elastic equations of motion can be reduced by transform techniques to a set of first-order differential equations. Those equations have to be solved subject to free surface boundary conditions and excitation by the source. This approach has been the basis of most purely numerical work for smoothly varying velocity distributions for both body waves and surface waves (see Aki and Richards, 1980). In models consisting of a number of uniform layers, these methods can be reduced to the manipulation of matrices. Refinements of the technique to remove numerical instability have been proposed by several authors, notably Gilbert and Backus (1966). With these modifications, or an alternative approach based on the reflection and transmission properties of a stratified model (see Kennett, 1983), it is possible to construct complete theoretical seismograms for realistic Earth models (see *Seismograms: Synthetic*).

A Homogeneous Medium

The disturbances associated with elastic waves can be described by specifying the Cartesian stress tensor τ_{ij}, the particle displacement vector **u**, and the associated strain tensor $e_{ij} = \frac{1}{2}(\partial u_i/\partial x_j + \partial u_j/\partial x_i)$.

In the presence of a body force f, the equation of motion relates the local gradient of the stress tensor components to the particle accelerations:

$$\frac{\partial \tau_{ij}}{\partial x_j} + \rho f_i = \rho \frac{\partial^2 u_i}{\partial t^2} \qquad (1)$$

where ρ is the local density. The stress and strain are linearly related by Hooke's law,

$$\tau_{ij} = \lambda C_{ijkl} e_{kl} = C_{ijkl} \frac{\partial u_k}{\partial x_l} \qquad (2)$$

using the symmetries of the elastic modulus tensor $C_{ijkl} = C_{jikl} = C_{ijlk} = C_{klij}$. In a general anisotropic medium, 21 independent moduli are needed, but the number reduces as further symmetry requirements are imposed. In transverse isotropy where the properties are uniform in a plane perpendicular to a specified symmetry axis, only five moduli are required. For full isotropy only the two Lamé moduli λ, μ are needed and

$$C_{ijkl} = \lambda \delta_{ij} \delta_{kl} + \mu(\delta_{ik}\delta_{jl} + \delta_{il}\delta_{jk}) \qquad (3)$$

μ is the shear modulus, and the bulk modulus $\kappa = \lambda + \frac{2}{3}\mu$.

When the elastic parameters have an arbitrary spatial variation, the equations for the displacement obtained by combining Eqs. 1-3 are still rather complex, but when the material is homogeneous they simplify significantly to

$$\rho \frac{\partial^2 u_i}{\partial t^2} = \frac{\partial}{\partial X_i}\left[(\lambda + \mu)\frac{\partial u_j}{\partial X_j} + \mu \frac{\partial u_i}{\partial X_j}\right] + \rho f_i \quad (4)$$

In the absence of sources, the three components of displacement are coupled but can be represented in terms of two simple classes of disturbance: P waves that are irrotational waves with wave speed $\alpha = [(\lambda + 2\mu)/\rho]^{1/2}$ and for which the particle motion in a plane wave solution is purely longitudinal, $\mathbf{u} = \hat{\mathbf{s}} \cdot u(t - \hat{\mathbf{s}} \cdot \mathbf{x}/\alpha)$, where $\hat{\mathbf{s}}$ is a unit vector in the propagation direction; and S waves that are equivoluminal with wave speed $\beta = [\mu/\rho]^{1/2}$. In this case the particle motion is transverse to the direction of propagation, $\mathbf{u} = \hat{\mathbf{n}} u(t - \hat{\mathbf{s}} \cdot \mathbf{x}/\beta)$ with $\hat{\mathbf{n}} \cdot \hat{\mathbf{s}} = 0$. The plane shear waves can be represented as a linear superposition of two such waves with orthogonal polarizations. It is conventional to take the polarization directions of these waves to lie in the vertical plane including the propagation direction (SV) and the horizontal direction mutually orthogonal to the propagation direction and $\hat{\mathbf{n}}_{sv}(SH)$.

The radiation from a point force was derived by Stokes in 1849 and consists of a longitudinal P-wave pulse with the same time dependence as the source with amplitude inversely proportional to the distance from the source point, followed later by a transverse S-wave pulse of similar character. Between these main P and S arrivals is a disturbance of intermediate character that decays more rapidly away from the source.

For applications in seismology an unbalanced force is not a suitable model for an earthquake. A general internal source can, however, be synthesized by superposition of force couples without moment (see Chapter 3 of Aki and Richards, 1980, or Chapter 4 of Kennett, 1983). The relative weights of the nine elementary dipoles and couples are specified by the moment tensor $M_{ij}(t)$ for the point source. For an earthquake with slip in direction $\mathbf{v}$ along a planar fault with normal $\mathbf{n}$,

$$M_{ij}(t) = M_0(n_i v_j + v_i n_j), \quad \mathbf{n} \cdot \mathbf{v} = 0 \quad (5)$$

a double-couple source.

A Homogeneous Half-Space

In a homogeneous medium P and S waves propagate independently, but as soon as a material surface is introduced they are coupled by their interaction in reflection. The most significant case for geophysical applications is the Earth's surface, which can be modeled to high accuracy by a free boundary unable to sustain any traction. In terms of a depth variable z, at the horizontal planar surface $z = 0$, the conditions for vanishing traction require

$$\tau_{xz} = \tau_{yz} = \tau_{zz} = 0 \quad \text{at } z = 0 \quad (6)$$

Plane SH waves with motion lying purely in the xy-plane are specularly reflected at such a free surface; however, the SV polarized waves couple with P on reflection. Thus when a plane P or SV wave impinges on the free surface, it will, in general, generate a reflected wave of the same type, with the same inclination to the vertical, and a converted wave. In order for the phase behavior to match on $z = 0$, the angles of inclination to the vertical for the P wave (i) and the SV wave (j) are related by a form of Snell's law.

$$\frac{\sin i}{\alpha} = \frac{\sin j}{\beta} \quad (7)$$

The S wave has a smaller inclination than the P wave since $\beta < \alpha$.

The associated reflection coefficients are determined by satisfying the traction conditions (Eq. 6). The particular forms are strongly dependent on the conventions used for the specific forms of the plane waves (see, e.g., Červený and Ravindra, 1975; Aki and Richards, 1980, for different choices).

The nature of the surface boundary condition of vanishing traction gives rise to strong coupling between the P and SV waves to the extent that once the inclination angle i exceeds around 30°, conversion to the other wave type on reflection is more efficient than reflection in the same type.

The behavior associated with plane waves is relatively simple, but additional effects become evident when one considers the way in which the radiation from a source interacts with the free surface. In an unbounded medium, the wave field just consists of direct P and S arrivals. The introduction of a free surface gives rise to additional reflected phases and a new type of wave, the Rayleigh wave, consisting of coupled P and S disturbances with amplitudes that decay exponentially away from the surface. This Rayleigh wave travels horizontally along the surface, with most of the energy concentrated near the surface. The theoretical problem of the excitation of the wave field in a half-space was first tackled by Lamb (1904), and his analysis has been extended by many later authors. Most methods of solution depend on the use of Fourier or Laplace transform techniques to reduce the problem to the excitation of plane or cylindrical waves, which then have to be reconstituted to generate the displacement field. A detailed treatment of the response due to a line source in a uniform half-space is given by Hudson (1980), and a number of ways of treating excitation by a point source are discussed in Chapter 6 of Aki and Richards (1980).

A concentrated source near the surface is an efficient generator of Rayleigh waves, and at moderate to large distances from the source these surface waves dominate the direct P and S waves in the surface displacement. Close to the source the wave field is rather complex, and the Rayleigh wave is not clearly established until a range somewhat larger than the depth of the source. Because the amplitude of the Rayleigh wave decays with depth, it becomes more difficult to excite with increasing source depth.

An Interface between Two Half-Spaces

When a plane P or S wave impinges on an interface between two homogeneous media, both reflected and transmitted waves are generated of the same wave type and with conversion. The boundary conditions that have to be satisfied by the wave field are the continuity of displacement and traction to ensure welded contact between the two half-spaces. Once again, in order to achieve continuity of phase, the angles of inclination to the normal to the interface for the reflected and transmitted waves must satisfy

$$\frac{\sin i_1}{\alpha_1} = \frac{\sin i_2}{\alpha_1} = \frac{\sin j_1}{\beta_1} = \frac{\sin j_2}{\beta_2} \qquad (8)$$

where the subscripts refer to the two different media. Equation 8 determines the trajectories of the reflected and transmitted waves, and their amplitudes are to be determined by matching the expressions for the displacement and traction on the two sides of the interface. The resulting analytical expressions for the reflection and transmission coefficients are rather complex and, as for the free surface case, depend on the conventions used for the specific forms of the plane waves (see, e.g., Cerveny and Ravindra, 1975; Aki and Richards, 1980).

For waves incident from a medium with lower velocities there is a limiting angle of incidence for P waves for which propagating transmitted P waves can exist in the lower medium. The P wave "critical angle" occurs when $i_2 = 90°$, so $i_1 = \sin^{-1}(\alpha_1/\alpha_2)$ and for larger angle of incidence there will be only reflected P waves but still-transmitted S waves. There will be a P disturbance established in the lower medium, but this evanescent wave decays exponentially away from the interface. If $\beta_2 > \alpha_1$, there is also a second angle $i_{s1} = \sin^{-1}(\alpha_1/\beta_2)$ at which transmission into propagating waves into the lower half-space ceases and only reflected P and S waves are produced. For S incident waves, the critical angle for S-wave transmission is $j_{c1} = \sin^{-1}(\beta_1/\beta_2)$. In addition to the P and S body waves, there is an analogue of the Rayleigh wave, called a *Stoneley wave*, in which there is exponentially decaying coupled P and S waves on both sides of the interface. Such Stoneley waves always exist for an interface between a fluid (which allows only P waves) and a solid, but the existence conditions for a solid-solid boundary impose rather restrictive conditions on the properties of the two materials.

An extensive study of the interaction of the elastic wave radiation from a concentrated source with an interface was made by Cagniard (1939). This work introduced a powerful technique for evaluating the contributions made by different types of waves to the displacement field, and it clearly demonstrated the importance of the sphericity of the wavefronts that emanate from the source. In addition to the expected reflected and transmitted waves, a new phenomenon is introduced when the material on the far side of the boundary has higher velocities. As already noted, beyond the P-wave critical angle there is no transmission of plane P waves into the lower medium. However, the effect of the curves wavefront is to produce a disturbance moving out along the interface with the P velocity in the lower medium and so outstripping the waves in the upper medium. The effect is that waves are radiated into the upper medium with an apparent velocity corresponding to the P-wave in the lower velocity medium, and a linear wavefront. When coupling between the wave types is allowed for, a rather complicated pattern of such "head waves" ensues, as illustrated in Fig. 1.

More General Stratification

For a medium consisting of a stack of uniform layers bordered by two uniform half-spaces, the

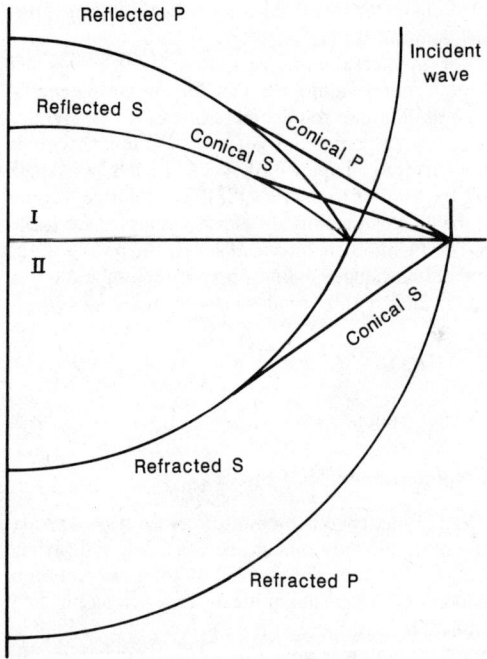

FIGURE 1. Reflection and refraction at an interface.

propagation within each uniform zone can be treated analytically. Two different approaches are often used to construct the response of a stack of layers in the frequency domain. In the first, attention is concentrated on the displacement and traction field. The displacements and traction at the top and bottom of a layer are connected by a propagator matrix, and, imposing continuity at interfaces, these propagators for the different layers are multiplied together to generate the propagator for the whole stack (see Sections 7.1 and 9.2 of Aki and Richards, 1980). The boundary conditions on the displacement and traction fields are now applied at the limits of the stack, and the resulting equations are solved for the unknown displacement field. For coupled P-SV waves, the solutions depend on 2×2 minors of the 4×4 propagator are evanescent matrices, and numerical instabilities can develop when P and S waves are evanescent within the stack. These instabilities can be avoided by working directly with the minors (see Gilbert and Backus, 1966).

The second approach to the response of the layer stack is to work directly with the reflected and transmitted waves. The reflection and transmission properties of the layer stack can be built up recursively by adding in the effect of one layer at a time (see Chapter 6, Kennett, 1983). This treatment avoids the numerical problems associated with the propagator matrices. For anisotropic media where we have three coupled wave types, this approach is particularly advantageous.

When we have a general stratified medium where the seismic parameters depend only on one coordinate (usually depth), we cannot use analytical methods. However, Alterman et al. (1959) showed how the propagation process can be handled in the frequency–wave member domain by using a first-order differential equation in depth for the displacement and traction components. Gilbert and Backus (1966) showed how the use of propagator matrices could systematize the application of the boundary conditions in such problems. Numerical solutions of these first-order differential equations have been extensively used for both body wave and surface wave calculations and for determining the free oscillation periods of a spherical model.

In a region where the seismic velocities increase continuously with depth, a "diving" wave can be generated. Snell's law for a continuous medium requires for P waves that

$$\frac{\sin i}{\alpha(z)} = \frac{\sin i_0}{\alpha_0} \qquad (9)$$

where the subscript 0 refers to the entry point. As the velocity α increases, the inclination i must increase until it reaches 90° and the trajectory is horizontal. The effect of wavefront curvature is to turn energy back with total reflection but a phase delay of 90°. The structure of the wave field in the neighborhood of such a turning level can be studied by using uniform asymptotic approximations due to Langer (see Section 9.4, Aki and Richards, 1980; Section 3.3, Kennett, 1983). Multiple reflections from the gradient zone can be very significant where there is a strongly reflecting overlying interface. The interference of many such reflections gives rise to a relatively strong wave, an "interference head wave," traveling with a speed a little greater than the seismic velocity just below the interface. Such waves persist to greater distances than the true head waves already discussed.

Guided Waves

In a stratified half-space, the presence of the free surface introduces a range of propagation processes that do not occur in unbounded media. Now we have reflection from the free surface interacting with reflection from the stratification beneath to give efficient, multiple, reflections, which can be important for both P and S waves.

Normally, in seismology, there is a zone of reduced velocities near the surface underlain at some depth by a relatively rapid increase to higher values. Examples are provided by sediments over basement rock, the Earth's crust overlying the upper mantle, and the upper mantle compared to the lower mantle. Multiple reflections at the boundaries of this lower velocity zone give rise to a complex interference pattern of guided waves carrying energy horizontally. An alternative viewpoint is that these waves can be regarded as a superposition of the free modes of the structure. The displacements associated with these modes have to satisfy the twin conditions of vanishing traction at the free surface and decay with depth (see Chapter 11, Kennett, 1983; Chapter 7, Aki and Richards, 1980). The simplest case is provided by the propagation of SH waves in a layer over a half-space first studied by Love in 1911. The modal solutions can only occur for certain combinations of frequency and wave number determined by the dispersion equation for the particular structure.

Guided near-surface SH waves in a general stratified half-space are now referred to as *Love waves*, and the dispersion equation is equivalent to requiring that the combination of frequency and wave number is such that a plane SH wave would be reflected back without change of amplitude and a phase increment of a whole number of cycles. There is a sequence of modes corresponding to the number of full cycles added to the phase that also determines the number of zero crossings with depth. Unlike the Rayleigh wave on a uniform half-space the apparent propagation speed of the Love waves varies with frequency. Similar dispersive behavior is shown by guided coupled P-SV waves that are commonly also referred to as *Rayleigh waves*, even though their character is rather different from the wave on a uniform half-space. The deepest part of the modal behavior is an exponential decay, but above this the displacements have an oscillatory character that becomes more rapidly varying with increasing wave number.

A shallow source in a stratified half-space will be very efficient at generating Rayleigh and Love waves, and as the range increases these waves tend to dominate the seismograms at the surface because the amplitudes decay only as the square root of the range rather than proportionally to the range as for body waves.

B. L. N. KENNETT

References

Aki, K., and P. G. Richards, 1980. *Quantitative Seismology: Theory and Methods*, 2 vols. San Francisco: W. H. Freeman, 932 p.

Alterman, Z., H. Jarosch, and C. L. Pekeris, 1959, Oscillations of the Earth, *Royal Soc. [London] Proc.* **252A**, 80-95.

Cagniard, L., 1939, *Réflexion et Refraction des Ondes Séismiques Progressives*. Paris: Gauthier-Villars.

Červený, V., and R. Ravindra, 1975, *The Theory of Seismic Head Waves*. Toronto: University of Toronto Press.

Ewing, W. M., W. S. Jardetsky, and F. Press, 1957, *Elastic Waves in Layered Medias*. New York: McGraw-Hill.

Gilbert, F., and G. E. Backus, 1966, Propagator matrices in elastic wave and vibration problems, *Geophysics* **31**, 326-332.

Hudson, J. A., 1980, *The Excitation and Propagation of Elastic Waves*. Cambridge: Cambridge University Press.

Kennett, B. L. N., 1983, *Seismic Wave Propagation in Stratified Media*. Cambridge: Cambridge University Press.

Lamb, H., 1904, On the propagation of tremors over the surface of an elastic solid, *Royal Soc. London Philos. Trans.* **203A**, 1-42.

Love, A. E. H., 1911, *Some Problems of Geodynamics*. Cambridge: Cambridge University Press.

Rayleigh, Lord, 1885, On waves propagated along the plane surface of an elastic solid, *London Math. Soc. Proc.* **17**, 4-11.

Cross-references: *Controlled Source Seismology; Deep Seismic-Reflection Profiling; Earthquake Seismology; Earth Structure: Global; Elasticity and Wave Propagation: Principles; Exploration Seismology; Finite-Difference Forward Modeling in Seismology; Free Oscillations of the Earth; Seismic Monitoring of Nuclear Explosions; Seismic Ray Theory; Seismic Wavefield Migration; Seismic Wave Scattering; Seismograms: Interpretation; Seismograms: Synthetic; Surface Waves.*

ENERGY BUDGET OF THE EARTH

The solid Earth can be considered as an enormous heat engine, where, through convection processes in the mantle, heat is transformed into mechanical energy that drives the plates, builds up mountains, etc. This section will deal with the global energy budget of the Earth. This includes the question of how the internal energy and the speed at which the heat engine was running have changed in geological time, i.e., the thermal history of the Earth. Four points must be considered: the sources and sinks of heat, the heat capacity of the Earth, the mechanisms of heat transport, and the initial thermal state. These points shall be discussed and synthesized into a theoretical model of the thermal history.

Heat Sources and Sinks

The most important energy source in the Earth is the decay of the four radioisotopes ^{238}U, ^{235}U, ^{232}Th, and ^{40}K. Their abundance depends strongly on rock type and is much lower in the mantle than in crustal rocks. The core is possibly devoid of these elements, although it has been speculated that it may contain some potassium. To estimate the total inventory of the Earth in these elements, we adopt the geochemical model of a chondritic Earth; i.e., the assumption that the composition of the bulk Earth is the same as that of meteorites called carbonaceous chondrites. This assumption appears reasonable for the "refractory" elements uranium and thorium; however, the Earth seems depleted in the moderately volatile element potassium compared with chondrites. The abundance ratio K:U is only about 10,000 in most crustal and mantle rocks, whereas the "chondritic ratio" is 70,000. Our best estimate of the bulk Earth abundance is about 20 ppb uranium, 75 ppb thorium, and 250 ppm potassium. About one fourth to one third of the total inventory resides in the continental crust. The thermal energy provided by the radioactive decay is calculated to 2×10^{13} W. The decay of these isotopes implies that they were more abundant in the past. This applies especially to ^{40}K and ^{235}U, which have shorter half-lives than the other two. Shortly after the formation of the Earth, 4.5 b.y. ago, the heat production was about five times higher than today, and 3.5 b.y. ago—the time where the geological record starts—it was three times more than today (see Fig. 1).

Other energy sources have been considered, but now it is believed that none of them has played a significant role during most of the Earth's history. Among them is the gravitational energy release during the separation of the metallic core from the mantle, which suffices to increase the temperature of the whole Earth by 2000 K, but which is assumed to have occurred as a catastrophic event in the earliest history of the Earth or during accretion. Further, there is the energy of tidal dissipation, which provides about 0.5×10^{13} W, but occurs mostly in shallow seas and not in the solid Earth.

The heat sink is the Earth's surface, from where the energy is lost by infrared radiation into space.

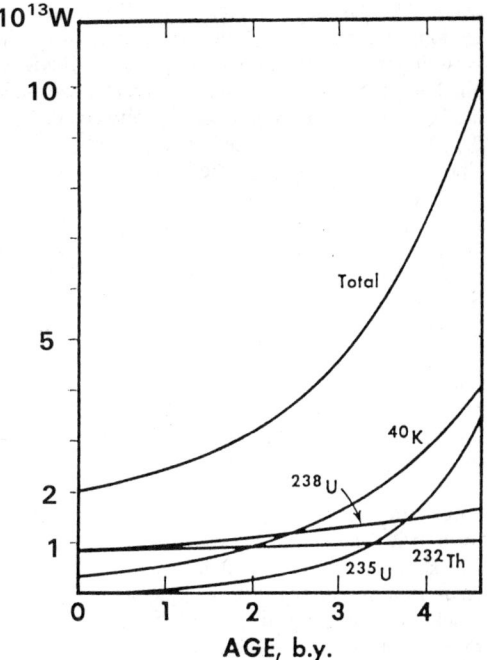

FIGURE 1. Variation of radioactive heat production in the Earth during geological time. Chondritic abundance of uranium and thorium and a K:U ratio of 10,000 is assumed (modified from Turcotte and Schubert, 1982).

Because the solar energy flux is 10^4 times larger than the flux from the interior, the surface temperature is independent of the internal heat flow and has probably remained fairly constant through most of geological time. The mean heat flow from within is 82 mW·m^{-2}, which amounts to a total heat loss from the Earth of 4.2×10^{13} W (see *Heat Flow in the Earth*). The ratio of heat production to heat loss is called the *Urey ratio*. If the assumption of the potassium-depleted chondritic composition of the Earth is correct, the Urey ratio is only about 0.5.

Heat Capacity

The ability of the Earth to store heat plays a big role in the energy budget. The specific heat capacity is the best-known thermal property of the Earth, being about 1250 J·kg·K for the mantle and 700 J·kg^{-1}·K^{-1} for the core. For the entire Earth the heat capacity is 6.4×10^{27} J·K^{-1}. If we consider that the Earth contracts slightly upon cooling, we can take the heat delivered by the compressional work into account and arrive at an effective heat capacity of 7.0×10^{27} J·K^{-1}. To illustrate this figure, one could assume that there were no heat sources at all. The body of the Earth could then supply the present heat flow for 5 m.y. before it cooled by 1°; during its whole history the temperature drop would amount to 900 K. As the present mean temperature of the Earth may be about 2500 K, it is conceivable that a significant part of the heat flow from within is due to cooling. The idea of a close equilibrium between heat production and heat loss, which prevailed for some time, must not necessarily be correct, and a Urey ratio of 0.5 might be possible.

Heat Transport

The characteristic time scale t for heat to be transported by conduction over a distance d is $t = d^2/\kappa$, where κ is the thermal diffusivity. If we take for d the depth of the mantle (3000 km) and for κ a value of 10^{-6} m^2·s^{-1}, which is thought typical for mantle rock, the time scale is 300 b.y., much larger than the age of the Earth. If heat would be transported by conduction alone, the thermal state of the Earth could not be in equilibrium. For example, the temperature in the deep interior had no effect on the surface heat flow. Heat transport by thermal convection is much more efficient. Here the characteristic time scale can be estimated as $t = d/u$, where u is the typical velocity of convection currents, say 3 cm/yr in the Earth's mantle. This yields 100 m.y., reasonably short compared with the age of the Earth. Since we know that mantle convection occurs, one can be fairly confident that the different parts of the Earth's body are in some kind of thermal equilibrium. Therefore, it suffices in a first approximation to consider only how the *mean* temperature changed during time, without worrying about details of the temperature distribution. However, we must assume that convection currents can traverse the entire mantle. If the upper and lower mantles convect separately, this simplification is not valid.

Obviously, the rate at which heat is brought from the deep interior to the Earth's surface must depend on the vigor of mantle convection. A measure for the vigor of convection is the Rayleigh number (see *Mantle Convection and Plumes*). Slightly different definitions are possible. The one best suited for our purposes is

$$\text{Ra} = \frac{\alpha g \rho T d^3}{\kappa \eta(T)} \quad (1)$$

where α is the thermal expansion coefficient, g is the gravity acceleration, ρ is the density, d is the depth of the mantle, η is the viscosity, and T is the mean mantle temperature. The Rayleigh number depends explicitly on the temperature; thus, if the mantle cools, Ra will be decreased. However, far more important is the dependence of the viscosity on temperature, which can be expressed by

$$\eta = A \exp \frac{E^*}{RT} \quad (2)$$

where E^* is an activation energy (see *Deformation*

of *Rocks and Minerals*). For mantle rock we can assume as a rule of thumb that the viscosity will increase by one order of magnitude when the temperature drops by 100°C. This has a far more pronounced effect on the Rayleigh number than the explicit temperature dependence.

A measure for the efficiency of the convective heat transport is the Nusselt number Nu, which is defined as the ratio of the actual heat flux Q to the (hypothetical) heat flow due to thermal conduction alone:

$$\text{Nu} = \frac{Q}{\lambda T d} \quad (3)$$

where λ is the thermal conductivity. In order to arrive at a simple model for the thermal history of the Earth, one has to parametrize the convective heat flux. This means that we ignore all the complexities of thermal convection, flow pattern, short-term fluctuations, etc., and restrict ourselves to a simple dependence of the Nusselt number on the Rayleigh number (which we hope to be at least approximately correct). Such a relation is usually expressed in the form

$$\text{Nu} \propto \text{Ra}^\beta \quad (4)$$

which applies when the Rayleigh number is far above its critical value, as it is the case for the Earth. As long as the viscosity is *not* strongly dependent on the temperature, the approximate validity of Eq. 4 with β near 0.3 is well documented by laboratory experiments, numerical calculations, and analytical theory. If viscosity does depend on temperature, things become more complicated. Nevertheless, Eq. 4 with $\beta \approx 0.3$ has been used by many workers to derive a thermal evolution model of the Earth. For variable viscosity convection a unique Nusselt-Rayleigh-number dependence no longer exists. Rather it must be specified how the Rayleigh number is varied. One way would be to increase the temperature at the top and bottom of the convecting layer by the same amount. The temperature difference and, therefore, the variation of the viscosity in the layer remain unchanged, but as the mean viscosity drops the Rayleigh number will rise. As laboratory experiments and numerical calculations indicate, the Nusselt number varies according to Eq. 4 with $\beta \approx 0.3$ (Fig. 2a) in this case, i.e., in the same way as it does in convection with temperature-independent viscosity. This was the rationale for using this parameterization for the Earth.

More applicable to a convecting planet is perhaps the second way to proceed. Here the temperature at the surface of the convecting layer is kept fixed, whereas the interior temperature changes. This implies a change of the Rayleigh number and of the viscosity differences within the convection layer at the same time. In this case the slopes of the Nusselt number versus Rayleigh number in the logarithmic diagram are no longer straight (Fig. 2b); i.e., Eq. 4 no longer applies. At high Rayleigh number the Nusselt number varies very weakly with the Rayleigh number. If these parts of the curve had to be fitted by Eq. 4, the exponent β would be between 0.0 and 0.1. This somewhat unexpected result can be understood when considering the stiff upper thermal boundary layer of the convection cell as the bottleneck for the heat release. Despite its stiffness, this layer moves, as the oceanic plates do, but its velocity is not controlled by the low viscosity in the interior of the convection cell. The rate of its movement is, according to the numerical experiments, determined by the viscosity at the upper boundary, which does not vary in the case of Fig. 2b since the surface temperature is fixed. If this latter case would be applicable to the convecting mantle, it would mean that the velocity of the lithospheric plates would be little affected by changes of the interior mantle temperature and viscosity. Because the Earth loses most of its heat by the process of seafloor spreading and subsequent cooling of the oceanic plates (see *Heat Flow in the Earth*), the global heat release would be little affected, too. If, however, the case of Fig. 2a is more applicable, the speed of plate motion and the heat loss would be strongly linked to the internal temperature via its control on the mantle viscosity. Most scientists, who have considered the thermal evolution of the Earth recently, have adopted this view (e.g., Davies, 1980; Spohn and Schubert, 1982). We will pursue both alternatives. A more detailed account can be found in Christensen (1985).

Initial Thermal State

Whether the Earth started out hot or cold depends on the mode of its accretion from the solar nebula. If the Earth accumulated from many dust-sized particles, virtually all the accretional energy was immediately radiated into space from the surface of the growing proto-earth, and it would have begun quite cold. However, today the view is strongly favored that the dust first accumulated into "planetesimals" growing in size, and the final formation of the Earth occurred in part by the impact of very large bodies. Here, a fraction of the accretional energy was deeply buried in the growing Earth and escaped the radiative reemission, which would imply a hot start.

Parametrized Thermal History Models

Starting point of a planetary thermal evolution model is the law of conservation of energy in the form

$$\frac{C \, dT}{dt} = H(t) - Q(T) \quad (5)$$

The change of the mean temperature T multiplied by the heat capacity C is obtained from the (im)balance

ENERGY BUDGET OF THE EARTH

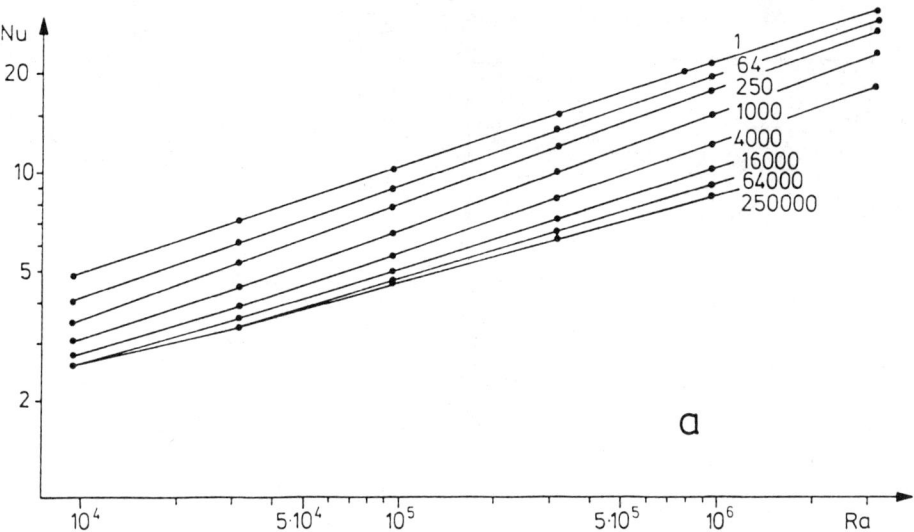

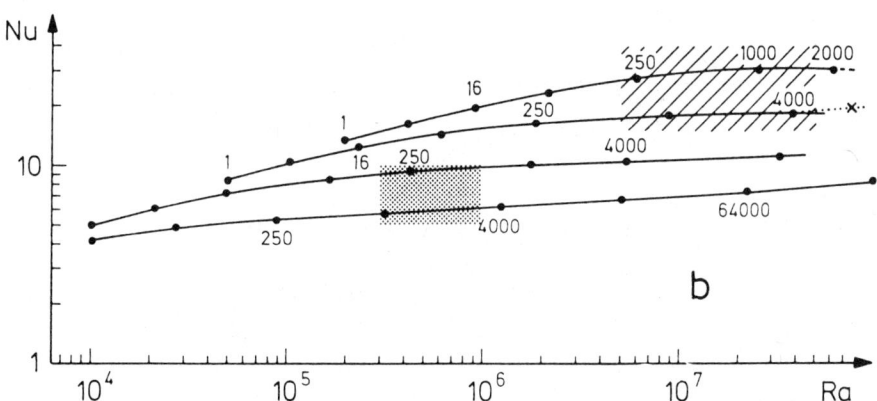

FIGURE 2. Dependence of the Nusselt number on the Rayleigh number in a logarithmic scale. Results from numerical calculations for convection with temperature-dependent viscosity are displayed. (*a*) The ratio between maximum and minimum viscosity is held constant on each line. The slope of the lines, which is equivalent to the exponent β in Eq. 4, is in the range 0.25–0.35. (*b*) The surface temperature (and surface viscosity) is held constant on each line. The viscosity ratio is indicated by the numbers at the data points. The hatched and dotted regions indicate the appropriate parameter range for present-day convection in the whole mantle and in the upper mantle, respectively.

between the time-dependent rate of radioactive heating H and the heat loss Q. We assume that Q is some function of T alone; this is a "parametrization" equivalent to the Nusselt–Rayleigh number dependency. Equations 1–5 can now be combined. Some constant of proportionality must be introduced in Eq. 4. To make things simple, we may combine this constant with all the other parameters, which do not interest us further, into a constant A in the final equation for a parametrized thermal evolution model:

$$\frac{C\,dT}{dt} = H(t) - AT^{1+\beta} \exp\frac{-\beta E^*}{RT} \quad (6)$$

Equation 6 is an ordinary differential equation for the mean temperature of the Earth (or its mantle) as a function of time. The constant A can be calibrated with the present heat flow from the Earth and the present mean temperature. Various concentrations of radioactive elements can be assumed, and H varies during time as in Fig. 1. Equation 6 may be integrated backward in time, starting from the present state; such integration avoids the guess for the initial temperature. If inappropriate parameters are chosen (e.g., a "wrong" amount of heat sources), this procedure may lead to unacceptable solutions, which, for example, have a temperature

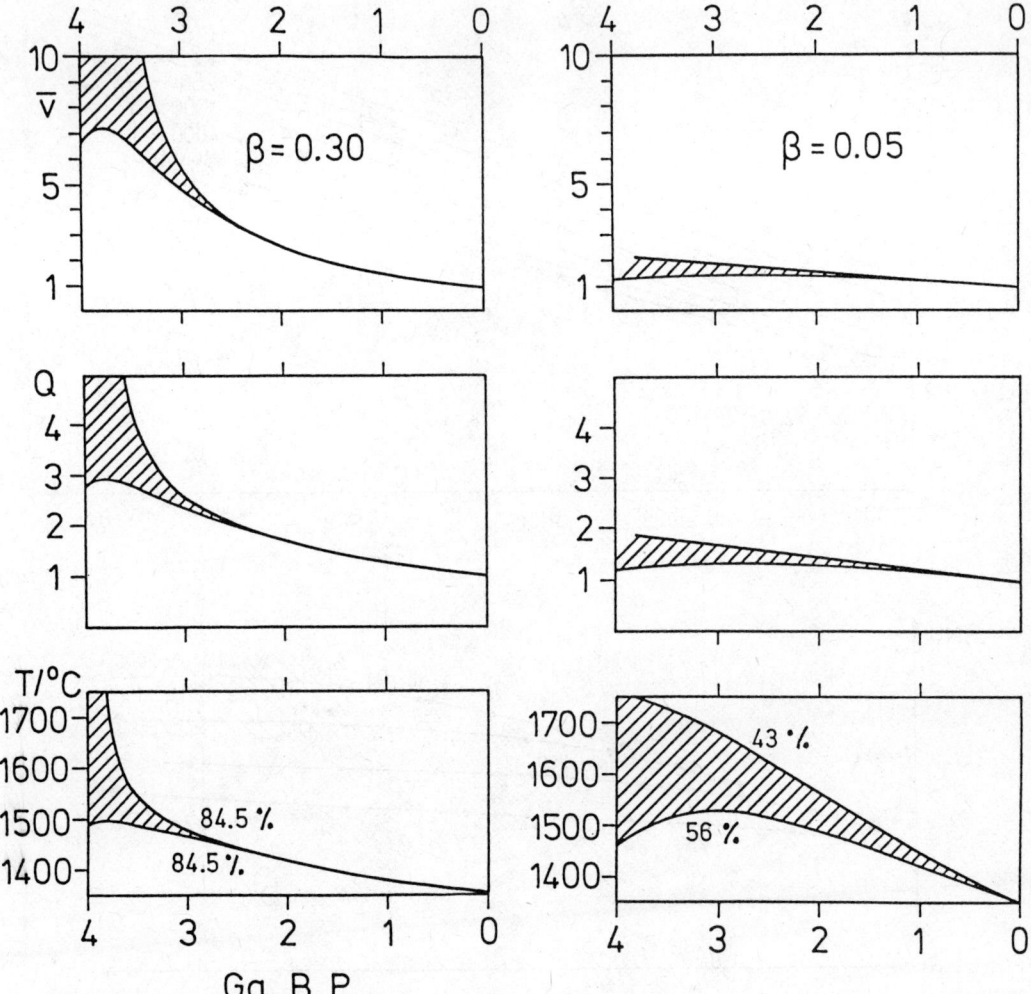

FIGURE 3. Results of thermal history models using two different parameterizations of the convective heat transport. In each column from top to bottom the variation of the mean plate velocity $\bar{v}$, of the heat flow Q, and of the temperature in the upper mantle below the lithosphere is shown. Velocity and heat flow are scaled to the present value. In the temperature diagram the limiting Urey ratios are also indicated.

singularity ($T = \infty$). By discarding such solutions, constraints can be put on the acceptable range for the rate of heating or, in other words, the Urey ratio. In Fig. 3 the range of permissible solutions is displayed for two models, where the exponent β in the Nusselt-Rayleigh-number relationship is taken as 0.30 and 0.05, respectively. Common to both kinds of models is that the Earth is cooling at least for the last 3 b.y. and perhaps for all its history. However, otherwise significant differences are obtained. The permissible range for the (present) Urey ratio is about 0.5 for low β, in good agreement with the geochemically inferred abundance of heat-producing elements in the Earth. With the "classical" β of 0.30 from constant-viscosity convection, the Urey ratio is about 0.85, which is too high, although there is some leeway due to uncertainties in various parameters (e.g., in E^*).

In the model with $\beta = 0.05$ only half of the heat flux is replenished by radioactive heat, which implies a high cooling rate. For the Archaean (about 3 b.y. ago) the mantle temperature would have been 180–350°C higher than at present. Some estimate of the paleotemperature can be obtained from the analysis of komatiites, which are very olivine-rich basaltic lavas that erupted almost exclusively in the Archaean. Their high olivine content implies a very high melting temperature. Comparing them with olivine-rich recent basalts, geophysicists have calculated a temperature difference of about 200–400°C.

In a thermal history model with $\beta = 0.3$ the temperature difference of about 130°C at 3 b.y. falls somewhat short of this figure.

In addition to the temperature history the variation of the surface heat flow and the average plate velocity $\bar{v}$ can be calculated (the latter from the relation $\bar{v} \propto Q^2$). With $\beta = 0.3$ the Archaean heat flow is determined to be about three times the present one, which goes parallel to the threefold increase in heat production. With the small value for β the difference is only a factor of 1.5. This implies that in the Archaean heat production and heat loss could have been in close equilibrium, but since then the gap between them widened, because the rate of heat generation dropped strongly, whereas the decrease in the heat flow was moderate. Through a study of Archaean metamorphic rocks it has been inferred that thermal gradients in the continental crust were not much higher than at present. Together with the high mantle temperatures indicated by the komatiites this has been called the *Archaean paradox*. In the low-β model this paradox would be explained. However, the continental heat flow may not be a good indicator of the overall heat loss in the past, which is assumed to occur predominantly in the oceans. The most dramatic differences between the high-β and the low-β models occurs in the inferred mean plate velocities: about five to ten times faster in the first case, and not more than twice as fast in the other case. Palaeomagnetism has been used to derive drift rates for the continental blocks during the last 2.5 b.y. There is no indication for a systematic decrease of the velocity during this period (Fig. 4).

Apparently the parametrized evolution model with a weak dependence of heat flow and plate velocity on the interior temperature leads to a thermal history that agrees with the independent evidence from geochemistry, geology, and paleomagnetism. The classical parametrized evolution model, assuming a strong coupling, does not perform equally well. However, it must be kept in mind that both models use highly simplifying assumptions and ignore some potentially important effects. For example, it is not clear if a high-viscosity boundary layer on top of a convection cell with temperature-dependent viscosity is already a good analogue for the Earth's lithosphere, where plastic, brittle, or elastic deformations also play a role. Another point is that the heat loss from the Earth could have been higher without increasing the mean velocity of surface plates, if there was a greater number of smaller plates in the past. Also, it must be kept in mind that the evidence on paleotemperature, heat flow, plate velocity, and Urey ratio is circumstantial and will perhaps need revision in the future. Thus the conclusions from these models must be taken with care. The most robust conclusion is that the Earth has been cooling for at least some billion years, and that whole Earth cooling provides a significant amount to the present heat flow.

Here only the simplest kind of parametrized thermal evolution model could be discussed. More elaborate models take into account the division of the Earth into several convecting subsystems. The division into separately convecting upper and lower mantles is disputed. In this case the insulating effect due to the lack of heat advection across the interface would allow almost any Urey ratio for the Earth,

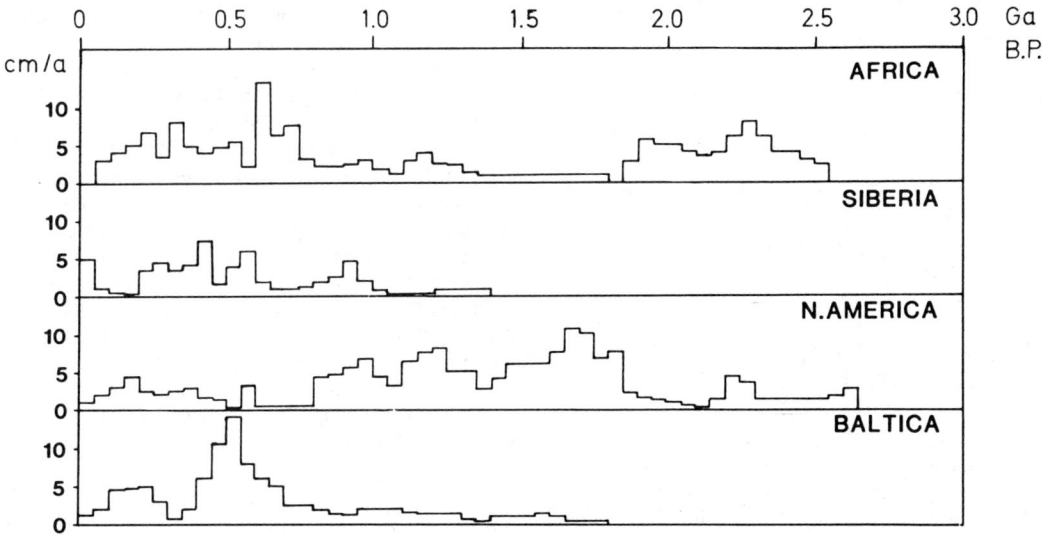

FIGURE 4. Minimum drift rates of four continental blocks with respect to the magnetic pole. (Adapted from Ullrich and Van der Voo, 1982)

depending on the initial state (McKenzie and Richter, 1981). An obvious division is that into mantle and core. Because the core contains probably no heat sources and constitutes only 20% of the Earth's heat capacity, it must not be explicitly considered in a model for the global evolution of the Earth. However, convection in the outer core is responsible for the Earth's magnetic field. The inner core probably grows by freezing pure iron-nickel from the outer core, leaving behind an alloy enriched in the lighter element(s). This may drive some kind of chemical convection. Obviously its vigor depends solely on the cooling rate and on the existence of a solid inner core. Thus the thermal history of the core is important for the presence and strength of the magnetic field in past times (Stevenson et al., 1983).

ULRICH CHRISTENSEN

References

Christensen, U. R., 1985, Thermal evolution models for the earth, *Jour. Geophys. Research* **90,** 2995–3007.
Davies, G. F., 1980, Thermal histories of convective earth models and constraints on radiogenic heat production in the earth, *Jour. Geophys. Research* **85,** 2517–2530.
McKenzie, D. P., and F. M. Richter, 1981, Parameterized thermal convection in a layered region and the thermal history of the earth, *Jour. Geophys. Research* **86,** 11667–11680.
Spohn, T., and G. Schubert, 1982, Modes of mantle convection and removal of heat from the earth's interior, *Jour. Geophys. Research* **87,** 4682–4696.
Stevenson, D. J., T. Spohn, and G. Schubert, 1983, Magnetism and thermal evolution of the terrestrial planets, *Icarus* **54,** 466–489.
Turcotte, D. L., and G. Schubert, 1982, *Geodynamics,* New York: John Wiley & Sons.
Ulrich, W., and R. Van der Voo, 1982, Minimum continental velocities with respect to the pole since the Archaean, *Tectonophysics* **74,** 17–27.

Cross-references: *Deformation of Rocks and Minerals; Earth and Moon: Origins; Earth's Core; Earth Structure: Global; Geothermal Energy; Heat Flow in the Earth; Mantle Convection and Plumes; Mantle Dynamics; Mantle Viscosity; Meteorites: Nature and Origin; Solar System: Origin.*

EQUILIBRIUM FIGURE OF THE EARTH

If the Earth were a fluid body, it would respond instantaneously to any stress, including stresses caused by the Earth's rotation, until it attains a state of zero stress. The figure that the Earth would assume if it were in such a state is called the *hydrostatic equilibrium figure* or, simply, *hydrostatic figure* or *equilibrium figure.* As we will show, the real Earth, however, is not in complete hydrostatic equilibrium but has small deviations from it. Since hydrostatic shape indicates a state of zero stress, any departures from this state are particularly interesting because they show the extent of available stresses in the interior of the Earth that can be invoked to explain any geophysical mechanisms that may be found or assumed to exist in the Earth's interior. For this reason, apart from its traditional appeal, the problem is particularly interesting to geophysicists.

The mathematical theory of hydrostatic equilibrium for the Earth, to the first order of small quantities, was originally developed by A. C. Clairaut in 1743. In 1885, R. Radau simplified the solution of Clairaut's differential equation by making an important substitution. The original purpose of the theory was to provide information about density distribution inside the Earth. It was found, however, that with the then-known data, the theory led to no discrimination between widely varying laws of density. But it did yield more accurate values of flattening for the Earth (assuming hydrostatic equilibrium) than were likely to be obtained by geodetic surveys. This stimulated further interest in the theory, and its development was extended to the second order by O. Callandreau in 1889 and G. H. Darwin in 1900. DeSitter (1938) modified the development and studied its actual application to the Earth. Subsequent applications of the second-order theory were made by Bullard (1948) and Jeffreys (1962).

With the advent of artificial Earth satellites and the consequent highly accurate determination of the second-degree harmonic coefficient of geopotential, it became possible to determine the actual flattening of the Earth directly from the external potential theory. The same coefficient, in combination with the precessional constant of the Earth, also yields accurate values of the Earth's polar moment of inertia; hence the hydrostatic theory could, for the first time, be used to yield the hydrostatic flattening independently. This provided to the geophysicists the long sought after opportunity of studying the departures of the actual Earth from its equilibrium state. Such studies were conducted by J. A. O'Keefe, S. W. Henriksen, M. Caputo, and M. A. Khan in the postartificial Earth satellite era. Since there is a fundamental difference in the pre- and postartificial Earth satellite applications of the hydrostatic theory, Khan (1969) revised and extended the second-order theory to suit readily the new applications and data types.

This article recounts the complete development of the theory. For the sake of brevity, however, some intermediate algebraic steps are omitted.

Theory of the External Field

The external gravitational potential U of a body symmetrical with respect to its equatorial plane and polar axis is (see, for example, Jeffreys, 1962)

$$U = \frac{GM}{r}\left(1 - \sum \left(\frac{a_e}{r}\right)^{2n} J_{2n}P_{2n}(\sin \phi)\right)$$

which, accurate to the square of flattening, reduces to

$$U = \frac{GM}{r}\left(1 - \left(\frac{a_e}{r}\right)^2 J_2 P_2(\sin \phi) - \left(\frac{a_e}{r}\right)^4 J_4 P_4(\sin \phi) - O(f^3)\right) \quad (1)$$

where

$P_{2n}(\sin \phi)$ = Legendre polynomials
a_e = equatorial radius of the body
r, ϕ = geocentric coordinates
J_{2n} = constant coefficients associated with Legendre polynomials
$O(f^3)$ = quantities of the order of the cube of flattening

U is the gravitational potential. The potential of gravity, V, is

$$V = U + \tfrac{1}{3}\omega^2 r^2 [1 - P_2(\sin \phi)]$$

where ω is the rate of rotation of the body. The condition for a surface to have a constant gravity potential, therefore, is

$$U + \tfrac{1}{3}\omega^2 r^2 [1 - P_2(\sin \phi)] = V = \text{constant} \quad (2)$$

Let the equation of an equipotential surface be

$$\frac{a_e}{r} = 1 + (f + \tfrac{3}{2}f^2)\sin^2 \phi - \tfrac{1}{2}f^2 \sin^4 \phi + O(f^3) \quad (3)$$

which in terms of the mean radius r_0 and the Legendre polynomials is expressible to the second degree as

$$r = r_0(1 + \alpha_2 P_2 + \alpha_4 P_4)$$
$$= r_0(1 - (\tfrac{2}{3}f + \tfrac{23}{63}f^2)P_2(\sin \phi) + \tfrac{12}{35}f^2 P_4(\sin \phi)) \quad (3a)$$

Substitution of Eqs. 1 and 3 or 3a in Eq. 2 yields, among other relations,

$$J_2 = \tfrac{2}{3}f - \tfrac{1}{3}f^2 - \tfrac{1}{3}m + \tfrac{2}{21}mf + O(f^3) \quad (4a)$$

$$J_4 = -\tfrac{4}{5}f^2 + \tfrac{4}{7}mf + O(f^3) \quad (4b)$$

where

$$m = \frac{\omega^2 a^3(1 - f)}{GM}$$

Equations 3a, 4a, and 4b will be used later.

Theory of the Internal Field

The condition of hydrostatic equilibrium for any point in the Earth's interior is

$$\frac{dp}{dr} = \rho \frac{dV}{dr} \quad (5)$$

where the pressure p and the density ρ are related to the point under consideration. Thus surfaces of constant V are also surfaces of constant p and ρ; i.e., the surfaces of equal density are equipotential surfaces. Let one such surface with a uniform density ρ' be expressed as

$$r = a'\left(1 + \sum \alpha_n P_n\right) \quad (6)$$

where α_n and ρ' are functions of a'. Let r_1 be the value of a' for the surface of constant density through an internal point. Then the potential $V(r_1)$ on this surface is the sum of potentials from (a) matter inside the shell of radius r_1, (b) matter outside the shell of radius r_1, and (c) rotational potential; i.e.,

$$V(r_1) = \frac{4}{3}\pi G \int_0^{r_1} \rho' \frac{\partial}{\partial a'}$$
$$\cdot \left(\frac{a'^3}{r} + \sum \frac{3}{2n+1} \frac{a'^{n+3}}{r^{n+1}} \alpha_n P_n\right) da'$$
$$+ \frac{4}{3}\pi G \int_{r_1}^{a} \rho' \frac{\partial}{\partial a'}$$
$$\cdot \left(\frac{3}{2} a'^2 + \sum \frac{3}{2n+1} \frac{r^n}{a'^{n-2}} \alpha_n P_n\right) da'$$
$$+ \frac{1}{2} \omega^2 r_1^2 (1 - \sin^2 \phi') \quad (7)$$

The mean density ρ_0 within the surface r_1 is

$$\rho_0 = \frac{3}{r_1^3} \int_0^{r_1} \rho' a'^2 \, da' \quad (8)$$

Since the second-order development of the hydrostatic theory becomes somewhat complicated and may tend to camouflage the correct structure of the development, I start by developing the first-order theory to illustrate clearly the structure of the problem and then extend it to the second order. The development of the first-order theory follows Jeffreys (1962).

Neglecting the difference between the geocentric and geodetic coordinates, which implies that V varies only radially, substituting the value of r from Eq. 6 in Eq. 7, and neglecting terms containing $n > 2$, including those containing P_2^2, yield

$$\frac{4}{3}\pi G \left(\frac{1 - \alpha_2 P_2}{r_1}\right) \int_0^{r_1} 3\rho' a'^2 \, da' + \frac{3}{5} P_2$$
$$\cdot \left\{\frac{1}{r_1^3} \int_0^{r_1} \rho' \, d(a'^5 \alpha_2) + r_1^2 \int_{r_1}^{a} \rho' \, d\alpha_2\right\}$$
$$+ \frac{1}{3}\omega^2 r_1^2 (1 - P_2) = F(r_1) \qquad (9)$$

Equation 9 indicates that the function on the left-hand side is constant for given r_1; i.e., V is a function of r only. Hence, the coefficient of P_2 must vanish on the equipotential surface; i.e.,

$$-\frac{\alpha_2}{r_1} \int_0^{r_1} \rho' a'^2 \, da' + \frac{1}{5}\left\{\frac{1}{r_1^3}\int_0^{r_1} \rho' \, d(a'^5 \alpha_2)\right.$$
$$\left. + r_1^2 \int_{r_1}^{a} \rho' \, d\alpha_2\right\} = \frac{\omega^2 r_1^2}{12\pi G} \qquad (10)$$

Multiply Eq. 10 by r^3 and differentiate; then the distinction between r and r_1 is no longer important. This gives

$$-\left(r^2 \frac{d\alpha_2}{dr} + 2\alpha_2 r\right) \int_0^r \rho' a'^2 \, da'$$
$$+ r^4 \int_r^a \rho' \frac{d\alpha_2}{da'} \, da' = \frac{5\omega^2 r^4}{12\pi G} \qquad (11)$$

Divide by r^4 and differentiate again:

$$\left(\frac{d^2\alpha_2}{dr^2} - \frac{6\alpha_2}{r^2}\right)\int_0^r \rho' a'^2 \, da'$$
$$+ 2\rho r^2 \left(\frac{d\alpha_2}{dr} + \frac{\alpha_2}{r}\right) = 0 \qquad (12)$$

Now substitute from Eq. 8:

$$\rho_0 \left(\frac{d^2\alpha_2}{dr^2} - \frac{6\alpha_2}{r^2}\right) + \frac{6\rho}{r}\left(\frac{d\alpha_2}{dr} + \frac{\alpha_2}{r}\right) = 0 \qquad (13)$$

This is Clairaut's differential equation.

Equation 11 is particularly interesting. For $r = a$, it becomes

$$-\left(a^2 \frac{d\alpha_2}{da} + 2\alpha_2 a\right)\int_0^a \rho' a'^2 \, da' = \frac{5\omega^2 a^4}{12\pi G} \qquad (14)$$

The mass of the whole body is

$$M = 4\pi \int_0^a \rho' a'^2 \, da' = \tfrac{4}{3}\pi a^3 \rho_0(a)$$

where $\rho_0(a)$ is the mean density of the mass bounded by a. Equation 14 then simplifies to

$$a\frac{d\alpha_2}{da} + 2\alpha_2 = -\frac{5\omega^2 a^3}{3GM} = -\frac{5}{3}m_e$$

where

$$m_e = \frac{\omega^2 a^3}{GM} \qquad (15)$$

But to the first order

$$m_e \simeq m = \frac{\omega^2 a^3 (1 - f)}{GM}$$

Thus

$$\frac{a}{\alpha_2(a)}\left(\frac{d\alpha_2}{da}\right)_a = -\frac{5m}{3\alpha_2(a)} - 2 \qquad (16)$$

where a as a subscript or in parentheses denotes the value of the quantity at $r = a$.

Let a new dependent variable be

$$\eta = \frac{d \log \alpha_2}{d \log r} = \frac{r \, d\alpha_2}{\alpha_2 \, dr} \qquad (17)$$

Then

$$\frac{d\alpha_2}{dr} = \frac{\eta \alpha_2}{r}$$
$$\frac{d^2\alpha_2}{dr^2} = \frac{\alpha_2}{r}\frac{d\eta}{dr} + \frac{\eta}{r}\frac{r}{\alpha_2}\frac{d\alpha_2}{dr}\frac{\alpha_2}{r} - \frac{\eta \alpha_2}{r}$$
$$= \alpha_2 \left(\frac{1}{r}\frac{d\eta}{dr} + \frac{\eta^2 - \eta}{r^2}\right) \qquad (18)$$

Substitute Eq. 18 in Eq. 17 and multiply the result by $r^2/\alpha_2 \rho_0$ to get

$$r\frac{d\eta}{dr} + \eta^2 - \eta - 6 + \frac{6\rho}{\rho_0}(\eta + 1) = 0 \qquad (19)$$

Differentiating both sides of Eq. 8, we get

$$\rho r^2 = \frac{1}{3}\frac{d}{dr}(\rho_0 r^3) \qquad (20)$$

which on simplification gives

$$\frac{\rho}{\rho_0} = 1 + \frac{1}{3}\frac{r}{\rho_0}\frac{d\rho_0}{dr} \qquad (20a)$$

Hence Eq. 19 becomes

$$r\frac{d\eta}{dr} + \eta^2 + 5\eta + 2\frac{r}{\rho_0}\frac{d\rho_0}{dr}(\eta + 1) = 0 \qquad (21)$$

Now consider

$$\frac{(d/dr)(\rho_0 r^5 \sqrt{1+\eta})}{\rho_0 r^5 \sqrt{1+\eta}} = \frac{1}{\rho_0}\frac{d\rho_0}{dr} + \frac{5}{r} + \frac{1}{2(1+\eta)}\frac{d\eta}{dr} \quad (22)$$

which gives

$$\frac{d\eta}{dr} = \frac{2\sqrt{1+\eta}}{\rho_0 r^5}\frac{d}{dr}(\rho_0 r^5 \sqrt{1+\eta})$$
$$- \frac{2}{\rho_0}\frac{d\rho_0}{dr}(1+\eta) - \frac{10(1+\eta)}{r} \quad (23)$$

Substitution of this in Eq. 21 yields

$$\frac{d}{dr}(\rho_0 r^5 \sqrt{1+\eta}) = 5\rho_0 r^4 \psi(\eta) \quad (24)$$

where

$$\psi(\eta) = \frac{1 + \tfrac{1}{2}\eta - \tfrac{1}{10}\eta^2}{\sqrt{1+\eta}} \quad (25)$$

Equation 25 is Radau's equation, as the substitution in Eq. 17 was conceived by R. Radau in 1885. The importance of the substitution lies in the fact that, for any reasonable law of density variation, the function $\psi(\eta)$ has the remarkable property of never departing from 1 by more than eight parts in 10^4. Then to an accuracy of this order,

$$\frac{d}{dr}(\rho_0 r^5 \sqrt{1+\eta}) = 5\rho_0 r^4$$

To the same degree of accuracy, the polar moment of inertia C is

$$C = \tfrac{8}{3}\pi \int_0^a \rho r^4\, dr$$

which, on integration by parts, yields

$$C = \tfrac{8}{9}\pi \int_0^a \left(3r^4\rho_0(r) + r^5\frac{d\rho_0}{dr}\right) dr$$

or

$$C = \tfrac{2}{3}Ma^2 - \tfrac{16}{9}\pi \int_0^a \rho_0 r^4\, dr \quad (26)$$

But the integral in Eq. 26 is

$$\int_0^a \rho_0 r^4\, dr = \frac{3}{20}\frac{Ma^2}{\pi}\sqrt{1+\eta(a)}$$

Thus

$$C = \tfrac{2}{3}Ma^2 - \tfrac{4}{15}Ma^2\sqrt{1+\eta(a)}$$

or

$$\eta(a) = \left(\frac{5}{2} - \frac{15}{4}\frac{C}{Ma^2}\right)^2 - 1 \quad (27)$$

But from Eq. 17 the value of η at $r = a$ is

$$\eta(a) = \frac{a}{\alpha_2(a)}\left(\frac{d\alpha_2}{dr}\right)_a$$

which, from Eq. 16, is

$$\eta(a) = -\frac{5}{3}\frac{m}{\alpha_2(a)} - 2 \quad (28)$$

It is the choice of the coefficient $\alpha_2(a)$ that links external potential theory to hydrostatic theory. Equations 3 or 3a, 4, and 5 of external potential theory are derived without any assumption about the density distribution inside the Earth. Hence, these equations are valid whether or not hydrostatic equilibrium exists in the Earth. But, if the Earth were in hydrostatic equilibrium, the exterior surface defined by Eq. 6 must match that defined by Eq. 3 or 3a. This should be intuitively clear: If the hydrostatic theory were true for the Earth, the surfaces defined by the theories of external and internal fields must be coincident at the outer boundary. To first order, Eq. 3a becomes

$$r = r_0(1 - \tfrac{2}{3}f_h P_2(\sin \phi)) + O(f^2)$$

and

$$\alpha_2 = -\tfrac{2}{3}f_h$$

where the subscript h refers to hydrostatic flattening. Consequently, from Eqs. 27 and 28,

$$\frac{C}{Ma^2} = \frac{2}{3} - \frac{4}{15}\sqrt{\frac{5}{2}\frac{m}{f_h} - 1}$$

or

$$f_h = \frac{5m}{2}\left(1 + \left(\frac{5}{2} - \frac{15}{4}\frac{C}{Ma^2}\right)^2\right)^{-1}$$

The first-order development illustrates clearly the structure of the hydrostatic equilibrium problem. The development of the second-order theory is somewhat complicated but follows exactly the procedure just outlined. Now all terms not greater than $O(f^2)$ must be retained, and the simplifying assumptions must be modified accordingly. The treatment we give follows deSitter (1938) and Khan (1969).

Development of the second-order theory becomes somewhat simpler if we choose as the independent variable the mean radius of a surface of equal density (deSitter, 1938). The potential V at any point (r, ϕ') within the Earth (Eq. 7), correct to the second order, can then be written as

$$V = \frac{4}{3}\pi G \left(3 \int_0^\beta \frac{\rho}{r} \beta^2 \, d\beta \right.$$

$$- \frac{2}{5} \left\{ \int_0^\beta \frac{\rho}{r^3} \frac{d}{d\beta} \left(\alpha' + \frac{2}{7}\alpha^2\right) \beta^5 \, d\beta \right.$$

$$+ \left. \int_\beta^1 r^2 \rho \frac{d}{d\beta}\left(\alpha' + \frac{16}{21}\alpha^2\right) \right\} P_2(\sin\phi')$$

$$+ \frac{12}{35} \left\{ \int_0^\beta \frac{\rho}{r^5} \frac{d}{d\beta} [\alpha^2 \beta^7] \, d\beta + \frac{32}{105} \right.$$

$$\left. \cdot \int_\beta^1 r^4 \rho \frac{d}{d\beta}(\beta^{-2}) \, d\beta \right\} P_4(\sin\phi') \bigg)$$

$$+ \frac{1}{3}\omega^2 r^2 [1 - P_2(\sin\phi')] \qquad (29)$$

where β is the mean radius of an arbitrary surface of equal density expressed in terms of the mean radius of the outer surface, and ρ is the density, expressed in terms of the mean density. The units of ρ differ from the previously used ρ'. In this notation, $\rho(\beta)$, the mean density within the surface β (corresponding to Eq. 8), is

$$\rho(\beta) = \frac{3}{\beta^3} \int_0^\beta \rho \beta^2 \, d\beta \qquad (30)$$

Thus for the surface $r = a$ (i.e., $\beta = 1$),

$$\rho(r) = \rho(1) = 1$$

The equivalent of Eq. 6 which defines a surface of equal density, also in this case an equipotential surface (Eq. 4), becomes

$$r = \beta(1 - (\tfrac{2}{3}\alpha' + \tfrac{4}{9}\alpha^2) P_2(\sin\phi')$$

$$+ \tfrac{12}{35}\alpha^2 P_4(\sin\phi')) \qquad (31)$$

where for convenience of algebraic treatment the quantity α' is introduced:

$$\alpha' = \alpha - \tfrac{5}{42}\alpha^2$$

In the second-order theory $\alpha'^2 = \alpha^2$. Substitution of Eq. 31 in Eq. 29 and the condition that the resulting equation must define an equipotential surface yield

$$\left(\alpha' + \frac{2}{7}\alpha^2 - \frac{1}{2}m(\beta)\right) \frac{3}{\beta^3} \int_0^\beta \rho \beta^2 \, d\beta$$

$$- \frac{3}{5\beta^3} \int_0^\beta \rho \frac{d}{d\beta}\left(\left(\alpha' + \frac{2}{7}\alpha^2\right)\beta^5\right) d\beta$$

$$+ \left(\frac{4}{7}\alpha - \frac{3}{5}\right) \int_\beta^1 \rho \frac{d}{d\beta}\left(\alpha' + \frac{16}{21}\alpha^2\right) d\beta$$

$$- \frac{4}{21}\alpha m(r_1) = 0 \qquad (32)$$

where

$$m(\beta) = \frac{\omega^2 \beta^3}{GM}$$

and

$$m(r_1) = \frac{\omega^2 r_1^3}{GM}$$

with β and r_1 denoting the radii of the same surface in different units. Also

$$9 \frac{\alpha^2}{\beta^3} \int_0^\beta \rho \beta^2 \, d\beta + \frac{3}{\beta^7} \int_0^\beta \rho \frac{d}{d\beta}[\alpha^2 \beta^7] \, d\beta$$

$$+ \frac{8}{3}\beta^2 \int_\beta^1 \rho \frac{d}{d\beta}(\beta^{-2}) \, d\beta$$

$$- \frac{6\alpha}{\beta^5} \int_0^\beta \rho \frac{d}{d\beta}$$

$$\cdot \left(\left(\alpha' + \frac{2}{7}\alpha^2\right)\beta^5\right) d\beta = 0$$

The treatment of Eq. 32 in the manner of Eq. 10 and the introduction of the variable η, which is now defined as

$$\eta = \frac{d \log \alpha'}{d \log \beta} = \frac{\beta}{\alpha'} \frac{d\alpha'}{d\beta}$$

yield

$$\left(\eta\alpha'\left(1 + \frac{4}{7}\alpha\right) - 3\alpha'\left(1 + \frac{2}{7}\alpha\right)\right) \frac{3}{\beta^3} \int_0^\beta \rho\beta^2 \, d\beta$$

$$- \frac{4}{21}\eta\alpha' m(r_1) + \frac{4}{7}\eta\alpha' \int_\beta^1 \rho \frac{d}{d\beta}$$

$$\cdot \left(\alpha' + \frac{16}{21}\alpha^2\right) d\beta + \frac{3}{\beta^5} \int_0^\beta \rho \frac{d}{d\beta}$$

$$\cdot \left(\left(\alpha' + \frac{2}{7}\alpha^2\right)\beta^5\right) d\beta = 0 \qquad (33)$$

which, on redifferentiation and treatment in the manner of Eqs. 11–20, yields the second-order counterpart of Eq. 21:

$$\beta \frac{d\eta}{d\beta} + \eta^2 + 5\eta - 2\zeta(1 + \eta) - \frac{4}{21}\zeta\xi = 0 \tag{34}$$

where

$$\zeta = -\frac{\beta}{\rho(\beta)} \frac{d\rho(\beta)}{d\beta} = 3\left(1 - \frac{\bar{\rho}}{\rho(\beta)}\right)$$

as yielded by the differentiation of Eq. 30 in a manner analogous to Eqs. 20 and 20a. Also

$$\xi = 7m(\beta)(1 + \eta) - 3\alpha(1 + \eta)^2 - 4\alpha$$

Like Eq. 11, Eq. 33 is particularly interesting. For $r = a$ (i.e., $\beta = 1$), it yields, after simplification,

$$\eta(a)f_h' = 3f_h' - \frac{6}{7}f_h^2 + \frac{4}{7}mf_h$$

$$- J_h\left(5 + \frac{10}{21}f_h + \frac{20}{21}m\right) \tag{35}$$

where $\eta(a)$ denotes the value of η for the outer surface on which α and α' also become $\alpha'(a) = f_h'$ and $\alpha(a) = f_h$; i.e., f_h denotes the value of hydrostatic flattening for the outer surface and J_h is the hydrostatic counterpart of $J = \frac{3}{2}J_2$.

By the procedure outlined in Eqs. 22 and 23, Eq. 34 yields the second-order counterparts of Eqs. 24 and 25; i.e.,

$$\frac{d}{d\beta}\rho(\beta)\beta^5\sqrt{1 + \eta} = 5\rho(\beta)\beta^4\psi(\eta) \tag{36}$$

where

$$\psi(\eta) = \frac{1 + \frac{1}{2}\eta - \frac{1}{10}\eta^2 + \frac{2}{105}\zeta\xi}{\sqrt{1 + \eta}}$$

Integration of Eq. 36 yields

$$\int_0^\beta \rho(\beta)\beta^4\psi(\eta) \, d\beta = \frac{1}{5}\rho(\beta)\beta^5\sqrt{1 + \eta}$$

or

$$\int_0^\beta \rho(\beta)\beta^4 \, d\beta = \frac{1}{5}\rho(\beta)\beta^5 \frac{\sqrt{1 + \eta}}{1 + \lambda}$$

$$= \frac{3}{20}\frac{M\beta^2}{\pi}\frac{\sqrt{1 + \eta}}{1 + \lambda}$$

which for the outer surface yields

$$\int_0^1 \rho(\beta)\beta^4 \, d\beta = \frac{3}{20}\frac{M}{\pi}\frac{\sqrt{1 + \eta(a)}}{1 + \lambda(a)} \tag{37}$$

where $1 + \lambda$ is the average value of the function $\psi(\eta)$ over the range of integration, and $\lambda(a)$ denotes the value of λ for the outer surface.

The polar moment of inertia C is given by the second-order counterpart of Eq. 26; i.e.,

$$C = \tfrac{8}{3}\pi \int_0^\beta \rho\beta^4 \, d\beta + \tfrac{2}{3}(C - A)$$

where A is the moment of inertia of the body around its equatorial diameter. Integration of this equation and substitution of Eq. 37 for the outer surface values yield

$$q = \frac{3}{2}\frac{C}{Ma^2} = 1 - \frac{2}{3}f_h + \frac{4}{9}f_h^2$$

$$- \frac{2}{5}\left(1 - \frac{2}{3}f_h + \frac{4}{9}f_h^2\right)\frac{\sqrt{1 + \eta(a)}}{1 + \lambda(a)} + J_{2h} \tag{38}$$

where a is now the equatorial radius.

Eliminating $\eta(a)$ from Eqs. 35 and 38, we get an explicit expression for f_h:

$$f_h = \frac{1}{3 - \eta_0}\left((5J_h + \delta_2)\right.$$

$$\left. + \frac{(5J_h + \delta_2)\delta_1}{3 - \eta_0} + \frac{25AJ_h^2}{(3 - \eta_0)^2}\right) \tag{39}$$

where

$$\eta_0 = \tfrac{25}{4}F^2 q'^2 - 1$$
$$q' = 1 - q$$
$$F = 1 + \lambda(a)$$
$$A = \tfrac{17}{14} - \tfrac{5}{42}\eta_0 - \tfrac{25}{3}F^2 qq'$$
$$\delta_1 = \tfrac{25}{3}F^2 q'J_h - \tfrac{4}{7}m + \tfrac{10}{21}J_h$$
$$\delta_2 = \tfrac{20}{21}mJ_h$$

The quantity δ_1 is of the order of f_h, and δ_2 is of the order of f_h^2.

Earlier, the consideration of the potential in free space of an ellipsoid of revolution with both polar and equatorial symmetry yielded Eq. 4a, which can be recast slightly to give

$$f = J_2(\tfrac{3}{2} + \tfrac{15}{28}m + \tfrac{9}{8}J_2) + \tfrac{1}{2}m + \tfrac{3}{56}m^2 \tag{40}$$

This equation is derived from external potential theory with no assumptions whatsoever as to the conditions existing in the Earth's interior. Hence, it should be valid whether or not hydrostatic equilibrium exists inside the Earth and should yield real flattening for observed value of J_2 and hydrostatic flattening for the hydrostatic value of J_2.

The polar moment of inertia is determined by

$$\frac{C}{Ma^2} = \frac{J_2}{H} = \frac{2}{3} q \qquad (41)$$

where

$$J_2 = \frac{C - A}{Ma^2} \quad \text{and} \quad H = \frac{C - A}{C}$$

The quantity J_2 is the coefficient of the second harmonic in the spherical harmonic expansion of geopotential and is directly determinable to a high degree of accuracy from the orbital motion of an artificial Earth satellite. The quantity H is also very accurately determined from the precession of the Earth's axis due to the torque exerted by the Moon and the Sun. Thus, in Eq. 39, if $\lambda(a)$ can be assigned an appropriate value, a knowledge of J_{2h} should yield f_h because the polar moment of inertia has already been determined from Eq. 41. But $J_{2h} \neq J_2$ for the Earth and thus is not known. We have reasoned earlier, however, that the solutions of external and internal potential theories must match at the outer boundary; i.e., Eqs. 39 and 40 must be satisfied simultaneously for the outer boundary. The correct value of f_h will then be obtained by this simultaneous solution.

Results

Shape of Equilibrium Figure of the Earth. The following data are adopted in the computation of f_h (Chovitz, 1981; Khan, 1983):

$$J_2 = 1082.63 \pm 0.01 \times 10^{-6},$$
$$H = 3273.66 \pm 0.07 \times 10^{-6}$$

Hence,

$$\frac{C}{Ma^2} = 0.330709 \pm 0.000007$$

$$m = 3449.80 \times 10^{-6}$$

$$\lambda(a) = 0$$

The resulting value of the hydrostatic flattening is

$$f_h = 3336.11 \pm 0.05 \times 10^{-6} = \frac{1}{299.75 \pm 0.005}$$

The corresponding value of hydrostatic J_2 is

$$J_{2h} = 1071.66 \pm 0.02 \times 10^{-6}$$

The value of actual flattening f corresponding to $J_2 = 1082.63 \pm 0.01 \times 10^{-6}$ is

$$f = 3352.81 \pm 0.03 \times 10^{-6} = \frac{1}{298.257 \pm 0.003}$$

and therefore

$$f - f_h = 16.7 \times 10^{-6}$$

The choice of $\lambda(a) = 0$ needs some justification. The value of $\lambda(a)$ lies between 1.3×10^{-4} (Jeffreys, 1962) and 1.6×10^{-4} (Bullard, 1948), as calculated from the known density distribution of the Earth. Thus, it may appear at first sight that the choice of λ is not compatible with second-order theory. However, choosing $\lambda(a) = 0$ instead of 1.3×10^{-4} results in f_h being greater by 0.6×10^{-6} only, whereas the difference $f - f_h = 16.7 \times 10^{-6}$. Hence our choices $\lambda(a) = 0$ and, consequently, $\psi(\eta) = 1$ do not affect the value of f_h to any significant degree (Khan, 1969).

Size of Equilibrium Figure of the Earth. The size of the hydrostatic Earth ellipsoid can be computed by assuming that it is volumetrically equivalent to the actual Earth. Thus, if a and b are the equatorial and polar radii of the actual Earth, and a_h and b_h are their hydrostatic counterparts, we have

$$a_h = a \left(\frac{1 - f}{1 - f_h} \right)^{1/3}$$

For $a = 6,378,137 \pm 2$ m and $b = 6,356,752 \pm 2$ m (corresponding to $f = 1/298.257$), we obtain

$$a_h = 6,378,101 \pm 2 \text{ m},$$
$$b_h = 6,356,823 \pm 2 \text{ m}$$

resulting in

$$a - a_h = 36 \text{ m}, \quad b - b_h = -71 \text{ m}$$

Gravity Field Referred to the Equilibrium Figure. As discussed earlier, since the hydrostatic figure is a figure of zero stress, it constitutes a geophysically meaningful reference figure for gravity anomalies for use in studying the internal state of the Earth. Since the gravity anomaly maps are normally available with reference to either the International Gravity Formula or Gravity Formula, 1967, or more recently, Geodetic Reference System, 1980, Khan (1976) computed the correction factors to convert these gravity anomalies to the equilibrium figure datum.

Note

A complete bibliography on the subject cannot be given due to space restrictions. Only selected references are given, which, however, contain references to other important work.

M. A. KHAN

References

Bullard, E. C., 1948, The figure of the earth, *Royal Astron. Soc. Monthly Notices, Geophys. Suppl.* **5,** 186-192.

Chovitz, B., 1981, 'Geodetic Reference System 1980', *EOS* **62,** 65-67.

deSitter, W., 1938, On the system of astronomical constants, *Astron. Inst. Netherlands Bull.* **8,** 213-229.

Jeffreys, H., 1962, *The Earth.* Cambridge: Cambridge University Press.

Khan, M. A., 1969, General solution of the problem of hydrostatic equilibrium of the earth, *Royal Astron. Soc. Geophys. Jour.* **18,** 177-188.

Khan, M. A., 1976, Some geophysical and geodetic contributions of satellite-determined gravity results, *Geophys. Surveys* **2,** 469-497.

Khan, M. A., 1983, Primary geodynamical parameters for the standard earth model, *Royal Astron. Soc. Geophys. Jour.* **72,** 333-336.

Cross-references: *Earth's External Gravity Field; Figure of the Earth; Geodesy: Physical; Gravity Fields: Implications for Planetary Interiors; Planetary Geodesy; Spherical Harmonic Analysis.*

EXPLORATION SEISMOLOGY

Exploration seismology is concerned primarily with prospecting for energy sources, chiefly oil and gas, but also coal and, more recently, geothermal reservoirs. The emphasis here will be on reflection methods as they are applied to the search for oil and gas, work that consumes about 95% of exploration budgets. This article will do no more than touch the surface of the vast field of exploration seismology, a subject about which hundreds of books have been written and to which whole journals are devoted. The standard texts cited in the references provide an entry into that huge literature.

The sedimentary basins in which oil and gas typically are found are made up of sedimentary layers, which in general differ from one another in rock density and seismic velocity. Seismic waves, generated artificially at the Earth's surface, propagate downward through the medium and are reflected or refracted at each discontinuity (Fig. 1). The reflected waves travel back to the surface, where they can be recorded and, ultimately, interpreted geologically. To obtain accurate geologic information about the subsurface from reflections, it is necessary to minimize insofar as possible distortions in the seismic records caused by unwanted signals such as multiple reflections, diffraction images, and out-of-plane ("sideswipe") reflections. Data can also be degraded by near-surface weathering effects and changes in elevation, lateral heterogeneity in velocity, discontinuous or irregular reflecting horizons, interference from other seismic waves generated in the course of the experiment (including surface waves and, occasionally, shear waves), and noise from wind and cultural activity. Enormous energy has gone toward devising methods of reflection exploration that will minimize these effects.

Seismic reflection surveying involves three parts: data acquisition, data processing, and data interpretation. We are concerned in this article primarily with the data acquisition and data processing, although, of course, every phase of reflection profiling is directed toward accurate geologic interpretation. In the sections that follow, we will describe routine methods of reflection seismology currently in use on land and at sea. A final section will deal with new developments and future directions.

Reflection Methods in Oil Exploration

While three-dimensional exploration (aerial coverage) has increased substantially in importance over the past several years, the basic principles are sufficiently similar to those of two-dimensional reflection profiling that we can ignore the added complexity of three-dimensionality. Full discussions of reflection techniques in two and three dimensions can be found in a number of standard texts, such as Waters (1987) or Dobrin (1976).

Nearly all seismic prospecting today is based on the common mid-point (CMP) method, frequently termed (inaccurately) common depth point (CDP) method within the exploration industry. While the method was first proposed in the 1930s and saw early use in the late 1950s, it was not until the advent of digital recording with a resulting huge increase in speed and reliability of data handling that the CMP method became an industry standard. The CMP method is overwhelmingly accepted because it provides ingenious solutions to a host of difficulties that had long plagued reflection profiling. A simplified CMP reflection spread is shown schematically in Fig. 2. Part of the power of CMP derives from its great redundancy of data. CMP uses multiple offsets centered about a common mid-point. The individual seismic traces of each source/receiver pair that share a common mid-point are grouped into CMP *gathers* (Fig. 2). Suitable time and waveform corrections are applied to each seismogram, after which the records are stacked (summed) together to form a single composite trace. In a typical 96-channel operation where distance spacing between sources is the same as that between receivers, the gathers will be 48 fold, meaning that 48 individual traces are summed to make a single "zero-offset" trace on a CMP stacked record section. [Strictly speaking, such a section is a *CMP-stacked section,* which mimics a zero-offset section but is not identical to it (Claerbout, 1985). For our purposes here, the distinction is not important.] The CMP

EXPLORATION SEISMOLOGY

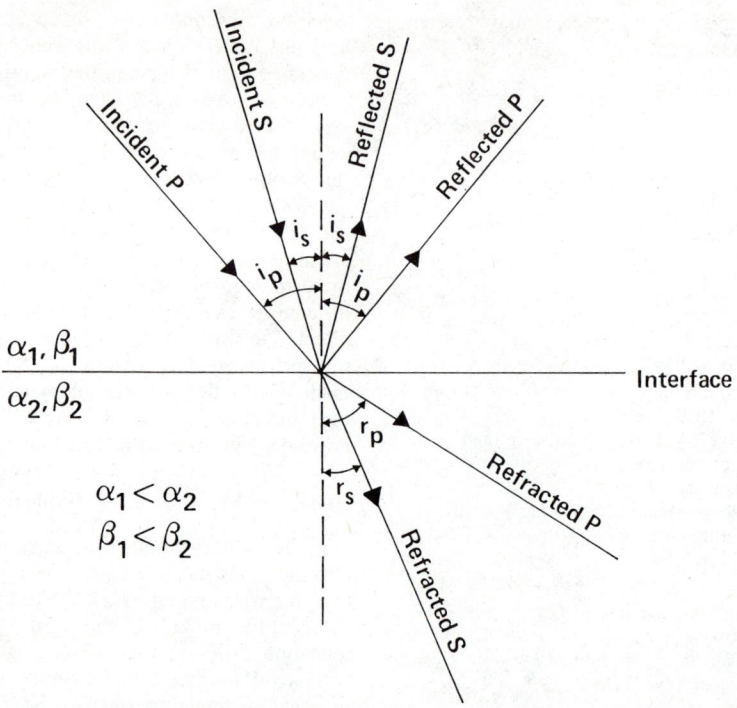

FIGURE 1. Schematic diagram of ray paths for P waves and S waves incident upon a horizontal discontinuity.

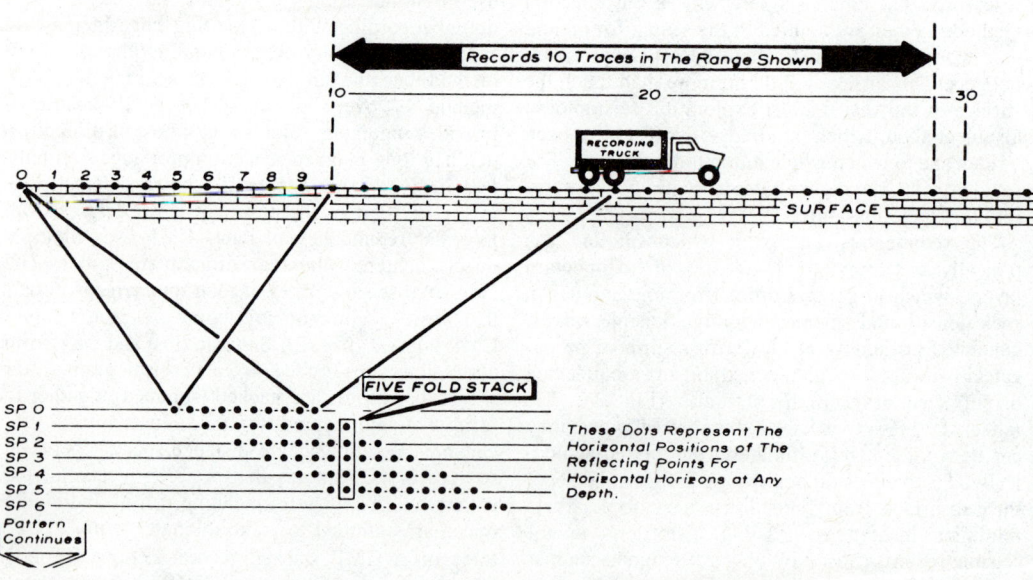

FIGURE 2. Schematic diagram of an idealized CMP reflection profile. This shot-receiver arrangement yields a five-fold stack. (From Waters, 1987, with permission)

method has two very important advantages: it eliminates or at least greatly reduces the effect of dip of the reflecting horizon on the determination of velocity; and it results in improved signal/noise ratio by cancelling effects of random noise and scattering events, multiple reflections, and surface waves.

Data Acquisition. Reflection seismic prospecting is concerned with Earth structures for depths less than about 10 km. Seismic sources on land are nearly always large truck-mounted ground vibrators (Vibroseis) or explosions set off in boreholes. Vibroseis signals are input to the ground by hydraulically vibrating a large baseplate on which most of the weight of the truck is supported. A standard Vibroseis signal has a duration of 12 seconds, during which interval the vibrator *sweeps* over a set frequency band, typically 10 Hz to 50 Hz. A single Vibroseis set-up will include multiple sweeps and the signals recorded for each receiver summed (stacked) in the field. Raw Vibroseis records are not interpretable and must be processed later to recover a seismogram resembling that caused by an impulsive source. Receivers are typically configured as 24, 48, or 96 channels connected by cable to a common recording unit. Between 12 and 48 geophones are connected in series to each channel and the signals electrically summed to provide the output for each receiver channel. The recording system is usually truck mounted, and signals from each receiver station are brought in with wires or telemetered. Preliminary data processing is done in the truck so that raw results can be displayed in the field and their quality monitored continuously.

Marine seismic sources in most common use today include airguns for deep prospecting and spark sources for shallow work. Airguns produce seismic signals through expansion of high pressure gas into a chamber; sparkers create a sound wave in the air. Although other sources are also employed, these two serve to illustrate the principles of marine seismic work. Explosive sources are almost never used in marine seismic studies because they kill fish and because they produce an unpredictable bubble effect. The advantage of airguns and sparkers is that they generate powerful pulses that are highly reproducible, can be repeated at intervals of several seconds, and can be enhanced by source arrays. For deep work, airguns are typically organized into arrays of several airguns, each with different chamber volumes to increase frequency bandwidth of the signal, and aligned in the water to produce constructive interference of the signals. Receivers are typically arrayed in *streamers* (sealed tubes) up to several kilometers in length that may contain up to several hundred receiver channels. Telemetry is commonly used for the larger multichannel systems. Receivers are piezoelectric transducers that detect pressure changes. Both receivers and airgun sources are set at a fixed depth underwater, typically about 13 m (Waters, 1987). Experiment organization is similar to that on land and is built around the CMP method. Expanding spread profiling is even more common at sea than it is on land because two ships (one the source and the other the receiver) steaming away from each other at the same speed will produce a near-vertical to wide-offset (expanding spread) CMP gather from which accurate subsurface velocity information can be obtained. Such expanding spread profiling is especially common in very deep crustal work (see *Deep Seismic-Reflection Profiling; Ocean Continent Transition: Structure*).

Data Processing. A flow chart of standard processing steps involved in obtaining a stacked seismic section using the CMP method is shown in Fig. 3. A complete description of each phase of the data analysis is given in many standard texts (e.g., Waters, 1987; Robinson and Treitel, 1980; Dohr, 1981; Nelson, 1983; Dobrin, 1976).

The following is a very brief summary of these basic procedures, broken down into twelve steps: demultiplexing, Vibroseis correlation, deconvolution, CMP sorting, static corrections, velocity analysis, normal moveout (NMO) correction, muting, gain recovery, stacking, filtering, and migration.

In a typical recording setup, the data are multiplexed. Multiplexing simply means that data from an N-channel receiver array will be recorded as a single data stream, with the first data sample of the first channel recorded first, the first data sample of the second channel recorded next, and so on until the first data sample of all N channels has been recorded. The cycle starts all over again with the second data sample of the first channel and continues until the last data sample of the Nth channel is recorded. *Demultiplexing* is simply the process of recovering the individual time-series for each channel from the multiplexed record.

If the seismic source is Vibroseis, the data have to be correlated with the original Vibroseis signal to produce a set of seismic records with waveforms equivalent to what would have been observed if the seismic source had been a delta function. (It turns out that things are not quite this simple. The actual seismic signal generated by Vibroseis depends not only on the control signal fed into the vibrator but upon the local surface conditions and rock type. Certain deconvolution operations, the details of which can be found in standard reflection seismology textbooks, are applied to the data to eliminate these effects.)

Deconvolution is a process of inverse filtering designed to undo "filtering" (by the Earth or by the seismic source) that has already taken place. Deconvolution has found its greatest application in marine work. In a way similar to Vibroseis correlation, deconvolution must be applied to marine records to compress the complicated airgun signal to an equiv-

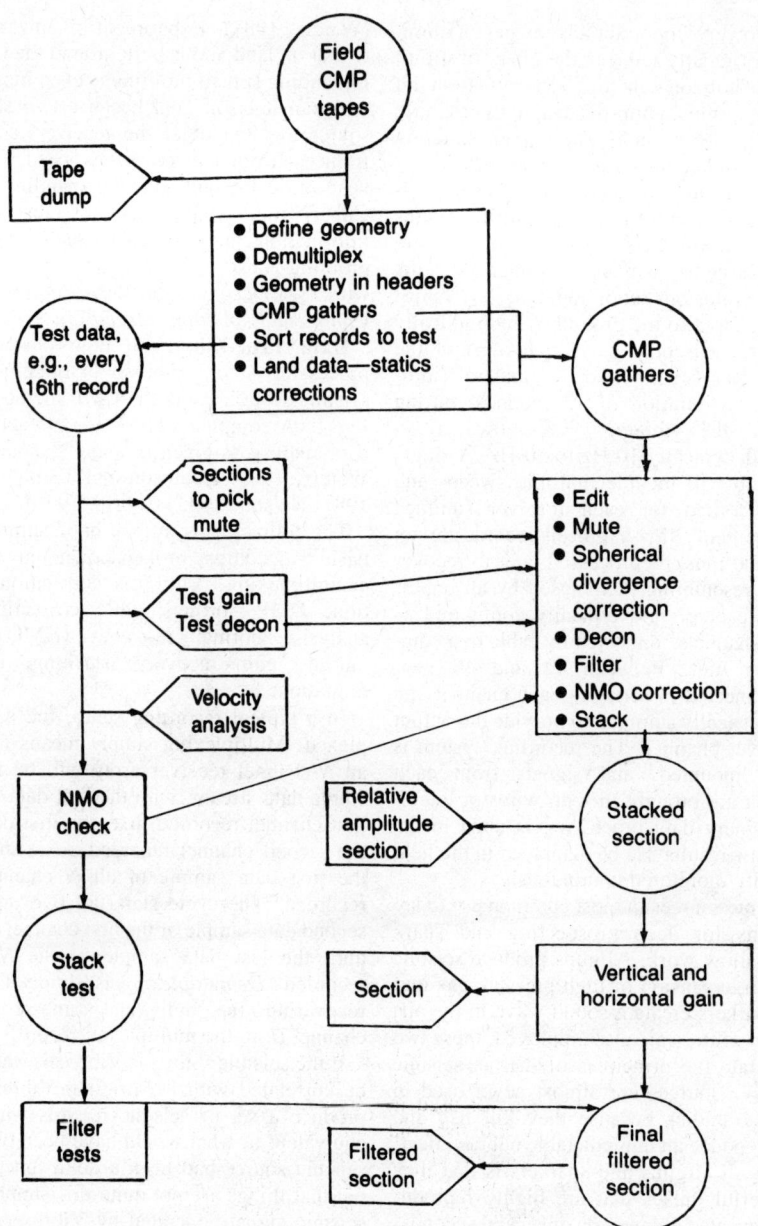

FIGURE 3. Flow chart of processing steps in CMP method up to a final stacked section (unmigrated). (From Nelson, 1983, with permission)

alent impulse source. Much more importantly for marine studies, however, are the strong resonances (reverberations or ringing) in the water layer that make the raw seismograms virtually uninterpretable. Deconvolution, properly applied to the marine data, greatly enhances the resolving power of the seismic section by suppressing these strong resonances. Deconvolution is most useful where reverberations are in shallow (i.e., water) layers, so that the sum of seismic source (e.g., airgun) effects and near-surface resonance can be treated *in toto* as a virtual source waveform.

Reverberation is simply a consequence of the fact that the water layer is bounded by media of very different acoustic impedance, so the water acts as a strong waveguide. In the 1950s the Geophysical

Analysis Group at MIT developed the method of *predictive deconvolution* to remove the so-called reverberation spike train (a reverberation wave train produced by a "spike" or unit pulse source at the water surface). The method of predictive deconvolution in practice is applied in different ways (for example, the first part of a seismic record can be used to predict and deconvolve later parts of the record), but here the term "predictive" is appropriate because the kinds of systematic reverberations encountered in a water layer can in fact be predicted.

All predictive deconvolution is based on the principle of minimum-delay. Indeed, most forms of deconvolution in reflection seismology are formulated to recover a minimum-delay wavelet from the data (see Robinson and Treitel, 1980). A minimum-delay (or minimum-phase) waveform is defined by Robinson and Treitel (p. 253) to be "the waveform that has the largest concentration of energy in the early part of the waveform in the class of all waveforms with the same magnitude spectrum." That is, it will be the most impulsive of all the possible wavelets. The concept of minimum-delay is extremely important for two reasons: (1) it permits a wide range of mathematical theories and techniques to be applied; and, (2) it is physically applicable to the Earth. It has been shown, for example, that pulses from explosions are minimum phase and that the filtering of such pulses by layered systems (such as water layers) is also minimum phase. This means that the *seismic records themselves* can be used to determine statistically the optimum inverse filter for removing unwanted source and reverberation effects from the seismic traces (see also *Seismic Signal Processing*).

At some time in the processing sequence, usually subsequent to demultiplexing, Vibroseis correlation, and/or deconvolution, the individual seismic traces are sorted into CMP gathers. In practice four principal types of trace gathers are used: common source, common receiver (geophone), common offset, and common midpoint (see Fig. 2). The terminology is self-explanatory and for the discussion here, we will be concerned only with the CMP gathers. For a 96-channel system, for example, the CMP gathers will typically consist of 48 traces when spacing between sources and receivers is the same. Properly adjusted traces of the CMP gather will sum constructively for primary reflections and destructively for other seismic signals that are out of phase.

The following processing steps, which lead to various kinds of time adjustments to the seismic traces are categorized either as statics corrections or as dynamic corrections. *Statics corrections* are applied as a constant time shift to the entire record; *dynamic corrections* vary with time along the record. Static adjustments are used primarily for land profiling where there may be changes in thickness and velocity of the weathering layer or in elevation. Such changes will produce time-delays of different magnitudes for different points on the surface so that each source or receiver site will have a time-delay related to near-surface conditions. In practice, statics corrections are generally made by applying time delays to the seismic records such that the sources and receivers appear to be adjusted downward to a fixed reference plane in the Earth, which is everywhere below the weathering layer. Statics corrections are usually applied in two stages: automatically near the start of general processing and at a later stage where more refined corrections may be added immediately before stack. Automatic statics corrections are typically applied on the basis of a grid of measured and assumed values for the effect of weathering and elevation. While such automatic corrections are sufficient for the early stages of data processing, they will not in general provide the best possible correction that can be gotten from the field data.

Dynamic corrections are any corrections that vary with time on the record. Included in these are velocity analysis and normal moveout (NMO) corrections, where NMO is simply the increase in travel time for a reflected wave as the distance increases between source and receiver. The correction for source-receiver offset changes with time along the seismogram and is a function of the velocity variation with depth and the length of the ray path. If average velocity as a function of depth is known or assumed in advance, NMO corrections can be applied simply on the basis of offset distance. If velocities are not known, they must be calculated from the slopes of the hyperbolic trajectories of primary reflections. A variety of methods can be used for this purpose, but most use statistical measures to compare signal coherence across primary records (e.g., a receiver gather) at various constant velocities to obtain the interval velocities (the velocities between reflecting horizons) that provide the best fit to the data. Accurate velocities are necessary not only for proper NMO corrections, but are also critical for seismic migration, which is carried out on the final zero-offset time section (see below) and for geologic interpretation. In some instances where velocity information is difficult to obtain, expanding-spread profiling may be done, where offset distances range from near-vertical to wide-angle, sufficient to determine velocity much more accurately than is possible with a standard receiver spread. NMO corrections, properly applied, will align the primary reflection events on the traces of a CMP gather regardless of offset distance, so that the reflections will add constructively when the traces are summed. Other arrivals on the records should interfere destructively, as the NMO time shifts will be valid only for primary reflections.

Muting is simply a procedure in which the amplitude of the signals on the seismogram is set equal to zero for those parts of the record that contain large-amplitude near-surface or airwave arrivals or are otherwise unacceptably noisy. Muting is done to assure that unuseable parts of any given seismic trace will not be added to the final stacked record.

Individual traces are commonly *energy normalized,* by which we mean the calculated total seismic energy (sum of the amplitudes squared) of each equal-length trace is scaled to be constant. This scaling is done by multiplying the amplitude of every point on each seismogram by an appropriate scaling factor. Energy normalization is designed to assure that the relative contributions of the individual traces to a stacked trace are roughly comparable. The assumption here is that signal/noise is constant trace to trace, but other factors such as gain setting and site response produce varying trace amplitudes. The process of energy normalization is in fact rather tricky, and has to account not only for noisy traces, varying offset distances, and other factors that have to be considered trace by trace, but it may also have to be done in such a way as to preserve true relative amplitudes (see discussion of amplitude preservation, below). In addition to energy normalization, the trace amplitudes are adjusted to compensate for spherical divergence of the seismic energy with distance from the source. This adjustment, which is applied equally to all traces, is performed simply by multiplying the amplitude of the trace, point by point, by the amplitude of a given gain function (Waters, 1987). In practice, the gain function is more than a simple compensation for spherical divergence, for other factors such as seismic attenuation are typically taken into account. Corrections for attenuation are determined empirically by examining the mean amplitudes of different segments of the seismogram.

One of the more important advances in reflection signal processing since the early 1970s has been the increased emphasis on the preservation of relative amplitude information. This has been particularly important in exploration for gas reservoirs, where it is recognized that large-amplitude signals, termed *bright spots,* can be very prominent indicators of such reservoirs (e.g., Robinson and Treitel, 1980). The large reflection amplitudes are due simply to the large impedance contrast between the gas and the surrounding rocks. A typical bright spot record is shown in Fig. 4. In addition to such spectacular amplitude features, it is also recognized now that more subtle amplitude variations may provide valuable information about whether structural traps (anticlines, fault offsets, and so on) in which oil may accumulate *actually* contain oil deposits. Amplitude preservation places important constraints on seismic processing, for it is no longer valid simply to adjust

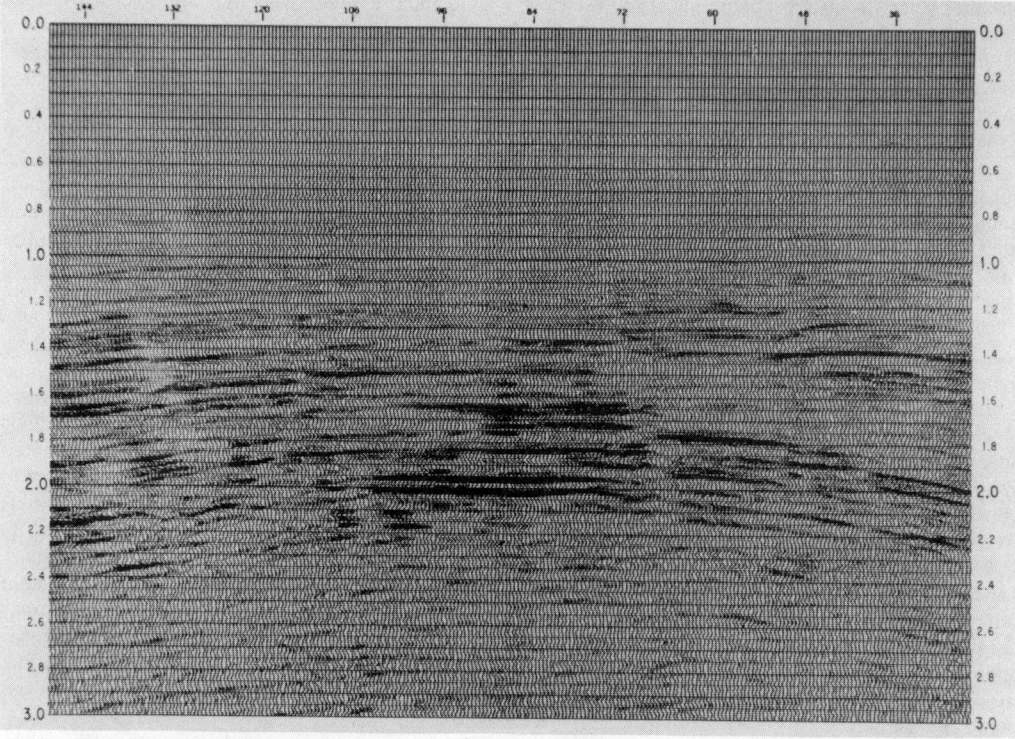

FIGURE 4. Relative amplitude stacked section showing a doublet "bright spot," (at about 2 seconds) probably imaging a site of trapped gas. (From Ware, 1977)

gains arbitrarily to optimize the signal-to-noise ratios and general appearance of the seismogram. Where bright spots and other amplitude-related factors are important for exploration, gain changes are typically limited to adjustments for spherical divergence or other effects for which there is a clear physical basis.

The appearance of individual seismograms or stacked records can commonly be enhanced by various kinds of filtering. In addition to deconvolution, which is a form of filtering and was discussed earlier, the seismic traces can be high-pass, low-pass, or band-pass frequency filtered to limit the range of frequencies on the record. All filtering now is done digitally, mostly in the frequency domain. Among its more common uses, high-pass filtering is used to suppress low frequency noise or shot-generated surface waves on the records. Low-pass filtering, on the other hand, may be used to eliminate spurious effects of spatial aliasing (insufficient density of source-receiver points along the profile) by retaining only the long wavelength information that is not aliased. Band-pass filtering finds application in a variety of circumstances. It is commonly desirable, for example, to record at a wide bandwidth and later band-pass the records as a means of reducing waveform distortion that occurs during removal of NMO corrections from wide-offset records (see Waters, 1987, for discussion of this and other problems of NMO distortions). At one time it was common to apply a series of narrow-band filters to the records to determine empirically the optimum band for observing reflections. In recent years, more objective approaches have been adopted using adaptive filters that can be designed from the data themselves and do not require operator intervention.

After the appropriate adjustments and corrections as outlined above have been applied to the individual traces of the CMP gathers, the records are stacked. It is common to examine the individual traces that make up the composite stacked trace for their coherence relative to a standard trace. Minor time shifts (statics corrections) may be applied to bring records into better coherence before stacking. If everything has been done correctly, the stacked record will have a substantially improved signal/noise ratio over any of the individual traces. The stacked CMP gather makes up a single *zero-offset* or *normal incidence* trace, equivalent to a common source and receiver point on the surface. The hypothetical reflection travel paths of zero-offset traces will be approximately normal incidence to the reflection interfaces, provided the reflecting plane does not have significant curvature.

The ensemble of zero-offset traces makes up the zero-offset record section. The vertical axis on these record sections is time, not depth, so while seismic time sections may have the appearance of a seismic depth section, they are not equivalent. Not only does the time axis not correspond to the depth axis, but more importantly, the reflecting points, which appear to lie directly beneath each recording point, are not in their proper geographic position. Thus, not only must the velocity information be added to produce a proper reflection cross section, but the reflection traces themselves must be "migrated" to their correct location in the section. Not surprisingly, this calculation is termed *seismic migration*. As defined by Robinson and Treitel (1980, p. 380), migration is "the process of constructing the reflector surface from the record surface." Seismic time records are migrated to correct for the fact that reflection horizons are generally not horizontal or even flat, and that they may be discontinuous. Migration collapses diffractions and compensates for dips and curvature of interfaces to give a correct representation of the subsurface. An example of migration applied to a two-dimensional synthetic seismic section is shown in Fig. 5. While the effects of migration are not always so startling as shown in Fig. 5, they are typically quite significant and few exploration surveys are considered complete until the records are migrated. Migration can be done either before or after stack. While better results can be obtained by migrating before stack, the computational costs are far greater.

A variety of computational methods have been developed for seismic migration and the subject continues to be an important area of active research. It is discussed in more detail in the entry *Seismic Wave Field Migration*. Some of the pitfalls associated with migration deserve mention, however. Migration requires accurate velocity-depth information. If velocities used in migration are incorrect, the position of the reflectors will be in error. The problem is particularly difficult when the velocities vary laterally. Out-of-plane reflections can introduce important errors in migrating two-dimensional seismic sections. These errors are most common where structures vary significantly in three-dimensions, and are manifest as incomplete collapse of diffractions and failure to resolve correct dips on interfaces. As emphasis in exploration has shifted increasingly to three-dimensional surveying, there has been an intensified effort to devise fast algorithms for three-dimensional migration. As yet, the computational requirements are so formidable that full three-dimensional migration is not likely to be standard for some time to come.

The final step in all seismic reflection exploration is interpretation. Technically, the seismologists' role ends with providing a migrated section, and our discussion of exploration seismology could end there. (Although, of course, many forms of filtering or other processes can be performed on the migrated record section.) The key to accurate interpretation now, as it has always been in the past, is seismic data that is as free as possible of distortions and artifacts. While it is far beyond the scope of this article to outline even the most elementary aspects of interpretation, some mention must be made of the

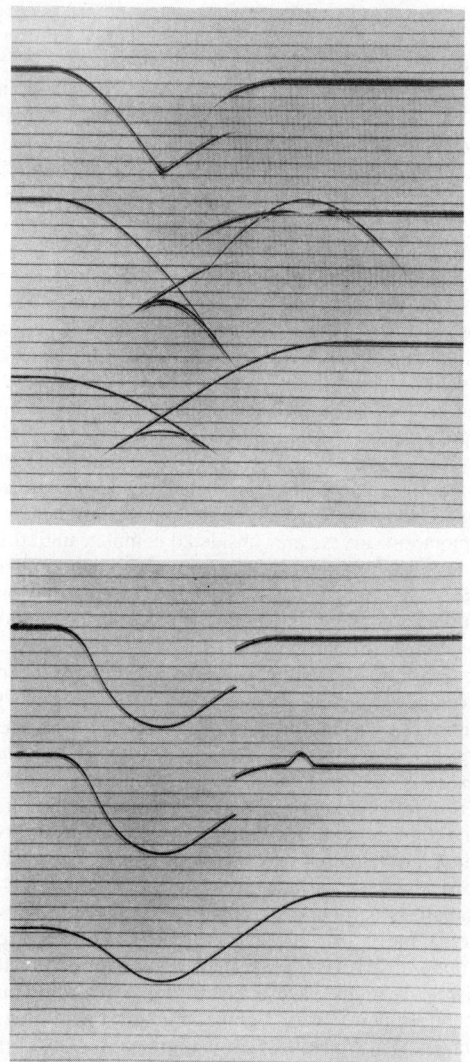

FIGURE 5. An example of seismic migration applied to two-dimensional synthetic seismic data. Upper section is unmigrated data, lower section is data after migration. (From Schneider, 1978)

developments in video graphics, an area of software development that has consumed the full time efforts of armies of programmers. The interactive three-dimensional color graphics displays now available in the oil industry have revolutionized the art of interpreting seismic information. Seismic sections and three-dimensional seismic blocks can be viewed interactively on color screens—rotated, enlarged, sliced in various ways, color modulated, time-shifted, filtered, and otherwise manipulated—such that the data can be more easily assimilated by the eye, and far more precise and comprehensive interpretations can be made of the results. The identification of key seismic features and the correlation with structures and lithologies, however, remains something of an art form that has resisted many efforts at automation.

New Techniques and Future Directions

The exploration industry spends very large sums of money to develop methods that will improve the prospects for discovering oil or gas. Among the major advances that have been made in the last decade or so must be included three-dimensional surveying, shear-wave profiling, and vertical seismic profiling. All are now in the standard arsenal of tools used by the exploration industry.

Three-Dimensional Studies. We have alluded earlier to three-dimensional seismic reflection surveys. As the search for oil has become increasingly difficult and the new geologic terranes less and less amenable to exploration by "standard" reflection profiling, the industry has turned increasingly to three-dimensional coverage. Three-dimensional seismic data are obviously of greatest importance in those terranes where structures are truly three-dimensional, and where linear profiles fail to give information about the nature and trend of structures out of the plane of the profile. In general, marine surveys can be organized for three-dimensional work without very significant changes in equipment or operating procedures. On land, however, and particularly in areas such as the overthrust belt of the western United States, the acquisition of three-dimensional data is difficult indeed. Instrument deployments over a grid can commonly be done only with helicopters or all-terrain vehicles. As Vibroseis trucks cannot travel off-road, the seismic sources are generally explosions. The problem of instrument deployment over a grid has been solved in part by the *seismic group recorders*, which are stand-alone units with cassette recorders for data acquisition. Communication with the recording units for turn-on at shot time is by radio. Cassette tapes are collected from each unit at the end of the day and replayed on field computers operated on-site in vans or trailers.

The processing of three-dimensional data is very similar to that for profile data, although of course the volume of data is increased by approximately n^2, where n is the number of observations along one side of the grid. True three-dimensional migration is still prohibitively expensive, typically requiring many days of computing time, but approximate three-dimensional migration is tractable. By the approximate method, two-dimensional slices are migrated along both orthogonal axes so that the number of seismograms to be handled is reduced from n^2 to $2n$ (Nelson, 1983).

The presentation of three-dimensional data is a

special problem, requiring the sophisticated high-resolution interactive color graphics display capabilities mentioned earlier. With these systems, the interpreter can examine the data from any angle, shift subsets of the volume relative to each other, display in colors according to amplitude or frequency, or carry out a host of other operations that aid in interpretation.

Shear-Wave Profiling. Another of the recent innovations is shear-wave profiling. Shear waves can in principle provide far more information than compressional waves, although to date they have been used mostly to supplement standard acoustic profiling. Modern Vibroseis shear-wave vibrators generate horizontally polarized (SH) shear waves. Recording techniques are very similar to those of standard CMP methods, but typically the frequencies used are about one-half those used for acoustic profiling. The reduced frequency is intended to maintain the same wavelength character of the seismic record: As velocity is simply the product of frequency times wavelength, and S-wave velocities are only about half those for P waves, comparable S wavelengths can be achieved only by decreasing frequency by a factor of about 2. Shear wave profiling finds its greatest application in situations where the S-wave and P-wave velocities do not have a simple relationship to one another. Different lithologies and lithological contacts will exhibit different S/P velocity ratios, for example, meaning that reflection coefficients for P and S waves will not in general be the same. Perhaps more importantly, the presence of pore fluids affects shear waves and compressional waves differently. These differences can be exploited to identify oil and gas deposits and even to distinguish hydrocarbons from water in porous rock.

Vertical Seismic Profiling (VSP). Vertical seismic profiling, which takes advantage of deep boreholes to make measurements within the earth, is widely used today to supplement conventional reflection profiling. Receivers, typically in multicomponent arrays, are deployed in one or more boreholes in the region of interest. Sources may either be on the surface or in other boreholes. There are several important advantages to VSP, including much decreased background noise, better coupling to the rock and therefore increased bandwidth, far greater ease in identifying events on the seismogram by measuring polarity of the incoming signals as they move along the array, and elimination of the unwanted effects of weathered layers (Wuenschel, 1976). The seismograms obtained give a view of the Earth both above and below the borehole receiver sites.

Tomographic Methods. Various tomographic methods have been developed both for standard reflection profiling (e.g., Bishop et al., 1985) and for cross-borehole tomography. Tomographic determination of reflector depth and seismic velocity in laterally varying media from reflection data as employed by Bishop et al. and others is formulated as a general inverse problem that produces a velocity depth model that minimizes traveltime residuals of selected reflection arrivals observed on unstacked seismic data. The method is very computationally intensive, however, and it remains in the experimental stage as far as its routine application is concerned.

VSP or borehole tomography requires deployment of seismic sources down one hole and receivers down the other. The resulting pattern of criss-crossing rays in the plane between the boreholes can be analyzed to determine velocity structure between the holes. While there is considerable interest in this technique, work by Menke (1984) demonstrates that only marginal resolution of horizontal velocity structure is possible by this method, although very good average horizontal velocities can be determined.

We have omitted mention of many important aspects of exploration seismology from this discussion. There are many excellent texts, however, that cover the field, some of which are listed in the references. While it is always risky to make predictions, it is certainly safe to say that the digital revolution in reflection seismology is far from over. Seismic images of the interior of the Earth are certain to become sharper and more accurate as the science keeps pace with and capitalizes on rapid technological advances in computer power, data storage, and video display systems.

DAVID E. JAMES

References

Bishop, T. N., K. P. Bube, R. T. Cutler, R. T. Langan, P. L. Love, J. R. Resnick, R. T. Shuey, D. A. Spindler, and H. W. Wyld, 1985, Tomographic determination of velocity and depth in laterally varying media, *Geophysics* **50**, 903–923.

Claerbout, J. F., 1976, *Fundamentals of Geophysical Data Processing.* New York: McGraw-Hill, 274p.

Claerbout, J. F., 1985, *Imaging the Earth's Interior.* Oxford: Blackwell Scientific Publications, 398p.

Dobrin, M. B., 1976, *Introduction to Geophysical Prospecting.* New York: Society for Exploration Geophysics.

Dohr, G., 1981, *Applied Geophysics,* 2nd rev. New York: Halsted Press, 231p.

Menke, W., 1984, The resolving power of cross-borehole tomography, *Geophys. Research Letters* **11**, 105–108.

Nelson, H. R., Jr., 1983, *New Technologies in Exploration Geophysics.* Houston, Tx.: Gulf Publishing Co., 281p.

Robinson, E. A., and S. Treitel, 1980, *Geophysical Signal Analysis.* Englewood Cliffs, N.J.: Prentice-Hall, Inc., 466p.

Schneider, W. A., 1978, Integral formulation for migration in two and three dimensions, *Geophysics* **43**, 49–76.

Ware, J. A., 1977, A simple upgrade for some bright spots, *Geophysics* **42**, 868-871.
Waters, K. H., 1987, *Reflection Seismology, A Tool for Energy Resource Exploration*, 3rd ed. New York: John Wiley and Sons, 527p.
Wuenschel, P. C., 1976, The vertical array in reflection seismology—Some experimental studies, *Geophysics* **41**, 219-232.

Cross-references: *Continental Crustal Structure; Controlled Source Seismology; Deep Seismic-Reflection Profiling; Elasticity and Wave Propagation: Principles; Elastic Waves in Homogeneous and Inhomogeneous Media; Finite-Difference Forward Modeling in Seismology; Ocean Continent Transition: Structure; Seismic Imaging; Seismic Signal Processing; Seismic Wavefield Migration.*

F

FAR-TRAVELED TERRANES

About one hundred anomalous regions ranging in size from 1000 km down to a few kilometers are embedded in the Earth's sea floor. These regions are typified by shallow water depth, thick crust, low upper crustal velocities, lack of clear magnetic lineations, and steep margins. They cover about 10% of the present-day ocean floor, with particular concentrations in the western Pacific and the Indian Oceans, as shown in Fig. 1. Typically these regions are well embedded within their plates and show no evidence for relative motions. Such relative motions can occur, however, when the plateaus move into the subduction zones where oceanic plates are consumed.

In this article we summarize ideas and evidence relevant to the fate of plateaus upon arrival at consuming plate boundaries, and their geological impact at these plate boundaries, the overriding plates, and the oceanic plates that carry them, including effects transmitted to the spreading ridges associated with these plates.

Oceanic Plateaus and Their Structure

Oceanic rises (Fig. 1) span a wide range of water depth and crustal thicknesses and may be classified on the basis of information obtained from drill holes, dredgings, multichannel seismic surveys, and core records: extinct arcs (e.g., Bowers Ridge, Aves Ridge), abandoned spreading ridges (e.g., the West Philippine Ridge), detached and submerged continental fragments (e.g., the Chatham Rise), anomalous volcanic piles (e.g., Iceland), hot spot traces (e.g., Cocos Ridge), and uplifted oceanic crust (e.g., the Nicoya complex in Central America).

Many of the large oceanic plateaus rise thousands of meters above their surrounding sea floor. Some, such as the Seychelles Bank, even rise above sea level, whereas the tops of others, such as the Ontong Java Plateau, are submerged 1500 to 2000 m below sea level. Most of the plateaus for which seismic refraction and gravity data are available range in thickness from 20 km to over 40 km, much more than usual oceanic crust.

Geological evidence from some plateaus shows continental affinities. For example, Precambrian granitic basement is exposed in the Seychelles Islands in the middle of the Indian Ocean. Granitic basement was also found in the Paracel Islands in the South China Sea, and dredging of the Agulhas and the Kerguelen plateaus yielded granitic rocks. Consequently, some plateaus or parts thereof are probably submerged continental fragments. The causes for their subsidence are not known. Other plateaus are of definite oceanic origin. For example, the Cocos and Carnegie ridges are probably the result of the continuous extrusion of basalts onto the Cocos plate, overriding an active hot spot. Other plateaus are of unknown nature. Drilling into the Ontong Java Plateau revealed a few meters of Lower Cretaceous vuggy basalt beneath a cover of more than 1 km of calcareous sediments, indicating shallow deposition since Early Cretaceous time. The nature of the rock underlying the volcanics remains unknown.

Some constraints on the nature of the enigmatic plateaus may be obtained from their crustal seismic velocities (Fig. 2): Much depends here on the interpretation of compressional velocity V_p of 6.0–6.3 km/s in the upper 5 km to 15 km in many of the plateaus. Nur and Ben-Avraham (1982) argue that these values of V_p and thickness, combined with the relief and gravity data, suggest a continental structure. Based on detailed velocity analysis it can be concluded that the central part of the Ontong Java Plateau shows continental affinity. By the same argument, the crust of the Manihiki Plateau, with structure like Ontong Java, is also similar to continental crust. One of the concerns is the lack of a reasonable continental source in the western Pacific. In the Indian Ocean, similar plateaus, proven to be continental fragments, originated from the breakup of Gondwana.

Migration and Consumption of Oceanic Plateaus

Because oceanic plateaus are well embedded in their moving plates, they are presumably fated to arrive at consuming plate boundaries—to become, at least in some cases, accreted allochthonous terranes. Paths of migration of terranes as oceanic plateaus have been suggested or inferred from paleomagnetic data obtained from rocks within terranes, or from plate motion reconstructions that are based on marine magnetic data. Paleomagnetic data are limited to providing the paleolatitude of a plateau, or a terrane, as a function of time, if rocks of appropriate age are present, and if these rocks have not been metamorphosed or otherwise transformed, so that their magnetic signature has been obliterated. Often there

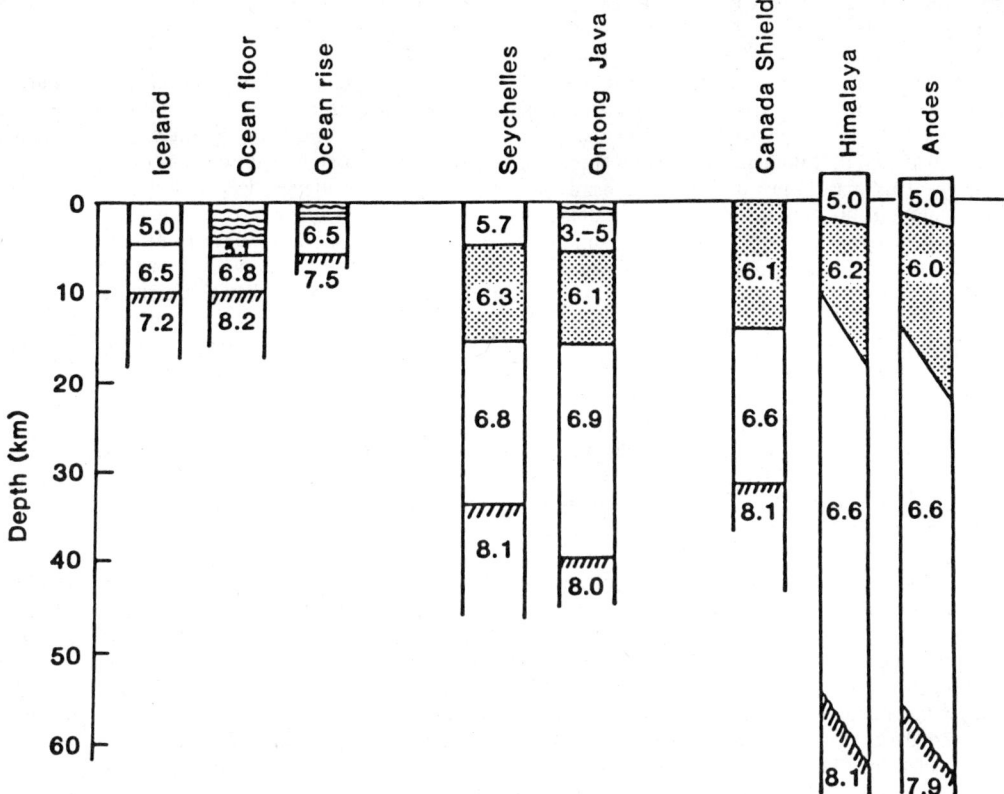

FIGURE 2. Comparison of some typical oceanic and continental crustal structures. Although morphologically similar, the Ontong Java Plateau is structurally dissimilar to Iceland or to typical oceanic crust. It is, however, remarkably similar to a typical shield structure, where compressional wave velocity in thick upper crust is 6.1 km/sec, and to the Seychelles Bank, which is known to be continental. Typical orogenic roots like those of the Himalayas and the Andes are even thicker. (After Nur and Ben-Avraham, 1978)

is also an ambiguity of polarity for periods in which reversals are abundant, so that it is not always possible to decide whether a terrane originated south or north of the paleo-equator. Nevertheless, partial or possible paths have been reconstructed for several important terranes such as Wrangellia and Salinia. Both came from the south and possibly the southern hemisphere. Similar results were obtained by others for terranes in Canada, east Siberia, Japan and China, as well as in other sectors of the Circum-Pacific. Multichannel seismic data have been used to infer the northward motion of the Bowers Ridge now located within the Bering Sea (Ben-Avraham et al., 1981).

FIGURE 1. Worldwide distribution of the present day's oceanic plateaus (hachured). Many of these plateaus may be continental fragments, in spite of their situation well within oceanic plates. Others are of different origins, such as extinct arcs and hot spot traces. (From Ben-Avraham et al., 1981; copyright © 1981 by the American Association for the Advancement of Science.)

A major question is the fate of oceanic plateaus as they move into otherwise normal oceanic subduction zones. Obviously due to their buoyancy, large continental masses tend to collide rather than subduct. On the other hand it has long been assumed that sediment accumulations and small oceanic rises are subducted with the normal oceanic crust in which they are riding. The fate of intermediate size plateaus is uncertain. Some believe, for example, that a sizeable and now extinct plateau, the counterpart of the Hess Rise in the Pacific, was carried with a now extinct plate into western North America where its subduction caused the Laramide orogeny. However, the extensive record of accreted terranes, including many oceanic ones attest to their resistance to subduction, even when they are rather small. For example, numerous accreted seamounts have been identified in central California (Jones et al., 1977). Monger and Price (1979) identified important accreted terranes in western Canada.

Several oceanic rises, such as the Nazca and Juan Fernandez Ridges off South America, and the Louisville and Marcus-Necker Rises in the western Pacific, are being consumed at present, either by collision or by subduction. These rises, which have already entered their respective subduction zones can be clearly associated with gaps in the seismicity of the downgoing slab. Often the seismic gaps are deep—hundreds of kilometers downdip from the collision. This may imply that stress transmission effects through the slab are responsible or that the gaps are associated with earlier consumption episodes of oceanic rises (Nur and Ben-Avraham, 1981).

Volcanic Gaps

Even more pronounced than seismic gaps are gaps in *volcanism* that are associated with these rises, such as the Nazca and Juan Fernandez ridges and the Benham Rise. Twenty-four gaps (Fig. 3) in the Circum-Pacific active volcanic rim were identified (McGeary et al., 1985). The extent of these gaps shows that over 30% of the Circum-Pacific oceanic crust is subducted without associated volcanism. Consequently the process that creates these gaps is significantly modifying the "normal" magmatic and possibly orogenic processes at convergent margins.

Figure 3 shows the close spatial association between the location of the Circum-Pacific volcanic gaps and the zones of collision or subduction of oceanic plateaus. The question is raised as to how these plateaus perturb the process of arc volcanism. The effect of a plateau may depend on characteristics inherent to the plateau such as its buoyancy, crustal thickness, age, and size, or on characteristics of the convergent system itself such as the rate of convergence, age of the plate, dip, length, depth of the slab, etc. The examination of the various characteristics for each volcanic gap and oceanic plateau by McGeary et al. (1985) reveal that the most likely cause of volcanic gaps is the perturbation in the supply of volcanic materials. This perturbation is probably tied to the different geochemical compositions and crustal thicknesses of oceanic plateaus, compared to normal oceanic crust. Normal oceanic crust and associated plate, when formed at a mid-ocean spreading ridge, are water rich. As this normal oceanic plate is subducted, it is hydrothermally altered. Estimates of the depth and extent of this alteration vary greatly, but the plate is generally thought to provide enough water through dehydration at depth to enhance partial melting in the asthenosphere with the resultant arc volcanism. In contrast, not enough water may be supplied by a oceanic plateau, with a different geochemistry and structure. In addition, subduction of water-rich sediments, especially trench sediment, is likely to be limited where a high topographic feature is subducted.

More specifically, three possibilities have been suggested for the effect of a plateau on the physico- chemical processes at depth, due to reduction in water supply:

1. The top of the plateau may be sheared off during subduction. If the top 2 or 3 km contain much of the water and/or hydrous minerals, this erosion may cause a significant decrease in the quantity of water available deeper for melting. This erosion may also remove water-rich sediments that may exist on the surface of the plateau.
2. The plateau remains intact through the subduction process, but doesn't initially contain the volume of hydrous minerals necessary to support the melting process needed for maintaining arc volcanism, in comparison with normal oceanic crust formed at spreading ridges.
3. A sufficient quantity of hydrous minerals exists in the plateau, but water may be released by dehydration reactions at a depth too shallow for melting within the mantle wedge.

All three possibilities require that the addition of volatiles from the subducted plate into the overlying mantle is required for partial melting in the wedge. On the other hand, if enough water is already in the mantle to cause partial melting, then the subduction of a ridge is not expected to significantly affect the deeper chemical processes. Because of the close correlation between oceanic rises and gaps, we conclude that the reduction in water supply is most likely responsible.

Volcanic Gaps Due to Oblique Consumption of Ridges

To gain more insight into the tectonic processes of gap formation we focus on the western margin of South America, which displays a remarkable and unusual variety of combinations of seismicity, volcanism, and morphology. The oceanic Nazca plate is consumed here from Colombia in the north to southern Chile, a process accompanied by volcanism on land and seismic activity along Benioff Zones to a depth of 600 km. However, clear anomalies are apparent: Where the Nazca ridge, towering more than 1500 m above the surrounding sea floor, collides with South America, the trench depth is greatly reduced, and volcanism stops. A 1500-km gap in active volcanism exists north of this point. In addition, the dip of the seismic plane here is anomalously shallow. A smaller volcanic gap of about 500 km is present further south, just north of the collision with the Juan Fernandez chain off the Chile coast.

Nur and Ben-Avraham (1981, 1982) have suggested that it is the obliqueness of the ridges that is responsible not only for the presence but also for the extent of the associated volcanic gap. If the ridge is oblique to the direction of plate motion, the point at which the ridge meets the continent moves along the boundary. Taking the plate velocity v, ϕ as the direction of plate motion relative to the ridge, the distance along the margin swept by the ridge portion consumed L, the velocity w at which the oblique ridge sweeps the plate boundary is

$$w = \frac{L}{t} = v \cdot \tan\phi$$

where t is the time increment over which motion is considered. For the Nazca Ridge, $v \approx 10$ cm/yr, $\phi \approx 45°$, $\tan\phi = 1$, and hence $w \approx 10$ cm/yr. Accordingly, the Nazca Ridge sweeps the consumption zone from north to south at a rate of 10 cm/yr, and would require about 15 my to sweep the length L of the 1500 km gap. The Juan Fernandez Ridge, with the same plate velocity $v = 10$ cm/yr but with $\phi \approx 15°$ and $\tan\phi = 0.3$ has a sweep velocity of only $w \approx 3$ cm/yr. Thus to sweep the length L associated with its gap, of about 500 km, would also require also about 15 my.

Figure 4 shows in cartoon fashion the process we envision. An elongated rise, oriented obliquely to the direction of plate motion, approaches the trench. Because of the buoyancy of the rise's light rocks the trench is erased, followed by a disruption of the downgoing slab. In this process some of the light material may be incorporated into the overriding plate as embedded fragments or crustal undercoating. The disrupted subduction leads to the temporary cessation of volcanic activity. Because of the oblique orientation of the rise relative to the movement of the oceanic plate, this disturbance migrates along the plate boundary. As the rise segment is consumed, subduction of the oceanic crust is reestablished, forming a new trench and renewing seismicity. This sequence of events may be responsible for the apparent flattening of the seismic active zone. Eventually, volcanic activity is renewed and perhaps offset seawards relative to the older volcanic zone, and normal subduction of the oceanic plate is restored, with trench, volcanism, and normally dipping seismic plane.

Entrapment Origin of Marginal Seas

The entrapment of a marginal part of a preexisting ocean by the formation of an island arc may also be related to the arrival of oceanic plateaus at trenches. Active back arc basins occur in only a very few locations on earth. There must be, therefore, a special setting in these places that allow back arc spreading to occur. Some marginal seas were obviously formed by rifting in the past. How spreading initiated in these seas and what made it stop remain major unsolved problems in geodynamics. For example, in the Shikoku Basin, spreading started about 30 myBP and ended 20 myBP. Several attempts to relate the initiation of spreading in this area to plate motions have been made but it is not clear why spreading stopped.

In most of the attempts to explain the origin of marginal basins, the irregularities of the sea floor have been ignored. However, the collision of plateaus with subduction zones could entrap pieces of the ocean basin by shifting the position of the subduction zones, particularly in low stress regimes. Several examples are reviewed below.

The Arctic Basin. It has been suggested that the development of the Arctic Ocean Basin involved plate accretion and the entrapment of a part of the proto-Pacific plate. As the northern Atlantic opened in the Jurassic, the Siberian and Russian platforms, which had been sutured together in the late Paleozoic, moved eastward around the Arctic while the North American plate moved westward. The moving continents consequently half closed off the ocean that separated Alaska-Chukotka from Siberia. The continuation of the circumpolar drift of continents in combination with northward motion of the Kula and Kolyma plates resulted in collision and suturing of Kolyma and Siberia in the Cretaceous, with multiple collisions and accretion of allochthonous terranes against the Pacific margin of Alaska-Chukotka resulting in orogenic deformation. Finally the collision and suturing of Alaska-Chukotka with the Eurasia plate isolated part of the Kula plate, creating a nucleus for the Arctic basin.

The Bering Sea. The Bering Sea is thought to be a marginal sea formed by the Aleutian Arc, which trapped a portion of the Kula plate. Prior to the formation of the Aleutian Arc in late Mesozoic or early Tertiary time, subduction probably took place along the present-day continental margin of the Bering Sea. As the arc formed, subduction shifted to the Aleutian Trench some time prior to the change in motion of the Pacific plate 42 my ago.

The formation of the arc may be related to the collision of oceanic plateaus with the Mesozoic subduction zone. At present the Bering Sea has three large oceanic plateaus and ridges: the Umnak Plateau, the Bowers Ridge, and the Shirshov Ridge. Refraction data from the Bowers Ridge and the Umnak Plateau indicate that a thickened crust is present beneath both. The Bowers Ridge is probably an extinct island arc. Multichannel seismic profiles reveal that there was a subduction zone on the northern side of the Bowers Ridge and that the Bering Sea margin was also a subduction zone. The Bowers Ridge, thus, must have moved in the past toward the Bering Sea margin.

It is possible that the Umnak Plateau, now situated between the Bering Sea margin and the Aleutian Ridge, also came from elsewhere. Thus, a possible scenario is that before formation of the Aleutian Ridge, the proto-Bowers Ridge and proto-Umnak Plateau moved into their present positions in the Bering Sea (Fig. 5). The collision of the Umnak Plateau with the then convergent Bering Sea margin

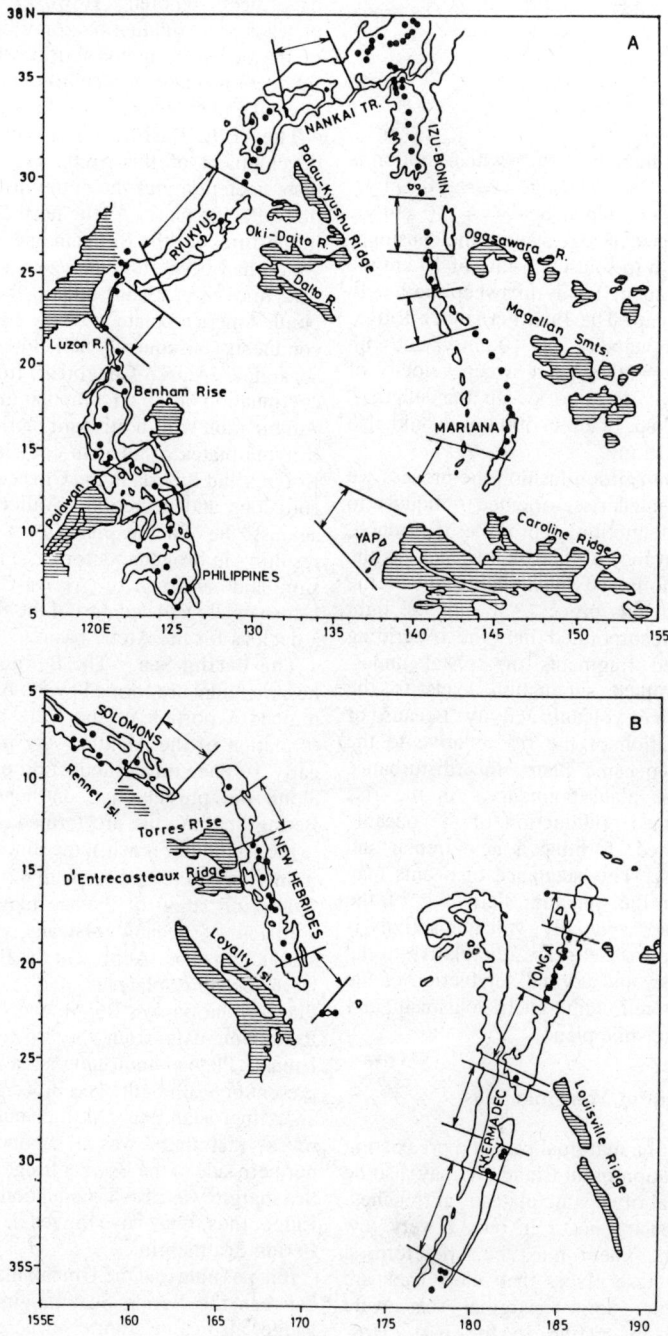

FIGURE 3. Maps showing locations of volcanic gaps and bathymetric rises. (*A*) The Philippine Sea region; (*B*) The Solomon, New Hebrides, Tonga, and Kermadec arcs; (*C*) South America; (*D*) Central America.

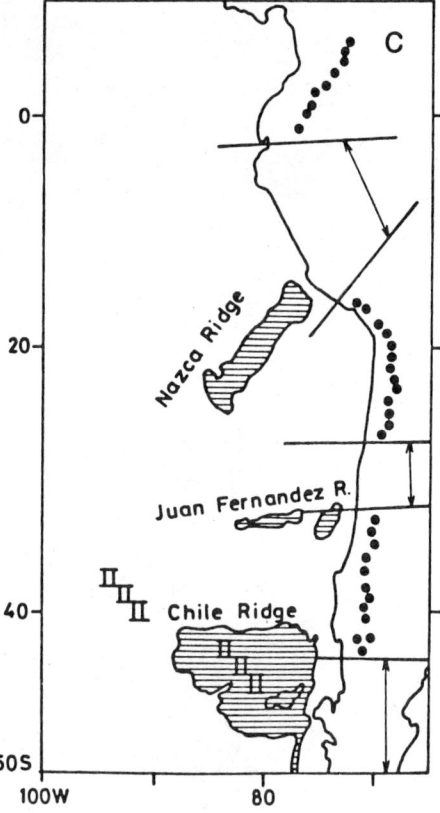

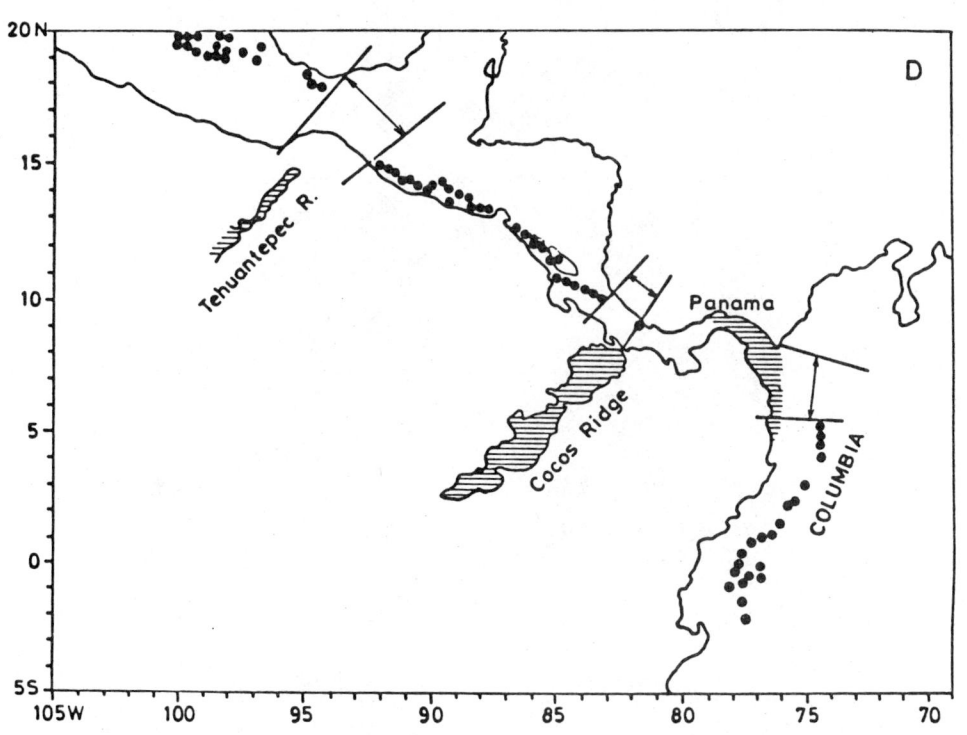

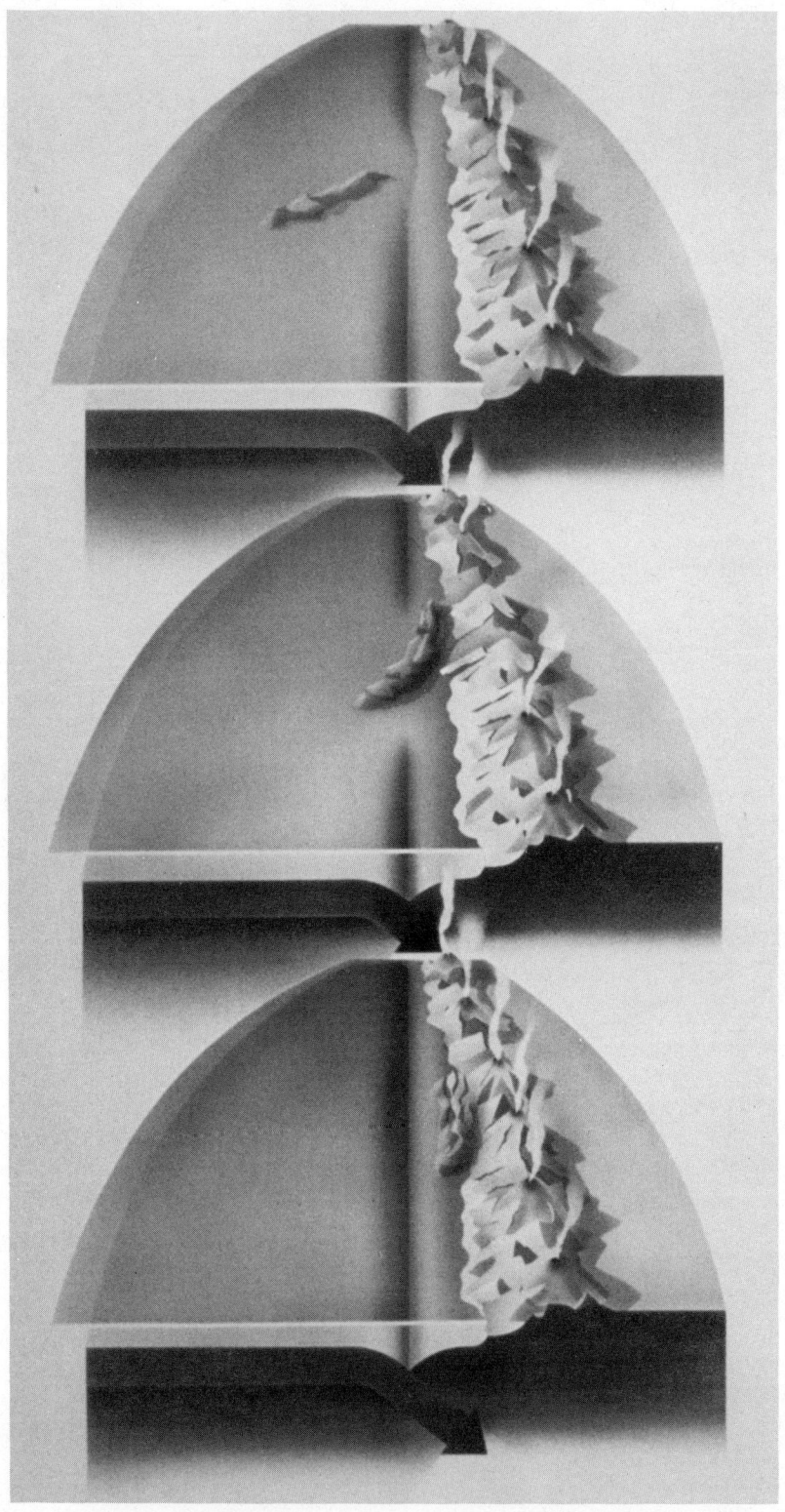

FIGURE 4. Cartoons depicting the consumption process of aseismic ridges. Top: Ridge is carried by oceanic plate to plate boundary. Middle: A volcanic gap develops as ridge segments are plastered onto the overriding plate. Bottom: Volcanic activity is resumed across the previous gap.

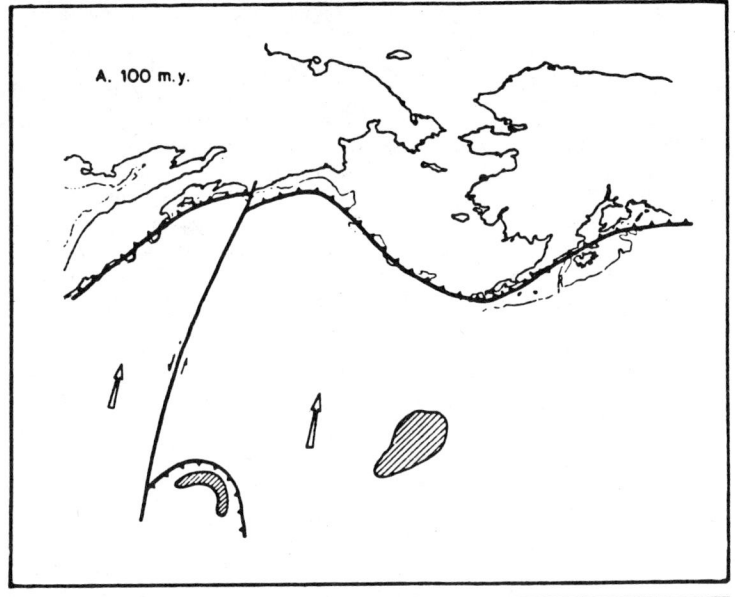

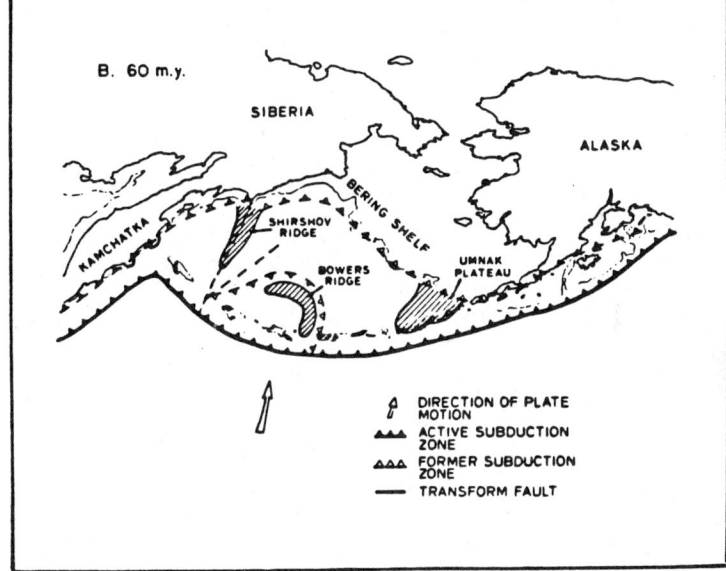

FIGURE 5. Conceptual model for the formation of the Aleutian ridge. Bowers ridge and Umnak plateau are thought to have come from the south with the Kula plate and Shirshov ridge to have formed in place. (A) Late Mesozoic; (B) Early Tertiary.

may have caused subduction to terminate and move southward, resulting in the formation of the Aleutian arc. Similarly, the Shirshov Ridge, separating the Aleutian and Komandorsky basins, could have been formed along a large transform fault that was active during the northward motion of the Kula plate or by rifting away from Kamchatka.

The Okhotsk Sea. Although the Sea of Okhotsk is underlain by continental crust, a calc-alkaline volcano-plutonic belt lay west and north of the present Sea of Okhotsk during late Cretaceous and early Cenozoic time, which has led to the proposal that a microcontinental Okhotsk block of unknown origin collided with the eastern margin of Eurasia during early Cenozoic time and caused the subduction zone to jump eastward, along the Kurile Ridge. The continental fragment, or possibly island arcs, lodged in the Okhotsk Sea and is now submerged;

hence a new marginal sea was formed. After the new subduction zone was established back-arc spreading may have occurred to form the Okhotsk abyssal plain (or the Kurile Basin).

The West Philippine Basin. The Philippine Sea is composed of two distinct units—the West Philippine Basin with WNW-trending magnetic anomalies and the eastern Philippine Sea with NS-trending magnetic anomalies. On the basis of symmetrical magnetic anomalies across the Philippine Ridge in the West Philippine Basin, it seems that the Central Basin Fault is actually an extinct spreading center. Based on this observation, Uyeda and Ben-Avraham (1972) have proposed that the Philippine Ridge was connected by a large transform fault to the Kula-Pacific Ridge and was migrating northward. Because of this transform fault, the Philippine Ridge was left behind while the other ridges descended beneath Japan. When the direction of motion of the Pacific plate changed at about 42 myBP, subduction started along the former transform fault because the sea floor west of the fault was much younger and therefore hotter and less dense than that to the east. The transform fault then became a subduction zone (now the Kyushu-Palau Ridge) and the Philippine Ridge became extinct. Thus, the part of the Kula plate west of the transform fault was trapped to become a marginal sea, the West Philippine Basin.

Ridge Offsets

In contrast with the direct roles of oceanic rises at active margins and the overriding plates, as discussed earlier, their role in the behavior of the oceanic plates in which they are embedded have not been explored very much. Two aspects are of particular interest here: the origins and behavior of leaky transform faults and the causes and mechanics of shifts and jumps of spreading ridge segments.

In this section we consider a process by which the consumption of an oceanic plateau or rise at a trench may cause shifts in spreading ridge associated with the oceanic plate that carries the rise. Consider a rise approaching the subduction zone in Fig. 6a. When the rise reaches the subduction zone, one of several things may happen: The rise may be sheared off the underlying subducting plate to become an accreted terrane (Coney et al., 1980), or it may be subducted together with the oceanic plate. Whether the process is accretion or subduction, it is reasonable to assume that when the rise is sufficiently large, its consumption may offer enough resistance, so that locally the subduction rate may be reduced until the rise is fully consumed. Such a slowdown has been demonstrated for the collision of India with Asia or suggested for the consumption of the Hess rise counterpart with western North America (Engebretson et al., 1982).

One question arises immediately: How is the local slowdown accommodated relative to the undisturbed motion of the surrounding oceanic plate? In the continental crust, differential horizontal motions are generally thought to be accommodated by shear faulting, involving thrust and strike slip motions along faults that tend to be oriented at 25° to 45° to the direction of maximum compression. Such a faulting pattern is rare in the oceanic crust. Instead, differential motion is most commonly accommodated by transform faults where ridge offsets are involved or leaky transform faults, where relative horizontal motion is involved beyond the ridge offsets. The shear motion along these faults is, unlike continental faults, in the direction of maximum compression, not at 25° to 45° to it. For reasons that have not as yet been identified, these faults in the oceanic crust do not obey the laboratory derived shear failure conditions of rocks. Hence for example the slight differential motion between the northeastern and the southeastern Atlantic is accommodated along the Azores leaky transform, which is oriented in the direction of plate motion.

We suggest that when the collision of an oceanic rise slows the subduction rate locally, the differential motion is accommodated along the two fracture zones that span the rise (Fig. 6c); these zones act as leaky transforms while the obstacle is present. As soon as the rise is consumed, the relative slip on the reactivated or leaky transform ceases, and the plate moves again as a rigid one.

We assume next that spreading at oceanic ridges is symmetrical, i.e., that material is added to each of the newly forming plates at an equal rate. Combining the three elements—symmetrical spreading at the ridge, reduced subduction rate where the oceanic rise is being consumed, and slip along fracture zones that border the rise—leads directly to the necessity that a gradual shift in the ridge's axis, away from the trench in which slip is slower and towards the opposite trench where slip is unchanged, must take place.

Together with this shift, the rate at which new crust is formed along this ridge segment must also lessen, leading to narrower magnetic anomalies. Figure 6 shows cartoonwise the progression of the proposed process: The oceanic plateau collides with the trench (6b), causing a slowdown of subduction where it is colliding, inducing relative slip across two fracture zones, which act as leaky transforms (6c), and causing a shift away of the ridge axis. While the collision process is in progress at the trench and the slowdown is maintained, this process continues (6c–e). Once the rise is consumed and subduction rate is reestablished (6f), a record of the disturbance remains frozen in the ocean floor, in the form of a ridge offset, associated with a domain of narrower magnetic anomalies and a progressive offset between magnetic anomalies. A domain of anomalous magnetic stripes is left not only in the plate that carried the plateau (right side in Fig. 6f), but also in its undisturbed counterpart (left side in Fig. 6f) plate, as shown in Fig. 6f.

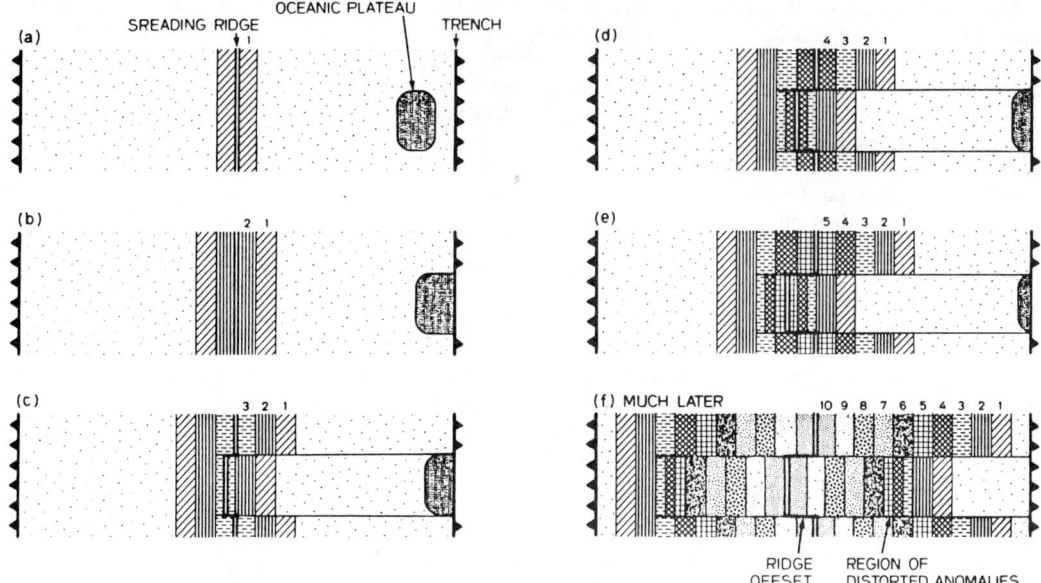

FIGURE 6. Sketches showing how the consumption of an oceanic plateau at a trench shifts the spreading ridge that carries the plateau.

The model proposed here, though not as yet tested, is amenable to rigorous testing. It predicts that the past consumption of some oceanic rises at active margins should be contemporaneous with associated ridge shifts, zones of narrower magnetic anomalies symmetrical about the spreading ridge, temporal activation of fracture zones as leaky transforms, and finally, a record of a contemporaneous gap in arc volcanism.

The Pacifica Continent

The accreted terranes in the Circum-Pacific have many origins. Of particular interest here are those with continental affinity. The accreted continental fragments could not have been formed in the open ocean; they had to be originally parts of large continental masses.

Three processes have been proposed to account for the presence of old continental crust in the Circum-Pacific belts: (1) separation, subsequent return and collision of continental slivers; (2) the transverse motion of irregular continental slivers along transform faults such as the San Andreas Fault; and (3) collision with and incorporation of island arcs. We envision a fourth process: the collision of continental fragments with the Circum-Pacific continents over at least the past few hundred million years. These fragments were part of a continental mass that disaggregated, in the manner of Gondwana and, currently, of Africa.

An example for the process of creation of continental fragments and their subsequent dispersal and accretion is provided by the history of the Indian Ocean. A complex pattern of rifting and ridge jumping produced a host of fragments and plateaus, including such fragments as Madagascar, the Seychelles Banks, the Kerguelen Plateau, and the Broken Ridge. Spreading eventually concentrated along the Carlsberg Ridge, which gradually propagated northward, finally separating the microcontinent of Arabia from Africa along the Red Sea. At present, while spreading along the Carlsberg Ridge-Red Sea axis continues and collision in the Himalayas and subduction in the Java Trench are well established, new rifting is developing in east Africa which could grow into a new spreading center. The east Africa rift has already led to the definition of the new, soon-to-be detached Somalia plate.

Analogously we suggested that the Circum-Pacific fragments were embedded in and moved with the major plates of the Pacific Ocean. The motion of such fragments, back to 190 myBP, may be reconstructed from the paleomagnetically derived motions of the Kula, Farallon, Phoenix, and Pacific plates (Larson and Chase, 1972; Hilde et al., 1977). By moving continental blocks from Alaska, western North America, Kamchatka, and Japan, with their corresponding plates back in time, we note that the various fragments, as they migrate toward their respective spreading ridges, also approach each other. It appears that by late Jurassic time (the earliest for which magnetic data is available) many of the fragments are on their way toward coalescing into a single continental mass. We suggest that such a mass could have actually been in existence before approximately late Permian times (Nur and Ben-Avraham, 1977).

We have proposed that such a continental mass, intact before early Mesozoic time, was broken up by the complex spreading activity that separated the Kula, Farallon, Pacific, and Phoenix plates. The fragments, divided into four groups associated with these plates, were then presumably carried toward subduction zones and eventually reached continental margins. Roughly speaking, the Kula fragments collided with Alaska and Eastern Siberia, the Farallon fragments with North America, and the Phoenix fragments with South America.

The oldest magnetic anomalies and sediments in the Pacific Ocean suggest that the continental fragments of Pacifica were already fully separated by late Jurassic or early Cretaceous times, approximately 120 myBP (Larson and Chase, 1972). It is quite likely that major continental collisions were already in advanced stages and that a large part of the Pacific Ocean had been swept clean at this time of continental fragments. By reconstructing the spreading junction, the early Pacifica may have been located to the northeast of today's Australia, either as a distinct mass in the middle of the Pacific Ocean, or more likely as part of Pangea. A hypothetical configuration of Pacifica as part of Pangea, sometime during or before Permian time, is shown in Fig. 7.

As pointed out by Nur and Ben-Avraham (1982) fragments of Pacifica-Gondwana have also played a major role in orogenic deformation of the Alpine-Himalaya chain, as suggested by numerous investigators for microplate and fragment incorporation. Many of these accreted fragments in the Alpine chain seem to have originated from Gondwana. The South European and Near Eastern blocks may have come from north Africa and Arabia, the Iran and Afghanistan blocks from east Africa, and the Tibet-Himalaya blocks from northern India, as illustrated hypothetically in Fig. 7. We also include Pacifica in this sketch, despite the vastly greater uncertainty of its location, to highlight the attractive possibility that the proposed Pacifica fragments may be the easternmost Gondwana fragments. Unlike more westerly parts of the northern rim of Gondwana, Pacifica faced a vast ocean; therefore its fragments may have been more widely spread and may have travelled much greater distances.

In summary, it seems possible that Gondwana's northern and eastern margins were repeatedly fragmented at least since Permian times. This fragmentation, which continues today, supplied numerous continental fragments to the oceanic plates, fragments that eventually have been or will be accreted by collision into active continental margins.

This generalization has interesting significance for our understanding of orogeny. As discussed already, the orogenic deformation in the northern Pacific rim may have been caused by collisions with accreted fragments and arcs and not just by the normal subduction of oceanic lithosphere. Now it appears that fragment collisions may also be responsible for much orogenic deformation in the Alpine-Himalaya chain. Consequently, both orogenic systems, the Alpine and the Circum-Pacific (excluding the Andes) may have similar mechanical origins: collisions with fragments. In the Andes, unlike the rest of the Circum-Pacific, there is no simple evidence at present of accreted terranes.

The notion of Pacifica and its breakup may help explain some of the similarities of faunas and rock sequences in widely separated locations in the mountain belts around the Pacific, and may tie in divergent paleobiogeographic data. It may explain the curious geological history of Wrangellia (Jones et al., 1977), where flood basalts and Triassic conti-

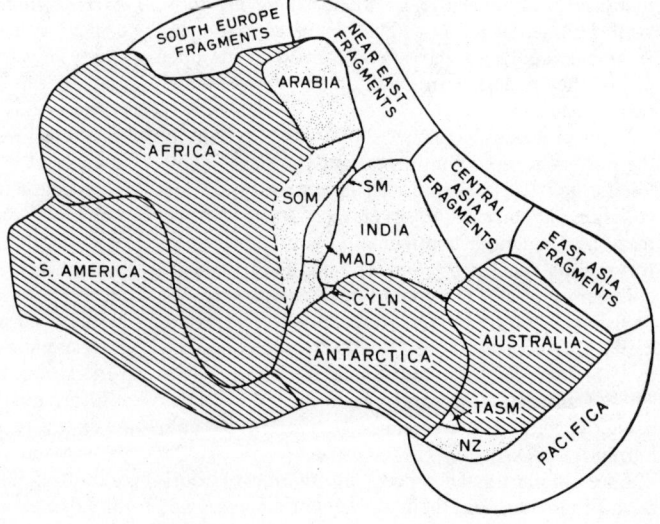

FIGURE 7. Speculation about the origin of many of the accreted allochthonous terranes in the Alpine and Pacific Mesozoic-Cenozoic orogenic belts. Possibly, the fragmentation of Gondwana, and particularly its margins, gave rise to a host of large and small fragments, which have become embedded in oceanic plates, moving toward subduction zones along the boundaries of other continental masses.

nental deposits are supplemented by strong evidence for continental rifting in an unknown location near the equator. The submergence of this fragment, and its subsequent emergence upon collision with North America, is almost a perfect example of the history we envision for the fragments of Pacifica.

Conclusion

We have reviewed some of the roles of oceanic rises in the behavior of the oceanic plate, subduction, volcanism, ridge mechanics and orogeny. Although many details remain to be sorted out and much data are still needed, it is quite clear that the anomalous rises including those that are now found in the oceans floors and those that existed in the past and have since been subducted or accreted, do exercise some control on lithospheric tectonics particularly at spreading or consuming plate boundaries. Some of the processes reviewed here are ripe at present for detailed research, going a step well beyond the large-scale model of plate tectonics.

AMOS NUR
ZVI BEN-AVRAHAM

References

Ben-Avraham, Z., A. Nur, D. Jones, and A. Cox, 1981, Continental accretion and orogeny: from oceanic plateaus to allochthonous terranes, *Science* **213**, 47-54.

Coney, P. J., D. L. Jones, and J. W. H. Monger, 1980, Cordilleran suspect terranes, *Nature* **288**, 329-333.

Cooper, A. K., M. S. Marlow, and D. W. Scholl, 1976, Mesozoic magnetic lineations in the Bering Sea marginal basin, *Jour. Geophys. Research* **81**, 1916-1934.

Engebretson, D. C., A. Cox, and G. A. Thompson, 1982, Convergence and tectonics: Laramide to Basin Range, *EOS* **63**, 911.

Hilde, T. W. C., S. Uyeda, and L. Kroenke, 1977, Evolution of the Western Pacific and its margin, *Tectonophysics* **38**, 145-165; **47**, 107-222.

Jones, D. L., N. J. Silberling, and J. W. Hillhouse, 1977, Wrangellia—a displaced terrane in northwestern North America, *Canadian Jour. Earth Science* **14**, 2565-2577.

Larson, R. L., and C. G. Chase, 1972, Late Mesozoic evolution of the western Pacific Ocean, *Geol. Soc. Am. Bull.* **83**, 3627-3644.

McGeary, S. E., A. Nur, and Z. Ben-Avraham, 1985, Spatial gaps in volcanism: the effect of collision or subduction of oceanic plateaus, *Tectonophysics* **119**, 195-221.

Monger, J. W. H., and R. A. Price, 1979, Geodynamic evolution of the Canadian Cordillera—progress and problems, *Canadian Jour. Earth Science* **16**, 770-791.

Nur, A., and Z. Ben-Avraham, 1977, Lost Pacifica continent, *Nature* **270**, 41-43.

Nur, A., and Z. Ben-Avraham, 1978, Speculations on mountain building and the lost Pacifica continent, *J. Phys. Earth Suppl.* **26**, S21-S37.

Nur, A., and Z. Ben-Avraham, 1981, Volcanic gaps and the consumption of aseismic ridges in South America, *Geol. Soc. America Memoir* **154**, 729-740.

Nur, A., and Z. Ben-Avraham, 1982, Oceanic plateaus, the fragmentation of continents and mountain building, *Jour. Geophys. Res.* **87**, 3644-3661.

Uyeda, S., and Z. Ben-Avraham, 1972, Origin and development of the Philippine Sea, *Nature* **240**, 176-178.

Cross-references: *Accretionary Tectonics: Examples from the North American Cordillera; Continental Collision Zones: Seismotectonics and Crustal Structure; Continental Crustal Structure; Deep Seismic-Reflection Profiling; Lithosphere, Oceanic: Formation and Evolution; Mantle Convection and Plumes; Ocean Continent Transition: Structure; Paleomagnetism and Plate Tectonics; Seafloor Spreading: Magnetic Evidence; Seismicity: Subduction Zone; Subduction Zones.*

FIGURE OF THE EARTH

That the Earth's surface resembles a sphere was recognized in ancient times. That it more closely approximates an oblate ellipsoid was deduced by Huygens and Newton and observationally confirmed in the eighteenth century. The ellipsoidal flattening is one part in 300, but the maximum deviation of the sea-level surface of the Earth from ellipsoidal shape is less than one part in 60,000. For many practical needs in mapping, surveying and engineering, assuming the figure of the Earth to be an oblate ellipsoid is sufficient. However, modern demands in both practical and scientific applications require spatial relationships known to several orders of magnitude better not only statically but as functions of time.

The Geoid

The determination of the figure of the Earth involves physical as well as geometric considerations. A point on the topographic surface is positioned by two horizontal (surface) coordinates and one vertical (elevation) coordinate. Measurement of elevation depends on the direction and intensity of gravity and is referred to sea level as the surface of zero elevation. A distinctive property of the sea-level surface, idealized over the oceans to remove fluctuations due to tides, currents, and other disturbing factors, is that it is equipotential; i.e., the direction of gravity is everywhere perpendicular to it. This surface is termed the *geoid* (so named by Listing in 1872) and is conventionally taken by geodesists as the "true" figure of the Earth. Thus, one definition of the geoid is: the equipotential surface of the Earth's gravity field that most closely approximates mean sea level. To the extent that "most closely approximates" and "mean" are vague, this definition lacks precision. But it can easily be made precise: For example, the geoid can be defined as that equipotential surface passing through the zero-elevation mark at a particular tide station. Because actual (as opposed to idealized) mean sea level varies from an equipotential surface by as much as 2 m globally, such a definition has an element of arbitrariness (to within a meter or so).

Geodetic nomenclature distinguishes between the Earth's *gravity* field and its *gravitational* field. The latter is affected only by the mass of (and its distribution within) the Earth and would apply, for example, to a satellite orbiting the Earth. The former pertains to a point on the Earth and thus is subject in addition to the centrifugal force caused by the Earth's rotation.

Because the external gravitational field of the Earth is a harmonic function, the gravitational potential V at a point on or above the surface can be expressed in terms of latitude ϕ, longitude λ, and radial distance from the Earth's center r, as

$$V = \sum_{n=0}^{\infty} \sum_{m=0}^{n} \left(\frac{1}{r}\right)^{n+1}$$
$$\cdot (A_{nm} \cos m\lambda + B_{nm} \sin m\lambda) P_{nm}(\sin \phi)$$

where $P_{nm}(\sin \phi)$ is an associated Legendre function of the first kind, and A_{nm}, B_{nm} are spherical harmonic coefficients that depend on the distribution of mass within the Earth. Taking the simplest possible example, if the Earth were a spherical body whose density is a function only of distance from the center of mass, it is well known that $V = GM/r$, where G is the universal gravitational constant and M is the mass of the Earth. This yields $A_{0,0} = GM$, and all other A_{nm}, B_{nm} are zero. On the surface r is a constant (the radius of the sphere); correspondingly, V there is a constant, and the sphere is an equipotential surface.

The potential W of the gravity field is given by

$$W = V + \tfrac{1}{3} \omega^2 r^2 [1 - P_{2,0}(\sin \phi)]$$

where ω is the rotational velocity of the Earth. Any fixed value for W in this equation will define an equipotential surface: The locus of all points (ϕ, λ, r) that satisfy that equation for an assigned set of A_{nm}, B_{nm}, ω, and the given value of W. The A_{nm}, B_{nm} settle the Earth's external gravitational field and remain constant (if shifts of mass with time are ignored). The geoid then can be found by establishing the position of a single point as belonging to that surface. From this vantage, the geodesist has two fundamental tasks in determining the figure of the Earth: to discover the correct A_{nm}, B_{nm}, which means determining the gravitational field, and to pick a proper point, which means determining a correct spatial scale.

The method described is most amenable to the analysis of observations of the perturbations of orbits of artificial satellites in the gravitational field of the Earth. Another type of data is the direct measurement of gravity itself on the surface. Historically, this well antedated the satellite era, and a formula conceived by Stokes in 1849 has been applied to resolve the geoid from this source. Consider a reference model (to be discussed later) commonly centered with the Earth. Then N, the difference in radial distance between the reference model and the geoid at a point on the latter, is given by the following surface integral over the unit sphere σ:

$$N = \frac{R}{4\pi\Lambda} \iint (\Delta g) S(\psi) \, d\sigma$$

where ψ is the spherical distance between the given point and the surface element $d\sigma$, $S(\psi)$ is an easily calculable function of ψ (see, e.g., Heiskanen and Moritz, 1967, p. 94), and R and Λ are mean global values for the radius of the model and surface gravity, respectively. The term Δg is called a *gravity anomaly*—the difference between measured gravity reduced to its value at the given point on the geoid and a corresponding value of gravity obtained by computation on the reference model. If gravity is known everywhere on the Earth's surface, then the position of any point of the geoid can be obtained directly from the integral.

There are other schemes for determining the geoid, but the two discussed here have been the most widely used. At present (1988) both the mean radius of the geoid and its variations from the reference model are known to within a meter. In regions of dense gravity coverage, the shape of the geoid can be determined to within a few centimeters. Measurements from radar altimeters on satellites of distances to the ocean surface are the principal source for enabling the geoid to be known over the oceans to less than a meter (see *Satellite Altimetry*).

Reference Models

Determination of distances and directions among networks of points is a basic geodetic activity. Performing these computations directly on the geoid is not feasible because, as can be seen from either the spherical harmonic representation or Stokes' formula, the geoid is not a mathematically tractable surface. As mentioned at the start, an Earth-centered oblate biaxial ellipsoid with optimally chosen parameters deviates from the geoid by no more than one part in 60,000. Methods of computation on this surface are well established, and any desired corrections to the geoid can be handled differentially. Thus the use of such an ellipsoid as a reference model for the figure of the Earth has been a traditional procedure for 150 years and still provides the basic reference surface for almost all types of geodetic investigations.

Let a and b designate the semimajor and semiminor axes of this model. The flattening f is defined by $(a - b)/a$. Customarily, the two parameters chosen to fix the figure geometrically are a and f. But the model must also possess physical properties to be applied to the gravity field, as in the definition of a gravity anomaly. The gravity field of the Earth depends on two components: the distribution of mass within the Earth, and centrifugal force due to the Earth's rotation. The latter can be specified by a single parameter, the Earth's rotational velocity ω. Because geodesy is primarily concerned with the surface and exterior gravity field, it can take advantage of the fact that this portion of the field can be uniquely determined without knowledge of the distribution of mass, provided only that the overall mass M is known and the surface is equipotential. Thus a standard reference model for the Earth can be defined by the parameters a, f, M, and ω.

One of the basic aims of geodesy has been to determine values for these (or equivalent) parameters that are closest in accord with reality: a has always been comparatively well determined from arcs of triangulation; ω depends on straightforward astronomical observations; however, for M and f, the direct observational material is relatively weak. Fortunately, these two can be replaced by quantities that are equivalent for the purpose intended and much better determined. Under the conditions stated for a reference model, the spherical harmonic formulation for the exterior gravitational potential becomes

$$U = \sum_{n=0}^{\infty} \left(\frac{1}{r}\right)^{n+1} A_{2n,0} P_{2n,0}(\sin \phi)$$

where U is termed the *normal* field. The first coefficient $A_{0,0}$ is equal to GM, and the second, $A_{2,0}$, is conventionally written as $-GMa^2 J_2$ for dimensional convenience. Since these assignments exhaust all the degrees of freedom available, all of the other coefficients are thereby decided. Both GM and the nondimensional coefficient J_2 can be determined from observations of space probes and artificial satellites to several orders of magnitude better than M and f. Thus the modern definition of the reference model is given by a, J_2, GM, and ω.

The relation

$$V = U + T$$

defines the *anomalous* field T, which consists of the remaining terms in the general spherical harmonic representation. Hence Δg, the gravity anomaly, corresponds to T, being the difference in gravity caused by the actual field and the normal (reference) field. (If the actual field is taken to be the gravity field W, then U is defined to include the centrifugal force term; thus T is the same in either case.)

The need for a universally accepted reference model as a basis for surveying, mapping, and scientific investigations was recognized early in this century. At the second general assembly of the International Union of Geodesy and Geophysics (IUGG) in 1924, values of a and f were officially approved. At the fourth IUGG general assembly in 1930, a formula for the determination of normal gravity (the magnitude of the gradient of U), γ, was sanctioned, based on chosen values of ω and γ_e, the value of γ at the equator (which can be employed as the remaining parameter). The so-called International ellipsoid and International gravity formula constituted the first internationally endorsed geodetic Earth reference model.

Observations on artificial Earth satellites, beginning in 1957, provided the means for spectacularly improving the accuracy of the reference model by yielding precise values of GM and J_2. The International Astronomical Union at its general assembly in 1964 recommended a set of a, GM, J_2, ω as the basis for a new reference model. In 1967, at its fourteenth general assembly, the IUGG adopted this model, termed the Geodetic Reference System 1967 (GRS 67). This was replaced by Geodetic Reference System 1980 (GRS 80) at the seventeenth general assembly of the IUGG, held in December 1979. Currently, GRS 80 represents the officially recognized reference model for the figure of the Earth. Table 1 gives the GRS 80 values of the parameters and the corresponding best range of estimates known at present. The "best" line refers to the last significant figure of the GRS 80 value. Thus, in the case of GM, the GRS 80 value is now (1988) acknowledged to be in error by about one part in 6×10^6. The range of values for ω represents not uncertainty but its periodic (primarily annual) fluctuation. The values of f and γ_e derived from the GRS 80 parameters are 1/297.257 and 9.780327 m · s^{-2}, respectively.

In the GRS 80 model, the mass of the atmosphere is included. This entails a correction of almost 1 mGal when computing γ at the surface. Also, the value of J_2 is premised on the basis that the model is strictly self-gravitating, meaning that it does not include the permanent tidal deformation due to the Sun and Moon. This latter effect adds about 9×10^{-9} to J_2.

Nonstandard reference models have also been proposed. The Earth's equator can be modeled by an ellipse, which introduces two additional parameters, one for the flattening and one for the orienta-

TABLE 1. GRS 80 Values and Best Range of Estimates

	a (m)	J_2	GM (m^3·s^{-2})	ω (rad s^{-1})
GRS 80	6,378,137	108,263 × 10^{-8}	3,986,005 × 10^8	7,292,115 × 10^{-11}
"best"	5 to 7	2.4 to 2.8	4.37 to 4.43	4.9 to 5.0

tion. (Equivalently, $A_{2,2}$ and $B_{2,2}$ in the spherical harmonic representation can be chosen.) The equatorial flattening amounts to one part in 90,000. Since there are other deviations of the geoid from the standard model of the same order of magnitude, the trade-off of improvement in the model versus its increased complexity does not seem worthwhile. Also proposed have been models which make assumptions concerning the Earth's density distribution. The one most discussed is hydrostatic equilibrium; i.e., interior equipotential surfaces are of constant density (see *Equilibrium Figure of the Earth*). Since the Earth is not in hydrostatic equilibrium, the resulting model does not approximate the geoid as well as the standard model and has not been seriously considered for geodetic purposes.

Time Variations

The distribution of mass within the Earth is not static but is subject to many forces that affect the gravity field and the geoid. These forces include oceanic and Earth tides, the shift of the axis of rotation with respect to the Earth's crust, periodic variations in rotational rate, postglacial rebound, and plate tectonic movements. Here we will touch only on the effect of these shifts on the parameters of the reference model. Short-term fluctuations, which average out, will not be considered.

Let us define by "significant" change in a parameter a relative amount not less than one order of magnitude smaller than the current accuracy of that parameter. There is no evidence at present of significant change in either a or GM. J_2 is known to four parts in 10^6; recent analyses of laser ranging observations to the laser-reflecting satellite *Lageos* (see *Satellite Laser Positioning*) indicate that this parameter is decreasing at the rate of 3×10^{-11} per year. The genesis of this is believed to be postglacial rebound. If it continues at the same rate, then in ten years the relative change will be a significant amount, 3×10^{-7}. For ω, tidal friction is the primary cause for a net decrease of about 2×10^{-12} rad · s^{-1} per century, a significant amount over that period with respect to the GRS 80 value of that parameter.

These considerations will markedly influence any future revisions of standard geodetic Earth reference models. It will not be sufficient to merely add significant figures to a set of parameters treated as constants, at least in the case of J_2 and ω. Future models for the figure of the Earth will have to take into account not only the static configuration of the Earth but also its dynamic behavior.

BERNARD H. CHOVITZ

References

Chovitz, B. H., 1981, Modern Geodetic Earth Reference Models, *EOS* (Am. Geophys. Union Trans.) **62,** 65–67.

Heiskanen, W., and H. Moritz, 1967, *Physical Geodesy*. San Francisco: W. H. Freeman, 364p.
International Association of Geodesy, 1967, Geodetic Reference System 1967, *Bull. Géod. Spec. Publ.*, 116p.
Moritz, H., 1984, Geodetic Reference System 1980, *Bull. Géod.* **58,** 388–398.
Torge, W., 1980, *Geodesy*, New York: W. de Gruyter, 254p.
Vanicek, P., and E. Krakiwsky, 1986, *Geodesy: The Concepts*, 2nd rev. ed. New York: North-Holland, 696p.

Cross-references: *Earth Orientation; Earth's External Gravity Field; Earth Tides; Equilibrium Figure of the Earth; Geodesy, Geometric; Geodesy: Historical Introduction; Geodesy Physical; Geodetic Reference Systems; Gravity Fields: Implications for Planetary Interiors; Planetary Geodesy; Very-Long-Baseline Interferometry (VLBI).*

FINITE-DIFFERENCE FORWARD MODELING IN SEISMOLOGY

Finite-difference wave equation techniques have assumed a prominent role in industrial and academic computational seismology, with finite-difference synthetic seismograms being used as a means of data interpretation and wave theory analysis, and finite-difference migration being routinely performed in seismic reflection data processing. Originally introduced to seismology for forward modeling by Alterman and Karal (1968) and for migration by Claerbout (1970), the use of finite-difference techniques has grown in importance in seismology with the increase in speed and memory of computers. The distinct advantage of difference methods is their ability to completely describe wave motion in media with any spatial variation of elastic properties. Their chief disadvantages are computational expense and gradual loss of accuracy. Aki and Richards (1980) have characterized the problems in which finite-difference methods are useful for forward modeling. Roughly speaking, they are most suited to problems in which total travel path is less than several hundred wavelengths and the scale of material heterogeneity (i.e., its approximate diameter) is 0.1–10 wavelengths (see Aki and Richards, 1980, p. 749).

Since the late 1960s the use of synthetic seismograms for interpreting field-recorded data in both exploration and earthquake seismology has become an increasingly important means of deciphering Earth velocity structure and earthquake source behavior. From the early, relatively simple Earth and fault models, the science has progressed to the investigation of the subtle complexities in fault mechanics and Earth structures. The use of synthetic seismograms for interpretation has spanned the range of Earth scales and seismic frequencies from the low-frequency response of the Earth's free oscillations to the high-frequency response of well log data. Finite-difference forward modeling of strong ground-motion records, fault rupture, long-period seismic surface waves, seismic refraction, reflection, and vertical seismic profile recordings has become common. The increase in seismic receiver density in all fields of observational seismology is allowing ever finer details of Earth structure to be examined, with the result that theoretically complete descriptions of wave motion in laterally invariant media are often inadequate for interpreting available data. Consequently, the use of finite-difference methods for calculating the Earth response in laterally varying media has grown in importance. Finite-difference synthetic seismograms have also proved invaluable in analyzing wave theory for laterally heterogeneous media, where theory is approximate or exact expressions are difficult to evaluate.

Exploration reflection seismology processing now customarily includes time migration of the common midpoint (CMP) stacked section to restore reflection events to their proper dip and location in the reflection image of the subsurface. Poststack migration is based on one of several approximations to the acoustic wave equation combined with an imaging principal. Finite-difference schemes are employed in several of the common migration algorithms. Computationally expensive finite-difference prestack depth and reverse-time migration are now also used as means of enhancing lateral resolution in reflection data from geologically complex areas. For a discussion of seismic migration techniques, see *Seismic Wavefield Migration*.

For general treatments of the finite-difference methods, see the elementary text by Gerald (1973) and the advanced texts by Mitchell and Griffiths (1980) and Richtmyer and Morton (1967). I will limit this discussion to *explicit* finite-difference methods used for calculating synthetic seismograms in media with laterally variable material properties. While implicit finite-difference schemes have been used for forward modeling (see Emerman et al., 1982; Marfurt, 1984), they are more widely used for seismic migration than for modeling.

The finite-difference scheme for the one- and two-dimensional acoustic wave equation will be discussed first. I will then discuss methods for two-dimensional elastic wave motion in media with any spatial variation in material properties. The one-dimensional equation is largely of pedagogical interest and is used as a simple means to introduce the numerical scheme. Currently, the bulk of interesting research applications in finite-difference forward modeling is in two-dimensional acoustic and elastic wave problems. Solutions of three-dimensional problems are feasible with modern supercomputers, and the extension of the methods presented here to three-dimensional geometries is straightforward.

Two-dimensional finite-difference schemes can be formulated in either a Cartesian system or a cylindrical system with azimuthal symmetry. The Cartesian system is useful for modeling geologic structures in which wave propagation is in a dip direction but has the disadvantages inherent in two-dimensional wave solutions: Spatially localized sources are line rather than point sources, being infinite out of the plane of propagation. Spatially localized sources in two dimensions exhibit a tail that decays as $1/\sqrt{t}$ rather than maintaining their impulsive character as in three dimensions. The cylindrical coordinate system allows for a true point source, preserving the source waveform. Only a few laterally varying Earth structures are close to being cylindrically symmetric, however, whereas many are locally two-dimensional.

The finite-difference scheme requires (1) a numerical approximation to the wave equation or the equations of motion for a homogeneous or inhomogeneous medium, (2) numerical approximations to boundary conditions along boundaries separating materials with different properties or at a free surface, (3) special boundary conditions at the periphery of the computational domain to simulate an infinite or semi-infinite medium, and (4) a means of initiating wave motion suitable for the modeling problem at hand (see Fig. 1). A particular finite-difference scheme is studied for stability and accuracy.

Acoustic Equation Approximations

I will illustrate the development of the one-dimensional and two-dimensional finite-difference numerical scheme, considerations of numerical stability and accuracy, and development of most boundary conditions with the acoustic wave equation. I develop the two-dimensional equation for a Cartesian system with x and z the horizontal and vertical coordinates (see Fig. 1). Material properties and wave propagation are constant in y. The acoustic constitutive law and equation of motion can be combined to give a linear second-order hyperbolic partial differential equation in pressure. Assuming constant density and a spatially varying bulk modulus, the wave equation for pressure P is

$$\frac{\partial^2 P}{\partial t^2} = c_0^2 \nabla^2 P \qquad (1)$$

where c_0 is the spatially variable acoustic wave velocity.

To develop a numerical scheme for the wave equation (1), discretize the one- or two-dimensional medium with $x = mh$ and $z = nh$, and let $t = l\,\Delta t$. The mesh size h is equal in both spatial directions in the two-dimensional geometry for convenience. We expand the pressure field in a Taylor's series about the location $(mh, nh, l\,\Delta t)$ in the $+x$ direction:

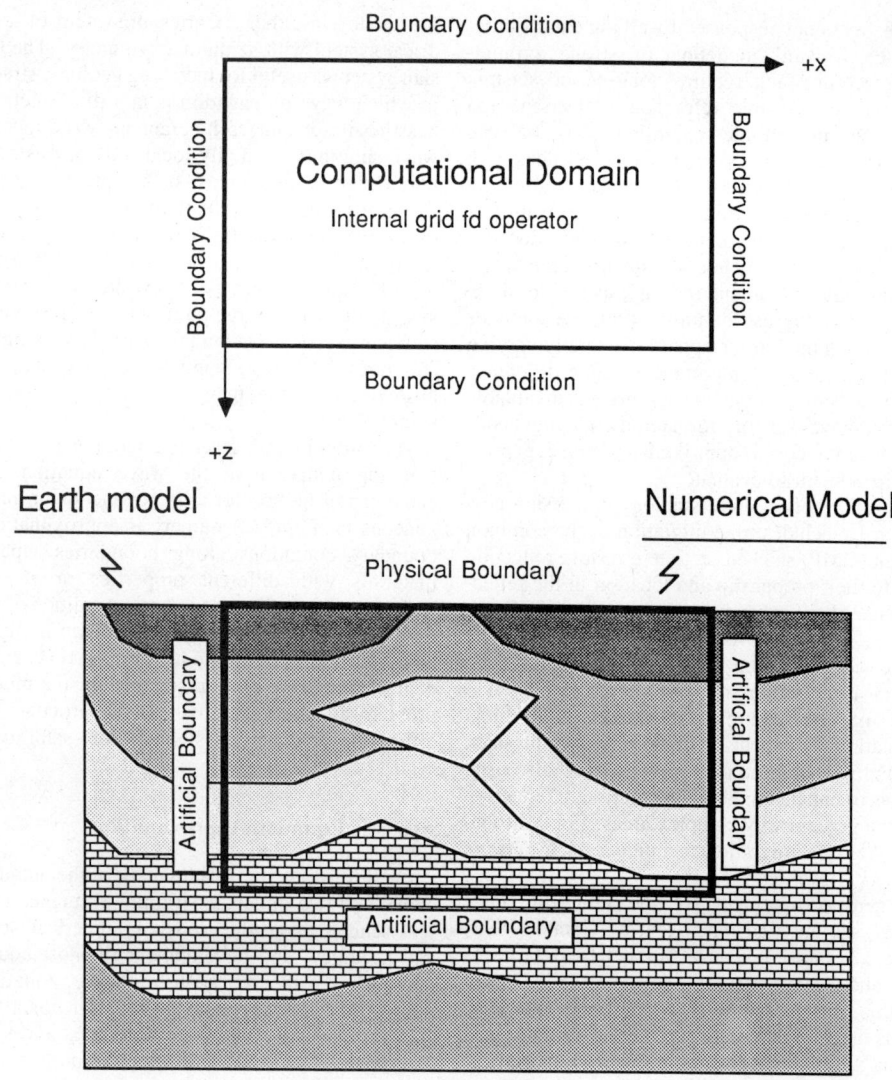

FIGURE 1. The two-dimensional Cartesian system for the finite-difference equations in the text. The numerical scheme consists of an internal grid finite-difference operator and numerical boundary conditions applied around the grid periphery. The numerical scheme is designed to simulate wave propagation in an Earth model with lateral velocity variation. The peripheral boundary conditions may be physical conditions or artificial conditions which are designed to make the computational domain appear infinite or plane layered outside of the region of interest.

$$P((m+1)h) = P(mh) + h\frac{\partial P(mh)}{\partial x} + \frac{h^2}{2}\frac{\partial^2 P(mh)}{\partial x^2}$$
$$+ \frac{h^3}{6}\frac{\partial^3 P(mh)}{\partial x^3} + O(h^4) \quad (2)$$

where $O(h^n)$ means terms that are the order of the nth power of h. Expanding the field variable in the reverse direction and summing the two Taylor's series provides a centered approximation to the second derivative, which is second-order accurate in the mesh size h:

$$\frac{\partial^2 P(m)}{\partial x^2} = \frac{P(m+1) - 2P(m) + P(m-1)}{h^2} + O(h^2)$$

We can similarly approximate the second time derivative. Approximating both the time and space derivatives gives the numerical scheme for the one-dimensional wave equation:

$$P(m, l + 1) = 2P(m, l) - P(m, l - 1)$$
$$+ c_0^2(m) \left(\frac{\Delta t}{h}\right)^2 (P(m + 1, l) - 2P(m, l)$$
$$+ P(m - 1, l)) + O(\Delta t^2, h^2)$$

The wave field at the $l + 1$ time level is isolated from the wave field at the l and $l - 1$ levels; hence this is an *explicit* method for solution of the one-dimensional wave equation. Initial values provided at $l = 0$ and $l = 1$ allow us to march the solution forward in time.

Extending this development to two spatial dimensions, we write the explicit scheme for the two-dimensional wave equation as

$$P(m, n, l + 1) = 2P(m, n, l) - P(m, n, l - 1)$$
$$+ c_0^2(m, n)\left(\frac{\Delta t}{h}\right)^2 (P(m + 1, n, l) + P(m - 1, n, l)$$
$$+ P(m, n + 1, l) + P(m, n - 1, l) - 4P(m, n, l))$$
$$+ O(\Delta t^2, h^2) \quad (3)$$

or, in a more compact notation,

$$D_{tt}P(m, n, l) = c_0^2(m, n)(D_{xx} + D_{zz})P(m, n, l)$$

where the operators D_{tt}, D_{xx}, and D_{zz} are the centered difference numerical equivalents of the continuous derivatives.

These expressions form computational stars or stencils for the numerical scheme in *x-t* or *x-z-t* space (see Fig. 2). Two time levels of the grid must be saved to update the solution. (The $l - 1$ level can be overwritten by the $l + 1$ level.) For this equation the calculation requires $2NM$ words for the field variables and NM words for the material velocities, where M is the number of nodes in the x direction and N is the number of nodes in z. Since the wave field is computed throughout the computational domain, the finite-difference method provides a continuous movie of the wave field as it evolves, in addition to the more conventional time series recordings along horizontal profiles and vertical sections through the model.

Numerical stability and the accuracy of explicit schemes are analyzed by assuming a medium with constant material properties, assuming a plane wave component as the field variable in the difference scheme, and demanding a bounded solution as the calculation advances in time. Substituting $P_0 \exp\{i(k_x mh - \omega l \Delta t)\}$ into the numerical approximation for the one-dimensional wave equation and rearranging gives the dispersion relation for the finite-difference scheme:

$$\sin^2\left(\frac{\omega \Delta t}{2}\right) = \frac{(c_0 \Delta t)^2}{h^2} \sin^2\left(\frac{k_x h}{2}\right) \quad (4)$$

where k_x is the wave number and ω is the angular frequency. For a given ratio of the time step and spatial increment a solution is stable if it is bounded as the calculation proceeds. Requiring that the frequency is a real quantity precludes the existence of exponentially growing components in the solution, implying that the left-hand side of the dispersion relation is real and less than or equal to 1 in magnitude. The resulting inequality is the stability criterion

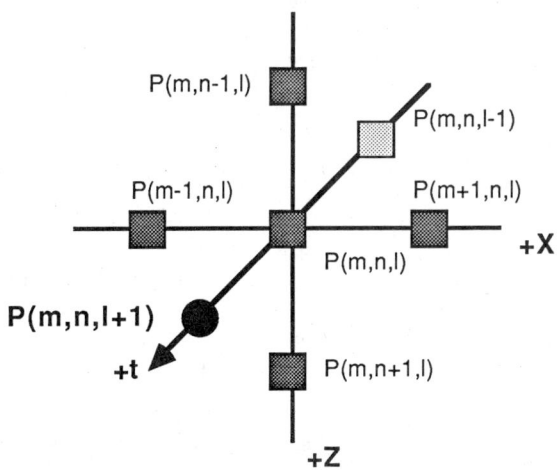

FIGURE 2. Finite-difference stencil for a two-dimensional $O(\Delta t^2, h^2)$ scheme. The pressure field P is advanced to the $l + 1$ time level using values from the l and $l - 1$ level.

$$1 \geq \frac{\Delta t\, c_0}{h} \quad (5)$$

For the two-dimensional acoustic Eq. 3 a similar analysis gives

$$\sin^2\left(\frac{\omega\, \Delta t}{2}\right) = \frac{(c_0\, \Delta t)^2}{h^2}\left[\sin^2\left(\frac{k_x h}{2}\right) + \sin^2\left(\frac{k_z h}{2}\right)\right] \quad (6)$$

with stability criterion

$$\frac{\Delta t\, c_0}{h} \leq \frac{1}{\sqrt{2}} \quad (7)$$

The number that bounds the ratio, $(\Delta t\, c_0)/h = p$, is referred to as the *Courant number* or the *stability limit*. The Courant number is interpreted physically as the fraction of the spatial increment that a wave can advance in a time step. For a spatially variable material velocity, the criterion used is the most conservative, i.e., the maximum velocity in the model.

The dispersion relation for the two-dimensional numerical scheme identifies inaccuracies inherent in simulating wave propagation through a continuous medium on a discrete mesh. Note the difference between the dispersion relation for the numerical scheme, Eq. 6, and the dispersion relation of the continuous differential equation:

$$\omega^2 = c_0^2(k_x^2 + k_z^2)$$

Waves in a perfectly elastic infinite continuous system are nondispersive, whereas waves in a discrete system are intrinsically dispersive. The material wave speed is c_0. Assume a plane wave propagating in the x direction. In Eq. 6, replace the quantity $k_x h$ with $2\pi/G$, where G is the grid points per wavelength. The phase velocity dispersion for the numerical waves is

$$\frac{c(G)}{c_0} = \frac{G}{\pi p}\sin^{-1}\left[p\sin\left(\frac{\pi}{G}\right)\right]$$

For large G, the normalized phase velocity is 1; the phase velocity of long wavelengths on the grid is the material velocity. For G approaching 2, the spatial Nyquist frequency, the phase velocity approaches the constant $c(G = 2)/c_0 = 0.707$.

The group velocity of the numerical waves is determined from Eq. 6 by rearranging for ω and differentiating with respect to k, for a wave traveling in x,

$$\frac{u(G)}{c_0} = \frac{\cos(\pi/G)}{(1 - p^2 \sin^2(\pi/G))^{1/2}}$$

For dense sampling, $u(G) \sim c_0$, whereas for $G = 2$, $u(G) = 0$. This dependence of numerical phase and group velocity on spatial sampling is referred to as *grid* or *numerical dispersion*. Grid dispersion is the physical result of short-wavelength waves sensing the discrete medium, in a manner analogous to the short-wavelength behavior of phonons sensing a crystal lattice. For densely sampled waves (i.e., long wavelengths) the dispersion relation for the numerical waves approaches that of the continuous waves. Long wavelengths do not sense the grid and accurately simulate waves in a continuous medium.

In Figs. 3 and 4 the phase and group velocities for two-dimensional waves are plotted as a function of spatial sampling for different propagation angles on the grid. Note from Eq. 6 and the figures that wave propagation is modeled most accurately for waves traveling at 45° to mesh coordinates. Waves traveling along the coordinate axes experience greater numerical dispersion at short wavelengths. The dependence of the phase and group velocities on the wave's orientation on the grid is referred to as *grid* or *numerical anisotropy*.

Grid dispersion for a given spatial sampling can be reduced by decreasing the truncation error of the approximations to the partial derivatives. For example, by expanding the $P(m, n, l)$ in a Taylor's series about $(m \pm 2)h$ and by using the expansions about $(m \pm 1)h$, one can approximate $\partial^2 P/\partial x^2$ to

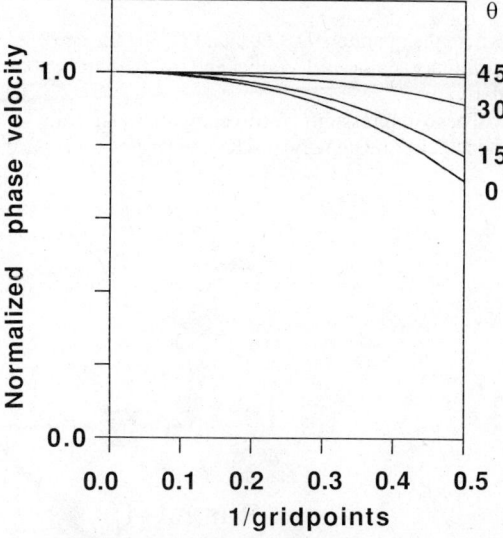

FIGURE 3. Normalized phase velocity dispersion of the $O(\Delta t^2, h^2)$ two-dimensional acoustic wave equation scheme (Eq. 3), for waves traveling at different angles to the grid. Waves that propagate in one of the coordinate directions, $\theta = 0°$, experience the greatest numerical dispersion. To accurately model a continuum requires spatial sampling of 10 gridpoints/wavelength. The curves were calculated for a scheme run at 100% of the stability limit.

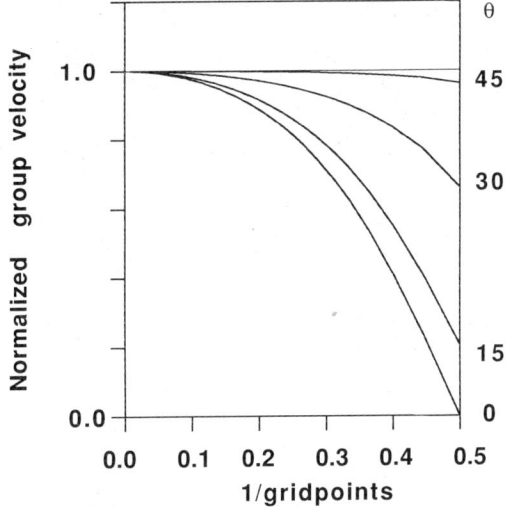

FIGURE 4. Normalized group velocity dispersion of the $O(\Delta t^2, h^2)$ scheme for waves traveling at different angles to the grid. Waves oriented in one of the coordinate directions, $\theta = 0°$, do not propagate when sampled at 2 gridpoints/wavelength. This condition corresponds to a standing wave on the finite-difference grid. Dispersion is least significant for waves oriented at $\theta = 45°$ to the grid. The curves were calculated for a scheme run at 100% of the stability limit.

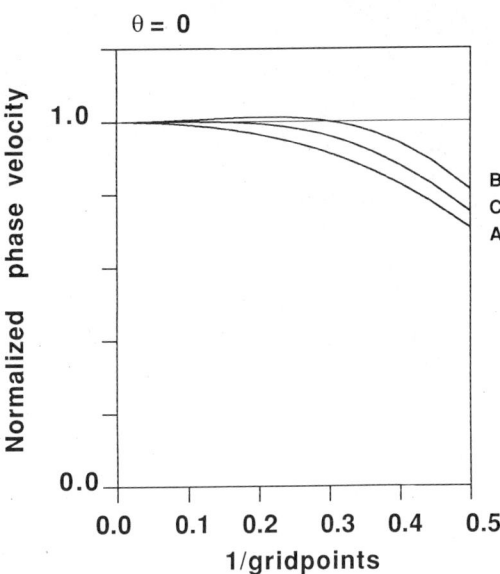

FIGURE 5. Comparison of normalized phase velocity dispersion of $O(\Delta t^2, h^2)$ and $O(\Delta t^2, h^4)$ schemes for the acoustic wave equation. **A** refers to the $O(\Delta t^2, h^2)$ scheme run at 99% of the stability limit, **B** the $O(\Delta t^2, h^4)$ scheme run at 98% of the stability limit, and **C** the $O(\Delta t^2, h^4)$ scheme run at 49% of the stability limit.

fourth-order accuracy in the spatial increment (see Abramowitz and Stegun, 1964, p. 884). In two dimensions the fourth-order-accurate Laplacian operator contains nine points rather than the five of the second-order-accurate approximation. Alford et al. (1974) have calculated the stability criterion and the dispersion relations for an acoustic wave scheme using a fourth-order-accurate approximation to the Laplacian and a second-order-accurate approximation to the temporal derivative. Their comparison of the dispersion of the $O(\Delta t^2, h^4)$ and the $O(\Delta t^2, h^2)$ schemes, reproduced in Figs. 5 and 6, indicates that waves are modeled accurately if they are sampled with ten or more grid points for the $O(\Delta t^2, h^2)$ scheme and five or more gridpoints for the $O(\Delta t^2, h^4)$ scheme. Both schemes exhibit grid dispersion and anisotropy at short wavelengths. The stability criterion for the $O(\Delta t^2, h^4)$ scheme is more restrictive than the $O(\Delta t^2, h^2)$ scheme (compare to inequality 7), being

$$\frac{\Delta t \, c_0}{h} \leq \left(\frac{3}{8}\right)^{1/2}$$

Dablain (1986) has further analyzed explicit finite-difference schemes of accuracy $O(\Delta t^2, h^2)$, $O(\Delta t^2, h^4)$, $O(\Delta t^4, h^4)$, and $O(\Delta t^4, h^{10})$ for the one-dimensional wave equation. Examining the discretization of the temporal derivatives alone, he concluded that the second-order-accurate numerical approximation

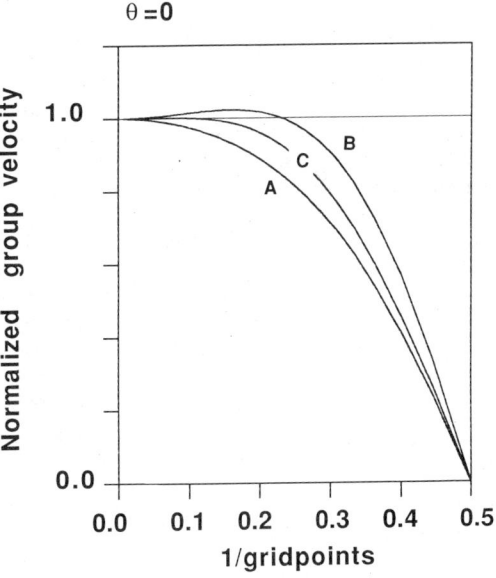

FIGURE 6. Comparison of normalized group velocity dispersion of $O(\Delta t^2, h^2)$ and $O(\Delta t^2, h^4)$ schemes for the acoustic wave equation; **A, B,** and **C** are labeled as in Fig. 5. The slightly high values of the group velocity in the **B** curve result in a precursory pulse (refer to Fig. 9). The $O(\Delta t^2, h^2)$ scheme requires sampling at 10 gridpoints/wavelength, whereas the $O(\Delta t^2, h^4)$ scheme requires sampling at 5.

results in phase advance, whereas examining the spatial derivatives alone, Dablain concludes that the second-order-accurate numerical approximation results in phase delay. For a scheme of similar accuracy in space and time, the effect of the spatial derivative dominates. Figure 7 compares the dispersion of $O(\Delta t^2, h^2)$, $O(\Delta t^2, h^4)$, and $O(\Delta t^4, h^{10})$ solutions of the one-dimensional wave equation in which the pulse bandwidth and the mesh size are identical (Dablain, 1986). The bandlimiting was chosen so that the maximum frequency in the pulse was sampled at 3 gridpoints/wavelength. Only the solution from $O(\Delta t^4, h^{10})$ scheme is adequately sampled, the other two solutions display how grid dispersion of different schemes affects the pulse shape.

The numerical inaccuracy due to the approximation for the second derivative can be understood by comparing the frequency domain representation of the second derivative with its numerical approximations for schemes with different order accuracy. The second spatial derivative has the wave-number representation $-k_x^2$. Figure 8 compares the exact expression to second-, fourth, and tenth-order-accurate finite-difference operators. The physical manifestation of the numerical inaccuracy can be seen in Fig. 9, where two-dimensional waves in a uniform medium have been calculated with $O(\Delta t^2, h^2)$ and $O(\Delta t^2, h^4)$ schemes. In these examples the mesh size used for the $O(\Delta t^2, h^4)$ calculation is twice that for the $O(\Delta t^2, h^2)$ calculation, giving pulses bandlimited for the same maximum frequency. Note the high-frequency precursor in the fourth-order simulation run at 98% of the stability limit. The precursor is attributable to the unphysically high value of the group velocity in the range $0.1 < 1/G < 0.25$ in Fig. 5. The pulses run with $O(\Delta t^2, h^2)$ and $O(\Delta t^2, h^4)$ schemes at 99%, and 49% of the stability limit traveling at 45° to the coordinate directions match the analytical solution very closely.

The dispersion and the stability analyses lead to the fundamental practical constraints in explicit finite-difference modeling. The stability criterion depends inversely on the highest material velocity, whereas the bandlimiting required to minimize numerical dispersion and anisotropy is proportional to the lowest material velocity; i.e.,

$$\Delta t \leq \left(\frac{rh}{\sqrt{n}\, c_{max}}\right)$$

whereas

$$f_{max} = \frac{c_{min}}{jh}$$

where n is the dimension of the problem, j is the minimum spatial sampling in gridpoints, and r is determined from the dispersion relation for a particular scheme (e.g., for an $O(\Delta t^2, h^2)$ scheme $j = 10$ and $r = 1$; for an $O(\Delta t^2, h^4)$ scheme $j = 5$ and $r = \sqrt{3/4}$.

The decrease in spatial sampling that is allowed by higher-order operators translates into greater efficiency in two-dimensional modeling. For example, while requiring approximately twice as many operations per node, a fourth-order scheme requires one quarter as many nodes as a second-order scheme for the same model and bandwidth (as in Fig. 9). The linear increase in computer cost in solving a higher-order operator is offset by the geometric decrease due to reduced spatial sampling. For spatial operators higher than fourth-order, the reduced number of gridpoints is balanced by the increased computational burden. Dablain (1986) confirmed that computer cost decreased marginally using an $O(\Delta t^4, h^{10})$ over an $O(\Delta t^2, h^4)$ scheme; however the memory required for the $O(\Delta t^4, h^{10})$ scheme is approximately two-thirds less. This is an important consideration when using fixed-memory machines, such as many of the supercomputers.

Determining higher-order schemes for spatial derivatives involves solving a tiresomely large system of equations developed from Taylor's series expansions similar to Eq. 2. Hale (pers. com.) has formulated this procedure for approximating any order derivative to any order accuracy as the solution of a Vandermonde system of equations (Golub and Van Loan, 1983, p. 119).

Elastic Equation Approximations

The acoustic equation is useful in many areas of seismic modeling and provides a simple system for understanding the numerical aspects of finite-difference modeling; however, a more realistic simulation of wave propagation in the Earth requires the inclusion of elastic effects. Again a two-dimensional Cartesian system is used to demonstrate the elastic equations. Numerically, the SH equation is similar to the acoustic equation (see Boore, 1970, 1972). Consequently I focus on the vector P-SV equations. In a two-dimensional Cartesian system (Fig. 1) the P-SV equations of motion are

$$\rho \frac{\partial^2 u}{\partial t^2} = \frac{\partial}{\partial x} \tau_{xx} + \frac{\partial}{\partial z} \tau_{xz}$$

$$\rho \frac{\partial^2 w}{\partial t^2} = \frac{\partial}{\partial x} \tau_{zx} + \frac{\partial}{\partial z} \tau_{zz} \qquad (8)$$

and the constitutive laws are

$$\tau_{xx} = (\lambda + 2\mu) \frac{\partial u}{\partial x} + \lambda \frac{\partial w}{\partial z}$$

$$\tau_{zx} = \mu \left(\frac{\partial u}{\partial z} + \frac{\partial w}{\partial x}\right)$$

$$\tau_{zz} = (\lambda + 2\mu) \frac{\partial w}{\partial z} + \lambda \frac{\partial u}{\partial x} \qquad (9)$$

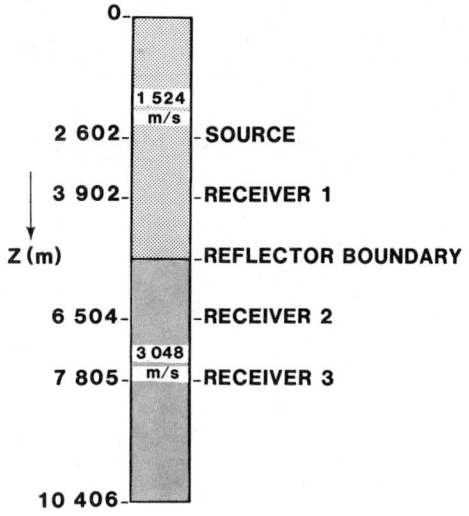
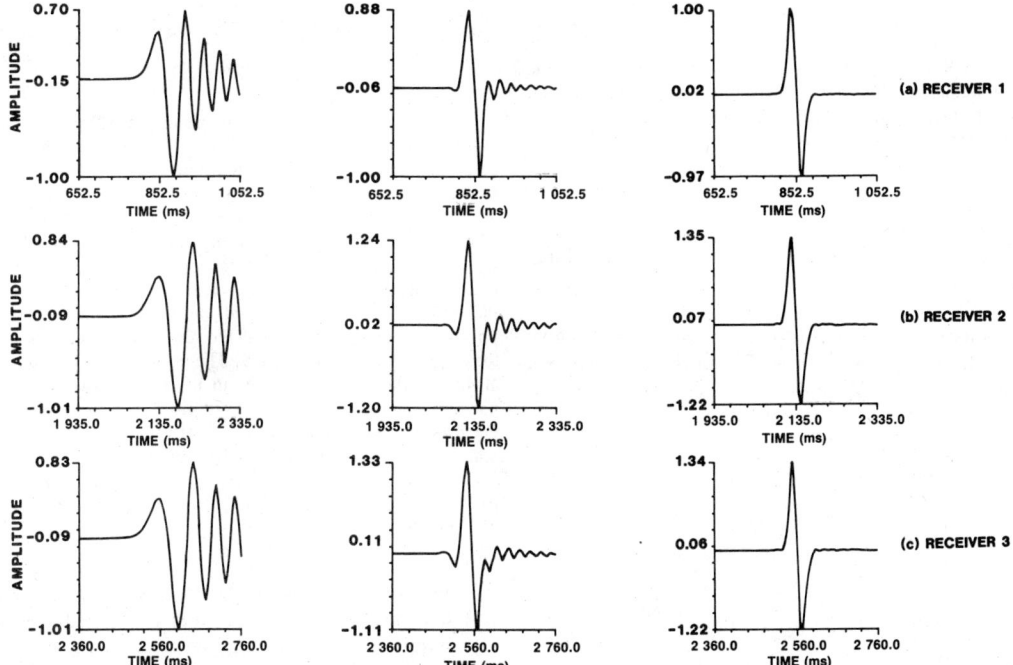

FIGURE 7. Coarse grid calculations for three one-dimensional finite-difference wave operators: the model is shown at the top, the signals from the three schemes at bottom: left $O(\Delta t^2, h^2)$; center $O(\Delta t^2, h^4)$; and right $O(\Delta t^4, h^{10})$. The first receiver is in a medium having half the velocity of the medium where the second and third receivers are located. The pulse shape changes continuously with the $O(\Delta t^2, h^4)$ scheme and is relatively stable with the other two. This suggests that most dispersion takes place in the low-velocity region in the $O(\Delta t^2, h^2)$ and $O(\Delta t^4, h^{10})$ schemes. (From Dablain, 1986, courtesy of Mobil Research and Development Corporation)

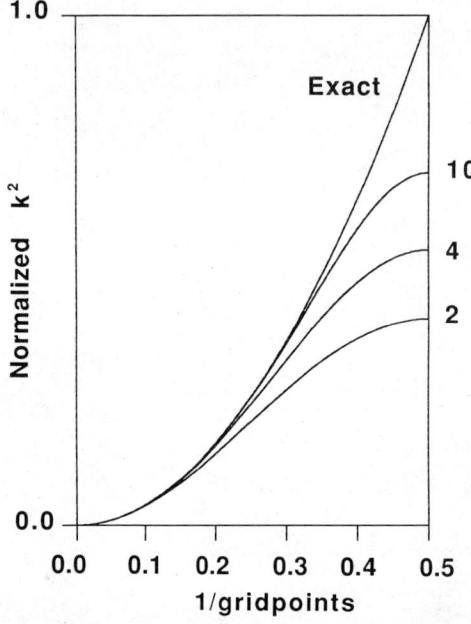

FIGURE 8. Frequency domain representation of the continuous second spatial derivative and its second- (2), fourth- (4), and tenth- (10) order-accurate finite difference approximations, plotted as a function of sampling.

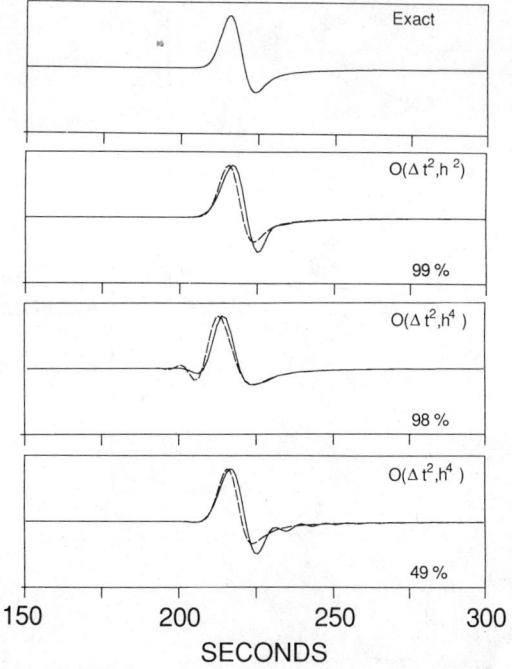

FIGURE 9. The effects of numerical dispersion on a pulse propagating in a whole space. The pulse has propagated 200 units from the source point in a medium with velocity 1 unit·s^{-1}. The top panel is an analytical solution. The second panel shows pulses from an $O(\Delta t^2, h^2)$ scheme calculated at 99% of the stability limit. In this and the following panels the solid line trace propagated in a coordinate direction, the dashed line trace propagated at 45° to the coordinate axes. The third and fourth panels show pulses from an $O(\Delta t^2, h^4)$ scheme run at 98% and 49% of the stability limit, respectively. The pulses that propagated at 45° to the coordinate axes in the second and bottom panels are most like the exact solution. The finite-difference sources were bandlimited at 10 gridpoints/wavelength and 5 gridpoints/wavelength for the $O(\Delta t^2, h^2)$ and $O(\Delta t^2, h^4)$ schemes, respectively. The highest-frequency energy in the finite-difference simulations has propagated 20 wavelengths.

where u and w are the displacement components in x and z, the τ_{ij} are the stresses, λ and μ are the Lamé parameters, and ρ is the density. The compressional velocity is given by $\alpha = \sqrt{(\lambda + 2\mu)/\rho}$, and the shear velocity by $\beta = \sqrt{\mu/\rho}$.

Equations 8 can be substituted into Eqs. 9 to give the elastic equations expressed in the displacement field:

$$\rho \frac{\partial^2 u}{\partial t^2} = \frac{\partial}{\partial x}\left(\lambda\left(\frac{\partial u}{\partial x} + \frac{\partial w}{\partial z}\right) + 2\mu\frac{\partial u}{\partial x}\right)$$
$$+ \frac{\partial}{\partial z}\left(\mu\left(\frac{\partial u}{\partial z} + \frac{\partial w}{\partial x}\right)\right)$$
$$\rho \frac{\partial^2 w}{\partial t^2} = \frac{\partial}{\partial z}\left(\lambda\left(\frac{\partial u}{\partial x} + \frac{\partial w}{\partial z}\right) + 2\mu\frac{\partial w}{\partial z}\right)$$
$$+ \frac{\partial}{\partial x}\left(\mu\left(\frac{\partial u}{\partial z} + \frac{\partial w}{\partial x}\right)\right) \quad (10)$$

Expressing Eq. 8 in particle velocity and differentiating Eq. 9 with respect to time gives a system of first-order, coupled, hyperbolic equations for particle velocity and stress. Equations 10 are coupled second-order hyperbolic equations for displacement. Both the first- and second-order systems have been used as the basis for explicit finite-difference schemes. A notable difference between the two systems of equations is that the second-order system has terms containing spatial derivatives of the material properties, which the first-order system lacks. The numerical schemes for the second-order elastic equations (10) require some averaging and differencing of the material properties. The form of the averaging and differencing determines the accuracy with which the implicitly determined stress field is modeled. Since the elastic equations support the propagation of compressional and shear waves, which couple at material property discontinuities through the stress field, an accurate representation of the stress field is necessary to correctly model elastic effects.

Several different numerical schemes exist for the second-order system of equations. Accuracy and stability are analyzed in a manner analogous to the acoustic equation. We examine a homogeneous material, assuming a solution of the form

$$\mathbf{u}(\mathbf{x}, t) = [u, w]e^{i(\mathbf{k} \cdot \mathbf{x} - \omega t)} = \mathbf{U}e^{i(\mathbf{k} \cdot \mathbf{x} - \omega t)}$$

Using Eq. 10, the definitions of the compressional and shear velocities, and substituting the numerical operators D_{tt}, D_{xx}, D_{zz}, and D_{xz} for their continuous counterparts gives

$$\mathbf{0} = \begin{bmatrix} (\alpha^2 D_{xx} + \beta^2 D_{zz}) - D_{tt} & (\alpha^2 - \beta^2)D_{xz} \\ (\alpha^2 - \beta^2)D_{xz} & (\beta^2 D_{xx} + \alpha^2 D_{zz}) - D_{tt} \end{bmatrix} \mathbf{U}$$

For a nontrivial solution the determinant of the coefficient matrix vanishes, providing a quadratic equation in D_{tt}. The quadratic equation yields the dispersion relation

$$D_{tt} = \tfrac{1}{2}(\alpha^2 + \beta^2)(D_{xx} + D_{zz}) \pm \tfrac{1}{2}(\alpha^2 - \beta^2)$$
$$\cdot \sqrt{(D_{xx} + D_{zz})^2 - 4(D_{xx}D_{zz} - D_{xz}D_{xz})}$$

with the two roots giving the compressional and shear wave dispersion relations. In the discrete system the dispersion relation depends on the form of the numerical operators D_{xx}, D_{zz}, and D_{xz} used to approximate the partial derivatives. Note that the numerical system dispersion relations for both wave types are functions of *both* material velocities when the second term of the radical is nonzero. In the continuous case the second term under the radical is identically zero, giving the dispersion relation for the continuous system:

$$\omega^2 = \alpha^2(k_x^2 + k_z^2) \quad \text{for compressional waves}$$
$$\omega^2 = \beta^2(k_x^2 + k_z^2) \quad \text{for shear waves}$$

Kelly et al. (1976) have published a frequently used second-order accurate formulation for the displacement equations, which I refer to as the U or displacement scheme. The difference stencil and the difference equations are given in Kelly et al. (1976). The stability criterion is

$$1 \geq \frac{\Delta t}{h} \sqrt{\alpha^2 + \beta^2}$$

(Alterman and Karal, 1968). The dispersion relation is given by

$$\sin^2\left(\frac{\omega \Delta t}{2}\right) = \frac{(\Delta t \alpha)^2}{2h^2} \left\{ \left(1 + \frac{\beta^2}{\alpha^2}\right)\left(\sin^2\left(\frac{k_x h}{2}\right)\right.\right.$$
$$\left.+ \sin^2\left(\frac{k_z h}{2}\right)\right) \pm \left(1 - \frac{\beta^2}{\alpha^2}\right)$$
$$\cdot \left[\left(\sin^2\left(\frac{k_x h}{2}\right) + \sin^2\left(\frac{k_z h}{2}\right)\right)^2\right.$$

$$\left. - 4\left(\sin^2\left(\frac{k_x h}{2}\right)\sin^2\left(\frac{k_z h}{2}\right)\right)\right.$$
$$\left.\left. - \frac{\sin^2(k_z h)\sin^2(k_x h)}{16}\right]^{1/2}\right\} \quad (11)$$

with the sign of the radical determining if the dispersion is for compressional or shear waves. The dispersion relations for this scheme are plotted in Fig. 10 for two different values of Poisson's ratio.

Madariaga (1976) introduced an $O(\Delta t^2, h^2)$ numerical scheme for the first-order coupled hyperbolic equations for modeling fault rupture. Virieux (1986) adapted it for general forward modeling. This scheme is used to solve for both the velocity components and the stress field (Eqs. 8 and 9). I refer to it as the V or velocity method. The grid is staggered in both space and time, with the stresses defined on one grid and the velocities defined on another. Solving the first-order system obviates the need for differencing or averaging material properties. By staggering the grid, the variables being updated are always centered between quantities used in the update. The finite-difference stencil and the difference equations are given in Virieux (1986). Analysis of stability and dispersion is similar to that for the U method. The stability criterion is

$$\frac{1}{\sqrt{2}} \geq \frac{\Delta t \, \alpha}{h}$$

and the dispersion relations are

$$\sin^2\left(\frac{\omega \Delta t}{2}\right) = \left(\frac{\Delta t \, \alpha}{h}\right)^2 \left(\sin^2\left(\frac{k_x h}{2}\right) + \sin^2\left(\frac{k_z h}{2}\right)\right) \quad (12)$$

for compressional waves and

$$\sin^2\left(\frac{\omega \Delta t}{2}\right) = \left(\frac{\Delta t \, \beta}{h}\right)^2 \left(\sin^2\left(\frac{k_x h}{2}\right) + \sin^2\left(\frac{k_z h}{2}\right)\right)$$

for shear waves (Collino, unpublished; see Fig. 11). Note that these dispersion relations are similar to that of the second-order accurate acoustic wave scheme (Eq. 6) but are scaled by the appropriate propagation velocity.

Both the second-order-accurate V and U schemes require sampling the shortest wavelength at 10 gridpoints/wavelength to minimize the effects of grid dispersion. In a constant-velocity medium the shear velocity wavelengths are always shorter than the compressional wavelengths. In a Poisson solid, for example, the ratio of compressional wavelength to shear wavelength is 1.73; in high-Poisson-ratio materials, such as in the sedimentary column or near the seafloor, the ratio can be much greater. If

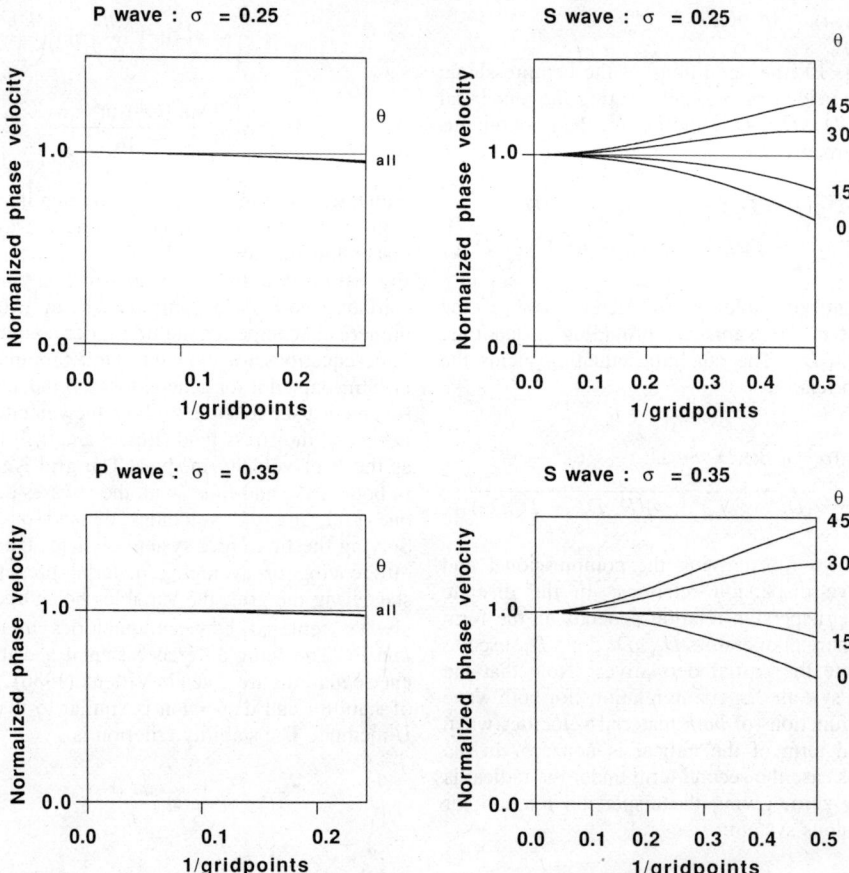

FIGURE 10. Compressional and shear wave phase velocity dispersion for the Kelly et al. (1976) U scheme for two values of Poisson's ratio shown for different propagation directions. The top panels are the compressional and shear velocity dispersion for $\sigma = 0.25$, with the minimum shear wave sampled at 2 gridpoints/wavelength. P-wave sampling is calculated to be consistent with S-wave sampling. The bottom panels are the dispersion for $\sigma = 0.35$, with the minimum shear wavelength sampled at 2 gridpoints/wavelength. Note that the shear wave dispersion increases with Poisson's ratio for constant sampling.

accurate shear wave modeling is desired, the source bandlimiting is chosen for proper sampling of the shortest shear wavelength.

Virieux's (1986) velocity-stress staggered grid scheme appears to have several advantages over the displacement scheme of Kelly et al. (1976). In contrast to the U scheme, the dispersion relation of the V scheme has no dependency on the ratio of shear velocity to compressional velocity, except implicitly through the time step (compare Eqs. 11 and 12 and Figs. 10 and 11). Because of the weaker dependence on the ratio of shear velocity to the compressional velocity, it is stable for mixed acoustic-elastic media and will accurately model compressional energy even if the shear waves are undersampled. A drawback to the V scheme is a somewhat more restrictive Courant number than the U scheme, approximately 18% smaller for Poisson solids.

The increase in computational cost for P-SV modeling over acoustic modeling can be dramatic; the solution requires more operations, the stability criterion is based on either the maximum compressional velocity or the sum of the maximum compressional and shear velocities, the bandlimiting is usually determined by the lowest shear velocity, and storage requirements are two to four times greater. It is important in P-SV simulation to reduce the spatial sampling necessary to accurately model shear waves. Higher-order approximations to the spatial derivatives of the velocity-stress equations and the displacement equations can be developed by using the methods of the preceding section (see Levander, 1988). The fourth-order-accurate P-SV schemes, like their acoustic counterparts, require minimum spatial sampling of 5 gridpoints/wavelength.

Storage required for a U algorithm is $7MN$ if the

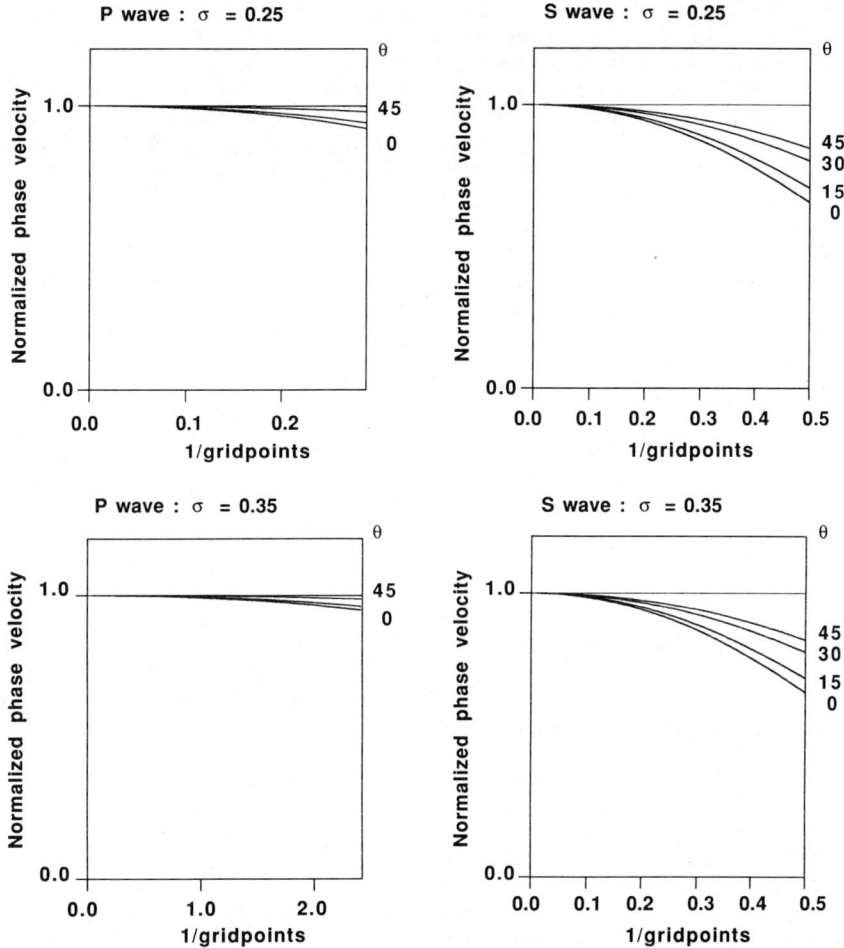

FIGURE 11. Compressional and shear wave phase velocity dispersion for the Virieux (1986) V scheme for two values of Poisson's ratio shown for different propagation directions. The top panels are the compressional and shear velocity dispersion for $\sigma = 0.25$, with the minimum shear wave sampled at 2 gridpoints/wavelength. The bottom panels are the dispersion for $\sigma = 0.35$. Note that the shear wave dispersion of the V scheme is much less sensitive to Poisson's ratio than is the shear wave dispersion of the U scheme (Fig. 10).

spatial averaging of the material properties is performed at each time step in the calculation, but storage can be as high as $11MN$ if the constants are computed at the beginning of the calculation and saved. Storage required for a V algorithm is $8MN$ or $9MN$.

Numerical Boundary Conditions

Boundary conditions are required at the edges of the computational grid and, depending on the finite-difference formulation, at material property boundaries within the model (refer to Fig. 1). Developing stable numerical boundary conditions often requires more effort than developing internal grid operators. Analysis of a difference scheme's stability including discrete boundary conditions is an inexact science, requiring formidable mathematics and, apparently, some degree of luck. Working with the one-dimensional convection equation, Trefethen (1983) has analyzed numerical boundary conditions in terms of the numerical group velocities supported by the grid. He classifies a numerical scheme including boundary conditions as unstable if the boundary condition permits (unphysical) group velocities with opposite polarity of the primary field. Although this is an encouraging result, its application to the analysis of two-dimensional systems is unclear.

Internal Grid Boundary Conditions

Considerable effort has gone into developing internal-grid boundary conditions for P-SV schemes at welded elastic-elastic interfaces (see Ilan et al.,

1975) and acoustic-elastic interfaces (Stephen, 1984). In the former case, the boundary conditions are continuity of displacement at the interface and stress in the plane of the interface. In the latter case, the boundary conditions are continuity of the normal component of velocity and normal stress across an interface. The numerical scheme is relatively simple to write where a material property boundary is aligned along one of the coordinate directions. In my opinion, the algorithmic clumsiness encountered when attempting to solve the explicitly satisfied internal boundary conditions for arbitrarily oriented interfaces makes the heterogeneous medium formulations, such as those presented in the preceding section, preferable. The heterogeneous medium operators implicitly satisfy internal welded-contact boundary conditions. In the comparisons of wave simulations between heterogeneous medium schemes and homogeneous medium schemes that have explicitly satisfied internal boundary conditions, the heterogeneous formulations model reflections and diffractions less accurately, by a few percent to as much as 20%, depending upon the formulation (Kelly et al., 1976; Kummer and Behle, 1982). The accuracy of the heterogeneous formulations is still a subject of research.

Boundary Conditions at the Grid Periphery

Both acoustic and elastic schemes require numerical boundary conditions at the edges of the computational domain. These can be either physical boundary conditions, for example the Earth's surface, or artificial conditions that attempt to make the finite model appear as if it were infinite or semi-infinite (see Fig. 1). The artificial conditions attempt either to satisfy the Sommerfeld radiation condition or to exploit some symmetry in the modeling problem.

Discrete physical boundary conditions, such as vanishing pressure and vanishing stress, are developed from their continuous analogs. The vanishing pressure condition for the acoustic wave equation is trivially solved, whereas the P-SV free surface conditions needed for either the U- or V-scheme elastic equations have required serious attention. In elastic wave problems we usually desire a solution calculated at the free surface, but a significant component of the wave field can stay near the surface as Rayleigh waves. An accurate approximation of the free surface condition is of considerable importance since the entire numerical scheme is only as accurate as the boundary conditions. The P-SV boundary conditions are most simply solved with a system of one-sided difference equations, or a centered scheme involving what is termed a "fictitious" row or column of gridpoints (see Fig. 12). The one-sided approximation is first-order accurate in space, and the centered approximation is second-order accurate in space but has a high-frequency instability (Ilan and Loewenthal, 1976). Neither condition is stable if the surface material has a large Poisson ratio. Ilan et al. (1975) and Ilan and Loewenthal (1976) analyzed the stability of these numerical free surface conditions and compared them with the stability of an approximation that makes use of higher-order approximation to the constitutive law. Their revised second-order-accurate composed approximation is stable for higher-Poisson-ratio surface materials than the one-sided or centered approximations, but is unstable for laterally heterogeneous surface structures. Vidale and Clayton (1986) have derived an implicit solution to the P-SV free surface condition for second- and fourth-order-accurate internal grid operators. They solve a simultaneous system of equations involving the free surface points and the grid row immediately beneath the free surface. This scheme is stable both for high values of Poisson's ratio and for laterally heterogeneous near surface structures (see Figs. 15, 16, and 17 in the Applications section).

The artificial boundary conditions at the edges of the computational domain designed to transmit energy off the grid are called *absorbing boundaries*. The physical problem is to keep waves from reflecting from the edge of the computational domain by satisfying a differential equation that makes the domain appear infinite. In one dimension an exact absorbing condition is the one-way wave equation

$$\frac{\partial P}{\partial t} \pm \frac{1}{c}\frac{\partial P}{\partial x} = 0$$

obtained by factoring the one-dimensional wave equation or by taking the differential form of the one-way wave equation dispersion relation.

An exact absorbing boundary condition for an acoustic wave equation in two or three dimensions requires a global frequency–wave-number domain expression, making the equation impractical for an explicit finite-difference scheme. Clayton and Engquist (1977) approximated the perfect absorbing condition for two-dimensional acoustic and P-SV waves. The Clayton-Engquist condition is a paraxial approximation to the wave equation designed for waves traveling toward a boundary. The order of the paraxial approximation determines the range of incidence angles absorbed by the boundary condition. For the acoustic wave equation the wave numbers of $+x$ traveling waves are given by

$$k_x = \frac{\omega}{c}\sqrt{1 - (k_z c/\omega)^2}$$

The horizontal wave number can be expanded in powers of $k_z c/\omega$ by using the Muir expansion (Claerbout, 1985) to first order as

$$k_x = \frac{\omega}{c}[1 + O((k_z c/\omega)^2)]$$

Centered approximation

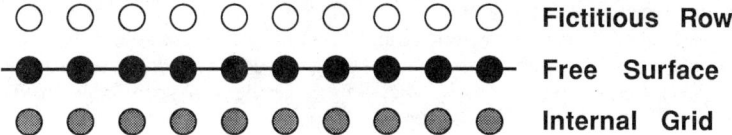

One-sided and composed approximations

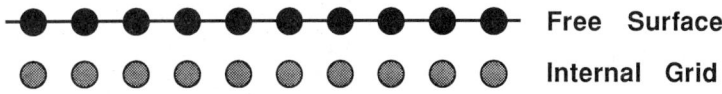

Implicit scheme

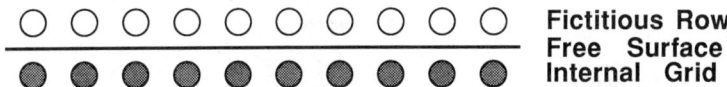

FIGURE 12. Nodes used in difference schemes for the free surface condition. Circles denote the mesh; the solid line denotes the location of the free surface. Top panel shows the mesh points used for the centered difference approximation. The free surface point is solved using the wave equation. The fictitious row and the first internal grid row are used to satisfy the boundary condition. Center panel shows the mesh points used for the one-sided difference scheme and the revised composed approximation. In the former a one-sided difference approximation to the free surface condition is used to extrapolate the internal grid values to the free surface. The one-sided scheme is first-order accurate in space. The composed approximation makes use of the wave equation to solve the free surface condition. The bottom panel shows the mesh points used for an implicit free surface difference scheme used for the U algorithm as described by Vidale and Clayton (1986).

and to second order as

$$k_x = \frac{\omega}{c}[1 - \tfrac{1}{2}(k_z c/\omega)^2 + O((k_z c/\omega)^4)]$$

These expressions have the differential forms

$$\frac{\partial P}{\partial x} + \frac{1}{c}\frac{\partial P}{\partial t} = 0$$

and

$$\frac{\partial^2 P}{\partial t \partial x} + \frac{1}{c}\frac{\partial^2 P}{\partial t^2} - \frac{c}{2}\frac{\partial^2 P}{\partial z^2} = 0$$

The first-order absorbing condition is identical to the one-dimensional one-way wave equation. Discrete formulations of either of these equations are used as boundary conditions at the $x = x_{\max}$ boundary of the grid. Similar boundary conditions are applied at other grid boundaries. The first-order condition absorbs waves at normal incidence, whereas the second-order condition absorbs waves at angles up to 30°. In both cases reflection from the boundary increases with incidence angle (see Clayton and Engquist, 1977, for reflection coefficients). Higher-order conditions can be derived. Clayton and Engquist (1977) also developed first- and second-order P-SV wave absorbing boundary conditions. Although the first-order equations work well, the second-order conditions appear to be unconditionally unstable for extreme changes in material properties along the boundary (Emerman and Stephen, 1983; Mahrer, 1986).

Numerous attempts at improving the original Clayton-Engquist conditions have met with mixed success. The most common tactic is to introduce a region around the periphery of the computational domain that contains a viscous damping term in the equation of motion. Israeli and Orzag (1981) have used a "Newtonian cooling," or frictional term, and an absorbing boundary term to improve absorption of outgoing two-dimensional acoustic waves. They call this a "sponge filter." They add a perturbational term consisting of a one-way wave equation scaled by a damping coefficient to the wave equation, applied in a narrow zone near the edge of the computational grid. The damping is increased in strength smoothly from zero at the interior of the finite difference grid to maximum strength at the boundary. For example, near $x = x_{\max}$, we write

$$\frac{\partial^2 P}{\partial t^2} = c^2 \nabla^2 P - \nu(x)\left[c\frac{\partial P}{\partial x} + \frac{\partial P}{\partial t}\right]$$

where $\nu(x)$ is the spatially variable damping coefficient such that

$$\nu(x) = 0; \quad x < x_{\max} - x_L$$
$$= \frac{A(n+1)(x - x_L)^n}{(x_{\max} - x_L)^n}$$

and x_L is the start of the smoothing region, $x_{\max}$ is the edge of the model region, and A is a constant and n is an integer. The term in the brackets can be a higher-order absorbing boundary condition than the first-order term used here. The intent of the term is to transmit waves in one direction ($+x$, in this case) with small damping and to severely damp waves in the opposite direction. The method works extremely well in acoustic propagation and is being applied to P-SV modeling.

If the modeling problem has a symmetry that can be exploited, the artificial edges of the grid boundary can be set up to satisfy Neumann (vanishing field derivative) or Dirichlet (vanishing field) boundary conditions. For example, if the modeling problem involves a vertically incident plane wave on a symmetric or periodic structure, the Neumann condition can be applied to the side boundaries to correctly model the response of the structure. Boore (1972) discusses this type of boundary condition more fully and presents an example of a planar SH wave incident on a symmetrical basin model.

Sources

Motion on the finite-difference grid is initiated exactly as in the continuous system by specification of initial conditions, boundary values, or an inhomogeneous term in the equation of motion. The most commonly used means of specifying an inhomogeneous term in the equation of motion is, paradoxically, to express it as a boundary value. The sources we are normally interested in are the two-dimensional point source, a planar body wave, or a pulse containing only a particular wave type, such as a surface wave mode. In all cases an overriding concern is to properly bandlimit the source to minimize grid dispersion in the solution. Bandlimiting is done in the wave-number or temporal frequency domains, depending on whether the source is specified as an initial value or a boundary value. In addition to the need for a bandlimited source, the numerical scheme also requires a source that is slowly varying, since the difference approximations to the derivatives of a rapidly varying function are inaccurate. The Gaussian function, its first derivative, or the Ricker wavelet is often used as the bandlimiting function. Boore (1972) has described a bandlimited delta function approximation that is also widely used.

An initial profile for the field variable provided at the first two time steps establishes either an initial displacement or an initial displacement and velocity. An analytic solution for a homogeneous or plane-layered medium is often used as the initial condition. Specifying the source in this fashion is useful for initiating a pulse consisting of a planar body wave or a surface wave mode. The initial wave profile is placed in a region of the finite-difference grid having the corresponding uniform or plane-layered material properties at time $t = 0$. The phase of the wave is advanced in time by $\omega \Delta t$ and supplied to the grid at time $t = \Delta t$ to establish the correct initial velocity. To construct an initial condition composed solely of a surface wave, one must make the frequency contain the surface wave dispersion relation for the layered medium. The initial displacements, consisting of the surface wave eigenfunctions, are calculated in the k_x-z domain and transformed to x-z. For a body wave initial condition the phase is advanced and the displacements are polarized for the appropriate body wave type. In Boore's (1972) review of finite-difference methods he shows an example of a plane SH wave initial condition used for input to an irregular sedimentary basin and an example of a Love wave pulse for a layer and half-space used as input to an irregular waveguide. The details of specifying a surface wave source can be found in Boore (1970) or Levander (1985).

An initial condition can occupy a significant fraction of the computational domain (see Fig. 13). Since this region must have uniform material properties for holding the analytically derived initial condition, this method can be expensive in computer memory. In particular, initiating a two-dimensional line source disturbance requires an excessively large area on the difference grid that must be homogeneous. A temporally smooth inhomogeneous term can be inserted at a single grid point, but the smoothness of the source function and the $1/\sqrt{t}$ decay of a two-dimensional point source can result in the source being active for a large part of the entire finite-difference calculation. Consequently, care must be taken to allow reflected energy to pass through the

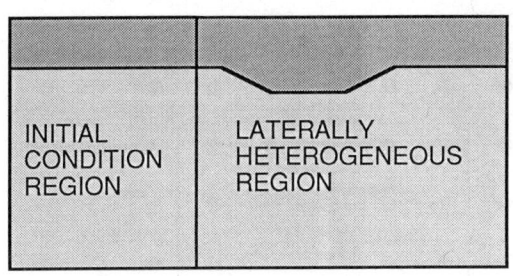

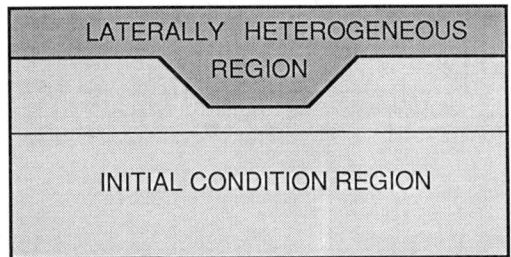

FIGURE 13. Use of initial conditions to initiate motion on the finite-difference grid. The initial condition region must be either homogeneous or plane layered. A surface wave pulse can be placed in the initial condition region of the top model or a planar body wave pulse in the bottom model. (Modified from Boore, 1972.)

source point; otherwise a diffracting point is created in the medium. Instead of using an initial condition or an inhomogeneous term specified at a single grid point, we can use the boundary value method to specify a two-dimensional point source or a planar body wave pulse. Alterman and Karal (1968) devised the means of extrapolating the source to a boundary a small distance from the source point for insertion into the larger finite-difference grid while allowing reflected energy to pass through the region. Kelly et al. (1976) clarified the concept. We divide the computation space into two regions: the source region S, which has constant material properties, and the exterior region E, which can have spatially variable properties. The two regions are separated by a surface, which in two dimensions is a curve C. In S the total field P_S is the result of the activation of the inhomogeneous term P_I and the field that is reflected back into S from the exterior region P_R. The term P_I satisfies the constant-velocity wave equation with the inhomogeneous term. The field reflected into S from the exterior region satisfies a homogeneous constant-velocity wave equation. The field in S along the inside of C is

$$P_S = P_I + P_R$$

The field in the exterior region, P_E, satisfies a homogeneous wave equation for a laterally variable velocity. Across C the exterior field and the interior field are continuous, as are the pressure or the stress in the plane of the curve. We write the field along C as

$$P_E = P_I + P_R$$

and

$$\mathbf{T}(P_E) = \mathbf{T}(P_I + P_R)$$

where $\mathbf{T}(P)$ is the traction due to P. Provided that the material properties are constant across C, the second condition will hold if the first does. We calculate P_I at C by using an analytical expression for the inhomogeneous term, and we supply P_I to the exterior region along C. Knowing P_I and calculating P_E with the finite-difference scheme allows us to calculate P_R at C. P_R can be propagated through S by using a separate finite-difference grid for S. $P_S = P_I + P_R$ is then supplied to C for the next time step in E. Note that the analytical solution is never used in the source region grid. One makes use of the analytically derived source function P_I along C to drive the grid E and the numerically derived source function P_R along C to drive the grid S. In the discrete case we actually use a zone of mesh points (see Fig. 14), with C defined as lying between mesh points. For a second-order-accurate spatial operator the stencil for the interior mesh point at C^- extends across C into region E, and the stencil for the exterior mesh point at C^+ extends across C into region S. Values at mesh points C^+ and C^- are used for calculation of both grids S and E. To calculate the source grid S, we supply the residual field to C^+ and C^- and calculate S from C^- inward. To calculate the exterior grid E, we supply the source terms plus the residual field to C^+ and C^- and calculate the grid E from C^+ outward. The zone is broadened to four grid points for use with a fourth-order-accurate spatial operator.

Vidale and Helmberger (1987) have extended use of the boundary value source insertion to include earthquake radiation patterns. They derived expressions to accurately calculate earthquake radiation patterns for strike-slip and dip-slip earthquake sources using quasi-near-field terms as part of the field supplied to the exterior grid.

Applications

To give the reader some flavor for the uses of finite-difference modeling, this section presents three model studies: an interpretation of strong ground-motion seismograms, forward modeling of the seismic reflection experiment, and analysis of the effects of waves scattered from a random medium on seismic reflection processing.

The first study is an interpretation of strong ground-motion records from the 1971 San Fernando earthquake. Using U-type $O(\Delta t^2, h^4)$ P-SV and SH schemes with fourth-order-accurate free surface

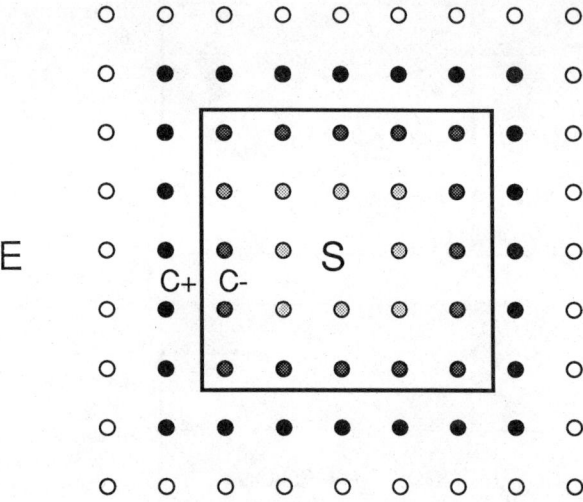

FIGURE 14. The boundary value method of inserting a spatially localized inhomogeneous term onto the finite-difference grid of an $O(\Delta t^2, h^2)$ scheme. The curve C in the continuous problem is extended to a zone two gridpoints wide from $C+$ to $C-$. S is the source region, E is the region external to the source. To calculate the reflected field in S at the $l + 1$ time step, the analytically determined source field for an infinite space is subtracted along $C+$ and $C-$ at the l time step and along $C-$ at $l - 1$. To insert the source into E, the analytically determined source values are supplied along $C+$ and $C-$ at the l time step, and to $C+$ at the $l - 1$ time step.

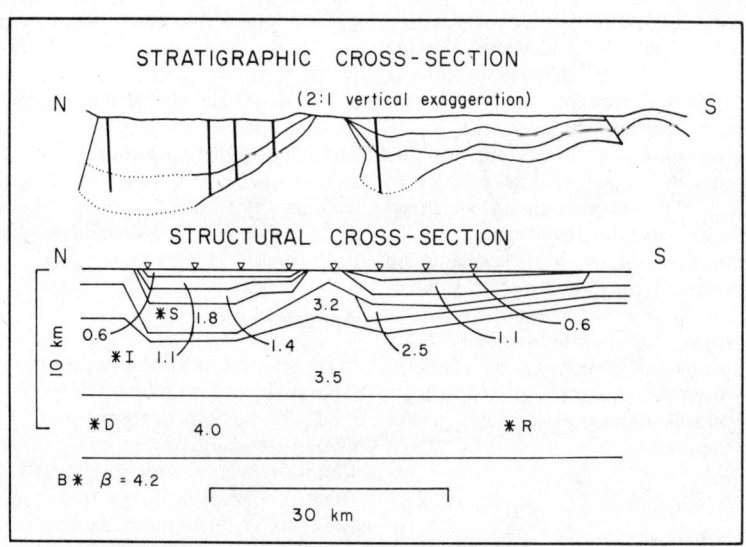

FIGURE 15. Structural cross section and finite-difference model of the San Fernando and Los Angeles Basins used to simulate strong ground-motion records from the 1971 San Fernando earthquake. (From Vidale and Helmberger, 1988)

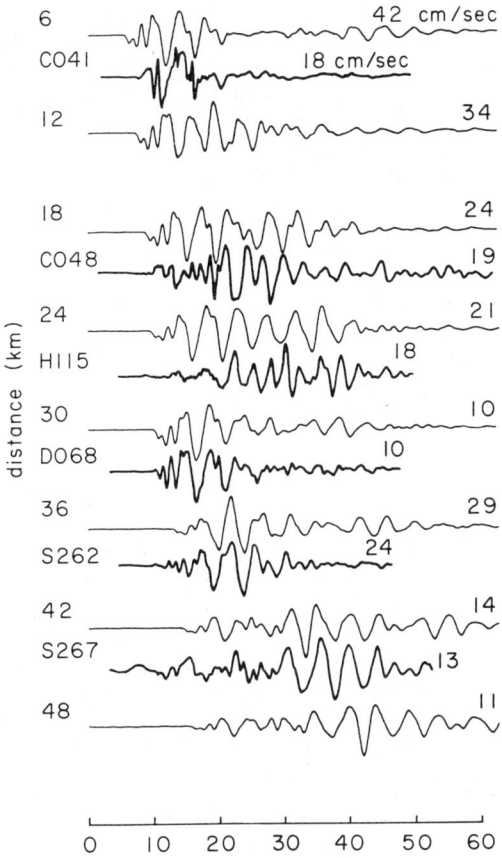

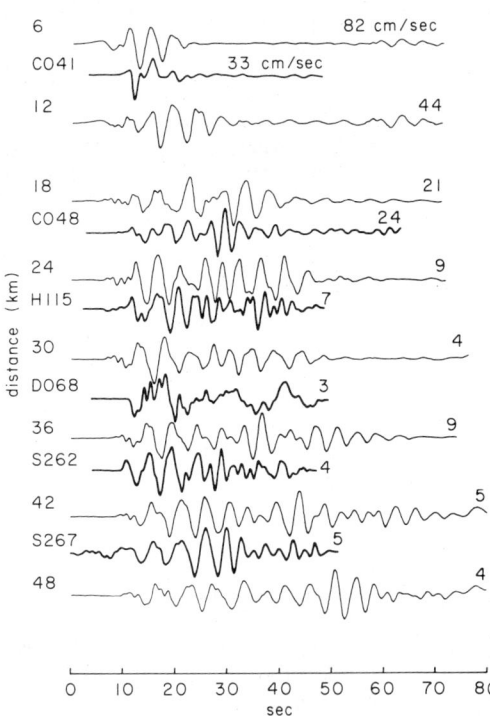

FIGURE 16. Transverse velocity records (dark lines) compared with finite-difference seismograms calculated from the model in Fig. 18. The numbers to the left are the source-to-receiver distance, the numbers to the right are the peak amplitudes on the traces in centimeters per second. The source was an impulsive line source with a strike-slip radiation pattern located at point D in Fig. 15. The data have been processed with an approximate line-to-point source filter and an attenuation filter. (From Vidale and Helmberger, 1988)

FIGURE 17. Vertical velocity records (dark lines) compared with vertical component finite-difference seismograms calculated from the model in Fig. 15. The source was an impulsive line source with a strike-slip radiation pattern located at point D in Fig. 15, inserted to the grid as in Fig. 14. The data have been processed as in Fig. 16. (From Vidale and Helmberger, 1988)

conditions, Vidale and Helmberger (1988) have computed the responses of earthquake sources at various depths in laterally inhomogeneous structures for comparison to strong ground-motion records from southern California. Figure 15 shows the laterally heterogeneous model of the San Fernando and Los Angeles Basins. The model was derived from well log data, refraction surveys, and geologic information. Seismograms for mixed strike-slip and dip-slip earthquake sources at B, D, I, and S were calculated. The synthetics of the transverse (SH) and vertical (P-SV) components of motion from source D are shown in Figs. 16 and 17. The synthetic data have been processed to remove the effects of the two-dimensional source tail and to approximately include the effects of attenuation. The agreement between the observations (heavy traces) and the synthetic seismograms (light traces) is very good for both transverse and vertical motion. Seismograms calculated for a plane-layered structure do not match the observations nearly as well as the finite-difference synthetics for the laterally varying structure. The authors conclude that the laterally varying structure is required to correctly model the observations and that the basin geometry is important in determining the falloff in peak velocity as a function of distance from the earthquake epicenter.

The second example is a large reflection seismology forward modeling problem. The model, 10 km by 3 km, consists of the complicated salt dome structure shown in Fig. 18 (Dablain, 1986). The author calculated an approximate exploding reflectors model using both $O(\Delta t^4, h^{10})$ and $O(\Delta t^2, h^4)$ approximations to the acoustic wave equation. The side and bottom boundaries had Israeli-Orzag sponge-filter boundary conditions. The source was bandlimited with a Ricker wavelet having a dominant frequency of 50 Hz. The $O(\Delta t^4, h^{10})$ calculation required 8.33-m mesh size to achieve 3 gridpoints/wavelength at the maximum frequency of 100

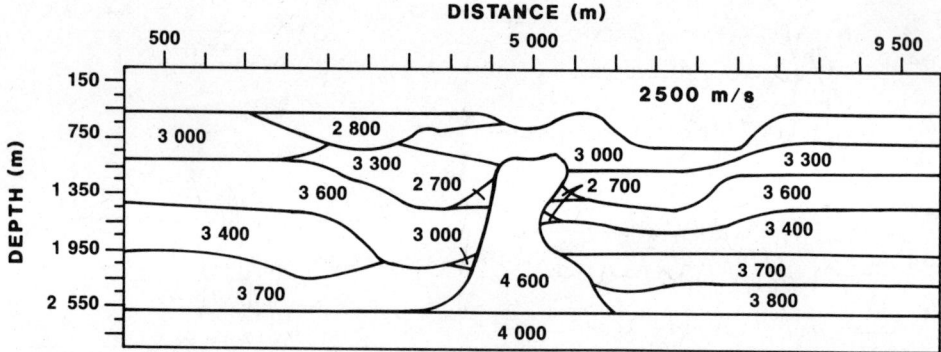

FIGURE 18. Acoustic wave exploration scale model of a salt dome in deformed sediments. The acoustic velocities of each layer are given in meters per second. Synthetic seismograms were calculated using exploding reflector sources (planar body wave pulses conforming to the boundaries, supplied as an initial condition.) (From Dablain, 1986, courtesy of Mobil Research and Development Corporation)

Hz; the $O(\Delta t^2, h^4)$ calculation required a 6.25-m mesh size for 5 gridpoints/wavelength. The $O(\Delta t^4, h^{10})$ solution required only 56% of the memory and 93% of the CPU time of the $O(\Delta t^2, h^4)$ solution. The difference between the two solutions was negligible; the $O(\Delta t^4, h^{10})$ solution is shown in Fig. 19. Dablain concludes that using the higher-order difference scheme results in substantial savings in computer memory for about the same cost in calculation. The exploding reflectors' response is used for comparison with a CMP stacked section.

The last example is a study of the effects of reflection seismic data processing on waves scattered from a medium with deeply buried random fluctuations in velocity (Gibson and Levander, 1988). The model, shown in Fig. 20, has a thin, smooth surface layer, two deep specular reflectors, and a zone of random velocity fluctuations between the two specular reflectors. The perturbed zone has a 5% root mean square (rms) random fluctuation about the mean velocity of 2400 m/s. The perturbed zone was spatially smoothed with a smoothing length corresponding to the wavelength of a 33-Hz signal. The authors calculated a 25-shot survey, using an $O(\Delta t^2, h^4)$ acoustic wave scheme having second-order absorbing boundary conditions at the sides and

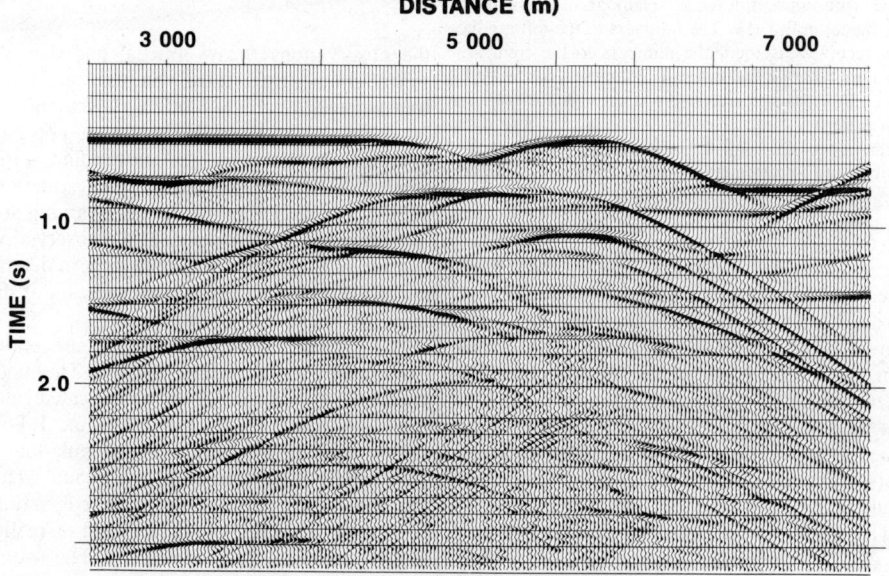

FIGURE 19. Exploding reflector synthetic seismograms calculated from the model in Fig. 18 with an $O(\Delta t^4, h^{10})$ acoustic wave finite-difference scheme. (From Dablain, 1986, courtesy of Mobil Research and Development Corporation.)

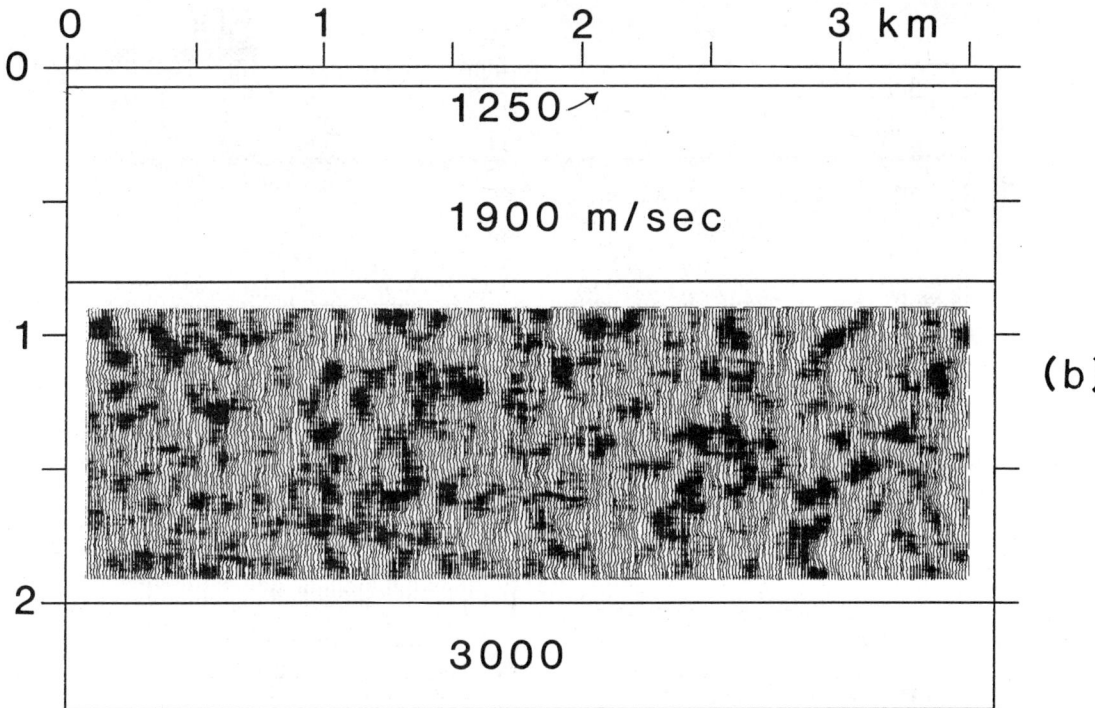

FIGURE 20. Finite-difference model designed to test the effects of CMP (common midpoint) stacking on events from a deeply buried zone of random velocity fluctuations between two specular reflectors. An $O(\Delta t^2, h^4)$ acoustic wave scheme was used to generate 25 shot records, which were subsequently processed as normal field data.

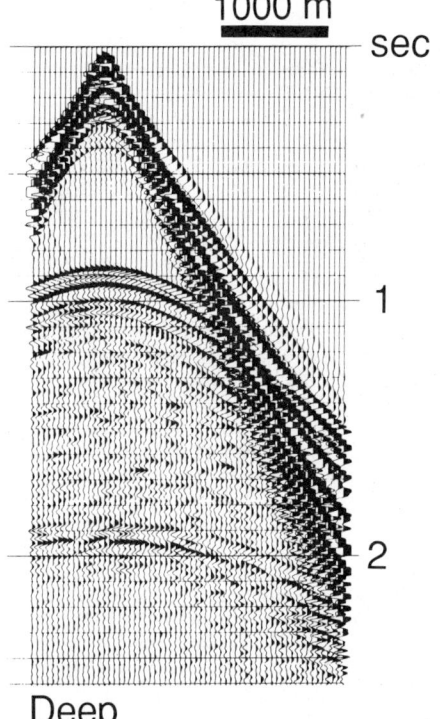

bottom and a free surface condition at the top of the grid. The data were summed over eight adjacent traces to simulate seismic group recording and were processed as a normal seismic reflection survey. An example of a single seismic shot record is shown in Fig. 21. The diffracted energy from the random zone appears between and beneath the two reflection events. Figure 22 is a migrated CMP stack of the synthetic data compared with a primaries-only section generated by convolution of the source pulse with the vertical path reflection coefficients calculated from the model. The diffractions from the random velocity fluctuations appear on the migrated section as a series of discontinuous horizontal events. The events have greater horizontal continuity than either the velocity perturbations in the random medium or the events on the primaries only section. The authors attribute this to the dip-filtering action

FIGURE 21. A finite-difference shot record from the model in Fig. 20. The zone of random velocity fluctuations produces a scattered wave field beginning immediately after the first specular reflection at 1.0 s and persisting after the second specular reflection. The data has been horizontally summed over eight grid points to simulate field group recording. The source was inserted into the thin surface layer using the technique in Fig. 14.

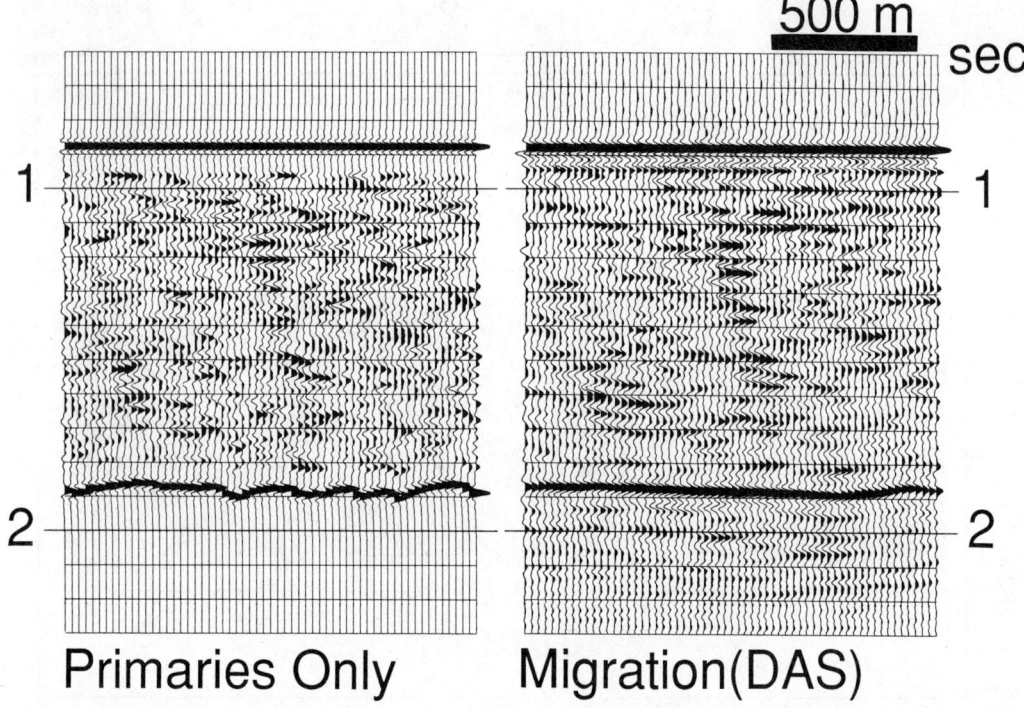

FIGURE 22. A CMP stack of 25 synthetic shot records generated for the model of Fig. 20 (right) and a primaries-only vertical incidence section generated from the model in Fig. 20 (left). The stack on the right is the result of reflection processing of the synthetic data including CMP stacking, deconvolution after stack (DAS), and time migration. The horizontally limited events appearing between the two specular reflections in the CMP stack have greater horizontal extent than the events in the primaries-only vertical incidence section.

of the CMP stacking process compounded by migration of the dip-filtered events.

Concluding Remarks

In this paper I have attempted to present the essentials of explicit finite-difference forward modeling methods and to discuss the properties of the modeling technique rather than review the large volume of research and applications of finite-difference modeling in the seismological literature. I have ignored the inclusion of anelastic effects, a topic of modern research. I have also ignored the algorithmic considerations that are involved in adapting difference methods to parallel computers, another area of research. The methods I have described are easily implemented on modern minicomputers and on the vector-oriented supercomputers.

ALAN R. LEVANDER

References

Abramowitz, M., and I. A. Stegun, 1964, *Handbook of Mathematical Functions*. New York: Dover Publications, Inc., 1046p.

Aki, K., and P. G. Richards, 1980, *Quantitative Seismology: Theories and Methods*, 2 vols. San Francisco: W. H. Freeman, 932p.

Alterman, Z!, and F. C. Karal, 1968, Propagation of elastic waves in layered media by finite difference methods, *Seismol. Soc. America Bull.* **58**, 367–398.

Alford, R. M., K. R. Kelly, and D. M. Boore, 1974, Accuracy of finite difference modeling of the acoustic wave equation, *Geophysics* **39**, 834–852.

Boore, D. M., 1970, Love waves in nonuniform wave guides: finite difference calculations, *Jour. Geophys. Research* **75**, 1512–1527.

Boore, D. M., 1972, Finite difference methods for seismic wave propagation in heterogeneous materials, in B. A. Bold, ed. *Methods in Computational Physics*, vol. 11, New York: Academic Press, 1–37.

Claerbout, J. C., 1970, Coarse grid calculations of waves in inhomogeneous media with application to delineation of complicated seismic structure, *Geophysics* **35**, 407–418.

Claerbout, J. C., 1985, *Imaging the Earth's Interior*. Oxford: Blackwell Scientific Publications, 398p.

Clayton, R., and B. Engquist, 1977, Absorbing boundary conditions for acoustic and elastic wave equations, *Seismol. Soc. America Bull.* **67**, 1529–1540.

Dablain, M. A., 1986, The application of higher-order differencing to the scalar wave equation, *Geophysics* **51**, 54–66.

Emerman, S. H., W. Schmidt, and R. A. Stephens, 1982, An implicit finite-difference formulation of the elastic wave equation, *Geophysics* **47**, 1521–1526.

Emerman, S. H., and R. A. Stephens, 1983, Comment on "Absorbing boundary conditions for acoustic and elastic wave equations," by R. Clayton and B. Engquist, *Seismol. Soc. America Bull.* **73**, 661–665.

Gerald, C. F., 1973, *Applied Numerical Analysis*. Reading MA.: Addison-Wesley, 340p.

Gibson, B. S., and A. R. Levander, 1988, Modeling and processing of scattered waves in seismic reflection surveys, *Geophysics* **53**, 466–478.

Golub, G. H., and C. F. Van Loan, 1983, *Matrix Computations*. Baltimore: The Johns Hopkins University Press, 447p.

Ilan, A., and D. Loewenthal, 1976, Instability of finite difference schemes due to boundary conditions in elastic media, *Geophysical Prospecting* **24**, 431–453.

Ilan, A., A. Ungar, and Z. Alterman, 1975, An improved representation of boundary conditions in finite difference schemes for seismological problems, *Royal Astron. Soc. Geophys.* **43**, 727–745.

Israeli, M., and S. A. Orszag, 1981, Approximation of radiation boundary conditions, *Jour. Comp. Physics* **41**, 115–135.

Kelly, K. R., R. W. Ward, S. Treitel, and R. M. Alford, 1976, Synthetic seismograms: a finite difference approach, *Geophysics* **41**, 2–27.

Kummer, B., and A. Behle, 1982, Second-order finite-difference modeling of SH-wave propagation in laterally inhomogeneous media, *Seismol. Soc. America Bull.* **72**, 793–808.

Levander, A. R., 1985, Finite difference calculations of dispersive Rayleigh wave propagation, *Tectonophysics* **113**, 1–30.

Levander, A. R., 1988, Fourth-order finite difference P-SV seismograms, *Geophysics* **53**, in press.

Madariaga, R., 1976, Dynamics of an expanding circular fault, *Seismol. Soc. America Bull.* **66**, 639–666.

Mahrer, K. D., 1986, An empirical study of instability and improvement of absorbing boundary conditions for the elastic wave equation, *Geophysics* **51**, 1499–1501.

Marfurt, K. J., 1984, Accuracy of finite-difference and finite-element modeling of the scalar and elastic wave equations, *Geophysics* **49**, 533–549.

Mitchell, A. R., and D. F. Griffiths, 1980, *The Finite Difference Method in Partial Differential Equations*. New York: Wiley, 272p.

Richtmyer, R. D., and K. W. Morton, 1967, *Difference Methods for Initial-Value Problems*, New York: John Wiley & Sons, 405p.

Stephen, R. A., 1984, Finite difference seismograms for laterally varying marine models, *Geophys. Jour. Royal Astron. Soc.* **79**, 185–198.

Trefethen, L. N., 1983, Group velocity interpretation of the stability theory of Gustafsson, Kreiss, and Sundstrom, *Jour. Comp. Physics* **49**, 199–217.

Vidale, J. E., and R. W. Clayton, 1986, A stable free-surface boundary condition for 2D elastic finite-difference wave simulation, *Geophysics* **51**, 2247–2249.

Vidale, J. E., and D. V. Helmberger, 1987, Path effects in strong motion seismology, in B. A. Bolt, ed., *Methods of Computational Physics*. New York: Academic Press, 267–319.

Vidale, J. E., and D. V. Helmberger, 1988, Elastic finite-difference modeling of the 1971 San Fernando earthquake, *Seismol. Soc. America Bull.* **78**, 122–141.

Virieux, J., 1986, P-SV wave propagation in heterogeneous media: velocity-stress finite-difference method, *Geophysics* **51**, 889–901.

Cross-references: *Deep Seismic-Reflection Profiling; Exploration Seismology; Seismic Diffraction; Seismic Imaging; Seismic Ray Theory; Seismic Signal Processing; Seismic Tomography; Seismic Wavefield Migration; Seismic Wave Scattering; Seismograms: Synthetic; Seismology: Physical Model Studies.*

FREE OSCILLATIONS OF THE EARTH

At very long periods, seismic signals may be represented more easily as the superposition of elastic-gravitational modes than as the sequential arrival of traveling waves. Because the Earth is a finite body, it will resonate as a whole only at discrete frequencies. This is apparent from Fig. 1, which shows the Fourier amplitude spectrum of a 30-hr-long accelerogram recorded at College, Alaska, following the Sumbawa, Indonesia, earthquake of 1977. To a first approximation, energy is concentrated within a number of peaks whose frequency depends upon the elastic and density structure of the interior and whose amplitude is controlled by the location of both source and receiver and by the radiation pattern of the earthquake. Physical attenuation acts to broaden each peak.

Ever since free oscillations were first unambiguously observed following the great Chilean earthquake of 1960, research in this area has followed two paths. One path has been to compile accurate measurements of free oscillation frequencies and attenuation rates. This task has been aided by improvements in instrumentation that have lowered the signal detection threshold and so allowed the use of recordings from smaller, more numerous earthquakes. New methods for analyzing time series are continually being introduced to glean additional information from high-quality digital recordings.

The second path has been to devise a theoretical framework for finding models of the interior that are consistent with these observations. Calculation of the free oscillations of a rotating, highly aspherical, anisotropic, and anelastic Earth is a tremendously difficult undertaking. Happily, the Earth's elastic properties deviate little from spherical symmetry and are almost uniformly isotropic. In this case, the problem is mathematically akin to those often encountered in quantum mechanics, and so the machinery of Rayleigh–Schrödinger perturbation theory may be brought to bear with great success. Given a rotating Earth model whose properties are small perturbations to spherical symmetry, it is

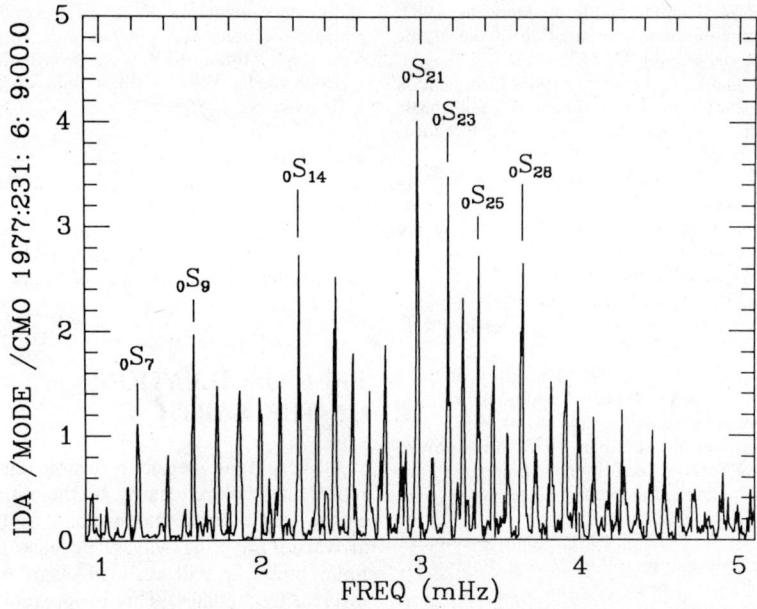

FIGURE 1. Spectrum of 30-hr vertical accelerogram recorded at IDA station CMO following the Sumbawa earthquake of 1977. Many prominent peaks are visible above the continuum of background noise and poorly excited overtones.

straightforward, if computationally laborious, to solve the forward problem of determining how such a model will oscillate. The inverse problem of determining what model or suite of models could have generated observed oscillations remains much more difficult, but progress here has also been made.

The next section contains an outline of what approximations have been necessary to make the general free oscillation problem tractable and how these suggest strategies for exploring the long-period spectrum and its significance for Earth structure.

The Forward Problem

An initial assumption made is that the Earth is disturbed from mechanical equilibrium by a fast-acting force, $\mathbf{f}(\mathbf{r}, t)$, which gives rise to an infinitesimal displacement $\mathbf{s}(\mathbf{r}, t)$. The linearized equation of motion (simply Newton's second law) satisfied by $\mathbf{s}(\mathbf{r}, t)$ is

$$\rho_0 \partial_t^2 \mathbf{s} + 2\rho_0 \mathbf{\Omega} \times \partial_t \mathbf{s} + \mathcal{H}\mathbf{s} = \mathbf{f} \quad (1)$$

where $\mathcal{H}$ is a linear integrodifferential operator containing all the forces that seek to restore the Earth to mechanical equilibrium. This equation must be solved subject to suitable boundary conditions on the displacement and its derivative as well as on traction at the outer surface and all internal discontinuities. (See Woodhouse and Dahlen, 1978, for a more complete formulation of the problem with correct boundary conditions.)

$\mathcal{H}$ includes all information on elastic and anelastic structure and incorporates the perturbations to density, gravitational potential, and stress associated with the deformation $\mathbf{s}(\mathbf{r}, t)$. For example, if one assumes that the rheology of the Earth is isotropic, $\mathcal{H}$ takes the form

$$\mathcal{H}\mathbf{s} = \rho_1 \nabla(\phi_0 + \psi) + \rho_0 \nabla \phi_1 \\ + \nabla[\rho_0 \mathbf{s} \cdot \nabla(\phi_0 + \psi)] - \nabla \cdot \mathbf{T} \quad (2)$$

where $\rho_1(\mathbf{r}, t)$, $\phi_1(\mathbf{r}, t)$, and $\mathbf{T}(\mathbf{r}, t)$ are the perturbations to density, gravitation potential, and stress away from the equilibrium conditions ρ_0 and ϕ_0. The centrifugal potential ψ is associated with rotation $\mathbf{\Omega}$.

Since the force acts over a short, finite time and then ceases, the motion of the Earth must consist of a superposition of simple harmonic oscillations of the form

$$\mathbf{s}(\mathbf{r}, t) = \mathbf{s}(\mathbf{r}) \exp[i\omega t] \quad (3)$$

After Eq. 3 is substituted into Eqs. 1 and 2, the equation of motion takes the form of an eigenvalue problem

$$\mathcal{H}\mathbf{s} + 2i\omega\rho_0 \mathbf{\Omega} \times \mathbf{s} = \omega^2 \rho_0 \mathbf{s} \quad (4)$$

for the operator $\mathcal{H} + 2i\omega\rho_0 \mathbf{\Omega} \times$. The eigenfunctions of this operator describe the shape of the oscillations; the eigenvalues, their corresponding frequencies. For a rotating, aspherical, anelastic

Earth, these eigenfunctions and eigenvalues are complex quantities.

Equations 1 and 4 are still a poor choice as starting points because of the complicated nature of $\mathcal{H}$. A better beginning is to search for a spherically symmetric, nonrotating, elastic, and isotropic (SNREI) model that is approximately correct and then find perturbed solutions to small perturbations of this model that incorporate more general effects. Ignoring rotation and assuming spherical symmetry, perfect elasticity, and isotropy for the moment, we further simplify the equation to

$$\mathcal{H}\mathbf{s} = \omega_d^2 \rho_0 \mathbf{s}$$

This problem may be converted to a system of Sturm–Liouville equations by separation of variables. The denumerably infinite solutions are grouped into two families.

The first family, the toroidal modes, has eigenfunctions of the form

$$\mathbf{s}(\mathbf{r}) = W(r) \, [-\hat{\mathbf{r}} \times \nabla_1 Y(\hat{\mathbf{r}})] \tag{5}$$

where $Y(\hat{\mathbf{r}})$ is a surface spherical harmonic of degree l and $\nabla_1 = \nabla - \hat{\mathbf{r}}(\hat{\mathbf{r}} \cdot \nabla)$ is the surface gradient operator on the unit sphere. Toroidal displacement fields are purely tangential and are therefore invisible on vertical accelerometers. They are denoted $_nT_l$ and classified by their radial order n and angular degree l. The fundamental mode branch corresponds to $n = 0$, and overtones correspond to $n > 0$. Toroidal modes of degree $l = 1$ are forbidden.

The second family, the spheroidal modes, has eigenfunctions of the form

$$\mathbf{s}(\mathbf{r}) = \hat{\mathbf{r}} U(r) \, Y(\hat{\mathbf{r}}) + V(r) \nabla_1 Y(\hat{\mathbf{r}}) \tag{6}$$

where $Y(\hat{\mathbf{r}})$ is a surface spherical harmonic of degree l as before. Spheroidal modes thus display both radial and tangential particle motion. They are denoted $_nS_l$ by their radial order and angular degree.

Any combination of the $2l + 1$ orthogonal spherical harmonics $Y_l^m(\theta, \phi)$, $-l \leq m \leq l$, that constitute the basis of a linear vector space may be substituted for $Y(\hat{\mathbf{r}})$ in Eqs. 5 and 6. The spherical coordinates θ and ϕ depend upon the choice of Cartesian axes, which is arbitrary in this spherically symmetric case. Once the coordinate system is fixed, however, the basis for the vector space is fixed as well. Each eigenfunctions $\mathbf{s}_m(\mathbf{r})$ of the spherical Earth may now be chosen to correspond to one spherical harmonic Y_l^m, and all modes of the same n, l will oscillate at the same eigenfrequency ω_d.

Not surprisingly, much of the nomenclature of the quantum mechanical formalism has been adopted as well. The condition that exists when two or more eigenfunctions have the same eigenfrequency is known as *degeneracy* and here is a consequence of the high degree of symmetry in the problem. Modes of a particular type (toroidal or spheroidal) sharing the same n and l are known collectively as a *multiplet*. Each mode within a multiplet is known as a *singlet*.

Gilbert and Dziewonski (1975) in their landmark study of the spherically averaged Earth solved the structure inverse problem for a suite of models that satisfied their observations of multiplet eigenfrequencies. Figures 2 to 6 are based upon the properties of one of their models, 1066A. Inversion schemes for the spherically averaged Earth often utilize the Rayleigh–Ritz variational method, which exploits the fact that eigenfunctions of models derived from body wave seismology or from smaller sets of free oscillation measurements are close to those of the spherically averaged Earth.

Because the attenuation mechanism is linear in the band of interest, anelasticity may be incorporated by adding a frequency-dependent imaginary component to the bulk and shear moduli. To express completely the properties of an anelastic Earth model, one must therefore specify the bulk and shear moduli at some reference frequency ω_r. The decay rate of a mode is described in terms of its quality factor Q, the fractional amount of energy lost per cycle. One consequence of introducing anelasticity is that the zeroth-order eigenfrequencies and eigenfunctions are no longer purely real, and the singlet eigenfunctions within a multiplet are not mutually orthogonal.

The second part of the forward problem, the extent to which each mode of oscillation will be excited, is determined by the nature of the imposed force in Eq. 1. For earthquakes, this term may be adequately represented by a body equivalent point force

$$\mathbf{f}(\mathbf{r}, t) = -\mathbf{M} H(t) \cdot \nabla \delta(\mathbf{r} - \mathbf{r}_0)$$

where $H(t)$ is the Heaviside step function and $\delta(\mathbf{r} - \mathbf{r}_0)$ is the Dirac delta function. $\mathbf{M}$ is a second-order symmetric moment tensor now computed routinely for moderate- and large-size earthquakes.

A multiplet's displacement for this slightly anelastic model is

$$\mathbf{s}(\mathbf{r}, t) = \sum_{m=-l}^{+l} \mathbf{A}_m \left[1 - \cos \omega_m t \, \exp\left[\frac{-\omega_m t}{2 Q_m}\right] \right]$$

where the amplitudes of singlet excitation

$$\mathbf{A}_m = \mathbf{M} : \boldsymbol{\epsilon}_m^*(\mathbf{r}_0) \, \mathbf{s}_m(\mathbf{r})$$

depend upon the moment tensor $\mathbf{M}$ and the strain ϵ evaluated at the source. Because of spherical symmetry, all $\omega_m = \omega_d$ and $Q_m = Q_d$.

Mode Properties

Figure 2 shows a dispersion diagram for the toroidal modes of model 1066A. High-frequency modes $_nT_l$ with $n \ll l$ are equivalent to Love surface

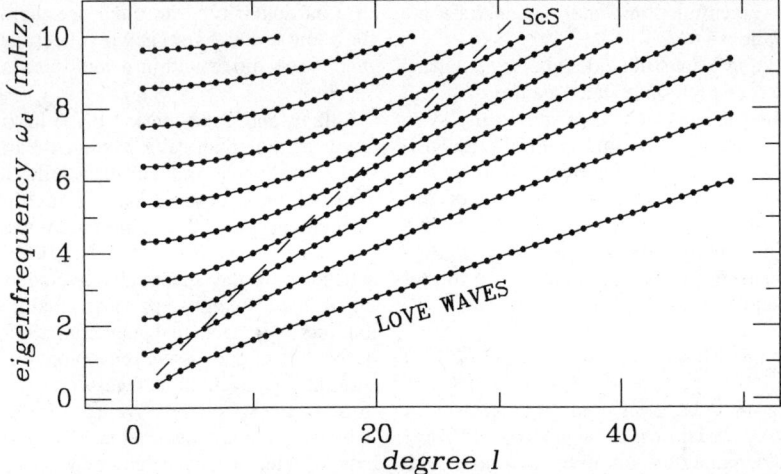

FIGURE 2. Dispersion diagram for toroidal modes of 1066A. Modes of the same radial order are connected with solid lines beginning with the fundamental branch to the lower right and proceeding through higher overtone branches to the upper left. Dashed lines group modes corresponding to body and surface waves of a given type. (Reprinted from Gilbert and Dziewonski, 1975 with permission of the authors.)

waves, and those with $n \gg l$, to $(ScS)_{SH}$ body waves. Note that as frequency increases, the density of modes in a given frequency band increases substantially. At frequencies much higher than 10 mHz, both the number of modes required to synthesize a seismogram and the coupling between such modes, an effect that is difficult to treat, render the modal superposition approach impractical, and one must revert to a traveling wave representation.

The dispersion diagram for the spheroidal modes is shown in Fig. 3. Additional "structure" in the upper left-hand corner of the diagram appears because the mantle is no longer completely decoupled from the liquid outer core as is the case with

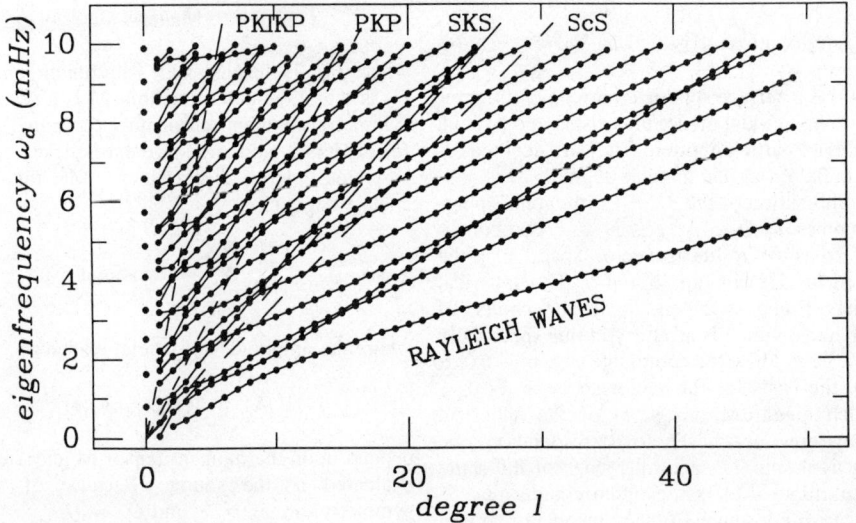

FIGURE 3. Dispersion diagram for spheroidal modes of 1066A. Modes of the same radial order are connected with solid lines beginning with the fundamental branch to the lower right and proceeding through higher overtone branches to the upper left. Dashed lines group modes corresponding to body and surface waves of a given type. (Reprinted from Gilbert and Dziewonski, 1975 with permission of the authors.)

toroidal modes. Regions within the diagram corresponding to body waves are labeled as in Fig. 2.

Each mode represented in Figs. 2 and 3 is sensitive to structure in different parts of the interior. Therefore, by choosing judiciously which particular mode or combination of modes to model observations of, one can target a specific region of the interior for study. Exactly how this may be done is the subject of generalized linear inverse theory. Parker (1977) contains an excellent review of the theory and includes an example of how this technique may be applied to a free oscillation problem.

Two cases illustrating how such a strategy may evolve are found in Figs. 4 and 5. Figure 4 shows energy density as a function of depth for three fundamental spheroidal modes. As the degree l increases, energy is concentrated toward the outer surface, and inspection of the shear and compressional velocity differential kernels (not shown) reveals that there is a corresponding increase in sensitivity to model changes at shallow depths. The fundamental modes are thus an excellent choice for studying upper mantle structure. By observing changes in properties along the fundamental mode branch, one can "peel back" the mantle layer by layer.

A different case is illustrated in Fig. 5. Here are three modes whose primary energy is in the core and are therefore suitable for studying the deep interior. Unhappily, these modes are excited less frequently, if at all, by earthquakes, and therefore our knowledge of the core is less detailed than of structure near the surface.

The lack of precise knowledge of the Earth's elastic structure complicates inferring details of the anelastic structure because the two problems are coupled. Mutual interference of modes, which results from lateral variations in the elastic structure, expresses itself as an apparent attenuation that is difficult to distinguish from physical attenuation, and only a poorly resolved pattern of anelasticity has emerged. Masters and Gilbert (1983) have succeeded in constructing several spherically symmetric models of Q, one of which is shown in Fig. 6b, that satisfy the data.

The spatial distribution of a mode's energy density and how that energy is partitioned between shear and compression give a clue to how fast that mode will attenuate. Generally speaking, the more a mode's energy density is concentrated in a region of low Q, the faster it will attenuate, and vice versa. Also because most dissipation takes place in shear motion, the greater the fraction of a mode's energy that is compressional, the less rapidly it will attenuate. The mode $_0S_{46}$'s energy density (Fig. 6a) is concentrated in the outer layers of the mantle where it attenuates ($Q_{0S_{46}} = 147$) rapidly. Mode $_6S_2$, on the other hand, has greater energy density in the high-Q core (Fig. 6c) and therefore attenuates ($Q_{6S_2} = 2900$) more slowly. Mode $_0S_0$ (not shown) is an almost pure compressional mode, which is reflected in its extremely high Q (5700).

Splitting

Rotation and any aspherical structure act to destroy the eigenfrequency degeneracy. (See Dahlen, 1980, for a thorough review of the topic.) What was a single line at the degenerate frequency ω_d splits into $2l + 1$ closely spaced lines. The eigenfunctions of the perturbed Earth are no longer functions of a single Y_l^m but now consist of linear combinations of eigenfunctions of the SNREI Earth model. If multiplets are spaced far enough apart in frequency, they may be treated as uncoupled, and this sum is over only the $2l + 1$ SNREI basis functions of the multiplet under consideration:

$$\mathbf{s}_j(\mathbf{r}) = \sum_{m=-l}^{+l} \alpha_j^m \mathbf{s}_m(\mathbf{r})$$

The expansion coefficients α_j^m and frequency perturbations $\delta\omega_j$ may be found by solving a Hermitian eigenvalue problem of the form

$$\sum_{m'=-l}^{+l} H_{mm'} \alpha_j^{m'} = \delta\omega_j \alpha_j^m \quad (7)$$

(Because $\mathbf{s}(\mathbf{r})$ has been expanded in a complete set of vector functions, the operator problem may be formally converted to a matrix equation; anelasticity is ignored here.) The matrix $\mathbf{H}$, termed the splitting matrix, has a block diagonal form and is generally much larger than shown in Eq. 7. By assuming multiplets are uncoupled, we may treat each $(2l + 1) \times (2l + 1)$ block independently. Each element $H_{mm'}$ is a linear functional of the perturbed model.

When aspherical structure is axisymmetric, the splitting matrix assumes a diagonal form, and the line spacing $\delta\omega_m$ is quadratic (Dahlen, 1968):

$$\delta\omega_m = \omega_d[a + mb + m^2 c] \quad (8)$$

where a, b, and c are constants that depend upon the multiplet under consideration. The term mb in Eq. 8 gives rise to a symmetric "Zeeman"-type splitting and is a consequence of the first-order effect of the Coriolis force. (See Fig. 7.) The constants a and c are due to axisymmetric aspherical and to second-order rotational effects. For nonaxisymmetric structure, the singlet frequency distribution is even more nonlinear.

When the individual lines of low-frequency modes can be resolved, splitting width measurements provide powerful constraints on aspherical structure. Recent work by Ritzwoller et al. (1986) and Giardini et al. (1987) have concentrated upon identifying, measuring, and modeling the splitting of low-harmonic-degree modes whose eigenfunctions are

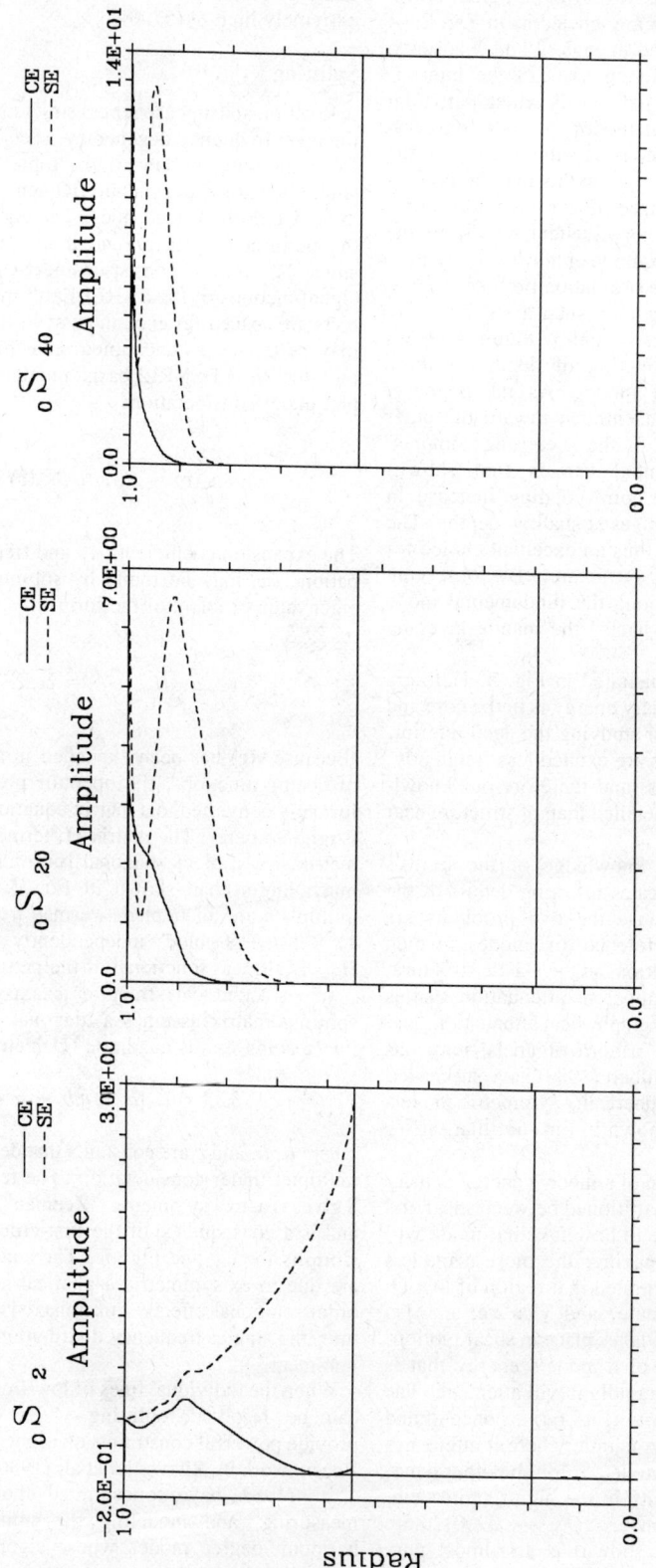

FIGURE 4. Compressional (solid line) and shear (dashed line) energy density for three fundamental spheroidal modes. As the degree l increases, energy is concentrated toward the surface, and the modes become sensitive only to upper-mantle velocity variations.

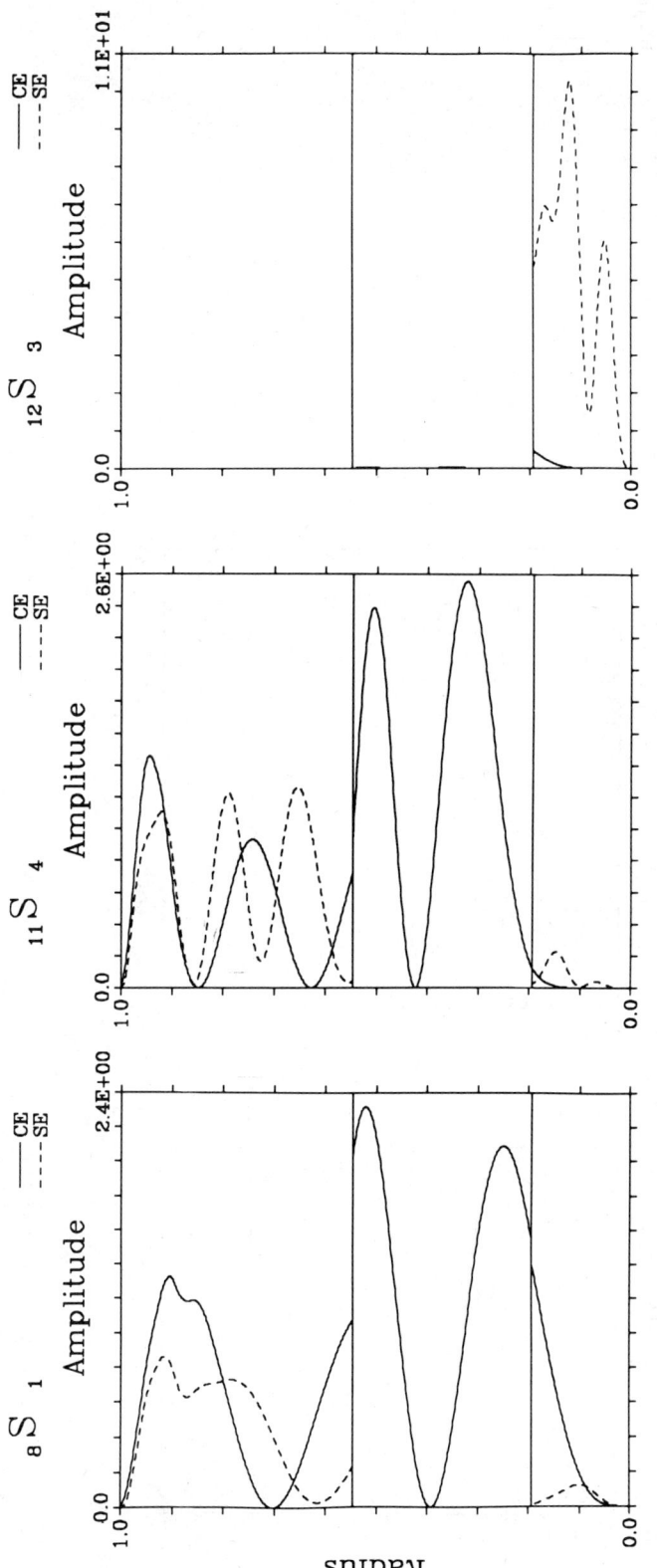

FIGURE 5. Compressional (solid line) and shear (dashed line) energy density for three core modes. Although their energy is distributed throughout the interior, these modes are more sensitive than most to structure in the core.

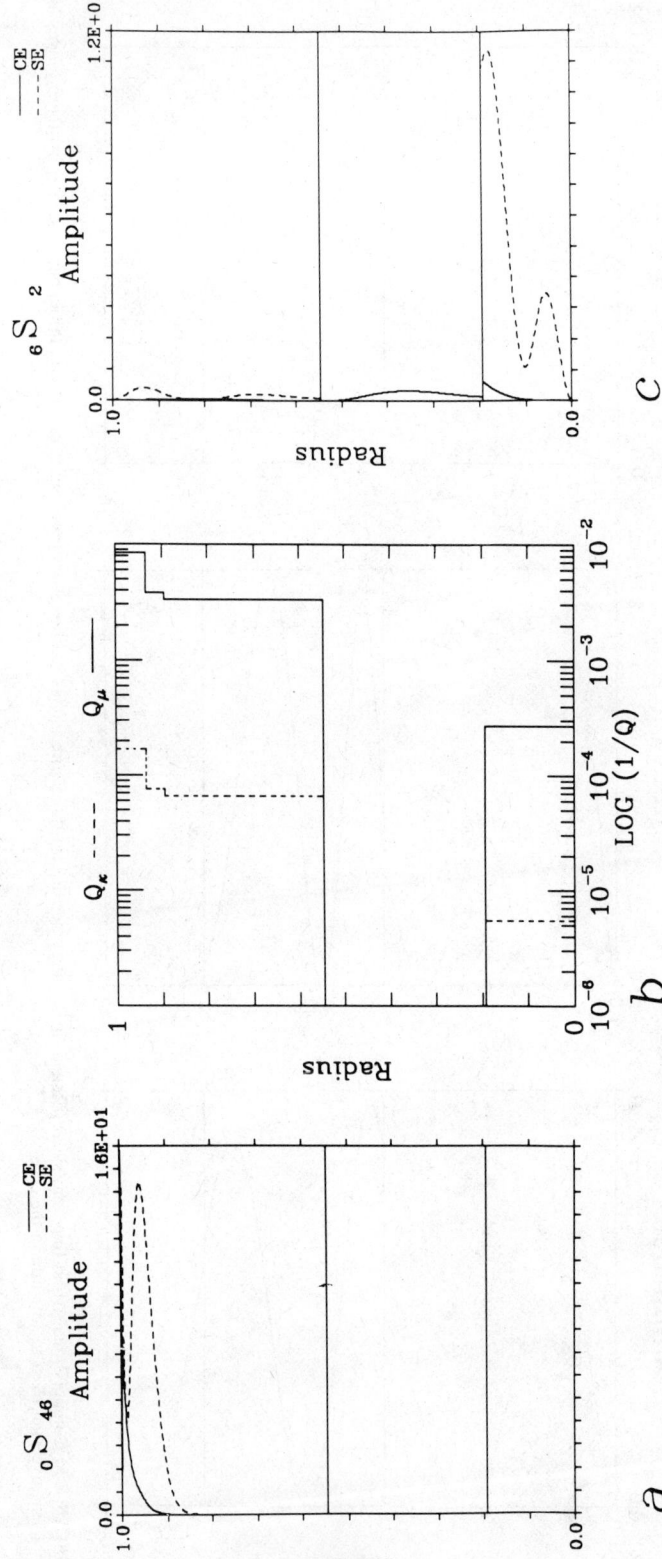

FIGURE 6. Energy density distribution for two modes (a, c) and an attenuation model (b). A mode whose energy is concentrated in the upper mantle, such as $_0S_{46}$, longer than a mode whose energy is concentrated in the core, such as $_6S_2$, will oscillate for much

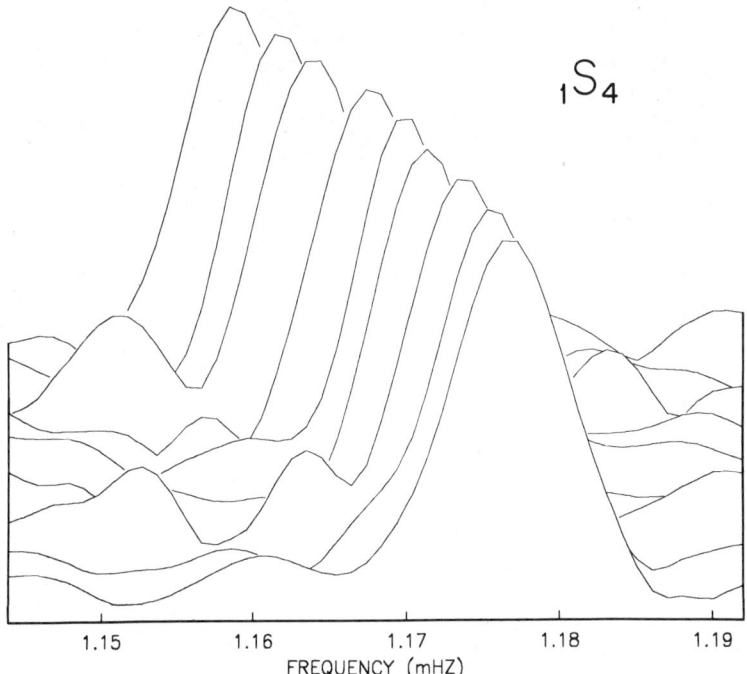

FIGURE 7. The splitting of $_1S_4$ is dominated by rotational effects, so that the singlet distribution is linear. For other modes of low angular degree, the splitting is nonlinear and provides valuable information on lateral structure. Spectra shown here are produced by a technique called stripping, which exploits knowledge of the singlet eigenfunction shape and earthquake moment tensor. (Reprinted from Ritzwoller et al., 1986 with permission of the American Geophysical Union.)

sensitive to structure within the core. The techniques employed usually take advantage of the low attenuation rates of the modes involved and that the source mechanism of the earthquake is known from higher-frequency information. For many low-degree multiplets, the actual splitting widths are affected little by the source, so that estimates of singlet resonance functions may be used to recombine spectra from more than one station and earthquake in order to enhance selectively the spectrum of an individual singlet at the expense of nearby peaks. Thus many singlet peaks are measured that would otherwise be unresolvable. The spectra shown in Fig. 7 are the product of such a procedure.

For the surface wave equivalent modes ($n \ll l$) of the fundamental and first few overtone branches, splitting is also important, but this potentially rich source of information about upper-mantle heterogeneity is difficult to measure. In these cases attenuation is so large compared with multiplet splitting widths that spectral lines of individual singlets cannot be resolved. The $2l + 1$ lines of the multiplet sum to give the appearance of a single peak rather than multiple peaks. For example, the splitting width of $_0S_{25}$ (Fig. 8) computed from the upper-mantle model M84A (Woodhouse and Dziewonski, 1984) is 16 μHz, but the peak is distributed over a band ≈ 100 μHz wide. Clearly then, one is reduced to making measurements of an entire multiplet rather than of individual singlets.

Jordan (1978) has constructed the lateral structure inverse problem in terms of multiplet observables by applying the method of moments used in quantum mechanics problems. To first order, the zeroth moment of a multiplet is independent of lateral structure, and the first moment is linear functional of lateral structure. The ratio of the two, which he termed the multiplet location, $\overline{\delta\omega}$, is easily measured.

In the limit of large l and heterogeneity that is both weak and of longer wavelength than the modes under consideration, each multiplet shifts its frequency by an amount $\overline{\delta\omega}$ that is proportional to the average of structure along that great circle P connecting source and receiver, namely,

$$\overline{\delta\omega} = \omega_d + \frac{1}{2\pi} \oint_P \delta\omega_{\text{local}}(\theta, \phi)\, ds$$

The function $\delta\omega_{\text{local}}$ is related to the phase velocity and may be expressed as an integral over the radius

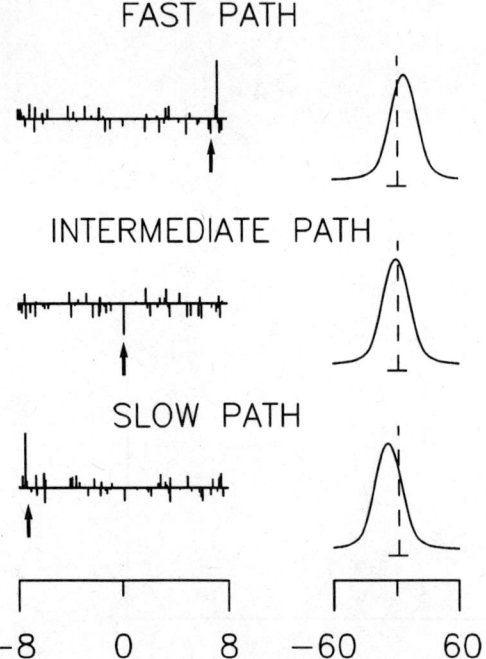

FIGURE 8. The singlet excitation of $_0S_{25}$ computed from model M84A for a receiver placed on fast, slow, and intermediate paths 115° from an explosive source is shown at left. In the extreme cases, the singlet spectrum is dominated by a single line at high or low frequency, and the multiplet peak, shown at right for a 20-hr record, is shifted relative to the SNREI frequency indicated here by a vertical dashed line. For the intermediate case, the singlets interfere to form a multiplet peak near the global average. Line segments beneath the multiplet peaks are 16 μHz wide, the width of the splitting shown at left. (Adapted from Davis and Henson, 1986.)

of a kernel function M_0 derived from an SNREI model:

$$\delta\omega_{\text{local}} = \int_0^d M_0(r)\, \delta m(r, \theta, \phi) r^2\, dr$$

where δm is the perturbation due to structure. If the Earth were spherically symmetric and varied radially as it does under (θ, ϕ), it would oscillate at the frequency $\omega_d + \delta\omega_{\text{local}}(\theta, \phi)$. A number of models of long-wavelength lateral structure have been obtained by regressing upon measurements of $\overline{\delta\omega}$. The result does not depend upon the source mechanism but does require a global coverage of great circle paths for a stable solution.

The major drawback in this approach is that, to this order, modal frequency shifts contain information only about even-parity structure (structure symmetric under reflection through the center of the Earth). This is a consequence of the selection rules in the formulas used to compute elements of the splitting matrix. In order to infer the properties of structure with odd parity, we must use information about modal amplitudes. By computing for each mode a fictitious frequency shift and perturbation in epicentral distance, Woodhouse and Dziewonski (1984) have introduced phase perturbations into their analysis and thus are able to draw inferences about both odd- and even-parity structure.

As models of lateral structure increase in resolution, additional physical effects will have to be taken into account. Masters et al. (1983) have demonstrated that the Coriolis force acts to couple some fundamental spheroidal and toroidal modes, and there are indications that coupling between the singlets of individual multiplets is important as well. These effects give rise to multiplets whose eigenfunctions are constructed from far more than the $2l + 1$ zeroth-order multiplet basis functions, and modes that are purely toroidal or spheroidal cease to exist. The increase in the amount of high-quality, digital seismic data and the continuing dramatic increases in computational capabilities will ensure that these problems become tractable in the near future.

PETER DAVIS

References

Dahlen, F. A., 1968, The normal modes of a rotating, elliptical Earth, *Royal Astron. Soc. Geophys. Jour.* **16,** 329–367.

Dahlen, F. A., 1980, Splitting of the free oscillations of the Earth, in A. M. Dziewonski and E. Boschi, eds., *Physics of the Earth's Interior.* Bologna, Italy: Soc. Italiana di Fisica.

Davis, J. P., and I. H. Henson, 1986, Validity of the great circular average approximation for inversion of normal mode measurements, *Royal Astron. Soc. Geophys. Jour.* **85,** 69–92.

Giardini, D., X.-D. Li, and J. H. Woodhouse, 1987, Three-dimensional structure of the Earth from splitting in free oscillation spectra, *Nature* **325,** 405–411.

Gilbert, F., and A. M. Dziewonski, 1975, An application of normal mode theory to the retrieval of structural parameters and source mechanisms from seismic spectra, *Royal Soc. London Philos. Trans. ser. A,* **278,** 187–269.

Jordan, T. H., 1978, A procedure for estimating lateral variations from low-frequency eigenspectra data, *Royal Astron. Soc. Geophys. Jour.* **52,** 441–455.

Masters, G., and F. Gilbert, 1983, Attenuation in the Earth at low frequencies, *Royal Soc. London Philos. Trans. Ser. A,* **308,** 419–422.

Masters, G., J. Park, and F. Gilbert, 1983, Observations of coupled spheroidal and toroidal modes, *Jour. Geophys. Research* **88,** 10285–10298.

Masters, G., T. H. Jordan, P. G. Silver, and F. Gilbert, 1982, Aspherical Earth structure from fundamental spheroidal mode data, *Nature* **298,** 609–613.

Parker, R. L., 1977, Understanding inverse theory, *Ann. Rev. Earth Planetary Sci.* **5,** 35–64.

Ritzwoller, M., G. Masters, and F. Gilbert, 1986, Observations of anomalous splitting and their interpretation in terms of aspherical structure, *Jour. Geophys. Research* **91,** 10203–10228.

Woodhouse, J. H., and F. A. Dahlen, 1978, The effect of a general aspherical perturbation on the free oscillations of the Earth, *Royal Astron. Soc. Geophys. Jour.* **53,** 335–354.

Woodhouse, J. H., and A. M. Dziewonski, 1984, Mapping the upper mantle: Three-dimensional modeling of Earth structure by inversion of seismic waveforms, *Jour. Geophys. Research* **89,** 5953–5986.

Cross-references: *Density Distribution in the Earth; Earth's Core: Structure; Earth Structure: Global; Elasticity and Wave Propagation: Principles; Inverse Theory and Methods: Seismology; Mantle, Lower: Structure; Mantle, Upper: Structure; Mantle Discontinuities; Mantle Viscosity; Seismic Anisotropy in the Earth; Seismic Attenuation: Observation and Measurement; Seismic Instrumentation; Spherical Harmonic Analysis.*

G

GEODESY: GEOMETRIC

Man has been interested in learning about the size and shape of the Earth for many centuries. Early Chinese imagined that the universe was an egg and the Earth the yolk inside. In early Greece, some speculated that the Earth was a flat disc but a spherical shape was the most widely accepted idea during the Greek era. The Greeks made efforts to determine the size of the Earth. The most famous of these ancient measurements was done by Eratosthenes in a well of Syene, an accomplishment for which he is called the founder of geodesy (see *Geodesy: Historical Introduction*). The telescope, logarithmic tables, and the method of triangulation developed in the seventeenth century contributed greatly to the foundation of geodetic science. The measurements of geodetic arc by the French Academy of Sciences in 1735 to determine the length of a meridian degree conclusively proved the Earth to be flattened ellipsoid, one of the milestones in the history of geodesy. In the eighteenth century, many famous scientists made important contributions to geodesy. Gauss (1775-1855) invented the least squares method and defined the geoid. He laid the theoretical foundations of geodesy and ushered it into its mature age. Laplace (1749-1827) devoted himself to celestial mechanics and tidal theory. Bessel (1784-1846) made the first accurate determination of the Earth's flattening.

In its early stage, geodesy was basically geometrical by nature, involving measurement of geometric quantities (angle and length) and employing geometric and trigonometric methods to study the geometric properties of the Earth (size, shape, and position). Physical geodesy came into being rather later, although it developed much faster.

The advent of modern computers, coupled with the development of space technology in the last three decades, has produced major breakthroughs in both geometric and physical geodesy.

Literally, geodesy means "dividing the Earth." The measurement and description of the Earth's surface was the province of classical geodesy as defined by Helmert in 1880. That definition no longer fully reflects the role of present-day geodesy. The scope of present-day geodesy may be seen in the structure of the International Association of Geodesy (IAG), with its five sections and various commissions and special study groups. For the period 1983-1987, the sections of IAG included: positioning, advanced space technology (used in geodesy), determination of the gravity field, general theory and methodology, and geodynamics.

Geodesy can be divided into a number of subdisciplines such as geometrical geodesy, geophysical geodesy, mathematical geodesy, and dynamical geodesy. With the advances of technology, new terms such as inertial geodesy, satellite geodesy, space geodesy, etc. have been introduced. Whereas these various divisions of geodesy are helpful for learning and understanding, too many subdisciplines can also cause unnecessary confusion. As a compromise, we suggest that it might be appropriate to divide geodesy into two subdisciplines: geometrical geodesy and geophysical geodesy. The former deals with positioning, including the size and shape of the Earth, and its temporal variation, while the latter includes studies of the gravity field of the Earth and its temporal variation.

It is difficult to set up a strict "boundary" between various subdisciplines of geodesy. For example, the shape of the Earth can be expressed geometrically by the semimajor axis and flattening of an ellipsoid; it can also be illustrated more precisely by the expression of the gravity potential as determined by geophysical geodesy. Geometrical and geophysical geodesy are, moreover, closely linked in the study of geodynamical problems. Therefore, the divisions of geodesy are rather artificial.

In geometrical geodesy the observations include quantities such as angles and lengths. Positioning (as expressed in a particular reference coordinate system) is the essence of geometrical geodesy. Positioning means to determine the coordinates of points on land, at sea, or in space with respect to an implied coordinate system. The study of the reference coordinate system is vitally important, since the coordinates of the same point are different in different reference systems. The coordinates of various points are not comparable unless they have been brought into the same reference system. The reference system could be global, regional, or local. Even the global systems set up by various techniques are different from each other. It is clearly preferable to have a single global Earth fixed reference system, so that positioning done anywhere in the world can be tied to a single system. Geodesists have pursued such a system for centuries, but it proved impossible before the advent of space techniques.

There are two classes of positioning methods. One is terrestrial and the other is celestial or extraterrestrial. The former is only able to yield direction differences or angles and differences in coordinates or point separations. Thus it can only provide *relative positioning*. Celestial methods link the terrestrial point with some celestial body so that the position of the point in some quasi-inertial reference system is known. Where the three dimensional coordinates of points are related to a global reference system is known as *absolute positioning*.

Astronomical optical observation is the most widely used conventional celestial method. Were it not for the vertical deflection and anomalies caused by atmospheric refraction, absolute positioning could have been accomplished long ago. Because of its complexity and because it is difficult to determine accurately, the vertical deflection has reduced the optical astronomical method to a secondary (although still important) role in geodesy. It could not replace the terrestrial method for positioning, but it can help to "control" the network and fix the orientation of the net.

The new space techniques also belong to the celestial method. The observed targets could be extragalactic radio sources, the moon, artificial satellites, or something else in the sky. The precise position of the source is known, either by a source catalogue, or by ephemerids, or as computed by dynamical methods. The space techniques are characterized by extremely high accuracy and independence of vertical line. For these reasons, as well as advantages such as ease of operation, space methods have revolutionized geodetic positioning.

Emphasis is put on absolute positioning in this article. In fact any method used for absolute positioning can also do relative positioning. One technique worth mentioning, which is especially used for relative positioning, is the inertial survey, based on utilizing the inertia of a mass to measure acceleration.

Positioning and Reference System by Conventional Techniques

By the conventional geodetic method, the three-dimensional coordinates of a point on Earth's surface are divided into horizontal and vertical components. The horizontal component is expressed in two-dimensional spheroidal coordinates. The vertical component is referred to as height. In classical geometrical geodesy, much emphasis is put on the 2-D spheroidal component. The horizontal and vertical components are measured by different surveying instruments, but all such instruments are adjusted by leveling devices so that the equipotential surface is implicitly involved in the observations. The plumb lines or verticals, which are normal to this surface, become the cardinal lines of this positioning method. The most important equipotential surface is the geoid, which in open ocean is simply mean sea level, and under the continents is the equivalent mean sea level. The geoid is a true physical surface and has a definite physical meaning. But its shape is too complicated to allow data to be easily referenced to it. In practice, a geometrically simple and regular surface must be taken for that purpose. Geodesists have adopted an ellipsoid as the main reference surface. The reference ellipsoid is a fictitious geometrical surface approximating the geoid but on which it is easy to carry out computations. Geodetic observations, which are intricately linked to the verticals, must be converted onto the reference ellipsoid to facilitate data processing.

Horizontal Positioning. The observables in horizontal positioning are angles and distances. Angles are observed by theodolite. Distances are measured by invar wires (in classical triangulation) or by electromagnetic, or microwave, or electro-optical distance measuring devices (in trilateration on traverse). Radar techniques can also be used, as in the Shoran and Hiran systems, but are not in common use. All of these instruments can be used only for relative measurements. Moreover, the stations between which distances or angles are to be observed must be mutually visible. (Shoran and Hiran systems are the exceptions). This condition means that the separation between two stations will be limited at most to some tens of kilometers. To link distant points together in a single system, additional intermediate points must be added. In this way, a geodetic network is formed. The layout of a geodetic net may take the form of triangulation, traverse, trilateration, or a combination of the three. Usually, the network is classified into one of four orders according to the accuracy of measurements, with first order the most accurate. Recent technological developments have necessitated an additional "zero order" classification, defined to be superior to first order.

The basic objects of the geodetic network have been (1) to establish a main network on which less precise observations may be based; and (2) in combination with astronomical and gravity observations to assist in determining the size and shape of the Earth.

The meridians and parallels of an ellipsoid constitute the reference system on which geodetic computation is carried out. The two coordinates are geodetic latitude and longitude. Angles observed between ground points and distances measured between ground points must be corrected to give corresponding angles and distances on the ellipsoid. If we are given the angles between ellipsoid points, the distance and azimuth of the normal section between at least one pair of these points, and the ellipsoidal coordinates of one point, then the latitudes and longitudes of the other points can be computed.

Computations carried out in this way are described as being done by Helmert's projection method. Conversely, if the coordinates of two points on the ellipsoid are given, the length and azimuth between these two points can be computed. In fact, these kinds of computations on the reference ellipsoid have been among the major tasks of classical geometrical geodesy.

Horizontal Positioning by Geodetic Astronomy. As we have already noted, we must know at least one azimuth and the ellipsoidal coordinates of one point to determine the ellipsoidal coordinates of all the points in a net. These qualities can be determined using astronomy.

Astronomic positioning is the oldest point positioning method. The observations are made by optical instruments, such as theodolite, astrolabe, zenith camera, etc. All the instruments contain leveling devices, hence the cardinal line of the observation is again the vertical, and the positions thus obtained are referenced to the geoid. Since the geoid is a nonmathematical surface, astronomic positions are independent of each other. The coordinates of a point are expressed as astronomic latitude and longitude, which differ from geodetic latitude and longitude. These differences constitute the two components of vertical deflection, where deflection is defined as the angle between the vertical and the spheroidal normal at the same point. Similar problems exist for other observables such as the astronomic azimuth. Again, because of the vertical deflection, this azimuth is different from the geodetic azimuth. An important relation between the longitude component of vertical deflection and the difference of astronomic and geodetic azimuth exists. This relation is the well-known Laplace's equation.

Astronomical latitudes, longitudes, and azimuths alone are of little value, since they do not relate one point to another with sufficient accuracy for the computation of distances and directions between the points. But when interconnected by geodetic positioning, they can be used to control the accumulation of systematic errors in the horizontal direction of a network and to orient the reference ellipsoid. In this way astro-geodetic positioning becomes absolute positioning.

Vertical Positioning. Vertical positioning is the process of determining height. Spirit leveling is the most accurate and widely used classical method for height measurement. Occasionally, trigonometric height is used as a substitute in areas with complicated terrain. The heights thus determined are the elevations above the geoid or mean sea level.

For some mapping purposes, especially in hydrology, the use of the geoid as a reference surface for the vertical coordinate is natural. In other cases, however, it is improper to take the ellipsoid as the reference surface for horizontal positions. If the horizontal component of a point is referred to one surface while the vertical component is referred to another, then the resulting incompatibility of the measurements means that no precise three dimensional positioning can be achieved.

When referenced to the ellipsoid the vertical component defines the ellipsoidal height, which is the distance above the ellipsoid measured along the ellipsoidal normal. To convert height above the geoid into ellipsoid height, the undulations of the geoid must be known. This is one of the intersections between geometrical and geophysical geodesy.

Geodetic Datums and the Reference Ellipsoid. The reference ellipsoid is a fictitious geometric surface. Its size, shape, center, and orientation can be arbitrarily defined. Any change of this definition will affect the positions of points as well as the amounts of vertical deflections and the undulations. The geodetic datum problem involves the definition of the reference ellipsoid.

A geodetic datum is defined as any numerical or geometrical quantity or set of such quantities that serve as a reference or base for other quantities. In geodesy two types of datums are considered: a horizontal datum that forms the basis for the computations of horizontal positioning, and a vertical datum to which elevations are referred.

The size and shape of the ellipsoid are represented by the semimajor axis and flattening. While these values can be derived by means of conventional geodetic networks, space techniques give much more accurate results. The origin of the ellipsoid is usually determined by a single astronomic point or by least squares. In the former, the geoid is taken to coincide with the ellipsoid at this point. In the latter, the summation of squares of the separation (between ellipsoid and geoid) at all points in a network are minimized. Similarly, the orientation of the ellipsoid to the geoid can also be determined at a single astronomic point, or by astro-geodetic orientation. In the latter case, a number of Laplace stations will be involved. The procedure reduces the sum of squares of the astro-geodetic deflections at all the Laplace stations to a minimum.

Thus, a horizontal datum consists of the longitude and latitude of an initial point (origin); an azimuth of a line (direction); the parameters (radius and flattening) of the ellipsoid selected for the computations; and the geoid separation at the origin. The reference ellipsoid defined in this way usually is not geocentric, and its rotation axis does not coincide with the maximum inertial axis, or the figure axis of the earth. Moreover, the datums defined by different countries or different data processing centers are commonly not equivalent. These datum differences can be removed by applying three rotations, three translations, and a scaling change, if there exist more than three common points for which coordinates in both datums are known. If two datums are separated by sea, then the equivalence between two datums will be very difficult to determine. Therefore, for conventional geodesy methods, there is no really precise global reference system.

Space Techniques for Positioning and Reference Coordinate System

In geometrical geodesy at present, the most widely used space techniques include Doppler positioning, satellite laser ranging (SLR), very long baseline interferometry (VLBI), and the global positioning system (GPS). In principle, lunar laser ranging (LLR) and some other techniques can also be used, provided technical or financial problems inherent in these techniques are overcome.

These new techniques belong to celestial methods, so that absolute positioning can be accomplished. They can also be used for relative positioning, and in most cases the accuracy of relative positioning is much higher than that of the absolute positioning done by the same technique.

With space techniques, the requirement of mutual visibility between adjacent stations is no longer a necessity, and intercontinental connections can be obtained directly. In principle, one does not need the dense network used in conventional positioning. There is no arbitrary defined ellipsoid involved in the data processing, and the observables have nothing to do with the geoid. The coordinates determined are those in the three dimensional Cartesian system. For conventional reasons, some people do use a reference ellipsoid, but it is not arbitrarily defined. The center of the ellipsoid must coincide with that of the Cartesian system and the rotation axis with the Z axis. The equator of the ellipsoid lies in XY plane. In fact, in this case, the ellipsoidal coordinates are computed from the directly determined Cartesian coordinates. Unlike the conventional geodetic datum, the Cartesian system inherent in any space technique is a global system. To maintain this global system (so that it can be available at any time) to detect possible global plate motions, and to monitor the irregularity of earth rotation (including polar motion), a global network, with a few stations on each major plate, is highly desirable.

A Common Principle for Space Techniques. A common theoretical principle exists for these diverse space techniques in positioning, although the detailed observing methods and data processing procedures may be completely different. In fact, this common principle is similar to that of optical astronomic observations.

Two different kinds of vectors are involved in space techniques: one is called "source vector," which embodies the position of the observing source; the other is "station vector," which represents the position of the station. These two kinds of vectors are expressed in two different reference coordinate systems, one being a celestial system and the other, a terrestrial system. The observables in space techniques may be time delay, Doppler counts, or other measured quantities. But the observables always provide a relationship between source and station vectors. In fact, all kinds of observables directly or indirectly correspond to the scalar product of source vector $\mathbf{O}$ and station vector $\mathbf{B}$. For instance, in astronomic observations, the cosine of zenith distance equals the scalar product of source and station vectors. In VLBI observation, the time delay divided by baseline length is no more than the scalar product of source vector and the baseline vector. In SLR the directly observed quantity is the distance $\mathbf{d}$ from station to satellite, so that if $\mathbf{B}$ is the geocenter vector of the station and $\mathbf{O}$ is the geocenter vector of the satellite, it is easy to prove that the following relationship exists for the scalar product $\mathbf{O} \cdot \mathbf{B}$

$$\mathbf{O} \cdot \mathbf{B} = (\mathbf{O}^2 + \mathbf{B}^2 - \mathbf{d}^2)/2$$

Therefore, the observable $\mathbf{d}$ is implicitly equivalent to the scalar product. Doppler counts represent differential distance, so that the vector relationships are similar to those of SLR.

Since $\mathbf{O}$ and $\mathbf{B}$ are expressed in different reference systems, to determine the scalar product, one must either convert $\mathbf{B}$ into celestial system or equivalently, bring $\mathbf{O}$ into terrestrial system. This can be accomplished through appropriate rotation matrix multiplication, where the matrices are astronomic parameters derived from information on precession, nutation, and Earth rotation. The scalar product (or the observables) link the three quantities $\mathbf{O}$, $\mathbf{B}$, and astronomic parameters together. Once we know any two of the three, the third one can be calculated, e.g., if $\mathbf{O}$ and $\mathbf{B}$ are given, one can determine the Earth rotation parameters, the precession, and/or nutation. By knowing $\mathbf{B}$ and the astronomic parameters, one can solve for $\mathbf{O}$. These are the objectives of astronomy and satellite dynamics. For geodetic positioning, one usually uses known $\mathbf{O}$ and astronomic parameters to determine $\mathbf{B}$.

In terms of the source vector, $\mathbf{O}$, some observed objects, such as the extra-galactic radio sources, have no motion at all. Thus the position of the object is fixed in the celestial system. Methods that make use of such sources are termed *geometrical*. If the observed sources are stars, which move slowly and their proper motions are known, then the position of the source at any observing epoch can be computed, in which case the method is called *kinematic*, although it is sometimes referred to, incorrectly, as geometrical. When the source is fast moving, such as an artificial satellite, and one knows only its approximate position at the observing epoch and the dynamical factors that cause the motion (such as perturbation forces), then the accurate position of the source $\mathbf{O}$ must be solved for simultaneously with $\mathbf{B}$. This method is referred as *dynamical*. If one uses only the approximate position of an object (artificial satellite, the Moon, or another planetary body) as given in an ephemeris, or if the observations at various stations are carried out simultaneously, then the method is also geometrical in nature.

The common sources of error for positioning by space techniques include:

1. Errors in **O**, for geometrical or kinematical methods, are the errors in the source catalogue or in the ephemeris; for the dynamical method, it is the error in the force model (perturbation).
2. Errors in astronomic parameters, including errors in precession, nutation, Earth rotation parameters, and other astronomic constants.
3. Errors in observables, including instrumentation errors (including frequency standard) and propagation errors.

Reference Systems of Different Techniques. Depending upon the forms of observables and **O**, the terrestrial system inherent in each technique may be different (see Table 1). If the observable is an angle or something equivalent, then no matter what form **O** takes, the calculated **B** and the corresponding terrestrial system must be two-dimensional. The techniques for which this applies include astronomic methods and optical (photographic) observations of satellites.

The other kind of observables is distance or distancelike, so that the determined **B** is three-dimensional. In this case, if **O** is also three-dimensional, the terrestrial system and **B** is truly three-dimensional, such as occurs for SLR, GPS, and Doppler positioning; if, however, **O** is two-dimensional as for VLBI source positioning, then the determined **B** is quasi three-dimensional. That is, only the relative 3-D coordinates can be obtained, and the center of the reference system is indeterminable.

In addition, the form of the terrestrial reference system still depends on the cardinal line of the observables. When the cardinal line is related to the plumb line, the inferred terrestrial system is astronomic, with coordinates given in astronomic latitude and longitude, which are affected by the vertical deflections. If the cardinal line has nothing to do with the plumb line, the terrestrial system is geodetic. Here the cardinal line means the direction of **B**. Table 1 summarizes these various concepts.

At present, it appears that VLBI is the most accurate (and most expensive) technique. But by this technique alone no geocenter coordinate system can be obtained. Satellite techniques are the only way to construct a geocentric coordinate system. At present, the best available accuracy for absolute positioning is at the centimeter level. At this accuracy, the global positioning system (GPS) is the most effective technique now available.

Difference between Various Terrestrial Systems. The X and Y axes of the geocentric coordinate system lie in the equatorial plane, and the Z axis is perpendicular to this plane. The Z axis is usually taken to be the maximum inertial axis or figure axis, which is quite close to the rotation axis of the Earth. Because of errors in determining polar motion, the Z axes determined by different techniques, and even by the same technique but for different observing data, are different. In addition, different techniques or even the same techniques by different data centers may entail different models and constants so that source errors also may not be the same. The inherent terrestrial systems will be differently affected. The differences between various systems can be expressed by seven transformation parameters: three translation parameters, three rotation parameters, and one scaling factor. Once these seven parameters are known, the coordinates of a point in one system can be transformed into another. By application of the appropriate transfor-

TABLE 1. Geodetic Reference Systems

Technique	Observing Source **O**	Station Position **B**	Observables	Terrestrial System
Astronomy (or optical observations of satellite)	Star (and satellite), 2-D coordinates in unit celestial sphere	Plumb line, 2-D spherical coordinates of the station	Angle	2-D astronomic system. No height information can be obtained
VLBI	Extragalactic radio source, 2-D coordinates in unit celestial sphere	Base line vector, relative 3-D Cartesian coordinates of two stations	Time delay, equivalent to distance	Quasi 3-D Cartesian coordinate system, center of the system is arbitrary
SLR, GPS, Doppler positioning	Satellite, Geocenter position of the satellite, 3-D Cartesian coordinates	Geocenter position of the station, 3-D geocenter Cartesian coordinates	Distance or equivalence	3-D geocenter Cartesian coordinate system

mation two different systems can be linked together. Clearly, however, it is preferable to have a combined single system that links all the existing terrestrial reference systems together. Such a system has been created and is termed the Conventional Terrestrial Reference System (CTRS). To tie all systems to CTRS, at least three colocation stations are needed, where colocation means that instruments of two or more techniques occupy the same station.

Temporal Variation of Positions

Geometrical geodesy is closely related to geodynamics. With measurement accuracies of a centimeter or better, the relationship becomes even closer. The positions of points on Earth's surface vary with time because of motions in the Earth's interior. If not corrected in the data processing, geodynamical motions will degrade the accuracy of positioning. On the other hand, information about temporal variations is contained in the observing data. This information may be recovered by proper data analyzing techniques and used in geodynamic studies.

Only contemporary movements (not the historical movements in the past millions of years) are of interest to geodesy. There are different kinds of crustal movements. Some are continuous, both in time and space, others are episodic. Some are secular (linear, slow motion, creeping), others are periodic or quasi-periodic. The periods range from seconds to centuries. Geometrical geodesy is concerned mostly with secular motions and some of the periodic motions. The most important motions include: plate tectonic motion, intraplate deformation, postglacial rebound, tides (both Earth tides and ocean tides), and wobble of the Earth's spin axis.

The magnitude of the wobble is the largest, about 10 m. But it affects the rotation of the Earth as a whole. Thus, only the absolute orientation of the network is affected, whereas the shape of the net unchanged. It does, however, have a serious effect on centimeter accuracy absolute positioning. If the wobble is not determined precisely, it introduces errors that will reduce positioning accuracy.

Some of the motions (such as tides and wobble) are better known than others (e.g., plate motion). Usually, the knowledge of the former is used to correct the observations or positions, whereas the information about the latter is supposed to be extracted from geodetic positioning. In either case, the time epoch at which the observation or position is obtained must be known.

Various data processing techniques can be used to separate different kinds of movements. The periodic components can be extracted by the spectral analysis method. The local and regional deformation (such as intraplate deformation) can be investigated by baseline length variation or changes in configuration. The study of the wobble of the Earth's spin axis and global plate motion must be closely related to the maintenance of the global reference system. Generally speaking, the magnitude of all these movements and deformations (except for the wobble and Earth tide) are quite small, of the order of centimeters. Only in recent years have their determinations become meaningful.

For geodetic positioning accurate to centimeters, as required for studies of Earth deformation, one must distinguish the "mean position" of a point from its "instantaneous" one. As far as periodic motions are concerned, the former refers to the position with the (short) periodic movements removed. Geodetic positioning should give this mean position as a final result. But the user must include the periodic terms to determine the true position at any instant in time. For secular motions, the quantity of interest is the rate of motion between points. Once this "proper motion" is determined, it can be corrected to obtain the instantaneous position.

SHENG-YUAN ZHU

References

Bomford, G., 1980, *Geodesy*, 4th ed. Oxford, Great Britain: Clarendon Press.

Groten, E., 1979, *Geodesy and the Earth's Gravity Field.* Bonn: Dümmler Verlag.

Marussi, A., 1985, *Intrinsic Geodesy.* Berlin, Heidelberg: Springer-Verlag.

Mueller, I. I., 1969, *Spherical and Practical Astronomy as Applied to Geodesy.* New York: Ungar.

Vanicek, P., and E. J. Krakiwsky, 1982, *Geodesy: The Concepts.* Amsterdam, New York, Oxford: North-Holland.

Cross-references: *Crustal Movements and Tectonic Deformation; Doppler Positioning: Satellite; Earth Orientation; Earth Tides; Equilibrium Figure of the Earth; Figure of the Earth; Geodesy: Historical Introduction; Geodesy: Physical; Geodesy: Satellite; Geodetic Ground Positioning; Geodetic Leveling; Geodetic Networks and Control Surveys; Geodetic Reference Systems; Global Positioning System (GPS); Map Projections in Geodesy; Planetary Geodesy; Satellite Altimetry; Satellite Laser Positioning; Very-Long-Baseline Interferometry (VLBI).*

GEODESY: HISTORICAL INTRODUCTION

He who understands the earth is a wise man, and he who understands the heavens is a sage. Knowledge is derived from the straight-line shadow, and that is derived from the right-angled joint. The combination of the right angle with numbers is what guides and rules the ten thousand things. (Needham, 1959, vol. 3, p. 23)

A tomb relief of the second century shows the Chinese deified king Fu-Hsi and his consort flaunting a carpenter's square and a quipu as their royal insignia "to govern all within the four seas." (Needham, 1954, vol. 1, p. 164, fig. 28)

The ancient Chinese thus described the fundamental role of astronomy and geodesy in understanding the universe and accorded to geodesy the role of establishing order on Earth. The word "geodesy" is Greek and means "partitioning the Earth surface." The carpenter's square and the quipu symbolize the basic geodetic tasks of localization and mensuration, of settling the queries "where?," "in what direction?," and "how far?"

Geodesy today is but one of many geosciences, but as the oldest and basic one it is the matrix from which the others split off in the course of time to develop their own ways. The history of geodesy thus suggests the picture of a variegated tree of knowledge with its roots in ancient time, nourished throughout its growth by the same urge to know and to know more. It is the story of looking for answers and finding new problems, searching for their solutions and running into even more new problems, all the while enlarging the scope of the subject matter to be absorbed.

Earliest Geodesy

Ancient artifacts and texts attest to the double aspect of geodesy: the wonder about the universe and the practical methods of surveying. Divine powers establish order in this mysterious and awesome world (Yü, the Great, in Chinese legends; the Almighty in the Bible, especially in "Job") and teach the mortals how to keep order and how to find their way in this world (a goddess is depicted teaching Sumerian and Egyptian kings how to find a north-south line, to lay out a base line, and to fix a cornerstone; Homer tells of divine guidance in navigation; Chinese immortals construct maps for the safety of travelers through unknown and terrifying countrysides).

The first professional geodesists mentioned are the Egyptian harpedonapts (rope stretchers and rope knotters, that is, surveyors and recorders). Their tools obviously were knotted ropes (familiar to us as nautical knots for a ship's speed per hour). The carpenter's square is mentioned in Chinese legends as a basic, multipurpose tool. The plumb line and the water level were generally known as well as the 3-4-5 right triangle. Ancient India knew and used many more "Pythagorean triangles," and Babylonia knew a rule for creating these number triples long before Pythagoras. Water depths were tested with a pole, also by the lead and line technique and by estimates of river sediments (Herodotus). Differential leveling made the ancient nilometers possible. For long distances, bematists, or royal mensors, counted their even steps. A day's journey on land and a day's sail (an etmal) at sea were distance units. Uncertain sailing distances were corrected by considering supporting or opposing winds, shapes of shore lines, and reliability of reporters. Directions were obtained from the sun and stars and the winds.

The oldest map we know of so far dates from the third millenium B.C.E. It depicts an area in the ancient empire of Ebla (near Aleppo) and even gives the ownership of fields. Other ancient maps range from local building plans to city layouts to imaginative world maps with a specific culture's hub as the known reference at the center: Babylon, or Jerusalem, Mecca, Khun-Lun mountains (north of Tibet), Meru or the Kailas mountains (Trans-Himalaya), or "the navel of the earth" at Delphi, Greece. A complete cadaster was maintained in ancient Egypt since at least 1300 B.C.E. for purposes of a fair tax assessment, and it was periodically adjusted after the annual inundations of the Nile.

In lieu of maps for sailors, the accumulated experiences of seafarers were collected in pilot books. The oldest extant pilot book, the *Periplous of Scylax* dates from the mid fourth century B.C.E. Ocean currents were well known and recorded as useful water lanes in the Mediterranean and in the Pacific (e.g., the Kuroshio current). Even water level differences at the ends of narrow channels (Homer's Scylla and Charybdis story) and at the entrance to the Mediterranean were well known, as were water discolorations in some regions to be watched as navigational aids.

The Spherical Earth

Speculations about the shape of the Earth took diverse fanciful forms from various myths to geometric solids of the Greeks. The Pythagorean sphere was only one of these speculations. It was trustfully accepted as the Earth model in the Western world, because of its aesthetic appeal as the most perfect geometric shape befitting the Pythagorean mystical notion of a perfect world. Several centuries elapsed before the emerging Greek scientific bent asked for a proof of this shape of the Earth, thus embarking on the scientific adventure of an open-ended search for the "Figure of the Earth."

Developments in China were different (Needham, 1959). Intense preoccupation with the revolution of the stars around a motionless pole, exemplifying the awesome role of the emperor and his officials revolving around him with the obligation to carry out the will of Heaven for cosmic order and unity, led to an early realization of partly and wholly invisible paths of circumpolar stars around and beneath the Earth. This preoccupation led further to the concept of concentric celestial and terrestrial spheres, "like a hen's egg and its yolk." It did not, however, lead to the study and use of a spherical Earth model for geodetic purposes. It had not done that for the Greek Anaximander either who was the first to recognize the full celestial sphere, yet retained a primitive cylindrical shape of the Earth with the inhabited world on its flat top. But the Chinese dichotomy between heaven and Earth was deliberate, distinguishing between different realms of achieving cosmic order, each in its own best way; those involved in the affairs of the Earth were specifically enjoined by Confucius to give priority to practical things of use to the government.

Here is a significant parting of geodetic ways between East and West. The driving purpose of the Chinese, confronted with a vast populated region in constant strife, was to collect detailed information about the countries to strengthen administrative and military power; to this end, detailed mapping in a simple rectangular grid on an inherently flat Earth was the most useful and efficient means, adequate for the next several centuries. The driving purpose of the Greeks, a seafaring people watching the effects of the Earth's curvature in daily navigation, was scientific curiosity for the Earth itself, leading to ever refined questions about its size and shape (Fischer, 1975) and the confounding problems of its mapping on flat paper.

The persistent Greek fascination with the generally accepted Pythagorean sphere evoked several attempts to prove this shape, among them several by Aristotle using astronomic or gravity arguments, and one by Archimedes, the "father of hydrostatics," using the concept of an ocean-covered Earth in equilibrium (the geodetic sea level or the "standard ocean" of later times). The curiosity about the size of the sphere brought some early estimates of the length of its circumference: 400,000 stadia, possibly by the Pythagorean disciple Archytas of Tarent, and 300,000 stadia, possibly by Dikaiarch (fourth century B.C.E). Eratosthenes (third century B.C.E.) is usually considered the "father of scientific geodesy" because he used measurements along the longest available, nearly meridional arc from Alexandria to Syene (now Aswan), in combination with the corresponding celestial arc measured with the sun dial at summer solstice. His result of 250,000 stadia was an improvement over his predecessors by an order of magnitude in accuracy. His method stayed essentially in use throughout the centuries into the modern satellite age.

The significance of his work, however, went far beyond computing the circumference by a superior method. This was only the first step toward the ultimate goal of constructing a map of the known world. The symbiosis of geography and geodesy was accumulating a mass of descriptive or quantified information from travel reports. Dikaiarch began to organize it with respect to a basic east-west reference line through known places on the Mediterranean coast. Eratosthenes' long arc added a north-south reference line, intersecting the first at Rhodes. His pairing the terrestrial and astronomical arcs established a conversion factor between terrestrially and astronomically determined distances, invaluable to utilize new information reported in either form. There was still the problem of the many types of stadia in use at the same time, comparable to our own jungle of feet, statute or nautical miles, meters, even different systems of geodetic coordinates. Hipparchus (190–125 B.C.E.) proposed therefore to use astronomical latitudes and longitudes also for terrestrial positions. This is the first uniform world coordinate system, an ancient precursor to our modern geodetic world systems.

The framework of Eratosthenes' map included a number of auxiliary parallels and meridians through important places. It was followed by a grid of straight-line meridians and parallels devised by Marinus of Tyre (first and second centuries C.E.) and then by Ptolemy's (127–151 C.E.) attempts at map projections. The problems of mapping a spherical surface onto a flat sheet of paper led to the development of cartography as a separate discipline.

Discounting a few isolated attempts at imitating Eratosthenes' arc measurement nothing of geodetic importance happened until the seventeenth century.

The Ellipsoidal Earth

In the highly intellectual ambience of seventeenth century Europe resplendent with famous scientists, scientific societies and academies, international scientific communication and cooperation, it was the Académie Royale des Sciences in Paris that became the hub for the next spectacular developments in geodetic history. The Académie was founded in 1666 for the purpose of improving maps and sailing charts, meaning specifically constructing a map of France and finding a method to determine the longitude at sea. A magnificent observatory was built just outside the lights and noise of Paris, and a super-accurate geodetic arc was planned to originate from there as the backbone of the map of France. The vast financial resources of Louis XIV and his ambition to outshine all of Europe backed this enterprise, attracting eminent scientists from all over and assembling the newest and best equipment available. Eratosthenes' approach of almost 2000 years earlier to an analogous project was the obvious model, and Jean Picard was asked to follow it in concept but update the procedures with the most modern techniques. Among these was triangulation, which had been developed from trigonometry in the sixteenth century by Gemma Frisius (Haasbroek, 1968), applied in limited form by Tycho Brahe, and shown by Willebrord Snellius to be usable for measuring a long meridional arc and extracting from that and astronomical latitudes at its endpoints the length of a degree in linear units à la Eratosthenes.

Therefore, he called himself modestly Eratosthenes Batavius.

Picard's arc (1669–1671) to determine the scale of the future map and the size of the Earth stretched from Malvoisine near Paris to Sourdon near Amiens and was carried out by triangulation, the newest astronomical techniques, and the utmost care for accuracy. The resulting size of the Earth was used by Isaac Newton in his contemporary studies of universal attraction.

The next step was the continuation of Picard's arc south to the Pyrenees by Dominique and Jaques Cassini, and north to Dunkerque by La Hire, to establish the standard meridian from border to border. The elation over the new capabilities, however, met with an upsetting puzzling paradox: the length of a degree calculated from these arcs seemed to decrease northward, questioning the adopted scale of the map as well as the age-old spherical model of the Earth; it seemed pointed toward the pole, egg-shaped. Such inference contradicted, however, Newton's deduction from his theory of gravitation that a rotating Earth would be flattened at the poles and bulged at the equator, grapefruit-shaped, which was further corroborated by pendulum observations of Jean Richer and others indicating less gravity toward the equator. While the Académie judged that the obtainable precision of surveys within France would produce a good mean value for a map scale but may not permit a definite decision about the shape of the Earth, Cassini claimed to see enough of a pattern in his work to draw up tables of systematic degree enlargements toward south.

Thus the famous international dispute between the French Earth elongators and the British Earth flatteners shaped up, to be resolved by the Académie's magnificent underwriting of the two expeditions to Peru (now Ecuador) in 1735 and to Lapland in 1736, far enough apart in latitude to decide the decrease or increase of a degree length toward north. The British won this one, to the delight of Voltaire who praised Maupertuis, the leader of the Lapland expedition, for having "flattened the earth and the Cassinis" (Todhunter, 1962), and scorned La Condamine, leader of the Peru expedition: "Vous avez trouvé par de long ennui ce que Newton trouva sans sortir de chez lui" [You have found by prolonged toil what Newton found without leaving his house]. (Newman, 1956, p. 268)

The acceptance of an oblate ellipsoid of revolution as the Earth model prompted the question of its size as well as its shape, i.e., two parameters versus one parameter. These parameters are either the semimajor axis a and the semiminor axis b, or more frequently, the semimajor axis a and the flattening $f = (a - b)/a$. Thus it takes at least two equations or two arcs to determine them, which started a frequently played international game of pairing arcs to calculate one's own value for these parameters, with as many different results. These differences were at first explained away by the faith that increasing accuracy capabilities would make them disappear so that the various arcs would fit smoothly onto one ellipsoid; but some astute observers such as Bouguer in Peru and Boscovich in Italy began to recognize an effect of topography on the survey measurements. Such possibility spread a gloom of inescapable uncertainty over all surveys so that Delambre (1822) suggested that one should be content with adopting approximate round numbers. Nonetheless, he and Méchain managed to get permission for another superaccurate arc measurement from Dunkerque to Barcelona as input into the establishment of a new international length unit, the meter, in lieu of the maze of current units. In 1791 the French National Assembly, on advice of the Académie, had proclaimed the meter as the ten millionth part of the meridional Earth quadrant through Paris. After the calculated relation of the toise to the meter was accepted into law in 1799, the length of the meter was fortunately represented by a prototype in the Archives and was thus removed from the uncertainties of remeasurements. Its historical connection with the figure of the Earth was then completely abandoned in the modern International System of Units (SI).

The intense geodetic activities in the seventeenth and eighteenth centuries give a seesaw picture of elated confidence in new sophisticated measuring capabilities (triangulation, astronomic techniques, accuracy precautions), puzzlement and irritation over the unexpected confusing results of their application (not a spherical Earth?), turning these into positive insights and new tasks (ellipsoid with two unknown parameters), confusion about differing results (unable to make them fit together), interference from the outside (Newton), admission that the outside was right (oblate versus prolate ellipsoid), discovery of the effect of topography on survey results, despondency over this effect as invalidating surveys, etc.

With the hindsight of later history one can discern here the kernels of a much broadened scope of geodesy arising from ostensible failures. The intricacies of triangulation and the inconsistencies of arc determinations led to theories of adjustment, the least squares method developed independently by Legendre and Gauss. The ellipsoidal Earth model sparked an ellipsoidal geometry and trigonometry to compute distances and positions on an ellipsoidal surface. The outside evidence for the shape of the Earth from the theory of gravitation and practical pendulum observations led to the absorption of this part of physics into a new branch of dynamical or physical geodesy, launched by Claude A. Clairaut's theorem in 1743, which links the flattening to latitudinal gravity variations. And the effect of topography on surveys was studied systematically, leading to the concept of an irregular Earth surface. There followed the consideration of underground mass distributions and theories of isostasy and of the interior of the Earth. Questions of a geophysical and geodynamical nature became part of the picture.

The Geoid

While the least squares adjustment procedures made peace with the unavoidable observational errors by distributing them, Gauss (1828, pp. 49–50) realized that large residuals beyond the allowance for observational errors must be caused by real features of the Earth, namely the irregular mass distributions all over, not just local ones. He wrote:

> ". . . the surface of the earth in a mathematical sense is nothing else but that surface which everywhere intersects the direction of gravity at right angles, and of which the surface of the ocean is a part. . . . The ideal ellipsoid of revolution would be the one where the computed directions of the normals best agreed with the astronomic observations. . . ." [in later terminology: the best fitting ellipsoid would be the one where the deflections of the vertical are minimized.]

Bessel (1837, pp. 269–270) elaborated further:

> ". . . All attractions together with the centrifugal force produce that surface to which the geodetic work refers;. . . it intersects the directions of these forces at right angles. . . . This condition, however, merely determines any of the surfaces which could be covered by a fluid in equilibrium; one must still decide which of these surfaces should be the mathematical surface of the earth. The choice . . . would be arbitrary if the earth were only a rigid body without an ocean. Since this, however, exists it is appropriate to adopt that one as the surface of the earth of which the ocean is a part. Imagine the earth covered by a net of channels connected with the ocean and filled by it, then the surface of the calm waters in them would coincide with the mathematical surface of the earth. . . ."

Bessel's theoretical discussion of the role of the ocean became very real in the early 1960s when geodesy, in anticipation of NASA's moon walk in project Apollo, reached out to the Moon (under the modified name as "selenodesy") and established there a first-order control network. There is no ocean and an arbitrary decision had to be made, indeed. The small crater Mösting A was chosen as the datum point because it was the most clearly defined feature.

Helmert (1884) systematized Gauss's and Bessel's ideas into a complete theory of the geopotential and included the study of its level (equipotential) surfaces into the realm of geodesy. The special level surface that contains the calm surface of the ocean (i.e., without winds, tides, currents) is called the geoid (coined by Listing in 1872); the calm surface of ponds and lakes would usually belong to other level surfaces. Helmert defined geodesy as "the science of surveying and mapping the earth's surface." He pointed out that the combined effects of attraction and rotation constitute a rule of formation for the shape of the Earth as a whole, which makes it possible to gain some knowledge of it from relatively small known parts. While Helmert's purpose was to fashion a consistent theoretical tool for the study of the Earth's surface, this tool, geopotential theory, proved much more powerful in the future developments than he had envisioned.

The broad enlargement of the scope of geodesy formed two distinct branches of geodesy, pursuing the same goal in very different approaches: the astro-geodetic or geometric geodesy handling the new insights in the traditional geometric way of terrestrial and astronomic measurements, versus the dynamic or physical geodesy concentrating on the Earth's gravity field.

The traditional way pursued the search for a "best fitting" world ellipsoid after Gauss's recipe by adopting first a preliminary ellipsoid with a starting point on it (together called a *geodetic datum*) in order to compute on it a triangulation network for a coherent accessible area, then pairing geodetic and astronomic positions where available to form "deflections of the vertical" (angles between the directions of the ellipsoid normals and the astronomical vertical), and then minimizing these by modifying the auxiliary geodetic datum in a least squares solution. In the process, two important insights were gained:

1. the regional or national practical needs for geodetic services could not wait for the elusive unique world datum to be used as a computational reference surface but had to be content with a more or less well-chosen auxiliary datum as a national datum. Thus, the Bessel 1841 ellipsoid is employed in China and Japan, the Clarke 1866 ellipsoid in North and Central America, the International 1924 ellipsoid (another unsuccessful candidate for a world ellipsoid) in Europe, etc.
2. The deflections of the vertical can be seen as the slopes of the irregular geoidal surface against the reference ellipsoid and thus be integrated to give the separation between these two surfaces (*geoidal heights*).

Such integration of the innocuously small deflections over a long distance can expose large, non-negligible geoidal separation of several hundred meters, as was shown first in 1939 on an east-west arc in the USSR, and then in the middle 1950s along the famous 100° meridional arcs that extended the European Datum throughout Africa, and the Northamerican Datum into Chile. From then on the distorting effect of such separations on the geodetic survey had to be studied systematically and corrected for. Contoured geoidal maps were then drawn up wherever possible, exposing the hidden three-dimensional content of the geodetic holdings, which had been considered to be two-dimensional up to then.

A specific bonus of interdisciplinary significance appeared when the first geoidal map of North America as a whole was constructed in 1957. To the delight of glaciologists one could discern from the contours a depression of the Hudson Bay region, caused by the Pleistocene ice load. Contour lines coincided with known hinge lines of the uplift and predicted where one might find their continuation (Fischer, 1959). This was one unexpected example of an interfacing between geodesy and another geoscience, so widespread and important today.

Conceptually, an important milestone was the distinction between the eighteenth century frantic and emotional search for the unique world ellipsoid seen as the focal concern of geodesy and the nineteenth century realization that the ellipsoid was a mere reference surface. The equatorial bulge deduced by Newton would appear as a bulge on a geoid map referred to a sphere, but not if referred to a more closely fitting ellipsoid. This fact played a role in a much later interdisciplinary paradox: mean sea level along Northamerica's north-south coasts seemed to rise toward south when measured by oceanographic techniques but not by geodetic leveling (Sverdrup et al. 1942). It could be established that the oceanographic Earth model, the "standard ocean" is a spherical concept (Fischer, 1977, Fig. 4), and thus all oceanographic meridional profiles exhibit the equatorial bulge.

The three-dimensional character of geoid maps made them immensely useful to visualize the irregular shape of the geoid in specific areas, to judge the suitability of adopted geodetic datums and suggest better fitting ones, and to devise another stab at a global solution by a new three-dimensional approach of minimizing geoidal heights. Among these pre-satellite world datums was the Vanguard Datum in 1959, the first operational world datum used to establish consistent coordinates of tracking stations for the Vanguard Satellite Project. The Tentative World Datum in 1959 incorporated gravity information (as explained below), and its refinement, the Mercury Datum of 1960, employed in NASA's Manned Space Flight Programs and several other government projects, made use of the first important geodetic return from the Vanguard satellite: a significant change in the value of the ellipsoidal flattening.

Physical or dynamic or gravimetric geodesy focused on Helmert's theory of the Earth's gravity field and its system of level surfaces. In practical application this meant determining the shape of the geoid from gravity observations (i.e., the deviation of the geoid from a reference ellipsoid). Stokes's Theorem of 1849 gave the theoretical basis for the shape of the geoid, provided geoid observations were available all over the world, had been made on the geoid or been referred to it, and there were no topographic masses outside the geoid. Since these conditions do not exist in the real world, much exertion went into devising corrections, assumptions, hypotheses, etc., soliciting geophysical and geodynamical assistance. The great advantage of the gravimetric over the geometric method lies in the global character of the force of gravity versus the local nature of triangulation; its drawback is its vulnerability caused by the lack of global data coverage. In a combination such as in the Tentative World Datum, global orientation is given by the one method (the gravimetrically derived Columbus Geoid) and distances or scale by the other.

Satellite geodesy covers both drawbacks. In its dynamic mode, it provides the long-wavelength features of the gravity field to fill in for uncertain distant regions, and in its geometric mode, it uses the satellite as another point in a long-distance network and thus produces global distances to connect conventional triangulation nets. Vice versa, a better knowledge of the size and shape of the Earth and its gravity field will improve the orbit calculations.

Marine Geodesy

The study of the geoid, approximately identified as the undisturbed sea level surface, could not be complete without considering the oceans, which comprise about 70% of the Earth's surface, and which since antiquity have played an integral part in the concept of the figure of the Earth. The apparent impossibility of making geodetic measurements in the open sea began to give way during the years of the early satellite age when unprecedented instrumentation such as electronic distance measurements aiding navigation and underwater techniques were developed.

In 1958, Maurice Ewing proposed the establishment of geodetic benchmarks on the ocean floor by transponders, the determination of their distances by underwater acoustic signals, and thus the formation of a trilateration network to be tied to the triangulation nets on land. In 1964, the First Marine Geodesy Symposium stressed the necessity of mapping the oceans as a prerequisite for utilizing the vast economic resources of the oceans, establishing international and territorial boundaries, and assisting geophysical (seismic, magnetic, gravimetric, etc.) surveys. Just as on land, geodesy in the oceans serves to establish where what is, how to get there, and how to recover and identify the same location again. This applies, e.g., to boundaries of leased exploration and exploitation areas in the sea, boundaries that will stand up in a dispute at court. It also applies to the delineation of the territorial waters and the continental shelf under the Law of the Sea. Bathymetric surveys and bathymetric charts of the ocean floor are part of the geodetic task as well as assistance to navigation and ship positioning and to oceanographic monitoring of mean sea level and other oceanographic phenomena.

Multifaceted Challenges

The value of the Earth's flattening as derived from the Vanguard satellite mission finalized the Mercury Datum, which was officially adopted for all government projects for the next ten years. The new Earth model was also used to update Crommelin's classical 1911 determination of the parallax and distance of the Moon. At the same time new radar capabilities produced direct measurements of the Moon's distance, and dynamically derived relationships in the Earth-Moon system corroborated the results. Also, a first-order geodetic control network was established on the Moon's surface, with the crater Mösting A as the datum point. Lunar maps followed.

The age-old goal of measuring the Earth's surface had been achieved; still unknown regions present no conceptual problem today, since the available techniques can provide whatever data should be needed, to the extent of available funding. It seemed geodesy was reduced to the never-ending service of the practical land surveyor who benefited, however, from advanced tools. But satellite geodesy brought new life to the old science by opening new vistas and challenges and revamping geodetic methods in a variety of novel ways, as described elsewhere in this volume.

Helmert had introduced the notion of the level surfaces of the gravity field into geodesy as a means to accomplish the old goal, but satellite geodesy went beyond Helmert's vision in exploring the whole external gravity field as an objective in itself, including it in the concept of the figure of the Earth. In the process, new requirements of expertise had to be absorbed too, such as celestial mechanics covering the Moon and the planets, new elaborate instrumentation and their periodic updating, and electronic, high-speed computation. Unprecedented capabilities and precision make new kinds of inquiries possible to serve other geosciences with the typical geodetic question of location. Such services may apply to the prediction of earthquakes and tsunamis, pole variations, plate tectonics and local crustal changes, sea level monitoring, etc. The feat of extreme precision, however, rebounded to undermine the results, as has been experienced so often in geodetic history; if tectonic deformations change the location or direction of the reference, what do you compare with what? And just as in previous experiences, exasperation is changed into new challenge: the time-dependence of phenomena must be considered. The scope of geodesy and the concept of the figure of the Earth is enlarged again; it becomes a four-dimensional research object with a four-dimensional reference frame in inertial space.

IRENE K. FISCHER

References

Bessel, F. W., 1837, Ueber den Einfluss der Unregelmässigkeiten der Figur der Erde auf geodätische Arbeiten und ihre Vergleichung mit den astronomischen Bestimmungen, *Astron. Nachr.* **329,** 269–272.

Delambre, J. B. J., 1822, *Grandeur et Figure de la Terre,* ed. G. Bigourdan, 1912, Paris: Gauthier Villars, 400p.

Fischer, I., 1959, The impact of the ice age on the present form of the geoid, *Jour. Geophys. Research* **64,** 85–88.

Fischer, I., 1975, The figure of the earth—Changes in concepts, *Geophys. Surveys* **2,** 3–54.

Fischer, I., 1977, Mean sea level and the marine geoid—An analysis of concepts, *Marine Geodesy* **1,** 37–59.

Gauss, C. F., 1828, *Bestimmung des Breitenunterschiedes zwischen den Sternwarten von Goettingen und Altona.* Goettingen: Vandenhoeck und Ruprecht, 48–50.

Haasbroek, N. D., 1968, *Gemma Frisius, Tycho Brahe, and Snellius and Their Triangulations.* Delft: The Netherland Geodetic Commission, 119p.

Helmert, F. R., 1884, *Die mathematischen und physikalischen Theorieen der Höheren Geodäsie.* Leipzig: Teubner, 612p.

Needham, J., 1954, 1959, *Science and Civilization in China.* London: Cambridge University Press.

Newman, J. R., ed., 1956, *The World of Mathematics*, vol. 1, p. 268.

Sverdrup, H. A., M. W. Johnson, and R. H. Fleming, 1942, *The Oceans, Their Physics, Chemistry and General Biology.* New York: Prentice Hall, 1087p.

Todhunter, I., 1962, *A History of Mathematical Theories*, p. 195.

Cross-references: *Earth Orientation; Equilibrium Figure of the Earth; Figure of the Earth; Geodesy: Geometric; Geodesy: Physical; Geodesy: Satellite; Geodetic Reference Systems.*

GEODESY: PHYSICAL

Earth sciences are concerned with the investigation of the state and the state change of the entire Earth's body, governed by an interplay between various physical and chemical processes and described in a space-time frame. Geodesy, as one section of Earth sciences, is concerned with the determination of both the shape of the Earth's surface and the Earth's (external) gravity field. The goal of physical geodesy in particular is the determination of the Earth's gravity field.

Gravity–Gravitation

Gravity results from a superposition of two effects: (1) a gravitational effect, which is, according to Newton's law of gravitation, caused by the mass distribution within the Earth's surface (as well as the small atmospheric mass and the mass distribution of celestial bodies), and (2) a centrifugal effect, which is caused by the rotation of the Earth.

This overly simplified picture cannot describe the considerably more complex reality in detail. Strictly speaking there is neither rest on Earth, nor outside the Earth, nor inside the Earth; but fortunately most of these movements are either very regular and can be modeled very accurately, or the movements are so slow that they can be disregarded within a small time frame. In any case, we assume that those effects can be modeled properly; they will not be considered here. However, the solid state formulation of the problem presented here should be considered with these reservations in mind.

If we remove the rotational effect, which can be handled very accurately, from gravity, we are left with gravitation. We know that the source of gravitation is the mass (density) distribution within the surface of the Earth. Consequently, were we to know the mass distribution of the entire Earth's body, its gravity field would be known both inside and outside the surface of the Earth. Since we do not have direct access to that kind of complete source information, we are forced to measure its effect in terms of various gravity field quantities. All these quantities are derived from a "mother function," called gravity potential, by elementary operations.

The gravity potential (like any other potential) is customarily described by the shape of surfaces of constant potential, called equipotential surfaces. There are an infinite number of equipotential surfaces, which in general are not parallel to each other. Among all equipotential surfaces one distinguished surface is of considerable concern to Earth sciences in general and to geodesy in particular: the geoid. It is the surface of constant gravity potential at mean zero level and coincides with the open ocean surface, considered at rest and not affected by the attraction of celestial bodies like Sun and Moon. It is the physical/mathematical surface of the Earth and serves as zero level for physical/geodetic height measurement. Therefore, most of the efforts in physical geodesy are concentrated on the geoid determination with steadily increasing accuracy (Heiskanen and Moritz, 1967).

Gravity Field Data

Surface Data. The most important gravity field quantities measured on the Earth's surface are gravity potential differences, determined by levelling in combination with gravity measurement; the modulus of the gravity vector, which is the vertical change of the gravity potential, determined by absolute and relative gravity measurement; the direction of the gravity vector, represented by the two parameters, latitude and longitude, and determined by geodetic-astronomical observations; and the components of the second order gravity gradient tensor, determined by torsion balance measurements and partly by vertical differential gravity measurements. All the quantities listed are straightforward derivatives of the gravity potential.

Satellite Data. Satellites move within the Earth's gravity field; therefore a satellite's orbit is controlled by the gravity field. Consequently, surveying satellite orbits yield data related to the Earth's gravity field in a somewhat more complicated way compared with the surface data listed earlier.

Some active satellites are used to monitor their distance from the surface of the ocean which, in an idealized way, can be regarded as the geoid. Consequently, such kinds of altimeter satellites provide a stream of data containing both implicit gravity field information through the orbit of the satellite, and explicit gravity field information through scanning the open ocean's geometry. That kind of altimeter data have provided an extraordinarily detailed and accurate picture of the ocean geoid (Rapp, 1982a, 1982b).

Another active satellite mission is designed to measure relative velocities between two satellites that are separated by a few hundred kilometers and orbiting at low altitudes of about 200 km. In this setup the gravity field information is hidden in the relative velocity between the two satellites. Such satellite-to-satellite tracking (SST) is expected to yield high-frequency information about the global gravity field (Colombo, 1984).

Finally, a satellite gradiometer mission is being designed, which is essentially a microscale SST setup, packed into one single shield. The measured quantities are the elements of the second order gravitational gradient tensor with an estimated precision of 10^{-11} s^{-2}. Satellite gradiometry is expected to map the very fine structure of the global gravity field with a resolution of about 100 km wavelength and an accuracy of about 3–4 mGal (Paik, 1981a, 1981b).

Gravity Field Models

According to potential theory the exterior gravity field of a body such as the Earth can be determined from surface data only without any assumption or knowledge about the mass distribution in its interior. The mass is the source, the gravity field its effect. Source information is very difficult to obtain, effect information is easily accessible. Therefore, it is not astonishing that effect studies dominate.

Effect Models. In applied sciences it is customary to refer all data to a model. In physical geodesy that model is an ellipsoid of revolution equipped with a gravity field of simple structure because in first order approximation the shape of the Earth is such an ellipsoid and the Earth's gravity field is well approximated by that ellipsoidal model gravity field. The simplicity of such a model is represented by the fact that only four parameters are required to completely define that model. The model is chosen such that its deviation from the actual figure of the Earth and its gravity field is minimum. The following model parameters are usually used (Moritz, 1984): $GM = 3,986,005 \cdot 10^8$ m^3 s^{-2}, the product of the gravitational constant and the total mass of the Earth; $J_2 = 108,263 \cdot 10^{-8}$, a dimensionless dynamical form parameter in terms of a second degree spectral coefficient of the Earth's gravity field, which is closely related to the flattening of the ellipsoid of revolution; $a = 6,378,137$ m, the equatorial radius, controlling the size of the ellipsoid; and $\omega = 7,292,115 \cdot 10^{-11}$ rad s^{-1}, the ellipsoid's constant angular velocity in agreement with the Earth's mean angular velocity.

The above listed parameters define the so-called "Geodetic Reference System 1980" (Moritz, 1984); its relative accuracy is estimated to be of the order of 10^{-7}.

Such an ellipsoidal model is very simple and can only describe the global feature of the shape of the Earth and its gravity field. But much more elaborate models can and have been designed. They are generally called "Earth Models" and represent the gravity field of the Earth in terms of a set of spectral coefficients (spherical harmonic coefficients), which are derived from surface and satellite data. The most recent Earth Models have been calculated by Rapp (1981) (around 33,000 coefficients), by Wenzel (1985) (around 40,000 coefficients), and by Rapp and Cruz (1986) (around 130,000 coefficients). These models are fully adequate to represent the Earth's gravity field on a global scale with a resolution of about 100 to 200 km wavelength.

Source Models. The steadily strengthening link between geodesy and geophysics requires effect information to be traced back to source information. Therefore, source models are now coming in fashion. As a first approximation source model we could consider a density model having radially changing but laterally constant mass density (see Dziewonski et al., 1975). Such a model can be described by a very few parameters and can therefore be compared with an ellipsoidal effect model. Recently considerable attempts were made mainly by the geophysical community to derive a high resolution three-dimensional mass density pattern by analyzing the travel time of earthquake waves (Dziewonski, 1984; Woodhouse and Dziewonski, 1984). This source Earth Model can be best compared to a high resolution effect Earth Models such as Rapp's (1981). It is my opinion that the determination of a very detailed source Earth Model will increasingly gain importance in the future. It cannot, should not, and will not be derived from seismic data only but rather from a combination of seismic and geodetic gravity field data.

Data Filtering

The resolution of a predicted signal depends on the data sampling rate. In physical geodesy that signal is the Earth's gravity field, which literally derives from the mass distribution within the Earth. The low-frequency part of the gravity field is caused by lateral density inhomogeneities in the mantle and probably even in the core and the medium to high-frequency part is caused by the mass distribution in the Earth's crust between the complex surface of the Earth and the surface of the upper mantle. Filtering out high-frequency constituents from a signal facilitates its prediction. Since the gravity field sampling rate cannot be made arbitrarily high, filtering plays a fundamental role in practical gravity field prediction problems (Forsberg and Tscherning, 1981; Forsberg 1984).

In particular, it is the filtering of the effects of shallow masses, called topographic-isostatic data reduction, based on the shape of the Earth's surface, which is available in terms of a digital terrain model and an adopted isostatic principle: the shape of the boundary surface between the upper mantle and the crust is controlled by a mass balance principle. According to the Airy/Heiskanen isostatic model, which is believed to represent the complex reality best, a mass surplus above zero level (continental masses) is compensated by a mass deficiency in the upper mantle, the result of a crustal "root" into the mantle; conversely, a mass deficiency below zero level (ocean masses) is compensated by a mass surplus, which is produced by crustal thinning, the result of upward bowing of the mantle/crust boundary surface. The current best estimate for the depth of the so-called compensation level (i.e., mantle/crust boundary surface corresponding to zero continental mass/zero ocean mass) for the Airy/Heiskanen model is around 34 km. Due to the strength of the crust, however, the mass balance is not 1:1; the mantle/crust boundary surface does not exactly mirror the shape of the solid Earth's surface; it is rather a blurred, smoothed image of the solid Earth's surface (Sünkel, 1985, 1986).

For the sake of simplicity the density of topographic masses is often assumed constant; more realistic topographic-isostatic models take into account lateral density variations. In any case, subtracting the effect of those topographic-isostatic masses from the observed gravity field data corresponds to a low-pass filtering process: the high-frequency constituents of the gravity field signal are removed to a great extent and, the signal is smoothed. The degree of smoothing depends on the accuracy and resolution of the digital terrain model of the Earth's surface, on the consideration of known topographic density variations, and to a minor degree on the chosen compensation model. Besides the filtering of the high-frequency part of the gravity field signal, which is due to the density distribution within the crust, there is also a low-frequency part for which filtering is both customary and useful: the model data reduction, which refers all data to either the adopted source or effect model.

The so obtained gravity field residuals would be exactly zero if the models used for data reduction would exactly represent reality. As a matter of fact, this is not the case because

1. a digital terrain model is only an approximation to the real Earth's surface,
2. the crustal density is usually assumed constant and differs as such up to about 30% from actual crustal density,
3. the adopted mantle/crust boundary is usually implied by an isostatic model which is again an idealization of reality,
4. Earth models are derived from geodetic and other data that are not available worldwide; data gaps, partly poor data distribution, and data noise affect the quality of the derived Earth model, and
5. gravity field data contain measurement noise due to incontrollable physical phenomena.

Therefore, gravity field residuals will still contain some power over the entire frequency range; however, compared to totally unreduced observations, the gravity field residuals will be generally very smooth.

Data Processing

As we have seen before, a variety of data are available that can be considered samples of the gravity field. The entire data set is heterogeneous, and the data are affected by unavoidable measurement noise. The problem of determining the gravity field is one of "inverting" the sampling process, a problem for which there is no unique solution. The gravity field is so complex that theoretically an infinite number of parameters would be needed to completely describe it; conversely, the data are finite in number, which means the problem is highly underdetermined.

In physical geodesy a virtue is made of necessity: from the multitude of possible solutions the one is chosen which is "best" in terms of approximation error minimization. Such a procedure, which can handle all heterogeneous gravity field quantities in a consistent way, is called least-squares collocation (Moritz, 1980).

Denoting the vector of residual gravity field quantities by l, its data signal variance-covariance matrix by C_{ll}, the residual gravity field quantity to be predicted by s, and the cross-covariance vector between predicted signal s and residual gravity field data by C_{sl}, least-squares collocation provides signal prediction by

$$s = C_{sl} C_{ll}^{-1} l$$

and the estimation of the error variance σ_s^2 of s,

$$\sigma_s^2 = C_{ss} - C_{sl} C_{ll}^{-1} C_{ls}$$

in a consistent way. Two facts are important to stress:

1. least-squares collocation can handle heterogeneous data sets consistently, provided that the statistical properties of the residual gravity field are available in terms of a proper covariance function. This function can be estimated from the given residual gravity field quantities.
2. using this covariance function to derive all variances and covariances that are contained in C_{ll}, C_{sl}, and C_{ss}, least-squares collocation provides the best possible linear estimate of any residual gravity field quantity among all conceivable linear estimation techniques. These properties make collocation superior over all existing gravity field estimation methods.

Collocation requires the set-up and solution of a linear system of as many equations as data are available with a fully occupied mapping operator. Since computational time required for solution of this system of equations increases as the third power of the number of equations (i.e., number of data), it is obvious that not even supercomputers are powerful enough to process more than about 10^4 data in a reasonable time. Consequently, the sophisticated collocation procedure is restricted to the processing of local and perhaps some regional data sets. A consistent handling of all available gravity field data in a single global collocation solution is not possible at present and apparently will remain so in the foreseeable future (Moritz, 1980; Tscherning, 1985).

But there are alternatives: taking only homogeneous subsets of the entire data set into consideration, integral formulas can be designed that avoid the expensive numerical inversion process. Somewhat loosely speaking, integral formulas contain an analytical inverse in terms of the integral kernel. That integral kernel, in turn, is the Green's function for a boundary value problem of physical geodesy: the determination of the structure of the gravity field from a homogeneous set of boundary (i.e., surface) data only. In mathematical terms it is a "non-linear, oblique derivative free boundary value problem," one of the mathematically most demanding problems which, after several decades of research, is still far from being completely solved (Hörmander, 1976; Sansò, 1977; Moritz, 1980).

In practical applications both integral formulas and the collocation method are merged, taking advantage of the simplicity of integral formulas and the power of collocation to process a heterogeneous data set. The input to that linear system is residual gravity field data, the output is any other residual gravity field quantity, for example a residual potential, any component of its residual gradient vector, etc., at any specific point in the exterior space. However, the predominantly estimated quantity is the residual gravity potential either at zero level or along the Earth's surface.

To obtain the total gravity field quantity, we have to reverse the filtering process described before and add the topographic-isostatic and Earth Model effect to the estimated residual gravity field quantity. If the effect of the reference ellipsoid is not added, then the obtained quantity is referred to that field and is called a disturbance quantity like gravity disturbance or vertical deflection. It is in particular the disturbed gravity field that is needed to reduce various geodetic measurements to its pure geometric content. This procedure—data filtering by removing the effect of the topography and its isostatic compensation, and the effect of the long-wavelength Earth model; processing of the so obtained residuals yielding predicted gravity field quantities at arbitrary positions; and adding the effect of the topography and its isostatic compensation and the effect of the long-wavelength Earth model to the predicted residual gravity field quantity—has become known as the "remove-restore" technique. It is applied worldwide in physical geodesy.

Results and Prospects

The variation of the gravity field can be described in several ways: in the function domain; in the frequency domain; in terms of the potential and/or its vertical and horizontal derivatives (gravity disturbance or anomaly, vertical deflection); in terms of global, regional, or local information and any combination thereof. Detailed results can be found in several subsequent articles of this volume. Therefore, we give here only an idea of the magnitude of observed surface/gravity field quantities.

The value of the disturbing potential is found to be in the range between -1000 and $+800$ m^2 s^{-2}, which corresponds to about -100 to $+80$ m geoidal height. Gravity disturbances are generally below ± 100 mGals, although considerably larger values have been observed, particularly along trenches and fault zones. Vertical deflections are predominantly well below $\pm 30''$ with a very few exceptions even up to $\pm 1'$, again observed in tectonically active areas.

The prediction accuracy of gravity field quantities depends on various parameters, including the smoothness of the gravity field in the area of investigation, the distribution and accuracy of available gravity field data in that region, and the availability of detailed topographical and geological information.

Results obtained in gravity field test areas certainly do not represent the global average, but they are a guideline for what can be achieved today in a well-observed area. Austria is such an area; let me therefore give some results that have been obtained from a recent gravity field determination: the relative geoidal height can be determined with an accuracy of ± 5 cm over a distance of 100 km, the gravity disturbance (anomaly) with an accuracy of a few mGals, and the vertical deflection with an accuracy of better than $\pm 1''$, even in the mountainous part of Austria (Sünkel, 1983; Sünkel et al., 1987)

The efforts to achieve such good results are considerable. In the future manpower will be replaced by satellite power, and future gradiometer missions may probably make most of the classical surface observations obsolete, providing detailed gravity field information worldwide with an accuracy comparable to present state of the art solutions within weeks instead of decades.

HANS SÜNKEL

References

Colombo, O. L., 1984, The global mapping of gravity with two satellites, *Netherlands Geodetic Commission, Publications on Geodesy*, vol. 7, no. 3, 253p.

Dziewonski, A. M., 1984, Mapping the lower mantle: Determination of lateral heterogeneity in P velocity up to degree and order 6, *Jour. Geophys. Research* **89** (B7), 5929-5952.

Dziewonski, A. M., A. L. Hales, and E. R. Lapwood, 1975, Parametrically simple Earth models consistent with geophysical data, *Physics of the Earth and Planetary Interiors* **10,** 12-48.

Forsberg, R., 1984, A study of terrain reduction, density anomalies and geophysical inversion methods in gravity field modelling, *Ohio State University Dept. Geodetic Sci. and Surveying Rept. 355*, 129p.

Forsberg, R., and C. C. Tscherning, 1981, The use of height data in gravity field approximation by collocation, *Jour. Geophys. Research* **86,** pp. 7843-7854.

Heiskanen, W. A., and H. Moritz, 1967, *Physical Geodesy*, San Francisco: Freeman, 364p.

Hörmander, L., 1976, The boundary value problems of physical geodesy, *Arch. Rat. Anal.* **62,** 1-52.

Moritz, H., 1980, *Advanced Physical Geodesy*. Karlsruhe: Wichmann Verlag, 499p.

Moritz, H., 1984, Geodetic reference system 1980, in *The Geodesist's Handbook 1984*, Vol. 58, No. 3. Paris: International Association of Geodesy, 388-398.

Paik, H. J., 1981a, Superconducting tensor gravity gradiometer for satellite geodesy and inertial navigation, *Jour. Astron. Sci.* **29** (1), 1-18.

Paik, H. J., 1981b, A spaceborne superconducting gravity gradiometer for mapping the Earth's gravity field, 1981 Int. Geoscience and Remote Sensing Symposium, Washington, D.C., June.

Rapp, R. H., 1981, The Earth's gravity field to degree and order 180 using Seasat altimeter data, terrestrial gravity data, and other data, *Ohio State Univ. Dept. Geodetic Sci. and Surveying Rept. 322*, 53pp.

Rapp, R. H., 1982a, A global atlas of sea surface heights based on the adjusted Seasat altimeter data, *Ohio State Univ. Dept. Geodetic Sci. and Surveying Rept. 333*, 63p.

Rapp, R. H., 1982b, A summary of the results from the OSU analysis of Seasat altimeter data, *Ohio State Univ. Dept. Geodetic Sci. and Surveying Rep. 335*, 19p.

Rapp, R. H., and J. Y. Cruz, 1986, Spherical harmonic expansions of the Earth's gravitational potential to degree 360 using 30' mean anomalies, *Ohio State University Dept. Geodetic Science and Surveying Rept. 376*, 22p.

Sansò, F., 1977, The geodetic boundary value problem in gravity space, *Mem. Accad. Naz. Lincei* **14,** 39-97.

Sünkel, H., 1983, Geoidbestimmung, Berechnungen an der TU Graz, 2. Teil, in *Das Geoid in Österreich*. Graz: Österreichische Kommission für die Internationale Erdmessung, Neue Folge, Band III, pp. 125-143.

Sünkel, H., 1985, An isostatic earth model, *Ohio State Univ. Dept. Geodetic Sci. Surveying Rept. 367*, 53p.

Sünkel, H., 1986, Global topographic-isostatic models, in *Mathematical and Numerical Techniques in Physical Geodesy*. Heidelberg: Springer-Verlag, 417–462.

Sünkel, H., N. Bartelme, H. Fuchs, M. Hanafy, W.-D. Schuh, and M. Wieser, 1987, The gravity field in Austria, in *The Gravity Field in Austria*. Graz: Österreichische Kommission für die Internationale Erdmessung, Neue Folge, Band IV, pp. 47–75,

Tscherning, C. C., 1985, Local approximation of the gravity potential by least squares collocation, in ed. K.-P. Schwarz, *Proceedings of the Beijing International Summer School in Local Gravity Field Approximation*. Alberta: University of Calgary, pp. 277–362.

Wenzel, H.-G., 1985, Hochauflösende Kugelfunktionsmodelle für das Gravitationspotential der Erde. Wiss. Arbeiten, *Fachrichtung Vermessungswesen der Universität Hannover*.

Woodhouse, J. H., and A. M. Dziewonski, 1984, Mapping the upper mantle: Three-dimensional modelling of Earth structure by inversion of seismic waveforms, *Jour. Geophys. Research* **89** (B7), 5953–5986.

Cross-references: *Absolute Gravity Measurements; Density Distribution in the Earth; Earth Orientation; Earth's External Gravity Field; Earth Structure: Global; Earth Tides; Equilibrium Figure of the Earth; Figure of the Earth; Gravity Fields: Implications for Planetary Interiors; Gravity and Isostasy; Mantle Convection and Plumes; Planetary Geodesy; Spherical Harmonic Analysis; Very-Long-Baseline Interferometry (VLBI)*.

GEODESY: SATELLITE

Geometric satellite geodesy is in its infancy compared to the classical geodetic disciplines. As with other space technologies, it is barely three decades old, tracing its beginnings to the first artificial Earth satellite, *Sputnik I*, which was launched on 4 October 1957. Though satellite geodesy is still in its infancy, its impact is pervasive.

The Transit, or Navy Navigation Satellite System (see *Doppler Positioning: Satellite*), was a direct outgrowth of observations of *Sputnik I*. Transit's prime purpose was to provide position updates for the inertial navigation equipment aboard the Polaris submarines. However, it was quickly recognized that the system provided a powerful new tool for geodetic measurements. In the early 1960s, the U.S. Defense Mapping Agency sponsored the development of the AN/PRR-14 Geoceiver. The Geoceiver was a special Transit receiver specifically designed to collect data for geodetic purposes. The new NAVSTAR, or Global Positioning System (see *Global Positioning System (GPS)*), promises to dramatically expand the geodetic measurement capabilities of the Transit system.

Relationship to Other Disciplines

As in any discipline undergoing rapid development, geometric satellite geodesy could not have progressed so rapidly without the concurrent development of other disciplines and technologies. Deserving special mention, geophysical satellite geodesy also developed rapidly; and human understanding and description of the Earth's gravitational field improved quickly. Also significant was the rapid development of the modern digital computer, which makes an astonishing computational power available to all.

The rapid improvement in knowledge of the gravitational field of the Earth was part of the very same process by which geodetic positioning capabilities were improved. In fact, in the early years of the Transit system, a process described as "pulling oneself up by one's own bootstraps" was employed. Measurements collected to determine geodetic positions over several years were reprocessed holding the geodetic positions fixed in order to improve the determination of the gravity field, which resulted in improved satellite orbit descriptions, which resulted in improved geodetic positions, which resulted in an improved determination of the gravity field, etc.

The Satellite Systems

Several satellite systems have been implemented specifically for navigation and positioning purposes. The positioning techniques employed have been varied. For example, Wild BC-4 cameras were used to photograph the balloon satellite PAGEOS against a star background. Such a system, however, encounters significant difficulty in automated data processing and hence has little promise of widespread use. Satellite laser ranging (SLR) to satellites such as *LAGEOS* and *Starlette* also is used for geodetic positioning purposes. However, the complexity, cost, and lack of portability argue against widespread use, although mobile SLR equipment is now being built. Even though there are significant cost and technological disadvantages to SLR systems, it can achieve some technical objectives better than other competing systems. Thus, SLR has proven very useful in an accurate determination of the size of the Earth and its fundamental gravitational coefficient.

Other satellite systems employing various ranging and direction-finding techniques have been used in the past but, for various reasons, are no longer used. The U.S. Army's SECOR (sequential collation of range) is an example of competitive accuracy at lower cost from other systems leading to its demise.

The Transit System

The Transit, or Navy Navigation Satellite system, has clearly become the most successful system for geodetic positioning because of its relatively low cost, portability, and ease of use of the equipment. *Doppler Positioning: Satellite* contains a more detailed description of the Transit system. A brief overview of the Transit system is given here in order to contrast it with the new Global Positioning System (GPS).

The Transit system typically consists of five or six satellites in approximately 1100-kilometer polar orbits. They transmit at frequencies near 400 Mhz and 150 Mhz. The CW (continuous wave) broadcast allows a Transit receiver to beat the received carrier against a receiver-generated (local) frequency, which is close to the satellite-generated frequency. The beat or difference frequency is composed of a component due to the difference between the transmitted and local frequency and the Doppler frequency, which is a measure of the rate of change of the slant range between the satellite and the receiver. By integrating (counting) this beat frequency over some time interval (typically on the order of thirty seconds), one obtains an integrated Doppler count. The integrated Doppler counts are measures of the change in range to the satellite over the integration interval. The satellites also transmit the orbital information describing the satellite position at each two-minute time mark by modulating the carrier phase with digital data.

Depending on the integration interval, about forty integrated Doppler counts will be obtained during a typical satellite pass. These, together with the orbital information from the satellite, are usually combined in an iterative least squares process to obtain a position fix. Because of geometric limitations, a single Transit pass can yield only a two-dimensional position solution (and an estimate of the difference in frequency between the transmitted and local receiver frequency).

The most widespread use of the Transit system is for obtaining position fixes for ships at sea. For this application, the two-dimensional nature of the solution is not a significant limitation, since the receiver antenna height is known. Of course, since the ship's velocity will affect the measured Doppler counts, the ship's velocity must be measured or estimated independently; and the accuracy of the position solution will depend on the accuracy of the velocity.

Transit Geodetic Positioning. For precise geodetic positioning, the two-dimensional nature of a single satellite pass can be overcome by remaining stationary and tracking many sucessive satellite passes. In addition, small error sources that were ignored in marine navigation applications are now carefully modeled or measured and removed. For marine navigation, integrated Doppler measurements are typically made to only a single cycle of precision. Geodetic receiver measurements are usually made to about 0.01 cycles. In a marine navigation receiver, the control of the Doppler integration interval is typically obtained from the recovery of the orbital message bit transitions. The time accuracy of these bit transitions will have a noise of about 30 microseconds, which results in fractional cycle errors in the Doppler measurements. Geodetic positioning receivers either control the integration interval or measure the time jitter.

Geodetic positioning also requires more care in processing than is typically used in marine positioning. For example, in marine positioning the integrated Doppler measurements are almost always processed as if they were independent, even though they are not. If the satellite is assigned an incorrect position at the end of a Doppler count such that a positive error in the measurement results, then that same position error will create a negative measurement error in the next Doppler count. Virtually all sources of error in the Doppler measurement have this characteristic, which results in a -0.5 correlation between adjacent measurements. When this correlation between the measurements is properly modeled, then a more accurate position results.

Treating the correlation between adjacent Doppler (range difference) measurements properly is completely equivalent to processing them as pseudorange or biased-range measurements. If the Doppler frequency is continuously integrated, then the set of measurements can be thought of as ranges with a bias equal to the negative of the initial range when the integration was commenced. If the initial range is included as one of the solution parameters, these "independent" range measurements will result in a solution identical with the solution of range difference measurements whose correlation has been properly modeled. The disadvantage of this approach is that the size of the solution vector is increased when the measurements are treated as biased ranges. Early literature often referred to these cumulative integrated Doppler counts as pseudorange measurements. More recently, the term *biased-range measurements* seems to be preferred to distinguish these from the GPS code measurements, which are also referred to as pseudorange measurements. GPS code measurements have the property that the range bias is equal to the receiver clock error, and thus the bias is the same for every satellite observed.

Transit Error Sources. The most significant error source of the Transit system is usually the orbital position ascribed to the satellite. The orbital messages obtained from the satellites are simply predicted positions based on tracking information that is as much as a day or two old. For the most precise absolute positioning, these "broadcast ephemerides" are usually replaced with "precise ephemerides" obtained from a network of stations that tracked the satellites concurrent with the observing receiver. With precise orbits obtained from the Naval Surface Weapons Center (NSWC), positions accurate to better than one meter can be obtained with a few days of data collected from one or more Transit satellites.

Since precise ephemerides are not instantly available, other methods of removing orbital inaccuracies have been widely used. The most popular of these is to locate one or more tracking receivers on local control points and to use the measurements from these receivers to adjust the satellite positions obtained from the broadcast ephemeris. Techniques that make use of knowledge of the gravitational field to assist in this process are usually referred to as "short arc" methods. Techniques that solve for a selected set of orbital bias parameters are usually referred to as "orbit relaxation" methods. Some methods may not fit neatly into either category. In a method employed by the author, orbital bias parameters (i.e., orbital height, along-track position, and cross-track position of the satellite) were included in the solution. Then the adjustment to the height was used together with the Earth's primary gravitational constant to adjust the along-course velocity of the orbit appropriate to its new modified orbital height. For separation distances of less than a few hundred kilometers, these differential positioning techniques can result in relative accuracies of a few tens-of-centimeters from one or two days of data collection.

The next largest source of error in the Transit system is the noise in the satellite and receiver clocks. For each satellite pass, the frequency of the local receiver clock is solved for as a nuisance parameter. However, it is not really constant over the typical twenty-minute duration of a satellite pass. It is the variation of both the local receiver frequency and the satellite frequency from a perfectly constant value that accounts for most of the residual error in the relative positions obtained. Little can be done to remove this error. If an extremely accurate frequency (e.g., a hydrogen maser) could be used for the receiver, the problem with the satellite oscillator would remain. The only promising method for removing this error source is to obtain simultaneous measurements from at least one other satellite at both the control and remote sites.

At the orbital height of the Transit satellites, it is not possible to always have two satellites simultaneously in view. However, hybrid schemes to achieve this new accuracy capability are possible. For example, tracking a Transit satellite and some other satellite from which integrated Doppler measurements can be made could yield new accuracies for geodetic positioning via the Transit system. The most logical satellite candidates for use in aiding the Transit system are the new GPS satellites. However, the geodetic positioning capabilities of the new GPS system are better than those of the Transit system; therefore, there is little incentive to include measurements to the Transit satellites with the measurements obtained from GPS satellites.

The Global Positioning System

A good description of the GPS system is given in *Global Positioning System (GPS)*, q.v. However, by comparing it to the Transit system, its strengths and weaknessess and why it shows such great promise for geodetic positioning should become more apparent.

The GPS satellites are in much higher orbits than the Transit satellites (about 20,000 kilometers rather than 1100 kilometers). This fact has two direct effects on geodetic positioning. First, the GPS satellites have orbital periods of one half of the sidereal day. Thus, the same satellite geometry repeats every day (only four minutes earlier). This creates a slight disadvantage in that the position solution is more susceptible to systematic solution biases. The second effect of the higher satellite altitude arises from the longer period of the satellite orbit. A single pass of a Transit satellite lasts for only about twenty minutes while a GPS pass might easily last for as long as five hours. The disadvantage of this is that, while a good receiver oscillator may be sufficiently characterized as having a constant frequency over twenty minutes, the same cannot be said for a five-hour data span. This disadvantage is overcome by having more than one satellite simultaneously in view. Multiple satellites in view allows for solution of the clock errors (or, more commonly, their removal by differencing of measurements).

The GPS satellites transmit on much higher frequencies than Transit satellites; thus, the ionosphere affects them significantly less. But the process of correcting for the ionosphere amplifies the noise much more for the GPS system.

The GPS carrier frequencies are biphase modulated with a pseudorandom code, which has the effect of spreading the energy over a wide spectrum and leaving negligible energy in the carrier. This means that, unlike the Transit system, we must first "reconstruct" the carrier before it can be tracked. The reconstruction is an automatic byproduct in those receivers that obtain code measurements by correlating a locally generated code sequence with that obtained from the satellite. Alternatively, the second harmonic of the carrier can be obtained by a wideband squaring of the received signal.

The earliest geodetic receivers used the squaring technique, although this is not the best method of obtaining the carrier. The process of squaring the signal has several disadvantages compared to the reconstruction of the carrier. The principal disadvantage of wideband squaring is that a significant loss in signal-to-noise ratio results, which makes it more difficult to track marginal signals. A second disadvantage of squaring is that one loses the information that is carried by the pseudorandom code and its modulation, which means the modulation cannot be used to accurately set and synchronize the receiver clocks. The pseudorandom code allows one to solve for the local clock quite easily to a few tens-of-nanoseconds, while accurate carrier processing requires clock accuracies of a few microseconds. Those receivers that square the received GPS signal require some process of external clock setting and synchronization for optimal accuracies. In addition, receivers that square the signal lose the satellite ephemeris information transmitted by the satellites, which means that they are dependent on an external source of orbital position information.

Up to this point, it has been tacitly assumed that GPS carrier phase measurements are the measurements of interest. However, the GPS system was designed as a navigation system that would use measurements of the pseudorandom code as a measure of the signal transit time (and hence range) from the satellite to the receiver. Two items combine to make the code measurements of little benefit to geodetic positioning.

The first of these is the reduced precision of the code measurement. It is about two orders of magnitude less precise than the carrier measurements. This lower precision arises from the fact that the code measurements are measures of the transit time and hence are scaled by the speed of light, while the carrier beat phase measurements are measures of the Doppler shift and are scaled by the relative velocity between the satellite and the observer.

The code measurements have a better geometrical strength than the carrier phase measurements, but this geometrical strength depends on the assumption that the satellite clocks are perfect. While the assumption of perfect satellite clocks is fine for a navigation receiver, it cannot be used when one is attempting to obtain high-precision geodetic results.

As mentioned earlier, one of the significant benefits of code measurements is that they can be used to synchronize the clocks of the receivers. The clocks can be solved for much more accurately using the code measurements because of their higher sensitivity to time. If one uses the code measurements to solve for the clock bias, then it need not be a parameter in the carrier phase solution.

GPS Geodetic Positioning

The methods used to process GPS measurements for geodetic positioning vary widely. The variation is, however, more in form than in function. Those people developing the geodetic GPS algorithms come from diverse backgrounds. Early developments were made by very-long-baseline-interferometry (VLBI) specialists. Other investigators came from a Transit system background. These diverse backgrounds have created positive tensions resulting from different viewpoints on the nature of the measurements. On the other hand, differences in terminology arise and often confuse the similarities and differences in approach. The different use of the term pseudorange was commented on earlier. Similarly, a number of terms have been used to describe the process of determining the number of whole cycles by which carrier phase measurements must be biased in order to transform them into range measurements. Those with a navigation background and a familiarity with the equivalent problem with Omega and Loran radio navigation systems usually refer to the process of determining the proper number of whole cycles as *lane resolution*. New terms have been used in the GPS context to describe the equivalent process as *ambiguity resolution* or *bias fixing*.

A fairly detailed description of the various processing schemes follows. It is the intent of this discussion to show that the processing methods, though very diverse, are largely equivalent and that the alternative approaches differ more in form than in substance.

The fundamental equation of carrier phase measurement can be expressed in the following form:

$$m_{i,j,k} = r_{i,j,k}/w_j + a_{i,k} - b_{j,k} - n_{i,j} \qquad (1)$$

where i designates the receiver site, j designates the satellite, and k designates the time epoch of the measurement. The carrier phase measurement is designated by m. It will be a fraction between zero and one for the first measurement; but, at subsequent epochs, it will also contain a whole value, since the received frequency is tracked and integrated. The slant range between the satellite and receiver at each epoch is designated by r. The wavelength of the satellite transmit frequency is represented by w and is used to scale the slant range into cycles. The clock phase of the receiver at each epoch is represented by a. Like m, the first value will be a fraction, but subsequent values may contain whole cycles as well. The clock phase of the satellite is represented by b and is similiar to a. The number of whole cycles by which the phase measurement needs to be biased to make it a true range measurement is represented by n. As long as carrier phase lock is maintained in the receiver this value will not change. Loss of lock will result in a new value for n. Multiple values of n can further complicate the processing, and some alternative methods of recovering loss of lock are discussed below.

As an example, let us assume that data is collected at one minute epochs for 30 minutes from five satellites at four sites. The result is a total of 600 measurements of the form of Eq. 1. Since Eq. 1 is nonlinear in the coordinates of the site position, it is typically expanded in a Taylor series around an estimated position and iterated until it converges. If one were to follow this procedure, a total of 12 position coordinates for the four sites, 120 independent values of receiver clock phase, 150 independent values of the satellite clock phase, and 20 values of whole-cycle biases would be found.

At first, it appears that there are a sufficient number of measurements, 600, to solve the set of resulting equations for the 302 parameters. However, the equations are rank deficient (i.e., not all the parameters can be separately resolved from the data). It becomes necessary to specify or constrain some parameters. One of the clocks (or an average of several clocks) must be selected as the master clock. It is not necessary that this clock be extraordinarily accurate, since any errors in this clock will alias into the other clocks directly and only extremely large errors can create a significant effect by distorting the time scale. If one of the satellite clocks is chosen as the master clock, then all of its associated clock phase values, b, would be removed from the solution parameters.

Because absolute positions are affected significantly by orbital biases, it is also typical to remove or constrain the coordinates of the master or control station from the solution. The resulting relative solution has the effect that most of the orbital errors are absorbed into the bias terms and removed from the solution. This modified set of equations can be solved directly but results in a very large matrix inversion. Most implementations reduce the matrix size by various methods.

Triple Difference Processing Methods. The geodetic solution method that most dramatically reduces the matrix size is usually referred to as the *triple difference* solution. If one subtracts the measurements obtained at each epoch from that of the next epoch, one obtains integrated Doppler measurements; and the values of the whole cycle bias, n, are cancelled from the resulting single difference measurements. The Doppler measurements at the reference site can then be subtracted from the corresponding measurements at the other sites. This double difference results in a cancellation of the satellite clock bias, b. Next, the double difference from a selected reference satellite can be subtracted from the double difference of the other satellites. This triple difference results in a cancellation of the receiver clock bias, a. (The order in which the differences were formed is not significant.)

The number of parameters that remain to be solved in the triple difference solution is only nine—the coordinates of each remote site. This solution also has the advantage that intermittent loss of lock does not unduly complicate the solution. It does not seem to be generally recognized that this much simpler solution is completely equivalent to the full solution described above—if the measurement-to-measurement correlation is properly modeled in the solution process.

The differencing of the measurements taken at one site from the measurements taken at other sites results in a $+0.5$ correlation of the combined measurements across sites. Similarly, the differencing of the measurements taken using one satellite from the measurements taken using other satellites results in a $+0.5$ correlation of the combined measurements across satellites. Finally, the differencing of the measurements from one time epoch with the measurements at the next epoch results in a -0.5 correlation of the measurements from one epoch to the next. It is generally true that positive correlation can be ignored without significantly affecting the solution. Ignoring positive correlation will, however, usually result in overstating the formal accuracy of the result. Ignoring negative correlation will almost always significantly affect the solution. There are a number of ways to properly account for the correlation in the solution process. One of the simplest of these is the whitening process described in Bierman (1977).

Once the coordinates of the sites have been determined, they can be back substituted into the original equations to determine the bias terms.

Many variations of the above process result in equivalent solutions. One of the more interesting variations eliminates the positive correlations across the satellites and sites. This is accomplished by forming the same differences as described above except that instead of selecting measurements from a specific site and measurements from a specific satellite to difference with the others, the average measurement across sites is removed from the corresponding measurement at all sites and the average measurement across satellites is removed from the corresponding measurement to all satellites. The -0.5 correlation from one epoch to the next still requires modeling.

Double Difference Processing Methods. The most common variation used to decrease the matrix size and still obtain at least a near-equivalent solution is the *double difference* method. This method involves the differencing across sites and satellites as described above, but the differencing across time epochs is not performed. Quite often the positive correlation, resulting from the selection of a specific site and specific satellite, is ignored; and this can compromise the solution slightly. Since the differences across time epochs are not performed, whole-cycle bias terms (composed of second differences of the original whole-cycle biases) remain and are solved for as part of the solution. Thus, the matrix size is larger than that of the triple difference method.

An additional disadvantage of the double difference method is that another bias term needs to be added to the solution every time a loss of lock occurs in the data. For this reason, it is common among those employing this algorithm to perform a preprocessing step to "repair" loss of lock in the data. This step is typically done by fitting the data or some form of the differenced data with a polynomial.

The most significant advantage of the double difference method is that the next processing step becomes obvious.

Whatever the method used to solve Eq. 1 above, once values for n have been obtained, it is standard practice to fix the value of each n (if the solution is "sufficiently close" to a whole value) to the nearest integer. Equation 1 then is reformed and becomes:

$$m_{i,j,k} + n_{i,j} = r_{i,j,k}/w_j + a_{i,k} - b_{j,k} \qquad (2)$$

In words, the whole-cycle bias is added to the measurements and only the clock biases remain. This equation is much stronger geometrically than Eq. 1 when short time spans of data are collected over short baselines. In this form, the attraction of the double difference algorithm is more apparent. Once the whole cycles have been determined, then the bias parameters are simply dropped from the solution and only the site coordinates remain. Similar processes are performed with the other algorithms described above; but, while they are usually quite simple, they are not so obvious as with the double difference algorithm.

GPS Error Sources

A number of problems not yet considered currently limit the positioning accuracy that can be obtained from the GPS system. These problems are now being addressed by many different institutions and government agencies. The solution of these problems will affect the future success of the GPS system. These problems are considered below.

Refraction Effects and Ambiguity Resolution. Ionospheric refraction can adversely affect the computed coordinates of the remote sites, particularly as the baseline distance between the control site and remote site is increased. The first step in overcoming this problem is to use a GPS receiver that tracks both the L_1 (1575.42 MHz) and the L_2 (1227.6 MHz) frequencies. Because the ionosphere affects these two frequencies differently, an ionosphere-free measurement can be generated by a linear combination of the measurements at the two frequencies. This works very well for all baseline lengths when solving Eq. 1.

Unfortunately, problems may arise when we attempt to fix the bias values n to whole integers and proceed to Eq. 2. The source of the problem is that, in order for the values of n to be exact whole cycles, the average ionospheric refraction error introduced in the range at the remote site must be identical to that at the control site. As the baseline distances are increased, the adequacy of this assumption becomes increasingly questionable. The only high accuracy methods for bias-fixing over long distances must involve either accurate modeling or precise measurement of the differences in the ionospheric range.

Sophisticated methods for using the GPS measurements to measure differences in the ionospheric range depend on obtaining accuracies of around two and one-half centimeters before bias-fixing can be accomplished.

Several suggestions have been made to deal with the bias-fixing problem at longer baseline distances. One suggestion has been to build a receiver designed to obtain maximum accuracy in the code measurements. This receiver would have a very precise calibration of the difference in range measured at the two frequencies, and the code measurements would be smoothed by the carrier measurements. It would appear that such a receiver might be able to measure the difference in average ionospheric range error at each satellite/site combination sufficiently well to allow whole-integer bias fixing. A simple, but unfortunately hypothetical, alternative would be to use a third frequency between L_1 and L_2 to solve the problem.

If precise orbits are separately available, then bias-fixing is simply not a problem when one observes the satellites over a long enough data span. If data is collected while the satellites travel virtually from horizon to horizon, the accuracy of the solution to Eq. 1 is essentially the same as that obtained from Eq. 2. It is only over shorter observation intervals that a significant accuracy improvement is made by stepping from Eq. 1 to Eq. 2. If one must also solve for the satellite orbits as part of the solution process then Eq. 2 remains much stronger than Eq. 1 even for long data spans.

Tropospheric refraction also can significantly affect the GPS relative position accuracy. Atmospheric models are normally used to reduce this error. It is generally agreed that the dry component of tropospheric refraction can be accurately modeled using the surface pressure. The wet component of tropospheric refraction is much more difficult to adequately model. Recent data, using water vapor radiometers to measure the refraction along the propagation path from the satellites, have suggested that surface humidity readings may make matters worse rather than better. Excellent results have been obtained by several investigators who have solved for the zenith tropospheric delay error as part of the solution. Some have allowed it to vary as a slowly changing stochastic variable while others have assumed a constant. The use of water vapor radiometers is encouraged for high accuracy geodetic positioning but the expense may prohibit their ordinary use. If making tropospheric parameters part of the solution proves practical then it will probably become the normal method for minimizing tropospheric errors.

Orbital Errors. The accuracy of the broadcast GPS orbit is such that baseline solutions will generally be limited to no better than one part in one million in each coordinate. Accuracies that are better by one or two orders of magnitude can be achieved if independent orbits are made part of the solution. Usually, this involves the use of three or four control stations, often referred to as fiducial sites, to collect data from very accurately known locations. As with the Transit system, some investigators propose orbit relaxation techniques or some hybrid method of improving the broadcast orbits rather than determining completely independent solutions. Each of these methods is yielding very good results now and undoubtedly will be improved in the future.

Selective Availability. One further problem needs to be addressed. The government intends to degrade the real-time navigational accuracy of the GPS system by implementing the "selective availability" option of the operational satellites. It appears that three actions will be taken that have the potential of affecting the geodetic use of the GPS system. These three items are: (1) corruption of the clock stability by some form of dithering; (2) corruption of the broadcast orbit; and (3) denial of access to the P code. The effect of each of these is assessed below.

The carrier phase will be affected by the clock dither. However, as long as the GPS receivers take essentially simultaneous measurements and maintain carrier phase lock then no adverse effect on relative position accuracy should result. This is because the normal process used to remove clock errors (either by direct differencing of measurements or by solution) effectively isolates clock performance from system accuracy.

The corruption of the broadcast orbit will stimulate wider use of post processing with independent orbits. Since highest geodetic accuracy already demands an independent orbit source, this is not seen as a significant limitation of the GPS system. In addition, indications are that the government may release precise orbit information after a delay of one or two weeks.

Since only the P code is present on the L_2 frequency, the implementation of selective availability means that no direct access to the L_2 frequency is possible. The only method of obtaining the carrier on the L_2 frequency is to implement some type of squaring receiver. Long distance geodetic baselines need to obtain carrier phase on both frequencies to make ionospheric corrections. Consequently, receivers used for geodetic positioning should contain squaring channels for the L_2 frequencies and correlation channels for the L_1 frequencies (so that accurate clock synchronization can be performed). Unfortunately, the squaring process doubles the carrier frequency and halves the wavelength, which means that the ambiguity resolution problem has become doubly difficult, which, in turn, shortens the baseline distances over which one can expect to properly resolve the ambiguities. It also argues for longer observation schedules on all baselines.

GPS Applications and Accuracies

All of the high accuracy applications of GPS positioning involve relative positioning where at least one receiver is located at a known reference station. The reference receiver allows the removal of common error sources that would otherwise dominate the solution. Three categories of application can be identified.

Long Distance Geodetic Positioning. For distances between 100 kilometers and several thousand kilometers, accuracies approaching one part in 10^8 have already been demonstrated. To obtain these accuracies it is generally necessary to use many hours of data collected from two-frequency receivers. Also, a precise post-fit orbit needs to be used rather than the broadcast ephemeris. In addition, it may be necessary to use water vapor radiometers to minimize tropospheric refraction effects.

Short Distance Engineering Surveys. For shorter distances, accuracies of from one to two parts per million of the baseline distances can be obtained using the broadcast orbits. For distances less than 20 to 30 kilometers, these accuracies can be obtained quickly even with single frequency receivers. For distances approaching 100 kilometers, it is necessary to remain on site for several hours or to use a two-frequency receiver in order to obtain these accuracies. This capability is unprecedented. As equipment prices come down, it can be expected to have a broad impact in the survey community.

Kinematic Surveying. When four or more satellites are simultaneously in view it is possible for a moving receiver, which maintains carrier phase lock, to solve for a new position at each epoch (a stationary reference receiver is still required). Accuracies in the vicinity of one centimeter can be obtained at each epoch. This capability is dependent upon having first determined the whole-cycle biases so that Eq. 2 can be employed. A number of methods of whole-cycle bias fixing specifically for this application have been proposed. Because kinematic surveying depends on having at least four satellites simultaneously in view, it can be expected to find much wider use when the GPS system becomes operational and many more satellites become available.

Conclusion

Work remains to develop and improve methods for each of the applications considered earlier. Nevertheless, the accuracy and speed with which the GPS system can be used to obtain geodetic positions is unprecedented. With the launch of additional satellites and full system operation, the ease of obtaining accurate positions will have entered a new era.

RONALD R. HATCH

References

Bierman, G. J., 1977, *Factorization Methods for Discrete Sequential Estimation*. New York: Academic Press, 49, 50.

Wells, D., 1986, *Guide to GPS Positioning*. Fredrickton, N.B. Canada: Canadian GPS Associates, University of New Brunswick.

Cross-references: *Doppler Positioning: Satellite; Earth Orientation; Earth's External Gravity Field; Equilibrium Figure of the Earth; Figure of the Earth; Geodesy: Geometric; Geodesy: Physical; Geodetic Reference Systems; Global Positioning System (GPS); Gravity Fields: Implications for Planetary Interiors; Satellite Altimetry; Satellite Laser Positioning; Spherical Harmonic Analysis; Very-Long-Baseline-Interferometry (VLBI).*

GEODETIC GROUND POSITIONING

The theory and technology of ground positioning (also called terrestrial geodetic networks, terrestrial horizontal geodetic networks, and horizontal control) dates directly from the fifteenth and sixteenth centuries with antecedents dating from Hellenic-Roman, and even Egyptian, eras. This development is an interesting story of the development of mathematics (plane and spherical trigonometry, plane and solid geometry, plane and solid analytical geometry, statistics, least squares, and calculus), Newtonian physics, positional astronomy, cartography, and surveying (mensuration and mapping). The mathematical principles of the data collection, instrumentation, and the adjustment-reduction for mapping purposes were well known by 1800.

Terrestrial geodetic surveying requires measuring distance and direction of lines between uniquely selected points on or near the Earth's topographic surface. By "direction" is meant the dihedral angle between a meridional plane and a plane containing two points on the Earth's surface and the center of the Earth. A meridional plane is any plane that is perpendicular to the Earth's equatorial plane and contains the Earth's polar axis (a straight line) in that plane. The direction of a survey line is called the *azimuth* or the *bearing* of the survey line. The azimuth is a clockwise angle from the meridional plane to the plane of the survey line. In most instances, the azimuth is reckoned from north, but in the United States, azimuths have been reckoned from south until institution of the North American Datum of 1983 (1986). If the direction is stated as a bearing, the direction is an acute dihedral angle from the meridional plane (either clockwise or counterclockwise) with reference to the cardinal directions of the compass (north, east, south, and west). The azimuth and bearing systems are compatible and commutable. If we designate the azimuth as α and the bearing angle as β, then the mathematical relationships are defined in Table 1.

Table 1 contains three alternatives for stating a bearing along a cardinal direction. The second and tertiary choices are of local usage. The use of "North," "East," "South," and "West" can be ambiguously interpreted. This ambiguity arises from legal, statutory, and diplomatic (treaty) usage. Therefore, it is recommended that these terms never be employed. Although bearings are employed regularly, it is recommended that azimuths be stated in all geodetic data, and the conversions (Table 1) be utilized only when user requested.

The geometrical concepts of the survey instrumentation and survey lines are based upon three-

Table 1. Mathematical Relations Between Azimuths and Bearings

	Clockwise from North		Clockwise from South	
α	β	Bearing	β	Bearing
0°–90°	$\beta = \alpha$	N β E	$\beta = \alpha$	S β W
90°–180°	$\beta = 180° - \alpha$	S β E	$\beta = 180° - \alpha$	N β W
180°–270°	$\beta = \alpha - 180°$	S β W	$\beta = \alpha - 180°$	N β E
270°–360°	$\beta = 360° - \alpha$	N β W	$\beta = 360° - \alpha$	S β E
0° or 360°	$\beta = 0°$	North / N 0° E / N 0° E	$\beta = 0°$	South / S 0° E / S 0° W
90°	$\beta = 90°$	East / N 90° E / S 90° E	$\beta = 90°$	West / N 90° W / S 90° W
180°	$\beta = 0°$	South / S 0° E / S 0° W	$\beta = 0°$	North / N 0° E / N 0° W
270°	$\beta = 90°$	West / N 90° W / S 90° W	$\beta = 90°$	East / N 90° E / S 90° E

dimensional geometry. The fundamental premise is that the mathematical definition of the Earth is a sphere. Although the currently employed reference figure is an ellipsoid of revolution about its minor axis, the ratio of the minor axis to the major axis is 99.66:100 (Clarke 1866 ellipsoid)—in other words, the Earth fails to be a sphere by about 0.4% (equatorial radius of about 6380 km). Therefore, the fundamental concepts for a sphere are equally applicable for the Earth ellipsoid. It can be demonstrated that the differences can be calculated and the observations corrected to satisfy ellipsoidal geometry. In many instances, low-order accuracy and small-area survey coverage (less than about 500 km^2) require only spherical trigonometric formulas for data reduction and mapping computations.

If we treat the Earth as a sphere, then every observed survey line between two points is a portion of a great circle and lies in a unique plane containing the two points and the sphere's (Earth's) center. A line from the Earth's center through a survey point is the *local vertical* through the survey point. Through any survey point an infinite number of planes (called vertical planes) also containing the Earth's center can be drawn. Therefore, there are an infinite number of vertical planes at a survey point, but only one of these planes will contain the second survey point. The arc of intersection of a vertical plane with the sphere is always a *great circle*.

The azimuth of a survey line is not normally measured directly but is calculated from the observed angles (dihedral) to other vertical planes whose azimuth is known. The reference azimuth (as it is called) may be the vertical plane to another point on the sphere or it may be the vertical plane containing the Sun or a star. Because the Sun and the stars continually move in the heavens, the azimuth must be instantaneously determined, which requires precise time. The azimuth at point A to point B, from a reference azimuth is

$$\alpha_{A \to B} = \alpha_{\text{Ref}} + \theta_{A, \text{Ref} \to B}$$

where α_{Ref} is the reference azimuth, $\theta_{A, \text{Ref} \to B}$ is the observed clockwise dihedral angle from the reference to point B, and $\alpha_{A \to B}$ is the azimuth from point A to point B. Furthermore, the azimuth from point B to point A usually satisfies

$$\alpha_{B \to A} = \alpha'_{A \to B} \neq \alpha_{A \to B} \pm \pi$$

The measurement of the dihedral angle $\theta_{A, \text{Ref } B}$ (generally called a *horizontal angle*) is performed with a *theodolite*. The design of the theodolite is intended to measure the dihedral angles at a point. The theodolite's vertical axis and the vertical through the survey point *coincide*, and the theodolite's horizontal circle is *perpendicular* to the theodolite's vertical axis. Therefore, the theoretical design rigorously conforms to three-dimensional geometrical (solid geometry) principles. Theodolites *do not exactly satisfy* the stated geometrical principles. However, a detailed analysis of the instrumental errors by Bessel and others has resulted in an observation program that minimizes the geometrical instrumental errors. For highly accurate work, those minor geometrical errors not corrected (eliminated) in the theodolite's observing procedure can be deleted by computational techniques.

The linear (great circle) distance between the survey points is "measured" by various techniques. Direct measurements utilize taping (bands, wires, tapes) or electro-optical technology to determine the distance. Electro-optical technology requires an accurate knowledge of the velocity of light in a vacuum and an accurate knowledge of the refractive index correction (a function of atmospheric gaseous mixture, water vapor, barometric pressure, and temperature). The electro-optical distancers measure distances from a few meters to hundreds of kilometers and have precisions (and repeatability) ranging from a few tens of meters to centimillimeters. The attainable accuracy and precision depends upon instrument design and electromagnetic frequency. It is the consensus that the current attainable (1986) *absolute outdoors* accuracy of geodetic distance (based upon the length standard definition) is 0.9 to 1.0 ppm. This limit is predicated on the limits of uncertainty to project a *suitable* "length standard" baseline for comparisons and the atmospheric modeling for the index of refraction. Taped distances (baselines) are accurate to no better than 1.0 ppm, because calibration laboratories have certified this limit in generating the special comparator for calibration.

The field observations for distances are usually not on the reference surface but on, or near, the topographic surface. So, field-measured distances must be converted from topographic lengths to *geodetic* lengths. The formulae require knowledge of the elevation of the survey point(s) above or below the Earth's mathematical surface, and the 'path' or 'curve' in the vertical plane that was actually measured.

An alternate "measurement" is the *indirect* method, in which the distance computed is based upon trigonometric or geometric relationships. This technique utilizes "triangulation" or "inverse" computations. The accuracy of this technique is a direct function of the accuracy of the quantities employed in the computation.

No mention has been made of survey network types. The reason for this omission is to emphasize that the geometrical principles of vertical planes—"great circles"—*are independent* of the survey network. The survey networks—*traverse, triangulation,* and *trilateration*—utilize identical geometrical concepts of "great circle geometry." The only difference is the "array" of survey stations established to provide geodetic positions on the Earth.

Triangulation is the procedure by which the angles of a series of triangles (either adjoining or overlapping) are observed, and the length of at least one triangle side is measured. Then, utilizing the law of sines (for plane triangles), the lengths of the other triangle sides are calculated. The use of plane triangles, when spherical trigonometry is the basis of physical and observational theory, is the result of the work of Adrien-Marie Legendre. The transformation function is known as *Legendre's Theorem*, and the derivation can be found in most nineteenth-century spherical trigonometry textbooks and pre-World War II geodesy textbooks. Because linear distances were the most difficult survey operation (economics, error source budgets, topography, etc.), triangulation was the most accurate method of providing geographic, or geodetic, positions for mathematical reference for mapping.

With the advent of electro-optical distancers, the trilateration network technique has been used. In this instance, the triangle's interior *plane angles* (called Legendre angles) are computed by the law of cosines (plane trigonometry). Trilateration has gained wide acceptance in construction and engineering, and in "small" surveying and mapping projects. The accuracy (commensurate with triangulation) and relatively low field survey cost (about one third that for triangulation) make the survey network a viable alternative. In the 1950s and 1960s, many geodesists did not consider trilateration comparable to triangulation. However, with conscientious planning and network design, trilateration is as accurate as, if not superior to, triangulation. The major (general) error source for pure triangulation is scale (length); for pure trilateration it is azimuth (angles).

If the triangulation and trilateration networks are plotted and if the observed measurements (distances and/or angles) are not indicated, it would be difficult to identify the type of geodetic network depicted. Both systems utilize triangles. Today a pure triangulation or trilateration network is rarely designed or observed. The combination of both distances and angles enhances the quality of the network.

The traverse is a geodetic network in which both lengths and dihedral angles are observed and is, geometrically, the weakest (error budget and error propagation) geodetic control network. In general, traverses are a polygon of *n* sides. Thus, in some European technical literature, traverses are referred to as polygometry. Considering all the forms of horizontal geodetic networks, traverses are the most-used form of geodetic network. The normally low-order accuracy requirements for a majority of general civil engineering and land surveying projects (about 50–100 ppm) justify traverse networks. However, national networks have been observed in those regions that cannot utilize triangulation or trilateration (e.g., rain forests). Individual traverses do not have the inherent geometrical strength comparable to triangulation and trilateration networks. The traverse can be employed to satisfy highly accurate geodetic requirements (i.e., 1–2 ppm). This accuracy is achieved by very special mensuration programs. In the United States this type of traverse is known as the precise transcontinental traverse (PCT or TCT).

The integrity of each type of network is a function of the instrumentation, observations, and methodology. To achieve a specific accuracy requires adherence to specific minimum criteria, which are more stringent for more accurate surveys. The analysis of instrumental errors and historical data provide the basis for minimum standards and specifications. These standards and specifications may seem conservative, which they are, but they result from careful study of historic information to ensure that the *minimally qualified* instrumentation *and* personnel *will satisfy* the minimum accuracy requirements. In the United States the Federal Geodetic Control Committee (FGCC) develops the survey specifications and requirements and evaluates instrumentation.

To achieve a specific survey accuracy requires adherence to all specification requirements. Some surveyors interpret the accuracy of a survey as a function only of the errors of misclosure resulting from the geometrical condition equations. This is unsatisfactory, because the survey network may contain compensating errors that may not be evident when the condition equations of misclosure are computed. However, if the condition equations of misclosure *and* the statistics of the data for each observed quantity are available, then the purported accuracy can be substantiated. As the required accuracy of a survey increases, additional observations are required to validate the result. For example, if one length and any other two elements of a triangle are observed (angles or lengths), then the remaining elements of the triangle are determined uniquely. However, there is no check of the "accuracy" of the results. If, however, additional triangle elements are measured, then there is verification of the results. For each additional element measured greater than the minimum number of elements required to determine uniquely the survey results, these addition elements are known as *redundancies* (*degrees-of-freedom* in statistics). No particular element is identified as a redundant element. The number of redundancies is the difference between the number of measured elements and the maximum number of elements that are required to determine uniquely the problem. For example, if all the interior angles and the three sides of a triangle are measured (only three are required to uniquely determine all the elements of the triangle), then the number of redundancies is three.

After survey networks have been observed, the data are reduced, adjusted, and published. Data reduction is the transformation of actual observations for instrumental, observational, and theoretical errors to the mathematical reference surface. The corrections for angle observations are (1) eccentricity of the theodolite's vertical axis from the survey point's vertical, (2) eccentricity of the observed survey point's signal vertical axis from the observed survey point's vertical, (3) lateral refraction (bending of the light ray between the theodolite and the signal from the true vertical plane between the two points), (4) reduction of angles to the ellipsoid (sometimes called the "reduction to sea level"), (5) instrumental errors, and (6) curvature of the verticals through the survey point and the observed survey point (called "curvature of the plumb line"). The corrections to observed distances include (1) instrumental corrections, (2) reduction from the topographic to the reference surface, (3) eccentric corrections, and (4) lateral refraction. In general, these corrections are referred to as systematic error corrections.

After the corrections have been applied, the geodetic positions (latitude and longitude) can be computed. The geodetic networks and the resulting geodetic positions provide a mathematical reference model upon which all types of mapping can be compiled. If there are redundant observations, then there are multiple mathematical models that result from the choice of the observations to compute the model. Statistically the *best model* incorporates all *valid* observations. Because precise observations contain small accidental errors, it is necessary to perform an adjustment that results in a unique mathematical model. The criteria of good adjustments are based on statistical and philosophical concepts. The statistical concepts are reasonably well understood and can be calculated (impartially). Although philosophical concepts are understood, they are often ignored because of their impreciseness. When an adjustment is performed, the corrections to the observations (called *residuals*) should be of the *same order of magnitude* as the precision (and accuracy) of the instrumentation and methodology employed in obtaining the original observations.

The second precept is that if other surveyors utilize equivalent, or lesser, instrumentation and procedures, they will discover *no significant* differences between their data and the previously adjusted results. If an adjustment fails to satisfy completely these two criteria, then an analysis is performed to determine the problems. If, after an in-depth evaluation of the observed data, there are no explicable reasons for the untoward discrepancies (residuals), then an analysis of the older geodetic network to which the new network has been connected should be analyzed for integrity. Old network integrity may be degraded due to unsuspected movement of the survey point's physical monument, tectonic or seismic actions, or unsuspected older network inaccuracies. Determination of the actual problem requires additional analysis that may require acquisition of nongeodetic information.

Adjustment of geodetic observations can employ either of two types of adjustments: *condition* equations or *observation* equations. Weighted least squares are employed in both types of adjustments. Basically, weighted least squares means that for a particular adjustment, the sum of the products of the square of the residuals with its associated weight is a *minimum*. Adjustment by condition equations requires identification of every independent geometrical condition in the network. The classical geometrical conditions for triangulation networks are angle (triangle), side, length, azimuth, and position closure (two equations—latitude and longitude). In the quadrilateral with two diagonals, there are four angle and five side equations that can be used. However, *only four* (at least one of each type of condition equation) are required to perform the adjustment. The four condition equations selected are *independent* of each other. If the equations are not independent, the solution of the resulting normal equations terminates when the redundant equation (redundant to one of the previously solved equations) is reached in the computation (called successive elimination of unknowns). In the traverse, there are only two types of conditions: azimuth and position closure. Trilateration networks have been adjusted by condition equations, but an alternate method is available, which is preferred.

The alternate method to adjust these networks employs observation equations. For all three network types, the observation equations for each line (one for distances and one for azimuth) are of the form

$$v(\alpha_{i \to j}) = A_{i,j}\delta\phi_i'' + A_{j,i}\delta\phi_j''$$
$$+ B_{i,j}(\delta\lambda_j'' - \delta\lambda_i'') + Z_i$$

$$v(s_{i,j}) = -\text{arc } 1'' \left\{ C_{i,j}\delta\phi_i'' + C_{j,i}\delta\phi_j'' \right.$$
$$\left. + D_{i,j}(\delta\lambda_i'' - \delta\lambda_j'') \right\} + Z_i$$

where

A, B, C, B = numerical coefficients
$\delta\phi_i''$, $\delta\phi_j''$ = corrections to the assumed latitude of points i and j (end points of the line) in arcseconds
$\delta\lambda_i''$, $\delta\lambda_j''$ = corrections to the assumed longitude of points i and j (end points of the line) in arcseconds
$v(\alpha_{i \to j})$ = residual of the azimuth $i \to j$
$v(s_{i,j})$ = residual of the distance between i and j
Z_i = orientation parameter at point i

This adjustment is frequently called the *adjustment by variation of coordinates*. Although this procedure was well known for several decades, it was not regularly employed until after the development of digital computers. One critical concern is retaining numerical precision, which is affected by computational roundoff and truncation. Presurvey planning and post-adjustment accuracy assessment utilize the same equations.

The various adjustment procedures (condition equations and/or observation equations) provide acceptable results. If a network is adjusted by either method (or a combination of the two methods) and employing appropriate statistical criteria (i.e., unbiased toward any adjustment method), analysis of the results reveals little difference between either method. The choice of adjustment is the preference and philosophy of the survey organization.

The results of terrestrial geodetic surveying is the geodetic position (geodetic latitude and longitude, a function of reference ellipsoid and datum). From these data, the mathematical reference system provides requisite information required by cartographers, engineers, geophysicists, planners, surveyors, and others. Each user has specific and unique requirements for these geodetic data. Therefore, geodetic networks must be designed (location and placement, instrumentation, observing program, etc.) to satisfy user requirements. Although regional and national geodetic networks satisfy general engineering and cartographic projects, an increasing number of special projects require long-distance measurements to hundredths of a millimeter and azimuth data accurate to about five-tenths of an arcsecond.

The attainable accuracy of geodetic networks depends on the *accuracy* and *precision* of the instrumentation employed and the observations obtained *and* on the existing geodetic network's integrity. When new networks are incorporated into old networks, the weaknesses and strengths of the old networks affect adjustment of the new network. In many instances, these effects usually are small, even "negligible," based upon only casual inspection of the adjustment results. Special analysis techniques can be employed to ascertain any untoward effects. In most instances, the differences result from scale (length) and meridional rotation (azimuth). One of the simple approaches to assess the effect is to perform a *minimally constrained adjustment* and a *rigorous adjustment*. In this procedure, identical observations with unchanged weights are adjusted as a free network (minimally constrained) and as a network constrained by ties to the adjacent networks. The differences between the two adjustments indicate the external effect exerted on the new observations. One method of indicating these changes is to graphically display the differences as vectors on a map of the network.

In more extensive regional networks, additional data, other than angles and distances, are observed. These data include astronomic azimuth and astronomic latitude and longitude. To obtain suitable results, one must carefully plan, execute, and properly adjust a horizontal network. The inclusion of astronomic azimuth and astronomic position or additional angles and distances requires technical analysis.

HERBERT W. STOUGHTON

Bibliography

Adams, O. S., 1915, Application of the Theory of Least Squares to the Adjustment of Triangulation, *U.S. Coast and Geod. Survey Spec. Pub.* 28.
Bomford, G., 1952, *Geodesy*, 2nd ed., 1962; 3rd ed., 1971; 4th ed., 1980. Oxford: Clarendon Press.
Clark, D., 1957, *Plane and Geodetic Surveying for Engineers*, two vols., (rev. J. Clendinning), 5th ed. New York: Frederick Ungar Publ.
Clarke, A. R., 1880, *Geodesy*. Oxford: Clarendon Press.
Crandall, C. L., 1907, *Text-book on Geodesy and Least Squares*. New York: John Wiley & Sons.
Durgin, C. M., and Sutcliffe, W. D., 1927, Manual of First Order Traverse, *U.S. Coast and Geod. Survey Spec. Pub.* 137.
Gossett, F. R., 1950, 1955, 1959, and 1978, Manual of Geodetic Triangulation, *U.S. Coast and Geod. Survey Spec. Pub.* 247.
Hirvonen, R. A., 1971, *Adjustment by Least Squares In Geodesy and Photogrammetry*. New York: Frederick Ungar Publ.
Hodgson, C. V., 1935, Manual of Second and Third Order Triangulation and Traverse, *U.S. Coast and Geod. Survey Spec. Pub.* 145
Hosmer, G. L., 1920, *Geodesy*. New York: John Wiley & Sons, 2nd ed., 1930/47.
Hotine, M., 1969, *Mathematical Geodesy*. Washington: Government Printing Office.
Leland, O. M., 1921, *Practical Least Squares*. New York: McGraw-Hill.
Merriam, M., 1884, *A Text-book on the Method of Least Squares*, 8th ed., 1911. New York: John Wiley & Sons.
Mikhail, E. M., 1976, *Observations and Least Squares*. New York: Harper & Row.
Mikhail, E. M., and Gracie, G., 1981, *Analysis and Adjustment of Survey Measurements*. New York: Van Nostrand Reinhold Co.
Mitchell, H. C., 1948, Definitions of Terms Used in Geodetic and Other Surveys, *U.S. Coast and Geod. Survey Spec. Pub.* 242.
Mussetter, W., 1941, Manual of Reconnaissance for Triangulation, *U.S. Coast and Geod. Survey Spec. Pub.* 224.
Rainsford, H. F., 1958, *Survey Adjustment and Least Squares*. (New York: Frederick Ungar Publ.
Reynolds, W. F., 1934, 1955. Manual of Triangulation Computation and Adjustment, *U.S. Coast and Geod. Survey Spec. Pub.* 138.
Richardus, P., 1966, *Project Surveying: General Adjustment and Optimization Techniques with Applications to Engineering Surveying*. Amsterdam: North-Holland.
Wright, T. W., 1884, *Adjustment of Observations*, 2nd ed., with J. F. Hayford, 1906. New York: Van Nostrand.

Cross-references: *Adjustment Methods; Figure of the Earth; Geodesy: Geometric; Geodesy: Physical; Geodetic Leveling; Geodetic Networks and Control Surveys.*

GEODETIC LEVELING

Geodetic leveling has been defined as leveling of a high order of accuracy, usually extended over large areas, to furnish accurate vertical control for surveying and mapping operations. Berry (1976, p. 137) states that

> a high order of accuracy (to be carefully distinguished from precision) is to be achieved only by use of a combination of a carefully designed and precisely constructed set of instruments (levels and rods), used by a competent and conscientious observer, in accordance with a proper observing routine, together with a data reduction system that applies appropriate corrections for all the physical and environmental situations that may affect the condition and/or calibration parameters of the instruments and observing conditions.

Geodetic leveling was traditionally restricted to using spirit leveling instruments (hence the term *spirit leveling*), which usually incorporate a curved cylindrical glass container, partially filled with fluids of low viscosity, e.g., alcohol/ether, so its bubble and associated telescope accurately determine a horizontal line of sight, i.e., a line of sight perpendicular to the gravity vector at the observation site.

The basic equipment used in geodetic leveling typically includes a geodetic leveling instrument and two leveling rods with graduated Invar strips. Auxiliary equipment may include thermometers or thermistors to measure the temperature of the leveling rods' Invar strips and atmospheric vertical temperature profiles.

Beginning in the 1950s, geodetic leveling instruments, incorporating some variation of a pendulum apparatus (compensator) to accurately determine a horizontal line of sight, became available for general use. These so-called self-leveling or automatic leveling instruments allow increased productivity while maintaining acceptable accuracy. The preferred terminology for both geodetic spirit leveling and geodetic leveling with compensator leveling instruments is high-accuracy differential leveling. In popular usage, the term leveling means differential leveling, unless another method of leveling is specified as part of the designation.

The fundamental observation in leveling is the elevation difference between two nearby points. The leveling instrument is set up midway between these points so its horizontal line of sight intersects graduations on two leveling rods set vertically at the points. From the readings on the leveling rods, observed elevation differences are determined.

When leveling between two widely separated points, it is necessary to set the leveling rods on intermediate temporary points, usually "turning pins," and accumulate a series of elevation differences until the overall elevation difference between permanently monumented points (bench marks) is determined. If proper instrumentation, observing procedures, and data reduction techniques are followed, the accuracy of these elevation differences can be sufficient to establish final heights of geodetic quality.

The National Geodetic Survey (NGS), a component of the Office of Charting and Geodetic Services, National Ocean Service, within the National Oceanic and Atmospheric Administration, U.S. Department of Commerce, is responsible for maintaining the National Geodetic Reference System (NGRS) in the United States. Within NGRS is a network of approximately 500,000 permanently monumented bench marks connected by geodetic leveling. These bench marks provide the basic vertical control for U.S. civilian surveying and mapping operations.

The observed elevation differences between these bench marks are processed through least-squares adjustments to determine orthometric heights referred to a common vertical reference surface. In this way, height values of all bench marks in the vertical control portion of NGRS are made consistent and can be compared directly to determine differences of elevation between bench marks in NGRS that may not be directly connected by lines of geodetic leveling.

The vertical reference surface in use in the United States, as in most parts of the world, approximates the geoid. The geoid was assumed to be coincident with local mean sea level at 26 tidal stations to obtain the Sea Level Datum of 1929 (SLD 29). Because heights of bench marks at other locations did not necessarily represent heights above mean sea level, and in order to avoid confusion when using bench mark heights referred to SLD 29 for surveying or mapping purposes, the datum was renamed (in 1973) National Geodetic Vertical Datum of 1929 (NGVD 29). The change was in name only; the same vertical reference system has been in use in the United States since 1929.

Geodetic leveling in the United States conforms to the Federal Geodetic Control Committee (FGCC) publication, *Standards and Specifications for Geodetic Control Networks* (FGCC, 1984). Other nations have adopted similar requirements. Geodetic leveling submitted to NGS for incorporation into NGRS must meet these FGCC criteria. Fulfillment of these criteria results in an assignment of a specific order and class to the bench marks in the survey. Order and class provide an overall assessment of the accuracy of the orthometric height of a bench mark. FGCC-acknowledged orders and classes for geodetic leveling are: first-order, class I; first-order, class II; second-order, class I; second-order, class II; and third-order.

An indication of the high accuracies attained by geodetic leveling is provided by the section-misclosure specifications in FGCC (1984). For example, the first-order, class I specification allows for a maximum of only 3 mm disagreement in elevation difference between forward- and backward-leveling between the same two bench marks, separated by a survey-route distance of 1 km. For third-order leveling, a maximum disagreement of 12 mm is allowed for a distance of 1 km.

Detailed requirements in FGCC (1984) include criteria for bench mark monumentation, instrumentation, calibration procedures, field procedures, office procedures, and proper ties to existing vertical control points. Inherent in geodetic leveling is the requirement for an absolute minimum of blunders and significant systematic errors. Both are controlled and minimized by proper field and office procedures (Zilkoski, 1983).

The following describes the corrections used to minimize systematic errors in geodetic leveling. The corrections are designated rod scale, rod temperature, level collimation, refraction, and magnetic, astronomic, and orthometric corrections. Procedures to minimize systematic errors are described in detail in Vaníček et al. (1980).

The length of the Invar (or equivalent) strip of a leveling rod used in geodetic leveling should be traceable to the National length standard. The length excess of an average rod meter is computed from the "actual minus nominal" length differences observed at several points along the rod. The correction is added with the resultant algebraic sign to the observed elevation difference:

$$C_r = De$$

where C_r = rod scale correction, D = observed difference of elevation for the section, and e = average length excess of the rod pair. (For C_r in mm, the units of D and e are typically m and mm/m, respectively.)

The length excess and index error are determined using a least squares adjustment of the observed "actual minus nominal" differences. The index error is the difference between the "imaginary" zero mark of the scale and the reference point on the bottom of the rod frame. The effect of index errors is minimized by appropriate field procedures; e.g., "leap-frogging" the leveling rods and using an even number of setups between the bench marks, or "matching" rod pairs with (nearly) equal index errors.

In 1980 the (U.S.) National Bureau of Standards developed an automated measurement system for the "detailed" calibration of geodetic leveling rods. The system incorporates a helium-neon laser interferometer to determine graduation locations on linear scales. The laser interferometer provides a length measurement referenced to either a fixture mounted to the foot piece of the rod or to any graduation on the rod scale itself. Accuracy (standard error) is better than ±50 micrometers.

The procedure to apply rod scale correction using detailed calibration data determines the standard length between the center of the rod foot plate, in line with the scale and each line graduation. The index correction (explained previously) is not needed when the detailed rod calibration is used.

The calibrated length of an average rod meter is determined for leveling rods for a standard temperature from the calibration and rod excess computations. The following procedure applies the rod temperature correction to the elevation difference between bench marks using the mean of the Invar temperatures observed at the beginning and end of a section. (A section is the interval between adjacent permanently monumented bench marks.) The correction is added with the resultant algebraic sign to the observed elevation difference for the section:

$$C_t = (t_m - t_s)DCE$$

where

C_t = rod temperature correction
t_m = mean observed temperature of the Invar strip
t_s = standardization temperature of the Invar strip
D = observed difference of elevation between the bench marks
CE = mean coefficient of thermal expansion per unit length per degree temperature of the rod pair

The units of D, t_m, and t_s must be consistent with the unit of CE. The unit of C_t is the same as the unit of D.

Collimation error is best minimized by field procedures. If sight lengths are balanced, i.e., $DS = 0$ and $SDS = 0$, where DS is the difference between backward and forward sight lengths at one setup (DS = backsight distance − foresight distance) and SDS is the accumulated DS for a section, the effect of the collimation error approaches zero. A well-adjusted instrument also minimizes this error without balancing sight lengths, although the collimation error of most leveling instruments changes slightly throughout the day as a result of changing temperature.

When SDS or the collimation error of the instrument is not zero, a correction is added with the resultant algebraic sign to the observed difference of elevation of each section:

$$C_c = -eSDS$$

where C_c = level collimation correction in mm, e = collimation error in radians × 1,000 or in mm/m, and SDS = accumulated difference in sight lengths for the section in m.

A simplified version of the model developed by the Finnish Geodetic Institute (Kukkamaki, 1939) is used to compute corrections to observed elevation differences of a section to minimize the effect of atmospheric refraction:

$$R = -10^{-5}a(s/100n)^2 tdW$$

where

R = refraction correction in mm for the section
a = 70
s = section length in meters
n = number of setups
t = "predicted" temperature difference in degrees Celsius between temperatures at 2.5 m and 0.5 m above the ground (upper temperature minus lower temperature)
d = difference of elevation for the section in units of half-cm
W = weather factor, which is 0.5 for 100 percent cloudy sky, 1 for 50 percent sunny sky, or 1.5 for 100 percent sunny sky

Holdahl (1981) provides a model for predicting temperature gradients for the conterminous United States from historic meteorological data, for use with the previous formula.

When temperature differences are observed at each setup, the atmospheric refraction correction is computed with the original Kukkamaki model (Kukkamaki, 1939) and added with the resultant algebraic sign to the observed elevation difference at a setup:

$$R = -10^{-5}a(S/50)^2 tD$$

where

R = refraction correction in mm
a = 70
S = sight length (instrument to rod) in m
t = observed temperature difference in degrees Celsius between temperatures observed at 1.3 and 0.3 m (upper minus lower temperature) at each setup
D = difference of elevation for the setup in half-cm

An additional source of error, caused by magnetic fields (including the geomagnetic field), is significant for certain compensator leveling instruments. For high-accuracy geodetic surveys, the use of instruments that are significantly influenced by the geomagnetic field is not recommended. Manufacturers of high-quality geodetic leveling instruments have changed their manufacturing materials and techniques so instruments are now routinely built that are not influenced by reasonably small magnetic fields.

Two magnetic calibration techniques have been used. The first involves the direct measurement, under laboratory conditions, of the influence of magnetic fields on leveling instruments. The second (empirical) method relies on the comparison of two or more surveys over the same leveling route, in which at least one of the surveys was performed with instruments not significantly influenced by magnetic fields, e.g., a spirit leveling instrument, and one or more surveys performed with the instrument for which a correction constant is to be determined. The magnetic correction to sections that are leveled with compensator instruments is:

$$M = AF \cos I \cos(a - D) s$$

where

M = magnetic correction in mm
A = magnetic constant, determined either in the laboratory or empirically, in mm/km/Gauss
F = intensity of the Earth's magnetic vector in Gauss
I = magnetic inclination
a = azimuth of the leveled section
D = declination, i.e., the azimuth of the Earth's magnetic vector
s = length of section in km

Magnetic error caused by the geomagnetic field has been found to be as much as 2 to 3 mm per km of north-south leveling observed with certain geodetic compensator leveling instruments.

The astronomic correction is applied to account for the effect of tidal accelerations due to the Moon and Sun on the Earth's equipotential surfaces. The astronomic correction is small, amounting, at most, to 0.1 mm/km, but it accumulates in the north-south direction. For example, the maximum accumulated astronomic correction between Spokane, Wash., and San Diego, Calif., is 7 cm. The astronomic correction is not usually applied to local leveling surveys, but it should be applied for systems of regional or continental extent. Development of the formulas involved in computing the astronomic correction is too lengthy for this article. Refer to Balazs and Young (1982).

The following formula is used to compute the approximate orthometric correction (based on normal gravity) to the observed difference in elevation of a section. The orthometric correction is applied to compensate for the fact that equipotential surfaces at different heights are not exactly parallel:

$$C_o = -2ha \sin 2p[1 + (a - 2b/a)(\cos 2p)] dp$$

where

C_o = orthometric correction
h = average height of the section
a = 0.002644
b = 0.000007
p = average latitude of the section
dp = latitude difference between the "begin-

ning" and "end" points of the section. (*dp* is positive when the "end" point is north of the "beginning" point.)

The unit of orthometric correction is the same as the unit of *h*. The orthometric correction is applied to the observed elevation difference of the section (elevation of the "end" point minus the elevation of the "beginning" point). The orthometric correction is not needed when geopotential numbers (explained later) are used.

By applying the orthometric correction and the other six corrections explained previously, a best estimate of observed, normal orthometric elevation differences, i.e., observed orthometric elevations based on normal (not observed) gravity values, is obtained. Gravity anomalies along leveling routes were not usually incorporated into height differences in the past, since, until the advent of modern gravimeters, it was not practical to measure gravity on a routine basis along leveling lines.

Observed, normal orthometric elevation differences have been used historically as input (observables) to least squares adjustments to determine adjusted ("final") normal orthometric elevation differences. By constraining the adjustment with appropriate normal orthometric heights from points in previously adjusted surveys, consistent adjusted normal orthometric heights are obtained. Adjusted normal orthometric heights referenced to the National Geodetic Vertical Datum of 1929 (NGVD 29) are the heights presently published and distributed by NGS.

Geopotential number is the difference between the potential of the geoid and the potential of the surface on which a particular point resides, and is expressed by:

$$C = \int_0^A g\, dn$$

where

C = geopotential number
0 = a particular point on the geoid
A = another point, connected to 0 by a line of leveling
g = average value of actual gravity between the successive bench marks
dn = elevation difference of the successive bench marks

Geopotential number is measured in geopotential units (gpu), where 1 gpu = 1 kgal meter = 1,000 gal meters. The use of geopotential numbers (instead of dynamic heights) was adopted by the International Association of Geodesy in 1955. Dynamic height differs from geopotential number only in scale, since division by a constant merely converts the geopotential number to a length.

The use of geopotential numbers (instead of orthometric heights based on normal gravity) to represent the heights of points of the National Geodetic Vertical Control Network is very advantageous. For example, geopotential differences between two points are independent of the route along which the leveling is run. Geopotential differences also provide an accurate measure of the "hydrostatic head" between points, i.e., if mean water surfaces of oceans or lakes represent a particular level surface, that surface has the same geopotential number at all locations. In the Great Lakes area of Canada and the United States, another vertical control system is in use by the Coordinating Committee on Great Lakes Basic Hydraulic and Hydrologic Data. This system, the International Great Lakes Datum of 1980, uses dynamic heights.

"True" orthometric heights are based on actual gravity values. In the next general readjustment of the National Geodetic Vertical Control Network, scheduled for completion in 1991, geopotential numbers, using actual gravity, will be computed for all bench marks. Geopotential numbers will then be converted to orthometric heights as part of the new adjustment associated with the North American Vertical Datum of 1988 (NAVD 88).

The Global Positioning System (GPS), incorporating NAVSTAR GPS satellites and ground-based GPS receivers, should revolutionize all geodetic surveying, including geodetic leveling, within the next few years. A point determined by GPS can be transformed into latitude, longitude, and *ellipsoidal* height, but the height is not directly related to an equipotential surface, i.e., the geoid, as is geodetic leveling. A knowledge of geoid heights (vertical differences between the geoid and a specific ellipsoid) is necessary to transform GPS-derived ellipsoidal heights to orthometric heights, which are the commonly used representation of geodetic heights.

It has been demonstrated that GPS can provide *differences* in ellipsoidal height at the same point to centimeter, or better, accuracy. This attribute has immediate geodetic applications to monitor, through repeated GPS surveys of the same points, areas undergoing significant vertical changes. The computation of the difference in GPS ellipsoidal height, measured at the same station at times differing by a few months or years, retains only the difference (over time) in geoid height at the point, which is usually negligible. In this way, the resulting difference in GPS ellipsoidal height provides information about vertical movement that is not significantly degraded by geoid height uncertainty.

Present determinations of geoid heights are not sufficient, in general, to provide geodetic-quality orthometric heights from GPS-derived heights. However, numerous applications can be satisfied. Tests in moderately severe terrain in both Germany and the United States have demonstrated that GPS heights, transformed to orthometric height differences using present geoid height determinations, were within 3–5 cm of geodetic leveling over lines averaging 14 km in length (Hannah, 1985).

GPS has already demonstrated that it can provide an economically attractive alternative for many applications that previously required traditional leveling techniques. GPS' advantages are certain to become even more pronounced in the future.

G. YOUNG

References

Balazs, E. I., and G. M. Young, 1982, Corrections applied by the National Geodetic Survey to precise leveling observations, *NOAA Technical Memorandum NOS NGS 34*, 12p.

Berry, R. M., 1976, History of geodetic leveling in the United States, *Surveying and Mapping* **36**(2) 137–153.

Federal Geodetic Control Committee, 1984, *Standards and Specifications for Geodetic Control Networks*, Rockville, Md.: National Geodetic Information Branch, NOAA, 29p.

Hannah, J., 1985, The global positioning system—the positioning tool of the future, *New Zealand Surveyor* **31**(3), 268–281.

Holdahl, S. R., 1981, A model of temperature stratification for correction of leveling refraction, *NOAA Technical Memorandum NOS NGS 31*, 27p.

Kukkamaki, T. J., 1939, Formeln und tabellen zur berechnung der nivellitischen refraktion, *Finnish Geodetic Institute Publication No. 27*, Helsinki, Finland, 18p.

Vaníček, P., R. O. Castle, and E. I. Balazs, 1980, Geodetic leveling and its applications, *Rev. Geophys. Space Physics* **18**(12), 505–524.

Zilkoski, D. B., 1983, Justification of FGCC vertical control specifications and procedures, *Technical Papers of the 43rd Annual Meeting*, ACSM-ASP Convention, Falls Church, Va.: American Congress on Surveying and Mapping, 128–143.

Cross-references: *Adjustment Methods; Figure of the Earth; Figure of the Earth: Equilibrium; Geodesy: Geometric; Geodesy: Historical Introduction; Geodesy: Physical; Geodesy: Satellite; Geodetic Ground Positioning; Geodetic Networks and Control Surveys; Geodetic Reference Systems; Geodetic Refraction; Global Positioning System (GPS); Gravity and Isostasy; Satellite Laser Positioning.*

GEODETIC NETWORKS AND CONTROL SURVEYS

Geodesy, as a science, draws on the knowledge and rigor of three basic disciplines, astronomy, mathematics, and physics, and in turn serves them and other Earth sciences through its unique discoveries. Geodesy is the study of the size and shape of the Earth, and thus supports a broad range of applications, from local surveys, to mapping of all kinds, to the study of the dynamics of the Earth as a whole, e.g., earthquakes and crustal movement.

Geodesy provides a foundational framework, a reference system of control points, required for many activities that demand precise locations or measurements on the Earth's surface. While no local survey by itself can be said to have any national significance, many surveys develop into regional networks, extending over large areas, and hence must be treated as national (and international) networks as a result of problems in the relationships between the different surveys. Accuracy standards have varied over time because of changing instrumentation and methods. Consequently, efforts have arisen to provide the essential systematic programming, rigorous standards, and continuity to maintain the geodetic networks necessary for economic development.

Before defining exactly what constitutes a geodetic network, let us examine the different requirements for spatial control. Such needs can be identified by user groups or by the various applications for the geodetic data. We find, in fact, that the two approaches are highly correlated.

Users of Geodetic Networks

To delineate the users of geodetic networks, we can consider the following example. An examination by the United States Office of Management and Budget (OMB 1973) on the use of horizontal geodetic networks revealed the following breakdown of users:

Federal agencies (47%)—mapping and charting, highway and facility construction, the operation and testing of defense weapon systems, and water resource studies and flood control projects.

State and local governments (17%)—planning and construction of streets and highways, regional and urban planning and development, water resources studies and flood control, and determination of boundaries and political subdivisions.

Industry and commerce (17%)—surveying and mapping projects, mineral exploration, transportation, utilities, and communications projects.

General public (19%)—college and university research and training projects and local surveys.

The geodetic networks of any country provide the control essential for mapping and charting programs. These programs must meet the expanding navigational needs as well as requirements for positioning boundaries and scientific studies. The political and economic significance of boundaries, with the increasing value of land and its resources, only serve to further accelerate the demands for spatial control information and systems to manage this information.

As our international involvement and outlook expands, we must reach toward the perennial goal of geodesists—a global geodetic reference framework, providing horizontal and vertical, or three-dimensional coordinates for national and international mapping and charting programs with the confidence that there will be no inconsistencies between the networks produced by individual countries. The International Association of Geodesy serves as a forum for geodetic research cooperation and information exchange throughout the world.

Defining a Geodetic Network

Now that we have established some of the requirements, we can proceed to define what constitutes a *geodetic network,* how one is established and maintained, and what the products or outputs of that effort are. We begin by defining the concept of a *geodetic reference system.* As defined by Epstein and Duchesneau (1984), a geodetic reference system is a set of marked points whose relative locations are known and which possess the following features:

A collection of permanently marked and maintained points.

Coverage of an extensive area.

A spatial relationship of known accuracy.

Relationships expressed in a common mathematical language or in a language translatable into other languages.

General availability of geodetic information.

Epstein and Duchesneau continue by distinguishing between local and global systems and the outputs that each system yields. As used here, local control refers to a reference system established for a specific site or project often with little or no emphasis on maintenance. A local control system serves two purposes: (1) to provide accurate measurement throughout the specific site, and (2) to create local compatibility for the site measurements. Thus, local control is site and project specific with its creation and maintenance tending to be short term.

A global system is one whereby spatial information products based on a geodetic network possess an additional and unique attribute. The individual products can be related to each other across all sites with a high degree of accuracy. We characterize this attribute as "universal compatibility." Such compatibility allows secondary and tertiary users to take two spatial information products that may have been produced by different individuals for totally unrelated primary purposes (two *local* control networks), such as one for the location of a river and one for an electric transmission line, and accurately depict their relative positions on a single map.

The existence of known mathematical relationships between elements of the geodetic network is vital to the notion of universal compatibility. By applying mathematical techniques to combine the observations taken in the individual surveys, a geodetic network is built up and expanded as new surveys are accepted and "fit" into the network. The usual or potential products of this mathematical processing are spatial coordinate(s) of the points within the individual surveys, measures of the accuracy of these coordinates relative to other points in the network, and improvements to the mathematical model representing the physical environment of the survey observations.

Datums, Standards, and Specifications

The mathematical model forming the basis of our computations includes the adoption of a "datum," a set of quantities to define a mathematical figure approximating the Earth (see *Figure of the Earth*), along with an origin and orientation of the network's coordinate system. Various datums are currently in use throughout the world having been derived to best fit the particular region.

After computation, the computed accuracies of the point coordinates are compared to "standards" in order to categorize the points. For example, when a horizontal point in a network is classified with a particular "order" and "class," the U.S. National Geodetic Survey certifies that the geodetic latitude and longitude of that point bear a relation of specific accuracy to the coordinates of all other points in the network. This relation is expressed as a distance accuracy, i.e., the ratio of the relative positional error of a pair of points to the horizontal separation of those points. The distance accuracy standards shown in Table 1 are used in the United States and are the responsibility of the Federal Geodetic Control Committee (FGCC 1984).

Although there is no internationally accepted set of geodetic standards, an effort is underway to develop such standards under the International Standards Organization.

The particular surveys are carried out according to "specifications"—recommended field instructions devised to most likely achieve the desired point standards. The specifications for a horizontal surveying technique known as "traverse" are an example of specifications recommended by the Federal Geodetic Control Committee (FGCC 1984), and are given in Table 2. Specifications are combinations of rules of thumb and studies of error propagation, based on experience, of how best to achieve

TABLE 1. Distance Accuracy Standards

Classification	Minimum Distance Accuracy
First-order	1:100,000
Second-order, class I	1: 50,000
Second-order, class II	1: 20,000
Third-order, class I	1: 10,000
Third-order, class II	1: 5,000

TABLE 2. Traverse Specifications

Order Class	First	Second I	Second II	Third I	Third II
Station spacing not less than (km)	10	4	2	0.5	0.5
Maximum deviation of main traverse from straight line (degrees)	20	20	25	30	40
Minimum number of bench mark ties	2	2	2	2	2
Bench mark tie spacing not more than (segments)	6	8	10	15	20
Astronomic azimuth spacing not more than (segments)	6	12	20	25	40
Minimum number of network control points	4	3	2	2	2

a desired level of quality. Unfortunately, there is no guarantee that a particular standard will be met if the associated specifications are followed. Therefore, specifications are usually designed with a safety factor such that slight departures from the strict specifications can usually be accommodated. The selection of standards and specifications are dictated by the purpose of the survey and the extent and accuracy of the existing geodetic network in the region of the survey.

Types of Geodetic Networks

The three primary types of geodetic networks—horizontal, vertical, and gravity—provide the common basis for all surveying and mapping operations to ensure a coherent product. These networks are interrelated in that the quantities derived in one network may be required for computations in another network. Also, a given station may be a point in one, two or all three geodetic networks.

Horizontal Geodetic Network. A horizontal geodetic network traditionally consists of the collection of established points and their interconnecting observations, latitudes and longitudes referenced to some adopted initial point (origin) and, employing a mathematical figure, an ellipsoid with its definition of size and orientation. The points within the horizontal geodetic network are subdivided according to their relative accuracy as determined by the mathematical evaluation of the survey observations as mentioned above. This evaluation includes some a priori knowledge of the potential for errors because of the observing techniques and the physical environment and the propagation of these errors to the final results.

On the North American continent, the horizontal geodetic network has just recently undergone a major redefinition effort to improve the coordinates of the reference points to evaluate the relative accuracies of these points, and thus the adequacy of the network, and to improve our knowledge of the figure of the Earth in North America. This redefinition known as the Readjustment of the North American Datum of 1983 (NAD 83) involved 250,000 horizontal points, 1,750,000 observations, the simultaneous solution of 900,000 mathematical equations, and significant international cooperation.

Vertical Geodetic Network. A vertical geodetic network consists of a series of lines of levels that have been interconnected in such a manner that closed loops or circuits have been formed. The resulting mathematical evaluation of these leveling observations provides heights for the individual marked points (usually called "bench marks") and an evaluation of the accuracy of these heights. The level surface to which these heights are referred is the *vertical datum*. The traditionally adopted datum for leveling operations is local mean sea level with a defined elevation of zero. For local surveys, when a sea-level connection is not available, an arbitrary datum may be adopted and defined in terms of an assumed elevation for some bench mark. An equipotential surface called the *geoid* is important to the definition of a vertical datum (see *Figure of the Earth*).

Like the redefinition effort for the horizontal network, the North American Vertical Datum is being redefined and will be completed in 1988 or 1989. It is known as the North American Vertical Datum of 1988 (NAVD 88). This massive effort, involving 620,000 bench marks, will provide an improved set of elevations, an evaluation of the adequacy and usefulness of the vertical geodetic network, improvements to our knowledge of the figure of the Earth, and also important, improved access to the geodetic data.

Gravity Networks. Points in the gravity network have gravity accelerations commonly expressed in milligals (10^{-5} m·s^{-2}) or microGals (10^{-8} m·s^{-2}) and classifications reflecting the measurement accuracies and site stabilities. The gravity values, established by absolute or relative observing techniques, contribute to our knowledge of the Earth and are required for computations of the other networks.

Special Geodetic Networks. In addition to the horizontal, vertical, and gravity networks, other special purpose networks exist as part of our geodetic framework. These networks include polar motion and Earth rotation networks to monitor variations in orientation and rate of the Earth's rotation and crustal motion networks to monitor shifts and subsidence.

Control Surveys

A *survey* can be defined as the orderly process of determining data relating physical characteristics of the Earth. In *geodetic surveys* we are establishing a metric relationship between physical points, and we account for the figure and size of the Earth in processing our measurements. The mathematical models for the Earth and physical influences affecting our measurements are used in processing our observations to obtain the desired relative quantities. We further define *control surveys* as being those surveys intended to establish or extend a geodetic network to serve as a reference for subsequent geodetic surveys.

Survey Techniques

Ever since the ancient Greeks first took an interest in the size and shape of the Earth, man has been attempting to determine the Earth's dimensions and the locations of features on its surface. During most of this time, the work has been carried out under the handicap of being confined to the Earth's surface. Because direct measurements of azimuth and distance could be made only between points that were intervisible, directly connected points could be no further away than the horizon. To measure longer distances and determine the relative positions of widely separated points, chains and networks of intermediate points, each one visible from its immediate neighbors, were established and linked together by measurements.

Traditional terrestrial control surveying with theodolites and geodimeters uses line-of-sight direction and distance measurements. This type of surveying technology has virtually reached its limits of accuracy, economy, and efficiency. The methods produce relative positioning accuracies up to one part in 10^5 of interstation distances over distances ranging up to 20 km to 30 km, which is generally adequate for local needs. However, traditional surveying methods are labor intensive and slow, and unfavorable error propagation seriously degrades accuracies over longer distances. The most serious limitation of traditional methods is the requirement for station intervisibility.

Inertial control surveying methods are much faster than the traditional methods and have the added advantage that stations need not be intervisible. Accuracy is in the 20 cm to 40 cm range over distances of 5 km to 1000 km. Adverse considerations are the relatively large bulk and cost of the equipment.

Doppler control surveying techniques, using the TRANSIT navigation satellites, produce 20 cm to 50 cm accuracies over distances of 5 km to 1000 km. The equipment is relatively compact and can be backpacked. Like inertial surveying, stations need not be intervisible, but unlike inertial methods, observing periods average about 2 days to acquire an adequate number of satellite passes.

For long (100 km to 2000 km) baseline measurements, satellite laser ranging produces results in the 3–8 cm accuracy range. Its use is limited by the length of time needed to acquire sufficient observations (1–3 weeks) and by the size and weight of the equipment.

Very-Long-Baseline Interferometry (VLBI)

After more than 15 years of development and refinement, geodetic VLBI has now proven to be a very powerful and cost-effective method of obtaining measurements that are vital to several aspects of geodesy and geophysics. VLBI is a system that employs multiple radio astronomy antennas to observe simultaneously signals from extragalactic radio sources, such as quasars. The time-tagged signals received at each antenna are cross-correlated to determine common signal arrival times that, together with the known speed of light and directions of the sources, can be used to determine the components of vectors connecting the observing stations. The lengths of vectors between stations separated by as much as several thousand kilometers can be obtained to 1–2 cm, accuracies achievable from observing periods of less than 1 day.

But VLBI provides much more than just highly precise distances. It also determines the orientation of the Earth with respect to the quasi-inertial coordinate system of radio sources. Repeated determinations at regular, frequent intervals provide a precise record of the Earth's orientation and rotational rate.

The accuracies that can be achieved by VLBI are unprecedented. Polar motion, which optical methods could determine to no better than 0.5 m to 1.0 m, is being determined to 3 to 5 cm in each component; and the true angular rotation of the Earth (UT1) is being obtained to better than 0.1 millisecond (Carter et al., 1985). In addition to being able to monitor the dynamic behavior of the Earth in support of geophysical as well as atmospheric research, these heretofore unattainable accuracies provide the basis of scale and orientation for our geodetic networks.

Global Positioning System (GPS)

The Navigation Satellite Timing and Ranging Global Positioning System (NAVSTAR GPS) is under development by the U.S. Department of Defense as a worldwide, all-weather navigation and timing system. When fully configured (1990), there will be at least 18 satellites, 3 in each of 6 evenly spaced orbital planes. The satellites will be maintained in near circular, 55° inclination orbits of 20,000 km radii (12-hour periods). This configuration is designed so that four to seven satellites will be visible from any point on the Earth at all times.

Using GPS, suitably equipped users will determine, instantaneously (or nearly so), their position and velocity with greater accuracy and at lower cost than with any other method previously available. The potential for revolutionizing control surveys, and many other aspects of our lives, should be obvious.

A significant collection of GPS articles related to control surveys can be found in the *Proceedings of the First International Symposium on Precise Positioning with the Global Positioning System* (NOAA 1985). Many authors have documented significant GPS capabilities to determine relative positions (distance and vector components) for control surveys. Such discussions of positioning capability often include reference to "absolute" positioning, that is, positioning relative to the defined origin of the coordinate system. Techniques have been devised, and will continue to be refined, for improved "static" and "dynamic" positioning.

Outlook for the Future

As the price continues to drop for GPS surveying equipment, there is little doubt that GPS will replace terrestrial methods for most control surveying. With GPS, and the ability to span distances of 100 km or more on each line, new geodetic networks in previously unsurveyed areas will initially have far fewer stations, and these will be at points more easily accessible than the hilltops so common in terrestrially established networks.

Network densification will also be accomplished by GPS, but it may be done only as the need arises for specific purposes, rather than as blanket coverage for all future needs. With the speed and economy of GPS geodetic positioning, and considering the cost and susceptibility to disturbance of permanent geodetic markers, the use of temporary markers may occur. Our notion of the geodetic network and its primary reference points must also now include those reference points in space—the quasar radio sources and the GPS satellites orbiting the Earth.

The concept of a geodetic network is undergoing radical change in light of the new surveying technologies. As geodesists work to refine our understanding and application of these technologies, they will discover additional ways to serve the scientific and informational needs of our society.

JOHN D. BOSSLER

References

Carter, W. E., D. S. Robertson, and J. R. Mackay, 1985, Geodetic radio interferometric surveying: Applications and results, *Jour. Geophys. Research* **90**(B6), 4577-4587.

Epstein, E. F., and T. D. Duchesneau, 1984, *The Use and Value of a Geodetic Reference System.* Orono, Me.: University of Maine.

Federal Geodetic Control Committee (FGCC), 1984, *Standards and Specifications for Geodetic Control Networks,* September. Rockville, Md.: National Geodetic Survey.

National Oceanic and Atmospheric Administration (NOAA), 1985, ed. C. Goad, *First International Symposium on Precise Positioning with the Global Positioning System, Proceedings,* April. Rockville, Md.: National Geodetic Survey.

Office of Management and Budget (OMB), 1983, *Report of the Federal Mapping Task Force on Mapping, Charting, Geodesy and Surveying,* July. Washington, D.C.: OMB.

Cross-references: *Adjustment Methods; Crustal Movements and Tectonic Deformation; Earth's External Gravity Field; Figure of the Earth; Geodesy: Geometric; Geodesy: Historical Introduction; Geodesy: Physical; Geodesy: Satellite; Geodetic Ground Positioning; Geodetic Leveling; Geodetic Reference Systems; Global Positioning System (GPS); Map Projections in Geodesy; Very-Long-Baseline Interferometry (VLBI).*

GEODETIC REFERENCE SYSTEMS

A geodetic reference system is made up of two components, (1) the astronomical and geodetic constants and theories used in the establishment of geodetic positions and orientations, and (2) the reference frame in which the positions and orientations are defined. Observations made to determine location or orientation on the surface of the Earth may be made by observing the orientation of a local reference vector with respect to an adopted realization of an inertial reference frame. This reference vector may be the direction of the local gravity vector, the radius vector from the center of mass of the Earth, or a baseline vector established by the location of two points on the Earth's crust. In the treatment of these observations it is necessary to assume that the positions of celestial sources defining the inertial system are known. The celestial sources might be stars; radio sources; the Moon, Sun, or planets; or artificial Earth satellites. The positions of these sources are given in a celestial reference frame at some instant in time. Astronomical and geodetic constants and theories are required to transform these positions to the instant at which the observations were made. By observing the relationship between the local reference vector and the vectors pointing to the celestial sources, the observer can determine the position in a local coordinate reference frame, which can then be transformed to a more general global reference frame if the necessary parameters are known.

Alternately, an observer may be able to determine

his location relatively with respect to some nearby sites whose locations are well defined in a geodetic reference frame. In this case, use of an astronomical coordinate system may not be required. He might observe the direction to some celestial source at this site as well as at the nearby sites with known locations and transform the differences in observed parameters to differences in station location. The observer might also be able to measure directly his distance from each of the terrestrial, previously defined, locations and position himself without need for a celestial frame. These relative measures then give the location of the observer in the reference frame in which the fiducial sites are located. In one case the observer makes use of celestial observations while in the second he might not.

Precise definitions of geodetic reference systems depend on the astronomical and terrestrial reference systems that are employed in the determination of the geodetic reference systems, as well as the procedures used to transform between them. In astronomy, reference systems seek to define coordinates or directions in quasi-inertial systems. These may be technique dependent reference systems, but procedures to transform between them continue to be developed. Terrestrial reference systems are being developed based on modern technology. This involves the choice of selected sites where the most precise instrumentation is available for the required observations. Colocations of multiple observing methods at these sites make it possible to establish the transformation procedures.

Astronomical and Geodetic Constants and Theories

The constants and theories that are part of any geodetic reference system may be divided into those that might be considered to be largely astronomical in nature and those that are considered geodetic. These areas may overlap, and every effort is made to ensure that the astronomical and geodetic constants and theories are consistent.

Astronomical constants and theories deal with the description of the locations of, or directions to, reference objects in a quasi-inertial reference frame. These include the description of precession, nutation, sidereal time, and Earth rotation as well as the dynamical theories involved in the description of the positions of artificial Earth satellites or solar system objects. Also included in this area are theories related to the intrinsic proper motions of stars and the assumption that quasars are at such very great distances that proper motions may be neglected.

The theories of *precession* and *nutation* describe the motion of the direction of the Earth's rotation vector in a quasi-inertial reference frame. This motion is caused by the gravitational attraction of solar system objects on the equatorial bulge of the Earth. Because the Earth is not truly spherical, these objects are able to exert a torque on the Earth causing its rotational axis to move in space. Precession refers to the slow motion of the axis in space by which the axis describes a cone with a half angle at the apex of approximately 23.5°. The Earth's axis completes one cycle of this motion in about 26,000 years. Nutation refers to the shorter periodic motions of the axis that are caused largely by the gravitational attraction of the Sun and the Moon. The largest of the nutational motions occurs with a period of 18.6 years and produces an elliptical motion of the rotational axis about the direction of the *mean pole* whose direction is given by the theory of precession. The periods of the nutational motions range from one day to 18.6 years, and the combination of all of the motions results in a complicated pattern of the direction of the rotational axis in space.

The constants describing precession and nutation are determined from astronomical observations. Before recent times these observations were of the directions and motions of stars. Since the advent of more modern technology, estimates of precessional constants are made by the study of the apparent positions of solar system objects as well as the stars. The constants describing the motion of the rotational axis depend on analytical theories of the motion of the Sun and Moon with respect to the Earth and on our theoretical knowledge of the distribution of the mass of the Earth and its viscosity. Based on our knowledge of the internal constitution of the Earth we are able to predict the motion of the Earth's axis and these can be checked using precise observations. The most current theories and constants are able to describe the direction of the rotational axis with an accuracy of about ± 0.001 seconds of arc.

The rotational motion of the Earth about its axis is described by *sidereal time* and the definition of *Universal Time*. Sidereal time refers to the angle measured along the equator between the direction of the origin of the right ascension system of the quasi-inertial reference frame (generally the *vernal equinox* or *first point of Aries*) and the direction of the local vertical. Greenwich sidereal time refers to the sidereal time at the meridian of Greenwich (longitude 0 degrees). The sidereal time that may be observed is called *apparent sidereal time*. Nutation causes the direction of the vernal equinox to undergo periodic variations that are easily determined. *Greenwich mean sidereal time* is equal to Greenwich apparent sidereal time corrected for the nutational motion of the equinox.

A numerical relationship exists between Greenwich mean sidereal time and Universal Time that effectively defines the Universal Time called *UT1*. This numerical relationship is based on an adopted motion for a fiducial direction located in the plane of the Earth's equator moving at a uniform rate equal to the mean motion of the Sun throughout the year. Astronomical observations of sidereal time are made that are related to UT1 using this theory. Then, the difference in time between UT1 and the civil time *UTC* (Coordinated Universal Time) may be determined. Because the Earth's rate of rotation is variable, the difference UT1 − UTC may not be determined theoretically as is the case with precession and nutation. This difference must be observed and values of the daily determinations are available from agencies such as the Central Bureau of the International Earth Rotation Service in Paris or the U.S. Naval Observatory. The UTC time scale is adjusted as needed to ensure that UT1 − UTC does not exceed +0.8 seconds of time.

In practice, an observer may determine the rotation angle of the Earth with respect to the origin of a quasi-inertial reference frame at some instant of UTC by using, first, the observed difference UT1 − UTC to find UT1; second, using the definition of UT1 to find Greenwich mean sidereal time; and finally, applying corrections to obtain Greenwich apparent sidereal time.

As is the case with the rotational angle, observations of the direction of the rotational axis with respect to the axis of the geodetic reference frame are also required. Because the Earth is not truly spherical the rotational axis will appear to move about with respect to the geodetic reference frame attached to the Earth with a period of about 435 days. This Chandler polar motion is due to the fact that the rotational axis is not aligned with the axis of the principal movement of inertia. Also the annual redistribution of atmospheric mass causes an annual motion of the rotational pole as seen by a terrestrial observer. These motions, which result in a difference of up to 0.5 seconds of arc in the directions of the axis of rotation and geodetic reference frame axis, must be observed and are tabulated in the publications of the International Earth Rotation Service. These corrections are given in the form of an angular offset between the Conventional International Origin (CIO) and the axis of rotation, the CIO being the direction of the rotational axis averaged over the years 1900 through 1906. It is used as an arbitrary origin to describe the motion of the rotational axis. The angular offset is given in two directions: x along the meridian of Greenwich and y along the meridian 90° W.

Also involved in the definition of quasi-inertial reference frames are the constants and theories devoted to the description of positions of solar system objects (particularly the Moon) and artificial Earth satellites. These include the gravitational constant G; the masses of the Sun, Moon, and planets; the dynamical shape of these objects; theories that describe the atmospheric drag on artificial Earth satellites; the gravity field; and the effect of solid Earth and ocean tides on satellites. All of these items are included in the formulation of ephemerides of the location of the object in a quasi-inertial reference frame. Modern technology has made it possible to account for all of these effects in the numerical integration of orbits with accuracies at the level of 0.001 seconds of arc.

Geodetic constants and theories are concerned with parameters describing the shape of the Earth, motion of crustal plates, the defining directions of geodetic reference frames, and local gravity fields. The Earth can be characterized as being generally ellipsoidal in shape having an equatorial semidiameter and flattening determined by geodetic measurements. Numerous estimates of these quantities have been made and one of these must be specified for each reference frame in use. The orientation of most geodetic reference frames is such that the direction of the north pole is assumed to be parallel to the direction of the CIO and that the zero of longitude is the point where the meridian of Greenwich intersects the equator. Not all geodetic reference frames use identical defining directions, and rotations based on intercomparisons of geodetic coordinates for the same site in different reference frames are employed to refer coordinates given in one frame to those given in another frame.

Since measurements are frequently made with respect to a local gravity vector, knowledge of the local gravity field is essential in transforming observations between reference frames. A number of gravity fields are in use for this purpose. Satellite geodesy, which has made possible the determination of coordinates with respect to the center of mass of the Earth with unprecedented accuracy, relies heavily on the adopted gravity field. Modern observations now indicate that motion of crustal plates must be taken into consideration in making precise geodetic measures. It can no longer be assumed that sites on the surface of the crust are not in motion with respect to the center of mass, and theories concerning the motion of the plates have been developed.

Astronomical and Geodetic Reference Frames

Reference frames refer to the lists of locations of or directions to specified objects or sites. Astronomical reference frames, for example, are defined by the directions to stars at some epoch, directions to distant quasars, and locations of solar systems objects or artificial Earth satellites. Geodetic reference frames are defined by specific locations on the Earth or reference directions assumed to be fixed to the Earth's crust.

Astronomical reference frames are defined by a reference plane (generally the plane of the Earth's equator) and a fiducial point (*vernal equinox*). The vernal equinox is the direction where the equatorial plane and the ecliptic plane defined by the plane of the Earth's orbit meet. The vernal equinox is specified because it is the direction of the Sun at the spring equinox when the Sun is moving from south to north in the sky viewed from the Earth. Although the term "coordinates" is often used, it may refer strictly to directions since some objects defining the astronomical reference frame are at essentially indeterminable distances. *Declination* is the angular distance above or below the equator measured along the great circle that passes through the orbit perpendicular to the equatorial plane. *Right ascension* is the angular distance from the equinox to the point where the great circle passing through the object, and perpendicular to the equator, intersects the equator. Alternately, coordinates may be specified in Cartesian coordinates if the distances may be measured with sufficient precision. The Cartesian coordinates refer to the equatorial plane and a fiducial direction in the plane, and the origin is generally the center of mass of the Earth.

Since precession and nutation act to move the position of the equinox in the sky, an epoch is assigned to the reference frame to specify the precise direction of the equator and equinox, for example at the beginning of the year 2000. For those reference systems describing the location of, or directions to, objects that have modelable motion, it is also necessary to specify an epoch, or the instant in time represented by the list of coordinates and motions.

The adopted astronomical reference frame is referred to as the FK5 system standing for the *Fifth Fundamental Catalog,* a system of stellar positions and motions obtained from the compilation of numerous astronomical observations and consistent with the most modern theories and constants. This frame is largely of interest to those dealing with observations of optical sources. Because the stars have individual motions (*proper motions*) with respect to a truly inertial frame, the proper motions are listed in the FK5 in order to achieve a quasi-inertial frame. The position and motions are given at a reference epoch (the year 2000.0) and the reference frame defined by the equator and equinox of the epoch 2000.0. Coordinates and proper motions are given in both right ascension and declination.

Because optical measures are hampered by atmospheric refraction, modern techniques capable of higher precision have made use of other astronomical reference frames related in some way to the FK5 system. Ephemerides of planetary and satellite positions obtained through numerical integration are given in the FK5 system of right ascension and declination. Positions of quasars obtained with radio interferometric techniques are also specified in the FK5 system. However since these quasars appear to be at such great distances, it is not necessary to model individual motions. For those quasars that appear to have measurable angular extent it may be necessary to model the structure of the source.

Geodetic reference frames are defined also with respect to a reference plane (the equator) and a fiducial direction in that plane. Since distances can be measured on the Earth, three coordinates can be given. Generally these can be in the form of Cartesian x,y,z coordinates or in the form of longitude, latitude, and height above a reference surface. The origins of these frames are in the vicinity of the center of mass of the Earth; the reference direction is in the direction of the meridian of Greenwich. The geodetic reference frames in use are chosen to meet specific requirements and to be useful in characterizing the measures made to meet those requirements. Local geodetic frames, for example, may be adequate to meet the requirements for precise geodetic coordinates in a certain region, but these may not be adequate for a global representation.

Different reference surfaces may also be employed depending on how the geodetic measures are made. Since measures are frequently made with respect to the direction of gravity, a reference surface defined by the surface with equal gravitational acceleration is often used as a reference and called a *geopotential surface.* The surface coinciding most nearly with the mean level of the oceans is called the *geoid.* Mathematical surfaces that are modeled to approximate these surfaces are called reference ellipsoids. They are characterized by an *equatorial semidiameter,* and a *flattening* defined by $f = (a - b)/a$ where b is the polar semidiameter and a is the equatorial semidiameter.

Astronomic latitude and longitude are measured using the direction of the plumb line as the local reference. Thus this measure is referred to the geopotential surface at the location of the observation. This observation can then be referred to a standard geoid or another geopotential surface if there is adequate knowledge of the local gravity field.

Geodetic latitude and longitude are measured with respect to a reference ellipsoid. Heights are measured with respect to the ellipsoid along the normal to the ellipsoid at the location of the observation. Coordinates are found from distances or direction measurements that are referred to the reference ellipsoid.

A *geodetic datum* is specified by a reference ellipsoid and the longitude, latitude, and height of an initial point. A variety of ellipsoids and datums are in use and procedures are available for transformation between them.

The BIH has compiled a directory of sites with extremely precise geodetic coordinates, which defines the BIH Terrestrial System (BTS). These coordinates are based on observations made using satellite laser ranging, lunar laser ranging, and very long baseline radio interferometry and serve as a standard for the definition of a terrestrial reference system.

Summary

As the precision with which astronomical and geodetic measurements can be made improves, the theories, constants, and reference frames are continuously improving. Current best estimates of these items are outlined in the *Project MERIT Standards* available from the U.S. Naval Observatory as U.S.

Naval Observatory Circular 167. Textbooks containing information on reference systems giving detailed procedures and constants are listed in the Bibliography.

DENNIS D. McCARTHY

Bibliography

Altamini, A., C. Boucher, and M. Feissel, 1988, *Directory of Sites Participating to the Realisation of the BTS for 1984*, Notes Internes Bureau International de l'Heure No. 5, available from the Bureau International de l'Heure, Paris, France.

Bomford, G., 1962, *Geodesy*. London: Oxford University Press.

Melbourne, W., R. Anderle, M. Feissel, R. King, D. McCarthy, D. Smith, B. Tapley, and R. Vicente, 1983, *Project MERIT Standards*, U.S. Naval Observatory Circular No. 167.

Mueller, I. I., 1969, *Spherical and Practical Astronomy as Applied to Geodesy*. New York: Frederick Ungar Publishing Co.

Woolard, E. W., and G. M. Clemence, 1966, *Spherical Astronomy*. New York: Academic Press.

Cross-references: *Earth Orientation; Earth's External Gravity Field; Figure of the Earth; Geodesy: Geometric; Geodesy: Historical Introduction; Geodesy: Physical; Geodesy: Satellite; Geodetic Ground Positioning; Geodetic Leveling; Geodetic Networks and Control Surveys; Global Positioning System (GPS); Map Projections in Geodesy; Planetary Geodesy; Satellite Laser Positioning; Very-Long-Baseline Interferometry (VLBI)*.

GEODETIC REFRACTION

Geodetic refraction is the term applied to cover the effects of the atmosphere on geodetic observations. An electromagnetic wave, which is the subject of the geodetic measurement, is affected by the inhomogeneities of the atmosphere's refractive index. When the wave traverses the atmosphere, the effects of variations in refractive index on its velocity and direction are the most significant; its amplitude, phase, and frequency are also affected. Thus the geodetic observations subject to refraction include horizontal and vertical angles, electronic distance measurements (EDM), leveling, laser ranges, Doppler and global positioning system (GPS) observations, and very long baseline interferometry (VLBI)—virtually all geodetic observations except those based on gravity.

It is not possible here to deal with all cases of geodetic refraction, but in four selected cases the refraction effect and its correction are covered. First, there is information on how refraction effects occur, the types of solution used in practice, and how to model the atmospheric boundary layer.

In the past, refraction has too often been regarded as a nuisance and relegated, in describing the methodology, to a single paragraph quoting the source of the correction model. The formulation of the correction has often been merely the most convenient available, selected without discrimination or understanding of its physical basis. However, the accuracy of geodetic measurements has increased dramatically in recent years, particularly in the past decade. Many observations now have an intrinsic accuracy of better than one part in a million. The instrumental accuracy has improved to the point where it has overtaken the refraction effects, so refraction has become the limiting factor in the accuracy of geodetic measurements. With this realization, refraction problems have been treated with greater care and have in recent years been the subject of substantial research efforts.

The magnitude and the variation of refraction depend on physical factors such as the frequency of the electromagnetic waves used in the measurement and the part of the atmosphere traversed. In terrestrial measurements (that is, between points on the Earth's surface), the waves must traverse the highly turbulent and variable boundary layer. For terrestrial measurements at optical frequencies, the most significant parameter is the atmospheric temperature. Temperature gradients control the bending of the waves, whereas the actual temperature and its variation along the path length determine the velocity. Particularly in the lower layers of the atmosphere, and for long lines, it may be difficult to model the corrections adequately.

Terrestrial measurements such as EDM at microwave frequencies are affected by atmospheric water vapor along the path, as well as temperature. The occurrence of water vapor, particularly close to the surface, is irregular and difficult to model.

In cases where the wave path traverses the whole depth of the atmosphere, observations at optical frequencies, such as laser ranging, are affected by temperature along the path, but microwave observations, for example Doppler satellite measurements, are affected by temperature and by water vapor. In this case the waves pass only briefly through the lowest layers of the atmosphere; the remainder of the atmosphere is "well-behaved" and can be modeled accurately. In addition, microwaves are subject to very substantial effects when traversing the ionosphere. Although affected by water vapor and ionosphere, which may be difficult to correct accurately, microwaves are unaffected by clouds. Microwave systems have the overwhelming advantage of all-weather capability.

Some geodetic techniques for correcting refraction are dependent on the dispersion qualities of the waves. In the neutral troposphere, optical waves are subject to dispersion but microwaves are unaffected. Conversely, in the ionosphere, optical waves are unaffected but microwaves are dispersive, with both the magnitude of the refraction effect and its dispersion dependent on the frequency. At higher frequencies, the refraction is smaller but so is the dispersion.

Types of Solution

The practical solutions applied to problems of atmospheric refraction can be classified as special observation techniques, modeling, instrumental solutions, and remote sensing of parameters. An example of a *special observation technique* is in precise leveling, where the lengths of backsights and foresights are made equal and lines of sight close to the surface are avoided. In this way the refractions on backsight and on foresight are very nearly equal and tend to cancel in the height difference Δh.

The classical example of *modeling* is the simple model for refraction in trigonometric heighting, which has been of great practical value over a period of at least 200 years. The model assumes that the curvature of the sightline in the vertical plane is constant. The radius of this curvature can be determined, and within limits it is nearly constant.

Instrumental solutions include development of instruments using dispersion as a measure of the total refraction. This has been applied for vertical angle measurements as well as for EDM. Using GPS as an alternative way of determining relative positions of points on the surface can be considered as an instrumental solution. When the relative positions of points 50 km apart are measured, the terrestrial measurements traverse the full 50 km close to the surface, in the irregular and turbulent lower atmosphere. The loss of accuracy over the distance is due largely to the atmospheric effects. Employing waves that connect two positions via satellites, the waves only pass through two relatively short sections of the difficult lower atmosphere, retaining the high relative accuracy. Hence, though the "instruments" are very elaborate, this procedure is an instrumental way of avoiding the refraction effect. We have the paradox that when relative positions of points are fixed more than about 50 km apart, it is more accurate to go all the way to satellites and back.

The best-known example of *remote sensing* for the refraction correction is the use of the water vapor radiometer (WVR), used to correct microwave measurements, particularly VLBI, for effects of atmospheric water vapor.

Water Vapor Radiometer (WVR)

Geodetic measurements are currently capable of measuring baselines of several thousand kilometers to accuracies of 2-5 cm. At present VLBI, using microwaves, measures baselines of this length. In the future, GPS measurements, with comparable accuracies, will extend to these distances. In order to achieve the full potential of these measurements, we must apply a water vapor correction, accurate to better than 2 cm.

Water vapor in the atmosphere is not a well-mixed constituent. It does not occur in a regular stratification, so accurate modeling does not appear possible. The water vapor content of the atmosphere must be measured directly so that the delay in arrival time can be modeled. A direct method of measuring the vapor is by passive microwave radiometry. The instrument detects the very faint microwave energy radiated by atmospheric water vapor.

There are two basic difficulties in the design of the radiometer. The power of the radiation is very small: The instrument has to deal with a noise power input of 10^{-19} watt. A large amplification is needed, with the result that there are problems of instrumental losses and instability. These problems are overcome by standardization at two points on the output scale, which is expressed as sky brightness temperature. The second problem is the presence of liquid water, which occurs as droplets in clouds. It causes dramatic changes in radiation but has very little effect on path delay. This problem is overcome by making the radiometer a dual-wave instrument.

Water vapor radiometers have been developed independently by several organizations. The present description refers to a set of WVRs designed and constructed at the Jet Propulsion Laboratory, Pasadena, California, to support VLBI experiments for the NASA Crustal Dynamics Project (Resch, 1984). The radiometers have two microwave channels, 20.7 and 31.4 GHz, to enable the precipitable vapor and liquid to be determined simultaneously. Each channel is fed by a corrugated horn antenna with a beam width of 7°. The radiometers are doubly stabilized, and two reference temperatures, 370 K and 315 K, are provided to correct for gain changes. Special calibration is necessary because the typical sky brightness range, between 20 and 80 K, does not lie between the reference temperatures.

The radiometer system is mounted on a simple alt-azimuth mounting, steerable to 1°, and there is a controller/interface module.

The WVR outputs two sky brightness temperatures. Developing the algorithm to convert these to a path delay involves a considerable amount of atmospheric modeling, followed by determination of the constants of the algorithm, by means of radiosonde data. The results with various radiometers have been carefully tested and analyzed. They indicate that the instruments are capable of determining the correction to an accuracy of ± 2 cm.

The WVR system appears to fulfill the requirements of accuracy, portability, and all-weather capability, but the cost, about $120,000 per unit, is high.

The Atmospheric Boundary Layer and Surface Layer

The layer of the atmosphere in which conditions are directly affected by the underlying surface is the boundary layer. Its depth is typically between 300 m and 2000 m, depending on conditions such as wind, surface roughness, and stability. The shallower *surface layer* has a depth of about 30 m. It is the layer in which various parameters, such as the vertical fluxes of heat, water vapor, and horizontal momentum, can be considered constant. Since all geodetic observations must pass through this layer for at least some part of their paths, the refraction here is especially significant. Extensive researches into atmospheric turbulence and heat balance in the surface layer have developed formulations for temperature gradients and other quantities, based on similarity theory. A valuable review of this subject related to geodetic applications is given by Webb (1984).

It is necessary to distinguish two distinct atmospheric conditions, stable and unstable, with an intermediate neutral state. When the surface is heated, convection carries heat upward in the form of turbulent eddies of air. The atmosphere is well-mixed, conditions are *unstable*, and the temperature gradient is negative. With a positive temperature gradient, the coldest, densest air is resting on the surface, with the next densest layer above, and so on. There is no tendency for the air to mix, and conditions are *stable*.

In unstable conditions, thermal upcurrents carry heat upward while cooler eddies sink between them. Temperatures and wind velocities fluctuate rapidly. Generally, however, for geodetic refraction the fluctuations are not modeled: The corrections deal with an atmosphere in which quantities are frozen at the mean of their fluctuating values.

The mechanisms of turbulence vary in different strata. The depths of the strata vary with time, depending on the value of a stability parameter, the Obukhov length L, which is proportional to the cube of the wind velocity. The whole structure of behavior expands and contracts with L. On a typical summer day with clear skies and wind velocity of 4 m · s^{-1}, L is about 30 m. Zone I is the stratum 0–1 m (0–0.03L), zone II is 1–30 m (0.03$L - L$), and zone III is above 30 m (L), up to the top of the boundary layer. If the wind velocity doubles to 8 m · s^{-1}, the zone depths increase by a factor of 8 (0–8, 8–240, >240).

The geodesist can use models for the gradients of temperature and water vapor. Expressions are available for these quantities, but they vary in different stability conditions and different vertical zones.

Zone I is generally below the range of geodetic observations. Zones II and III contain most terrestrial geodetic observations. In zone II the temperature gradient is given by

$$\frac{dT}{dz} = \left[\frac{H^2 T}{(C_p \rho)^2 g}\right]^{1/3} z^{-4/3} \qquad (1)$$

where H is the upward heat flux (flow) due to convection in W · m^{-2}, C_p is the specific heat, at constant pressure, of the air, ρ is the density of the air, and g is the acceleration due to gravity. Substituting for constant or slowly varying parameters, we obtain the gradient

$$\frac{dT}{dz} = -0.027\, H^{2/3} z^{-4/3}$$

The gradient depends upon powers of the height z and the upward heat flow H. It is possible to measure H, but it involves sensitive measurements with specialized equipment. Alternatively, it is possible to estimate it, based on considerations of the heat balance. The net radiation R_{net} reaching the ground surface is made up of four components:

$$R_{net} = S_d + L_d - S_v - L_u \qquad (2)$$

where S_d is downward shortwave radiation received directly from the Sun, L_d is downward long-wave radiation from clouds and sky, S_u is upward shortwave radiation, which depends on the albedo of the surface, and L_v is upward long-wave radiation, directly dependent on the surface temperature. Since there must be a balance of heat at the surface, the heat loss must equal R_{net}. Omitting some very small components, we get

$$R_{net} = H + G + E \qquad (3)$$

where G is the heat conducted into the ground, E is the latent heat loss in vaporization of surface moisture, and H is, as before, the upward heat flux carried by convection. The required quantity is H. From Eqs. 2 and 3,

$$H = S_d + L_d - S_u - L_u - G - E \qquad (4)$$

The terms on the right side of Eq. 4 can be calculated or estimated on the basis of simple meteorological observations. Alternatively, R_{net} may be measured with a net radiometer so that H is derived more directly:

$$H = R_{net} - G - E$$

The temperature gradient takes up different forms, depending on the height above the surface. At the top of zone II, above about 0.7L, the gradient of potential temperature intermittently diminishes to zero, for periods of up to a few minutes, before returning to the value of Eq. 1. In zone III the gradient is zero and temperatures are quiescent, except for isolated thermal plumes. These columnar plumes extend right to the top of the boundary layer. At this level there is an increase of temperature (an inversion), and above the interface, the free atmosphere, with its stratification unaffected by the surface.

There are equations, similar to Eq. 1, for the

gradient of specific humidity in the unstable surface layer and analogous equations for the gradients of temperature and humidity in the stable surface layer.

The model described has been applied as the basis for refraction corrections in the case of vertical angles, EDM, leveling, and network adjustment by the ratio method (see references in Angus-Leppan, 1984; Brunner, 1984).

Refraction in Trigonometric Heighting

In 1810, an officer of the British army visiting Katmandu in Nepal observed vertical angles to the magnificent snow-covered peaks of the Himalayas to the north. He was astonished to find that the peaks appeared to move up and down by nearly 1000 ft through the day. This was an effect of atmospheric refraction and its daily variation. The principle of trigonometric heighting is that a theodolite, set up on a station of known height, observes a vertical angle to a second station of unknown height but known distance. The height of the second station can be calculated from simple trigonometry, but it was learned early that corrections are needed for the curvature of the Earth and curvature of the line of sight due to refraction.

It can be shown from physics that the curvature F of a line of sight in a vertical plane is given (Bomford, 1971, p. 263) by

$$F = 16.5 \frac{P}{T} (0.033 + dT/dz) \quad (5)$$

where P is the atmospheric pressure in millibars, T is the temperature K, and z is the vertical height in meters. If this curvature is constant, then the well-known *coefficient of refraction*, k, defined as the ratio of the radius of the Earth to the radius of the line of sight, is related to F by

$$k = \frac{FS}{\theta}$$

where S is the distance and θ is the angle subtended at the center of the Earth by S. Usually k is determined by reciprocal vertical angle observations, that is, forward and back observations along the same line. The conditions for observing vertical angles are well known: They should be observed around the midday period (10 to 15 hr), in sunny conditions, and the sightlines must not be "grazing rays"; that is, they must be well above the surface. Then an average value of k ($k = 0.13$ is often used) may be adopted.

To calculate the height difference h, first correct each observed angle B for Earth curvature and refraction:

$$B' = B + (1 - k)\theta$$

Then

$$h = S \sin B' \sec\left(B' + \frac{\theta}{2}\right)$$

That the coefficient of refraction is almost constant under the conditions specified is understandable from the information on the atmosphere outlined in the previous section. The rules for observing ensure unstable conditions (near midday and sunny) with most of the sightline in zone III (not a grazing ray) where the gradient is constant; in fact the potential temperature gradient is zero. Substituting the value of the potential temperature gradient in Eq. 5 and taking into account the small portion of the sightline close to the theodolite that passes through a numerically larger gradient, we obtain a value of 0.13 for the coefficient of refraction k.

The foregoing procedure is very practical and consistently produces satisfactory results when moderate accuracy is sufficient. Modifications, in order of increasing accuracy, are set out below:

1. Adopt a constant value, say 0.13, for k.
2. Determine a local value for k. This is achieved by measuring reciprocal vertical angles in the optimum time period but not necessarily on the same day. On the lines where angles are reciprocal, the effect of k cancels out. On other lines, the local value, determined from the reciprocal angles, is used.
3. Observe reciprocal angles at exactly the same time at both ends. This improves accuracy considerably.
4. Shorten lines of sight. The refraction propagates as the square of the distance, so shorter lines bring increased accuracy. Several recent investigations have shown that motorized high-precision traversing, with precise angle and EDM measurements and careful target centering, can produce results comparable in accuracy to precise leveling. Lines of sight are reduced to 100 or 200 m.

Further refinements that can be introduced in trigonometric heighting are modeling of the atmospheric surface layer or determining refraction by using dual or multiple wavelengths. Modeling of the atmosphere to determine the temperature gradient field, which is then integrated along the line, has been advocated and tested by several investigators. The results have been promising, but this approach does not appear to have been adopted.

The simplicity of the "coefficient of refraction" approach makes it very attractive. Generally, surveyors follow the criteria for the optimum observing conditions so slavishly that they are unaware of what happens under other conditions. Variations of the temperature gradient are the chief cause of variability in the refraction. The gradient varies rapidly with height above the surface, particularly in the surface layer. Like the temperature, the gradient fluctuates rapidly, but even smoothed values vary greatly with time. For example, the shadow of a cloud passing over can cause a large and sudden jump in refraction. In the layer at about the height of a theodolite, 1.5 m, the coefficient of refraction could typically vary from -10 to $+5$, from dawn to midday. At a height of 2.5 m, the variation would be approximately halved.

Two-wavelength instruments, which measure the angular dispersion, represent the instrumental solution. A recent paper reviewed the designs and instruments developed by eight different groups (Williams and Kahmen, 1984). Although there are several different approaches, none of the instruments has passed the prototype stage. In most cases, although laboratory tests show a high precision, the results under field conditions are disappointing. The basic problem is that the measured quantity, the dispersion, is very much smaller than the refraction, which is to be determined from dispersion. So dispersion needs to be measured to a high accuracy, about 1/100 of an arcsecond. Under field conditions, the turbulence of the atmosphere makes it almost impossible to achieve this accuracy. In addition, dispersion causes the paths of the two waves to separate, to pass through different eddies, and to decorrelate. This puts limits on accuracy and length of lines used. In the future, it may be possible to choose the optimum instrumental design from the range of possibilities and to incorporate a microcomputer. Multiple observations could then be taken automatically and filtered to produce results of the required precision and reliability.

Height differences, measured over a survey network, need to be adjusted. Usually only the heights are adjusted, but it would be more appropriate to consider the coefficient of refraction as an unknown. This has been thoroughly tested by Hradilek (1980) in a long series of investigations. The most successful model assumes that k is constant at one station for each set of angles measured within a short period. If observations are taken at a different time, another parameter is introduced. The method gives good precision and has the additional advantage that it is not necessary to measure every line from both directions, as it would be with the reciprocal angle method.

Leveling Refraction

In typical daytime sunny conditions the error due to vertical refraction on a 50-m leveling sightline may be 1 mm. At night it might be 1 or 2 mm, but with opposite sign. In an extreme case, it has been observed in windless conditions over a surface of snow and ice to be as high as 8 mm. If such errors accumulated along a line of levels, the results would be of little value.

In leveling *on the flat*, the refraction on the leveling foresight and backsight are nearly equal and cancel almost exactly in the height difference h.

In leveling *up a slope*, the foresight is closer to the ground and refraction is larger, causing the observed height differences to be too small. The effect is greatest on gentle slopes, where long sightlines are still possible. The error may amount to 2 to 3 cm per 100 m of height difference. If conditions are symmetrical, the height lost going up will be regained coming down, and the error will not show up as a misclosure of the leveling circuit.

Another systematic but hidden error accumulates on north-south slopes. Here the sunny slope facing the equator is heated more than the shady slope; the heat flux and, hence, the refraction are higher, causing an error that accumulates over long north-south lines. The effect is quite variable, between 5 and 100 cm per 1000 km of the north-south component of a leveling line. In leveling circuits, if the topography on north-south segments is similar, the error will not show up in circuit misclosures. The effect is to give sea level, as depicted by leveling, an apparent downward tilt toward the equator.

The first serious investigations in leveling refraction were in 1938 by Kukkamaki, who derived a form of correction using a model for temperature gradient. Constants for the model came from the most comprehensive set of data available for a station in southern England. Taking these very mild conditions as typical, it appeared justifiable to neglect refraction.

It was 40 years before geodesists realized that these conditions were not representative. Conditions in most of the United States and in many parts of the world outside Central Europe result in far higher values of refraction. The development of better models for the surface layer and the preparations for the readjustment of the North American Vertical Datum (NAVD 88) led to theoretical and practical developments in the United States, Canada, Europe, and Australia.

For the NAVD 88, Holdahl used the Kukkamaki model for refraction and the surface layer model outlined earlier to derive corrections for refraction. Using meteorological data sets, he was able to derive parameters for correcting "historical leveling," that is, leveling carried out in the past. The U.S. National Geodetic Survey investigated and compared various formulations of the refraction correction. The conclusion was that none of the methods can predict all the variations, but most can eliminate some 80% of the effect. The various methods give similar results. It is perhaps surprising that general prediction of temperature gradient from models leads to slightly more accurate corrections than measurements of the temperature gradient at the time of the leveling.

Angus-Leppan derived a more rigorous form of the correction from first principles and proposed applying a simple form of heat balance relationship for estimating the gradient.

As a result of the investigations in the late 1970s, it is safe to conclude that corrections for refraction should be applied in precise leveling.

Another means of investigating refraction effects in leveling is through the statistical approach. It is easily shown that leveling observations are correlated and contain systematic and random errors. Much of the correlation between observations in the same vicinity is due to refraction. Remmer has advocated eliminating refraction effects by adjustment. In his observation equation he includes terms with unknown constants in a form that depends on the second and fourth derivatives of the temperature gradient and on the propagation of refraction proportional to the square of the sight length.

When the method is applied to the Finnish geodetic leveling, results show that the Kukkamaki correction should be multiplied by 3, that the loop closures are improved significantly, and the correlation is reduced effectively to zero. However, the method has the serious flaw that it assumes that the derivatives of temperature are invariant with time and position. This assumption is far from the truth.

Geodetic leveling has remained unchanged in principle for close to two centuries. To speed up the operation, we need longer sightlines, but refraction limits the sight length. Initiatives have been taken, notably by the U.S. National Geodetic Survey, to develop a new leveling system. It will almost certainly involve a precise height-traversing system, carried on vehicles, with automated readings and some means of measuring refraction. Currently the system of "refraction by reflection" devised by Angus-Leppan is being investigated.

Practical measures to deal with refraction, in order of increasing complexity, are

1. The widely applied observing method of keeping backsight and foresight equal and avoiding sightlines close to the ground.
2. Shortening sightlines, which is effective because the refraction varies as the square of the sightline length. This means that halving the length will reduce refraction by a factor of 4 and should also reduce the irregularities in refraction by a similar factor. However, this also slows down the operation and increases costs.
3. Applying the refraction correction using either a temperature model or actual temperature differences measured in the field. This correction should be applied in all first-order leveling.
4. In the future, with the use of new instrumental systems still under development, it will be possible to progress faster, without the loss of accuracy.

Finally, there is a technique that provides a very precise and secure elimination of refraction, at a cost of some additional time in the field. This technique combines two sets of observations in the leveling run. One has sight lengths s, and the other $2s$, with common change points. It is possible to arrange the observing procedure effectively, using one level and two leveling rods, and moving them only in the forward direction between sights. From the observations, the refraction on each sight length can be determined and eliminated to a high precision.

Electronic Distance Measurement

The EDM technique uses measurements of the time of propagation of light or microwaves to determine the distance between points. It requires the velocity of the waves, which depends upon the refractive index of the medium. In practice, an EDM instrument determines a distance σ_r in relation to its internal *reference refractivity* N_r.

The reduction ΔS of the EDM observation σ_r to the required chord distance S is given by

$$\Delta S = 10^{-6} \int_0^S (N - N_r)\, dx \quad (6)$$

where N is the actual refractivity and the integration with respect to x is performed along the length of the line. Complete equations and examples are in Angus-Leppan and Brunner (1980). Normally, for the actual refractivity, the mean of the meteorological values observed at the end points is taken as representative of the conditions along the whole line. Using the mean of the temperature, pressure, and humidity values at the two end points, we obtain the refractivity N_m:

$$N_m = \frac{1}{T}(0.2696 N_{go} P - 11.25 e)$$

where N_{go} is the group refractivity, a function of wavelength, for standard atmospheric conditions, and P, T, and e are the atmospheric pressure, temperature, and partial water vapor pressure. Under these conditions, Eq. 6 for the EDM correction becomes

$$S = 10^{-6} S(N_m - N_r)$$

When the atmospheric condition is neutral, the gradient of potential temperature is zero, and the mean value of end point measurements gives the correct reduction. This occurs in the hours soon after sunrise and around sunset, so these times are the most favorable for EDM.

At other times, if the highest accuracy is sought, an additional correction is needed. Microwave measurements are not considered here because they should not be used when seeking the highest accuracies. For light waves, the temperature gradient as modeled by theory in the surface layer can be applied to correct the distances. This correction reaches a maximum of nearly 4 ppm during the daytime. It can reach values of 10 ppm or more on a windless night with clear sky, and these conditions should generally be avoided for EDM.

If a distance is measured with two different wavelengths, the difference between the measurements should, in principle, give a measure of the refraction. In practice, because the difference is only a small fraction of the refraction, about 1/20, it must be measured with great accuracy. This instrumental solution has formed the basis for some very accurate instruments, capable of accuracies of better than 1 ppm. These solutions have been applied successfully, for example in crustal deformation studies. One instrument has been taken beyond the prototype stage. It is very specialized and provides only the last digits of the distance to a high precision, but it is unsuitable for normal EDM. The method suffers from some problems. Over longer distances the two waves become decorrelated and travel along different paths, which limits the range and precision, and there is a problem of self-modulation in the instruments.

In a totally different approach to distance observations, the *ratio method* adopts as the basic observation the ratio between two distances measured at the same time and from the same station. These observations can then be used in calculating the control network. There are several forms of solution, but results in all cases indicate that it is capable of increasing the accuracy of measurements by at least an order of magnitude. One proponent advocates applying corrections, including the correction for nonneutral conditions, before calculating with the ratios. Another proposes an equivalent solution that solves for a scale parameter at each station. The practical disadvantage of the method is that most lines need to be measured from both ends, nearly doubling the amount of observing time.

Refraction in Satellite Doppler Observations

This section illustrates the refraction correction for satellite observations. It deals specifically with the tropospheric correction for Doppler satellite positioning, but it is easy to see how this is related to the corrections for ranging, VLBI, or the various forms of GPS observation. The formulation is based on the work of Hopfield (1969) and deals with the effects of the neutral atmosphere (the "troposphere") on radio waves. For these wavelengths, the atmosphere is nondispersive. A correction is also required for the ionosphere. Since the radio waves are dispersive here, the correction uses the two-frequency method, which removes first-order ionospheric effects.

Apparent range of the satellite is given by $\int n \, ds$ along the wave path. The tropospheric effect on the Doppler shift is the difference between the shift in a vacuum and the apparent Doppler shift, given by

$$\Delta f = -10^{-6} \frac{f}{c} \frac{d}{dt} \int N \, dr \quad (7)$$

where F is the signal frequency, C is the velocity of light, $N = 10^6(n - 1)$ is the refractivity, where n is the refractive index, and r is the distance from the Earth's center. What is required is a profile of N that has the correct value N_0 at the surface, that gives the correct value of the integral at the zenith, and that also has, at any geographic location, the correct value of dN/dr near the surface. This last condition is to ensure that the integral gives the correct values for wave paths at low elevation angles.

The integral was evaluated first for the dry component of the atmosphere, the first term in the equation of Smith and Weintraub:

$$N = 77.6 \frac{P}{T} + 77.6 \left(4810 \frac{e}{T^2} \right) \quad (8)$$

where P and e are the total pressure and the water vapor pressure in the atmosphere (mbars) and T is the temperature in kelvins.

Assuming that the vertical temperature gradient is constant, the profile of the dry refractivity N_d is given by

$$N_d = N_{0d} \left[\frac{T_0/a - Z}{T_0/a} \right]^u \quad (9)$$

$$= K_d (Z_d - Z)^u \quad (10)$$

where N_{0d} is the value at the surface of the first term in Eq. 8, T_0 is the temperature at the surface, a is the temperature gradient, Z is the height above sea level, u is $g/R - 1$, in which R is the gas constant, and K_d and Z_d are parameters that are determined for the refraction correction.

The index u in Eq. 9 is a function of the gradient. If u is a positive integer, Eq. 7 can be integrated in closed form. The assumption that $u = 4$, corresponding to a gradient of $-6.8°C/km$, yields a good approximation for the shape of the profile, including the gradient dN/dz near the surface. The height parameter Z_d represents the height at which the refraction effect falls to zero.

Although there is little theoretical basis for the analogy, the water vapor content is assumed to have the same form as the dry component in Eq. 10. This assumption yields the "two-quartic tropospheric refractivity profile." Z_d and the corresponding parameter for the wet component, Z_w, are determined so as to make the modeled values of $N \, dZ$, at the zenith, agree with values calculated by using data from balloon ascents. Z_d is about 43 km and varies with latitude, whereas Z_w, whose value is not critical, is fixed at 12 km. The parameters K_d and K_w are then determined for each time and locality, depending upon the observed surface conditions. The resulting corrections are accurate to about 2%.

After a problem of round off errors was discovered, Yionoulis, in 1970, developed a series formulation for the correction. In 1978, Black presented a greatly simplified algorithm that still retained the required precision. The correction can be calculated to about ±5 cm, but differences between the model and the real atmosphere give rise to far greater errors, particularly for low elevations. The model for the water vapor component is approximate, but is generally satisfactory because the water vapor pressure is low and occurs in the lower regions of the atmosphere. Thus water vapor contributes only about one tenth of the correction.

The effect of tropospheric refraction on a position fixed by Doppler is to make the station appear closer to the satellite than it actually is. The magnitude is about 18 m for a satellite whose orbit passes close to the zenith; it increases to 100 m for an orbit whose maximum elevation is 10°. A lower cutoff of 10° elevation is recommended for satellite tracking.

Besides modeling errors, there are other error sources. The geometrical effect, due to the curvature of the wave path through the troposphere, is generally neglected. It is very small for high elevations, but increases to 3 m at 1°. The temperature profile in the model starts with the observed temperature, extending upward with a constant gradient of a (-6.8 K/km). However, in the boundary layer, temperature gradients differ significantly from -6.8, causing the modeled temperatures to differ from actual temperatures through the whole profile. This is the boundary layer effect, which varies with time of day and which may amount to several centimeters. Horizontal changes of atmospheric temperature and humidity cause errors of similar magnitude.

Tropospheric effects in observations of satellite laser ranging, VLBI, and GPS are related to those in Doppler observations. The present VLBI baselines and the future long GPS baselines call for improvements in the accuracy of the correction. Improved modeling for the dry component and application of corrections for the error sources mentioned can improve the accuracy of the correction, but the water vapor component remains a difficulty. The solution may be WVR, as described earlier.

P. V. ANGUS-LEPPAN

References

Angus-Leppan, P. V., 1984, Refraction in geodetic levelling, in F. K. Brunner, ed., *Geodetic Refraction, Effects of Electromagnetic Wave Propagation through the Atmosphere*. Berlin: Springer Verlag, 163–180.

Angus-Leppan, P. V., and F. K. Brunner, 1980, Atmospheric temperature models for short-range EDM, *Canadian Surveyor* **34**(2), 153–165.

Bomford, G., 1971. *Geodesy*. London: Oxford University Press, 731p.

Brunner, F. K., 1984, Modelling of atmospheric effects on terrestrial geodetic measurements, in F. K. Brunner, ed., *Geodetic Refraction, Effects of Electromagnetic Wave Propagation through the Atmosphere*. Berlin: Springer Verlag, 143–162.

Hopfield, H. S., 1969, Two-quartic tropospheric refractivity profile for correcting satellite data, *Jour. Geophys. Research* **74**(18), 4487–4499.

Hradilek, L., 1980, Evaluation of refraction by geodetic methods, in E. Tengstrom and G. Teleki, eds., *Symp. on Refractional Influences in Astrometry and Geodesy Proc.* Upsala: International Astronomical Union, 191–193.

Resch, G. M., 1984, Water vapor radiometry in geodetic applications, in F. K. Brunner, ed., *Geodetic Refraction, Effects of Electromagnetic Wave Propagation through the Atmosphere*. Berlin: Springer Verlag, 53–84.

Webb, E. K., 1984, Temperature and humidity structure in the lower atmosphere, in F. K. Brunner, ed., *Geodetic Refraction, Effects of Electromagnetic Wave Propagation through the Atmosphere*. Berlin: Springer Verlag, 85–142.

Williams, D. C., and H. Kahmen, 1984, Two wavelength angular refraction measurement, in F. K. Brunner, ed., *Geodetic Refraction, Effects of Electromagnetic Wave Propagation through the Atmosphere*. Berlin: Springer Verlag, 7–32.

Cross-references: *Doppler Positioning: Satellite; Geodesy: Geometric; Geodesy: Satellite; Geodetic Ground Positioning; Geodetic Leveling; Geodetic Networks and Control Surveys; Global Positioning System (GPS); Satellite Altimetry; Satellite Laser Positioning; Very-Long-Baseline Interferometry (VLBI).*

GEOMAGNETIC FIELD: ASYMMETRIES

It is now well accepted that, on occasion, the geomagnetic field reverses its polarity, the two states being referred to as *normal* (the present state) and *reverse* polarity. Over the past few years there have been several studies that have concluded that the normal and reverse polarity fields differ in respects other than a simple change of sign. In other words, if we were to take the geomagnetic field during a time of reverse polarity and simply reverse the direction of the field everywhere, we would not have a normal polarity field. Without question we would have a field very similar to a normal polarity field, but the suggestion is that it would differ in certain characteristics. Such differences are referred to as *asymmetries* in the polarity states of the geomagnetic field.

There may of course be asymmetries in the geomagnetic field that are not related to the different polarity states. For example, asymmetries between the northern and southern hemispheres (Merrill and McElhinny, 1977): Possible explanations for such asymmetries include the variations in topography and/or temperature at the core-mantle interface. However, as will be discussed later in the article, it is crucial to our understanding of the generation of the field to know whether genuine asymmetries in the polarity states do exist, whereas the existence of other forms of asymmetry is not of such fundamental importance for an understanding of the geodynamo. Not surprisingly, therefore, this particular form of asymmetry in the geomagnetic field has attracted more attention in the literature than other asymmetries. This article concentrates quite specifically on asymmetries in the polarity states.

Evidence for Polarity Asymmetries

It is not the intent of this section to provide a comprehensive coverage of all the investigations which have concluded that a polarity asymmetry of some form exists. Instead the aim is to provide an overview of the evidence and to cover the types of asymmetry that have been suggested in the literature.

Polarity Bias. Over the years an enormous amount of effort has been put in to determine a reliable chronology for reversals as far back as possible. Such a chronology is now quite well determined for the past 170 Ma from analyses of marine magnetic anomalies (see *Magnetostratigraphy* and *Seafloor Spreading: Magnetic Evidence*). For examples of such time scales, see Heirtzler et al. (1968), LaBrecque et al. (1977), Ness et al. (1980), Lowrie and Alvarez (1981), Harland et al. (1982), or Berggren et al. (1985). As the reversal time scale developed, it became clear that the character of the reversal pattern has changed markedly with time. The most obvious change occurred about 118 Ma ago, when the field apparently locked in to a normal polarity state and remained that way for the next 35 Ma, after which reversals again occurred. Similarly, there are other periods, as in the late Paleozoic, during which the field was of reverse polarity most of the time. This phenomenon of long intervals of time in which there were no polarity changes is referred to as *polarity bias*; it has been recognized for some time now (e.g., see McElhinny, 1971; Irving and Pullaiah; 1976).

This situation where the field apparently locks in to one particular polarity is an extreme bias. If the field, while not locked in to one particular polarity, were to spend, on average, substantially more time in one polarity than the other, this too would be a polarity bias, but one of a nonextreme form.

The existence of polarity bias has been taken to represent a distinct asymmetry between the two polarity states, the field apparently exhibiting a clear preference for one polarity state over the other. A further point of interest is that bias during the late Paleozoic was for the reverse polarity, but during the Cretaceous it was for the normal polarity, suggesting that the bias itself has been a function of time. Cox (1981) has attributed polarity bias to variations in the physical properties of the lower mantle with a spatial distribution that can be described by an odd zonal harmonic (i.e., a spatial distribution that has a northern-southern hemisphere asymmetry).

Relative Stabilities of the Two Polarity States. The intervals of time between reversals (excluding times of extreme polarity bias, as discussed above) may be modeled as a gamma distribution. Thus if x represents an interval of time between reversals, the probability density $P(x)$ of x is

$$P(x) = \frac{1}{\Gamma(k)} (k\lambda)^k x^{(k-1)} \exp(-k\lambda x) \quad (1)$$

where the true mean length τ of the intervals is

$$\tau = \frac{1}{\lambda} \quad (2)$$

If the true mean length for normal polarity intervals is different from the true mean length for reverse polarity intervals, then an asymmetry exists that may be considered merely as a nonextreme polarity bias. Although a physical interpretation for the parameter λ is readily apparent through Eq. 2, it is not so simple to provide a physical interpretation for the parameter k (often referred to as the shape parameter in statistical literature). For a discussion of the gamma distribution and interpretation of the parameters, see *Geomagnetic Reversal Sequence: Statistical Structure*.

Phillips (1977) analyzed the Heirtzler et al. (1968) time scale and concluded that, during the Cenozoic, k for the normal polarity intervals was consistently larger than k for the reverse polarity intervals. He interpreted k in terms of the stability of the field (against a further reversal) immediately after a reversal, and so concluded that the normal polarity field has exhibited a greater stability immediately following a reversal than has the reverse polarity field. Lowrie and Kent (1983) have analyzed the LaBrecque et al. (1977) and Lowrie and Alvarez (1981) time scales and have also concluded that an asymmetry in k exists. However, they note that 57 very small coherent "wiggles" in the records may represent very short polarity intervals, and these wiggles have not been included in the polarity time scales. On reanalyzing the time scale with these events added as polarity intervals, they still concluded that an asymmetry exists, but that the added short events alter the sense of the asymmetry between the normal and reverse states.

Time-averaged Structure of the Field. The geomagnetic field at the Earth's surface is very complex and continually varies with time in a complicated way. This variation with time is referred to as *secular variation* (see *Geomagnetic Secular Variation: Theory; Geomagnetic Secular Variation: Direction and Intensity*; and *Paleomagnetic Secular Variation*). Because of this continual variation it would be totally unrealistic to make an observation at one point in time during a normal polarity interval, compare this with a similar observation at another point in time during a reverse polarity interval, and conclude that an asymmetry exists because the observations differ. When one looks for asymmetries in the structure of the field it is important to perform very careful time-averaging to obtain that which is truly characteristic of the field. Because of the complexity of the field, it is also important to obtain as wide a spatial coverage as possible in order to determine this structure reliably. If the time-averaged structure of the normal polarity field differs discernibly from that of the reverse polarity field, then an asymmetry in polarity states does exist.

By far the most common way to analyze the geomagnetic field is through the use of spherical harmonics (see *Geomagnetic Field Analysis*; further information may be found in many mathematics texts). With regard to the use of spherical harmonics in geomagnetism, a full account may be found in Chapman and Bartels (1940, 1962), and an introduction and account suitable for this particular article may be found in Merrill and McElhinny (1983). It turns out that the time-averaged field can be modeled using only low-degree zonal harmonics. These zonal harmonics are harmonics that are symmetrical about the Earth's spin axis. Hence, although the field changes with latitude, it does not change with longitude. The lower-degree harmonics are probably better known by the terms dipole (degree 2), quadrupole (degree 3), octupole (degree 4), etc. For example, the zonal harmonic of degree 3 is a quadrupole positioned at the center of the Earth and, because it is zonal, oriented along the Earth's spin axis. The geomagnetic field at the Earth's surface is dominantly that of a dipole.

Because the time-averaged field does not vary with longitude, it is possible to obtain an effective time-averaged field by averaging paleomagnetic observations of the field's inclination (well distributed over time) within latitudinal bands. These mean values may then be analyzed to determine the structure of the field: In this instance, this implies determining the relative proportions of the zonal harmonics that make up the time-averaged field. This was performed by Merrill and McElhinny (1977) for data covering the last 5 Ma. From their analysis they concluded that, at least over the last 5 Ma, the structure of the time-averaged reverse polarity field has been discernibly different from that of the time-averaged normal polarity field. In essence, they concluded that the time-averaged reverse polarity field has proportionately had a larger quadrupole and octupole content than has the time-averaged normal polarity field.

Intensity of the Field. Even if the geomagnetic field were precisely that of an axial dipole, the intensity of the field would vary over the surface of the Earth. For example, the intensity at the pole would be twice that at the equator. Consequently, paleointensities (see *Paleomagnetic Field: Intensity*) cannot be compared directly but must be referred to the strength of the source producing the field. The field is, in detail, very complex, but at the surface of the Earth it is dominantly dipolar. One method of comparing paleointensities involves the use of a virtual dipole moment (VDM). One calculates a VDM by assuming that the field is solely dipolar, then uses the observed paleoinclination to calculate the assumed position of the observation relative to the orientation of this dipole, and then uses this information together with the paleointensity to calculate the moment (i.e., strength) of the dipole. Naturally, because secular variation affects the field intensity at any given locality, it is also important to obtain a good temporal and spatial distribution of VDMs for reliable comparison of the normal and reverse polarity fields.

McFadden and McElhinny (1982) analyzed all available VDMs for the past 5 million years. They were able to model the observations by assuming that the true dipole moment (TDM) varies such that if sampled at random times a truncated Gaussian distribution is obtained. This truncation is taken to represent the lowest dipole moment associated with a stable polarity field. A VDM is then a TDM plus a Gaussian distributed error (caused by the nondipole field and its secular variation) with standard deviation proportional to the TDM. Comparison showed that the models for the normal and reverse polarity fields were statistically discernible only in that the truncation point for the normal polarity TDMs was larger than that for the reverse polarity TDMs.

Roberts and Shaw (1984) determined paleointensities and paleomagnetic directions from 106 late Tertiary lavas in eastern Iceland. The mean paleointensity values for the normal and reverse polarity fields do not differ, but they suggest an asymmetry whereby the normal polarity field maintained its magnitude over a wider range of directions than did the reverse polarity field. This form of asymmetry is similar to the different truncation values (of the TDM distributions) that were observed by McFadden and McElhinny (1982), but it is of the opposite sense.

Pesonen and Halls (1983) determined 54 paleointensity values in late Precambrian Keweenawan rocks around Lake Superior and found that the reverse polarity paleointensities were, on average, 40% higher than their normal polarity counterparts. This apparent difference disappears when VDMs are calculated. The difference might be explained by invoking plate movement between the normal polarity observations and the reverse polarity observations. However, evidence at one locality of possibly three reversals poses problems for this interpretation, and they concluded that an asymmetry in polarity states must have existed.

Theoretical Considerations and Implications

Merrill and McElhinny (1977), Merrill et al. (1979), and Merrill and McElhinny (1983) have discussed the theoretical aspects and implications of asymmetries in the polarity states, and reference should be made to these papers for further information. Central to any dynamo theory is the magnetic induction equation

$$\frac{\partial \mathbf{B}}{\partial t} = K\nabla^2 \mathbf{B} + \nabla \times (\mathbf{v} \times \mathbf{B}) \quad (3)$$

where $\mathbf{B}$ is the magnetic field, t is the time, K is the magnetic diffusivity, and $\mathbf{v}$ is the fluid velocity field in the Earth's outer core. The velocity field must also satisfy the Navier-Stokes equation, which, for an incompressible fluid, is

$$\rho\left(\frac{\partial}{\partial t} + \mathbf{v} \cdot \nabla\right)\mathbf{v} + 2\rho(\mathbf{\Omega} \times \mathbf{v})$$
$$= -\nabla P + \eta\nabla^2\mathbf{v} + \mathbf{F} + (\mathbf{J} \times \mathbf{B})$$

where P is the pressure, ρ is the density, η is the viscosity, $\mathbf{\Omega}$ is the Earth's angular rotation velocity, $\mathbf{J}$ is the electric current density, and $\mathbf{F}$ is the gravitational body force. The term $\mathbf{J} \times \mathbf{B}$ is the Lorentz body force and is the only term in which the magnetic field enters the Navier-Stokes equation. For conditions in the core we have

$$\nabla \times \mathbf{B} = \mu_0 \mathbf{J}$$

where μ_0 is the permeability of free space. Thus,

$$\mathbf{J} \times \mathbf{B} = (\nabla \times \mathbf{B}) \times \frac{\mathbf{B}}{\mu_0}$$

Therefore the sign of $\mathbf{B}$ does not enter into the Navier-Stokes equation. Furthermore each term in Eq. 3 is an odd function of $\mathbf{B}$, so the sign of $\mathbf{B}$ is irrelevant in the magnetic induction equation. Thus we have the important result that if a particular $\mathbf{B}$ field is a solution to the dynamo problem, then another solution is obtained simply by reversing the sign of $\mathbf{B}$ everywhere. Thus, although dynamo theory does not predict that reversals will necessarily occur, it does predict that, if they do occur, the two states should have identical time-averaged properties: There should be no asymmetries in the polarity states.

Any asymmetries that exist must therefore result from boundary conditions and/or initial conditions. Merrill et al. (1979) have suggested a standing field hypothesis and a transition probability hypothesis as possible candidates for explaining asymmetries. By definition, the geodynamo starts with some initially weak magnetic field (almost always assumed to have originated outside the Earth's core), and this field is greatly magnified through appropriate movement of electrically conductive core material. Once the dynamo begins to act, this initial field is no longer required. However, if it is still present today it would be magnified by the dynamo processes, and its manifestation at the Earth's surface would be an example of a "standing magnetic field." It would almost certainly lead to some asymmetry, but it must be asked whether the hypothesized asymmetry would be discernible at the Earth's surface. Thermoelectric currents could produce a "standing magnetic field," but the magnitude of this field is very small (Merrill et al., 1979; Merrill and McElhinny, 1983). Alternatively, some weak solar field early in the Earth's history may have been the initial seed field for the dynamo, and because this field cannot now reach the Earth's core (because of magnetospheric processes) no asymmetry at all would be expected.

The transition probability hypothesis effectively states that there is more than one convection state that can produce the geodynamo. Asymmetry is hypothesized to occur by the ad hoc assumption that the normal and reverse polarity states are distributed differently between the possible convection states. This distribution may change with time, so this hypothesis provides a possible explanation for reports that the character of the asymmetry has changed during the Cenozoic. In this case the "asymmetry" reported by Merrill and McElhinny (1977) might simply reflect that the last 5 million years is insufficient time over which to average. However, this hypothesis is contrived and there is considerable disagreement between the analyses reputed to show how the alleged asymmetries have changed during the Cenozoic.

A possible alternative explanation is that all reported asymmetries are data artifacts, and this explanation will now be considered.

Examination of the Evidence

Polarity Bias. As previously discussed, when considering asymmetries in the structure or intensity of the field, it is crucial for one to obtain a proper time-averaging with good spatial coverage. When looking at a phenomenon such as polarity bias, one is interested only in knowing when the reversals occurred; the global details of the field are unimportant. Thus, conclusions drawn with regard to polarity bias (or relative stabilities of the two polarity states) should be somewhat more robust than those for structure or intensity.

McFadden and Merrill (1984), using statistical tools developed by McFadden (1984), analyzed the Ness et al. (1980) time scale from the present back to 83 Ma and the Cox time scale in Harland et al. (1982) from 118 Ma back to 165 Ma. In neither of these intervals do the data give any reason to reject the hypothesis of a common mean length for normal and reverse polarity intervals. In other words, there is no evidence for any real polarity bias right up to, and immediately after, the long normal polarity interval from 118 to 83 Ma. Although there is some evidence (Irving and Pullaiah, 1976) of a few short, irregularly spaced reverse periods within this interval, the evidence is weak and none of these periods appears in any of the recent marine magnetic anomaly time scales. Furthermore, the rate at which reversals occur (λ in Eqs. 1 and 2) slowly decreases toward zero going into the long normal polarity interval and slowly increases from zero from 83 Ma to the present.

Consequently, a consistent interpretation is as follows. The process producing reversals (but which does not generate the field itself) gradually slowed down from 165 Ma until it stopped about 118 Ma ago. The field then had to remain in whatever polarity state it occupied at the time, the probability for each polarity being 0.5. The process causing reversals then restarted about 83 Ma ago and has gradually sped up since then.

Thus it would appear there is no robust evidence for polarity bias being a real phenomenon.

Relative Stabilities of the Two Polarity States. McFadden and Merrill (1984) have also analyzed the time scales for differences in k between the normal and reverse polarity sequences. Previous analyses had not properly taken into account the lack of precision with which k can be estimated, and it was basically this point that led to suggestions of asymmetry. As an example, Figure 1 shows an analysis of the Ness et al. scale using a sliding window to move down the data sequence (see McFadden, 1984 for a discussion of the problems with using sliding windows). The immediate visual impact is that there is a very clear difference. However, it is recognized that several very short intervals have probably been missed in the records, so Fig. 2 is an analysis of exactly the same sequence but with one very short normal interval added from 33.40 to 33.41 Ma. The difference caused by this one interval to a sequence of nearly 200 intervals is quite striking and indicates the perils of concluding that an asymmetry exists on the basis of estimated k values. After a careful statistical investigation, McFadden and Merrill (1984) concluded that the data do not support rejection of the null hypothesis that the normal and reverse polarity sequence share a common value of k.

An examination of the recent marine magnetic

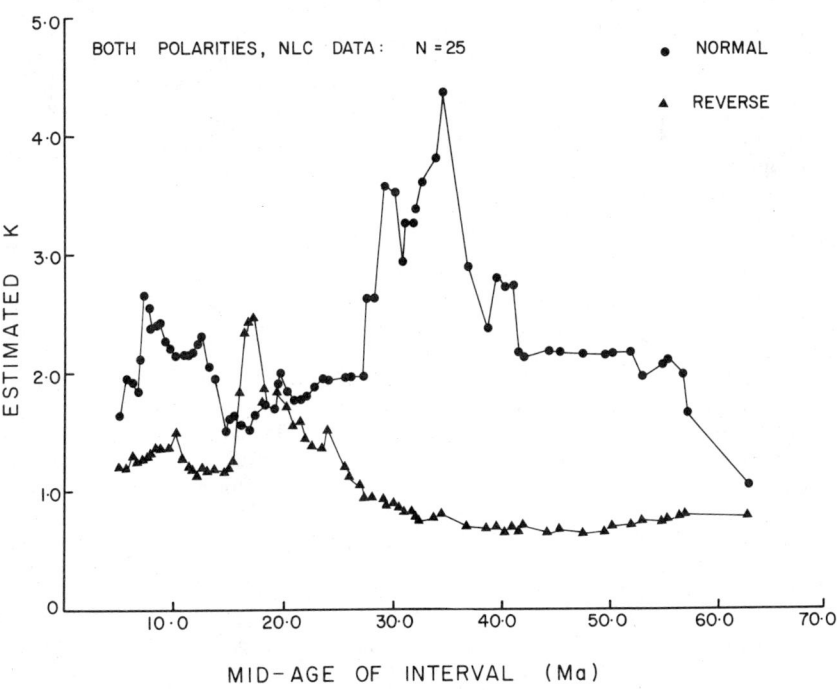

FIGURE 1. Estimates of k for normal and reverse polarities for the Ness et al. (1980) time scale. Sliding window covers 25 intervals of each polarity and shifts by one interval of each polarity each time. (After McFadden and Merrill, 1984).

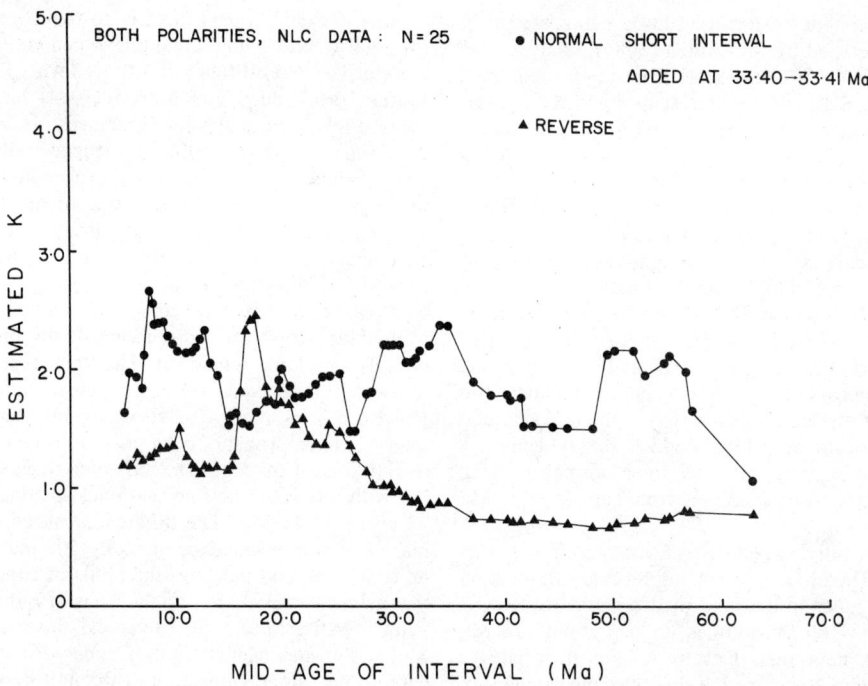

FIGURE 2. As for Fig. 1, but with a single short interval added from 33.40 to 33.41 Ma. (After McFadden and Merrill, 1984).

anomaly time scales shows their structures to be sufficiently similar that both this conclusion and that for polarity bias are independent of which scale is used.

Intensity of the Field. As noted by McFadden and McElhinny (1982), the estimated value of the truncation point for the distribution of TDMs is critically dependent on a few observations at the low end of the range of VDMs. Thus, a single reliable observation of a relatively low VDM with normal polarity would alter McFadden and McElhinny's conclusion that the truncation point for the normal polarity TDMs is discernibly larger than that for the reverse polarity TDMs. Consequently, that conclusion cannot be considered as robust.

According to their own calculations, Roberts and Shaw (1984) were only able to reject the hypothesis of a common intensity at the 75% level of confidence (at most). If anything, their observations indicate symmetry rather than asymmetry.

The conclusion of Pesonen and Halls (1983) suffers because the observations were tightly grouped together in space. Consequently, their evidence cannot be used as robust evidence for an asymmetry in the polarity states. Furthermore, if one considers their observed scatter of paleointensities rather than just the mean values, one cannot reliably reject the hypothesis of a common paleointensity for the two polarity states.

Time-averaged Structure of the Field. The paleomagnetic data used to obtain the time-averaged magnetic field come primarily from lava flows on continents and islands and from deep-sea sedimentary cores. For the most part, the latter provide inclination data only, although this is less serious than it may initially seem because the time-averaged field does appear to be axially symmetric. However, there are considerable problems in obtaining a time-averaged field from these data, and these problems are magnified if one attempts to determine second-order effects such as asymmetries. Because of the very limited absolute paleointensity data, one is, at best, forced to analyze data that provide information on direction only. The spatial distribution of data is often quite poor, particularly in the southern hemisphere. In addition, the data are poorly distributed in time. Finally, rock magnetic and other effects can give rise to erroneous estimates of the odd-degree harmonics, particularly the octupole term.

It now appears likely that the alleged asymmetries in stability and intensity are due to data artifacts. In both cases these artifacts arose because the statistical parameters used to delineate the asymmetries were unstable in the sense that they were significantly affected by small changes in the data. Given the problems mentioned above with the time-averaged data, it is possible that the statistical parameters used to test for differences between the reverse and normal polarity states are also not robust, a possibility currently under investigation by the authors of this article.

Conclusion

During the last decade, paleomagnetic analyses of temporally and spatially distributed data have improved considerably. For example, one can now say with considerable confidence that the paleomagnetic field averaged over a few million years is rarely, if ever, properly characterized by a geocentric axial dipole field, as is commonly assumed in paleomagnetic studies.

With the improvement in paleomagnetic data and analyses have come hypotheses that there are asymmetries between the reverse and normal polarity states. However, recent detailed examination of the data for these hypothesized asymmetries suggests that there is not at present any robust evidence for them. Although the possibility of asymmetries still exists, at this stage it seems reasonable to accept the simple hypothesis of symmetry.

Acknowledgments

This article is published with the permission of the Director, Bureau of Mineral Resources, Geology and Geophysics.

P. L. McFADDEN
R. T. MERRILL
M. W. McELHINNY

References

Berggren, W. A., D. V. Kent, J. J. Flynn, and J. A. van Couvering, 1985, Cenozoic geochronology, *Geol. Soc. America Bull.* **96,** 1407-1418.

Chapman, S., and J. Bartels, 1940, *Geomagnetism*, vols. 1 and 2; 2nd ed., 1962. Oxford: Oxford University Press.

Cox, A. V., 1981, A stochastic approach towards understanding the frequency and polarity bias of geomagnetic reversals, *Physics Earth and Planetary Interiors* **24,** 178-190.

Harland, W. B., A. V. Cox, P. G. Llewellyn, C. A. G. Pickton, A. G. Smith, and R. Walters, 1982, *A Geologic Time-Scale*. Cambridge: Cambridge University Press, 131 p.

Heirtzler, J. R., G. O. Dickson, E. M. Herron, W. C. Pitman III, and X. LePichon, 1968, Marine magnetic anomalies, geomagnetic field reversals, and motions of the ocean floor and continents, *Jour. Geophys. Research* **73,** 2119-2136.

Irving, E., and G. Pullaiah, 1976, Reversals of the geomagnetic field, magnetostratigraphy, and relative magnitude of paleosecular variation in the Phanerozoic, *Earth-Sci. Rev.* **12,** 35-64.

LaBrecque, J. L., D. V. Kent, and S. C. Cande, 1977, Revised magnetic polarity time scale for Late Cretaceous and Cenozoic time, *Geology* **5,** 330-335.

Lowrie, W., and W. Alvarez, 1981, 100 million years of geomagnetic polarity history, *Geology* **9,** 392-397.

Lowrie, W., and D. V. Kent, 1983, Geomagnetic reversal frequency since the Late Cretaceous, *Earth and Planetary Sci. Letters* **62,** 305-313.

McElhinny, M. W., 1971, Geomagnetic reversals during the Phanerozoic, *Science* **172,** 157-159.

McFadden, P. L., 1984, Statistical tools for the analysis of geomagnetic reversal sequences, *Jour. Geophys. Research* **89,** 3363-3372.

McFadden, P. L., and M. W. McElhinny, 1982, Variations in the geomagnetic dipole 2: statistical analysis of VDMs for the past 5 million years, *Jour. Geomagnetism and Geoelectricity* **34,** 163-189.

McFadden, P. L., and R. T. Merrill, 1984, Lower mantle convection and geomagnetism, *Jour. Geophys. Research* **89,** 3354-3362.

Merrill, R. T., and M. W. McElhinny, 1977, Anomalies in the time-averaged paleomagnetic field and their implications for the lower mantle, *Rev. Geophys. Space Physics* **15,** 309-323.

Merrill, R. T., and M. W. McElhinny, 1983, *The Earth's Magnetic Field: Its History, Origin and Planetary Perspective*. London: Academic Press, 401p.

Merrill, R. T., M. W. McElhinny, and D. J. Stevenson, 1979, Evidence for long-term asymmetries in the Earth's magnetic field and possible implications for dynamo theories, *Physics Earth Planetary Interiors* **20,** 75-82.

Ness, G., S. Levi., and R. Couch, 1980, Marine magnetic anomaly time scales for the Cenozoic and Late Cretaceous: A precis, critique and synthesis, *Rev. Geophys. Space Physics* **18,** 753-770.

Pesonen, L. J., and H. C. Halls, 1983, Geomagnetic field intensity and reversal asymmetry in late Precambrian Keweenawan rocks, *Geophys. Jour. Royal Astron. Soc.* **73,** 241-270.

Phillips, J. D., 1977, Time variation and asymmetry in the statistics of geomagnetic reversal sequences, *Jour. Geophys. Research* **82,** 835-843.

Roberts, N., and J. Shaw, 1984, The relationship between the magnitude and direction of the geomagnetic field during the Late Tertiary in Eastern Iceland, *Geophys. Jour. Royal Astron. Soc.* **76,** 637-651.

Cross-references: *Geomagnetic Field Analysis; Geomagnetic Polarity Reversals: Observations; Geomagnetic Polarity Reversals: Theory and Models; Geomagnetic Reversal Sequence: Statistical Structure; Geomagnetic Secular Variation: Direction and Intensity; Geomagnetic Secular Variation: Theory; Paleomagnetic Field: Intensity; Paleomagnetic Secular Variation; Seafloor Spreading: Magnetic Evidence.*

GEOMAGNETIC FIELD: ELEMENTS

The geomagnetic field is a vector field that varies in space and time. The field may be characterized by two vectors: **F**, the magnetic intensity, and **B**, the magnetic induction. In free space, **F** differs from **B** only by a constant factor μ_0, the permeability of free space. That is,

$$\mathbf{B} = \mu_0 \mathbf{F}$$

In the old electromagnetic units (emu) $\mu_0 = 1$ and was nondimensional, so **B** and **F**, expressed in gauss, were indistinguishable. This system led to units of electrical resistance and electromotive force that were inconveniently small. To overcome this situation, most countries have adopted the International System of Units (SI), which is based on the meter, kilogram, and second (MKS). In SI units the permeability of free space is equal to $4\pi \times 10^{-7}$ kg·m coulombs^2 which makes **F** and **B** not only very different in magnitude, but also gives them different units.

The International Association of Geomagnetism and Aeronomy (IAGA), which encourages research in geomagnetism, adopted a resolution at the IAGA Scientific Assembly in Kyoto, Japan, in 1973 (Alldredge, 1973) that recommended the adoption of SI units for geomagnetic studies. Among other things, the resolution recommended that values of the geomagnetic "field" be expressed in terms of the magnetic induction **B** (SI unit tesla (T) = W·m^{-2}). This choice of using **B** to express values of the geomagnetic field made the transition to the new units easy because $1\,T = 10^4$ Gauss, and 1 nT = 1 gamma (10^{-5} Gauss), a unit frequently used in the past.

Basic Elements

At any point 0 the vector magnetic field **B** can be specified in alternative ways (Fig. 1). One method is to give the magnitude of the vector **B** (B) and the two angles D and I. D is the deviation of **H**, the horizontal component of **B**, from the northward horizontal direction; I is the deviation of **B** from **H**. D is positive if eastward, and I is positive if downward. D is called the *declination* or compass variation, and I is called the dip or *inclination*. The magnetic meridian plane at any point is the vertical plane defined by the direction of the magnetic lines of force at that place.

Another method is by giving H, D, and I, where H is the positive magnitude of **H**. This method was most often used before the advent of nuclear, atomic, and flux gate magnetometers, because these elements were the easiest to measure with the classical instruments. Yet another way of specifying **B** is by its northward and eastward horizontal components X and Y and its vertical component Z, which is taken as positive if downward. The intensities (magnitudes) of B, H, X, Y, and Z are expressed in nanoteslas; D and I are expressed in degrees and minutes of arc or in radians, if required in interconnecting equations given here. The seven quantities B, H, D, I, X, Y, and Z are called *magnetic elements*; H, X, Y, and Z are also referred to as *magnetic components*. These quantities are the most-used magnetic elements and components, but the vector **B** can be resolved into other special elements as may be required in special problems.

The magnetic elements are interconnected by the following equations:

$$H = B \cos I \qquad Z = B \sin I = H \tan I$$
$$X = H \cos D \qquad Y = H \sin D$$
$$X^2 + Y^2 = H^2 \qquad X^2 + Y^2 + Z^2 = H^2 + Z^2 = B^2 \quad (1)$$

Three independent elements are required to specify **B**; any of the others can then be determined from Eq. 1.

Slow changes in the Earth's magnetic field, from sources in the core of the Earth, are called *secular variations*. These variations are studied by examining the time rate of change (indicated by a dot over the element) of some of the magnetic elements. From Eq. 1 we get

$$\dot{H} = \dot{B} \cos I - \dot{I} B \sin I$$
$$\dot{Z} = \dot{B} \sin I + \dot{I} B \cos I = \dot{H} \tan I + \dot{I} H \sec^2 I$$
$$\dot{X} = \dot{H} \cos D - \dot{D} H \sin D$$
$$\dot{Y} = \dot{H} \sin D + \dot{D} H \cos D$$

Here $\dot{B}$, $\dot{H}$, $\dot{X}$, $\dot{Y}$, and $\dot{Z}$ are usually expressed in nanoteslas per year, and $\dot{D}$ and $\dot{I}$ are expressed in radians per year. In the study of secular variation, instantaneous derivatives are not usually used, but estimated for intervals from one to five years.

The distribution of the magnetic elements is best illustrated by magnetic charts on which lines are drawn through points at which a given magnetic element has the same value. Special names have been assigned to the lines for several of the elements: *isogonic* lines refer to D, *isoclinic* lines refer to I, and *isodynamic* lines refer to H, X, Y, Z, and B.

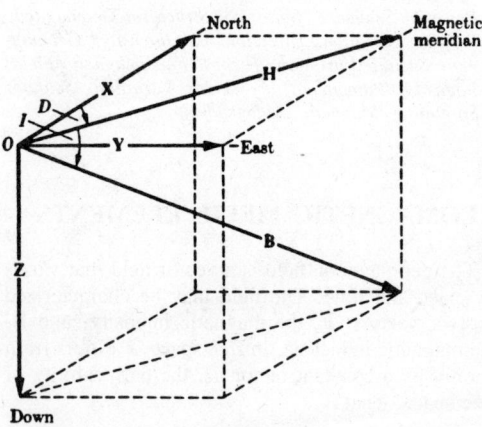

FIGURE 1. The geomagnetic force B, its rectangular components X, Y, and Z, and the elements H, D, and I.

The secular variations or long-period changes of the geomagnetic field are usually indicated on world or national charts by *isoporic* lines, which are the contours of equal rate of change.

For nautical, aeronautical, and surveying purposes, the isogonic or D charts are most important. Years ago, world isogonic charts were normally published by several countries at five-year intervals, and component charts were published usually at 10-year intervals. These charts were handdrawn directly from adjusted observations of the elements. Recently IAGA has been determining International Geomagnetic Reference Field (IGRF) models every five years, and all elements are now charted at this interval. These models are derived from the observed elements by using spherical harmonic analysis, which is described later.

The magnetic poles are defined as the points where $I = \pm 90°$, $H = 0$, and the magnetic equator is defined by the points where $Z = 0$, $I = 0$. The isogonic lines converge at both the magnetic poles and the geographic poles.

Other Derived Elements of the Geomagnetic Field

Magnetic elements measured at observatories and repeat stations, as well as by satellites become voluminous. As described in this volume (see *Geomagnetic Field Analysis*), a good way to organize and systemize the great volume of magnetic element information is to determine from the data, by least squares, the spherical harmonic coefficients g_n^m and h_n^m in the following expression for the geomagnetic potential:

$$V = a \sum_{n=1}^{N} \sum_{m=0}^{n} \left(\frac{a}{r}\right)^{n+1} P_n^m(\theta)$$
$$\cdot (g_n^m \cos m\phi + h_n^m \sin m\phi)$$

In this equation, $P_n^m(\theta)$ is the associated Legendre polynomial of degree n and order m, θ is the colatitude, ϕ is the east longitude, a is the radius of the Earth, N is the truncation degree, and r is the desired radius for the calculated potential. In a sense, g_n^m and h_n^m become rather basic elements of the geomagnetic field, because X, Y, and Z are obtained by taking the gradient of V (Chapman and Bartels, 1940). In a similar manner, $\dot{g}_n^m$ and $\dot{h}_n^m$ can be obtained. In precise analyses of the field, the Earth's oblateness is taken into account. This is done by converting X and Z components measured in the geodetic coordinate system to corresponding geocentric components.

Once the spherical harmonic coefficients of the field have been obtained, other interesting and useful quantities can be obtained, such as the magnitude of the dipole moment of the Earth,

$$M \cong [(g_1^0)^2 + (g_1^1)^2 + (h_1^1)^2]^{1/2} \quad (2)$$

and the axis of the centered dipole, or the spectrum of the magnetic energy outside the Earth,

$$E_n = \frac{2\pi a^2}{\mu_0} \frac{n+1}{2(n+1)} \sum_{m=0}^{n} [(g_n^m)^2 + (h_n^m)^2]$$

(McDonald and Gunst, 1968).

At the IAGA Scientific Assembly in Prague in 1985, a new IGRF field model (IAGA Division I, Working Group 1, 1985) was adopted, which gives a value of 30,438 nT for M (Eq. 2). The axis of the centered dipole for this model emerges in the northern hemisphere at 79.0° latitude and 70.9° west longitude.

Coordinate Systems Based on the Geomagnetic Field

The axis of the centered dipole is almost as fundamental for geomagnetism as the axis of rotation of the Earth is for geography. Because of this, A. Schmidt, in 1891, introduced geomagnetic coordinates as described by Chapman and Bartels (1940). In this system the axis of the centered dipole is taken as the polar axis. The coordinates of a point on the Earth are given by the geomagnetic colatitude of the point measured from this new polar axis, and the geomagnetic longitude is measured eastward from the meridian half-plane bounded by the polar axis and containing the geographical South Pole. The geomagnetic coordinates of a point can be computed from its geographical coordinates by spherical trigonometry.

Geomagnetic time at a station is a useful quantity defined by the angle between the geomagnetic meridian through the station and the one opposite to that through the Sun; it can therefore be found by computing the geomagnetic longitudes of the station and of the Sun at any instant. At stations in middle and low latitudes the geomagnetic time differs only slightly from the geographic local time.

Soon after the first satellite was launched, scientists had difficulty in attempts to map the intensities of trapped particles. The geomagnetic field can be thought of as consisting of continuous magnetic lines of force that indicate the path a positive magnetic pole (if such existed) would take if released in the field. McIlwain (1961) described a new two-coordinate system, later improved by Stone (1963), which was useful in organizing measurements along lines of the magnetic field. One coordinate is taken as B, the magnitude of the magnetic field at point A. The other coordinate is a magnetic shell parameter $L = f(B, I)$, where I is the integral invariant

$$I = \int_A^{A'} \left(1 - \frac{B_l}{B}\right)^{1/2} ds$$

where ds is the differential path length along the line of force connecting point A with its conjugate point

A', and B_l is the magnitude of the magnetic field along the line of force.

The magnetic shell parameter L, not defined in detail here, retains most of the desirable properties of I, organizes the measurements along lines of force, is approximately constant along lines of force, and is the analogue of the equatorial radius of a magnetic shell in a dipole field. B and L form the primary spatial coordinates for an intercomparison of artificial satellite studies of trapped particles.

Magnetic Indices

As indicated earlier, all of the geomagnetic elements vary with time. Variations with periods of several years and longer are thought to originate in the core of the Earth. Most shorter-period variations are from sources in the magnetosphere and ionosphere with corresponding induced parts in the Earth. The Sun is no doubt the ultimate cause of most of these variations.

Over the years, a variety of geomagnetic indices (Akasofu and Chapman, 1972) have been used to describe these variations. One of the most-used indices is the K index, which is intended to be a measure of solar plasma flow disturbances. A local K index (integer value) from 0 to 9 is assigned for each 3-hr interval of Greenwich time; thus, in all there are eight K indices for each Greenwich day. The elements H and D recorded on a magnetogram at each station are examined for each 3-hr interval, and an estimate is made of the range r during the interval, excluding solar-quiet-day variation, lunar variation, and solar flare effects. Each observatory uses a quasilogarithmic scale for converting these estimated r values into K values. The mean of the K values from 12 selected observatories is denoted by K_p (a global K index) and is expressed in a scale of thirds:

$$0_0, 0+, 1-, 1_0, 1+, \cdots, 9_0$$

Efforts have been made to establish indices that are related to some magnetospheric parameters. The D_{st} index gives the average (in longitude) depression of H in low latitudes in units of nanoteslas. It is proportional to the total kinetic energy of the particles injected into and trapped in the Van Allen belt, provided that the particles are distributed symmetrically with respect to the dipole axis (Akasofu and Chapman, 1972). Hourly values of the D_{st} index are normally provided.

The AE, AL, and AU indices were devised to obtain the intensity of the auroral electrojet in units of nanoteslas (Davis and Sugiura, 1966). These indices are based on the H records from a set of auroral-zone magnetic observatories, which is uniformly distributed in longitude. The records are superposed, and the upper envelope gives the AU index, the lower envelope gives the AL index, and the distance between the upper and lower envelopes gives the auroral electrojet index AE. Because digital records are used to determine these indices, nearly unlimited resolution can be given, but for practical purposes only 2.5-min value are given (Akasofu and Chapman, 1972).

LEROY R. ALLDREDGE

References

Akasofu, S.-I., and S. Chapman, 1972, *Solar-Terrestrial Physics*. Clarendon Press: Oxford, 901 p.

Alldredge, L. R. (ed.), 1973, *IAGA Bulletin 35*, International Union of Geodesy and Geophysics, Paris: IUGG Publication Office.

Chapman, S., and J. Bartels, 1940, *Geomagnetism*, vols. 1 and 2. Oxford: The University Press.

Davis, T. N., and M. Sugiura, 1966, Auroral electrojet activity index AE and its universal time variations, *Jour. Geophys. Research* **71**, 785-801.

IAGA Division I, Working Group 1, 1986, International geomagnetic reference field revision 1985, *Geophysics* **51**, 1020-1023.

McDonald, K. L., and R. H. Gunst, 1968, Recent trends in the Earth's magnetic field, *Jour. Geophys. Research* **73**, 2057-2067.

McIlwain, C. E., 1961, Coordinates for mapping the distribution of magnetically trapped particles, *Jour. Geophys. Research* **66**, 3681-3691.

Stone, E. C., 1963, The physical significance and application of L, B_0, and R_0 to geomagnetically trapped particles, *Jour. Geophys. Research* **68**, 4157-4166.

Cross-references: *Geomagnetic Field: Measurement; Geomagnetic Field, Main: Theory; Geomagnetic Field Analysis; Geomagnetism: Historical Introduction; Spherical Harmonic Analysis*.

GEOMAGNETIC FIELD: MEASUREMENT

Elements of the Geomagnetic Field

The Earth's magnetic field at any point is a vector quantity with both magnitude and direction. The total intensity of the vector **F** acts in the plane of the magnetic meridian. Declination D (the angle between the magnetic meridian and true north) and Inclination I (the angle between **F** and the horizontal) define the direction of the vector. In the meridian plane **F** may be resolved into a horizontal component **H** and a vertical component **Z**. It is frequently convenient to express these quantities in relation to three fixed orthogonal axes, true N-S (x), true E-W (y) and the vertical (z). In this context the horizontal component **H** may be resolved into a north component **X** and an east component **Y**.

The magnetic elements, illustrated in Fig. 1, are related by the following equations:

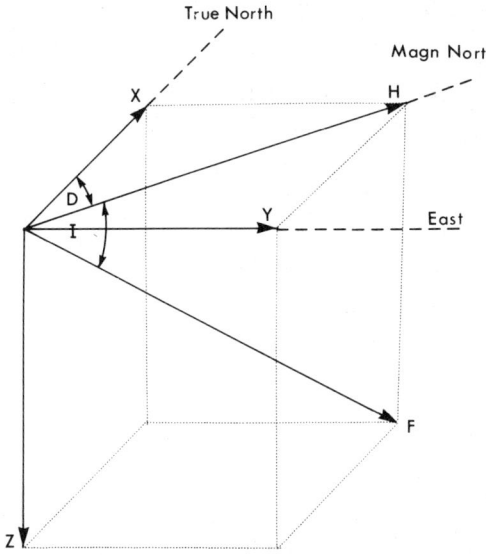

FIGURE 1. The geomagnetic elements.

$$F^2 = X^2 + Y^2 + Z^2$$
$$F^2 = H^2 + Z^2$$
$$X = H \cos D$$
$$Y = H \sin D$$
$$H = F \cos I$$
$$Z = F \sin I$$
$$\text{Tan } I = Z/H$$
$$\text{Tan } D = Y/X$$

Z and I are by convention reckoned positive to the north of the magnetic equator. Declination east of true north is assumed to be positive. Three elements suffice to describe the magnetic vector, e.g., D, **H**, **Z**; **X**, **Y**, **Z**; D, I, **F**, etc. Vector field intensities are expressed in nanotesla (nT) units. Directions are usually given in degrees and minutes, or one-tenth minutes.

Description of the Field and Measurements

Main Field. The Earth behaves as a great magnet. Its magnetic field extends from the core out into space where it creates the tear-shaped cavity in the solar wind known as the magnetosphere. A major part of the Earth's field (>90%), the main field, is of internal origin. The field is sustained by electric currents induced in the conductive liquid core as a consequence of slow convective motions within the core. These motions cause a gradual change in the pattern of the surface field. At a fixed measurement site this appears as a progressive change in the field intensity and direction, known as the secular variation. Figure 2 shows the secular variation at Eskdalemuir observatory revealed by the annual mean values of **H**, D, and **Z**.

Because of this change there is a continuing requirement to update magnetic charts and mathematical models of the field. Declination charts are produced at five-year intervals and charts of the other elements every ten years. The production of charts and models requires a continuing input of worldwide absolute, or near absolute, data. A small but essential proportion of these data is obtained from a network of about 160 permanent magnetic observatories that monitor the field on a continuous basis and standardize their records by regular absolute measurements. The bulk of the data is obtained from uncoordinated land, marine, aeromagnetic, and satellite surveys. The preparation of data for chart production involves the reduction of measurements, made at different times and locations, to a common epoch. This process requires accurate knowledge of the secular variation, which is primarily derived from observatory measurements.

The most effective method of collecting globally distributed data, within a short period of time, is by means of a satellite placed in near-earth, near-polar orbit. In the past the accuracy of field models has been limited by the uneven geographical distribution of the data and because many surface measurements are affected by the presence of localized magnetic features in the Earth's crust. Satellite measurements are an improvement in both of these respects.

During the period 1965–1967 the Polar Orbiting Geophysical Observatory (POGO) series of satellites provided the first global coverage of scalar measurements, which contributed to the 1965 main field model. More recently MAGSAT (October, 1979–June, 1980) demonstrated the capability of making vector and scalar measurements in a low orbit, the 1980 International Geomagnetic Reference Field (IGRF) main field model being based entirely on MAGSAT data.

Time Variable Field. Temporal and spatial variations superimposed on the main field are manifestations of the external time variable field and the crustal field. Temporal variations, revealed by continuous magnetic records, are attributable to magnetospheric and ionospheric currents. Some variations exhibit well-defined periodicities, others are seen as irregular disturbances. The dynamic range of the time variable field is extremely large extending from decades (the solar cycle) to milliseconds (ELF pulsations). Amplitudes are similarly variable ranging from about 5×10^3 nT (intense magnetic storm) to less than 10^{-2} nT (Pcl micropulsations).

The study of the magnetosphere has been dramatically stimulated since the International Geophysical Year (IGY) 1957–1959 by spaceborne measurements and multidisciplinary campaigns involving the simultaneous measurement of parameters associated with the magnetic variations. Ground-based magne-

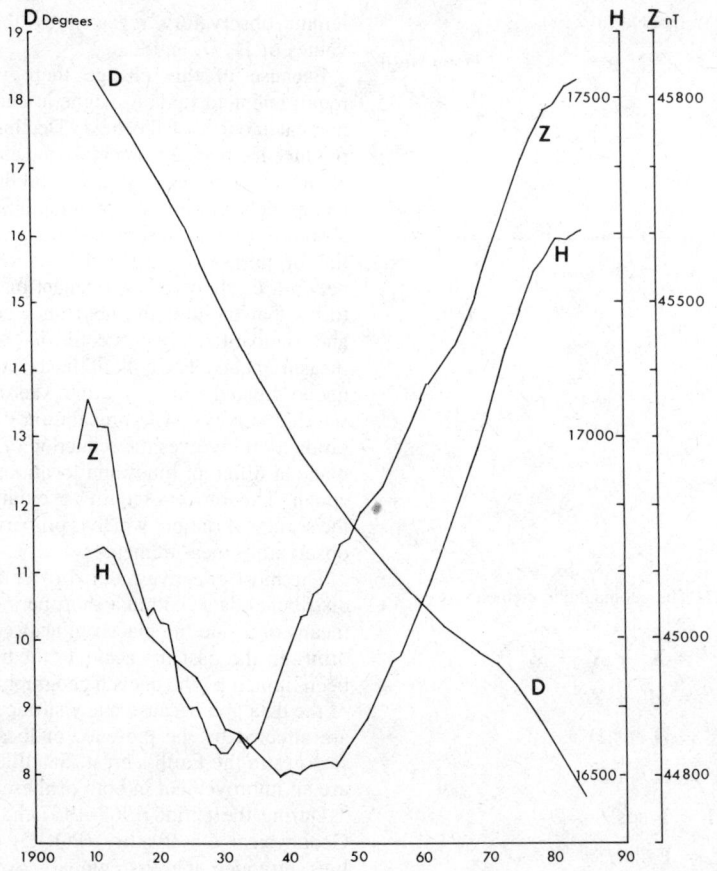

FIGURE 2. The secular variation in H, D and Z at Eskdalemuir magnetic observatory, 1907–1984.

tometer arrays recording three components of the field without absolute control were extensively deployed during the period of the International Magnetospheric Study (IMS) 1976–1979.

Crustal Field. Spatial variations in the pattern of the surface field, revealed by local and regional survey measurements, reflect the differing magnetizations of crustal rocks and are thus the surface signature of the underlying geology. Low-level aeromagnetic surveys are used to delineate areas of magnetic anomaly for selective investigation by ground-based or marine geophysical methods and possible resource exploitation. Surveys flown at higher altitudes and satellite measurements in near-earth orbit do not resolve the short wavelength anomalies and so provide information about large scale geological and tectonic features. Crustal studies require the removal of the estimated main field and the time variable field from the collected data.

Rocks acquire their magnetization under the influence of the ambient magnetic field at the time of their formation. Paleomagnetic measurements of the natural remanent magnetization (NRM) (q.v.) of rock samples, which define the magnetic vector, extend our knowledge of the Earth's field back into geological time. Paleomagnetism has provided valuable evidence in support of continental drift, field reversals, and plate tectonic theory.

Induced Field. We have seen so far that any magnetic measurement contains contributions from the main field, the external field, and the crustal field. One further but less tangible contribution is particularly relevant in the field of modern geomagnetic measurements. Varying external magnetic fields induce currents in conductive rocks, thereby generating secondary magnetic fields. The recorded time variable field at a measurement site therefore, combines an inducing and an induced component, so that the measurement contains information related to the physical properties of the underlying rocks and geological structure.

The geomagnetic deep sounding (GDS) method (refer to *Magnetotelluric Method: Fundamental Concepts*) exploits this induced field by deploying an array of magnetometers, recording three field components without absolute control, within the survey region. Data obtained from the array are compared with data obtained from a base station outside the survey area over a broad band of frequencies. GDS is based on the idea that a normal transient uniform field has no vertical component, and that if a vertical component is observed, it is necessarily produced by a lateral conductivity discontinuity. Gough-Reitzel variometers, fluxgate magnetometers, and torsion photoelectric variometers have been successfully used for this type of application.

The depth of penetration of an electromagnetic wave into the Earth is dependent on its wavelength and the resistivity of the ground. Long period waves penetrate deep into the lithosphere and beyond, while rapid variations achieve relatively shallow penetration. Magnetotelluric (MT) measurements exploit this effect to obtain a depth versus apparent resistivity sounding by means of surface measurements of the magnetic and electric fields. Three orthogonal magnetic components (X, Y, Z) are recorded together with two electric field components (E_x and E_y) in several discrete frequency bands. The natural time variable field conventionally provides the signal source, although artificial signal sources have been used at the higher frequencies. MT data are typically recorded in the range 5×10^2 Hz to 5×10^{-3} Hz. Induction coils and Superconducting Quantum Interference Devices (Squids) are the preferred sensors because signal amplitudes are extremely small in parts of this waveband. Electric field measurements are made by telluric lines and sensitive amplifiers.

Two innovations have greatly improved the quality of MT data in recent years: remote reference and real-time data processing. The remote reference technique involves the simultaneous recording of X (or Y) together with E_y (or E_x) at a site some distance from the measurement site. This recording facilitates discrimination between coherent source field signals and incoherent electromagnetic noise. In-field real-time data processing allows examination of the sounding curve while recording is still in process. The operator is thus able to gauge when sufficient measurements have been made within each frequency band and assess the suitability of the site for measurement purposes. This is a great improvement over older types of instrumentation where there was often a delay of weeks between data collection and calculation of the final results.

Magnetic Instruments

Instruments have been developed for making measurements at all parts of the geomagnetic spectrum. The choice of instrument depends on the nature of the measurement, the frequency and amplitude range of the variation, the required measurement resolution, etc. Other practical considerations involve such factors as power consumption, size, and cost.

Instruments may be classified in different ways. One basic distinction is that between magnetometers and variometers. The former are used to make absolute measurements, and the latter are used to monitor variations of the field without reference to absolute field magnitude. A second distinction is that between scalar and vector measurement. A scalar magnetometer measures the magnitude of the field without regard to direction; a vector instrument measures the field intensity in a prescribed direction (usually the sensor axis). Vector measurements can be made by scalar instruments using precisely aligned bias coils to create auxiliary fields. Vector measurements require a coordinate frame of reference, e.g., the measurement of declination requires knowledge of true North, and measurement of Z by fluxgate magnetometer requires the sensor axis to be vertical. Making accurate vector measurements from a mobile platform, subject to pitch, roll and yaw, poses real problems that can only be solved by sophisticated methods, e.g., referencing the sensor to an on-board inertial navigation system or an attitude sensing (or control) system.

The most commonly used instrument classification is probably that based on the physical principle invoked in the measurements. Five classes are considered in this review: (1) torsion magnetometers, (2) induction magnetometers, (3) saturable core magnetometers, (4) resonance magnetometers, and (5) cryogenic magnetometers.

Torsion Magnetometers. When a magnet is suspended (or balanced) in a magnetic field it adopts an equilibrium position in which the couple exerted on the magnet due to the field is compensated by a restoring couple due to the suspension (or to gravity). A small change in the field causes the magnet to rotate into a new position of equilibrium. The resulting small movement is conventionally amplified by optical lever and photographically recorded. Instruments based on this principle have provided the basis for continuous records at observatories for nearly 150 years (Laursen and Olsen, 1971).

Several field portable instruments of this type have been developed for temporary deployments. The best known of these are probably the Askania variograph and the relatively inexpensive Gough-Reitzel variometer (Gough and Reitzel, 1969). The latter instrument has been widely used in magnetometer arrays for conductivity and magnetospheric investigations.

The torsion photoelectric magnetometer (Marianiuk et al., 1978) utilizes the Bobrov pattern quartz variometer as a null detector. Any small magnet rotation caused by a change in the ambient field is optically detected, producing a current that is fed back to create a field at the variometer opposing the change. This pattern of instrument preserves the excellent qualities of the variometer (low temperature coefficient, good long-term stability) and provides a voltage output suitable for input to a data logger. Drifts as low as 2 nT per year have been reported when the instrument is deployed in a temperature stabilized observatory environment. Field portable versions of the instrument have proved suitable for use in conductivity arrays and MT studies.

Torsion (and balanced) magnetometers played an important role as absolute observatory instruments from the mid-nineteenth century until the development of the proton vector magnetometer. Declination is still measured by observation of a suspended magnet at about 75% of the world's geomagnetic observatories. Two secondary standard instruments of this type, the Quartz Horizontal Magnetometer (QHM) and Balance Magnetometrique Zero (BMZ) also remain in current use at a number of observatories.

Induction Magnetometers. In this category of instrument the distinction between an absolute instrument and a variometer is implicit in the measurement method. Absolute measurements of a slowly changing field are made with a rotating coil, while rapid variations are monitored by a static coil. Both methods involve the detection of the electromotive force (emf) induced by the changing flux linking the coil.

When a small coil is rotated about a diameter axis aligned parallel to F, no flux linkage occurs and zero emf is detected. Inclination is then defined by the rotation axis. This principle was the basis of the classical earth inductor, an observatory instrument that is no longer in use but that had a reported 0.1' measurement accuracy. Attempts have been made to modernize this instrument, replacing the manually rotated coil with a pneumatically driven rotor and using orthogonally positioned stator coils to sense the misalignment between F and the rotor spin axis.

Static induction coils, with high permeability cores, are widely used for pulsation and MT measurements. When a basic coil is used, variations in the ambient field produce a signal proportional to the first derivative of the field. Simple coils have limited bandwidths, and knowledge of the frequency and phase characteristics of the coil is a prerequisite for interpreting the records obtained using these devices. The use of feedback flux (Clerc, 1971) has been an important innovation. In this configuration, natural variations of the field are compensated by current feedback to an auxiliary winding on the coil, which generates a self-balancing flux. When the induction coil is used as a null detector in this way, the coil characteristics are no longer important and a flat response can be obtained over a four-decade frequency bandwidth.

Saturable Core Magnetometers. The fluxgate magnetometer was developed in the 1940s for use in antisubmarine warfare. Shortly after World War II it was adopted for airborne geophysical surveys. The measurement depends on the variable permeability of ferromagnetic core material when subjected to an AC excitation that drives the core through saturation around the B/H hysteresis loop (Primdahl, 1979). Various sensor geometries have been developed. Probably that most commonly used is the Vacquier sensor, which comprises two high permeability cores wound with series-connected, antiparallel excitation coils. Both cores are contained within a common signal winding and an auxilliary feedback winding. When an AC excitation is applied, both cores are driven into saturation during a significant part of each half cycle. In the presence of an ambient field, saturation occurs earlier during one half cycle than during the next, causing a signal at the second harmonic of the excitation frequency to be induced in the signal winding, which is sensitively related to the magnetic field component parallel to the sensor axis.

Figure 3 shows a schematic diagram of a fluxgate magnetometer employing second harmonic detection and negative feedback. The signal voltage is amplified by a tuned amplifier and rectified by the phase sensitive detector. The detector output is low-pass filtered and DC amplified to provide current drive to the feedback winding in such a sense as to cancel the ambient field. The fluxgate sensor operates as a null detector and the feedback current is a measure of the field intensity.

Since the fluxgate magnetometer is a vector instrument the output voltage is attitude dependent. This feature is exploited in the fluxgate theodolite, which uses the fluxgate as a zero-field detector to measure D and I (Trigg, 1970).

The inherent advantages of the fluxgate are its simplicity, reliability, small size, and modest power consumption. Its disadvantages include zero-offset, thermal drift, relatively poor resolution, and a restricted bandwidth. Recently introduced instruments with ring core sensor geometries appear to offer improved performance in all these respects.

Thin film magnetometers, which utilize the transverse hysteresis characteristic of a thin Fe-Ni film are similar to the fluxgate magnetometer, but offer an improved low noise, low power performance with a superior bandwidth (Irons and Schwee, 1972).

Resonance Magnetometers. Resonance magnetometers monitor the precession of atomic particles in the presence of an ambient magnetic field. The best known instrument in this class is the proton precession magnetometer (PPM). When an atom or nucleus with magnetic moment M and angular momentum G is placed in a steady field H_o, M precesses about H_o at a frequency f_p that is directly proportional to the scalar magnitude of the field.

$$f_p = \gamma_p |H_o| = (|M|/|G|) \cdot H_o$$

The quantity γ_p, the gyromagnetic ratio of the proton is an immutable atomic constant. If the Larmor frequency f_p is measured accurately, the magnitude of the field H_p is derived in absolute terms. In 1960

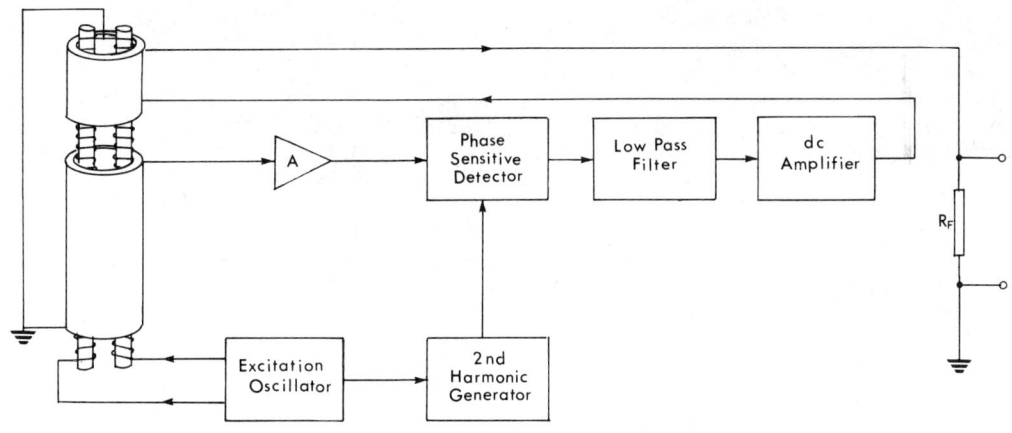

FIGURE 3. Schematic diagram of a second harmonic fluxgate with total negative feedback.

the International Association of Geomagnetism and Aeronomy (IAGA) adopted a value for γ_p of $(2.6753 = 0.00002) \cdot 10^{-1}$ Hz $\cdot$ nT^{-1} for protons in water.

The PPM sensor consists of a sample bottle, typically containing water or kerosene, wound with a polarizing coil. When current is applied to the coil a strong axial field H_T is established normal to H_o creating a large equilibrium magnetization directed along H_T. When the polarizing field is rapidly removed the protons in the sample precess about H_o with coherent phase, inducing a signal at the Larmor frequency in the coil. The precession is damped by interaction between neighboring atoms causing the signal to decay into noise within 2–3 s. In the majority of modern PPMs the signal frequency is multiplied and then counted during a preset time interval (Fig. 4).

Given a suitable choice of multiplier N and gating period τ, the measured value can be displayed in nT or $^1\!/_{10}$nT units.

The PPM is suitable for all applications requiring absolute measurement of the field. The sensor is robust and insensitive to attitude and temperature. Even in the adverse circumstances of a towed marine or aeromagnetic deployment, the noise level is < 1nT.

One disadvantage of the PPM, particularly for aeromagnetic survey, is its finite polarizing time, which makes any measurement discontinuous. The Overhauser/Abragam magnetometer provides continuous precession, which is achieved by the use of an organic free radical in solution that introduces free electrons into the sample. When subjected to the radio frequency (rf) radiation at the electron resonance frequency of the salt, the spin energy of the electrons is transferred to the proton spin of the solvent, thereby sustaining continuous precession at the Larmor frequency.

Alkali vapor magnetometers (Bloom, 1962) provide a continuous frequency output at the frequency of electron resonance of the alkali atom. Since the gyromagnetic ratio for an electron is

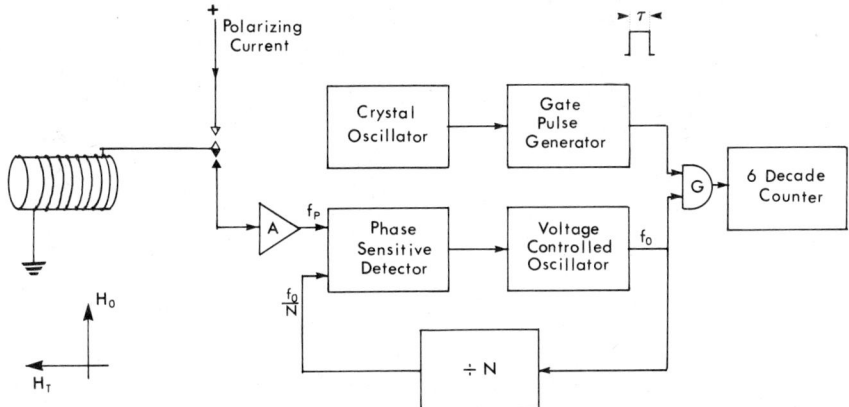

FIGURE 4. Schematic diagram of a proton precession magnetometer. The phase sensitive detector controls the frequency at the VCO output such that $f_p = f_0/N$.

considerably larger than for a proton, this type of instrument has an enhanced sensitivity. Three types of instrument have been developed: the metastable helium, rubidium, and cesium magnetometers. All three instruments make use of optical pumping and monitoring techniques.

Optical pumping involves the excitation of alkali vapor atoms within an absorption cell by means of a light beam of collimated, filtered, and circularly polarized resonance radiation. This pumping serves to polarize the sample, creating a non-zero equilibrium magnetization in the $^2S_{1/2}$ ground states of the alkali atom. In this pumped condition the optical transmission through the cell is maximum. The application of a small rf field at the Larmor frequency normal to H_o redistributes the ground state population. Two consequential optical effects are observable. The transmitted light intensity in the direction of H_o is diminished, and the intensity of a beam normal to H_o is modulated at the Larmor frequency. Two patterns of sensor have been developed that utilize these effects.

The locked oscillator magnetometer, Fig. 5a, monitors the transmission of the pumping beam directed along H_o and locks the Larmor signal, derived from the voltage controlled oscillator, to the peak of the resonance.

The self-oscillator, Fig. 5b, monitors the intensity modulation of the cross beam and feeds back an amplified, current limited, signal at the Larmor frequency to the $H1$ coil, suitably phase shifted to sustain continuous oscillation. When the sensor axis

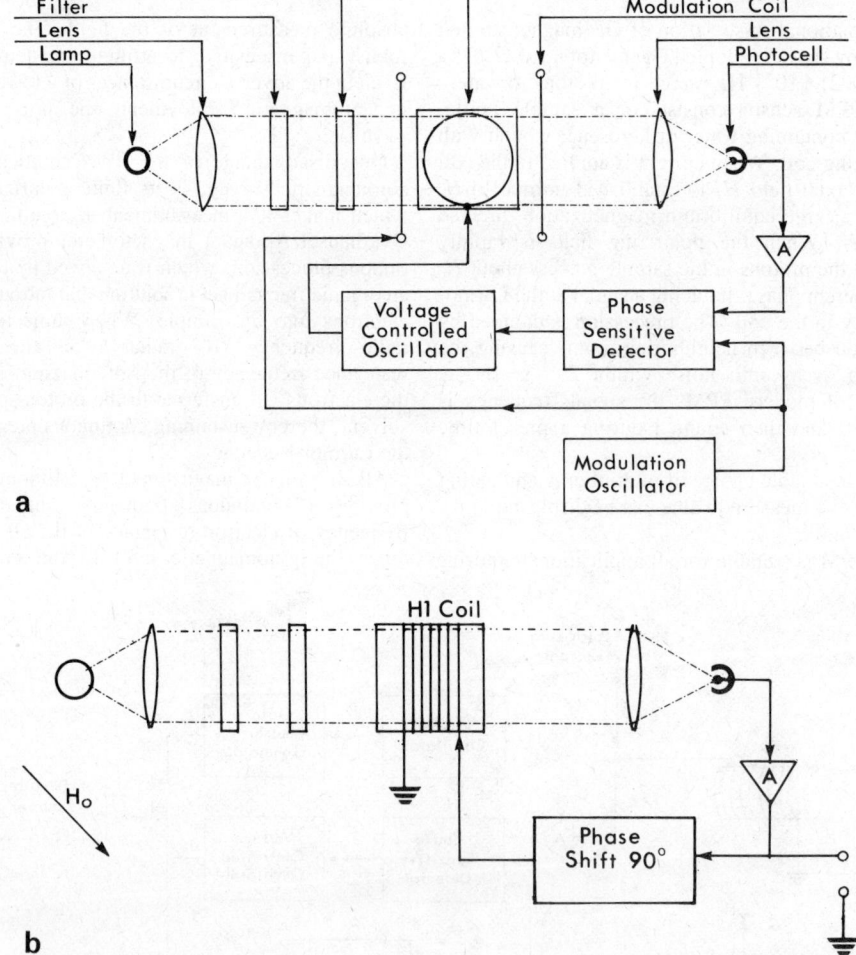

FIGURE 5. (a) Schematic diagram of a locked oscillator magnetometer. The modulation oscillator imposes a small modulation on the H_0 field. At phase lock the amplified signal is the second harmonic of the modulation frequency. When H_0 changes, a signal at the fundamental is detected with a consequent adjustment of the VCO frequency, which thus tracks the Larmor frequency. (b) Schematic diagram of the self-oscillator magnetometer.

is orientated at 45° to the ambient field H_o the light beam provides both the pumping and monitoring components. The sensor is only suitable for fixed attitude operation because null signal zones occur whenever the optical axis is aligned parallel or normal to H_o. More sophisticated configurations using multiple sensors are needed to minimize or eliminate signal null zones for aeromagnetic and satellite use.

Gradiometer configured resonance magnetometers (Hood, 1981) are increasingly being used for geophysical surveys. One of the many advantages of this mode of operation is that it provides common mode rejection of the time variable field, thus considerably simplifying the reduction of survey data for mapping the crustal field.

Cryogenic Magnetometers. The cryogenic magnetometer or Squid is the most sensitive magnetometer presently available (Goree and Fuller, 1976). Its operation depends on various physical effects associated with superconductivity. When a superconducting ring is cooled below a critical temperature it becomes a perfect conductor. The magnetic flux inside the ring ϕ_i is trapped and remains constant even when the external flux ϕ_x changes because of the presence of induced supercurrents, carried by electron pairs, circulating within the ring that cancel the flux variation. As the field increases, the net circulating current also increases until it reaches a critical value I_c, at which point the ring becomes resistive again and flux adjustment occurs.

The flux inside the ring ϕ_i cannot assume any arbitrary value but is quantitized in units of flux quantum $\phi_o = h/2e$, $(2.07 \times 10^{-15}$ Webers), where h is Planck's constant and $2e$ is the charge carried by paired electrons. In a continuously varying field ϕ_i changes discontinuously by an indeterminate number of quanta $n\phi_o$ each time I_c is exceeded. If a suitably designed weak link is introduced into the ring then I_c is reduced to a level such that $n = 1$ and only $\pm\phi_o$ adjustments are possible, and the response of the Squid to a periodic change of external flux is hysteretic (Fig. 6a). The small energy dissipation involved in traversing the loop ABCD is detectable and forms the basis of operation of most rf Squids.

A simplified schematic diagram of the rf Squid is shown in Fig. 6b. The voltage developed across the LC tank coil, which is loosely coupled to the Squid, is a linear function of the rf current drive as long as the net circulating current is below I_c. When a flux quantum enters or leaves the ring, the tank circuit energy is abruptly reduced by the energy dissipated in traversing the hysteresis loop. The tank coil voltage is amplified and the rf detected, producing a voltage at the detector output that is periodic as a function of the external field, Fig. 6c.

For a given datum field and level of rf drive the signal input to the phase sensitive detector is the second harmonic of the audio-frequency (af) modulation (ω) imposed on the field. Any departure of the ambient field from this level produces a signal input at the fundamental frequency, which is rectified and fed back to create a self-balancing flux at the Squid, thus locking the Squid to the datum field level. The Squid functions as a null detector and the feedback current is a measure of the departure of the ambient field from the datum level.

All the critical components are enclosed within a super-conducting shield that stabilizes the background field to 1 part in 10^{10}. Input to the Squid is effected via a flux transformer. The magnetometer is a vector instrument with high sensitivity (10^{-6} nT $\cdot$ $HZ^{-1/2}$) and broadband (dc $-$ 10^7Hz) characteristics. The DC Squid, which has similar characteristics, is slightly more sensitive.

Geophysical applications (Clarke, 1983) include MT sounding, geodynamic measurement (volcanomagnetic, seismo-magnetic, and tecto-magnetic), palaeomagnetic measurement, aeromagnetic gradiometric measurements, etc. that exploit its high sensitivity. The logistical problems involved in the supply of liquid helium have been cited as its principal disadvantage. Reported Squid use in such unlikely environments as the sea bottom and in boreholes suggest that modern cryogenic technology has gone some way towards solving this problem.

Classes of Measurement. Table 1 summarizes the principle characteristics and use of the various instruments. There has been a considerable rationalization of sensor use during the past decade. Alkali vapour magnetometers, which appeared to hold great promise in the 1960s, are now rarely used for surface measurements. Torsion variometers and induction coils have been given a new lease of life by the application of negative feedback. Fluxgate magnetometers and torsion photoelectric variometers are steadily replacing the classical magnetographs at observatories where the fluxgate theodolite/proton magnetometer combination is increasingly being used for absolute measurement. The high sensitivity of the Squid has been exploited in a number of geophysical applications. At the present time its high cost and the logistical problems involved in its operation seem likely to mitigate against its widespread use.

Conclusion

The main thrust of geomagnetic measurement is three pronged: directed towards an improved understanding of the main field, magnetospheric/ionospheric coupling, and electromagnetic induction in the crust and oceans. It seems unlikely that there will be any dramatic shift in emphasis during the next decade. The pattern of data collection will inevitably change because of the success of MAGSAT, and the current technological advances in the measurement-related fields of data logging, processing, and communication. Main field and crustal measurements will almost certainly be made by a regular

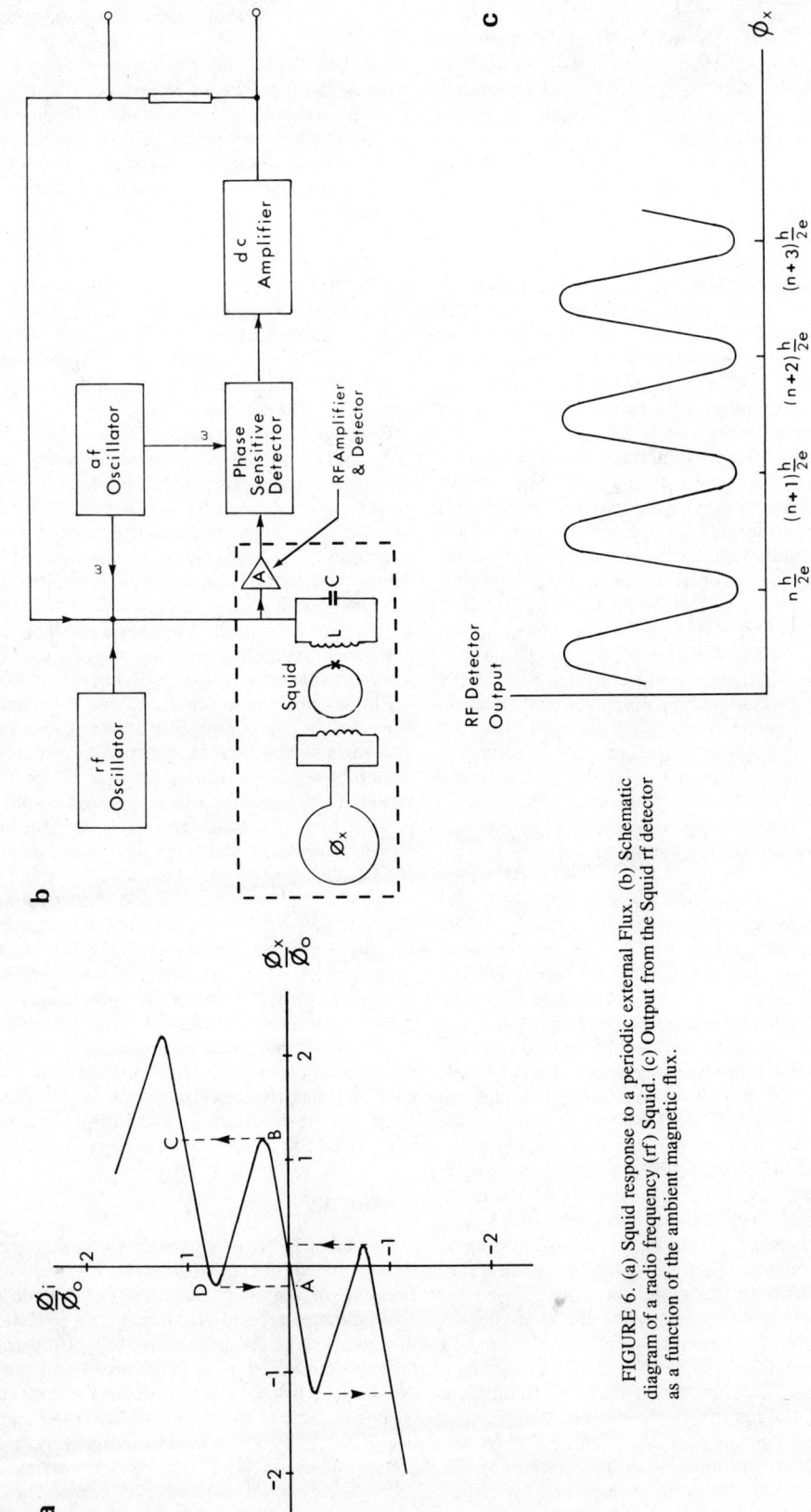

FIGURE 6. (a) Squid response to a periodic external Flux. (b) Schematic diagram of a radio frequency (rf) Squid. (c) Output from the Squid rf detector as a function of the ambient magnetic flux.

TABLE 1. Use and Characteristics of Geomagnetic Instruments

Class	Instrument	Mode	Area of Use[a]	Sensitivity	Output
Torsion	Torsion variometer	vector	1,10	<0.5 nT	photographic
	Gough-Reitzel	vector	6,7	<5 nT	35mm film
	Photo-electric	vector	1,6,7,8	$<10^{-2}$ nT	voltage
Induction	Earth inductor	vector	1,10	0.1^1 min	voltage
	Induction coil	vector	6,8	$10^{-2}-10^{-4}$ nT	voltage
Saturable Core	Fluxgate	vector	1,2,3,4,5,6,7,9,10	$0.5-10^{-2}$ nT	voltage, digital
	Fluxgate theodolite	vector	1,2	0.1 min	voltage
	Thin film	vector	8,9,10	$10^{-1}-10^{-2}$ nT	voltage
Resonance	Proton precession	scalar	1,2,3,4,5	$1-10^{-1}$ nT	display, digital
	Continuous proton	scalar	4,9	$10^{-1}-10^{-2}$ nT	frequency, digital
	Alkali vapor	scalar	1,3,4,5,6,9	$10^{-2}-5.10^{-4}$ nT	frequency, digital
Cryogenic	Squid	vector	8,9,10	$10^{-5}-10^{-6}$ nT	voltage

[a] The figures 1 to 10 indicate use within the following classes of measurement: 1 = observatories, 2 = land survey, 3 = marine survey, 4 = aeromagnetic survey, 5 = satellites, 6 = magnetospheric arrays, 7 = conductivity array, 8 = magnetotelluric, 9 = geophysical, 10 = paleomagnetic.

succession of survey satellites. Some rationalization of the pattern of surface stations monitoring the field may be anticipated to support the change and also provide real time data for solar terrestial monitoring.

A. J. FORBES

References

Bloom, A. L., 1962, Principle of operation of the rubidium vapour magnetometer, *Applied Optics* **1**, 61–68.
Breiner, S., 1981, Magnetometers for geophysical applications, in H. Weinstock and W. C. Overton, Jr., eds., *Squid Applications to Geophysics*, Tulsa, Ok.: Society of Exploration Geophysics, 3–12.
Clerc, G., 1971, Contribution a l'optimisation des capteurs destines a la mesure du champ magnetique. Ph.D. dissertation, Universite Paris.
Foner, S., 1981, Review of Magnetometry, *IEEE Trans.* **Mag-17**(6), 3358–3363.
Goree, W. S., and M. Fuller, 1976, Magnetometers using RF-driven Squids and their applications in rock magnetism and palaeomagnetism, *Rev. Geophysics Space Physics* **14**, 591–608.
Gough, D. I., and J. S. Reitzel, 1969, A portable three-component magnetic variometer, *Jour. Geomagnetism and Geoelectricity* **19**(3), 203–215.
Hood, P., 1981, Aeromagnetic gradiometry—a superior geological mapping tool for mineral exploration programs, in H. Weinstock and W. C. Overton, Jr., eds., *Squid Applications to Geophysics*. Tulsa, Ok.: Society of Exploration Geophysics, 72–76.
Irons, H. R., and L. J. Schwee, 1972, Magnetic thin film magnetometers for magnetic field measurements, *IEEE Trans.* **Mag-8**, 61–65.
Laursen, V., and J. Olsen, 1971, Classical methods of geomagnetic observations, in *Encyclopedia of Physics/Geophysics*, vol. 49/3, Geophysics III. Springer Verlag, Berlin-Heidelberg-New York: Springer Verlag.
Marianiuk, J., A. Gnoinski, and A. Szymanski, 1978, Recording of geomagnetic field elements with a photo-electric magnetometer, *Pol. Acad. Sci. Inst. Geophysics Pub.* **C-5**(125), 57–75.
Parkinson, W. D., 1983, *Introduction to Geomagnetism.* Edinburgh: Scottish Academic Press.
Primdahl, F., 1979, The fluxgate magnetometer, *Jour. Physics—E: Soc. Instrum.* **12**, 241–253.
Serson, P. H., 1973, Instrumentation for induction studies on land, *Physics Earth Planetary Interiors* **7**, 313–332.
Stuart, W. F., 1972, Earth's field magnetometry, *Rep. Prog. Physics* **35**(8), 803–881.
Trigg, D. H., 1970, A portable D and I magnetometer, *Dept. Energy Mines and Resources, Earth Phys. Branch, Rept. 70-3*, Canada.

Cross-references: *Aeromagnetic Surveying; Geomagnetic Field: Elements; Geomagnetic Field, Main: Theory; Geomagnetic Field Analysis; Geomagnetic Measurement Techniques and Surveys; Geomagnetism: Historical Introduction; Satellite Magnetic Measurements.*

GEOMAGNETIC FIELD: WESTWARD DRIFT

The westerly drift of the Earth's magnetic field was first described by Edmund Halley in 1692. Halley noted that many of the changes in the field that had been observed in the previous century, such as the shift in the position of no variation from Cape d'Agulhas to the meridian of St. Helena (about 23° in 90 years), could be accounted for by a slow westward rotation of the internal part of the Earth. Halley's suggestion of a westward drift has since proved to be remarkably foresighted as magnetic field features have continued to move westward during the succeeding three centuries. Furthermore, as a result of theoretical calculations, the source of the field is now widely attributed to dynamo action in the Earth's fluid core with part of the core bodily moving westward carrying features of the field along

TABLE 1. Estimated Westward Drift of the Earth's Magnetic Field

Drift Velocity (degree/yr)	Field Element	Approximate Epoch	Source
0.25	Point of zero declination	1600–1690	Halley (1692)
~0.1	Region of maximum declination	1550–1750	van Bemmelen (1899)
0.18	Nondipole field	1907–1945	Bullard et al. (1950)
0.32	East component of secular variation	1907–1945	Bullard et al. (1950)
0.11	Spherical harmonic components $m = 1, n = 1$	1829–1885	Carlheim-Gyllenskold (1896)
0.26	$m = 1, n = 2$	1829–1885	Carlheim-Gyllenskold (1896)
0.30	Eccentric dipole	1830–1950	Vestine (1952)
−0.02	Canadian vertical field		Whitham (1958)
0.13	Spherical harmonic components up to $n = 6$ at core/mantle boundary	1965	Richmond (1969)
0.18	Six null flux curve normals on core/mantle boundary	1965	Booker (1969)
0.08	Eight touch points on core/mantle boundary	1945–1985	Author's unpublished calculations

with it as part of the dynamo process. Alternatively the westward drift may be associated with hydromagnetic wave propagation inside the core, again as part of the dynamo process.

The movements of a great variety of magnetic features, both at the Earth's surface (Skiles, 1970) and at the core's surface, have been documented (Table 1). The predominant direction of motion of nearly all of these features during the last 400 years has been westward. Typical rates of movement have been a few tenths of a degree per year. Several of the magnetic features have also changed latitude while a few have moved eastward. Most of the features studied have lifetimes of a few hundreds of years at most. This time span is significantly less than the time needed to drift once around the world. So field changes are caused by the growth, evolution, and decay of magnetic features, along with an overall westward drift in their positions, coupled with a decrease in intensity of the dipole moment of the Earth's field.

One approach to analyzing magnetic data from different localities around the world is to draw charts of the magnetic components and to plot the longitudinal movement of various features through time. The earliest estimates of westward drift (Table 1) were made in this way. Figure 1 illustrates this approach for the zero declination (D = 0) line. We can see in Fig. 1 how lines of D = 0 tend to drift westwards and also how they tend to form and to die out. For example, the zero declination line over South America in 1550 drifts rapidly westwards across the Caribbean through the seventeenth and eighteenth centuries, but moves more steadily across the South Pacific to die away around 1750 at A. A new area of D = 0 arises over Persia around 1800. This region of D = 0 expands in the following years to produce a westward drift across the Near East, but northward drift across the Caspian Sea and no drift over the Himalayas at B. Over Western Europe a D = 0 line is seen to have drifted westward before disappearing at C around 1700. Eastward drift is observed at D over Canada, slowing to no drift at the present day. The South Atlantic, area E, has experienced steady westward drift for 400 years, continuing the changes first noted by Halley. The overall impression of westward drift gained from charts is of turbulent motion similar to that of the atmosphere as seen in meteorological maps of cyclones and anticyclones. The typical length scale of the magnetic variations of several thousand kilometers is frequently taken to indicate that the westward drift is caused by the external layers of the core rotating more slowly than the inner core.

A second approach to estimating drift rates is to match the secular change with the longitudinal field gradient. Whitham (1958) has used this approach and noted an area of slow eastward drift in Canada (see Table 1).

An alternative, very popular, mathematical method of analyzing magnetic data is to represent the geomagnetic field through a spherical harmonic expansion. An advantage of this approach is that global drift rates can be calculated directly from the spherical harmonic coefficients and their changes with time (Table 1). It is found both from these

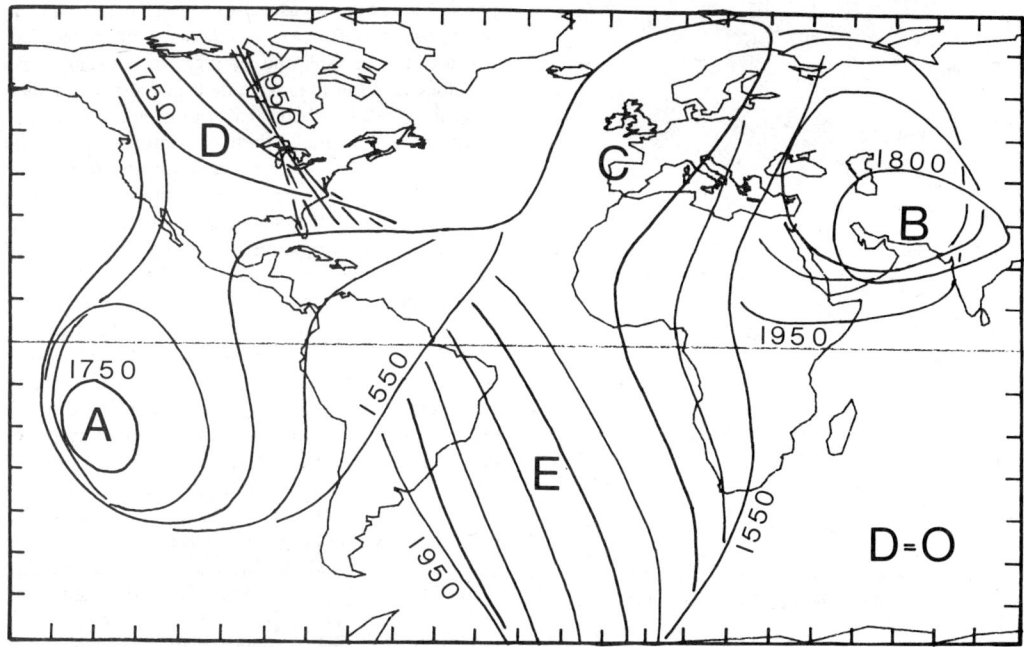

FIGURE 1. Locations of the zero declination line at 50-year intervals between A.D. 1550 and A.D. 1950 for selected regions to illustrate steady westward drift (E), eastward and no drift (D), the appearance of new magnetic features (B) and their disappearance (A and C).

global drift rate estimates and from analyses of particular magnetic features that the drift rate has varied with time.

Vestine (1952) suggested that temporal changes in the westward drift rate were associated with fluctuations in the length of the day through a mechanism involving transfer of angular momentum between the core and mantle. Subsequent studies have been less successful in correlating drift rate changes with the irregular fluctuations of the Earth's rotation. A particularly pronounced change in secular variation occurred in 1969 when the second time derivatives of the north, east and vertical magnetic field components suddenly changed all over the world. These sudden changes are interpreted as originating within the Earth rather than being associated with external field variations, which are generally of lower amplitude. Westward drift rate, which had been decreasing during the 1960s, increased again at the time of the 1969 impulse. A clear link between the 1969 magnetic change and the Earth's rotational fluctuations is eagerly being sought.

During the last few decades the establishment of permanent magnetic observatories at some 200 locations worldwide, with continuous monitoring of all three components of the field, has provided enough information to allow calculations of the field within the Earth's mantle through downward continuation of the magnetic potential. Much interest presently lies in estimating the field at the core/mantle boundary and in determining core motions from the field variations. Figures 2 and 3 illustrate the effect of downward continuation on the position of the magnetic equator ($I = 0$). The magnetic equator does not follow a line of latitude but oscillates about the geographic equator partly on account of the tilt of the geomagnetic dipole axis away from the Earth's spin axis. We see in Fig. 2 how in 1950 the magnetic equator ran through the south of Brazil and then across the Atlantic to the Sahara. We can also see in Fig. 2 how the magnetic equator has been drifting westward at some 0.4° per year. In particular the position at which the magnetic equator inter-

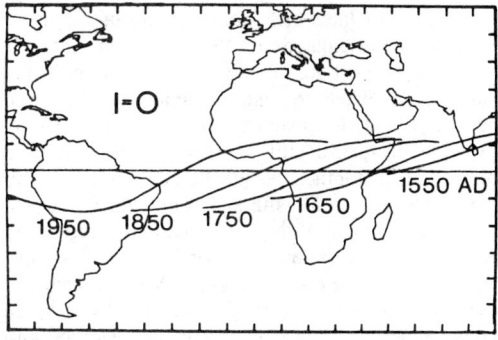

FIGURE 2. Locations of the magnetic equator at 100-year intervals between A.D. 1550 and A.D. 1950.

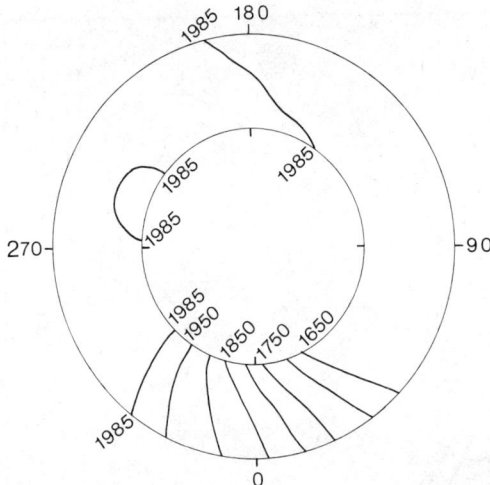

FIGURE 3. Equatorial cross section of the Earth showing the position of the magnetic equator and its variation with depth in the mantle for A.D. 1985 and at 50-year intervals for earlier epochs beneath Africa and the equatorial Atlantic.

sected the geographic equator has moved westward from the Seychelles in 1550 to cross the Greenwich meridian in 1870 and on to its present position in the western Atlantic.

Figure 3 gives another view of these changes. It shows a cross section through the equator of the Earth. The position of the magnetic equator in this cross section is marked for 1985 along with its position, beneath the Atlantic, at earlier epochs. In Fig. 3 we can see the same general overall movement of the magnetic equator at the core-mantle boundary as at the Earth's surface, but we can also see a number of differences in detail. For example, the magnetic equator crossed the Greenwich meridian around 1700 at the core's surface rather than in 1870, as at the Earth's surface. Also we can see how the drift rate of the magnetic equator was slower at the core's surface.

Table 1 tabulates drift rates of various magnetic features as calculated at the core's surface. Of particular interest are the movement of the null-flux curves (lines where the radial magnetic field component vanishes). The magnetic equator of Fig. 3 is an example of such a null-flux line. If the core is a perfect conductor then the null-flux curves move with the conducting fluid, although only the horizontal motion orthogonal to the curves can be determined. Hide (1986) has pointed to the topological significance of touch points where the horizontal magnetic field is tangential to null-flux curves. These touch points appear to have moved only very slowly with time (Table 1).

Palaeomagnetic and archaeomagnetic studies hold out the potential for extending the geomagnetic record through measurements of the remanent magnetization of rocks, sediments, and archaeological artifacts. Lavas and pottery, for example, can "memorize" the magnetic field at the time of their cooling. If undisturbed lava or brick samples can be collected and dated, then a pattern of ancient geomagnetic changes can be built up from their stable remanence directions and intensities. Dating the age of a remanent magnetization, however, presents many difficulties especially in the case of the detrital remanence of sediments. Tracking the movement of magnetic features across the Earth's surface through comparison of the palaeomagnetic records from one locality to another has proved to be extremely unreliable so far. A more promising approach to investigating ancient drift through natural remanence studies involves analyzing direction changes at individual localities where ordered palaeomagnetic or archaeomagnetic data are available. Westward drift tends to produce clockwise magnetic field looping (sensu Bauer, 1896) while eastward drift leads to an anticlockwise motion (Runcorn, 1959). Archaeomagnetic, sediment, and lava flow records all point to anticlockwise field looping and hence eastward drift at earlier times (Thompson, 1982). The most reliable of these records are probably those of the thermoremanences of archaeological materials, especially those from Britain, France, and the Ukraine, all of which point to eastward drift around 1000 years ago.

R. THOMPSON

References

Bauer, L. A., 1896, On the secular variation of a free magnetic needle, *Phys. Rev.* **3**, 34–48.

Bullard, E. C., C. Freedman, H. Gellmann, and J. Nixon, 1950, The westward drift of the Earth's magnetic field, *Royal Soc. London Philos. Trans.* **A243**, 67–92.

Booker, J. R., 1969, Geomagnetic data and core motions, *Royal Soc. (London) Proc.* **A309**, 27–40.

Carlheim-Gyllenskold, V., 1896, Sur la forme analytique de l'attraction magnétique de la terre, *Astron Iakt. Stockholm* **5**, 1–36.

Halley, E., 1692, On the cause of the Change in the Variation of the Magnetic Needle; with an Hypothesis of the Structure of the Internal Parts of the Earth, *Royal Soc. London Philos. Trans.* **17**, 470–478.

Hide, R., 1986, The Earth's differential rotation, *Royal Astron. Soc. Quart. Jour.* **27**, 3–20.

Richmond, A. D., 1969, Relation of the westward drift of the geomagnetic field to the rotation of the earth's core, *Jour. Geophys. Research* **74**, 3013–3018.

Runcorn, S. K., 1959, On the theory of the geomagnetic secular variation, *Ann. Geophys.* **15**, 87–92.

Skiles, D. D., 1970, A method of inferring the direction of drift of the geomagnetic field from palaeomagnetic data, *Jour. Geomagnetism and Geoelectricity* **22**, 441–461.

Thompson, R., 1982, A comparison of geomagnetic secular variation as recorded by historical archaeomagnetic and palaeomagnetic measurements, *Royal Soc. London Philos. Trans.* **A306**, 103–112.

Van Bemmelen, W., 1899, Die Abweichung der Magnetnadel, Beobachtungen, säcular-variation Wert- und Isogonensystemebis zur Mitte des XVIII Jahrhunderts, *Batavia R. Magn. Met. Obs.*, Suppl. to vol. 21 of observations.

Vestine, E. H., 1952, On variations of the geomagnetic fluid, fluid motions, and the rate of the Earth's rotation, *Natl. Acad. Sci. (USA) Proc.* **38**, 1030–1038.

Whitham, 1958, The relationships between the secular change and the non-dipole fields, *Canadian Jour. Physics* **36**, 1372–1396.

Cross-references: *Archaeomagnetism; Core-Mantle Coupling; Detrital Remanent Magnetization (DRM); Geomagnetic Field, Main: Theory; Geomagnetic Field Analysis; Geomagnetic Secular Variation: Direction and Intensity; Geomagnetic Secular Variation: Theory; Geomagnetism: Historical Introduction; Paleomagnetic Secular Variation; Thermoremanence.*

GEOMAGNETIC FIELD, MAIN: THEORY

The theory of geomagnetism goes back to Gilbert's book *De Magnete*, published in 1600, in which he proposed that the Earth itself is a magnet like a lodestone. The idea that the geomagnetic field originates from the ferromagnetism of the Earth's material prevailed until the early years of this century when it became clear that the Curie temperature of ferromagnetic materials is exceeded a few tens of kilometers inside the Earth's crust and that an unreasonably high magnetization of the upper crust would be required. Even before that time the phenomenon of secular variation was difficult to reconcile with the hypothesis of remanent magnetism.

Numerous other hypotheses about the origin of geomagnetism have been explored and disregarded (Rikitake, 1966) except for the dynamo hypothesis originally put forward in 1919 by Larmor to explain the magnetic field of sunspots. According to the dynamo hypothesis, motions within the outer liquid-iron core of the Earth can amplify a magnetic field from arbitrarily small amplitudes and sustain it at the observed finite strength against ohmic dissipation. The feasibility of this mechanism in a simply connected region with nearly uniform electrical conductivity seemed to be doubtful, especially after Cowling was able to demonstrate in 1934 that axisymmetric fields cannot be generated by the dynamo mechanism. Since the geomagnetic field is axisymmetric in first approximation, Cowling's theorem led to a search for alternative mechanisms. It was not until 1958 that Backus and Herzenberg demonstrated independently that dynamos are indeed possible in spheres of uniform conductivity. Since that time dynamo theory has developed rapidly, and it has become generally accepted that a geodynamo operates in the liquid outer core of the Earth. The main remaining problem is the nature of this process. No detailed theory predicting the form of the observed geomagnetic secular variation, strength of the magnetic field inside the core, and the mechanism of geomagnetic reversals is yet available. Theoretical research in this direction is progressing, however, and a brief outline of some of the main elements of this research is given here.

Kinematic Dynamo Theory

The description of the dynamo process is based on Maxwell's equations in the magnetohydrodynamic approximation, in which the displacement current is neglected, and on Ohm's law:

$$\nabla \cdot \mathbf{B} = 0, \quad \frac{\partial \mathbf{B}}{\partial t} = -\nabla \times \mathbf{E}, \quad \frac{1}{\mu} \nabla \times \mathbf{B} = \mathbf{j},$$

$$\mathbf{j} = \sigma(\mathbf{E} + \mathbf{v} \times \mathbf{B}) \quad (1)$$

where **B** is the magnetic flux density, **E** is the electric field, **j** is the current density, σ is the electrical conductivity, and **v** is the velocity vector. Ohm's law has been written in the form valid for a moving conductor. Accordingly Eqs. 1 are invariant with respect to a Galilei transformation; in other words, a system moving with constant velocity **V** with respect to the original system of reference, Eqs. 1 are valid when **E** and **B** obey the relationships

$$\mathbf{B}' = \mathbf{B}, \quad \mathbf{E}' = \mathbf{E} + \mathbf{V} \times \mathbf{B}$$

where the primed quantities refer to the new system. It is convenient to eliminate **E** and **j** from Eqs. 1 and to write

$$\frac{\partial}{\partial t} \mathbf{B} + \nabla \times (\lambda \nabla \times \mathbf{B}) = \nabla \times (\mathbf{v} \times \mathbf{B}) \quad (2)$$

where $\lambda \equiv (\sigma\mu)^{-1}$ is the magnetic diffusivity. Because of the absence of ferromagnetic effects in the Earth's core, the permeability μ is usually replaced by the permeability μ_0 of a vacuum. From the form of Eq. 2 it is obvious that $\nabla \cdot \mathbf{B} = 0$ remains satisfied if this equation is satisfied by the initial conditions.

Since the Earth's outer core can be regarded as an incompressible fluid in first approximation, Eq. 2 can be simplified further by the use of equations $\nabla \cdot \mathbf{v} = 0, \nabla \cdot \mathbf{B} = 0$,

$$\left(\frac{\partial}{\partial t} + \mathbf{v} \cdot \nabla\right) \mathbf{B} - \lambda \nabla^2 \mathbf{B} = \mathbf{B} \cdot \nabla \mathbf{v} \quad (3)$$

For simplicity λ = const. has been assumed. This form of the equation of magnetic induction is revealing since it resembles the heat equation with the magnetic field stretching term $\mathbf{B} \cdot \nabla \mathbf{v}$ acting as a heat source. In order that growing solutions of Eq.

3 exist, $|\mathbf{B} \cdot \nabla \mathbf{v}|$ must exceed $|\lambda \nabla^2 \mathbf{B}|$ at least in some region of the core. The ratio of the two terms is described by the magnetic Reynolds number

$$R_m \equiv \frac{VL}{\lambda}$$

where V is a typical velocity and L is a typical length scale, which in the case of the Earth's core is usually identified with the core radius r_o. The necessary condition for a dynamo—that the Reynolds number must exceed a value of the order unity—has been derived rigorously by Backus and Childress for a fluid sphere of constant magnetic diffusivity:

$$R_m \equiv \frac{Vr_o}{\lambda} > \pi \quad \text{for dynamo action} \quad (4)$$

In Childress's formulation V refers to the maximum velocity, whereas in Backus's formulation $V(\pi r_o)^{-1}$ is a measure of the maximum rate of strain. It should be emphasized that Eq. 4 holds with respect to a rotating system since Eq. 3 is invariant not only with respect to Galilei transformations but also to transformations to rotating systems.

Cowling's theorem demonstrates that the necessary condition Eq. 4 is far from being sufficient. The theorem states that axisymmetric or two-dimensional growing solutions of Eq. 3 do not exist. The original proof by Cowling has been extended by many authors and now includes the case of Eq. 2 for a compressible fluid.

The simplest dynamo is the disk dynamo, as shown in Fig. 1. A weak initial magnetic field $\mathbf{B}_0$ intersects a metal disk rotating with angular velocity Ω, and an electromotive force (emf) is generated between the axis and rim of the disk:

$$U = \Omega \int_0^{s_0} B_z s \, ds \quad (5a)$$

When the stationary circuit with resistance R and inductivity L indicated in the figure is closed with the help of two brushes, a current I flows and obeys the differential equation

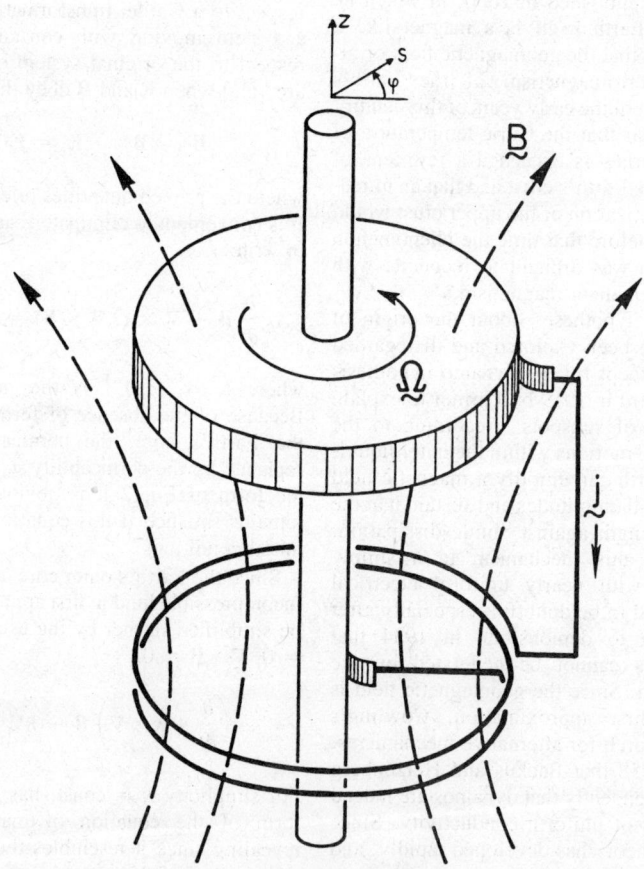

FIGURE 1. Sketch of a disk dynamo.

$$L\frac{dI}{dt} = U - RI \quad (5b)$$

The current I induces a magnetic flux F intersecting the disk, which is given by

$$F \equiv 2\pi \int_0^{s_0} B_z s\, ds = MI \quad (5c)$$

where M is the mutual inductivity between circuit and disk. The solution of Eqs. 5 given by

$$I = I_0 \exp\left\{\frac{1}{L}\left(\frac{1}{2\pi} M\Omega - R\right)t\right\}$$

indicates that the current I, and thus the magnetic field, is growing exponentially if the dynamo condition

$$\frac{M\Omega}{R} > 2\pi$$

is satisfied. The exponential growth will be modified as soon as the magnetic torque has grown sufficiently to affect the angular velocity Ω.

The disk dynamo is a better example for technical dynamos than it is for the dynamo operating in the Earth's core, where a multiple connected distribution of conductivity cannot be expected. It can be proven that a simple form of motion, such as that of the disk dynamo, cannot drive a dynamo in a sphere if λ is a function of the radial coordinate only. More generally, the toroidal theorem states that growing solutions of Eq. 2 do not exist if $\mathbf{v}$ and λ can be written in the form

$$\mathbf{v} = \nabla \times \mathbf{r}\psi, \quad \lambda = \lambda(|\mathbf{r}|)$$

An analogous theorem holds for the planar case, $\mathbf{v} = \nabla \times \mathbf{k}\psi$, $\lambda = \lambda(\mathbf{k} \cdot \mathbf{r})$, where $\mathbf{v}$ is confined to planes with normal unit vector $\mathbf{k}$.

It is evident from the foregoing discussion that growing solutions of Eq. 3 require a complex spatial structure. For the sphere this structure can be elucidated by separation of the magnetic field into its axisymmetric and nonaxisymmetric components

$$\mathbf{B} = \nabla \times A\hat{\boldsymbol{\varphi}} + B_\phi \hat{\boldsymbol{\varphi}} + \mathbf{b} \quad \text{with } \overline{\mathbf{b}} = 0$$

where A describes the meridional component of $\mathbf{B}$, $\hat{\boldsymbol{\varphi}}$ is the unit vector in the azimuthal direction, and the azimuthal average is indicated by a bar. A similar decomposition can be made for the velocity field

$$\mathbf{v} = \nabla \times \psi\hat{\boldsymbol{\varphi}} + U\hat{\boldsymbol{\varphi}} + \mathbf{u} \quad \text{with } \overline{\mathbf{u}} = 0$$

The axisymmetric component of Eq. 2 can now be written in the form

$$\left[\frac{\partial}{\partial t} - \lambda\left(\nabla^2 - \frac{1}{r^2 \sin^2\theta}\right)\right]A = \overline{\mathbf{u} \times \mathbf{b}}|_\phi \quad (6a)$$

$$\left[\frac{\partial}{\partial t} - \lambda\left(\nabla^2 - \frac{1}{r^2 \sin^2\theta}\right)\right]B_\phi \quad (6b)$$
$$= \nabla \times (\overline{\mathbf{u} \times \mathbf{b}})|_\phi + \nabla$$
$$\times (\hat{\boldsymbol{\varphi}} U \times (\nabla \times \hat{\boldsymbol{\varphi}} A))|_\phi$$

where λ is assumed to be constant and θ is the colatitude. For simplicity we have also assumed $\psi \equiv 0$ because axisymmetric meridional circulations are usually very small compared with the differential rotations described by U. From Eq. 6a it can be seen that the interaction between fluctuating components of magnetic and velocity fields is necessary to generate the meridional field described by A. If the notation of mean-field magnetohydrodynamics (Krause and Rädler, 1980) is followed, this effect is also called the α-effect, since it can be shown that under certain circumstances the relationship $\overline{\mathbf{u} \times \mathbf{b}} = \alpha \overline{\mathbf{B}}$ holds.

In Eq. 6b two terms on the right side are capable of amplifying the magnetic field B_ϕ. The first term is analogous to the term on the right side of Eq. 6a and describes the α-effect. Dynamo solutions for which the second term on the right side of Eq. 6b is relatively unimportant are thus called α^2-dynamos. The stretching of poloidal field lines into the azimuthal direction becomes important in situations with strong differential rotation. In those cases the second term on the right side of Eq. 6b dominates, and the growing solutions of Eqs. 6 are called $\alpha\omega$-*dynamos*.

The mutual amplification of mean components of the magnetic field by fluctuating components, and vice versa, or of toroidal components by poloidal components, and vice versa, is characteristic for the dynamo process. It is complicated by the structure of the velocity field. Efficient dynamos are usually obtained only for finite values of the helicity h, which is defined by

$$h \equiv \overline{\mathbf{v} \cdot \nabla \times \mathbf{v}}$$

and provides a measure of the screw sense of the velocity field. For further details on the dynamo process, see Moffatt (1978), Parker (1979), and Krause and Rädler (1980).

Energetics of the Geodynamo

For a prescribed stationary velocity field, Eq. 2 admits solutions with an exponential time dependence, of which the growing ones are called *dynamos*. As the magnetic field grows, the action of the Lorentz force on the velocity field can no longer be neglected. The work done by the velocity field against the Lorentz force balances out the energy consumed by ohmic heating. Thus, the sustenance of a finite-amplitude magnetic field against ohmic dissipation requires a source of mechanical energy. A lower bound of about 10^8 W on the ohmic heating can be obtained by a calculation of the minimum dissipation corresponding to the observed geomagnetic field, while an upper bound of about 10^{13} W is given by the observed geothermal heat flux. A realistic value is likely to be within the range 10^{11} to 10^{12} W.

Numerous sources of energy for the geodynamo have been considered in the history of dynamo theory. The major ones that cannot be excluded right away are the precession of the Earth, thermal heat sources, and compositional buoyancy. The latter source, first proposed by Braginsky in 1963, seems to be the most likely possibility, since the turbulence induced by precession appears to be marginal and since the Carnot efficiency of the heat sources is relatively low (Gubbins and Masters, 1979). Compositional buoyancy is generated by the growing solid inner core. Iron and, perhaps, nickel crystallize at the inner core–outer core boundary while lighter elements stay in solution. The decrease in density caused by the increase in the concentration of light elements provides a source of potential energy that can drive convection in the outer core. Since the outer core contains about 10% light elements by weight, most likely sulfur and oxygen, the buoyancy provided by this process appears to be more than adequate to drive the geodynamo.

The compositional buoyancy mechanism has the advantage that it could explain the apparent absence of a dynamo in the case of Venus (Stevenson et al., 1983). Since Venus resembles the Earth in most respects, the absence of a measurable magnetic field (see *Planetary Magnetic Fields*) has posed a puzzle. The somewhat higher pressure at the Earth's center seems to have allowed the inner core to form, whereas Venus will reach this stage of its thermal history much later.

Estimates for the age of formation of the inner core of the Earth range from 1 to 2 b.y. Since evidence for geomagnetism exists for earlier times in the paleomagnetic record, thermal buoyancy is likely to have played a role in driving the geodynamo in the Earth's early history. In this respect the Earth resembles the major planets for which thermal buoyancy appears to be the only viable source of energy to drive the planetary dynamo.

The way in which compositional buoyancy drives convective motions in the outer core is not well understood. It is even possible that the outer part of the outer core is stably stratified, although the strong stratification originally proposed by Higgins and Kennedy has long been refuted. Another peculiar property of the solidification process at the inner core is the so-called mushy zone in which liquid and solid phases coexist (see *Earth's Core*). There even seems to be evidence that the inner core as a whole is not entirely solid and that its high anelasticity can be attributed to a partially molten state. But these details have little influence on the theory of the geodynamo in its current state. Since the form of motions in the outer core is primarily determined by Coriolis and Lorentz forces, the theory of geomagnetism has proceeded without distinguishing between thermal and compositional buoyancy.

Some typical values of parameters used in the theory of the geodynamo are given in Table 1.

Magnetohydrodynamic Theory of the Geodynamo

The theory of the geodynamo requires the inclusion of the equations of motion for the rotating liquid core for two reasons. First of all, constraints on the velocity field used in the dynamo equation are necessary, since rather arbitrary velocity fields are capable of generating fields that resemble the observed geomagnetic field. The velocity fields driven by geophysically realistic forces share special properties, such as the approximate alignment of the vorticity with the axis of rotation, that makes them useful in the interpretation of certain features of the observed magnetic field. The second reason for including the equations of motion stems from the necessity of taking into account the Lorentz force, without which an equilibrium amplitude of the magnetic field cannot be determined.

The basic equations to be solved are usually written in the following forms:

$$\left(\frac{\partial}{\partial t} + \mathbf{v} \cdot \nabla\right)\mathbf{v} + 2\mathbf{\Omega} \times \mathbf{v}$$
$$= -\nabla \pi - \mathbf{g}\alpha\theta + \nu\nabla^2\mathbf{v} + \frac{1}{\rho\mu}(\nabla \times \mathbf{B}) \times \mathbf{B}$$

(7a)

$$\nabla \cdot \mathbf{u} = 0 \quad (7b)$$

$$\left(\frac{\partial}{\partial t} + \mathbf{v} \cdot \nabla\right)\theta + \mathbf{v} \cdot \nabla T_s = \nabla^2\theta \quad (7c)$$

TABLE 1 Properties of the Earth's Core Relevant to Dynamo Theory

Angular velocity	$\Omega = 72.5 \times 10^{-6}$ s^{-1}
Inner and outer radius of the liquid core	$r_i = 1217$ km, $r_O = 3486$ km
Mean density and coefficient of thermal expansion	$\rho = 10^4$ kg $\cdot$ m^{-3}, $\alpha = 10^{-5}$ K^{-1}
Thermal and magnetic diffusivities	$\kappa = 7 \times 10^{-6}$ m$^2 \cdot$ s^{-1}, $\lambda = 1.6$ m$^2 \cdot$ s^{-1}
Rough estimate for the kinematic viscosity	$\nu = 10^{-6}$ m$^2 \cdot$ s^{-1}

$$\left(\frac{\partial}{\partial t} + \mathbf{v} \cdot \nabla\right)\mathbf{B} - \mathbf{B} \cdot \nabla\mathbf{v} = \lambda\nabla^2\mathbf{B} \quad (7d)$$

$$\nabla \cdot \mathbf{B} = 0 \quad (7e)$$

where $\mathbf{g}$ is the force of gravity, ν is the kinematic viscosity, Ω is the angular velocity of rotation, π is the dynamic pressure divided by the density, and θ is the deviation from the potential temperature distribution T_s. The latter describes the difference between the temperature distribution for the static solution of the problem and the adiabatic temperature distribution with the same temperature at the inner core boundary. As mentioned before, Eqs. 7 remain virtually unchanged if T_s and θ are interpreted as concentrations of light elements. In that case the coefficient of thermal expansion α and the thermal diffusivity κ need to be reinterpreted as well. In writing Eqs. 7, we have assumed constant material properties and used other simplifications to simplify the problem without losing physically important features.

Since λ/ν is roughly 10^6 in the Earth's core, the main sink of mechanical energy is ohmic dissipation, and it is generally believed that the term $\nu\nabla^2\mathbf{u}$ is unimportant except close to the rigid boundaries of the liquid core. Without this term the important J. B. Taylor constraint can be derived: After integrating the azimuthal component of Eq. 7a over a coaxial cylindrical surface inside the liquid core, one must conclude that an accelerating differential rotation occurs unless the condition

$$\iint [(\nabla \times \mathbf{B}) \times \mathbf{B}]_\phi s \, d\phi \, dz = 0 \quad (8)$$

is satisfied for every distance s from the axis. In deriving Eq. 8, we used the fact that the inflow across the cylindrical surface is balanced by the outflow, with the result that the contribution from the Coriolis force term vanishes. Different strategies have been designed to accommodate the Taylor constraint, Eq. 8. In Braginsky's (1978) model Z the friction of the viscous Ekman layer at the solid surface of the sphere provides a balance on the right side of Eq. 8. But this effect is small such that the constraint of Eq. 8 becomes nearly satisfied by a poloidal field that is extended parallel to the axis of rotation except for the region close to the boundary. Some authors have considered the kinematic dynamo problem and have included condition 8 as the only constraint imposed by the equations of motion. It is generally found that Eq. 8 significantly increases the magnetic Reynolds number for dynamo action.

It is instructive to consider Eqs. 7 in the limit of a vanishing magnetic field. In that case the dominant terms are the Coriolis force and the pressure gradient, the balance of which,

$$2\Omega \times \mathbf{v} = -\nabla\pi \quad (9)$$

is called the *geostrophic balance*. The curl of Eq. 9 together with Eq. 7b gives rise to the Proudman-Taylor theorem

$$\Omega \cdot \nabla\mathbf{v} = 0 \quad (10)$$

which states that steady motions of a rotating inviscid fluid must not depend on the coordinate in the direction of the axis of rotation. Convection in a sphere requires, however, a finite radial velocity component and therefore cannot satisfy condition 10. As a result, nonmagnetic solutions of Eqs. 7 exhibit a time dependence in the form of a drift in the prograde azimuthal direction. In addition, they are characterized by small length scales in the directions perpendicular to the axis of rotation, whereby viscous friction is increased. A sketch of the motion at the onset of convection in a homogeneously heated sphere is shown in Fig. 2.

Because of the small viscosity in the Earth's core, the Lorentz force offers a more effective release of the rotational constraint expressed by condition 10 than does viscous friction. The analysis of the onset of convection in the presence of an azimuthal magnetic field shows that the scale of the preferred mode of convection increases dramatically as the magnetic field strength is increased and the drift rate decreases. The dimensionless parameter governing this effect is the Elsasser number

$$\Lambda \equiv \frac{B_0^2}{2\Omega\mu_0\rho\lambda}$$

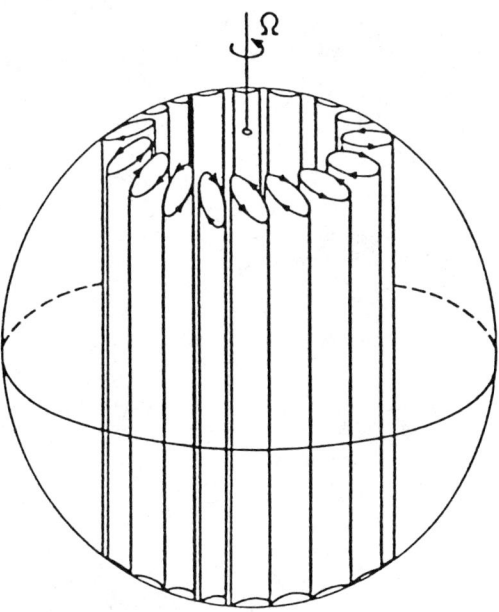

FIGURE 2. Sketch of convection columns in a rotating, internally heated sphere for Rayleigh numbers close to the critical value.

which describes the ratio between Lorentz and Coriolis forces. When the magnetic field becomes very strong, it exerts an inhibiting influence on convection. It turns out that the minimum Rayleigh number for the onset of convection is achieved for $\Lambda \simeq 1$. Eltayeb and Roberts suggested in 1970 that this condition is likely to determine the equilibrium strength of the magnetic field inside the Earth's core. Based on the values of Table 1, this argument gives a not unreasonable field strength of 1.5×10^{-3} T. For major planets, however, this argument could not be applied, since the poloidal field alone would exceed this equilibrium value. Numerous other ideas for the balance governing the field strength in various planets have been studied. Here we refer only to the theoretical approaches of Busse (1976) and Stevenson (1983).

For the solution of Eqs. 7 it is convenient to introduce the following general representation of the solenoidal vector fields **v** and **B**:

$$\mathbf{v} = \nabla \times (\nabla \times \mathbf{r}\Phi) + \nabla \times \mathbf{r}\psi,$$

$$\mathbf{B} = \nabla \times (\nabla \times \mathbf{r}h) + \nabla \times \mathbf{r}g$$

The r-components of the curl and of the curlcurl of Eq. 7a and the r-components of Eq. 7d and of its curl offer four equations for the scalar fields Φ, ψ, h, and g. These fields are uniquely determined when the condition is imposed that their average over any surface $r =$ const. vanishes. Stress-free boundary conditions are usually assumed at the inner and outer boundaries of the core. In the geophysically interesting case of high rotation rates the presence of a rigid boundary causes modifications of the solution in the thin Ekman boundary layers. Since the outside of the core is insulating, g must vanish at the boundary, and h is determined by the condition that the outside poloidal field is a potential field.

The fact that the preferred velocity field of convection is symmetric with respect to the equatorial plane as shown in Fig. 2 implies that the solutions **B** of Eq. 7a can be separated into two classes: a quadrupolar class that has the same symmetry with respect to the equatorial plane as the velocity field, and a dipolar class that exhibits the opposite symmetry:

dipolar class: h is symmetric, g is antisymmetric with respect to the equatorial plane.

quadrupolar class: h is antisymmetric, g is symmetric with respect to the equatorial plane.

In both cases the Lorentz force is compatible with the symmetry of the velocity field. The observed geomagnetic field is primarily a dipolar field and so are most of the other planetary magnetic fields, but from a basic point of view quadrupolar dynamos are not less likely. The more complex structure of their poloidal field is balanced by the simpler form of the zonal component. The latter is symmetric with respect to the equatorial plane, whereas it is antisymmetric in the dipolar case. In the work of Cuong and Busse (1981) only the differential rotation induced by convection causes a preference of dipolar modes.

The observed geomagnetic field includes a substantial contribution from quadrupolar terms. This property is hardly surprising, since the motions inside the core are turbulent and since the symmetry property exhibited in Fig. 2 does not hold inside the cylindrical surface touching the inner core at the equator. Moreover, the thermal conditions at the core surface may not be spherically homogeneous because of the presence of convection flow in the Earth's mantle. In the present state of the geodynamo theory, it is inappropriate to take these secondary effects into account, and research has thus been focused on the simplest solutions of Eqs. 7 in spherical shells. In a series of papers Braginsky (1978) has developed an analytic approach based on the assumption of a large axisymmetric toroidal field, which has allowed him to use the inverse of the magnetic Reynolds number (R_m^{-1}) as the parameter for a perturbation expansion. The solution of Eqs. 7 becomes much simpler if the geometry of the spherical shell is replaced by a cylindrical annulus. Numerical computations for a spherical shell basically yield results similar to the annulus results (Cuong and Busse, 1981). Numerical models for turbulent dynamos in spherical shells have also been developed for applications to the solar convection zone (see, e.g., Glatzmaier, 1984).

Time Dependence of the Geodynamo

From observations of the secular variation and from the paleomagnetic evidence for reversals of the direction of the geomagnetic field, it is clear that the mechanism of the geodynamo is not steady. Since there is no reasonably complete model of the geodynamo, the various features that may lead to the observed time dependence cannot be derived from a basic model. Thus, the theories of secular variation and of reversals rely on simplifying assumptions (see *Geomagnetic Secular Variation: Theory; Geomagnetic Polarity Reversals: Theory and Models*). Eventually, however, the observed secular variation and the increasing body of data on the reversal process are likely to be important for discriminating between different models of the geodynamo.

The time dependence of the nonaxisymmetric part of the magnetic field is a result of the basic theory discussed in the preceding section. Even without a magnetic field, the convection motion exhibits a time dependence in the form of a prograde or eastward drift. In the presence of a magnetic field, this drift is decreased substantially because the Lorentz force balances the ageostrophic part of the Coriolis force. The order of magnitude of the drift corresponds roughly to the observed westward drift of the nondipole part of the geomagnetic field. But since there is likely to be a zonal flow of the liquid core relative to the mantle of a similar order of magnitude, no direct comparison between theoretical and observed drift rates can be made without further knowledge about the zonal flow in the core.

In contrast to the solar magnetic field, the geomagnetic field does not reverse in periodic intervals. Since the time scale of reversals is at least an order of magnitude larger than the time scale of about 2×10^4 yr for the diffusive decay of Earth's magnetic field, the geodynamo is basically a steady dynamo. But like many nonlinear dynamic systems, it exhibits an irregular time dependence that leads to reversals in random intervals. A simple example for such behavior is the coupled disk dynamo (Rikitake, 1966). Two dynamos of the form shown in Fig. 1 are coupled in that each circuit is attached to the other disk. After the torque of the Lorentz force has been taken into account, a periodic time dependence, including reversals, can be observed. The basic reason for the possibility of reversals is the property that Eqs. 5 or 7 are invariant with respect to the sign of the magnetic field. For any solution of Eqs. 7 there is thus another solution with the opposite sign of **B**. The mechanism that causes a reversal, however, is not yet clearly identified.

F. H. BUSSE

References

Braginsky, S. I., 1978, Nearly axially symmetric model of the hydromagnetic dynamo of the earth, *Geomagnetism and Aeronomy* **18**, 225-231.
Busse, F. H., 1976, Generation of planetary magnetism by convection, *Physics Earth and Planetary Interiors* **12**, 350-358.
Cuong, P. G., and F. H. Busse, 1981, Generation of magnetic fields by convection in a rotating sphere I, *Physics Earth and Planetary Interiors* **24**, 272-283.
Glatzmaier, G. A., 1984, Numerical Simulations of Stellar Convective Dynamos. I. The Model and Method, *Jour. Comp. Physics* **55**, 461-484.
Gubbins, D., and T. G. Masters, 1979, Driving Mechanisms for the Earth's Dynamo, *Advances in Geophysics* **21**, 1-50.
Krause, F., and F.-H. Rädler, 1980, *Mean-field Magnetohydrodynamics and Dynamo Theory*. Oxford: Pergamon.
Moffatt, H. K., 1978, *Magnetic Field Generation in Electrically Conducting Fluids*. Cambridge: Cambridge University Press.
Parker, E. N., 1979, *Cosmical Magnetic Fields*. Oxford: Clarendon Press.
Rikitake, T., 1966, *Electromagnetism and the Earth's Interior*. Amsterdam: Elsevier.
Stevenson, D. J., 1984, The energy flux number and three types of planetary dynamo, *Astron. Nachr.* **305**, 257-264.
Stevenson, D. J., T. Spohn, and G. Schubert, 1983, Magnetism and thermal evolution of the terrestrial planets, *Icarus* **54**, 466-489.

Cross-references: *Core-Mantle Coupling; Earth's Core; Geomagnetic Polarity Reversals: Theory and Models; Geomagnetic Secular Variation: Theory; Geomagnetism: Historical Introduction; Magnetohydrodynamic Waves within the Earth.*

GEOMAGNETIC FIELD ANALYSIS

The geomagnetic field observed about the Earth is the vector sum of several different components:

1. The *main* field, which is generated by electrical currents in the liquid core of the Earth and possibly by leakage currents into the deep mantle
2. The *crustal* or *anomaly* field, which results from direct magnetization of permeable material in the outer layers from the present main field, as well as remanent magnetization from ancient fields
3. The *external* field, produced by electrical currents in the ionosphere and magnetosphere, and modulated by the solar wind and the interplanetary field
4. The *induced* field, from electrical currents in the oceans and the conducting layers of the crust and mantle produced by flux changes from the external field, and possibly by changes in the core field.

The analysis of data must include consideration of all of these fields. Prior to satellite surveys, there was generally no distinction made between such sources. The word "geomagnetic" then meant the "regional" or main field as opposed to the anomaly field, and the induced and external current contributions were assumed to be negligible. This writer adopts the view that "geomagnetic" includes all of the above. The purpose of the analysis is to identify and model all sources while retaining the historical emphasis on the main field which, near the Earth's surface, averages over 95% of the total signal.

The method to be used is an iterative adjustment to coefficients describing these sources so as to minimize the weighted sum squares of the difference between the observations of the field and those values computed from such coefficients.

Computing the Static Field

Because the conductivity of the atmosphere is small below about 90 km altitude, the vector magnetic field **B** from the surface to this level is considered to be curl free ($\nabla \times \mathbf{B} = 0$). Thus it may be uniquely derived from a scalar potential function $B = -\nabla V$ satisfying Laplace's equation ($\nabla^2 V = 0$). One solution of this equation for **V** is in terms of spherical harmonic functions (see *Spherical Harmonic Analysis*). In theory **V** is the sum of an internal portion $\mathbf{V}_i$ and external portion $\mathbf{V}_e$. However, in practice only the first harmonic of V_e is determined.

Representation of the field using other functions such as multipoles, dipoles, and current loops is possible, but has no special advantage over spherical harmonics and will not be developed herein (see Langel, 1987).

The three field components B_θ, B_ϕ, and B_r in spherical coordinates are computed by taking the gradient of the potential function. For example, the value of B_r is computed from the internal potential by the expression

$$B_r = \sum_{n=1}^{n^*} (n+1) \left(\frac{a}{r}\right)^{n+2} \sum_{m=0}^{n} (g_n^m \cos m\phi + h_n^m \sin m\phi) P_n^m(\theta)$$

The g_n^m and h_n^m are called Gauss coefficients and are usually expressed in nanoteslas (nT), which are equivalent to the older unit gamma (γ). The range of n omits the monopole term g_0^0 and approximates the infinite series by truncation at n^*.

Because the magnetic elements (e.g., X, Y, Z) are measured on the surface of the oblate Earth in local topocentric coordinates, they do not correspond exactly to the spherical components just described. The directions of X and Z differ in angle from their geocentric counterparts by up to 0.2^0. To ignore this difference produces errors in field models of about 100 nT. Computations also should take into account the changes in Earth radius with latitude (e.g., 6378 km at the equator to 6357 km at the poles) and the fact that altitude is usually measured in the vertical, not in the direction toward the Earth's center.

For comparison with observed data the X and Z components are thus computed by a rotation from geocentric to topocentric as

$$X = -B_\theta \cos \alpha - B_r \sin \alpha$$

$$Z = -B_r \cos \alpha + B_\theta \sin \alpha$$

The Y component is equal to B_ϕ. The angle α is the small difference in direction between a line to the center of the Earth and one perpendicular to the spheroid.

The spatial structure of this field is an important consideration in the analysis. As can be seen in Fig. 1, the dipole or $n = 1$ term dominates the spectrum. Plotted is the spatial spectrum of the geomagnetic field for each degree of the spherical harmonic potential function in units of nT^2 at an average satellite altitude of 420 km (see *Satellite Magnetic Measurements*). This spectrum is computed from the expression

$$W_n = (n+1) \left(\frac{a}{r}\right)^{2(n+2)} \sum_{m=0}^{n} [(g_n^m)^2 + (h_n^m)^2]$$

It also represents the energy density multiplied by $2\mu_0$ for each degree (Mauersberger, 1956), where μ_0 is the permeability of free space. As can be seen

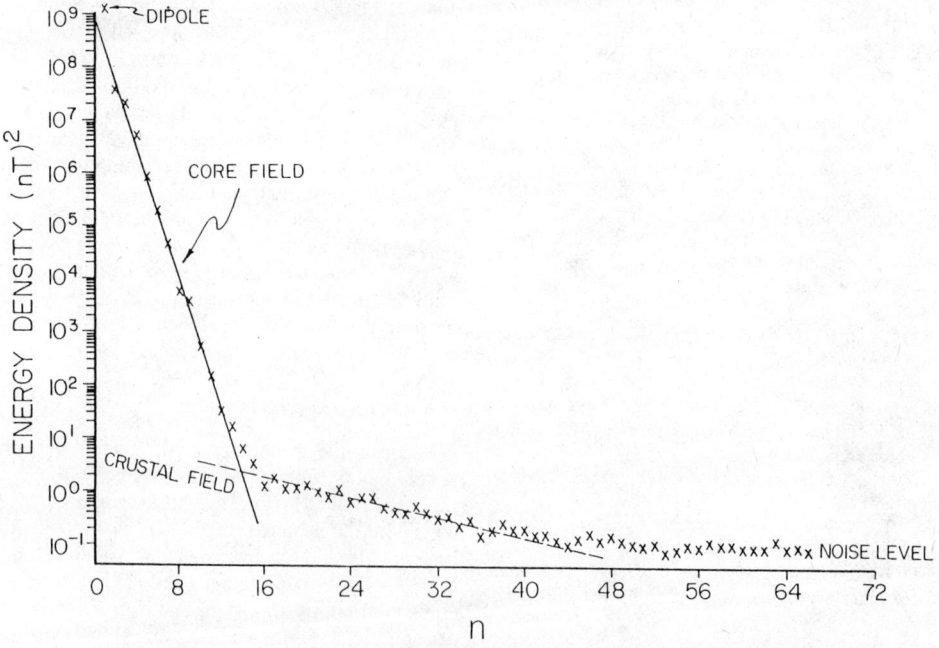

FIGURE 1. Spatial energy density spectrum of the geomagnetic field as seen by *Magsat* at an average altitude of 420 km (units are nT^2).

here, the core components have a steep spectrum and dominate the field until $n = 14$. Above this degree the crustal components predominate until the signal drops into the noise above about $n = 50$. From aeromagnetic survey data we know that anomalies exist with scale sizes that would produce points on such a plot well beyond $n = 400$ if suitable data were available.

Meyer et al. (1983) have shown that the crustal components contribute to the lower part of the total spectrum at a power comparable to that above $n = 14$. Conversely, it is likely that there are observable core components that change with time above this degree.

The international group that provides standard reference models (Peddie, 1986) has attempted to offer coefficients that only represent the main field. However, all such models adopted to date make no attempt to separate the crustal component or contributions from induced subsurface currents, and do not provide any reference to the state of the external field. The recently adopted definitive International Reference Fields are limited to $n = 10$ and thus omit some of the known core components.

Effect of External and Induced Fields

Although data may be discarded from analysis if taken during magnetically disturbed intervals, there are systematic variations that may yet be included in results if this is done. Currents continue to flow in the ionosphere and magnetosphere even during magnetically quiet intervals, and the times of low activity are very rare. At any given time there are contributions from the ionospheric and magnetospheric currents of the order of a few tens of nanoteslas at mid-latitudes, up to 200 nT near the dip equator, and up an order of magnitude larger in auroral regions.

The best way to make sure that such sources do not affect the internal coefficients is to model and remove their contribution from observed data or to include solutions for their coefficients in the same analysis as for the internal field. Indeed, if satellite data are used in the analysis, even the condition of a curl-free space is no longer true. To date, no one has adequately included such complexities in modeling, but instead the problem is minimized by selecting data with no obvious disturbance.

Present practice is to select data taken during magnetically quiet conditions using the global index of magnetic activity K_p, and to delete data taken near noon when the ionospheric currents are strongest if the data are not from the auroral ovals.

Induction from external sources is likely small if averaged over several years. Studies such as those of Yukutake and Cain (1979) for the $n = 1$ terms have shown it is related to the known changes in the external field. Figure 2 compares estimates of the year-to-year variation of the axial dipole components g_1^0, both internal and external, the ring current index Dst, magnetic activity, and sunspot number. Due to the changes seen in the external dipole term $(g_1^0)^e$, the $n = 1$ external coefficients should be included in any analysis to minimize possible errors in determining the internal dipole. However, there is no way to distinguish an internal component produced by the core from one produced by induction. Although the figure only plots the dipole, it is likely that higher-degree modes are also induced from the external field, perhaps to levels higher than that of the dipole.

Present practice is to attempt to model and remove the variations in $(g_1^0)^e$ by using Dst. This quantity is a rough index of the changes in the combined external and internal first-degree harmonics as observed near the equator, though to an arbitrary base line. After the effect is estimated and removed, the analyses may still include terms for the constant level of the external dipole. Even during quiet conditions this level can be a few tens of nanoteslas from zero.

Observational Factors

Besides these geophysical considerations, there are also instrumental and practical considerations. For example, it is very expensive to obtain vector data due to the complexity of very accurately determining the attitude or orientation of the observing platform (see *Satellite Magnetic Measurements*). To obtain a high accuracy in components requires directions accurate within a few seconds of arc. That is, 1-nT error transverse to a 57,300-nT field would give an angle of 1/57,300 rad or 0.01^0 of arc. By an effort that cost about $1 million in 1980, NASA was able to determine the attitude of the *Magsat* spacecraft instruments for the seven-month data collection interval to better than 0.4' of arc. This survey resulted in observed field components with accuracies of about 8 nT. Aircraft and shipboard surveys of the vector field have been less successful. Aircraft surveys have not been able to achieve vector accuracies better than about 150 nT.

However, instruments measuring the absolute magnitude of the vector have been developed that are inexpensive and accurate (e.g., to 1 nT). Aircraft or satellite surveys can be highly accurate in the scalar field, especially if the instruments are placed on tethers or booms away from the fields generated by the vehicle. The most accurate survey of the scalar field was made by the *POGO* spacecraft (1965–1971) with accuracies approaching 1 nT.

Data distribution is as important as data accuracy in producing good models of the geomagnetic field. From recent satellite work it is known that most of the uncertainties in the past field analyses using only surface data were due to the sparseness of data and the failure of schemes to "update" survey data to a single epoch. It is now understood that accurate

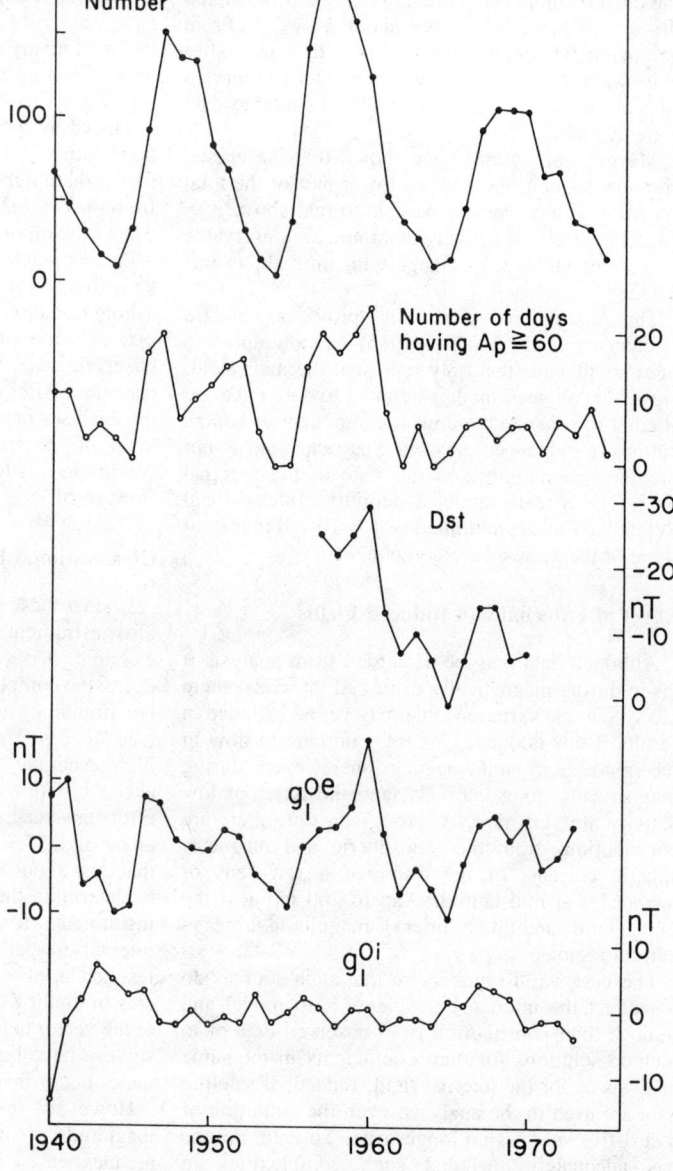

FIGURE 2. Comparison of the axial dipole field with variations in other indices. From the top down are shown the annual values of the relative sunspot number, number of days when the magnetic activity index Ap exceeded 60, annual Dst values, and the filtered external and internal axial dipole components computed from magnetic observatory annual means.

modeling of the geomagnetic field can only be done by the use of a nearly complete coverage of the Earth with a vector satellite survey.

At this writing, the theory and testing of techniques to best deal with sparse data are incomplete. Those attempting to estimate internal fields from planetary flybys have the most difficult problem of analysis and thus have made the most sophisticated attempts to determine the internal harmonics from the least data (Connery et al., 1982). Variants of the general least squares techniques have been introduced by P. Fougere, L. Shure, D. Gubbins, and R. Langel. Some of these techniques add other constraints on the models or ensure smoothness of the results that reflect assumptions for their particular purposes. Other techniques are yet under development.

Procedures in Truncated Least Squares

The original analyses by Gauss and others late in the nineteenth century (Chapman and Bartels, 1940) were performed on data read from the grid points of handdrawn contour maps of survey data. Such maps were constructed at a given "epoch" or year (e.g.,

1835.0) by making a separate analysis for secular change and "updating" survey observations to that epoch. More recent analyses, such as those developed by Cain et al. (1967), made a simultaneous adjustment on the Gauss coefficients and their time derivatives (secular change terms) using only the input data. The spherical harmonic expansions were thus used as spatial interpolating functions. The mathematics used for solving for the coefficient adjustments are given in the entry on spherical harmonic analysis. This method employs a nonlinear expansion of the usual least squares equations into a first-order Taylor's series about an initial model so as to accommodate both vector component data, which can be analyzed without such an expansion, and data such as the scalar or angular observations, which cannot be expressed as linear combinations of the Gauss coefficients.

When different types of data are combined into one least squares solution, it is necessary to construct a cost function of the form

$$\sum_i w_i (O - C)^2 = \sum_i w_i \delta^2$$

where O represents the observed component, C the computed component, and w some constructed form of weights. The differences $\delta = O - C$ for each observation are called *residuals*. Although it is not necessary for the minimization procedures that all observations be in the same units, one useful way of combining angular data with those of components has been to effectively convert the angular residuals into nanoteslas. In an analysis utilizing the observations of D, I, H, X, Y, Z, and F, the sums have been formed of $H_c\delta D$, $F_c\delta I$, δH, δX, δY, δZ, and δF, where H_c and F_c are computed from the prior estimated field coefficients.

Combining data from different sources, one constructs weights by including factors that are related to the error of observations. There is a divergence of opinion of the best way for such a combination in geomagnetic analyses, since the accuracy of an observation itself is not the most significant source of noise, because the truncation level falls far short of describing the local anomaly field. For example, data taken by a ship-towed magnetometer generally show much smaller δ's than do land observations because the sensor is more distant (by the ocean depth) from the very local crustal anomalies. Satellite data are taken at sufficient altitude that the very small scale anomalies are completely filtered out.

Past practice has been to use weighting factors in the fitting equations that tend toward minimum variance. That is, factors such as $1/\sigma^2$ are used for a given class of data, where σ is the standard deviation of those observations from a model. Because the presence of magnetic anomalies makes any residual distribution non-Gaussian, the choice of such weighting factors tends to be qualitative. The use of the concept of root mean square (rms) deviation of data from a given magnetic model as the criterion of the quality of that model is quite dependent on the cutoff or limit of data. For such reasons surface data taken in the presence of high crustal noise (e.g., 200 nT rms), if weighted accordingly, have little effect on a model of the field when combined with satellite data that have residuals under 10 nT.

Scalar data alone can be used to generate a model, but the resulting field tends to have angular errors at low latitudes due to a phenomenon not completely understood, but which has been labeled as the *Backus ambiguity* (Backus, 1970) or the *transverse error*. Numerical tests have shown that models made using a small sample of scalar data can have errors up to several hundred nanoteslas in the vertical field near the magnetic equator. The zonal, or $m = 0$, terms are the least well-determined. Further, as the data span a larger range in radius, the error is reduced. This aspect likely results from the fact that near the magnetic equator in nearly curl-free space ($\nabla \times \mathbf{B} = 0$) the vertical gradient of the north-south field contains information on the north-south gradient of the vertical field. Since the main field is generally almost north-south near the equator, a thicker shell of scalar field observations has the effect of having observations of the vertical field.

The best results, however, come from also having vector observations of the field, or at least those components that will give a measure of its direction, especially near the magnetic equator. Having vector data elsewhere has less effect on correcting a model determined by scalar data. Even sparse surface vector observations can contribute significantly to a model derived from scalar data, provided some means can be found to mitigate the effects of crustal anomalies.

Theory requires that there be a continuum of data over a sphere for a perfect determination of the field. Such a requirement is never met in practice. Some refinements have been developed by Schmitz and Cain (1983) to alleviate the problems of uneven data distribution. The data are first selected by area to obtain the most even distribution possible, and then additional weighting factors are applied that depend on the resultant areal density. Otherwise, the coefficients that are determined are unstable as the value of n^* changes (Merrill and McIlhinny, 1983). This phenomenon is a kind of spatial "aliasing" similar to the distortions in linear spectra caused by power at frequencies beyond the truncation level.

Secular Change

The main impediment to an accurate geomagnetic field model is the fact that the field undergoes constant change. The axial dipole term g_1^0 was about 29,850 nT in 1985 and decreasing by about 25 nT per year. Secular change in some areas is as high as 200 nT per year (see *Geomagnetic Secular Variation: Direction and Intensity*). Thus, if data are collected for more than a week and model accuracies of a few nanoteslas are sought, it is essential in analysis to take into account such changes.

As noted earlier, analyses have been done both by making a separate determination of secular change or by combining the secular change determination with the spatial determination. In an early combined computation (Cain et al., 1967), the Gauss coefficients were expanded to second degree in time by using only survey observations. This technique did not take into account the additional information that many of the observing points were fixed in position and only varied in time. Magnetic observatory data, which were taken in as noisy a crustal environment as any survey observation, could not be weighted as strongly as their 1-nT accuracy indicated without adversely affecting the spatial field coefficients. Two different developments have used this added information. In the technique of Schmitz and Cain (1983), the values of $\dot{X}$, $\dot{Y}$, and $\dot{Z}$ were added to the cost function. In the development by Langel (1987), "biases" or correction factors were included in the analysis. Such biases were the predetermined average deviations between a field model truncated to $n = 13$ and the data taken over a number of years. The attempt is designed to correct the observatory data for the crustal components and thus be able to give them much higher weights. To date, however, such correction factors are known only approximately, since they change with analyses done over different time spans.

In the combined technique, there is no theoretical basis for the best relative weighting of secular change data and survey observations in the cost function. Schmitz and Cain (1983) used weights for the $\dot{X}$, $\dot{Y}$, and $\dot{Z}$ data based on the estimated accuracy of their determination, but then also used other arbitrary factors, depending on the outcome of the analysis.

Integral Extensions

Due to present limitations on the size and speed of computing systems, the least squares approach described here has not been applied beyond $n = 29$. Other techniques have been used to obtain the points shown in Fig. 1 beyond $n = 29$. To produce these higher-degree terms, residuals of satellite vertical field observations B_r were formed from an $n = 29$ model and then averaged in cells of about 3^0 in latitude and longitude. Then, a small modification of the Gauss-Legendre quadrature technique (see *Spherical Harmonic Analysis*) was applied, and the coefficients residual to those of the initial least squares set were obtained up to $n = 66$. However, such determinations are limited to the noise in the data and other approximations so that present results become increasingly uncertain above $n = 40$. In principle there is no reason such techniques could not be applied to much higher degrees given adequate data.

Summary

The determination of the geomagnetic field structure is the subject of a continuing research effort in spite of its origins over a century ago. As the data become increasingly accurate and numerous, the results continue to provide new insights into the physical sources as well as challenges in numerical analysis.

JOSEPH C. CAIN

References

Backus, G. E., 1970, Non-uniqueness of the external geomagnetic field determined by surface intensity measurements, *Jour. Geophys. Research* **75**, 6337-6341.

Chapman, S., and J. Bartels, 1940, *Geomagnetism*. Oxford: Clarendon Press.

Cain, J. C., S. J. Hendricks, R. A. Langel, and W. V. Hudson, 1967, A proposed model for the International Geomagnetic Reference Field—1965, *Jour. Geomagnetism and Geoelectricity* **19**, 335-355.

Connery, J. E. P., M. H. Acuna, and N. F. Ness, 1982, Voyager 1 assessment of Jupiter's planetary magnetic field, *Jour. Geophys. Research* **86**, 3623-3627.

Langel, R. A., 1987, Main Field, in J. Jacobs, ed. *Geomagnetism*, Chap. 3., New York: Academic Press.

Mauersberger, P., 1956, Das Mittel der Energiedichte des geomagnetischen Hauptfeldes an der Erdoberfläche und seine säkulare Änderung, *Beiträge zur Geophys.* **65**, 207-215.

Merrill, R. T., and M. McIlhinny, 1983, *The Earth's Magnetic Field*, International Geophysics Series 32, New York: Academic Press.

Meyer, J., J.-H. Hufen, M. Siebert, and A. Hahn, 1983, Investigations of the internal geomagnetic field by means of a global model of the Earth's crust, *Jour. Geophys.* **52**, 71-84.

Peddie, N. W., 1986, International geomagnetic reference field revision 1985, *Geophysics* **51**, 1020-1023.

Schmitz, D. R., and J. C. Cain, 1983, Geomagnetic spherical harmonic analyses 1. Techniques, *Jour. Geophys. Research* **88**, 1222-1228.

Yukutake, T., and J. C. Cain, 1979, Solar cycle variations in the first-degree harmonic components of the geomagnetic field, *Jour. Geomagnetism and Geoelectricity* **31**, 509-544.

Cross-references: *Geomagnetic Field: Asymmetries; Geomagnetic Field: Elements; Geomagnetic Field: Westward Drift; Geomagnetic Field, Main: Theory; Geomagnetic Secular Variation: Direction and Intensity; Geomagnetic Secular Variation: Theory; Geomagnetism: Historical Introduction; Spherical Harmonic Analysis.*

GEOMAGNETIC MEASUREMENT TECHNIQUES AND SURVEYS

When the earliest measurements of the Earth's magnetic field were taken several hundred years ago, the modern science of geomagnetism began. Before then, although the magnetic compass had already been invented and we had long made invaluable use of it to tell direction, our picture of the magnetic field that surrounds the Earth and makes the compass work was very unclear. Since then, as more and more geomagnetic measurements have been taken, and as ever more accurate magnetic charts and mathematical models have been derived from them, our picture of geomagnetism has become much clearer.

Because of geomagnetic measurements and surveys we now know that the Earth's magnetic field is a composite of fields from several distinct sources. The major part, which we call the main field, is believed to be generated by electric currents deep inside the Earth, probably in the liquid core. Another part, the crustal field, is caused by magnetic rocks in the crust. A third part, the external field, is caused by electric currents in the upper atmosphere and beyond. We know that the intensity of the main field ranges from about 24,000 nT (nanotesla) in part of South America to nearly 62,000 nT in the polar regions. The intensity of the crustal field averages only about 200 nT, but exceeds 1000 nT in a few places. In contrast to the crustal field, which is relatively stable, and to the external field, which is mostly quite transient, the main field undergoes long-term change, called secular variation. The secular variation presently exceeds 100 nT per year in several locations. Although the secular variation in a particular region may be nearly constant for many years, it can also change rapidly as shown in Fig. 1, which shows the magnetic declination at Agincourt, Canada, from 1900 to 1970. The rate of change of declination at Agincourt held fairly steady at about $-0.07°$ per year during the period from 1902 to 1932, but changed, in approximately two years, to about $+0.01°$ per year, and remained at that value for more than 10 years. Changes in the rate of change such as these are not predictable at present, mainly because a usable quantitative theory explaining the cause of the main field and secular variation has not yet been developed.

Generally speaking, there are two kinds of magnetic surveys. In one, measurements of one or more of the elements of the field, usually only the field's total intensity, are taken at closely spaced points, perhaps no more than 1 km apart. The results of this kind of survey are used to study the local geology and are often instrumental in the search for petroleum and other minerals. (See *Aeromagnetic Surveying*.) In the other kind, measurements of the field's intensity and direction are taken at more widely spaced points. The results of this kind of survey are used chiefly for magnetic charting and modeling and for basic research.

Magnetic charts and mathematical models are the principal means of conveying information on the space-time structure of the geomagnetic field. Magnetic charts have been compiled and issued, mainly by those nations with strong maritime interests, for hundreds of years. The chart of magnetic declination, also called the isogonic or compass-variation chart, has been especially important because mariners and other navigators have long relied on it to convert compass readings to true headings. Mathematical models have become more important in recent years as the use of computers has increased. Particularly noteworthy are the models that make up the International Geomagnetic Reference Field (IGRF) (e.g., Peddie, 1982), a complete mathematical description of the Earth's magnetic field from 1945 to 1990. The mathematical form of most geomagnetic models, including those of the

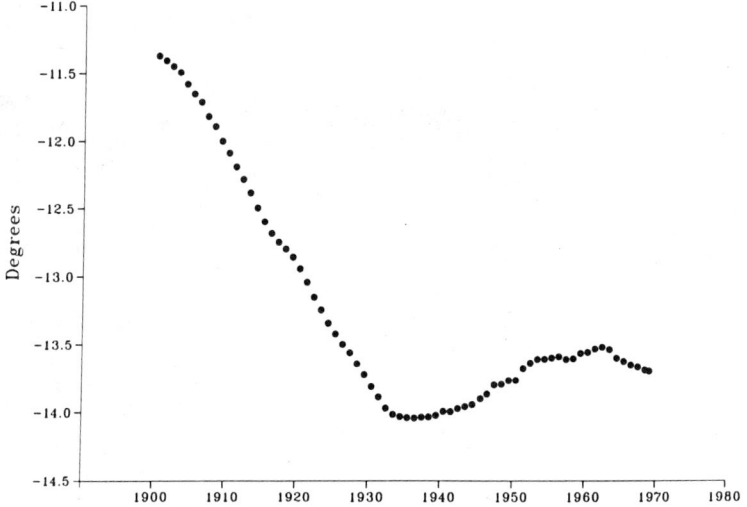

FIGURE 1. Magnetic declination at Agincourt, Canada. Negative values indicate west declination.

IGRF, is the series of internal solid spherical harmonics. (See *Geomagnetic Field Analysis*.)

Models and charts usually include information on the expected secular variation so that they will be applicable for several years. On charts this information appears as isopors, or lines of equal annual change. Models and charts must be revised periodically, normally every five or ten years. Naturally, the quality of the revisions depends on the continuing availability of new measurements.

In this article we will explore some of the more important kinds of instruments that are used to measure the geomagnetic field, see how they are used, and learn about the kinds of geomagnetic surveys that are undertaken. We will see that geomagnetic equipment has evolved over the years from simple mechanical devices with permanent magnets, requiring a great deal of time and effort to operate, to solid-state electronic devices that automatically take measurements every few seconds. We will also see that geomagnetic surveys have evolved from the difficult and dangerous land and sea expeditions of the eighteenth and nineteenth centuries to the global satellite surveys of today.

Geomagnetic Instruments

Instruments that are used to measure the Earth's magnetic field are called magnetometers. Before World War II, most types of magnetometers used either permanent magnets or induction coils. We will briefly discuss three of these: the classical magnetometer, the Earth inductor, and the QHM. More complete descriptions of these as well as other older types of magnetometers are given by Chapman and Bartels (1940, 1962), Fleming (1939), and Hazard (1930). Most types of magnetometers in general use today for survey work were developed during or after World War II and utilize electronic circuitry. We will briefly discuss the proton-precession magnetometer, the fluxgate magnetometer, and the optically-pumped magnetometer. Great care must be taken in the construction of magnetometers to see that only nonmagnetic materials are used to make their nonactive parts.

Although there are many kinds of magnetometers, each can be classified as either a variations, absolute, or relative instrument. A variations instrument, also called a variometer, records the short-term variations in the field, but of itself provides no information on the field's absolute intensity or direction. Magnetometers of the other two types are capable of measuring the absolute intensity or direction, but only the absolute magnetometer provides a truly absolute measurement of the magnetic field. The relative magnetometer requires occasional calibration, which is done by comparing its measurements with those of an absolute magnetometer.

Measurements with pre-World War II instruments are laborious and time-consuming. The time required to take a complete set of measurements, usually of the elements D (declination), I (inclination or dip), and H (horizontal intensity), may exceed one hour. (See *Geomagnetic Field: Elements*.) The classical magnetometer and the Earth inductor are two examples of absolute instruments. Although now seldom used in field survey work, they are still used at many magnetic observatories.

For more than a century the classical magnetometer, which can measure D and H, was the standard instrument for determining the intensity of the field (see Fig. 2). For the D measurement the instrument serves essentially as a very accurate compass. The orientation of a small bar magnet, suspended from a gold fiber, is determined optically and compared with a reference direction. H is measured using the method of oscillations and deflections invented by Poisson and perfected by Gauss in the early 1800s. The H measurement takes almost one hour and is difficult to do. Only the most skilled observers achieve probable errors of less than 10 nT.

The Earth inductor, used to measure I, consists essentially of a coil of wire connected by means of a commutator and brushes to a sensitive galvanometer (see Fig. 3). It operates thus: While the Earth inductor's coil is rotated about one of its diameters by means of a hand crank, the orientation of the coil is carefully adjusted until the galvanometer indicates that no net current flows in the coil. The value of I

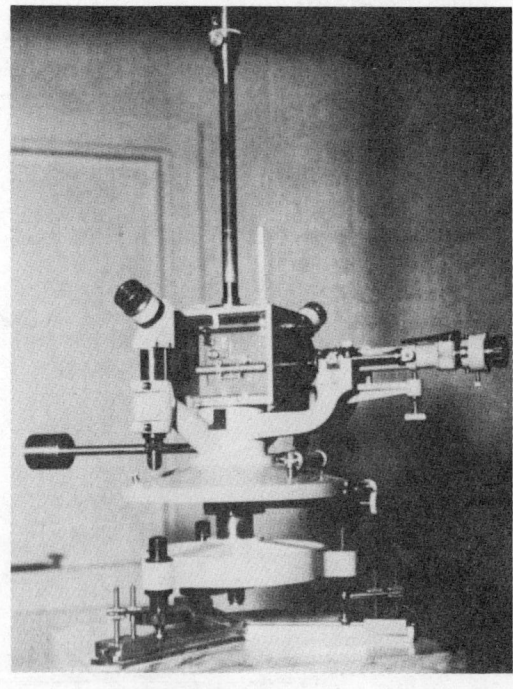

FIGURE 2. Classical magnetometer, designed for absolute measurements of magnetic declination and horizontal intensity.

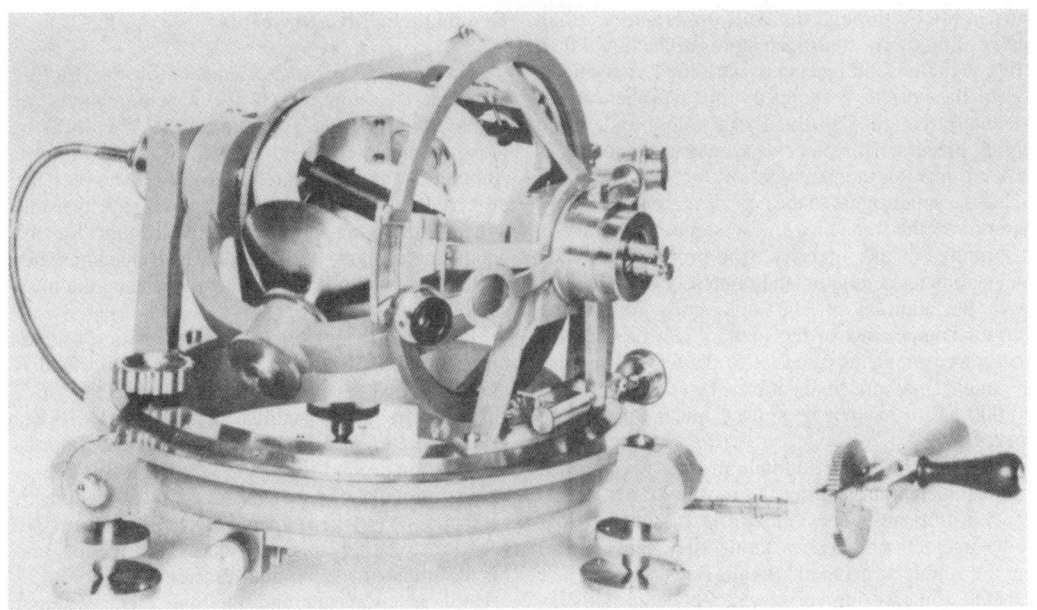

FIGURE 3. Earth inductor, used for measuring inclination or dip.

is then obtained as the angular difference between this final orientation of the coil's rotation axis and the horizontal plane, as determined by a spirit level.

The QHM (quartz-fiber horizontal-intensity magnetometer) is an example of a relative instrument (see Fig. 4). It consists essentially of a vertical tube containing a small magnet suspended from a quartz fiber and an optical system for determining the horizontal direction of the magnet. The observer rotates the tube, causing the fiber to twist and the direction of the magnet to become deflected from that of the horizontal magnetic field. The amount of deflection indicates the horizontal-field intensity. Periodic calibration is required because the magnet's magnetic moment and the fiber's elasticity change with time.

The fluxgate magnetometer, which was developed during World War II for the aerial detection of submarines, is widely used today both for survey work and at many magnetic observatories. It makes use of the variation of magnetic permeability with intensity of magnetization. The sensor, which consists of a magnetically saturable core wound by several coils, measures the component of the field in the direction of its axis. Often three sensors are arranged so that each is perpendicular to the other two. If the orientation of this three-component sensor is determined, the instrument provides information on the total field vector. Because fluxgate measurements can be taken in rapid succession, information on rapidly varying fields can be obtained.

The proton precession magnetometer, developed in the 1950s and used extensively today, measures the total intensity of the field. Its sensor comprises a coil of wire surrounding a small sample of a liquid that is rich in hydrogen, such as water or kerosene. A measurement is begun by passing an electric

FIGURE 4. Quartz-fiber horizontal-intensity magnetometer (QHM).

current briefly through the coil, producing a strong artificial field. The hydrogen atoms in the liquid then align with this field instead of with the Earth's field. When the current is stopped, the hydrogen atoms gradually realign with the Earth's field and, in so doing, precess. This precession, whose frequency is proportional to the strength of the Earth's field, induces a small signal in the coil. Electronic circuitry determines the frequency of the signal and converts it to units of field intensity. The proton precession magnetometer is reliable and relatively inexpensive. With the addition of Helmholz coils for nulling various components of the field, it can be used for vector measurements (see Fig. 5). However, because it cannot measure intensities of less than about 20,000 nT, it has not been used much for satellite surveys.

The rubidium-vapor and helium optically pumped magnetometers are among the most recently developed instruments. They make use of the fact that some interactions between atoms and light depend on the ambient magnetic field. Because they are capable of measuring very weak magnetic fields (as low as 10 nT), they are used extensively for satellite surveys.

Geomagnetic Measurements

Measurements of the elements of the Earth's magnetic field are taken for a wide variety of purposes. Modeling and charting of the local or global geomagnetic field are two of the main ones. Another is basic research into, for example, the properties of the Earth's crust and mantle, possible dynamo mechanisms that could account for the geomagnetic field, and the physical characteristics of the ionosphere, the magnetosphere, and the local geology.

Magnetic Observatories. A magnetic observatory is a laboratory where the geomagnetic field is monitored and recorded continually. The first such observatories were established in the early 1800s. Today there are nearly 200 permanent observatories in operation around the world (see Fig. 6). Observatories provide for the necessary calibration of magnetic survey instruments. Data from them are vital for updating magnetic charts and models and for the study of secular and transient variation.

Until a few years ago the optically recording magnetograph was used at nearly all observatories (Fig. 7). The magnetograph employs variometers,

FIGURE 5. Sensor and coils of proton vector magnetometer. The sensor is at the center. The inner set of four coaxial coils, with axis vertical, is for nulling Z during the measurement of H. The outer pair of coaxial coils, with axis horizontal, is for nulling H during the measurement of Z. Neither coil system is energized during the F measurement.

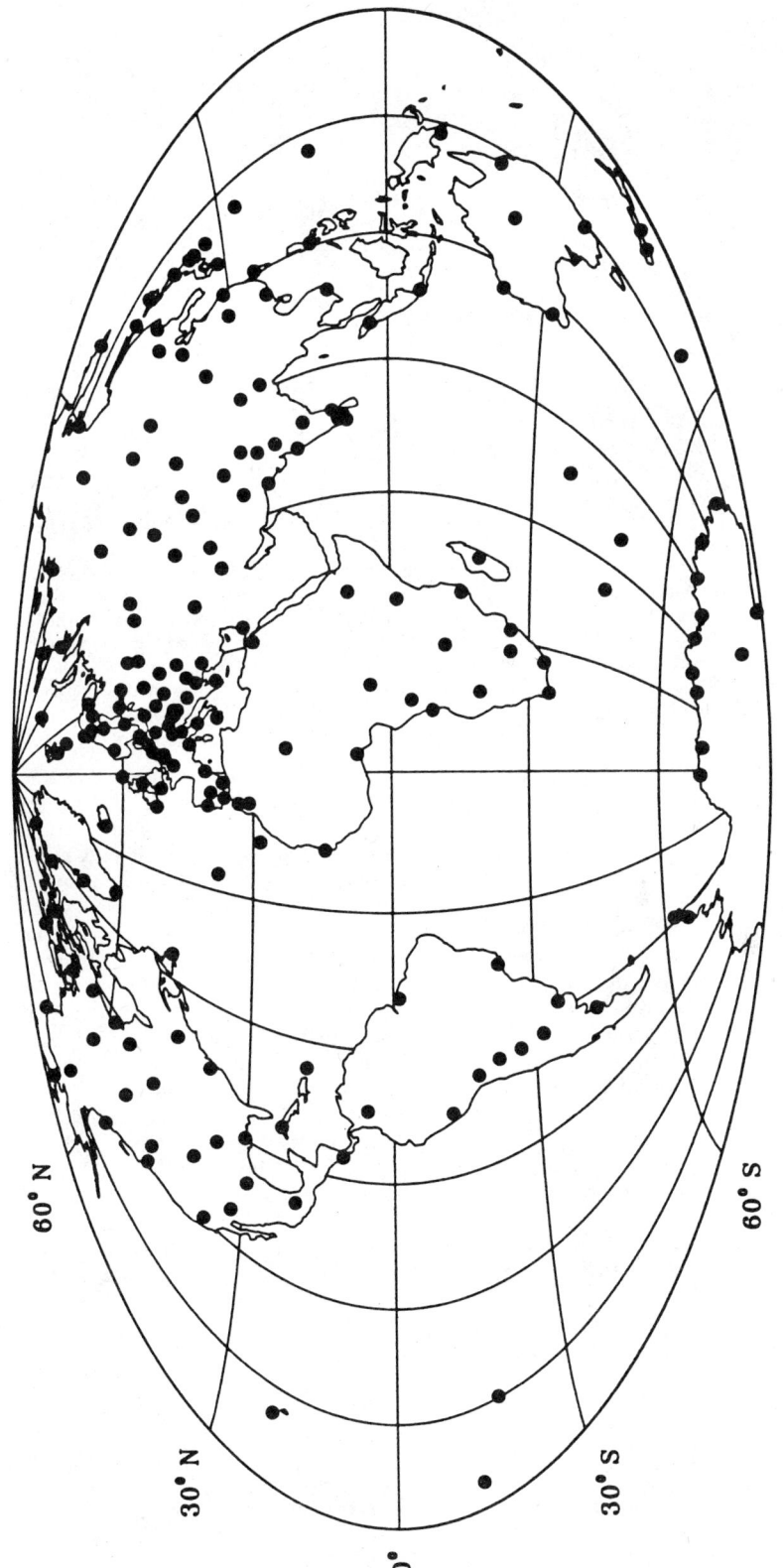

FIGURE 6. Magnetic observatories in operation in 1985.

FIGURE 7. Optically recording magnetograph. At left is the recorder, which contains a slowly rotating cylindrical drum to which photographic paper has been attached. Next to the recorder are the recording and time-flash lamps. To the right are the H, D, and Z variometers. Each variometer contains a carefully oriented magnet to which is connected a small mirror. A beam of light from the recording lamp is reflected from each mirror to the photographic paper. The three reflected beams move in response to changes in the Earth's magnetic field and record traces on the paper. The mirror next to the D variometer reflects hourly flashes from the time-flash lamp, putting time lines on the record. Each variometer is equipped with a Helmholz coil for determining its sensitivity or scale value.

each of which has a small permanent magnet that is free to rotate about a vertical or horizontal axis. Usually three variometers are used, each one carefully oriented to measure changes in D, H, or Z (vertical intensity). The field variations are recorded as traces on photographic paper of light beams reflected off small mirrors attached to the magnets. Each variometer is calibrated by determining the scale value of its trace, usually with a Helmholz coil, and the baseline of its trace, with periodic absolute measurements at a nearby location.

These systems are gradually being replaced by electronic systems capable of digital recording and responding more quickly to rapid field variations. Typically, the modern system employs three-axis fluxgate magnetometers, which are programmed to measure the field components every few seconds. The measurements are stored in random-access memory or on magnetic tape and are periodically transmitted by telephone line or satellite to a central office.

Directional stability of the sensors is critical for producing accurate observatory data. The piers upon which the sensors are mounted sometimes tilt erratically for many months after being constructed. During this time, more frequent absolute measurements may be necessary to keep track of the baselines. Pier instability is a chronic problem at high-latitude observatories built on permafrost.

Surface Surveys. For hundreds of years, measurements taken on land and sea were our only source of information on the spatial structure of the geomagnetic field. Among the earliest of such measurements are those of magnetic declination taken by mariners to correct their compass readings during voyages. In 1698, Sir Edmund Halley, the Astronomer Royal, undertook the first voyage whose main purpose was magnetic survey, and used the resulting data in compiling the first world isogonic (declination) chart. Figures 8–10 show where surface measurements were taken during the periods 1900–1930, 1930–1955, and 1955–1985. The coverage attained by surface surveys was quite uneven, with the polar regions almost devoid of measurements. The density of measurements, in terms of number of measurements per 5-degree geographic quadrangle, ranged from more than 800 in Western Europe, Japan, and parts of North America and Australia, to fewer than 100 in much of Africa and southern Asia.

The Carnegie Institution of Washington was responsible for a great deal of the survey work accomplished in the early part of the twentieth century. It sponsored ten cruises of the survey ships *Galilee* and, its successor, the *Carnegie* until 1929 when the *Carnegie* was destroyed by fire at Apia, Samoa. It also obtained measurements at more than 6000 land stations until 1944. The 1935–1955 period witnessed ambitious surveys of Europe and the Soviet Union but little survey activity at sea. Ocean surveying increased during the 1955–1985 period mainly as a result of the deployment of the Soviet Union's survey ship *Zarya* in 1956. However, except

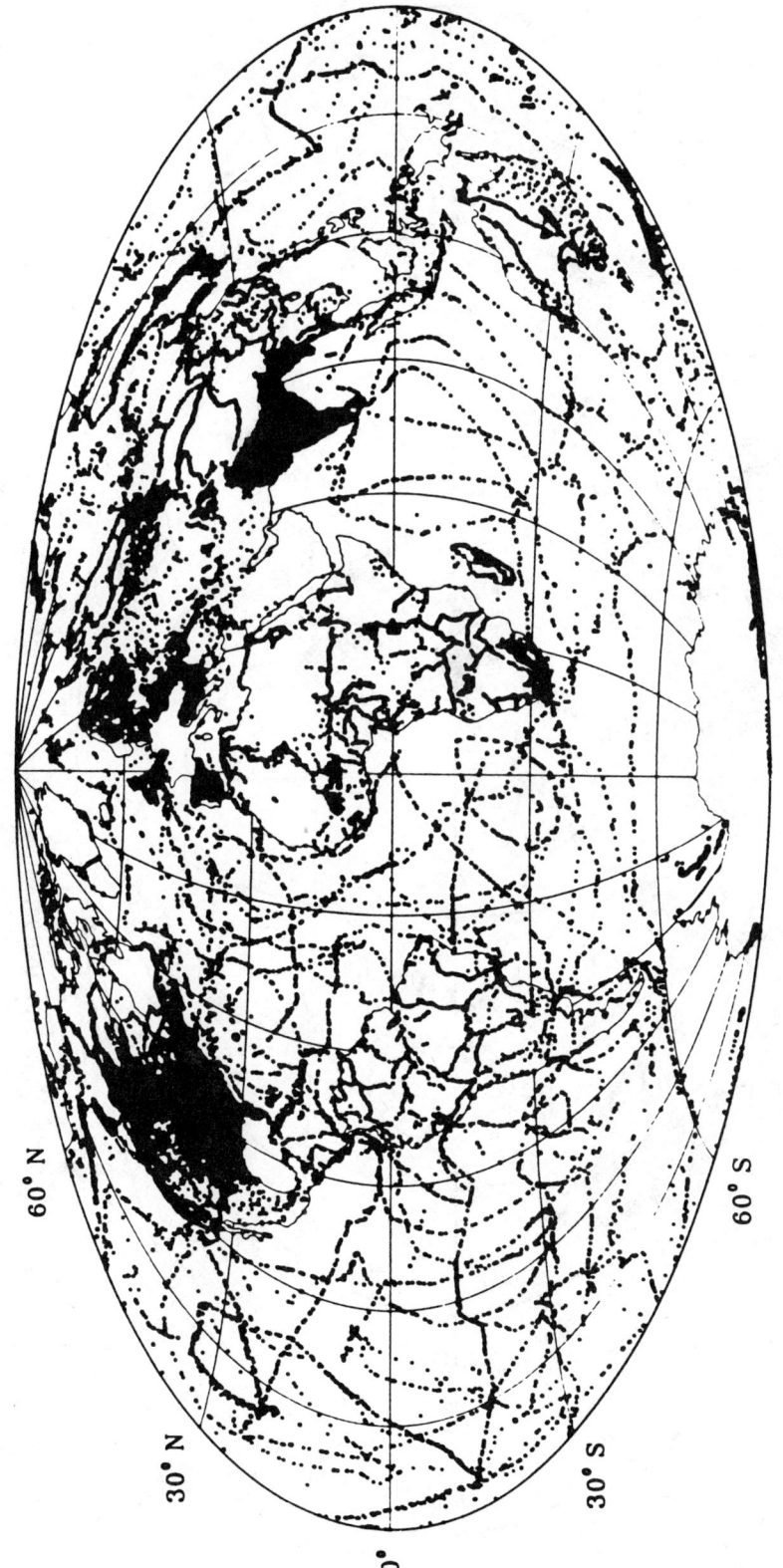

FIGURE 8. Land and marine surveys, 1900–1930.

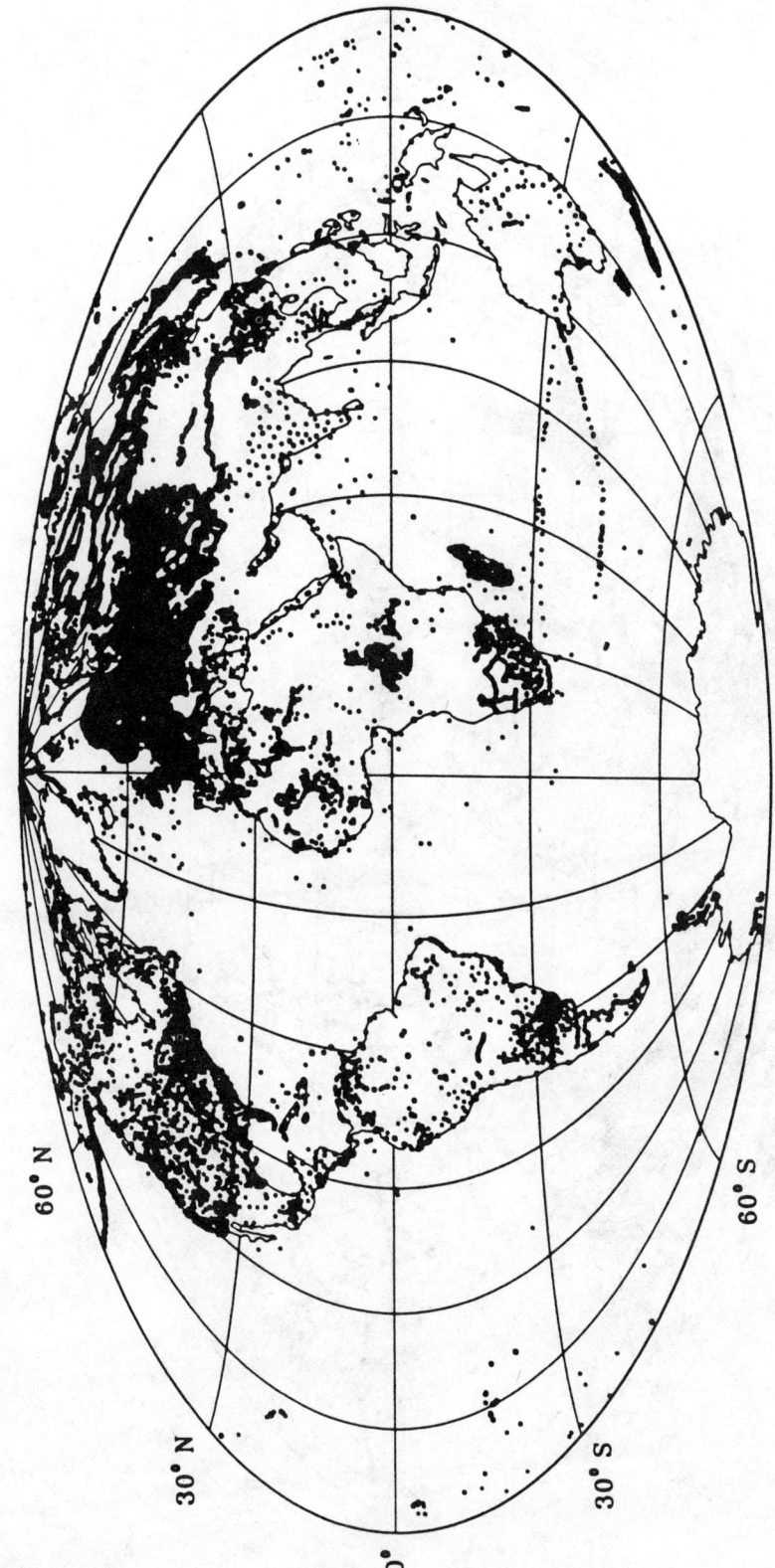

FIGURE 9. Land and marine surveys, 1930–1955.

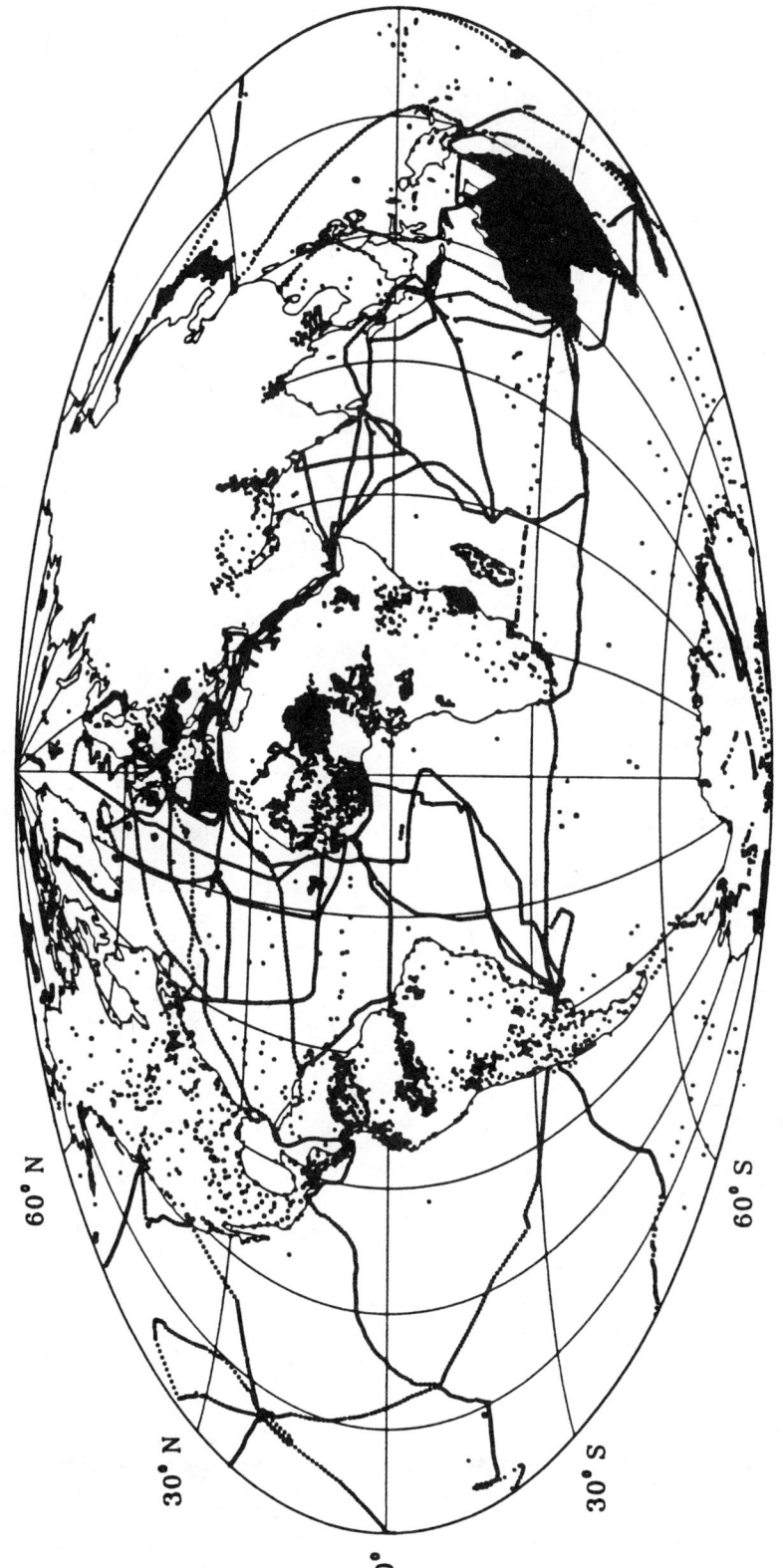

FIGURE 10. Land and marine surveys, 1955–1985.

for a fortunate increase in survey activity in and around Antarctica, surface survey activity declined during this latest period chiefly because of the advent of long-range aerial and satellite surveying.

Repeat Surveys. To improve the accuracy of their magnetic charts and models, many countries undertake magnetic repeat surveys to augment the secular-variation information available from local and nearby magnetic observatories. Repeat survey measurements are taken at repeat stations, which are permanently marked locations that have been carefully selected as magnetically undisturbed and likely to remain so for many years. Repeat stations are usually reoccupied about every three to five years. During an occupation, which typically lasts several days, many sets of measurements are taken. The measurements may then be reduced, or corrected for any transient variations that occurred, using information from the nearest magnetic observatory or a variation instrument set up temporarily near the station.

Aerial Surveys. During World War II, great advances were made in the speed, range, and payload of aircraft. These advances led to a great increase in the use of aircraft for geophysical surveying after the war. The most ambitious aerial magnetic surveys have been those of Canada and Scandinavia conducted by the Dominion Observatory of Canada and the worldwide Project MAGNET surveys conducted by the U.S. Navy (Fig. 11). These surveys have made use of three-component fluxgate magnetometers mounted inside the aircraft and scalar-intensity instruments, such as proton-precession or metastable-helium magnetometers, having sensors located at the end of a boom or tether outside the aircraft.

One of the fundamental problems associated with aerial vector surveying is the maintenance of an accurate reference direction. A directional error of only 1 minute of arc may cause errors in the field-component measurements of 15 nT. In the Canadian system, the magnetometers are mounted on a gyro-stabilized platform to isolate them from the accelerations of the aircraft. Directional reference is obtained from a gyroscope mounted on the stable platform. The drift of the gyroscope is monitored by means of celestial observations with a periscopic sextant, which is also stabilized.

Because the fluxgate sensors are located inside each aircraft, their readings are affected by the aircraft's own magnetic field. Suitable corrections are determined from time to time by calibration flights, in which the aircraft is flown in different directions over a place where the field vector is well known, usually a magnetic observatory.

Satellite Surveys. The only geomagnetic surveys that are truly global have been accomplished by artificial satellites equipped with magnetometers. The first such satellites were launched in the 1960s by the Soviet Union and by the United States as part of the World Magnetic Survey, an international program of cooperation in the study of geomagnetism (Zmuda, 1971). The most successful of these early satellite surveys were the Soviet Union's Kosmos-49 survey and the United States's OGO-2, -4, and -6 surveys (also called POGO, for Polar Orbiting Geophysical Observatory).

The magnetometers used in the early surveys were capable of measuring only the field's total intensity. It was hoped that spherical harmonic analysis of the resulting measurements would yield models that not only would accurately represent the total intensity but also the vector field. As it turned out, the resulting models suffered from a systematic inaccuracy, mainly in the Z field component near the magnetic equator. This phenomenon, which was dubbed "perpendicular-error effect" by Lowes (1975), who first explained it, affected mainly certain subseries of the spherical harmonics, which were identified by Backus (1970). Naturally, this problem served as a powerful impetus for the development of a satellite capable of measuring the complete field vector.

The dream of a satellite survey of the vector field was finally realized nearly a decade after the last POGO mission, when the United States launched the MAGSAT (MAGnetic SATellite) spacecraft on October 30, 1979. MAGSAT orbited the Earth for seven and a half months. It carried a three-component fluxgate vector magnetometer and a cesium-vapor scalar magnetometer, which together measured the field strength and vector components with expected error of less than 2 nT and 6 nT, respectively. Two star cameras, a sun sensor, and a pitch-axis gyroscope provided the necessary information on the fluxgate sensors' orientation. Measurements were stored in onboard recorders and periodically transmitted to receiving stations on the ground. The MAGSAT mission is considered the most successful geomagnetic survey ever accomplished. Several proposed successors to MAGSAT were being considered when this article was written. (See *Satellite Magnetic Measurements*.)

Magnetometer Arrays. A set of magnetometers deployed in a region for simultaneous recording is called a magnetometer array. The resulting array data provide a detailed picture of the local magnetic field and its time variations and may reveal much about the local geology, including the electrical conductivity and material composition and discontinuities. Magnetometer arrays have been used effectively in studies of the subduction zones occurring at the boundaries of tectonic plates and the piezomagnetic effect associated with earthquakes.

Magnetic Pole Surveys. The north and south magnetic poles are those two locations on the Earth where the average (undisturbed) magnetic field is exactly vertical. As the geomagnetic field undergoes secular variation the locations of the magnetic poles change. Their locations were first determined in the

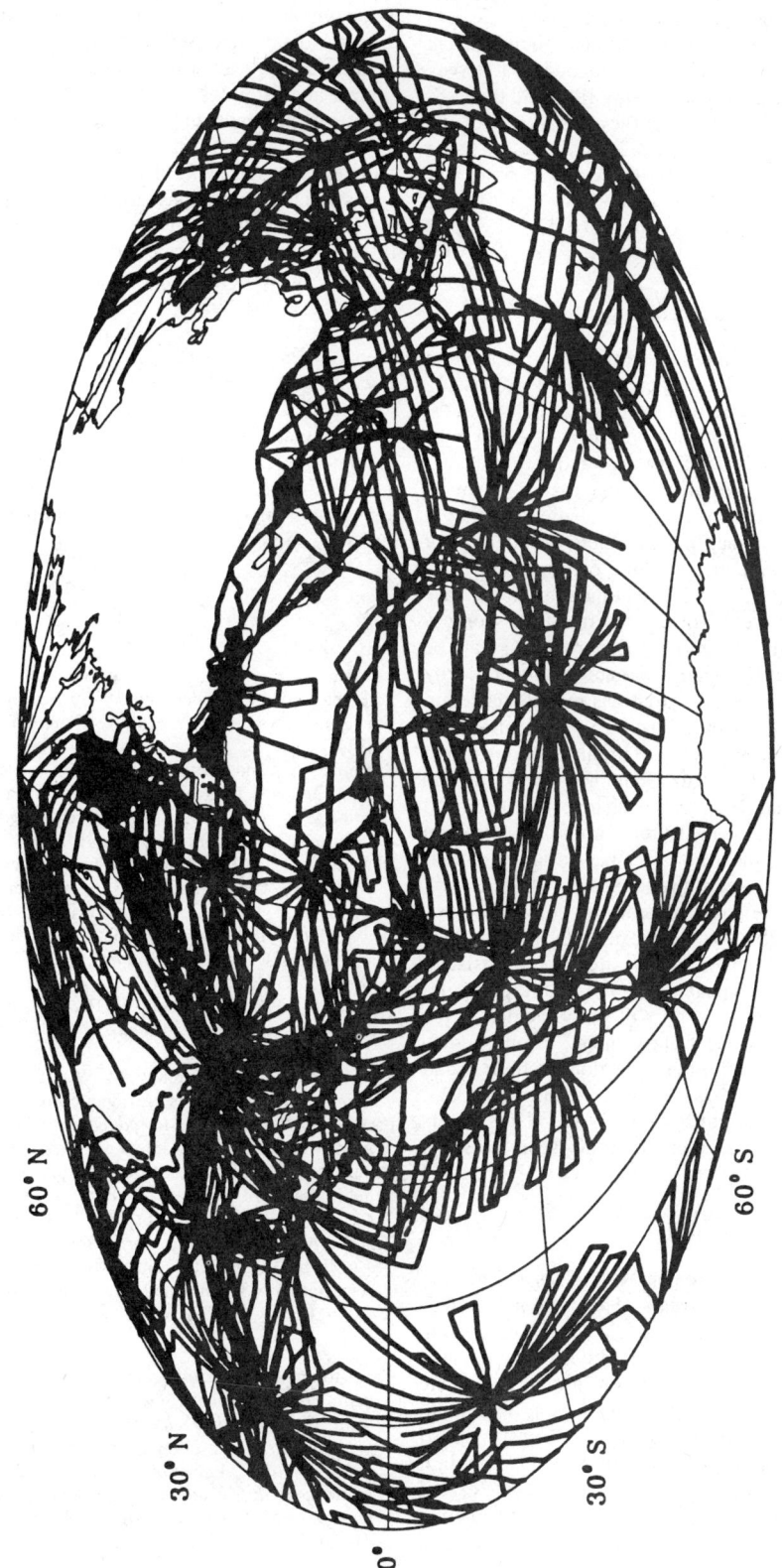

FIGURE 11. Aerial surveys, 1953–1985.

early 1800s. Since then, dozens of special surveys have been undertaken to determine their current location. Most recently Canada conducted a survey of the north magnetic pole in May 1984 and the U.S. Navy's Project MAGNET surveyed the area of the south magnetic pole in 1982 and 1985. These surveys found that the north magnetic pole was located at 77.0°N, 102.3°W in 1984 and that the south magnetic pole was located at 62.5°S, 139.2°E in 1985.

Data Processing and Archiving

The raw measurements resulting from a modern magnetic survey often undergo several processing steps, including correction for external-field disturbance, reduction to a common epoch, calculation of mean values, conversion of units, and reformatting. Correction for external-field disturbance involves estimating the portion of each measurement that is caused by the external field and then subtracting that value. Some components of the external field may be estimated from mathematical models. In the case of regional surveys, other external-field components, generally the more transient ones, can be estimated either by using information from a nearby magnetic observatory or variations instrument set up near the survey, or by adjusting the measurements so that they are consistent at the intersections of the flight lines. Reduction to a common epoch or date involves estimating the field change that occurred between the date of measurement and the common date, usually by means of a secular-variation model, and adding or subtracting the change. If necessary, at some stage the measurements are converted from the original units (volts, for example) to equivalent field-intensity units and reformatted to be consistent with the results of related surveys or the particular requirements of those who will study the data.

A system of World Data Centers, established in the 1950s, provides for the collection, archiving, and distribution of geophysical data. Geomagnetic data, in particular, are archived in the World Data Centers located in Boulder, Colorado, in Moscow (USSR), and in Kyoto (Japan).

Summary and Comment

Geomagnetic measurement techniques and surveys have evolved over the years. The early magnetometers were simple optomechanical devices employing permanent magnets. They required great skill and effort in their operation and were not very accurate by today's standards. Great advances in instrumentation were made during and after World War II. Today's magnetometers are easier to use, permit more rapid sampling of the field, and provide digital electronic storage of data. Advanced instrumentation has greatly increased the speed with which measurements can be made. Surveys that formerly required data gathering over several days at one location have given way to broad-scale surveys by aircraft and satellites, in which measurements take only a few seconds and are electronically recorded and transmitted to a distant receiving station. Thus, geomagnetic surveying has evolved from labor-intensive land expeditions having narrow geographic scope, through the great ocean surveys of the early 1900s, to extensive aerial surveys undertaken after World War II, and finally to today's truly global surveys by satellite.

The Earth's magnetic field, like its weather, may be part of a system that is so chaotic that it will always defy long-term forecast. If so, we will continue to need new geomagnetic measurements. Until a quantitative theory is discovered that permits accurate forecast of the field for many years ahead, it is crucial that we maintain the magnetic observatory system and continue to carry out regular global magnetic surveys.

NORMAN W. PEDDIE

References

Backus, G. E., 1970, Non-uniqueness of the external geomagnetic field determined by surface intensity measurements, *Jour. Geophys. Research* **75,** 6337–6341.

Chapman, S., and Bartels, J., 1940, *Geomagnetism.* 2 vols., 2nd ed., 1962. Oxford: Clarendon Press, 1049p.

Fleming, J. A., 1939, *Terrestrial Magnetism and Electricity.* New York: McGraw-Hill, 794p.

Hazard, D. L., 1930, Directions for magnetic measurements, U.S. Coast and Geodetic Survey Serial No. 166, Washington, D.C.: Government Printing Office, 129p.

Lowes, F. J., 1975, Vector errors in spherical harmonic analysis of scalar data, *Geophys. Jour. Royal Astron. Soc.* **42,** 637–651.

Matsushita, S., and Campbell, W. H., eds., 1967, *Physics of Geomagnetic Phenomena,* 2 vols. New York: Academic Press, 1398p.

Merrill, R. T., and McElhinny, M. W., 1983, *The Earth's Magnetic Field.* London: Academic Press, 401p.

Parkinson, W. D., 1983, *Introduction to Geomagnetism.* Amsterdam: Elsevier, 433p.

Peddie, N. W., 1982, International geomagnetic reference field: the third generation, *Jour. Geomagnetism and Geoelectricity* **34,** 309–326.

Zmuda, A. J., ed., 1971, *World magnetic survey 1957-1969,* IAGA Bulletin No. 28, Paris: International Association of Geomagnetism and Aeronomy.

Cross-references: *Aeromagnetic Surveying; Geomagnetic Field: Elements; Geomagnetic Field: Measurement; Geomagnetic Field, Main: Theory; Geomagnetic Secular Variation: Direction and Intensity; Geomagnetic Secular Variation: Theory; Geomagnetism: Historical Introduction; Magnetotelluric Method: Fundamental Concepts; Satellite Magnetic Measurements.*

GEOMAGNETIC POLARITY REVERSALS: OBSERVATIONS

In 1906 the French physicist Bernard Brunhes discovered some volcanic rocks magnetized almost exactly in the direction opposite to that of the Earth's present magnetic field. He also noticed that some clays that had been heated and baked by the flow were likewise reversely magnetized. At the beginning of the century paleomagnetism was more a subject of scientific curiosity than an established method in the earth sciences. However, it was already known then that when molten volcanic rocks cool and solidify, the magnetic minerals in them acquire a magnetization in the direction of the geomagnetic field and that bricks and pottery similarly acquire a magnetization along the ambient field when cooling from high temperatures in a furnace. From his observations Brunhes then concluded that the Earth's magnetic field must have reversed its polarity in the past. This is the first documented suggestion for reversals of the geomagnetic field.

Although later accepted by other scientists, at the time this idea attracted little attention, probably because geophysicists were not ready to accept such an unexpected property of the Earth's field. It was only in the early 1960s, when the development of the K/Ar isotopic dating method made possible to date relatively young rocks with sufficient accuracy, that Cox, Doell, and Dalrymple (1964) could establish a one-to-one correlation between the magnetic polarity and the age of rocks of different nature and geographical origin.

With the discovery and interpretation of the marine magnetic anomalies the total evidence for reversals became overwhelming. Although the reversed polarity of a few particular rocks may be explained by a complex process known as self-reversal (see *Magnetic Self-Reversal*), whereby a rock can acquire a magnetization opposite to that of the ambient field, today no doubt can be raised about the reality of this major aspect of the geomagnetic field.

Studies of reversals have proceeded along two separate lines. One has been to analyse the frequency of reversals using different mathematical methods and polarity scales of increasing precision in order to reveal statistical aspects of the reversal phenomenon. The other approach has been to try to obtain the sequence of changing field directions, and sometimes intensities, recorded in a sequence of rocks formed during the time spell of a geomagnetic reversal, in order to understand the geometry of the transitional field. Both approaches are important to impose crucial constraints on the modeling of the reversal mechanism and ultimately on the mechanisms of the terrestrial dynamo itself.

This article is concerned only with the second aspect of the study of the reversal process. Although in principle very simple, it is not easy to obtain detailed records of reversals because of the relatively very short duration of the phenomenon (estimates vary between 2000 and 20,000 years). Nevertheless, considerable effort over the last two decades has resulted in a number of detailed records from a wide variety of rock sequences including sediments, intrusions, and volcanic rocks. The widely different rock magnetic properties of these sequences might be expected to determine, to a certain extent, some of the characteristics of the paleomagnetic records. Indeed, controversies, some of which will be discussed later, exist among the different authors on the interpretation of the differences observed in records from different lithologies.

Early Records and the Nondipole Character of the Transitional Fields

The main concern of the early work on geomagnetic reversals has been to ascertain whether the geomagnetic field retained its dipolar nature during a reversal. Of the two simplest possible models for obtaining a reversal, one is a decrease in the main dipole field so that the transitional field is dominated by the nondipole field. Alternatively one could imagine that the dipole simply "turns over." In the first case the virtual geomagnetic pole paths* of the same transition from different sites would have no reason to coincide. On the contrary, in the second strong similarities among these paths are expected. Although these simple models do not yield any idea on the underlying physics of the problem they are useful because they provide a theoretical record of field directions.

Considering the present lack of paleomagnetic records of reversals from the Southern Hemisphere, it is somewhat surprising that one of the first detailed studies of geomagnetic reversals was reported in 1962 from the Stormberg sequence of lava flows of Triassic-Jurassic age in South Africa.

In this sequence, the lowermost flows are reversely magnetized and the uppermost ones are normally magnetized with a transition zone in between where the declination and inclination change progressively. A careful examination of the rock-magnetic properties of the different flows showed no possibility of self reversal, so that most probably the observed intermediate directions resulted from a reversal of the Earth's field in Triassic-Jurassic times. It was also found that the paleofield strength decreased during the transition by a factor of four to

---

*The virtual geomagnetic pole is defined as the pole of the centered dipole field, which would give the observed direction of magnetization at the site of observation. Although the transitional fields are not dipolar, this representation is widely used because it allows easy comparison of the behavior of the transitional field at different times and different places.

five, which is considerably larger than the maximum factor of two expected from a simple 180° rotation of a dipole.

Nevertheless, because the virtual geomagnetic pole (VGP) paths of many of the earliest studied polarity transitions all happened to lie along the same longitudinal band, more or less centered on the Indian Ocean, many later studies assumed a dipolar structure for the transitional field with a decrease of the dipole strength during the directional changes. In one of the earliest models, developed by Creer and Ispir in 1970, the results are explained in terms of two dipoles, the main one of which is the usual axial dipole, and the second, a nonaxial dipole of lesser strength. During the transition the two dipoles reverse at different times.

Subsequent work, however, has not sustained the hypothesis of a dipolar transitional field. First, the existence of two preferential longitudinal bands for the VGP paths has been suggested, then the accumulation of the results showed that this hypothesis was not correct either. For instance, the paths relative to the lower and upper Jaramillo transitions reported from a study of deep-sea cores do not at all match those situated over the Indian Ocean. Later it was realized that no preferential sector exists and the only common characteristic was a tendency for paths from the same geographical region to coincide.

Several detailed records were obtained in the early 1970s by different authors in different countries, using lava sequences, marine sediments, or intrusions. All these studies have documented the existence of intermediate paleofield directions and confirmed the drop in the paleofield strength associated with the directional changes. Finally, in 1976, Hillhouse and Cox obtained a detailed record of the Matuyama-Brunhes transition from Lake Tecopa, California, and compared its VGP path to the record of the same transition which had been obtained by N. Nijtsuma from marine sediments in the Boso Peninsula in Japan. The paths are quite different from each other (Fig. 1) definitively proving that the field was not dipolar during the transition. Although a very recent work by J. P. Valet, L. Tauxe, and D. V. Clark indicates that the paleomagnetic directions of the original work are largely overprinted, this conclusion has been proved correct by all the following studies.

Standing and Flooding Field Mechanisms for Reversals

The recognition of the nondipolar nature of the transitional field has been a major step in the study of geomagnetic reversals and led naturally to investigations of whether the transitional field basically resembles the usual nondipole field after collapse of the main dipole, or whether it arises from time dependent terms related to the reversal.

J. Hillhouse and A. Cox used the results of other authors, according to whom the nondipole field can be described as the sum of a westward drifting and a standing component, and the estimate of 4600 years for the duration of a reversal, obtained from records in deep-sea cores. They calculated that, if the usual nondipole field were basically unchanged during a polarity reversal, its westward drifting component should cause large longitudinal swings well visible

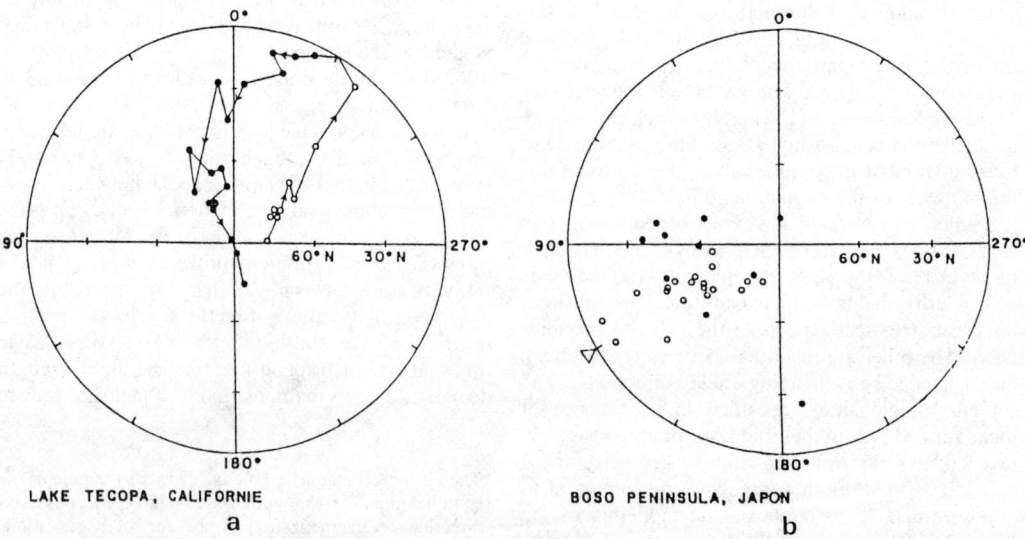

FIGURE 1. Comparison of VGP paths for the Brunhes-Matuyama transition recorded at Lake Tecopa and Japan. The paths do not overlap showing that the transitional field is not dipolar. Successive poles from Japan have not been connected, but the crossover at the equator is marked by an open triangle. (From Hillhouse and Cox, 1976)

in the records. The absence of such a characteristic in the Lake Tecopa record where the transitional directions are confined along a great circle, led Hillhouse and Cox to suggest that a standing component of the field is largely predominant during the main part of the transition, while the dipole field decreases to zero and regenerates with opposite polarity. This theory has been called the *standing field* model for reversal.

A completely different conclusion was reached by K. A. Hoffman in 1977 who suggested that the reversals originate in a well-localized region of the core and subsequently "floods" into other regions. The transitional field originates from time dependent components associated with this process in the core. Using a simple model, he calculated that at some point of the reversal the transitional field at the surface of the Earth will become vertical (upward or downward depending on the sense of the reversal) at any site of observation. In other words, the VGP paths pass near the site of observation or near the antipode to the site ('near' or 'far sided' path). Both the standing field and the flooding field models are described in the entry *Geomagnetic Polarity Reversals: Theory and Models*. In 1981, Hoffman pointed out that the standing field and the flooding field models predict different field behavior associated with sequential reversals that is theoretically distinguishable. Specifically, if a standing field predominates during the transition and remains unchanged for times that are longer than the interval separating two successive reversals, the intermediate field geometries associated with two sequential reversals should be identical, i.e., independent of the sense of transition. On the contrary widely different intermediate directions are expected if a flooding field controls the reversal process. This step has been important because it has provided a guideline for the analysis and classification of reversals.

In this respect one may, for instance, examine the VGP paths associated with four different reversals, three of which are successive, recorded in an upper Miocene marine section at Potamida in western Crete, which have been reported by Valet, Laj, and colleagues in a series of papers in 1981–1984. The main results, shown in Fig. 2, is that the north-VGP paths associated with the two (R-N) transitions, and the south-VGP path for the middle (N-R) transition are identical, within noise limits. Thus the directional field variations at the site during this (N-R) transition is identical to that associated with the R-Ns except for a change in sign. The south-VGP path for the upper (N-R) reversal is different, about 60° of longitude away from the other ones. Although the paths are not situated along the site's meridian or antipodal to it, as predicted by the original flooding model, this situation is exactly opposite to a standing field situation that requires identical transitional field along with sign, but is consistent with a flooding model. Furthermore, the variation of the field intensity as a function of the latitude of the VGP closely matches the predictions of this model.

An investigation of two sequential (R-N) and (N-R) reversals of early Pliocene age from lava sequences in Kauai, reported by S. W. Bogue and R. S. Coe in 1982, was initially interpreted in terms of a standing field. Indeed the two paths, although ill defined because of the limited number of transitional directions, seem identical and close to the site. However, the record of the transitional paleointensity obtained more recently for the (R-N) reversal is quite asymmetric, with an abrupt drop at the beginning of the transition and a much smoother intensity recovery that begins while the field direction is still transitional (Fig. 3). This point is important because a plot of the field intensity versus angular deviation for any particular standing field model is symmetrical, so that the standing field model cannot explain the asymmetric intensity pattern observed in the paleomagnetic data from Kauai. Bogue and Coe reinterpreted their data in terms of a particular zonal flooding model (see *Geomagnetic Polarity Reversals: Theory and Models*).

These two studies clearly favor the flooding field model, but conflicting evidence has been obtained from two independent studies of the Olduvai and Jaramillo transitions. In one case, the (N-R) upper Olduvai and the (R-N) lower Jaramillo transitions were studied by Clement and Kent (1984) in deep-sea sedimentary cores from the Crozet Basin in the southern Indian Ocean. The transitional VGPs are somewhat constrained in longitude and are roughly centered on the 180° meridian. As they fall more than 90° away from the site longitude (60°E) they can both be classified as far sided. These results are thus consistent with the hypothesis that a standing field has persisted across both these reversals.

In the other case, Herrero-Bervera and Theyer (1986) studied both the lower and the upper reversals of the Olduvai and Jaramillo chrons in deep-sea sediments in north-central Pacific. Although the records are characterized by a large scatter, the two (R-N) paths are predominantly near sided and the two (N-R) ones predominantly far sided. Thus the back-to-back reversals have well separated paths, a situation that cannot be reconciled with a standing field model. On the contrary, although the paths are not precisely antipodal, the situation is readily described with a flooding field model. Although it is possible that some of the characteristics of the southern Indian Ocean records may result from remanence acquisition phenomena and part of the scatter in the North Pacific records from incomplete removal of secondary magnetizations, the two studies cannot presently be reconciled, illustrating one of the difficulties in these studies.

To summarize, even though the application of Hoffman's test to sequential reversals leads in certain cases to a somewhat confusing situation, the clearest

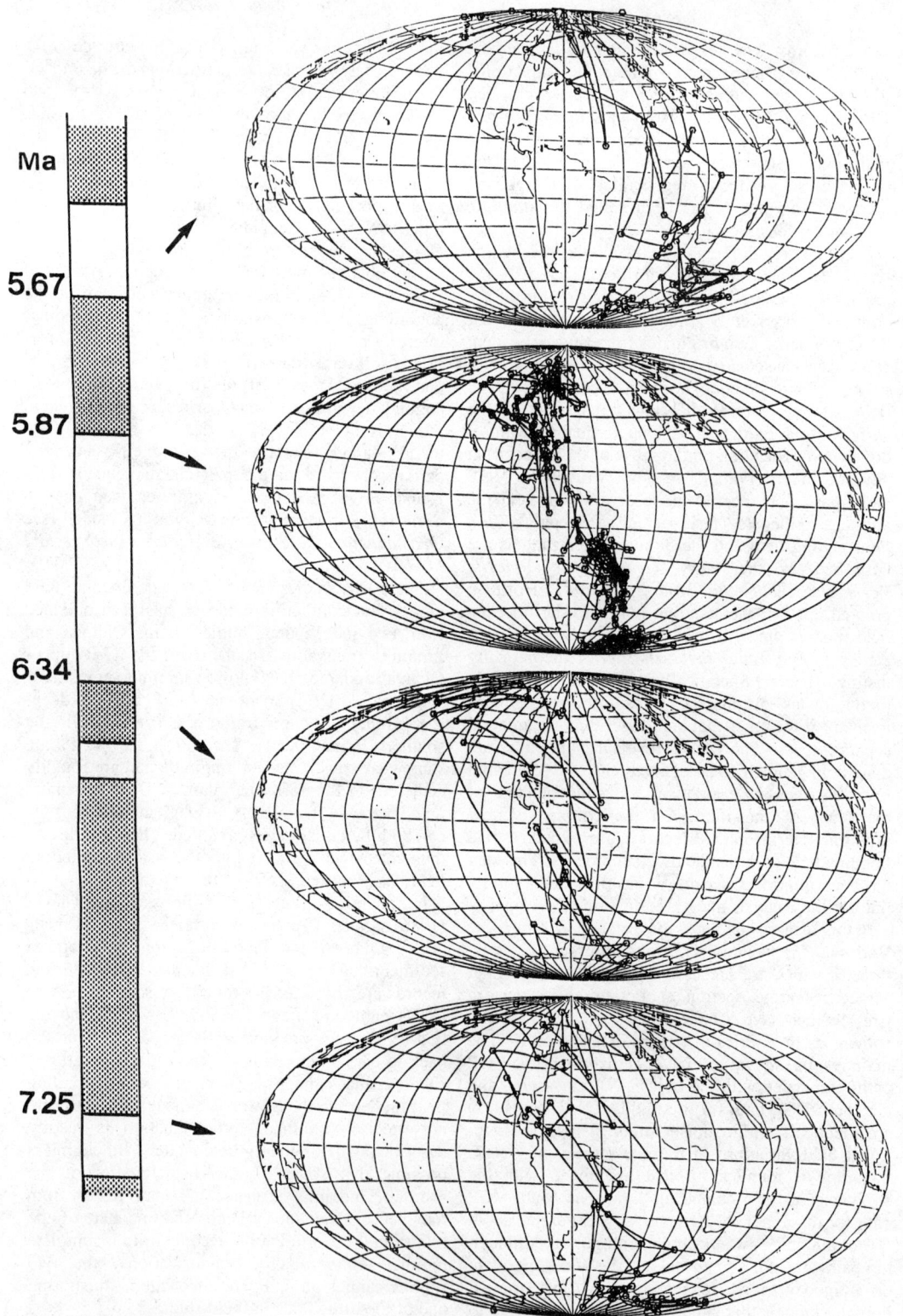

FIGURE 2. VGP paths of four polarity transitions in Crete (north VGPs for the R-N and south VGPs for the N-R transitions). The VGP paths of the three lower transitions are identical within noise limits, while the upper one differs by 60â. (After Valet and Laj, 1984)

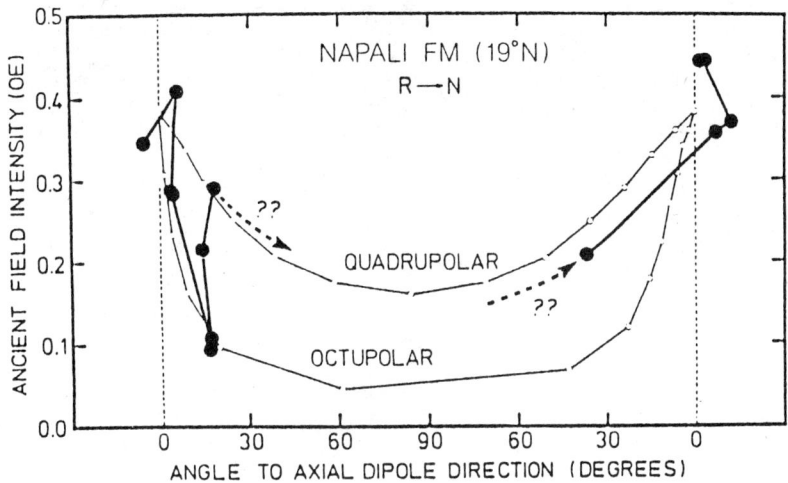

FIGURE 3. Comparison of paleointensity variations in the Napali (R-N) transition zone with that predicted by quadrupolar or octupolar zonal flooding. The initial decrease is consistent with an octupolar symmetry, the smooth recovery with an octupolar one. (From Bogue and Coe, 1984)

results favor the flooding model. Additional evidence in this sense comes from the observation, discussed later, that in many volcanic and sedimentary records the VGP paths undergo large directional deviations that require at least a superposition of a standing and a drifting component. Of course it cannot be excluded that the standing field process might be active in some cases.

The assumption that the nondipole field is basically unchanged during a polarity reversal is incorrect, as can easily be judged from the existence of the sequence of invariant transitional configurations observed in Crete. An additional and fundamental argument showing that the nature of the transitional field is basically different from the usual nondipole field is the occurrence of the rapid fluctuations of transitional field directions recently reported in both volcanic and sedimentary records. Furthermore, Merrill and McElhinny (1983) have shown that the nondipole components change sign when the axial dipole components reverse indicating that the nondipole field itself must undergo some transitional process.

The Geometry of the Transitional Field

In addition to the studies of the reversal mechanisms, the symmetry of the transitional field, which is basic to any interpretation of the reversal records, has been the object of many investigations. In examining all the records available at the time, Hoffman and M. Fuller pointed out in 1979 that the VGP paths of the (R-N) reversals from mid-latitudes in the Northern Hemisphere tend to be predominantly in the hemisphere centered on the site longitude while the (N-R) paths show a tendency to lie in the hemisphere opposite the site. This tendency is particularly marked when the five paths for the Brunhes-Matuyama reversal available at the time are considered (Fig. 4). If plotted as observed in geographical coordinates, no obvious correlations are observed among them (Fig. 4a). If, however, each path is plotted with respect to the site longitude (a plot called common site longitude) (Fig. 4b), the paths appear much more constrained in longitude, especially in the first half of the transition. This is strong evidence that axisymmetric terms dominate the harmonic contents of the transitional field.

This observation also led K. A. Williams and M. Fuller to develop in (1981) a mathematical model in which the dipole energy is partitioned among low order zonal terms during the reversal. This model predicts a relationship among the harmonic contents and the inclination record of a reversal and the latitude of the site of observation. With suitable partition coefficients it produced a synthetic record for inclination that closely matched the one observed for the Matuyama-Brunhes reversal.

Evidence for low order zonal transitional terms has been reported by several authors. Clement and Kent (1984) studied the Brunhes-Matuyama transition in several marine cores latitudinally distributed from 45°N to 33.5°S and obtained a plot of the duration of the reversal versus core site latitude, which is in good agreement with the prediction of the Williams and Fuller model assuming zonal octupolar flooding (Fig. 5). Bogue and Coe (1984) obtained a fit of the intensity data from Kauai assuming a flooding model whose exact symmetry, although axial, undergoes some changes through the reversal. Indeed the results (Fig. 3) fit neither the octupolar nor the quadrupolar models, but rather exhibit some characteristics of

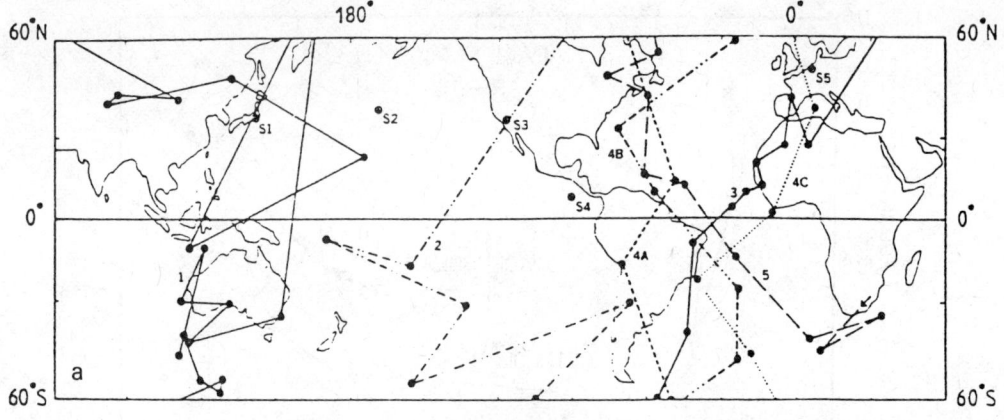
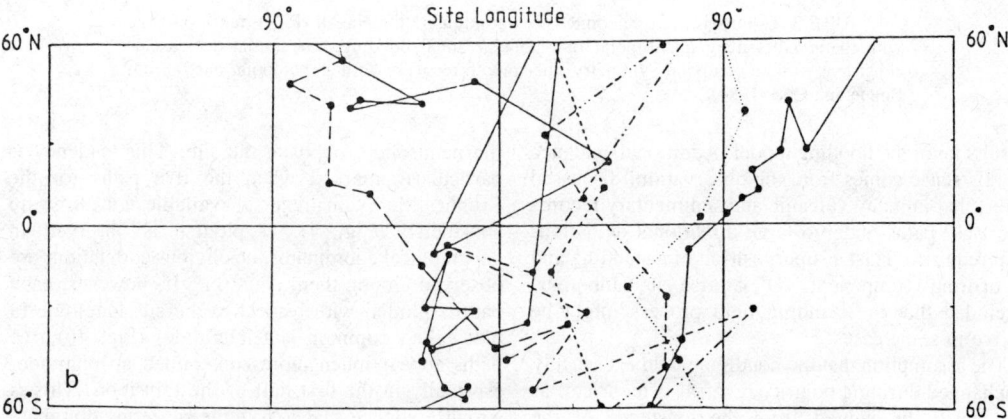

FIGURE 4. VGP paths for Brunhes-Matuyama transitions in (a) geographical and (b) common site longitude. The longitude confinement suggests an axisymmetric transitional field. 1. Boso Peninsula; 2. north central Pacific; 3. Lake Tecopa; 4. east equatorial Pacific; 5. Bruggen. (From Fuller et al., 1979)

both successively. The large intensity drop at the beginning of the transition can readily be interpreted in terms of zonal octupolar symmetry, while the smooth intensity recovery is more easily described by a zonal quadrupolar model. The rapid intensity recovery just after the initial drop may thus be interpreted as a brief return of the field to the initial reverse state after an aborted attempt to reverse its polarity by flooding with an octupolar symmetry.

Examination of other records, however, has shown that in most cases the transitional field is not purely axisymmetric, a situation that can be described by a nonsymmetric flooding model also developed by Hoffman in 1981.

The Cretan records for instance, are situated about 90° east or west of the site depending on the sense of the transition and thus cannot be reconciled with a simple zonal symmetry, which requires the paths to lie close to the site meridian or opposite to it. The large longitudinal confinement of these records is, however, significant of some degree of symmetry of the reversal process. Other evidences for asymmetry are found in many other sedimentary records.

A specific test using successive reversals, recently published by F. Theyer and colleagues, has shown that the addition of a small sectoral component is necessary to reconcile the flooding model with the records of the Olduvai and Jaramillo transitions. And the mentioned investigation of the four Olduvai and Jaramillo transitions has confirmed that large fluctuations exist in the directional records, so that the reversals appear to be controlled by both zonal and nonzonal terms with different temporal characteristics. For the Olduvai onset, for instance, the path is initially roughly 90°W of the site, then moves about 90°E and finally to a normal polarity along a near-sided path (Fig. 6). Also the path of the onset and the termination transitions, although longitudinally separated, are not antipodal.

Large asymmetries are also observed in almost all

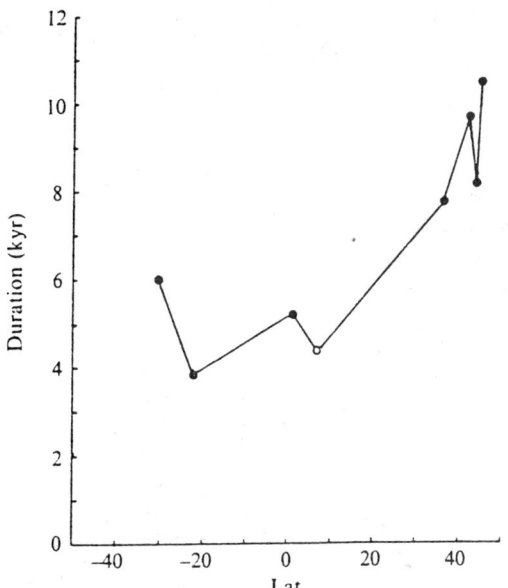

FIGURE 5. Latitudinal dependence of the duration of the Brunhes-Matuyama transition in different marine cores. (From Clement and Kent, 1984b)

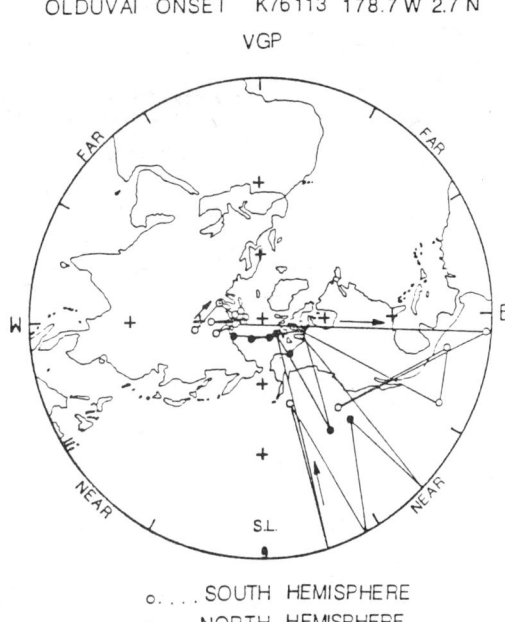

○.... SOUTH HEMISPHERE
●.... NORTH HEMISPHERE

FIGURE 6. VGP paths for the Olduvai onset transition in central Pacific showing largely nonaxisymmetric fields. (From Herrero-Bervera and Theyer, 1986)

the volcanic records and particularly in the very detailed record of a Miocene transition recently obtained at the Steen's Mountain by E. Mankinen, M. Prevot, and colleagues. The first part of the transition has a (R-N) path, which is somewhat confined in longitude, at least in the mid and low latitudes, about 150°–180° away from the site. The rest of the record is a normal-transitional-normal "rebound," in the authors' terminology, which clearly is not related to an axisymmetric transitional field (Fig. 7).

Volcanic reversal records do not appear to have the same properties of longitudinal confinement apparent on many sedimentary records. Some of the axisymmetric characteristics of the latter might result largely from the process of acquisition of magnetization by a sedimentary rock. A volcanic record is a series of spot readings of the vector field. Sediments, on the other hand, provide records that are naturally smoothed by the process of acquisition of magnetization. If the time constant associated with the locking-in of the magnetization is great compared to the time constant associated with the reversing field, the resulting record might be not only smoothed but also seriously altered.

It is certainly possible that in some of the published sedimentary records, smoothing has resulted in a simple structure of the paths not related to the changing field directions. However in many cases the records contain well-established fine-scale structures that definitely prove that the sediments have recorded fast transient properties of the transitional field, and yet the records are largely constrained in longitude: the Cretan records (Fig. 2) are in this respect a good example. The differences between volcanic and sedimentary records do not, in my opinion, all result from a simple smoothing present in the sedimentary records. Records of the same transition obtained from the two lithologies will certainly be of help in this matter, as well as records from high-deposition rate sedimentary formation, which should allow time resolution in a range similar to volcanic rocks and eventually lead to a reconciliation of the sedimentary and volcanic records.

Intensity of the Field during Polarity Transitions

The accurate determination of the paleofield intensity is quite delicate and time consuming from an experimental point of view. These difficulties have resulted in a limited number of reliable intensity records, although such records of the intensity changes during a transition are extremely important for the understanding of the reversal mechanisms of reversal of the Earth's dynamo.

Absolute paleointensities can only be obtained from volcanic rocks. However many of the published intensity records have been obtained from sedimentary sequences and sometimes from intrusions. Intensity records derived from sediments not only depend on the many spurious factors that may seriously obliterate the magnetic record but also

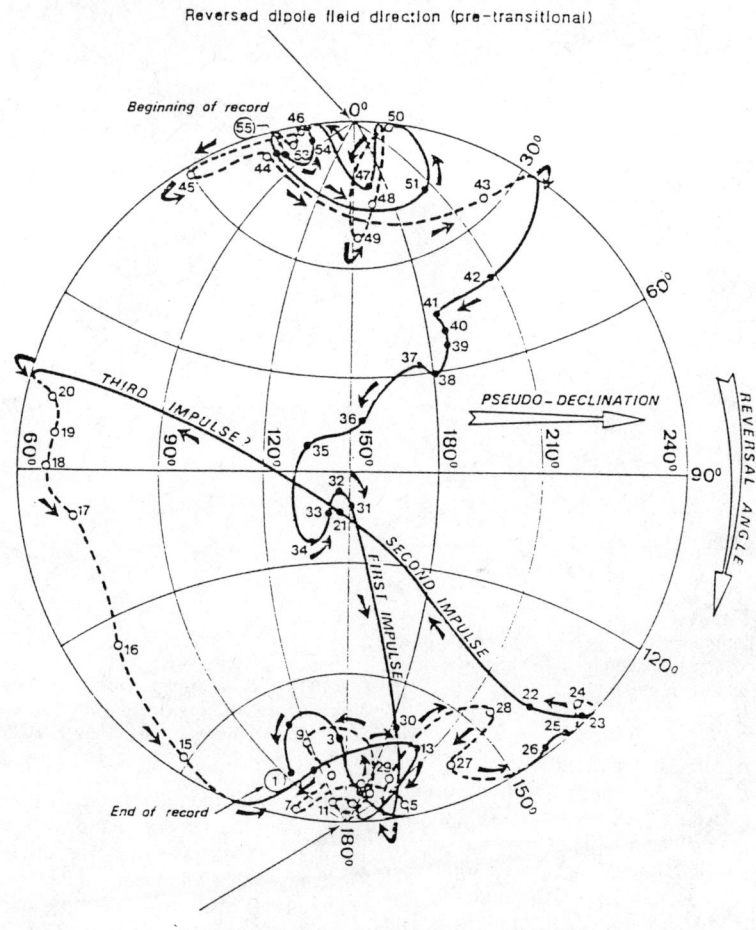

FIGURE 7. VGP paths for the Steen's Mountain Miocene reversal. (See Prévot et al., 1984 for details.)

necessarily demand a normalizing factor. Methods have been developed to deal with these difficulties and to obtain reliable records after selection of suitable sequences. I, among others, am convinced of the overall accuracy of these methods and thus of the reliability of some of the published sedimentary records. Other authors, however, consider that probably only very approximate determinations of paleofield behavior can be obtained from sedimentary sequences. The question is open to debate. Records obtained from intrusions are in many respects similar to those obtained from sediments, although the primary magnetization is a thermoremanence, because individual samples record a time-averaged field. Difficulties in interpretation arise from uncertainties regarding the time interval during which the magnetization was acquired, which can only be estimated from considerations of the shape of the body and its cooling history.

A great majority of the records suggest an overall reduction of field strength down to 10–30% of its nontransitional value. Only occasionally and in early records such a decrease has not been observed.

As early as 1973, Opdyke and colleagues (Opdyke et al. 1973) (Fig. 8) reported a record of the Lower Jaramillo transition in which the period of reduced magnetization intensity coincides with that for the directional changes (no normalization was attempted at the epoch). In other cases, the Tatoosh intrusion or the records from Crete for instance, the intensity decrease appears to last longer than the change in direction. The intensity, declination, and inclination records of the Tatoosh intrusion obtained by Dodson and colleagues are reported in Fig. 9. The results from Kauai discussed above similarly indicate that the drop in intensity lasts longer than the directional change and also that the collapse and recovery phases may not be symmetric.

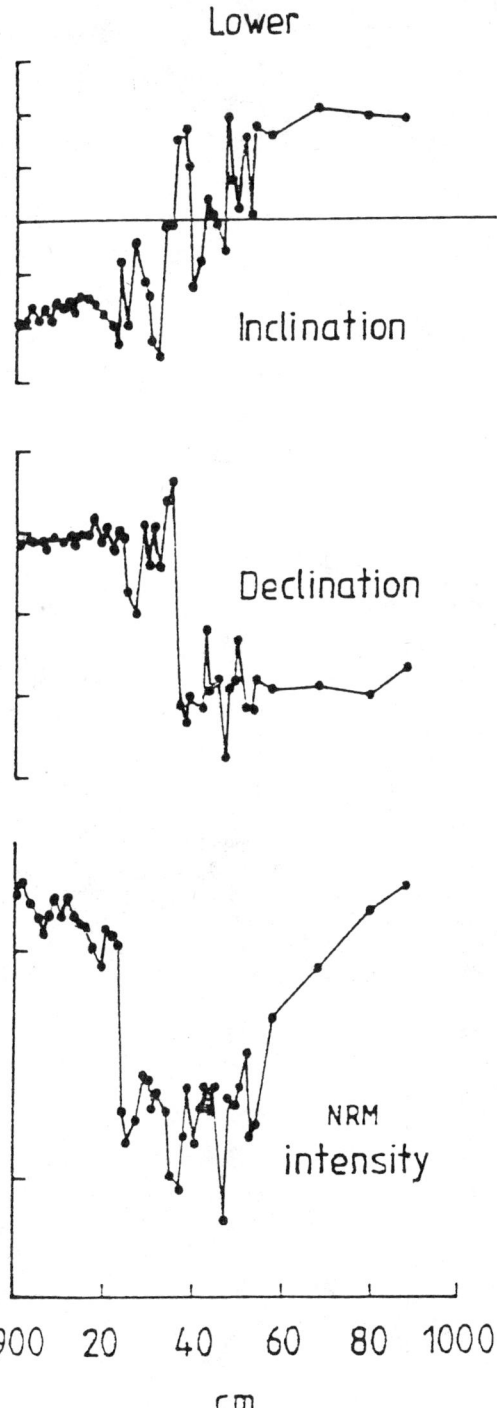

FIGURE 8. Record of the inclination, declination, and intensity for the lower Jaramillo transition. The decrease of the intensity of magnetization of the sediments appears to coincide with the directional changes. (From Opdyke et al., 1973)

In 1975, J. Shaw reported evidence for a significant increase of field intensity in the middle of the transition, while the VGP is close to the equator (Fig. 10). Interestingly, the suggestion of strong intermediate intensities had already been made by R. L. Wilson in 1971, as a result of statistical analysis of over 3500 specimens from 1500 flows. The presence of a high intensity episode during a transition has also been reported in the Steen's Mountain reversal record. In this case, however, the episode of high intensity values occurs in the period of time that separates the two phases of the directional history, during which a normal dipole direction has been reestablished. This period thus corresponds to a temporary reestablishment of a nontransitional dipole configuration. The intensity decreases again during the following normal-transitional-normal "rebound" and finally attains a value higher than the pre-transitional one, but not significantly different from the expected average Miocene field. The significantly lower pretransitional intensity most probably results from a long-term decrease in dipole strength before the reversal, while the large intensity fluctuations observed after the transition could reflect instabilities of the reestablished dipole configuration.

A completely independent evidence of field intensity decrease associated with the Brunhes-Matuyama reversal has been obtained from measurements of the abundance of cosmogenic ^{10}Be in marine sedimentary cores obtained by G. Raisbeck and colleagues. The abundance of ^{10}Be is expected to increase by a factor of two to three as a result of the reduced field intensity, which allows a significantly increased flux of high-energy particles to reach the Earth's atmosphere. The measurements, which were made using a Tandetron accelerator mass spectrometer facility, show that the ^{10}Be concentration increases by a factor of 2 for transitional samples. Interestingly, the interval of increased ^{10}Be is significantly longer than the time taken for the reversal in inclination, suggesting that the intensity low might last longer than the directional changes. Also, a well-marked minimum is present in the ^{10}Be profile, implying a significant decrease in production. Although the magnitude and duration of this minimum might be largely controlled by bioturbation of the sediment, this phenomenon is reminiscent of the previously mentioned evidence of high intensity occurring midway through a reversal. However, for the Brunhes-Matuyama transition no effect of this kind has been reported so far in a paleomagnetic study.

Recent Developments

Recent developments in the study of geomagnetic reversals are characterized not only by very accurate and complete records but also by the fact that particular characteristics of the records have been directly

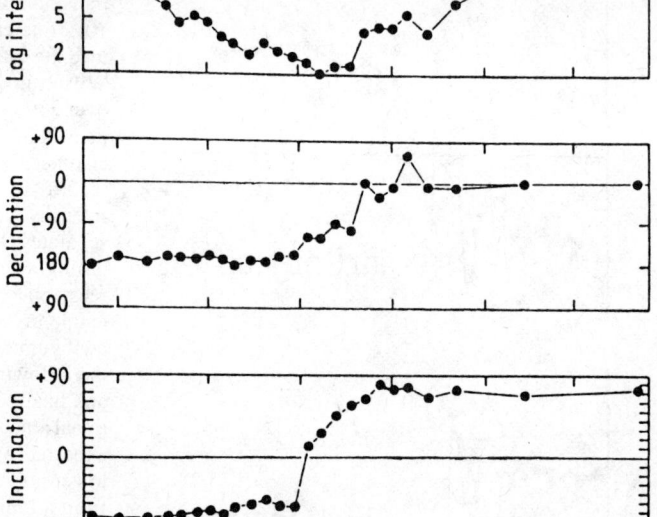

FIGURE 9. Records of the inclination, declination, and intensity for the Tatoosh intrusion reversal plotted against perpendicular distance from the reversal isothermal plane. The intensity decrease appears to last longer than the directional changes. (From Dodson et al., 1978)

related to particular processes in the core. In this respect the full vector record of the Steens Mountain reversal is quite significant of this new trend in reversal studies (Fig. 11). Although its time resolution, based on the assumption of continuous and constant rate of extrusion, which is difficult to assess on a time scale of less than 1000 years, might not be as precise as claimed by the authors, this record yields very convincing evidence of very rapid vectorial field changes during the reversal. These impulses have been interpreted in terms of transport of field lines, under frozen flux conditions, and upwelling of the toroidal field. This interpretation implies a significant increase in the level of turbulence within the core during a reversal.

In a recent investigation of a reversal from Crete, Valet and colleagues also obtained evidence of a significant increase of rapid field direction fluctuations, which were interpreted in terms of increased core turbulence, as in the Steens Mountain record. On another time scale, long term oscillations of field direction are present with the same time constant both before and through the reversal (Fig. 12), suggesting that the secular variation, at least in its

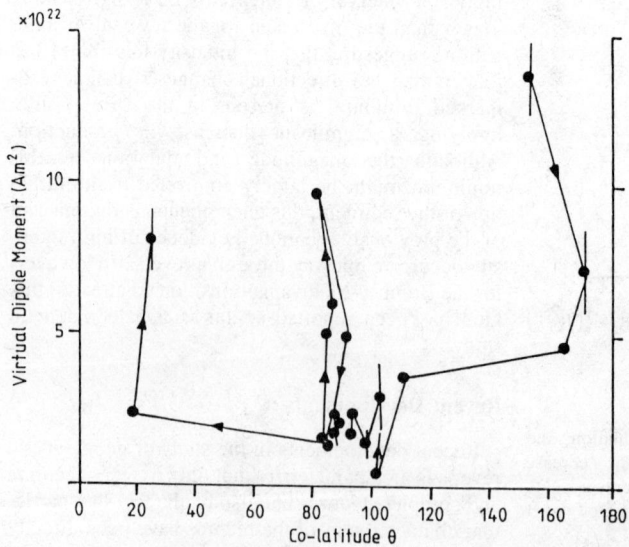

FIGURE 10. Values of the virtual dipole moment (vdm) against colatitude during an R-N reversal in Iceland. The strong increase in vdm occurs for equatorial VGPs. (From Shaw, 1975)

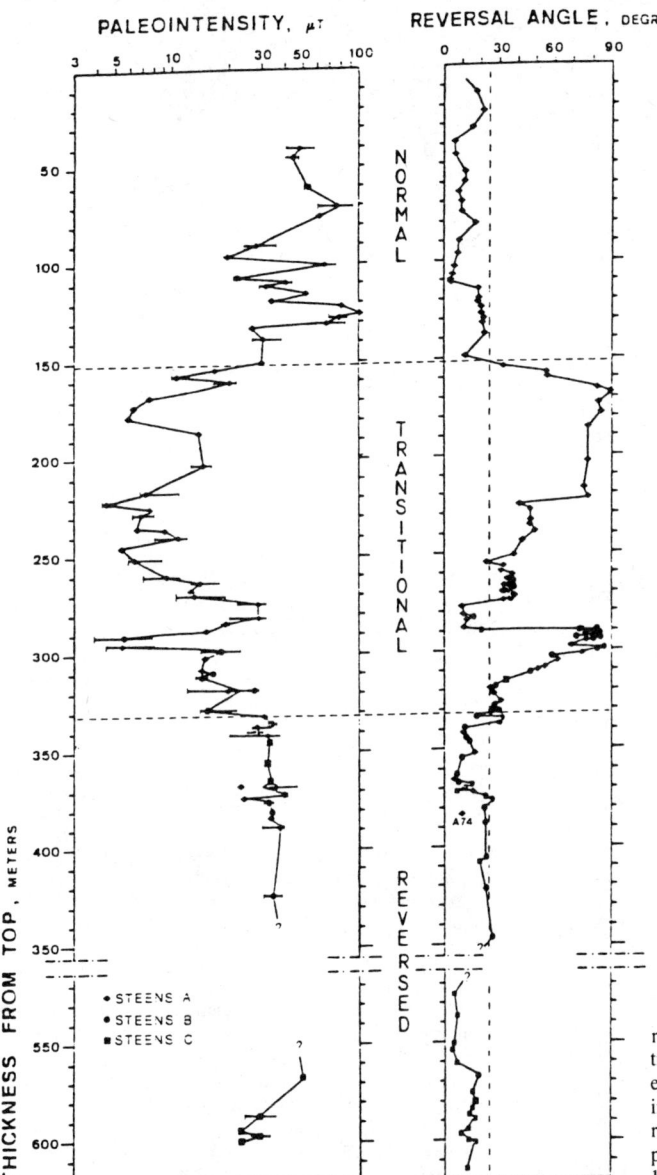

FIGURE 11. Plot of the paleointensity and reversal angle for the Steen's Mountain transition. The large intensity recovery at the end of the first phase is evident. Large intensity fluctuations are apparent after the reversal, while low values are present in the pretransitional period. (From Prévot et al., 1984)

longer period terms, is not affected by the reversal. When examined in the light of a recent theoretical prediction this observation implies that the frozen flux approximation might be valid over periods of time of the order of the duration of a reversal.

Finally, records from Australian basalts reported by Hoffman support the hypothesis of an intermediate field state from which both successful and unsuccessful reversal may originate. They also show that several aborted attempts involving the same intermediate field geometry may precede a successful polarity transition.

Thus, the records of polarity reversals of the geomagnetic field obtained in the last years have provided important insight into the mechanisms of the Earth's dynamo. The first big step has been the recognition that the transitional field is not dipolar. Then attempts were made to characterize an average field configuration and possible reversal mechanisms. In some cases evidence for axial symmetry was obtained but nonaxisymmetric terms are present in most of the very detailed records. The most recent records are interpreted in terms of precise processes in the core and, in turn, theoretical papers have tried

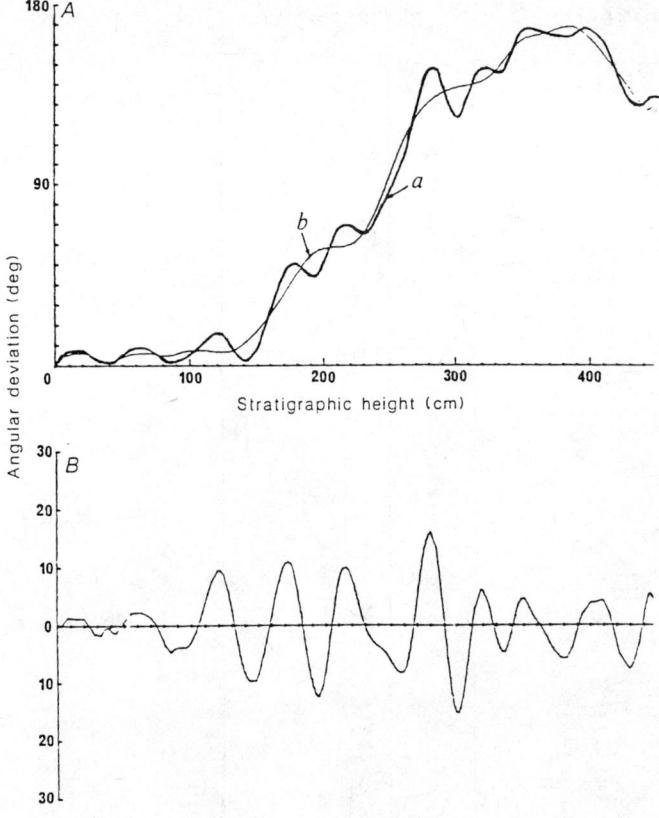

FIGURE 12. A. Angular deviations from the prereversal directions plotted against stratigraphic height (a) after smoothing with cubic splines 27 knots, (b) same with 15 knots. B. Periodic oscillations obtained by subtracting curve (b) from curve (a). (From Valet et al., 1986)

to explain particular features of these records. The study of reversal has thus evolved in a short time from a period in which observations and theory were completely separated to the present situation where progress is to be expected from their interaction.

CARLO LAJ

References

Bogue, S. W., and R. S. Coe, 1984, Transitional paleointensities from Kauai, Hawaii and geomagnetic reversal models, *Jour. Geophys. Research* **89(B2)**, 1049-1055.

Clement, B. M., and D. V. Kent, 1984, Latitudinal dependency of geomagnetic polarity transitional durations, *Nature* **310**, 488-491.

Cox, A., R. R. Doell, and G. R. Dalrymple, 1964, Reversals of the Earth's magnetic field, *Science* **144**, 1537-1543.

Dodson, R., J. R. Dunn, M. Fuller, I. Williams, H. Ito, V. A. Schmidt, and M. Wu Yu, 1978, Paleomagnetic record of a late tertiary field reversal, *Geophys. Jour.* **53**, 373.

Fuller, M., I. Williams, K. A. Hoffman, 1979, Paleomagnetic records of geomagnetic reversals and the morphology of the transitional field, *Rev. Geophysics Space Physics* **17**, 172-203.

Herrero-Bervera, E., and F. Theyer, 1986, Non axisymetric behaviour of Olduvai and Jaramillo polarity transition recorded in north-central Pacific deep-sea sediments, *Nature* **322**, 159-162.

Hillhouse, J. W., and A. Cox, 1976, Brunhes-Matuyama polarity transition, *Earth and Planetary Sci. Letters* **29**, 51-64.

Jacobs, J. A., 1984, *Reversals of the Earth's Magnetic Field*. Bristol: Hilger.

Merrill, R. T., and M. W. McElhinny, 1983, *The Earth's Magnetic Field*. London: Academic Press.

Opdyke, N. D., D. V. Kent, and W. Lowrie, 1973, Details of magnetic polarity transitions recorder in a high deposition rate deep-sea core, *Earth and Planetary Sci. Letters* **20**, 315-324.

Prevot, M., E. A. Mankinen, C. S. Gromme, and R. S. Coe, 1984, The Steens Mountain polarity transition II. Field intensity variations and discussion of reversal models, *Jour. Geophys. Research* **89**, 6285-6292.

Shaw, J., 1975, Strong geomagnetic fields during a single Icelandic polarity transition, *Royal Astron. Soc. Geophys. Jour.* **40**, 345-350.

Valet, J-P., and C. Laj, 1984, Invariant and changing transitional field configuration in a sequence of geomagnetic reversals, *Nature* **311**, 552-555.

Valet, J-P., C. Laj, and P. Tucholka, 1986, High resolution sedimentary record of a geomagnetic reversal, *Nature* **322**, 27-32.

Cross-references: *Geomagnetic Polarity Reversals: Theory and Models; Geomagnetic Reversal Sequence: Statistical Structure; Magnetostratigraphy; Paleomagnetic Field: Intensity; Paleomagnetism and Plate Tectonics; Seafloor Spreading: Magnetic Evidence.*

GEOMAGNETIC POLARITY REVERSALS: THEORY AND MODELS

It has long been observed in the paleomagnetic record obtained from rocks that the Earth's field, which is generated in the fluid-conducting outer core, possesses two possible polarity states: normal and reverse. Such a situation brings forth the intriguing question as to how the magnetic field manages to change from one polarity to the other. Over the past quarter century, several ideas and models have been advanced regarding the actual mechanism. A review of the variety of approaches to geomagnetic reversal found in the literature is presented here. Comprehensive works on our current knowledge about polarity transitions may also be found in Merrill and McElhinny (1983) and Jacobs (1984).

Predictions from Theory

Since, in the final analysis, any relevant and complete model of the geomagnetic reversal process must be consistent with hydromagnetic theory, we start with a brief discussion along theoretical grounds. First, two polarity states of the geomagnetic field are expected. Indeed, the hydromagnetic equation

$$\frac{\partial \mathbf{H}}{\partial t} = k_m \nabla^2 \mathbf{H} + \nabla \times (\mathbf{v} \times \mathbf{H}) \quad (1)$$

which governs changes in the vector magnetic field $\mathbf{H}$ associated with a fluid of magnetic diffusivity k_m and velocity field $\mathbf{v}$, is "blind" to the sign of $\mathbf{H}$. That is, either sign of the magnetic field can be associated with a given fluid-mechanical state.

Further, it is seen in Eq. 1 that the time rate of change in $\mathbf{H}$ is due to two terms. Insight regarding the physical nature of each term comes from considering extreme, or limiting, cases. For example, forcing the velocity field $\mathbf{v}$ to zero eliminates the second term on the right, leaving the vector diffusion equation. Given the absence of fluid motion, then, the magnetic field will decay in a manner consistent with this now simplified equation. In contrast, the second term on the right describes the interaction of the velocity field with the magnetic field. For the extreme case of infinite conductivity (i.e., $k_m = 0$), the first term on the right is eliminated. Analysis of the resulting equation indicates that magnetic flux lines need to be "frozen" into the fluid, are thus constrained to move precisely with it. Moreover, they may be stretched in so doing. Locally in the core such "field transport" can either enhance or degrade the magnetic field strength.

Since the mechanism of magnetic field reversal must be consistent with Eq. 1, it must involve a process in the fluid outer core that incorporates flux diffusion and/or flux transport. However, since the extreme cases just considered are not strictly relevant to the fluid outer core, both phenomena to some degree must be at work. With this in mind, we now review models of geomagnetic reversal. Several approaches will be considered in this entry: theoretical, phenomenological, statistical, and analogical. Ultimately, a given model cannot be considered relevant unless it can somehow be tested. In this regard, each model discussed need be associated with particular predictions regarding geomagnetic field behavior. Whenever possible, such predictions will be addressed in light of available paleomagnetic transition data.

Approaches to Geomagnetic Reversal

Electromechanical Analogs. One approach to modeling the geomagnetic reversal process involves the investigation of electromechanical analogs to the geodynamo whose workings can be quantitatively analyzed by way of simple physics. The first such study was made by T. Rikitaki in 1958, who investigated the characteristics of two identical, coupled-disk (homopolar) dynamos (Fig. 1) and found the system to have the capacity to undergo reversal of the induced magnetic field. Each electrically conducting disk, when rotated by an applied torque in an external magnetic field, produces a current that passes along a wire wound around the shaft (see Fig. 1). Each winding is then a source of a magnetic field. Under certain conditions, the current output of this two-disk system can be made to oscillate unstably, ultimately changing sign spontaneously and, hence, causing reversal of the induced magnetic field. Analyses of this, as well as other electromechanical systems (see Jacobs, 1984), have shown field reversals to be a common characteristic; however, the manner in which reversal is achieved is often not obviously compatible with paleomagnetic observation. Indeed, the relevance of each component constituting such systems as an analog to the fluid dynamical nature of the Earth's core is not obvious.

Theoretical and Phenomenological Models

Free Decay. A simple class of reversal models is one in which the initial stage of the process is

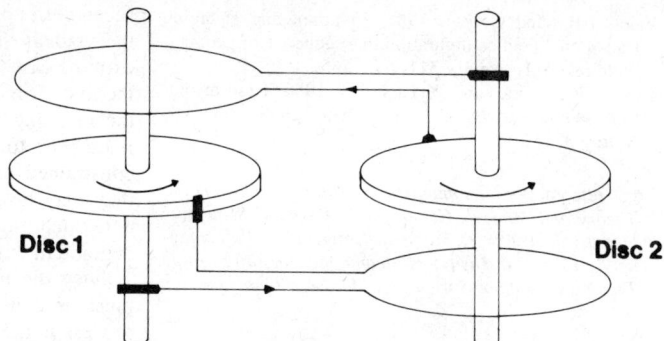

FIGURE 1. Two coupled disk (homopolar) dynamos, an electromechanical system whose output can show reversals. (After Rikitaki, 1958)

dominated by diffusive decay of the magnetic field. Such an approach is based on Cowling's theorem (1934), which points out that poloidal magnetic fields symmetric about a given axis cannot be maintained by fluid motion that happens to be symmetric about the same axis. Specifically, during times when the fluid core's velocity structure becomes largely symmetric about the rotation axis, T. Nagata and later F. E. M. Lilley have argued that the geomagnetic field would then decay to zero. This suggestion makes several implicit predictions regarding paleomagnetically observed transitional field behavior. With regard to field geometry, a dipolar configuration is expected to be maintained, indeed enhanced, while the field freely decays. This follows from the fact that the time constant associated with free diffusive decay of a given field component depends on its characteristic wavelength; the dipole term thus possesses the longest duration. Hence, a paleomagnetically recorded free decay transition at any site should show a field direction that becomes increasingly closer to that expected from an axial dipole source as the intensity decreases toward zero. However, although paleomagnetic transition records often do show an initial intensity decrease prior to any significant directional change, movement of the field vector is always observed to occur while the field strength is still far from zero. Also inconsistent is our best estimate of the time scale associated with free decay of the dipole field. On the order of 10,000 yr, this value appears to be too long to account for reversals. Indeed, our best estimates for the duration of *complete* reversal events from paleomagnetic records span several thousands of years, yet typically do not exceed this estimate of the free decay time. However, the determination of the free decay time constant for the core field is sensitive to the value of conductivity of the fluid, a value not precisely known.

Consistent with the free decay approach to reversals is the observation that reversing fields sometimes appear to be more complex during the final phase of the process than during the initial phase. For the case of an assumed free decay, full polarity may only be restored after a velocity structure in the core is developed that is sufficiently asymmetric about the rotation axis to again generate a magnetic field. Such an asymmetric velocity field must produce, at least initially, an asymmetric magnetic field configuration.

It may further be argued that subsequent to a free decay event either polarity sense may be generated. This contention that not all reversal attempts are successful is consistent with the observation of geomagnetic excursions or apparent aborted field reversals.

Cox's Probabilistic Model. In an attempt to explain the stochastic nature of reversals, Cox (1968) proposed that polarity transitions may occur when nondipole fields of generally opposite sign to the dipole term momentarily dominate the most significant source regions within the core (Fig. 2). This situation, Cox argued, may occur when the intensity of the dipole, assumed to vary in a regular, oscillatory manner, is low and the intensity of the nondipole field, assumed to randomly fluctuate with changes in the distribution of cyclonic convection cells, is high. At the time of such a reversal-causing instability, the newly predominant "reverse" flux is stretched and transported through the core while the earlier transported flux (of opposite sign) decays by diffusion.

Dipolar Reversals. The notion that polarity reversal may be characterized by a transition field that remains dipolar throughout the process has been explored. Examples of proposed models envision changes of polarity to be either the result of the interplay of discrete, geocentric dipole sources, which have the ability to reverse independently, or the result of simple rotation of the main dipole source. Such models neither are based on strict hydromagnetic theory nor argued from the standpoint of the nature of the core. In general, dipolar models of geomagnetic reversal were constructed as a first attempt to interpret particular lone records of paleomagnetic transitions that appeared to be associated with rather smoothly varying directional field behavior.

Other models to be discussed here, which attempt to simulate paleomagnetic observations of the

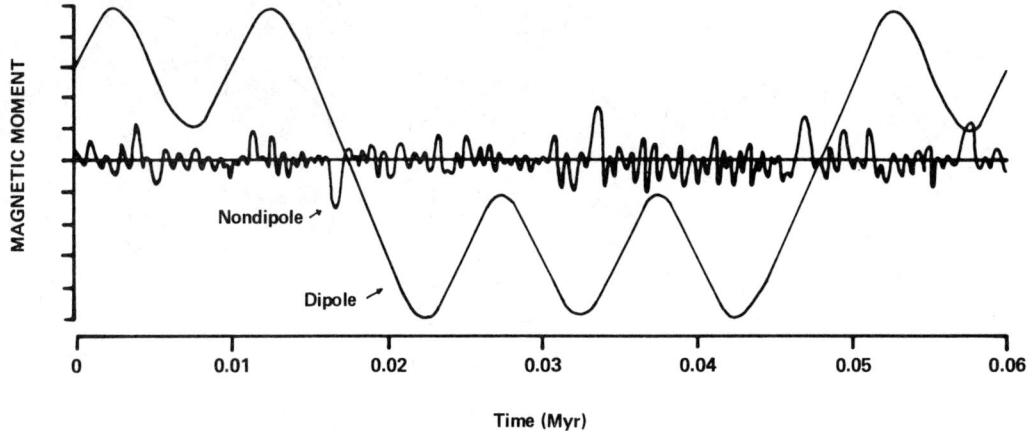

FIGURE 2. Probabilistic reversal model in which the axial dipole field undergoes periodic variations in strength between strong and weak values while more rapid, less regular variations in the nondipole field take place. Reversal occurs when an opposing nondipole field possesses a characteristic strength larger than the dipole. (After Cox, 1968)

reversal process, are based on the contention that transition fields are largely nondipolar. At present, available data clearly favor this particular view. Indeed, it is possible to test for dipole versus nondipole dominance of transitional fields, provided that multiple paleomagnetic records are available from widely distant sites. Such is accomplished by comparing the paths of the virtual geomagnetic pole (VGP) (see *Geomagnetic Polarity Reversals: Observations*) as each migrates from one polarity to the other. If a given transition field is controlled by the dipole term, then all sites over the globe will experience directional field changes associated with essentially the same VGP path. If, on the other hand, the VGP paths from different recording sites are largely dissimilar, it can be concluded that the transition field must have been more complex and of a higher order than dipole. In 1976, J. Hillhouse and A. Cox applied this test to two reversal records associated with the Matuyama-Brunhes transition, one obtained from Japan and the other from California, and found the VGP paths to be widely dissimilar. These authors concluded that the Matuyama-Brunhes transition field must have been dominated by nondipole components. Additional records of this reversal from sites scattered over the Northern Hemisphere (see Fig. 4a in *Geomagnetic Polarity Reversals: Observations*) further support this basic, but significant, observation.

Standing Field Model. Hillhouse and Cox (1976) proposed that the smoothly varying transitional field behavior sometimes seen in the paleomagnetic record may be due to a process by which the dipole field, along with a fraction of the nondipole field, decays through zero and then regenerates in the opposite orientation. The nondipole fraction that does not take part in this process is assumed by Hillhouse and Cox to remain invariant, thus providing a higher-order and more complex field geometry midway through such a reversal. This suggestion is paleomagnetically testable, provided that the nonreversing portion of the nondipole field remains basically unchanged for a time sufficient to span successive transitions. If so, then sequential paths of the VGP associated with one transition sense would be expected to retrace those paths associated with the other transition sense.

The Parker-Levy Approach: Inspiration for "Flooding" Models. Recall that for a free decay reversal, the process of flux transport is significant only during the final, or regenerative, phase. In contrast, the theoretical approach to reversals proposed by Parker (1969) and further elaborated by Levy (1972) involves significant flux transport early in the process and significant diffusive effects during the final phase. As with Cox's model, the Parker-Levy approach is based on the assumption that both nonuniform rotation (velocity shear) as well as cyclonic convection are significant aspects of the core's fluid motions. The interaction of cyclones with the existing toroidal magnetic field produces poloidal field components through the Coriolis effect. According to Parker and Levy, it is the precise introduction and removal of bursts of cyclonic activity at distinct latitudinal bands within the core that makes polarity reversal achievable. Following the generation of a "reverse" magnetic field within some latitudinally defined region, flux diffusion causes the reversal to spread, or "flood," axisymmetrically throughout the core, thus completing the polarity transition (Fig. 3).

The Parker-Levy models make quite definite predictions concerning the configurational characteristics of transition fields. In 1977, K.A. Hoffman pointed out that Parker-Levy reversals are characterized by a transfer of magnetic energy from that of

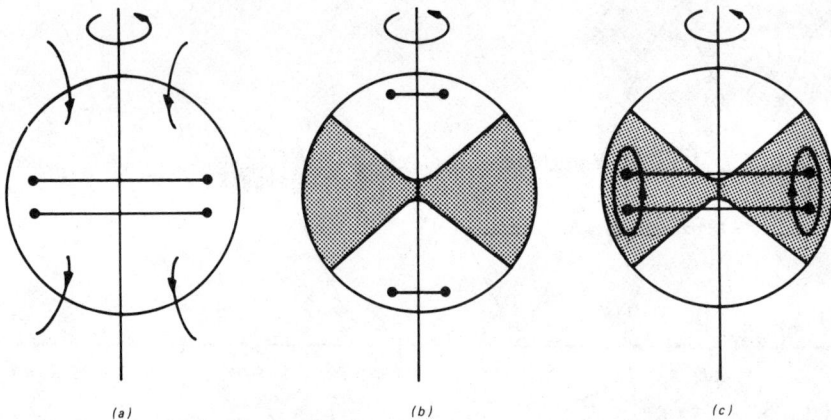

FIGURE 3. One possible scenario of core reversal in which the dipole field is assumed to be a consequence of cyclonic convection. In this case (a) the dipole is maintained by rings of cyclones at low latitudes. (b) A large burst of cyclones at high latitudes (again shown as rings) gives rise to opposing toroidal flux at low latitudes. (c) A return to low-latitude cyclonic conditions then twists the toroidal field into a poloidal field of opposite sign to the original dipole. (After Levy, 1972b)

the full polarity dipole to higher-order (i.e., nondipole), axisymmetric terms, and again back to the dipole. Thus, in contrast to simple free decay models, directional field movement during Parker-Levy polarity transitions at essentially any site on the Earth's surface must be dominated by swings in the north-south meridional plane (i.e., zonal field behavior). Returning for a moment to the free decay approach, R. Hide argued in 1981 for zonal field behavior from the standpoint of Cowling's theorem. However, due to the characteristic time dependence of diffusive decay on the geometrical complexity of a given field component, axisymmetric terms of order higher than dipole are simply not expected to be significant during free decay.

Hoffman (1977) tested existing transitional paleomagnetic data against both the Nagata-Lilley approach as well as the Parker-Levy models. What was found was that the majority of reversal records—which happened to correspond to reverse-to-normal (R-N) transitions and which were obtained from northern latitudes—were associated with VGP paths that reside on the hemisphere of the globe centered about the longitude of the recording site (the so-called near side). Such an observation, Hoffman argued, is consistent with transition fields dominated by axisymmetric terms and with the predictions of the Parker-Levy approach. Hoffman and Fuller (1978) then categorized predicted VGP paths in terms of being either near-sided or far-sided for sites in both hemispheres (North and South) and for both transition senses (R-N and N-R) associated with flooding-type reversals that initiated at high and at low core latitudes (Fig. 4). What Hoffman and Fuller showed is that the location in the core of the onset of axisymmetric reversals (i.e., those initially associated with a particular core latitudinal band) is paleomagnetically distinguishable, provided that records were available from distant sites in both hemispheres.

The Matuyama-Brunhes reversal, for which the largest number of records is available, indicates transitional field characteristics consistent with a Parker-Levy flooding reversal process that initiated either at equatorial or southern latitudes of the core but not at northern latitudes. Williams and Fuller (1981) expanded on the axisymmetric approach to field reversals by partitioning the magnetic energy of the main (dipolar) field into the three lowest-order zonal components (g_2^0, g_3^0, g_4^0) during the transition and then into a main field of opposite polarity. By varying the respective ratios of these energy terms, Williams and Fuller were able to effectively simulate several existing time logs of inclination for the Matuyama-Brunhes reversal (Fig. 5). The Williams-Fuller approach points out the strong relationships among the zonal geometric content of the transition field, the location of the site of observation on the earth's surface, and the temporal paleomagnetic signature of inclination.

S. W. Bogue and R. S. Coe (1984) also have expanded on the early axisymmetric flooding approach by constructing a spherical magnetic source region in the core. Specifically, the source region consists of thin latitudinal "disks" each of which are uniformly magnetized along the rotation axis. A polarity transition is accomplished first by the reversal of magnetization within one disk, and then the progressive reversal of adjacent disks to both the north and the south.

With the testing of these axisymmetric models came the realization that reversing fields do not appear to be purely axisymmetric. Indeed, nearly all records show some degree of nonzonal field behavior

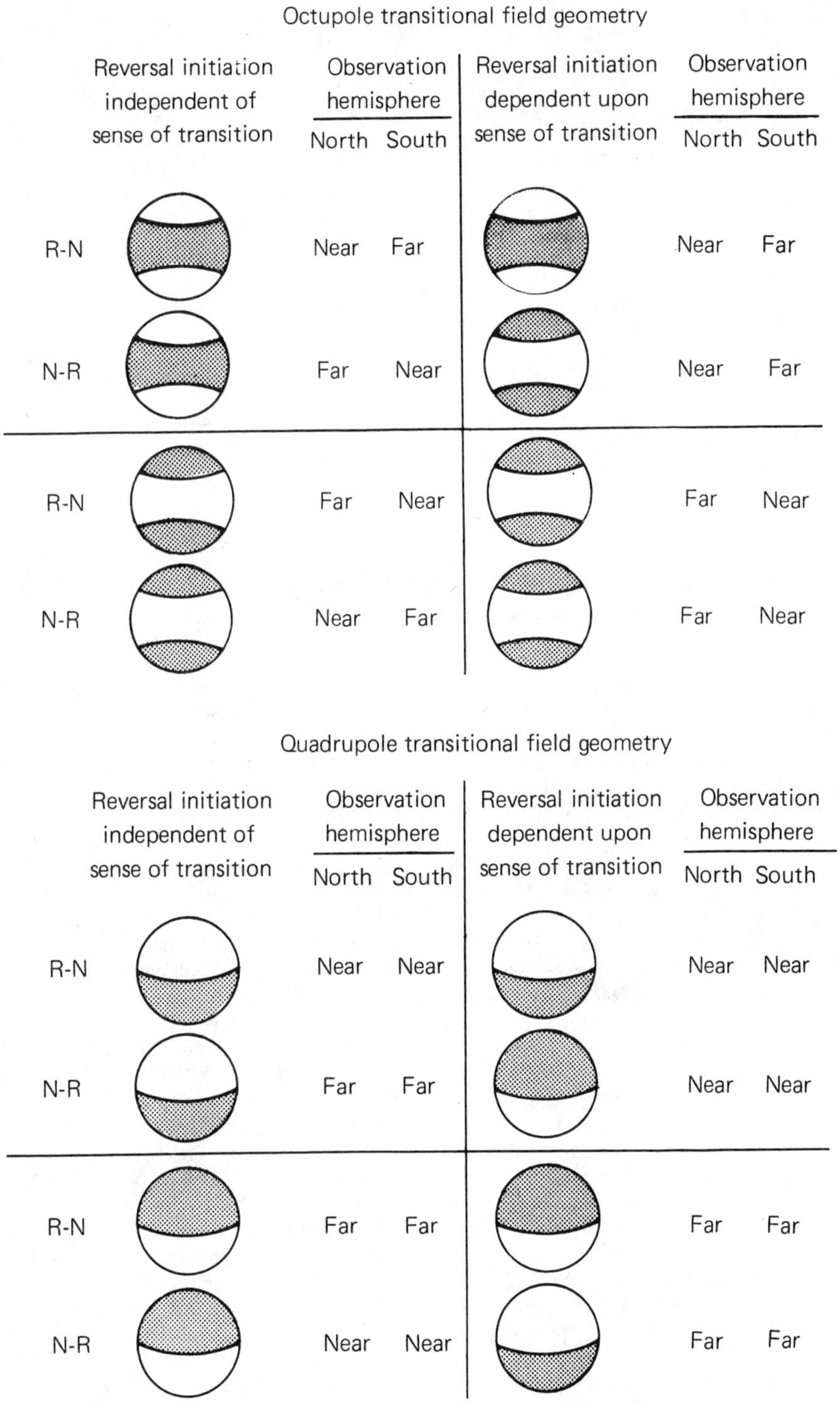

GEOMAGNETIC POLARITY REVERSALS: THEORY AND MODELS

Octupole transitional field geometry

Quadrupole transitional field geometry

FIGURE 4. Zonal flooding models and the predictions of paleomagnetic directional path behavior associated with each case. Reversal initiation (darkened regions) is considered at equatorial latitudes of the core (producing octupole transition fields) as well as at high latitudes of each hemisphere (producing quadrupole transition fields). Systematic reversal mechanisms that are dependent and independent of transition sense are seen to produce identical and antipodal directional paths, respectively, for successive reversals. (After Hoffman and Fuller, 1978)

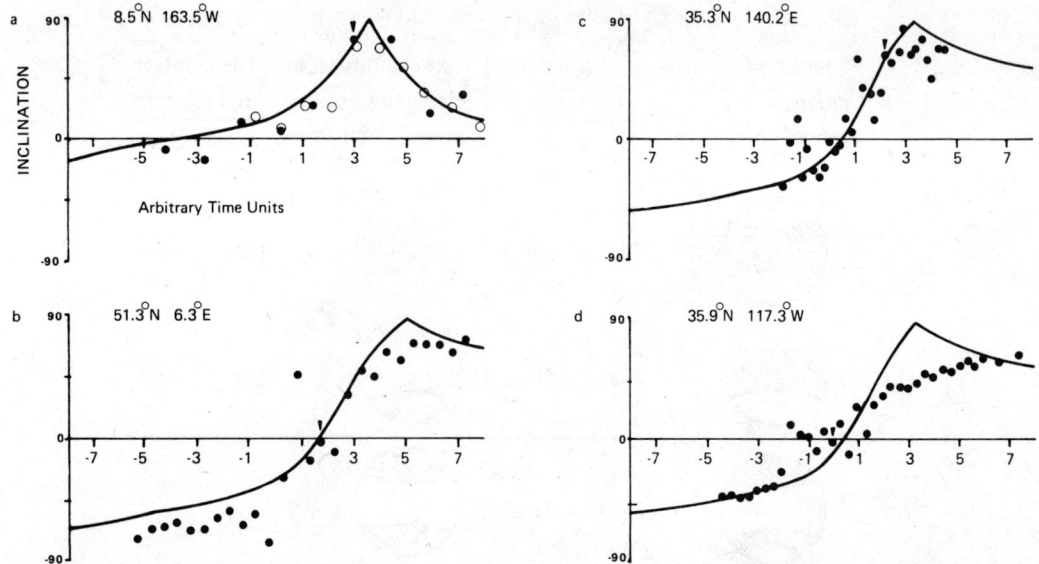

FIGURE 5. Inclination data and predictions of a zonal harmonic model for records of the Matuyama-Brunhes reverse-to-normal transition obtained from sedimentary sections at the indicated geographic coordinates. (After Williams and Fuller, 1981)

during transitions and some of them appear to be dominantly nonaxisymmetric. In 1979, Hoffman expanded the flooding approach to incorporate nonaxisymmetric terms to reversing fields. This phenomenological approach assumes a localized region within the core to be the initiation point for a given transition. Subsequent to the start of the process, both north-south and east-west flooding are assumed to take place until the magnetic flux emanating from the entire core is reversed. For the purpose of first-order modeling, a simple cylindrical surface composed of lines of dipoles is employed as the magnetic source region of the core. A. D. Kaiser and K. L. Verosub (1985) have since attempted to construct a more realistic source region, being spherical and incorporating latitudinal rings of radial dipoles. Both nonaxisymmetric models have been able to roughly simulate directional behavior associated with several records of the Matuyama-Brunhes reversal. Figure 6 summarizes the evolution of flooding models as has been applied to polarity transitions.

Frozen-flux Reversals. The possibility that reversals are events of sufficiently short duration to be able to approximate the process by neglecting diffusive effects was suggested in 1982 by E. R. Benton and explored by Gubbins and Roberts (1983). If this is a valid approximation, then the problem of field reversal reduces to one of fluid dynamics in which magnetic flux is transported with the fluid. However, field behavior during a frozen-flux reversal is strongly constrained. For example, the total magnetic flux (neglecting sign) leaving the surface of the core is conserved. More rigorously, the number of null flux contours on the core's surface (separating regions of oppositely signed emerging flux) is conserved, as is the amount of flux that emanates from each defined region. Such frozen-flux constraints, Gubbins and Roberts (1983) argue, make a purely axisymmetric reversal impossible. These authors also point out, however, that it is unlikely that the conductivity of the core is sufficiently large to confidently eliminate the need to consider diffusive effects throughout an entire reversal.

A point of controversy that has recently arisen is whether the paleomagnetic record contains evidence that the frozen-flux approximation is valid during portions of polarity transitions. A case in point is the observation of what may be rapid swings in field direction during at least some reversals for which records are available. Prévot et al. (1985) argue that the presence of these rapid "geomagnetic impulses" indicates that the fluid velocity field within the core increases during reversals. Yet an apparent stop-and-go jerkiness observed for several recordings also suggests that the frozen-flux approximation may be totally inapplicable during more quiescent intervals within a given transition. Hence, it is possible that reversals may contain both intervals of time in which flux transport is the dominant process as well as intervals of time when flux diffusion is the controlling factor.

Olson's Helicity Approach. Noting that the basic requirement for dynamo action in a turbulent core is that the mean helicity, that is, the average correlation between velocity and vorticity, be greater

AXISYMMETRIC

LINE OF DIPOLES

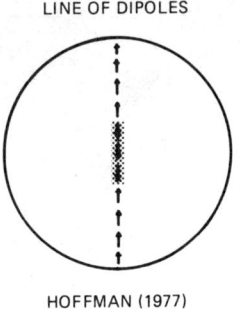

HOFFMAN (1977)
HOFFMAN & FULLER (1978)

DIPOLE RINGS

BOGUE & COE (1984)

NON-AXISYMMETRIC

CYLINDRICAL SHELL

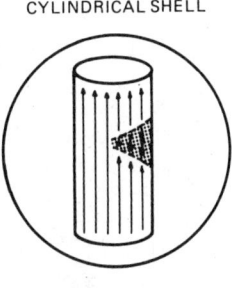

HOFFMAN (1979)

RADIAL DIPOLES

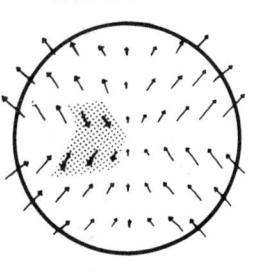

KAISER & VEROSUB (1985)

FIGURE 6. Phenomenological magnetic source regions in the core associated with proposed flooding models.

than a critical value, Olson (1983) points out that this condition applies to the magnitude of helicity, but not to its sign. Helicity can be positive or negative, the sense affecting only the symmetry of the generating magnetic field. Hence, change the sign of helicity and rapid reversals of the dipole field can occur. Olson (1983) shows that two classes of polarity transitions are possible (Fig. 7): *Component reversals* involve reversal of the dipole field without reversal of the toroidal field within the core. This type of reversal is argued to be the result of a single, long-term fluctuation in core helicity. *Full reversals*, on the other hand, involve reversal of both the dipole and toroidal fields and are the result of a series of short-term fluctuations in core helicity. How these two types of reversals might be distinguished paleomagnetically is not known. In support of this approach, Olson shows that the time interval required to complete such a reversal can be as short as 7500 yr, reasonably consistent with paleomagnetic observations.

Systematics of the Reversal Process

The Analysis of Frequency. An understanding of the statistical nature of geomagnetic reversal has steadily developed from the analysis of the pattern of marine magnetic anomalies (see *Geomagnetic Reversal Sequence: Statistical Structure* and *Geomagnetic Field: Asymmetries*). In particular, there may be a rather simple underlying structure to the variation in reversal frequency over geologic time such that smooth increases or decreases in rate occur over several tens of millions of years. McFadden and Merrill (1984) argue that the median length of a polarity interval (to first order) has smoothly decreased since the beginning of the Cenozoic (Fig. 8). Notwithstanding this element of systematics, the reversal triggering process can be shown to be stochastic and, according to McFadden and Merrill, one that does not distinguish between normal and reverse polarities. If this is indeed so, then, for one, the behavioral characteristics of the actual mechanism of reversal also may be expected to be independent of transition sense.

There do exist paleomagnetic transition data compatible with this contention. Valet and Laj (1984), reporting on a series of reversals recorded in exposed marine sediments on Crete, note that within a particular time duration spanning a few polarity intervals, reversal paths associated with transitions of opposite sense are antipodal. Such a

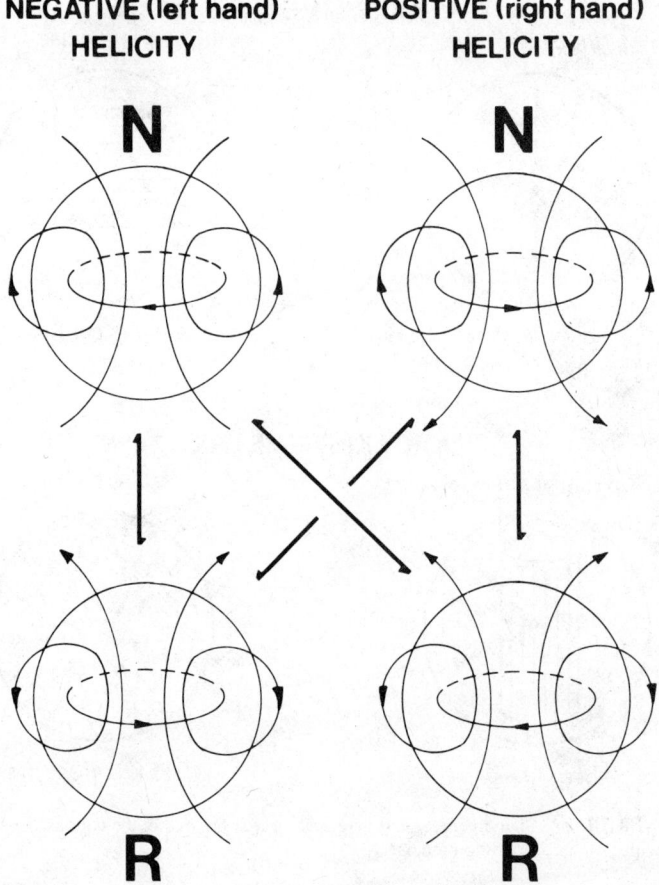

FIGURE 7. Schematic representation of *component* reversals (diagonal paths) and *full* reversals (vertical paths) associated with the helicity approach to transitions. (After Olson, 1983)

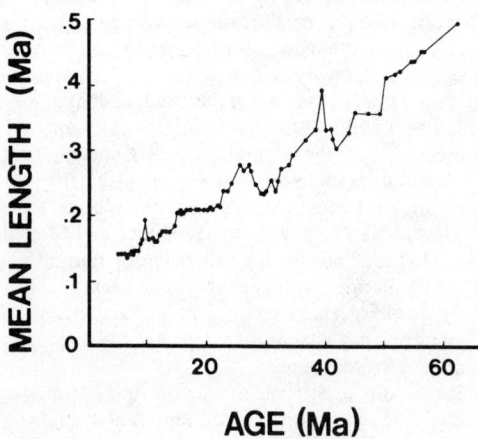

FIGURE 8. Variation in the mean length of polarity intervals since the end of the Cretaceous. (After McFadden and Merrill, 1984)

finding is expected when the reversal process is spatially blind to the initial polarity of the field. However, not all reportings of directional behavior during sequential reversals fit this simple picture.

What Triggers Reversals? The fact that the reversal process basically appears to be a random event provides a most profound constraint as to the actual physical mechanism. As we have seen, the free decay models of Nagata and Lilley, the probabilistic model of Cox (1968), and the cyclonic convection models of Parker (1969) and Levy (1972) all involve a triggering mechanism compatible with the random component of the process. In Olson's (1983) approach, reversal triggering involves a fluid mechanical disturbance that acts to change the sign of core helicity. Further, Olson offers a physical basis for such events: Briefly, the core's net helicity is seen to vary rapidly in response to fluctuations in the level of turbulence produced by two competing energy sources, namely, thermal convection and segregation of the inner core. Olson further points out that in each hemisphere helicity generated by heat loss at the core-mantle boundary should have the opposite sign compared with helicity generated by energy release at the outer core–inner core boundary. The random variation in the rates of these two processes, according to Olson, can cause the net helicity to change sign occasionally, thus initiating a reversal.

McFadden and Merrill (1986) argue from statistical grounds that the data indicate the existence of some physical triggering process whose energy source is effectively *independent* of the source powering the main geomagnetic field. They then propose two possible models, compatible with the data, that identify the independent energy source that causes the necessary perturbation of the existing core field:

Model I: The assumption here is that core convection is driven primarily by cooling at the core-mantle boundary and that the reversal-causing perturbation is an occasional less dense, "warmer blob" rising from the inner core boundary due to the freezing of the outer core.

Model II: Here core convection is driven primarily by the chemical and buoyancy differences resulting from freezing at the inner core boundary and that the occasional reversal-causing perturbation is a more dense, "cooler blob" derived from cooling at the core-mantle boundary.

In contrast to these triggering mechanisms that arise from internally induced disturbances of the core fluid, the suggestion has been made that reversals may arise from external phenomena, particularly, collisions with extraterrestrial objects (Muller and Morris, 1986). In this scenario, large meteoritic impacts on the Earth's surface are considered to be capable of raising a significant amount of particulate matter into the atmosphere, causing an effect similar to a nuclear winter. It is further argued that such a situation would alter the Earth's climate and ultimately affect the mechanical motion of the core fluid. Muller and Morris (1986) propose that abrupt accelerations of the fluid produced in this manner may be all that is needed to initiate a polarity transition. Confirmation of such a mechanism, however, rests strongly not only on temporal correlations between geomagnetic reversals and such phenomena as crater formation, microtektite levels, changes in temperature, and biota extinctions, but also on a reasonable explanation for the observed long-term variation in reversal frequency (Fig. 8).

It seems fitting that we end our discussion with this recent, rather controversial suggestion. Clearly, research into the geodynamo and its capacity to undergo reversal remains a lively topic of geophysical research, even after some three decades of effort. Moreover, the scope of the problem is now becoming clear. Fueled in large part by observations of both today's field as well as the paleofield, our knowledge about the workings of the geodynamo has steadily evolved. What is now beginning to emerge is a wholistic approach to the reversal process, one which incorporates both the fluid-mechanical as well as thermodynamic state of the earth's deep interior.

Acknowledgments

The preparation of this article was partially supported by the National Science Foundation (grant #EAR84-17307 and #EAR87-08524).

KENNETH A. HOFFMAN

References

Cox, A., 1968, Lengths of geomagnetic polarity intervals, *Jour. Geophys. Research* **73**, 3247-3260.

Gubbins, D., and N. Roberts, 1983, Use of the frozen flux approximation in the interpretation of archaeomagnetic and paleomagnetic data, *Geophys. Jour. Royal Astron. Soc.* **73**, 675-687.

Hoffman, K. A., 1977, Polarity transition records and the geomagnetic dynamo, *Science* **196**, 1329-1332.

Hoffman, K. A., and M. Fuller, 1978, Transitional field configurations and geomagnetic reversal, *Nature* **273**, 715-718.

Jacobs, J. A., 1984, *Reversals of the Earth's Magnetic Field*. Bristol: Hilger.

Levy, E., 1972, Kinematic reversal schemes for the geomagnetic dipole, *Astrophys. Jour.* **171**, 635-642.

McFadden, P. L., and R. T. Merrill, 1984, Lower mantle convection and geomagnetism, *Jour. Geophys. Research* **89**, 3354-3362.

McFadden, P. L., and R. T. Merrill, 1986, Geodynamo energy source constraints from paleomagnetic data, *Physics Earth and Planetary Interiors* **43**, 22-33.

Merrill, R. T., and M. W. McElhinny, 1983, *The Earth's Magnetic Field*. London: Academic Press.

Muller, R. A., and D. E. Morris, 1986, Geomagnetic reversals from impacts on the earth, *Geophys. Research Letters* **13**, 1177-1180.

Olson, P., 1983, Geomagnetic polarity reversals in a turbulent core, *Physics Earth and Planetary Interiors* **33**, 260-274.

Parker, E. N., 1969, The occasional reversal of the geomagnetic field, *Astrophys. Jour.* **158**, 815-827.

Prévot, M., E. A. Mankinen, C. S. Grommé, and R. S. Coe, 1985, How the geomagnetic field vector reverses polarity, *Nature* **316**, 230-234.

Valet, J.-P., and C. Lai, 1984, Invariant and changing transitional field configurations in a sequence of geomagnetic reversals, *Nature* **311**, 552-555.

Williams, I. S., and M. Fuller, 1981, Zonal harmonic models of reversal transition fields, *Jour. Geophys. Research* **86**, 11657-11665.

Cross-references: *Geomagnetic Polarity Reversals: Observations; Geomagnetic Reversal Sequence: Statistical Structure; Magnetostratigraphy; Paleomagnetic Field: Intensity; Paleomagnetism and Plate Tectonics; Seafloor Spreading: Magnetic Evidence.*

GEOMAGNETIC REVERSAL SEQUENCE: STATISTICAL STRUCTURE

Early in the 1960s, developments in the K-Ar dating method made it possible to date young volcanic rocks with reasonable precision. Consequently, it was possible to undertake systematic studies attempting to define a polarity time scale using joint magnetic polarity and K-Ar age determinations on young lavas. As data rapidly became available, it was established that rocks of the same age had the same polarity of magnetization, thereby helping to confirm that the observed reversals of magnetization were indeed due to reversals of the geomagnetic field itself. Quite early, it became obvious that these geomagnetic reversals did not occur at regular intervals: some of the observed polarity intervals were nearly a million years in length, and some were as short as 0.1 Ma. Furthermore, there did not appear to be any regular pattern to these different lengths. This led to the suggestion that there is a random component in the reversal process and, therefore, to interest in the statistical structure of the geomagnetic reversal sequence.

It has proven extremely difficult to solve the geodynamo problem—i.e., to determine just how the geodynamo operates. Determination of the statistical structure of the geomagnetic reversal sequence provides insight for the long-term dynamic process and in so doing can provide powerful constraints on geodynamo models.

When one examines a time sequence such as this, it is crucial that the observed events be ordered correctly, and this (together with other difficulties) has led to insurmountable problems with the land-based geomagnetic polarity information. As one goes further back in time, the absolute error in dating a rock soon becomes larger than the length of the shorter polarity intervals. Thus, because the land-based information comes from scattered rocks rather than a single continuous sequence, the ordering of events is effectively indeterminate for events closer together than the dating error. As a result, the length of the reliable land-based polarity time scale is too short for reliable estimation of the statistical structure of the geomagnetic polarity sequence.

It turns out that the most useful data source for this type of analysis comes from marine magnetic anomalies (see *Seafloor Spreading: Magnetic Evidence*), and a reversal chronology is now quite well determined for the past 165 Ma. The absolute error in the age assigned to individual reversals will still often be greater than the length of the shorter intervals, but the continuous nature of the data source means that the ordering of events will be correct. Furthermore, because of the way ages are assigned in the marine magnetic anomaly time scales, the lengths of the intervals between reversals will be about right.

There are still problems with the marine magnetic anomaly time scales. As we go further back in time, the record will naturally become more degraded, so the time scale from 120 Ma back to 170 Ma is less reliable than that from the present back to 80 Ma. Furthermore, because of subduction of the seafloor, any record prior to about 170 Ma has been destroyed.

A major problem with marine magnetic anomalies is that it is fairly easy to miss very short intervals, particularly if a short interval appears between two relatively long intervals of the opposite polarity. As might be expected, when determining the statistical structure of a sequence, it is quite a serious matter to miss out several of the very short intervals. Obviously it is impossible to overcome such data deficiencies entirely, but it is possible to consider this particular problem from a theoretical point of view and, to a large extent, to take it into account in the interpretation.

Relevant Probability Distributions

It has been established that individual reversals take only a few thousand years to complete, which is quite a short period of time relative to the average interval between reversals. Thus it is a reasonable approximation to assume that, relative to the time constants of interest, the reversals themselves are instantaneous. It has already been noted that there appears to be a random component in the reversal process, and so it is sensible to look for a probabilistic description. Consider an interval of time Δx short enough that the probability of having two reversals in Δx is negligible, but with probability of $\lambda \Delta x$ of having a single reversal in Δx. This is then a Poisson process, and it is a simple matter to show that the probability density $P(x)$ of interval lengths x between reversals is given by

$$P(x)\, dx = \lambda e^{-\lambda x}\, dx \tag{1}$$

The mean length μ of the interval lengths is the expectation of x, $\langle x \rangle$, and is given by

$$\mu = \langle x \rangle = \frac{1}{\lambda}$$

so λ is just the mean rate at which reversals would occur. The variance of interval lengths $\text{Var}(x)$ is

$$\text{Var}(x) = \frac{1}{\lambda^2}$$

It is possible to test whether the observed distribution of interval lengths is compatible with the distribution predicted by Eq. 1. Initially it was concluded that there was indeed compatibility.

however, after more testing, it was suggested that there were too few very short intervals in the observed.

Following on from this, a general gamma process was considered, with the probability density of interval lengths given by

$$P(x) \, dx = \frac{1}{\Gamma(k)} (k\lambda^*)^k x^{(k-1)} \exp(-k\lambda^* x) \, dx \quad (2)$$

$\Gamma(k)$ is the gamma function of k, given by

$$\Gamma(k) = \int_0^\infty z^{k-1} e^{-z} \, dz$$

If k is an integer, then $\Gamma(k)$ is the factorial function of $k - 1$; i.e., $\Gamma(k) = (k - 1)!$. The mean length μ^* is again given very simply by

$$\mu^* = \langle x \rangle = \frac{1}{\lambda^*}$$

so λ^* is the mean rate at which reversals would occur with a gamma process. The variance of interval lengths is now

$$\mathrm{Var}(x) = \frac{1}{k(\lambda^*)^2}$$

The observed distribution of interval lengths is compatible with that predicted by the gamma distribution (Eq. 2). Consequently, this distribution has been assumed for most analyses of the geomagnetic reversal sequence. The best-fit gamma distribution to the polarity intervals has $k > 1$, and it will be assumed in the following discussion that $k \geq 1$.

As may be seen from Eqs. 1 and 2, a Poisson process is in fact a gamma process with $k = 1$. However, as we will show, it is a very special case.

Figure 1 shows the probability density $P(x)$ for $k = 1$ (Poisson), 2, 3, and 4.

Implications of a Gamma or Poisson Process

Consider the general probability of a reversal occurring in an interval Δx. Let x be the interval of time since the last reversal (i.e., no reversals have occurred in the interval 0 to x), and let $f(x) \, \Delta x$ be the probability that a reversal will occur in the interval x to $(x + \Delta x)$. The function $f(x)$ is then

$$f(x) = \frac{P(x)}{1 - \int_0^x P(t) \, dt} \quad (3)$$

with $P(\cdot)$ given by Eq. 2.

With a gamma process $f(x)$ is zero for $x = 0$ and approaches $k\lambda^*$ asymptotically as x increases. The larger the value of k, the slower $f(x)$ increases. This is shown in Fig. 2, where $f(x)/k\lambda^*$ is plotted against $kx\lambda^*$ for $k = 1, 2, 3$, and 4. Thus the probability of having a reversal in an interval Δx immediately after a reversal is zero (thereby reducing the number of short intervals that will occur) and gradually rises (with an asymptotic limit of $k\lambda^* \Delta x$) until another reversal occurs. In other words, a reversal inhibits the occurrence of a future reversal, this inhibition decreasing with time since the last reversal. Thus a gamma process exhibits a "memory," in that the probability for a future event is a function of what has happened in the past. However, the probability is a function only of the time since the last reversal; the memory does not extend farther back than this.

As already noted, a Poisson process is a gamma process with $k = 1$ (and $\lambda = \lambda^*$). However, if we put this into Eq. 3, we get $f(x) = \lambda$ (as already stated), and this is no longer a function of x (see also Fig. 2). Consequently, the Poisson process is a very

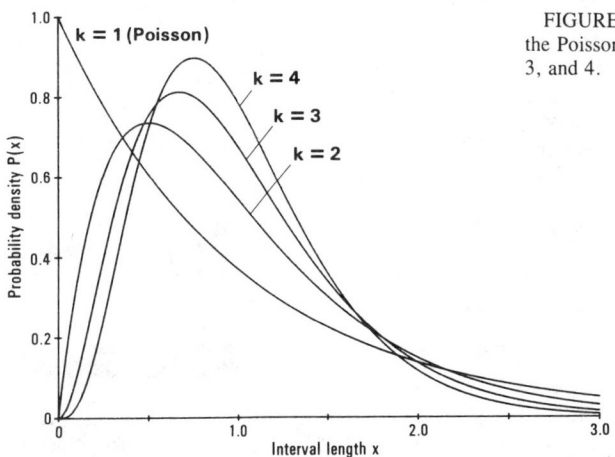

FIGURE 1. Probability densities of interval lengths for the Poisson and gamma processes for $k = 1$ (Poisson), 2, 3, and 4.

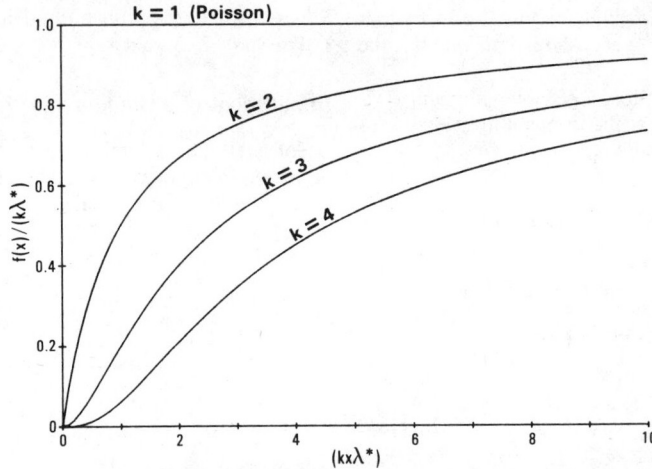

FIGURE 2. Illustration of the memory (inhibition) in a gamma process ($k = 2, 3,$ and 4) and the lack of memory in a Poisson process ($k = 1$).

special case of a gamma process: A Poisson process does not exhibit any memory.

Is the Process Gamma or Poisson?

From the previous section it follows that for a full understanding of the geodynamo it is important to distinguish between a Poisson process and a general gamma process. It has already been noted that there may be too few very short intervals in the observed distribution for a Poisson process, but that the observed distribution is compatible with a gamma process. Thus it may initially appear as an open-and-shut case in favor of a general gamma process, but a little investigation raises problems with this.

If the sequence is analyzed as a general gamma process to estimate the parameters μ^* (or λ^*) and k, it is found that from the present back to 80 Ma, k is about 1.25 (with 95% confidence limits of 1.02 and 1.55), not much different from unity (i.e., a Poisson process). Furthermore, as already noted, the major problem with marine magnetic anomaly time scales is that the very short intervals are easily missed: Thus, we should expect that there were more very short intervals in the actual reversal sequence than appear in the marine magnetic anomaly time scales.

Consider a sequence of normal and reverse polarity intervals $R_1N_1R_2N_2R_3N_3R_4N_4$, with N_2 being a very short interval. Assuming that this interval is missed, the time scale will have the sequence $R_1N_1R_cN_3R_4N_4$, where R_c is a concatenation of the intervals R_2, N_2, and R_3 (see Fig. 3). Consider now a sequence of intervals from a Poisson process, but concatenate the first n intervals into a single interval I_1, the next n intervals into the single interval I_2, etc. The resulting sequence of interval lengths $I_1I_2I_3I_4 \cdots$ will have a distribution given precisely by Eq. 2 with $k = n$, but the process is not a gamma process.

Obviously, with the transformation from the actual geomagnetic sequence to the marine magnetic anomaly time scale, there is no regular concatenation of intervals, as in the previous paragraph. What happens instead is that very short intervals are preferentially missed, and these misses occur in a random manner through the sequence. It may be shown that if the original process were Poisson, then the resulting distribution of interval lengths will be very similar to that predicted by Eq. 2 with $k > 1$, even though again the process is not a gamma process. The similarity is such that it would require an extremely long sequence to be able to distinguish that the distribution is not precisely that given by Eq. 2 and with the geomagnetic sequence there is no possibility of making such a distinction. Let A be the proportion of intervals in the marine magnetic anomaly time scale that agree with the actual intervals in the polarity sequence (i.e., they are not the concatenation of two or more actual polarity intervals). The best-fit gamma distribution then has λ^* (now the apparent reversal rate) given by

$$\lambda^* = \frac{\lambda}{2 - A} \quad (4)$$

where λ is the actual rate of reversals (from the Poisson process), and k given by

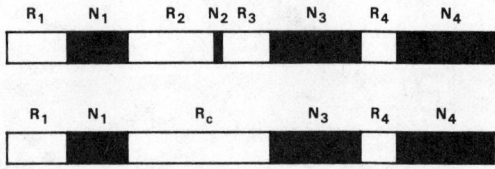

FIGURE 3. Effect of missing an interval: the reverse polarity intervals R_2 and R_3, and the normal polarity interval N_2 are concatenated in the single interval R_c.

$$k = \frac{(2-A)^2}{2-A^2} \quad (5)$$

Having estimated k and λ^*, we may estimate A by inverting Eq. 5 to give

$$A = \frac{2 - \sqrt{2k(k-1)}}{1+k} \quad (6)$$

so that λ may be estimated.

Direct support for this interpretation is available. Comparison of detailed studies of Icelandic lavas (age 2 to 14 Ma) with the marine magnetic anomaly time scale gives almost precisely the same ratio (λ/λ^*) as is obtained from the marine magnetic anomaly time scale alone by using Eqs. 4, 5, and 6. Also, in the stacked marine magnetic profiles for the present back to 80 Ma, there are 57 "tiny wiggles" that have not been included in the marine magnetic anomaly time scales. Although it is unlikely that all of these represent very short intervals that have been missed, many of them probably do. Analysis of the marine magnetic anomaly time scale with these wiggles included as very short intervals gives results that are consistent with this interpretation.

Although the time taken for a reversal to occur is short compared with the typical interval between reversals, it is finite, so the process cannot be truly Poisson. However, in practical terms this is unimportant, so the process may be thought of as "essentially Poisson."

Polarity Asymmetries

For a more complete discussion of polarity asymmetries, see *Geomagnetic Field: Asymmetries*. It is sufficient here to note that the data give no reason to reject the hypothesis that there is no difference between the normal and reverse polarity sequences. Consequently, the polarity sequence can be analyzed simply as a sequence of intervals from a single process, without regard to the actual polarity of the interval.

Nonstationarity of the Reversal Process

At the present time the reversal rate is about 4.5 Ma^{-1}, 40 Ma ago it was about 2 Ma^{-1}, and from about 83 Ma ago back to about 119 Ma ago there were no reversals, the field maintaining normal polarity. Clearly the rate at which the reversal process occurs has not been constant through time. This means that λ^* (or λ) should really be written as a function of time, $\lambda^*(t)$. Although this complicates the analysis, it is still quite possible to do, particularly if the function is simple. From the present back to about 83 Ma ago the structure of the nonstationarity is fit surprisingly well by

$$\lambda^*(t) = \alpha + \beta t$$

with the maximum likelihood estimates $\hat{\alpha}$ and $\hat{\beta}$ of α and β given by $\hat{\alpha} = 4.41$ Ma^{-1} and $\hat{\beta} = -0.051$ Ma^{-2}.

If the 57 tiny wiggles previously mentioned are included in the time scale, then this linear approximation is no longer acceptable, but $\lambda^*(t)$ still varies smoothly from its present value down to zero at about 83 Ma ago. Similarly, $\lambda^*(t)$ increases smoothly (but not monotonically) from about 119 Ma ago back to about 170 Ma ago (when the record terminates).

Periodicities in the Reversal Rate

There have been several suggestions that certain periodicities exist within the geomagnetic reversal sequence either as harmonic variations in the reversal rate or as peaks in the reversal rate. This implies that the rate λ is given by

$$\lambda = \phi(t) + H_1 \cos(\omega_1 t + \theta_1) + H_2 \cos(\omega_2 t + \theta_2)$$
$$+ \cdots + D_1 \psi(\rho_1) + D_2 \psi(\rho_2) + \cdots \quad (7)$$

where $\phi(t)$ is the long-term variation already mentioned under nonstationarity of the process. The function $\psi(\rho_i)$ is used here to represent regular peaks and may be thought of as a Dirac comb with period ρ_i (and appropriate phase) convolved with a fairly narrow Gaussian. As yet, there has been no formal specification of any of the H_i or the D_i.

A Poisson (or gamma) process with a constant λ produces intervals with varying lengths in a random manner. If a filter designed to extract spectral information is applied to such a sequence, it will indicate the presence of spectral components, even in the case when each of the H_i and D_i in Eq. 7 is zero and $\phi(t)$ is a constant. With a Poisson process all frequencies will be present, the expectation amplitudes decreasing with increasing period. However, if only a short sequence is observed, it will not be possible to have all frequencies within that sequence, and consequently certain frequencies (chosen at random) will appear to stand out. Naturally it is tempting to assign nonzero values of H_i to the observed values of ω_i that appear to stand out. A similar situation occurs with regard to short sequences and apparent periodic peaking, leading to apparent nonzero values of D_i. Because of this, before concluding that a genuine periodicity exists (i.e., that there actually is a nonzero H_i or D_i), very careful testing must be performed to show that the observed amplitude at that period is discernibly larger than might be expected from random processes. Furthermore, because the particular frequency ω_i or period ρ_i has appeared in the sequence and, due to the random nature, it will not always appear in sequences of that length, the testing must be performed conditional upon observing that particular ω_i or ρ_i.

To date, all suggestions of periodicities have come about as a result of the application of a filter to the polarity sequence, but none of these has been backed up with testing conditional upon the observed periodicity. It is certain that if any genuine periodicities do exist, the amplitudes are small. For the time being, it is probably sensible to take a conservative line and assume that all the H_i and D_i are zero.

Conclusions

The simplest statistical structure apparently consistent with the available data is that the reversal process is essentially Poisson, without distinction between normal and reverse polarities. In our records several of the very short intervals appear to have been missed, producing a time scale with interval lengths that are gamma distributed with k about 1.25. The rate of the process has varied with time, the rate reducing smoothly from around 4 Ma^{-1} about 165 Ma ago to zero about 119 Ma ago, remaining at zero until about 83 Ma ago, and then increasing smoothly to around 4.5 Ma^{-1} in the recent past. During the time that the reversal process was in abeyance (119 to 83 Ma ago), the polarity was normal, but presumably there was a 50% chance of it being in either polarity.

Knowledge of the statistical structure of the reversal sequence is very important in the assessment of geodynamo models. Clearly, any truly viable model must include the mechanism for reversals, and this mechanism must produce a Poisson process. It must be possible for the rate of this process to vary smoothly, while remaining Poisson, and it must be possible for the reversal process to be held in abeyance for a few tens of millions of years while not bringing the geodynamo itself to a halt.

Because of seafloor subduction, we have an ordered sequence of geomagnetic polarity intervals for less than 4% of the Earth's history (i.e., to about 170 Ma). However, from the land-based record we do know that reversals occurred for most of that time. There is also evidence that there were other occasions when the reversal process was held in abeyance. Naturally it is tempting to suggest that throughout most of the Earth's magnetic history the statistical structure of the geomagnetic sequence has been similar to that of the past 170 Ma, and the rather meager evidence available to us is not inconsistent with this idea.

Acknowledgment

This article is published with the permission of the Director, Bureau of Mineral Resources, Geology and Geophysics.

P. L. McFADDEN

References

Berggren, W. A., D. V. Kent, J. J. Flynn, and J. A. van Couvering, 1985, Cenozoic geochronology, *Geol. Soc. America Bull.* **96**, 1407-1418.

Cox, A. V., 1968, Lengths of geomagnetic polarity intervals, *Jour. Geophys. Research* **73**, 3247-3260.

McFadden, P. L., 1984, Statistical tools for the analysis of geomagnetic reversal sequences, *Jour. Geophys. Research* **89**, 3363-3372.

McFadden, P. L., and R. T. Merrill, 1984, Lower mantle convection and geomagnetism, *Jour. Geophys. Research 89*, 3354-3362.

McFadden, P. L., R. T. Merrill, W. Lowrie, and D. V. Kent, 1987, The relative stabilities of the reverse and normal polarity states of the Earth's magnetic field, *Earth and Planetary Sci. Letters* **82**, 373-383.

Cross-references: *Geomagnetic Field, Asymmetries; Geomagnetic Polarity Reversals: Observations; Geomagnetic Polarity Reversals: Theory and Models; Magnetostratigraphy; Paleomagnetism and Plate Tectonics; Seafloor Spreading: Magnetic Evidence.*

GEOMAGNETIC SECULAR VARIATION: DIRECTION AND INTENSITY

What Constitutes Secular Variation?

The secular variation of the geomagnetic field denotes the gradual change with time of the Earth's magnetic field, measured on a timescale of months to thousands of years. Among the longest records we have of direct observations of the geomagnetic field are those from London and Paris dating from the mid sixteenth century. The record at Greenwich, London is illustrated in Fig. 1. It was fortunate that when the earliest measurements were being made in Europe, the amplitude of the secular variation was unusually large. If the study of terrestrial magnetism had developed in Australia, for example, where directional changes have been much smaller (Fig. 2), the secular variation would have passed unnoticed for much longer. An account of the discovery of secular variation is given in *Geomagnetism: Historical Introduction*.

The magnetic field observed on the Earth's surface is due predominantly to the field generated within the Earth's liquid core, the so-called main field. Lesser contributions arise from external electric currents in the ionosphere and magnetosphere, currents induced within the Earth by external field variations, and the remanent and steady-state induced magnetization of crustal rocks. In much of the literature, little distinction is made between the secular variation of the total field and that of the main (core) field. The latter is sometimes called the *true secular variation* to avoid this ambiguity. Because the frequency spectra of main field variations and external-plus-induced field variations overlap, isolation of true secular variation poses a difficult problem.

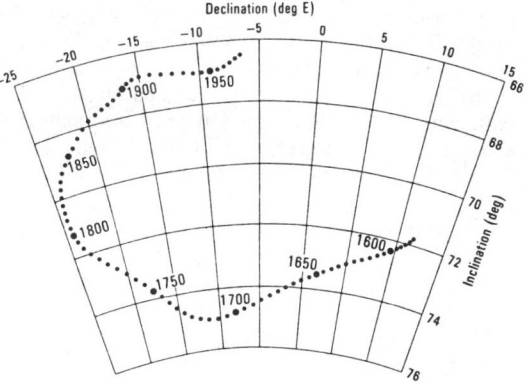

FIGURE 1. Direction of the Earth's magnetic field at Greenwich, London, interpolated at 5-yearly intervals using cubic spline fits to declination and inclination observations made at, or in the vicinity of Greenwich. (After Malin and Bullard, 1981)

Magnetic signals from the core propagating outward become attenuated as a result of the finite electrical conductivity of the mantle, particularly the lower mantle. The degree of attenuation increases with frequency. It is generally accepted that the lower bound for the time constant of identifiable signals from the core is in the range of 3.7 yr to 13 yr. However, if geomagnetic "jerks" are indeed secular variation impulses of internal origin then these bounds may have to be revised downwards substantially. See the section on secular acceleration impulses.

On the other hand, most external field variations and associated transient induction effects (excluding the small seasonal influences described below) have time constants of much less than a year. For example, the dominant external field contribution is the 24-hour solar diurnal variations (Sq), with an average amplitude of about 50 nT and a maximum amplitude of typically 200 nT at the geomagnetic equator. Hence it is usually assumed that external and transient effects average to zero in annual mean values of the magnetic field. Actually it is preferable to use only night-time values on magnetically undisturbed days in order to minimize the bias in all-day mean values caused by the long-term effects of magnetic storms. Differences between these two types of annual means can be several nannotesla, which is more than the annual change over much of the Earth's surface. The secular variation is taken to be the difference between successive annual means.

There are several long period, seasonal, contributions to the external field:

1. a semimonthly lunar modulation of Sq, amounting to about 40 nT at the magnetic equator and lessening towards the poles;
2. a 27-day quasi-periodicity in geomagnetic activity corresponding to the synodic (apparent) rotation period of the Sun;
3. a weak semiannual periodicity, associated with alternation of the sense of the interplanetary magnetic field borne by the solar wind;
4. annual variations, with amplitudes of a few nT;
5. a biannual modulation of the amplitude of diurnal variations, with amplitudes of 1-2 nT;
6. an 11-year periodicity, associated with solar cycle induced variations in geomagnetic activity, with typical amplitude 10 nT;
7. a clustering of spectral lines around 21.4 yr close to the double solar-cycle, the origin of which is uncertain.

Finally there is some evidence for 60-year variations of the field, but these are more likely to be of internal origin (see the section on the spectrum of secular variation).

With only 200 or so permanent magnetic observatories distributed irregularly over the globe, our

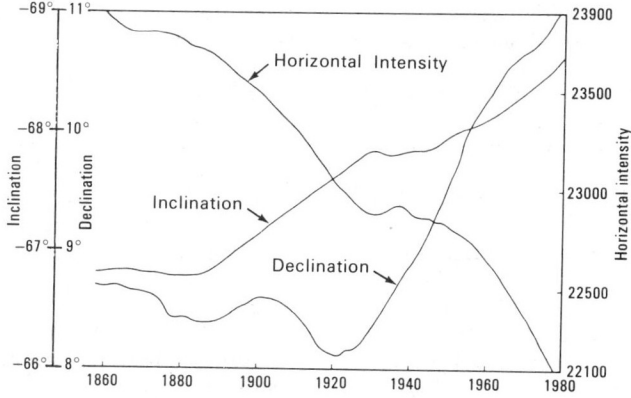

FIGURE 2. Secular variation at Toolangi magnetic observatory, including data from Melbourne corrected to Toolangi. (After Parkinson, 1983)

knowledge of the spatial change of the secular variation has been restricted essentially to a 10^3 km scale or greater. This is illustrated by global spherical harmonic models of the field which show a pattern of positive and negative focii of continental size and larger, superimposed on a trend dictated by the decay of the dipole moment (see Fig. 3).

At the very low frequency end of the spectrum, paleomagnetic records (see the next section) show continuous variations of the main field with characteristic times of up to 10^4 years, and amplitudes similar to those observed during the last few hundred years. The upper limit for the time constant of natural secular variation of the geodynamo is generally thought to be about 10^5 years, which is not far short of that for polarity reversals (the mean polarity interval has decreased linearly during the last 86 My and is now about 250 000 yr).

There have been suggestions of extremely long time-constant variations in the geomagnetic field (for example in the strength of the dipole moment) arising from external forcing factors of astronomical origin, and which would therefore be associated with climatic variations. Furthermore, on a geological timescale, changes in remanent magnetization of the crust and evolution of the internal structure of the Earth will also influence the geomagnetic field. Such effects are not considered to be part of the secular variation.

Determination of the Secular Variation

The most accurate information about the secular variation is provided by the global network of about 200 permanent magnetic observatories that record three or more components of the field at least every minute. Additional, but less accurate information comes from regional networks of magnetic repeat stations. These are permanently marked points, where observations of the field are made for a few days at intervals of usually every 2 to 5 years. Satellites potentially provide the ultimate means of monitoring the main field and its secular variation. The first satellite equipped to measure the vector field was MAGSAT, which operated from 30 October 1979 to 11 June 1980. The MAGSAT mission led to a major improvement in our knowledge of the geomagnetic field, but was too short lived to be of much value in determining the secular variation. To date there has not been a successor to MAGSAT.

To determine the nature of the secular variation over longer timescales we must rely on paleomagnetic results from sequences of lava flows, rapidly deposited sediments, and baked materials and artifacts (see *Archaeomagnetism* and *Paleomagnetic Secular Variation*). Unaltered materials that acquired their magnetic remanence during cooling often preserve a good record of both the direction and the intensity of the ambient field at the time of cooling. The detrital remanent magnetization of sediments [see *Detrital Remanent Magnetization (DRM)*] is less reliable as a recorder of the geomagnetic field and seldom permits determination of the ancient field intensity. Despite this, marine and lacustrine sediments do provide a vital source of information about the long-term behavior of the geomagnetic field because of their wide spatial and temporal coverage.

On a very long (geological) timescale it is possible to estimate the magnitude of the secular variation as a function of latitude from the scatter in groups of paleomagnetic observations. This estimation can be used to discriminate between possible models for the secular variation, e.g., the relative importance of dipole and nondipole fluctuations. *Paleomagnetic Secular Variation* provides further information on this topic.

Description of the Secular Variation

The most widely used global models of the secular variation are provided by the International Geomagnetic Reference Field (IGRF). This set of spherical harmonic models of the scalar magnetic potential of the field is updated every 5 years (see *Geomagnetic Field Analysis*). The 1987, fifth generation, revision of IGRF adopted by the International Association of Geomagnetism and Aeronomy (IAGA Division I Working Group 1, 1987) defined models of the field at 5-yearly epochs from 1945.0 to 1985.0 and a secular variation model for the interval 1985.0 to 1990.0. These models are expressed as spherical harmonic Gauss coefficients and their annual changes (listed for IGRF 1985 in Table 1). The secular variation model is truncated at degree and order 8 (equivalent to an equatorial wavelength of about 5000 km), which is the limit of resolution of the available data. For intermediate epochs prior to 1985 the field, and hence the secular variation, is defined by linear interpolation of the Gauss coefficients between neighboring 5-year epochs. IGRF epoch models from 1945 through 1980 have been classified as "Definitive," in the sense that no further revision is anticipated and are referred to as DGRF models. Various smoothing procedures have been proposed to avoid the sudden change in secular variation inherent in the DGRF models at 5-yearly epochs, although none of these have been adopted by IAGA.

In principle, it should be necessary to determine the main field once, and thereafter determine only the secular variation at regular intervals. In practice, the errors inherent in measuring the secular variation are such that regular remeasurements of the main field are required. For example, by the early 1970s it was found that observed values of the field differed from forward extrapolation of IGRF 1965.0 by some hundreds of nannotesla over much of the Earth's surface, an error that was unacceptably large for many applications.

World charts of the secular variation (displaying

GEOMAGNETIC SECULAR VARIATION: DIRECTION AND INTENSITY

TABLE 1. International Geomagnetic Reference Field, 1985

n	m	g_n^m	$(\dot{g}_n^m)$	h_n^m	$(\dot{h}_n^m)$
1	0	−29877	(23.2)		
1	1	−1903	(10.0)	5497	(−24.5)
2	0	−2073	(−13.7)		
2	1	2045	(3.4)	−2191	(−11.5)
2	2	1691	(7.0)	−309	(−20.2)
3	0	1300	(5.1)		
3	1	−2208	(−4.6)	−312	(5.3)
3	2	1244	(−0.6)	284	(2.3)
3	3	835	(0.1)	−296	(−10.8)
4	0	937	(0.1)		
4	1	780	(−0.6)	233	(3.8)
4	2	363	(−7.8)	−250	(2.2)
4	3	−426	(−1.4)	68	(2.5)
4	4	169	(−6.8)	−298	(0.9)
5	0	−215	(1.3)		
5	1	356	(0.1)	47	(0.1)
5	2	253	(−1.5)	148	(−0.2)
5	3	−94	(−3.2)	−155	(−0.1)
5	4	−161	(0.1)	−75	(0.6)
5	5	−48	(−0.1)	95	(0.0)
6	0	52	(1.4)		
6	1	65	(−0.3)	−16	(−0.4)
6	2	50	(1.7)	90	(−1.1)
6	3	−186	(0.6)	69	(−0.8)
6	4	4	(0.0)	−50	(−2.3)
6	5	17	(0.9)	−4	(−0.5)
6	6	−102	(1.2)	20	(−0.1)
7	0	75	(0.2)		
7	1	−61	(−0.6)	−82	(0.2)
7	2	2	(−0.5)	−26	(1.0)
7	3	24	(0.8)	−1	(1.1)
7	4	−6	(1.0)	23	(1.9)
7	5	4	(0.4)	17	(0.3)
7	6	9	(−0.5)	−21	(0.2)
7	7	0	(−0.1)	−6	(0.9)
8	0	21	(0.7)		
8	1	6	(0.0)	7	(0.1)
8	2	0	(0.3)	−21	(−1.0)
8	3	−11	(0.4)	5	(0.1)
8	4	−9	(−0.3)	−25	(−0.8)
8	5	2	(−0.3)	11	(0.2)
8	6	4	(0.1)	12	(−0.8)
8	7	4	(−0.5)	−16	(−0.1)
8	8	−6	(−0.8)	−10	(1.3)
9	0	5			
9	1	10		−21	
9	2	1		16	
9	3	−12		9	
9	4	9		−5	
9	5	−3		−6	
9	6	−1		9	
9	7	7		10	
9	8	2		−6	
9	9	−5		2	
10	0	−4			
10	1	−4		1	
10	2	2		0	
10	3	−5		3	
10	4	−2		6	
10	5	5		−4	
10	6	3		0	
10	7	1		−1	
10	8	2		4	
10	9	3		0	
10	10	0		−6	

Note: The Gauss coefficients of degree n and order m are in nT, and their time derivatives (in parentheses) are in nT/yr.

"isoporic" contour lines of equal annual change of elements of the field) for epoch 1980.0 are illustrated in Fig. 3. The pattern for the vertical component (Z) is dominated by a large negative isoporic focus over the western Central Atlantic, a lesser negative focus over Irian Jaya, and a broad positive focus over the Southern Ocean centered below South America. The maximum rate of decrease of Z is about 190 nT yr^{-1} at the center of the western Central Atlantic focus. Over the Pacific basin the secular variation of all the geomagnetic elements is subdued, and the nondipole field is also particularly weak. This 'Pacific nondipole low', or 'dipole window' is discussed later.

It is evident from Table 1 that the secular variation field is not as strongly biased toward the low degree spherical harmonic terms as is the main field, which is a reflection of the more complicated pattern of the secular variation.

The distribution over the surface of the Earth of the secular variation in total field (F) for IGRF 1985.0 is illustrated in Fig. 4a. Truncation of the IGRF secular variation model to degree and order 8 has the effect of clipping the two extreme tail ends of the distribution, but does not lead to any substantial change in the form of the distribution. The annual change in F ranges from −120 nT yr^{-1} to 50 nT yr^{-1}, with a strong negative bias resulting from the decay of the dipole moment (Fig. 4b). The positive bias in the histogram for the nondipole field (Fig. 4c) illustrates how the nondipole field is growing as the dipole field is decaying. The effect of the Pacific nondipole low is to produce a strongly peaked distribution of the secular variation of the nondipole field and a corresponding bulge near zero in the histogram of the total field.

In fitting curves to sequences of archaeointensity data some authors have implied changes in F of 0.3–0.5% per year (i.e., around 200 nT yr^{-1}) averaged over 200 years. Clearly, such rates are not characteristic of the present-day secular variation. For example, a secular variation in F of 100 nT yr^{-1} is presently exceeded over only 4.8% of the Earth's surface. Furthermore, most isoporic focii are not fixed but drift gradually with time. The largest feature of the nondipole field is the Mongolian high

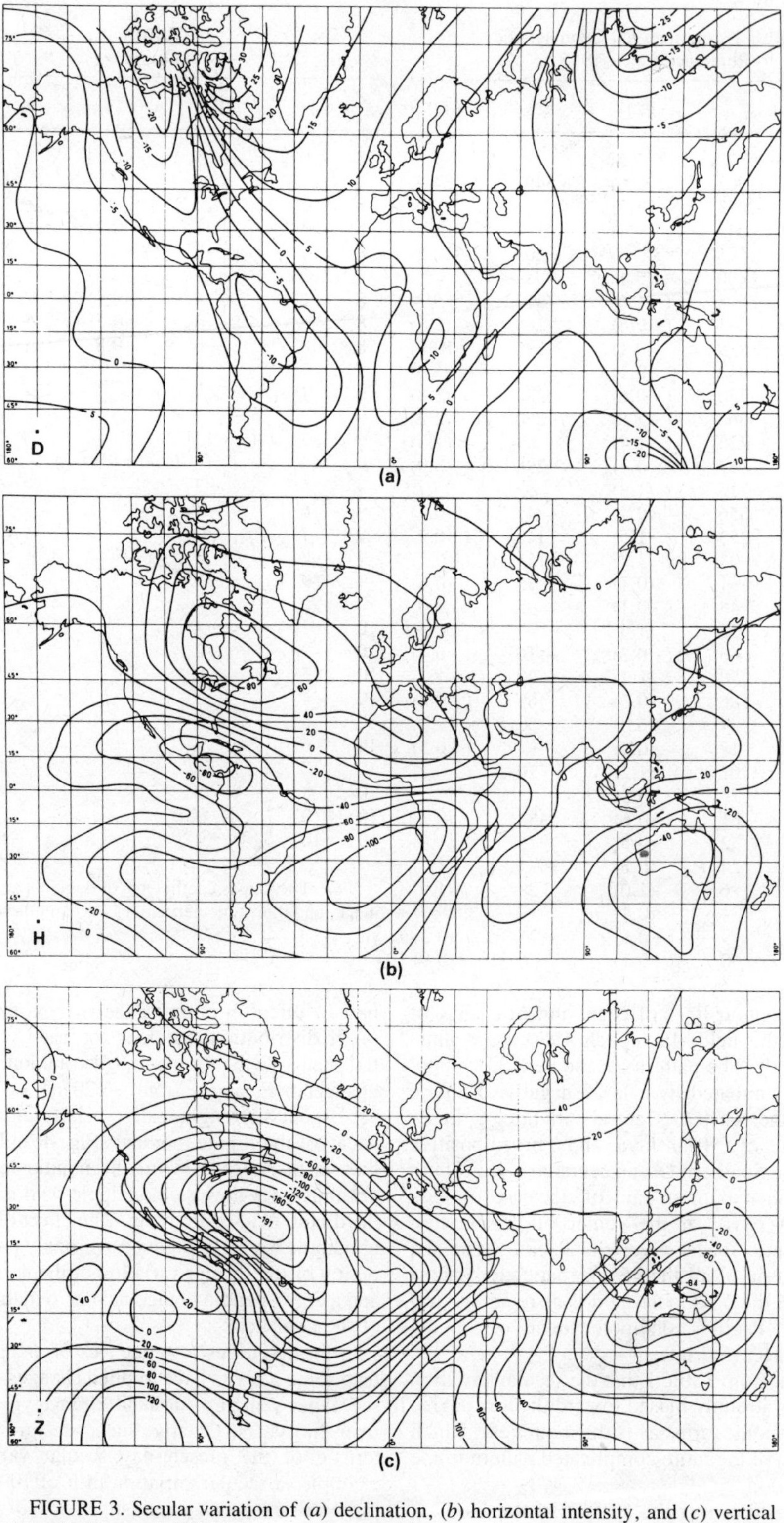

FIGURE 3. Secular variation of (*a*) declination, (*b*) horizontal intensity, and (*c*) vertical intensity for epoch 1980.0, based on the world magnetic chart model of Barker et al., 1981. Declination and intensity contour intervals are 5 minutes of arc per year and 20 nT yr^{-1} respectively. (After Parkinson, 1983)

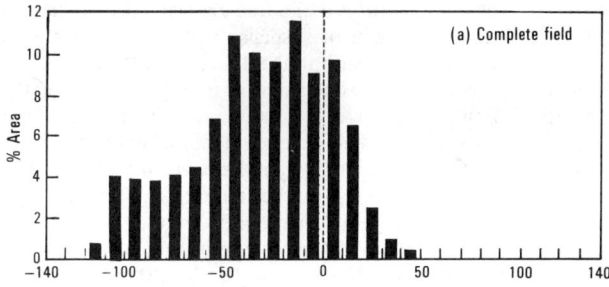
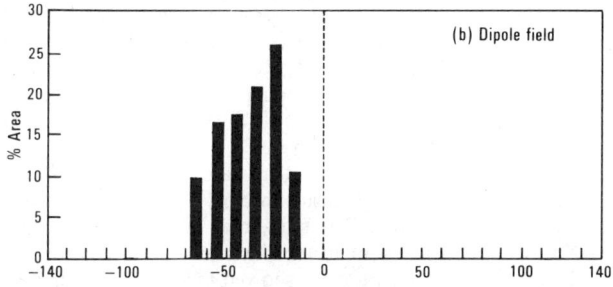
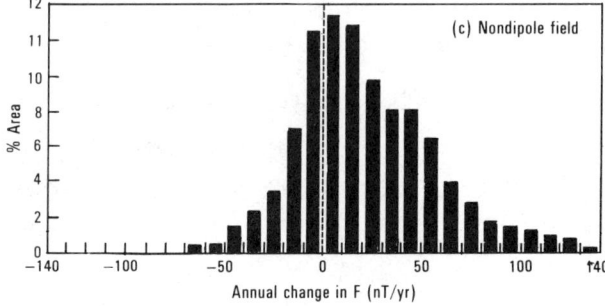

FIGURE 4. Spatial distribution over the Earth's surface of the secular variation in total intensity, F, of (a) the whole field, (b) the dipole field, and (c) the nondipole field for IGRF 1985. The ordinate shows the percentage of the Earth's surface, summed for 1° by 1° tessera, for a given range of annual change in F.

(see Fig. 7), which has grown at a mean rate of 53 nT yr^{-1} for the last 400 years. During the twentieth century the peak value of the secular variation has been typically 150 nT yr^{-1}.

Spectrum of Secular Variation

Long series of observatory data show a well-defined annual cyclicity that clearly has an external origin. One of the first attempts to use series of annual means to separate true secular variation from other phenomena was made by Fisk in 1931. His work suggested the presence of an 11-year component with an amplitude of about 10 nT. In 1947 Vestine and co-workers advocated the use of low degree polynomials to model the true secular variation, a precedent that has been followed largely ever since. The character of residuals from such polynomial fits to long sets of observatory annual means convincingly confirmed the existence of an 11-year cyclicity associated with the sunspot cycle.

Fourier and maximum entropy spectrum analyses of sequences of observatory data have revealed several additional spectral peaks. Periodicities of 2 years, 6.7 years, and 22 years (the double sunspot cycle) have been claimed but have yet to gain general acceptance and be identified reliably with physical origins. There is growing evidence for a concentration of power in the geomagnetic spectrum at periods around 60 years. This cyclicity has been detected in both time and frequency domain analyses of observatory data although the available records are not long enough to establish the phenomenon beyond doubt. An origin for a 60-year geomagnetic signal has yet to be identified, but may well be internal. Support for this view comes from a similar periodicity (57.5 yr) in decade fluctuations in the length of day. The two phenomena could both stem from exchange of angular momentum between the core and the mantle. J. I. Braginskiy has endeavoured to account for the 60-year signal in terms of torsional, Alfvén-type magnetohydrodynamic waves in the core.

Records of the secular variation on a thousand-year timescale, obtained from archaeomagnetic and paleomagnetic studies, have failed to reveal any discrete worldwide periodicities. Individual records do display clearly resolved spectral peaks, but the differences in periods between records from various parts of the world appear to be larger than can be accounted for by inaccuracies in the time scales and limitations in resolution of the analytical techniques. Despite this, there appears to be a clustering of spectral peaks at periodicities of 400–600 years (generally attributed to growth and decay of the nondipole field), at periodicities of a few thousand years (generally attributed to drift of the nondipole field), and at periodicities of many thousands of years (generally attributed to dipole variations).

Westward Drift

Louis Agricola Bauer (1865–1932) made extensive use of plots of declination vs. inclination for different places around the world. These plots, which effectively illustrate the motion of the tip of the magnetic vector, have come to be known as Bauer diagrams. Bauer (1895, p. 37) concluded, "In consequence of the secular variation of geomagnetism, the north end of a free magnetic needle viewed from the center of suspension of the needle moves on the whole earth in the direction of the hands of a watch." This conclusion is true over much of the Earth's surface (e.g. London, Fig. 1), although there are regions, notably the southern Indian Ocean where Bauer had no data, where the sense of looping of the magnetic vector is either anticlockwise or indeterminate. The global pattern of looping is illustrated in Fig. 5 which shows virtual geomagnetic pole (VGP) paths for sites at 10° by 10° grid points. (VGP paths map the secular variation at a site in terms of the motion of the northern dip pole of an equivalent geocentric dipole, and are similar in shape to Bauer plots for sites remote from the poles.) Archaeomagnetic and paleomagnetic records indicate that the sense of looping of the magnetic vector alternates irregularly on a timescale of thousands to tens of thousands of years.

Clearly westward (eastward) precession of the dipole axis will result in clockwise (anticlockwise) looping on a Bauer diagram provided one is not observing from inside the circle of precession. In 1959 S. K. Runcorn showed that westward drift of the nondipole field produces clockwise looping on a Bauer diagram for most plausible distributions of

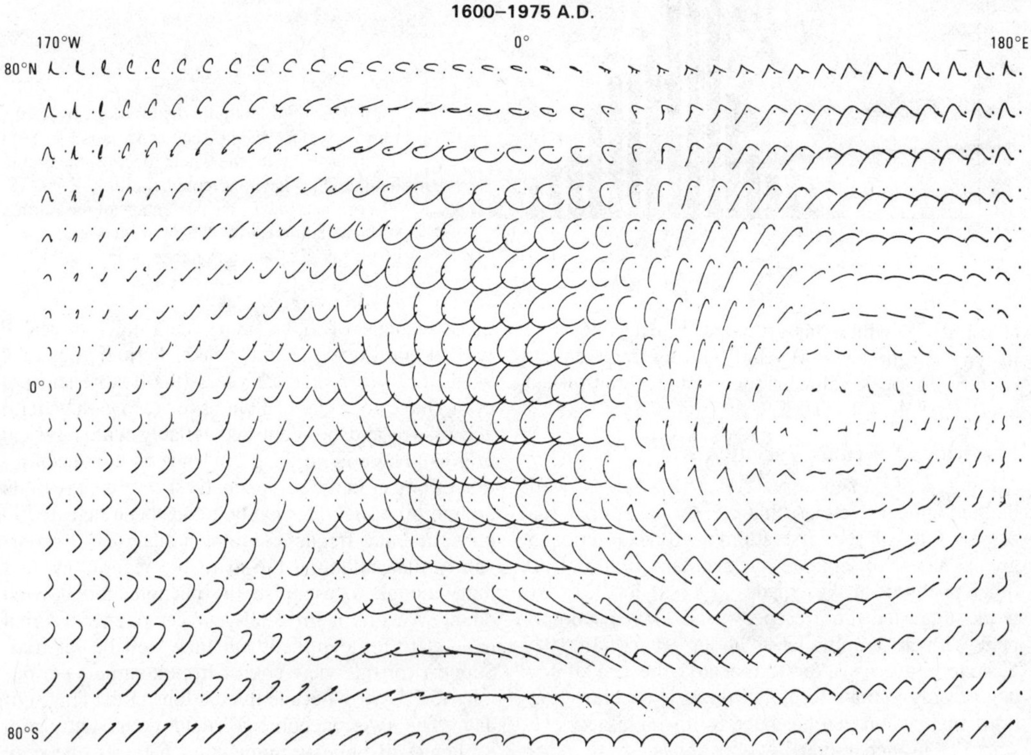

FIGURE 5. Virtual geomagnetic pole paths for A.D. 1600 to A.D. 1975 on a 10° by 10° grid, based on least squares cubic spline fits. All paths display clockwise looping except those in the southern Indian Ocean, which loop in an anticlockwise direction. Clockwise looping in the European region is particularly well defined. (After Thompson and Barraclough, 1982)

spherical harmonic coefficients of the magnetic potential (Runcorn's Rule). However, attempts at modeling the field in terms of dipole or current loop sources in the outer regions of the core have highlighted the fact that certain configurations, particularly for sources that oscillate in strength, lead to violations of Runcorn's Rule. Thus the existence of westward drift cannot be demonstrated uniquely from isolated secular variation records.

During recent times each of the dipole, quadrupole, and octupole components of the field as well as the position of the eccentric dipole (i.e., the best fitting off-centered dipole) have drifted westward, although the movement of the dipole axis has been very much smaller than that of the others. There is good evidence that the westward drift rate fluctuates on a decade timescale, and that the rate has increased between 1970 and 1980, as illustrated in Fig. 6. This point is relevant to the discussion that follows on the 1969/1970 geomagnetic jerk.

Longitudinal drift of the nondipole field is most simply explained in terms of a small difference in average rotation rate between the crust and the outer regions of the core. (There is an intuitive assumption here that the nondipole field is mainly influenced by fluid motion in the outer layer of the liquid core.) The basic idea was first hypothesized by Edmond Halley in 1683 (Halley conceptualized a set of revolving concentric magnetized shells within the Earth), and pursued in detail by Bullard et al. (1950). However, a net differential rotation between the core and the crust has yet to be clearly demonstrated, and the possibility put forward by Raymond Hide in 1966 that westward drift results from the propagation of waves in the core cannot be discounted. Furthermore, westward drift of the nondipole field does not necessarily imply net westward flow of core fluid, and the alternation of clockwise and anticlockwise looping of the magnetic vector on a thousand-year timescale is difficult to explain in terms of rapid changes in differential motion between the crust and the core.

The topic of westward drift of the geomagnetic nondipole field is dealt with in detail in *Geomagnetic Field: Westward Drift*.

Standing and Drifting Components

If the secular variation was caused entirely by westward drift, it could be determined simply from the present-day variation of the field around circles of latitude. However, in most cases estimates based on this principle have proved to be inaccurate. During the last few hundred years, for which we have adequate records, growth and decay of features of the nondipole field have been much more prominent than westward drift (Fig. 7). For example, the Mongolian anomaly increased by 50% during the period 1829 to 1980.

Some features of the secular variation show an obvious westward drift whereas others appear essentially fixed and only grow and decay. This can be seen in Fig. 7 where the African anomaly has drifted westwards (at about $0.28°/\text{yr}^{-1}$) and the North American and Mongolian anomalies are more-or-less stationary. This led Yukutake and Tachinaka (1968) to model the secular variation in terms of two steady components—one standing and the other drifting westward at $0.29°/\text{yr}^{-1}$. Yukutake demonstrated later that a better fit with the observations is obtained if the two components are allowed to vary with time, or, alternately, an additional steady component drifting eastward is introduced. In the latter case, the westward drifting component dominates the slower eastward drifting one. Models of the radial component of the magnetic field at the core-mantle boundary for 6 epochs since 1715 A.D., developed by Bloxham and Gubbins (1985), likewise show large standing features, together with features that migrate westward, and a few features that migrate eastward.

Dipole and Nondipole Contributions

Separation of the geomagnetic field into dipole (i.e., the best fitting geocentric dipole) and nondipole (the residual) parts is mathematically simple and is convenient for depicting the global and regional behavior of the field respectively. However, there is no compelling physical reason for doing this. Justification is generally based on

1. the large size of the dipole field—about 90% of the total field at the Earth's surface;
2. a notion that the secular variation has two distinct timescales—many thousands of years, which characterizes the global behavior of the field, and

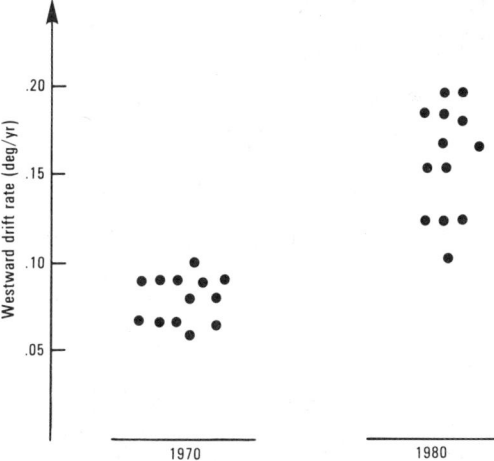

FIGURE 6. Various estimates of the westward drift rate of the geomagnetic field for 1970 and 1980. Estimates have been spread out laterally to avoid overlap. (After Gire et al., 1984)

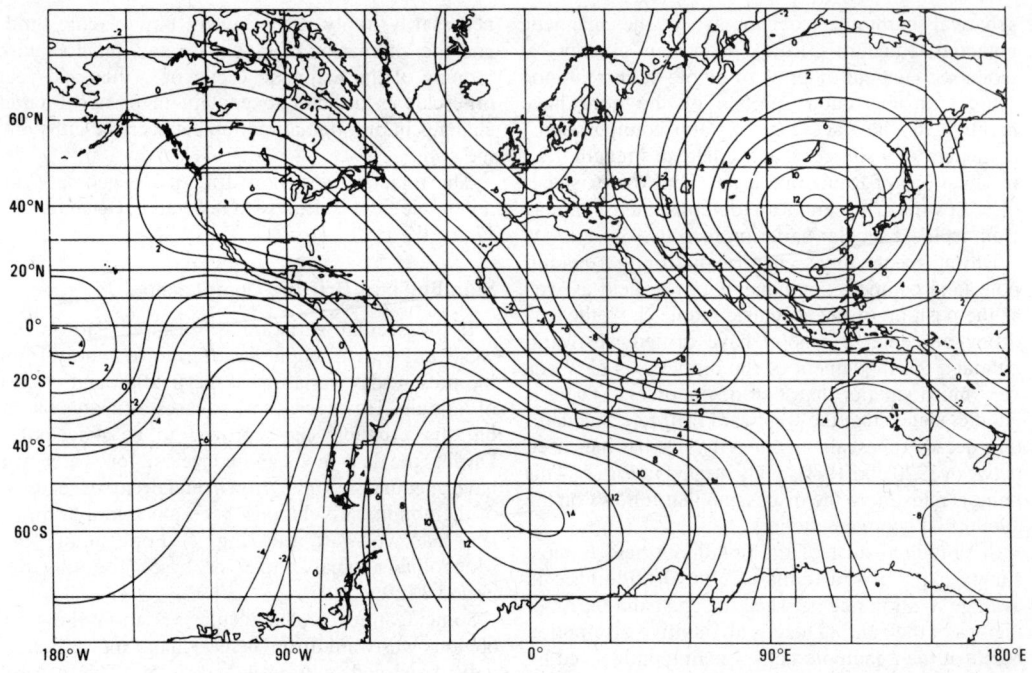
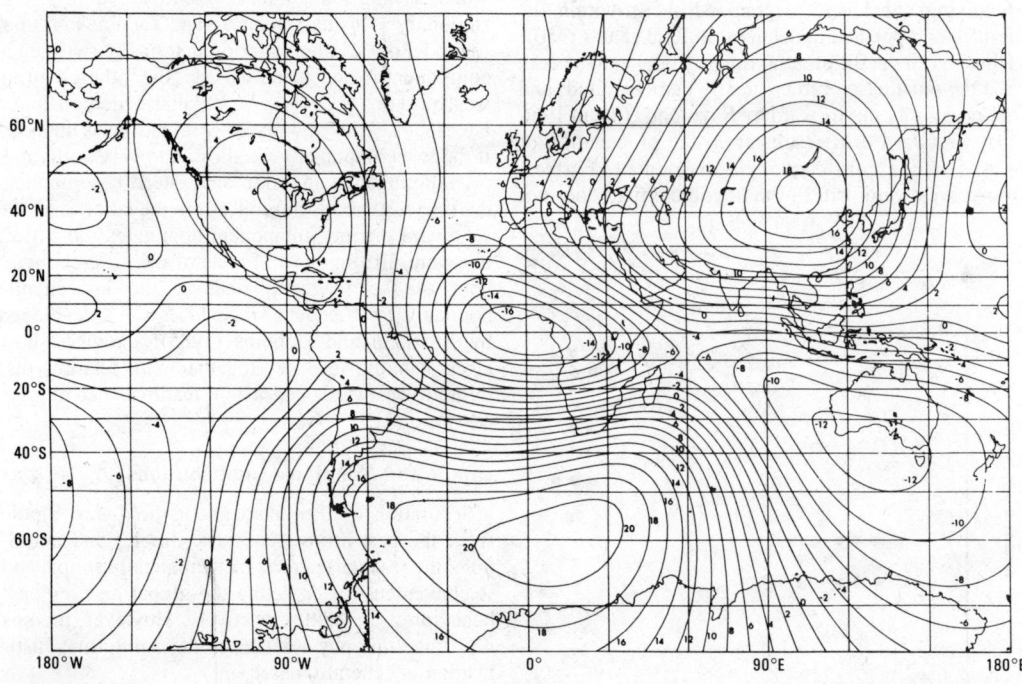

FIGURE 7. The vertical component of the nondipole field for (a) 1829 (from Yukutake and Tachinaka, 1968), and (b) 1980. Contours are in units of 2000 nT. (Figure reproduced by courtesy of D. R. Barraclough and Geomagnetism Unit, British Geological Survey, Edinburgh.)

hundreds to thousands of years, which characterizes the nondipole field and regional variations; and
3. the geomagnetic (dipole) axis is generally confined to within a few degrees of the Earth's rotation axis and coincides with it when averaged over a sufficient interval of time (typically a few million years).

The weakness of the first argument is that our perception of the geomagnetic field is very much a result of our particular position as observers on the Earth's surface. If observations were made at the core-mantle boundary we would see the field as essentially nondipolar (Fig. 8). Conversely, as we go further from the Earth, the field becomes increasingly dipolar because of preferential attenuation of the higher harmonics.

The second argument is also weak as the timescale distinction is an inevitable consequence of the longer free diffusive decay times of lower harmonics of the core field. For example, for a core conductivity of 10^5 S m^{-1} the free decay relaxation times for the lowest order radial modes are 5000 yr ($n = 1$), 2460 yr ($n = 2$), 1483 yr ($n = 3$) and 970 yr ($n = 4$), where n is the degree of the surface spherical harmonic (McFadden et al., 1985). For a core conductivity of 10^6 S m^{-1} these times are an order of magnitude greater. Thus, regardless of the nature of the process responsible for generating the geomagnetic field, we should expect dipole and nondipole variations to have different timescales with an intermediate region where spectral overlap occurs. Guesses as to the form of the geomagnetic power spectrum have indeed shown a dip separating the supposed frequency ranges of dipole and nondipole contributions. However, a composite power spectrum that is based on data from observatories, on archaeomagnetic data, and on paleomagnetic data from rapidly deposited sediments shows a monotonic increase of spectral power density with increasing period up to periods of about 10^4 yr (Fig. 9). There is a suggestion of an intermediate peak in power at periods around 60 years.

The last argument remains the strongest, as it suggests that there is something fundamental about the geomagnetic dipole arising from the constraint imposed on the geodynamo by the Earth's rotation.

There is increasing evidence that the time-averaged geomagnetic field on a million-year timescale is axisymmetric but not exactly dipolar (see *Geomagnetic Field: Asymmetries*). This evidence implies that there are very long-lived zonal contributions to the nondipole field. Suggestions for the origin of these include topographic irregularities at the core-mantle boundary and thermoelectric and electrochemically driven current systems near the core-mantle boundary.

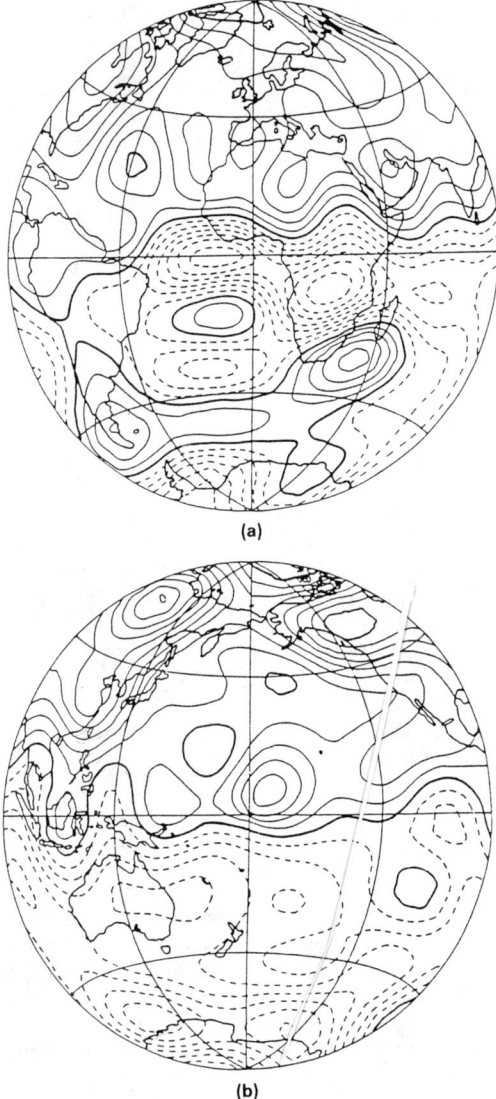

FIGURE 8. A model of the radial field at the core-mantle boundary. The contour interval is 100 nT. Solid (dashed) contours represent flux into (out of) the core. The thicker lines are null-flux lines. (After Bloxham and Gubbins, 1985)

Secular Variation of the Dipole Field

The moment and orientation of the geomagnetic dipole are determined from a spherical harmonic analysis of worldwide observations. In terms of the first degree spherical harmonic Gauss coefficients, the moment of the geocentric axial dipole is:

$$M_a = \frac{4\pi}{\mu_0} r^3 g_1^0 \text{ in SI units,}$$

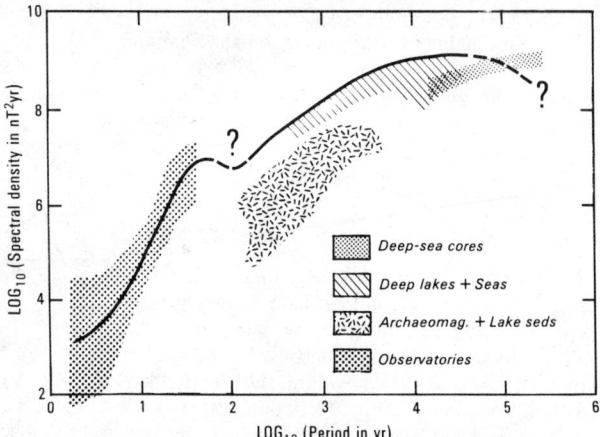

FIGURE 9. Geomagnetic power density spectrum, based on observatory, archaeomagnetic, and paleomagnetic data.

where r is the radius of the Earth and $\mu_0 = 4\pi\ 10^{-7}$ H m^{-1}. Gauss coefficient g_1^0 is the strength of the axial dipole field at the equator in nT. The moment of the inclined geocentric dipole is:

$$M_i = \frac{4\pi}{\mu_0} r^3 \sqrt{(g_1^0)^2 + (g_1^1)^2 + (h_1^1)^2}.$$

The orientation of the inclined geocentric dipole, with north pole (i.e., the south geomagnetic pole) at latitude (λ) and east-longitude (ϕ) is:

$$\operatorname{Tan} \lambda = \frac{g_1^0}{\sqrt{(g_1^1)^2 + (h_1^1)^2}};\ \operatorname{Tan} \phi = \frac{h_1^1}{g_1^1}.$$

Only since the mid-nineteenth century have there been sufficient observations to permit reliable spherical harmonic analysis of the field. Barraclough (1978) has compiled 264 published analyses, for epochs from 1550 onward. Data for the earlier epochs were supplemented with archaeomagnetic inclination and intensity data in order to obtain uniform data sets. Table 2 gives the mean values of the parameters of the inclined geocentric dipole for each epoch, based on analyses selected as follows:

only analyses taken to degree 4, or greater were used;
the 1985 revision of IGRF was used for 1945.0 onward;
for 1800 to 1942, mean values were taken of M, λ, and ϕ for all the models in Barraclough's compilation;
for epoch 1780, Braginskiy's analyses were used in preference to the two nineteenth century analyses;
prior to 1780, only analyses with dipole moments falling within the interquartile mean (i.e., the mean of the median half of the data, interpreted loosely, see below) were used, in an attempt to eliminate the less accurate analyses (the assumption being that M is the parameter most likely to vary smoothly).

In the table, N is the number of values averaged; for interquartile means this is followed in parentheses by the total number of analyses for that epoch. After 1800, the distinction between interquartile means and averages ceases to be important, which is an indication of the improved quality of the later models. If five or more values were averaged, the standard deviations are given in the table in parentheses; otherwise the range limits are given in square brackets.

The dipole parameters are plotted in Fig. 10. Solitary analyses for 1770 and 1820, which give dipole moments inconsistent with the trend of M, have been omitted. Crosses joined by dashed lines are for the 50-year epoch models of Thompson and Barraclough (1982), which are based on a robust cubic spline smoothing technique applied to historical data in B. P. Veinberg and V. P. Shibaev's catalog supplemented with 19 archaeomagnetic data for the earlier epochs. The thick solid lines from 1910 to 1950 are for the fifth-degree spherical harmonic models computed at 2-yearly intervals by Hodder (1981) using observatory annual means only. Points that show large transient departures from this solid line, e.g. at 1932 and 1937, are unlikely to represent dipole behavior.

Error bars in Fig. 10 show only the variability between different spherical harmonic models for the same epoch. The inaccuracies of the models will certainly be larger, particularly for earlier epochs. Even with identical data sets, the type of analysis alone can have a surprisingly large effect on the dipole parameters. For example, Hodder's analyses taken to degree 5 and 6 give dipole moments differing by up to 1%, and geomagnetic pole longitudes varying between $\pm 1.5°$.

Up to the mid-nineteenth century the dipole axis drifted westward fairly uniformly at a mean rate of 0.14 deg yr^{-1} (period 2570 years), while its tilt (colatitude) increased steadily from about 3° in 1550 A.D. to about 11.5° in 1830–1840. For the first half

TABLE 2. Parameters of the Geomagnetic Dipole (Inclined, Geocentric)

Epoch 1	Dipole Moment ($\times 10^{22}$ A m^2)	g_1^0 (μT)	Geomagnetic North Pole Colatitude (°)	Longitude (°E)	N
1990	7.835	−29.761	10.81	289.02	1
1985	7.871	−29.877	11.02	289.09	1
1980	7.906	−29.992	11.19	289.24	1
1975	7.938	−30.100	11.31	289.53	1
1970	7.972	−30.220	11.41	289.82	1
1965	8.004	−30.334	11.46	290.15	1
1960	8.024	−30.411	11.47	290.51	1
1955	8.049	−30.507	11.44	290.21	1
1950	8.067	−30.571	11.51	291.10	1
1945	8.084	−30.634	11.48	291.10	1
1942	8.03 (.04)	−30.450 (159)	11.43 (.53)	292.6 (2.2)	7
1937	8.10 [±.00]	−30.625 [±.015]	12.06 [±.28]	295.2 [±1.2]	2
1935	8.08	−30.662	11.37	290.1	1
1932	8.11 [±.00]	−30.665 [±.045]	12.07 [±.34]	294.9 [±1.3]	2
1925	8.15	−30.892	11.37	290.0	1
1922	8.07 (.11)	−30.574 (.409)	11.44 (.09)	291.1 (.9)	5
1915	8.22	−31.176	11.42	290.1	1
1910	8.25	−31.246	11.55	291.7	1
1905	8.29	−31.423	11.45	290.0	1
1900	8.27 [±.02]	−31.331 [±.081]	11.46 [±.06]	292.0 [±.08]	2
1890	8.36	−31.718	11.32	294.8	1
1885	8.39 (.04)	−31.790 (.175)	11.41 (.18)	291.2 (1.0)	22
1880	8.45 (.20)	−31.669 (.041)	11.55 (.19)	292.3 (1.0)	5
1860	8.60	−32.576	11.57	296.1	1
1850	8.47 [±.02]	−32.111 [±.111]	11.52 [±.22]	295.6 [±.4]	2
1845	8.49 [+.04] [−.04]	−32.186 [+.145] [−.089]	11.33 [+.11] [−.10]	295.6 [+.41] [−.71]	4
1842	8.51 [+.06] [−.05]	−32.273 [+.179] [−.176]	11.45 [+.12] [−.13]	294.8 [+.17] [−.35]	3
1840	8.52	−32.233	12.14	294.4	1
1835	8.54 [+.06] [−.08]	−32.296 [+.188] [−.240]	12.00 [+.41] [−.57]	297.2 [+2.1] [−1.3]	3
1830	8.55	−32.422	11.38	298.7	1
1829	8.47	−32.077	11.70	295.2	1
1820	8.81	−33.381	11.57	292.9	1[a]
1800	8.61 (.06)	−32.710 (.250)	10.80 (.15)	301.0 (.7)	6
1780	8.73 [±.00]	−32.230 [±.00]	10.34 [±.04]	303.7 [±.2]	2[b]
1770	8.19	−31.185	10.01	302.5	1[a]
1750	8.84	−33.646	10.08	305.4	1(3)
1700	9.00 [±.00]	−34.450 [±.031]	8.31 [±.17]	314.6 [±1.7]	2(6)
1650	9.18 [±.00]	−35.231 [±.039]	6.98 [±.30]	322.3 [±3.4]	2(6)
1600	9.36 [±.00]	−36.060 [±.00]	5.37 [±.01]	330.3 [±3.9]	2(6)
1550	9.54	−36.840	3.09	334.1	1(3)

Note: Figures in parentheses are standard deviations; figures in brackets are range limits; N is the number of analyses averaged, followed in parentheses by the total number if different.
[a] Omitted from Figure 10.
[b] Braginskiy's analyses only.

of the twentieth century, the axis remained relatively static, with north geomagnetic pole longitude around 291° E, and tilt close to 11.5°. Since 1960, the axis has resumed its westward drift, at about 0.12°/yr^{-1}, whereas the tilt has dropped back to 11°.

The view expressed in literature of the 1960s, that the movement of the geomagnetic axis is dominantly a westward precession with virtually no variation in colatitude, is unfounded (Fig. 11). The movement of the dipole axis has tended to slow down over the last 400 years and went through a quiescent phase during the first half of this century.

Variation of the Earth's Dipole Moment

The most striking characteristic of the dipole variation is its rapid, near-linear decay in strength (Fig. 10). Prior to the twentieth century the dipole moment decayed at a rate of approximately 0.0031 $\times 10^{22}$ A m^2 yr^{-1}, i.e., 3.2% per 100 years for 400

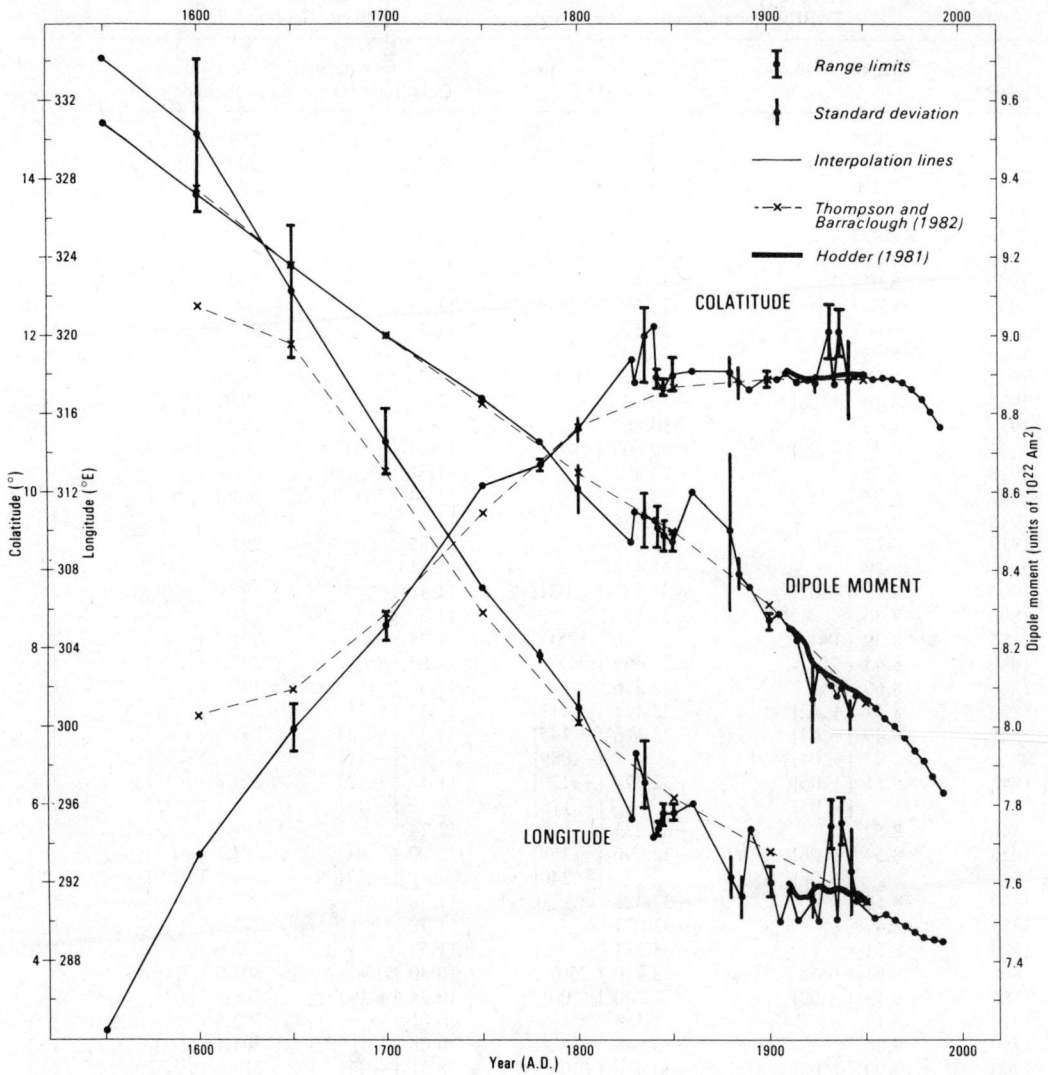

FIGURE 10. Secular variation of the geomagnetic dipole, based on mean values of spherical harmonic analyses (Table 2). Colatitude and longitude are for the north geomagnetic pole. Data for epochs 1770 and 1820 are omitted. Error bars with squared ends denote range limits; plain error bars denote standard deviations (5 or more values).

years. (Note that absolute intensity measurements were not started until 1832 and that dipole moments before this are based on archaeomagnetic measurements and extrapolation). Around 1900 A.D. the decay rate increased, and for the last 30 years it has been about 0.0048×10^{22} A m² per year (5.8% per 100 years).

The average value of M over longer periods of time (hundreds to thousands of years) can be estimated from worldwide archaeointensity measurements (see *Archaeomagnetism*). Results indicate that M has been decreasing steadily since reaching a maximum of approximately 11.5×10^{22} A m² around 600 B.C. (Fig. 12), and that between 50,000 and 15,000 years ago, the dipole was generally 50% weaker than during the last 10,000 years. It has been speculated that if the decay rate during the last 400 years were to persist, the demise of the Earth's dipole moment would occur approximately 2500 years from now, with a 50% probability of triggering a polarity reversal. However, since the dipole moment has been recovering from a historical high this speculation carries little weight.

Paleointensity determinations on ancient rocks give the mean value of the dipole moment as 8.7×10^{22} A m² during the last 5 million years and show that this has not changed substantially during the last 3 billion years. Thus it is probable that the geodynamo was established early in the lifetime of the Earth in a form similar to that of today.

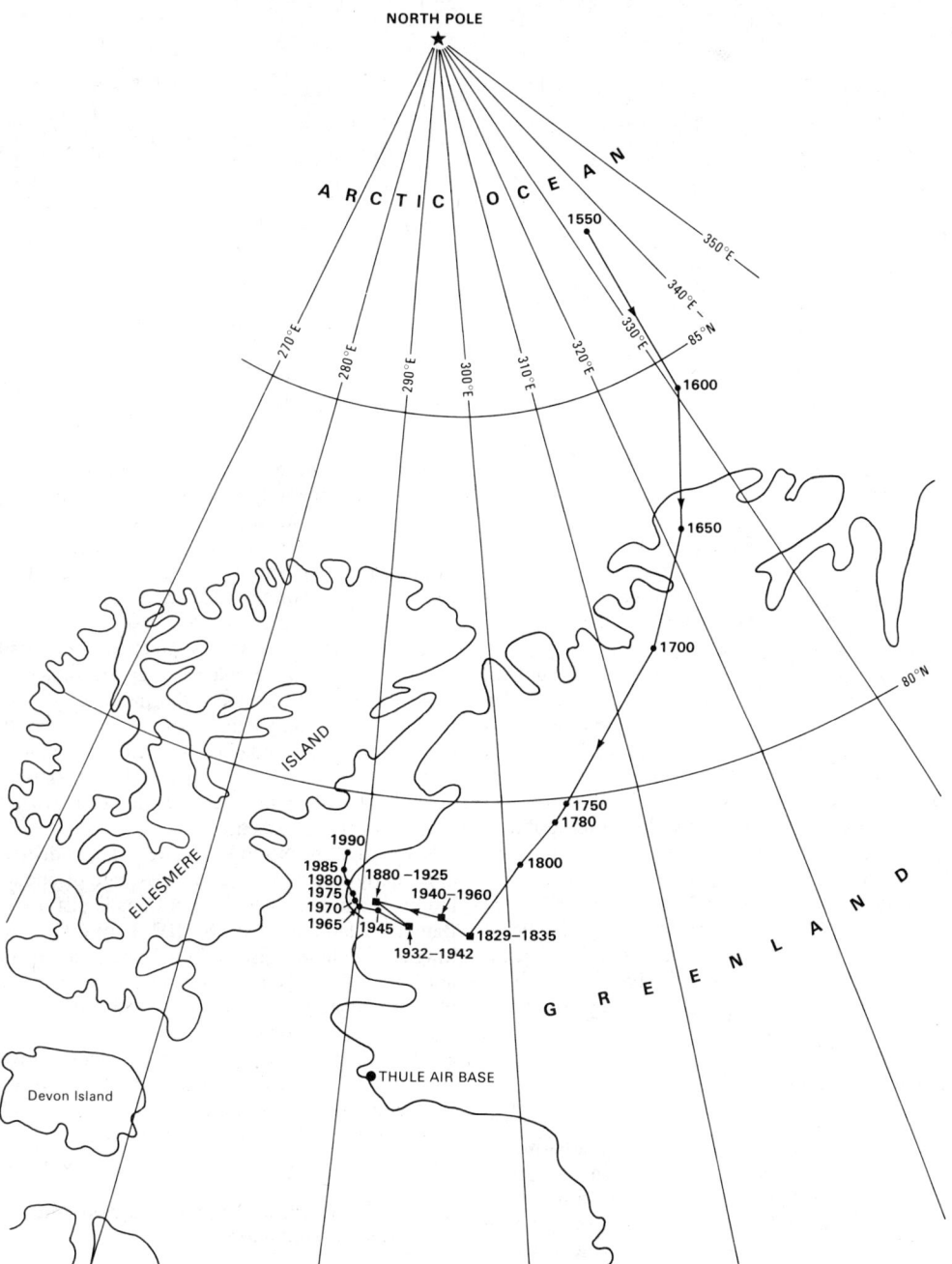

FIGURE 11. Movement of the north geomagnetic pole since A.D. 1550. Solid circles denote single epoch data; squares denote coordinates obtained by averaging over the stated intervals. The 1985 revision of IGRF is used from 1945 onward.

The Pacific Nondipole Low

Spherical harmonic models and charts of the geomagnetic field covering the last 200 years show that the Pacific Ocean is characterized by persistently low secular variation. Furthermore, the field over the Pacific region is more markedly dipolar in character than over the Earth as a whole. For this reason, the phenomenon has been termed the *Pacific nondipole low*, or *dipole window*.

Paleomagnetic data from sequences of lava flows from islands in the Pacific, primarily Hawaii,

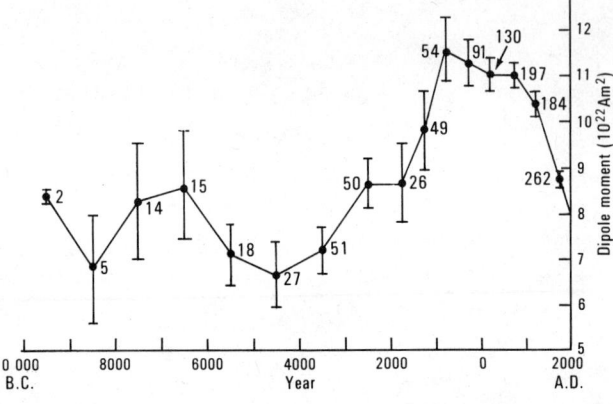

FIGURE 12. Global mean dipole moments with 95% confidence limits obtained from archaeomagnetic data. The figures denote the number of analyses per 500 year interval post 2000 B.C. or per 1000-year interval pre-2000 B.C. (After McElhinny and Senanayake, 1982)

likewise display abnormally low directional dispersion, leading to the suggestion that low secular variation in the region is a permanent feature. The hypothesis has attracted considerable attention since lateral heterogeneity of the geomagnetic field on this scale would provide a constraint on dynamo models and could be diagnostic of major inhomogeneities in the internal structure of the Earth.

There now appears to be general concensus that the lava flow sequences studied were not extruded uniformly through time, and the time intervals sampled paleomagnetically were too short to give a representative measure of the secular variation. Subsequent measurements made on lavas from the Society Islands and Easter Island do not show abnormally low directional scatter, i.e., secular variation, although results from radiocarbon-dated lava flows from Hawaii spanning the interval 200–31,000 years B.P. yield angular dispersions lower than predicted by global models of the secular variation. The Pacific nondipole low is now generally believed to be a transient characteristic of the geomagnetic field, albeit a long-lived one.

Secular Motion of the Magnetic Poles

Unlike the geomagnetic poles, which are determined from the coefficients of a spherical harmonic analysis of the global field, the north and south magnetic poles have a physical manifestation, namely the principal points on the Earth's surface (remote from localized magnetic anomalies), where the field is exactly vertical (inclination = $\pm 90°$).

During the nineteenth and early part of the twentieth century, the search for the magnetic poles assumed an importance little short of that for the geographic poles, due partly to the belief that a knowledge of their location was important for global navigation. A flag was first placed at the North Magnetic Pole by Commander James Clark Ross in 1831. Ross's ambition was to take the same flag to the South Magnetic Pole, but his attempt to do so in 1840–1841 was thwarted at an estimated distance of 250 km from the pole. The honor of first reaching the South Magnetic Pole went to Australians T. W. Edgeworth David and Douglas Mawson, and Scotsman Alistair Mackay who reached their estimated mean position of the pole (72° 25′ S, 155° 16′ E) on the Antarctic Plateau in Victoria Land on 16 January 1909.

As both parties knew, it is practically impossible to reach the magnetic poles as they are subject to a rapid diurnal motion under the influence of externally generated magnetic fields. During a day they trace out an approximately elliptical path with axes varying in length from less than 10 km on extremely quiet days to many hundreds of kilometers when the field is highly disturbed.

Since 1832 the North Magnetic Pole has drifted in a northerly direction at an average rate of about 11 km yr^{-1}, and is now located in Canada just north of Bathurst Island at 77.0° N, 102.3° W. The South Magnetic Pole has also been drifting in a generally north-northwesterly direction, at an average speed of approximately 9 km yr^{-1} since 1841. It drifted off the Antarctic continent into the Southern Ocean in about 1960 and is presently in the vicinity of 65° 20′ S, 138° 10′ E, about 150 km NNW of the French Antarctic base of Dumont d'Urville. These differences in the secular motion of the magnetic poles reflect regional variations in the nondipole field.

Secular Acceleration Impulses (Jerks) and Geophysical Implications

The concept of impulses in the secular acceleration (i.e., very rapid variations in the second derivative of the field) was introduced by J. B. Walker and P. L. O'Dea in 1952. By assigning them an internal origin, S. K. Runcorn in 1955 and K. L. MacDonald in 1957 were both able to estimate the electrical conductivity of the lower mantle. However, it was later demonstrated that most of these impulses stemmed from solar cycle effects.

Ideally an impulse is manifested as an instantaneous change in the slope of the first time derivative (i.e., secular variation) of one or more elements of the field, with a corresponding step change in the second derivative, and a Dirac delta function change in the third derivative. In practice, observatory annual mean values are used to monitor the secular variation, so it is difficult to tie down the duration of observed impulses. Shorter time averages, e.g., monthly means, inherently have a higher scatter due to imperfect removal of external disturbances, and do not necessarily lead to better time resolution.

Around 1969/1970 a sharp change in slope of the secular variation showed up in annual mean values from a large number of observations (Fig. 13). It was particularly noticeable at European observatories in the easterly (Y) component of the field, for which the 11-year solar cycle signal is weak. The effect was initially noted by H. Mizuno in Japan, V. Courtillot, J. Ducruix, and J. L. Le Mouël in France, and by S. R. C. Malin, D. R. Barraclough, and B. M. Hodder in Britain, and has become known as the "geomagnetic jerk."

The question as to whether the jerk is a distinct geomagnetic phenomenon is currently under debate. One body of opinion, championed by the French group, claims that the 1969/1970 jerk was worldwide in character, synchronous over the globe, of internal origin, and also that there have been additional jerks in 1910 and about 1850 (Courtillot et al., 1978). The opposing camp contends that the discontinuities in the second derivatives are a figment of the type of analysis applied, namely the fitting of low order polynomials to data on each side of an inferred jerk; hence there is no reason to suppose that the observations represent anything other than natural fluctuations of a continuously varying signal (Alldredge, 1984, 1985). Alldredge argues that there are insufficient grounds to suppose that a worldwide impulse occurred in 1969–1970, and that the rapid changes in secular acceleration observed simultaneously in Europe and Alaska at that time can be explained as an external effect caused by the auroral electrojet.

There is now an extensive literature on the subject of the 1969/1970 geomagnetic jerk, and earlier jerks around 1910 and possibly around 1845. The reader is referred to Courtillot and Le Mouël (1984) and Alldredge (1984, 1985) who review the literature and present, respectively, the arguments for and against the existence of jerks. The existence of geomagnetic jerks of internal origin has far reaching geophysical ramifications.

Mantle Conductivity. The estimated duration of the 1969/1970 jerk observed at the Earth's surface is 1 to 2 years, i.e., considerably shorter than

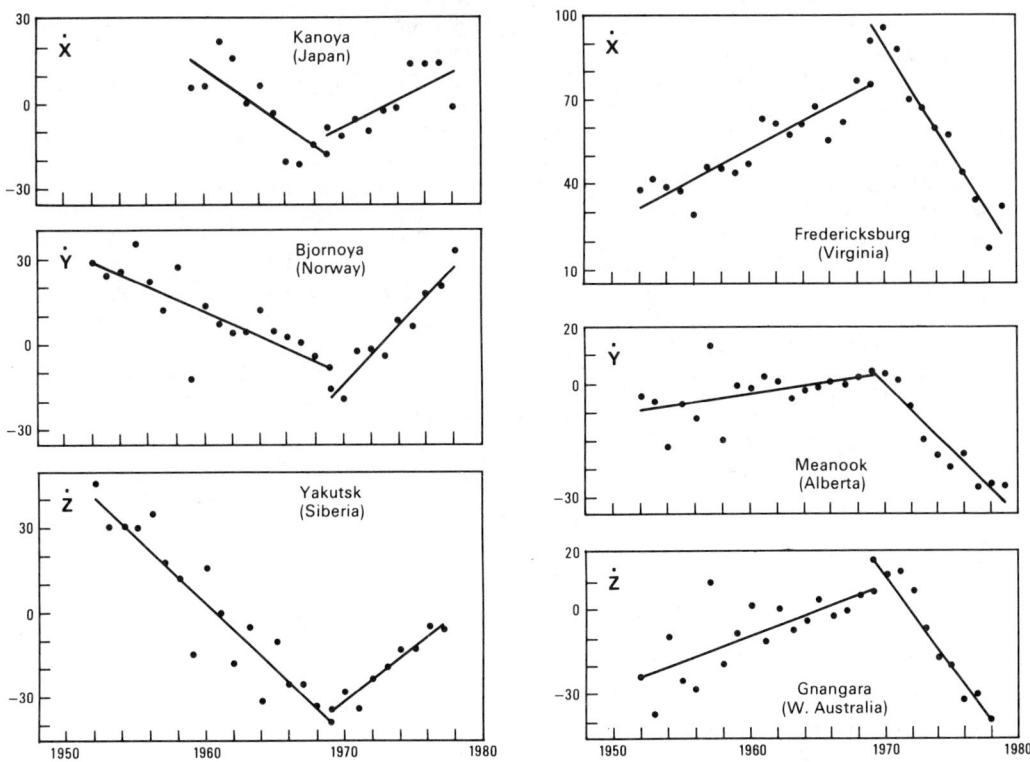

FIGURE 13. Examples of linear segment fits to the secular variation of the magnetic elements (differences between annual means) at some observatories. (After Madden and Le Mouël, 1982)

previous estimates of the lower cutoff for magnetic signals from the core. This shortened estimate has been used to infer a substantial downward revision of the electrical conductivity of the mantle, to no more than a few hundred S m^{-1} in the lower mantle. G. Backus pointed out in 1983 that some earlier analyses were over-simplistic and failed to allow for harmonic mode mixing (i.e., the resultant signal observed at the Earth's surface is a mixture of the different harmonics of the source impulse, each having its own characteristic, Gaussian smoothing and delay times after passing through the mantle). The effect of this is to underestimate the conductivity of the mantle. Backus shows that for reasonable conductivity distributions, the ratio of the smoothing time to the delay time must lie between 0.3 and 0.8. Thus for a smoothing time of about 1 year, as claimed for the jerk, the delay time must be 1 to 3 years.

Correlation with Westward Drift Rate. There was a decreasing trend in the westward drift rate for about 20 years prior to the 1969/1970 jerk, followed by a subsequent increase. There is also a suggestion that a significant change in westward drift rate occurred about the time of the 1910 jerk. It should be noted, however, that there is considerable variability between various estimates of the westward drift rate, and also between estimates for different features of the field (as illustrated in Fig. 6). Hence it is difficult to be specific about when the acceleration of the drift occurred. Based on these drift-rate changes it has been inferred that the sources of the jerk are associated with angular momentum transfer in the outer regions of the core. This conclusion, combined with the claimed global synchroneity of the jerk, has been taken to indicate that the westward drift is due to rotational effects rather than to hydromagnetic waves propagating in the core. Efforts during the last few years to model fluid motions at the surface of the core, using the frozen flux approximation, likewise indicate that a change in character of fluid motions occurred around 1970.

Correlation with Fluctuations in the Length of Day. A third phenomenon that may be closely associated with the jerk is the correlation between long-term decade fluctuations in the length of day, i.e., the rotation rate of the Earth, and the secular variation (notably the westward drift rate). Such a relationship was first suggested by Vestine in 1953. A striking example of the correlation with the westward drift rate for European declination data is shown in Fig. 14. However, it should be noted that other estimates of the westward drift rate, for example for the position of the eccentric dipole, do not show such a clear correlation with the length of day. Courtillot et al. (1978) also argue that the proposed jerks around 1850, 1910, and 1969/1970 coincide with minima in the Earth's rotation rate (peaks in the excess length of day).

Decade fluctuations in the length of day appear to be caused by transfer of angular momentum between the core and the mantle. Electromagnetic coupling between the mantle and fluid motions in the core is generally accepted as the causal link with geomagnetic variations. The effects of viscous drag appear to be too small to be important, although coupling via topographic irregularities at the core-mantle boundary may play a role.

The correlation evident in Fig. 14 has been interpreted in two ways. Le Mouël, Courtillot, and coworkers argue that increased convection imparts a westward torque to the outer shell of the core and an equal and opposite torque on the main body of the core. With low electrical conductivity in the mantle, the magnetic signal is rapidly propagated, whereas there is a 10-year time lag before electromagnetic coupling between the main body of the core and the mantle leads to an observable increase in rotation rate. The adoption of this concept to model the amplitude and phase relationships between the two signals indicates that the thickness of the outermost layer of the core responsible for the westward drift is of the order of 100 km, and that the toroidal and radial velocities within the shell are about 3×10^{-4} m s^{-1} and 10^{-6} m s^{-1} respectively.

On the other hand Vestine, Runcorn, and more recently Backus have each taken the view that increases in westward drift rate lag behind decreases

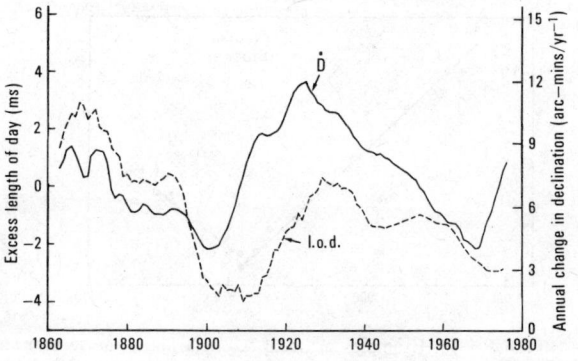

FIGURE 14. Secular variation of declination in Paris (5-year means, solid line, right-hand scale) and the excess length of day (dashed line, left-hand scale). The latter is proportional to the angular velocity of the Earth. (After Le Mouël and Courtillot, 1981)

in rotation rate by approximately 13 years. It is argued that increased convection in the core leads to reduction in the angular velocity of its outer shell, which is rapidly transmitted to the mantle via electromagnetic coupling. A high electrical conductivity in the lower mantle is required so that the magnetic signal takes approximately 13 years to propagate to the surface.

The latter interpretation is somewhat simpler in terms of the angular momentum exchange processes involved and is now more widely accepted than the former.

Debate regarding the reality of geomagnetic jerks of internal origin is currently at a critical stage. Both camps have drawn opposite conclusions from the same data. Given the strong evidence for jerks in European data, researchers have proceeded confidently to identify corroborative evidence from worldwide observatories. However, if the density of observatories in Europe was as low as in many parts of the world, the 1969/1970 jerk might not have been recognized so readily, if at all. As pointed out by Alldredge, the arguments concerning the relationship between the length of day and the secular variation are largely independent of the outcome of this debate.

Conclusion

An understanding of the secular variation has long been recognized as one of the most important problems in geomagnetism. Indeed, Albert Einstein ranked the origin (and hence, indirectly, the behavior) of the geomagnetic field among the five outstanding problems in physics. Despite the wealth of observational data that we now have, the problem remains as a major challenge.

Acknowledgments

I thank D. Gubbins, P. L. McFadden, and R. T. Merrill for their constructive comments. This paper is published with permission of the Director, Bureau of Mineral Resources.

C. E. BARTON

References

Alldredge, L. R., 1984, A discussion of impulses and jerks in the geomagnetic field, *Jour. Geophys. Research* **89**, 4403-4412.

Alldredge, L. R., 1985, More on the alleged 1970 geomagnetic jerk, *Physics Earth and Planetary Interiors* **39**, 255-264.

Barker, F. S., D. R. Barraclough, and S. R. C. Malin, 1981, World magnetic charts for 1980—spherical harmonic models of the geomagnetic field and its secular variation, *Royal Astron. Soc. Geophys. Jour.* **65**, 525-533.

Barraclough, D. R., 1978, Spherical harmonic models of the geomagnetic field, *Inst. Geol. Sci. Geomagnetic Bull.* No. 8.

Bauer, L. A., 1895, On the secular motion of a free magnetic needle. II, *Rev. Physics* **3**, 34-48.

Bloxham, J., and D. Gubbins, 1985, The secular variation of Earth's magnetic field, *Nature* **317**, 777-781.

Bullard, E. C., C. Freedman, H. Gellman, and J. Nixon, 1950, The westward drift of the Earth's magnetic field, *Royal Soc. London Philos. Trans.* **A243**, 67-92.

Courtillot, V. and J.-L. Le Mouël, 1984, Geomagnetic secular variation impulses, *Nature* **311**, 709-716.

Courtillot, V., J. Ducruix, and J.-L. Le Mouël, 1978, Sur une accéleration récente de la variation séculaire du champ magnétique terrestre, *Acad. Sci. Paris Comptes Rendus* **D287**, 1095-1098.

Gire, C., J.-L. Le Mouël, and T. Madden, 1984, The recent westward drift rate of the geomagnetic field and the body drift of external layers of the core, *Annales Geophys.* **2**, 37-46.

Hodder, B. M., 1981, Geomagnetic secular variation since 1901, *Royal Astron. Soc. Geophys. Jour.* **65**, 763-776.

IAGA Division I Working Group 1, 1987, The International Geomagnetic Reference Field revision 1987, *Jour. Geomagnetism and Geoelectricity* **39**, 773-779.

Le Mouël, J.-L., and V. Courtillot, 1981, Core motions, electromagnetic core-mantle coupling and variations in the Earth's rotation: new constraints from geomagnetic secular variation impulses, *Physics Earth and Planetary Interiors* **24**, 236-241.

McElhinny, M. W., and W. Senanayake, 1982, Variations in the geomagnetic dipole: the last 50,000 years, *Jour. Geomagnetism and Geoelectricity* **34**, 39-51.

McFadden, P. L., R. T. Merrill, and M. W. McElhinny, 1985, Non-linear processes in the geodynamo: palaeomagnetic evidence, *Roy. Astron. Soc. Geophys. Jour.* **83**, 111-126.

Madden, T., and J.-L. Le Mouël, 1982, The recent secular variation and the motions at the core surface, *Roy. Soc. London Philos. Trans.* **A306**, 271-280.

Malin, S. R. C., and E. C. Bullard, 1981, The direction of the Earth's magnetic field at London, 1570-1975, *Roy. Soc. London Philos. Trans.* **A299**, 357-423.

Parkinson, W. D., 1983, *Introduction to Geomagnetism.* Scottish Academic Press, distributed by Elsevier, 433 p.

Thompson, R., and D. R. Barraclough, 1982, Geomagnetic secular variation based on spherical harmonic and cross validation analyses of historical and archaeomagnetic data, *Jour. Geomagnetism and Geoelectricity* **34**, 245-263.

Vestine, E. H., 1953, On variations of the geomagnetic field, fluid motions, and the rate of the Earth's rotation, *Jour. Geophys. Research* **58**, 127-145.

Yukutake, T., and H. Tachinaka, 1968, The non-dipole part of the Earth's magnetic field, *Tokyo Univ. Earthquake Research Inst. Bull.* **46**, 1027-1074.

Cross-references: *Earth Orientation; Geomagnetic Field: Asymmetries; Geomagnetic Field: Measurement; Geomagnetic Field: Westward Drift; Geomagnetic Field, Main: Theory; Geomagnetic Field Analysis; Geomagnetic Secular Variation: Theory; Geomagnetism: Historical Introduction; Magnetohydrodynamic Waves within the Earth; Paleomagnetic Secular Variation; Spherical Harmonic Analysis.*

GEOMAGNETIC SECULAR VARIATION: THEORY

A major part of the geomagnetic field is considered as being generated through hydromagnetic processes in the Earth's liquid core, which consists mostly of iron. Since the electrical conductivity of iron is high, electric currents are induced in the core by motions of the liquid iron in the presence of the magnetic field. The electric currents induced in this way recreate the magnetic field. Accordingly, time variation in the geomagnetic field is regarded as a manifestation of the fluid motions in the core, and the geomagnetic secular variations are understood within a frame of the generation process of the geomagnetic field dynamo process (see *Geomagnetic Field, Main: Theory*).

There is yet no unanimous agreement regarding the dynamo process of maintaining the geomagnetic dipole field. Some require intense toroidal fields (the field confined within the core *without* the radial component) more than ten times stronger than the poloidal fields (the field *with* the radial component and observable at the Earth's surface). Others claim that the toroidal fields need not be stronger than the poloidal fields. The former process is called the *strong field model*, the latter the *weak field model*. According to models, interpretation of the secular variations differs.

Geomagnetic Secular Variations

Axisymmetric Field. One of the most pronounced characteristics of the geomagnetic secular variations is the change in the dipole field (see *Geomagnetic Secular Variation: Direction and Intensity*). At present the dipole moment is decreasing at a rate of about 6% per century. Its rate of decrease has not been constant for the past hundred years, but subject to a year-to-year variation of about a 60-yr period. Paleomagnetic investigations indicate that the dipole field has changed its intensity significantly on a much longer time scale (Merrill and McElhinny, 1983). The dipole moment seems to have been a maximum about 1500 years ago, about 1.5 times as intense as the present one, whereas it was nearly the same as the present one about 8000 years ago. It is also well known that the dipole field has changed its polarity frequently over the geologic past.

Westward Drift. Westward drift is another aspect of the geomagnetic secular variation (see *Geomagnetic Field: Westward Drift* and Yukutake, 1979). The nonaxisymmetric field is known to have been drifting westward with a velocity of approximately 0.3° per year over the past several hundred years. Variations in declination, for example, show a clear westward drift. The longitude of an intersection point of an agonic line, where the declination becomes zero, with the equator is found to be drifting westward with a velocity of 0.28° per year since 1700. The westward drift is also seen in the phase shift of harmonic components of the geomagnetic potential expanded in a spherical harmonic series. Phase angles of harmonics such as sectorial terms, for which the degree (n) is equal to the order (m), decrease linearly with time, giving westward drift velocities of 0.24° per year for $n = m = 2$, for example.

It is not true, however, to say that the whole pattern of the geomagnetic field is drifting westward. Significant parts are standing at the same location over several hundred years. Such a nondipole anomaly of the positive vertical component that covers most of the Eurasian continent with its center near Mongolia, is an example of the standing field. Besides the westward-drifting component, eastward-drifting fields are supposed to exist. When archeomagnetic data going back to 2000 years ago are examined, a pronounced eastward drift is seen in the time variation of the equatorial dipole component with a velocity of 0.3° per year.

Short-Period Phenomena. The geomagnetic field is also subjected to shorter-period fluctuations. Variations of 20 to 30 years are observed at many places. These decade variations are considered to be of core origin and have complicated spatial structures.

A disputed question has been whether impulsive changes exist in the secular variation of the core origin. The phenomenon *secular change impulse* found in earlier dates has been confirmed to be the variation related to solar activity. Recently, although not identical, a rapid phenomenon called *geomagnetic jerk* was found. It is an impulsive change in the second time derivative of declination—in other words, a discontinuous change in the rate of change, which occurred in Europe around 1970. Although there are still arguments about its extraction process, many are inclined to believe this to be of core origin.

Theory of the Westward Drift

Generation of the nonaxisymmetric field and its drift are considered to be deeply involved in the dynamo process. In the strong field model, electromagnetic coupling between the core and the mantle (see *Core-Mantle Coupling*) is considered to make the outer part of the core rotate more slowly than the mantle. Then the magnetic field rotating with the outer core is observed as drifting relative to an observer at the Earth's surface. If a hydromagnetic wave is excited and propagates along the toroidal field, it can be a drifting field (see *Magnetohydrodynamic Waves within the Earth*). In the weak field model, on the other hand, the rotating field around the Earth's rotation axis is obtained as a result of the dynamo process to maintain the axial dipole.

Equations. In order to understand secular variations, we have to solve the induction equations together with the Navier-Stokes equation and a thermodynamic equation. Let **B** be the magnetic flux density, σ the core conductivity, and μ the core magnetic permeability. Then the induction equation becomes

$$\frac{\partial \mathbf{B}}{\partial t} = \text{curl}(\mathbf{v} \times \mathbf{B}) + \eta \nabla^2 \mathbf{B} \qquad (1)$$

where $\eta = 1/\sigma\mu$ is the *magnetic diffusivity*. The vector **v** represents fluid velocity in the coordinate system fixed to the mantle. We define the magnetic Reynolds number R_m, a dimensionless parameter, as $R_m = \sigma\mu UL$, where L and U are typical length scale and magnitude of velocity, respectively. The Navier-Stokes equation becomes

$$\frac{\partial \mathbf{v}}{\partial t} + (\mathbf{v} \cdot \nabla)\mathbf{v} + 2\mathbf{\Omega} \times \mathbf{v}$$

$$= -\frac{1}{\rho}\nabla P + \nu\nabla^2\mathbf{v} + F_B + F_\alpha \qquad (2)$$

$$F_B = \frac{1}{\rho\mu} \text{curl } \mathbf{B} \times \mathbf{B}$$

where $\mathbf{\Omega}$ is the angular velocity of the mantle, ρ is the density, P is the pressure, ν is the kinematic viscosity, F_B is the Lorentz force, and F_α is the buoyancy force.

The thermodynamic equation to be solved differs according to the model. When the thermal origin is considered as a driving force, the equation becomes

$$\frac{\partial T}{\partial t} + \chi\nabla^2 T = Q \qquad (3)$$

where T is temperature, χ is thermal diffusivity, Q is a heat source. Subsequently, we have

$$F_\alpha = \alpha(T - T_0)g$$

where α denotes the thermal expansion coefficient and g the gravitational acceleration.

When a compositional convection is considered instead, where lighter material released at the surface of the inner core floats up and the heavier iron sinks from the bottom of the mantle (see *Earth's Core*), Eq. 3 changes and we have (Braginskiy, 1978)

$$\frac{\partial C}{\partial t} + \mathbf{v} \cdot \nabla C + \text{div } \mathbf{i}^c = Q^c$$

where C represents density inhomogeneity that causes buoyancy, $\mathbf{i}^c$ is the diffusion flux, and Q^c is the source of enhanced density.

When σ is high, we have $R_m \gg 1$, and the diffusion term expressed by the second term on the right side of Eq. 1 can be ignored. Then Eq. 1 reduces to

$$\frac{\partial \mathbf{B}}{\partial t} = \text{curl}(\mathbf{v} \times \mathbf{B})$$

This implies that the magnetic flux passing through an enclosed surface is preserved. This phenomenon is well known as a *frozen field:* The magnetic lines of force move around with fluid particles as if they were frozen into the fluid particles.

Bullard Model. Using a strong field model, Bullard and others (1950) considered the westward drift of the magnetic field to be a manifestation of the relative rotation of the outer core to the mantle by assuming that the magnetic lines of force were frozen in the conducting outer core.

The Earth's rotation has an overwhelming effect on the fluid motion in the core. However, the dipole field cannot be maintained by rotational motion alone. A certain type of convectional motion is necessary, whether thermal convection or compositional.

Suppose that the convection occurs in a rotating system, and the angular momentum of each fluid particle is conserved. Then a rising fluid particle is decelerated as it floats up so as to conserve its angular momentum, whereas a sinking fluid particle is accelerated. This causes a shear motion inside the core; the inner part of the core rotates faster than the outer part. Through interaction with the dipole field, the shear motion creates an intense toroidal field of about 100 gauss, which has only an azimuthal component, eastward in the Northern Hemisphere and westward in the Southern Hemisphere. The toroidal field thus induced is associated with poloidal electric currents flowing in the meridional planes, as shown in Fig. 1*a*.

Since the mantle is weakly conducting, part of the electric current leaks out into the mantle. Interacting with the dipole field, the current exerts an accelerating torque on the mantle. This can be balanced by an electromagnetic torque to be generated when a relative motion exists between the outer core and the mantle. If the mantle rotates faster than the outer core crossing the dipole field, the electric currents are induced as shown in Fig. 1*b*. Interaction of the currents with the dipole field exerts a decelerating torque on the mantle.

Using a simple core model consisting of a sphere and a spherical shell of equal volume, Bullard and others (1950) investigated the electromagnetic coupling between the core and the mantle and found the torque on the mantle to be

$$N = k\sigma_m a^5 B_0^2 \left[(\omega_b - \omega_a)\frac{b^2}{a^2} - (\omega_m - \omega_a) \right]$$

provided conductivity of the mantle (σ_m) is small in comparison with that of the core (σ_c). Here k is a geometrical factor, a is the radius of the conducting core, b is the radius of the spherical surface dividing

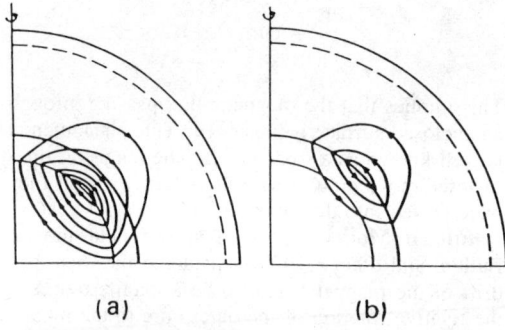

FIGURE 1. (a) Electric currents associated with the toroidal field originated from the deep interior of the core. Interacting with the dipole field, the currents exert a torque on the mantle to accelerate its rotation. (b) Electric currents generated by a relative motion between the mantle and the outer part of the core. The currents exert a decelerating torque on the mantle to balance with the accelerating torque generated by the currents from the deep interior.

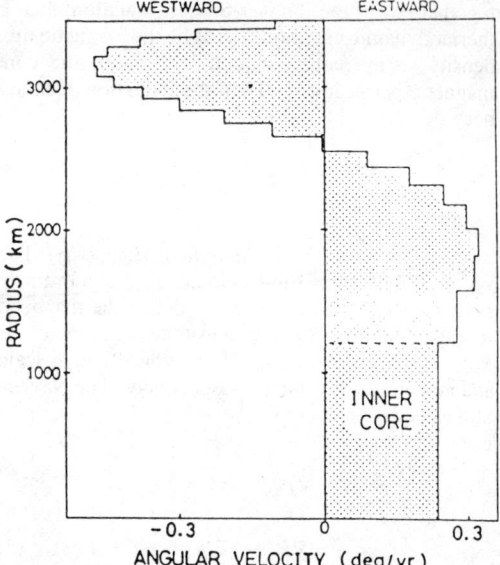

FIGURE 2. Velocity profile for the multilayered core model, in which the electromagnetic force balances with change in the angular momentum of each layer caused by convectional fluid motion. (From Watanabe and Yukutake, 1975)

the core into two equal volumes, B_0 represents magnetic flux density due to the dipole, and ω_m, ω_a, and ω_b are angular velocities of the mantle, the outer core, and the inner core, respectively.

For the rotation of the mantle to be steady, the following relation should hold:

$$\omega_m - \omega_a = \frac{b^2}{a^2}(\omega_b - \omega_a)$$

Since the angular velocity of the inner part of the core is larger than that of the outer core ($\omega_b - \omega_a > 0$), we have $\omega_m - \omega_a > 0$. The outer core rotates more slowly than the mantle; namely, the outer part of the core moves westward as a whole relative to the mantle. For the multilayered structure of the core, the velocity profile is obtained as shown in Fig. 2 (Watanabe and Yukutake, 1975). If the magnetic fields of complex patterns are either frozen in the uppermost part of the core or generated by motions in the layers near the surface of the core, they are observed to drift westward from an observer at the surface of the Earth.

A shear motion is derived by a somewhat different approach. Braginskiy (1978) considers that a magnetic buoyancy force due to the curvature of the toroidal field and a buoyancy force due to an inhomogeneous density distribution (Archimedean force) are balanced by the Coriolis force, which leads to the conclusion that the inner part of the core rotates faster than the outer part.

Hydromagnetic Waves. In a conducting liquid permeated by a magnetic field, the Alfvén wave, a hydromagnetic wave, is excited and propagates along the magnetic lines of force. Therefore it is natural to consider that hydromagnetic waves are excited in the Earth's core. In analogy to Rossby waves in meteorology and oceanography, R. Hide has proposed propagation of a magnetic wave modified by the Earth's rotation to explain the observed westward drift. Existence of this type of wave has been shown in a heuristic way by D. J. Acheson.

Let v and b be the disturbance velocity and the magnetic field of a small quantity in a constant field B. Suppose v and b change in time as $e^{i\omega t}$. When σ is infinite and ν is zero, leading terms of Eq. 1 give

$$\omega b \sim \frac{vB}{L} \quad (4)$$

where L is a characteristic length scale. Similarly, the Navier-Stokes equation, Eq. 2, becomes

$$\omega v \sim \Omega v + \frac{bB}{\rho\mu L} \quad (5)$$

Substituting b of Eq. 4 into Eq. 5, we obtain the magnitudes of the main terms in Eq. 5:

$$\omega^2 v \sim \omega\Omega v + \frac{V_A^2 v}{L^2} \quad (6)$$

where $V_A = B/\sqrt{\rho\mu}$ is the Alfvén wave velocity. When the magnetic field is weak, then V_A is small and Eq. 6 gives

$$\omega \sim \Omega$$

namely the mode oscillating with a diurnal period,

which is called the *inertial mode*. For the extremely slow mode, $\omega \ll 1$, the left side of Eq. 6 can be ignored, and we have

$$\omega \sim \frac{V_A^2}{\Omega L^2} \quad (7)$$

This is sometimes called the *magnetostrophic mode*, which is a magnetic mode modified by the Coriolis force. For the mode for which the Coriolis force is ineffective, we simply have

$$\omega \sim \frac{V_A}{L}$$

namely the Alfvén wave.

If the toroidal field B is 100 G, $V_A = 10$ cm · s^{-1} since $\rho \sim 10$ g · cm^{-3}. $\Omega = 0.7 \times 10^{-4}$ s^{-1}. If we take $L \sim 10^8$ cm, we have $\omega \sim 10^{-10}$ s^{-1}. This is the right order of magnitude with the angular velocity of the magnetic field estimated from the westward drift velocity (0.3° yr^{-1}). Braginskiy has examined the case where buoyancy (Archimedean) force is operating simultaneously and has obtained a composite wave called the MAC (Magnetic, Archimedean, and Coriolis) wave.

The wave theory is attractive in that the observed velocity of the westward drift is explained naturally by the frequency of the magnetostrophic mode, provided that the intensity of the toroidal field is about 100 G. Therefore, the idea that the westward drift is caused by propagation of magnetostrophic waves along the toroidal field superposed on the nonuniform rotation of the core is acquiring wider support.

However, the wave theory is not completely consistent with observation. As is seen in Eq. 7, the frequency of the magnetostrophic mode depends on the square of the length scale of the wave. This implies that the phase velocity is highly dispersive, and a wave with a shorter wavelength or a larger wave number propagates faster. The observation, on the other hand, indicates that the drift velocity of the geomagnetic field is rather uniform for different degrees (n) and orders (m) of the spherical harmonics, around 0.25° yr^{-1}. Even if there exists a dispersive relationship, it is very weak and the lowest degree and order ($n = m = 1$) field drifts the fastest, 0.30° yr^{-1}.

The second problem concerns the direction of propagation. The characteristic equation for the wave frequency gives not only positive but also negative frequency, implying that the magnetostrophic mode propagates both east and west. Until recently the magnetic field was believed to drift westward only. Therefore theoretical efforts have been concentrated on explaining how the westward-drifting field is selectively excited, although the recent suggestion that eastward-drifting fields are also likely to exist seems to relax the constraint on the unidirectional propagation so far assumed. The solution has been sought in the stability problem, which makes a westward mode the first to destabilize.

Solving a linear stability problem, I. A. Eltayeb and J. Kumar have examined the onset of thermal convection in the presence of a magnetic field for a self-gravitating fluid sphere rotating with a constant angular velocity. Heat sources are assumed to be distributed uniformly inside the sphere, and the magnetic field has only the azimuthal component to increase its intensity linearly with distance from the axis of rotation.

When the magnetic field is weak, rotation and the magnetic field both exert a stabilizing effect. However, when the magnetic field strength is increased until the Lorentz force balances with the Coriolis force, the magnetic field facilitates convection. If the magnetic field is further increased, it works to suppress the convection again. Hydromagnetic waves first destabilized depend on the field strength. For a weak field, waves are excited that propagate eastward with large azimuthal wave numbers (m), whereas for a stronger field, waves are generated that propagate westward with small wave numbers ($m = 1$ or 2). For such a strong field as the Lorentz force balances with the Coriolis force, the westward drift velocity becomes approximately $V_A^2/\Omega d$, where d is characteristic length scale, such as the radius of the sphere. This velocity is equivalent to that previously derived for the magnetostrophic mode (Eq. 7).

Weak Field Model. A weak field model also predicts westward drift of the nonaxisymmetric field. One of the weak field models is a turbulent dynamo. When turbulence has nonzero helicity (i.e., $\mathbf{v} \cdot$ curl $\mathbf{v} \neq 0$), the electric field $\mathbf{E}$ can be induced parallel to the impressed magnetic field $\mathbf{B}$, thus $\mathbf{E} = \alpha \mathbf{B}$. This phenomenon is called the α-effect, which is in noticeable contrast with the normal induction process, where the induced electric field is orthogonal to the magnetic field (i.e., $\mathbf{E} = \mathbf{v} \times \mathbf{B}$). Due to this α-effect the electric currents flow parallel to the inducing magnetic field. Consequently, where the inducing field is poloidal, toroidal fields are induced. Interaction of turbulent motions exerts another α-effect on the toroidal fields and generates a poloidal field of original type. This process is called the α^2-dynamo (Krause, 1977).

Assuming a form $e^{i(\omega t + m\varphi)}$ for the nonaxisymmetric field, S_1 and A_1 in Fig. 3, Krause has solved dynamo equations to determine eigenvalues for nondimensional parameters, $R_\alpha = \sigma\mu\alpha r_0$, $R_\omega = \sigma\mu\omega r_0^2$, and obtained similar magnitudes of R_α for the fields of A_0 (axial dipole), S_0 (axial quadrupole), S_1 (equatorial dipole), and A_1 (field asymmetric to the equator with $m = 1$, such as the field expressed by $P_2^1(\theta)e^{i\varphi}$ in terms of spherical harmonics). This implies that these types of fields are almost equally excitable. R_ω's determined for A_1- and S_1-type

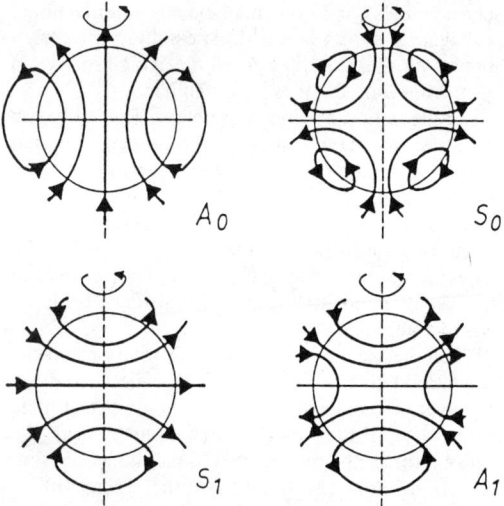

FIGURE 3. Types of fields excited by α^2-dynamo. (From Krause, 1977)

dynamos become negative. This means that these types of fields rotate westward. Krause estimated the drift velocity to be 0.036° per year (rotational period 10,000 yr) for the equatorial dipole S_1. It is about an order of magnitude smaller than the observation. This velocity seems to be a natural consequence of the weak field model. Even in a strong field model, the magnetostrophic mode would give a similar small drift velocity for such a weak strength of the toroidal field as 10 G.

Variations in the Axisymmetric Field

Oscillation of the dipole field is an important feature of the secular variations. An extreme case of the oscillation is polarity reversals of the dipole in geologic times. This should be interpreted as one aspect of the dynamo process. In fact, the $\alpha\omega$-dynamo is characterized by oscillations of the dipole field (ac dynamo), although the α^2-dynamo is characterized by a steadiness of the dipole (dc dynamo). An $\alpha\omega$-dynamo model based on turbulent motions has been applied to a realistic Earth model by H. Watanabe in which the effect of the Lorentz force on the α-effect is taken into consideration and a reversal process of the dipole field has been demonstrated, which is accompanied by a polarity change in the axial quadrupole.

A large-amplitude variation, about 50% of the present dipole moment, over the past 8000 years is supposed to be caused by perturbation of the main dynamo process. However, no successful theory has been presented yet.

About a 60-year variation is remarkable not only in the dipole component but also in other components. However, the theory is still in its infancy. Braginskiy (1984) has investigated the induction process in a magnetic diffusive layer about 30 km thick near the core surface. He first considers that the advection of axisymmetric poloidal fields by axisymmetric meridional flows in the layer is responsible for the dipole oscillation. The meridional flows are assumed to be induced so that the azimuthal component of the Coriolis force may balance with the inertial force for the azimuthal flow caused by a change in the Earth's rotation with a 60-year period. When a steady dipole field is assumed as the inducing field, whose intensity is the same as the present field, only a variation of less than 10^{-2} of the observed has been obtained. Subsequently, instead of advection of the axisymmetric field, Braginskiy suspects the existence of intense non-axisymmetric fields in the layer to create the observed variation through interaction with nonaxisymmetric flows.

Decades Variation

Main Stream of the Core as a Source of Decades Variation. When the fluid of the core moves fast in comparison with the magnetic diffusion time, during which the magnetic field decays through ohmic dissipation, the core fluid can be regarded as a perfect conductor and the frozen-flux approximation can be applied to explain the behavior of the magnetic flux. Roberts and Scott (1965) have estimated the dynamical time scale τ_u to be 30 years, during which fluid motions can substantially alter the field by redistributing the magnetic lines of force. This contrasts with the magnetic free decay time τ_B, which is 12,000 years, which implies that the diffusion term is less important in the induction equation (Eq. 1) with a magnetic Reynolds number ($R_m = \tau_B/\tau_u$) as large as 400.

Since the conductivity of the mantle is as low as 10^2–10^3 S · m^{-1} at the bottom of the mantle, we may regard the mantle as an insulator. Therefore if the frozen-flux approximation is applicable in the core, we can infer the core motions from the movement of the magnetic lines of force at the surface of the core. Although the viscosity of the core is not well known (10^{-2}–10^9 stokes), the viscous boundary layer cannot be thicker than 10 km. If concentration of the electric currents occurs in the boundary layer, the tangential component of the magnetic field becomes discontinuous across the layer. However, the radial component of the magnetic flux density (B_r) should be continuous through the layer. Therefore with another boundary condition on the velocity whose radial component should vanish at the boundary, we have

$$\frac{\partial B_r}{\partial t} + \nabla_s(B_r \mathbf{v}_s) = 0 \qquad (8)$$

where

$$\nabla_s(B_r\mathbf{v}_s) = \frac{1}{r\sin\theta}\left[\frac{\partial}{\partial\theta}(\sin\theta \cdot B_r v_\theta) + \frac{\partial}{\partial\varphi}(B_r v_\varphi)\right]$$

Using this relationship, A. B. Kahle and others have estimated fluid velocity ($\mathbf{v}_s$) at the surface of the core from the magnetic field (B_r) and its secular variation ($\dot{B}_r$) continued downward from the Earth's surface to the core-mantle boundary.

Although it is not easy to obtain B_r and $\dot{B}_r$ accurately at the core surface by downward continuation, a more serious problem exists—nonuniqueness of the solution. Backus (1968) has demonstrated the existence of an infinite number of eligible flows that satisfy Eq. 8. The term $\mathbf{v}_s B_r$ in Eq. 8 is expressed as follows:

$$\mathbf{v}_s B_r = \nabla_s\phi - \mathbf{n}\times\nabla_s\psi \qquad (9)$$

where $\mathbf{n}$ is the unit vector normal to the core surface. Backus shows that no matter how ψ in Eq. 9 is chosen, the resulting $\mathbf{v}_s$ satisfies Eq. 8 and therefore ψ is indeterminate except at null flux points where $B_r = 0$. At null flux points, Eq. 8 becomes

$$\frac{\partial B_r}{\partial t} + \mathbf{v}_s\cdot\nabla_s B_r = 0$$

If B_r and $\dot{B}_r$ are given, $\mathbf{v}_s$ is uniquely determined from the gradient of B_r normal to the null flux curve. The fluid velocities across the null flux curves have been examined by J. R. Booker, E. R. Benton, and K. A. Whaler, and predominance of the westward drift has been obtained. If the tangential component of the electric field $\mathbf{E}_s$ is known by any means at the surface of the core, Backus claims that the fluid motion can be determined uniquely by

$$\mathbf{E}_s = -B_r\mathbf{v}_s$$

This method has been employed by Muth and Benton (1981), equating $\mathbf{E}_s$ to a vector potential of the secular variation.

There are some efforts being made to solve the nonuniqueness problem by imposing some a priori constraints. D. Gubbins assumes that the motion is purely toroidal and steady to obtain not only the westward drifting azimuthal velocity but also the meridional flow. J. L. Le Mouël and others assume that the motion is of sufficiently large scale that it can be expressed by low-degree terms of spherical harmonics and of geostrophic type. Then they conclude that both toroidal and poloidal motions are likely to exist. They note that an acceleration of an azimuthal toroidal motion occurred sometime during the period that includes the time of the *geomagnetic jerk*. The results obtained by different authors are still far from being in agreement. C. V. Voorhies assumes a steady motion, and obtains a large scale combined toroidal-poloidal circulation.

Boundary Layer Phenomena. Although the nature of the boundary layers near the surface of the core has not yet been clarified, fluid motions in the layers play a crucial role in modifying the magnetic field created in the main body of the core into the observed form at the Earth's surface. Three layers are now considered near the core-mantle boundary: the viscous boundary layer, the magnetic diffusive layer, and the stratified layer due either to concentration of lighter elements or to thermal stratification.

As mentioned already, the viscous boundary layer (Ekman layer) is thinner than 10 km if a reasonable value of viscosity is assumed for the core. In the presence of a magnetic field the Ekman layer is modified into a Hartman-type layer (Acheson, 1978), of which the thickness is given by

$$\delta_H = \frac{\delta_E}{[C + (1 + C^2)^{1/2}]^{1/2}}$$

$$\delta_E = \left(\frac{\nu}{2\Omega}\right)^{1/2}, \quad C = \frac{V_A^2}{2\Omega\eta}$$

This indicates that the magnetic field reduces the thickness of the boundary layer. The detailed nature of the layer has not been investigated, but a meridional flow is assumed to be induced in association with a nonuniform azimuthal flow in the main volume of the core (Braginskiy, 1978).

The importance of the magnetic diffusive layer has been pointed out by Tough and Roberts (1968) and Braginskiy (1984). Through the layer, the magnetic field originated in the interior region of the core is adjusted to continue to the field in the mantle. Braginskiy considers that the short-period secular variations, including the 60-year period variation of the dipole field, are produced within this layer, and estimates the thickness of the layer to be 36 km from the skin-depth formula for the 60-year period.

Braginskiy has proposed a dynamo model, called model Z, in which the magnetic lines of force are aligned almost parallel to the rotation axis in the core. They are sharply bent in a boundary layer where concentration of electric currents occurs and continue to the field in the mantle. The layer thickness of the current concentration is estimated to be less than 100 km.

The third type of the layer, which is characterized by stable stratification, has not been well established yet. Argument about the stable core was initiated by Kennedy and Higgins (1973), who examined the melting temperature gradient of iron under high pressure. They concluded that the melting temperature gradient was less than that of the adiabatic in the core and that no convectional motions were possible except for a limited region near the inner-outer core boundary. It is now believed that, although two temperature gradients are close to each other, the melting temperature gradient is slightly steeper and convection can occur. However, there is a possibility of a stable density-stratified layer near the core-mantle boundary with a thickness of 100 to 1000 km, provided the core-mantle boundary cools more slowly than the core interior.

Stratification of the core is also suggested from a completely different point of view. The core does not consist of pure iron but contains 10 to 20% lighter elements, such as sulfur, silicon, oxygen, carbon, or hydrogen. As the inner core solidifies, which consists of almost pure iron, lighter materials are released at the inner core boundary and rise to the top surface of the core to form a stably stratified layer (Braginskiy, 1984). D. Loper and P. H. Roberts estimate its thickness to be about 70 km, whereas Braginskiy gives 20 km.

Dynamics in this layer is supposed to be similar to that in the ocean and the atmosphere in many aspects, such as excitation of Rossby waves or baroclinic waves (Braginskiy, 1984). The difference is that they are modified by the existence of a magnetic field. Yukutake considers that the standing part of the geomagnetic field is generated through interaction of stratified fluid motions in this layer with the dipole field, whereas the drifting field is produced by interaction of convectional motions with a toroidal field in the deep interior. Whaler has examined recent secular variations and concluded that the fluid motions near the core surface are most likely to be toroidal, namely stratified.

Although core motions are not determined uniquely from secular variation data alone, it is obvious that the dynamics in the core is influenced by the presence of the magnetic field, and the geomagnetic data no doubt provide important clues to clarify the hydromagnetic phenomena in the core.

TAKESI YUKUTAKE

References

Acheson, D. J., 1978, Magnetohydrodynamic waves and instabilities in rotating fluids, in P. H. Roberts and A. M. Soward, eds., *Rotating Fluids in Geophysics*. London: Academic Press, 315-349.

Backus, G. E., 1968, Kinematics of geomagnetic secular variation in a perfectly conducting core, *Royal Soc. London Philos. Trans.* **263A**, 239-266.

Braginskiy, S. I., 1978, The geomagnetic dynamo, *Izvestiya, Acad. Sci. USSR. Physics Solid Earth* **14**, 659-668 (E).

Braginskiy, S. I., 1984, Short-period geomagnetic secular variation, *Geophys. Astrophys. Fluid Dynamics* **30**, 1-78.

Bullard, E. C., C. Freedman, H. Gellman, and J. Nixon, 1950, The westward drift of the earth's magnetic field, *Royal Soc. London Philos. Trans.* **243A**, 67-92.

Kennedy, G. C., and G. H. Higgins, 1973, The core paradox, *Jour. Geophys. Research* **78**, 900-904.

Krause, F., 1977, Mean-field electrodynamics and dynamo theory of the Earth's magnetic field, *Jour. Geophysics* **43**, 421-440.

Merrill, R. T., and M. W. McElhinny, 1983, *The Earth's Magnetic Field—Its History, Origin and Planetary Perspective*. London: Academic Press, 401p.

Muth, L. A., and E. R. Benton, 1981, On the frozen flux velocity field at the surface of earth's core necessary to account for the poloidal main magnetic field and its secular variation, *Physics Earth and Planetary Interiors* **24**, 245-252.

Roberts, P. H., and S. Scott, 1965, On analysis of the secular variation. I. A hydromagnetic constraint: theory, *Jour. Geomagnetism and Geoelectricity* **17**, 137-151.

Tough, J. G., and P. H. Roberts, 1968, Nearly symmetric hydromagnetic dynamos, *Physics Earth and Planetary Interiors* **1**, 288-296.

Watanabe, H., and T. Yukutake, 1975, Electromagnetic core-mantle coupling associated with changes in the geomagnetic dipole field, *Jour. Geomagnetism and Geoelectricity* **27**, 153-173.

Yukutake, T., 1979, Review of the geomagnetic secular variations on the historical time scale, *Physics Earth and Planetary Interiors* **20**, 83-95.

Cross-references: Core-Mantle Coupling; Earth Orientation; Earth's Core; Earth Structure: Global; Geomagnetic Field: Westward Drift; Geomagnetic Field, Main: Theory; Geomagnetic Field Analysis; Geomagnetic Secular Variation: Direction and Intensity; Geomagnetism: Historical Introduction; Magnetohydrodynamic Waves in the Earth; Paleomagnetic Secular Variation; Satellite Magnetic Measurements; Spherical Harmonic Analysis.

GEOMAGNETISM: HISTORICAL INTRODUCTION

Geomagnetism, the study and use of the magnetic field of the Earth, is one of the oldest branches of science and can be said to have begun with the discovery that a piece of lodestone (a naturally occurring, and usually naturally magnetized, form of magnetite, Fe_3O_4), when freely suspended or supported in some manner, always came to rest in a roughly north-south direction. The fact that lodestone attracts small pieces of iron was widely known by about the middle of the first millennium B.C., but its directional property in the Earth's magnetic field was not discovered until much later.

This article is a roughly chronological sketch of the most significant events in the history of geomagnetism from the discovery of this directive property until the end of the nineteenth century. For the present century no attempt has been made to be at all comprehensive and the choice of topics is no doubt strongly influenced by my own interests.

The Magnetic Compass in China

The discovery of the directivity of lodestone appears to have first been made in China. Needham (1962) has discussed the evidence in great detail and considers that the extant texts point to the directivity having been discovered by Chinese geomancers in about the first century A.D. Geomancy has been described as "the art of adapting the residences of the living and the tombs of the dead so as to cooperate and harmonize with the local currents of the cosmic breath." It involved the use of a diviner's board consisting of two plates: one square and the other circular, the latter being able to rotate on a pivot located at the center of the square plate. Both plates were marked with astronomical and divinatory signs and the circular plate also contained a representation of the stars of the constellation now known as Ursa Major (the Great Bear). The tail of the Great Bear acted as a form of pointer. At some time about the first century A.D., the circular disc was replaced with a model of the Great Bear, resembling a Chinese spoon and it was discovered that, if this spoon was made from lodestone, it would point to the south, one of the most important points in the heavens from the point of view of geomancy. In order for this device, known as the south-pointing spoon, to work, the base plate and the lodestone spoon had to be highly polished; even so no great sensitivity or accuracy could have been attained.

To make directional measurements with any accuracy it was necessary to replace the lodestone spoon with an elongated piece of magnetized iron or steel and to support it in a more sensitive way. From an analysis of Chinese texts that refer to lodestones and to needles, Needham considers that magnetized needles first came into use at some time between the fourth and sixth centuries A.D. The first application of such compasses to navigation probably occurred about A.D. 1000, since there is a reference to a south-pointing fish that Needham dates at about A.D. 1030. This compass probably used a piece of iron in the shape of a fish that was floated on water contained in a shallow dish.

The first detailed description of a magnetic compass occurs in a book with the somewhat unlikely title of *Dream Pool Essays* (*Meng Qi Pi Tan*) by Shen Kua, written about 1088. In fact, three types of compass are described: one in which the needle was supported by attaching it to a floating reed, another in which the needle was suspended by means of a single thread of new silk, and a third in which it was balanced on a sharp edge. From the early twelfth century onward there are many Chinese references to the use of the magnetic compass (usually some form of floating-needle instrument) in navigation.

In the book cited above, Shen Kua also states quite clearly that the compass needle does not, in general, point exactly N-S. This recognition of the magnetic declination predates European knowledge of the phenomenon by some 400 years. From a study of geomantic literature and of the various circles on the diviner's board, Needham claims that the Chinese geomancers were aware of the declination as e. as the middle of the ninth century A.D. A formal definition of magnetic declination is given in *Geomagnetic Field: Elements*.

The Magnetic Compass in the West

The first non-Chinese reference to anything geomagnetic is a description of the navigational use of the compass by Alexander Neckam in his treatise *De Naturis Rerum* written in 1190 (almost exactly 100 years after Shen Kua's *Dream Pool Essays*). This reference is followed, during the next few decades, by several brief mentions of the compass in western European literature, that are mostly of a metaphorical nature (see Needham, 1962 for further details).

Arabic sailors were using the compass in navigation by about 1230, but whether theirs was an independent discovery or derived from the Chinese or from western Europe is not known. Needham (1962) gives details of Arabic knowledge of magnetism in early medieval times.

In August 1269, Pierre de Maricourt, or Petrus Peregrinus, was fighting in the army of Charles of Anjou, King of Naples and Sicily, that was laying siege to the town of Lucera in southern Italy. His military duties were evidently not too burdensome, since he found time to write a lengthy letter to his friend and neighbor Sygerus de Foucaucourt on the subject of the magnet. This letter is, in fact, one of the finest contributions to natural science produced during the Middle Ages and is the first western treatise on magnetism. Petrus describes how to find the positions of the poles of a spherical piece of lodestone, recognizes that the two poles are of different kinds (north and south), describes the attraction of unlike poles for one another and the repulsion between like poles, and describes two improved forms of magnetic compass. One of these used a small lodestone secured inside a small box and thus enabled to float in a vessel of water. The other instrument used a magnetized iron needle attached to a vertical axle free to turn in pivots at its upper and lower ends. The addition of a sighting arrangement enabled these two instruments to measure azimuths of distant objects more accurately than the crude compasses then in use. Harradon (1943) gives a translation of the letter of Petrus Peregrinus.

The Discovery of Magnetic Declination in Europe

The fact that the magnetic compass does not in general point exactly N-S seems to have dawned gradually upon European compass makers and users. During the fifteenth century Nuremberg was the center for the manufacture of pocket sundials. To be of use these instruments had to be correctly oriented with respect to the true N-S direction, and an obvious way of doing this was to incorporate a small magnetic compass. By about 1450, many of these compasses had two marks, one indicating true north and the other displaced by several degrees. The sundial makers had become aware that the compass needles did not point truly and attempted to compensate for this. Whether they realized that they had discovered a fundamental property of the geomagnetic field is, of course, quite another matter. There is, in fact, some evidence that this deviation from true north was ascribed to defects in the needles or in the lodestones used to magnetize them.

Before the discovery of these compensated pocket sundials, it was long believed that Columbus had been the first westerner to observe the magnetic declination, during his first voyage in 1492. Mitchell (1937) studied the surviving versions of the journals of all Columbus's voyages and concluded that Columbus may have discovered that the declination (of which he was aware before setting out on his first voyage) varied with position on the Earth's surface. The journals are so vague and have been edited so extensively that this conclusion is by no means certain.

It is, however, certain that by 1529, the spatial variation of the declination was not only recognized but was beginning to be put on a quantitative basis. In that year the French sailors Jean and Raoul Parmentier made several declination measurements in the South Atlantic during a voyage from France to Sumatra. The observations cover a range of almost 10° in declination.

The sixteenth century was the Golden Age of maritime exploration and observations of the magnetic declination were made and recorded by many of the explorers. Particularly noteworthy are the series of observations made by the Portuguese explorer João de Castro during his voyages to and from Goa and along the west coast of India and in the Red Sea. During his explorations along the Indian coast, de Castro made the first observation of a local magnetic anomaly, on the small island of Chaul, and correctly ascribed the discrepant results to the magnetization of the rocks in the vicinity. Harradon (1944) gives translations of the relevant parts of de Castro's journals.

Fewer observations were made on land and these were almost always at sea ports. An exception was Georg Hartmann's observation of the declination in Rome in about 1510, which is the earliest known land magnetic observation (Harradon, 1943).

The Discovery of Magnetic Inclination

Hartmann was also the first to realize that the direction of the geomagnetic field is not, in general, parallel to the Earth's surface. He noted, but did not accurately measure, the angle of inclination or dip at Nuremberg in about 1544 (Harradon, 1943). For a definition of magnetic inclination see *Geomagnetic Field: Elements*.

The first accurate measurement of inclination was made by a London instrument maker, Robert Norman, probably at Radcliffe, in London, in 1576. Norman used an instrument that he had designed specially for this experiment: the prototype of the dip-circle. He published a description of his instrument and the results of the observations made with it in the first book to be devoted to geomagnetism, *The Newe Attractive*, first published in 1581. From his results Norman concluded that the point to which his dip needle was oriented lay below the horizon. This conclusion directed attention towards the Earth, rather than the heavens, as the possible seat of the force that controls a magnetized needle.

William Gilbert: The Earth a Great Magnet

William Gilbert, physician to Queen Elizabeth I, placed all previous work in geomagnetism in a modern scientific context in his masterly treatise *De Magnete*, which has been described as the first modern scientific monograph. In this book, published in 1600, Gilbert not only summarized what was then known about the geomagnetic field but also described his own extensive experiments in the subject. Gilbert fashioned pieces of lodestone into spheres (*terellae* or little Earths) and, using small iron needles, investigated how the magnetic field was distributed over their surfaces. Since these distributions were, in all cases, remarkably similar to what was known about the distribution of the geomagnetic field over the Earth's surface, and, in particular, since they exhibited the phenomenon of dip, Gilbert concluded that the Earth itself is a great magnet. It was, therefore, the Earth itself that controlled the compass needle and the dip needle, and a study of geomagnetism might be able to provide information about the interior of the Earth. Thus did geomagnetism become part of geophysics; indeed, since geophysics itself hardly existed at this time it might be truer to say thus was geophysics born. Gilbert's contributions to geomagnetism have been discussed in detail by Chapman and Bartels (1940).

The Discovery of Secular Variation

The next major geomagnetic discovery was that the Earth's magnetic field changed with time. William Borough had made a series of declination observations at Limehouse in London in 1580 and 1581. These observations indicated that the declination in 1580 was 11° 19′ east. Edmund Gunter, Professor of Astronomy at Gresham College, repeated Borough's observations in 1622, as nearly as possible at the same place, and found the declination to be 5° 56.5′ east. Gunter, however, suspected an error in either his own or Borough's results and did not follow up the matter. His successor, Henry Gellibrand, is credited with the discovery of the time variation of the geomagnetic field. In 1634 Gellibrand made a series of observations at Deptford (where Gunter in 1622 had found the declination to be 6° 15′ east) which gave a value for the declination of 4° 05′ east. Gellibrand was sufficiently confident of his own and Gunter's observations to ascribe this difference to a real phenomenon (now known as the secular variation) rather than to errors of observation. He duly published his results and conclusions in 1635 in his book *A Discourse Mathematical on the Variation of the Magnetical Needle.*

The Earliest Magnetic Charts

The fact that magnetic north and true north differed and that this difference varied from place to place and with time meant that the magnetic compass alone was not the total answer to the navigator's need for a direction-finding instrument. Towards the end of the sixteenth century and during most of the seventeenth several efforts were made to solve this problem. The first step was to collect existing data. The first attempt at this was made by Simon Stevin in his book *De Havenfinding* published in 1599. Since the secular variation had not yet been discovered, Stevin did not consider it necessary to record the dates of the observations in his list. A more comprehensive collection of declination observations was included in the book *Magnes sive De Arte Magnetica* by Athanasius Kircher first published in 1641. Lists of declination values were, however, of little value to the navigator. What the navigator needed was some kind of cartographic representation of the distribution of declination at a particular time. There are references to at least two attempts at providing this kind of description, one by the Spanish Chief Pilot Alonzo de Santa Cruz in about 1536 and another about the middle of the sixteenth century by Cristoforo Borri and Martin Martinus. The earlier of these seems to have consisted merely of declination values entered on a map; the other may have been some crude kind of contour chart. No copies of either chart appear to have survived.

The first declination chart to have survived, and the first to be widely known and used, is the famous Atlantic Chart of Edmond Halley, published in 1701. Halley, later the second Astronomer Royal and famous for his comet, included among his diverse scientific activities an abiding interest in geophysics. Indeed, his first recorded scientific endeavor was an observation of the magnetic declination at his home in London when he was 16 years old. From a study of what little was known about the secular variation, Halley came to the conclusion that a major part of this change could be accounted for by assuming that the geomagnetic field was drifting slowly westward relative to the Earth's surface. Halley also put forward a theory to explain this westward drift. His model involved two pairs of magnetic poles, one pair fixed relative to the Earth's crust and the other attached to an internal part of the Earth that Halley assumed to be moving relative to the surface layers. In order to test and refine his model Halley realized that he needed more data. With the energy and perseverance that characterized much of his scientific work, Halley persuaded the British government to provide him with a specially built ship, a pink named the *Paramore*, in which to make a voyage in the Atlantic Ocean measuring the magnetic declination at all convenient points en route. Halley's first voyage, which lasted from October 1698 to July 1699, was not a complete success and he made a second between September 1699 and September 1700 during which he reached a latitude of 52° 24′S before having to turn back because he was surrounded by icebergs.

Using the data collected during his voyages, Halley compiled his Atlantic Chart, which depicts the declination by means of contour lines. With the possible exception of the now lost chart of Borri and Martinus mentioned earlier, this was not only the first magnetic chart but also the first use of contour lines to represent the spatial distribution of any variable. A year later, having collected further data from other sources, Halley produced a second chart, the so-called World Chart, on which the contour lines are extended to cover the Indian and western Pacific Oceans. So famous and widely known did these charts become among seamen that contour lines were for many years known as Halleyan curves and it is probably true to say that Halley's main claim to fame during the rest of his life and for the century following his death was as the inventor of the magnetic chart.

Halley was well aware that his charts would need periodical revision because of the effects of secular variation and several such revisions were made at irregular intervals during the next 150 years. In 1858 the British Admiralty produced the first of their series of World Magnetic Charts, which have been revised at regular intervals since then. The latest, compiled in conjunction with the U.S. Navy, are for epoch 1985.

As scientific interest in geomagnetism increased, charts of the other geomagnetic elements were also produced. The first inclination chart, covering southeastern England, was produced by William Whiston in 1721. The first global inclination chart was compiled by Wilcke and published in 1768. The earliest attempt at producing a chart of the intensity of the geomagnetic field was the map of the total intensity for northern South America for 1799 produced by Humboldt.

The Discovery of Magnetic Storms and the Daily Variation

The first suspicion that the geomagnetic field varied on time scales shorter than those that characterize the secular variation resulted from observations made by Tachard and his fellow missionaries in Thailand in 1685. The systematic study of these rapid variations began when George Graham, a leading London clock and instrument maker, constructed a very sensitive and accurate declinometer. In 1722 Graham discovered, from a long series of careful observations with his new instrument, that the magnetic needle was continually in motion with time scales extending from a few minutes to several hours. Some of these variations were irregular whereas others had a pattern that repeated from day to day. Andreas Celsius made similar observations in Uppsala and confirmed Graham's results. In 1747 Celsius's assistant Hiorter discovered a connection between the irregular variations and the aurorae: strong disturbances of the magnetic needle were always accompanied by aurorae and vice versa. A particularly significant aspect of this work was the close collaboration between Graham and the Uppsala group that showed that the irregular variations (later named magnetic storms by Humboldt) occurred simultaneously in London and in Uppsala and that the regular (diurnal) variations depended on local time. As we shall see, from this initial collaboration between two observers developed the worldwide network of magnetic observatories that exists today.

Humboldt and Measurements of Geomagnetic Intensity

The latter half of the eighteenth century saw an upsurge in scientific voyages of exploration and discovery, of which the three voyages of James Cook are probably the most famous. Most of these voyages included in their observational program measurements of the magnetic declination and, often, of the inclination.

Extensive series of magnetic observations on land were still lacking, however. Humboldt had made observations at several places in Europe and, between 1799 and 1804, he made many more during his extensive journeys in South and Central America. As important as the geographical spread of Humboldt's observations was his inclusion of measurements of the intensity of the field along with those of declination and inclination. See *Geomagnetic Field: Elements* for definitions of the intensity elements.

Graham had, in 1723, made observations of the total intensity of the field by measuring the period of oscillation of a dip needle, but experimental difficulties prevented him from achieving any great accuracy. In 1769 Mallet measured the period of oscillation of a horizontal needle at several places in northern Russia and concluded that they showed no evidence for a spatial variation in the horizontal intensity. Borda made observations with a dip needle during an expedition to the Canary Islands in 1776 and concluded that the total intensity did not change with position. During La Pérouse's ill-fated voyage of 1785-1788, de Lamanon had observed the oscillations of a dip needle and his results indicated that the total intensity increased with latitude on either side of the equator. Although it is probable that de Lamanon's preliminary results were read before the French Academy in 1787, they were not published. Similar observations made by de Rossel during d'Entrecasteaux's voyage of 1791-1794 in search of La Pérouse, which led to the same conclusion, were not published until 1808.

Thus Humboldt's observations, published in 1805 in his account of his American travels, were the first published evidence for the increase of the total intensity with latitude. Humboldt considered this result to be the most important of his American expedition, though he was scrupulous in giving due credit to de Lamanon and de Rossel when he summarized his results later in *Kosmos*.

All these measurements of the intensity of the geomagnetic field were relative and were based on the assumption that the magnetization of the oscillating needle had not varied during the series of observations. Humboldt chose to express all his total intensity values in terms of the smallest value that he had observed, at Micuipampa in Peru on the magnetic equator where the inclination was zero. This value was arbitrarily assigned the value 1.0 and Humboldt's became the accepted unit of relative intensity for the next three decades.

Humboldt and Magnetic Observatories

Humboldt was also interested in the time variations of the geomagnetic field, in particular the irregular variations that he named *magnetische Störungen* or magnetic storms. In order to study this phenomenon he set up a magnetic observatory in Berlin in 1806 where, with J. Oltmanns, he made a series of observations at hourly and half-hourly intervals between May 1806 and June 1807. As a result of these observations, Humboldt suggested that observations be made at other places, simultaneously with those in Berlin, to continue the investigations of Graham, Celsius, and Hiorter on the occurrence of storms at different places. Unfortunately politics intervened (Humboldt was employed at the time as a diplomat) and the project did not come to fruition until 1828. In that year Humboldt had a second observatory built in Berlin (in the garden of Abraham Mendelssohn-Bartholdy, the father of the composer). In October 1828 simultaneous observations were made in Berlin and down a mine at Freiberg in Saxony. In 1829, observatories at Paris and Kazan' joined the program, making observations on predetermined days known as *term-days*.

During a visit to Russia in 1829 Humboldt was able to persuade the authorities there to set up a chain of magnetic observatories which by 1835 numbered

11 extending from Arkhangel'sk in the north to Peking in the south and from St. Petersburg in the west to Sitka in the east. A total of about 20 observatories in various parts of the world contributed to this network (or *verein* as Humboldt called it) at various times between 1829 and 1835.

Gauss: Intensity Measurements and the Göttingen Magnetic Union

In 1828, at a conference on the natural sciences held in Berlin, Humboldt met the great mathematician Carl Friedrich Gauss. Humboldt demonstrated to Gauss the instruments in his magnetic observatory and this seems to have stimulated Gauss to interest himself in geomagnetism. By 1832 Gauss had devised a method for measuring the intensity of a magnetic field in absolute units and, with Wilhelm Weber, had constructed instruments to implement the method. Gauss and Weber also built and equipped a magnetic observatory at Göttingen that was operating by 1833, and by 1835 the responsibility for coordinating the term-day observations had been assumed by Gauss and Weber. They had rather more success than Humboldt in collecting and publishing the term-day observations from the various contributing observatories and the Göttingen Magnetic Union expanded gradually over the next 5 years. It grew to include the British observatories at Greenwich and Dublin and the Royal Society was asked by Humboldt in his famous letter to the President, the Duke of Sussex, to give consideration to establishing magnetic observatories in some of the British colonies.

British Colonial Observatories

Humboldt's suggestion was acted upon with remarkable speed and James Clark Ross, during his Antarctic expedition of 1839–1843, established observatories at St. Helena, the Cape of Good Hope, and Hobart in Tasmania. Other observatories were built at about the same time at Toronto, Singapore, Simla, and Madras. It was therefore natural that, when the Göttingen Magnetic Union became defunct in 1841, the Royal Society took over the responsibility for specifying the term-days and coordinating the observations.

From these beginnings—the collaboration between Graham and Celsius, the magnetic unions of Humboldt and of Gauss and Weber at Göttingen, and the creation of the British colonial observatories—stems the worldwide network of almost 200 magnetic observatories that exists today.

Gauss and Spherical Harmonic Analysis

Gauss made another fundamental contribution to geomagnetism with his introduction, in 1839, of spherical harmonic analysis as a technique for mathematically modeling the Earth's magnetic field.

Using values of the declination, inclination, and total intensity interpolated from charts of these three elements at 84 regularly spaced points on the Earth's surface, Gauss was able to derive values for the first 24 coefficients in the spherical harmonic expansion of the geomagnetic field. He was also able to show that, within the uncertainties in his data, the main geomagnetic field had no sources external to the Earth's surface. He thus made quantitative Gilbert's assertion of over 300 years before. The technique of spherical harmonic analysis has been used many times since to model not only the main geomagnetic field but also the secular variation and some of the more rapid variations. It is now an essential preliminary stage in the production of global magnetic charts. Details of this technique are given in *Spherical Harmonic Analysis*.

Ross and the Magnetic Poles

Mention has already been made of James Clark Ross in connection with the establishment of the British colonial observatories. Ross's involvement with geomagnetism probably began when he sailed as midshipman with the Royal Navy expedition led by his uncle John Ross in 1818, which went in search of the Northwest Passage. During the next 10 years the younger Ross took part in four more Arctic expeditions, all under the command of Parry. In 1829 John Ross made another attempt to find the Northwest Passage, this time in command of a privately-funded expedition and with his nephew as second-in-command. The expedition was forced to spend four winters on the shores of the Boothia Peninsula in northern Canada, in the vicinity, according to the magnetic charts of the time, of the North Magnetic Pole (the place where the inclination was 90°). A temporary magnetic observatory was set up near the ship and its results indicated that the pole was indeed quite near, probably within sledging distance. Thus, in May 1831, a party led by James Ross set out to cross the peninsula and try to find the position of the pole by direct observation. On June 1st, their dip circle indicated an inclination of 89° 59.5′, Ross marked the spot with a cairn and took possession of the North Magnetic Pole in the name of King William IV.

During his Antarctic expedition of 1839–1843 Ross tried to complete the double by visiting the South Magnetic Pole. Although he was able to estimate its position from observations made on board ship, icebergs and mountains prevented him from setting foot on the South Magnetic Pole.

Sabine and Magnetic Surveys

Soon after his return from the Antarctic expedition, Ross was involved in the first magnetic survey of the British Isles. This survey, the first national survey to be organized anywhere, took 4 years (1834–1838) to complete and seems to have been instigated by Edward Sabine, who was also responsible for the reduction and analysis of the observations.

Sabine was an indefatigable organizer of scientific projects and was particularly good at collecting and analyzing the data from them. He was largely responsible for the processing of the data from the British colonial observatories and from the observatory at Kew. He also made a comprehensive collection of global survey observations made during the first half of the nineteenth century and produced a series of magnetic charts based on these data.

Sabine and Solar-Terrestrial Physics

In 1852 Sabine discovered the statistical connection between the occurrence of magnetic storms and the number of spots on the Sun, thus establishing that the storms were triggered from outside rather than inside the Earth. This discovery was a direct result of his wife's translation of Humboldt's *Kosmos* into English. Mrs. Sabine worked from proof sheets sent by Humboldt and included in these were Heinrich Schwabe's first results of his studies of the time dependence of the number of sunspots. When Sabine saw these data he realized that there was a striking similarity between the variation with time of the number of sunspots and the way the number of magnetic storms observed at the British colonial observatories had changed over the years. Some further quantitative work then established the statistical relationship between the two phenomena, which can be regarded as the beginning of solar-terrestrial physics.

Photographic Recording at Magnetic Observatories

The term-day observations at observatories, involving as they eventually did reading the instruments by eye every five minutes for 24 hours or longer, were enormously tedious and labor intensive. Also, in Britain and its colonies, it was not considered fitting that such work be performed on Sundays. The answer was some way of making the magnetometers self-recording and a prize was offered by the British government in 1845 for the first person to successfully devise such a technique. By the following year two claimants had come forward: Charles Brooke, a London surgeon and amateur inventor, and Francis Ronalds, the superintendent of the Kew magnetic observatory. Both Brooke and Ronalds used photographic techniques to produce a permanent record of the variations of the geomagnetic field. Largely through the support of the Astronomer Royal and Director of the Royal Greenwich Observatory, George Biddell Airy, the prize of £500 was awarded to Brooke and the Greenwich magnetic observatory was equipped with Brooke's self-recording magnetographs in 1847. Kew observatory, however, was equipped with a development of the Ronalds magnetograph, which came into operation in 1852. Ronalds eventually received a consolation prize of £250 from the government. Self-recording instruments were gradually installed at most magnetic observatories and the tedium of reading the instruments was replaced by that of scaling the magnetograms.

One of the first new phenomena discovered with the help of photographic recording was a class of regular variations with periods of several seconds to a few minutes. These were first found on the Kew magnetograms for August and September 1859 by Balfour Stewart. They were given their present name, pulsations, by van Bemmelen in 1908 and have since been classified into various types on the basis of their period and morphology. Pulsations have proved to be a fruitful source of information about the magnetosphere.

Daily Variations and the Ionosphere

The data derived from the magnetograms were used in studies of magnetic storms and the daily variations. Canton, in 1759, had discovered that the daily variation in declination in London was greater in summer than in winter. The more extensive data from magnetic observatories now established that all elements of the geomagnetic field were affected by storms and showed daily variations. A detailed analysis of the magnetograms showed that, besides the daily variation discovered by Graham, which has a period of 24 hours, there is also a much smaller variation with a period of a lunar day. This variation was first recognized by Kreil in 1850 using data from the magnetic observatory at Prague. The problem of the source of these diurnal variations was as yet unsolved. In 1882 Balfour Stewart suggested that Sq (the solar daily variation on magnetically quiet days) was caused by electric currents flowing in the upper atmosphere, which must, therefore, be an electrical conductor. This hypothesis, the first suggestion that something like the ionosphere might exist, was partially confirmed in 1889 by Arthur Schuster who used spherical harmonic analysis to show that the majority of Sq arises from sources external to the Earth. Schuster also found a smaller part of internal origin, which he correctly ascribed to electric currents induced in the Earth by the external source.

The Twentieth Century

We have now arrived at the beginning of the twentieth century and lack of space and a superabundance of interesting topics prohibit anything more than a very brief glance at a selection of the events that seem to be among the most significant. I shall group these into three categories: international collaboration, data collection, and theoretical work.

The first important program of geomagnetic studies that involved international collaboration was the first International Polar Year which took place, following a suggestion by Carl Weyprecht, between 1882 and 1883. The work covered primarily geomagnetism and meteorology. Eleven countries established 14 temporary magnetic observatories, 12 in the Arctic and 2 in the southern hemisphere (at Cape Horn and South Georgia).

Fifty years later (1932–1933), the second International Polar Year encompassed a much wider program of geophysical studies and a much wider geographical coverage. Once again, geomagnetism was featured strongly. North of 55° N latitude 16 permanent and 24 temporary magnetic observatories were established and several midlatitude and equatorial stations also contributed.

It had been planned to organize a third International Polar Year 25 years after the second. By that time, however, the subject of geophysics had greatly expanded, and the third Polar Year became the International Geophysical Year (IGY) of 1957–1958. This triumph of international scientific cooperation included, in addition to geomagnetism, many other geophysical disciplines such as oceanography, glaciology, seismology, and auroral studies. Many magnetic observatories were established especially for the IGY. Although some of these were closed down at the end of 1958, many remained in operation for longer, and some are still operating. Another lasting legacy of the IGY is the system of discipline-oriented World Data Centers established to organize the archiving and dissemination of the large quantities of data produced by the IGY observing programs.

Turning now to data collection and, in particular, to the surveying of the main geomagnetic field worldwide, the first truly global surveying program began with the formation of the Department of Terrestrial Magnetism of the Carnegie Institution of Washington. Between 1905 and 1944 magnetic surveys were conducted on land and at sea that, taken together, constituted the nearest approach to world-wide coverage that had yet been achieved. The oceanic work was particularly innovative. It began with three cruises using a chartered brigantine, the *Galilee*, between 1905 and 1908. By the following year, a specially built nonmagnetic yacht, the *Carnegie*, was ready for service. Between 1909 and 1929, the *Carnegie* made seven cruises covering almost 300,000 miles and including four circumnavigations of the Earth. All the major ocean areas were surveyed, many for the first time. This extremely valuable surveying program ended tragically when the *Carnegie* caught fire while refuelling in the harbour of Apia, Samoa, on 29 November 1929 and was destroyed.

A British nonmagnetic ship, the *Research*, was built but was never commissioned because of the Second World War. In 1956, the USSR launched the nonmagnetic schooner *Zarya* and this vessel has provided many valuable observations during the past 30 years, though her sphere of operation has been restricted in recent years to near-coastal waters.

In 1945 a modified Lancaster bomber, *Aries*, of the Royal Air Force made several long-distance flights over Greenland and the Canadian Arctic. The main purpose was to develop aerial navigational techniques at high latitudes and, for this reason, several different types of magnetic compass were carried. The flights clearly showed the potential of aircraft for making long-range magnetic surveys.

This potential was realized by the U.S. Navy, which established a program of long-range aeromagnetic surveying in 1951. Known as Project MAGNET, the program became operational in 1953 using a Neptune aircraft. Since then, with several changes of the aircraft and equipment, Project MAGNET has covered several million miles and has surveyed most of the ocean areas of the world as well as large regions over the continents. Canada and Japan have also conducted extensive aeromagnetic surveys during the past 25 years, though with a more restricted coverage than Project MAGNET.

In addition to these ship and airborne surveys of the complete magnetic vector, there have been an increasing number of surveys of the total intensity of the geomagnetic field made by towing magnetometers behind ships and aircraft during the last three decades. These surveys are particularly useful in studies of crustal magnetic anomalies (see *Aeromagnetic Surveying; Magnetostratigraphy; Seafloor Spreading: Magnetic Evidence*).

The first artificial satellite, *Sputnik 1*, was launched during the IGY. Not only was it now possible to make truly global surveys of the main geomagnetic field in a short time, it was also possible to investigate how the field behaved at large distances from the Earth. Satellites have been used to explore the magnetosphere and how it responds to the solar plasma that continually streams past it and they have also improved our knowledge of the distribution of the field over the Earth's surface.

Sputnik 3 in 1958 and *Vanguard 3* in 1959 both carried magnetometers to measure the main field, but they were restricted in both accuracy and ability to collect data over the whole Earth. The first complete and accurate global satellite survey was made by the Polar Orbiting Geophysical Observatories (POGOs) between 1965 and 1971.

However, because of difficulties in controlling and measuring the attitude of a satellite, the POGO satellites measured only the total intensity of the field. By 1979 these difficulties had largely been overcome and *Magsat*, launched in October of that year, measured the strength and direction of the field for seven months and provided the most accurate description of the main geomagnetic field that we have yet been able to obtain (see *Satellite Magnetic Measurements*).

Finally, turning to theories of the origins of the various parts of the geomagnetic field, we have seen that, by the end of the nineteenth century, it was known that the main field originated within the Earth, that magnetic anomalies were caused by crustal rocks, that the daily variations had their source outside the Earth, and that magnetic storms were somehow connected with sunspots.

Theories of the origin of the main field have considered permanent magnetization, the field of rotating charges, systems of currents at various depths within the Earth and the hypothesis that a magnetic field might be an intrinsic property of a rotating body. This last suggestion was put forward by Blackett in 1947 and is of interest despite the fact that, when tested experimentally, the hypothesis had to be rejected. One of the tests involved the development of a very sensitive magnetometer that was later used extensively in the early work on rock magnetism and palaeomagnetism. The presently accepted theory of the origin of the main field involves a self-exciting dynamo that causes electrical currents to flow in the liquid outer core of the Earth. The first suggestion of such a dynamo was made by Sir Joseph Larmor in 1919, though it was not successfully applied to the Earth until the 1940s (see *Geomagnetic Field, Main: Theory*).

The statistical connection between magnetic storms and sunspots turned attention to the possibility that the Sun might be emitting streams of charged particles. In 1919 Lindemann pointed out that such streams must be electrically neutral, containing an equal number of positively and negatively charged particles (a plasma in modern terminology). Chapman and Ferraro used this assumption as the basis of the first quantitative theory of magnetic storms, which involved the impact of such a plasma stream on the geomagnetic field. Their analysis showed that a hollow cavity would be formed in the stream and that the geomagnetic field would be confined within it. The existence of this cavity, named the magnetosphere by Gold in 1959, has been confirmed by satellite observations and the Chapman-Ferraro theory has been revised in the light of these data.

These advances in geomagnetism are just a few that have been made during the present century. Although much has been done and our knowledge of the Earth's magnetic field has greatly increased since the time of Shen Kua and Petrus Peregrinus, we are still a long way from a complete understanding of this important geophysical phenomenon.

DAVID R. BARRACLOUGH

References

Balmer, H., 1956, *Beiträge zur Geschichte der Erkentniss des Erdmagnetismus*. Arrau: Sauerländer.

Chapman, S., and J. Bartels, 1940, *Geomagnetism*, chap. 26. Oxford: Clarendon.

Harradon, H. D., 1943, Some early contributions to the history of geomagnetism, *Terr. Magn. Atmos. Electr.* **48**, 3-17, 79-91, 127-130, 197-199, 200-202.

Harradon, H. D., 1944, Some early contributions to the history of geomagnetism, *Terr. Magn. Atmos. Electr.* **49**, 185-198.

Harradon, H. D., 1945, Some early contributions to the history of geomagnetism, *Terr. Magn. Atmos. Electr.* **50**, 63-68.

Merrill, R. T., and M. W. McElhinny, 1983, *The Earth's Magnetic Field: Its History, Origin and Planetary Perspective*, chap. 1. London: Academic.

Mitchell, A. C., 1932, Chapters in the history of terrestrial magnetism, *Terr. Magn. Atmos. Electr.* **37**, 105-146.

Mitchell, A. C., 1937, Chapters in the history of terrestrial magnetism, *Terr. Magn. Atmos. Electr.* **42**, 241-280.

Mitchell, A. C., 1939, Chapters in the history of terrestrial magnetism, *Terr. Magn. Atmos. Electr.* **44**, 77-80.

Mitchell, A. C., 1946, Chapters in the history of terrestrial magnetism, *Terr. Magn. Atmos. Electr.* **51**, 323-351.

Nelson, J. H., L. Hurwitz, and D. G. Knapp, 1962, *Magnetism of the Earth*, chap. 6. U.S. Coast and Geodetic Survey Publication 40-1.

Needham, J., 1962, *Science and Civilisation in China*, Vol. 4, *Physics and Physical Technology. Part I: Physics.* Cambridge: Cambridge University Press, 229-334.

Parkinson, W. D., 1983, *Introduction to Geomagnetism*, chap. 6. Edinburgh: Scottish Academic Press.

Cross-references: *Aeromagnetic Surveying; Archaeomagnetism; Geomagnetic Field: Elements; Geomagnetic Field: Measurement; Geomagnetic Field: Westward Drift; Geomagnetic Field, Main: Theory; Geomagnetic Field Analysis; Geomagnetic Measurement Techniques and Surveys; Geomagnetic Secular Variation: Direction and Intensity; Geomagnetic Secular Variation: Theory; Magnetostratigraphy; Paleomagnetism and Continental Drift: The Early Years; Paleomagnetism and Plate Tectonics; Satellite Magnetic Measurements; Seafloor Spreading: Magnetic Evidence; Spherical Harmonic Analysis.*

GEOTHERMAL ENERGY

A vast amount of naturally occurring heat is stored in the Earth and is transferred to the surface both by conduction (diffusion) and mass convection of magma and water. In the upper 10 km of the crust, the amount of geothermal energy stored above surface temperature is on the order of 10^{27} joules, equivalent to roughly 50,000 times the energy in the ultimately recoverable oil and gas resources of the world. However, most of this heat is too deeply buried and too dispersed to be tapped for electrical power production or direct-heat uses. Geothermal energy is exploitable only where it occurs in high concentrations at accessible depths, in a manner analogous to concentration of other natural resources such as ore deposits and petroleum reservoirs. Geothermal *resources* are defined as the fraction of the stored thermal energy that might be extracted economically at some future time, assuming reasonable advances in technology and economic favorability.

Nature and Distribution of Geothermal Systems

Many geothermal systems occur in regions where young magmatic intrusions in the upper crust provide sources of heat by cooling and crystallization. Favorable plate tectonic settings for magma generation and associated geothermal systems are along subduction zones and oceanic spreading centers, in continental rift zones and regions of crustal extension, and at intraplate hot spots (Fig. 1). Geothermal systems on continents commonly are located in areas of silicic magmatism because SiO_2-rich magma tends to form large intrusive bodies that lodge in the crust. Less viscous basaltic magma, on the other hand, rises comparatively rapidly and dissipates much of its thermal energy as volcanism at the Earth's surface. In oceanic environments, however, the amount of basaltic magma stored at shallow crustal depths may be large enough to support geothermal systems, as in Hawaii and Iceland.

Thermal energy conducted away from a cooling intrusion heats meteoric water circulating to depths of a few kilometers through pores and fractures of the surrounding rock and parts of the crystallized intrusion itself. In the process of hydrothermal convection, the hot, lower density water rises buoyantly upward and it commonly is discharged at the surface as thermal springs and geysers. Convecting water, either as liquid or steam, is the working fluid that concentrates diffuse heat and transports it to shallow depths. Some hydrothermal systems have only feeble hot spring activity at the surface because of capping by low-permeability rock that prevents free leakage of water and heat upward. The Salton Sea field in southern California is a well-known example of a blind geothermal system where the cap rock is an impermeable rock formation of regional extent. In other cases, the cap rock is produced by self-sealing of rock pores and fractures resulting from precipitation of silica or calcite from thermal waters or the formation of clays and zeolites.

Figure 2 shows a simplified temperature-depth profile of a hydrothermal-magmatic system. A high conductive gradient exists between the intrusion and the base of the overlying hydrothermal system (line CD). Upward-flowing water in a nearly isothermal convection zone is represented by line BC. The upper limit of the isothermal zone is determined by the boiling-point curve of water (line AB). A spectacular hydrothermal-magmatic system having more than 10,000 identified geysers, hot springs, and fumaroles is found in Yellowstone National Park. These thermal features occur within a 600,000-year-old volcanic caldera beneath which a large, partially molten body is inferred to exist at depths greater than 5 km. Fluid temperatures within the hydrothermal system likely reach 350° to 450°C, and the convective heat flow of the hot-spring system is at least 1800 mW/m^2 averaged over 2500 km^2. Calculations indicate that crystallization and cooling of about 0.1 km^3 of rhyolitic magma per year would be sufficient to furnish the heat discharged by the hot-spring system. The average temperature gradient between the magmatic heat source and the hydrothermal convection system is estimated to be 700° to 1000°C/km. In comparison, heat flow and temper-

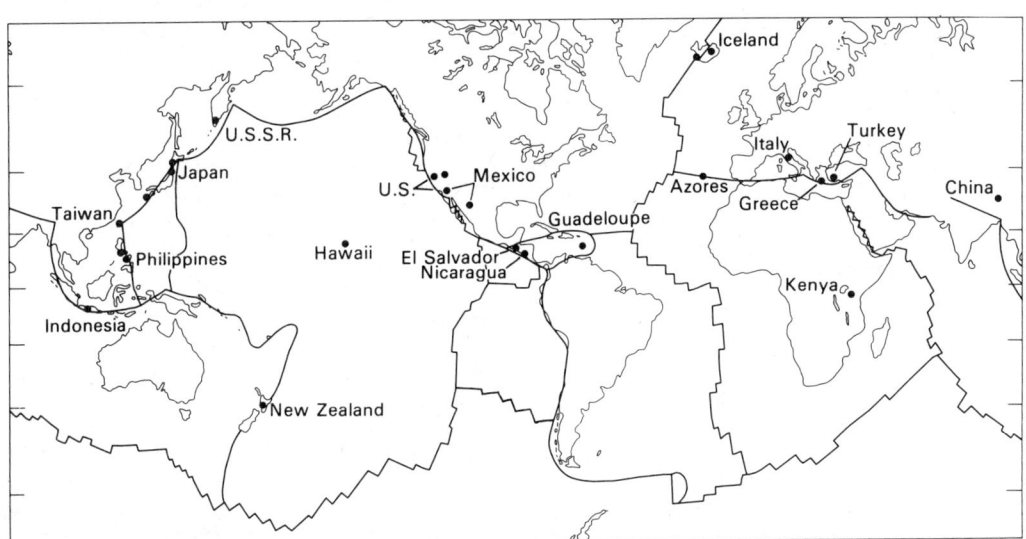

FIGURE 1. Map showing major plate boundaries and locations of geothermal systems developed for electrical power.

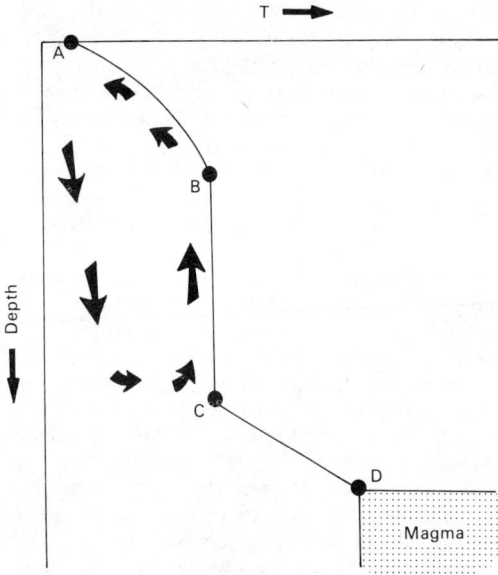

FIGURE 2. Schematic temperature-depth profile of hydrothermal-magmatic system. Arrows indicate downward flow of cool ground water and rise of heated water. Line AB is boiling-point curve of water; line BC is isothermal zone; line CD is conductive gradient between magmatic heat source and hydrothermal convection system. (From Fournier and Pitt, 1985)

ature gradient in nonmagmatic conductive regimes typically range from 40 mW to 100 mW/m^2 and 15° to 50°C/km, respectively. [*Note:* The Yellowstone hydrothermal system is within a national park and is not available for commercial power development so that its outstanding natural thermal features can be preserved.]

Most of the total thermal energy associated with a cooling igneous intrusion is stored in the magma itself and in surrounding *hot dry rock*. If heat could be mined directly from magmatic bodies, the recovered energy would be considerable. However, technical obstacles involved in accurately locating shallow magma bodies and drilling in an extremely hostile chemical and physical environment make such energy recovery very difficult. Geothermal systems characterized by hot dry rock contain much heat but lack the permeability and water of hydrothermal convection systems. Extracting thermal energy from hot dry rock involves creating reservoir permeability by hydraulic fracturing and then pumping water down one borehole and up a second in order to extract heat from the interconnecting fractured zone with minimal loss of water to surrounding rocks. At an experimental site near Valles caldera in the Jemez Mountains of New Mexico, zones of hydraulically fractured granite have been connected with two boreholes, and thermal water has been recovered at the surface. However, the process of developing hot-dry-rock reservoirs is far from routine.

In addition to concentrating thermal energy, hydrothermal activity can form ore deposits that represent *fossil* geothermal systems in which the thermal energy has been largely dispersed. These deposits include epithermal veins of gold, silver, antimony, and mercury in the upper kilometer or so and porphyry-copper and molybdenum deposits at deeper levels. Along oceanic spreading ridges, submarine hot springs have been observed that are actively precipitating massive sulfide ores on the ocean floor. Hydrothermal systems also may be responsible for low-grade metamorphism. At the Salton Sea field for example, analysis of drill core has shown that metamorphic transformations to greenschist facies are actively taking place as hot, saline (>250,000 ppm dissolved solids) brines of the geothermal system interact at elevated temperature and pressure with sediments of the Colorado River delta.

Not all geothermal systems are directly related to specific young igneous intrusions in the upper crust. Meteoric water may circulate to depths of several kilometers along favorable structures such as faults and fracture zones, becoming heated as temperature increases with depth and then rising convectively toward the surface as thermal springs. An example of this type of geothermal environment is the Basin and Range province of the western United States, a region of high conductive heat flow where extensional faulting enhances ground-water circulation and basaltic magmatism moves heat into the lower crust. In general, hot springs are more abundant in areas of elevated regional heat flow and high geothermal gradient, such as the Basin and Range province, and circulating fluids attain higher temperature at shallower depths, compared to areas of low to moderate heat flow (such as the eastern United States).

Geothermal systems also may occur in deep sedimentary basins that have normal to moderately high geothermal gradients (25°C/km or greater) and aquifers of sufficient porosity, permeability, and size for production of thermal water. This type of lower temperature geothermal resource is utilized for space and agricultural heating in the Paris basin, France, and in the Pannonian basin, Hungary. In some sedimentary basins, pore waters are *geopressured.* That is, fluid pressures are greater than that of the water column (which is hydrostatic) and approach lithostatic as fluids bear a large fraction of the weight of the rock overburden. An example of geopressured geothermal systems is found in the northern Gulf of Mexico Basin beneath parts of Texas, Louisiana, and Mississippi. The pore waters are trapped in thick sequences of rapidly accumulating marine sediments and are unable to attain hydrostatic equilibrium because impermeable clay barriers isolate more porous reservoirs. With increasingly deep burial, the trapped waters become heated to higher temperatures. Dissolved methane is present in the geopressured fluids of the Gulf Coast and constitutes a source of energy about equal in magnitude to the thermal fluid. Although the geopressured resource of the Gulf Coast is large, competitive development at this time (1988) is hampered by the lower temperatures, great depths (3 km to 7 km), and uncertain producibility of reservoirs.

Utilization of Geothermal Energy

At present, the best targets for electrical-power production are hydrothermal convection systems because they have both high temperature at depths within reach of current drilling technology (to about 3 km) and sufficient water or steam to transport thermal energy to the surface. In the most common type of hydrothermal convection system, the fluid phase in rock pores and fractures is liquid water. During exploitation, hot pressurized water enters a drill hole, and part of it flashes to steam. The water-steam mixture flows up the well to the surface where the steam and water commonly are separated mechanically at a pressure of 500 kPa to 700 kPa. The steam is then routed to turbines to generate electricity, and the unwanted water usually is injected back into the ground. In some cases, a second stage of decompression (flashing) is used to separate steam for lower pressure turbines. In the much less common vapor-dominated systems, steam in open fractures and channels in the reservoir rock is the predominant fluid phase at depth. Vapor-dominated systems occur where the heat supply is very high and recharge of cooler water from surrounding rocks is low enough that the steam zone is not swamped by the influx of liquid. Within the steam zone, which generally is capped by impermeable rock and shallow ground water, temperature and pressure are relatively constant with depth because of the low weight of steam. Although rare, such systems are economically desirable because of the relative simplicity of producing dry steam at the wellhead without the need to separate and dispose of large volumes of water. Vapor-dominated systems are developed for electrical power at The Geysers in northern California, three fields in Italy (Larderello, Travale, and Monte Amiata), Matsukawa in Japan, and Kamojang in Indonesia.

Various distributions of temperature with depth are possible in hydrothermal convection systems (Fig. 3). For comparison, lines A–D show a range of conductive gradients (8° to 75°C/km) typically found in the upper 2 km of the crust (assuming a constant value of rock thermal conductivity). In the convective hot-water system at Wairakei, New Zealand, exemplified by curve G, liquid water at 260°C occurs in an isothermal zone at depths greater than 600 m; at shallower depths and lower pressures, a two-phase mixture of liquid and steam closely follows the boiling-point curve of pure water. A geothermal system such as the Salton Sea field has temperatures above the boiling-point curve for pure water because of the effect of high salinity (curve H). In some hydrothermal systems, water flowing rapidly to the surface does not lose much heat by conduction from the margins of the systems, and the deep temperature is not much higher than that of boiling springs discharging at the surface (curve E). In a vapor-dominated system (curve F), temperature near 240°C is determined by the maximum enthalpy of saturated steam; beneath the vapor fraction, temperature increases in a hydrostatically pressured zone of liquid water. Systems having temperature distributions like curves F, G, or H are the best exploration targets.

Hydrothermal systems with temperatures less than about 150°C are not currently economical for generating electricity in a turbo-generator supplied directly with steam separated from geothermal fluids. However, it is feasible to use lower temperature geothermal water to heat a secondary, low-boiling-point fluid (e.g., freon or isobutane) that in turn drives turbines; this method is called the binary method. In addition, water at temperatures of approximately 40°C to 100°C can be used for nonelectrical (direct) purposes such as space heating, industrial drying, and aquacultural and agricultural applications.

Geothermal energy was first developed for electrical power in 1904 from the vapor-dominated Larderello field in Tuscany, Italy, where power production continues today. In 1959, New Zealand was the first country to build a commercial plant utilizing a hot-water system. Improvements in the late 1970s and early 1980s in binary and flash technology now permit economic use of hot-water systems that have high salinity or temperatures lower than 150°C. Geothermal development has been increasing rapidly since the early 1970s; the annual growth rate of electrical capacity from 1978 to 1984 was 16%. In 1986, 17 countries generated 4,733 MW$_e$ from geothermal resources (Table 1), about 0.1% of electrical generation from all modes. The United States is the world's major geothermal producer; its capacity is dominated by production at The Geysers field in northern California. This large, vapor-dominated system has been developed since 1960, and the 1986 capacity of 19 separate power plants was nearly 1800 MW$_e$, enough to satisfy the needs of the San Francisco/Oakland metropolitan area. The total developed geothermal capacity of the United States, 2006 MW$_e$ in 1986, is only a fraction of the nation's potential geothermal resource. The resource contained in 52 hydrothermal convection systems (with reservoir temperatures greater than 150° and at depths less than 3 km) is estimated to be 220×10^{18} joules, which is equivalent to 23,000 MW$_e$ for 30 years (Muffler, 1979).

Worldwide use of geothermal resources for direct-

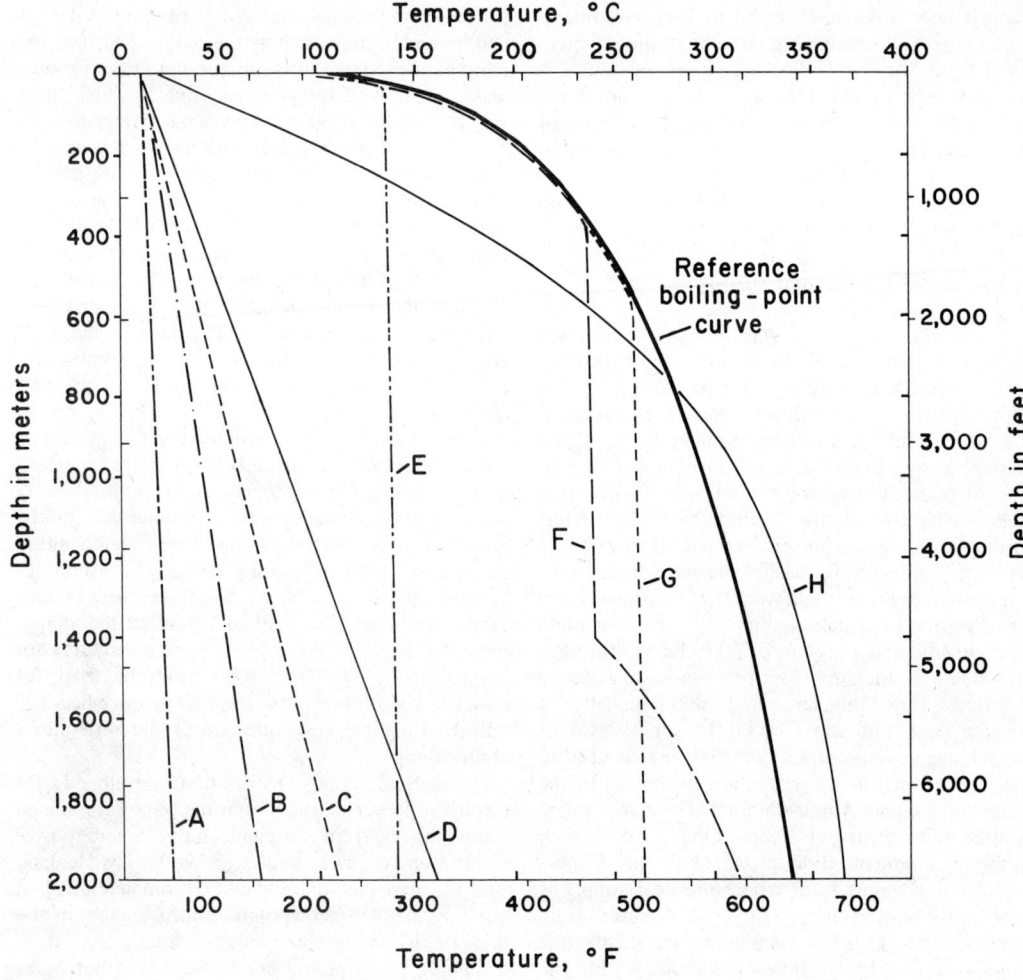

FIGURE 3. Various distributions of temperature with depth in hydrothermal convection systems. See text for detailed discussion of lettered curves. (Modified from White, 1973)

heat applications was approximately 7100 MW (thermal) by 11 countries in 1984, with Japan and Hungary having the greatest development (Table 2). Iceland utilizes geothermal energy extensively for district heating that involves about 75% of the population. In the Paris basin of France, water that is 60° to 100°C is pumped from depths of 1.5 km to 2 km, and 40 geothermal plants provide heat for about 500,000 people. District heating systems, which employ a network of pipes to transport geothermal fluids from a centralized well field to individual users or blocks of users, constitute the major portion of direct-heat usage. Other uses include bathing (especially in Japan), greenhouse and soil heating, fish farming, animal husbandry, and industrial drying. Low-temperature geothermal resources are widespread throughout the world and their use has great potential for growth.

The environmental impact of geothermal development generally can be satisfactorily managed. Gaseous components of geothermal fluids, such as H_2S, are controlled by chemical scrubbers and limited venting of steam to the atmosphere. Spent brines commonly are injected back into geothermal fields to prevent contamination of surface waters and to maintain reservoir pressures, but careful treatment is required to prevent plugging of wells or rock formations by chemical precipitation. Microearthquakes are known to be induced by fluid extraction and injection but are not considered to be a significant hazard. In some areas, land subsidence is monitored as fluid withdrawal increases. When steam is released directly to the atmosphere, the noise can be reduced by silencers. Careful siting of power plants and routing of pipelines is important where scenic values are threatened. Not all these environmental factors attend development at every geothermal field. Moreover, geothermal energy can be

TABLE 1. Worldwide Electrical Capacity from Geothermal Plants (as of 1986)

Country	Megawatts (electrical)
United States	2006
Philippines	894
Mexico	645
Italy	504
Japan	215
New Zealand	167
El Salvador	95
Kenya	45
Iceland	39
Nicaragua	35
Indonesia	32
Turkey	21
China (incl. Taiwan)	15
Soviet Union	11
France (Guadeloupe)	4
Portugal (Azores)	3
Greece	2
TOTAL	4733

Source: Data from DiPippo (1986).

TABLE 2. Countries Making Significant Direct Use of Geothermal Energy in 1984

Country	Megawatts (thermal)
Japan	2686
Hungary	1001
Iceland	889
Soviet Union	402
China	393
France	365
United States	339
Italy	288
Romania	251
New Zealand	215
Turkey	166
Other	142
TOTAL	7137

Source: Data from Gudmundsson (1985).

produced and used on-site, thus eliminating problems associated with mining and transportation.

Geochemical and Geophysical Exploration Techniques

Because geothermal systems are not always delineated completely by surface features, a variety of exploration techniques are employed to determine the configuration of subsurface reservoirs. Drilling deep holes is expensive and risky, so most geothermal exploration first involves shallow heat-flow and temperature-gradient holes and surface geochemical and geophysical techniques.

The compositions of hydrothermal fluids are controlled principally by temperature-dependent chemical reactions between thermal fluids and reservoir rocks. Assuming that chemical equilibrium occurring at depth is preserved when fluids ascend rapidly to the surface, the subsurface temperature within a geothermal reservoir can be estimated from the chemistry of surface springs or shallow well waters. Many different chemical and isotopic reactions are possible geothermometers; the most commonly used are based on the solubility of SiO_2, the cation ratios Na/K and Na/K/Ca, and the fractionation of oxygen isotopes between water and dissolved sulfate ($SO_4^=$). Chemical geothermometers greatly increase the chances of finding a high-temperature hydrothermal system before drilling and are applied as standard exploration tools worldwide. Geochemical methods also are useful for identifying vapor-dominated zones, recognizing underground mixing of hot and cold waters, and deciphering reservoir processes such as underground boiling and movement of cooler injected water into producing reservoirs.

Hydrothermal systems usually are more electrically conductive because of the effects of high temperature, salinity of thermal fluids, and mineral alteration than are surrounding volumes of rock. Geoelectrical techniques are used to map the lateral and vertical extent of hydrothermally related resistivity lows (i.e., conductivity highs). Direct-current methods measure the resistance of a volume of rock to the passage of electrical current applied to the ground and are useful to depths of 1 km to 2 km. Time-domain electromagnetic techniques analyze decay of a transient magnetic field induced underground by current flowing through a wire loop on the surface. Magnetotelluric methods measure naturally occurring magnetic fields and electrical currents over a range of frequencies including signals in the audio band. These types of geophysical techniques are sensitive to the presence of thermal fluids; other geophysical methods, such as gravity and aeromagnetic techniques, are useful for investigating subsurface rock compositions and crustal structures that may control the occurrence of magma bodies and geothermal systems.

Seismic techniques also help to characterize geothermal systems. In seismic refraction and reflection experiments, the paths of seismic waves generated by controlled explosions are analyzed to delineate faults and layered structures in both sedimentary and volcanic environments. In teleseismic *P*-wave investigations, the arrivals of *P* waves from distant earthquakes are timed in order to locate zones of low velocity (*P*-wave delays) inferred to be partially molten rock or magma chambers in the crust. This technique has been applied widely in the United States. Magma chambers of horizontal and vertical dimensions greater than about 5 km, the resolution limit of most teleseismic experiments, have been interpreted to underlie volcanic fields at Long Valley, The Geysers-Clear Lake area, and

Coso in California, Yellowstone National Park, and Roosevelt Hot Springs in Utah. In the Cascade Range of Washington, Oregon, and California, magmatic bodies beneath major volcanoes have not been detected as coherent, upper crustal features by teleseismic and other geophysical studies. There, magma reservoirs appear to be small, deep, and/or complex bodies.

Summary

Geothermal systems related to young igneous intrusions in the upper crust can be divided into hydrothermal convection, magma, and hot dry rock systems. Hydrothermal convection systems that are not related to specific young intrusions occur where meteoric water circulates deeply along faults and fractures. Some deep sedimentary basins of high porosity and permeability contain thermal waters that can be extracted for nonelectrical uses. Geopressured geothermal systems occur in deep sedimentary basins where thermal fluids cannot attain hydrostatic equilibrium. Hydrothermal convection systems are exploited worldwide for the production of electricity; 4733 MW_e were produced by 17 countries in 1986. Various geochemical and geophysical exploration techniques are employed to discern the subsurface extent of thermal fluids and magma.

M. GUFFANTI

References

DiPippo, R., 1986, Geothermal power plants—worldwide status 1986, *Geothermal Res. Council Bull.* 15(11):9-18.

Fournier, R. O., and A. M. Pitt, 1985, The Yellowstone magmatic-hydrothermal system (USA), in C. Stone, ed., *1985 International Symposium on Geothermal Energy.* Geothermal Resources Council, Davis, Calif., 319-327.

Gudmundsson, J. S., 1985, Direct uses of geothermal energy in 1984, in C. Stone, ed., *1985 International Symposium on Geothermal Energy.* Geothermal Resources Council, Davis, Calif., 19-29.

Muffler, L. J. P., ed., 1979, Assessment of geothermal resources of the United States—1978, *U.S. Geol. Survey Circular 790,* 163p.

White, D. E., 1973, Characteristics of geothermal resources, in P. Kruger and C. Otte, eds., *Geothermal Energy: Resources, Production, Stimulation.* Stanford, Calif.: Stanford University Press, 69-94.

Cross-references: *Continental Drilling: Ultradeep; Energy Budget of the Earth; Heat Flow in the Earth.*

GLOBAL POSITIONING SYSTEM (GPS)

The NAVSTAR global positioning system (GPS) is the successor to the Navy navigation satellite system (NAVSAT) (see *Doppler Positioning: Satellite*). It was developed to provide instantaneous, global, three-dimensional position and velocity for navigation to an accuracy higher than that possible with the NAVSAT system, and to provide the data at any time rather than once per hour or so. During the course of the development of the GPS system, scientists devised ways of exploiting the transmitted signals for precise geodetic and geophysical applications. The high precision is obtained by methods related to radio interferometry (see *Very Long Base Line Interferometry (VLBI)*) methods in the sense that signals received at two sites are compared to determine the relative positions of the sites. For short distances between sites, GPS data can provide an accuracy equivalent to VLBI data at considerably less cost and complexity in equipment and computation. But for distances exceeding 100 km or so, errors in knowledge of the positions of the GPS satellites begin to degrade the results. At distances of thousands of kilometers it is no longer possible to observe GPS satellites simultaneously at the pair of sites. Thus the GPS and VLBI systems are complementary for positioning over a range of distances. Furthermore, the VLBI system provides precise orientation of GPS-derived results through its determination of the direction of the Earth's spin axis and its angular rate (see *Earth Orientation*).

In some applications, it is also necessary to know the position of sites with respect to the center of the Earth, which cannot be determined by the VLBI system. Although the GPS system does provide positions in an Earth-centered frame, high precision in Earth-centered coordinates is obtained by laser satellite observations. Therefore the laser system is also complementary to the GPS system in providing Earth-centered coordinates for selected sites (see *Satellite Laser Positioning*).

During the course of the development of the GPS system, demonstrations (and operational use of some applications) have been performed showing its applicability for satellite, aircraft, vehicle, and ship navigation, for geodetic positioning, and for precise crustal motion measurements. Approximate accuracies obtained are as follows:

Application	Accuracy
General navigation	100 m absolute, instantaneous
Precise navigation	20 m absolute, instantaneous
Geodetic position	1 m absolute after 12-24 h
Relative position	1 cm per 100 km site separation

These values should not be considered definitive, because the accuracy achieved depends on many factors. The general navigation value is set by national security requirements, which could change. The precise relative positioning accuracy could be better or worse, depending on methods and equipment used. There is another category of relative navigation for a moving receiver that could be at the 1-cm to 10-m level, depending on methods used and operating conditions.

History of the GPS System

The Navy navigation satellite system has proven highly successful for ship navigation and for geodetic applications. Tens of thousands of receivers are aboard commercial and military ships providing positions at about hourly intervals. Tens of thousands of geodetic positions have been determined worldwide, strengthening and extending national geodetic networks, providing orientation and an Earth-centered origin for the networks, and connecting the networks and isolated islands to a global system. However, hourly two-dimensional positions are not adequate for aircraft navigation, and improved accuracy was required for many applications.

Several concepts were considered for the basis of an improved navigation system. The NAVSTAR GPS finally evolved to best meet the needs of the U.S. Department of Defense. A joint project office was established under Air Force leadership with representation from all interested Department of Defense and civilian agencies. Authorization was received for a test program that included maintenance of six satellites in 12-h orbits and the development of various types of receivers. Navigation is based on the measured time of arrival of signals from four satellites. The six satellites allowed testing of the system in limited geographical regions during a period of about 2 h each day. Following the fully successful tests of the system in all proposed modes of operation, authorization was given for deployment of 18 satellites (plus spares) to provide instantaneous positions globally at any time of day.

System Description

Satellite Constellation. The operational GPS system includes at least 18 satellites in nearly circular 12-h orbits (20,000 km above the Earth). Three satellites in each of six orbit planes very nearly repeat their ground tracks each day, the time of equator crossing occurring 4 min earlier each day. Satellites launched during the test program at an orbital inclination of 63° will be included in the operational system during the balance of their useful lifetimes. All new satellites are at an orbital inclination of 50°. At least 4 of the 18 satellites are visible at an elevation angle above 10° at any time or location.

Transmitted Signals. The satellites transmit signals for navigation purpose at coherent L-band frequencies of 1227.6 MHz (L_2) and 1575.42 MHz (L_1). Two frequencies are used in order to permit correction of the time of arrival of the signal for the effect of ionospheric refraction, which is inversely proportional to the square of the frequency. Each satellite broadcasts its own position in real time from a prediction that has been uploaded in memory by a master control station on the ground, and each also broadcasts an almanac giving an approximate orbit and the health status for all satellites. The ephemeris, almanac, and clock corrections are transmitted at both frequencies by phase modulation at a 50 bits per second (bps) rate. The measurement of the time of arrival of reference signals (handover words) transmitted at 6-s intervals is accomplished by cross-correlation of a unique code for each satellite generated in the receiver with the corresponding code modulated on the signals transmitted by the satellite. A precision (P) code modulates both frequencies at a 10.23-MHz rate, the code repeating each week. Because of the difficulty in correlating this long code sequence in a blind time-frequency search, the L_1 signal is also modulated by a clear acquisition (C/A) code. The C/A code is available to all potential users of the GPS system; arrangements for access to the precision code must be made with the Department of Defense. A more complete discussion of the signal is given by Spilker (1980).

Measurement Types

When the transmitted code is correlated with the corresponding code generated on the ground, the difference in the time of emission and time of receipt of the 6-s time marks, multiplied by the speed of light, gives an estimate of range to the satellite. In addition to tropospheric and ionospheric effects on the propagation time, the range estimate will have a large bias due to errors in the receiver clock. The measurement is therefore referred to as a *pseudorange*. Pseudorange measurements to four satellites along with the broadcast satellite positions allows the computation of the error in the receiver clock as well as the position of the observer. The pseudorange measurements may be based on C/A or P code measurements. In the case of the P code, measurements will normally be made at both the L_1 and the L_2 frequencies. The vacuum pseudorange can be inferred from the L_1 and L_2 measurements (approximately) as

$$R_v = 2.54 R_1 - 1.54 R_2 \qquad (1)$$

Pseudorange measurements at each frequency made with an omnidirectional antenna have a precision on the order of 1 m when made on the P code and on the order of 10 m when made on the C/A code. Note from Eq. 1 that the precision of the vacuum range will be about four times poorer than that of the original L_1 and L_2 precisions as a result of the ionospheric refraction correction. Once the code correlation has been made, the receiver also tracks the carrier signal, allowing measurements of phase or Doppler to be made. Since the L_1 and L_2 wavelengths are only 19 cm and 24 cm, respectively, the phase measurements can be made to a precision of the order of 2 mm with an omnidirectional antenna. Phase measurements may be differenced over a time interval: If the integer number of carrier cycles received during the time interval is also counted, the result, called the *Doppler count*, can be converted to a difference in the range to the satellite over the time interval by

$$R(i+1) - R(i) = \frac{cN}{f}$$

where N is the number of integer and fractional cycles measured, c is the velocity of light, and f is the transmitted frequency. In practice, it is difficult to measure the phase directly; instead, the phase measurements are made on the difference between the received frequency and a ground reference frequency f_g that has been offset $(f_t - f_g)$ from the transmitted frequency f_t. The range difference is then

$$R(i=1) - R(i)$$
$$= \frac{c(N - (f_t - f_g)(t(i=1) - t(i)))}{f}$$

the sign of the result depending on whether the reference frequency is offset above or below the transmitted frequency.

Methods of Computing Position

Methods of computation of the position of the observer are based on either one of two underlying principles, with a number of variations in actual operations. The major distinction in the methods is whether the pseudorange measurements are unique, such as in the case of P code measurements, or ambiguous, such as in the case of phase measurements of the carrier signal. In the latter case the range measurements will be in error by an integral number of carrier wavelengths (a different integer for each satellite and each station). In either case, the position of the observer introduces three unknowns into the problem and, unless the observer is equipped with a hydrogen maser clock synchronized to GPS time, the error in the observer's clock introduces at least one more unknown. Variations in the solution are introduced to

1. Accommodate a moving observer, with or without additional data such as accelerometer outputs
2. Minimize the effects of random errors by the use of redundant observations
3. Optimize accuracy by balancing random error reduction obtained by introducing clock models for the receiver against errors caused by unmodeled clock variations
4. Minimize the effects of GPS satellite position and clock errors
5. Accommodate asynchronous observations for those receivers that do not make simultaneous observations of four or more GPS satellites.

For uncommon pseudorange measurements, the fundamental equations to be solved based on pseudorange measurements $R(j)$, $j = 1$ to 4 for satellites are

$$(x(j) - x(0))^2 + (y(j) - y(0))^2 + (z(j) - z(0))^2$$
$$+ (ct(0))^2 = (r(0))^2$$

The solution is ambiguous, but normally easily resolvable with only a very crude knowledge of the location of the observer. The effect of satellite geometry on the strength of solution is expressed in terms of quantities such as the geometric dilution of precision (GDOP). GDOP is simply the square root of the trace of the covariance matrix for unit weight of observation. Other terms, such as position dilution of precision (PDOP) omit clock variance from the trace. As noted earlier, variations in the solution are made if the observer's position $(x(0), y(0), z(0))$ is time-varying and/or if a clock model is introduced. Satellite position and clock errors can be minimized by making simultaneous measurements at a known location and either using the data from the two stations simultaneously in a solution for the unknown site with respect to the known site or simply offsetting the solution for the unknown site by the displacement calculated by using the data from the known site.

If the pseudorange measurements are ambiguous by an integer number of carrier signal wavelengths, additional parameters must be introduced into the solution and additional measurements must be made to allow the determination of all the unknowns. This technique is utilized for the determination of the relative position of two or more sites since the ambiguity cannot be resolved in the presence of GPS position and clock errors, which significantly exceed a wavelength, unless the effects of these error sources are suppressed by treating them as common mode errors at two or more sites. If the receiver maintains phase lock on each satellite over a sufficient interval of time (an hour or two), only one additional parameter need be introduced for each satellite observed by each station:

$$(x(ji) - x(0))^2 + (y(ji) - y(0))^2 + (z(ji)$$
$$- z(0))^2 + (ct(0i))^2 + (wn(0j))^2 = (R(ji))^2$$

where w is the wavelength of the signal on which the i phase measurements are made and $N(0j)$ is the number of cycle ambiguities for the jth satellite at the station. It is assumed here that the station position is fixed. Usually $N(0j)$ is treated as a continuous variable in a preliminary solution and then, if the $N(0j)$ are found to be near integers, they are set to the nearest integers in a final solution. Departures from integer values increase as base lines between sites increase because the effects of GPS position and clock errors have different projections on the lines of site from distant stations; if single-frequency data are used, then differences in ionospheric effects also increase as the site separation increases. Note that without further constraint, a solution with both a clock epoch error and an ambiguity error will be singular. The clock at one of the stations is usually used as a master clock so that the singularity can be removed by setting the clock error of the master clock to zero. Also note that the satellite position used in the calculations is a function of the clock error at the station so that if the station clock error exceeds a microsecond, the functional dependence of the interpolated satellite position must be considered in the solution. Lastly, if the cycle count is not continuous, then the ambiguity is different for each observation becoming $N(0ji)$, thus increasing the geometry requirements for the resolution of the ambiguity, but not affecting the precision of the solution if the ambiguities are resolvable.

Types of Receivers

Receivers currently available differ in

1. Degree of simultaneity of observations of different satellites
2. Whether measurements are made only on C/A code or on both C/A and P
3. Whether measurements are made on pseudorange and/or phase
4. Whether the satellite almanacs, ephemeris, and clock equations are recovered with the observations or must be obtained after the observation campaign

At the extremes of simultaneity, receivers may sequence through two frequencies for each of four or more satellites or have multiple channels to provide simultaneous observations. Quasi-simultaneity can also be achieved by multiplexing the signals so that the switching among the signals is so rapid that simultaneity is effectively achieved. Measurement on the C/A signal alone precludes L_2 measurements, since the L_2 channel does not have the C/A code, which means that the ionospheric correction cannot be made. However, if the base line between the stations is very short, perhaps 10 km or so, the precision of single-frequency solutions could be higher than that of dual-frequency solutions because the ionospheric errors may be common while the random error of the measurement is degraded by the ionospheric correction algorithm. Pseudorange measurements are invaluable for the determination of the position of an isolated receiver, but cannot compete in accuracy with the relative position precision achievable from phase measurements at two sites. Readout of satellite ephemeris and clock errors are vital for real-time operations, but data obtained after the operation are adequate for most geodetic and geophysical operations.

Applications

The impact of the global positioning system on the commercial, scientific, and public sectors will continue to grow over the years as the system reaches operational status and costs of receivers continue to drop. Shipping will be able to rely on satellites for navigation in straits and harbors as well as in the open oceans; aircraft will have a global system obtained more cost-effectively, mean sea level can be established inland by comparison of the GPS positions with gravimetric data, and crustal motions are being monitored with GPS systems. The LANDSAT mapping satellites have been provided with real-time positions with satellite-borne receiver-navigators, and the TOPEX oceanographic satellite will have still more precise GPS receivers aboard which, with radar altimeter data, will permit the mapping of ocean currents. Receivers and map displays being developed for the motoring and sailing public will bring direct benefits of the GPS system to everyone.

RICHARD J. ANDERLE

Reference

Spilker, J. J., Jr., 1980, GPS signal structure and performance characteristics, *Global Positioning System*. Washington, D.C.: Institute of Navigation, 29-54.

Cross-references: *Crustal Movements and Tectonic Deformation; Doppler Positioning: Satellite; Geodesy: Geometric; Geodesy: Satellite; Geodetic Reference Systems; Satellite Altimetry; Satellite Laser Positioning; Very-Long-Baseline Interferometry (VLBI)*.

GRAVITY ANOMALIES: INTERPRETATION

Many geodetic and geophysical processes and computations require, as input data, quantities that represent gravitational variations caused by the Earth's internal heterogeneous mass distribution. These quantities are known as *gravity anomalies*. Several types of gravity anomalies may be computed. Each has a different interpretation. The differences between the various kinds of gravity anomalies depend primarily on how the topographic masses are treated in the computational process used to obtain the gravity anomalies.

The three most commonly used gravity anomaly types are the free-air gravity anomaly, the Bouguer gravity anomaly, and the isostatic gravity anomaly. Other varieties of gravity anomalies have been proposed, for example, the Rudzki inversion and Helmert condensation gravity anomalies, but these are seldom used. A few of the infrequently used kinds of gravity anomalies are mentioned in some of the geodetic references listed at the end of this article. The discussion in this article is limited to the three commonly used gravity anomalies.

The ensuing discussion of gravity anomalies and their interpretation applies primarily to continental land areas. (Modifications needed for application to normal oceanic areas are given parenthetically.) Nonstandard methods of computation and interpretation apply in some special cases. For example, gravity anomaly patterns over such structures as oceanic islands and subduction zones will require somewhat different approaches to interpretation than gravity anomaly patterns over normal continental or oceanic areas. Also, special formulas must be used for computation of gravity anomalies when gravity measurements are made in unusual locations, such as lake surfaces and glacial ice caps. Appropriate formulas for gravity anomaly computation in various situations are given at the end of this article.

Construction of an Earth Model

In order to convey a clear and basic understanding of the meaning and correct interpretation of free-air, Bouguer, and isostatic gravity anomalies, we begin by constructing a model of the Earth. Start with a simple theoretical model void of irregular mass distribution. Then define and add elements to the simple initial model to create a final model whose mass distribution duplicates that within the real Earth. Design all elements of the model such that they can be used to define and interpret the gravity anomalies. The end product of model development is Eq. 5, which expresses gravity at the physical surface of the Earth as the sum of the gravitational elements used to construct the final model. Equation 5 and subsequent modifications to it are used as a basis for discussion of the gravity anomalies.

Define a biaxial oblate ellipsoid of revolution to be the simple theoretical model of the Earth. Place the center of this ellipsoid at the center of gravity of the Earth, and make the minor axis of this ellipsoid coincide with the mean rotation axis of the Earth. Rotate this ellipsoid about its minor axis with the same angular velocity as the Earth's rotation. Fill the interior of this ellipsoid with mass. The total mass contained within the ellipsoid is equal to the total mass contained within the real Earth. Distribute the mass inside of this ellipsoid such that it is both rotationally symmetric and symmetric with respect to the equatorial plane. Thus, the lateral mass distribution within this ellipsoid is uniform. Finally, the surface of this ellipsoid should approximate the sea-level surface of the real Earth, called the *geoid*, as closely as possible.

The theoretical ellipsoidal model just described is usually called the *normal ellipsoid*. Since it contains mass, it must generate a theoretical or normal external gravity field. This normal gravity field depends only on the size, shape, total mass, and rate of rotation of the normal ellipsoidal model. Exactly four independent parameters are required to completely define the external normal gravity field. The simplest set of parameters that can be used is (1) the length of the semimajor axis of the ellipsoid, (2) the length of the semiminor axis of the ellipsoid, (3) the total mass contained within the ellipsoid, and (4) the angular velocity of the ellipsoid's rotation. The external normal gravity field varies only with latitude and distance from the center of the ellipsoid. Because of the rotationally symmetric mass distribution within the ellipsoid, there is no dependence on longitude.

Geodesists have derived analytical expressions that enable computation of normal gravity at any point in space. The most commonly used expression for computing normal gravity *on the surface of the ellipsoid*, γ, is the *normal gravity formula*. The normal gravity formula for the normal ellipsoid of the Geodetic Reference System 1980 (GRS80) is

$$\gamma = 978032.7(1 + 0.0053024 \sin^2 \phi$$
$$- 0.0000058 \sin^2 2\phi) \text{ mgal} \qquad (1)$$

where γ = normal gravity on the surface of the ellipsoid and ϕ = geodetic latitude. Values of γ on the GRS80 ellipsoid range from 978032.7 mgal on the equator to 983218.6 mgal at the poles. In connection with gravity anomaly computation for geophysical prospecting applications, γ is sometimes called the *latitude correction*.

The constants that appear in Eq. 1 are computed from the four parameters that define the GRS80 normal ellipsoid. Since Eq. 1 is limited in accuracy to 0.1 mgal, sometimes more accurate expressions are sometimes used to compute normal gravity.

The milligal (mgal) is the traditional and most convenient unit for expressing gravity anomaly magnitudes in many applications. In geophysical prospecting work, the gravity unit (gu) is sometimes used. For the most accurate work, the microgal (μgal) is often used. None of these are SI units. The SI unit for gravity is meters per second per second ($\text{m} \cdot \text{s}^{-2}$). Appropriate equivalences are given here for reference purposes.

$$\mu\text{gal} = 10^{-6} \text{ gal} = 10^{-8} \text{ m} \cdot \text{s}^{-2}$$
$$\text{gu} = 10^{-4} \text{ gal} = 10^{-6} \text{ m} \cdot \text{s}^{-2}$$
$$\text{mgal} = 10^{-3} \text{ gal} = 10^{-5} \text{ m} \cdot \text{s}^{-2}$$
$$\text{gal} = \text{cm} \cdot \text{s}^{-2} = 10^{-2} \text{ m} \cdot \text{s}^{-2}$$

The next step in model construction is to add

elements to the normal ellipsoid that convert this simple initial model into a final model that duplicates the real world situation. The first element to be added elevates normal gravity from the surface of the ellipsoid to the true physical surface of the real Earth. This operation can then be accomplished by subtracting the free-air reduction F from normal gravity on the ellipsoid surface:

$$F = \frac{\partial g}{\partial h} h \qquad (2)$$

where $\partial g/\partial h$ = vertical gradient of real gravity, g, and h = orthometric elevation of the physical surface of the Earth above sea level.

A second-order term can be added to Eq. 2 to yield greater accuracy when h is large.

The variable h in Eq. 2 is defined to be the conventional orthometric elevation above sea level. It is the distance between a point on the physical surface and sea level, the geoid, measured along the curved plumb line that passes through the point on the physical surface. It may seem that the variable actually needed is the elevation of the physical surface above the ellipsoid. For technical reasons, to be explained later, the distance between sea level, the geoid, and the ellipsoid is immaterial to the present discussion and may be neglected without introducing any appreciable error. Most elevations published in the Western Hemisphere are orthometric elevations.

Since the intent is to develop elements that convert from the simple theoretical model to a model that corresponds to the real world, the vertical gradient of real gravity, $\partial g/\partial h$, should be used in Eq. 2. Unfortunately, correct values for $\partial g/\partial h$ are seldom known. Therefore it is common practice to substitute the vertical gradient of normal gravity, $\partial \gamma/\partial h$, for $\partial g/\partial h$ in the free-air reduction. Introducing the approximation

$$\frac{\partial \gamma}{\partial h} \approx \frac{\partial g}{\partial h}$$

we may then write

$$F = \frac{\partial \gamma}{\partial h} h \qquad (3)$$

The numerical value of $\partial \gamma/\partial h$ depends on the values of the four parameters that define the normal ellipsoid. For all modern ellipsoids, including the GRS80 ellipsoid,

$$\frac{\partial \gamma}{\partial h} = 0.3086 \text{ mgal} \cdot \text{m}^{-1} \qquad (4)$$

The value of $\partial \gamma/\partial h$ given by Eq. 4 is generally applied as a constant, thereby neglecting a very slight dependence on latitude.

The free-air reduction is valid only in free space outside of all mass. Thus, when the free-air reduction is used to elevate, or lower, gravity, it must be assumed that no mass exists over the distance h. However, by definition, no normal mass exists outside of the normal ellipsoid's surface. Therefore, use of F to elevate gravity upward from the normal ellipsoid surface is completely valid at this point in the argument.

The next three elements to be introduced express the gravitational effects, at the physical surface of the Earth, of differences between the actual nonuniform mass distribution within the real Earth and the laterally homogeneous mass distribution within the simple normal ellipsoidal model.

The most obvious difference between the simple normal ellipsoidal model and the real Earth is that the latter has topographic land masses that extend above sea level (and oceanic basins where the rock surface is below sea level). The normal ellipsoidal model is void of any topographic (or bathymetric) features. Therefore, the next step in model construction involves adding the topographic masses (and ocean basins) to the simple ellipsoidal model. The added topographic masses must have the same nonuniform mass distribution as the real topographic masses. Let the *total* gravitational attraction of these topographic masses at a point on the physical surface be A_T.

(In oceanic areas, the mass represented by A_T must be subtracted. The seawater filling the ocean basins is considerably less massive than the "rock" material within the normal ellipsoidal model. For oceanic areas, A_T is the difference in gravitational attraction at the ocean surface between that of the uniform normal "rock" mass and that of the uniform density seawater mass within the ocean basins. For simplicity, this effect will be referred to as the *bathymetric mass deficiency*.)

The second mass-related element to be introduced corrects an imbalance created by adding (or subtracting) A_T. The normal ellipsoid was defined as containing the total mass of the real Earth. Adding topographic masses above the ellipsoid (and subtracting mass below the ellipsoid surface in ocean basins) may have altered the total mass of the Earth. Therefore, it is now necessary to subtract mass under the continents (and add mass under the oceans) such that, in the end, the correct total mass of the Earth is restored. The justification for this operation rests on the well-accepted principle of isostasy. According to the principle of isostasy, the total mass contained within vertical columns of unit cross section between the physical surface (or ocean surface) of the Earth and the so-called depth of compensation, 60–100 km, is constant. Thus, the excess masses contained in the topography of land areas are balanced or compensated by a corresponding mass deficiency at depth within the Earth (and the bathymetric mass deficiency in the ocean basins is compensated by a mass excess at depth). Thus, subtracting the gravitational attraction of the compensating mass deficiency in land areas effectively removes the mass added in the topography. (In oceanic areas, the gravitational attraction of the compensating mass excess is added.) Let the gravitational attraction of the compensating mass deficiencies (or excesses) at a point on the physical surface (or ocean surface) be A_C.

The final mass element represents the only remaining difference between the real Earth and the normal ellipsoidal model. Namely, the real Earth has laterally nonuniform mass distribution beneath sea level. Let the gravitational effects at the physical surface (or ocean surface) of the differences between the nonisostatic heterogeneous mass distribution beneath sea level (or ocean floor) of the real Earth and the laterally homogeneous mass distribution within the normal ellipsoid be A_I. This operation merely redistributes masses. It does not change the total mass of the Earth.

It is now possible to write an equation expressing gravity at a point on the physical surface of the Earth in land areas (or ocean surface in water areas) as a function of the model parameters just discussed. The equation is

$$g = \gamma - F + A_T - A_C + A_I \quad (5)$$

where g is the gravity at the physical surface (or ocean surface).

Equation 5 applies for land areas. (The signs of A_T and A_C are reversed for application in oceanic areas.) Except for any approximations used in F, Eq. 5 is exact.

Free-Air Gravity Anomaly

Transposing γ and F from the right side to the left side of Eq. 5 gives

$$g + F - \gamma = A_T - A_C + A_I \quad (6)$$

The left side of Eq. 6 is the definition of the free-air gravity anomaly Δg_F:

$$\Delta g_F = (g + F) - \gamma \quad (7)$$

Equation 7 represents Δg_F in its conventional form. The gravity g measured at the physical surface of the Earth is lowered to sea level, the geoid, by means of the free-air reduction F. Then normal gravity γ computed at the ellipsoid surface is subtracted. Although $g + F$ applies at the geoid and γ applies at the ellipsoid, Δg_F is generally considered to apply at the geoid surface. (If g is measured on the ocean surface, then $F = 0$ and Eq. 7 reduces to $\Delta g_F = g - \gamma$.)

Use of the free-air reduction F to lower measured gravity from the physical surface of the Earth to sea level requires the assumption that no mass exists above sea level. In reality, the topographic masses are actually present. The topographic masses are simply ignored when F is used to compute Δg_F.

Since real measured gravity is being lowered to the geoid by application of F, Eq. 2 must be used for F in order to obtain a correct reduction. That is, the vertical gradient of real gravity $\partial g / \partial h$ is required in order to obtain completely accurate results. However, because the value of $\partial g / \partial h$ is seldom known, it is common practice to use Eq. 3 for F. When the vertical gradient of normal gravity, $\partial \gamma / \partial h$, is used to compute the conventional free-air gravity anomaly, the resulting Δg_F at the geoid is only an approximate rather than a rigorously correct value.

To circumvent this approximation problem, we must define the free-air gravity anomaly in a different way. Rewrite Eq. 7 as

$$\Delta g_F = g - (\gamma - F) \quad (8)$$

In this form, known as the modern interpretation, normal gravity computed at the ellipsoid surface is elevated over the distance h by using the free-air reduction. The resulting upwardly reduced value $\gamma - F$ is subtracted from gravity measured at the physical surface, g, to yield the surface free-air gravity anomaly that applies at the physical surface of the Earth.

Because no normal mass exists outside of the ellipsoid surface, normal gravity is literally elevated through free space in the modern interpretation. Therefore, the no-mass condition for use of the free-air reduction is rigorously satisfied. Since normal gravity is being elevated, it is completely correct to use the vertical gradient of normal gravity in F, and Eq. 3 is exactly right for this application. The only assumption involved is that $\partial \gamma / \partial h$ is constant everywhere. No appreciable error is introduced by this assumption in most practical cases. Consequently, the surface free-air gravity anomaly computed by Eqs. 8 and 3 involves no important approximations or assumptions and, hence, is essentially correct.

If Eq. 3 is used in both cases, the conventional free-air gravity anomaly value computed by Eq. 7 will be identical in value to the surface free-air gravity anomaly value computed by Eq. 8. However, this value will be rigorously correct at the physical surface, but only an approximation at the geoid. (In oceanic areas, the conventional and surface free-air gravity anomalies are the same.)

The proper interpretation of the free-air gravity anomaly can be deduced by comparing Eqs. 6 and 7 to obtain

$$\Delta g_F = (A_T - A_C) + A_I \quad (9)$$

The free-air gravity anomaly is seen to include the

total gravitational effects of the heterogeneous topographic masses (or the bathymetric mass deficiency), A_T, the gravitational effects of the compensating mass deficiencies (or excesses), A_C, and the gravitational effects of nonisostatic mass distribution irregularities beneath sea level, A_I.

It is convenient to divide A_T into two parts. The first element of A_T will be referred to as the gravitational effect of gross topography. The second part of A_T is the gravitational effect of mass distribution irregularities in the topography that are caused by rock density variations.

The contribution of each of these two elements to A_T can be visualized by referring to a topographic model. Let the average density of the topographic masses be σ_T. Then suppose that the topographic masses are formed from rock matter having the constant density σ_T. Call this model *gross topography*. Since the assumed average density of the gross topography is a constant, the gravitational attraction of the gross topography at a point on the physical surface of the Earth depends upon the elevation of that point; that is, it depends upon the thickness of nearby gross topography—the thicker the gross topography, the greater the mass, and the larger the gravitational attraction. In mountainous areas, where there are short-wavelength elevation variations having significant amplitudes, the gravitational attraction of the gross topography must be characterized by short-wavelength variations having significant amplitudes. Conversely, in flat plains regions, the gravitational attraction of gross topography is nearly constant from point to point because the elevations are nearly constant.

The gravitational effects of rock density variations within the topography are superimposed on the gravitational effects of the gross topography. This second element of A_T can be visualized as the difference in gravitational attraction between that of the real topographic masses having nonuniform mass distribution and that of the gross topographic model that has uniform density distribution. (In oceanic areas, the signs of A_T and A_C in Eq. 9 are reversed. The bathymetric mass deficiency generates only a gross topographic effect. Short-wavelength variations in nearby ocean bottom topography having significant amplitudes generate short-wavelength variations in A_T having significant amplitudes. On the other hand, A_T is nearly constant over abyssal plains. The gravitational effects of density variations in the rock masses underlying the ocean basins are included in A_I).

The gravitational effects of gross topography are very important in free-air gravity anomaly interpretation. These effects, however, must be discussed in conjunction with the effects of isostatic compensation. Therefore, it is necessary to examine the combined contribution to free-air gravity anomalies of $A_T - A_C$, where A_T includes only gross topographic effects.

Isostatic compensation of the topographic masses exists only on a regional basis. That is, individual mountain peaks and valleys having a small lateral extent are not compensated locally by complementary mass deficiencies or by excesses at depth. However, an entire mountain range will be compensated by mass deficiencies at depth on an overall average basis.

Because of the regional character of isostatic compensation, surface mass variations due to short-wavelength features of gross topography are not neutralized by any compensating mass variations at depth. Thus, A_C tends toward zero for short-wavelength features of gross topography. Consequently, the full gravitational effects of the short-wavelength variations in gross topography included in A_T are reflected in the free-air gravity anomaly. In other words, the free-air gravity anomalies are highly correlated with local elevation changes. Free-air gravity anomalies tend to have large positive values on top of mountain peaks and large negative values on the floors of deep, narrow valleys.

On the other hand, the long-wavelength gross topographic variations are likely to be largely neutralized by compensating mass variations at depth, and the combined contribution of $A_T - A_C$ in this case tends toward zero. In other words, on a regional basis, there is very little correlation between free-air gravity anomalies and gross topography. The average free-air gravity anomaly over an entire mountain range should be, at most, only slightly positive.

Topographic variations of intermediate wavelengths are likely to be partially compensated.

Some departures from complete regional isostatic balance do exist, and these departures can contribute to regional free-air gravity anomaly magnitudes. As an example, regionally negative free-air gravity anomalies over the Canadian shield are generally attributed to a geologically recent removal of a glacial cover at the end of the most recent ice age. These regionally negative free-air gravity anomalies reflect a condition of overcompensation; that is, there is too much compensating mass deficiency at depth. The compensating mass deficiencies now present are appropriate to balance a "topographic" ice load mass excess at the surface that no longer exists. Various other mechanisms can also affect the degree of isostatic compensation that is present.

Superimposed on the gravitational effects of gross topography and its isostatic compensation are the gravitational effects of nonisostatic mass distribution irregularities beneath sea level, A_I, and rock density variations within the topography, included in A_T. The gravitational effects of these irregularities can be quite small. For example, salt domes, anticlines, and some other structures of interest to geophysical prospecting typically generate gravity anomaly variations ranging from a few milligals down to a few tenths of a milligal. In other cases, however, mass distribution irregularities can have a significant effect on gravity anomalies. An extreme example is the mid-continent gravity high in the central United States where subsurface masses of anomalous density generate free-air gravity anomaly variations that approach 100 mgal. Interpretation can be complicated by the fact that anomalous density masses having a large lateral extent may be partially or completely compensated.

All elements that contribute to free-air gravity anomaly magnitudes must be considered in their interpretation (e.g., short-wavelength variations in gross topography, longer-wavelength variations in gross topography and the degree of isostatic compensation, and variations in rock density both above and below sea level). For example, it is incorrect to conclude that free-air gravity anomalies will be uniformly negative at points below average terrain levels due to short-wavelength topographic effects alone. Numerous examples exist of positive free-air gravity anomalies in valleys where anomalous mass distribution effects are important.

The largest known free-air gravity anomalies occur atop mountain peaks on islands in deep ocean basins. For example, free-air gravity anomalies exceeding +600 mgal are known on Mauna Loa in Hawaii. This mountain is one of the highest in the world when the elevation of its summit above the ocean floor is considered. The largest negative free-air gravity anomalies probably occur in conjunction with deep oceanic trenches and subduction zones. Although inadequate measured data is available to enable confirmation of a maximum value, it is probably a few hundred milligals.

Free-air gravity anomalies exceeding 100 mgal are uncommon except in mountainous areas, over oceanic islands and seamounts, and over areas of disturbed structural characteristics such as subduction zones. The worldwide average magnitude of free-air gravity anomalies as estimated by the square root of the variance of the magnitudes is about ±40 mgal. The worldwide average of all free-air gravity anomalies should be very close to zero.

The best source for viewing free-air gravity anomaly magnitudes on a worldwide basis is the free-air gravity anomaly map of the world compiled by Carl Bowin of the Woods Hole Oceanographic Institute. The map is published both in colored form and, at a larger scale, as a black-and-white atlas. The map and atlas were published in 1982 as the Geological Society of America Map and Chart Series MC-45 and MC-46.

The free-air gravity anomaly is very useful for many geodetic applications. It is easily computed. It is a gravitational parameter, valid either at the Earth's surface or along the geoid, that truly represents the actual mass distribution within the Earth. Its computation requires no assumptions about Earth structure and does not cause appreciable changes in the Earth's external gravity field. The indirect effect, discussed later, is negligible.

The free-air gravity anomaly is infrequently used for geophysical or geological interpretation of land areas. The principal reason is that the short-wavelength, topographically generated variations in the free-air gravity anomaly tend to completely mask the generally smaller variations due to internal mass and density distribution. For the same reason, large-scale, free-air gravity anomaly maps of continental land areas are quite rare. (Contrarily, geophysical interpretations of free-air gravity anomalies and free-air gravity anomaly maps are common over oceanic areas. One reason for this is that the detailed, accurate ocean depth data required for Bouguer gravity anomaly computation is usually lacking. Fortunately, except over major bathymetric features, such as seamounts and fracture zones, short-wavelength topographic variations in the free-air gravity anomaly tend to be somewhat less pronounced in oceanic areas due to the thick, uniform water layer between the ocean surface and the ocean floor.)

Bouguer Gravity Anomaly

There appears to be a definite need for a type of gravity anomaly that is free of short-wavelength topographic effects. The easiest way of obtaining such a form is simply to remove the topographic masses completely by computation. The result of such an operation is the Bouguer gravity anomaly.

Topographic mass removal, required for Bouguer gravity anomaly computation, is accomplished in two steps. The first step is subtraction of the Bouguer plate. The second step, application of the terrain correction, rectifies certain errors committed in the first step.

The Bouguer plate is a right circular cylinder having an infinite radius and a thickness h equal to the orthometric elevation of the point P on the physical surface where the Bouguer gravity anomaly is being computed. The axis of the cylinder is vertical and passes through P. The upper surface of the Bouguer plate passes through P. The lower surface is tangent to sea level, the geoid.

The gravitational attraction at P of the topographic mass contained within the Bouguer plate is given by the simple Bouguer reduction A_b:

$$A_b = 2\pi k \sigma h \qquad (10)$$

where

A_b = gravitational attraction at P of the mass within the Bouguer plate
π = 3.14159 . . .
k = gravitational constant
σ = density of the mass contained within the Bouguer plate, assumed to be constant
h = orthometric elevation of P above sea level

The most commonly used value for the density of the topographic mass within the Bouguer plate is $\sigma = 2.67\ \text{g} \cdot \text{cm}^{-3}$. This value is supposed to represent the average density of all masses above sea level. It is probably too low in many cases, but usually works reasonably well to eliminate the major portion of short-wavelength topographic effects. For a number of special geophysical interpretation purposes, other, more exact, density values can be used in the Bouguer reduction. If $\sigma = 2.67\ \text{g} \cdot \text{cm}^{-3}$ is adopted in the Bouguer reduction, then Eq. 10 becomes

$$A_b = 0.1119h\ \text{mgal} \cdot \text{m}^{-1}$$

The computed gravitational attraction of the Bouguer plate is always larger than the actual gravitational attraction of the topographic masses, because the flat upper surface of the Bouguer plate does not conform exactly to the irregular topographic surface of the Earth, particularly in mountainous areas. Therefore, a correction term that accounts for the gravitational effects of mass, or lack of mass, lying between the actual physical surface of the Earth and the plane upper surface of the Bouguer plate is necessary, except in flat plains regions where the actual topographic surface is essentially planar. The correction required is known as the terrain correction A_t.

There are two situations to be considered: (1) the physical surface of the Earth is below the upper surface of the Bouguer plate, or (2) the physical surface of the Earth is above the upper surface of the Bouguer plate. It is necessary to restore the mass above the physical surface mistakenly subtracted in the Bouguer plate in case (1) and to remove the mass above the upper surface of, and therefore not included within, the Bouguer plate in case (2). The gravitational attractions of both residual masses are computed to yield the terrain correction that is subtracted from the gravitational attraction of the Bouguer plate.

Computation of terrain correction values is rather laborious when conventional methods are used. Hundreds of average elevation values must be estimated for surface areas of various sizes, called *compartments*, in the vicinity of the point P, called the *computation point*, for which a terrain correction value is desired. The average elevation values are estimated manually from height contours contained on large-scale topographic maps. Actual field measurements may be necessary in the immediate vicinity of the computation point. Then differences between the estimated average elevations for each compartment and the elevation of the upper surface of the Bouguer plate are entered into standard formulas, and the gravitational attraction at the computation point P of mass lying between the physical surface of the Earth and upper surface of the Bouguer plate is computed for each compartment. The gravitational attractions for all compartments are summed up to obtain the final terrain correction value. A density value of $2.67\ \text{g} \cdot \text{cm}^{-3}$ is generally used for the mass contained within each compartment. For best accuracy, the terrain correction computation should encompass all areas within a radius of 167 km from the computation point. The contribution of areas beyond the 167-km radius is negligible. However, because of the labor involved, some terrain correction computations have encompassed smaller areas.

The advent of digital elevation data bases has significantly reduced the labor involved in terrain correction computations. By using digital elevation data bases, all average elevation estimations and the subsequent gravitational attraction computations are accomplished automatically on high-speed computers. However, for best accuracy, it is still necessary to work the area within a few kilometers of the computation point manually.

The magnitude of the terrain correction is correlated with the ruggedness of the topography nearby the computation point. In extremely rugged areas, terrain correction values can exceed 100 mgal. Typical values in most mountainous regions are generally less than 50 mgal. In flat plains areas, the terrain correction is negligible.

Let the combined gravitational attraction of the Bouguer plate and the terrain correction be A_B, the complete Bouguer reduction. Then

$$A_B = A_b - A_t \qquad (11)$$

Now subtract A_B from both sides of Eq. 6:

$$g - A_B + F - \gamma = (A_T - A_B) - A_C + A_I \qquad (12)$$

The left side of Eq. 12 is the definition of the Bouguer gravity anomaly Δg_B:

$$\Delta g_B = (g - A_B + F) - \gamma \qquad (13)$$

The meaning of the definition of Eq. 13 is that the gravitational attraction of the topographic masses is subtracted from gravity measured at the physical surface of the Earth. That is, these masses are removed completely. Then the free-air reduction is applied to lower $g - A_B$ through free space from the physical surface of the Earth to sea level, the geoid. Finally, normal gravity computed on the ellipsoid surface is subtracted from $g - A_B + F$.

If the terrain correction is included in A_B, the Δg_B computed by Eq. 13 is called the *complete* Bouguer gravity anomaly. If, on the other hand, the terrain correction is omitted from A_B, the Δg_B computed by Eq. 13 is called the *simple* Bouguer gravity anomaly. Unfortunately, Bouguer gravity anomaly terminology has not been standardized, and other names are occasionally encountered that denote the Δg_B with or without the terrain correction.

It is often said that the terrain correction is always positive in land areas. This statement applies in the context that A_t always has the effect of increasing gravity g at the computation point. This can be seen by inserting Eq. 11 into Eq. 13.

The Bouguer gravity anomaly defined by Eq. 13 is conventional; that is, it is valid at the geoid surface. The approximation problems pertaining to the computation of F, noted during the discussion of the conventional free-air gravity anomaly, apply also to the conventional Bouguer gravity anomaly.

A modern definition of the Bouguer gravity anomaly is possible. In this case, Eq. 13 becomes

$$\Delta g_B = (g - A_B) - (\gamma - F)$$

Here, the free-air reduction is used to reduce normal gravity upward over the elevation h. The resulting value $\gamma - F$ is subtracted from gravity measured at the physical surface minus the gravitational effects of the topographic masses $g - A_B$. In this form, the Bouguer gravity anomaly is valid in free space at the location of the physical surface prior to topographic mass removal and is free of approximations in F. Although this modern interpretation of the Bouguer gravity anomaly is more rigorous than the conventional interpretation, the modern form is not widely known and, consequently, has seen little use.

In some geophysical prospecting applications, Bouguer gravity anomalies that apply to some intermediate reference elevation are computed. In other words, F and A_B are applied only partially or not at all. The resulting form is called an *incomplete* Bouguer gravity anomaly. It is useful only for interpretation of structural features within the uppermost parts of the Earth. (In oceanic areas, Bouguer gravity anomaly computation is handled differently. The Bouguer reduction is used to *add* mass to the ocean basins such that the density of the seawater, about 1.027 g $\cdot$ cm^{-3}, is raised to a standard rock density, preferably the average of the rock masses that underlie the ocean floor. Since information on actual rock density is generally lacking, the standard value of 2.67 g $\cdot$ cm^{-3} is most often used for oceanic Bouguer gravity anomaly computation. In this case, the density factor in the Bouguer plate computation, Eq. 10, becomes $\sigma = 2.67 - 1.027 = 1.643$ g $\cdot$ cm^{-3}, and

$$A_b = 0.06886d \text{ mgal} \cdot \text{m}^{-1}$$

where A_b is the gravitational attraction of added mass within the Bouguer plate and d is the depth of the water. Although the terrain correction is needed in A_B to completely eliminate short-wavelength bathymetric variations when significant ocean bottom relief exists, it is seldom applied in oceanic areas. The usual reason for terrain correction omission is that ocean bottom relief data is generally inadequate to enable its computation. In oceanic areas, A_B is applied in Eq. 13 with reversed sign.

The proper interpretation of the Bouguer gravity anomaly can be deduced by comparing Eqs. 12 and 13 to obtain

$$\Delta g_B = (A_T - A_B) - A_C + A_I \quad (14)$$

The term $A_T - A_B$ is the difference between the gravitational attraction of the real topographic masses, and the gravitational attraction of the topographic masses as modeled by the Bouguer plate, the terrain correction, and a constant value for rock density. If the constant density used to compute A_B closely matches the average density of the real topographic masses, then gross topography and, consequently, the gravitational attraction of gross topography are removed completely. In this case, $A_T - A_B$ includes only the gravitational effects of rock density variations within the real topography. Since all gravitational effects of gross topography have been eliminated, the gravitational effects of short-wavelength variations in topography must be absent from the complete Bouguer gravity anomaly. In other words, the *complete* Bouguer gravity anomaly has no short-wavelength variations due to local topography. *Simple* Bouguer gravity anomalies, however, are correlated with the terrain correction. For example, the simple Bouguer gravity anomaly atop a high mountain peak will appear as a pronounced short-wavelength, negative maximum value in comparison to surrounding values.

If the constant density used to compute A_B does not closely match the average density of the real topography, some small, short-wavelength, topographically caused variations may remain in the computed complete Bouguer gravity anomalies. This residual effect can be eliminated empirically by finding the density value that, when used in A_B, yields the minimum local correlation between elevation and computed Bouguer gravity anomaly. This procedure, called *density profiling*, is attributed to L. L. Nettleton. It can be used for both simple and complete Bouguer gravity anomalies.

The full negative gravitational effects of the compensating mass deficiencies, A_C, are contained in the Bouguer gravity anomalies given by Eq. 14. This negative influence is not counteracted in Δg_B computation by any positive gravitational effects of the gross topography because the gross topography has been removed by the Bouguer reduction. Thus, the Bouguer gravity anomalies must be highly correlated with the compensating mass deficiencies. The amount of compensating mass deficiency present depends upon regional elevation levels since compensation is a regional effect. Consequently, the Bouguer gravity anomalies are inversely correlated with regional elevation—the higher the average elevation, the more negative the Bouguer gravity anomaly. Thus, it is often said that the Bouguer gravity anomalies are smoothed mirror images of the topography. Regional Bouguer gravity anomaly magnitudes also depend partly on the degree of isostatic compensation that is present.

(In oceanic areas, Eq. 14 reads

$$g_B = (-A_T + A_B) + A_C + A_I \quad (15)$$

Short-wavelength gravitational variations due to ocean bottom topography should be eliminated if the constant density used to compute A_B closely matches the actual density of the mass that forms the ocean floor. Since the bathymetric mass deficiency $-A_T$ is essentially eliminated by adding A_B, the full positive gravitational effects of the compensating mass excesses, A_C, are contained in the Bouguer gravity anomalies given by Eq. 15, and these Bouguer gravity anomalies must be highly correlated with the compensating mass excesses—the larger the average ocean depth, the more positive the oceanic Bouguer gravity anomaly).

Thus, Bouguer gravity anomaly magnitudes are strongly tied to average elevation levels (or average ocean depths). As a worldwide average,

$$\Delta g_B \simeq -0.1\bar{h} \quad (\Delta g_B \simeq +0.07\bar{d})$$

where

Δg_B = approximate Bouguer gravity anomaly magnitude, mgal
$\bar{h}$ = average elevation level, m
$\bar{d}$ = average ocean depth, m

Bouguer gravity anomalies are affected by anomalous mass distribution in the same way as are free-air gravity anomalies. However, for complete Δg_B, the correlation with elevation appears only as a smoothly varying inverse regional effect. In other words, the term A_C in the context of Eq. 14 is a largely predictable, smoothly varying function. All short-wavelength variations generated by gross topography are absent. Consequently, the Bouguer gravity anomalies are sensitive and valid indicators of density variations above sea level, $A_T - A_B$, and mass distribution irregularities below sea level, A_I. Regional trends in Bouguer gravity anomalies, such as those due to inverse correlation with regional elevation, are frequently removed by computation before Bouguer gravity anomalies are used to interpret local structures.

A good understanding of Bouguer gravity anomaly magnitudes and variations is best achieved by viewing some of the numerous Bouguer gravity anomaly maps that have been published for continental areas. An outstanding example is the Bouguer gravity map of the United States published in 1982 by the Society of Exploration Geophysicists.

The Bouguer gravity anomalies are very well suited for Earth structure interpretation and are widely used for geological and geophysical exploration and for other scientific geophysical work. Specific procedures for applying Bouguer gravity anomaly data to interpret geological and geophysical structures and conditions are generally quite straightforward. However, a discussion of these procedures is beyond the scope of this article. Some guidelines and leads to appropriate sources of information may be found in the references.

The Bouguer gravity anomaly is a poor geodetic tool. Elimination of the topographic masses significantly changes the total mass of the Earth, perceptibly changes the external gravity field of the Earth, and generates a large indirect effect. Consequently, Δg_B is not a valid geodetic parameter. Geodetic uses of Bouguer gravity anomalies are limited to interpolation and extrapolation operations involved in estimating average gravity anomaly values.

(A definition of Bouguer gravity anomalies, in oceanic areas, that differs from the one given here may be encountered. The concept involved in this alternative definition is that the primary purpose of gravity anomaly computation is a reduction of the measured values to sea level. Since an ocean surface gravity measurement is made at sea level, no such reduction is necessary according to this concept and, therefore, Bouguer gravity anomalies at sea are defined to be identical to free-air gravity anomalies at sea. Although this alternative definition has a logical basis, it unnecessarily restricts oceanic gravity anomaly options and is incompatible with the concepts advanced in this article.)

Isostatic Gravity Anomalies

It may be advantageous to obtain a gravity anomaly form that retains the desirable properties of the Bouguer gravity anomaly, such as freedom from correlation with local topographic variations, but lacks features of Δg_B that may be viewed as undesirable, such as the strong inverse correlation with regional topography and the distortion of the external gravity field due to topographic mass elimination. The isostatic gravity anomalies provide just such an alternative.

To compute an isostatic gravity anomaly, the topographic masses are removed exactly as is done in computing the complete Bouguer gravity anomaly. The masses removed by the Bouguer reduction are then reinserted into the Earth at depth in such a way that the compensating mass deficiencies are neutralized. The mass insertion must be accomplished in accordance with some accepted model of the isostatic compensation mechanism. The most frequently used isostatic theories are the Pratt-Hayford theory and the Airy-Heiskanen theory. Several plausible models can be developed based on each of these two theories. The gravitational attraction of the reinserted masses is computed and added to the Bouguer gravity anomaly. The computation process is quite laborious since it requires estimation of average elevations for surface areas of various sizes over the entire Earth. Tables and maps of isostatic effects, available for some isostatic models, do simplify the process somewhat. (In oceanic areas, the mass inserted into the ocean basins by A_B is taken from the compensating mass excess at depth. The gravitational attraction of the compensating mass excess at depth is computed according to some isostatic theory and subtracted from the Bouguer gravity anomaly.)

Let the gravitational attraction of the masses reinserted at depth in accordance with some isostatic model be A_m. Then add A_m to both sides of Eq. 12. The result is

$$g - A_B + A_m + F - \gamma$$
$$= (A_T - A_B) - (A_C - A_m) + A_I \quad (16)$$

The left side of Eq. 16 is the definition of the isostatic gravity anomaly Δg_I:

$$\Delta g_I = g - A_B + A_m + F - \gamma \quad (17)$$

The meaning of Eq. 17 is that the gravitational attraction of the topographic masses is subtracted from gravity measured at the physical surface of the Earth, using A_B. Then the gravitational attraction of these masses reinserted at depth in accordance with some isostatic model is added, using A_m. The free-air reduction F is applied to lower $g - A_B + A_m$ through free space to sea level, the geoid. Finally, normal gravity computed at the ellipsoid surface is subtracted.

Equation 17 is the conventional isostatic gravity anomaly valid at the geoid. A modern interpretation is also possible, but has not been used. Equation 17 omits a small term that is generally included in isostatic gravity anomaly computations, namely, the curvature correction. This term accounts for the departure of the flat Bouguer plate from the curved Earth surface. The maximum value of the curvature correction is about 1.5 mgal. (In oceanic areas, for Eq. 17, the signs of A_B and A_m are reversed.)

The interpretation of the isostatic gravity anomaly can be deduced by comparing Eqs. 16 and 17 to obtain

$$\Delta g_I = (A_T - A_B) - (A_C - A_m) + A_I \quad (18)$$

The term $A_T - A_B$ is identical in meaning and interpretation to the same term contained in the Bouguer gravity anomaly (Eq. 14). Interpretation of the term $A_C - A_m$ requires a careful analysis. If the isostatic model accurately represents the isostatic mechanism *and* a condition of complete regional isostatic equilibrium exists, then the computed gravitational attraction of the reinserted masses A_m should be essentially equal to the gravitational attraction of the actual masses required to neutralize the compensating mass deficiencies A_C. In this case, $A_C - A_m$ is zero, and Eq. 18 reduces to

$$\Delta g_I = (A_T - A_B) + A_I$$

Thus, in its purest form, the isostatic gravity anomaly represents only the gravitational attraction of density variations in the topography, $(A_T - A_B)$, and nonisostatic mass distribution irregularities beneath sea level, A_I. It is not subject to any kind of correlation with topography.

If the isostatic model is accurate but a regional isostatic imbalance exists, then $A_C - A_m \neq 0$ and Eq. 18 applies. In this case, the magnitude of $A_C - A_m$ is a measure of the isostatic imbalance that is present. Consequently, the isostatic gravity anomaly should accurately reflect prevailing isostatic conditions. However, it is not always easy to distinguish whether a particular isostatic gravity anomaly denotes either anomalous mass distribution, or isostatic imbalance, or both.

It is most likely that the adopted isostatic model does not exactly reproduce the real-world isostatic mechanism, but just gives a reasonable approximation. In this case, $A_C - A_m$ is affected by modeling parameters. In fact, use of different isostatic models yields slightly different values for the isostatic gravity anomalies. The model dependency of the computed isostatic gravity anomalies must be taken into account in geophysical interpretations. In addition, as was discussed in connection with Bouguer gravity anomalies, the term $A_T - A_B$ may also be somewhat model dependent.

As suggested by the foregoing, isostatic gravity anomalies have a number of subtle and elusive properties that make proper interpretation a rather complicated matter. For comprehensive discussions, see the excellent article by Simpson et al. (1986) and the references by G. P. Woollard cited there. (In oceanic areas, Eq. 18 reads

$$\Delta g_I = (-A_T + A_B) + (A_C - A_m) + A_I$$

The interpretation is similar to that of Eq. 18 for land areas.)

Because of the laborious computation process required to obtain isostatic gravity anomalies, values of Δg_I exist for only a small fraction of the sites where gravity has been measured. Consequently, isostatic gravity anomaly statistics are limited, and no worldwide average magnitude has been estimated. However, in an overall sense, isostatic gravity anomalies certainly have the smallest magnitudes of all common gravity anomaly types.

Isostatic gravity anomaly magnitudes exceeding 100 mgal are known to occur on some oceanic islands, probably indicating that such islands are largely uncompensated mass distribution anomalies. Elsewhere, Δg_I values of 100-mgal amplitudes appear to be quite rare. Most isostatic gravity anomalies are less than 50 mgal.

Isostatic gravity anomaly maps are less frequently produced than Bouguer gravity anomaly maps over land areas or free-air gravity anomaly maps over oceanic areas. However, viewing isostatic gravity anomaly maps is the best way to obtain an understanding of typical magnitudes and anomaly distributions. A good example of such a map is the isostatic gravity anomaly map of the United States, published by the NOAA Geophysical Data Center.

In the past, the isostatic gravity anomaly was used occasionally for geodetic applications. It is a valid gravitational parameter that gives a reasonably accurate expression of the Earth's mass distribution. Its isostatic model dependency does not invalidate a geodetic application provided that the adopted isostatic model is used consistently. However, its computation is laborious and does require that Earth structure assumptions be made, the external gravity field is changed slightly more than it is for free-air gravity anomalies, and the indirect effect, though small, is significant. Consequently, the isostatic gravity anomalies are little used today for geodetic purposes.

Isostatic gravity anomalies continue to be used for some types of geophysical applications, especially to study isostatic conditions and to analyze regional geological and geophysical structures. However, they are seldom used for geophysical prospecting work.

Some Additional, Mostly Theoretical, Considerations

In order to provide simplicity and clarity, several conceptually simple models have been proposed and discussed in this article. However, the real world is characterized by an almost infinite variety of conditions and structures, some of which may be more complex than the modeled situation. No simple model can accurately represent every mass distribution actually encountered in nature. For rigorous accuracy, gravitational attraction at a point should be obtained by integration over all mass elements close enough to that point to have an appreciable contribution. Nonetheless, the models presented are generally valid for the purposes of conveying a basic understanding of the principles involved.

Gravity anomalies, in reality, are purely gravitational quantities. Gravity is the vector sum or resultant of the Newtonian gravitational force generated by the Earth's masses and the centrifugal force of the Earth's rotation. Only the magnitudes of these forces are considered in gravity anomaly computation. The centrifugal force component of the magnitude of gravity g measured at the physical surface of the Earth is virtually identical to the centrifugal force component of the corresponding normal gravity magnitude γ computed at the ellipsoid. When the difference between g and γ is taken during gravity anomaly computation, therefore, centrifugal force cancels out completely, and only gravitational differences between g and γ remain.

When gravity anomalies are computed, γ must be obtained by using a formula compatible with the reference system of measured gravity g. Most modern gravity measurements are referred to the International Gravity Standardization Network of 1971 (IGSN71), a worldwide gravity measurement reference system having an absolute accuracy of 0.1 mgal or better. A few modern gravity measurements are referred to more recent national or local reference systems which are tied directly to recent, high-accuracy, absolute gravity measurements. These systems have an absolute accuracy on the order of 0.01–0.02 mgal. Examples of geodetic reference systems that can be used to compute normal gravity values compatible with the IGSN71 and other modern absolute gravity measurement systems are the Geodetic Reference System 1967 (GRS67) and the Geodetic Reference System 1980 (GRS80). The international gravity formula of 1930 is incompatible with the IGSN71 or other modern absolute systems.

An accuracy of 0.1 mgal or better can be attained easily for modern land gravity measurements. In order to retain an accuracy of 0.1 mgal or better in the computed free-air gravity anomaly, the accuracy of the elevation value used in the free-air reduction must be 0.3 m or better. This can be seen by applying Eq. 4 as follows:

$$(0.3086 \text{ mgal} \cdot \text{m}^{-1})(0.3 \text{ m}) = 0.1 \text{ mgal}$$

An elevation accuracy of 0.3 m can be obtained only by spirit leveling or, possibly, satellite positioning. Elevation data at points where gravity is measured is frequently determined by less accurate means, such as interpolation from topographic maps and barometric surveying methods. The accuracy of such elevation data is not likely to be much better than about 3 m. A 3-m error in elevation translates to a 1-mgal error in the computed free-air gravity anomaly. Consequently, unless the elevation data used in the free-air reduction is determined by very precise methods such as spirit leveling or satellite positioning, free-air gravity anomaly accuracy on land is not likely to be much better than about 1 mgal. In like manner, Bouguer and isostatic gravity anomaly accuracies are also limited by elevation reliability. (The best accuracy that can be attained for gravity measured aboard ships on the ocean surface is probably about 1–2 mgal. Since $F = 0$ for ocean surface gravity measurements, the accuracy of the computed free-air gravity anomaly is the same as the accuracy of the gravity measurement. However, Bouguer and isostatic gravity anomalies computed in oceanic areas are affected by errors in oceanic depth values.)

The error incurred by using Eq. 3 instead of Eq. 2 to lower gravity to the geoid surface can be as large as 0.1 mgal $\cdot$ m^{-1}, but usually is much less. Since values for $\partial g/\partial h$ are seldom known, it is difficult to assess the magnitude of the error for any particular gravity anomaly computation. Similarly, it is difficult to evaluate the error generated in the gravity anomaly by assuming that $\partial g/\partial h$ is constant over the elevation h. Because of these difficulties, many geodesists prefer to apply the modern interpretation.

Strictly speaking, the elevation value used in the modern interpretation (Eq. 8) to obtain a free-air gravity anomaly valid at the physical surface of the Earth should be normal height rather than the conventional orthometric elevation. Use of the orthometric elevation instead of normal height in Eq. 8 does not cause appreciable error in topographically flat areas, but can falsify the surface free-air gravity anomaly by about a milligal in rugged mountainous terrain. However, in the latter case, this error is most likely no larger than the error due to uncertainty in determination of the elevation value itself. In most practical cases, therefore, the type of elevation used need not be a major concern.

Certain geodetic applications that use the surface free-air gravity anomaly computed by Eq. 8 require a correction term that is approximated by the conventional terrain correction A_t. For this reason, terrain-corrected free-air gravity anomalies, $\Delta g_F + A_t$, have seen increasing use in geodesy. Terrain-corrected free-air gravity anomalies are sometimes referred to as *Faye anomalies*, but this terminology is by no means standardized.

Some other geodetic operations, such as least squares collocation prediction procedures, can be enhanced by removing topographically generated gravitational effects from the free-air gravity anomalies before the operation is carried out, and then restoring them afterward. The topographically generated gravitational effects are generally determined empirically, and their removal from free-air gravity anomalies yields a form similar, but not identical, to the Bouguer gravity anomalies. This form is ideal for prediction and certain other procedures, but, as is true for all Bouguer-type gravity anomalies, it does not represent the total mass effects of the Earth because some of the topographic effects have been eliminated. Consequently, topographic effects must be restored at the end to ensure correct geodetic results in subsequent operations.

Despite its planar geometry, the Bouguer plate, combined with the terrain correction, is a perfectly acceptable model to use for computation of the gravitational effects of nearby topographic masses. The Bouguer plate and terrain correction are effective to evaluate the gravitation of topographic masses that lie within a radius of 167 km from the point where the gravity anomaly is being computed. Although the Bouguer plate has an infinite radius, masses within the Bouguer plate that are more than 167 km from the computation point contribute nothing since the attraction of these masses is almost entirely horizontal (gravitational attraction is vertical by definition). Further, within a radial distance of 167 km from the computation point, the Bouguer plate departs only slightly from the Earth's nearly spherical surface, and the gravitational effects of this departure can be computed accurately, if desired, by using the small curvature correction.

Spherical formulas can be used to compute the gravitational effects of actual distant topographic masses that lie atop the Earth's curved surface outside a 167-km radius from the Bouguer gravity anomaly computation point. In general, however, the gravitational effects of distant topography change very, very slowly between adjacent computation points. Within any surface area of moderate size, the gravitational effects of distant topography constitute only a minute random component of the regional trend of the Bouguer gravity anomalies. Since the major purpose of the Bouguer reduction for Bouguer gravity anomaly computation is to eliminate short-wavelength gravity anomaly variations due to nearby short-wavelength variations in gross topography, the small gravitational effects of distant topography are usually ignored in Bouguer gravity anomaly computations. This omission does not invalidate the Bouguer gravity anomaly for its intended applications, namely, local or regional Earth structure analysis, and gravity anomaly interpolation and extrapolation.

In isostatic gravity anomaly computation, on the other hand, distant topography *and* its isostatic compensation lying beyond 167 km from the computation point all the way to the antipodes *must* be included in order for the computed isostatic gravity anomaly to truly and consistently represent the Earth's total internal mass distribution. This is absolutely essential if the isostatic gravity anomalies are to be used for geodetic applications or for regional isostatic analysis. Spherical formulas are used to compute the combined gravitational effects of distant topography and its isostatic compensation for isostatic gravity anomalies. These distant effects are added to the nearby effects where topographic and isostatic effects are separately computed. The Bouguer plate and terrain correction are perfectly valid for modeling nearby topographic effects in isostatic gravity anomaly computation.

More rigor can be given to gravity anomaly definition by a closer examination of the geoid and ellipsoid relationships. The geoid is an equipotential surface, that is, a surface of constant gravity potential. Let the constant gravity potential of the geoid be W_0. The potential W_0 is generated by the real masses inside the Earth.

The surface of the normal ellipsoid is defined to

be an equipotential surface. Let the gravity potential of the normal ellipsoid surface be U_0. The potential U_0 is generated by the mass contained within the normal ellipsoid.

By definition,

$$U_0 = W_0 \qquad (19)$$

Equation 19 is preserved, except for the indirect effect, when gravity anomalies are computed. In the gravity reduction process used to compute conventional gravity anomalies, all topographic mass is either removed entirely (e.g., Bouguer gravity anomalies) or reinserted inside the geoid (e.g., free-air and isostatic gravity anomalies). This process is known as *regularization* of the Earth.

The mass movement that occurs during free-air gravity anomaly computation is not obvious because, when Eq. 7 is used to compute Δg_F, the topographic masses are simply ignored as F is applied to lower g to the geoid. Nonetheless, a mass transfer does take place. What happens is that the topographic masses are condensed vertically into a surface layer of infinitesimally small thickness that lies on the underside of the geoid. This mass transfer is an exact consequence of a special gravity reduction known as *Helmert's second condensation reduction*. The free-air gravity anomaly is a close approximation of the exact gravity anomaly form given by the Helmert reduction.

Thus, the reduced value of measured gravity (e.g., $g + F$ for the conventional free-air form) applies on the surface of the geoid, which now contains all the mass of the real Earth. The ellipsoid contains all the mass of the normal ellipsoidal Earth model. Both the geoid and ellipsoid have the same gravity potential. Therefore, the magnitude of $g + F$ on the geoid is exactly comparable to the magnitude of γ on the ellipsoid. Thus, it is entirely reasonable and proper to take the difference between values specified at two different points in space to obtain the gravity anomaly, a quantity that expresses gravitational differences between total masses within the geoid and total masses within the ellipsoid. It should now be clear that the geoid-ellipsoid separation can be ignored when writing Eq. 5.

Similar arguments apply to justify the validity of the surface free-air gravity anomaly. For example, the potential of the equipotential surface of real gravity that passes through the point on the physical surface of the Earth where gravity is measured is exactly equal to the potential of the equipotential surface of normal gravity that passes through the point to which it is elevated by F if normal height is used in F. Thus, the magnitude of g measured on the physical surface is exactly comparable to the magnitude of $\gamma - F$.

The gravity anomaly is perfectly suited for most geodetic and geophysical applications. In some special applications, such as dynamic problems involving gravitational effects on orbits and trajectories, it is more desirable to use a gravity parameter computed as the difference between real and normal gravity both given at the same point in space. This gravity parameter is known as the *gravity disturbance*. The gravity disturbance at the geoid is related to the free-air gravity anomaly by the relation

$$\delta g = \Delta g_F - 0.3086 N$$

where

δg = gravity disturbance, mgal
Δg_F = free-air gravity anomaly, mgal
N = height of geoid above ellipsoid, m

The mass transfer that takes place during gravity anomaly computation has a so-called indirect effect. The position of the geoid in space is changed slightly. The moved geoid is called the *cogeoid*. A correction for the indirect effect is necessary for the most accurate geodetic applications. The correction, applied directly to the computed gravity anomalies, is

$$\delta = +0.3086 \delta N$$

where δN = movement of the geoid (m), and δ = correction for indirect effect (mgal). The indirect effect is very large for Bouguer gravity anomalies, small but appreciable for isostatic gravity anomalies, and negligible for free-air gravity anomalies.

There is one, and only one, normal ellipsoid of revolution that contains the same mass as the real Earth, whose surface has the same potential as the geoid and whose surface matches the geoid surface in position as closely as possible. Geodesists call this the *mean Earth ellipsoid*.

The best normal ellipsoids developed as of this writing are close approximations to the true mean Earth ellipsoid. But, in general, for any modern normal ellipsoid, there will be a small difference between the masses enclosed by the geoid and ellipsoid and/or a small difference in gravity potential between the geoid and ellipsoid surface. These differences give rise to a zero-degree harmonic in the gravity anomalies. The practical significance of the zero-degree harmonic is that the worldwide average of the free-air gravity anomalies will not be exactly zero. Fortunately, the zero-degree harmonic is probably small enough to be neglected for most practical applications.

Gravity anomaly values that are applicable at the geoid or physical surface of the Earth also can be computed by using an infinite series of solid spherical harmonics and geopotential harmonic coefficients determined from artificial Earth satellite dynamics and/or global surface gravity data. A commonly used formula is

$$\Delta g(r, \phi, \lambda)$$

$$= \frac{km}{r^2} \sum_{n=2}^{\infty} \sum_{m=0}^{n} \left\{ \left(\frac{a}{r}\right)^n (n-1)(\overline{C}_{nm} \cos m\lambda \right.$$

$$\left. + \overline{S}_{nm} \sin m\lambda) \overline{P}_{nm}(\sin \phi) \right\} \quad (20)$$

where

- k = gravitational constant
- m = total mass of the Earth
- a = length of semimajor axis of the normal Earth ellipsoid
- r = geocentric radius to the point on the geoid or physical surface where Δg is being computed
- $\overline{C}_{nm}, \overline{S}_{nm}$ = fully normalized geopotential harmonic coefficients
- $\overline{P}_{nm}(\sin \phi)$ = fully normalized associated Legendre function
- ϕ, λ = geocentric latitude and longitude of the point where g is being computed
- n, m = summation indexes (n = degree, m = order)

In Eq. 20 the even zonal coefficients (e.g., $\overline{C}_{20}$, $\overline{C}_{40}$) must be the differences between the geopotential harmonic coefficient values for the real Earth and the corresponding values for the normal ellipsoid.

The Δg computed by Eq. 20 is generally considered to be a free-air gravity anomaly. Since the summation is always truncated at some finite values for n and m, the gravity anomaly computed by Eq. 20 is an average value. The degree of generalization depends upon the maximum values of n and m in the truncated series. For example, if the series is truncated at $n = m = 180$, the Δg computed by Eq. 20 represents about a $1° \times 1°$ average value.

There is another way to deduce gravity anomalies in oceanic areas. The location of the geoid with respect to the mean Earth ellipsoid can be determined over oceans from satellite-borne radar altimeter measurements in conjunction with orbital and ellipsoidal geometry. These geoid heights can be converted to free-air gravity anomalies representing $1° \times 1°$ and larger surface areas by using methods such as inverse Stokes integrals and least squares collocation. Improved satellite altimetry missions may enable future recovery of oceanic gravity anomalies with greater resolution.

Most geodetic applications are simpler when conventional gravity anomalies, computed on the geoid, are used instead of surface gravity anomalies, computed at the physical surface of the Earth. The problem is that reduction of gravity to the geoid by conventional methods almost always requires assumptions and approximations that cause undesirable errors in the gravity anomalies applicable at the geoid.

An alternative procedure, suggested by the Swedish geodesist A. Bjerhammar, might be advantageous in some cases. First, compute the surface free-air gravity anomalies. These values will be wholly accurate. Then lower these surface free-air gravity anomalies to the geoid by some method of analytical continuation. This process is known as *downward continuation*. Two possible methods are an inversion of the well-known Poisson integral formula for upward continuation, and least squares collocation.

Unfortunately, downward continuation of force fields is an ambiguous process. There are infinitely many free-air gravity anomaly distributions along the geoid that, when analytically continued upward, will yield the same surface free-air gravity anomaly distribution at the physical surface of the Earth.

The free-air gravity anomalies computed at the geoid surface by downward continuation are correct in the sense that, upon upward continuation, they will exactly reproduce the given surface free-air gravity anomalies. They may even be very close to the true values of the free-air gravity anomalies along the geoid, but there is no rigorous method to verify just how close to the truth they really are.

Downward continuation is not a mass-transferring operation. Application of Poisson's integral requires that no mass exist over the area of its application. Therefore, free-air gravity anomalies computed at the geoid surface by downward continuation may not be strictly comparable to those computed by conventional gravity reduction techniques. The most straightforward current application of downward continuation is to reduce airborne gravity measurements through free space to the geoid or physical surface of the Earth.

A completely accurate solution for gravity anomaly values applicable at the geoid may not be possible without a detailed knowledge of the Earth's interior mass distribution, and many geodesists will continue to prefer the surface gravity anomalies even though applications procedures tend to be complex. This problem is of less importance for geophysical interpretation applications and can probably be overlooked safely in most cases.

Conventional Gravity Anomaly Computation Formulas

The gravity anomaly computation formulas (Table 1) included in this section enable computation of conventional free-air and Bouguer gravity anomaly formulas in different types of environments. To obtain isostatic gravity anomalies, add the term A_m, computed in accordance with some isostatic model, to the Bouguer gravity anomaly.

The free-air reduction F used in these formulas is always based on Eqs. 3 and 4. All Bouguer gravity anomaly formulas give simple Bouguer gravity anomalies. To obtain complete Bouguer gravity anomalies, add the terrain correction to all formulas.

TABLE 1. Gravity Anomaly Computation Formulas

Location of Gravity Measurement	Free Air	Simple Bouguer
Land		
At physical surface	$\Delta g_F = g + 0.3086h - \gamma$	$\Delta g_B = \Delta g_F - 0.1119h$
Underground in cave or tunnel	$\Delta g_F = g + 0.2238u + 0.3086(h - u) - \gamma$	$\Delta g_B = \Delta g_F - 0.1119h$
Ocean		
At surface	$\Delta g_F = g - \gamma$	$\Delta g_B = \Delta g_F + 0.06886d$
Submerged, e.g., in submarine	$\Delta g_F = g - 0.2225s - \gamma$	$\Delta g_B = \Delta g_F + 0.06886d$
Bottom, i.e., on ocean floor	$\Delta g_F = g - 0.2225d - \gamma$	$\Delta g_B = \Delta g_F + 0.06886d$
Lake		
At surface; both lake surface and lake bottom above sea level	$\Delta g_F = g + 0.3086K - \gamma$	$\Delta g_B = \Delta g_F - 0.04191b - 0.1119(K - b)$
At surface; lake surface above sea level; lake bottom below sea level	$\Delta g_F = g + 0.3086K - \gamma$	$\Delta g_B = \Delta g_F - 0.04191K - 0.06999(K - b)$
At surface; both lake surface and lake bottom below sea level	$\Delta g_F = g + 0.3086K - \gamma$	$\Delta g_B = \Delta g_F - 0.1119K + 0.06999b$
Bottom; i.e., on lake floor; both lake surface and lake bottom above sea level	$\Delta g_F = g + 0.08382b + 0.3086(K - b) - \gamma$	$\Delta g_B = \Delta g_F - 0.04191b - 0.1119(K - b)$
Bottom; lake surface above sea level; lake bottom below sea level	$\Delta g_F = g + 0.08382b + 0.3086(K - b) - \gamma$	$\Delta g_B = \Delta g_F - 0.04191K - 0.06999(K - b)$
Bottom; both lake surface and lake bottom below sea level	$\Delta g_F = g + 0.3086K - 0.2248b - \gamma$	$\Delta g_B = \Delta g_F - 0.1119K + 0.06999b$
Glacial Icecap		
At surface; bottom of ice above sea level	$\Delta g_F = g + 0.3086c - \gamma$	$\Delta g_B = \Delta g_F - 0.03843t - 0.1119(c - t)$
At surface; bottom of ice below sea level	$\Delta g_F = g + 0.3086c - \gamma$	$\Delta g_B = \Delta g_F - 0.03843c - 0.07347(c - t)$

Source: These formulas were compiled from material furnished by the DoD Gravity Library.

TABLE 2. Explanation of Symbols Used in Table 1

Δg_F	free-air gravity anomaly
Δg_B	Bouguer gravity anomaly
g	measured (observed) gravity; always positive
h	elevation of physical surface of the Earth; positive if above sea level and negative if below sea level
u	depth of underground measurement below physical surface of the Earth; always positive
d	depth of ocean; always positive
s	depth of underwater measurement below ocean surface; always positive
b	depth of lake bottom below lake surface; always positive
K	elevation of lake surface; positive if above sea level and negative if below sea level
c	elevation of ice surface; positive above sea level
t	thickness of glacial ice; always positive
γ	normal gravity

TABLE 3. Density Factors Used in Formulas in Table 1

Substance	Density, σ (g · cm^{-3})	Bouguer Plate Factor ($2\pi k\sigma$)
Freshwater	1.0	0.04191
Salt Water	1.027	0.04304
Ice	0.917	0.03843
Land	2.67	0.1119
Land Minus Freshwater	1.67	0.06999
Land Minus Salt Water	1.643	0.06886
Land Minus Ice	1.753	0.07347

k = gravitational constant = 6.67×10^{-8} cm^3 · g^{-1} · s^{-2}
$\pi = 3.14159\ldots$

One case not included in the formulas pertains to gravity measurements made at some altitude above the Earth's surface in an aircraft. Conventional gravity reductions tend to give unsatisfactory results in this situation. The recommended procedure is to compute a "surface" free-air gravity anomaly at the altitude of the measurement by using Eq. 8. Then this value should be lowered to the geoid or physical surface by downward continuation procedures.

The principle that is applied to compute gravity anomalies in certain unusual situations has not yet been introduced in this article. When gravity is measured underwater or below the physical surface of the Earth, the gravitational attraction of the overlying masses always must be treated when gravity anomalies are computed. The gravitational attraction of the overlying masses is determined by using the Bouguer plate with appropriate density factors. For example, when gravity is measured on the ocean floor at a depth d below the ocean surface, the gravitational attraction of the mass within the seawater lying above the ocean bottom measurement is given by the Bouguer plate formula (Eq. 10) in the form

$$A_b = 2\pi k\sigma d \tag{21}$$

where σ is now the density of salt water, 1.027 g · cm^{-3}. Putting this value into Eq. 21 along with appropriate values for π and the gravitational constant k gives

$$A_b = 0.04304d \tag{22}$$

The reduction to obtain the free-air gravity anomaly now proceeds as follows. Remove the seawater mass. Compute its gravitational attraction by using Eq. 22. Removal of mass lying above the measurement point causes gravity to increase at that point. Therefore Eq. 22 is applied with a positive sign. Next, apply the free-air reduction F to elevate the gravity measurement from the bottom to sea level through free space. Since gravity is being moved outward, F is applied with a negative sign. Finally, reinsert the seawater mass that is now below the measurement elevated by F, thereby increasing gravity again. Compute the gravitational attraction of the reinserted seawater mass by using Eq. 22 and apply with a positive sign. Thus,

$$F = +0.04304d - 0.3086d + 0.04304d$$
$$= -0.2225d$$

Free-air reductions for other underwater or underground measurements are obtained by similar principles and appropriate density factors.

To obtain the Bouguer gravity anomaly from the free-air gravity anomaly, remove the gravitational attraction of land, water, and/or ice masses above sea level and/or bring the density of water or ice masses below sea level up to the density of land. The gravitational attraction values are computed by using the Bouguer plate and appropriate density factors.

Suggestions for Further Reading

Classical treatment of gravity anomaly computation procedures and more extended discussions of many of the formulas given in this article may be found in the following:

Heiskanen, W. A., and H. Moritz, *Physical Geodesy*, W. H. Freeman and Company, San Francisco, 1967, Chapter 3.

Tsuboi, C., *Gravity*, George Allen and Unwin, London, 1983, Chapter 3.

Dehlinger, P., *Marine Gravity*, Elsevier Scientific Publishing Co., Amsterdam, 1978, Chapter 6.

Heiskanen, W. A., and F. A. Vening Meinesz, *The Earth and Its Gravity Field*, McGraw-Hill Book Company, Inc., New York, 1958, Chapter 6.

The literature dealing with use of Bouguer gravity anomalies for geophysical exploration is extensive. The following are recommended as starters:

Dobrin, M. B., *Introduction to Geophysical Prospecting*, 3rd edition, McGraw-Hill Book Co., 1976.

Nettleton, L. L., *Gravity and Magnetics in Oil Prospecting*, McGraw-Hill Book Co., 1976.

Paterson, N. R., and C. V. Reeves, Applications of gravity and magnetic surveys: The state of the art in 1985, *Geophysics* **50**, No. 12, 1985, pp. 2558-2594.

For further information and an excellent reference list pertaining to interpretation of free-air gravity anomalies over oceanic areas, see:

Watts, A. B., D. P. McKenzie, B. E. Parsons, and M. Roufosse, The relationship between gravity and bathymetry in the Pacific Ocean, *Geophysical Journal of the Royal Astronomical Society* **83**, 1985, pp. 263-298.

A comprehensive treatment of isostatic gravity anomaly interpretation is given by:

Simpson, R. W., R. C. Jachens, and R. J. Blakely, A new isostatic residual gravity map of the conterminous United States with a discussion of the significance of isostatic residual anomalies, *Journal of Geophysical Research* **91**, 1986, pp. 8348-8372.

G. P. Woollard published many classic articles dealing with interpretation of gravity anomaly data. Several of his more important works are listed in the references of Simpson et al. given above.

A very readable textbook that treats geodetic applications of gravity anomalies is:

Vanicek, P., and E. Krakiwsky, *Geodesy: The Concepts*, 2nd edition, North Holland, New York, 1986.

All of the preceding citations contain extensive reference lists.

LUMAN E. WILCOX

Cross-references: *Absolute Gravity Measurements; Density Distribution in the Earth; Earth's External Gravity Field; Geodesy: Physical; Gravity Anomalies: Statistical Analysis; Gravity and Isostasy; Spherical Harmonic Analysis.*

GRAVITY ANOMALIES: STATISTICAL ANALYSIS

Gravity anomalies can be defined in different ways (see *Gravity Anomalies: Interpretation*). For this discussion we adopt the following definition:

$$\Delta g = g_0 - \gamma \quad (1)$$

where g_0 is an observed value of gravity reduced to the geoid (or reference equipotential surface) and γ is the theoretical or normal value of gravity at a point on the ellipsoid. The theoretical value can be precisely determined by knowing the parameters of the reference ellipsoid (see *Figure of the Earth*) and the latitude of the point. The values of the anomalies depend on what parameters are adopted for the ellipsoid. The definition in Eq. 1 is the classical definition of anomalies (Heiskanen and Moritz, 1967); alternative definitions are possible.

The type of anomaly we have depends on the procedure used to reduce the observed value of gravity to the geoid. The most common reduction procedures are called free-air, Bouguer, and isostatic. Each procedure makes some assumption on the treatment of the mass between the geoid and the point at which gravity was measured. These reduction procedures lead to quantities called free-air, Bouguer, and isostatic anomalies.

The quantity defined in Eq. 1 is called a *point gravity anomaly* because it is based on a point measurement of gravity. Mean gravity anomalies can be defined for some specific area A in the following way:

$$\overline{\Delta g} = \frac{1}{A} \iint_A \Delta g \, dA \quad (2)$$

In practice, the actual evaluation of Eq. 2 is carried out by recognizing that gravity values are available at discrete points and not as a continuous function.

Since many gravity anomalies have been determined over much of the surface of the Earth, it is natural to discuss such anomalies in a statistical sense. We first look at the descriptive behavior of gravity anomalies and consider numerical values that convey information on the anomalies. Second, we might consider anomalies in some probabilistic sense and try to describe the behavior of gravity anomalies in this context. Much controversy exists about regarding gravity anomalies in a stochastic way because there is only one anomaly at a given point and only one "Earth" for which gravity anomalies are being determined. Considering the existence of strong anomaly structure (or patterns) in various geographic areas, we cannot think of anomalies as a true stochastic (random) process. However, we can still use statistical tools to describe properties of the anomalous gravity field (Moritz, 1980, Section 38).

The magnitude of gravity anomalies depends on the geographic location of the point and on the height of the point in land areas or on the depth of the ocean in water areas. Free-air anomalies are generally positive at high elevations. The largest positive anomaly is near 700 mgal (1 mgal = 10^{-5} m · s^{-2}) on the island of Hawaii at an elevation of 4113 m. The most negative free-air anomaly is about -360 mgal over the Puerto Rican Trench. Free-air anomalies show a correlation with elevation. Such correlations are clear in an area where there is substantial elevation difference between the observation stations. The consideration of this elevation correlation is quite important when interpolating or predicting free-air anomalies at points where the elevation is substantially different from that of the given points.

The global mean free-air anomaly is within 0.5 mgal of being zero. This is because of the appropriate choice of gravity formula (i.e., γ) parameters. The root mean square free-air anomaly is about ± 32 mgal. This value varies greatly from one area to another. For example, large areas of the Pacific Ocean have anomalies that are about ± 10 mgal, and some trench areas have anomalies of about -200 mgal.

Bouguer gravity anomalies are negative in land areas. At elevations of 2000 m it is not unusual to see Bouguer anomalies of -350 mgal. Bouguer anomalies show a regional correlation with elevation as opposed to the local correlation shown by free-air anomalies. The higher the regional elevations, the more negative is the Bouguer anomaly. Because Bouguer anomalies are smoother, in a local sense, than free-air anomalies, they are easier to interpolate or predict.

Isostatic gravity anomalies depend on the type of isostatic hypothesis used and parameters adopted for the isostatic reduction. Isostatic anomalies show much smoother variations than free-air anomalies although their magnitude is similar.

Our discussion to this point has dealt with point anomalies. Mean anomalies have been defined in Eq. 2. Such values are associated with areas of different sizes (A). Typical global collections of mean anomalies exist for $1° \times 1°$ equiangular blocks where there is a total of 64,800 blocks. The linear dimension of the $1°$ block is about 111 km $\times$ 111 km at the equator. The maximum and minimum $1° \times 1°$ anomalies are 340 mgal and -270 mgal, respectively. The root mean square $1° \times 1°$ anomaly is ± 28 mgal. A set of $1°$ equal area anomalies may be derived from the $1°$ equiangular set. A histogram showing the distribution of these equal area anomalies is shown in Fig. 1 with an interval of 6 mgal. A theoretical normal distribution curve is plotted. A goodness-of-fit test would show that the $1°$ anomalies are not normally distributed.

As noted earlier, we often apply statistical procedures to gravity anomaly data. Quantities such as a covariance function or a correlation function can be computed. Consider functions $g(x)$ and $h(x)$ defined on a line of length T. Then the covariance function is defined as (Bendot and Piersol, 1980)

$$C_{gh}(s) = \lim_{T \to \infty} \int_0^T (g(x) - \mu_g)(h(x + s) - \mu_h) \, dx$$

(3)

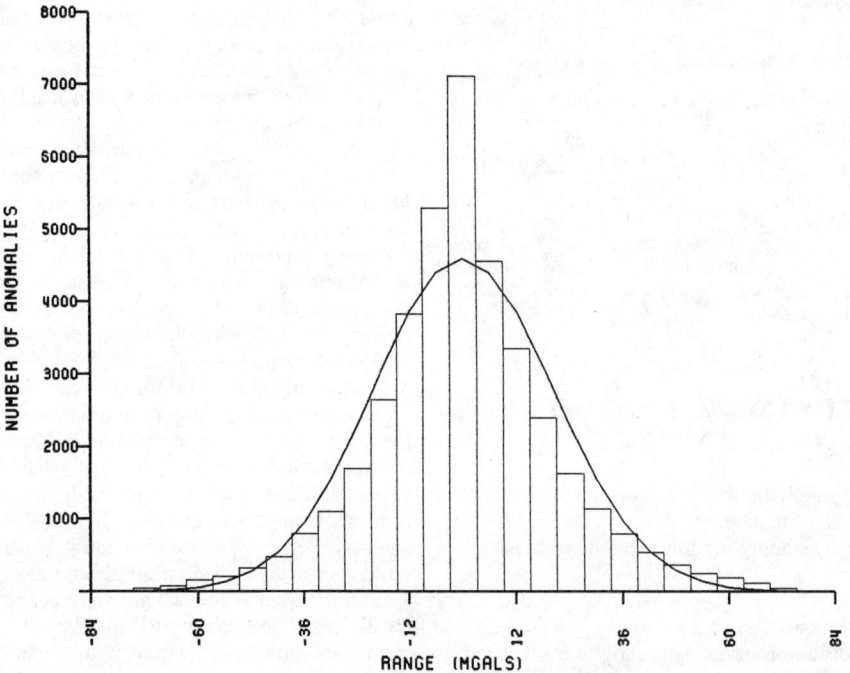

FIGURE 1. Histogram of 39,046 $1°$ equal-area free-air anomalies and the theoretical normal distribution based on a variance of $(20.3 \text{ mgal})^2$. Some extreme values in the complete data set have been deleted from this plot. The class interval of the histogram is 6 mgal.

where μ_g and μ_h are the mean value of the functions over the line T. The *autocovariance* function for one of the functions would be computed as follows:

$$C_{gg}(s) = \lim_{T \to \infty} \frac{1}{T} \int_0^T (g(x) - \mu_g)$$
$$\cdot (g(x + s) - \mu_g) \, dx$$

If μ_g and μ_h are zero, $C_{gh}(s)$ and $C_{gg}(s)$ yield the cross-correlation function and the autocorrelation function (Schwarz, 1984). We might consider $g(x)$ to be gravity anomalies and $h(x)$ to be some other gravimetric quantity, such as a deflection of the vertical, or a nongravimetric quantity, such as topographic elevation.

In practice we deal with functions defined over a finite-length line. In this case the cross-covariance function is defined as

$$\hat{r}_{gh}(s) = \frac{1}{T - s} \int_0^{T-s} g(x)h(x + s) \, ds \quad (4)$$

If the functions are defined at discrete intervals Δx in the range $0 \le k \le x \le (N - 1) \Delta x$, the *discrete cross-correlation function* is

$$\hat{r}_{gh}(k \, \Delta x) = \frac{1}{N - k} \sum_{i=0}^{N-1-k} g(i \, \Delta x) h((i + k) \, \Delta x) \quad (5)$$

The maximum k is usually restricted to $N/2$.

The functions calculated from Eqs. 4 or 5 represent one-dimensional averages over a specific line where the variable is the spatial separation. If we were given data, anomaly or otherwise, in a two-dimensional case, it is possible to consider a two-dimensional correlation function. Let there be N data points of equal spacing in the x- and y-directions. Then the correlation function that depends on r steps in the x-direction and s steps in the y-direction is

$$\hat{r}_{gh}(r, s) = \frac{1}{(N - r)(N - s)}$$
$$\cdot \sum_{i=1}^{N-r} \sum_{j=1}^{N-s} g(i, j) h(i + r, j + s),$$
$$r, s = 0, 1, \ldots, m, \quad m = \frac{N}{2}$$

$$(6)$$

The estimates of the covariances obtained from Eqs. 5 or 6 may depend on several factors, including data length, geographic area, and direction of profile (if not two-dimensional data). To obtain correlation functions that may be more representative, we often carry out an averaging process. This process may be carried out by averaging various covariance functions determined with local data (i.e., data given in a restricted region) or by attempting to estimate a global covariance function from globally distributed data. This discussion has been carried out in a planar context. Analogous discussion for data on a sphere is described next.

Consider our two functions given on the surface of a sphere where coordinates are polar distance θ and longitude λ. Then a global correlation function could be defined as a function of the distance between points P and Q as

$$C_{gh}(P, Q) = \frac{1}{4\pi} \int_{\lambda=0}^{2\pi} \int_{\theta=0}^{\pi} \frac{1}{2\pi} \int_0^{2\pi} g(\theta, \lambda) h(\theta', \lambda')$$
$$\cdot \sin \theta \, d\theta \, d\lambda \, d\alpha \quad (7)$$

where α is the azimuth. Moritz (1980, p. 283) refers to the process represented by Eq. 7 as a rotation group average. An alternative general notation for the covariance between two functions is

$$C_{gh}(P, Q) = M(g(P)h(Q)) \quad (8)$$

where M is an averaging operator defined by Eq. 7 in a global case or by a suitable average for products in a local area. In most gravimetric computations the average value of $g(P)$ and $h(P)$ is assumed or forced to be zero. The variable in Eqs. 7 and 8 is the separation between points P and Q.

A covariance function defined by Eqs. 7 or 8 is said to be isotropic (independent of azimuth) and homogeneous (independent of position). Such a covariance function would be obtained from stationary data. In reality, gravity anomalies are not stationary since they depend on geographic location. However, in making the isotropic, homogeneous assumption, useful covariance functions can be found. Local covariance functions computed from Eq. 8 would be isotropic, and they would be representative of a specific geographic location. A nonisotropic or anisotropic covariance function is one that depends on direction. Such functions can be computed, but they are not often used in practice.

Figure 2 shows the form usually taken by local one-dimensional anomaly covariance functions. A local covariance function can be characterized by (Moritz, 1980) (1) variance C_0, (2) correlation length ξ, where $C(\xi) = C_0/2$, and (3) curvature parameter χ. The curvature parameter is related to the curvature κ of the covariance function at the origin:

$$\chi = \frac{\kappa \xi^2}{C_0} \quad \text{or} \quad \chi = \frac{\xi^2 G_0}{C_0}$$

where G_0 is the horizontal anomaly gradient variance.

K.-P. Schwarz and G. Lachapelle studied gravity data in Canada to arrive at values of C_0, ξ, and G_0

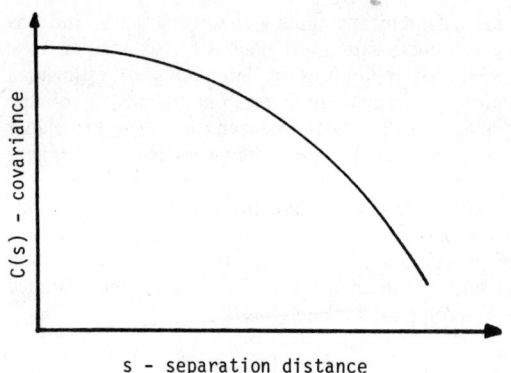

FIGURE 2. Generic form of a local anomaly covariance function. $C(s)$ is the covariance given as a function of the separation distance s.

for various regions. Average values, after a spherical harmonic reference field to degree 30 was removed, were $C_0 = 460$ mgal^2, $G_0 = 200$ E^2 (where 1 E = 0.1 mgal/km = 10^{-9} s^{-2}), and $\xi = 52$ km (28 arc minutes). Both C_0 and G_0 are quite sensitive to the geographic region. Goad and others have studied covariance functions for terrain-corrected Bouguer anomalies in the United States.

Global anomaly covariance functions take the general form shown in Fig. 3. The computation of such a function is usually done through mean gravity anomalies given in elements on the sphere corresponding to 1° or 5° blocks. An example of a global $C(\psi)$ is given in Tscherning and Rapp (1974). On the basis of the data then available C_0 was 1795 mgal^2. More recent estimates give $C_0 = 1000$ mgal^2.

Two-dimensional covariance functions have been computed by Forsberg (1984) and Schwarz (1984), among others. An example of such a function, taken from Kearsley, is shown as Fig. 4. If the anomaly field were isotropic, the covariance plot would be concentric circles. Because the covariance contours in Fig. 4 are elongated in a certain direction, the covariances (and the anomaly field) are considered to be anisotropic.

The discussion to this point has considered covariance or correlation functions in the spatial domain. We next look at representation in the frequency domain. We start with a Legendre series representation of the global anomaly covariance function (Heiskanen and Moritz, p. 256):

$$C(\psi) = \sum_{l=2}^{\infty} c_l P_l(\cos \psi) \tag{9}$$

where ψ is the spherical distance separating the points of interest, c_l are called anomaly degree variances, and P_l are Legendre polynomials of degree l. Given a set of c_l values or a model for c_l (Tscherning and Rapp, 1974), it is possible to estimate c_l values if a global covariance function can be defined.

For applications in a smaller area, flat-area approximations can be made so that Fourier analysis techniques can be applied. Spectral density functions can be determined by computing the Fourier transform of the correlation functions. From Eq. 3 (taking μ_g and μ_h zero) we have for the cross-spectral density function (Schwarz, 1984) at frequency u:

$$S_{gh}(u) = \int_{-\infty}^{\infty} C_{gh}(s) e^{-i2\pi us} \, ds \tag{10}$$

where s is the distance. The autospectral density function or power spectrum is

$$S_{gg}(u) = \int_{-\infty}^{\infty} C_{gg}(s) e^{-i2\pi us} \, ds \tag{11}$$

Values of $S_{gg}(u)$ are somewhat analogous to the c_l

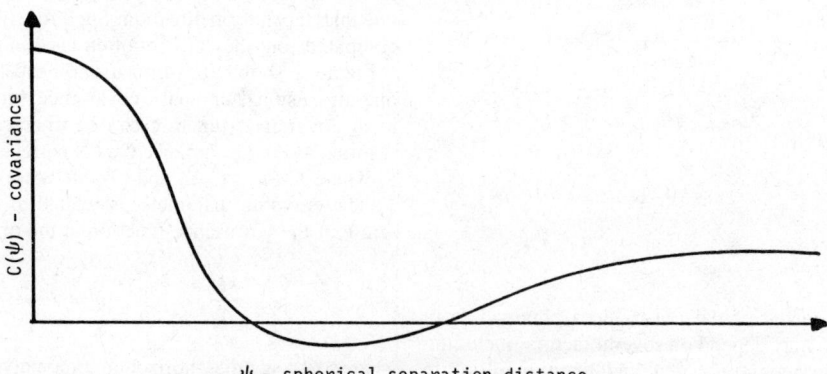

FIGURE 3. Generic form of a global anomaly covariance function. $C(\psi)$ is the covariance given as a function of the spherical separation distance ψ.

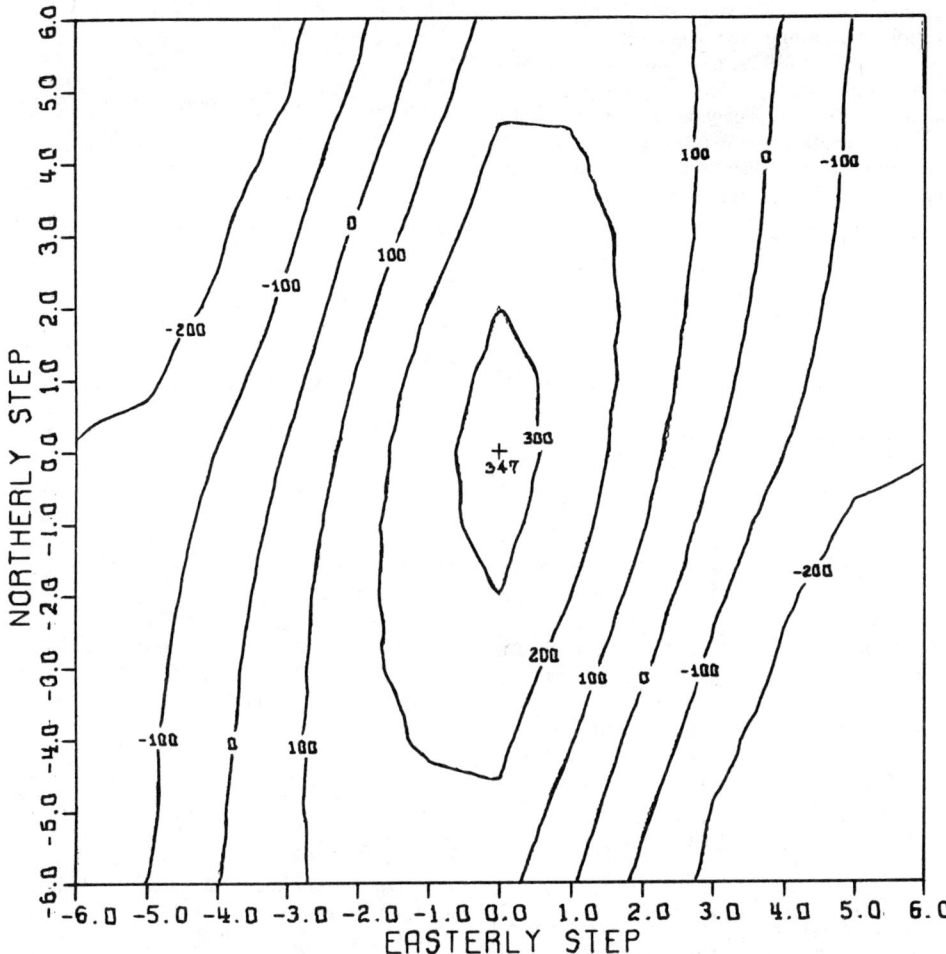

FIGURE 4. A two-dimensional covariance function with contours in mgal². The covariance function is given as a function of northerly and easterly steps where each step is approximately 9 km.

values in Eq. 9. The evaluation of Eqs. 10 or 11 can be done through one- or two-dimensional (fast) Fourier transforms. Forsberg (1984) showed a link between the two estimates valid for all but low-harmonic ($l < 10$) degrees at which a flat Earth approximation is valid. Specifically

$$c_l = \frac{1}{2\pi} \frac{l + \frac{1}{2}}{R^2} S_{gg}(\omega_l) \qquad (12)$$

where ω_l is a circular frequency (wave number) that has a relationship with the degree l, depending on the number of data points in one direction of the grid and the data spacing. Forsberg (1984), Rapp (1986), and Vassiliou and Schwarz (1987) have used this procedure of performing a Fourier analysis on locally gridded data to obtain high-degree anomaly degree variance (or the spectrum of the gravitational potential) information.

Numerous studies have been carried out to develop models for c_l or $S_{gg}(\omega)$. A very simple model for c_l, implied by the potential decay predicted by Kaula, is

$$c_l = \frac{192}{l + 1.5} \text{ mgal}^2 \qquad (13)$$

Tscherning and Rapp (1974) suggested

$$c_l = \frac{A(l - 1)}{(l - 2)(l + 24)} s^{l+2}$$

where $A = 425.28$ mgal² and $s = 0.999617$. Values of A can be changed to tailor the model to a local region. Moritz (1980) discusses the limitations of models such as Eqs. 12 and 13 and suggests more complex models. J. Nigi and V. Dimri discuss various power spectral density models and consider implications concerning covariance functions.

Heller and Jordan (1979) have suggested a multicomponent model for c_l. They postulate that the anomalous gravity field is generated by a white noise on a shell of depth D with respect to the Earth's surface. If σ_T^2 is the disturbing potential degree variance, they show that on the surface of the sphere of radius approximating the Earth,

$$c_{l,D} = \frac{(2l + 1)(l - 1)^2 D^2 (2R - D)^2 \sigma_T^2}{R^4 (2R^2 - 2DR + D^2)} \cdot \left(\frac{1 - D}{R}\right)^{2l} \quad (14)$$

Heller and Jordan postulate five such shells and calculate the D and σ_T values associated with each shell by fitting to known covariance functions. The total c_l value is then the sum of the degree variances implied by each shell. The model represented by Eq. 14 is called an *attenuated white noise model*. Parameters for the model can be determined on a global scale or on a local basis.

The statistical analysis of gravity anomaly data is easily extended to other quantities that depend on the Earth's gravity field. The results of these analyses give insight into the behavior of the gravity field and provide specific information that may be used to estimate gravimetric quantities on the basis of different kinds of observations.

RICHARD H. RAPP

References

Bendat, J. S., and A. G. Piersol, 1980, *Engineering Applications of Correlation and Spectral Analysis*. New York: Wiley.
Forsberg, R., 1984, Local covariance functions and density distributions, *Ohio State Univ. Dept. Geod. Sci. Surveying Rept. 356*.
Heller, W., and S. Jordan, 1979, Attenuated white noise statistical gravity model, *Jour. Geophys. Research* **84**(B9), 4680-4688.
Heiskanen, W., and H. Moritz, 1967, *Physical Geodesy*. San Francisco: W. H. Freeman.
Moritz, H., 1980, *Advanced Physical Geodesy*. Karlsruhe, FRG: Herbert Wichmann Verlag.
Nigi, J., and V. Dimri, 1977, Three-dimensional statistical gravity disturbance model, *Royal Astron. Soc. Geophys. Jour.* **48**, 187-196.
Rapp, R. H., 1985, Gravity anomalies and sea surface heights derived from a combined GEOS-3/Seasat altimeter data set, *Jour. Geophys. Research* **91**(B5), 4867-4876.
Schwarz, K. P., 1984, Data types and their spectral properties, in *Local Gravity Field Approximation Proc.* Alberta, Canada: University of Calgary Division of Surveying Engineering.
Tscherning, C. C., and R. H. Rapp, 1974, Closed covariance expressions for gravity anomalies, geoid undulations, and deflections of the vertical implied by anomaly degree variance models, *Ohio State Univ. Dept. Geod. Sci. Rept. 208*.
Vassiliou, A. A., and K. P. Schwarz, 1987, Study of the high-frequency spectrum of the anomalous gravity potential, *Jour. Geophys. Research* **92**(B1), 609-617.

Cross-references: *Absolute Gravity Measurements; Figure of the Earth; Geodesy: Physical; Gravity Anomalies: Interpretation; Gravity Fields: Implications for Planetary Interiors; Gravity and Isostasy; Spherical Harmonic Analysis.*

GRAVITY FIELDS: IMPLICATIONS FOR PLANETARY INTERIORS

Variations in the gravity field of a planet require variations in mass distribution. These variations may be either in surface loads, interface location, or material density. Variations in mass distribution require differences in shear stress in the planet's interior, because there are variations in the gravitational attraction of the planet proportionate to the variations in mass distribution. The magnitude of the variations are limited by the finite strength of the rocks constituting a planet. The yield stresses of rocks not near melting are on the order of one kilobar (10^8 pascal). The magnitude of variations in mass distribution among planets would be expected to vary inversely with the magnitude of gravity g among planets. Thus in the Earth the maximum free air gravity anomaly is perhaps 120 mgals (0.12 cm/sec^2), hardly 1/8000 of the total attraction, while in the Moon, which has one-sixth the gravity, the maximum is around 300 mgal: about 1/500 of the total. If the maximum shallow stresses in the two bodies were the same, the maximum gravity anomaly in the Moon would be six times the Earth's, rather than 2.5 times. Hence the Moon is closer to equilibrium than the Earth by the physically meaningful measure of shearing stress magnitudes.

The interpretation of gravity anomalies can be described as having at least four levels. (1) The anomalies can be explained as arising from static irregularities in mass distribution. (2) These irregularities can arise from a variety of causes: compositional variations (as in the classic Airy and Hayford isostatic compensations); thermal variations (as prevail in the oceanic lithosphere); or variations in the surface topography. (3) The various causes of density irregularities require different ongoing processes in the Earth: the compositional variations require petrological differentiation; the thermal variations require some sort of thermal convection (a broad term that means the flow of matter with heat); and the topographic variations may arise from either the chemical or the physical processes. (4) These ongoing processes in turn indicate different evolutionary patterns in the planet's interior.

Gravity data alone are quite ambiguous, so other data must be employed for a useful interpretation at any of the four levels. Most commonly employed for interpretation of short wavelength variations, necessarily of shallow origin, are geologic evidence of rock types and seismological evidence of density variations at depth. However, the longer the wavelengths, the greater the stresses for a given amplitude, and the deeper may be the sources. Hence the less the interpretation can be constrained by geology and reflection seismology, the more physical considerations arising from tectonophysical models or thermal convection theory must be used. *Gravity and Isostasy* discusses the considerations associated with interpreting gravity variations arising from the crust and lithosphere. This article deals with the broader and deeper variations in the gravity field, which necessarily involve dynamic considerations. It thus is auxiliary to *Mantle Convection and Plumes* and *Mantle Dynamics*.

This article first discusses the Earth, next the Moon, and then the other planets. The discussion concerning the Earth follows Kaula (1980) and Richards and Hager (1986). The lunar discussion follows Kaula et al. (1986). The primary reference for variations among the terrestrial planets is Phillips and Lambeck (1980).

The Earth

Since plate tectonics became accepted around 1967, it has been recognized that the suboceanic lithosphere is the boundary layer of a mantle convective system, upon which ride the continents. As a consequence of this reality, many phenomena can be treated as functions of the aging of the lithosphere and its interaction with the subjacent asthenosphere. The problem areas have become essentially twofold: (1) the physics of marginal zones (which may be 1000 km wide) between tectonic plates: deformation, earthquake occurrence, heat flow, petrology, etc.; and (2) the nature of mantle convection: its energy sources and constitutive relationships, and the resulting planforms, evolving in time. These two areas have no sharp boundary between them, particularly since the subducted material associated with convergent plate margins certainly has a strong influence on the pattern of mantle convection.

The main contemporary question related to mantle convection is the extent to which it is heterogeneous. Isotopic constraints require separation of sources for long periods, in some cases 2,000 my or more, on a variety of scales, ranging from continent vs. ocean to the source regions of volcanoes hardly 10 km apart. The long length scale heterogeneities have some relevant seismological evidence. In the radial direction, there is a strong gradient in density, totalling about 0.3 gm/cm^3, near depth 670 km. In addition, reflections at a depth around 220 km have been reported by some seismologists. In the lateral direction, there are differences in S-wave velocity between continents and oceans to perhaps 400 km depth—most pronounced, faster velocities under old cratons—as well as indications of subducted slabs to 700 km depth.

For length scales of more than 100 km and time scales of more than 100 my—hardly 2% of the Earth's lifetime—the parts of the mantle participating in what drives plate tectonics are essentially fluid. Hence, larger volume elements are smeared out and lose their identity. The lower limit on the size of such elements is quite uncertain. There likely exist under continents keels of 200 km–300 km depth that are depleted in lithophiles and high in Mg: Fe ratio that have remained undisturbed for 1,000 my or more, but for the bulk of the mantle the maintenance of long-wavelength heterogeneity must entail an inhibition to large-scale mixing from a compositional stratification, most obviously at the 670 km depth, but possibly at the 220 km. However, the only rocks that clearly sample a large reservoir are the midocean ridge basalts; the rocks from which they must be separated for 2,000 my could come from "plums" of 10 km extent just as well as a much larger body that cannot maintain its identity in the mantle flow system on the longer time scales.

It is thus desirable to seek evidence of lower mantle heterogeneities and of the extent to which they are connected to the upper mantle. The most significant advance in recent years has been the identification of lateral heterogeneities from seismic "tomography" (Clayton and Comer, 1983; Dziewonski, 1984). With the appropriate viscous model of the mantle, these heterogeneities can explain a large part of the low degree harmonics of the gravity field, subject to a free parameter that is the ratio of density to seismic velocity (Hager et al., 1985).

This identification of deep heterogeneities finally gives some corroboration for much of the variations in gravitational potential; heretofore, a major part of the long wavelength variations has been necessarily attributed to something unknown in the lower mantle.

The sources of the Earth's gravity field can be categorized as sixfold:

Lithospheric strength	5%
Crustal isostasy	5%
Thermal isostasy	10%
Plate tectonics	20%
Surface loads	5%
Deep heterogeneities	50%

The percentages are rough estimates of the contributions to the mean square of the anomalous potential, or geoid height (see Richards and Hager, 1988).

By *lithospheric strength* is meant the flexural rigidity of the lithosphere: the support of loads, surficial or internal, by an elastic rheology. That flexural rigidity can persist for a long time is evidenced by gravity anomalies associated with geologically ancient features, such as the midcontinent high: a feature 1200 km long and 80 km wide that has a peak anomaly of 50 mgals and stretching from the western tip of Lake Superior down to the Missouri River. The limited strength of rocks confines support by flexural rigidity to shorter wavelengths of less than 200 km.

Crustal isostasy means the classically hypothesized support of topography (and hidden loads as well) by compositional variations, which may be either lateral variations in crustal composition (Pratt isostasy) or variations in the depth to the crust-mantle interface (Airy isostasy). Crustal isostasy is the dominant mode of support of the continents. Since this isostatic compensation is more than 30 km deep in some places, it still is a perceptible contributor to the geoid at wavelengths less than 600 km.

By *thermal isostasy* is meant the support of oceanic features associated with the cooling, spreading oceanic lithosphere. Since this compensation extends some 125 km deep (Parsons and Sclater, 1977) and the oceans are broad, this effect contributes to rather long wavelengths, up to 5,000 km.

By *plate tectonics* is meant effects arising from the density differential of lithosphere that is part of plate tectonics and dynamics associated therewith: most prominently, the subducted slabs, which penetrate to 700 km or deeper.

By *surface loads* are meant the recent glacial load transfer, some 9000 years ago, and the associated response of the mantle. This transferred load still makes a perceptible contribution over a range of wavelengths, because the isostatic backsurge of the mantle is incomplete.

By deep heterogeneities are meant the density inhomogeneities between 670 and 2900 km deep, which only in recent years have been detected by seismic analyses. These inhomogeneities are undoubtedly related to deep convective systems. A mantle of the viscosities inferred from postglacial rebound would flow in response to them, although possibly at rates more than an order of magnitude slower than some oceanic tectonic plates at the surface. The inhomogeneities also must include undulations of the core-mantle boundary. However, they cannot lie in the fluid outer core itself: if they did, they would drive flows much faster than inferred from the magnetic field generated by the geodynamo. Undulations of the core-mantle boundary of a few hundred meters amplitude are not only predicted by viscous mantle models from the seismic density inhomogeneities (Hager et al., 1985), but also inferred by the nutation of the Earth in response to torques exerted by the Moon and the Sun (Gwinn et al., 1986).

The Moon

The Moon differs radically from the Earth in that it has one global lithosphere, some 800 km thick. It also has a crust, about 60 km thick, composed largely of calc-aluminous rocks called anorthosite, a simpler differentiate than characterizes the Earth's continental crust. The surface regions of this crust are called terrae. The principal variations in the surface, covering about 20%, are darker basaltic lava flows called maria. The basins in which these maria lie appear to be caused by impacts into the crust. The more recent, less than 4000 my old, have well-defined margins caused by these impacts. The ringed maria range up to 1000 km in diameter, the most prominent example being Mare Imbrium.

The Moon's gravity field, while having larger variations than the Earth in anomalous acceleration, indicates a quieter interior in the sense of stress-difference. The most prominent features of the Moon's gravity field are positive features that are associated with ringed maria greater than 400 km in diameter. The ages of the maria, as estimated from radiochronology in two and crater counts in all, are 4,000 to 3,200 my. This correlation is consistent with a Moon whose outer parts have been cooling off for most of the last 4,000 my but whose deep interior has warmed up. The correlation of positive gravity anomalies with ringed maria apparently arises from the Moon having had an interior active enough at the times of impact that the weakening of the lithosphere caused by the impacts led to their being sites of convective upthrust and volcanic outpourings. As on the Earth, the volcanism must be small in mass and heat compared to the upthrust below it. Well-defined craters smaller than 300 km are associated with negative anomalies; these impacts were too small or too late to stimulate internal responses.

Aside from the anomalies associated with ringed maria and craters, the generally mild gravity field of the Moon indicates isostatic compensation of the topography at a depth of less than 100 km. This compensation arises from variations that may be either thickness of the anorthositic layer or Mg:Fe ratio in the upper mantle—probably both. This quiescent state is consistent with other indicators that the outer regions of the Moon differentiated a crust very early, which stabilized by 4,400 my ago and has steadily cooled ever since.

Venus

Venus is the planet most similar to the Earth, being 20% smaller in mass and 5% smaller in radius. Hence its mean density, corrected for pressure, is only 2% less than the Earth's, requiring that it have a similar bulk composition. Plausible models of planet formation indicate that it is virtually certain that Venus has differentiated an iron core like the Earth. The absence of a magnetic field is most likely caused by Venus being sufficiently smaller than the

Earth that it has not differentiated a solid inner core, and hence lacks this energy source for a geodynamo. Venera lander data on surface rock composition also indicate that Venus has differentiated a crust, the main difference being a moderately higher MgO:FeO ratio, as would be expected if the Venerean crustal rocks were products of crystallizations from higher temperature partial melts. The topography of Venus differs markedly from the Earth's. For 90% of the surface, there is a gentle undulation of ~0.5 km about a modal level. The remaining 10% is comprised of plateaus some kilometers high, the most prominent being Aphrodite, stretching along the equator some 4000 km; Ishtar, near the north pole, smaller but higher; and Beta Regio, on the equator 4000 km west of Aphrodite, about 1000 km in extent. Most markedly lacking is a connected ridge system like the Earth's ocean rises. If Venus had the same density of heat sources and transferred internal heat to the surface in the same manner as the Earth, its topographic system analogous to the ocean ridges would rise about 1.4 km above the modal plane and be about 40,000 km long. The absence of such a system indicates that plate tectonics—the coming to the surface of the boundary layer of mantle convection—is absent on Venus. This absence suggests that Venus is totally covered by a continental crust, from which there is negligible recycling to the mantle.

For features more than 4,000 km in extent, the variations in Venus's gravity field are smaller than those of the Earth. However, for smaller features they are greater. The most striking difference from the Earth's field is that the gravity field has a marked positive correlation with the topography at all wavelengths. The ratio of the gravity to topography indicates that if this correlation arises from isostatic compensation, the depth of the compensation must exceed 100 km for most positive features; for the prominence Beta Regio, as deep as 400 km. Given the high surface temperature of 740° K on Venus, these depths are much too great for compensation by passive compositional isostasy, as prevails under the Earth's continents. Hence the compensation must arise from dynamic processes: mantle convection, probably complicated by interaction with a crustal layer much thicker than the Earth's.

Venus's field is much simpler than the Earth's, for which we listed six qualitatively different phenomena above. On Venus, at the wavelengths observable, only one of the phenomena is necessarily significant; another is plausible; and two are possible, but unlikely. The high surface temperature utterly precludes changes in ice loading and lithospheric strength as perceptible contributors. There may be a "plate tectonic" contribution associated with Ishtar, in the region of a convective downstream. There also could conceivably be some crustal isostasy if there are significant lateral variations in upper mantle radioactive heat sources and hence temperatures. The one process that must be dominant is thermal isostasy; this is dictated by the great depths of compensation inferred. The correlation with topography is imperfect enough, however, that there very well could be some contribution from deep heterogeneities, which perhaps may be generated by hot instabilities at the core-mantle boundary, the reverse of the instabilities known to be prominent in the Earth's mantle, the cold subduction zones.

Mars

Mars, like the Moon, appears to have one coherent lithosphere completely surrounding the planet. Hence the final heat transfer outward must be almost entirely by conduction through this lithosphere. Mars differs in that it evidences mainly tensional tectonics, most markedly the great Vallis Marineris along the equator. This tensional character implies a thermal history that is predominantly a warming, with relatively late formation of the core, less than 2000 my ago. The ages of major lava flows inferred from crater counts are less than 1000 my. However, the outer parts appear to have cooled enough for the lithosphere to be more than 200 km thick. A consequence of this thick lithosphere is the confinement of volcanism to a few major features. The Tharsis region with its crowning peak Mons Olympus (27 km high) is the greatest topographic irregularity on any of the planets.

The Tharsis region also dominates the irregularities in the gravity field. Its most prominent feature is a peak of more than 300 mgals over Mons Olympus, more than three times as great as the next greatest anomaly. The long wavelength variations in the Martian Field have a magnitude about four times as great as predicted from the Earth on the equal stress implication assumption. These variations also show a positive correlation of about 0.5 with the topography. If the field is assumed to arise from isostatic compensation, and if the Tharsis region is excluded, then the mean depth of compensation is about 100-150 km. However, the Tharsis region must either have a much deeper compensation—more than 300 km—or lithospheric stress differences greater than one kilobar. Probably the support of Tharsis is a combination of these processes. The evidence of a warming thermal history of Mars suggests that there remain significant radioactive heat sources at depth, which generate sufficiently strong convection to provide a significant fraction of the support of Tharsis: a dynamic compensation with an effective depth of more than 400 km. A considerable part of the support could also be by an elastic lithosphere about 200 km thick, with deviatoric stresses of a few hundred bars. This strong, thick lithosphere would not be inconsistent with plausible temperature gradients in Mars, and would also tend to confine the surface manifestations of a convective system to relatively few regions, such as the Tharsis Plateau and Vallis Marineris.

Summary

The four terrestrial bodies for which gravimetry exists demonstrate a heterogeneity in their gravity fields comparable to other properties. This heterogeneity reflects a diversity of thermal and compositional evolution. The Earth is certainly the most complicated of the four bodies, and perhaps has examples of all the mechanisms that produce variations in the gravity field. The six mechanisms listed in the section on the Earth are, however, not only somewhat subjective in their selection, but their terminology is strongly influenced by circumstances on the Earth. A more general scheme would employ different combinations of at least four properties:

depth: shallow, deep
density cause: thermal, compositional
rheology: elastic, viscous
motion: dynamic, static

In reality, there normally are not only gradations between these extremes, but also sub-categories. Thus "shallow" includes both "crustal" and "surficial"; "elastic" and "viscous" are idealizations of actual rheology; while "dynamic" can vary from "steady state" to "transient." The categories are also essentially independent; the only gross incongruity would be elastic rheology with steady-state dynamics.

Hence the six categories of sources for the Earth can be restated as:

1. lithospheric strength: shallow static support of loads of all kinds by elastic rheology
2. crustal isostasy: shallow static support of loads of all kinds by compositional variations
3. thermal isostasy: shallow support of loads by thermal variations
4. plate tectonics: density variations primarily thermal and secondarily compositional, supported by primarily viscous, secondarily by elastic rheology, associated mainly with the dynamics of mantle convective down flows
5. surface loads: the transient, viscous flow in the mantle in response to the transfer of the water load from the ice sheets (primarily Laurentide) to the oceans in recent geological time
6. deep heterogeneities: density variations in the mantle deeper than 670 km that are some combination of thermal and compositional variations, supported by the viscous flow of the mantle in an essentially steady-state manner.

Phrased in this manner, the contributors to the Earth's gravity field do not seem as comprehensive, and the gravity fields of other planets not necessarily interpretable as simple variants of the Earth. For example, the longer wavelength variations in the mild gravity field of the Moon may reflect deep heterogeneities that could be elastically, rather than viscously, supported.

The prominences of Venus's gravity field a confined to much fewer regions than the Earth, and because of the high admittance ratio of gravity topography, probably reflect thermal features of viscous mantle convective system that is essential steady state for gravity interpretation, but probab time-varying on a scale of 100 my. It is simpler think of these features as thermal, but their smal ness makes it possible that they arise from a lon; term compositional differentiation of the Venere. interior. The dominant blandness of most of Venus gravity field probably reflects isostatic compens tion—by some unknown combination of variatio in crustal composition and interface depth.

Mars is similar to Venus in that its prominenc are even more confined to a few features, suggesti suppression of manifestations of mantle convectic by a thick surface layer. But in the case of Mars th layer must be thermotectonic rather than compos tional: a thick, strong lithosphere. Both Mars a Venus could have significant contributions to t gravity field from deep heterogeneities not close associated with topography, but perhaps with inst bilities at the core mantle boundary.

Gravity fields are of value in studying the interio of planets because they "see deep": i.e., refle density irregularities at all depths. But this san property makes them ambiguous in interpretatio and, of all data pertaining to a planetary interio their analysis most requires a context generated b other data and models.

WILLIAM M. KAUL

References

Clayton, R. W., and R. P. Comer, 1983, A tomograph analysis of mantle heterogeneities from body wave trav times, *EOS (Am. Geophys. Union Trans.)* **62**, 776.

Dziewonski, A. M., 1984, Mapping the lower mantl determination of lateral heterogeneity in P velocity to degree and order 6, *Jour. Geophys. Research* **8** 5929-5952.

Gwinn, C. R., T. A. Herring, and I. I. Shapiro, 198 Geodesy by radio interferometry: studies of the forc nutations of the Earth, 2, interpretation, *Jour. Geophy Research* **91**, 4755-4765.

Hager, B. H., R. W. Clayton, M. A. Richards, R. Comer, and A. M. Dziewonski, 1985, Lower man heterogeneity, dynamic topography, and the geoi *Nature* **313**, 541-545.

Kaula, W. M., 1980, Material properties for man convection consistent with observed surface fields, *Jou Geophys. Research* **85**, 7031-7044.

Kaula, W. M., M. Drake, and J. W. Head, 1986, T Moon, in J. A. Burns and M. W. Matthews, eds., *Sat lites,* Tucson, Az.: University of Arizona Press, 58 628.

Parsons, B., and J. G. Sclater, 1977, An analysis of t variation of ocean floor bathymetry and heat flow w age, *Jour. Geophys. Research* **82**, 803-827.

Phillips, R. G., and K. Lambeck, 1980, Gravity fields of the terrestrial planets: long-wavelength anomalies and tectonics, *Revs. Geophys. Space Physics* **18**, 27–76.

Richards, M. A., and B. H. Hager, 1988, The Earth's geoid and the large-scale structure of mantle convection, in S. K. Runcorn, ed., *The Physics of Planets*, New York: Wiley, 247–272.

Cross-references: *Absolute Gravity Measurements; Continental Lithosphere; Earth Orientation; Earth's External Gravity Field; Earth Structure: Global; Figure of the Earth; Geodesy: Physical; Gravity Anomalies: Interpretation; Gravity Fields: Statistical Analysis; Gravity and Isostasy; Lithosphere: Mechanical Properties; Mantle Convection and Plumes; Mantle Dynamics: Mantle Viscosity; Planetary Geodesy.*

GRAVITY AND ISOSTASY

If the Earth were a perfect fluid, its exterior surface and all interior density interfaces would assume the shape of an ellipsoid of revolution in hydrostatic equilibrium with the combined forces of gravity and centrifugal acceleration due to rotation. The Earth is not a perfect fluid, however, and there exist notable departures from the hydrostatic density distribution, the most obvious being surface topography. If the Earth were a perfectly rigid body, it could statically maintain its topography and any buried density anomalies indefinitely. But the Earth is also not perfectly rigid, and it has been known for over a century that the excess mass of surface topography must be correlated with density deficiencies at depth in order to maintain topography over geologically significant time periods.

It first became apparent toward the close of the eighteenth century that elevated regions must be underlain by compensating low density. In 1749 P. Bouguer obtained a value for Newton's gravitational constant G, from which one could determine the mean density of the Earth, by measuring the deflection of a plumb line caused by the excess mass of the mountain Chimborazo in Ecuador. The small deflection of the vertical implied a value for G that proved to be grossly underestimated compared to Cavendish's more accurate number obtained in the laboratory with a torsion balance. About 50 years later, during the survey of India, J. Pratt and G. B. Airy proposed specific models for the density compensation beneath mountains that was responsible for the observation that the deflection of a plumb line was only one third of the expected value, given the calculated excess mass of the topography. The term "isostasy" was coined by C. E. Dutton in 1889 to refer to the condition of compensation of elevation variations such that the stress state is hydrostatic below some depth within the Earth. Accordingly, a topographic feature is said to be *isostatically compensated* when the deficiency of mass buried below the topography exactly equals its excess mass. The *compensation depth* is the shallowest depth below which the stress state induced by the topography and its compensation is hydrostatic.

Compensation Models

Various models have been suggested as potential ways in which the Earth compensates its topography. These models can be classified as *local* or *regional*, depending on the spatial extent of the compensating mass relative to the topography. For the local mechanisms, the compensation for an elevated point is confined to a vertical column directly beneath it, which in turn implies that shearing stresses between columns must vanish to allow independent vertical motion in response to changes in surface load. The state of isostasy is most commonly expressed as the requirement that the total mass in all straight, vertical columns above the compensation depth be a constant, although F. A. Dahlen has recently pointed out several alternative definitions of isostasy, such as equality of mass in conical columns on a spherical Earth, minimum deviatoric stress in the compensating layer, or equality of pressure below the compensation depth. All four definitions allow hydrostatic stress below the compensation depth and zero shear stress between vertical columns, but differ in the potential field produced by the compensated topography. For the first three definitions, the most realistic ones, the differences in geoid or gravity anomaly are small and only appear at extremely long wavelengths. Regional isostatic mechanisms involve lateral as well as vertical distribution of the compensation with no requirement that shear stresses vanish. Individual vertical columns will not, in general, have equal mass above the depth of compensation. Only after horizontal and vertical averaging does the total mass deficit in the compensation equal the total mass excess in the topography.

Local Compensation. The first models of isostasy put forth by Airy and Pratt are both examples of local compensation. Airy proposed a thin crust floating on a dense substratum with compensation achieved by variations in crustal thickness inversely proportional to the land elevation (Fig. 1). The condition for local equality of mass in vertical columns requires a simple relationship between the height of the topography h at a given point and the excess thickening of the crust w immediately beneath:

$$w = \frac{\rho_0 h}{\rho_m - \rho_0} \qquad (1)$$

where ρ_0 is the density of the topography and ρ_m is the density of the mantle. The compensation depth lies just below the deepest crustal root.

According to Pratt's hypothesis, the crust extends

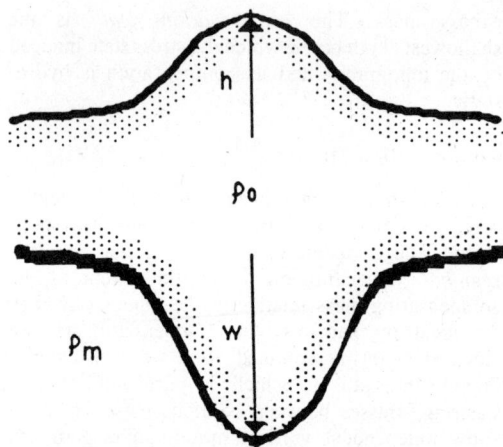

FIGURE 1. Airy isostasy. Mountains are buoyed up by thickened crustal roots extending into the denser mantle beneath.

to a uniform depth Z_P with density in each vertical column inversely proportional to the height of the topography h (Fig. 2). The condition for local equilibrium is

$$\Delta\rho = \rho_0 \frac{h}{Z_P + h} \quad (2)$$

in which ρ_0 is the density in a crustal column at sea level and $\Delta\rho$ is the density deficiency relative to ρ_0 for the column rising to height h. In Pratt's scheme, the compensation depth is at the constant depth Z_P.

There exists a continuum of local compensation models between those of Pratt and Airy in which variations in crustal thickness are combined with density changes in order to bring about equality of mass in vertical columns. A familiar example is the support for the elevation difference between continents and the seafloor. Continents stand higher because their crust is both lighter (average density

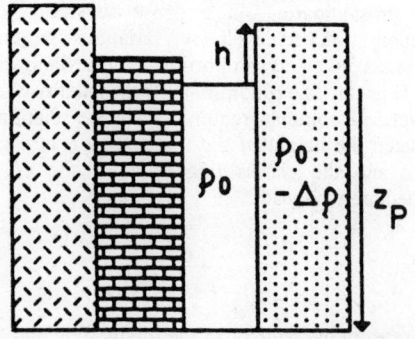

FIGURE 2. Pratt isostasy. Changes in elevation are caused by variations in density in independent crustal columns extending to the compensation depth Z_P.

of 2670 kg/m^3) and thicker (average thickness of 35 km) than oceanic crust (average density of 2800 kg/m^3 and average thickness of 6 km).

E. V. Artyushkov has criticized the local compensation model as being unrealistic in its anisotropy. It requires that Earth materials have no strength to support vertical shear tractions between crustal columns, but have infinite strength to maintain horizontal stresses due to lateral density changes. For an isotropic material with a topographic feature supported by Airy's mechanism, for example, there would be a tendency for the mountain and its compensation to spread out laterally in the absence of other stresses on the system. The magnitude of the horizontal deviatoric stress is of the order of the excess load associated with the topographic variations. The ability of rocks to support such stress over geologic time scales depends on the strain rate of the system and illustrates the necessity of discussing isostasy within the context of realistic models of lithospheric rheology.

Regional Compensation. The development of regional compensation models has been motivated both by the observation that gravity anomalies reveal many topographic features not in local equilibrium and by the desire to incorporate more realistic material behavior. For example, a striking case of regional compensation is found in the Hawaiian Islands. Large positive gravity anomalies over the summits of the Hawaiian volcanoes show that the compensating mass cannot be localized immediately beneath the loads. Gravity lows flanking the volcanoes mark the lateral extent of the compensation. Furthermore, with the advent of plate tectonics, the recognition of a relatively rigid Pacific Plate drifting as a unit on a more mobile asthenosphere suggests that the strength of the Pacific Plate has regionally distributed the weight of the Hawaiian volcanoes over a broad area.

The most popular model for regional compensation assumes the upper surface of the Earth behaves like a thin elastic plate overlying a fluid. The excess weight of a topographic feature causes the plate to displace the fluid so that, as in Airy's model, buoyancy forces support the load, but in this case the displaced fluid extends over an area much broader than the feature itself. The distribution of the compensation is maintained by the flexural rigidity of the elastic plate, D, which is proportional to the cube of the elastic plate thickness T_e. For a very thick plate (limit $T_e \to \infty$), the flexure is imperceptible, and we would say that the feature appeared uncompensated on a perfectly rigid Earth. For an extremely thin plate (limit $T_e \to 0$), the elastic plate model reduces to Airy's mechanism.

The relationship between the deflection of the elastic plate w at $\mathbf{x} = (x, y)$ and the load $P(\mathbf{x})$ acting on the plate is described by the thin-plate equation

$$D\nabla^2 w(\mathbf{x}) + (\rho_m - \rho_0)gw(\mathbf{x}) = P(\mathbf{x})$$

in which ρ_m and ρ_0 are the density of the materials below and above the plate and g is the gravitational acceleration (Fig. 3). The solutions to this differential equation consist of combinations of damped periodic functions with the exact form depending on the boundary conditions (e.g., continuous or fractured plate) and load geometry (e.g., point loads, lineated topography, circular symmetry, arbitrary two-dimensional loads). Commonly encountered modifications to this equation include the addition of a term for lateral thrusts acting on the plate or the incorporation of time derivatives to account for viscoelastic behavior.

In accord with the predictions from the elastic plate model, the seafloor around the Hawaiian Islands is flexed into a characteristic moat followed by a low-amplitude arch. The elastic plate thickness consistent with the bathymetry and gravity anomalies is about 25 km.

Using Gravity to Study Isostasy

Because the essential feature of isostasy is buried density deficiencies beneath elevated regions, gravity data are most commonly used in studies of isostasy. From observed free-air gravity anomalies, the theoretical attraction of the topographic mass is removed to yield Bouguer gravity anomalies. If the topography near the gravity station is relatively smooth, it is common to calculate the "simple" Bouguer anomaly by approximating the attraction of the topography Δg_{topo} with the Bouguer slab formula

$$\Delta g_{topo} = 2\pi \rho_0 G h$$

If full topographic corrections are made, as would be necessary in mountainous terrain, the anomaly is called the *complete* Bouguer anomaly.

In regions of appreciable topography, the principal signal in the Bouguer gravity field is that from the buried compensation, and, accordingly, Bouguer gravity is strongly negative in elevated regions. Although Bouguer gravity anomalies alone cannot unambiguously determine the depth and distribution of the compensating mass due to the nonuniqueness in the gravity inverse problem, they can distinguish among competing hypotheses.

Methods employed for incorporating gravity into studies of isostatic compensation include forward modeling to compare the predicted anomaly from assumed bodies to the observed gravity, inverse methods for deriving features of the compensation from the gravity anomaly under a minimum of assumptions, and admittance techniques for studying the statistical properties of compensation over large areas. Algorithms used for forward modeling and direct inversion of gravity data are similar to those employed in gravity field interpretation for other purposes, such as geophysical prospecting. The admittance method, which is used in conjunction with forward modeling and direct inversion, relies on correlations between gravity and topography, and therefore is almost exclusively reserved for isostatic and some geodynamic applications.

Forward Modeling. Direct modeling of gravity anomalies is probably the most used of the available techniques and also the most abused because it is extremely difficult to account for the nonuniqueness in the density distributions consistent with the data. The typical procedure in forward modeling is to calculate the predicted distribution of compensating mass beneath the observed topography, assuming a specific compensation mechanism, such as those given by Eqs. 1 or 2. The gravity anomaly produced by this compensating mass can be calculated on a digital computer, even for irregular bodies, using the formulas of Talwani and Ewing (1960) or Parker (1972). Talwani's method computes the gravity anomaly by adding the contributions from a finite number of polygonal laminae that define the anomalous mass. Parker's technique sums Fourier transforms of increasing powers of the surface enclosing the anomalous mass to produce the gravity anomaly. Parker's technique is best suited for undulating density interfaces with relief small compared to the depth of burial. Talwani's method is suitable for even shallow bodies extending to great depth, but is

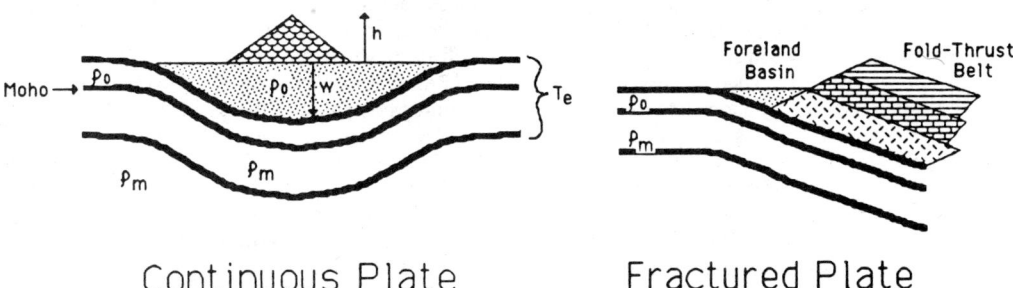

FIGURE 3. Elastic-plate compensation. The upper diagram shows compensation by flexure of a continuous elastic plate. The lower diagram shows compensation by flexure of an elastic plate with a free end beneath the load. In both cases, the Bouguer gravity anomaly is caused by warping of the Moho discontinuity embedded in the elastic plate.

computationally time-consuming for complicated models. Free parameters in the compensation model (i.e., crustal thickness, density contrasts, flexural rigidity, etc.) are then adjusted until the calculated gravity field matches the observed Bouguer gravity anomaly.

The drawback of the forward modeling approach is that an infinite number of mass distributions can be found consistent with a particular gravity anomaly, even under the assumption of isostasy. Often the number of free parameters in the compensation model approaches the number of truly independent pieces of information in the gravity field. Forward modeling in isostatic studies is best used when independent evidence from other data sources in support of a particular compensation mechanism is so compelling that an exercise in parameter fitting is justified. Examples of geophysical observations pertinent to studies of isostasy include the configuration of buried density interfaces from seismology, lateral variations in thermal regime from heat flow data, and predictions of the response of lithospheric rocks to loading from laboratory-derived flow laws.

Inverse Methods. Most "inverse methods" for interpreting gravity anomalies are in fact techniques for determining a finite number of parameters in a restricted model given observations of gravity. Therefore, they differ from the forward modeling described above only in the sense that one employs mathematical or graphical means to rapidly converge on the values for the unknown parameters in the model using the gravity data. The solutions, therefore, only have meaning if, as in the forward modeling approach, independent evidence corroborates the appropriateness of the chosen model.

A few of the inverse methods do attempt to allow for the nonuniqueness in the solutions by choosing to produce properties common to all solutions or extremal models that bound permissible solutions. Unfortunately, the gravity inverse problem is so highly nonunique that we can conclude only a few properties of the isostatic compensation unambiguously from the Bouguer gravity anomaly alone without restrictive assumptions. The integral of the Bouguer gravity anomaly is a measure of the total compensating mass, which in turn should be equal in magnitude but opposite in sign to the total excess mass in the topography. Theoretically, therefore, integrals of Bouguer gravity anomalies could be used to establish whether features are indeed perfectly compensated. In practice, it is often difficult to remove accurately regional trends from deep sources unrelated to isostasy and to choose the proper limits of the area over which the integration is performed.

Several formulas exist for bounding the depth to the compensating mass, given only the gravity anomaly. They rely on the fact that short-wavelength features in the gravity anomaly are strongly attenuated at increasing distance from the source region. Therefore, the shortest-wavelength contribution to the gravity field at the surface places a limit on the maximum depth to the top of the compensating body. The extent to which the bounds provide useful information for studies of isostasy depend upon the compensation mechanism and the accuracy with which one can remove any gravity anomalies due to variations in lithology near the surface.

Admittance Technique. In regions of complicated topographic loads that extend over large areas, the time and expense required for forward modeling of the gravity data can be prohibitive. In 1970, L. Dorman and B. T. R. Lewis introduced a more efficient means of extracting information related to isostatic compensation by applying methods of time series analysis to topography and gravity data. Following their formulation, the linear part of the relationship between the topography h and the Bouguer gravity anomaly Δg can be written as a two-dimensional convolution

$$\Delta g(\mathbf{x}_0) = \int dx \int dy \; q(|\mathbf{x}_0 - \mathbf{x}|) \; h(\mathbf{x}) + n(\mathbf{x}_0) \quad (3)$$
$$= q * h + n \quad (4)$$

in which q is the unknown isostatic Green's function and n represents the noise due to Bouguer gravity anomalies unrelated to isostasy. The dependence of q on

$$|\mathbf{x}_0 - \mathbf{x}| = \sqrt{(x_0 - x)^2 + (y_0 - y)^2}$$

is based on the assumption that the isostatic compensation for a point load would cause symmetrical anomalies beneath it. If the isostatic anomalies are uncorrelated with n, we may recover the function q directly from observations of Δg and h. This is most easily accomplished in the wave-number domain, because if the Earth's curvature can be neglected, Eq. 4 can be Fourier-transformed to obtain

$$\Delta G(\mathbf{k}) = Q(\mathbf{k})H(\mathbf{k}) + N(\mathbf{k}) \quad (5)$$

in which $\mathbf{k}$ is the two-dimensional wave number. The isostatic admittance Q is the wave-number domain representation of the gravity anomaly from the compensation of a point topographic load. Interpretation of $Q(\mathbf{k})$ or its spatial counterpart $q(|\mathbf{x}|)$ by either direct modeling or formal inversion is greatly simplified compared to the problem of dealing with two two-dimensional gravity and topography maps. Most investigators have chosen to directly compare the observed Q to theoretical admittances functions $\tilde{Q}$ corresponding to the linear part of preferred local and regional compensation mechanisms. A few studies have directly inverted the admittance using spectral expansion or linear programming techniques to recover the vertical structure of density or density gradients responsible for the compensation.

Despite the expediency of using the isostatic

admittance to investigate isostasy on a regional scale, some assumptions involved in the derivation of Eqs. 3-5 are not always justified. For example, Eq. 3 assumes a linear relationship between Bouguer gravity and topography. Most compensation mechanisms, such as those of Pratt, Airy, and the elastic plate, are nonlinear, although the second-order terms are only appreciable in areas of strong relief. In addition, the form of the equations assumes azimuthal independence of the Earth's response to loading. This condition is violated for fractured-plate models, which reasonably approximate the situation near lithospheric-plate boundaries where so much of the Earth's topography is created. Perhaps the greatest restriction in the admittance method is the requirement that the compensation mechanism be uniform over the entire region of study. On the continents, in particular, it is difficult to find large enough tectonic provinces over which this condition is met so that the mid- to long-wavelength admittance estimates are recovered. For all of these reasons, the most successful applications of the admittance method have been for oceanic features formed in midplate regions.

Results from Studies of Isostasy

So far, we have discussed isostatic compensation as a problem that involves the determination of a density distribution subject to the constraints of mechanical equilibrium. The more interesting problem, however, is to incorporate information of isostatic compensation with models of Earth kinetics provided by plate tectonics and with models of rock deformation provided by laboratory experiments to obtain a more complete picture of how certain forms of isostatic compensation are achieved and maintained. Once we understand how compensation depends on the physical state of the lithosphere and tectonic history, it then becomes possible to use gravity anomalies to predict the temperatures and state of stress in the lithosphere, calibrate laboratory-derived flow laws at geologic strain rates, and speculate on the tectonic processes responsible for the formation of topographic features.

Oceanic Isostasy. Topographic features on the ocean floor can be roughly categorized into two groups on the basis of isostasy and causality. The first category consists of long-wavelength slopes from the shallow midocean ridges to the deep abyssal plains and more localized bathymetric swells that frequently surround active hotspots. The elevation and slopes of the midocean ridges are successfully explained by a thermal plate model in which the subsidence of the seafloor as it moves away from the ridge is controlled by cooling as the plate ages. Thus midocean ridges stand high due to thermal expansion, and the topography is the consequence of the compensation. This thermal isostasy is a variation of Pratt's mechanism with the density variations $\Delta\rho$ in the vertical crustal columns being a function of the anomalous temperature ΔT via the coefficient of thermal expansion α:

$$\Delta\rho(x, z) = -\rho_0 \alpha\, \Delta T(x, z)$$

For topography compensated by this mechanism, observations of elevation differences and geoid or gravity anomalies can be used to constrain thermal models of the lithosphere.

Young hot-spot volcanoes usually cap 500–2000-km-wide areas of shallow seafloor lying 1 to 2 km above the depth for a normal lithosphere of that age. The subsidence of these swells downstream from hot spots and the elevated heat flow along the swells leave little doubt that these swells also have a thermal origin. Gravity and geoid data provide a direct measure of the depth to the thermal anomaly, which in turn constrains theories of the formation of the swells and mechanisms of heat transfer between the asthenosphere and the lithosphere. The compensation depth for swells surrounding the Hawaiian and Marquesas Islands lies in the mid- to lower lithosphere. Heat from the interior cannot be conductively transported through the lithosphere to depths of 45–70 km rapidly enough to explain the several-million-year rise time of the swells. It appears that we must invoke some physical process associated with hot-spot volcanism that convectively thins the lithosphere.

The second class of topographic features on the seafloor includes oceanic plateaus, seamounts, and islands formed by excess outpouring of basalt. Studies of the isostatic compensation of these features, using both forward modeling and admittance techniques, have established the mechanical nature of oceanic lithosphere on geologic time scales. Gravity and geoid data from both shipboard surveys and satellite altimeters demonstrate rather conclusively that volcanic constructs are compensated by the elastic plate mechanism with plate thickness controlled by the depth to a particular isotherm θ_e at the time of loading (Watts et al., 1980). The rapid transition from high strength to very low strength at a certain temperature is consistent with the exponential dependence of the strength of olivine $\Delta\sigma_0$ on temperature θ as determined by empirical flow laws from laboratory experiments:

$$\Delta\sigma_0 = \left[\frac{\dot{\epsilon}}{A} \exp(E/R\theta)\right]^{1/n} \qquad (6)$$

in which $\dot{\epsilon}$ is strain rate, E is the activation energy for ductile creep, R is the gas constant, and A and n are empirically determined constants. At depths above the isotherm θ_e in the lithosphere, the strength $\Delta\sigma_0$ exceeds the bending stresses involved in elastic plate flexure for processes proceeding at geologic strain rates of 10^{-15} to $10^{-16}\,\mathrm{s}^{-1}$. M. K. McNutt and

H. W. Menard used observations of flexure of the oceanic lithosphere at the outer rise seaward of subduction zones along with the thermal model of B. E. Parsons and J. Sclater to demonstrate that θ_e approximately corresponds to the 600°C isotherm. This result is in surprisingly good agreement with the predictions of Eq. 6 despite the extrapolation of this laboratory flow law over 10 orders of magnitude in strain rate.

The correspondence between the depth to θ_e and the thickness of the elastic plate compensating oceanic topography is so convincing that studies of oceanic isostasy have been used to probe the tectonic and thermal history of the ocean basins. For example, Watts et al. (1980) have used the estimates of elastic-plate thickness determined from gravity anomalies to classify oceanic features as having formed either near oceanic ridge crests (low T_e) or off-ridge on older, stiffer lithosphere (larger T_e). Thus gravity information can be used to assign approximate ages to unsampled seamounts, ridges, and plateaus. Since the elastic-plate thickness T_e measures the depth to the 600°C isotherm, H. W. Menard and M. K. McNutt cite the reduction in elastic plate thickness beneath hot-spot islands and seamounts as determined from geoid and gravity anomalies as evidence that hot spots thin the oceanic lithosphere enough to elevate the 600°C isotherm.

The simple elastic-plate model ignores the component of viscous response from the lithosphere. Bodine et al. (1981) discuss the time dependence of the elastic thickness of the lithosphere from the perspective of the rheological model derived from rock mechanics. Immediately after a loading event, such as a volcanic eruption, the Earth's surface rapidly sinks for a period of 10,000 years or so as the viscous lower lithosphere reestablishes hydrostatic equilibrium. For the next 1 m.y., the load slowly sinks an additional amount as relaxation of elastic stresses at the base of the plate thin the effective elastic thickness by about 5 km. Finally, the elastic-plate thickness stabilizes at 35 to 40 km for old oceanic lithosphere because the creep rates at these depths according to Eq. 6 are so low that it would require an addition 200 m.y. to relax the elastic stresses in the lowest 1 km of the plate. It is possible that the period of viscous response of the lithosphere can greatly exceed 1 m.y. if heat is injected into the base of the elastic plate in the process of lithospheric thinning by a hot spot.

Continental Isostasy. The state of our knowledge of isostatic mechanisms is not nearly so advanced on the continents as in the oceans. Interpreting gravity anomalies on the continents in terms of general mechanisms of isostasy is complicated for a number of reasons. To begin with, lithology varies rapidly. Simple models based on laboratory experiments with olivine may not apply. For example, at the temperatures and pressures representative of the continental crust, quartz is an order of magnitude weaker than olivine. Second, continental rocks are overprinted by numerous thermal events. The state of isostasy in any region will be a complicated function of the thermal and load histories. Third, topography above sea level erodes. It is hard to separate changes in compensation due to processes driven at depth from those due to removal of the surface load itself (isostatic rebound). Finally, the modes of topographic formation on the continents range from simple volcanism to large-scale plate collisions involving the entire lithosphere. The method of compensation may depend on the degree to which the lithosphere is deformed in the process of forming the load.

Although the observations of continental isostasy still defy an unambiguous interpretation, it appears that stable continental basement that has not undergone a major regional metamorphic event within the last 100 m.y. or so will respond to applied loads in the manner of an elastic plate, with thickness increasing with the square root of time since the last major thermal event. Significant elastic-plate thicknesses have been found for shield areas that have rebound isostatically after the melting of Pleistocene glaciers (e.g., 90 km for Fennoscandia) and for intercratonic sedimentary basins (e.g., 50 km for the Michigan basin). Flexure of the continental lithosphere where it underthrusts orogenic belts created by continent-continent collision is better modeled with the fractured-plate boundary condition (Fig. 3). From gravity anomalies over the mountain range and their associated foreland basins, estimates of elastic-plate thickness for continental lithosphere have been obtained ranging from 30 km for Hercynian basement rocks to more than 90 km for Precambrian shields. The analysis of gravity anomalies in orogenic belts is complicated by the presence of loads other than just the topography involved in the isostatic balance of the plate. These "hidden" loads include ophiolites obducted in the process of ocean closing (Karner and Watts, 1983) or the excess mass of the subducted slab itself as imaged by seismic tomography (Sheffels and McNutt, 1986). The connection between effective elastic plate thickness and plate age (as measured by K-Ar geochronology of basement rocks) is not so systematic as in the oceanic regime. Although the maximum elastic-plate thickness for plates of a given age increases with age, McNutt and co-workers (1988) infer from gravity anomalies that some ostensibly old plates appear very weak due to mechanical failure of the lithosphere when subjected to large bending stresses.

Summary

The concept of isostasy began as a way to explain the reduced attraction of mountain masses, but now principally used to determine lithospheric response to loading for the purpose of geodynamic and rheological modeling. Gravity and geoid anomalies have been the primary observational constraints in establishing the validity of the elastic-plate model of isostasy. The rigidity of the elastic plate supporting topographic features varies spatially with the lithology and the temperature of the plate at the time the load was formed. Observations of isostasy can be used to probe the rheological and thermal structure, as well as the tectonic history, of the lithosphere.

MARCIA McNUTT

References

Bodine, J. H., M. S. Steckler, and A. B. Watts, 1981, Observations of flexure and the rheology of the oceanic lithosphere, *Jour. Geophys. Research* **86,** 3695-3707.
Karner, G. D., and A. B. Watts, 1983, Gravity anomalies and flexure of the lithosphere at mountain ranges, *Jour. Geophys. Research* **88,** 10449-10477.
McNutt, M. K., M. Diament, and M. G. Kogan, 1988, Variations of elastic plate thickness at continental thrust belts, *Jour. Geophys. Research* **93,** 8825-8838.
Parker, R. L., 1972, The rapid calculation of potential anomalies, *Geophys. Jour. Royal Astron. Soc.* **31,** 447-455.
Sheffels, B., and M. K. McNutt, 1986, The role of subsurface loads and regional compensation in the isostatic balance of the Transverse Ranges, California: Evidence for intracontinental subduction, *Jour. Geophys. Research* **91,** 6419-6431.
Talwani, M., and M. Ewing, 1960, Rapid computation of gravitational attraction of three-dimensional bodies of arbitrary shape, *Geophysics* **25,** 203-225.
Watts, A. B., J. H. Bodine, and N. M. Ribe, 1980, Observations of flexure and the geological evolution of the Pacific Ocean basin, *Nature* **283,** 532-537.

Cross-references: *Figure of the Earth; Gravity Anomalies: Interpretation; Gravity Anomalies: Statistical Analysis; Gravity Fields: Implications for Planetary Interiors; Lithosphere: Mechanical Properties; Mantle Convection and Plumes; Mantle Dynamics; Mantle Viscosity.*

H

HEAT FLOW IN THE EARTH

Heat flow in the Earth, from its hot interior to its relatively cool exterior, is the primary energy flow responsible for the dynamic nature of our planet. Global heat loss is about 4×10^{13} W, or an average of about 80 mW m^{-2}. This energy flux is over two orders of magnitude less than the energy received by the Earth from the Sun, but as the solar energy is mainly reradiated into space, it has little effect on temperatures deeper than a few meters in the Earth. Effectively, transfer of solar energy controls the temperature of the Earth's surface, internal heat controls the Earth's internal temperatures. About 80% of the Earth's present internal thermal energy is derived from radioactive decay of long-lived isotopes, primarily ^{232}Th, ^{238}U, ^{40}K, and ^{235}U (listed in order of decreasing importance). The remaining 20% is primordial heat of Earth formation and differentiation (Turcotte and Schubert, 1982). Most of this heat is lost through the Earth's surface: A small fraction is converted to other forms of energy that drive tectonism, magmatism and other internal dynamic processes (Bott, 1982). Three heat transfer mechanisms are thought to be effective: (1) lattice (or phonon) conduction, (2) convection (sometimes called advection), and (3) radiation (or photon conduction).

Over most of the Earth's surface heat is lost by conduction, and heat transfer is governed by Fourier's Law of Conduction. To a first approximation, heat flows vertically from the Earth, and can be represented by the relationship

$$q = K(\delta T/\delta z) \qquad (1)$$

where heat flow, q, is given by the product of rock thermal conductivity, K, and $\delta T/\delta z$ is the rate of increase of temperature, T, with depth, z (by convention z is defined as positive downward). Conductive heat flow, q, is the parameter commonly quoted with reference to terrestrial heat flow and is the usual quantity measured by geophysicists. In stable regions, this conductive thermal regime is thought to extend to the base of the lithosphere at depths of about 100 to 200 km.

In localized regions at the Earth's surface, during tectonic deformation of the lithosphere, and below the base of the lithosphere, heat flow is thought to be dominated by convection in which heat is transferred by physical movement of Earth material across a temperature difference (i.e., rock, c ascending and descending fluids, such as water c magma). Heat flow by convection is given by

$$q = \rho c v (T_1 - T_2) \qquad (2)$$

where q is the heat flow in the direction of flow, ρ c and v are the density, specific heat and velocity c the flowing medium, respectively, and $T_1 - T_2$ the temperature difference in the flow. (*Note:* Thi expression must be modified for flow in a porou medium, such as water flow in an aquifer, but th basic parameters and principle are the same). Th threshold velocity at which convection become more efficient than conduction depends upon materi thermal properties (K, ρ and c), but for most terres trial heat flow problems this velocity is of the orde of a few centimeters per year, or about 10^{-9} m s$^-$ Convection can be driven by thermally induce buoyancy forces (free convection), as in the circu lation of seawater through young oceanic crust (se below) or in mantle and core convection (see *Mantl Convection and Plumes*). Convection may also b externally driven (forced convection), as in regiona gravity-driven groundwater flow and tectonic defor mation of the lithosphere (see below). A significar portion of the Earth's heat loss is convected to th surface at midocean ridges, and convection i thought to be the dominant heat transfer process a depths greater than 100 to 200 km (below the bas of the lithosphere).

Radiative heat transfer becomes increasingl effective at high temperatures and can be defined b the same relationship as conductive heat transfer (Eq 1) by replacing the term for thermal conductivity K, by an equivalent and strongly temperature-depen dent radiative conductivity. The role of radiative hea transfer in the Earth is poorly defined. Experiment show that in single crystals of minerals like olivine which are thought to be common in the upper mantle radiative heat transfer becomes efficient relative t phonon conduction at temperatures in excess of abou 500°C. However, grain (or crystal) boundarie probably scatter photons and reduce the efficiency o radiative heat transfer. The unknown efficiency o radiative heat transfer in the lower lithosphere create uncertainties in the geotherm at depth, but as radia tive heat transfer, like conduction, does not involv the physical movement of material, its effects can b quantitatively estimated by using an effective (conductive plus radiative) thermal conductivity for the conductive lid of the lithosphere.

Measurement of Terrestrial Heat Flow

The only quantity that can be directly measured related to heat flow in the Earth is the near surface heat flow. Conductive heat flow is usually determined by making separate measurements of the thermal gradient, $\delta T/\delta z$, and the thermal conductivity, K, of the rocks in which the gradient is measured, and calculating heat flow from Eq. 1.

On land, gradients are usually measured using an electrical thermometer in boreholes, 100 m or more in depth, below the transient effects of surface temperature variations and below the groundwater infiltration zone. Thermal conductivities are generally measured by steady-state thermal-comparitor techniques or transient heating techniques on core samples or drill-cutting samples from the boreholes (A. E. Beck, *Techniques of Measuring Heat Flow on Land,* in Lee, 1965; Kappelmeyer and Haenel, 1974). At sea, and in some deep lakes, gradients are measured in the upper few meters of sediment using a string of electrical temperature sensors on a sediment-penetrating probe. The same probe may include a heating element for in situ thermal conductivity measurements, or conductivities may be measured on sediment cores (M. G. Langseth, *Techniques of Measuring Heat Flow through the Ocean Floor,* in Lee, 1965). Stable thermal conditions at depth in the oceans and in deep lakes allow heat flow determinations to be made over a much shallower depth interval at these sites than for land measurements. A number of corrections may be applied to the heat flow measurements to account for local thermal effects (J. C. Jaeger, *Application of the Theory of Heat Conduction to Geothermal Measurements,* in Lee, 1965; Kappelmeyer and Haenel, 1974; Bunterbarth, 1984).

Occasionally estimates of convective surface heat transfer can be made by measuring the discharge rate and temperatures of flowing hot springs, through airborne infrared measurements, or from chemical analyses of discharging groundwater. This convective flux is an important part of the global heat flow budget, as discussed below: However, because of the difficulties of accurately measuring its magnitude and its extremely localized nature, unfortunately it is not generally included in large-scale heat flow maps.

An additional parameter essential for understanding heat flow in the Earth is the intrinsic radiogenic heat production of Earth materials. This parameter is usually measured in the laboratory on approximately 1 kg samples by measuring the radioactive isotope abundances through their natural gamma decay and using the isotope half-lives and decay energy to convert abundances to heat production (Rybach, 1971). The heat producing isotopes are isotopes of Th, U and K (see earlier), which are incompatible elements (not readily incorporated into mineral lattices). They tend to be concentrated in lower melting-point silicic igneous rocks, although their abundances do not closely follow major element abundances.

The main measured parameters in heat flow are given in Table 1 together with typical ranges of these parameters (see also Sass, 1971).

TABLE 1. Major Parameters Measured in Heat Flow Determinations

Parameter	Working Unit	Typical Range
Vertical temperature gradient	°C/km (mK/m)	10–50
Thermal conductivity	W/m/K	1–4
Heat production	$\mu W/m^3$	0–8
Heat flow	mW/m^2	0–125

Heat Flow and Internal Temperatures

Regionally, near surface heat flow varies from zero or near zero to more than 100 mW m^{-2}. The low limit of regional heat flow is imposed by the requirement that temperature must increase with depth (the Earth is losing heat), but the geothermal gradient can be reduced to almost zero over the descending portions of convection systems. The upper limit of regional heat flow (typically about 125 mW m^{-2}) is probably controlled by melting in the crust: Higher heat flow requires super-solidus temperatures in the crust or uppermost mantle that typically result in magmatic convection and local redistribution of heat. Locally, a much larger range of heat flow is measured, either associated with redistribution of heat by groundwater on land, seawater convection through oceanic crust, and/or magmatism.

The temperature profile within the Earth, or geotherm, is constrained by the surface heat flow, but, as the effectiveness of different heat transfer mechanisms at depth and the distribution of heat production in the Earth are somewhat uncertain, it must be constrained at depth by indirect evidence and theoretical considerations. A global average geotherm is shown schematically in Fig. 1. The temperature increases rapidly in the upper 100 km or so as heat is conducted through the lithosphere. In the mantle below the lithosphere, an absence of evidence for widespread melting, seismic, and other geophysical evidence, indicates that heat is transferred by solid-state convection at temperatures just below the solidus, and the geotherm is less steep. Variations in the geotherm in this zone are related to the details of mantle convection (see *Mantle Convection and Plumes*). Seismic data reveal topography on the core-mantle boundary, which is inter-

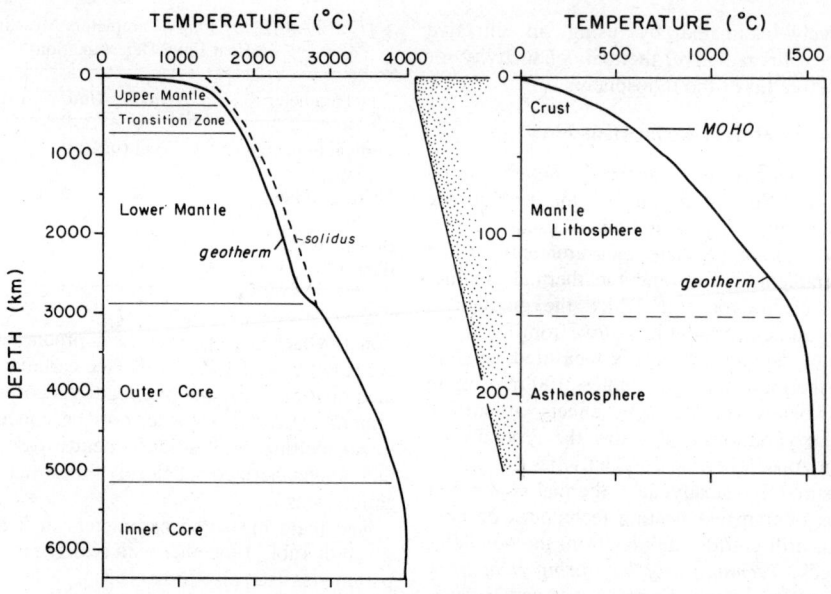

FIGURE 1. Global average temperature profile (geotherm) in the Earth (left) and in the crust and uppermost mantle (right). (After Stacey, 1977)

preted to indicate a dynamic thermal interchange between the core and mantle and the presence of a thermal boundary layer with a relatively rapid temperature increase near the base of the lithosphere. The seismic deduction that the outer core is liquid constrains temperatures in the outer core to be super-solidus. Temperatures within the core are based upon the energy balance within the core and seismic evidence for a solid (subsolidus temperature) inner core (Stacey, 1977). Apart from convective instabilities in the mantle (thermal plumes), major departures from the global average geotherm are expected only in the uppermost thermo-mechanical layer of the earth, the lithosphere, and the thermal structure of this zone is the subject of the remainder of this discussion.

The magnitudes and major modes of heat loss from the Earth are summarized in Table 2 (for additional details see Sass, 1971 and Sclater et al., 1980, 1981). Almost 75% of the global heat loss is through the oceans, approximately 85% of which is from the creation of new oceanic lithosphere. In contrast, over 60% of the 25%–30% of the global heat loss from the continents is generated by radioactive decay in the crust. Just as continents and oceans are fundamentally different with respect to crustal thickness, bulk composition, and mean age, there are fundamental differences in their thermal regimes.

TABLE 2. Heat Loss of the Earth

Component	Value
Heat loss through continents	1.2×10^{13} J/s
Heat loss through oceans	3.1×10^{13} J/s
Total	4.2×10^{13} J/s
Heat loss by hydrothermal circulation	1.0×10^{13} J/s
Heat lost in plate creation	2.6×10^{13} J/s
Mean heat flow	
Continents	50 mW m^{-2}
Oceans	100 mW m^{-2}
Worldwide	84 mW m^{-2}
Convective heat transport by surface plates[a]	~65% heat loss
Radioactive decay in crust	~17% heat loss

Source: After Sclater et al., 1981.
[a]Includes lithospheric creation in oceans and magmatic activity in continents.

Oceanic Heat Flow

When heat flow was first measured in the oceans in the early 1950s it was expected that oceanic heat flow would be much lower than continental heat flow because it was known that thin, basaltic oceanic crust is much less rich in the heat-producing isotopes than thick, granitic continental crust. Early data indicated that oceanic heat flow was approximately equal to continental heat flow. However, we now know that mean oceanic heat flow is approximately 70% higher than mean continental heat flow. Clearly, therefore, most of the heat from the Earth's interior is lost through oceanic crust, and this heat loss is part of the process of seafloor spreading through the creation of new oceanic crust at midocean ridges (see *Pa-*

leomagnetism and Plate Tectonics; Lithosphere, Oceanic: Formation and Evolution).

Heat is conducted upward at the midocean ridges as the mantle rises and partially melts due to pressure release. These melts continue to rise and eventually cool to form new oceanic crust with a very high near-surface geothermal gradient and heat flow. Newly formed oceanic lithosphere moves approximately symmetrically away from the midocean ridge and cools by conduction to the surface. As it cools, the geothermal gradient in the lithosphere and surface heat flow decrease. General cooling in the lithosphere-asthenosphere system is accompanied by thermal contraction and an increase in the mean density of the system, which is manifested by isostatic sinking of the ocean floor. These features of the thermal development of new oceanic lithosphere are shown in Fig. 2.

Cooling of oceanic lithosphere is essentially one-dimensional, and the thermal characteristics of all oceanic lithosphere is primarily dependent upon its age (i.e., it is independent of spreading rate and distance from the midocean ridge). A variety of models of lithospheric development associated with cooling have been proposed, all of which explain the basic observable thermal and elevation characteristics of oceanic lithosphere, and contrasting examples of which are shown in Fig. 2B. None of the models, however, explains all known features of oceanic lithosphere development.

There are two major departures in the observed parameters from the simple model predictions. The first departure is in the measured heat flow. Observed heat flow is generally much lower and more erratic than predicted, as shown in Fig. 3A. Heat flow data from all oceans are in reasonable agreement with the theoretical cooling curves for older oceanic lithosphere (greater than 10 to 70 Ma), and sonar and

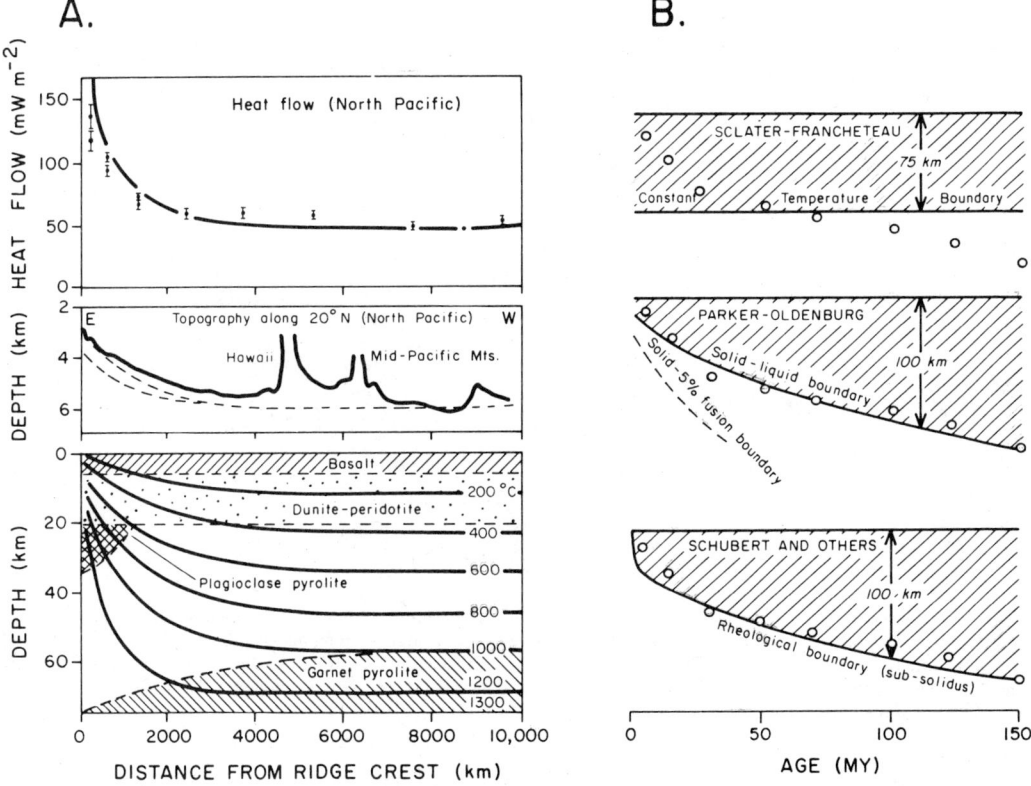

FIGURE 2. Thermal development of oceanic lithosphere. (A) Lower: isotherms and mineralogy in oceanic lithosphere as it cools and moves away from the midocean ridge with a horizontal scale appropriate to the northern Pacific. Middle: topography along 20°N (solid curve) compared with theoretically predicted topography (dashed curves) for models including only thermal expansion (upper) and both thermal expansion and mineral phase transitions (lower). Upper: Observed heat flow averages with error bars (symbols) and calculated heat flow (solid curve). (B) Models of the cooling oceanic lithosphere suggested by different workers. Upper: constant thickness (plate) model in which temperatures are perturbed relative to a constant-temperature lower boundary; Middle: increasing thickness model in which the lower boundary of the lithosphere is defined by the solidus temperature; and Lower: increasing thickness model in which the base of the lithosphere is defined by a subsolidus rheological constraint. Circles show the estimated thickness of the lithosphere based upon surface wave seismic data. (From Bott, 1982)

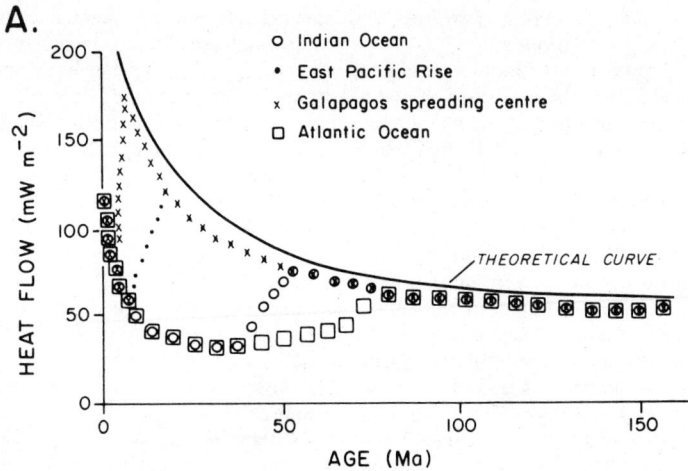

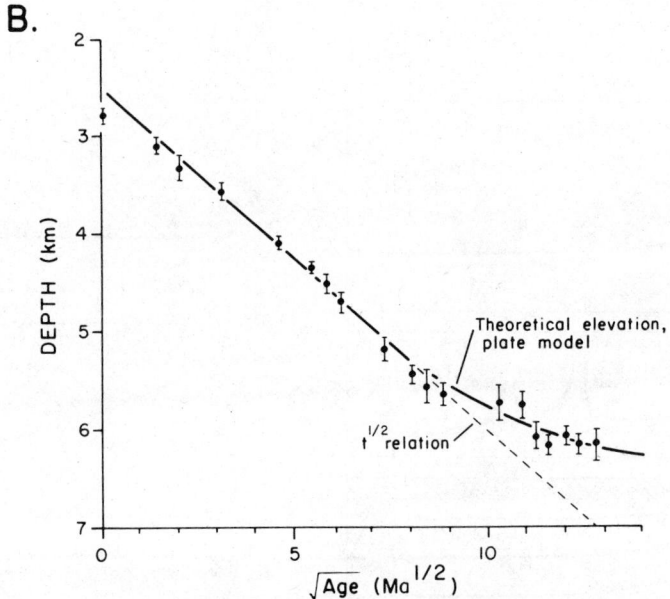

FIGURE 3. Deviation of oceanic heat flow and depth from predictions of simple cooling models. (A) Composite heat flow versus ocean-floor age for the main oceans. Heat flow near ocean ridges (young ages) is depressed due to the effects of hydrothermal circulation. (B) Mean regional depths of the north Pacific Ocean versus the square root of ocean-floor age. Solid curves shows depth predicted by 125 km thick constant-thickness plate model; $t^{1/2}$ relation shows depth predicted by simple unrestricted cooling half-space model. (From Bott, 1982)

seismic observations of the sea floor indicate that this agreement is achieved when a coherent layer of sediment first covers the igneous oceanic crust. Where sedimentation rates are high, observations and predictions agree in relatively young crust; where sedimentation rates are low, agreement is not achieved until much older crust. With the exception of a few very high heat flow measurements, observed heat flow is always much less than predicted by the models in very young (less than 5 Ma) crust.

Measured heat flow is much less than predicted by the models, especially near midocean ridges, because much of the heat is removed from young oceanic crust by the convection of seawater through the crust, as illustrated in Fig. 4A. Measured heat flow is low, because only the conductive component of the total heat loss is measured, and, by redistribution of heat flow, the convecting seawater results in large areas of recharge with low heat flow and very small areas of discharge with high heat flow. This convection

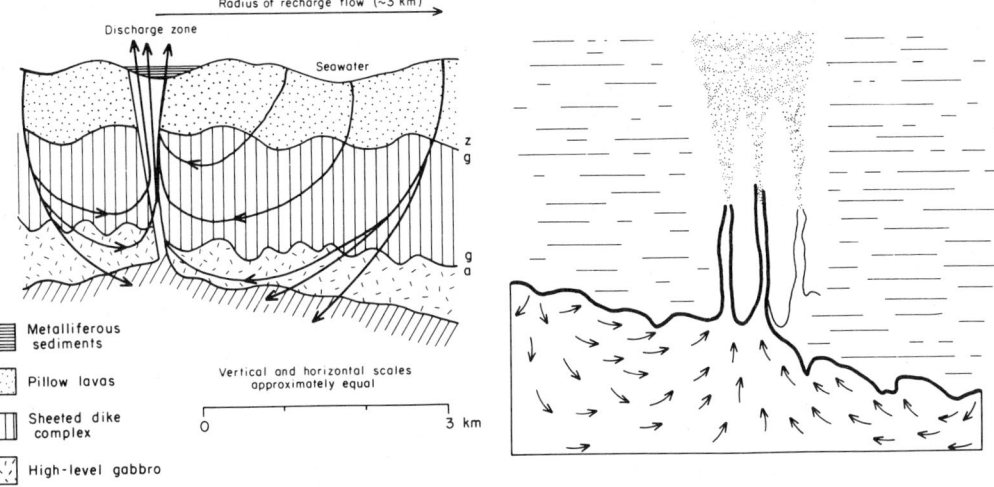

FIGURE 4. Examples of hydrothermal circulation in young oceanic crust. (A) Circulation pattern through crust at midocean ridge. If buoyant ascending heated seawater is not mixed below the surface it exits as a black smoker. (B) Black smokers, the discharge vents for very hot (> 300°C) hydrothermal circulation near midocean ridges. The black smoker chimneys are built from precipitates of the cooling metalliferous effluent. The "black smoke" results from further precipitation as the effluent mixes with cold seawater. (From Anderson, 1986)

system is best developed at the midocean ridges where the discharging submarine hot-springs produce spectacular black-smoker chimneys rich with metalliferous effluent (Fig. 4B) and remarkable local biological communities (Anderson, 1986). Access for seawater to the igneous oceanic crust is eventually cut off by the buildup of sediments on the ocean floor, and as soon as an impermeable layer of sediments is developed, all heat is lost by conduction, and observed and model predictions of heat flow are in good agreement.

The second departure from predictions of the simple thermal models is the flattening of the ocean floor with increasing age in contrast to continuously increasing depth with increasing age predicted by the unconstrained cooling models (Fig. 3B). Unfortunately the predicted heat flow variations for old ocean floor where the observed depths deviate from predictions cannot be resolved with available data, but it is likely that heat flow also does not continue to decrease indefinitely with increasing ocean-floor age. The thickness of the oceanic lithosphere does not thicken indefinitely with age and several models have been proposed to explain this behavior.

The simplest model used to explain the thermal behavior of older oceanic lithosphere is the model of a cooling constant-thickness plate with a fixed temperature at its base. This model successfully predicts the observable parameters associated with aging of the ocean floor, but is difficult to justify through geological arguments. Other models are based on a maximum thickness of the lithosphere controlled by heat input to the base of the lithosphere by shear heating, small-scale convection, radiogenic heat production in the upper mantle, or mantle plumes impinging on the base of the lithosphere.

A dramatic example of the probable effect of a mantle plume interacting with the lithosphere is the Hawaiian Swell and the string of volcanic islands that run along its axis. This swell, some 1000 km wide and 5000 km long, slopes gently west from the Hawaiian Islands and has a small heat flow anomaly (5 to 10 mW m^{-2}) relative to other ocean floor of the same age (Fig. 2A). The simplest explanation of this behavior is that the age of the Pacific plate is reset to a younger age as it passes over the Hawaiian plume or hot spot: Heat input from the mantle plume thins the lithosphere and the cooling process starts again as the oceanic lithosphere moves laterally westward away from the plume. Another well-defined example of a plume or hot spot is Iceland, where the effects of the plume are superimposed on a midocean ridge. Numerous other smaller mantle plumes probably cause similar, but less pronounced, deviations from the simple cooling behavior of oceanic lithosphere.

Oceanic lithosphere is returned to the mantle at subduction zones (see *Subduction Zones*). Cool sinking lithosphere in these zones convects heat downward, and surface heat flow is significantly reduced, sometimes to effectively zero. Melting and magmatism associated with the descending plate reverses this trend, however, and heat flow is high in zones of active volcanism behind subduction zones (volcanic arcs). The depression of isotherms associated with subduction, and the dual low/high heat flow signature of a typical subduction zone are illustrated schematically in Fig. 5. Active, or very recent volcanism must be present for the high heat flow zone to exist, and the high heat flow is clearly associated with upward advection of heat by magmas. In zones where arc volcanism is no longer active (e.g., northern central Andes), or has moved back from the trench (e.g., U.S. Cascades), heat flow is low.

If the stress field is favorable, subduction zones can also be associated with extension and the creation of new oceanic lithosphere in zones of back-arc spreading. Sea-floor spreading in these zones tends to be less well organized than at midocean ridges, but their thermal structure is essentially identical to that of their midocean counterparts.

Continental Heat Flow

Compared with the relatively short life (< 200 Ma) and rapid turnover of oceanic lithosphere, continental lithosphere plays a relatively passive role in plate tectonics, with a rate of creation and destruction at least an order of magnitude slower than the creation and destruction of oceanic lithosphere. It is chemically much more heterogeneous than oceanic lithosphere, particularly with respect to the heat-producing isotopes. Unlike oceanic lithosphere, which is all created in association with an apparently uniform thermal process and which does not experience major tectonism, apart from the effects of mantle plumes, before destruction, continental lithosphere typically experiences a variety of tectonic and thermal events of differing magnitudes during its lifetime. The transient thermal processes operating in continental lithosphere are conceptually similar to the processes operating in oceanic lithosphere, but their magnitudes are much less predictable, and their signal is mixed with thermal variations associated with the chemical heterogeneity of the continents.

Measurements of continental heat flow are plotted in Fig. 6 as a function of the tectonothermal age, the age of the last major tectonic or magmatic event in the region of each measurement site. Comparison of the continental data with oceanic data (Figs. 6, 2, and 3) suggests that the continental data follow a similar trend to the oceanic data as a function of "age," but with more scatter in the continental data, especially as "age" increases. Detailed inspection of individual data sets indicates more fundamental differences between the two settings, however.

Continental tectonothermal settings of Cenozoic age include zones of extension, compression, and volcanism. High heat flow is generally measured in zones of extension as heat is advected upward by magmas and by the general upward movement associated with thinning of the lithosphere in response to extension. The advective processes in these zones are similar to advection in midocean ridges, and there is good evidence that in many cases they are the precursors of midocean ridges (see *Continental Rifting: Types and Rates*).

Similar zones of high heat flow are associated with general magmatism in the continents in all tectonic settings, and the dominant process operating in these

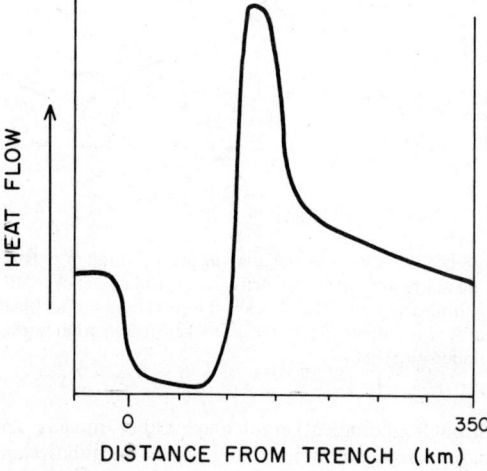

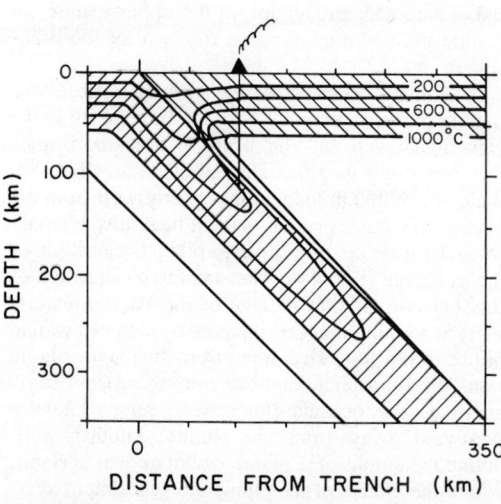

FIGURE 5. Schematic thermal structure associated with subduction. Upper: heat flow profile across zone. Low heat flow between trench and volcanic arc is associated with the cold descending plate; high heat flow in the arc and back-arc zone is associated with rising magmas. Lower: Depression of isotherms associated with descending slab. (See also Anderson, 1986.)

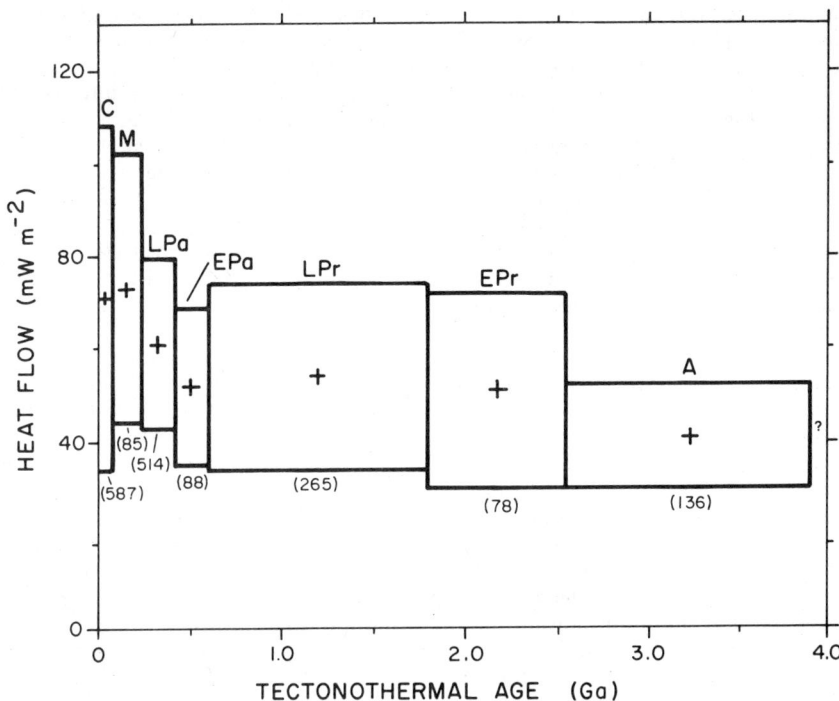

FIGURE 6. Observed continental heat flow versus tectonothermal age (age of the last major tectonic or magmatic event at the measurement sites). Data are grouped in geological age ranges: C: Cenozoic, M: Mesozoic, LPa: late Paleozoic, EPa: early Paleozoic, LPr: late Proterozoic, EPr: early Proterozoic, A: Archean. Crosses are plotted at the mean heat flow and midpoint of the age range for each group. Box widths indicate age range; box heights indicate ± one standard deviation of the data about the mean. Numbers in parentheses indicate number of data in each group. (From Morgan and Sass, 1984)

zones is upward advection of heat by ascending magmas. High heat flow has been predicted, but not observed, in association with shear heating on continental transform faults. Locally, heat in continental crust is redistributed by convecting groundwater, somewhat analogous to the convection of seawater through young oceanic crust. On continents the fluid motions are commonly driven by hydraulic gradients associated with variations in the water table elevation in addition to flow driven by thermal buoyancy. On a regional scale there seems to be an approximate upper limit to the average heat flow of about 125 mW m^{-2}. At this level of heat flow, near-solidus temperatures are indicated near the Moho, and higher regional heat flow averages may be prevented by buffering of maximum temperatures in the crust by melting.

When the source of heating ceases to operate, continental lithosphere appears to cool and subside in a similar manner to oceanic lithosphere. However, it is less easy to demonstrate this behavior in the continents. Yellowstone National Park (the Yellowstone caldera), Wyoming, in the western United States is perhaps the clearest example of the effects of a mantle plume in continental lithosphere, and this hot spot has a trace that extends west-southwest along the eastern Snake River Plain from Yellowstone. Like the midocean ridges, elevations decrease along the Snake River Plain away from the heat source (at Yellowstone) at a rate consistent with a cooling lithosphere model in which the age of cooling increases approximately linearly with distance along the plain. Cooling and subsidence appear to cease in this example at a cooling age of approximately 20 Ma, when the heat flow has decreased to the high background heat flow of the western United States. In place of the large constructional basaltic volcanoes associated with the Hawaiian plume, basaltic volcanism associated with the Yellowstone/Snake River Plain system heats the crust causing massive silicic eruptions, and locally makes the crust more basic and more dense, resulting in a topographically depressed plain relative to the surrounding terrain. A second good example of cooling and subsidence in a continental setting is found in some sedimentary basins of extensional origin, in particular, basins on passive continental margins. Sediments in these basins generally record a history of decreasing subsidence rate with time, consistent with cooling and subsidence models of the lithosphere. Subsidence is amplified in these basins by the loading effects of the sediments, and high heat flow in the early stages of basin formation can be recorded in temperature-dependent processes, such as argon loss, fission-track annealing, or hydrocarbon maturation, in the older sedimentary rocks deposited in the basin.

Subduction zones on continental margins have similar thermal structure to subduction zones in completely oceanic settings, as shown schematically in Fig. 5. The contrasting adjacent hot/cold thermal regimes in these zones are preserved in paired metamorphic belts in older terranes, with blueschist grade metamorphism marking the high pressure, low temperature conditions over the subducting slab between the trench and subduction-related volcanism, and amphibolite grade metamorphism recording the moderate pressure, high temperature crustal conditions associated with the volcanism.

Similar, but more complex thermal effects to subduction zones are found in zones of continental collision such as the Alpine-Himalayan belt (see *Continental Collision Zones: Seismotectonics and Crustal Structure*). Where simple or imbricate thrusts result in the underthrusting of one continental mass beneath another, the net movement of material in the lithosphere is downward, resulting in downward advection of heat and low surface heat flow. Heating of the resulting thickened crust in these zones commonly results in magmatism, however, which causes high surface heat flow. Thus, zones of continent-continent collision may be expected to be associated with complex interspersed high and low heat flow regions. Unfortunately, however, because of the logistical problems of measuring heat flow in these zones and redistribution of heat by flowing groundwater in the resulting rugged topography, data defining this pattern are very sparse.

A final complication to continental heat flow in regions of recent tectonothermal activity is the effects of advection of heat by sedimentation and erosion. Sedimentation results in a net downward movement of material, advecting heat downward and depressing surface heat flow; erosion has the opposite effect, advecting heat upward and increasing surface heat flow. These effects can extend the time period for thermal relaxation of continental lithosphere (the time for thermal equilibrium to be attained after tectonothermal activity) as they result in advection of heat in the lithosphere as the lithosphere responds isostatically to its changing geotherm. Examples of possible surface heat flow recovery paths are shown in Fig. 7 for a variety of initial tectonothermal disturbances and the effects of sedimentation and erosion.

Large scatter in heat flow in continental regions of young tectonothermal age (Cenozoic and Mesozoic in Fig. 6) is thus a result of a mixture of different complex thermal processes with different magnitudes and signs operating in these zones. Relative to oceanic data, however, there is still a large scatter in the continental data in regions of older tectonothermal age (greater than about 250 Ma), and this scatter reflects the chemical heterogeneity of continental crust.

In general, continental heat flow has three significant components: (1) Heat loss from the deep interior of the Earth; (2) heat generated within the continental lithosphere; and (3) advection within the lithosphere associated with tectonothermal activity. The third component should be insignificant in regions with tectonothermal ages of Paleozoic or older, and work by F. Birch, R. F. Roy, E. R. Decker, and D. D. Blackwell in the late 1960s showed that for measurements in major silicic plutons the first two components can be separated. These workers discovered linear relationships between surface heat flow and surface heat production for silicic igneous sites in different thermal settings. The heat flow intercept of these relationships, or reduced heat flow, gives the surface heat flow for zero surface heat production, and is interpreted to be the heat flow from below the zone of upper crustal enrichment in the heat-producing isotopes. The slope of the linear relationship has the units of length, and is interpreted to be a scaling parameter for the depth distribution of heat production in the crust. A. H. Lachenbruch has shown that for the relationships to survive the effects of differential erosion, heat production should decrease exponentially from the surface with an exponential decrement defined by this scaling parameter (see Morgan, 1984).

The linear heat flow-heat production relationship has now been extended to high-grade metamorphic terranes in addition to silicic igneous terranes, although the quality of the relationship is poorer in the metamorphic terranes. Relationships have been established for seventeen regions, or heat flow provinces, covering all continents except Antarctica. In provinces with young tectonothermal ages, the reduced heat flow values show a wide range, correlating with the style of tectonothermal disturbance (e.g., see Fig. 7). In provinces with tectonothermal ages Paleozoic or older, however, the reduced heat flow is remarkably constant, with a mean and standard deviation of 27 ± 4 mW m^{-2} ($n = 10$; Morgan, 1984). Therefore the deep component of continental heat flow in stable regions appears to be approximately constant, and the scatter in surface heat flow is primarily a function of lateral variations in crustal heat production. As shown in Fig. 6, the means and standard deviations of heat flow data from Paleozoic and Proterozoic provinces are insignificantly different, consistent with their having similar heat flow at depth and similar ranges in crustal heat production. Mean heat flow for Archean age sites is lower than for younger sites, and the scatter in these data is significantly less. Examination of heat flow-heat production relationships for Archean terranes indicates that reduced heat flow in these provinces is insignificantly different

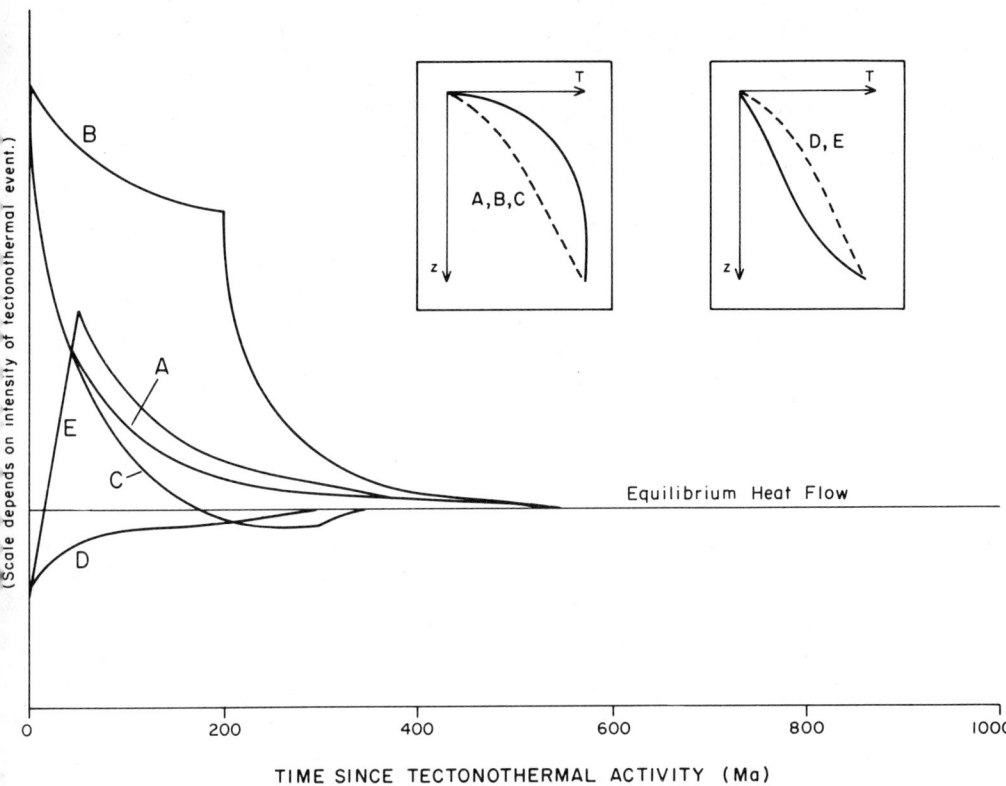

FIGURE 7. Schematic examples of surface heat flow relaxation following continental tectonothermal activity. Insets [in] upper right show initial geotherms (temperature, T, versus depth, z) by solid curves relative to the dashed curve of [th]e equilibrium geotherm: initial high heat flow on the left (rifting, midplate magmatism, hot spot, subduction, or colli[si]on-related magmatism); initial low heat flow on the right (subduction or compression without magmatism). A: Cooling [wi]thout sedimentation or erosion. B: Cooling with erosion for 200 Ma. C: Cooling with sedimentation for 300 Ma. D: [R]eheating without sedimentation or erosion. E: Reheating with rapid erosion for 50 Ma. (From Morgan and Sass, 1984)

[fr]om that in younger stable terranes, but that high [he]at production crust is much rarer in Archean [ter]ranes than in younger terranes. This conclusion is [co]nsistent with geochemical evidence that Archean [cr]ust is significantly statistically lower in incompat[ib]le elements, including Th, U, and K, than younger [cr]ust (Morgan, 1984).

Models of variations in crustal heat production in [st]able continental terranes, with an approximately [un]iform deep component of heat flow, indicate that [st]able continental lithosphere has a range in thick[ne]ss, higher heat production crust resulting in thinner [lit]hosphere than low heat production crust (Fig. 8). [Th]ere is great uncertainty in the geotherm in the [lo]wer lithosphere and in the best thermal criterion [fo]r the base of the lithosphere, but regardless of these [pa]rameters, variations in upper crustal heat produc[ti]on causes a spread in geotherms at depth which is [pr]obably equivalent to a difference between [m]inimum and maximum thicknesses of the stable [co]ntinental lithosphere of approximately a factor of [tw]o.

Paleo-Heat Flow

Variations in surface heat flow for the modern Earth are summarized schematically in Fig. 9 relative to the global mean heat flow. Earlier in Earth history this global mean must have been significantly higher as the abundances of the heat producing isotopes within the Earth were higher and the component of heat flow associated with secular cooling of the Earth was larger (Fig. 10). As heat loss from the Earth is closely related to tectonic and magmatic activity, the mode by which this extra heat was lost from the Earth is closely related to the tectonic style of the early Earth.

The largest fraction of the global heat budget lost from the modern Earth is lost in association with the creation of new oceanic lithosphere. The record of seafloor spreading preserved beneath the current oceans indicates that the rate of creation of new oceanic lithosphere has not been constant over the last 150 Ma. A peak in seafloor spreading occurred in the late Cretaceous, approximately 75 Ma ago,

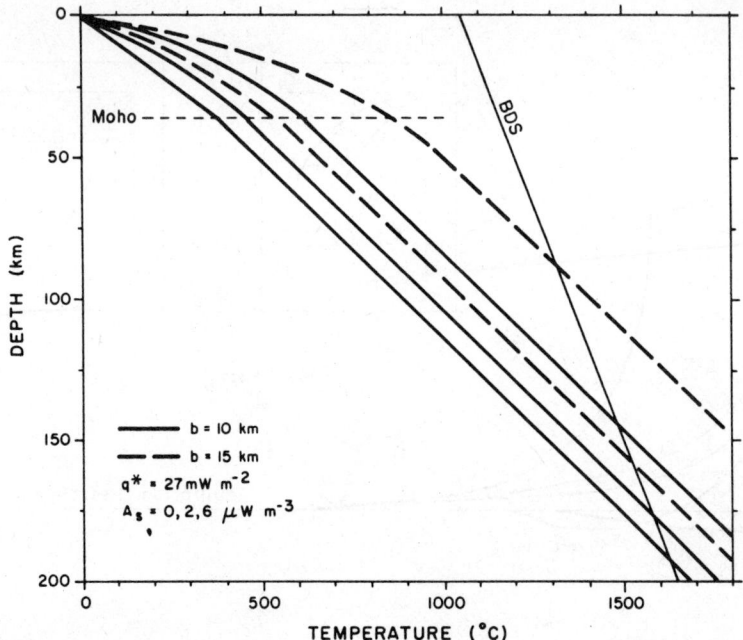

FIGURE 8. Hypothetical geotherms for stable continental lithosphere with a range in crustal heat production. Geotherms correspond to a single reduced heat flow, q^*, of 27 mW m^{-2}, crustal and mantle thermal conductivities of 2.5 and 3.4 W m^{-1} K^{-1}, respectively, and exponentially decreasing heat production with depth. Solid curves correspond to an exponential scaling length, b, of 10 km and to surface heat production values, A_s, of 0, 2 and 6 μW m^{-3} for the lower, middle and upper curves, respectively. Dashed curves correspond to a b of 15 km, and A_s of 2 and 6 μW m^{-3} for the lower and upper curves, respectively. BDS is an approximation of the basalt dry solidus, and the intersection of the geotherms with the BDS gives a maximum thickness for the lithosphere. (From Morgan and Sass, 1984)

associated with the opening of the Atlantic. This spreading peak was probably associated with a peak in mean oceanic heat flow of about 125 mW m^{-2}, and a peak in mean global heat flow of about 100 mW m^{-2}, relative to the present mean global value of about 80 mW m^{-2} (Sprague and Pollack, 1980). Therefore, much of the additional heat available in the early Earth was probably lost by an accelerated rate of oceanic lithosphere creation, either through seafloor spreading at a faster mean rate than at present, or through smaller plates and greater ridge length than at present. Hot young oceanic lithosphere subducts more slowly than cold old lithosphere in present-day plate tectonics, suggesting that the early Earth was covered by many small plates moving slowly: If Archean heat flow was three times that of the present, twenty-seven times as much ridge would have been required to lose this heat, assuming the same balance between continental and oceanic heat-loss mechanisms in the Archean and modern Earth (Hargraves, 1986).

Other heat loss mechanisms may have also been significant in the early Earth. The lack of preservation of crust prior to 3.8 Ga, about 100 Ma, or abo[ut] one lithospheric cooling time after the last heav[y] meteorite bombardment of the Earth suggests th[at] this bombardment disrupted the stabilization of ear[ly] continents, and may have caused a constant[ly] changing pattern of heat loss through plate conve[c]tion and magmatism. More heat may have been lo[st] in the early Earth by conduction into the base [of] continental lithosphere, and calculations indicate th[at] the present stable reduced heat flow can be appro[x]imately doubled to about 55 mW m^{-2} before [an] unacceptable amount of melting is predicted in t[he] lower crust. As discussed earlier, however, it [is] unlikely that peak heat flow values were higher [in] the Archean than at present, as peak heat flow [is] buffered even today by crustal melting, although ve[ry] high heat flow may have been more common a[nd] widespread in the early Earth.

Tectonic processes may have been slight[ly] different in the early Earth from modern processe[s] and it has been suggested that ancient oceanic cru[st] was thicker than modern crust in association wi[th] more efficient advection of heat at ancient midocea[n]

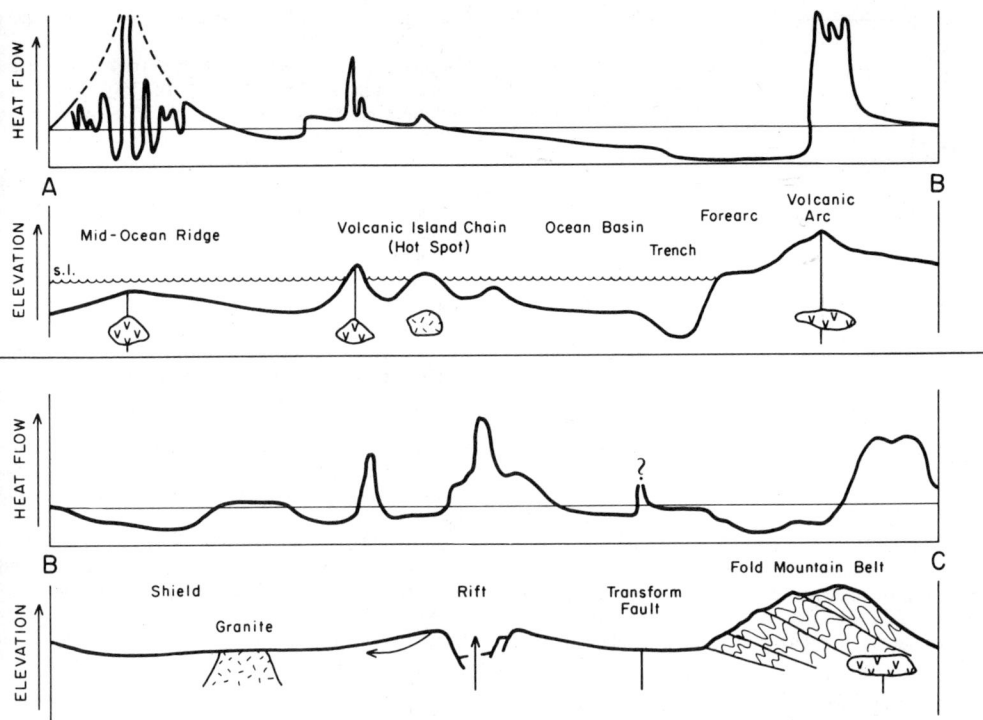

FIGURE 9. Schematic heat loss associated with different tectonic regimes. Thin horizontal line on heat flow plots indicates global mean heat flow. Wavy line (s.l.) on upper tectonic style plot indicates sea level. *Note:* Heat flow variations shown only qualitatively; tectonic style plots not to scale. Features shown from A to B: High irregular heat flow associated with hydrothermal convection at midocean ridge; reheating of oceanic lithosphere with local magmatic advection of heat associated with volcanic island chain (hot spot); stable low heat flow in ocean basin; low heat flow over descending slab and in fore-arc region of subduction zone; high heat flow associated with magmatic advection in volcanic . Feature shown from B to C: Generally low heat flow over shield with higher heat flow over granite rich in heat ducing isotopes; local heat flow high in region of groundwater discharge (hot spring); groundwater flow shown by ow); high heat flow with local magmatic advection of heat in continental rift zone; possible local heating associated h transform fault (unproven); low heat flow associated with underthrusting in fold mountain belt, and high heat flow ere magmatism occurs in same.

ges. The occurrence of komatiites, very high-temperature magnesium-rich lavas in Archean enstone belts suggests a more dynamic mantle rmal regime in the Archean, although the dating Archean age diamonds, which require the conditions of a relatively cool geotherm for formation, licates at least some local "cool" spots in the chean upper mantle. The problem in understanding the thermal regime of the early Earth lies finding data to constrain the processes in operation rather than in finding mechanisms capable of plaining the additional heat loss.

The statistical chemical difference (low heat oduction) between surviving Archean continental ist and younger crust may also be related to higher it loss during the Archean. As shown in Fig. 8, v heat production crust has a cooler geotherm than h heat production crust and is mechanically onger by virtue of its lower temperature and requires more heating before it can be melted. If reduced heat flow was generally significantly higher in the Archean than at present, possibly high heat production crust was selectively reworked during Archean orogenesis, and low heat production crust was selectively preserved.

Concluding Remarks

Much progress has been made in understanding the thermal regime of the Earth during, and in association with, the development of the concepts of plate tectonics. In particular our understanding of the thermal development of oceanic lithosphere, as well as of the components of continental heat flow has greatly improved. More study is required to constrain the details of heat loss associated with continental tectonism and magmatism, and the thermal process(es) associated with flattening of the ocean

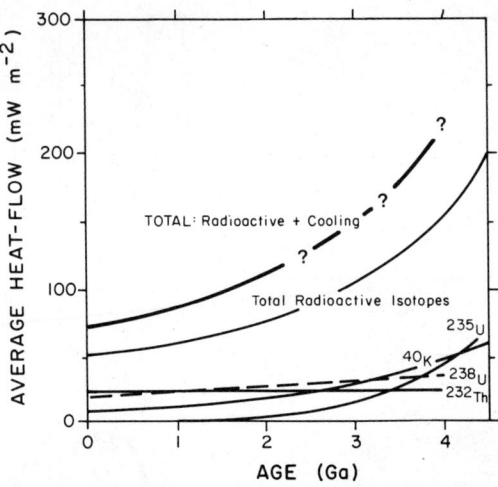

FIGURE 10. Global heat budget versus geologic time. ^{235}U, ^{40}K, ^{238}U and ^{232}Th indicate curves for heat flow generated by each individual isotope. Total Radioactive Isotopes curve indicates combined heat flow generated by isotopes. TOTAL curve indicates heat flow from isotopes plus secular cooling. (After Turcotte and Schubert, 1982)

floor with age. Through study of these processes and through study of the tectonothermal processes recorded in ancient continental rocks, we may gain more insight into the thermal regime of the early Earth.

PAUL MORGAN

References

Anderson, R. N., 1986, *Marine Geology*. New York: Wiley, 328p.

Bunterbarth, G., 1984, *Geothermics*. Berlin: Springer-Verlag, 144p.

Bott, M. H. P., 1982, *The Interior of the Earth*, 2nd e New York: Elsevier, 403p.

Hargraves, R. B., 1986, Faster spreading or greater rid length in the Archean? *Geology* **14**, 750-752.

Kappelmeyer, O., and R. Haenel, 1974, *Geothermic* Berlin: Gebrüder Borntraeger, 238p.

Lee, W. H. K., 1965, *Terrestrial Heat Flow*. Washingto D.C.: American Geophysical Union (Geophysic Monograph Series), **8**, 276p.

Morgan, P., 1984, The thermal structure and therm evolution of the continental lithosphere, *Physics a Chemistry of the Earth* **15**, 107-193.

Morgan, P., and J. H. Sass, 1984, Thermal regime of t continental lithosphere, *Jour. Geodynamics* **1**, 143-16

Rybach, L., 1971, Radiometric Techniques, in R. Wainerdi and E. A. Uken, eds., *Modern Methods Geochemical Analysis*. New York: Plenum Press, 27 318.

Sass, J. H., 1971, The Earth's heat and internal temper tures, in I. G. Gass, P. J. Smith, and R. C. L. Wilso eds., *Understanding the Earth*. Sussex: Artemis Pre 81-88.

Sclater, J. G., C. Jaupart, and D. Galson, 1980, The he flow through oceanic and continental crust and the he loss of the earth, *Rev. Geophysics and Space Physics* 269-311.

Sclater, J. G., B. Parsons, and C. Jaupart, 1981, Ocea and continents: Similarities and differences in mechanisms of heat loss, *Jour. Geophys. Research* **8** 11335-11552.

Sprague, D., and H. N. Pollack, 1980, Heat flow in t Mesozoic and Cenozoic, *Nature* **285**, 393-395.

Stacey, F. D., 1977, *Physics of the Earth*, 2nd ed. N York: Wiley, 414p.

Turcotte, D. L., and G. Schubert, 1982, *Geodynamic* New York: Wiley, 450p.

Cross-references: *Energy Budget of the Earth; G thermal Energy; Lithosphere, Oceanic: Formation a Evolution; Magmatic Processes; Mantle Convection a Plumes; Mantle Dynamics; Mantle Viscosity.*

INVERSE THEORY AND METHODS: SEISMOLOGY

In the past 20 years seismologists have come to recognize a fundamental fact: Analysis of seismological data to infer properties of the Earth's interior always involves the implicit or explicit solution of what we now call an *inverse problem*. This article discusses what inverse problems are and reviews various methods for solving inverse problems in seismology.

Solution of an inverse problem is required to interpret any measurements that are indirect. Indirect measurements are best understood by contrast with a direct measurement. If we measure the length L of a bar of steel with a meter stick, we are making a simple direct measurement. Now suppose we mount ultrasonic transducers on this bar of steel and measure the transit time of a P wave through this bar as time τ. We would all say that the P-wave velocity of the bar was L/τ, but we may not recognize that we are actually implicitly solving a trivial inverse problem in the process. τ *indirectly* measures the P-wave velocity, because to convert τ to velocity we have to measure L as well and calculate L/τ.

Interpretation of real seismic data differs little, in principle, from this example. The basic measurements of seismology are arrival times and amplitudes of different phases. These data are controlled by the elastic and inelastic properties of the Earth's interior, so we immediately expect that there should be some way to use these data to infer the elastic and inelastic properties of the Earth. Inversion is a formal way to make such inferences. In principle, the idea differs little from calculating L/τ. The difference is in the complexity introduced by the three-dimensional geometry of the real Earth, and the problem posed by the need to handle data sets that are often large but yet are universally inadequate.

Major Concepts

The zeroth-order requirement for solution of any inverse problem is that before we can hope to solve the inverse problem, we have to be able to solve the "forward problem." This basically means we have to understand the physical process that produced the observations well enough to make a reliable mathematical model of the process. That is, given some function $\hat{m}(\mathbf{r})$ that describes some physical property of the Earth (e.g., P wave velocity, S wave velocity, density) as a function of position $\mathbf{r}$, we assume we can calculate

$$\hat{d}_i = F_i[\hat{m}(\mathbf{r})] \quad (1)$$

where $\hat{d}_i$ is a predicted value for datum i based on the model $\hat{m}$. d_i is what we measure, and m is what we wish to know. F_i is a functional whose existence implies that if we know m exactly, we could predict the data perfectly.

Equation 1 is the basis of the simplest class of inverse solutions known as *forward modeling*. Forward modeling involves the use of educated guesses to derive an $\hat{m}$ that fits all the observable data according to some defined measure of goodness of fit (e.g., a chi-square test). Forward modeling has merit in two circumstances: (1) problems where F_i is an extremely difficult calculation (e.g., modeling of complex seismic reflection sections) and (2) problems for which auxiliary information cannot be easily incorporated in a formal inverse scheme (gravity models are a good example).

Forward modeling is attractive because of its simplicity. However, its fundamental weakness is that one may find a model that fits the data, but one generally has little idea how reliable that model is. Every inverse problem has the following aspects that need to be addressed.

Existence: Does any model fit the data?
Uniqueness: Can the data uniquely constrain the model?
Construction: How can we find a solution?
Appraisal: How well do the data constrain the model?

The first two are considered esoteric by many and are, unfortunately, largely ignored. Existence is self-explanatory. Uniqueness in the context above is an idealized concept more akin to uniqueness proofs for boundary value problems. The classic example in seismology is the proof by Gerver and Markushevich (1966) that in the absence of low-velocity zones in a radially symmetric Earth, travel time curves can be uniquely inverted for seismic velocity as a function of radius. Much of the effort in the solution of a problem often goes to the third aspect, construction. The fundamental difference between construction by an inverse procedure compared to forward modeling is that the data are used directly to construct a solution. Finally, appraisal is the most significant feature of a good inverse procedure. It answers the question of how good the solution one constructs really is.

A Canonical Problem

To illustrate the principles of inverse theory, consider a special, but practical, inverse problem. In recent years exploration geophysicists have begun to make extensive use of a technique called *vertical seismic profiling* (VSP). The geometry of a VSP is illustrated in Fig. 1. A receiver is positioned at a series of m depths, z_i, in a borehole where we measure the corresponding first arrival travel time t_i. (Real VSPs are used for considerably more, but this is the data of interest here.) The inverse problem is to use these travel time measurements to infer the P-wave velocity as a function of depth, z, in the vicinity of the borehole. We assume velocity varies only in the z-direction.

Analytic Solutions

The VSP travel time problem introduced above has a trivial analytic solution. If $u(z)$ is the slowness of the rocks along the borehole, then the travel time to some depth z_0 is

$$t(z_0) = \int_0^{z_0} u(z)\, dz \qquad (2)$$

That is, $t(z_0)$ is the antiderivative of $u(z)$.

Thus it follows directly from the fundamental theorem of calculus that

$$u = \frac{dt}{dz} \qquad (3)$$

This is the simplest analytic formula possible for an inverse problem. To estimate u, we only have to differentiate the travel time curve.

The original seismological inverse method, known as the *Herglotz-Wiechert (HW) integral* is the classical example of an analytic inverse solution. Under an assumption that the Earth is radially symmetric, the HW integral provides a recipe for estimating seismic velocity as a function of radius directly from the travel time curves. Application of the HW integral led to the earliest models of deep Earth structure. Conceptually the HW integral is essentially the same as Eq. 3. The form of the solution, however, is much more complex because of the need to correct for refraction effects that we assumed were not present in our VSP problem.

Another major analytic inverse method that has developed in recent years is inverse scattering techniques. Inverse scattering techniques seek to estimate acoustic impedance as a function of position within the Earth directly from a reflection seismic section (Newton, 1981). The problem has been solved under an assumption that the medium is horizontally stratified, but solutions for more realistic Earth models that are three-dimensional remain to be solved.

In most respects an analytic solution is the ultimate solution for any inverse problem. An analytic solution, however, is not always ideal for two reasons. First, to make the problem tractable, one must make assumptions about Earth structure that are not always justified. For example, with VSP travel times one assumes horizontal variations in velocity are negligible. This may not always be true. Second, analytic solutions in their simplest form always ignore two features of real data: (1) finiteness and (2) measurement errors. All seismological data sets are finite, even though they are often extremely large. Literal solutions based on equations like Eq. 3, however, require specification of an infinitely densely sampled curve $t(z)$, which is never possible in real life. Furthermore, if we tried refining $t(z)$ in our example VSP problem by progressively finer sampling, we would come to an interesting paradox. Suppose we sampled at intervals of Δz and calculated by a simple finite difference

$$u(z_0) = \frac{t(z_0 - \Delta z) - t(z_0)}{\Delta z}$$

As $\Delta z \to 0$, we could find the solution degrading into a random sequence of points. The reason is that real data contain measurement errors that are amplified to greater percentages as $\Delta z \to 0$. Similar problems are the plague of any analytic solution that makes no concession to data errors. The solution is clear. One has to fit a curve through the data that is not required to pass through every point. A variety of methods could be chosen to do so, but there is no clear choice, a symptom of the universal cancer of

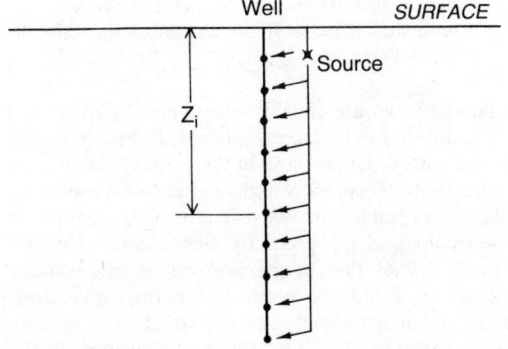

FIGURE 1. Schematic illustration of vertical seismic profiling geometry. The wavefield is recorded at a series of depths, z_i, by a borehole receiver using a surface source.

inversion. The finiteness of real data sets and measurement errors always combine to make any solution fundamentally nonunique and limit the ultimate resolving power of the data.

Inversion Using Discrete Parameter Models

Without doubt the most commonly used class of inverse solutions in seismology is the class of solutions that construct Earth models based on a finite number of parameters. This approach predates concepts of "inversion" by at least 50 years. Classic refraction analysis techniques assume an Earth structure made up of a finite stack of homogeneous layers. This classical analysis is an inverse problem. The data are a set of observed travel times. The model consists of the velocities of the layers. The construction technique, however, is specialized to the refraction problem. "Discrete inverse theory" (Menke, 1984) is a generalized method of solution construction for problems where the Earth model is blindly assumed to consist of a finite number of parameters. Discrete methods also provide good methods for appraising errors within the limitations imposed by the assumed parameterization.

To obtain an understanding of discrete methods, let us again consider the VSP travel time problem. Now, however, we will cast the problem as a discrete inverse problem by *assuming* that the rocks near the borehole consist of a set of n homogeneous layers of unknown slowness u_j ($j = 1, 2, \ldots, n$). In real life, the layer boundary depths might be chosen from auxiliary well log data, but for simplicity here we assume the layers are a constant thickness Δz. The problem is to determine the u_j from a set of m observed first-arrival travel times, t_i, measured with the receiver at depths z_i. The travel times are functions of the u_j, so we write

$$t_i = F(u_1, u_2, \ldots, u_n) \qquad (4)$$

where F is the simplest possible form of Eq. 1:

$$t_i = \sum_{j=1}^{n} G_{ij} u_j \qquad (5)$$

where

$$G_{ij} = \begin{cases} \Delta z & \text{if } j \Delta z \leq z_i \\ z_i - j \Delta z & \text{if } j \Delta z < z_i \leq (j+1) \Delta z \\ 0 & \text{if } (j+1) \Delta z < z_i \end{cases}$$

That is, t_i is a *linear* function of the unknowns u_j. Equation 4 can thus be written as a matrix equation

$$\mathbf{Gu} = \mathbf{t} \qquad (6)$$

which is simply an alternate notation for Eq. 5.

Consider first a special case. Suppose we choose the receiver depths so that $(z_0)_i = i \Delta z$. In this case $m = n$ and Eq. (6) has an obvious solution

$$\mathbf{u} = \mathbf{G}^{-1}\mathbf{Gu} = \mathbf{G}^{-1}\mathbf{t} \qquad (7)$$

where $\mathbf{G}^{-1}$ denotes the inverse matrix of $\mathbf{G}$. It is not hard to show that Eq. 7 yields

$$u_i = \begin{cases} \dfrac{t_i - t_{i-1}}{\Delta z}, & i = 2, 3, \ldots, n \\[1em] \dfrac{t_i}{\Delta z}, & i = 1 \end{cases}$$

That is, the solution, Eq. 7, for this special case is equivalent to a finite difference form of the analytic solution Eq. 3. This should not be surprising because even though Eqs. 7 and 3 may appear quite different they still are both seeking a common solution. (For a deeper understanding of this relationship see Lanczos, 1961.) This relationship is shown graphically as the exactly determined solution in Fig. 2.

When $m \neq n$, there is a similar solution to Eq. 7, called the *pseudoinverse* solution. The pseudoinverse is the generalized form of $\mathbf{G}^{-1}$ for nonsquare matrices and has some remarkable mathematical properties (see, e.g., Ben-Israel and Greville, 1974; Lanczos, 1961; Lawson and Hanson, 1974). For any $m \times n$ matrix $\mathbf{G}$, there exists a unique $n \times m$ pseudoinverse matrix $\mathbf{G}^+$ (see Lawson and Hanson, 1974, for a description of how $\mathbf{G}^+$ can actually be calculated) that can be used to "solve" Eq. 6 as follows:

$$\hat{\mathbf{u}} = \mathbf{G}^+\mathbf{Gu} = \mathbf{G}^+\mathbf{t} \qquad (8)$$

where the parallel to Eq. 7 is clear. In fact, it can be shown that if $m = n$ and $\mathbf{G}$ is full rank, $\mathbf{G}^+ = \mathbf{G}^{-1}$, and we revert to the special case described above. Normally, however, $m \neq n$ and we recognize three distinct cases.

1. A problem is *overdetermined* if $m > n$ and $\mathbf{G}$ has rank n (Menke, 1984, p. 48). In this case, $\mathbf{G}^+\mathbf{G} = \mathbf{I}$ and Eqs. 7 and 8 are almost identical. This particular case can be approached from a completely different direction through the method of least squares. That is, when $m > n$ there are more equations than unknowns, so one cannot expect all m equations to be compatible. Of all possible solutions Eq. 8 is the unique one that minimizes the root mean square misfit among all these equations:

$$\|\mathbf{t} - \mathbf{Gu}\| = \left[\sum_{i=1}^{m}\left(t_i - \sum_{j=1}^{n} G_{ij} u_j\right)^2\right]^{1/2} \qquad (9)$$

2. If $m < n$, $\mathbf{G}$ is said to be *underdetermined*. Underdetermined problems suffer from a funda-

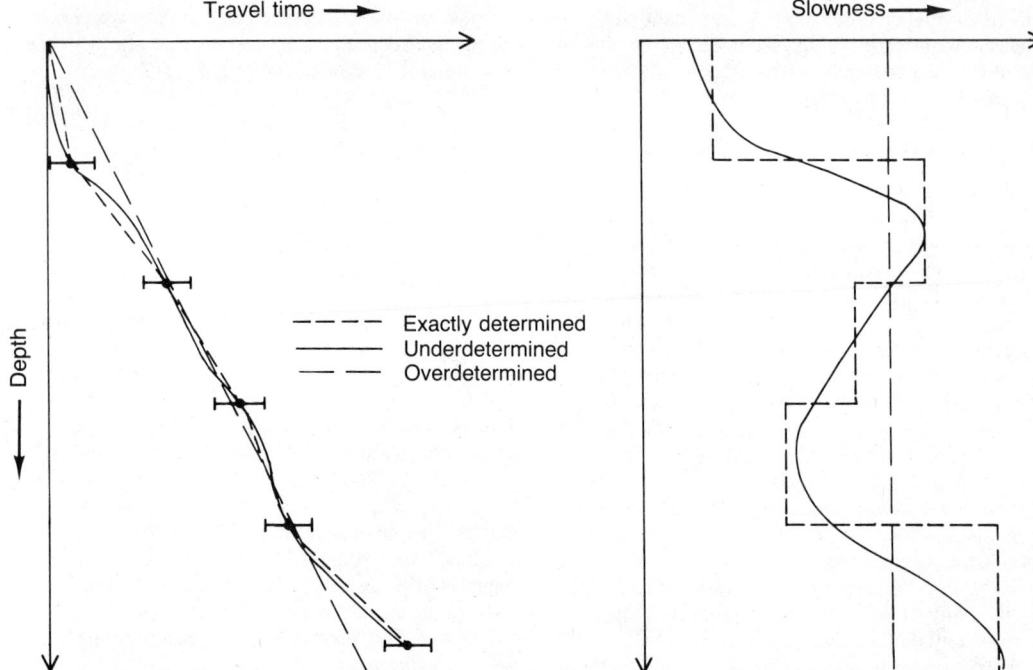

FIGURE 2. Types of solutions in discrete inverse problems. The left graph shows a set of five travel time measurements with a set of error bars. The right graph shows three types of solutions that could be generated from these data. The exactly determined case estimates one layer velocity for each data point; the underdetermined case determines a smooth curve solution; and the overdetermined case assumes the whole interval is characterized by a single velocity. The left-hand graph also shows the way each of these solutions fits the observations and how they implicitly interpolate between data points.

mental lack of information, since $m < n$ implies there are more unknowns than equations. Equation 6 is then inherently ambiguous because there are infinitely many solutions that fit all the data perfectly. The solution defined by Eq. 8 is still unique in this case, however, because it finds the "smallest model." That is, Eq. 8 is the solution that minimizes $\|\mathbf{u}\|$, where $\| \ \|$ is as defined in Eq. 9.

3. Most real seismological inverse problems are *mixed*. That is, they contain elements of overdetermined and underdetermined problems. Overdetermination is a consequence of data redundancy, and underdetermination implies a fundamental lack of information. For mixed problems the pseudoinverse solution is the unique one that simultaneously satisfies the least square criterion of overdetermined problems and the smallest model criterion for underdetermined problems.

The fundamental rule of inversion is that for any real problem the solution is always fundamentally ambiguous. This ambiguity comes from two sources: (1) the data are always inadequate in some respect, and (2) the data are always inaccurate to some degree. This nonuniqueness has two consequences. The first is that in a sense any solution is arbitrary. A particular solution is special only if it has some well-defined properties. The pseudoinverse is attractive for this reason, and this is one of the reasons it is widely used. The reader must recognize, however, that there are an infinite number of alternative "generalized inverses" $\mathbf{G}^{-g}$ that could also be used to solve Eq. 6 as follows:

$$\hat{\mathbf{u}} = \mathbf{G}^{-g}\mathbf{G}\mathbf{u} = \mathbf{G}^{-g}\mathbf{t} \quad (10)$$

For a review of some of the common choices in geophysics see Menke (1984). For a more mathematical treatise see Ben-Israel and Greville (1974).

The second consequence of the inherent ambiguity of inverse problems is that asking what "the" solution of a problem is is a meaningless question. Instead, the proper question is, how good is the solution? Proper applications of inverse methods focus on error analysis. The purpose of the analysis is to ascertain whether features found in some particular solution are real or an artifact of the choice of $\mathbf{G}^{-g}$ and/or random errors. For discrete problems, the former is routinely appraised by the resolution matrix

$$\mathbf{R} = \mathbf{G}^{-g}\mathbf{G}$$

e name is derived from the relation of **R** to
olving kernels that are encountered in the continuous theory described below. The basic idea, however, can be obtained by comparing Eqs. 8 and
. For full rank, overdetermined problems $\mathbf{G}^+\mathbf{G} =$
When the data constrain the model well, a reasonable choice for $\mathbf{G}^{-g}$ can be judged by how closely
approximates an identity matrix. Common practice
publications using discrete methods is to publish
:omplete listing of the elements of **R** or, at least,
diagonal elements.

The other major aspect of discrete inversion error
praisal is estimating the potential impact of
:asurement errors. The observed data t_i can be
oken down as

$$t_i = \tilde{t}_i + e_i$$

iere $\tilde{t}_i$ is the "true" datum and e_i is the measurement error. Normally one assumes e_i is a sample
om a zero-mean random process. Then the uncertainty in the solution of Eq. 10 can be appraised by
e covariance matrix

$$\mathbf{C}_u = \mathbf{G}^{-g}\mathbf{C}_d(\mathbf{G}^{-g})^T \qquad (11)$$

iere T denotes a matrix transpose and $\mathbf{C}_d$ is the
ta covariance, which is the multivariant generalition of the variance discussed in any introductory
itistics course (see Menke, 1984, for further
tails). Common practice in seismology is to
blish only the "standard errors" of a solution,
iich are the square roots of the diagonal elements
$\mathbf{C}_u$. For problems with three or fewer parameters,
can be presented in the form of error ellipsoids.
is practice is common for earthquake locations.

Discrete inverse techniques are widely used in
smology. Earthquake hypocenter location is an
ierently discrete inversion problem. Travel time
d surface wave inversions for crustal structure
nost universally apply discrete inverse methods.
r an excellent review see Thurber and Aki (1987).
Before continuing, we summarize the strengths
d weaknesses of discrete methods. The biggest
ength of discrete methods is the relative ease with
iich a solution can be calculated. This factor is
;nificant in many problems for two reasons. First,
lving the forward problem and calculation of the
efficients of the equivalent of **G** in Eq. 6 can
metimes be a formidable calculation of its own.
• make the problem tractable at all may require
iking the number of parameters n smaller than one
ght like. The second factor is that most important
ismological problems are nonlinear. That is,
uations equivalent to Eq. 4 cannot be reduced to
e simple matrix form of Eq. 6. The usual solution
thod in seismology for such equations is to use a
iuss-Newton method. (I avoid describing the
tails of the Gauss-Newton method to avoid confu-
in. For an excellent description see Lee and

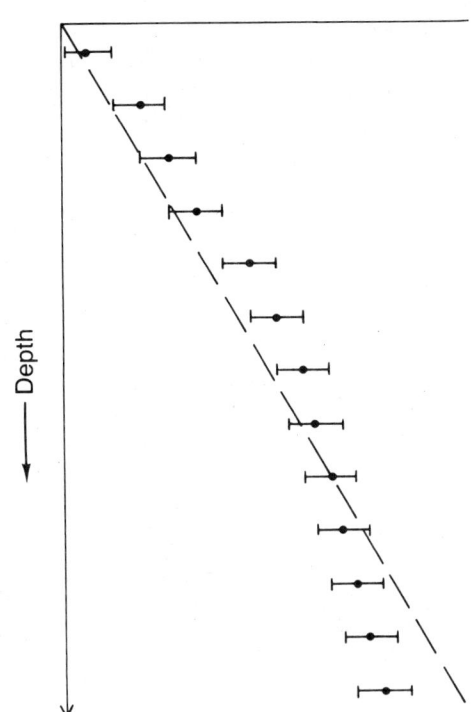

FIGURE 3. The pitfall of overdetermined, discrete parameter solutions. This graph is comparable to the left graph of Fig. 2. A single layer solution for this set of data would yield the broken straight line, with the error bars shown, the fit to the data is in a gray area where it almost works. In this case, the obvious thing to do is use two layers instead of just one. With other seismological inverse problems, such a simple solution may not be so obvious.

Stewart, 1981.) This presents two problems from a computational standpoint. (1) We are required to calculate $d_i - \hat{d}_i$ (Eq. 1) and partial derivatives $\partial F_i/\partial X_j$ (X_j is one of n discrete parameters). Both are usually calculated numerically and require at least $n + 1$ solutions of the forward problem for each datum. (2) The Gauss-Newton method is iterative and requires repeated solutions of equations of a form similar to Eq. 10. This can present an enormous computational problem and drives one to choose an overdetermined discrete problem with n made as small as possible.

The major weakness of discrete methods is that it is not always clear how the result is influenced by the a priori parameterization. A simple demonstration of this problem based on the VSP problem can be seen from Figs. 2 and 3. For this simple problem an overdetermined least squares solution (very coarse layers) can be shown to amount to fitting a series of lines to the data over the depth interval defined by each layer. Figure 3 shows how one such fit might work. Error bars are shown to emphasize that the data contain information that fundamentally cannot be extracted because the model was parameterized too coarsely. This deficiency in the result would be easy to detect for the VSP problem. Normally, however, it is very difficult to detect such a problem even when one strongly suspects it exists.

Continuous Inverse Theory

Continuous inverse theory combines the best features of analytic solutions and discrete methods. As such, most seismologists would agree that it is theoretically the best approach, in general, to follow. As a practical matter, however, implementing a continuous procedure can prove computationally intractable and may require assumptions as difficult to justify as the a priori parameterization in a discrete procedure.

The distinguishing feature of continuous inverse theory is recognition that the model, $m(\mathbf{r})$, is a function that requires, in principle, an infinity of parameters to be completely represented, whereas data sets are always finite and inaccurate.

I will again introduce the concepts of the continuous theory through the VSP travel time problem and will limit the discussion to a formulation due to Backus and Gilbert (1968, 1970). For an understandable, more general development see Parker (1977). Equation 2 gives the specific form of Eq. 1 that is our starting point. For reasons that will be apparent momentarily it is necessary to rewrite Eq. 2 as

$$t_i = \int_0^L G_i(z) u(z) \, dz \quad (12)$$

where L is conveniently chosen as the deepest receiver depth, $G_i(z) = H(z_i - z)$, where H denotes the Heaviside functions, and the G_i are fixed functions called *data kernels*.

The basic idea of the Backus-Gilbert formulation is to examine linear estimators of u of the form

$$\hat{u}(z_0) = \sum_{i=1}^m a_i(z_0) t_i \quad (13)$$

Given Eq. 12, however, it follows that

$$\hat{u}(z_0) = \int_0^L R(z, z_0) u(z) \, dz$$

where

$$R(z, z_0) = \sum_{i=1}^m a_i(z_0) G_i(z)$$

The weights $a_i(z_0)$ in Eq. 13 are, in a sense, entirely arbitrary. However, if the weights could be chosen such that $R(z, z_0)$ approaches the Dirac delta function, $\delta(z - z_0)$, $\hat{u}(z_0)$ would become a perfect estimator of $u(z_0)$. Because the data are finite however, this is never possible. (If it were, we would return to the analytic solution of Eq. 3.) Instead, we must be content to work with some R that is concentrated as much as possible around the target depth z_0. There are a variety of "deltaness criteria" for determining R. A general, unified theory of "deltaness criteria" appears in a rather difficult paper by Backus (1970). The most important ones, however, are closely connected with the most common class of generalized inverses (Eq. 10) used in discrete problems.

Error appraisal in continuous problems centers around the function R. The other player in determining errors in an estimate of u is measurement error. The standard error of the estimates (Eq. 13) is

$$\sigma_{u(z)} = \left[\sum_{i=1}^m \sum_{j=1}^m a_i a_j (C_d)_{ij} \right]^{1/2} \quad (1)$$

where $(C_d)_{ij}$ are the coefficients of the data covariance as given in Eq. 11. Backus and Gilbert (197) prove that resolution as defined by some "deltaness criteria" and the standard errors of the solution (Eq. 20) have a trade-off relationship. That is, decreasing the width of the peaked function R (increasing resolution) cannot occur without a corresponding increase in statistical uncertainties, and vice versa.

Figures 4 and 5 give two concrete examples of this effect. Figure 4a shows a typical set of resolving functions one might calculate for a Backus-Gilbert solution of the VSP travel time problem. The parameter θ is used to generate a family of related solutions with varying resolving power. As θ increases, the resolving function widens dramatically. Figure 4 shows the corresponding values of σ. As θ increases σ always decreases.

Figure 5 shows results of a more complete analysis of a closely related, but different, problem. The figure represents a resolving power map of a set of earthquake arrival times from the Coso geothermal area, California. These data were inverted by a Backus-Gilbert procedure for P-wave slowness as a function of depth by Pavlis (1982). This procedure represents one way to reduce the volume of information from Fig. 4. Only widths of the peaks of the resolving functions are determined by a formula related to the "deltaness criteria." These "resolution lengths" are contoured. The results show in one picture how resolution and standard errors trade of in this solution. The result is that we can see the solution is relatively well resolved down to a depth of about 7 km, but rapidly becomes poor below that depth.

The great advantage of continuous methods is that a very complete error appraisal is possible. However, they have one overriding downfall; often a solution

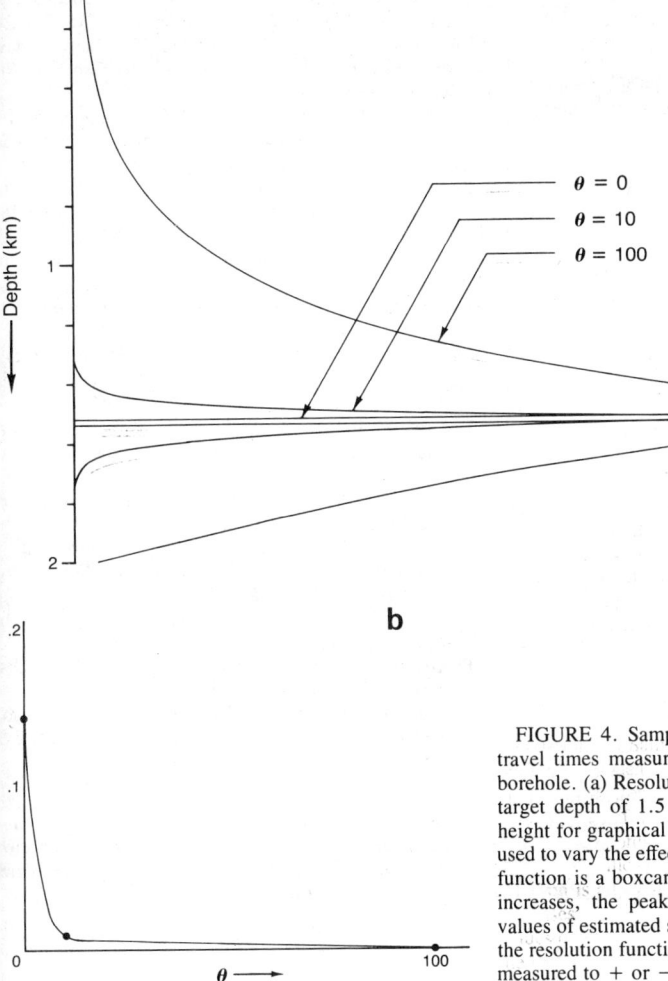

FIGURE 4. Sample resolution characteristics for VSP travel times measured at 10-m intervals in a 2-km deep borehole. (a) Resolution functions for three solutions for a target depth of 1.5 km. Curves are normalized by peak height for graphical purposes. θ is a parameter that can be used to vary the effective peak width. The $\theta = 0$ resolution function is a boxcar function with a width of 10 m. As θ increases, the peaks broaden. (b) Shows corresponding values of estimated statistical uncertainties associated with the resolution functions shown in (a) when travel times are measured to $+$ or $-$ 1 ms.

almost impossible to compute, and a complete or analysis is out of the question. A case in point three-dimensional travel time inversions. Finding olution requires m^2 integrations on a three-dimennal grid followed by inversion of an $m \times m$ atrix. The problem becomes computationally possible before m can be made large enough to tain reasonable resolution.

actical Guidelines

In this article I have chosen to group inverse ocedures used in seismology into three classes: (1) alytic solution, (2) discrete methods, and (3) ntinuous methods. If each class is broken down in tail, one finds a virtually infinite variety of lutions. This may be bewildering, but it really is ly a reflection of the fundamental ambiguity of any al inverse problem. If you find yourself lost in the inversion jungle, I offer the following practical advice.

1. If an analytic solution is available, use it. Bear in mind, however, that real data are inadequate and inaccurate. Any analytic solution that does not bow to this fact of life should be viewed with suspicion.
2. If no analytic solution exists, a continuous solution is the next best approach, provided such a technique can be made computationally tractable.
3. In many real problems the only alternative is a discrete parameter solution. Be aware of the inherent limitations imposed by the driving need to discretize the problem and be creative in appraising the influence of the often arbitrary parameterization.

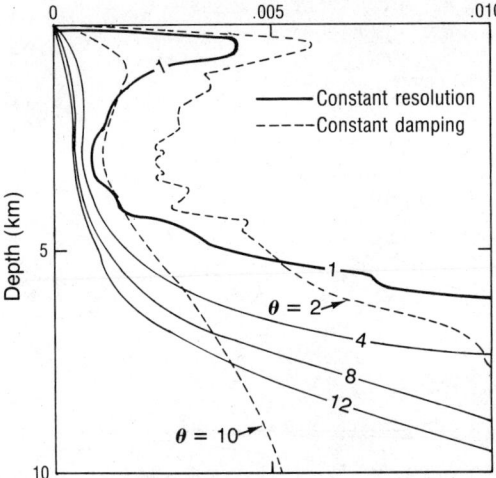

FIGURE 5. Example of complete set of trade-off curves. These curves show trade-off characteristics for a set of earthquake arrival times inverted for P-wave velocity structure. The horizontal axis is comparable to the vertical axis in Fig. 4. The solid lines are contours of a measure of the peak width of resolving function like those in Fig. 4. The dashed lines show corresponding values for a family of solutions with a constant θ.

Above all, no matter what solution one chooses to focus on, the inversion analyst must keep one fact in mind. Finding *a* solution means nothing. The focus of a good analysis must lie with appraising the nonuniqueness of the solution.

GARY L. PAVLIS

References

Backus, G. E., 1970, Inference from inadequate a inaccurate data, II, *Natl. Acad. Sci. Proc.* **65**, 281-2

Backus, G. E., and F. Gilbert, 1968, The resolving pov of gross earth data, *Royal Astron. Soc. Geophys. Jou* **16**, 169-205.

Backus, G. E., and F. Gilbert, 1970, Uniqueness in inversion of inaccurate gross earth data, *Royal S London Philos. Trans. ser. A*, **266**, 123-192.

Ben-Israel, A., and T. N. E. Greville, 1974, *Generaliz Inverses: Theory and Applications*. New York: Wile

Gerver, M., and V. Markushevich, 1966, Determinati of a seismic wave velocity from the travel-time curv *Royal Astron. Soc. Geophys. Jour.* **11**, 165-173.

Lanczos, C., 1961, *Linear Differential Operators*. N York: Van Nostrand.

Lawson, C. H., and R. J. Hanson, 1974, *Solving Le Squares Problems*. Englewood Cliffs, N.J.: Prenti Hall.

Lee, W. H. K., and S. W. Stewart, 1981, Principles a applications of microearthquake networks, *Advances Geophysics*, vol. 23. New York: Academic.

Menke, W., 1984, *Geophysical Data Analysis: Discr Inverse Theory*. New York: Academic.

Newton, R. G., 1981, Inversion of reflection data layered media: a review of exact methods, *Royal Astro Soc. Geophys. Jour.* **65**, 191-215.

Parker, R. L., 1977, Understanding inverse theory, *A Rev. Earth Planetary Sci.* **5**, 35-64.

Pavlis, G. L., 1982, *Progressive Inversion*, Ph.D. Dis University of Washington, Seattle.

Thurber, C. H., and K. Aki, 1987, Three-dimensio seismic imaging, *Ann. Rev. Earth Planetary Sci.* 115-139.

Cross-references: *Earthquakes: Location Techniqu Earth Structure: Global; Finite-Difference Forwa Modeling in Seismology; Seismic Imaging; Seism Tomography; Seismic Wavefield Migration.*

L

LITHOSPHERE: MECHANICAL PROPERTIES

The lithosphere is the strong layer of the Earth that allows the plates to move as coherent units. Its mechanical properties govern the response of the crust to the underlying convective processes that drive plate tectonics and may also dictate the style of deformation in the vicinity of plate boundaries. Normally consisting of both crust and uppermost mantle layers, the mechanical properties of the lithosphere are governed by temperature, composition, and state of stress. In the last few years, there has been a remarkable convergence toward a coherent picture of the nature of the lithosphere that is based on such diverse approaches as laboratory studies of rock deformation, earthquake depths and focal mechanisms in a variety of intraplate settings, the evolution of sedimentary basins, the topographic response to loading of the lithosphere by the formation of seamounts and volcanic islands, the shape of oceanic trenches and outer rises, and the statistical relationship between topography and gravity. Although it is has long been known that the surface layers of the Earth are strong enough to support large loads without complete, local isostatic compensation, our understanding of the factors controlling the rigidity of the lithosphere in different tectonic settings is now approaching the point where we can make useful quantitative predictions.

Rheology of the Lithosphere

In some settings, particularly in continents, there may actually be two or more lithospheres, i.e., the lithosphere may comprise alternating strong and weak layers. Laboratory studies of rheology show that at low temperatures rocks deform in a brittle fashion. The yield strength increases with pressure and is greater under horizontal compression than under extension (Fig. 1). The uppermost part of the crust may be quite weak and incapable of supporting much stress. With increasing depth, the yield strength is large enough that the lithosphere will respond as an elastic plate to loading. The flexural response of the lithosphere to an applied vertical load depends, however, on whether there are any applied horizontal loads. These can bring the lithosphere close to the yield point, thinning the elastic core of the plate, so that an additional applied load results in greater deformation than would occur in an unstressed state. At higher temperatures, the rock undergoes a transition to ductile behavior. At some depth within the Earth, the temperature becomes great enough that the mode of deformation switches to ductile flow. The stress accompanying ductile deformation decreases with further increases in temperature and increases with increasing strain rate. Thus, for a given strain rate, a stress envelope for the lithosphere can be defined (Fig. 1). Stratification of the lithosphere with one or more weak crustal asthenospheres sandwiched between strong layers is possible because the pressure and temperature of the brittle-ductile transition depends on the composition and mineralogy of the rock. Olivine retains strength to temperatures as high as 1000°C at high strain rates, while typical crustal rocks yield ductilely at much lower temperatures. With a thick crust or sufficiently steep temperature gradient, the upper crust should be in the brittle regime, the lower crust should yield ductilely at low stresses, and the top of the

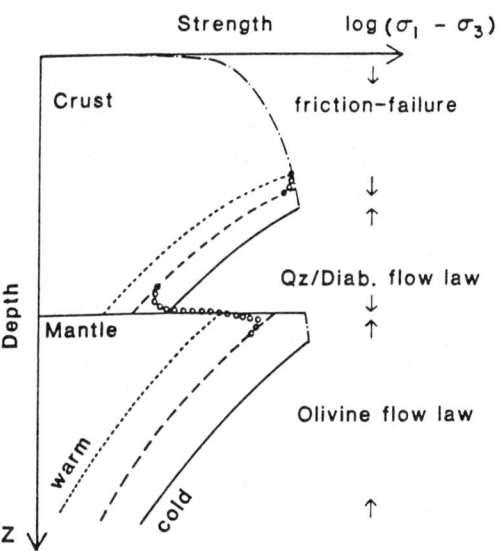

FIGURE 1. Schematic illustration of the variation of the mechanical strength of the lithosphere as a function of depth for high (warm), low (cold) and normal (dashed line) geothermal gradients. Brittle failure may or may not occur in the mantle portion of the lithosphere, depending on the thickness of the crust and the geothermal gradient. The apparent strength or stress level in the regions governed by creep or flow also is a function of the strain rate. (After Chen and Molnar, 1983)

olivine-rich mantle should again be within the strong, brittle regime. Eventually, with increasing depth, the mantle will also pass into the ductile regime at the top of the mantle asthenosphere. If the continental crust is internally stratified in composition, with quartz-rich rocks near the surface and plagioclase and pyroxene-rich rocks at greater depth, then more than one weak crustal layer is possible. (For more information on the rheological behavior of rocks, see *Deformation of Rocks and Minerals* and *Brittle Phenomena* in this volume).

A weak, intracrustal layer may serve as a detachment surface for normal faults, such as in the Basin and Range province; it may allow delamination of the mantle lithosphere from the crust as has been suggested for the Colorado plateau and in general may be responsible for thin-skin tectonics, such as the extensive underthrusting of one crustal layer beneath another in the southern Appalachians or the extensive folding of the upper crust in the Zagros region (see *Thin-Skin Tectonics*). It is less likely to occur in the oceanic lithosphere, because the brittle-ductile transition will not take place within the thin oceanic crust unless there is a steep thermal gradient and very high heat flow. However, the fact that there are ophiolite sequences that do not include an ultramafic layer suggests that it is possible to detach the upper oceanic crust from the mantle. The existence of two scales of deformation within the Central Indian Ocean Basin, one a pattern of faulting and block rotation on a 5 to 15 km scale and the other a 100 to 300 km wavelength undulation in the seafloor also indicates that there may be a weak layer in the lower oceanic crust. Water or hot fluid circulation within the crust could aid deformation and create weak layers under temperature and pressure conditions that would ordinarily be associated with elastic behavior.

In many situations, the mechanical response of the lithosphere can be successfully represented by the simple model of an elastic plate overlying an inviscid asthenosphere. The bending of the plate in response to an applied surface or subsurface load is then dependent on a single parameter, the flexural rigidity. The flexural rigidity, D, is related to the thickness of the elastic plate, h, by the expression

$$D = \frac{Eh^3}{12(1 - v^2)}$$

where E and v are Young's modulus and Poisson's ratio. Although this model is clearly an oversimplification of the true rheological behavior, so that the apparent elastic thickness cannot be directly interpreted in terms of depth to the asthenosphere, it provides an adequate description of many geological features such as the shape of sedimentary basins. If earthquakes or fault structures are observed, for example, then a more complicated rheological behavior can be inferred. If only the shape of the topographic response to the load is known, then the lithospheric response is normally described in terms of the effective elastic thickness, recognizing the limitations of both the data and the model.

Oceanic Lithosphere

The effective elastic thickness of the oceanic lithosphere increases roughly in proportion to the square root of the age of the seafloor (Fig. 2). This pattern is expected if the base of the mechanical lithosphere is controlled by an isotherm that deepens as the crust and upper mantle cools as it moves away from the midocean spreading centers. If the elastic thickness is interpreted simply as depth to base of the lithosphere, then the governing isotherm is in the 400° to 500°C range, which may be an underestimate for several reasons. First, the oceanic crust may be so weak that it contributes little to the rigidity. Second, the flexural rigidity is often estimated from the flexure and degree of local isostatic compensation associated with loading the plate by the formation of volcanic seamounts. Once formed, the seamount and flexure are in equilibrium and presumably frozen into the lithosphere as it continues to cool, so the apparent rigidity is assumed to describe the lithosphere at the time of formation of the seamount rather than at the current age. However, some reheating and softening of the lithosphere from mantle hotspots probably accompanies most seamount formation, so there may be a systematic bias towards underestimating the rigidity. The elastic thickness is also estimated from the bending of plates before subduction at oceanic trenches. This flexure is accompanied by many earthquakes whose cumulative seismic moment indicates that a large fraction of the deformation is nonelastic, so that the apparent thickness of the elastic core of the lithosphere must be significantly less than the depth to the asthenosphere. If the elastic thickness is corrected for these effects, the apparent transition temperature from seamount loading and bending of the plate at trenches is about 600° to 700°C, in agreement with the apparent growth of the plate in the immediate vicinity of the ridge axis. See also *Lithosphere, Oceanic: Formation and Evolution* for further discussion of the question of lithosphere thickness.

The depth of intraplate earthquakes provides a constraint on the depth to the brittle-ductile transition, although the strain rate dependence of the transition must be kept in mind. The maximum depth of well-located events beneath the outer rise before old oceanic lithosphere subducts is 45 to 50 km. A similar maximum is found for the larger events near the island of Hawaii and for intraplate events in the Central Indian Basin and near the Azores-Gibralter diffuse plate boundary. The maximum depth of intraplate events increases roughly in proportion to the square root of the age of the lithosphere.

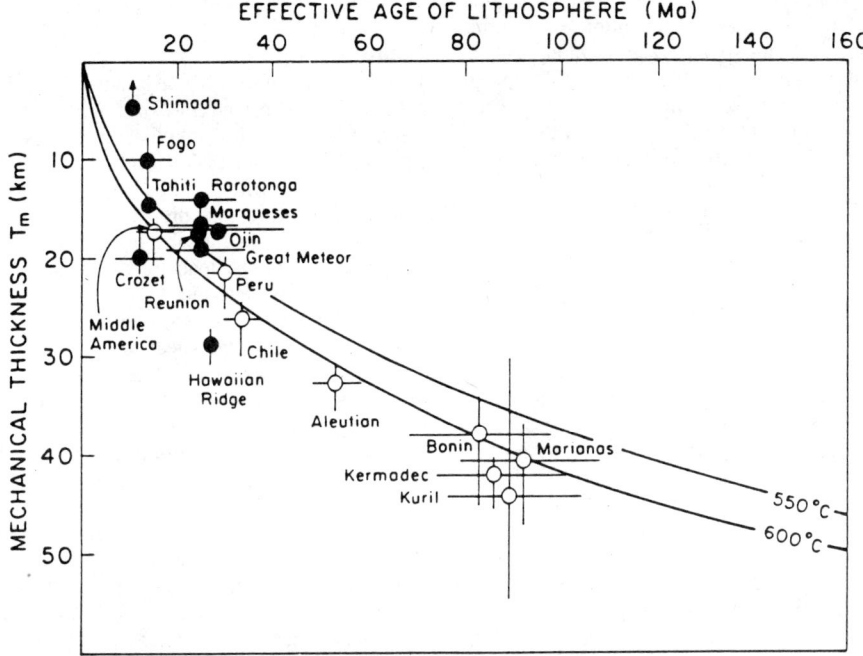

FIGURE 2. Apparent elastic thickness of the oceanic lithosphere as a function of the effective age of the plate (after McNutt, 1984). The age is corrected for the effects of reheating the oceanic lithosphere by hot spots, and the elastic thickness is corrected for the effects of exceeding the yield stress. Isotherms are based on a cooling plate model with bottom boundary at 125 km.

Continental Lithosphere

Within continents, there is a wide range of apparent elastic thicknesses. In extensional environments, the effective elastic thickness may be a few km or less. In the Los Angeles basin, a pull-apart basin that is part of the ridge-transform system of the Gulf of California that extends onto land, the response is indistinguishable from completely local isostasy, equivalent to zero flexural rigidity. In the development of the Anadarko basin in another extensional setting, the apparent thickness was initially very small, but gradually increased with time to about 10 or 15 km as sediments continued to be deposited, presumably because the lithosphere was cooling and becoming more rigid after the active extension ceased. In the Basin and Range province in the United States the pattern of uplift and subsidence accompanying the Borah Peak earthquake of 1983 suggests an elastic plate thickness of only a few km. It is possible, however, that this estimate reflects only the rigidity of an upper crustal layer that begins a few km beneath the weak surface layer; the total rigidity of the entire lithosphere may be greater. Deformation associated with loading and unloading of the glacial Lake Bonneville in the eastern Basin and Range suggests elastic thicknesses on the order of 20 km. Not all extensional regimes are characterized by very low flexural rigidities. Effective elastic thicknesses of the East African rift system range from 20 to 40 km, although the total amount of extension in East Africa is very small. The depth extent of earthquakes is consistent with these differences in rigidity. In California and the Basin and Range provinces, well-located events are shallower than 10 to 15 km, while in East Africa, earthquakes in the vicinity of the rifts occur as deep as 35 km. In these extensional environments, the apparent rigidity seems to be controlled both by the thermal state of the lithosphere and by the relatively easy movement allowed on normal faults. The blocks between faults may be stiff, but the overall response in terms of isostatic state may resemble local isostasy if the level of stress required for slip on the faults is low. Clearly, however, the large regions of lowest rigidity are those with the highest heat flow.

In intraplate regions of the continents, rigidity is greater than in the extensional regions. The nature of the subsurface loads responsible for the subsidence of the Michigan and Williston basins is not known, but several investigators have suggested that their formation is triggered by the passage of the lithosphere over hot spots. Whatever the trigger may have been, the basins have subsided gradually with the apparent elastic thickness estimated from the shape of depositional surfaces increasing from about

30 km to about 75 km with a time constant of about 200 Ma. Again, the thermal evolution of the lithosphere following hot-spot passage is thought to be the controlling process for plate rigidity. In eastern Australia, there has been a progression of volcanic activity from north to south over the last 30 to 40 Ma, with very recent activity in the southeast. This pattern suggests the passage of Australia over a hot spot. In keeping with the model of lithospheric thinning by hot-spot activity, followed by a long period of cooling and thickening, the apparent elastic thickness of southeastern Australia is about 25 km, increasing to about 40 km in the northeast. The effective elastic thickness of most of the rest of Australia is on the order of 100 km.

In continent-continent convergence zones, the shapes of foreland sedimentary basins associated with collisional mountain belts such as the Himalayas, Alps, and Appalachians indicate that the apparent elastic thickness of continents can exceed 100 km. In these regions, it appears that subsurface loads other than the weight of the mountains themselves are partially responsible for maintaining the depression of the plate after collision ceases (see *Continental Collision Zones: Seismotectonics and Crustal Structure*). Continental collision zones are analogous to oceanic trenches, so it may be that portions of the subducted mantle lithosphere remain attached to the continental lithosphere and act as buried loads (Fig. 3). Because the state of stress and the rheological stratification of the continental lithosphere affects the flexural lithosphere, the apparent elastic thickness does not mark the depth to a particular isotherm. Overall, the greater maximum rigidity of the continental lithosphere strongly supports the hypothesis that continental roots may extend much deeper than the base of the oceanic lithosphere.

Coherence Between Gravity and Topography

Most of the estimates of continental rigidity described earlier are based on the shapes of sedimentary basins and horizons. Another approach based on the statistical relationship between topography and gravity anomalies is applicable in a wider range of tectonic settings. If a topographic load, such as a volcanic edifice, is applied to the surface of an elastic plate, it will flex the plate providing some compensation for the load by downward displacement of density interfaces, such as the Moho, beneath the load. The Bouguer gravity anomaly, having corrected for the effects of the surface topography itself, represents the attraction of lateral, subsurface variations in density, such as those created by the flexural response to surface loads. The flexure thus produces a Bouguer anomaly that is directly correlated to surface topography. Similarly, an applied subsurface load, such as a volcanic intrusion or a density anomaly associated with reheating of the lithosphere, will deflect the plate, creating a correlated topographic signal. At wavelengths long compared to the characteristic flexural wavelength of the elastic plate, both surface and subsurface loads will be fully compensated (local isostatic compensation) producing similar relationships between topography and gravity anomalies. In this case, the coherence, or square of the correlation coefficient, between topography and gravity will be close to unity. At short wavelengths, the lithosphere supports the loads so that neither surface nor subsurface loads produces a compensating deflection of the plate. In this case, the relationship between the gravity anomalies produced by subsurface loads and topography produced by surface processes will be incoherent unless surface and subsurface loads are systematically correlated. In the wavenumber

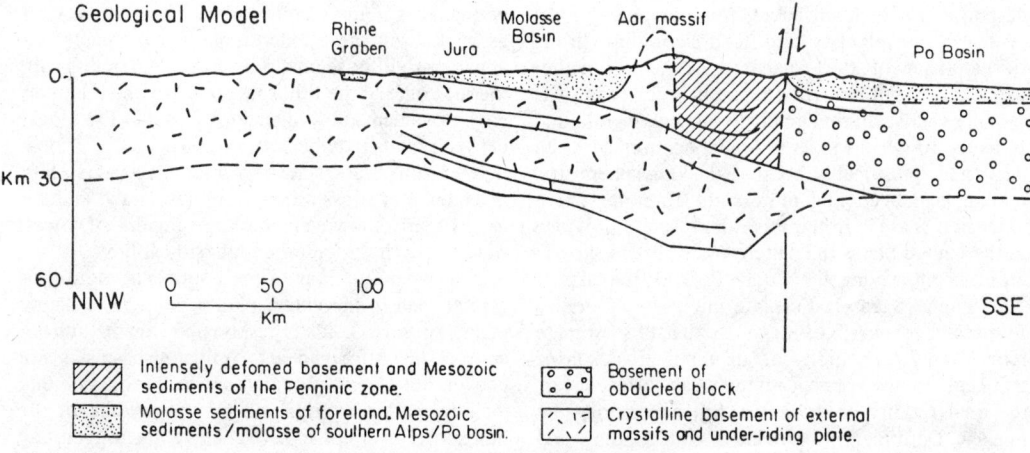

FIGURE 3. Schematic model of the flexure of the lithosphere in a continental collision zone (after Karner and Watts, 1983). In this cross section across the Alps, the Molasse Basin represents a foreland basin that forms on the down-flexed, subducting plate. The width of the basin depends on the flexural rigidity of the underthrust plate.

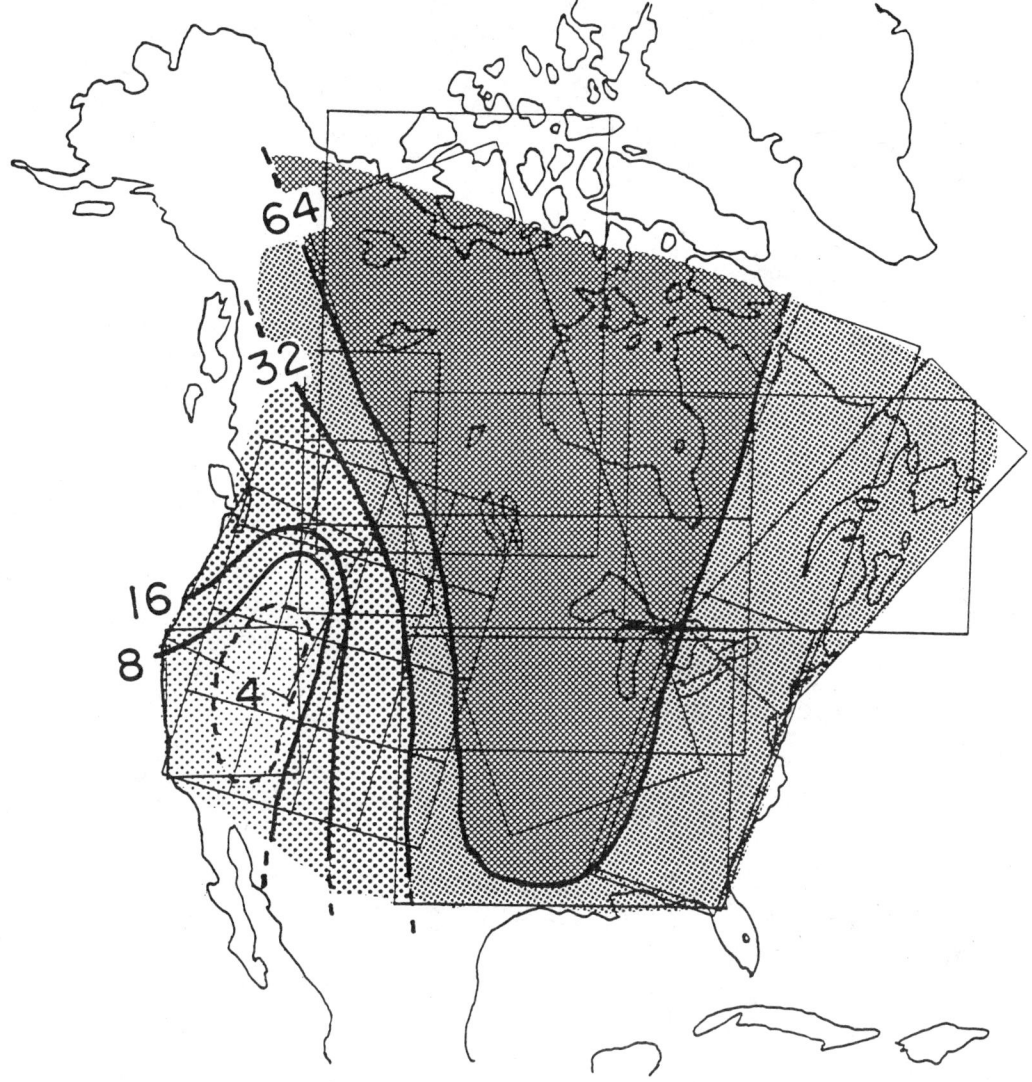

FIGURE 4. Contour map of variations in effective elastic thickness within the North American continent (after Bechtel and Forsyth, in preparation). Elastic thicknesses (in km) are based on the variation in wavelength between a coherent and an incoherent relationship between topography and Bouguer gravity anomalies. Rectangular regions represent some of the subareas analysed for coherence.

domain, the transition between a coherent and incoherent relationship between topography and Bouguer gravity anomaly occurs at around the characteristic flexural wavelength, from which the apparent elastic thickness can be deduced.

Maps of the variations in apparent elastic thickness in North America, Australia, and much of Africa and Europe have been produced using the technique of determining the coherence between gravity and topography as a function of wavelength. In each continent, much of the stable continental interior shields and platforms approach or exceed 100 km in effective elastic thickness. In North America (Fig. 4), the rigidity decreases to between 32 and 64 km along the eastern and Gulf coasts, areas of more recent tectonic activity. Extremely low values of rigidity are found in the center of the Basin and Range province, with apparent elastic plate thickness approaching 4 km and the transition between coherent and incoherent topography and gravity occurring between 70 km and 200 km. This means that individual basins and ranges are largely supported by stresses within the lithosphere, but that the dominant 200 km wavelength Bouguer anomaly signal corresponds to compensated topography. There is also some azimuthal dependence of the coherence, indicating that slip on faults may play an important role in achieving isostatic compensation.

Conclusions

Although the actual rheological behavior of the lithosphere can be quite complex, to a first approximation, the mechanical response of the lithosphere to loads can be represented by the flexure of an elastic plate. The effective elastic thickness of the oceanic lithosphere is primarily a function of the temperature structure and thus varies systematically with the age of the seafloor. The effective elastic thickness of the continental lithosphere varies from a few km in extensional tectonic provinces with high heat flow to about 100 km in stable continental interiors.

DONALD W. FORSYTH

References

Chen, W.-P., and P. Molnar, 1983, Focal depths of intracontinental and intraplate earthquakes and their implications for the thermal and mechanical properties of the lithosphere, *Jour. Geophys. Research* **88**, 4183–4214.

Forsyth, D. W., 1985, Subsurface loading and estimates of the flexural rigidity of continental lithosphere, *Jour. Geophys. Research* **90**, 12623–12632.

Karner, G. D., and A. B. Watts, 1983, Gravity anomalies and flexure of the lithosphere at mountain ranges, *Jour. Geophys. Research* **88**, 10449–10474.

Kirby, S. H., 1983, Rheology of the lithosphere, *Rev. Geophysics Space Physics* **21**, 1458–1487.

Kirby, S. H., and A. K. Kronenberg, 1987, Rheology of the lithosphere: Selected Topics, *Rev. Geophysics* **25**, 1219–1244.

Kuo, B. Y., D. W. Forsyth, and E. M. Parmentier, 1986, Flexure and thickening of the lithosphere at the East Pacific Rise, *Geophys. Research Letters* **13**, 681–684.

McNutt, M. K., 1984, Lithospheric flexure and thermal anomalies, *Jour. Geophys. Research* **89**, 11180–11194.

McNutt, M., 1987, Lithospheric stress and deformation, *Rev. Geophysics* **25**, 1245–1253.

Watts, A. B., J. H. Bodine, and M. S. Steckler, 1980, Observations of flexure and the state of stress in the oceanic lithosphere, *Jour. Geophys. Research* **85**, 6369–6376.

Wiens, D. A., and S. Stein, 1985, Implications of oceanic intraplate seismicity for plate stresses, driving forces, and rheology, *Tectonophysics* **116**, 143–162.

Cross-references: *Continental Collision Zones: Seismotectonics and Crustal Structure; Continental Crustal Structure; Continental Drilling: Ultradeep; Continental Lithosphere; Crustal Movements and Tectonic Deformation; Deformation of Rocks and Minerals; Earthquakes and Crustal Deformation; Earthquake Seismology; Gravity and Isostasy; Lithosphere, Oceanic: Formation and Evolution; Mantle, Upper: Structure; Mantle Convection and Plumes; Mantle Dynamics; Mantle Viscosity; Seismicity: Intraplate; Stress in the Earth's Lithosphere; Thin-Skin Tectonics.*

LITHOSPHERE, OCEANIC: FORMATION AND EVOLUTION

The structure of the oceanic crust and lithosphere is most striking in its gross uniformity across the oceans. While many parts of the ocean basins, such as seamounts and plateaus, are anomalous, most of the seafloor is underlain by a surprisingly homogeneous crust which implies that the processes acting to form this structure are similar throughout the world's oceans. These processes can be divided between those that contribute to the formation of the lithosphere and those that contribute to its evolution. The former take place at the mid-ocean ridges and involve a complex interaction of magmatic, tectonic, and hydrothermal phenomena. As the lithosphere ages these phenomena become less important and conductive cooling takes over as the dominant process acting beneath the seafloor. It is this regime that we shall term the evolutional phase of the seafloor spreading.

A discussion of the structure and certainly the evolution of the oceanic lithosphere would have been pointless as little as thirty years ago. Although the term lithosphere had been introduced by Joseph Barrell in 1914 to describe isostatic compensation mechanisms, the idea that the Earth beneath the oceans might consist of a young, rigid, and evolving carapace was not anticipated. Even as late as 1950, when the major geological expeditions from the Scripps Institution of Oceanography, Woods Hole Oceanographic Institution, Columbia University, and the University of Cambridge, began to put to sea, the composition of the seafloor and the thickness of the oceanic crust were entirely unknown. Our level of knowledge of that seafloor and lithosphere has changed dramatically since that time. Plate tectonics, which was largely developed during the 1960s, forced geologists and geophysicists to view the seafloor as an assemblage of rigid plates that could transmit stresses over enormous distances. The lithosphere floats upon a much more poorly defined plastic or fluid asthenosphere. Decades of objections to continental drift and seafloor spreading, which were based on well-reasoned ideas about the strength of rocks, disappeared. The oceanic lithosphere is now known to be formed along the midocean ridges and rises within extraordinarily localized volcanic zones. This lithosphere spreads away from these axes and is cooled by conductive and convective processes. The cooling plates increase in thickness and density and sink slowly to abyssal depths. The oceanic crust, which constitutes the upper portion of the lithosphere, is fully formed within the first few hundreds of thousands of years of its generation, most probably by cooling of a substantial crustal magma chamber. Although phenomenological descriptions of many of these seafloor processes that affect the evolution of the oceanic lithosphere are now known, the dynamic systems that control the aggregation of melt and the growth and translation of the lithosphere remain as problems for future generations of marine geologists.

An elucidation of the structure of the oceanic crust and uppermost mantle lagged well behind the study of the continental crust in the first half of the twentieth century. The seismic structure of the continents was initially studied using local and regional recordings of earthquakes in Europe. Similar experiments in the ocean were precluded simply because of an insufficient density of islands to support seismic observatories. Early studies of the gravity field in the oceans revealed that the field over the oceans, except for the deep trenches, was very similar to that measured on continents. The principle of isostasy required that oceanic and continental crustal columns be in hydrostatic equilibrium at some depth which, for the sake of argument, will be taken as 80 km. Given the low density of the water in the oceans, isostasy simply required that the light upper layers of the continental crust be replaced with substantially denser material in the oceanic crust. Sir Harold Jeffreys argued that the oceanic crust must be quite thin and that the composition of the rocks should be intermediate between granite, which characterizes the upper portion of the continents, and the supposed mantle dunite composition. Very few samples of the oceanic crust were available to test this hypothesis and islands were composed of rocks ranging from basaltic (Hawaii) through andesitic (Tonga) to granitic (Fiji) compositions.

Prior to 1950, the only seismic method that was successful in outlining the structure of the oceanic crust and uppermost mantle was the analysis of surface waves from earthquakes that propagated across an expanse of ocean basin. Surface waves are confined to the outer portions of the Earth with little seismic displacement at depths in excess of a few hundred kilometers. Generally, the compressional and shear velocities increase as a function of depth so that the lower frequency surface waves, with oscillations that extend to great depth, arrive before the higher frequency portions of the wavetrain. This phenomenon is generally termed normal dispersion and is easily observed for earthquakes that have propagated across continental structures. Early work with surface waves that crossed ocean basins indicated that the degree of dispersion was a great deal less than on continents, and this was largely explained by either reducing the thickness of the granitic layer in the oceanic crust to something like 10 km with high velocity dunite below or by eliminating the crust altogether at the seafloor.

Maurice Ewing went to sea for the first time in 1935 to introduce explosion or seismic refraction methods for the study of the seafloor. The early years of refraction were plagued by technical problems and a general lack of sea time. The war interceded and the study of the ocean basins was delayed until the late 1940s. Seismic refraction studies of the seafloor were, by this time, being pursued not only by Ewing, by this time at Columbia, but by the Scripps Institution of Oceanography and Cambridge University (Menard, 1986). Dr. Russell Raitt, at Scripps, reported the first complete seismic sections of the thin oceanic crust off southern California at a meeting of the Geological Society of America in El Paso in 1949. He expanded on this initial success with many new measurements on Scripps expeditions that were mounted to the north and south Pacific in the period 1950-1953. Raitt (1963) published these and other available results years later in support of his model of the average oceanic crust (Table 1).

TABLE 1. Raitt's Seismic Sections Results

Layer	Velocity (km/s)	Thickness (km)
2	5.07 ± 0.63	1.71 ± 0.75
3	6.69 ± 0.26	4.86 ± 1.42
4	8.13 ± 0.24	—

Layer 1, which is not included in this table, consists of seafloor sediments that are variable in thickness as well as velocity. The uniformity of the velocity of the third layer was striking as was the total thickness of the oceanic crustal column. Any models for the generation of the oceanic crust must necessarily account for these impressive uniformities.

Subsequent surface wave studies indicated that the explosion seismology model of the oceanic crust was consistent with the average structure proposed by Raitt. Earlier investigators had not carried their analyses to frequencies high enough to be affected by the thickness of the water column and the composition of the thin crust. In other cases, scientists had simply disregarded the effect of the overlying fluid and used methods that had been developed for studying the structure of continents.

The choice of a layered velocity structure to describe the oceanic crust arose from the observation that the arrival times, when plotted as a function of distance between the source and receiver, generally fell along several straight lines. From the slopes of these straight lines seismologists could infer the velocities of discrete layers in the Earth. The intercepts of these straight lines with the time axis provided information on the delays within each layer and permitted the thicknesses of the layers to be simply calculated. Although the net result of this approach to the problem was a velocity—depth function that contained as many homogeneous layers as straight line segments in the travel time plot, it is clear that the homogeneity of the observed layers and the existence of discrete interfaces between them are artifacts of the analysis. Raitt (1956) stated:

Calculation of the velocity-depth structure from the observed travel times was based on a simplified model of layers of constant velocity as determined from the travel-time plots. . . . This model represents a simplification of the actual situation, where velocities probably vary continuously with depth and possibly with horizontal distance, but it is a rough approximation and provides a basis for comparison of results.

The layer solution approach of analysis of oceanic as well as continental seismic data was used well into the 1970s. The Layer 2 of Raitt was subdivided into Layers 2A, 2B, and 2C by Houtz and Ewing (1976). 2A was characterized by a very low velocity for basalts and was largely confined to the young lithosphere. Some researchers found evidence that a high velocity layer might exist at the lowest point in the crust.

A new approach to the analysis of marine refraction data became available with the development of techniques for computing synthetic seismograms. Not only must these synthetic seismograms match the observed travel times, but a reasonable match must also be achieved to the observed amplitudes and phases. Travel time inversion techniques were also improved to allow a less restricted set of velocity models. The confluence of these two new approaches has led to a greatly modified view of the nature of the oceanic crust (Spudich and Orcutt, 1980). Layer 2 is presently viewed as a region of very high velocity gradient although the precise form of this velocity increase is still not known. The replacement of individual layers with a gradient in velocity is reasonable given that the increase in velocity is almost certainly associated with increasing lithostatic pressure and the closing of pores. There is growing evidence from modern seismic refraction experiments using seafloor receivers and towed hydrophone arrays that a very high gradient or even a discontinuity separates the upper gradient zone from the more homogeneous central crust or Layer 3. Layer 3 is generally characterized by very low velocity gradients with the result that the classical layer solution methods provided quite a good picture of the structure. Little evidence persists to support the idea of a high velocity basal layer (3b) immediately above mantle (see Fig. 1).

Early in the 1960s, Harry Hess, Russell Raitt, George Shor, and others discovered that the uppermost mantle was anisotropic. This anisotropy is usually attributed to olivine crystals that have been aligned by tectonic strains in the uppermost mantle. More recently, studies of travel times, P-wave polarizations and shear wave splitting have led to hypotheses of upper crustal anisotropy. Measurements of crustal and mantle anisotropy have been made at a site in the South Pacific where the directions of anisotropy were found to be at right angles (Shearer and Orcutt, 1985). Upper mantle anisotropy is likely associated with the alignment of

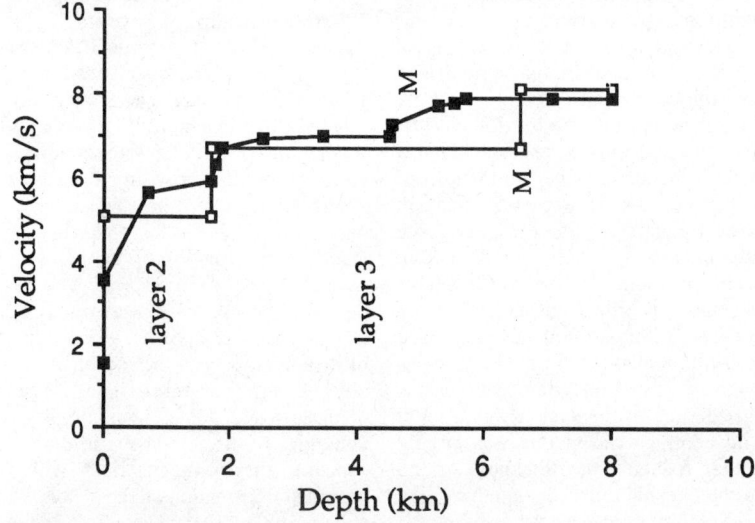

FIGURE 1. A comparison of the compressional-wave velocity structure of the oceanic crust determined from the use of synthetic seismograms (solid squares), with the average layered structure (open squares) given by Raitt (1963). Note that the upper crust (layer 2) has high velocity gradients in the modern structure, while the lower crust (layer 3) is similar in the two models. The Moho (M) and other boundaries in the crust have been shown by the modern techniques to be gradational transitional zones rather than sharp boundaries. The difference in the total thickness of the crusts is not significant.

olivine crystals in the spreading direction while in the upper crust anisotropy is associated with cracks perpendicular to the spreading direction that form through faulting in young crust. The directions of anisotropy are normal to one another and future measurements will undoubtedly settle this issue.

A number of heroic efforts at long range explosion profiles have been mounted over the years in the Gulf of Mexico and in the Pacific and Atlantic oceans. These profiles in the Pacific and the Atlantic provided information on the compressional velocity structure of the uppermost mantle. In general, all these profiles failed to detect a significant velocity reversal at depths greater than 65 km that might be associated with a boundary between the lithosphere and asthenosphere. Most profiles support the proposition that velocity increases at a depth of about 55 km to values in excess of 8.4 km/s. Generally, the derived velocity profiles show alternate high and low velocities below the Moho although these variations can as easily be ascribed to small heterogeneities along the path of propagation. The velocity gradient below the Moho is so slight that the conduct of refraction experiments that rely upon increasing velocities to return energy to the surface are not particularly effective at delineating lithosphere structure.

Rather recently, analogies have been drawn between ophiolites and the oceanic lithosphere. The current ophiolite hypothesis postulates that ophiolites are segments of ocean crust that have been emplaced on land. This theory is generally accepted although the origin of the oceanic structures, whether normal ocean crust, back-arc basin crust, or island arc material, is almost certainly tied to processes associated with subduction zones (e.g., Moores, 1982). In spite of a lack of understanding of the origin of these ophiolites, it is generally agreed that the rocks have an oceanic origin and that the ophiolites can be used to learn about the largely inaccessible ocean crust. Although the stratigraphy of these ophiolites is very variable, the uppermost section consists of oceanic pillow basalts adjacent to sedimentary rocks with marine origins. The pillow basalts are underlain by a set of feeder dikes termed *sheeted dikes*. These dikes are generally characterized by a single chilled margin, which apparently arose as the central dike was subsequently intruded in the spreading process. The sheeted dikes are, in turn, juxtaposed upon gabbros in the central part of the ophiolite. These isotropic gabbros give way to more mafic assemblages characterized by cumulate textures. These mafic cumulates are underlain by ultramafic cumulates that are, in turn, underlain by harzburgites, which are probably typical of the uppermost mantle. The ophiolite model has been quite successful in explaining many aspects of the oceanic lithosphere and crust and the structures that are predicted are frequently testable with remote, geophysical methods. Spudich and Orcutt (1980) compared a detailed seismic study of young Pacific crust with velocities of samples from the well-studied Bay of Islands ophiolite suite in Newfoundland (Christensen and Salisbury, 1975). Although the overall thicknesses of the seismic and reconstructed ophiolite differed, the overall stratigraphy and velocities of the ophiolite compared very favorably with the detailed seismic measurements. In broad outline, the upper oceanic crust characterized by a high gradient in velocity corresponds to the pillow basalts and sheeted dikes while Layer 3 is associated with the gabbros. In the oceans the Mohorovičić discontinuity is a smooth transition. This transition can be related to the increasing ultramafic cumulates at the base of the ophiolite. The oceanic crustal rock suite is frequently attributed to differentiation and fractionation of a molten, central crustal magma chamber. This picture is enormously simplified, however, in that a molten chamber will convect vigorously, and crystal settling to form the cumulates at the base of the crust simply cannot occur.

The Formation of the Oceanic Crust

The study of the formation of the oceanic crust and uppermost mantle has included seismic studies, petrological studies of basalts dredged from the seafloor, geological studies of ophiolites, bathymetric and submersible observations, and numerical modeling. These studies have resulted in a remarkably consistent, primarily two-dimensional, picture of seafloor spreading. The major elements of this model include the upwelling and subsequent partial melting of mantle material in response to the spreading of the ocean plates; the separation and rise of the mafic melt leaving behind a depleted ultramafic uppermost mantle; the further differentiation of the melt in a shallow (lower crustal) magma chamber; the eruption of the magma to form seafloor basalts; the faulting and deformation associated with the extensional tectonics, and the cracking and cooling of the crust by seawater. These processes do not act independently, but all interact and affect each other. However, for convenience, we shall divide the discussion between magmatic, tectonic, and hydrothermal processes.

Magmatic and Volcanic Processes. The magmatic processes under mid-ocean ridges begin in the mantle where the upwelling material undergoes adiabatic decompression, partial melting, and melt aggregation. The anomalously high temperatures and melting result in an uppermost mantle that dramatically attenuates seismic signals; seismic experiments have thus far failed to yield reliable details about the distribution of melt. The results of gravity surveys nearly always require a low density mantle extending over a relatively broad region. The portion of the low density region that is actually melt (as opposed to merely an anomalously hot solid) cannot be determined from gravity measurements alone, and the nature of the upper mantle beneath the ridges remains a fundamentally important avenue for future research.

From petrological and geochemical studies it is known that the melt and residual solid remain in equilibrium until they begin to segregate at depths from 15 km to 40 km. The mafic melt rises to form the crust, leaving behind a primarily harzburgite uppermost mantle. It is these harzburgites that tectonically deform, resulting in a preferred orientation of olivine crystals to form the widely observed mantle anisotropy.

The rising mafic magma apparently undergoes fractionation at very shallow depths. This and the presence of isotropic and cumulate gabbros in ophiolite suites have long been cited as evidence that a magma chamber must exist in the lower oceanic crust beneath mid-ocean spreading centers (e.g., Moores, 1982). Theoretical studies, which are based on numerical modeling of a conductively cooled lithosphere, have yielded predictions concerning the size, shape, and lifetimes of crustal magma chambers. For slow spreading ridges such studies predict a small ephemeral chamber, while ridges spreading at intermediate to fast rates may have extremely large ($>$ 10 km in half-width) and steady-state magma chambers (Sleep, 1975). Even when hydrothermal cooling of the upper crust is included in models, the magma chambers remain surprisingly large, though deeper.

Early seismic studies seemed to confirm the presence of magma chambers under the axes of various segments of the East Pacific Rise (e.g., Orcutt et al., 1975). Some of these experiments put loose upper bounds on the width of the magma chambers, upper bounds that appeared wide enough to be consistent with the numerical and geological models discussed earlier. Under the more slowly spreading Mid-Atlantic Ridge, the experiments indicated that magma chambers were absent or very small. While analyses of seafloor basalts appear to require at least a temporary or fluctuating permanent magma chamber, the seismic results tend to confirm thermal models that indicate the crust in the Atlantic is too cool to support a permanent magma chamber. Resolving this contradiction will be difficult, for the rugged topography of the Mid-Atlantic Ridge makes detection of a small magma chamber extremely difficult.

Thus, as of 1981 the seismic experiments were largely in agreement with the conductively cooled models. Of particular interest were the large magma chambers suggested for intermediate- to fast-spreading ridges. A different view of the seismic structure under the East Pacific Rise began to emerge as a number of investigators argued that seismic observations showed that magma chambers were either extremely small or absent. This view led some to question if long-lived magma chambers can exist under any spreading center. However, the most recent and convincing experiments have confirmed that magma chambers are present under substantial portions of the ridge (McClain et al., 1985). Detrick et al. (1987) have shown that a fairly continuous reflector exists under several hundred kilometers of the East Pacific Rise. In some areas this reflector displays a phase reversal that would indicate that it marks a decrease in velocity, as would be expected at a magma chamber lid. However, these recent studies show that these magma chambers (magma defined as largely molten rock) are narrow ($<$ 3 km half-width), even under the fastest ridges. They may also fluctuate in size (e.g., Macdonald, 1982). The disagreement between the seismic results and most numerical models indicates that the cooling of the magma chamber is far more efficient than the models predict, perhaps because of the dynamics of the magma chamber and/or hydrothermal cooling.

Whatever the characteristics of the magma chamber, it is tapped periodically because of the continued spreading at the axis. The result is the intrusion of magma into the upper crust and extrusion of basalts at the surface. The intrusion probably occurs along narrow conduits that solidify into dikes. With ongoing extension the dikes are rifted apart and intruded by new magma, which results in the sheeted dike units observed in most ophiolites (e.g., Moores, 1982).

From bathymetric and petrological studies there is growing evidence that magma upwelling and volcanism does not occur uniformly in time or along the strike of the axis. Instead, it may be that magma rises through the mantle at isolated loci along the axis. In some cases these loci may be hot spots (Vogt and Johnson, 1975), but it now appears that even normal segments of mid-ocean ridges exhibit systematic variations. Francheteau and Ballard (1983) noted that the East Pacific Rise axis displays long-wavelength (100–200 km) variations in depth that may be tied to magma supply. The more elevated portions of the ridge exhibit a broad axial cross section whereas the deeper segments have a narrow cross section as if they were deficient in magma. Ken Macdonald, Jeff Fox and their other colleagues have suggested that the flow of magma along the axis away from loci can result in overlapping spreading centers (OSCs) if two opposite flows are misaligned.

Morphological, structural, and petrological variations have been well documented along the strike of the Mid-Atlantic Ridge. In general these variations are correlated with the distance from a fracture zone and are usually attributed to the anomalous cooling resulting from the placement of the spreading center adjacent to older, colder lithosphere. The suppressed temperatures result in increased depths, in anomalous basalts, and in an anomalously thinned crust.

Tectonism and Faulting at the Mid-Ocean Ridges. That the mid-ocean ridges are extensional zones where two plates diverge is one of the pillars of plate tectonic theory, and the resulting seismicity provided major supporting evidence for the theory. Earthquakes under the midocean ridges are invariably shallow (< 10 km) and usually small, reflecting the thinness of the "brittle" lithosphere, and hence the elevated geotherms of the spreading centers. The details of the thermal structure have a profound influence on the morphology and tectonics of the ridges, affecting the balance between magma supply and extensional deformation.

The most critical control on the thermal structure beneath various segments of the mid-ocean ridges is the spreading rate. At slower rates, ridges are usually characterized by a substantial rift valley. On the Mid-Atlantic Ridge the valley is as deep as 3 km. As spreading rate increases, the topography becomes more subdued and at the fastest spreading rates (> 100 mm/yr) the ridges are generally characterized by a small axial high (for a more thorough review, see Macdonald, 1982).

Numerous models considering the balance between the supply of magma to the ridge and the mechanics of rifting have been advanced to explain the presence or absence of rift valleys. Control of seafloor topography by the thermal structure of the ridges is a common thread in these models. Hence, factors that influence the thermal structure should also influence the topography. A subtle variation on this theme was first advanced by Vogt and Johnson (1975). Pointing out that magma flow along an axis would be dammed by a fracture zone, they suggested that "downstream" segments of the ridge might be magma starved, while the upstream segment would have excess magma. The slow-spreading Reykjanes Ridge exhibits a topographic high that may be related to magma flowing south from the Iceland hotspot. Similarly, the upstream side of the Blanco Fracture Zone (the Juan de Fuca ridge) displays a topographic high while the Gorda Rise to the south has a prominent rift valley. A virtually identical geometry can be seen at the Tamayo Fracture Zone on the East Pacific Rise.

The importance of thermal structure is also seen in the faulting observed on the ridges. Earthquakes on the slow spreading Mid-Atlantic extend into the mantle, while on faster ridges they may be restricted to the upper crust. The topographic expressions of the faults on the ridges also reflect this difference. On the Mid-Atlantic Ridge the throws on faults are large and provide the actual mechanism for forming the rift valley, while on ridges with intermediate to fast spreading rates the throws are smaller. The latter may be responsible for the formation of the horst and graben "abyssal hills" observed in the Pacific.

The Nature of the Hydrothermal Circulation in the Crust. Simple seafloor spreading models must recognize the cooling of the oceanic crust and the lithosphere as the major geological process acting under the oceans. Conductive cooling models were amazingly successful at predicting the decrease in heat flow and increase in depth observed as the lithosphere aged. The models failed, however, in young seafloor where the heat flow was too variable and the depths too deep compared to the thermal predictions. These observations confirmed a growing realization that hydrothermal circulation made a major contribution to the cooling of the young oceanic crust (Lister, 1972). With the discovery of "black smokers" on various segments of the midocean ridges, the importance of hydrothermal circulation was well established. These high temperature hot springs have a tremendous heat flux associated with them, and they may account for a substantial portion of the Earth's total heat flow. Their exact contribution depends upon the largely unknown geometry and longevity of the hydrothermal circulation.

There has been considerable speculation on the nature of the crustal permeability and hydrothermal circulation associated with the high temperature springs at the rise axis. Mathematical models are governed by the need to make the models tractable, and two general geometries have been considered. In one, water circulation is confined along "slots," or fractures running parallel to the spreading axis. The alternative geometry is one of pervasive cracking. The high temperatures of the fluids as they exit the seafloor require the former geometry; a pervasive permeability would diffuse the high temperatures. The nature of the permeability in the recharge and source regions of the hydrothermal systems are more poorly constrained. Recent results have shown that the seismic velocities in the vent zone tend to be somewhat higher than adjacent off-axis areas (McClain et al., 1985). The decrease in velocity as the crust ages is probably the result of pervasive fissuring; such fissures have been observed from submersibles and may form the recharge zone of fluids entering the system (Macdonald, 1982).

Seismic results also suggest that the porosity in the lower crust (that is, Layer 3) must be very low, even in very young crust. If water penetrates into the lower crust, it must do so by conduits that are unresolved by seismic means. Isolated faults reaching into the lower crust could provide the mechanism for such deep penetration. It is this problematical deep circulation that must cool the axial magma chambers; for their narrowness cannot be explained by hydrothermal circulation restricted to the uppermost crust.

The balance between hydrothermal circulation, volcanism, and tectonism remains an important question in the earth sciences. What is emerging in the last five years is a growing appreciation for the episodicity and along-strike variations of ridge processes. Led by new technology, most notably the multibeam and sidescan sonar systems (SeaBeam and SeaMarc), these hitherto second order effects are being recognized as the major question marks in understanding the creation of the oceanic lithosphere.

Evolution of the Oceanic Lithosphere

As the oceanic lithosphere is spread away from the spreading center where it was created, it is reasonable to expect that it may evolve as it cools and interacts with seawater. The most obvious changes are the increasing depth of the seafloor and the increasing thickness of sediments. Whether these changes are accompanied by significant changes in structure, composition, and mechanical properties of the lithosphere has been an important question in marine geophysics.

In discussions of the evolution of the oceanic lithosphere one must be cautious in defining the term *lithosphere;* at least three definitions can be made. They are: (1) that portion of the crust and upper mantle above the seismic low velocity zone (LVZ); (2) that part of the Earth that behaves as a brittle and elastic solid rather than plastically; and (3) the portion of the Earth that spreads with the tectonic plate. The distinction is not trivial; in the Pacific, for example, it is only the uppermost crust that behaves in a brittle manner at the ridges, yet the entire crust and some portion of the mantle must move with the oceanic plate. We shall try to avoid this confusion in the discussion that follows.

To a first order, the composition and structure of the crust and upper mantle are set at the mid-ocean ridges. Tectonism, as revealed by intraplate faulting and earthquakes is rare. Ongoing volcanism, resulting in seamounts, is probably the most dramatic process occurring away from the ridges and is poorly understood. Seamounts may cover as much 5-10% of the seafloor, and interest in seamounts as an important departure from plate tectonic theory is growing. However, for the remaining 90% of the seafloor, it appears that magmatic activity has ceased. Hence, for most of the oceanic crust, only hydrothermal processes are expected to continue as the crust ages. Because conductive cooling models (see below) are so successful at predicting the topography and heat flow in the ocean basins, it is clear that hydrothermal circulation is not a major contributor to the thermal budget away from the ridges.

Evolution of the Crust. The coexistence of basalt and seawater on the seafloor ensures that the composition of the basalt will change as it undergoes alteration. That this alteration will yield a systematic evolution of at least the upper crust has been documented from geochemical and petrological studies of seafloor rocks. For the geophysicist, alteration is most significant in that it affects the physical properties, such as seismic velocity and density, in the crust. Perhaps the best documented age dependence for crustal structure is the increase in velocities of the uppermost crust as the lithosphere ages to 70 Ma (Houtz and Ewing, 1976), which has been generally attributed to compaction and sealing of the pores in the crust. Changes in upper crustal velocities begin as soon as the crust is formed at the spreading axis. It appears that fissuring near the spreading axis causes a substantial decrease in velocity as the crust ages over thousands of years. On the East Pacific Rise as much as a 20% decrease in velocity in the top 500 m of the crust has been reported within 6 km of the Rise axis (McClain et al., 1985). However, within 16 km the process appears to reverse itself and velocities increase as the cracks are filled with alteration products.

The most controversial aspect of crustal evolution has been the reported thickening of the oceanic crust. Several workers showed that crustal thickening appeared to extend to some 40 million years, and argued that the probable mechanism for such thickening was the penetration of water into the uppermost mantle and the subsequent alteration of peridotite to form serpentinite. Serpentinization would lower the mantle velocities to crustal values, and the Moho would then become a hydration boundary rather than a petrological one. This model contradicts the accepted model for the ocean crust and has met with substantial resistance. Many workers did not accept the possibility that large-scale water penetration could occur. Christensen and Salisbury (1975) found similar crustal thickening but preferred a model in which ongoing igneous intrusion was responsible.

A number of experiments seemed to confirm at least localized crustal thickening. Brian Lewis and his students at the University of Washington found evidence for a seismic low velocity zone at the base of the crust under the northern Cocos plate. The low velocity zone appeared to thicken with age out to about 8 Ma, which they attributed to deepening serpentinization as the isotherms retreated with cooling. Such a serpentine lower crust might also explain the overall uniformity of crustal thicknesses; the ultimate crustal thickness would be controlled by pressure regulated cracking rather than the igneous processes at the rise axis.

In order to examine this question more closely, James McClain and Carl Atallah at the University of California at Davis reinvestigated the statistical evidence for crustal thickening in the Pacific. They found that while average thickness of the crust increases, the increase is very slight (.34 km) and can probably be attributed to localized, rather than general volcanism or serpentinization. The average Pacific crust does not increase in thickness, and crustal thickening cannot be used as evidence that serpentine is a major component of the lower crust.

Evolution of the Lithosphere as a Whole. The general increase in the depth of the seafloor as it ages has long been attributed to the cooling of the lithosphere. Assuming various conductively cooled models, a number of workers were able to reproduce the gross features of the depth increase as well as a corresponding decrease in heat flow. In keeping with the concept of plate tectonics, these models were derived for a conductively cooled slab. Alternative models, which include a cooling half-space, were also proposed. All of the models either implicitly or explicitly predicted depths that were proportional to the square root of age ($\sqrt{t}$) and a heat flow proportional to the inverse of the square root of age ($1/\sqrt{t}$) (see Forsyth, 1977; Parsons and Sclater, 1977). When averaged, seafloor depths and heat flow measurements fit the conductive cooling models extremely well with two exceptions: First, near the spreading axes, hydrothermal circulation was substantial enough to make the ridges too deep and the heat flow too variable. Second, for seafloor that was older than about 80 Ma, the seafloor depths ceased to increase as rapidly as the models predicted (Parsons and Sclater, 1977). Similar results were found when measuring the geoid height over the aging ocean lithosphere. A linear relationship between height and age was predicted from cooling plate models and found to hold to an age of about 80 Ma. Beyond that age, the geoid began to flatten out, an effect correlated with the flattening of the seafloor bathymetry.

While many of these studies argued about the exact mechanism for the anomalously shallow older seafloor, the essential reason is that the lithosphere ceases to undergo thermal contraction as it ages, which implies that the heat supplied to the lithosphere from below a certain depth equals the flow of heat flowing through the surface. This conclusion led many researchers to accept the plate cooling model; that is, the conductively cooled layer of the Earth has a finite thickness, with mantle convection supplying heat from below. Many would call this layer the lithosphere, and the bathymetric and geoid data allow estimation of the lithospheric thickness. Values range from 50 km to about 100 km.

If the lithosphere, as defined earlier, is that part of the Earth that travels with the oceanic plates, it may have a constant thickness. Most other commonly used definitions for lithosphere, such as depth to the low velocity zone (LVZ), brittle-ductile transition, or elastic thickness, share the trait that they all depend on temperature. They must change as the suboceanic mantle ages and cools. Therefore, it is likely that these alternatively defined lithospheres must thicken with age.

The study of earthquake source mechanisms is useful in determining the stress field in the lithosphere as it ages. The tensional axis associated with intraplate earthquakes is generally aligned with or somewhat oblique to the direction of spreading in young (<35 Ma) lithosphere. The ridge push forces associated with the elevated ridges does not appear to control the stress field. Rather, the thermoelastic stresses that arise with cooling are felt to be responsible. Earthquakes in the oceans in areas other than subduction zones are generally very shallow. In young lithosphere, earthquakes appear to occur above the 800°C isotherm calculated with half-space, conductive cooling models. Tensional earthquakes, which are associated with the spreading process, are apparently confined to a very narrow band about the neovolcanic zone where the lithosphere is created.

Shear wave velocities are particularly sensitive to temperature and the existence of partial melt, and studies have showed that shear body waves exhibit a systematic increase in velocity as they pass through mantle with increasing age. Similarly, surface wave studies reveal an increase in velocities with age. Forsyth (1977) argues that the depth to the top of the low velocity zone increases from less than 25 km at the ridge to 60 km within ten million years. Beyond that age, the increase is slower, reaching a maximum of 90 km. He attributes the lithosphere-asthenosphere boundary to the solidus of upper mantle peridotites.

Another measure of lithospheric thickness is the effective elastic thickness determined from flexural studies associated with island loading and flexure caused by the subduction process. Compilations of such estimates from lithosphere with different ages showed a systematic increase in the lithospheric thickness. In both studies, thickness went from near zero at the spreading axis to about 50 km. This change is substantially slower than estimates for the seismic boundary but roughly parallel estimates of the 500°C to 600°C isotherms. The difference between the seismic and flexural studies may arise from the inherently different responses of the Earth. In addition, each technique relies on different and artificial assumptions of layering. That the results should be different should not be surprising.

The discussion of the evolution of the lithosphere leaves a number of important questions unanswered. First, the contribution of off-axis hydrothermal circulation to the thermal and chemical budget of the lithosphere is not well understood. Even more fundamental are questions about the nature of the oceanic lithosphere and its boundary with the underlying asthenosphere. Finally, the ocean basins exhibit major departures from simple plate tectonic theory, such as seamounts and intraplate earthquakes. How these come about are exciting avenues for ongoing research.

JAMES S. McCLAIN
JOHN A. ORCUTT

References

Christensen, N. I., and M. H. Salisbury, 1975, Structure and composition of the lower oceanic crust, *Rev. Geophysics and Space Physics* **13**, 57-86.

Detrick, R. S., P. Buhl, E. Vera, J. Mutter, J. Orcutt, J. Madsen, and T. Brocher, 1987, Multi-channel seismic imaging of a crustal magma chamber along the East Pacific Rise, *Nature* **326**, 35-41.

Forsyth, D. W., 1977, The evolution of the upper mantle beneath mid-ocean ridges, *Tectophysics* **38**, 89-118.

Francheteau, J., and R. D. Ballard, 1983, The East Pacific Rise near 21°N and 20°S: Inferences for along-strike variability of axial processes of the mid-ocean ridge, *Earth and Planetary Sci. Letters* **64**, 93-116.

Houtz, R. E., and J. Ewing, 1976, Upper crustal structure as a function of plate age, *Jour. Geophys. Research* **81**, 2490-2498.

Lister, C. R. B., 1972, On the thermal balance of a mid-ocean ridge, *Royal Astron. Soc. Geophys. Jour.* **26**, 515-535.

McClain, J. S., J. A. Orcutt, and M. Burnett, 1985, The East Pacific Rise in cross section: A seismic model, *Jour. Geophys. Research* **90**, 8627-8640.

Macdonald, K. C., 1982, Mid-ocean ridges: Fine Scale tectonic, volcanic and hydrothermal processes within the plate boundary zone, *Ann. Rev. Earth Planetary Sci.* **10**, 155-190.

Menard, H. W., 1986, *The Ocean of Truth*. Princeton, N.J.: Princeton University Press, 353 p.

Moores, E. M., 1982, Origin and emplacement of ophiolites, *Rev. Geophysics and Space Physics* **20**, 735-760.

Orcutt, J. A., B. Kennett, L. Dorman, and W. Prothero, 1975, Evidence for a low-velocity zone underlying a fast-spreading rise crest, *Nature* **256**, 475-476.

Parsons, B., and J. G. Sclater, 1977, An analysis of the variation of ocean floor bathymetry and heat flow with age, *Jour. Geophys. Research* **82**, 803.

Raitt, R. W., 1956, Seismic refraction studies of the Pacific Ocean basin, *Geol. Soc. America Bull.* **67**, 1623-1640.

Raitt, R. W., 1963, The oceanic rocks, in *The Sea*, vol. 3, M. N. Hill, ed. New York: Wiley-Interscience, pp. 85-102.

Shearer, P., and J. Orcutt, 1985, Anisotropy in the oceanic lithosphere-theory and observations from the Ngendei seismic refraction experiment in the south-west Pacific, *Geophys. Jour. Royal Astron. Soc.* **80**, 493-526.

Sleep, N. H., 1975, Formation of the oceanic crust: Some thermal constraints, *Jour. Geophys. Research* **80**, 4037-4042.

Spudich, P. K. P., and J. A. Orcutt, 1980, Petrology and porosity of an oceanic crustal site: Results from waveform modeling of seismic refraction data, *Jour. Geophys. Research* **85**, 1409-1433.

Vogt, P. R., and G. L. Johnson, 1975, Transform faults and longitudinal flow below the midoceanic ridge, *Jour. Geophys. Research* **80**, 1399-1428.

Cross-references: *Continental Crustal Structure; Continental Lithosphere; Deep Seismic-Reflection Profiling; Earthquake Mechanisms and Plate Tectonics; Far-Travelled Terranes; Lithosphere: Mechanical Properties; Magmatic Processes; Mantle, Upper: Structure; Mantle Convection and Plumes; Mantle Dynamics; Ocean-Continent Transition; Paleomagnetism and Plate Tectonics; Seafloor Spreading: Magnetic Evidence; Seismic Anisotropy in the Earth; Seismicity: Mid-Ocean Ridge; Stress in the Earth's Lithosphere; Surface Waves.*

LUNAR PALEOMAGNETISM

One of the major surprises of the Apollo program was the discovery that many lunar samples carried a substantial remanent magnetization, evidence that the Moon may have had an inherent magnetic field sometime in its past. The detection of lunar surface fields of up to several hundred gammas, which are in general associated with the younger basin ejecta deposits, also suggested a magnetic history for the Moon. Because the returned lunar samples represent loose ejecta and not bedrock, paleomagnetic studies on lunar samples have focused on determining the intensity of the ancient magnetizing fields, rather than the direction relative to the present lunar rotational poles.

Before lunar sample magnetism can be evaluated, an understanding of the origin of their natural remanent magnetization (NRM) (q.v.) is necessary. Among the Apollo samples, the mare basalts and the highland melt rocks are most likely to have acquired a primary thermoremanent magnetization (TRM) (see *Thermoremanence*) when they were formed. If this has been preserved over the 3 to 4 billion years since that time, then the intensity of the ancient magnetizing field may be recovered.

Alternating field (AF) demagnetization analyses (see *Demagnetization*) have been carried out for a large number of lunar samples. Their interpretation is complicated because of the variety of carriers that may be in the lunar samples, but some definitive patterns have been identified. For instance, a correlation exists between the petrologically-determined cooling rates of the mare basalt samples and the amount of demagnetization they suffer when exposed to weak alternating fields, i.e., the least resistant samples were found to have the slowest cooling rates, which is caused by the formation of larger multidomain grains with the slower cooling rates. The regolith breccias have generally high mean destructive fields (MDFs), suggesting that the carriers are predominantly fine. As a group, the melt breccias are the most consistent, with MDFs of close to 10 mT, which is typical of coarser grain behavior (see *Rock Magnetism*).

Thermal demagnetizations (Johannes, this volume) on a limited number of lunar samples all indicate that the NRM is carried by a magnetic phase with a Curie point (see *Curie Temperature*) close to that of iron (770°C). Thermomagnetic analyses by Nagata et al. (1972) also indicated low-Ni iron as the principal magnetic phases in lunar rocks.

Whereas different NRM results have occasionally been obtained from subsamples of the same lunar sample, for the most part different subsamples have yielded similar behavior, even when the studies were made in different laboratories. Therefore, it is not sensible to ascribe the major variation seen between samples to differences in measuring technique or to contamination picked up after the samples left the Lunar Receiving Laboratory.

Mare Basalts

Most available lunar paleomagnetic data are from mare basalts. These results are particularly important because the mare basalts in general have the simplest surface history of the lunar samples. Two basic types of AF demagnetization curves for NRM are seen in lunar basalts. One has a roughly linear decrease on logarithmic plots of magnetization vs. alternating field, and the other has a sharp initial decrease followed by more gradual decrease in higher AF fields, which gives the curve a concave upward appearance. The difference between the two curves is caused by the differing amounts of soft magnetization, which is lost in the lowest AF steps. There appears to be some correlation between the degree to which the NRM and IRM_s (saturation remanence—the maximum remanence that a sample can carry, measured after exposure to a strong field) demagnetization curves are concave upward, which implies control due to the inherent AF demagnetization response of the sample.

Those samples that have the concave upward curves frequently show changes in direction as a result of the low field demagnetization. It therefore appears that the NRM is multicomponent and the AF demagnetization is successfully separating a "soft" component. In some samples the more resistant component exhibits directional stability, but in others the sample shows little or no convergence to a stable direction. Those samples that have more linear AF demagnetization curves tend to exhibit a higher degree of directional stability.

Breccias

The demagnetization characteristics of NRM in the breccias is highly variable. At one extreme are the strong but highly viscous NRMs of many regolith breccias, which cannot be distinguished by AF demagnetization from viscous remanence (see *Viscous Remanent Magnetization (VRM) and Viscous Magnetization*) acquired in the Earth's field. At the other extreme are the highly stable NRMs of many melt breccias, which may afford the best opportunity to determine a paleomagnetic record of the Moon, apart from the mare basalts.

The regolith breccias have magnetic properties similar to the lunar soils from which they are derived during small impact events. They contain abundant fine iron particles that include stable single domain and finer superparamagnetic particles that are magnetically viscous (see *Rock Magnetism*). Their magnetization can be hard against AF demagnetization, but is less resistant to thermal treatment. Indeed, thermal demagnetization to only a few hundred degrees centigrade reduces the remanence by about an order of magnitude in many of the regolith breccias. Since the magnetic material in these samples is very unstable upon heating, it has not been possible to carry out reliable thermal demagnetizations to high temperature so the NRM of the samples is enigmatic. A possible exception is sample 15498, in which the recrystallized matrix appears to carry stable NRM.

Melt breccias are the crystallization products from impact melts. They are particularly important for studies of NRM because they afford rocks that acquired a primary thermomagnetic remanence. The presence of varying amounts of clasts in these breccias suggests that in some parts may not have exceeded the Curie point of iron during assemblage.

Among the various clast-poor rocks, the *Apollo 16* collection dominates. As a group, the samples show some directional stability of NRM upon AF demagnetization. The NRM is less readily demagnetized in these samples than is IRM_s. The clast-bearing melt breccias include more *Apollo 16* material and a substantial representation from *Apollo 17*, which includes samples collected from large boulders derived from a geologic formation representing basin ejecta. The behavior of these samples is rather similar to that of the clast-free melt breccias; the NRM shows some directional stability and is usually harder than the IRM_s. The clast-rich melt breccias include those samples in which the evidence for melting and recrystallization is limited, although recrystallization temperatures are estimated to be in the range of 800°C. Therefore, where evidence of recrystallization is present, NRM would have been acquired at least as a partial TRM. The *Apollo 14* (Fra Mauro) annealed breccias fall into this category.

Interpretation of NRM of the Apollo Samples

Most lunar samples clearly have more than one component of NRM. In particular, many have a soft component that might have been acquired as isothermal remanence (IRM) in a field of about 1–2 mT. Some samples are so magnetically soft that even if they have a stable component of NRM, the effect of the soft phase dominates and the stable components are obscured. However, many samples have remanence components stable against AF demagnetization to high fields. Moreover, other samples are stable against thermal demagnetization to close to the Curie point.

During processing and return from the Moon, lunar samples may have been exposed to fields of as much as 1 mT. Sawing may have induced further contamination, as could the process of sealing the bags in which the subsamples were placed. Finally there is the possibility of acquisition of VRM in the geomagnetic field.

A number of processes take place on the lunar surface that could give rise to components of NRM. These include irradiation, lunar diurnal temperature cycling, and shock effects. Irradiation effects have been demonstrated to be inefficient in comparison with TRM in generating remanence. Since the fields in which it occurs on the present lunar surface are very small, it is not likely to account for any part of the measured NRM.

The possibility that the lunar diurnal temperature cycle might generate a stable NRM has been suggested. In its simplest form, this remanence would be acquired by samples that contain superparamagnetic and near-superparamagnetic material at room temperature. Such magnetic phases might become magnetized as they are thermally cycled below the maximum of the lunar diurnal cycle of about 150°C. However, if they were collected from the lunar surface at temperatures approaching the diurnal maximum, they would be thermally demagnetized at that time in the weak present surface field of the Moon. Any lunar diurnal cycle magnetization should readily demagnetize in the laboratory at temperatures below 200°C.

Impact-related shock has also been called upon to account for the NRM of the Apollo samples. In the absence of a magnetic field, when a sample is shocked it will be demagnetized. The degree of demagnetization depends upon the strength of the shock. Moreover, shock demagnetizes the softer magnetization preferentially. These effects dominate in the range of a few hundred megapascals (a few kilobars). As higher shock ranges are reached, the rock experiences residual heating effects, which give rise to partial or complete thermal demagnetization of the rock. In the presence of a field these same effects will give rise to remanent magnetization. However, there is an important caveat; the field must be strong to produce any important NRM in the low shock range. If the field is small, the effect will still be to demagnetize any preexisting remanence. In the range of shock that gives important residual heating effects, the situation is quite different. TRM and even partial TRM are efficient mechanisms of magnetization even if the inducing fields are weak, so that measurable NRMs could be generated in the presence of a coherent external field.

Additional NRM could be generated if the shock produces any other new magnetic phase in the affected rock. Cisowski and others in 1973 demonstrated changes in the remanence carrying ability of lunar soil with shocks in the range of thousands of megapascals. Of particular importance was the production of fine iron in the glass. Such iron could become strongly magnetized if formed in a significant field.

T. Gold and S. Soter in 1976 suggested that strong local fields may be produced by compressing the solar wind or lunar surface remanent field during impact events. But to maintain this field while the heated samples cool and acquire a TRM, as the younger basin ejecta apparently has, does not seem feasible. In 1984, L. L. Hood and A. Vickery demonstrated theoretically that only weak fields can be expected to result from effects relating to impact-produced plasmas. Indeed, if large fields are created during impact events, the craters themselves should have substantial anomalies associated with them. However R. P. Lin in 1979, after an analysis of subsatellite electron reflectance data, concluded that this was not indicated in lunar surface field patterns. Also pervasively shocked lunar samples are not in general strongly magnetized.

The characteristics of TRM in single domain, pseudosingle domain, and multidomain magnetite are well established (see *Thermoremanence*). Less is known about TRM in ferromagnetic iron and nickel iron phases, but the experiments that have been carried out in connection with the lunar studies suggest that it conforms broadly with TRM in the ferrimagnetic materials found in terrestrial rocks. Thus weak field TRM is a stable type of magnetization, particularly in fine magnetic particles. In fine particles, TRM is blocked near the Curie point, if the particle size is large compared to the superparamagnetic to single domain transition size. On the other hand, if the magnetic carrier is multidomain, the magnetization is softer against AF demagnetization and can be thermally unblocked well below its Curie temperature. Such characteristics are found in a number of the more strongly magnetized mare basalts and recrystallized and melt breccias, implying that their NRMs are primarily of thermal origin.

Paleointensity of Lunar Fields

The most successful methods of paleointensity determination are based on the acquisition of laboratory TRM. Unfortunately, they obviously require the heating of each studied sample, which often produces irreversible magneto-chemical changes in the remanence carriers. Interactions between adjacent magnetic phases with different Curie temperatures may also contribute to difficulties in interpreting laboratory TRM, as suggested by Pearce and others in 1976.

Paleointensity determinations employing dual stepwise heatings to determine alternately NRM lost and partial TRM gained in a laboratory field have been made on lunar samples with only limited success. However Walton in 1983 contended that such two-stage results may overestimate the paleointensity of the ancient field due to progressive destruction of fine grained magnetic carriers and proposed an additional heating at each temperature step. Such an experiment has yet to be accomplished on any lunar sample.

Given the difficulties in heating the lunar samples without bringing about irreversible changes in the magnetic phases, attempts have been made to use other techniques. Shaw in 1974 developed a paleointensity method that utilizes anhysteretic remanence (ARM—a remanence acquired during application of both an alternating and a steady biasing field) to monitor changes brought about in the sample by the heating necessary to induce the laboratory TRM. A single heating is used, and the AF demagnetization of NRM and TRM are compared so that the ratios of the two remanences can be obtained incrementally. This method, like the dual heating method, has the virtue that an attempt to detect irreversible changes is made.

Two methods that do not require heating each specimen have been developed specifically for lunar samples. These methods are both normalization methods, so that a separate calibration experiment is needed to determine the absolute intensity. Thus they initially give only relative intensities but not absolute intensities without the additional heating experiment.

The ARM normalization method proposed by A. Stephenson and D. W. Collinson in 1974 consists of an experiment somewhat analogous to the dual heating method, except that the magnetization used is ARM rather than TRM. Thus the sample is first AF demagnetized and then given an ARM in a weak field. The ratio of NRM lost to ARM gained over particular ranges of alternating field strength permits one to normalize the NRM in terms of ARM. The method can be calibrated by a heating experiment in which the ratio of ARM to TRM is established for a range of fields. A number of determinations have been carried out and a degree of internal consistency between different samples has encouraged some confidence in the method.

The second method employed by Cisowski et al. normalizes the AF demagnetized NRM with AF demagnetized IRM_s. This method has the virtue of being applicable to many samples and again does not involve heating of each sample. Calibration has now been carried out by heating a small number of lunar samples to give a TRM in various fields. The IRM_s used in the calibration was measured after heating in which the TRM was acquired. A weakness of this IRM_s normalization method is that it gives no direct indication of whether the NRM is a primary TRM or some other form of remanence.

Figure 1 summarizes the results of paleointensity estimates as a function of sample age, using the KTT and ARM methods. Sample 62235, a highland melt rock, gave results of about 0.1 mT (1.0 oersted) by two of these methods. Results from both older and younger samples suggest a peak value for the lunar field at about 3.8 AE.

Figure 2 presents lunar sample paleointensities that were obtained by using the IRM_s normalization method as a function of age of the sample. Samples whose IRM_s demagnetization curves are dominated by low coercivity carriers are not plotted since the calibration heatings indicated that such samples cannot be used to distinguish between fields of less than a few hundredths of a millitesla. A period of high magnetic field intensity for the Moon is indicated between about 3.6 and 3.8 AE with a maximum value of several hundredths of a millitesla (tenths of an oersted). Such fields are far higher than present fields on the lunar surface, but about an order of magnitude smaller than those suggested by the other paleointensity methods.

Origin of Lunar Magnetism

A number of models for the origin of ancient lunar fields have been hypothesized. The most tenable of these are models involving the amplification or generation of magnetic fields during impact events, the recording of an external (e.g., solar) field by the

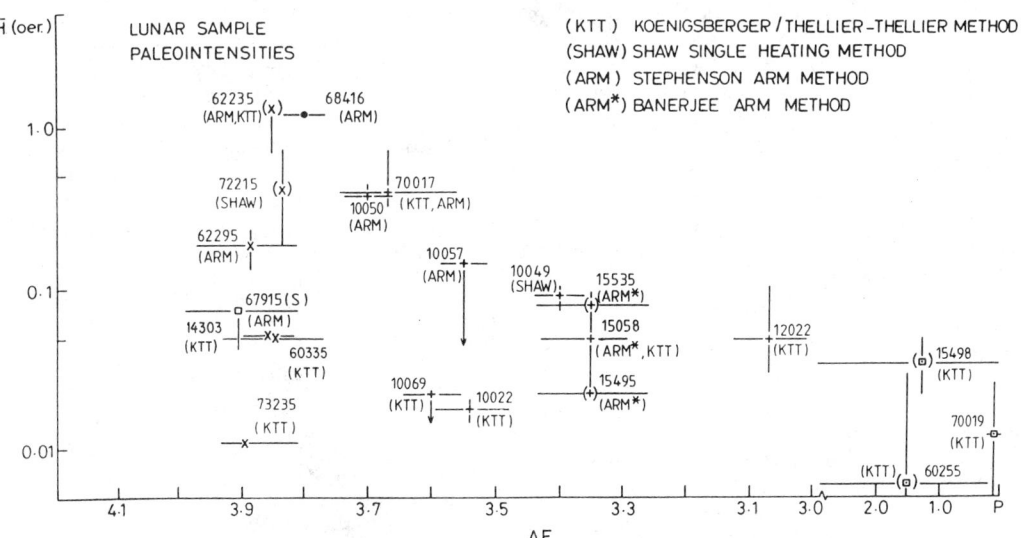

FIGURE 1. A compilation of paleointensity versus radiometric age for 21 lunar samples via the dual heating (KTT) and ARM methods. 1.0 oersted (oer.) = 0.1 millitesla.

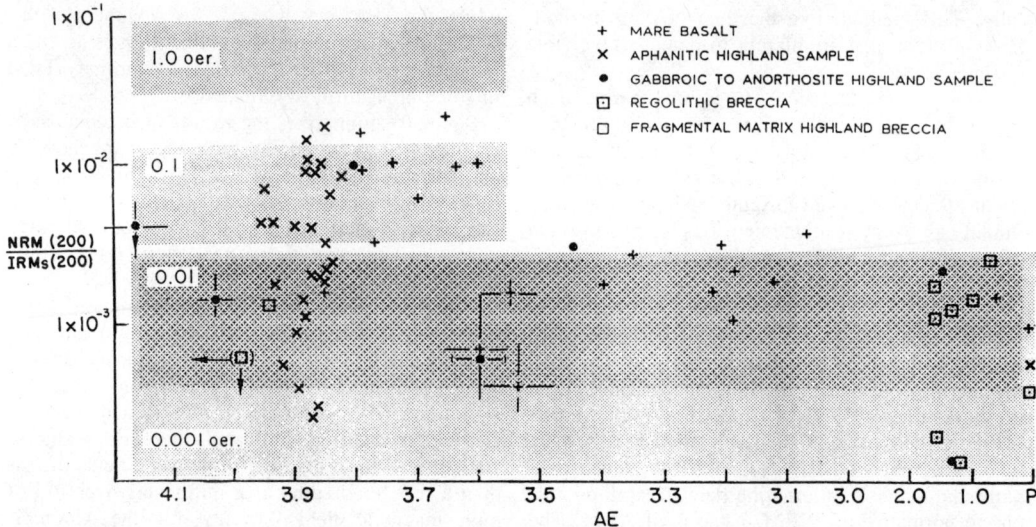

FIGURE 2. Paleointensity vs. radiometric age for 67 lunar samples via the IRM_s normalization method. Field calibrations are based on laboratory heatings of 5 lunar samples. 1.0 oersted (oer.) = 0.1 millitesla.

Moon during accretion or cooling, and generation by an internal dynamo.

Impact generated transient field models are less appealing because mare craters appear to have no systematic magnetic expression. Also samples containing pervasive mineralogical shock effects are not in general among the most strongly magnetized samples, when normalized to their remanence carrying potential, as shown in Fig. 2. Moreover, the restricted age range of the strongly magnetized material also argues against impact phenomena being the sole cause of the magnetization; impacts are an ever present aspect of lunar history.

External field models do not seem to provide a helpful explanation of lunar fields of planetary scale. They require substantial inducing fields late in the evolution of the solar system, if they are to explain the magnetization of the lunar samples, and of the sources for the presently observed lunar surface fields. They also encounter a profound difficulty with a magnetostatic theorem stated by S. K. Runcorn in 1975, i.e., that an external homogeneous field would magnetize the Moon in a manner to leave a detectable dipole moment. Such a lunar dipole magnetization is not observed by satellite measurement. While inhomogeneous external field models do not face this difficulty, they and indeed all remanent field models face another possibly fatal difficulty in explaining why the strongly magnetized material has a restricted age range, with the oldest samples not among the most strongly magnetized.

The remaining possibility is an ancient lunar planetary field of internal origin. S. K. Runcorn in 1983 argued that the directions of magnetization of various mare surfaces and basin ejecta units, as derived from Apollo subsatellite data, can be interpreted in terms of a reversing lunar dipole field that periodically changed its position relative to the lunar crust. This lunar "polar wander" implies that a molten lunar core, capable of sustaining a hydrodynamic dynamo, existed at least until Imbrian times (about 3.8 AE). However Runcorn's interpretation of the lunar surface fields was challenged by Hood et al. (1984), as other geophysical data does not allow for the existence of a lunar metallic core with a radius of greater than 400 km.

The new calibration of the IRM_s normalization technique, which suggests peak fields in Imbrian times in the range of only hundredths rather than tenths of millitesla, makes a lunar dynamo originating from such a relatively small core somewhat more tenable. The closer proximity of the Moon to the Earth and the terrestrial field in early lunar history might also make the inception of a lunar dynamo a more likely occurence, as was suggested by E. Levy in 1979 for the moons of the giant planets.

M. FULLER
S. M. CISOWSKI

References

Cisowski, S. M., D. W. Collinson, S. K. Runcorn, A. Stephenson, and M. Fuller, 1983, A review of lunar paleointensity data and implications for the origin of lunar magnetism, (Proc. Lunar and Planetary Sci. Conf., 13th) *Jour. Geophys. Res.* **88,** A691–A704.

Fuller, M., 1974, Lunar magnetism, *Rev. Geophysics and Space Physics* **12,** 23–70.

Gose, W. A., and R. F. Butler, 1975, Magnetism of the Moon and meteorites, *Rev. Geophysics and Space Physics* **13**, 189–193.

Hood, L. L., C. P. Sonett, and L. J. Srnka, 1984, Lunar magnetism, *Nature* **307**, 661–662.

Hood, L. L., and S. M. Cisowski, 1983, Paleomagnetism of the Moon and meteorites, *Rev. Geophysics and Space Physics* **21**, 676–684.

Mason, B., and W. G. Melson, 1970, *The Lunar Rocks.* New York: Wiley-Interscience, 179p.

Nagata, T., R. M. Fisher, and F. C. Schwerer, 1972, Lunar rock magnetism, *The Moon* **4**, 160–186.

Pearce, G. W., G. S. Hoye, D. W. Strangway, B. M. Walker, and L. A. Taylor, 1976, Some complexities in the determination of lunar paleointensities, (Proc. 7th Lunar Sci. Conf.) *Geochim. Cosmochim. Acta* **40**, suppl. 7, 3271–3297.

Taylor, S. R., 1975, *Lunar Science: A Post-Apollo View,* New York: Pergamon, 372p.

Cross-references: *Curie Temperature; Demagnetization; Earth and Moon: Origins; Magnetic Properties of Minerals; Natural Remanent Magnetization (NRM); Planetary Magnetic Fields; Rock Magnetism; Thermoremanence; Viscous Remanent Magnetization (VRM) and Viscous Remagnetization.*

LUNAR SEISMOLOGY

Lunar Seismographs and Locations

As part of the Apollo manned lunar landings, which ended in 1969, seismographs were placed at six sites on the Moon. In addition, Apollo astronauts carried out shallow seismic refraction experiments at several landing sites. Recording of seismic data was terminated in September 1977.

The landing sites of the Apollo lunar landing missions were restricted to the front side of the Moon so there were uncertainties in the location and travel times of detected seismic events.

The seismographs placed on the Moon contained three orthogonal matched long-period seismometers (with resonant periods of 15 seconds) and a vertical component short-period seismometer (a resonant period of 1 second). Because the ground noise was very low compared to that of the Earth the seismographs could be operated at very high magnifications. The long-period seismometers could detect vibrations of the lunar surface in the frequency band from 0.004 to 2 Hz whereas the short-period seismometer covered the band from 0.05 to 20 Hz. Ground motions as small as 0.3 nanometers at maximum sensitivity could be detected.

Character of Seismic Signals

The lunar seismographs detected between 600 and 3000 events every year of their operation. Three basic types of seismic events were recorded: (1) events caused by the impact of lunar landing modules, booster rockets, and meteorites; (2) shallow focus (< 100 km) moonquakes, and (3) deep moonquakes with focal depths ranging from 800 to 1000 km.

Typical seismic signals recorded by the lunar seismographs are shown in Fig. 1. Perhaps the most striking feature of the recorded seismograms is their exceedingly long duration. Signals lasting as long as 4 hours were observed from the large artificial impacts, such as the third stage of the Saturn boosters. Several other features are worthy of note. The signals are emergent in onset and the sequence of body and surface wave phases typical of terrestrial seismic signals is absent or, at best, very weakly developed.

The long duration of the lunar seismic signals indicates that the efficiency of transmission of seismic energy in the lunar interior is very high. Attenuation of elastic energy can be described by the quality factor Q where $2\pi Q$ is the fractional loss of elastic energy for one cycle of vibration. A high value of Q indicates low attenuation and analyses of lunar seismic signals indicates values of Q ranging from 2000–4000, which are an order of magnitude larger than values observed at comparable depths on the Earth. It is probable that, in the absence of volatiles, frictional losses induced by the transmission of seismic waves is much less than noted on the Earth.

The natural moonquakes are very small, having estimated magnitudes of less than 2 or so. Translating this to an annual average seismic energy

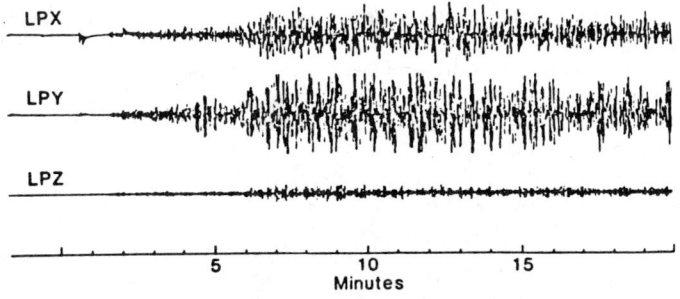

FIGURE 1. Typical moonquake seismograms.

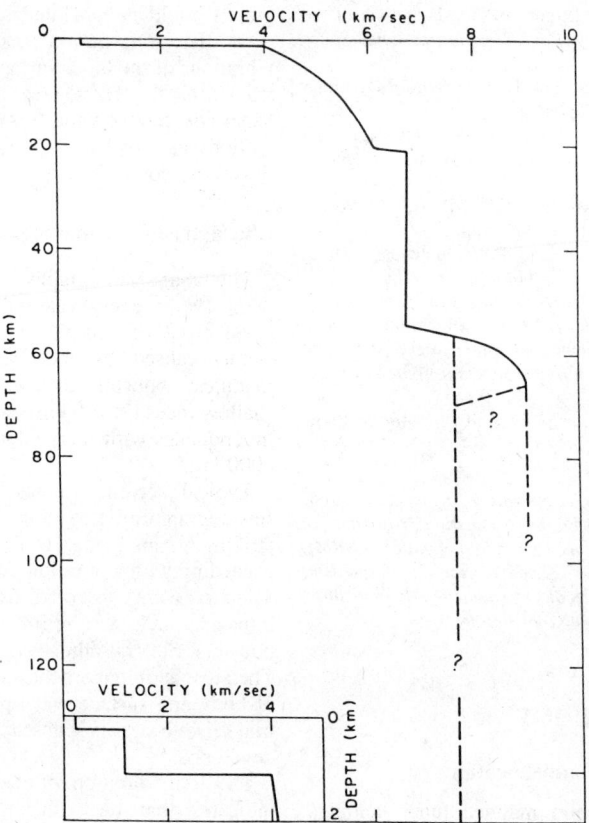

FIGURE 2. Seismic velocity structure of the Moon.

release gives a value of approximately 1×10^8 joules, which is about ten orders of magnitude less than that observed on the Earth.

Another aspect of the moonquakes is that, except for a small number of shallow focus events, the moonquakes originate deep within the Moon at depths ranging from 600 to 1000 km and are very strongly correlated with a component of the Earth tidally induced shear stresses acting on the Moon. The inferred tidal stresses are small, however, consistent with the size of the moonquakes and argue against any pronounced thermal convection within the Moon that could produce higher stresses within the Moon's outer shell. On the Earth, deep focus earthquakes are associated with lithospheric slabs that sink to great depths being driven by a global convection system. One thus concludes that the outer shell of the Moon is cold and rigid and much more tectonically stable when compared to the Earth.

Velocity Structure and Composition of the Lunar Interior

The velocity model for the Moon that best satisfies all the artificial impact data is shown in Fig. 2. In the upper 1 km of the Moon the seismic velocity increases abruptly to a value of 4.0 km/sec and then gradually increases to 6.0 km/sec at a depth of 20 km where it increases abruptly to a value of 6.7 km/sec. Between a depth of 20 km and a depth of 55 km (taken to be the base of a lunar crust) the velocity remains constant. In the depth range of 55 to 60 km the velocity increases to a relatively constant value of 8 to 9 km/sec to a depth of 1000 km or so but there are some uncertainties in the interpretation. The presence of a central core could not be determined from the existing seismic data.

The velocity structure of the lunar crust is compatible with an anorthositic composition such as those found in the rocks forming the lunar highlands. At greater depths the seismic velocities are compatible with those inferred for olivine-pyroxene mixtures or olivine rich compositions.

ROBERT L. KOVACH

Bibliography

Anderson, D. L., and R. L. Kovach, 1972, The lunar interior, *Physics Earth and Planetary Interiors* **6**, 116–112.

Kovach, R. L., and J. S. Watkins, 1973, Apollo 17 seismic profiling—Probing the lunar crust, *Science* **180,** 1063-1064.

Latham, G., M. Ewing, J. Dorman, D. Lammlein, F. Press, N. Toksoz, G. Sutton, F. Duennebier and Y. Nakamura, 1971, Moonquakes, *Science* **174,** 687-692.

Toksoz, M. N., F. Press, K. Anderson, A. Dainty, G. Latham, M. Ewing, J. Dorman, D. Lammlein, G. Sutton, F. Duennebier, and Y. Nakamura, 1972, Lunar crust: Structure and composition, *Science* **176,** 1012-1016.

Toksoz, M. N., N. R. Goins, and C. H. Cheng, 1977, Moonquakes: Mechanisms and relation to tidal stresses, *Science* **196,** 979-981.

Cross-references: *Earth and Moon: Origins; Earthquake Seismology; Elasticity and Wave Propagation: Principles; Seismic Instrumentation; Seismic Wave Scattering.*

M

MAGMA PHYSICS

Molten rock, with or without crystals and bubbles, is *magma*. Most magmas are highly polymerized silicate solutions with relatively minor amounts of other constituents (Table 1). They are rheologically unusual in their extremely strong dependence of viscosity on temperature during crystallization. From the first sign of melting (solidus) to the disappearance of the last crystal (liquidus), viscosity can easily vary by a factor of 10^{15}, which is perhaps the largest physical property variation within the Earth for any given temperature drop ($\sim 100°C$). It is this single property that makes the behavior of magma so varied and its physical analysis so interesting.

Volcanoes erupt magma as lava and pyroclastic debris and furnish abundant evidence for its varied state. The roots of volcanoes, exposed in deeply dissected terranes, on the other hand, show solidified bodies (plutons), sometimes containing spectacular sortings of crystals and intricate textural patterns due to differential movements of highly fluid magma and crystals at various stages of crystallization. It is evidence such as this that has continually brought petrologists to the study of magma through analysis of its production, extraction, ascension, behavior in chambers, and eruption. These evolutionary stages delineate the magmatic life cycle (Fig. 1).

These problems are singular, and complex, in their intimate coupling to the phase equilibria and chemical evolution of magma. The study of magma from a purely chemical viewpoint is an old and richly developed facet of igneous petrology. The main goal of that work has been to understand the observed diversity of the igneous rocks. These studies have implied and surmised physical processes that have often become generally accepted without even the most cursory understanding of their physical fundamentals. And as fruitful as this approach has been, the fundamentals of the principal evolutionary stages are inaccessible to this chemical means of analysis. It has thus become increasingly apparent that the true meaning of chemical studies can only be revealed through a strongly coupled knowledge of the physical and chemical evolution of magma.

The aim here is to provide an introduction to the fundamentals of some of the significant physical aspects of magmatic behavior. After a brief historical perspective, the basics of magma rheology are discussed and followed by the basic equations, which are scaled and discussed in terms of their fundamental meaning in terms of magma physics. The essentials of this research are then introduced with a special emphasis on the newest developments, namely, the style and vigor of convection in magma chambers and the distribution, sorting, and sedimentation of crystals during convection.

History

There are two principal early actors in the physics of magma as appreciated today: Reginald Aldworth Daly (1871–1957) and Frank Fitch Grout (1880–1958) (Fig. 2). Daly was a professor at Harvard University and Grout was a professor at the University of Minnesota. Both were petrologists, both were keen field men, both wrote an excellent text, and both had clear insight into magmatic processes. This similarity ended in personality and style of research. Daly was strong-minded, domineering, and clearly crusty. Grout was formal and reserved, yet warm and gentle. He refused to accept the presidency of the Geological Society of America because his department chairman had not been offered the same. Grout had many students (e.g., F. J. Pettijohn), whereas Daly worked essentially alone, but with the careful editorial hand of his admired wife, whom he called his inspiring fellow worker, though she was not a geologist. Daly's worldwide broad interests and imagination allowed him to spawn such novel ideas as sea-level fluctuation by postglacial rebound and erosion of submarine canyons by turbidity currents. Grout's mind traveled more through igneous, sedimentary, and metamorphic rocks, and inventing and elucidating novel physical processes.

The idea of magmatic stoping, whereby magma essentially mines its way upward, had been enunciated before Daly got hold of it, but it is he who

TABLE 1. Representative Compositions of Magma (mass %)

	Basaltic	Silicic
SiO_2	50	70
Al_2O_3	16	13
CaO	12	2
MgO	7	1
FeO	6	2
Fe_2O_3	4	1
Na_2O	3	4
K_2O	1	5

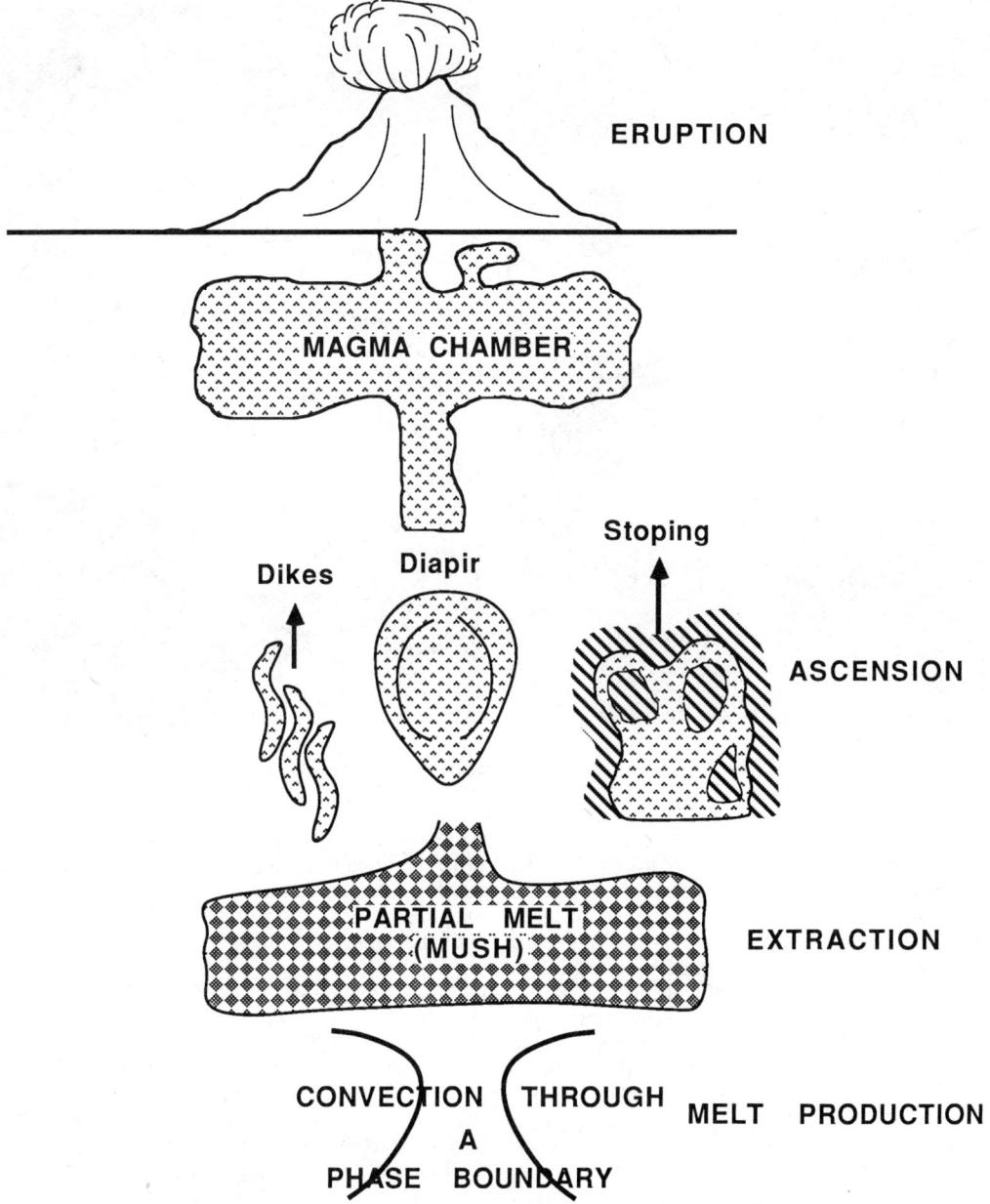

FIGURE 1. The magmatic life cycle: Melt produced by convection through a phase boundary is extracted to form upward-moving dikes, diapirs, or stoping bodies. Magma collects in the near-surface, forming a magma chamber in which further cooling and crystallization give rise to a spectrum of chemical compositions displayed in volcanic ejecta.

relentlessly pushed its acceptance through careful field work and a quantitative evaluation of its basic mechanics. He investigated the role of thermal stress in loosening roof rock, and he evaluated the settling rate of stoped blocks through magma. Stoping was (and is) not well accepted as an important and common means of magma transfer, and Daly was spurred by a strong desire to refute doubters. His calculations showed him what physical properties were important and the dearth of such information available. Daly's text, for its day, contains a vast amount of crucial physical property data and is virtually a model of what an igneous petrology text should be. His balance of recognizing critical field areas, worldwide, illustrating fundamental igneous processes is masterful.

FIGURE 2. Pioneers in physical research of magma: R. A. Daly (left), F. F. Grout (right).

Grout believed in Daly's ideas of stoping, for he had seen good field evidence for it, but he also sometimes saw flow of the evidently heated and softened adjacent wall rock. From this evidence he quantitatively developed the idea of igneous diapirism, whereby upward transfer of magma is accommodated by a return or downward flow of roof and wall rock. His quantitative, fairly well scaled experiments established the important features of this model. Much earlier in his Yale dissertation of 1918, however, Grout showed sterling insight in his enunciation of the mechanics of two-phase convection in magma. The actual term *two-phase convection* had been used earlier (1911) by Daly in describing gas-driven convection of Hawaiian lava lakes, but Grout extended it to the more general case of convection driven by sinking crystals. He also realized that cooling along a tall vertical wall would create great buoyancy and strong convection. And, perhaps still more important, he saw that increasing crystallinity increases viscosity to the point that near 50% solids the magma locks and becomes essentially infinitely viscous.

In their time, Daly had the dominant name. This was a measure of his personality, global virtuosity, and dogged attempt to quantify igneous petrology. Grout's ideas were more specific to magmatic processes; they were exact, quantitative, and true to the test of time. Careful study of their faces (Fig. 2) will reveal their style and even means of magmatic transfer (i.e., gentle diapirism and brittle stoping). They started and gave physical igneous petrology its initial impetus.

Rheology of Magma

At a concentration of about 50% (v), solids in a suspension begin to touch and form a rigid, interlocking network. In order to shear this fluid, the solids must locally separate to move past one another, and this simple feature makes this congested fluid dilatant. The fluid expands upon shearing and effectively stoppers its container. Near this point of critical crystallinity magma changes from a highly viscous mush into a dilatant solid; viscosity increases essentially without limit (Fig. 3).

At very small concentrations A. Einstein showed that viscosity increases linearly with solid concentration:

$$\frac{\mu}{\mu_0} = 1 + 2.5N$$

where N is the fractional concentration of solids and μ_0 is the crystal-free viscosity. This small effect is of little consequence to magma dynamics; for to

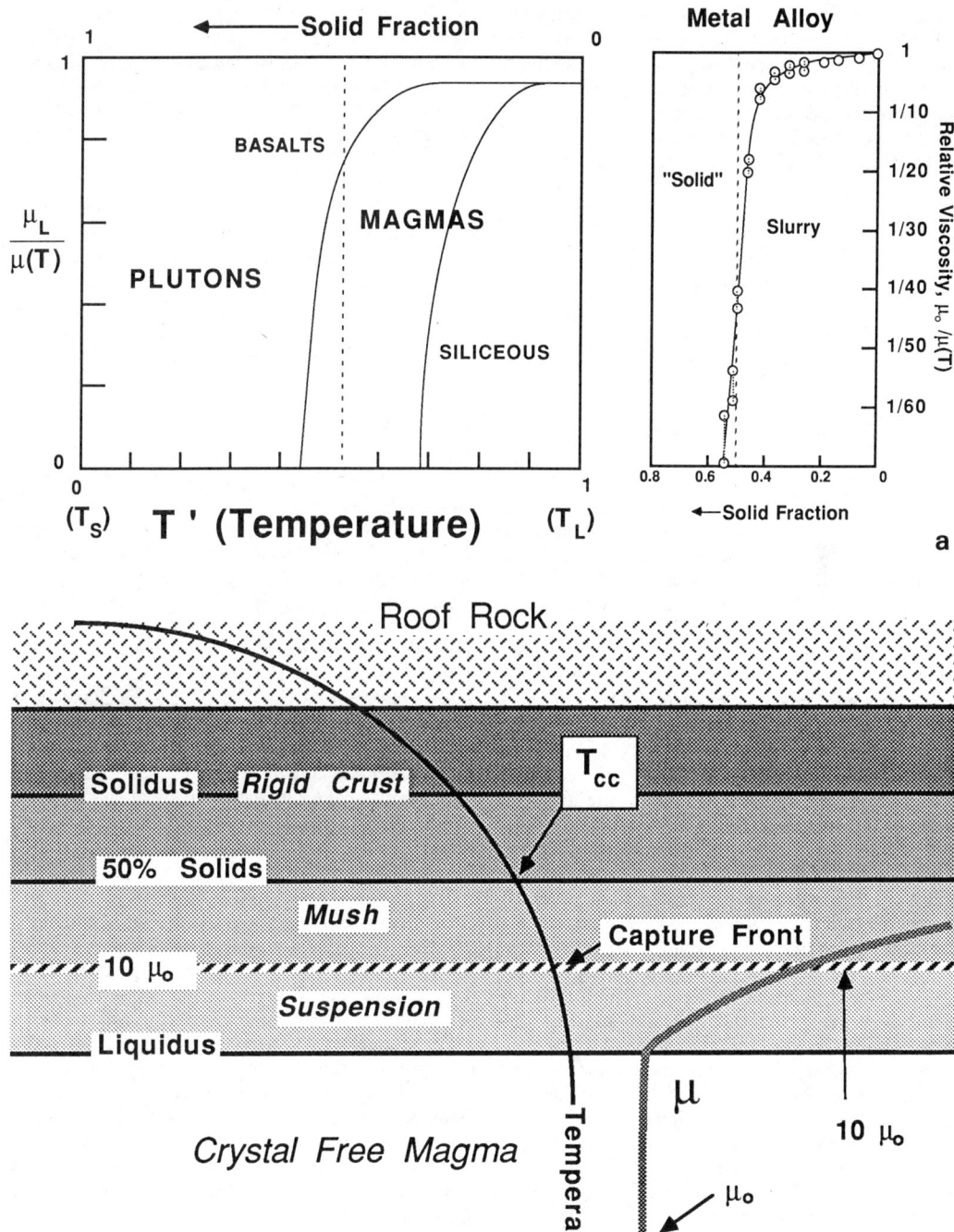

FIGURE 3. (a) The variation of magmatic viscosity, relative to the crystal-free state, decreasing temperature, and increasing concentration of solids. μ_L is the (constant) crystal-free viscosity and $\mu(T)$ is that at any lower temperature. Beyond a crystallinity of ~50% (v) in basalts and ~30% in siliceous magmas, the magma is too viscous to transport. A very similar variation in viscosity is found in metals. (b) The rheological divisions of the thermal boundary layer at the roof of a crystallizing magma (see text).

cause a tenfold increase in viscosity the fractional volume of solids must increase to the impossible limit of 3.6. Over the whole range of crystallinity from liquidus to critical crystallinity, the general variation of viscosity has been characterized by a wide range of formulae, of which the simplest, but still useful, is Roscoe's for uniform spheres:

$$\frac{\mu}{\mu_0} = \left(1 - \frac{N}{N_m}\right)^{-2.5}$$

where N_m is the concentration of solids at maximum packing. Viscosity increases tenfold when $N/N_m \simeq 0.6$, which for $N_m = 0.5$ is at 30% solids, close to that observed.

This strong variation in viscosity occurs not only within the relatively small temperature range that is defined by the solidus and liquidus, but over only the (approximately) upper half of this range (i.e., ~100°C). This profound change in viscosity separates crystallizing magma into a rigid, immobile part, where the solid concentration exceeds about 50%, and a mobile part, with strongly varying viscosity, near the margins, where cooling and crystallization occur.

This critical interplay between crystallinity and mobility is best shown by the complete absence of lavas erupted containing more than about 50–60% phenocrysts (large, subterranean-grown crystals). The exact point of critical crystallinity depends very much on the silica content of the magma: the higher the viscosity of the interstitial liquid, the lower the point of critical crystallinity. In low-silica (~45%, m), syrupy basaltic or ultramafic liquids, critical crystallinity approaches 60% solids, whereas in high-silica (~70%, m), highly viscous granitic liquids it is much smaller, perhaps 25–30%. The exact value depends on, in addition to liquid viscosity, the size range and abundance of crystals in the suspension.

The exact point of critical crystallinity also depends on the applied shear stress or strain rate. At large strain rates (e.g., eruptive processes) the point is at large crystallinities, but at low strain rates, such as that due to convection, it is at the point where the viscosity increases tenfold, which is at $N/N_m = 0.6$. That is, convection is partitioned in variable viscosity flows to an essentially isoviscous region wherein viscosity increases by no more than about tenfold. Thus the liquidus-solidus region is characterized by three principal rheological zones (Fig. 3b): (1) *rigid crust* extending out to $N/N_m \simeq 1$ or $N \simeq 0.50$; (2) a highly viscous *mush* zone $1 \leq N/N_m \leq 0.6$; and (3) a *suspension* zone over $0.6 \leq N/N_m \leq 0$.

The excellent natural example of this overall character is found in the drilling of Hawaiian lava lakes. Ponding of lava in natural depressions forms lakes up to 100 m deep and several kilometers wide. A rigid crust forms by cooling from the top and slowly thickens with time (Fig. 4). Rotary, spindle-style viscosity measurements have been made by drilling through the crust and inserting the viscometer directly into the underlying magma. The crust acts as rigid, drillable rock until crystallinity has decreased to about 50% (where $T \simeq 1065$°C), beyond which the drill stem can be easily pushed downward by hand into the mushy magma. Drilling and the weight of the drill stem itself provide a high local shear stress, apparently approximately equivalent to that of eruption, and the so-defined critical crystallinity is large. Careful examination of the recovered cores, however, suggests that, in terms of the natural flow of the magma itself, the critical crystallinity is much smaller, perhaps 20–25% ($T \simeq 1120$°C), which delimits the suspension zone.

All of what has been said so far implies that magma is a Newtonian fluid, and without crystals it probably is, but suspensions are well known to possess yield strengths and are therefore clearly non-Newtonian. Magmas always contain crystals and, at least in the crystallizing portions, most surely have yield strengths that depend on the spectrum of crystal size and abundance. That magma has a yield strength means that heavy crystals cannot sink from the liquid unless they are sufficiently large to exceed the yield strength. Crystal settling, or *fractionation*, is the single most universally held process by which magmatic chemical composition can be changed. It is how basalts are made into more granitic compositions. A yield strength allows only the larger, earlier formed crystals to settle, and this yield strength may have a fundamental effect on magmatic evolution. Overall, this effect is yet to be reckoned with.

Equations

In the most general sense, the full set of equations needed to describe a fluid as rheologically complex as magma rarely yields an exact meaningful solution. Experiments and theory on simpler fluids, though still having strongly variable viscosity, have shown that magmatic flow regimes may be dynamically separated into three regions: (1) high viscosity, rigid, and thermally conductive wall, (2) thin thermal boundary layers where viscosity varies strongly and convective heat transfer is approximately equivalent to conduction, and (3) a hot interior of low and constant viscosity where any motion would allow convection to dominate conduction.

The field equations of a constant-viscosity fluid are meaningful here because they describe the bulk of the fluid, and these solutions can be matched to boundary results involving variable viscosity and conductive heat transfer in the wall rock. The Boussinesq approximation is employed, whereby the fluid density is taken as constant everywhere except in the buoyancy term of the momentum equation. That is, for conservation of mass the equation of continuity reduces to

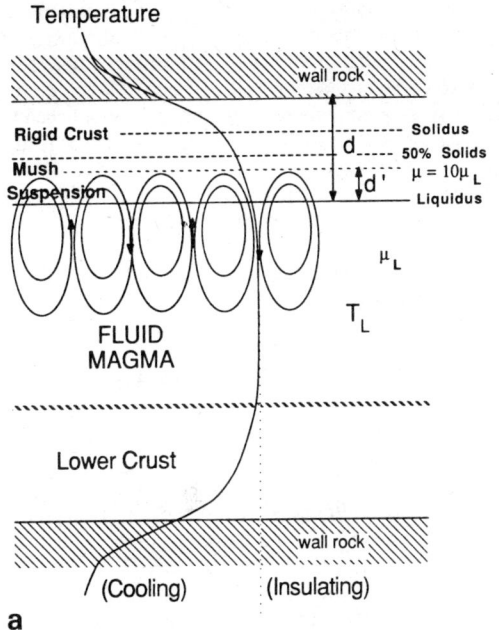
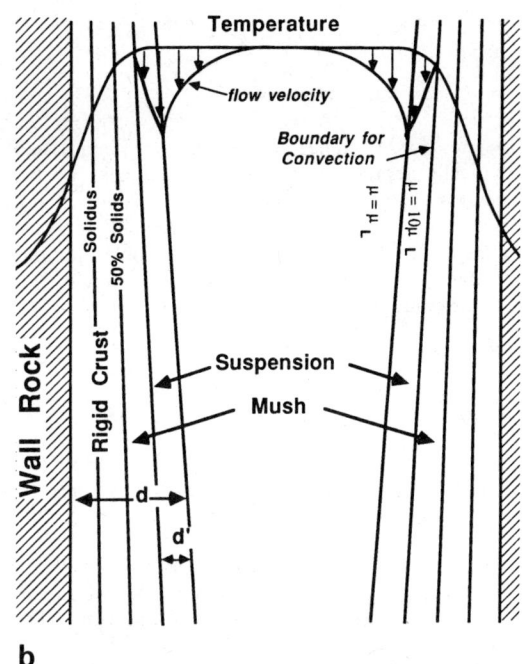

FIGURE 4. Styles of convection in sheetlike (a) and tall (b) magma chambers. (a) Convection is driven by a thin, cool region (suspension zone) forward of the advancing mush zone. The floor region is convectively stable because of the complete absence of heating from the base. Only steady convection is dependent on the thickness of the layer itself. (b) Convection in tall chambers is also driven by the suspension zone, but the characteristic length scale for convection is now the height of the chamber itself.

$$\nabla \cdot \mathbf{V} = 0 \quad (1)$$

Conservation of energy gives the heat equation for temperature (T),

$$\frac{\partial T}{\partial t} + \mathbf{V} \cdot \nabla T = K \nabla^2 T \quad (2)$$

where $\mathbf{V}$ is velocity, the fluid thermal diffusivity K is assumed constant, and a source term, including the possible effect of heat of crystallization and viscous dissipation, is ignored. Conservation of momentum gives the well-known Navier-Stokes equations

$$\frac{\partial \mathbf{V}}{\partial t} + (\mathbf{V} \cdot \nabla)\mathbf{V} = -\frac{1}{\rho} \nabla P + \nu \nabla^2 \mathbf{V} - \alpha g \, \Delta T \quad (3)$$

where the pressure P is a dynamic or flow pressure, g is gravity, α is the coefficient of thermal expansion, ν is kinematic viscosity ($= \mu/\rho$), and ΔT is the departure of temperature from a basic static (e.g., conductive) state.

Complete solution of this set of equations, without approximations, is notoriously difficult. Most problems, however, have features that allow sensible approximations to be made, which amounts to dropping or simplifying certain terms, and solution is readily possible. The means through which such approximations can be judged as sensible or not is *scaling*.

Scaling of these equations renders all terms dimensionless, but even more important it yields equations where each differential term is more or less of the same numerical size. The magnitude or importance of each term can then be judged solely by the magnitude of the dimensionless group or amplitude term in front of each differentiation. Moreover, solutions to the scaled equations hold for any problem having similar dimensionless parameters. This is the *principle of dynamic similarity*.

Although the dimensional scales used are often obvious, experience and physical intuition soon demonstrate the most direct and revealing means of scaling. For this set of equations, a convenient set of characteristic scales are

distance $\quad x_i = L_i x_i^*$

time $\quad t = \dfrac{L^2}{\nu} t^*$

velocity $\quad V = \dfrac{K}{L} V^*$

pressure $\quad P = \dfrac{\rho\nu K}{L^2} P^*$

temperature $\quad T = \Delta T T^*$

where, for example, L has units of length and typically represents the characteristic size of the magma chamber, whereas x^* is a variable without dimensions. These scales are all reasonably intuitive except that for pressure, which is only obvious through experience in analyzing viscous flows.

Substituting these into Eqs. 1–3 gives

$$\nabla \cdot \mathbf{V} = 0 \quad (4)$$

$$\text{Pr}\,\frac{\partial T}{\partial t} + \mathbf{V} \cdot \nabla T = \nabla^2 T$$

$$\frac{\partial \mathbf{V}}{\partial t} + \frac{1}{\text{Pr}}(\mathbf{V}\cdot\nabla)\mathbf{V} = -\nabla P + \nabla^2\mathbf{V} - \text{Ra}\,T \quad (5)$$

where for convenience all the asterisks have been dropped; recall that all variables are free of dimensions, as are all terms in the three equations. The equations are characterized by two nondimensional groups:

Prandtl number $\quad \text{Pr} = \dfrac{\nu}{K}$

Rayleigh number $\quad \text{Ra} = \dfrac{\alpha g\,\Delta T\,L^3}{\nu K} \quad (6)$

The importance of any term in these equations is fully measured by either Pr, Ra, or unity, which is the amplitude of terms not containing Ra or Pr, and all solutions are completely characterized by Ra and Pr. Magmas are highly viscous fluids with small, nearly constant thermal diffusivities, so Pr is thus very large ($\sim 10^4$ or greater). It is well known that if Pr $\gtrsim$ about 10, its effect on any solution is the same as for Pr $\to \infty$. Under this condition of very large Pr, this set reduces to

$$\nabla \cdot \mathbf{V} = 0 \quad (7)$$

$$\mathbf{V} \cdot \nabla T = \nabla^2 T \quad (8)$$

$$\frac{\partial \mathbf{V}}{\partial t} = -\nabla P + \nabla^2\mathbf{V} - \text{Ra}\,T \quad (9)$$

and all solutions are independent of Pr and dependent only on Ra.

The physical meaning of these equations is now fairly clear. Equation 7 is unchanged, still clearly conservation of mass. Equation 8 shows an intimate balance between convective and conductive heat transfer, and since conduction is only important around the edges of the body, where heat is going in or out, it states that the heat transfer is dominated by effects in this part of the flow, i.e., through thermal boundary layers. Equation 9 appears more complex than the others, but the only real physics in it is represented by the last two terms, a balance of viscous effects and thermal buoyancy. The first term is simply a monitor of whether the flow is accelerating, and, since viscous flows rapidly reach a steady velocity state (in time $\sim L^2/\nu$), this term can usually be dropped. The second, or pressure, term appears at the request of the other two and by itself is passive in thermally driven flows, unlike for flow through a pipe where the balance is between the second and third terms.

Rare lava flows of essentially molten magnetite [a compound of magnetite (Fe_3O_4) and ulvospinel ($FeTiO_3$)] have been observed running from at least one South American volcano. These metallic melts, like the Earth's core, have a small Pr (<1), and the essential balance in Eq. 5 is between fluid inertia (second term) and buoyancy (last term). All solutions now depend on Pr and Ra and combinations thereof; the quotient Ra/Pr is the Grashof number (Gr) and the product Ra Pr is sometimes called the Boussinesq number (Bo).

For systems that have no intrinsic horizontal temperature gradients, convection occurs only if Ra exceeds some critical value (Ra_c), which, depending on boundary conditions, is commonly in the range 150–2000. Systems heated or cooled from the side are unstable for all nonzero Ra, for equipotential surfaces can never be isothermal surfaces.

Typical magmas are characterized, aside from crystallization effects, by the Rayleigh number, which is a measure of the strength of induced buoyancy, through heating and cooling, relative to viscous drag retarding motion. All characteristics (i.e., velocity, heat transfer, boundary layer thickness) of the flow will be solely functions of Ra. To see this, consider the fundamental balances as described above in terms of the interaction of the cool, negatively buoyant boundary layers with the rest of the fluid (Fig. 4).

In dimensional terms, convection is a balance between viscous and buoyant forces; that is,

$$\mu\nabla^2\mathbf{V} \simeq \rho\alpha\,\Delta T\,g$$

In terms of fundamental scales, where the surfaces, to begin with, are stress free, viscous drag acts throughout the fluid L to impede buoyancy concentrated in the boundary layers of thickness δ.

$$\frac{\mu V}{L^2} \simeq \rho\alpha\,\Delta T\,g\left(\frac{\delta}{L}\right) \quad (10)$$

where the last quantity allows for the fact that buoyancy exists only in the thermal boundary layers (Fig. 4). Solving for velocity gives

$$V \simeq \frac{\rho\alpha\,\Delta T\,g\delta L}{\mu} \quad (11)$$

The temperature equation (Eq. 8) furnishes a second relation for velocity. That is, in dimensional terms

$$\mathbf{V} \cdot \nabla T \simeq K \nabla^2 T$$

and since convection (first term) scales with a length L, whereas conduction through the boundary layer (second term) scales with δ,

$$V \frac{\Delta T}{L} \simeq K \frac{\Delta T}{\delta^2} \qquad (12)$$

Solving for V and setting it equal to Eq. 11 gives

$$\frac{L}{\delta} \simeq \left(\frac{\alpha g \, \Delta T \, L^3}{\nu K} \right)^{1/3} = \mathrm{Ra}^{1/3} \qquad (13)$$

which is a measure of the heat transferred by convection relative to that for pure conduction under the same ΔT. That is, for conduction in the (imaginary) static system, the heat flux (per unit area) is

$$q_{cd} \simeq K_c \frac{\Delta T}{L}$$

whereas for the convecting system the heat flux is that due to conduction across the thin thermal boundary layers,

$$q_{cv} \simeq K_c \frac{\Delta T}{\delta}$$

where K_c is thermal conductivity.

The ratio q_{cv}/q_{cd} is a pure (dimensionless) number called the Nusselt number (Nu). That is,

$$\mathrm{Nu} \equiv \frac{q_{cv}}{q_{dc}} = \frac{L}{\delta} \qquad (14)$$

and then, from Eq. 13,

$$\mathrm{Nu} \simeq \mathrm{Ra}^{1/3} \qquad (15)$$

If the boundaries are not stress-free but rigid, the dominating drag acts over the length scale δ (i.e., $\mu V/\delta^2$ in Eq. 10) and the exponent becomes 1/5. And if the boundaries are rigid with buoyancy not modulated by the factor δ/L in Eq. 10, as perhaps with internal heating, the exponent becomes 1/4. A multitude of experiments determined to measure this exponent generally agrees, at large Ra, with these scalings and commonly finds an exponent of approximately 1/3.5. These approximations (Eq. 15) can be made equalities by inserting a constant of proportionality, which is of order unity (i.e., $c \simeq 0.15$–0.50) and can only be found from experiment or an analytical solution. Or, in general,

$$\mathrm{Nu} = c \, \mathrm{Ra}^b \qquad (16)$$

In flows where Ra is small ($\sim 10^3$), the boundary layer approximation does not hold and b is closer to unity.

Convection velocity is easily found from Eqs. 14 and 12:

$$V \simeq \frac{K}{L} \mathrm{Ra}^{2/3} = \frac{K}{L} \mathrm{Ra}^{2b} \qquad (17)$$

The rate of convection increases as Ra^{2b}. Multiplying Eq. 17 by L/ν allows the Reynolds number (Re, a measure of inertial to viscous effects) for the flow to be found:

$$\mathrm{Re} \simeq \mathrm{Pr}^{-1} \mathrm{Ra}^{2b} \qquad (17a)$$

or, using Eq. 16,

$$\mathrm{Re} \simeq \mathrm{Pr}^{-1} \mathrm{Nu}^2 \qquad (17b)$$

Since $\mathrm{Pr} \gg 1$, the magnitude of Nu is of paramount importance in deciding Re.

The boundary layer thickness is given by Eq. 13:

$$\delta \simeq L \, \mathrm{Ra}^{-1/3} = L \, \mathrm{Ra}^{-b}$$

which decreases with increasing Ra, making the outward (and inward) heat flux larger.

The important lesson here is that all information pertaining to the flow in a magma chamber relies in some way on an accurate knowledge of Ra. The basic functional forms of the relationships hold for any magmatic convection, although the exponent will change, as will the definition of Ra.

For a shallow (relative to width) horizontal layer heated from below, Ra is as defined by Eq. 6, but for a tall, narrow chamber, or convection due to sidewall cooling, the length scale in Ra is distance along the wall (Fig. 4). If the flow (either magmatic or hydrothermal) is through a porous material, Ra depends on the permeability K_p:

$$\mathrm{Ra} = \frac{\alpha g K_p \, \Delta T \, L}{\nu K}$$

and, consequently, only on length L to the first power.

The opportunity of knowing the spatial distribution of temperature and velocity is lost through a scaling analysis, which depends on global averages, but its ease, versatility, and accuracy far outweigh this disadvantage.

Scaling and the ensuing parametric analysis are especially valuable in studying systems as complex as magmas, where physical properties may vary strongly with temperature and the relevant scales may initially be in question.

Magmatic Longevity—Heat Transfer

Whether magma is ascending, stewing in a chamber, or in eruption, its life is its heat. Cooling kills magma. Thus to understand the magmatic life cycle, its rate of cooling or transfer of hotness to its surrounding is of paramount importance.

The rate of change of mean temperature (T) is directly proportional to Nu, and cooling can therefore be linked directly to internal or external convection. That is, for conservation of energy

$$C_p V_0 \frac{dT}{dt} = -Q_T$$

where C_p is specific heat, V_0 is volume, t is time, and Q_T is the total heat away from the body. (The effects of heat sources or sinks can also be included.) From the definition of Nu in Eq. 14, we obtain

$$Q_T = Q_{cd} \text{Nu} = K_c \frac{T - T_w}{L} \text{Nu} \quad (18)$$

where T_w is the wall rock temperature far from the body. Combining this with Eq. 18, with account taken of the area of the body (A), gives

$$\frac{dT}{dt} = -\left(\frac{K_c}{\rho C_p}\right)\left(\frac{A}{V_0}\right)(T - T_w) \text{Nu} \quad (19)$$

As might be expected, cooling is proportional to the conductivity of the wall rock, the area-to-volume ratio of the body, the temperature contrast, and Nu; it is inversely proportional to the heat capacity of the body. Once the shape, initial temperature, and wall rock of the body are chosen, the change in mean temperature is easily found, providing a proper value of Nu is known. If, as in Eq. 16, Nu also depends on the temperature contrast, through Ra, the solution is not as easy, but the expression can still be expanded and solved analytically. The key obstacle is a proper choice of Nu or Ra.

Convective Style and Ra Magnitude

A layer heated from below can be maintained at any level of vigor by appropriately choosing the rate of heating, regardless of rheology. The buoyancy driving the flow comes both from the heating at the base and from the cooling at the top, such that the net effect is a force couple across a lever arm of length L, the layer thickness. The motion is, mechanically speaking, essentially a thermally pumped flywheel.

The rate of heat transfer *from* the layer is given by Eq. 16, but, because an equivalent amount of heat is continually being added through the base, the mean temperature, as described by Eq. 19, remains constant. Put another way, if Ra is taken to be based on L and Eq. 16 is used in Eq. 19 to give the temperature history of the layer, it will cool at a rapid rate. Even when the hot layer is insulated on all sides other than the top, convection and cooling appear dominated by the length scale L. This is so because cool fluid, sinking from the roof, is steadily replaced by deeper, warmer fluid. The net result is that the tank convects as if it were heated from below or from within. But because no heat is replaced, convection diminishes rapidly and overall is transient and fleeting.

Magma chambers are most commonly perceived as unheated bodies of magma cooled on all sides. There is certainly clear and persuasive evidence that replenishment occurs, but it is nowhere near the requisite volume needed to act as anything more than a short-term pulse of heat. What, then, is the vigor of convection? How big is Nu (or Ra)? What is the temperature history?

Maximum Nu. An unheated layer can transfer heat through the top, say, no faster than if it is kept thermally well mixed and of uniform viscosity for the duration of cooling. That is, the layer is always of spatially uniform (although changing in time) temperature, and heat must be conducted away through the infinite overburden. The maximum rate of heat transfer relative to that of a stagnant, conducting layer that began cooling at the same instant, is easy to deduce. A stagnant layer has, in effect, a thermal boundary layer on each side of the upper (and lower) boundary; the temperature at the boundary just after initiation of cooling is, for equivalent thermal properties, the average of the two initial temperatures.

In marked contrast, a thermally well mixed body has only a single boundary layer, residing in the wall rock, and the boundary temperature is always the temperature of the magma itself. Except for the doubling of the very early boundary temperature, the two problems are identical in terms of the heat transfer in the overburden. Doubling the early contact temperature doubles the earliest heat flux. Thus the maximum Nu, as here specifically defined, is 2.0, the full temperature history is straightforward to calculate, and it is also observed experimentally. The actual calculation gives

$$\text{Nu} = 2 \frac{1 - \sqrt{\pi} \eta e^{\eta^2} \text{erfc}(\eta)}{1 - e^{-1/\eta^2}}$$

where $\eta = \sqrt{4Kt}/L$ is dimensionless time and L is the full-layer thickness. As $\eta \to 0$, Nu $\to 2$ and the overall variation with η is given by Fig. 5, which shows Nu decreasing to near unity when $\eta \simeq 0.5$. The maximum rate of heat transfer from this thermally perfectly well mixed body is only twice as great as if the body were stagnant and cooling purely by conduction.

An effective Rayleigh number necessary to supply the requisite heat flow from the well-mixed body can be estimated by using Eq. 16. With, for example, $c = 0.15$ and $b = \frac{1}{3}$, Ra = $(\text{Nu}/0.15)^3$, and values of Ra calculated as such are given along the right axis of Fig. 5. Ra decreases from a maximum of about

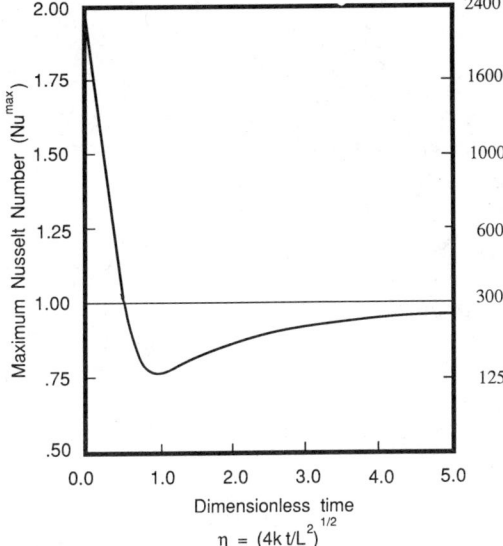

FIGURE 5. The maximum heat transfer in terms of a special Nusselt number (Nu) as a function of time (dimensionless) for a layer of thickness L perfectly well mixed thermally. On the right axis is the effective Rayleigh number necessary to maintain this heat flux.

2400 and drops below about 300 for $\eta >$ about 0.5, beyond which point the connecting layer cools more slowly than the same layer by time-dependent conduction. This specially defined Nu cannot strictly be used in Eqs. 14–16, which assume a different basic state, and so these Ra are only illustrative.

This estimate of an effective Ra is in sharp contrast to that found by directly using the common definition of Ra, as in Eq. 6, where for $L = 1$ km, $\alpha = 5 \times 10^{-5}$ °C^{-1}, $g = 10^3$ cm/s^2, $\nu = 10^3$ cm^2/s, $K = 10^{-2}$ cm^2/s, $\Delta T = 10$°C, Ra $= 5 \times 10^{13}$ and Nu $\simeq 4 \times 10^3$. If such an enormous rate of heat transfer were maintained and no heat were added, virtually any magmatic body would solidify in a geologically very short time. That is, from Eq. 19, $dT/dt \sim$ Nu. The fact that all erupted magmas are never observed to be superheated reflects the ease of large Ra convection to evacuate any potential superheat. Such convection maintains or buffers these systems at their convective liquidus, where Ra is small and the effects of crystallization dominate convection.

The slow, small Ra rate of cooling of magma can be checked by comparing such calculations against the observed rate of cooling of Hawaiian lava lakes (Fig. 6). The well-mixed body cools much faster than the lava lakes, which more closely follow the cooling of a stagnant layer without overburden but with latent heat. Applying Eq. 19 to the core region of the lava lake suggests that, since the rate of decrease of mean temperature is very small, the rate of convective heat transfer must also be small. This rate of cooling is much slower than that [see curve THS (1986) of Fig. 6] calculated based on Eq. 16, with Ra dependent on L^3, where L for a lava lake may be 14–100 m, and Ra $\simeq 10^8$–10^{13}.

Taken altogether, these calculations and observations suggest that the rate of magmatic heat transfer is small and Ra is small when the effects of solidification (e.g., nucleation and strongly variable viscosity) dominate convection. But what is the form of convection? Could the body be truly stagnant?

Style of Convection. There is little doubt that some form of convection takes place in such sheets of magma, but it is certainly not the usual, layer-filling cellular type so common to one's imagination. When an otherwise insulated layer is cooled from above, a cold, dense boundary layer grows downward from the upper surface. When it is thick enough, the layer drops away as plumes descending into the deeper, hotter regions of the layer. The boundary layer is destroyed and begins to reestablish itself. The descending plumes are warmed and eventually thermally mix with the lower fluid. If the layer is thin enough, the plumes may reach the floor, forming a cool stagnant basal layer where further cooling may also occur.

This instability is unusual in that it must continually reestablish itself and thus, of necessity, always remains near some crucial value of Ra. Both the initial and final states are isothermal and stable, and convection is merely a transitory linking process. The Rayleigh number describing this transient motion is quite different, for the fundamental length scale is not L but the thermal boundary layer thickness, which is proportional to $(Kt)^{1/2}$. A host of studies, both experimental and analytical, show that for this type of convection the product of the usual Ra (based on L—i.e., Ra$_L$) and the dimensionless time for the onset of convection to the power $\frac{3}{2}$ is a constant.

$$\mathrm{Ra}_L t_L^{3/2} = \frac{\alpha g \, \Delta T \, L^3}{\nu K} \left(\frac{Kt}{L^2}\right)^{3/2} = \mathrm{Const.}$$

or

$$\mathrm{Ra}_t \equiv \frac{\alpha g \, \Delta T (Kt)^{3/2}}{\nu K} = \mathrm{Const.}$$

The full-layer thickness L drops out and is insignificant, being replaced by the thermal boundary layer thickness

$$d = E(Kt)^{1/2} \qquad (20)$$

E is a constant (e.g., $\sqrt{\pi}$, 3, 4) that depends on the exact definition of d in terms of, for example, 80%, 90%, or 99% of the full temperature variation. Then

$$\mathrm{Ra}_d \equiv \frac{\alpha g \, \Delta T \, d^3}{\nu K} \qquad (21)$$

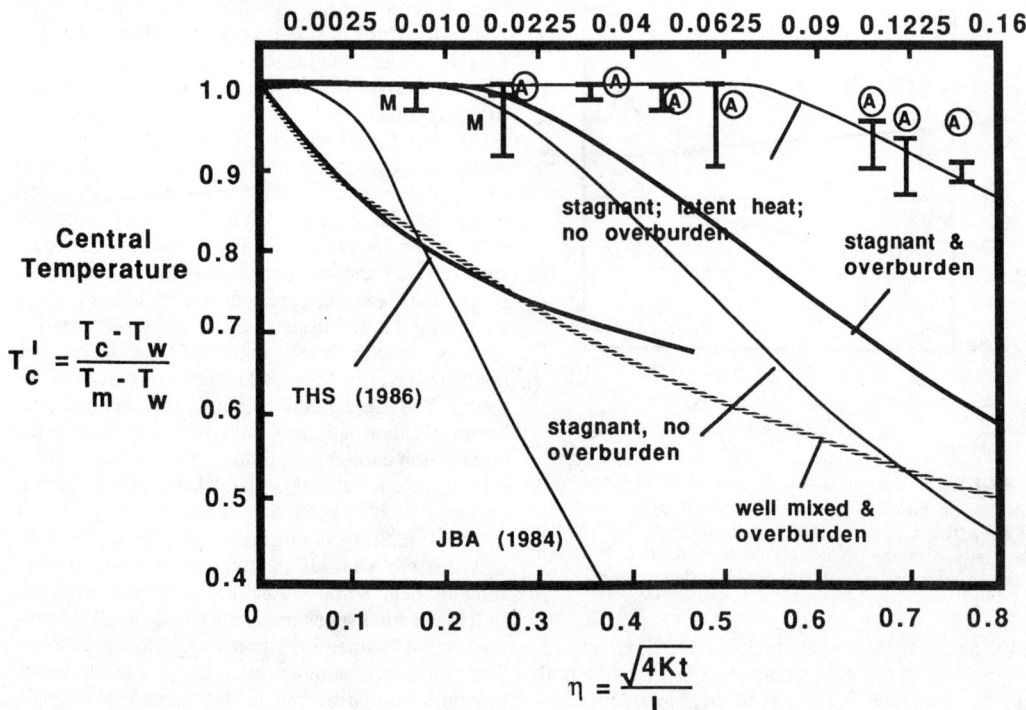

FIGURE 6. The central temperature (dimensionless) of a layer of thickness L with time (dimensionless) for thermally well mixed and stagnant layers with and without overburden and latent heat. The two curves marked JBA and THS represent, respectively, experiments and theory for well-mixed cooling. The data bars shown across the top are from Hawaiian lava lakes Makaopuhi and Alae. The closest agreement is with the stagnant layer.

The inherent time dependence of the boundary layer thickness makes both Ra_t and Ra_d also time-dependent. There is no escape from this feature; it is a fundamental characteristic of this instability.

The critical value of Ra_t, beyond which convection sets in, is generally about 200, but this value depends somewhat on the exact nature of cooling at the boundary, and thus there is some time before the interior temperature feels the full effect of convective cooling. This lag in cooling is clear in the experimental results shown in Fig. 6 [see curve JBA (1984)].

Most of this discussion implicitly assumes a constant-viscosity fluid, but the same general process also occurs in fluid whose viscosity varies as strongly with temperature as that of magma. An important difference is that the full thickness of the cool, dense, and highly viscous boundary layer is not itself unstable; only the outermost part (suspension zone of Figs. 3 and 4), inward of the point of critical crystallinity is unstable. If the rigid crust grows downward too fast, it will continually overtake incipient fluid instabilities, thus stifling convection. Hence, there is competition between the growth of instabilities and their capture by the growing crust. If the crust, although rigid, were itself not stable and could plunge back into the magma, the effect of variable viscosity would be minimized and the temperature would follow the well-mixed model. This foundering of crust happens during the earliest phase of lava lake solidification, but shortly thereafter a stable crust forms and the interior temperature changes slowly. In crustal chambers there is no reason to expect foundering unless the wall rock roof melts and does not form a competent attachment structure, which apparently has happened for some large mafic bodies such as Bushveld in South Africa.

Crust growth and instability capture further modify Ra in a predictable fashion. If the crust is taken to grow at velocity V, then, judging from Eq. 20, an appropriate length scale for d in Eq. 21 is K/V and

$$Ra_V \equiv \frac{\alpha g \, \Delta T \, (K/V)^3}{\nu K}$$

When V is large, Ra_V is small and no convection occurs. In time V decreases and Ra_V increases to the point that convection sets in. Decreasing V with time reflects thickening of the crust and loss of initial magmatic heat, reducing the upward heat flux, which restricts the energy transfer available to drive convection. It is therefore doubtful that the vigor of convection ever becomes any greater than it is shortly after the onset of instability.

Magma bodies have a finite amount of energy that they lose in a fashion controlled by conduction in the wall rock. This gives the magma a much longer thermal life than if it cooled without wall rock and at the mercy of convection within the body itself. But can the heat transfer in the wall rock be increased by any means? Yes. Vigorous hydrothermal circulation in the wall rock can greatly speed heat transfer, over pure conduction, and increase Nu significantly. Hydrothermal systems set up in response to heat from the magma, but, because this takes some time, there is a significant time lag between solidification of the body and the enhancement of Nu. The amplified cooling rate is most effective in the later stages of cooling, when the body is beyond critical crystallinity and dynamically dead.

Crystal Distribution

A collection of rocks from any volcano will show a range of chemical compositions whose explanation is the raison d'être of igneous petrology. Every crystal has a composition different than its magma, and separation of liquid and solid gives a differentiated or fractionated magma. Any number of compositions can be produced by this method of crystal fractionation; without any exaggeration, it is the central tenet of igneous petrology. If we accept that some form of such separation takes place, it is intriguing to investigate the distribution of crystals in a dynamically evolving magma chamber.

Consider a general convective flow whose fluid elements follow paths defined by a stream function $\psi(x, z)$ (x horizontal, positive to right and z vertical, positive upward, both from middle of base of a rectangular chamber). The horizontal and vertical velocity components are given by, respectively,

$$U_x \equiv \frac{\partial \psi}{\partial z} \quad \text{and} \quad U_z \equiv -\frac{\partial \psi}{\partial x}$$

which by definition satisfy the continuity equation (Eq. 4).

If the solid particles are neutrally buoyant and disperse, their paths will be given by this same stream function. But since they generally tend to sink or rise, their stream functions will differ. Given that gravity provides a single, vertical velocity component, the horizontal particle velocity can only be

$$V_x = U_x \tag{22}$$

but the vertical particle velocity is modified by convection

$$V_z = -V_g + U_z \tag{23}$$

where V_g is the settling velocity (i.e., downward and negative) in still fluid. It is immediately clear that Eqs. 22 and 23 do not satisfy the continuity equation unless $V_g = 0$, and thus the particle stream function is distinct from that of the fluid alone. Continuity is satisfied, however, by defining a particle stream function as

$$\phi \equiv V_g X + \psi$$

The paths of particles in a convecting fluid are found simply by inserting the fluid stream function and tracing out the positions of constant values of ϕ. Contours of ω forming a closed loop indicate complete particle retention. At this point, no assumptions have been made about the nature (i.e., rheology) of the fluid or the function V_g, which could involve the effects of crystal concentration. It is assumed, however, that the redistribution of particles does not affect the flow itself.

The single nondimensional parameter that determines particle behavior is

$$S \equiv \frac{V_g}{U_0}$$

where U_0 is the magnitude of convective velocity. There are four possible ranges of S:

1. $S \simeq 0$: The particles are neutrally buoyant and follow the fluid elements. $V_g = 0$, $V_x = U_x$, and $V_z = U_z$.
2. $S \gg 1$: Either the settling (rising) velocity is very large or the convective velocity very small, but the fluid motion has no effect on the particle. $V_z \simeq V_g$ and $V_x \simeq 0$.
3. $S = 1$: The particle is barely supported in the strongest part of the flow; eventually all particles will settle out.
4. $0 < S < 1$: There is strong competition over most of the flow between the particle and fluid stream functions. The distribution and possible retention of crystals strongly depend on the nature of the flow.

Figure 7 shows distributions for sinking and floating particles in a single cellular flow. The greatest retention is in the strong part of the flow, relative to the particle velocity. Sinking crystals are retained in the central portion, and floating crystals at the side walls of the chamber. Magma containing two solid phases (e.g., olivine and plagioclase), where one sinks and the other floats, could isolate each phase with such a flow field. As convection

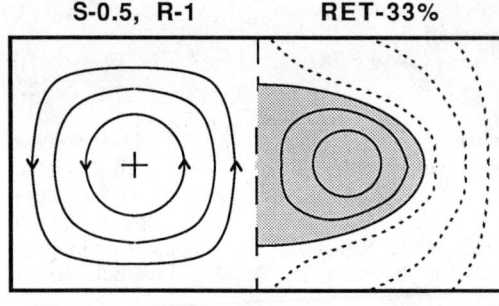

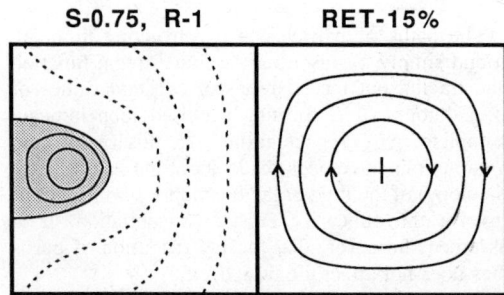

FIGURE 7. Crystal retention in convecting magma. The arrowed lines show fluid stream lines, and all other lines show particle path lines. Particle retention occurs within the stippled region. *Top:* Heavy particles are retained in the central section, which actually continues across the center line, and for $S = 0.5$, 33% (v) of the crystals are retained; the chamber aspect ratio $R = 1$. *Bottom:* Light particles are retained along the margins of the system, and when $S = 0.75$, there is 15% retention.

weakens or S increases due to crystal growth, layers rich in each phase might be formed at the top and bottom of the chamber. Since any eruption of magma will evacuate the less viscous part of the chamber, it is possible that the erupted magma may show a more or less evolved chemical composition, depending on the density relations between solids and liquid.

Present Trends

Magma physics is the study of the physicochemical origin and evolution of molten rock. Its origin in the early part of the century was headed by R. A. Daly and F. F. Grout, whose ideas, although still important, suffered in their detailed development because of the unavailability of techniques and results still to be developed in heat transfer and fluid mechanics. The boom of engineering and applied mathematics in the last 30 years has provided a wealth of fundamental concepts upon which present-day magma physics has been founded in the last 20 years. H. R. Shaw is without question the founder in this respect.

The problems of magma physics are generally so complex (e.g., change of phase, variable viscosity, multiphase flows) that without the heuristic and scaling techniques of fluid mechanics there would normally be little progress. Experimental results have revealed powerful simplifying principles, such as separation of flows of strongly varying viscosity into an active region of near-uniform viscosity and a highly viscous, rigid region, which acts as part of the container. The basics of chemical and mechanical sedimentation in convecting magmas are steadily revealing a veritable garden of interesting results. Magma physics now enjoys the position of making fundamental contributions to the sciences to which it is itself indebted.

BRUCE D. MARSH

Bibliography

Hubbert, H., 1986, The intrusion of fluid mechanics into geology, *Jour. Fluid Mechanics* **173**, 557–594.

Marsh, B. D., 1987, Magmatic processes, *Revs. Geophysics* **25**, 1043–1053.

Marsh, B. D., and M. R. Maxey, 1985, On the distribution and separation of crystals in convecting magma, *Jour. Volc. Geotherm. Research* **24**, 95–150.

Shaw, H. R., 1965, Comments on viscosity, crystal settling and convection in granitic magmas, *Am. Jour. Sci.* **263**, 120–152.

Spera, F. J., D. A. Yuen, and D. V. Kemp, 1984, Mass transfer rates along vertical walls in magma chambers and marginal upwellings, *Nature* **310**, 764–767.

Cross-reference: *Magmatic Processes.*

MAGMATIC PROCESSES

Primary Role of Heat Transfer

Magma is a mobile mixture of liquids, crystals, gas bubbles, and debris that intrudes within the Earth or erupts as lava. The principal process controlling the transformation of hot magma into cold rock is heat loss. The transformation may involve additional interactive processes that result in variations in magma during cooling. But not all magmas "begin" with the same temperature and/or constitution. Thus any inventory of magmatic processes must also account for the generation of magmas with different initial temperatures and compositions.

Magma generation is a manifestation of planetary heat loss and largely results from the internal convection of the Earth's mantle, which drives the plate tectonic cycle. As with the solidification of magma, the generation of magma involves a number of interactive processes that cause variations to occur. This article describes the processes of magma generation and the subsequent transformation of magma into igneous rock, and illustrates their interaction. A common theme of many of the interactions is that instabilities develop, magmatism itself being a consequence of thermal instability in an otherwise orderly planetary interior.

Tectonic Setting

Naturally occurring magmas are thermally and chemically uniform neither in space nor time. Certain broad associations of types of magma with specific plate tectonic settings do exist. For instance, tholeiitic magmas erupted on accreting plate margins at mid-ocean ridge spreading centers are hot (1200–1300°C). This contrasts with the cooler (1100–1200°C) calc-alkaline magmatism at convergent plate margins above subduction zones. This contrast is systematic and encourages the view that the thermal regime and perhaps stress states of tectonic settings control magmatic effusions. Nevertheless within specific tectonic areas, at neighboring eruptive centers, or even at single volcanoes, the variations of magmas in space or time are often as profound as the variations that characterize different provinces. These variational patterns may occur or repeat in ways quite difficult to reconcile with simple control by gross thermal regime. Thus additional controls must cause the observed variations and we focus on the processes rather than on the plate tectonic synthesis. In this respect it is important to recognize that planets without plate tectonics (e.g., Moon, Mercury, Mars) also have interesting ranges of magmatic types. This reinforces the need to concentrate on process rather than tectonics.

Outline of Magma Genesis

Magma generation requires (1) source material, (2) a fusion process, and (3) a plumbing system to collect and deliver the magma produced.

Variations in the product magma can be introduced by perturbations in the processes controlling any of these three aspects of magma generation (Fig. 1). The nature of the variations produced (and observed) requires special comment. The physical state of material can be changed during magmatic evolution. But magmas also have compositions and compositional evolution that indicate the operation of control processes. It is almost not an exaggeration to say that magma composition is the most central issue. The discovery of and understanding of the role of various physical processes causing compositional variation in igneous rocks has been a major concern in petrology for the past 100 years. As we shall see, the physical processes that operate are not entirely divorced from the chemical state during the evolution. Physical process and chemical state are mutually interactive. Thus compositional variation is not only the most commonly observed marker of magmatic evolution, it is also a determinant of that evolution.

Source Region Processes

The constitution of the source region of magma has a direct bearing on magma production. At some level the processes of planetary accretion and primordial differentiation determine the general character and range of variation of magmatic source regions. The peridotitic nature of the upper mantle source of basalt magma was established early in Earth history, yet the upper mantle remains a heterogeneous and dynamically evolving mass. Solid state convection driven by the release of the Earth's internal heat is the primary cause of this dynamism. Most melting in the Earth's interior occurs as the result of pressure release in convective updrafts. Hot buoyant, solid material decompresses during ascent and partially melts. The extra buoyancy and fluidity of the liquid phase leads to its segregation from the residual crystalline feedstuff. Because the separated magmatic liquid and residual crystals are of different compositions, chemical differentiation is produced by the partial melting process. The volume and chemistry of the magma depend on whether the particular source parcel has suffered previous episodes of melt extraction leading to depletion in fusible components and whether the source parcels have been enriched in fusible components by the incorporation of melts produced elsewhere. The latter may occur as a result of injection and trapping of magma interrupted in transit from some other source parcel to the surface; it may also occur if mantle convection remixes solidified lavas back to the heated depths for another round of convective upwelling. For instance, the heterogeneous bands seen in Fig. 1 as (2) or (4) could be subducted and deformed ancient oceanic crust. Thus the previous magmatic activity of the source region can be a strong determinant of the character of magma produced.

Fusible components and geochemically incompatible elements (incompatible from the point of view of occupying the crystal structure of the major minerals present) may also be enriched or depleted by the agency of fluids rich in H_2O, CO_2, CH_4, SO_2, etc. The process of changing the chemistry of a rock parcel is called metasomatism whether the agency is fluid migrating along microscopic pores or silicate liquid forming macroscopic injections. That metasomatism does occur is a conclusion drawn from geochemical study of pieces of the Earth's mantle in the form of xenoliths entrained in erupting basalt and kimberlite. The isotopic signatures in basalt and preserved zonation of mineral chemistry in xenoliths suggest that metasomatism has often immediately predated eruption. It may even have provided the instability leading to melting and eruption. The relative importance of various liquid and vapor agents of metasomatism is an active area of current inquiry.

Once heterogeneities become established (for

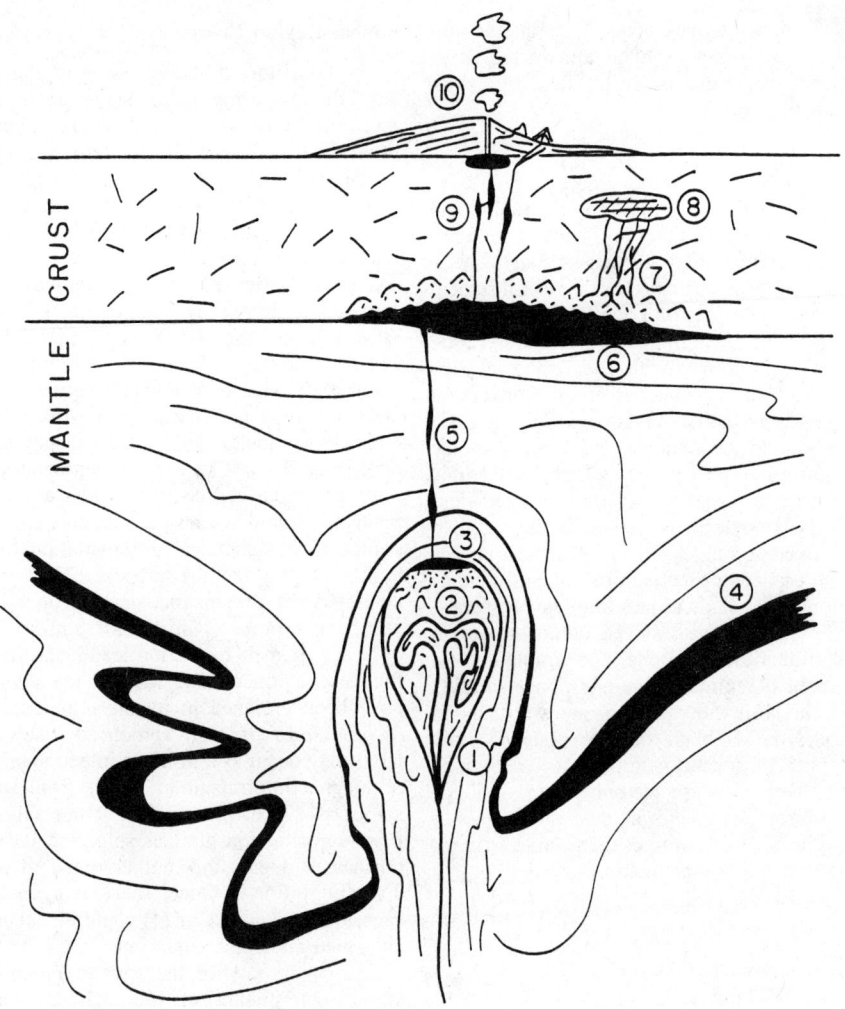

FIGURE 1. Overview of magma generation.
1: Rising solid state diapir; 2: Deformed heterogeneity; 3: Percolated melt; 4: External mantle heterogeneity; 5: Elastic melt-filled crack; 6: Magma pond at base of crust; 7: Partial crustal fusion and contamination of melt pond; 8: Secondary magma body; 9: More crack propagation; 10: Volcano.

instance by accretion, previous magmatic activity, or metasomatism) mantle convection can further process the materials. Contrasting substances may be juxtaposed and eventually blended, producing new compositions. Thus, although convection is responsible for producing heterogeneities by triggering magmatic differentiation, it may also act to erase them by mixing them back together. Mixing in mantle convection might be thought to act on the scale of the parcels in motion (kms and up). However sufficient stretching occurs in the convective motion that equant blobs of km dimension may be reduced to ribbons of cm-m dimension in selected regions (compare (3) and (4) in Fig. 1). Thus mixing by convection can be locally effective enough that diffusion on the grain-size scale (mm-cm) can complete the homogenization process. However for source region enrichment by convective mixing to have impact on magma character, the material need only be admixed on a scale less than that sampled in the melting process. Fine scale homogenization, for example by diffusion, is superfluous.

Fusion Processes

As mentioned earlier, most magmatism in the Earth's mantle is thought to result from adiabatic pressure-release melting (see Fig. 2). (Adiabatic signifies that the process occurs without addition or evolution of heat. In the case of large convective parcels of Earth materials, thermal conduction is simply too slow to allow significant heat transfer to

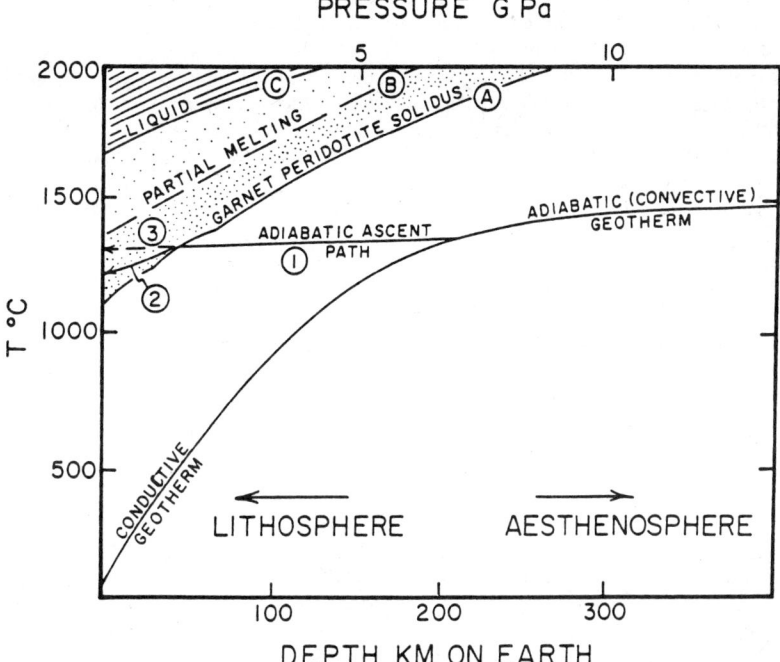

FIGURE 2. P-T of magma generation.
A: Solidus for garnet peridotite; describes the initial melting curve for the upper mantle. B: Solidus for refractory material such as harzburgite. C: Liquidus for mantle; complete melting curve.
1: Adiabatic ascent path of parcel. 2: Pressure release adiabatic melting of garnet peridotite. 3: Failure of harzburgite of same initial temperature as (2) to melt.

Heavy curve gives an average geothermal temperature distribution. Within the lithosphere, temperatures rise rapidly with depth and heat transport is primarily by conduction. Within the aesthenosphere the average temperature distribution is adiabatic because the effective viscosity is low enough to allow solid state convection to be the principal mode of heat transport. Regions of ascending parcels will be warmer than average; descending parcels will be cooler. In regions around mid-ocean ridges for example, the lithosphere will be thinner and convective updrafts from the aesthenosphere can intersect the solidus and partially melt.

or from the parcel.) Compressed materials have energy stored in them in the same way a spring can store energy. Work needs to be done in order to compress them. If the confining pressure is relaxed, then this energy can be recovered. The energy of compression is released during ascent and decompression and may drive the fusion process. For partial melting to occur, the solid material at depth must have been sufficiently warmed to intersect the pressure-temperature condition of liquid stability during ascent. Material insufficiently warmed to reach the melting regime will simply ascend along an adiabatic P-T path without melting. Ascent paths are illustrated in Fig. 2. The temperature distribution along an adiabat leads to cooling during ascent if the ascent is sufficiently slow to be approximately reversible in the thermodynamic sense. Although energy of compression is released, which would tend to raise T in the absence of other effects, it is more than compensated by the P-V expansion work done on the surrounding medium. Thus a slight temperature drop is sustained by ascending convective parcels of Earth materials (see (1) in Fig. 2). Ascending parcels that are hot enough to enter the melting regime sustain larger temperature drops per km of rise because more energy of compression is absorbed in the fusion process and because more P-V work is done on the surroundings by the expansion caused by melting (see (2) in Fig. 2). Consider two parcels that start at the same temperature at the same depth but one of which is chemically too refractory to enter the melting regime. Adiabatic ascent produces the apparently paradoxical result that the partially molten material is cooler than the unmolten material (compare (2) and (3) in Fig. 2). The paradox is, of course, easily resolved by reference to the effect of material composition, specifically here the inventory of fusible components. Again this highlights the importance of source region composition, and its variation, upon magmatic fusion processes.

The adiabatic fusion process associated with convective movements of the planetary interior derives its energy from the dissipation of the planet's internal heat. This heat may be primordial from accretion, from large scale early differentiation of the core, or it may be from long term radioactive decay. In any case, the parcels that ascend and melt are driven by the larger picture of planetary thermal and gravitational energy dissipation. The energy which leads to the fusion is not supplied locally to the melting parcel at the time of fusion. Rather it is recovered at that time after being supplied earlier by slow heating of larger volumes of deep, compressed planetary interior. The energy content of parcels that rise to melt is probably not dramatically different from that of their initial neighbors. They only need be slightly higher than their neighbors in order for large parcels to become buoyant, to rise, and to melt. Buoyancy may be triggered by small heterogeneities in radioactivity or chemistry. Once ascent begins, no further input of energy is needed to drive the fusion as it occurs.

Fusion may also occur, however, by local energy input at the time of fusion. Although few examples are known, it is conceivable that geochemical concentration of radioactivity could be so intense as to lead in special circumstances to local partial fusion. Likewise local partial fusion associated with shearing is known to occur within crustal mylonite zones, and viscous shear dissipation may play a limited role in the magmatism associated with subduction.

Pressure release melting is usually considered to be adiabatic because of the difficulty in transferring heat to or from an ascending parcel. But radioactivity and shearing can overcome this difficulty by depositing their energy directly within the parcel. Radioactive decay will not be important in the ascent time scale unless some enormous geochemical anomaly exists. But internal deformation of the parcel may be a non-trivial mechanism for energy dissipation in aid of fusion. A limit to the possible magnitude of this effect can be roughly found from consideration of the gravitational potential energy released during ascent. If one assumes that all the gravitational potential energy is dissipated by viscous stirring of the materials involved in the convection, then the contribution to the fusion process may be substantial. The energy per gram of parcel could easily be as much as the heat of fusion. However the partition of this energy between the parcel and disturbed surroundings is not well known. It will not be better known until mantle rheology as a function of stress, temperature, and pressure is better characterized. Future research is needed in this area.

A spectacular example of fusion driven by energy input at the time of volcanism is provided by Jupiter's innermost Gallilean satellite, Io. Viscous dissipation by tidal forcing in Jupiter's gravitational field is apparently responsible for the volcanic plumes observed by the *Voyager I* space mission. It has also been suggested that tidal dissipation may have been important in early lunar volcanism when the Moon was only slightly beyond the Earth's Roche limit.

Perhaps the most important type of magmatism that arises through local energy addition occurs by the emplacement of hot, mantle-derived magma within or at the base of the crust. The cooling of basic magma can lead to substantial partial fusion of crust, resulting in calc-alkaline and silicic magmatism. Continental silicic volcanism may be a thermal consequence of basalt underplating [see Fig. 1 parts (6), (7), and (8)]. For instance the rhyolite ash flows of the Island Park caldera and the Yellowstone region may be viewed as a precursory phase to the propagation of adjacent Snake River Plain basaltic volcanism to the northeast. Current and past Snake River Plain basaltic volcanism was preceded by silicic volcanism, which suggests that basalt ponds within or below, and partially fuses, the crust before it breaks through to erupt. Likewise the calc-alkaline volcanism on continental crust above Benioff zones may be driven by the interaction, both thermal and chemical, between basaltic magma and crust. The plagiogranites of ocean ridge systems may be, in part, the result of partial fusion of (altered) crust by off-axis volcanism.

Even though we can enumerate and give examples of a variety of fusion processes, it is important to appreciate the leading role of adiabatic pressure-release melting. Not only can that process draw upon the vast internal energy reserves of the planet, but it can provide the energy for the secondary magmatism just mentioned. Considering the ubiquity of pressure-release volcanism in time and space, we may conclude that the Earth's energy budget has been sufficient to maintain the Earth's temperature hovering about the melting point since primordial differentiation. In a sense the Earth's magmatic differentiation is still in progress. However changes in the potency of the pressure-release process in generating hot magma may be seen in the geological record. The komatiitic volcanism of the Archaean period of Earth's history erupted with temperatures up to $\sim 1600°C$, which is considerably in excess of the hottest lavas erupted today $\sim 1300°C$. This difference suggests a secular cooling of the earth of $\sim 100°$/billion years since the Archaean.

Geochemical Consequences of the Fusion Process

The compositional difference between liquids produced during partial fusion and residual crystals promotes chemical fractionation as these phases are separated from one another. By definition, the incompatible elements are enriched in the mobile liquid phase because they are incompatible in the structures of the residual crystals. As it happens, the most potent radioactive heat-producing nuclides ^{40}K, ^{238}U, ^{235}U, and ^{232}Th are moderately-to-strongly incompatible in the olivine, pyroxenes, garnet, spinel, and plagioclase thought to comprise the upper mantle. Partial melting may then be expected to strip source material of its internal heat sources. Insofar as the magmas ascend, often to the crust, we expect the heat producing elements to be strongly concentrated upward in a differentiated Earth. These expectations might be confounded if, as has been occasionally suggested, K becomes compatible in high pressure sulfides segregating to the Earth's core, or if liquids are not buoyant compared to the crystals from which they are produced by partial fusion. For example, high liquid compressibilities may render silicate liquid densities higher than those of residual crystals. The early differentiation of the Earth to great depths may have been influenced by this relationship of high liquid density, but it could not have affected small planets like the Moon where the gravitational field is insufficient to cause the required internal compressions. Thus, small planets not only have a larger surface/volume ratio to effect heat loss but they also cannot conserve their heat sources by burying them, which undoubtedly promotes the observed rapid thermal extinction of the small terrestrial planets in the solar system (with the exception of Io mentioned above).

More subtle geochemical fractionation effects result from variations in the total amount of melting sustained by a parcel and the extent to which equilibrium is maintained between crystals and liquid. It is easy to appreciate that the more a parcel is melted, the more refractory the residue will become. The possible extent of melting is largely determined by the initial internal energy of a parcel during adiabatic ascent or by the amount of heat input in the other melting processes.

The extent to which equilibrium is maintained between liquid and crystals is largely determined by the rate of removal of liquid from crystalline residue. In the limit of no removal of liquid and very slow heating, diffusion between and within crystals (and liquid) can maintain full chemical equilibrium. For crystal sizes in the range mm-cm and chemical diffusion constants in crystals at magmatic temperatures in the range 10^{-10}–10^{-14} cm^2/sec, equilibration times may be as long as a million years or more. However, diffusion constants are appreciably faster within magmatic silicate liquids (10^{-5}–10^{-9} cm^2/sec). Thus we may expect liquids to be chemically in equilibrium only with the surfaces of residual crystals on a time scale of years or less. Let us assume that the cycling time between distinct magmatic pulses at individual eruptive centers is related to the rate at which liquid separates from residue. For basaltic magmas these times are on the order of years to thousands of years, which leads to the expectation that surface equilibrium of melt with residual crystals is common but that bulk chemical equilibrium with the residue is unlikely to be the rule.

To explore the geochemical consequences of these contrasting equilibration modes of partial fusion (determined by the rapidity of the extraction rate compared to the diffusion rate), we can use models of either *equilibrium* or *fractional* partial fusion respectively. These models show that the major element evolution of liquids produced by equilibrium partial fusion is continuous in composition and temperature. The depletion of incompatible elements from the crystals is moderated by the continued presence of liquid. By contrast, during fractional partial fusion, where only transient surface equilibrium with crystals is realized by the liquid before separation, the major element evolution of liquids produced is discontinuous in composition and temperature. Depletion of incompatible elements from the crystalline residue is more complete with fractional partial fusion than with equilibrium fusion because each increment of melt leaves the system with its burden and cannot moderate subsequent depletions. Intermediate possibilities with a variety of complex consequences can occur when the mechanical segregation of liquid from the residue is incomplete as partial fusion proceeds. The conclusion is that the physical details and the rates of the fusion and segregation process may have significant impact on the chemistry of magmas produced.

A final important aspect of the fusion process is the pressure at which it occurs. Pressure variations can change mineral stabilities with the result that melting at different pressures of the same bulk composition can lead to very different melting products if the residue mineralogy is different. For example, melting of lherzolite is pressure sensitive. At 25 kb garnet is the stable aluminous phase in the crystalline assemblage and melting in this regime can produce alkaline basalts with enrichment of the light rare earths with respect to the heavy rare earths, a characteristic garnet fractionation signature. By contrast, melting at 5–9 kb in the plagioclase stability field can produce tholeiites with little relative rare earth fractionation except for Eu depletion, a characteristic plagioclase fractionation signature. Examining mineral stability relations as a function of pressure is a continuing enterprise of experimental petrology laboratories.

Plumbing Processes

We have considered the source regions and the fusion processes responsible for magmatism, yet many mechanical and chemical adventures remain before the magmatic drama is exhausted. Magma must be separated from residue and transported. It may be ponded, interact with its surroundings, partially solidify, deform and differentiate during flowage, suffer immiscibility, mix with cognate batches in the plumbing system, inflate a chamber, convect in conjunction with chemical and thermal diffusion, become charged with groundwater, accumulate crystals, exsolve vapor, and possibly even erupt before completing the transition to igneous rock. This list of complicating processes is formidable, but it has a first among "equals." Historically the chemical fractionation resulting from the separation of crystals precipitated upon cooling has been given the most detailed attention in order to understand magma evolution. This approach is perhaps best illustrated in the work of Norman L. Bowen. Nevertheless the plethora of additional processes deserves recognition.

Segregation and Transport. Partial fusion processes eventually produce completely liquid masses, requiring that a separation of liquid from residual crystals has occurred. Percolation of melt along grain interstices may be the initial stage of this process, driven either by buoyancy or directed stress. An impermeable barrier may collect the percolating fluid. Alternatively liquid-crystal surface forces or tectonic shears may promote spontaneous segregation of melt into tabular dike or veinlike bodies that can migrate considerably faster than melt occupying a network of capillaries. The intimacy of contact between liquid and crystals is reduced somewhat by the segregation to veinlets and for geochemical purposes may mark the initiation of isolation of magma from the residue. The migration rate is key. Percolation rates, although geologically rapid (can be m/year or more), are not so rapid as to short-circuit diffusive exchange between the crystal surfaces and the liquid. As a result if liquid percolates through regions of varying geochemical character, important exchange may occur between magma and porous crystalline reservoirs. Isotopic as well as chemical exchange may occur with the effect that the magma may bear a geochemical signature of the last region through which porous flow occurred rather than the signature of the actual parcel that generated the melt. On the other hand, once segregation to dikes or veinlets occurs, the migration rates may approach meters/day. These rates would preclude all but the most superficial exchange between magma and crystalline host medium.

Crystal Fractionation. At any point during a magma's subsequent history, heat loss will lead to crystallization. As with partial fusion, partial crystallization is a potent mechanism for introducing chemical fractionation because of the difference in composition between liquid and precipitating crystals. As with partial fusion, the pressure at which crystallization occurs dictates the crystalline assemblage and mineral proportions subtracted during cooling. The series of residual liquid compositions encountered during cooling is called the liquid line of descent and this evolutionary path may be sensitive to pressure. Many common rock associations such as alkali olivine basalt →hawaiite →benmoreite →mugearite →trachyte are thought to arise as serial liquids in a line of descent produced by fractionation of olivine, pyroxene, and feldspar crystals. If two or more crystals participate in the fractionation, residual liquids cannot cross in composition the array representing mixtures of those crystals. The composition plane between olivine, plagioclase and augite is one such barrier (called a thermal divide) to chemical fractionation. Tholeiites and alkaline basalts lie on different sides of this plane, known as the olivine gabbro thermal divide. Thus tholeiites and alkaline basalts cannot occupy a single liquid line of descent at low pressures where olivine, plagioclase, and augite form the stable crystallization assemblage. Neither can be parental to the other at low pressure.

As with partial fusion, crystallization may proceed either with bulk chemical equilibrium maintained among all phases or with only surface equilibrium maintained between crystals and liquid. As with partial fusion, greater differentiations are possible during fractional (surface equilibrium) crystallization as opposed to equilibrium (bulk) crystallization. However discontinuities do not occur in the fractional crystallization process as they may in the fractional fusion process.

Understanding and enumeration of the variety of crystal fractionation mechanisms has been an abiding and central concern of several generations of igneous petrologists. Reconciliation of bulk rock chemical information with mineral chemistry data and experimental results has been the principal avenue of inquiry. Major elements are most often treated with the aid of saturation (or liquidus) diagrams whereas trace elements are most often treated with partitioning relations and the Rayleigh distillation equation or more sophisticated derivatives.

For a sequence of liquids along a line of descent there should be a complementary sequence of crystalline extracts. If the fractionation occurs during transit, these crystals may be spread throughout the plumbing system. They may coat the walls of dikes if they are precipitated there. However, if they are precipitated within the magma then hydrodynamic sorting may influence their distribution and eventual deposition. In narrow conduits, flowage tends to move crystals in suspension away from the areas of high liquid shear near the solid margins. This flowage differentiation can lead to liquid segregation toward the margins and crystal concentration to the middle of dikes, a common feature of hypabyssal sheets.

If the magma has an opportunity to pond, then gravitational settling or flotation may segregate the crystals from the liquid body. Crystals collected in this way form cumulus sequences that are often layered. The layering seen in solidified intrusions, historically has been ascribed to such a process, but this interpretation has proven to be controversial. Whether the layering has a gravitational or other origin, as long as it is stratigraphic, then study of cumulus sequences holds an important key to magmatic processes. They hold a record of changes in a magma body during its evolution. For this reason the cumulus sequences of layered intrusions found in the geological record provide a natural laboratory for testing crystal fractionation hypotheses for explaining the variations in eruptive lavas.

Ponding. Buoyant rise of percolating fluid and propagating magma-filled cracks may be halted by impermeability, by nonbuoyancy, or by a free-slip surface. Any of these conditions may cause magma ponding as melt piles up against the barrier. For instance, imagine an interface between crust and mantle. If the mineralogy changes so that percolating fluid no longer wets the grain edges to form connecting tubules, the crust's base is effectively impermeable. Migration would proceed to the base of the crust and percolating melt would collect there. Should the crustal assemblage have significantly lower density, for instance by virtue of a higher proportion of feldspar, the melt may no longer be buoyant enough to rise further but may instead occupy the horizon of neutral buoyancy and perhaps spread laterally, mechanically decoupling the solid strata. The free-slip underside of this sill-like body then forms an effective trap for dikelike bodies propagating upward as elastic cracks. These considerations are undoubtedly important in causing sill-like ponding whether at the base of the crust or at any other suitable horizon. Ponding may at first glance be thought to promote uniformity in a series of incremental melts collected by aggregation and mixing. However large magma bodies have a greater thermal power to contaminate their magmas than do small dispersed magmas.

Assimilation. Magma out of thermal equilibrium with its surroundings has the capacity to corrode and assimilate country rock. Ponded bodies of hot magma have more thermal power whereas thin tabular bodies have more effective surface/volume through which to absorb contaminants. Corrosion of country rock will cool the magma. Magma energy raises the temperature of the surroundings and must supply the heat of fusion and solution to partially fuse and absorb material. Most magmas are initially heated little if at all above their melting interval because they are generated as partial fusion products themselves. Superheat of magma is generally not a large part of the energy budget for assimilation. Energy must mostly be derived from cooling and crystallization of the magma. Thus the power of magma to assimilate is usually limited to absorbing only a small fraction of its mass in foreign material. Nevertheless this limited power may be capable of geochemically interesting consequences, especially if there are interesting materials to be assimilated.

The digestion of continental crust is a common occurrence in flood tholeiites erupted through continental crust, possibly after some residence at the base of the crust. Assimilative contamination often does not produce straightforward geochemical mixing arrays between magma and contaminant, partly because the crystallization induced in the process provides additional components of chemical variation. The resulting process is called assimilative fractional crystallization. But complications also result from the selectivity of the corrosion process. In a partial fusion of potential assimilant, the most fusible constituents will be preferentially extracted. For instance it has been suggested that basaltic magma can preferentially mobilize the incompatible-element-rich interstitial material from large volumes of country rock in crust or mantle without undue thermal expense (wall rock reaction process). Another mode of selectivity in assimilation processes is encountered in the absorbtion step. Elements of different ionic charge and radius diffuse at very different rates through silicate melt. Likewise the compatibility of various elements in melt of differing composition and structure is governed by ionic radius and charge considerations. Silicic melts formed from crustal fusion by basic melt may preferentially deplete comingled basic liquid in alkalies by diffusive exchange, to satisfy melt structure considerations. Likewise diffusive exchange between alkali-rich crustal fusion products and basic melts may lead to preferential enrichment of more mobile K over Na in the basic melt. The assimilation and absorbtion process can be nonintuitive and complex.

Immiscibility. Additional chemical diversity in magmas may be introduced by the agency of liquid phase separation. Stable immiscibility of liquid pairs is known to occur in many silicate systems at low temperature. It is a common petrographic feature of the groundmass of lunar and terrestrial basalts. Many alkaline igneous rocks contain ocellar structures that may preserve immiscible melt fractions. The immiscibility of silicate, sulfide, and carbonate liquids is also clear. However the importance of this process in causing magma diversity is controversial. By the time demonstrable cases occur, the rock system is usually too well crystallized to be able to mobilize the pairs as viable separate magmas. However a more important role for immiscibility may be in keeping distinct the silicic partial fusion products of crustal anatexis and the basic magmas causing the partial fusion.

Mixing. In the context of ponding, several interesting magmatic processes may occur within the chamber. Separate magma batches may enter the chamber with various temperatures, compositions, and hence, densities. Convection and mixing may ensue so that the chamber mixing process reduces the range of initial magmas that can be recovered.

On the other hand magma chambers can contribute to the diversity of magmas through the mixing process. Consider the periodic addition of the same magma type to a chamber that is cooling and producing a crystal fractionation liquid line of descent that is generally not a linear compositional array. The recharge of the chamber with the same initial liquid can produce a linear mixing array that generates compositions not found on the normal liquid line of descent. This process can lead to the appearance of seemingly anomalous crystals in the mixtures, compared to normal crystal fractionation. This feature appears to be quite common in mid-ocean ridge basalt petrogenesis.

Boundary Layers. Ponded bodies of magma may initially have sharp interfaces of temperature and composition with the surroundings. The sharpness of this interface deteriorates with time by diffusive processes. Thermal conduction reduces the marginal temperature gradient and also leads to marginal crystallization. The chemical perturbation in the liquid caused by this crystallization will be ameliorated in time by chemical diffusion. If the thermal and chemical shocks produce buoyancy differences and hence convection in the chamber, then the margin is also a boundary for the viscous dissipation of the momentum of moving magma parcels. Associated with the transport of heat, mass, and momentum are boundary layers (or zones over which the principal gradients of the properties occur) with thicknesses scaling as the relevant diffusivity. These diffusivities (i.e., D = chemical diffusion coefficient, κ = thermal diffusivity, and ν = kinematic viscosity) increase in the order mass < heat < momentum in jumps of roughly 10^4 or more, giving boundary layer thicknesses that may differ by more than a factor of 100. For instance, a chemical perturbation ($D \sim 10^{-6} cm^2/sec$) propagates a shorter distance into the chamber than a thermal disturbance ($\kappa \sim 10^{-2} cm^2/sec$); and both these are likely to be embedded in the stagnant marginal layer not participating in the general pattern of chamber circulation ($\nu \sim 10^2 cm^2/sec$).

However if density differences arise from the chemical and thermal effects, then these boundary layers may not be isolated from flow but may be the cause of it through buoyancy. Consider a chamber's vertical side margin through which heat is lost and against which crystallization is proceeding. Cool liquid should tend to sink along the wall due to thermal contraction. However the liquid adjacent to the wall may be modified in composition as well as temperature by the crystal fractionation process. In basic magma systems undergoing calcalkaline fractionation trends, the residual liquid can be chemically buoyant even though thermally contracted. Thus there will be two flow regimes. Closest to the wall ascending flow will be driven by chemical buoyancy whereas farther into the magma descending flow will be driven by thermal contraction. Advective flow and collection of these various boundary layer fractions may be a novel form of magma chamber differentiation. It is crystal fractionation, but the liquids are separated from the crystals rather than having the crystals separate individually from the liquid by gravitative settling or flotation.

Doubly Diffusive Convection. Given the disparity in the diffusion rates of heat and chemical components, both of which influence liquid density, an unusual mode of convective dissipation of heat and mass differences may occur in magma chambers. Special initial conditions are required. For instance consider the introduction of hot, but chemically dense, magma into the bottom of a chamber occupied by cooler, but chemically buoyant magma. Eventually the chamber should approach a condition of uniform magma, but the path by which the initial contrasts are dissipated is complex.

The lower hot layer will impart heat and mass to the upper layer across the interface which initially separates them; but more heat than mass is transported because of the differing diffusivities. As a result the base of the upper layer becomes buoyant and begins to break up into rising warm parcels. As the parcels proceed into the cool upper layer they lose their heat and approach the same temperature as the upper layer. But because the parcels initially acquired some densifying mass from the lower layer, the newly cooled parcel is denser than the surroundings and settles back towards the initial interface, where the process can be repeated. The result is the establishment of a convective layer between the two original masses.

At the same time the base of the upper layer was heated and chemically densified, the top of the lower layer was cooled and made slightly less chemically dense by mass transfer. Thus parcels should also start sinking from the initial interface and come rapidly to thermal equilibrium with the surroundings. But the mass transfer causes the thermally equilibrated sinking parcel now to be buoyant and to proceed back up toward the interface. Thus the original interface is actually replaced by a double layer, each part internally convecting separately and separated from the main masses of the initial layers by sharp boundary layers. Eventually each of the layers can spawn additional convecting layers above and below.

This phenomenon of stratification into a series of internally convecting layers is one manifestation of doubly diffusive convection. Other modes are possible but have as the common property that two controls on buoyancy are present that have unequal diffusivities. The example discussed is the "layer" mode and has received attention as a possible explanation for the cause of some forms of layering and chemical differentiation in layered intrusions. As bizarre as the process sounds, it is well established to occur in the ocean, for instance beneath melting icebergs. However it is still a very controversial process in magmatic evolution because the physical parameters that control the phenomenon are not as well known as for aqueous solutions. (Also the phenomenon is difficult to observe in progress!) The effects of strong thermal and chemical viscosity variations is unclear. Also in magma systems the presence of crystals and the complications caused by crystallization during heat transfer have yet to be fully sorted out. Elucidating this problem and related ones involving magma circulation through cumulus crystal piles is a current focus of fluid mechanics research in magmatic systems.

Soret Effect. There has been considerable recent renewed interest in the thermal diffusion process (Soret effect) in connection with convective boundary layers in magma chambers, especially those coupled with crystallization processes. Previously these diffusion processes had largely been rejected on the basis that temperature contrasts driving chemical diffusion ought to decay before significant chemical mass transport and differentiation can occur because heat diffuses so much more rapidly than mass in magmatic systems. Recent calculations taking into account the complications of coupling convection to the process, especially doubly diffusive convection, have suggested the possibility that this stricture may be relaxed. However laboratory experiments showing the actual chemical fractionation produced in Soret fractionation are difficult to reconcile with the chemical fractionations observed in magmatic rocks other than in the most limited of circumstances.

In Situ Crystal Accumulation. The growth of crystals as a consequence of the loss of heat through a chamber's margins leads to crystal accumulation. Much recent attention has been given to the possibility that accumulative crystals may have nucleated and grown in situ in the layers in which they are eventually found rather than being put in those layers from elsewhere by gravity or a flow process. This could help explain the apparent buoyancy conundrums such as coaccumulation of olivine (which should sink) and plagioclase (which should float) in the same layers of some fossil magma chambers. Likewise puzzling features of adcumulus texture, comb layering, and chain structure may be less mysterious in the context of in situ growth. The escape of the intercumulus liquid back to a chamber whose margins are undergoing in situ fractionation has geochemical consequences not easily understood by the normal models of crystal fractionation. For instance, chambers infected with a return of in situ boundary layer crystallization liquid may show some of the characteristic chemical variations shown in partial melting models, which produces an ambiguity in the geochemical interpretation of effusive lavas. Although simple partial melting is easily distinguishable from simple crystal fractionation in its geochemical consequences, realistically complex physical models may make the processes less clearly distinguishable on geochemical grounds.

Volatile Transfer. The agency of C–O–H–S fluids in metasomatism of the source region has already been mentioned. Undoubtedly volatile fluids can also play a role in modifying magma composition and properties when directly introduced or expelled from the magma itself.

Of perhaps more interest from the physical point of view is the role of volatile fluids as propellants during explosive eruption of magma. When magma absorbs constituents such as H_2O, for instance from contact with country rock charged with deep circulating hydrothermal fluids, it does so at pressures greater than the vapor pressure of the constituent in the magma. Magma can absorb until the vapor pressure equals ambient. However, as magma ascends and/or decompresses, the vapor pressure may exceed the total pressure with the result that effervescence of gas bubbles occurs. The solubility of gas in the magma depends on pressure and magma composition. Fluid basaltic magmas of low gas content may relieve their gas overpressure with mild vesiculation upon eruption. However viscous silicic magmas containing water up to their high solubility limits may have a much less innocuous release. Once a conduit is established from such a magma body to the surface, low pressure venting and eruption of the upper portions may relieve the pressure on the lower portions causing the gas exsolution front to advance downward. The expansion of the gas-charged magmatic foam into the previously evacuated parts of the chamber and perhaps then out through the collapsing roof can lead to catastrophic explosive eruptions. The 1980 Mt. St. Helens eruption appears to have been triggered by the depressurization of the magma chamber when part of the edifice sloughed away. Once the magma chamber began to unload, gas exolution sustained the eruption and discharged much fragmented magma in the form of ash. The magnitude of volcanic catastrophe is largely a function of the involvement of volatile constituents.

Emplacement and Solidification. Because heat loss is the primary agent of solidification, the state of solidified magma will depend most importantly upon cooling history. In the limit of rapid cooling (10–10^3°C/sec), as in the quenching of erupted magma against water, dense glass may be formed. Crystallization, effervescence, assimilation, and flow differentiation are precluded in such an abbreviated evolution. In the limit of slow cooling (10^{-8}–10^{-12}°C/sec), as in plutonic rocks of orogenic belts, coarse-grained, crystalline, foliated rocks are produced that often merge gradationally into their surroundings. The insulating properties of country rock around plutons cause a protracted cooling history in which magma thoroughly crystallizes, differentiates, recrystallizes, deforms in response to orogenic stress, and interacts with its surroundings.

The properties of solidified magma do not simply form a series of linear combinations between those of the two limits of rapid and slow cooling history, because of the variety of ways in which the secondary processes including crystallization, vesiculation, assimilation, deformation etc., can be amalgamated and interact with heat loss upon emplacement. The secondary processes may also interact with one another to give new combinations.

Consider whether gas discharge occurs before or after substantial crystallization. Early gas discharge may produce magmatic foams or pumice whereas late gas discharge may produce vug-filled crystalline lavas. Continuous gas discharge may produce neither, or either, or both. The extent of the gas effervescence and the extent to which the gas causes explosive foams, or is trapped as discrete vesicles, or freely escapes, depends on magma composition among other factors. The key aspects of magma composition are gas content and proportion of ingredients such as SiO_2, which produce viscous, polymerized liquids. Sticky liquids are more apt to foam if charged with gas. Furthermore, gas discharge can influence subsequent evolution of a magma batch. Loss of H_2O-rich gas causes a magmatic residue to stiffen in several ways. H_2O tends to depolymerize melt structure and therefore makes magma more fluid. Loss of H_2O reverses this process and stiffens the residual melt phase. H_2O in solution tends to increase the solubility of crystalline phases. Exsolution of H_2O may then trigger crystal precipitation if saturation boundaries are shifted by the discharge. The presence of crystals and gas bubbles in magma may transform a Newtonian fluid rheology into a Bingham plastic with a finite yield strength and larger effective viscosity. And finally gas exsolution may contribute to the magma's heat loss and stiffening simply through the thermally activated increase of viscosity of silicate liquids with cooling.

Magmatism is completed upon magma solidification, unless of course tectonic recycling reincorporates the material into the planetary thermal engine through subduction or orogenesis.

Summary

Magmatism is primarily an expression of excess planetary heat dissipation. Many magmatic processes derive from contrasts in rates of transport processes. Adiabatic pressure-release melting occurs because advection can dissipate heat more rapidly than conduction. Magma ponding occurs in part because of contrasts in physical properties controlling transport rates such as permeability, density, and effective viscosity of layers of country rock. Many magma chamber processes are driven by the difference in the rates of heat and mass transport. And finally explosive eruptions are driven by the slowness of magma to dissipate the mechanical energy of vesiculation by viscous flow compared to the rapidness of chemical exsolution in supplying expansion work. Interaction among the transport processes is an important aspect of magmatism. Magmatism will cease when a planet's internal energy is insufficient to trigger the instabilities that result from the imbalances in transport processes leading to thermally extinct objects such as the Moon and Mercury whose thermal evolution is now amagmatic.

DAVID WALKER

Bibliography

Anderson, D. L., 1984, The Earth as a planet: Paradigms and paradoxes, *Science* **223**, 347-355.
Bowen, N. L., 1928, *The Evolution of Igneous Rocks*. Princeton, N.J.: Princeton University Press.
Basaltic Volcanism Study Project, 1981, *Basaltic Volcanism on the Terrestrial Planets*. New York: Pergamon.
Grove, T. L., and R. J., Kinzler, 1986, Petrogenesis of andesites, *Ann. Rev. Earth Sci.* **14**, 417-454.
Hargraves, R. B., ed., 1979, *Physics of Magmatic Processes*. Princeton, N.J.: Princeton University Press.
Hildreth, E. W., 1980, Gradients in silicic magma chambers; implications for lithosphere magmatism, *Jour. Geophys. Research* **86**, 10153-10192.
McBirney, A. R., and R. M. Noyes, 1979, Crystallization and layering of the Skaergaard Intrusion, *Jour. Petrology* **20**, 487-554.
Stolper, E. M., D. Walker, B. H. Hager, and J. F. Hays, 1981, Melt segregation from partially molten source regions: the importance of melt density and source region size, *Jour. Geophys. Research* **86**, 6261-6271.
Walker, D., 1983, New developments in magmatic processes, *Rev. Geophys. Space Phys.* **21**, 1372-1384.
Watson, E. B., 1982, Basalt contamination by continental crust: some experiments and models, *Contr. Mineralogy and Petrology* **80**, 73-87.
Yoder, H. S., Jr., ed., 1979, *The Evolution of the Igneous Rocks, 50th Anniversary Perspectives*. Princeton, N.J.: Princeton Univ. Press.

Cross-references: Earth and Moon: Origins; Earthquakes: Volcanogenic; Earth Structure: Global; Energy Budget of the Earth; Geothermal Energy; Heat Flow in the Earth; Lithosphere, Oceanic: Formation and Evolution; Magma Physics; Mantle Convection and Plumes; Mantle Dynamics; Mantle Viscosity; Mineralogic Phase Transitions in the Earth; Subduction Zones.

MAGNETIC ANALYSIS OF ROCK FABRIC

A rock has a uniform fabric when the geometrical and spatial distributions of its constituent grains show no directional preferences. Many geological processes active during the formation or subsequent deformation of a rock lead to statistically preferential alignment of grain shapes or crystal axes. The ensuing anisotropic fabric is a combination of planar and linear elements, one of which may be strongly dominant. A nonrandom fabric results in directional dependency of the physical properties of the rock, including its magnetic behavior. Magnetic anisotropy can be measured by a variety of techniques that are as accurate and much faster than direct petrofabric observations.

Nonrandomness of the grain distribution in a rock generally affects both the nonferromagnetic matrix-forming grains, which constitute the bulk of the rock, and the weak concentration (often much less than

1% by volume) of ferromagnetic minerals dispersed throughout the matrix. The responses of the matrix minerals and the ferromagnetic fraction are both inherently present; in some cases special techniques allow separation of their individual contributions. Often, unless the ferromagnetic fraction is extremely dilute, its anisotropy dominates that of the rock because of the intrinsically strong magnetic properties of ferromagnetic minerals.

Magnetic Anisotropy of Minerals

The physical theories of diamagnetism, paramagnetism and the different classes of ferromagnetism, and the magnetic behavior of minerals most commonly found in rocks are described extensively in several excellent textbooks (Stacey and Banerjee, 1974; O'Reilly, 1984 and elsewhere in this volume). Only a cursory description relevant to magnetic fabric analysis is given here.

The magnetic anisotropy of a mineral arises from (1) the magnetostatic energy associated with its intrinsic magnetization (M_s) and the grain shape and (2) the quantum-mechanical preferential alignment of the intrinsic magnetization with particular crystallographic directions. Grain-shape dependency of magnetic anisotropy is important in only a few ferromagnetic minerals that have a strong spontaneous magnetization: magnetite ($M_s = 480 \times 10^3$ A · m^{-1}), the titanomagnetite solid solution series, and maghemite ($M_s = 430 \times 10^3$ A · m^{-1}). Only a small departure from equidimensional shape is necessary in minerals with high susceptibility (k) for shape-dependent anisotropy to be more important than magnetocrystalline anisotropy. For example, in magnetite an axial ratio of only 1.2:1 is enough for this condition to be realized. In rocks that contain magnetite as the main magnetic mineral, the magnetic anisotropy is determined primarily by the degree of preferential alignment of the magnetite grain shapes.

Hematite is an important ferromagnetic mineral, but its spontaneous magnetization (around 2×10^3 A · m^{-1}) and its susceptibility (around 10^{-2}) are comparatively weak. Magnetostatic energy is proportional to the square of M_s and is therefore some 50,000 times weaker in hematite than in magnetite for the same grain shape. However, hematite has a strong magnetocrystalline anisotropy; at room temperature the susceptibility in the crystal basal plane is about two orders of magnitude greater than along the hexagonal c-axis. The ferrimagnetic form of pyrrhotite also has a strong magnetocrystalline anisotropy with susceptibility in the basal plane much greater than along the c-axis. The spontaneous magnetization is comparatively strong (around 90×10^3 A · m^{-1}) and may result in shape-dependent magnetic anisotropy within the crystalline basal plane (Fuller, 1963). However, the magnetic anisotropy of rocks containing only hematite or pyrrhotite as predominant magnetic mineral is determined primarily by the degree of preferred alignment of basal planes of these minerals.

The matrix-forming minerals have weak magnetic properties that are comparatively difficult to measure. Consequently, precise information about their magnetic anisotropy is very sparse. Calcite is diamagnetic, with susceptibility -1.38×10^{-5} along the mineral c-axis and -1.24×10^{-5} perpendicular to it. Magnetic susceptibility anisotropy can develop through crystallographic preferred orientation in a calcite rock. This development is only likely to contribute significantly to the anisotropy of pure marbles and limestones that are exceptionally free of paramagnetic or ferromagnetic mineral phases. A high degree of alignment of calcite crystallographic axes is necessary for a measurable diamagnetic susceptibility anisotropy.

Weak magnetic anisotropy of magnetocrystalline origin is also present in several paramagnetic minerals, especially some rock-forming phyllosilicates (e.g., micas, clay minerals) and inosilicates (e.g., pyroxenes, hornblende). For example, susceptibilities of order 10^{-4} and large differences between maximum and minimum values are known for single crystals of hornblende, in which the direction of maximum susceptibility is along the crystallographic c-axis. In general, however, there is a paucity of reliable data for susceptibility and its anisotropy in the matrix-forming silicate minerals.

It is evident that the magnetic fabric of a rock depends on the relative concentrations of the constituent minerals as well as on their preferred alignments. The strongly magnetic ferromagnetic fraction is frequently a very dilute dispersion in a paramagnetic or diamagnetic matrix. It alone determines the magnetic fabric determined by remanent magnetization methods, and, unless its concentration is extremely low, its presence tends to dominate low-field susceptibility observations.

Methods of Magnetic Fabric Analysis

The laboratory methods of estimating magnetic fabric fall into two categories. The first involves measurement of the directional dependence of remanent magnetization, which is an expression of the anisotropy of the small ferromagnetic fraction. The second category embraces anisotropic effects induced in the presence of an applied magnetic field, and includes methods that depend on the anisotropy of magnetic susceptibility in low fields or the energy of magnetization in high fields.

Anisotropy of Remanent Magnetization. In the early 1960s paleomagnetists were concerned that the direction of natural remanent magnetization (NRM) in anisotropic rocks might be deflected from the paleomagnetic field direction. Laboratory studies showed that the anisotropy of isothermal remanent magnetization (IRM), induced in strong fields, was often much more pronounced than the anisotropy of low-field magnetic susceptibility (AMS) measured in the same sample (Fuller, 1963). This is because the IRM method analyzes only the ferromagnetic fraction in the rock, whereas AMS results also include contributions from paramagnetic and diamagnetic minerals. The paramagnetic matrix of rocks containing hematite, pyrrhotite, or very low concentrations of magnetite exerts a dilution effect that reduces the magnitude of the anisotropy of susceptibility (Fuller, 1963).

The anisotropy of anhysteretic remanent magnetization (ARM) provides an alternative method of analyzing rock magnetic fabric. However, ARM only affects the fraction of the magnetic mineralogy with coercivities less than the peak alternating field used to create the ARM. Because the peak field usually is less than 0.1 T, the low-coercivity magnetite fraction of a rock will be activated predominantly. Although the acquisition of IRM is a nonlinear function of the applied magnetic field, IRM may be a more suitable remanence type because it can be produced in fields up to several teslas, which affect most of the ferromagnetic fraction in a rock. ARM and IRM are comparatively strong remanences and can be measured with greater precision than can AMS. However, the remanent magnetizations are carried by single-domain and pseudo-single-domain grains, whereas AMS in low fields may be controlled by superparamagnetic or large multidomain grains. In rocks containing both magnetite and hematite, medium-field ARM may be dominated by the comparatively soft magnetite fraction, but high-field IRM will have a strong contribution from the hematite; in AMS results both of these ferromagnetic components may be superposed on a paramagnetic component. The individual subfabrics may be acquired by different mechanisms or at different times in the rock's history. Therefore, separation of the anisotropy components and a clear understanding of the rock magnetic mineralogy are necessary for correct geological interpretation.

Anisotropy of Magnetic Susceptibility. In an anisotropic medium the magnetization produced by an applied field is, in general, not parallel to the field. The components of induced magnetization (M_i) and applied field (H_j) are linearly related in weak fields:

$$M_i = k_{ij} H_j$$

The second-rank symmetric susceptibility tensor k_{ij} is represented geometrically by a triaxial ellipsoid. The magnitudes and directions of the principal axes of susceptibility ($k_1 > k_2 > k_3$) are given by the eigenvalues and eigenvectors of k_{ij}, respectively. Several experimental methods allow measurement of the susceptibility tensor elements k_{ij}. Most notable are the susceptibility bridge, spinner magnetometer, and torsion magnetometer methods.

In an alternating current, susceptibility bridge circuit two arms of the bridge consist of identical coils (Fig. 1a). Within each coil the current produces an alternating field parallel to the coil axis. When a sample is introduced into one of the coils, its permeability alters the inductance of this arm of the bridge, disturbing the balance. The out-of-balance voltage is proportional to the sample susceptibility along the field direction (coil axis). If an anisotropic rock sample is rotated about an axis perpendicular to the coil axis, the output voltge is sinusoidal, with phase and amplitude determined by the grain alignment in the plane normal to the rotation axis. By rotating about each of three orthogonal axes in turn, sinusoidal outputs are obtained that enable reconstruction of the susceptibility ellipsoid and determination of the magnitudes and directions of its principal axes. These define a magnetization ellipsoid (Fig. 1b), the radius of which in any direction is proportional to the magnetization that can be produced in that direction. Note that the field required to generate this magnetization must be applied in a different direction, because applied field and the resulting magnetization are not parallel in an anisotropic medium. Susceptibility is proportional to the magnetization component along the field direction; therefore, the susceptibility and magnetization ellipsoids are not equivalent. The relationship between susceptibility and magnetization (or magnitude) ellipsoids is discussed more fully by Hrouda (1982) and Collinson (1983).

Consider the rotation of a sample about one of its reference axes. The induced magnetization has components M_1 and M_2 given by

$$M_1 = k_{11}H_1 + k_{12}H_2 = k_{11}H \cos \theta + k_{12}H \sin \theta$$
$$M_2 = k_{21}H_1 + k_{22}H_2 = k_{21}H \cos \theta + k_{22}H \sin \theta$$

where $H_1 = H \cos \theta$ and $H_2 = H \sin \theta$ are the components of the applied field (H) along the sample axes. The magnetization induced in the direction of H is

$$M_H = M_1 \cos \theta + M_2 \sin \theta$$

The susceptibility in the direction of the field is

$$k_H = k_{11} \cos^2 \theta + k_{22} \sin^2 \theta + 2k_{12} \sin \theta \cos \theta$$
$$= \tfrac{1}{2}[(k_{11} + k_{22}) + (k_{11} - k_{22})\cos 2\theta$$
$$+ 2k_{12} \sin 2\theta] \qquad (1)$$

Similar expressions are obtained for rotations about the other two axes. Thus a sinusoidal curve is obtained in each plane of measurement, which

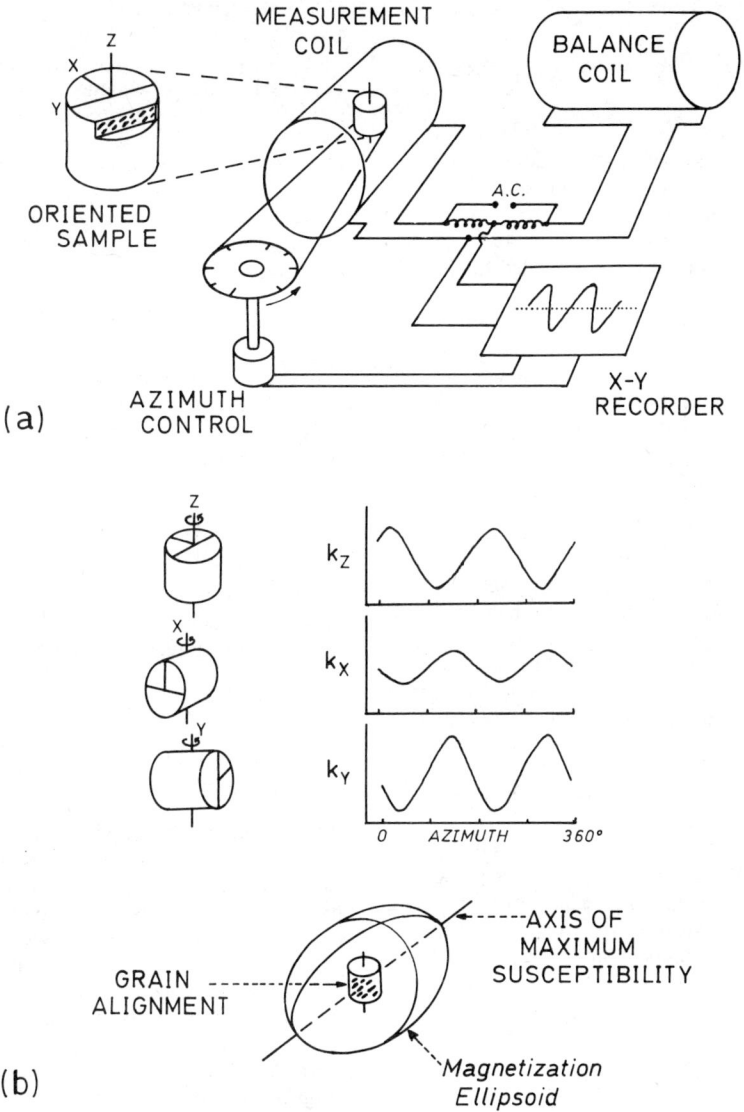

FIGURE 1. (a) Schematic diagram of an alternating current, susceptibility bridge. (b) The amplitude and phase of the sinusoidal signal generated by rotating an oriented sample about each of three orthogonal reference axes in turn are combined to describe a triaxial ellipsoid of magnetic anisotropy.

changes through two cycles for each cycle of rotation of the sample. Each curve is sampled at discrete intervals, usually 45°, so the number of measurements exceeds the number of unknowns. The overdefined susceptibility tensor elements can be optimally derived by least squares matrix manipulation.

In the spinner magnetometer method the sample is continuously rotated about each reference axis in turn. The induced signal changes with twice the rotational frequency; its amplitude and phase give the $(k_{ii} - k_{jj})/2$ and k_{ij} ($i \neq j$) terms, as in Eq. 1. Thus, two quantities are determined for each rotation, and the combination of results from the three measurement planes exactly determines the six unknown susceptibility elements k_{ij}.

In the torsion magnetometer method the sample is suspended from a fine fiber perpendicular to a magnetic field of low to medium intensity. Usually an alternating field is employed to nullify the torque resulting from interaction with the remanent magnetization of the sample. The torque (T_3) exerted by the applied field on the magnetization induced in an anisotropic sample is

$$T_3 = \mu_0(M_1H_2 - M_2H_1)$$

$$= \tfrac{1}{2}\mu_0[(k_{11} - k_{22})\sin 2\theta - 2k_{12}\cos 2\theta]\,H^2 \quad (2)$$

An analogous measurement procedure is used as for the susceptibility bridge method, and the overdefined k_{ij} are obtained by matrix analysis. Because the k_{ii} terms always occur as differences, an independent measurement of one of them is needed for complete analysis.

Energy of Magnetization in High Fields. High-field analysis of magnetic anisotropy is carried out with a torquemeter method, similar to that described above, except that the sample is suspended in a strong field between the poles of a powerful electromagnet. Equation 2 shows that the torque exerted on anisotropic paramagnetic and diamagnetic grains increases proportionally to H^2. However, in ferromagnetic materials the susceptibility is field-dependent, and in high fields magnetic saturation (M_s) is achieved. In this condition a ferromagnetic grain can still exhibit anisotropy because of directional differences in the internal demagnetizing field (H_d).

The internal demagnetizing field results from the distributions of magnetic poles on surfaces of the particle oblique to the magnetization (M), to which it is proportional:

$$H_d = -NM$$

The proportionality constant N is the self-demagnetizing factor. It depends on particle shape and can be computed for ideal ellipsoidal forms. If an ellipsoidal particle with principal axes $A > B > C$ is magnetized to saturation along a direction in the AB-plane at an angle θ to the A-axis, the magnetostatic energy is given by

$$E = \frac{\mu_0}{2}(N_a M_a^2 + N_b M_b^2)$$

$$= \frac{\mu_0 M_s^2}{2}(N_a \cos^2\theta + N_b \sin^2\theta)$$

The torque exerted on the particle by the strong magnetic field needed to produce the saturation magnetization is

$$T = \frac{-dE}{d\theta} = \frac{\mu_0 M_s^2}{2}(N_a - N_b)\sin 2\theta \quad (3)$$

Separation of Ferromagnetic and Paramagnetic Anisotropy Components. The torque that is experienced by an anisotropic rock sample in a uniform magnetic field contains contributions from all the constituent grains. High-field torque magnetometry provides a way of determining and comparing the anisotropies that are carried by the ferromagnetic and paramagnetic fractions. Examination of the dependence of torque on the square of the applied field (Fig. 2) shows a linear relation for the paramagnetic fraction, whereas a saturating contribution results for the ferromagnetic content. The combined curve shows a change in gradient at the ferromagnetic saturation field H_s. If two complete high-field torque analyses are made in fields stronger than the saturation field ($H_2 > H_1 > H_s$), then at a given angle θ of the torque curve the amplitudes are given by

$$T_1 = A_f + k(\theta)H_1^2, \quad T_2 = A_f + k(\theta)H_2^2 \quad (4)$$

where A_f is the contribution from the saturated ferromagnetic component and $k(\theta)$ is the paramagnetic susceptibility as in Eq. 2. The simultaneous solutions of Eq. 9 are

$$k(\theta) = \frac{T_2 - T_1}{(H_2^2 - H_1^2)}, \quad A_f = \frac{T_1 H_2^2 - T_2 H_1^2}{H_2^2 - H_1^2}$$

After separation of the ferromagnetic and paramagnetic amplitudes at an appropriate number of measurement angles θ, the principal axes of the corresponding magnetic anisotropy ellipsoids can be completely determined from Eqs. 2 and 3.

Some Results of Magnetic Analysis of Rock Fabric

Sedimentary Rocks. Several different forces can influence the final orientation of a nonequidimensional grain in a sediment or sedimentary rock. The most important effects in the production of sedimentary fabric arise at or near the sediment-water interface. On contact of a settling grain with the bottom, the mechanical couple exerted by gravity rotates the grain so that the longest dimensions align as closely as possible with the depositional surface. This planar alignment is expressed in the magnetic fabric as a magnetic foliation.

During free settling, a ferromagnetic grain is also subject to the orienting couple of the ambient geomagnetic field on its remanent magnetization, which can produce a statistical alignment and depositional remanent magnetization. However, if the water is in motion, hydrodynamic forces may overwhelm the magnetic couple, deflecting the particle away from the magnetic field direction (Fig. 3). During laminar flow transport, the stable attitude of an elongate grain is with its longest axis perpendicular to the current. After deposition the hydrodynamic forces rotate the grain about the fulcrum of its bottom contact so as to present the smallest area to the moving water. This rotation produces a statistical alignment of grain long axes with the current direction. The magnetic fabric is characterized by a lineation within the magnetic foliation plane. Under sedimentary conditions the magnetic lineation is

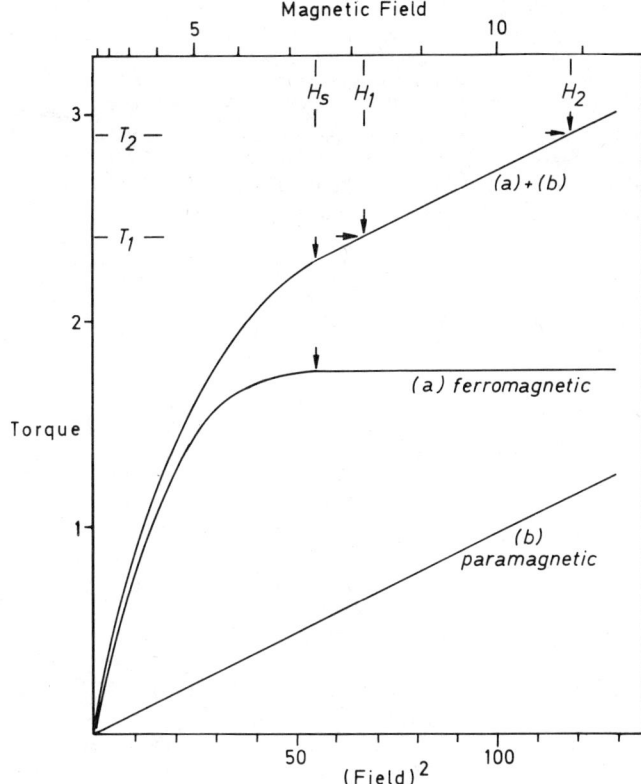

FIGURE 2. Plots of torque against the square of magnetic field show (a) a saturating effect for ferromagnetic grains and (b) a linear relationship for paramagnetic minerals. The two components in the composite curve can be separated by analysis of the torque in two fields, H_1 and H_2, stronger than the field H_s required to saturate the ferromagnetic magnetization.

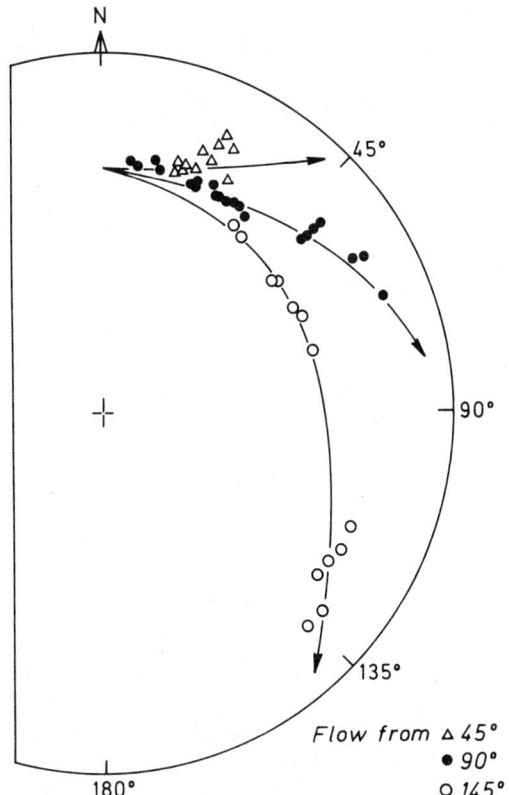

FIGURE 3. Lower hemisphere stereographic projection of maximum susceptibility axes of sediments redeposited in currents of different flow strengths and directions. With increasing velocity of flow the axes diverge from the magnetic field direction toward the flow azimuth (After Hamilton and Rees, 1970).

usually more weakly expressed than the magnetic foliation. The characteristics of sedimentary magnetic fabrics have been established in laboratory redeposition experiments and confirmed by measurements on modern unconsolidated sediments and undeformed, indurated sedimentary rocks that contained macroscopic indicators of paleocurrent direction (Hamilton and Rees, 1970).

If the water flow is turbulent, particles at the sediment-water surface can be reoriented; the sediment may be scoured, eroded, and redeposited elsewhere. If the new transportation process involves rolling of the grains along the bottom, a magnetic fabric may result in which the magnetic lineation is normal to the current. The rolling mechanism may also cause a magnetic lineation normal to the flow direction in turbiditic deposits. Observations of magnetic anisotropy in modern sediments and in thoroughly mixed limestone beds show that bioturbation apparently has negligible effects on the magnetic fabric. However, compaction of the sediment before induration enhances the magnetic foliation.

In summary, the relationships between magnetic foliation and lineation and the corresponding depositional processes are well established. Magnetic fabric analysis can quickly provide parameters that are of value in sedimentological studies, and are useful in the determination of paleocurrent directions in undeformed sediments.

Igneous Rocks. The physical processes of grain alignment in sediments are not greatly affected by the low viscosity of the water. In igneous rocks the viscosity of the fluid molten rock probably plays an important role in determining the fabric. When ellipsoidal objects are transported in a slow-moving viscous fluid, the motion involves end-over-end tumbling. Thus the magnetic fabric measured in an indurated lava must reflect the attitudes of the individual grains when the lava finally stiffened to the extent that further flow became impossible. Available evidence suggests that the grain orientation mechanism in volcanic rocks is not very effective. The magnetic fabrics are more weakly expressed than in sedimentary rocks, but generally have a foliation near to the flow plane (Hrouda, 1982). The magnetic lineation has been reported to be perpendicular to the known flow direction, but in some cases was parallel to it (Hrouda, 1982). The magnetic anisotropy of basaltic lavas may originate while the lava is flowing, but the stresses associated with later shrinking, cracking, and the formation of basaltic columns may be able to modify the pattern.

The magnetic anisotropy of plutonic rocks is highly variable. Where the anisotropy relates to flow conditions, the inefficiency of the grain orientation

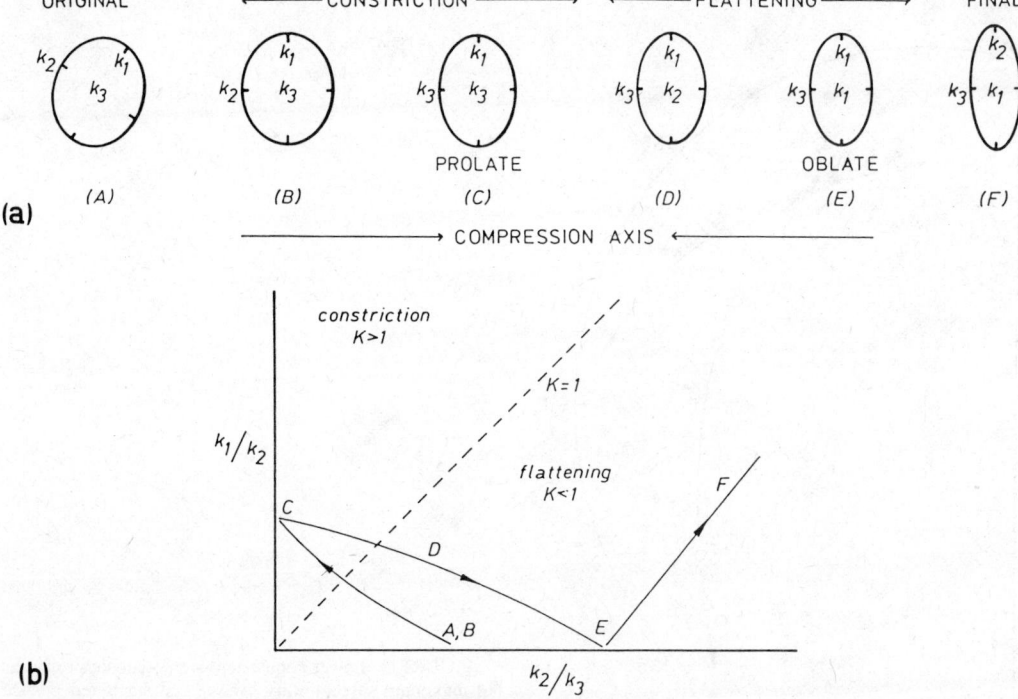

FIGURE 4. (a) Change of orientations of principal axes ($k_1 > k_2 > k_3$) and shape of the magnetic anisotropy ellipsoid accompanying compressional deformation (after Graham, 1966). (b) The same sequence (A–F) of ellipsoid shape changes is plotted qualitatively as axial ratios, in the same way as a structural geological deformation diagram.

process produces weak magnetic fabrics, as observed in some granites (Hrouda, 1982). However, in some plutonic rocks magnetite may crystallize in intergranular voids and may reflect the preexisting fabric of the host silicate structure. The mimetic magnetic fabric then exhibits a high degree of anisotropy (Hrouda, 1982).

The results of magnetic fabric studies in igneous rocks have sometimes proved to be more ambiguous than in sedimentary rocks. Nevertheless, the magnetic anisotropy constitutes a useful additional petrological parameter for analyzing the mode of emplacement of some igneous rocks.

Metamorphic and Deformed Rocks. Rocks that have suffered metamorphism of high enough grade to enable the growth of a secondary magnetic mineralogy can be expected to exhibit a mimetic magnetic fabric that reflects the conditions under which the new mineral grew. For example, the magnetic fabric of deformed oolitic limestones in the Swiss Alps is carried by magnetite that crystallized during a greenschist facies metamorphic event following a schistosity-forming deformation. The mimetic magnetic fabric reflects the anisotropy of the ellipsoidally shaped ooids (Kligfield et al., 1982).

The most interesting and promising results of magnetic fabric investigations have been found in rocks that have undergone ductile deformation but only low-grade or no metamorphism. The grains of ferromagnetic mineral are interpreted to behave as rigid objects that rotate without individual change of shape in a ductile matrix. The magnetic anisotropy

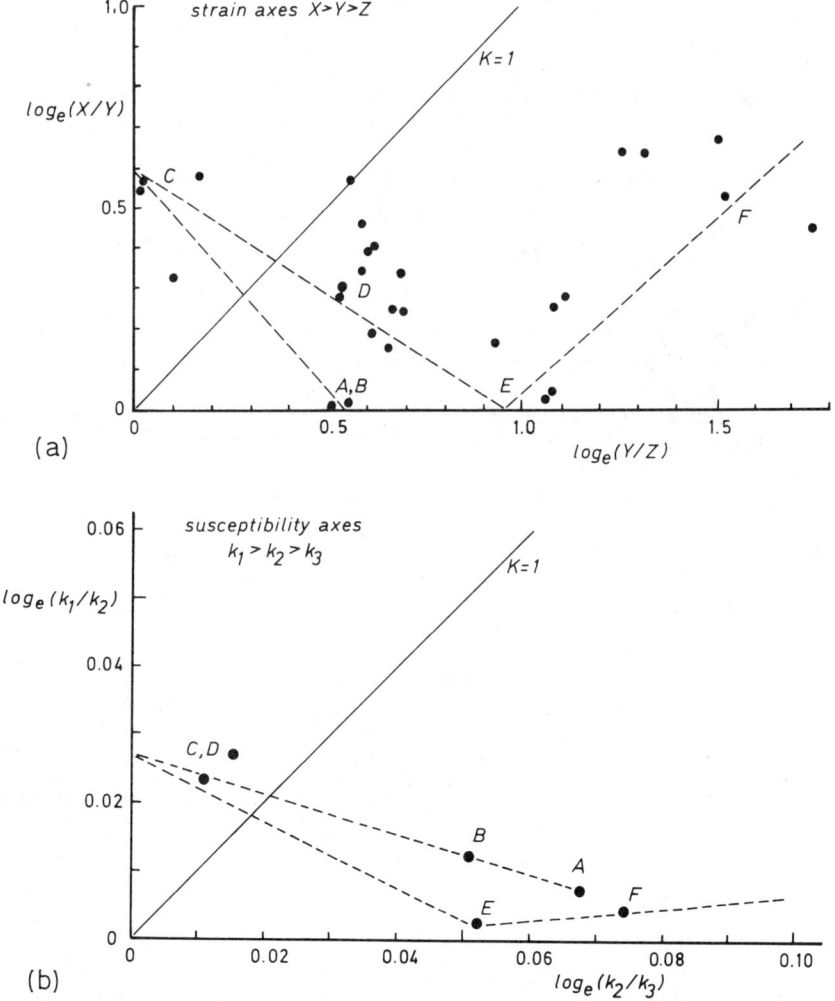

FIGURE 5. (a) Deformation plot of principal strain axes ($X > Y > Z$) observed for successive stages of progressive deformation in the Maritime Alps, France. (b) Analogous plot of average ratios of magnetic susceptibility principal axes ($k_1 > k_2 > k_3$) at the same deformation stages. The letters A–F identify the same shapes of strain or magnetic ellipsoid as in Fig. 4(a).

ellipsoid responds to the deformation in an analogous manner to the strain ellipsoid. J. W. Graham (1966), in a study of the anisotropy of magnetic susceptibility of various deformed sedimentary rocks from the Appalachian Mountains, showed that the principal susceptibility axes remained aligned with the corresponding principal strain axes. The shape of the AMS ellipsoid changed qualitatively in the same manner as a strain ellipsoid (Fig. 4a). Graham postulated that these changes occurred by a mechanism of plastic deformation of the unconsolidated sediments rather than by ductile deformation of the indurated rock. The rocks were not subjected to microfabric analysis, which would be the case in a modern structural geological study. The analogy between magnetic and strain ellipsoids is also fulfilled by ductile deformation of the rocks; the magnetic susceptibility ellipsoid changes shape progressively with increasing strain. If the axial ratios of the magnetization ellipsoid are plotted akin to those of the strain ellipsoid, a "deformation path" results (Fig. 4b) similar to that seen in progressive ductile deformation.

The progressive deformation of Permian red beds in the Maritime Alps of southern France can be followed by using deformed reduction spots as strain indicators (Fig. 5a). Kligfield et al. (1981) investigated the response of the magnetization ellipsoid to progressive deformation in these rocks. Undeformed red beds showed typical sedimentary strain and magnetization ellipsoids, flattened slightly in the bedding by compaction. The superposition of horizontal tectonic shortening caused the development of elongate pencil structures in the rocks. The associated magnetization ellipsoid was prolate parallel to the pencils, along the bedding-cleavage intersection direction. With increasing ductile deformation a tectonic flattening was superposed on, and eventually replaced, the original compactional fabric; the magnetization ellipsoid became oblate, with minimum axis perpendicular to the cleavage. The deformation path derived from strain measurements was found to have a magnetic equivalent (Fig. 5b), which was derived from the quantitative relationship between natural strain and the magnetic anisotropy acquired during progressive deformation of these Permian red beds. Similar relationships between strain and magnetic fabric have been found in other rock types, but there is no universal relationship.

Conclusions

When any event in the history of a rock produces preferential grain alignments, the paramagnetic and ferromagnetic grains are oriented anisotropically. This results in a magnetic fabric that usually has similar or identical properties to that of the matrix and is related to the causative event. Several laboratory techniques allow the measurement of rock magnetic fabric with a high degree of precision. The magnetic anisotropy of rocks has been used as a fundamental tool in petrofabric analyses, ranging from determinations of paleocurrent directions in sedimentary rocks to investigations of strain in deformed rocks. The magnetic results do not replace other techniques of sedimentological or structural geological analysis, but they constitute an additional, compatible data source that has a potential usefulness that has not yet been fully appreciated.

WILLIAM LOWRIE

References

Collinson, D. W., 1983, *Methods in Rock Magnetism and Palaeomagnetism*. New York: Chapman and Hall, 503p.

Fuller, M. D., 1963, Magnetic anisotropy and palaeomagnetism, *Jour. Geophys. Research* **68**, 293–309.

Graham, J. W., 1966, Significance of magnetic anisotropy in Appalachian sedimentary rocks, in J. S. Steinhart and T. J. Smith, eds. *The Earth beneath the Continents, Geophys. Monogr. Ser.*, vol. 10, Washington, D.C.: American Geophysical Union, 627–648.

Hamilton, N., and A. I. Rees, 1970, The use of magnetic fabric in palaeocurrent estimation, in S. K. Runcorn, ed., *Palaeogeophysics*. New York: Academic Press, 445–464.

Hrouda, F., 1982, Magnetic anisotropy of rocks and its application in geology and geophysics, *Geophys. Surveys* **5**, 37–82.

Kligfield, R., W. H. Owens, and W. Lowrie, 1981, Magnetic susceptibility anisotropy, strain and progressive deformation in Permian sediments from the Maritime Alps, *Earth and Planetary Sci. Letters* **55**, 181–189.

Kligfield, R., W. Lowrie, and O. A. Pfiffner, 1982, Magnetic properties of deformed oolitic limestones from the Swiss Alps: the correlation of magnetic anisotropy and strain, *Eclogae Geol. Helvetiae* **75**, 127–157.

O'Reilly, W., 1984, *Rock and Mineral Magnetism*. Glasgow: Blackie, 220p.

Stacey, F. D., and S. K. Banerjee, 1974, *The Physical Principles of Rock Magnetism*. Amsterdam: Elsevier, 195p.

Cross-references: *Chemical Remanent Magnetization; Curie Temperature; Demagnetization; Detrital Remanent Magnetization (DRM); Magnetic Domains; Magnetic Properties of Minerals; Natural Remanent Magnetization (NRM); Paleomagnetism: Orogenic Belts; Rock Magnetism; Rock Magnetism: Measuring Techniques and Apparatus; Thermoremanence; Viscous Remanent Magnetization (VRM) and Viscous Remagnetization.*

MAGNETIC DOMAINS

The existence of a paleomagnetic record testifies to the ability of magnetic minerals in rocks to retain their remanent magnetizations over geologic time. In the earlier days of rock magnetism and paleomagnetism, it was thought that the stable components of magnetization always resided in single-domain grains. However, it is now recognized that, due to their small size and scarcity, single domains are not likely to be the major carriers of stable remanence in many rocks. The prevailing view today is that much of the characteristic remanent signal is carried by larger grains, which, for energetic reasons, are subdivided into two or more magnetic domains.

For an original remanent direction to survive, an arrangement of domains must remain stable in two ways. First, the domain wall (the transition region between two adjacent domains) must be so strongly locked into its remanent position that it can withstand various perturbing influences, such as thermal activation at ambient or slightly elevated temperature, moderate changes in pressure, and changes in the Earth's field direction. Second, the fundamental domain structure, defined by the number of domains and the overall geometrical style in which walls are arranged in the particle, must remain stable with time, despite the influence of the effects noted above. In short, the domain structure that accompanies the initial remanent state must resist many changes of physical environment during the history of the rock. Thus, magnetic domain structure lies at the heart of magnetic stability criteria upon which paleomagnetism has been founded.

This article presents a brief discussion of the fundamental principles of magnetic domain theory. First, the energies that govern domain structure are discussed. Although these discussions usually refer to ferromagnetic materials, they can be generalized to the more complex but more geologically important ferrimagnets (see *Magnetic Properties of Minerals*). Simple domain models are presented, and domain patterns that are observed on several natural phases are described in view of these models. Although domain models have been refined considerably by various workers, those presented here are based on older, classical models for the sake of greater mathematical simplicity. Nevertheless, the classical models provide much physical insight into the domain problem. Next, we will discuss the temperature dependence of domain structure and how it may influence acquisition of thermal remanent magnetization (TRM) and viscous magnetization (VM). Finally, domain studies and domain theory are discussed in view of the behavior of the fine particles that are critical to paleomagnetism.

Energies

A magnetic domain consists of a volume that is spontaneously magnetized along a single direction. As implied by its name, a single-domain particle is uniformly magnetized along a single direction throughout its entire volume. There is potential energy associated with the termination of magnetic "free poles" at the grain's surface, and this energy is equal to the work required to assemble the

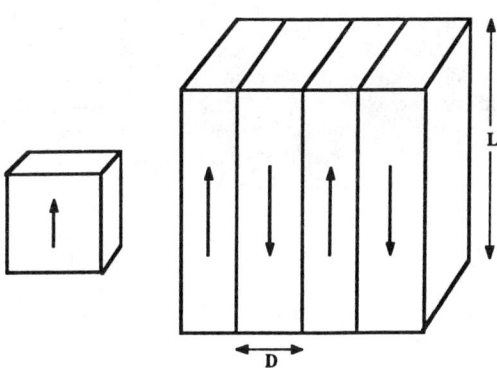

FIGURE 1. A single-domain particle (left) and a multi-domain particle (right) that is subdivided by 180° domain walls.

magnetic dipoles (the atomic moments) at the surface of a particle with a given shape. With increasing particle size, this magnetostatic energy rises in proportion to the grain volume, until a single-domain configuration is no longer energetically favorable. Thus when the grain exceeds a certain critical size, it breaks up into several domains to reduce the surface energy (Fig. 1) rather than remaining homogeneously saturated.

It will be shown that the number of domains, n, within a particle is a function of the particle size, so small particles of a given material contain fewer domains than larger grains with the same shape. This size variation depends upon the intrinsic magnetic properties of the material at hand, the state of internal stress, and the temperature. At a given grain size, the domain configuration of minimum energy represents a balance between the surface magnetostatic (or, in some cases, surface anisotropy) energy and the total energy of the domain walls necessary to produce the configuration. The wall energy is largely determined by two competing energies: (1) the anisotropy energy of magnetocrystalline or stress origin that determines the directions of minimum energy ("easy" directions) along which the spontaneous magnetization is aligned, and (2) the exchange energy, which tends to align adjacent spins and which provides the physical basis for spontaneous magnetization. These energies are discussed here (see *Magnetic Properties of Minerals*).

Exchange Energy. The origin of ferromagnetism was originally postulated by Weiss in terms of a molecular field that causes the adjacent spins in a ferromagnetic material to be aligned. Later, the Weiss molecular field model was shown to be a classical, phenomenological description of the quantum-mechanical exchange force. Calculations by Heisenberg demonstrated that the exchange energy between two adjacent spins S_i and S_j is given by

$$E_{ex} = -2J\mathbf{S}_i \cdot \mathbf{S}_j = -2JS_iS_j \cos\theta_{ij}$$

where J is the exchange integral of the given material. The sign of the exchange integral determines whether parallel or antiparallel spins yield the lowest energy state. J is negative in most materials, so antiparallel spins are favored; there is a mutual cancellation of neighboring spins and no spontaneous magnetization. In certain other materials J is positive, giving rise to the relatively rare phenomena of spontaneous magnetization and permanent magnetization. Above absolute zero, thermal agitation disturbs the alignment between neighboring spins in a ferromagnet, so the spontaneous magnetization M_s decreases with temperature. The temperature at which the aligning effect of the exchange force is overwhelmed by the randomizing effect of thermal agitation is the material's Curie temperature. Note that the exchange energy depends only on the angle between adjacent spins, and not on the orientation of the spins with respect to the crystal lattice. Hence, ferromagnetism and spontaneous magnetization can exist even in noncrystalline (amorphous) materials.

Magnetocrystalline Anisotropy. Magnetocrystalline anisotropy arises from the coupling between the spin and the orbital motion of the electron. Therefore, work is required to rotate the spins away from the "easy" direction determined by the spin-orbit coupling. This type of anisotropy is known as *magnetocrystalline anisotropy*, since it gives rise to easy and hard directions of magnetization along given crystallographic directions that depend upon the material. In hexagonal crystals, the work required to rotate the spontaneous magnetization away from the c-axis through an angle θ is

$$E_{\text{cryst}} = K_0 + K_1 \sin^2 \theta + K_2 \sin^4 \theta + \cdots$$

Thus when $K_1 > 0$ and $K_2 > -K_1$, the c-axis is the easy direction of magnetization in a hexagonal substance.

For a cubic material, such as magnetite, the energy associated with a particular orientation of M_s is

$$E_{\text{cryst}} = K_0 + K_1(\alpha_1^2\alpha_2^2 + \alpha_2^2\alpha_3^2 + \alpha_1^2\alpha_3^2)$$
$$+ K_2(\alpha_1^2\alpha_2^2\alpha_3^2) + \cdots$$

where α_1, α_2, α_3 are the direction cosines of the angles between M_s and the crystal axes [100], [010], and [001], respectively, and where K_0, K_1, and K_2 are the magnetocrystalline anisotropy constants for the given material. Higher powers usually are neglected, and often K_1 so exceeds K_2 in magnitude that only the term in K_1 is needed for most calculations. The signs of K_1 and K_2 will determine which crystallographic directions are "easy" and "hard" for the material under consideration (see Table 1). Easy directions in magnetite, for which $K_1 < 0$ and $K_2 < 0$, thus lie along $\langle 111 \rangle$.

Stress Anisotropy. Due to spin-orbit coupling, the lattice of a ferromagnetic material will spontaneously strain below the Curie temperature with respect to its "ideal" dimensions ordinarily determined by the crystallographic structure and interatomic spacing. The inverse of this phenomenon is that the direction of spontaneous magnetization can be rotated through application of stress or, equivalently, by accumulation of internal strain. The magnetoelastic energy density that results in a cubic material when a stress σ is applied is

$$E_{\text{stress}} = -3/2\lambda_{100}\sigma(\alpha_1^2\gamma_1^2 + \alpha_2^2\gamma_2^2 + \alpha_3^2\gamma_3^2)$$
$$- 3\lambda_{111}\sigma(\alpha_1\alpha_2\gamma_1\gamma_2 + \alpha_2\alpha_3\gamma_2\gamma_3$$
$$+ \alpha_3\alpha_1\gamma_3\gamma_1)$$

where α_1, α_2, α_3 are the direction cosines of M_s with respect to [100], [010], and [001], respectively, γ_i are the equivalent direction cosines of σ, and λ_{100} and λ_{111} are the magnetostriction constants of the material. Here, the sign of tensile stress is assumed to be positive. For the simplest case of isotropic magnetostriction where $\lambda_{100} = \lambda_{111}$ is a valid approximation,

$$E_{\text{stress}} = \tfrac{3}{2}\lambda\sigma \sin^2 \theta$$

θ being the angle between M_s and σ. In this simple case, the minimum energy configuration will correspond to M_s being parallel to σ when $\lambda\sigma > 0$ and to M_s being perpendicular to σ when $\lambda\sigma < 0$.

Magnetostatic Energy. The work required to assemble magnetic dipoles into a given configuration is the *magnetostatic energy*. Consider a material whose spontaneous magnetization is constant in magnitude everywhere and is uniformly directed within each domain. That is, div $M_s = 0$ and grad $M_s = 0$ (except within domain walls). The self-

TABLE 1. Directions of Easy, Medium, and Hard Magnetizations in a Cubic Crystal, Determined by Magnetocrystalline Anisotropy

K_1	+	+	+	−	−	−
K_2	$+\infty$ to $-9K_1/4$	$-9K_1/4$ to $-9K_1$	$-9K_1$ to $-\infty$	$-\infty$ to $9\lvert K_1\rvert/4$	$9\lvert K_1\rvert/4$ to $9\lvert K_1\rvert$	$9\lvert K_1\rvert$ to $+\infty$
Easy	$\langle 100 \rangle$	$\langle 100 \rangle$	$\langle 111 \rangle$	$\langle 111 \rangle$	$\langle 110 \rangle$	$\langle 110 \rangle$
Medium	$\langle 110 \rangle$	$\langle 111 \rangle$	$\langle 100 \rangle$	$\langle 110 \rangle$	$\langle 111 \rangle$	$\langle 100 \rangle$
Hard	$\langle 111 \rangle$	$\langle 110 \rangle$	$\langle 110 \rangle$	$\langle 100 \rangle$	$\langle 100 \rangle$	$\langle 111 \rangle$

energy of the surface magnetic poles is given by the surface integral $\int \phi \sigma \, ds$, where σ is the surface free pole density at a point and where ϕ is the magnetic potential due to the free poles at this point. In general, both σ and ϕ are functions of their position on the surface. Magnetostatic calculations for magnetized grains with irregular shapes and irregular distributions of domain walls are therefore very cumbersome and require numerical methods.

However, we can gain much physical insight into this problem by calculating the self-energy of a semi-infinite, magnetized body, whose domain walls are arranged in a periodic manner. In this case, σ is also periodic, so the potential ϕ can be obtained analytically with a Fourier series approach. For the regular array of planar domains contained by a semi-infinite platelet with net moment, shown in Fig. 2, the magnetostatic energy per unit surface area is

$$E = 2\pi M_s^2 L \left(\frac{M}{M_s}\right)^2 + \frac{8M_s^2 L}{\pi^2 g} \sum_{n=1}^{\infty} \frac{1}{n^3}$$
$$\cdot [1 - \exp(-2\pi n g)] \sin^2\left[\frac{n\pi}{2}\left(1 + \frac{M}{M_s}\right)\right]$$

where $M = M_s(D_1 - D_2)/(D_1 + D_2)$ and $g = L/(D_1 + D_2)$ (Kooy and Enz, 1960).

When $L \gg D_1 + D_2$ and $D = D_1 + D_2$ so that $M = 0$ (zero moment), Eq. 1 greatly simplifies to

$$E = \frac{16 M_s^2 D}{\pi^2} \sum_{n:\text{odd}}^{\infty} \frac{1}{n^3} = 1.705 M_s^2 D$$

which is Kittel's (1949) well-known result for a regular array of uniformly spaced domains in a semi-infinite platelet. Thus the magnetostatic energy drops in proportion to the width of the domains.

Magnetic minerals in rocks usually occur as nearly equant grains rather than as thin platelets. Nevertheless, the magnetostatic energy given by the simple Kittel (1949) model for a subvolume containing N domains within a semi-infinite platelet is nearly equal to that obtained for a finite, cubic particle having the same thickness and number of N domains by using rigorous and much more complicated calculations (Rhodes and Rowlands, 1954; Halgedahl, 1987). Thus, the simple Kittel calculation yields a convenient expression for the magnetostatic energy in closed form that can be applied without serious error to the nearly equant grains in rocks.

Energy and Width of the Domain Wall. A *domain wall* is the transition region where the spontaneous magnetization changes direction from one domain to the next. The two most important energies that affect the domain wall are the dominant anisotropy energy, of magnetocrystalline or stress origin, and the exchange energy. The wall does contribute a small amount of magnetostatic energy, but this effect is neglected here. Were the magnetocrystalline energy acting alone, the spontaneous magnetization vector would change abruptly from one domain to the next, so domain walls as we know them would not exist. However, an abrupt transition involves a large expenditure of exchange energy. For this reason, the spins rotate gradually across the wall, giving the wall a finite width.

The energy and width of a 180° Bloch wall that separates two adjacent domains whose magnetizations are antiparallel to each other can be roughly estimated as follows. It is assumed that spins are always parallel to the plane of the wall and that the angle between each adjacent spin in the walls is a constant (Fig. 3). The exchange energy for a pair of atoms with the same spin S is

$$E_{ex} = -2JS^2 \cos \phi \cong -2JS^2\left(1 - \frac{\phi^2}{2}\right)$$

for small ϕ. For simplicity, we assume a simple cubic crystal structure with an atom at each corner of a unit cell of edge a, with the plane of the wall parallel to a cube face. Let the wall be N atoms thick. Therefore, that part of the exchange energy that depends upon the angle is given by

$$E_{ex} = \frac{JS^2 \phi^2 N}{a^2} \quad \text{(per unit area of wall)}$$
$$= \frac{JS^2 \pi^2}{Na^2} \quad \left(\text{when } \phi = \frac{\pi}{N}\right)$$

The anisotropy energy per unit area of wall is about KNa. At stable equilibrium, the derivative of the total energy is zero; thus,

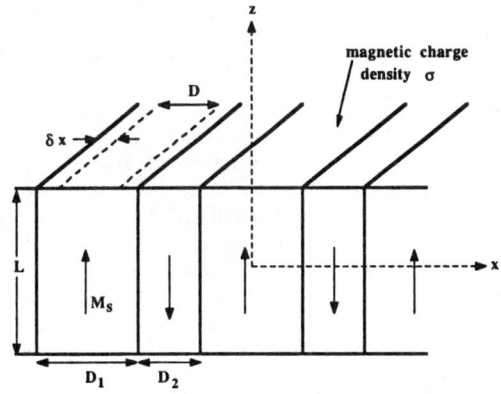

FIGURE 2. The semi-infinite platelet model analyzed by Kooy and Enz (1960) to determine the magnetostatic energy of a periodic distribution of domain walls, whose displacements yield a net magnetization $M = M_s(D_1 - D_2)/(D_1 + D_2)$ per unit volume. δx = the displacement of the wall from the uniformly spaced configuration of zero moment; σ = the density of surface magnetic "charge."

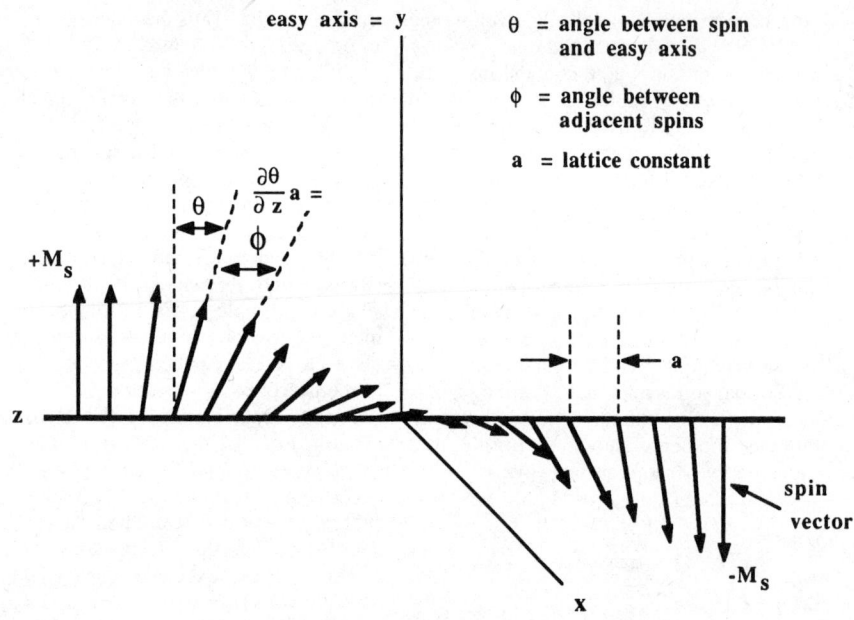

FIGURE 3. A 180° Bloch wall.

$$\epsilon_w(180°) \cong 2\left(\frac{JS^2\pi^2 K}{a}\right)^{1/2}$$

$$\delta_w(180°) \cong \left(\frac{JS^2\pi^2}{Ka}\right)^{1/2}$$

According to this simple model, the wall energy is minimized when the anisotropy and exchange contributions are equal. We also see that the exchange energy tends to broaden the wall. Whereas the anisotropy energy serves to narrow it.

For stable equilibrium, the torque exerted on a spin by the anisotropy field, which tends to restore the spins in the wall toward an easy direction, must be exactly balanced by an opposing torque from exchange. For this reason, the angle between spins in the wall actually must increase as the spins rotate into the hard direction. A more correct treatment of the domain wall accounts for the change of the angle between adjacent spins as a function of position within the wall (Lilley, 1950). For uniaxial anisotropy, Lilley's calculations give

$$\epsilon_w = 4(AK)^{1/2}$$

where $A = JS^2/a$ (the exchange constant) for a simple cubic lattice, but is equal to other multiples of JS^2/a for other types of cubic lattices (see Kittel, 1949). In ferrimagnetic materials, the exchange constant is a complex function of the individual exchange interactions between the various magnetic sublattices.

Wall energy calculations for cubic materials, though analogous in principle to the uniaxial case, are much more complicated, particularly when the magnetoelastic energy in the wall is accounted for. For details, the reader is referred to Lilley (1950). Values of the wall energy have been calculated by Moskowitz and Banerjee (1979) and by Moskowitz (1980) for magnetite and for $x = 0.6$ titanomagnetite by Lilley's method. These calculations give $\epsilon_w = 0.8$ erg·cm^{-2} for $x = 0.0$ (magnetite) and 0.42 erg·cm^{-2} for $x = 0.6$. However, a fair estimate of ϵ_w in cubic materials such as magnetite and titanomagnetite is provided by the uniaxial Eq. 2. The major source of error in calculating ϵ_w for magnetite and titanomagnetite lies in our uncertain knowledge of the exchange constant A in these geophysically important materials rather than in the particular method used for calculations. Also, the magnitude of internal stress is usually unknown. Because the Ti-rich titanomagnetites are quite stress-sensitive, wall energies that are calculated for stress-free crystals may be in error. A summary of parameters important to domain calculations is given in Table 2 for magnetite, one of the most important magnetic carriers in sedimentary and many igneous rocks, and for $x = 0.6$ titanomagnetite, the initial magnetic phase in unoxidized oceanic basalts.

Domain Structures and Their Dependence upon Grain Size

The style in which domain walls terminate at the crystal surface is largely determined by the dominant anisotropy energy in the crystal. Domain wall

TABLE 2. Important Parameters for Magnetite and $x = 0.6$ Titanomagnetite

	T_c (°C)	M_s (emu/cm^3)	K_1 (erg/cm^3)	λ_{111}	λ_{100}	A (erg/cm)	ϵ_w (erg/cm^2)	δ (cm)
Fe$_3$O$_4$	575	480	-1.36×10^5	78×10^{-6}	-20×10^{-6}	6.7×10^{-7}	0.8	1.4×10^{-5}
Fe$_{2.4}$Ti$_{0.6}$O$_4$	150	130	-0.40×10^5	90×10^{-6}	—	3.3×10^{-7}	0.42	1.03×10^{-5}

Notes: T_c, M_s, K_1, λ_{111}, λ_{100} for magnetite taken from Syono (1965). Corresponding values for $x = 0.6$ estimated from Syono's data by linear extrapolation. A (exchange constant), ϵ_w (domain wall energy), and δ (wall width) from Moskowitz and Banerjee (1979) and Moskowitz (1980). ϵ_w and δ calculated for the extended medium.

patterns that we observe on the surface of cubic materials thus may be quite different from those found on materials that are magnetically uniaxial. Furthermore, distinctly different styles of surface domain structures occur in uniaxial materials, depending on whether the anisotropy is strong or weak with respect to the magnetostatic term.

Both cubic and uniaxial magnetic minerals occur in rocks. For example, pure magnetite is the classic cubic spinel, and its domain structure has a cubic style. Pyrrhotite, whose crystal structure is pseudo-hexagonal, is magnetically uniaxial due to its strong, first-order magnetocrystalline term. The Ti-rich titanomagnetites in ocean floor basalts have a cubic spinel crystal structure. However, Ti-rich titanomagnetites are so highly magnetostrictive that they may be controlled by residual stresses that, magnetically speaking, make them behave in a uniaxial manner. For this reason, we will discuss several prototypical domain models for cubic and uniaxial materials, whose counterparts are also observed on natural magnetic phases in rocks.

Uniaxial Structures. The simplest type of domain structure consists of a series of uniformly spaced, planar, 180° Bloch walls, as in Figs. 1 and 2. If we assume that the magnetostatic energy can be well-approximated by a thin platelet model, the total energy per unit area of crystal surface is

$$E = 1.705 M_s^2 D + \frac{\epsilon_w L}{D}$$

where L is the grain thickness. It is clear from this expression that a reduction of magnetostatic energy favors the addition of walls, whereas the energy required to add walls will oppose further subdivision. This fundamental principle drives such an "open" structure toward its minimum energy state, no matter how intricately subdivided the pattern of walls at the surface may become. After minimizing the total energy with respect to D, one obtains the familiar half-power law originally presented by Kittel (1949) for the dependence of domain width upon crystal thickness:

$$D = \frac{1}{M_s} \left(\frac{\epsilon_w L}{1.705} \right)^{1/2}$$

If we assume that the magnetostatic energy of a cubic particle of thickness L can be well-approximated by Kittel's model, then the number of domains is

$$n = M_s \left(\frac{1.705 L}{\epsilon_w} \right)^{1/2}$$

Thus large grains contain more domains than small grains with the same shape, given the same material. With increasing crystal thickness, uniaxial materials develop a remarkable variety of surface domain structures that reduce magnetostatic energy more effectively than a simple "Kittel-like" array of planar domains. The fundamental style of these surface patterns is determined by the relative strength of the magnetostatic energy with respect to the dominant anisotropy energy of magnetocrystalline or stress origin.

Materials in which $2\pi M_s^2/K > 1.0$ will form prism-shaped closure domains bounded by 90° walls, as in Fig. 4a. Because the magnetic flux of this arrangement is completely enclosed within the specimen, there is no magnetostatic energy from surface poles as in the Kittel-like arrangement addressed earlier. However, anisotropy is expended within the closure domains, where M_s is oriented perpendicular to the easy axis. For a given crystal thickness L the equilibrium domain width D is

$$D = \left(\frac{2L \epsilon_w}{K} \right)^{1/2}$$

so that, again, D varies with the half-power of grain thickness. Note, however, that $\epsilon_w \propto K^{1/2}$ to a first approximation. Therefore, $D \propto K^{-1/4}$. Domain widths in uniaxial "low-anisotropy" materials bounded by simple closure domains consequently respond to a rise in anisotropy energy by growing narrower. This trend is just the opposite to what occurs for an "open" flux arrangement (Fig. 2), in which the domains become broader when the anisotropy energy, and hence wall energy, increase.

The completely closed structure is predicted to give way to a much more complicated surface arrangement of closure and reverse spike domains

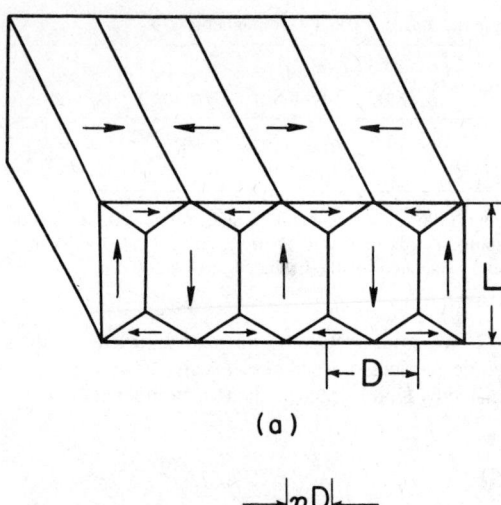

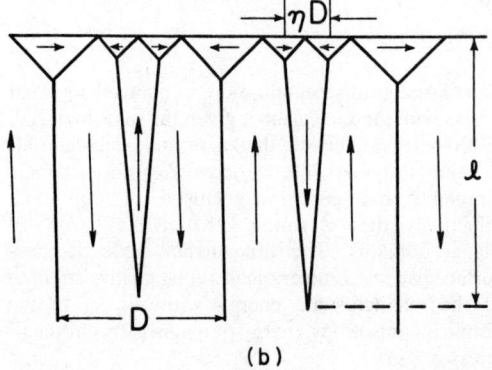

FIGURE 4. (a) Prototypical domain structure in a uniaxial material with "low"-anisotropy ($2\pi M_s^2/K > 1.0$), in which the magnetic flux is completely closed at the surface by closure domains. (b) Elaborate surface domain structures which can develop in thick crystals of low anisotropy, uniaxial materials.

when the grain thickness exceeds

$$L_c = \frac{27\pi^2 \epsilon_w M_s^2}{\mu^2 K^2}$$

where $\mu = 1 + 2\pi M_s^2/K$ (Fig. 4b).

A very different style of surface pattern is favored when $2\pi M_s^2/K < 1.0$. In this case, the anisotropy energy far outweighs the magnetostatic energy, so particles cannot afford to expend anisotropy energy in surface closure domains. Instead, uniaxial crystals that exceed a certain critical thickness develop wavy walls (Fig. 5) that accomplish a more intimate mixture of surface free poles while still permitting M_s to lie along the easy axis. According to calculations by Szymczak (1968), the wavy structure becomes more energetically favorable than the Kittel-like structure above a critical domain width D_c and critical particle thickness L_c given by

$$D_c = \frac{3.94\epsilon_w}{\mu M_s^2}$$

$$L_c + \frac{1.705 M_s^2 D_c^2}{\epsilon_w} \cong \frac{26.5\epsilon_w}{(M_s\mu)^2}$$

where $\mu = 1 + 2\pi M_s^2/K$. Szymczak's analyses predict that the domain width $D \propto L^{2/3}$ in uniaxial crystals that are thick enough to support the wavy walls.

With increasing thickness, surface domain structures on uniaxial crystals with "high anisotropy" (i.e., $2\pi M_s^2/K < 1.0$) become increasingly more elaborate and display wavy walls that alternate with rows of reverse spikes. Very ornate, branching structures eventually develop on the surfaces of very thick crystals. All of these intricate surface patterns are superficial, however, and do not represent the domain structure within the bulk of the crystal. Penetration of the waviness into the crystal depths would result in excessively high total wall energies. The waviness consequently dies out away from the surface. Thus, although complex patterns may appear on the basal plane of high-anisotropy uniaxial materials, a simple array of 180° walls is found parallel to the easy axis within the body of the grain.

Cubic Materials. Materials that are governed by cubic magnetocrystalline anisotropy may contain prism-shaped closure domains at the surface, with 180° walls within the crystal. Surface closure domains in materials for which $\langle 100 \rangle$ are easy directions, such as iron, are bounded by 90° walls, as in Fig. 4a. Cubic materials store no anisotropy energy within these closure domains, since M_s is oriented along an easy axis at the surface. That closure domains require magnetoelastic energy to be expended, however, owing to their being deformed by the underlying "principal" domains.

In materials such as magnetite, whose easy directions lie along $\langle 111 \rangle$, the closure domains are bounded by 71° and 109° walls and magnetizations within the closure domains are oriented with a normal component to the surface (Fig. 6). Comparison of the energies associated with the closure domains in magnetite reveals that magnetostatic energy far outweighs the magnetoelastic contribution. For a "platelet" of magnetite as in Fig. 6, the equilibrium domain width is

$$D = \left(\frac{1}{M_s \cos \phi}\right)\left(\frac{\epsilon_w L}{1.705}\right)^{1/2}$$

$$\cong \frac{3.0}{M_s}\left(\frac{\epsilon_w L}{1.705}\right)^{1/2}$$

where $\phi = 71°$ or $109°$. Magnetites that contain a well-developed closure structure therefore should contain about one third the number of domains that

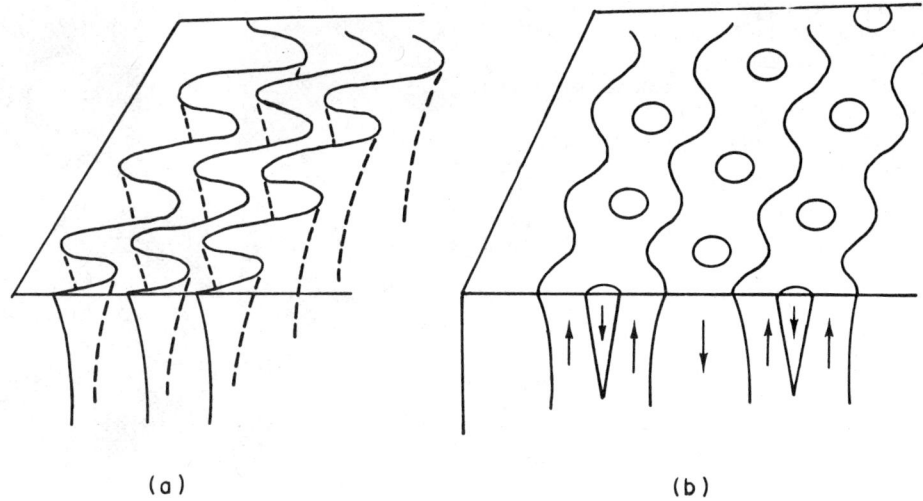

FIGURE 5. (a) Wavy walls at the surface of a "high"-anisotropy ($2\pi M_s^2/K < 1.0$) material. Note that adjacent undulations are "in phase" with each another. (b) Wavy walls and reverse spikes that can develop at the surface of thick crystals with "high" anisotropy. Note that adjacent undulations are now 180° out of phase.

would be expected in the absence of closure domains.

Domain Patterns on Natural Magnetic Phases

In the field of rock magnetism, domain patterns have been studied most extensively on pyrrhotite, magnetite, and titanomagnetite. The technique used to image domain walls with the greatest success has been the Bitter technique. In preparation for observations, samples are first mechanically polished with diamond compounds to achieve a smooth, shiny surface. Next, one must remove the topmost layer of material, which is in a strained state as a result of mechanical polishing. Successful removal of this strained layer is vital, since surface strain usually produces a thin layer filled with finely spaced "maze" domains that bear no resemblance to the domain structure of undisturbed, bulk material. Several methods have been utilized by various workers to remove this strained layer, including thermal annealing at high temperature in vacuum, chemical polishing in hot acid solution, electrolytic polishing, ionic polishing, and Syton polishing. Single magnetite crystals are amenable to all these treatments, since they are chemically stable at high temperatures (i.e., 800°C) in high vacuum and are sufficiently conductive as to be electrolytically polished. However, the latter two methods are the most effective on titanomagnetite samples and on rocks. Due to their tendency to unmix, titanomagnetites cannot be treated at high temperature where crystal reordering occurs unless the oxygen fugacity is carefully controlled and the titanomagnetites are highly stoichiometric. The individual magnetic grains in rocks of greatest geologic interest cannot be electrolytically polished, owing to their being electrically insulated by the surrounding silicates. Nor can the magnetites in rocks always be thermally annealed with success, since they may undergo an unacceptable degree of chemical change due to reactions with other minerals.

The technique that has been used to prepare strain-free surfaces on rocks with greatest success is the

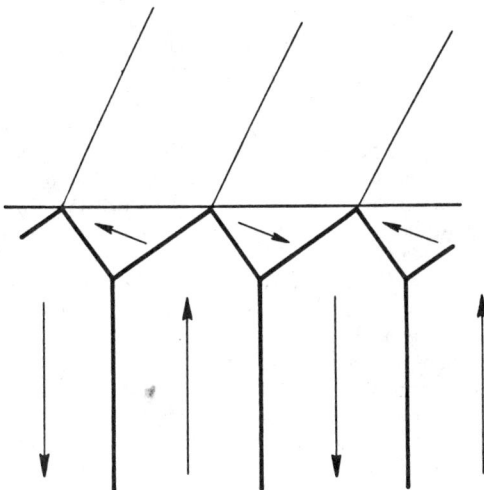

FIGURE 6. Prototypical domain structure in a cubic material with easy directions along $\langle 111 \rangle$, such as magnetite. Flux is partially closed at the surface by closure domains bounded by 71° and 109° domain walls. 180° walls are shown deeper within the crystal.

ionic polishing technique developed by Soffel and Petersen (1971). In this method, the mechanically polished surface is bombarded with ionized air or argon that has been accelerated across several thousand volts.

Good results have also been obtained simply by using a slurry of colloidal silica (Syton) for the final polishing step. Though Syton polishing is a well-known technique in the magnetic materials industry, it has been neglected by rock magnetists until recently.

After adequate surface preparation, the walls are observed with the Bitter method, as follows. A drop of magnetic colloid is placed on a cover slip, which is inverted and pressed firmly against the sample's surface. The strong field gradients associated with domain walls preferentially attract the colloid, so the walls appear as dark lines when viewed in reflected light.

Bitter patterns observed on a large multidomain grain of natural magnetite are shown in Fig. 7. This pattern indicates a network of 180°, 71°, and 109° walls, as we would expect in magnetite dominated by cubic magnetocrystalline anisotropy with $K_1 < 0$.

Domain patterns on several pyrrhotite grains are shown in Fig. 8. Owing to its very high magnetocrystalline anisotropy, low sensitivity to stress, and pseudohexagonal structure, pyrrhotite requires only mechanical polishing to yield patterns of 180° walls, as in the figure. Domain widths in pyrrhotite are nearly proportional to $L^{1/2}$, as one would expect for a simple arrangement of lamellar domains. The D-$L^{1/2}$ dependence in pyrrhotite gives a single-domain boundary size near 1.5 μm and $\epsilon_w \cong 2.5$ erg·cm^{-2} (Soffel, 1977; Halgedahl and Fuller, 1983).

Despite their cubic crystallographic structure and

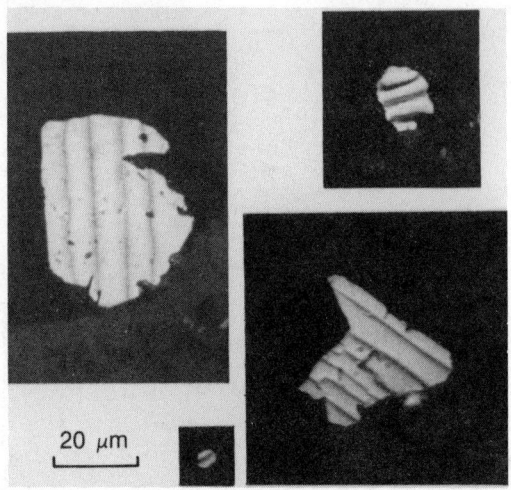

FIGURE 8. Typical domain patterns on natural pyrrhotite particles, indicating 180° walls and uniaxial anisotropy.

cubic magnetocrystalline anisotropy, titanomagnetites ($Fe_{3-x}Ti_xO_4$) rich in titanium yield domain patterns that suggest an underlying stress control. Unlike magnetite, titanomagnetites with x above about 0.4–0.5 rarely display patterns of 180°, 71°, and 109° walls. Instead, typical patterns on Ti-rich compositions are complex and are not characteristic for a dominant cubic magnetocrystalline anisotropy. In many cases such patterns consist of wavy walls and reverse spike domains, as on the $x = 0.6$ titanomagnetite particle in Fig. 9. This type of pattern clearly indicates a strong uniaxial anisotropy, for which $2\pi M_s^2/K < 1.0$, and an easy axis of magnetization that is largely perpendicular to the surface of view. Most likely, such patterns are caused by long-range stress that overrides the relatively weak magnetocrystalline control in Ti-rich compositions (Appel and Soffel, 1984, 1985). Whether this stress is intrinsic to the rock and is important in controlling remanence and coercivity, or whether this stress is an artifact of the surface preparation remains an unresolved problem.

In rare instances, very simple domain patterns are observed on high x-value titanomagnetites, such as that shown in Fig. 10. A plot of the number of domains as a function of grain diameter for $x = 0.55$ titanomagnetite grains with simple patterns is shown in Fig. 11, after work by Soffel (1971). As expected, $D \propto L^{1/2}$, to a good approximation. These data indicate a single-domain boundary $L \cong 0.6$ μm and a wall energy $\epsilon_w \cong 1.0$–1.5 erg·cm^{-2} (depending upon the value of M_s assumed) for this composition. However, even these simple patterns may reflect fairly high internal stresses, since the average number of domains observed for any given size is about one half the number predicted for the unstressed state (Halgedahl, 1987).

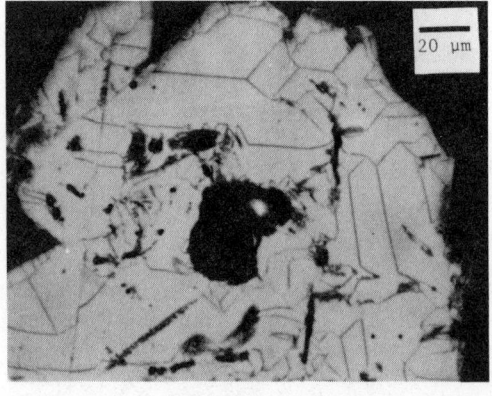

FIGURE 7. Bitter patterns on the surface of a large natural magnetite grain. The network of 180°, 109°, and 71° walls indicates the dominance of magnetite's cubic magnetocrystalline anisotropy.

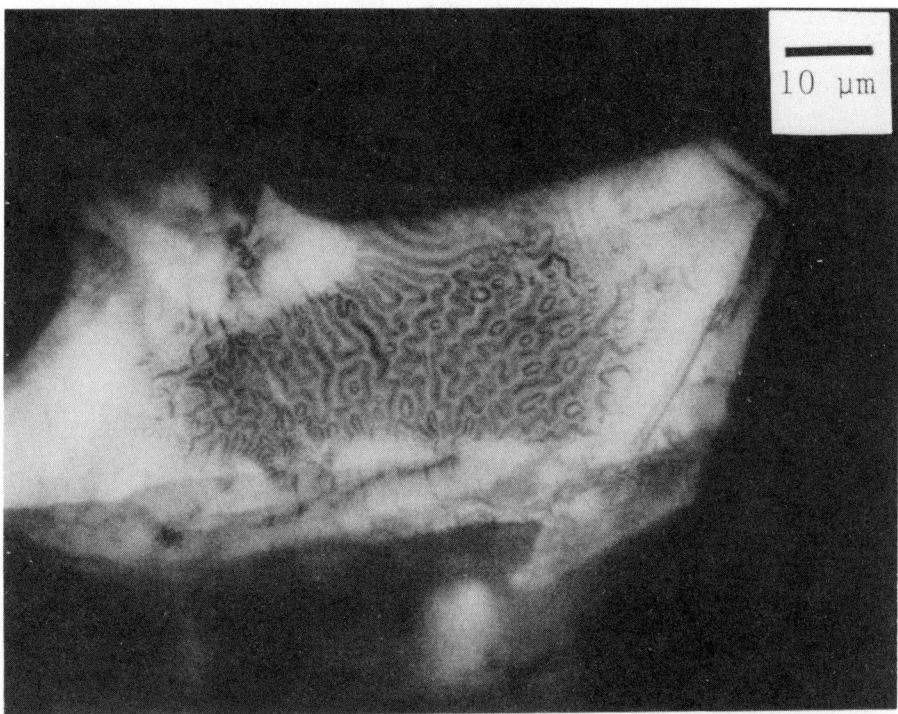

FIGURE 9. Wavy walls and reverse spikes (ovals) on the surface of a natural titanomagnetite grain whose composition is near $Fe_{2.4}Ti_{0.6}O_4$ ($x = 0.6$). This type of pattern is most likely generated by internal stress that, in this grain, yields an easy axis of magnetization that is normal to the surface of view.

Interestingly, plots for the number of domains observed as a function of grain size in the nearly equant grains in rocks invariably are scattered about the $L^{1/2}$-curve that best fits the data. Some of this scatter is undoubtedly attributable to variations in grain shape, bulk composition, and internal wall pinning energies. However, due to the following two effects, much of this scatter would occur even in the most ideal assemblages of grains, whose shapes and anisotropy energies were invariant throughout the population. First, real particles (as opposed to the semi-infinite plates that are so amenable to theoretical treatment) can only contain an integral number of walls. In a sense, the energies associated with adding or subtracting a wall are "quantized." As a result, the same number of domains give a ground state for particles that span a range of size. Simple domain theory applied to finite grains thus predicts that the N-versus-L plot will be intrinsically "scattered" in the horizontal sense (Halgedahl, 1987; Moskowitz and Halgedahl, 1987). Second, scatter in the vertical sense will originate from the various local energy minimum (LEM) states that grains can occupy. When in a LEM state, a grain contains a different number of domains than for an absolute energy minimum (Moon and Merrill, 1985). LEM states originate from the work associated with marching a wall into the grain when other walls, which must become subsequently rearranged, are present. For these reasons, the number of domains observed as a function of grain size in real particles

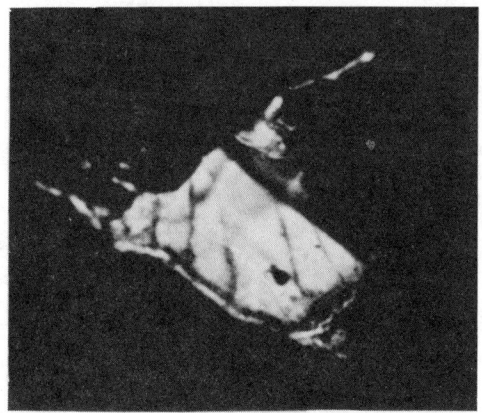

FIGURE 10. Simple domain pattern on a particle of $x = 0.6$ titanomagnetite, which is most simply interpreted as a series of 180° walls.

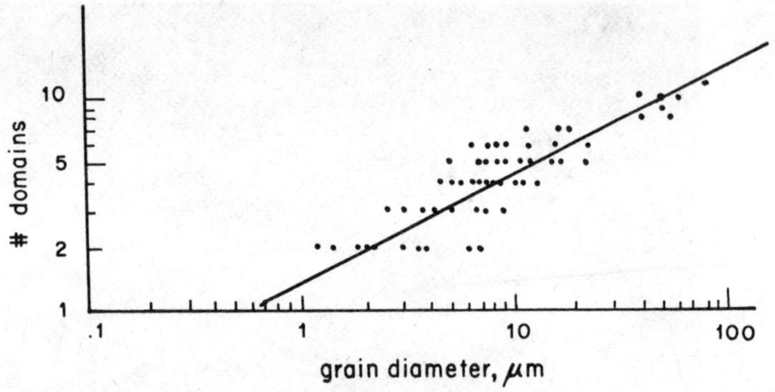

FIGURE 11. The number of domains observed as a function of grain size on particles of natural $x = 0.55$ titanomagnetite, which display simple domain patterns similar to those in Fig. 10 (after Soffel, 1971).

carries an intrinsic scatter. Nevertheless, the data can be fit to the $L^{1/2}$-curve predicted by idealized models to a very good approximation for the purposes of estimating wall energy and the single-domain boundary.

Thermal Variation of Domain Structure

Domain structure is expected to undergo fundamental changes with temperature. These changes are triggered by thermal variations in both the magnetostatic energy, which promotes the addition of walls, and the domain wall energy, which prevents unlimited subdivision. To predict how the domain spacing and the number of domains will vary with temperature, it is convenient to express the temperature dependence of the fundamental energies in terms of the reduced spontaneous magnetization $m(T) = M_s(T)/M_{s0}(T_0)$, in which the reference temperature T_0 is assumed to be room temperature here. To a good approximation, the exchange constant A and the dominant anisotropy constant K may be expressed as powers of the spontaneous magnetization at T:

$$\frac{A(T)}{A(T_0)} = [m(T)]^p, \quad \frac{K(T)}{K(T_0)} = [m(T)]^q$$

(see, for example, Chikazumi, 1964 for a development of these power laws). In magnetite, for example, the first magnetocrystalline anisotropy constant K_1 varies approximately with M_s^8 (Fletcher and Banerjee, 1969). In pure ferromagnets, A varies approximately with M_s^2. The thermal dependence of the exchange constant in a ferrimagnet is difficult to predict, due to the contribution of the sublattice magnetizations; however, we expect A to vary between M_s^1 and M_s^3.

It is most instructive to consider the thermal evolution of domain structure in the simple "open" arrangement of 180° Bloch walls shown in Fig. 1b. Similar reasoning can be applied to other domain structures. When normalized to their values at room temperature, the domain width and the number of domains for the planar arrangement vary with temperature according to

$$\frac{D(T)}{D(T_0)} = [m(T)]^{(p+q)/4 - 1},$$

$$\frac{n(T)}{n(T_0)} = [m(T)]^{1-(p+q)/4}$$

For this simple array of planar walls the number of domains will either increase or decrease with heating according to whether the wall energy decays more or less rapidly than the magnetostatic energy. That is, walls will be added with heating if $p + q > 4.0$; walls will be lost with heating if $p + q < 4.0$; the number of domains will remain constant if $p + q = 4.0$. The inverse trend applies when the sample cools (Halgedahl, 1987; Moskowitz and Halgedahl, 1987). This model is quite simple and it assumes that the configuration of absolute minimum energy is always accessible. Nevertheless, it highlights the physical phenomena that control the variation of domain structure with temperature.

Simple, Kittel-like domain patterns have been observed at high temperatures on uniaxial ferrites, such as barium ferrite (Kojima and Goto, 1962), magnetoplumbite (Gemperle et al., 1963), and Sr-Al hexaferrites (Florescu and Rosenberg, 1970). Domain widths in these three materials increase with temperature, due to the less rapid thermal dependence of wall energy with respect to that of the magnetostatic energy.

The style of surface domain patterns may undergo a thermal hysteresis in certain materials. For example, polycrystalline pyrrhotite displays uncomplicated patterns of straight, 180° walls after af (alternating field) demagnetization at room temperature. Heating to a certain critical temperature transforms these straight walls into a more complex, meandering pattern that is retained upon cooling. The simple pattern can be restored once again by af demagnetization. The critical temperature for the domain pattern transition corresponds to the blocking temperature of TRM (Halgedahl and Fuller, 1981). Thus, TRM acquisition may involve dramatic changes in domain geometry, as well as blocking of walls by the defects that intersect them.

Little is known about the thermal dependence of

domain structure in magnetic phases of paleomagnetic interest. Magnetite, whose closure domains allow some flux leakage, is expected to add walls with rising temperature, since ϵ_w is governed by the thermally sensitive magnetocrystalline energy ($K_1 \propto M_s^8$). High-temperature observations on magnetite have yet to be reported, however.

Recent efforts have been directed toward studying domain patterns on $x = 0.6$ titanomagnetite at high temperature. Because the Curie point of this composition is relatively low (150°C), patterns are observable in the temperature range where TRM is acquired, using special high-temperature colloids. Surprisingly, domain patterns on this composition exhibit few changes with heating (Metcalf and Fuller, 1986; Halgedahl, 1987). This behavior suggests that the dominant anisotropy in $x = 0.6$ decays much more gently with temperature than the magnetocrystalline energy of magnetite. Possibly, this observation lends further support to stress control in Ti-rich compositions (Halgedahl, 1987). Whether it has a magnetocrystalline or stress origin, however, this anisotropy remains sufficiently strong at high temperatures to allow some otherwise multidomain grains to remain in virtual saturation when they acquire TRM, with only small reverse nuclei pinned at their surfaces (Metcalf and Fuller, 1986). Thus TRM in fine particles of Ti-rich titanomagnetites may be attributable to single-domain-like grains that fail to nucleate mature domain structure, as Halgedahl and Fuller (1983) have suggested.

Domains, Domain Walls, and the Grain-Size Dependence of Magnetic Behavior

It is well known that the remanence-carrying efficiency and coercivity increase as grain size decreases toward the single-domain boundary. These trends with grain size occur in a wide variety of materials, ranging from those of industrial interest to the magnetic phases in rocks. Ultimately, the magnetic "hardness" and hence the ability to acquire remanence are rooted in two phenomena: the pinning of domain walls by intervening defects, and the nucleation of domain walls at defects. Though somewhat beyond the scope of this article, these two processes will be discussed in the light of what domain pattern observations have revealed about their impact upon magnetic behavior.

Moving walls can be pinned by defects before they reach a position of absolute minimum energy. Once pinned, the walls require some external influence, such as an applied field or a rise in temperature, to become dislodged. This phenomenon is commonly referred to as *bulk* wall pinning, and the total pinning force upon a wall is approximated by

$$\frac{\partial(\epsilon_w S_w)}{\partial x}$$

where S_w is the wall's area, which is composed of magnetic material, and where x is normal to the wall. Defects include strained (stressed) regions, regions whose composition deviates from stoichiometry, dislocations, concentrations of point defects, holes, nonmagnetic inclusions, and shape irregularities at the grain surface. As a general rule, the width of a defect must be about equal to the width of the wall to have maximum pinning effect. If the wall's area remains constant, pinning originates from a gradient in the wall's intrinsic energy density—that is, from $\partial(AK)^{1/2}/\partial x$. On the other hand, holes, nonmagnetic inclusions, and irregularities in grain shape change the wall's area S_w. In this second case, a field is necessary to drive the wall past the region where its area (and hence total energy) is lowest.

The motion of domain walls in a grain of $x = 0.6$ titanomagnetite during hysteresis is illustrated in Fig. 12. Nucleation is energetically easy in this grain, since walls appear well before the field is removed. Furthermore, wall pinning is fairly weak, since the walls move readily in fields of a few tens of oersteds. By contrast, a surface defect in the pyrrhotite grain illustrated in Fig. 13 has a high pinning strength, since 525 Oe are necessary to drive the wall past the defect during hysteresis.

The process of wall pinning has long been familiar to workers in magnetics. Yet, we still await a pinning theory that satisfactorily accounts for the grain-size dependence of such parameters as H_c (coercivity), J_r/J_s (saturation remanence–saturation magnetization), and TRM from first principles. Much of this difficulty arises from having to account for a random spatial distribution of defects, with a distribution of pinning strengths, that the wall will encounter in its motion. A second complication arises from the spatially dependent internal field, which varies according to how the walls are disposed throughout the grain. Walls do not act independently of one another, so a comprehensive multidomain theory is a "many-bodied" problem. Also, a multidomain theory must consider that the number of walls is a function of grain size. For these reasons, models for the behavior of grains that contain walls rapidly become complicated and may require a statistical approach.

Nucleation processes have received much attention from the magnetics industry in the quest for very hard magnetic materials. According to micromagnetic calculations (e.g., Brown, 1963) the condition for spontaneous nucleation (that is, the initial creation) of a wall is

$$|\mathbf{H}_{app} + \mathbf{H}_{demag}| \geq \frac{2K}{M_s}$$

Here, $\mathbf{H}_{app}$ is the applied field, $\mathbf{H}_{demag}$ is the demagnetizing field, and $2K/M_s$ is the anisotropy field, all at the nucleation site. Perhaps owing to the reduced effect of the exchange interaction, as well as the

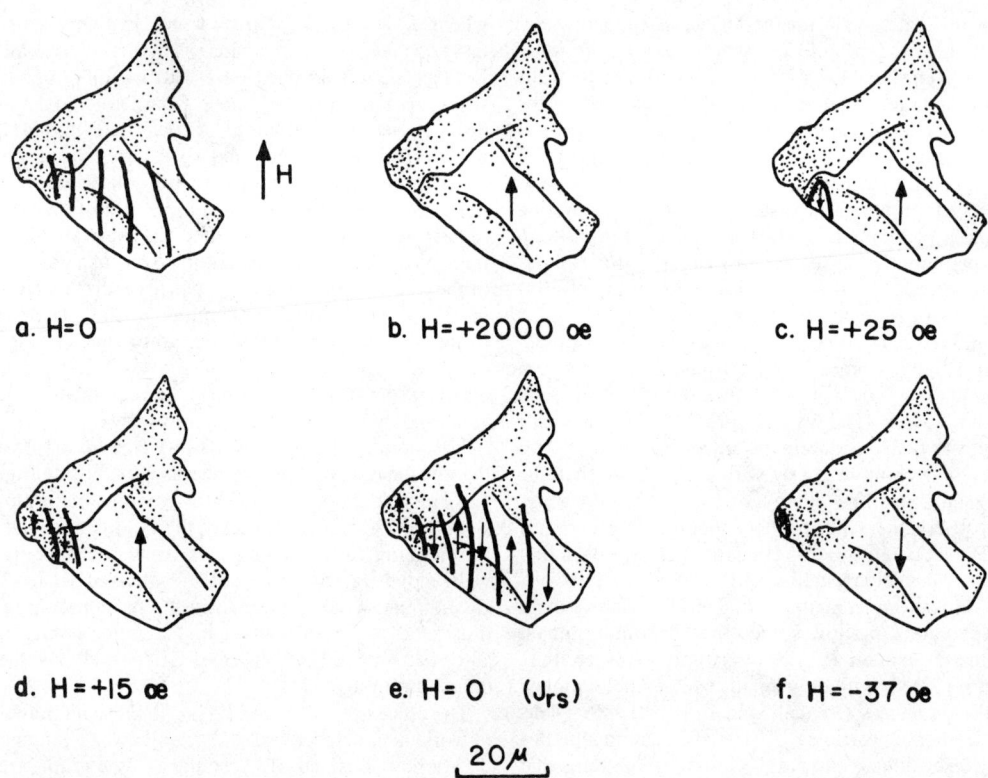

FIGURE 12. Domain patterns observed on an $x = 0.6$ titanomagnetite grain during part of its hysteresis loop. This grain is the same as that shown in Fig. 10. Nucleation of a reverse domain always occurs in the lower left-hand portion of the grain when the field has been reduced to about 25 Oe (c). Additional walls nucleate and move into the particle shortly before the field is reduced to zero (e). A reverse field of only a few tens of Oe is sufficient to achieve a state of near saturation (f), indicating that wall pinning is weak in this grain.

presence of shape irregularities, most nucleation sites occur at grain surfaces. Thus, when the local demagnetizing field at the site is too weak to rotate the local spins away from the easy direction, a reverse field must be applied before a wall can form.

Somewhat paradoxically, it follows that multidomain particles, which may be magnetically "soft" when domain walls are present, actually can be quite "hard" due to nucleation difficulties. Once saturated, particles of high anisotropy-field materials (i.e., $2K/M_s > 1.0$) can remain single-domain until a back-field of sufficient magnitude is applied to produce walls. For example, nucleation fields in Co_5Sm can be thousands of oersteds.

Yet another seeming paradox is that those sites that are most favorable for nucleation (i.e., where either $2K/M_s$ is anomalously low or where $\mathbf{H}_{demag}$ is unusually strong) can increase the remanence and magnetic hardness in two ways. First, although reverse nuclei may actually form at these favorable sites, the submicroscopic domain walls may remain trapped within the site due to magnetostatic pinning by the strong, local demagnetizing field, or due to the anomalously low wall energy at the site with respect to that of the bulk material. Particles may remain in a virtually saturated condition in this manner, despite the presence of many trapped nuclei at their surfaces. Intense back-fields may be required to expand these nuclei into the grain. Second, given that a mature domain structure has managed to evolve, the surface defects then may pin large-scale walls, for the same reasons.

Consequently, two types of nucleation "events" may control magnetic behavior in some materials: the true nucleation of fresh walls at sites that satisfy Brown's condition, and the unpinning and subsequent expansion of submicroscopic nuclei that have been previously trapped within defects. In either case, grains may remain in the saturated, or virtually saturated, state after exposure to a strong field and require a reverse field to trigger development of energetically favorable domain walls. In such particles, the microscopic coercivity may be controlled either by the nucleation field required to expand reverse nuclei into the grain or by the bulk pinning of fully expanded walls by intervening defects, depending on whichever is greater (Becker, 1976).

Recent domain studies have demonstrated the

importance of nucleation in fine-particle behavior of some natural magnetic minerals. Domain studies of $x = 0.6$ titanomagnetite and pyrrhotite grains during hysteresis reveal that otherwise multidomain grains of these two materials appear to remain as metastable single domains in the saturation remanent state (Figs. 13 and 14). The probability that particles fail to nucleate walls in saturation remanence is found to increase as the grain size decreases, in a manner similar to the size dependence of J_r/J_s in these two materials. Walls are nucleated with the application of a back-field, and the nucleation field of pyrrhotite grains rises appreciably as the grains become smaller. It is usually impossible to observationally discriminate the nucleation of a fresh wall from the unpinning of a submicroscopic, preexisting nucleus that was previously trapped at the surface. Nevertheless, the observations suggest that nucleation processes can provide a straightforward mechanism for explaining the grain-size dependence of hysteresis parameters (and, possibly, of TRM) in certain natural materials. This may be understood from a theoretical standpoint, since the probability that a grain will contain a "favorable" nucleation site, which simultaneously fulfills Brown's condition and has a low "trapping" strength, will decrease as the grains become smaller (Halgedahl and Fuller, 1980, 1983).

The acquisition of TRM by multidomain grains will be influenced by the intrinsic thermal variation of the wall pinning forces, the internal field, and the number of domains and domain wall geometry (see *Thermoremanence*). Thermal reequilibration of the domain state has received little attention, yet this phenomenon may have a strong impact upon TRM blocking. Imagine a preexisting series of domain walls in stable equilibrium as the grain cools through a certain temperature interval. Next, imagine that a wall is suddenly lost with further cooling, when the grain assumes the next stable domain state. The loss of a wall will cause an abrupt change in the internal field. In response, the remaining walls must reshuffle into new positions. Unless strong wall pinning prevents this reordering, the moment carried by the grain prior to the transition will be unblocked.

This model suggests the concept of a "domain blocking temperature" T_{db} at which the final domain transition occurs as the grain cools toward room temperature. To first order, this temperature is

$$\frac{n_0 \pm 1}{n_0} = [m(T_{db})]^{1-(p+q)/4}$$

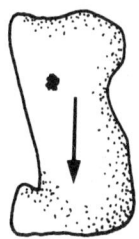

a. H = −2000 oe

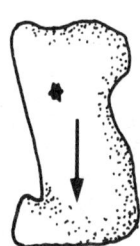

b. J$_{rs}$

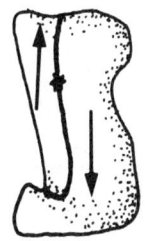

c. H = +60 oe

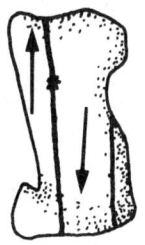

d. H = +300 oe

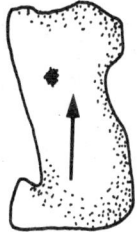

e. H = +525 oe

FIGURE 13. Domain patterns observed on a pyrrhotite grain during part of its hysteresis cycle. Note that this grain appears to remain saturated at saturation remanence (b) and requires a back-field to trigger nucleation of a wall (c). The wall becomes tightly pinned at a defect, indicated by the pit on the surface. This defect has a strong pinning strength, since the wall remains locked until 525 Oe are applied (d, e).

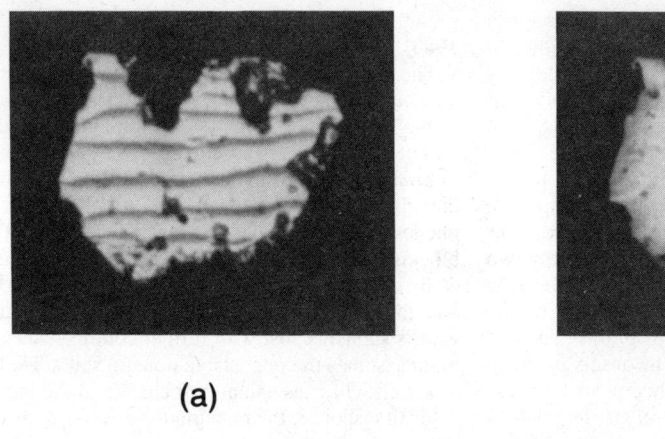

(a) AF Demag ($\tilde{H}$ = 1 KOe) (b) J_{rs} (15 KOe)

FIGURE 14. Large pyrrhotite grain that is clearly multidomain after af demagnetization, but which fails to nucleate domain walls in the saturation remanent state after exposure to 15 kOe.

Since $n_0(T_0) \propto L^{1/2}$, the average domain blocking temperature for small grains with few walls will be higher than that for large grains with many walls. If domain blocking is related to TRM blocking, it follows that small particles with few walls can have higher TRM blocking temperatures than can much larger grains, at least on average. This trend applies to all materials, regardless of the specific thermal variation of the internal energies, except for materials whose domain state is independent of temperature (Halgedahl; 1987; Moskowitz and Halgedahl, 1987). However, both TRM blocking experiments on well-sized assemblages and domain pattern studies as a function of temperature are needed to test this model.

S. L. HALGEDAHL

References

Appel, E., and H. C. Soffel, 1984, Model for the domain state of Ti-rich titanomagnetites, *Geophys. Research Letters* **11,** 189-192.

Appel, E., and H. C. Soffel, 1985, Domain state of Ti-rich titanomagnetites deduced from domain structure observations and susceptibility measurements. *Jour. Geophysics* **56,** 121-132.

Becker, J. J., 1976, Reversal mechanism in copper-modified cobalt-rare-earths, *IEEE Trans. Magn.* **MAG-12,** 965-967.

Brown, W. F., 1963, *Micromagnetics*. New York: Wiley, 143 p.

Chikazumi, S., 1964, *Physics of Magnetism*. New York: Wiley.

Fletcher, E. J., and S. K. Banerjee, 1969, High temperature dependence of single crystal anisotropy constants of titanomagnetite (abstract), *EOS (Am. Geophys. Union Trans.)* **50,** 132.

Florescu, V., and M. Rosenberg, 1970, Temperature dependence of domain structure in strontium alumino-ferrites single crystals, *Japan Jour. Appl. Physics* **9,** 217-223.

Gemperle, R., E. V. Shtolts, and M. Zeleny, 1963, The temperature dependence of the domain structure of magnetoplumbite, *Phys. Stat. Sol.* **3,** 2015-2028.

Halgedahl, S. L., 1987, Domain pattern observations in rock magnetism: progress and problems, *Physics Earth and Planetary Interiors* **46,** 127-163.

Halgedahl, S., and M. Fuller, 1981, The dependence of magnetic domain structure upon magnetization state in polycrystalline pyrrhotite, *Physics Earth and Planetary Interiors* **26,** 93-97.

Halgedahl, S., and M. Fuller, 1983, The dependence of magnetic domain structure upon magnetization state with emphasis upon nucleation as a mechanism for pseudo-single-domain behavior, *Jour. Geophys. Research* **88,** 6505-6522.

Kittel, C., 1949, Physical theory of ferromagnetic domains, *Rev. Mod. Phys.* **21,** 541-583.

Kojima, H., and K. Goto, 1962, Temperature dependence of domain width in thin layers of $BaFe_{12}O_{19}$, *J. Phys. Soc. Japan Jour.* **17,** supp. B-1, 201-203, (Int. Conf. on Magnetism and Crystallography Proc., 1961, v. 1).

Kooy, C., and U. Enz, 1960, Experimental and theoretical study of the domain configuration in thin layers of $BaFe_{12}O_{19}$, *Philips Research Rept.* **15,** 7-29.

Lilley, B. A., 1950, Energies and widths of domain boundaries in ferromagnetics, *Philos. Mag.* **41,** 792-813.

Metcalf, M., and M. Fuller, 1986, Domain observations of titanomagnetites from room temperature to Curie point and the nature of thermoremanent magnetism in fine particles, *Nature* **321**, 847-849.

Moon, T., and R. T. Merrill, 1985, Nucleation theory and domain states in multidomain magnetic material, *Phys. Earth and Planetary Interiors* **37**, 214-222.

Moskowitz, B. M., 1980, Theoretical grain size limits for single-domain, pseudo-single-domain and multi-domain behavior in titanomagnetite (x = 0.6) as a function of low-temperature oxidation, *Earth and Planetary Sci. Letters* **47**, 285-293.

Moskowitz, B. M., and S. K. Banerjee, 1979, Grain size limits for pseudo-single domain behavior in magnetite: implications for paleomagnetism, *IEEE Trans. Magn.* **MAG-15**, 1241-1246.

Moskowitz, B. M., and S. L. Halgedahl, 1987, Theoretical temperature and grain-size dependence of domain state in x = 0.6 titanomagnetite, *Jour. Geophys. Research* **92**, 10,677-10,682.

Rhodes, P., and G. Rowlands, 1954, Demagnetising energies of uniformly magnetised rectangular blocks, *Leeds Philos. Soc., Sci. Sec. Proc.* **6**, 191-210.

Soffel, H., 1971, The single domain-multidomain transition in natural intermediate titanomagnetites, *Zeitschr. Geophysik* **37**, 451-470.

Soffel, H., 1977, Pseudo-single-domain effects and single-domain multidomain transition in natural pyrrhotite deduced from domain structure observations, *Jour. Geophysics* **42**, 351-359.

Soffel, H. C., and N. Petersen, 1971, Ionic etching of titanomagnetite grains in basalts, *Earth and Planetary Sci. Letters* **11**, 312-316.

Syono, Y., 1965, Magnetocrystalline anisotropy and magnetostriction of Fe_3O_4-Fe_2TiO_4 Series—with special application to rock magnetism, *Japan. Jour. Geophysics* **4**, 71-143.

Szymczak, R., 1968, The magnetic structure of ferromagnetic materials of uniaxial symmetry, *Electron Technology* **1**, 5-43.

Cross-references: *Chemical Remanent Magnetization; Curie Temperature; Demagnetization; Magnetic Properties of Minerals; Magnetic Self-Reversal; Natural Remanent Magnetization; Rock Magnetism; Rock Magnetism: Measuring Techniques and Apparatus; Thermoremanence; Viscous Remanent Magnetization (VRM) and Viscous Remagnetization.*

MAGNETIC MEASUREMENTS BY SATELLITE—See SATELLITE MAGNETIC MEASUREMENTS.

MAGNETIC PROPERTIES OF MINERALS

The intrinsic magnetic properties are those that are determined by the species of magnetic atom in the compound and the arrangement of the atoms in space in the crystal lattice. The minerals of the Earth's crust have a ferrimagnetic spin structure. The properties of the magnetic structure may be described in terms of the observables: spontaneous magnetization (M_s), magnetocrystalline anisotropy constants (K, K_1, K_2, etc.), and the magnetostriction constants (λ_{100}, λ_{111}, etc.), which are, in effect, statistics describing the state of the population of magnetic atoms in dynamic equilibrium with temperature. At the Curie point (see *Curie Temperature*) the observables vanish along with the spin structure.

The Titanomagnetites

The titanomagnetites share the spinel crystal structure with the large body of ferrite materials, many of which (such as γ-Fe_2O_3) have useful applications. The building block of the spinel structure is shown in Fig. 1. The nearest neighbors of an oxygen ion are a tetrahedrally sited (A) cation (shared with three other oxygens) and three octahedrally sited (B) cations (each shared with five other oxygens). The spinel formula unit is therefore $A_{1/4}B_{1/2}O$ or AB_2O_4. The nearest neighbors (anions) of A and B sites are also shown in the figure.

Spontaneous Magnetization. The magnetic moments of the A and B site cations are coupled via the oxygen ions (a "superexchange interaction"). The result in many spinel oxides is an antiparallel alignment of the magnetic moments of the A and B

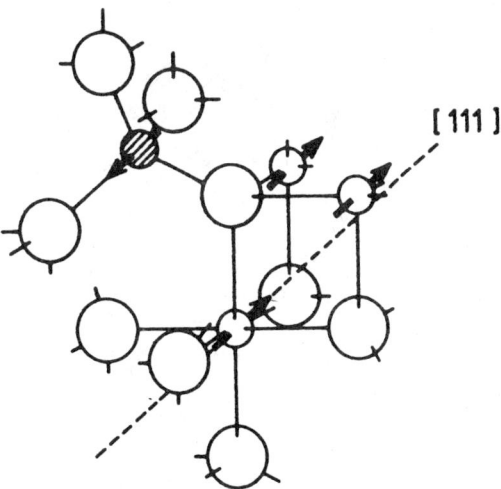

FIGURE 1. The building blocks of the spinel structure. Each oxygen ion (large sphere) is surrounded by a tetrahedral cation site (small hatched sphere, which is shared with three other anions) and three octahedral cation sites (each shared with five other anions). The tetrahedral and octahedral sites become the A and B sublattices of a ferrimagnetic spin structure. The electronic orbitals of the cations interact with the negatively charged nearest-neighbor anions to produce magnetocrystalline anisotropy ($K_1 < 0$ shown) and linear magnetostriction.

sublattices (a ferrimagnetic spin structure). The spontaneous magnetization $M_s(T) = M_B(T) - M_A(T)$, where M_A and M_B are the (temperature-dependent) sublattice magnetizations.

The spontaneous magnetization (measured as saturation magnetization) depends on the distribution of cation species, each with their particular magnetic moment, between the two sublattices. The structural formula of model magnetite is $Fe^{3+}[Fe^{3+}Fe^{2+}]O_4^{2-}$, indicating that the A sublattice is populated by Fe^{3+} ions and the B sublattice by an equal mixture of Fe^{3+} and Fe^{2+}. The magnetic moments of the two species are 5β and 4β, respectively, the moments coming from the five and four uncompensated electron spins of Fe^{3+} and Fe^{2+} (the Bohr magneton $\beta = 9.27 \times 10^{-24}$ A·m^2). The resultant of the perfectly aligned low-temperature spin structure is then 4β per formula unit or, expressed in macroscopic units, 96 A·m^2·kg^{-1}. The structural formula of ulvospinel is universally taken as $Fe^{2+}[Fe^{2+}Ti^{4+}]O_4^{2-}$. Ti^{4+} has no magnetic moment, so the low-temperature magnetization is very small. The model moments of intermediate compositions $Fe_{3-x}Ti_xO_4$ ($0 < x < 1$) and, in particular, TM60 ($x = 0.6$)—the analog of the magnetic phase in new submarine crust—depend on the distributions of Fe^{2+} and Fe^{3+} between the sublattices, Ti^{4+} being assigned to octahedral sites. Figure 2 shows a selection of experimental values of saturation magnetization for the titanomagnetite solid solution. The Fe^{3+}-Fe^{2+} cation distribution is evidently variable. One suggestion, that the distribution reflects the thermal history and corresponds to the last elevated temperature at which there was time for equilibrium to be reached (analogous to the "freezing in" of TRM (see *Thermoremanence*), has yet to be demonstrated experimentally. Perhaps the relaxation time for the redistribution of Fe^{2+} and Fe^{3+} rapidly becomes very long below a "blocking temperature," thus preventing equilibrium being reached with moderate laboratory temperatures in accessible times. At room temperature, a typical range for the saturation magnetization of TM60 is 20–30 A·m^2·kg^{-1}; for Fe_3O_4 it is about 92 A·m^2·kg^{-1}. The saturation magnetization of the series may be approximated to $92(1 - 1.25x)$ A·m^2·kg^{-1} at room temperature.

The Curie Point Temperature (also see *Curie Temperature*). The exchange forces organize the spin structure; thermal agitation disorganizes it. As the temperature rises from absolute zero, the spontaneous magnetization falls monotonically for ferro-

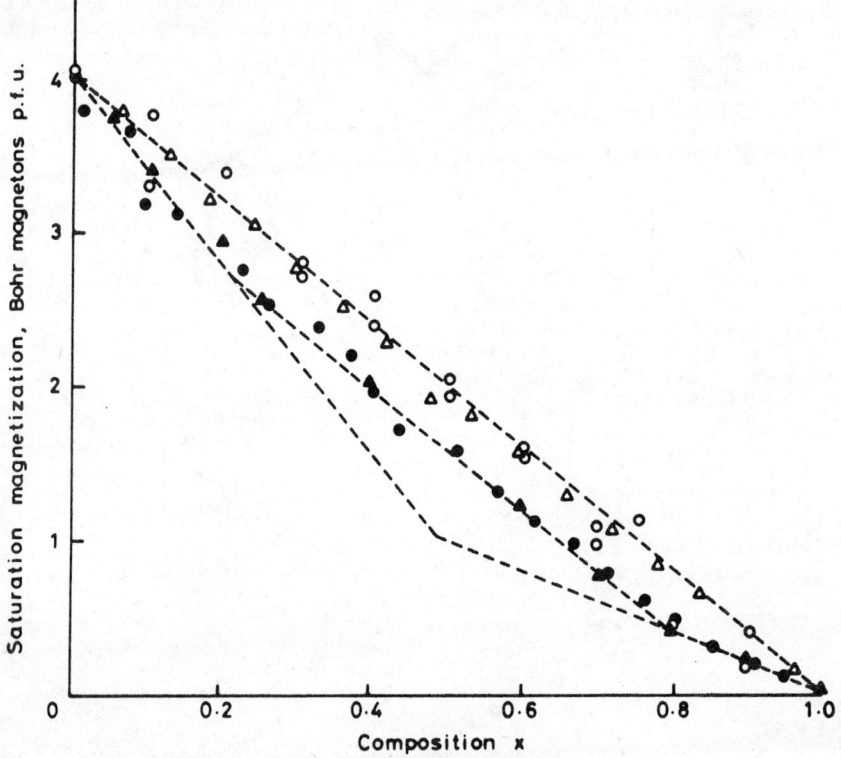

FIGURE 2. A problem in crystal chemistry. The saturation magnetization has a simple relation to the distribution of cations between the tetrahedral and octahedral sublattices of the spinel structure. Titanomagnetite of a given composition, but different provenance, has evidently a variable cation distribution. The source of this variation still has to be resolved.

magnets, but for ferrimagnets the magnetization may rise and fall as temperature rises. The unambiguously defined Curie point at temperature T_c is a fiction of physical models; experimentally T_c must be defined conventionally. According to physical models, T_c depends on some kind of product of the concentrations of the magnetic ions on the two sublattices of a spinel ferrimagnet. T_c is thus only weakly dependent on the distribution of the cation population between the sublattices and is much more strongly dependent on the composition.

In the titanomagnetite solid solution, T_c falls approximately linearly with composition from 575°C ($x = 0$) to $-153°C$ ($x = 1$). The fall is readily explicable in terms of a dilution of the AB interaction by nonmagnetic Ti^{4+} and replacement of the strongly interacting Fe^{3+} by weakly interacting Fe^{2+}. T_c seems very strongly dependent on degree of high-temperature nonstoichiometry (z, the oxidation parameter lying between 0 and 0.1). The empirical dependence is $T_c(x, z) = T_c(x, 0) + 700z$. The Curie point of TM60 may therefore lie between 150 and 220°C. It is not clear how the present physical models for T_c can account for this variation.

Other nonmagnetic diluents (Al^{3+}, Mg^{2+}) also reduce T_c; the application of hydrostatic pressure increases T_c ($\sim 10^{-8}$ K·Pa^{-1}).

Magnetocrystalline Anisotropy. The exchange interactions establish the spin structure (ferromagnetism, ferrimagnetism, etc.). Crystal anisotropy governs the orientation of the spin axes with respect to the crystal structure (e.g., whether the spins lie along the [111] or [100] axes). The two effects do not necessarily operate independently [at the Morin transition in hematite (see the third section of the article) a change in crystal anisotropy produces a change in spin structure]. The principle source of magnetocrystalline anisotropy in the spinel structure is believed to arise from "single-ion" effects. A chain of interactions links the spin (and hence the magnetic moment) of each ion with the crystallographic directions. Due to its motion about the positively charged nucleus, an orbiting electron experiences a magnetic field. The energy of the electron depends on the orientation of its magnetic moment with respect to this field, hence spin-orbit coupling. The orbit is, in turn, coupled to the lattice by the interaction of the negatively charged orbitals with the negatively charged nearest-neighbor cations (Fig. 1). The electronic orbitals are not in general spherically symmetrical. There will therefore be a torque on the extended lobes of the orbitals until the lobes lie in some symmetrical position with respect to the electric field due to the anions.

The symmetries of the electric fields at the two cation sites in the spinel structure are different, and the same cation may discover different preferred orientations of its orbitals in the two sites. Different ions have different electronic structures and therefore different preferred orientations in the same symmetry field. Because the magnetic moments of the cations are coupled together by exchange, the preferred magnetization axis of the population as a whole is the resultant of the torques experienced by individual cations. The expression for the angular dependence of the anisotropy energy (per unit volume) in cubic symmetry is

$$E_{\text{anis}} = E_0 + K_1(\alpha_1^2\alpha_2^2 + \alpha_2^2\alpha_3^2 + \alpha_3^2\alpha_1^2) + K_2\alpha_1^2\alpha_2^2\alpha_3^2 + \cdots$$

where α_1, α_2, and α_3 are the direction cosines of the magnetization vector with respect to the principal cubic axes. The size of K_1, the first-order anisotropy constant, gives the magnitude of the anisotropy, and the sign of K reveals which axis is preferred. $K_1 > 0$ indicates a minimum in the stored energy when the magnetization lies along [100], and $K_1 < 0$ a minimum along [111] (neglecting the effect of K_2).

In the titanomagnetites, four contributions to K_1 will come from Fe^{2+} and Fe^{3+} in the A and B sites. Fe^{3+} is believed to make a positive contribution in A sites and a negative contribution in B sites, the size of each contribution simply being proportional to the concentration of Fe^{3+} on each sublattice. The contributions are temperature dependent, falling monotonically as temperature rises. Fe^{2+} is more complicated, the size and sign of the contribution being both concentration and temperature dependent. The complicated behavior of Fe^{2+} is believed to underlie the temperature dependence of K_1 of near-Fe_3O_4 titanomagnetites (Fig. 3). Notable features are the zero points in K_1, where the spontaneous magnetization is strong but only weakly linked to the crystal lattice, and the large values of K_1 at low temperature.

Magnetostriction–Strain Anisotropy. If the lobes of the cation orbitals experience a torque when they are pulled from the symmetrical minimum energy positions in the crystal field due to the anions, the lobes exert a concomitant torque on the anions. The anions are not absolutely fixed in their positions in the crystal and will be repelled until a new balance is reached with the restoring force of the chemical bonds. The crystal will change dimensions as the direction of the magnetization vector is changed. This phenomenon is *linear magnetostriction*. For a cubic system, the fractional change in length along an axis defined by direction cosines β_1, β_2, and β_3, when the magnetization has direction cosines α_1, α_2, and α_3, is

$$\frac{\delta\ell}{\ell} = \tfrac{3}{2}\lambda_{100}(\alpha_1^2\beta_1^2 + \alpha_2^2\beta_2^2 + \alpha_3^2\beta_3^2 - \tfrac{1}{3}) + 3\lambda_{111}(\alpha_1\alpha_2\beta_1\beta_2 + \alpha_2\alpha_3\beta_2\beta_3 + \alpha_3\alpha_1\beta_3\beta_1)$$

where λ_{100} and λ_{111} are the saturation magnetostriction constants for the [100] and [111] directions. If

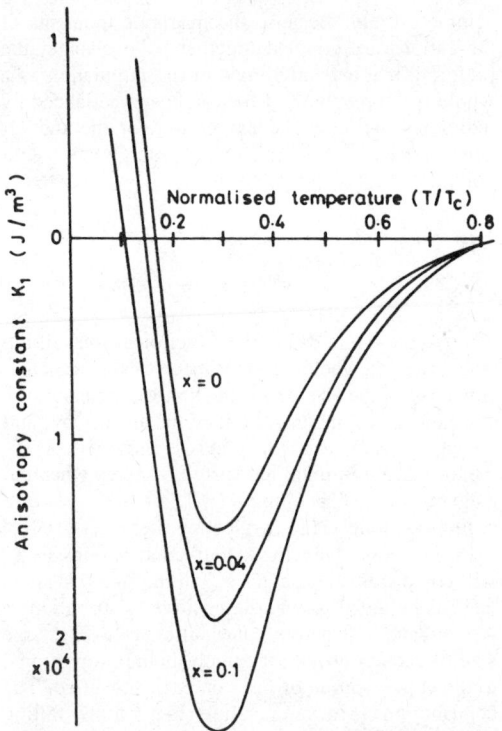

FIGURE 3. The first magnetocrystalline anisotropy constant K_1 for Fe-rich titanomagnetites. The behavior near T_c is relatively simple. Compensation points occur at lower temperatures. Such complex properties can lead to complex magnetization process behavior such as the rapid decay of remanence when cooling or heating through a compensation point.

the magnetostriction is taken to be isotropic, $\lambda_{100} = \lambda_{111} = \lambda_s$ and $\delta l/l = \frac{3}{2} \lambda_s (\cos^2 \theta - \frac{1}{3})$, where $\delta\ell/\ell$ is the strain along a direction making an angle θ with the magnetization vector.

If a constant stress σ is applied along a random axis in the magnetostrictively isotropic material to define the $\theta = 0$ direction, the variation in stored energy per unit volume, due to the variation in $\delta\ell$ along $\theta = 0$ as the magnetization vector rotates, is $\sigma\delta\ell/\ell = -\frac{3}{2} \lambda_s \sigma \cos^2 \theta$. This is an induced anisotropy—strain anisotropy—in which the stress-defined axis is the preferred axis if $\lambda_s \sigma > 0$, i.e., the combination of a compression and negative λ_s or a tension and positive λ_s.

Magnetostriction and magnetocrystalline anisotropy are related; a highly anisotropic ion is also a highly magnetostrictive one. λ_{111} lies in the range $+(80 - 100) \times 10^{-6}$ for $x = 0$ to $x = 0.68$ at room temperature, λ_{100} varying from -20×10^{-6} to $+170 \times 10^{-6}$. The available data show strong temperature dependencies but no points of zero magnetostriction, although they may well occur, probably in λ_{100} for the appropriate composition (low x).

Changing the composition influences all material properties, including the Curie temperature. The temperature dependencies of M_s, K_1, and λ of a suite of related compositions are best compared at the same "magnetic isotherm" by plotting against the ratio T/T_c. Thus, although Fe^{2+} is an anisotropic ion, it also weakens the exchange interaction, reduces T_c, and at, say, room temperature $|K_1|$ may actually *fall* as Fe^{2+} concentration increases. Compositional changes in material properties also produce complex temperature dependencies in magnetization process parameters, which may also be resolved by recasting temperature as T/T_c.

In the magnetization process, a material with high M_s is more strongly coupled to an applied magnetic field. A high $|K_1|$ or λ indicates a strong coupling of M_s to the lattice or direction of an applied stress. The ratios $|K_1|/M_s$ and $|\lambda|/M_s$ are "hardness" parameters—high values of the ratios lead to a "hard" material (in either the monodomain or multidomain states (see *Magnetic Domains*). Figure 4 shows the variation of these hardness parameters for the titanomagnetite solid solution at room temperature. The progressive potential for strain anisotropy as the Fe^{2+} concentration increases is striking.

The Titanomaghemites

The titanomaghemites are nonstoichiometric spinel oxidation products of the titanomagnetites. Oxidation may be effected by adding oxygen to the lattice or removing metals from it. The structure becomes progressively cation deficient with respect to the stoichiometric metal: oxygen ratio of 3:4. The resultant open structure of the titanomaghemites is metastable and collapses to a more stable configuration on application of hydrostatic pressure or elevated temperatures. In the special case of maghemite, $Fe^{3+}[Fe^{3+}_{5/3} \square_{1/3}]O_4$, the application of 150 bar (1.5×10^7 Pa) at 0°C or a temperature of, typically, 350°C at atmospheric pressure "inverts" the structure to hematite α-Fe_2O_3. The inversion of the general titanomaghemite produces a mixture of phases, the important magnetic constituent often being near in composition to magnetite.

Spontaneous Magnetization and Curie Point Temperature. The variation of spontaneous magnetization with the oxidation parameter z can be interpreted in terms of the evolving cation distribution under certain assumptions (e.g., about the location of $\square$, in addition to that of Ti^{4+} in octahedral sites only). The cation distribution can then be related to the maghemitization mechanism at the atomic level (e.g., in terms of the relative availability for oxidation of Fe^{2+} on tetrahedral or octahedral sites). Some room temperature saturation magnetization data are shown in Fig. 5.

A fall in room temperature spontaneous magnetization due to maghemitization will not be universal. In addition to the complications suggested by the

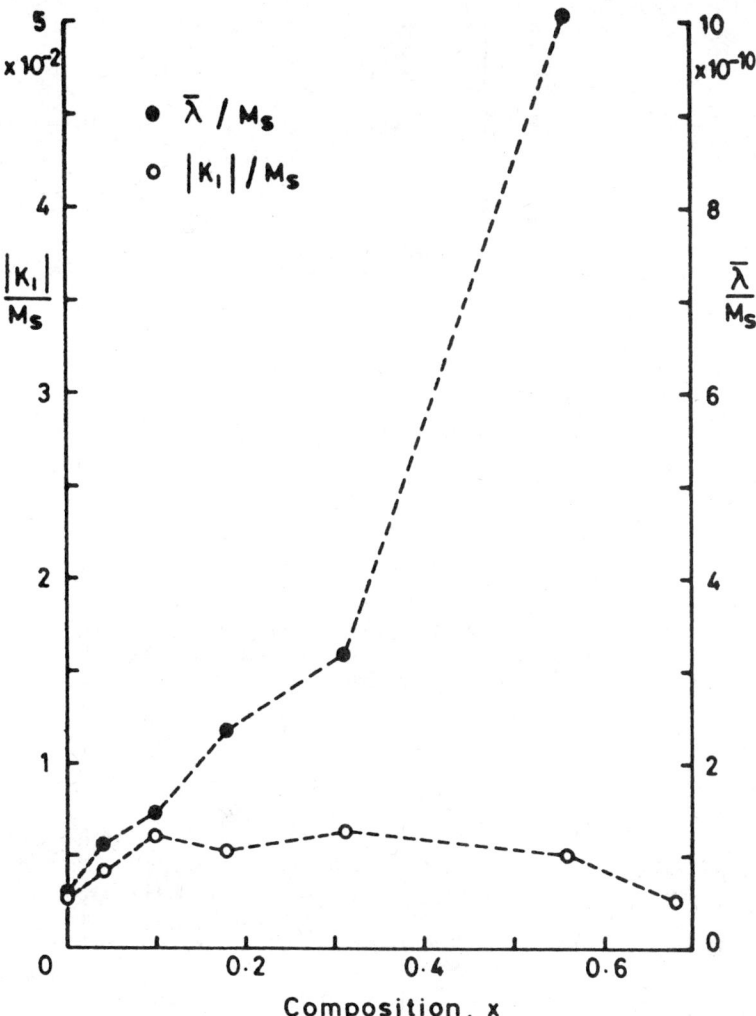

FIGURE 4. The "magnetic hardness" parameters $|K_1|/M_s$ and $\bar{\lambda}/M_s$ at room temperature for members of the titanomagnetite solid solution series. The potential for barriers to magnetization change to be increasingly dominated by strain in Fe^{2+}-rich compositions is evident.

aluminum-substituted TM60 of Fig. 5, an additional factor is the universal rise in T_c as maghemitization proceeds. An example is shown in Fig. 6. In the extreme case, titanomagnetites with T_c below room temperature and, therefore, a zero spontaneous magnetization at room temperature will become magnetic and increase in spontaneous magnetization as T_c rises through room temperature (and acquire chemical remanence; see *Chemical Remanent Magnetization*). The Curie point rises because the increasing concentration of strongly interacting Fe^{3+} offsets the dilution by the increasing □ concentration.

The inversion of titanomaghemites at moderately elevated temperatures finds magnetic expression in the variation of saturation magnetization with temperature. An example is shown in Fig. 7. The inversion product contains an Fe-rich spinel phase near to Fe_3O_4 in composition, and the ratio of magnetization after and before inversion (M_F/M_I) is significantly greater than unity. Such thermomagnetic behavior is a *signature* of maghemitization.

No data exist for K_1 or λ as a function of degree of maghemitization. Experience suggests a general fall in the two properties as the concentration of Fe^{2+} falls, although the variation at some specific temperature, especially near the Curie point, may well be more complex.

Hematite and the Hematite-Ilmenite Solid Solution Series

Hematite (α-Fe_2O_3) crystallizes in the corundum structure. This structure consists of a hexagonal

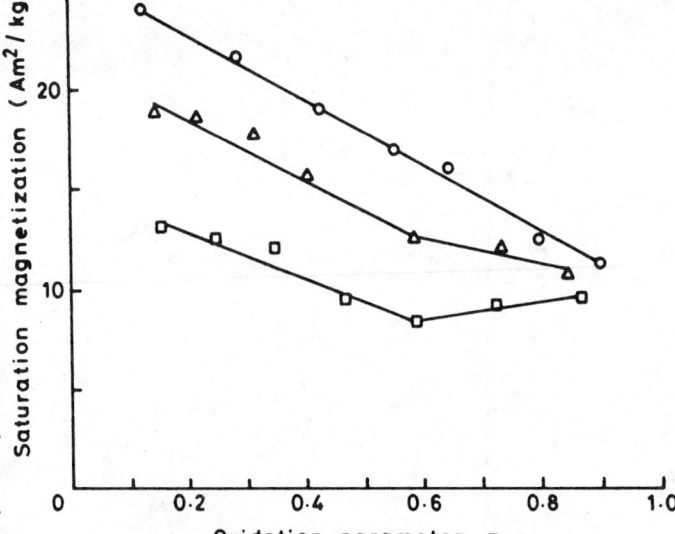

FIGURE 5. The variation of room-temperature saturation magnetization with composition for maghemitized TM60 (upper line) and two related compositions in which part of the Fe^{3+} population is replaced by Al^{3+}. The fall in M_s with degree of maghemitization is relevant to the change with time of the natural remanent magnetization of submarine basalts. Changes in microstructure (grain cracking) during maghemitization may also be a major factor in this context.

close-packed oxygen lattice that encloses interstitial sites of sixfold symmetry (octahedral sites), two thirds of which are occupied by cations (Fig. 8). An antiferromagnetic spin arrangement follows from such a crystal structure. The ilmenite ($FeTiO_3$) structure is essentially similar, with alternate cation layers occupied by the two different species of cation. The hematite-ilmenite solid solution $Fe_{2-y}Ti_yO_3$ ($0 < y < 1$) may have ordered or disordered distributions of the Fe^{3+}, Fe^{2+}, and Ti^{4+} among the cation layers.

Spontaneous Magnetization and Curie Point Temperature. Although it is in essence an antiferromagnet, hematite has a spontaneous magnetization at room temperature and above and often also at low temperatures. The antiferromagnetism is expressed through a magnetic susceptibility (χ_{HF}). The subscript HF refers to "high field" and distinguishes this intrinsic susceptibility from the microstructure-dependent susceptibility of ferri- and ferromagnets below T_c. The latter is a magnetization process parameter. Both types of susceptibility contribute to the low-field induced magnetization in hematite. $\chi_{HF} = 25 \times 10^{-8}$ $m^3 \cdot kg^{-1}$ and is, to a good approximation, constant in hematite samples of different provenances and particle size. It is also relatively independent of temperature, consistent with being the "perpendicular susceptibility" of an antiferromagnet. By contrast, the spontaneous magnetization is very variable, values of ferromagnetic saturation magnetization between 0.2 and 0.5 $A \cdot m^2 \cdot kg^{-1}$ being reported.

The origin of the spontaneous magnetization is a "canted" spin structure. The alternate cation layers are the A and B sublattices of an antiferromagnet. When M_A and M_B lie in the plane of the layers, they are not exactly antiparallel, thus producing a small resultant M at right angles within the cation layer. If M_A and M_B lie at right angles to the plane of the layers (the magnetocrystalline anisotropy constant by

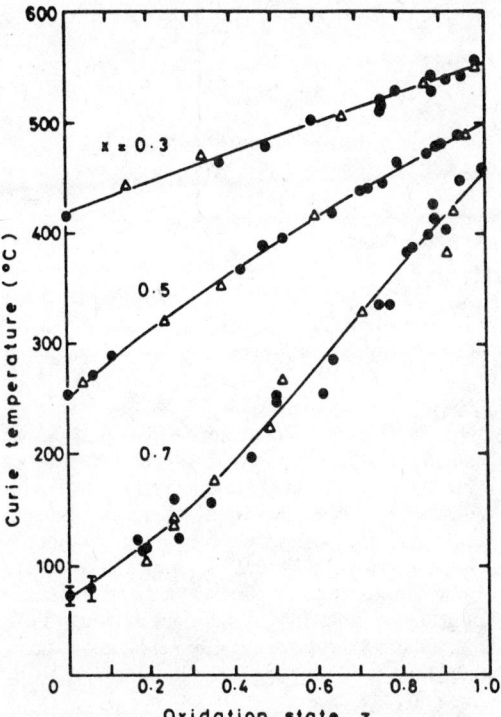

FIGURE 6. Variation of Curie point temperature with degree of maghemitization, with initial titanomagnetite composition as a parameter. Such data find use in the determination of degree of maghemitization of submarine basalts.

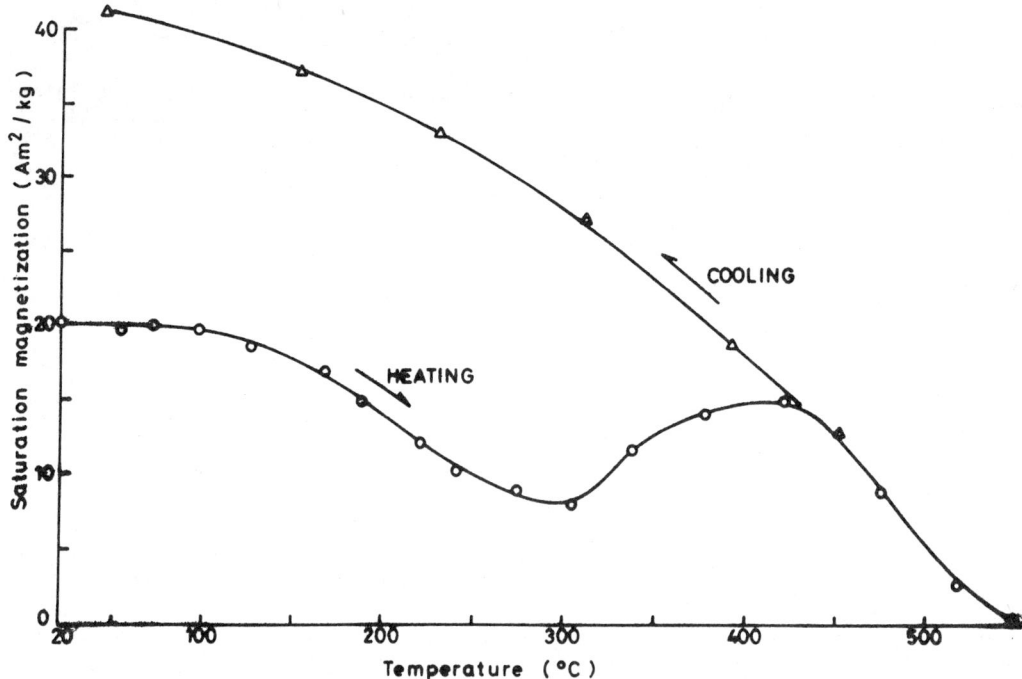

FIGURE 7. The irreversible variation of saturation magnetization during heating a vacuum-encapsulated maghemitized TM60 ($z = 0.29$). The rise in M_s at about 300°C indicates a structural change ("inversion") that produces a strongly magnetic phase near Fe_3O_4 in composition.

implication having changed sign), M_A and M_B are exactly antiparallel and no spontaneous magnetization exists. The spin structure switches, at the Morin transition, from being parallel to being at right angles to the layers on cooling below room temperature. The spontaneous magnetization is then lost.

In addition to the "weak ferromagnetism" associated with the canted spin structure, the presence of a variable "defect moment" has been hypothesized. The precise definition of what might constitute a defect moment has yet to be worked out. Because the intrinsic ferromagnetism of hematite is so weak compared to that of associated iron oxides such as γ-Fe_2O_3 and Fe_3O_4, the presence of undetected quantities of the latter oxides in hematite samples provides the simplest explanation for reductions in saturation magnetization observed after some hematite samples are heated. Such a simple explanation does not preclude the existence of variable moments arising from perturbations of the spin structure by induced anisotropies or yet more exotic mechanisms.

The spontaneous magnetization of the hematite-ilmenite solid solution is variable. Quenched material may have a near-random distribution, each cation layer containing $(y/2)Fe^{2+}$, $(1-y)Fe^{3+}$, and $(y/2)Ti^{4+}$ per formula unit and weak ferrimagnetism due to the almost identical magnetic sublattices. The ordered arrangement, perhaps approached in slowly cooled material, may have layers containing $(1-y)Fe^{3+}$ and yTi^{4+} that alternate with layers containing $(1-y)Fe^{3+}$ and yFe^{2+}. The low-temperature saturation magnetization would be 4β per

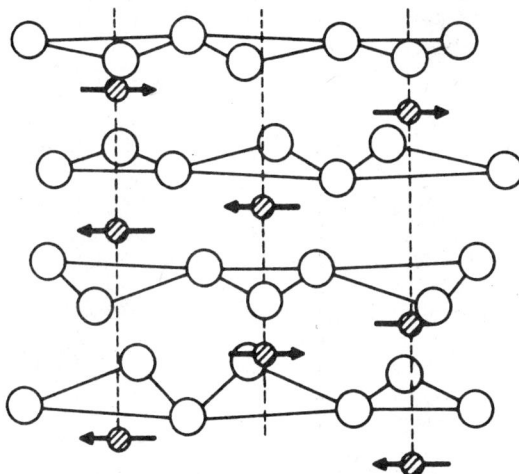

FIGURE 8. The corundum structure of hematite. The cations lie between the close-packed oxygen layers, the electrostatic energy being reduced by the indicated distortions to the ideal structure. The magnetic structure above the Morin transition is shown, although the small canting angle between the magnetic sublattices, which leads to the weak ferromagnetism, cannot be shown on the scale of the figure.

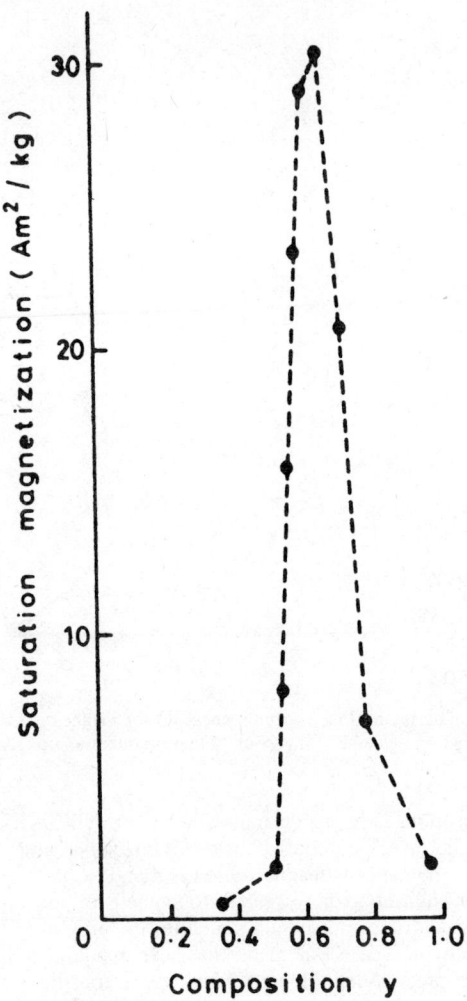

FIGURE 9. The compositional variation of saturation magnetization at room temperature of 9.8 μm size samples of the hematite-ilmenite solid solution $Fe_{2-y}Ti_yO_3$ ($0 < y < 1$).

formula unit, the spins lying in the basal plane, like high-temperature α-Fe_2O_3. This strong ferrimagnetism cannot persist up to $y = 1$ (ilmenite) because the Ti^{4+}-rich layers become progressively nonmagnetic and leave the Fe^{2+}-rich layers with nothing with which to couple by exchange interactions. The ferrimagnetic spin structure breaks down, in preference to an antiferromagnetic structure resulting from exchange between the Fe^{2+}-rich next-nearest layers. The transitions from the spin structure of α-Fe_2O_3 to that within the solid solution near the $y = 0$ end of the series and the transition to the spin structure of $FeTiO_3$ near the $y = 1$ end are no doubt complex.

Some room temperature saturation magnetization data for the series are shown in Fig. 9.

The Curie point of Fe_2O_3 ($y = 0$) lies at 675°C.

T_c may vary nonlinearly in the range $0 < y < 0.3$, although no data are available. In the range $0.3 < y < 1$, $T_c = 622 - 655y$°C.

Magnetocrystalline Anisotropy and Magnetostriction. For a crystal with uniaxial symmetry the angular dependence of the energy of the axis of the spin structure is

$$E_{anis} = E_0 + K_{U1} \sin^2 \theta + K_{U2} \sin^4 \theta + \cdots$$

where θ is the angle between the axis of the spin system (antiferromagnetic in the present context) and the unique symmetry axis. Neglecting K_{U2}, we can see that when $K_{U1} > 0$, the unique axis is preferred, and when $K_{U1} < 0$, the unique plane has the lowest energy. The unique plane itself may not be isotropic. The basal plane of the corundum structure has hexagonal symmetry, so a term $K_3 \sin^6 \theta \cos^6 \phi$ must be added.

As for cubic spinel oxides, contributions to K_{U1} come from the single-ion anisotropy of Fe^{3+} in octahedral symmetry (K_{si}) and also Fe^{2+} when $y > 0$. The symmetry of the structure now allows anisotropy to arise from the magnetostatic interactions between the magnetic dipole moments of the Fe^{3+} and Fe^{2+} ions (K_{dip}). Thus, $K_{U1}(T) = K_{si}(T) + K_{dip}(T)$. K_{si} and K_{dip} have opposite signs and different temperature dependences. It seems clear that dipolar interactions would favor the spins lying in the basal planes, which suggests that K_{si} dominates at low temperature and falls faster with rising temperature than K_{dip} does. At the Morin transition temperature $|K_{si}| = |K_{dip}|$; the lowering of the transition temperature with compositional change ($y > 0$) and reduction in particle size indicates modification of one, or both, contributions or the addition of further contributions. In pure bulk hematite the transition lies at about -10°C. About 0.3 mole Ti^{4+} suppresses the transition.

The orientation of the weak ferromagnetism lying in the (unstrained) basal plane above the transition is governed by K_3 (about 0.05 to 20 J·m^{-3}, several orders of magnitude less than the K_1 of Fe^{3+}-bearing spinel oxides). The magnetostriction coefficient of hematite at room temperature is about 10^{-5}, more or less typical of Fe^{3+}-containing oxides. Because M_s is so small, hematite is only weakly coupled to an external field. The "hardness parameter" $|\lambda|/M_s$ is therefore one or more orders of magnitude bigger than the titanomagnetites, and $|K_3|/M_s$ is several orders smaller.

How the magnetostriction and the crystal anisotropy constant K_3 change with composition in the solid solution series is not known, although these properties will become progressively dominated by the increasing concentration of anisotropic Fe^{2+}.

The Pyrrhotites

The pyrrhotites $Fe_{1-x}S$ ($0 < x < 0.13$) crystallize in the NiAs structure (Fig. 10). Stoichiometric FeS sublattice magnetizations lying in the basal planes of the structure. Thus, troilite is antiferromagnetic, with $\chi_{HF} = 20 \times 10^{-8}$ m$^3 \cdot$kg^{-1}. Fe$_7$S$_8$ would have a model ferrimagnetic low-temperature moment of 4β per formula unit (32 A$\cdot$m$^2 \cdot$kg^{-1}) (from the structural model above), assuming moments of the S ions cancel. Experimental room temperature values are about 20 A$\cdot$m$^2 \cdot$kg^{-1}.

Low-temperature hexagonal pyrrhotites exhibit ferrimagnetism below T_c and above the "γ-transition" (Fig. 11). At room temperature the stable state is antiferromagnetism, although the ferrimagnetism of the T_γ to T_c zone can be preserved by quenching. The metastable ferrimagnetism can be removed by annealing just below T_γ. This latter property would serve to distinguish the peak-type thermomagnetic curve of a rapidly cooled low-temperature hexagonal pyrrhotite from that of an inverting titanomaghemite (Fig. 7). A typical saturation magnetization value

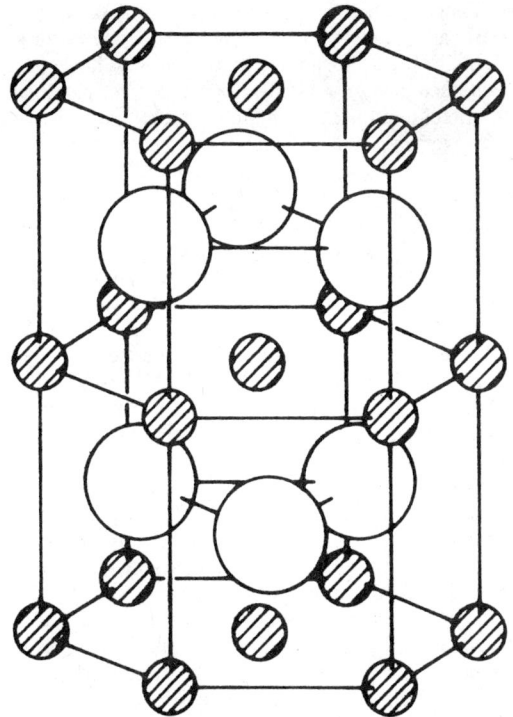

FIGURE 10. The NiAs structure of stoichiometric troilite FeS. In nonstoichiometric monoclinic pyrrhotite Fe$_7$S$_8$ the cation layers (hatched circles) are believed to be alternatively filled and partly filled. The adjacent layers have opposing and unequal magnetizations that lie in the basal planes (like hematite, Fig. 9).

(troilite) has the six sites within the body of the hexagonal prism occupied by S^{2-} and sites on the surface occupied by Fe^{2+}, both ions in six-fold coordination with nearest neighbors. Nonstoichiometric pyrrhotites are cation deficient; the vacancies are believed to be ordered below the β transition (about 310°C) and random above. The valency of the iron may remain unchanged as the degree of nonstoichiometry increases, unlike cation-deficient titanomagnetite. In this case the average sulfur valency is less than 2, and the (ordered) structural formula unit for Fe$_7$S$_8$ (monoclinic pyrrhotite) would be $[Fe_4^{2+}][Fe_3^{3+} \Box](S_8^{2-} h_2^+)$, in which the alternate cation layers are shown in parentheses and h^+ represents a positive hole associated with the S ions. In more nonstoichiometric (low-temperature hexagonal) pyrrhotites, Fe$_9$S$_{10}$ and possibly Fe$_{10}$S$_{11}$ and Fe$_{11}$S$_{12}$, the ordered arrangement may consist of layers on which one quarter of the sites are vacant interleaved with sequences of one or more filled layers.

Spontaneous Magnetization, Curie Point, Crystal Anisotropy, and Magnetostriction. Like hematite, the spin structure of the pyrrhotites consists of ferromagnetic layers with opposing

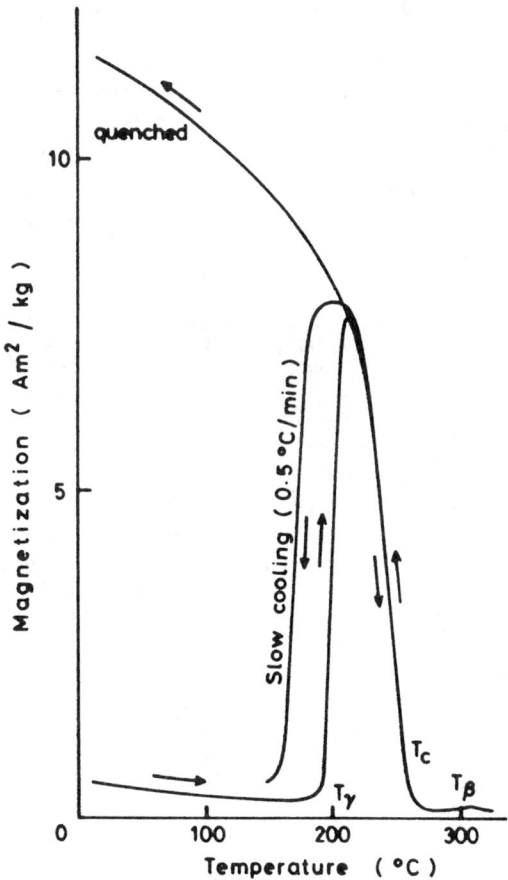

FIGURE 11. The variation of spontaneous magnetization of low-temperature hexagonal pyrrhotite, Fe$_9$S$_{10}$, showing the dependence of the thermomagnetic curve on the history of the specimen and the rate of heating and cooling. The "peak-type" curve should be distinguishable from the inversion of a titanomaghemite (Fig. 8).

for the room temperature quenched-in ferrimagnetism of Fe_9S_{10} is about $10 \text{ A} \cdot \text{m}^2 \cdot \text{kg}^{-1}$.

The Curie point of Fe_7S_8 lies at about 300°C, at or slightly below the β-transition. The Curie points of Fe_9S_{10}, etc., lie from 270 to 210°C, the γ-transitions from about 220 to 180°C.

Unlike α-Fe_2O_3, the spontaneous magnetism of Fe_7S_8 can exist normal to the basal plane. The uniaxial anisotropy constant describing the coupling of the magnetization to the basal plane is about 10^5 $J \cdot m^{-3}$. The anisotropy within the basal plane has a pronounced triaxial component. The magnetostriction coefficient may be as high as 10^{-5}.

Other Minerals

Goethite. The spin structure axis of antiferromagnetic goethite (α-FeOOH) lies parallel with one edge of the orthorhombic unit cell. The origin of the weak ferromagnetism shown by some goethite samples is unknown, but vanishes at the same temperature as the antiferromagnetism (60–170°C). Observed saturation magnetizations lie in the range 10^{-3} to $1 \text{ A} \cdot \text{m}^2 \cdot \text{kg}^{-1}$.

Iron. Iron is ferromagnetic with a T_c of 770°C. The saturation magnetization at room temperature is $220 \text{ A} \cdot \text{m}^2 \cdot \text{kg}^{-1}$. The cubic anisotropy constants are $K_1 = 4.7 \times 10^4 \text{ J} \cdot \text{m}^{-3}$ and $K_2 = 1.5$–3.0×10^4 $J \cdot m^{-3}$ at room temperature; $\lambda_{100} = 2.5 \times 10^{-6}$ and $\lambda_{111} = -19 \times 10^{-6}$.

W. O'REILLY

Bibliography

Fuller, M. D., 1970, Geophysical aspects of palaeomagnetism, *CRC Critical Revs. Solid State Sci.* **1**, 137–219

Hellevege, K.-H., ed., 1982, *Landolf-Börnstein, Numerical Data and Functional Relationships in Science and Technology. Group V: Geophysics and Space Research. Volume 1: Physical Properties of Rocks*, Subvolume b. Berlin, Heidelberg, New York: Springer-Verlag, 600 p.

O'Reilly, W., 1984, *Rock and Mineral Magnetism*. Glasgow: Blackie, New York: Chapman and Hall, 220 p.

Rumble III, D., ed., 1976, *Oxide Minerals*. Washington, D.C.: Mineralogical Society of America, 463 p.

Ward, J. C., 1970, The structure and properties of some iron sulphides, *Rev. Pure Appl. Chem.* **20**, 175–206.

Cross-references: *Chemical Remanent Magnetization (CRM); Curie Temperature; Demagnetization; Magnetic Domains; Magnetic Self-Reversal; Natural Remanent Magnetization (NRM); Rock Magnetism; Rock Magnetism: Measuring Techniques and Apparatus; Thermoremanence; Viscous Remanent Magnetization (VRM) and Viscous Remagnetization.*

MAGNETIC SELF-REVERSAL

One of the most exciting results of early paleomagnetic studies was that many rocks carry a remanent magnetization direction antiparallel (or nearly antiparallel) to that of the present day Earth's magnetic field. Assuming that the rocks have not been tectonically rotated, there are two possible interpretations of these findings: field reversals or self-reversals. In a field reversal, the rocks record the polarity of the ancient geomagnetic field during a time when the Earth's field had the opposite polarity to the present day field. Alternately, in a self-reversal, the reversal of magnetization is due to some intrinsic property of the magnetic minerals in the rocks such that the rocks acquire a remanence that is antiparallel to (or reverse of) the Earth's field.

There is now overwhelming evidence that the Earth's field has reversed its polarity periodically throughout geologic history and that most rocks having reversed remanence directions do indeed record such field reversals. However, there are also a number of studies where the phenomenon of self-reversal is well documented. During a paleomagnetic investigation, then, it is important to know whether or not self-reversal can be excluded as a remanence-controlling mechanism.

In the most general case, a material is self-reversing if, when cooled from above its Curie temperature (T_c) in an applied field, it acquires a remanent moment that is opposite in direction to that of the applied field.

Self-Reversal Mechanisms

Néel (1955) described several possible mechanisms for self-reversal. These mechanisms can be grouped into the following three general classes.

1. The first mechanism involves the magnetic coupling of two phases in a sample. Here, the phases form at high temperature and are already present in the sample when it acquires its remanence. In order for a sample to acquire a reverse remanence, the two phases present must have different Curie temperatures, and they must be negatively coupled. Let us consider a sample containing phase A and phase B, such that $T_c(A) > T_c(B)$. During cooling in an applied field, phase A becomes magnetized first and acquires a remanence parallel to the field. With further cooling, the Curie temperature of phase B is reached. Because of the negative coupling between the two phases, phase B acquires a remanence that is the reverse of phase A and thus the reverse of the applied field. When the sample cools to room temperature (or the temperature of interest), the measured remanence is a combination of the remanent moments of phase A and phase B. Therefore, if the magnetization of phase B is the more intense of the two, the total measured remanence will be reverse of the applied field.

This reversal mechanism is responsible for most of the well documented cases of self-reversal in natural rocks and minerals, and it has been studied in detail in laboratory experiments using both natural and synthetic minerals.

2. The second mechanism involves development of reverse remanence in a sample at some time after acquisition of the initial remanence. This type of reversal requires that changes in the original mineralogy take place at a temperature lower than its Curie temperature. These changes include cation ordering of the original magnetic phase or growth of a second magnetic phase by either exsolution or oxidation.

In the case of cation ordering, reversal can occur if the intensity of magnetization of the sample's sublattices changes as a result of ordering. For example, in disordered ferrites, the B (octahedral) sublattice magnetization is more intense than that of the A (tetrahedral) sublattice. Depending on the bulk composition of the ferrite, it is possible that with ordering the A sublattice magnetization will become the more intense one. Since the A and B sites are antiferromagnetically (negatively) coupled, a reversal of remanence would occur in this situation.

In the case of growth of a second magnetic phase, reversal would occur if the second phase is negatively coupled to the original phase, and if the magnetization of the second phase becomes stronger than that of the original phase.

Several natural examples have been reported where this second self-reversal mechanism seems to have operated. However, it is impossible in most cases to duplicate in laboratory experiments the conditions that the rocks would have been subjected to during their cooling history. In these cases, the existence of this type of self-reversal mechanism can not be verified directly (so-called nonreproducible self-reversals).

3. The third mechanism for self-reversal is found in materials exhibiting Néel's N-type behavior. Here, the phase is homogeneous, but the sublattice magnetizations, while having similar saturation values (though in opposing directions, due to the antiferromagnetic coupling between sublattices), vary differently as a function of temperature. In this situation, there will be some temperature below T_c, called the compensation temperature, at which the intensity of the magnetization of the sublattices will be equal and opposite in direction, so that there will be no macroscopic magnetization of the sample. Below that temperature, one sublattice will have the more intense magnetization, and above that temperature the other sublattice will be more intense. Thus, when cooling from above the Curie temperature in an applied field, the direction of the overall remanent magnetization of the sample will switch from being parallel to the applied field above the compensation temperature to reverse of the field below it. The existence of N-type behavior has been well documented in synthetic Li–Cr ferrites. Reported examples of similar behavior in natural materials, on the other hand, are extremely rare and not well documented.

Interaction Mechanisms. Where two phases are coupled magnetically to produce a self-reversal, two types of interactions between the phases are possible: exchange interactions and magnetostatic interactions. Exchange interactions are spin-dependent electrostatic interactions between electrons on neighboring atoms. Although this type of interaction is quite strong (the exchange energy between neighboring atoms is approximately 1,000 times stronger than the magnetostatic energy), it operates only over very short distances. Thus, phases coupled by exchange interactions must be intimately intermixed on a very fine scale. In contrast, magnetostatic interactions, though very weak, operate effectively on the macroscopic scale (i.e., between coexisting, discrete mineral grains or between adjacent exsolution lamellae). Magnetostatic energy is that energy associated with the demagnetizing field of a phase. The intensity of the applied field that is necessary to suppress self-reversal is much higher where exchange interactions are operating than where magnetostatic interactions are involved (see Stacey and Banerjee, 1974 for a more detailed discussion of magnetostatic and exchange interactions). The results of the majority of well-documented cases of self-reversal indicate that exchange interactions are the most important means of magnetically coupling two phases to produce self-reversal. In fact, Stacey (1963) has argued that the stringent constraints on grain geometry and magnetization necessary for magnetostatic interactions between grains to be stronger than the applied field make it unlikely that this is an important mechanism for producing self-reversals.

Partial Self-Reversal. The concept of partial self-reversal is used to denote rock samples whose overall remanence is normal, but that contain one (or more) mineral phase that is self-reversed and one (or more) phase that is normal. In a rock having a partial self-reversal, the intensity of the rock's normal remanence is diminished due to the presence of the self-reversed phase. In another rock that has the same minerals present but in different proportions, it is possible to get a complete self-reversal if the remanence of the self-reversed phase is stronger than that of the normal phase.

Ilmenite-Hematite Minerals

In minerals of the ilmenite-hematite solid solution series, two different types of self-reversal have been recognized. Ferrian ilmenite samples (those that have intermediate compositions between the end members) can acquire a reproducible, intrinsic self-reversal (one that can be mimicked in laboratory experiments). Titanohematites (those that have hematite-rich compositions), on the other hand, can acquire a nonreproducible self-reversal (one that can not be reproduced in the laboratory).

Ferrian Ilmenites. Ferrian ilmenites having compositions between approximately $Ilm_{48}Hem_{52}$ and $Ilm_{75}Hem_{25}$ are the best known and most thoroughly studied examples of naturally occurring self-reversing materials. Nagata and coworkers first discovered the self-reversing nature of these minerals in a dacite pumice from Mt. Haruna in Japan (Nagata et al., 1953). More recently, other occurrences, all from dacite volcanic rocks, have been found.

Since the initial studies of Mt. Haruna ferrian ilmenites, much detailed work on synthetic ferrian ilmenites has been devoted to elucidating the mechanism responsible for the self-reversal found in these minerals. It is clear that Néel's first reversal mechanism, where two phases are present prior to the acquisition of remanence and are coupled antiferromagnetically, operates in this system. Ferrian ilmenites have been shown to acquire reverse remanence even when exposed to large applied fields, indicating that the coupling between the two magnetic phases is probably through exchange interactions. In one extreme example, a ferrian ilmenite sample with the composition $Ilm_{48}Hem_{52}$ acquired a reverse remanence in applied fields up to 1.27×10^6 A/m (16,000 Oe). Although the results of early magnetic studies were most reasonably interpreted assuming that two phases are present, the samples appeared to be homogeneous down to the resolution of the analytical techniques used (x-ray diffraction, reflected light microscopy, electron microprobe analyzer, and, more recently, scanning electron microscopy). Ishikawa and Syono (1963) called the second phase the X-phase, because they were never able to observe it directly. The X-phase was believed to form during cation ordering and was envisioned as small scale disordered regions surrounding ordered regions. In order to explain the results of their magnetic experiments, Ishikawa and Syono developed the following criteria for the X-phase. It must (1) be enriched in iron relative to the bulk of the sample (i.e., have a higher Curie temperature than the bulk), (2) have a disordered Fe-Ti distribution, (3) be metastable during long-term annealing, and (4) be antiferromagnetically coupled to the ordered regions.

Recent transmission electron microscope (TEM) studies have provided direct evidence of the existence of the X-phase. TEM observations have shown that ilmenite-hematite minerals develop transformation-induced cation-ordered domains (as distinct from magnetic domains) and domain boundaries when cooled through the ordering transformation temperature. The domain boundaries act as the X-phase during acquisition of reverse remanence. Three of the four restrictions on the character of the X-phase (enrichment in iron, disordered Fe-Ti distribution, and metastability during long-term annealing) are satisfied when domain boundaries are equated with the X-phase. The restriction that the X-phase be antiferromagnetically coupled to the ordered regions is neither substantiated nor contradicted by equating transformation-induced domain boundaries with the X-phase. However, antiferromagnetic coupling is a necessary condition for the acquisition of reverse remanence.

Titanohematites. The results of several studies indicate that titanohematite compositions in the ilmenite-hematite solid solution series may carry a nonreproducible self-reversed remanence. Examples of such self-reversals have been reported in granitic rocks from the Sierra Nevada of California, the Bergell Massif of Switzerland and the Piedmont of Georgia, as well as the earlier reported occurrences from metamorphic and igneous rocks of the Grenville Province of New York and Quebec. As the name implies, it is not possible to reproduce a reverse remanence in these samples in laboratory experiments. The operating mechanism for nonreproducible self-reversal has yet to be determined. In addition, although the circumstantial evidence is credible, it is impossible to demonstrate directly that these samples acquired a self-reversed remanence during their cooling history.

The ilmenite-hematite minerals responsible for this nonreproducible self-reversal all come from slowly cooled, relatively deep-seated rocks. In most localities, the ilmenite-hematite phase is pervasively exsolved, consisting of a complex intergrowth of ferrian ilmenite and titanohematite lamellae. The reverse remanence is presumed to reside in the titanohematite phase, since the ferrian ilmenite phase is paramagnetic at room temperature. The exact mechanism responsible for this type of self-reversal is not known. However, the mechanism is thought to involve a magnetic coupling between two phases.

Magnetite-Ulvöspinel Minerals

Natural examples of self-reversing magnetite-ulvöspinel minerals all come from basaltic lavas. In all reported cases, the magnetite-ulvöspinel phase is an intermediate titanomagnetite having a composition in the range between approximately $Mag_{60}Usp_{40}$ and $Mag_{20}Usp_{80}$ and has undergone some degree of oxidation (or maghemitization), either low-temperature, high-temperature or both. These phases are more properly called titanomaghemites. Unlike the ilmenite-hematite minerals, where one reversal mechanism is responsible for all the known examples of reproducible self-reversals (i.e., those known from dacitic volcanic rocks and their synthetic analogues), all three of Néel's reversal mechanisms have been invoked to explain self-reversals found in magnetite-ulvöspinel minerals.

Negative magnetic coupling between two phases has been called upon to explain reverse remanent magnetizations induced during laboratory heat treatments of both natural and synthetic samples. In general, experiments run at moderate temperatures (between approximately 300° and 450°C) for short periods of time (less than a few hours) are necessary to produce self-reversed remanences. In longer duration, higher temperature experiments, samples acquire normal remanent magnetizations. Although never observed directly, the two magnetically coupled phases are thought to be a relatively highly oxidized titanomagnetite (titanomaghemite) phase (having a higher T_c) and an unoxidized (or relatively unoxidized) titanomagnetite (having a lower T_c). In long duration or high temperature experiments, little or no unoxidized material remains, and a normal remanence is acquired. Magnetostatic interactions are thought to be the coupling mechanism between the two phases, since self-reversal is usually suppressed except at very low applied fields. In one case, however, it was found that self-reversal occurred (although only at a very low temperature, approximately $-200°C$) even when the applied field was 2.07×10^5 A/m (2600 Oe). This result suggests that the coupling of the phases, at least in this one sample, is through an exchange interaction.

Néel's first reversal mechanism is responsible for many of the examples of self-reversal produced in the laboratory. However, in those remanence acquisition experiments where the degree of oxidation changes during the course of the experiment, a variation of Néel's second mechanism may operate. In this case, although the original titanomagnetite may be above its Curie temperature at the temperature of the experiments, the titanomaghemite that forms may be below its Curie temperature.

The results of a few studies of natural samples have been interpreted as resulting from N-type magnetic behavior (Néel's third reversal mechanism) of titanomaghemites. Rather than having a true compensation temperature where the spontaneous magnetization goes to zero at a temperature lower than T_c, these natural samples have a minimum point in their magnetization curves. It has been suggested that this minimum point is due to the inhomogeneous nature of the materials, but it is not clear whether it is the titanomaghemite phase that is supposed to be inhomogeneous or the rock as a whole. In the vast majority of these samples, the minimum point in magnetization occurs below room temperature. It is only in a very few samples where laboratory induced remanences are reversed at room temperature.

Based on theoretical considerations, it has been suggested that self-reversal in impure titanomagnetites (those containing cations such as Al, Mg, or Mn) could result from cation ordering of phases of appropriate composition. This mechanism requires that the initial Fe-Ti distribution be disordered, with cation ordering occurring only after the acquisition of the initial remanence (Néel's second mechanism). Results from experimental studies suggest that this mechanism may not operate, since there is a significant degree of Fe-Ti ordering even in quenched titanomagnetites. This type of self-reversal is nonreproducible, making it impossible to prove that any natural titanomagnetite samples have self-reversed as a result of this mechanism.

In spite of the fact that there are numerous examples of titanomaghemites that acquire reverse remanent magnetizations in laboratory experiments, it is not clear how important this phenomenon is in nature. There are only a few studies where there is good evidence that the natural remanent magnetization is self-reversed or partially self-reversed.

CHARLES A. LAWSON

References

Gorter, E. W., and Schulkes, J. A., 1953, Reversal of spontaneous magnetization as a function of temperature in LiFeCr spinels, *Phys. Rev.* **90**, 487–488.

Heller, F., and Petersen, N., 1982, Self-reversal explanation for the Laschamp/Olby geomagnetic field excursion, *Physics Earth and Planetary Interiors* **30**, 358–372.

Ishikawa, Y., and Syono, Y., 1963, Order-disorder transformation and reverse thermoremanent magnetism in the FeTiO$_3$-Fe$_2$O$_3$ system, *Jour. Phys. Chem. Solids* **24**, 517–528.

Lawson, C. A., G. L. Nord, E. Dowty, and R. B. Hargraves, 1981, Antiphase domains and reverse thermoremanent magnetism in ilmenite-hematite minerals, *Science* **213**, 1372–1374.

Nagata, T., S. Akimoto, and S. Uyeda, 1953, Self-reversal of thermoremanent magnetism of igneous rocks (III), *Jour. Geomagnetism and Geoelectricity* **5**, 168–184.

Néel, L., 1955, Some theoretical aspects of rock magnetism, *Adv. Physics* **4**, 191–242.

Stacey, F. D., 1963, The physical theory of rock magnetism, *Adv. Physics* **12**, 45–133.

Stacey, F. D., and S. K. Banerjee, 1974, *The Physical Principles of Rock Magnetism.* Amsterdam: Elsevier, 195p.

Uyeda, S., 1958, Thermoremanent magnetism as a medium of paleomagnetism, with special reference to reverse thermoremanent magnetism, *Japan. Jour. Geophys.* **2**, 1–123.

Cross references: *Curie Temperature; Magnetic Domains; Magnetic Properties of Minerals; Natural Remanent Magnetization (NRM); Rock Magnetism; Thermoremanence* Vol. IVA: *Exsolution; Solid Solution.* Vol. IVB: *Magnetic Minerals; Order-Disorder; Spinel Group.*

MAGNETOHYDRODYNAMIC WAVES WITHIN THE EARTH

The fluid core of the Earth is electrically conducting and it is permeated by a magnetic field that makes the propagation of waves, generally called magnetohydrodynamic (MHD) waves, possible. These waves were predicted in the 1940s by Hannes Alfvén, who proved theoretically that the magnetic field lines in a perfectly conducting fluid behave like elastic strings frozen to the medium. It is precisely this elasticity that makes MHD waves possible.

There are several reasons why the study of wave propagation in the core can add insight into the MHD of the core and the origins of geomagnetism. Theoretically, these determine features of core waves that can be put into correspondence with the observations, e.g., the spectrum of core frequencies, the spatial structure of the waves that are most easily excited to large amplitudes, and the rate of drift of the waves about the axis of rotation of the Earth. By choosing the theoretical model whose wave properties fit the observations best, one can hope to learn much about the state of the core. To take a simple analogy, information about the size and constitution of a bell results from an analysis of the sound it emits when struck. Second, destabilizing forces such as buoyancy may introduce instabilities. Such instabilities are likely to be significant ingredients of the dynamo mechanism that maintains the geomagnetic field. By studying waves and their instabilities, we may hope to add to our understanding of the geodynamo process. Third, competing theories of the geomagnetic dynamo could be more easily evaluated if the time scale of the torsional oscillations of the core could be reliably determined and compared with observation. Torsional waves are investigated here. They are similar to the original Alfvén wave, and from that simple starting point we begin our review.

According to linearized theory, unbounded plane waves are characterized by variables that are proportional to $\exp(i\mathbf{k} \cdot \mathbf{r} - i\omega t)$, where t is time, ω is the wave frequency, $\mathbf{r}$ is the position vector, and $\mathbf{k}$ is the wave-number vector. The frequency of plane Alfvén waves in an incompressible fluid of density ρ_0 permeated by a magnetic field $\mathbf{B}$ is $\pm\omega_A$, where $\omega_A^2 = (\mathbf{k} \cdot \mathbf{V}_A)^2$; here $\mathbf{V}_A = \mathbf{B}(4\pi\rho_0)^{-1/2}$ is the Alfvén velocity. These waves travel in the direction of the magnetic field with velocity $\pm\mathbf{V}_A$. Bounded volumes can also execute eigenoscillations of Alfvén type. Their frequency can be estimated by replacing $|\mathbf{k}|^{-1}$ by a characteristic linear dimension L of the volume: $\omega \sim \omega_A \sim V_A / L$. Taking $B \sim 10^2$ G as characteristic of the Earth's core, we have $\omega_A \sim 10^{-7}$ s^{-1}, corresponding to a period of $T = 2\pi/\omega \sim$ 2 years. We have used $L \sim R_1 / \pi \sim 10^3$ km, where R_1 is the core radius. In what follows, we shall frequently, as we have done here, make simple estimates based on plane wave solutions, implicitly supposing that such estimates are reasonable, i.e., would be confirmed by generalizations to the more realistic geometry.

The Earth's daily rotation, with angular velocity $\Omega = 0.729 \times 10^{-4}$ s^{-1}, gives rise to Coriolis forces that are highly significant to core dynamics. It also imparts an elasticity to an incompressible fluid, making another type of wave motion possible. The frequency of these "inertial waves" is $\pm\omega_i$, where $\omega_i^2 = (2\mathbf{\Omega} \cdot \mathbf{k})^2 / k^2$; see, for example, Chandrasekhar (1961). Closed cavities of fluid can execute inertial eigenoscillations of frequency 2Ω or less.

The frequency of plane Alfvén waves as modified by the presence of rotation is $\pm\omega_\pm$, where

$$\omega_\pm^2 = \omega_A^2 + \frac{1}{2}\omega_i^2 \pm \left[\left(\omega_A^2 + \frac{1}{2}\omega_i^2\right)^2 - \omega_A^4\right]^{1/2}$$

see, for example, Chandrasekhar (1961). If $\omega_i \ll \omega_A$, the Alfvén frequency is only weakly split by rotation, but in core conditions $\Omega \gg \omega_A$ and two different types of wave arise. One is the inertial wave weakly distorted by the magnetic field: $\omega_+ \approx \omega_i \sim \Omega$; the other is a slow wave with frequency $\omega_- \approx \omega_A^2 / \omega_i$. If $B \sim 10^2$ G and $\omega_i \sim \Omega$, then $2\pi/\omega_- \sim 10^3$ years.

All waves propagate on some background state, whose character is crucially important in determining their properties. Some assumptions are here inevitable. We will suppose that a dynamo of "strong field" type operates in the Earth's core, and that therefore a large toroidal field B_ϕ is present that dominates the poloidal field $\mathbf{B}_p$ known to exist from observations made at the Earth's surface. In accordance with the recent dynamo model of Braginsky (1978), it is assumed that $B_\phi \sim 10^2$ G. In a strong field dynamo, the fluid velocity $\mathbf{V}$ possesses a large zonal component V_ϕ. It is just this rapid nonuniform rotation that generates the large field B_ϕ. The driving mechanism of the Earth's hydromagnetic dynamo (the geodynamo, as we shall call it) is convection resulting from the buoyancy force, created by a small excess of the actual fluid density ρ over the equilibrium density ρ_0. This "Archimedean force" (as we shall term it) is proportional to the acceleration due to gravity $\mathbf{g}$ and to the relative density excess, or (as we shall call it) the "condensation," $C = (\rho - \rho_0) / \rho_0$. In what follows, we shall assume the core conductivity to be $\sigma = 4 \times 10^5$ ohm$^{-1} \cdot$ m^{-1}, corresponding to a magnetic diffusivity of $D = 2$ m$^2 \cdot$ s^{-1}.

The complete fields $\mathbf{V}$, $\mathbf{B}$, and C include both the background and the waves. They can be subdivided through their different time scales. Characteristic of changes in the background are T_0 and τ_D ($\sim T_0$). Here $T_0 \approx 0.9 \times 10^4$ years is the observed period of oscillations in the geomagnetic dipole moment, and $\tau_D \sim L^2 / D \approx 1.6 \times 10^4$ years is the ohmic diffusion time. Our prime interest here is the large-scale MHD processes with periods significantly shorter than T_0; these processes can be observed and are distinguished clearly from the background, which we have taken to be a strong field dynamo. The waves that can arise in a weak field dynamo were considered by Busse (1975) and Soward (1979b). Various types of oscillations can arise when magnetoconvection is excited in a rotating fluid. Relevant references and reviews may be found in Braginsky (1964), Roberts and Soward (1972), Hide and Stewartson (1972), Acheson and Hide (1973),

Acheson (1978), Roberts and Loper (1979), Soward (1979a,b), Eltayeb (1981), Fearn and Proctor (1983), and Braginsky (1984), to name but a few.

In what follows, we shall employ both spherical coordinates (r, θ, ϕ) and cylindrical coordinates (z, s, ϕ), where $z = r \cos \theta$, $s = r \sin \theta$, and Oz is parallel to Ω; $\mathbf{1}_c$ is the unit vector in the direction in which coordinate c increases; the subscript p denotes meridional projections (e.g., $\mathbf{B}_p = \mathbf{1}_z B_z + \mathbf{1}_s B_s$). We often use $\sim$ as an order of magnitude sign; this should be translated as "up to a factor 1.5-2-3," depending on context. The sign $\approx$ is employed to signify an accuracy of $\sim 10\%$.

MAC Waves

In order to study MHD convection in the core, motions are referred to a coordinate frame rotating with the solid Earth, and the Boussinesq approximation is adopted. The resulting equations are

$$\partial_t \mathbf{V} + \mathbf{V} \cdot \nabla \mathbf{V} = -\nabla P + \Sigma \mathbf{F} \quad (1a)$$

$$\partial_t \mathbf{B} + \mathbf{V} \cdot \nabla \mathbf{B} = \mathbf{B} \cdot \nabla \mathbf{V} + D \nabla^2 \mathbf{B} \quad (1b)$$

$$\partial_t C + \mathbf{V} \cdot \nabla C = Q^C - \nabla \cdot \mathbf{J}^C \quad (1c)$$

and $\nabla \cdot \mathbf{V} = \nabla \cdot \mathbf{B} = 0$. Here $\partial_t = \partial / \partial t$ and, in a similar way later, $\partial_r = \partial / \partial r$, etc.; the pressure P includes the centrifugal force and is divided by ρ_0; $\Sigma \mathbf{F}$ is the sum of the body forces, consisting of the Coriolis force $\mathbf{F}^\Omega = 2 \mathbf{V} \times \mathbf{\Omega}$ (where $\mathbf{\Omega} = \Omega \mathbf{1}_z$), the Archimedean force $\mathbf{g} C$, the magnetic force $\mathbf{F}^B = \mathbf{B} \cdot \nabla \mathbf{B} - \nabla B^2 / 2$, and the viscous force $\mathbf{F}^\nu = \nu \nabla^2 \mathbf{V}$, where ν is the kinematic viscosity. In this section and the next, $\mathbf{B}$ stands for the magnetic field divided by $(4\pi\rho_0)^{1/2}$. It therefore has the units of velocity, 1 cm · s^{-1} corresponding to 11.7 G; it coincides with $\mathbf{V}_A$.

The condensation C is created by inhomogeneities of both temperature and composition, but in the Earth's core it is the compositional part of C that dominates (Braginsky, 1964). The two terms on the right-hand side of Eq. 1c measure respectively the source of condensation and the convergence of its flux. The relevant molecular diffusivities are very small: The temperature diffusivity is $\kappa^T \sim 10^{-5} D$, and the compositional diffusivity $\kappa^C \sim 10^{-10} D$ is quite negligible.

Let $\mathbf{V} = \overline{\mathbf{V}} + \mathbf{V}'$, $C = \overline{C} + C'$, etc., where the background quantities carry an overbar and a prime is used to distinguish deviations from the background. Small oscillations are governed by the linearized form of system 1a–1c. Its solution presents a formidable problem in the general case, and one is compelled to use idealized models. The simplest case is that of the plane wave traveling across a uniform background. We suppose that

$$\overline{\mathbf{V}} = \mathbf{1}_y \overline{V}_y, \quad \overline{\mathbf{B}} = \mathbf{1}_y \overline{B}_y, \quad \nabla \overline{C} = \mathbf{1}_z \overline{C}_z, \quad \mathbf{g} = \mathbf{1}_z g_z$$

where y is the Cartesian substitute for ϕ. We assume that all barred quantities are constant and that all primed quantities are proportional to $\exp(i \mathbf{k} \cdot \mathbf{r} - i\omega t)$; for these quantities $\partial_t \to -i\omega$ and $\nabla \to i \mathbf{k}$.

The simple and direct way of describing the physical nature of MAC waves is to write down and comment on the linearized equations that govern them. We have

$$-\overline{\omega}^2 \mathbf{\eta} = -i \mathbf{k} P' + i \overline{\omega} \, 2\mathbf{\Omega} \times \mathbf{\eta}$$
$$+ \mathbf{1}_z \eta_z \omega_\alpha^2 - \mathbf{\eta} \omega_A^2 \Delta^{-1} \quad (2a)$$

$$\mathbf{B}' = \mathbf{\eta} i \omega_A \Delta^{-1}, \quad C' = -\eta_z \overline{C}_z \quad (2b,c)$$

and $\mathbf{k} \cdot \mathbf{\eta} = 0$, where $\omega_A = \mathbf{k} \cdot \overline{\mathbf{B}}$, and $\Delta = 1 + Dk^2 / \overline{\omega}$. The fluid particle displacement $\mathbf{\eta} = i \mathbf{V}' / \overline{\omega}$ is introduced here in place of $\mathbf{V}'$, where $\overline{\omega} = \omega - \mathbf{k} \cdot \overline{\mathbf{V}}$ is the Doppler-shifted frequency of the wave, i.e., its frequency in the frame moving with mean velocity $\overline{\mathbf{V}}$. The magnetic pressure is absorbed into P', so $\mathbf{F}_B' = i(\mathbf{k} \cdot \overline{\mathbf{B}}) \mathbf{B}'$. The viscous force is neglected because $\nu k^2 \ll \Omega$. Since $\overline{\omega} \ll \Omega$, the inertial term is also small but is retained here for clarity. Dissipation is neglected in Eq. 2c. Because we assume that the stratification is unstable, we use the Archimedean frequency, $\omega_\alpha = (-g_z \overline{C}_z)^{1/2}$, in Eq. 2a rather than the better-known Brunt-Väisälä frequency $N = (g_z \overline{C}_z)^{1/2}$.

The magnetic force is antiparallel to the displacement $\mathbf{\eta}$ and imparts an elasticity to the medium, whereas the Archimedean force is destabilizing. The inertial term in Eq. 2a may be neglected. The force balance in the direction parallel to $\mathbf{\Omega}$ determines P'. In the absence of inertial forces, the velocity required to create the Coriolis force that balances the net remaining force perpendicular to $\mathbf{\Omega}$ arises spontaneously and instantaneously. Magnetic, Archimedean, and Coriolis forces (together with the pressure gradient) are therefore in perpetual balance as the waves propagate and for this reason are called MAC waves. Frequency enters the equations governing these quasi-static waves only because the magnetic and Archimedean forces are proportional to the displacement of the fluid from its equilibrium position, whereas the Coriolis force is proportional to its velocity. Since $\mathbf{k} \cdot \mathbf{\eta} = 0$, it follows from Eq. 2a that

$$\omega = \omega_d \pm \left[(\omega_A^2 - \Gamma) \frac{\omega_A^2}{\omega_i^2} \right]^{1/2} - i Dk^2 f \quad (3)$$

(Braginsky, 1964). Here $\omega_d = k_y \overline{V}_y$; for this model, $\Gamma = \omega_{\alpha \perp}^2$, where $\omega_{\alpha \perp}^2 = \omega_\alpha^2 (\mathbf{k} \times \mathbf{1}_z)^2 / k^2$. If $\omega_\alpha = 0$ and $D = 0$, then Eq. 3 gives the frequency $\overline{\omega} = \omega_A^2 / \omega_i$ of the slow MHD waves that arise at fast rotation rates. They may be referred to for short as MC waves. The addition of the Archimedean force converts them to MAC waves that grow in amplitude when $\Gamma > \omega_A^2$. The ohmic decay term added to

Eq. 3 is assumed to be small (i.e., $Dk^2 \ll \bar{\omega}$); the coefficient f depends on ω_A^2 and Γ and is ~ 1. The MAC wave that arises in an infinite plane layer of thickness L_z must satisfy the condition $V_z' = 0$ at both boundaries. It can be expressed as a linear combination of the unbounded plane waves just considered, with $k_z = \pm \pi / L_z$ and with the same frequency as Eq. 3.

The infinite model just described is an oversimplification. Many attempts have been made to solve more realistic models. The first study of MC waves in a sphere was that of Hide (1966), who considered a thin spherical shell of fluid and assumed that $D = 0$, $\bar{V} = 0$, and $\bar{B}_\phi$ is constant. He established the existence of fast waves with $\omega \sim \omega_i$ and slow waves with $\omega = \omega_A^2 / \omega_i$. The latter had an eastward-propagating phase velocity. Hide also stressed the significance of slow waves for the geomagnetic secular variation.

Braginsky (1967) derived the MAC wave spectrum for a thin torus of rectangular cross section $L_z \times L_s$. It was assumed that $D = 0$, that $\omega_\alpha^2 (= -g_z \bar{C}_z)$ is constant, and that $\bar{B}_\phi(s)$ and $\bar{V}_\phi(s) = s\bar{\zeta}(s)$ are independent of z. The wave amplitude is proportional to $\exp(im\phi - i\omega t)$, where m is an integer, so $k_\phi = m/s$ appears in place of k_y and $\omega_A^2 = m^2 \bar{B}_\phi^2 / s^2$. The expression for the frequency has the form of Eq. 3 with $k_z = \pi / L_z$ and $k_s = \pi / L_s$, but some additional terms are present that allow for the curvature of the magnetic field lines and a small inhomogeneity in the background:

$$\omega_d = m\bar{\zeta} - 2m\zeta_B,$$

$$\Gamma = \omega_{\alpha\perp}^2 + 2\Omega \left(\frac{k_z^2}{k^2}\right) s\partial_s(\zeta_B - \bar{\zeta}) \quad (3a,b)$$

where $2\Omega\zeta_B = \bar{B}_\phi^2 / s^2$. The second term in ω_d imparts a westward direction to the wave propagation relative to the fluid.

Similar results can be deduced from a local stability analysis of the axisymmetric background; see Acheson (1978), Roberts and Loper (1979), and Fearn and Proctor (1983). In this approach, small-scale waves are considered that are locally plane: $k_p \gg L^{-1}$ and $k_\phi \sim L^{-1}$. If $\mathbf{g} = \mathbf{1}_r g_r$ and $\nabla \bar{C} = \mathbf{1}_r \bar{C}_r$, then $\omega_\alpha^2 = -g_r \bar{C}_r$ and, if it is assumed that $Dk^2 \ll \bar{\omega}$, Eqs. 3, 3a, and 3b are valid, with $\omega_{\alpha\perp}^2 = \omega_\alpha^2 k_\theta^2 / k^2$. The condition $\Gamma > \omega_A^2$ for instability can sometimes be satisfied even in the absence of buoyancy. For example, if $\omega_\alpha = 0$ and $\nabla \bar{\zeta} = 0$, "field-gradient instability" is possible if $\partial_s (\bar{B}_\phi^2 / s^{m+2}) > 0$. It seems plausible that Eqs. 3, 3a, and 3b are qualitatively correct also for large-scale wave motion with $k \sim L^{-1}$. The few global solutions, obtained mainly by numerical methods, confirm this expectation.

Analytical solutions of the linearized form of Eq. 1 are possible for $\bar{V} = 0$, $\bar{B}_\phi / s = $ constant. Soward (1979a) used this simplification to construct an analytic model for the plane layer $0 \leq z \leq L$; he assumed $\omega_\alpha = -g_z \bar{C}_z$ is constant, and retained all diffusivities. He confirmed Eq. 3a for MAC waves, and could also study other wave motions in this convenient cylindrical model. Malkus (1967) had previously also used the simple $\bar{V} = 0$, $\bar{B}_\phi / s = $ constant model with $D = 0$ to study waves in a rotating fluid sphere. The constancy of $\omega_A = m\bar{B}_\phi/s$ allowed him to reduce the governing equation to a simple form that he could solve analytically and satisfy the boundary condition $V_r' = 0$ on the surface of the sphere. In this way he obtained the fast and slow wave spectra. The slow waves have frequencies $\sim \omega_A^2 / \Omega$ and travel both eastward and westward. Malkus (1967) also noted that when $D \neq 0$ a thin boundary layer is associated with the waves, which determines their ohmic decay.

The equation governing dissipationless waves riding on a general axisymmetric background $\bar{V}_\phi$, $\bar{B}_\phi$, $\bar{C}$ is of the form

$$M\boldsymbol{\eta}_p + A\boldsymbol{\eta}_p = \frac{\omega}{m} K\boldsymbol{\eta}_p \quad (4)$$

(Braginsky, 1967). Here $\boldsymbol{\eta}$ is the displacement of a fluid particle from the position it would have occupied on the circular trajectory it would have described in the absence of the wave; M, A, and K are self-adjoint operators corresponding to the action of magnetic, Archimedean, and Coriolis forces. This theory was generalized by Braginsky and Roberts (1975) to include magnetic dissipation. Unfortunately, explicit solutions of Eq. 4 for sufficiently general background states are lacking. Special cases nevertheless demonstrate the possibility of MAC waves with both eastward and westward phase velocities, and that these waves grow in amplitude for a sufficiently top-heavy density distribution (i.e., for a sufficiently large $|\nabla \bar{C}|$).

The direction of wave propagation is much discussed in the geophysical literature in connection with the observed westward drift ζ_W of the geomagnetic field. In this context, the important role played by the fluid velocity $\bar{V}_\phi = s\bar{\zeta}$ inherent to the dynamo mechanism must be stressed. Some average of $\bar{\zeta}$ is included in the wave velocity relative to the mantle. Moreover, a large shear gradient $|s\partial_s \bar{\zeta}| \gg \omega_A$ hampers the creation of large-scale waves. This was demonstrated by Braginsky (1980) by means of a crude "duct" model in which ω_A and ω_α are constant and the shear velocity $\bar{V}_y(x)$ is large. A large shear $\bar{\zeta}(s)$ an order of magnitude larger than ζ_W was obtained by Braginsky (1978) in the dynamo model he named "model Z." He found that the angular velocity was directed to the west for $s < s_0$, where $s_0 = 0.73 R_1$, whereas for $s > s_0$ it was eastward and had a much smaller magnitude. A velocity $\bar{V}_y(x)$ of a similar form was therefore assumed in the duct model (Braginsky, 1980). It was found that wavelengths in the x-direction were small in the region where $|\partial_x \overline{\mathbf{V}}_y|$ is large; MAC waves of rather large wavelength in the x-direction tend to occupy the region where $|\partial_x \overline{\mathbf{V}}_y|$ is small. Both directions of wave propagation are possible. It is natural to suppose that MAC waves develop in the Earth's core mainly within an equatorial belt.

Fearn and Proctor (1983) investigated numerically the effect of a nonuniform rotation $\overline{\mathbf{V}}_\phi$ on waves in a rotating sphere containing a field with only a $\overline{\mathbf{B}}_\phi$ component and in the presence of a buoyancy force. They demonstrated that the critical Rayleigh number for convection increases with increasing $|\overline{\mathbf{V}}_\phi|$.

The complicated pattern of MAC waves in the Earth's core is still unclear. The background determines the waves, but, as stressed by Roberts and Soward (1972), the finite-amplitude MAC waves also influences that background. In fact, the MAC wave problem is an integral part of the full geodynamo problem. At present we have only a sketch of a linear theory, where large-scale waves can grow in amplitude. It seems natural to assume that their nonlinear interaction with one another and with the smaller-scale motions, the wave-background interaction, and Joule dissipation, will all stabilize the waves. As a result, MAC waves of rather large amplitude will be established. These MAC waves are an essential part of the generation mechanism of the geodynamo, and they reveal themselves in the observed variations of the geomagnetic field.

Torsional Oscillations

In one class of motions in the Earth's core, the otherwise influential Coriolis force is precisely balanced and smaller forces can make themselves felt. These "geostrophic" motions, $\tilde{v}_\phi = s\tilde{\zeta}(s, t)$, are independent of z, so the Coriolis forces $\mathbf{1}_s 2\Omega \tilde{v}_\phi$ they create are exact gradients that are balanced by pressure gradients $\nabla \tilde{p}(s)$. Such motions do not interact with the field $\overline{\mathbf{B}}_\phi$ to induce a further field, but they stretch the field lines of $\mathbf{B}_p$ in the ϕ-direction, and the resulting elasticity provides the restoring force for torsional oscillations (TO) of Alfvén type (Braginsky, 1970). The frequencies of the TO are $\omega_k \sim \overline{\mathbf{B}}_p L^{-1}$; for $\overline{\mathbf{B}}_p \sim 0.3$ cm s^{-1} (corresponding to 3 G) and $L \sim 10^3$ km, we have $T_k = 2\pi/\omega_k \sim 60$ years. Each cylindrical shell, $s = $ constant, oscillates in a TO with angular velocity $\tilde{\zeta}(s, t)$ under the action of inertial, magnetic ("elastic") restoring forces, friction between the core and mantle, and also some exciting force that compensates for the frictional losses of energy. The uniformity in the z-direction of the total azimuthal force is brought about by a very weak secondary flow, $\tilde{\mathbf{v}}_p \sim (\omega/\Omega)\tilde{\mathbf{v}}_\phi$, which creates a Coriolis force $2\Omega\tilde{\mathbf{v}}_s$ that evens out $\tilde{\mathbf{v}}_\phi$.

The equation governing TO was considered by Braginsky (1970, 1980, 1984) and by Roberts and Soward (1972). It has the form

$$-\omega^2 \mathcal{P}\tilde{\zeta} = \partial_s(\mathcal{K}\partial_s\tilde{\zeta}) + i\mathcal{F}(\tilde{\zeta} - \tilde{\zeta}_M)$$

and is supplemented by the equation governing changes in the angular velocity $\tilde{\zeta}_M$ of the mantle

$$-\omega^2 \mathcal{I}_M \tilde{\zeta}_M = -i \int \mathcal{F}(\tilde{\zeta} - \tilde{\zeta}_M)\, ds$$

where $\mathcal{I}_M$ is the moment of inertia of the mantle. The coefficients $\mathcal{P}$, $\mathcal{K}$, and $\mathcal{F}$ are functions of s that allow for the inertia of the fluid, the magnetic elasticity, and the core-mantle friction. To make progress, one must select a simple model and calculate $\mathcal{K}(s)$, which is proportional to the average of B_s^2 over the cylindrical shell of radius s. The following simple model was assumed by Braginsky (1980, 1984): the regions $s < R_2 = 0.35R_1$ and $s > s_b = 0.8R_1$ were taken to be rigid, while for $R_2 < s < s_b$ it was assumed that B_s^2 is constant and equal to $\mathbf{B}_f^2$. Here R_2 is the radius of the solid inner core; the fluid adjacent to it is carried with it by the frozen-in magnetic field. The rigidity of the equatorial belt, $s > s_b$, is ascribed to the large amplitude of the MAC waves it contains. Expressions for the electromagnetic friction between core and mantle were included, but the origin of the exciting force remained obscure; it cannot be the Archimedean force, because it is orthogonal to $\tilde{v}_\phi$. To circumvent this difficulty, Braginsky assumed that turbulent eddies in the core transport angular momentum in the s-direction toward regions of large $\tilde{\zeta}$. This mechanism is well known in astrophysics and geophysics and is called *negative viscosity*. It results here in the replacement of B_f^2 by a complex constant $B_f^2(1 + i\nu^-)$. The unknown constants B_f^2 and ν^- were obtained by assuming that the TO are steady and by fitting their calculated period to the observed period. It was found that $B_f = 2.1$ G for $T_1 = 65$ years. According to the model calculations, the frequencies of the first three modes, $k = 1, 2,$ and 3, are such that $\omega_k = k\omega_1$, approximately. Because the resonance condition $\omega_3 = \omega_1 + \omega_2$ is satisfied, the excitation of the third mode by nonlinear coupling of the first and second modes is made easier.

Core-mantle friction couples the TO to mantle oscillations, which reveal themselves as variations in the length of the day (l.o.d.); these variations are directly observed.

The Geomagnetic Data and Their Implications

The geomagnetic secular variations (SV) are complicated in nature. It is possible to sort them by their periods in the following way: short-period variations on time scales comparable with $T_k < 10^2$ years, intermediate-period variations on time scales comparable with $T_\alpha \sim 10^3$ years, long-period variations on time scales comparable with the fundamental period $T_0 \approx 0.9 \times 10^4$ years, and even longer periods associated with the evolution of the Earth and its core.

The short-period geomagnetic SV are the subject of a review by Braginsky (1984). They have small amplitudes, only about 1% of the corresponding background fields. They have been observed by direct measurements, principally in the twentieth century. One can see peaks in their spectrum near periods of 60 years and 30 years. Geomagnetic and l.o.d. variations are closely linked through the TO. The l.o.d. data are more suitable for determining the spectrum, for they are more precise and are based on observations over a longer interval of time. The observed l.o.d. variation can be represented to high accuracy by a combination of a slow trend and three oscillations, with periods T_1, $T_2 = T_1/2$, and $T_3 = T_1/3$. The first oscillation started close to 1860 and experienced a slow but significant decay; the second oscillation remained almost stationary, and the third was always rather weak. The amplitudes of the first two oscillations are ~ 1 ms. Through the TO model, this leads to an estimate of $\sim 1°$ for the angular displacement $\tilde{\zeta}/\omega$ of fluid during the oscillation. This motion carries the magnetic field with it, thus creating the geomagnetic SV. Calculations based on the TO model indicate that this mechanism is unable to give a complete description of the observed SV, though it creates effects of the right order of magnitude. One must therefore presume the existence of another, as yet unknown, mechanism (or mechanisms) of short-period SV. Direct evidence for the TO in the Earth's core is provided by the variations in the l.o.d., while the geomagnetic SV serves only to confirm the existence of some process acting on the time scales $\sim T_1$ and T_2. One may hope that the elucidation of the complicated pattern of short-period SV and the discovery of all the mechanisms underlying them will eventually provide a geomagnetic demonstration of the existence of TO and make it possible to estimate some significant parameters of the geodynamo.

The slower geomagnetic SV are studied by the methods of archeomagnetism and paleomagnetism (cf. Creer et al., 1983). The available data are numerous but incomplete, their inaccuracies being comparable with the SV amplitudes themselves. Despite this, they provide us with extremely important information about the geomagnetic field in the past.

The variation in the mean geomagnetic dipole moment $M(t)$ during the past 10^4 years has been determined by a few authors, using an appropriate sliding time average of the archeomagnetic field strengths. The curves obtained resemble a one-wave sinusoid superimposed on a constant value M_0. The period of the sinusoid is approximately 0.9×10^4 years, though values of between 0.7×10^4 years and 10^4 years have also been favored. The average value M_0 is close to 8×10^{25} G·cm^3, its value today; the amplitude of this "fundamental" oscillation is about $0.3 M_0$. Earlier data are scanty, but they do not contradict the existence of the same fundamental oscillation, in which $M_0(t)$ and $T_0(t)$ vary very slowly.

Even the naked eye can discern variations in the archeo- and paleomagnetic curves of field elements from different geographic localities and of characteristic time scales of $\sim 10^3$ years. A spectral analysis of these curves reveals peaks that correspond to periods in the interval 600–1800 years, but larger and smaller periods are also mentioned. Archeomagnetic curves are crude and available only for short intervals of time, usually not more than 2×10^3 years. Only the order of magnitude of the periods can therefore be trusted; the spectra must be regarded as tentative.

A reliable analytic model of the surface geomagnetic field $\mathbf{B}(t, \theta, \phi)$ over a large interval of time may become possible by combining the entire sets of data available with the constraints imposed by the theory of the geodynamo. The theory can guide the choice of a suitable analytical model, one that contains certain adjustable parameters with which to fit the data sets, the number of these parameters being very much less than the number of data. The data available today and in the foreseeable future are numerous but too rough and incomplete to determine by themselves the function $\mathbf{B}(t, \theta, \phi)$. One may hope, however, that it will become possible to fit the parameters of a model to sufficient high accuracy, provided that the model itself is in accord with reality. As a rough attempt, we may mention the tentative parametric model of Braginsky (1974), where the field over the last 2000 years was represented by a set of waves traveling both east and west. For a model to be deemed adequate, it must describe the field over time intervals comparable with T_0, and it must rest securely on a well-developed theory of the geodynamo. In this way, we may hope, despite the scarcity of observational data, to obtain the parameterized field model and to confirm the geodynamo theory. At present only the qualitative agreement between the observed time scales (≈ 60 years, ≈ 30 years, $\sim 10^3$ years, and $\approx 10^4$ years) and those anticipated on theoretical grounds can be pointed to with satisfaction.

Concluding Remarks

The geodynamo is a complicated auto-oscillating system having several different time scales. The MAC waves and the TO are essential facets of its activity.

Numerical investigations of the stability of rotating fluids containing magnetic fields reveal several types of oscillation beside the MAC waves and the TO, particularly those that evolve on diffusion time scales. Some of these can be excited by inhomogeneities in the magnetic field and arise even when the fluid is stably stratified. It is, however, hard to say what MHD waves exist in the core besides MAC waves and TO.

The molecular diffusivities κ^T and κ^C are extremely small, and the corresponding Rayleigh number of the core is extremely large; the geodynamo works far beyond the marginal convective state. It was shown by Braginsky (1964) that, under these conditions, perturbations having the form of small platelike cells, very narrow in the s-direction and stretched out in the z- and ϕ-directions, grow rapidly. They evolve into small-scale turbulent eddies that produce intense mixing of heat and composition. For example, for a mixing length $l^t \sim 50$ km, the turbulent diffusivity, D^t, is about 2×10^4 cm$^2 \cdot$s^{-1}, even when the eddy velocities have the rather modest value of $v^t \sim 10^{-2}$ cm$\cdot$s^{-1}. Probably all unstable small-scale motions in the main body of the core are merged together through their nonlinear coupling and form part of a fully developed turbulence. Because of this turbulence, $\kappa \sim D^t \sim D$ effectively. The long-period waves ($T > T_0$) are not independent; they are intimately linked to the principal convective processes at work in the core. In contrast, MAC waves and TO are clearly distinguished from the background as independent entities. Their frequencies are mainly determined by the mean magnetic field, though dissipative processes, including turbulent transport, may also be significant.

According to Braginsky (1984), special short-period disturbances localized near the core-mantle must be expected, but they have not yet been confirmed by observation.

An identification of the observed intermediate-period secular variations with MAC waves requires that $B_\phi \sim 10^2$ G; this estimate of B_ϕ corroborates the view that the geodynamo is of strong field type. The characteristic large parameter of such a dynamo is $\mathcal{R}^{1/2} \sim B_\phi / B_* \sim B_* / B_p$, where $B_* = (2\Omega D 4\pi\rho_0)^{1/2} \approx 20$ G (cf. Braginsky, 1978). True, $\mathcal{R}$ is not large ($\mathcal{R}^{1/2} \sim 5$), but it satisfies qualitatively the demands of the strong field theory.

We can probably conclude that MAC waves and TO are MHD waves that actually exist in the Earth's core. Their reality is supported by the rough correspondence between theoretical estimates and observational data. The uncertainty expressed in this conclusion and the frequent use of the "$\sim$" sign in this review are dictated by the present state of the subject. Fortunately, the MHD of the Earth's core is now being developed rapidly, and we may hope that in the near future will add enough certainty to allow most of the "$\sim$" signs to be replaced by "$\approx$".

S. I. BRAGINSKY

References

Acheson, D. J., 1978, Magnetohydrodynamic waves and instabilities in rotating fluids, in P. H. Roberts and A. M. Soward, eds., *Rotating Fluids in Geophysics*. London: Academic Press, 315-349.

Acheson, D. J., and R. Hide, 1973, Hydromagnetics of rotating fluids, *Rept. Progr. Physics* **36**, 159-221.

Braginsky, S. I., 1964, Magnetohydrodynamics of the Earth's core, *Geomagnetism and Aeronomy* **4**, 898-916, Engl. trans., 698-712.

Braginsky, S. I., 1967, Magnetic waves in the Earth's core, *Geomagnetism and Aeronomy* **7**, 1050-1060, Engl. trans., 851-859.

Braginsky, S. I., 1970, Torsional magnetohydrodynamic vibrations in the Earth's core and variations in day length, *Geomagnetism and Aeronomy* **10**, 3-12, Engl. trans., 1-8.

Braginsky, S. I., 1972, Analytical description of the geomagnetic field of past epochs and determination of the spectrum of magnetic waves in the core of the Earth, *Geomagnetism and Aeronomy* **12**, 1092-1105, Engl. trans., 947-957.

Braginsky, S. I., 1974, Analytical description of the geomagnetic field of past epochs and determination of the spectrum of magnetic waves in the core of the Earth. II. *Geomagnetism and Aeronomy* **14**, 522-529, Engl. trans., 441-447.

Braginsky, S. I., 1978, Nearly axially symmetric model of the hydromagnetic dynamo of the Earth, *Geomagnetism and Aeronomy* **18**, 340-351, Engl. trans., 225-231.

Braginsky, S. I., 1980, Magnetic waves in the core of the Earth, II, *Geophys. Astrophys. Fluid Dynam.* **14**, 189-208.

Braginsky, S. I., 1984, Short-period geomagnetic secular variation, *Geophys. Astrophys. Fluid Dynam.* **30**, 1-78.

Braginsky, S. I., and P. H. Roberts, 1975, Magnetic field generation by baroclinic waves, *Royal Soc. London Proc.*, **A347**, 125-140.

Busse, F. H., 1975, A model of the geodynamo, *Royal Astron. Soc. Geophys. Jour.* **42**, 437-459.

Chandrasekhar, S., 1961, *Hydrodynamic and Hydromagnetic Stability*. Oxford: University Press, 652p.

Creer, K. M., P. Tucholka, and C. E. Barton, eds., 1983, *Geomagnetism of Baked Clays and Recent Sediments*. Amsterdam: Elsevier, 324p.

Eltayeb, I. A., 1981, Propagation and stability of wave motions in rotating magnetic systems, *Physics Earth and Planetary Interiors* **24**, 259-271.

Fearn, D. R., and M. R. E. Proctor, 1983, Hydromagnetic waves in a differentially rotating sphere, *Jour. Fluid Mech.* **128**, 1-20.

Hide, R., 1966, Free hydromagnetic oscillations of the Earth's core and the theory of the geomagnetic secular variation, *Royal Soc. London Philos. Trans.* **A259**, 615-650.

Hide, R., and K. Stewartson, 1972, Hydromagnetic oscillations in the Earth's core, *Rev. Geophysics Space Physics* **10**, 579-598.

Malkus, W. V. R., 1967, Hydromagnetic planetary waves, *Jour. Fluid Mech.* **28**, 793-802.

Roberts, P. H., and D. E. Loper, 1979, On the diffusive instability of some simple steady magnetohydrodynamic flows, *Jour. Fluid Mech.* **90**, 641-668.

Roberts, P. H., and A. M. Soward, 1972, Magnetohydrodynamics of the Earth's core, *Ann. Rev. Fluid Mech.* **4**, 117-153.

Soward, A. M., 1979a, Thermal and magnetically driven convection in a rapidly rotating fluid layer, *Jour. Fluid Mech.* **90**, 669-684.

Soward, A. M., 1979b, Convection driven dynamos, *Physics Earth and Planetary Interiors* **20**, 134-151.

Cross-references: *Core-Mantle Coupling; Earth's Core; Geomagnetic Field: Asymmetries; Geomagnetic Field: Westward Drift; Geomagnetic Field, Main: Theory; Geomagnetic Secular Variation: Direction and Intensity; Geomagnetic Secular Variation: Theory.*

MAGNETOSTRATIGRAPHY

In the early 1960s scientists proved that the Earth's magnetic field reverses itself and thus provides a worldwide and synchronous marker. It was further shown that these reversals occurred at random intervals, suggesting that if a pattern in a stratigraphic sequence could be tied to a standard pattern, calibrated with respect to time, that the sequence could be used for dating the stratigraphic section. Thus was born the field of magnetostratigraphy.

Since the application of magnetostratigraphic analysis relies to a very large extent on the accurate dating of the polarity history of the magnetic field, a brief account of the development of the time scale models is provided in the following section. Then we proceed to a beginner's guide to magnetostratigraphic methodology followed by a description of three possible applications of magnetostratigraphic techniques to solve geologic problems.

Development of the Geomagnetic Reversal Time Scale

The Geomagnetic Reversal Time Scale (GRTS) is a chronometric calibration of the record of geomagnetic polarity history. Its development has been an iterative process, synthesizing data from a variety of sources such as marine magnetic anomaly records, magnetostratigraphic records, dates derived from the decay of radioactive elements and age information available from the record of biotic evolution.

In order to establish the relative pattern of magnetic polarity intervals, scientists rely on two fundamentally different records of geomagnetic polarity history: the marine magnetic anomaly record and the magnetostratigraphic record. Such records are generated in a manner analogous to a low fidelity magnetic tape recorder. The "tape" in this analogy is either the oceanic crust created and magnetized at the great chains of mid-ocean ridges or stratigraphic sequences consisting of sediments or volcanic layers.

Carrying the tape recorder analogy further, the "magnetic tape head" for oceanic crust is at the mid-ocean ridge, where molten rock first solidifies, then cools through its Curie temperature at which point it acquires a permanent magnetization known as thermal remanent magnetization or TRM. Subsequently, the magnetized layer undergoes alteration from the circulation of seawater through the hot crust and acquires a chemical remanent magnetization or CRM. The strongly magnetized layers of newly created oceanic lithosphere are gradually carried away from the ridge by the process of seafloor spreading and, as the magnetic field switches polarity, forms lineated bands of reversely and normally magnetized crust. The oceanic lithosphere thus generates a magnetic field that alternately adds to and subtracts from the Earth's field, creating the lineated magnetic anomalies measured by ship-towed magnetometers (Fig. 1). This marine magnetic anomaly record carries the polarity history since the early Jurassic and has a resolution of about 40,000 years. It is, however, subject to distortions caused by changes in the rate of seafloor spreading and must be calibrated with respect to absolute time.

The magnetostratigraphic record of polarity history comprises sediments or volcanics that were magnetized at the time of deposition or eruption. Stratigraphic sections generally do not span great lengths of time and can be subject to abrupt changes in accumulation rates (similar to a sticky tape recorder). However, stratigraphic records can provide a much higher resolution picture of the details of polarity history and thus can be used to confirm and supplement the record provided by the marine magnetic anomalies.

Chronometric calibration of the magnetic polarity history is not as easy as it might at first appear. The most straightforward approach would be to date marine basalts recovered by drilling at clearly identifiable anomalies; such dates are often unreliable owing to the extensive alteration history of oceanic crust. The calibration of the GRTS, then, has relied on dates that have been correlated to the geomagnetic reversal pattern.

The development of the GRTS involves the parallel endeavors of establishing the geomagnetic reversal pattern itself as well as calibrating it with respect to time. The first time scales were based on determinations of both age and polarity for many basaltic units from around the world. When stacked up against age, a pattern began to emerge of the polarity history of the Earth's magnetic field (see column A of Fig. 2). Subsequent time scales, which were based on marine magnetic anomalies, relied on the matching of the anomaly pattern to that of the basalt record. One reversal boundary was selected (the beginning of the Gauss) as a calibration point in addition to the anomaly centered over the ridge (zero age). By assuming a constant spreading rate for a profile taken in the South Atlantic, a time scale was then constructed for the last 80 million years (column B of Fig. 2). Recent time scales are syntheses of data from a wide range of sources, hence are compromises that are undoubtedly close to being "true" but are continually undergoing modification.

The basic template of the geomagnetic reversal pattern comes from magnetic anomaly data compiled from many different places in the ocean (column F of Fig. 2). The pattern, therefore, will not be seen exactly reproduced in any ocean but represents a

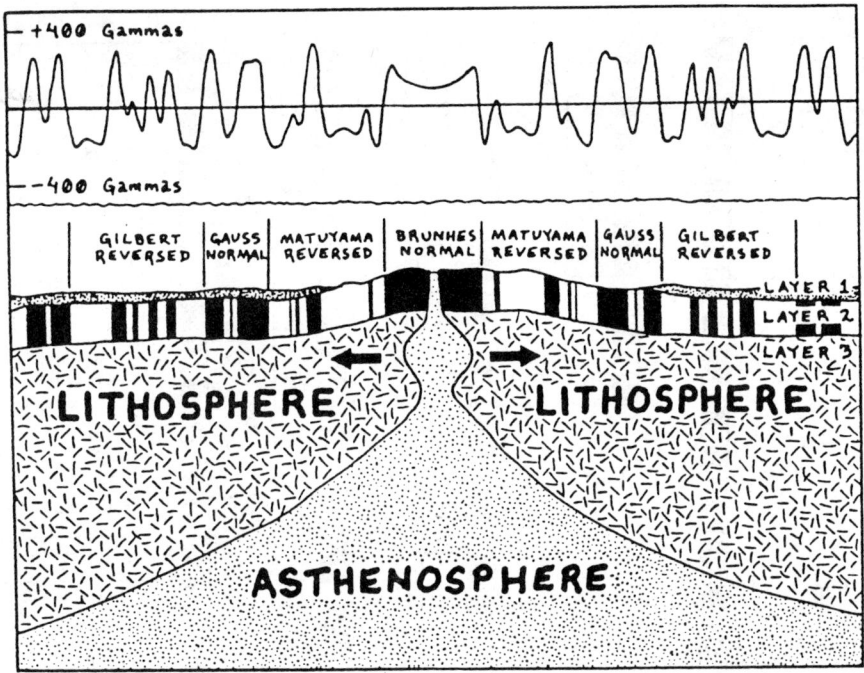

FIGURE 1. Basalts generated at the ridge crest spread away to form lineated magnetic anomalies. Black units symbolize crust formed during periods of normal polarity and white, during reversed. This highly magnetic layer alternately adds to and subtracts from the Earth's field resulting in the anomalies observed at the sea surface. (From Cox and Hart, 1986).

best-guess as to what the relative durations of the polarity intervals were. For calibration into time, models also draw upon a large data set, attempting to minimize differences among the various data points. For the youngest part of the time scale, calibration relies on a technique similar to that used for the first time scale: determining age and polarity of many continental basalts. This rather straightforward method of assigning dates to the reversal pattern is not useful for the sequence older than about 4 to 5 million years because the inherent error in absolute age estimates exceeds the average interval of time in any one polarity state, obscuring the fine structure of the time scale pattern. For the earlier part of the record, dates are tied to the geomagnetic reversal pattern by two methods. Dates on units within a stratigraphic section may be tied to the time scale either by matching the pattern of reversals within the stratigraphic sequence to that of the time scale or by the less direct method of determining the biostratigraphic age of the strata including the dated unit. Several biostratigraphic systems have been tied to the reversal pattern through careful combined magneto-biostratigraphic studies and provide an indirect link to the GRTS. A recent time scale calibration is illustrated by the solid line in Fig. 3. Ironically, this highly sophisticated and integrated time scale model agrees with the early model, which assumed constant spreading rate on a single profile (labeled HDHPL on Fig. 3) to within about 10%.

Another element involved in the history of the GRTS is the development of a terminology capable of uniquely and simply referring to a particular time interval. The evolution in terminology is shown in Fig. 2. The first time scales (e.g., column A) were divided into "epochs" averaging about one million years. The first four of these were named after scientists contributing to early research in magnetism and geomagnetism. Thus the present epoch, Brunhes, was named for a French scientist who was among the first to establish the existence of reversely magnetized rocks and postulate that the field had undergone a reversal. Epochs were then numbered starting with Epoch 5. These epochs were punctuated by "events" of shorter duration named after the place they were first established. Thus the "Olduvai event" was discovered at Olduvai Gorge, home of the early makers of stone tools some 2 my ago. Since magnetic anomalies cannot be construed as time units per se and the dating of isolated basalt units cannot resolve polarity intervals beyond about 5 my, it made sense to turn to stratigraphic sequences for defining earlier epochs of polarity history; hence Epochs 6 through 23 were defined on the basis of the polarity sequences found in deep sea sediment cores (columns C through E of Fig. 2). These epochs were correlated to the then current time scale. Epoch 11 was originally correlated to Anomaly 5 of the time scale shown in Column B.

In 1974, a great synthesis of magnetostratigraphic

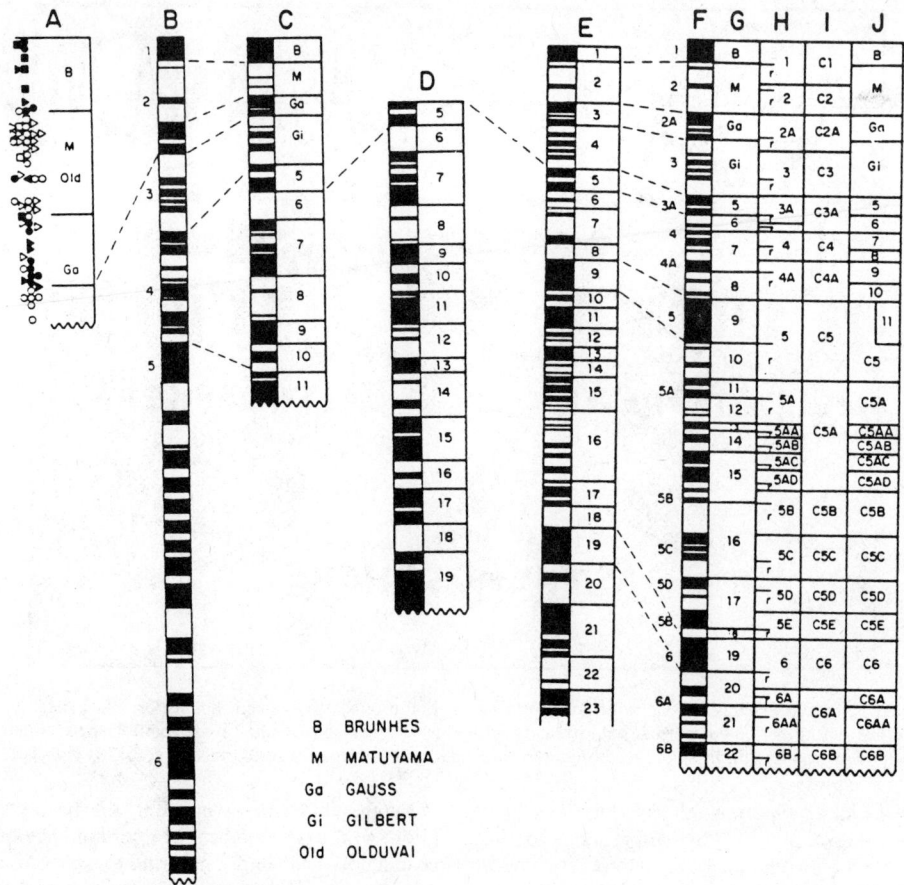

FIGURE 2. The development of the Geomagnetic Reversal Time Scale (GRTS). Column A: Polarity "epochs" based on polarity and K-Ar dates of lava flows (Cox et al., 1964). Column B: Reversal pattern based on marine magnetic anomalies (Heirtzler et al., 1968). Column C: Polarity "epochs" defined on deep-sea sediment cores and correlated to the anomaly pattern (Foster and Opdyke, 1974). Column D: Extension of the "epoch" system in sediment cores and correlated to revised anomaly pattern shown in F (Opdyke et al., 1974). Column E: Completion of the Neogene epoch system based on Pacific piston cores (Theyer and Hammond, 1974). Column F: A revised polarity history based on detailed surveys of marine magnetic anomalies (Talwani et al., 1971, and Blakely, 1974), with numbering scheme of LaBrecque et al., 1977). Column G: Correlation of the epochs derived from sediment cores (C-E) to the anomaly pattern (Ryan et al., 1974). Column H: The Chron terminology of Cox, 1982. Column I: Chrons of LaBrecque et al., 1983. Column J: Chron usage of Berggren et al., 1985.

and biostratigraphic data resulted in a correlation of these epochs to a revised magnetic anomaly pattern (Column F) as shown in Column G. Note that Anomaly 5 was tied to Epoch 9 as opposed to Epoch 11. There has been an evolution in terminology since the 1974 synthesis. Complications arose from a variety of areas. First, because the terms *epoch* and *event* had previously understood connotations in stratigraphy, the subcommission on stratigraphic nomenclature recommended that the terms be dropped in favor of *chron* and *sub-chron*. Second, the correlation of Chron (Epoch) 9 to Anomaly 5 of the 1974 synthesis is now thought to be in error (e.g., Berggren et al., 1985). Because of the confusion resulting from this revision as well as the need to extend the terminology to the Paleogene and earlier parts of the time scale, several alternative systems were proposed (Columns H through J). The systems shown in Columns H and I are similar in that they define chrons in terms of the correlative anomaly numbering scheme. The prime disadvantage of that shown in H is the likelihood of confusion with the former chron system, which is thoroughly entrenched in the literature. For example, Chron 9 can be understood in three ways: as correlative to Anomaly 5 (Column G), as correlative to Anomaly 4A some 1.5 million years younger (Column J), or as correlative to Anomaly 9, some 20 million years older (extension of system in Column H). Column J is the most recent time scale. This column further complicates

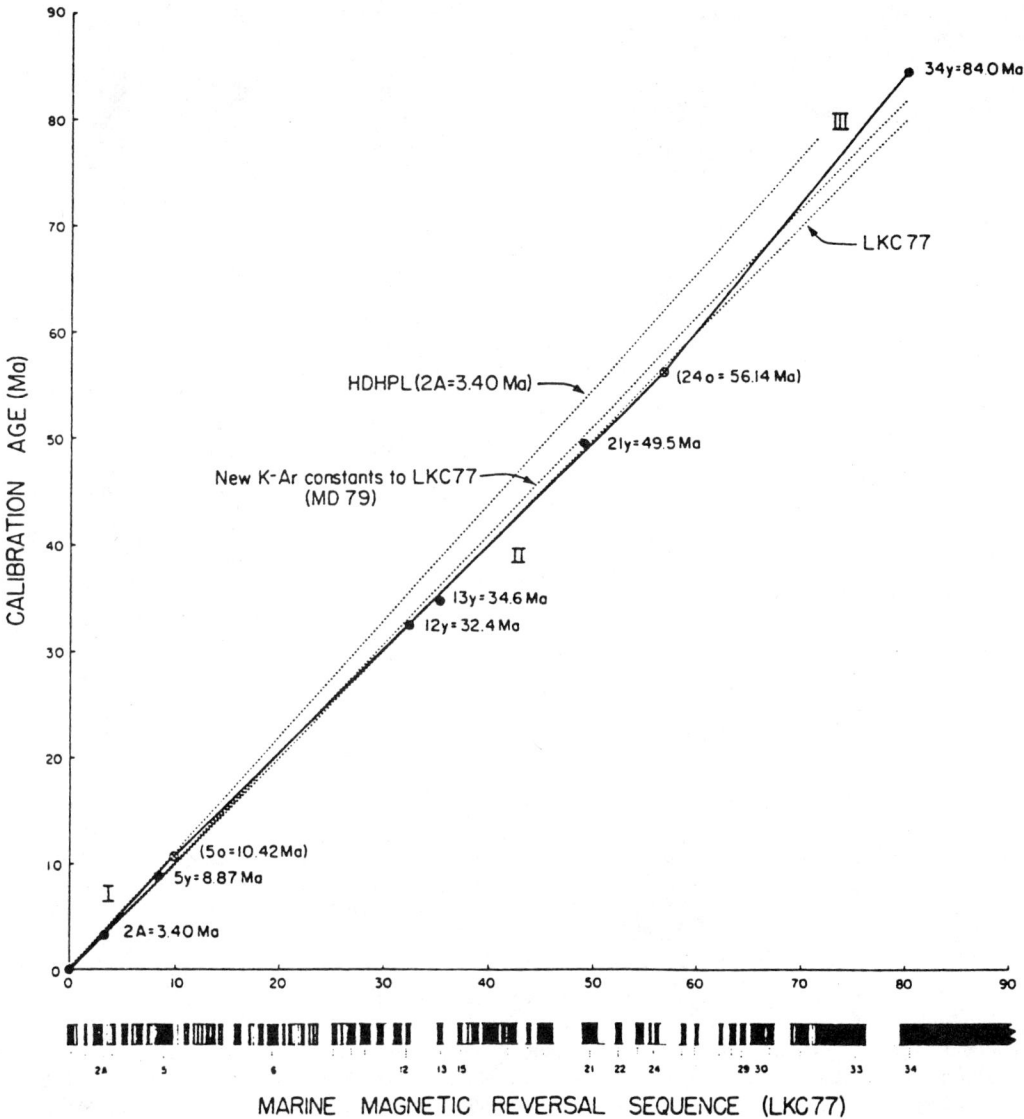

FIGURE 3. A recent time scale calibration (Berggren et al., 1985). The reversal sequence is that compiled by LaBrecque et al. (1977) and is the same as that shown in Column F of Figure 2. The solid circles are dates correlated to the reversal pattern. Open circles with xs are inferred changes in seafloor spreading rates generating the reversal pattern. Shown for comparison are two other time scale models; HDHPL is that shown in Column B of Fig. 2. The two earlier time scales were recalculated using revised radioactive decay constants for consistency with the most recent calibration.

the issue by suggesting the new correlation of Chrons 1 through 11 (defined in column C) to Anomaly 5 and then adopting a system similar to that shown in Column I. For the purposes of simplicity and clarity, I prefer the use of the system shown in Column I.

Methods

The basic techniques involved in magnetostratigraphy are illustrated in Fig. 4. To the left is a lithostratigraphic description of a sedimentary section with stratigraphic height shown in meters and any additional biostratigraphic or absolute age data available to aid in ultimate correlation to the GRTS. Samples for paleomagnetic analysis are taken from discrete horizons either by drilling out cores or as blocks. These are fully oriented before removal from the ground or core section. Whenever possible, multiple samples are taken from each level to test for reproducibility. The paleomagnetic samples are then prepared into specimens that will fit into the

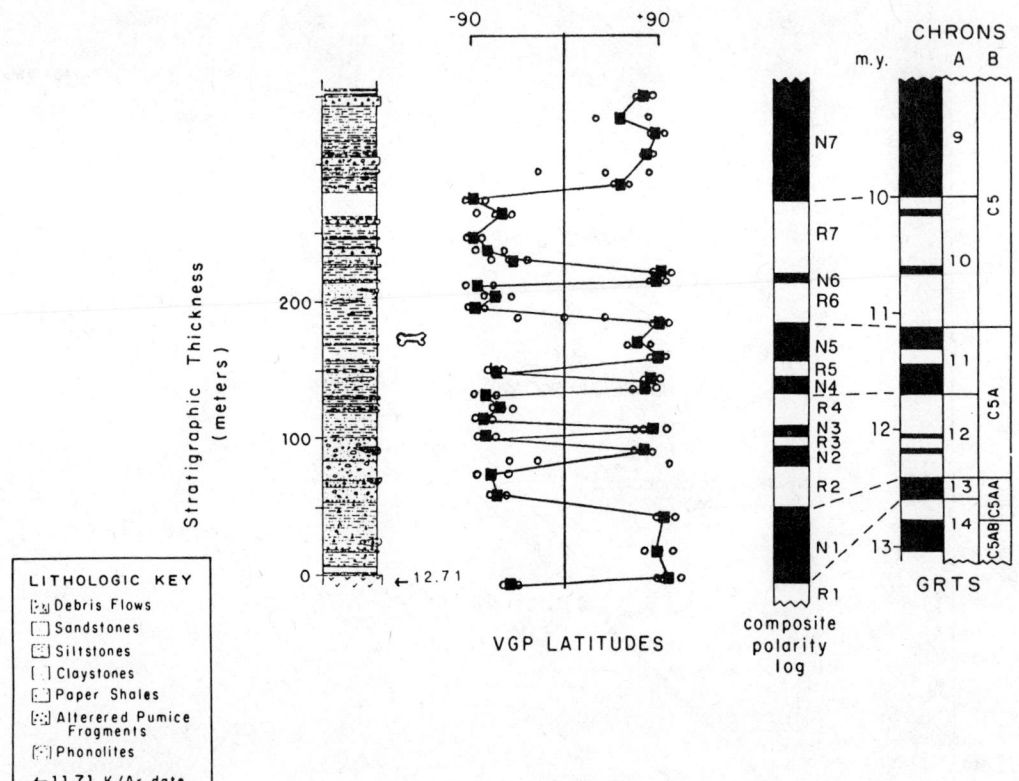

FIGURE 4. A hypothetical magnetostratigraphic study. To the left is a lithologic description of the section. The virtual geomagnetic pole (VGP) latitudes of individual specimens are plotted as small circles. Statistically nonrandom data averaged from each level are shown as squares connected by lines. The polarity interpretation is based on the nonrandom VGP averages (black is normal and white is reversed). The magnetostratigraphy is correlated to the GRTS aided by the date on an underlying volcanic flow unit and is based on pattern matching.

standard holders of magnetometers (typically cubes or cylinders measuring roughly one inch in diameter). Specimens are then subjected to the standard paleomagnetic cleaning techniques as discussed elsewhere (see *Demagnetization*), in order to isolate the primary direction. Once the reliability of the data has been established, the primary vector is isolated from each specimen. A convenient way to represent the directional information is as the latitude of the virtual geomagnetic pole (VGP; the location of the pole that would generate the observed direction). Since certain levels may contain highly scattered data, only sites with nonrandom data should be used for polarity interpretations. Thus, individual specimen data are shown as small circles and statistically reliable averages are shown as squares. Polarity interpretations can be made on the basis of the reliable VGP latitudes. Here, normal zones (positive latitudes) are indicated by black, as is the paleomagnetic convention. Correlation of the magnetostratigraphy to the time scale requires some independent age information and a convincing match to the pattern of the GRTS.

Before discussing applications, however, some mention must be made of the limitations of the method. Effective use of magnetostratigraphy requires a record that faithfully relects a sufficient number of sequential reversals for reliable correlation to the GRTS. The time resolution of the technique is therefore limited by the fidelity of the recording mechanism itself as well as the frequency of reversals.

The recording mechanism involves aspects related to the mode of magnetization (the "tape head") and the process of stratigraphic accumulation (the "tape drive"). Rocks may become magnetized in a variety of ways (see *Rock Magnetism* and *Natural Remanent Magnetization*) and magnetizations may become superposed resulting in very complex magnetic behavior. Also, stratigraphic accumulation may be sporadic, severely limiting the reliability of any correlation to the GRTS.

Since stratigraphic sections are generally of limited duration and/or completeness, the frequency of reversals plays an important role. When reversals are frequent, a very complete record is necessary in order to reveal the true pattern of reversals. When reversals are infrequent, a very long record is necessary in order to obtain a record with a sufficient number of polarity zones to establish a unique match to the time scale pattern.

Various Applications of Magnetostratigraphy

Despite these limitations, magnetostratigraphic data are the basis for a wide variety of useful geological applications. The primary information provided by magnetostratigraphy is temporal. Applications therefore involve the use of the time control provided either for the purpose of calibration in terms of absolute age, correlation of strata from place to place or calculation of rates of geologic processes.

Perhaps the best known application is for the calibration of stratigraphic sequences in terms of "absolute" time. Although the GRTS is inherently an ordinal time scale, so many absolute dates have been tied to it that it has long been considered to provide absolute age information at least as reliable as any single date based on radioactive decay. Certainly, the precision of correlation within the framework of the GRTS can be quite high. The identification of the same reversal boundary within two different sequences allows us to say that the two horizons are the same age to within perhaps 10,000 years or better. Therefore, various biostratigraphic systems have been calibrated in terms of the GRTS, allowing a comparison of events within, say the mammalian world (such as the evolution of the genus Homo), to events recorded in the deep ocean (such as ice ages). An example of this application is shown in Fig. 4. The fossil horizon, indicated by a bone symbol next to the lithologic section, falls within the upper part of Chron C5A and has an age of about 11.2 Ma (depending on the choice of time scales). Ultimately, this line of research will allow the integration of information on the evolution of various plants and animals with details of such processes as might have some controlling influence such as climate, tectonics, and even possible collisions of the Earth with large extraterrestrial bodies.

Magnetostratigraphic data have also been used to provide information about the rates of geologic processes such as sediment accumulation, basin subsidence, or tectonic rotations. Because every reversal boundary constitutes a date, rates can be calculated over a variety of time scales, from the shortest polarity intervals in the time scale (40,000 years) to the length of the entire section. Also, changes in rate can be pinpointed with a resolution comparable only to the best biostratigraphic systems (eg. coccoliths).

A third possible application of magnetostrati-

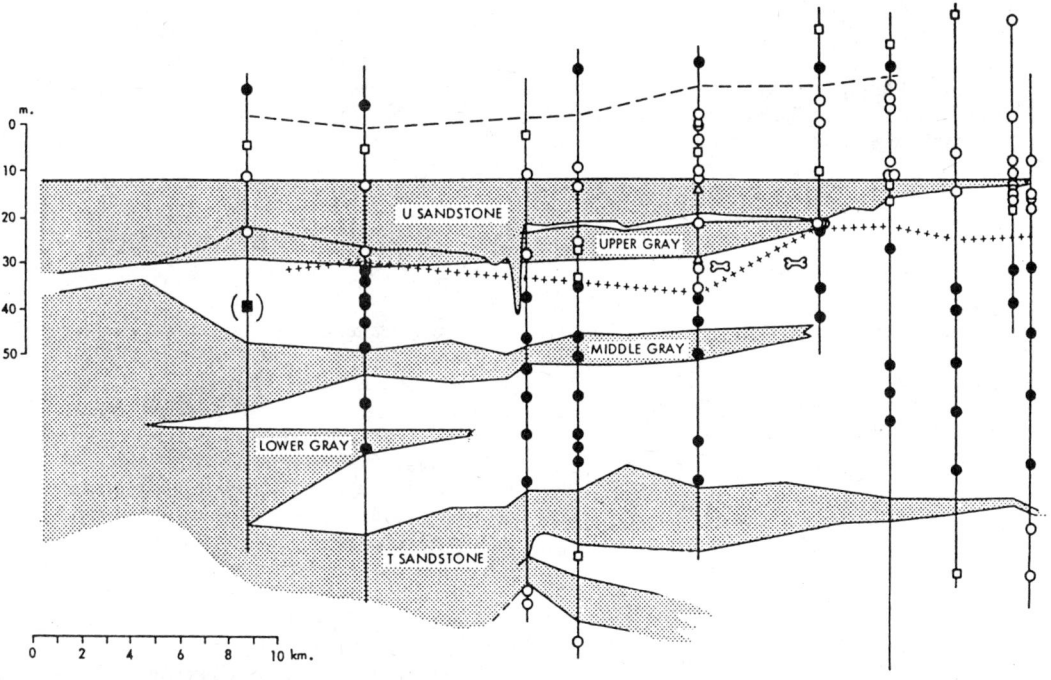

FIGURE 5. Illustration of the use of magnetostratigraphy for providing an isochronous horizon within a sedimentary basin. Paleomagnetic sections define discrete polarity units whose boundaries are isochrons. In the figure, the two bones represent fossil localities that are clearly different ages, despite the fact that they occur at the same position relative to a regional lithologic marker unit (labelled U Sandstone). (Figure from Behrensmeyer and Tauxe, 1982).

graphic data is to provide a traceable isochronous horizon within a single sedimentary basin. Such an application would be useful in sequences without other traceable units such as volcanic horizons or events such as abrupt evolutionary change. Parallel magnetostratigraphic sections may reproduce the same reversal boundary, which can then be correlated to provide an isochronous surface independent of lithology. This potentially powerful application is limited by the fact that paleomagnetic properties of rocks are subject to diagenetic change and great care must be taken in order to demonstrate that the record is "primary" and reversal boundaries are indeed isochrons and not artifacts.

An example of isochron tracing is shown in Fig. 5. Neighboring sections through a sequence of fluvial sediments established an isochronous horizon, allowing the reconstruction of paleoenvironments on the ancient flood plain. The two bones represent hypothetical fossil localities that are clearly different ages, despite the fact that they occur at the same position relative to lithologic marker units. Thus, magnetostratigraphic data can be used to improve temporal resolution on a finer scale than either lithostratigraphy or biostratigraphy.

Concluding Remarks

Magnetostratigraphy has been applied as a tool for correlation since the first time scales were developed in the 1960s. Each year brings a refinement to the time scale, the methodology, and/or an innovative application. In this article, I have briefly described three ways of utilizing magnetostatigraphic data. These are just a few of the possibilities. Magnetostratigraphy, born during the revolution in the earth sciences of the 1960s, has now come of age.

LISA TAUXE

References

Behrensmeyer, A. K., and L. Tauxe, 1982, Isochronous fluvial systems in Miocene deposits of northern Pakistan, *Sedimentology* **29,** 331-352.
Berggren, W. A., D. V. Kent, J. J. Flynn, and J. A. Van Couvering, 1985, Cenozoic geochronology, *Geol. Soc. America Bull.* **96,** 1407-1418.
Blakely, R. J., 1974, Geomagnetic reversals and crustal spreading rates during the Miocene, *Jour. Geophys. Research* **79,** 279-300.
Cox, A. V., 1982, Magnetostratigraphic time scale, in W. B. Harland, ed., *A Geologic Time Scale.* Cambridge, England: Cambridge University Press, 63-84.
Cox, A. V., and R. B. Hart, 1986, *Plate Tectonics: How It Works.* Palo Alto, Ca.: Blackwell Scientific Publications.
Cox, A. V., R. R. Doell, and G. B. Dalrymple, 1964, Reversals of the Earth's magnetic field, *Science* **144,** 1537-1543.
Foster, J. H., and N. D. Opdyke, 1970, Upper Miocene to recent magnetic stratigraphy in deep-sea sediments, *Jour. Geophys. Research* **75,** 4465-4473.
Heirtzler, J. R., D. C. Dickson, E. M. Herron, W. C. Pitman, III, and X. Le Pichon, 1968, Marine magnetic anomalies, geomagnetic field reversals and motions of the ocean floor and continents, *Jour. Geophys. Research* **73,** 2119-2136.
LaBrecque, J. L., D. V. Kent, and S. C. Cande, 1977, Revised magnetic polarity time scale for Late Cretaceous and Cenozoic time, *Geology* **5,** 330-335.
LaBrecque, J. L., et al., 1983, Contributions to Paleogene stratigraphy in nomenclature, chronology and sedimentation rates, *Paleogeography, Paleoclimatology, Paleoecology* **42,** 91-125.
Opdyke, N. D., L. H. Burckle, and A. Todd, 1974, The extension of the magnetic time scale in sediments of the Central Pacific Ocean, *Earth and Planetary Sci. Letters,* **22,** 300-306.
Ryan, W. B. F., M. B. Cita, M. Dreyfus Rawson, L. H. Burckle, and T. Saito, 1974, A paleomagnetic assignment of Neogene stage boundaries and the development of isochronous datum planes between the Mediterranean, the Pacific and Indian Oceans in order to investigate the response of the world ocean to the Mediterranean "salinity crisis," *Riv. Ital. Paleont.* **80,** 631-688.
Talwani, M., C. C. Windisch, and M. G. Langseth, 1971, Reykjanes ridge crest: a detailed geophysical study, *Jour. Geophys. Research* **76,** 473-517.
Theyer, F., and S. R. Hammond, 1974, Cenozoic magnetic time scale in deep-sea cores: completion of the Neogene, *Geology* **2,** 487-492.

Cross-references: *Curie Temperature; Demagnetization; Geomagnetic Polarity Reversals: Observation; Geomagnetic Polarity Reversals: Theory and Models; Geomagnetic Reversal Sequence: Statistical Structure; Natural Remanent Magnetization (NRM); Paleomagnetism and Continental Drift: Historical Introduction; Paleomagnetism and Plate Tectonics; Rock Magnetism; Seafloor Spreading: Magnetic Evidence; Thermoremanence.*

MAGNETOTELLURIC METHOD: FUNDAMENTAL CONCEPTS

The magnetotelluric (MT) method has evolved over the last several decades as a versatile geophysical tool for both exploration and regional studies. The technique employs the propagation of extremely low frequency (10^{-4}-10^3 Hz) natural electromagnetic energy from distant transient sources in the ionosphere and magnetosphere (Fig. 1) to probe the resistivity structure of the Earth's interior from depths of hundreds of meters to many hundreds of kilometers. An example of a transient disturbance used in such studies is shown in Fig. 2. This particular event was simultaneously recorded at observatories around the globe.

The use of this technique is quite analogous to the way in which seismologists use signals from natural earthquakes to probe the acoustic properties of the Earth. Interest in mapping the bulk resistivity of geologic materials at depth stems from the fact that

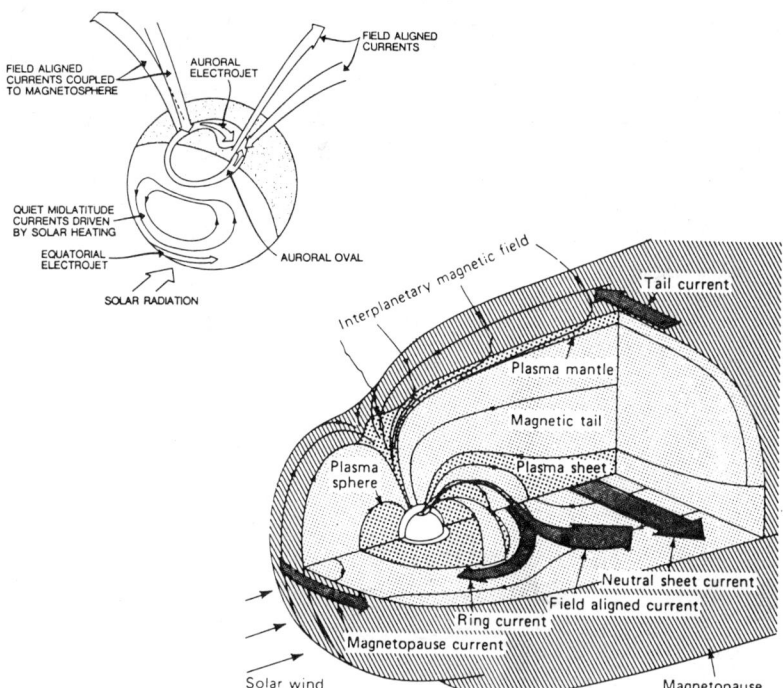

FIGURE 1. Sources of natural electromagnetic signals in the magnetosphere and ionosphere. The lower figure shows the overall distortion of the Earth's magnetic field in the solar wind, along with the principal electric current systems deep in the "magnetosphere" (from T. A. Potemra). The top left figure illustrates the penetration of currents originating from deeper in the magnetosphere that flow along magnetic field lines to lower altitude (approx. 110 km) in the auroral zone (65° to 75° latitude), where they then flow horizontally as broad current sheets in the ionosphere. Also shown in this figure are ionospheric currents at lower latitude: the quiet day Sq variations at mid latitudes, and the equatorial electrojet (from A. D. Richmond). All of these phenomena are time dependent, and "map" to magnetic field fluctuations at the Earth's surface.

this parameter is greatly affected by the ambient temperature of the host material and by the presence and composition of electrolytic fluids contained in cracks and pores. Thus, knowing the electrical resistivity as a function of depth can often place significant bounds on the physical state and processes in the Earth's interior.

The MT method is finding wide application to petroleum exploration and, to a limited degree, the search for geothermal, mineral, and groundwater resources. Moreover, since one of the factors affecting the electrical conductivity of geologic materials is temperature, the MT method is also finding increased application to studies in basic science, ranging from research on the thermal/mechanical evolution of the lithosphere to studies that characterize the roots of major volcanic complexes.

An MT survey involves the measurement of orthogonal electric and magnetic field variations at single sites sequentially or at a number of sites simultaneously (Fig. 3). The electric field is measured quite simply as the voltage difference between two metal electrodes driven into the ground about a hundred meters apart. Early geophysicists (before the development of the MT method) used these voltages to monitor naturally induced earth currents, which they called "telluric" currents (from the Latin word "tellus," the earth).

The magnetic field can be measured with a variety of sensors. One popular device is the induction coil, where one measures the electromotive force (voltage) induced by a time-varying magnetic field in a multi-turn coil of wire wound on a large-diameter air-core, or wound in the form of a long solenoid on a high-permeability core of ferrite or laminated ferromagnetic alloy. Other popular magnetic field sensors include flux-gate magnetometers and superconducting quantum interference devices (SQUIDs). In some applications, an effective approach is simply to measure the deflection of a suspended magnet due to changing torques produced by fluctuations in the Earth's magnetic field.

Louis Cagniard coined the word "magnetotellurics" to represent a geophysical method that uses simultaneous measurements of the Earth's magnetic and electric (telluric) field components (Fig. 4). He essentially employed the theory from classical

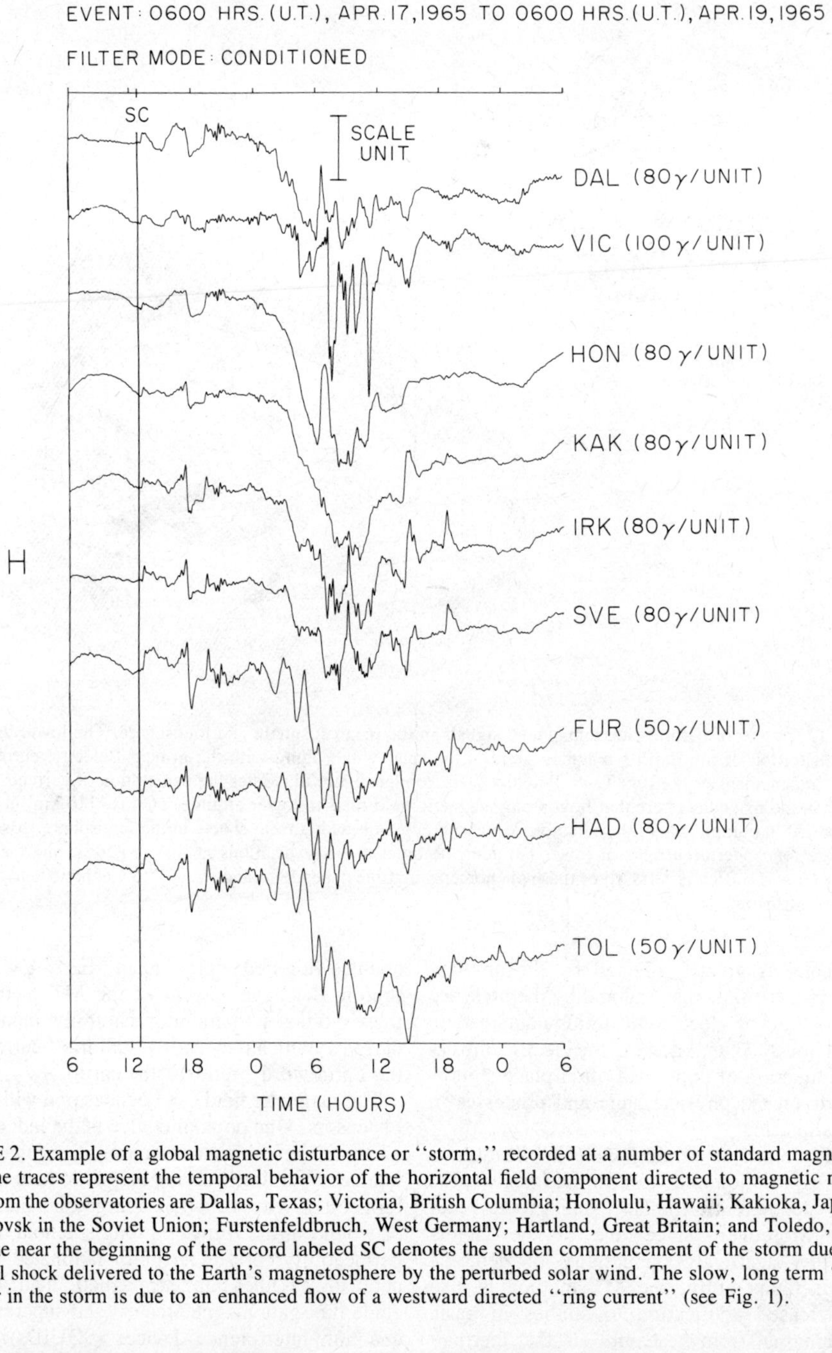

FIGURE 2. Example of a global magnetic disturbance or "storm," recorded at a number of standard magnetic observatories. The traces represent the temporal behavior of the horizontal field component directed to magnetic north. From top to bottom the observatories are Dallas, Texas; Victoria, British Columbia; Honolulu, Hawaii; Kakioka, Japan; Irkutsk and Svedlovsk in the Soviet Union; Furstenfeldbruch, West Germany; Hartland, Great Britain; and Toledo, Spain. The vertical line near the beginning of the record labeled SC denotes the sudden commencement of the storm due to the first mechanical shock delivered to the Earth's magnetosphere by the perturbed solar wind. The slow, long term "recovery" phase later in the storm is due to an enhanced flow of a westward directed "ring current" (see Fig. 1).

physics of the propagation of electromagnetic waves in a linear homogeneous isotropic medium. In such a medium an electromagnetic wave propagates such that the electric field **E** and the magnetic field **H** are orthogonal (recall that **E** × **H** is the Poynting vector, or the power flow along the direction of propagation). In addition, the ratio of electric to magnetic field intensity is a characteristic measure of the electromagnetic properties of the medium, often called the *characteristic impedance* and denoted by Z. In free space, $Z = (\mu/\epsilon)^{1/2}$, but in the more general case of a lossy medium such as the Earth, $Z = (i\omega\mu/\sigma)^{1/2}$, where σ is the total conductivity given by $\sigma = \sigma_{dc} + i\omega\epsilon$, as described later.

The fundamental principle on which the method is based involves the concept of *skin depth* or the *depth of penetration* of electromagnetic energy in a lossy electrical medium (a class into which all geologic

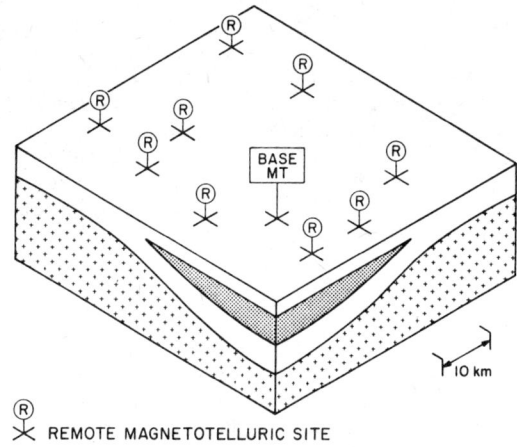

® REMOTE MAGNETOTELLURIC SITE

FIGURE 3. Schematic illustrating the deployment of magnetotelluric sites in the vicinity of a geologic structure to be studied.

at increasingly longer periods, one can obtain a measure of the electrical conductivity of the Earth's interior at progressively greater depths.

In order to emphasize the close connection between the MT method employed by geophysicists and the classical electromagnetic theory of physicists and engineers, we consider Maxwell's equations in MKSA units. We employ uppercase letters and lowercase letters to discriminate between vector fields in the frequency domain [e.g., $\mathbf{E}(\omega)$] and their counterparts in the time domain [e.g., $\mathbf{e}(t)$], respectively. In general, we do not indicate the time or frequency dependence explicitly, so the reader will have to be cautious in noting the above convention implicitly. Accordingly, Maxwell's equations in a homogeneous linear isotropic medium can be expressed in time as

$$\nabla \times \mathbf{e} = \frac{-\partial \mathbf{b}}{\partial t} \quad (1)$$

$$\nabla \times \mathbf{h} = \mathbf{j} \quad (2)$$

$$\nabla \cdot \mathbf{e} = \frac{\rho}{\epsilon}$$

$$\nabla \cdot \mathbf{b} = 0$$

materials fall). For a medium of given conductivity, lower-frequency signals propagate to greater depths than do higher-frequency signals. Therefore, by estimating the characteristic impedance of the Earth

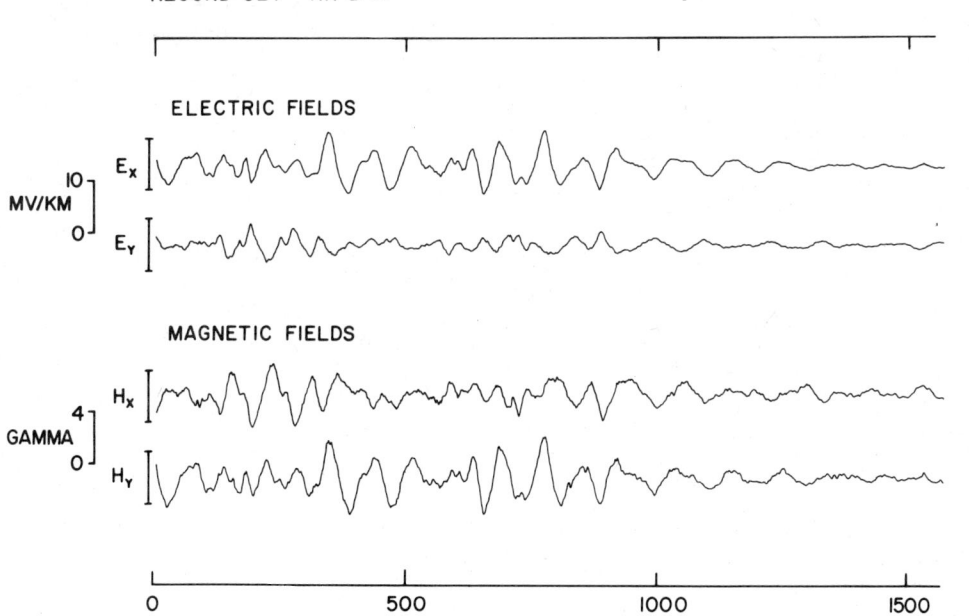

FIGURE 4. An example of a magnetotelluric data set from Krusivik, Iceland. The data are band pass filtered from 10 to 100 s, and the signals represent broad band pulsation activity. The electric fields are shown for the horizontal components directed to the North (E_x) and East (E_y) and the magnetic fields are shown for the North (H_x) and East (H_y) directions. Note the high temporal correlation between the electric North component, E_x, and the magnetic East component, H_y. Similarly there is good agreement between the electric East component, E_y, and the magnetic North component, H_x. These sets are known as Cagniard pairs.

with the constitutive relations $\mathbf{b} = \mu\mathbf{h}$, $\mathbf{d} = \epsilon\mathbf{e}$, and $\mathbf{j} = \sigma\mathbf{e}$. These equations relate the electric field intensity $[\mathbf{e}(t)$ in $V \cdot m^{-1}]$, the magnetic flux density $[\mathbf{b}(t)$ in $W \cdot m^{-1}$ or T], the magnetic field intensity $[\mathbf{h}(t)$ in $A \cdot m^{-1}]$, the electric current density $[\mathbf{j}(t)$ in $A \cdot m^{-2}]$, and the volumetric charge density $[\rho(t)$ in $C \cdot m^{-3}]$. In addition, we have used a generalized form for Ohm's law:

$$\mathbf{j} = \sigma_{dc}\mathbf{e} + \frac{\partial \mathbf{d}}{\partial t}$$

where $\mathbf{d}$ is the displacement field or electric flux density. Hence, $\mathbf{j}$ represents the *total* electric current density consisting of both the dc conduction current term $\sigma_{dc}\mathbf{e}$ and the displacement current term $\partial\mathbf{d}/\partial t$. Following a fairly well established practice, we denote the *total* conductivity σ as the sum of the dc conductivity and the displacement admittivity, such that

$$\sigma = \sigma_{dc} + i\omega\epsilon$$

where σ_{dc} is the low-frequency conductivity in $S \cdot m^{-1}$, and ϵ is the volumetric permittivity in $F \cdot m^{-1}$, which is related to the relative permittivity or the dielectric constant κ by $\epsilon = \kappa\epsilon_0$, where ϵ_0 is the permittivity of free space, approximately $1/36\,\pi \times 10^{-9}$ F $\cdot$ m^{-1}.

Work in the late 1940s and early 1950s by Rikitake, Kato, and Kikuchi in Japan, by Tikhonov in the Soviet Union, and by Cagniard in France resulted in the MT method as we know it today. Cagniard's paper had an enormous impact on the science because it formulated the relation between electric and magnetic field variations within a framework of clearly stated assumptions; namely he argued that the electromagnetic fields from distant sources in the magnetosphere and ionosphere could be modeled by plane waves and that the Earth could be modeled by horizontal layers having different resistivities. He pointed out that these assumptions are valid for many cases of geological and geophysical interest, but nowhere did he imply that these assumptions are universally true for all cases, which they are not, particularly when the source is relatively close to the observer and nonplanar, and when the local geology is laterally complex.

A One-Dimensional Wave (or Helmholtz) Equation. To illustrate several of the basic concepts underlying the Cagniard method, consider the case of a plane wave propagating in the $+z$-direction having its electric field polarized in the $+x$-direction. Making use of the constitutive relations, Maxwell's equations, Eqs. 1 and 2, reduce to

$$\frac{\partial e_x}{\partial z} = \frac{-\mu \partial h_y}{\partial t} \quad (3)$$

and

$$\frac{\partial h_y}{\partial z} = -\sigma e_x \quad (4)$$

Assuming that the electromagnetic field components have a simple harmonic time dependence of the form $\exp(i\omega t)$ (or that more complicated time series can be reduced to such a form through Fourier analysis), then we can expand the respective electric and magnetic field components as

$$e_x(t) = E_x \exp(i\omega t)$$

and

$$h_y(t) = H_y \exp(i\omega t)$$

The two time-domain expressions, Eqs. 3 and 4, can now be written in terms of their frequency domain coefficients such that

$$\frac{\partial E_x}{\partial z} = -i\omega\mu H_y \quad (5)$$

and

$$\frac{\partial H_y}{\partial z} = -\sigma E_x \quad (6)$$

where we have factored out the time dependent term $\exp(i\omega t)$. Upon taking the second derivative of Eq. 5 with respect to z, we obtain $\partial^2 E_x/\partial z^2$ on the left, and we can substitute Eq. 6 for $\partial H_y/\partial z$ on the right side of the new expression, thus obtaining

$$\frac{\partial^2 E_x}{\partial z^2} = i\omega\mu\sigma E_x \quad (7)$$

which is simply the lossy Helmholtz or wave equation for E_x, dependent on only the spatial variable z. Usually z is measured in the vertical direction in geophysics (positive downward) and Eq. 7 is termed a one-dimensional (1-D) relationship.

For a wave propagating in the positive (vertically downward) direction, a solution of Eq. 7 has the form

$$E_x^0 \exp(-\gamma z) \quad (8)$$

where $\gamma = (i\omega\mu\sigma)^{1/2}$, and we take the appropriate root such that

$$\gamma = (1 + i)\left(\frac{\omega\mu\sigma}{2}\right)^{1/2}$$

$$= \alpha + i\beta$$

where at low frequency $\alpha = \beta = (\omega\mu\sigma/2)^{1/2}$. Since $\exp[-(\alpha + i\beta)z] = \exp(-\alpha z)\exp(-i\beta z)$, the electric field is therefore the product of two spatial factors:

$$E_x = E_x^0 \exp(-\alpha z) \exp(-i\beta z) \quad (9)$$

Depth of Penetration. The first term represents a monotonic exponential damping with depth such that the field decreases to $1/e$ of its value at a depth z_e when the argument $\alpha z_e = 1$ or when $z_e = (2/\omega\mu\sigma)^{1/2}$. This depth is termed the *skin depth* or the *depth of penetration* in classical physics and is denoted by

$$\delta = \left(\frac{2}{\omega\mu\sigma}\right)^{1/2} \quad (10)$$

Typical values for the frequencies employed in geophysical field surveys vary from 10^{-4} to 10^3 Hz. Thus at a frequency of 0.01 Hz (or a period of 100 s), and for a conductivity of the earth of .01 S · m^{-1} (or a resistivity of 100 Ω · m), we obtain a depth of penetration, from Eq. 10, of approximately 50 km. If instead of a period of 100 s, one used 0.01 s (100 Hz), then the depth of penetration would be reduced to 0.5 km—the depth of penetration goes as the square root of the period of the signal.

Phase Velocity. The second spatial factor in Eq. 9 represents a harmonic dependence with depth that, when combined with the time factor $\exp(i\omega t)$, results in a wave traveling vertically downward in space-time coordinates such that any given phase front follows a relation of the form

$$\omega t - \beta z = \text{constant} \quad (11)$$

Differentiating Eq. 11 with respect to time, we can solve for the phase velocity

$$v_p = \frac{\partial z}{\partial t} = \frac{\omega}{\beta} = \left(\frac{2\omega}{\mu\sigma}\right)^{1/2}$$

For the same case as the one initially considered ($\rho = 100$ Ω · m, $T = 100$ s), a typical phase velocity would be 3.2 km · s^{-1}, which is clearly much lower than, for example, the phase velocity of electromagnetic energy in free space (approximately 3×10^5 km · s^{-1}). The corresponding "wavelength" of a 100-s signal traveling at a velocity of 3.2 km · s^{-1} would be approximately 320 km. Thus, since the depth of penetration is only 50 km, the wave is sharply attenuated before it penetrates the Earth by even a significant fraction of a wavelength. The behavior of typical signals with depth is illustrated in Fig. 5.

The Characteristic Impedance. Assuming a solution to the Helmholtz equation, Eq. 7, for the electric field of the form of Eq. 8, one can derive a consistent expression for the magnetic field by substituting Eq. 8 into the left side of Eq. 5. Upon taking the derivative with respect to z, one can rearrange the expression to obtain the characteristic impedance of a homogeneous medium

$$Z = \frac{E_x}{H_y} \quad (12)$$

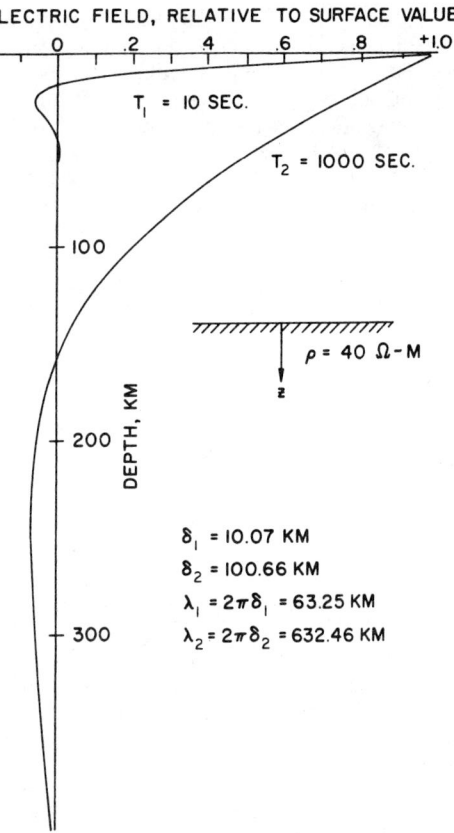

FIGURE 5. The attenuation with depth of the electric field strength for two signals, one at 10 s, the other at 1000 s, in a homogeneous earth of 40 Ω · m. The depth of penetration at 10 s is $\delta_1 = 10$ km, and the depth of penetration at 1000 s is $\delta_2 = 1000$ km.

which in turn can be expressed in terms of various parameters of interest:

$$Z = \frac{i\omega\mu}{\gamma} \quad (13a)$$

$$= \left(\frac{i\omega\mu}{\sigma}\right)^{1/2} \quad (13b)$$

$$= (i\omega\mu\rho)^{1/2} \quad (13c)$$

where our symbol for denoting the resistivity, $\rho = 1/\sigma$ (and having the units ohm · m), is not to be confused with the same symbol when used to denote charge density. Expression 12 states that, in a homogeneous medium, the electric and magnetic field components for a propagating plane wave are intimately related through the medium's characteristic impedance. Expressions 13a through 13c state that upon determining the characteristic impedance, one can then determine other parameters of the medium, such as its conductivity or its resistivity.

Cagniard (1953) was the first to recognize that simultaneous measurements of natural electric and magnetic field variations could be analyzed into appropriate frequency contributions, and, upon forming the ratios of orthogonal component pairs (e.g., E_x and H_y, or E_y and $-H_x$), could be used in Eq. 12 to determine the impedance of the Earth. The relation between the impedance and the actual resistivity of the medium (Eq. 13) could then be factored to yield

$$\rho_a = \frac{1}{\omega\mu} |Z|^2 \qquad (14a)$$

or, in terms of the actual electric and magnetic field components measured at the Earth's surface, one could determine a measure of the depth-averaged resistivity

$$\rho_a = \frac{1}{\omega\mu} \left|\frac{E_x}{H_y}\right|^2 \qquad (14b)$$

The subscript a denotes that the left-hand term in Eq. 14 is an "apparent" resistivity, since, for a heterogeneous Earth, a particular value calculated from this relation may represent some type of volumetric average of two or more "true" resistivities. For a simple two-layered Earth, for example, the skin-depth phenomenon will cause observations at short period to lead to apparent resistivities asymptotic to the true resistivity of the upper layer, whereas at very long periods the skin depth will be very large and the apparent resistivity will asymptotically approach the true resistivity of the deeper layer (Fig. 6). The range of periods over which the transition of ρ_a from ρ_1 (the resistivity of medium 1) to ρ_2 (the resistivity of medium 2) occurs is characteristic of the depth to the interface between the two media. Thus an MT "sounding" in such a setting will yield three essential pieces of information on the structure: ρ_1, ρ_2, and the depth to the interface between the two layers.

Such information is often of great value in a reconnaissance survey of a broad sedimentary basin. Because sediments tend to have a relatively high porosity (10–20%) and to be saturated with groundwater having a relatively high concentration of ions, their resistivity is typically quite low (5–20 $\Omega \cdot$ m). In contrast, the crystalline basement that underlies the sediments tends to have a much lower porosity (less than 1%), which results in far fewer connected pathways along which ions (the major charge carriers for the electric current) can flow. Consequently, crystalline and metamorphic basement materials tend to have a much higher resistivity. Thus, as a first-order approximation, a sedimentary basin can, in some sense, be characterized as a single layer of low-resistivity sediments (5–20 $\Omega \cdot$ m), that is underlain by a highly resistive crystalline basement (10^3–10^4 $\Omega \cdot$ m). Determining the depth to basement is often a key objective in many geophysical studies, a parameter that is readily determined for the idealized case considered here.

Another parameter readily determined from MT studies is the depth to a conductor. This parameter

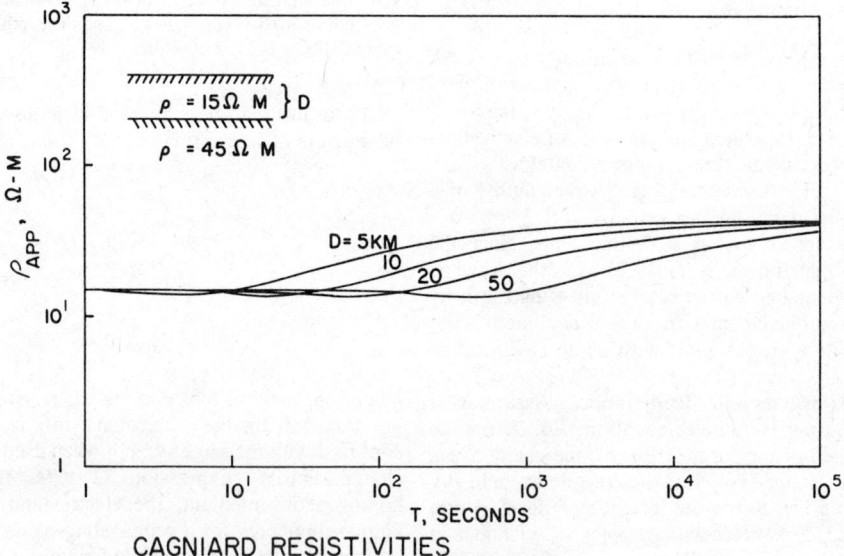

FIGURE 6. Apparent resistivity as a function of period for the simple case of two layers; the first layer has a resistivity of 15 $\Omega \cdot$ m and the second layer has a resistivity of 45 $\Omega \cdot$ m. Each curve corresponds to a different depth to the interface separating the two layers. Note that when d is greater, the inflection in the apparent resistivity curve shifts to longer periods, as one would expect from the depth of penetration phenomena.

is often the objective in regional surveys when large-scale thermomechanical processes at depth in the lithosphere or asthenosphere are investigated. In tectonically active areas, where the surface thermal gradients may be quite high (50–60°C · km^{-1}), temperatures on the order of 1000°C may be expected at depths of only 15–20 km. Because at such high temperatures the host rock itself becomes conductive, such thermal regimes can be confirmed and their distribution mapped by MT surveys. For example, whereas the low-temperature conductivity of the basement at depths of 1–2 km is usually limited by conduction in pore fluids and has a resistivity of 10^3–10^4 Ω · m, at temperatures of 1000°C semiconduction mechanisms in the bulk material itself cause the resistivity of even crystalline rock to drop to values of only a few ohm-meters or less.

Thus, MT surveys in active tectonic areas often reveal a resistive upper crust (10^3–10^4 Ω · m) underlain by a zone having a resistivity significantly less than 10 Ω · m. In some cases, temperatures are so high that materials can be sustained in a molten state, so the resistivity may drop to less than an ohm-meter. Therefore zones containing significant concentrations of melt are also an appropriate target for MT surveys.

On the other hand, not all conductors in the crust may indicate abnormally high temperatures, but their presence might have important geological implications, nevertheless. For example, in some regions large-scale compressive stresses may have resulted in the overthrusting of blocks of crystalline basement on top of a sedimentary section that was originally at the surface. Many workers think that some of the conductive features (~ 10 Ω · m) detected at depths of 10 km or more beneath some of the less active tectonic areas of the world represent slices of sediments that were entrapped during such basement overthrusting in the distant past.

Clearly, detecting conductive zones in the crystalline basement is extremely interesting for several geologic reasons, and determining the depth, thickness, and lateral configuration of such features is a primary objective of many regional MT surveys.

Magnetotelluric Field Relations for a Plane-Layered Earth

General. The MT method involves determining the electrical properties of the Earth's interior through the analysis of natural magnetic and telluric (electric) field variations at its surface. Cagniard developed the original theory of MT fields for a plane-layered Earth employing concepts from the theory of the propagation of electromagnetic plane waves in a dissipative medium that is linear, homogeneous, and isotropic. As mentioned in the previous section, in such a medium an electromagnetic wave propagates so that the electric and magnetic field vectors are orthogonal and so that the ratio of electric to magnetic field intensity is a characteristic measure of the electromagnetic properties of the medium; this parameter is often called the characteristic impedance.

If indeed the Earth is homogeneous and isotropic, then the true resistivity of the Earth is related to the characteristic impedance through the relation

$$\rho = \frac{1}{\omega\mu} |Z|^2 = \frac{1}{\omega\mu} \left|\frac{E_x}{H_y}\right|^2 \quad (15)$$

where ρ is the resistivity in Ω · m, ω is the radian frequency, and μ is the permeability.

For a horizontally layered Earth (or 1-D Earth), Eq. 15 becomes an *apparent* resistivity:

$$\rho_a = \frac{1}{\omega\mu} \left|\frac{E_x}{H_y}\right|^2 \quad (16)$$

which is frequency dependent. For example, if the Earth locally consists of two layers, the finite depth of penetration of the fields would cause Eq. 16 to be asymptotic to the resistivity of layer 1 at short periods and asymptotic to the resistivity of layer 2 at long periods.

Moreover, because of the symmetry of the problem, estimates of the characteristic impedance for either a homogeneous or a *layered* Earth do not depend on rotation of the measuring axes in the horizontal plane, so that the north and east electric components are related to the orthogonal magnetic components through the linear equations

$$E_x = ZH_y \quad \text{and} \quad E_y = -ZH_x$$

Each of these equations is a statement of the condition that, at a particular period or frequency, an *electric* field component is linearly related to its orthogonal *magnetic* field component through a single-valued complex scalar transfer function. This relation is a direct analog from simple filter theory. The output signal of a filter (E_x) is linearly related to the input signal (H_y) multiplied by the frequency response of the filter (Z). A comparison between the engineering viewpoint and the geophysical viewpoint is illustrated in Fig. 7.

Equation 16 was formulated for the first time by Cagniard (1953) and as a result is known as the *Cagniard relation*. The conditions under which Eq. 16 is valid are called the *Cagniard conditions*, which state that the incident electromagnetic fields are plane waves at the Earth's surface and that the Earth consists of horizontal plane layers. It was Cagniard's clever combination of characteristic impedance, apparent resistivity, and depth of penetration that allowed him to develop the MT method into a quantitative tool for the field geophysicist.

General Magnetotelluric Relations for an *n*-Layered Earth. Consider a plane wave propagating vertically in a medium consisting of n

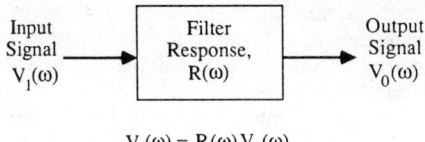

$$V_1(\omega) = R(\omega)V_0(\omega)$$

The Conventional Engineering Viewpoint

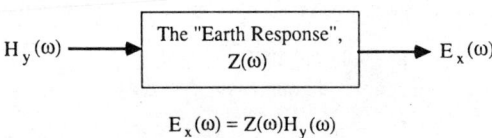

$$E_x(\omega) = Z(\omega)H_y(\omega)$$

The Geophysical Viewpoint

FIGURE 7. Comparing the conventional engineering view of the input and output relation for a simple filter to the geophysical view showing the relation between the electric field and the magnetic field connected through the surface impedance.

horizontally plane layers (Fig. 8). The electric field in each layer can be represented as a solution to the lossy Helmholtz equation

$$\frac{\partial^2 E_{xj}}{\partial z^2} = i\omega\mu\sigma_j E_{xj} \quad (17)$$

where j denotes the jth layer. We consider the *source* of the signal as being at some position above the surface, and assume that the wave propagates only in the z-direction. In each layer, however, two waves will generally exist, one wave transmitted from above and traveling vertically downward in the $+z$ direction, and another wave reflected from interfaces below and traveling upward in the $-z$-direction. The electric field in either case is polarized in the horizontal (x) plane.

Solutions to Eq. 17 therefore consist of the sum of these two modes in the jth layer:

$$E_{xj} = A_j \exp(+\gamma_j z) + B_j \exp(-\gamma_j z) \quad (18)$$

where $\gamma_j = (1 + i)(\omega\mu\sigma_j/2)^{1/2}$, σ_j being the conductivity of the jth medium.

A magnetic field consistent with Eq. 18 is obtained from Faraday's law. In the 1-D case considered here, the magnetic field in the jth layer is simply

$$H_{yj} = -\frac{1}{Z_j}[A_j \exp(+\gamma_j z) - B_j \exp(-\gamma_j z)] \quad (19)$$

where we have used the general definition for the characteristic impedance,

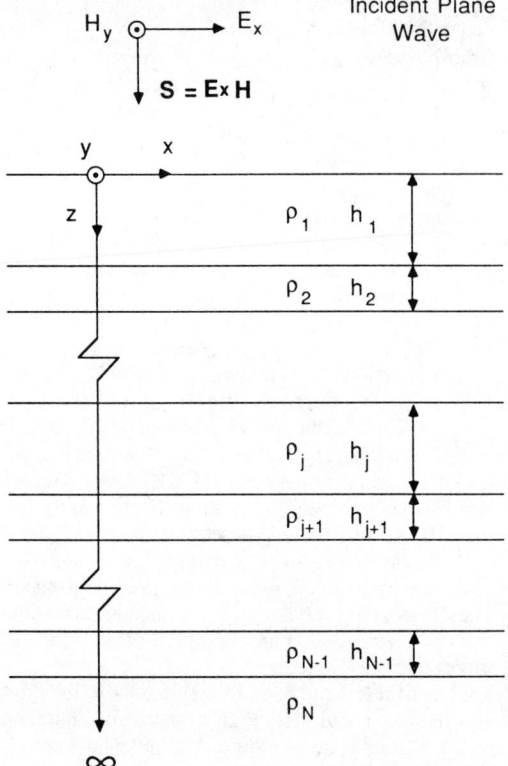

FIGURE 8. A plane layered model for the earth. Each layer has a resistivity ρ_j, and a thickness h_j. The lowermost medium, ρ_n, extends to infinite depth.

$$Z_j = \frac{i\omega\mu}{\gamma_j}$$

Both E_x and H_y are continuous across any horizontal interface. With reference to a local coordinate system having an origin at the bottom of the jth layer, the ratio of Eq. 18 to Eq. 19 at $z = 0$ is equal to the *surface* impedance of the ($j + 1$)st layer Z^{j+1}. (Because of possible reflections from below, the *surface* impedance Z^{j+1} of a layer, denoted by a superscript, is generally different from the layer's *characteristic* impedance Z_{j+1}, denoted by a subscript.)

At the *top* of the jth layer (having a thickness h_j), $z = -h_j$, so the ratio of Eq. 18 to Eq. 19 leads to the surface impedance $Z^j = E_{xj}/H_{yj}$, or

$$Z^j = Z_j \frac{[Z^{j+1} + Z_j \tanh(\gamma_j h_j)]}{Z^{j+1} \tanh(\gamma_j h_j) + Z_j} \quad (20)$$

It is easy to see that because $\tanh \theta$ goes to 0 as θ goes to 0, and to 1 as θ goes to infinity, Z^j approaches Z_j when the jth layer is electrically "thick" ($\gamma_j h_j \gg 1$ or $h_j/\delta_j \gg 1$), and Z^j

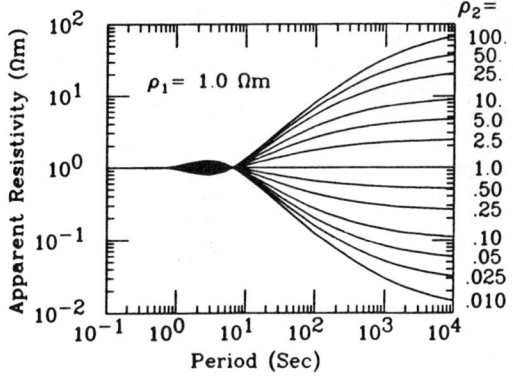

FIGURE 9. Master curves for apparent resistivity for the case of a 1000 m thick surface layer having a resistivity of 1 Ω · m underlain by a second layer having the various resistivities shown along the right hand side of the figure.

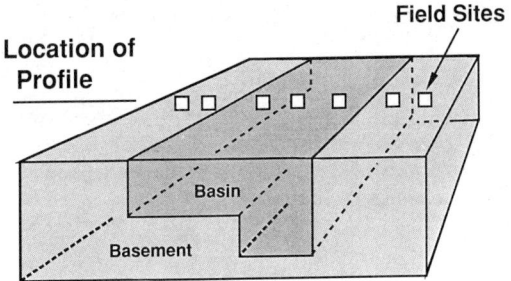

FIGURE 10. An example of a 2-D geologic model; in this case, an asymmetric sedimentary basin.

approaches Z^{j+1} when the jth layer is electrically "thin" ($\gamma_j h_j \ll 1$ or $h_j/\delta_j \ll 1$).

Wait (1962) generated a recursion relation by noting that at the deepest interface the surface impedance equals the characteristic impedance, or $Z^n = Z_n$. Using Eq. 20, he showed that a surface impedance could then be calculated at the top of the $(n-1)$st layer (Z^{n-1}), and from there at the top of the $(n-2)$nd layer (Z^{n-2}), etc., continuing progressively to the surface ($j = 1$, and Z^1).

The Two-Layered Impedance. As an example of an MT response function, consider that of a simple two-plane-layered model. The surface impedance at the top of the first layer is seen from Eq. 20 to be

$$Z^1 = Z_1 \left[\frac{Z_2 + Z_1 \tanh(\gamma_1 h_1)}{Z_2 \tanh(\gamma_1 h_1) + Z_1} \right]$$

An example showing how the response varies when we keep ρ_1 and h_1 fixed at 1 Ω · m and 1 km, respectively, and vary ρ_2 is shown in Fig. 9.

Of particular interest are the asymptotic limits at short and long periods. At short periods, layer 1 becomes electrically *thick* (i.e., $\gamma_1 h_1$ or h_1/δ_1 approaches infinity). In this case the hyperbolic tangent function approaches unity and $Z^1 = Z_1$, or $\rho_a = \rho_1$. In the alternative case, when layer 1 becomes electrically *thin*, or when $\gamma_1 h_1$ or h_1/δ_1 approaches zero, the surface impedance equals the impedance of the lower medium, $Z^1 = Z_2$, or $\rho_a = \rho_2$. This behavior should be clear in Fig. 9.

The Effect of 2-D and 3-D Lateral Heterogeneities

In regions where the Earth has a structure more complicated than simple plane layers, the coupling between electric and magnetic fields is more complicated than the 1-D case considered by Cagniard (1953). As an example, measurements often are made in elongated, but narrow sediment-filled valleys (Fig. 10). If we can neglect the variation of conductivity along the axis of the valley and adopt a coordinate system such that the y-axis is parallel to the strike of the valley, the x-axis is perpendicular to the strike, and the z-axis is directed vertically downward, then to first-order we can assume that $\sigma = \sigma(x, z)$. In such a case, we say that the electrical fabric is two-dimensional (2-D). In other cases, the structure is more complicated and one must consider the finite dimensions of the structure in all directions, the so-called three-dimensional (3-D) problem in which $\sigma = \sigma(x, y, z)$. In reality all geophysical problems are three-dimensional, and the success we have in modeling such features with simple 1-D or even 2-D models depends largely on the structure we are studying, and, in some cases, on the accuracy that we expect from the interpretation of our data.

The Two-Dimensional Case. For a strictly 2-D geometry illuminated by a plane wave source field, Maxwell's equations decouple into two distinct modes. Consider, for example, a situation (e.g., Fig. 10) where a conducting structure varies in the x- and z-directions, but is uniform along the strike in the y-direction. In other words, $\sigma = \sigma(x, z)$. Because of this symmetry, fields do not vary in the y-direction regardless of polarization; hence all derivatives with respect to y vanish. With this condition two classes of relations are present. One class is that in which H_x, H_z, and their derivatives are related to E_y only (the electric field component parallel to strike) and its derivatives. This class is the *E-polarization*, *E-parallel*, or *transverse electric* (TE) mode.

The other class of relations is one in which E_x, E_z, and their derivatives are related to H_y only (the magnetic field component parallel to strike) and its derivatives. This class is the *H-polarization*, *H-parallel*, or *transverse magnetic* (TM) mode.

Each of these modes can be solved quite independently from the other for a given 2-D conductivity distribution such that, with suitable approximations, in the TE mode, E parallel to strike depends only on H perpendicular to strike, and in the TM mode, E perpendicular to strike depends only on H parallel to strike. In other words, the impedance relation decouples into two modes represented by

$$E'_x = Z'_{xy} H'_y \quad \text{and} \quad E'_y = Z'_{yx} H'_x$$

where the prime indicates that the measuring axes x and y are aligned parallel and perpendicular to the strike of the 2-D lateral inhomogeneity. This special orientation (x', y') is called the *principal coordinate system*, and Z'_{xy} and Z'_{yx} are called the *principal impedance values*.

In matrix notation, the coupling between $\mathbf{E'}$ and $\mathbf{H'}$ can be expressed as

$$\begin{bmatrix} E'_x \\ E'_y \end{bmatrix} = \begin{bmatrix} 0 & Z'_{xy} \\ Z'_{yx} & 0 \end{bmatrix} \begin{bmatrix} H'_x \\ H'_y \end{bmatrix} \quad (21a)$$

or

$$\mathbf{E'} = [\mathbf{Z'}] \cdot \mathbf{H'} \quad (21b)$$

where $[\mathbf{Z'}]$ is often termed the *principal impedance tensor*.

For arbitrary orientations of the measuring axes away from the principal directions, one finds, as might be expected, a linear coupling of *each* electric component to *both* magnetic components. This coupling may be seen by a simple rotation of the new measuring coordinates away from the principal directions through an angle measured in the clockwise sense from north looking down. In matrix form, the effects of such a coordinate rotation on a vector $\mathbf{X'}$ can be expressed as

$$\begin{bmatrix} X_x \\ X_y \end{bmatrix} = \begin{bmatrix} \cos\theta & \sin\theta \\ -\sin\theta & \cos\theta \end{bmatrix} \begin{bmatrix} X'_x \\ X'_y \end{bmatrix}$$

or

$$\mathbf{X} = [\mathbf{R}] \cdot \mathbf{X'}$$

where $\mathbf{X'}$ represents the vector elements in the original coordinate system, $\mathbf{X}$ represents the vector elements in the rotated coordinate system, and $[\mathbf{R}]$ is the rotation operator. Upon left-multiplying both sides of Eq. 21b by $[\mathbf{R}]$ and noting that $\mathbf{E} = [\mathbf{R}] \cdot \mathbf{E'}$ and $\mathbf{H'} = [\mathbf{R}^{-1}] \cdot \mathbf{H}$, we obtain

$$\mathbf{E} = [\mathbf{R}][\mathbf{Z'}][\mathbf{R}^{-1}] \cdot \mathbf{H}$$

or

$$\mathbf{E} = [\mathbf{Z}] \cdot \mathbf{H}$$

where $[\mathbf{Z}] = [\mathbf{R}][\mathbf{Z'}][\mathbf{R}^{-1}]$. Thus, the new tensor elements expressed in terms of the original principal values (Z'_{xy}, Z'_{yx}) and the rotation angle θ are

$$Z_{xx} = \frac{Z'_{xy} + Z'_{yx}}{2} \sin 2\theta$$

$$Z_{xy} = Z'_{xy} - (Z'_{xy} + Z'_{yx}) \sin^2 \theta$$

$$Z_{yx} = Z'_{yx} - (Z'_{xy} + Z'_{yx}) \sin^2 \theta$$

$$Z_{yy} = -\frac{Z'_{xy} + Z'_{yx}}{2} \sin 2\theta$$

From inspection of these tensor elements several properties appear obvious. First, we have the invariant properties with rotation: $Z_{xx} = -Z_{yy}$ and $Z_{xy} - Z_{yx} = $ constant, for *all* rotation angles. In addition, during rotation through 180° the off-diagonal elements (Z_{xy} and Z_{yx}) each go through *one* maximum and *one* minimum, whereas the diagonal elements (Z_{xx} and Z_{yy}) each go through *two* minima. Hence, for a purely 2-D situation each of the two diagonal elements goes through zero every time the rotating coordinate system passes through a principal direction.

The values of the off-diagonal tensor elements (Z'_{xy}, Z'_{yx}) in the principal coordinates are used for estimating the principal resistivity values (ρ'_{xy}, ρ'_{yx}), one of which is maximum and the other minimum.

The Three-Dimensional Case. *General.* Although it has been recognized for many years that lateral heterogeneities in the electrical properties of the Earth are likely to cause severe departures of MT response functions from their plane-layered counterparts, only recently have developments in computer modeling algorithms led to a better understanding of the kinds of distortions associated with 3-D structures (see Vozoff, 1987, and Hermance, 1983). For the most part, these studies have focused on the response of relatively localized features, such as cubes or rectangular prisms or simple combinations of such structures (Fig. 11).

On the other hand, many actual MT surveys are concerned with the distortion effects introduced by 3-D distributed structures having a more regional scale. Of particular interest to this latter problem are the thin-sheet techniques, because they allow us to study a variety of practical problems. For example, many MT measurements are made in elongated sedimentary basins that, though relatively homogeneous internally, often develop irregular structures along strike. A simple example is shown in Fig. 12. To consider such a case in detail, however, would

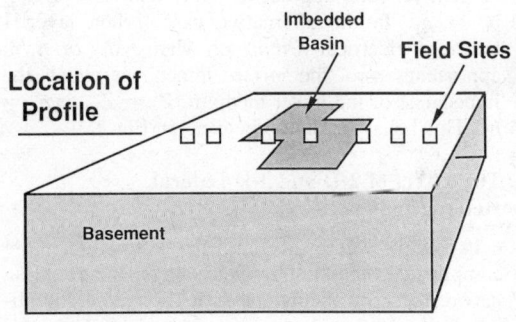

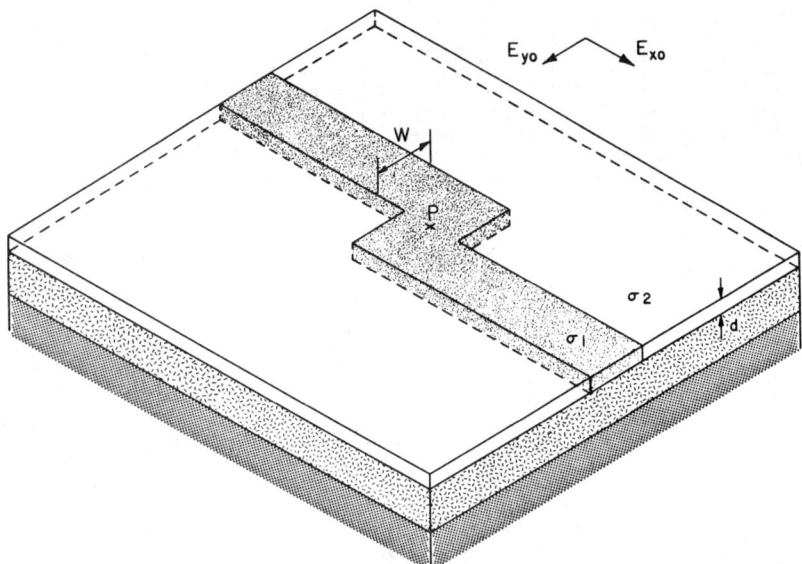

FIGURE 12. A simple theoretical model of an offset sedimentary basin. The basin is a lateral discontinuity having a width W, and conductivity σ_1 imbedded in a thin sheet having a constant thickness d. The coordinate system is such that x increases toward geographic north, and y increases toward geographic east. Material outside the basin has conductivity σ_2 extending to infinity in the horizontal direction.

tax even the largest computer. Nevertheless, significant inroads into understanding this problem can be made if one invokes some simplifying assumptions, which can be done if one has a fundamental feeling for the physics that controls the problem at hand.

The basic phenomenon that determines the electromagnetic response of a case such as Fig. 12 is associated with the fact that broad-scale telluric current sheets are induced in surficial conducting layers over large areas of the Earth's surface. The currents tend to be preferentially channeled along the axes of conducting structures—in the case illustrated here, it is a sedimentary basin. The dominant effect on MT fields at long periods will be due to the distortion of the telluric current system as it flows along or across the axis of the basin and is modulated in strength and direction by changes in the basin structure. Since the distortion of the telluric fields is dominated by local geometrical effects rather than by local induction effects, this class of phenomena is often termed *current channeling*.

In Fig. 12 the sedimentary basin is represented by a conducting zone having a width W, a thickness d, and a conductivity σ_1. The important feature of this model is the lateral offset in the strike of the basin, which is characteristic of many such structures in nature. In the case considered here, the basin is laterally offset a distance equal to its width W. For the actual model we consider, this surface layer has a constant thickness of 500 m, whereas the sedimentary basin has a total conductance (depth-conductivity product) of 100 S (Ω^{-1}), and the region on either side has a total conductance of 10 S. We simulate the response of this sheet as if it were underlain by an infinitely resistive crust extending to a depth of 10 km, beneath which lies a superconductor. The nature of the deep structure that we employ is based on the fact that in many areas of the world, sedimentary basins are underlain by broad-scale conductive features within the Earth's crust at depths of 10 km or so. These features are seen in parts of Europe, Africa, the USSR, and western North America, particularly in the Basin and Range Province. Because of the ubiquitous occurrence of this type of crustal structure, we will treat a highly idealized version of such a case. We say "highly idealized" because usually the crystalline basement is not infinitely resistive; typical values are more like 10^3–10^4 $\Omega \cdot$ m. In addition, the real Earth is not superconducting at a depth of 10 km. Typical resistivities for those regions where the Earth *does* become highly conducting by geophysical standards are more like 0.1 to 2.0 $\Omega \cdot$ m.

Determining the distortion of telluric fields within the thin sheet reduces to a 2-D solution of Laplace's equation for the potential field in the horizontal coordinates x and y. At the extreme low-frequency limit, one simply solves for a 2-D potential function $U(x, y)$, usually numerically, observing the proper continuity conditions on the normal current density and the tangential electric field at all internal boundaries. The electric field components at infinity (E_x^0, E_y^0) serve to "drive" the electric currents in the thin sheet.

A region in the interior of our model is illustrated in Fig. 13 (the actual model extends to ± 100 km

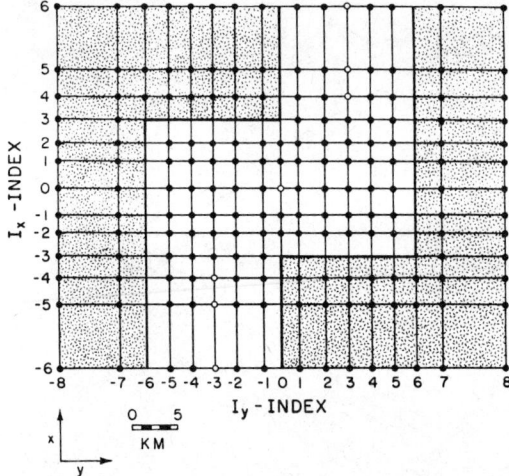

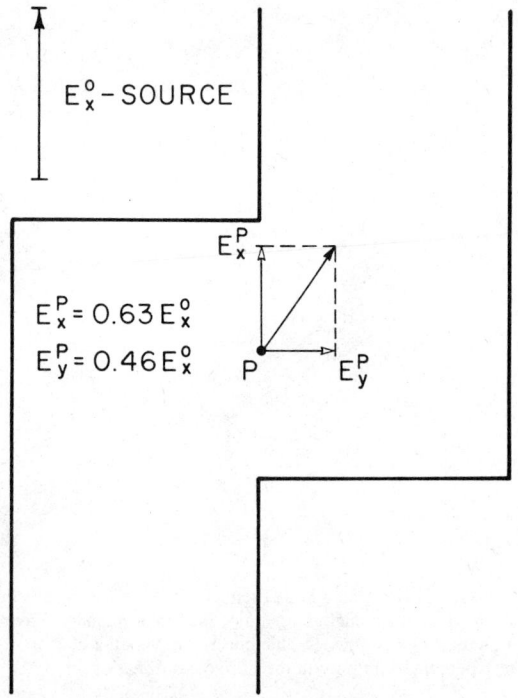

FIGURE 13. A plan view showing the 2-D mesh of nodal lines used to simulate the basin offset. The basin (conductivity = σ_1) is contained within the heavier lines bordered by the textured pattern. Only a portion of the model's interior is shown. The outer boundaries of the model are actually ± 100 km from the origin.

from the origin at the center of the offset). The width of the basin is 15 km and is contained within the textured pattern indicated in the figure. The electric field components at each node (E_x, E_y) were determined from the potential field ($U(x, y)$) at nearest-neighbor nodes using centered differences.

Telluric Transfer Coefficients for the Offset Basin. The basic character of telluric field distortion is illustrated by considering the fields at the origin $P(0, 0)$, located at the center of the basin (see Fig. 13). We first consider the effects of a source at infinity, E_x^0, of unit amplitude polarized in the north (or x) direction. As seen in Fig. 14, two horizontal electric field components are generated at the point $P(0, 0)$ such that $E_x^p = 0.63 E_x^0$ and $E_y^p = 0.46 E_x^0$. If, on the other hand, we have a source as shown in Fig. 15, which consists of a unit amplitude field at infinity, E_y^0, polarized in the east (or y) direction, the two horizontal electric field components observed at the point $P(0, 0)$ would be $E_x^p = 0.086 E_y^0$ and $E_y^p = 0.24 E_y^0$. Clearly, for the model presented here, if both sources were present, then at point $P(0, 0)$, $E_x^p = 0.63 E_x^0 + 0.086 E_y^0$, and $E_y^p = 0.46 E_x^0 + 0.24 E_y^0$.

More generally, the field components at an arbitrary point P are related to the reference source or field components at infinity through a relation having the form

$$E_x^p = T_{xx} E_x^0 + T_{xy} E_y^0$$

and

$$E_y^p = T_{yx} E_x^0 + T_{yy} E_y^0$$

FIGURE 14. The vector electric field components at the point $P(0, 0)$ due to a unit amplitude electric field E_x^0 at infinity having the polarization and magnitude shown for reference in the upper left portion of the figure.

This expression can be reduced to the matrix form

$$\mathbf{E}^p = [\mathbf{T}] \cdot \mathbf{E}^0$$

where

$$[\mathbf{T}] = \begin{bmatrix} T_{xx} & T_{xy} \\ T_{yx} & T_{yy} \end{bmatrix}$$

if often termed the *telluric transfer matrix*.

If the source field vector $\mathbf{E}^0$ rotates in direction through 360°, the tip of the telluric field vector at a point generally traces out an ellipse. In Fig. 16 we illustrate the telluric field ellipses observed at various points within the sedimentary basin referred to a unit amplitude, circularly polarized telluric field at infinity. We call these ellipses *compensated* telluric ellipses, since they are calculated with reference to undistorted values for the field far from the effects of the lateral heterogeneity.

Note, for example, the behavior of the major axes of the telluric ellipses that align themselves along the preferred direction of current flow within the offset region. Moreover, as one gets far from the disturbing effects of the offset along the linear segments of the basin, the magnitudes of the major axes approach the diameter of the circle circum-

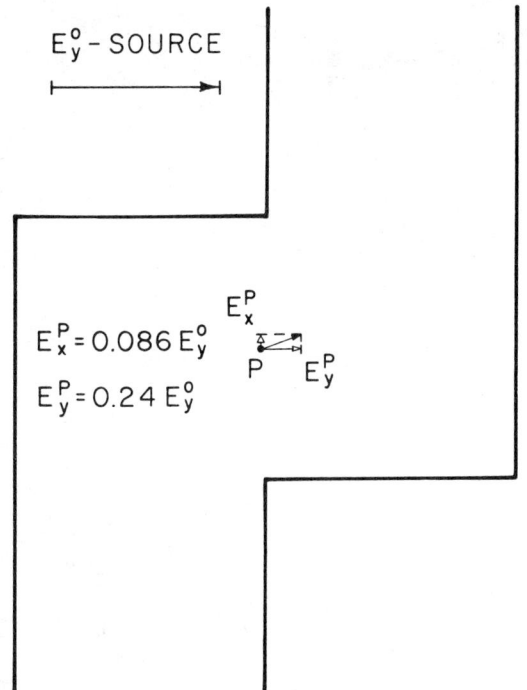

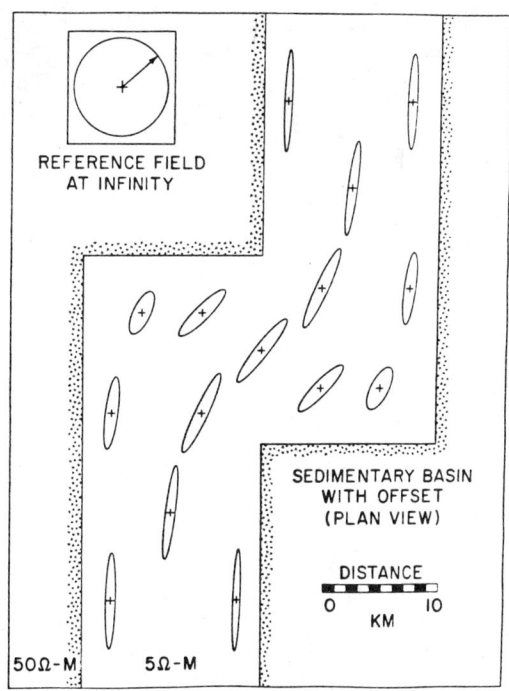

FIGURE 15. The vector electric field components at the point $P(0, 0)$ due to a unit amplitude electric field E_y^0 at infinity having the polarization and magnitude shown for reference in the upper left portion of the figure.

FIGURE 16. Plan view showing the telluric field ellipses at a number of representative points within the offset basin. These figures represent the trace of the tip of the electric field vector at a specific point referenced to a unit amplitude circularly polarized field at infinity. Shown in the upper left corner is the magnitude of the circularly polarized reference field at infinity.

scribed by the reference field at infinity. This behavior is a reflection of the continuity of the electric field component tangential to or parallel to electrical discontinuities.

On the other hand, the *minor* axes of the ellipses for the linear segments of the model are tending to approach one tenth the magnitude of the major axes. This result is a reflection of the boundary condition on the continuity of the normal component of the current density ($J_n = \sigma E_n$). Since J_n is continuous, E_n, the electric field perpendicular to strike within the basin, tends to be reduced by a factor equal to the resistivity contrast between the basin (5 Ω · m) and the surrounding material (50 Ω · m).

Magnetotelluric Parameters for the Offset Basin. The telluric distortion phenomenon described above can be recast in terms of the distortion of conventional MT parameters. The basic concept we are stressing in this model is that, at long periods, telluric currents induced in surficial layers over large regions are locally distorted through current channeling in distributed conductors, such as this elongated sedimentary basin. We argue that the primary induction process, i.e., the generation of telluric currents in the thin sheet, is determined primarily by a driving impedance that, in general, will represent the properties of the thin sheet averaged over a large area. For the model considered here, we assume that the lateral dimensions of the basin (W) are small compared with the lateral dimensions of the entire sheet (which extends to infinity). Hence, the average properties of the thin sheet over its entire dimension will be dominated by the medium outside the basin.

We assume that at infinity (i.e., beyond the effects of perturbations from the basin itself) the following isotropic relations between the electric field and the magnetic field components pertain:

$$E_x^0 = Z^0 H_y^0 \tag{22a}$$

and

$$E_y^0 = -Z^0 H_x^0 \tag{22b}$$

where the surface impedance Z^0 is some plane-layered response function to be specified. We also assume that the plane-layered structure beneath the thin sheet is horizontally uniform everywhere. This is not a strict requirement for our modeling concept (outside of ρ being infinite directly beneath the thin sheet), but is an assumption made to simplify the interpretation of our results in the present application.

With the above restrictions in mind, we may

substitute Eq. 22a and 22b for E_x^0 and E_y^0, respectively, in relations 26a and 26b and obtain

$$E_x^p = T_{xx}Z^0H_y^0 - T_{xy}Z^0H_x^0 \quad (23a)$$

and

$$E_y^p = T_{yx}Z^0H_y^0 - T_{yy}Z^0H_x^0 \quad (23b)$$

With a simple redefinition of coefficients (e.g., $Z_{xx} = -T_{xy}Z^0$), the relations in Eqs. 23a and 23b can be rearranged to the matrix expression

$$\mathbf{E}^p = [\mathbf{Z}] \cdot \mathbf{H}^0$$

where

$$[\mathbf{Z}] = \begin{bmatrix} Z_{xx} & Z_{xy} \\ Z_{yx} & Z_{yy} \end{bmatrix} \quad (24)$$

is known as the impedance transfer matrix or tensor. If, as is true in many cases, the horizontal magnetic field at infinity, $\mathbf{H}^0$, is approximately equal to the horizontal magnetic field at the point P, $\mathbf{H}^p$, then $[\mathbf{Z}]$, as represented by Eq. 24, becomes the conventional impedance matrix. Hence, for the present discussion we will assume that the impedance matrix is given by the operation

$$[\mathbf{Z}] = [\mathbf{T}] \cdot [\mathbf{Z}^0] \quad (25)$$

where the impedance at infinity has the matrix form

$$[\mathbf{Z}^0] = \begin{bmatrix} 0 & Z^0 \\ -Z^0 & 0 \end{bmatrix} \quad (26)$$

It is useful to introduce the terminology that $[\mathbf{Z}]$ is the "point impedance," $[\mathbf{Z}^0]$ is the "driving impedance," and $[\mathbf{T}]$ is the "transfer matrix." For this discussion, we choose a driving impedance at infinity of $Z^0 = (26.4 \pm i\,288.5) \times 10^{-6}\,\Omega$. This value is compatible with an apparent resistivity of $3.2\,\Omega \cdot m$ at a period of 300 s (see Fig. 17) and corresponds to a point on the right-hand descending branch of the plane-layered response curves (Fig. 17).

The Point Impedances. As described earlier, a telluric transfer matrix $[\mathbf{T}]$ is now calculated for each nodal point P in our thin sheet that relates the electric field at a particular point to a unit amplitude reference or source field at infinity. Then, using Eq. 25, we calculate an approximation to the point impedance, $[\mathbf{Z}] = [\mathbf{T}] \cdot [\mathbf{Z}^0]$, where, in the present case, the elements of the driving impedance are given by Eq. 26. Assuming appropriate values for the coefficients in Eq. 26 and employing numerically determined values for the elements of $[\mathbf{T}]$ allow one to use Eq. 25 to generate values of the impedance at each nodal point over the surface of the model.

Rotated Impedance Figures. In Fig. 18 we show plots of the rotated impedance elements (Z_{xy} and Z_{xx}

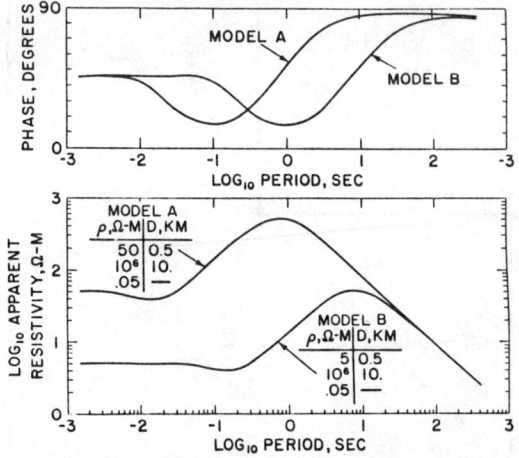

FIGURE 17. The equivalent plane layered MT response (phase and the apparent resistivity as a function of period) for two one-dimensional models: Model A represents the case for a site outside, and at infinite distance from, the sedimentary basin. Model B represents the case of a site within the sedimentary basin, but where the basin extends to infinity in all directions. The depth to the bottom of the respective layers is indicated in the inset tables by D in kilometers and the resistivity of each layer ρ is given in ohm-meters ($\Omega \cdot m$).

at a series of representative points within the interior of the sedimentary basin. At points away from 3-D effects, the rotated diagonal impedance (Z_{xx} or Z_{yy}) attains the shape of a symmetric, four-lobed, cloverleaf figure. On the other hand, at points where 3-D effects are severe, the symmetric four-lobed figure degrades into a strongly asymmetric form.

As we attempt to understand the nature of the distortions associated with the basin offset illustrated in Figs. 16 and 17, we should keep in mind what we would expect for the *unperturbed* parameters. In the absence of any lateral variations whatsoever (i.e., for a strictly 1-D structure), (1) the telluric ellipse at all points should degenerate to a circle of unit radius, (2) the off-diagonal impedance elements should be equal and constant in all directions, and (3) the diagonal elements should go to zero. On the other hand, for a strictly 2-D lateral variation, such as is appropriate for the linear segments of the basin far from the effects of the basin offset, (1) telluric ellipses should be polarized parallel to the strike of the basin such that the semimajor axis has a unit amplitude and the semiminor axis has a value of 0.1, (2) the diagonal elements of the rotated impedance matrix should tend to be symmetrical cloverleafs, and (3) off-diagonal elements of the impedance matrix should be two-lobed, lying parallel to the strike of the basin. The departure of the actual parameters in Figs. 16 and 17 from these unperturbed values is a measure of the 3-D distortion imposed by the basin offset.

One can use these results to evaluate where the

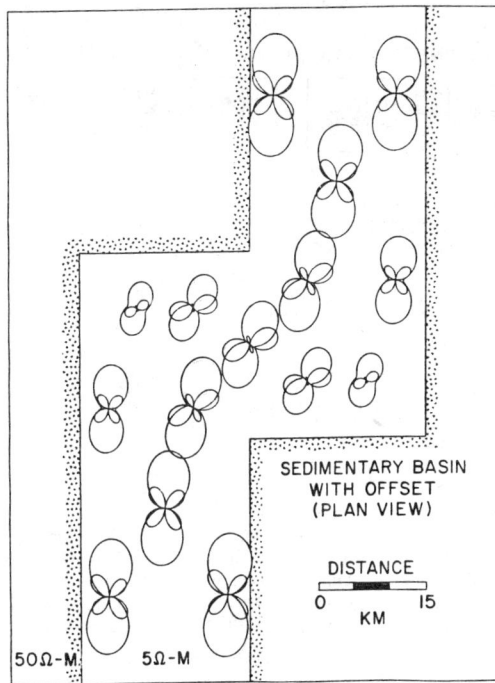

FIGURE 18. Rotated impedance elements at a number of representative points in the interior of the basin. In all cases, Z_{xy} appears as a two-lobed figure and Z_{xx} appears as a four-lobed figure.

effects of the basin offset become small. At distances outside the offset of only a fraction (2.5 km) of a characteristic dimension of the perturbation ($W = 15$ km), the response parameters are already tending to their unperturbed values. At the bottommost row of nodes ($I_x = -6$) or the topmost row of nodes ($I_x = +6$) in Fig. 13, the distortion of the parallel electric field component is less than 5%. Hence, according to the model described here, it seems that if one is a characteristic dimension away from a structural heterogeneity, then its electrical perturbations may be minimal.

Data Acquisition

The Magnetotelluric Field System. Although one of the earliest recognized advantages of the MT method was that, in principle, it allowed one to deduce the gross layered structure of the Earth from measurements at a single site, it very soon developed that more than one site was usually occupied in a survey in order to obtain lateral control on the structural interpretation. Today, MT experiments invariably involve multiple sites occupied simultaneously or in sequence.

Selecting the type of field system to be deployed needs to recognize the nature of the experiment to be performed. Broadband MT surveys covering the spectrum from 1000 Hz to 10,000 s are the most demanding in terms of acquiring and archiving results. A block diagram illustrating the major components of a modern MT/magnetic variation field system is shown in Fig. 19. This particular system is built around a DEC PDP-11/73 microcomputer, which serves as the central processor for acquiring, analyzing, displaying, and archiving all data gathered in the field (Fig. 20). Other minicomputers, even personal computers, can be readily adapted to serve the same functions. In the following section, the various components of the system sensors, signal conditioning electronics, microcomputer, and display devices are described in more detail.

Long-period induction coils can be readily used to monitor frequency bands from 300 s to 30 Hz, and audio MT induction coils can be used to monitor from 50 s to 1000 Hz. In addition, one could use three-component, ring-core flux-gate magnetometers to monitor the magnetic field from periods of dc to 10 s (a comparison of various sensors is shown in Fig. 21). Alternatively, cryogenic (SQUID) magnetometers may be used throughout this frequency range. The electric field sensors are $Cu-CuSO_4$ or $Pb-PbCl_2$ porous pot electrodes buried 20 cm to 30 cm in the earth spaced from 50 m to 200 m apart. These electrodes are then connected to the recording system using single conductor, or coaxial "electric lines."

Signal Conditioning. Electric field signals are preamplified (see Fig. 19) with a low-noise instrumentation amplifier with a bias adjust circuit for nulling the self-potential of the ground. Magnetic field signals from induction coil sensors are preamplified with a high-quality differential amplifier. (Note that preamplification is not normally required for signals from the SQUID or ring-core flux-gate magnetometers.)

Each channel passes through a commercial active bandpass filter (e.g., Ithaco model 4111) consisting of four-pole high-pass and four-pole low-pass stages in cascade. The cutoff frequencies of each stage are selectable in 1–10 decade increments over the period range 0.01 to 10^4 s. These filters bandlimit the signals to prevent aliasing problems during analog-to-digital (A/D) conversion, and to permit a total data set to be limited to a reasonable number of samples (usually 4096).

Notch filtering is supplied at 60 and 180 Hz, and postamplification is performed when necessary. In addition, comparators are often used in each amplification stage to detect amplifier saturation. Such a condition can be signaled both to the system operator (via light-emitting diodes) and to the microcomputer (via a hardware interrupt) so that data acquisition can be aborted and amplifier gains can be adjusted.

The Data Acquisition System. A microcomputer such as the DEC LSI-11/73 (see Fig. 20) can be used for real-time data acquisition and for the display and analysis of data in the field. The system

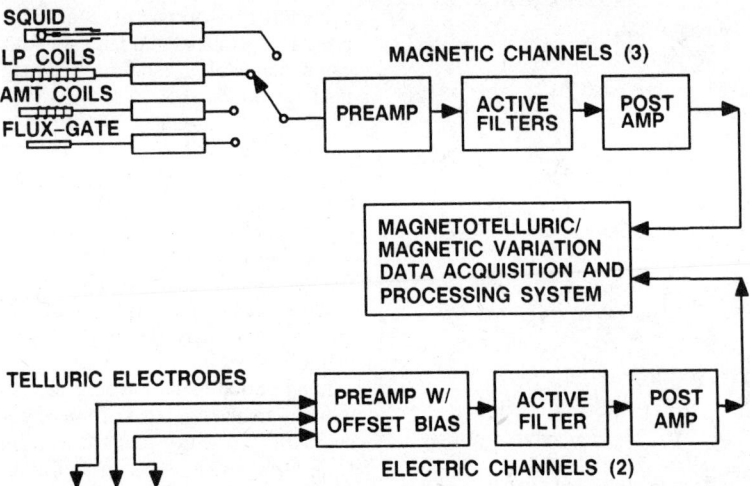

FIGURE 19. An example of a magnetotelluric signal conditioning subsystem showing at the top of the figure possible magnetic field sensors and the corresponding magnetic signal conditioning channels (usually 3 identical channels). The electric or telluric field sensor (grounded electrodes) and the corresponding signal conditioning channels (usually 2) are shown in the lower portion of the figure. Possible magnetic field sensors could be SQUIDS, long period induction coils (wire solenoids wound on high permeability mu metal cores, also known as "search coils"), audio magnetotelluric induction coils, and flux gate magnetometers.

device consists of a hard (Winchester) disk plus a floppy disk drive. This unit is a reliable direct access device that allows one to run a real-time operating system such as DEC RT-11 under field conditions. A nine-track magnetic tape unit permits archiving data and computational results. A graphics terminal and a small hard-copy printer permit the system operator(s) to monitor and control operations.

Data acquisition is often performed with an 8- to 16-channel, 12-bit A/D converter and a programmable, crystal-controlled clock synchronized to standard radio time so that several systems can be operated simultaneously in the field. An adequate maximum data conversion rate is approximately 20,000 samples per second on each channel. Upon command from the operator, a multichannel MT data set can be read into the microcomputer memory and held for immediate display and analysis, or it can be archived. Alternatively, data can continually stream into memory in a wraparound manner while the computer itself tests for the occurrence of significant bursts of signal or an "event." In this mode, a data window centered around the occurrence of the triggering event can be captured and stored for follow-up analysis.

Analysis of captured data is usually accomplished immediately with the computation of auto and cross spectra of the multichannel data, and the calculation of relevant MT parameters. The RT-11 operating system, for example, has a foreground/background capability that permits concurrent analysis of one data set and real-time acquisition of another. This capability, plus the fast floating-point hardware, make data analysis in quasi real time quite feasible.

Computational results, and raw data if desired, can be archived to either floppy disks, magnetic tape, or tape cartridge; further interpretation (e.g., data inversion) is usually performed later.

Display. Several types of display devices are normally available as part of the field system; for example, for low-frequency monitoring, one might employ a six-channel Gould strip-chart recorder, whereas for high-frequency monitoring, one might use two dual-beam Phillip's oscilloscopes. Both types of devices can display either real-time analog data or, through use of an eight-channel D/A converter, data that are being held in computer memory can be recalled and plotted. In addition, time series and data parameters at various stages of processing can be inspected on a graphics monitor (e.g., a GraphOn GO-230).

Multichannel Data Processing

Most modern MT surveys are concerned with analyzing data where several variables (E_x, E_y) are mutually and simultaneously coupled with several other variables (H_x, H_y) through the matrix coefficients of the impedance tensor

$$[Z] = \begin{bmatrix} Z_{xx} & Z_{xy} \\ Z_{yx} & Z_{yy} \end{bmatrix}$$

Following closely the concepts developed by statisticians for dealing with multivariate spectral analysis, one can develop a procedure for optimizing the estimate of a tensor element if one is provided with many independent data sets or closely spaced

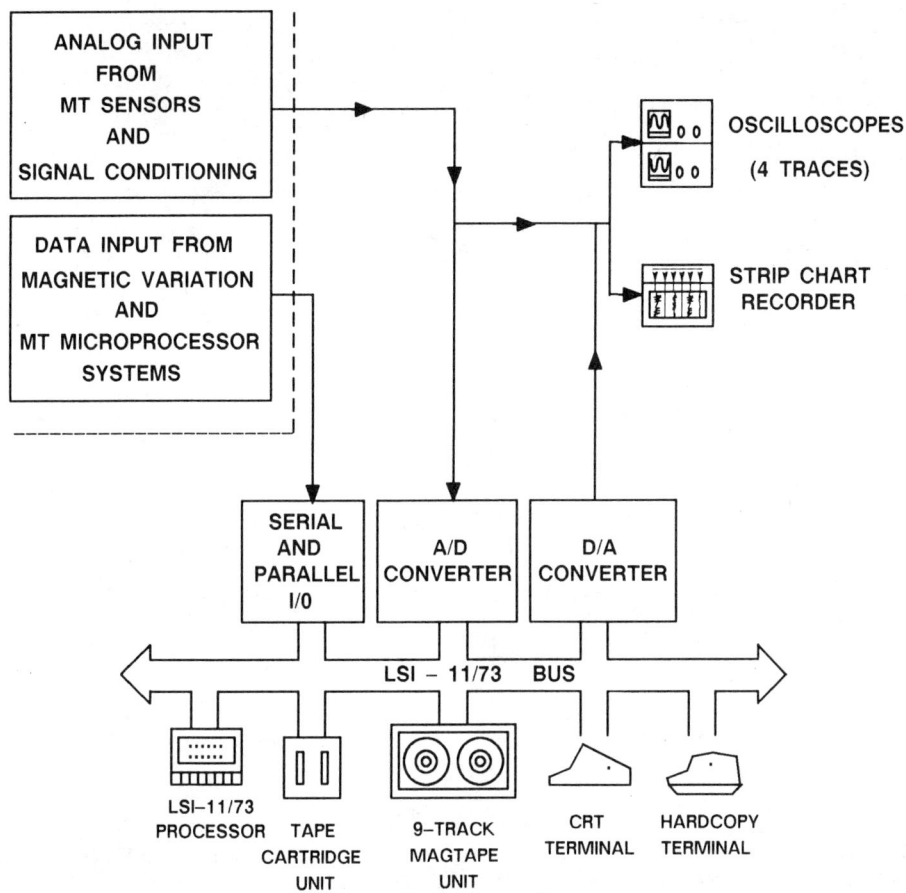

FIGURE 20. The magnetotelluric and magnetic variation data acquisition and processing system. The top left hand box represents analog data input from the system shown in Fig. 19. Also shown within the dashed line in the upper left hand corner is a box representing digital data from previous experiments which could have been archived on a variety of media for later processing or reprocessing. Data acquisition and processing is performed under control of a mini or microcomputer.

frequency samples. As an example, for the expression $E_x = Z_{xx}H_x + Z_{xy}H_y$, one can define the best estimate of Z_{xx} and Z_{xy} in a way that minimizes the mean of the squared differences between the measured electric field component E_{xj}^{meas} and the electric field predicted from the measured magnetic field components through the relation

$$E_{xj}^{\text{pred}} = Z_{xx}H_{xj}^{\text{meas}} + Z_{xy}H_{yj}^{\text{meas}}$$

where the subscript j denotes the jth measured data set. The difference between the measured and predicted electric fields is

$$R_j = E_{xj}^{\text{meas}} - E_{xj}^{\text{pred}}$$

Multiplying this quantity by its complex conjugate and averaging over the various independent samples leads to a sum of the squared residuals

$$S^2 = \sum R_j R_j^* = \sum (E_{xj} - Z_{xx}H_{xj} - Z_{xy}H_{yj}) \cdot (E_{xj}^* - Z_{xx}^*H_{xj}^* - Z_{xy}^*H_{yj}^*) \quad (27)$$

where the sum is over the independent data sets or over each component of the Fourier coefficients of the measured time series in a given frequency band. The values Z_{xx} and Z_{xy} that minimize Eq. 27 can be determined by least squares such that the derivatives of S^2 with respect to the real and imaginary parts of Z_{xx} and Z_{xy} are zero, respectively, thus leading to the simultaneous equations

$$\langle E_x H_x^* \rangle = Z_{xx} \langle H_x H_x^* \rangle + Z_{xy} \langle H_y H_x^* \rangle \quad (28a)$$

$$\langle E_x H_y^* \rangle = Z_{xx} \langle H_x H_y^* \rangle + Z_{xy} \langle H_y H_y^* \rangle \quad (28b)$$

where the angle brackets denote the ensemble

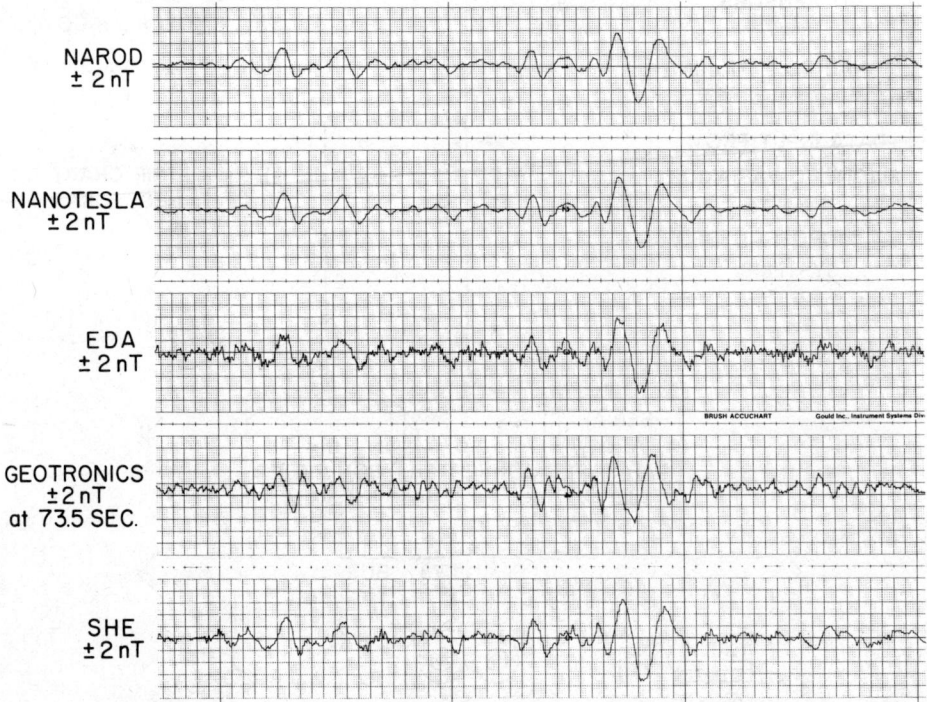

FIGURE 21. A comparison of the output of five different magnetometers running simultaneously. The top two traces represent data from two modern ring core flux gate sensors. The middle trace represents a recording from a conventional flux gate magnetometer. The Geotronics trace is from an induction coil, and the bottommost trace is from a superconducting SQUID. Note the improvement in data quality of both the upper two traces for the ring core sensors, compared with the middle trace for the conventional EDA flux gate magnetometer, which shows a higher degree of noise "jitter".

average, which might, for example, consist of frequency-band averages of discrete Fourier coefficients as approximations to smoothed cross-power spectral estimates for different data sets. The terms Z_{xx} and Z_{xy} are, in a sense, average estimates of the tensor elements over the smoothing bandwidth. One may solve these two equations for Z_{xx} and Z_{xy}. A similar set of equations can be set up and solved for Z_{yx} and Z_{yy}. After the tensor elements for our *measuring* coordinate system are calculated, the last step in estimating MT parameters is usually to determine the *principal* coordinate system, which is best done by rotating the tensor elements and determining the angle that minimizes the diagonal elements of the corresponding rotated impedance tensor.

We must remember that [Z] is an "optimal" estimate in the context of our analysis—it may not be a *true* estimate if our noise model is not correct or if our analysis procedure leads to biased results. It is well known, for example, that whereas noise on the E channels has virtually no effect on an analysis using relations of the form Eq. 28, the contribution of noise on the H channels has the effect of *biasing* the estimate of the multivariate coupling coefficients to values that are systematically lower than the *true* values. Such concerns are the topic of current research.

One way to circumvent such an effect is by making two simultaneous but independent measurements of the H channels using two parallel sets of sensors; one set is placed at some remote location and serves as a reference. Such "remote reference" techniques lead to estimates of [Z] that, for some sources of noise, are much less biased than the simple least squares estimate represented by relation 28. Remote reference methods are finding wide application in modern MT surveys (see Clarke et al., 1983).

Another significant advance toward improving data quality is in the area of robust data analysis, which is currently under active development (see Chave and Booker, 1987).

Fundamental View of the Magnetotelluric Inverse Problem

A Simple Example: Inversion of 1-D MT Data.

Assume that an MT experiment has been performed in a relatively flat lying, plane layered environment, and produced the data shown in Fig. 22a. The objective of an MT interpretation is to derive a model for the Earth whose response "fits" the observed data in some optimal sense. This is known as the *inverse* problem, since one begins with an observed response and attempts to determine the causative model; as opposed, for example, to the conventional *forward* problem in physics and engineering where one starts with a model and determines its response. In the inverse problem, one usually begins their analysis with an initial "first guess" model (known as the *a priori* model) whose response exhibits a substantial misfit with the actual observations (Fig. 22b). The objective of the inversion procedure is to systematically perturb the parameters of the initial model in such a way that the misfit is minimized (Fig. 22c). In most cases the goodness of fit is judged by minimizing the root mean square (rms) residual between the response of the final model and the actual observed data. If one has N data points at periods T_i (where $i = 1$ to N), then the rms residual R_{rms} is

$$R_{rms} = \left\{ \frac{1}{N} \sum_i [\rho_a^i(\text{data}) - \rho_a^i(\text{model})]^2 \right\}^{1/2}$$

where $\rho_a^i(\text{data})$ is the observed data and $\rho_a^i(\text{model})$ is the response predicted by a trial model. The summation is generally performed over all periods T_i ($i = 1$ to N). The idea is to perturb the parameters of the initial model in such a way that its response at each period T_i shifts (say by $\Delta\rho_a^i$, as shown in Fig. 22b) so as to better approximate the idealized situation (Fig. 22c).

To see this more clearly, consider the data in Fig. 23a, which is to be modeled by a simple two-layered Earth having an upper layer of resistivity ρ_1 and thickness h_1, and a lower layer having infinite resistivity ($\rho_2 = \infty$). Referring to Fig. 23b, we want to perturb the initial model in such a way that its response shifts by a measure $\Delta\rho_a^i$ at each period in order to better "fit" the data. In other words, we want to construct a perturbation to the model in "model space" such that its response better fits the data in "data space."

As viewed from the perspective of model (or parameter) space (Fig. 23c), we can look upon this adjustment as a change from an initial set of values ($\rho_1(\text{old})$, $h_1(\text{old})$) to a new set of values ($\rho_1(\text{new})$, $h_1(\text{new})$) such that the direction and magnitude of this change in parameter space results in the desired change in data (or response) space (Fig. 23b). In other words, for the simple two-parameter case

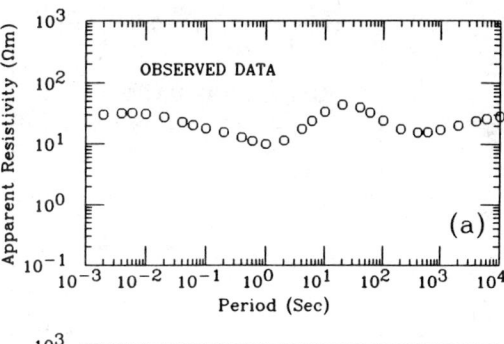

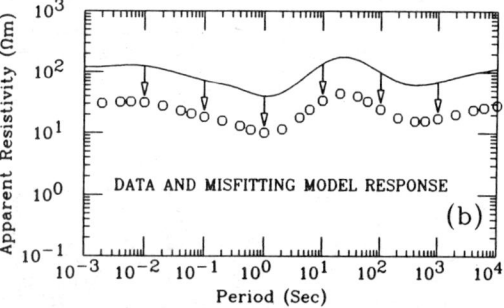

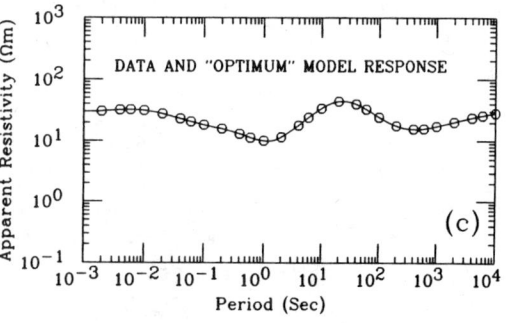

FIGURE 22. (a) An example of observed data to be inverted into a plane layered model. (b) An example of observed data shown as the open points and a model response that misfits the data. The objective is to perturb the model such that the response moves in the direction of the arrows. (c) The observed data with the fit of an optimum model response.

considered here, we want a given perturbation in ρ_1 and h_1 to result in a better-fitting response function ρ_a^i over a range of periods in data space.

Providing that the initial model response is close to the new model response, we can expand the response function in the vicinity of the initial model as a first-order Taylor's series at each period:

$$\rho_a^i(\text{new}) = \rho_a^i(\text{old}) + \Delta\rho_1 \left(\frac{\partial \rho_a^i}{\partial \rho_1}\right) + \Delta h_1 \left(\frac{\partial \rho_a^i}{\partial h_1}\right)$$

(29)

where

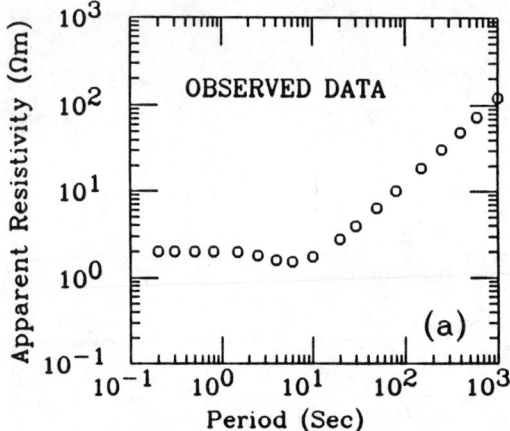

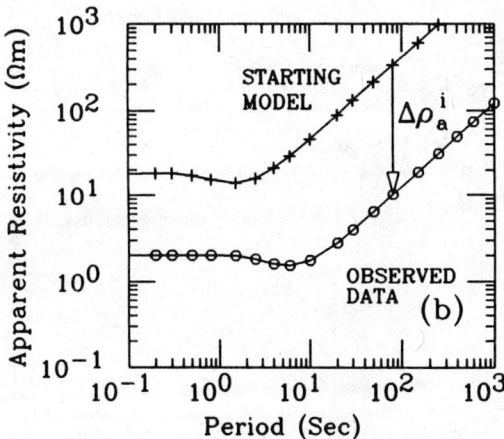

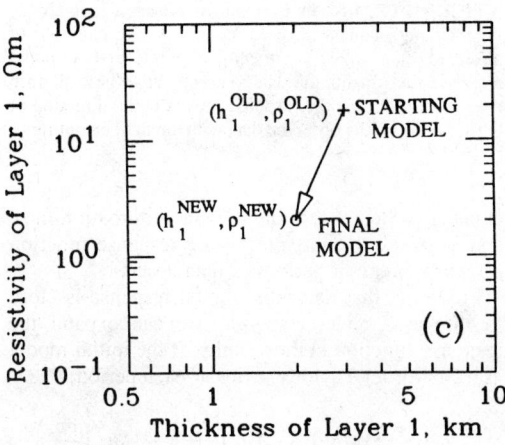

FIGURE 23. (a) An example of observed data. (b) The misfit of the reponse of a starting model from the observed data showing the perturbation in the response parameter $\Delta\rho_a^i$. (c) The perturbation in model space from a starting model to a new final model.

$$\Delta\rho_1 = \rho_1(\text{new}) - \rho_1(\text{old})$$

$$\Delta h_1 = h_1(\text{new}) - h_1(\text{old})$$

By subtracting $\rho_a^i(\text{old})$ from either side of Eq. 29 and defining

$$\Delta\rho_a^i = \rho_a^i(\text{new}) - \rho_a^i(\text{old}) \quad (30)$$

we obtain

$$\Delta\rho_a^i = \Delta\rho_1 \left(\frac{\partial \rho_a^i}{\partial \rho_1}\right) + \Delta h_1 \left(\frac{\partial \rho_a^i}{\partial h_1}\right) \quad (31)$$

For a given initial model, one can determine the derivatives $\partial\rho_a^i/\partial\rho_1$ and $\partial\rho_a^i/\partial h_1$ analytically or, using finite differences, numerically. The problem thus becomes one of determining the proper perturbations in parameter space (Fig. 23c) to generate the appropriate perturbations $\Delta\rho_a^i$ at each period in data space so that the initial model response is moved closer to the desired final model response. This adjustment is usually done iteratively for the MT problem because of the strong nonlinear dependence of ρ_a^i on the model parameters.

We assume that through some means we can calculate the appropriate perturbations in parameter space $(\Delta\rho_1, \Delta h_1)$ so that the initial model response at each period T_i ($i = 1$ to N) is modified by an amount $\Delta\rho_a^i$ ($i = 1$ to N) that causes the response of the new model to optimally "fit" the observed data. If we were able to precisely fit each data point, then we could define

$$\rho_a^i(\text{new}) \equiv \rho_a^i(\text{data})$$

so that Eq. 30 becomes

$$\Delta\rho_a^i = \rho_a^i(\text{data}) - \rho_a^i(\text{initial})$$

At each period we "know" the value of $\Delta\rho_a^i$, so we can use Eq. 31 along with the known derivatives $\partial\rho_a^i/\partial\rho_1$ and $\partial\rho_a^i/\partial h_1$ to calculate the required perturbations in the parameters $\Delta\rho_1$ and Δh_1. In principle, we only need two independent estimates of $\Delta\rho_a^i$ to solve for $\Delta\rho_1$ and Δh_1 by simultaneous equations. Often, however, we have many data points (N) and a smaller number of model parameters (M). In the present example $M = 2$. We therefore have an overdetermined set of equations (Eq. 31) to solve for $\Delta\rho_1$ and Δh_1—a classic problem in least squares estimation.

The Method of Least Squares; Global Minimization. In general, we cannot expect that any given set of values for $(\Delta\rho_1, \Delta h_1)$ is going to precisely fit the data everywhere (i.e., for all T^i, $i = 1$ to N). We might achieve this fit at one or several data points, but our theoretical model will not likely fit the data [i.e., $\rho_a^i(\text{data}) \neq \rho_a^i(\text{new})$] precisely at all other periods due to (1) errors in the data, (2) the strong nonlinear dependence of ρ_a^i on the model parameters, or (3) errors in the underlying assumptions regarding the model. We define the "error" e_i between our predicted response in data space [ρ_a^i(new)] and the actual observed data as

$$e_i = \rho_a^i(\text{data}) - \rho_a^i(\text{new}) \quad (32)$$

A useful measure of the goodness of fit of the model to the data is given by the sum of the squared values of this error,

$$S^2 = \sum_i (e_i)^2, \quad i = 1 \text{ to } N \quad (33)$$

from which one can determine the rms error

$$e_{\text{rms}} = \left(\frac{S^2}{N}\right)^{1/2}$$

Often, however, one simply employs S^2 to investigate how well various models fit the data. For example, S^2 is used in Fig. 24a to illustrate how well models with various values of (ρ_1, h_1) fit the data in Fig. 23a. It is quite clear that a prudent choice of ρ_1 and h_1 will minimize S^2; this choice is the optimal least squares solution. As illustrated in Fig. 24b, what we seek is a systematic method to perturb a starting model (which generally has a high S^2) into a least squares solution (the model with minimal S^2). Clearly, S^2 will be minimized when

$$\frac{\partial S^2}{\partial(\Delta\rho_1)} = 0 \quad (34)$$

and

$$\frac{\partial S^2}{\partial(\Delta h_1)} = 0 \quad (35)$$

In other words, we wish to solve the following equations for $\Delta\rho_1$ and Δh_1:

$$\frac{\partial}{\partial(\Delta\rho_1)}\left\{\sum_i\left[\Delta\rho_a^i - \Delta\rho_1\left(\frac{\partial\rho_a^i}{\partial\rho_1}\right)\right.\right.$$
$$\left.\left. - \Delta h_1\left(\frac{\partial\rho_a^i}{\partial h_1}\right)\right]^2\right\} = 0 \quad (36)$$

$$\frac{\partial}{\partial(\Delta h_1)}\left\{\sum_i\left[\Delta\rho_a^i - \Delta\rho_1\left(\frac{\partial\rho_a^i}{\partial\rho_1}\right)\right.\right.$$
$$\left.\left. - \Delta h_1\left(\frac{\partial\rho_a^i}{\partial h_1}\right)\right]^2\right\} = 0 \quad (37)$$

Performing the differentiation in Eq. 36 with respect to $\Delta\rho_1$ and the differentiation in Eq. 37 with respect to Δh_1, respectively, gives

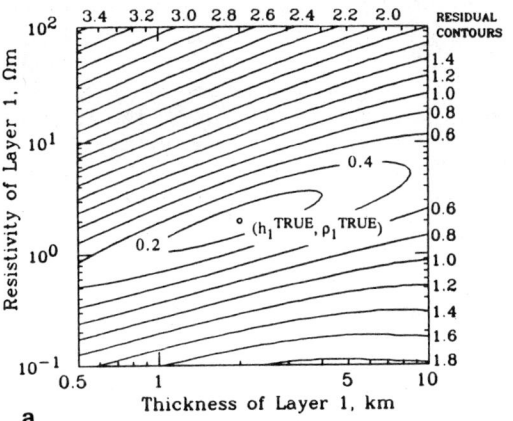

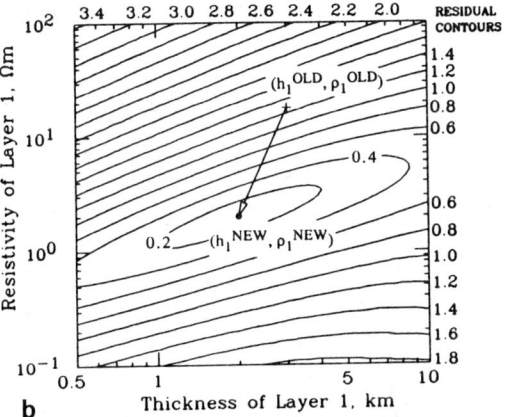

FIGURE 24. Contours of the squared residuals between the observed data and the response of a theoretical two layered model for various combinations of the thickness (km) and the resistivity ($\Omega \cdot$ m) of layer 1. (b) The perturbation of a starting model along the arrow to the new model. The objective of the inversion is to minimize the normalized residual between the observed data and the new model.

$$\Delta\rho_1\left[\left(\frac{\partial\rho_a^i}{\partial\rho_1}\right)\left(\frac{\partial\rho_a^i}{\partial\rho_1}\right)\right]$$
$$+ \Delta h_1\left[\left(\frac{\partial\rho_a^i}{\partial h_1}\right)\left(\frac{\partial\rho_a^i}{\partial\rho_1}\right)\right] = \frac{\Delta\rho_a^i\partial\rho_a^i}{\partial\rho_1} \quad (38)$$

and

$$\Delta\rho_1\left[\left(\frac{\partial\rho_a^i}{\partial\rho_1}\right)\left(\frac{\partial\rho_a^i}{\partial h_1}\right)\right]$$
$$+ \Delta h_1\left[\left(\frac{\partial\rho_a^i}{\partial h_1}\right)\left(\frac{\partial\rho_a^i}{\partial h_1}\right)\right] = \frac{\Delta\rho_a^i\partial\rho_a^i}{\partial h_1} \quad (39)$$

where summation over i is implicit in both equations following the Einstein convention of repeated indices.

These two equations represent the normal equations to be solved for the unknown parameters $\Delta\rho_1$ and Δh_1. They have resulted from implementing the least squares conditions Eqs. 34 and 35 on the sum of the squared errors, Eq. 33. This method is the conventional least squares or Gauss-Newton method. Because of the strong nonlinear dependence of the response function on the model parameters, one has to solve for $\Delta\rho_1$ and Δh_1 iteratively.

Matrix Formulation of the Least Squares Method. We now represent the model parameters and the response perturbations by column vectors, respectively, so that

$$\Delta\mathbf{x} = [\Delta\rho_1 \quad \Delta h_1]^T$$

and

$$\Delta\mathbf{y} = [\Delta\rho_a^1 \quad \Delta\rho_a^2 \quad \cdots \quad \Delta\rho_a^N]^T$$

where the superscript T denotes the matrix transpose. We next define the matrix of partial derivatives, $\mathbf{A} = [A_{ij}] = [\partial\rho_a^i/\partial x_j]$, termed the *Jacobian* or the *sensitivity matrix*. Thus Eq. 31 may be written in the following matrix form:

$$\Delta\mathbf{y} = \mathbf{A}\,\Delta\mathbf{x} \tag{40}$$

where

$$A_{i1} = \frac{\partial\rho_a^i}{\partial\rho_1} \tag{41a}$$

and

$$A_{i2} = \frac{\partial\rho_a^i}{\partial h_1} \tag{41b}$$

Equation 32 can be expressed as

$$\mathbf{e} = \Delta\mathbf{y} - \mathbf{A}\cdot\Delta\mathbf{x} \tag{42}$$

Thus Eq. 33 becomes

$$\mathbf{e}^T\mathbf{e} = [\Delta\mathbf{y} - \mathbf{A}\,\Delta\mathbf{x}]^T[\Delta\mathbf{y} - \mathbf{A}\,\Delta\mathbf{x}] \tag{43}$$

Expanding Eq. 43 leads to

$$(\Delta\mathbf{y}^T - \Delta\mathbf{x}^T\mathbf{A}^T)(\Delta\mathbf{y} - \mathbf{A}\,\Delta\mathbf{x}) =$$
$$\Delta\mathbf{y}^T\Delta\mathbf{y} - \Delta\mathbf{x}^T\mathbf{A}^T\Delta\mathbf{y} - \Delta\mathbf{y}^T\mathbf{A}\,\Delta\mathbf{x}$$
$$+ \Delta\mathbf{x}^T\mathbf{A}^T\mathbf{A}\,\Delta\mathbf{x} \tag{44}$$

For minimum error

$$\frac{\partial[\mathbf{e}^T\mathbf{e}]}{\partial(\Delta\mathbf{x})} = 0 \tag{45}$$

which leads to

$$-2\mathbf{A}^T\,\Delta\mathbf{y} + \mathbf{A}^T\mathbf{A}\,\Delta\mathbf{x} = 0 \tag{46}$$

Rearranging and dropping the common factor of 2, we obtain

$$\mathbf{A}^T\mathbf{A}\,\Delta\mathbf{x} = \mathbf{A}^T\,\Delta\mathbf{y} \tag{47}$$

which one can readily show is equivalent to Eqs. 38 and 39. The solution to Eq. 47 is given by

$$\Delta\mathbf{x} = (\mathbf{A}^T\mathbf{A})^{-1}\mathbf{A}^T\,\Delta\mathbf{y} \tag{48}$$

and is known as the Gauss-Newton solution in matrix form. Although we have used only a two-parameter problem as an example, the relations represented by Eqs. 40 through 48 are valid for the general M-parameter problem (where $M < N$ for overdetermined problems). It often happens, however, that the matrix $\mathbf{A}$ is ill-conditioned, so its least squares inverse $(\mathbf{A}^T\mathbf{A})^{-1}\mathbf{A}^T$ is singular and special steps have to be taken to assure that the elements of $\Delta\mathbf{x}$ are stable. Moreover, it is sometimes instructive to systematically explore the manner in which various $\Delta\mathbf{x}$ and $\Delta\mathbf{y}$ are coupled together. This has led workers to employ more generalized methods to study the inverse problem.

Generalized Inverse Theory. Over the past 15–20 years, dramatic advances have been made in the field of geophysical generalized inverse theory (e.g., Menke, 1984). The 1-D MT inversion problem (i.e., where the resistivity varies only with depth) has been studied extensively. The most popular approaches involve damped least squares iterative methods. In contrast to the rich literature on 1-D MT inversion methods, the literature is very sparse on 2-D MT inversion techniques and practically nonexistent for 3-D inverse methods. Nevertheless, any of these so-called inverse problems can be set up as the solution of a matrix expression having the form of Eq. 40, where $\Delta\mathbf{y}$ is a vector containing the difference between the observed data point and the response of the starting model ($\Delta y_i = y_i - y_i^0$). The unknown vector $\Delta\mathbf{x}$ must be determined; it contains the difference between the "true" model parameters $\mathbf{x}$ and the starting parameters $\mathbf{x}^0$ ($\Delta x_j = x_j - x_j^0$). The matrix $\mathbf{A}$ is referred to by various workers as the Jacobian, partial derivative, or the sensitivity matrix and has elements given by $A_{ij} = \partial y_i^0/\partial x_j$. For the simplest cases where $\mathbf{A}^{-1}$ exists (for example, if $\mathbf{A}$ is square and nonsingular), the problem is reduced to calculating $\Delta\mathbf{x}$ according to $\Delta\mathbf{x} = \mathbf{A}^{-1}\,\Delta\mathbf{y}$, and our new model will be $\mathbf{x} = \mathbf{x}^0 + \Delta\mathbf{x}$.

On the other hand, for most cases where a simple form for $\mathbf{A}^{-1}$ does not exist (such as for an $N \times M$ matrix where $N \neq M$, and when $\mathbf{A}$ may be ill-conditioned), one can use a generalized inverse operator $\mathbf{A}^+$ to calculate $\Delta\mathbf{x}$, such that

$$\Delta\mathbf{x} = \mathbf{A}^+\Delta\mathbf{y} \tag{49}$$

Here we employ the generalized inverse method to determine a least squares solution by minimizing the error $\|e\|^2 = \|\Delta y - A \Delta x\|^2$.

Singular Value Decomposition. Lanczos (1958) describes a method commonly known as *singular value decomposition* (SVD), which allows one to calculate the generalized inverse of an arbitrary matrix in a very simple and economical way by eigenvalue analysis. The SVD decomposes the Jacobian matrix **A** to the form

$$A = U\Lambda V^T$$

where **U** is an orthogonal $N \times N$ data eigenvector matrix (i.e., the whole data space is spanned by the columns of **U**), **V** is an orthogonal $M \times M$ model parameter eigenvector matrix (i.e., the model parameter space is spanned by the columns of **V**), and Λ is an $N \times M$ diagonal matrix, whose elements λ_i are nonnegative, ordered in decreasing magnitude and referred to as the *singular values*.

The generalized inverse of **A** can then be written as

$$A^{\div} = V_p \Lambda_p^{-1} U_p^T$$

where Λ_p^{-1} is a diagonal matrix whose diagonal elements are $1/\lambda_i$, where $i = 1$ to p (p being the rank of the matrix **A** or the number of nonzero singular values). Equation 49 is then written as

$$\Delta x = V_p \Lambda_p^{-1} U_p^T \Delta y \qquad (50)$$

which is known as the *generalized inverse solution*.

Model and Data Resolution. By left-multiplying both sides of Eq. 50 by V_p^T and introducing the transformations

$$\Delta \varphi_p = V_p^T \Delta x, \qquad \Delta \psi_p = U_p^T \Delta y$$

we can modify Eq. 50 to

$$\Delta \varphi = \Lambda^{-1} \Delta \psi \qquad (51a)$$

or, in index notation,

$$\Delta \phi_i = \frac{\Delta \psi_i}{\lambda_i} \qquad (51b)$$

where, for convenience, we have dropped the subscript p.

Thus $\Delta \varphi$ is the mapping of the true model parameter corrections onto the independent model parameter eigenvectors in the model parameter space; i.e., $\Delta \phi_i$ is the correction term along the ith model parameter eigenvector ϕ_i. If all the real parameters (**x**) were independent, then $\phi_i \equiv x_j$, where i and j may or may not be the same. In other words, it may turn out for a particular problem that, for example, $\phi_1 = x_3$, $\phi_2 = x_5$, $\phi_3 = x_1$, etc. More commonly, however, ϕ_i is a linear combination of some or all of the real parameters (**x**).

Similarly, $\Delta \psi$ is the mapping of Δy onto the independent data eigenvectors ψ in the data space. The vectors φ and ψ are referred to as the *eigenparameters* and *eigendata*, respectively, and **x** and **y** are referred to as the *real parameters* and *real data*, respectively. The advantage of these transformations is that each of the members of Eq. 51 is decoupled, and, therefore, each eigenparameter can be solved for independently.

Determining the Relative Importance of Various Model Parameters. A number of approaches are available to assist the data interpreter in determining which parameters of a model are "important" in the analysis. One straightforward way is to look directly at the various terms in the Jacobian matrix itself ($A_{ij} = \partial y_i/\partial x_j$). The largest coefficients represent those data that best constrain a given model parameter. In other words, a large A_{ij} indicates that a small perturbation in the model parameter results in a large change in the theoretical response function; thus observed data of that type constrain the model parameter quite closely.

However, because many of the Jacobian terms are often coupled together both columnwise (along the data columns y_i) and rowwise (across the model rows x_j), it is sometimes instructive to decompose the Jacobian into a more fundamental, decoupled form as described in the last section. Such an eigenvalue decomposition can be used as a basis for classifying parameters as important and nonimportant, based on the amplitude of the singular value corresponding to each $\Delta \phi_i$. Obviously the first eigenparameter (ϕ_1) is the most important (or most sensitive) combination of the real parameters. For equal perturbations in each of the ϕ_i (i.e., $\Delta \phi_i$), the largest change in the eigendata will be in $\psi_1 = \psi_1^0 + \lambda_1 \Delta \phi_1$, because λ_1 is the maximum singular value. Conversely, the last eigenparameter ϕ_p (p being the number of nonzero singular values) is the least important eigenparameter. Some examples, illustrating these and other approaches for the 1-D and 2-D case, respectively, are given in Pedersen and Hermance (1986) and Eysteinsson (1988).

Concluding Remarks

The contribution that Louis Cagniard made to the field of magnetotellurics can be appreciated by recalling statements made in the classic treatise of Chapman and Bartels in 1940: "The simultaneous occurrence of magnetic and earth-current disturbances is the best established fact about earth-currents," however, ... "as to quantitative relations between the two vectors, the earth current density and the disturbance of the magnetic force, the evidence is much less conclusive (p. 435)." In other words, in the early 1940s, electric field variations were a curiosity, but after 1953, with the publication of Cagniard's classic paper, they became a quantitative tool.

Following Cagniard's paper, developments were systematic, though at times episodic. Progress was somewhat stalled when workers began using the MT method in geological environments far more complicated than the simple plane-layered basin structures that were intended to be studied by Cagniard. Two- and three-dimensional effects caused cross-coupling between the various electromagnetic field components that was first perceived as "noise." Recognizing the tensor quality of the MT impedance and separating observed parameters into their appropriate modes or principal directions led to a dramatic improvement in the quality of data estimates.

Determining the nature of the transfer function that couples together the telluric and the magnetic field variations is the fundamental problem in processing MT data. In general, this transfer function has a tensor character, and one is faced with the problem of obtaining reliable estimates of tensor elements that are stable from data set to data set and that vary smoothly with frequency. The objective of data processing is to develop analysis techniques that minimize, or at least identify, both the random and the systematic effects of noise on determining reliable estimates of the fundamental response parameters. Remote reference techniques and robust data analysis have made substantial contributions toward addressing these concerns.

Developments in instrumentation have played, and will continue to play, as dramatic a role in improving the MT technique as have refined data processing methods. System noise, the ultimate constraint on data quality, is being pushed down to its theoretical limits, and researchers are continuing to find creative ways to circumvent the residual effects that still remain.

Obtaining the basic data in the field, however, is only one component of the complex montage of the MT method; one ultimately needs to interpret their results in terms of realistic physical models of the Earth. At present, interpretational procedures to generate such models are lagging our ability to routinely obtain high-quality data. But this imbalance will not likely persist for long. The combination of increased computer power and advances in generalized inverse theory will, in the future, make spectacular inroads into the application of these methods to studying complicated geological structures. Just as the ability to do quantitative 2-D modeling led to a resurgence of interest in the MT method in the 1970s, 3-D modeling using powerful new supercomputers will lead to dramatic refinements in the method in the 1990s.

Acknowledgments

This article was developed through funds from the following agencies to Brown University: Department of Energy, Office of Basic Energy Sciences Grant DEFG02-87ER13665, and Office of Naval Research Contract N00014-85-K-0685.

JOHN F. HERMANCE

References

Bendat, J. S., and A. G. Piersol, 1986, *Random Data, Analysis and Measurement Procedures*, 2nd ed. New York: Wiley, 566p.

Cagniard, L., 1953, Basic theory of the magnetotelluric method of geophysical prospecting, *Geophysics* **18**, 605–635.

Chapman, S., and J. Bartels, 1940, *Geomagnetism, The International Series of Monographs on Physics*. Oxford: Clarendon Press.

Chave, A. D., and J. R. Booker, 1987, Electromagnetic Induction Studies, *Rev. Geophysics* **25**(5), 989–1003.

Clarke, J., T. D. Gamble, W. M. Goubau, D. A. Koch, and R. F. Miracky, 1983, Remote-reference magnetotellurics: equipment and processing, *Geophys. Prosp.* **31**, 149–170.

Eysteinsson, H., 1988, *The Inversion of Two-Dimensional Magnetotelluric and Magnetic Variation Data*, Ph.D. Dissertation, Brown University.

Hermance, J. F., 1983, Electromagnetic induction studies, *Rev. Geophysics and Space Phys.* **21**(3), 652–665.

Lanczos, C., 1958, Linear systems in self-adjoint form, *Amer. Math. Monthly* **65**, 665–679.

Menke, W., 1984, *Geophysical Data Analysis: Discrete Inverse Theory*. London: Academic Press, 260p.

Pedersen, J., and J. F. Hermance, 1986, Least squares inversion of one-dimensional magnetotelluric data; an assessment of procedures employed by Brown University, *Surveys in Geophysics* **8**, 187–231.

Vozoff, K., 1987, *Magnetotelluric Methods*. Geophysics Reprint Series 5. Tulsa, Okla.: Society of Exploration Geophysicists, 763p.

Wait, J. R., 1962, Theory of magneto-telluric fields, *Jour. Research* **66D**(5), 509–541.

Cross-references: *Continental Lithosphere; Controlled Source Electromagnetic Mapping of the Crust; Earth Structure: Global; Heat Flow in the Earth; Mantle Convection and Plumes; Mantle Dynamics; Thin-Skin Tectonics.*

MANTLE, LOWER: STRUCTURE

The vast portion of the Earth extending from below the seismic velocity discontinuity at a depth of 670 km to the 2886-km deep outer core boundary is called the lower mantle. The only radial subdivision of this thick layer is in the lowermost 200 km, where the D″ region, a zone of distinctive properties, overlies the core-mantle boundary (CMB). To first order, the lower mantle can be characterized as an adiabatic, homogeneous layer of the Earth in which material properties such as density and elastic velocities vary systematically with depth due to gravitational self-compression. The composition of the lower mantle is not well known, but is believed to consist primarily of magnesiowustite $(Mg,Fe)O$ with a halite structure, and silicates $(Mg,Fe)SiO_3$ with a perovskite structure. Seismologically derived models for the radial variation of the material properties of this region have changed very little since the classic works in 1939 of Jeffreys, Gutenberg, and Richter. In fact, with the exceptions of the D'' region and the 670-km discontinuity, recent radial models for the lower mantle have velocities within 1% of those in models available in 1939. Despite this, much attention is focused on improving our understanding of the lower mantle, for the small lateral velocity deviations from the reference Earth models have tremendous significance as signatures of dynamic processes in the interior.

It is generally agreed that thermal convection is the primary mechanism that the Earth adopts to remove heat from the core and mantle. However, the configuration of this convection system is one of the primary unresolved problems in the earth sciences, with the critical issue being whether or not there is material transport across the 670-km discontinuity. If the lower mantle is compositionally different from or sufficiently more viscous than the upper mantle, convection will occur in a layered system of cells rather than in large convection cells extending throughout the mantle. Recent seismological and geodynamical analyses have radically revised the role of the lower mantle in this dynamic system. The D'' region is now believed to be a major thermal boundary layer in the Earth, resulting from an 800° to 1000° K temperature increase from the lowermost mantle to the outermost core. This boundary layer may be thermally unstable, shedding thermal plumes that rise through the mantle. There is evidence that subducting oceanic lithosphere does penetrate into the lower mantle and that lower mantle viscosities are not high enough to inhibit deep-seated convection. Recent three-dimensional models of lateral velocity variations in the lower mantle indicate a complex, presumably nonsteady state, dynamic configuration. The lines of evidence supporting this new image of a dynamic lower mantle will be summarized later.

Radial Structure of the Lower Mantle

Seismological methods provide the best constraints on the properties of the deep mantle. Seismologists have derived a series of radially symmetric models of variations of elastic velocities, density, and anelastic properties in the lower mantle. The most recent model, PREM, (Dziewonski and Anderson, 1981) is shown in Fig. 1. This Earth model was constructed to satisfy normal mode periods, travel time observations and the moment of inertia of the

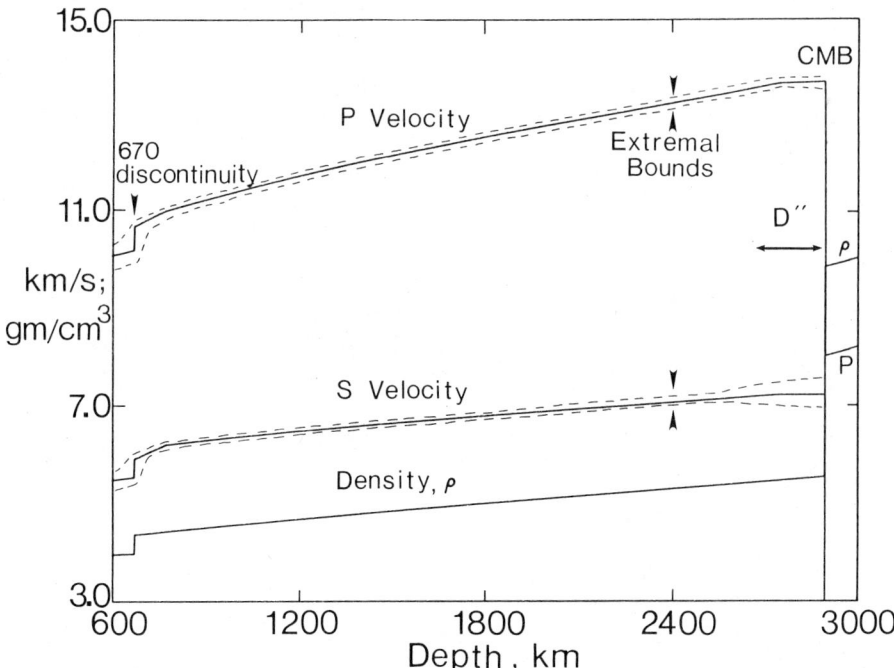

FIGURE 1. Variation of seismic velocities and density through the lower mantle for Earth Model PREM of Dziewonski and Anderson (1981). The extremal bounds indicate the confidence interval on the velocities that are consistent with observed travel time data. The D'' region is the lowermost 200 km of the lower mantle overlying the core-mantle boundary (CMB).

planet. Superimposed on the velocity variations are confidence intervals based on global travel time constraints. These extremal bounds convey the resolution of any radially symmetric Earth model compatible with the travel time data, but lateral variations at a given depth are not necessarily bracketed by the bounds. The PREM model suggests the presence of a steep velocity gradient below the 670-km discontinuity and a decreased gradient in the D" region. The first feature is supported by detailed P and S waveform modeling by Walck and Grand and Helmberger. The decrease in velocity gradient within D" is required to match the travel time slope for P waves observed at distances greater than 90°. A similar change in velocity gradient near the base of the mantle existed in the early models of Gutenberg and Jeffreys, and led Bullen to identify the D" region as chemically inhomogeneous. The spreading of extremal bounds on the velocities in D" reflects an increase in scatter of travel times produced by radially and laterally varying structure near the base of the mantle. However, through the bulk of the lower mantle there is little evidence for radial inhomogeneity.

Many body wave travel time and apparent velocity studies have suggested the presence of small scale discontinuities at various depths, such as 1050 km, in the lower mantle; however, none of these has been demonstrated to be of global extent. The effects of near source structure and lower mantle lateral heterogeneity complicate the interpretation of many of the proposed small scale features. Detection of radial velocity discontinuities with less than 1% increases in velocity is probably below the threshold of available seismic data, so minor layering may have been missed; however, the data do preclude the existence of larger radially symmetric discontinuities everywhere except in the D" region.

Many diverse seismic models have been proposed for the D" layer, including models with up to 3% velocity discontinuities, strong negative velocity gradients just above the core, or strong positive velocity gradients just above the core (see Fig. 2). Cleary (1974) reviewed the existing models based on travel time analysis, core-reflected phases (PcP), and waves diffracted around the core. While almost all models show general decreases in P and S wave velocity gradients, as in model PREM, there has been substantial debate over whether the velocities actually begin to decrease with depth above the CMB. This issue is critical to the interpretation of the thermal interaction between the core and mantle.

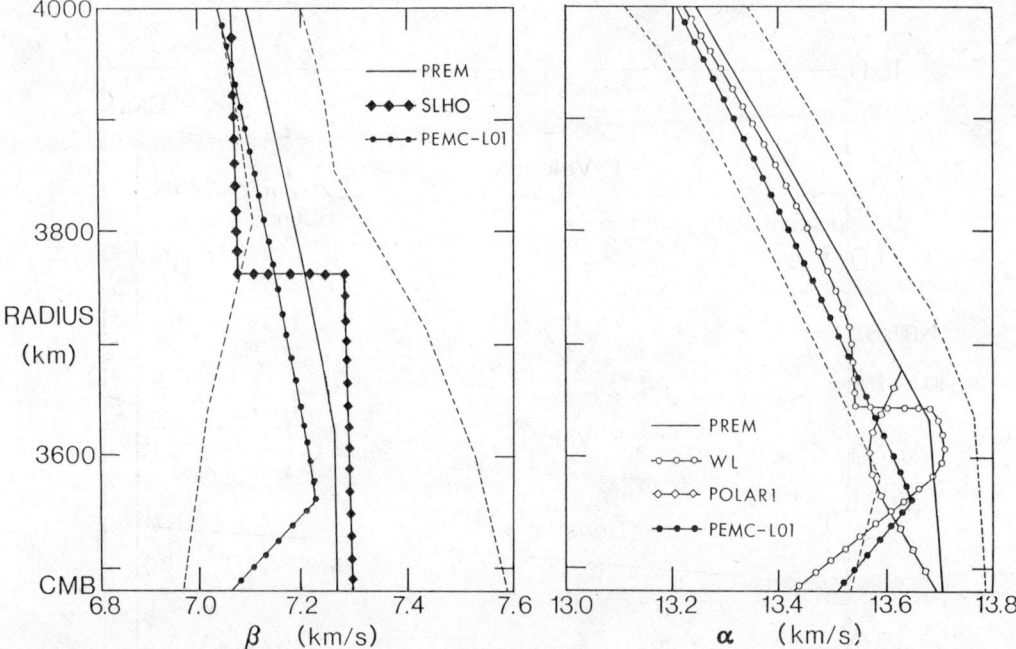

FIGURE 2. S-wave (left) and P-wave (right) velocity models for the D" region. The PREM model is representative of the smooth velocity structures obtained from gross Earth data sets of body wave travel times and free oscillation periods. Model Pemc-L01 (Doornbos and Mondt, 1979) has a negative velocity gradient in the lowermost 75 km of the mantle required to satisfy diffracted P and S waves. The POLAR1 model was derived from P wave observations near the core shadow boundary at 95°. The discontinuity models SLHO and WL were proposed on the basis of lower mantle triplication observations. The light dashed curves indicate the extremal bounds on the radially symmetric models of D" consistent with travel time data.

Strong velocity decreases have been suggested on the basis of anomalous PcP reflections and more convincingly on the basis of low apparent velocities of diffracted P and S waves. More recent quantitative waveform analysis of PcP and diffracted waves has resulted in models with milder negative velocity gradients or slight positive gradients in the lowermost 100 km of D″ (e.g., Doornbos and Mondt, 1979). This quantitative modeling has established that waves diffracting along the core are sensitive to structure throughout D″, and complete synthesis of the waveforms is needed to account for frequency dependent effects. The majority of the available data indicate the presence of at least a mild velocity decrease with depth into the D″ layer, particularly observations of high frequency diffracted S and P wave energy at large distances into the core shadow zone that would not be observed otherwise.

More complicated models for the D″ region have recently been proposed on the basis of detailed body wave analysis. A series of array studies of P waves traversing D″ has indicated the presence of a sharp 1.5%–3.0% P-wave velocity discontinuity about 180 km above the core (e.g., Wright et al., 1985). This interpretation is based on an abrupt change in the P-wave travel time curve near 87°. One of the proposed P wave discontinuity models is shown in Fig. 2. The model is radically different from the smooth PREM model. The strong negative velocity gradient below the discontinuity is similar to the gradient in the models obtained from diffracted waves, but if the discontinuity reflects compositional stratification, the impact on D″ dynamic models will be profound. Even more complex D″ structure is suggested by the POLAR1 model Ruff and Helmberger obtained in 1982 by analysis of short period P waves near the onset of the core shadow at 95°. The diversity of proposed P velocity models may reflect the limited sampling of the laterally heterogeneous D″ layer in each study, in which case these structures are only approximate "local" models. However, the different techniques used to generate the models are also intrinsically sensitive to different aspects of the actual mantle structure.

An S velocity discontinuity was proposed by Lay and Helmberger (1983), with a 2.75% increase 280 km above the CMB. They found direct evidence for this structure in the form of a reflected signal arriving ahead of the core reflection, ScS. Four separate regions have been shown to have an S velocity discontinuity in D″, suggesting the possibility that this is a global structure. This model is also radically different from PREM (Fig. 2). The S-wave discontinuity is about 100 km shallower than the P-wave discontinuity, but much of this discrepancy may be attributable to the difference in velocity gradients below the discontinuities. Recent modeling has shown that a negative velocity gradient below the S-wave discontinuity is needed to match diffracted S-wave observations. Additional work will establish whether the D″ velocity discontinuities are global features. The variation in the seismic models in Fig. 2 should be kept in mind when drawing conclusions about the dynamics of the lower mantle.

The depth variation of anelastic properties of the lower mantle has also been modeled by seismic methods. It is generally recognized that body waves are less attenuated in passing through the lower mantle than in passing through the upper mantle. Radial models of the quality factor, Q, which is inversely proportional to the amount of anelastic loss, reflect this, with frequency independent Q models giving relatively high Q values of 300–500 throughout the lower mantle except for a low Q zone in D″. A recent frequency dependent Q model by Anderson and Given allows signals with periods less than 1 sec to have almost purely elastic transmission (infinite Q) through the lower mantle, while still producing Q values near 400 for long period waves. The existence of a low Q zone at the base of the mantle for body wave periods has been questioned by Mula and is not compatible with observations of high frequency diffracted signals.

Lateral Heterogeneity of the Lower Mantle

With the exception of the D″ region, there is little evidence for radial structure in the lower mantle; however, there is ample evidence for moderate lateral heterogeneity in this region. Numerous travel time and array studies have indicated the presence of large scale heterogeneities in the lower mantle. Efforts to constrain the dimensions of particular heterogeneities have indicated 2% lateral variations over 1000 km scale lengths. While these lateral variations are much smaller than the 10% variations observed in the uppermost mantle, body waves with long pathlengths through the lower mantle can accumulate appreciable anomalies. Several efforts have been made to invert large travel time data sets directly for the three-dimensional configuration of the lower mantle velocity variations (e.g., Dziewonski, 1984).

The three-dimensional heterogeneity inversions suggest a high level of perturbations just below the 670-km discontinuity and within the D″ region. The low order spherical harmonic components of this heterogeneity have 1.0–1.5% velocity variations in D″, 3 to 4 times greater than those found in the central mantle. Perhaps the most profound conclusion drawn from the first generation of three-dimensional models for the lower mantle is that the velocity heterogeneity is not regularly ordered (as would be expected for a steady state convecting system) but is both chaotic and asymmetric, suggesting a nonsteady state dynamic system.

A particularly interesting lateral heterogeneity in the lower mantle has been detected along the downward extension of those oceanic slabs that have seismicity to depths of 670 km. Creager and Jordan

(1984) have shown that tabular high velocity structures extend to at least 1000 km depth along the trajectory of the subducted slabs. They interpret this as evidence that the slabs penetrate into the lower mantle, refuting the idea of strictly layered convection. Detailed raytracing calculations indicate that thermal models of the subducted lithosphere closely match the observations as long as the slabs are allowed to penetrate. Alternate interpretations, such as thermally coupled convection in which a downwelling current is induced in the lower layer by the cold downwelling current in the upper layer or near source slab anisotropy have been suggested, but these alternatives have not been shown to match the data as well as the slab penetration model. The current generation of three-dimensional global inversion models do not have sufficient resolution to confirm or refute the slab penetration model, but they do exhibit broad regions of high velocity material extending throughout the lower mantle beneath the circum-Pacific rim.

While the global inversions indicate very long wavelength lateral variations in D'', many scales of heterogeneity are indicated by different seismic waves. Long period diffracted signals indicate large-scale variations in the velocity structure of D''; scattered short period precursors to PKP phases indicate lateral heterogeneities with 40 to 150 km scale lengths with a mean velocity variation of about 1%; and PKP amplitude anomalies indicate regions with varying degrees of heterogeneity in D''. Detailed modeling of ScS phases indicates the presence of either localized regions with high velocity gradients in D'' or anisotropy in D''. These combined lines of evidence for strong heterogeneity in D'' mandate caution when interpreting subtle features in radially symmetric velocity structures like those in Fig. 2. There is particular need for caution when interpreting average models derived from diffracted signals, which have long horizontal transits through the heterogeneous D'' region.

In addition to the heterogeneity observed within D'', there is evidence for topographic relief on the CMB itself. Anomalies in PKKP phases (P waves once reflected off the inside of the CMB boundary) have been attributed to backscattering from the underside of the CMB. Topography of a few hundred meters is sufficient to produce the scattered arrivals in all but a few cases. Greater relief, of up to 10 km has been suggested by recent three-dimensional models of the outer core obtained by inverting large PKP and PcP travel time data sets. The amount of inferred topography trades off with the heterogeneity attributed to the D'' region, however, excess ellipticity of the CMB may also be required to account for anomalous splitting of normal modes sensitive to the outermost core. Topography on the CMB and heterogeneity within D'' may influence convection in the outer core as well as within the lower mantle.

Dynamics of the Lower Mantle

The velocity heterogeneity indicating lower mantle slab penetration is an obvious manifestation of lower mantle dynamics, but in fact all of the lateral heterogeneity in the lower mantle can be associated with deep convective processes, because velocity heterogeneity must result from thermal or compositional variations, which in turn will have associated density variations. Over the long time scales operating in the Earth, any density heterogeneity must drive viscous flow. One of the most striking confirmations of the relationship between velocity heterogeneity deep in the mantle and the Earth's dynamics is the successful prediction of long-wavelength components of the Earth's geoid by density heterogeneity in the lower mantle inferred from the three-dimensional seismic models. At spherical harmonic degrees 2 and 3, geoid lows are associated with high velocity regions in the lower mantle, whereas geoid highs are associated with slow lower mantle. Long wavelength temperature differences on the order of $\pm 20°C$ could produce the observed low order velocity heterogeneity.

The dynamic response calculations for the low order lower mantle heterogeneity also predict dynamically supported topography of the CMB with a total excursion of about 3 km. Shorter wavelength heterogeneities such as those associated with thermal boundary layer instabilities should produce even larger dynamic topography, which may account for the PKKP scattering and is important for theories of geodynamo excitation and decade fluctuations in the length of day.

The importance of lower mantle viscous flow has been enhanced as estimates of the lower mantle viscosity have decreased. It is now believed that there is little increase in viscosity from the upper mantle to the lower mantle or at most a moderate viscosity increase of a factor near 30 to 100. If viscosity increases further with depth, the effect will be to slow the convection velocities in the deepest mantle but not to prevent convection. A viscosity increase with depth in the lower mantle thus provides a mechanism for preserving primordial mantle material in an undifferentiated state, as is required by geochemical observations (Davies, 1984).

The D'' region is intimately associated with lower mantle dynamics because it is believed to be a thermal boundary layer arising from a strong temperature contrast between the core and mantle (e.g., Jeanloz and Richter, 1979). Estimates of the temperature at the base of the mantle extrapolated from the upper mantle assuming adiabatic conditions combined with estimates of the outer core temperature extrapolated from the melting temperature for iron at the inner core boundary lead to an estimated temperature drop across the CMB of 800–1000° (Stacey and Loper, 1983). This temperature increment reduces the viscosity at the base of D″ relative to the overlying mantle by a factor of more than 10^4. This boundary layer, which is perhaps 75 km thick, will inevitably become unstable, producing deep mantle plumes with about 20 km dimensions. These plumes may rise all the way through the mantle, producing hotspots. Also, the strong shear flow in the thin layer at the base of D″ may produce seismic velocity anisotropy that would affect diffracted waves. A critical factor in these calculations is the assumption that the D″ region is not compositionally stratified. If the evidence for seismic discontinuities at the top of D″ ultimately confirms the presence of a global layer, the dynamical structure of D″ may prove to be much different, perhaps with a double boundary layer system.

The strong lateral heterogeneity of D″ led Jordan to speculate in 1979 that D″ is compositionally stratified, analogous to the lithosphere, where lateral compositional gradients between oceans and continents persist despite dynamic convection processes. Hofmann and White (1982) have suggested that the D″ region is a storage cachement for chemically distinct material maintained by density differences, possibly including subducted oceanic crust, that eventually becomes unstable and rises in plumes. An interesting difference between D″ and the lithosphere is that the former boundary is hottest and softest on its surface, whereas the latter is coolest and most rigid on its surface.

Whatever the detailed nature of the D″ thermal boundary layer proves to be, heating from below is a critical aspect of the thermal convection system in the lower mantle. Whole mantle convection models incorporate this heat flux, as well as internal heating from radioactive decay in modeling the deep mantle dynamics. However, most thermal calculations have been conducted for simplified steady state geometries, which may not be applicable to the actual mantle situation. The conceptual model advanced by Davies (1984), which has heterogeneous blobs of material throughout the lower mantle but concentrated toward the lowermost mantle, appears most compatible with the range of seismic observations to date.

THORNE LAY

References

Cleary, J. R., 1974, The D″ region, *Physics Earth and Planetary Interiors* **9**, 13-27.

Creager, K. C., and T. H. Jordan, 1984, Slab penetration into the lower mantle, *Jour. Geophys. Research* **89**, 3031-3049.

Davies, G. F., 1984, Geophysical and isotopic constraints on mantle convection: an interim synthesis, *Jour. Geophys. Research* **89**, 6017-6040.

Doornbos, D. J., and J. C. Mondt, 1979, P and S waves diffracted around the core and the velocity structure at the base of the mantle, *Astron. Soc. Geophys. Jour. Royal* **57**, 381-395.

Dziewonski, A. M., 1984, Mapping the lower mantle: Determination of lateral heterogeneity in P velocity up to degree and order 6, *Jour. Geophys. Research* **89**, 5929-5952.

Dziewonski, A. M., and D. L. Anderson, 1981, Preliminary reference Earth model, *Physics Earth and Planetary Interiors* **25**, 297-356.

Hofmann, A. W., and W. M. White, 1982, Mantle plumes from ancient oceanic crust, *Earth and Planetary Science Letters* **57**, 421-436.

Jeanloz, R., and F. M. Richter, 1979, Convection, composition and thermal state of the lower mantle, *Jour. Geophys. Research* **84**, 5497-5504.

Jordan, T. H., 1979, Structural Geology of the Earth's Interior, *Natl. Acad. Science (USA) Proc.* **76**, 4192-4200.

Lay, T., and D. V. Helmberger, 1983, A lower mantle S wave triplication and the shear velocity structure of D″, *Royal Astron. Soc. Geophys. Jour.* **75**, 799-838.

Stacey, F. D., and D. E. Loper, 1983, The thermal boundary layer interpretation of D″ and its role as a plume source, *Physics Earth and Planetary Interiors* **33**, 45-55.

Wright, C., K. J. Muirhead, and A. E. Dixon, 1985, The P wave velocity structure near the base of the mantle, *Jour. Geophys. Research* **90**, 623-634.

Cross-references: *Core-Mantle Coupling; Earth's Core: Structure; Earth Structure: Global; Free Oscillations of the Earth; Mantle Convection and Plumes; Mantle Discontinuities; Mantle Dynamics; Mantle Viscosity; Seismic Tomography.*

MANTLE, UPPER: STRUCTURE

When Bullen defined his nomenclature for the layering of the Earth in 1940, he split the upper part of the mantle into two regions: layer B beneath the Moho, where seismic velocities vary relatively little as a function of depth, and layer C, from 400 to 1000 km depth, which was thought to be characterized by high gradients in V_p and V_s. Together, layer B and C formed what was commonly known as the Upper Mantle. In the 1960s, evidence accumulated that there is a second important worldwide discontinuity at a depth of 650 km or thereabouts, and it has become more practical to define this discontinuity as the lower boundary of the Upper Mantle. As such the term will be used in this section.

Most of what we know about the structure of the Upper Mantle has come from seismological observations. The upper part of the Mantle is the region where the properties vary most drastically as a function of depth. In fact, there are only two fields that, when measured at the surface of the Earth, provide insight into the depth dependence of physical characteristics in the Mantle: the seismic and the electromagnetic wavefield. In practice, seismic wave velocities at depth can be measured with a precision that is more than one order of magnitude better than the electrical conductivity.

For many years, seismologists have considered the Earth as a body with spherical symmetry, in which the seismic velocities and the density could be specified as a function of depth or radius only. The primary reason for this simplification was lack of sufficient data to reveal lateral heterogeneity. The primary apology was that symmetric Earth models served a very useful role, and continue to do so at present, in earthquake location and source mechanism determinations and in petrological modeling. This role is possible because the deviations from spherical symmetry are generally of second order, due to the strong influence of the pressure on the mineralogy of the rocks. From Fig. 1, in which the velocity-depth structure for a large number of regions is plotted, we see that the P-velocity in the Upper Mantle grows from 8 km/s at the Moho to almost 11 km/s at 700 km depth, an increase of some 35%, which is more than the horizontal component of the velocity gradient. The change between different geophysical provinces, over horizontal distances comparable to the same interval of 700 km, rarely exceeds 0.6 km/s or about 7%. The last in a continuing series of symmetric Earth models is the Preliminary Reference Earth Model (PREM, Dziewonski and Andersen, 1981). The PREM velocities in the Upper Mantle are shown in Fig. 2. The model shows several features that are believed to be worldwide.

Common Features of Upper Mantle Models

We have compiled a number of recently published Upper Mantle models for several regions (Table 1). This set of models is not homogeneous, and some care must be exercised in interpreting it. Models based on body waves are sensitive to the fine structure of discontinuities, and may sample both S- and P-wave velocities. Most body wave studies published in the last decade make use of synthetic seismograms, either qualitatively for short period waves or with some degree of waveform fitting for long period waves, which are less affected by scattering (but have a reduced resolution). Provided phases are correctly identified, the body wave travel times may give fairly accurate depth estimates to discontinuities, with a precision of some 10 km. Misidentifications may, however, lead to much larger errors. The fact that potentially dipping or curved structures are approximated by a plane-layered Earth adds another uncertainty, which is difficult to estimate.

Surface wave studies, on the other hand, give a very accurate averaged S-velocity, but are insensitive to the fine detail. Only very long period fundamental modes penetrate with useful resolution beyond 150 km depth. Most of the surface wave models compiled in Table 1 are therefore based on higher mode phase velocities.

The Low-Velocity Layer (LVL). The existence of a layer with a velocity decrease with depth was originally proposed by Gutenberg and definitely confirmed with surface wave dispersion studies by Press, Ewing, Dorman, and others around 1959. Because of the inherent ambiguity in the interpretation of body wave travel times in the presence of a low-velocity layer, surface waves are needed to study this feature. However, it should be kept in mind that the depth resolution of surface waves is poor, and that only the S-velocity can be determined. In general, it is possible to deliniate the top of the LVL with an accuracy of a few tens of kilometers and to find the drop in S-velocity in this channel. Only very low frequency (< 5 mHz) surface waves are able to penetrate into the rock below the LVL.

A large number of surface wave studies has shown the LVL to be a worldwide phenomenon for S

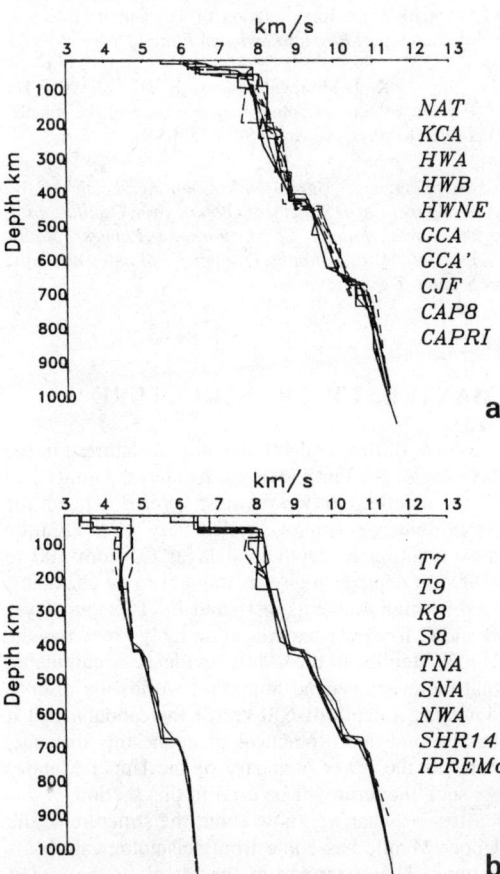

FIGURE 1. A compilation of a large number of regional models for compressional velocity V_P (a and b) or shear velocity V_S. (b) shows that, despite regional deviations of up to 10%, there is a strong increase of velocity with depth in the Upper Mantle. All models possess strong velocity gradients near 400 and 670 km (Figure courtesy B. L. N. Kennett).

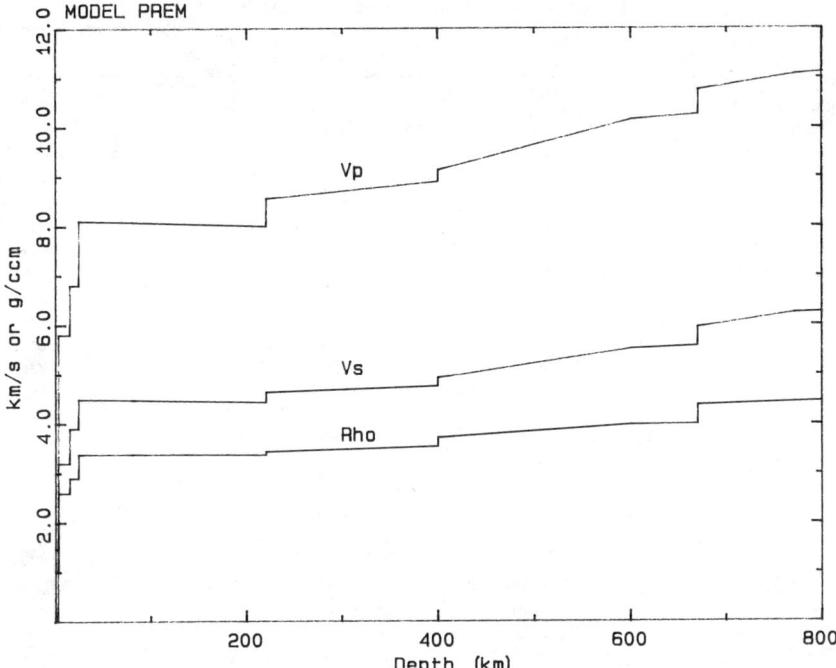

FIGURE 2. Velocities and density of the Preliminary Reference Earth Model (Dziewonski and Anderson, 1982) in the Upper Mantle.

waves, with the possible exception of the oldest shields. On the other hand, the LVL is most pronounced under very young oceanic lithosphere, such as under the East Pacific Rise. A compilation of LVL observations in different regions is given in Table 2. The depth of the LVL is here taken as the onset of a negative velocity gradient with depth.

Although attenuation of the shear modulus of the rock may start already far below the solidus, velocities in the LVL are so low in some areas that partial melting must be implied. It has been suggested that the LVL marks the boundary where the mantle is effectively decoupled from the lithosphere. In view of the widely different time scales involved in seismic and tectonic movements, however, a strict identification of the LVL with the bottom of the mechanically rigid lithosphere is unwarranted.

The Lehmann Discontinuity. Many seismologists have reported reflections or conversions from a discontinuity somewhere between 190 and 250 km depth. Early detailed studies on this feature in Europe and the western part of the United States were done by Lehmann, and it is after her that the discontinuity is often named. Anderson (1979) reviewed the evidence for the worldwide occurrence of this discontinuity, and proposed that it is due to a change in chemistry. A density difference would also imply that the Lehmann discontinuity marks a minor thermal boundary layer, which would explain the kink that is observed around this depth in pyroxene-derived geotherms.

The observations typically indicate a 3-4% jump in seismic velocities. The depth to this discontinuity varies considerably from region to region. Only very few of the recent models in Table 1 show a first order discontinuity that could qualify as the Lehmann discontinuity (Table 3). Some models, such as T7 for the western United States or NWA for the NW Atlantic have no distinct Lehmann discontinuity and could not satisfy the data if one was included.

This raises some doubt as to the worldwide occurrence of the Lehmann discontinuity as an Upper Mantle feature with a common cause and common characteristics. Since a velocity contrast of 3-4% is also observed for lateral heterogeneities, one cannot rule out that waves, scattered from more disorganized heterogeneities, have been misinterpreted.

Models without a Lehmann discontinuity do in general show a strong gradient in velocity to 400 km depth. Such gradients have been observed in western North America (T7), western Russia (KCA), and western Canada. However, other regions (central United States, Japan-Kurile arc and NE Australia) show weak or even negative gradients between the LVL and the 400 km discontinuity. Evidence for such deep-seated heterogeneities comes also from travel time differences of up to 15 seconds in ScS waves, and S-wave station anomalies (Jordan, 1979).

The 400 km Discontinuity. This discontinuity is also known as the 20° discontinuity, after a sudden change in slope of travel time curves of P-waves near 20°, where the slowness drops from more than 12

TABLE 1. Regionalized Upper Mantle Models

Model	Region	Reference	Data
Continental			
CAPRI	Central Australia	Leven, PEPI, v38, 9, 1985	Body waves
EB	East African Rift	Nolet & Mueller, Tect, v84, 151, 1982	Body and surface waves
GCA	W Mexico	Walck, GJRAS, v76, 697, 1984	Body waves
K8	NW Eurasia	Given & Helmberger, JGR, v85, 7183, 1980	Body waves
KCA	W Russia	King & Calcagnile, GJRAS, v46, 407, 1976	Body waves
M7	W Europe	Nolet, JG, v43, 265, 1977	Higher modes
S8	NE U.S.	Burdick, JGR, v86, 5926, 1981	Body waves
SCSH1	Scandinavia	Dost, Tect, v128, 289, 1986	Higher modes
SNA	N America, Canada	Grand & Helmberger, GJRAS, v76, 399, 1984	Body waves
T7	Western US	Burdick & Helmberger, JGR, v83, 1699, 1978	Body waves
TNA	Western US, Pacific	Grand & Helmberger, GJRAS, v76, 399, 1984	Body waves
USNE, USSW	NE and SW U.S.	Cara, GJRAS, v57, 649, 1979	Higher modes
WB	East Africa	Nolet & Mueller, Tect, v84, 151, 1982	Body and surface waves
WEPL1	W European platform	Dost, Tect, v128, 289, 1986	Higher modes
Oceanic			
ARC-TR	W Pacific, Kuriles	Fukao, GJRAS, v50, 621, 1977	Body waves
EPR-B	East Pacific Rise	Wielandt & Knopoff, JGR, v87, 8631, 1982	Surface waves
FBI-LVZ	Fiji Plateau	Frohlich et al., GJRAS, v50, 185, 1977	Body waves
NWA	NW Atlantic	Grand & Helmberger, JGR, v89, 11465, 1984	Body waves
PAC	Pacific	Cara, GJRAS, v57, 649, 1979	Higher modes

Note: The following journal abbreviations have been used: GJRAS = *Geophys. J. Roy. Astr. Soc.*, JGR = *J. Geophys. Res.*, PEPI = *Phys. Earth Plan. Int.*, Tect = *Tectonophysics*, JG = *J. of Geophysics*.

TABLE 2. The Low Velocity Layer (LVL)

Model	Depth	Lid velocity	Minimum velocity
V_s models			
EB	78	4.43	4.09
EPR-B	10	4.59	4.15
NWA	125	4.75	4.46
PAC	45	4.56	4.14
PREM	24	4.49	4.42
SCSH1	100	4.54	4.46
SNA	175	4.77	4.63
TNA	50	4.35	4.29
US-NE	85	4.75	4.47
US-SW	45	4.43	4.05
WB	55	4.52	4.47
WEPL1	100	4.52	4.28
V_p models			
ARC-TR	85	8.23	8.10
GCA	20	7.90	7.70
K8	140	8.34	8.04
PREM	24	8.11	7.99
T7	65	8.05	7.70
WB	55	7.97	7.69

TABLE 3. The Lehmann Discontinuity

Model	Depth	ΔV_p	ΔV_s
CAPRI	190–210	8.30–8.60	
EB	220		4.41–4.59
PREM	220	7.99–8.56	4.42–4.64

sec/degree at 18° to 10 sec/degree at 22°. This phenomenon was first reported by Byerly in 1926. In the 1930s, Birch and Bernal proposed that this indicates a transition zone with high pressure transformations of silicate minerals, and in 1939 Jeffreys suggested that the travel time curves near 20° could be explained by a discontinuity in mantle velocities at 413 km depth. The depth at which this discontinuity is reported from recent seismological observations remains fairly close to this value. The average depth for the body wave models compiled in Table 4 is 401 km with a standard deviation of 11 km, and extremes of 385 and 420 km.

Although precursors to P'P' have been identified as reflections from the underside of this discontinuity, these are somewhat scattered and rather weak, and reliable conversions have not been observed, which indicates that the transition is not sharp, at least not with respect to the wavelengths of seismic body waves (≈ 10 km). Numerical experiments show that the synthetic modelling of long period seismograms is too insensitive to discriminate between a first order discontinuity and a velocity increase spread over a depth interval of some 20 km. The sharpness shown by many of the models in Table 4 is therefore misleading.

The Transition Zone. A minor first or second order discontinuity has been reported in some studies near 500 km depth, but is absent in others. Again, it has been shown that waveform modelling of long period body waves is too insensitive to recognize this feature. However, to fit the observations of very high

TABLE 4. The 400 km Discontinuity

Model	Depth	ΔV_p	ΔV_s
ARC-TR	385–400	8.75–9.25	
CAPRI	406	8.80–9.33	
EPR-B	466		4.85–5.14
FBI-LVZ	375–395	8.40–9.60	
GCA	390	8.82–9.25	
K8	410–420	8.84–9.24	
KCA	420	8.66–9.27	
M7	340–400		4.68–4.98
PREM	400	8.90–9.13	4.76–4.93
SNA, TNA, NWA	405		4.78–5.00
S8	395–400	8.81–9.24	
T7	392–395	8.80–9.24	

TABLE 5. The 650 km Discontinuity

Model	Depth	ΔV_p	ΔV_s
ARC-TR	640–680	10.14–10.80	
CAPRI	610–645	9.80–10.55	
GCA	660	10.36–10.65	
K8	664–675	10.22–10.66	
KCA	690	10.38–10.80	
PREM	670	10.26–10.75	5.57–5.94
SNA, TNA, NWA	659		5.50–5.91
S8	650–655	10.25–10.76	
T7	665–669	10.38–10.80	

order Rayleigh modes, a gradient had to be introduced near 500 km in model M7. In this model, the gradient in V_s jumps quite suddenly from $5 \times 10^{-4} s^{-1}$ at a depth of 450 km to $4.5 \times 10^{-3} s^{-1}$ at 520 km.

The 650 Discontinuity. In 1965, Niazi and Anderson observed the changes in the slowness of P waves recorded with the Tonto Forest array in Arizona, for earthquakes with epicentral distances between 17° and 20°, in great detail. This was probably the first convincing observation of what is now called the *650 discontinuity* (sometimes also named *670 discontinuity*). This major feature in the upper Mantle has been observed worldwide (Table 5). The average depth for the models in Table 5 is 662 km, with a standard deviation of 17 km. The most notable outliers are central Australia (≈ 625 km in model CAPRI) and western Russia (690 km in model KCA), but all other models lie in the 650–670 km range, and when outliers are removed we find 662 ± 6 km.

Underside reflections appear as precursors to P'P' of sometimes remarkably high amplitude, and converted waves have been reported as well. The discontinuity is probably fairly sharp with respect to the seismic wavelength. No known univariant phase transition is able to reproduce the observed high reflection coefficients (Lees et al., 1983), and the fact that these coefficients show a high variability between different observations indicates that focussing and defocussing of the seismic waves, induced by a slight warping of the 650 discontinuity, is responsible for the observed amplitudes, not an abnormally high velocity contrast. If the velocity contrast for the PREM model is taken, reflection coefficients are 7% for P, 8% for S.

Anelasticity. The anelasticity of the Earth, represented by the intrinsic $Q_\mu(\mathbf{r})$ and $Q_\kappa(\mathbf{r})$ is very difficult to measure. Although a frequency dependence of Q is required on theoretical grounds, measurements are not precise enough to establish this correspondence, and for frequencies higher than about 1 Hz, Q may be approximately constant. The Q values for PREM are representative for global averages in the Upper Mantle and listed in Table 6. Considerable local variations exist in $Q_\mu(\mathbf{r})$. The radial normal modes are the only type of seismological data that are not consistent with a zero bulk loss, but the depth at which this loss is made is undetermined, and a $Q_\kappa < \infty$ in the Upper Mantle is not necessarily required. The LVL is also a region of low Q.

The Constitution of the Upper Mantle

Because of a combination of inadequate and sometimes conflicting data, both from seismology and from laboratory experiments, there is no full consensus about the nature of the Upper Mantle. Some crucial data on the velocities of seismic waves in high pressure compounds of likely mantle rocks are still unavailable.

The petrological models that have been proposed start from the plausible assumption that the composition of the primitive mantle must have been such that it yields a basaltic magma upon melting and then predict the mineral assemblages formed by the remaining residue as a function of depth or pressure from laboratory experiments. Such models are far from unique, and there is some freedom to adjust them to satisfy seismological or other constraints.

Most geophysicists believe the upper mantle to be largely peridotitic, such as the pyrolite model advocated by Ringwood (1975). These models are characterized by olivine [$(Mg,Fe)_2SiO_4$] as the dominant mineral, with roughly equal amounts of orthopyroxene [$(Mg,Fe)SiO_3$], clinopyroxene [diopside, $MgCaSi_2O_6$, and jadeite, $NaAlSi_2O_6$], and garnet [$CaMg_2Al_3Si_3O_{12}$].

However, as early as 1922, Goldschmidt proposed

TABLE 6. Anelasticity in PREM

Depth (km)	Q_μ	Q_κ
3–80	600	57822
80–220	80	57822
220–670	143	57822
>670	312	57822

TABLE 7. Mineral Assemblages

	pyrolite	piclogite
Olivine Ol	57	16
Orthopyroxene Opx	17	3
Clinopyroxene Cpx	12	46
Garnet Gt	14	37

an Earth consisting of a 120-km thick silicate layer on top of an eclogite mantle. Despite incessant attempts to discredit the eclogite mantle hypothesis, it is still around and its most recent advocate is D.L. Anderson (1984). He argues that the volume of the lunar crust, which represents 10% of the Moon, contrasts significantly with the 0.5% found for the Earth. If the Earth's mantle is as well differentiated as that of the Moon, this would result in a basalt layer with a thickness of several hundred kilometers. This layer is unstable, since basalt transforms to eclogite at a depth of several tens of kilometers. The high density (3.5 gr/ccm) eclogite sinks and accumulates at the 650 discontinuity, resulting in a mantle consisting of depleted peridotite on top of an eclogitic ("piclogite") layer. Both the Lehmann discontinuity and the 400 km discontinuity have been proposed as possible boundaries between these two sections. The model composition of the pyrolite (Ringwood, 1975) and piclogite (Bass and Anderson, 1984) mantle assemblages is given in Table 7.

In view of the worldwide occurrence of the 400- and 650-discontinuity, it is evident that a petrological model must not only fit the general velocity increase with depth, but also the sudden changes associated with these transitions. In laboratory experiments, the following phase transitions have been observed in the minerals listed in Table 7:

Olivine. At pressures roughly corresponding to the 400 km discontinuity, α-olivine transforms to β-modified spinel. According to Ringwood (1975), this transition would be spread out over a 30 km depth interval and most geophysicists consider this phase transition responsible for the seismologically observed discontinuity, even though the observed velocity increase (from 8.82 to 9.24 km/s in PREM) is small in comparison with laboratory measurements at room conditions, and would fit only a 40% Ol content. However, there are considerable uncertainties in the pressure derivatives of the shear modulus for the α and β phase, needed to extrapolate from room conditions to ambient pressure and temperature, and a pyrolite composition with about 60% olivine cannot be ruled out. A piclogitic mantle with only 16% Ol, on the other hand, predicts a velocity jump which is far too low.

β-spinel will transform to γ-spinel at about 500 km, with only a small change in physical properties. At depth corresponding to the 650 discontinuity, γ-spinel transforms to perovskite. The properties of perovskite are not known sufficiently to model the magnitude of the velocity jump. Spinel may also combine with stishovite to produce ilmenite, but this transition does not create a first or second order velocity discontinuity.

Pyroxenes. Opx transforms to a garnet-like phase, majorite, starting at about 300 km depth. This transition becomes really important between 350 and 400 km where the transformed Opx forms a garnet solid solution assemblage (garnetite). This transformation may be responsible for the high velocity gradients between 300 and 400 km, but the transformation is not sudden enough to satisfy seismological models of the 400 km discontinuity.

Cpx is stable to much higher pressures because it is hard to substitute Ca^{2+} for Al^{3+}, required for the transformation to majorite. This transition would give a high velocity gradient in the 400–600 km depth range.

For the upper 200 km of the mantle, the seismological data seem to point to a pyrolite-type composition. The disagreement between proponents of the pyrolite and piclogite mantle focuses mainly on the transition region between the 400 and 650 km discontinuity, and the nature of these discontinuities themselves (Weidner, 1985; Bass and Anderson, 1984). The $\alpha \rightarrow \beta$ transition is obviously an attractive candidate for the 400 km discontinuity. In a piclogite mantle, this discontinuity must be explained as a chemical boundary. On the other hand, reasonable assumptions for the elastic constants of majorite lead to rather high velocities near 500 km for a pyrolite mantle, and the high gradient between 400 and 600 km could be fitted with the Cpx$\rightarrow$Gt transition in the piclogite model.

On the basis of the presently available data, it is not possible to reject one model in favor of the other. Both models suffer the shortcoming that they are devised for a spherically symmetric Earth, and none of the models has a truly satisfactory explanation of the Lehmann discontinuity and the variability in its observations. There are, however, two models for the Upper Mantle that take lateral variability explicitly into account. These are Jordan's tectosphere model, and Vlaar's model of extensive lithospheric doubling.

The Tectosphere Model. As shown elsewhere in this volume (see *Lithosphere, Oceanic: Formation and Evolution*), the oceanic lithosphere can be modeled very well as a thermal boundary layer. Jordan (1979) points to a fundamental difficulty in extending this model to the old cratonic parts of the continents. In oceans, increasing water depth maintains isostatic balance when the lithosphere cools and its density increases. Continued cooling of the lithosphere of continental cratons would result in lithospheric thicknesses of up to 300 km. There is seismic evidence, notably from surface wave higher modes and ScS waves, that the oldest cratons exhibit a distinct, presumably cool, structure to this depth. To maintain isostatic balance, Jordan proposed that the old shields possess a "root," which consists primarily of peridotite, depleted in basaltic constituents by a melting process (10–20%). The depleted residue would be lighter, such that depletion compensates for the temperature effects. A similar model has been proposed by Oxburgh and Parmentier. This model gives a satisfactory explanation of those seismological observations that correllate strongly with the surface tectonics.

Lithospheric Doubling. In 1976, Vlaar and Wortel showed that only the oldest parts of the oceans subduct as deep as the 650 km discontinuity, whereas lithosphere with an age younger than ± 70 my will come to gravitational equilibrium *within* the Upper Mantle. Vlaar (1983) proposes that the opening and closing of fairly young oceanic basins has been a quite general phenomenon in plate tectonic processes of the recent past. Consequently, the Upper Mantle is expected to be littered with remnants of lithospheres that have been subducted quasi-horizontally, often including an oceanic ridge that may act as a mantle hotspot. The subducted lithosphere not only retains a specific petrological identity but remains a thermal marker for tens of million years. This theory is the only one for the constitution of the Upper Mantle that includes an explanation for the often large variability of the Lehmann discontinuity and the large scatter of seismic energy.

Lateral Heterogeneity

It is beyond dispute that much of the large scale heterogeneity reflected in models of the lithosphere and low velocity layer correlates with surface geology. Regions with tectonic activity have thin lithospheres and low V_s in the LVL (cf. models EB, US-SW, TNA, EPR-B). Lithospheric thickness and V_s is larger than average in shield areas (SCSH1, SNA). In oceanic areas, lithospheric thickness increases from very low values close to the ridge (EPR-B) to 125 km (NWA) and has been shown to correlate very well with the age of the ocean floor. These inferences are visible from the results of regional studies in Table 2, and corroborate a number of studies based on the great circle measurements of low frequency surface waves (for references, see Woodhouse and Dziewonski, 1984).

Limitations in available data and computing power have for a long time forced seismologists to construct their models for lateral heterogeneity on the basis of an assumed scheme of regionalization. The tomographic method has been applied in a number of early studies by Aki, Dziewonski, Romanowicz and others (see Nolet, 1987) without a priori assumptions but with rather limited resolution.

However, the growth in computing power as well as the accumulation of digital data from new global networks has brought a significant improvement in the precision and resolution of tomographic models. We now possess a number of regional and global Upper Mantle models in three dimensions that are independent of any a priori regionalization (Table 8). The global models listed are based on surface wave, normal mode data or even complete low-frequency seismograms. On a regional scale in the continents, body wave travel time delays are able to provide a very much increased resolution.

On the long end of the wavelength scale, model H1 shows that the global heterogeneity of the Earth has a strong degree 2 component (apart from the ellipticity) that correllates with the geoid. Models M84C and H4 agree with this observation and place the heterogeneity in the Upper Mantle.

The analysis of the more recent tomographic models listed in Table 8 confirms the results on Upper Mantle heterogeneity from regional studies: shields are fast, ridges are slow. Upper Mantle heterogeneity extends to at least 300 km and possibly deeper. This is illustrated fairly well by the global transect for model M84C in Fig. 3. The African continent has a root of high S velocity that extends to about 300 km. High velocities are also found in the western part of the Pacific and near the Brazilian passive margin, but although these velocity anomalies extend beyond the bottom of the lithosphere, they are much more limited in depth. The Atlantic, the young part of the Pacific and the deep part of the Indian Ocean beneath the Carlsberg ridge show low S velocities. These results clearly indicate that ridge systems are not merely "gaps" created by tensional forces that pull the lithospheres apart.

TABLE 8. Large Scale 3D Studies of the Upper Mantle

Model	Region	Data	Reference
H0	U.S.	*P*-delays	Romanowicz, GJ, v57, 479, 1979.
H1	Global	Normal modes	Masters et al., Nature, v298, 609, 1982.
H2	Eurasia	Surface waves	Feng and Teng, JGR, v88, 2261, 1983.
H3	S. California	*P*-delays	Humphreys et al., GRL, v11, 625, 1984.
M84C	Global	Full waveforms	Woodhouse and Dziewonski, JGR, v89, 5953, 1984.
H4	Global	Surface waves	Nataf et al., JGR, v91, 7261, 1986.
H5	Indian Ocean	Surface waves	Montagner, GRL, v13, 145, 1986
EUR85A	Europe	*P*-delays	Spakman, Geol. Mijnb., v65, 145, 1986.

Note: Models without an original name have been designated H0, H1, ...

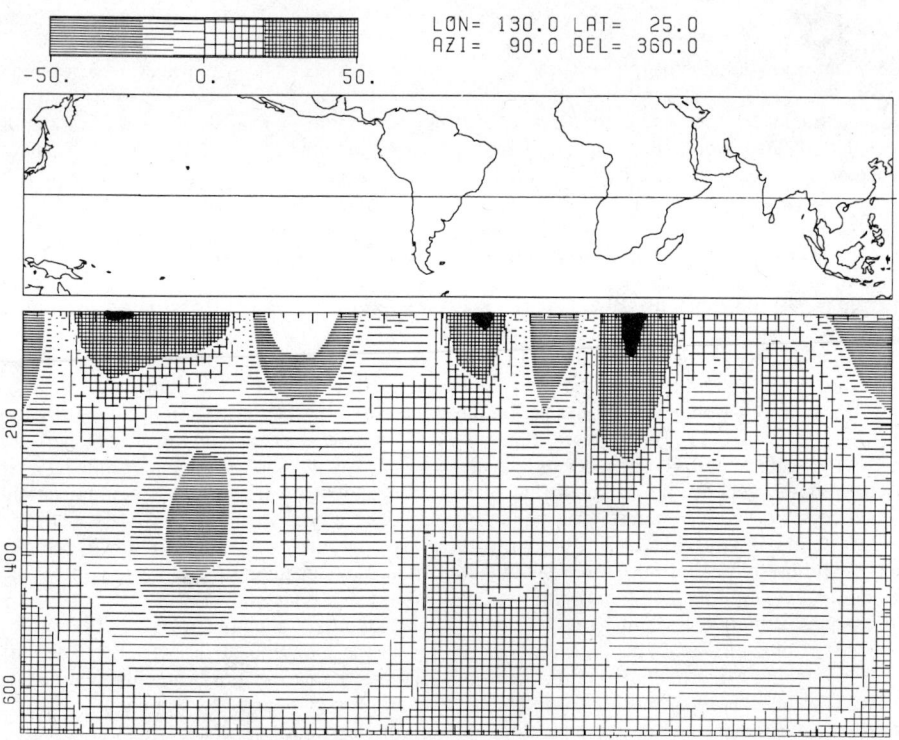

FIGURE 3. A global transect through model M84C showing deep lateral heterogeneity in the Upper Mantle. The legend gives the deviation in V_S in promilles with respect to the PREM model.

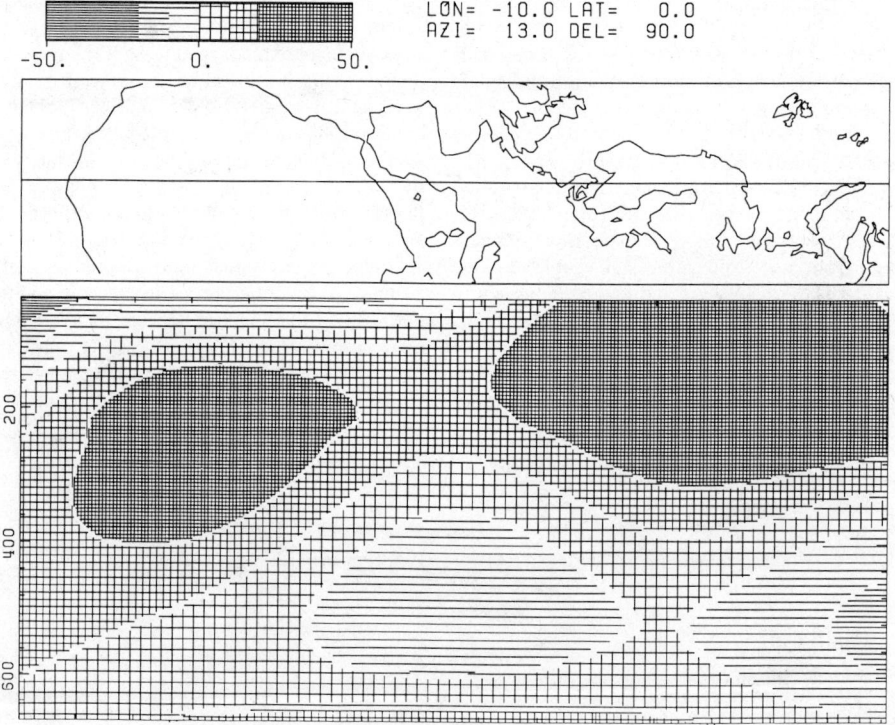

FIGURE 4. M84C shows deep, high velocity roots beneath the African and the Scandinavian Shield. The legend gives the deviation in V_S in promilles with respect to the PREM model.

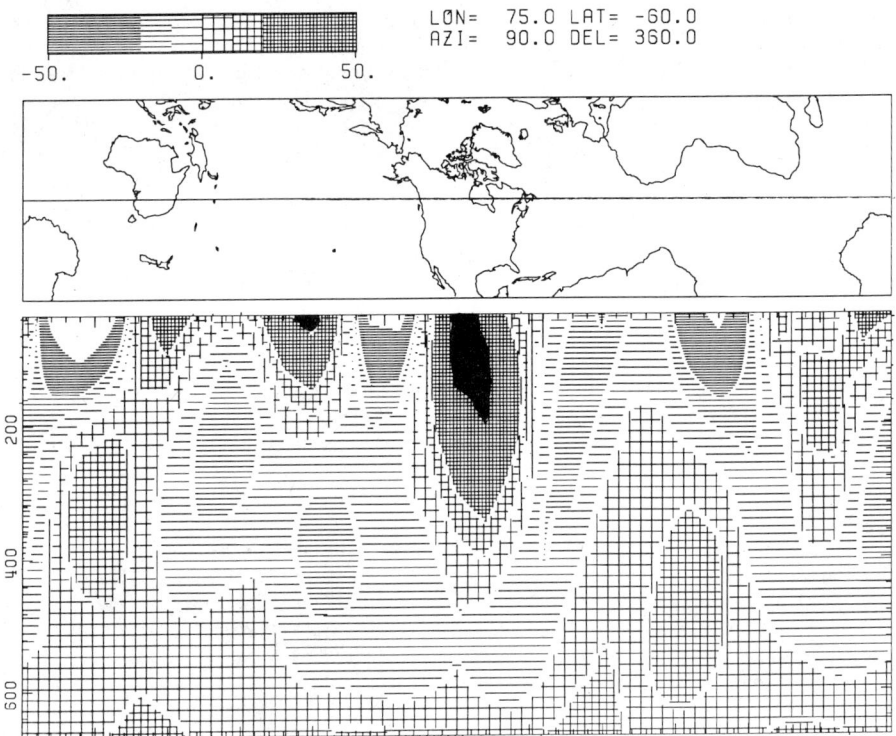

FIGURE 5. In M84C, the 300-km root beneath the Canadian Shield shows a much larger velocity deviation than the West Pacific lithosphere that subducts near the New Hebrides. The legend gives the deviation in V_S in promilles with respect to the PREM model.

Either the deeper regions are strongly affected by the surface rifting or the low velocity areas indicate a causal relationship between rifting and upper mantle conditions.

Figure 4 gives another illustration of the extent of deep heterogeneities. The Upper Mantle beneath the Scandinavian shield and the Barentz sea, as well as the deeper part of the Upper Mantle under west Africa, all show high velocities. The complexities in the Mediterranean area are not well resolved.

M84C has also a very deep root (300 km) with a high anomaly in S velocity ($>5\%$) under the North American shield (Fig. 5), which supports Jordan's hypothesis that the oldest cratons can withstand tectonic forces because they have a cold, strong root deep into the Upper Mantle. However, a look at the high-resolution models in Table 8, such as H3 and EUR85A, shows that not all features at depth are clearly correlated with surface tectonics.

The Fate of the Subducting Slab in the Upper Mantle

The subduction process involves the descent of cold and (generally) dense lithosphere into the mantle. In the mantle the subducted lithosphere is reheated by thermal conduction, adiabatic compression, shear heating and by latent heat of phase changes. The relevance of the subduction process for the structure of the Upper Mantle stems from the considerable magnitude of the velocity and density perturbations that are induced and maintained by subduction. For example, the low temperatures in the subducting slab cause anomalies in seismic velocity that may amount to 3-5% of the ambient mantle structure. Through the temperature-dependence of the slab's rheology the relatively cold descending lithosphere is capable of supporting higher stresses than the surrounding mantle, which is a factor of primary importance in understanding why intermediate-depth and deep earthquakes only occur in subducted lithosphere.

The fate of the subducting slab in the Upper Mantle is part of the problem of the depth extent of mantle convection. Current ideas on mantle convection can be simplified and summarized in terms of two models, the whole-mantle convection model and the layered-mantle convection model (see Peltier and Jarvis, 1982, for a review). In the latter model the Upper Mantle is assumed to convect separately from the deeper mantle, whereas such a separation is not assumed in the whole-mantle model. Attempts to discriminate between the two models focus on the question whether or not there is a barrier to slab penetration (into the lower mantle) at a depth approximately coinciding with the 650-discontinuity. Such a barrier, if it exists, would obviously play an important role in the physical and chemical evolution of the Upper Mantle.

To be able to assess the perturbing effects of subduction we briefly consider the initial presubduction structure of the subducting lithosphere (as subducting lithosphere is predominantly of oceanic nature we will concentrate on subduction of oceanic lithosphere): The temperature distribution in oceanic lithosphere resembles that of a thermal boundary layer with an error function type of depth-dependence. The surface (ocean bottom) temperature is near 0° C and the asymptotic value is between 1200° C and 1400° C. Partial melting and differentiation of mantle material underneath the spreading ridge gives rise to a chemically zoned lithosphere with probably, below the basaltic oceanic crust, residual harzburgite and lherzolite. As it spreads away from the ridge axis the lithosphere thickens by cooling.

Subduction Zone Seismicity. The location and geometry of subducted lithosphere is traditionally inferred from the Wadati-Benioff zones, the inclined zones of intermediate-depth and deep earthquakes (see Isacks and Molnar, 1971). From numerous studies of the seismicity in all subduction zones, the following observations appear to be quite general:

1. The maximum focal depth of earthquakes is about 680 km. (In several subduction zones, however, the maximum depth is considerably less.)
2. The focal mechanisms of deep earthquakes very consistently indicate downdip compression.
3. The pattern of seismic energy release in deep earthquake zones often shows a minimum at about 300–350 km and a relative maximum in the depth range of about 500–680 km.

The most straightforward explanation of these observations is that the subducting slab meets strong resistance at the depth of about 650 to 700 km, which prevents it from penetrating into the lower mantle. A phase change with a strongly negative slope of the Clapeyron curve ($dP/dT < 0$), a high viscosity contrast and a chemical boundary would be possible causes for such a resistance. Alternatively, the observations are considered in the light of thermal assimilation of the subducting lithosphere.

In early analyses (around 1970) both McKenzie and Griggs suggested that the maximum depth of seismic activity is controlled by a critical temperature above which earthquakes can not be generated. Although their studies lent qualitative support to this hypothesis, their analyses were based on limited data and hampered by great uncertainties in various geophysical parameters. Vlaar and Wortel established that the maximum depth of seismicity in subduction zones increases with the age of the descending lithosphere. Later, in 1982, Wortel used the by that time available information on the oceanic lithosphere (age distribution and age-dependence of the thermal structure) to show that the maximum focal depths in subduction zones are indeed in agreement with the existence of a critical temperature (T_{cr}), for which he found $T_{cr} = (0.54 \pm 0.04).T_m$, where T_m is the composite mantle solidus as determined by Stacey in 1977 (see Fig. 6). In the meantime, it was shown by O'Connell and by Richter that the second observation does not provide a strong constraint on the presence of a barrier.

Whereas the first observation was shown to be in agreement with thermal assimilation, the combination of the first and the third observation (absolute cessation of seismic activity at about 680 km depth and a relative maximum in seismic energy release in the 200 km depth range above this depth) are often cited as strong evidence against thermal assimilation and in favor of a barrier to slab penetration. The model recently proposed by Wortel (1986), however, appears to account for both the cut-off and the relative maximum in energy release. By incorporating the interaction of depth-dependent forces and rheology into the model it turns out that the seismicity observations can be explained by thermal assimilation. A barrier to slab penetration or a special role played by a phase change at this depth is not required. In conclusion, it seems no longer warranted to dismiss slab penetration into the lower mantle on the basis of observations regarding deep earthquakes in subducted slabs.

Aseismic Continuation of Subducted Slabs. The critical temperature T_{cr} for earthquake generation is several hundreds of degrees Celsius below the

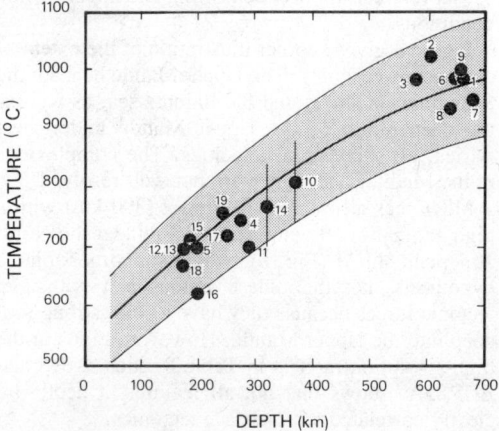

FIGURE 6. Minimum temperatures at depths of deepest earthquakes in subduction zones, indicated by dots. 1. Kuriles, 2. Northern Honshu, 3. Izu-bonin, 4. Ryukyus, 5. Sumatra, 6. Java, 7, 8. Tonga, 9. Kermadec, 10, 11. New Zealand, 12, 13, 14. Chile, 15. North-central Peru, 16, 17. Central America, 18. Alaska, 19. Aleutians. The thick solid curve represents T_{cr} as a function of depth, with its uncertainty range (stippled area).

ambient mantle temperature. Even at the depths of the deepest earthquake foci thermal modelling indicates a temperature difference of at least 500–600°C. This means that beyond the location of the deepest earthquakes in a slab we may expect to find a continuation (with some unknown geometry) with lower temperatures than the ambient mantle.

Jordan made a direct search for such a aseismic continuation in 1977. He studied travel time anomalies (for S waves) on the unit focal sphere for a deep event in the Kuriles-Kamchatka arc (Sea of Okhotsk). From this study and from similar studies (with P waves) by Creager and Jordan, in the same arc and in the Japanese and Mariana subduction zones, it was concluded that slab penetration of at least 300 km below the 650-discontinuity is required to fit the travel time data (see Creager and Jordan, 1986). The dip of the aseismic continuation was found to be somewhat greater than that of the seismically active part in the Upper Mantle. From their data and their thermal model Creager and Jordan also determined an effective thermal coefficient of velocity $\partial V_p/\partial T$ of -0.5 m/s/K if Upper Mantle phase changes are not included in the slab model, and of -0.4 m/s/K if they are. These values are consistent with laboratory measurements.

A direct illustration of the discrepancy between the maximum depth of a seismic zone and the penetration depth of the corresponding subducted slab is given in a tomographic study of the Upper Mantle structure beneath the Mediterranean region by Spakman (1986). Figure 7 displays his results for the structure beneath the Hellenic arc and the Aegean Sea, with a high-velocity slablike anomaly extending all the way to the bottom of the investigated region. The thickness of the anomaly is not to be identified

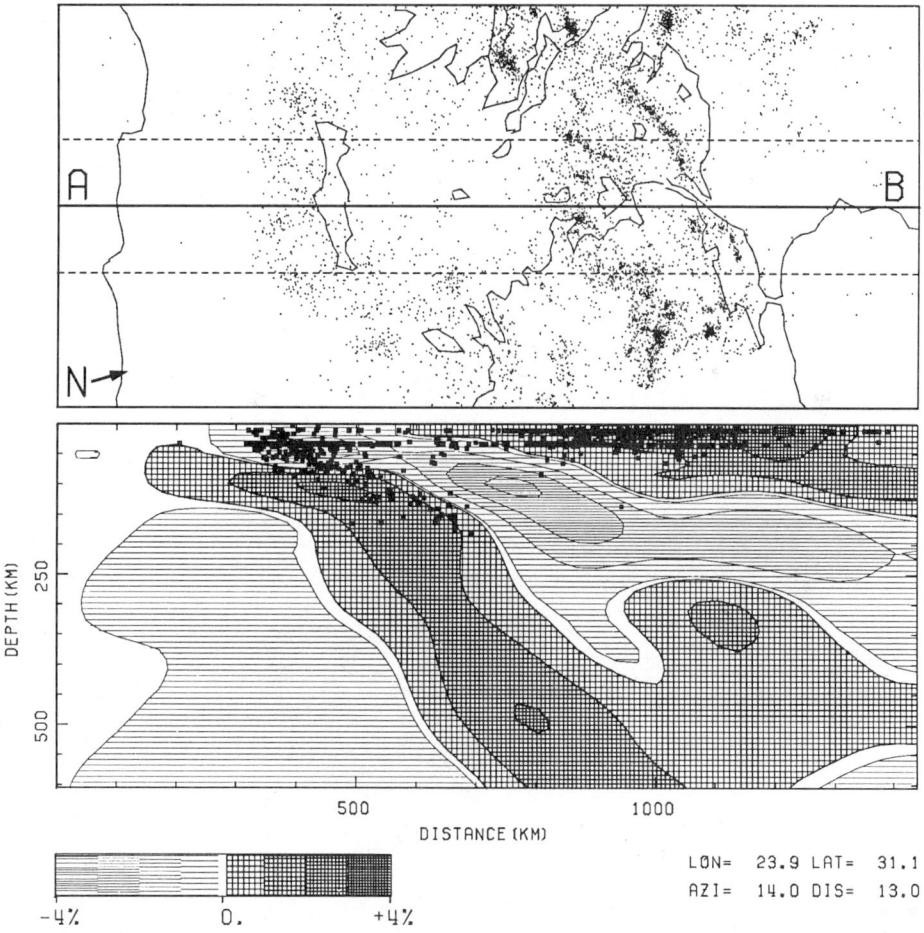

FIGURE 7. A cross section through the Aegean Upper Mantle, along great circle segment A (North Africa)—B (Europe). Dots in the upper part of figure indicate epicenters of earthquakes reported by PDE for the years 1964–1984. The hatching in the lower part of the figure indicates inferred V_P anomalies measured in percents with respect to Jeffreys-Bullen model. Dots indicate the projections on the plane of the cross section of the hypocenters of events located between the two dashed lines shown in upper part of the figure. (Figure courtesy of W. Spakman)

with the thickness of the subducted slab. The penetration of cold material also cools the surrounding mantle, which tends to smear out the seismic velocity anomaly.

In the studies cited earlier the subducted lithosphere was older than 100 my (presubduction age). For the thick, cold, and dense lithosphere of this age, slab penetration below the 650-discontinuity appears to be a possibility. Subducted lithosphere with a presubduction age less than, say, 90–100 my may well settle in the Upper Mantle. Vlaar (1983) suggested that very young lithosphere (< 30 my), if it is subducted, resides in the Upper Mantle immediately below the overriding plate.

Disintegration of Subducting Lithosphere. It is generally thought that oceanic lithosphere subducts coherently. In fact, this was implicitly assumed in the present discussion so far. Several authors, however, have formulated hypotheses in which disintegration of the subducting lithosphere plays a central role (e.g., Ringwood, 1975, 1982; Hofmann and White, 1982). A key element in these hypotheses are the intrinsic densities of the originally chemically zoned lithosphere and the resulting nonuniform distribution of density in a cross section of a descending slab.

Ringwood (1982) presents an extensive discussion of phase changes in subducted lithosphere. In particular, he draws attention to the density (and the density contrast with respect to the mantle) as a function of depth for the various "layers" of the chemically zoned subducting slab, that is (in his model) for the basaltic crust, the residual harzburgite, the residual lherzolite, and the depleted pyrolite. For the lower part of the lithosphere (depleted pyrolite) the density contrast with respect to the surrounding pyrolite mantle is very small. The negative buoyancy forces are concentrated in the colder upper part of the lithosphere. Ringwood (1982) suggests that this may cause the lower lithosphere to be stripped off gradually and resorbed in the Upper Mantle. Furthermore, after subduction, the former oceanic crust transforms into eclogite, which remains denser than pyrolite at all depths. After several phase changes, however, the former harzburgite layer becomes buoyant with respect to the mantle at depths below the 650 discontinuity. In Ringwood's hypothesis the consequences of this dynamic situation are as follows: the relative buoyancy of the former harzburgite below about 600 km forces the lower part of the slab into compression. Buckling and severe deformation are likely to occur, the net result being the formation of a large "megalith" of relatively cool but deformed oceanic

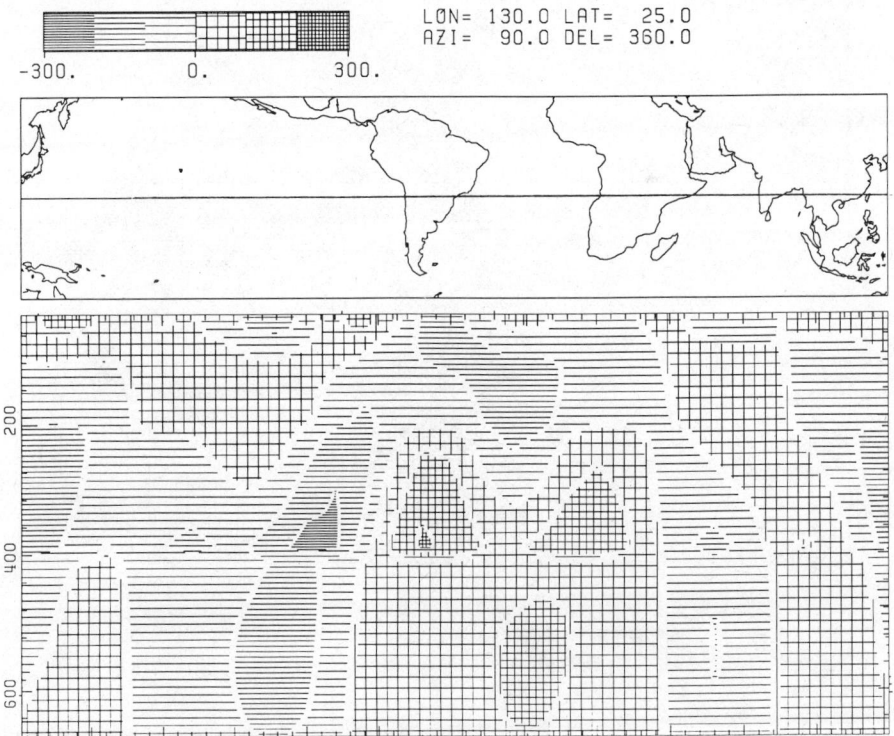

FIGURE 8. The anisotropy parameter ξ, defined as $(V_{SH}/V_{SV})^2 - 1$ is indicative for vertical flow ($\xi < 0$) or horizontal flow ($\xi > 0$). This global transect for model H4 shows, among other things, a horizontal flow pattern in the Pacific LVL and a vertical flow pattern near the South American and Indonesian subduction regions. The legend gives $\xi \times 10^3$.

lithosphere at the base of the Upper Mantle or just below it. It should be noted that the postulated buckling phenomena are not "visible" in the seismic activity of deep earthquake zones.

Hofmann and White (1982) developed a somewhat related hypothesis. They emphasize the high density of the former oceanic crust after it is transformed into eclogite. Unlike Ringwood they expect the eclogite to separate from the lithosphere and to sink to a certain level deep in the mantle, which may be the core-mantle boundary.

The nature of mantle perturbations arising from slab disintegration along these lines would be very much different from that of the thermally induced anomalies considered earlier. Whereas the latter have a transient nature and disappear after prolonged reheating, the anomalies arising from intrinsic chemical density differences would be gradually accumulating.

The fundamental dynamical problems associated with this class of hypotheses, however, are yet to be investigated. A first step into this direction was made by Gurnis (1986) who performed numerical experiments on the effects of chemical density differences on convective mixing in the mantle. So far, his conclusion is that it is unlikely that intrinsic density of subducting oceanic lithosphere can have much effect on the dynamics of the mantle and therefore on the chemical evolution of the mantle.

Seismic Anisotropy in the Upper Mantle

Although the existence of seismic anisotropy in the oceanic lithosphere is undisputed (see Crampin et al., 1984, for a review), the existence of seismic anisotropy in the LVL or even deeper regions is more questionable. Love and Rayleigh phase velocities have been reported as inconsistent for oceanic regions unless about 3% transverse isotropy was assumed. The PREM model shows this kind of anisotropy to 200 km depth. Olivine is anisotropic, and if crystal orientation is governed by large scale strain fields, we may expect some consistency in the direction dependence of seismic velocities. This anisotropy should not necessarily be of the transverse type. Indeed, there is petrological evidence that the (fast) a-axes of olivine crystals will cluster around the flow direction.

Tanimoto and Anderson, who mapped the direction dependence of Rayleigh phase velocities at very low frequencies, detected a good correlation between fast phase velocities and the direction of mantle return flow as predicted from kinematic considerations. Model H4, which explicitly included transverse isotropy in the model paramaterization, shows that $V_{SV} > V_{SH}$ for the youngest and oldest oceanic Upper Mantle between depths of 200 km and 400 km (indicative of vertical flow), but that intermediate oceanic Upper Mantle has $V_{SH} > V_{SV}$ (Fig 8). However, most of the evidence for seismic anisotropy in or below the LVL is still fragile. If it is important, it may have a profound influence on the results of studies that now interpret seismic data in terms of lateral heterogeneity alone.

G. NOLET
M.J.R. WORTEL

References

Andersen, D. L., 1979, The deep structure of continents, *Jour. Geophys. Research* **84**, 7555-7560.

Anderson, D. L., 1984, The Earth as a planet: paradigms and paradoxes, *Science* **223**, 347-355.

Bass, J. D., and D. L. Anderson, 1984, Composition of the Upper Mantle: geophysical tests of two petrological models, *Geophys. Research Letters* **11**, 237-240.

Crampin, S., E. M. Chesnokov, and R. G. Hipkin, 1984, Seismic anisotropy—the state of the art II, *Royal Astron. Soc. Geophys. Jour.* **76**, 1-16.

Creager, K. C., and T. H. Jordan, 1986, Slab penetration into the lower mantle beneath the Mariana and other island arcs of the Northwest Pacific, *Jour. Geophys. Research* **91**, 3573-3589.

Dziewonski, A. M., and D. L. Anderson, 1981, Preliminary reference Earth model, *Physics Earth and Planetary Interiors* **25**, 297-356.

Gurnis, M., 1986, The effects of chemical density differences on convective mixing in the Earth's mantle, *Jour. Geophys. Research* **91**, 11407-11419.

Hofmann, A. W., and W. M. White, 1982, Mantle plumes from ancient oceanic crust, *Earth and Planetary Sci. Letters* **57**, 421-436.

Isacks, B., and P. Molnar, 1971, Distribution of stresses in the descending lithosphere from a global survey of focal mechanism solutions of mantle earthquakes, *Rev. Geophysics Space Physics* **9**, 103-174.

Jordan, T. H., 1979, Structural geology of the Earth's interior, *Natl. Acad. Sci. (USA) Proc.* **76**, 4192-4200.

Lees, A. C., M. S. T. Bukowinski, and R. Jeanloz, 1983, Reflection properties of phase transition and compositional change models of the 670 discontinuity, *Jour. Geophys. Research* **88**, 8145-8159.

Nolet, G., ed., 1987, *Seismic Tomography, with Applications in Global Seismology and Exploration Geophysics*. Dordrecht: Reidel, 387p.

Peltier, W. R., and G. T. Jarvis, 1982, Whole mantle convection and the thermal evolution of the earth, *Physics Earth and Planetary Interiors* **29**, 281-304.

Ringwood, A., 1975, *Composition and Petrology of the Earth's Mantle*. New York: McGraw-Hill, 618p.

Ringwood, A., 1982, Phase transformations and differentiation in subducted lithosphere: implications for mantle dynamics, basalt petrogenesis, and crustal evolution, *Jour. Geology* **90**, 611-643, 1982.

Spakman, W., 1986, Subduction beneath Eurasia in connection with the Mesozoic Tethys, *Geol. Mijnbouw* **65**, 145-153.

Vlaar, N. J., 1983, Thermal anomalies and magmatism due to lithospheric doubling and shifting, *Earth and Planetary Sci. Letters* **65**, 322-330.

Vlaar, N. J., and M. J. R. Wortel, 1976, Lithospheric aging, instability and subduction, *Tectonophysics* **32**, 331-351.

Weidner, D. J., 1985, A mineral physics test of a pyrolite mantle, *Geophys. Research Letters* **12**, 417-420.

Woodhouse, J. H., and A. M. Dziewonski, 1984, Mapping the Upper Mantle: Three dimensional modeling of Earth structure by inversion of seismic waveforms, *Jour. Geophys. Research* **89**, 5953-5986.

Wortel, R., 1986, Deep earthquakes and the thermal assimilation of subducting lithosphere, *Geophys. Research Letters* **13**, 34-37.

Cross-references: *Continental Lithosphere; Deep Earthquakes; Deep Seismic-Reflection Profiling; Earth Structure: Global; Lithosphere: Mechanical Properties; Lithosphere, Oceanic: Formation and Evolution; Mantle, Lower: Structure; Mantle Convection and Plumes; Mantle Discontinuities; Mantle Dynamics; Mantle Viscosity; Ocean-Continent Transition: Structure; Seismic Anisotropy in the Earth; Seismicity: Subduction Zone; Seismic Tomography; Surface Waves.*

MANTLE CONVECTION AND PLUMES

The term "mantle convection" refers to the slow, but organized, subsolidus creeping motion in the mantle, driven by buoyancy forces. These forces arise primarily because of thermal expansion and contraction associated with horizontal temperature differences, although compositionally and mineralogically produced density differences contribute in certain environments.

The subject of mantle convection developed in step with the larger subject of geodynamics. Like so many topics in Earth sciences, particularly those in which direct observation is impossible, it was first regarded as outlandish, then passed through a phase of intense debate about its existence, and then on to the present phase, in which it has become a fully accepted paradigm. Indeed, it is now recognized as the primary physical process driving the evolution of the solid Earth. It is important to remember that this elevated status did not come about overnight. It required a couple of generations of research activity on the mechanics of thermal convection, plus several timely breakthroughs in geophysics, most notably the plate tectonic revolution. Even so, we do not have anything approaching a complete picture of convection in the mantle, and, in particular, its vertical structure is still largely undetermined. Curiously, this lack of definitive information does not appear to have jeopardized the paradigm in the least; if anything, it seems to be a source of strength, allowing individuals to conceive of it in ways that suit their own applications.

Arthur Holmes was perhaps the first prominent Earth scientist to recognize the importance of convection in Earth history, both as a source of power for tectonic activity and as a means of removing heat from the deep interior. It was Alfred Wegener's continental drift proposal that brought the idea to the forefront, and over the ensuing decades it was championed by a small minority of visionary geophysicists such as S. K. Runcorn and J. Verhoogen. The beginning of universal acceptance came with the advent of plate tectonics. Plate motions require an enormous mass transfer into and out of the mantle. At the present time, on average, 3 km^2 of oceanic lithosphere is formed each year at spreading centers, and approximately an equal amount is subducted back into the mantle at trenches. Accompanying this circulation is an advective heat transfer amounting to nearly 3×10^{13} W, about 75% of the Earth's estimated heat loss. It was immediately recognized that the process of subduction, in which relatively cold slabs sink under their own negative buoyancy to at least 670 km depth, is an extremely efficient mechanism for removing heat from the interior. In this way plate tectonics can be viewed as the surface expression of a convective heat engine operating in the mantle.

Aside from providing the driving force for plate tectonics, there are three major reasons why convection has become accepted generally. First, our knowledge of mantle rheology indicates that subsolidus creep readily occurs at mantle temperatures (see *Mantle Viscosity*). Second, the concept of convection has provided an explanation for recently acquired data on mantle structure and composition coming from such diverse fields as seismology, gravity, mineral physics, and geochemistry, which would otherwise appear to be quite unrelated. And finally, the theory of thermal convection, applied to the mantle using the best available estimates for critical parameters, indicates that the mantle should be in a state of vigorous, fully developed convective motion.

Convective Instability

A quantitative description of mantle convection begins with an analysis of conditions necessary for motion, which also serves to define the important dimensionless parameters. The prototype problem is the onset of convective motion in a layer of viscous fluid heated from below and cooled from above, so-called Rayleigh-Benard convection. As a first approximation, we idealize the mantle as homogeneous with uniform properties except for density differences as they contribute to buoyancy, and ignore all compressibility effects. These simplifications constitute the *Boussinesq* approximation, which is the starting point for the description of mantle convection. Later, we shall relax some of these restrictions. Using the Boussinesq approximation, conservation of linear momentum, mass, and heat can be written as

$$\rho_0 \frac{d\mathbf{u}}{dt} = -\nabla P + \rho \mathbf{g} + \mu \nabla^2 \mathbf{u} \quad (1)$$

$$\nabla \cdot \mathbf{u} = 0$$

$$\rho_0 C_p \frac{dT}{dt} = k\nabla^2 T + H \quad (2)$$

In these equations **u** is the velocity of the material, t is time, P is pressure, $\mathbf{g} = -g\hat{z}$ is acceleration of gravity, and T is temperature. Material properties are dynamic viscosity μ, specific heat at constant pressure C_p, thermal conductivity k, heat source concentration H (due to radioactive decay), density ρ, and spherically averaged density ρ_0. The terms in the momentum balance are the inertia, pressure, gravity, and viscous forces, respectively. Terms in the heat equation are the rate of change of heat content, heat conduction, and heat production by radioactive decay, respectively.

The hydrostatic pressure variation can be removed by writing pressure and density as the sum of a *static* part and a *dynamic* part:

$$P = P_0 + P_1, \quad \rho = \rho_0 + \rho_1 \quad (3)$$

Density variations are related to temperature variations according to

$$\rho_1 = -\rho_0 \alpha T \quad (4)$$

where α is the thermal expansion coefficient. Substituting Eqs. 3 and 4 into Eq. 1 yields the Navier-Stokes equation of motion

$$\rho_0 \frac{d\mathbf{u}}{dt} = -\nabla P_1 + \rho_0 \alpha g \hat{z} T + \mu \nabla^2 \mathbf{u} \quad (5)$$

It is instructive to now introduce dimensionless variables. The basic length, time, and temperature scales are the convecting layer depth D, the thermal diffusion timescale $D^2/K = \rho_0 C_p D^2/k$, where K is thermal diffusivity, and ΔT, the temperature difference between bottom and top boundaries. Then Eqs. 5 and 2 become

$$\frac{1}{Pr} \cdot \frac{d\mathbf{u}'}{dt'} = -\nabla' P' + Ra_T \hat{z} T' + \nabla'^2 \mathbf{u}' \quad (6)$$

$$\frac{dT'}{dt'} = \nabla'^2 T' + \frac{Ra_T}{Ra_H} \quad (7)$$

the primes denoting dimensionless variables. There are three dimensionless parameters, the *Prandtl number*

$$Pr = \frac{\nu}{K}$$

the ratio of kinematic viscosity ($\nu = \mu/\rho$) to thermal diffusivity, and two *Rayleigh numbers*

$$Ra_T = \frac{\alpha g D^3}{K\nu} \Delta T \quad (8)$$

and

$$Ra_H = \frac{\alpha g D^5}{kK\nu} H \quad (9)$$

The Prandtl number for the mantle is astronomical—approximately 10^{23}—which makes inertial effects strictly negligible. The Rayleigh numbers are dimensionless measures of buoyancy forces, arising from the vertical temperature gradient and heat production by radioactive decay, respectively. The character of mantle convection is largely determined by the magnitude of these parameters.

The conditions for onset of convective motion are determined by considering the following thought experiment. Suppose that the temperature difference between bottom and top boundaries of a fluid layer is increased steadily with time, but slowly enough so that the interior temperature profile remains essentially in conductive equilibrium. The vertical density gradient becomes increasingly negative, and hence the fluid layer becomes progressively less stable. Eventually the adverse density gradient becomes large enough so that small disturbances in the fluid are accelerated by buoyancy forces sufficiently to overcome the stabilizing influence of viscous and thermal diffusion. Instability sets in the form of spatially periodic cellular flow, called *Rayleigh-Benard convection*. The horizontal or *planform* depends on the Prandtl number, boundary conditions, and amount of internal heating. Allowable symmetric planforms are those shapes capable of filling a planar surface, including hexagons, equilateral triangles and rectangles, with *two-dimensional rolls* being the limiting case of a long rectangle. Planforms commonly realized in laboratory experiments are shown in Fig. 1.

The *critical Rayleigh number*, which defines the condition of *neutral stability*, can be found by introducing small-amplitude disturbances, along with the basic (undisturbed) conductive temperature profile into Eq. 6 and 7. This results in a set of homogeneous linear equations for each *normal mode* of the disturbance, with the eigenvalue being the critical Rayleigh number. The simplest case relevant to mantle convection is instability of a layer with free-slip (zero shear stress) boundaries. In the presence of basal heating only ($Ra_H = 0$), the critical Rayleigh number can be found analytically to be

$$Ra_c = \frac{(\pi^2 + a^2)^3}{a^2}$$

where a, the wave number of the basic periodicity, is related to Δ, the cell aspect ratio, by $\Delta = \pi/a$. The *minimum* critical Rayleigh number occurs for $a = \pi/\sqrt{2}$ and is approximately 657. Thus, convection in the mantle is expected if $Ra > 10^3$. Conditions in the mantle, though, are far more complex than the homogeneous fluid layer considered here, but

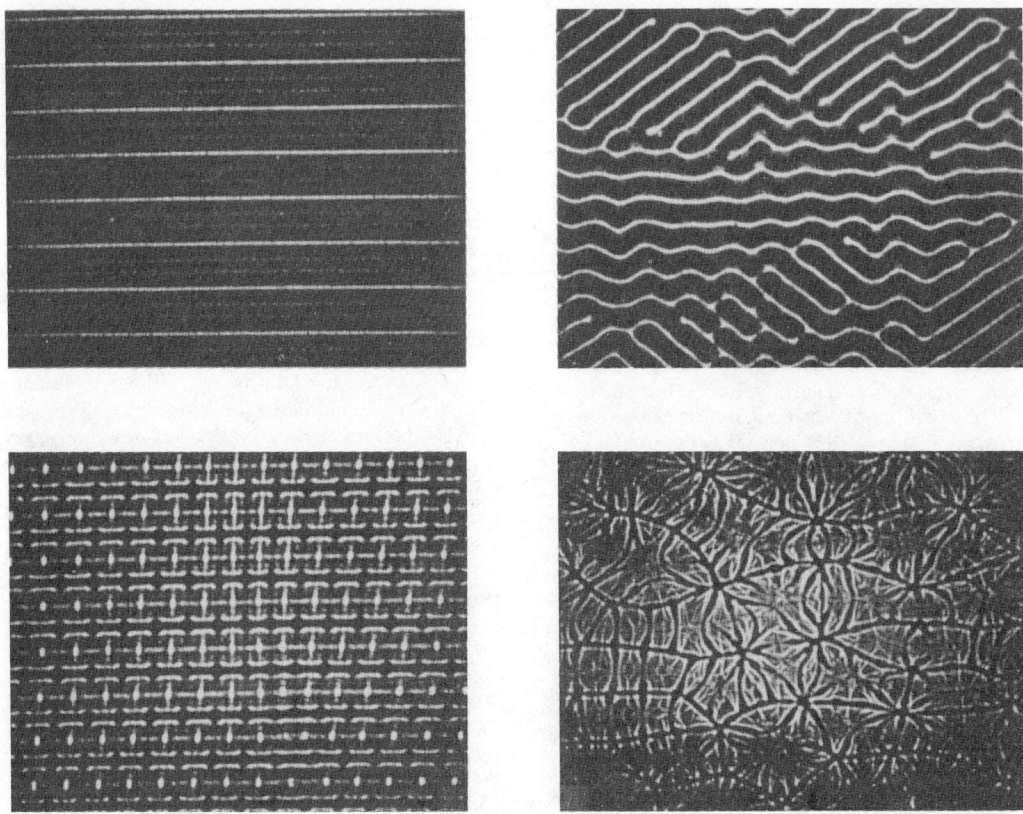

FIGURE 1. Shadowgraph images of Rayleigh-Benard convection planforms in a high Prandtl number fluid: (top left) two-dimensional rolls at $Ra_T = 2 \times 10^4$; (top right) zigzag instability of aspect ratio 1.4 rolls at $Ra_T = 3 \times 10^4$; (bottom left): bimodal convection at $Ra_T = 4 \times 10^4$; (bottom right) spoke pattern convection at $Ra_T = 2.4 \times 10^5$. (From Busse and Whitehead, 1971)

stability analyses of this type done for a wide variety of geometries (including spherical shells), boundary conditions, and modes of heating all give critical Rayleigh numbers of the same order of magnitude, which gives us some confidence that it represents a useful criterion for judging the stability of the mantle.

The mantle Rayleigh number can be crudely estimated by taking $\Delta T \approx 2000\ K$, the sum of the superadiabatic temperature rise across the lithosphere, 1300 K, plus 700 K distributed deeper in the mantle, along with $\alpha = 1 \times 10^{-5}\ K^{-1}$, $g = 10$ m · s^{-2}, $K = 10^{-6}$ m^2 · s^{-1}. We then need values for the convecting layer depth and the average effective viscosity. Classical studies of postglacial rebound indicate $\nu \approx 3 \times 10^{17}$ m^2 · s^{-1}, which may represent the upper-mantle average, with the whole-mantle average considerably greater (see *Mantle Viscosity*). There are only two reasonable choices for the depth of the convecting layer: the 670-km discontinuity, which nearly coincides with the deepest Wadati-Benioff zone seismicity, or the whole-mantle depth, 2886 km. The implications of these alternatives will be discussed later. For now, we simply estimate the Rayleigh number in each case. For whole-mantle convection, the foregoing parameters substituted into Eq. 8 give $Ra_T \approx 2 \times 10^7$, whereas for upper-mantle convection, $Ra_T \approx 2.3 \times 10^5$. Using reasonable estimates for the rate of internal heat generation by radioactive decay in Eq. 9 gives $Ra_H \approx (10-100)\ Ra_T$.

Regardless of which parameter values are chosen, the implication is the same: The mantle Rayleigh number is far beyond the critical value, and vigorous, fully developed convection is expected. Convection in the Earth occurs because of high interior temperatures, which provide negative buoyancy for the lithosphere and allow for a finite, subsolidus creep viscosity, and also because of the large physical dimensions of the mantle.

Convection at High Rayleigh Number

Since the Rayleigh number for the mantle is far greater than the critical value, it is useful to follow the changes in structure of thermal convection in a homogeneous viscous fluid for the insight it brings before we address the question of subsolidus convection in the Earth. Figure 2 shows contours of temperature and stream function from numerical calculations using finite-difference representations of Eqs. 6 and 7 in the infinite Prandtl number limit. The geometry is a two-dimensional cylindrical annulus, which incorporates the curvature and connectedness of the spherical mantle without requiring resolution of the flow in the third dimension. At low Rayleigh numbers (less than about 10^5) the equilibrium structure consists of steady-state cellular flow with the *aspect ratio* (cell width normalized by layer depth) dependent on the mode of heating. Purely basal heating produces cells with aspect ratio near unity, and a combination of basal plus volumetric heating (as in the mantle) produces elongated large aspect ratio circulation (elongated cells). The four-cell circulation that is shown in Fig. 2 for $Ra_T = 3.2 \times 10^4$ and $Ra_H = 4Ra_T$ is equivalent to a spherical harmonic degree $l = m = 2$ circulation in a spherical annulus, and corresponds to the dominant wave number in present-day plate tectonics, mantle heterogeneity deduced by seismic tomography, and the nonhydrostatic geoid. There are good reasons for believing this mode to be the dominant, or "energy-containing," mode in mantle convection.

As the Rayleigh number increases to 10^5 and higher, convective velocities become large, and advection of heat dominates heat conduction everywhere except within thin *thermal boundary layers* adjacent to the inner and outer surfaces and within narrow *thermal plumes* or sheets located between cells. These layers contain nearly all the temperature variations, and their buoyancy drives the circulation. The velocity field, however, does not become localized, as indicated by the more-or-less evenly spaced streamlines.

A *global-scale* velocity field driven by *local-scale* thermal anomalies is the hallmark of high Rayleigh number convection with uniform viscosity. Convection in the mantle is characterized by structures analogous to those in the numerical simulations. The lithosphere is the surface thermal boundary layer, and subducted slabs are the geophysical equivalent of the negatively buoyant thermal plumes. Similarly, the mantle geotherm closely approximates the temperature distribution found in high Rayleigh number convection, including strongly superadiabatic boundary layers at the surface and at the core-mantle boundary. The rest of the mantle is relatively well mixed by convection, with an isentropic (adiabatic) variation of temperature, except perhaps in the transition zone, where there may be an additional superadiabatic thermal layer.

Another property of high Rayleigh number convection can be seen in Fig. 2. At Rayleigh numbers in excess of 10^5, *boundary layer instabilities* repeatedly appear as perturbations in the global-scale circulation. In a uniform viscosity medium they grow from both boundary layers, travel along with the speed of the global circulation, and are eventually destroyed by mixing. The steady states characteristic of low Rayleigh numbers are replaced, in this regime, by nonlinear oscillations about a time-averaged mean state. As the Rayleigh number reaches and exceeds 10^6, the strength of the recurring instabilities increases to the point where they are able to disrupt the general circulation. The flow becomes a chaotic pattern of spawning and merging cells with pulsations in global properties such as average velocity and heat transport. In this regime the lifetime of an individual convection cell is only about as long as the turnover time for material within that cell. This is a special case of turbulent flow called *thermal turbulence*, in which random fluctuations are produced by nonlinearities in advection of temperature. It is characterized by lack of symmetry, rapidly varying flow pattern, and large fluctuations in kinetic energy and heat output. Since the Rayleigh number based on the mantle depth is about 10^7, it is virtually ensured that thermal turbulence is present if the whole of the mantle convects as a single layer.

The case for thermal turbulence in convection limited to the upper mantle is less clear, but it too becomes more likely when effects of temperature-dependent viscosity are included. The theoretical arguments just given indicate that convection in the mantle is likely to be thermally turbulent, and this prediction agrees with the complexity of mantle structure as revealed by seismic tomography and mantle-derived isotopic patterns, the rapid appearance of new plates, and the disappearance of old ones. Indeed, those aspects of the geologic record that have an episodic character, such as long-term sea level and growth of continental crust, may eventually be interpreted as a result of pulsating thermal turbulence in mantle convection.

Convection Statistics: Boundary Layer Theory

It is possible to circumvent some of the complexity of time dependence by focusing attention on the *time average* properties of mantle convection and then using these statistics for constructing a "zero-order" model. Many of the important aspects of convection at high Rayleigh numbers can be found in the dynamics of the thermal boundary layers. Referring to Fig. 2, it is evident that the interior or core of each cell is virtually isothermal, and all significant temperature variations are localized in thin boundary layers wrapped around the core. The driving force for this flow is concentrated in the rising and sinking plumes that form where two horizontal boundary layers merge and detach. Note that in this case the term "plume" denotes localized temperature variations; because this flow has uniform viscosity, the velocity field is *not* similarly localized. A small parcel circulating in the thermal boundary layer is

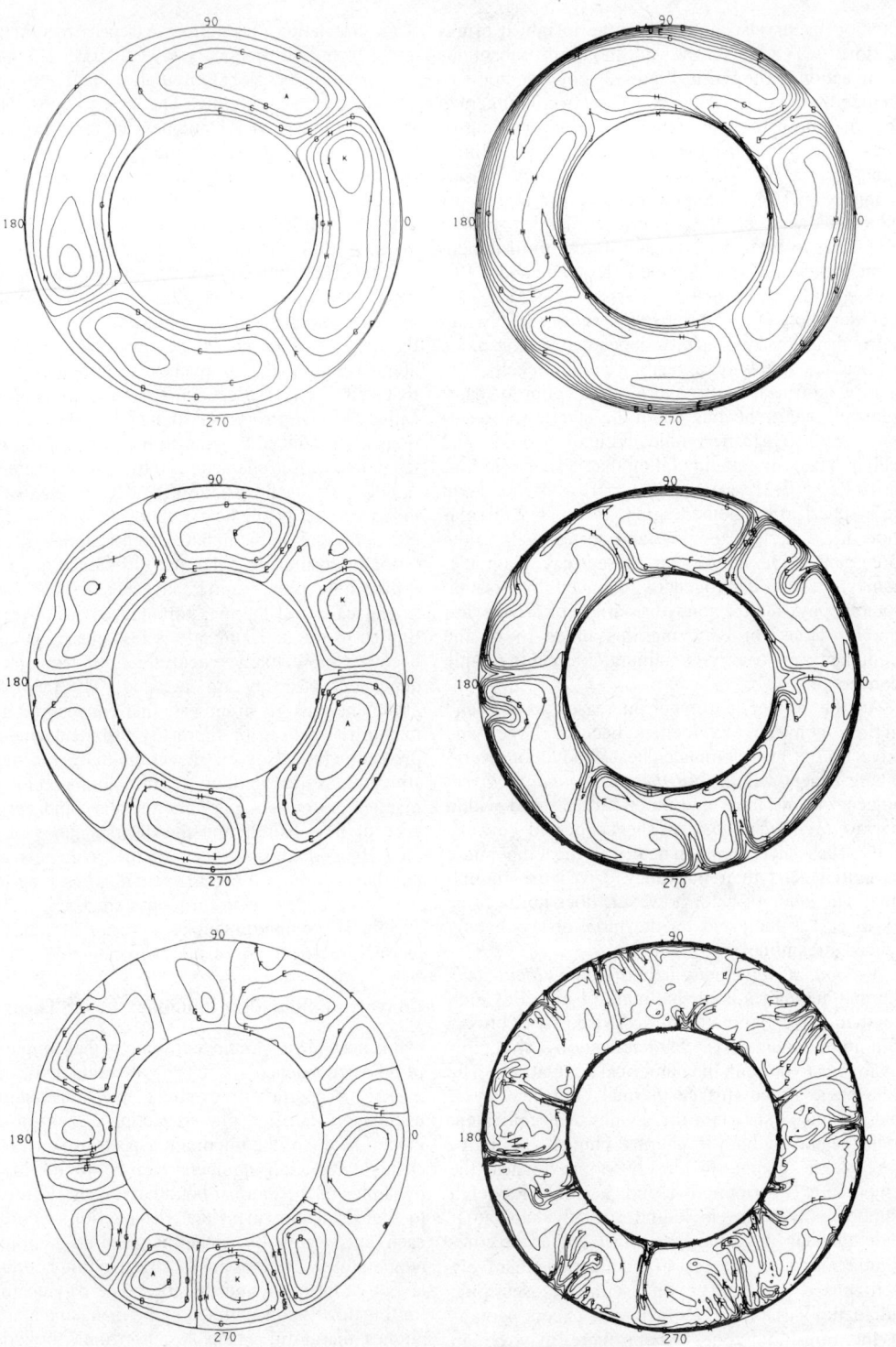

FIGURE 2. Streamlines (left) and isotherms (right) from numerical simulations of two-dimensional convection in a cylindrical annulus with a radius ratio equal to the whole mantle: (top) four-cell convection at $Ra_T = 3.2 \times 10^4$, $Ra_H = 4Ra_T$; (middle) pulsating six-cell convection with traveling boundary layer instabilities at $Ra_T = 10^5$, $Ra_H = 4Ra_T$; (bottom) thermally turbulent convection at $Ra_T = 10^6$, $Ra_H = 4Ra_T$. (Courtesy B. Travis)

alternatingly heated along the lower boundary and cooled at the upper surface and advected into the ascending and descending sheets. All the driving force for this circulation is concentrated in the boundary layer regions, particularly in the vertical sheets. The negative or positive buoyancy in each sheet exerts a torque on the fluid, and the couple provided by adjacent sheets drives the cellular flow between them. The relative strength of rising and sinking sheets is determined by the ratio of basal to surface heat flux. If this ratio is small, the surface boundary layer and the sheets that detach from it will dominate the lower boundary layer and its instabilities, and the flow is driven primarily by the torques provided by the sinking sheets. This situation is present in the mantle, where it is estimated that the core provides only 10–30% of the surface heat flow and sinking slabs provide the major driving force. In addition, mantle convection is fully three-dimensional. The descending flow in subduction zones is indeed sheetlike, but the evidence from seismic tomography and hot spots is that ascending flow in the mantle is grossly columnar, except immediately beneath ridges where it is constrained to be sheetlike by the geometry of seafloor spreading.

The characteristics of the thermal boundary layers can be estimated by using simplified, local stability arguments. These arguments are based on the premise that the thermal layers develop up to the point at which they become gravitationally unstable. This argument is not strictly correct, but it has proven to be useful for extracting the basic statistics of thermal convection. We define a local or *boundary layer Rayleigh* number based on the boundary layer thickness δ and the temperature drop δT across it:

$$\text{Ra}_\delta = \frac{\alpha g \delta^3 \, \delta T}{K\nu} \quad (10)$$

and we assert that equilibrium boundary layers are characterized by a *critical local Rayleigh* number Ra_c of about 10^3. Combining Eqs. 10 and 8 gives

$$\frac{\delta}{D} \approx \left(\frac{\text{Ra}_c}{\text{Ra}_T}\right)^{1/3}$$

The *Nusselt number*, the dimensionless ratio of convective heat transfer q to the equivalent conductive heat transfer, is

$$N = \frac{qD}{k \, \Delta T} \approx \frac{D}{\delta} \approx \left(\frac{\text{Ra}_T}{\text{Ra}_c}\right)^{1/3} = c\text{Ra}_T^{1/3} \quad (11)$$

Precise experimental and numerical determinations of heat transfer give a power-law relationship similar to Eq. 11, usually with a coefficient c close to 0.1. The time scale of convection can be estimated by the time to conduct heat through the boundary layer to a depth δ, which must be equal to the transit time across the surface of the cell,

$$t \approx \frac{0.2 \delta^2}{K} \approx \frac{0.2 \, D^2}{c^2 \, K \, \text{Ra}_T^{2/3}} \quad (12)$$

With $\Delta T = 2000 \, K$, $k = 3.3 \, \text{W} \cdot \text{m}^{-1} \cdot K^{-1}$, $K = 10^{-6} \, \text{m}^2 \cdot \text{s}^{-1}$, $D = 2.9 \times 10^6 \, \text{m}$, and $\text{Ra}_T = 2 \times 10^7$ as before, Eq. 11 predicts a Nusselt number of 27, a surface heat flux of $62 \cdot \text{W} \cdot \text{m}^{-2}$, and an average lithosphere thickness of 107 km, the latter two very close to the observed average values. With the same parameters, Eq. 12 predicts a transit time of 72 my at the surface, which is within 20% of the mean age of the oceanic lithosphere. All of this has been derived by assuming whole-mantle convection. If the transition zone breaks mantle convection into two layers, as some geophysicists think, then the preceding formulas do not apply, since they are derived for convection in a homogeneous fluid. Two-layer convection is still a rather unexplored phenomenon; it is not simply the linear sum of convection in two homogeneous layers, because of the complex interactions that can occur along the dividing interface (see the discussion of transition zone effects).

The Structure of Mantle Convection

Convection in a homogeneous fluid with uniform viscosity serves as the starting point in understanding subsolidus convection, but convection in the Earth's mantle is significantly more complex than this simple model, largely because of its heterogeneous rheology and structure. Subsolidus thermal convection in a medium with nonuniform rheology and structure is an area of active research, and we do not yet have a complete picture of all the facets of this difficult problem. We do know that flow in the mantle is strongly affected by variable viscosity, compression effects (particularly polymorphic phase changes), and radial variations in bulk composition. We have a fair appreciation of how these affect thermal convection individually, but there is still no clear picture of how they act together in the mantle.

The subsolidus creep rheology of the mantle is poorly constrained from either observation or theory (see *Mantle Viscosity*). Postglacial rebound studies have given the canonical value of 10^{21} Pa $\cdot$ s, which is presumably most representative of the upper-mantle average beneath the lithosphere. Its variation with temperature and pressure is usually expressed in terms of the Ahrrenius formula

$$\mu = \mu_0 \exp\left(\frac{E + pV}{RT}\right) \quad (13)$$

Creep measurements on olivine (see *Deformation of Rocks and Minerals*) indicate the activation energy E is 100 kcal $\cdot$ mol^{-1} or larger, and the activation volume V may exceed 8 cm$^3 \cdot$ mol^{-1}. Formula 13 predicts a factor of 10 rise in viscosity for each approximate 100 K drop in temperature and several orders of magnitude increase with depth through the pressure range of the upper mantle. Furthermore, there is evidence for a viscosity increase of a factor of 3–30 across the transition zone associated with phase transitions, and the dependence on pressure may be responsible for a low-viscosity asthenosphere channel beneath the lithosphere, coinciding with the seismic low velocity. There may also be low viscosity in the layer D'' at the base of the mantle, a consequence of heat flux from the outer core. Large variations in viscosity result in the distinctive style of mantle convection, which is dominated by the lithosphere and the constraints of plate motions in the upper mantle and by interaction with the core in the lower mantle. In that context, the transition zone may define the region where the flow adjusts from one regime to the other.

The most important effects of temperature-dependent viscosity occur in the thermal boundary layers. Cold boundary layers—the oceanic lithosphere being the primary one—are stabilized by their high viscosity, whereas hot boundary layers are destabilized. This distinction is reflected in the difference in local Rayleigh number (Eq. 10) for cold versus hot boundary layers, when the viscosities differ by several orders of magnitude. In the lithosphere, localized instabilities are inhibited by the high viscosity; the only instabilities with finite growth rates are long wavelength, involving the underlying mantle, in which the lithosphere buckles and collapses to form an isolated subduction zone. In contrast, hot boundary layers become unstable through formation of smaller-scale plumes and diapirs, originating in the regions of lowest viscosity.

The difference in behavior between hot and cold thermal boundary layers provides a qualitative explanation for the gross structure of mantle convection as expressed in surface tectonics. The majority of the driving force resides in the highly viscous subducted slabs and their aseismic extensions, where the greatest lateral temperature contrasts occur. They descend as widely spaced, nearly two-dimensional sheets in order to minimize internal deformation. Return flow occurs mostly in the regions far from subduction zones and contributes less to the driving force than slabs do. Return flow with significant amounts of positive buoyancy occurs as smaller-scale thermal plumes derived from instability of a hot boundary layer. Plumes that reach the lithosphere without being absorbed and mixed by the global circulation form hot spots and the related topographic rises. The source of mantle plumes is most probably the D''-layer at the base of the mantle, which is a complex thermochemical boundary layer formed by chemical differentiation and heat flux from the core. An alternative source region is the transition zone, which could generate plumes if the mantle is stratified in bulk composition and heat from the lower mantle is conducted to the upper mantle across the 670-km discontinuity. Mid-ocean ridges form where large tensional stresses develop in the lithosphere, as dictated by the pattern of subduction, their location perhaps influenced by the presence of midplate hot spots that create zones of weakened lithosphere. Ridge systems that migrate rapidly, in response to changing stress fields, necessarily draw material from the uppermost mantle only, and the flow beneath these spreading centers is primarily pressure-driven. Ridge systems that are more stationary, such as the Mid-Atlantic Ridge, are able to entrain mantle material from greater depths, including the hot boundary layer. This phenomenon is reflected in high density of hot spots associated with ridges of this type.

The pattern of global circulation in the mantle is highly asymmetric and time-variable, as expected in high Rayleigh number convection. The mode of time dependence, as revealed by present-day plate motions and plate reconstructions, is qualitatively the same as in the time-variable convection simulations shown in Fig. 2. Plates are destroyed by the retrograde motion of pairs of subduction zones, as is occurring now along the Pacific margin. Retrograde motion of subduction zones leads to subduction of former spreading centers. New plate boundaries are created by instabilities of the lithosphere; the process in the numerical simulations is instability of the top boundary layer, leading to formation of a new descending sheet. In the numerical calculations the top boundary layer dominates as a consequence of the flow being driven mostly by internal heating; in the mantle the same is true, and there it is further augmented by the effects of temperature-dependent viscosity.

In addition to influencing the structure of mantle convection, temperature-dependent viscosity affects the long-term evolution of the Earth by controlling the rate of change of heat loss from the deep interior. When an isoviscous fluid cools by convection, the decrease in the rate of heat loss is nearly in linear proportion to the decrease in temperature of the fluid, relative to the surface temperature. With variable viscosity, however, small decreases in absolute temperature result in large increases in viscosity, which, according to Eq. 11, produces large decreases in heat loss. Because subsolidus viscosity is so sensitive to temperature, a decrease in average mantle temperature of 100 K reduces the Rayleigh number by a factor of 10 and the heat transfer by a factor of about 2. This effect can be made quantitative by using a global energy balance consisting of the heat equation (Eq. 2) averaged over the entire mantle, with Eq. 11 representing the surface heat loss (assuming a rheological law similar to Eq. 13 for the dependence of viscosity on mean mantle temperature), and by introducing the radioactive decay formulas of mantle heat-producing isotopes into the heat source term. The result is a *thermal history model* that can be integrated backward in time, starting from present conditions, to give an estimate of the thermal regime in the Earth's early history. Results of these calculations indicate that the Earth was most likely hotter in the past by a few hundred degrees on average. The main inference, according to this model, is that convection was once much more vigorous because of lower average viscosity. This would imply a large reduction in the time scale of every major Archean geodynamical process, relative to the present, by roughly a factor of 10. The evidence on this point is equivocal, although most geodynamicists support the general conclusion of thermal history models that the Earth is slowly cooling and that the loss of thermal energy is as important as radioactive heat production in fueling mantle convection.

The Role of the Transition Zone

Of all the contentious issues in mantle convection, none is so enduring and so important as the influence of the transition zone between 400 and 800 km depth. The seismicity cutoff near 670 km indicates that brittle deformation terminates in the transition zone, which in turn implies that slabs are either prevented from entering the lower mantle or that they do so with a ductile rheology. In either case it is not possible to use seismicity as a tracer for mantle flow beneath the transition zone. Recently, global seismic tomography and travel time anomaly studies in Benioff zones have demonstrated that high-velocity material is present in the lower mantle beneath subduction zones and that the core-mantle boundary is depressed beneath these anomalies, as expected under regions of descending flow (see *Earth Structure, Global* and *Mantle, Lower: Structure*). Thus there is good evidence for coupling between upper- and lower-mantle flow, on the largest scales, across the transition zone. This coupling may be by slab penetration directly into the lower mantle or indirectly, with separate upper- and lower-mantle convection systems and with descending flow aligned along the same radius in each layer. The first alternative is *penetrative convection*, with whole-mantle convection being a limiting case. The second alternative is called *two-layer convection*.

Most of the increase in seismic velocity and density through the transition zone is due to polymorphic phase changes from olivine to spinel structure near 400 km and from spinel to postspinel phases, such as perovskite, near 670 km. It is now understood that these phase changes are by themselves not sufficient to arrest descending slabs. The principal effect of phase transformations is the additional buoyancy arising from deflection of the equilibrium phase boundary. In cold slabs the exothermic olivine-spinel transformation is elevated by about 40 km, providing additional negative buoyancy. The spinel-perovskite transformation, however, is now understood to be endothermic, which adds positive buoyancy to those slabs that extend to 670 km, canceling at least part of the negative buoyancy force provided by the olivine-spinel reaction. Consequently the sum of all phase transformation effects in the transition zone is thought to be of secondary importance compared to thermal and compositional buoyancies.

Both theory and experiments indicate that the vertical structure of thermal convection is extremely sensitive to even slight variations in bulk composition with depth. This sensitivity demonstrates the importance of possible compositional variations through the transition zone in determining the style of mantle convection. Differences in bulk composition due to silica or iron enrichment in the lower mantle may account for as much as one third of the density increase across the 670-km discontinuity— equivalent to the lower mantle being 2–3% denser than the upper mantle, after correcting for phase changes. This density increase is comparable to the relative density excess in the slab due to thermal contraction, $\alpha\delta T$, where δT represents the average slab-mantle temperature contrast. The behavior of subducted slabs in the presence of a compositionally produced density increase is illustrated by numerical simulations in Fig. 3 and similarly conceived laboratory experiments in Fig. 4. Virtually no slab penetration occurs when the compositional density jump across the transition zone exceeds 4%; with a 3% jump slabs penetrate deeply into the lower layer, although they may be arrested before reaching the core-mantle boundary. This is the regime of penetrative convection. A compositional density jump of much less than 3% would be unstable to slab penetration and would lead to mixing between upper and lower mantles unless accompanied by *strongly* endothermic phase changes at the same depth. Based on what is known about the 670-km discontinuity, penetrative convection is the most likely style of upper-lower-mantle interaction if the transition zone contains a slight change in composition, and whole-mantle convection is expected if the transition zone contains only phase transformations. The third alternative is strict two-layer convection without significant slab penetration, illustrated in Fig. 5. In two-layer convection, the flow in one layer is not coupled to the flow in the other by direct mass transfer. Rather, the coupling occurs indirectly through viscous stresses and by heat conduction across the transition zone. Mass transfer between layers can occur by entrainment where the discontinuity is particularly distorted, as in Fig. 5, but at a rate which is too small to homogenize the mantle over the age of the Earth. Two-layer convection is accompanied by a double thermal boundary layer in the transition zone.

To summarize, there are three geodynamically plausible regimes for the largest scale of mantle convection: (1) whole-mantle convection, with the

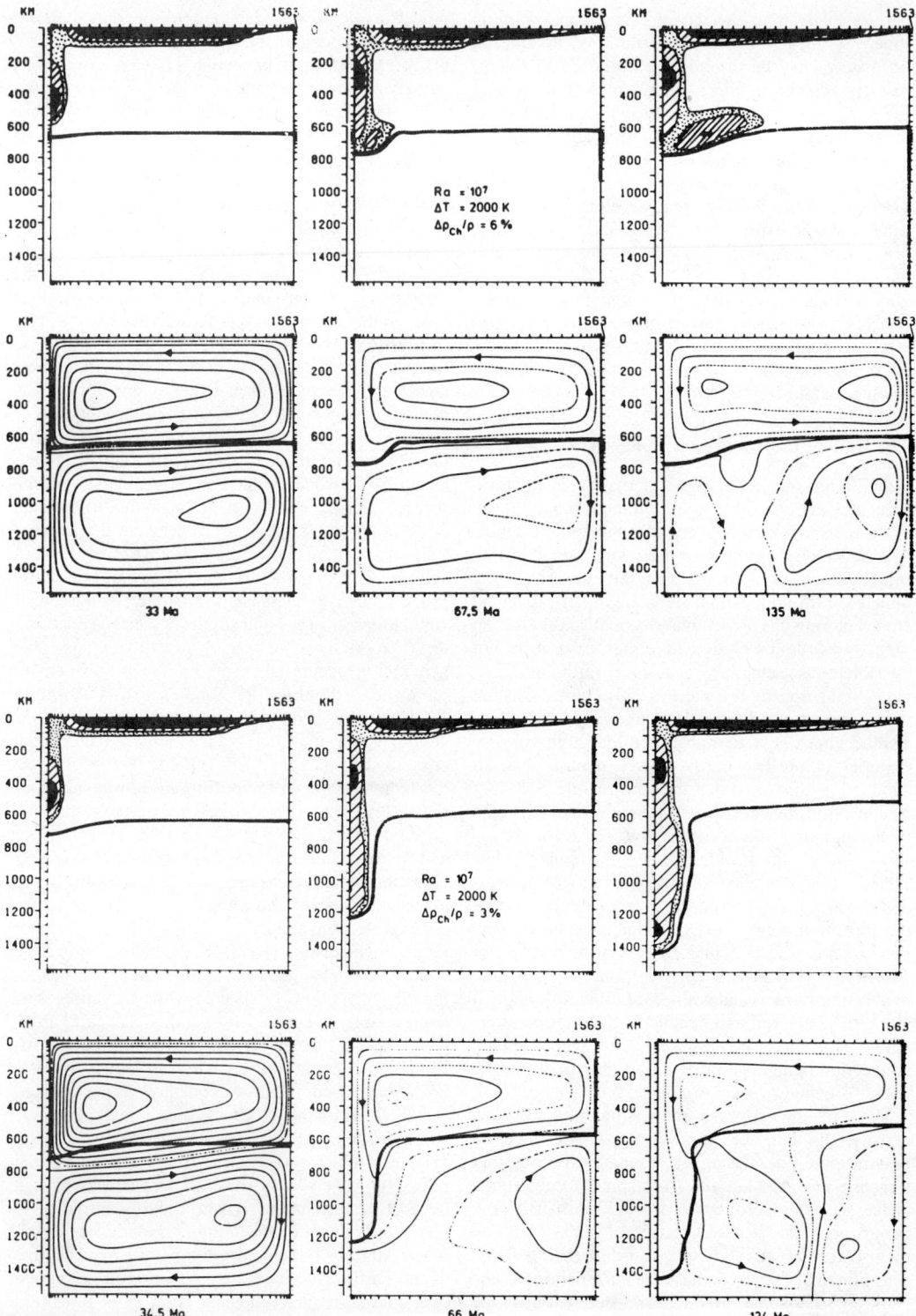

FIGURE 3. Numerical simulations of the interaction of a subducting slab with a compositional boundary at the 670-km discontinuity, showing slab structure (shaded according to viscosity) and streamlines: (top) deflection of slab by a 6% density jump; (bottom) penetrative convection with a 3% density jump. (From Christensen and Yuen, 1984)

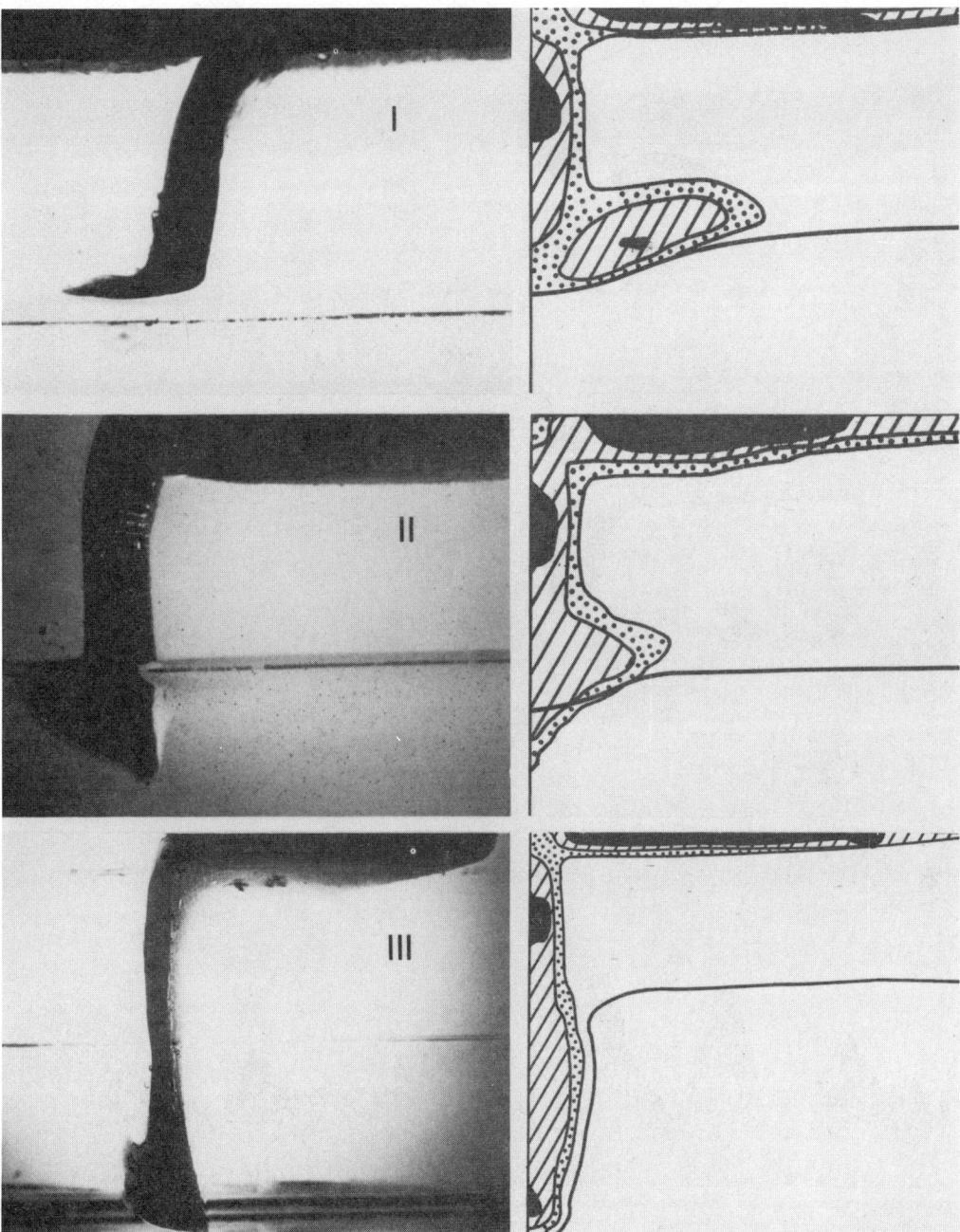

FIGURE 4. Laboratory experiments on slab penetration across a density discontinuity in a temperature-dependent viscous sucrose solution (left) compared with numerical simulations (right) by Christensen and Yuen (1984): (top) slab deflection by strong discontinuity; (middle) shallow penetration by intermediate strength discontinuity; (bottom) deep penetrative convection across a weak discontinuity. (Courtesy C. Kincaid)

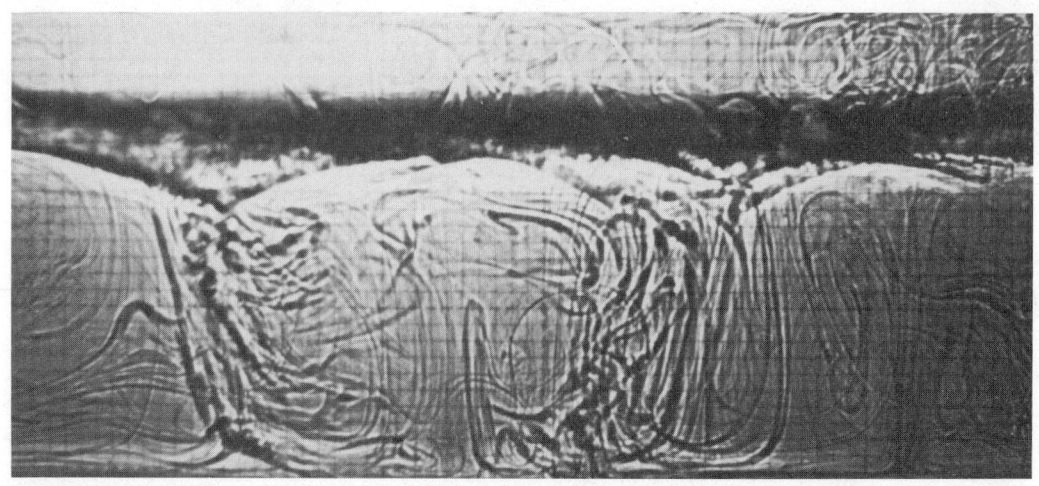

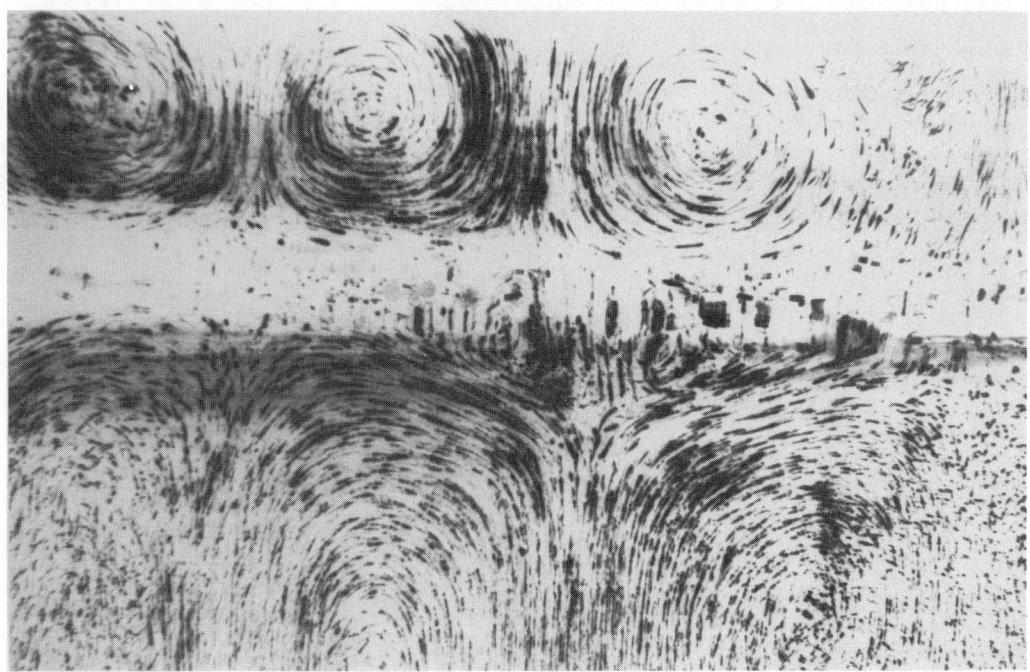

FIGURE 5. Shadowgraph image (*top*) and laser streak image (*bottom*) showing the vertical structure of two-layer convection in a temperature-dependent viscous sucrose solution, at $Ra_T \approx 10^6$. Note the complex transition zone structure and the schlieren of material entrained across the interface.

transition zone consisting of phase changes only; (2) penetrative convection across the transition zone, consisting of phase changes plus a slight change in composition; and (3) convection in two separate layers, divided by the transition zone in which large changes in composition occur. Arguments can be made in support of each of these alternatives, although at present no argument is conclusive. The final resolution of this issue probably must wait until the nature of the transition zone is better understood.

Mantle Plumes

In addition to the global circulation that is responsible for plate tectonics, there exist secondary flows that give rise to intraplate tectonics. Perhaps the most striking examples of these are hot spots and their associated topographic swells, the best examples being large oceanic rises surrounding Hawaii, Iceland, Cape Verde, Bermuda, and other oceanic islands. Attention has traditionally been focused on

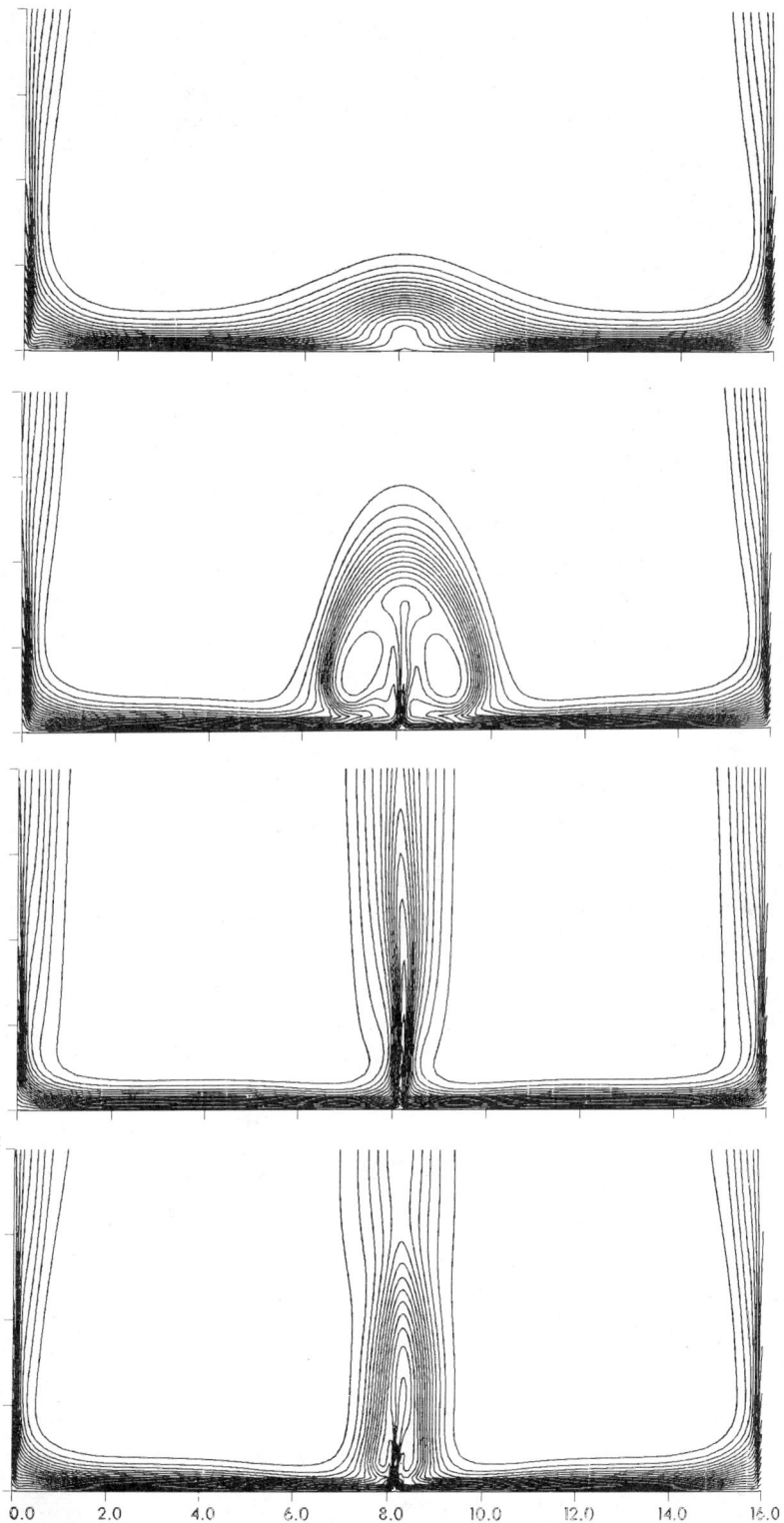

FIGURE 6. Isotherms from a numerical simulation of diapir and plume formation in the D''-layer above the core-mantle boundary. Frames are at 15 my intervals.

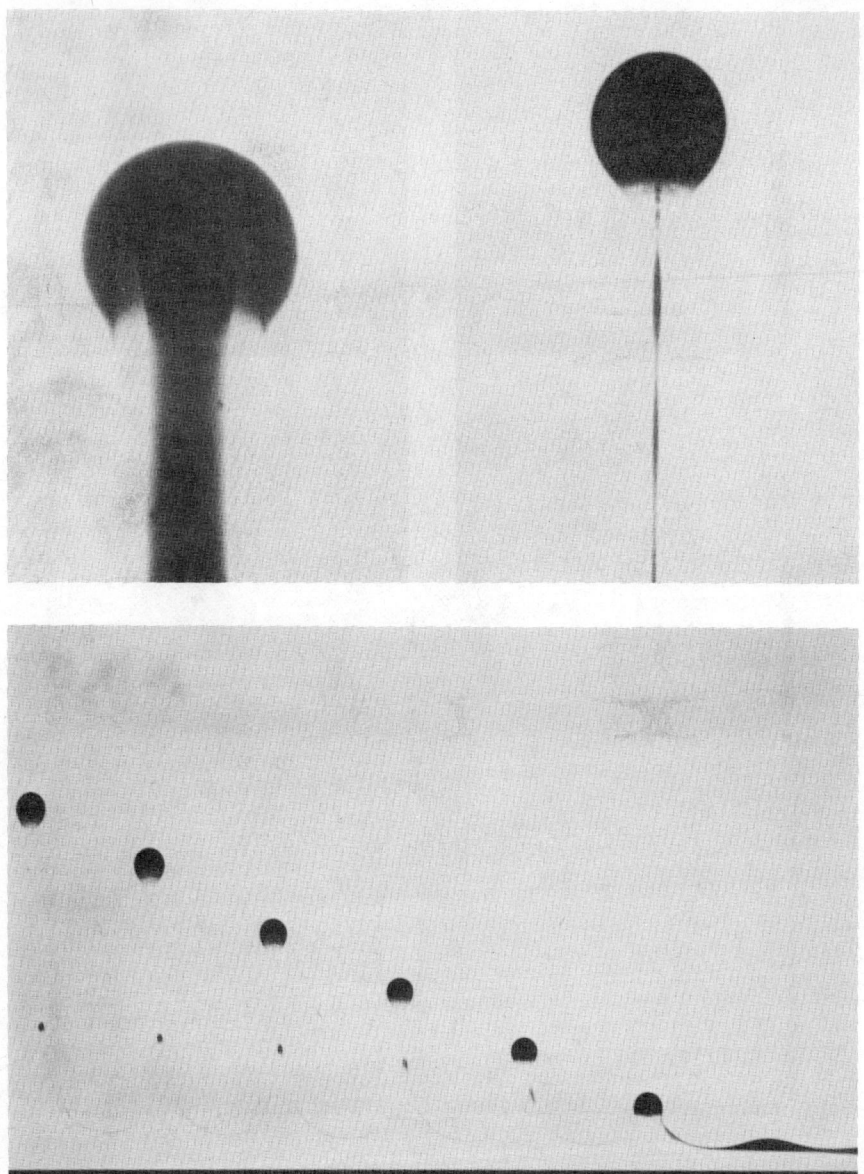

FIGURE 7. The morphology of buoyant plumes in a viscous fluid: (top left) an isoviscous plume in a stationary fluid; (top right) a low-viscosity plume in a highly viscous, stationary fluid; (bottom) a low-viscosity diapir chain formed by relative motion between plume source and surrounding fluid.

volcanic hot spots; only relatively recently has the significance of the swells been recognized. Hot spots are remarkable for their apparent independence from plate tectonic motions. They are nearly stationary with respect to plate motions and also with respect to each other, for time intervals in excess of 100 my. The Hawaiian Ridge is a nearly continuous hot-spot track, with volcanism recurring on a 1-my time scale. Most other hot spots are less regular, and indeed most hot-spot tracks appear to be composed of discrete segments. In the Atlantic, some hot spots have little or no recognizable track, either because the plate above them is nearly stationary or because they were formed in a single episode.

Evidence of the geodynamical origin of hot spots comes from the character of their swells—broad topographic rises 1-2 km high and typically 1000 km across. The correlation of geoid and topographic heights, plus the presence of anomalously high heat flow, indicates that swells mark reheated and thermally expanded lithosphere. The heat flux needed to maintain the swells of the currently active

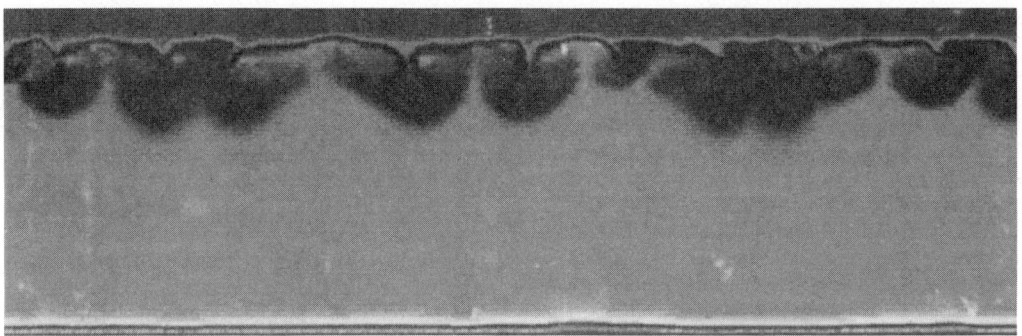
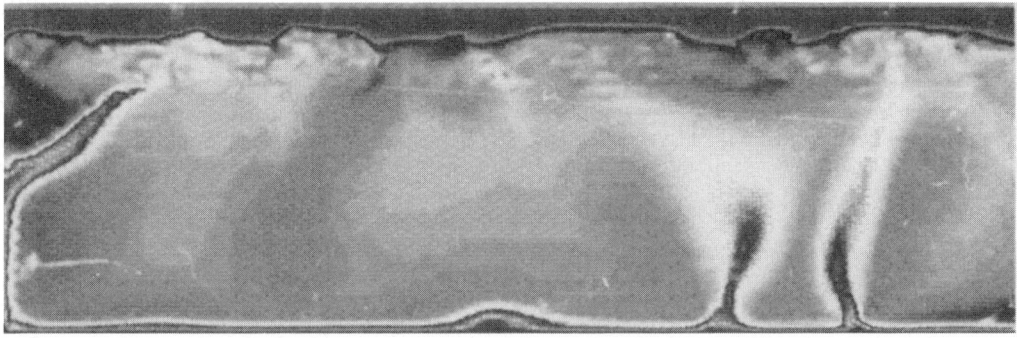

FIGURE 8. Grey scale temperature fields from numerical simulation of the interaction of thermal plumes with the lithosphere in the mantle with both temperature- and pressure-dependent rheology: (top) small-scale convection in low-viscosity asthenosphere channel precedes plume formation in lower thermal boundary layer; (bottom) plumes erupt from lower boundary layer and further erode lithosphere. (Courtesy G. Schubert and C. Anderson)

hot-spot population is approximately 3×10^{12} W, nearly 10% of the Earth's total heat loss. Furthermore, the large size of the major swells demands a source at depth in the mantle. These facts, plus geochemical differences from normal mid-ocean ridge basalts, indicate that hot spots form from a localized, hot upwelling or plume. The swell represents the area engulfed by the plume as it stagnates within the lithosphere, at depths of 60–100 km.

Until seismic tomography can provide very high resolution images of mantle structure with which to test the plume hypothesis, there is no choice but to rely on inferences drawn from a combination of fluid-mechanical models and indirect geophysical data. We know from fluid mechanics that plumes originate from instabilities of hot thermal boundary layers, which restricts their source in the mantle to either the region just above the core-mantle boundary or the transition zone. The process of plume initiation at the base of the mantle by boundary layer instability is illustrated by the sequence in Fig. 6. At the leading edge of the plume a nearly spherical *diapir* forms by drawing low-viscosity material along the core-mantle boundary. The diapir inflates to the critical size necessary for separation (approximately 400-km diameter) and ascends, leaving behind a trailing conduit. The conduit acts as an umbilical line connecting the leading diapir to its source region through which low-viscosity heated material rises as in pipe flow. Two possible morphologies of fully developed plumes during ascent through the mantle are shown in Fig. 7 (top). When the background medium is fixed with respect to the plume source, the conduit remains continuous. Often the conduit supports propagating *solitary waves*. This configuration may serve as a model for plumes beneath stationary or slowly moving plates, where the global circulation is negligible. However, under rapidly moving plates, the global circulation can disrupt the conduit, producing the *diapir chain* in Fig. 7 (bottom). The ordered intrusion of a diapir chain into the lithosphere results in a track, which appears continuous if the diapirs are closely spaced, and discontinuous otherwise. These are the basic characteristics of hot spots beneath fast-moving plates such as the Pacific.

Thermal plumes incident from below reheat the lithosphere either by direct intrusion or by exciting secondary scale instabilities in the low-viscosity asthenosphere. Figure 8 shows a numerical simulation of plumes formed from a hot boundary layer eroding high-viscosity lithosphere, with a low-viscosity asthenospheric channel beneath. Both direct intrusion and secondary flows are evident.

Concluding Remarks

The study of mantle convection over the past two or three decades has given us a fragmentary picture of the processes that are important in the Earth's internal heat engine, plus plausible and physically defensible alternative models with which to view it. Mantle convection is unquestionably useful as a theory in the general sense, but at the present stage of development it is rarely so useful for specifics. We do not yet understand enough about mantle convection to fully explain the origin of particular structures on the Earth in terms of convection theory, in most cases. A solid conceptual framework has been built, but the details that would make it a practical tool are, for the most part, still missing. Advances in other areas of geophysics and geochemistry, combined with further appreciation of the phenomenon of convection itself will eventually provide the missing pieces of the picture if we have the persistence and the imagination to find them.

PETER OLSON

References

Busse, F. H., and J. A. Whitehead, 1971, Instability of convection rolls in a high Prandtl number fluid, *Jour. Fluid Mechanics* **47,** 305.

Christensen, U., and D. A. Yuen, 1984, The interaction of a subducting lithospheric slab with a chemical or phase boundary, *Jour. Geophys. Research* **89,** 4389-4402.

Davies, G. F., 1980, Thermal histories of convective earth models and constraints on radiogenic heat production in the earth, *Jour. Geophys. Research* **85,** 2517-2530.

Jarvis, G. T., and W. R. Peltier, 1984, Mantle convection as a boundary layer phenomenon, *Royal Astron. Soc. Geophys. Jour.* **68,** 389-427.

Loper, D. E., 1985, A simple model of whole mantle convection, *Jour. Geophys. Research* **90,** 1809-1836.

Olson, P., and H. Singer, 1985, Creeping plumes, *Jour. Fluid Mechanics* **158,** 511-531.

Oxburgh, E. R., and D. L. Turcotte, 1978, Mechanisms of continental drift, *Rept. Prog. Physics* **41,** 1249-1312.

Peltier, W. R., ed., 1989, *Mantle Convection.* New York: Gordon and Breach.

Schubert, G., 1979, Subsolidus convection in the mantles of terrestrial planets, *Ann. Rev. Earth Planetary Sci.* **7,** 289-342.

Silver, P. G., R. W. Carlson, and P. Olson, 1988, Deep slabs, geochemical heterogeneity, and the large-scale structure of mantle convection: Investigation of an enduring paradox, *Ann. Rev. Earth Planet. Sci.* **16,** 477-541.

Cross-references: *Core-Mantle Coupling; Deformation of Rocks and Minerals; Earth's Core; Earth Structure, Global; Energy Budget of the Earth; Heat Flow in the Earth; Mantle, Lower: Structure; Mantle, Upper: Structure; Mantle Dynamics; Mantle Viscosity; Seismic Tomography.*

MANTLE DISCONTINUITIES

According to seismic data, the mantle includes layers where wave velocities are slowly varying functions of depth and relatively thin anomalous zones of pronounced changes in the elastic parameters. The anomalous zones are often called *discontinuities*. A discontinuity can be characterized by the depth and thickness of the anomalous layer; P-velocity, S-velocity, and density contrasts; and by the fine velocity structure. The parameters of a discontinuity can be expected to vary laterally. An extremely sharp discontinuity is often termed first-order one. Precise knowledge of the seismic properties of a discontinuity may help to understand its origin and identify the mantle material. For this reason, seismic data on the mantle discontinuities play an important role in the current discussions of the physics, chemistry, and dynamics of the mantle. The review of seismic data on the properties of the mantle discontinuities would be incomplete without a short description of relevant seismic techniques, their advantages and limitations.

The bulk of the presently available seismic data on the mantle discontinuities comes from observations of refracted and overcritically wide angle reflected body waves at distances up to about 4000 km. A typical example of the P-wave travel-time data in this distance range is shown in Fig. 1. The branch A, which is observed as a first arrival at distances less than 2200 km, corresponds to the waves penetrating depths up to about 300 km. This line is followed by the triangle (triplication) formed by the lines AB, BC and CD, and corresponds to the well-known 400 km discontinuity. The amplitudes of the record corresponding to the branch BC in the vicinity of the point C are often very large due to the focusing effect of the discontinuity, and this effect facilitates observations of this branch in many upper mantle studies. The branch BC corresponds to overcritical (wide-angle) reflections, while undercritical reflections may form a continuation of it toward shorter epicentral distances. The undercritical reflections, however are too weak to be routinely observed. The triplication ABC in Fig. 1 is followed by the triplication CDEF, which corresponds to the other well-known discontinuity located near the 650-km depth.

In its turn, the line A in Fig. 1, if subjected to a more detailed study using densely spaced seismometer arrays on long-range profiles, appears to break up into several refraction (overcritical reflection) lines (for a review see, e.g., Fuchs and Vinnik, 1982), and thus the mantle at depths less than 300 km may contain a number of discontinuities. The reader must be warned, however, that lateral inhomogeneities in the lithosphere cause many complications in the records of long-range refraction profiling. Up to now this complexity is poorly

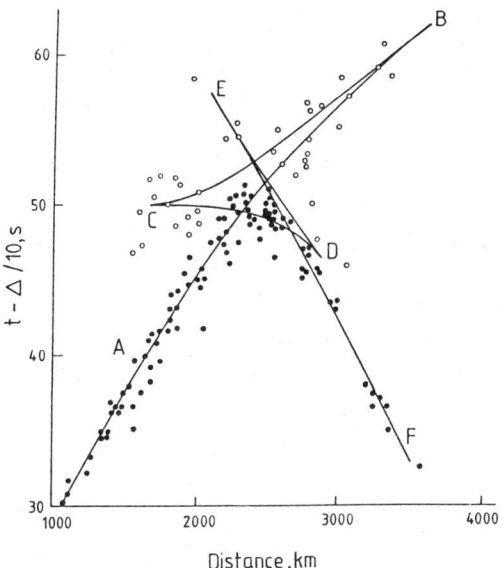

FIGURE 1. Reduced travel times of P first arrivals (solid circles), of later arrivals (open circles) and the corresponding travel-time curve for the East-European platform (From Vinnik et al., 1983).

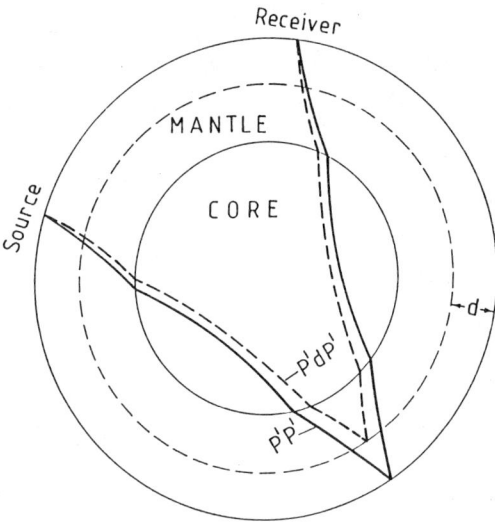

FIGURE 2. Ray paths of P'P' and P'dP'.

understood and can be responsible for some dubious features in the resulting depth-velocity models.

Many data sets like that shown in Fig. 1 were inverted for the mantle velocity structure. Often the inversion of travel-time data was complemented by the more advanced methods of calculating synthetic seismograms and then fitting them to the observed ones. In spite of the developments in the techniques, the observations of refracted waves are still strongly limited in their accuracy and resolving power. This point can be illustrated by the data shown in Fig. 1. To determine the depth of the corresponding discontinuity one should fix positions of the points D and E as accurately as possible. The seismic phase DE, however, interferes with the other strong phases, and this complicates the matter. In practice, the errors of positioning points D and E can be of the order of a second in time and a few hundred kilometers in distance. Such errors are sufficient to change the depth of the discontinuity by a few tens of kilometers and this alone may explain the scatter in the depth estimates reported in the literature. Similarly, refracted and wide-angle reflected waves are rather insensitive to the sharpness of a discontinuity. First-order discontinuities in the published mantle models are usually not demanded by the data but preferred due to their simplicity. A limited lateral resolution is the other deficiency of the method: refracted waves penetrating to a depth of 600 km return to the Earth's surface at a distance of about 3000 km from the epicenter thus precluding observations of the lateral variations in the deep structure of the Earth on a scale less than, say, 1000 km.

The views on the mantle velocity structure were much affected by the observations of the so-called underside reflections (see, e.g., Engdahl and Flinn, 1969). These phases were identified as precursors to the strong short-period reflections from the Earth's surface. The ray-paths of an underside reflection (P'dP') and of the corresponding surface reflection (P'P') are shown in Fig. 2 (d is for a depth of the reflector). The observation of a near-vertical reflection means that the thickness of the discontinuity is less than about a quarter of a wavelength. A typical wavelength of P'dP' is about 10 km and thus the corresponding layer thickness should be less than 3 km. Several authors presented evidence that many of the reported precursors, and especially those with a relatively small lead time, were formed not by the reflections at the mantle discontinuities but by the random wave scattering near the Earth's surface. The earliest precursors to P'P' avoided explanation in terms of the near-surface scattering but J. Cleary suggested that they could be formed by the wave scattering at the inside of the core-mantle boundary. Thus a reliable identification of the underside reflections among other phases presents the major problem in this direction of seismic research.

Some important data on the mantle discontinuities were obtained from observations of waves converted in the mantle either from P to S or from S to P. The time difference between the arrivals of the converted phase and the corresponding primary wave depends upon the depth of the converting discontinuity and can be used to determine this depth. The amplitude of the converted phase depends on the S-velocity contrast across the discontinuity and is practically

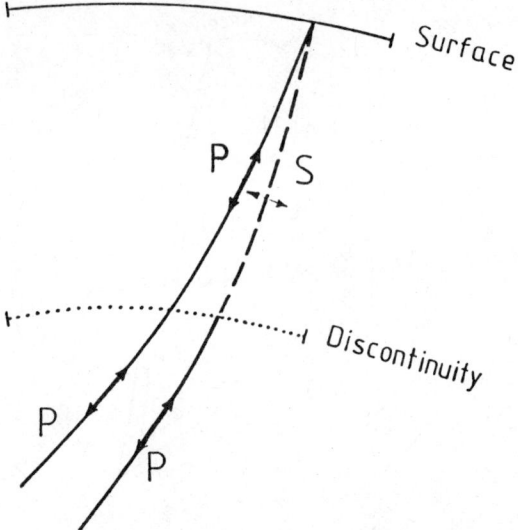

FIGURE 3. Ray path of the P to S converted phase.

independent of the P-velocity and density contrasts. The sharpness of the discontinuity affects the amplitude although the dependence is weaker than in the case of a near-vertical reflection.

Figure 3 shows a ray path of a teleseismic Ps phase which is formed by conversion from P to S beneath a seismograph station. The Ps phases are recorded in the P-wave coda with amplitudes of the order of a few percent of the P-wave amplitude; to detect such weak phases a special signal-processing procedure is required. This procedure is, in principle, similar to the conventional velocity filtering commonly performed in seismic array experiments but instead of an array of sensors, use is made of the array of seismic sources and the apparent velocity as the unknown parameter is replaced by the depth of conversion. The technique has been applied to the records of a few seismograph stations, and the Ps phases corresponding to the mantle discontinuities in the 400–700 km depth range were consistently observed (Vinnik et al., 1983). The reported accuracy of the depth determinations was of the order of a few kilometers; the lateral resolution was of the order of a few hundred kilometers.

The phases converted from S to P are observed as precursors to S either on the records of deep local events or on the teleseismic records in the epicentral distance range around 80° (see, e.g., Sacks and Snoke, 1977; Bock and Ha, 1984; Faber and Müller, 1984). These observations do not require a complicated technique of record processing but, in many records, the true converted waves are mixed with other phases.

After this brief discussion of the techniques we may proceed to the data on the mantle discontinuities.

In the uppermost mantle the observations on long-range profiles often reveal a discontinuous increase of the P-velocity with depth from about 8.0–8.2 km/s to 8.3–8.5 km/s at depths between approximately 50 km and 90 km. The discontinuity was found especially often in the lithosphere of the old continental platforms where some authors even attempted to map its depth. A. Hales suggested that this discontinuity could be produced by the spinel-to-garnet phase transformation in the ultramafic mantle material. The pressure of transition from spinel peridotite to garnet peridotite is known to be sensitive to chemical composition, which could explain the variations of depth of the discontinuity.

The observations on long-range profiles provide evidence of some other discontinuities between the top of the mantle and a 300 km depth. The resulting high- and low-velocity layers sometimes can be traced along the profiles many hundreds of kilometers long. A laminated structure was found, for example, in the mantle of the old Siberian platform, although the most plausible (garnet pyrolite) composition if combined with a suitable geotherm predicts a monotonous increase of the P-velocity with depth in this depth range. Some authors claim observations of (undercritical?) reflections from the upper-mantle discontinuities, which may imply that at least some of the discontinuities are relatively sharp. In general, the nature of the laminated structure of the uppermost mantle is enigmatic, but one can think about anisotropy among the other possibilities.

Among the discontinuities of the uppermost mantle the one at a depth around 200 km is the best known. It was observed in Europe, Asia, Australia and North America. Almost everywhere it shows up as the bottom of a zone of low P velocities. The most likely value of the P-velocity at the underside of the discontinuity is about 8.7 km/s, the most likely P-velocity contrast is about 0.3 km/s. In Siberia where the system of observations was exceptionally favorable for deep structural studies the data suggest that the discontinuity changes its depth from approximately 200 km under the young West-Siberian platform to 270 km under the old Siberian platform. The observations of this discontinuity in Australia were discussed by Leven et al. (1981). They concluded that neither a phase transformation nor a compositional change could explain this feature, but it could be interpreted in terms of azimuthal anisotropy in a thin zone of possible decoupling of the continental lithosphere from the underlying mantle.

Little is known about the discontinuities in the depth range between approximately 300 km and 400 km since the corresponding seismic phases are usually hidden in late arrivals. The phases corresponding to this depth range can be observed as first arrivals if the seismic sources are located in the upper mantle. A. Hales and his colleagues exploited this opportunity in Australia and reported observations of a discontinuous P-velocity increase at a depth of 325 km.

The strongest velocity contrasts are known to exist in the mantle transition zone. The top of the zone is marked by the discontinuity which is well pronounced in almost every set of seismic refraction data (see example in Fig. 1). The published estimates of depth of this discontinuity are scattered between approximately 370 km and 450 km, though my own experience suggests that the depth of this discontinuity is probably the same in such widely separated regions as the North of Europe and the West of North America and does not deviate from 400 km by more than 10 km. This effective depth corresponds to a middle point of the transition. The P- and S-velocity contrasts at the discontinuity are around 0.5 km/s and 0.2–0.3 km/s, respectively.

The observations of the short-period phases P'400 P' are extremely rare, implying that the normal thickness of the transition is more than 3 km. An accurate determination of this thickness is, however, difficult. The published estimates vary between 0 km and 100 km partly due to ambiguity of data and partly due to real lateral variations of sharpness of the discontinuity. The observations of the phases converted from P to S at the 400 km depth imply that the thickness of the discontinuity is less than 30–50 km.

The discontinuity at a 400 km depth can be explained by the phase transition from olivine to β-spinel (see, e.g., Jeanloz and Thompson, 1983). A broad agreement between the properties of the discontinuity and those of the phase transformation in olivine is the main reason to believe that olivine or, more precisely, its high-pressure modifications, dominate in the mantle transition zone. The pressure of the transition depends on temperature, and the depth of the discontinuity may indicate temperature in the mantle. The sharpness may depend on the MgO/FeO ratio in olivine.

In the depth interval between 500 km and 550 km some authors observed a discontinuity with a P-velocity contrast of about 0.2 km/s. This discontinuity is present in some of the published depth-velocity models and is missing in others. The observations of the P to S converted phases indicate that the discontinuity is weak or absent in some regions (e.g., north of Europe) but is very well pronounced, for example, in the Rio Grande rift area.

As a rule, the most pronounced discontinuity is the one located near 650-km depth and which separates the upper and the lower mantle. Understanding the nature of this discontinuity is important for constraining the models of composition, dynamics, and thermal regime of the Earth. A phase transformation alone would most likely permit whole-mantle convection while a change in chemical composition could result in a two-layered convective system.

Sometimes the observations of the P'650 P' phases have been regarded as evidence of sharpness of the discontinuity and this, in turn, was taken to imply a

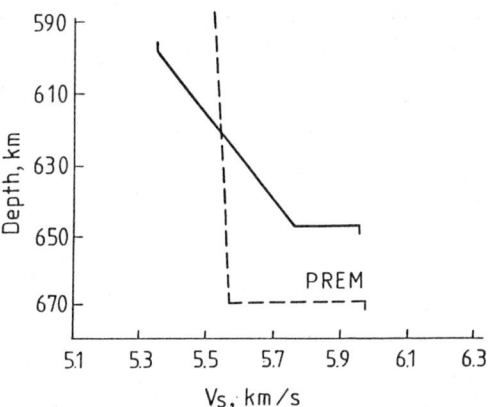

FIGURE 4. S-velocity model of the 650-km transition inferred from the observations of P to S converted phases at NORSAR (solid line). The distribution of the S-velocity in the model PREM by Dziewonski and Anderson is shown for comparison (dashed line).

chemical change. The strength of this evidence, however, seems to be somewhat overestimated (see Muirhead, 1985). First of all, the observations of P'650 P' are consistent with Cleary's hypothesis of random wave scattering at the inside of the core-mantle boundary. What can be equally important, these phases were observed only on a small portion of records, and the majority of inferred reflections occurred under Antarctica and surrounding seas. Thus, even if the hypothesis of reflections is in general correct, the reported observations may represent a structure of a regional rather than world-wide extent.

A result that was inferred from the observations of the P to S converted phases is presented in Fig. 4. The model, which is consistent with the seismic data obtained at the Norwegian Seismic Array (NORSAR), incorporates a gradient layer 50 km thick and a sharp discontinuity at the 650-km depth. There are some indications that the gradient layer can be weakly anisotropic. The sharp discontinuity in the model could be replaced by an equivalent gradient layer ~10 km thick with an S-velocity contrast of about 0.2 km/s. In some locations a sharp discontinuity is almost definitely absent while the gradient layer can be ~100 km or more thick.

Understanding the nature of the 650-km discontinuity could be greatly improved by observations of its properties in anomalous regions such as subduction zones or zones of upgoing mantle flow. The questions that could be addressed here are as follows: Do the subducting lithospheric slabs penetrate through the discontinuity into the lower mantle or not? Is there any discrepancy between the depths of the deepest earthquakes in the subduction zones and the normal depth of the discontinuity? Is there anything special in the properties of the 650-km discontinuity in the regions of upgoing mantle flow?

The most likely answer to these questions seems to be "yes" but more definitive studies will be required in the future.

There are indications of the presence of discontinuities in the mantle below 650 km. These discontinuities are comparatively weak, however, and the related seismic data are often somewhat controversial due, primarily, to the lateral heterogeneity of the Earth's interior.

In general, the mantle discontinuities present a fascinating and challenging subject in seismology. The studies of their properties are often hampered by a lack of resolving power in the presently available data and techniques. Progress in this field depends on the developments in digital seismology and seismic instrumentation.

L. P. VINNIK

References

Bock, G., and J. Ha, 1984, Short-period S to P conversion in the mantle at a depth near 700 km, *Royal Astron. Soc. Geophys. Jour.* **77,** 593-615.

Engdahl, E. R., and E. A. Flinn, 1969, Seismic waves reflected from discontinuities within the earth's upper mantle, *Science* **163,** 177-179.

Faber, S., and G. Müller, 1980, S-phases from the transition zone between the upper and lower mantle. *Seismol. Soc. America Bull.* **70,** 487-508.

Fuchs, K., and L. P. Vinnik, 1982, Investigation of the subcrustal lithosphere and asthenosphere by controlled source seismic experiments on long range profiles, in G. Palmason, ed., *Continental and Oceanic Rifts.* Washington, D.C.: American Geophysical Union and Geological Society of America (Geodynamics Series) **8,** 81-89.

Jeanloz, R., and A. B. Thompson, 1983, Phase transitions and mantle discontinuities, *Rev. Geophysics Space Physics* **21**(1), 51-74.

Leven, J. H., I. Jackson, and E. A. Ringwood, 1981, Upper mantle seismic anisotropy and lithosphere decoupling, *Nature* **289,** 234-239.

Muirhead, K., 1985, Comments on 'Reflection properties of phase transition and compositional change models of the 670-km discontinuity' by Alison C. Lees, M. S. T. Bukowinsky, and Raymond Jeanloz, *Jour. Geophys. Research* **90,** 2057-2059.

Sacks, I. S., and J. A. Snoke, 1977, The use of converted phases to infer the depth of the lithosphere-asthenosphere boundary beneath South America, *Jour. Geophys. Research* **82,** 2011-2017.

Vinnik, L. P., R. A. Avetisjan, and N. G. Mikhailova, 1983, Heterogeneities in the mantle transition zone from observations of P-to-SV converted waves, *Physics Earth and Planetary Interiors* **33,** 149-166.

Cross-references: *Continental Lithosphere; Earth Structure, Global; Mantle, Lower: Structure; Mantle, Upper: Structure; Mantle Convection and Plumes; Mantle Dynamics; Seismicity: Subduction Zone; Seismology: History.*

MANTLE DYNAMICS

With the development and confirmation of the theory of plate tectonics in the 1960s, it became clear that the Earth's surface is quite mobile on time scales of tens of millions of years or more. That this implies that the mantle underlying the Earth's lithospheric plates is also mobile was also quickly recognized. It was seen that some form of mantle convection, previously widely regarded as a speculative mechanism for the speculative theory of continental drift, must be associated with the moving plates. However, the form of this convection, and the connection between it and the moving plates, has been the subject of a great deal of debate. More recently, the debate has been joined by geochemists, whose evidence for several types of ancient (1-3 Gyr) chemical heterogeneities in the mantle has been difficult to reconcile with the relatively rapid stirring presumed to accompany mantle convection.

Many ideas have been considered. It has been proposed that plates ride relatively passively on large mantle convection cells, that the heavy subducted lithosphere pulls its attached plate along behind it and sinks into a largely passively mantle, and that plates slide off the flanks of the mid-ocean ridges. It has been assumed that mid-ocean ridges are sites where anomalously hot mantle material wells up, and that the mantle rising under ridges is indistinguishable from mantle elsewhere. It has been suggested that plates are largely decoupled from the underlying mantle by a low-viscosity layer, that subducted lithosphere is also decoupled from the surrounding mantle, functioning mainly as a mass source, and that the plates and mantle are tightly coupled. It has been argued that the upper mantle, above 670 km depth, flows separately from the lower mantle, and that a single flow system penetrates the whole mantle. It has been suggested that there are narrow ascending plumes of hot mantle that are not directly related to plates and plate-scale flow, that they originate at the core-mantle boundary, and that they originate at a putative thermal boundary layer at 670 km depth. It has been supposed that much of the heat transport through the upper mantle is accomplished by modes of convection smaller in horizontal scale than the plates, such as plumes or upper mantle rolls or cells, with the flow associated with the plates a sort of large-scale background.

On the chemical side, it has been proposed that the mantle has one, two and three chemically distinct layers and that there may be some chemical stratification, but not in sharply defined layers. The mantle has been characterized as remarkably well mixed and as remarkably heterogeneous. Heterogeneities have been supposed to come from internal layering in the mantle, from interactions with the core, from reinjection of continental crustal material, from detached continental lithosphere, from subducted oceanic crust and lithosphere and to be residuals from the formation of the Earth; here, at least, the possibilities are not all mutually exclusive. It has been argued that subducted lithosphere and/or crust is so heavy it sinks to the bottom of the mantle, that it is light and stays near the top, that it is buoyant only in the transition zone and accumulates there, and that the density differences are not sufficient to prevent it from being carried around by the convecting mantle.

One picture from this profusion of ideas has gradually gained empirical and theoretical support, with some recent results strongly favoring it over its main rival. In this picture, there is large-scale flow associated with the plates that penetrates through most or all of the mantle, and there are plumes ascending through this flow from, at, or near the core-mantle boundary. A moderate viscosity increase with depth, by perhaps two orders of magnitude, reduces velocities in the deepest mantle by about one order of magnitude. Chemical heterogeneities from various sources, but mainly from subducted lithosphere, are stirred through the mantle, but some survive for several billion years before being sampled at the surface.

This picture is simple enough, but it is not immediately obvious that it is consistent with observations, especially with the geochemical evidence for ancient heterogeneities. Many details remain to be quantitatively tested, but it has so far survived a number of important tests, whereas the layered mantle model, its main rival, seems to be faring poorly in these tests. Therefore, this article concentrates on explaining and justifying this general model as apparently the best current candidate for how the dynamics of the mantle work. A somewhat complementary view is discussed in *Mantle Convection and Plumes*.

Physical Processes

Here we will discuss the physical relationship between the moving lithospheric plates at the surface and the convecting mantle underneath. This discussion involves looking at the mechanical and thermal coupling between them and at various possible modes of mantle convection that are not directly related to the moving plates.

Coupling Between Plates and Mantle Convection. Mantle convection was proposed as a mechanism of continental drift by Arthur Holmes in 1928, and continued to be regarded favorably by those disposed to take the concept of continental drift seriously. However, when the theory of plate tectonics was developed in the 1960s, the picture that emerged at the Earth's surface did not look much like the conventional view of a convecting fluid: simple, regular cells with nearly symmetrically diverging and converging zones, smooth variations of surface velocity with position and, if there is time-dependence, slow, smooth changes. Instead, the plates have a large range of sizes, irregular shapes, uniform (angular) velocities within plates and sharp changes in velocity at plate margins, symmetrical divergence but completely asymmetrical convergence (only one plate subducts), and occasional sudden changes in plate direction or spreading center location (see *Paleomagnetism and Plate Tectonics*). Each of these properties of plate geometry and kinematics is indicative of the mechanical properties of the lithosphere, the strong, but brittle cool thermal boundary layer near the Earth's surface that the plates comprise. In other words, it seems that the surface flow pattern of our dynamic mantle is controlled by the mechanical properties of the lithosphere rather than by the mechanical properties of the more fluid underlying mantle.

The lithosphere's defining property is its strength, and it derives its strength from its relatively low temperatures. This strength then tends to prevent the cool thermal boundary layer from sinking under its excess weight. Experiments with fluids whose viscosity varies strongly with temperature indicate that only the lower part of the thermal boundary layer, which has a viscosity no more than two or three orders of magnitude greater than the underlying mantle, will be mobile enough to detach and partake in any independent flow in the underlying fluid. This observation limits the mobile part of the thermal boundary layer to that which is less than about 300°C cooler than the mantle interior. In other words, the lithosphere extends from the surface to about the 1100°C isotherm. This range includes about 85% of the negative buoyancy of the thermal boundary layer. Only about 15% of the negative buoyancy included in the range from about 1100°C to 1400°C will have some freedom to generate independent flow under the lithosphere.

Together, these features imply that the lithosphere largely controls the locations of upwelling and downwelling at the top of the mantle: upwelling occurs under spreading centers, downwelling occurs at subduction zones, and very little of either occurs in between. In fact, the lithosphere may have a strong influence on the whole structure of convection in the mantle. This inference is supported by some numerical convection models illustrated in Fig. 1. Figure 1a shows convection in a fluid of uniform viscosity with a free-slip (zero shear stress) top surface and with uniform heating throughout, as would result from radioactivity in the mantle. This convection is of the conventional type, with cell widths comparable to the layer depth and relatively smooth variations in surface velocity. Figure 1b shows convection in a fluid with a temperature-dependent viscosity, so that the top thermal boundary layer has viscosities of up to 100 times that in the interior. This stiff layer is kept mobile, like the Earth's lithosphere, by imposing a uniform horizontal velocity on the top

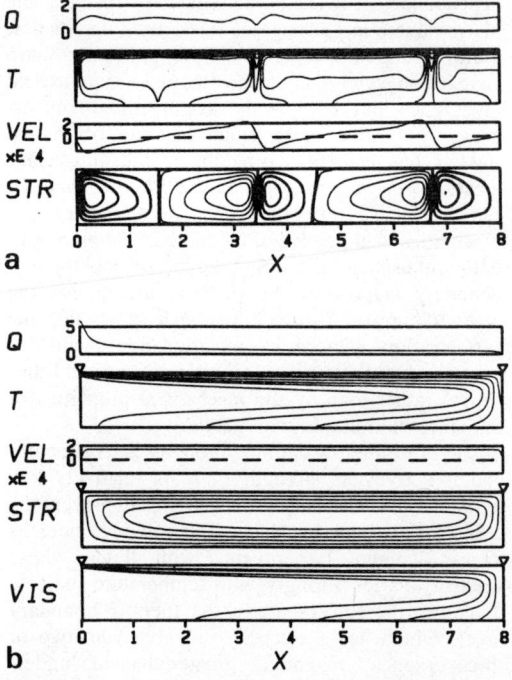

FIGURE 1. Comparison of convection without and with a simulated mobile lithospheric plate. In each part, Q denotes the surface heat flux, T denotes the internal thermal structure, represented as isotherms, VEL denotes the surface horizontal velocity, STR denotes streamlines, which show the instantaneous flow direction and VIS denotes viscosity, shown contoured. Units are dimensionless. (a) shows convection with a free-slip (zero shear stress) top surface and a constant viscosity fluid. (b) shows convection with an imposed top surface velocity profile (scaled from a plate velocity of about 15 mm/yr) and a temperature-dependent viscosity with a factor of 100 total variation (highest at the top). Sides and bottom are free-slip and insulating; heating is internal and uniform; Rayleigh number, based on heat flux, is 10^6; Peclet number in part (b) is 150.

surface, scaled from observed plate velocities. Thus this model incorporates both aspects of the lithosphere discussed earlier, and the result is a single, long, stable convection cell. Upwelling and downwelling are confined to the ends of the box, where the reflecting boundary conditions imply spreading and (symmetric) subduction.

The flow in Fig. 1b is driven largely by the sinking cool boundary-layer fluid, since the imposed surface velocity in this model is relatively slow (corresponding to a plate velocity of about 15 mm/yr). In this respect it is like the convection in Fig. 1a; the main effect of the "lithosphere" has been to change the structure of the flow. Similarly, we can expect that in the mantle the subsiding lithosphere, which comprises most of the thermal boundary layer, provides the main driving force within the mantle.

A Decoupling "Low Viscosity Zone"? It is likely that the viscosity of the mantle goes through a minimum between about 100 and 200 km depth because of the competing effects of temperature and pressure. The rapid rise in temperature in the first hundred kilometers or so causes a rapid drop in viscosity, but the temperature gradient then drops from 10–20°C/km to about 0.3°C/km in the underlying mobile mantle. The effect of the steadily increasing pressure is to increase viscosity, and this increase is probably dominant below the shallow thermal boundary layer. Estimates based on the rebound of the Earth's surface after the removal of ice-age glaciers and lakes tend to confirm this, suggesting a minimum viscosity of 10^{19} Pa · s (10^{20} poise), compared with an average upper mantle viscosity of about 10^{21} Pa · s. These observations are confined to continental regions. In oceanic regions, indirect observations must be used: There is a more pronounced minimum in seismic shear wave velocities in this depth range under oceans, most probably caused by the same combination of temperature and pressure effects. Thus there may be a more pronounced viscosity minimum, especially near oceanic spreading centers where high mantle temperatures rise nearer to the surface. See *Mantle Viscosity* for more detail.

The probable existence of this "low viscosity zone" has led some to postulate that it acts as a lubricating layer that tends to decouple the plates from the underlying mantle. However, two things argue against taking this idea too literally. First, calculations have shown that a layer only 100 km thick or less under a plate thousands of kilometers wide would have to have a viscosity four or five orders lower than the underlying mantle before the viscous drag from the moving plate was prevented from penetrating through into the deeper mantle. Such a pronounced drop in viscosity may not be implausible under younger oceanic lithosphere, although it has not been convincingly argued either way, but it is unlikely under older oceanic lithosphere and under stable continental regions, so there should be significant coupling there. Secondly, and more importantly, the subducted lithosphere is traced by deep earthquake zones to much greater depths, so the underlying mantle will still be driven by this dense sinking material. In effect, both the plates and the deeper mantle below the low viscosity zone will be driven by the same sinking boundary layer, so they will tend to move together whether the coupling between them is strong or weak.

One extreme model, which was discussed soon after plate tectonics was formulated, had the "return flow," which complements the moving surface plates, returning through the low viscosity zone. It seems quite unlikely that the viscosity minimum is so low as to permit this. Another popular view has been that the return flow would occur between the low viscosity zone and the 670-km seismic discontinuity, which has been supposed by many to be a barrier to mantle flow. Arguments against the latter view will be given later. The effect of a moderate low viscosity zone in plausible models of mantle dynamics has not really been quantitatively evaluated, but the above arguments indicate that it will not have a first-order effect on the structure of mantle flow.

Transport of Heat by the Plates. A discussion of the relationship between the plates and mantle convection should also consider thermal aspects. The plates may not look much like the surface pattern of "conventional" convection, but their motion accounts for the transport of a considerable amount of heat, as will now be demonstrated, and is apparently largely driven by body forces resulting from temperature differences that are associated with the subducted lithosphere. Plate motions therefore constitute thermal convection in the general sense.

Plates form at spreading centers and then cool, thicken, and subside as they drift away, so that they are typically 2–3 km deeper below sea level when they reach a subduction zone. The subsidence is caused by thermal contraction as the plate cools by conduction to the surface, which accounts quite well for the existence of the mid-ocean ridges, which are the dominant topography of the sea floor. The thermal role of the lithosphere, like any thermal boundary layer, is to bring heat to the surface and there to lose it, thus accomplishing the transfer of heat from the interior of the convecting mantle though the top surface.

The amount of heat transferred in this way can be estimated from the amount of subsidence. If the thickness from which heat is lost is L, and the average temperature drop is $\Delta T/2$, where ΔT is the temperature difference between the surface and the mantle interior, then the change in heat content per unit surface area of plate is

$$H = \frac{\rho C_P L \Delta T}{2} \qquad (1)$$

where ρ is density and C_P is specific heat. The thermal contraction caused by this heat loss is

$$h = \frac{\alpha L \Delta T}{2} \qquad (2)$$

where α is the volume coefficient of thermal expansion. Eliminating $L\Delta T$ from equations 1 and 2 gives

$$H = \frac{\rho C_P h}{\alpha} \qquad (3)$$

If the average areal rate at which sea floor is created (and destroyed) is A, then the total heat flow due to this lithospheric cooling is

$$Q = \frac{A\rho C_P h}{\alpha} \qquad (4)$$

Typical values of the quantities in equation 4 are $A = 3$ km^2/yr, $\rho = 3300$ kg/m^3, $C_P = 750$ J/kg K, $h = 3$ km and $\alpha = 3 \times 10^{-5}$ K^{-1}, which yield $Q = 2.5 \times 10^{13}$ W. The heat flow actually observed through the ocean floor is about 3×10^{13} W. Thus, taking account of the approximations in this estimate, it is clear that the moving plates account for a large proportion of the heat removed from the mantle. (A further 1.2×10^{13} W of heat emerges from the continents, but only half or less of this comes from the mantle. The balance is generated in the crust by radioactivity.) In other words, the moving plates are the dominant mode of convection transferring heat to the Earth's surface.

The Depth Extent of Flow Associated with the Plates. This aspect of mantle convection has been the most controversial. A widely held view has been that the 670-km seismic discontinuity is the site of a small change in mantle composition, with the lower mantle being intrinsically denser by a few percent, implying that there was no flow across this level, and the density change is usually taken to be the mechanism preventing such flow.

The main reasons for this assumption are probably that deep earthquake zones, which are interpreted to trace the path of cool subducted lithosphere, cease abruptly at 670-km depth in several widely separated places, and the stress orientations deduced from the seismic waves emitted by these deep earthquakes indicate the deeply penetrating lithosphere encounters resistance from below (the lithosphere is in down-dip compression). One possible inference is that the subducted lithosphere encounters an impenetrable barrier (the density increase). However, an equally plausible inference is simply that there is increased resistance, perhaps due to increasing viscosity or to phase transformations in this vicinity (see *Mineralogic Phase Transitions in the Earth*), but it is not sufficient to stop the further penetration of the lithosphere. The cessation of earthquakes is also plausibly explained by the occurrence of phase transformations, which would tend to relieve stresses and possibly remove the material from the brittle regime. In fact, the proposed increase in intrinsic density at 670 km has never been convincingly demonstrated. There is a substantial density increase, but a large part of it is clearly due to phase transformations (see *Mineralogic Phase Transitions in the Earth*).

Many arguments have been proposed to settle this controversy, but none has succeeded. Recently, however, a new one has been demonstrated that seems likely to be conclusive. If there were a barrier to convection at 670-km depth, then heat would have to be transferred across it by conduction. Since the radioactivity in the upper mantle is sufficient to account for only a small fraction of the surface heat flow, nearly as much heat would have to flow into the base of the upper mantle as flows out the top. This means that there would be a hot thermal boundary layer at the base of the upper mantle that would be nearly as pronounced as the cool thermal boundary layer at the top. Thus hot, buoyant material would be generated at the base at about the same rate as cool, negatively buoyant material is generated at the top. When this material rose to the base of the lithosphere, it would generate topography that would be comparable to that due to the cooling lithosphere, since the buoyancy flux would be comparable in magnitude, though of opposite sign.

The effect is illustrated in Fig. 2. Figure 2a shows a numerical convection model heated from below. This model has two diverging "plates," the result of a temperature-dependent viscosity and an imposed horizontal surface velocity profile (plotted as "VEL"). In fact the spreading center between the plates is migrating in the model, simulating real spreading center migration, which results from symmetric spreading combined with unequal plate speeds (the spreading center velocity is the average of the vector velocities of the two plates). The model was started from a steady-state model with a stationary spreading center in the middle of the box. Thus, initially, there was a column of buoyant fluid rising under the spreading center while columns of cool fluid descended at the ends of the box. As the spreading center migrated, the rising column of hot fluid trailed behind, so it is no longer centered under the spreading center. The resulting topography is illustrated in Fig. 2b: It is completely asymmetric about the spreading center. In fact, the spreading center is no longer a topographic high. This result is obviously completely inconsistent with the actual topography of the seafloor, in which spreading centers are consistently at the crests of the mid-ocean rises (to the point where the terms are used interchangeably), and the mid-ocean rises are the dominant seafloor topography, as already noted. The model in Fig. 2 predicts gravity and geoid anomalies that are also inconsistent with observations, and this is true also when the spreading center is stationary.

The model shown in Fig. 2 is but one example of what must be a general phenomenon. In fact, the result could have been anticipated. We noted earlier that the subsidence of the ageing seafloor is well accounted for by the cooling lithosphere, which implies that the underlying mantle is relatively passive, in the sense that there are no large temperature variations that would produce topography at the surface. Since buoyant material rising from a 670-km boundary layer would clearly constitute "active" mantle material, there can be no such boundary layer.

If, on the other hand, the fluid is heated predominantly from within, then the buoyancy flux from the base is reduced in proportion and the fluid more

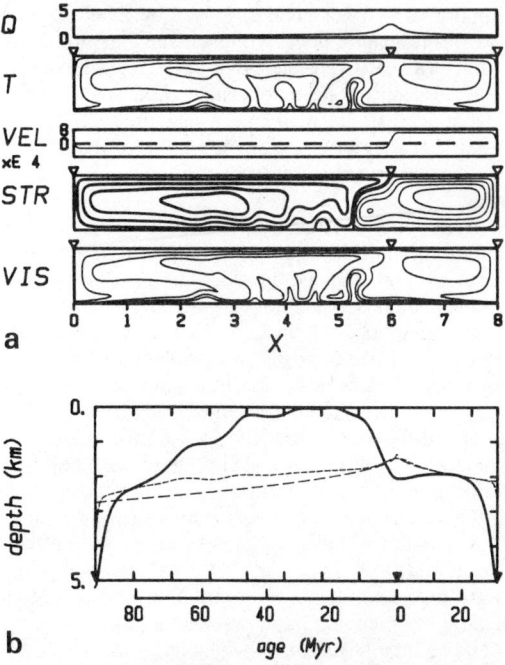

FIGURE 2. (a) Convection in a base-heated fluid under a simulated migrating spreading center. The surface velocity profile is imposed, and the spreading center has migrated from the center of the box ($x = 4$) at a speed corresponding to about 15 mm/yr. The viscosity is temperature-dependent, the dimensionless quantity varying from 0.2 to 5; it is near 1 in the interior of the fluid. The high Rayleigh number (10^6), low viscosity and lower velocities deep under the left "plate" have allowed the bottom boundary to become unstable: The irregularities are due to hot blobs of fluid detaching from this boundary layer. The Peclet number is 600, corresponding to a plate velocity of about 60 mm/yr. (b) Topography from the model in Fig. 1a (solid curve), scaled to the mantle. The topography is due mainly to the hot fluid rising from the base. The migrating spreading center no longer coincides with the topographic high because the hot rising fluid is no longer centered under it. Also shown are the predicted topography due to a simple conductive cooling model of the top boundary layer (long-dashed) and the topography due to the actual top thermal boundary layer (short-dashed).

closely approximates the passive fluid assumed in the standard theory of seafloor subsidence. Numerical models, not shown here, have shown that the predicted topography then conforms reasonably well with what is observed.

While one might argue that an internal boundary layer could be tolerated at some deeper level in the mantle, such that most of the surface heat flux is generated above it and it would therefore generate a smaller buoyancy flux, which might be tolerable, there is no independent evidence for such an interface within the mantle. The straightforward conclusion is that there is no boundary to flow within the mantle, so that the flow associated with the plates penetrates to the core-mantle boundary and any hot thermal boundary layer would be located there. The latter possibility will enter the following considerations.

Mantle Plumes. So far we have been concerned with plates or with mantle flow directly associated with the plates. Now we turn our attention to some postulated modes of mantle convection not directly associated with the plates.

Mantle plumes are thought to be narrow columns or pipes of hot material rising from deep in the mantle under volcanic hotspots. They are discussed in more detail in *Mantle Convection and Plumes*. Here we focus only on their buoyancy flux and the topography they generate. The discussion is analogous to the above estimate of oceanic heat flux from mid-ocean rise topography.

Volcanic hotspots such as Hawaii or Iceland typically have a broad, low rise around them. In the case of Hawaii, the rise, or "swell" is about 1000 km across and about 1 km high. It is well established that the hotspots, of which there are 50 or more, tend to move only very slowly relative to each other (no faster than about 10 mm/yr), whereas plates move somewhat faster (50–100 mm/yr). The result of the rapid motion of the Pacific plate over the inferred Hawaiian hotspot source is that a trail of volcanic islands and seamounts has been left on the Pacific plates, which stretches to the northwest for thousands of kilometers (the Hawaiian-Emperor seamount chain). In order for the swell surrounding Hawaii to persist, it must be renewed at the southeastern end. Assuming, as is usual, that the swell is due to buoyant material rising under the lithosphere at Hawaii, the rate at which new topography is raised in the swell must be a measure of the buoyancy flux in the underlying plume. The velocity of Hawaii over the plume is about 100 mm/yr, so every year a strip this wide and 1000 km long must be elevated by about 1 km: New topography is created at a rate of about 10^8 m^3/yr or 3 m^3/s. Similar results are obtained for the several other major Pacific hotspots. On the other hand, the same calculation for hotspots on other plates, most of which are moving much more slowly relative to hotspots, give much smaller values. For example, for Cape Verde the flux is less than 0.2 m^3/s and even for Iceland, which has a swell perhaps 2000 km wide and 3 km high, the flux is only about 1.6 m^3/s.

If the buoyancy flux under all hotspots is added up, the result is less than about 30 m^3/s. From this we can deduce the plume heat flux, in a way similar to the derivation of Eq. 4. If the buoyancy flux is S, then

$$Q = \frac{\rho C_P}{\alpha} S \quad (5)$$

With the values used earlier, we get $Q = 2.5 \times 10^{12}$ W for the plume heat flux. This is less than 10% of the global heat flow (4.2×10^{13} W).

Although the estimates used here are fairly rough, it is clear that plumes cannot be transporting a large fraction of the heat flow through the mantle. Thus it cannot be argued that the buoyancy from the putative boundary layer at 670 km is being transported in the plumes.

Plumes were originally suggested by J. Tuzo Wilson, and the idea was later taken up by Jason Morgan, who proposed that they were coming from the core-mantle boundary. The heat flow emerging from the core can be roughly estimated, for example, by estimating the adiabatic temperature gradient and the thermal conductivity, or by using the rate at which the Earth's interior is believed to be cooling (about 100°C per billion years), and the result of both estimates is about 3×10^{12} W. While these estimates have considerable uncertainty, their similarity to the plume heat flux indicates that the idea that the plumes are coming from the core-mantle boundary is quite plausible. Thus, heat emerging from the core would generate a thermal boundary layer at the base of the mantle, and the resulting buoyant material would rise as plumes, eventually to be manifest at the surface as hotspots.

Other Small-Scale Modes of Mantle Convection. Other modes of mantle convection with horizontal scales less than typical plate sizes have been suggested, including upper mantle cells or rolls and boundary layer instabilities at the base of the lithosphere. Both of these modes were proposed to explain the fact that old seafloor does not subside as rapidly as predicted by simply assuming conductive cooling to the surface. An earlier motivation was that oceanic heat flux also seemed to decline less rapidly than the conductive cooling model predicted, but this turned out to be an artifact resulting from the neglect of hydrothermal convection in young oceanic crust, which caused the heat flux from young crust to be underestimated. It also now seems that the "flattening" of seafloor subsidence is adequately explained by the effects of hotspot tracks and their associated residual swells. Thus the empirical motivation for these modes can be questioned, but there are also strong, direct arguments against them being major modes of mantle convection.

The upper mantle cells were envisaged as extending down to 670 km depth and having a comparable horizontal extent. If they were under a fast plate, the preferred geometry would be rolls with axes aligned with the direction of plate motion. If they were to stop the subsidence of old lithosphere, they would have to transport as much heat to the base of that lithosphere as emerges at the surface, i.e., at least 40 mW/m^2 or about half the global average flux. This hypothesis fails for the same reason that the large-scale upper mantle mode discussed earlier fails: It would generate too much topography. We would expect topographic anomalies of 1 km or more amplitude under old oceanic lithosphere. At first thought it might seem that the residual swells along hotspot tracks might be reinterpreted as due to the presence of this mode of convection, and in fact there has been some discussion of this point. However, upper mantle cells provide no obvious explanation for the relatively sudden onset of the swells near the hotspots, nor for the near stationarity of the hotspots themselves.

A boundary layer instability at the base of the lithosphere has also been proposed to accomplish the same purpose. The idea is that the lowermost lithosphere is mobile enough to become unstable and fall away. The earlier discussion suggests that perhaps 15% of the negative buoyancy of the lithosphere might be available for this mode. The instability would start when a local critical Rayleigh number was exceeded as the lithosphere thickened and cooled. The effect would be to prevent the lithosphere from thickening any further and, it has been argued, thereby to prevent the further subsidence of the lithosphere. The flaw in this argument is that it ignores the underlying mantle, and in particular the cool material that is dropping off the lithosphere.

The preferred mode of instability would again be rolls aligned in the direction of plate motion, but a different mode, illustrated in Fig. 3, can demonstrate the problem. The model in Fig. 3a has a uniform velocity imposed at the top, as in the earlier models, but now the fluid has a uniform viscosity to enhance the tendency of the top boundary layer to drop away from the surface. (This uniform viscosity is necessary in a two-dimensional model because the instability tends to be suppressed by the shearing in the main flow. The preferred mode of "longitudinal" rolls is intrinsically three-dimensional and so much more difficult to model.) It is evident in Fig. 3a that the thickening boundary layer becomes unstable and drops blobs of cool material into the interior. The topography resulting from this model is shown in Fig. 3b. The main feature in the topography is a series of large depressions over the falling blobs. These depressions are much larger in amplitude than can be tolerated by observed seafloor topography. Furthermore, although the instability stops the boundary layer from thickening, it does not stop the surface from subsiding, because the influence of the cold material is still being felt at the surface, even after it has subsided into the interior. The first-order effect of the instability is merely to exchange the vertical positions of warm and cool material. Thus this mode of convection does not accomplish what it was supposed to have, and instead it produces unacceptably large depth anomalies. It can be concluded that if the instability exists (and it is quite plausible that it does), then it can be responsible for transporting only a small fraction of the heat flux emerging at the ocean floor. In other words, it can only be a minor mode of mantle convection.

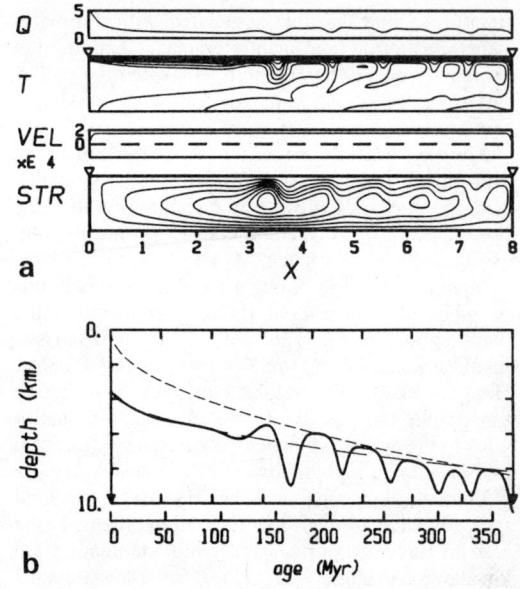

FIGURE 3. Effect of a top boundary layer instability on topography. (a) Convection in an internally heated, constant viscosity fluid with the top velocity profile imposed. The initial state was that shown in Fig. 1: With the viscosity variations eliminated, the top boundary layer is unstable, dropping cool blobs into the interior. Rayleigh number is 10^6 and Peclet number is 150. (b) Topography from this model (solid curve). The short-dashed curve is the topography predicted from simple conductive cooling of the top boundary layer, using the model parameters, while the long-dashed curve is of the same form (depth proportional to square root of age) but empirically fitted to the model topography (as is done with the seafloor topography). It is clear that the general subsidence continues after the instability develops, and its main effect is to generate unacceptably large depressions over the descending blobs.

Viscosity Increase with Depth. Our knowledge of the viscosity of the mantle has changed a lot in the past two decades. It used to be thought by many that the viscosity was so high that mantle convection would be impossible, although studies in the 1930s of postglacial rebound clearly indicated an upper mantle viscosity of about 10^{21} Pa · s. The advent of plate tectonics and further studies of postglacial rebound and of the Earth's rotation led to the view that the mantle's viscosity is very uniform. This view now seems to have been an over-reaction to the earlier view, with some recent studies of postglacial rebound and rotation suggesting that the lower mantle may have a viscosity one or two orders of magnitude higher than the upper mantle. An independent constraint has come from the study of the geoid near subduction zones, which indicates a fairly sharp increase in viscosity by one to two orders of magnitude in the transition zone (see *Mantle Viscosity*). The vertical resolution is not yet very good, but the picture suggested by these results is of a smooth increase in viscosity caused by the effect of increasing pressure, with one or more sharp increases through the transition zone due to phase transformations there. The total variation through the depth of the mantle may be up to three orders of magnitude, with a substantial portion of that increase occurring in the transition zone. This picture would be consistent with our limited knowledge of the effects of temperature, pressure, and phase transformations on the relevant materials, which would indicate, if anything, a larger increase in viscosity with depth.

An increasing viscosity with depth is likely to have some important effects on mantle dynamics. Convective velocities would be slower at depth, but not in simple proportion to the viscosity increase: a decrease by about one order of magnitude would be likely, providing an immediate explanation for the slow relative velocities of hotspots at the surface if they are due to plumes rising from the base of the mantle. It also would slow the stirring of chemical heterogeneities and increase their residence times in the mantle, as will be discussed in the next section.

Summary: The Dynamics of the Plate-Mantle System. The picture that emerges from the above discussion is fairly simple. The mantle convects as a single layer and with two modes of flow for which there is clear evidence: a large-scale circulation arising from the cool, stiff top boundary layer and small-scale plumes arising from a bottom boundary layer. The plates, being pieces of the top boundary layer, are an integral part of the large-scale circulation; in the terms used earlier, they are strongly coupled to the large-scale flow. The large-scale circulation, including the plates, accounts for about 90% of the heat transport out of the mantle. The plumes seem to occur fairly independently of the large-scale flow, although there is a tendency for them to cluster near spreading centers and a better correlation with regions of the deep mantle inferred, from recent three-dimensional seismic velocity models and from the geoid, to be relatively warm. The plumes transport less than 10% of the heat emerging from the mantle. Other modes of convection may exist, such as a boundary-layer instability of the lower lithosphere, but they must be minor, and no compelling evidence for them has been presented. A moderate increase in viscosity with depth, suggested by some independent evidence, would explain the low relative horizontal velocities of plumes and would have important effects on the persistence of chemical heterogeneities in the mantle, as we shall now see.

Stirring and Persistence of Chemical Heterogeneities

If the physical constraints preclude significant chemical layering in the mantle, then attention is focussed on how to reconcile geochemical observations of ancient mantle heterogeneities with a mantle that is stirred throughout its depth. Some recent work has shown that stirring in this kind of flow proceeds in surprising ways, which are not like stirring in turbulent, multiscale flows such as occur in the ocean and atmosphere.

Geochemical Constraints. Some of the key geochemical evidence is summarized in Fig. 4, which shows plots of the various isotopic ratios measured in young oceanic basalts, which are derived fairly directly from the mantle. Each numerator isotope has a radiogenic contribution, that is, it is produced by radioactive decay of a parent isotope. These plots thus contain information about chemical heterogeneity and about age, since the abundance of a radiogenic isotope depends on both the abundance of the parent and on time.

In each plot a field is identified as MORB: mid-ocean ridge basalt. The remaining fields are from oceanic island basalts (OIB), and specifically from islands and seamounts in hotspot chains. Generally, the range of variation of MORB is less than the range of OIB. It is pointless to get into semantic arguments about whether MORB is "very" uniform or not. Some have argued that MORB is uniform or "well-mixed" while OIB is not, implying that they form distinct populations with distinct sources, but there does not seem to be any clear demarcation between the populations. In fact the restricted range of MORB in these plots is to some extent an artifact of the fact that some on-ridge hotspots, notably Iceland, are counted as OIB rather than MORB. The difference in ranges is probably connected to the fact that melting is the norm for material rising under a spreading center, so that MORB tends to be a fairly random sample, whereas melting is clearly abnormal away from spreading centers, so OIB must be a biased sample.

The data in Fig. 4a spread out in a fan away from the lower left corner, labelled "DM" for "depleted MORB". This implies that there must be at least three types of source in the mantle that are separated on the plot at least as much as the points DM, A and B. The intervening points could be due to mixtures of these source types or they could have distinct source types of their own. Taking account of isotopes of other elements, such as helium, argon and xenon, some authors have identified as many as five source types or end members. A direct implication is that a simple mantle model comprising two uniform layers cannot explain even the data in Fig. 4a: At least one of the layers would have to be heterogeneous to accommodate more than two source types, so the problem of the survival of ancient heterogeneities is not avoided by invoking layers, even aside from the strong geophysical arguments against layering given above.

Large apparent ages are exhibited in several ways, most obviously in Fig. 4 by the positive correlation evident in Fig. 4c. Isochrons plot on this diagram as

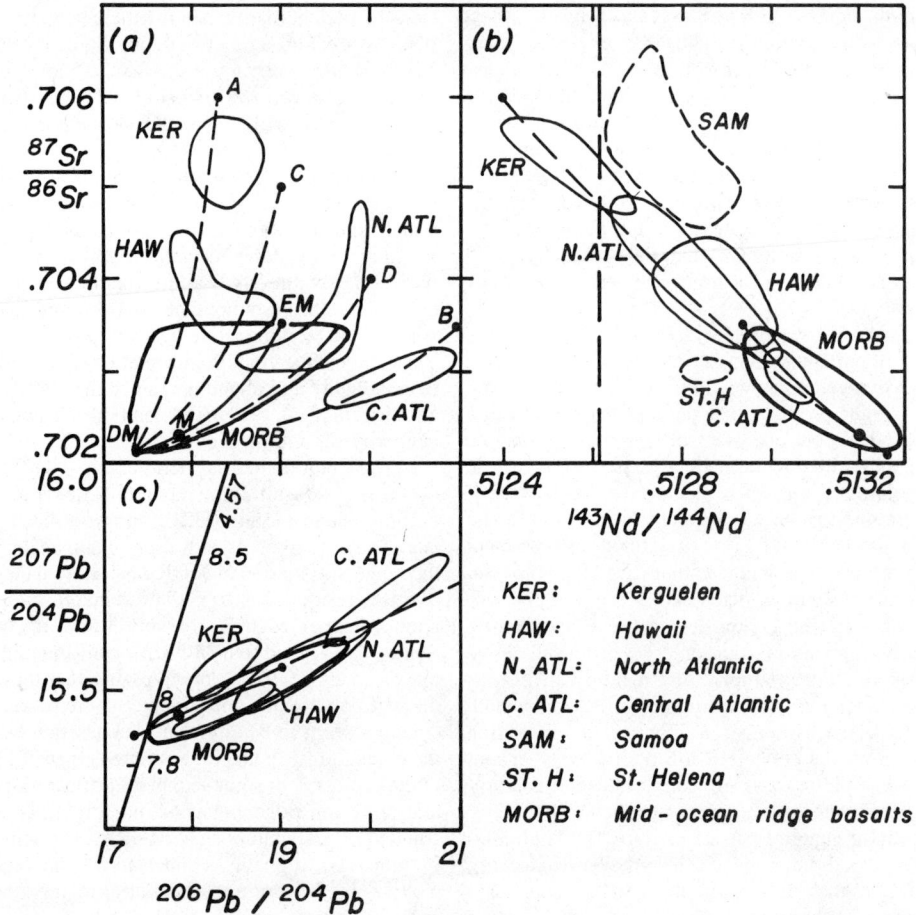

FIGURE 4. Summary of some measured isotopic variations in oceanic rocks derived from the mantle, including midocean ridge basalts (MORB) and oceanic island chain basalts (but not island arc basalts). Dashed curves are estimated mixing lines between representative compositions. Solid curve is a mixing line between depleted mantle (DM) and a typical enriched mantle (EM), with an estimated mean at M. In (b), the vertical dashed line is the mean composition of chondritic meteorites. In (c), the sloping line is the 4.57 Gyr meteorite isochron, marked with values of $\mu = {}^{238}U/{}^{204}Pb$.

straight lines, with age directly related to slope. The apparent age of the MORB field on this plot is about 1.6 Gyr, while individual island groups give apparent ages ranging from 1 to 3 Gyr. The term *apparent age* is used here because the data might be derived from mixtures of several sources, and this can blur the ages or even give unphysical apparent ages. However, other isotopic systems indicate comparable apparent ages, and so it is fairly clear that they represent some kind of mean age of the mantle source. Since the basalts from which these data were obtained are all relatively young (a few million years), these old ages must correspond to some much earlier event in the mantle source.

Several possible origins for these mantle heterogeneities have been suggested: subducted lithosphere (in which we can distinguish oceanic crust, the depleted mantle underlying the oceanic crust, and marine sediments, which are mainly of continental origin), detached continental lithosphere, separate mantle layers, and "primitive" mantle (i.e., mantle which has remained distinct since soon after the formation of the Earth).

It is clear that heterogeneities have been introduced by subduction of lithosphere. Detached continental lithosphere is a plausible source, but it has not been clearly established. A continental crustal component has been suggested by some OIBs with high values of $^{87}Sr/^{86}Sr$ and $^4He/^3He$, both of which are characteristically high in the continental crust. Separate mantle layers are incompatible with the geophysical arguments given above.

Primitive mantle was originally suggested on the basis of some basalts (continental and oceanic) whose ratio $^{143}Nd/^{144}Nd$ was the same as that in primitive (chondritic) meteorites, but the occurrence of these values in the basalts now seems to be coincidental in the light of more abundant data. Other evidence for primitive material comes from the occurrence of low values of ^{4}He/^{3}He in some OIBs (notably Hawaii), which are interpreted as reflecting a lower level of degassing of these sources, so that the radiogenic ^{4}He component is proportionately less. However, the relationship between "less degassed" and "primitive" in terms of other, refractory isotopic systems is not a direct one at all, especially since very little is known about how volatiles are lost from the mantle and there are virtually no constraints on the relative proportions of more and less degassed mantle. Anomalous ratios of ^{129}Xe/^{130}Xe have also been interpreted in terms of a primitive mantle component, since ^{129}Xe is a daughter of ^{129}I, which has a half life of only 17 Myr, so it should have become "extinct" very early in the Earth's history. If we take the relatively few Xe data at face value, then they certainly constitute a primitive signature. We can conclude that there are indications that some mantle sources have remained distinct since very early in the Earth's history, and some sources (not necessarily the same ones) have been less degassed than most. The proportions of the mantle which these source types comprise is unclear.

To summarize the observations, isotopic (and therefore chemical) heterogeneities in the mantle are clearly established. There are at least three and possibly many more source types. Heterogeneities as old as 2–3 Gyr are clearly established, and some may be virtually as old as the Earth (i.e., about 4.5 Gyr). Some portions of the mantle have not been degassed as much as others. Many heterogeneities are probably derived from subducted lithosphere, and some may be derived from detached continental lithosphere or from mantle which has been processed very little since very early in earth history.

Stirring of Heterogeneities. At a velocity of, say, 30 mm/yr, the vertical transit time of material through the depth of the mantle (about 3000 km) is 100 Myr. Thus a typical overturn time of the mantle is a few hundred million years. Furthermore, the Earth's interior was probably warmer in the past, because of higher radioactivity, and this would have meant lower viscosities and faster convection. Estimates from thermal history models suggest perhaps twice the heat flow and four times the velocities at the end of the Archean era, 2.5 Gyr ago, and perhaps several hundred transits since the formation of the Earth. It is not obvious that mantle heterogeneities from early in Earth's history could have survived to the present.

Some simple numerical stirring experiments are illustrated in Fig. 5. In Fig. 5a, a simple two-cell, steady circulation carries a passive heterogeneity, comprising the square of dots in the upper center, into a stagnation point of the flow. There the heterogeneity is stretched into two parts, but the parts are connected by a very thin "tendril" which has such a small volume of the heterogeneity in it that no tracer points plotted within the central part. Elsewhere, in the upper corners, the heterogeneity remains substantially thicker. This stagnation point stretching process is the most drastic stirring event undergone by heterogeneities.

If the flow is time-dependent, rather than steady, some different phenomena come into play. An example is shown in Fig. 5b. The initial flow is like that in Fig. 5a, but the boundary between the cells oscillates to the left and right, with each cell alternately growing and shrinking. Only the corners of the box are drawn, so as not to obscure the tracers near the edge. A sequence is shown, marked by the number of transit times that have elapsed. An initial circular heterogeneity is shown in the top left, initially within one of the cells. In the early part of the flow it is simply sheared out within that cell, but after 15 transit times some of it has passed through a stagnation point and been stretched into a very long tendril. In this experiment, tracers were used to mark the perimeter of the heterogeneity, rather than the interior, and new tracers were added as necessary to display the tendrils regardless of how thin they became. Thus, the number of tracers is not a measure of the volume of fluid, as it was in Fig. 5a. Note that at 15 transit times, the main part of the heterogeneity is in the right-hand cell, which means it is being sheared in the opposite sense to the initial shearing in the left-hand cell. After 22 transit times, the tendril has become very long, thin, and convoluted, passing completely around the edge of the box. Yet there is still a portion of the heterogeneity, shown inset, which has a maximum thickness, θ_{max}, 10–20% of the initial diameter, θ_0.

What has happened in this experiment is that the tendril, once formed, is so long that it is frequently carried through a stagnation point and stretched further. On the other hand, the thickest part of the heterogeneity is only occasionally carried near a stagnation point, as is evident from the fact that it survived over 10 transit times before forming a tendril. Furthermore, when the thick part of the heterogeneity passes into the other cell, the sense of shearing reverses and it is partially "unmixed." The result is that a significant fraction of the heterogeneity remains relatively thick and compact, while the rest is drawn out into extremely thin tendrils.

It can be shown that the length of the tendrils increases exponentially with time, and their average thickness decreases exponentially with time. Thus this part of the heterogeneity is stirred into fine streaks, which will be unlikely to be separately sampled by melting near the Earth's surface. On the other hand, the thickest part of the heterogeneity undergoes simple shearing, which is often reversed in sense in time-dependent flow, so its thickness decreases only inversely with time, and at an average rate substantially less than the rate within a steady cell. The results of several such numerical experi-

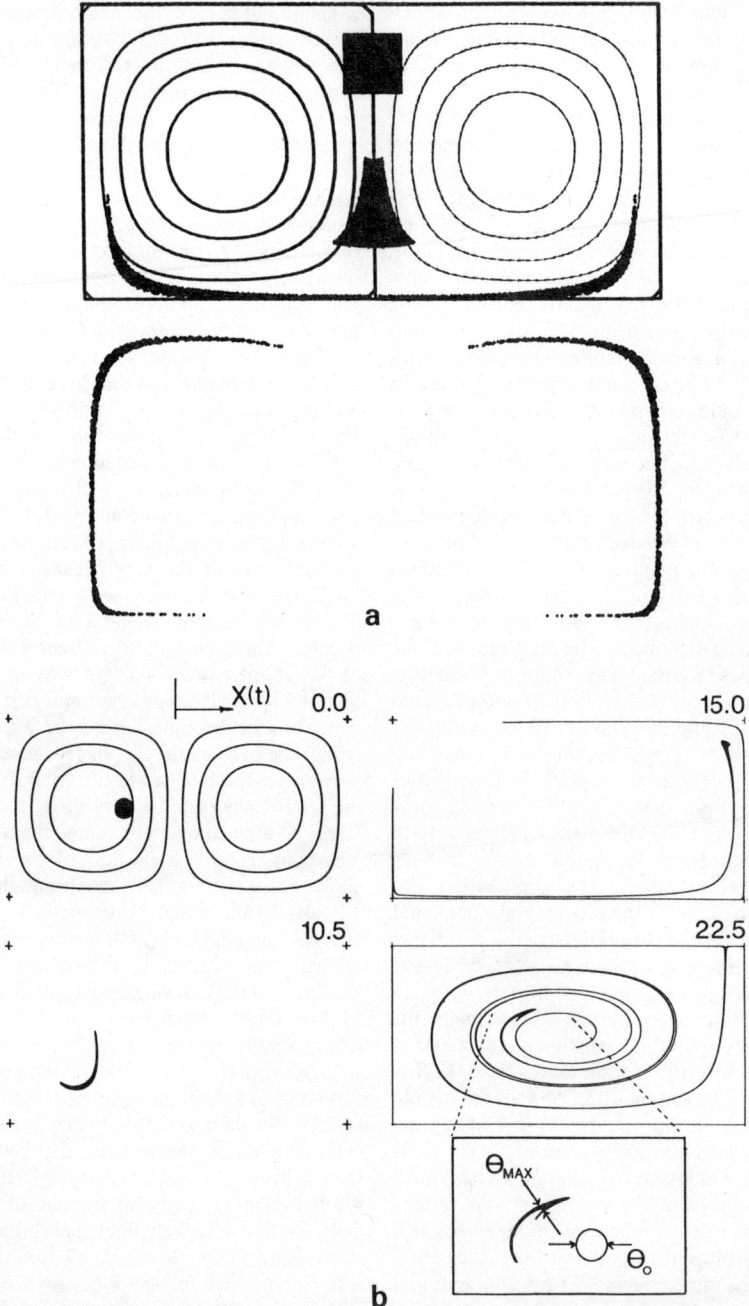

FIGURE 5. (a) Stirring of an array of passive tracers advected into a stagnation point of a simple flow, shown as streamlines. The initial array of 2500 tracers (black square) is carried downward and shown at several successive times. For the last time, the tracers are shown below the box, so they can be distinguished from the sides of the box. The two tracer groups are joined by a very thin tendril in which no tracers have plotted. (b) Stirring of a passive blob by an unsteady flow. The flow is like that in (a), but the boundary between the cells oscillates to the left and right, the cells growing and shrinking. Streamlines are shown only in the first frame. The times of successive frames are shown by the number of transit times (the time for a vertical transit at a typical velocity). The tracers are here used to mark the perimeter of the blob, with tracers added as necessary, so that thin tendrils can be resolved. Only the corners of the box are marked, so as not to obscure tracers. After 22.5 transit times, very thin and convoluted tendrils extend around the box, but a relatively thick portion of the original blob remains (inset). θ_0 denotes the initial blob diameter and θ_{max} denotes the maximum thickness of the stirred blob.

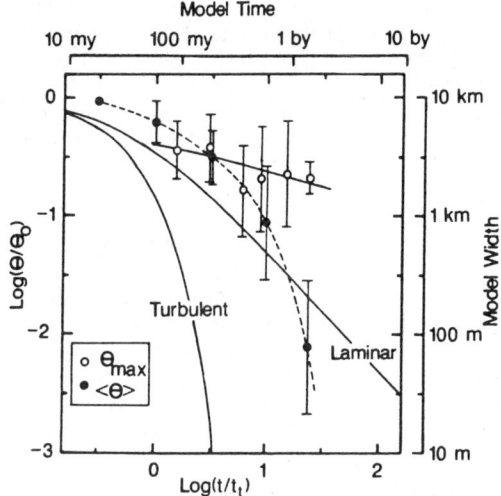

FIGURE 6. Relative thickness of stirred blobs as a function of time. Solid curves show the predicted average blob thickness for laminar mixing (i.e., simple-shear stirring) and turbulent mixing (i.e., multiscale stirring). Solid circles show the average blob thickness from several numerical experiments: The initial decrease in average thickness is due to simple shearing within a single cell, while the later, more rapid decrease sets in when the tendril scale becomes larger than the cell scale, approximating the conditions for multiscale stirring. The open circles show the variation of maximum blob thickness, which is much slower because it is due to simple shearing with partial "unmixing" as the sense of shear reverses. Approximate scales for whole-mantle mixing are included.

ments are summarized in Fig. 6, with scalings to whole-mantle flow. It is evident that the thick parts can persist for billions of years with significant thicknesses, while the tendrils, once formed, are stirred down to the point where they will be unresolvable within less than a billion years.

These results depend on there being only one dominant scale of flow in the mantle. If there were strong small-scale flow, there would be many more stagnation points and the probability of thick heterogeneities being carried into them would be much higher. The simple shearing elsewhere in the fluid might also be more effective in dispersing heterogeneities.

Flows in low-viscosity fluids (strictly speaking, in high Reynolds number flows) like the ocean and atmosphere often become turbulent, which means that a large-scale flow spontaneously breaks down into multiscale flow, including very small scales. Thus stirring is much more efficient in this kind of flow. In fact, stirring in which the average thickness decreases exponentially has become known as *turbulent mixing*, while the simple shearing that occurs within a cell is known as *laminar mixing*. However, it seems that a better terminology would be *multiscale stirring* and *simple-shear stirring*, since multiple scales can occur in laminar flows. We have also distinguished stirring, which causes the intermingling of fluids, from mixing, which requires diffusion to actually homogenize the stirred fluid.

The finding that flows of smaller scale than the plates are secondary in the mantle is thus very important. In fact, the only clearly identified secondary flow is the plumes. These have heat and mass fluxes an order of magnitude smaller than that of the large-scale flow. Furthermore, the upward flow is strongly concentrated, which means that the compensating downward flow will involve very low velocity gradients. Thus we may expect that only those heterogeneities that actually pass close to a plume will be significantly stirred. This would not be true, for example, for a space-filling system of rolls or other-shaped cells, such as those that might be due to the instability of the lower lithosphere, superimposed on the large-scale flow: all mantle that passed into this zone would be more rapidly stirred within it.

This topic is by no means thoroughly studied. The effects of plumes, low-amplitude small-scale flows, and of three-dimensionality all need to be quantitatively investigated. The above results and discussion have illustrated that intuition, or experience based on different types of flow, may be poor guides to the stirring behavior of mantle flow. They have also demonstrated that this stirring behavior is consistent with the observations that were summarized in the last section, to the extent that it has so far been quantified.

Mantle Residence Times. In addition to considering whether heterogeneities will persist within the mantle, we must also consider the likelihood that they will be sampled at the surface. This likelihood has two aspects: the probability that a given piece of mantle will pass near enough to the surface to be sampled (usually by melting), and the melting process itself, especially the scale of melting.

The main sampling process is melting under spreading centers. This occurs to a depth of perhaps 30–50 km and the melting zone probably has a comparable width. The largest degrees of partial melting, up to about 30%, occur in a much narrower zone. Mixing in the melt zone ought to be greatly enhanced over mixing in the solid state because of both faster diffusion and magma mixing. Nevertheless, observations reveal significant variations along ridge axes down to less than 10-km length scales. Thus, heterogeneities that are stirred down to scales of the order of 1 km or less will presumably not be resolvable in zones of substantial melting.

In zones where the degree of partial melting is small, which may be true at some slow-spreading ridges and hotspots and will be true on the margins of more pronounced melting zones, any heterogeneities with a lower solidus temperature, which means a greater propensity for melting, will be melted preferentially. In this way, some heterogeneities may be more selectively sampled, and finer scales of heterogeneity may be evident. In fact, substantial heterogeneities have been observed in single "off-ridge" volcanic seamounts whose dimensions are only of the order of 1 km.

Oceanic crust is about 6 km thick and is usually believed to represent about a 20% partial melt of the mantle ascending under spreading centers, which implies that melting extends to at least 30 km depth, with lesser amounts of partial melting probably extending significantly below that, say to 50 km. Mantle material that rises above this level produces melt that may be sampled at the surface, and the residue will have its radiogenic clocks reset.

What is the average frequency with which a given piece of mantle will be sampled in this way? Or, what is the average residence time of mantle material? A simple estimate can be obtained from the areal rate of seafloor spreading (about 3 km^2/yr), the density of upper mantle material (3300 kg/m^3) and the mass of the mantle (about 4×10^{24} kg). The mantle is then processed at a rate of about 5×10^{14} kg/yr, so it will take about 8 Gyr to process the whole mantle. Thus at present rates, a given piece of mantle would have a residence time of about 8 Gyr, on average.

This estimate is enough to establish that the residence time is very long, but for times so long we must take into account the likely changes in mantle heat flux, convection velocity, and depth of melting. The heat flux has probably decreased with a half life of about 2 Gyr, tracking the decay of the U, Th and K heat-producing isotopes, and convection velocities vary as approximately the square of heat flux. Melting depth is harder to estimate but not critical to the answers: Mantle temperature has probably varied by less than 20%, but melting depth may have been hundreds of kilometers in the early Earth. Taking these variations into account, we get mean mantle ages in the range of 1–2 Gyr: The mean age is the average time since the mantle material was processed at a spreading center, and is the generalization of the residence time in the steady state. These results are comparable to the observed apparent ages discussed earlier, though perhaps a bit low.

The effect of an increase in viscosity with depth, such as was suggested earlier, is not so easily calculated, but some numerical models have revealed that a moderate increase of 2–3 orders of magnitude in viscosity will increase the mean age by up to about 50%. A more dramatic effect is obtained for material initially in the deep, higher viscosity mantle: some small fraction of this, perhaps 5%, may only now have been brought into the shallower, more mobile mantle.

Another possible influence would be if parts of the subducted lithosphere, such as the oceanic crust component, were denser than normal mantle and tended to sink to the bottom. This speculation has been proposed as a possible way of achieving long residence times in the mantle. Numerical models reveal that for the likely density contrasts the effect is not strong (there is a moderate tendency for the heavy material to be in the lower part of the flow, and residence times are enhanced by less than about 20%).

It thus seems that the average apparent age of the mantle inferred from isotopic observations can be adequately accounted for by estimates of rates of processing of the mantle at spreading centers, present and past, taking into account the effects of the likely increase in viscosity with depth and the possibility that some subducted material is slightly denser than normal mantle. Even the evidence from He and Xe isotopes for some mantle having survived with a distinct signature since very early in Earth history is explicable by this type of model.

Summary: Heterogeneities in a Convecting Mantle. The results described here indicate that material injected into the convecting mantle will be carried around for about 2 Gyr, on average, before it is brought near enough to the surface to melt and produce a chemical signature at the Earth's surface. Furthermore, some of the injected material will have survived as a reasonably compact and discrete entity, so that its geochemical signature may differ significantly from the average. Some material initially at the bottom of the mantle may survive essentially for the age of the Earth. Thus there is no evident conflict between the whole-mantle convection model and the observations of mantle heterogeneities with apparent ages in the range 1–3 Gyr, with suggestions of some material having survived from nearly the beginning of the Earth.

Conclusion

At this time a reasonably simple picture of mantle dynamics seems to have satisfied the physical and chemical constraints, to the extent that quantitative tests have been made. In this picture, mantle flow is dominated by plate-scale flow, with the plates as an integral part, which accounts for perhaps 90% of the heat and mass fluxes through the mantle. Plumes, originating at the core-mantle boundary, transport most of the balance of the heat, and produce hotspots, with accompanying topographic swells, at the Earth's surface. Chemical heterogeneities are introduced mainly by subducted lithosphere, but also possibly from other sources, and some of these survive mantle stirring for long enough to yield recognizable chemical signatures when they return to the near-surface and are sampled, which is, on average, about 2 Gyr later, in accord with observations. A small amount of deep mantle material may have survived from the earliest times.

Mantle layering, and in particular separate upper mantle and lower mantle layers, seems to be precluded by the prediction that the hot boundary layer at the base of the upper layer would produce large topographic anomalies that are not observed. Other modes of mantle flow are possible, such as an instability at the base of the lithosphere, but no convincing evidence for such modes has yet been demonstrated. In particular, such small-scale modes do not account for the slower-than-predicted subsidence of old ocean floor, and their lack of obvious topographic expression limits them to a minor role, if they exist at all. The old ocean floor topography seems to be explained by residual swells along hotspot tracks.

There are many possible further tests to which the model described here can be subjected. With full, three-dimensional spherical models now becoming feasible on supercomputers, detailed comparisons of physical predictions should be made with the Earth's topography, gravity and geoid, three-dimensional seismic structure and Wadati-Benioff deep seismic zones. Stirring by three-dimensional flows and the effects of secondary modes of flow, including plumes, need to be studied. More specific modeling of isotopic and trace element heterogeneities, addressing particular chemical systems, source types and source sizes should be done in the context of this type of flow model.

Finally, two general observations. Mantle convection, for a long time widely regarded as the domain of theoreticians unconstrained by the real world, seems at last to have matured to the point where there are many empirical tests available to constrain the models. In the debates that have raged and will continue around the topics discussed here, one must be careful to distinguish something that doesn't seem possible but has not been quantitatively tested, from something that has been demonstrated to be inconsistent with observations. The former indicates the need for further work, while the latter, if upheld, can rule a hypothesis out of contention.

GEOFFREY F. DAVIES

References

Allegre, C. J., T. Staudacher, and P. Sarda, 1987, Rare gas systematics: formation of the atmosphere, evolution and structure of the earth's mantle, *Earth Planetary Sci. Letters* **81**, 127–150.

Allegre, C. J., and D. L. Turcotte, 1985, Geodynamic mixing in the mesosphere boundary layer and the origin of oceanic islands, *Geophys. Research Letters* **12**, 207–210.

Crough, S. T., 1983, Hotspot swells, *Ann. Rev. Earth Planetary Sci.* **11**, 165–193.

Davies, G. F., 1984, Geophysical and isotopic constraints on mantle convection: an interim synthesis, *Jour. Geophys. Research* **89**, 6017–6040.

Davies, G. F., 1987, Role of the lithosphere in mantle convection, *Jour. Geophys. Research* **93**, 10451–10466.

Davies, G. F., 1987, Ocean bathymetry and mantle convection I. Large-scale flow and hotspots, *Jour. Geophys. Research* **93**, 10467–10481.

McKenzie, D. P., 1983, The earth's mantle, *Sci. American* **249**, 66–78.

McKenzie, D. P., and F. Richter, 1976, Convection currents in the earth's mantle, *Sci. American* **228**.

Gurnis, M., and G. F. Davies, 1986, Mixing in numerical models of mantle convection incorporating plate kinematics, *Jour. Geophys. Research* **91**, 6375–6395.

Hager, B. H., R. W. Clayton, M. A. Richards, R. P. Comer, and A. M. Dziewonski, 1985, Lower mantle heterogeneity, dynamic topography and the geoid, *Nature* **313**, 541–545.

Hofmann, A. W., 1984, Geochemical mantle models, *Terra Cognita* **4**, 157–165.

Cross-references: *Core-Mantle Coupling; Deep Earthquakes; Density Distribution in the Earth; Earthquake Seismology; Earth Structure: Global; Energy Budget of the Earth; Far-Travelled Terranes; Figure of the Earth; Heat Flow in the Earth; Mantle, Lower: Structure; Mantle, Upper: Structure; Mantle Convection and Plumes; Mantle Discontinuities.*

MANTLE VISCOSITY

It is a fundamental assumption of the thermal convection hypothesis of continental drift that the mantle of the Earth must deform as a viscous fluid when it is subjected to a shear stress of sufficient temporal duration. It is furthermore true that this hypothesis can be construed as valid only if the viscosity of the mantle "fluid" has a value close to that which may be computed on the basis of simple scaling considerations that follow from the power-law relations that are obeyed by thermal convection at high Rayleigh number. This required value may thereby be shown to be close to 10^{21} Pa · s. An important means whereby the internal consistency of the convection hypothesis of continental drift may be assessed then involves measurement of the viscosity of the mantle. If one can show, either by direct laboratory measurement of the stress-strain relation for materials of mantle composition in the appropriate range of pressure and temperature or through analysis of geophysical data not related to the convection process itself that the viscosity of the mantle has this value, then the convection hypothesis might be considered verified. Although great progress has been made in direct laboratory investigation of the creep law for polycrystalline mantle materials, the experimental results must inevitably be extrapolated over a great many orders of magnitude in strain rate, the strain rates obtaining in the laboratory being very much in excess of those that are characteristic of the thermal convection process.

Because of this difficulty with the interpretation of laboratory-derived creep measurements, the inference of mantle viscosity on the basis of the interpretation of large-scale geophysical observations continues to play an extremely important, indeed a dominant, role in the science. There are two classes of data that have been employed to make such inferences. The first class, and the one that has played the most important role in the historical development of the subject, consists of observations related to the phenomena of glacial isostatic adjustment. Data in this class derive from four different kinds of geophysical observation, each of which represents a distinct signature of the planet's response to the glaciation–deglaciation cycle that has been a dominant process in the geological system over the past 2 million years of Earth history, the part of this history called the Pleistocene geological epoch. Observations of relative sea-level histories, constrained by ^{14}C dating, in the time since the last glacial maximum, which occurred approximately 18,000 years ago, constitute the first and most important kind of data in this class. Closely associated with such relative sea-level observations are observations of free-air gravity anomalies that are associated with present-day regions of land emergence in the areas that were once ice covered, notably the northern part of the North American continent centered on Hudson Bay, the northwestern part of Europe surrounding the Gulf of Bothnia, and the part of the Antarctic continent west of the Transantarctic Mountains. Complementing these relative sea-level and free-air gravity data are two observations of properties of the Earth's rotational state that are sufficiently strongly influenced by the glacial rebound process that they may be extracted from measurements made with modern space-based geodetic techniques, such as very long baseline radio interferometry (VLBI) (see *Very-Long-Baseline Interferometry*) and satellite laser ranging (SLR) (see *Satellite Laser Positioning*). These data consist of the so-called nontidal component of the acceleration of the Earth's rate of axial rotation and the true secular drift of the pole of rotation with respect to the surface geography. Since these four observations are not related to the thermal convection process, to the extent that they may be invoked collectively or individually to constrain the viscosity of the mantle, one may then employ them to verify the internal consistency of the convection hypothesis.

The second class of data that may be invoked to constrain this crucial dynamical parameter has become accessible due to the successful application of the techniques of seismic tomography to the construction of global models of the lateral heterogeneity of seismic velocities in the mantle (see *Earth Structure, Global*). To the extent that the observed lateral heterogeneity may be ascribed to lateral temperature changes, these data may be simply employed to construct models of the lateral heterogeneity of mantle density and thus direct images of the mantle convective circulation itself. Since the motion of the mantle fluid is forced entirely by this lateral density heterogeneity, given the variation of viscosity the rate of flow may be directly predicted. In particular, the pattern of flow at the Earth's surface may be predicted and compared to the observed flow, the viscosity being adjusted until an acceptable fit to the surface plate velocities is obtained. Other geophysical observations may also be predicted on the basis of the same tomographic images of the internal density field by using the same theoretical models as are employed to predict plate velocities. The main additional data that have been analyzed in this way consist of nonhydrostatic geoid height anomalies, as representative of the planet's large-scale gravitational field, and the topography of the core-mantle boundary (CMB). The former data are delivered by analyses of the orbits of artificial Earth satellites, whereas the latter have been a further by-product of the analysis of seismic data.

This article gives examples of each kind of geophysical data used to constrain mantle viscosity and illustrates how the constraints are obtained with reference to the theoretical models employed to invert the data. Separate sections focus on the glacial isostatic adjustment and mantle convection/seismic tomography analyses. The final section summarizes the results of these analyses and provides a perspective on the contribution of mantle viscosity analyses to understanding the convection process.

Mantle Viscosity from Analyses of Glacial Isostatic Adjustment

The past decade of research on mantle viscosity has considerably helped us understand how the Earth's internal structure influences each signature of the isostatic adjustment process. It has also proven crucial to reconciling some controversies. Each controversy arose due to a failure to comprehend how elements in the radial elastic structure of the planet contribute to inferences of the radial viscous structure based upon analyses of glacial "rebound" data. The way in which these two elements of Earth structure interact during the rebound process can be easily understood with the theoretical model developed to explain these observations.

The Linear Viscoelastic Field Theory for Glacial Isostasy. The small-amplitude creeping flow gravitationally induced in the Earth in response to surface ice-sheet loading satisfies the following Laplace transform domain forms of the equations of momentum balance and continuity and perturbation of the gravitational potential:

$$\nabla \cdot \tau - \nabla(\rho_0 g_0 \mathbf{u} \cdot \mathbf{e}_r) - \rho_0 \nabla \phi_1 - g_0 \rho_1 \hat{\mathbf{e}}_r = 0 \tag{1a}$$

$$\rho_1 = -\rho_0 \nabla \cdot \mathbf{u} - \mathbf{u} \cdot (\partial_r \rho_0) \hat{\mathbf{e}}_r \tag{1b}$$

$$\nabla^2 \phi^1 = 4\pi G \rho_1 \tag{1c}$$

These equations have been linearized in small deviations from a background hydrostatic equilibrium configuration (ρ_0, p_0, ϕ_0) that satisfies

$$\nabla p_0 = -\rho_0 g_0 \hat{e}_r \quad (2a)$$

$$\nabla^2 \phi_0 = 4\pi G \rho_0 \quad (2b)$$

In Eq. 1 τ is the stress tensor, **u** is the displacement vector, ϕ_1 is the perturbation of the gravitational potential, ρ_1 is the density perturbation, and G is the gravitational constant. In general, ϕ_1 is the sum of two parts, ϕ_2 and ϕ_3, that are, respectively, the potential of the externally applied gravitational force field (the load) and the potential due to the internal redistribution of mass forced by the load-induced deformation.

In the Laplace transform domain of the variable s, the stress-strain relation that relates τ to the strain tensor **e** is that for a three-dimensional Maxwell solid with no bulk dissipation; it has the form

$$\tau_{ij} = \lambda(s) e_{kk} \delta_{ij} + 2\mu(s) e_{ij} \quad (2)$$

in which the modulii are

$$\lambda(s) = \frac{\lambda s + \mu K/\nu}{s + \mu/\nu}$$

$$\mu(s) = \frac{\mu s}{s + \mu/\nu}$$

Clearly in the limit $s \to \infty$ ($t \to 0$), the modulii reduce to the usual elastic Lamé parameters, and the constitutive relation Eq. 2 reduces to Hooke's law. In the opposite limit $s \to 0$ ($t \to \infty$), we have $\lambda(s) \to K$ and $\mu(s) \to s\nu$, so the constitutive relation effectively reduces to that for a Newtonian viscous fluid with molecular viscosity ν. In this model the bulk modulus is $K = \lambda + 2\mu/3$. The problem of glacial isostasy is to solve Eq. 1 for the ice-sheet-loading-induced deformations of the Earth and, by fitting predictions of the model to observations, to infer $\nu(r)$ in the interior of the planet.

It is crucial to our implementation of the so-called principle of correspondence, which is used to obtain the time-domain solution of Eq. 1 from that in the Laplace transform domain, that the quasi-static momentum balance equation, Eq. 1a, contain the second term on the left-hand side. This describes the extra body force to which the system is subject because it is in a state of hydrostatic prestress before the surface load is applied. Because of this term, effectively nonadiabatic density discontinuities in the planetary interior introduce extra internal buoyancy when deflected from their equilibrium positions. As well as employing this general theory to infer the viscosity of the mantle, we can also use it to infer the presence of any such internal buoyancy-producing discontinuities. We will not discuss the technical details of how the field equations are solved; instead we focus on the most important results that have been obtained through application of the theory. A complete review of the mathematical methods is found in Peltier (1982, 1985).

Mantle Viscosity and Postglacial Sea Levels. The primary data base for the study of mantle viscosity consists of relative sea-level histories over the past 10,000–15,000 years since the melting of the major ice sheets began about 18,000 years ago. The observed variation of sea level at any location is obtained by measuring the height above or depth below present-day sea level of a sequence of relict beach horizons of age determined by application of ^{14}C dating techniques. Typical examples of such data are shown in Fig. 1. At sites well within the ice-sheet margins, such as the Ottawa Islands in present-day Hudson Bay, the oldest beach is found at the greatest height (>100 m) above the present level of the sea, and the uplift of the region (glacial rebound) has clearly preceded at an exponentially decreasing rate through time. At sites near the ice-sheet margin, such as Boston, the relative sea-level histories are strikingly nonmonotonic, with initial emergence followed by submergence that is ongoing today. Further south along the U.S. East Coast at Clinton, Connecticut, the sea-level history is one of monotonic submergence, whereas at a very distant site like Recife, Brazil, the sea-level record is marked by the appearance of a raised beach about 6000 years ago, after which the coast has been continuously uplifted, but at a rate that has been a decreasing function of time.

For measuring mantle viscosity, the most useful sea-level records are from sites that were once ice covered. Since the disintegration of the ice sheet over Canada (centered on Hudson Bay) was essentially complete about 6000 years ago, it is clear from the fact that the land has continued to rise out of the sea since that time that the Earth's mantle is not Hookean elastic. If it were, then deformation would have ceased with the cessation of melting. We seek to determine $\nu(r)$ by fitting the predictions of our viscoelastic field theory to the sea-level data. Using a reasonably well-constrained a priori model for the glaciation history, we predict local sea-level histories by solving an integral equation that directly predicts the time-dependent separation of the geoid and the surface of the solid Earth. This integral equation (e.g., Peltier, 1982) follows simply by application of the constraint that the meltwater produced by glacial disintegration be distributed over the surface of the global ocean in such a way that the ocean surface remains equipotential.

Figure 2 shows some results from one such sequence of analyses for three sea-level sites in Canada that were once ice covered and for three along the U.S. East Coast that were beyond the margin of the Laurentian ice sheet. Comparisons between theory and observation are shown for three different models of the internal viscoelastic layering that differ from one another only in lithospheric thickness, with the elastic structure fixed to that of the seismically realistic model 1066B of Gilbert and Dziewonski (1975). The models all have an upper-

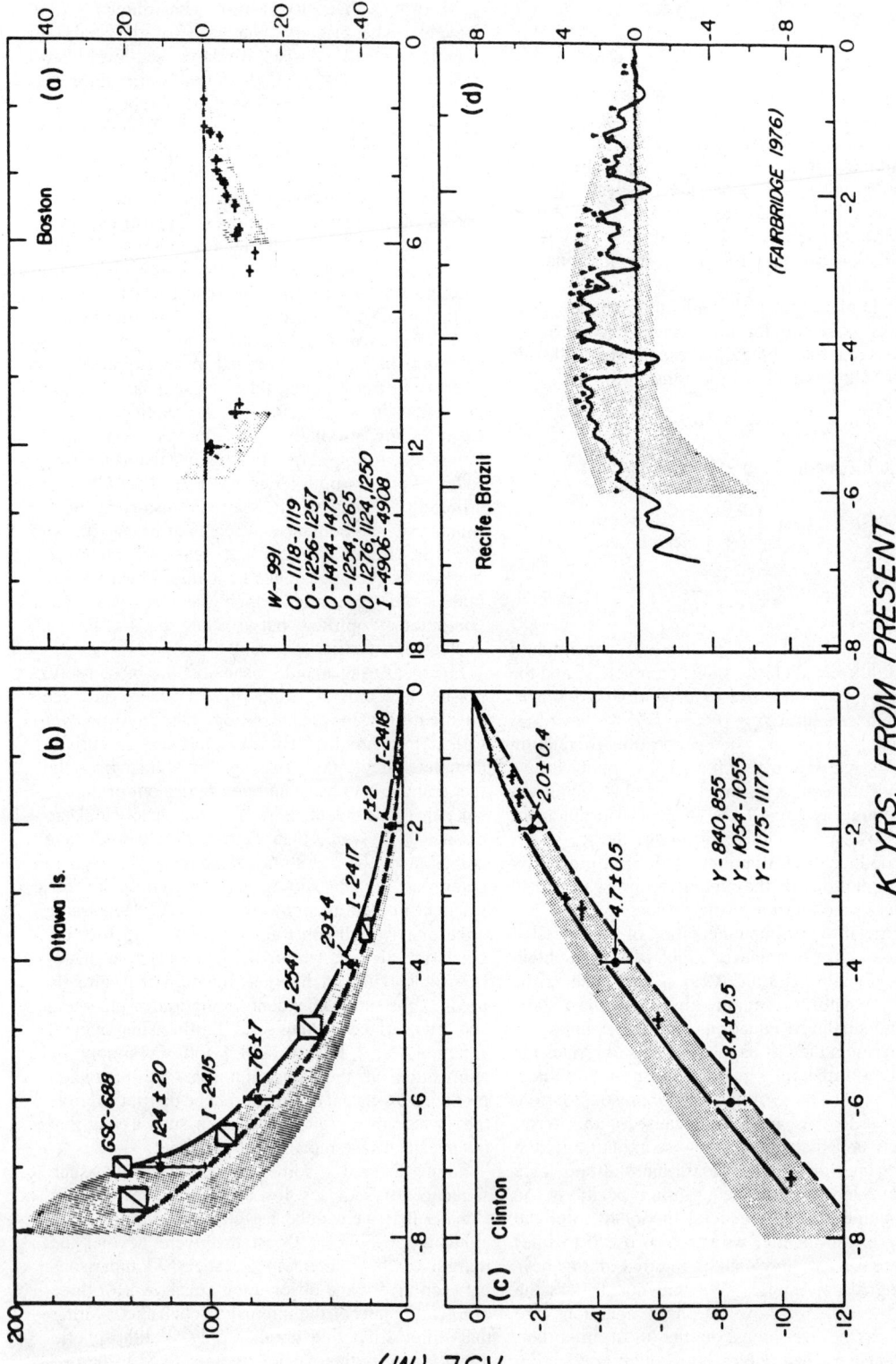

FIGURE 1. Four examples of radio-carbon-controlled relative sea-level curves from sites located at various distances from the Laurentian ice sheet, which last covered Canada 18000 years ago. The Ottawa Islands are located in Hudson Bay near the center of rebound, Boston is located near the margin of the ice sheet at Laurentide maximum, Clinton, Connecticut, is south of the margin on the U.S. East Coast, and Recife, Brazil, on the east coast of South America, is well removed from any major location of glaciation.

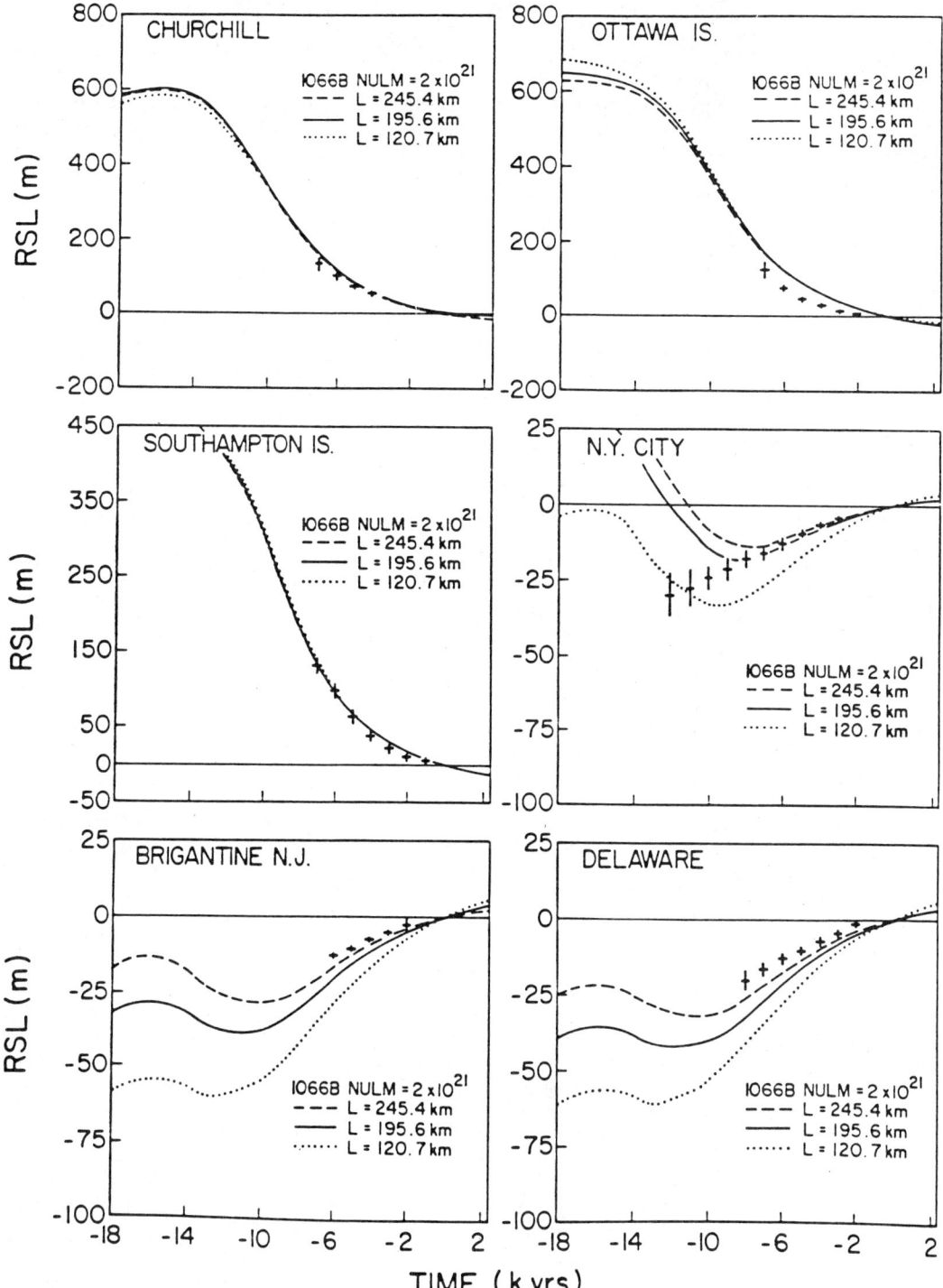

FIGURE 2. Comparisons of observed relative sea-level data with theoretical predictions at six North American sites for three Earth models, which differ from one another only in lithospheric thickness. Data from sites within the margins of Laurentian ice (Churchill, Ottawa Islands, Southampton Island) are not influenced by lithospheric thickness variations and are well fit by a model that has an upper-mantle viscosity of 10^{21} Pa · s and a lower-mantle viscosity of 2×10^{21} Pa · s. Data from beyond the margin (New York City, Delaware, Brigantine, N.J.) are quite sensitive to lithospheric thickness. The apparent preference of these data for very thick lithosphere is due in part to imperfections in the deglaciation history assumed for Laurentia.

mantle viscosity of 10^{21} Pa · s to a depth of 670 km and a lower-mantle viscosity of 2×10^{21} Pa · s. Inspection of these comparisons shows that the "interior" data are insensitive to variations of lithospheric thickness L and are generally well fit by the model with very modest increase of viscosity with depth. Any further increase of lower-mantle viscosity induces significant misfits to the relative sea-level data. These data strongly constrain the upper-mantle viscosity close to the value $\nu_{LM} = 10^{21}$ Pa · s and require 10^{21} Pa · s $\leq \nu_{LM} \leq 10^{22}$ Pa · s, with the preferred value for ν_{LM} close to 2×10^{21} Pa · s. Exterior data provide similarly useful constraints on lithospheric thickness. The preferred value for mantle viscosity is near that required by the thermal convection hypothesis of continental drift and seafloor spreading. As we discuss later, other signatures of the rebound process may be invoked to confirm this deduction.

Mantle Viscosity, $\dot{J}_2$, and the Nontidal Acceleration of Planetary Rotation. Given the mass of the ice sheets whose melting caused the variations of sea-level discussed in the last section ($\sim 2 \times 10^{19}$ kg for the Laurentian ice sheet, $\sim 0.6 \times 10^{19}$ kg for the Fennoscandian ice sheet, and $\sim 0.7 \times 10^{19}$ kg for the West Antarctic ice sheet), it should not be too surprising to learn that the disintegration of these large ice masses induced substantial changes in the Earth's rotation, causing both the rate of rotation to change and the rotation pole itself to wander with respect to the surface geography. Even though the ice sheets have long since disappeared, their influences on these two elements of planetary rotation are still evident today because of the slow viscous flow in the mantle induced by the associated gravitational imbalance. For several years it has been increasingly recognized that the nontidal acceleration of rotation inferred from analyses of ancient eclipse data was due, as originally suggested by R. H. Dicke, to the influence of deglaciation. This datum has now been more accurately inferred through the use of laser ranging data to the *LAGEOS* satellite, which has been shown to require a present-day secular variation of J_2 (the degree-2 zonal coefficient in the spherical harmonic expansion of the Earth's gravitational potential field) in an amount near $(-3.5 \pm 0.3) \times 10^{-11}$ yr^{-1}.

The theoretical prediction of $\dot{J}_2$ or, equivalently, of the nontidal acceleration of rotation $\dot{\omega}_3/\Omega$, since the two are linearly related, takes the form

$$\frac{\dot{\omega}_3}{\Omega} = -\frac{2m_e a^2}{3C} \dot{J}_2$$

$$= -\frac{I^R_{33}}{C}\left[D_1 f(t) + \sum_{j=1}^{M} \Gamma_j^2 \frac{d}{dt}(f * e^{-S_j t})\right]$$

in which C is the Earth's axial component of inertia, I^R_{33} is the axial inertia perturbation that would be produced by the surface ice and water load if the Earth were rigid, $D_1 = 1 + k_2^E$, where k_2^E is the elastic surface load Love number of degree 2, and $f(t)$ is a function that describes the history of glaciation and deglaciation (equal to 1 at glacial maximum and 0 during interglacial periods), and r_j and s_j are the amplitudes and inverse relaxation times, respectively, of the M normal modes of viscous gravitational relaxation required to synthesize the time dependence of the Love number k of degree 2. In such calculations the load history $f(t)$ is taken to be a periodic sawtooth function of period 10^5 yr with a linear rise of duration 9×10^4 yr and a 10-kyr linear collapse. This function is reasonably well constrained by $^{18}O/^{16}O$ concentration data from deep-sea sedimentary cores that are proxy for past fluctuations in the volume of continental ice. Figure 3 shows a sequence of predictions of the *LAGEOS* observation for a suite of models, all of which (again) have 1066B elastic structure and which differ from one another only in their viscosities beneath a depth of 670 km (labeled ν_{LM}). Calculations are shown for models including one (Laurentia), two (Laurentia + Fennoscandia), and all three (+ Antarctica) ice sheets. With the upper-mantle viscosity fixed at 10^{21} Pa · s, we require 2.7×10^{21} Pa · s $\leq \nu_{LM} \leq 4.4 \times 10^{21}$ Pa · s in order to fit the observed $\dot{J}_2$ if the observation is taken to be represented by the value of Yoder et al. (1983), and somewhat lower if one accepts the reduced value obtained by Rubincam (1984), which is probably closer to the truth. This value is consistent with the result from the sea-level data discussed in the last section. Although the $\dot{J}_2$ observation can also be fit with an extremely high value for the lower-mantle viscosity (near 10^{23} Pa · s), this possibility is ruled out by the sea-level data.

Mantle Viscosity and the Anomalous Gravitational Field. In all three regions of the Earth that were once ice covered, the surface remains somewhat depressed below its height of gravitational equilibrium. In consequence, if one subtracts from the local vertical component of the gravitational acceleration the value for an equilibrium reference spheroid at the same latitude, one finds a negative anomaly in each of these locations. Maps of these anomalies for Laurentia and Fennoscandia are shown in Fig. 4. Over Hudson Bay the peak anomaly Δg is such that $-40 \leq \Delta g \leq -30$ mgal, for Fennoscandia the range is $-20 \leq \Delta g \leq -15$ mgal (when appropriate corrections are made for larger-scale bias), and for Antarctica (not shown) the anomaly is very near that for Fennoscandia. Our theoretical prediction of the free-air anomaly uses a Green function for this signal of the form

$$G(\theta, t) = \frac{g}{m_e} \sum_{l=0}^{\infty} [(l+2) - (l-1)k_l] P_l (\cos \theta)$$

(3)

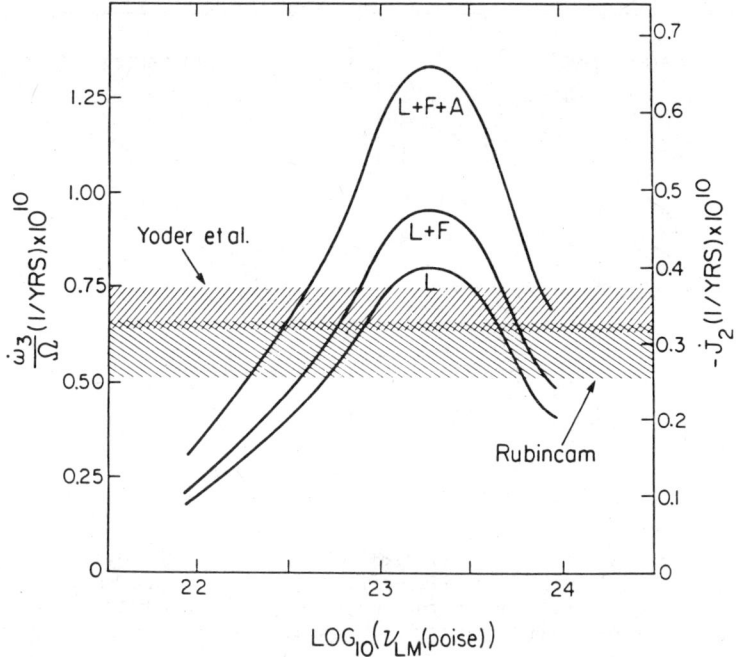

FIGURE 3. Predictions of present-day nontidal acceleration of rotation (and $\dot{J}_2$) due to Pleistocene glaciation and deglaciation for models that include only one ice sheet (Laurentia: L), two ice sheets (Laurentia + Fennoscandia: L + F), and three ice sheets (Laurentia + Fennoscandia + Antarctica: L + F + A). Predictions are shown as a function of lower-mantle viscosity ν_{LM}, with upper-mantle viscosity $\nu_{UM} = 10^{21}$ Pa · s fixed. The satellite laser ranging observation is shown as either one or the other of the discrepant estimates of Yoder and others from the Jet Propulsion Laboratory of the California Institute of Technology and of Rubincan of the Goddard Space Flight Center of NASA.

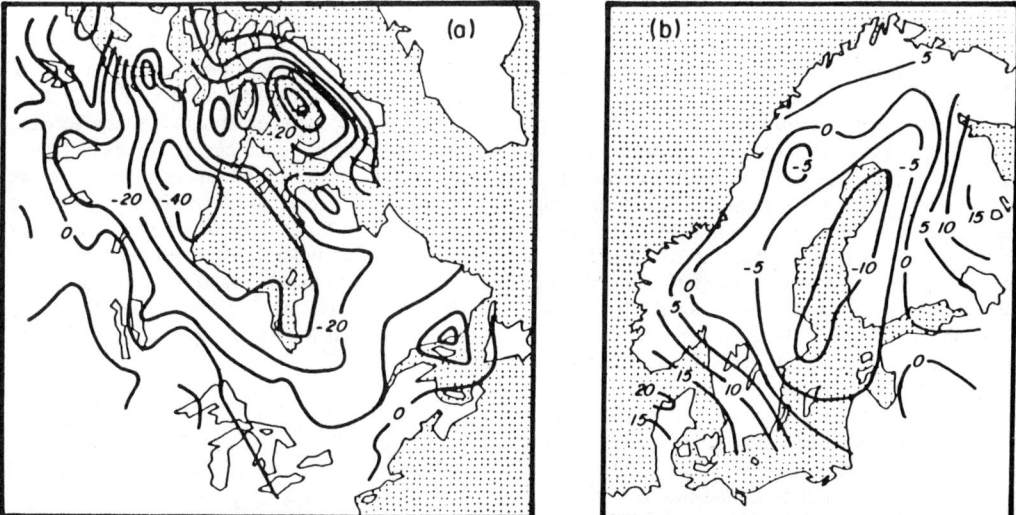

FIGURE 4. Free-air gravity anomalies based upon surface pendulum observations for (a) Laurentia and (b) Fennoscandia.

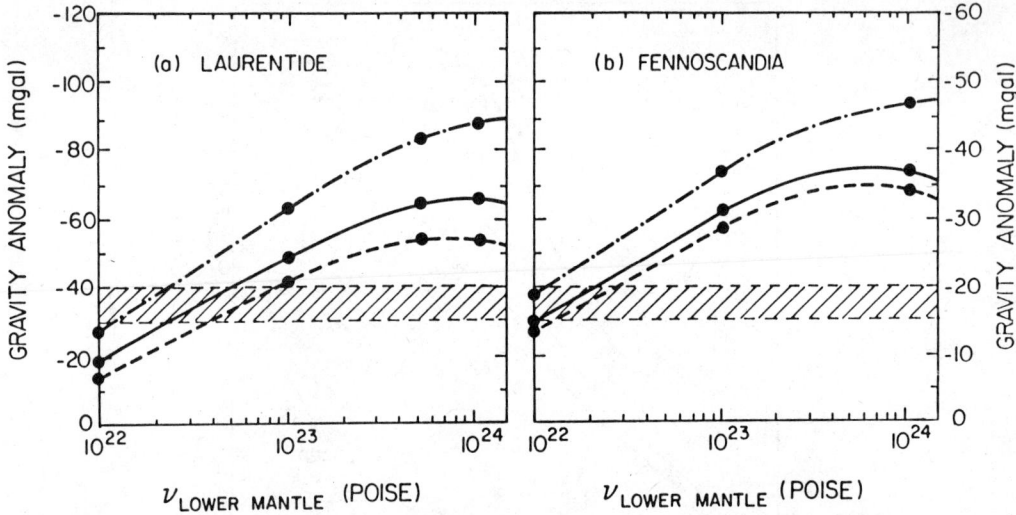

FIGURE 5. Predicted peak present-day free-air gravity anomalies for (a) Laurentia and (b) Fennoscandia as a function of lower-mantle viscosity ν_{LM} (in poise rather than Pa · s; 10^{20} P = 10^{21} Pa · s), with the upper-mantle viscosity ν_{UM} = 10^{21} Pa · s fixed. The observed peak anomaly is shown as the hatched region on each plate, and theoretical predictions are shown for three different assumptions concerning the history of glaciation and deglaciation as discussed in the text.

in which k_l is the surface load Love number of degree l. To predict Δg, we simply convolve this Green function with the space- and time-dependent surface load. Because the observation provides an absolute measure of the current degree of gravitational disequilibrium, these calculations are particularly sensitive to the degree of gravitational disequilibrium that existed before deglaciation began; they are also extremely sensitive to the existence of any internal buoyancy-generating interfaces in the mantle.

With the history of loading again constrained by $^{18}O/^{16}O$ data from deep-sea cores, and 20 prior 10^5-yr glacial cycles employed in the calculation, Fig. 5 compares the observed free-air gravity anomaly with the predicted value (for circular disk load models of the ice sheets) as a function of the mantle viscosity beneath 670-km depth. The upper-mantle viscosity is again fixed at 10^{21} Pa · s, and the 1066B structure is employed to constrain the elastic component of the model. Results are shown for both Laurentia and Fennoscandia. Three calculations are shown on each plate: The dashed-dotted line is the prediction based on the assumption that isostatic equilibrium prevails initially; the dashed line includes the influence of initial disequilibrium but assumes that the ice-sheet radius remains fixed while the volume fluctuates; and the most accurate prediction, the solid line, incorporates the effect of simultaneous volume and radius variations under the assumption that the ice sheet maintains a plastic profile at all times. The lower-mantle viscosity preferred by the data again agrees with that deduced on the basis of the previously discussed relative sea-level fluctuations and $\dot{J}_2$ observation.

Mantle Viscosity and Deglaciation-Induced True Polar Wander. The last of the main signatures of glacial isostasy that can be invoked to constrain the internal viscoelastic structure of the planet is the *polar wander*, which is presently occurring at a rate of $(0.95 \pm 0.15)°$ per 10^6 yr along the $(76 \pm 5)°$ meridian toward Greenland (see Fig. 6, where the secular drift, measured with the photozenith tube data of the International Latitude Service, is displayed). This effect can be explained as a memory of the planet of the last deglaciation event of the current ice age, although previously, following work by Walter Munk and Roger Revelle, it was believed that true polar wander could be produced only if the surface load was actively varying.

The mathematical form of the correct theoretical prediction of polar wander due to surface loading by ice sheets is

$$\frac{\dot{\omega}_j}{\Omega} = \frac{\Omega}{A\sigma_0} I^R_{j3} \left[D_1 \dot{f}(t) + D_2 f(t) \cdot + \sum_{i=1}^{M-1} E_i \frac{d}{dt}(f * e^{-\lambda_i t}) \right] \quad (4)$$

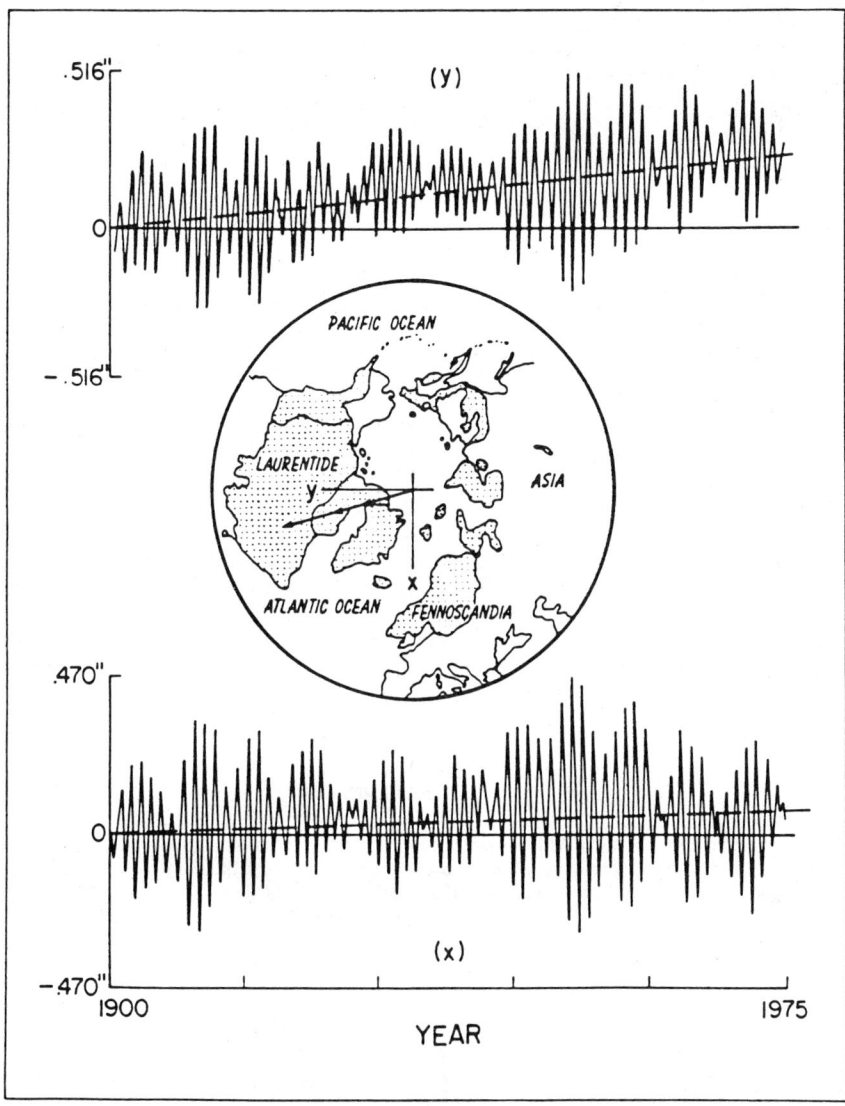

FIGURE 6. The motion of the rotation pole relative to the surface geography in a coordinate system with origin at the CIO (conventional international origin) with x- and y-axes as shown on the inset polar projection. The dominant oscillatory motion is due to the interference between the 14-month Chandler and 12-month annual wobbles. It is superimposed upon a slow secular drift of the pole in the direction shown on the inset polar projection at a rate near $(0.95 \pm 0.15)°$ my^{-1}. This secular drift is explicable as due to the glacial isostatic adjustment process. The data are from the International Latitude Service.

where D_1, D_2, E_i, and λ_i are defined in Peltier (1982). The main difference between the mathematical structure of the polar wander solution, Eq. 4, and that for the nontidal acceleration, Eq. 3, is that the history-dependent (convolution) term in the present case involves a sum over $M - 1$ modes rather than the M modes that govern the evolution of $\dot{J}_2$. This difference arises because when the surface load is zero ($f = 0$), $\dot{\omega}_{1,2}/\Omega$ differs from zero only because the centrifugal and isostatic adjustment contributions to the forcing fail to cancel. In fact, for a homogeneous Earth model that has only one mode

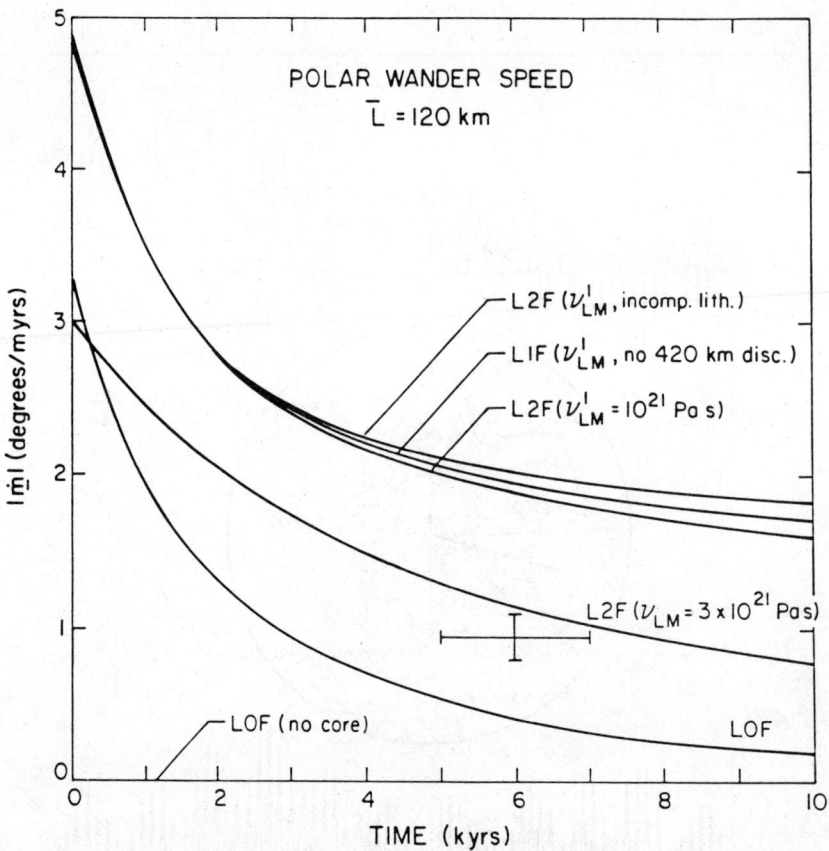

FIGURE 7. Polar wander speed predictions for a sequence of Earth models that differ only in their radial viscoelastic structures as discussed in the text. The observed polar wander speed is shown as the cross centered upon the time of 6 kyr, which corresponds to the present day in the model of glaciation and deglaciation employed to make the theoretical predictions.

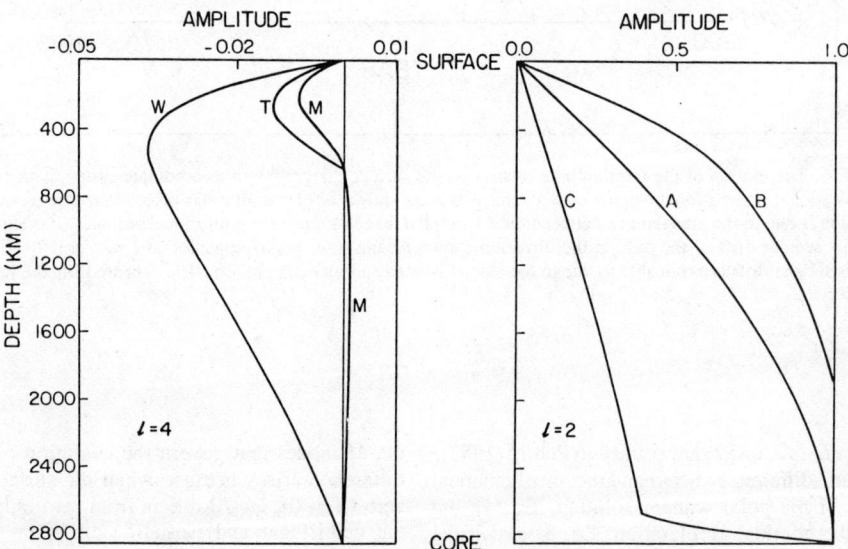

FIGURE 8. Surface divergence kernels (left plate) and core-mantle boundary topography kernels (right plate) for the range of different convection models discussed in the text.

(a) WHOLE MANTLE FLOW

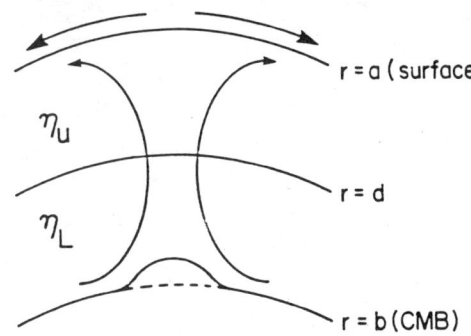

LAYERED FLOW

(b) MECHANICAL COUPLING   (c) THERMAL COUPLING

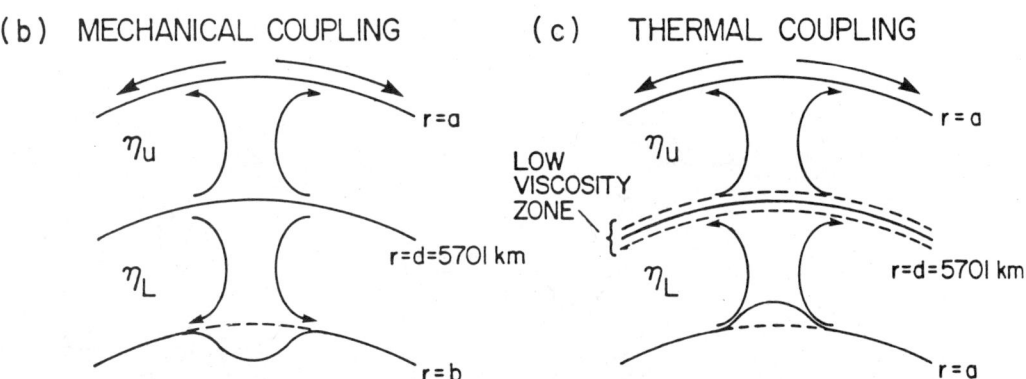

FIGURE 9. Schematic diagrams illustrating the connection between surface divergence and core-mantle boundary topography for the three different versions of the convection model discussed in the text.

of relaxation, the contributions do exactly cancel, and such a model therefore cannot explain the present-day observed wander of the rotation pole. The nonzero polar wander speed is therefore carried by a term involving fewer (by 1) exponential time constants than are required to describe the isostatic adjustment process itself. Figure 7 shows a series of polar wander speed predictions over a period of 10 kyr following the last of seven 100-kyr pulses of glaciation and deglaciation for a sequence of Earth models that differ only in their radial viscoelastic structures. The almost identically zero prediction for model L0F (no core) confirms that, for a homogeneous model of the internal viscoelasticity, no polar wander is predicted to occur after melting ceases. The prediction for model L0F demonstrates the effect of including a high-density inviscid core in the model, and this effect sharply increases the speed prediction. Adding additional radial density heterogeneity in the model further increases the speed prediction, so that with the mantle viscosity everywhere fixed to the value of 10^{21} Pa · s the predicted speed exceeds the observed polar wander speed (shown as the cross) by almost a factor of 2. Increasing the lower-mantle viscosity by a factor of 3, however, lowers the speed into the range required by the data. This radial viscosity structure is the same as that shown to be required by the three previously discussed signatures of the isostatic adjustment process. Given that all of the data of glacial isostasy appear to be reconciled by this same radial viscosity variation, we will demonstrate the extent to which this structure is acceptable to the convective circulation itself.

Mantle Viscosity from Analyses of the Convective Circulation Using Data from Seismic Tomographic Imaging

Whenever the motion of a viscous fluid is characterized by an extremely small value of the Reynolds number (Re = UL/η, where U and L are characteristic velocity and length scales, respectively, and η is the kinematic viscosity; i.e., $\eta = \nu/\rho$), then the influence of inertial forces on the motion is completely negligible compared with the influence of viscous forces. The equation for the conservation of momentum may then be written, assuming the fluid to be incompressible and Newtonian, in the form

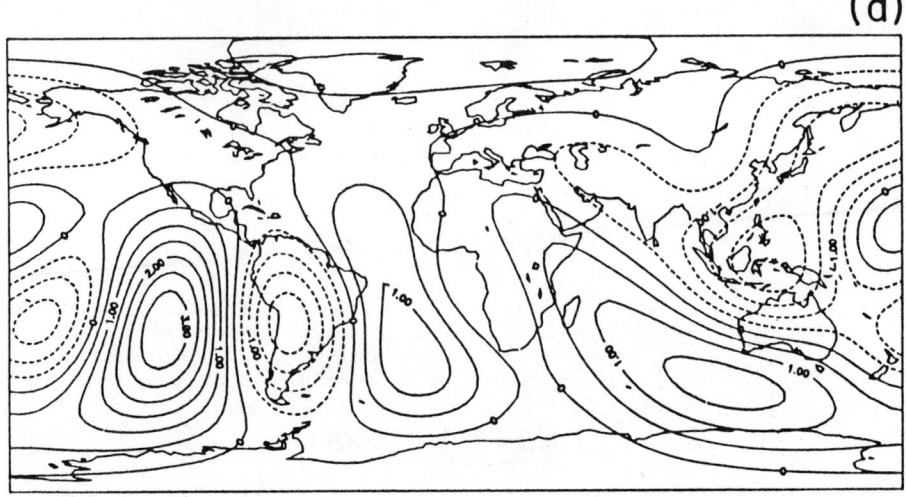

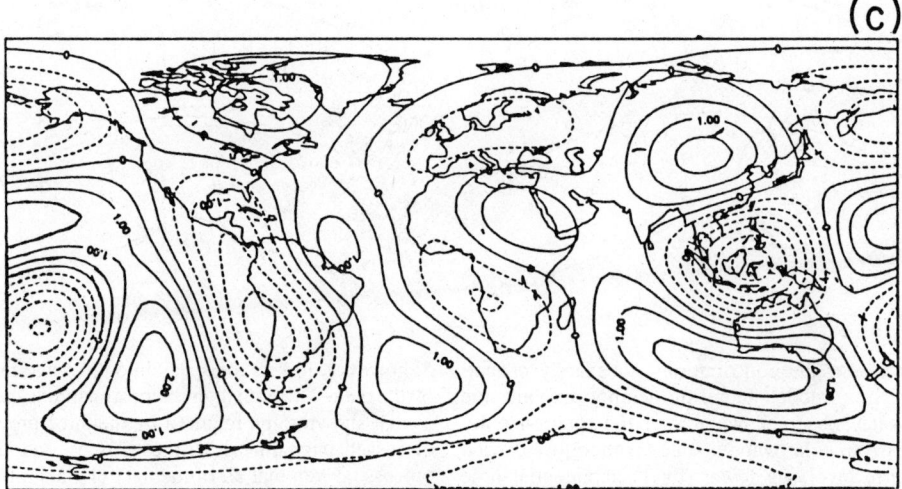

FIGURE 10. Surface divergence in the range of spherical harmonic degree $1 \leq l \leq 5$: (a) the observed divergence deduced on the basis of surface plate velocities, (b) the predicted divergence using a whole-mantle convection model, (c) the predicted divergence using the layered convection model with mechanical coupling, and (d) the predicted divergence using the layered convection model with thermal coupling.

$$\rho_0 \mathbf{g}_1 + \rho_1 \mathbf{g}_0 - \nabla p + \nu \nabla^2 \mathbf{u} = 0 \quad (5)$$

where, as before, ρ_1 and $\mathbf{g}_1$ and p are the perturbations of density, gravitational acceleration, and pressure to a hydrostatic equilibrium configuration due to the motion. Because incompressibility is assumed, the conservation of mass constraint reduces to the form

$$\nabla \cdot \mathbf{u} = 0 \quad (6)$$

and thus the velocity field $\mathbf{u}$ may be represented in terms of a poloidal-toroidal decomposition as

$$\mathbf{u} = \nabla \times \Lambda p + \Lambda q \quad (7)$$

in which p and q are respectively the poloidal and toroidal generating scalars. Following Forte and Peltier (1987), use of Eq. 7 in Eq. 5 reduces the momentum conservation equation to

$$\left[\frac{d^2}{dr^2} + \frac{2}{r}\frac{d}{der} - \frac{l(l+1)}{r^2}\right] p_l^m(r) = \frac{g_0}{\nu} \frac{(\rho_1)_l^m(r)}{r} \quad (8)$$

in which $p_l^m(r)$ and $(\rho_1)_l^m(r)$ are the depth-dependent amplitudes in the spherical harmonic expansions of

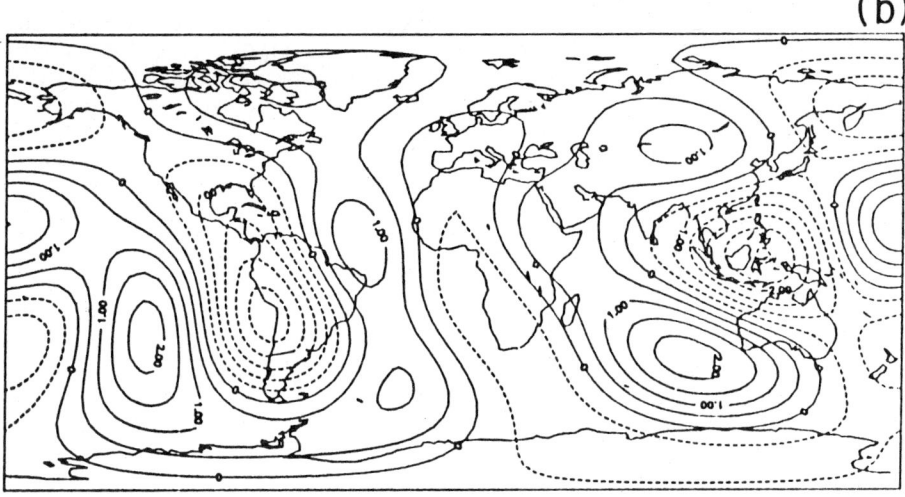

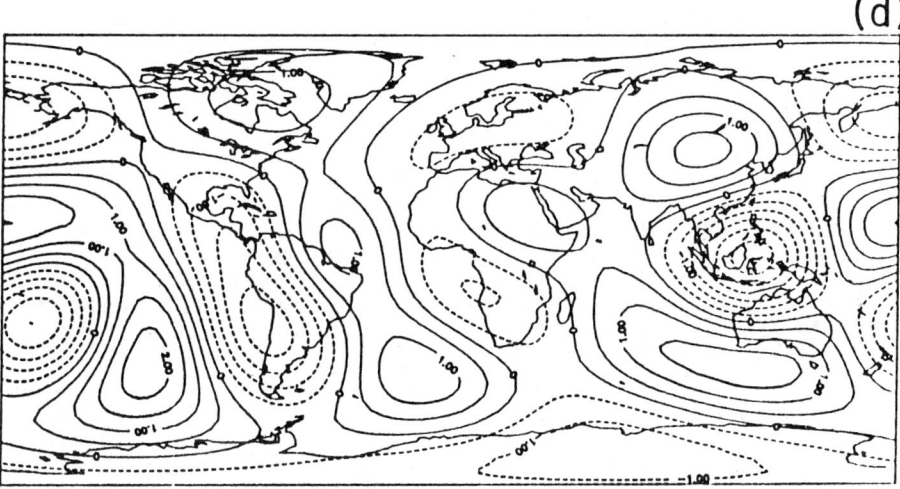

FIGURE 10. (*Continued*)

the poloidal scalar and lateral density heterogeneity, respectively. For this simple hydrodynamic model the corresponding equation for q_l^m is homogeneous, demonstrating that the buoyancy force associated with nonzero ρ_1 does not couple into toroidal motion. Equation 8 demonstrates that if $(\rho_1)_l^m$ is known, then p_l^m and, using Eq. 7, the velocity field can be predicted. Since the prediction of p_l^m given $(\rho_1)_l^m$ clearly depends upon the viscosity ν, which appears in the denominator of the right-hand side of Eq. 8, we may vary ν so as to achieve a best fit to p_l^m (and thus **u**) and thereby measure the value of the viscosity.

The importance of the results achieved by seismic tomographic imaging techniques is that they provide implicit models for $(\rho_1)_l^m(r)$ that can be invoked to make the new method of inferring mantle viscosity work. Just as mantle viscosity can be inferred by using isostatic adjustment data, in which several different signatures of the glacial rebound process may be invoked to constrain this variable, so can the convective circulation itself, as described by Eq. 6. Several Earth properties related to this dynamical process may be separately invoked as constraints on ν.

The three most important Earth properties that have been examined consist of nonhydrostatic geoid anomalies, surface plate velocities, and CMB topography. The theoretical predictions of these quantities take the forms

$$(Ge)_l^m = \frac{3}{(2l+1)\bar\rho} \int_b^a G_l\left(r; \frac{\nu_L}{\nu_U}; d\right)(\rho_1)_l^m(r)\, dr \tag{9a}$$

$$(\nabla_H \cdot \mathbf{u})_l^m = \frac{g_0}{\nu_u} \int_b^a \delta_l(r; \nu_L/\nu_U; d)(\rho_1)_l^m(r)\, dr \tag{9b}$$

$$(\delta b)_l^m = \frac{1}{\rho_0 - \rho_0} \int_b^a B_l\left(r; \frac{\nu_L}{\nu_U}; d\right)(\rho_1)_l^m(r)\, dr \tag{9c}$$

These forms clearly show that both geoid height anomalies and CMB topography are insensitive to the absolute magnitude of the viscosities in the upper and lower regions of the mantle. They depend only on the *ratio* ν_L/ν_U of viscosities in the two layers. Only the surface divergence datum is sensitive to the absolute value of viscosity, as shown clearly in Eq. 9b by the appearance of the preintegral factor g_0/ν_U. Furthermore, each of these observables depends on the internal lateral density heterogeneity of the mantle in a different manner, since the kernels G_l, S_l, and B_l have different radial structures. Figure 8 illustrates in plates (a) and (b) a series of S_l kernels for $l = 4$ and a series of B_l kernels for $l = 2$, respectively, where $l = 4$ and $l = 2$ are the spherical harmonic degrees at which the surface divergence and CMB topographies have their maximum amplitudes.

The divergence kernels shown on plate (a) are for three different two-layer viscosity models. The first model, the kernel for which is denoted W, is a whole-mantle flow model that has the interface separating the two layers at 670 km depth and a factor of 18 increase in viscosity from the upper mantle to the lower mantle. For this model the boundary condition on vertical velocity at the interface between layers is simple continuity, such as to deliver a nonzero mass flux across this interface. The remaining two models, whose kernels are denoted by T and M on plate (a), correspond to different versions of the layered convection hypothesis; that labeled M representing the mechanically coupled model for which the radial velocity is zero at the 670-km depth but for which no significant radial shear in horizontal velocity obtains, and that labeled T represents the thermally coupled model for which the radial velocity is zero at 670-km but for which there is strong radial shear of tangential velocity through a low-viscosity sublayer centered at this depth. As illustrated in Fig. 9, these two versions of the layered convection hypothesis respectively correspond to layered flow in which cold plumes overlie hot plumes, and vice versa (M), and in which cold plumes overlie cold plumes and hot plumes overlie hot plumes (T). The kernel shown for model M is that which delivers the best fit to the surface divergence data, and this fit is achieved with a viscosity increase across the 670-km depth by a factor of 200. The kernel shown for model T is identically zero below the 670-km depth, and thus lower-mantle viscosity does not influence the surface divergence field at all, a consequence of the assumed decoupling of the motion across the thin low-viscosity zone at the layer interface.

Figure 10 illustrates the quality of the fits to the observed surface divergence in the degree range $1 \le l \le 5$ [plate (a)], which are obtained when the above-described three convection models are employed to fix the divergence kernels. The models of the internal lateral heterogeneity of density employed to fix the internal buoyancy in the models were obtained by Adam Dziewonski and John Woodhouse at Harvard University, based upon the application of seismic tomographic imaging techniques. One significant adjustment was made to those objective models for present purposes: The anomalously cold part of the upper-mantle lateral heterogeneity was replaced by that assumed to be associated with down-going slabs beneath regions of surface convergence. The anomalously warm fraction of the upper-mantle lateral heterogeneity was held fixed, and the lower-mantle lateral heterogeneity was untouched for purposes of these analyses. Plate (b) of Fig. 10 shows the best fit for the whole-mantle model W, plate (c) that for layered model M, and plate (d) that for layered model T. Although the variance reduction achieved with W significantly exceeds that for either M or T, the main difference between the best-fit viscosity models obtained with these three different versions of the convection hypothesis concerns the value of the viscosity required in the upper mantle to achieve reasonable fits to the observed plate velocities. For model W, $\nu_U = 2.2 \times 10^{21}$ Pa $\cdot$ s, whereas for layered models M and T we require $\nu_U = 0.3 \times 10^{21}$ Pa $\cdot$ s and $\nu_U = 0.56 \times 10^{21}$ Pa $\cdot$ s, respectively. It is crucial to note that ν_U for model W exceeds the upper-mantle viscosity required to fit postglacial rebound data, whereas ν_U for either of the layered convection models is less than that required to reconcile glacial isostatic adjustment observations. Since laboratory data on the transient creep of polycrystalline materials of upper-mantle composition demonstrate that the rheology is strain hardening, the long time scale process of thermal convection must be governed by a viscosity higher than that which controls the shorter time scale postglacial rebound process. On the basis of the above analysis, it is clear that in the context of a two-layer viscosity model only the whole-mantle form of the convection hypothesis is compatible with this a priori rheological constraint.

This inference based upon mantle viscosity stratification that the convective circulation is not layered is reinforced by consideration of a second seismic-

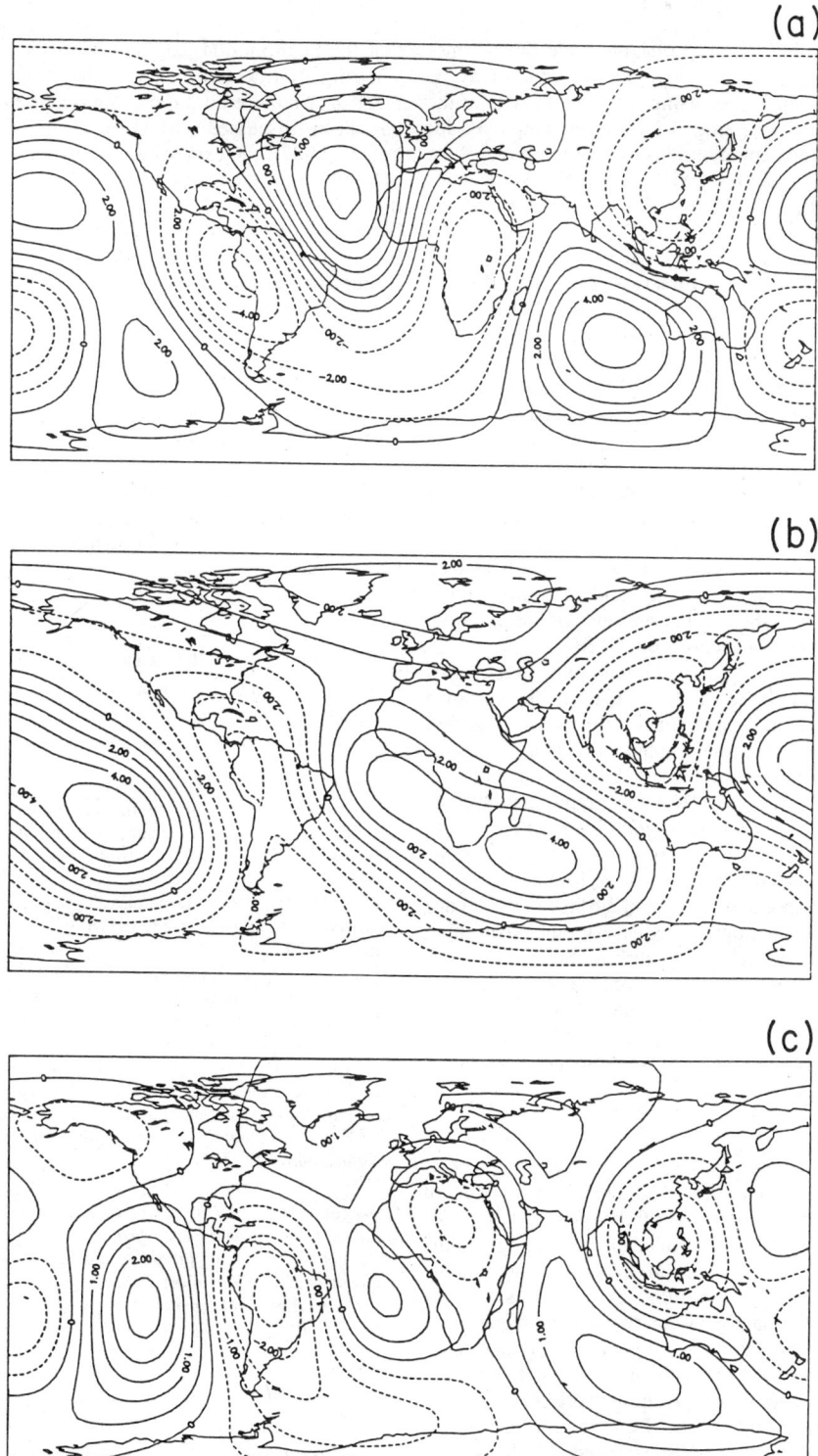

FIGURE 11. Core-mantle boundary topography in the range of spherical harmonic degree $2 \leq l \leq 4$: (a) a recent model of the observed topography on the core mantle boundary in this same range of wavelengths, (b) the predicted core-mantle boundary topography based upon the whole-mantle convection model, and (c) the surface divergence of the plate velocities in the same range $2 \leq l \leq 4$.

ally observable property of large-scale Earth structure, namely the CMB topography, which may be predicted from Eq. 9c using the same models of the internal density heterogeneity as were employed for the divergence computations discussed above. Plate (b) of Fig. 8 shows CMB topography kernels for three different Earth models, which are all of the "whole-mantle convection" variety (see also *Core-Mantle Coupling*). The kernel marked A is for an isoviscous mantle, B is for a two-layered mantle with boundary at 670-km depth and a factor of 18 increase of viscosity, and C is the CMB topography kernel that has uniform viscosity everywhere except in a thin layer below a depth of 2700 km and continuing to the CMB in which viscosity is reduced by a factor of 10^3. The thin low-viscosity layer in the latter model is meant to represent the low viscosities that obtain in the hot lower boundary layer that is observed seismologically as the D" layer. Inspection of these kernels demonstrates that CMB topography is somewhat enhanced in the two-layer model with high lower-mantle viscosity and somewhat more strongly diminished by the presence of the thin low-viscosity layer adjacent to the CMB.

The seismically observed topography on the CMB is shown in plate (a) of Fig. 11 in the degree range $l = 2$–4. Plate (c) shows for comparison the observed velocity divergence field in this same range of spherical harmonic degree. The correlation between seismically observed CMB topography and surface divergence is clearly very good everywhere except in the North Atlantic. Furthermore the sign of the correlation is such that regions of low CMB topography correspond to regions of surface convergence and thus to regions in which anomalously high density material is sinking, just as one would expect in the whole-mantle convection scenario. Plate (b) shows the predicted CMB topography based upon Eq. 9c. The fit to the observed topography obtained with the whole-mantle convection model is rather encouraging.

Conclusions

This article has reviewed the main sources of information on viscosity of the planetary mantle. These sources consist of data pertaining to the phenomena of postglacial rebound and mantle convection. Although there is some discrepancy between the mantle viscosity profiles required to reconcile these two types of geophysical data, these discrepancies are not extreme and are themselves useful, since they may be employed to make important dynamical inferences. For example, the differences in upper-mantle viscosity required by whole-mantle and layered-mantle versions of the convection hypothesis compared with that delivered by inversion of glacial rebound data argued strongly for the validity of the whole-mantle convection scenario. The article does not discuss the further discrepancy

that the plate motion data appear to require strong viscosity stratification, whereas the isostatic adjustment data require an almost isoviscous mantle. As has been discussed elsewhere (e.g., Forte and Peltier, 1987), the explanation of this discrepancy is probably connected with the purely poloidal form of the dynamical model that delivers Eq. 9. Since the surface velocity field has 50% of its kinetic energy in torroidal flow, the purely poloidal model is clearly inadequate. This inadequacy is probably responsible for the preference of such models for strong viscosity stratification and thus for their rejection of the viscosity profile delivered by the isostatic adjustment constraints. In order to construct a dynamical model capable of fitting surface divergence, CMB topography, and geoid height observations with the same almost isoviscous model preferred by isostatic adjustment observations, explicit incorporation of the influence of chemical heterogeneity in the near-surface region in the convection model will probably be required.

W. R. PELTIER

References

Forte, A. M., and W. R. Peltier, 1987, Plate tectonics and aspherical earth structure: the importance of poloidal-toroidal coupling, *Jour. Geophys. Research* **92**, 3645–3679.

Gilbert, F., and A. M. Dziewonski, 1975, An application of normal mode theory to the retrieval of structural parameters and source mechanisms from seismic spectra, *Royal Soc. London Philos. Trans. Ser. A*, **278**, 187–269.

Peltier, W. R., 1982, Dynamics of the ice-age earth, *Advances in Geophysics* **24**, 1–146.

Peltier, W. R., 1985, Mantle convection and viscoelasticity, *Ann. Rev. Fluid Mech.* **17**, 561–608.

Rubincam, D. P., 1984, Postglacial rebound observed by LAGEOS and the effective viscosity of the lower mantle, *J. Geophys. Res.* **89**, 1077–1087.

Yoder, C. F., J. G. Williams, J. O. Dickey, B. E. Schultz, B. J. Eanes, and B. D. Tapley, 1983, Secular variation of earth's gravitational harmonic J_2 coefficient from LAGEOS and nontidal acceleration of earth rotation, *Nature* **303**, 757–762.

Cross-references: *Deformation of Rocks and Minerals; Earth Orientation; Earth Structure, Global; Energy Budget of the Earth; Figure of the Earth; Gravity and Isostasy; Heat Flow in the Earth; Lithosphere: Mechanical Properties; Magmatic Processes; Mantle Convection and Plumes; Mantle Dynamics; Seismic Attenuation: Observation and Measurement.*

MAP PROJECTIONS IN GEODESY

Principles of Map Projection

A map projection is a systematic representation of all or part of the surface of a round body, especially the Earth, on a flat surface. This representation usually includes lines delineating meridians of longitude and parallels of latitude, but they may be omitted, depending on the purpose of the map. Since this transformation cannot be achieved without distortion, the cartographer must choose the characteristic that is to be shown accurately at the expense of others, or a compromise of several characteristics. If the map portrays a continent or larger part of the Earth, distortion will be visually apparent. If the region is small, distortion may be barely measurable using any of the more appropriate projections. There is literally an infinite number of map projections that can be devised, and several hundred have been published, but most are rarely used novelties. Most projections can also be infinitely varied by choosing different points or lines on the Earth to define the projection.

No matter how carefully the projection is chosen, the inherent distortion is increased not only by errors in placement of information on the map, but also because maps are commonly plotted and printed on paper, which is dimensionally unstable. Variations in atmospheric humidity can easily cause dimensional changes greater than those resulting from the choice between common projections for large-scale topographic quadrangle maps. Furthermore, the dimensional changes vary with direction on a given sheet. The use of stable plastic bases for maps is recommended for precision work. On large-scale maps, accurate determination of distances is facilitated by rectangular grid overprints, because the grid expands with the paper, and point positions are determined by reference to the nearest grid lines.

It cannot be said that there is one "best" projection for mapping in general, or even for a given application, unless the parameters chosen are artificially constricting. The characteristics normally considered in choosing a map projection are either preservation of local shape or preservation of all areas.

1. *Shape.* Many of the common and most important projections, such as the Mercator and the Lambert Conformal Conic, are *conformal* (or *orthomorphic*), in that normally the relative local angles about every point on the map are shown correctly. (On a conformal map of the entire Earth there are usually one or more "singular" points, for example the poles of the Mercator, at which local angles are still distorted.) Although a large region must still be shown distorted in shape, its small features are shaped essentially correctly. An important result of conformality is that the local scale in every direction around any one point is constant. Because local angles are correct, meridians intersect parallels at right angles on a conformal projection, just as they do on the Earth. The linear and area scales are generally too great or too small throughout the map, but they are correct along certain lines, depending on the projection. Nearly all large-scale maps of the U.S. Geological Survey and other mapping agencies throughout the world are now prepared on conformal projections.

2. *Area.* Many map projections, such as the Albers Equal-Area Conic, are designed to be *equal-area* (or *equivalent*), so that a coin of any size on one part of the map covers exactly the same area of the actual Earth as the same coin on any other part of the map. Shapes, angles, and scale must be distorted on most parts of such a map, but usually some parts of an equal-area map do retain these characteristics correctly, or very nearly so. Whereas conformality applies on a point or infinitesimal basis, an equal-area map projection shows areas correctly on a finite, in fact mapwide basis. No projection can be both conformal and equal-area.

Some map projections, such as the Azimuthal Equidistant, are neither equal-area nor conformal, but linear scale is correct along all lines radiating from the center, along all meridians, or following other special patterns. No map projection can show scale correctly throughout the map. In addition, compromise projections, usually restricted to small-scale mapping, are used because they balance distortion in scale, area, and shape.

Projections are often classified by the type of surface onto which the Earth may be mapped. The concept of cylindrical or conic projections, such as the Mercator or Lambert Conformal Conic, respectively, involves placing a cylinder or cone around a globe, projecting the map features onto the new surface, and then unrolling this map surface. If the axis of the cone or cylinder coincides with the polar axis of the globe, the projection has equally spaced straight meridians, parallel on the cylindrical projections and converging on the conics. The meridians are intersected at right angles by the parallels of latitude. The latter are straight on the cylindrical and concentric circular arcs on the conic projections. The spacing of the parallels is seldom actually based on geometric projection, but rather on mathematical formulations. There is normally no distortion along the line of tangency of the map surface to the globe.

A plane tangent to the globe at a pole leads to polar azimuthal projections, such as the polar Stereographic, with the parallels mapped as arcs of concentric circles and meridians as equally spaced radii of the circles. The point of tangency is free of distortion. Scale remains constant along each parallel of latitude on a regular cylindrical, conic, or polar azimuthal projection, but it changes from one latitude to another. Directions of all points are correct as seen from the center of an azimuthal projection.

If the cylinder or cone is secant instead of tangent to the globe, the projection conceptually has two lines instead of one that are free of distortion. A secant plane can provide a line rather than a point of no distortion only for a conformal projection (the Stereographic). Wrapping the cylinder about a meridian leads to transverse projections. By placing a plane tangent to the Equator instead of a pole, equatorial aspects of azimuthal projections result. Tilting the cylinder, cone, or plane to relate to another point on the Earth leads to an oblique projection, and the meridians and parallels are not the straight lines or circular arcs they are in the normal aspect. The lines of constant scale are correspondingly rotated.

The Earth as an Ellipsoid

For maps at scales smaller than 1:5,000,000 and covering regions with an area larger than that of the United States, the Earth may be considered a perfect sphere. The distortions resulting from mapping the Earth onto a flat surface are much greater at these scales than the slight additional corrections needed to compensate for the fact that the Earth is not a perfect sphere. For large-scale planimetric and topographic maps at a scale of 1:250,000 or larger, the distortions resulting from projecting the round Earth onto a flat plane are less than the half-percent shape distortion resulting from treating the Earth as a sphere instead of an ellipsoid.

For precision large-scale mapping, the Earth is taken to be an oblate ellipsoid of revolution, usually referred to as an ellipsoid or spheroid. This mathematical figure is formed by rotating an ellipse about its minor axis. As discussed in greater detail in *Figure of the Earth,* the flattening of the ellipse in the case of the Earth is about 1 part in 300. There have been numerous measurements of the two axes of the ellipsoidal Earth, and over a dozen of these representations of the "figure of the Earth" are currently used by one or more countries (Table 1). The axes of ellipsoids measured earlier than 1966 are not necessarily coincident with the axes of the Earth but are shifted by 100–300 m to fit better a particular portion of the Earth. The coordinates of this fit comprise a datum. Large-scale mapping of North America is based on the North American Datum (NAD) 1927; the ellipsoid used is the Clarke 1866. The International Ellipsoid, adopted by the International Union of Geodesy and Geophysics (IUGG) in 1924, is used in much of the world, although it is individually adjusted to various datums. The WGS (World Geodetic System) 72, GRS (Geodetic Reference System) 80, and WGS 84 are satellite-determined figures with axes coinciding with those of the Earth. The GRS 80 was adopted internationally by IUGG in 1980, and some mapping agencies in North America are in the process of shifting to the NAD 1983, which utilizes this ellipsoid. There is also a move toward adopting the WGS 84, the global equivalent of NAD 83.

Coordinates on the Ellipsoidal Earth

Longitude is measured from the prime meridian of Greenwich, counting in both directions from 0° to 180°, east longitude being plus and west, minus. All meridians on the ellipsoid are identical ellipses, each in its own plane, and the longitude is the angle that the plane of the meridian passing through a given point makes with the prime meridian. Regular or geographic latitude is measured as the angle that a perpendicular to a given point on the surface of the ellipsoid makes with the plane of the Equator, 0° to 90° north (+) or south (−) of the Equator. This latitude is not the same as the geocentric latitude, except at the Equator, a pole, or on the sphere. The length of a degree of latitude on the ellipsoid increases slightly away from the Equator, and may be calculated as follows:

$$1° \text{ lat.} = \pi a (1 - e^2)/[180°(1 - e^2 \sin^2\phi)^{3/2}]$$

where a is the semimajor axis of the ellipsoid (see Table 1), e is the eccentricity of the ellipsoid, and ϕ is the latitude. To convert the flattening f (Table 1) to e,

$$e^2 = 2f - f^2$$

TABLE 1. Some Official Ellipsoids in Use Throughout the World

Name	Date	Equatorial Radius, a meters	Polar Radius b meters	Flattening f	Use
GRS 80	1980	6,378,137	6,356,752.3	1/298.257	Newly adopted
WGS 72	1972	6,378,135	6,356,750.5	1/298.26	NASA; U.S. Dept. of Defense
Australian	1965	6,378,160	6,356,774.7	1/298.25	Australia
Krasovsky	1940	6,378,245	6,356,863.0	1/298.3	Soviet Union
Internatl.	1924	6,378,388	6,356,911.9	1/297	Remainder of world
Clarke	1880	6,378,249.1	6,356,514.9	1/293.46	Most of Africa; France
Clarke	1866	6,378,206.4	6,356,583.8	1/294.98	North America; Philippines
Bessel	1841	6,377,397.2	6,356,079.0	1/299.15	Central Europe; Chile; Indonesia
Airy	1830	6,377,563.4	6,356,256.9	1/299.32	Great Britain
Everest	1830	6,377,276.3	6,356,075.4	1/300.80	India; Burma; Pakistan; Afghanistan; Thailand

Note: Some of these ellipsoids (Clarke 1880, Airy, and Everest) are also used in some regions with various modified constants. WGS 84 differs from GRS 80 by 1 mm in b; a is the same; above digits apply to both.

The length of a degree of longitude on the ellipsoid decreases to zero at the poles using the following formula:

$$1° \text{ long.} = \pi a \cos \phi / [180°(1 - e^2 \sin \phi)^{1/2}]$$

The shortest distance between two points on a spherical Earth is along a great circle, which is the intersection of the surface of the sphere with the surface of a plane passing through the two points and the center of the Earth. This curve is fairly simple to describe mathematically. On the ellipsoid, however, the shortest distance is a geodesic, a very complicated curve unless the two points fall along the same meridian or along the Equator. As a result, map projections that are azimuthal or that show great circles as straight lines, when applied to the sphere, often lose some of their properties when applied to the ellipsoid, except in an approximate form. The projections below are described as they relate to the ellipsoid. Generally these descriptions also apply to the sphere, but the spherical form may have additional characteristics as well. The formulas for the various projections discussed are not given here, since most are lengthy when applied to the ellipsoid. The forward (lat/long to rectangular coordinates) and inverse forms, with worked examples and other references, are given in Snyder (1982, 1987).

Conformal Map Projections for Large-Scale Maps

Mercator Projection. Best known of all, the Mercator projection (Fig. 1) was presented by Gerardus Mercator for navigational purposes in 1569. On this projection rhumb lines, or lines of constant bearing or direction, are plotted straight. This use of the projection for maps of regions away from the Equator is justifiable, but the Mercator is generally unsuitable as a world map for geographical purposes, although it is often so used.

It is a normal cylindrical projection, with the cylinder conceptually tangent to the Equator. Lines of constant scale follow the parallels of latitude, all of which are straight and run parallel to the Equator, except for the poles, which are at infinity. The scale on the Mercator increases away from the Equator. The projection is conformal, and it is recommended for large-scale conformal mapping of regions bordering the Equator. Even though the scale is

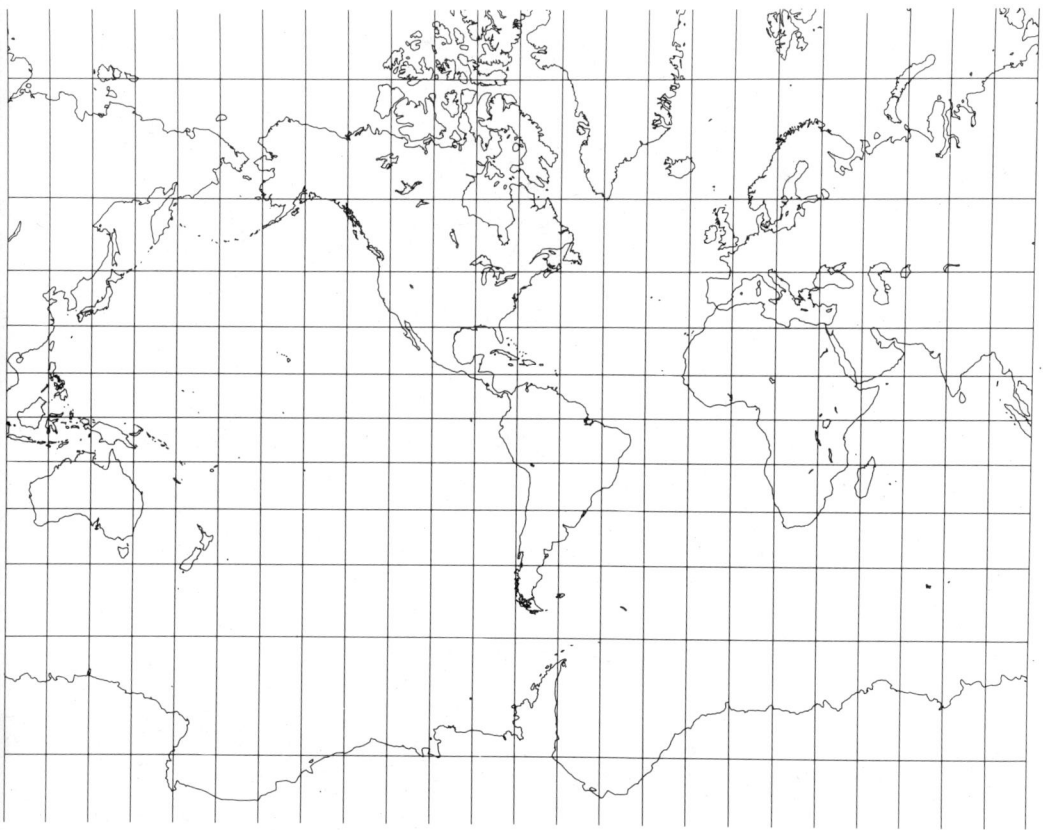

FIGURE 1. Mercator projection.

commonly made correct along the Equator, it may be shifted to two standard parallels that are symmetrically N and S of the Equator.

Transverse Mercator Projection. Rotating the cylinder of the Mercator so that it is conceptually tangent along a meridian of the ellipsoid leads to the very important conformal projection called the Transverse Mercator (Fig. 2). The central meridian, the Equator, and each meridian 90° from the central meridian are straight lines. All other meridians and parallels are complex curves. The central meridian remains at a constant scale, usually reduced from the stated map scale to balance errors in measurement over the rest of the map. The lines of constant scale are nearly straight lines parallel to the central meridian, at which distortion is zero, except for the scale reduction.

The projection is recommended for conformal mapping of regions principally north to south in extent. It was developed for the sphere by Lambert in 1772, but Gauss and later Krüger developed the mathematics for the ellipsoidal form, which is sometimes called the Gauss-Krüger projection. It is used more than any other projection for large-scale topographic mapping in numerous countries. It is also the basis of the Universal Transverse Mercator (UTM) projection and grid system, consisting of 60 zones, each 6° of longitude wide and extending from lats. 80° S to 84° N. The zones are numbered east from the 180th meridian. Each UTM zone has its own central meridian with a reduced scale set at 0.9996 of the stated map scale.

Oblique Mercator Projection. A cylinder may be placed around a sphere representing the Earth so that it is tangent along a great circle that is neither a meridian nor the Equator. Such a wrapping is not quite possible with the Earth as an ellipsoid, but the Oblique Mercator (Fig. 3) may be thus conceptually projected for conformal mapping of a region chiefly extending along this oblique central line. Nearly all meridians and parallels are complex curves. Here the lines of constant scale run nearly parallel with the central line. There are several ways of adapting the Oblique Mercator to the ellipsoid, although none is ideal: If there is perfect conformality, as is the case with Hotine's frequently used adaptation and most others, the central line does not remain at a precisely constant scale.

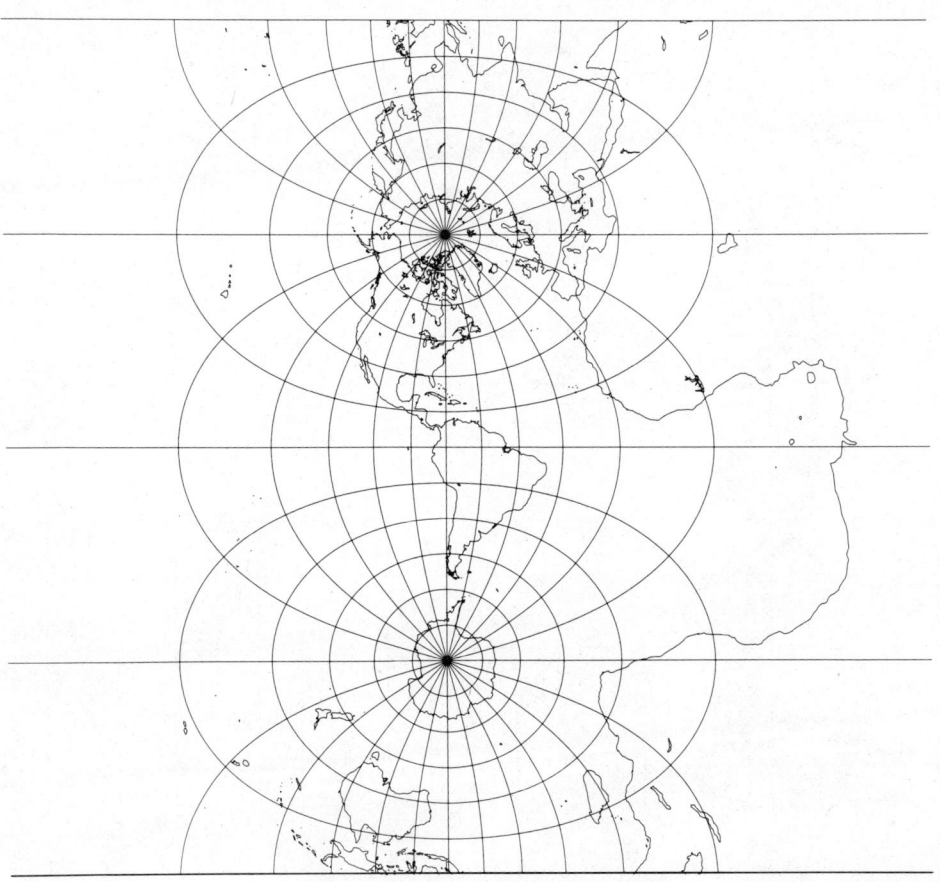

FIGURE 2. Transverse Mercator projection.

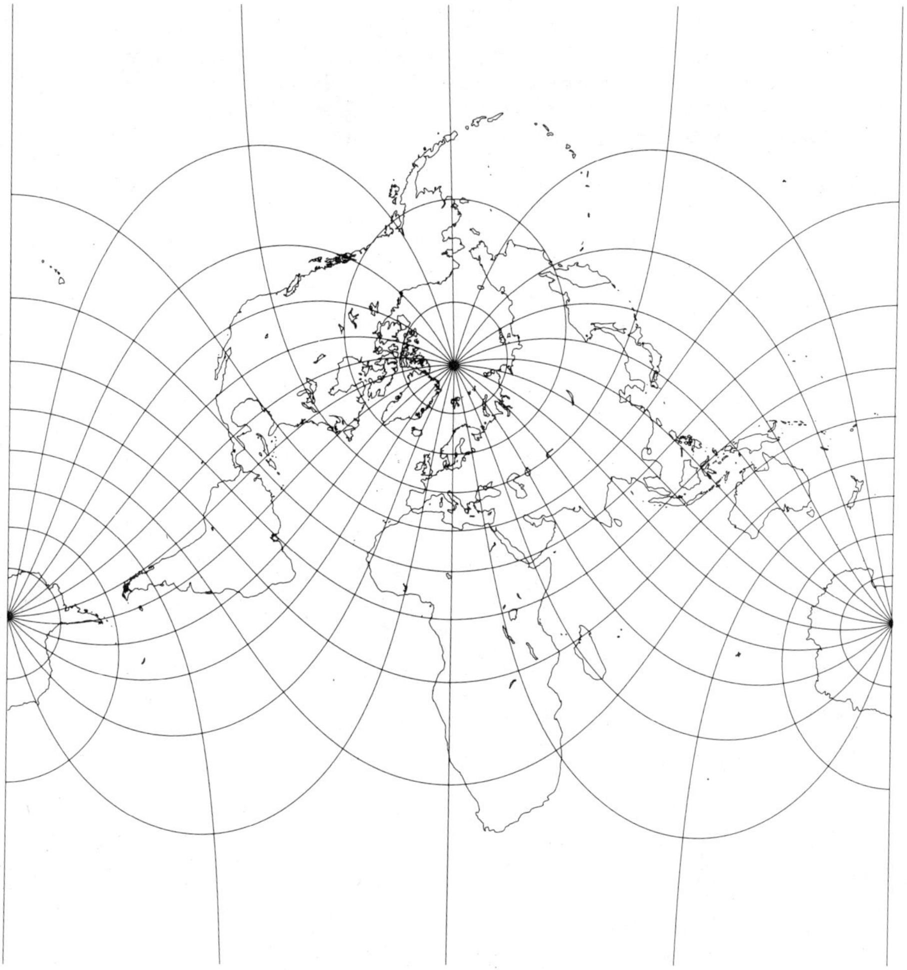

FIGURE 3. Oblique Mercator projection.

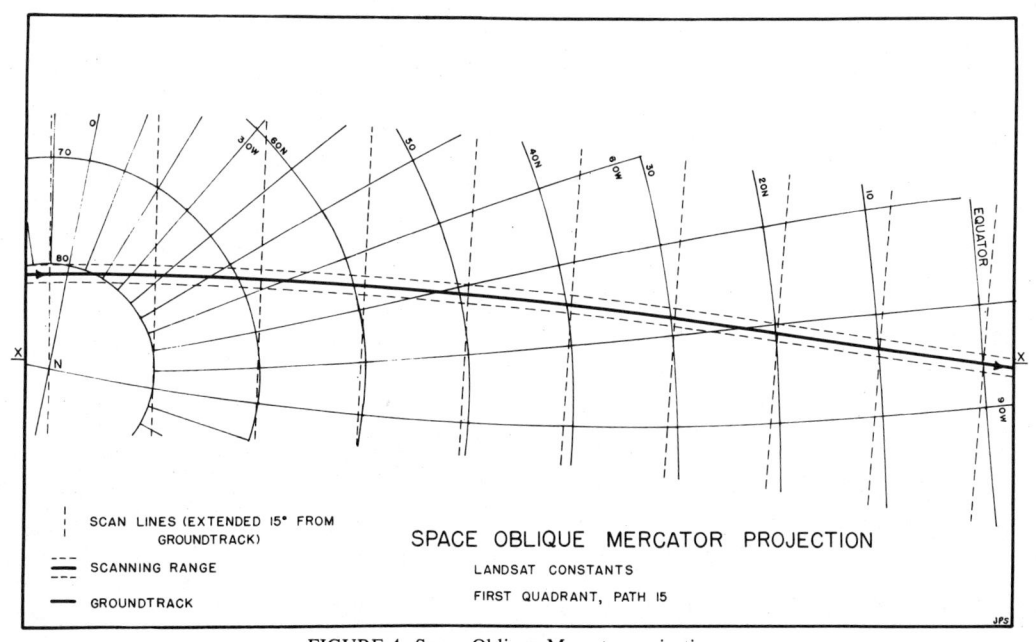

FIGURE 4. Space Oblique Mercator projection.

Space Oblique Mercator. One of the most recent and complicated projections was designed during the 1970s specifically for the continuous mapping of imaging from satellites such as Landsat. Called the Space Oblique Mercator (Fig. 4) and developed by Colvocoresses and Snyder of the U.S. Geological Survey, it is a basically conformal projection with the satellite groundtrack held at true scale. Because of the relative motion of Earth and satellite, the groundtrack is an almost sinusoidal curve on the map. The parameter of time is thus fundamental to this projection.

Lambert Conformal Conic Projection. Presented by Lambert in 1772, the Lambert Conformal Conic projection (Fig. 5) shows parallels as concentric circular arcs and meridians as equally spaced radii of those circles. One pole is at the center of the circles, while the other pole is at infinity. The paral-

FIGURE 5. Lambert conformal conic projection, standard parallels 20° and 60° N lat.

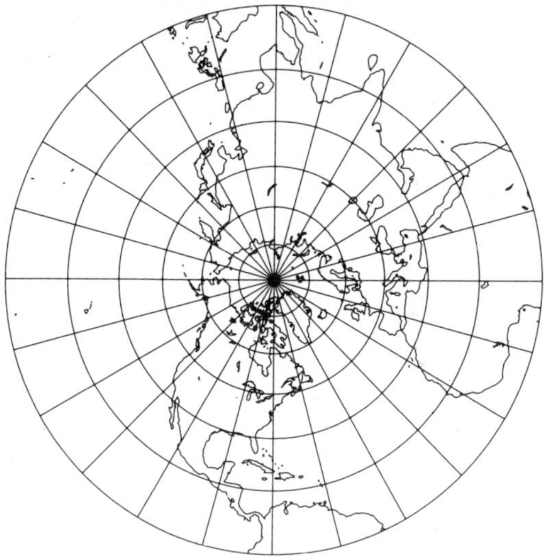

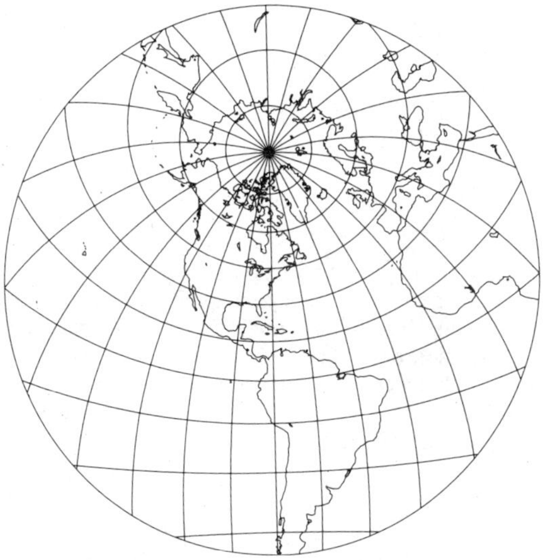

lels are more closely spaced between the (normally) two standard parallels, which have no distortion of area or scale as well as of shape.

Because scale is constant along any given parallel of latitude, the projection is recommended for regions predominantly E to W in extent at latitudes other than the Equator and poles. The Lambert is used second to the Transverse Mercator for much of the large-scale mapping throughout the world. The regular Mercator projection is its limiting form if the standard parallels are made symmetrical about the Equator; the polar Stereographic is the limiting form at the polar extremes.

Stereographic Projection. For larger-scale maps of polar regions, the Stereographic projection (Fig. 6) is commonly used. The oblique or equatorial aspect is also recommended for essentially circular or square regions centered away from either pole. The Stereographic is conformal in both spherical and ellipsoidal forms. The spherical form is a perspective and azimuthal projection of the sphere onto a tangent or secant plane. The point of perspective lies on the opposite surface of the globe. For the ellipsoid, the projection as normally used is not quite perspective, and, if a pole is not the center, it is not quite azimuthal.

All meridians and parallels on any aspect are straight lines or circular arcs, and the lines of constant scale are circles (nearly so for the ellipsoid) centered on the projection center. True scale may be at this center or along one of the concentric circles. The polar aspect, used for over 2000 years, displays straight equally spaced meridians radiating from the central pole, and the parallels of latitude are circular arcs centered on the pole, with their spacings gradually increasing away from the pole.

Equal-Area Map Projections for Large-Scale Maps

Cylindrical Equal-Area Projection: Normal, Transverse, and Oblique Aspects. The equal-area counterparts of the Mercator projection and its transverse and oblique aspects are the various forms of

FIGURE 6. Stereographic projection: (*top*) polar aspect; (*bottom*) oblique aspect, centered at 40° N lat.

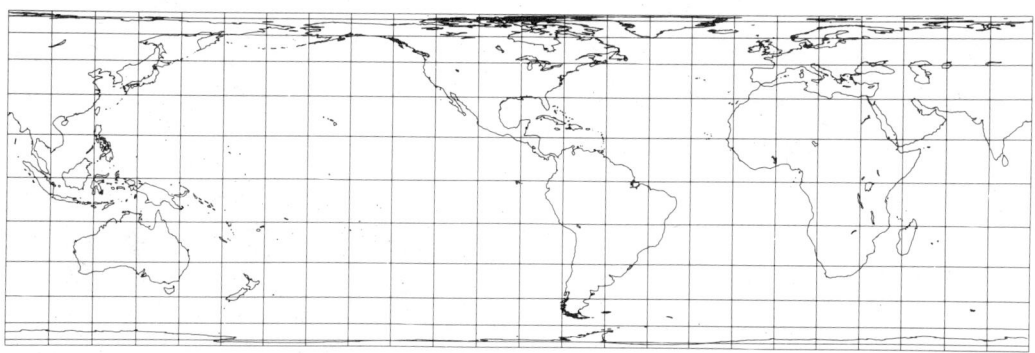

FIGURE 7. Cylindrical equal-area projection, normal aspect. (Transverse and oblique aspects bear same relationship to this aspect that Figures 2 and 3 bear to Figure 1.)

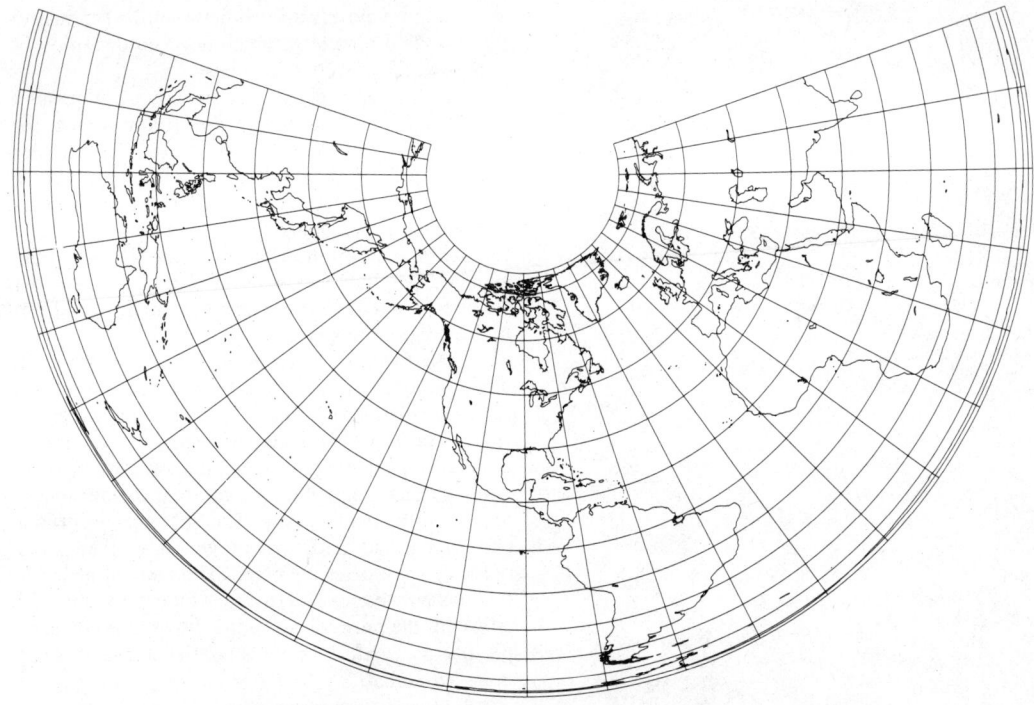

FIGURE 8. Albers equal-area conic projection, standard parallels 20° and 60° N lat.

the Cylindrical Equal-Area projection (Fig. 7), first presented by Lambert in 1772. On the normal aspect, the parallels of the Cylindrical Equal-Area become closer together with increased distance from the Equator, but the meridians remain equidistant, just as they are on the Mercator. The lines of constant scale are parallel to the Equator or other central line, or nearly so. The projection is recommended for equal-area mapping of regions predominantly lying along the Equator (normal aspect), a meridian (transverse), or oblique "great circle" (oblique), but it has only recently been developed in the ellipsoidal form (Snyder, 1985, 1987) and has been almost unused for large-scale mapping.

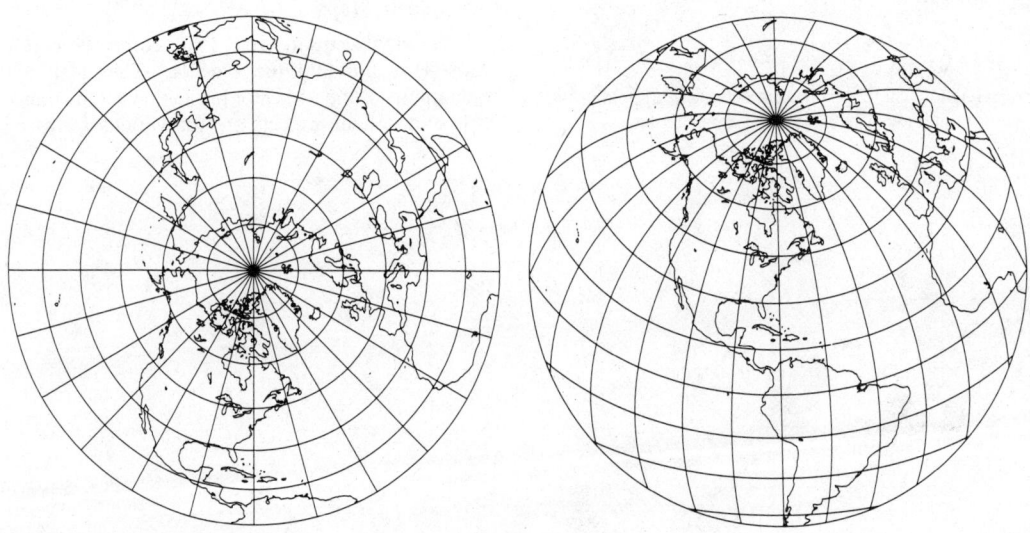

FIGURE 9. Lambert azimuthal equal-area projection: (*left*) polar aspect; (*right*) oblique aspect, centered at 40° N lat.

Albers Equal-Area Conic Projection. Albers developed the equal-area counterpart of the Lambert Conformal Conic (Fig. 8) in 1805. The meridians are equally spaced radii of the concentric circular arcs representing the parallels, and there are normally two standard parallels, free of all distortion, just as on the Lambert. The parallels are, however, spaced farther apart between the standard parallels than they are beyond, and both poles are circular arcs. For a map of the United States, both projections appear similar; only careful measurements (or the label) indicate a difference.

Because meridians intersect parallels at right angles, it may at first be thought that there is no angular distortion. It exists, however, for any angle other than that between a meridian and parallel, except at the standard parallels. Scale is constant in a given direction along any given parallel; therefore, the Albers is recommended for equal-area mapping of regions predominantly E–W in extent. The normal Cylindrical Equal-Area and the Polar Lambert Azimuthal Equal-Area are limiting forms.

Lambert Azimuthal Equal-Area Projection. Although rarely used in the ellipsoidal form, the Lambert Azimuthal Equal-Area projection (Fig. 9), also presented in 1772, is commonly used in the spherical form and is available in the ellipsoidal, as the equal-area counterpart of the Stereographic. The polar aspect shows meridians as radii of the concentric circles representing parallels, but the parallels are more closely spaced as the distance from the center increases, the opposite of the Stereographic spacing. The meridians and parallels of the oblique and equatorial aspects are complex curves. The projection is recommended for equal-area mapping of a region circular or square in shape.

Miscellaneous Projections

Polyconic Projection. Regularly used for topographic mapping of the United States until the 1950s and for coastal mapping until about 1900, the Polyconic projection (Fig. 10), developed about 1820 by Hassler of the U.S. Survey of the Coast, is simple to construct and has correct scale along the central meridian and along all parallels. Only the central meridian is free of distortion. The remainder of the projection is neither equal-area nor conformal. The parallels of latitude are circular arcs with radii equal to the length of the element of a cone tangent

FIGURE 10. Polyconic projection.

at the particular parallel (the basis of the name polyconic). The projection is not recommended for any new maps in view of other projections available.

Other Projections. Other ellipsoidal forms of projections have been developed, sometimes custom-made for a special application. Examples include recent conformal map projections for New Zealand and Alaska designed to provide lines of constant scale that follow more nearly the irregular shape of these two regions. The distortion on such a projection for the region involved is substantially reduced by the use of a complex-algebra polynomial (Reilly, 1973; Snyder, 1987). The availability of digital computers makes it possible to use projections such as these that would previously have been impractical.

JOHN P. SNYDER

References

Maling, D. H., 1973, *Coordinate Systems and Map Projections*. London: George Philip & Son, 255p.

Richardus, P., and R. K. Adler, 1972, *Map Projections for Geodesists, Cartographers, and Geographers*. Amsterdam: North-Holland, 174p.

Reilly, W. I., 1973, A conformal mapping projection with minimum scale error, *Survey Review* **22**(168), 57-71.

Snyder, J. P., 1982, *Map Projections Used by the U.S. Geological Survey*, U.S. Geol. Survey Bull. 1532, 313p.

Snyder, J. P., 1985, The transverse and oblique cylindrical equal-area projection of the ellipsoid, *Assoc. of Am. Geographers Annals* **75**(3), 431-442.

Snyder, J. P., 1987, *Map Projections: A Working Manual*, U.S. Geol. Survey Prof. Paper 1395, 383p.

Cross-references: *Figure of the Earth; Geodesy: Geometric; Geodetic Networks and Control Surveys; Geodetic Reference Systems.*

METEORITES: NATURE AND ORIGIN

A thunderous explosion was heard from the sky, it was a big star almost as bright as the moon . . . falling into a garden. . . . The star was found aglow at the bottom of a pit about three feet deep. . . . A warm round fist-sized rock was dug out. Its color and density were similar to those of iron. . . .
from *Notes from Dreamy Creek*
by Kuo Shen [(A.D. 1064) Sung dynasty]

As witnessed by this astute Chinese observer almost a thousand years ago, all terrestrial planets including the Earth are under a constant barrage of fascinating solid objects from space. Those objects that survive the passage through the atmosphere and reach the Earth's surface are *meteorites*. A few thousand meteorites, ranging in size from less than a gram to several tons, have been collected either as accidental *finds* or as purposefully recovered *falls*

after observing the object's entry. The most important locale is the Antarctic, where perhaps half of the known meteoritic samples have been found. The meteorites are important because they keep relatively undisturbed records of conditions of the early solar system. In particular, they may be samples of the building material from which the terrestrial planets were accreted.

The major meteorite types and frequency of falls are listed below. The percentage of total observed falls that belong to each type according to Dodd (1981) is given in parentheses.

Stones (95.7%)
  chondrites (87.4%)
    —ordinary   high iron (abbreviated as H type)
                    low iron (L and LL)
    —carbonaceous (C)
    —enstatite (E)
  achondrite (8.3%)
Irons (3.2%)
Stony irons (1.2%)

The most abundant meteorites are stones, of which chondrites dominate. The name *chondrite* comes from the ubiquitous presence in them of millimeter-sized subspherical masses of silicates called *chondrules,* which most likely crystallized from rapidly cooled melt droplets. Chondrites are agglomerate rocks consisting of diverse components featuring a variety of chondrules set in a fine-grained matrix. This distinct texture can be seen in Figs. 1 and 2. Some chondrites contain carbonaceous material including complex organic compounds, hence they are termed *carbonaceous chondrites.* Others are rich in the mineral enstatite ($MgSiO_3$)

FIGURE 1. Allende carbonaceous chondrite. Notice the fusion crust (darker patches) on both sides of the lower surface. The large white object in the middle is an inclusion rich in refractory elements, in which extinct ^{26}Al was found. Some of the smaller, round light-colored objects are chondrules. The fine-grained matrix forms the grayish black background which is rich in volatiles, organic compounds, and water. (*Courtesy of Jim H. Chen*)

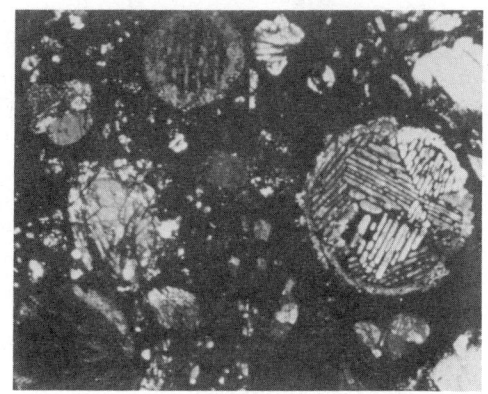

FIGURE 2. Thin section of type H5 chondrite Faith through transmitted light under a petrographic microscope. The opaque parts are the matrix. The round light-colored objects are chondrules. (*Photograph courtesy of Arizona State University*)

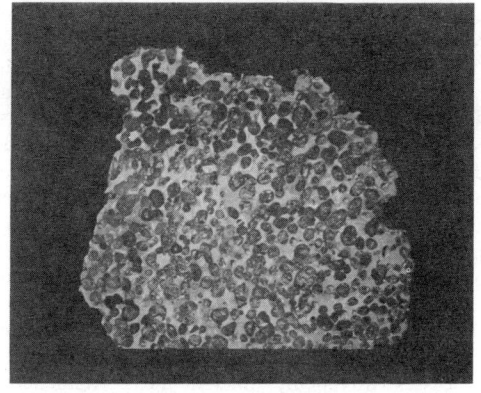

FIGURE 3. Stony-iron meteorite Springwater. The darker parts are olivine crystals (stone), which are embedded in the bright reflective Fe-Ni metal alloy (iron). (*Photograph courtesy of Arizona State University*)

forming the *enstatite chondrite* group. The remaining chondrites, which are the most common meteorites, are called *ordinary chondrites*. They can be further divided into the high- and low-iron varieties. In addition to the above scheme, chondrites can be subdivided into six petrologic types numbering from 1 to 6 in increasing order of the severity of the metamorphic effects suffered in early solar system heating events. For example, in type 6 the heating caused nearly complete recrystallization, homogenization of the chemical composition between different olivine crystals, coarsening of grain sizes, and the obliteration of the chondrule-matrix texture.

Stony meteorites without chondrules are called *achondrites,* which are extremely diverse. Many resemble terrestrial and lunar basalts. Some are almost monomineralic cumulates; others are breccias. The genesis of most achondrites seems to involve melting processes. Iron meteorites consist primarily of metallic Fe and Ni alloys with minor sulfides and phosphides. Some contain silicate inclusions. Several subtypes, based on either texture or chemistry, exist. An example of stony iron meteorites consisting of comparable amounts of silicates and Fe-Ni metal is shown in Fig. 3. Irons and stony-irons crystallized from melts. The segregation between immiscible silicate and metallic liquids probably figured prominently in their genesis.

The closer the elemental abundance patterns of a meteorite approaches that of the Sun, the more *primitive* it is considered to be, since meteorites are thought to have been derived ultimately from the same material that formed the Sun through successive stages of chemical differentiation. Using this criterion we tend to believe that type 1 carbonaceous chondrites is the most primitive meteorite type since its abundance pattern closely resembles that observed in the solar atmosphere (Fig. 4). The exceptions are H, C, N, O, and noble gases that are clearly too volatile to be quantitatively incorporated into rocks. The other types of chondrites show subtle differences from this composition. One example is that the abundance of volatile elements such as Tl and Pb decreases with increasing petrological types reflecting their different thermal histories. The nonchondrites have abundance patterns grossly different from solar abundances. They are thus termed *differentiated* since presumably they are the products of chemical differentiation of some material whose composition was originally much closer to solar composition. Magmatic processes in genesis of many of them obviously did much of the differentiation.

Isotopic methods have yielded a wealth of

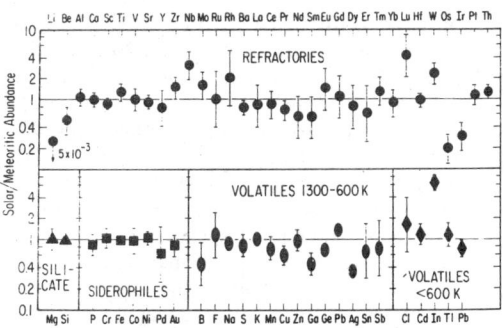

FIGURE 4. Comparison between solar and C1 meteoritic abundance (normalized to Si). Only solar Li shows large depletion, which is the result of thermal nuclear destruction inside the Sun after its formation. Eight elements show offsets outside the nominal errors, which are probably caused by uncertainties in the solar abundance determination. The abundances of the remaining 51 elements agree independent of their diverse volatility and chemical affinity. (After Anders and Ebihara, 1982)

constraints on the nature of meteorites (Kerridge and Matthews, 1987). Long-lived radio-nuclides such as ^{238}U have been used to determine the age of the meteorites. It was found that the ages cluster around 4.5 billion years with very few exceptions. Therefore, meteorites are much older than the oldest terrestrial rocks and most of the lunar rocks. They are thus good recorders for the earliest period of solar system evolution. The old ages also imply that they probably resided on bodies that were at most a few hundred kilometers in size, because one would expect that on larger bodies, the thermal evolution time is long so that most rocks would have registered younger ages due to later planetary geological processes driven by internal heat. Radioactive nuclides with half-lives much shorter than the age of the solar system are extinct today. However, it was discovered that several such extinct nuclides were present in some meteoritic samples when they first crystallized (Fig. 1). One of these extinct nuclides, ^{26}Al, has a half-life of only 0.7 million years. Therefore, meteorites had to solidify within a few million years of the synthesis of some ^{26}Al inside stars, otherwise the ^{26}Al would have all decayed by the time the meteorites solidified. The solar system could not have been much older than the meteorites, and the age of the meteorites essentially gives the age of the system. Furthermore, the ^{26}Al was quite abundant, hence its decay could have provided a strong but transient heat source for planetary objects in the early solar system.

The oxygen isotopic composition of different types of meteorites, the Moon and the Earth are different implying that the early solar system was isotopically heterogeneous. This heterogeneity provides a way to test the genetic relationship between different types of meteorites and to trace their origin. For example, if the source of certain meteorites is Mars then their oxygen isotopic composition must be the same as that of Mars. Unfortunately, so far we do not have precise enough data for other planets to apply this potentially powerful tool.

Where do the meteorites originate? Several resemble the lunar rocks returned from the *Apollo* missions, thus most probably came from the Moon. A rare group of achondrites (commonly abbreviated as SNC using the initials of its prominent members) displays young radiometric ages thus is thought to have come from late heating events on planet-sized bodies. Mars is the most likely candidate. However, the vast majority of meteorites probably have come from relatively small parent bodies perhaps no larger than a few hundred kilometers. Cooling rate studies and the study of pressure-sensitive mineralogical or chemical indicators are consistent with this estimate. We also know from studying how long the meteorites have been exposed to cosmic rays that they have been shielded from the irradiations for most of their life times hence must have been inside bodies at least about a kilometer in size. The only solar system solid

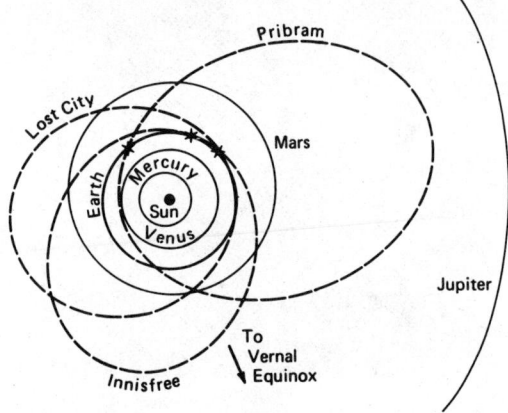

FIGURE 5. Preentry orbits of three ordinary chondrites are similar to those for the earth-crossing asteroids and have aphelions near the asteroid belt supporting the asteroidal origin of meteorites. However, the orbits of many short period comets also resemble these. (From Wood, 1979)

bodies of this size range are asteroids and cometary nuclei.

Another clue of the origin of meteorites comes from the orbits of meteorites before entry. The preentry orbits of three ordinary chondrites have been determined from simultaneous photographic observations by widely separated cameras. The aphelions are within or near the asteroid belts (Fig. 5). Furthermore, the reflectance spectra of several types of meteorites matches those of several types of asteroids (Fig. 6). For these reasons, the asteroidal origin of meteorites is favored. However, cometary nuclei remain a possibility, especially for carbonaceous chondrites whose water and volatile rich composition is reminiscent of the icy dust from comets.

To summarize, currently the evolutionary history

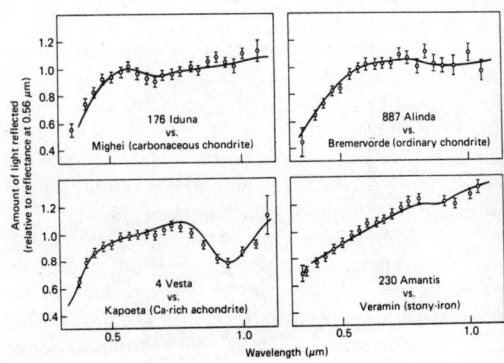

FIGURE 6. The reflectance spectra of four different types of asteroids (dots with error bars) can be well fitted by laboratory spectra of meteorite powders (solid curves) (From Wood, 1979).

of meteorites is thought to involve the following steps:

1. Different components in chondritic meteorites formed from solar nebula condensates under diverse physio-chemical conditions at different locales. Some presolar interstellar dusts may have also been mixed in. These early materials experienced an extremely complicated history of heating and cooling causing various chemical fractionation effects.
2. These diverse components assembled themselves into meteorite parent bodies (asteroidal?). Certain transient heat sources (^{26}Al?) caused some to metamorphose and others to differentiate through melting. But most of the actions were over within the first one hundred million years of 4.55 billion years ago. These bodies presumably were mostly consumed during the accretion of terrestrial planets.
3. The leftover bodies stayed dormant except for occasional mutual collisions, which caused impact reworking and produced shock features often found in meteorites. These bodies gradually ground themselves to pieces. The pieces were perturbed into earth-crossing orbits and eventually fell. The interval between breaking into meter-sized bodies and fall is of the order of 10^7 to 10^8 years.

TYPHOON LEE

References

Anders, E., and M. Ebihara, 1982, Solar system abundances of the elements, *Geochim. Cosmochim. Acta* **46**, 2363.

Dodd, R. T., 1981, *Meteorites, A Petrologic-Chemical Synthesis*. Cambridge, U.K.: Cambridge Univ. Press, 368p.

Kerridge, J. F., and M. S. Matthews, eds., 1988, *Meteorites and the Early Solar System*. Tucson, Az.: University of Arizona Press.

Wasson, J. T., 1974, *Meteorites: Classification and Properties*. Berlin: Springer-Verlag, 316p.

Wood, J. A., 1979, *The Solar System*. Englewood Cliffs, N.J.: Prentice-Hall, pp.95–111.

Cross-references: *Absolute Age Determination: Radiometric; Earth and Moon: Origins; Paleomagnetism: Meteorites; Solar System: Origins.*

MINERALOGIC PHASE TRANSFORMATIONS IN THE MANTLE

The forces responsible for both convection and vertical chemical stratification in the Earth are derived from a combination of the gravitational and the thermodynamic potentials. Gravity alone would force the iron to magnesium ratio to increase steadily with depth and heavy elements such as uranium to be depleted from the crust. In fact, we find that uranium is highly enriched in the crust and the iron to magnesium ratio is higher in the crust than in the mantle. The chemical potential is responsible for partitioning both uranium and iron in melts. Since the resulting liquid is less dense than the residual solid, the liquid rises to the surface under the influence of gravity. Similarly, the minerals in the cooler subducting portions of the mantle undergo phase transitions at depths, which differ from the ambient mantle owing to the reduced temperatures. Since the densities of the two phases are different, the gravity driven subduction is either enhanced or retarded depending on whether the denser phase in the cooler region transforms at lower or higher pressures. In fact, the density differences due to phase transformations dominates that due to thermal expansion.

Seismic data provide the primary information for defining the variation of chemical composition within the Earth. Since virtually all minerals that exist in the upper mantle undergo phase transformations with depth, we must expect that much of the mantle's seismic signature reflects the effects of these phase transformations. Those minerals that are the most abundant will produce the greatest effects, because acoustic velocities of the aggregate are dependent on the volume per cent of the particular mineral. The seismic structure will exhibit a discontinuity for transformations that have a substantial velocity increase over a narrow transition pressure interval or a steep velocity gradient if the transition occurs over a broad pressure interval. Low velocity zones will be created if the mantle is partially molten over some depth interval. Thus, in order to define the chemical state of the mantle, it is necessary to understand the nature of all possible phase transformations including their pressure, temperature, and chemical dependence, along with the physical properties of all phases. Then their effects on the acoustic velocities and density can be properly defined.

Experimental Techniques

Figure 1 illustrates estimates of the pressure and temperature as a function of depth in the Earth. Also illustrated are the compressional wave seismic velocities illustrating the different regions of the Earth's interior. Birch (1952) demonstrated that the material of the mantle must either change in composition or in phase or both in the region between 200 km and 1000 km depth. Ringwood made great strides in defining the possible phase transformations in this region, but because the pressure capabilities of experimental equipment were quite limited initially, he had to rely on investigating chemical analogues to silicates. The research in the last two decades has been spurred by developments in high pressure technology where pressures equal to those at the

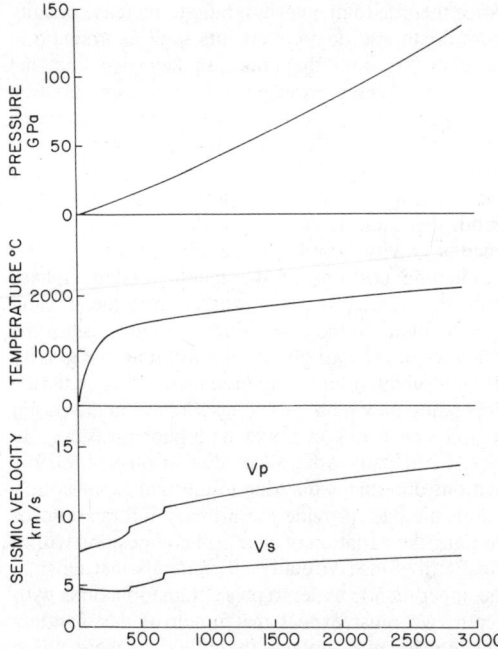

FIGURE 1. Pressure, temperature, and acoustic velocities as a function of depth in the Earth's mantle.

center of the Earth have been achieved for very small samples, and pressures equivalent to those of the lower mantle have been achieved in well controlled experiments for relatively large samples.

The diamond anvil high pressure cell allowed major breakthroughs in high pressure physics and geophysics. The principle of the diamond anvil is quite simple. A sample is sandwiched between the truncated tips of two diamonds and the diamonds are squeezed together. Since diamond is strong, large stresses can be generated and since the area is small, a relatively small force is required to generate these high stresses. A schematic of such a system is illustrated in Fig. 2. The sample can be heated by focusing a high powered laser on the specimen. The results of the elevated pressure and temperature can be evaluated by X-ray analysis of the recovered material. Using this technique, Liu (1976) demonstrated that the magnesium silicates transform to the perovskite structure at depths of about 700 km. Further transformations at higher pressures have not been discovered, suggesting that most materials of the lower mantle are in the perovskite crystal structure.

In addition to outlining the general stability fields of the different phases, the diamond anvil system is useful to provide spectroscopic information and X-ray data at elevated pressures since the diamonds are transparent to such radiation. However, the diamond cell cannot provide detailed phase equilibria data nor high quality recovered samples. The temperature gradients may approach one thousand degrees per micron in the high temperature experiments, which will create a chemical potential gradient and force chemical heterogeneity in the sample on a very small scale. Furthermore, the samples are generally about 10 microns thick and often heavily deformed due to nonhydrostatic stresses.

Fortunately, as the diamond anvil cell was breaking high pressure records, the Japanese laboratories were continuing to develop large volume high pressure systems. Such systems, which now can simulate the conditions of the top of the lower

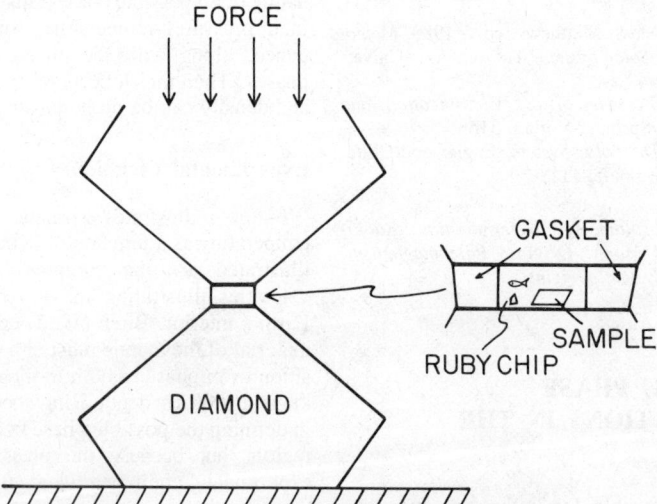

FIGURE 2. Diamond anvil high pressure cell. The high pressure face of the diamonds may have a diameter of 0.5 mm and the gasket thickness is typically 0.1 mm. The sample assembly illustrated includes a gasket with the sample in a fluid that creates an hydrostatic pressure field. The fluorescence of the ruby chip serves as a pressure calibrant.

mantle, provide a much more uniform thermodynamic environment for the sample. While the sample volumes are still only of the order of cubic millimeters, these samples are three to four orders of magnitude greater than those of the diamond anvil cell. Thus, it is possible to produce sufficient quantities of the high pressure phases to analyze the properties on recovered specimens, often single crystals.

The high pressure system developed by Kawai is illustrated in Fig. 3. Pressure is generated by a uniaxial ram. In the jaws of the ram are the guideblocks that house a hardened steel sphere which, upon closing, creates a cubic cavity. This stage is the first of a two stage system. The second stage is created by forming an oversized cube consisting of eight smaller tungsten carbide cubes (the hardest metal available). These cubes are constructed so as to leave a small octahedral cavity when they are assembled. The sample is located in a ceramic pressure medium inside this cavity. As the jaws of the press close, the forces are concentrated onto the sample cavity. Temperature can be increased by introducing an electrical conductor in the sample chamber creating an electrical furnace. Conditions have been attained that melt the mantle minerals at pressures corresponding to those of the lower mantle. The maximum pressure for these systems is limited by the strength of the material of the second stage. These systems owe their success to a design feature that uses the force to support the second stage anvils so that they do not break at lower pressures.

Phase Relations in Mantle Systems

The mineralogy within the mantle is a reflection of the pressure, temperature, and composition. Oxygen, the most plentiful element, is sufficiently abundant as to completely oxidize most cations. Of the cations, magnesium is considered the most

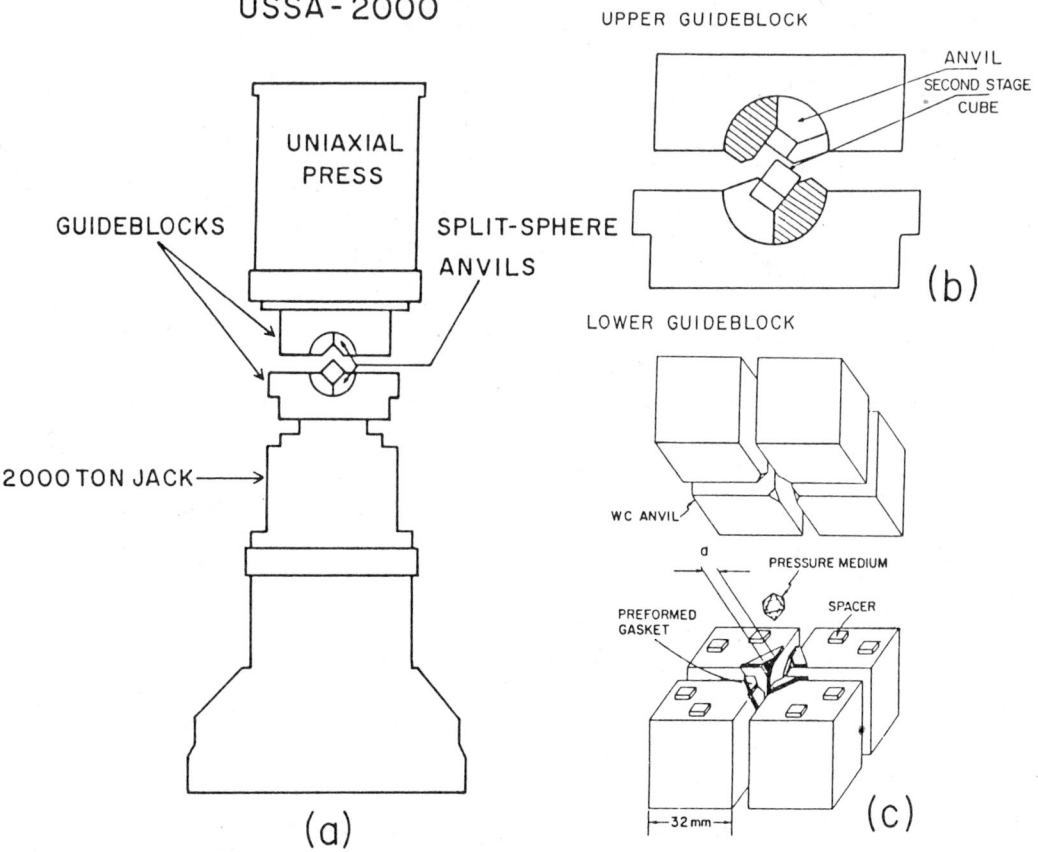

FIGURE 3. Large volume high pressure system. The press illustrated here produces a force of 2000 tons. The guideblocks (b) squeeze together on a cubic hole and the cube that fits in the hole (c) is composed of eight anvils, which have truncated corners. Inside this octahedral cavity is the pressure medium. Sample volumes are typically several cubic millimeters.

abundant with silicon a close second. Iron, the third most abundant cation, is about a factor of ten more scarce. Calcium and aluminum, which are present in about equal amounts, are reduced in numbers by yet another factor of two from iron. Sodium, the most plentiful monovalent element, is about another factor of five lower in abundance than aluminum.

The chemical formula for most mineral phases that will be found in the mantle can be written as:

$$(AO)_i(B_2O_3)_j(C_2O)_k(SiO_2)$$

where A are the divalent cations including Fe^{2+}, Mg, and Ca; B are the trivalent cations such as Fe^{3+}, Al, and Cr; and C are the monovalent cations such as Na and K. The phase relationships in such a complicated chemical system depend on the chemical potential of each component in each phase. However, for these systems, we can separately evaluate two systems: a low silica system where $i = 2$ and a high silica system with $i = 1$. These two groups do not interact appreciably due to the limited solubility of most cations in the phases dominated by magnesium and iron.

Most models of the mantle conclude a value of i intermediate between these two systems. If the earth was produced with a similar elemental abundance of a type I carbonaceous chondrite and this material has been reduced sufficiently to produce a metal phase (iron but no silicon) which formed the core, then the value of i is about 1.2. The pyrolite model, which has been proposed by Ringwood, has a value for i of 1.4. These models will contain phases belonging to both the high silica system and the low silica system.

The low silica phases include olivine, modified spinel, and spinel. A is either Mg or Fe^{2+}, and both j and k are effectively zero. The phase assemblage for this portion of the mantle's material will thus reflect the pressure and temperature and the Fe/Mg ratio but it will be insensitive to the abundance of other components. The high silica phases include pyroxene, garnet, ilmenite, and perovskite. These phases will include Mg and Fe^{2+}, as well as Ca, Al, Na, Cr, K, and Fe^{3+}. The mineralogy of the silicon rich portion of the mantle will depend on the abundance of all of these elements.

The Low Silica System. The silica poor system represents about 60% by volume of a pyrolitic mantle. The restrictive chemistry of this system simplifies the phase relations and increases the possibility of these transformations occurring over a narrow depth range. The only significant chemical variable is the Fe/Mg ratio. Figure 4 displays the phase relations in this system as a function of depth and of composition for an assumed temperature gradient. The temperature of 1600°C corresponds to that at 650 km depth and the temperature is assumed to vary with depth along an adiabatic gradient, as defined by Brown and Shankland (1981). Such a temperature gradient would be expected in a convecting mantle but the absolute level remains somewhat uncertain.

Figure 5 illustrates the per cent of the different phases as a function of depth for Mg/(Mg + Fe) of 0.85. Initially, olivine is the stable phase. At about 350 km depth, the olivine begins to transform to spinel. The compositions of the two phases will not be the same and can be inferred from the position of the curves in Fig. 4. At about 380 km depth, olivine,

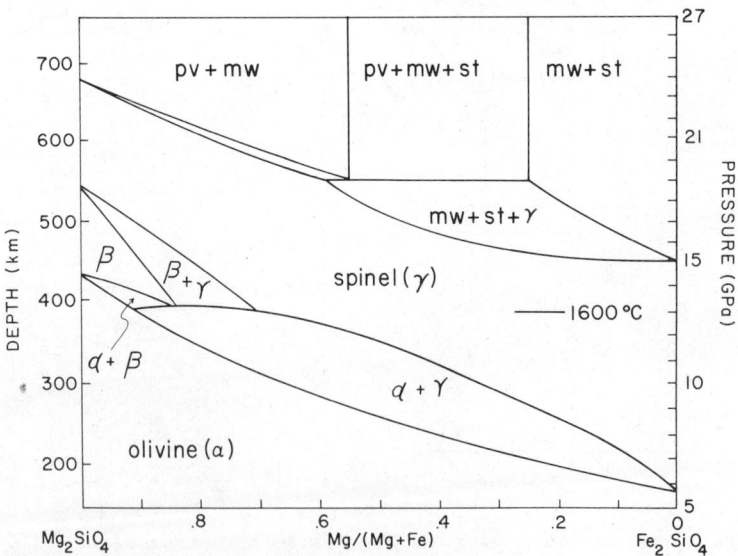

FIGURE 4. Phase relations in the system Mg_2SiO_4-Fe_2SiO_4 as a function of composition and depth. The curves correspond to an adiabatic temperature gradient that is 1600° at 650 km depth.

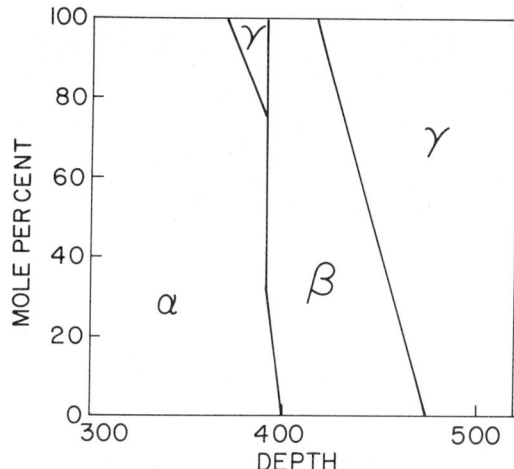

FIGURE 5. The percent of the phases, olivine, modified spinel, and spinel, as a function of depth for the composition indicated in Fig. 4.

spinel, and modified spinel all coexist. Slightly deeper, only olivine and modified spinel coexist and the olivine content has discontinuously decreased. Even deeper, the olivine completely disappears, modified spinel yields to spinel, which in turn yields to perovskite and magnesio-wustite.

The topology of the phase relations in Fig. 4 has been demonstrated in several laboratories; however, the details are still very much in question and are the focal points of several current research programs. For example, the pressure, temperature, and compositions where olivine, spinel, and modified spinel coexist are not well defined. Furthermore, the bulk of this phase diagram is based on data that was gathered for temperatures several hundred degrees lower than indicated on the figure.

The perovskite producing phase transformation in Fig. 4 has only recently been studied in large volume high pressure equipment with low thermal gradients. While the spinel to perovskite plus magnesio-wustite transformation is a divariant reaction, it appears to be very narrow in pressure. Experiments to date suggest that this transformation may occur over a sufficiently narrow region as to produce a seismic discontinuity only a few kilometers wide.

Figure 6 illustrates the effect that different mantle temperatures would have on the low pressure portion of the phase diagram. Higher temperatures generally increase the depth at which the phase transformation takes place. Thus, in a subducting region, where the temperatures are anomalously low, the phase transformation will occur at a shallower depth. The added density of the high pressure phase will add to the sinking forces. The temperature dependence of the transformation to perovskite has not been included on this figure as only the end member reactions have been studied. In this case, Ito and Yamada (1982) demonstrated that increasing temperature actually decreases the depth at which the transformation takes place. Thus, this transformation will work against whole mantle convection. As to whether or not such

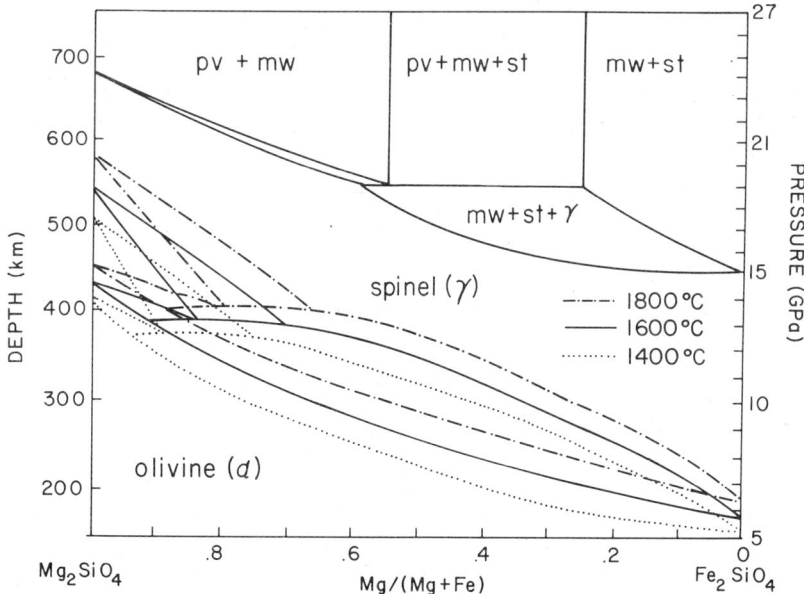

FIGURE 6. Phase relations in the system Mg_2SiO_4-Fe_2SiO_4 as a function of pressure for several adiabatic temperature gradients. The indicated temperatures correspond to the temperature at 650 km depth. The perovskite forming transition is not changed due to lack of sufficient data.

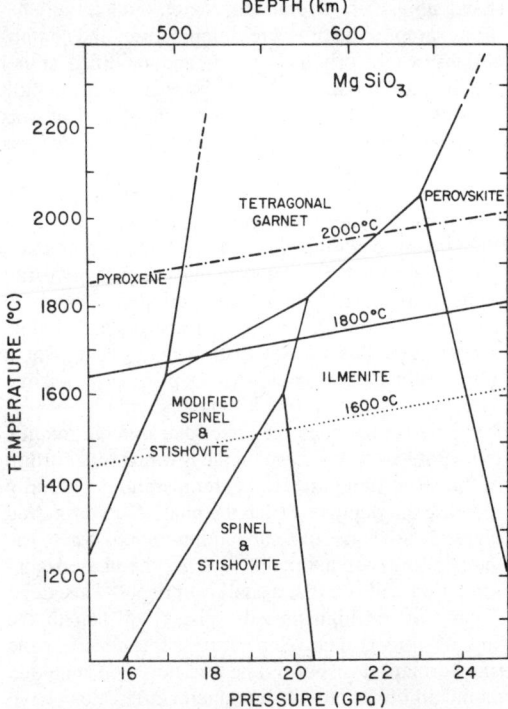

FIGURE 7. The phase stability fields for MgSiO$_3$ as a function of pressure and temperature. The curves represent different adiabatic temperature gradients possible in the Earth's mantle. (Modeled after Sawamoto, 1987)

a phase boundary forms a barrier to convection is not yet resolved.

The High Silica System. The silica rich phase relations are more complex both because there are more stable phases that are possible in the Earth's mantle for even simple systems and because the chemistry is "dirty." The silica rich phases provide a home for most of the minor elements and these elements further complicate the stability regions of the phases. Figure 7 illustrates the stability fields as described by Sawamoto (1987) for MgSiO$_3$ as a function of pressure and temperature. Also illustrated are the estimated geotherms used in the previous discussion. Given the uncertainties in the mantle temperature, we see that, even for the pure magnesium end-member, the possible mantle phases include pyroxene, modified spinel, spinel, stishovite, garnet, ilmenite, and perovskite.

Figure 8 demonstrates the effect of adding Al to this system. This phase diagram, which is patterned after Akaogi et al. (1987) illustrates the solid solution of pyrope garnet (the extreme right end member) and majorite (the left end member) along possible geotherms. If we consider the composition at 80, we see that pure pyroxene (MgSiO$_3$) coexists with a garnet of 20 mole per cent majorite at the very bottom of the diagram. At 450 km depth, the garnet composition is almost 70 mole per cent majorite. The proportion of garnet continually grows along this curve until the garnet has the same composition as the starting material. At this point, only garnet exists in the assemblage. The garnet-pyroxene curve appears to be fairly temperature insensitive. Deeper than 500 km, the garnet composition moves toward pyrope with modified spinel plus stishovite representing the far left side mineralogy as the stability pressure of pyroxene has been exceeded. In this region the proportion of garnet is shrinking as the solid-solution curve moves to the right. As with the olivine phase diagram, many aspects of these curves are still not certain. The curves representing the

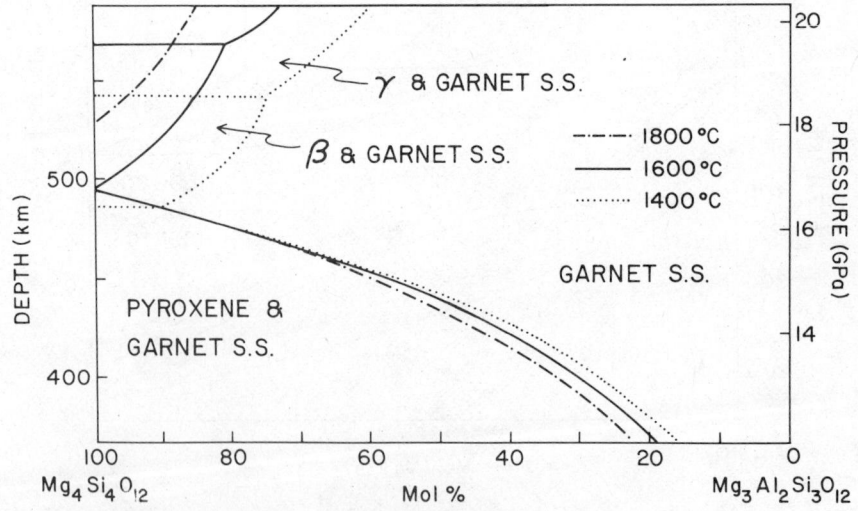

FIGURE 8. Phase boundaries in the enstatite (Mg$_4$Si$_4$O$_{12}$) pyrope (Mg$_3$Al$_2$Si$_3$O$_{12}$) system as a function of depth for different adiabatic temperature gradients. The temperatures correspond to the temperature at 650 km depth.

equilibrium of the garnet solid-solution and the spinel phase has not been experimentally determined.

In contrast to the silica poor system, phase transitions in this system will be distributed over a considerable depth. The pyroxene to garnet transition occurs at a fixed depth in Fig. 7, but when aluminum is added as in Fig. 8, the transition occurs over a depth range in excess of 100 km. The aggregate effect of sodium, calcium, and iron is probably to increase the stability of the garnet phase; however, the specific effects of adding these elements are not yet defined.

Seismic Signature of Phase Transitions

Associated with phase transformations are changes in physical properties, including the acoustic velocities, which are observable via seismic studies. If two phases have different acoustic velocities, then the transformation will cause a discontinuity in seismic velocity if the transformation occurs over a very small depth range and a steep velocity gradient if the transformation occurs over a wider depth range. The amount of the effect is proportional to the difference in velocity between the two phases and the volume per cent of the total that these phases represent.

Using Brillouin spectroscopy, in the laboratory of Weidner, the acoustic velocities of all of the significant mantle phases have been determined at room pressure and temperature. These results demonstrate that the modified spinel structure has very similar acoustic velocities to the spinal structure. Thus, this transformation will not be associated with a seismic discontinuity. Also, spinel plus stishovite is very similar to ilmenite. Most of the other transformations are capable of producing seismic signatures.

The details of the phase transformation become important in defining their effect on the seismic velocity. An example is given in Fig. 9 showing acoustic velocities expected from the olivine to modified spinel transformation as the composition is changed relative to the topology of the phase diagram. Curves A through E represent increases in iron content in that the compositions approach closer and closer to the point where olivine coexists with both modified spinel and spinel. For curves A and B the olivine transforms directly to modified spinel. The two phase region is seen as a distributed velocity increase. Curves C through E represent compositions where the olivine first transforms to spinel until the pressure obtains the required value (12 GPa) for the stability of modified spinel. At this point, the velocity jumps discontinuously because there is a discontinuous decrease in the amount of olivine remaining in the assemblage. Thus, the shape of the velocity increase changes dramatically with a change in the details of the phase relations or compositions. Even though the seismically inferred discontinuity includes very steep gradients as well as discontinu-

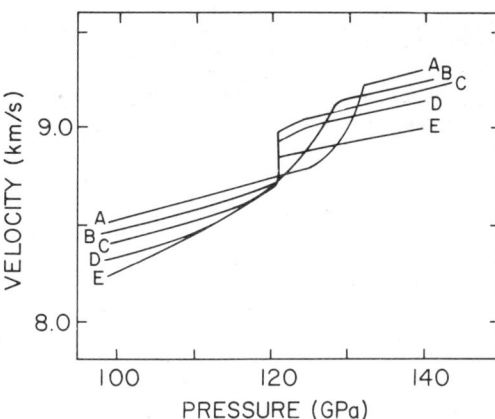

FIGURE 9. Compressional velocity as a function of pressure for the region corresponding to the 400 km discontinuity. The different curves represent different amounts of iron relative to the position of the univariant phase boundary. (From Weidner 1986).

ities owing to the seismic resolving power, these different models will yield different values for the size of the discontinuity.

Owing to the sharpness of the phase boundaries, the low silica minerals can produce seismic discontinuities as olivine transforms to the modified spinel phase and as spinel transforms to perovskite plus magnesio-wustite. The silica rich phase transformations are broadened by the presence of aluminum and other minor elements. Thus, with the possible exception of the perovskite producing transformation, these phase changes will produce steep velocity gradients but not discontinuities. It is very appealing to associate the 400 km discontinuity with the olivine to modified spinel transformation and the 670 km discontinuity with the perovskite producing transformation. Steep velocity gradients in the transition zone may then be related to the distributed transformations. Uniform mantle compositions with 40-70% olivine can thereby provide a seismic signal that agrees with the observed. Refinements of the mantle composition await a better understanding of the effects of pressure and temperature on the acoustic velocities of these high pressure phases.

Melting

The major chemical features of the Earth's crust have been produced by melting. As experimental techniques have progressed to higher pressures and temperatures, the results have raised questions regarding the features of the Earth's mantle. Answers that are both surprising and intriguing have been found, although the full impact of these new studies have not been completely assimilated into our description of the Earth.

One important and rapidly growing body of knowledge concerns the density of silicate liquids at high pressures. In recent experiments, Rigden et al. (1984) have been able to measure the density of silicate liquids up to megabar pressures using shock wave techniques. They have found the mantle melts may behave quite differently from crustal melts. Whereas crustal melts tend to be less dense than the solid residue (i.e., the melts are buoyant), at mantle pressures the buoyancy is greatly reduced and in some cases, the liquid is denser than the solid indicating that the liquid will sink relative to the remaining crystals. As a result, magmas generated deeper than 200 km or so will be much more difficult to separate from the parent body as the buoyant forces become very small or even negative. Melting experiments at high pressures by Ohtani, Takahashi, Ito, Scarfe, Herzberg, and Sawamoto have demonstrated that pressure plays a very important role on the chemical composition of the melt in equilibrium with the residual solid. At higher pressures, garnet rather than olivine becomes the last phase to melt. It is further demonstrated that a peridotitic melt can be produced from a chondritic source with either garnet or perovskite as the residual solid, which suggests that it is possible to produce an olivine rich upper mantle by partially melting a chondritic mantle leaving a silica rich lower mantle. Of course, such a model does not proceed without dispute as Kato et al. (1987) demonstrate that such a process would greatly affect the rare-earth element abundances.

The chemical homogeneity of the Earth's mantle is thus still an open question. The seismic structure can be produced by such a model. Still, it seems that the upper mantle is silica poor relative to chondritic sources. Mantle wide melting provides a mechanism for chemically layering a chondritic mantle in a manner consistent with our current views of the upper mantle. On the other hand, abundances of trace elements appear to be inconsistent with such a process. The development of new experimental capabilities adds to the excitement as we anticipate that most of the remaining questions can be answered in the near future. Acoustic velocities for all high pressure phases have now been determined at room pressure and temperature. We can anticipate that these properties will be determined at pressures and temperatures appropriate to the mantle. More detailed phase relations will soon be determined. With these data, precise seismic models could be made for any model chemical composition. Added to this will be more information on melt chemistry, density, and viscosity. Then, dynamics of early mantle evolution can be evaluated along with the chemical effects. Knowledge of trace element partition coefficients between various phases including liquids is essential to unravel the chemical record, which is imprinted on rocks we can examine.

DONALD J. WEIDNER
GABRIEL GWANMESIA
VIRGINIA HANIFORD
ANNE REMSBERG

References

Akaogi, M., A. Navrotsky, T. Yagi, and S. Akimoto, 1987, Pyroxene-garnet transformation: Thermochemistry and elasticity of garnet solid solution and applications to a pyrolite mantle, in M. H. Manghnani and Y. Syono, eds., *High Pressure Research in Mineral Physics*, Geophysical Monograph 39. Washington, D.C.: American Geophysical Union, 251–260.

Birch, F., 1952, Elasticity and constitution of the Earth's interior, *Jour. Geophys. Research* **57**, 227–286.

Brown, J. M., and T. J. Shankland, 1981, Thermodynamic parameters in the Earth determined from seismic profiles, *Royal Astron. Soc. Geophys. Jour.* **31**, 200–208.

Ito, E., and H. Yamada, 1982, Stability relations of silicate spinels, ilmenites, and perovskites, in S. Akimoto and M. H. Manghnani, eds., *High Pressure Research in Geophysics*. Tokyo: Center for Academic Publications, 405–419.

Kato, T., T. Irifune, and A. E. Ringwood, 1987, Majorite partition behavior and petrogenesis of the Earth's upper mantle, *Geophys. Research Letters* **14**, 546–649.

Liu, L., 1976, The high pressure phases of $MgSiO_3$ *Earth and Planetary Sci. Letters* **31**, 200–208.

Rigden, S. M., T. J. Ahrens, and E. M. Stolper, 1984, Densities of liquid silicates at high pressures, *Science* **226**, 1071–1074.

Sawamoto, H., 1987, Phase diagram of $MgSiO_3$ at pressures up to 24 GPa and temperatures up to 2200°C: Phase stability and properties of tetragonal garnet, in M. H. Manghnani and Y. Syono, eds., *High Pressure Research in Mineral Physics*, Geophysical Monograph 39. Washington, D.C.: American Geophysical Union, 209–220.

Weidner, D. J., 1986, Mantle model based on measured physical properties of minerals, in S. K. Saxena, ed., *Advances in Physical Geochemistry*, vol. 6, *Chemistry and Physics of Terrestrial Planets*. New York: Springer-Verlag.

Cross-references: *Continental Lithosphere; Earth Structure, Global; Mantle, Lower: Structure; Mantle, Upper: Structure; Mantle Convection and Plumes; Mantle Discontinuities; Mantle Dynamics; Seismic Properties of Rocks; Seismic Tomography.*

N

NATURAL REMANENT MAGNETIZATION (NRM)

Definition

Almost all rocks in nature, i.e., terrestrial and lunar rocks and meteorites that so far have actually been examined, possess permanent magnetization of a certain magnitude. A few exceptions are pure biogenic rocks, such as pure limestones, which do not contain ferromagnetic or ferrimagnetic minerals.

The permanent magnetization of natural rocks, which have never been artificially magnetized, is called the *natural remanent magnetization* (NRM). Natural rocks in populated areas often have an artificially produced remanent magnetization superimposed on the original NRM. To deal with NRM, therefore, the superimposed artificial remanent magnetization should be removed or separated.

Bearers of NRM

In this article, both ferro- and ferrimagnetic minerals are generally called magnetic minerals, as distinguished from paramagnetic minerals such as olivine and pyroxene.

Magnetic components of most terrestrial rocks are titanomagnetites, the chemical composition of which is expressed by $x\text{Fe}_2\text{TiO}_4(1-x)\text{Fe}_3\text{O}_4$, where Fe_3O_4 corresponding to $x = 0$ is the stoichiometric magnetite while Fe_2TiO_4 corresponding to $x = 1$ is the stoichiometric ulvospinel. In some rocks, particularly in oxidized rocks, maghemite ($\gamma - \text{Fe}_2\text{O}_3$) or titanomaghemite ($\gamma - x\text{FeTiO}_3 \cdot (1-x)\text{Fe}_2\text{O}_3$) is the main magnetic component. In sulfur-rich rocks, pyrrhotite (FeS_{1+x}, $1/10 < x < 1/7$) often plays the principal magnetic characteristics.

The magnetic components of lunar rocks and most meteorites, except some carbonacious chondrites, are Fe-Ni-Co metallic alloys. In some carbonaceous chondrites, magnetite is the main magnetic component by itself or together with Fe-Ni-Co metals.

In some particular terrestrial rocks, other magnetic minerals such as the substituted magnetites chemically expressed by $(\text{Mn},\text{Fe})_3\text{O}_4$, $(\text{Mg},\text{Fe})_3\text{O}_4$ are the main or auxiliary magnetic components. In phosphorus-rich meteorites, the Fe-Ni phosphides such as $(\text{Fe},\text{Ni})_3\text{P}$ play a significant magnetic role.

These magnetic minerals generally have the magnetic domain structure to minimize the total energy consisting of the atomistic exchange energy within the crystal lattice and the magnetostatic energy among individual magnetic constituents. As far as the volume of a magnetic mineral is small enough, its total energy maintains the minimum value by forming a single-domain structure in which the spontaneous magnetization of elemental crystal lattices is uniformly oriented. If a magnetic mineral is large enough, the mineral has a multidomain structure consisting of a number of magnetic domains so that the sum of the exchange energy of individual domains and the reduced magnetostatic energy among them becomes the minimum.

Natural rocks (including lunar rocks and meteorites) contain both or either of single-domain magnetic grains and multidomain grains. A single-domain magnetic grain behaves like a magnetic dipole of stable permanent magnetization in an applied magnetic field so long as the field is not sufficiently strong to reverse the dipole to the opposite direction. NRM borne by single-domain magnetic minerals is therefore very stable. (See *Rock Magnetism*, *Magnetic Domains*, and *Magnetic Properties of Minerals*.)

Origins of NRM

Isothermal Remanent Magnetization (IRM). Magnetic materials and natural rocks containing magnetic minerals acquire a remanent magnetization, $\mathbf{J}r(H)$, after applying a static magnetic field (H) and then completely removing the field at a constant temperature (T). The remanent magnetization thus acquired is called the *isothermal remanent magnetization* (IRM). The direction of IRM agrees with that of $\mathbf{H}$, and the intensity of IRM, $\mathbf{J}r(H)$, is an increasing function with an increasing H. With an increase of H, however, $\mathbf{J}r(H)$ approaches asymptotically toward a finite saturated value, $\mathbf{J}r(\infty) = \mathbf{J}_R$, which is called the saturated IRM. When no special remark is given, the temperature, T, for acquisition of $\mathbf{J}r(H)$ and $\mathbf{J}_R$ customarily means the room temperature or the atmospheric temperature.

A heavy DC electric current accompanying a lightning discharge to rocky Earth's surface often causes IRM of the surface rock by an associated strong pulse of magnetic field around the lightning path. This kind of NRM caused by the IRM acquisition process is widely observed in mountain areas, in particular, in basaltic mountains (see *Rock Magnetism*).

Thermoremanent Magnetization (TRM). An igneous rock that is cooled down from an elevated temperature, T, to the atmospheric temperature, T_0, in the presence of a weak magnetic field, $\mathbf{H}$, acquires a relatively intense and stable remanent magnetization, $\mathbf{J}r(T,H)$. The direction of $\mathbf{J}r(T,H)$ is in agreement with that of $\mathbf{H}$ and its intensity is approximately proportional to H, provided that H is about 10 Oe or less and T is fixed. The remanent magnetization thus acquired is called the *thermoremanent magnetization* (TRM) acquired by cooling from T in a magnetic field $\mathbf{H}$.

Terrestrial igneous rocks were formed by cooling down from a high temperature molten state in the presence of the geomagnetic field in historic or geologic age. Therefore, igneous rocks generally possess TRM, the direction and intensity of which were subjected to the geomagnetic field when the igneous rocks were cooled down. The direction and intensity of NRM of igneous rocks thus acquired represent those of the geomagnetic field in situ at the time when the rocks were formed.

TRM reaches its saturation value for temperature beyond the highest *Curie point*, θ_c, of the rock-forming magnetic minerals, so long as $\mathbf{H}$ is kept invariant. The saturated TRM is NRM of most igneous rocks. TRM acquired by cooling from T^* to T^{**} in temperature, where $T^{**} < T^* < \theta_c$, in $\mathbf{H}^*$ of a magnetic field, where no magnetic field for temperature ranges higher than T^* and lower than T^{**}, is the *partial thermoremanent magnetization* (PTRM). PTRM acquired by cooling from T^* to T^{**} in $\mathbf{H}$ can be demagnetized by heating to T^*. An addition law holds for PTRMs. Namely, PTRM acquired by cooling from T^* to T^{***} in $\mathbf{H}$ is equal to the sum of PTRM acquired by cooling from T^* to T^{**} and that acquired by cooling from T^{**} to T^{***}, where $T^* > T^{**} > T^{***}$.

A superposition of PTRMs often appears in nature. Metamorphic rocks that were thermally metamorphosed in different times in respectively different geomagnetic fields are typical examples of such a case. (See *Thermoremanence* and *Curie Temperature*.)

Chemical Remanent Magnetization (CRM). In the course of a chemical formation of magnetic minerals from nonmagnetic (i.e. paramagnetic) materials in the presence of a magnetic field, $\mathbf{H}$, the newly created magnetic mineral acquires a remanent magnetization called the *chemical remanent magnetization* (CRM). The direction of CRM is parallel to that of $\mathbf{H}$ and its intensity is approximately proportional to H as far as H is of the order of 10 Oe or less.

As the newly created magnetic minerals in the early formation stage are small enough to be of a single-domain structure, their CRM is relatively intense and stable. In nature a chemical deposition of hematite ($\alpha - Fe_2O_3$) and maghemite ($\gamma - Fe_2O_3$) in sediments owing to a chemical dehydration process such as

$$2FeOOH \rightarrow Fe_2O_3 + H_2O$$

in the presence of the geomagnetic field is believed to be the origin of NRM of redbeds, which are widely spread over the Earth's surface. [See *Chemical Remanent Magnetization (CRM)*.]

Detrital Remanent Magnetization (DRM). An aggregation of sediments of detritus such as ocean-bottom or lake-bottom sediments in nature has a remanent magnetization. Laboratory experiments show that the sediments of magnetic grains having the spontaneous magnetization of single-domain or TRM produced by their sedimentation through water in the presence of a magnetic field, $\mathbf{H}$, are statistically aligned along the field direction, and that magnetic grains in a resting (or still-standing) wet mass of sediments in the presence of $\mathbf{H}$ also are statistically aligned along the direction of $\mathbf{H}$. The remanent magnetization of a coagulated mass of sediments thus produced is called the *detrital remanent magnetization* (DRM).

DRM of ocean-bottom sediments is widely examined for the purpose of paleomagnetic research of deep-sea sediments. (See *Detrital Remanent Magnetization (DRM)* and *Paleomagnetism: Deep-Sea Sediments*.)

NRM of Rocks in Geophysical Problems

NRM of terrestrial rocks is very stable, in many cases, being altered very little even during a period of more than several tens of million years, provided that no considerable chemical change has taken place in the rocks during the geologic period concerned. A stable NRM of terrestrial rock can therefore be considered as a fossil of the geomagnetic field at the locality and at the time the rocks were formed. In paleomagnetism, NRM data of rocks of known ages are the basis for research of geophysical problems such as *continental drift*, seafloor spreading, and regional crustal movements.

A remark to be made in connection with paleomagnetism with the aid of NRM of rocks may be the magnetic viscosity owing to the coexistence of very fine magnetic minerals in the rocks. Very fine magnetic grains of single-domain structure are effectively subjected to the thermal agitation at temperature T. In a weak magnetic field such as the geomagnetic field, very fine magnetic grains of single-domain tend to be aligned along the field direction, the growth rate with time (t) of magnetization along the field direction being approximately proportional to $(1 - \exp(-t/\tau))$, where τ is the relaxation time dependent on temperature T and the volume of magnetic grains and others. This kind of magnetization is called *viscous magnetization*.

If the magnetic field is removed, the acquired viscous magnetization tends to diminish with a decay form proportional to $\exp(-t/\tau)$. If τ is very large, say 10^2 years for example, no measurable viscous magnetization appears in the geomagnetic field during the usual laboratory experimental time, say 10 days. In the geologic time scale, however, the viscous magnetization acquired during a very long period, say several tens of million years, can amount to a detectable magnitude. NRM thus acquired through the viscous magnetization process is called *viscous remanent magnetization* (VRM).

In paleomagnetic research, the VRM component superimposed on a stable NRM component representing a paleomagnetic field should be carefully checked, and if any VRM component coexists, it must be removed. (See *Viscous Remanent Magnetization (VRM) and Viscous Magnetization* and *Seafloor Spreading: Magnetic Evidence*.)

T. NAGATA

Bibliography

McElhinny, M. W., 1973, *Paleomagnetism and Plate Tectonics*. Cambridge: Cambridge Univ. Press, 358 p.

Nagata, T., 1961, *Rock Magnetism*. Tokyo: Maruzen Co., 339 p.

Cross-references: *Chemical Remanent Magnetization; Curie Temperature; Demagnetization; Magnetic Domains; Magnetic Properties of Minerals; Magnetic Self-Reversal; Magnetostratigraphy; Rock Magnetism; Rock Magnetism: Measuring Techniques and Apparatus; Thermoremanence; Viscous Remanent Magnetization (VRM) and Viscous Remagnetization.*

NORMAL MODES—See FREE OSCILLATIONS OF THE EARTH

O

OCEAN–CONTINENT TRANSITION: STRUCTURE

The map of the global distribution of earthquake epicenters (Fig. 1) clearly divides the margins of continents into two major categories. Margins that border the Pacific are characterized by a large number of earthquakes and are termed active margins. Also included in this category are the northern margin of the eastern Mediterranean, the northeast margin of the Indian Ocean, and the eastern margins of the Caribbean and Scotia Seas. Conversely, the margins of Africa, Australia, and the seaward margin of India, as well as the majority of the margins of both sides of the Atlantic Ocean, are virtually devoid of earthquakes and constitute passive margins.

At active margins two crustal plates are moving with respect to each other. At most active margins the predominant component of motion is one of convergence. The oceanic plate is being thrust under the continental or Island Arc plate and subduction is said to be taking place. Oceanic crust generated at the crests of mid-ocean ridges is being consumed at an equal rate at plate margins by subduction and the size of the Earth remains the same. At some active margins, particularly in the vicinity of southern California, the oceanic and continental plates are mostly slipping by each other in transform motion, rather than converging. We will not discuss the ocean continent boundary, where the motion is largely transform, for two reasons. First, even at these boundaries some convergence takes place and features of subduction exist here. Second, the deep structures of transform margins are relatively poorly known.

According to the theory of plate tectonics, continents that constitute the visible part of the lithosphere drift over the asthenosphere and collide to form larger continents. Later in time these continents resplit and divide into smaller continents, and the process goes on. All passive margins represent fossil splitting zones of larger supercontinents. Consequently, for each passive margin there is another conjugate passive margin—the two being joined where the split occurred. Conjugacy of the Atlantic margins is relatively easy to establish. The conjugacy of the East African, Indian, Australian, and Antarctic margins is more difficult to establish precisely, although they were all together once and constituted the Gondwana supercontinent. The splitting apart motion had extensional as well as shear components, depending on the azimuth of the line along which the split took place. Passive margins, formed where the splitting was mainly extensional, have been studied the most and we will concentrate on them in this article.

Geophysical Techniques

The structure of the ocean continent transition has been probed by a variety of geophysical and geological techniques and observations. Seismicity studies have played a very large role in determining the structure of active margins and are described more fully in *Seismicity: Subduction Zone*. Of course, passive margins are characterized by far less seismicity. Earthquakes, such as the 1886 Charleston earthquake that took place close to the U.S. East Coast, represent an infrequent example of seismicity near a passive margin. Some inferences have been drawn from such seismicity about the deep structure, but considerable uncertainty is associated with these inferences.

Seismic refraction studies have provided the most detailed information about the crustal structure of ocean basins and continents. They have also provided information about the ocean continent transition zones, but these zones have been difficult to study for a variety of reasons, and earlier results have been lacking in consistency as well as in accuracy. Newer techniques involving multichannel seismic detectors and very closely spaced shots (using large airguns as sound sources) developed by the oil prospecting industry for seismic reflection work promise to obtain refraction and wide-angle reflection results with far greater accuracy than was possible earlier. The Expanding Spread Profiles (ESPs) in which two ships, one carrying a large sound source and the other towing a multichannel seismic detector, both steaming in opposite directions from a common midpoint, have considerable advantages over conventional two ship refraction profiles. Using a common midpoint rather than shooting from either end provides some geometrical advantages in the interpretation of the data. The very large data density allows advanced techniques of seismic data processing to be applied to these measurements with the ultimate hope of obtaining much better velocity resolution and imaging of the subsurface. Recordings that faithfully preserve signal

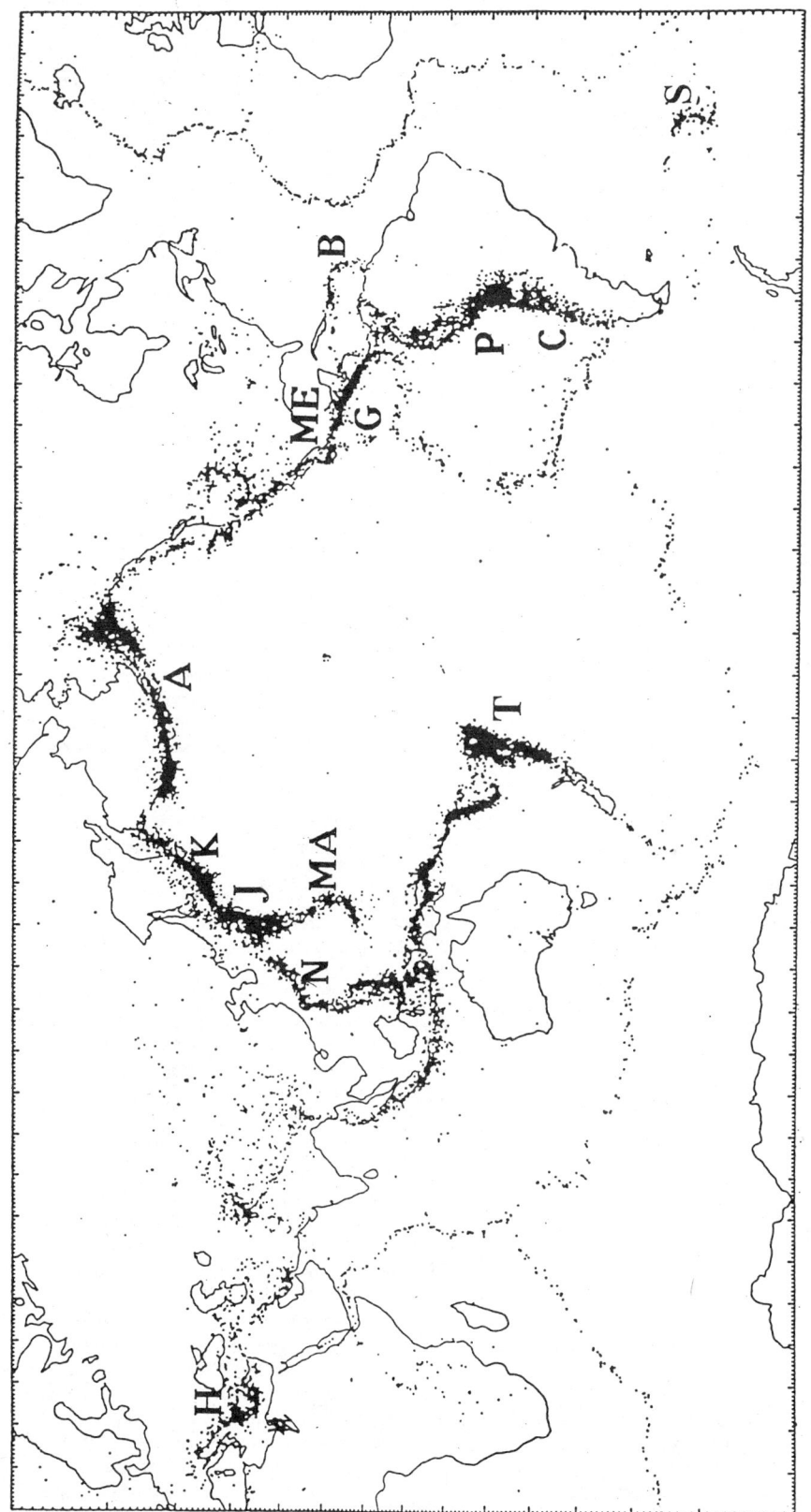

FIGURE 1. Global distribution of earthquake epicenters for 1961–1967 (Barazangi and Dorman, 1969). Subduction zones lie at the continental margins associated with earthquake activity. They are designated by letters as follows: A = Aleutian; B = Barbados; C = Chile; G = Guatemala; H = Hellenic; J = Japan; K = Kuril; MA = Marianas; ME = Mexican; N = Nankai; P = Peru; S = Scotia; T = Tonga. Passive margins (bordering the Atlantic, south of India, and around Australia) are margins not associated with earthquake activity.

amplitudes will make inversion schemes much more valuable. ESPs have been shot over several passive margins, notably off the U.S. East Coast, off Hatton Bank in the northeast Atlantic, in the Mediterranean Sea, in the Biscay area, and in the Norwegian Sea.

Multichannel seismic reflection work has yielded very important detailed information about the very shallow part of the crust, notably the sedimentary layers of interest to the oil and gas industry. Numerous reflection studies have been made over the continental margins, especially in the shallow water areas of the continental shelves. Reflection studies also hold great future promise for unravelling the deep structure of continental margins. Needed improvements include much more powerful sound sources so that deep penetration can be achieved. Larger towed multichannel receiving arrays will be particularly useful because they will allow determination of velocities from wide angle reflections. Such arrays can be attained either by constructing physically larger arrays or by creating wide synthetic apertures by towing arrays from two or more ships. (It must be mentioned, however, that the highly deformed rock strata in subduction zones pose formidable challenges to the seismic reflection method).

Gravity measurements first made by Vening-Meinesz aboard submarines first located the Indonesian and Caribbean subduction zones. Continuous gravity measurements from surface ships added considerable detail to the knowledge of the gravity field over the oceans. With present instrument accuracies the shorter wavelength components of the Earth's gravity field are determined accurately by gravity measurements, but the longer wavelength components are better expressed in undulations of the geoid. The satellite altimetric methods have greatly increased our ability to resolve geoid elevations and thus obtain the long wavelength gravity field. As far as the deep structure of the ocean continent boundary is considered, the gravity field, with few exceptions, has been used in a passive mode. It has not been inverted to directly yield structural detail, but rather has been used to verify that structures derived by other methods are not in conflict with observations of the gravity field. The fact that small density inhomogeneities may persist to great depths within the Earth has made it particularly difficult to obtain structures only on the basis of the gravity field.

Measurements of the Earth's magnetic field have shed light on certain aspects of the ocean continent transition. The continuation of the striped pattern of magnetic anomalies observed on ocean ridges and basins beyond deepsea trenches underneath their landward walls has confirmed that oceanic basement is subducted under and beyond the trench. The existence of the familiar striped pattern over back-arc regions confirms that the basement there was formed by seafloor spreading and makes it possible to obtain the age of the back-arc opening. Prominent magnetic anomalies also exist over passive margins, but they are not yet completely understood. We discuss them later.

Observations of heat flow emanating from the ocean floor has enabled some constraints to be placed on deep structures, especially in active margins. We will consider these, briefly, later. Magnetotelluric measurements have also been used to obtain values of electrical conductivity of rocks buried under the margins and therefore also place some constraints on deep structures. These have, however, been made in only a very few areas.

Active Margins

Geologists and solid earth geophysicists have emphasized different features in describing subduction zones. Taken together, Fig. 2 (ocean on the right side) and Fig. 3 (ocean on the left side) are sketches that demonstrate most of the features of subduction zones.

Just seaward of the deepsea trench, a topographic bulge attributed to elastic bending of the plate undergoing subduction is seen in the oceanic plate. An accretionary wedge of sediments culminating in an outer-arc high is formed by sediments from the deep basin being offscraped onto the landward plate. Heavy black lines represent décollement faults found within the accretionary wedge. Sediments derived from the island arc are ponded between the accretionary wedge and the island arc to form a fore-arc basin. On the landward side of the volcanic arc a back-arc basin is developed in some subduction zones. This back-arc basin is floored by basement created by seafloor spreading. Heat flow is high in the back-arc basin, in contrast to the generally low heat flow observed in the vicinity of the trench.

Seismic observations clearly demonstrate the presence of an oceanic plate, with high Q and high seismic velocities, that is subducting to great depths. Earthquake source mechanisms indicate the state of stress. Near the topographic bulge referred to earlier, the plate is under extension. The stresses are also extensional within the plate at some depth. However, it is the interplate thrust earthquakes where the two plates are tightly coupled to each other that produce the largest and the most damaging earthquakes.

We do not in any way mean to imply that all subduction zones are identical in nature. In fact, there are major differences in subduction zones. Studies presently focused on establishing these differences and explaining them should lead to improved understanding of the subduction processes. We briefly describe some of these differences and attempt to explain them. A major difference is in the presence or absence of back-arc spreading. The Peru, Chile, Japan, Kuril, and Aleutian trenches are not associated with presently active back-arc spreading, while the Marianas and Scotia subduction zones

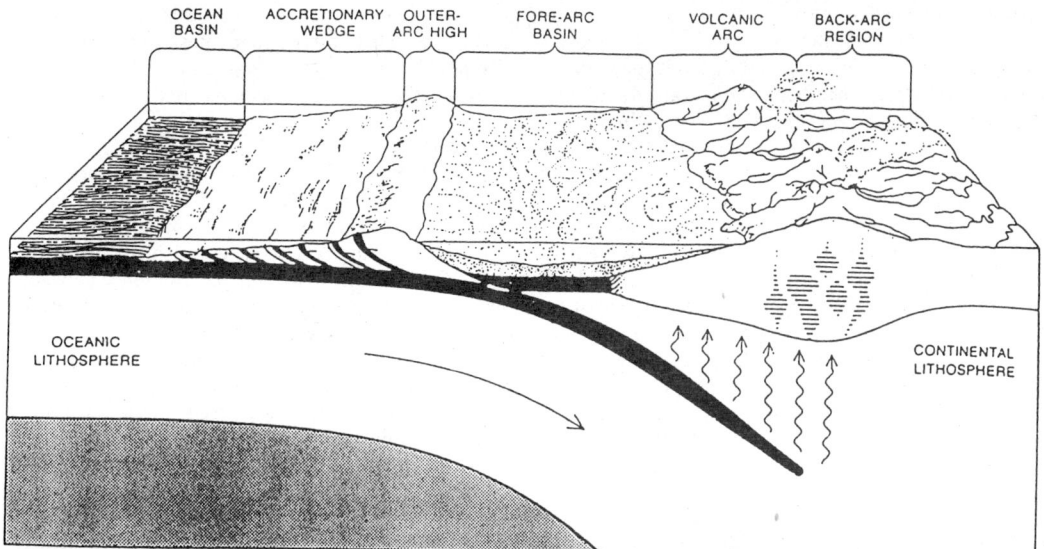

FIGURE 2. Geologists' view of subduction zones. Oceanic lithosphere (crust and upper mantle) is subducted under continental lithosphere. Fragments of oceanic crust and sedimentary rock scraped from the subducted plate form an accretionary wedge and an outer-arc high. Next comes a fore-arc basin, which accumulates sediments from the adjacent elevations, and then a volcanic arc, the most characteristic feature resulting from subduction. Some of the magma rising from the subduction zone solidifies in the crust. The back-arc region behind the volcanoes may show convergence (such as crustal faulting and folding) or divergence (such as crustal thinning and subsidence). Heavy black lines denote decollement faults within the accretionary wedge. (From Burchfiel, 1983)

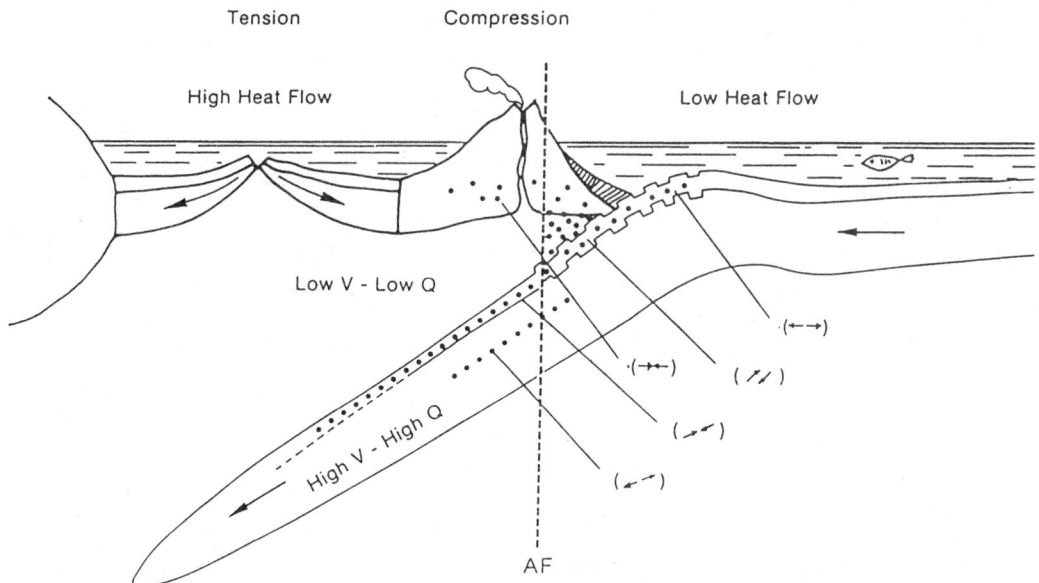

FIGURE 3. Geophysicists' view of idealized subduction zone. Dots - earthquake foci; heavy arrows - plate motions; Q - seismic quality factor; arrows in parentheses indicate earthquake source mechanism. (From Uyeda, 1984)

possess active back-arc spreading. This major difference manifested in the presence or absence of presently active back-arc spreading led Uyeda (1984) to define two "end member" subduction zones—the high stress Chilean-type and the low stress Marianas-type. The sketches in Figs. 4 and 5 summarize the major points of difference between the two subduction end members. The Chilean-type subduction zones are characterized by well developed Cordilleran-type mountain ranges, which are topped by calc-alkaline andesitic strato-volcanoes. These mountain belts contain copper porphyry deposits. The Marianas-type subduction zones have less massive island arcs but abundant basaltic volcanic piles overlying oceanic basement crust. Copper deposits are of the massive sulfide types. The stresses are compressive for the Chilean-type zones behind the trench while the back-arcs of the Marianas-type exhibit extensional stresses. The Chilean-type oceanic plate is subducted at a shallow angle, has a pronounced topographic bulge seaward of the trench, and the subduction is associated with truly great interplate thrust earthquakes. The Marianas-type oceanic plate, on the other hand, is subducted at a steep angle, the topographic bulge is poorly developed, and truly great earthquakes do not occur. The landward walls of Marianas-type trenches, particularly the Tonga trench, are devoid of sediments, whereas the Nankai and Barbados troughs have abundant sediments on the landward side.

A particularly interesting contrast exists in the amount of sediment on the landward wall of two neighboring segments of the Middle America Trench. The northern segment, the Mexican Trench, is of the Chilean-type and possesses abundant sediments. The southern segment, the Guatemala Trench, is of the Marianas-type and is practically devoid of sediment.

Uyeda has called attention to the striking difference in the rates of uplift of shorelines between the two types of subduction zones. Chilean-type shorelines rise at rates of a few mm/yr, while the Marianas-type shorelines rise at much smaller rates.

The simplest way to explain the difference between the two styles of subduction is to attribute them to the difference in mechanical coupling between the subducting and overriding plates. Where the coupling is strong, the oceanic plate bulges upwards seaward of the trench and strong interplate thrust earthquakes take place as the plate dips down at a shallow angle. Large amounts of sediments are scraped off to form massive accretionary prisms and the shorelines rise rapidly. Compressional stresses behind the trench give rise to mountain building and porphyry copper mineralization is associated with aerial volcanic activity taking place under horizontal compressional crustal stresses. This subduction is the Chilean type.

Where the coupling is weaker, the seaward bulge is poorly developed, strong interplate thrust earthquakes do not take place, and the oceanic plate bends down at a steeper angle. Sediments, especially indurated sediments, tend to move down with the subducting plate. Thus major accretionary prisms are not formed and the shore line is not rapidly pushed up. Extensional stresses behind the trench are

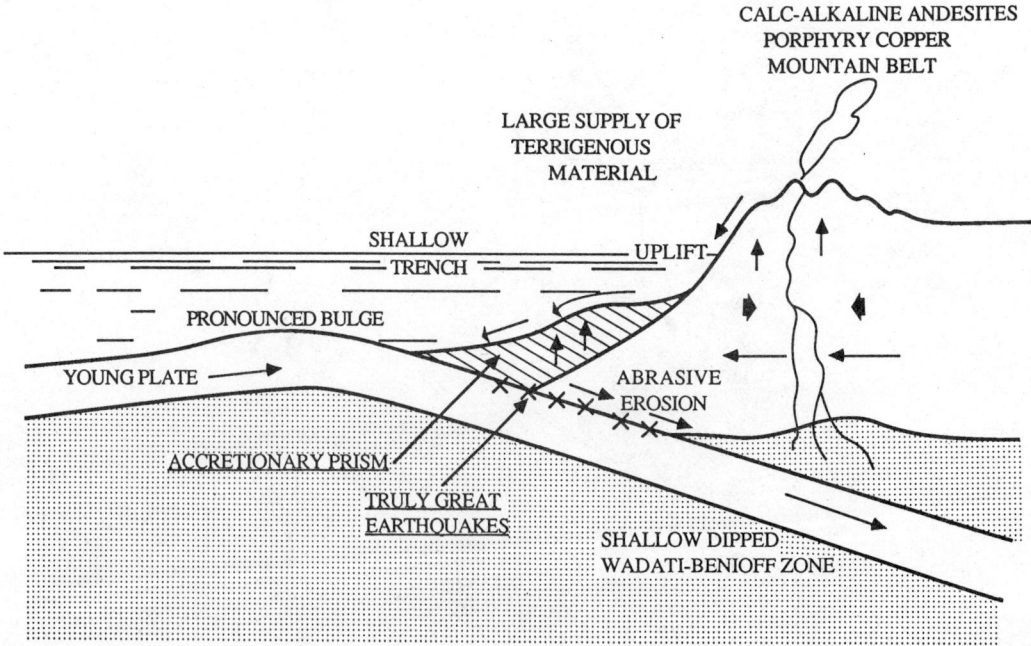

FIGURE 4. High stress (Chilean-type) subduction zone.

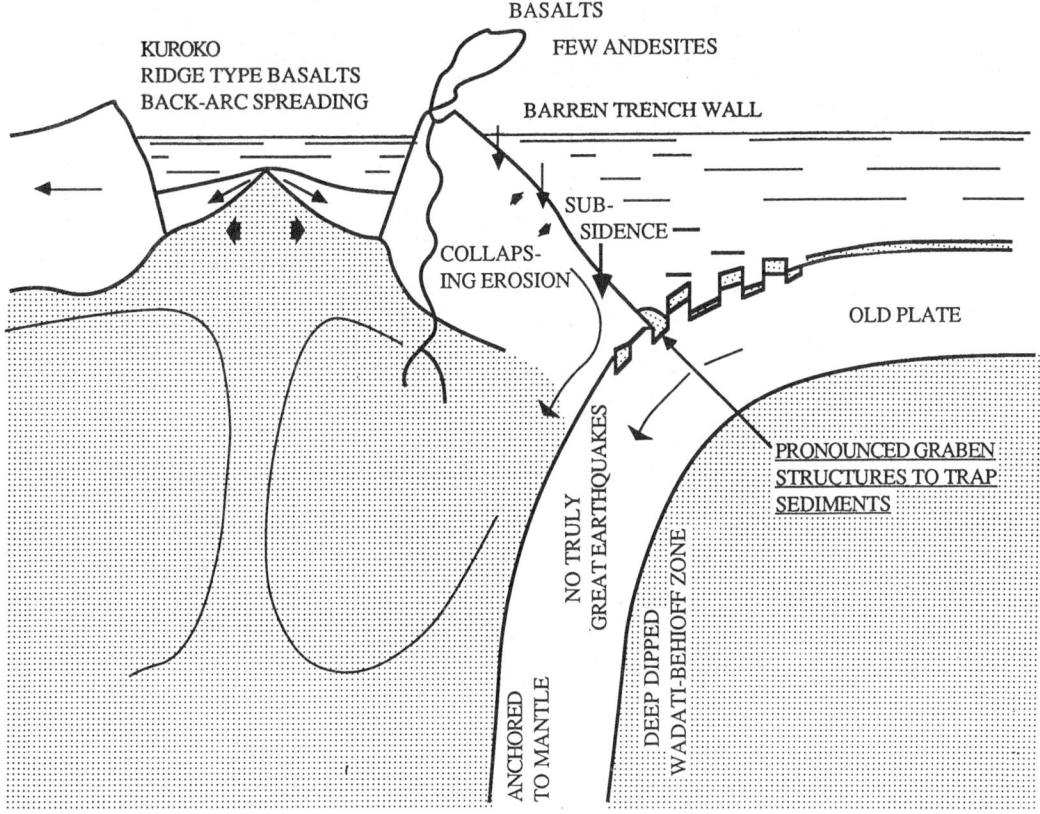

FIGURE 5. Low stress (Marianas-type) subduction zone.

associated with back-arc spreading and high heat flow. The generation of basaltic crust in seafloor spreading involves the implacement of massive sulfide deposits. This subduction is the Marianas type.

Various explanations have been given for differences in mechanical coupling. One explanation attributes the differences in coupling to the differences in the nature of the interface between the plates. Where the interface is lubricated with fluids or magmas, the resistance to subduction is small and low-stress subduction of the Marianas-type results. However, where the interface is rugged rather than smooth and lack of lubrication allows the plates to get locked, high-stress subduction of the Chilean-type occurs.

Another explanation suggests that as plates get older they get cooler and heavier as they move away from the spreading centers, giving rise to rapid sinking and oceanward retreat of a relatively decoupled Marianas-type oceanic plate. More buoyant younger plates, on the other hand, tend to be more strongly coupled and have to be forced down by the overriding plate in Chilean-type subduction. This explanation is supported by the observation that in the western Pacific the ocean plates tend to be older, giving rise to Marianas-type subduction while in the eastern Pacific the ocean plates tend to be younger with attendant Chilean-type subduction.

While the factors previously cited undoubtedly affect the nature of the subduction process, perhaps the most elegant explanation for the differences in the subduction process has been given by Uyeda (1984). He believes that the degree of mechanical coupling is controlled by plate motions and is determined by the vectorial relative motion between the continental overriding plate and the hinge line of the subducting oceanic plate. This hinge line is the position of the trench. Let us first assume the approximation that the hinge line stays fixed relative to the asthenosphere. This assumption is not bad because a slab subducted deep into the asthenosphere should be hard to displace sideways. Analysis of present global plate motions indicate that in the cases of the Marianas-type subduction, the overriding plate O is retreating (with velocity V_o in Fig. 6) causing back-arc spreading to occur. In the opposite case of Chilean-type subduction, the South American, Japanese, or Bering Sea plates are moving toward the hinge line of the Peru and Chile, Japan and Aleutian trenches, respectively. In some cases the hinge line may not be fixed. In the case of the

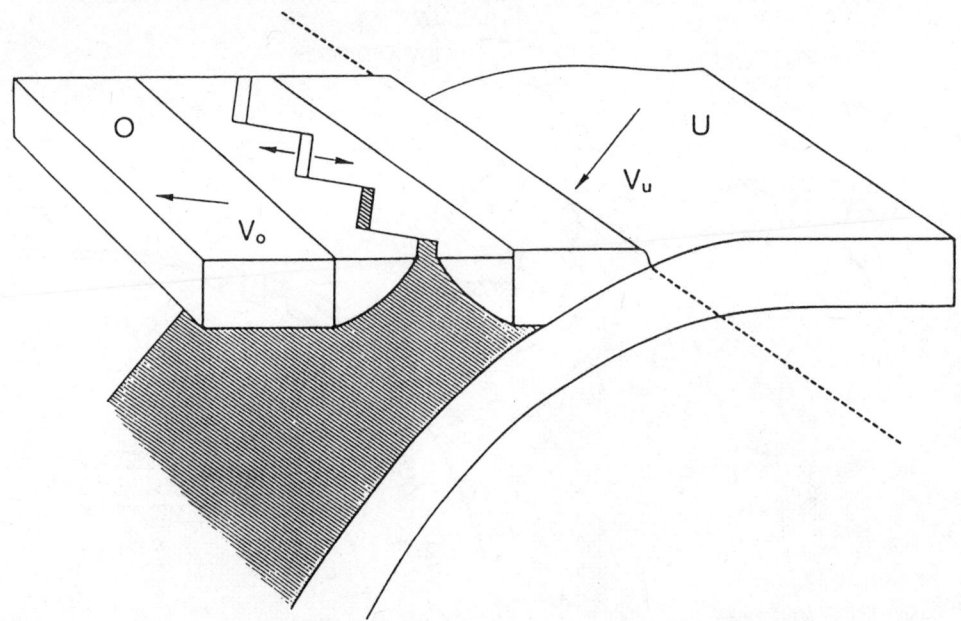

FIGURE 6. Schematic diagram of subduction geometry. (From Uyeda, 1984)

Scotia Arc, for example, the relative motion for the Marianas-type subduction is caused by a retreat of the hinge line (also known as a rollback of the trench).

Although Uyeda's explanation probably established the most plausible framework to understand differences in modes of subduction, other factors can also play a part, two of which were mentioned earlier—the nature of the interface between the two plates and the age of the oceanic plate. Still other factors have been proposed. For instance, it has been suggested that when buoyant features (such as sea mounts) in the ocean plate come into contact with the overriding plate, these features tend to choke the subduction process and actually bring it to a halt. Subduction may start again at a nearby but different locality. It has also been observed that trench depths vary systematically with age of the neighboring basement and that subduction rates are systematically greater with greater trench depths. These effects are generally attributed to lower temperatures and greater densities of older plates. Another interesting observation is the relationship between the maximum depth of the subducting plate and its age. Younger plates are shallow and older plates, which have had time to move further, are deeper. A subject of considerable interest is the presence or absence of the accretionary wedge. The supply of sediments (axial as well as transverse) and mechanisms for removal of sediments play an important role in determining the volume of accreted sediments.

Finally, like other plate tectonics phenomena, subduction is ephemeral and changing. Subduction stops as when choked by buoyant features in an oceanic plate and then can restart. Moreover, the mode of subduction can change from one type to the other. In the Japan region the subduction has changed from a Marianas-type to a Chilean-type. Uyeda suspects that even the Chilean subduction zone, which is the model for its type, may at some time in the past have been the Marianas-type of subduction.

Passive Margins

Major tectonic activity stopped at most passive margins a long time ago in contrast to active margins, where it is going on today. Subsidence has been a ubiquitous phenomenon at passive margins, and the crustal rocks often lie buried at great depths below a thick cover of sediments. In spite of the fact that the structure of continental margins has been investigated for a long time, it is still poorly known for most margins. A principal difficulty has been the presence of the thick sedimentary cover, which has been difficult to penetrate by geological and geophysical methods. Conversely, the mechanism by which the thick cover of sediments was deposited and the subsidence of the crust that accompanied it have been the subject of extensive studies and have led to the development of theories of the detailed origin of passive margins.

By the late 1950s the first order structure of the oceans and continents had been established mainly on the basis of seismic refraction studies. The oceanic crust is thin and the continental crust attains thicknesses of a few tens of kilometers. Seismic measurements generally gave more uncertain results under the margins but it was imagined that the thin crust in the oceans gradually increased in thickness under passive continental margins to attain continental thicknesses.

Gravity calculations in which the gravity effects of the density variations associated with an assumed crustal structure under the passive margins were matched with observations supported the idea of a gradual decrease in crustal thickness from continent to ocean. Arguments based on gravity work were to some extent circular in that continuity of crustal layering was assumed from ocean to continent as was the fact that the subcrustal density remained constant. If the crustal layering and subcrustal density under the margin was different on either side, gravity calculations would not necessarily yield a smoothly changing crustal thickness. However, seismic results, did support gravity results. Indications of seaward thinning of the continental crust or landward thickening of the oceanic crust were often encountered in seismic refraction measurements. However, some reservations must be expressed regarding the early seismic refraction measurements. One reservation concerns velocities of about 7.2–7.5 km/sec that were often encountered under passive continental margins but were considered too high for either continental or oceanic crustal velocities and too low for subcrustal velocities. There was a tendency among early investigators to try to force the data away from these anomalous velocities so that the layers with these velocities may actually be more numerous than have been reported.

The presence of the M-discontinuity at the base of the crust has been detected both in the ocean and on the continent from continuous seismic reflection measurements but seldom under the margin. One of the few measurements of this discontinuity by reflection is in the Antarctic; it was made offshore Wilkes Land and is shown in Fig. 7.

The best determinations of the crustal structure of the passive margins has come from the use of Expanding Spread Profiles (ESPs), which utilize wide-angle reflection as well as refraction measurements supplemented by continuous multichannel seismic reflection profiles (preferably utilizing long towed receiving arrays, thereby attaining wide receiving apertures). Such measurements have been made in a number of areas including the North Biscay margin, the Mediterranean Sea, the Norwegian margin, and the northwest Australian margin. We have chosen to illustrate passive margin structures along two lines, one across the U.S. margin off New Jersey over the Baltimore Canyon Trough (Figs. 8 and 9) and the other across the margin off Hatton Bank—a submarine feature in the northeast Atlantic lying between Iceland and Ireland (Fig. 10).

Figure 8 shows a line drawing interpretation of a reflection line off New Jersey, which shows the sedimentary section. Figure 9 shows a section obtained by the LASE project, including ESP results. Individual ESP lines are shown in the inset. The section in Figs. 8 and 9 are nearly coincident. They are shown separately to emphasize the additional information obtained by the LASE ESPs. The layering in the shallow part of Fig. 9 is not substantially different from that derived earlier along the reflection line in Fig. 8. The ESP results, however, do indicate that immediately below the prograding carbonate bank lower seismic velocities exist, supporting the notion that sediments occupy the poorly mapped part of the crustal cross section. If that is true, sediments extend down to the great depth of almost 20 km! Although the LASE results represent by far the best crustal structure determination off the U.S. East Coast, even these results suffer from considerable uncertainty. The top of the layer under the margin with seismic velocities lying between 7.1 km/sec and 7.5 km/sec is fairly well determined from reflection data, but its bottom is considerably more uncertain both with regard to position as well as the underlying velocity. Oceanic basement determined by seismic reflection is shown by scalloped lines. Although it has been extended considerably westward (Fig. 8) it is never really seen under the carbonate bank, and the more conservative westward limit for oceanic basement in Fig. 9 is more credible. We note that the velocity of 7.2 km/

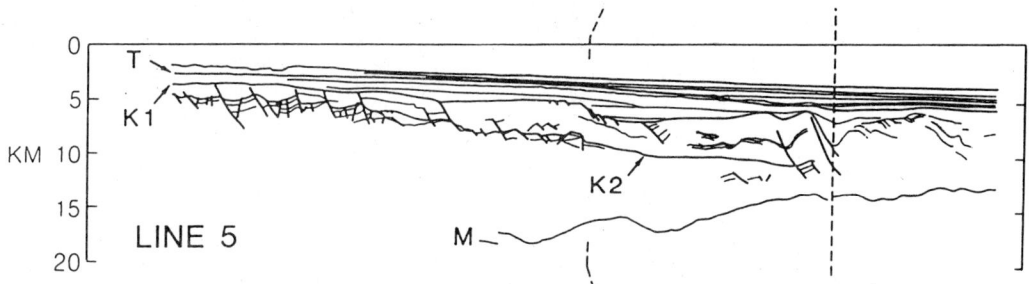

FIGURE 7. Tracing of seismic reflection record off Wilkes Land (Antarctica) showing the landward shallowing off the M-discontinuity under a passive margin. It has not been possible to trace the M-discontinuity as clearly below other passive margins. K2, K1 and T represent major unconformities. K2 is associated with the onset of rifting, K1 follows the continental breakup and T is a Tertiary event (from Eittreim and Smith, 1987).

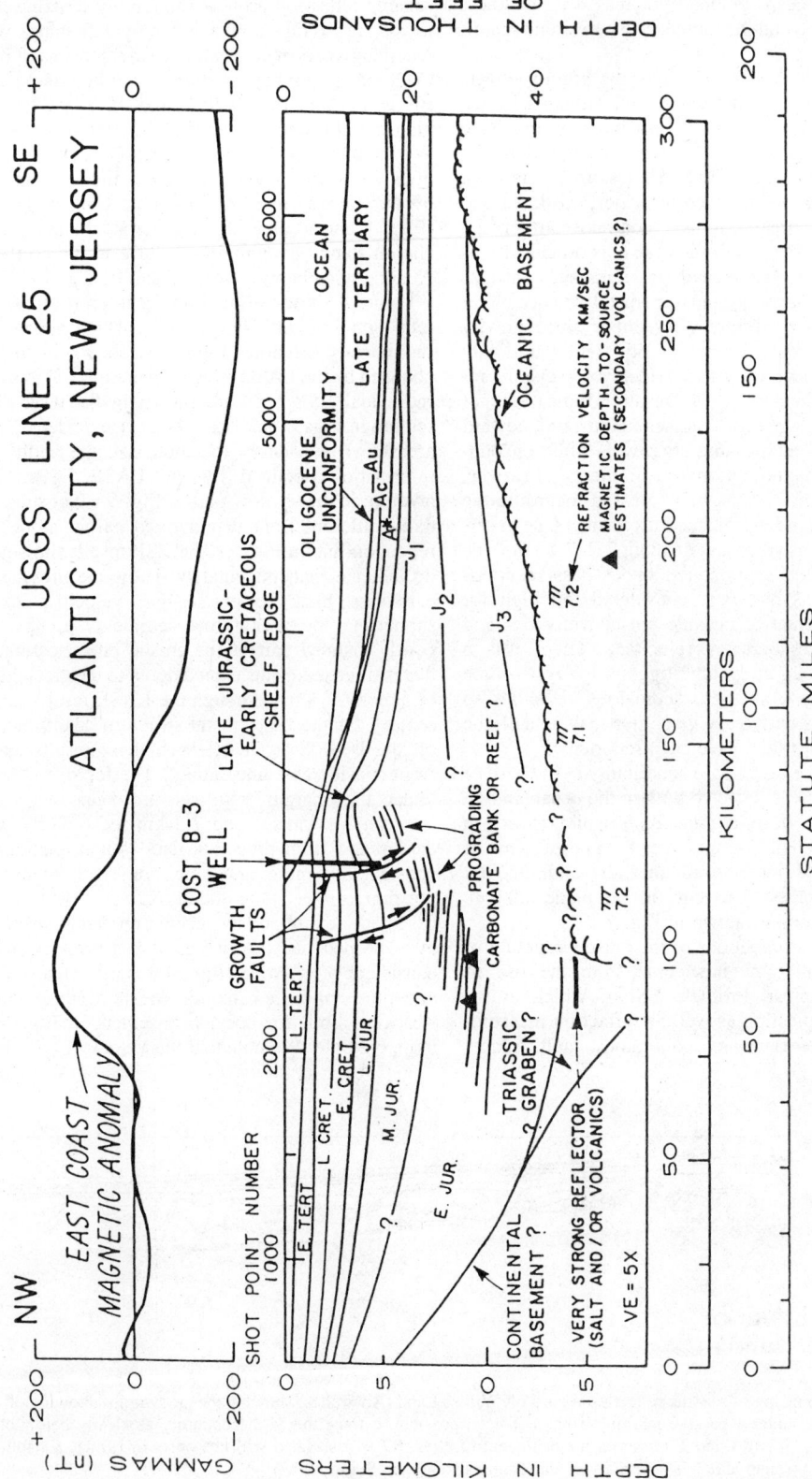

FIGURE 8. East Coast Magnetic Anomaly and crustal cross section obtained from line drawing interpretation of seismic reflection line shot by the U.S. Geological Survey across the Baltimore Canyon Trough. Deeper horizons were not obtained by seismic reflection. (From Grow, 1980)

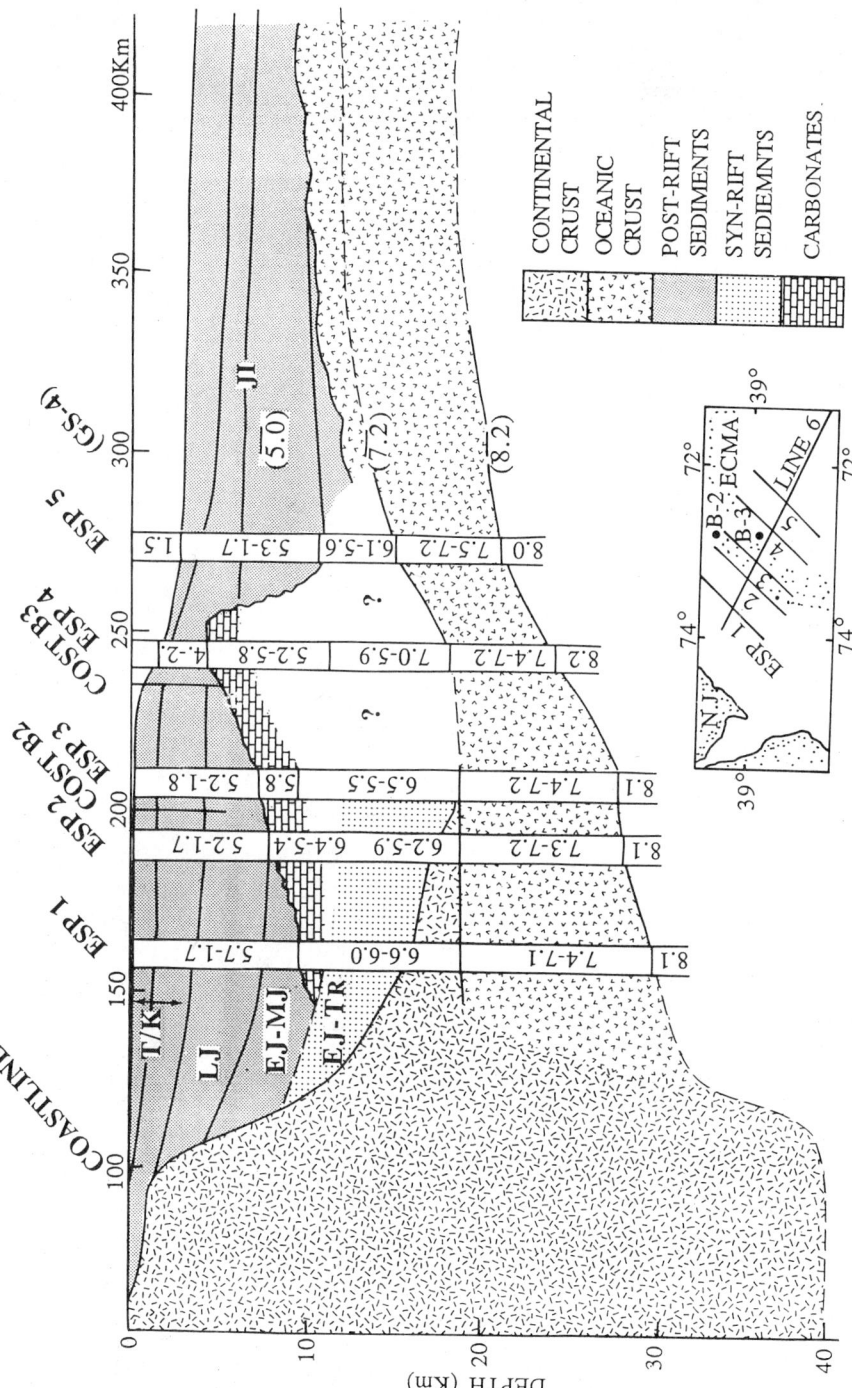

FIGURE 9. Crustal cross section across the Baltimore Canyon Trough obtained at the same location as the section in Fig. 8 by the LASE Group, which used Expanding Spread Profiles to obtain deep structure based on seismic wide-angle reflections and refractions. The layer with velocity about 7.2 km/sec appears to be underplating the entire passive margin crust. (From LASE Study Group, 1987).

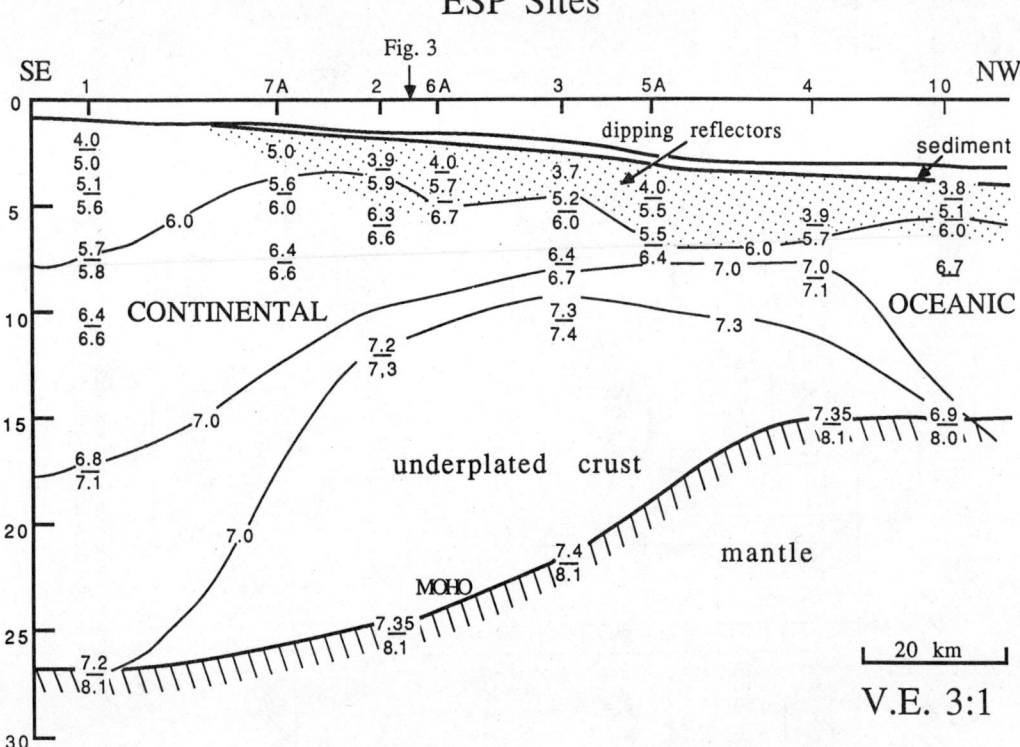

FIGURE 10. Crustal section across the Hatton Bank in the NE Atlantic shows structure similar to the LASE section across the Baltimore Canyon Trough in Fig. 9. (From White et al. 1987)

sec (see GS-4, Fig. 9) is too high for typical oceanic crust and the interpretation of this area as oceanic rests in part on the scalloped appearances of an overlying reflector, which is interpreted as oceanic basaltic basement. Accepting the 7.2 km/sec layer as oceanic, the LASE section raises some very interesting questions. The boundary between continental and oceanic crust would lie over the shallow part of the continental shelf. However, all earlier estimates had placed this boundary much further to the east—over the continental slope or beyond. Second, the continental crust appears to thin very abruptly and a wide zone of thinned continental crust, which most theories of the formation of passive margins demand, would seem to be absent.

The results obtained over the Hatton Bank margin are very similar. In Fig. 10, if we imagine the layer with velocity exceeding 7.0 km/sec (including the area indicated as underplated crust) to be oceanic crust, then the oceanic layer extends to areas considered continental; and continental crust (if defined as the layer with velocity under 7.0 km/sec) thins considerably under the margin, in places being barely 1 km thick.

Most current theories involve the lithospheric stretching model of McKenzie (1978). This model, involving rifting of the continental crust, starts with extension and thinning. The pure-shear model, which is currently most popular (Fig. 11, top) invokes listric normal faulting in the brittle upper crust above a décollement zone at about 10 km depth. These listric normal faults have been best observed in the Biscay margin (off France and Spain). Ductile stretching takes place below this zone. Wernicke (1981) has proposed an asymmetric model in which the detachment fault, instead of representing the brittle ductile transition, cuts through the entire lithosphere (Fig. 11, bottom). The two models will lead to differences in the nature of the conjugate passive margins that are ultimately formed. They will be symmetrical in the first case but asymmetrical in the Wernicke case. In any event, necking or thinning of the crust takes place. As hot asthenospheric material rises, isostatic subsidence takes place since the asthenospheric material has a higher density than the crustal material it replaces. The process of extension and rifting continues until the hot asthenospheric material finally breaks through and the process of drift starts with the movement apart of the two edges of the old continents. As they move apart they subside because of thermal cooling and consequent contraction as well as the loading of

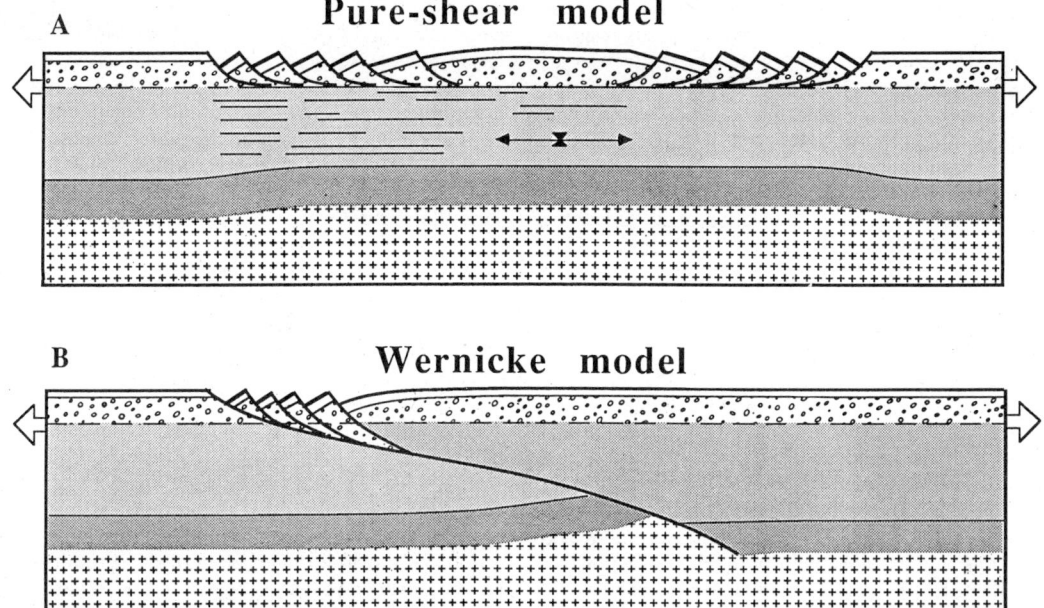

FIGURE 11. The pure shear mode (McKenzie, 1978) and the asymmetrical model (Wernicke, 1981) for extension of the continental extension. One or the other of these models is believed to precede continental breakup, which leads to the formation of a passive margin. (From Lister et al., 1986).

sediments from respective parts of the continent adjacent to each edge. This general scheme has been applied to examine the passive margins in a number of areas including the Bay of Biscay and the margin off Nova Scotia, Canada. The amount of initial stretching is estimated either from the motion on listric normal faults or on the observed attenuation of the continental crust. Various factors such as the subsidence history of the margin, the elastic properties of the lithosphere (including in some cases the rheological properties of the lower crust), and the degrees of stretching are related to obtain a consistent picture of the evolution of these margins.

However, to apply the previous concepts to the LASE and to the Hatton Bank lines requires some reexamination, or at least some modification, of the earlier concepts. Major amounts of listric normal faulting are not observed for either one of the sections, although it is possible for the Baltimore Canyon Trough line that faulting in the basement has simply not been resolved in the reflection lines. (Normal faulting is observed in the continental basement under neighboring Georges Bank). The biggest problem revolves around the presence of the layer with velocity about 7.2 km/sec, which appears to be continuous from the ocean to the continent with seeming disregard for the ocean continent boundary. The apparently very small thickness of the continental crust under the margin or perhaps even its complete disappearance in the Baltimore Canyon are also puzzling. However, the LASE study group reconciles these data to the existing concepts of continental margin formation by interpreting the 7.2 km/sec layer as being formed by basaltic magma generated within the asthenosphere during extension and rifting of the lithosphere. The melt rises through the lower lithosphere and intrudes the lower continental crust or solidifies below it. This process, which has been termed underplating, could form a large part of the observed lower crust. This explanation, which applies equally well to the Hatton Bank line, is not an unreasonable suggestion but it does makes the task of ever resolving the precise nature and location of the ocean continent boundary and of explaining the East Coast Magnetic Anomaly (see below) very difficult. It also creates difficulties in estimating the thinning of the continental crust and therefore of making quantitative tests of the theories of passive margin formulation.

In conclusion we examine two other aspects of passive continental margins: the slope magnetic anomalies and the seaward dipping reflectors. Both have been studied to obtain clues about the deep structure.

A prominent magnetic anomaly, the East Coast Magnetic Anomaly (ECMA) exists off the U.S. East Coast, generally lying over the continental slope. Where the shelf has prograded considerably seaward, the ECMA lies over the shelf as in the Baltimore Canyon Trough area (Fig. 8). Similar slope anomalies exist over a number of continental slopes, for example, off the coasts of Newfoundland,

Ireland, South Australia, Antarctica, and NW Africa. The prominence of the slope anomalies suggest that they are genetically related to the formation of passive margins and should be helpful in constraining interpretation of seismic data. The most popular explanation of the ECMA is that it is caused by the magnetization contrast between oceanic and continental crust. We see in Fig. 9 that the western margin of the ECMA lies near the location of ESP2 and the eastern margin lies between ESP3 and ESP4. However, the crustal cross section does not show a sharp ocean continent basement edge in this area. A prominent normal fault known as the East Coast Boundary Fault (ECBF) has been deduced from the LASE data in the vicinity of ESP2 (but is not shown in Fig. 9). This fault is coincident with the western margin of the ECMA. If this fault were to be coincident with the magnetic edge, and if oceanic crust that is reversely magnetized at a low angle of inclination were to lie east of it, an anomaly would result that would appear similar to the ECMA but would lie considerably to the west of the observed position of the ECMA. An alternative explanation would involve a highly magnetized body with its western edge coincident with the ECBF and its eastern edge coincident with the eastern margin of the ECMA. The present state of resolution of the seismic data makes it difficult to choose between differing explanations for the ECMA. A more satisfactory reconciliation between magnetic and seismic data will, in the future, lead to a better understanding of the passive margin off the U.S. East Coast. South of Australia a slope anomaly is readily understood as being associated with a normal fault with large displacement. Normally magnetized continental crust lies on the landward side of the fault. The crust on the other side lies at much greater depth and its origin is not known with certainty, but because of its large depth and weak magnetization it does not contribute significantly to the magnetic anomaly. Thus, the south Australian anomaly can be attributed to an edge effect and the magnetic and seismic data do not conflict. The unanswered question is whether the south Australian margin anomaly, the ECMA as well as all other slope anomalies, are all genetically related and have a common origin. If that is so, what is that origin and can it be corraborated from seismic data?

Other phenomena associated with the ocean continent transition are the so-called seaward-dipping reflectors. They have the following characteristics: (1) the dip on the reflectors is seaward and individual reflectors have an arcuate shape, (2) there is a down-section and down-dip increase in dip that results in an overall wedge-shaped appearance with reflectors diverging seawards, and (3) there is no evidence of bottom-set aspect to the reflectors. Figure 12 shows seaward dipping reflectors in the Voring Plateau off of Norway. A seafloor spreading type magnetic anomaly is associated with these reflectors and they have been interpreted to have been formed by subaerial seafloor spreading that took place at the inception of continental separation and drift. Other interpretations of these reflectors also invoke a large

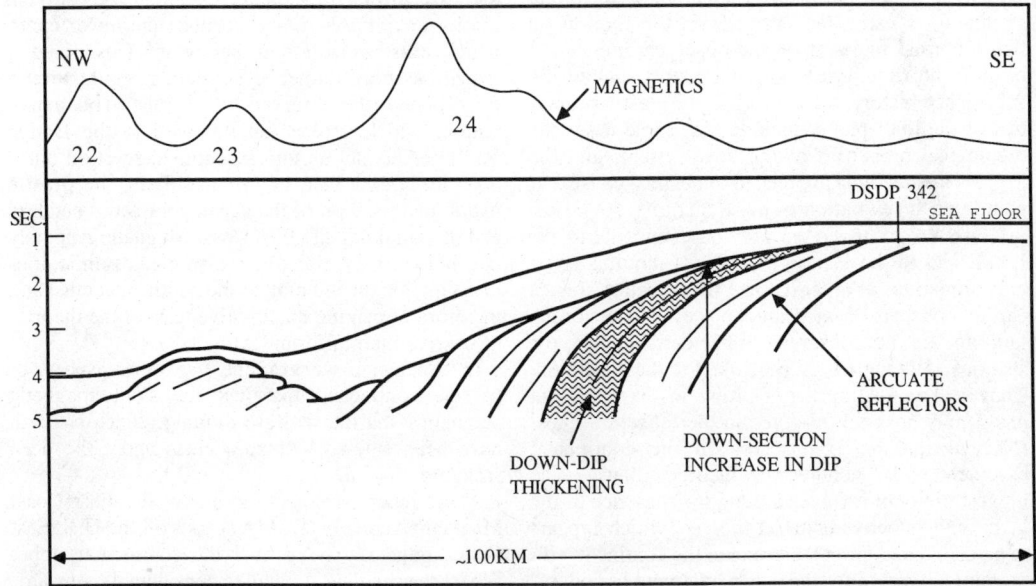

FIGURE 12. Schematic drawing of crossing of outer Voring Plateau showing the seaward dipping seismic reflectors believed to characterize the top part of the crust at the time of continental breakup. A magnetic anomaly profile is plotted above the section. (From Mutter et al., 1982)

amount of volcanic activity but just before the inception of opening rather than just after. These characteristic seaward dipping reflectors have been observed at a number of passive margins, including the U.S. East Coast, the Hatton Bank margin, and off the Antarctic. Their importance lies in their being associated with the beginning of the opening of a new ocean and therefore in their being able to identify the opening event. Since these reflectors are deeply buried it is not always easy to map them and one has to be careful not to identify other varieties of seaward dipping reflectors as those that characterize early opening.

MANIK TALWANI

References

Barazangi, M., and J. Dorman, 1969, World seismicity map compiled from ESSA Coast and Geodetic Survey epicenter data, *Seismol. Soc. America Bull.* **59,** 369-380.

Burchfiel, B. C., 1983, The continental crust, *Sci. Am.* **249,** 130-144.

Eittreim, S. L., and G. L. Smith, 1987, Siesmic sequences and their distribution on the Wilkes land margin, in *The Antarctic Continental Margin: Geology and Geophysics of Offshore Wilkes Land*, Earth Science Series vol. 5A. Houston: Circum Pacific Council for Energy and Mineral Resources, 15-43.

Grow, J. A., 1980, Deep structure and evolution of the Baltimore Canyon Trough in the vicinity of the East No. B-3 well, in P. A. Scholle, ed., *Geological Studies of the East No. B-3 Well, United States Mid-Atlantic Continental Slope Area*, U.S. Geol. Survey Circ. 833, 117-132.

LASE Study Group, 1986, Deep Structure of the U.S. East Coast passive margin from large aperture seismic experiments (LASE), *Marine Petrol. Geol.* **3,** 234-242.

Lister, G. S., M. A. Etheridge, and P. A. Symonds, 1986, Detachment faulting and the evolution of passive continental margins, *Geology* **14,** 246-250.

McKenzie, D. P., 1978, Some remarks on the development of sedimentary basins, *Earth and Planetary Science Letters* **40,** 25-32.

Mutter, J. C., M. Talwani, and P. Stoffa, 1982, Origin of seaward-dipping reflectors in oceanic crust off the Norwegian margin by "subaerial sea-floor spreading," *Geology* **10,** 353-357.

Uyeda, S., 1984, Subduction zones: Their diversity, mechanism and human impacts, *GeoJournal* **8,** 381-406.

Wernicke, B., 1981, Low-angle normal faults in the Basin and Range province: Nappe tectonics in an extending orogen, *Nature* **291,** 645-648.

White, R. S., G. K. Westbrook, S. R. Fowler, G. D. Spence, P. J. Barton, M. Joppen, J. Morgan, A. N. Bowen, C. Prestcott, and M. H. P. Bott, 1987, Hatton Bank (northwest U.K.) continental margin structure, *Royal Astron. Soc. Geophys. Jour.* **89,** 265-272.

Cross-references: *Earth Structure, Global; Lithosphere, Oceanic: Formation and Evolution; Mantle, Upper: Structure; Seismicity: Subduction Zone; Seismicity and Plate Tectonics; Subduction Zones.*

P

PALEOMAGNETIC FIELD: INTENSITY

In principle the determination of the intensity of the paleomagnetic field is simple. Nagata (1943) demonstrated that for constant magnetic fields of intensity less than 100 μT, the strength of a thermoremanent magnetization (TRM) acquired in the field is proportional to the field intensity. By measuring the strength of some natural thermoremanent magnetization (NRM) acquired in an ancient field (BN) and then giving the same sample a laboratory TRM in a known laboratory field (BL) it is possible, in principle, to calculate the ancient field intensity from the following equation:

$$\frac{BN}{BL} = \frac{NRM}{TRM} \qquad (1)$$

The present geomagnetic field intensity varies from about 30 μT at the equator to about 75 μT at the poles, and provided that BN is not much different from the present day value, then Nagata's relationship holds and Eq. 1 is valid.

Unfortunately Eq. 1 can become invalid for a number of other reasons. When the sample is heated in the laboratory to form the TRM, changes can occur to the magnetic minerals and their ability to acquire a TRM. Over geological time the NRM that originally formed in the sample may be partially erased or overprinted by reheating (e.g., during metamorphism or burial) or by the acquisition of a later chemical remanent magnetization (CRM). It is possible that during the initial cooling of a sample, alteration of the magnetic minerals may occur at temperatures below their Curie temperature. If this alteration occurs then the sample will not acquire a simple TRM but a thermochemical remanence (TCRM) instead. A further complication is that all NRMs decay with time, some more than others, and acquire a viscous remanent magnetization (VRM) in the present day field direction. Clearly any change in the NRM or TRM acquisition or the use of an NRM that is not a single component thermoremanent magnetization will invalidate Eq. 1. It is important to detect and, if possible, remove the effects of alteration and to check, by continuous monitoring of the NRM direction, that we are dealing with a single component magnetization.

Paleomagnetic Field Intensity Methods

A number of techniques have been applied to archaeological samples but are not suited to geological material because of the severe alteration that often occurs to rock samples. Generally two techniques are commonly used to investigate the paleomagnetic field intensity. The first was developed by Thellier and Thellier (1937) and used progressive thermal demagnetization and remagnetization to investigate the sample in detail. The second was developed by Shaw (1974) and used alternating field (af) demagnetization combined with a single heating to form a laboratory TRM. Both

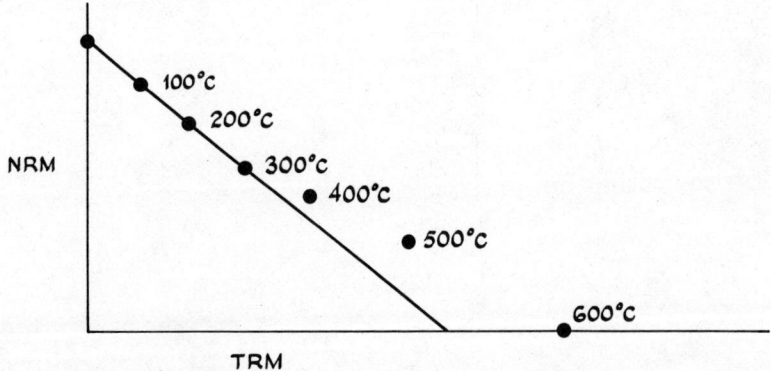

FIGURE 1. Graphical representation of a Thellier field intensity determination. This sample shows alteration above 300°C and no evidence of VRM at low temperatures.

methods have rigorous checks for alteration of the magnetic minerals during the laboratory heating. They both rely on being able to identify an unaltered part of the laboratory TRM for comparison with an equivalent part of the NRM in Eq. 1.

The *Thellier method* uses the fact that the NRM can be gradually removed by heating to higher and higher temperatures and at the same time the TRM can be gradually installed. A plot of NRM removed against TRM gained, using demagnetization/remagnetization temperatures as the parameter, should be a straight line with slope BN/BL. Because of VRM the low temperature region may not fall on the line and because of high temperature chemical alteration of the magnetic minerals, the high temperature region may depart from the straight line (Fig. 1).

The value of BN is calculated from the straight line portion of the graph, thus avoiding the effects of VRM and chemical changes during the laboratory heating. A modification of the technique can be employed to check for a gradual linear chemical alteration. In such a case the BN/BL plot would still give a straight line but with an "incorrect" gradient. To detect such alteration involves, after each temperature step, repeating the previous lower temperature step.

The *Shaw method* uses af demagnetization to gradually remove the NRM by the application of stepwise increasing alternating magnetic fields (af demagnetization). The sample is then given an artificial remanence called an anhysteretic remanent magnetization (ARM1), which is installed by applying a constant magnetic field while the peak alternating field is slowly reduced to zero. ARM1 is then af demagnetized in the same steps as the NRM. A full TRM is then given to the sample by heating above the Curie temperature and af demagnetized in the same way. Finally ARM2 is installed and similarly af demagnetized. ARM1 and ARM2 are used to monitor the magnetic characteristics of the sample. A plot of ARM1 against ARM2, using the peak af value as a parameter, should be a straight line of slope 1 if there has been no change in the magnetic minerals. Chemical alteration that occurred while the sample was being given the TRM can be observed as regions of the ARM1/ARM2 plot that do not fall on a straight line of slope 1 (Fig. 2). It is often possible to isolate an unaltered region that does have a slope of 1. Any such region can be used to compare the NRM/TRM values and obtain a ratio for BN/BL. The main advantage of the Shaw method is that, because only one heating is required, the experiment can be performed considerably faster than the standard Thellier method.

Kono (1978) proposed a variation of the Shaw method that allowed for the correction of limited amount of linear thermal alteration. In such cases the ARM1/ARM2 plot produces a linear slope that is slightly greater or slightly less than 1. The ARM1/ARM2 slope can then be used to correct for the

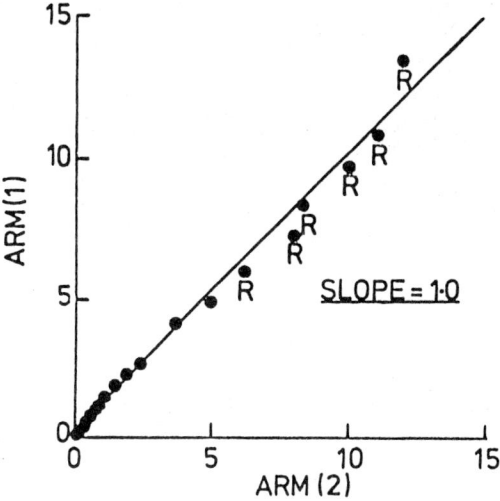

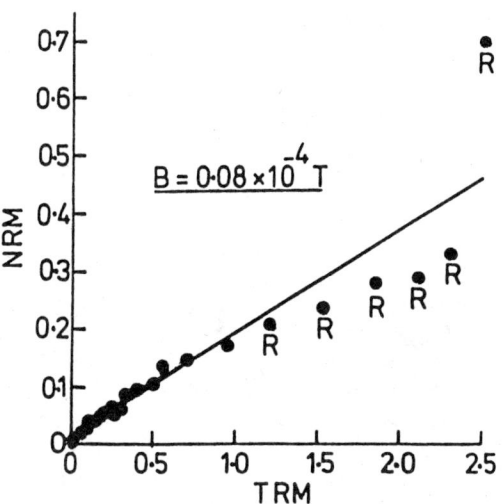

FIGURE 2. Graphical representation of a Shaw field intensity determination. Points that do not fall on a line of Slope 1 on the ARM1/ARM2 plot are marked R and rejected. The NRM/TRM points corresponding to the same AF demagnetization values are similarly rejected and the NRM/TRM ratio is calculated from the unaltered values. In this case the TRM field was 0.5×10^{-4} T.

thermal alteration in the NRM/TRM graph in the following manner

$$\frac{BN}{BL} = \frac{NRM}{TRM} \times \frac{ARM2}{ARM1}$$

This step was a first toward using samples that altered during heating. Rolph and Shaw (1985) extended this work to include curved ARM1/ARM2 graphs. To ensure that only single domain grains

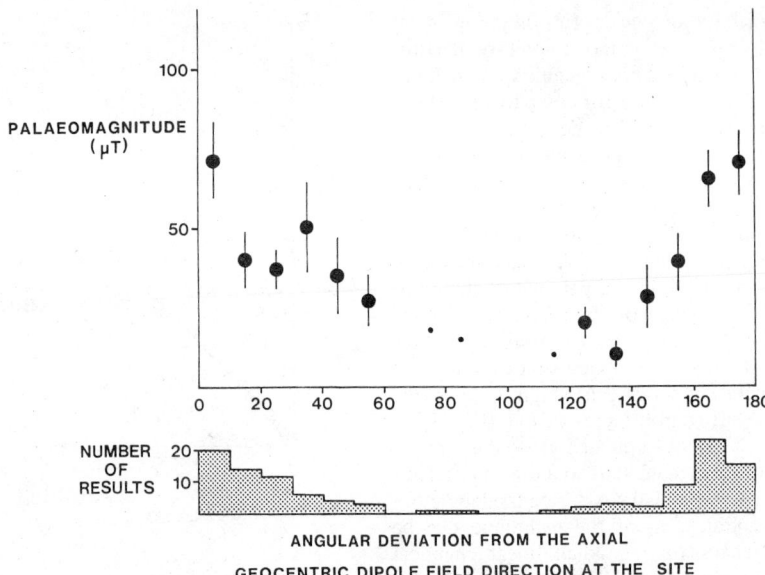

FIGURE 3. Average paleomagnetic field intensity results from Icelandic basalts extruded over the past 12 Ma.

were being investigated, this method of correction is limited to data derived above a 100 mT peak alternating field. This modification proved to be very successful when applied to modern lavas, the results agreeing well with magnetic observatory measurements.

The two techniques (Thellier and Shaw) have been extensively compared with good agreement (Senanayake et al., 1982).

Paleomagnetic Intensity over the Past 5 Ma

There have been a large number of field reversals over the past 5 Ma, the average frequency of reversal is about 5/Ma. In order to discover if the geomagnetic field changed its intensity during reversals, a number of early field intensity determinations were made using samples that had been magnetized at some time during a field reversal, with intermediate paleomagnetic directions between normal and reversed. Lawley discovered that the intermediate field was generally weaker than the normal or reversed fields. This finding has been confirmed by statistical analysis of the magnetization of Iceland's basalts and by additional field intensity measurements. Generally the average intermediate field intensity in Iceland falls to less than 20% of the average normal or reversed intensity value of 55 μT (Fig. 3).

Although it is possible to construct an average picture of geomagnetic field behavior by using a large collection of lava flows, it is very difficult to examine a single field reversal in detail because it is uncommon for a large number of lavas to be extruded during a field reversal. However there have been attempts to examine a single reversal in this way (Prevot et al., 1985; Shaw, 1975, 1977). In every case the intermediate field intensity was weak but in two cases a peak was observed in the intensity of the intermediate field (Fig. 4) without any significant associated change in field direction.

We can express a field intensity in terms of the magnetic moment of an axial geocentric dipole that would be required to produce the field. A large number of field intensities covering the past 5 Ma have been expressed in this way and used to produce a time average virtual dipole moment (VDM) for this period of about 9×10^{22} Am^2 (McFadden et al.,

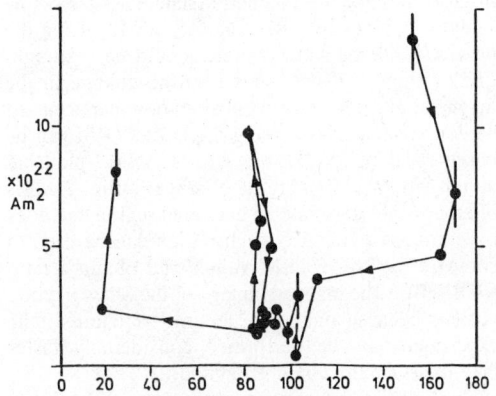

FIGURE 4. Plot of VDM against paleomagnetic pole colatitude. These results are from a reversal recorded in Icelandic lavas. The field moves from the reversed state to the normal state but appears to hesitate, grow and then decay while in an intermediate configuration.

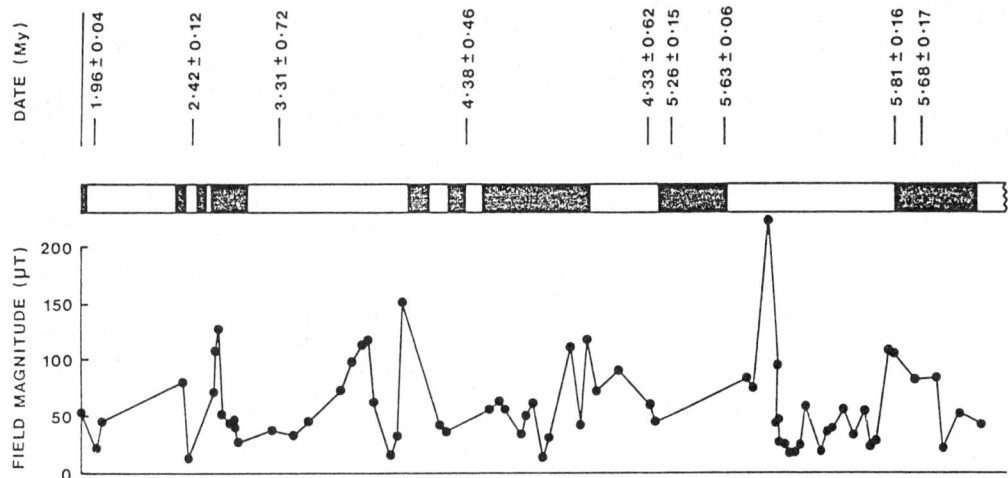

FIGURE 5. Time ordered field intensity results from Iceland and the corresponding polarity time scale (dark areas represent normal polarity).

1982; Roberts et al., 1984). This is a little larger than the present day value of about $8 \times 10^{22} Am^2$.

By assuming that, on average, lavas are extruded at regular intervals it is possible to observe changes in paleomagnetic intensity with time (Fig. 5). The effect of these changes on the magnetic record contained in sea floor basalts can be observed as small scale but widely observed and correlated changes in the amplitude of the marine magnetic anomaly record (Shaw et al., 1982; Cande et al., 1974).

Paleomagnetic Intensity beyond 5 Ma

The accurate determination of the intensity of the paleomagnetic field becomes more difficult as the age of the samples increases. Problems of secondary magnetizations, chemical alteration, and decay of the original remanence all increase with age and consequently only a small number of measurements have been made on samples older than 5 Ma. However, there is general agreement that the paleomagnetic field intensity during the Carboniferous gave an average VDM of only $1 \times 10^{22} Am^2$ (Rolph et al., 1985; Carmichael, 1967) but that at about 1000 Ma BP it produced a VDM close to the present day value of $8 \times 10^{22} Am^2$, which it maintained for the rest of the Proterozoic.

Pesonen and Halls (1983) carried out a detailed field intensity study on Precambrian Keweenawan rocks with ages of 1000 to 1200 Ma BP. These rocks had previously been measured for paleomagnetic directions and the average reverse field direction was not, as we would expect, antiparallel to the average normal field direction. The field intensity values also showed some asymmetry in that the average reversed field intensity was 40% higher than the average normal field intensity. The same direction and intensity asymmetry has been observed in the 1300 Ma Gardar lavas and it would appear that the source of the geomagnetic field during this period was not a simple axial geocentric dipole.

Clearly it is important that we understand the configuration of the paleomagnetic field if we are to use it to investigate the movement of the continents in the past. Paleomagnetic field intensities provide additional information that can be used to constrain models of continental evolution. They also help our understanding of the processes within the liquid core of the Earth that produce the geomagnetic dynamo, the source of the paleomagnetic field.

JOHN SHAW

References

Cande, S. C., and J. L. Labreque, 1974, Behaviour of the Earth's paleomagnetic field from small scale marine magnetic anomalies, *Nature* **247**, 26-28.

Carmichael, C. M., 1967, An outline of the intensity of the paleomagnetic field of the Earth, *Earth Planetary Sci. Letters* **3**, 351-354.

Kono, M., 1978, Reliability of palaeointensity methods using alternating field demagnetisation and anhysteretic remanence, *Royal Astron. Soc. Geophys. Jour.* **54**, 241-261.

McFadden, P. L., and M. W. McElhinny, 1982, Variations in the geomagnetic dipole, 2: Statistical analysis of VDMs for the past 5 million years, *Jour. Geomagnetism and Geoelectricity* **34**, 163-189.

Nagata, T., 1943, The natural remanent magnetism of volcanic rocks and its relation to geomagnetic phenomena, *Tokyo Univ. Earthquake Res. Inst. Bull.* **21**, 1-196.

Pesonen, L. J., and C. H. Halls, 1983, Geomagnetic field intensity and reversal asymmetry in the late Precambrian Keweenawen rocks, *Royal Astron. Soc. Geophys. Jour.* **73**, 241-270.

Prevot, M., E. A. Mankinen, R. S. Coe, and C. S. Gromme, 1985, The Steens Mountain (Oregon) geomagnetic polarity transition. 2. Field intensity variations and discussion of reversal models, *Jour. Geophys. Research* **90**, 417-448.

Roberts, N., and J. Shaw, 1984, The relationship between the magnitude and direction of the geomagnetic field during the Late Tertiary in Eastern Iceland, *Royal Astron. Soc. Geophys. Jour.* **76**, 637-651.

Rolph, T. C., and J. Shaw, 1985, A new method of palaeofield magnitude correction for thermally altered lavas and its application to Lower Carboniferous lavas, *Royal Astron. Soc. Geophys. Jour.* **80**, 773-781.

Senanayake, W. E., M. W. McElhinny, and P. L. McFadden, 1982, Comparisons between the Thellier's and Shaw's palaeointensity methods using basalts less than 5 million years old, *Jour. Geomagnetism and Geoelectricity* **34**, 141-163.

Shaw, J., 1974, A new method of determining the magnitude of the paleomagnetic field, *Royal Astron. Soc. Geophys. Jour.* **39**, 133-141.

Shaw, J., 1975, Strong geomagnetic fields during a single Icelandic polarity transition, *Royal Astron. Soc. Geophys. Jour.* **40**, 345-350.

Shaw, J., 1977, Further evidence for a strong intermediate state of the paleomagnetic field, *Royal Astron. Soc. Geophys. Jour.* **48**, 263-269.

Shaw, J., P. Dagley, and A. E. Mussett, 1982, The magnitude of the paleomagnetic field in Iceland between 2 and 6 Myr ago, *Royal Astron. Soc. Geophys. Jour.* **68**, 211-218.

Thellier, E., 1937, Recherche de l'intensite du champ magnetique dans le passe, *Premier resultato. Ann. de l'institut de phy. du Globe, Univ. Paris* **15**, 179-184.

Cross-references: *Archaeomagnetism; Chemical Remanent Magnetization (CRM); Curie Temperature; Demagnetization; Geomagnetic Field, Main: Theory; Magnetic Domains; Magnetic Properties of Minerals; Natural Remanent Magnetization (NRM); Paleomagnetic Secular Variation; Rock Magnetism; Thermoremanence; Viscous Remanent Magnetization (VRM) and Viscous Remagnetization.*

PALEOMAGNETIC SECULAR VARIATION

Historic secular variation (HSV) and paleomagnetic secular variation (PSV) describe the general spatial and temporal variability of the Earth's internal magnetic field during periods of stable magnetic polarity. HSV describes the historic field variability for just the last 10^2 years, while PSV describes the prehistoric field variability over timescales of 10^2 years or longer. High-resolution studies of HSV and PSV provide the data base for evaluating both models of the Earth's core-dynamo process, which is the primary source of the internal magnetic field, and the axial-dipole hypothesis, which is a cornerstone for plate tectonic reconstructions.

Whereas HSV is estimated from direct measurements of the Earth's magnetic field, PSV is estimated from proxy measurements of the field derived from the paleomagnetic study of archaeological materials, unconsolidated sediments, and rocks. The paleomagnetic methods used to recover PSV data have been summarized recently by Tarling (1983) and Collinson (1983). They note that quite different methods are normally used to recover PSV vector and paleointensity estimates. Therefore, PSV vector data sets usually do not have associated paleointensity estimates and vice versa. (One remarkable exception to this rule is the archeomagnetic record from Bulgaria, noted in Figs. 1 and 2, that was developed by Kovacheva (1983) and Kovacheva and Zagniy (1985).)

The purpose of this paper will be to describe the *vector* paleomagnetic secular variation (PSV) of the Earth's magnetic field; it will provide an overview of PSV data sources, methods of PSV analysis, long-term characteristics of PSV, and models for PSV behavior. Special attention will be paid to the relationship between PSV and HSV, the evidence for long-term stationarity of PSV, and the characteristics of PSV that may be useful in dynamo studies. A more lengthy summary of PSV can be found in Merrill and McElhinny (1983), and a summary of recent PSV and dynamo studies can be found in Lund and Olson (1987).

Paleomagnetic Secular Variation Data

PSV data come from a wide variety of paleomagnetic studies that can be delineated on the basis of the type of sediment or rock measured, the degree of detail in stratigraphic sampling, and the degree of radiometric-age control for each study. This detail usually separates PSV studies into three types: (1) studies of late-Quaternary-aged sequences of unconsolidated sediments, lava-flows, or archeological materials, which can be radiocarbon-dated in detail, and which are sampled in sufficient detail to resolve PSV waveform information; (2) studies of older sediment or lava-flow sequences that have waveform information but no detailed age-control; and (3) studies of any-aged rock or sediment sequences that have poor within-sequence age control and no waveform information (uncorrelated or random data). The first type of PSV study can be used for a full spectrum of time-series analyses (waveform, spectral, or statistical analyses); the second type of study can be used for waveform and statistical analyses; the third type of study is only suitable for statistical analysis. Examples of high-resolution late-Quaternary PSV data sets from Europe and North America are shown in Figs. 1-3. An example of a statistical PSV data set is shown in Fig. 4 (page 880).

The materials normally used for detailed paleomagnetic studies of PSV are archeologic kilns and fire pits, lava flows, and lake or marine sediment sequences. Each of these materials has inherent advantages and disadvantages for the accurate

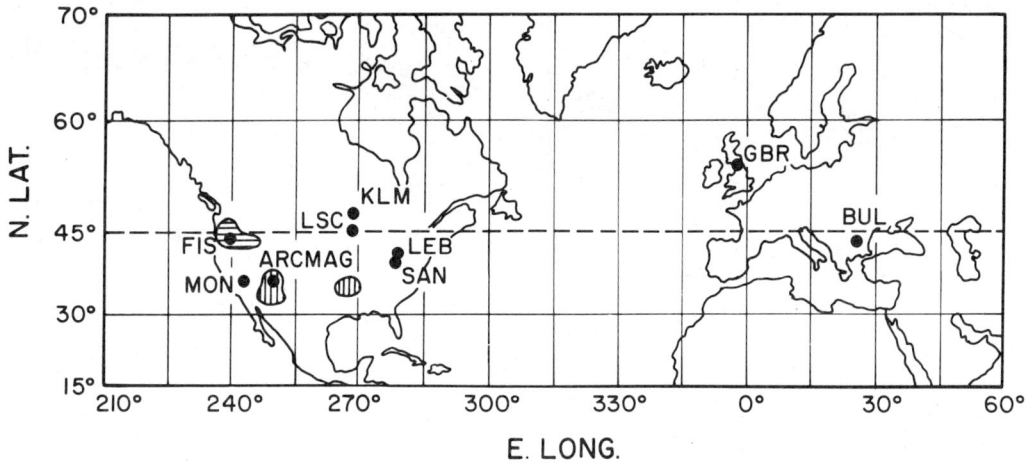

FIGURE 1. Sampling sites of paleomagnetic records displayed in Figs. 2 and 3. ARCMAG is a composite record that summarizes archeomagnetic data from Arizona, Colorado, and Arkansas (vertical hachured regions), and lava-flow data from the Pacific Northwest (horizontal hachured region). FIS = Fish Lake, Oregon; MON = Mono Lake, California; LSC = Lake St. Croix, Minnesota; KLM = Kylen Lake, Minnesota; LEB = Lake LeBoeuf, Pennsylvania; SAN = Sandy Lake, Pennsylvania; GBR = composite lake-sediment record from Great Britain; BUL = composite archeomagnetic record from Bulgaria.

recording of PSV, and the accumulation of paleomagnetic records from all three materials in parallel is necessary to properly develop regional high-resolution PSV records. Lava flows and archeological materials have the advantage that their NRM is normally a TRM that is acquired over time intervals of less than a few days. They, therefore, retain a truly "instantaneous" record of PSV. The primary disadvantage of archeological materials is their scarcity prior to about 2000 ybp. The primary disadvantage of lava flows is the difficulty in finding sufficient radiocarbon-dated flows within a region to develop a long-duration, composite PSV record; only two such studies are currently available in the whole world (western USA, Champion, 1980; Hawaii, Holcomb et al. 1986). The primary advantage of sediment sequences is their potential to provide continuous, high-resolution PSV records far back in time from many sites around the world. The primary drawback to sediments, however, is the lower resolution of the DRM/PDRM recording process (see *Detrital Remanent Magnetization [DRM]*) due to some degree of inherent smoothing of the PSV signal during remanence acquisition near the sediment/water interface. In most high-resolution sediment records, the smoothing interval can be estimated to be less than 20–50 years in duration, but further study is necessary to better establish the role of smoothing in the DRM/PDRM acquisition process. A second drawback to sediments is the difficulty in recovering paleointensity estimates from their paleomagnetic records.

An added complexity associated with all PSV data is the limited extent to which they can be compared to HSV data. This difficulty in correlation is due primarily to the fact that (1) PSV data, most of which comes from a few localized Northern Hemisphere sites, does not have the broad spatial (global) sampling distribution of HSV data, (2) the inherent vector resolution of PSV data (2° to 4° α_{95} at best) is significantly lower than the resolution of HSV data (typically 1° α_{95} or better), and (3) the radiometric ages associated with PSV records have relatively large errors (ca. ± 100 years). PSV records derived from sediments have the added disadvantage, noted previously, of not recovering instantaneous estimates of secular variation due to inherent DRM/PDRM smoothing. Because of these differences, there may be few occasions where HSV and PSV data can be compared in detail. Even so, we can hope that analysis of PSV data will yield characteristics that relate, in some way, to observed HSV characteristics.

Special note should be made of paleomagnetic excursions, which are anomalous PSV fluctuations defined by virtual geomagnetic poles (VGPs) located more than 45 degrees away from the geographic pole. It is clear that excursions do occur; it is not usually clear, however, what is their waveform morphology, or whether some excursions are really artifacts of field/laboratory measurement errors. Excursions have been primarily identified in Brunhes-aged rocks (0–730,000 ybp), but they have also been noted in older rocks. Most excursions are difficult to correlate because of uncertainties in their age estimates. Two of these excursions, however, are better constrained, the Mono Lake excursion (ca. 28,000 BP, Fig. 3) and the Cobb Mountain excursion (ca. 1.1 my BP). The future possibility of recovering waveform information in different

PALEOMAGNETIC SECULAR VARIATION

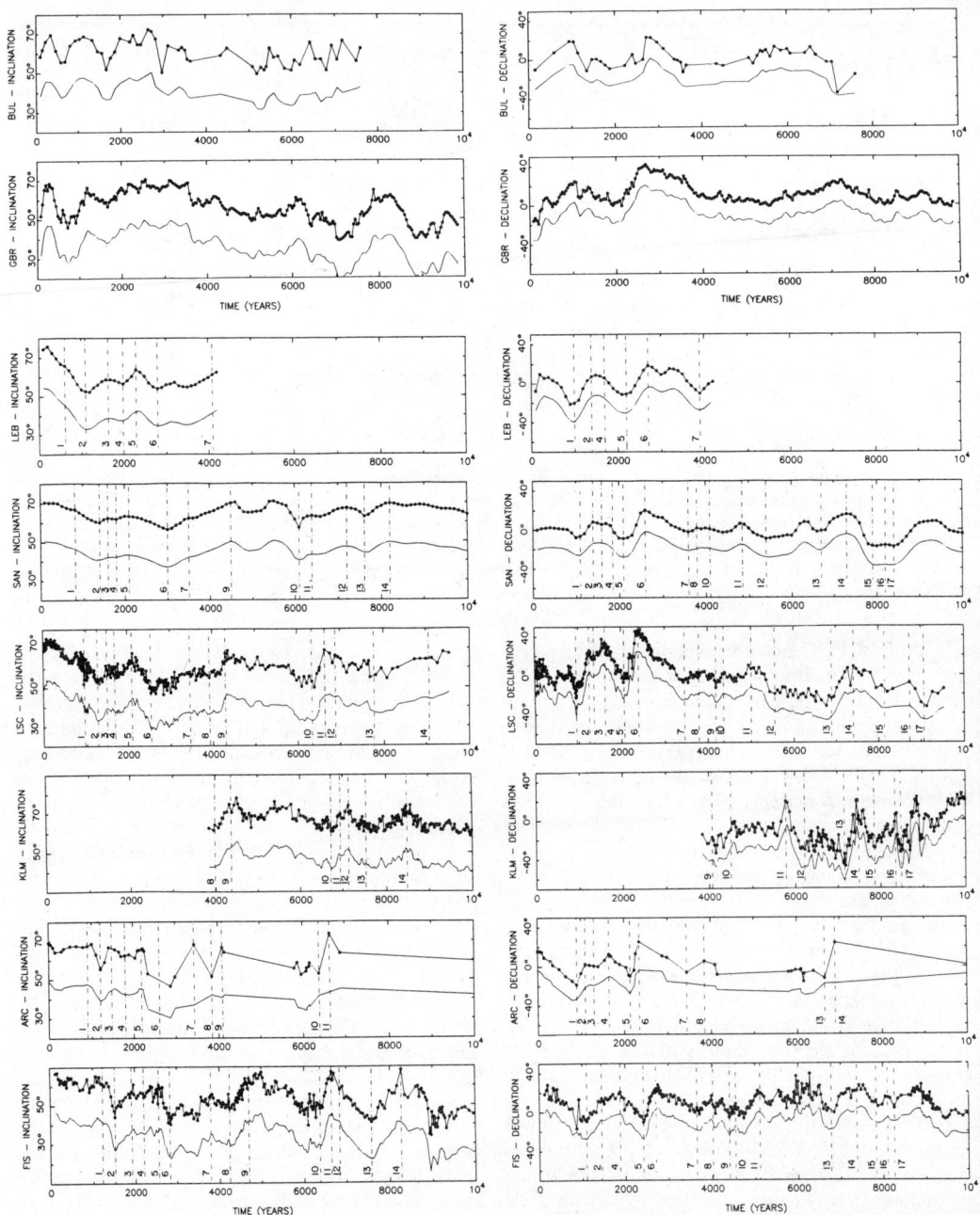

FIGURE 2. Selected Late-Quaternary Paleomagnetic secular variation records from North America and Europe. (*left*) Inclination records, (*right*) declination records. FIS = Fish Lake, Oregon (Verosub et al., 1986); ARCMAG = composite archeomagnetic record from the western United States (Champion, 1980; Sternberg, 1983); KLM = Kylen Lake, Minnesota (Lund and Banerjee, 1985); LSC = Lake St. Croix, Minnesota (Lund and Banerjee, 1985); SAN = Sandy Lake, Pennsylvania (King, 1983); LEB = Lake LeBoeuf, Pennsylvania (King, 1983); GBR = Great Britain composite record (Turner and Thompson, 1981); BUL = Bulgaria composite archeomagnetic record (Kovacheva, 1983).

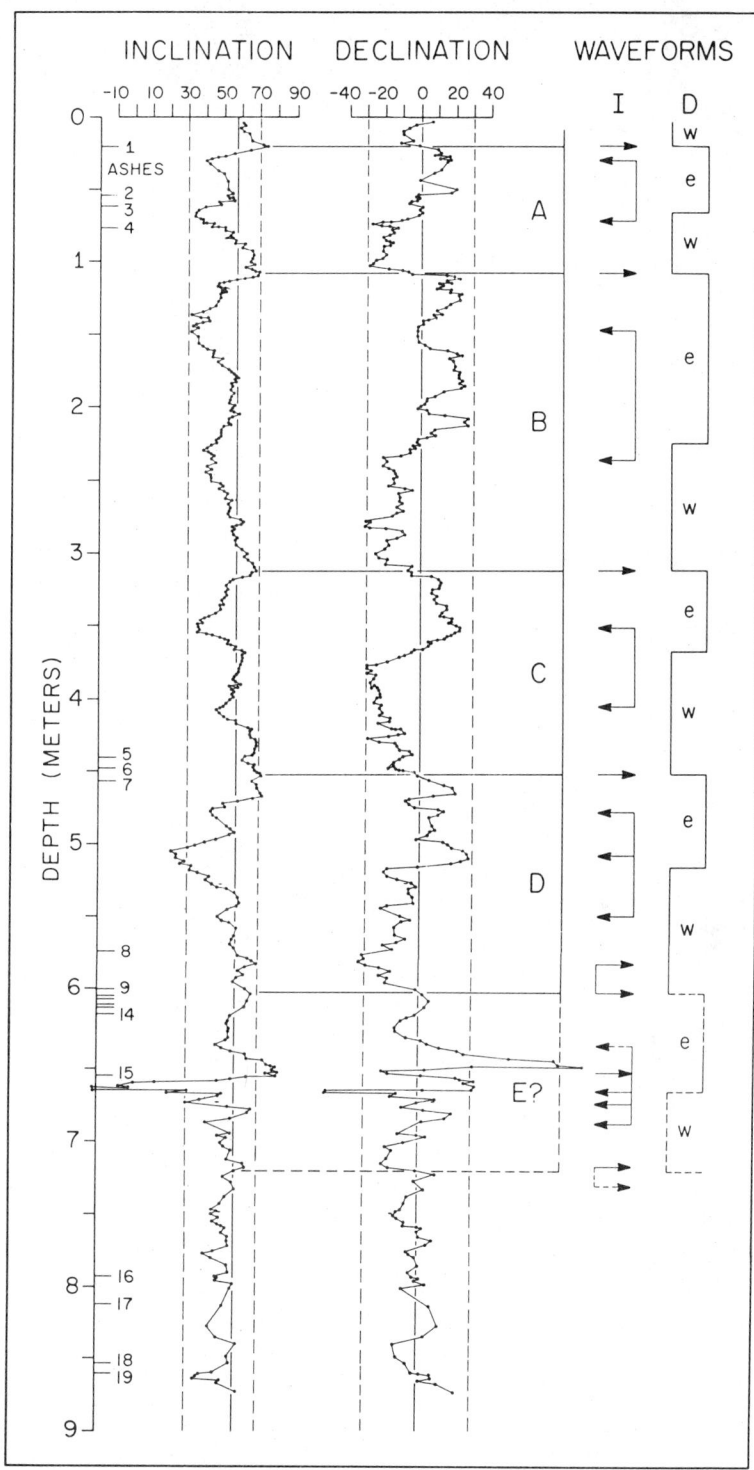

FIGURE 3. Late-Quaternary Paleomagnetic record From Mono Lake, California. This figure summarizes the paleomagnetic data recovered by several paleomagnetic studies of Mono Lake: Denham and Cox (1971), Denham (1974), Liddicoat and Coe (1979), Lund et al. (1988). The record contains several repetitions of a complex waveform (labeled A-E?), and the Mono Lake Excursion (waveform E?). (After Lund et al., 1988.)

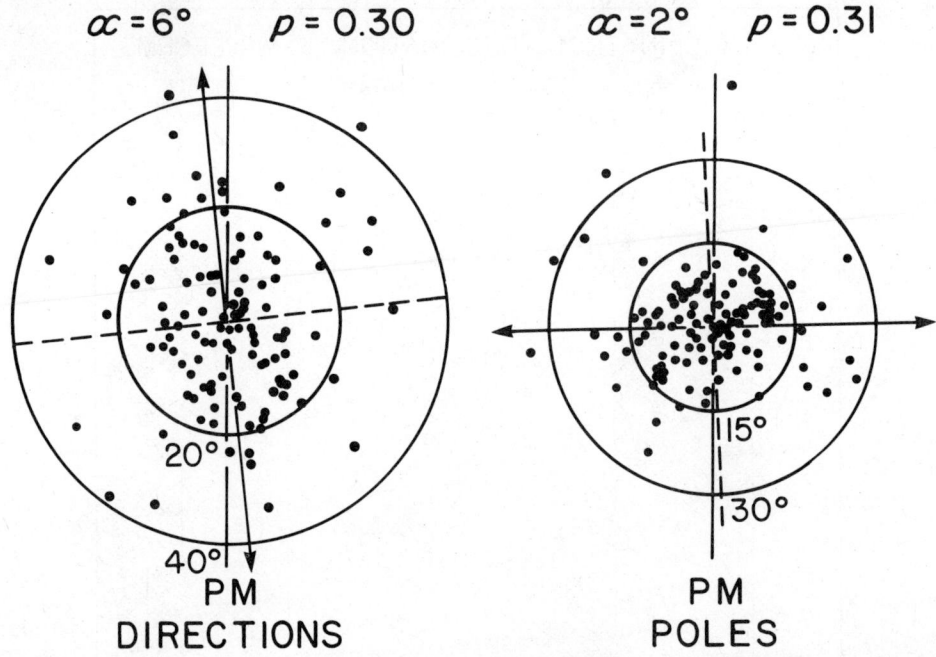

FIGURE 4. Statistical distribution of field vectors and virtual geomagnetic poles (VGPs) in Plio-Pleistocene Rocks from Equatorial Africa. These results, from Brock (1971), illustrate the non-Fisherian distribution of field vectors and VGPs from a geographic region. Shape statistical analysis (Brock, 1971) indicates that these data sets are both elliptical in distribution ($P = 0.0$ for circular distribution) with the ellipse major axes aligned ($\alpha = 0$ for perfect alignment) to the site longitude (PM). (After Brock, 1971)

geographic regions from these two excursions is an exciting prospect to improve our understanding of the importance and source of excursions. Only when excursions are considered within the context of the related PSV can the cause of excursions be considered in detail, in particular the possibility that excursions are aborted reversals (see also *Geomagnetic Polarity Reversals*) rather than large-amplitude PSV.

Time-Series Analysis of Paleomagnetic Secular Variation

The conceptual framework for analysis of PSV data is based on the fact that HSV records are too short in duration to adequately characterize the temporal field variability, while PSV records are too scattered spatially to adequately describe the prehistoric spatial field variability. We can hope, however, to identify spatial and short-duration temporal components of the HSV that may relate to long-term temporal components of the PSV. We can also attempt to improve the spatial estimates of PSV by, perhaps, appealing to analogs in HSV. Such a coordinated analysis of PSV and HSV can perhaps answer the key unresolved question in secular variation studies: what is the mapping function between the observed spatial and temporal variations of the geomagnetic field? Only with a coherent view of the total spatio-temporal variability of the historic and prehistoric geomagnetic field can we properly evaluate models of the core-dynamo process, which is the source of the field variability.

Analysis of PSV data uses a variety of time-series and modeling techniques in order to delineate the spatial and temporal characteristics of PSV. Time-series techniques that can be applied to PSV data will be considered in this section; modeling techniques will be considered in the next section. Time-series techniques are classified into three broad categories: waveform studies, spectral analysis, and statistical analysis. Each of these techniques has unique advantages for characterizing a particular type of PSV data and thereby providing a point of comparison with HSV data.

Waveform analysis. PSV waveforms, which reflect distinctive temporal variations in the paleomagnetic field at a single site, are often delineated within high-resolution paleomagnetic records (e.g., Figs. 2 and 3). Waveform comparisons (1) within individual paleomagnetic records, (2) between records from different sites, and (3) between PSV and HSV records, where they overlap in time, provide one method for characterizing the temporal aspect of secular variation. Comparisons of the amplitudes and the phase-relationships of PSV waveforms with their counterparts(?) in global maps of the present-day spatial field may provide one method for characterizing the spatio-temporal mapping of secular variation. Below we discuss five different types of comparative waveform analysis that document distinctive PSV characteristics, all of which are correlative with HSV waveform characteristics.

The first type of waveform comparison, between PSV records for the last 500 years (e.g., Lake St. Croix [Figs. 1 and 2]; British lacustrine PSV records [Figs. 1 and 2]; Japanese or British archeomagnetic records) and HSV records from the same sites (estimated from spherical harmonic analyses), indicates that only the largest-amplitude, longest-period (ca. 500 years) HSV waveforms can be correlated with the PSV records. This limited correlation is due to the fact that PSV records only resolve waveforms more than about 4° in amplitude and a few hundred years in duration thus limiting the potential for detailed HSV and PSV correlations. Even so, it is certain that high-resolution PSV records can accurately continue the long-term temporal variation hinted at in HSV records.

The second type of comparison, between different high-resolution PSV records from the same region (e.g., all European and Russian records; all North American records), indicates that PSV waveforms, with periods longer than a few hundred years, can routinely be correlated over about 3000 km. For example, Figs. 1 and 2 show well-dated Holocene PSV records from six different sites in North America spread over 40° of longitude (~3000 km). It is readily apparent that several I and D waveforms, numbered for clarity, can be correlated among these records. (That is not to say that the records are identical, but variations in sampling-rate and signal/noise ratios or errors in data acquisition and analysis probably can explain most of the differences in single records.) Turner and Thompson noted in 1981 a similar scale of coherence in the correlation of Holocene PSV records from Europe.

A similar comparison between PSV records from different geographic regions, is much more problematic. For example, the comparison of North American and European PSV records (see Fig. 2) indicates that there is no simple correlation between records from the two regions that preserves phase-relationships or long-term trends in the vector data. It thus appears that straightforward PSV vector correlations break down beyond ≤ 5000 km. Thompson (1984) noted a similar scale of spatial coherence in the correlation of HSV waveforms. He determined that HSV could be broken down into about 9 different regions over the Earth's surface. Within each region HSV waveforms were broadly correlative, but between regions the patterns were significantly different.

A third type of comparison, between PSV waveforms within individual paleomagnetic records, has occasionally identified distinctive waveforms that seem to recur every 2400–3500 years (e.g., Lake St. Croix, Mono Lake). The PSV record from Mono Lake, shown in Fig. 3, illustrates this recurrence pattern. At least 4 recurrences (A–D) of basically the same complex waveform can be noted. With each recurrence, the waveform is slightly altered; however, the general pattern persists for more than 15,000 years. It is likely that this distinctive waveform has evolved out of the Mono Lake excursion (waveform E? in Fig. 3). The *persistence* of a distinctive waveform that recurs over 3000 year intervals for at least 15,000 years argues strongly for some type of long-term memory within the core-dynamo process.

A fourth type of comparison, more speculative in its application, can be suggested between temporal PSV waveforms and spatial HSV waveforms. For example, the waveform amplitudes and phase-relationships in Holocene and Pleistocene PSV records from mid-latitude sites in the Northern and Southern Hemispheres typically display $\pm 20°$ variation in I, $\pm 30°$ variation in D, and a 90° out-of-phase relationship between I and D. The range of I and D variation in the present-day spatial geomagnetic field (e.g., epoch 1975 Internal Geomagnetic Reference Field) at mid latitudes is almost identical to the range of PSV. Also, a similar 90° out-of-phase relationship can be found in the nondipole foci identified regionally in the present-day spatial field. Such comparisons may provide the basis for developing a spatio-temporal mapping of the geomagnetic field variability. However, an important exception to the similarity of temporal PSV and spatial HSV occurs within the Mono Lake excursion (Fig. 3), where the inclination deviates 65° and the declination deviates 90° from the axial-dipole expectation. There is no analog for this anomalous behavior in the present-day field. An important, but unanswered, question is how do excursions relate to normal PSV?

A final type of waveform comparison is based on the characteristic circularity (looping) of field vectors or VGPs in both HSV and PSV. Such behavior was suggested by Skiles in 1970 to be an indicator of westward (or eastward) drift (see the following section, Models for Paleomagnetic Secular Variation). Clockwise looping, normally associated with westward drift, has been noted consistently in the North American, European, Australian, and Argentinian PSV records over the last 8000 years. In contrast, counterclockwise looping, normally associated with eastward drift, has been noted predominantly in older PSV records from Europe and North America. The correlation between observed looping and drift is unfortunately, however, not unique. Nevertheless, systematic looping behavior does appear to be an important characteristic of the long-term field.

A summary of distinctive PSV waveform characteristics follows:

1. PSV waveforms are normally greater than 4° in amplitude and a few hundred years in duration; smaller-amplitude or shorter-duration waveforms cannot normally be resolved due to sampling problems or analytical errors.
2. PSV studies for the last 500 years can accurately recover waveforms comparable with the long-period HSV waveforms and continue them beyond the time range of historic measurement.
3. PSV waveforms with durations of a few hundred to a few thousand years can be routinely correlated between sites up to about 3000 km apart; correlations over distances greater than 3000 km are not straightforward; an analog for these characteristics can be found in the correlation of HSV waveforms.
4. Holocene PSV waveforms from midlatitude sites of the Northern and Southern Hemisphere show typical non-axial-dipole variations of about $\pm 20°$ in inclination and $\pm 30°$ in declination; there is some preference for a 90° out-of-phase relationship between I and D components of these waveforms. Quite comparable non-axial-dipole amplitude variations occur spatially in the present-day field; also, historic non-dipole foci have a similar 90° out-of-phase relationship between I and D.
5. Circularity of Holocene PSV vector waveforms is predominantly clockwise, while late Pleistocene circularity is predominantly counterclockwise.

These characteristics envisage a prehistoric magnetic field comparable to the present-day field in its spatial and temporal variability, although the Holocene field may be subtly different from the Pleistocene field. Patterns of PSV are regionally coherent, but perhaps unrelated between regions. Occasionally, the regional pattern of PSV may be dominated by a distinctive, periodically recurring waveform that may persist for at least 10^4 years. Anomalously large-amplitude waveforms (excursions) may also occur on occasion, but they are probably regional in extent and transient in time.

Spectral Analysis. Spectral analysis describes the frequency content of PSV over timescales of 10^2–10^6 years. Traditional ideas suggest that PSV should be a band-limited random process. That is, that the spectral power of PSV should markedly diminish beyond some cutoff period on the order of 10^4 years. PSV records longer than the cutoff period should then be stationary in a statistical sense, and have an average field-vector direction that is constant through time for a given site. The axial-dipole hypothesis, a cornerstone of plate tectonic reconstructions, assumes that the PSV process is stationary and that each site's average field vector (during intervals of normal polarity) satisfies the formula

$$\tan I = 2 \tan \lambda,$$
$$D = 0°,$$

where λ is the site paleolatitude. One important goal of PSV studies is to test the validity of the axial-dipole hypothesis.

Spectral analysis also describes the characteristic distribution of spectral power within the PSV process and identifies whether there are distinct frequency bands within the continuous spectrum that have higher-than-average spectral power. Knowledge of the PSV power spectrum and its potential changes in time and space is critical for better understanding the relationship between PSV and the core dynamo process that generates it.

The primary limitation in recovering detailed estimates of the PSV spectrum is the quality of age estimates associated with the PSV records. Most records are dated using radiocarbon methods that have systematic errors with respect to 'true' time; these errors can only be estimated for the last 7000 years. Also, radiocarbon-age estimates will have random errors on the order of 10^2 years or worse, thus causing an added uncertainty to the spectral resolution.

Despite this limitation, important and convincing spectral estimates of late-Quaternary PSV have been recovered from several important regions around the world: Europe, North America, the Far East, Australia, South America. Together, results from these areas provide an important long-term view of the PSV spectrum that is summarized in Fig. 5 and compared with HSV spectral estimates.

The various spectra clearly indicate that the largest-amplitude secular variations occur at periods greater than 10^2 years (beyond the range of HSV). An apparent interval of relatively low spectral power near 10^2 years, which marks the boundary between PSV and HSV spectra, may be an artifact of the spectral process or may be due to the fact that Earth's magnetic field intensity was relatively lower prior to 2000 ybp. The longest PSV records usually display a significant decrease in spectral power for periods greater than 10^4 years, which may indicate the occurrence of a cutoff period that should be present if PSV is stationary in a statistical sense. Spectral analyses of lower-resolution PSV records (Barton, 1982, Fig. 5), however, sometimes display high spectral power in the interval of 10^4–10^5 years. Unfortunately these records don't contain normal PSV, and their reality as recorders of longer-term PSV is in doubt. If these results can be verified in high-resolution PSV records, they would indicate that the PSV process is much less stationary than we now presume.

Spectral analysis of the high-resolution PSV records has identified two general bands of spectral power below the cutoff-period (Fig. 5), a long-period

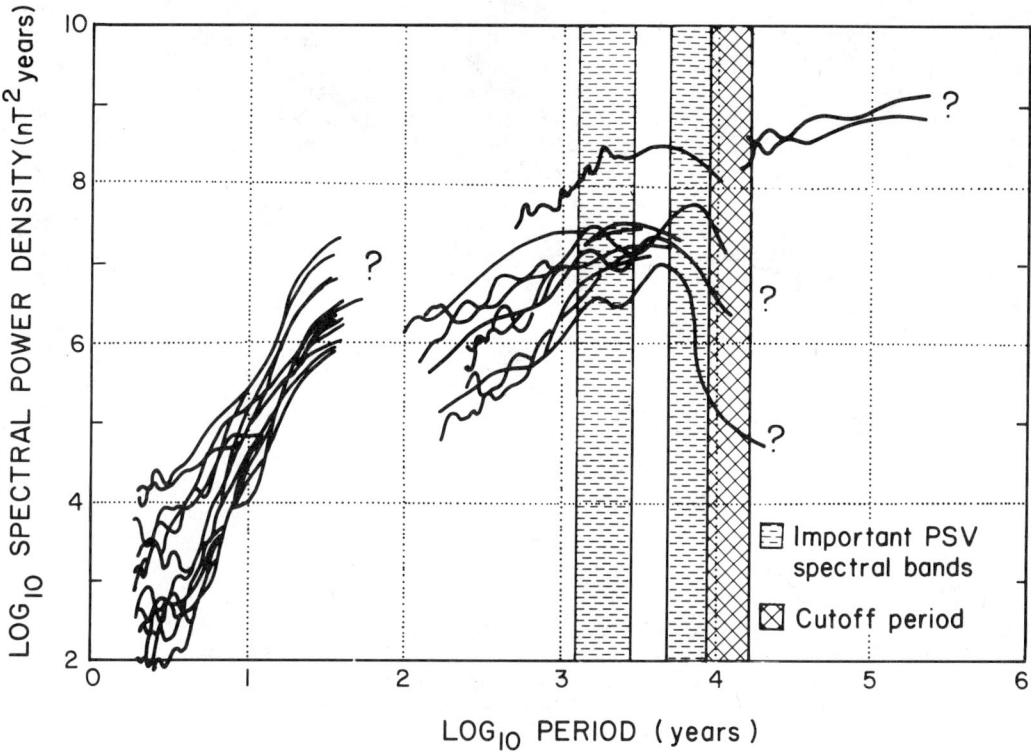

FIGURE 5. Summary of spectral analyses of historic and paleomagnetic secular variation records. Several spectral analyses, primarily from Barton (1983) with additions from Lund and Banerjee (1985) and Liddicoat and Lund (1983), are stacked here to illustrate the probable long-term average spectrum of the magnetic field secular variation. Most of the spectral power occurs at periods greater than 10^2 years and may display a cutoff period on the order of 10^4 years. See the text for further discussion. (After Barton, 1982)

interval near 7000–10,000 years and a shorter-period interval between 500–3500 years. The long-period band may be due to dipole-like behavior and the shorter-period band may be due to nondipole field behavior. The shorter-period band is distinctive in that the spectral peaks almost all appear to be multiples of the largest spectral peak, about 2400–3000 years. The periods associated with these two spectral bands may vary significantly between different locations on the Earth's surface and may also change through time. Under such circumstances, the average global spectral power for 10^4-year or longer intervals, may not show these spectral bands.

A summary of distinctive PSV spectral characteristics follows:

1. The dominant spectral power associated with HSV and PSV occurs at periods greater than 10^2 years, periods that both HSV and PSV studies are able to routinely resolve.
2. Most high-resolution PSV records display a significant drop in spectral power for periods greater than about 10^4 years (cut-off period), which suggests that the long-term secular variation process is band-limited and stationary over intervals greater than 10^4 years.
3. Some low-resolution PSV records point to the possibility of large-amplitude waveforms with periods greater than 10^4 years; if this possibility is validated, then long-term secular variation may only be quasi-stationary.
4. Distinct spectral bands have been noted at periods near 2500 years and 9000 years; the shorter-period band may be related to long-term nondipole activity.

Most of these spectral characteristics have no temporal analog in HSV because of the short time span of historic measurement. One might, however, attempt to relate these long-term temporal spectral characteristics to the spatial spectrum associated with the present-day field. The present-day spatial field due to core sources has a cutoff near spherical harmonic degree 12, and the spectral power decreases quickly from a maximum at harmonic degree 1 (dipole terms). Only spatial terms of about degree 6 or lower have vector amplitudes large enough to be recorded in PSV. Therefore, one might hypothesize that the long-term PSV spectrum must be due to temporal variations of spatial components of the present-day field with spherical harmonic degree ≤ 6.

Statistical Analysis. The aspect of PSV that is easiest to measure is its statistical behavior averaged over some interval of time. For this reason, statistical properties of PSV, within time windows on the order of 10^5 or 10^6 years, were the first PSV characteristics to be compared spatially, and are still the only PSV characteristics that can be compared spatially on a global scale. Such comparisons provide the strongest evidence relating to stationarity of the field, the axial-dipole hypothesis, and the truly global extent of PSV characteristics.

Statistical study of PSV follows two very different paths on the basis of sampling frequency and age control of the paleomagnetic measurements. In the first approach, paleomagnetic field directions in undated rock sequences are measured under the assumption that the age difference of successive rock units is large compared to the longest period of PSV (ca. 10,000 years?). Each data point is therefore assumed to be an independent random value picked from the frequency-band-limited PSV process. Data sets from small regions, averaged over 10^5 or 10^6 years, are then statistically analyzed and compared with some global model of the expected statistical behavior. The second approach is to measure radiocarbon-dated paleomagnetic sequences where the sampling interval is less than the shortest period of PSV (about 30 years). It is not often feasible to find sequences with such short time spacing, but useful information can be obtained with sample intervals up to 250 years. Statistics are estimated from equispaced time series derived from the dated paleomagnetic records. This method permits spatial comparison of statistical parameters averaged over much shorter time intervals, on the order of 1000 to 50,000 years.

Statistical analyses of the probability distributions of both field vectors and their equivalent VGPs from single sites indicate that neither distribution is typically Fisherian (Fisher, 1953; spherical analog of the normal distribution). For example, Brock's 1971 results from equatorial Africa (Fig. 4) show that both field vectors and VGPs tend to have somewhat elliptical distributions. Engebretson and Beck (1978) summarized the statistical parameters normally used to characterize the shape of each distribution. It is probable that shape statistics vary systematically as a function of latitude (and longitude?) and future studies of shape statistics may provide important added characteristics of the Earth's long-term PSV.

Currently the two statistical parameters most often measured in PSV studies are the ΔI anomaly, which is the site mean inclination (I) minus the expected axial-dipole field inclination, and the angular dispersion associated with a site's vector (or equivalent VGP) variation. The global pattern of the ΔI anomaly estimates how well the axial-dipole hypothesis, the cornerstone of plate tectonic reconstructions, fits the actual geomagnetic field behavior. The global pattern of angular dispersion estimates the degree of dipole wobble and the latitudinal variation in relative strengths of the dipole and nondipole fields.

The ΔI anomaly was perhaps first quantified by Wilson in 1970 who noticed that the average paleomagnetic pole positions associated with individual geographic regions (e.g. Australia, Europe, North America) were always farther from the sampled region than the known geographic pole. This offset, termed the *far-sided effect*, is due to paleomagnetic inclinations that are systematically lower than their axial-dipole expectation. Merrill and McElhinny (1977) have determined the global ΔI anomaly for the last 5 my and their results are plotted in Fig. 6a. Their analysis indicates that the ΔI anomaly has persisted for Plio-Pleistocene time and that it appears to be zonal, that is any site along a line of latitude will have the same magnitude of ΔI anomaly. Other workers have noted that similar ΔI anomalies have existed for longer periods, up to at least 100 my.

Analysis of angular dispersions has established that, like the ΔI anomaly, a distinctive zonal pattern of amplitude variation with latitude (Fig. 6) exists for this parameter. The magnitude of the angular dispersion and its latitudinal variation are related to the relative contribution of dipole and nondipole components to the total PSV. McFadden and McElhinny (1984) have presented an exciting new model (Fig. 6) for the 5 my-average latitudinal distribution in VGP dispersion, which only assumes that the nondipole field intensity varies linearly with the associated dipole intensity variation in latitude. This model accurately predicts the observed long-term latitudinal distribution of dispersion. Lund noted in 1985, however, that the average VGP dispersion for the last 30,000 years is significantly lower in the Northern Hemisphere than the 5-my average, which suggests that VGP dispersion may only be stationary for intervals greater than 10^5 years.

Statistical studies of angular dispersion have commonly referred to angular dispersion as *paleosecular variation*. This term is unfortunate, for the angular dispersion is only one of several statistical parameters that can characterize PSV, and statistical analysis is only one of several more general methods for the study of PSV. The continued use of the term paleosecular variation for studies of angular dispersion is not recommended.

A summary of PSV statistical characteristics follows:

1. Field-vector and VGP distributions derived from PSV records at single sites are both elliptical in shape rather than Fisherian.
2. The time-averaged paleomagnetic field measured

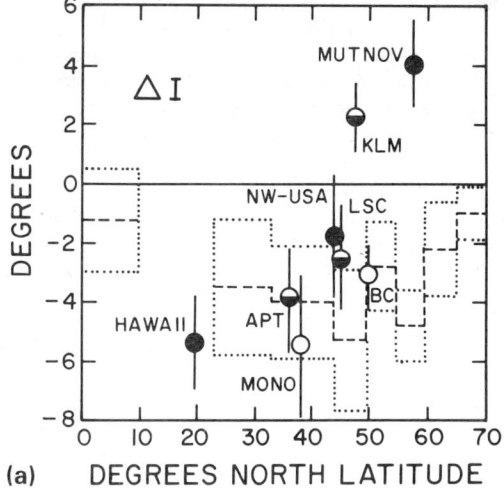

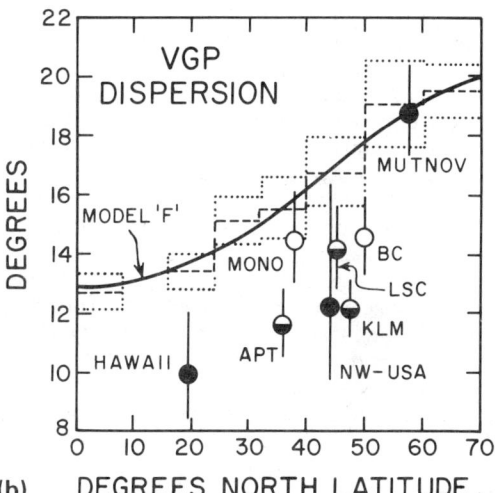

FIGURE 6. Zonal variation of the ΔI Anomaly and VGP dispersion. (a) The average variation of the ΔI anomaly for the last 5 my is shown by the dashed line (Lee and McElhinny, 1984); the ΔI anomalies of selected late-Quaternary PSV records (dots; Lund, 1985) are shown for comparison. (b) The average variation of VGP dispersion for the last 5 my is shown by the dashed line (McFadden and McElhinny, 1984) and model F (solid line) was determined by McFadden and McElhinny (1984) to fit the observed dispersion; the VGP dispersions for selected late-Quaternary PSV records (dots; Lund, 1985) are shown for comparison. (After Lund, 1985)

at a single site is not exactly axial-dipolar; a systematic offset in site inclinations, termed the ΔI anomaly, which varies systematically with site latitude is present.

3. The ΔI anomaly is present today and has persisted for at least the last 100 my; its level is apparently stationary for the last 35 my.

4. The angular dispersion of the PSV also varies with latitude in a systematic manner; this variation is related to the relative importance of dipole and nondipole contributions to PSV.

5. Model 'F' of McFadden and McElhinny (1984) accurately fits the long-term pattern of dispersion; there is evidence, however, that the patterns of dispersion for the last 3×10^4 years is different from the 5-my average, suggesting that stationarity may exist for this parameter only over timescales greater than 10^5 years.

These characteristics clearly indicate that the long-term prehistoric field is almost, but not exactly, axial-dipolar in its average behavior. Systematic errors in paleolatitude of up to perhaps 5° are typical. Even so, these nondipole contributions appear to be stationary over timescales of 10^4 years or longer. It is also apparent, however, that some parameters of PSV, such as angular dispersion, may not be stationary over time intervals less than 10^5 years. Such an observation may be related to the evidence for subtle long-term changes in late-Quaternary PSV waveforms and the possibility of significant spectral power in the 10^4–10^5-year time window noted in some low-resolution PSV records. Taken together, these characteristics point to a field that is only loosely stationary; some parameters, such as average inclination, may be near (but never at) their axial-dipole expectation within time intervals of 10^4 years, while other parameters, such as angular dispersion, may require averaging over 10^5 years or longer to reach apparent stationarity.

Models for Paleomagnetic Secular Variation

An alternative method for the analysis of PSV is to develop models for the observed field variability. Such models may be conceptual in nature with their primary purpose being to qualitatively estimate the style of variability that a potential source of PSV might generate, or the models may be more quantitative (essentially mathematical-simulation models), with their primary purpose being to replicate observed PSV. The mathematical-simulation models may be developed from more conceptual models, with sources that may have some basis in reality, or from purely empirical (unrealistic) inputs. A third group of models that characterize the actual magnetohydrodynamic process that generates the Earth's core field are beyond the scope of the present discussion.

Conceptual Models. In the past, various conceptual models have been proposed to qualitatively explain characteristic features of the observed HSV. Three of the models that have been discussed most often are (1) dipole wobble, (2) westward (or eastward) drift of the total nondipole field, and (3) standing and drifting nondipole fields. These conceptual models have also been called upon to qualitatively explain specific components of the PSV or to justify and physically explain specific mathematical-simulation models of PSV.

Dipole wobble has been suggested as one component of HSV and PSV based on the presence of an $11.5°$ offset in the present-day dipole field, its persistence during historic time, and the indication from paleomagnetic data that the field is normally *axial*-dipolar over time scales of 10^4 years or longer. Therefore, the average dipole field direction must have moved prehistorically and perhaps it has 'wobbled' irregularly to produce the *axial*-dipole average.

The paleomagnetic evidence for dipole wobble is problematic because of nonuniqueness. PSV at a single site is really the non-axial-dipole variation (dipole wobble plus true nondipole variation) at that site. The proportion of dipole wobble versus true nondipole contributions to PSV can only be assessed by analyzing globally distributed paleomagnetic data. Merrill and McElhinny (1977) carried out such an analysis of Northern Hemisphere archeomagnetic data for the last 2000 years and suggested that a significant dipole wobble component does exist. Dipole wobble contributions have also been estimated from the statistical analysis of angular dispersion; however, the proportion of dipole wobble depends upon the proposed model of dipole-wobble variability.

Westward drift has been suggested as an important element of HSV based on the observation that temporal changes in nondipole field components at the Earth's surface are due primarily to the westward drift in time of the spatial nondipole field components. The cause of westward drift has been related to differential rotation of the fluid outer-core, where the field is generated, versus the overlying lithosphere. The importance of westward drift is complicated by the fact that some areas of the Earth have exhibited eastward drift and other areas have exhibited no drift at all during historic times.

Paleomagnetic evidence for westward drift comes from a variety of PSV observations, few of which are unique. The circularity of PSV data has long been associated with westward or eastward drift of the paleomagnetic field, although other non-axial-dipole variations could produce the same effect. The recurring waveforms and related nondipole(?) spectral band (2400–3000 year period and multiples), noted earlier, may indicate westward (or eastward) drift of a complex nondipole waveform that changes very slowly in time compared to the time it takes for the waveform to drift entirely around the Earth (2400–3000 years). In such a model, similar waveform and spectral characteristics should be noted at all sites along a line of latitude. The Holocene waveform comparisons noted earlier do not, however, appear to be compatible with this model. Regional PSV comparisons within Europe and North America separately display waveform correlations that are consistent with a westward drift model; however, no simple correlation can apparently be made between the two regional data sets. Such a correlation between North American and European PSV records is necessary if westward (or eastward) drift is a predominant aspect of PSV.

Standing-nondipole field components have been proposed to improve the fit of drifting-nondipole-field components to the total HSV. If truly present, their origin might be related to standing components of fluid-flow near the core-mantle boundary caused by roughness of the boundary surface. The presence of standing nondipole components in the paleomagnetic record, however, is very difficult to evaluate because of problems of nonuniqueness and the uncertainties of spatial PSV behavior. To the extent that standing nondipole components might produce nonzonal components of I and D, their importance must be below the level of noise associated with parametric statistical analyses of long-term PSV (average I, D; ΔI; angular dispersion), for all of these parameters are apparently zonal in their spatial distribution. However, in the study of late-Quaternary PSV waveforms, standing-nondipole sources have been suggested as reasonable (but nonunique) alternatives to westward drift to explain the observed waveform variability within individual paleomagnetic records.

Mathematical Simulation Models. Models that are more quantitative have also been applied to HSV. Spherical harmonic models separate the field into dipole and multipole components and may include secular-change coefficients for predicting short-term temporal variations. The primary drawback to spherical harmonic models is their lack of relevance to the underlying physical causes of the Earth's internal magnetic field. Radial-dipole and current-loop models have occasionally been used as alternatives to spherical harmonic models. These models are more appealing in that their sources may mimic that part of the core process associated with nondipole foci observed at the Earth's surface.

Spherical harmonic models are hard to apply to PSV because of the inherent timing uncertainties associated with PSV data and because of the poor spatial distribution of most PSV data. Time-averaged PSV statistical parameters, such as the ΔI anomaly and vector dispersion, however, are more amenable to spherical harmonic analysis. For example, the ΔI anomaly can be modeled by an axial-dipole with added quadrupolar and octupolar components; the long-term changes in ΔI can then be modeled as changes in the quadrupole/octupole amplitude ratio (see Merrill and McElhinny [1983] for a more detailed discussion).

Radial-dipole models, with either standing or drifting sources, have been applied to individual high-resolution PSV records, as well as to statistical PSV records. An example of a drifting-radial-dipole model for the Lake St. Croix PSV record (Figs. 1

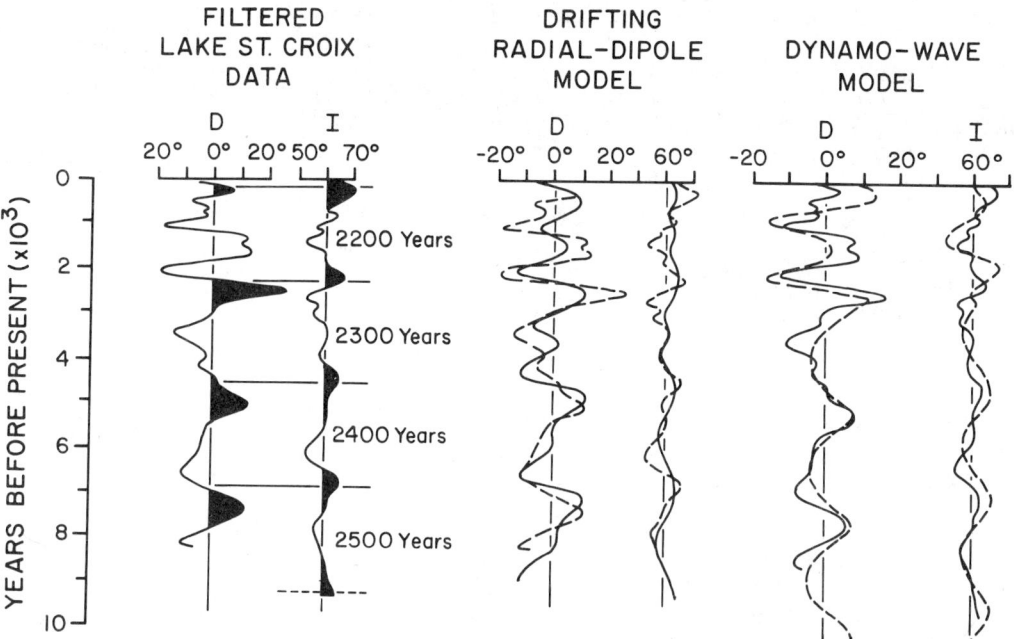

FIGURE 7. Mathematical-simulation models of the paleomagnetic secular variation record from Lake St. Croix. The paleomagnetic record from Lake St. Croix (Figs. 1 and 2) can be modeled by both (a) drifting radial-dipole models (Lund and Banerjee, 1985) and (b) dynamo-wave models (Olson and Hagee, 1987). Both models can fit the data equally well; the drifting radial-dipole model, however, predicts similar PSV at all sites near Lake St. Croix's latitude; the dynamo-wave model requires only a more regional coherence in PSV.

and 2; Lund and Banerjee, 1985) is shown in Fig. 7. Two drifting radial dipoles plus an axial dipole are able to model almost all of the observed variability at Lake St. Croix for the last 9000 years. Unfortunately, more complicated standing-radial-dipole models could also fit the data. But, these models would require more sources in order to fit the characteristic phase-relationships (see the previous section, Paleomagnetic Secular Variation Data) of the Lake St. Croix PSV data. The drifting-radial-dipole model for Lake St. Croix predicts similar PSV behavior for other sites on Lake St. Croix's latitude; the standing-radial-dipole models will only produce regional coherence. As has been noted before, the correlations in the section on paleomagnetic secular variation data, if validated with other similar studies, would argue against drifting-radial-dipole models for PSV. Conversely, the complex recurrent waveforms noted in several PSV records are very difficult to model with standing-radial dipoles due to the number of sources required and the detailed timing of recurrent intensity variations that each source must maintain relative to the other sources.

A new and appealing model of PSV, based on dynamo waves in the Earth's outer core, has been developed recently by Olson and Hagee (1987). In principle, their model successfully mimics several elements of the observed PSV: waveform phase-relationships and amplitude, waveform circularity, preferred westward drift, repeating waveforms. Figure 7 shows the result of their model for the observed Lake St. Croix PSV. It is apparent that the dynamo-wave model does just as good a job of fitting the observed variability as the drifting radial-dipole model. The dynamo-wave model, however, only requires regional coherence in the waveform correlations. This dynamo-wave model is an example of future interactions that will be possible between the PSV studies discussed here and dynamo-related studies of the Earth's core.

Conclusions

Paleomagnetic secular variation (PSV) studies provide a long-term view of the Earth's internal magnetic field variability that is unavailable from historic secular variation (HSV) studies. Analyses of PSV records have shown that the paleomagnetic field behavior is quasi-stationary over timescales of 10^4 years, although some aspects of the PSV may not be stationary until 10^5 years or longer. The time-averaged field appears to be essentially axial-dipolar, although inclinations are persistently lower than the axial-dipole expectation. This effect can cause perhaps 5° errors in paleolatitude calculations for plate tectonic reconstructions. On occasion, anomalously large-amplitude excursions of the PSV have occurred; the relationship of these excursions to normal PSV is unclear.

Comparative analyses of PSV and HSV indicate that the present-day field is probably indicative of the prehistoric spatial field variability. However, the dominant temporal field variations occur at timescales of 10^2 years or longer, beyond the range of HSV. Future coordinated analyses of HSV and PSV records, that keep in mind the fact that HSV and PSV are two related views of the same spatiotemporal secular variation process, should be able to produce a more coherent view of the Earth's total magnetic field variability.

Acknowledgements

This work has been supported by NSF Grants EAR-8212930, EAR-8509019, and EAR-8610001 to Steve P. Lund.

STEVE P. LUND

References

Barton, C. E., 1982, Spectral analysis of paleomagnetic time series and the geomagnetic spectrum, *Royal Soc. London Philos. Trans.* **A306**, 203.

Barton, C. E., 1983, Analysis of palaeomagnetic time series-techniques and applications, *Geophys. Surveys* **5**, 335-368.

Brock, A., 1971, An experimental study of palaeosecular variation, *Royal Astron. Soc. Geophys. Jour.* **24**, 303.

Champion, D. E., 1980, Holocene geomagnetic secular variation in the western United States: implications for the global geomagnetic field, *U.S. Geol. Survey Rept. Open file ser. 80-824*, 314.

Collinson, D. W., 1983, *Methods in Rock and Paleomagnetism: Techniques and Instrumentations.* London: Chapman and Hall.

Denham, C. R., 1974, Counterclockwise motion of paleomagnetic directions 24,000 years ago at Lake Mono, California, *Jour. Geomagnetism and Geoelectricity* **26**, 487.

Denham, C. R., and A. V. Cox, 1971, Evidence that the Laschamp Polarity Event did not occur 13,300-30,400 years ago, *Earth and Planetary Sci. Letters* **13**, 181.

Engebretson, D. C., and M. E. Beck, Jr., 1978, On the shape of directional data sets, *Jour. Geophys. Research* **83**, 5979-5982.

Fisher, R. A., 1953, Dispersion on a sphere, *Royal Soc. London Proc.* **A217**, 295.

Holcomb, R., D. Champion, and M. McWilliams, 1986, Dating recent Hawaiian lava flows using paleomagnetic secular variations, *Geol. Soc. America Bull.* **97**, 829-839.

King, J., 1983, Geomagnetic secular variation curves for northeastern North America for the past 9000 years, Ph.D. diss., University of Minnesota, Minneapolis.

Kovacheva, M., 1983, Archeomagnetic data for Bulgaria and southeastern Yugoslavia, in K. M. Creer, P. Tucholka, and C. E. Barton, eds., *Geomagnetism of Baked Clays and Recent Sediments*. Amsterdam: Elsevier, 106-109.

Kovacheva, M., and G. Zagniy, 1985, Archeomagnetic results from some prehistoric sites in Bulgaria, *Archeometry* **27**, 179-184.

Lee, S. H., and McElhinny, M. W., 1984, Estimates of the time-variation of the non-dipole part of the time-averaged paleomagnetic field (in press).

Liddicoat, J. C., and R. S. Coe, 1979, Mono Lake geomagnetic excursion, *Jour. Geophys. Research* **84**, 261.

Liddicoat, J. C., and S. P. Lund, 1983, A high resolution record of secular variation from Quaternary sediments from Mono Lake, California, *EOS Am. Geophys. Union Trans.* **64**, 685.

Lund, S. P., 1985, A comparison of the statistical secular variation recorded in some late Quaternary lava flows and sediments, and its implications, *Geophys. Research Letters* **12**, 251-254.

Lund, S. P., and S. K. Banerjee, 1985, Late Quaternary paleomagnetic field secular variation from two Minnesota Lakes, *Jour. Geophys. Research* **90**, 803-825.

Lund, S. P., and P. Olson, 1987, Historic and paleomagnetic secular variation and the Earth's core dynamo process, *Rev. Geophys.* **25**, 917-928.

Lund, S. P., J. Liddicoat, K. Lajoie, T. L. Henyey, and S. Robinson, 1988, Paleomagnetic evidence for long-term (10^4 year) memory and periodic behavior in the Earth's core dynamo process, *Geophys. Research Letters* **15**, 1101-1104.

McElhinny, M. W., and R. T. Merrill, 1975, Geomagnetic secular variation over the past 5 my, *Rev. Geophysics Space Physics* **13**, 687.

McFadden, P. L., and M. W. McElhinny, 1984, A physical model for palaeosecular variation, *Royal Astron. Soc. Geophys. Jour.* (in press).

Merrill, R. T., and M. W. McElhinny, 1977, Anomalies in the time-averaged paleomagnetic field and their implications for the lower mantle, *Rev. Geophysics Space Physics* **15**, 309.

Merrill, R. T., and M. W. McElhinny, 1983, *The Earth's Magnetic Field, Its History, Origin and Planetary Perspective*. New York: Academic Press.

Olson, P., and V. L. Hagee, 1987, Dynamo waves and paleomagnetic secular variation, *Royal Astron. Soc. Geophys. Jour.* **88**, 139-159.

Skiles, D. D., 1970, A method of inferring the direction of drift of the geomagnetic field from paleomagnetic data, *Jour. Geomagnetism and Geoelasticity* **22**, 441.

Sternberg, R. S., 1983, Archeomagnetism in the Southwest of North America, in K. M. Creer, P. Tucholka, and C. E. Barton, eds. *Geomagnetism of Baked Clays and Recent Sediments*. Amsterdam: Elsevier.

Tarling, D. H., 1983, *Paleomagnetism: Principles and Applications in Geology*. London: Chapman and Hall.

Thompson, R., 1984, Geomagnetic evolution: 400 years of change on planet Earth, *Physics Earth and Planetary Interiors* **36**, 61-77.

Turner, G. M., and R. Thompson, 1981, Lake sediment record of the geomagnetic secular variation in Britain during Holocene times, *Geophys. Jour. Royal Astron. Soc.* **65**, 703-725.

Verosub, K. L., P. J. Mehringer, Jr., and P. Waterstraat, 1986, Holocene secular variation in western North America: paleomagnetic record from Fish Lake, Harney County, Oregon, *Jour. Geophys. Research* **92**, 3609-3623.

Wilson, R. L., 1970, Permanent aspects of the earth's non-dipole magnetic field over Upper Tertiary times, *Royal Astron. Soc. Geophys. Jour.* **19**, 417.

Cross-references: *Archaeomagnetism; Geomagnetic Field: Westward Drift; Geomagnetic Secular Variation: Direction and Intensity; Geomagnetic Secular Variation: Theory; Paleomagnetic Field: Intensity.*

PALEOMAGNETIC TIME SCALE—See MAGNETOSTRATIGRAPHY

PALEOMAGNETISM: DEEP-SEA SEDIMENTS

Applications

Many types of deep-sea (or pelagic) sediments, recovered by ocean drilling or cropping out on land, are efficient recorders of the direction of the geomagnetic field at the time of deposition of the sediment. These records have been used to quantify the relative rotation of rock bodies, on a scale that ranges from thrust sheets to entire lithospheric plates, by comparing paleomagnetic pole positions of the same age from one rock body to another. Perhaps more importantly, these records have been a major source of information on the polarity history of the geomagnetic field.

Geomagnetic polarity reversals are globally synchronous phenomena and the reversal process is rapid in a geologic time-frame (10^3 to 10^4 years). Therefore the reversal record in sediments provides a high resolution global correction tool, which can enhance the stratigraphic resolution provided by the fossil record, correlate biozonations from contrasting paleoenvironments, and monitor rapid paleoceanographic events such as stable isotope shifts. In addition, intervals of normal and reversed polarity (zones) in pelagic sections can be correlated to biozonations and to the marine magnetic anomaly record to give us our best estimates of the age of the ocean floor. The reversal record in sediments (magnetostratigraphy) has become a standard stratigraphic correlation technique for Cenozoic pelagic sediments drilled at sea or cropping out on land. The magnetostratigraphic correlation method is particularly applicable to pelagic sediments not only because they are often reliable recorders of the paleomagnetic field, but also because the relatively slow, constant sedimentation rates facilitate the recognition of polarityzone patterns that can be correlated to the geomagnetic reversal time scale (GRTS). For Mesozoic and older sediments, this correlation technique is not yet widely used due to uncertainty in the correlation of key biozonations to the GRTS.

Reliable geomagnetic paleointensity information from deep-sea sediments is more elusive than directional data. If the remanence is purely depositional in origin (DRM) and the sediment can be completely deaggregated, laboratory deposition can theoretically determine the relationship between DRM intensity and applied field. In practice, relative paleointensity determinations are often estimated by normalizing the natural remanent magnetization (NRM) intensity using a parameter such as saturation remanent magentization (SIRM), initial susceptibility or anhysteretic remanent magnetization (ARM). These techniques are based on the assumption that the magnitude of the normalizing parameter is proportional to the concentration of magnetic carriers in the sediment. As SIRM and initial susceptibility tend to overemphasize the role of the coarser grains, the ARM intensity is the preferred normalizing factor.

Characteristics of the Magnetization

The magnetization of most rocks and sediments is a composite of a number of magnetization components acquired at various times during their history. Demagnetization techniques using alternating fields, heating, or chemical methods are designed to "clean" the sample magnetization and resolve the primary component, that associated with sediment deposition.

Fortunately some deep-sea sediments have very simple magnetic properties, exhibiting a single magnetization component carried by ferrimagnetic titanomagnetite. It has been proposed that these sediments acquire their magnetization by the mechanical rotation of (titano) magnetite grains into line with the prevalent geomagnetic field in the bioturbated upper centimeters of soft sediment. This acquisition process is referred to as depositional remanent magnetization (DRM). The depth (below the sediment/water interface) and size of the "lock-in zone" in which the grains lose their mobility will depend on grain size, magnetic mineralogy, sedimentation rate, and the extent of bioturbation. Estimates of this depth for pelagic sediments are generally a few tens of centimeters, which correspond to a few times 10^4 yrs, at typical pelagic sedimentation rates.

Sediments of this category are excellent recorders of the direction of the geomagnetic field at a time close to the time of deposition and there are numerous examples where studies from deep-sea sediments have provided high resolution magnetostratigraphies. It has been suggested that the magnetite grains that carry the magnetization are entirely detrital in origin. However, the Curie temperatures often indicate the presence of almost pure magnetite (Fe_3O_4). This indication is inconsistent with a totally detrital source as titanomagnetites with significantly lower Curie temperatures are characteristic of mid-ocean ridge basalts and igneous rocks exposed on land. It has recently been discovered that several marine organisms, including bacteria and chitons, precipitate nearly pure magnetite as an integral part of their metabolism (Blakemore, 1975; Kirschvink and Lowenstam, 1979). Magnetite crystals of probable bacterial origin have been observed in pelagic sediments by using transmission electron microscopy (Petersen et al., 1986). These typically octahedral or parallelepiped shaped magnetite grains are usually 0.05 to $0.1 \mu m$ across and are within the single domain grain size range, which is optimal for retention of magnetic remanence. Magnetite of bacterial origin may form at a few tens of centimeters below the sediment-water interface at the transition from iron-oxidizing to iron-reducing conditions (Karlin and Levi, 1985). It appears that nearly pure magnetitie of biogenic origin is a major carrier of magnetic remanence in deep-sea sediments (Petersen et al., 1986). The possibility of authigenic magnetite forming during subsequent diagenesis should not be discounted, and secondary magnetite has been associated with hydrocarbon migration during the deformation of some limestones. Certain rock magnetic studies in pelagic sediments have indicated two distinct phases of magnetite, a dominant low titanium phase of probable bacterial/biogenic origin and titanomagnetite of probable detrital origin.

Not all deep-sea sediments are efficient recorders of the geomagnetic field at the time of deposition. The lack of a primary magnetization may be due to the magnetic mineralogy or grain size of the detritus, to adverse diagenetic conditions, which result in the alteration of primary magnetite, or to the growth of authigenic magnetic mineral phases.

It has been observed, particularly in the unfossiliferous Pacific clay facies, that the primary magnetic record degrades at a few meters depth below the sediment/water interface. The boundary between the primary and unstable magnetic records often coincides with the Late Pliocene onset of Northern Hemisphere Glaciation, and it has been suggested that a change in eolian detritus is responsible. The reduced grain size of magnetite detritus in the pre-Late Pliocene sediments (due to less intense atmospheric circulation) may account for their ability to acquire viscous remanence (VRM) and for the masking of the primary magnetization (Yamazaki and Katsura, in press).

In some cases the degradation of the primary magnetization may be due to diagenetic alteration of detrital or biogenic magnetic minerals and/or growth of authigenic magnetic minerals. In oxidizing diagenetic conditions, magnetite may alter to maghemite (γ-Fe_2O_3), which may carry either a secondary chemical remanent magnetization (CRM) or a VRM (Kent and Lowrie, 1974). Henshaw and Merrill (1980) have suggested that secondary ferromanganese oxides and oxyhydroxides (such as the jacobsite ($MnFe_2O_4$) solid solution series) are often important authigenic magnetic phases. Ferrimagnetic goethite (α-FeOOH) and the less common lepidocrosite (γ-FeOOH) can be precipitated directly from seawater and will dehydrate during diagenesis to hematite (α-Fe_2O_3) and maghemite (γ-Fe_2O_3) respectively. Authigenic hematite is an important contributor to the magnetization of many red pelagic sediments, and goethite may sometimes be its precursor. There is evidence from red pelagic limestones exposed on land that hematite growth can occur within a few meters of the sediment water interface (Channell et al., 1982). Geothite and hematite can also form as a late low-temperature oxidation product of more primary Fe phases such as sulfides. Hematite, geothite, maghemite, and magnetite can coexist in some red pelagic limestones exposed on land, but very often the magnetization is dominated by magnetite with a minor contribution from authigenic hematite (Lowrie and Heller, 1982).

In areas of high sedimentation rate and/or restricted circulation, the burial of organic matter results in reducing diagenetic conditions and the formation of iron sulfides. These sulfides may form in part by the reduction of primary magnetite. The metastable iron sulfides mackinawite (FeS) and griegite (Fe_3S_4) are capable of carrying magnetic remanence, as are some compositions of the more stable pyrrhotite ($Fe_{1-x}S$). The secondary magnetic remanence carried by these authigenic sulfides is rarely useful because it cannot be associated with a temporally defined depositional or diagenetic event.

Conclusions

Biogenic and detrital magnetite are the principal carriers of primary magnetization in deep-sea sediments. However, the fidelity of the paleomagnetic record is a function of detrital and biogenic sources and of diagenetic conditions. If conditions are reducing, primary magnetite may be reduced to sulfide. In highly oxidizing conditions, maghematization of magnetite and growth of secondary magnetic minerals may occur, which can obscure the primary magnetization.

J. E. T. CHANNELL

References

Blakemore, R., 1975, Magnetic Bacteria, *Science* **190**, 377-379.

Channell, J., R. Freeman, F. Heller, and W. Lowrie, 1982, Timing of diagenetic haematite growth in red pelagic limestones from Gubbio (Italy), *Earth and Planetary Sci. Letters* **58**, 189-201.

Henshaw, P. C., and R. T. Merrill, 1980, Magnetic and chemical changes in marine sediments, *Rev. Geophysics Space Physics* **18**, 483-504.

Karlin, R., and S. Levi, 1985, Geochemical and sedimentological control of the magnetic properties of Hemipelagic sediments, *Jour. Geophys. Research* **90**, 10373-10392.

Kent, D. V., and W. Lowrie, 1974, Origin of magnetic instability in sediment cores from the central North Pacific, *Jour. Geophys. Research* **79**, 2987-3000.

Kirschvink, J. L., and H. A. Lowenstam, 1979, Mineralization and magnetization of Chiton teeth: paleomagnetic, sedimentologic, and biologic implications of organic magnetite, *Earth and Planetary Sci. Letters* **44**, 193-204.

Lowrie, W., and F. Heller, 1982, Magnetic properties of marine limestones, *Revs. Geophysics Space Physics* **20**(2), 171-192.

Petersen, N., T. Von Dobeneck, and H. Vali, 1986, Fossil bacterial magnetite in deep-sea sediments from the South Atlantic Ocean, *Nature* **320**, 611-615.

Yamazaki, T., and I. Katsura, 1989, Magnetic grain size and viscous remanent magnetization of pelagic clay, *Jour. Geophys. Research* (in press).

Cross-references: *Chemical Remanent Magnetization (CRM); Curie Temperature; Demagnetization; Detrital Remanent Magnetization (DRM); Magnetic Properties of Minerals; Magnetic Domains; Magnetic Properties of Minerals; Natural Remanent Magnetization (NRM); Rock Magnetism; Rock Magnetism: Measuring Techniques and Apparatus; Seafloor Spreading: Magnetic Evidence; Thermoremanence; Viscous Remanent Magnetization (VRM) and Viscous Remagnetization.*

PALEOMAGNETISM: METEORITES

Meteorites represent the most primitive bulk material available for scientific study. The discovery by Stacey and Lovering in 1969 that meteorites can possess remanent magnetizations of probable extraterrestrial origin suggested that these ancient objects might provide information about magnetic fields in the early history of the solar system. Paleomagnetic studies on meteorites have since focused on determining the intensity of the ancient magnetizing fields. These efforts have been hindered by the complex interacting magnetic mineralogies and textures found in many meteorites and by irreversible changes that these mineral assemblages often undergo upon heating, even in vacuum.

Magnetic contamination of meteorites and their components by exposure to stray magnetic fields, particularly from hand-held magnets, can be a serious problem, as meteorites are commonly identified by their sometimes intense magnetic attraction. Because their remanent magnetization can be severely altered or even destroyed by exposure to magnets, suspected meteorites should never be identified by this technique. The large numbers of meteorites discovered on the ice sheets in Antarctica since 1969 are particularly valuable for magnetic remanence studies, since particular care has been taken to preserve their pristine condition. Contamination by the Earth's field due to atmospheric heating is less of a problem since the stable magnetization of only the outer 1 mm (i.e., the fusion crust) is affected in stony meteorites.

Interpretation of the paleointensity results for meteorites is dependent on assumptions about the nature of their parent bodies, because the source of the magnetic field might be either of internal (e.g., liquid core dynamo) or external (e.g., solar field) origin. Evidence for the existence of substantial magnetic fields of external origin are important to understanding the early evolution of the solar system. Inductive heating of celestial bodies by an early solar wind magnetic field, angular momentum transfer by magnetic fields from an initially rapidly spinning sun, and metal-silicate fractionation due to high field gradients in the solar nebula, are hypothesized processes for which primitive meteorite paleointensities might provide supporting evidence.

Magnetic Classification of Meteorites

Meteorites have historically been classified into two types, metal rich (NiFe) irons and silicate rich stones, with the stones further subdivided into chondrites or achondrites, depending on whether they contain millimeter-sized once-molten spherical bodies called chondrules. Composite types, called stony irons, represent a minor group of meteorites. The subclassification of chondrites is in part based on their iron content, with the sequence of LL (amphoteric), L (hypersthene), H (bronzite), and E (enstatite) chondrites corresponding to increasingly higher concentrations of metallic nickel iron (NiFe). C (carbonaceous) chondrites all contain varying amounts of magnetite (Fe_3O_4), as well as metal in the more metamorphosed types. Chondrites other than carbonaceous and enstatite types are generally referred to as ordinary chondrites. Except for the composite stony irons and a unique subclass called the ureilites, achondrites have very low NiFe contents compared to chondrites.

Because of this trend in metallic iron content and an opposing trend of higher Ni content in the metal of L and LL chondrites, a combination of saturation magnetization (I_s) (see *Rock Magnetism*) and thermomagnetic measurements is sufficient to classify stony meteorites into their various subtypes (Fig. 1). The use of thermomagnetic data to determine meteorite type follows from the recognition that two NiFe alloys dominate the high field magnetic properties of most meteorites. The alpha phase, called kamacite, is a Ni poor (Ni <8 wt%) alloy with a body centered cubic (bcc) crystal structure and an irreversible thermomagnetic curve, with a Curie temperature (T_c, see *Curie Temperature*) increasing from about 740°C to 770°C with decreasing Ni content. The gamma phase, called taenite, is a Ni rich alloy with a face centered cubic (fcc) structure, and a T_c varying from below 100°C for 25-30 wt% Ni, to over 600°C for >60 wt% Ni. Fine intergrowths of kamacite and taenite, called plessite, are commonly found in chondrites. Kamacite, both as discrete grains and as intergrowths in plessite, transforms to the gamma phase,

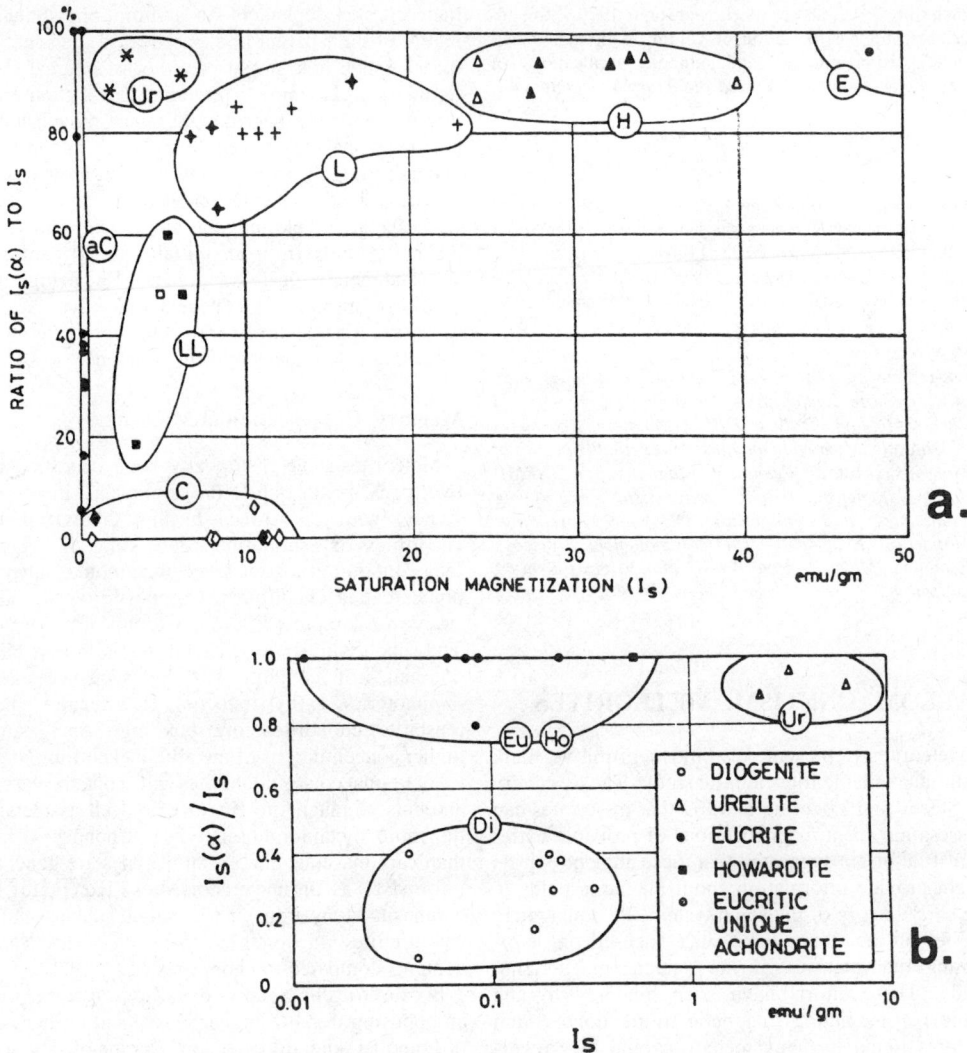

FIGURE 1. (a) Classification of chondrites and achondrites via magnetic measurements. (b) Classification of Antarctic achondrites using magnetic measurements. (From Nagata, 1980)

taenite, at high temperature. It is this transformation, along with a delayed regression to the alpha state on cooling, which is responsible for the irreversible thermomagnetic curves of many kamacite- and plessite-bearing meteorites. An identification of the kamacite thermomagnetic curve with its characteristic thermal hysteresis allows for an estimate of the relative contribution of the Ni poor alpha phase to the total magnetization of the meteorite, and hence the relative Ni content of the component NiFe phases. A near total absence of kamacite in carbonaceous chondrites distinguishes them from the LL and L chondrites (Fig. 1).

Natural remanent magnetization (NRM) (q.v.) intensities for meteorites range over more than four orders of magnitude and correlate with magnetic susceptibilities (Fig. 2) and the variation in magnetic grain content shown in Fig. 1. Although some meteorites are prone to acquire viscous remanence (VRM) (see *Viscous Remanent Magnetization (VRM) and Viscous Magnetization*) in the Earth's field, VRM acquired since falling to Earth is unlikely to alone account for the considerable NRM intensities seen in many meteorites. Instead thermal (TRM, see *Thermoremanence*) or chemical (CRM, see *Chemical Remanent Magnetization*) remanence acquired in the presence of ancient extraterrestrial fields, or some combination of both, is judged as the most likely source of remanence in the more stably magnetized meteorites. Shock magnetization (SRM) is a possible source of remanence for meteorites showing mineralogical evidence of one or more

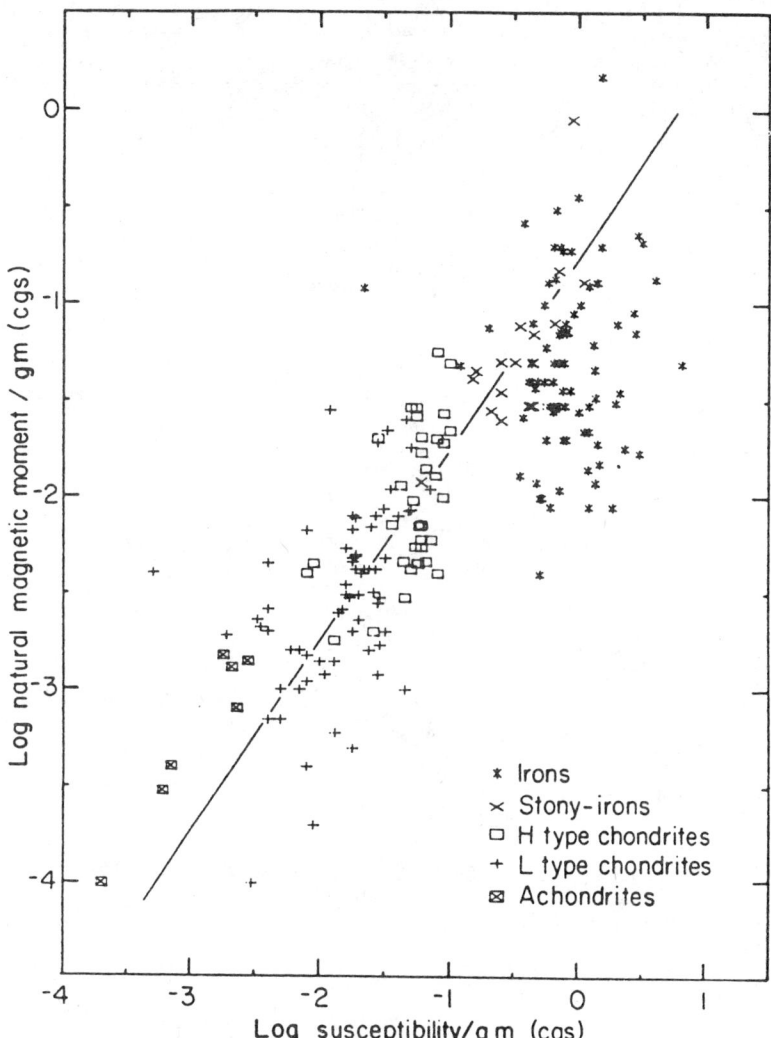

FIGURE 2. NRM vs. susceptibility for various types of meteorites, from a compilation of measurements from Russian collections by Herndon et al., 1972. (From Levy and Sonett, 1978)

shock events. However the presence of an ambient field seems necessary if an SRM of significant magnitude is to be acquired. Reduction of preexisting NRM through shock demagnetization is more likely, particularly in recent (<1 by old) shock events.

Paleointensity Studies on Meteorites

Carbonaceous Chondrites. The meteorite that has undergone the most thorough magnetic analyses to date is the Allende carbonaceous chondrite, which fell in northern Mexico in 1969. Its petrologic type (3) indicates that it experienced an intermediate degree of heating (about 300°C) during or since assembly (Type 2 meteorites contain the least metamorphosed components, while Type 6 contain the most). Its radiometric age is about 4.5 by, although it may have experienced a reheating episode about 1 by later.

Allende's high field thermomagnetic properties are dominated by taenite and possible magnetite ($T_c =$ 590–610°C). However its remanence carrying ability resides primarily in unidentified low temperature phases ($T_c < 320°C$). These phases may include NiFe and magnetic sulfides such as pyrrhotite. High resistance to alternating field (af) demagnetization (see *Demagnetization*) and directional stability and coherence indicate that the low temperature portion of the NRM resides in fine magnetic grains and represents a thermal, chemical, or detrital remanence (DRM, see *Detrital Remanent Magnetization*) acquired uniformly throughout the meteorite.

Dual heating paleointensity determinations, which

TABLE 1. Dual Heating Paleointensity Results on Meteorites

Meteorite	Sample	Temp. (°C)	Field (mT)	Workers
Carbonaceous Chondrites				
Orgueil (C1)	bulk	20-120	(0.067)	S.K. Banerjee and R.B. Hargraves, 1972
Murchinson (C2)	bulk	20-90	(0.018)	
Allende (C3)	bulk	20-130	0.109	
Allende (C3)	bulk	30-150	0.093-0.131	R.F. Butler, 1972
Allende (C3)	bulk	20-330	0.079-0.33 (median-0.123)	T. Nagata and M. Funaki, 1983
	matrix	20-330	0.19-0.30	
	chondrules	20-330	0.07 and 0.13	
Allende (C3)	chondrules	160-300	(0.20-0.30)	M. Lanoix et al., 1977; 1978
		350-450	(0.30 and 0.70)	
Leoville (C4)	bulk	20-350	0.097	T. Nagata and N. Sugiura, 1977
Ordinary Chondrites				
Yam. 74160 (LL7)	bulk	30-580	(>0.0054)	T. Nagata and M. Funaki, 1982
ALHA 77260 (L3)	bulk	20-170	(0.0185)	
		20-270	(0.005)	
Dagley Dns. (L4)	bulk	20-470	0.068	T. Nagata and N. Sugiura, 1977
Fukutomi (L5)	bulk	20-400	0.010	
Seminole (H4)	bulk	20-350	0.039	
Yonozu (H4,5)	bulk	20-470	(0.018)	
Bjurbole (L4)	matrix	20-400	(0.025)	N. Sugiura and D.W. Strangway, 1982
	chondrule	20-585	(0.06)	
Prairie Dog (H)	bulk	120-300	(0.137)	M. Westphal and H. Whitechurch, 1983
Bath (H)	bulk	50-300	(0.170)	
Ochansk (H)	bulk	60-360	(0.188)	
Mooresfort (H)	bulk	60-400	(0.069)	
Pultusk (H)	bulk	50-300	(0.048)	
Indio Rico (H)	bulk	60-120	(0.080)	
Oakley Stone (H)	bulk	120-400	(0.120)	
ALHA 76009 (L6)	chondrules	30-360	(0.047 & 0.091)	T. Nagata and M. Funaki, 1981
Enstatite Chondrites				
Abee (E4)	clast	74-228	(0.61)	N. Sugiura and D.W. Strangway, 1981
	matrix	56-216	(0.70)	
Indarch (E4)	bulk	20-385	(1.78)	N. Sugiura and D.W. Strangway, 1982
Achondrites				
Yam. 7308 (7307)	bulk	20-240	0.007	T. Nagata, 1979
Shergotty	bulk	20-138	0.002	S.M. Cisowski, 1986
		362-598	<0.001	

Note: Values in parentheses are either described as less reliable by the experimenters, in conflict with other paleointensity estimates, or derived from samples heated in air.

allow for a comparison of NRM lost at various heating steps to partial thermoremanence (PTRM) gained in a laboratory field, give results centering on 0.1 millitesla (mT) for bulk samples of Allende (Table 1). Thus the meteorite may have been metamorphosed or assembled in a magnetic field similar in intensity to the Earth's. However nonlinear behavior was observed for these experiments at temperatures beginning at 135-170°C indicating that irreversible magnetochemical changes became significant at this point. The monitoring of high field magnetic properties during initial heating suggests that more subtle alteration in Allende may be induced at even lower temperatures. Interactions between low temperature magnetic phases may additionally complicate the interpretation of these paleointensity estimates.

Dual heating paleointensity results as high as 1.6 mT (later revised to 0.7 mT) have been reported for large chondrules (greater than 2 mm in diameter) taken from the Allende meteorite (Table 1). Unlike the matrix and bulk samples, the strongly magnetized chondrule remanence persists up to temperatures of 590°C, and has been attributed to magnetite or taenite. Since the texture of chondrules suggests a high temperature of origin, the remanence carried by chondrules could represent a TRM acquired before consolidation of the parent meteorite. Potential sources for this intense remanence include a very strong field of the early T-Tauri stage sun or lightning in the planetary nebula, with the latter mechanism relating directly to the formation of the chondrules.

The large and intensely magnetized Allende chondrules were not oriented with respect to one another as they were removed from the meteorite by a previous investigator. Sugiura and Strangway (1985) found that smaller oriented chondrules from Allende reveal a weak directional coherency, suggesting an overprint by a low temperature thermochemical event. However, Wasilewski (1981) had already demonstrated that these smaller chondrules have much weaker NRM intensities when normalized to their total remanence carrying ability (IRM$_s$, saturation isothermal remanence), as compared to the large chondrules. Paleointensity experiments suggest that they carry only a low temperature remanence of about 0.1 mT similar to the results on the bulk samples (Table 1).

Until the high paleofield values attributed to the large chondrules can be verified by studying similar in situ chondrules, a possibility exists that their remanences may have been reset by postsampling contamination in stray magnetic fields. If these high field magnetizations are verified by future studies of Allende chondrules, this would imply a significant decay in strength of the magnetizing field between the time of chondrule formation and parent body assemblage.

Other carbonaceous chondrites show NRMs with less stability to af demagnetization than Allende, suggesting that their magnetic carriers are coarser grained. A paleointensity value again of about 0.1 mT (Table 1) was obtained at temperatures below 300°C for the Leoville meteorite of the same (C3) petrologic type as Allende. Paleointensity determinations on less metamorphosed Murchison (C2) and Orgueil (C1) meteorites produced somewhat lower field values (Table 1). ^{129}I–^{127}I ratios on these two carbonaceous chondrites indicate that their magnetite grains formed from about 4.3 by to 4.5 by ago.

A fundamental problem in the interpretation of paleointensity results for carbonaceous chondrites is whether their low temperature magnetizations could persist for billions of years. This is particularly true for the C1 and C2 meteorites, since their NRMs are mostly unblocked at temperatures below 125°C. If their NRM carriers are coarse multidomain or extremely fine near-superparamagnetic grains of magnetite or metal, their magnetic relaxation times might be considerably less than the assumed ancient age of magnetization.

Ordinary and Enstatite Chondrites. In general, radiometric age dating shows that most ordinary chondrites were formed not long after the 4.53 by age of the solar system. However a large proportion (about 1/3) have experienced significant reheating since their initial formation, probably as a result of shock events on their parent bodies. Thus, though many ordinary chondrites, along with carbonaceous chondrites, had the potential for recording information about magnetic fields in the early solar system, many others would have been magnetically reset in much later times.

An understanding of the magnetization record of shock-heated meteorites is complicated by a number of metallurgical and magnetic transitions and transformations that their component NiFe grains undergo in response to heating and cooling. Rapid cooling of taenite, as would likely follow shock heating, may produce martensite, the crystal-distorted form of alpha NiFe that is metastable, particularly at higher temperatures. Because the transformation to martensite is capable of producing stable remanence even in large grains, as demonstrated by Wasilewski in 1974, an assessment of the role of martensite in the remanent magnetization of shocked meteorites is important.

The NRMs of many E, H, and L chondrites decay rapidly when exposed to AF demagnetization, with samples often displaying order of magnitude decreases in intensity at alternating fields of less than 5 mT. In contrast, the NRMs of many of the measured LL chondrites show moderate to high resistance to the AF demagnetization technique, probably due to the presence of plessite, or tetrataenite.

Tetrataenite, an ordered crystal form of taenite with Ni content between 48 and 57 wt%, has been recently recognized as an important magnetic mineral in slowly cooled (unshocked) ordinary chondrites. Its potential for contributing to a meteorite's remanent magnetization lies in its extremely high coercivity and resistance to AF demagnetization, due to its high crystalline anisotropy (see *Rock Magnetism*). Dual heating paleointensity determinations on two tetrataenite-bearing chondrites (Yamato 74160 and ALH 77260—Table 1) give minimum paleofield values of only 0.005 mT for the "hard" component of magnetization. However these results were reported with reservation, as some transformation of the tetrataenite to ordinary taenite will occur during stepwise heating, while the thermal nature of the tetratanite NRM is not assured.

Dual heating paleointensity experiments on five other H and L chondrites estimated the strengths of paleofields to be several hundredths of a millitesla (Table 1). Out of these five, only Fukutomi and Yonuzu, which gave the lower paleofield estimates, were demonstrated to have a high resistance to AF demagnetization. Two of the others, Dalgety Downs and Seminole, display mineralogical evidence of moderate shock and reheating since their formation. The third, Bjurbole, may contain plessite and/or tetrataenite as the remanence-carrying phases. Thus the NRM might be of (thermo-) chemical origin, and the validity of the paleointensity determinations may be questionable.

Dual heating paleointensity estimates on an additional seven H chondrites, with the highest paleofield values >0.1 mT, are also listed in parentheses in Table 1. These experimental results are suspect as the heatings were done in air. Since metallic iron is highly susceptible to oxidation at even moderate temperatures, destruction of magnetic carriers in the indicated temperature range is probable and is suggested by a loss of PTRM for many of the sequential heating steps. Destruction of PTRM acquisition potential could account for these significantly higher paleofield values as compared to the Seminole and Yanozu H chondrite paleofield values, which were obtained through vacuum heatings.

Detailed magnetic studies on the enstatite chondrites Abee and Indarch, and an Antarctic L chondrite, ALHA 76009, illustrate the difficulties in interpreting paleointensity results from chondrites. Paleointensity determinations of 0.6–0.7 mT were obtained for a clast and matrix sample of the 4.5 by old Abee meteorite (Table 1). These results must be considered with reservation because of the complex nature of the magnetic remanence in this specimen. The principal carrier of stable remanence appears to be cohenite (FeC_3), which is microscopically intergrown and thus magnetically interacting with kamacite. These properties, along with a much lower paleofield value (0.033 mT) obtained from a single heating paleointensity determination on a bulk sample of Abee by Brecher and Ranganayaki (1975), make it difficult to interpret these results on clast and matrix samples. An even higher paleofield value of 1.78 mT was obtained from another E chondrite, Indarch (Table 1). Again evidence of cohenite and other exotic minerals as prime remanence carriers, and the confinement of the NRM to low coercivity phases suggests caution in the interpretation of paleointensity result from this E chondrite.

In ALHA 76009, 13 measured chondrules had stable but random directions of magnetization, although the matrix samples showed evidence of a magnetic overprinting event. Again paleointensity estimates on chondrules of this L6 meteorite (Table 1) cannot be accepted unequivocally because of uncertainty about the nature of the remanence and evidence of interaction effects in the NRM lost vs. PTRM gained curves.

Achondrites. Achondrites represent meteoritic material derived from a chemically differentiated parent body. Therefore their paleofield strength might relate to an internal dynamo mechanism with a conductive core rather than to an interplanetary field thought to be recorded in chondrites. Radiometric ages for most achondrites are more than 4 by, indicating that their magnetizations may be nearly as old as the assumed age of chondrite remanent magnetism.

Achondrites are petrologically similar to lunar samples, with eucrites and diogenites somewhat analogous to monomict breccias (clasts and matrix produced from the same parent rock) derived from basaltic and ultramafic source rocks, respectively. Howardites, mesosiderites, and aubrites are analogous to lunar polymict breccias (clasts and matrix derived from a variety of source rocks). In general achondrites have much lower NiFe metal contents than ordinary and enstatite chondrites (Fig. 1), with saturation magnetization (I_s) values comparable to those of lunar igneous rocks. Exceptions to this generalization are the ureilites, highly shocked ultramafic achondrites with abundant low-Ni metal, and the pallasite and mesosiderite stony irons. A few achondrites appear to represent initially unbrecciated igneous rocks. These include the shergottites, nakhlites, and chassignites, whose less than 1.5 by crystallization ages set them apart from all other meteorites.

Achondrites are often characterized by a considerable directional stability and resistance to AF demagnetization, with variable proportions of kamacite, taenite, and possibly magnetic chromite as the remanence-carrying magentic phases. Because of their relative scarcity and complications in their thermal demagnetizing characteristics, only one dual heating paleointensity result has been reported on a common type of achondrite, indicating a paleofield of less than 0.01 mT (Table 1). This single estimate has been supplemented by additional paleofield estimates by Nagata (1980b) on six achondrites collected in Antarctica. The method substituted anhysteretic remanent magnetization (ARM) for laboratory heating and gave upper limits of from 0.003 to 0.009 mT for five samples, consistent with the dual heating result. A sixth sample, one of two unbrecciated diogenites included in the group, gave an upper limit value of 0.024 mT.

Two ureilites (Kenna and Goalpara) exhibit strong NRMs, which comprise a considerable fraction of their total remanence-carrying ability (NRM/IRM_s values >0.1). These intense magnetizations may indicate a strong (about 0.1 mT) nebular magnetizing field, as suggested by Brecher and Fuhrman (1979), but contrast with the 0.0089 mT result reported for an Antarctic ureilite studied by Nagata (1980b). Their concave downward AF demagnetization curves (resistant only at low alternating fields) are reminiscent of those of intensely magnetized and metamorphosed mafic and ultramafic Archaean terrestrial rocks, which do not reflect the Earth's paleofield intensity. This magnetic analogy with similarly behaving terrestrial samples suggests caution in the interpretation of the ureilites' sometimes strong magnetization.

Shergottites, chassignites, and nakhlites, collectively referred to as SNC meteorites, represent a unique class of achondrites with crystallization ages much less than those of other extraterrestrial samples. Because their crystallization ages may range to as young as 340 my, they must have been derived from a fairly large planetary body that has had volcanic activity in the recent past. The presence of partially oxidized titanomagnetite as a principal magnetic carrier makes these meteorites more similar in magnetic character to terrestrial samples than to other extraterrestrial specimens. Their derivation from a geologically active parent body, their oxidized state, and a large number of geochemical characteristics, including noble gas and nitrogen compositions similar to those reported by the Viking spacecraft, have led to the contention that the SNC meteorites were derived from the planet Mars following a major impact event.

A dual heating paleointensity experiment on a subsample of the Shergotty meteorite revealed a low temperature remanence component acquired in a field of about 0.002 mT and a high temperature (greater than 362°C) component apparently acquired in a much weaker field (Table 1). The high temperature remanence is more likely to reflect the ambient field present when this meteorite was shocked and ejected from its parent body and is consistent with the weak surface field thought to exist on Mars.

Irons. Iron meteorites are capable of carrying directionally stable and AF-resistant NRM apparently due to the subdivision of their magnetic fabric by intergrown kamacite and taenite phases. However, Brecher and Albright (1977) demonstrated that the TRM acquired by iron meteorites is not directly proportional to the strength of the applied field and that the direction of remanence is mainly controlled by the crystallographic orientation of the intergrown phases. This finding, combined with their greater susceptibility to heating effects while entering the Earth's atmosphere, makes it unlikely that any reliable information concerning paleofields can be deduced from iron meteorites.

STANLEY M. CISOWSKI

References

Brecher, A., and L. Albright, 1977, The thermoremanence hypothesis and the origin of magnetization in iron meteorites, *Jour. Geomagnetism and Geoelectricity* **29**, 379-400.

Brecher, A., and M. Fuhrman, 1979, The magnetic effects of brecciation and shock in meteorites: II. The ureilites and evidence for strong nebular magnetic fields, *Moon and Planets* **20**, 251-263.

Brecher, A., and R. P. Ranganayaki, 1975, Paleomagnetic systematics of ordinary chondrites, *Earth Planetary Sci. Letters* **25**, 57-67.

Gose, W. A., and R. F. Butler, 1975, Magnetism of the Moon and meteorites, *Rev. Geophysics Space Physics* **13**, 189-193.

Herndon, J. M., M. W. Rowe, E. E. Larson, and D. E. Watson, 1972, Magnetism of meteorites: a review of Russian studies, *Meteoritics* **7**, 263-284.

Hood, L. L., and S. M. Cisowski, 1983, Paleomagnetism of the Moon and meteorites, *Rev. Geophysics Space Physics* **21**, 676-684.

Levy, E. H., and C. P. Sonett, 1978, Meteorite magnetism and early solar system magnetic fields, in T. Gehrels, ed., *Protostars and Planets*. Tucson, Ariz.: University of Arizona Press, 516-532.

Nagata, T., 1979, Meteorite magnetism and the early solar system magnetic field, *Physics Earth and Planetary Interiors* **20**, 324-341.

Nagata, T., 1980a, Magnetic classification of Antarctic meteorites, *11th Lunar Planet. Sci. Conf. Proc.* 1789-1799.

Nagata, T., 1980b, Paleomagnetism of Antarctic achondrites, *National Inst. Polar Research, Memoirs, Spec. Issue 17*, 233-242.

Ramdohr, P., 1973, *The Opaque Minerals in Stony Meteorites*. Amsterdam: Elsevier, 245p.

Sugiura, N., and D. W. Strangway, 1985, NRM directions around a centimeter-sized dark inclusion in Allende, 15th Lunar Planet. Sci. Conf. Proc., in *Jour. Geophys. Research* **90**, C729-C738.

Wasilewski, P. J., 1981, New magnetic results from Allende C3(V), *Physics Earth Planetary Interiors* **26**, 134-148.

Wasson, J. T., 1985, *Meteorites, Their Record of Early Solar-System History*. New York: W. H. Freeman, 267p.

Cross-references: *Chemical Remanent Magnetization (CRM); Curie Temperature; Demagnetization; Earth and Moon: Origins; Magnetic Properties of Minerals; Meteorites: Nature and Origin; Natural Remanent Magnetization (NRM); Rock Magnetism; Rock Magnetism: Measuring Techniques and Apparatus; Thermoremanence; Viscous Remanent Magnetization (VRM) and Viscous Remagnetization.*

PALEOMAGNETISM: OROGENIC BELTS

Paleomagnetic laboratory measurements are designed to define geomagnetic field vectors as functions of space and time. When paleomagnetism first developed as a technique in the 1950s, the principal use intended for these vectors was to trace the record of the geomagnetic field directions for a given continent. Granted the assumption of a geocentric coaxial dipole field for the geological past, these directions could then be translated into paleopole positions and apparent polar wander paths (APWPs) by holding the continent fixed in today's geographic reference frame. It took very little time, however, to recognize that the reciprocal use of the paleomagnetically determined vectors would yield many (and often more interesting) applications: Instead of holding the continent fixed and letting the vectors trace an apparent path, it rapidly became customary to prescribe the vector system as "fixed" while letting the continent trace a path as a function of time. The latter approach has the advantage that it corresponds better to our current conceptual framework of geodynamics, in which continental drift and lithospheric plate movements are described with respect to a stationary rotation axis and latitude grid. It remains a possibility that the rotation axis moves as well (true polar wander), but for Earth-bound comparisons such movement is immaterial since *relative* positions of points or areas on the globe remain unchanged during true polar wander.

One can expand on the above-mentioned idea of holding the vector system "fixed" by simply substituting the notion that the orientation of the vector (for a given point in space and for a given time) is precisely prescribed. In other words, one predicts the vector for a latitude and longitude pair at a particular geological time. If an observation deviates from this prediction, one can attribute the inconsistency to three possible origins: (1) the latitude-longitude pair or, more broadly, the spatial orientation and location of our starting point is incorrect; (2) the geological time used for the prediction is in error; or (3) the whole premise of the methodology, including the assumption of a geocentric coaxial dipole field and validity of the analytical laboratory techniques, is flawed. These three points form the very essence of paleomagnetic analysis. In general, the success of the paleomagnetic methodology in terms of independently and consistently documenting plate tectonic movements has shown that to first-order approximation (i.e., within 5° to 10°) possibility (3) usually can be ruled out. Thus, interpretations of paleomagnetic measurements generally center on points (1) and (2), that is, space and time.

This comparative predictor-observation technique can be applied at all scales, ranging from the dimensions of continents to those of microplates, allochthonous terranes, thrust sheets, fold limbs, slump folds, pebbles, or even individual grains. Paleomagnetism in orogenic belts is simply a description of the application of general paleomagnetic methodologies to problems in tectonically active zones with scales much smaller than that of a continent and generally larger than that of an outcrop. As already mentioned, the application involves a comparison between a set of observations and a predictor, which is based on a reference framework extrapolated from a neighboring craton or block, with space and time as the common variables. Statistical techniques are used to validate inferences drawn from this comparison and to render the conclusions statistically significant.

Methodology

Paleomagnetic techniques, when applied to rocks in orogenic belts, can yield solutions to well-defined questions ("equations") that have only a single variable. It is in principle impossible to use a paleomagnetic direction from a mountain belt to determine both the (unknown) age of magnetization and the displacement of the terrane in which the rocks are found. Although this fact appears to be a priori self-evident, examples can be found in the literature where age of magnetization and displacement (or rotation) are both unknown, and such studies obviously have a diminished value. Furthermore, because of the axial symmetry of the geomagnetic (dipole) field model, generally only north-south displacements can be resolved; east-west components of convergence or divergence cannot be determined. An exception to this rule is found when a sequence of paleopoles can be determined for two areas being compared, provided that there is independent evidence that the areas did not move with respect to one another *during* the acquisition of the associated magnetizations. An example of such a case is provided by the APWPs of cratonic Europe and North America during the late Carboniferous–early Cretaceous interval. Fitting of the two APWP segments for this interval results in a (paleomagnetically) rather precise determination of a subsequent longitudinal relative rotation of 36° ± 5° that occurred when the Atlantic Ocean opened. Despite the predominantly east-west direction of spreading, this amount accurately matches the (more precisely known) width of the North Atlantic obtained from continental coastline fitting (see *Paleomagnetism and Continental Drift: Historical Introduction* and *Seafloor Spreading: Magnetic Evidence*).

A single paleomagnetic pole determination, however, leaves the relative paleolongitude undetermined, so for ease of comparison paleomagneticians use the basic directional parameters, inclination and declination, more often than paleopoles. In essence, inclinations are used to derive paleolatitude comparisons (relative north-south displacements), and declinations are used for the assessment of relative rotations. A further advantage of this decomposition of the total magnetic vector into two angles is that for stratified rocks the inclination with respect to bedding is generally independent of subsequent tectonic deformation and rotation; thus in theory the inclination and declination angles can be treated as independent variables. In practice, measurement and orientation error, secular variation of the geomagnetic field (q.v.), and scatter of the magnetic directions due to imperfect recording of the paleofield result in a statistical distribution of directions in space that can only be treated in terms of independent inclinations and declinations to first approximation.

Because of the factors that produce a scatter in observations, statistical analysis is required to determine mean directions and associated confidence limits (see McElhinny, 1973). Derivations of such confidence limits, generally taken at the 95% probability level, can be obtained for parameters such as inclination (I), declination (D), paleolatitude (λ), and paleopole location. Given a radius of the cone of 95% confidence around a mean direction (α_{95}), the following relationships hold:

$$\Delta I = \alpha_{95}, \quad \Delta D = \sin^{-1}\left(\frac{\sin \alpha_{95}}{\cos I}\right),$$

$$\Delta \lambda = \tfrac{1}{2} \alpha_{95}(1 + 3 \sin^2 \lambda)$$

where Δ denotes the 95% confidence limit around a mean value, implying that statistically the true mean is thought with 95% probability to lie within the range of the mean plus or minus the confidence limit.

When paleomagnetic directions from a possibly displaced or rotated terrane are compared with those from a neighboring craton (Beck, 1980), the declination difference between an expected or reference direction (with subscript r) and an observed direction (subscript 0) indicates the amount of rotation (R), whereas the paleolatitude difference (L) indicates north-south relative displacement:

$$R = D_0 - D_r, \quad L = \lambda_0 - \lambda_r$$

Positive R indicates clockwise rotation, positive L southward displacement. The observed and expected directions each have confidence limits; thus the differences have confidence limits as well, given by

$$\Delta R = (\Delta D_0^2 + \Delta D_r^2)^{1/2}, \quad \Delta L = (\Delta \lambda_0^2 + \Delta \lambda_r^2)^{1/2}$$

$$\Delta \lambda_r = A_{95}, \quad \Delta \lambda_0 = \tfrac{1}{2} \alpha_{95}(1 + 3 \sin^2 \lambda_0)$$

where A_{95} is the cone of confidence about a mean pole, in contrast to α_{95}, which is the cone of confidence about a mean direction. Minor statistical refinements have recently been proposed by Demarest (1983).

Time Considerations

The realization that rocks can have magnetizations that are ancient yet many millions of years younger than the rocks themselves has increased dramatically (e.g., McCabe et al., 1983). Such remagnetizations can be caused by ancient diagenetic or chemical events, heating and burial episodes followed by cooling and uplift, or slow viscous acquisition during long periods of a single polarity of the geomagnetic field (see *Chemical Remanent Magnetization*; *Thermoremanence*; and *Viscous Remanent Magnetization and Viscous Remagnetization*). Because of the typically long periods during which the remagnetizations are acquired, they are generally represented by well-clustered directions that probably average out the effects of secular variation, dipole wobble (q.v.), or even long-term nondipole fields (see *Geomagnetic Field: Asymmetries*; *Geomagnetic Field: Elements*; *Geomagnetic Field: Measurement*; *Geomagnetic Field: Westward Drift*; *Geomagnetic Field, Main: Theory*; *Geomagnetic Field Analysis*).

At first thought to be a mere nuisance, these remagnetizations can also be used as a dating tool. The causative event can be dated, provided that sufficient field tests, such as the fold test, conglomerate test, or contact test (McElhinny, 1973), can be applied to constrain the age of magnetization, or if for the area studied an APWP of sufficient quality is already available so that the paleopoles of these remagnetizations can be compared with the reference APWP. It appears, for instance, that widespread remagnetization occurred within the North American craton during late Carboniferous–early Permian time, and this remagnetization has been speculatively related to widespread fluid migrations during the Alleghenian orogeny (Oliver, 1986).

In general, however, it cannot be stressed enough that if sufficient information about the age of magnetization is absent, then interpreters of paleomagnetic directions should be circumspect about the possibility of remagnetization before drawing conclusions about large-scale tectonics.

Microplates and Displaced Terrains

Present-day plate tectonics involves about eight major plates and a few microplates, such as the Juan de Fuca Plate west of the Oregon-Washington coast. Microplates should have, by definition, motion parameters that are different from those of adjacent plates, a thickness sufficient to comprise the crust and a lithospheric part of the upper mantle, and ideally a stable interior that is not deforming. Although not microplates today, Corsica and Sardinia are ideal examples of ancient microplates, since they moved independently from Africa and Europe when they rotated counterclockwise during the Miocene: They were of sufficient size to assume that lithospheric mantle moved with the rotating crustal blocks, and they have a fairly rigid Hercynian crust that shows Alpine and younger deformation only at rifts and along the former microplate margins.

Displaced terrains, on the other hand, may have been transported as part of larger plates rather than independently, their basal detachment may be inside the crust, and they are often internally deformed during displacement. A good example is the series of terranes collectively called Wrangellia in the western United States and Canada (Coney et al., 1980). These terranes display a similar lower Mesozoic stratigraphy, have similar paleomagnetic signatures, and have traveled a very large distance before being incorporated into the North American continent. One problem with the Wrangellia paleomagnetic data is that the polarity of the available results is unknown; thus the possibility of very large rotations precludes a determination of the sign of the paleolatitude (i.e., Northern or Southern Hemisphere). However, regardless of this ambiguity, the minimum displacement of Wrangellia is thought to be about 3000 km.

The Mediterranean-Himalaya-Indonesia chain and the western North American Cordilleran chain are the best-studied continental areas where microplates and displaced terranes have been documented, primarily through paleomagnetic techniques (Van der Voo and Channell, 1980). These techniques have been successful in the two chains for different reasons. In the Mediterranean, microplates have been documented through the use of declinations that show significant and mainly counterclockwise rotations in Spain, Portugal, Corsica, Sardinia, the Alps, and Italy, whereas the paleolatitudes do not deviate from those predicted by extrapolating from the European craton. In the American Cordillera, on the other hand, both declinations and inclinations have been found to deviate significantly from those predicted by extrapolation from the North American craton, indicating generally northward displacements as well as, for the most part, clockwise rotations.

Thrust Sheet and Other Rotations

At the margins of the cratons, thrust sheet movements may have been considerable, such as in the Laramide Rocky Mountains and Alleghenian Valley and Ridge overthrust belt of North America, and it is logical to expect that during such movement some rotation of the thrust sheets may have occurred. Paleomagnetic data indeed support this statement, but more interesting is the question of whether such rotations are in any way systematic. Eldredge et al. (1985) have examined the available paleomagnetic data from curved mountain belts in order to test S. Warren Carey's oroclinal model. As originally defined, an orocline is a now-curved mountain belt that was originally straight(er). Carey's model implicitly includes bending of deeper crust (and possibly even some lithospheric upper mantle) in addition to rotation of supracrustal rocks, but this cannot be tested paleomagnetically, nor is the model very compatible with what we know about the structure of the crust and mantle under overthrust belts. Paleozoic rotations in Hercynian Europe, on the other hand, may indeed have involved deeper crust or upper mantle and are clearly seen in the paleomagnetic declinations.

It appears that in many now-curved overthrust belts, systematic rotations have occurred that correlate with the structural trends, implying that these belts originally were straighter than they are today (Eldredge et al., 1985). For several belts, the mechanical cause of the paleomagnetically determined rotations has been sought in the effects of buttressing or indentation (e.g., Tapponnier et al., 1982).

Paleomagnetic Determinations of Structural Tilt

In the previous two sections, the rotations discussed were implicitly confined to movements within a horizontal plane; i.e., the rotations occurred about a vertical axis. When rotations occur about a horizontal axis, it is preferable to speak of structural tilt. Tilts can be documented just as readily by paleomagnetic comparisons between observations and a predictor as can rotations or latitudinal displacements. The problem is that, unless these tilts are assumed to be about a perfectly horizontal axis, the answers are not unique. In reality it is likely that tilts are often parts of plunging fold systems, i.e., settings in which rotations and tilts are combined.

Paleomagnetic determinations of tilt have been most useful in settings where stratified rocks were not available or where the original horizontality of the stratification was in doubt. Examples include studies of Basin and Range plutonic rocks of Jurassic-Cretaceous age, host to copper porphyry ores, where paleomagnetic directions have been observed that deviate by more than 90° from those expected for the area. These deviations have been interpreted as due to structural tilts of large magnitude (Geissman et al., 1982), an interpretation strengthened by the observed tilts of Tertiary ignimbrite flows in the area. Obviously, such a conclusion has important implications for ore exploration methods in the district.

In summary, paleomagnetic directions in vector space can be, and have been, used for many interesting applications that involve the tracing of movements (lithospheric plate motions, displacement of terranes, structural rotations or tilts) through geologic time. In addition, paleomagnetic directions may be overprinted on rocks because of chemical or thermal events; when there is a reliable set of reference directions available for the area of these rocks, the events can be dated paleomagnetically. As such, paleomagnetic studies are making important contributions to our knowledge of the world's orogenic belts.

ROB VAN DER VOO

References

Beck, M. E., 1980, Paleomagnetic record of plate-margin tectonic processes along the western edge of North America, *Jour. Geophys. Research* **85**, 7115-7131.

Coney, P. J., D. L. Jones, and J. W. H. Monger, 1980, Cordilleran suspect terranes, *Nature* **288**, 329-333.

Demarest, H. H., 1983, Error analysis for the determination of tectonic rotation from paleomagnetic data, *Jour. Geophys. Research* **88**, 4321-4328.

Eldredge, S., V. Bachtadse, and R. Van der Voo, 1985, Paleomagnetism and the orocline hypothesis, *Tectonophysics* **119**, 153-179.

Geissman, J. W., R. Van der Voo, and K. L. Howard, 1982, A paleomagnetic study of the structural deformation in the Yerington district, *Am. Jour. Sci.* **282**, 1042-1109.

McCabe, C., R. Van der Voo, D. R. Peacor, C. R. Scotese, and R. Freeman, 1983, Diagenetic magnetite carries ancient yet secondary magnetizations in some Paleozoic sedimentary carbonates, *Geology* **11,** 221–223.

McElhinny, M. W., 1973, *Paleomagnetism and Plate Tectonics.* Cambridge, U.K.: Cambridge University Press, 358 p.

Oliver, J., 1986, Fluids expelled tectonically from orogenic belts: their role in hydrocarbon migration and other geologic phenomena, *Geology* **14,** 99–102.

Tapponnier, P., G. Peltzer, A. Y. Le Dain, R. Armijo, and P. Cobbold, 1982, Propagating extrusion tectonics in Asia: new insights from simple experiments with plasticine, *Geology* **10,** 611–616.

Van der Voo, R., and J. E. T. Channell, 1980, Paleomagnetism in orogenic belts, *Rev. Geophysics Space Physics* **18,** 455–481.

Cross-references: *Accretionary Tectonics: Examples from the North American Cordillera; Far-Traveled Terranes; Geomagnetic Polarity Reversals: Observation; Geomagnetic Polarity Reversals: Theory and Models; Magnetostratigraphy; Natural Remanent Magnetization (NRM); Paleomagnetism and Continental Drift: Historical Introduction; Paleomagnetism and Plate Tectonics; Rock Magnetism; Seafloor Spreading: Magnetic Evidence; Thermoremanence; Viscous Remanent Magnetization (VRM) and Viscous Remagnetization.*

PALEOMAGNETISM AND CONTINENTAL DRIFT: HISTORICAL INTRODUCTION

Paleomagnetic studies (studies of the Earth's ancient magnetic field as recorded in geological deposits) were crucially important in reviving interest in continental drift and challenging orthodox geological theories in Britain and North America in the 1950s and early 1960s. Many who took up paleomagnetism in the 1950s and helped reshape its techniques and aims came to it from geophysics or physics rather than from more traditional geological specialties. Often these researchers had some familiarity with drift, read widely in the drift literature and interpreted their results as evidence for drift.

Paleomagnetic phenomena had been noticed in the nineteenth century in baked clays, lava flows, and even pottery. For example, it was observed that beds were magnetized roughly in accord with the Earth's magnetic field. This phenomenon was attributed to the alignment of iron in the molten lava with the Earth's magnetic field, which was locked in as the lava cooled. This magnetization seemed to be stable but some samples had an orientation reversed from the Earth's present poles.

Paul L. Mercanton, a meteorologist at the University of Lausanne, suggested in 1926 that paleomagnetic studies might bear on drift. He proposed comparing present and former locations of the Earth's poles by a worldwide survey that might "corroborate in an unexpected manner the hypothesis of great displacements of the Earth's axis of rotation defended by A. Wegener" (p. 190). His plea for this project to the International Union of Geodesy and Geophysics was unheeded. Paleomagnetism was a marginal specialty and drift was a highly controversial program. A combination of the two apparently had little appeal to most geologists. Three decades later, an alliance between paleomagnetic studies and drift was to be forged.

A Promising Past: The DTM Group

E. A. Johnson, an electrical engineer and physicist in the Department of Terrestrial Magnetism (DTM) at the Carnegie Institution, in the late 1930s developed a spinner magnetometer to measure weak magnetic fields. He and his co-workers believed that the device would further one of the major aims of the DTM, the collection of data on fluctuations in the geomagnetic field in order to elucidate its origin. The magnetometer was so sensitive that it could even be used on sedimentary rocks, which had a much weaker paleomagnetism than igneous rocks but which, unlike igneous rocks, could be dated by traditional geological methods. They applied their apparatus to sediments and varved clays, some of which dated back 20,000 years. They found some divergences from the current magnetic field but were uncertain of the stability of their samples. The DTM group were enthusiastic over these developments: perhaps one could reconstruct the Earth's magnetic history and use it as a method for dating rocks. The program was, however, put aside on the outbreak of World War II.

In the immediate postwar years, the Carnegie group, which included Johnson and J. W. Graham, renewed their studies but now with a specific aim: to evaluate conflicting theories of the origin of the Earth's magnetic field. The physicist W. M. Elsasser postulated an internal origin, i.e., that geomagnetism arose from processes inside the Earth. In the late 1940s the English geophysicist E. C. Bullard developed a sophisticated version according to which geomagnetism was produced in a manner analogous to a self-exciting dynamo by the formation and circulation in the Earth of thermal convection currents. The English physicist P. M. S. Blackett countered with a "fundamental" theory. In 1947 he became intrigued by observations of the rapidly rotating star 78 Virginius, which indicated that it had a magnetic field. He speculated that magnetism might be a property of all rotating matter: the geomagnetic field was generated simply by the Earth's rotation. Johnson argued that if the "fundamental" theory be correct, ancient magnetic poles determined by paleomagnetic studies should coincide with the present ones, assuming that the Earth's axis of rotation had not changed appreciably. His bid for paleomagnetism to be the arbiter between these theories gave his research agenda and his technical skills new direction and importance.

Johnson and his co-workers (1948) refined their instrument and tested samples dating back 1,000,000 years. There were only slight variations from the present field. He concluded that the magnetism of his samples was stable and, on their evidence, that the Earth's magnetism was generally stable. His results were consonant with a "fundamental" theory but he needed to trace further the prehistory of the Earth's magnetic field before drawing firm conclusions: after all, even a million years is an eyeblink in geological history. The farther back these measurements were pushed, the greater their importance for evaluating theories of the origin of the Earth's magnetic field but the more concern there might be about the validity of methods, techniques, instruments, and results. Paleomagnetics was far from being a "black box," that is, a set of procedures and instrumentation the principles of which are no longer controversial, which can be utilized by semiskilled technicians and the output of which is treated as unproblematic. Discussions of instruments and methods were common in many of the early papers reporting paleomagnetic results. Researchers in the 1940s and 1950s confronted many practical and theoretical difficulties. There was no accepted explanation of the Earth's magnetic field. The techniques of paleomagnetism were not well-developed; for example, there were no established sampling methods or agreed mathematical procedures for reducing results or indicating limits of error. Even for igneous rocks, there was no *prima facie* reason why stability should be presumed for rocks formed millions of years ago and folded, elevated, eroded, and altered in other ways. Might not reversely-magnetized rocks be evidence for the instability and unreliability of paleomagnetism itself?

Graham (1949) addressed the trustworthiness of paleomagnetism as a guide to Earth history in reporting measurements on sedimentary rocks dating back approximately 200 mya. He suggested two tests that were quickly accepted. One was applicable to folded strata: if rock magnetism be stable, the orientation of the strata should not coincide with the Earth's field but should more or less follow the contours of the folds. The other test was applicable to deposits that contained pebbles or other extraneous matter: if the magnetization of the strata be stable, the inclusions, which presumably had been formed earlier and possessed their own magnetization, should be randomly oriented with respect to the overall orientation of the strata.

Graham also proposed an additional aim for paleomagnetic studies which soon gave them pride of place in revitalizing drift. Not only might it be possible to trace the history of the geomagnetic field and discriminate between theories of geomagnetism, one could answer "questions regarding large-scale movements of the crust, such as continental drift and polar wandering" (p. 160). The DTM program, however, soon ran aground on technical and institutional reefs. The Carnegie group (and most other American paleomagnetists) began to focus on such problems as reversals and self-reversals and the nature of paleomagnetism rather than on its application to drift. Graham himself became increasingly concerned with and frustrated over the mechanisms of magnetization and reversals, matters which seemed of less relevance to the aims of the DTM. By the mid- to late 1950s, Graham's studies of magnetostriction in some types of rocks, conducted in association with J. R. Balsley and A. F. Buddington, led him to the view that little, if any, reliable information about the geomagnetic field in the past might be garnered from paleomagnetic studies. Only lengthy investigations of seemingly highly complex phenomena might determine if there were any relationship between rock magnetism measured in the laboratory and paleomagnetism; a negative answer was expected. This pessimistic view, which was shared by other U.S. researchers marked a redirection from the larger questions that had motivated the earlier work at the DTM: the Earth's magnetic history, the origin of geomagnetism, and drift. After Johnson's departure, Graham's position at the DTM became increasingly precarious: his superior, Merle Tuve, disapproved of the direction and rate of progress of Graham's research and discouraged him from pursuing it. Ironically, the larger program begun at the DTM was, in the hands of researchers in Britain, then beginning to bear intriguing fruit.

English Directionalists and Theories of Geomagnetism

A small group of physicists and geophysicists in England by the mid-1950s applied studies of rock magnetism to polar wandering and drift. Their approach stressed the determination of the direction—the location—of the ancient magnetic poles by measurements on rocks from different geologic periods and the treating of "normally" and "reversely" magnetized rocks as equivalent. These "directionalists," associated with the universities of Cambridge, London, and Newcastle-upon-Tyne, gained preeminence in this area. Their early claims on behalf of drift were greeted with skepticism. They then devoted much effort to refining and defending their techniques, gathering additional data and developing corroborating lines of evidence. By the early 1960s paleomagnetism had pushed drift back into the limelight. Their starting point was the same as the Carnegie group's: the competing Blackett and Elsasser-Bullard theories.

Blackett conceived a laboratory test for his theory: if a massive object is rotated, it should generate a measurable field, provided extraneous magnetic fields could be damped. By 1950 he had developed a suitable astatic magnetometer. His hopes were dashed: no measurable field was generated. His results agreed instead with the views of Bullard and S. K. Runcorn. Runcorn had studied geophysics at Cambridge under Bullard and then had gone to Blackett's department at Manchester as an assistant lecturer. Soon after his arrival, he heard Blackett lecture on his theory. Runcorn preferred Bullard's internal theory. He carried out his own experimental program involving measurements in mineshafts of changes in the Earth's magnetic field, which yielded results opposed to fundamental theories.

Blackett (1952) did not give up hope for his theory. The astatic magnetometer was "admirably suited to the measurement of very weakly magnetized geological specimens, for instance, certain sedimentary rocks" and he called attention to the work of the Carnegie group, which he described as being of "extreme interest." His enthusiasm was fired by the application of his device to elucidate the "prehistory" of the Earth's magnetic field. If ancient magnetic poles corresponded to the present ones (and roughly with the Earth's axis of rotation), this would be evidence for his fundamental theory. On the other hand, convection currents in the Earth would presumably be unstable over geological time and should give rise to apparent "wandering" of the magnetic poles or, perhaps, nondipolar fields. He was not oblivious to drift. About 1950 he began to read himself into geology and into the drift literature. He seems quickly to have become convinced that drift had occurred.

Blackett and his group soon applied his instrument to rock magnetism. Now based in London, they had found by 1953 that, contrary to previous studies, there seemed to be a systematic discrepancy between the modern poles and those calculated from British sedimentary rocks from Cambrian and more recent periods. This could have been simply explained had Blackett surrendered his fundamental theory. His group instead proposed that England had rotated some 34° and moved toward the North Pole (Clegg et al., 1954). This rotation did not necessarily entail drift, since it could reflect a general shifting of the crust about the poles, but data from other land masses were needed. They seem to have anticipated that their unexpected findings and startling conclusion would draw criticism. They dwelt on the design and use of the instrument and the reliability and stability of the rock samples. Such concerns are often a feature of the introduction of new techniques and instrumentation, especially when these are used to further relatively heterodox ideas. Much time and effort is focused on the techniques and instruments themselves before their output is accepted as neither artifacts nor "noise."

Runcorn and his group, located in the Department of Geodesy and Geophysics at Cambridge and later

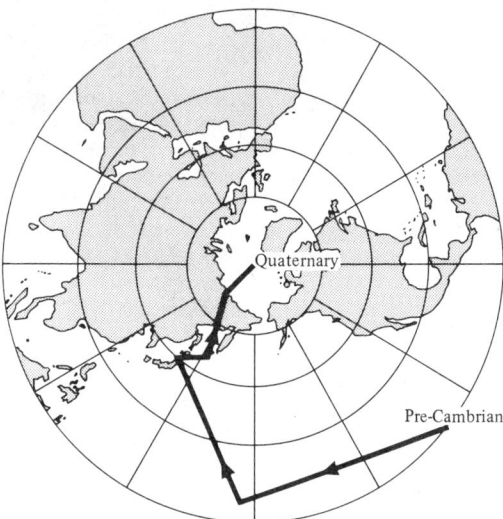

FIGURE 1. One of the first polar-wandering paths, constructed about 1954 by the Runcorn group. The orientation of samples from Britain dating from the Precambrian to the modern period were calculated, plotted, and connected in a "smooth" curve. The representation of the data in this form is striking and suggestive.

in the Physics Department at Newcastle-upon-Tyne, accepted Blackett's techniques and data but rejected his interpretation. His group built a magnetometer after Blackett's design and in 1951 started measuring rock magnetism, no doubt hoping to find support for an internal theory. They tested samples from major geological periods going back to the Precambrian. Their data were consistent with Blackett's. They shared his presumption that the magnetic poles were roughly aligned with the poles of rotation but there agreement ended. A northerly drift and clockwise rotation of England could not be ruled out, but they proposed (Creer et al., 1954) that the Earth's poles had moved with respect to land masses, not that the land masses had moved relative to one another, and they represented their data in graphic form as polar-wandering paths (see Fig. 1)

Runcorn, like Blackett, felt it essential to defend paleomagnetic techniques, not just his interpretation of the results they yielded. Both Blackett and Runcorn were in part responding to the contention of Louis Néel in 1951 that rock magnetism might be altered by chemical and physical processes. Néel, a French physicist who later became a Nobel laureate for his research on magnetic phenomena, had touched on a sore point: the phenomenon of reversed magnetism, which was problematic for both fundamental and internal theories of the Earth's magnetism, especially the former. A reversal of the Earth's field implied a reversal of the Earth's direction of rotation! If reversals were due to properties of rocks themselves, as Néel suggested, then how sure a guide was paleomagnetism to past geography? A Japanese group (Nagata et al., 1953) soon reported spontaneous reversals in pumice from an active volcano. This finding seemed to fit with one of Néel's mechanisms and to undercut the assumption of the stability of paleomagnetism central to the directionalists' agenda. Only a few other instances were soon reported but it was not until the 1960s that it was agreed that most reversals were to be explained by reversals of the Earth's field. In the interim, the phenomenon could constitute an objection to drawing sweeping conclusions from paleomagnetic studies. Jan Hospers, a Dutch geologist and geophysicist, also took issue with polar wandering data and interpretations. He had received the Ph.D. from Cambridge and was aware of possible pitfalls in the techniques. His work at Cambridge, which helped draw the attention of Runcorn and others to the importance of reversals, pertained to a study of Icelandic lavas containing bands of normally and reversely magnetized rock. He claimed that his samples gave a maximum polar wander of 5° or 10°—far less than that found by Runcorn in his studies. Despite his negative stance toward directionalists, his work contributed to their agenda. R. A. Fisher, the celebrated statistician, at Runcorn's urging had developed methods for the treatment of paleomagnetic measurements and these were first applied in Hospers's study. Furthermore, that Hospers's samples were reversed almost exactly 180° from the present poles supported the assumption of an axial dipole rather than a nonaxial or nondipolar geomagnetic field.

Runcorn in 1955 discussed possible mechanisms for rock magnetism, replied to arguments that it could be due to causes other than the Earth's magnetic field, and described Graham's tests for sample stability. All this preceded his remarks on the interpretation of results. He discounted drift as an explanation although he admitted that it could not yet be excluded. He also questioned the value of paleoclimatic data used as evidence for drift. He plainly expected further data from other locations to tell in favor of polar wandering. Later in 1955 he published results of studies he had begun in 1953 on North American sedimentary rocks. There was rough agreement between the poles he calculated for these samples and the curve for British rocks of the same approximate age: additional evidence for his interpretation and, inferentially, evidence for his model of geomagnetism and against Blackett's.

The Alliance of Paleomagnetism and Drift

Edward Irving at the Australian National University gave support to and charted new directions for the Blackett group. Irving had been an assistant to Runcorn and a postgraduate student at Cambridge before going to the new Department of Geophysics established by J. C. Jaeger at the A.N.U. Irving, despite his Cambridge connections, favored Blackett's interpretation. He thought that his own paleomagnetic studies pointed not only to relative motion between the continents and the magnetic poles but also to relative motion between the continents themselves, i.e., drift. In 1956 Irving spelled out clearly and concisely the major contentions of and the research agenda for paleomagnetic investigations relating to drift that were to be pursued by both the Runcorn and Blackett groups over the next few years.

Irving assumed a coincidence between the magnetic and the rotational axes. If this were correct, the magnetic latitude of a deposit on the Earth's surface would correspond to its ancient geographical latitude. The paleolatitudes so determined did not correspond to modern ones, indicating that large-scale crustal movement had taken place. Using the measurements reported by others as well as his own, he concluded that Europe, North America, India, and Australia all seemed to have undergone displacement with respect to the present poles. These movements could be depicted as polar wandering curves by plotting successive locations of the poles with respect to each continent. He developed two additional claims. First, paleoclimatic evidence should agree with paleomagnetic data. A more established specialty could thus be used to help legitimate a new one. For each of the four landmasses considered, there were found no major discrepancies between the paleomagnetic and the available paleoclimatic data. Second, the polar wandering path for each of these landmasses were inconsistent with those for the others and this constituted evidence for drift. If the path of the pole is determined from, for example, European samples, and it is assumed that the crust of the Earth had moved as a whole relative to the poles, this clashed with both the paleoclimatic and paleomagnetic evidence for the other regions. The continents had moved with respect to the magnetic (and rotational) poles but they had not moved as a unit—they had moved relative to one another. Some form of drift had taken place.

The London group pursued these same lines. They found, for example, that if a fundamental theory of geomagnetism is assumed, India had apparently moved north 40° and rotated about 25°. Since India's calculated pole positions did not coincide with those of Britain or America, it must have drifted relative to them. Blackett did not abandon his fundamental theory in response to these results. Instead, he officially shifted the aims of that research. He connected rock magnetism with drift, citing Graham on this point, and suggested that his own techniques were of value not only for the history of the Earth's magnetic field, but also "to settle the long-debated and highly controversial problems of continental drift and polar wandering." The measurements from England and India, taken together, eliminated explanations in terms of a simple shifting of the crust as a whole relative to the poles. Moreover, a northward movement of India of some 7,000 km, resulting in India's crushing into Asia and throwing up the Himalayas, was in accord with drift. Finally, the paleoclimatic arguments previously made for drift could be turned to advantage. If paleoclimatic and paleomagnetic data agreed, they would be mutually supporting, would support drift, and would not conflict with his fundamental theory nor with the more limited claim that the rotational and magnetic poles were always roughly parallel. From the mid-1950s, drift was the preferred view of Blackett's group.

Runcorn switched his position in 1956 and accepted Blackett's view that drift was needed. In a paper published early that year, he had reaffirmed his opinion that the discrepancies between polar positions given by American rocks and British rocks were minor and that both sets of data could be accounted for by polar wandering without drift. Now, on rechecking his data, he rejected polar wandering as a sufficient explanation. He found a small but systematic difference between the sets of polar wandering curves. The North Pole for America seemed to have shifted steadily to the west (see Fig. 2).

Given his assumptions that the Earth at any one time had only one set of magnetic poles and that these were aligned with the poles of rotation, he thought he could match up the polar wandering curves only by postulating that North America had been displaced about 24° since the Triassic (about 213 mya). He saw this as a powerful new argument for drift and warned that "the geophysical objections to continental drift must be re-examined."

Drift had become central to both the Blackett and the Runcorn groups. Paleomagnetism gave new evidence for large-scale horizontal motion; such traditional arguments for drift as paleoclimatic ones could serve to verify paleomagnetic conclusions. The audience was no longer limited to physicists and geophysicists. Runcorn's group swiftly established ascendancy in directionalist studies and their work by the early 1960s formed part of more general geological discussions.

Runcorn and his co-workers tried to dispel skepticism concerning their techniques, results, and revised interpretations. In a beautifully orchestrated series of six papers in the Royal Society's *Philosophical Transactions* for 1957, the soundness of each link in the chain from instrument to sample to interpretation was argued for. The first dealt with the design and theory of the astatic magnetometer, the techniques of its operation, and the mathematical treatment of the instrument readings. In the second, Irving and Runcorn discussed the theory of rock magnetism, especially the question raised by the phenomenon of reversed polarity. They claimed that mechanisms other than reversals of the geomagnetic field to explain reversely-magnetized rocks were complex and usually invoked special characteristics of particular sorts of rocks that had not been observed in the field. In contrast, the view that reversed rock magnetism was caused by reversed geomagnetism when the rocks were formed appealed to the same mechanism as for "normally" magnetized rocks and thus gave a "simple and general explanation." Irving in the third paper used this preferred mechanism to explain the origin of magnetism of the samples tested and concluded that their measured paleomagnetic orientation corresponded to that of the Earth at the time they were formed. The fourth and fifth papers discussed at an empirical level the problems of stability and reversals in specific samples. The final paper delivered the interpretation. Runcorn had originally summed up in favor of polar wandering without drift. However, after being convinced of drift, in the galley proofs he completely redrafted the conclusions to be drawn from the series. The published version interpreted the data in terms of multiple polar-wandering curves that coincided if large-scale drift were accepted.

Runcorn in 1959 addressed a more general audience in the pages of *Science*. He built a case for the value and reliability of rock magnetism, sketched the more striking findings of his group, and summarized the support given the drift interpretation of paleomagnetic results by paleoclimatology. His conclusion was unambiguous: drift had occurred. He also contended that it was occurring now: he cited

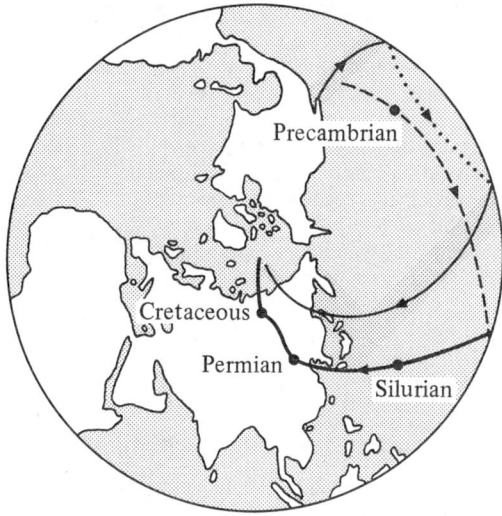

FIGURE 2. Polar-wandering curves for North America (solid and dashed line) and Europe (solid and dotted line). The curves, Runcorn argued, should coincide if the Earth at any one time had only one pair of poles. They could be made to coincide if it were assumed that the two continents had been joined before the Permian and had since drifted apart.

movement along the San Andreas Fault, which was of the same magnitude as demanded by his interpretation of paleomagnetic results. That there was no mechanism, excepting possibly convection currents, that explained these movements was irrelevant: they had occurred and were occurring even if geologists could not explain them.

Paleomagnetism, Drift, and the Geological Community

The Runcorn and Blackett approach of interpreting paleomagnetic results in terms of drift and using drift arguments to buttress their paleomagnetic work accelerated from 1956. The stream of publications relating paleomagnetism to drift conveyed the impression of a promising and progressing line of research. It not only attracted new recruits into the small community of directionalists and the attention of paleomagnetists working on other problems, but also stirred interest outside this subspecialty. Three symposia, in 1956, 1958, and 1960, illustrate different stages of reaction to the novel paleomagnetic evidence. Each of the three was explicitly concerned with a reexamination of drift. Each was stimulated in part by the directionalist studies of Blackett's and Runcorn's groups. Each generally regarded paleomagnetic studies as powerful evidence in favor of drift, with or without polar wandering. Each had a different composition and institutional setting.

Carey arranged a symposium on drift for 1956 at the University of Tasmania. Directionalist studies were seen to offer a way out of the impasse over drift although recognition of their potential was not unmixed with scepticism in respect of the interpretations offered by Runcorn and Blackett. Irving (1958, p. 53) concluded that although a case for drift had not yet been firmly established, "... there are many paleomagnetic data which are strikingly consistent with certain aspects of the drift hypothesis." In the opening address the venerable Chester Longwell had suggested that the paleomagnetic results could thus far be explained in terms of polar wandering without drift and expressed a critical attitude toward drift on other grounds. His view had altered slightly by the end of the symposium and he singled out paleomagnetism as significant (1958, p. 357): "... the divergence between ancient polar locations ... seems too consistent to be accidental. Readers will see much merit in Irving's analysis and will regard rock magnetism as a highly promising source of information on ancient geographies." He judged that the paleomagnetic data did not constitute "firm proof" of drift: more data must be gathered and the assumptions underpinning the techniques and conclusions scrutinized. The tenor of the symposium in respect to paleomagnetism could be described as guarded: it is promising but more work is required before the claims of its practitioners can be recognized. The printed *Proceedings* were, however, dominated by Carey's case for expansionism and this may have overshadowed for readers the discussions of paleomagnetism.

Alberta, poles away from Hobart, provided the setting in 1958 for a symposium on polar wandering and drift. Geologists and geophysicists from all over the world had been drawn to Alberta by the Canadian oil boom. Many joined the Alberta Society of Petroleum Geologists and this society sponsored the symposium. This symposium (Raasch, 1958), unlike its Hobart counterpart, was not an academic exercise. The aim was to gain familiarity with the work of the directionalists and to evaluate it with respect to the problems and issues of concern to the members of the Alberta Society, not to break new ground in either paleomagnetism or global theorizing. Few of the participants mentioned the economic aspects of polar wandering or drift but all were aware of their possible relevance to petroleum geology. Both polar wandering and drift had a direct bearing on paleoclimatology. A knowledge of ancient climatic zones would be of enormous use in oil exploration. The results of directionalist studies were greeted with enthusiasm; paleomagnetism "presents a new kind of evidence, capable of rendering a decision" on polar wandering and drift, which in turn bore on paleoclimatology.

Polar wandering and drift were featured at a symposium held as part of the 1960 meeting of the Society of Economic Paleontologists and Mineralogists. Interest centered on paleomagnetism and its interpretation in terms of drift and polar wandering. A. C. Munyan (1963, p.3), the organizer, captured this well: "Probably no other single line of investigation has done so much ... to advance the cause of continental shifting and polar wandering." Runcorn recapped the assumptions behind and difficulties inherent in paleomagnetism, including a fairly technical discussion of statistical methods. Most of the participants seem to have been less interested in paleomagnetism *per se* than in how their own repertoires of techniques and lines of evidence reinforced—and were reinforced by—the conclusions of Blackett and Runcorn. These ranged from traditional paleobotanical approaches to paleoclimatology and glaciology to the plotting of deep-focus earthquakes.

Paleomagnetism and Drift in the Early 1960s

By 1960 what had begun with Runcorn and Blackett as a technique to test physical theories had yielded results in respect to drift, which attracted the attention in other problem fields and specialties. The diffusion of the work of the directionalists into other specialties not only stimulated research on topics traditionally associated with drift, it also promoted the development of other techniques, methods, and research projects. Drift enjoyed a favorable press in the late 1950s and early 1960s and for this, as Munyan suggested, paleomagnetism was largely responsible. Nonetheless, there were still reservations not only about drift with or without polar wandering, but also about the techniques and interpretations of the directionalists. Most geologists, unless they were themselves specialists in paleomagnetism, were unlikely to have read the relevant papers or to have a first-hand familiarity with the new techniques, instrumentation, and so on. There was no direct, unequivocal demonstration that rock magnetism as measured in the laboratory provided a trustworthy, stable record of the Earth's magnetic field. Blackett, Runcorn, and most other workers in the area were not impressed by such objections. Naturally they had little sympathy with hints that their techniques might be faulty. One can conceive of complex mechanisms that might account for divergent magnetic orientations of a few rocks but to admit that these were widespread would be to undermine their own investment in paleomagnetic studies, which were predicated on an "imprinting" of the Earth's magnetic field as the cause of rock magnetism. The rationale for the program pursued by Allan Cox and Richard Doell bore precisely on this point: to collect enough samples from different locations and different geological eras to establish whether reversals were a general feature of the Earth's magnetic field or of the rocks themselves. Their approach was complemented by the experiments of P. M. Stott and F. D. Stacey in 1959 and 1960, which indicated that the concern of Graham and others over magnetostriction were groundless, at least for the rocks commonly used in paleomagnetic pole determinations. As for a demonstration of the validity of the paleomagnetic data and a drift interpretation of it, that was not possible in isolation but various lines of corroborating evidence from other specialties could be adduced—although these too were equivocal or inconclusive.

We can reconstruct choices for geologists: if one's confidence in paleomagnetic techniques were sufficiently strong (or if one were predisposed to drift on other grounds), then one could follow the argument leading from certain assumptions and data and conclude in favor of drift despite well-known objections. If one had little confidence in directionalist studies or if one were predisposed against drift or both, then it might be preferable to adopt complex mechanisms of rock magnetism or to challenge other assumptions or techniques of paleomagnetism and maintain the relative stability of continents. Or, one could adopt a wait-and-see attitude. The last option was probably the most popular. Directionalist studies did not cause a sudden, massive reversal of opinion about drift. They had, however, contributed significantly to breaking the impasse that had existed in the drift debate during the 1940s and early 1950s.

Allan Cox and Richard Doell's 1960 influential review article in the *G.S.A. Bulletin* reflects enthusiasm for the potential use of paleomagnetism in settling the question of drift, reservations about the data and interpretations offered thus far, and the desirability of more—and more reliable—data. Cox and Doell were not directionalists. Both had studied geophysics at Berkeley and were working under the auspices of the U.S. Geological Survey on a project to establish a time scale of geomagnetic reversals. This may have contributed to the influence of their review: they were experts in paleomagnetism but could be considered "hostile witnesses" on the question of the support given to drift by the directionalist approach. Their conclusions must be seen in the context of a general review of the whole field, not selected parts of it, and the global body of data collected, not selected measurements. Their concern was not to establish the merits of global theories but to give a sober, dispassionate defense of the techniques of paleomagnetism, upon which they themselves relied. In fact, Cox had come to the view as a graduate student that drift was probably true but thought that there was not much to be gained by openly endorsing it. Paleomagnetism was a reliable tool for the reconstruction of past geographies; misgivings on that score should be laid to rest.

They did not accept the case for drift in the Mesozoic and early Tertiary nor for drift or polar wandering in the late Tertiary and after. However, in their opinion the body of data cited by Runcorn and his group from the Carboniferous and Permian in Europe and North America was internally quite consistent and "... indicates a magnetic field configuration vastly different from the present. ... [And] these results constitute a strong case for polar wandering" (p. 760). The Australian data for these same periods were even more impressive: they "constitute evidence for a relative displacement of Australia with respect to North America and Europe which cannot be ignored" (p. 762). What was needed "to test properly the hypotheses of large-scale continental drift" (p. 763) in times more recent than the Permian was to eliminate ambiguities in the existing data by the collection of new samples from well-dated strata in a broader range of geographical locations. Only a year later, they expressed a more positive attitude toward the growing body of paleomagnetic evidence purportedly showing that drift had occurred: not only was there now "little doubt that the magnetic pole has wandered," if one made "reasonable assumptions about the Earth's field in the past, it is difficult to explain all of the presently available paleomagnetic data without invoking continental drift" (Doell and Cox, 1961, p. 302).

Directionalists continued to pursue their agenda, with some success, in the early 1960s. A month before Cox and Doell's review appeared, Blackett gave his own survey of the field. He focused not so much on the data or the techniques, but on alternative interpretations of the data. He ruled out as improbable processes other than the Earth's magnetic field as the source of rock magnetism: the more data amassed, the more improbable such alternatives became. He considered whether the Earth's crust might have shifted as a whole with respect to the magnetic poles but argued that polar wandering was insufficient since polar-wandering curves did not coincide unless drift was also invoked. Next, he took up the suggestion that the Earth's magnetic field had not always been an axial dipole field. Against this he offered both a theoretical argument; namely, that none of the extant theories of the Earth's magnetism are consistent with a nondipole field, and a phenomenological argument: to embrace a nondipole field is to accept that the present alignment of the magnetic and rotational poles is purely coincidental and atypical. By a process of exhaustion, drift was the most plausible theory for explicating rock magnetic data.

Blackett and Cox and Doell were not in complete agreement about the interpretation of data but the disagreement was one of degree. Paleomagnetism by the early 1960s was becoming a "black box" and perhaps this was a more remarkable achievement of paleomagicians than the revival of drift. Directionalist paleomagnetics was becoming a routinized, unproblematic set of instruments and techniques into one end of which samples were fed and from the other end pole positions were spewed out. This came about through a complex process of negotiation and the resolution of disagreements over a series of theoretical, empirical, and methodological issues, which ranged from geology into physics, chemistry, and mathematics. Consider such a minor technique as the "cleaning" of samples. By the late 1950s paleomagicians as a matter of routine exposed their samples to either heat or an alternating-current magnetic field. The purpose was to remove "extraneous" magnetization, which might have been acquired since the formation of the rock and which overlay or disguised its true, original, permanent magnetization. Practitioners might have every confidence that this technique did not alter the "original" magnetism; outsiders who did not have first-hand familiarity with the technique might well be dubious. Every step from sample collection to plotting poles could be analyzed in similar terms; every step involved a welter of empirical and theoretical assumptions. These were potential sources of divergent interpretation and disagreement. Paleomagnetists eventually gained acceptance for their instruments and techniques. Perhaps, in part, this acceptance was aided by practioners' prestige and credibility inside and outside the field. However, what formed the substance of the emerging consensus were the internal consistency of their results, consistency with results obtained by other methods in other fields, the provision of a coherent theoretical account of the instrumentation and techniques and their relation to the phenomena they supposedly measured, the provision of theories of geomagnetism and how these related to paleomagnetism, and, of course, the explication of their results in terms of a global theory. The construction of the paleomagnetics black box was complete when this consensus was established. This was a significant accomplishment for the agenda initiated just over a decade earlier; as Bullard later remarked: "The clarity which was finally achieved in the interpretation of paleomagnetism should not obscure the complexity and difficulty of the route by which it was attained ... the surprising thing is that by 1960 the case was substantially complete" (Bullard, 1975, p. 14).

By the early 1960s directionalist paleomagnetic studies had evolved from a minor problem-field on the fringes of geophysics and physics to an emergent specialty of recognized importance for the evaluation of rival global geological theories. Paleomagicians might accomplish what other specialists had been unable to accomplish: the breaking of the stalemate among drift and its rivals. Both Runcorn and Blackett were "outsiders" in respect to geological specialties and much of geophysics. They were not completely ignorant of the theories, data, methods, and problems central to other problem fields and specialties, nor were they unaware of widespread opposition to drift. Nonetheless, these matters seemed to have weighed far less heavily for them than the apparent value of drift in explicating their results—in solving their problems—within their problem field. However, they sought to publicize their work among geologists; to demonstrate that their techniques, results, and interpretations met some notional minimal standards in the problem field, the discipline, and other disciplines; to show that there was compatibility between the data and theories of their problem field and others, and to show that paleomagnetism and drift could be of use in addressing central problems of other specialties. Blackett, Runcorn, and their colleagues thereby broadened their arguments to assemble a more general case for drift. Had they failed to do so, their approach may have stagnated or been ignored.

The early relationship between paleomagnetic directionalist studies and drift was a symbiotic one. Drift offered a rationale. If drift were rejected, then the patterns, the divergent polar-wandering paths, elicited from the scatter of apparent ancient pole positions were meaningless; perhaps researchers would have turned to other phenomena or perhaps this field of investigation would have reverted to a marginal problem field of little consequence. Drift also served a heuristic role, i.e., it suggested ways in which the directionalists could develop their agenda. At a theoretical level, drift afforded an interpretation of diverging polar curves in terms of relative motions of land masses and motivated Runcorn's work on convection currents. At an empirical level, drift directed attention to the employment of paleoclimatic data as a cross check on and reinforcement of paleomagnetic data. At a "technique" level, drift offered guidelines as to appropriate locations and approximate dates of deposits that might be studied. Conversely, directionalist research provided fresh arguments for drift.

There were, of course, agendas in paleomagnetism other than that of the directionalists. The work of Néel and others on mechanisms of paleomagnetism and reversals formed part of more general investigations of magnetism. Cox and Doell's agenda of developing a geological time scale owed much to Cox's concern with geochronology and their familiarity with the young-rock dating research at Berkeley. Marine geologists collected data on seafloor magnetic anomalies as part of a general data-gathering exercise. Within a few years, some of these other strands of paleomagnetic research, especially reversal age-scales and seafloor magnetics, together with the new theories of seafloor spreading and plate tectonics, resulted in the "modern revolution in the earth sciences." It is a moot question whether directionalist studies alone or even in concert with paleoclimatology and paleobiography would have been sufficient to lead to widespread acceptance of drift. The line of research pioneered and sustained by Blackett, Runcorn, and their associates played a pivotal role among land-based geologists in bringing drift out of the doldrums in which it existed in the 1940s and 1950s. It sparked renewed interest in drift, which transcended the specialty in which they worked. It gave the impression that drift was progressing empirically and theoretically. However, it was not the directionalist studies alone, but a confluence in the mid-1960s of those with other theoretical and empirical developments that produced a fundamental reorientation and a global geology that incorporated drift.

H. E. Le GRAND

References

Blackett, P. M. S., 1952, A negative experiment relating to magnetism and the earth's rotation, *Royal Society Trans.* **A245,** 309-370.

Bullard, E. C., 1975, The emergence of plate tectonics: A personal view, *Ann. Rev. Earth Planetary Sci.* **3,** 1-30.

Clegg, J. A., M. Almond, and P. H. S. Stubbs, 1954, The remanent magnetism of some sedimentary rocks in Britain, *Philos. Mag.* ser. 7, **45,** 583-598.

Cox, A., and R. R. Doell, 1960, Review of paleomagnetism, *Geol. Soc. America Bull.* **71,** 645-768.

Creer, K. M., E. Irving, and S. K. Runcorn, 1954, The direction of the geomagnetic field in remote epochs in Great Britain, *Jour. Geomagnetism and Geoelectricity* **6,** 163-168.

Doell, R. R., and A. Cox, 1961, Paleomagnetism, *Adv. Geophys.* **8,** 221-313.

Graham, J. W., 1949, The stability and significance of magnetism in sedimentary rocks, *Jour. Geophys. Research* **54,** 131-167.

Irving, E., 1958, Rock magnetism: A new approach to the problems of polar wandering and continental drift, in S. W. Carey, ed., *Continental Drift: A Symposium*. Hobart: University of Tasmania Geology Department, 24-61.

Johnson, E. A., T. Murphy, and O. W. Torreson, 1948, Prehistory of the earth's magnetic field, *Jour. Geophys. Research* **43,** 349-372.

Longwell, C. R., 1958, Epilogue in S. W. Carey, ed., *Continental Drift: A Symposium*, Hobart: University of Tasmania Geology Department, 356-358.

Mercanton, P., 1926, Inversion de l'inclinaison magnétique terrestre aux ages géologiques, *Terrestrial Magnetism and Atmospheric Electricity* **31,** 187-190.

Munyan, A. C., ed., 1963, *Polar Wandering and Continental Drift*. Tulsa, Okla.: Society of Economic Paleontologists and Mineralogists Spec. Pub. No. 10.

Nagata, T., S. Akimoto, and S. Uyeda, 1953, Origin of reverse thermo-remanent magnetism of igneous rocks, *Nature* **172,** 630.

Néel, L. E. F., 1951, L'inversion de l'aimantation permanente des roches, *Ann. Géophysique* **7,** 90-102.

Raasch, G. O., ed., 1958, Polar wandering and continental drift, a symposium, *Alberta Soc. Petroleum Geologists Jour.* **6,** 139-178.

Bibliography

Glen, W., 1982, *The Road to Jaramillo: Critical Years of the Revolution in Earth Science*. Stanford, Calif.: Stanford University Press, 459p.

Le Grand, H. E., 1988, *Drifting Continents and Shifting Theories*. Cambridge: Cambridge University Press.

Takeuchi, H., S. Uyeda, and H. Kanamori, 1970, *Debate About the Earth*. San Francisco: Freeman Cooper & Co., 281p.

Wood, R. M., 1985, *The Dark Side of the Earth*. London: George Allen & Unwin, 246p.

Cross-references: *Curie Temperature; Demagnetization; Detrital Remanent Magnetization (DRM); Geomagnetic Field: Westward Drift; Geomagnetic Field, Main: Theory; Geomagnetic Polarity Reversals: Observations; Geomagnetism: Historical Introduction; Magnetic Properties of Minerals; Magnetic Self-Reversal; Magnetostratigraphy; Natural Remanent Magnetization (NRM); Paleomagnetism and Plate Tectonics; Rock Magnetism; Rock Magnetism: Measuring Techniques and Apparatus; Seafloor Spreading; Magnetic Evidence; Thermoremance; Viscous Remanent Magnetization (VRM) and Viscous Remagnetization.*

PALEOMAGNETISM AND PLATE TECTONICS

Between 1962 and 1968, a rapidly growing body of geological and geophysical data caused most earth scientists to accept dramatic changes in their model of how the Earth works and to switch from a *fixist* to a *mobilist* view of horizontal motions of the Earth's surface. Central to this revolution was evidence from paleomagnetism, which is the study of the ancient magnetic memory of rocks. In this article we attempt to sketch some of the history of this revolution, with an emphasis on paleomagnetic data. We then turn to some of the key elements of and later modifications to plate tectonic theory.

Hypotheses related to plate tectonics were known long before the 1960s. The most famous of these is the theory of continental drift proposed by Alfred Wegener in 1912. According to Wegener, the continents were once joined in a single supercontinent, termed *Pangaea*, and have been moving continually apart since Pangaea's breakup. In support of continental drift Wegener presented several powerful arguments including evidence that parts of the Earth known to be made of solid rock behaved like fluids over long intervals of time, evidence for fundamental differences between continental crust and oceanic crust, and the excellent fit of the coastlines of different continents (e.g., the well-known fit of the east edge of South America to the west edge of Africa as illustrated in Fig. 1). Wegener also argued that the reconstructed continents brought together provinces of distinctively similar geology or fossils.

Wegener's theory also had serious flaws, however. He envisioned the continents plowing through the oceanic crust, which many geophysicists thought unlikely because of the strength of the oceanic crust. He overestimated the rate of continental drift because he underestimated the age of Pangaea's breakup. He also tried to explain several phenomena that we now know are unexplainable by his theory. Perhaps most important, however, was that Wegener could not convincingly explain what force moved the continents. The lack of a reasonable driving mechanism caused many geophysicists to reject continental drift. Wegener's invocation of the far too weak pole-fleeing force (*Polfluchtkraft*) was inadequate. Although many earth scientists found his theory of great interest, almost no one thought that the data justified accepting it, and many strongly rejected it.

Paleomagnetism

Paleomagnetic data from the continents provided an important new test of continental drift that was independent of prior geological and geophysical data. Many rocks are reliable recorders of the direction of the magnetic field at the time a rock is formed. These magnetic directions can be determined by collecting oriented samples from outcrops of rocks and then measuring their magnetization in the laboratory using sensitive magnetometers. When averaged over tens of thousands of years, the magnetic field is to a good approximation that of a geocentric dipole aligned with the Earth's axis of rotation. Thus the time-averaged magnetic declination, which is the horizontal projection of the magnetic field (i.e., what a compass measures), points to true north. Moreover the time-averaged magnetic inclination, which is the angle the magnetic field makes with horizontal, varies predictably with latitude. The magnetic field points down at the North Pole, is horizontal at the equator, and points up at the South Pole (Fig. 2). The precise relationship between inclination of the magnetic field and latitude is given by the equation $\tan I = 2 \tan \lambda$ where I is magnetic inclination and λ is latitude. By simple calculation the location of the ancient North Pole or spin axis can be determined from the paleomagnetic declination and inclination. These ancient poles are termed *paleomagnetic poles*. Paleomagnetic poles can be determined from rocks of many different ages. A time sequence of paleomagnetic poles is termed an *apparent polar wander* (APW) *path* (Fig. 3).

Paleomagnetic investigations during the 1950s by British workers (P. M. S. Blackett, K. Runcorn, and E. Irving) showed that as the age of rocks being studied increased, the paleomagnetic poles determined from ancient rocks from different continents diverged from the present pole and from each other. Earth scientists were initially divided on the implications of these results. Some geophysicists believed the data provided strong support of the hypothesis of continental drift. Other geophysicists were more cautious and suggested that the magnetic field in the distant geologic past was more complex than the magnetic field over the past few tens of millions of years. A third group of geophysicists had an even more cautious point of view. They felt that invoking continental drift to explain the data introduced too many adjustable parameters, and therefore the whole field of paleomagnetism was of dubious value. In summary, paleomagnetic directional data provided new evidence favoring continental drift. These data were instrumental in persuading a significant minority of earth scientists to favor continental drift, but were unconvincing to most.

Another path of paleomagnetic research was the study of polarity reversals. Early this century scientists found that some rocks are naturally magnetized in a direction opposite to the present direction of the Earth's magnetic field. These observations can be explained by two key alternative hypotheses: (1) that some minerals become naturally magnetized in a direction opposite to the prevailing magnetic field, or (2) that the Earth's magnetic field used to point in a direction opposite to that of today. During the late 1950s and early 1960s the study of the cause of reversely magnetized rocks accelerated greatly, led by Allan Cox, Richard Doell, and Brent Dalrymple in California and by Ian McDougall and Don Tarling in Australia. The results of this work were two-fold: First, it was established by paleomagnetic study of the radiometrically dated rocks from sites widely distributed around the world that the principal cause of reversely magnetized rocks is that the Earth's past magnetic field often was of the polarity opposite of that today. The process by which the field flips from one polarity to another is termed a geomagnetic polarity *reversal*, a process that is thought to take about 4,000–10,000 years to occur. Second, the history of reversals over the past several million years was accurately determined (Fig. 4a), showing

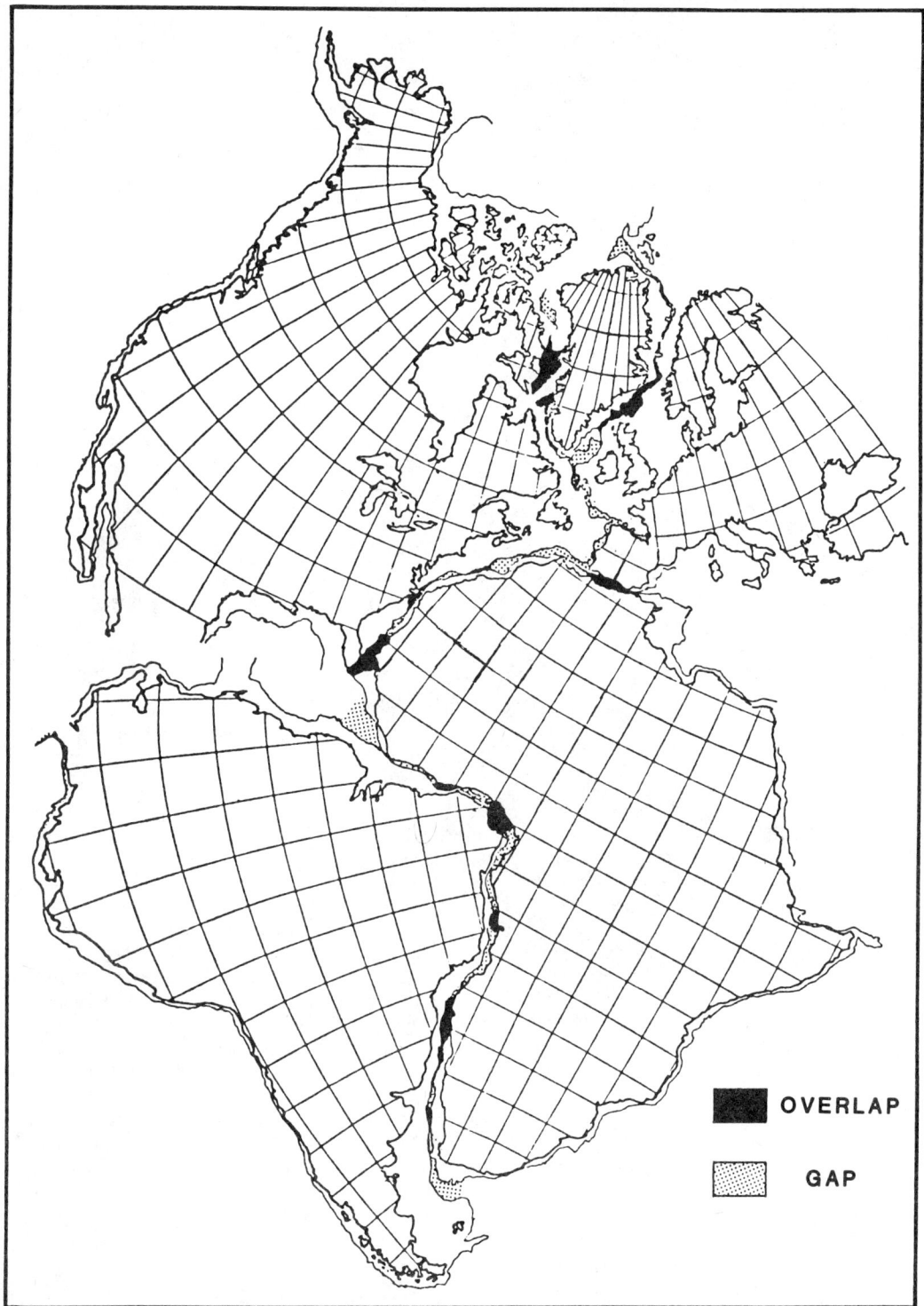

FIGURE 1. Reconstruction by E. Bullard, J. E. Everett, and A. G. Smith of the relative positions of the circum-Atlantic continents about 200 Ma. The fit is excellent but not perfect. The black areas show areas of overlap, and the stippled areas show small gaps in the reconstructions.

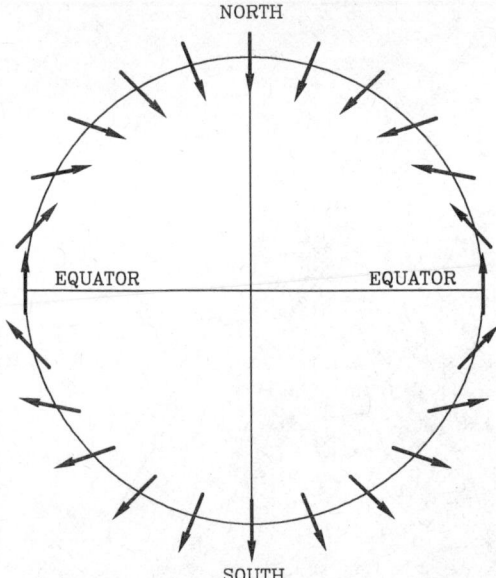

FIGURE 2. Inclination of the time-averaged geomagnetic field predicted by the axial geocentric dipole hypothesis during an interval of normal magnetic polarity, as has been the case over the past 730,000 years. This cartoon shows the Earth as it would be seen by a distant observer lying in the Earth's equatorial plane. Thus the Earth's north pole is at the top of the cartoon, and the south pole is at the bottom. The arrows illustrate the direction (but not the magnitude) of the time-averaged magnetic field at the surface of the Earth. Note that the field points straight down at the north pole, has a downward component throughout the northern hemisphere, is horizontal at the equator, has an upward component in the southern hemisphere, and straight-up at the south pole. During intervals of time when the geomagnetic field was reversed, the time-averaged field would be precisely opposite to the directions shown here.

that reversals have typically occurred at irregular intervals of ~200,000–2,000,000 years, although much shorter intervals are suspected and much longer intervals (~30 million years) are documented further back in the geologic past.

Seafloor Spreading

The proof of geomagnetic reversals and the accurate determination of their history appear unrelated to the question of continental drift. Surprisingly, however, these two discoveries proved to be essential tools in the demonstration of seafloor spreading and continental drift, and furthermore provided the accurate timing needed for a useful, quantitative theory of plate tectonics. Among the unusual features of mid-ocean ridges are the pronounced high in the intensity of the magnetic field over the ridge-axis and the flanking magnetic highs and lows (Fig. 4b), which form a pattern elongated

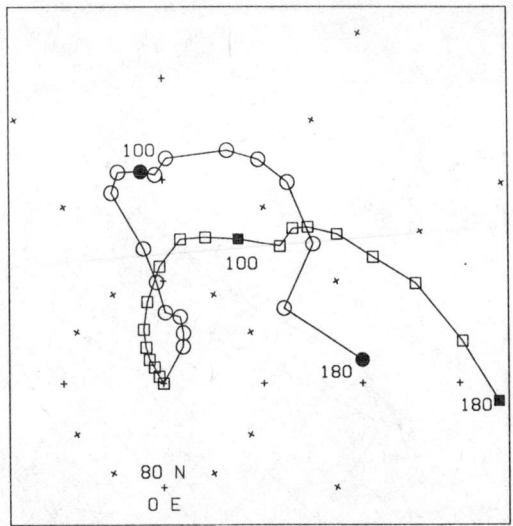

FIGURE 3. The observed apparent polar wander path (determined by C. G. A. Harrison and T. Lindh) of North America (circles) is compared to an apparent polar wander path of North America predicted if the hotspots have been fixed relative to the spin axis (squares). Poles are given every 10 my from 20 to 180 Ma. The 100 Ma and 180 Ma poles are solid; the rest are open. Predicted poles are determined from the North America-hotspot motion model of W. J. Morgan. The radii of the 95% confidence circles of the observed poles range from 3° to 10°. Differences between observed and predicted poles, when significant, imply motion of the hotspots relative to the spin axis. (From Gordon, 1987)

parallel to the ridge axis. These magnetic highs and lows are observed through magnetic measurements near the sea surface made from ships and low-flying airplanes. In 1963 Fred Vine and Drummond Matthews, and (independently) Lawrence Morley proposed a fantastic explanation for the origin of these magnetic highs and lows. Their hypothesis can be considered to be a corollary of two hypotheses: (1) The geomagnetic field reverses polarity and (2) as proposed by Harry Hess and Robert Dietz, the seafloor spreads, new material wells up along the axis of a mid-ocean ridge, and then is transported laterally away to make room for more seafloor to be created. (A similar model had earlier been proposed by Arthur Holmes. His model differed in one important way: seafloor spreading was distributed over a wide area, unlike the model of Hess and Dietz who proposed that spreading occurred in a narrow zone at the axis of the mid-ocean ridge.)

Thus new seafloor created at the ridge axis since the most recent geomagnetic reversal is *normally* magnetized (i.e., parallel to the present magnetic field), whereas seafloor created before the most recent reversal (but after the reversal preceding it) is *reversely* magnetized (i.e., antiparallel to the present

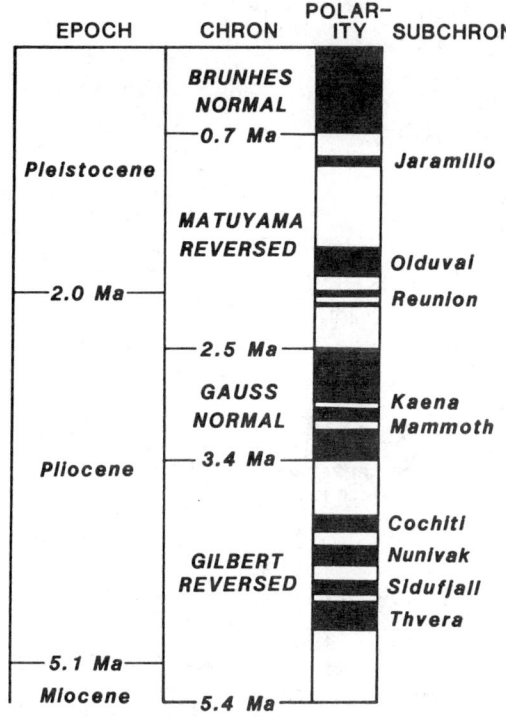

FIGURE 4a. Geomagnetic reversal time scale determined from potassium-argon dating and paleomagnetic analysis of oriented rocks collected on land. The ancient geomagnetic field was of normal polarity (i.e., the same polarity as today) during the intervals shown in black, and of reversed polarity (i.e., the opposite of today) during intervals shown in white. Polarity intervals vary greatly in length. The basic unit for describing a time interval of dominantly one magnetic polarity is termed a *polarity chron*, which in practice generally refers to intervals with lengths ranging from hundreds of thousands of years to millions of years. Shorter intervals, which may occur entirely within chrons, are termed *polarity subchrons*, and typically have lengths of about ten thousand years to hundreds of thousands of years. For example, here the Jaramillo, Olduvai, and Reunion subchrons are brief intervals when the geomagnetic field was of normal polarity that occur within the Matuyama revised polarity chron, during which the geomagnetic field was dominantly of reversed polarity. (After Cox and Hart, 1986)

magnetic field). With further increases in the age of the seafloor, alternating bands of normally and reversely magnetized seafloor are predicted by their model (Fig. 4b). Their hypothesis also predicts that for many latitudes and orientations of spreading ridges, the magnetic highs and lows are bilaterally symmetric or nearly symmetric with respect to the axis of the mid-ocean ridge (Fig. 4b). Normally magnetized oceanic crust usually adds to the present magnetic field causing a magnetic high, whereas reversely magnetized oceanic crust usually subtracts from the present magnetic field causing a magnetic low. The resulting pattern of highs and lows, termed *anomalies*, can be compared to the geomagnetic polarity reversal time scale determined independently from dated rock samples on land (Fig. 4a). If the rate of seafloor spreading is constant, then the distance of any mapped anomaly from the axis of the ridgecrest can be equated with seafloor age, and the hypothesis can be rigorously tested (Fig. 4a and b).

Observed magnetic anomalies across spreading ridges and refinements in the reversal time scale soon provided convincing confirmation of seafloor spreading at rates that are approximately constant in time at any one site, but vary around the world from ~10–160 mm/year. Linear patterns of magnetic anomalies occur over most of the Earth's seafloor and careful correlation of the pattern worldwide, combined with age data obtained by drilling into the seafloor, has permitted most of the Earth beneath the sea to be accurately dated with ages as old as ~170 Ma. Magnetism and the dating of the seafloor allows the reconstruction of past positions of continents. For example, we know that in the past India and Asia were separate continents, that India later collided with Asia, and seems to be continuing to collide today, causing the formation and continuing uplift of the Himalayan Mountains.

Plate Tectonics

The key idea connecting seafloor spreading to plate tectonics was J. Tuzo Wilson's concept of transform faults. Wilson noted that the Earth's main zones of deformation or mobile belts take three forms: mountains (including deep sea trenches), mid-ocean ridges, and strike-slip faults with large horizontal displacements. He noted the puzzling tendency of an individual zone of deformation to end abruptly when it is traced along its longest horizontal dimension. He proposed that these individual features are not isolated, but together form a continuous network that divide the Earth's surface into several large rigid plates. The abrupt ends of individual features are really the points where one type of mobile belt is transformed into another. A key transformation occurs at the ends of the strike slip faults, termed *transform faults*, that connect separate segments of the mid-ocean ridge. At these intersections the horizontal slip or shear of the transform fault is changed into an extension across the mid-ocean ridge segment, where the seafloor spreads.

We now recognize about a dozen rigid plates divided by narrow boundaries where horizontal deformation is concentrated (Fig. 5). Most plate boundaries are neatly marked by seismicity (Fig. 6; cf. Fig. 5). The plates are composed of the strong upper layer of the mantle and the overlying layer of crust, which together are termed the *lithosphere*.

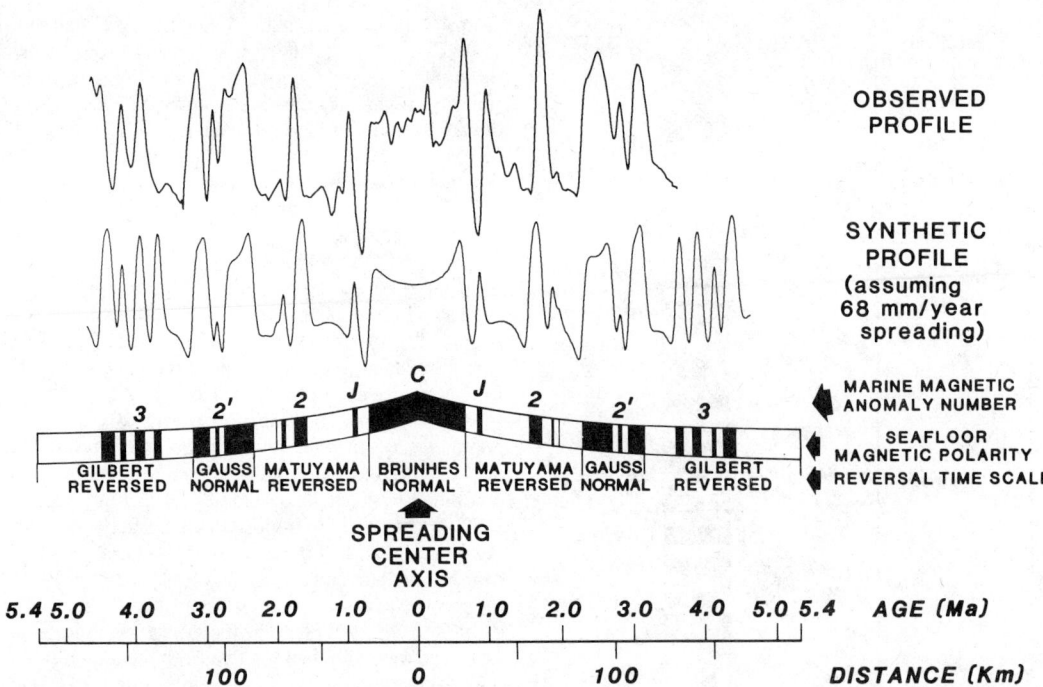

FIGURE 4b. The link between magnetic anomalies observed on profiles crossing seafloor spreading centers and the geomagnetic polarity time scale determined from paleomagnetic studies of radiometrically dated rock samples on land (Fig. 4a). At the top is an observed magnetic anomaly profile that crosses the Southeast Indian Ridge in the southeast Indian Ocean between Tasmania and Antarctica. The ups and downs of the profile reflect increases and decreases in the intensity of the magnetic field measured by a proton precession magnetometer towed behind an oceanographic research vessel. The variation in field strength amounts to only about 1% of the total field intensity. Just below the observed profile is a synthetic magnetic anomaly profile computed from the geomagnetic reversal time scale shown in Fig. 4a assuming a constant rate of seafloor spreading of 68 mm per year, which is equivalent to 68 km per million years. In general, magnetic "highs" are predicted to occur over seafloor that is normally magnetized and magnetic "lows" are predicted to occur over seafloor that is reversely magnetized. The agreement between predicted and observed anomalies is good, although the fit could be slightly improved by varying the spreading rate slightly as a function of age. Beneath the synthetic anomaly is a cartoon cross section of the seafloor and its assumed magnetic polarity. The spreading center axis is an axis of bilateral symmetry; seafloor age at the axis is 0 and increases monotonically in both directions with increasing distance from the spreading center axis. Portions of the seafloor colored black are normally magnetized and white regions are reversely magnetized. The portions of the seafloor formed during named magnetic chrons are shown and are intended to be easy to compare to the geomagnetic reversal time scale shown in Fig. 4a.

Beneath the lithosphere is the softer, more easily deformable layer of the mantle termed the *asthenosphere* (Fig. 7). The assumption that the plates are rigid provides a simple, yet powerful, quantitative theory of surface tectonics, first worked out for flat plates on a plane by Wilson, and given greater rigor and extended to a spherical Earth by Dan McKenzie, Robert Parker, and Jason Morgan. The motion between plates is typically determined from three types of data: plate separation rates from magnetic anomalies across spreading ridges, direction of plate motion shown by the azimuths of transform faults dividing plates, and the orientation of slip vectors calculated from ground motions caused by earthquakes along plate boundaries. The geometrical precision and rigor provided by the assumption of plate rigidity, combined with the accurate timing provided by magnetic anomalies and precisely dated magnetic reversals, allow horizontal rates of motion to be determined accurately over most of the Earth's surface.

As was recognized from its inception, plate tectonics is not the ultimate tectonic theory. Several workers have instead compared it to the simplicity, elegance, and explanatory power of the Bohr model of the atom, which was an important stepping stone in the development of the modern theory of quantum physics. Plate tectonics explains why the seafloor is much younger than the continents, why earthquakes occur in narrow belts and are in some places confined to depths shallower than a few tens of kilometers, and in others extend to depths of many hundreds of

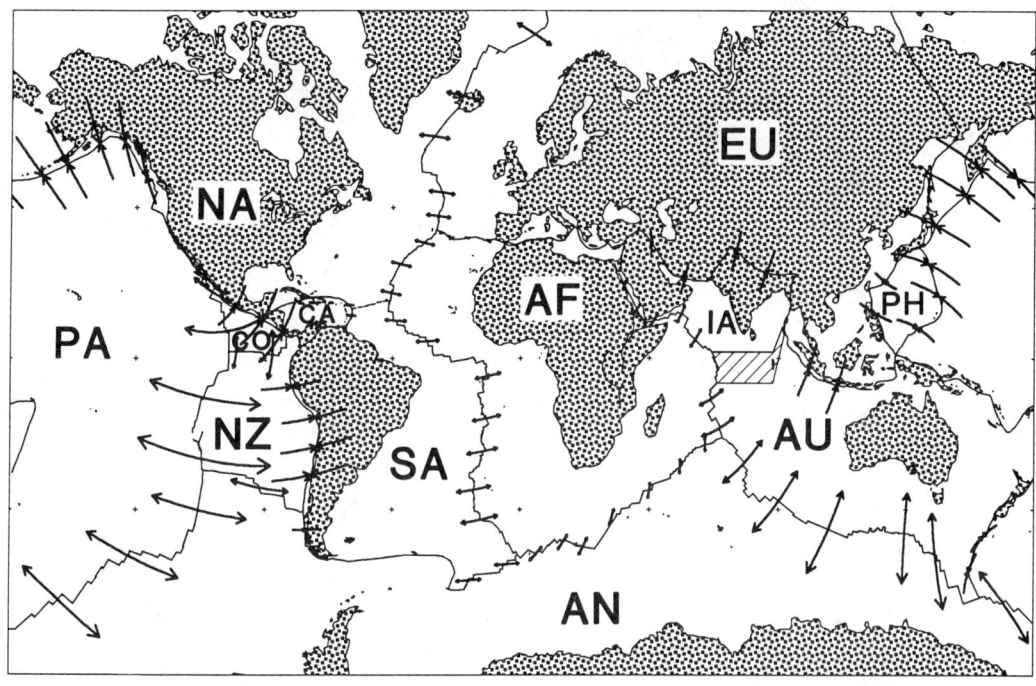

FIGURE 5. Current plate motions from a preliminary version of the NUVEL-1 plate motion model of C. DeMets, R. G. Gordon, D. F. Argus, and S. Stein. The length of each arrow is proportional to relative plate speed. Black lines are used to show the interconnected network of plate boundaries that divide the following plates: North America (NA), South America (SA), Africa (AF), Eurasia (EU), Australia (AU), Antarctica (AN), Pacific (PA), Philippine (PH), Nazca (NZ), Cocos (CO), Caribbean (CA), and Indo-Arabia (IA). (In the final version of the NUVEL-1 model, India and Arabia were treated as separate plates with slow relative motion of 2 mm/yr.) Inward facing pairs of arrows show plate convergence, usually at subduction zones associated with deep sea trenches (Fig. 7), but also at sites of active continent-continent collision and mountain building, such as occurs in the Himalayan Mountains. Outward-facing arrows show plate divergence, usually along the world's mid-ocean ridge system. The region with diagonal rules located between the Indo-Arabian and Australian plates is interpreted as a diffuse plate boundary (see Fig. 10 for details).

kilometers. It explains why faults slip in the direction they do, how the great fracture zones and rifts of the seafloor formed, and why the deep sea trenches occur where they do. It also explains why many volcanos occur in narrow belts and why many of the belts are concentrated around the rim of the Pacific. It provides the basis for understanding the major variations of depth and heat flowing out of the seafloor, and the basis for understanding the distribution and cause of the Earth's major mountain belts.

Also, like the Bohr atom, it has some clear limitations. Although the theory works well in describing deformation in the oceanic lithosphere, it works less well in describing deformation within the continents. When plate boundaries cross continents, they splay out into broad zones of deformation hundreds or thousands of kilometers wide (See Fig. 6, in particular the western United States and the seismic zones north of the Alpine-Himalayan belt in Eurasia). In contrast, the zone of significant deformation at submarine spreading ridges is thought to be 20 km or less in width. Similarly the zone of active faulting on submarine transform faults is thought to be a few km or less in width, consistent with the narrow submarine seismic zones shown in Fig. 6. Plate tectonics is nevertheless pivotal in understanding the tectonics of continents: plate motion models provide the boundary conditions that distributed deformation within the continents must satisfy.

Refinements to the Seafloor Spreading Model

Asymmetric and Oblique Spreading. Although plate tectonics does not require seafloor spreading to be symmetric (i.e., crust is produced at equal rates on both sides of a spreading ridge) or orthogonal (i.e., ridges and magnetic lineations are orthogonal to transform faults and the direction of plate motion), they typically are nearly symmetric and nearly orthogonal. In some places, however, anomaly patterns are asymmetric about the ridge axis, showing that crust is created at unequal rates on the

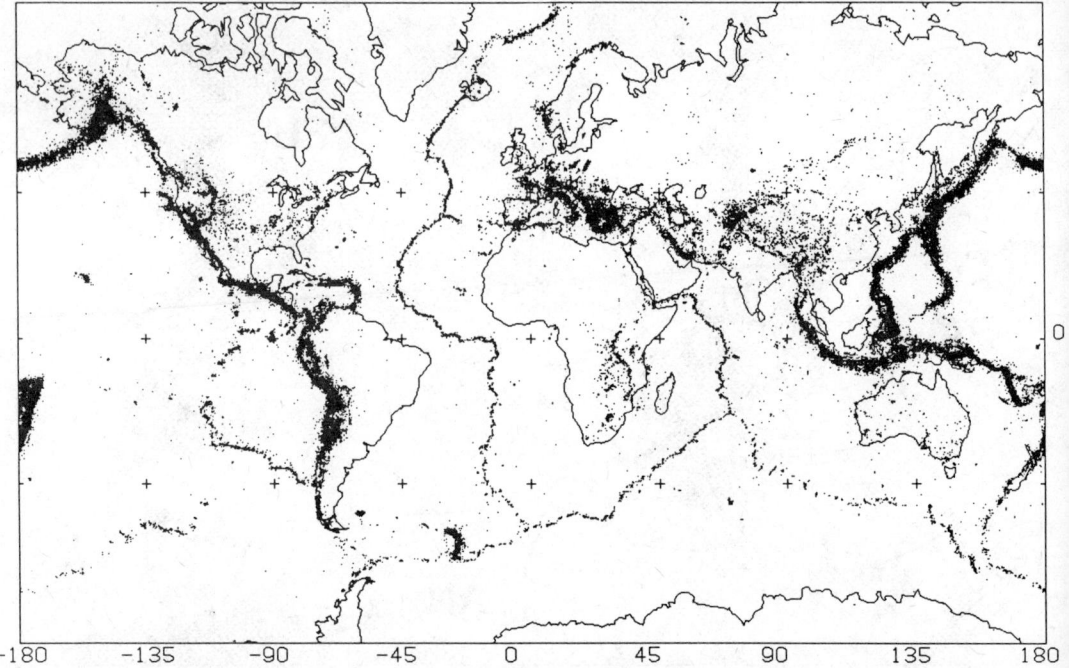

FIGURE 6. The worldwide distribution of earthquakes from 1963 to 1987 that occurred at depths of 100 km or less. Each black dot represents an earthquake epicenter from the earthquake data file of the National Geophysical Data Center. The belts of frequent earthquakes coincide closely with the plate boundaries shown in Fig. 5, which was made using exactly the same projection as this figure. Earthquake belts along the midocean ridge system are narrow, whereas plate boundaries that pass through continents, such as in the western United States, tend to have wide zones of earthquakes. (Figure courtesy of C. DeMets.)

two sides of the spreading center, a process termed *asymmetric spreading* (Fig. 8). It has also been observed that some anomalies, while forming parallel to ridges, are not perpendicular to the direction of relative motion (Fig. 8). This phenomenon, termed *oblique spreading*, is observed only for spreading slower than about 30 mm/year.

Propagating Rifts. Analysis of isolated magnetic profiles across spreading centers in some places suggests that part of the anomaly sequence is missing from one side of the ridge crest, but is duplicated on the other. The first, and simplest, explanation for this observation is that the spreading center "jumped," i.e., a new seafloor spreading rift formed while the old rift ceased spreading. However, detailed studies of adjacent closely spaced magnetic profiles across spreading centers have shown that the age when the spreading center jumped is diachronous. Thus the new rift did not form everywhere simultaneously, but increased its length with time. These newly forming rifts, termed *propagating rifts* by Richard Hey, who first recognized them, form adjacent to a preexisting spreading center. The existence of propagating rifts suggests that the configuration of some plate boundaries varies with time, possibly in response to changing conditions (e.g., change in direction of plate motion) that favor a different location or different orientation of a spreading center.

A new rift can grow in several ways. In one limiting case, once formed it would instantaneously spread at the full separation rate of the plate-pair it divides. The old rift would immediately cease spreading and the two rifts would not coexist (Fig. 9a). Alternatively, as shown by magnetic lineations and seismicity in several regions, some time interval may be needed for the new ridge to attain the full separation rate between the two plates. During this interval two spreading centers coexist, the sums of their spreading rates equal the total separation rate of the major plates they divide. Between the two coexisting spreading centers lies a tectonically independent overlap region, which may be a rigid microplate (Fig. 9b), or may instead be deformed by extension, shear, or both.

Modifications to Plate Tectonics

Deformation of Lithospheric Plates. It is important to recognize that the key plate tectonic assumption of rigidity only applies to horizontal motions. It is widely recognized that there are significant widespread vertical motions of the lithosphere

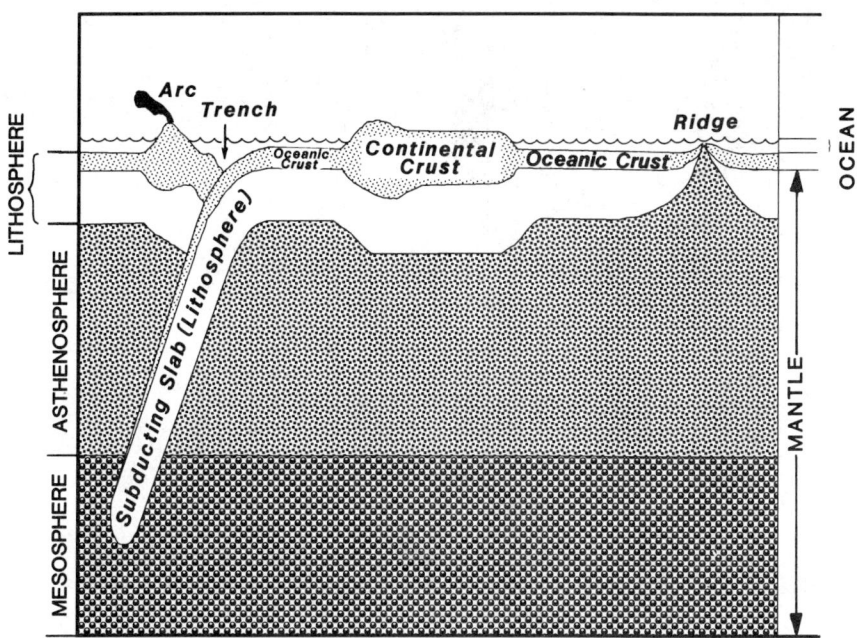

FIGURE 7. Cartoon cross section through the Earth showing plate tectonic layering, and contrasting with an older scheme of layering based on discontinuities in seismic velocities. Plate tectonic layering is labeled on the left edge of the figure, whereas simplified seismic velocity layering is labeled on the right edge of the figure. Rigid, cold lithosphere slides over the soft asthenosphere until it encounters a trench, where it sinks or subducts. The continent does not plow through the oceanic crust, as in the model of continental drift, but rides passively with the oceanic crust and subcrustal lithosphere as part of a single plate. (After Cox and Hart, 1986)

including the subsidence with age of the lithosphere away from the spreading ridges, uplift associated with large volcanos, uplift of broader regions that have experienced episodes of high heat-flow, and subsidence and uplift from flexure associated with loading of the lithosphere by seamounts, sedimentary wedges, etc. Thus, there is much evidence that the lithosphere is deforming.

There is evidence, moreover, for horizontal deformation of the lithosphere other than at conventionally defined plate boundaries. Maps of shallow global seismicity, such as Fig. 6, show that many

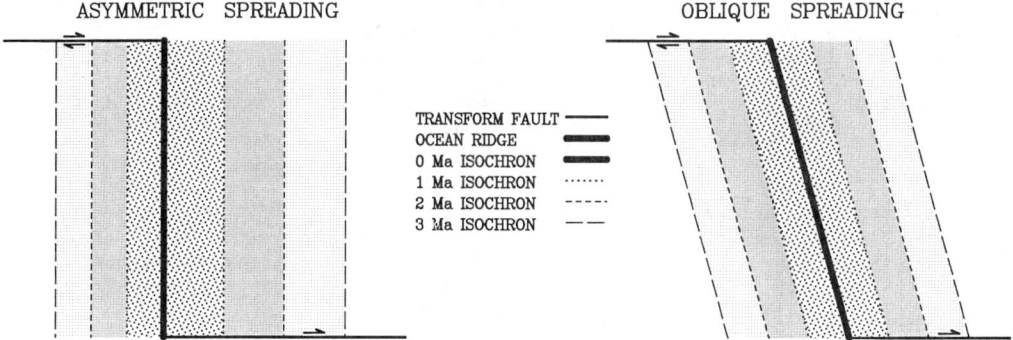

FIGURE 8. Two departures from orthogonal, symmetric seafloor spreading that have been observed in many places on the Earth. On the left panel, spreading is orthogonal (the strike of the spreading ridge and magnetic isochrons is orthogonal to the strike of the transform faults, which are parallel to the direction of relative plate motion) but asymmetric: more seafloor is accreted to the plate to the right of the spreading center than on the plate to the left of the spreading center. On the right, spreading is symmetric, but oblique (i.e., not orthogonal): the spreading ridge intersects the transform fault at an angle greater than 90°. Isochrons, and the magnetic lineations that mark them, parallel the spreading ridge and are also not at right angles to the transform fault.

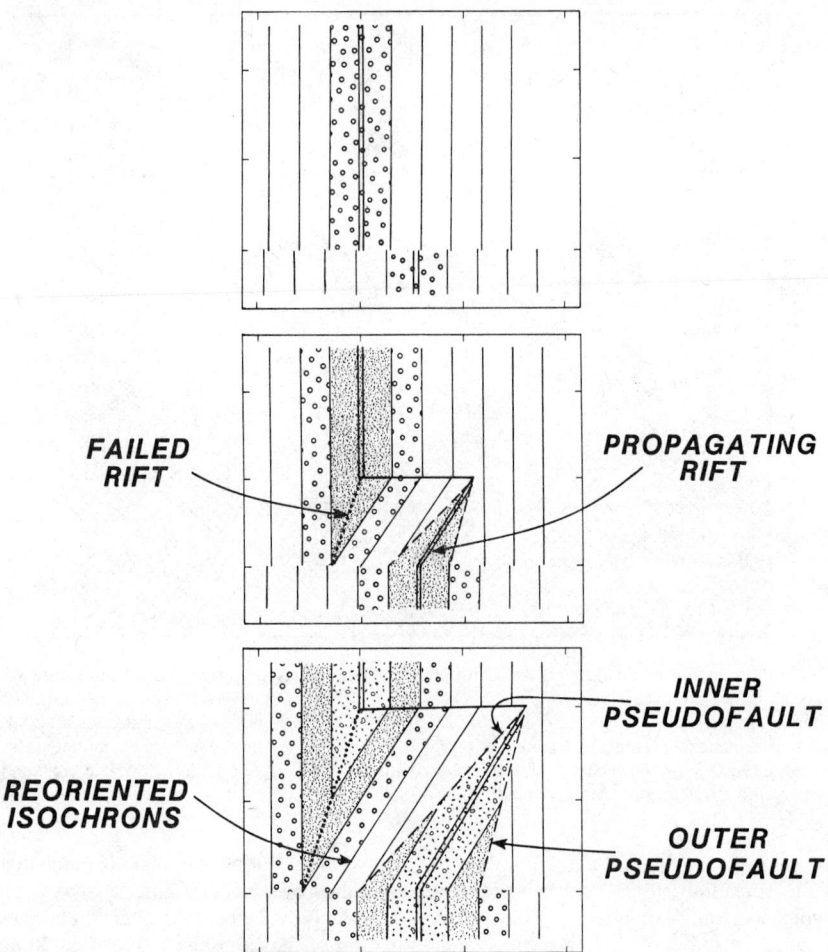

FIGURE 9a. Schematic evolution of the geometry of a propagating rift where the old (failing) rift ceases spreading immediately. In the earliest time frame (top) is shown two spreading center segments offset by a transform fault. Seafloor formed during the prior one million years is shown by a pattern of small open circles. In the next time frame (center), one million years later, the spreading center in the bottom of the panel has increased in length. In this model the dying rift ceases spreading as soon as the propagating rift grows enough to overlap it. Thus seafloor originally moving with the plate to the right is immediately shifted to the plate to the left. In the bottom panel, after an additional one million years, the propagating rift has grown further at the expense of the dying rift. Pseudofaults, which mark a discontinuity in seafloor age, are the boundaries between the seafloor formed at the old rift and seafloor formed at the new (propagating) rift. Isochrons located between the failed and propagating rift change orientation relative to their original strike.

earthquakes occur within the interiors of supposedly rigid plates (although most earthquakes occur on plate boundaries). This observation violates the rigid plate assumption underlying plate tectonics, and it has suggested some modifications to plate tectonic theory including the following: (1) The rigid-plate hypothesis may be only approximately valid and plate interiors deform pervasively and slowly with perhaps some concentration of deformation on several old zones of weakness. Seismicity distributed throughout a plate would suggest pervasive intraplate deformation. (2) Alternatively, motion between plates may not always be confined to a narrow boundary zone and in some places may be taken up in wide zones (up to 1000 km or more). In contrast to pervasive intraplate deformation, large areas of lithosphere outside the plate boundary zone would be aseismic and (nearly) rigid.

For example, it seems reasonable to treat the relatively aseismic lithosphere between the Mid-Atlantic Ridge and the Wasatch Mountains of Utah as part of the North American plate. Any deformation across it would reasonably be termed *intraplate deformation*. In contrast, the portion of North

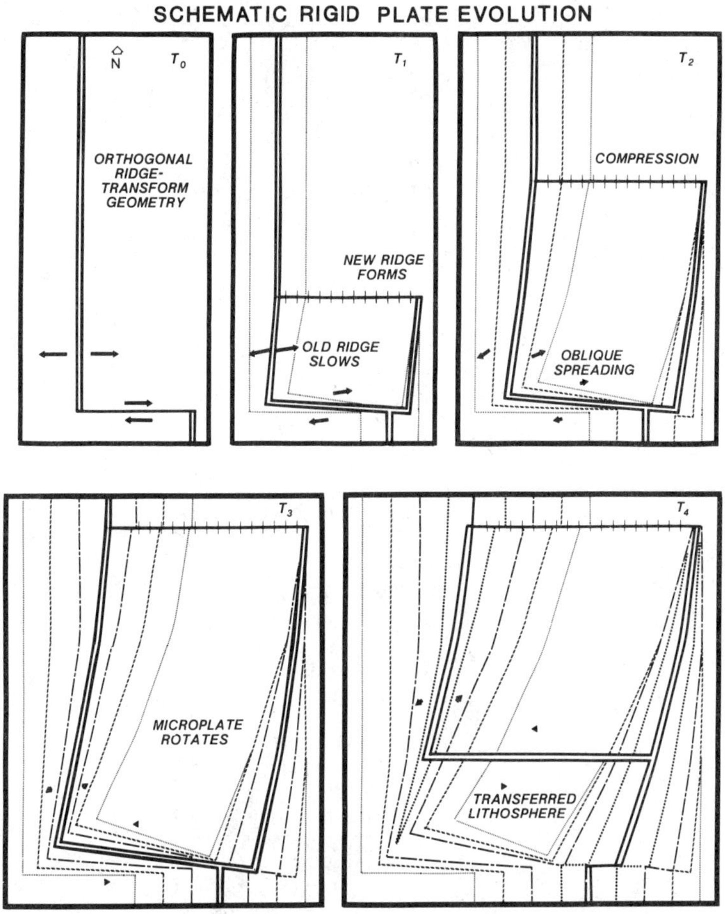

SCHEMATIC RIGID PLATE EVOLUTION

FIGURE 9b. A possible geometry resulting from rift propagation when the dying ridge does not cease spreading immediately, resulting in two coexisting spreading ridges. The region between the propagating and dying rifts moves with neither the major plate on the left nor the major plate on the right. Instead it forms a tectonically independent region that can reasonably be regarded as a microplate. Successive stages of evolution are illustrated from panel T_0 (earliest) to T_4 (latest). As the geometry evolves, the microplate rotates, the ridges change orientation, the northern edge of the microplate becomes a convergent or contractional boundary, and the original transform fault evolves into an obliquely divergent boundary. (From Engeln et al., 1988)

America west of the Wasatch Mountains, including the Basin and Range Province, the California Coast Ranges, and San Andreas and related fault systems, seems more sensibly regarded as part of a wide plate boundary zone taking up the motion between the Pacific and North American plates.

Intraplate Deformation. Plate tectonics provides tools for understanding and quantifying deformation away from plate boundaries. The rate and direction of intraplate deformation can be determined through geological, geodetic, and seismological observations; however, plate motion models can also be used to predict the location and magnitude of intraplate deformation. Plate motion models use plate circuit closure (i.e., the motions within a closed plate circuit sum to zero if plates are rigid) to tie motions from three or more plate boundaries into a plate circuit. Because these models use many data, they may provide the accurate description of plate motions needed to quantify intraplate deformation.

For example, if all motion between the Pacific and North American plates were taken up along the San Andreas Fault in central California (Fig. 10a), then an accurate plate motion model would predict a rate and direction of motion at the San Andreas consistent with geologic and geodetic observations of motion across the fault. However, the predicted rate and direction differs significantly from that observed along the San Andreas (Fig. 10b), suggesting that additional deformation occurs east of the fault, west of the fault, or both. Some, but not all, of this discrepancy is accounted for by the NW-SE exten-

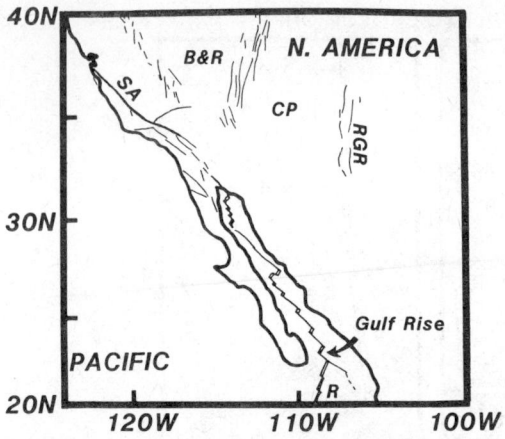

FIGURE 10a. Simplified tectonic map of part of the western United States and Mexico and the eastern Pacific, showing elements of the Pacific-North America plate boundary including the following: the San Andreas Fault (SA), the Basin and Range Province (B&R), the Colorado Plateau (CP), the Rio Grande Rift (RGR), and the Gulf Rise. Also shown is part of the Rivera plate (R). Note the location of the Gulf Rise, which is a short but important spreading center forming part of the boundary between the Pacific and North American plates. In the region near the Gulf Rise (but north of the Rivera plate) the Pacific-North America boundary appears to be narrow, but becomes much wider and diffuse to the north (as is also suggested by Fig. 5), possibly extending from offshore California through the Basin and Range province, Colorado Plateau, and Rio Grande Rift. (From DeMets et al., 1987)

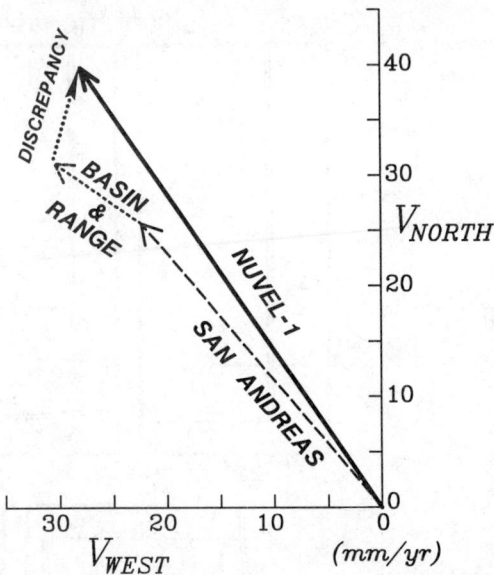

FIGURE 10b. Linear velocity vectors showing the observed and predicted relative motions at 36°N along the San Andreas Fault in central California. The predictions of the Pacific-North America Euler vector of the NUVEL-1 global plate motion model (solid line) are compared to motion observed along the San Andreas Fault (long-dashed line) and corresponding to extension within the Basin and Range province (short-dashed line). The sum of the San Andreas and Basin and Range velocity vectors disagrees with the motion predicted from the NUVEL-1 plate motion model suggesting that additional significant deformation occurs within the diffuse plate boundary dividing the Pacific and North American plates. The effect of this additional deformation on plate motion predictions along the San Andreas fault is estimated by the dotted velocity vector, which is the summed San Andreas and Basin and Range *discrepancy vector*. The discrepancy vector has a small component parallel to the San Andreas fault, suggesting that minor strike-slip motion parallel to the San Andreas fault occurs on other faults. The component of the discrepancy vector perpendicular to the San Andreas fault is larger and suggests convergence or contraction perpendicular to the fault, as might be taken up in part by the uplift of the California Coast Ranges. (After DeMets et al., 1987)

sion of about 10 mm/year extension within the Basin and Range Province (as estimated by J. B. Minster and T. Jordan), and significant motion may also occur along near-coastal faults in California.

In contrast to well-documented, intracontinental deformation, the limit to the rigid-plate approximation when applied to oceanic lithosphere is less clear. Deformation of most oceanic plates appears to be less than the 3–4 mm/year resolution of the types of plate motion data described above, but there are exceptions. The oceanic lithosphere sandwiched between the Pacific plate and the northwestern margin of the United States is deforming internally, as shown by the nearly uniform distribution of seismicity throughout the plate and the usual pattern of magnetic anomaly stripes. New techniques for detailed mapping of the seafloor and magnetic anomalies should provide data accurate enough to permit new, more rigorous tests of the lower limit of the rigid-plate approximation for oceanic lithosphere.

Diffuse Plate Boundaries. Typically, horizontal deformation is concentrated within a few kilometers of a plate boundary; however, the motion between two plates is sometimes taken up within a wider zone, as reflected by the extensive normal faulting in the East African rift valleys and in western North America. However, diffuse boundaries within oceanic lithosphere are rare. An outstanding example of extensive deformation of oceanic lithosphere that may reasonably be regarded as a diffuse India-Australia plate boundary lies in the equatorial Indian Ocean extending from the Central Indian Ridge eastwards to the Sumatra Trench (Fig. 11). Evidence for the deformation of lithosphere in the central Indian Ocean includes lithospheric folding and faulting and the most intense concentration of oceanic seismicity unrelated to a conventionally defined plate boundary. Plate motion data from spreading ridges in the Indian Ocean give a model

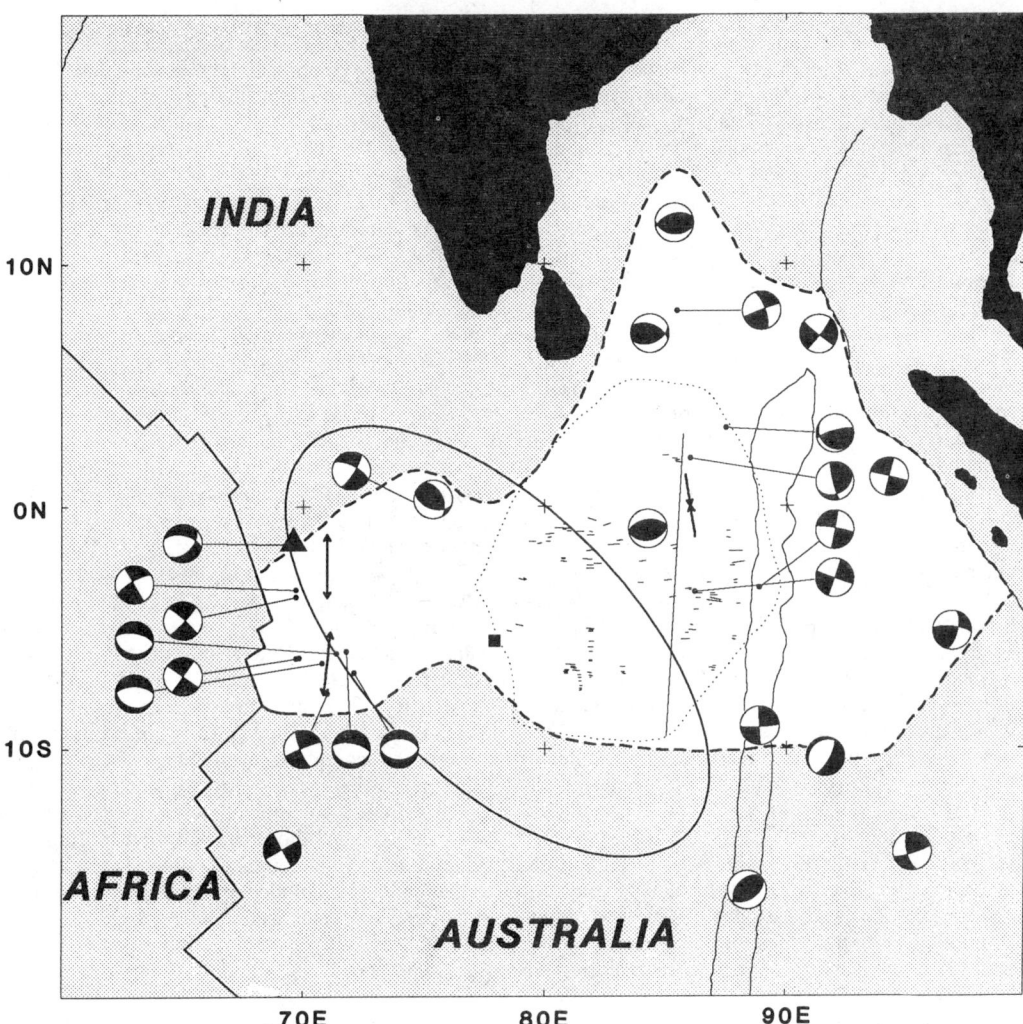

FIGURE 11. Model for current deformation of central Indian Ocean lithosphere. A diffuse boundary (white area) dividing the assumed-rigid Australian and Indian plates. The Euler pole (solid square) describing their motion and its 95% confidence region are shown. The Australian plate moves counterclockwise relative to the Indian plate. Thin lines within the diffuse boundary are faults with significant basement offsets. Earthquake focal mechanisms and other deformation data suggest several distinct strain or stress provinces within the diffuse plate boundary. Thin arrows show two average tensional axes for nine focal mechanisms near the Central Indian Ridge, and an average compressional axis for four mechanisms near the Ninetyeast Ridge. (From R. G. Gordon, C. DeMets, and D. Argus, manuscript in press, *Tectonics*)

that predicts rates and directions of deformation within the diffuse boundary zone consistent with those inferred from geological and seismological data. Thus, accurate plate motion models based on the assumption of plate rigidity are useful for locating regions where modifications to simple, rigid-plate models are required, and for providing quantitative models of diffuse deformation.

Absolute Plate Motions, True Polar Wander, and Paleomagnetic Euler Poles

Plate tectonics, strictly speaking, describes the motion between plates but does not describe the motion of any plate relative to a frame of reference external to any one plate. To determine such motions, termed *absolute* motions, further assumptions must be made. The most popular reference frame assumes that long-lived vigorous volcanic sources, termed *hotspots*, are fixed relative to one another and presumably to the deep mantle. Examples of hotspots include the hotspot that created the Hawaiian Islands and now lies under the big island of Hawaii, and the hotspot beneath Iceland. A second class of models of absolute plate motions are based on various assumptions about mantle convection and about mechanical coupling between the lithosphere and asthenosphere. These include absolute motion models determined by assuming the lithosphere (the solid Earth's outermost layer of strength) exerts no net-torque on the asthenosphere (the weak layer underlying the lithosphere). It is encouraging that various models give plate motions that are similar, the differences between models typically being about 10 mm/year, whereas the fastest plates move 80-90 mm/year. These models of absolute plate motion have been important in assessing what moves the plates. For example, models of absolute plate motions show that plates that are being subducted (i.e., the plate being underthrust at the trench as shown in the left-hand side of Fig. 7) tend to move much faster than plates that are not being subducted. This observation has led to the important conclusion that the negative buoyancy of the subducting slabs (Fig. 7) attached to the rapidly moving plates is the cause of the rapid motion.

The concept of absolute plate motions is related to a fundamental problem in paleomagnetism: the kinematic interpretation of apparent polar wander (APW) paths. Early paleomagnetic studies established that with increasing age the ancient paleomagnetic pole diverged from the present North Pole, suggesting *true polar wander* (TPW), the motion of the entire solid Earth relative to the spin axis. Later data showed that APW paths from different continents diverged from one another, which reflected plate motion. With the acceptance of plate tectonics, the concept of TPW is not needed to explain the paleomagnetic observations, but the question of whether TPW has an observable effect on APW paths has not been convincingly resolved. Paleomagnetic and plate motion data have recently improved enough that TPW can be examined more rigorously. There are two complementary approaches to exploring this problem.

The first focuses on the comparison of APW paths to plate motion relative to an independent deep mantle or *absolute* reference frame, described above. For example, differences between the locations of observed paleomagnetic poles and paleomagnetic poles predicted by assuming the hotspots have been fixed relative to the paleomagnetic axis show that the hotspots have moved relative to the paleomagnetic axis (Fig. 3). If the hotspots are a reliable reference frame, and if the paleomagnetic axis reliably tracks the spin axis, then this motion shows that TPW has been large enough to affect APW paths. No consensus on the fixity or mobility of hotspots has emerged, but recent work suggests that the hotspot and no-net-torque reference frames agree well, and that the inferred motion of the paleomagnetic pole relative to the deep mantle depends only weakly on how absolute motions are estimated. We think that the main conclusion is that TPW over the past 70 Ma was small, but a shift of 10-15° may have occurred between 100 and 70 Ma. However, this view is not universally accepted. Further paleomagnetic work is needed to test this hypothesis and to narrow the timing and rate of this shift.

An alternative approach is to assume that TPW is negligible and to exploit the plate motion information available in APW paths, which consist of long, gently curved segments termed *tracks* linked by short segments with sharp curvature termed *cusps* or *hairpins*. A few years ago it was thought that paleomagnetic data provided information on only two of the three components of the rotation that describes plate motion. Once a plate is correctly oriented in paleolatitude and in paleodeclination, its paleolongitude is indeterminate because of the axial symmetry of the time-averaged geomagnetic field. Recently, however, it was realized that a time-sequence of paleomagnetic poles might give all three components of plate motion if APW tracks correspond to intervals when the direction of plate motion was constant. APW tracks would thus tend to lie along small circles, the center of which is termed a *paleomagnetic Euler pole* (Fig. 12).

This new approach has given several useful results. Whereas plates bearing large continents are now moving slowly (less than ~20 mm/year), paleomagnetic Euler pole analysis shows that North America and Eurasia moved rapidly (60-100 mm/year) before 140 Ma. Gordon et al. (1984) interpreted this result as follows: During intervals of rapid motion the continents were attached to oceanic lithosphere that was being subducted at some distance from the continent. Rapid motion stopped when the oceanic lithosphere had been consumed by subduction and the continent, which is too buoyant to be subducted, collided with another continent or with a major oceanic island arc system.

Concluding Remarks

The current view of a mobile Earth has now been widely accepted for about twenty years. Plate motions over the past few millions of years are well described, and motions over many tens of millions of years (or longer) are described, but less completely and less accurately. Although at a slower pace than 10 to 20 years ago, studies of plate motion continue to be fruitful, in large part because the plate tectonic model is quantitative and predictive. There are now well-established quantitative relationships between plate tectonics and many other geophysical and geological observations and processes (e.g., heat flow, subsidence, and spreading ridge morphology) too numerous to discuss here. A few of the outstanding problems that remain in plate motion studies include the following: (1) accurate determination of the history of subduction and strike-slip motion around the margin of the Pacific ocean, (2) determination of reliable and accurate limits on the rate of motion between hotspots, (3) determination of reliable and accurate limits on the rate and distribution of intraplate deformation, and (4) determi-

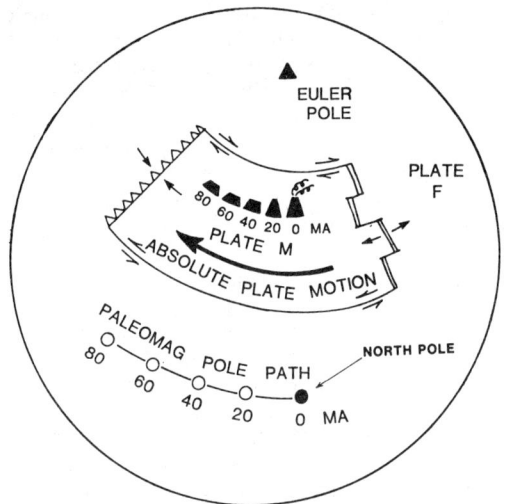

FIGURE 12. Cartoon illustrating the key assumptions underlying paleomagnetic Euler pole analysis. Plate F has been fixed relative to the lower mantle for the past 80 my, while Plate M has been moving about a fixed Euler pole relative to the lower mantle for the past 80 my. Transform faults between Plate F and Plate M lie on small circles centered on the Euler pole of the motion of Plate M relative to Plate F. The absolute Euler pole of Plate M coincides with its Euler pole for motion relative to Plate F. Each point on Plate M has moved along a small circle centered on this Euler pole. A hotspot fixed relative to the lower mantle is shown beneath Plate M; the island and seamount chain formed on Plate M lies on such a small circle centered on the same Euler pole. The direction of plate motion is the direction from the youngest volcano toward older volcanos. Similarly, paleomagnetic poles recording the ancient location of the spin axis lie on a small circle centered on the Euler pole. The direction of plate motion is the direction from the youngest paleomagnetic pole (the solid circle depicting the present spin axis) towards older paleomagnetic poles. (From Gordon et al., 1984)

nation of whether significant TPW has occurred. Advances in these problems should lead to improved understanding of whether and how plate convergence rates are reflected in deformation and mountain building along the edges of continents, and to a more complete understanding of what forces drive and resist plate motion. Recent and current work has improved the understanding of the tectonics of continental regions poorly described by plate tectonics. Promising directions of future research include space-technology-based measurements of plate motion over short time scales (i.e., years) and measurements of plate motion over long time scales through new paleomagnetic data and methods of analysis. Such promising investigations, coupled with many other important directions of research in earth science, may eventually lead to a more comprehensive theory of global tectonics on a mobile Earth.

Acknowledgments

We thank S. O. Schlanger and C. DeMets for helpful comments.

RICHARD G. GORDON
GARY ACTON

REFERENCES

Cox, A., 1973, *Plate Tectonics and Geomagnetic Reversals*. San Francisco: W. H. Freeman, 702 p.

Cox, A., and R. B. Hart, 1986, *Plate Tectonics: How it Works*. Palo Alto, Calif.: Blackwell Scientific Publications, 392 p.

DeMets, C., R. G. Gordon, S. Stein, and D. F. Argus, 1987, A revised estimate of Pacific-North America motion and implications for western North America plate boundary zone tectonics, *Geophys. Research Letters* **14,** 911-914.

Engeln, J. F., S. Stein, J. Werner, and R. Gordon, 1988, Microplate and shear zone models for oceanic spreading center reorganizations, *Jour. Geophys. Research* **93,** 2839-2856.

Glen, W., 1982, *The Road to Jaramillo*. Stanford, Calif.: Stanford University Press, 459 p.

Gordon, R. G., 1987, Polar wandering and paleomagnetism, *Ann. Rev. Earth Planetary Sci.* **15,** 567-593.

Gordon, R. G., A. Cox, and S. O'Hare, 1984, Paleomagnetic Euler poles and the apparent polar wander and absolute motion of North America since the Carboniferous, *Tectonics* **3,** 499-537.

Menard, H. W., 1986, *The Ocean of Truth*. Princeton, N.J.: Princeton University Press, 353 p.

Takeuchi, H., S. Uyeda, and H. Kanamori, 1967, *Debate about the Earth*. San Francisco: Freeman, Cooper, and Co., 281 p.

Cross-references: *Geomagnetic Polarity Reversals: Observations; Lithosphere, Oceanic: Formation and Evolution; Magnetostratigraphy; Mantle Convection and Plumes; Mantle Dynamics; Natural Remanent Magnetization (NRM); Paleomagnetism and Continental Drift: Historical Introduction; Rock Magnetism; Seafloor Spreading: Magnetic Evidence; Seismicity and Plate Tectonics.*

PHOTOGRAMMETRY

Photogrammetry is the theory and practice of deriving geometric information about an object from geometric measurements on pictures of the object. The object may be as small as a grain of sand or as large as a planet; the picture may have been taken at any wavelength from that of hard X-rays to meter-long radio waves. With rare exceptions, the pictures are photographs; usually, they are black and white photographs (b/w) taken in the optical range (roughly, between 0.4 and 0.8 micrometers) and used as positive or negative transparencies. (The following explanations therefore deal with this most commonly encountered situation and can be assumed to apply without significant differences generally [e.g., color photographs are treated much the same as b/w photos]. Where special conditions create significant differences in photogrammetric practice, these will be mentioned.)

Physical Interpretation

The basic principles of photogrammetry are the same as those governing human vision. Physically, a camera or other receptor is substituted for the eye, a photographic emulsion is substituted for the retina, and an optical/mechanical apparatus or a computer is substituted for the optical nerves and the brain. The brain interprets the two-dimensional pattern on the retina of one eye as the visualization of a flat object without depth and at an unknown distance; usually, the brain has enough prior knowledge of the nature of the object (e.g., the width of a street or the height of a person in the scene) that it can also make some inferences about the object's absolute size and distance. *Monoscopic photogrammetry* uses exactly the same kind of information to infer geometric properties of an object from single photographs. In binocular vision, the brain uses the same information as that from monocular vision. In addition, it uses the known distance between the eyes, the known angle of convergence of the line of sight (as derived from the muscular tension used in focusing the two eyes on the same point) and the perceived differences in the patterns on the two retinas to derive as full information about the object's metric properties as the imperfect nature of vision will allow. *Stereoscopic photogrammetry* is an exact analog but is more accurate: the distance between cameras can be made as great or small as desired to provide the most favorable angle of convergence of line of sight, and the imperfections of the lens systems and other uncertainties can be taken care of better.

A photograph can be thought of as having been created by a bundle of light-rays. Each ray proceeded from an external point on the object, through a central point of the lens onto a point on the emulsion. This point becomes visible when the photograph is developed. By a fundamental law of optics, a light-ray proceeding from a point on the emulsion through the central point will then pass to the corresponding point on the object. In a first approximation, the rays may be considered straight lines, the lens may be considered free of all aberrations, the photograph free of distortion and of infinite resolution, etc. One can then attempt to reconstruct the object by (1) actually passing rays back through the photograph; (2) represent a ray by a combination of rods, linkages, and oculars in such a way that the geometric relationship between photograph, lens center, and viewer is that of object, center, and photograph; or (3) represent a ray by a mathematical formula and use the measured coordinates of points on the photograph, together with the formula, to describe the object mathematically. (In the troposphere, the straight-line approximation is usually satisfactory for radio waves over any distance and for infrared and optical radiation over distances of a few hundred meters. Outside the troposphere, it is usually satisfactory for infrared and optical radiation at any distance but is not satisfactory for radio waves longer than a few cm wavelength. X-rays travel in straight lines.)

Mathematical Interpretation

Light-rays can be represented mathematically by straight lines, in a first approximation. The equation of a straight line in three dimensions can be written in several ways. One is $\mathbf{z} = a\mathbf{x} + b\mathbf{y}$. Here $\mathbf{z}$, $\mathbf{x}$, and $\mathbf{y}$ are points on the straight line; a and b are scalars such that $a + b = 1$. The point $\mathbf{z}$ is always a point on the image of an object; $\mathbf{x}$ can be taken to be a point on the object and $\mathbf{y}$ is then the center of perspective (usually taken as the center of the lens system in terrestrial photogrammetry) through which all rays from the object pass. The components of $\mathbf{z}$ are always known, by measurement, in some coordinate system fixed in the image.

The problems of photogrammetry are of two kinds: either the location $\mathbf{x}$ of the camera is known and the points $\mathbf{y}$ on the object must be determined; or the locations $\mathbf{y}$ of points on the object are known and the location $\mathbf{x}$. In either case, of course, the coordinates $\mathbf{z}$ of points on the image are known by measurement. Topographic photogrammetry usually contains both problems. The first problem is solved first, by photographing terrain on which points (the $\mathbf{y}$'s, called "ground control points) of known location have been marked and the equation is solved for the camera's coordinates $\mathbf{x}$. Three ground control points are sufficient. Once the $\mathbf{x}$'s of all positions of the camera have been determined, the equations are solved for the coordinates $\mathbf{y}$ of all points on the ground that are of interest and that appear on more than one image. It is not necessary to have three ground control points for each individual position of the camera. A simple count of the number of unknowns and number of equations in each case shows that (1) three separate $\mathbf{y}$'s suffice to determine the position $\mathbf{x}$ of the camera; (2) the values of $\mathbf{z}$ for five points on the ground, measured on two different images, suffice to determine the coordinates $\mathbf{x}'$ and $\mathbf{y}'$ in the coordinate system of one of the images (but not, of course, in the coordinate system of the whole space); and (3) the location $\mathbf{y}$ of a point on the ground can be determined from the measurements $\mathbf{z}$ on two different images for which the camera locations $\mathbf{x}$ are known. The procedure is, therefore, as follows. Ground control points are put in by surveying. The region is photographed so that each point on the ground appears on at least two images. The coordinate systems of all images are determined relative to that of a particular image or, more often, to an arbitrary system. This step is, in effect, the stitching together of all the separate images into one large image whose coordinate system is still not related to that of the terrain. The coordinates of all points of interest on the ground are determined in this arbitrary system. Finally, the arbitrary system is related to the ground system by introducing the ground control points.

To the equations given are usually added corrections for distortion of the lens system, atmospheric refraction, shrinkage of the emulsion, etc. There may be condition equations added because there are more data than those given by ground control points and image measurements alone: e.g., data on the orientation of the camera, location of the camera (of the aircraft or satellite, actually), etc. Note that the preceding equations apply to problems in 3-space. Many geophysical problems are basically 2-space problems; the object to be reconstructed photogrammetrically is flat or there is no interest in y_3. Traditionally, the equations given have been applied as such only when the number of points to be determined is small and the points are widely separated. The procedure is then called "aerotriangulation" and is limited by the number of equations that can be solved economically on the computer. If, as is the case in topographic mapping, the points are so close as to form almost continuous lines, the equations (or, one might say, the rays the equations simulate) have represented by rods and/or optical trains connecting the physical points z and x to the points y in a model. Within the past two decades, electronic computers and computing methods have made it possible to apply numerical methods even to mapping, so that the mechanical/optical method need not be used.

Techniques

Three categories of substantially different photogrammetric techniques can be distinguished using the object's size as criterion.

1. Microscopic photogrammetry
    objects of molecular size to about 1 cm
    distance less than 30 cm (very-short-range photogrammetry)
    short to medium focal-length cameras
    object moved for stereoscopy
    examples: crystal structure, deformation of grains in rock

2. Macroscopic photogrammetry
    objects 10 cm–100 metres or more in size
    distances of 3 m to 100 m (short-range photogrammetry)
    cameras of short-medium focal lengths
    object or camera may be moved for stereoscopy, or rigidly-connected, paired cameras may be used
    examples: geomorphology of stream banks, glacial marks

3. Megascopic photogrammetry
    objects of kilometer size to planetary size
    distances of greater than 1 km (long-range photogrammetry)
    cameras of medium (20 cm) or longer focal length
    camera moved for stereoscopy
    examples: crustal deformation in small regions; lunar-surface mapping

Topographic photogrammetry is a subdivision of megascopic photogrammetry. It concerns mapping of a region of the surface of the Earth or similar body. It is classified as *terrestrial* if the photographs were taken from the ground or as *aerial* if they were taken from the air. The techniques are quite different but the former is very little used any more. *Nontopographic photogrammetry*, sometimes called *close-range photogrammetry*, concerns all other kinds of photogrammetry.

Data

Data come from the following sources: (1) measurements on photographs or optical images produced from photographs; (2) measurements on or known information about the geometry of the object; (3) measurements (calibration of the receiving and recording equipment (e.g., camera, antenna, circuitry, emulsion); (4) measurements from auxiliary equipment (e.g., rangefinder, altimeter, thermometer, refractometer); (5) tables and specifications (e.g., standard atmospheric tables, indices of refraction, aberrations of lens systems). See Table 1.

Errors and Their Removal (Calibration)

The principal error introduced by a camera is distortion of the image by the lens system. Most of this is proportional to radial distance from the optical

TABLE 1. Cameras for Nonmicroscopic Photogrammetry

	Macroscopic		Megascopic	
Focal length (mm)	60	120	89	210
Field of view	—	—	121.5	46
Size of image (mm)	60 × 60	90 × 120	228 × 228	c.200
Maker	Hasselblad	Zeiss (Oberk.)	Zeiss (Jena)	Wild
Model	MK79 (Biogon)	TMK-12	LMK-9	BC-4

axis; some is perpendicular to the radial direction at each point. It can be reduced by applying corrections given by the manufacturer of the camera or by determining the corrections oneself, e.g., by photographing a pattern of known geometry. Some instruments used later in the measuring process eliminate calibration by making the measurements through the same lens system as was originally used for taking the photographs, thus cancelling the original distortions. Other aberrations such as coma create some systematic errors that may be removed by calibration but mostly cause random errors that limit the accuracy attainable. During development, the emulsion shrinks and warps; some of this error can be removed by applying corrections based on manufacturer's data or on experiment. Greater accuracy can be achieved by taking the photograph through a glass plate on which a regular, rectangular pattern has been etched. The photograph then shows not only the object but also the pattern (*reseau*). Development distorts both object-image and reseau-image in the same way and corrections to the first can be determined by measurements on the second.

Errors are introduced during photography if the incoming rays cannot be considered as following straight lines or if the object is moving with respect to the camera during exposure. If only a few points are involved, the problem may first be solved assuming straight rays and corrections then applied, based on the known refractive index of the medium and on the determined geometry. If a complete model of the object is to be constructed, a special device may have to be used to make the correction during measurement or an analytical plotter used. Spatial filtering has been used to correct for relative motion of image and object; techniques peculiar to the circumstances of photographing have also been used.

Measuring systems introduce systematic errors; the errors introduced by the optical and mechanical components can be reduced by substituting patterns of known metric geometry for the photographs and using the deviations of the measurements from the known values as corrections to later measurements on photographs. Some errors occur because each point of the object is represented in the final image not by a point but by a blurry patch. Camera, photographic process, and measuring instrument all contribute. The errors appear random if measurements are made by a human operator. Some of the total error can be removed by calibration if the images are of clearly defined points on the object and if the cross-hairs or other reference-mark in the measuring instrument is moved to the measuring position by an automatic-centering device rather than by hand. The results obtained using plotting instruments that convert the measurements into graphic or digital representations of the object contain further errors introduced by the mechanical and/or electrical connection between the measuring engine and the plotter proper. Whether these errors can be reduced depends on the type of instrument involved and a general rule cannot be given. Plotting instruments in which a large part of the plotting is governed by a computer introduce errors because the equations used are truncated and because the numbers are rounded off during computation. The nature of the errors and the method of reducing them depends on the instrument used, the equations and program for computation, and often also on the nature of the problem.

Monoscopic Photogrammetry

A ray from a specific point on the object creates an image of that point on the recording medium. However, any point on the ray, acting as a source, could have created the same image point. Hence, each image point on a picture could have originated from any one of an infinitude of points in object space. Without more information, a single photograph gives only directions, in a camera-fixed coordinate system with x, y-plane in the focal plane. The orientation of this coordinate system with respect to one fixed at the observer is determined first. The direction of a point on the object with respect to the camera-fixed system is then determined from the ratios y/x and $[\sqrt{(x^2 + y^2)}]/f$, where f is the focal length and x, y are coordinates measured on the photograph. The two systems of direction are then combined to give the direction of point on object with respect to observer-fixed system.

In some applications, only directions are needed. Most celestial objects outside the solar system are located by two angular coordinates only. The orbits of artificial satellites and of planets have been determined from directions found by monoscopic photogrammetry. In other applications, the distance or size and shape of the object are wanted. The particular circumstances of the problem dictate how much can actually be determined about the object's metric qualities.

Stereoscopic Photogrammetry

The photogrammetric analog of binocular vision is stereoscopic photogrammetry. Starting with a pair (stereoscopic pair) of photographs showing the same object from different points of view, the two pictures are placed as nearly as possible in the same relative positions they occupied when taken, but at a convenient scale. (The pictures used may be reduced in size to fit the type of equipment being used.) This step is known as *relative orientation* and requires that at least five identifiable points on the object be viable in both pictures. In the second step, the scale and position of the model with respect to the photographs are determined by identifying three or more points of known location in the model with the corresponding points in the stereoscopic image. This process is called *absolute orientation*. In the third step (plotting), the shape and (scaled) size of the object are determined and either drawn as contour lines or profiles on drafting material or are stored as strings of digits on magnetic tape or disk. See the descriptions in Fig. 1.

Analytical Photogrammetry

If information is wanted only about distances and directions between a few isolated points on the object and not about the object as a whole (as will often be the case in geophysical studies), the complicated and expensive equipment used to draw contours or profiles is not needed. Only a simple measuring engine and a computer are needed.

After all the measurements have been made and the auxiliary data are available (distortions of the camera lens system, shrinkage of the emulsion or other medium during processing, refractive effects, etc.), the basic photogrammetric equations are linearized by taking the differentials of the quantities involved. This changes the equations to the form $d\mathbf{z} = a\, d\mathbf{x} + b\, d\mathbf{y}$ or, simply, $d\mathbf{z} = \mathbf{C}\, d\mathbf{X}$. These are the observation equations. To them must be added the equations (3) for transforming from the coordinate system in the object to the coordinate systems in pictures and cameras and the condition equations (2) which state that corresponding points in different

---

TECHNIQUE 1: Operator does all positioning of photographs, shifting each photograph until corresponding points match.

Operator observes a projected, composite image of the photographs (double-projection, direct-viewing equipment).
   General Characteristics: Pictures are projected to form a composite image on a flat surface. This image is viewed through special spectacles to present a three-dimensional image to the operator, who makes measurements on this image.
   Advantages: concept simple; equipment simple and less costly than that used in other techniques but more versatile; model may be formed from three or more pictures at one time.
   Disadvantages: accuracy somewhat lower than other types; not as easily adaptable to automatic or semi-automatic operation.
   Typical Varieties: Multiplex (Fig. 2); Balplex (Bausch and Lomb); Photocartograph VI (Nistri).

Operator observes the photographs directly (mechanical/optical equipment).
   General Characteristics: each of two pictures is projected to a different eye, giving the observer the impression of seeing the object directly. Measurements are made on this apparent object.
   Advantages: concept simple; high accuracy; easily automated.
   Disadvantages: equipment usually costly; only one pair of pictures can be accommodated at one time; difficult to adapt to unusual scales or types of pictures.
   Typical Varieties: Stereoplanigraph C8 (Zeiss); Autograph A8 (Wild); PG2 (Kern); Planimat (Zeiss/Oberkochen) (Fig. 3)

TECHNIQUE 2: Operator moves photographs only for measurement (analytical plotters).

Operator identifies corresponding points but equipment moves photographs.
   General Characteristics: Relative and absolute orientation done automatically and governed by computer; contouring or profiling may also be done by computer but is usually governed or monitored by operator.
   Advantages: almost unlimited versatility; some varieties are less expensive than those of type 1b; easily operated; mathematical principles simple; very fast in operation.
   Disadvantages: complex circuitry; completely automatic varieties extremely expensive; only one pair of pictures can be worked on at a time.
   Typical Varieties: UNAMACE (Bunker-Ramo); AP/C4 (OMI); U.S. 1 (Bendix); APPS-IV (Autometric) (Fig. 4)

Operator makes measurements directly on individual photographs (analytical photogrammetry).
   General Characteristics: The photographs are measured individually. The measurements and other data are given to a computer which then calculates all required results such as distances and directions between points on object or coordinates of points on object. Usually only a few (less than 1000) points are involved.
   Advantages: direct control over what is done; extremely versatile and accurate.
   Disadvantages: least efficient (accuracy aside) per point plotted.

FIGURE 1. Techniques of stereoscopic photogrammetry.

FIGURE 2. Multiplex—general view.

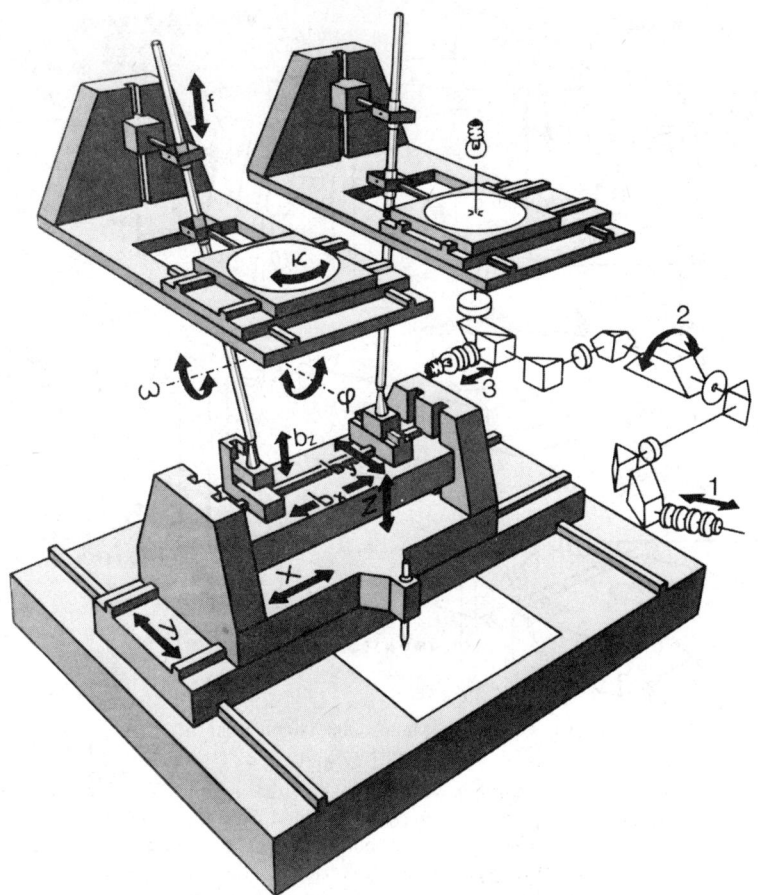

FIGURE 3. Mechanical and optical design of Planimat.

pictures correspond to the same point on the object. There may also be condition equations imposed by known distances between points on the object, known distances between cameras or locations of cameras, etc. All of these equations can usually be collected into a single set of linear equations, $z = Gr$. The particular conditions of the problem determine the exact structure of the rectangular observation matrix G, but it is often strongly diagonal with relatively few nonzero elements.

Applications

There are no typical applications of photogrammetry to geophysics. The methods used vary according to whether the atmosphere, hydrosphere, or crust, the magnetic, thermal, or gravitational field is being studied, and whether the investigation is taking place in the laboratory or in the field. A few examples will be given, but the principal factor involved is the investigator's ingenuity.

Cartography. The most common (although by no means the earliest) application of photogrammetry is to map making. In the region to be mapped, surveyors determine the coordinates of a suitable number of points and mark these points so that they will be visible on aerial photographs. An airplane then flies over this region and photographs it; the photographs are spaced so closely that adjacent pairs form stereoscopic pictures. The pictures are developed, placed in accurate measuring engines, and the measurements converted by the standard techniques described previously either to coordinates (of the desired features on the ground) or to plots of those features (the coordinates of the previously surveyed points being used to establish the coordinate system). The accuracy is increased if the aircraft's position is also known (through on-board instruments, by using satellite-aided navigation, etc.). If the region to be mapped is extraterrestrial, then a preliminary survey on the ground is not possible. One then uses data on the picture-taking satellite's position in orbit to provide the coordinate system.

Classically, measurements have been considered to be between points on pictures, with the observer being relied on to find the points. This concept was

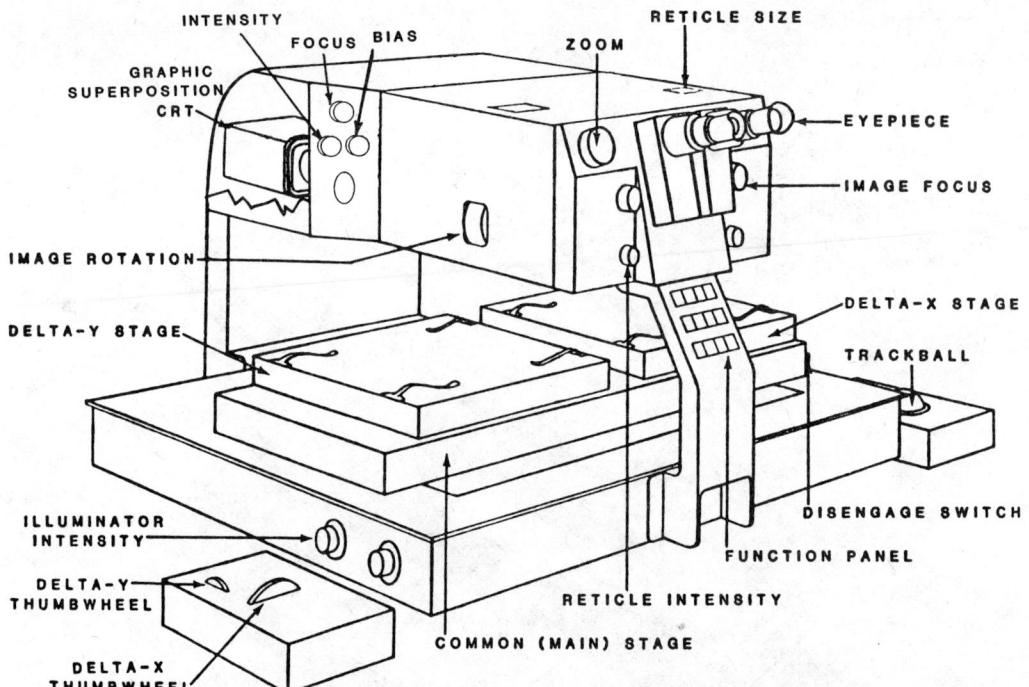

FIGURE 4. APPS-IV Model 4000 (Autometric, Inc.)

generalized when automatic measuring engines were introduced. These machines found corresponding points in stereoscopic pairs by defining points as the centers of small patches and correlating a patch on one picture with a patch on the other. When large computers became generally available, the concept was further generalized to allow points to be defined precisely on the coarse-grained pictures taken of other planets' surfaces by television, thereby also allowing camera-introduced errors to be reduced or eliminated. The present trend is to use the concept of a point as a patch wherever computing facilities are adequate, because greater precision (although not necessarily greater accuracy) is usually obtained.

Wind Waves. Study of the form and history of individual waves by photogrammetry is difficult and best carried out in a laboratory by kinematographic photography of waves in a glass-walled tank. A single camera may be sufficient, or a sequence of cameras may be lined up to get a continuous picture of the wave. Measurements can be made on a monoscopic, reasonably precise measuring engine. The waves generated have, however, characteristics of swell rather than wind waves. More commonly, wind waves are studied in the aggregate and by photographing from a fixed position or from an aircraft. One technique used for determining the sea state involves photographing the surface from a downward-pointing camera aboard an aircraft on a sunny day. Each patch of the water's surface acts as a tiny mirror, and some of these patches send the light to the camera. The slope of a mirror can be determined from the coordinates of the image on the photograph, the orientation of the camera, and the direction of the sun. The data are reduced using elementary trigonometric formulae, and the topography of the surface is inferred from the slopes of the reflecting patches. In a more recent technique, the aerial photograph is placed on an optical bench and illuminated by monochromatic, coherent light. The transmitted light is gathered by a lens system and focused onto a photographic plate. The developed plate shows the Fourier transform of the original scene. This plate is placed in a microdensitometer and the wavelengths present measured. These wavelengths can be converted to wave height and sea state (with some necessary assumptions).

Catastrophic Crustal Changes. Rock slides, mud slides, avalanches, etc., are usually best studied using aerial photographs of the region. The photographs can be reduced using the standard methods of aerial mapping and volumetric inventorying. In North America, there is a good chance that suitable photographs will also be available of the region before the event, e.g., the eruption of Mt. St. Helens.

Mechanics of rocks. Most photogrammetric work is done in the laboratory. In one class of methods, the carefully polished surface of a rock is photographed through an optical microscope. The deformations of individual crystals in the rock are then measured on a measuring engine and these can be converted into strains and stresses applied during the history of the rock. The photogrammetry needed is elementary, but the mechanical theory is complex.

Crustal Deformation. Study of crustal deformation uses the standard techniques of topographic photogrammetry. Photographs are usually taken from aircraft using common types of aerial cameras and are measured on stereographic measuring engines. The data are best reduced using block adjustment for the control points. A model of the surface can be reconstructed using any common stereographic plotting instrument. But to be useful, suitable photography at two widely separated epochs must be available. In studying the mechanics of local subsidences caused by withdrawal of groundwater or the meandering of streams, photographs taken at monthly or yearly intervals may be adequate, depending on the rate of change of the topography. Conversely, if movement along faults is being monitored, a decade or century may have to elapse before movement is great enough to be reliably determined photogrammetrically. The deciding factors are cost and politics. Most of North America has been photographed at scales such that changes of 5 cm per year horizontally will be able to be determined reliably from photographs taken after 2000. Smaller annual changes can be determined in some regions, e.g., Salt Lake City, using photographs taken in another ten years or so. If one waits long enough, the geophysical changes will be greater than the photogrammetric errors in the models that are constructed. The waiting period can be shortened by (within limits) increasing the scale and the number of photographs. This, however, greatly increases the cost of an aerial survey, and these could well be greater than the cost of a limited traverse or leveling, or even of installing strain gauges and extensometers. If the region is politically important, e.g., if it is densely populated and subject to earthquakes or subsidence, small movements in short periods of time must be determined, and photogrammetric methods will not be adequate.

SOREN HENRIKSEN

Cross-references: *Crustal Movements and Tectonic Deformation; Geodesy: Geometric; Geodesy: Historical Introduction; Map Projections in Geodesy; Geodetic Ground Positioning; Planetary Geodesy.*

PLANETARY GEODESY

The objectives of planetary geodesy are to determine the gravity, topography, and rotation of the planets. These measurements are put to two basic uses. In a somewhat pragmatic vein, the geodetic parameters are useful for navigational purposes: The motion of a spacecraft in orbit around a planet, relative to features of interest on the planetary surface, depends on the gravity, topography, and rotation of the planet. On the other hand, these same geodetic parameters provide the only remotely accessible information on the internal mass distribution in the planet.

The methods used to obtain geodetic information on the planets are quite diverse. The earliest estimates of planetary masses came from observing the orbital motions of natural satellites. For a small satellite, the size of the orbit (semimajor axis a) and the mean orbital angular velocity (mean motion n) are related to the mass of the primary body M via $GM = a^3 n^2$, where G is the gravitational coupling constant. Through the gravitational constant $G = (6.6726 \pm 0.0005) \times 10^{-20}$ km$^3 \cdot$ s$^{-2} \cdot$ kg^{-1} is only known to an accuracy of roughly one part in 10^4, the product GM is known for many of the planets to an accuracy of a few parts in 10^7.

Departures from spherical symmetry in the gravity field of a planet will change the shape and/or orientation of the orbits of satellites. Detailed observations of natural satellites (and rings) yield estimates of the orbital precession rates. These natural probes have to some extent revealed the low-degree gravity harmonics of Jupiter, Saturn, Uranus, and Neptune. The limiting factor in most natural satellite observations is the ~1-arcsec positional accuracy attainable with optical techniques.

Artificial satellites of other planets are much too small to be observed optically, but the Doppler shift in the radio communication links induced by the relative motions of the satellite and the ground tracking station can be accurately measured. Accuracies of a few millimeters per second are often available and have allowed fairly detailed mapping of the gravity fields of the Earth, Moon, Mars, and Venus. For Earth gravity measurements these Doppler observations are supplemented by laser ranging to specially constructed geodetic satellites. Accuracies of a few centimeters in range allow very accurate determinations of the long- and intermediate-wavelength gravity anomalies.

The earliest determinations of the sizes and shapes of the planets relied on optical telescopic measurements of the dimensions of the projected disk. Much better accuracy is attainable by timing the duration of stellar occultations. Better still are the results of spacecraft occultations and radar altimeters. For nearby planets, Earth-based radar ranging to the planetary surface yields good results.

Estimates of the rotational states of the planets have also come from a variety of sources. Obviously, the rotational states of the Earth and Moon have been known fairly well for centuries, though recent technological improvements have helped reveal subtle variations. For the other planets, initial estimates of rotation rates and axial orientations came from optical telescopic observations of the motion of discrete features across the observed disc. Though the method worked well for Mars, Jupiter, and

Saturn, the initial results obtained this way for Mercury and Venus were wrong.

A small solid body can maintain whatever shape it acquired during formation. However, for sufficiently large bodies (or for sufficiently weak material), the mutual gravitational attraction of the constituent parcels of mass will cause the body to assume an essentially spherical form. Thus, in the absence of tidal or rotational distortions, the gravitational potential at an external point will depend only on the total mass of the planet and neither the gravity, nor the topography will convey any information about the internal density distribution. However, a rotating hydrostatic body will be distorted by an amount that depends on the radial density variation $\rho(r)$ and the spin rate ω.

A dimensionless parameter that indicates the relative importance of rotational distortion versus the gravitational tendency toward sphericity is $q = \omega^2 R^3/GM$. Table 1 gives current estimates for all of these parameters for the bodies we will be discussing later. It appears that, in terms of the importance of rotation, the planets can be divided into two basic categories. Saturn, Jupiter, Uranus, and Neptune are rapid rotators ($q > 10^{-2}$), whereas Venus, Mercury, and the Moon are slow rotators ($q < 10^{-4}$). The Earth, Mars, and Pluto are in an intermediate regime where rotation is important but not completely dominant.

In many applications it is convenient to describe the topography and gravitational potential of a planet in terms of spherical harmonic functions:

$$H(\theta, \phi) = R \sum \sum H_{nm} \Lambda_{nm}(\theta, \phi)$$

$$G(r, \theta, \phi) = \frac{GM}{R} \sum \sum G_{nm} \left(\frac{r}{R}\right)^{-(n+1)} \Lambda_{nm}(\theta, \phi)$$

where (r, θ, ϕ) are radius, latitude, and longitude, and $\Lambda_{nm}(\theta, \phi)$ is a complex surface harmonic of degree n and order m. Each such harmonic is a product of a latitude-dependent associated Legendre function $P_{nm}[\sin(\theta)]$ and a longitude-dependent complex exponential $e^{im\phi}$. The real and imaginary parts of the harmonic coefficients are customarily denoted C_{nm} and S_{nm}, respectively, and the zonal harmonic ($m = 0$) coefficients are frequently reported in terms of $J_n = -C_{n,0}$. If R and M are chosen to be the mean radius and total mass respectively, the degree-zero harmonic coefficients are both equal to 1. If the center of mass is chosen as the coordinate origin, the first-degree gravity coefficients are identically zero.

For rotating bodies in hydrostatic equilibrium, the only nonzero coefficients are the even-degree zonals, $J_2, J_4, J_6, \ldots$, and these coefficients form a rapidly decreasing series with $J_{2n} = O(q^n)$. This high degree of symmetry (symmetry about the rotation axis and the equatorial plane) makes it possible to reliably determine the low-degree harmonics of topography and gravity for Jupiter, Saturn, Uranus, and Neptune from relatively few observations. No such symmetry applies to nonhydrostatic planets, and their topographic and gravitational variance spectra can only be determined from detailed observations over the entire surface.

For a spherical body with internal density variations described by

$$\rho(r, a, b) = <\rho> \sum \sum A_{nm}(r)\Lambda_{nm}(a, b)$$

where $<\rho> = 3M/4\pi R^3$ is the mean density, the gravity harmonics are

$$G_{nm} = \frac{3}{2n + 1} \int_0^1 A_{nm}(x) x^{n+2} \, dx$$

with $x = r/R$. As a special case, the gravitational effect of topography in a surface layer of density $\rho_s = \tau<\rho>$ can be approximated by $G_{nm} = 3\tau H_{nm}/(2n + 1)$.

This formula conveys two basic ideas: If the surface layers have low density, the topography has relatively little gravitational impact, and, for a given density, long-wavelength topography has the greatest effect.

For a fluid planet of uniform density, the relation-

TABLE 1. Geodetic Parameters of the Planets

Planet	Mass (km³·s⁻²)	Radius (km)	Period (s)
Mercury	22,031.8 ± 1.0	2,440. ± 1.	5,067,020. ± 430.
Venus	324,858.15 ± 0.17	6,051.47 ± 0.03	20,996,800. ± 1,200.
Earth	398,600.434 ± 0.002	6,371.01 ± 0.02	86,164.090 ± 0.004
Moon	4,902.799 ± 0.003	1,737.53 ± 0.03	2,360,591.554 ± 0.005
Mars	42,828.3 ± 0.1	3,389.92 ± 0.04	88,642.663 ± 0.002
Jupiter	126,712,767. ± 5.	69,946. ± 6.	35,729.70 ± 0.05
Saturn	37,929,100. ± 2,400.	58,300. ± 6.	38,364.12 ± 0.05
Uranus	5,783,900. ± 3,800.	25,456. ± 60.	61,200. ± 720.
Neptune	6,809,000. ± 14,000.	24,650. ± 100.	64,200. ± 300.
Pluto	908. ± 42.	1,123 ± 4.	551,850. ± 20.

ship between rotation, gravity, and topography is particularly simple. To lowest order in q, we have

$$-G_{2,0} = J_2(g) = \frac{q}{2}, \quad -H_{2,0} = J_2(t) = \frac{5q}{6}$$

These values are maximal in the sense that, for a given value of q, a hydrostatic body whose density increases with depth will have gravity and topography harmonics less than these limits. In fact, for a hydrostatic body, the mean moment of inertia I is related to the gravitational oblateness via

$$\frac{I}{MR^2} = \frac{2J_2}{3J_2 + q}$$

It is primarily by just such comparisons of the gravity, topography, and rotation that we can make inferences about the composition and internal structure of the Jovian planets. On the other hand, for a nonhydrostatic body the values of the low-degree harmonics will not bear any simple relationship to the rotation rate.

This article discusses the planets in order of their distance from the Sun. The results summarized here are the work of many people. Because of space limitations, however, specific references have only rarely been cited. Comprehensive bibliographies of recent work are given in Bills and Synnott (1987) and in articles cited in the references.

Mercury

The primary sources of geodetic information for Mercury are occultations and Doppler tracking data from the flybys of the *Mariner 10* spacecraft and Earth-based radar observations.

Rotation. Prior to 1965 the rotation period of Mercury was believed to equal the orbital period of 87.9686 days. Radar observations of the surface revealed that the actual rotation period was (59 ± 5) days. It was quickly realized that these error bounds enclose the 58.6457-day period corresponding to a 2:3 commensurability between the spin and orbital periods. Analysis of *Mariner 10* images provides the best estimate of the rotation period: (58.6461 ± 0.005) days. Solar tidal torques tend to despin the planet toward a Sun-synchronous state. The present resonance is stable only if the solar torque acting on the planet's permanent equatorial ellipticity exceeds the tidal torque. Stability of resonance requires that $Q(B - A)/C > 10^{-7}$, where $1/Q$ is a measure of frictional dissipation. Even if Q were as low as 10, this stability only requires that $(B - A)/C > 10^{-8}$. As we shall see, the actual value far exceeds this limit.

The slow rotation of Mercury may explain its lack of natural satellites. The tidal torque exerted on a satellite depends on the angular rate at which the satellite moves across the planetary surface. For a planet rotating as slowly as Mercury, the orbits of all but the most distant satellites would have decayed to the point of an eventual impact onto the planet.

Gravity. Because of its lack of satellites, the earliest estimates of the mass of Mercury relied on observations of the perturbations it produced in the orbit of Venus. The perturbations are small, but by combining a long series of optical angular observations and a shorter but more precise series of radar range measurements, investigators determined that $GM = (22030 \pm 55)$ km$^3 \cdot$s^{-2} in 1971. The close flybys of the *Mariner 10* spacecraft allowed much better estimates of the mass and the second-degree harmonics: $GM = (22031.8 \pm 1.0)$ km$^3 \cdot$s^{-2}, $J_2 = (81 \pm 23) \times 10^{-6}$, $C_{2,2} = (0 \pm 4) \times 10^{-6}$, $S_{2,2} = (13 \pm 10) \times 10^{-6}$. Assuming that librations about the resonant spin state are fully damped, the a priori expectation is that $S_{2,2} = 0$. These gravity harmonics suggest that the normalized equatorial moment difference $(B - A)/C$ is roughly 10^{-5}, two to three orders of magnitude larger than is required for stability of the spin resonance.

Topography. Because of its smallness and angular proximity to the Sun, Mercury is a difficult object to study telescopically. Nevertheless, classical estimates of the radius have been surprisingly accurate, giving (2440 ± 7.5) km. More recent radar altimetry measurements suggest a mean radius of (2440 ± 1) km with a significant equatorial ellipticity: $C_{2,2} = (15 \pm 5) \times 10^{-5}$, $S_{2,2} = (6 \pm 5) \times 10^{-5}$. The radar data are confined to the equatorial region (6°S to 12°N) and thus allow no estimate of J_2.

Discussion. The mean density of Mercury is (5.4262 ± 0.0073) Mg$\cdot$m^{-3}. The dominant source of error in this value is uncertainty in the mean radius. The high density is generally accepted as evidence of a significant enrichment in iron, and some calculations suggest that it may be composed only of those materials that condensed at temperatures near that of metallic iron condensation in the cooling of the solar nebula. The moment of inertia is unknown observationally, but theoretical calculations suggest a range of $0.325 < I/MR^2 < 0.394$, corresponding to fully differentiated and nearly homogeneous models, respectively. The spin orbit resonance produces significant longitudinal temperature variations ($\sim 60°$C). If the entire silicate mantle is in conductive equilibrium with these boundary conditions, differential thermal expansion alone could produce second-harmonic topographic variations of $C_{2,2} = 6.8 \times 10^{-5}$ and gravity harmonics of $C_{2,2} = 7.5 \times 10^{-6}$. In each case this is about half of the observed value.

Venus

The primary sources of geodetic information on Venus are the *Pioneer* Venus orbiter spacecraft and radar ranging to the surface.

Rotation. The rotation of Venus is anomalous in several respects. It is slow, retrograde, and the axial orientation is unusual. When radar observations of the surface first revealed a retrograde rotation with a period of (247 ± 5) days, it was realized that this value was close to the 243.1650 days required for a synodic spin-orbit resonance with the Earth. If Venus were in such a resonance, it would present the same face toward the Earth at each inferior conjunction. However, more recent measurements indicate that the period is (243.019 ± 0.014) days, which seems to exclude the resonance. On the other hand, the atmospheric angular momentum is a large enough fraction (1.6×10^{-3}) of the solid-body angular momentum that a 40% decrease in atmospheric superrotation (over a period of perhaps decades) would allow the resonance to survive. Whether or not a resonance with Earth is involved, the current rotation rate of Venus almost certainly reflects a near balance between the dominant solar torques: a positive atmospheric thermal tidal torque and a negative solid-body tidal torque. The retrograde rotation may also be explained as a result of tidal evolution. A further peculiarity of the rotation is that, whereas the tidal evolution should drive the spin axis very close to the orbit normal, it is actually observed to be closer to the invariable pole of the Earth-Venus system.

Gravity. The earliest estimates of the mass of Venus come from observations of its perturbations of the orbits of Mercury and the Earth: In 1895 Newcomb estimated that $GM = (326{,}400 \pm 1100)$ km$^3 \cdot$s^{-2}. With more recent observations of planetary perturbations, a value of $GM = (324{,}860 \pm 80)$ km$^3 \cdot$s^{-2} was obtained in 1971. The best current estimate comes from analysis of Pioneer Venus Orbiter (PVO) Doppler tracking data. The preferred value at present in $GM = (324{,}858.15 \pm 0.17)$ km$^3 \cdot$s^{-2}. These same data have allowed estimates of higher-degree harmonics of the gravity field. The overall shape of the gravitational variance spectrum is similar to that of other solid planets. A notable difference between the Earth and Venus concerns the degree of correlation between gravity and topography. On Venus the correlation is significantly positive for all harmonic degrees greater than 2. On Earth, the correlation does not become significantly positive until degree 10 or so. This difference presumably reflects a difference in the internal viscosity stratification on the two planets. Whereas the viscosity of the Earth's mantle increases with depth, the Venusian mantle viscosity appears to be more nearly constant.

The second-degree harmonics reveal that the axis of least inertia is inclined to the rotation axis by about 3°. This presumably reflects a free wobble analogous to the Chandler wobble on Earth. Though the angular amplitude is about 10^5 times larger on Venus, the extremely slow rotation and nearly spherical mass distribution on Venus suggests that the energy requirements for maintaining the two wobbles are roughly comparable.

Topography. The earliest telescopic estimates of the radius of Venus yielded values in the vicinity of (6120 ± 8) km, which is now known to represent the cloud tops. The best current information comes from the PVO altimeter. The mean radius is (6051.47 ± 0.03) km. The mean density is (5.2448 ± 0.0005) Mg·m^{-3}, 4.8% less than the density of the Earth. Most of this difference is due to the greater self-compression in the more massive Earth. The remainder has been variously attributed to differences in bulk composition, state of oxidation, basalt/eclogite ratio, or some combination thereof. Though the total range of elevations is nearly the same as on Earth, the frequency distribution is very different. Over 60% of the surface area of the planet lies at elevations within 1 km of the mean. The variance spectrum of the topography is, however, very similar to the Earth, both in terms of a high-degree spectral decay $O(n-2)$, and in terms of a low-degree deficit relative to the extrapolated trend of the high-degree harmonics.

Earth

A great deal more is known about the Earth than any of the other planets. The material covered here is merely an attempt to view the Earth from a planetary perspective.

Gravity. The mass of the Earth is best determined from the motions of artificial satellites, monitored via laser ranging measurements. A recent estimate is $GM = (398{,}600.434 \pm 0.002)$ km$^3 \cdot$s^{-2}. The low-degree harmonics are also well determined: $J_2 = (1082.622 \pm 0.004) \times 10^{-6}$, $J_3 = (-2.5332 \pm 0.0007) \times 10^{-6}$, and $J_4 = (-1.631 \pm 0.007) \times 10^{-6}$. Hydrostatic values of these parameters can be calculated from the known rotation period and seismically determined radial density variations: $J_2 = 1074.467 \times 10^{-6}$, $J_3 = 0$, and $J_4 = -3.363 \times 10^{-6}$. Higher-degree harmonics have been estimated from combinations of satellite tracking data, ocean surface altimetry, and land surface gravimetry. In 1980, R. H. Rapp has estimated harmonic coefficients up to degree 180. The variance spectrum of the nonhydrostatic potential has the approximate form $\sigma_n(G) = 10^{-5}/n^2$.

Topography. Whereas the continental topography of the Earth has been known with reasonable accuracy for many decades, it is only in the last two decades that a reasonably clear picture of suboceanic topography has emerged. The mean radius of the Earth is (6371.01 ± 0.02) km. The mean density is thus (5.5148 ± 0.0005) Mg·m^{-3}. The distribution of elevations is strongly bimodal, reflecting a strong continental versus oceanic dichotomy. The variance spectrum of the nonhydrostatic topography is similar in form to the gravitational spectrum, but the low-degree harmonics do not correlate well.

Rotation. The sidereal rotation period of the Earth is (23.934469 ± 0.000001) hr. The quoted uncertainty represents real variations of ± 0.005 s in the length of day observed since A.D. 1800. In addition to these seemingly random variations, some of which apparently involve angular momentum exchange between the atmosphere and the solid body of the Earth, there is a more regular tidal deceleration of 2.4×10^{-3} s per century, which involves angular momentum exchange with the lunar orbit.

Moon

Rotation. On the basis of a long series of observations, Cassini, in 1693, deduced three empirical laws respecting lunar rotation: (a) the Moon rotates eastward, about a fixed axis, with constant angular velocity and in a period equal to one sidereal month of $(27.32166150 \pm 0.00000006)$ days, (b) the inclination of the Moon's axis of rotation to the ecliptic remains constant, and (c) the Moon's rotation axis and orbit normal both precess about, and remain coplanar with, the Earth's orbit normal. It has been subsequently shown that this rotational state minimizes the rate of tidal energy dissipation. The primary geodetic interest is in determining the extent of departures from Cassini's laws. The most successful observational technique has involved laser ranging to reflectors left on the lunar surface. These range measurements are now routinely performed with an accuracy of 2–3 cm, and have revealed a rich spectrum of librations, wobbles, and nutations. The amplitudes and periods of these motions provide valuable constraints on the internal mass distribution and the degree of internal energy dissipation. Among the valuable constraints obtained this way are the moment difference ratios: $(C - A)/B = (631.69 \pm 0.13) \times 10^{-6}$ and $(B - A)/C = (228.02 \pm 0.10) \times 10^{-6}$.

Gravity. The best present estimate of the lunar mass and low-degree harmonics comes from a combination of *Lunar Orbiter 4* Doppler signatures and lunar laser ranging data: $GM = (4902.799 \pm 0.003)$ km$^3 \cdot$s^{-2}, $J_2 = (202.2 \pm 1.2) \times 10^{-6}$, $C_{2,2} = (22.30 \pm 0.13) \times 10^{-6}$, and $S_{2,2} = (0.017 \pm 0.006) \times 10^{-6}$. These values, in conjunction with the rotationally determined moment difference ratios, provide an estimate of the mean moment of inertia: $I/MR^2 = 0.3905 \pm 0.0023$.

This same data set has been used, in conjunction with considerable additional low-altitude Doppler data, to obtain a degree-16 harmonic model. The variance spectrum of the gravity is similar in form to that of the Earth, and the correlation with topography is rather poor.

Topography. It is somewhat ironic that the topography of the Moon, our nearest planetary neighbor, is less well known that of Venus or even Mars. The near-side topography is known, to some extent, from Earth-based photogrammetry. The equatorial regions ($\pm 15°$ latitude) are very well known from laser altimeters on *Apollos 15, 16*, and *17*. The limb zone is fairly well known from a long series of stellar occulations. The remainder of the far side is essentially unknown.

B.G. Bills and A.J. Ferrari have used essentially all of the available data to estimate a degree-12 harmonic model of the topography. The mean radius is $R = (1737.53 \pm 0.03)$ km. This implies a mean density of (3.3440 ± 0.004) Mg$\cdot$m^{-3}. This value, in combination with the mean moment of inertia constraint, suggests an internal density stratification greater than could be produced by hydrostatic pressure. A model with an Fe/Ni core 300 km in radius, surrounded by a mantle with density 3.36 Mg$\cdot$m^{-3}, and topped by a crust 70 km thick with a density of 2.85 Mg$\cdot$m^{-3}, is consistent with these constraints. The first-degree harmonics imply that the center of figure is displaced from the center of mass by (1.98 ± 0.06) km. This offset may be due to a thickening of the crust on the far side. The topographic variance spectrum is also similar to the Earth's.

Mars

Rotation. The rotation period of Mars has been known with impressive accuracy for a long time. By 1666 Cassini had determined a period of just over 24 hr. In 1886, Wislicenus used observations of the time of central meridian crossing of Syrtis Major, spanning the interval from 1659 to 1881, to derive a period of $(24.6229597 \pm 0.0000036)$ hr. Radio tracking data from the *Viking* landers give the best current estimate: $(24.62296194 \pm 0.00000056)$ hr. Exchange of mass between the atmosphere and the seasonal polar caps will produce annual and semiannual variations in the rotation period with an amplitude of roughly 0.04 s.

Gravity. The earliest reliable estimates of the mass of Mars were obtained from observations of the orbital motions of its satellites Phobos and Deimos. In 1878, Hall estimated $GM = (42,900 \pm 70)$ km$^3 \cdot$s^{-2}. The best current estimate is based on Doppler data from the *Mariner 4* flyby: $GM = (42,828.3 \pm 0.1)$ km$^3 \cdot$s^{-2}. Subsequent orbiters have helped delineate lateral variations. G. Balmino et al. have estimated harmonics up to degree 18 from the orbital motions of *Mariner 9* and the *Viking* orbiters: $J_2 = (1960.45 \pm 0.04) \times 10^{-6}$, $C_{2,2} = (-54.733 \pm 0.006) \times 10^{-6}$, and $S_{2,2} = (31.395 \pm 0.006) \times 10^{-6}$. Assuming that the mean moment of inertia is $I = 0.365 MR^2$, hydrostatic values for the low-degree zonals are $J_2 = 1817.70 \times 10^{-6}$ and $J_4 = 8.1 \times 10^{-6}$. The variance spectrum is displayed in Fig. 1. Mars is notable in terms of having the largest known gravity anomalies of any planet. Mars is similar to Venus in terms of the good correlation between gravity and topography, but has affinities with the Moon in terms of the relative amplitudes of the observed gravity relative to that predicted from uncompensated topography.

Topography. The approximate size and shape of Mars was established quite early by telescopic observations. Classical estimates of the polar and equatorial radii are summarized by de Vaucouleurs (1964): $R_p = (3350 \pm 20)$ km, $R_e = (3375 \pm 20)$ km. The best current estimates of Martin topography come from Earth-based radar (restricted to $\pm 25°$ latitude) and spacecraft data (principally occultations). B.G. Bills and A.J. Ferrari have used essentially all of the pre-*Viking* data to estimate spherical harmonic coefficients to degree 16. The mean radius is $R = (3389.92 \pm 0.04)$ km. The corresponding mean density is (3.9335 ± 0.0004) Mg·m^{-3}. The center of figure is displaced from the center of mass by (2.50 ± 0.07) km toward Tharsis.

Jupiter

Rotation. The mean rotation period of Jupiter is known from observations of the periodic modulation in decametric radiation associated with the magnetic field. The period is (9.924917 ± 0.000014) hr. Observations of cloud features in the belts and zones reveal zonal wind velocities up to 150 m·s^{-1}. The zonal wind structure is quite stable and may be a surface manifestation of a more deep-seated differential rotation. The departures from solid-body rotation considerably complicate the connection between gravity harmonics and internal density distribution.

Gravity. The earliest estimate of the mass of Jupiter was made by Newton in 1686 from observations of the orbit of Callisto. He found $GM = 124,400,000$ km^3·s^{-2}. The best present estimate is based on Doppler data acquired during the encounters with *Pioneers 10* and *11* and *Voyagers 1* and *2*: $GM = (126,712,767 \pm 5)$ km^3·s^{-2}. These encounters have also provided information on the low-degree zonal harmonics of the gravity field: $J_2 = (14,736 \pm 1) \times 10^{-6}$, $J_4 = (-587 \pm 5) \times 10^{-6}$, and $J_6 = (31 \pm 20) \times 10^{-6}$.

Topography. The size and shape of Jupiter are best determined from occultations of *Voyagers 1* and *2*, and *Pioneers 10* and *11* spacecraft. The polar, equatorial, and mean radii at a 1-bar pressure level estimated from these observations are $R_p = (66,854 \pm 10)$ km, $R_e = (71,492 \pm 4)$ km, and $R = (69,946 \pm 6)$ km. The resultant mean density is (1.3248 ± 0.0004) Mg·m^{-3}. This low density implies a composition dominated by H and He. In 1980, W.B. Hubbard et al. estimated a composition of roughly 5 Earth masses of "rock," 15 Earth masses of "ice," 60 Earth masses of He, and 240 Earth masses of H.

Saturn

Rotation. The rotation period of Saturn is most reliably determined from observations of nonthermal radio emissions linked to the magnetic field of the planet. The best current estimate is (10.6567 ± 0.0019) hr. This presumably is the average rotation period of the deep interior. Significant zonal winds (up to 500 m·s^{-1}) make the rotation period of the visible surface a strong function of latitude.

Gravity. The first mass determination for Saturn (in 1831) was based on observations of orbital motions of Titan, the largest of its satellites, and gave a value of $GM = (37,948,000 \pm 43,000)$ km^3·s^{-2}. The best present estimate is based on a combination of Doppler data from *Pioneer 11* and the orbital motion of five of Saturn's natural satellites. The resulting value is $GM = (37,929,100 \pm 2400)$ km^3·s^{-2}. The low-degree zonal harmonics were also estimated from the same data: $J_2 = (16,479 \pm 18) \times 10^{-6}$, $J_4 = (-937 \pm 38) \times 10^{-6}$.

Topography. The best estimates of the size and shape of Saturn come from *Voyager* radio occultation data. The polar and equatorial radii at a 1-bar pressure level are $R_p = (54,365 \pm 10)$ km and $R_e = (60,268 \pm 4)$ km. The mean radius at this level is $(58,300 \pm 6)$ km. The corresponding mean density is (0.6848 ± 0.0003) Mg·m^{-3}. A density this low implies that the composition must be predominantly hydrogen and helium with very little admixture of heavier elements. Using the mass, radius, and low-degree harmonics of the gravity field as constraints, Hubbard and Stevenson (1984) have estimated that Saturn consists of roughly 5 Earth masses of "rock," 15 Earth masses of "ice" (mostly methane, ammonia, and water), and 10 Earth masses of helium, with the remaining 65 Earth masses composed of hydrogen.

Relative to an equipotential surface whose shape is determined by the observed gravity harmonics and mean rotation period, the observed 100-mbar pressure surface has an excess oblateness of (122 ± 11) km produced by the eastward zonal winds.

Uranus

Gravity. Not long after its discovery by Herschel in 1781, estimates of the mass of Uranus became available from observations of the orbits of its satellites. In 1849, J.C. Adams obtained a value of $GM = 6,320,000$ km^3·s^{-2}. The best current estimate of the mass is likewise based on natural satellite observations: $GM = (5,783,900 \pm 3800)$ km^3·s^{-2}. Though the satellites provide the best mass estimates, precession of the rings, as monitored via stellar occultations, is the best source of information about the low-degree zonal harmonics of the gravity field: $J_2 = (3.349 \pm 0.005) \times 10^{-3}$, $J_4 = (-3.8 \pm 0.9) \times 10^{-5}$.

Topography. The size and shape of Uranus are best known from stellar occultations. Observations from five separate occultations suggest polar and equatorial radii at the 1-bar pressure level of $R_p = (25,046 \pm 60)$ km and $R_e = (25,662 \pm 60)$ km.

Thus, the mean radius is $R = (25{,}456 \pm 60)$ km, and the mean density is (1.2545 ± 0.0096) Mg·m^{-3}. Though compositional models are not as well constrained for Uranus as for Jupiter and Saturn, it is at least clear that the mass fraction of hydrogen and helium in Uranus is much smaller than for the larger planets.

Rotation. Telescopically, Uranus is almost entirely featureless. Thus, estimates of the rotation period have relied, until very recently, on spectroscopic determinations in the Doppler shift in reflected sunlight. These observations have yielded a rotation period of (15.6 ± 0.8) hr. Observations of nonthermal radio emission and in situ magnetic field measurements by the *Voyager 2* spacecraft have refined estimates of the mean rotation period to (17.0 ± 0.2) hr. Spacecraft imaging results suggest significant (100 m·s^{-1}) zonal winds.

Neptune

Rotation. Neptune has a considerably higher visual contrast than Uranus, and optical tracking of cloud features in the mid-latitudes has revealed a rotation period of (17.83 ± 0.08) hr. Though this period pertains only to the latitudes at which the clouds were observed, if the atmospheric dynamics of Neptune is similar to that observed on Jupiter and Saturn, the rotation period of the mid-latitudes should be close to that of the interior of the planet.

Gravity. Anomalies in the orbital motion of Uranus lead Adams and Leverrier in 1846 independently to predict the existence of another planet beyond Uranus. These predictions not only led to the actual discovery of Neptune but also provided reasonably accurate prediscovery mass estimates. Within a year of Neptune's dicovery, B. Peirce had obtained a refined mass value from observations of the orbital motions of Triton: $GM = (6{,}848{,}000 \pm 37{,}000)$ km^3·s^{-2}. The best current estimates of the mass and the oblateness also rely on Triton's orbital motion: $GM = (6{,}809{,}000 \pm 14{,}000)$ km^3·s^{-2}, $J_2 = (4.3 \pm 0.3) \times 10^{-3}$.

Topography. The size and shape of Neptune are best determined from stellar occultation observations. From an occultation of 15 June 1983, estimates have been obtained for equatorial and polar radii at the 1-bar level of $R_e = (24{,}830 \pm 100)$ km and $R_p = (24{,}280 \pm 100)$ km. The resulting mean radius is $R = (24{,}648 \pm 100)$ km, which implies a mean density of (1.627 ± 0.023) Mg·m^{-3}.

Discussion. Until very recently, the rotational period had been uncertain enough that internal density modeling was rather inconclusive. Hubbard et al., using a period of 15 hr, claim that the gravity analysis of A.W. Harris and the occultation derived topography suggest that Neptune is essentially homologous to Uranus. On the other hand, French et al. (1985), using the same gravity data and essentially the same topography data, but favoring the longer rotational period of 18.2 hr, claim that Neptune is the least centrally condensed of all the giant planets. The recent period estimate of Smith (1985) seems to support the latter conclusion.

Pluto

Rotation. Pluto is far too small to allow direct telescopic observation of the rotation. The first estimates of the rotation period came from photometric detection of a $\sim 20\%$ variation in brightness with a period of (6.390 ± 0.003) days. Using a longer time span of observations, investigations improved this value by 1974 to (6.3874 ± 0.0002) days. At about the same time it was shown that a long-term increase in the amplitude of the light curve accompanied by a decrease in the mean brightness was best explained by a large obliquity $(90 \pm 40)°$. The sense of the rotation was not known until after the discovery of the satellite Charon (Christy and Harrington, 1978). The satellite has an orbital period of (6.3871 ± 0.0002) days, which is very nearly coincident with Pluto's rotation period. If the satellite is in a synchronous, equatorial orbit, the obliquity is $(118.5 \pm 1.0)°$. Thus, its rotational state is similar to Uranus in that it is retrograde and has its spin axis nearly in the orbital plane.

Gravity. The discovery of Neptune helped explain most of the anomalies in the orbit of Uranus, but some small residual anomalies remained. This led to predictions that yet another planet existed, beyond Neptune. The search for this object eventually led to the discovery of Pluto. However, contrary to the Neptunian case, the body that was found proved to be much too small to account for the anomalies that originally motivated the search. Most prediscovery estimates suggested a body at least as massive as the Earth. The first indication that the mass was much less than this came from considerations of the telescopic brightness, which, assuming a plausible range of albedos, yields maximum radii more Moonlike than Earthlike. The issue was not clearly resolved until the discovery of a satellite allowed a direct estimate of the combined mass of Pluto and Charon. The semimajor axis of the orbit is $(19{,}130 \pm 300)$ km; this value, in conjunction with the above-quoted orbital period, allows the mass of the system to be estimated as $GM = (908 \pm 42)$ km^3·s^{-2}.

Topography. The radius of Pluto is not particularly well determined at present. Analysis of the observations of several Pluto-Charon occultation and transit events suggests that the sum of the radii of Pluto and Charon are (1123 ± 4) km and (600 ± 6) km respectively. Little, if anything, is known of the shape of Pluto or Charon, though the preeclipse light curve suggests either quite extreme albedo variations or an irregular shape.

Discussion. The best current estimates of the mass and radius of Pluto suggest a density of $(1.99$

± 0.15) Mg·m⁻³. Pluto is thus very rock-rich, with a rock/(rock + ice) ratio of 0.7–0.8.

BRUCE G. BILLS

References

Anderson, J. D., 1975, Planetary geodesy, *Rev. Geophysics* **13**, 274–275.

Bills, B. G., and S. P. Synnott, 1987, Planetary geodesy, *Rev. Geophysics* **25**, 833–839.

Bretagnon, P., 1982, Integration constants and mean elements for all the planets, *Astron. Astrophysics* **108**, 69–75.

Christy, J. W., and R. S. Harrington, 1978, The satellite of Pluto, *Astron. Jour.* **83**, 1005–1008.

Davies, M. E., V. K. Abalakin, M. Bursa, T. Lederle, J. H. Lieske, R. H. Rapp, P. K. Seidelman, A. T. Sinclair, V. G. Teifel, and Y. S. Tjuflin, 1986, Cartographic coordinates and rotational elements of the planets and satellites, *Celest. Mech.* **39**, 103–113.

De Vaucouleurs, G., 1964, Geometric and photometric parameters of the terrestrial planets, *Icarus* **3**, 187–235.

Ferrari, A. J., and B. G. Bills, 1979, Planetary geodesy, *Rev. Geophysics* **17**, 1663–1677.

Hubbard, W. B., 1984, *Planetary Interiors*. New York: Van Nostrand Reinhold.

Hubbard, W. B., and D. J. Stevenson, 1984, Interior structure of Saturn, in T. Gehrels and M. S. Matthews, eds., *Saturn*. Tucson, Ariz.: University of Arizona Press, 47–87.

Sjogren, W. L., 1983, Planetary geodesy, *Rev. Geophysics* **21**, 528–537.

Smith, B. A., 1985, Ground-based observations of Uranus and Neptune using CCD instruments, *Soviet Astron.* **19**, 29–32.

Cross-references: *Earth and Moon: Origins; Earth Orientation; Geodesy: Physical; Gravity Fields: Implications for Planetary Interiors; Photogrammetry; Solar System: Origins.*

PLANETARY MAGNETIC FIELDS

Magnetic fields are generated by electric currents. Currents can flow in the highly electrically conducting fluid cores of the planets, which convect and which may be in differential rotation, i.e., they may rotate with different periods in different regions. These electric currents dissipate energy and the energy contained in the magnetic field would be drained by this dissipation if some source did not put energy back into the magnetic field. This energy source is the same source that drives the motion of the electrically conducting fluid, and it must have sufficient vigor to sustain the magnetic field against dissipative losses. The overall system of energy sources, convective motions, electric currents, and magnetic field generation is called a *planetary dynamo*.

Many energy sources have been proposed to drive planetary dynamos: primordial heat left over from planetary accretion, radioactive heating, precessional stirring, latent heat of fusion, and gravitational settling. The last two sources occur when a constituent of a solution, such as iron, solidifies, releasing its latent heat of fusion, and sinks to the center of a planet releasing gravitational energy and stirring the liquid. The nature of the energy source is not deemed to be so important as its strength, or more properly, its efficiency in producing the required fluid motions to produce the planetary magnetic field.

Four basic equations of physics govern the interrelationship of electrical charge, electric currents, and magnetic and electric fields. These equations are called Maxwell's laws. In an electrical resistor, current is usually directly proportional to the applied electric field. In a magnetized flowing conductor, an additional term arises because currents flow due to the motion of the conductor. The relationship between the current, the electric and magnetic fields, and the flow velocity is called the Ohm's law. The Ohm's law can be combined with the four Maxwell's laws to give what is known as the *dynamo equation*. This equation relates the change in magnetic field due to the resistive decay of the currents and the regeneration of the field due to the motion of the fluid. In the absence of motion in the conducting fluid, the magnetic field will decay with a time scale of a few tens of thousands of years for the Earth's dynamo to a few millions of years for the Jovian dynamo.

Solutions to the dynamo equation are known that produce magnetic fields given a specific velocity field. These solutions are called kinematic solutions because the velocity is specified, a priori. A self-consistent magnetohydrodynamic (MHD) dynamo in which the velocity and magnetic fields are self-consistently derived has not yet been achieved, but computer simulations of self-consistent MHD dynamos appear to be close. Studying planetary magnetic fields helps us in this regard because we can observe the varying responses of planetary dynamos to varying boundary conditions, such as rotation rate and heat flux, and hence infer more about the nature of the dynamo process itself.

Electrical currents are also found in the various plasmas of the solar system. These plasmas, or electron-ion gases are highly electrically conducting. The material inside the sun, the solar atmosphere, the expanding solar wind and the gases in the high upper atmospheres of all the planets are plasmas. These currents provide their own magnetic fields, which add to those generated by currents within the planets. In many instances these magnetic fields equal or exceed that due to the internal sources in the region accessible to spacecraft. These currents can cause difficulties in the precise determination of the planetary contribution.

Finally, there are currents at the atomic level in solid materials. Our usual exposure to such magnetic fields is in the form of artificial magnets but nature too magnetizes materials. This natural remanent magnetization preserves a record of the magnetic environment of the material at the time of its cooling. The study of magnetized terrestrial rocks and lunar samples has led to improved understanding of the magnetic history of both the Earth and the Moon.

The strong magnetic fields of some planets, e.g., that of the Earth and of Jupiter, can lead by various processes to the energization of charged particles and to the trapping of intense radiation belts. These charged particles, in turn, generate radio waves that can be detected far from the planet. It was through the study of radio waves from Jupiter that its magnetic field was first discovered and characterized. Because of various asymmetries in these radio emissions, it is possible to determine from their periodic modulation the rotation rate of planets. It is generally supposed that this rotation rate represents the rotation rate of the interior generation region deep inside the planets, which is not otherwise measured from spacecraft.

The instrument for measuring magnetic fields is called a magnetometer. Several different types of magnetometers are in use but the one most commonly employed on planetary spacecraft is the flux gate magnetometer, which is simple, light, and rugged. Magnetic fields in space are measured in gammas, Gauss, and Teslas where 1 Tesla equals 10^4 Gauss or 10^9 gammas. The Tesla is the standard unit of the International System of Units and is gradually becoming more and more prevalent in its usage. The magnetic field on the surface of the Earth at the equator is about 31 μT (one millionth of a Tesla). In the outer reaches of the Earth's magnetosphere, it is about 100 nT (1 nT equals one billionth of a Tesla). Outside the Earth's magnetosphere in the solar wind at 1 AU, the magnetic field strength is about 6 nT. The magnetic field strength on the surface of Jupiter is about 1 mT (one thousandth of a Tesla), but quite varied because of the high harmonic content of the surface field of the planet.

The two dynamo driven fields for which we have extended observations, that of the Earth and that of the Sun, periodically reverse. The Sun does this fairly regularly every 11 years. The Earth does this less regularly over time spans of hundreds of thousands of years. We do not have data that demonstrate that any planetary fields have reversed or are in the process of reversing. However, one interpretation of lunar paleomagnetic data is that the ancient lunar magnetic fields did reverse. For the planets, we have but a snapshot in time. We have to take them as they are today, trusting that their present properties are typical of their state over a more extended period.

The planets are treated below in order of their distance from the Sun with the exception of the Earth's moon, which we defer to last. We will not treat either the Earth or meteorites, leaving them to authors of other articles in this volume. Those interested in pursuing any of the topics covered in this article in further depth are referred to the bibliography at the end of this article.

Mercury

Mercury is the smallest of the terrestrial planets (R_m = 2440 km), intermediate between the Earth's moon and Mars in size. It rotates more slowly than the Moon, rotating with a period of 59 days compared to the Moon's 28-day period. It is heavily cratered like the Moon and apparently has a very ancient crust. Mercury differs from the lunar case in that its rotational period of 59 days is not the same as its orbital period, and hence Mercury does not always keep the same face to the Sun. It is much denser than the Moon, 5.4 g $\cdot$ cm^{-3}, compared to the lunar 3.3 g $\cdot$ cm^{-3}. It travels in an elliptical orbit of moderately high eccentricity (e = 0.206) and inclination to the ecliptic (i = 7°). Furthermore, Mercury appears to have an active dynamo in its core. The Moon does not. Recently, sodium and potassium in Mercury's atmosphere have been observed spectroscopically from Earth. It is possible that the sodium and potassium at Mercury are due to outgassing from the interior of Mercury and is possibly an indication that Mercury is an active, dynamic planet despite its outward resemblance to the Moon.

Mercury has been visited by only one spacecraft, *Mariner 10*, which made three passes by the planet from March 1974 to March 1975, of which only the first and the third passes were suitable for studying planetary magnetism. On the first pass, the spacecraft crossed the dark side of the planet, approaching within 723 km of the surface. The maximum field strength observed was nearly 100 nT at closest approach. The characteristics of the data were very similar to those that would be obtained on a pass through the terrestrial magnetosphere but were on a far smaller scale here. Figure 1 shows a sketch of the Mercury magnetosphere and its interaction with the solar wind, the rapidly expanding, ionized, outer atmosphere of the Sun. As at Earth there is a bow shock behind which the interplanetary magnetic field is compressed while the solar wind flowing out from the Sun is suddenly slowed down and heated as it supersonically reaches the planet. Behind the shock, another boundary is encountered, the magnetopause. This boundary defines the region that is shielded by the planetary magnetic field from the solar wind. The third Mercury pass was also behind the planet but closer to the planetary surface and at higher latitudes. The maximum magnetic field was over 400 nT. Otherwise the features were similar to those observed on the first pass.

The magnetic measurements of *Mariner 10* were strongly influenced by the current systems flowing

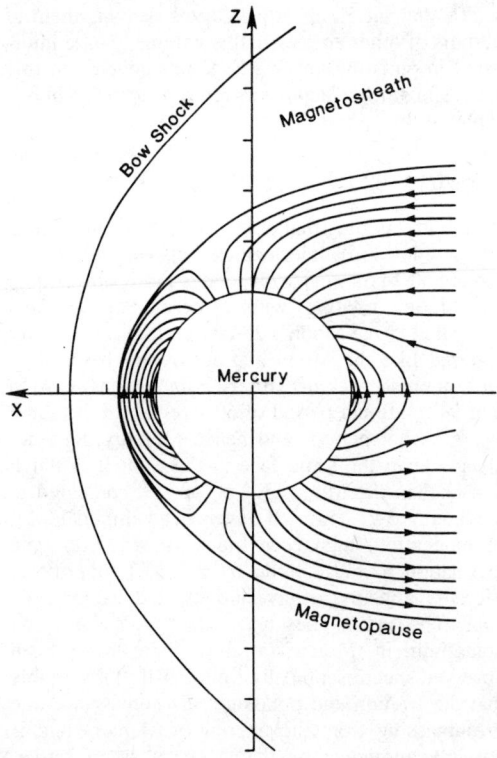

FIGURE 1. Sketch of the magnetic field lines in the noon-midnight meridian of the magnetosphere of Mercury. The bow shock slows, heats, and deflects the supersonic solar wind around the planetary magnetosphere. The magnetosheath is the name given to this shocked solar wind flow. The magnetopause is the boundary between this flowing plasma and the planetary magnetic fields. The solar wind interaction with the Earth, Jupiter, Saturn, and Uranus has this same basic form. The most noticeable difference between these magnetospheres is the relative size occupied by the planet. Mercury occupies the largest fractional volume of its magnetosphere and Jupiter the least.

on the magnetospheric boundaries. This influence can be seen in Fig. 1, which shows how much the field lines in the night side magnetosphere of Mercury differ from a dipolar configuration. One must take external conditions into account when attempting to derive the planetary moment. No solar wind ion data were obtained from *Mariner 10*. The solar wind properties were deduced from electron measurements, a less accurate technique. Further, the two *Mariner 10* passes through the Mercury magnetosphere surveyed only a limited portion of the planet. In order to survey accurately the magnetic moments of the internal field, a low altitude polar orbit is required. As a result of this lack of data the model inversions are not unique and the derived moments are controversial. The derived magnetic moments range from 1.8×10^{12} T · m^{-3} to 6×10^{12} T · m^3. There is general agreement that the observed magnetic field is generated by an active dynamo. The observed magnetic fields do not have the characteristics expected if they were induced by the interaction of the solar wind, and they appear to be too strong to be caused by natural magnetic remanence of the crust of Mercury.

Venus

Of the terrestrial planets, Venus is the closest to the Earth and in size almost a twin to the Earth (1 $R_v = 6051$ km). However, Venus rotates much more slowly than the Earth; one sidereal day on Venus equals 243 Earth days. This is about one quarter of Mercury's rotation rate. All else being equal, from dimensional considerations we would expect the magnetic field in a dynamo to be proportional to the rotation rate and the core radius. Since the magnetic moment is proportional to the cube of the radius times the field strength, we obtain a scaling proportional to the rotation rate and the radius of the core to the fourth power. Applying this scaling law to Venus and assuming a core radius of 0.5 R_v, we would predict a moment of 1.9×10^{13} T · m^3, which is 2×10^{-3} of that of the earth and corresponds to a surface field of 86 nT. Thus if Venus had an internal dynamo, we would expect a weak but measureable magnetic field. However, spacecraft would have to probe to low altitudes to observe it.

In December 1962, *Mariner 2*, the first successful mission to another planet, carried a flux gate magnetometer but at its closest approach distance of 6.6 R_v detected no evidence for a planetary magnetic field. This observation put an upper limit of 0.05 of that of the Earth's on the Venus magnetic moment. In October 1967 *Mariner 5* flew much closer to Venus coming within 1.7 R_v of the center of the planet and *Venera 4* carried a magnetometer to within 200 km of the surface. These data lowered the upper limit on a planetary magnetic moment to 10^{-3} of that of the Earth.

Venera 9 and *10* were placed into orbit about Venus in October 1975. While they revealed that Venus had a magnetic tail pointing in the antisolar direction in many respects similar to that of Mercury and of the Earth, in the final analysis they also revealed no evidence for an intrinsic magnetic field. The Veneras were followed by the *Pioneer Venus* orbiter in December 1978. This spacecraft had a controllable distance of closest approach (i.e., periapsis) and could probe deep down into the upper atmosphere. It also detected no evidence for a planetary magnetic field and lowered the upper limit of any intrinsic field to being less than 8×10^{10} T · m^3 or less than 10^{-5} of the terrestrial magnetic moment. The implication of such a small intrinsic magnetic field is that Venus has a stably stratified fluid core with no dynamo action, which is possible because Venus is slightly smaller than the Earth and has lower internal pressures as its center.

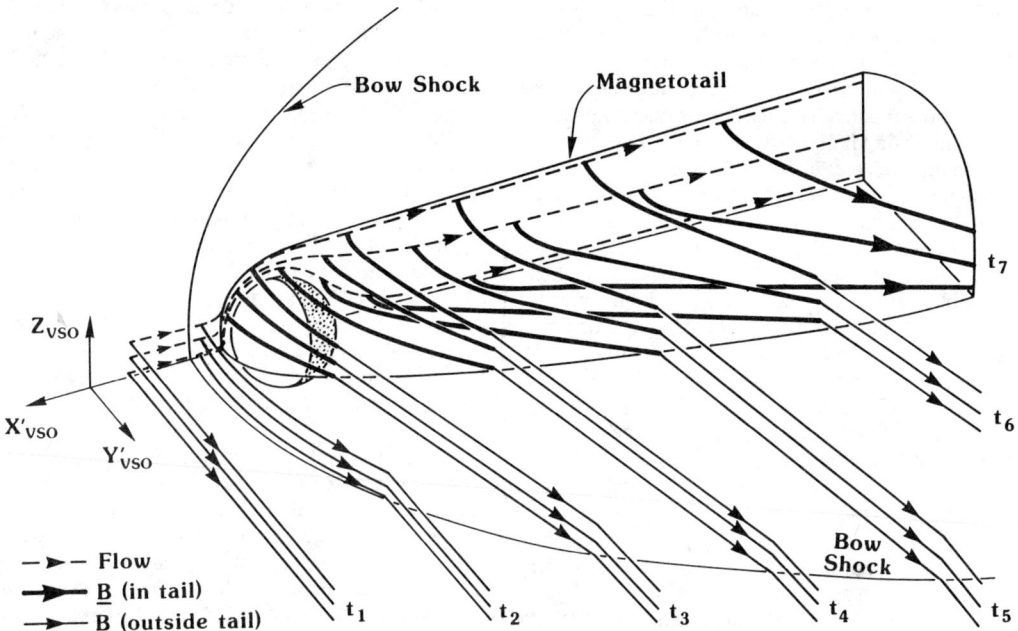

FIGURE 2. Sketch of the distortion in the magnetic field caused by the solar wind interaction with Venus. At time t_1 the magnetic field lines have not encountered the planet and are undistorted. At time t_2 they have just passed through the bow shock and entered the magnetosheath. The solar wind carrying the field lines is slowed in the magnetosheath and the field lines bend. As the field lines approach more closely to the planet at time t_3, the field lines closest to the planet, where the plasma is moving most slowly, are distorted the most. The flowing plasma picks up additional ions from the neutral atmosphere close to the planet and slows down even more in this region. The net result as shown at time t_4 and later is a "taillike" configuration of magnetic field lines, which are most distorted for those field lines that passed at one time closest to the planet.

The many Venus missions have allowed us to derive a moderately sophisticated understanding of the interaction of the magnetized solar wind with Venus. Figure 2 shows this interaction schematically. Magnetic field lines are carried to the planet by the solar wind. As they pass through the extended upper atmosphere of Venus, they pick up ions due to various processes, such as photoionization and charge exchange. This addition of mass slows down the plasma near the planet while the ends of the field lines are carried downstream by the solar wind. The net result of this process is an extended "induced" magnetotail. The direction of the tail field is governed by the position of the observation point and the direction of the interplanetary field transverse to the flow. This process is the same as that which forms the tail of a comet.

Mars

Mars is intermediate between the Earth and Mercury in size (radius of Mars is 3390 km) and has a spin period only slightly longer than 1 day. If internal conditions were similar to those in the Earth and Mercury, we might expect a magnetic moment intermediate between that of the Earth and Mercury. However, it is clear that the magnetic moment of Mars is at most equal to, but probably very much less than, that of Mercury. It is also quite evident that the interior of Mars is different from that of the Earth and Mercury, if only in the size of the core. The mean density of Mars is 3.9 g · cm^{-3} in contrast to Mercury's 5.4 g · cm^{-3}. Estimates of the core size range from 1500 to 2000 km. Mars appears to be less seismically active than the Earth.

Even though Mars has been visited by many spacecraft, only a few of these have carried magnetometers. *Mariner 4* in 1965 was the only American Mars spacecraft to make magnetic measurements. Later, in November and December 1971, *Mars 2* and *3* were placed into orbit by the USSR. *Mars 2* and *3* came within only 1500 km of the surface at periapsis. *Mars 5* was placed into orbit in February 1974. It came within only 1800 km of the surface at closest approach.

Mariner 4 came close enough to Mars on its flyby trajectory to detect a planetary bow shock but it did not detect any unambiguous evidence for a planetary magnetic field. *Mars 2* and *3* were placed into Mars orbit on November 27 and December 2, 1971 with quite different orbits. *Mars 2* had a periapsis altitude of 1300 km and an apoapsis altitude of 28,000 km.

Mars 3 had a similar periapsis and an apoapsis of 212,000 km. *Mars 2* never achieved inertial stabilization and continued to rotate about the Mars-Sun line thus compromising its measurements. The Soviet investigators have used these data as evidence for an intrinsic Martian magnetic field. However, the orientation of the claimed magnetic field is not unambiguous from these data and is model dependent. Suffice it to say that alternate explanations have been proposed for all the claimed pieces of evidence for a field of intrinsic origin. Thus, it is most probable that the present data establish but an upper limit on the Martian magnetic moment. This upper limit ranges from about 10^{-5} to 3×10^{-4} of the Earth's magnetic moment depending on the author. Indirect means of deducing any planetary magnetic moment, such as from the location of the planetary bow shock, have been equally unsatisfactory.

There are several possible explanations why a Martian magnetic dynamo may have stopped. The correct explanation, if needed, may have to wait until we have gained more knowledge of the Martian interior through seismology and chemical analyses. It is possible that Mars had an ancient planetary dynamo. If so we would expect to find remanent magnetization in the Martian rocks which in turn should allow us to deduce the properties of this ancient dynamo.

Jupiter

Jupiter is the largest of the planets (1 R_j = 71,400 km) and the most rapidly rotating, with a period of 09 h 55 m 29.7 s. For several decades it also has been known to possess a magnetic field because of its polarized radio emissions. Jovian radio emissions were first detected at Megahertz frequencies by Burke and Franklin in 1955. Later synchrotron radiation was detected due to energetic particles gyrating in Jupiter's strong magnetic field. Jupiter's interior is far different from the interiors of the terrestrial planets. The planet consists mainly of hydrogen and helium. The pressure in the interior of Jupiter is so great that it is expected that the deep interior is metallic hydrogen. The transition from the liquid molecular state to the metallic state is thought to occur at about 0.71–0.75 R_j. It is within this metallic liquid that Jupiter's dynamo is thought to act.

Pioneer 10 flew by Jupiter in December 1973 in a prograde sense passing within 2.9 R_j and *Pioneer 11* flew by Jupiter in December 1974 in a retrograde sense passing within 1.6 R_j. *Voyager 1* arrived on March 5, 1979 passing within 4.9 R_j of the planet's center and *Voyager 2* on August 20, 1979 passing within 10.1 R_j. These spacecraft revealed a planetary magnetic field that had a dipole moment of 1.55 $\times 10^{20}$ T·m³ or close to 20,000 times larger than that of the Earth. This field had strong harmonic content. The dipole to quadrupole to octupole ratio was 1.00 to 0.24 to 0.21 compared to the 1.00 to 0.14 to 0.10 for Earth. The tilt angle of the magnetic dipole axis to the rotation axis is about 9.7° compared to the terrestrial 11.4°.

The Jovian magnetosphere contains four large satellites, Io, Europa, Ganymede, and Callisto, and many small ones. All satellites contribute some mass to the Jovian magnetosphere, which becomes ionized and spun up to high velocities in the rotating Jovian magnetic field. Io has active volcanoes which contribute greatly to the amount of mass in the magnetosphere. This rapidly rotating, added mass greatly distorts the magnetic field beyond Io creating a so-called magnetodisk. This distortion is so great that the internal planetary field contributions can be accurately determined only inside the orbit of Io at 6R_j.

The magnetosphere of Jupiter is immense. The magnetopause stands off the solar wind in front of Jupiter at a distance of about 5 million km. Its tail appears to stretch away from the sun all the way to Saturn. There has been some speculation that Io has an intrinsic magnetic field but no firm evidence that it or any other Galilean satellite actually has an active internal dynamo.

Saturn

Saturn is the second largest planet with a radius of 60,330 km and a rotation period of 10 hr 39 min. On September 1, 1979, *Pioneer 11* reached Saturn passing within 1.4 Saturn Radii R_s of the center of the planet. *Pioneer 11* was soon followed by *Voyager 1*, on August 22, 1980, passing within 3.1 R_s of the center of the planet and by *Voyager 2* on August 26, 1981 passing within 2.7 R_s of the center of the planet. The results obtained were quite surprising, based on our experience at other planets.

The first surprise is the magnitude of the field. The dipole moment of 4.6×10^{18} T·m³ corresponding to an equatorial surface field of about 20 µT is much less than the 100 µT expected. Secondly, the radial dependence falls much closer to the inverse cube law expected of a dipolar field than the Jovian field. In fact, the ratio of the quadrupole moment to the dipole moment for Saturn is 0.07 compared with the terrestrial ratio of 0.14. The simplest explanation of this harmonic purity is that the conducting core of Saturn is much smaller relative to the radius of the planet than the terrestrial core. The third surprise was the tilt of the dipole moment. The terrestrial dipole moment is tilted at an angle of 11.4° to the Earth's rotation axis. The Jovian dipole is tilted about 10° to the rotation axis but the dipole moment of Saturn is inclined less than 1° to the rotation axis. *Voyager* made a close flyby of Titan, Saturn's largest moon and found no evidence of an intrinsic magnetic field.

The surprising nature of the Saturn magnetic field is believed to be possibly caused by the formation of helium rain in the interior of the planet. Under

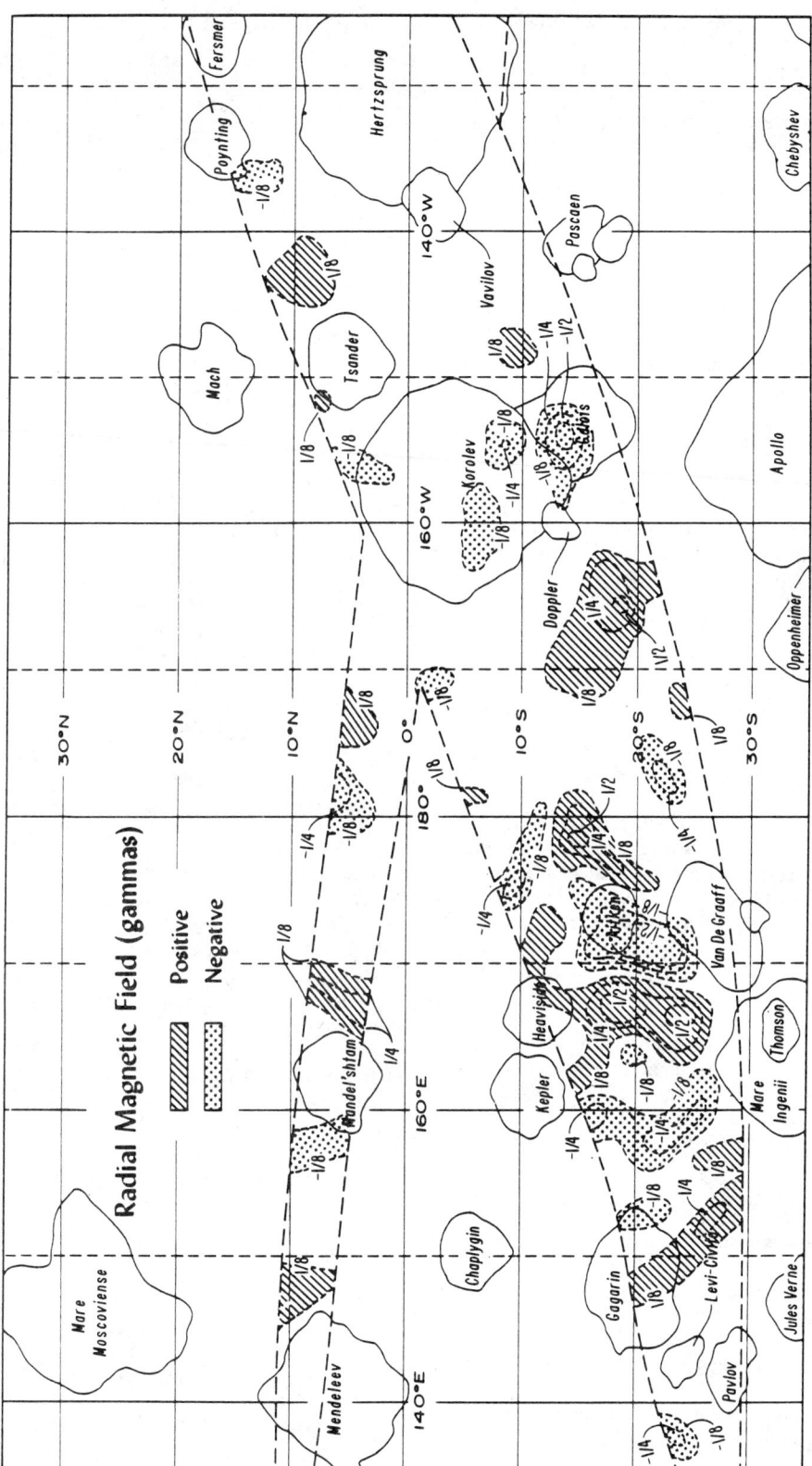

FIGURE 3. The radial component of the lunar magnetic field mapped by the *Apollo 15* and *16* subsatellites from 135° E to 125° W at altitudes between 65 km and 100 km. The limited coverage of the lunar surface resulted from the fact that the mapping could be done only when the Moon was in the geomagnetic tail together with the limited life time of the two satellites. The upper swath was obtained by *Apollo 16*; the lower broader swath was obtained by *Apollo 15*.

the temperatures and pressures that are thought to be present in the interior of Saturn, helium is thought to condense in the hydrogen fluid. This condensation into raindrops releases heat and the sinking of the raindrops releases gravitational energy. This process is thought to reduce the size of the dynamo region, which reduces the strength of the magnetic field and decreases the relative strength of the higher harmonics at the planetary surface because of their more rapid radial fall off. The axial alignment of the magnetic field is thought to be associated with attenuation of the nonaxisymmetric components by the differential rotation of highly conducting material in the core. Still, Saturn's field is unique and somewhat mysterious, and was made all the more so by the Uranian observations.

Uranus

Uranus is twice as far from the Sun as Saturn with a mean heliocentric distance of 19.2 AU. It has an equatorial radius of 25,900 km, less than half of that of Saturn. It has a mean density of 1.25 g · cm^{-3} and rotates with a period of about 16.8 hours. Uranus has the largest rotational axis inclination of any of the planets, 97.9°. Its satellite and rings orbit the planet in planes almost orthogonal to the ecliptic plane. At present, the interior structure is not well understood but continued analysis of the *Voyager 2* data are improving this situation. Presently favored are three layer models of rock, ice, and gas.

An intrinsic magnetic field at Uranus was generally expected in order to explain the intense Lyman α emissions observed there. The expectation was that these emissions were auroral phenomena that derive their energy from the solar wind. On January 24, 1986, *Voyager 2* encountered Uranus passing within 107,000 km of the center of the planet. In the last few days before the encounter, radio emissions were seen. A planetary blow shock and magnetopause were crossed and *Voyager 2* entered a planetary magnetosphere. The solar system had one more surprise for those studying planetary magnetism. The dipole moment of Uranus was not tilted at about 11° to the spin axis as at Jupiter and the Earth or near zero as at Saturn but close to 60° to the spin axis. The surface field was Earthlike near 23 µT corresponding to a magnetic moment of 3.9 × 10^{17} T · m^3 or slightly more than 50 times that of the Earth.

Prior to the encounter the speculation had been that any Uranian magnetic field would be generated either in an iron-rich core deep inside the planet and have a weak surface field of about 10 µT or be generated in a water ice mantle and have a surface field about an order of magnitude greater. The observed field sits nearly between these two limits. There was no prior speculation, nor any clear present explanation, for the large tilt angle.

Neptune and Pluto

There is no present observational evidence for or against magnetic fields at Neptune and Pluto. Neptune could quite possibly be similar to Uranus because it is thought to be similar in size and composition to Uranus. Pluto appears to be too small for dynamo action.

The Moon

The Earth's moon is the most accessible of all the small bodies in the solar system and hence we have more information about the properties of the Moon than of any other solar system body, except the Earth. The first lunar magnetic measurements were made in September 1959 by *Luna 2*. Later, in April 1966, *Lunar 10* carried a magnetometer in Lunar orbit. No lunar magnetic effects were detected. In July 1967, *Explorer 35* was placed into lunar orbit with an apoapsis of 5.4 lunar radii (R_L) and a perilune of 1.4 R_L. No lunar dipole field was detected but compressional disturbances were seen at the edge of the lunar wake, which were interpreted to be due to the deflection of the solar wind by magnetized regions on the lunar surface. This interpretation was later confirmed by magnetic measurements on the subsatellites put into low altitude lunar orbit by the *Apollo 15* and *16* astronauts. The regions that were supposed to be magnetized because they corresponded to solar wind deflections were directly sensed to have magnetic fields. Furthermore, the returned lunar samples were found to be magnetic. Figure 3 shows a map of the radial component of the lunar magnetic field observed by the *Apollo 15* and *16* subsatellites.

The Moon clearly does not have an active planetary dynamo today, but the simplest explanation of the magnetization of the lunar surface is that it once did. It seems likely that the Moon once had an internal dynamo. Studies of the intensity of magnetization of the returned lunar samples suggest that this field was strongest about 4 billion years ago and weakened considerably over the next billion years.

Synopsis

Mercury, the Earth, Jupiter, Saturn, and Uranus all have active planetary dynamos today. Neptune may have. Venus and the Moon do not, although they both may once have had one. Mars probably does not have an active dynamo, but this question is not completely settled. There are still many mysteries left about planetary magnetism. We still do not have a good model of a planetary dynamo and we still have many observational gaps in our knowledge of the planets.

C. T. RUSSELL

References

Burke B. F., and K. L. Franklin, 1955, Observations of a variable radio source associated with the planet Jupiter, *Jour. Geophys. Research* **60**, 213-217.

Dessler, A. J., ed., 1983, *Physics of the Jovian Magnetosphere*. New York: Cambridge University Press.

Gehrels, T., and M. S. Matthews, eds., 1984, *Saturn*. Tucson, Ariz.: University of Arizona Press.

Hunten, D. M., L. Colin, T. M. Donahue, and V. I. Moroz, eds., 1983, *Venus*. Tucson, Ariz.: University of Arizona Press.

Merrill, R. T., and M. W. McElhinny, 1983, *The Earth's Magnetic Field: Its History, Origin and Planetary Perspective*. New York: Academic Press.

Russell, C. T., 1987, Planetary magnetism, in J. A. Jacobs, ed., *Geomagnetism*. New York: Academic Press.

Stevenson, D. J., 1983, Planetary magnetic fields, *Rep. Prog. Physics* **46**, 555-620.

Cross-references: Earth and Moon: Origins; Earth's Core; Geomagnetic Field, Main: Theory; Natural Remanent Magnetization (NRM); Paleomagnetism: Meteorites; Planetary Geodesy; Rock Magnetism; Solar System: Origins.

PLUMES—See MANTLE CONVECTION AND PLUMES.

RADIOMETRIC AGES—See ABSOLUTE AGE DETERMINATION: RADIOMETRIC.

REFLECTION SEISMOLOGY—See EXPLORATION SEISMOLOGY.

ROCK MAGNETISM

The past three decades have witnessed a new paradigm, the plate tectonics paradigm, in Earth sciences. The record of the Earth's magnetic field stored in rocks played a major role in the establishment of this paradigm. The study of rock magnetism is the study of the magnetic properties of rocks and, as a corollary, the study of magnetic minerals in those rocks. Excellent books on the subject are Nagata (1961), Stacey and Banerjee (1973), and O'Reilly (1984).

All rocks and minerals carry an induced magnetization described by

$$M = \chi H$$

where M is the magnetization, H is the external magnetic field present, and χ, a second-order tensor, is the magnetic susceptibility. If $\chi < 0$ for a material, the material is called *diamagnetic*; if $\chi > 0$, the material is *paramagnetic*. Halite (NaCl) is an example of a diamagnetic mineral, and fayerite (Fe_2SiO_4) is an example (at room temperature) of a paramagnetic mineral.

In addition to an induced magnetization, many minerals and rocks have a "permanent magnetization." More properly this permanent magnetization is called *remanent magnetization* (RM). It is defined as that magnetization present in a zero external magnetic field. Although there are many types of RM, the most common forms are given in Fig. 1. An RM of about 10^{-2} to 10^{-3} A · m^2 · kg^{-1} would be considered to be strong in rock magnetism, and an RM of about 10^{-6} or less would be considered weak. Basalts commonly exhibit moderate to strong RM, whereas limestones commonly exhibit weak RM.

Natural Remanent Magnetization (NRM): the RM acquired by a sample (rock or mineral) under natural conditions.

Thermal Remanent Magnetization (TRM): the RM acquired by a sample during cooling from a temperature above the Curie or Néel temperature to room temperature. The term partial TRM refers to the RM acquired over a subinterval of this temperature range.

Chemical Remanent Magnetization (CRM): the RM acquired during any chemical change (including grain growth) of that sample.

Thermal CRM (TCRM): the RM acquired during cooling and chemical alteration.

Detrital (or depositional) Remanent Magnetization (DRM): the RM acquired by sediments formed by grains settling out of a fluid.

Post-Depositional Remanent Magnetization (Post-DRM): the RM acquired by the sample from physical processes acting after deposition (e.g., bioturbation or compaction).

Viscous Remanent Magnetization (VRM): the RM acquired over a long time in an external magnetic field when all other factors (e.g., sample chemistry and temperature) are held constant.

Isothermal Remanent Magnetization (IRM): the RM acquired over a short time in an external magnetic field (that is usually strong). Although there is no agreed upon definition of "short time" versus "long time" used in this definition and in the VRM definition, in practice "short time" often refers to a few seconds and "long time" to anything longer than this.

Anhysteretic Remanent Magnetization (ARM): the RM acquired when the peak value of an alternating magnetic field is decreased from a large value to zero in the presence of a weak constant magnetic field. (This definition differs slightly from that used in magnetic tape research in which the strength of the steady field is linearly proportional to the peak value of the alternating field.)

FIGURE 1. Some common types of remanent magnetization.

For an example of one form of RM, consider a basalt lava flow erupted on land. Although the final crystallization temperature of the basalt depends on such factors as the amount and kind of volatiles present, it often will be around 1100°C, far above the Curie and Néel temperatures of all known magnetic minerals at atmospheric pressures, and hence there can be no RM until further cooling occurs. In general, basalts contain numerous magnetic grains that are randomly oriented. In this case the rock is magnetically isotropic, and during subsequent cooling it will acquire a magnetization parallel to (or occasionally antiparallel; see *Magnetic Self-Reversal*) the external magnetic field present. Sometimes this magnetization, thermal RM (TRM) in this case, is remarkably stable at room temperature. For example, in numerous instances one cannot alter the direction of the TRM in a basalt sample by subjecting it to a magnetic field in an arbitrary direction that is more than a thousand times the magnitude of the field in which the sample cooled. The TRM formed in this case essentially was acquired when the rock formed and is called *primary magnetization*. Any RM forming subsequently, say viscous RM (VRM), is called *secondary magnetization*. Usually, the primary magnetization is of interest to paleomagnetists because it records the Earth's magnetic field when the rock formed.

Because physical systems prefer to occupy minimum energy states, a basalt sample containing many magnetic grains effectively should exhibit no RM if equilibrium conditions are present. In a zero external field the rock should be demagnetized: The magnetic moments of individual grains should be randomly oriented. The very existence of remanence in many rocks illustrates that one is dealing with a nonequilibrium phenomenon. A sample can be demagnetized in various ways. Two of the most common methods are *thermal demagnetization*, in which a sample is cooled from temperatures above the highest Curie or Néel temperatures present in a zero external magnetic field, and *alternating field (af) demagnetization*, in which the sample is tumbled randomly in an initially very high alternating magnetic field that is slowly decreased to zero (see *Demagnetization*).

To understand remanence and other magnetic properties of rocks, one must understand the properties of the individual minerals that constitute rocks. Permanent magnetization exists because certain individual atoms and ions exhibit a permanent magnetic moment (primarily due to an electron spin moment). Exchange energy, or coulombic electric field energy associated with electron overlap, tries to make the magnetic moments of adjacent atoms (or ions) parallel (ferromagnetic coupling) or antiparallel (antiferromagnetic coupling); see Mattis, 1965. In contrast, thermal energy tries to produce randomization of the magnetic moments. At high enough temperatures, the thermal energy dominates and there is no magnetic ordering in minerals. For those materials that have net magnetic moments associated with individual atoms or ions, magnetic ordering first occurs (by definition) at the Curie temperature or Néel temperature (see *Curie Temperature*). Consequently, no spontaneous magnetization can exist above these temperatures. Strictly speaking, the Curie temperature refers to the situation when ferromagnetic coupling between adjacent *magnetic* atoms occurs, and the Néel temperature refers to the situation when there is antiferromagnetic coupling between adjacent *magnetic* atoms. (It is important to note that "adjacent magnetic atoms" often differ from "adjacent atoms.") Iron exhibits ferromagnetic order below its Curie temperature of 1043 K at 1 atm pressure. Ilmenite ($FeTiO_3$) exhibits antiferromagnetic coupling below its Néel temperature of 40 K. Magnetite (Fe_3O_4) exhibits ferrimagnetic order below its Néel temperature of 853 K: Antiferromagnetic coupling occurs between unequal sizes of adjacent magnetic moments (associated with the ferrous and ferric ions). Iron is the most common substance carrying the remanence in samples returned from the Moon, and iron oxides, hydroxides, and sulfides are the most common compositions of terrestrial minerals that carry a remanence (see *Magnetic Properties of Minerals*).

Although, typically, basalts are magnetically isotropic, other rocks sometimes exhibit magnetic anisotropy because of lineation or foliation in the rock. Individually, all magnetic minerals are magnetically anisotropic. That is, all minerals have certain directions, referred to as *easy axes*, along which they prefer to be magnetized. This anisotropy is the underlying cause of remanence. Easy axes can be thought of as energy minimum states in which the magnetization prefers to reside. Energy must be supplied to change the magnetic direction in a crystal from one easy axis to another or to reverse the direction of magnetization by 180°. This follows because the magnetic moment must be rotated through some hard anisotropy direction. The origin of this anisotropy on a quantum-mechanical level involves factors such as electron spin-orbit coupling, dipole-dipole interaction, and ionic distortion. Because these factors are typically too complicated to describe the anisotropy at a level useful for most applications, a phenomenological division of anisotropy is often used. Anisotropy is subdivided into magnetocrystalline anisotropy, stress or magnetostrictive anisotropy, and shape anisotropy. In addition, on occasion there is exchange energy anisotropy associated with exchange energy coupling between two different mineral phases.

Magnetocrystalline anisotropy is the magnetic anisotropy associated with an ideal stoichiometric, infinite size mineral. Usually it is described by one or two anisotropy constants. For example, a typical relationship describing the magnetocrystalline anisotropy energy E_M for a uniaxial anisotropy situation is

$$E_M = K \sin^2 \theta_1$$

where K is the anisotropy constant and θ_1 is the angle between the magnetization and the easy axis ($\theta_1 = 0$). Clearly, the larger K is, the larger the anisotropy is, and the greater the energy required to reverse the magnetization over the energy barrier ($\theta_1 = \pi/2$).

The magnetic forces acting between atoms or ions affect the lattice spacing. Conversely, any stress, whether of external or internal origin, changes the lattice spacing and hence the strengths of the magnetic interactions. Magnetostrictive anisotropy originates from internal stress, as would be associated, say, with a crystal dislocation. Like magnetocrystalline anisotropy, its origin is understood in principle (e.g., Mattis, 1965), but in practice experimental values must be used. An equation for uniaxial magnetostrictive anisotropy energy E_S is similar to that for magnetocrystalline anisotropy:

$$E_S = \lambda \sin^2 \theta_2 \qquad (1)$$

where λ is the magnetostrictive anisotropy constant and θ_2 denotes the angle between the magnetostrictive easy axis ($\theta_2 = 0$) and the magnetization. Because θ_2 in Eq. 1 is often different from θ_1 and because magnetocrystalline and magnetostrictive anisotropy are both always present in actual situations, the magnetization will usually not lie along either the magnetocrystalline or magnetostrictive easy axes.

The origin of shape anisotropy is illustrated in Fig. 2. The largely dipole magnetic moments associated with atoms are illustrated by arrows. At these boundaries of the grain where the magnetization is not parallel to the boundary, there will be a net bound magnetic charge (illustrated by a + or − in Fig. 2; mathematically, bound charge exists wherever the divergence of the magnetization is nonvanishing). The existence of this bound charge produces an internal magnetic field. This internal field will be less when the magnetization is along the long axis of a grain (Fig. 2A) than when it is perpendicular to the long axis (Fig. 2B). Hence, if all other anisotropy energies were negligible, a grain would prefer to be magnetized along its long axis, producing a shape anisotropy.

In addition to the foregoing anisotropy energies, the external magnetic field also produces a directional preference. If μ is the magnetic moment of a uniformly magnetized grain (μ is the magnetization times the grain volume), the energy E_{ex} associated with the external magnetic field H is

$$E_{ex} = -\mu \cdot H$$

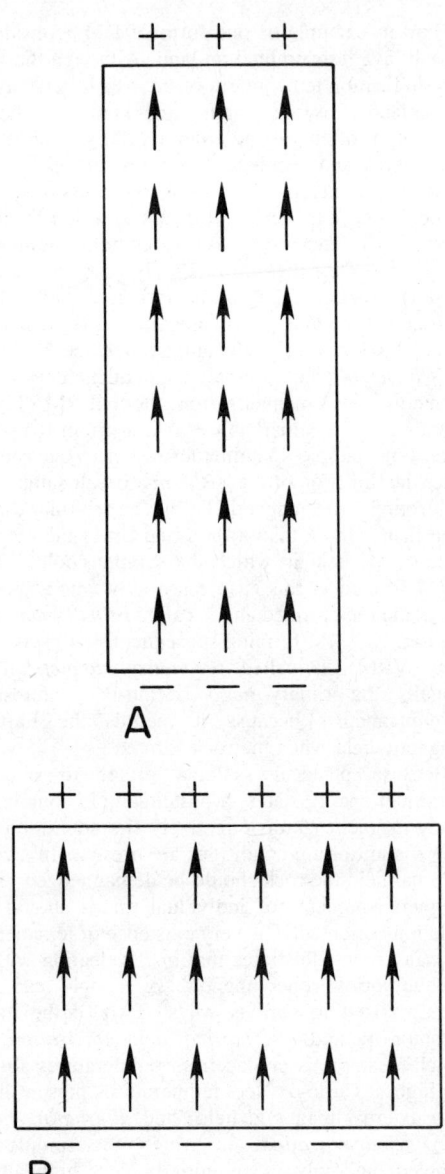

FIGURE 2. The origin of magnetic shape anisotropy. Bound magnetic charge occurs wherever the divergence of the magnetization is nonzero. The internal magnetic field, or the demagnetization field, is larger in 2B than 2A.

The energy minimum state is found mathematically by minimization of the total energy (exchange energy, the anisotropy energies, and the energy associated with the external magnetic field) as a function of the direction of the magnetization.

Shape, magnetostrictive, and magnetocrystalline anisotropies usually decrease rapidly (more rapidly than does the saturation magnetization) as the temperature is increased from room temperature to the Curie or Néel temperature of rock-forming minerals. This implies that the energy barriers separating minimum energy states typically rapidly increase with decreasing temperature. These changes cause the magnetization to become "locked into" a sample during cooling. The magnetization direction will follow any changes in direction of the external magnetic field just below the Curie temperature. The energy barriers between easy directions grow on further cooling, resulting in the "locking-in" of the magnetic direction. This locking-in process typically occurs over such a small temperature range that one refers to a *blocking* temperature, below which the magnetization direction is invariant (more rigorously one must speak in terms of relaxation times; see *Thermoremanence*). Variations in mineralogy, defect structures, and shapes of magnetic grains in rocks typically result in a spectrum of blocking temperatures in an actual sample. These considerations are key ingredients in explanations of the origin of thermal RM (TRM) and other remanent magnetizations.

A further complication in rock magnetism concerns the domain state of a sample. In small grains (e.g., a spherical magnetite grain with diameter 0.05 μm) the magnetization is essentially uniform throughout. Such grains are called *single-domain* grains. However, the internal magnetic field energy (resulting from the bound magnetic charge) in a uniformly magnetized grain increases as the square of the saturation magnetization times the grain volume. Large grains can lower this internal magnetic field by becoming multidomain grains. Multidomain grains contain regions called domains in which the magnetization is (nearly) uniform, separated by transition regions referred to as domain walls (see *Magnetic Domains*). The magnetization in domains is typically along easy axes, and, roughly speaking, the larger the number of domains in any given grain, the lower the internal magnetic field energy. However, balanced against this condition is an increase in domain wall energy that occurs because some of the magnetic moments in domain walls must be directed along hard anisotropic directions. Minimization of all the energies involved can only be done reasonably on a computer and, even then, only with simplifying assumptions. Comparison of calculated domain structures with observed domain structures shows only fair agreement (see *Magnetic Domains*).

Consider what happens during grain growth of a single magnetic grain. When the grain is small, the magnetization is uniform throughout. However, when the grain becomes large, it has an inclination to become multidomain. For example, consider a sphere of magnetite that has grown to a diameter of 0.2 μm. At this size a two-domain state is found to be energetically preferred, but the grain actually may be in a single-domain state. Nucleation of a domain wall (which requires energy) must occur to transform the grain from a single-domain to a multidomain state. Because of such complications, rock magnetists often characterize the properties of grains as *superparamagnetic* (small uniformly magnetized grains that do not carry a remanence but can affect such properties as susceptibility), single-domain (SD), pseudo-single-domain (PSD), and multidomain (MD). Pseudo-single-domain grains are intermediate in size between SD and MD grains. For example, the 0.2-μm magnetite grain would be classified as a PSD grain; in most ensembles of such grain, some grains would be found to be uniformly magnetized while others would contain domain walls. Because of problems associated with nucleation and denucleation of domain walls, there is no precise size dividing a PSD grain from an MD grain. Nevertheless, for spherical magnetite the PSD size range falls roughly between diameters of 0.08 to 15 μm or so.

Classification into SD, PSD, or MD grains is usually made on the basis of such magnetic measurements as saturation magnetization (SM), saturation RM (SRM), bulk coercive force, remanent coercive force, and susceptibility. As an example, the change in the ratio of SRM to SM with grain size is considered here. SM M_s is measured in the presence of an external magnetic field that is strong enough to saturate the sample. The RM associated with that field is M_{rs}. Hence, $M_{rs}/M_s < 1$. This ratio decreases from values around $\frac{1}{2}$ for SD grains to values 0.01 or lower in very large MD grains. SD and PSD grains are believed to be more stable than large MD grains, and rocks containing SD and PSD grains are believed to be better recorders of the external magnetic field than are large MD grains when the rock formed.

It usually is the stability of the remanence with time that is of prime interest to the Earth scientist. Techniques involving partial demagnetization are often used to infer time stability. For example, partial thermal demagnetization involves heating the sample to some temperature T_1 that is lower than the Curie or Néel temperature T_c (of the minerals present) and cooling the sample in a zero external field. The magnetic moments in those grains with blocking temperatures less than T_1 will be randomized by this process. The process can be repeated to a new temperature T_2, where $T_2 > T_1$, until T_c has been reached (and then the sample has been completely thermally demagnetized). Usually, but not always (particularly when chemical remanent magnetization might be involved), samples that are most resistant to thermal demagnetization are believed to be most stable with time. Similarly, partial af demagnetization can be used by employing peak alternating fields of different initial sizes. In this case, stability with time is inferred from stability with respect to alternating fields. Rocks that are stable with respect to both af and thermal demagnetization are usually the best bets to use to obtain the primary magnetization. More complete description of demagnetization techniques are found in O'Reilly (1984) (see also *Demagnetization*).

Because all demagnetization tests can, and on occasion do, fail, paleomagnetists and rock magnetists often use a variety of criteria obtained from geological mapping to decide whether the primary magnetization has been obtained after demagnetization. For example, consider a sedimentary bed that has undergone folding. If the magnetization is parallel throughout the bed, it cannot be a primary magnetization because the folding event occurs after the bed was formed. These and other so-called field tests are described more completely in McElhinny (1973) and Merrill and McElhinny (1983).

For the purist, the puzzles as to why and how minerals and rocks obtain, preserve, and alter their magnetic properties are sufficient to justify the study of rock magnetism. However, most Earth scientists are interested in rock magnetism as a tool to solve other problems. One important example of this interest is to use the intensity and direction of the primary magnetism to study the history of the Earth's magnetic field (see Merrill and McElhinny, 1983; Jacobs, 1984). For example, it is now known that the Earth's magnetic field reverses polarity. This opens up new questions, such as how does this occur? In turn, in some cases the history of the Earth's magnetic field (e.g., the magnetic field reversal chronology) is known so well that the magnetic record in rocks can be used for dating rocks.

Perhaps the most well-known use of rock magnetism is in tectonics. Some of the most important evidence of plate tectonics, displaced terranes, etc., comes from the use of paleomagnetic poles and seafloor magnetic anomalies (McElhinny, 1973).

Sometimes the presence of some exotic property of a magnetic mineral can be used for stratigraphy purposes, and often the knowledge of the Earth's magnetic history is useful in the correlation of geology formations. The magnetic properties of lunar samples has provided one useful way of classifying some lunar rocks, although the mystery as to how these samples became magnetized is not yet well resolved (Merrill and McElhinny, 1983).

In addition to the use of magnetic anomalies in tectonics, magnetic anomalies are sometimes useful in mineral exploration. For example, often the burning of coal seams or the spontaneous combustion of hydrocarbons results in the production of strongly magnetized minerals. Magnetic anomalies associated with the RM produced in this way (and sometimes with the induced magnetization) can be useful in locating the associated ore bodies.

Although the presence of magnetic anisotropy in rocks is usually not desirable in most applications, it can be very useful when paleocurrents or paleostress directions are desired. For example, because of magnetic shape anisotropy, a current flow that produced a preferred orientation of grains in rocks can sometimes be detected by measuring the magnetic anisotropy of susceptibility or the anisotropy associated with a laboratory-produced RM.

Rocks often exhibit a spectrum of blocking temperatures, which on occasion can be used to extract the temperature history of rocks. For example, consider a baked zone of sediments adjacent to igneous rocks. If a given sedimentary rock in this baked zone were heated to some temperature T_0, then the magnetization it contains in minerals with blocking temperatures less than T_0 will be reset by this heating. In this example, magnetic measurements in sediment in a baked zone provides a mechanism by which estimates of the temperature in that zone can be made.

Finally, it now seems likely that on occasion some organisms have the ability to sense the magnetic field for navigation purposes. At least some of these organisms, including some bacteria, fish, and birds, appear to have magnetic particles that are used to sense the external magnetic field (Jacobs, 1984).

Thus the uses of rock magnetism are numerous and varied; one expects such uses to increase in the future.

RONALD T. MERRILL

References

Jacobs, J., 1984, *Reversals of the Earth's Magnetic Field.* Bristol, U.K.: Adam Hilger.

Mattis, D. G., 1965, *The Theory of Magnetism.* New York: Harper and Row.

McElhinny, M. W., 1973, *Paleomagnetism and Plate Tectonics.* London: Cambridge University Press.

Merrill, R. T., and M. W. McElhinny, 1983, *The Earth's Magnetic Field: Its History, Origin and Planetary Perspective.* London: Academic Press.

Nagata, T., 1961, *Rock Magnetism* 2nd ed. Tokyo: Maruzen.

O'Reilly, W., 1984, *Rock and Mineral Magnetism.* New York: Chapman and Hall.

Stacey, F., and S. K. Banerjee, 1973, *The Physical Principles of Rock Magnetism.* Amsterdam: Elsevier.

Cross-references: Chemical Remanent Magnetization; Curie Temperature; Demagnetization; Magnetic Domains; Magnetic Properties of Minerals; Magnetic Self-Reversal; Natural Remanent Magnetization (NRM); Rock Magnetism: Measuring Techniques and Apparatus; Thermoremanence; Viscous Remanent Magnetization (VRM) and Viscous Remagnetization.

ROCK MAGNETISM: MEASURING TECHNIQUES AND APPARATUS

The magnetic properties of rocks are mostly caused by the presence of ferrimagnetic minerals (natural ferrites) in the rocks. The ferrimagnetic properties observed in rocks depend on the minerals present, on their concentration, and grain size. The values of the magnetic parameters of rocks vary over several orders; from weakly magnetic, sedimentary, metamorphosed, and igneous rocks to strongly magnetic, basic, igneous, and metamorphosed rocks, or to pure magnetite and other ores. All these rocks or ores are the object of research into magnetic properties and, therefore, the instruments used to measure these properties have to be highly sensitive, accurate, and have a large dynamic range. It is technically quite difficult to satisfy these dynamic conditions.

The following quantities are usually measured directly in rock magnetism and paleomagnetism:

saturated magnetic polarization, SMP, J_{so}
saturated remanent magnetic polarization, SRMP, J_{rso}
isothermal remanent magnetic polarization, IRMP, J_r
natural remanent magnetic polarization, NRMP, J_n
magnetic polarization induced by field H, IMP, $J(H)$
initial magnetic susceptibility, κ
coercive force, H_c
Curie temperature, T_c

These quantities are also studied under high pressures and temperatures, and their changes due to temperature, external A.C. and D.C. fields, etc., are investigated. Most of the other parameters in rock magnetism are derived from these directly measurable quantities.

The purpose of this paper is to acquaint the reader briefly with the methods used to measure these parameters, and with their basic applications; references in which each of the methods is described in detail are also given. Some of the methods can be used to measure a single quantity, others to measure two or more. The instruments used to measure magnetic properties are, therefore, usually divided either with regard to the parameter they measure, or with regard to the physical principle involved (A.C., D.C. methods). Recently, these methods have been divided into dynamic and static (depending on whether the samples are required to move or not during measurement), and methods founded on the effect the force of a magnetic field has on the measured sample.

The instruments most frequently used in static methods are astatic magnetometers, fluxgate magnetometers, superconducting magnetometers (SQUID), A.C. bridges for measuring the initial susceptibility, and oscillographs for measuring the magnetic characteristics. Those most often used in dynamic methods are ballistic magnetometers, vibrating-sample magnetometers, and spinner magnetometers. The instruments used in methods founded on the effect of the force of a magnetic field on the measured sample are magnetic balances and torsion magnetometers.

Figure 1 shows the mean ranges of the individual instruments for measuring the remanent magnetic polarization (RMP), the magnetic susceptibility (MS) and the induced (or saturated) magnetic polarization (IMP), $I = I(H)$. Figure 2 shows the possible uses of the individual types of instruments for measuring the remanent magnetic polarization (RMP), the natural RMP (NRMP), the saturated magnetic polarization (SMP), the magnetic susceptibility (MS), the coercive force (CF), the Curie point T_c, changes of RMP with temperature $[J_r(T)]$, changes of SMP with temperature $[J_s(T)]$, changes of H_c with temperature $[H_c(T)]$, changes of IMP with magnetic field $[J(H)]$, changes of susceptibility with temperature $[\kappa(T)]$, changes of susceptibility with magnetic field $[\kappa(H)]$, changes of RMP with hydrostatic pressure P $[J_r(P)]$, changes of RMP with uniaxial pressure p $[J_r(p)]$ and changes of susceptibility with pressure $[\kappa(P), \kappa(p)]$.

Static Methods

Astatic Magnetometer. This instrument belongs to the classical instruments used in rock magnetism and paleomagnetic laboratories. Nevertheless, the astatic magnetometer is still being used in measuring medium and strongly magnetic rocks, and is now undergoing intensive development and modernization.

The basic element of the astatic magnetometer is the astatic system that usually consists of two (sometimes of three or more) horizontal, firmly connected magnets oriented antiparallel. This astatic system is suspended on a torsion fiber (Fig. 3). The position of the system is such that the axes of the magnets are in the plane of the magnetic meridian. The magnetic field of the measured sample does not act equally on the separate magnets, equilibrium is lost, and the system rotates through an angle determined by the torsion of the fiber and the action of the magnetic field on the sample. Three fundamental positions of the sample relative to the astatic system are used in measuring with the astatic magnetometer: first Gauss position (in the plane of the lower magnet), second Gauss position (underneath the system), and As' position (in the plane passing through the center of the astatic system. See Fig. 3). The limiting sensitivity of an astatic magnetometer is determined by the Brownian motion that is close to the value 2×10^{-9} A · m. However, this sensitivity cannot be achieved practically because the astatic system is also affected by external, inhomogeneous, disturbing magnetic fields whose noise is estimated at about 2×10^{-8} A · m. The sensitivities of the best astatic magnetometers with feedback are now coming close to this value.

The first classical analysis and optimization of the astatic magnetometer was presented by Blackett (1952); this was improved upon by As (1960), and the most comprehensive review of astatic magne-

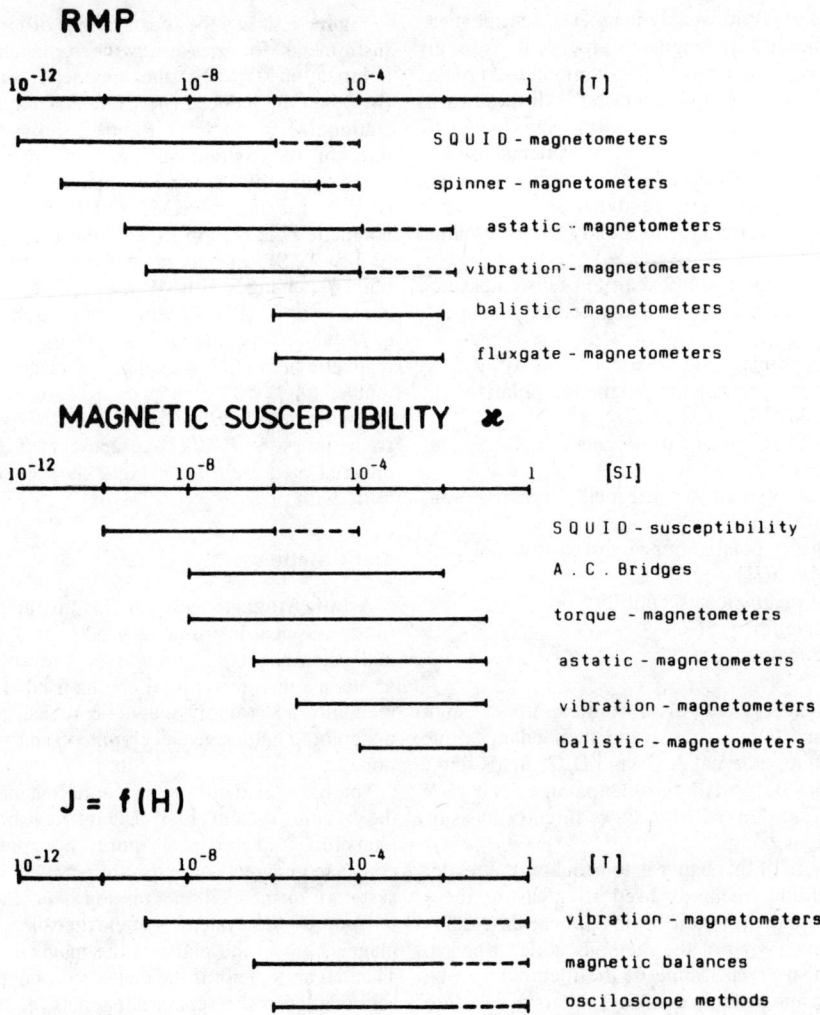

FIGURE 1. Review of mean ranges of instrument types for measuring magnetic parameters.

tometers can be found in Collinson's book (1983). The computer-controlled astatic magnetometer for fully automated measurements, as presented by Fujiwara and Yoshida (1979), is interesting; its disadvantage is the relatively long time required for the measurement (as much as 25 min) at a sensitivity of 6×10^{-8} A · m.

Since astatic magnetometers do not require a moving sample, they are frequently combined with nonmagnetic furnaces for measuring the temperature dependence of the RMP or with nonmagnetic pressure chambers for measuring the RMP under hydrostatic or uniaxial pressure (Kropáček et al., 1975).

Fluxgate Magnetometers. The increase in the sensitivity of fluxgate magnetometers to 0.1 nT made it possible to use them in rock magnetism. The principle of the method is discussed in *Geomagnetic Field: Measurement*.

For the purposes of rock magnetism, either a back-to-back type of probe or a single probe is used. These magnetometers can be used for fast measurements of irregular rock samples even in the field.

In the other variant two probes are differentially connected and, e.g., in connection with a solenoid can be used to measure the hysteresis curves of strongly and intermediately magnetic rocks.

Fluxgate magnetometers are not particularly sensitive, and their use is restricted to measuring intermediately to strongly magnetic rocks in the field, in situ. Some of the magnetometers of this type are described by Collinson et al. (1967).

SQUID Magnetometers. This type began to be developed after 1970; the instrument is founded on

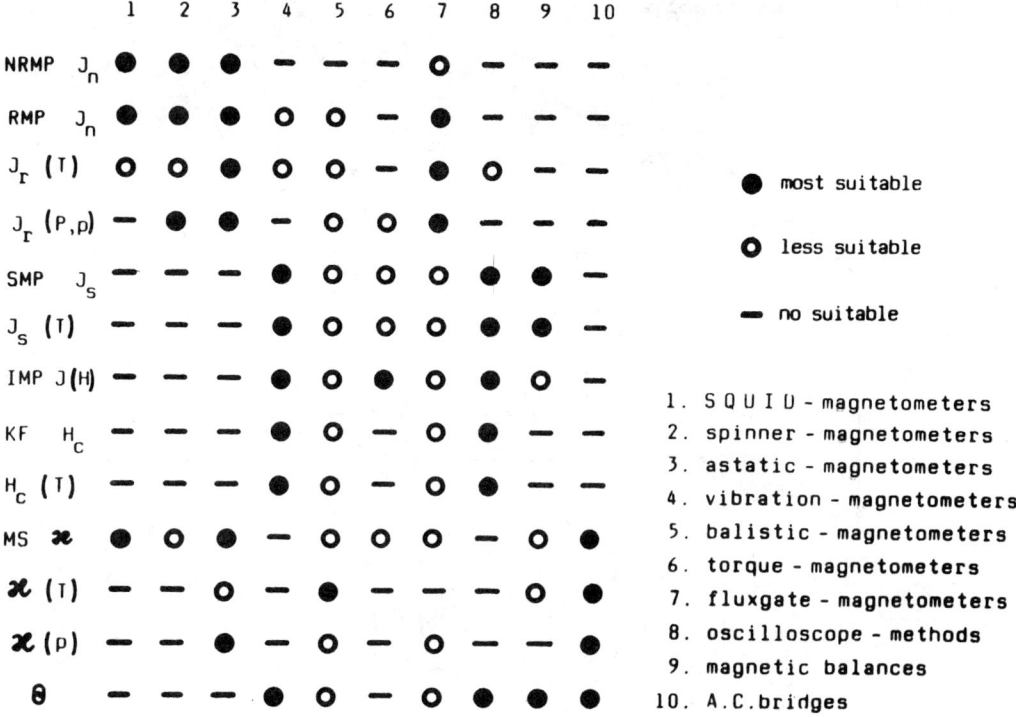

FIGURE 2. Types of instruments used to measure the relevant magnetic parameters.

Josephson's effect. SQUID stands for superconducting quantum interference device. The principle of this method is discussed in *Geomagnetic Field: Measurement*.

The block diagram of the SQUID magnetometer for measuring the remanent magnetization with the feedback is shown in Fig. 4. The magnetic field is not sensed directly by the superconducting ring, but by a pick-up coil that is tightly coupled with the superconducting ring. The sensor contains yet another winding, which is part of the high-Q-tuned circuit of the rf oscillator. This winding is weakly coupled with the sensor. The magnitude of the controlled oscillator amplitude is such that the SQUID is quantized under both polarities of the current, guaranteeing that the flux quantization is phase-shifted for the positive and negative polarity of rf excitation when the rf field is superimposed on the field of the pick-up coil. A phase-sensitive detector is used to evaluate this shift, which is superimposed on the exciting frequency. The output of this detector is the voltage proportional to the superimposed field, which represents the change in the magnetic field being measured (Fig. 5).

The above implies that the dependence of the output voltage of the phase detector on the measured field is nonlinear, and the voltage depends on the rf field of the exciting source. That is why a feedback is included in the system by means of generating a compensation field, and the measuring process described above is used as a zero indicator. As in astatic magnetometers, a zero indicator method guarantees a higher accuracy and dynamic range. There are usually two pick-up coils for measuring one component, and SQUID magnetometers are designed either to measure just one component, or to measure two or three simultaneously.

These magnetometers are now the most sensitive, their rate of measurement is very high (about 1–2 minutes per sample); their disadvantage is that liquid helium and permalloy or superconducting screening have to be used against the external magnetic noise. In combination with a source of a weak magnetic field they are also suitable for measuring the initial susceptibility and its anisotropy.

Bridges for Measuring Magnetic Susceptibility. A.C. bridges are now used most frequently for measuring the initial magnetic susceptibility and its anisotropy. Their greatest advantage is that the measurement is independent of the RMP value of the sample; their disadvantage is that the values of the magnetic susceptibility are affected by electrically conducting samples.

The principle of the measurements is based on the change of the magnetic induction generated by inserting the sample into a coil, $B = \mu_0 (H + M)$, where μ_0 is the permeability of vacuum, H the external magnetic field, and M the magnetic polar-

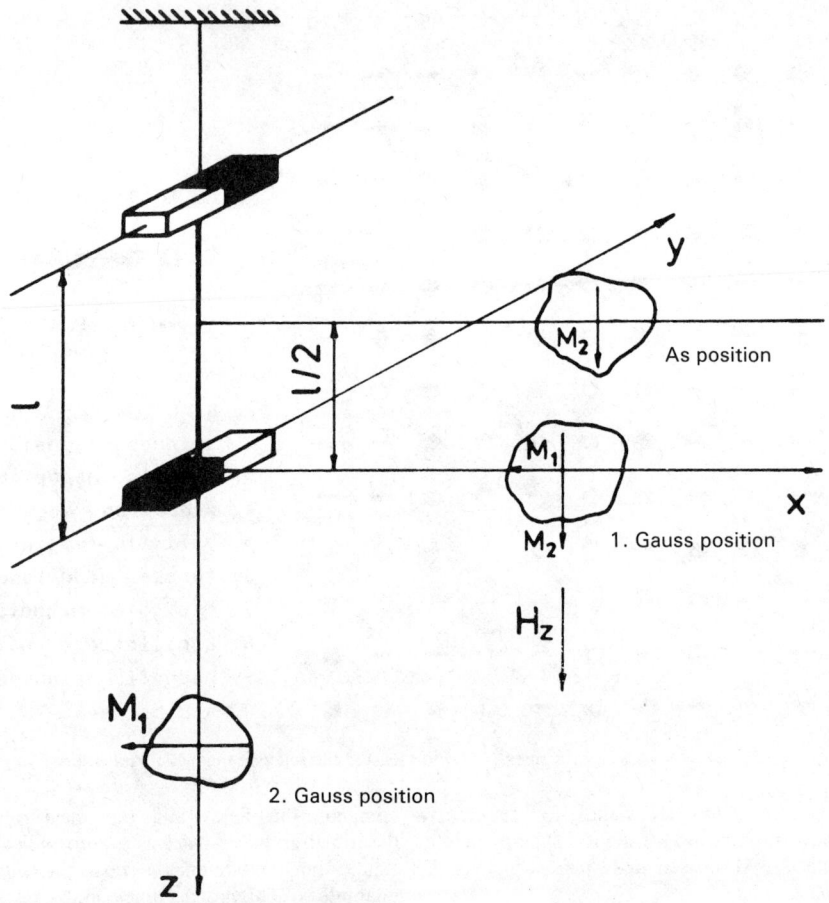

FIGURE 3. Three fundamental positions for measuring with the astatic magnetometer.

ization induced in field H, i.e., $M = \kappa H$, where κ is the susceptibility. Therefore, $B = \mu_0 H(1 + \kappa)$.

The change in the inductance of the coil due to inserting a sample of susceptibility κ_s into it can be expressed as

$$L = \mu_0 N^2 A(\kappa_s - \kappa_a)/L,$$

where N is the number of turns of the coil, A its area, κ_a the susceptibility of air, and L the length of the coil.

The bridge connection is usually used for measuring the susceptibility κ of rocks whose values are between 10^{-7} and 10^{-1} SI.

A.C. bridges are also used to determine the Curie point of rocks by means of Hopkinson's effect. Most A.C. bridges measure directional susceptibility along the coil axis. Since the susceptibility of a magnetically linear anisotropic medium is in general a symmetric tensor of the second order with six independent components, the susceptibility tensor can be determined if a sufficient number of measurements of directional susceptibilities is made.

A special version of A.C. bridges are kappameters designed for field measurements; the pick-up coil in these instruments is designed to pick-up the susceptibility of the whole halfspace formed by the measured rock in situ. These portable instruments may also be used in some cases for tentative measurements of the susceptibility of rock samples and drill cores.

Oscillographic Methods of Measuring Magnetic Characteristics. These methods also belong to the methods based on magnetic induction. The measured sample is placed either in a solenoid, or in an electromagnet with a C core, supplied by alternating current, which generates a time-variable magnetic field. The sample is inserted into the pick-up winding. The magnetizing field is proportional to the current flowing through the primary winding, however, the voltage across the pick-up secondary winding is directly proportional to the time variation of the induction flux, i.e., the time variation of the induction $\partial B/\partial t$. Consequently, there are no particular difficulties in obtaining a record of the magnetizing field as a function of time, whereas to obtain

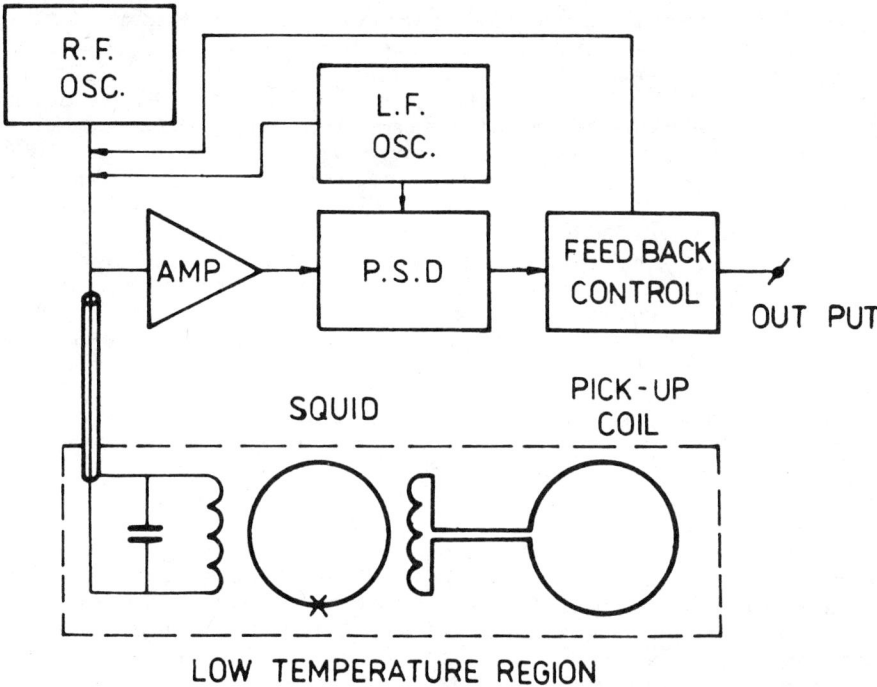

FIGURE 4. Block diagram of the SQUID magnetometer.

a record of the time variation of the sample's induction the voltage across the secondary winding has to be integrated. Oscillographs are used to display hysteresis loops of fast processes (over 10 Hz). Oscillographic methods are employed in measuring dynamic hysteresis curves without the possibility of obtaining the curve of the initial (primary) magnetic polarization and, moreover, one cannot, in general, decide what frequency for measuring the hysteresis curve can still be considered quasistatic, i.e., in its consequences equivalent to static magnetization. In this respect, it depends on the way in which the phenomena of magnetic viscosity of the substance itself are manifest, etc.

A successful instrument that operates on this principle was designed and built by Likhite et al. (1965); in this instrument samples containing magnetite can be saturated by an electromagnet generating a maximum exciting field of 0.36 T in a 12-mm gap 50×25 mm in cross section. The signal-

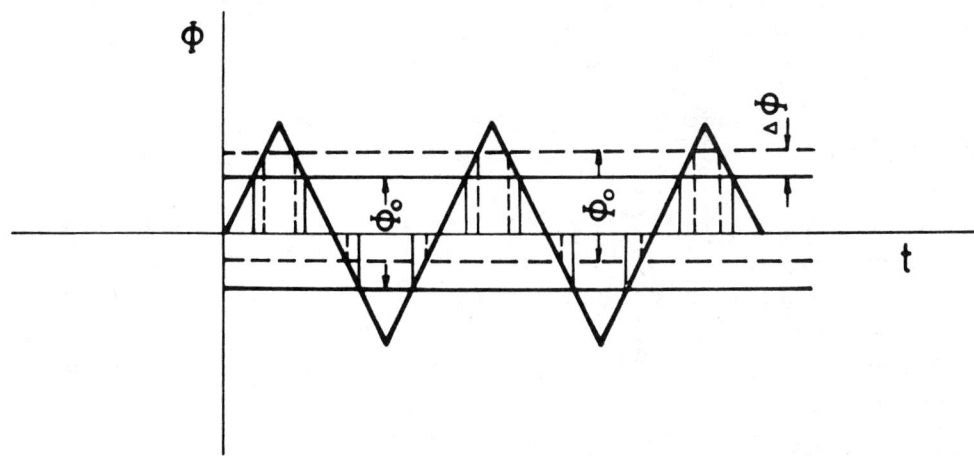

FIGURE 5. The effect of the variations of the measured field on the phase shift of field quantizing.

to-noise ratio is limited by the uncompensated harmonic frequencies in the exciting magnetizing field of the electromagnet. In a magnetic field of 0.24 T, for example, the signal-to-noise ratio is about 10. This type of instrument is also suitable for measuring hysteresis curves under low and high temperatures.

Dynamic Methods

Ballistic Magnetometer. Earlier, ballistic magnetometers were used very frequently in magnetic laboratories. Now they are being used rarely because of their low sensitivity, because the measurement is lengthy, and because the data cannot be digitized. The description of this method can be found in Collinson (1983).

The ballistic method is now practically used only for measuring magnetic properties under pressure and, instead of the sample moving from one coil to the other, the coil moves over the sample.

Spinner Magnetometers. The principle of the spinner magnetometer for measuring the remanent magnetic polarization is in generating an A.C. voltage by the sample rotating in the pick-up system, which consists either of pick-up coils, or a fluxgate. Together with the sample-generated voltage, a reference voltage $u_r(t)$ is generated with the aid of which either the amplitude and phase or the amplitudes of the cosine and sine components are measured. We then speak of measuring in the polar or component mode. To be able to determine the vector J_r in full, therefore, it is necessary to measure with the sample at least in two positions by rotating the sample about two mutually perpendicular axes. Measurements in six positions are mostly used, and these yield four measurements for each component.

The block diagram of the spinner magnetometer is shown in Fig. 6. Jelinek (1966) presented an optimum design of the pick-up coils for suppressing the Johnson noise. This problem was also treated by a number of other authors. The latest version of the magnetometer designed by Jelinek, the JR-4, has an RMP sensitivity of 4 pT.

The first design of a fluxgate spinner magnetometer was presented by Foster (1966). As opposed to the earlier antiparallel configuration, the present spinner magnetometers with fluxgate pick-ups use a sensor with only one fluxgate and multilayer magnetic screening. Many designs have been summarized, e.g., in Collinson (1983). The author also gives the methods of calibration and analysis of the effects of sample inhomogeneities.

New trends in developing spinner magnetometers should also be mentioned. Molyneux (1971) presented a method of processing the signal by direct digitization of an analogue output by means of one hundred and twenty-eight 12-bite samples which are repeatedly summed into the memory of the processor over the chosen period of integration. These data are Fourier analyzed, and the first harmonic is separated. The appropriate samples are then obtained by phase comparison with a reference signal. The fully digitized processing improves the stability especially as regards long time constants of very weak samples.

Another very interesting spinner magnetometer was designed by Kono et al. (1981), in which the sample rotates about two mutually perpendicular axes in the pick-up system (in the same way as in the A.F. demagnetizing device), and all three components of the vector of remanent magnetic polarization are determined simultaneously by means of harmonic analysis. It was found that the optimum ratio of revolutions is 4:7. The magnetometer is suitable for measuring intermediate and strongly magnetic rocks.

Vibrating-Sample Magnetometer. These instruments have now replaced ballistic magnetometers. The principle of the vibrating-sample magnetometer is similar to that of the spinner magnetometer, the rotation of the sample being replaced by its vibration. Moreover, the pick-up system is located in a controlled homogeneous magnetic field, which is stationary relative to the pick-up system. In a nonzero external magnetic field the pick-up system also indicates the induced magnetic polarization as a function of the external magnetic field, making it possible to measure the complete magnetic charac-

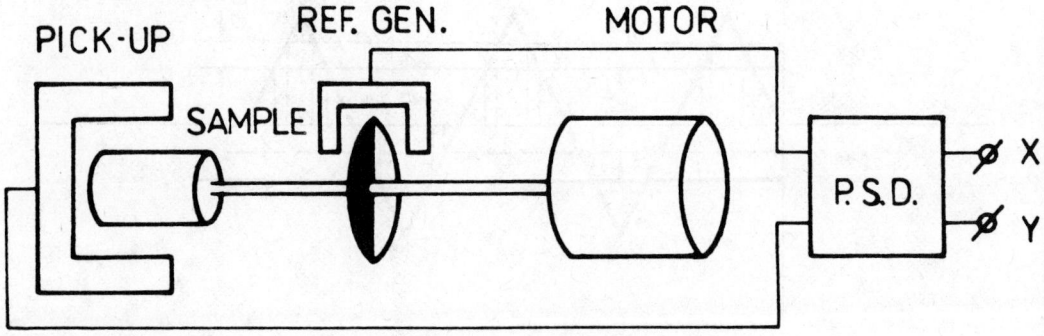

FIGURE 6. Block diagram of the spinner magnetometer.

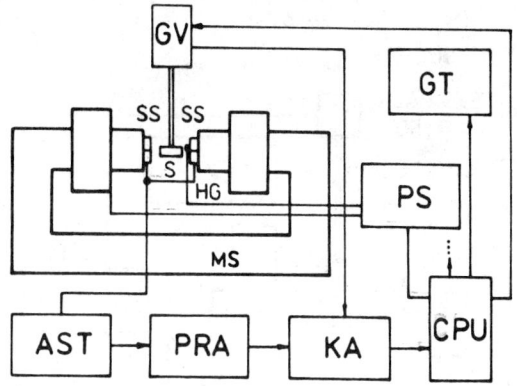

FIGURE 7. Block diagram of the vibrating-sample magnetometer.

optimum designs of the pick-up system of the vibrating-sample magnetometer. A precise analysis of the design of this system must be based on the topology of the magnetic field in the neighborhood of the sample, and the finite dimensions of the system and sample must be considered (Zelinka et al., 1984).

The block diagram of a vibrating-sample magnetometer is shown in Fig. 7 (Zelinka et al., 1984). Sample S is set in harmonic motion perpendicular to the applied external magnetic field H. The signal is picked up by four pick-up coils L1-L4. The working magnetic field is generated by the pole shoes of the magnetic circuit MS. The latter is feedback controlled by power supply PS. The intensity of the magnetic field is measured by a Hall generator HG. Sample S is set in harmonic motion by vibration generator GV. The signal of the sensing system SS (coils L1-L4) is modified by an asymmetric transformer AST, a pre-amplifier featuring high internal resistance PRA and a correlation analyzer KA, whose reference signal is derived from the sample's motion by means of an optoelectronic sensor. The correlation analyzer and the power supply are connected to the processor CPU. Communication with the operator is via the graphical terminal GT. Figure 8 shows an example of the output of this vibration-sample magnetometer.

teristic, as well as individual points (RMP, SMP, coercive force). In view of their design, vibrating-sample magnetometers, do not have as high a sensitivity as spinner magnetometers, however, the possibility of measuring in a homogeneous magnetic field enables them to be used in a number of other applications.

A whole series of pick-up system configurations is known, a number of which have been described by Collinson (1983). Many authors have presented

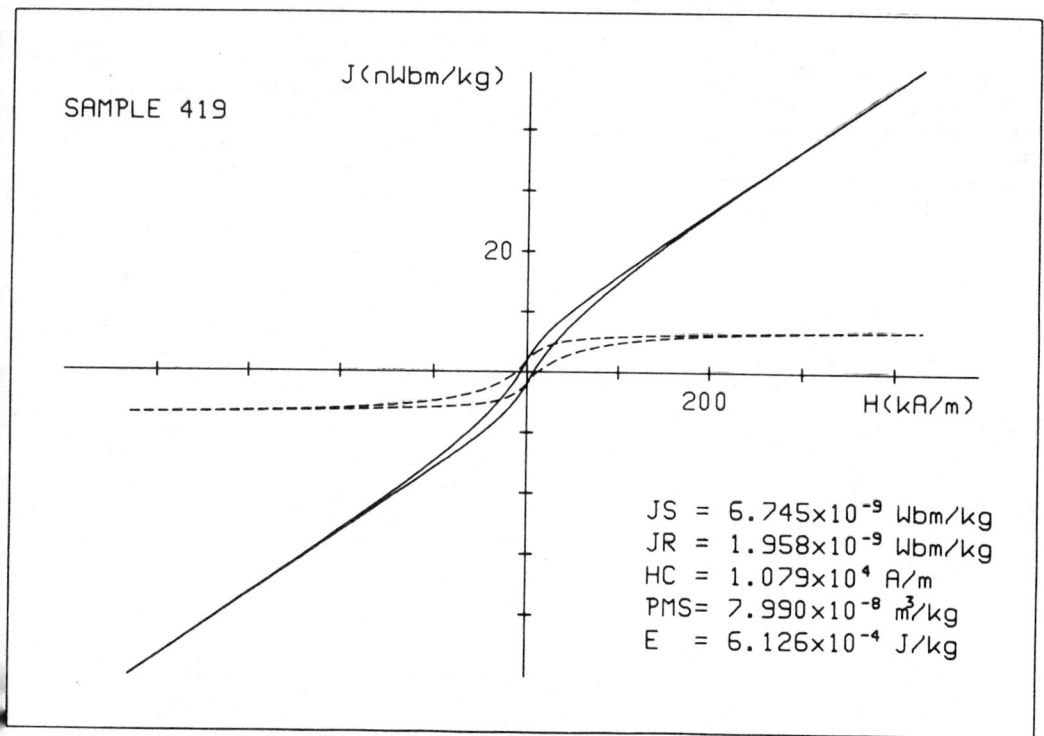

FIGURE 8. Example of output of processed data measured with the vibrating-sample magnetometer. Sample: phonolite.

Methods Based on the Effect of the Force of the Magnetic Field Acting on the Measured Sample

Magnetic Balances. The principle of magnetic balances is founded on measuring the force of an inhomogeneous magnetic field acting on the measured sample. In general, an inhomogeneous magnetic field $\mathbf{H}$ in vacuum acts on a sample whose magnetic moment is $\mathbf{M}$ with the force $\mathbf{F} = \text{grad}\,(\mathbf{M} \cdot \mathbf{H})$. If the sample is small, volume dV, its magnetic moment is $\mathbf{J}\,dV$, where $\mathbf{J} = d\mathbf{M}/dV$, and if the magnetic circuit is such that the inhomogeneous field only acts in one direction (x), the force acting on the sample is $dF_x = \mu_0 \mathbf{J}(\partial \mathbf{H}/\partial x)dV$. The measurement methods, based on this equation, can be divided into two groups, depending on whether the method is based on the above differential equation directly, or whether the equation is used in its integrated form:

$$F_x = \mu_0 \mathbf{J} \int_{(V)} (\partial \mathbf{H}/\partial x)dV$$

The first group includes methods used to measure samples that are so small that the intensity of the field $\mathbf{H}$ can be considered constant in direction and magnitude practically within their whole volume V. The second group consists of methods used to measure larger samples in which $\mathbf{H}$ is not constant within the whole volume of the sample, and the equation has to be integrated. The methods of the first group with small samples are mostly used in rock magnetism; a more detailed description of these methods can be found in Parry's study (1967).

Most magnetic balances are based on the design presented by Weiss and Foëx (1911), in which a light horizontal balance beam is suspended on bifilar fibers V and able to move only in the direction of the balance beam length K. The sample being measured (Fig. 9) is placed at the end of the beam, at point A,

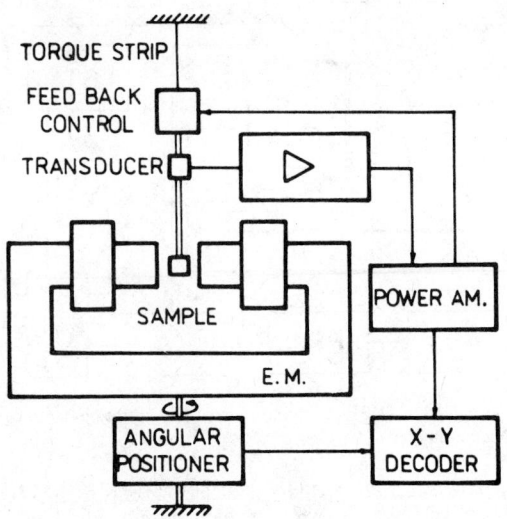

FIGURE 10. Block diagram of the torsion magnetometer.

between the poles of electromagnet M. As soon as the magnetic field is generated, the sample is deflected from its original position and this deflection is compensated by the electrodynamic effect of coils C1 and C2. Whereas a constant current flows through coil C2 fixed to the balance beam, the current in the stationary coil changes and acts to return the sample to its original position. The current in coil C1 is then the measure of the force with which the magnetic field acts on the sample.

Torsion Magnetometers. The torsion magnetometer is an instrument used to study two kinds of magnetic anisotropy of rocks: the anisotropy of magnetic susceptibility, and the anisotropy of saturated magnetic polarization (SMP).

The theory of measurement with the torsion magnetometer has been developed by a number of authors, including King and Rees (1962). The diagram in Fig. 10 will be used to describe the principle of the torsion magnetometer. A sample, placed in a homogeneous magnetic field, is rotated due to the field's action so that its potential energy in the field is minimal. The torque acting on the sample $T = -dW(\theta)/d$, where W is the potential energy of the sample in the magnetic field, and θ the angle of rotation of the sample relative to the direction of the magnetic field. The energy of the sample in the interval of saturation is $W = \frac{1}{2} \mu_0 V J_s^2 N_d(\theta)$, where J_s is the saturated magnetic polarization, V the volume of the magnetic fraction in the sample, and $N_d(\theta)$ the average demagnetization factor in the direction of angle θ. The angular deflection of the sample is recorded by the position sensor, amplified and fed back to the actuator, which returns the sample to its initial position. The output signal is the voltage proportional to the compensation torque,

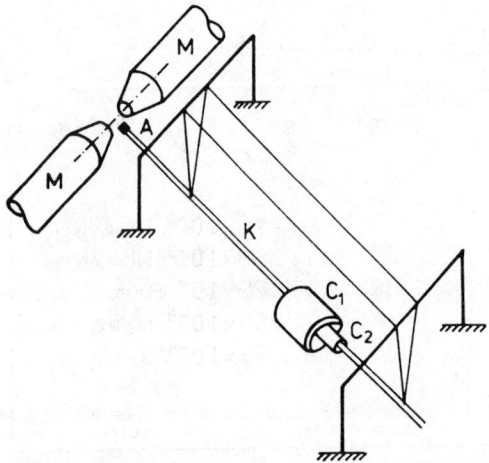

FIGURE 9. Block diagram of the Weiss-Foex magnetic balance.

i.e., $T(\theta)$. The sensitivities of torsion magnetometers are around 3×10^{-13} m^{3}Kg^{-1}. Noise sources are mostly of a mechanical nature and are discussed by King (1967).

Measurement Methods and Instruments for Measuring Magnetic Parameters under High Hydrostatic and Uniaxial Pressure

In recent years, the research into the effect of high pressures (hydrostatic as well as uniaxial) on the magnetic properties of rocks has become an important part of the study of magnetic properties of rocks. As already demonstrated, the measurement of the RMP of rocks is specific in that it requires a high sensitivity of the instruments which, moreover, must be nonmagnetic. These requirements do not change even when high pressures are applied but, on the contrary, are combined with the requirement of high-strength nonmagnetic material.

Modified ballistic magnetometers with mobile pick-up coils are used in studying the changes of magnetic properties under pressure. In these instruments the change of magnetic flux is caused by the displacement of the pick-up coils. Magnetometers of this type have been designed and described by Nagata and Kinoshita (1965) and Jelenska (1970). Another device for measuring the RMP under hydrostatic pressures is the nonmagnetic chamber, placed together with the sample in an astatic magnetometer. In this case either direct pressurization is used and the measurement is made only along one axis, or a detachable pressure chamber is used and the RMP is measured along three axes. The design of some of the chambers can be seen in Fig. 11 (Kropáček et al. 1975). This method was already used earlier by Kawai and Sawaoka (1967) and Carmichael et al. (1969). The minimum distance between the astatic system and the sample is determined by the external dimensions of the pressure chamber, which depend on the size of the sample measured and on the strength of the material. The sensitivity of the system cannot, therefore, be increased by increasing the size of the sample, because this would require the external dimensions of the chamber to be increased substantially, which would in turn increase the distance between the astatic system and sample. However, higher sensitivity can be achieved if the external dimensions of the chamber do not come to bear, i.e., if the sensor is located in the immediate neighborhood of the sample inside the pressure chamber. For this purpose it is then best to use a spinner magnetometer in which the sample rotates at a constant angular velocity. The fixed pick-up coils can be placed inside the pressure

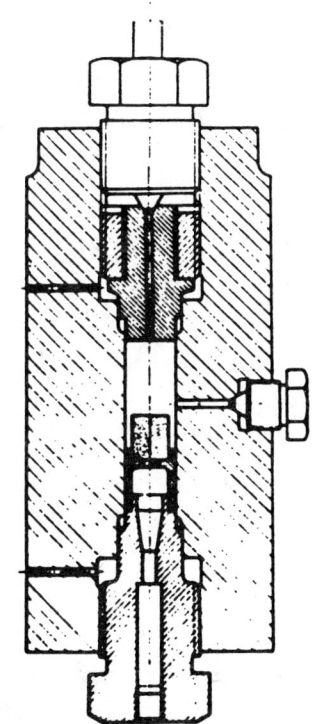 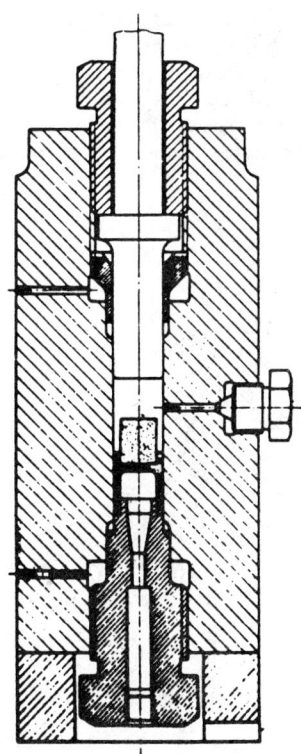

FIGURE 11. Examples of the design of (*a*) a nonmagnetic chamber up to 4 kb; (*b*) a pressure-tight chamber up to 10 kb.

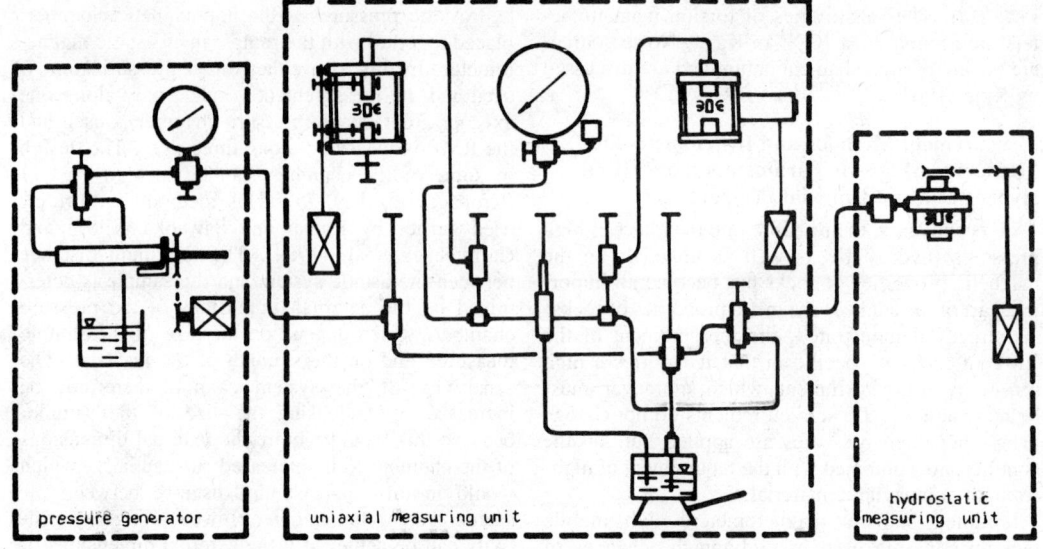

FIGURE 12. Block diagram of the magnetometric pressure system.

chamber and, consequently, only the problems related to the constant frequency of rotation of the sample in a high-pressure medium need be considered (increase in the viscosity of the medium with pressure, pressure-tight rotational bushings). A spinner magnetometer of this type was designed and described by Kropáček et al. (1975); the maximum permissible hydrostatic pressure inside the instrument is 300 MPa and its sensitivity is 30 nT.

Based on this spinner magnetometer in the hydrostatic pressure chamber, a spinner magnetometer was constructed for measuring under uniaxial pressure with rotating pistons, in which the signal is again generated in fixed pick-up coils by the sample rotating at a constant angular velocity with the pistons of the nonmagnetic press. The maximum pressure acting on the samples is 100 kN and the sensitivity of the apparatus is again 30 nT. The next step in this development is a uniaxial press with vibrating coils sensing the magnetic field in the sample along the direction in which the uniaxial pressure is acting. The block diagram of the whole

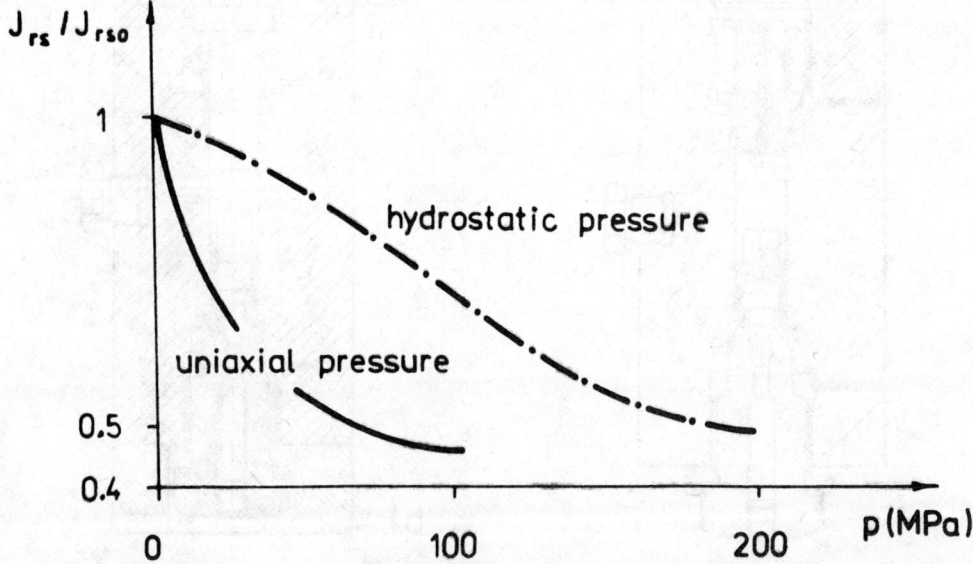

FIGURE 13. Comparison of the effect of uniaxial and hydrostatic pressure on changes of saturated RMP.

magnetometric device is shown in Fig. 12 (Podroužková et al., 1983). A comparison of the effect of hydrostatic and uniaxial pressure on the same sample, i.e., of $J_r(P_{hydro})$ and $J_r(P_{uniax})$, is presented in Fig. 13.

The purpose of this paper was to characterize briefly the methods and apparatuses currently used in rock magnetism research. With a view to the limited space available, the separate methods could not be discussed in more detail. Nor was attention concentrated on some of the special types of instruments used to measure the magnetic properties of very small grains of minerals, and on other types of magnetic apparatuses. We refer to the work by Collinson (1983), which presents a well-founded analysis of methods and instruments used.

T. ZELINKA
V. KROPÁČEK
J. TAUER

References

As, J. A., 1960, Instruments and measuring methods in palaeomagnetic research, *Mededel. Verhandel. K.N.M.I.* **78**, 1-56.

Blackett, P. M. S., 1952, A negative experiment relating to magnetism and the Earth's rotation, *Royal Soc. London Philos. Trans.* **A245**, 309-370.

Collinson, D. W., 1983, *Methods in Rock Magnetism and Palaeomagnetism*. London-New York: Chapman and Hall.

Collinson, D. W., K. M. Creer, and S. K. Runcorn, eds., 1967, *Methods in Palaeomagnetism*. Amsterdam: Elsevier, 609p.

Foster, J. H., 1966, A palaeomagnetic spinner magnetometer using a fluxgate gradiometer, *Earth Planet. Sci. Letters* **1**, 463-467.

Fujiwara, Y., and M. Yoshida, 1979, A computer controlled astatic magnetometer, *Rock Magnetism and Palaeogeophysics* **6**, 143-145.

Goree, W. S., and M. D. Fuller, 1976, Magnetometers using r.f.-driven squids and their application in rock magnetism and palaeomagnetism, *Rev. Geophysics Space Physics* **14**, 591-608.

Jelenska, M., 1970, Effect of uniaxial compression on magnetization of rocks, *Acta Geophys. Polonica* **18**, 171-183.

Jelinek, V., 1966, A high sensitivity spinner magnetometer, *Studia geophys. et geod.* **10**, 58-77.

Jelinek, V., 1973, Precision a.c. bridge set for measuring magnetic susceptibility of rocks and its anisotropy, *Studia geophys. et geod.* **17**, 36-48.

Kawai, N., and A. Sawaoka, 1967, Some techniques on magnetic measurements under hydrostatic pressure, *Ann. Progress Report of Palaeogeophysics Research in Japan*, pp. 151-164.

King, R. F., and A. I. Rees, 1962, The measurement of the anisotropy of magnetic susceptibility of rocks by the torque method, *Jour. Geophys. Research* **67**, 1565-1572.

Kono, M., Y. Hamano, T. Nishitani, and T. Tosha, 1981, A new spinner magnetometer: principles and techniques, *Royal Astron. Soc. Geophys. Jour.* **67**, 217-227.

Kropáček, V., Z. Podroužková, and O. Štěrba, 1975, The measurement of the magnetic properties of rocks subjected to high hydrostatic pressure, *High Temperatures-High Pressures* **7**, 559-562.

Likhite, S. D., C. Rahdhakrishnamurty, and P. W. Sahasrabudhe, 1965, Alternating current electromagnet type hysteresis loop tracer for minerals and rocks, *Rev. Sci. Instrum.* 1558-1564.

Molyneux, L., 1971, A complete result magnetometer for measuring the remanent magnetization of rocks, *Royal Astron. Soc. Geophys. Jour.* **24**, 429-433.

Nagata, T., and H. Kinoshita, 1965, Studies on piezomagnetism /I/ magnetization of titaniferous magnetite under uniaxial compression, *J. Geomag. Geoelectr.* **17**, 121-135.

Parry, J. H., 1967, Principles of magnetic balances, D. W. Collinson, K. M. Creer, and S. K. Runcorn, eds., *Methods in Palaeomagnetism*. Amsterdam: Elsevier, 431-437.

Zelinka, T., P. Hejda, and V. Kropáček, 1984, Vibrating-sample magnetometer for measuring magnetically weak materials, *Tesla Electronics* **17**, 35-43.

Cross-references: *Curie Temperature; Demagnetization; Detrital Remanent Magnetization (DRM); Magnetic Domains; Magnetic Properties of Minerals; Magnetic Self-Reversal; Natural Remanent Magnetization (NRM); Rock Magnetism; Thermoremanence; Viscous Remanent Magnetization (VRM) and Viscous Remagnetization.*

ROTATION OF THE EARTH—See EARTH ORIENTATION.

S

SATELLITE ALTIMETRY

A basic need for planetary exploration is the measurement and description of surface topography. This information is of fundamental importance for studies of ocean, ice, and land by a broad set of disciplines. The satellite altimeter is an instrument that can make a global measurement of the surface topography.

A satellite altimeter measures the round trip travel time of a short electromagnetic pulse sent from an orbiting spacecraft toward the planetary body and reflected off its surface. This measurement estimates the height of the satellite above the planet's surface. The altitude of the spacecraft relative to a reference ellipsoidal surface of the planet can be determined independently using orbit determination procedures. The difference between the altimeter measurement and the orbit range estimate is the height of the topography relative to the reference ellipsoid in geodetic coordinates.

While simple in concept, the practical application of altimetry is difficult because of accuracy requirements of less than a meter from a nominal altitude of 1000 km above the surface. The descriptions of land or ice surfaces that vary several kilometers need accuracies of less than a meter to be useful for most applications. Mean sea level, which is closely related to the geoid, varies about 100 m and requires submeter accuracy. The temporal variability of sea level, which can be directly related to ocean circulation, changes about a meter and requires subdecimeter accuracy. Eventually, scientists will seek information about secular fluctuations in topography that relate to changes in the climate and the dynamics of the planet's interior. These variations can be less than a decimeter over a few decades and will demand accuracies of a few millimeters or less.

Altimeter systems have become increasingly accurate over the past two decades in order to meet the application demands. The first missions on *Apollo* and *SKYLAB* in the early 1970s had precisions of about one meter and absolute accuracies of several meters. In contrast, the joint U.S./French *Ocean Topography Mission* (TOPEX/Poseidon) sponsored by NASA and the Centre National d'Etudes Spatiales (CNES), that is scheduled for the early 1990s, plans to achieve an instrument precision of 2 cm and an absolute accuracy of 10 cm over the ocean (Born et al., 1984). Satellite altimeters have already provided significant new scientific information about the oceans, ice, and land on Earth, as well as the surfaces of the Moon and Venus.

The Altimeter System

Satellite altimeter systems are composed of three principal components. The instrument and spacecraft provide the primary measurement of the height of the satellite above the surface. Precise satellite tracking and orbit determination procedures are necessary for determining the height of the spacecraft relative to the reference ellipsoid. Ancillary measurements are required to correct for altimeter range delays over the path of the pulse: They can be made from the same spacecraft but often are not. In addition to these components, post processing of the data is very important. For some applications the errors are too large to directly examine the signal of interest. Many processes have been successfully derived from the data through the development of analysis techniques for separating signal from noise.

The Instrument. There are a variety of altimeter instruments, including single beam nadir pointing systems, experimental multiple beam methods and scanning synthetic aperture techniques. This article discusses only single beam nadir pointing altimeters. Multibeam systems are in the early stages of development. Scanning synthetic aperture radar altimeters, which have the potential for obtaining high spatial resolution topography measurements from space, have been recently described by Elachi (1987).

Nadir pointing altimeters have two modes of operation (see Fig. 1). In a *beam limited altimeter*, there is very little dispersion of the electromagnetic pulse and the effective size of the footprint on the surface is quite small. A beam limited system is advantageous over rough terrain as found on land or ice. In a *pulse limited* system the beam dispersion is large. In order to limit the size of the surface footprint in this system the pulse length is made very short. Pulse limited systems have worked extremely well over the oceans, where the bigger footprint averages the surface wavefield and accurately measures the large scale and low frequency sea level variations.

Beam limited systems are often built with laser technology, whereas pulse limited systems are commonly built with microwave radars. Both microwave and laser altimeters have been used on satel-

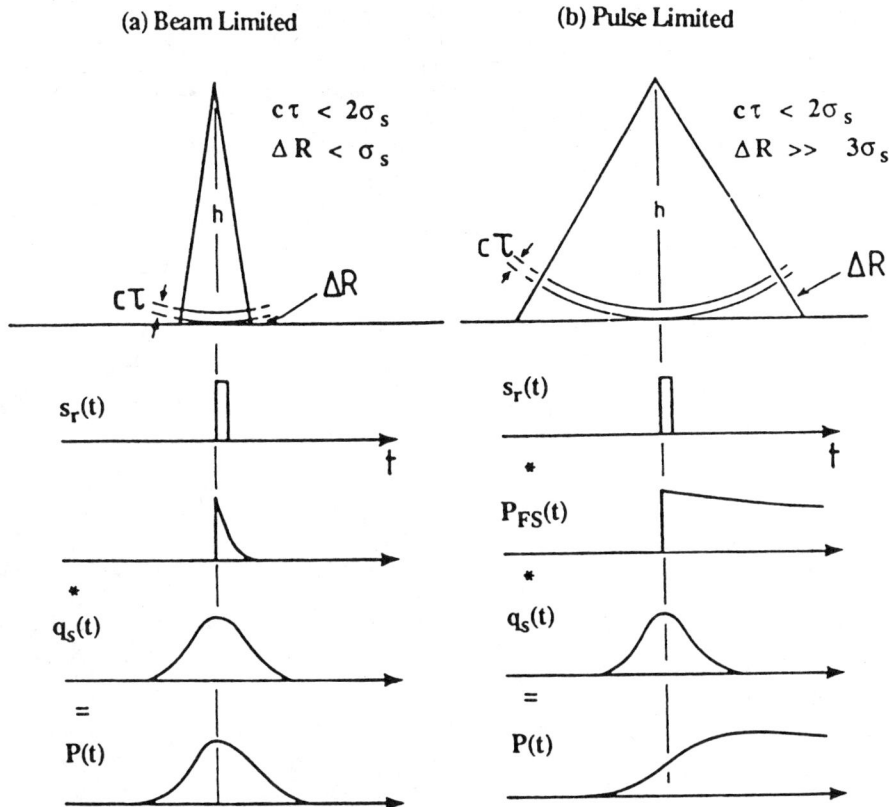

FIGURE 1. The operation of a radar altimeter and its interactions with the surface. (From Rapley et al., 1987)

lites. The advantage of a laser altimeter is the size of the beam limited footprint on the surface (< 100 m) compared to the larger pulse limited footprint of the microwave sensor (>1 km). The microwave sensor has the advantage of being able to penetrate most atmospheric conditions.

A theoretical description of how the altimeter beam reflects from the surface has been described by Barrick and Lipa (1985). In a beam limited system the return power is a convolution of the point target response and the probability distribution of the scattering facets on the surface. In a pulse limited system a convolution with the antenna response is required in addition to the beam limited response in order to adjust for the angular dependence of the larger footprint.

Microwave Sensors. The design of a microwave altimeter is constrained by spacecraft antenna sizes that force the beam to be relatively wide. This restriction is overcome by operating in a pulse limited mode. Another design constraint is the pulse frequency. The pulse width is inversely proportional to the bandwidth. On earth, this bandwidth is restricted by international agreements on acceptable transmission frequencies for satellites. Altimeters have been built to operate with a central frequency in the Ku (12.5–18 GHz) or C (4–8 Ghz) bands. At these frequencies the available bandwidth permits an effective surface footprint of a few kilometers.

The power of the radar echo is measured in a sequence of bins or *timing gates*. The spectrum of the received power measured by the timing gates is called the *return waveform*. The surface facets within the footprint modulate the return echo causing a low signal to noise ratio in the return of an individual pulse. To overcome this difficulty, several pulses are averaged to yield a smooth return waveform. The satellite height is usually estimated as the time of the half power point in the return waveform (see Fig. 2). A detailed discussion of the operation of pulse limited altimeters has been presented by Chelton et al. (1989).

The shape of the returned waveform varies considerably with the type of surface. Typical waveforms for a variety of conditions over land, ice, and ocean are shown in Fig. 2. Over relatively smooth surfaces, such as the ocean, the return power from the surface footprint quickly increases and then gradually decreases resulting in a well defined estimate of the height of the satellite. Return waveforms from land or ice are more difficult to interpret (Martin et al., 1983).

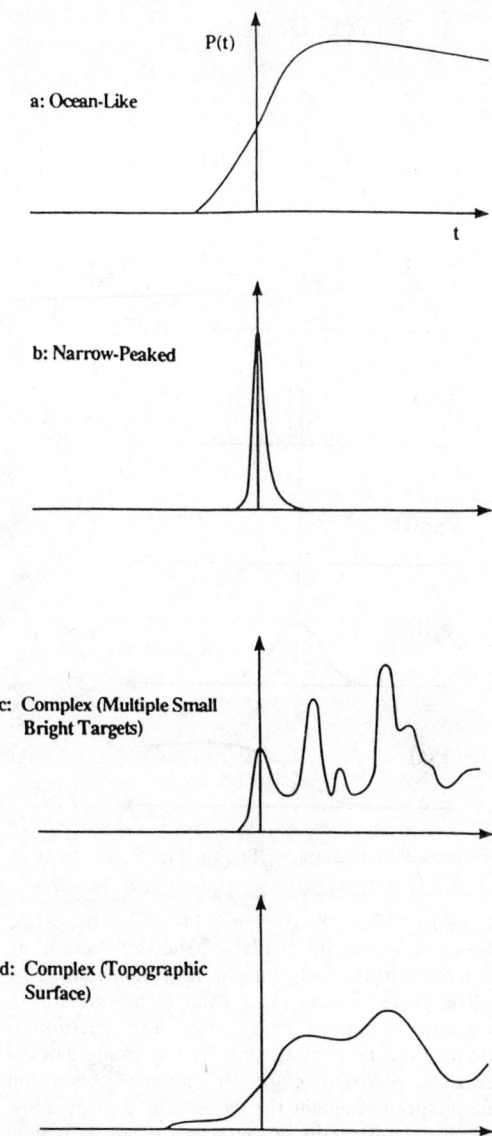

FIGURE 2. Sample return waveforms over a variety of land, ice and water conditions. (From Rapley et al., 1987)

The point target response and the antenna response for the radar beam do not change, therefore variations in the return waveform are related to changes in the probability distribution of the scattering facets within the footprint. This relationship permits estimates of averaged waveheight over the ocean. The total power return reflects the roughness of the surface, which can be related to wind speed over the ocean and the dielectric constant over land.

Since radar pulses travel at the speed of light, a few tenths of nanoseconds in travel time is the same as a few centimeters of precision in the range estimate. Because the number of receiver timing gates are limited to a range of a few meters and surface topography can vary by several kilometers, the opening of the timing gates to collect the return pulse must be set in some predetermined fashion. This is the job of the *tracker* subsystem. Tracker techniques vary depending on the application, however, they all operate on the preceding return waveform. If the receiver is not turned on at the proper time, the tracker loses its lock on the transmitted pulse. In order to regain lock on the return echo the tracker must change its procedure and search for the return pulse. Over the ocean the estimated return time of the next pulse is so accurate that it is often used for the range estimate without examining the actual return waveform to determine the travel time. Over land and ice the individual return waveforms are retracked computationally in post processing to determine the range accurately (Martin et al., 1983).

The design of a specific altimeter will vary depending on its application. For example, the TOPEX altimeter, which is interested in studying the mean and varying ocean surface, will average several hundred pulses over 128 timing gates to get a precise estimate of the sea surface and its wave field. The TOPEX tracker will be based on the fact that the probability distribution of the scattering facets is reasonably well known for ocean waves. In contrast, an altimeter for land and ice topography might average only 20 pulses in 24 gates because of the reduced accuracy requirements over land and ice compared to the ocean and the need for more flexible tracking because of the variety in possible return waveforms over land and ice (See Fig. 2). Some systems have multiple modes to account for the different types of surfaces on a planet. The European Space Agency's Earth Resources Satellite-1 (ERS-1) will have separate operating modes and trackers for ice and ocean.

Laser Sensors. The major difference between a beam limited laser system and a pulse limited microwave sensor is the much smaller beamwidth, which results in a less biased estimate of the height of the satellite over rough topography and/or sloping terrain as is commonly found over land or ice. In addition, the smaller footprint increases the accuracy in the horizontal position measurement of the primary return echo in relation to the spacecraft nadir point. Similar to microwave systems, laser instruments have timing gates and tracker systems to measure the return echo. However, the short high powered pulse of the laser and the reduction in surface scattering effects within the smaller footprint result in a high signal-to-noise ratio for individual pulses. Consequently, return waveforms are usually not averaged. The overall precision of the laser altimeter is on the order of 10 cm for present systems.

While laser altimeters were experimented with successfully during the Apollo missions to the Moon, there have been no dedicated satellite laser altimeters. Several have been proposed for the future and are now undergoing tests on aircraft. These include missions to measure topography on the Earth, Mars, and the Moon.

Spacecraft Requirements. Satellite altimeters place a number of requirements on the spacecraft. Most important of these are pointing accuracy and mass to area ratio. The spacecraft pointing must be accurate for the position measurement, which is most critical with laser systems over land or ice. In addition, the pointing must be held to within a degree to insure adequate power return in the microwave echo over the ocean.

The friction of the atmosphere on the spacecraft affects its altitude. The forces of atmospheric drag are not well understood at the altitudes of the altimetric satellites (800 Km) and are parameterized in orbit determination models. The amount of drag on a spacecraft is a function of the mass-to-volume ratio of the spacecraft, the shape of the spacecraft and the altitude. The drag on a spacecraft is inversely proportional to the mass-to-area ratio.

Orbit Determination. The estimation of the spacecraft altitude relative to a reference ellipsoid is a critical part of satellite altimetry. The orbit is determined by fitting an incomplete set of satellite tracking data with an integration of the equations of motion. Satellite tracking is done most frequently from Earth with either lasers or electronic tracking systems. Satellite-to-satellite tracking has been successful on an experimental basis and should become routine when a satellite global positioning system is in place. Integration of the equations of motion requires a detailed knowledge of the force field. Over the past two decades, satellite geodesy has provided increasingly accurate gravity models of the Earth and planets as the number of satellites increase. Deceleration forces on the spacecraft, such as atmospheric drag, are determined in a parametric fashion because of imperfect knowledge of the physics. The projected goal of 14 cm rms absolute accuracy in satellite altitude for TOPEX gives an indication of the resolution that orbit determination procedures will soon achieve.

Path Length Corrections. The travel time of the altimeter pulse is delayed by refraction as it travels through the atmosphere. Path length corrections can be determined by independently measuring the constituents of the atmosphere that most significantly affect the refraction. For a microwave sensor orbiting the Earth, the dry atmosphere, tropospheric water content, and electron density of the ionosphere cause path length corrections of order 2.5 m, 10–50 cm, and 1–15 cm, respectively.

Surface Interactions. The composition and the structure of the reflecting surface can have a dramatic effect on the return time of the altimeter echo. For example, over the ocean the return power is focused in wave troughs but scattered at wavecrests resulting in a negative bias in the range estimate that can be over 10 cm. Post measurement corrections must be made to the height measurement to adjust for this effect. The overall power return varies considerably because of the surface composition and roughness.

Techniques for Data Analysis. A number of techniques have been developed for processing altimeter data. Many methods improve the altimeter accuracy by effectively removing various sources of noise in the data by spatial or temporal filtering. Instrument noise at the shortest wavelengths is effectively removed with a median filter. Residual errors in the determination of the orbit have been found to be concentrated in very long wavelengths of once per revolution or more. Usually a *crossover adjustment* is performed on several intersecting tracks by removing those low order polynomials, which minimize the differences in the altimeter data where ascending and descending orbits cross.

There are some time varying effects on topography that are often well known and desirable to remove from the altimeter data in order to study other phenomena. Solid earth and ocean tides represent periodic phenomena that are reasonably well known on Earth and can be nominally removed from the measurements using model output in order to study nontidal oceanographic processes. One of the important science goals of oceanographic altimetric missions is to derive improved estimates of global ocean tides. Tidal variations on the earth amount to a signal of about 10 cm rms for the solid earth and 30 cm rms for the ocean. The barometric response to atmospheric pressure is often removed with the inverse barometer approximation in order to study the low frequency change in sea level. Variations in sea level pressure in terms of the barometric response have a rms of less than 10 cm.

Applications

The first altimeter to fly in space was a laser system on *Apollo 15*. This and subsequent flights on *Apollos 16* and *17* were used to examine a very limited region of the lunar surface. In 1974 *Skylab* flew the first microwave altimeter over the Earth in a proof of concept mission (McGoogan et al., 1974). While the system accuracy was several meters, analysts were able to distinguish real geophysical spatial variations in the sea surface. The first dedicated altimeter spacecraft was *GEOS-3*. This system was a geodetic mission to examine the mean shape of the sea surface. It flew from 1975 through 1978. A number of useful scientific results were obtained from *GEOS-3* and many are described in the special journal publication edited by Stanley (1979). In 1978 the *Pioneer Venus Orbiter* began observations of Venus with a radar altimeter system and some initial results have been described by Pettingill et al. (1980).

The first altimeter specifically designed to meet the rigorous accuracy restrictions for ocean science flew on the *SEASAT* spacecraft in 1978. The overall accuracy of the system was less than a meter and the instrument precision was less than 10 cm. Several of the many useful analyses from these measurements have appeared in the special journal publications edited by Bernstein (1982) and Kirwan et al. (1983). Review articles describing the recent results of analyses of *GEOS-3* and *SEASAT* have been presented by Marsh (1983), Brown and Cheney (1983), and Douglas et al. (1987).

In March 1985, the U.S. Navy launched the *GEOSAT* spacecraft to make altimeter measurements over the ocean for geodetic and oceanographic purposes. It is the first such system to deliver continuous global observations for more than a year. Early analyses from this system are described in the April–June, 1987 issue of the *Johns Hopkins APL Technical Digest* (Vol. 8, No. 2).

The observations from these systems have been used for applications over land, ice, and water. In some cases data from a single system were found useful for all three applications.

Land. Topographic data is a fundamental measurement for many disciplines of Earth science including hydrology, ecology, and botany, as well as, geology, geodesy, and geophysics. Unfortunately, global topographic data sets with sufficient resolution and precision for most applications are not available for any planet, including the Earth. To overcome this deficiency, satellite altimeters have been successfully demonstrated over land and are now being developed or proposed for a number of planets.

The first altimeter that successfully returned topographic information over land were laser systems flown on *Apollo*. These data were not continuous track measurements but included several thousand individual pulses at nominal separations of 30 km or more on the surface. The measurements were used to calibrate metric camera photographs.

The first microwave altimeter to fly over land was the *Pioneer Venus Orbiter* (*PVO*) radar altimeter. This system was constrained by the highly elliptical orbit of *PVO* and a spinning spacecraft. Consequently, the footprint averaged about 100 km. Nevertheless, investigators were able to map 93% of the surface of Venus at a coarse resolution of 100–200 km and a vertical accuracy of less than 200 m.

A few investigators have studied overland data from the *GEOS-3*, *SEASAT*, and *GEOSAT* ocean altimeter missions. They have shown that these data can be used over smoother land features and inland water bodies with a precision of about a meter. The most extensive analyses have been presented by Rapley et al. (1987). Desert areas have been found to have return waveforms very similar to the ocean, and dune heights and desert zonation have been successfully identified. Some large land areas, such as Australia, have been mapped and the data have been found to be in good agreement with higher resolution surveys done from aircraft and on the surface.

Ice. The most basic question of glaciology concerns the growth of the polar ice sheets. Understanding the ice mass balance would provide important knowledge concerning the global water budget and the climate. Despite years of polar research, field measurements have not provided adequate coverage to determine whether ice sheets are receding or growing. Satellite altimeter systems can provide routine topographic mapping of the major ice sheets and supply this information.

Data from *GEOS-3*, *SEASAT*, and *GEOSAT* have been used to successfully make preliminary maps of the Greenland and Antarctic sheets up to 72° North and South latitude respectively (Zwally et al., 1983). Estimates of growth can now be considered using data from *GEOSAT* and *SEASAT*. Thomas et al. (1985) has recommended accuracies of less than 0.5 m vertically and 10 m horizontally in order to make routine estimates that would enable useful estimates of the ice mass balance.

Ocean. The most successful applications of radar altimeters have been in the study of the oceans. Three dedicated missions (*GEOS-3*, *SEASAT*, and *GEOSAT*) have measured the ocean surface in the past decade and a half. Included among the many achievements derived from these data are: the mapping of a global mean sea surface at a spatial resolution of 12 km and submeter vertical accuracy (see Fig. 3); the global mapping of sea level variability; regional estimates of large scale, low frequency sea level change at subdecimeter accuracy; and global estimates of surface wind speed and wave heights. Altimeter studies of the ocean include both oceanographic and geophysical studies. The former focus on studies of mean and variable ocean circulation processes as well as properties of the sea surface, whereas the latter utilize the relationship between the geoid and the mean sea surface to explore the structure of the Earth beneath the sea.

Oceanographic Studies. The ocean provides a reservoir of heat for the atmosphere because of its much larger thermal capacity. Determining how and where heat is transported in the ocean is of fundamental importance to a further understanding of climate change. Conventional oceanographic field measurements are inadequate to monitor the global ocean circulation; a satellite system is required.

At periods of several days and longer most unforced ocean currents are maintained by a balance between the Coriolis force and horizontal pressure gradients. In this *geostrophic balance* the flow is perpendicular to the pressure gradient. Near the ocean surface, horizontal pressure gradients are directly reflected in the sea surface slope. Altimeters with subdecimeter precision represent an unprecedented capability for oceanographers to measure the geostrophic circulation and gain a better under-

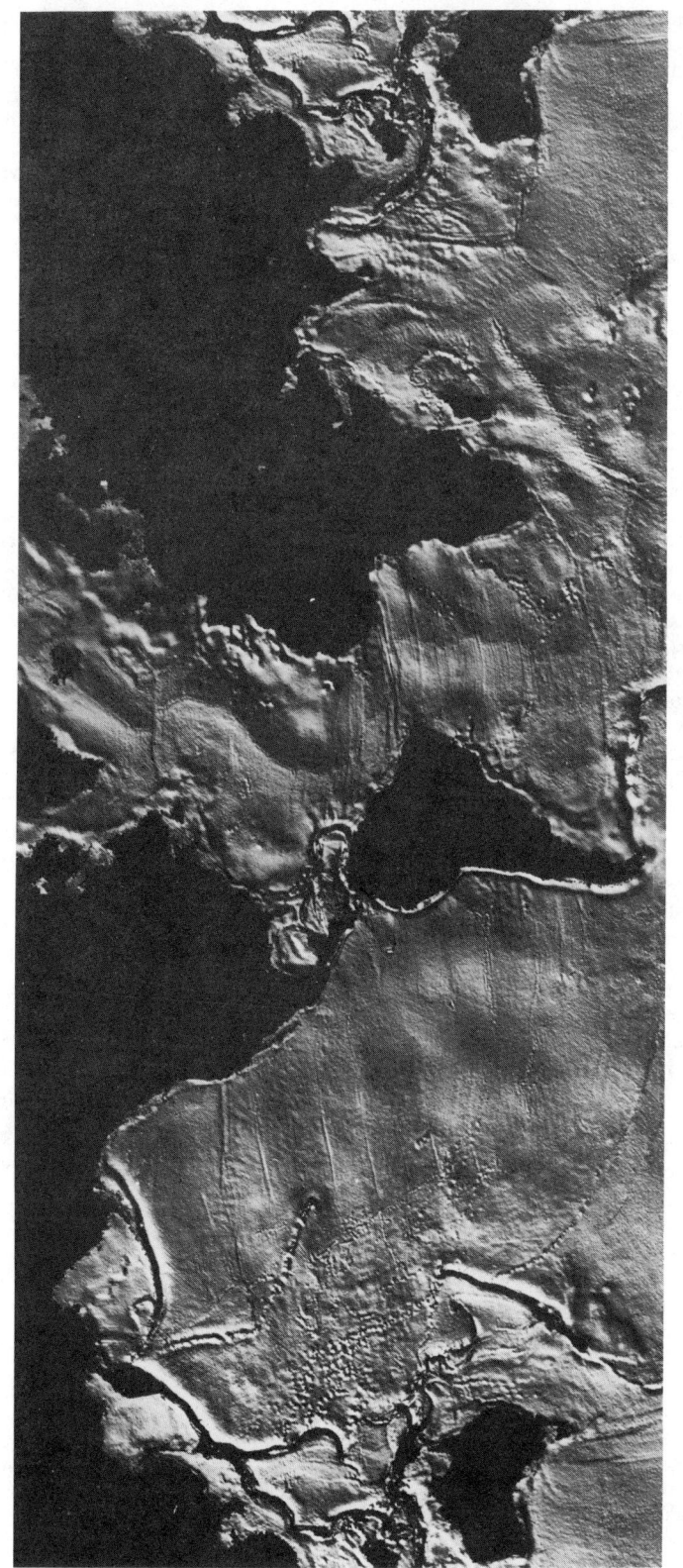

FIGURE 3. The mean sea surface from a combination of *SEASAT* and *GEOS-3* altimeter data.

standing of the surface transport. The general circulation of the ocean varies on time scales of decades. Appropriately placed ship transects can measure the deep transport relative to an undetermined geopotential surface with routine hydrographic surveys. The sea surface slope determined from altimetry can provide the necessary boundary condition and permit an estimate of the absolute circulation of the ocean throughout the water column. Converting these concepts into quantitative measurements is one of the major goals of ocean studies with satellite altimeters.

Initial altimeter studies of the ocean with *GEOS-3* and *SEASAT* developed many useful techniques for deriving oceanographically useful quantities from altimeter data. Major achievements included: global estimates of sea level variability, qualitatively correct estimates of the contribution to the mean sea surface from the long term mean ocean circulation, and regional estimates of large scale sea level change. The flight of the Navy's *GEOSAT* altimeter has provided the next step in this development. This system has been already used to make very precise estimates of global sea level variability over a one year period (Koblinsky, 1988). Regional analyses of large scale sea level change in the equatorial Pacific over a 2 year period have demonstrated the first quasi-synoptic view of the sea level response to anomolous equatorial wind forcing during the El Niño-Southern Oscillation phenomena (Miller et al., 1988).

Geophysical Studies. Satellite altimeters have provided the first global view of the marine geoid. Figure 3 shows a mean sea surface derived from *SEASAT* and *GEOS-3* data. The mean sea surface is a combination of the height of the geoid and the dynamic sea level resulting from the long-term mean ocean circulation. Height variations in the geoid are more than two orders of magnitude larger than the dynamic sea level. This large signal to noise ratio has been exploited by geophysicists, who use the mean sea surface as a close approximation of the geoid.

The features in the mean sea surface with the largest vertical range are long wavelength (>3000 km) undulations that are predominantly maintained by mantle-wide convection. These fluctuations in the altimeter derived mean sea surface compare quite well with independent estimates made from global gravity models and they have been removed from the mean sea surface shown in Fig. 3. Short wavelength features are accentuated in the mean sea surface in Fig. 3 through the use of computer generated artificial light shading. The large contrast in density between the ocean and the Earth leads to a strong correlation between the sea floor topography and the mean sea surface at these wavelengths. This relationship has enabled scientists to find many new features in the marine bathymetry such as seamounts and fracture zones that had been previously unexplored with ship surveys.

Shorter wavelength features in the mean sea surface also reflect density anomalies within or just beneath the lithosphere. The altimeter data has enabled detailed studies of fracture zones. Other studies have searched for evidence of short scale convection in the mantle. Around deep ocean trenches, the mean sea surface has yielded new information about lithospheric flexure.

Future Missions

The success of satellite altimetry has led to a number of important new missions that are scheduled for the next decade. The European Space Agency (ESA) on *ERS-1* plans to fly an altimeter in near polar orbit over the Earth to examine the oceans and polar ice sheets in 1990. A joint U.S. and French mission intends to fly a new U.S. altimeter system of unprecedented accuracy and an experimental French altimeter in the TOPEX/Poseidon mission in 1992. Also for the early 1990s a team of U.S. scientists are developing a laser altimeter to be used on NASA's Mars Observer Mission. A number of altimeters have been proposed for projects in the mid-1990s, including follow-up satellites to *ERS-1* and *GEOSAT*, a two dimensional Land and Ocean Radar Altimeter (LORA) for NASA's polar orbiting *Earth Observing System*, and perhaps a mission to map the surface of the Moon. In addition, NASA's Magellan project will use a synthetic aperture radar to map the surface of Venus in unprecedented detail in the early 1990s.

CHESTER J. KOBLINSKY
JAMES G. MARSH

References

Barrick, D. E., and B. J. Lipa, 1985, Analysis and interpretation of altimeter sea echo, in B. Saltzman, ed., *Satellite Oceanic Remote Sensing*. Orlando, Fla.: Academic Press, 61–100.

Bernstein, R. L., ed., 1982, SEASAT Special Issue I: Geophysical Evaluation. *Jour. Geophys. Research* **85,** 3173–3438.

Born, G. H., C. Wunsch, and C. A. Yamerone, 1984, TOPEX: Observing the oceans from space, *EOS (Am. Geophys. Union Trans.)* **65,** 433–434.

Brown, O. B., and R. E. Cheney, 1983, Advances in satellite oceanography, *Rev. Geophysics Space Physics* **21,** 1216–1230.

Chelton, D. B., E. J. Walsh, and J. L. MacArthur, 1989, Pulse compression and sea level tracking in satellite altimetry, *Jour. Atmos. Oceanic Technol.*, in press.

Douglas, B. C., D. C. McAdoo, and R. E. Cheney, 1987, Oceanographic and geophysical applications of satellite altimetry, *Rev. Geophysics Space Physics* **25,** 875–880.

Elachi, C., 1987, *Spaceborne Radar Remote Sensing: Applications and Techniques*. New York: IEEE Press, 255p.

Kirwan, A. D., T. J. Ahrens, and G. H. Born, eds., 1983, SEASAT Special Issue II: Scientific Results, *Jour. Geophys. Research* **88**(C3), 1529-1952.

Koblinsky, C. J., 1988, GEOSAT vs. SEASAT, *EOS* **69**, 44, 1026.

McGoogan, J. T., L. S. Miller, G. S. Brown, and G. S. Hayne, 1974, The S-193 radar altimeter experiment, *IEEE Proc.* **62**, 793-803.

Marsh, J. G., 1983, Satellite altimetry, *Rev. Geophysics Space Physics* **21**, 10955-10964.

Martin, T. V., H. J. Zwally, A. C. Brenner, and R. A. Bindschadler, 1983, Analysis and retracking of continental ice sheet radar altimeter waveforms, *Jour. Geophys. Research* **88**, 1608-1616.

Miller, L., R. E. Cheney, and B. C. Douglas, 1988, GEOSAT altimeter observations of Kelvin waves and the 1986-87 El Niño, *Science* **239**, 52-54.

Pettingill, G. H., E. Eliason, P. G. Ford, G. B. Loriot, H. Masursky, and G. E. McGill, 1980, Pioneer Venus radar results: altimetry and surface properties, *Jour. Geophys. Research* **85**, 8261-8270.

Rapley, C. G., M. A. J. Guzkowska, W. Cudlip, and I. M. Mason, 1987, An exploratory study of inland water and land altimetry using SEASAT data, *ESA Contract Report 6483/85/NL/B1*.

Stanley, H. R., ed., 1979, The GEOS-3 Project, *Jour. Geophys. Research* **84** (B8), 3779-4079.

Thomas, R. H., R. A. Bindschadler, R. L. Cameron, F. D. Carsey, B. Holt, T. J. Hughes, C. W. M. Swithinbank, T. M. Williams, and H. J. Zwally, 1985, Satellite remote sensing for ice sheet research, *NASA Technical Memorandum TM-86233*, 40p.

Zwally, H. J., R. A. Bindschadler, A. C. Brenner, T. V. Martin, and R. H. Thomas, 1983, Surface elevation contours of Greenland and Antarctic ice sheets, *Jour. Geophys. Research* **88**, 1589-1596.

Cross-references: *Doppler Positioning: Satellite; Earth Orientation; Figure of the Earth; Geodesy: Satellite; Global Positioning System; Satellite Laser Positioning.*

SATELLITE DOPPLER POSITIONING—See DOPPLER POSITIONING: SATELLITE.

SATELLITE LASER POSITIONING

Satellite laser ranging (SLR) is one of the space methods developed since the mid-1960s to accurately determine the position of a spacecraft and its orbit and to determine positions on the Earth's surface. A laser range observation is a time and a distance (range) from the laser tracking system on the Earth's surface to a spacecraft in orbit. The observation is obtained by measuring the time taken by a short pulse of light to travel to the spacecraft and return to the Earth (Fig. 1).

The principal components of a laser ranging system are a laser transmitter, which generates a short pulse of laser light; a laser receiver, which collects the photons of the returning laser pulse; and a time interval counter that measures the time between the pulse leaving the transmitter and returning to the receiver. In addition, the system requires a cooperative target, that is, a satellite equipped with laser retro-reflectors. The specifications of the laser ranging system operating at the NASA Goddard Space Flight Center in Greenbelt, Maryland, as of March 1987, follow:

Wavelength	532	nm
Power-energy/pulse	100	millijoule
Pulse repetition rate	10	pps
Pulse width	100	picosecs
Divergence	0.1	mrad
Receive telescope aperture	75	cm
Receive sensitivity	60	photons
Accuracy (single shot)	1	cm
Precision (single shot)	0.8	cm

The accuracy of the better laser tracking systems of the mid-1980s is about 1 cm and improvements continue to be made. Over the last two decades laser systems have improved two orders of magnitude. Approximately 15 spacecraft in orbit carry laser retro-reflectors, but only a few are regularly tracked by laser systems. Of these, the *Lageos* (Laser Geodynamics Satellite) and *Starlette* spacecraft are the most frequently tracked. Both spacecraft are spherical in shape and studded with laser retro-reflectors. The specifications and the orbital elements of these two spacecraft are shown in Table 1.

Laser systems are operated by, and from, many countries, and at various accuracy levels. These systems generally operate in a loosely organized network that routinely track the *Lageos, Starlette* and other spacecraft. Figure 2 shows the laser tracking systems that were cooperating with the National Aeronautics and Space Administration (NASA) in early 1987 in a program of measurement of motion of the Earth's crust. The specifications and performance of these systems are shown in Table 2. In addition to those shown in Fig. 2, several other countries were planning to join this network in the next few years. Laser tracking systems are also operated by the USSR and other East European nations.

Analysis Approach

Two basic approaches can be employed in the analysis of laser tracking data for determining station positions—the dynamical method and the geometrical method. The former utilizes the motion of the spacecraft and requires knowledge of the forces acting on the satellite, or an ephemeris of the spacecraft. The latter uses the geometry of the relationship between the spacecraft and tracking stations and

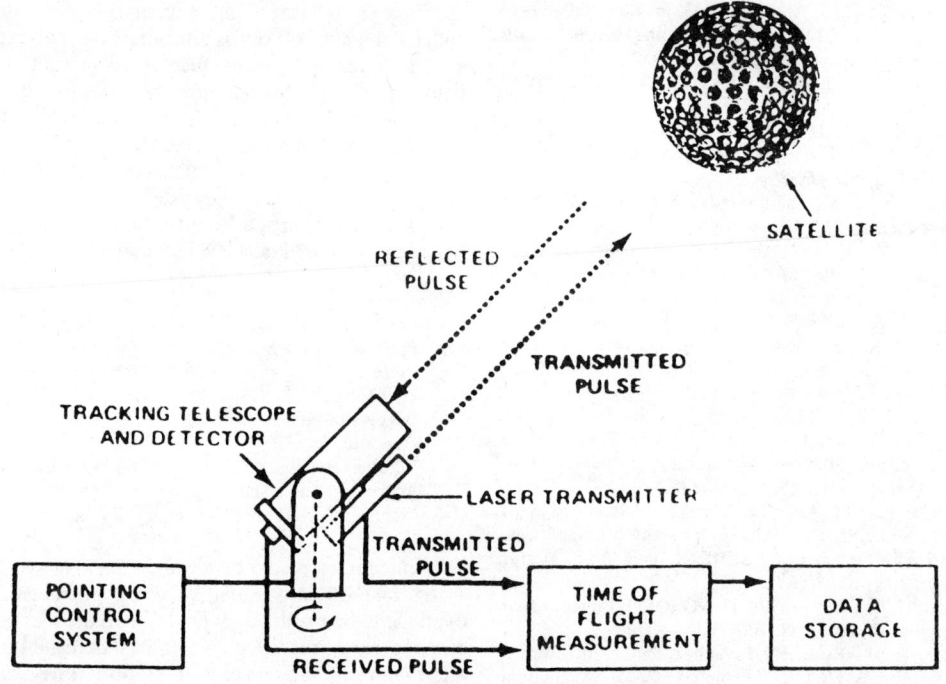

FIGURE 1. Concept of laser ranging station.

requires simultaneity of data from two or more sites. The former approach is the method most commonly used because simultaneity of data from several sites is rarely obtained in some regions because of poor weather conditions, and also because it limits the distance of separation between the tracking stations for a given height of a satellite above the Earth's surface. The dynamic method is also a much more powerful method and permits the estimation of many other quantities in addition to the position of the laser tracking stations, such as, the Earth's gravity field and tidal parameters. However, the dynamic approach is more complex and generally requires more computational effort.

The basis of the dynamical approach is the determination of the orbital motion of the spacecraft simultaneously with the relative positions of the tracking stations. In addition, it is generally necessary to determine the rotation of the Earth (polar motion and rotation rate) because Earth rotation is a form of variation in station position. The model parameters typically required for precision orbit

TABLE 1. Descriptions and Orbital Elements of the *Lageos* and *Starlette* Spacecraft

	Lageos	*Starlette*
Shape	sphere	sphere
Size	60 cm diam.	24 cm diam.
Weight	407 kg	47 kg
Retro-reflectors	422 fused silica 4 germanium	60
Area-to-mass ratio	6.95×10^{-4} m^2/kg	9.54×10^{-4} m^2/kg
Semimajor axis	12,265 km	7328 km
Inclination	109.8°	49.8°
Eccentricity	0.004	0.00
Perigee height	5858 km	806 km
Apogee height	5958 km	1108 km
Node rate	0.343° per day	−3.950° per day
Perigee rate	−0.214° per day	3.308° per/day

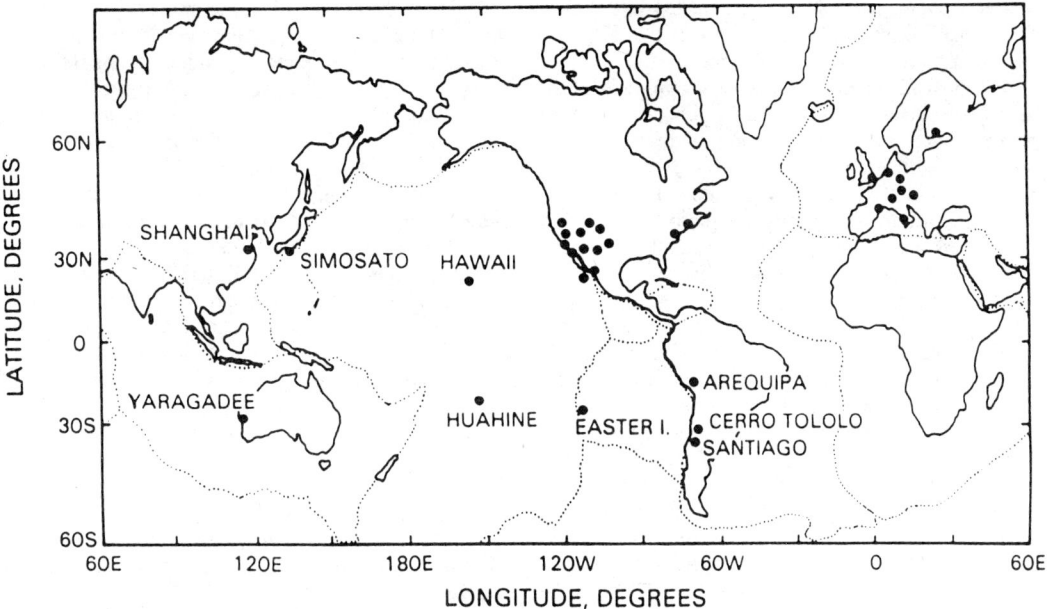

FIGURE 2. Global laser tracking network.

determination are as follows:

6 orbit parameters (a, e, i, w, W, M; or $x, y, z, \dot{x}, \dot{y}, \dot{z}$)
1 or 2 drag parameters
solar radiation parameters
gravity field [$C(1m), S(1m)$]
GM (gravitational constant * earth's mass)

lunar and solar gravity
planetary gravity (Venus, Mars, Jupiter, Saturn)
earth and ocean tides (gravity)
earth tides (body displacement tides)
ocean loading
station coordinates (lat, long, ht; or X, Y, Z)

These parameters define the forces acting on the

TABLE 2. The Global Laser Network, 1987

Location	System	Operating Agency	Accuracy
Greenbelt, Maryland	Moblas 7	NASA	1 cm
Quincy, California	Moblas 8	NASA	1 cm
Monument Peak, Calif.	Moblas 4	NASA	1 cm
Mazatlan, Mexico	Moblas 6	NASA	2 cm
Ft. Davis, Texas	MLRS	U. Texas	2 cm
Mt. Haleakala, Hawaii	Hollas	U. Hawaii	2 cm
Huahini, Fr. Polynesia	TLRS 2[a]	NASA	2 cm
Arequipa, Peru	SAO 1	U. Peru	10 cm
Easter Is.	TLRS 2[a]	NASA	2 cm
Yarragadee, Australia	Moblas 5	Dept. of Sci. & Tech.	2 cm
Orroral Val., Australia	NLRS	Dept. of Nat. Mapping	3 cm
Simosato, Japan	—	Simos. Hydro. Obs.	3 cm
Shanghai, China	—	Shanghai Obs.	10 cm
Wettzell, F.R. Germany	—	IFAG	3 cm
Kootwijk, Netherlands	—	U. of Delft	15 cm
Matera, Italy	SAO 2	CNR	5 cm
Grasse, France	—	CNES	3 cm
Graz, Austria	—	Inst. f. Weltraumf.	2 cm
Herstmonceaux, England	—	Roy. Greenwich Obs.	2 cm
Zimmerwald, Switzerland	—	Inst. f. Geodasie	3 cm
Jerusalem, Israel	Moblas 2	Israeli Space Agency	5 cm

[a]Beginning in June 1987, TLRS 2 spends approx. six months at each site each year

spacecraft. Normally some are adjusted in the orbit determination process, for example the air drag perturbation on a low orbiting spacecraft.

A critical factor in the dynamical approach to the analysis of laser tracking data for geodesy and orbit determination is the length of the orbital arc, the time span over which the observations are to be analyzed as part of a continuous orbit. In general, the shorter the orbital arc the more sensitive the orbit and other parameters are to the quality and quantity of data, and the less sensitive these parameters are to errors in the adopted force model that determines the motion of the satellite. For long orbital arcs it is the reverse. The orbital parameters are more sensitive to the force model and less sensitive to the observational data. Consequently, if one is primarily concerned with determining the forces acting on the spacecraft then the analysis is usually conducted using longer orbital arcs. However, if precise orbit position is required, as in the case for station positioning, then a shorter arc would probably be used, in part because it requires less information about the forces acting on the satellite. Thus, as the orbital arc length increases the quality of the fit of the orbit to the data generally degrades, i.e., the rms of fit goes up. Figure 3 shows a schematic of the contribution of various error sources to the position error of a spacecraft as a function of orbital arc length.

The major perturbations of a spacecraft orbit are caused by the Earth's gravity field, the Sun and Moon's gravity, earth and ocean tides, air drag, and solar radiation pressure. These forces move the satellite out of the Keplerian ellipse, the orbit that the spacecraft would follow if it were orbiting about a simple point mass and in the absence of the Sun or the Moon, or an atmosphere. In addition to these forces, several other effects introduce errors into the orbit determination process, including the tracking station coordinates, the polar motion and Earth rotation, and solid body tides. For all of these forces and effects there are models that enable their perturbing influence to be computed, but if the quality of the laser data is at the few-centimeter level then their knowledge is inadequate except for the shortest orbital arc. In these situations some of these parameters will be adjusted simultaneously with the orbit and the station coordinates. Typically, these parameters are the air drag, solar radiation pressure forces, and the polar motion and Earth rotation, all of which are difficult to predict.

For the very high accuracy data presently available, the definition of the coordinate system in which the tracking stations are located becomes important. The tectonic plate motion that each of the stations is undergoing causes the station coordinates to change with time. These motions can exceed 10 cm per year and for very long orbital arcs, or for studies involving data over a long period of time (a year or more) these motions need to be included. For most purposes the motions of the stations can be adequately represented by a relative motion model that describes the tectonic behavior of each of the plates in terms of its motion about a rotation pole. The introduction of such motions into the station coordinates can improve the orbit determination of the spacecraft. The station coordinates also help define the position of the pole of rotation of the Earth. If the relative locations of each of the laser tracking stations is known, then the Earth's pole of rotation can be

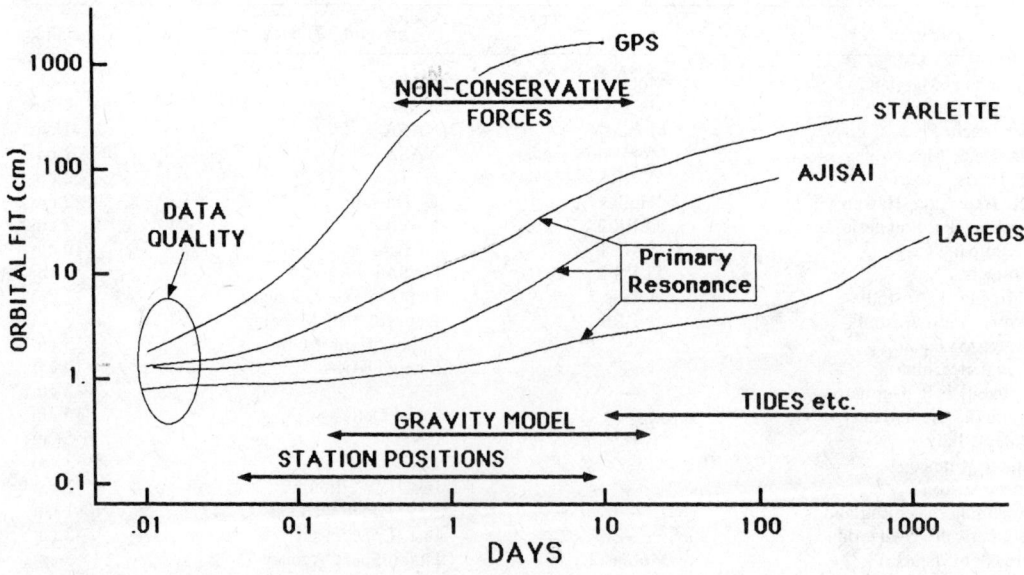

FIGURE 3. Schematic of error sources and orbit error.

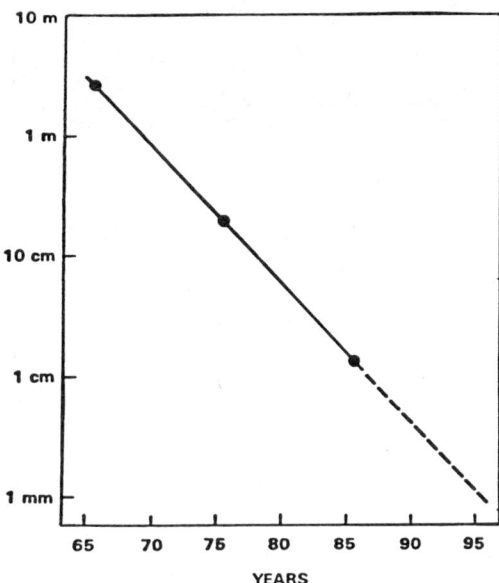

FIGURE 4. Accuracy of laser data over two decades.

defined as the point on the Earth's surface about which these stations are instantaneously rotating, since the angular distance from the pole to the station is the station's colatitude (by definition). If a model of the Earth's rotation is adopted then it must imply the correct instantaneous latitudes of the tracking stations. Alternatively, if the data permits, the position of the Earth's pole of rotation can be recovered from the data simultaneously with the orbit parameters. This approach is normal since a priori knowledge of the pole position to sufficient accuracy is unlikely to be available.

Results

Laser tracking of the *Lageos* spacecraft has continued since the launch in 1976 and the quantity and quality of the data obtained by the laser network has steadily improved. The accuracy of the data over the past two decades is shown schematically in Fig. 4, indicating that there has been an order of magnitude improvement every 10 years.

Dynamical methods of data analysis have predominated the interpretation of these data with orbital studies being based on short orbital arcs of just a few days to arc lengths of a thousand days and more. The primary purpose of all these analyses has been the measurement of present-day motions of the Earth's crust caused by tectonic and earthquake activity, the measurement of the motion of the Earth on its axis, and the basic understanding of all the forces affecting the motion of one body (*Lageos*) orbiting another body in a planetary system. The quality of the orbital fits of the *Lageos* data for various arc lengths is shown in Table 3. The typical number of parameters adjusted in the orbital solution is also shown, and the orbital fits assume that the station coordinates have been independently and accurately determined at the few-centimeter level.

The typical procedure for determining accurate station coordinates from laser data is (1) the selection of a suitable arc length, (2) the determination of the orbit from the data within an orbital arc, (3) the adjustment of the station coordinates, polar motion and any other model parameter, (4) the combining of the results from a single arc with similar arcs to form a robust solution, and (5) the back substitution of these parameters into a second iteration. (The second iteration does not usually change parameters such as station positions but it does correct the orbit and is therefore necessary for precision orbital studies.)

The results from several years of *Lageos* tracking have provided time histories of the position changes of over twenty laser tracking stations on at least five tectonic plates with most of these having over four years of data. The relative changes in position between three pairs of these stations is shown in Fig. 5. This figure shows the change in geodesic distances between Greenbelt, Maryland, and Arequipa, Peru (6167 km); between McDonald Observatory, Texas, and Hawaii (5311 km); and between Yaragadee, Australia, and Wettzell, Germany (13,264 km). The geodesic is defined as the shortest distance along the reference ellipsoid between the two sites, and does not involve the height of the tracking station. The geodesic baselines are therefore independent of the individual height values. Also shown in Fig. 5 is the observed rate of change of the geodesic and its standard deviation, and the rms scatter of the individual yearly values about the straight line. The

TABLE 3. Orbital Fits to *Lageos* Laser Data of 1 cm Noise

Arc Length	Typical Parameters Adjusted	Orbital Fit to Data (rms)
30 mins	6 orbit	1 cm
2 days	6 orbit + 1 drag	3 cm
30 days	6 orbit + 2 drag + rad. pressure	5 cm
100 days	6 orbit + 8 drag + rad. pressure	15 cm
1000 days	6 orbit + ~40 drag + rad. pressure	~30 cm

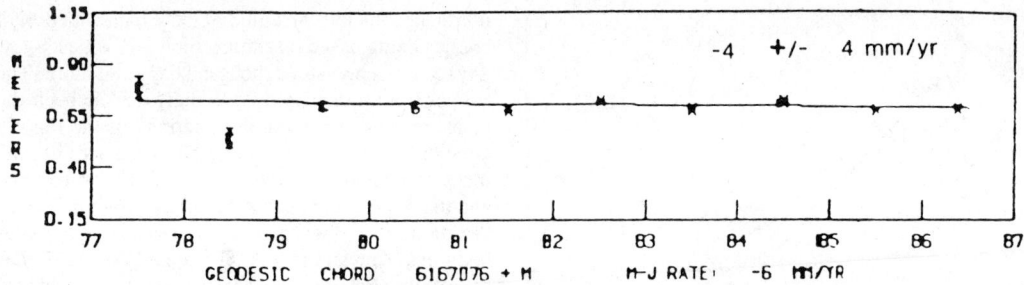

Fig 5a GREENBELT, MARYLAND TO AREQUIPA, PERU

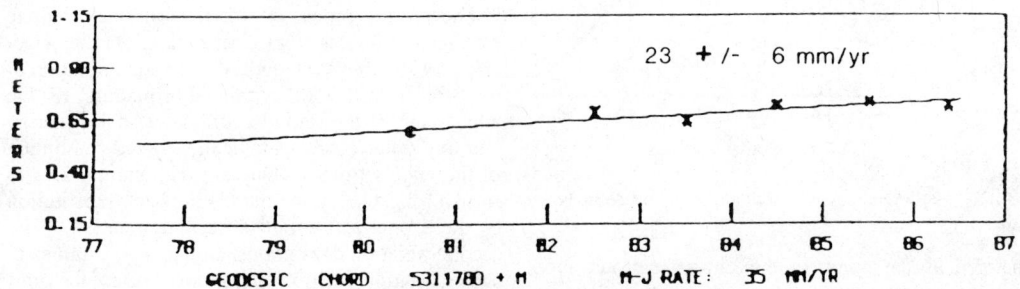

Fig 5b McDONALD, TEXAS TO HAWAII

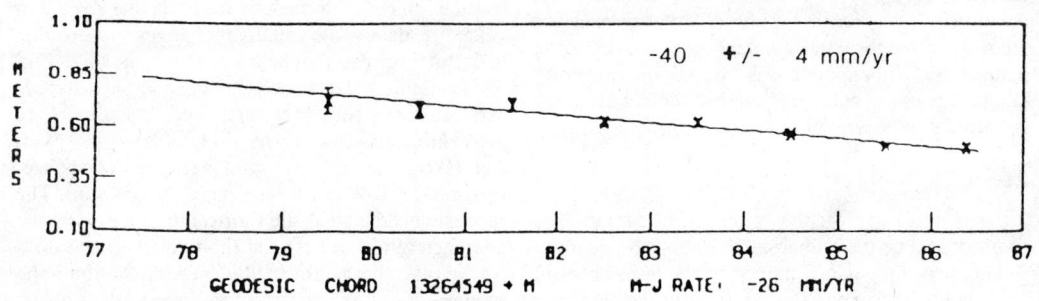

Fig 5c YARAGADEE, AUSTRALIA TO WETTZELL, GERMANY

FIGURE 5. Observed tectonic plate motion between laser sites.

former is a measure of the tectonic motion occuring between the sites, and the latter is a measure of the quality of the yearly baseline values. Each measurement is the result of differencing two station positions, each of which was obtained from the data collected at the particular site during a given year. In the solutions shown in Fig. 5 the data from all the stations in the network were combined into 30-day orbital arcs (1 month) and twelve orbital arcs were combined to produce each annual solution. Table 4 summarizes the results obtained on twenty-eight baselines connecting one or more sites on the North American (NA), South American (SA), Pacific (PAC), Indo-Australia (AUS), and Eurasia (EUR) plates.

Table 4 enables comparison of the observed plate motions and the expected motions based on geological data, such as seafloor magnetic anomalies. The general agreement between the observed and the geological rates is very good and suggests that

TABLE 4. Summary of Observed Motions on 28 Baselines

Baseline (plate)	Data Span (yrs)	Obs. Geodesic Rate (mm/yr)	Geologic Rate (mm/yr)
McDonald Obs. (NA)			
Yaragadee (AUS)	8	−70 +/− 10	−71
Greenbelt (NA)	8	12 +/− 9	0
Quincy (NA)	5	10 +/− 14	0
Mon. Peak (PAC)	5	19 +/− 9	41
Hawaii (PAC)	7	23 +/− 6	35
Wettzell (EUR)	8	35 +/− 31	21
Arequipa (SA)	8	−8 +/− 3	−11
Yaragadee (AUS)			
Greenbelt (NA)	8	−80 +/− 8	−89
Quincy (NA)	8	−80 +/− 6	−83
Mon. Peak (PAC)	6	−89 +/− 6	−104
Hawaii (PAC)	7	−93 +/− 7	−103
Wettzell (EUR)	8	−40 +/− 4	−26
Simosato (EUR)	5	−56 +/− 5	−77
Arequipa (SA)	8	65 +/− 15	61
Greenbelt (NA)			
Quincy (NA)	6	3 +/− 8	0
Mon. Peak (PAC)	6	18 +/− 6	16
Hawaii (PAC)	7	19 +/− 8	17
Wettzell (EUR)	8	21 +/− 22	21
Simosato (EUR)	5	−43 +/− 34	−6
Arequipa (SA)	10	−4 +/− 4	−6
Quincy (NA)			
Hawaii (PAC)	6	2 +/− 6	8
Arequipa (SA)	6	4 +/− 9	−12
Mon. Peak (PAC)			
Hawaii (PAC)	6	1 +/− 5	0
Arequipa (SA)	6	35 +/− 9	43
Hawaii (PAC)			
Simosato (EUR)	5	−73 +/− 20	−99
Arequipa (SA)	?	79 +/− 7	66
Wettzell (EUR)			
Simosato (EUR)	5	−68 +/− 24	0
Arequipa (SA)	8	20 +/− 13	21

present-day tectonic motions are essentially the same as the average motions over the last ten million years. Because of the locations of most of the sites in reasonably stable areas, nearly all these lines can be expected to show most or even all of the plate motion occuring between them. In California, however, where some of the laser sites are in actively deforming areas, the observed motions are significantly different from the expected geological rates. Figure 6 shows the rate of change of the geodesic between Monument Peak in southern California, west of the San Andreas fault, and Quincy in northern California, east of the fault. According to the geological prediction the motion between these two sites should be about 54 mm/year. The observed value is 23 ± 5 mm/year. Figure 6 is a plot of the baseline solutions derived from the monthly station coordinates for the period late 1981 to September 1986. If it is assumed that the geological rate is also the present-day plate motion velocity between North America and Pacific, as indicated by the motions shown in Fig. 5c and Table 4, then we are forced to conclude that additional motion, or deformation, is taking place further to the east of Quincy and/or further to the west of Monument Peak, for example, off shore in the Pacific ocean. The slower apparent motion being observed by laser ranging across the San Andreas is supported by ground based measurements of the U.S. Geological Survey on geodimeter networks close to the fault, who are observing about 20 mm/year, and also by very long baseline interferometer measurements between Monument Peak and Quincy who detected around 30 mm/year between 1983 and 1985. The question of where this additional motion is taking place is of considerable importance to understanding the earthquake risk in California today.

The tectonic motions of the stations are derived from the horizontal positions of the stations. The heights of the stations above a reference ellipsoid are

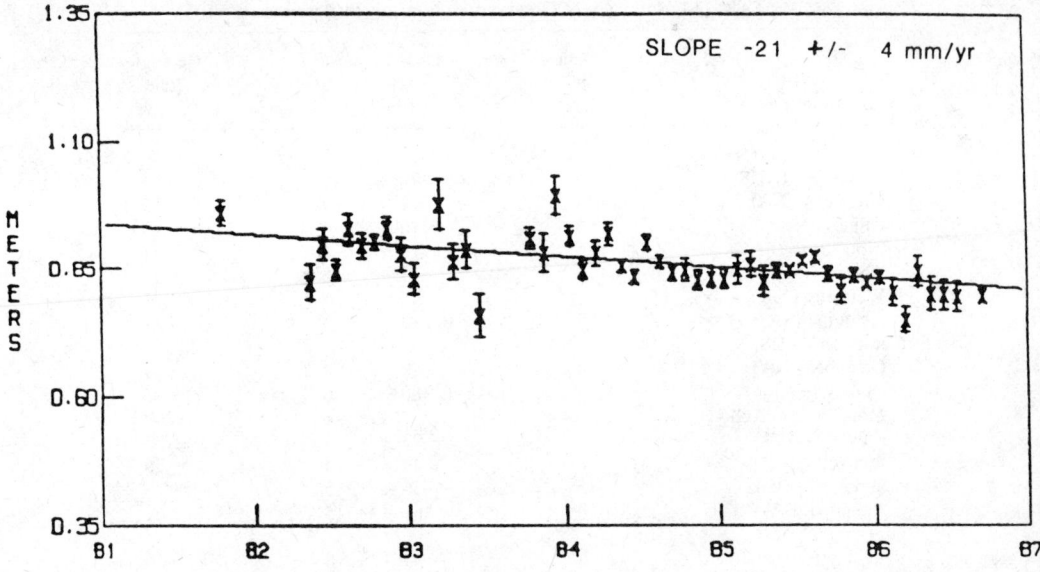

FIGURE 6. Geodesic distance from Monument Peak to Quincy.

also derived from the data simultaneously with the horizontal positions and provide a measure of the contemporary vertical changes. In comparison with horizontal motion, vertical motion is believed to be, on average, an order of magnitude smaller. Further, there is no basic hypothesis for vertical motions equivalent to plate tectonics for horizontal motion so there is no expected pattern or signature of vertical changes to be sought after. The expectation for most sites is no vertical motion. If changes are observed it is difficult for them to be ruled out as "observational error" unless they are extremely large. Figure 7 shows the annual height values observed at the Greenbelt, Maryland, and Arequipa, Peru, laser stations. These are the two longest data sets presently available and show stability at the few-centimeter level from year to year. Both data sets suggest the existence of very long-term trends but at neither site

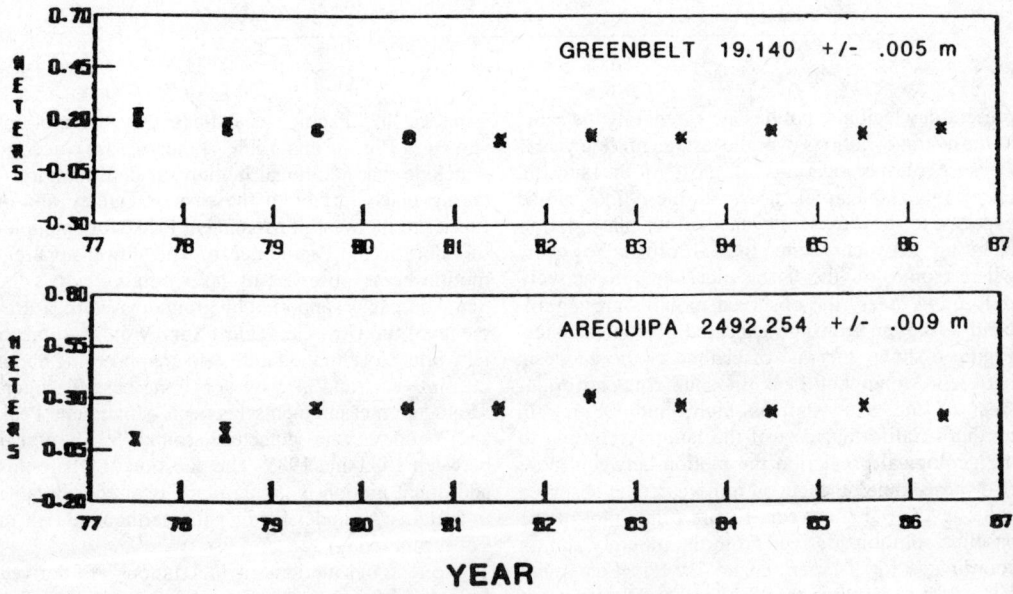

FIGURE 7. Annual laser station heights at Greenbelt, Maryland, and Arequipa, Peru.

is the evidence overwhelming. At Greenbelt the curve appears to be slightly concave upwards while at Arequipa there is an appearance of concave downwards; but in both cases it is caused almost exclusively by the data in 1977 and 1978. Exclusion of these data removes nearly all of both trends and in both cases it is clear that no major vertical motions occurred during the 10-year period. Incidentally, any change at Greenbelt above the centimeter level would be surprising considering its location in the stable center of the plate. At Arequipa, however, high in the Peruvian mountains and close to the subduction zone off the Peruvian coast, significant changes in elevation might be expected.

Future Trends

Technological developments in progress in the late 1980s will bring satellite laser ranging systems to approach the millimeter accuracy level by the mid-1990s, or before. Two-color systems will permit the almost complete removal of tropospheric effects on the range measurement and the overall limitation on system accuracy will become the spacecraft construction, distribution of the corner reflectors on the spacecraft, and knowledge of its attitude and rotation.

Modeling of the orbital dynamics will contine to improve, particularly in the areas of gravity and tides, and a better understanding will be developed about the various nonconservative forces that perturb the spacecraft movements. If similar improvements are made during the second decade after the launch of *Lageos* as were made during the first decade then orbital positioning of the spacecraft in its orbit will reach the 1-cm level for orbital arcs of a month and ground positioning of the tracking stations will approach the millimeter level.

A second *Lageos*, *Lageos 2,* a joint U.S./Italian project, is expected to be launched in the early 1990s into a similar size orbit to *Lageos 1* but with an orbital inclination near 50 degrees. This second *Lageos* will provide a different geometry for the laser stations, a different sensitivity to orbit modeling errors, and thereby strengthen the recovery of the geodetic parameters in joint solutions with data from *Lageos 1*. The *Lageos 2* spacecraft will be almost identical to *Lageos 1* in shape, size, weight, and construction. A second *Starlette* satellite is also being planned by the French for launch in the late 1980s. A third *Lageos* is also being discussed.

Another laser tracking system presently under design is the spaceborne laser which it is planned will be carried aboard the Earth Observing System (EOS) polar platform in the mid-1990s. The spaceborne system is a laser on the satellite with corner cube reflectors on the ground. Two-color range measurements will be made to an array of retroreflectors on the ground from which the relative coordinates of all the reflectors can be determined. This concept is planned to be used as a densification scheme within a large network monitored by ground based lasers or VLBI.

DAVID E. SMITH

Bibliography

Minster, J. B., and T. H. Jordan, 1978, Present-day plate motions, *Jour. Geophys. Research* **83,** 5331-5354.

Special Issue on Lageos Scientific Results, *Jour. Geophys. Res.* **90** (B11), Sept. 1985, 9217-9438.

Special Issue on Satellite Geodynamics, *IEEE Trans. on Geoscience and Remote Sensing* **GE-23,** July 1985, 4.

Cross-references: *Doppler Positioning: Satellite; Earth Orientation; Geodesy: Satellite; Global Positioning System (GPS); Paleomagnetism and Plate Tectonics; Satellite Altimetry; Seafloor Spreading: Magnetic Evidence; Very-Long-Baseline Interferometry.*

SATELLITE MAGNETIC MEASUREMENTS

The geomagnetic field at and near the Earth's surface stems from sources in the Earth's core, in the Earth's crust, and from currents external to the Earth in the ionosphere and magnetosphere (see *Geomagnetic Field: Elements*). Knowledge of this field is a primary source of information about these source regions. For logistical reasons, traditional measurements of the magnetic field at magnetic observatories and with sea, air, and land surveys, are unable to adequately cover the entire globe in a uniform manner. One of the principal contributions to geomagnetism of satellite magnetic field measurements has been to make available a high density global distribution of data from which a geographically uniform subset can be extracted.

The region external to the Earth, in which currents flow, extends from the lower ionosphere, at about 100 km altitude, thousands of kilometers into space (Fig. 1). With data only below the ionosphere, it is not possible to map the locations and characteristics of these currents. To a large degree, satellites have succeeded in carrying out such a mapping. In this article we are concerned only with the volume within an altitude of about 2000 km, because measurements from this region are pertinent to the study of the solid Earth.

Measurements of the geomagnetic field from Earth-orbiting artificial satellites were among the first space science investigations, beginning with *Sputnik 3* in May, 1958. Near-Earth magnetic field measurements have since been carried out by many satellites from several countries, as summarized in Table 1. This article summarizes the technology and history of these satellites and some highlights of research results.

About Satellites

Before specifically discussing magnetic field measurements it is useful to understand something

THE MAGNETOSPHERE

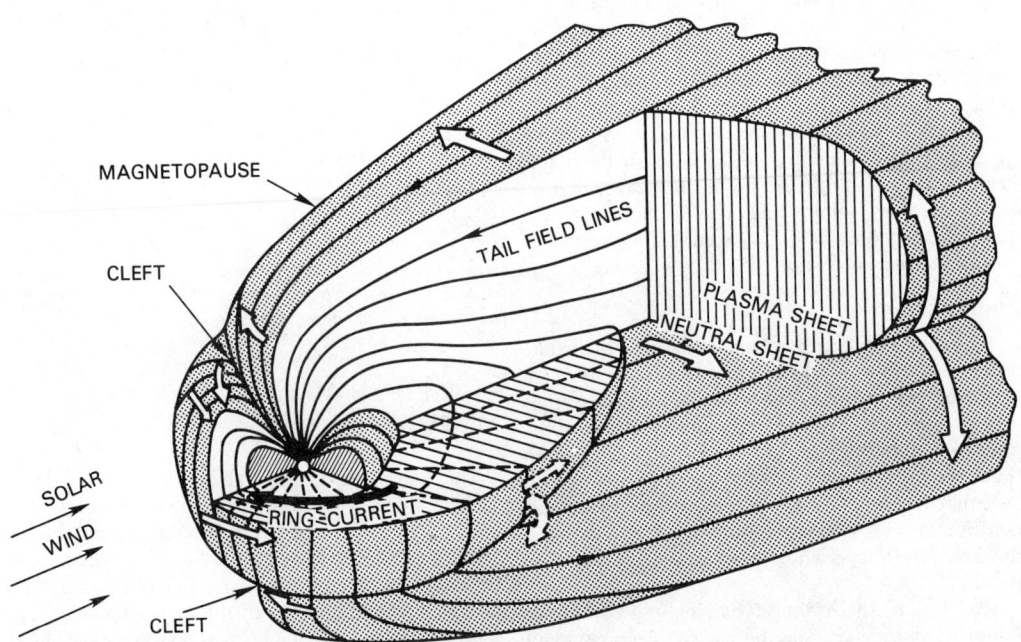

FIGURE 1. Artist's conception of the magnetic environment of the Earth: the magnetosphere.

about the nature of satellite orbits. Referring to Fig. 2, an Earth-orbiting satellite traverses an elliptical path with the center of mass of the Earth located at one focus of the ellipse. As the satellite traverses this path, its altitude above the Earth's surface varies due to the orbital ellipticity (or eccentricity); the closest and farthest points from the Earth's surface are called perigee and apogee, respectively. At near-Earth altitudes (200–2000 km), the time for one full orbit, or the orbital period, varies from about 90 to 200 minutes. The angle of the plane of the satellite path, with respect to the Earth's equatorial plane, is the inclination, i, and the intersection of the orbital plane and the equatorial plane, at the point where the satellite is going north, is the ascending node. To a first approximation, the orbit geometry remains fixed in inertial space, while the Earth rotates beneath. The projection of the satellite position onto the surface of the Earth is called the ground track, or subsatellite track, and this path over a period of time is a measure of the amount of coverage or amount of the globe from which data may be acquired.

Figure 2 illustrates that data are only acquired at latitudes up to the value of the inclination. The latitude coverage is 100% at $i = 90°$. The dependence of coverage on inclination is illustrated in Fig. 3, which shows ground tracks from the *Magsat* satellite with an inclination of 97.15°, and Fig. 4,

from the *OSTA-1* mission with an inclination of 38°. *Magsat* gave near global coverage, whereas *OSTA-1* only acquired data below 38° latitude. (An inclination greater than 90° means that the rotational component of the satellite direction is opposite the Earth's direction of rotation, called a retrograde orbit; orbits that are not retrograde are called prograde.)

In addition to the latitude range of the satellite, two other aspects of data coverage are important—data spacing in longitude and in local time. Data spacing in longitude depends on the way the subsatellite tracks move in longitude. For example, it is possible to choose the orbital parameters so that the tracks repeat after a fixed number of orbits, leaving large gaps in longitude coverage, or so that the tracks never repeat. Data spacing in longitude is a complicated function of satellite altitude, orbit ellipticity, and inclination. Figure 5 shows the longitude change between ascending nodes for a circular orbit at inclination 90° as a function of satellite altitude. For example, at an altitude of about 380 km the longitude of successive equator crossings changes by 23°. As long as an integral multiple of this change does not equal an integral multiple of 360°, the tracks will not exactly repeat from day to day and the coverage in longitude will become denser as time goes on.

Another factor in spacecraft surveys is that because

TABLE 1. Spacecraft Obtaining Near-Earth Magnetic Field Measurements

Satellite	Inclination	Altitude Range (km)	Dates	Instrument	Approximate Accuracy (nT)	Coverage
Sputnik 3	65°	226–1881	5/58–6/58	Fluxgate	100	USSR
Vanguard 3	33°	510–3750	9/59–12/59	Proton	10	near ground station
1963-38C	Polar	1100	9/63–1/74	Fluxgate	unknown	near ground station
Cosmos 26	49°	270–403	3/64	Proton	unknown	whole orbit
Cosmos 49	50°	261–488	10/64–11/64	Proton	22	whole orbit
1964-83C	90°	1040–1089	12/64–6/65	Rubidium	22	near ground station
OGO-2	87°	413–1510	10/65–9/67	Rubidium	6	whole orbit
OGO-4	86°	412–908	7/67–1/69	Rubidium	6	whole orbit
OGO-6	82°	397–1098	6/69–7/71	Rubidium	6	whole orbit
Cosmos 321	72°	270–403	1/70–3/70	Cesium	unknown	whole orbit
Triad	Polar	750–832	9/72–present	Fluxgate	about 200	near ground station
Magsat	97°	325–550	11/79–5/80	Fluxgate and cesium	6 / 3	whole orbit
DE-2	Polar	309–1012	8/81–2/83	Fluxgate	about 100	whole orbit
ICB-1300	81°	825–906	10/81–?	Fluxgate	>75	part orbit
AUREOL-3	82.5°	408–2012	9/81–?	Fluxgate	>150	part orbit
Hilat	82°	800	6/83–?	Fluxgate	about 200	near ground station
DMSP F-7	Polar	835	11/83–1/88	Fluxgate	>1000	whole orbit
Polar Bear	Polar	1000	1/86–present	Fluxgate	about 200	near ground station

the Earth is aspherical, or nonsymmetric, the plane of a satellite orbit precesses very slowly in inertial space. The rate of this precession depends on the orbital geometry, i.e., apogee, perigee, inclination, ellipticity. Since the Earth's rotation brings each longitude under the orbit plane, the data are globally well distributed. However, all observations at one latitude may have nearly the same local time for an extended period of time. In fact, the orbital parameters may be chosen so that the orbital plane remains fixed in its relation to the Sun and hence fixed in local time. Such an orbit is called sun-synchronous. The *Magsat* spacecraft was in sun-synchronous orbit in the dawn-dusk meridian plane of the Earth. Figure 6 shows the daily change in local time of the ascending node of the orbital plane as a function of inclination and of altitude. For prograde orbits the local time is decreasing with time; the rate of decrease becomes smaller as the satellite altitude increases and as the inclination increases. The local time of a satellite in a 90° inclination orbit changes about 4 minutes a day.

The capability to store and transmit data is of major importance to achieving optimal coverage. In the absence of onboard data storage, data acquisition can only occur at those times when the satellite transmits data directly to a ground receiving station, called "real time" data acquisition. The total time available for data acquisition at near-Earth altitudes, as a satellite passes by a ground receiving station, is generally from 3 to 15 minutes. Relying on real time data acquisition severely limits the actual coverage capability. To overcome this limitation a storage device, e.g., a tape recorder, must be flown onboard the spacecraft. Data are then recorded and, at a convenient time, are played back at an accelerated rate to a receiving station.

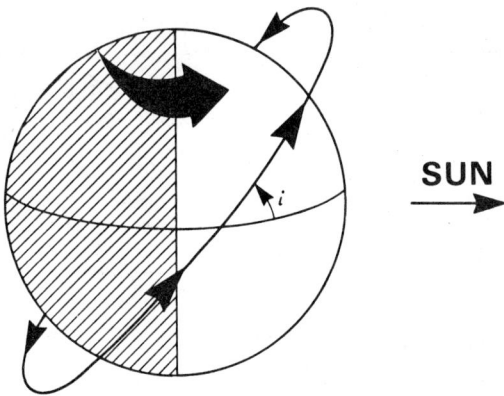

FIGURE 2. Illustration of the path of a spacecraft at inclination, i, in orbit around the rotating Earth.

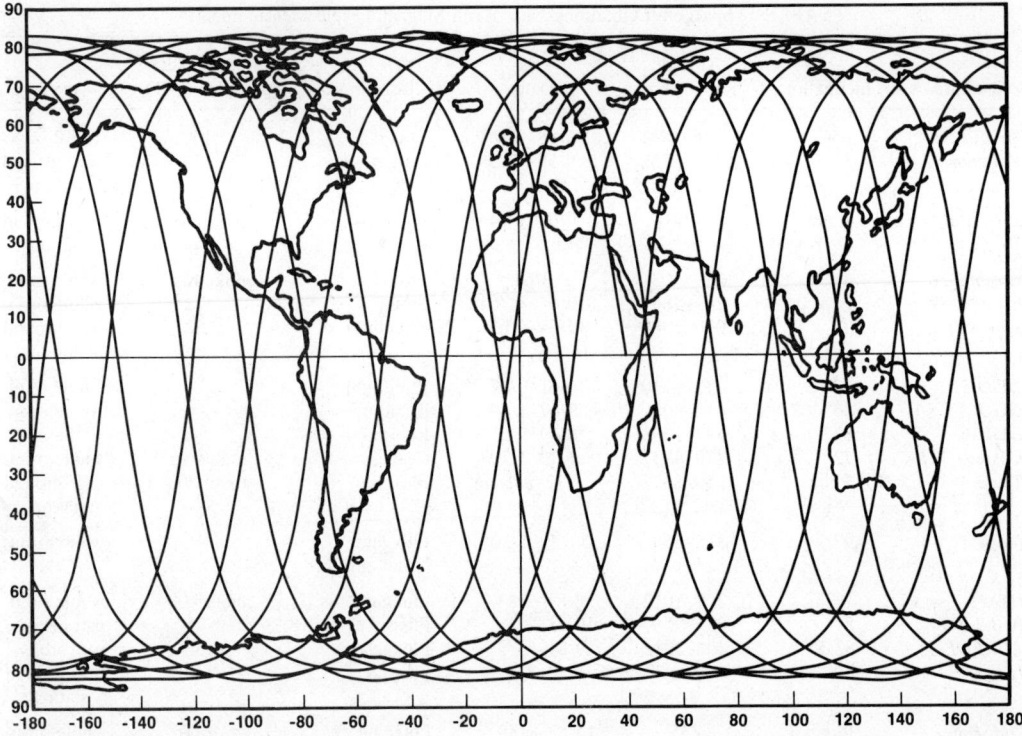

FIGURE 3. Ground track of the *Magsat* spacecraft for 24 hr. Apogee was 550 km, perigee 325 km and inclination 97.15°.

Instrumentation

Because the geomagnetic field is a vector field, a complete measurement consists of three vector components or, equivalently, the field magnitude and direction. The following summarizes some of the important features of instruments used on near-Earth spacecraft.

Component measurements on satellites have generally been made with fluxgate magnetometers. The heart of a fluxgate magnetometer is one or more fluxgate sensors (see *Geomagnetic Field: Measurement*). Such sensors have a very wide dynamic range but, to improve the noise performance of the instruments, the range of measurements is typically limited to ±1000 to ±2500 nT. To extend the total instru-

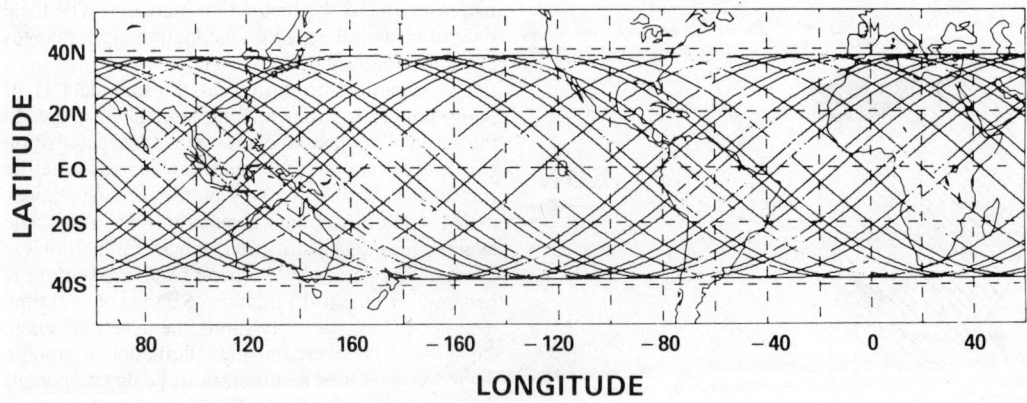

FIGURE 4. Ground track of shuttle on OSTA-1 mission: circular orbit with 38° inclination and 262 km average altitude. (Courtesy of P. D. Lowman, Goddard Space Flight Center).

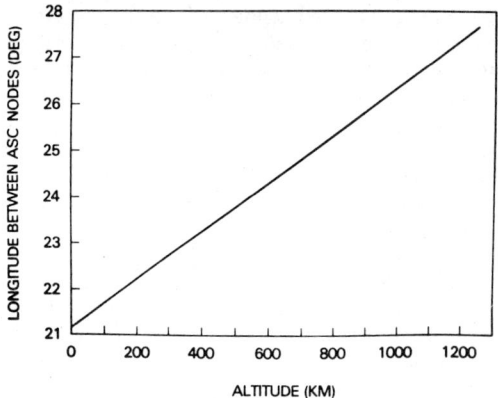

FIGURE 5. Change of the longitude between successive ascending nodes (i.e., north-going equator crossings) as a function of satellite altitude for a circular orbit. (Courtesy of K. A. Vance, Goddard Space Flight Center)

the coil is added to that measured by the fluxgate sensor in order to provide a measurement of the total ambient field.

Measurement of the field magnitude only, called a scalar measurement, has been accomplished with two types of magnetometers: the proton precession magnetometer and the alkali-vapor magnetometer.

Proton-precession and alkali-vapor magnetometers are absolute instruments because their output frequencies depend only on atomic constants. In practice, a poor optical alignment or the wrong electronic phase shift can introduce errors of several nT into the alkali-vapor magnetometer. In contrast the fluxgate magnetometer is not an absolute instrument; it requires calibration and is subject to changes in zero level and sensitivity, e.g., caused by physical changes in the coils from changing temperature or aging of electronic components.

Magnetometers onboard spacecraft measure in situ fields at the sensor location. Since the spacecraft is moving (about 7 km/sec for near-Earth spacecraft), these fields change with position and time, so when a finite counting time is necessary, as for the proton-precession and alkali-vapor magnetometers, the field readout from the instrument is an average of the ambient field over the counting time. It is thus desirable to minimize this time as much as possible since nonlinearities in the variation of the field over the counting time interval produce an average value that will not in general correspond to the field value in the middle of the counting interval.

ment range to the ±64,000 nT needed to measure the Earth's field, precisely known currents are added to the coils surrounding each sensor. The current strength is adjusted so that the magnetic field from the coil cancels most of the ambient field, allowing the sensor to function in the desired range. The electronic systems that generate these currents are called "offset generators." The field generated by

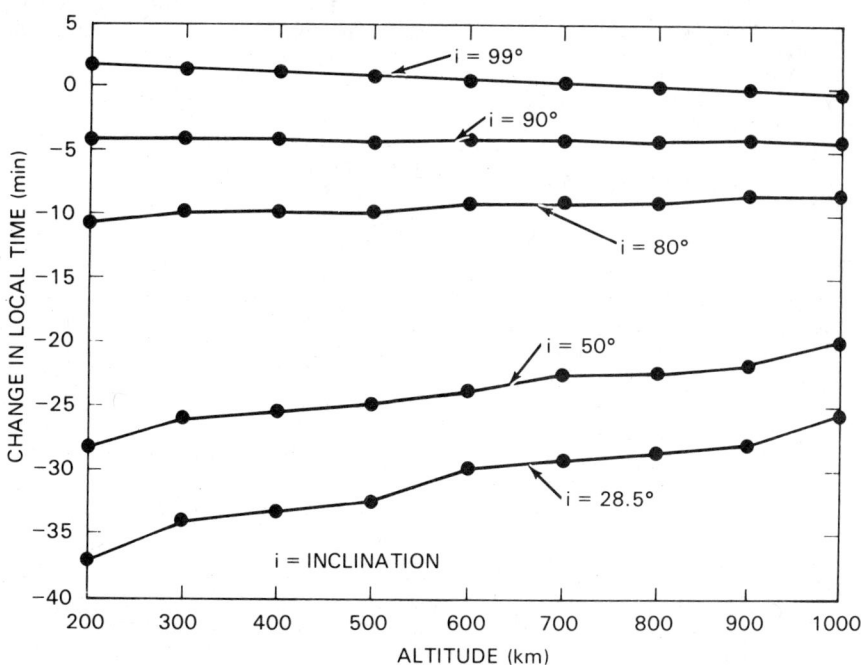

FIGURE 6. Daily change of the local time of the ascending node of an Earth orbiting satellite as a function of altitude and inclination. Circular orbit is assumed. (Courtesy of K. A. Vance, Goddard Space Flight Center)

Spacecraft Design

A major problem in spacecraft design is magnetic interference from the spacecraft itself. Any magnetic fields from the spacecraft that are above the noise level of the instrument constitute a noise source. Ideally, the magnetometer sensors will be sufficiently separated from magnetic sources to eliminate spacecraft fields as a major noise source. This separation is typically accomplished by placing the magnetometer sensors at the end of a nonmagnetic boom, while most of the electronics are placed in the spacecraft proper.

To minimize the required boom length, care must be taken in the spacecraft design and construction to use materials that are minimally magnetic and to route wires so that the current geometry causes field cancellation rather than addition.

These steps have not been possible on all spacecraft. Consequently the noise level of the resulting data is dominated by the spacecraft fields, a situation which is particularly troublesome when field generating subsystems are periodically turned off and on, resulting in a changing spacecraft field.

In order to measure the direction of the geomagnetic field, one must know the directions of the magnetometer axes. If the field along one axis is of magnitude 50,000 nT, then a 20 arc-second error in knowing the magnetometer orientation results in a 5 nT error in measuring the field perpendicular to that axis. Scalar magnetometers easily measure the field magnitude to 1–2 nT. Comparable component measurements require knowing the spacecraft attitude, and the magnetometer orientation relative to the spacecraft, to within a few arc-seconds. To date this has only been accomplished by the *Magsat* spacecraft. On *Magsat* the spacecraft attitude was measured with two star trackers and a precision sun-sensor and the orientation of the magnetometer relative to the star cameras was measured with an onboard optical system.

Orbit Accuracy

It is not only important to measure the field accurately and in a clean environment but also to know the location of the measurement. Satellite positions and/or velocities are not measured continuously. Rather, measurements are acquired at discrete times and then used to derive model ephemeris parameters. Two sources of error exist: errors in the measurements and errors or shortcomings in the model. The technology in both areas has advanced considerably over the years. One way of looking at the effect of position error is to ask what the field difference is between the actual position and the estimated position. By far the largest differences are due to the gradients in the main geomagnetic field, which can be estimated. Table 2 summarizes the global maxima and means of the various field gradients. For example, if one wants a maximum error of 1 nT, then, because the maximum vertical gradient in B and B_r is -28 nT/km, the spacecraft position must be known to within 36 m vertically.

History

Table 1 lists spacecraft that have made significant contributions to our understanding of the near-Earth geomagnetic field. The very first satellite magnetic field measurement was accomplished by a triaxial fluxgate magnetometer onboard *Sputnik 3*. The instrument was mounted in a gimballed fashion so that it could be reoriented in flight. One axis was maintained along the ambient field by reorienting the instrument until the fields measured by the other axes were zero. The axis parallel to the field gave the field magnitude; the position of the gimbal gave the orientation of the spacecraft relative to the field; i.e., the magnetometer was used to measure the spacecraft attitude, a common practice on many spacecraft. Spacecraft fields were high on *Sputnik 3* and the coverage was limited to the Soviet Union.

The *Vanguard 3* spacecraft acquired the first really accurate measurements of field magnitude from space. Because it carried no onboard storage device, however, data were acquired only in the vicinity of receiving stations, in this case in central Australia, South America, South Africa, and southern California.

The U.S. Navy satellite *1963-38C* was magnetically stabilized to within about 6° of the ambient magnetic field by a large permanent magnet. Its fluxgate magnetometers gave useful data only on the field transverse to the permanent magnet. These data

TABLE 2. Global Maxima and Means of Field Gradients

Component	Maximum Gradient (γ/km)			Mean of Gradient Magnitude (γ/km)		
	Vertical	Along Track	Cross Track	Vertical	Along Track	Cross Track
B_r	−28.0	−13.3	6.8	14.9	5.3	2.0
B_θ	18.0	−6.5	23.4	7.6	3.0	1.7
B_ϕ	8.4	−2.0	23.3	2.6	0.7	2.1
B	−28.0	−6.1	5.7	18.4	2.5	1.6

provided the first evidence for the presence of transverse magnetic fields caused by field aligned currents in the auroral belt. The Navy *Triad* satellite, operative since 1972, carried a triaxial fluxgate that obtained higher quality data than *1963-38C* and thoroughly mapped the characteristics of the fields due to the field aligned currents. Other uses of the data from these two satellites were minimal for several reasons. Neither had really adequate attitude determination; neither carried an absolute instrument to calibrate the fluxgate; and neither had onboard data storage so the resulting coverage was quite limited.

As can be seen from Table 1, until *Magsat* other surveys were performed with either proton-precession or alkali-vapor magnetometers that measured only the field magnitude. Lack of onboard recording devices limited the coverage of *Vanguard 3* and of *1964-83C*, whereas the *Cosmos*, *POGO*, and *Magsat* satellites all carried tape recorders and achieved full orbit coverage. Spacecraft fields were well above the noise level for *Sputnik 3, 1963-38C, Cosmos 49,* and *1964-83C*. An additional error source in the *Cosmos 49* data arose because the time assigned to the data was uncertain to ±0.5 second.

The first survey to combine near polar inclination, onboard data storage, and high measurement accuracy was conducted by the *OGO-2, -4,* and *-6 (POGO)* satellites that operated between 1965 and 1971. *Magsat*, launched in October, 1979, was the first and, to date, only satellite to survey the vector components of the field with high accuracy.

Measurement of field aligned currents, begun with *1963-38C* and *TRIAD*, has been carried on by a series of satellites with fluxgate magnetometers: *DE-2, ICB-1300, AUREOL-3, Hilat, DMSP F-7,* and *Polar Bear*. Most of these have also acquired data complimentary to the magnetic field data, i.e., measurements of particle precipitation, auroral imaging or electric fields. However these magnetic field data, not being absolute, have, in general, not been useful for solid Earth studies. The exception is that field magnitude data from *DE-2* has been used in modeling the Earth's main field.

Magsat

Magsat has set the standard for accurate vector measurements from near-Earth spacecraft. For a complete description of this spacecraft see Langel et al., 1982.

The *Magsat* spacecraft (Fig. 7) consisted of two independent modules—an instrument module and a base module. The base module contained the normal supporting systems including the solar array/power system, the attitude control system, the telemetry/transponder communications system, and redundant tape recorders to capture all of the *Magsat* data continuously. The instrument module contained the spacecraft attitude determination system, an attitude transfer system (ATS) for determining the vector magnetometer axis relationship to the spacecraft attitude and the vector and scalar magnetometers located on the instrument platform at the end of the magnetometer boom.

The scalar magnetometer was of the Cesium vapor type, accurate to about 1.5 nT. The fluxgate vector magnetometer performed flawlessly except for a gradual drift (about 20 nT over the mission lifetime), which was removed by calibration against the Cesium vapor magnetometer. The vector magnetometer consisted of three fluxgate sensors mounted on a ceramic structure; its construction was designed so as to minimize the effects of thermal expansion and contraction. The instrument accuracy, after inflight calibration, is estimated to be within ± 3 nT in each axis.

Knowledge of the magnetic field orientation to a total system accuracy of better than 20 arc-seconds was critical to the determination of the vector field components. This accuracy requirement dictated that the two star cameras and the ATS optical heads be mounted on a temperature controlled graphite-epoxy optical bench. Two mirrors were mounted on the back of the vector magnetometer which, in turn, was located on the end of a 6-m boom. These mirrors provided reflected beams of collimated light for accurate magnetometer axis determination by the ATS. In addition, measurements were made by a precision sun sensor mounted on the vector magnetometer and by a pitch gyro mounted in the spacecraft. The star camera measurements were combined with the sun sensor/gyro measurements to provide a cross check and to allow for interpolation of attitude between star camera tracks. Instrument and attitude accuracies, together with the other principal errors, are summarized in Table 3.

Results

Satellite data have been the basis for the most accurate models of the Earth's main field, have mapped for the first time crustal fields in the 400–3000 km wavelength scale, and have provided both a synoptic and detailed picture of ionospheric and magnetic field-aligned currents. (For magnetic field models see *Geomagnetic Field Analysis*.)

The main field of the Earth undergoes a long term, or secular, variation that can amount to as much as 0.5% per year in some localities. Thus, to represent the global geomagnetic field accurately at any given epoch, worldwide measurements must be made at times near that epoch, a feat only achieved by satellite observations, partially by *Cosmos 49* and, most notably, by the *POGO* and *Magsat* satellites with their onboard tape recorders and polar orbits. Models based only on scalar data have proved to have large errors in the vertical and east-west components at

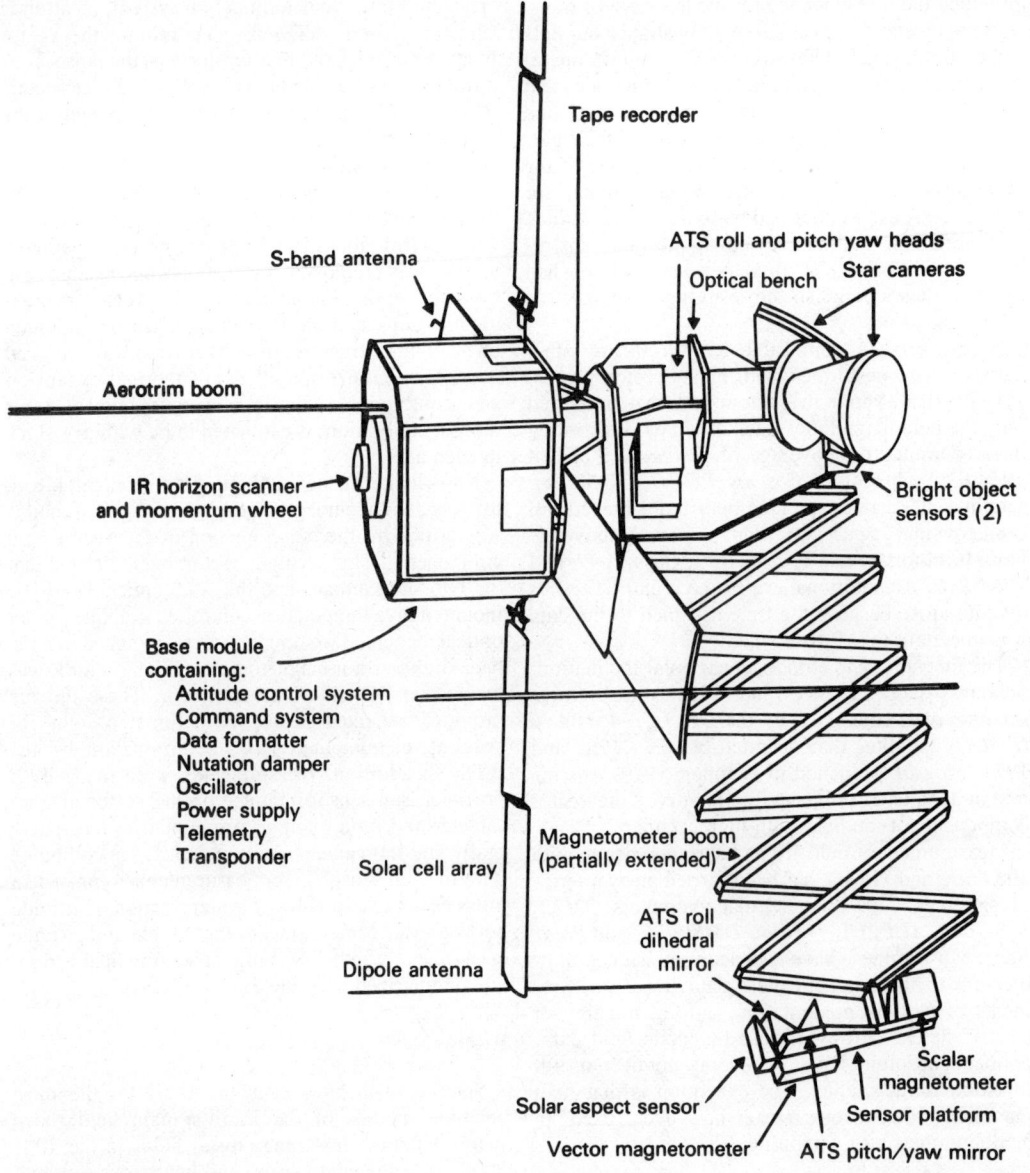

FIGURE 7. Pictorial diagram of the *Magsat* spacecraft. (From Langel et al., 1982; copyright © 1982 by American Geophysical Union)

and near the geomagnetic equator. Models based on vector *Magsat* data are not subject to these errors and are thus the most accurate models available for any epoch. They are estimated to describe the field at 1980 to an accuracy better than 20 nT for the main field at the surface of the Earth. At the same time they incorporate the first definitive determination of the long wavelength fields resulting from external sources. *Magsat* models have given new impetus to, and are being used in, studies of the field at the core-mantle boundary and studies of the fluid motions causing that field.

A crustal magnetic anomaly is the residual field after estimates of the core and external fields have been subtracted from the measured field. An anomaly map is a contour map of the measured average anomaly field at the altitude of the data. Techniques for preparing an anomaly map from satellite data include selecting suitable quiet-time data, removing the best estimate of the fields not originating in the Earth's crust, and averaging data at the appropriate resolution. Anomaly maps have been derived from aeromagnetic and shipborne data for many years and have been used in the construction of geologic/

TABLE 3. Magsat Error Budget

Error Source	Scalar (nT)	Vector (nT)
Instrument	1.5	3.0
Position and time errors	1.0	1.0
Digitization noise	0.5	0.5
Attitude error (20″ at 50,000nT)	—	4.8
Spacecraft fields	0.5	0.5
r.s.s. (root sum square)	1.96	5.8

geophysical models of the crust. These investigations have mainly concentrated on the very localized anomalies associated with small-scale geologic features and localized mineralization. However, in the past few years there has been an increased interest in studies of the broad-scale anomalies that appear in regional compilations of aeromagnetic and shipborne data. Satellite anomaly maps are of recent origin and describe only the very broadest scale anomalies, i.e., with characteristic size greater than 200 km. Aeromagnetic and shipborne anomaly maps have usually been interpreted assuming a flat Earth and a constant ambient field over the region of interest. Because of the extremely large scale of satellite-derived anomalies, in many analyses both of these assumptions become invalid, thus necessitating development of new analysis techniques.

Originally, it was thought impossible to detect fields of crustal origin in satellite data. However, while analyzing data from the *POGO* satellites, Regan et al. (1975) discovered that the lower altitude data contained separable fields caused by crustal anomalies, thus opening the door to a new class of investigations. Figure 8 provides a comparison between a *POGO*-derived anomaly map and upward-continued aeromagnetic data from western Canada. The two maps are in substantial agreement, except in the southwestern corner where the satellite map is suspect, demonstrating further both the reality and crustal origin of the anomalies.

The basic anomaly maps are only a starting point for interpretation. To maximize their usefulness they must be transformed to a common altitude and to the anomalies that would be present if the Earth's field had the same inclination everywhere (reduction to common inclination). This reduction is accomplished by assuming that the anomaly source is due to an array of dipoles, at or just below the Earth's surface, with moments aligned along the main field as estimated from a model. A least-squares solution then estimates the dipole magnitudes that best reproduce the measured anomaly field. If it is assumed that the thickness of the magnetic crust is constant over the entire region, and that the magnetization varies laterally but not with depth, then contouring the dipole moments gives the corresponding lateral variation of crustal magnetization, called the equivalent magnetization. Such a map for the United States is shown in Fig. 9. On this map are labeled some well-known geologic features of the United

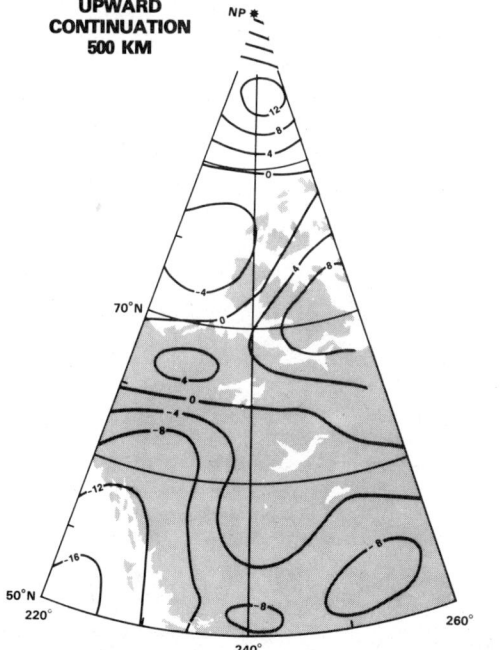
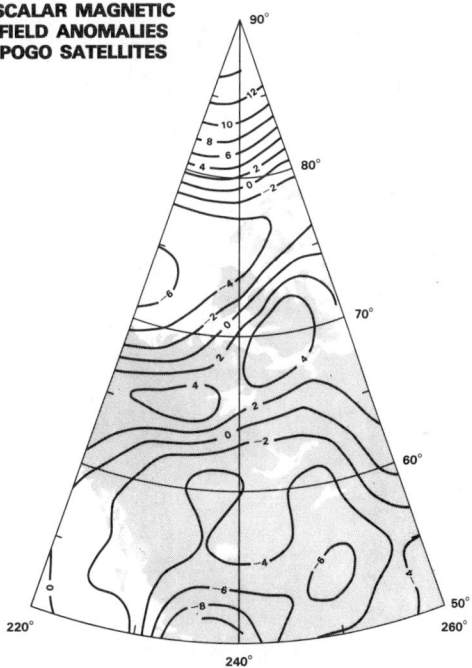

FIGURE 8. Comparison of magnetic anomaly maps from upward continued aeromagnetic data and from data from the *POGO* spacecraft. Units are nanoteslas (nT).

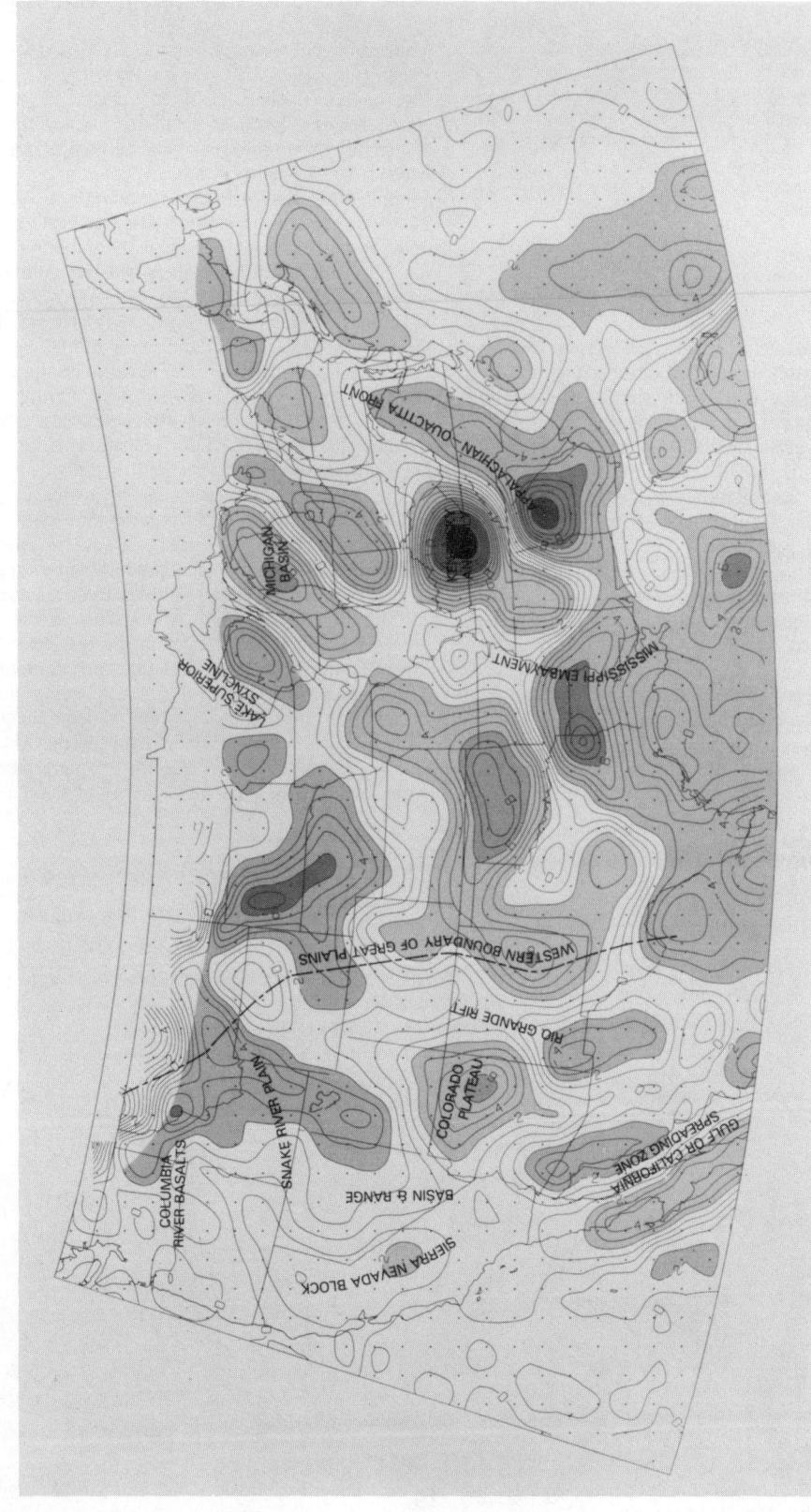

FIGURE 9. Relative magnetization map of the United States derived from *POGO* data. Units are EMU/cm^3 × 10^4. Some geologic features are indicated. (From Mayhew, 1982; copyright © 1982 by American Geophysical Union)

States, which have an apparent signature in the computed magnetization. For example, the Colorado Plateau and Michigan Basin both correspond to relatively high magnetization while the Rio Grande Rift and Mississippi Embayment, and its northward extension through the United States, which is thought by many to be a failed rift, are both reduced in magnetization. The Kentucky Anomaly shows no apparent surface expression but does correspond to a gravity anomaly and an apparent thickening of the lower crust as indicated by seismic refraction data.

Crustal models have now been made for many of the magnetic anomalies found in the satellite data. For continental crust, such models include Curie isotherm mapping, models of the structure and magnetization of the Kentucky Anomaly and of similar features near Bangui, Africa, and Kursk, USSR. For the oceans, models include several subduction zones, oceanic plateaus, and the entire North Atlantic.

As already noted, the existence of field-aligned currents was discovered and mapped with satellite data. These currents constitute a direct connection between the ionosphere and the magnetosphere. Figure 10 summarizes the large scale morphology of these currents as presently known. These currents circle the geomagnetic pole in two regions; arbitrarily, the poleward region is called "Region 1" and the equatorward region "Region 2." Region 1 currents are flowing into the ionosphere at morning local times and out of the ionosphere at evening local times. Region 2 currents flow in the opposite direction. When current densities in Regions 1 and 2 are equal, the ionospheric closure is along the magnetic meridian. This condition does not always prevail resulting in current flow across meridional lines. In the midnight region the situation is more complicated because the patterns overlap into a three-region flow, and it is not possible to distinguish Region 1 and Region 2. This basic pattern is present during all magnetic conditions, except that the regions become wider and move equatorward during more magnetically disturbed times.

What the Future May Hold

Given the present state of knowledge, the basic requirements for measurements of the core (main) and crustal fields are only partly compatible. For crustal fields the need is for increased spatial resolution while maintaining global coverage. Measurements at a given altitude need be acquired only once, although repeat missions are desirable for increased confidence in the results. This approach leads naturally to low altitude (<160 km) missions. Such a mission is useful for external and main field studies in that it provides a "snapshot" of those fields, but mapping of the temporal variation of the external and main fields would require innumerable such missions and is better accomplished by a higher altitude

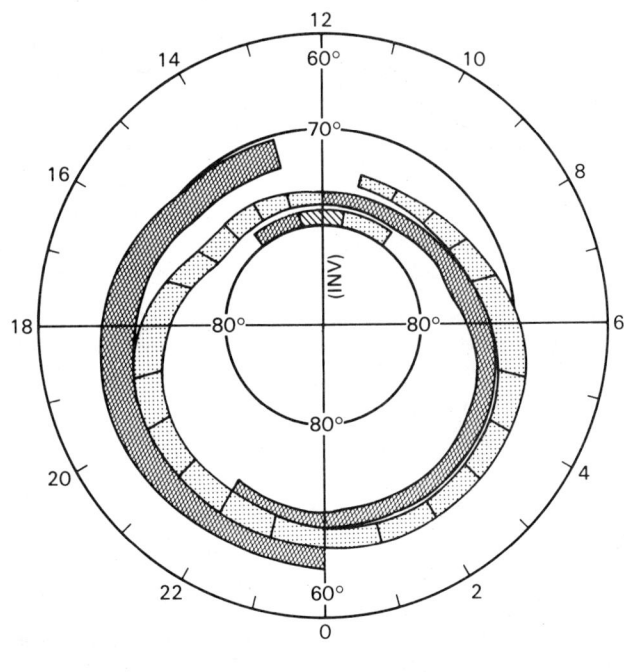

FIGURE 10. Pictorial summary of the distribution and flow directions of large scale field-aligned currents, based on data from the *Triad* spacecraft during periods of weak magnetic disturbance. Coordinates are magnetic local time (MLT) and invariant latitude (Λ). The hachured area shown between 1130 and 1230 MLT, just equatorward of 80°, indicates that current flow directions are often confused. (From Iijima and Potemra, 1976; copyright © 1976 by American Geophysical Union)

TABLE 4. Description of Proposed/Suggested Magnetometer Satellite Missions

Name	Orbit	Duration	Space Platform	Scientific Purpose
Electric and Magnetic Field Explorer (EMF)	400–500 km Polar	One year	Scout launch, similar to Magsat w/o altitude transfer system	Main field secular variation studies, crustal field studies, also electric fields and particles
MFE/Magnolia	600 km 85° inclination	Five years	Ariane launch, U.S. and French magnetometers	Main field, secular variation studies, also electric fields and particles
Tethered Satellite System (TSS)	100–150 km from shuttle 175–125 km above Earth STS orbit (Polar)	Less than 24 hours	Attached to STS by a 100 to 150 km tether; magnetometer on a small boom	High resolution crustal magnetic field profiling; proof of concept study
Polar Orbiting Geomagnetic Survey (POGS)	760 km Polar	1 year (design)	Atlas launch. Fluxgate on boom	Main field charting
Geomagnetic Autonomous Shuttle Launched Probe (GASP)	300 km	1–2 months	STS launch Fluxgate, proton or alkali vapor	Main field charting
Earth Observation System, Eos	800 km Polar	Decades	Polar platform subsystem; magnetometer attached by a boom to Eos	Main field and secular variation studies
Attached Payload	250 km 28° inclination	Decades	Space station subsystem; attached by boom	Crustal field studies; contribute to main field

(between 500 and 1000 km) mission with as long a lifetime as is practical.

These considerations have led to the mission concepts that are summarized in Table 4. At present the only approved mission is the Polar Orbiting Geomagnetic Survey (POGS). POGS and the Geomagnetic Autonomous Shuttle Launch Probe (GASP) are U.S. Naval Oceanographic Office (USNOO) missions to meet the requirements for their charts of the main field. The measurements proposed are not of the same accuracy as Magsat or of the other missions proposed, but will nevertheless be of use in modeling the Earth's main field. The MFE/Magnolia mission, a joint U.S. (NASA) and French (CNES) concept, is under study and studies are proposed for the other missions in the table.

In the long term, main field investigations require a "permanent" observatory in space. The measurements being proposed as an integral part of the Earth Observing System (EOS) mission, commencing in the mid-1990s would be a large step in that direction. Such measurements will literally revolutionize the study of geomagnetism.

A crustal field survey would be part of the objectives of the EMF mission. The tether concept is proposed as a means of achieving finer resolution of the crustal field. Tethered measurements would be complementary to those obtained by polar orbiting spacecraft and are proposed both as a space shuttle experiment and as part of the attached payload of the space station.

ROBERT A. LANGEL

References

Iijima, T., and T. A. Potemra, 1976, Field aligned currents in the dayside cusp observed by Triad, *Jour. Geophys. Research* **81**, 5971–5979.

Langel, R., G. Ousley, J. Berbert, J. Murphy, and M. Settle, 1982, The Magsat mission, *Geophys. Research Letters* **9**, 243–245.

Mayhew, M. A., 1982, An equivalent layer magnetization model for the United States derived from satellite altitude magnetic anomalies, *Jour. Geophys. Research* **87**, 4837–4845.

Regan, R. D., J. C. Cain, and W. M. Davis, 1975, A global magnetic anomaly map, *Jour. Geophys. Research* **80**, 794–802.

Bibliography

APL Technical Digest, 1980, **1**, (3) whole issue.

Cain, J. E., 1971, Geomagnetic models from satellite surveys, *Rev. Geophysics Space Physics* **9**, 259–273.

Langel, R. A., 1979, Near-Earth satellite magnetic field measurements: A prelude to Magsat, *EOS (Am. Geophys. Union Trans.)* **60**, 667–668.

Langel, R. A., 1982, Results from the Magsat mission, *APL Tech. Digest* **3**, 307–324.

Langel, R. A., 1982, The magnetic Earth as seen from Magsat, Initial Results, *Geophys. Research Letters* **1**, 239–242.

Langel, R. A., 1985, Introduction to the special issue: A perspective on Magsat results, *Jour. Geophys. Research* **90**, 2441–2444.

Potemra, T. A., 1985, Field-aligned (Birkeland) currents, *Space Sci. Rev.*, **42**, 295–311.

Webster, W. J., Jr., P. T. Taylor, C. C. Schnetzler, and R. A. Langel, 1985, The magnetic field of the Earth: Performance considerations for space-based observing systems, *IEEE Trans. on Geoscience and Remote Sensing*, **GE-23**, 541–551.

Zmuda, A. J., ed., 1971, *World Magnetic Survey 1957–1969*, IAGA Bulletin 28. Paris: IUGG Publication Office.

Cross-references: *Geomagnetic Field: Asymmetries; Geomagnetic Field: Elements; Geomagnetic Field: Measurement; Geomagnetic Field, Main: Theory; Geomagnetic Field Analysis; Geomagnetic Measurement Techniques and Surveys; Planetary Magnetic Fields.*

SEAFLOOR SPREADING: MAGNETIC EVIDENCE

The period following World War II saw a burgeoning of marine geophysical research and by 1960 a massive amount of data had been acquired dealing with marine bathymetry, thickness variation of the soft sediment layer (or seismic layer 1) overlying suspected igneous ocean crust (seismic layer 2), the linear midoceanic ridges, trenches near island arcs and continental margins, as well as the submerged oceanic islands with erosion-cut flat tops called guyots. A large number of these new observations were integrated into the seafloor spreading hypothesis suggested independently in 1962 by H. H. Hess and R. S. Dietz. The two chief cornerstones of the hypothesis were the continuous creation by linear spreading of new igneous ocean crust at the eponymous *spreading centers* and their equally inexorable destruction at the *subduction zones* located at the edges of the oceanic plates, the latter consisting not only of the newly formed ocean crust but also that part of the underlying mantle that lay above the asthenosphere.

Although the seafloor spreading hypothesis was the work of geologists alone, the confirmation of the hypothesis came chiefly from geophysicists, among whom the earliest to contribute were the geomagnetists. First came the establishment of the geomagnetic reversal timescale, which aided in explaining

This is Publication No. 1108 of The School of Earth Sciences, Department of Geology and Geophysics, University of Minnesota.

linear marine magnetic anomalies invoking seafloor spreading. Next, it was shown that the reversal stratigraphy is also recorded in vertical sections of soft sediments of layer 1. The magnetic age of the bottommost sediments was shown to increase laterally, away from the spreading center thus confirming again, albeit indirectly, the seafloor spreading process. Finally, with the advent of ocean drilling programs under the aegis of the Deep Sea Drilling Program (DSDP) and Ocean Drilling Program (ODP) it became possible to obtain a detailed rock magnetic stratigraphy of that part of the igneous ocean crust that is the source of marine magnetic anomalies. This rock magnetic stratigraphy is most easily explained on the basis of the present advanced seafloor spreading model with its attendant requirement of time- and depth-dependent alteration of rocks.

This article discusses mainly the first and most important magnetic evidence of seafloor spreading: linear marine magnetic anomalies. These anomalies with amplitudes of 50–100 nT and extending over hundreds of kilometers in near-straight lines had been discovered by R. G. Mason and A. D. Raff in the late 1950s. Cross-correlation of many such positive and negative anomalies observed along nearby shiptracks led to the discovery of *magnetic stripes* of positive or negative polarity. At that time they caused a great deal of excitement, especially since their conventional explanation required strike-slip faulting over hundreds of kilometers, assuming that the magnetic anomaly was caused by a contrasting magnetization induced by the present geomagnetic field in a region representing different magnetic susceptibilities caused by a lithologic contrast. It was known from the work on continental magnetic anomalies, however, that magnetic anomalies could also be caused partly or wholly by the natural remanent magnetization (NRM) acquired by a rock as a fossil magnetization during its formation at a previous geological time. Vine and Matthews (1963) and Morley (1981, but written originally in 1963) assumed the latter to be the case and, with the additional assumptions of seafloor spreading and geomagnetic reversals, claimed that a corollary of the hypothesis would be the generation of linear marine magnetic anomalies. With the advent of direct drilling of the ocean crust it became possible to attempt a direct test of the Vine-Matthews-Morley hypothesis by (1) age dating of the ocean crust as a function of distance from the spreading center and (2) measuring the NRM of the drilled igneous ocean crust. As often happens in science, the ocean drilling program and further detailed data acquisition and analyses of marine magnetic anomalies have not resulted in a tame confirmation of the original Vine-Matthews-Morley hypothesis but have raised new questions about the nature of the rifting process at a spreading center, the large amplitude of the central anomaly over the youngest ocean crust, anomalous skewness and amplitude modulation of the linear anomalies elsewhere in the ocean basins and the so-called Quiet Zones of diminished amplitude of the anomalies over some parts of the oldest crust. These topics will be touched upon later insofar as they relate to the magnetic evidence of seafloor spreading. In the same way, the latest observations and ideas about the magnetic stratigraphy of the igneous source layer of the marine magnetic anomalies will be discussed because the discussion will lead to a more comprehensive picture of the seafloor spreading process than was known in 1963.

Marine Magnetic Anomalies

W. Glen in his 1982 book, *The Road to Jaramillo* (Stanford University Press), and Menard (1986) have chronicled the history of the acceptance of the seafloor spreading hypothesis, and the critical journal articles related to the magnetic evidence for seafloor spreading have been collated with annotated comments by Cox (1973). Refer to these three volumes for a fuller account of the explanation of linear marine magnetic anomalies. What amazes most readers, however, is the fact that knowledge about the so-called magnetic stripes were in the public domain since the 1958 work of R. G. Mason but it took five years for a tentative explanation to be published by F. J. Vine and D. H. Matthews and it took another two years before the work of Vine and J. T. Wilson firmly established a mechanism for the origin of the marine magnetic anomalies. Both W. Glen and H. W. Menard have gone into this discussion in some detail. An explanation along the Vine-Matthews-Morley model required the acceptance of three critical assumptions, all of which were major departures from contemporary thinking:

1. New seafloor is created at oceanic spreading centers by thermal convection and it spreads symmetrically on either side of the linear spreading centers.
2. At the same time that the above occurs, the predominantly dipolar geomagnetic field undergoes sporadic reversals, about once in a million years.
3. Upon cooling, the new igneous crust acquires a thermoremanent magnetization (TRM) at the spreading center parallel to the ambient dipolar field. The reversals of the field produce, therefore, alternate linear stripes of magnetization in the igneous crust, which are perceived at the sealevel as linear marine magnetic anomalies.

In addition to these novel (for 1963) assumptions, there is another hidden internal assumption about the origin of the anomalies, which is given in the third assumption listed above. When the spatial variation of the present field is subtracted from the observed anomaly, the two remaining causes are (a) induced magnetization ($= \Delta\chi \cdot B$, where $\Delta\chi$ = difference in magnetic susceptibility between two rock types and B = ambient geomagnetic field) and (b) natural remanent magnetization (NRM, which could be thermoremanent magnetization or TRM in the newly created igneous crust of basalt and gabbro). Since both these sources are ordinarily present in igneous rocks, a knowledge about their ratio (called Koenigsberger Ratio, $Q_n = (\Delta\chi \cdot B)/\text{NRM}$) is needed before anomaly modeling can be carried out. The Vine-Matthews-Morley hypothesis for the origin of marine magnetic anomalies assumes that $Q_n \gg 1$ for the oceanic magnetic source layer and hence the alternately magnetized linear regions of the crust, parallel to the spreading center axis, were the only sources for the observed magnetic stripes. In 1963 it required considerable courage to assume that this was the case for the source layer of the marine magnetic anomalies, but later studies of samples of oceanic crust have confirmed it to such a large extent that high values of Q_n have become one of the requirements for proving oceanic provenance for ophiolites (Levi et al., 1978), ancient oceanic crustal rocks now uplifted (or obducted) and available as continental samples.

For the confirmation of seafloor spreading through magnetic evidence, independent proof of the second assumption above is required and was provided by the 1963 results of A. Cox working with R. R. Doell and G. B. Dalrymple and I. McDougall working with D. H. Tarling. These authors have provided potassium-argon radiometric ages of volcanic rocks of known magnetic polarities from California and Hawaii that show that irrespective of provenance, rocks of the same age show similar polarity magnetizations, either normal, i.e., parallel to the present earth's field or reversed, i.e., antiparallel to it. This evidence was critical to the proof of the field-reversal hypothesis, which maintained that oppositely magnetized rocks with high values of Q_n acquired their NRM due to 180° reversals of the dipolar geomagnetic field and not due to self-reversal, i.e., the acquisition of NRM antiparallel to the ambient field due to physico-chemical interactions between two magnetic phases in the rock.

The 1963 work of the above-mentioned scientists covered less than the last 4 million years (m.y.) but later work, especially that of I. McDougall in 1977 on Icelandic basalts has extended the radiometric dating of recent geomagnetic polarity reversals to the last 9 m.y. From the early age dating work F. J. Vine, in 1966, was able to date the first few magnetic anomalies at the Juan de Fuca ridge in the northeast Pacific, which led to the calculation of the rate of seafloor spreading during the most recent polarity reversals. Vine could then assume the rate to be constant and date the anomalies up to 10 m.y. ago by extrapolation. Heirtzler et al. (1968) compiled anomalies worldwide and assumed a constant spreading rate for the S. Atlantic Ocean to calculate the ages of major anomalies up to 80 m.y. ago. They then showed that similarly constant, but different in magnitude, values of spreading rates for North and South Pacific can be used to obtain a good match in the ages of all the anomalies from the three oceans. The work of Heirtzler et al. constituted firm magnetic evidence for seafloor spreading but questions remained regarding many aspects of marine magnetic anomalies and these are by no means mere details because continuing work on these problems has aided in a thorough understanding of the seafloor spreading process.

As examples of these second order but critical problems we can mention the unusually large central anomaly for the present geomagnetic epoch (named after Brunhés); a small minimum or axial magnetic low in the center of the larger central anomaly; a rapid, exponential, decrease of anomaly amplitudes in the first 10 m.y. in many of the world's spreading centers; an apparent change in reversal frequency at about 40 m.y.; an increase in anomaly amplitude in the Mesozoic M-sequence of anomalies; a very large decrease in the amplitude in the Jurassic and Cretaceous Quiet Zones; and an anomalous skewness seen in the shape of the anomalies even after subtracting the contribution to skewness made by latitude variation of the geomagnetic field. Solutions to these problems have been suggested by those working in magnetic anomaly modeling, rock magnetism, structural geology and petrology. They have resulted in an increased understanding of ridge crest tectonics and hydrothermal alteration processes. The conclusions are not always incontrovertible but in a few cases, conclusions of different workers seem to be converging on unique explanations.

The large central anomaly at the spreading centers seems to be a case where various workers are now in agreement that the increase in amplitude in the rift valley over the elevated spreading center is not due to shallowness of the magnetic source layer but due to relatively rapid low temperature oxidation (halmyrolysis) of the pillow lava pile to brownstone facies. For the first 1 m.y. the NRM of the fresh fine-grained titanomagnetite ($Fe_{2.4}Ti_{0.6}O_4$) decays exponentially from a peak value of, say, 50 A/m to half of this value in 0.6 m.y. corresponding to a titanomaghemite composition characterized by its cation-deficient spinel crystallographic structure. Raymond and LeBrecque (1987) follow Harrison (1981) in thinking that a second rate process is operative between 1 m.y. and 5 m.y. when the exponential half-life of the NRM is 3 m.y. There is an even slower rate of decay seen between 5 m.y. and 20 m.y., after which the average NRM, according to a compilation made by V. Bleil and N. Petersen, appears to increase, although these investigators find that the magnetic Curie temperature (a measure of the degree of low temperature oxidation suffered by titanomagnetite) continues to increase at a very slow rate from 20 m.y. to 140 m.y.

Most authors suggest that the above changes in NRM of the oceanic magnetic source layer, and duly reflected in the sea surface anomaly amplitudes, simply reflect a low ($\sim 0°C$) temperature oxidation process at the spreading center. During the first m.y. the finest grains are attacked while after that hydrothermal alteration of coarser grains at elevated ($\sim 150°-200°C$) temperatures takes place by convection through cracks and veins in the magnetic carapace. In the last phase (20 m.y. to 140 m.y.) oxidation continues extremely slowly by diffusion (of iron ions), again, at $0°C$.

There are two problems with the above description of alteration associated with the seafloor spreading process. The first is that both sea surface anomaly profiles and the more recent observations of global intermediate wavelength magnetic anomalies from the space by *Magsat* satellite show that beyond about 20 m.y. anomaly amplitudes slowly increase in intensity, reaching a maximum during 80-100 m.y. (the Cretaceous Quiet Zone). If seafloor alteration took place in the form of loss of iron from titanomagnetite to seawater, the attendant oxidation of titanomagnetite to titanomaghemite should cause a monotonic decrease in NRM intensity of the ocean crust, not an increase as required by the anomaly observations referred to above. V. Bleil and N. Petersen offer an ingenious explanation by postulating preferential loss of iron ions from the tetrahedral sites of the titanomaghemite spinel structure after 20 m.y. Since the saturation magnetization of a spinel is given by the difference in magnetic spin moments of oppositely directed iron ions in tetrahedral and octahedral sites, this would indeed cause an increase in spontaneous magnetization and if the grains were small enough to be uniformly magnetized (or single domain), the NRM of the ocean crust and the amplitude of marine magnetic anomalies would both increase with age after 20 m.y. However, this type of unusual oxidation is not known to occur commonly, and, what is more relevant, Raymond and LaBrecque (1987) have suggested a much simpler explanation. During the first 10 m.y. after a piece of ocean crust is formed, low temperature oxidation and hydrothermal alteration can decrease the original NRM which was a thermoremanent magnetization (TRM) and a new magnetization is acquired which is a chemical remanent magnetization (CRM) parallel to one or more different orientations of the geomagnetic field. The net CRM will be a function of the polarity history of a specific 10 m.y. window of time and by trial and error, Raymond and LaBrecque have shown that a combination of a loss of TRM to 16% of the original value and a gain of CRM which grows to four times the value of the remaining TRM can simulate successfully the observed anomaly amplitude modulation from zero to 180 m.y. This period includes not only the Cretaceous Quiet Zone increase in amplitude but also the modulations of the Mesozoic M-sequence anomalies.

A second characteristic of the anomalies not addressed by the alteration model outlined above is the 1 km-wide dip or low seen in deeptow magnetometer profiles in the exact center of the large central anomaly, where the axial graben is located, as discovered in the East Pacific Rise at 19°30'S latitude by S. P. Miller and K. C. Macdonald in 1985. These scientists suspected an elevated magma chamber under the axial graben to be responsible for the presence of a shallow Curie point isotherm which, in turn, could lead to the observed axial magnetic low. An alternative explanation depends on the low being a magnetic expression of an episodic bow-form volcanic edifice due to periodic fissuring (Tivey and Johnson, 1987). In either case, it is obvious how the fine details of marine magnetic anomalies can help define complexities of the seafloor spreading process.

An additional, and final, example of this kind of tectonic study through a thorough examination of the magnetic anomalies is the anomalous skewness referred to earlier. The "normal" component of skewness results, superimposed on the observed squarewave pattern of anomalies (due to alternate blocks of positive and negative magnetization), because of the latitude-dependent contribution by the geomagnetic field to the anomaly. Anomalous skewness is displayed most directly when magnetic lineations are replotted after reducing it to the geomagnetic pole where the magnetic field is purely vertical. S. C. Cande and D. V. Kent in 1976 suggested one explanation for anomalous skewness based on sloping isotherms, which will be discussed in the last section because the chief conclusion pertains to the thickness of the magnetic source layer, which is attended to in that section. Another explanation alluded to by Cande and Kent is rotation of oceanic crustal blocks along historic faults at the spreading center, when each degree of rotation away from the vertical produces a degree of anomalous skewness. This is not a favored model because of the need to postulate very large (35°) rotations to explain observed skewnesses. In 1978, Cande suggested yet another explanation based on a decreasing intensity of the geomagnetic field and/or undetected multiple short reversals of the field prior to a stable reversal of polarity. This explanation, however, has a somewhat ad hoc flavor to it. The CRM model of Raymond and LaBrecque can also explain anomalous skewness. Such theoretically derived skewnesses are shown to be variable with time depending, as it does, on the polarity history and thus providing a simple explanation of the observed variations.

If the CRM model of Raymond and LaBrecque gathers support from future studies, its corollaries regarding ocean crust magnetization and petrologic history must also be appreciated. As these authors point out, their model of CRM acquisition along a new ambient field direction is explicitly dependent on such a mechanism proposed earlier for the oceanic crust by Smith and Banerjee (1986). This apparently flies in the face of other studies, the earliest among them being that of a 1972 study on oxidized rims of dredged pillow basalts by M. Marshall and A. Cox, who found that low temperature oxidation produced a CRM parallel to the original TRM and not a divergent field direction. Ö. Özdemir and D. J. Dunlop found similarly in 1985 that laboratory oxidation of titanomagnetite to titanomaghemite merely degrades the intensity of the prior TRM but does not produce a new CRM along a field applied perpendicular to the prior TRM. C. A. Raymond and J. L. LaBrecque consider these results to be at variance with the observations of G. M. Smith and S. K. Banerjee who studied the cores from the (as yet) deepest hole (no. 504B) in the ocean crust. But the variance is only apparent because Smith and Banerjee suspect second generation CRM to be present in the basaltic dike layer where the magnetic mineral was not titanomaghemite but magnetite, presumed to have been formed in response to moderate temperature hydrothermal alteration in a convective regime near the spreading center. By inference, the good match between the predicted (by Raymond and LaBrecque, 1987) and observed anomaly amplitude modulations suggest that the CRM-imparting process is the formation of new magnetite at moderately elevated temperatures and not titanomaghemite at 0°C. The former is a true second phase formation while the latter is a topotactic transformation that allows the original oxygen structure to remain intact. Recent laboratory studies of CRM formation due to exsolution of magnetite in heated samples of pillow lavas from DSDP hole 504B confirm that the new CRM diverges from the original TRM.

Furthermore, very small single domain grains such as those used in the study of Özdemir and Dunlop will not acquire a divergent CRM but that larger multidomain grains will. Therefore, even if the chemical change is low temperature oxidation or titanomaghemitization, a new CRM could grow, as Raymond and LaBrecque require in their model, provided the grains are multidomain. It is pleasing to conclude this section with the realization that interdisciplinary study of the peculiarities of marine magnetic anomalies (or "anomalies of anomalies") by theoretical geophysicists, rock magnetists, and petrologists indeed has a tremendous potential for a comprehensive understanding of seafloor geology and geophysics.

Magnetic Source-layer of the Ocean Crust

In this article we set out to show how a study of marine magnetic anomalies led to first, a proof of the seafloor spreading hypothesis and second, a detailed picture of tectonics and alteration processes operating at shallow levels of the oceanic igneous crust at spreading centers. As examination of the Vine-Matthews-Morley hypothesis has continued further, it has resulted in a desire to constrain the hypothesis further by determining the depth, thickness, and composition of the magnetic source-layer. This, in turn, has resulted in the delineation of the depth- and time-dependent alteration processes at the spreading center. Here we provide a few examples.

Vine and Matthews (1963) postulated a 20 km thick source-layer bounded at the bottom of the Curie point isotherm of magnetite. If 20 km is the thickest source-layer suggested up to now, the suggestion for the thinnest layer (500 m) came from a 1971 modeling of strike-parallel undulations of the central anomaly at Reykjanes ridge by Talwani et al. (1971). Harrison (1981) has provided a tabulation of all the data published up to 1979 that address this question and a thorough review of the data. Similar reviews were also published by Johnson (1979) and Banerjee in 1984. In more recent times M. A. Mayhew has addressed the question from an analysis of both sea surface and satellite anomaly data.

There is a convergence of views that the magnetic source-layer of the ocean crust, from Brunhés age to the Jurassic, has to be thicker than 500 m as suggested originally by M. Talwani and his co-workers in 1971. It is the details of such variation of NRM intensity, metamorphic grade, and petrology and magnetic mineralogy with depth that are subject to discussion and disagreement at this stage. A recent contribution to the discussion has been made by Smith and Banerjee in their 1986 article, which studies the upper 1 km of the 6 m.y. old igneous ocean crust at DSDP hole 504B in the eastern equatorial Pacific just south of the Costa Rica rift. This study, for the first time, provides NRM measurements of nearly 300 m of an in situ basaltic dike swarm that underlies the uppermost basaltic pillow lava layer (500 m) and a transition zone (200 m) of mixed pillows and dikes riven with signs of penetrating hydrothermal circulation (Lister, 1977). Although this study is only the first of an in situ dike layer, its mean (arithmetic) NRM intensity of 1.4 A/m is not very much less than the mean value of 5.5 A/m of the uppermost pillow lava layer, an early contender for the sole source according to the 1971 model of Talwani, Windisch, and Langseth.

The NRM intensities of dikes and underlying gabbro from ophiolite suites (obducted and subaerially exposed ancient ocean crust) and DSDP drill cores from near fracture zones have been measured by a few authors and the data have been summarized and analyzed by Banerjee in 1984. Although such samples are not pristine examples of the ocean crust, their NRM values are high enough ($\sim$ 0.5 A/m) so as to be contenders for inclusion in a source model, in particular for the older oceanic crust whose strongly magnetic uppermost pillow lava layer (a 500 m "icing") will have been thoroughly altered and demagnetized.

Another approach has been taken by anomaly modelers who have looked at anomaly characteristics such as variations in the width of transition zones (from normal to reverse polarity) as a function of age, changes in reversal frequency with age, anomalous skewness (referred to earlier), and inversion of deep-tow magnetometer surveys over lineated anomalies. The previously cited articles by Johnson (1979) and Harrison (1981) deal with this problem as does the article by Blakely (1983). It appears that deep-seated greenschist facies rocks (basalts and gabbro) do contribute to the anomalies observed at the sea surface and at satellite altitudes.

Further work involving more deep drillholes in the ocean crust will settle the issue of the depth and thickness of the oceanic magnetic layer. In the process we are learning important facts about the nature of metamorphism associated with the seafloor spreading process.

Future Directions

Marine magnetic anomalies were once explained by hypothetico-deductive reasoning based on the seafloor spreading process. Today the peculiar characteristics of the anomalies observed at different heights and direct sampling of the ocean crust are being used as a powerful combination to understand tectonic processes such as faulting, block rotation, oblique spreading, and rift propagation. We are also learning rapidly about the depth- and time-dependent alterations on the seafloor caused by hydrothermal circulation. It is an exciting time to be associated with studies related to marine magnetics.

Acknowledgment

I thank the National Science Foundation for support through grants OCE-8516175 and EAR-8707428. I am particularly indebted to C. Raymond and J. LaBrecque for an opportunity to see the preprint of their 1987 article.

SUBIR K. BANERJEE

References

Blakely, R. J., 1983, Statistical averaging of marine magnetic anomalies and the ageing of oceanic crust, *Jour. Geophys. Research* **88**, 2289-2296.

Cox, A., ed., 1973, *Plate Tectonics and Geomagnetic Reversals*. San Francisco: W. H. Freeman.

Harrison, C. G. A., 1981, Magnetism of the oceanic crust, in C. Emiliani, ed., *The Sea*, vol. 7. New York: Wiley, 219-240.

Heirtzler, J. R., G. O. Dickson, E. M. Herron, W. C. Pitman III, and X. LePichon, 1968, Marine magnetic anomalies, geomagnetic field reversals, and motions of the ocean floor and continents. *Jour. Geophys. Research* **73**, 2119-2136.

Johnson, H. P., 1979, Magnetization of the oceanic crust, *Rev. Geophysics* **17**, 215-226.
Levi, S., S. K. Banerjee, S. Beske-Diehl, and B. Moskowitz, 1978, Limitations of ophiolite complexes as models for the magnetic layer of the oceanic lithosphere, *Geophys. Research Letters* **5**, 473-476.
Lister, C. R. B., 1977, Qualitative models of spreading-center processes, including hydrothermal penetration. *Tectonophysics* **37**, 203-218.
Menard, H. W., 1986, *The Ocean of Truth.* Princeton, N.J.: Princeton University Press.
Morley, L. W., 1981, An explanation of magnetic banding in ocean basins, in C. Emiliani, ed., *The Sea*, vol. 7. New York: Wiley, 1717-1719 (originally written 1963).
Raymond, C. A., and J. L. LaBrecque, 1987, Magnetization of the oceanic crust: Thermoremanent magnetization or chemical remanent magnetization? *Jour. Geophys. Research* **92**, 8077-8088.
Smith, G. M., and S. K. Banerjee, 1986, Magnetic structure of the upper kilometer of the marine crust at Deep Sea Drilling Project Hole 504B, eastern Pacific Ocean, *Jour. Geophys. Research* **91**, 10337-10354.
Talwani, M., C. C. Windisch, and M. G. Langseth, 1971, Reykjanes Ridge crest: A detailed geophysical study, *Jour. Geophys. Research* **76**, 473-517.
Tivey, M. A. and H. P. Johnson, 1987, The central anomaly magnetic high: Implications for ocean crust construction and evolution, *Jour. Geophys. Research*, in press.
Vine, F. J., and D. H. Matthews, 1963, Magnetic anomalies over oceanic ridges, *Nature* **199**, 947-949.

Cross-references: *Chemical Remanent Magnetization; Curie Temperature; Detrital Remanent Magnetization; Geomagnetic Polarity Reversals: Observations; Geomagnetic Reversal Sequence: Statistical Structure; Natural Remanent Magnetization (NRM); Paleomagnetism: Deep-Sea Sediments; Paleomagnetism and Continental Drift: Historical Introduction; Paleomagnetism and Plate Tectonics; Rock Magnetism; Thermoremanence; Viscous Remanent Magnetization (VRM) and Viscous Remagnetization.*

SEISMIC ANISOTROPY IN THE EARTH

Why Seismic Anisotropy?

Seismic anisotropy is the direction-dependent nature of propagation velocities of seismic waves. Natural minerals usually have some crystallographic structure. Under the physical conditions of high temperature and high pressure in the Earth's deep interior, rocks undergo slow plastic deformation. Through the deformation process, crystallographic axes of minerals are realigned to a particular direction due to uniaxial tectonic stress. Rocks may thus exhibit petrofabric structure, which in turn will produce seismic anisotropy on a macroscopic scale. The Earth's deep interior is supposed to be necessarily anisotropic, and in fact mantle nodules of upper-mantle origin usually display distinct seismic anisotropy. Thus, seismic anisotropy can be viewed as a characteristic property of the Earth.

Lithospheric thicknesses of seismological models for the upper mantle are primarily controlled by Rayleigh and Love wave dispersion at periods shorter than 50 s. If dispersion curves of Rayleigh and Love waves at these periods are affected differently by seismic anisotropy, then the joint inversion of Rayleigh and Love waves for lithosphere/asthenosphere structure produces results quite different from those that would be obtained if the constituent materials were isotropic. Thus, isotropic inversion for dispersion data from an anisotropic Earth might be largely misleading, and invoking seismic anisotropy for seismic modeling is crucial.

Seismic Wave Propagation in an Anisotropic Medium

Seismic anisotropy is defined as direction-dependence of the stress-strain relation of elasticity. The stress-strain relation in an anisotropic medium is the generalized Hooke law, which, in the Cartesian coordinates, has the form

$$\sigma_{ij} = \lambda_{ijkl}\, e_{kl} \qquad (i,j = 1,3) \qquad (1)$$

where summation over repeated indices is implied, σ_{ij} are components of the stress tensor, e_{kl} is the strain tensor, and λ_{ijkl} is the fourth-order tensor of elasticity, called the elastic constant. The strain tensors are defined as

$$e_{kl} = \frac{1}{2}\left(\frac{\partial u_l}{\partial x_k} + \frac{\partial u_k}{\partial x_l}\right)$$

where x_k is a reference coordinate and u_l is an elastic displacement.

Because of symmetry of the stress and strain tensors and thermodynamic constraints in Eq. 1, only 21 independent elastic constants are required to describe the general stress-strain relations. If the elastic medium displays certain kinds of symmetry, the stress-strain relations in Eq. 1 are further simplified. If the medium exhibits symmetry about three orthogonal planes, the number of independent elastic constants reduces to nine and the medium is described as orthotropic. If the medium is axisymmetric, there are only five independent elastic constants and the medium is described as hexagonal. If the direction of the symmetry axis is vertical, the medium can be described as transversely isotropic. If the medium is hexagonally symmetric with respect to two orthogonal axes, it is isotropic with only two independent elastic constants (Lamé's constants).

There are three principal consequences of seismic anisotropy: (1) direction-dependence of wave propagation velocity, (2) polarization-dependence of wave propagation velocity as found in S-wave splitting, and (3) polarization anomaly that is described as anomalous particle motions. Consequence (1) in a horizontal plane is recognized as azimuthal anisotropy in seismology. Consequence (3) occurs since the particle motion planes in an anisotropic medium are, in general, neither parallel nor perpendicular to the wave propagation direction, whereas those planes in an isotropic medium are either parallel or perpendicular to the wave propagation direction.

In an anisotropic and homogeneous medium, there always exist three independent body waves with orthogonal planes of particle motions, called quasi-compressional waves (qP) and quasi-shear waves (qSV and qSH). The wave propagation directions of qP, qSV, and qSH waves are not generally perpendicular to their respective wavefronts, except for directions within the planes of symmetry of elasticity.

To a first-order approximation about the differences between isotropic and anisotropic elastic constants, azimuthal variations of "squares" of the three-body wave velocities within the X_1-X_2 plane are expressed as follows (Crampin, 1977):

$$\rho V_P^2 = A + B_c \cos(2\theta) + B_s \sin(2\theta)$$
$$+ C_c \cos(4\theta) + C_s \sin(4\theta) \quad (2)$$
$$\rho V_{SH}^2 = D - C_c \cos(4\theta) - C_s \sin(4\theta)$$
$$\rho V_{SV}^2 = F + G_c \cos(2\theta) + G_s \sin(2\theta)$$

where

$$A = \frac{3(C_{11} + C_{22}) + 2C_{12} + 4C_{66}}{8}$$

$$B_c = \frac{C_{11} - C_{22}}{2}$$

$$B_s = C_{16} + C_{26}$$

$$C_c = \frac{C_{11} + C_{22} - 2C_{12} - 4C_{66}}{8}$$

$$C_s = \frac{C_{16} - C_{26}}{2}$$

$$D = \frac{C_{11} + C_{22} - 2C_{12} + 4C_{66}}{8}$$

$$F = \frac{C_{44} + C_{55}}{2}$$

$$G_c = \frac{-C_{44} + C_{55}}{2}$$

$$G_s = C_{45}$$

ρ is the density. θ is the azimuth measured from the X_1- to the X_2-axis, and V_P, V_{SH}, and V_{SV} are phase velocities of qP, qSH and qSV waves. qP and qSH are polarized nearly parallel to, and qSV nearly perpendicular to the horizontal plane. Correspondences between λ_{ijkl} and C_{mn} are as follows. Subscripts 1~6 of C_{mn} correspond to pairs (1,1), (2,2), (3,3), (2,3), (1,3), and (1,2) of subscripts of λ_{ijkl}, respectively. Thus,

$$C_{11} = \lambda_{1111}$$
$$C_{22} = \lambda_{2222}$$
$$C_{33} = \lambda_{3333}$$
$$C_{44} = \lambda_{2323}$$
$$C_{55} = \lambda_{1313}$$
$$C_{66} = \lambda_{1212}$$
$$C_{12} = \lambda_{1122}$$
$$C_{13} = \lambda_{1133}$$
$$C_{23} = \lambda_{2233}$$

In seismology C_{mn} rather than λ_{ijkl} is common notation.

Figure 1 shows an example of directional variations of qP-, qSH-, and qSV-wave velocities within three planes of symmetry of single crystal olivine, believed to be the major component of the Earth's upper mantle, where the elastic constants are from laboratory experiments. A distinct feature is the 4θ variation (namely, four-lobed variation) with azimuth for the qSH-wave velocity within any plane of symmetry, and 2θ variation for qP- and qSV-wave velocities with azimuth (two-lobed variation), coincident with the trigonometric series of Eq. 2. This is the general character for velocity variation in olivine-rich aggregates, and we tentatively adopt this as a working model for the following discussion of seismic anisotropy in the Earth's interior.

The first-order approximation of azimuthal dependence of surface wave phase velocities was given in a form quite similar to Eq. 2 with some second-order corrections (Smith and Dahlen, 1973).

Rayleigh and Love waves in an anisotropic medium are coupled to each other and hence not independent. They are called generalized surface waves in this sense. A new notational convention was proposed for generalized surface waves: FG (fundamental mode generalized surface waves), 1G (first higher-mode generalized surface waves), 2G and so on, in an order of slowness (Crampin, 1977). However, for simplicity we use traditional terms, Rayleigh and Love waves, in this article, although in a rigorous sense the terms are not the same as for an isotropic medium.

We can define a magnitude of azimuthal anisotropy as the fastest velocity minus the slowest velocity divided by an average of the extreme veloc-

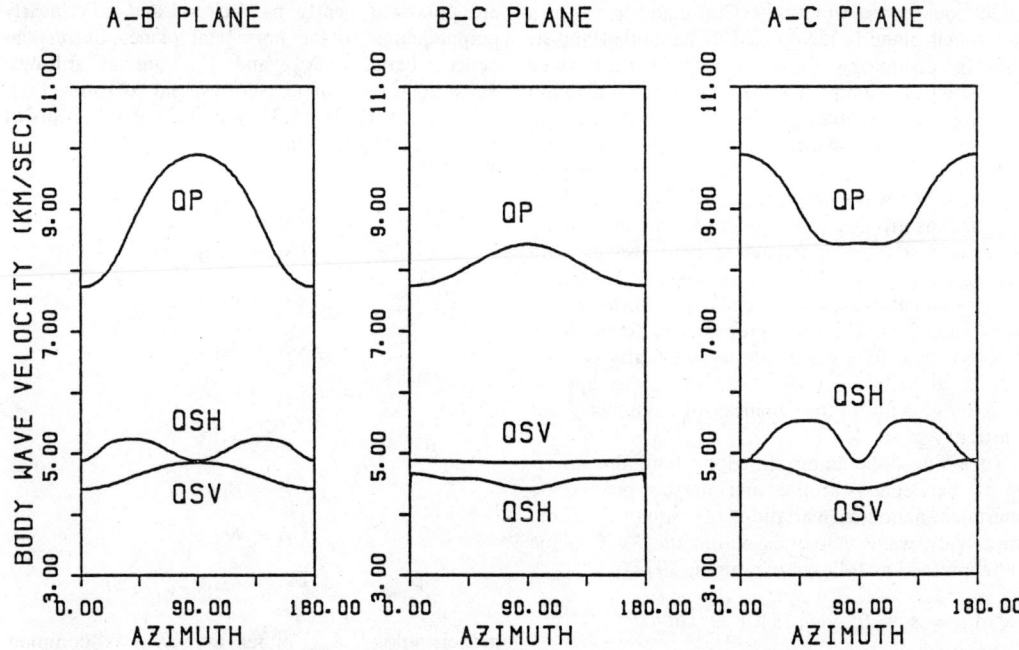

FIGURE 1. Variations of qP, qSH, and qSV wave velocities within planes of symmetry of single crystal olivine. The labels A-C, B-C, and A-C planes denote symmetry planes that include a- and b-axes, b- and c-axes, and a- and c-axes, respectively.

ities. Thus, the magnitude of the azimuthal anisotropy for qSV waves is twice a root of mean squares of coefficients of 2θ terms of the trigonometric series.

Seismic Sources in an Anisotropic Medium

Another aspect of anisotropy is its effects on calculating seismic moment. The source moment tensor M_{ij}, equivalent for a plane shear dislocation in an anisotropic medium, has the simple form (Kawasaki and Tanimoto, 1981)

$$M_{ij} = - D_0 S_0 \; \lambda_{ijkl} \; b_k \; s_l \qquad (3)$$

where

$$D = D_0 \, (b_1, b_2, b_3)$$
$$S = S_0 \, (s_1, s_2, s_3)$$
$$b_1^2 + b_2^2 + b_3^2 = 1$$
$$s_1^2 + s_2^2 + s_3^2 = 1$$

The dislocation vector **D** represents a constant discontinuity D_0 of displacement across the surface of the dislocation loop. The axial vector **S** is perpendicular to a dislocation surface with an absolute value S_0 equal to an area bounded by a dislocation loop. Since M_{ij} is symmetric because of moment equilibrium and symmetry conditions of elastic constants, M_{ij} can be transformed into a diagonal form $\mathbf{M}^0$ with diagonal elements of M_a, M_b, and M_c through rotation of reference coordinate axes. A trace of the moment tensor is not zero in general, whereas it is always zero for the isotropic case. $\mathbf{M}^0$ has the form

$$\mathbf{M}^0 = \begin{bmatrix} M_a & 0 & 0 \\ 0 & M_b & 0 \\ 0 & 0 & M_c \end{bmatrix}$$

where $M_a > M_b > M_c$.

Source moment tensors have been considered as follows. $\mathbf{M}^0$ can be decomposed as

$$M^0 = V \begin{bmatrix} 1 & 0 & 0 \\ 0 & 1 & 0 \\ 0 & 0 & 1 \end{bmatrix} + (M_b - M_c) \begin{bmatrix} 1 & 0 & 0 \\ 0 & 0 & 0 \\ 0 & 0 & -1 \end{bmatrix}$$
$$+ 2 \, (V - M_b) \begin{bmatrix} 1 & 0 & 0 \\ 0 & -\frac{1}{2} & 0 \\ 0 & 0 & -\frac{1}{2} \end{bmatrix} \qquad (4)$$

where V is equal to $(M_a + M_b + M_c)/3$. The first term on the right side of Eq. 4 indicates volumetric change at the seismic focus. The second term denotes a double-couple component. The last term is called the Compensated Linear Vector Dipole (CLVD). CLVD was interpreted to exhibit pair dislocations under uniaxial tectonic stress or a sudden change of shear modulus in a finite volume in the presence of an external strain field. Thus, above transformation clarifies that a plane shear dislocation in an anisotropic medium is no longer equivalent to the double couple but, in general, to the sum of three orthogonal dipole forces with different magnitudes M_a, M_b and M_c.

Imagine a pure thrust fault occurring at a 40-km depth with a strike direction of $10°E$, a dip angle of $45°$ to the west, a fault area of S_0, and a constant dislocation of D_0 over the fault. Then, a dislocation vector **D** and an axial vector **S** are given by $\mathbf{D} = D_0(q, 0, q)$ and $\mathbf{S} = S_0(q, 0, -q)$, respectively, where q is $1/\sqrt{2}$. The X_1-axis is aligned to the east, X_2 to the north, and X_3 vertically. From Eq. 3, we have

$$M_{11} = -(C_{11} - C_{33})S_0 D_0/2$$

$$M_{22} = -(C_{12} - C_{23})S_0 D_0/2$$

$$M_{33} = -(C_{13} - C_{33})S_0 D_0/2$$

$$M_{ij} = 0 \quad (i \neq j)$$

Taking values of C_{ij} from the preliminary reference Earth model (PREM) of Dziewonski and Anderson (1981) at the source depth, we obtain

$$M_{11} = -70.1 S_0 D_0 \text{ GPa}$$

$$M_{22} = 3.6 S_0 D_0 \text{ GPa}$$

$$M_{33} = 65.5 S_0 D_0 \text{ GPa}$$

M_{ij} can be simply decomposed as follows:

$$\begin{bmatrix} -70.1 & 0 & 0 \\ 0 & 3.6 & 0 \\ 0 & 0 & 65.5 \end{bmatrix} = -0.3 \begin{bmatrix} 1 & 0 & 0 \\ 0 & 1 & 0 \\ 0 & 0 & 1 \end{bmatrix}$$

$$-61.9 \begin{bmatrix} 1 & 0 & 0 \\ 0 & 0 & 0 \\ 0 & 0 & -1 \end{bmatrix}$$

$$-7.9 \begin{bmatrix} 1 & 0 & 0 \\ 0 & -\frac{1}{2} & 0 \\ 0 & 0 & -\frac{1}{2} \end{bmatrix}$$

where the common multiplier, $S_0 D_0$ GPa, is omitted.

Thus, this thrust faulting possesses a volumetric component of $\sim 1\%$ and a CLVD of $\sim 13\%$ of the magnitude of the double-couple component. The double couple clearly does not correspond to a plane shear dislocation. A radiation pattern of body waves is no longer of a quadrant type. Figure 2 displays a radiation pattern of qP waves from the thrust faulting, calculated with Eq. 3 and the far-field approximation of anisotropic Green's tensor. The deformation of the radiation pattern is not large in this case, though it sometimes becomes drastic, depending on the relative orientation of anisotropy and fault geometry.

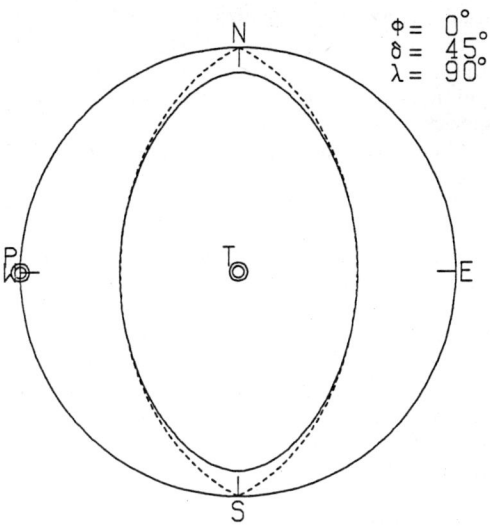

FIGURE 2. A radiation pattern of qP-wave first motion directions on the lower focal hemisphere by the equal-area projection. The strike direction, dip angle, and slip direction are N0°E, 45° to the west, and 90° (a pure reverse faulting), respectively. The slip direction is measured clockwise from the strike direction on the fault plane. The hypothetical seismic source is located at 40 km depth of PREM (Dziewonski and Anderson, 1981). The qP-wave first motion directions are calculated in a consistent manner with the equivalent body forces given by Eq. 3 in the text and the far-field approximation of anisotropic Green's tensor.

Recently, Dziewonski and Woodhouse (1983) developed a centroid moment tensor (CMT) inversion method with the transversely isotropic Earth model PREM. They considered the double-couple component of CMT solutions to correspond to a plane shear dislocation. They attributed a residual portion to nonplane shear dislocation terms: e.g., a tensile dislocation and multiplicity of rupture processes. However, their interpretation is inconsistent with their assumption of an anisotropic Earth model. If an earthquake occurs in an anisotropic region and an anisotropic moment tensor inversion is performed in a consistent fashion, CMT solutions for a shear dislocation necessarily involve non-double-couple components. Unfortunately, so far the source anisotropy problem has been little studied from this viewpoint.

Seismological Observations of Large-Scale Anisotropy

By assuming that the Earth is isotropic and radially symmetric, seismologists have made simplified analyses of Earth's deep structure. However, evidence for large-scale anisotropy has been steadily accumulating. We summarize the observations of the large-scale anisotropy in the Pacific Ocean. However, for the limited space available, only a few typical investigations will be cited for respective subjects.

P_n Velocity Anisotropy. P_n velocity anisotropy is usually recognized as a 2θ variation with azimuth of P_n refraction arrivals in the oceans, propagating immediately below the Moho discontinuity.

Hess (1964) first recognized P_n velocity anisotropy of 8.0–8.6 km·s^{-1} ($\sim 7\%$) in the northeast Pacific for ocean floor with ages of 20–80 Ma. The fastest direction of the P_n velocity was east-west and parallel to trends of fracture zones in an adjacent area, which were presumably fossil traces of the paleoplate spreading. The slowest direction was perpendicular to the fossil plate spreading direction. Recently, P_n velocity anisotropy of 7.9–8.6 km · s^{-1} ($\sim 8\%$) was measured in the northwest Pacific Ocean by ocean bottom seismometer (OBS) experiments mainly of longshots of ~ 1500-km distances (Shimamura et al., 1983), where ocean floor ages are greater than 130 Ma.

In the Pacific Ocean area, many other experiments have been operated to observe the P_n velocity anisotropy (e.g., see *Lithosphere, Oceanic: Formation and Evolution*). All of the observations available in the literature commonly suggest that the magnitude of the P_n velocity anisotropy is 0.3–0.6 km · s^{-1} (4–8%) and that the fastest directions are coincident with fossil directions of paleoplate spreading, as illustrated in Fig. 3.

SV_n Velocity Anisotropy. One of a few observations of very weak SV_n velocity anisotropy is ~ 4.65 km · s^{-1} in the Ngendei area reported by Shearer and Orcutt (1986), in contrast to the strong P_n velocity anisotropy of 7.95–8.4 km · s^{-1} ($\sim 6\%$) within the same experiment area.

Azimuthal Anisotropy of Rayleigh Waves. Azimuthal anisotropy of $\sim 2\%$ of 2θ variation with azimuth of Rayleigh waves for periods of 20–125 s was first discovered by Forsyth (1975) in an area covering the East Pacific rise and the Nazca Plate with ocean floor ages of 0–40 Ma.

It is now possible to map azimuthal anisotropy of Rayleigh waves both for periods longer than 100 s (Tanimoto and Anderson, 1985) and shorter than 100 s (Suetsugu and Nakanishi, 1987; Nishimura and Forsyth, 1988). The fastest directions for shorter period Rayleigh waves are parallel to the fossil spreading directions, just as for P_n velocity anisotropy.

Azimuthal anisotropy of P_n velocity and shorter period Rayleigh waves requires lithospheric azimuthal anisotropy. Azimuthal anisotropy of $\sim 2\%$ for Rayleigh waves extending to periods of 250 s would also require asthenospheric anisotropy.

Small-Degree Azimuthal Anisotropy of Love Waves. One controversy of large-scale anisotropy has been the very weak (less than 1%) azimuthal anisotropy of Love waves in the Pacific ocean (e.g., Nishimura and Forsyth, 1985). Though azimuthal anisotropy of Love waves would be more difficult to recognize than anisotropy of Rayleigh waves because

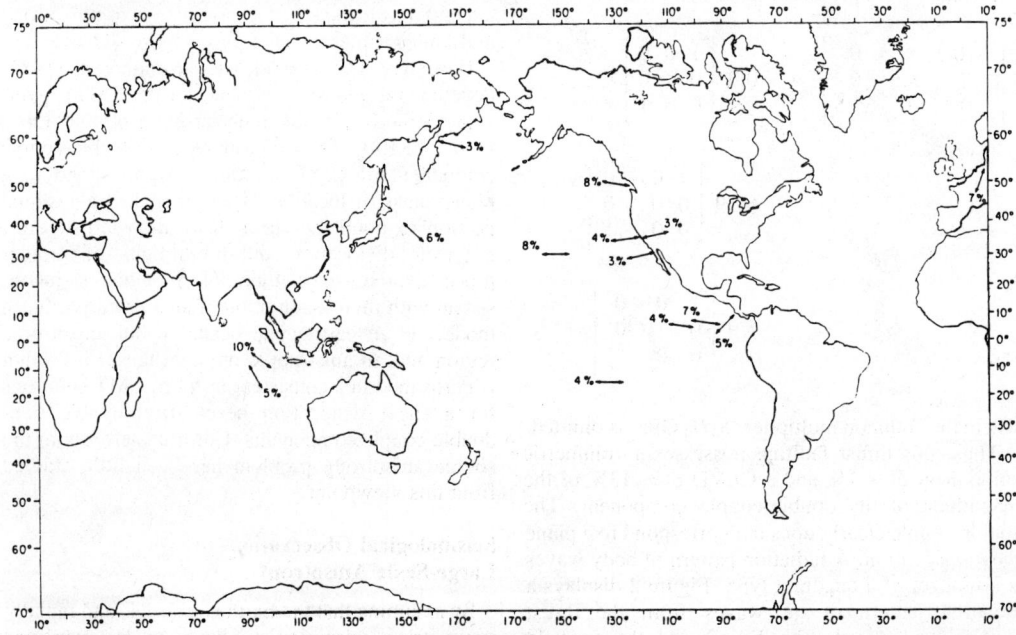

FIGURE 3. A summary of P_n velocity anisotropy in the Pacific Ocean. A direction of a bar centered over a station indicates the fastest direction of P_n velocity. A numeral is a magnitude of anisotropy. (From Christensen, 1984)

of a 4θ variation with azimuth in the olivine-rich upper mantle, azimuthal anisotropy of Love waves could be considered to be very weak and probably less than 1% if it exists at all.

ScS Polarization Anisotropy. ScS polarization anisotropy denotes arrival time differences (splitting) between two core-reflected ScS-waves vertically traveling in the mantle with two mutually orthogonal directions of linear particle motions.

An extensive search was first made by Ando (1984) in the circum-Pacific area, yielding a measure of the seismic anisotropy within the oceanic plates descending below the seismic stations. He observed that ScS-waves, polarized parallel to the fossil spreading direction of descending oceanic plates, arrived earlier, by ~1 s, than ScS waves polarized in the orthogonal direction. In Fig. 4, solid bars denote particle motion directions for the faster component of split ScS waves, and the magnitudes of the bars are proportional to arrival time differences.

SH-SV Polarization Anisotropy of Surface Waves. SH-SV polarization anisotropy is the discrepancy between V_{SH} and V_{SV} inverted separately, assuming the upper mantle to be isotropic but with different S-wave velocities, V_{SH} and V_{SV}, for Love and Rayleigh waves, respectively. This kind of surface wave inversion is called *separate inversion*.

In the Pacific Ocean region (see Table 1), the SH-SV polarization anisotropy was found to be ~0.15 km·s^{-1} in the outer 125 km (Forsyth, 1975), and ~0.1 km·s^{-1} in the lithosphere (Mitchell and Yu, 1980). Recently, SH-SV anisotropy of 3–4% was obtained for fundamental to second higher mode Rayleigh waves (e.g., Leveque and Cara, 1985). This multimode data suggest that the anisotropy may extend to depths greater than 200 km in the Pacific Ocean. The SH-SV discrepancy of a few percent was also measured for great-circle surface waves and from Earth's free oscillation data even by transversely isotropic inversion.

Thus, V_{SH} obtained from Love wave data by separate inversion is always larger than V_{SV} obtained from Rayleigh data (e.g., Mitchell and Yu, 1980). However, since the magnitude of the SH-SV anisotropy depends on the depth distribution of the anisotropy, the results cited cannot be directly compared with each other.

Other Works. P-wave arrival time data from ~300 shallow focus events to ~1000 seismic stations were inverted to refine P-wave travel times and station corrections (Dziewonski and Anderson, 1983). The station corrections involved 2θ terms, reflecting global azimuthal anisotropy in the upper mantle. Figure 5 displays the global distribution of the P-wave travel time residuals as a function of azimuth. Bars centered over stations indicate the slowest directions with lengths proportional to the magnitude of the residuals. The general tendency of the fastest P-directions, which are perpendicular to directions of the bars in this figure, in the western part

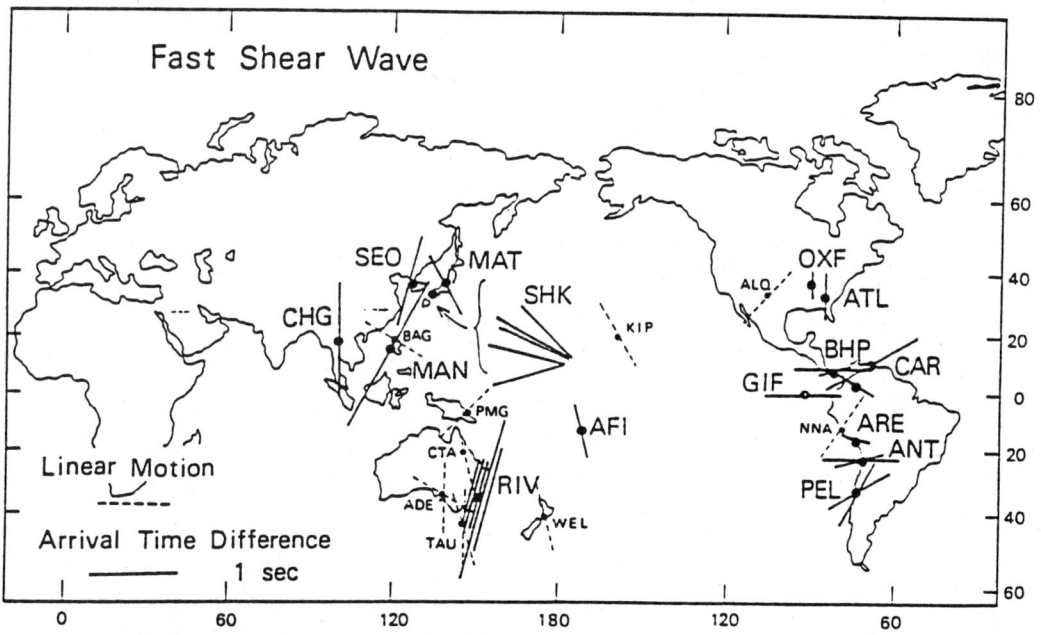

FIGURE 4. ScS polarization anisotropy. A direction of a bar indicates a direction of a particle motion of faster split ScS-waves, vertically traveling in the Earth's mantle. A magnitude of the bar is proportional to an arrival time difference between the faster and the slower split ScS-waves. (From Ando, 1984)

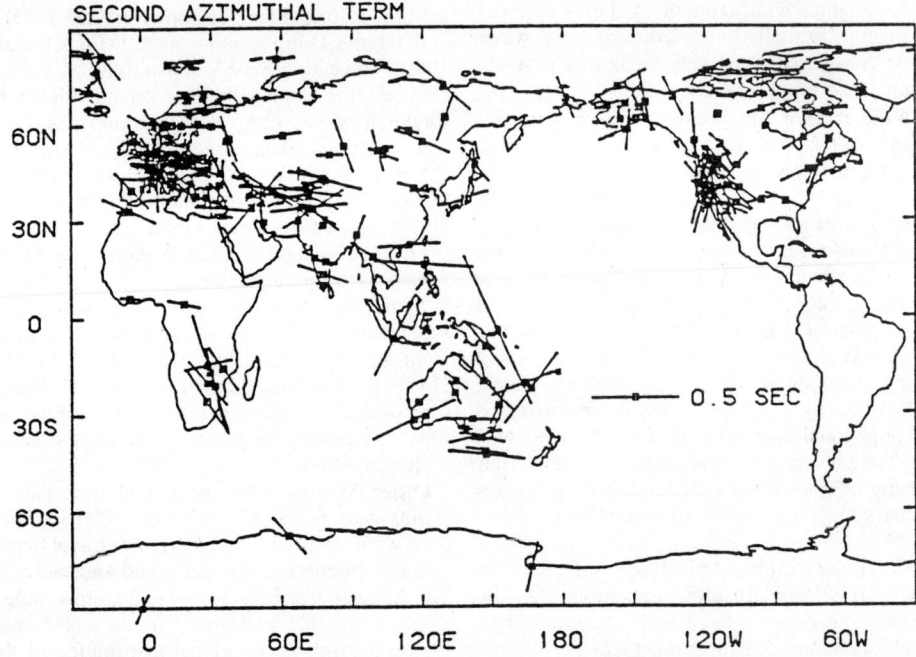

FIGURE 5. Global spatial distribution of station corrections of P-wave travel times. Bars centered over stations indicate the slowest directions of 2θ variations of the P-wave travel time residuals. (From Dziewonski and Anderson, 1983)

of the circum-Pacific region to resemble that of the ScS polarization anisotropy (Fig. 4) suggests a common root of these two independent observations.

Seismic Anisotropy beneath Continents and Island Arcs. Most observations of seismic anisotropy in continents have been made by P_n refraction arrivals. For an example, the P_n velocity anisotropy found was ~3% in northern Britain and the United States (Bamford et al., 1979). Fuchs (1983) recently presented a three-anisotropic-layer model of continental subcrustal lithosphere in southern Germany to explain P_n velocity anisotropy in Europe, suggesting that the anisotropy was caused by the preferred orientation of olivine due to plastic flow under continents. The P_n velocity anisotropy of 7.7–8.1 km·s^{-1} (~5%) was found even in a marginal sea, the Sea of Japan (Okada et al., 1978).

A three-dimensional anisotropic picture of P-wave velocities was drawn within the Pacific and Philippine sea plates descending under the Japan Islands (Hirahara and Ishikawa, 1985). It was found that the fastest direction of the P-wave velocity was perpendicular to the strike direction of the magnetic stripes of the ocean floor in the adjacent area, but unfortunately the estimate was not statistically significant.

Tanimoto and Anderson (1985) mapped the worldwide distribution of azimuthal anisotropy of longer-period (100–250 s) Rayleigh waves. Though the patterns of the azimuthal anisotropy in the oceanic environments were stable, those in the continents displayed strong dependence on cutoff angular order, suggesting a more complex tectonic signature beneath the continents.

Extensive-dilatancy anisotropy due to aligned cracks, commonly reported in the continental crust, is certainly a major focus of seismic anisotropy study. However, we omit it here because it is outside the scope of lithospheric/asthenospheric anisotropy and mantle convection flow.

A New Era of Anisotropy Study

In the early 1980s, a new era of anisotropy studies emerged with improved quantity and quality of seismic data, inversion theories, and models. A new Earth model, PREM, was developed from a transversely isotropic inversion that assumed transverse isotropy for the outer 220 km of the upper mantle (Dziewonski and Anderson, 1981). It was demonstrated that the introduction of anisotropy could yield quite novel results, with a notably less pronounced low velocity zone (LVZ) than previous isotropic inversions. More recently, azimuthal anisotropy has been emphasized as a possible tracer of mantle convection flow.

We now address two basic questions: (1) What kinds of petrofabric anisotropy are consistent with seismic observations of large-scale anisotropy? (2) What kinds of mantle flow mechanism do the petrofabric anisotropy suggest? In response to the first

TABLE 1. Large-Scale Seismic Anisotropy in the Pacific Ocean

Type of Anisotropy	Typical Observation	Reference
P_n velocity anisotropy	7.9–8.6 km.s^{-1}	(1)
S_n velocity anisotropy	<0.1 km.s^{-1}	(2)
Azimuthal anisotropy of		(3)
Rayleigh waves at 30–100 s	2–3%	
at 200 s	~2%	(4)
Love waves at 30–100 s	<1%	(5)
ScS polarization anisotropy	~1 s	(6)
SH-SV anisotropy	~0.1 km.s^{-1} in lithosphere	(7)

References: (1) Shimamura et al. (1983), (2) Shearer and Orcutt (1986), (3) Nishimura and Forsyth (1988), (4) Tanimoto and Anderson (1985), (5) Nishimura and Forsyth (1985), (6) Ando (1984), and (7) Mitchell and Yu (1980).

question, the following is an attempt to synthesize the seismic anisotropy observations into a seismological portrayal of mantle flow in the oceanic upper mantle. For simplicity and clarity we focus only on items listed in Table 1.

Fortunately, the trade-off between seismic anisotropy and lateral heterogeneity is small (Suetsugu and Nakanishi, 1987). We can refer to seismic anisotropy almost independent of lateral heterogeneity. However, seismic anisotropy seems to share the same root with the lateral heterogeneity in the Earth's deep interior. Both are believed to be dynamically maintained by mantle flow (see *Mantle Dynamics; Mantle Convection and Plumes*). Thus, we must keep in mind that seismic anisotropy is just one aspect of the mantle flow, and lateral heterogeneity should be incorporated to obtain a more complete seismological picture of the mantle convection flow.

Azimuthal Anisotropy as a Tracer of Mantle Flow

All observations of P_n velocity anisotropy in the Pacific Ocean display overall agreement with the fastest P_n velocities parallel to paleoplate spreading directions. The same holds for the azimuthal anisotropy of shorter-period Rayleigh waves. Figure 6 shows fossil plate spreading directions (top) as inferred from the magnetic anomalies and directions of the present-day absolute plate motions (bottom) (Minster and Jordan, 1978). Figure 7 shows patterns of the fastest directions of shorter-period Rayleigh waves. Resolution is poor for data at a period of 33 s. But patterns for periods of 59 and 91 s show excellent agreement with the patterns of the fossil plate spreading (Fig. 6, top).

Great-circle Rayleigh wave dispersion data for periods of 100 to 250 s could be inverted to simultaneously map worldwide distributions of lateral heterogeneity and azimuthal anisotropy (Fig. 8) in the upper mantle by the spherical harmonic expansion method (Tanimoto and Anderson, 1985). The azimuthal anisotropy obtained was ~2% on average in the Pacific Ocean area. It exhibited little correlation with the paleoplate spreading directions (Fig. 6) but a surprisingly strong correlation with return flow

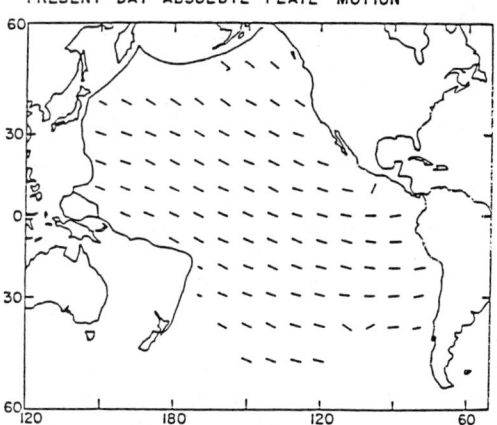

FIGURE 6. Fossil plate spreading directions inferred from magnetic anomalies (top) and absolute plate spreading motion directions (bottom). (From Nishimura and Forsyth, 1988)

Axial Symmetry of Elasticity of Oceanic Upper Mantle

Extensive studies by Christensen and his associates (see also *Seismic Properties of Rocks*) provide petrographic data on the preferred orientation of olivine and orthopyroxene of ultramafic rocks from the ophiolite complexes of the Bay of Islands, Newfoundland, and Oman, which were once a part of the paleooceanic upper mantle that has been obducted onto the continental margin. They discovered a striking tendency for the a-axis of olivine and the c-axis of pyroxene in ultramafic rocks to be aligned parallel to the paleoplate spreading direction, and the b-axis of olivine and the a-axis of pyroxene to be vertical, parallel to the sheeted dykes.

One of the prominent gross elastic features of the ultramafic rocks is axial symmetry of velocities of qP, qSH and qSV waves with the direction of the olivine a-axis, parallel to the spreading direction of the paleoceanic plate. The other notable feature is the uniform spatial variation of faster S-wave velocity. The faster S waves correspond to qSH waves.

Additional evidence was presented for the axial symmetry of elasticity (Estey and Douglas, 1986). If elastic constants for single crystal olivine, pyroxene, and garnet are mixed by Voigt-Reuss-Hill averages with the same volumetric modal ratios of the three components as those in the pyrolite model (Ringwood, 1975) and with the above mentioned mineral alignments found in the ultramafic rocks, the resultant assemblage exhibits approximate axial symmetry of elastic properties, with the olivine a-axis parallel to flow.

Anisotropic Inversions for Upper Mantle Structure

Forsyth (1975) first attempted to invert Rayleigh wave dispersion data for upper-mantle SV-wave velocities and Love wave dispersion data for SH-wave velocities in the east Pacific Ocean by the separate inversion. He obtained a thickness of ~ 80 km for oceanic lithosphere of 10–50 Ma. A similar separate inversion was applied to obtain the average thickness of 40–70 km for ages of 20–100 Ma (Mitchell and Yu, 1980) in the Pacific Ocean.

In response to the criticism that the separate inversion might be misleading, a transversely isotropic inversion method was developed in a self-consistent manner (Dziewonski and Anderson, 1981). The transverse isotropy was introduced into the outer 220 km of the upper mantle, and a new Earth model, PREM, was derived with a less pronounced LVZ.

The transversely isotropic inversion was applied to regional data of shorter-period (< 100 s) surface wave dispersions in the Pacific Ocean (Regan and Anderson, 1984). A new picture was drawn for the oceanic upper mantle (Fig. 10), where the average thickness of the lithosphere was ~ 45 km, far thinner

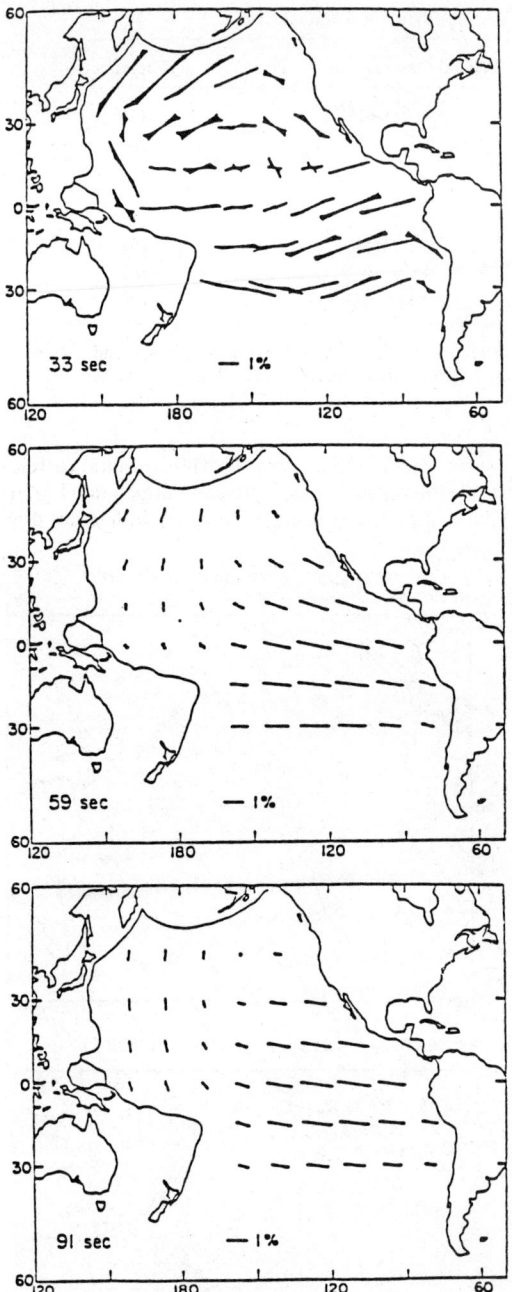

FIGURE 7. Direction of bar denotes fastest direction of Rayleigh waves of periods indicated. (From Nishimura and Forsyth, 1988)

directions (Fig. 9) derived from kinematic considerations (Hager and O'Connell, 1979), suggesting the existence of the return flow in the mesosphere in a low viscosity channel. Thus, the azimuthal anisotropy of P_n and Rayleigh wave velocities appears to indicate flow directions.

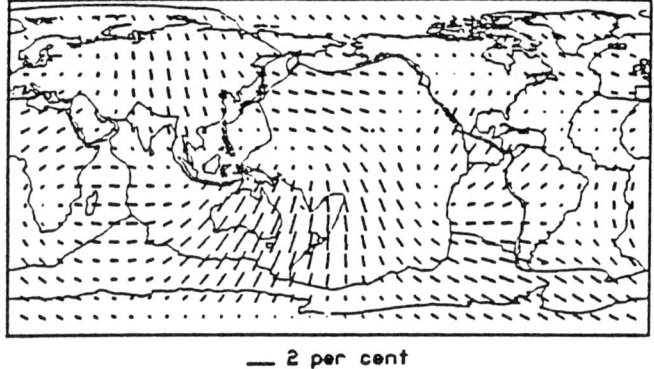

FIGURE 8. Azimuthal anisotropy of Rayleigh waves of a period of 200 s. The lines indicate the fastest directions of phase velocities. The line length is proportional to the magnitude of azimuthal anisotropy. (From Tanimoto and Anderson, 1985)

than those obtained previously from isotropic and separate inversions. The thin oceanic lithosphere is in better agreement with lithosphere thickness based on nonseismic definitions, for example, by seamount loading (Watt et al., 1980). The results from transversely isotropic inversion had the LVZ with age-dependent anisotropy and S-wave velocities of 4.3–4.4 km·s^{-1}, higher than those from previous isotropic surface wave inversions. For most of the oceanic region, SH-wave velocities were higher than SV-wave velocities in the LVZ, except for the East Pacific rise, suggesting ascending flow there.

Quasi-azimuthally anisotropic modeling of the oceanic upper mantle was attempted by Kawasaki (1986), who invoked the axial symmetry of elasticity as one of his basic frameworks. Resulting models had a thin and azimuthally anisotropic lithosphere of ~45 km for average Pacific Ocean of the ages of 20–100 Ma. The degree of azimuthal anisotropy in the oceanic lithosphere and asthenosphere of Kawasaki's model was quite similar to that of ultramafic rocks from the ophiolite complex. The azimuthally anisotropic model could successfully explain the overall features of the seismic observations listed in Table 1, including the very small azimuthal anisotropy of Love waves. Thus, his study

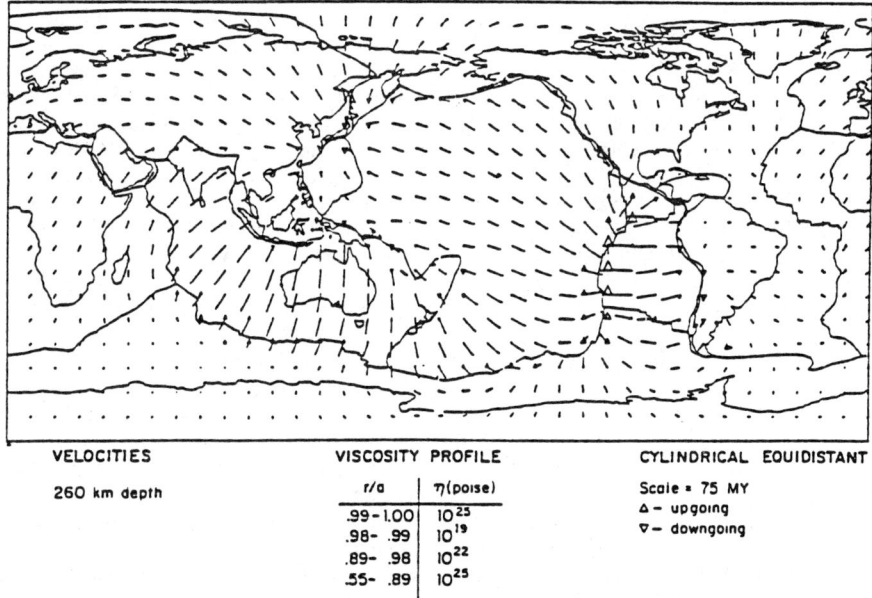

FIGURE 9. Flow lines at 260-km depth for the kinematic model of large-scale flow in the upper mantle. (From Hager and O'Connell, 1979)

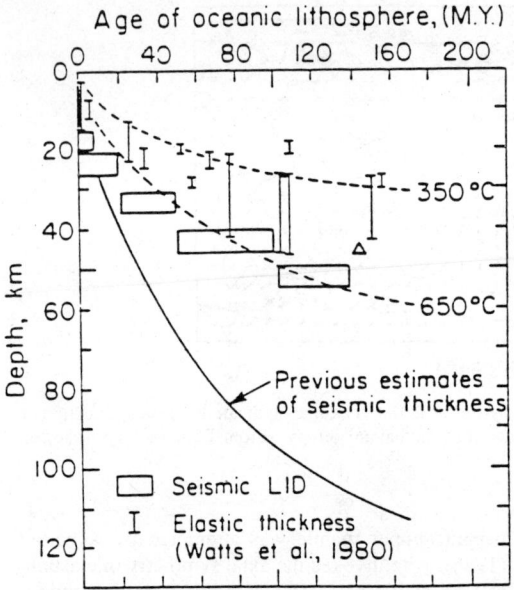

FIGURE 10. The thickness of the oceanic lithosphere determined by anisotropic surface wave inversion. The upper and lower edges of open boxes indicate the thickness without and with the oceanic crust. (From Regan and Anderson, 1984)

seems to have confirmed that the ultramafic rocks from the ophiolite complex could be the constituent materials throughout the oceanic upper mantle above the 220-km discontinuity.

Thin Lithosphere or Thick Lithosphere?

Thin oceanic lithosphere (~45 km) was suggested in the Pacific Ocean by anisotropic modeling. On the other hand, high-frequency P_n arrivals were observed at long distances of ~1500 km by ocean bottom seismometers (OBS) in the northwest Pacific Ocean (e.g., Shimamura et al., 1983), where ocean floor ages exceed 100 Ma. These observations suggested thick oceanic lithosphere or no LVZ, inconsistent with the thin oceanic lithosphere models. Which is right? What can remove this discrepancy?

Schubert et al. (1976) presented a coupled thermal and mechanical solid-state model of the oceanic lithosphere and asthenosphere by a numerical experiment that includes temperature-dependent thermal conductivity and nonlinear shear deformation, assuming that dislocation motion is the controlling mechanism for mantle flow in the asthenosphere. One of the important consequences of their numerical experiment is a steady increase of body wave velocities in the LVZ with the ocean floor ages, suggesting a less distinguishable LVZ in old oceans.

Another possibility for the discrepancy is that the oceanic upper mantle could be a highly non-Poisson solid. Because dislocation motion mainly accompanies shear deformation, only S-wave velocities could be effectively decreased, so that surface wave inversion would tend to produce a well-developed LVZ.

Though these various factors seem partially to resolve the discrepancy, we do not have a complete solution to the problem at present. More laboratory data are required on the elastic properties of deformed aggregates of constituent minerals under the conditions of mantle flow.

Dislocation Motion or Partial Melting?

It has been widely accepted that the LVZ under oceans corresponds to the intersection between the solidus of the composite minerals and the temperature profile in the upper mantle. Seismological observations, including anisotropy, have been interpreted in this vein.

Schubert et al. (1976) pursued the other possibility that mantle flow occurs through dislocation motions with no partial melting. Partial melting, if present at all, would be restricted to young oceans. Actually, if partial melting is a dominant cause of the LVZ and of seismic anisotropy, it appears to be very difficult to explain the lithospheric anisotropy (P_n velocity anisotropy and azimuthal anisotropy of shorter-period Rayleigh waves), because partial melting would be frozen in the lithosphere. Seismic observations are consistent with dislocation motions and can be explained without invoking partial melting in the LVZ, though they do not completely rule out the possibility of partial melting.

Seismic observations indicate that the petrofabric anisotropy associated with the mantle flow under the ocean is the same as that found in ultramafic rocks from the ophiolite complex. Thus, the relative orientation of crystallographic axes of olivine and pyroxene in ophiolite complex could be the petrofabric morphology of the mantle flow due to the dislocation motion.

Unfortunately, we do not have laboratory experiments of plastic flow due to dislocation motions under simple shear to give us an insight on how this relative orientation is relevant to mantle flow. Such experiments could be one of the urgent research subjects of solid-Earth geophysics.

Other relative orientations of crystallographic axes of olivine and pyroxene could dominate under different temperature and pressure conditions for stress-controlled dynamic recrystallization in the asthenosphere or deeper parts of the Earth. If other orientations do occur, seismic anisotropy in the asthenosphere or deeper Earth could be much different from what we have seen for the lithosphere and shallower part of the asthenosphere. This question can not be resolved from existing data.

Seismic anisotropy in the oceanic upper mantle appears to be emerging from obscurity, both petrologically and seismologically, in a consistent fashion. The mantle flow mechanism, the understanding of which is one of the goals in geophysics, seems to be near at hand. The relationship of seismic anisotropy to the mechanisms for mantle flow is one reason why it is so important to solid-Earth geophysics.

ICHIRO KAWASAKI

References

Ando, M., 1984, ScS polarization anisotropy around the Pacific Ocean, *Jour. Physics Earth* **32**, 179-196.

Bamford, D., M. Jentsch, and C. Prohdehl, 1979, P_n anisotropy studies in northern Britain and the eastern United States, *Royal Astron. Soc. Geophys. Jour.* **57**, 397-429.

Christensen, N. I., 1984, The magnitude, symmetry and origin of upper mantle anisotropy based on fabric analyses of ultramafic tectonites, *Royal Astron. Soc. Geophys. Jour.* **76**, 89-111.

Crampin, S., 1977, A review of the effects of anisotropic layering on the propagation of seismic waves, *Royal Astron. Soc. Geophys. Jour.* **49**, 9-27.

Dziewonski, A. M., and D. L. Anderson, 1981, Preliminary reference Earth model, *Physics Earth and Planetary Interiors* **25**, 297-356.

Dziewonski, A. M., and D. L. Anderson, 1983, Travel times and station corrections for P waves at teleseismic distances, *Jour. Geophys. Research* **88**, 3296-3314.

Dziewonski, A. M., and J. H. Woodhouse, 1983, An experiment in systematic study of global seismicity: centroid-moment tensor solutions for 201 moderate and large earthquakes of 1981, *Jour. Geophys. Research* **88**, 3247-3271.

Estey, L. H., and B. J. Douglas, 1986, Upper-mantle anisotropy: A geophysically consistent model, *Jour. Geophys. Research* **91**, 11393-11406.

Forsyth, D. W., 1975, The early structural evolution and anisotropy of the oceanic upper mantle, *Royal Astron. Soc. Geophys. Jour.* **43**, 103-162.

Fuchs, K., 1983, Recently formed elastic anisotropy and petrological models for the continental subcrustal lithosphere in southern Germany, *Physics Earth and Planetary Interiors* **31**, 93-118.

Hager, B., and R. O'Connell, 1979, Kinematic models of large-scale flow in the Earth's mantle, *Jour. Geophys. Research* **84**, 1031-1048.

Hess, H., 1964, Seismic anisotropy of the uppermost mantle under oceans, *Nature* **203**, 629-631.

Hirahara, K., and Y. Ishikawa, 1984, Travel time inversion for three-dimensional P-wave velocity anisotropy, *Jour. Physics Earth* **32**, 197-218.

Kawasaki, I., 1986, Azimuthally anisotropic model of the oceanic upper mantle, *Physics Earth and Planetary Interiors* **43**, 1-21.

Kawasaki, I., and T. Tanimoto, 1981, Radiation patterns of body waves due to the seismic dislocation occurring in an anisotropic source medium, *Seismol. Soc. America Bull.* **71**, 37-50.

Leveque, J.-J., and M. Cara, 1985, Inversion of multimode surface wave data: evidence for sub-lithospheric anisotropy, *Royal Astron. Soc. Geophys. Jour.* **83**, 753-773.

Minster, J. B., and T. H. Jordan, 1978, Present-day plate motions, *Jour. Geophys. Research* **83**, 5331-5354.

Mitchell, B. J., and G.-K. Yu, 1980, Surface wave dispersion, regionalized velocity models, and anisotropy of the Pacific crust and upper mantle, *Royal Astron. Soc. Geophys. Jour.* **63**, 497-514.

Nishimura, C. E., and D. W. Forsyth, 1985, Anomalous Love-wave phase velocities in the Pacific: sequential pure-path and spherical harmonic inversion, *Royal Astron. Soc. Geophys. Jour.* **81**, 389-408.

Nishimura, C. E., and D. W. Forsyth, 1988, Rayleigh wave phase velocities in the Pacific with implications for azimuthal anisotropy and lateral heterogeneity, *Royal Astron. Soc. Geophys. Jour.* **94**, 479-501.

Okada, H., T. Moriya, T. Masuda, T. Hasegawa, S. Asano, K. Kasahara, A. Ikami, H. Aoki, Y. Sasaki, N. Hurukawa, and K. Muramatsu, 1987, Velocity anisotropy in the Sea of Japan as revealed by big explosions, *Jour. Physics Earth* **26**, Suppl., S491-S502.

Regan, J., and D. L. Anderson, 1984, Anisotropic models of the upper mantle, *Physics Earth and Planetary Interiors* **35**, 227-263.

Ringwood, A. E., 1975, *Composition and Petrology of the Earth's Mantle*. New York: McGraw-Hill, 618p.

Schubert, G., C. Froidevaux, and D. A. Yuen, 1976, Oceanic lithosphere and asthenosphere: thermal and mechanical structure, *Jour. Geophys. Research* **81**, 3525-3540.

Shearer, P. M., and J. A. Orcutt, 1986, Compressional and shear wave anisotropy in the oceanic lithosphere—The Ngendei seismic refraction experiment, *Royal Astron. Soc. Geophys. Jour.* **87**, 967-1004.

Shimamura, H., T. Asada, K. Suyehiro, T. Yamada, and H. Inatani, 1983, Longshot experiments to study velocity anisotropy in the oceanic lithosphere of the northwestern Pacific, *Physics Earth Planetary Interiors* **31**, 348-362.

Smith, M. L., and F. A. Dahlen, 1973, The azimuthal dependence of Love and Rayleigh wave propagation in a slightly anisotropic medium, *Jour. Geophys. Research* **78**, 3321-3333.

Suetsugu, D., and I. Nakanishi, 1987, Regional and azimuthal dependence of phase velocities of mantle Rayleigh waves in the Pacific ocean, *Physics Earth and Planetary Interiors* **47**, 230-245.

Tanimoto, T., and D. L. Anderson, 1985, Lateral heterogeneity and azimuthal anisotropy of the upper mantle: Love and Rayleigh waves 100-250 s, *Jour. Geophys. Research* **90**, 1842-1858.

Watts, A. B., J. H. Bodine, and M. S. Steckler, 1980, Observations of flexure and the state of stress in the oceanic lithosphere, *Jour. Geophys. Research* **85**, 6396-6376.

Cross-references: *Deformation of Rocks and Minerals; Earth Structure: Global; Free Oscillations of the Earth; Mantle, Upper: Structure; Mantle Convection and Plumes; Mantle Dynamics; Seafloor Spreading: Magnetic Evidence; Seismic Attenuation: Observation and Measurement; Seismic Properties of Rocks; Surface Waves.*

SEISMIC ATTENUATION: OBSERVATION AND MEASUREMENT

The amplitude of seismic waves decreases with increasing distance from earthquake, explosion, and impact sources. How this amplitude decrease occurs, how rapidly it occurs, and how it depends on frequency of the seismic waves is fundamentally important to the efforts to describe Earth structure and seismic sources.

Seismic attenuation and its variation with location within the Earth are useful for determining the type and state of the rocks and minerals composing the Earth. In addition to providing information on a physical property, research in seismic attenuation has also been strongly motivated by more practical problems. One problem has been the prediction of the size of ground motion due to probable earthquakes in different regions. The frequency content and decay with distance of this strong ground motion is an important input to the design of earthquake resistant structures and to disaster planning. Another problem has been to estimate the size and detectability of underground nuclear tests. The monitoring of the threshold test ban treaty has depended mainly on the estimates of explosion size from the amplitude of elastic waves radiated by the explosion and recorded at large distances by seismographs (see *Seismic Monitoring of Nuclear Explosions*).

Ways in Which Seismic Waves Attenuate

The attenuation of seismic waves is due to three effects: geometric spreading, intrinsic attenuation, and scattering attenuation.

Geometric Spreading. *Geometric spreading* is simply the energy density decrease that occurs as an elastic wavefront expands. In a homogeneous Earth of constant velocity and density, the geometric spreading of a seismic body wave is proportional to the reciprocal of the distance between source and receiver. In the real Earth, velocity and density vary strongly with depth and less so laterally. Given a model of this variation, however, the geometric spreading of a body wave can be easily calculated (see *Seismic Ray Theory*).

Intrinsic Attenuation. *Intrinsic attenuation* is energy lost to heat and internal friction during the passage of an elastic wave. The microscopic mechanisms of intrinsic attenuation have been described in several different ways, including the resistive properties of oscillator models of the atoms in crystalline lattices, the movement of interstitial fluids between grain boundaries and cracks, and the frictional sliding of cracks (Jackson and Anderson, 1970).

Scattering Attenuation. *Scattering attenuation* is not true energy loss in this sense. Elastic energy is not converted into heat but is redistributed into angular directions away from the receiver or converted into wave types arriving in different time windows at the receiver. Scattering takes place by reflection, refraction, and conversion of elastic energy by irregularities in the medium. These irregularities are discontinuous or rapid variations in the velocity and/or density of the medium. In the Earth, the most important discontinuities are the approximately planar layers in the crust and the spherically symmetric boundaries such as the core-mantle boundary and Mohorovičić discontinuity (Moho). In addition, there are rapid variations of velocity and density throughout the Earth, occurring over variable scale lengths, with distributions varying strongly with depth. In the crust and uppermost mantle, variations in velocity and density can be particularly strong in the lateral as well as the vertical direction.

Linear Intrinsic Attenuation

Stress-Strain Hysteresis. Intrinsic attenuation can be best illustrated by the time behavior of the stress-strain relation. A stress is a vector force per unit area applied to a solid. The strain is a nondimensional measure of the deformation of the solid due to the applied stress, such as the change in a length element divided by the original length. In a perfectly elastic solid, stress σ is directly proportional to strain ϵ by an elastic modulus M such that $\sigma = M\epsilon$. All real solids, however, are anelastic or attenuative. A cycle of increasing and decreasing stress does not produce a perfectly proportional increase and decrease in strain. Instead a hysteresis occurs (Fig. 1), in which the cycle of the strain response lags the applied stress in time. The area enclosed by the hysteresis curve is a measure of the energy lost to heat and internal friction. During the stress cycle associated with the passage of a seismic wave, the energy lost to this internal friction is not available to deform adjacent regions of the solid just ahead of the wavefront.

From the hysteresis curve, one can see immediately that the stress-strain relation cannot be described by a simple constant of proportionality. A more complicated relation is required to describe the time lag of strain, involving the time history of the applied stress. A simpler form of the relation can be stated by Fourier-transforming the stress-strain

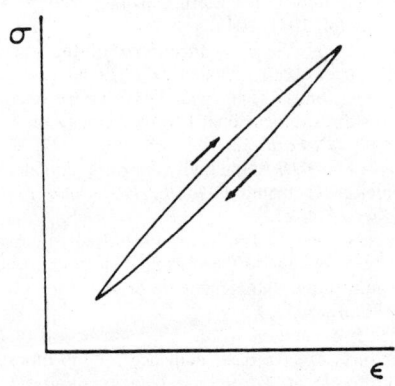

FIGURE 1. A hysteresis curve for a cycle of stress applied to a solid with intrinsic attenuation.

relation in time into the frequency domain. In this form the Fourier transform of stress at a given frequency is proportional to the Fourier transform of strain by an elastic modulus $\hat{M}(\omega)$, This modulus must be a complex number to describe the phase lag of strain. It must also be frequency dependent because the phase lag of strain depends on the time history of the stress, the shape of the hysteresis curve changing with different load histories.

A feature of the complex modulus is that its real part will be smaller at zero or very low frequency and larger at infinite or very high frequency. That is, there will be an instantaneous response of strain to the applied stress, which is smaller than the eventual equilibrium response after longer time. The trend of the frequency dependence can be inferred from the time lag of strain. The upper and lower limiting values of the real part of the modulus can be determined by observing a sequence of hysteresis curves for increasing frequencies of monochromatic loads. This frequency dependence of the real part of the modulus implies that the propagation of a stress pulse will be dispersive, with higher frequencies traveling faster than lower frequencies.

Superposition in the Linear Domain. The linear domain of intrinsic attenuation is restricted to small strains of 10^{-6} or less. In this linear domain the stress-strain relation can be described by a linear differential equation

$$L_1 \sigma(t) = L_2 \epsilon(t) \qquad (1)$$

where L_1 and L_2 are any linear combinations of operators of the type $\partial^n/\partial t^n$ or $\int dt^n$. A superposition principle is valid in the linear domain, in which the total strain due to separately applied stresses or forcing functions can be calculated simply by summing the solutions to the differential equation for each separate stress or forcing function. This operation makes it possible to find solutions to the seismic wave equation for a complicated forcing function, such as an earthquake, by convolving the earthquake forcing function with the solution for a simple forcing function, such as a vector point force. In the frequency domain, these convolutions over time history of applied stress are in the form of multiplications, in which the time information is contained in the phase of complex quantitites.

Q and Complex Velocity. Because simple mechanical systems, composed of springs and dashpots, and simple electric circuits also obey linear equations of this form, there are analogies between the quantities describing these systems and quantities in the stress-strain relation. For example, strain behaves like voltage and stress like current in electric circuits. Similar to the resonance phenomena in circuits and mechanical systems, a Q can be defined by the average energy W per cycle divided by the energy lost or work done per cycle, ΔW:

$$Q = \frac{W}{\Delta W}$$

Large Q's imply small energy loss, small Q's imply large loss. Q is a measure of the area contained in the hysteresis loop of a stress-strain cycle. A complex elastic modulus $\hat{M}(\omega)$ can be found by Fourier-transforming Eq. 1:

$$\hat{M}(\omega) = \frac{\tilde{L}_2}{\tilde{L}_1}$$

where $\tilde{L}_2$ and $\tilde{L}_1$ are the Fourier-transformed operators.

Plane waves of frequency ω and propagating in the $+x$-direction can be defined by the phasor $e^{[i\hat{k}x - \omega t]}$, where $\hat{k}$ is a complex wave number $\omega/\hat{c}$ and $\hat{c}$ is a complex velocity defined from the local density ρ and complex modulus $\hat{M}$:

$$\hat{c} = \sqrt{\frac{\hat{M}}{\rho}}$$

From the average energy density and loss per cycle of a complex plane wave it can be shown that $Q =$ Re $\hat{M}/$Im $\hat{M}$. It is usually more convenient to report the reciprocal of this quantity (Q^{-1}), which represents the usually small perturbations to perfect elasticity.

$Q^{-1}(\omega)$ **Absorption Band.** Since $\hat{M}$ depends on frequency, Q also depends on frequency. Zener (1960) has described the frequency-dependent effects on an elastic modulus of a solid having a single characteristic time for the relaxation of stress. A Zener relaxation gives a Q^{-1} that is peaked at the frequency $f = 1/2\pi\tau$, where τ is the relaxation time. A distribution of relaxation times τ can be constructed to give a Q^{-1} having a variable dependence on frequency. The function $Q^{-1}(\omega)$ is called the *relaxation spectrum*. In the Earth and in many solid materials, the relaxation spectrum is observed to be slowly varying and nearly constant over a broad band of frequencies. In this case, a distribution of relaxations is needed to describe near-constancy of Q. Figures 2a,b show how the distribution of Zener relaxations can produce a Q that is nearly constant with frequency over a broad band. Note in Fig. 2b that the real part of the elastic modulus varies continuously with frequency over the relaxation spectrum from its lower, relaxed, limit to its upper, unrelaxed limit. Once the limits of an absorption band are specified, it is not possible to have an arbitrarily high Q^{-1} at any frequency because the total modulus defect $[\hat{M}(\infty) - \hat{M}(0)]$ is related to the distribution of Im $\hat{M}(\omega)$ over the absorption band. Although this dispersion in elastic moduli had long been known and predicted from the theory of linear viscoelasticity, it was only first recognized in

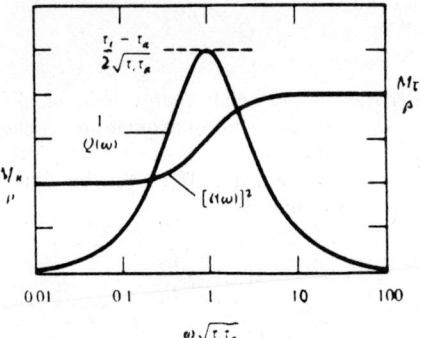

FIGURE 2(a). A single Zener relaxation peak in Q^{-1} centered at radian frequency $\omega = 1/\sqrt{\tau}$. $\tau = \sqrt{\tau_\epsilon \tau_\sigma}$, the geometric mean of the relaxation time for stress and the relaxation time for strain, respectively. The elastic modulus varies from the relaxed limit M_R at zero frequency and the unrelaxed limit M_U at infinite frequency.

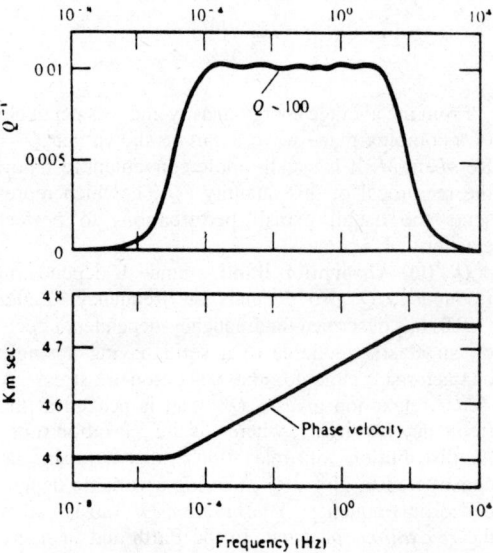

FIGURE 2(b). An absorption band in Q^{-1} obtained from the superposition of 12 different Zener relaxations. The dispersion of phase velocity is obtained from $v = \mathrm{Re}\sqrt{M(\omega)/\rho}$. (After Liu et al., 1976)

seismology when velocity models determined in the low-frequency band from the normal modes of the Earth (0.0001 to 0.01 Hz) were compared with velocity models determined in a high-frequency band (0.1 to 10 Hz) from body waves. The models were found to differ (Fig. 3), and the difference was found to agree with the amount of dispersion predicted from average Q models of the Earth (see *Earth Structure, Global*).

Velocity Dispersion. Another, more subtle, effect of this velocity dispersion can be seen in the propagation of pulses as body waves. A stress disturbance that propagates from its point of initiation as a symmetric delta function in time gradually evolves into an asymmetric pulse (Fig. 4). High frequencies, traveling faster than low frequencies, are preferentially loaded into the front of the pulse. Infinite frequency arrives earliest at the front of a pedestal before the first visible onset of the pulse. In practice, the pedestal cannot be observed above Earth noise, but occasionally a well-designed experiment having very broadband data can observe the pulse asymmetry. An earthquake or explosion source will have a pulse shape, which itself is more complicated than a delta function in time. The shape of the source pulse must be known or removed in order to determine the changes in shape due to attenuation.

Scattering Attenuation

Scattering of elastic waves is a large subject and is beyond the scope of this article (see *Seismic Wave Scattering*). It is important, however, to briefly consider some of the biasing effects of scattering on measurements of intrinsic attenuation. At frequencies that are so low that wavelengths are much larger than linear dimensions of a heterogeneity, the effects of scattering can usually be neglected for the propagation of body waves. Very low frequencies are in the domain of Rayleigh scattering, in which the energy loss in the forward direction is proportional to the inverse fourth power of frequency. At sufficiently low frequency, the medium behaves like an equivalent medium, having properties that are an average of small-scale heterogeneities.

At higher frequencies, those in which the wavelength is on the order of the scale length of heterogeneities, the presence of the heterogeneities can profoundly alter the propagation of the wave field. For wavelengths much smaller than the scale length of the heterogeneities, it is possible in some situations to calculate the effect of scatterers by using ray theory. The distribution and shape of the scatterers, however, must first be known or assumed.

Stochastic Dispersion. The most complicated domain in which to perform calculations is where wavelength is on the order of the scale length of the heterogeneity. In this domain, fully numerical calculations, such as finite-difference solutions of the elastic wave equations, are often the only way of predicting the effects of scattering on the wave field. A more simplified experiment is to calculate the effects of scattering in a one-dimensional medium consisting of thin planar layers in which the velocity in each layer is assigned randomly. This experiment simulates the effect of stratigraphic attenuation of a medium (Fig. 5a,b) in which plane layering varies rapidly within a wavelength. A prediction of such experiments is that body waves will exhibit a stochastic dispersion in which high-frequency energy is transferred into the coda following the first several cycles. This stochastic dispersion can be seen in

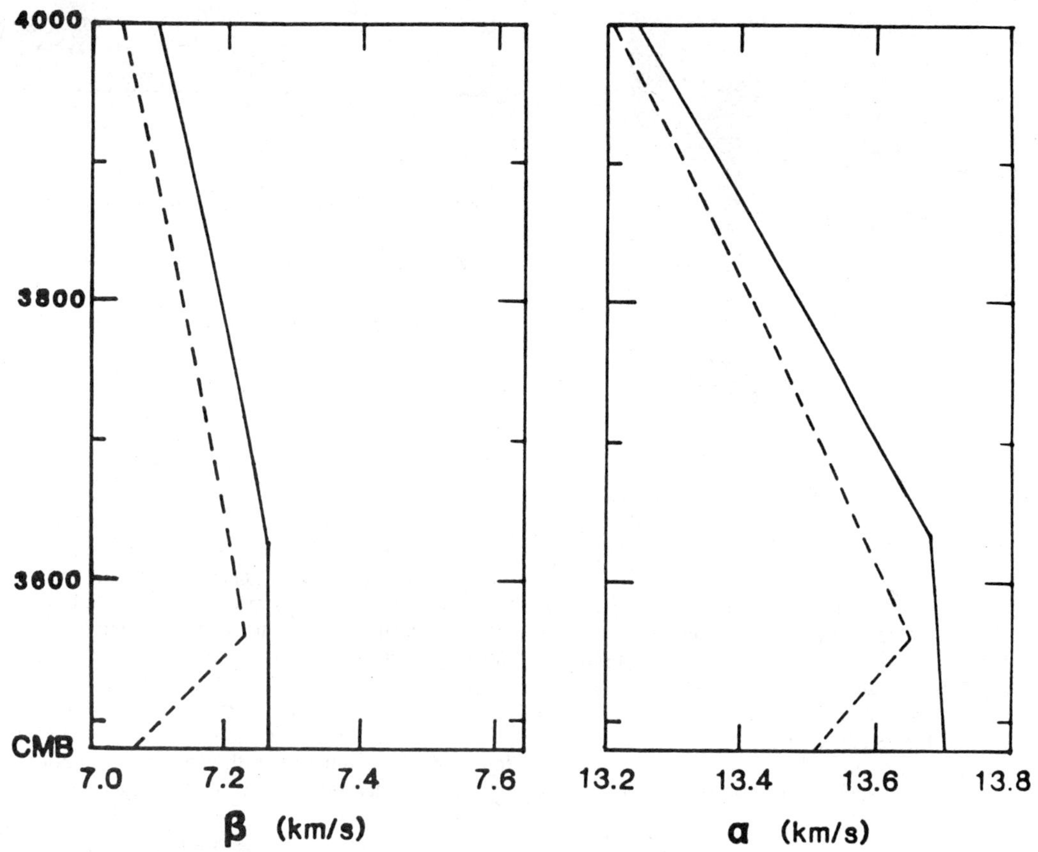

FIGURE 3. S-velocity profile (left) and P-velocity profile (right) in the lowermost 520 km of the Earth's mantle. The solid lines are for an earth model referenced to a frequency of 1 Hz.; the dashed lines are for an earth model referenced to a frequency of 0.005 Hz.

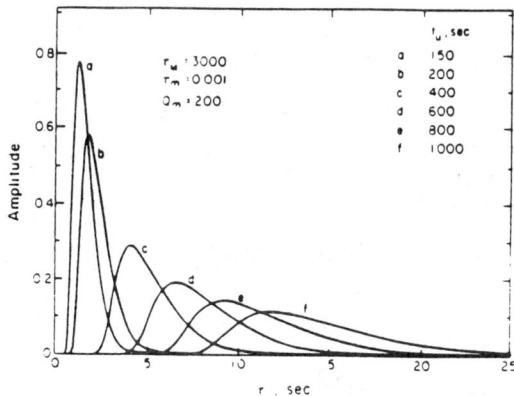

FIGURE 4. Evolution of a delta function with increasing travel time in a solid having intrinsic attenuation specified by an absorption band. The maximum Q^{-1} is 1/200; the lower and upper corner frequencies of the absorption band are $f = 1/\tau_V$ and $f = 1/\tau_m$, respectively. The pulses are referred to the high-frequency, unrelaxed arrival time t_u in each case; i.e., $\tau = t - t_u$, where $t_u = x/V_u$. (After Minster, 1978)

seismic data (Fig. 6a,b), suggesting that scattering may have some biasing effects on measures of intrinsic attenuation. In spectral measures taken over a narrow time window, different results can be obtained, depending on the length of window analyzed. Pulse measurements such as width and rise time may also be biased because higher-frequency energy has been transferred out of the pulse into the coda. Note in Fig. 6b that stochastic dispersion is in the opposite sense to the dispersion of intrinsic attenuation shown in Fig. 2a,b. High frequencies travel slower than low frequencies. A transition in the sense of dispersion may occur around 1–2 Hz in Fig. 6b.

Mean Free Paths. Another useful concept in evaluating the importance of scattering is to examine the mean free path resulting from scattering versus that resulting from intrinsic attenuation (see *Seismic Wave Scattering*). For scattering, the mean free path is the distance a wave travels before it is scattered. For intrinsic attenuation the mean free path is the distance traveled before amplitude decreases by e^{-1}. In seismograms observed on the Moon, the mean

SEISMIC ATTENUATION: OBSERVATION AND MEASUREMENT

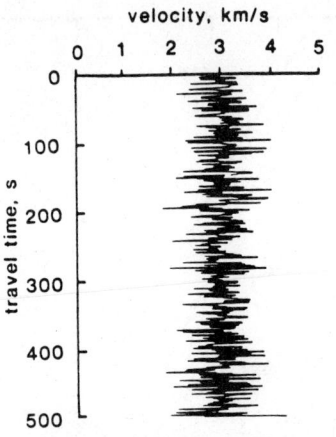

FIGURE 5(a). Velocity profile having 500 layers with velocities Gaussian distributed about a mean of 3.0 km·s^{-1} with a standard deviation of 0.25 km·s^{-1}. (After Richards and Menke, 1983)

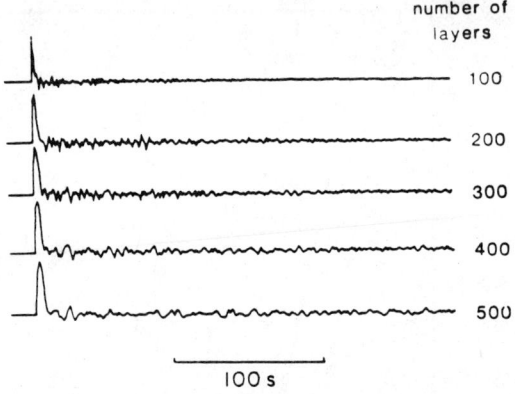

FIGURE 5(b). Transmission response of a medium similar to that shown in Fig. 5a. The incident wave is a delta function. (After Richards and Menke, 1983)

free path due to scattering is much shorter than that for intrinsic Q throughout a wide frequency band of common seismographs. The wave field can only be approximately calculated by a diffusion process. In the Earth, the mean free path of scattering is larger than that due to attenuation in most of the seismic band. Only beginning around 10–20 Hz do waves in the Earth's crust and uppermost mantle seem to have a scattering mean free path approximately equal to the mean free path of intrinsic attenuation. In this frequency band, the loss of energy with increasing range is primarily due to scattering rather than intrinsic attenuation. Body waves that sample the deep interior of the Earth are sufficiently coherent across large arrays of seismometers that their direction of approach agrees closely with that predicted for a radially symmetric Earth. The small variations in the recorded wavefront of 1 Hz and lower-

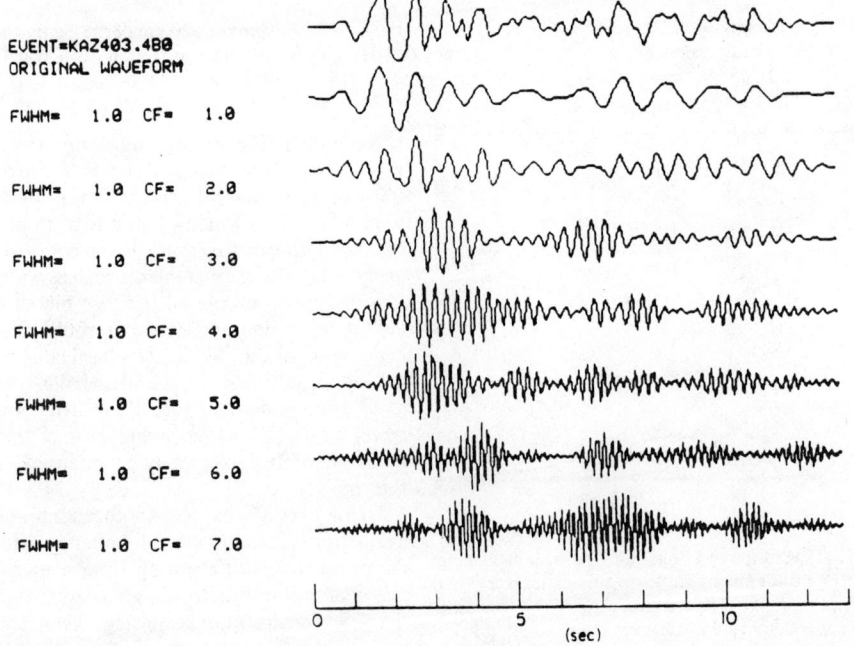

FIGURE 6(a). A teleseismic P wave recorded by a seismograph in Norway from an underground nuclear explosion at the east Kazakh test site of the Soviet Union. The original waveform has been filtered by a sequence of narrowband filters having center frequencies from 1 to 7 Hz. (After McLaughlin and Anderson, 1987)

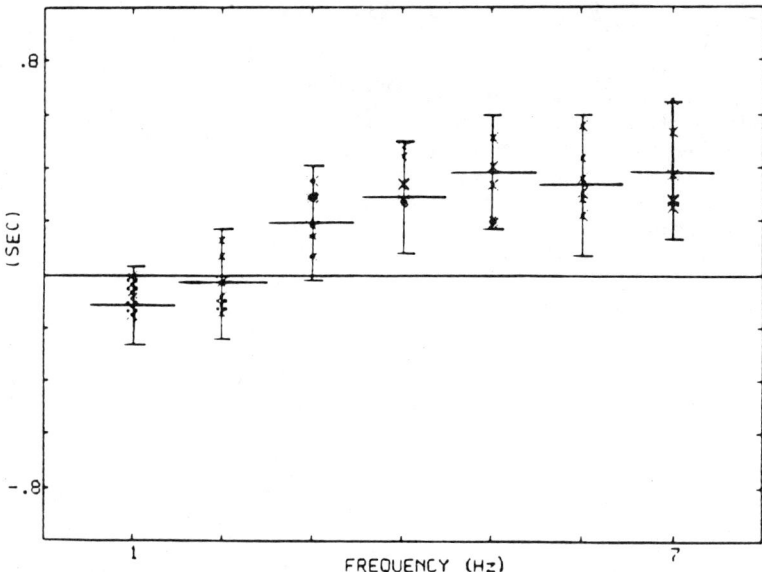

FIGURE 6(b). The temporal energy centroid of bandpass-filtered P waves from east Kazakh explosions recorded in Norway. Note that the higher-frequency energy is shifted to a later time relative to the lower-frequency energy. (After McLaughlin and Anderson, 1987)

frequency body waves can be explained by heterogeneities that are concentrated in the Earth's uppermost mantle and crust.

The Measurement of Body Wave Attenuation

Forward Modeling. Measurement of seismic attenuation is the art of measuring the absence of something in the data. For example, a certain amplitude is expected for a body or surface wave at a distant station based on measurements taken very close to the source. Alternatively, a certain difference in amplitude of seismic waves is expected between two stations at different distances. In both cases, the difference between observed and expected amplitudes or amplitude ratios is attributed to attenuation. Measurements of amplitude may be taken directly from seismograms or from their Fourier transforms in the frequency domain. To measure the attenuation, we must predict its effects from a model and vary the parameters of the model to fit the observed amplitude or amplitude ratio. The effects of intrinsic attenuation can be simply obtained by allowing elastic moduli and/or propagation velocities to become complex. Elastic boundary conditions, reflection and transmission at boundaries, travel times, and amplitudes are calculated exactly as in a nonattenuating solid but with elastic moduli and associated velocities analytically continued to complex values. This step of analytic continuation of real moduli to complex moduli is the same, whether one wishes to predict the waveform of a body wave or surface wave or a spectrum of free oscillations. For each particular wave type, the only differences in the calculations are details associated with choice of integral transforms, harmonic expansions, and asymptotic approximations. These choices are exactly the same as those made for calculations in a nonattenuating Earth. The size of the imaginary part of elastic moduli, parameterized by the value of Q as a function of depth and frequency, is chosen to match an observed waveform, spectrum, amplitude ratio, or spectral ratio.

The Attenuation Operator for Body Waves. As an example of these procedures, consider an experiment with body waves. The effects on a body wave of source radiation, geometric spreading, reflection-transmission, and intrinsic attenuation are most conveniently expressed in the frequency domain by a product of complex functions. The complex $g(\mathbf{x}\ \omega)$ spectrum of a body wave propagating from a point $\mathbf{x}_0$ to a receiver at $\mathbf{x}$ is

$$g(\mathbf{x}, \omega) = G_0(\mathbf{x}, \mathbf{x}_0)\tilde{A}(\omega) \qquad (2)$$

The function G_0 incorporates a product of reflection-transmission coefficients, reverberations at source and receiver, geometric spreading, and source radiation pattern. The function $\tilde{A}(\omega)$ is defined by

$$\tilde{A}(\omega) = e^{[i\omega\hat{T}(\omega)]}$$

where $\hat{T}(\omega)$ is a complex travel time obtained by integrating the reciprocal of complex velocity along a ray or normal to the wavefront of the body wave:

$$\hat{T}(\omega) = \int \frac{ds}{\hat{c}(\omega)}$$

For body waves, the dominant effect of attenuation on amplitude and phase is given by $\tilde{A}(\omega)$. The effects of attenuation on reflection-transmission coefficients and geometric spreading, which have been lumped into G_0, are much smaller and can be neglected unless the attenuation is very large (Q is very small). For $Q \gg 1$, $\tilde{A}(\omega)$ can be rewritten as

$$\tilde{A}(\omega) = \exp\left[-\frac{\omega t^*(\omega)}{2}\right]$$

$$\cdot \exp\left\{i\omega\left[\operatorname{Re}\tilde{T}(\infty) - \frac{H[t^*(\omega)]}{2}\right]\right\}$$

where

$$t^*(\omega) = \int \frac{Q^{-1}}{\operatorname{Re}\hat{c}(\omega)}\,ds$$

The effect on amplitudes of attenuation is contained in the factor $\exp[-\omega t^*(\omega)/2]$. Over a band in which dispersion is relatively weak, $\operatorname{Re}\hat{c}(\omega)$ may be replaced by a velocity c_0 for some reference frequency contained in the band. The dispersive effect of attenuation is contained in the factor $\exp\{i\omega[\operatorname{Re}\hat{T}(\infty) - \frac{1}{2}H[t^*(\omega)]]\}$. The operator H is a Hilbert transform. In a band of frequencies in which Q and t^* are nearly constant

$$H[t(\omega)/2] = \frac{\ln(\omega/\omega_0)}{\pi} t^*$$

where ω_0 is a reference frequency contained in the band. The value of $\hat{T}(\infty)$ need not be known and can be replaced by some reference time taken from an Earth model or table of travel times appropriate for the phase being analyzed.

Most measurements of attenuation reported in seismic literature attempt to measure only the amplitude effect of attenuation through the term $\exp[-\omega t^*(\omega)/2]$ measured from the spectral shape of body waves. There are basically two types of experiments commonly reported: matching of (1) spectral decay rates and (2) spectral ratios. In experiment (1) a shape for the source spectrum $|\tilde{A}(\omega)|$ is assumed, usually a flat level followed by decay at ω^{-2} above a corner frequency (Fig. 7). The additional decay observed at high frequencies in data spectra is taken as a measure of t^* in $\exp[-\omega t^*(\omega)/2]$. In experi-

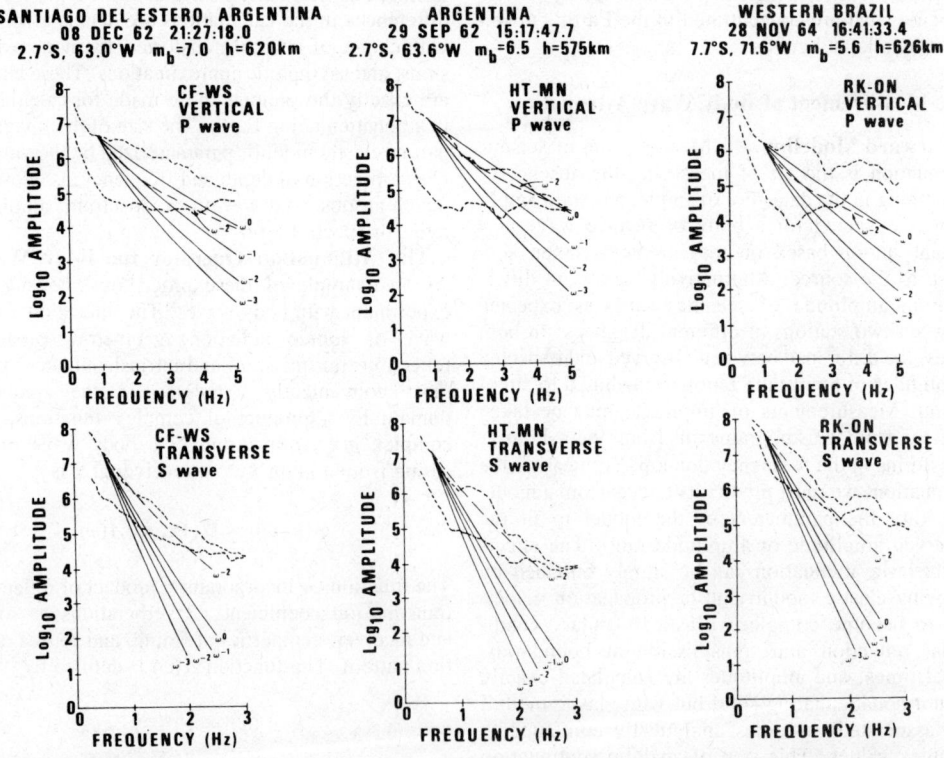

FIGURE 7. Spectra of body waves from deep focus earthquakes in South America recorded by seismographs in North America. The upper dotted lines are signal, the lower are noise. The spectral decay for $t_P^* = 1$ and $t_S^* = 4$ cannot fit the observed spectra, even assuming a white spectrum (ω^0) for the earthquake source spectra. (After Der et al., 1982)

ment (2), a ratio of spectra is observed, in which the source spectrum is assumed to approximately cancel. If the phases analyzed are recorded at the same receiver and are incident at nearly the same angles, then the source and receiver crustal reverberations will also approximately cancel. Both types of experiments usually apply some type of smoothing to the spectra to remove biasing effects of spectral holes caused by interfering crustal multiples, source complexities, scattering, and multipathing that are not included in the simple propagation model.

In the last decade, it has been more common to attempt to model the dispersive effects of attenuation as well as the amplitude effects in experiments based on the simple propagation model of Eq. 2. This modeling can be done in either the frequency domain or the time domain. In the frequency domain the phase of $\tilde{A}(\omega)$ is reported. In the time domain, an inversion is made for the time domain $A(t)$ function, which, when convolved with the time function describing the effects of the source and elastic structure matches, observed waveforms. Figures 8 and 9 illustrate the results in the spectral and time domains of determining the operator $A(t)$ needed to convert a broadband P wave into a broadband S wave. The uncertainties in the path-integrated parameter t^* shown in Fig. 9 are determined from a range of spectral shapes for an earthquake source, including differences in source spectra radiated as either P or S (see *Seismic Source: Theory*). Since t^* measures only the path-integrated effect of attenuation, many such experiments for different ray paths, bottoming at a range of different depths, are needed to construct a model of Q as a function of depth.

Free Oscillations. Measurements of attenuation in the low-frequency band of the free oscillations of the Earth face are conducted in the frequency domain by observing the width of the individual resonance peaks associated with each mode. These measurements face special problems associated with the broadening produced by lateral heterogeneity of elastic Earth structure. This heterogeneity splits the

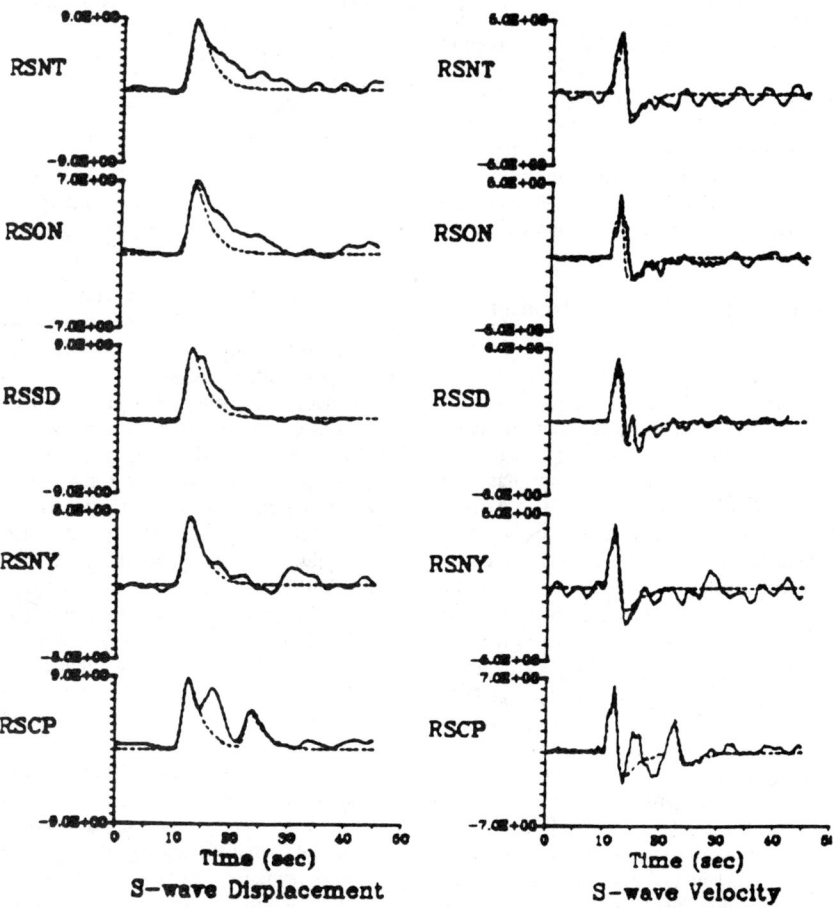

FIGURE 8. Broadband S waves observed by seismographs in North America from a deep focus earthquake beneath the Sea of Okhotsk. Dashed waveforms are best fits to the source function convolved with an attenuation operator obtained from comparison of P and S waveforms. (After Choy and Cormier, 1986)

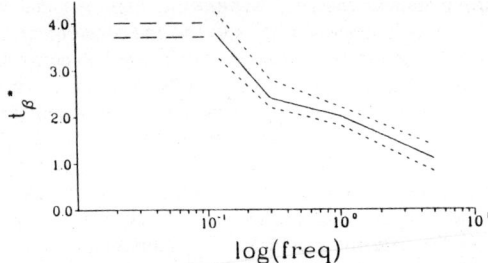

FIGURE 9. Constraints on the path-integrated attenuation t_S^* obtained from the broadband S waveforms shown in Fig. 8. The short dashed lines indicate bounds obtained from a range of source-time functions consistent with observed P waveforms. The long dashed lines at low frequency are from limits suggested by other studies. (After Choy and Cormier, 1986)

degenerate modes of a radially symmetric Earth, making a set of modes that would have the same frequency in a radially symmetric Earth have slightly different frequencies. The slightly different frequencies of the split modes may not be easily resolved in the data spectra and can be confused with the broadening of a single resonance peak of a mode caused by attenuation. Work is currently in progress both in predicting and correcting for the effects of mode splitting caused by lateral heterogeneity as well as in improving techniques of data analysis for measuring the widths of resonance peaks (see *Free Oscillations of the Earth*).

Interpretation of Attenuation Measurements in the Earth

Shear versus Bulk Attenuation. Although plausible mechanisms for defects in bulk moduli have been found in both laboratory measurements and analytic models of specific attenuation mechanisms, measurements on real data find that bulk dissipation in the Earth is small and, in most cases, can be neglected. Intrinsic attenuation occurs almost entirely in shear, associated with lateral movement of lattice defects, grain boundaries, and/or fluids rather than with changes in material volume. For attenuation purely in shear in a Poisson solid, $Q_P = \frac{9}{4} Q_S$, and the parameter for path-integrated attenuation of S waves, or t_S^*, is approximately $4 t_P^*$. Most experiments confirm these values. There is a suggestion, however, that Q_P tends to approach Q_S and $t_S^* < 4 t_P^*$ at frequencies higher than 1 Hz. These observations should not necessarily be taken as evidence of bulk attenuation, because it has been shown that as the effects of scattering increase at higher frequencies the apparent Q_S tends to approach the apparent Q_P when they are measured from pulse widths or spectra taken over short time windows bracketing the pulse. The assumption of attenuation mainly in shear can aid in separating the effects of scattering from intrinsic attenuation in body wave pulses.

Frequency Dependence. Similar to the discovery of the dispersive effect of intrinsic attenuation, its frequency dependence only began to be appreciated as the measurements taken in widely separated frequency bands, such as free oscillations and body waves, began to be compared. One of the earliest observations was reported by Gutenberg in 1958, who noted that in order for the phase PKPPKP to be observable its t_P^* must decrease with frequency as ω^{-1}. This observation would imply that Q_P^{-1}, averaged along the ray path, also decreases with frequency as ω^{-1}. Later, when the results of attenuation measurements determined from free oscillations and body waves in the 0.0001–0.1 Hz band began to be compared with observations of body wave spectra in the 1–10 Hz band, it became apparent that even under the assumption of a white source spectrum that an increase in Q with frequency was necessary to explain the amplitude of spectra in the 1–10 Hz band.

Thermal Activation. This frequency dependence has been interpreted in terms of physical mechanisms of attenuation that are thermally activated. In these mechanisms, the high-frequency corner f_c is tied to a relaxation time τ_1, where $f_c = 1/2\pi\tau_1$. The time τ_1 depends on temperature and pressure as follows:

$$\tau_1 = \tau_0 \exp\left(\frac{E^* + PV^*}{RT}\right) \quad (3)$$

As an example of such a mechanism, imagine a standard linear oscillator model of the movement of a lattice defect within a potential well. Temperature and pressure affect the shape of the potential well and, hence, the spring constant and frequency of oscillation of the defect within the well. In Eq. 3, E^* and V^* are activation energy and volume, respectively, associated with the movement of Avogadro's number of defects over their respective potential barriers. The higher time constant τ_2 (or low-frequency cutoff) of an absorption band is also assumed to be similarly affected, temperature and pressure acting to slide the absorption band through a band of frequencies. A typical width to expect for the relaxation spectrum of the mantle is about five orders of magnitude in frequency, $\tau_2/\tau_1 = 10^5$ (Anderson and Given, 1982). A simplified model of an absorption band with depth in the Earth's mantle is shown in Fig. 10. The movement of the absorption band toward lower frequencies (longer periods) in the mantle below 400-km depth is consistent with the type of behavior shown in Fig. 9 for the t_S^* measured from the body waves of an earthquake. The difference in the location of the absorption band with respect to the band of seismic frequencies is consistent with models of the temperature and pressure profiles of the Earth's mantle for specific

SEISMIC ATTENUATION: OBSERVATION AND MEASUREMENT

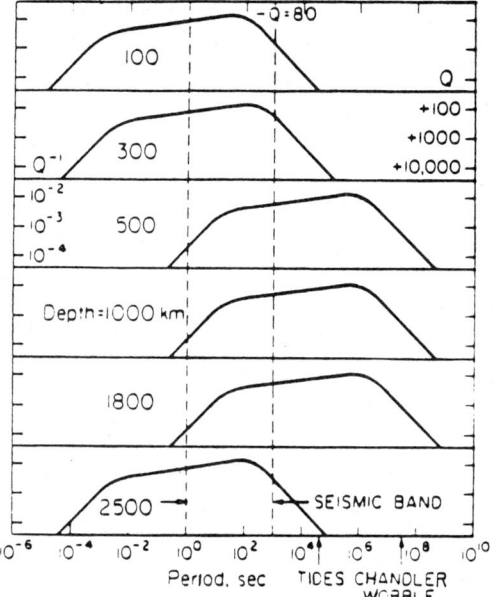

FIGURE 10. The seismic relaxation spectrum as a function of depth and frequency in the Earth's mantle as proposed by Anderson and Given (1982).

values of E^* and V^* (Fig. 11). In the uppermost mantle, a rapid increase in temperature with depth can rapidly change the location of the absorption band with respect to the seismic band. Depending on the values chosen for E^* and V^*, the slower temperature and pressure changes in the mid- and lower mantle makes the location of the relaxation spectrum more stable. Regional variations in attenuation in the Earth's mantle also agree well with presumed thermal variations in at least a qualitative sense. Tectonically, active regions having radiogenically younger crust are more attenuating than the mantle underlying inactive regions, such as continental shields. The shape of the frequency dependence across the seismic band seems to remain similar in different regions, although the Q at a given frequency is lower for a tectonic region than for a shield region (Fig. 12). It has been observed the variations in travel time correlate with variations in apparent attenuation observed from spectral content. Fast travel times of body waves correlate with spectra enriched in high frequency (low attenuation), and slow travel times correlate with spectra depleted in high frequency (high attenuation). These correlations appear to be consistent with thermal activation, in which the dispersive effect of attenuation acts jointly with variations in the high-frequency corner of the mantle relaxation spectrum to produce the observed variations in travel time and frequency content. Deep chemical differences between shields and young continents and oceans, however, have also been suggested (see *Mantle, Upper: Structure*). In this case, regional variations in attenuation are as likely to be affected by differ-

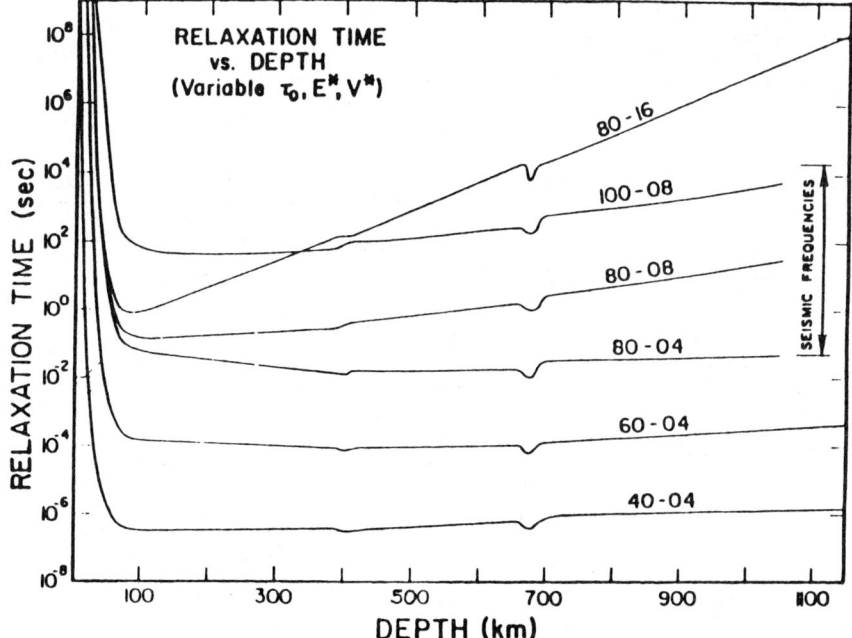

FIGURE 11. Possible behavior of the relaxation time τ_1 that controls the high-frequency corner of the relaxation spectrum in the Earth's mantle. A dependence on temperature and pressure is assumed in the form $\tau_1 = \tau_0 \exp[(E^* + PV^*)/RT]$. The variation in τ_1 is shown for different pairs of E^* and V^*

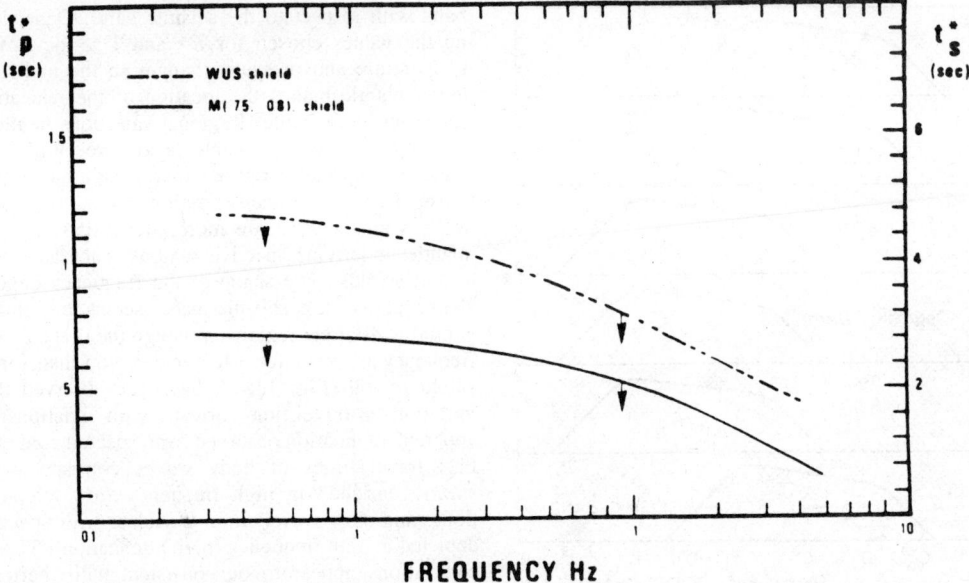

FIGURE 12. Frequency dependence of t_P^* and t_S^* proposed by Der et al. (1982) for shallow focus sources. The WUS-shield curve is for mixed western United States-shield paths. The solid curve is for pure shield paths. There will, in general, be a slight distance as well as frequency dependence of the path-integrated attenuation factors.

ences in composition and activation constants as by the temperature profile. The regional variation of attenuation complicates the design of experiments for searching for frequency dependence, in that additional criteria are needed for sorting data into separate sets of regionally uniform data.

Strain Dependence. Laboratory measurements of Q in rocks find a dependence on strain beginning at strains of about 10^{-6} (Fig. 13). The strain dependence decreases with confining pressure. The Q is also strongly dependent on moisture and interstitial fluids between cracks and grain boundaries. These observations are consistent with a physical mechanism of frictional sliding of cracks. Unlike viscoelastic relaxations, which are representative of all linear mechanisms, frictional sliding of cracks is an inherently nonlinear mechanism. The anelastic effects of frictional sliding cannot be calculated by substitution of complex moduli. A relation between stress and strain is required that includes

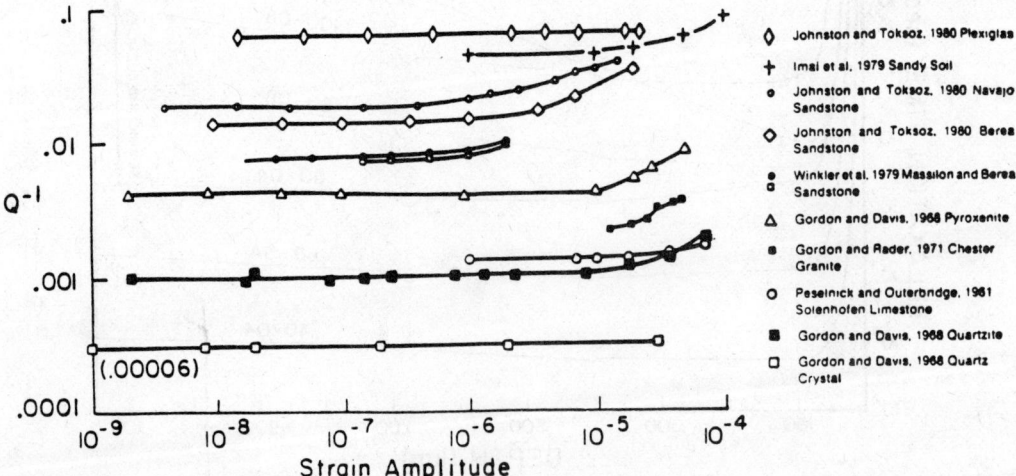

FIGURE 13. Strain dependence of Q^{-1} for a variety of rocks and solids observed in the laboratory, usually at atmospheric pressure. (After Stewart and Toksöz, 1983)

the local orientation and magnitude of strain as well as the history of the applied stress.

Q measurements on the porous and cracked rocks in the Earth's crust agree with the predictions for the strain and pressure dependence of Q obtained from the theory of Herzian contact friction. Many data have been fit to the law

$$Q^{-1} = \frac{k \zeta \epsilon}{P^{4/3}}$$

where k is a function of the elastic parameters of the rock matrix, ζ is the crack density, ϵ is the strain amplitude, and P is the confining pressure. Effects on waveforms can be approximately calculated by assuming linear propagation over short range, using Q values determined from the average local strain over the range interval. The failure of linear superposition and the transfer of energy between frequencies are not included in this approximate calculation, but the results seem to agree with more exact calculations in the 10^{-4}–10^{-5} strain regime. Estimates of when nonlinear effects occur in a particular data set may be made from calculating the strain associated with the seismic wave being analyzed. A rough estimate can be obtained by assuming that the wavefront is a plane wave and dividing the particle velocity by the propagation velocity. For example, the particle velocity of body waves observed in strong ground motion recordings from 0 km to 10 km from the hypocenter of a $m_b = 6$ earthquake are typically about 10 cm·s^{-1}. If the body wave propagates at 3 km·s^{-1}, the strain observed at the strong ground-motion site is roughly 10 cm·s^{-1} divided by 3×10^5 cm·s^{-1}, or $\epsilon = 3.3 \times 10^{-5}$. This value is likely to be in the nonlinear regime of surficial rocks having open cracks.

Conclusions

The intrinsic attenuation of seismic waves in the Earth has been found to be consistent with loss mechanisms that are thermally activated. The observed regional and frequency dependencies of seismic Q agree with the expected lateral variations in a geotherm having a rapid temperature increase in the upper 400 km of the mantle, followed by a slower variation in the mid- and lower mantle. High seismic velocities correlate with regions of low attenuation; low seismic velocities correlate with regions of high attenuation. Most measurements are consistent with losses in shear rather than bulk deformations.

The existence of lateral heterogeneity in the elastic properties of the Earth complicates the measurement of its anelastic properties. The longer scale lengths of heterogeneity can split modes of free oscillation and focus and defocus body waves and surface waves. Shorter scale lengths scatter seismic energy, broaden the waveforms of body waves, and redistribute energy into different time and azimuthal windows. The intensity of heterogeneity in percent fluctuation of velocities and densities is higher at shorter scale lengths and at shallower depths in the Earth's crust and upper mantle. There is still a need for experiments that determine finer details of how the distribution of heterogeneity changes with depth and lateral location in the Earth as well as for experiments that seek to separate the relative contribution of scattering versus intrinsic anelasticity to the attenuation of body waves.

Laboratory experiments find a transition from linear to nonlinear anelastic behavior at strains on the order of 10^{-6}. The observed strain dependence of Q and its dependence on pressure agree with a mechanism of frictional sliding of cracks. In situ experiments, close to earthquakes or artificial seismic sources, have not yet been designed so that the transition to nonlinear behavior can be carefully observed. It is still unknown how and at what strain levels linear superposition begins to break down.

Although a consensus has been reached on the major features of intrinsic attenuation in most of the Earth's mantle and outer core, this is less true of other regions of the Earth. Definitive experiments are still needed for the distribution of Q in the lowermost 400 km of the mantle. A unifying concept is needed to explain attenuation in the inner core, which supports high-Q free oscillations rich in shear deformation, but which strongly attenuates high-frequency body waves. High-frequency body waves and waveguide modes in the Earth's crust suggest Q_S values of 1000 or more, while laboratory measurements on crustal rocks often measure a Q_S of several hundred. Corrections for the frequency dependence of viscoelastic relaxations may provide the answer to some of these problems, but there is still a need for independent verification of this hypothesis with laboratory data.

VERNON F. CORMIER

References

Anderson, D. L., and J. W. Given, 1982, The absorbtion band Q model for the Earth. *Jour. Geophys. Research* **87**, 3893–3904.

Choy, G. L., and V. F. Cormier, 1986, Direct measurement of the mantle attenuation operator from broadband P and S waveforms, *Jour. Geophys. Research* **91**, 7326–7342.

Cormier, V. F., 1982. The effect of attenuation on seismic body waves, *Seismol. Soc. America Bull.* **72**, S169–S200.

Der, Z. A., T. W. McElfresh, and A. O'Donnell, 1982, An investigation of regional variations and frequency dependence of anelastic attenuation in the United States in the 0.5-4 Hz. band, *Royal Astron. Soc. Geophys. Jour.* **69**, 67–100.

Gutenberg, B., 1958, Attenuation of seismic waves in the Earth's mantle, *Seismol. Soc. America Bull.* **48**, 269–282.

Jackson, D. D., and D. L. Anderson, 1970, Physical mechanisms of seismic attenuation, *Rev. Geophysics and Space Physics* **8**, 1–63.

Liu, H.-P., D. L. Anderson, and H. Kanamori, 1976, Velocity dispersion due to anelasticity: implications for seismology and mantle composition, *Royal Astron. Soc. Geophys. Jour.* **47**, 41-58.

McLaughlin, K. L., and L. M. Anderson, 1987, Stochastic dispersion of short-period body waves due to scattering and multipathing, *Royal Astron. Soc. Geophys. Jour.* **89**, 933-964.

Minster, J. B., 1978, Transient and impulse responses of a one-dimensionally linearly attenuating medium—I. Analytical results; II. A parametric study, *Royal Astron. Soc. Geophys. Jour.* **52**, 479-524.

Minster, J. B., 1980, Anelasticity and attenuation, in A. M. Dziewonski and E. Boschi, eds., *Physics of the Earth's Interior, Proc. Enrico Fermi Int. School Phys.* Amsterdam: Boschi, North-Holland, 152-212.

Nowick, A. S., and B. S. Berry, 1972, *Anelastic Relaxation in Crystalline Solids*. Orlando, Fla.: Academic Press, 677p.

Richards, P. G., and W. Menke. 1983, The apparent attenuation of a scattering medium, *Seismol. Soc. Am. Bull.* **73**, 1005-1021.

Stewart, R. R., M. N. Toksöz, and A. Timur, 1983, Strain dependent attenuation: observations and a proposed mechanism, *Jour. Geophys. Research.* **88**, 546-554.

Toksöz, M. N., and D. H. Johnston, eds., 1981, *Seismic Wave Attenuation*. Tulsa, Okla.: Society of Exploration Geophysicists, 459p.

Zener, C., 1960, *Elasticity and Anelasticity of Metals*. Chicago, Ill.: The University of Chicago Press.

Cross-references: *Earth Structure: Global; Earthquake Seismology; Elasticity and Wave Propagation: Principles; Elastic Waves in Homogeneous and Inhomogeneous Media; Free Oscillations of the Earth; Mantle Viscosity; Seismic Diffraction; Seismic Monitoring of Nuclear Explosions; Seismic Properties of Rocks; Seismic Ray Theory; Seismic Source: Observations; Seismic Source: Theory; Seismic Wave Scattering; Seismology: Physical Model Studies; Surface Waves.*

SEISMIC DIFFRACTION

Diffraction and Scattering

The words "diffraction" and "scattering" are often used to describe the same wave phenomena. Initially "diffraction" was reserved for wave motion beyond an obstacle that has cut off portions of the advancing wavefront. Later the meaning of the word was widened to also describe the nongeometrical scattering effects by an irregular interface or surface. Some have explained the difference as follows: Let $\mathbf{u}^{in}$ represent the incident wave field. The interaction of $\mathbf{u}^{in}$ with the interface results in a total field $\mathbf{u}^t$. Then

$$\mathbf{u}^t = \mathbf{u}^{in} + \mathbf{u}^{sc}$$

where $\mathbf{u}^{sc}$ is the scattered field. On the other hand, one can try and predict the scattering by geometrical ray theory, and the predicted total field is denoted $\mathbf{u}^{ge}$. Then

$$\mathbf{u}^t = \mathbf{u}^{ge} + \mathbf{u}^{di}$$

where $\mathbf{u}^{di}$ is the diffracted field.

Although it is nearly impossible to give a complete classification of diffraction patterns, it is possible to identify some of the more common types of diffraction that are of seismological interest. Five situations will be considered in the following text. Some of these are associated with a one-dimensional velocity structure. These problems are quantitatively well understood and well documented (see Aki and Richards, 1980; Chapman and Orcutt, 1985). Other phenomena are associated with laterally varying structures. They currently form an area of active research in theoretical seismology.

Fresnel Shadow

Diffraction by the edge of an opaque screen (edge diffraction) is a classical problem in optics that can serve as a basis for studying seismologically more relevant problems. In optics a Fresnel shadow is formed behind the screen (Fig. 1). The wave amplitude begins to decrease near point F where the first Fresnel zone in the plane of the screen lies just off the screen, and the amplitude is reduced by a factor of 2 at the geometrical shadow boundary (Fig. 2).

Two types of complications can be expected when replacing the screen by a seismic diffractor: (1) The diffractor geometry may be different from the screen, and (2) the boundary conditions at the diffractor must

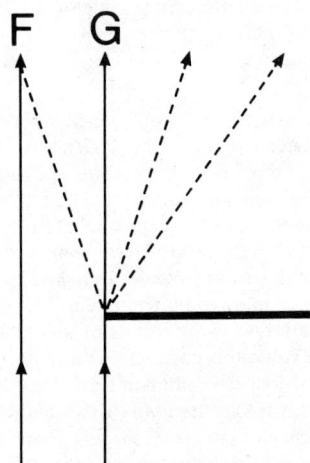

FIGURE 1. Plane wave incident on semi-infinite opaque screen. A Fresnel shadow is formed behind the screen: (solid lines) geometrical rays; (dashed lines) diffracted rays. For a receiver at distance d beyond the screen, the edge of the first Fresnel zone in the plane of the screen for wavelength λ is defined by the distance $r = d + \frac{1}{2}\lambda$. In F the first Fresnel zone lies just off the screen.

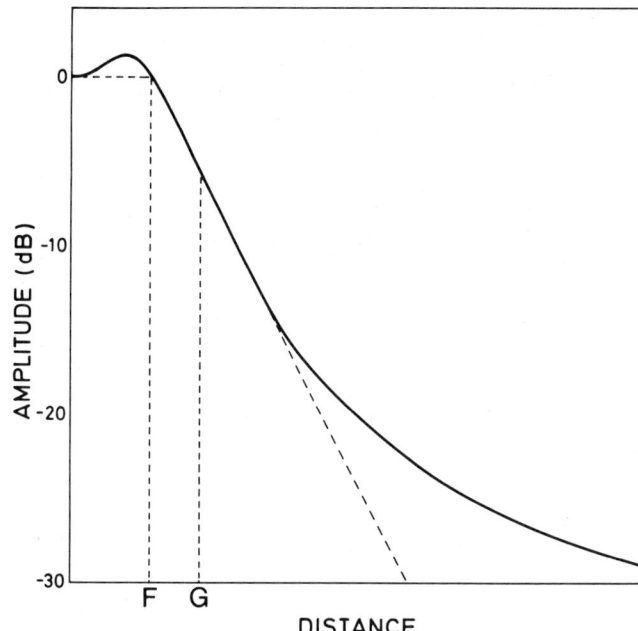

FIGURE 2. Diffracted wave amplitude variation into the shadow zone. Geometrical shadow boundary is at G. Effective shadow boundary is near F: (solid line) Fresnel shadow; (dotted line) asymptotic behavior of diffracted waves around the core.

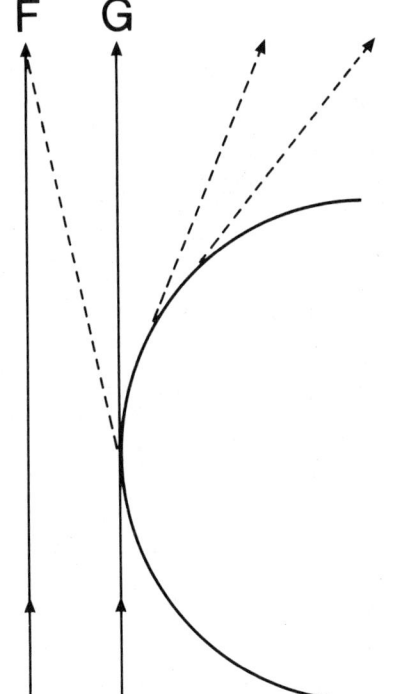

FIGURE 3. Diffraction of plane wave around the core. In a homogeneous mantle the ray trajectories are straight lines. The effect of dissecting the wavefront by the core would be a Fresnel shadow; the additional scattering of the wave interacting with the boundary results in a nearly exponential amplitude decay in the deep shadow.

be taken into account. Thus, although seismic diffraction by the edge of a crack implies a diffractor geometry similar to the screen, point (2) causes the seismic problem to be much more complicated. Achenbach and co-workers (1982) have made a detailed analysis of this problem using a method due to Keller in 1962, called the *geometrical theory of diffraction*.

Diffraction Around an Object

Diffraction around the Earth's core constitutes one of the best-documented examples of this type (Fig. 3). One of the reasons is that the diffracted waves are well observed because the core forms a shadow zone for both P and S waves. It forms a shadow for S waves because they do not penetrate the liquid core, and it forms a shadow for P waves because the core acts as a low-velocity zone for P waves. It is not necessary for the diffractor to be bounded by a low-velocity zone. However, with a high-velocity zone there will be no shadow, and the diffracted wave will not be a first arrival.

The significance of Fresnel-type diffraction to this problem can be investigated by replacing the core by an opaque screen and deducing the position of the first Fresnel zone and the amplitude decay of diffracted waves into the shadow. This approximation would at best be short-period, since for relatively short wavelengths the spherical geometry of the core is less important. Indeed, Phinney and Cathles (1969) found the effective boundary of the Fresnel shadow to be in reasonable agreement with short-period data of diffracted P wave. However, the

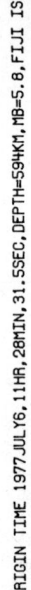
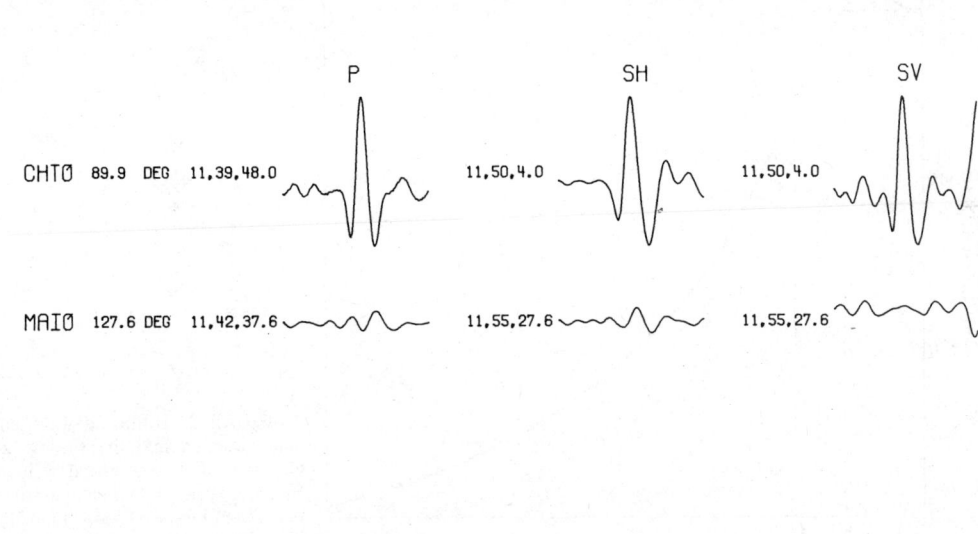

FIGURE 4. P, SH, and SV long-period waves from an event near Fiji Islands, recorded at station MAIO near 90° epicentral distance, and the diffracted waves at station CHTO near 130°. Length of traces is 2 min. The geometrical shadow boundary of the core is near 100°. The amplitude decay of diffracted P, SH, and SV is different. The diffracted SV wave has become too small to be observed.

observations do not show the amplitude decay of a Fresnel shadow.

Most observations of diffraction into the shadow of the core are of long-period waves. Figure 4 shows records of P, SV, and SH waves at a seismic station near the shadow boundary of the core and at another station some distance into the shadow. Not only are the deduced decay rates in disagreement with a Fresnel shadow, the rates for P, SH, and SV are also mutually different. These observations attest to the significance of wave interaction with the boundary. This problem will be discussed in the next section.

Diffraction at a Plane Boundary

For diffraction to occur at a plane boundary there must be a velocity increase with depth above the (assumed horizontal) velocity discontinuity. Diffraction characteristics are then very similar to those observed in the shadow of the spherical core (Fig. 5). The similarity is to be expected, since a simple "Earth-flattening transformation" relates the velocity-depth profiles in the two geometries such that kinematic features of the wave field are unchanged (for more details, see Aki and Richards, 1980, pp. 363–365). Thus, constant velocity in the spherical layers of Fig. 3 corresponds to a velocity increase with depth in the plane layers of Fig. 5.

Chapman and Orcutt (1985) showed that the so-called WKBJ approximation accounts for the nongeometrical part of the amplitude decay into the shadow by a factor that precisely reproduces the Fresnel part of the shadow, i.e., the effect of the dissection of the wavefront by the velocity discontinuity. It does not account for scattering due to the wave interacting with the boundary. It appears that this interaction is most sensitive to the velocity structure above the boundary; it provides a means to infer the velocity structure from diffracted wave observations. Doornbos and Mondt (1979), among others, have investigated the various effects in some detail.

The diffracted field here can be taken to represent the backscattering from the interface. Backscattering is generally quantified by means of reflection coefficients. These coefficients are usually consistent with a WKBJ approximation of the wave functions.

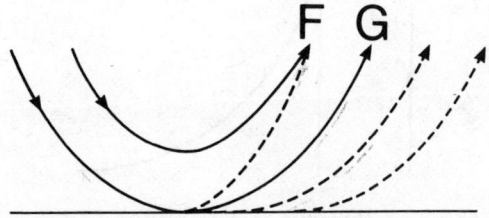

FIGURE 5. Diffraction at a plane boundary. The wave velocity above the boundary increases with depth. The geometry of the boundary and the ray trajectories in this figure and in Fig. 3 are related through an Earth-flattening transformation.

However, this approximation is invalid near a turning point of the ray, which is the region of interest here. The problem has been circumvented in two ways (see the references for the details): (1) The structure is divided into homogeneous layers (in which there are no turning points) and the reflectivity method is used to obtain a generalized reflection coefficient. (2) The WKBJ approximation of the wave functions is replaced by the uniformly asymptotic Langer approximation. Formulas for reflection coefficients are modified accordingly. The effect of velocity structure enters these formulas through the argument $\omega\tau$ of the vertical wave functions. For a flat Earth model,

$$\tau = \int_{z_0}^{z} \left(\frac{1}{v^2 - p^2}\right) dz$$

where v is the P- or S-wave velocity, p is the ray parameter, z_0 is the turning point, and z is the level at which the wave function is to be evaluated (i.e., at the interface). The dominant contribution to the diffracted wave involves ray parameters with a turning point above the interface. The method in its original form was valid only for velocity gradients that were not too strong. Kennett (1983), and Chapman and Orcutt (1985) discuss various extensions and alternatives to handle more complicated structures.

Figure 2 includes a typical result with a nearly exponential amplitude decay in the deep shadow. Such a result for diffraction around the core and a homogeneous mantle was already obtained by Scholte (1956), by introducing spherical wave reflection coefficients. The computed amplitude decay for a homogeneous mantle is much stronger than for Fresnel diffraction. However, it is also stronger than what has been inferred from many observations, particularly for SH. The discrepancy suggests that the S velocity just above the core-mantle boundary is not constant but decreases with depth.

Diffraction at a Caustic

A *caustic* is a point of focusing of rays (i.e., it is characterized by the vanishing of the geometrical spreading) in a one-dimensional model $dX/dp = 0$. This condition exists, for example, at the far end of a shadow zone caused by a low-velocity layer (Fig. 6). One of the best-known examples of such a caustic is created by P waves refracted through the Earth's core (PKP). This caustic extends throughout the mantle. Caustics are associated with some other core phases as well.

In 1939, H. Jeffreys used an Airy function to approximate the amplitude-distance variation of the wave field near the caustic in a one-dimensional model:

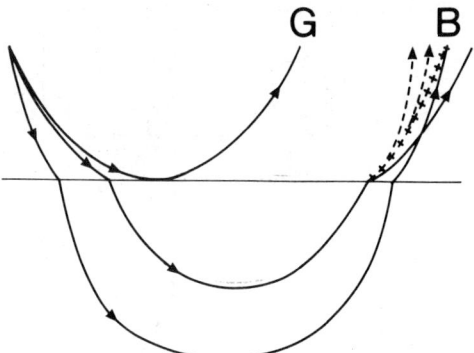

FIGURE 6. Shadow between G and B due to a velocity decrease with depth across the boundary. B represents a caustic. The caustic in this model extends throughout the upper layer (crosses). Diffracted waves into the shadow near B are described by an Airy function.

$$A(x) \sim \text{Ai}\left(-2^{1/3}\omega^{2/3}(x - X_B)\left(\frac{d^2X}{dp^2}\right)_B^{-1/3}\right) \quad (1)$$

where p_B and X_B are the ray parameter and distance at the caustic. Chapman and Orcutt (1985) showed that the variation according to Eq. 1 is precisely reproduced by the WKBJ seismogram. Since the caustic shadow is not directly associated with the cutting off of the wavefront by a boundary, wave interaction effects are less important and the WKBJ seismogram is expected to reliably account for the diffraction. For the same reasons the diffraction at caustics in two- or three-dimensional media can be modeled by an appropriate generalization of ray methods, such as the WKBJ or Maslov method or the method of Gaussian beams. For further information on these methods, see *Seismic Ray Theory*.

Figure 7 (originally from Jordan's Ph.D. thesis in 1972) shows short-period PKP waves at the caustic near 143° and at shorter distances into the "shadow zone." In this distance range the first-arriving train of waves is precursory to the refracted/reflected wave through the inner core. The amplitudes of the precursors are much larger than expected for diffracted waves at an Airy caustic. It is now generally accepted that most of the precursors are not directly associated with this diffraction. Presumably they are generated by scattering at or above the core-mantle boundary. This phenomenon belongs more appropriately to the next section.

Diffraction by Topographic Relief

Some striking features on seismic sections and on earthquake seismograms are thought to be caused by diffraction due to topographic relief of the Earth's surface and internal discontinuities (Fig. 8). The radiation pattern of the diffracted waves depends on the geometry of the diffractor, but a simple relation-

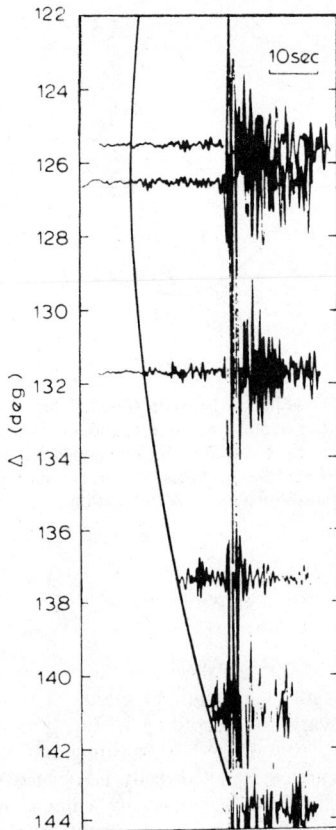

FIGURE 7. PKP records from an earthquake in the Java Sea assembled by Jordan and later also shown by Haddon and Cleary. The phase refracted through the inner core (PKIKP) is lined up in the records. The precursors to PKIKP cannot be explained by Airy diffraction.

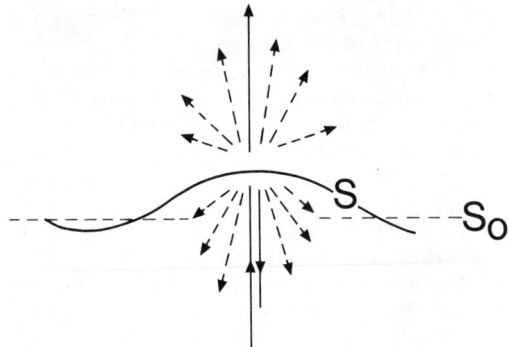

FIGURE 8. Scattering by an irregular boundary (S): (solid lines) incident wave, and scattered waves for an auxiliary plane boundary (S_0); (dashed lines): diffracted waves due to irregularity of the boundary S.

ship exists only in a first (Born) approximation. Ray methods will not account for the radiation pattern. The application of complete three-dimensional finite-difference or finite-element methods to these problems can, at least in principle, account for the diffraction effects, but perhaps the most straightforward way to account for the diffraction is to generalize the Kirchhoff integral. The original Kirchhoff integral for Fresnel diffraction involves field quantities of acoustic waves, and the integral is taken over the dissected wavefront. Trorey (1970) modeled diffraction due to an irregular reflector by taking the secondary sources for scattered waves to coincide with the reflector. The Kirchhoff integral is modified accordingly.

A further generalization is based on the elastodynamic representation theorem. Displacement $\mathbf{u}_S$ and traction $\boldsymbol{\sigma}_S$ on the irregular surface or interface S are taken to be the secondary sources of scattered waves. These quantities are to be solved from a system of coupled integral equations. A similar theorem has been applied to the problem of scattering by a perfectly conducting surface in electrodynamics, and the application was termed the *extended boundary condition method*, the *null field method*, or the *extinction theorem method* (an introduction to scattering applications in this field can be found in Brown, 1985). The terminology used to identify the method emphasizes that the scattered field inside the perfectly conducting body must cancel the incident field to ensure that the total field is zero. The elastodynamic counterpart of this problem is that of a rigid surface, $\mathbf{u}_S = 0$, and applying the extinction theorem to the empty space above S in which there is no field leads to a vector integral equation for $\boldsymbol{\sigma}_S$. On the other hand, the geophysically more relevant case of a free surface corresponds to $\boldsymbol{\sigma}_S = 0$; the theorem then leads to a vector equation for $\mathbf{u}_S$. For a solid-solid interface the extinction theorem can be applied to the opposite sides of the interface, leading to two vector equations for $\mathbf{u}_S$, $\boldsymbol{\sigma}_S$. Finally, the secondary source field for a solid-liquid interface consists of the normal displacement and traction and of the tangential displacement components on the solid side of the interface. These four quantities are determined by four integral equations.

A solution for the secondary source field must be obtained numerically in all of the above-mentioned cases. If $\mathbf{u}_S$ and $\boldsymbol{\sigma}_S$ are "not too large," it is instructive to try a solution in the form of a perturbation series:

$$\mathbf{u} = \sum_{n=0}^{\infty} \mathbf{U}^{(n)}, \qquad \boldsymbol{\sigma} = \sum_{n=0}^{\infty} \mathbf{S}^{(n)}$$

The zeroth-order term gives the conventional reflection/transmission coefficients for a plane interface, the first-order term includes the Born approximation, and the higher-order terms account for multiple scattering. For a detailed account of these series

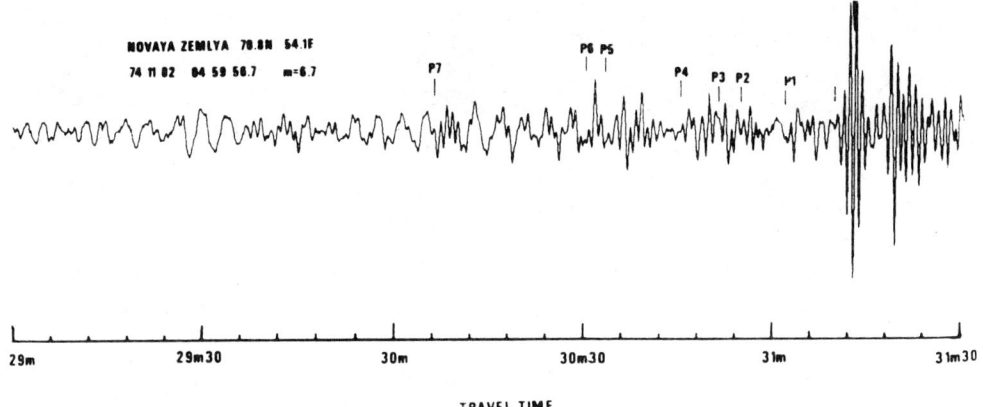

FIGURE 9. Record of the LASA array in Montana (no longer operational), shown by Chang and Cleary. The record is an array beam in the time interval of PKKP from a Novaya Zemlya explosion at distance about 60° from the array. Precursor phases are numbered P1 through P7. Not only the precursors but also the main phase in this record is thought to be caused by backscattering due to topographic relief.

solutions refer to Doornbos (1988). The Born approximation was obtained previously by B. L. N. Kennett in 1972. The approximation gives a linear relationship between the wave-number spectrum of the diffractor and that of the diffracted wave field: Let the topography of the boundary S be measured by a function $f(x, y)$ on the reference plane S_0, and let an incident monochromatic plane wave with wave-number components k_x^0, k_y^0 be represented by

$$\mathbf{u}^{in}(k_x, k_y) = \mathbf{U}^0 \delta(k_x - k_x^0)\, \delta(k_y - k_y^0)$$

Then for the first-order scattered wave

$$\mathbf{U}^{(1)}(k_x, k_y) = F(k_x - k_x^0, k_y - k_y^0) \mathbf{P}(k_x, k_x^0, k_y, k_y^0) \mathbf{U}^0$$

where $F(k_x, k_y)$ is the wave-number spectrum of $f(x, y)$.

Diffraction phenomena of this type are especially important for the Earth's most striking discontinuities such as the surface, the sea bottom, and the core-mantle boundary. Diffraction by topography of the sea bottom represents a significant source of coherent noise in marine seismic surveys, as demonstrated by Larner and co-workers (1983). The precursors to the core phases PKP at epicentral distances less than about 140° (Fig. 7) may be forward-scattered waves due to topography of the core-mantle boundary, although these precursors can also be explained by other forward-scattering models. This hypothesis was first published by Haddon (1972). An application of scattering theory to this problem is included in Doornbos (1988). Topography of the core-mantle boundary would also backscatter waves arriving as precursors to PKKP, and sometimes these have indeed been observed. Usually the observed precursors are weak, but an exceptional example (taken from a publication by Chang and Cleary, 1978) suggesting strong back scattering is shown in Fig. 9.

DURK J. DOORNBOS

References

Achenbach, J. D., A. K. Gautesen, and H. McMaken, 1982, *Ray Methods for Waves in Elastic Solids*. London: Pitman Books, 251 p.

Aki, K., and P. G. Richards, 1980, *Quantitative Seismology*. San Francisco: Freeman Co., 932 p.

Brown, G. S., 1985, A comparison of approximate theories for scattering from rough surfaces, *Wave Motion* **7**, 195-205.

Chang, A. C., and J. R. Cleary, 1978, Precursors to PKKP, *Seismol. Soc. America Bull.* **68**, 1059-1079.

Chapman, C. H., and J. A. Orcutt, 1985, The computation of body wave synthetic seismograms in laterally homogeneous media, *Rev. Geophysics* **23**, 105-163.

Doornbos, D. J., 1988, Multiple scattering by topographic relief with application to the core-mantle boundary, *Royal Astron. Soc. Geophys. Jour.* **92**, 465-478.

Doornbos, D. J., and J. C. Mondt, 1979, Attenuation of P and S waves diffracted around the core, *Royal Astron. Soc. Geophys. Jour.* **57**, 353-379.

Haddon, R. A. W., 1972, Corrugations on the mantle-core boundary or transition layers between inner and outer cores? *Am. Geophys. Union Trans.* **53**, 600.

Keller, J. B., 1962, Geometrical theory of diffraction, *Optical Soc. America Jour.* **52**, 116-130.

Kennett, B. L. N., 1983, *Seismic Wave Propagation in Stratified Media*. Cambridge: Cambridge University Press.

Larner, K., R. Chambers, M. Yang, W. Lynn, and W. Wai, 1983, Coherent noise in marine seismic data, *Geophysics* **48**, 854-886.

Phinney, R. A., and L. M. Cathles, 1969, Diffraction of P by the core: A study of long period amplitudes near the edge of the shadow, *Jour. Geophys. Research* **74**, 1556-1574.

Scholte, J. G. J., 1956, On seismic waves in a spherical earth, *Koninkl. Nederlandse Meteorol. Inst. Publ.* **65**, 1–55.

Trorey, A. W., 1970, A simple theory for seismic diffractions, *Geophysics* **35**, 762–784.

Cross-references: *Earthquake Seismology; Earth's Core: Structure; Earth Structure: Global; Elasticity and Wave Propagation: Principles; Elastic Waves in Homogeneous and Inhomogeneous Media; Mantle, Lower: Structure; Seismic Attenuation: Observations and Measurement; Seismic Ray Theory; Seismic Wave Scattering; Seismograms: Interpretation.*

SEISMIC IMAGING

Seismic imaging is the set of methods that obtain images of the Earth using observed seismograms as inputs. The images are well-defined representations of the Earth's interior, for instance, density, viscoelastic parameters, or "reflectivity." Three traditional methods of imaging are migration, linearized inversion, and tomography, and a new method is nonlinear inversion.

The most widely used methods of seismic imaging are migration techniques that obtain "reflectivity" images of the Earth using reflection seismograms. Reflection seismograms record ground motions corresponding to reflected seismic waves, i.e., seismic waves that have traveled down from a seismic source into the Earth to a discontinuity in the elastic parameters where they are reflected upward to the Earth's surface. Migration uses a knowledge of the kinematics of wave propagation to find where the reflections were generated in the Earth (see *Seismic Wave Field Migration*). One main pitfall is that the image does not represent a precise physical quantity but the reflection strength ("reflectivity"). A second pitfall is that a velocity model of the Earth describing the kinematics of wave propagation is required in order to obtain the image. Also, oversimplifications, such as assuming the Earth is a liquid or that seismic waves are not refracted while propagating in the Earth, are traditionally made to speed the computations.

Closely related to migration are the linearized inversion methods of seismic imaging. These methods seek an Earth model that predicts the observed seismograms. They overcome the first pitfall of migration in that they yield a precise physical quantity such as density or the elastic parameters. However, it is well known that, like migration, they will not obtain, but rather require, a kinematical description of the Earth. Furthermore, they frequently contain the same oversimplifications as migration. Another pitfall is that the seismograms are *not* linearly related to the Earth properties, and so the results of linearized inversion are inaccurate.

The linearized inversion methods can use either reflection or transmission seismograms, where transmission seismograms record waves that do not suffer reflection but are transmitted through the Earth. When transmission seismograms are used, the linearized inversions update the kinematical model, and when reflection seismograms are used, they obtain the reflector model. Note that travel time curves of reflections are known to resolve the kinematical model (e.g., using "velocity analysis"), so linearized inversions do have deficiencies.

A third imaging method related to the linearized inversion methods using transmission seismograms is tomography. The delays and advances of the transmitted seismic waves indicate decreases and increases in the average velocity of the transmitted wave. Tomographic methods project these decreases and increases back along the path of wave propagation to obtain a kinematical model of the Earth (wave-speed model). As with migration and linearized inversion, tomography requires a beginning kinematical model and it is frequently based on the same assumptions as the linearized inversion methods. It thus has the same pitfalls.

Nonlinear inversion is an alternative that theoretically overcomes the pitfalls of the traditional methods. Nonlinear inversion seeks the Earth model (values for the elastic parameters at each point of the Earth's interior) whose predicted seismograms best fit the observed seismograms with no linearity assumptions or oversimplifications. This task is difficult because seismograms depend critically on many factors (precise configuration of sources and receivers, near-surface effects, physics of propagation, etc.), which need to be modeled realistically.

The least squares criterion of goodness of fit leads to the simplest computations. This article shows that the problem can be solved by using an iterative method. Each iteration involves the solution to the wave equation with the actual sources in the current medium, the solution to the wave equation backwards in time with the current residuals (difference between observed and currently predicted seismograms) as sources, and the correlation at each point of space of the two wave fields thus obtained.

The Forward Problem

The *forward problem* attempts to predict the seismograms corresponding to a given Earth model.

Let $\mathbf{x}$ denote a point in the Earth's interior and ξ a point on the Earth's surface. We describe an Earth model using the density $\rho(\mathbf{x})$ and elastic stiffness $c^{ijkl}(\mathbf{x})$, and assume a seismic wave source (vibrator) on the Earth's surface. The ith component of displacement at time t of a particle whose rest position is $\mathbf{x}$ is denoted by $\vec{u}(\mathbf{x}, t)$. The displacement field $\vec{u}(\mathbf{x}, t)$ is uniquely defined by the system of equations (see *Elasticity and Wave Propagation: Principles*)

$$\rho(\mathbf{x}) \frac{\partial^2 \vec{u}^i}{\partial t^2}(\mathbf{x}, t) - \frac{\partial}{\partial x^j}\left(c^{ijkl}(\mathbf{x}) \frac{\partial \vec{u}^k}{\partial x^l}(\mathbf{x}, t)\right) = 0$$

$$n^j(\xi) c^{ijkl}(\xi) \frac{\partial \vec{u}^k}{\partial x^l}(\xi, t) = \tau^i(\xi, t)$$

$$\vec{u}^i(\mathbf{x}, 0) = 0$$

$$\frac{\partial \vec{u}^i}{\partial t}(\mathbf{x}, 0) = 0 \qquad (1)$$

where $n^j(\xi)$ is the outward unit normal at surface point ξ and the time coordinate t belongs to the interval $[0, T]$.

Given the variables $\rho(\mathbf{x})$ and $c^{ijkl}(\mathbf{x})$, the displacement field $\vec{u}^i(\mathbf{x}, t)$ is computed by solving equation system 1 with a numerical method, such as finite differences (e.g., Kosloff et al., 1984). Once the field $\vec{u}^i(\mathbf{x}, t)$ is known for any $\mathbf{x}$, the seismograms, $\vec{u}(\xi, t)_{cal}$, can be obtained by simply extracting its surface values.

The Inverse Problem

Let $\vec{u}^i(\xi, t)_{obs}$ be the observed displacements. The aim of least squares inversion is to obtain the Earth model for which the misfit function

$$E = \int_0^T dt \int_S dS(\xi) \, [\vec{u}^i(\xi, t)_{cal} - \vec{u}^i(\xi, t)_{obs}]^2$$

is a minimum. The misfit function E is not a quadratic function of the parameters describing the Earth (density and elastic stiffness), because the calculated displacements do not depend linearly on them. Thus, the minimization of E is not an easy problem. Ideally, fully nonlinear methods such as Monte Carlo methods (i.e., methods, based on random generation of models until some model with a sufficiently low value of the misfit function is obtained) should be used. But present day computers are too slow, and Monte Carlo methods cannot be used for realistic problems. Gradient iterative methods have the advantage that a few iterations can be performed on present day computers, but the convergence rate is so slow that it is not possible, in general, to iterate long enough to achieve convergence.

In fact, the strongest nonlinearity of the problem relates the calculated seismograms to the *long wavelengths* of the medium; and if these long wavelengths are given by an ad hoc estimation, then gradient methods perform fairly well. The problem has been *linearized*.

The simplest gradient method is the method of steepest descent: The current model is updated by a finite amount in the direction (in the model space) of steepest descent of the misfit function at the position of the current model. The analytical development required to obtain the directions of steepest descent for the density and elastic stiffness is lengthy but not difficult (Tarantola, 1987); it uses the concepts of functional analysis instead of matrix algebra because the problem is better cast as that of obtaining the *functions* $\rho(\mathbf{x})$ and $c^{ijkl}(\mathbf{x})$ rather than finite-dimensional representations of these functions. The result is an iteration of the steepest descent method (Tarantola, 1984) given by

$$\rho(\mathbf{x})_{n+1} = \rho(\mathbf{x})_n - \alpha_n \int_0^T dt \, \frac{\partial \vec{u}^i}{\partial t}(\mathbf{x}, t)_n$$
$$\cdot \frac{\partial \bar{u}^i}{\partial t}(\mathbf{x}, t)_n$$

$$c^{ijkl}(\mathbf{x})_{n+1} = c^{ijkl}(\mathbf{x})_n - \alpha_n \int_0^T dt \, \frac{\partial \vec{u}^i}{\partial x^j}(\mathbf{x}, t)_n$$
$$\cdot \frac{\partial \bar{u}^k}{\partial x^l}(\mathbf{x}, t)_n \qquad (2)$$

In these equations, $\vec{u}^i(\mathbf{x}, t)_n$ denotes the currently predicted field obtained by extrapolating waves *forward* in time using the known seismic sources. The field $\bar{u}^i(\mathbf{x}, t)_n$ is named the *missing field* and is defined as the field obtained by extrapolating *backward in time* null final conditions in the current medium, using the current data residuals as sources. The constant α_n is small enough to ensure a decrease of the misfit function value from the current iteration to the next.

More precisely, the field $\vec{u}(\mathbf{x}, t)_n$ is the solution of the (forward) time extrapolation problem

$$\rho(\mathbf{x})_n \frac{\partial^2 \vec{u}^i}{\partial t^2}(\mathbf{x}, t)_n$$
$$- \frac{\partial}{\partial x^j}\left(c^{ijkl}(\mathbf{x})_n \frac{\partial \vec{u}^k}{\partial x^l}(\mathbf{x}, t)_n\right) = 0$$

$$n^j(\xi) c^{ijkl}(\xi)_n \frac{\partial \vec{u}^k}{\partial x^l}(\xi, t)_n = \tau^i(\xi, t)$$

$$\vec{u}^i(\mathbf{x}, 0)_n = 0$$

$$\frac{\partial \vec{u}^i}{\partial t}(\mathbf{x}, 0)_n = 0$$

and the field $\bar{u}^i(\mathbf{x}, t)_n$ is the solution of the (backward) time extrapolation problem

$$\rho(\mathbf{x})_n \frac{\partial^2 \bar{u}^i}{\partial t^2}(\mathbf{x}, t)_n - \frac{\partial}{\partial x^j}\left(c^{ijkl}(\mathbf{x})_n \frac{\partial \bar{u}^k}{\partial x^l}(\mathbf{x}, t)_n\right) = 0$$

$$n^j(\xi) c^{ijkl}(\xi)_n \frac{\partial \bar{u}^k}{\partial x^l}(\xi, t)_n = \vec{u}^i(\xi, t)_n - \vec{u}^i(\xi, t)_{obs}$$

$$\bar{u}^i(\mathbf{x}, T)_n = 0$$

$$\frac{\partial \bar{u}^i}{\partial t}(\mathbf{x}, T)_n = 0$$

(notice that there are *final* time conditions instead of *initial* time conditions).

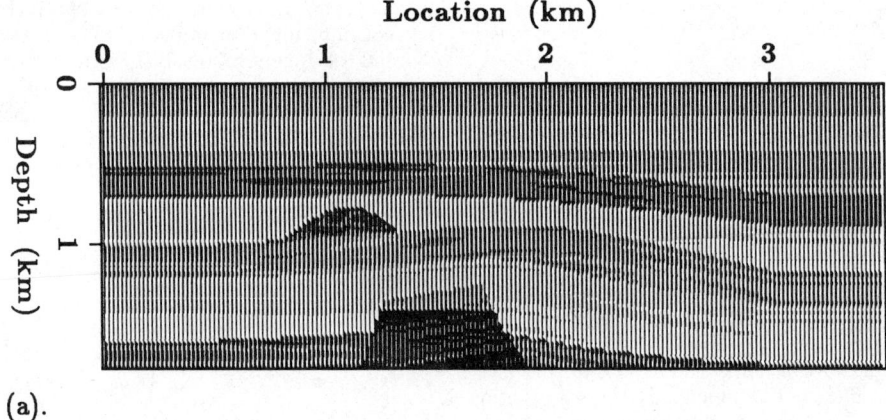
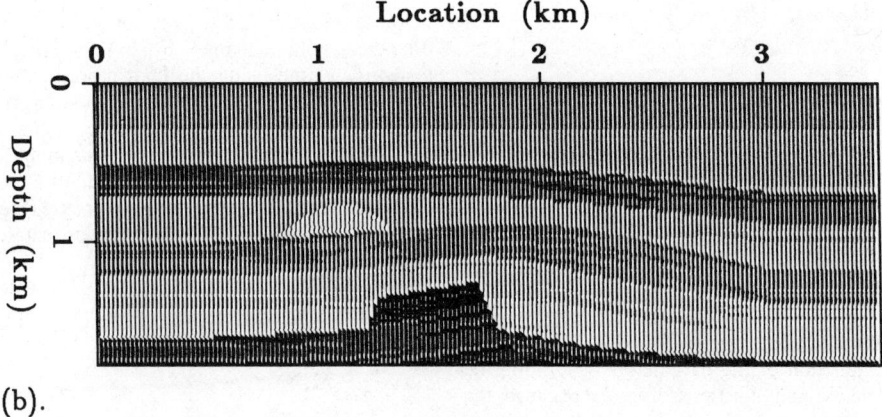
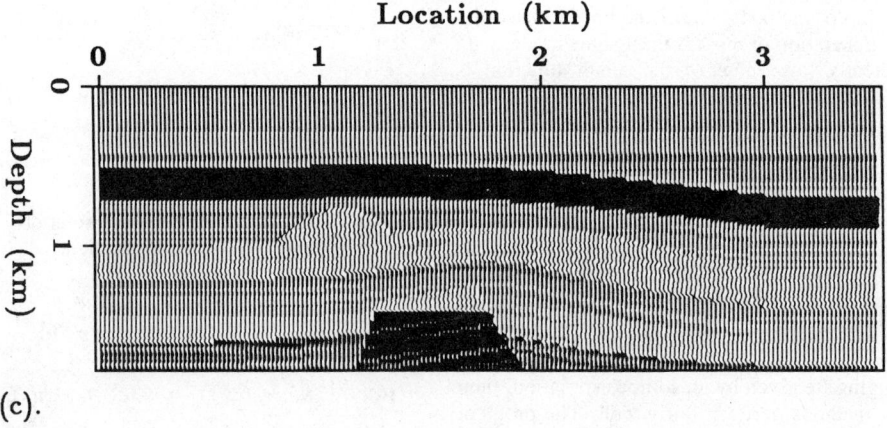

FIGURE 1. Horst and reef model used to generate the synthetic data shown in Figs. 2 and 3: (a) P-wave velocity; (b) S-wave velocity; (c) density.

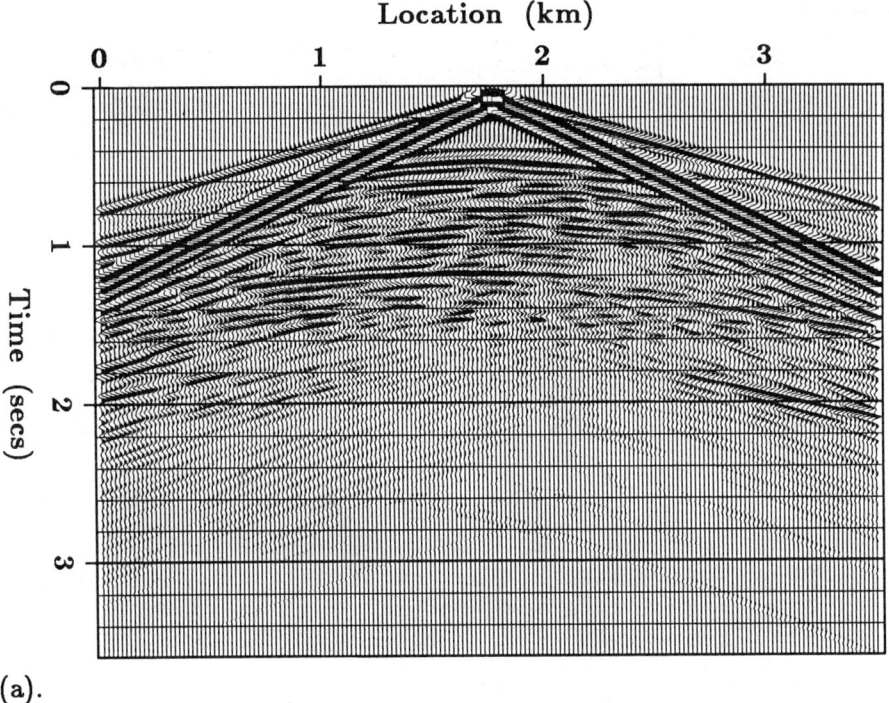

(a).

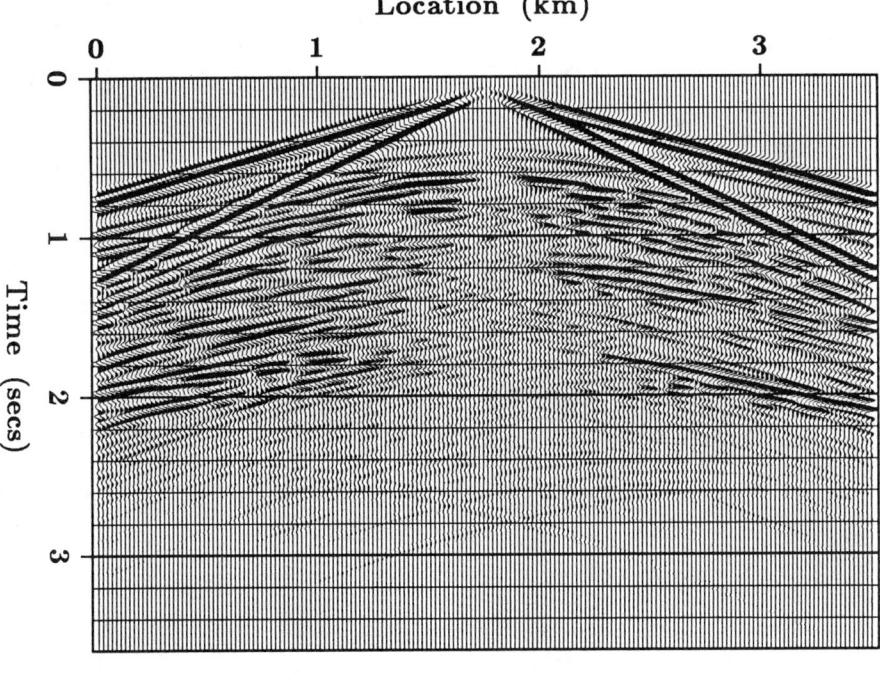

(b).

FIGURE 2. One of the split spread shot profiles used in the inversion. It is due to a vertical source located at 1.76 km and was generated by elastic finite differences with the model shown in Fig. 1. This synthetic data is noise free: (a) vertical component; (b) horizontal component.

The intuitive interpretation of Eq. 2 is that if there is a time correlation between the velocity of the missing field and the velocity of the current field, a density diffractor must be added to the model to decrease the misfit; if there is a time correlation between the strain associated with the missing field and the strain associated with the current field, the elastic stiffness must be modified to decrease the misfit.

With this method, nonlinear inversion of seismic reflection data requires computations that are quite similar to those realized today in accurate prestack migration. They are more expensive because the process is iterative, and the full elastodynamic wave equation is used to model seismic waves instead of the less realistic acoustic wave equation.

Although the equations given here are for three-component data, the restriction to vertical component data is straightforward.

Numerical Results

The following is an example developed by Mora (1988). The medium is assumed isotropic, and the compressional wave (P-wave) velocity

$$\alpha(\mathbf{x}) = \sqrt{\frac{\lambda(\mathbf{x}) + 2\mu(\mathbf{x})}{\rho(\mathbf{x})}}$$

and the shear wave (S-wave) velocity

$$\beta(\mathbf{x}) = \sqrt{\frac{\mu(\mathbf{x})}{\rho(\mathbf{x})}}$$

have been used instead of the elastic stiffness $c^{ijkl}(\mathbf{x})$ ($\lambda(\mathbf{x})$ and $\mu(\mathbf{x})$ denote the Lamé parameters, which are related to the elastic stiffness by $c^{ijkl} = \lambda \delta^{ij} \delta^{kl} + \mu[\delta^{il}\delta^{jk} + \delta^{ik}\delta^{jl}]$). This leads to only minor changes in the equations (Mora, 1987a; Tarantola, 1987).

To simplify computations, we use an absorbing surface boundary condition instead of a more realistic free surface boundary condition. Furthermore, the size of the velocity and density perturbations has been deliberately chosen to be small (only a few percent relative to a linear with depth initial model) so that the inverse problem would be more linear and hence convergence more rapid. Hence, the results will show how it is possible to do inversions when the initial velocity is known fairly accurately, which is often the case where detailed velocity analyses were carried out. The results also illustrate that inversion can obtain the S-wave velocity as well as the P-wave velocity if the observed seismograms contain S-wave events such as mode conversions, direct S waves, and S-S reflections as well as the usual P-wave events.

The synthetic data were calculated from the model shown in Fig. 1 by a finite-difference approximation to the elastic wave equation (a version of Kosloff et

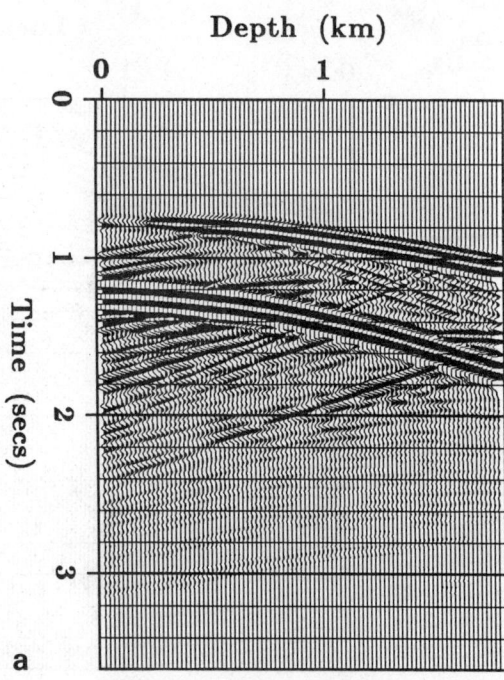

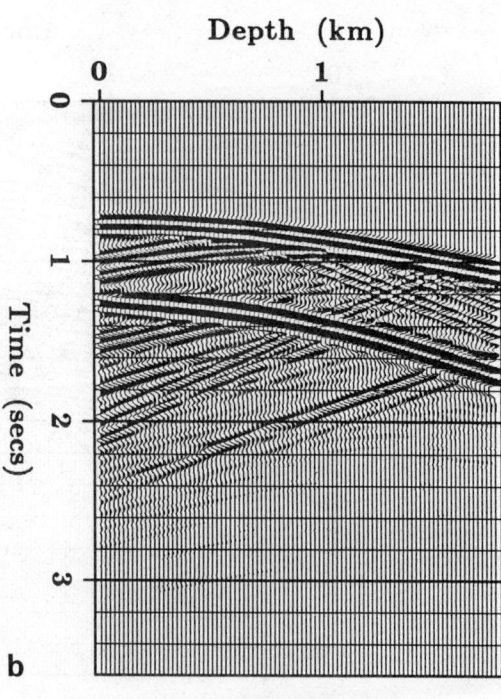

FIGURE 3. One of the offset VSPs used in the inversion. It corresponds to a well located at 0.0 km due to a vertical source located at 2.26 km. It was generated by elastic finite differences with the model shown in Fig. 1. This synthetic data is noise free: (*a*) vertical component; (*b*) horizontal component.

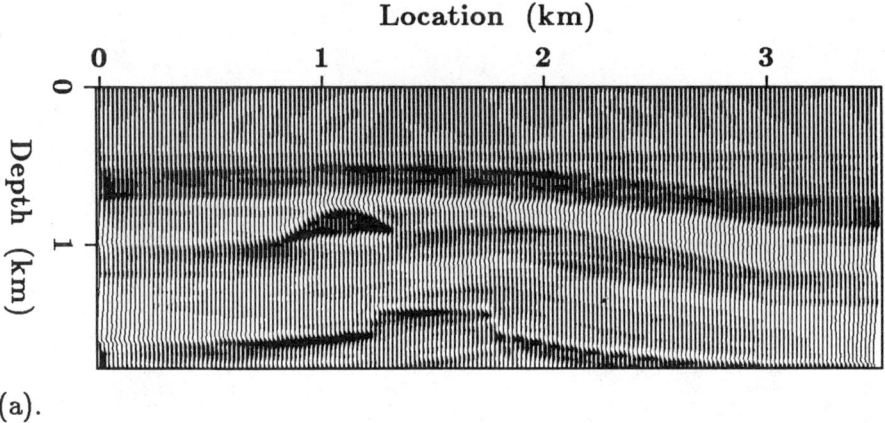

(a).

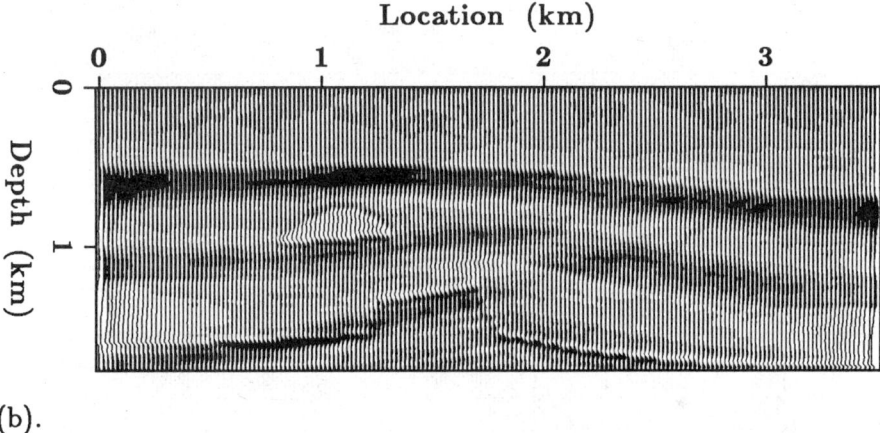

(b).

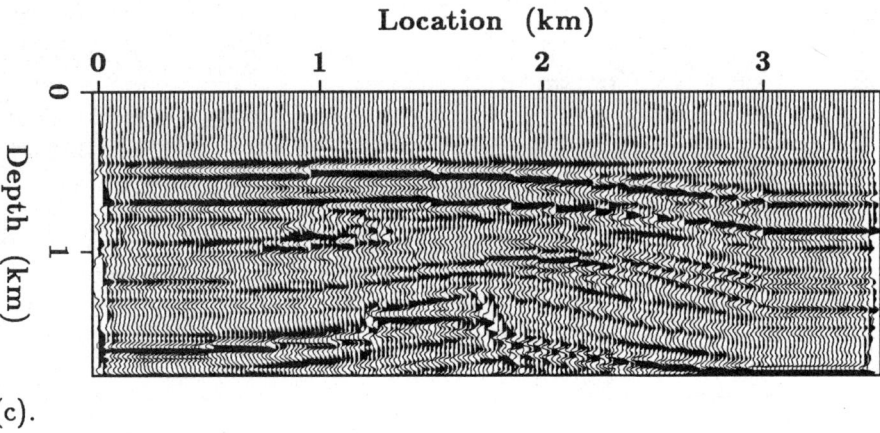

(c).

FIGURE 4. Inversion results after 10 iterations: (a) P-wave velocity; (b) S-wave velocity; (c) density. Note that these plots are not at the same scale as Fig. 1.

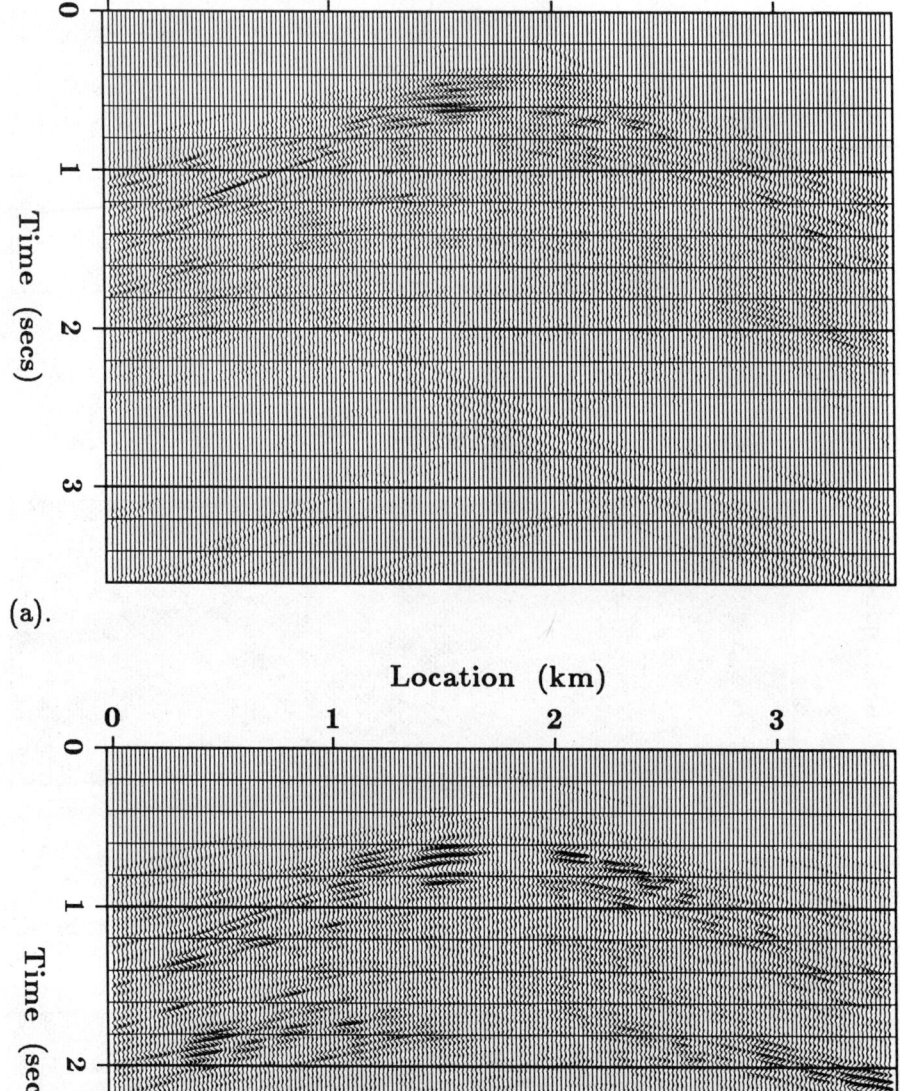

FIGURE 5. Unmatched part of the shot profile of Fig. 2 (i.e., the residual) after 10 iterations of the inversion algorithm. It is plotted at the same scale as Fig. 2: (a) vertical component; (b) horizontal component.

al., 1984, with space rather than Fourier domain spatial derivative operators). This model depicts a fairly complex geology that may be encountered in a sedimentary basin consisting of a horst structure and limestone reef complex amidst a multitude of larger layers containing fine layer structure. It is parameterized by the three isotropic elastic parameters (P- and S-wave velocity and density) on a square grid with a grid spacing of 20 m.

Nine shot simulations were carried out, with shots spaced every 0.5 km starting at -0.24 km (just to the left of the displayed part of the model). A source wavelet with a similar band limitation to real seismic wavelets was used (a fourth derivative of a Gaussian curve with a fundamental frequency of 20 hz). The surface geophone array was fixed and covered the 3.52-km extent of the model. There were also two wells at 0.0 km and 3.52 km that recorded vertical seismic profiling (VSP) data for these shots. The geophone spacing on the Earth's surface and down the wells was the same as the grid spacing. Two-component receivers were used (P-SV geophones), and a vertical force was applied. A representative shot profile due to a shot located at 1.76 km is shown in Fig. 2, and a typical offset VSP recorded in the well at 0.0 km due to the shot at 2.26 km is shown in Fig. 3.

The inversion was carried out with a conjugate gradient algorithm. The same method of elastic finite differences was used to compute both the forward modeled wave field $\vec{u}^i$ and the back-propagated residual wave field $\bar{u}^i$. The data recorded for all nine shots were used simultaneously in the inversion. This included nine two-component shot profiles, nine two-component offset VSPs from the well located at 0.0 km, and nine two-component offset VSPs from the well located at 3.52 km. The data set contains both reflected and transmitted waves. The reflections are produced by the short-wavelength perturbations in the P- and S-wave velocities and density, so reflections will help resolve the high-wave-number part of the model. By comparison, the transmitted waves (direct waves in the VSPs) are most affected by long-wavelength velocity anomalies that cause travel time delays and advances, so they tend to help resolve the low-wave-number part of the model. Note that the limitation that reflections tend to resolve mainly high wave numbers may be overcome in principle. This can be achieved by splitting the inversion algorithm into two components that perform migration-like operations and reflection-tomographic-like operations and applying some different scaling to the components (Mora, 1989a).

The inversion result for P-wave velocity, S-wave velocity, and density is shown in Fig. 4. Note that the P- and S-wave velocity results are close to the true model, whereas the density result is only a high-wave-number approximation of the true density model. The reason is that the low-wave-number part of the P- and S-wave velocities could be resolved by

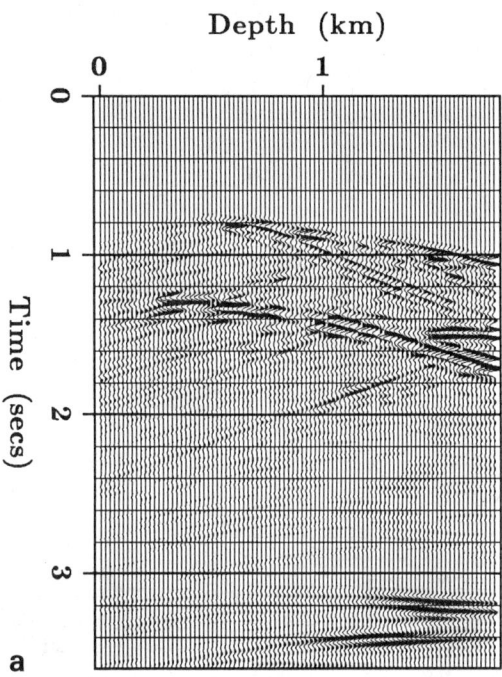

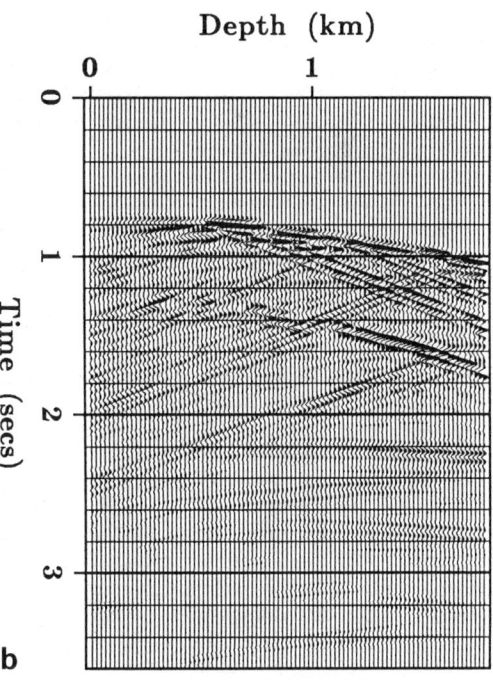

FIGURE 6. Unmatched part of the VSP of Fig. 3 (i.e., the residual) after 10 iterations of the inversion algorithm. It is plotted at the same scale as Fig. 2: (a) vertical component; (b) horizontal component.

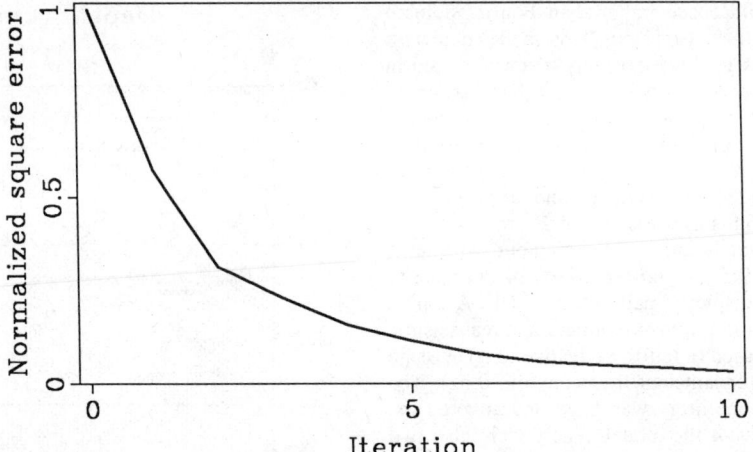

FIGURE 7. The normalized sum of the square error (misfit function) as a function of iteration number.

travel time anomalies in the direct waves, whereas low-wave-number density fluctuations have little effect on the seismic waves and so cannot be resolved. Notice that, just as in the true model, the flat spot at the top of the horst (gas-water contact) can be seen on the P-wave velocity result but does not exist on the S-wave velocity result. This difference illustrates that there is good resolution between the P- and S-wave velocities. Also observe that the the reef can be seen in the inversion results and has the correct polarity of P- and S-wave velocity perturbations. Figures 2 and 3 show the data residuals (unmatched part of the wave field) after ten iterations for the gather and VSP, respectively, shown in Figs. 1 and 2. A graph of the square error sum (misfit) as a function of iteration is shown in Fig. 7, indicating that the amount of mismatched energy in the seismic wave field steadily decreases as the least squares iterations proceed.

Comments

Waves propagate in a three-dimensional Earth, and, even if the geology is invariant in the direction perpendicular to the survey line, accurate modeling of amplitudes can be obtained only by using three-dimensional models. Finite-difference approximations to the three-dimensional wave equation are possible but expensive with modern vector computers. The new generation of parallel computers will probably make the task feasible, as demonstrated by the three-dimensional calculations of elastic waves in the whole earth of Mora (1989b). For the time being, some ad-hoc corrections in the observed amplitudes may account for part of the effect of using a two-dimensional modeling theory for inversions. But correct inversion is highly dependent on forward modeling, and this limitation could be major. The method presented here is just now beginning to be applied to real data (Mora, 1987b; Crase et al., 1989), but considerable research and testing remains. With increasing computer power it will probably replace traditional methods.

PETER MORA
ALBERT TARANTOLA

References

Crase, E., A. Pica, M. Noble, and A. Tarantola, 1989, Robust elastic nonlinear inversion of seismic waveforms: application to real data, submitted to *Geophysics*.

Kosloff, D., M. Reshef, and D. Loewenthal, 1984, Elastic wave calculations by the Fourier method, *Seismol. Soc. America Bull.* **74,** 875–891.

Mora, P., 1987a, Nonlinear 2D elastic inversion of multi-offset seismic data, *Geophysics* **52,** 1211–1228.

Mora, P., 1987b, Nonlinear elastic inversion of real data, Soc. Exploration Geophysicists 57th annual meeting and exposition, Oct. 11–15, 1987, Proceedings.

Mora, P., 1988, Elastic wavefield inversion of reflection and transmission data, *Geophysics* **53,** 750–759.

Mora, P., 1989a, Inversion = migration + tomography, rev., *Geophysics* **54.**

Mora, P., 1989b, Wave-equation modeling 3D anisotropic waves in the Earth, AGU Fall meeting, Dec. 5–9, 1988, Trans., *EOS*, 70.

Tarantola, A., 1984, The seismic reflection inverse problem, in F. Santosa, Y. H. Pao, W. Symes, and Ch. Holland, eds., *Inverse Problems of Acoustic and Elastic Waves*. Philadelphia: SIAM.

Tarantola, A., 1987, *Inverse Problem Theory. Methods for Data Fitting and Model Parameter Estimation*. Amsterdam: Elsevier.

Cross-references: *Controlled Source Seismology; Deep Seismic-Reflection Profiling; Elasticity and Wave Propagation: Principles; Elastic Waves in Homogeneous and Inhomogeneous Media; Exploration Seismology; Finite-Difference Forward Modeling in*

Seismology; Inverse Theory and Methods: Seismology; Seismic Diffraction; Seismic Ray Theory; Seismic Signal Processing; Seismic Tomography; Seismic Wavefield Migration; Seismic Wave Scattering; Seismology: Physical Model Studies.

SEISMIC INSTRUMENTATION

All but the simplest seismic instruments are designed to record the time history of ground motion. For discussing such instruments, it is useful to break them down into three parts (although these may be combined in an actual instrument). First there is a sensor, which converts ground motion into some other quantity, usually an electrical one; this sensor is called a seismometer, or a geophone in exploration seismology. Next, the sensor output is modified, usually by amplifying some frequencies and suppressing others; very often such modification is done partly within the sensor itself. Third, the modified output is recorded, along with a time scale (and in most cases absolute time). With the growth first of analog and then of digital electronics the systems used for the signal modification and recording have increasingly often been adapted from devices built for other purposes, rather than being purely seismological inventions. As a consequence, seismic instrument design has become less and less the province of the seismologist and more and more that of the electrical engineer.

Seismic Instrumentation: History treats the historical development of seismometry: Figure 1 summarizes the development quantitatively by showing the magnification (record/ground displacement) curves for different early instruments. Figure 2 shows similar curves for several more modern instruments using analog recording, and Fig. 3 shows the equivalent amplitude response curves for digitally-recording instruments. For digital instruments, of course, the concept of magnification must be replaced with the digital units for given ground motion.

These figures show that there has been a steady push toward higher and higher gain, limited only by ground noise (see *Seismic Noise*); the one exception has been instruments to record strong ground shaking. With the increase in gain came a split into short-period and long-period sensors, for recording at frequencies 1–10 and 0.01–0.1 Hz respectively; seismic systems were designed to filter out the high microseism noise in the intermediate-period band 0.1–1 Hz. Very recently the increased dynamic range of recording has made possible high-gain responses that are the same over a broad band of frequencies. Another trend in instrument design has been the increasing use of smaller and lighter sensors, partly for easier installation, and partly to take advantage of the reduced noise in a borehole.

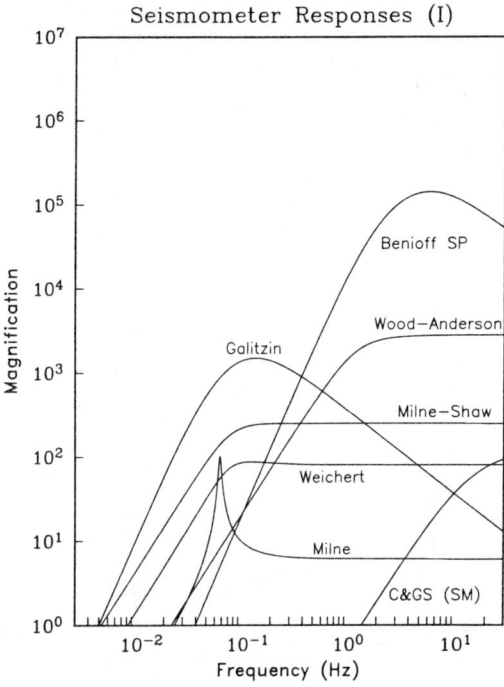

FIGURE 1. Magnification curves for seismic systems of special historical interest. Milne's instrument was the first designed specifically to record unfelt earthquake waves; the Milne-Shaw was developed from it. The Weichert instruments were among the first with adequate damping, and the 125-kg model was widely used. Galitzin's instrument was the first to use electromagnetic sensors and recorders. The Wood-Anderson instrument, the first designed for recording small local earthquakes, is the basis of the local magnitude scale. The Benioff short-period served as the basis for many later teleseismic instruments. The C&GS instrument was the first useful strong-motion recorder.

Sensors

Almost all seismometers contain a "proof mass" (of mass m) that is loosely coupled to the instrument frame and measure the relative displacement of the two. This displacement is due to the inertia of the mass, and these sensors are called "inertial." The frame is usually assumed to move with the ground, an approximation that can break down at high frequencies (Krohn, 1984) or in soft materials (Prothero, 1984). The sensor is designed so that the mass can move only in one direction (usually but not always vertically or horizontally [Melton, 1976]). For a simple mass-spring system, the force exerted on the mass by the frame is the sum of a restoring force $-kq$ and a velocity-dependent damping force $-d\dot{q}$, where q is the displacement of the mass from its equilibrium position, measured relative to the frame, and k and d are instrumental constants.

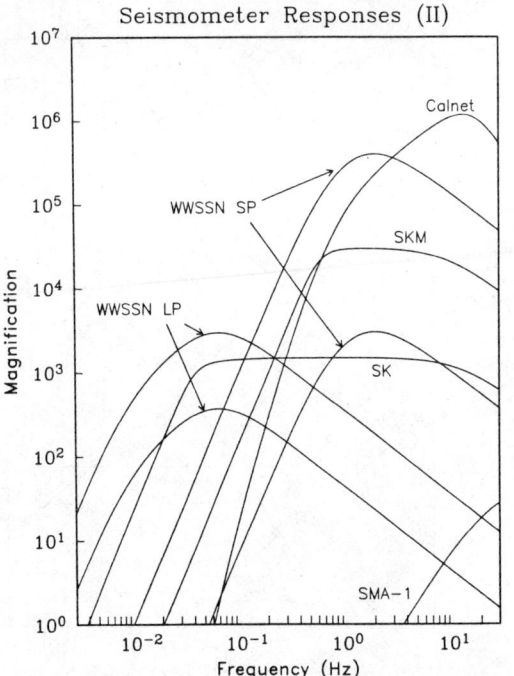
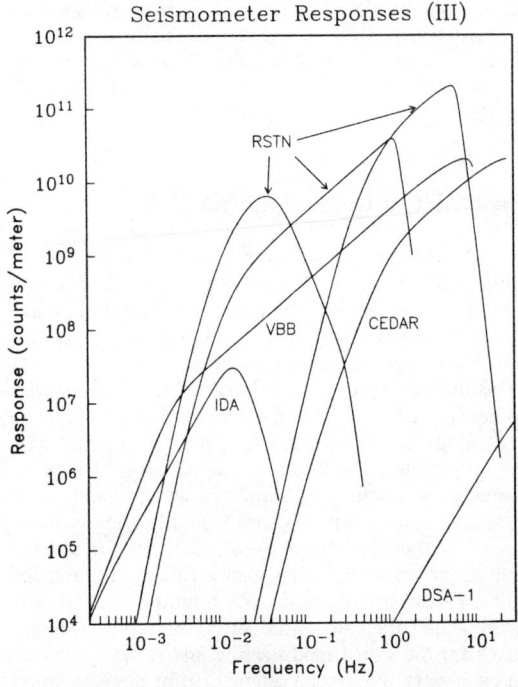

FIGURE 2. Magnification curves for seismic instruments with analog recording in use around 1970. The WWSSN curves show the range of short-period (SP) and long-period (LP) magnifications used in this global network. The Kirnos instruments (SK and SKM) are used in the Soviet Union. The Calnet curves show the magnification on instruments of the local network in central California. The SMA-1 curve shows the typical response of a modern analog strong-motion recorder.

FIGURE 3. Response curves (vs. ground displacement) of several digitally-recording seismic systems. (For a plotting scale of 0.1 mm/count, a response of 10^4 counts/meter would be a magnification of 1). The IDA instruments were an early global digital network, designed for ultra-long-period recording. The RSTN is a recent system, recorded in three bands. The CEDAR curve shows the response of the digital local network for Southern California. The curve marked VBB shows the response of a system that uses feedback to obtain a uniform response (flat to ground velocity) over a wide range of frequencies. The DSA-1 curve is the response of a digital strong-motion system.

Relative to inertial space the displacement of the mass is $q + x$ where x is the ground motion; Newton's law then becomes $-kq - d\dot{q} = m(\ddot{q} + \ddot{x})$ or

$$\ddot{q} + 2\gamma\omega_0\dot{q} + \omega_0^2 q = -\ddot{x}$$

which is usually called the indicator equation, $\omega_0 = (k/m)^{1/2}$ being the natural frequency and $\gamma = d/2(km)^{1/2}$ the damping factor. If $\gamma \ll 1$ (underdamped) the mass responds primarily to motions with frequencies near ω_0, so usually γ is designed to be near 1 ($\gamma = 1$ is critical damping).

The period of a seismometer ($T_0 = 2\pi/\omega_0$) is often used to describe it as short-period, long-period, or intermediate-period, the frequency ranges being the ones given earlier for seismic waves. Until recently the best records of seismic waves with a given frequency would be obtained using a sensor with about the same frequency, and a long-period seismic system would be both one intended to record long-period waves and one that used a long-period seismometer. The increasing use of feedback (see later) has made the performance of a seismometer less dependent on its period.

The indicator equation shows that an inertial sensor responds to ground acceleration; at periods greater than several hundred seconds, the response may also include the effects of tilt and changes in gravity (Gilbert, 1980; Rodgers, 1968). For ground motions with frequencies much lower than ω_0, we have $q \approx -\ddot{x}/\omega_0^2$, and the displacement of an inertial sensor looks like ground acceleration, leading it to be termed an accelerometer. For motions with frequencies much higher than ω_0, $q \approx x$ and the sensor could be called a displacement meter. These are only approximations; no inertial sensor can give a perfect reproduction of ground motion.

Much of the mechanical ingenuity displayed in seismometer design has gone into making the restoring force exactly proportional to mass motion. This proportionality is simply achieved in instruments with high ω_0 (short-period seismometers) through elastic elements, either a simple helical spring or a strip in torsion. Conventional springs are too stiff to give a small value of ω_0, so, for long-period seismometers, other mass suspensions are needed. For vertical instruments the preferred solution has been the LaCoste spring suspension (Melton, 1971), which can provide (in principle) an arbitrarily small k; for horizontal instruments the horizontal pendulum ("garden gate") suspension does the same thing by using a small part of gravity for the restoring force. (Both of these require the mass to move in a circle, so that q is strictly speaking an angle, but for q small the indicator equation remains valid.)

In the earliest seismometers the mass motion was magnified mechanically and recorded directly. All modern instruments use a transducer to convert mass displacement q into some other quantity. Of all the displacement transducers available (Agnew, 1986), the most commonly used in seismometry is the moving-coil electromagnetic transducer introduced by Galitzin, in which the motion of a wire coil in a magnetic field generates a voltage V proportional to mass velocity: $V = G\dot{q}$. The constant of proportionality G is usually called the "generator constant," with dimensions $V\text{-}s/m$ or N/A. (The Benioff variable-reluctance transducer gets the same result in a slightly different way). These systems have the advantage of needing no external power and the disadvantages of decreasing response at lower frequencies and the large mass of the permanent magnets. Most recent designs for teleseismic seismometers employ an active transducer, which modulates an electrical input signal to produce an output proportional to mass displacement. Both capacitive and inductive transducers (Agnew, 1986) have been used.

Whereas in many cases the limits on what a seismometer records are set by local ground noise (see *Seismic Noise*), there are also fundamental limits on what an inertial sensor can detect. These limits arise from the damping applied to the mass: Random fluctuations in the damping element (whether a surrounding fluid or a resistor) cause random forces on, and displacements of, the mass. The randomness of the forces means that they appear equivalent to random ground accelerations (a flat power spectrum of ground acceleration). The mass displacement must be such that the mass is in thermal equilibrium with its surroundings. It may be shown that these conditions imply a noise level of $8\omega_0\gamma k_B T/m$, where T is the temperature and k_B is Boltzmann's constant; at room temperature $k_B T = 4 \cdot 10^{-21}$ J. For light short-period seismometers the resulting noise level can be above the lower limits of ground noise unless $\gamma \ll 1$ (Melton, 1976).

Amplification and Filtering

In the earliest seismic systems, amplification was carried out using levers, either mechanical or optical. Much greater gain and a more flexible response are possible if a galvanometer is used to convert the electrical output of a seismometer back into a displacement of the galvanometer mirror. Like the moving-coil transducer the galvanometer requires no external power, and it also has much less noise at low frequencies than any other amplifier (Melton, 1976). Its major disadvantage is fragility. The conventional procedure has been to connect the galvanometer to an electromagnetic seismometer through a resistor network: The theory for the response of such a system is given by Chakrabarty and Choudhury (1964) and is summarized in Willmore (1979). By suitable choice of constants it is possible to obtain a wide range of responses (e.g., the WWSSN and SK curves in Fig. 2 and the Benioff and Galitzin curves in Fig. 1).

As the cost, power consumption, and noise levels of electronic amplifiers have fallen, they have become widely used to process the voltages from a seismometer. All amplifiers become noisier at low frequencies, so they are used most easily in short-period systems (e.g., networks for local-earthquake studies). Such amplifiers possess a much wider range of responses than the classical galvanometric systems do.

Many designs for seismometers now use feedback, in which the output of the mass-position transducer is amplified, filtered, and put into a system that converts it into an additional force on the seismometer mass. This procedure also allows a wide range of responses to be obtained (Plešinger, 1984): For example, by applying a force proportional to $\ddot{q}$, the apparent mass of the instrument may be increased and the period lengthened. One major advantage of feedback is that the response is determined by electronic components, and thus can be more stable than a purely mechanical system. A second advantage is that damping applied through feedback does not introduce the thermal noise that passive damping must. With more use of smaller, lighter instruments, such feedback is required to keep noise levels acceptably low.

Recording

So many techniques are used for recording that it is also hard to summarize adequately. One special aspect should be noted: To a degree unequaled by other geophysical observations, seismological data gathering has a combined need for detailed time resolution (because of the frequencies of seismic waves) and for continuous recording (because of the infrequent occurrence of earthquakes). This combination has always exerted a strong pressure toward compactness, and continues to bedevil instrument designers.

Until very recently nearly all seismic systems wrote visible records (hence the term *seismograph*). The most popular early techniques were mechanical recording, with a stylus scratching a trace on soot-covered paper, and optical recording, with a light spot moving over photographic paper or film. Smoked-paper recording, originally used to get low friction (important in mechanical instruments), remains popular for its reliability under rugged conditions. Photographic recording allows the use of optical levers for magnification, and is, of course, especially suited for use with galvanometers; it has been the most widely used. Recording several traces in parallel on film (Melton, 1981) was popular at one time because of its compactness. Active electronics can be used to drive a relatively heavy recording pen and thus record with ink on conventional paper or with a heated stylus on heat-sensitive paper.

All these visual recording techniques are now being replaced by nonvisual recording. Such recordings can draw on developments in audio and computer technology and have become increasingly appropriate since the digital computer has joined the eye of the seismologist as the major tool for analyzing a seismic record. A number of systems for recording on magnetic tape have been devised, but only digital systems have been widely used. In these systems, the voltage from the amplifier and filter is sampled at regular intervals and converted into a number; the resulting sequence of numbers is then recorded. The important specifications of a digital recorder are the sampling interval Δt, the voltage corresponding to a change of one unit, and the ratio of largest to smallest number that can be recorded. This last specification, termed the dynamic range, is often expressed in decibels (dB), which are $20 \log_{10}$ of the ratio. A digital system will record inaccurately signals with significant energy above the Nyquist frequency $(2\Delta t)^{-1}$ or that are bigger than the largest number or smaller than one unit. The frequencies of seismic waves are so low that it is easy to sample often enough, but adequate dynamic range is more difficult to get.

Many techniques exist for storing the digital sequence once it has been produced (Goldstein, 1984). In this area technology is advancing so rapidly that generalizations quickly become obsolete. The most notable advance in recent years has been the development of portable low-power systems, generally using microprocessor control. Some of the first systems were developed for use in ocean-bottom seismology (Prothero, 1984) but have since become commercially available for use on land as well. Such systems have made digital measurements possible in the field and have enormously expanded the amount of high-quality data available in local earthquake studies and controlled-source seismology.

Whatever recording technique is used should also make it possible to determine the absolute time of the data, at least to within 0.1 second. For long a major problem in seismic recording, this task has become relatively straightforward with the advent of crystal-controlled clocks and broadcast and satellite time signals (Kamas and Howe, 1979); with the use of the latter, accuracies of better than 0.001 second are possible. In networks for recording local earthquakes (Lee and Stewart, 1981), it has been more customary to solve the timing problem by telemetering all the signals to a central recorder, which also simplifies the difficulties of handling large masses of data.

Seismic Arrays

The techniques described earlier (especially for older seismographs) aim to maximize the signal-to-noise ratio by selecting and rejecting frequencies. Another way to do this is to combine the outputs of spatially separated seismometers: If the signal is the same on separate instruments but the noise is not, then addition of outputs will enhance the signal-to-noise. Such spatial arrays can also determine the apparent speed of propagation and direction of arrival of a seismic wave.

Arrays of seismometers are routine in exploration seismology (where they are called "spreads"): These are usually set along a line. In earthquake seismology arrays have been fewer and have been used largely for seismic discrimination between earthquakes and explosions (see *Seismic Monitoring of Nuclear Explosions*). All such arrays have had areal coverage to accept waves from any azimuth; their sizes have ranged from a few km for the "Geneva-type" arrays to 200 km for the Large Aperture Seismic Array (LASA). Such arrays have in fact been shown to increase signal-to-noise and hence detectability of small events, although the optimal size and configuration remain areas of study (Douglas, 1982; Filson, 1975).

Special Topics

The recording of ground motion in the meizoseismal area has generally been done with a special class of instruments known as strong-motion recorders. These have a very short free period (about 0.04 sec) so that they act as accelerometers. They thus record directly the quantity of greatest interest to earthquake engineers: Recording acceleration also minimizes the dynamic range required.

Rather than measuring seismic waves with an inertial sensor, we may instead use a strain seismometer (or strainmeter), which responds to the changes in distance between nearby points (deformation) during seismic waves. None of the many strainmeter designs that have been developed (Agnew, 1986) have been useful for more than limited seismological purposes. Another alternative to the inertial sensor is to measure ground tilts directly using a gyroscope as a reference, but no available gyroscope is stable enough to be seismologically useful.

DUNCAN CARR AGNEW

References

Agnew, D. C., 1986, Strainmeters and tiltmeters, *Rev. Geophysics* **24**, 579-624.

Chakrabarty, S. K., and S. N. R. Choudhury, 1964, Response characteristics of electromagnetic seismographs, *Seismol. Soc. America Bull.* **54**, 1445-1458.

Douglas, A., 1982, Two problems in forensic seismology twenty years on, *Royal Astron. Soc. Quart. Jour.* **23**, 26-44.

Filson, J., 1975, Array seismology, *Ann. Rev. Earth Planetary Sci.* **3**, 157-181.

Gilbert, F., 1980, An introduction to low-frequency seismology, in A. M. Dzeiwonski, ed., *Physics of the Earth's Interior—International School of Physics "Enrico Fermi."* Amsterdam: North-Holland, 41-81.

Goldstein, C. M., 1984, Computer-based information storage technologies, *Ann. Rev. Inf. Sci. Tech.* **19**, 65-96.

Kamas, G., and S. L. Howe, 1979, *Time and frequency users manual. U.S. National Bureau of Standards Special Pub.* **559**.

Krohn, C. E., 1984, Geophone ground coupling, *Geophysics* **49**, 722-731.

Lee, W. H. K., and S. W. Stewart, 1981, Principles and applications of microearthquake networks, *Adv. Geophys. Suppl. Ser.* **2**, 1-293.

Melton, B., 1971, The LaCoste suspension—principles and practice, *Royal Astron. Soc. Geophys. Jour.* **22**, 521-543.

Melton, B., 1976, The sensitivity and dynamic range of inertial seismographs, *Rev. Geophysics Space Physics* **14**, 93-116.

Melton, B., 1981, Earthquake seismograph development: A modern history—Part 2, *EOS (Amer. Geophys. Union Trans.)* **62**, 545-548.

Plešinger, A., 1984, Analysis and optimization of wideband force-balance seismometer responses, *Studia Geoph. et Geodet.* **28**, 67-81.

Prothero, W., 1984, Ocean-bottom seismometer technology, *EOS (Am. Geophys. Union Trans.)* **65**, 545-548.

Rodgers, P. W., 1968, The response of the horizontal pendulum seismometer to Rayleigh and Love waves, tilt, and free oscillations of the earth, *Seismol. Soc. Amer. Bull.* **58**, 1384-1406.

Willmore, P. L., 1979, *Manual of Seismological Observatory Practice*. Boulder, Colo.: World Data Center A: Report SE-20.

Cross-references: *Controlled Source Seismology; Earthquake Seismology; Exploration Seismology; Seismic Instrumentation: History; Seismic Monitoring of Nuclear Explosions; Seismic Noise; Seismograms: Interpretation; Strong Motion Seismology.*

SEISMIC INSTRUMENTATION: HISTORY

The problem in detecting ground motions is to find a place to stand from which to make observations. Devices for this purpose fall into two categories: those in which the relative positions of an inertial mass only loosely connected to the moving ground are observed, and those in which the varying separations (strains) of two fixed points are measured.

Devices whose purpose is primarily to determine that an earthquake has occurred are called *seismoscopes*. Those that make some sort of record of the motion, usually accompanied by an indication of the time of occurrence, are called *seismographs*. The recording produced is called a *seismogram*. An instrument that converts the ground motion into some sort of signal that is proportional to the ground displacement or strain or one of their derivatives is called a *seismometer* and is usually a part of a seismograph.

Seismoscopes

The earliest known seismoscope was built by Chang Heng in China about A.D. 132. Eight bronze dragons on the surface of a hollow vessel held balls in their mouths so arranged that when an earthquake occurred, one (or more) of the balls fell into the mouth of one of eight toads (Fig. 1). It is believed that within the vessel was an inertial pendulum that knocked the ball from the dragon's mouth when an earthquake occurred. The drop of the ball would mark not only the occurrence of an earthquake but also the direction to its source on the assumption, rarely correct, that the principal motion was a step away from the source of the waves.

Early seismoscopes were intended primarily to indicate that an earthquake had occurred. In 1703 J. de la Haute Feuille designed a seismoscope where mercury would spill from a central reservoir into cups around its periphery if the ground moved. It was hoped that this instrument would predict the

FIGURE 1. Chang Heng's seismoscope.

location of future large shocks by the detection of foreshocks, but it did not do so.

In 1845 Robert Mallet attempted to measure the time of transmission of seismic waves generated by explosions by observing the deformation of the surface of a pool of mercury as the waves shook it. The method lacked sufficient sensitivity to give accurate results.

Luigi Palmieri built a more sophisticated seismoscope in Italy on the same principle in 1855. Sloshing of the mercury closed an electrical circuit stopping a clock to indicate the time of the event and starting a drum on which a pencil was pressed whenever the electrical current was complete, giving a measure of the duration of the shaking. In 1856 he added a vertical component instrument in which a mass was suspended by a spiral spring over a basin of mercury. Raising the basin of mercury closed an electrical circuit.

Pendulum seismoscopes have been in use since the eighteenth century. At first, the motion was watched by an observer directly or with the aid of a microscope. In 1751 Andrea Bina added a simple recording device consisting of a tray of fine sand into which a pointer on the pendulum dipped, tracing out the motion of the pendulum bob. Some modern horizontal-component strong-motion seismometers are built on the same principle, although the recording system has been improved. In 1796, A. Filomarino, Duca della Torre, attached a hair to the pendulum so that, on its being disturbed, it released the balance wheel of a clock, allowing the time of the event to be determined. (A seismoscope built by G. Wagoner in Japan in 1880 used the same principle to stop a clock.)

In 1841, James D. Forbes in Scotland designed and shortly thereafter built several inverted-pendulum seismographs. The pendulum was mounted on a stiff wire that returned the pendulum to its rest position after it was disturbed. By varying the length of the wire, the period of the pendulum could be altered. A record was made by a pencil attached to the pendulum. Only two earthquakes were recorded by Forbes's seismograph, although several dozen were felt in the area where it was set up. However, J. D. Forbes appears to have been the first, in 1844, to discuss the mathematical theory of the pendulum as a recorder of seismic waves.

Inertial Seismographs

The simple pendulums initially used to detect earthquakes lacked sensitivity, and it was difficult to get a soft enough suspension to prevent the suspended mass from following the slower parts of the ground motion. Generally, more earthquakes were actually felt at any location than were recorded by these early seismoscopes. The first successful true seismograph was built by Filippo Cecchi in Italy in 1875. It recorded the motions of two simple pendu-

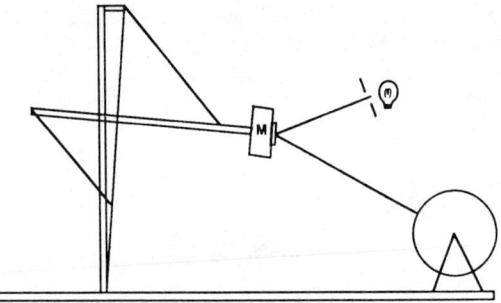

FIGURE 2. Principle of the Zollner-suspension seismograph. The axis of rotation of the pendulum is tilted at an angle with the vertical.

lums, one swinging N-S and the other E-W, with their motions magnified three times by a thread and pulley apparatus. A vertical component incorporated a mass suspended by a spiral spring. Unfortunately, the magnification of this instrument was too small for it to record any but the strongest earthquakes.

Much better instruments were built after the advantages of tilting the pendulum's axis of rotation were realized. A pendulum suspended so that its axis of rotation makes a small angle with the vertical is very sensitive to horizontal ground movements and to tilt. Such instruments were first built to study deflections of the vertical due to the lunar and solar tides by L. Hengler in 1832, A. Gerard in 1851, and M. F. Zollner in 1869 (Fig. 2). The free period of such a pendulum can be made as long as desired by reducing the angle of the axis of rotation with the vertical. Hengler's, Gerard's, and Zollner's instruments were not used as seismometers, but they were known to others who recognized their potential. Thomas Gray in Japan was familiar with Gerard's pendulum; and Zollner's was copied by Ernst von Rebeur-Paschwitz, who in 1884 built a tiltmeter on this principle, which in 1889 made the first recording of a distant earthquake. At first, it was believed that such instruments responded primarily to the tilt of the ground; but eventually it was recognized that they were equally sensitive to the horizontal ground movement at a right angle to the axis of rotation of the pendulum.

Another important improvement copied from the Zollner instrument was the use of a light spot reflected from a mirror on the suspended mass to record the motion. Use of the light spot had two important advantages over the scribers used previously: first, it eliminated any friction between the recording medium and the scriber, which tended to distort the recorded motion, and for very small motions to suppress it altogether; and second, it allowed significant amplification by the use of a long optical lever arm.

John Milne was most influential in developing a practical seismograph. He was appointed to the

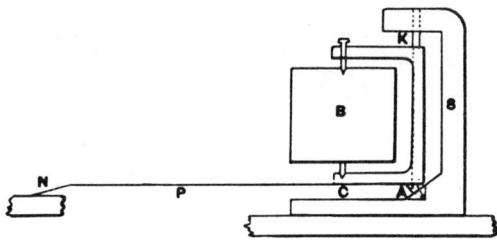

FIGURE 3. Principle of Ewing's horizontal-component pendulum seismometer (Milne, 1899).

Faculty of the Imperial College of Engineering in Tokyo starting in 1876. There, initially in association with Thomas Gray and J. Alfred Ewing, he experimented with a variety of pendulum instruments for recording ground motions. By 1881, Gray, Ewing, and Milne had developed a system of three pendulums whose motions were recorded on a strip of smoked paper advanced by a clockwork mechanism. The time of an earthquake was noted by tripping an electromagnet that caused inkpads on the hands of a clock to press on a dial. The ground motion was amplified by a factor of six through the mechanical lever of a long recording arm (Fig. 3). This instrument incorporated the three essential features of all useful seismographs: it produced a seismogram whose trace deflection was roughly proportional to the ground motion; the motions were amplified so that small motions could be studied; and the exact time of the event was preserved. The step from a simple strip of paper for recording ground motion and a separate time record to recording a spiral trace on a rotating drum with time marks coordinated with the seismic events was relatively simple and soon accomplished. Many modern instruments differ from this simple beginning largely in improvements in sensitivity, in control of the breadth of frequency spectrum recorded, and in the accuracy of reproduction of ground motion as a function of time. Beginning with the earthquake of 3 November 1880, useful seismograms of Japanese earthquakes were obtained.

One of Milne's most important contributions was the establishment of a network of about 40 seismographs distributed worldwide. He returned from Japan to England in 1895 and established a seismic observatory at Shide on the Isle of Wight. Through the Seismological Committee of the British Association for the Advancement of Science, of which he was secretary, he encouraged the installation of seismographs of his own design at observatories all over the world, but largely in British Empire countries. These observatories forwarded their results to Milne at Shide, and he summarized the findings in the annual reports of the Seismological Committee, beginning the system of seismological world data centers.

Before the 1880s seismology depended almost completely on subjective descriptions of what was felt during the earthquake and on examination of damage. After the development of the seismograph, the science gradually shifted over a period of 30 or 40 years to dependence primarily on seismograms as a source of information on what actually took place. The bulk of the report written by Andrew C. Lawson and associates of the 1906 California earthquake details the effects of the earthquake, but 15 of the 40 charts in the Atlas volume were copies of seismograms from stations all over the world. After the 1880s, scientists studying earthquakes were able to examine the actual motions of the ground. At first these recordings were crude. They were of low magnification, and were limited in the range of frequencies recorded, especially at low frequencies, and they sometimes introduced serious distortions, especially at very low amplitudes. As time passed, better and better instruments were developed, and the number of observatories providing detailed, accurate seismograms increased.

To reproduce ground motion accurately, the free period of a simple pendulum seismograph must be longer than the principal periods of the ground motion. Long periods are easier to achieve with a horizontal than with a vertical component instrument. Vertical component pendulums were built at first with the inertial mass supported directly from above with a coiled spring. Gray increased the period by attaching the spring between the center of mass and the axis of rotation, and astatized it by adding a mercury reservoir whose motion increased the moment as the arm tilted down and decreased it as the arm moved up. Ewing improved this instrument by attaching the restoring spring below the axis of the lever arm, causing the point of suspension to move toward and away from the vertical axis as the ground moves down and up, thereby altering the restoring force on the pendulum in such a manner that the equivalent free period is increased. Modern instruments utilize a supporting spring so wound that unstretched it would have zero length. Such a spring was first developed by Lucien J. R. La Coste in 1934 for measuring variations in the force of gravity. The free period of the pendulum can also be increased by arranging the spring so that its direction of pull makes an angle with the vertical (Fig. 4).

The only damping of the motion in the earliest seismographs was provided by the solid friction of the recording stylus on the recording paper. This type of damping allowed the pendulum to oscillate after the initial motion ended and distorted the recorded trace in other ways, including the loss of very small motions. A very sharp, clear trace was obtained by the use of a hair rubbing on smoked paper. Newer methods of recording have yet to improve upon this method for sharpness of the trace.

The introduction of viscous damping by Emil Wiechert in 1898 was an important advance in seismograph design. Wiechert used a dash pot,

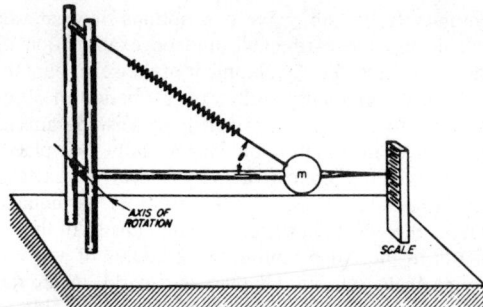

FIGURE 4. By using a diagonal spring, the free period of the seismometer pendulum is increased (Howell, 1959).

which has the undesirable feature that the force is not a linear function of the motion it is opposing. Better damping can be provided by immersing a vane in a cup of oil. If the vane moves parallel to its surface, the resistive force is proportional to the velocity of motion.

Wiechert is also noted for his design of horizontal component inverted-pendulum seismographs using very large masses of a ton or more and mechanical registration (Fig. 5). By using a very large mass, his instruments were affected very little by the friction of the stylus on the recording paper, and clear records with substantial amplification were readily obtained. With large masses and weak restoring springs, a long free period for the pendulums was obtained, allowing Wiechert's seismographs to record a wide spectrum of seismic signals, accurately reproducing ground motions.

Seismographs involving large masses are difficult to construct and maintain. High sensitivity with more easily controlled damping was achieved by B. Galitzin in Russia in 1906. Galitzin mounted a coil of wire on the inertial mass in such a manner that a magnet fixed to the earth moved with respect to it, generating an electrical current used to rotate a galvanometer coil. The movement of a light spot reflected from a mirror mounted on the galvanometer coil was recorded on a rotating drum of photographic paper. Damping was provided in two ways. A copper vane moving in a magnetic field produced eddy currents that resisted the motion, and the reaction of the main electrical current on the inertial-mass coil provided damping that could be adjusted to any degree required. Very large magnification is available from the mass ratio of the inertial to the galvanometer coil and the optical-lever arm of the galvanometer light spot. Such instruments differ from direct recording seismographs in that the electrical system is responsive to the velocity of the ground with respect to the inertial mass rather than to its displacement. Velocity sensitivity is sometimes considered an advantage inasmuch as the square of the velocity is a measure of the kinetic energy of the moving ground. Electromagnetic seismographs approach a flat response with frequency for seismic-wave energy above the natural frequency of the seismometer suspension.

Anderson and Wood (1925) designed a highly successful seismograph in 1922 damped only by the relative motion of a copper inertial mass placed in the magnetic field of a magnet attached to the ground (Fig. 6). The pendulum has a free period of 0.8 seconds and a gain, obtained by mounting a mirror on the inertial mass, of 2800. For the low frequencies predominant in earthquake waves, this seismograph acts as an accelerometer.

To maximize the sensitivity of an electromagnetic seismometer, in 1930 Benioff (1955) introduced a variable reluctance transducer as a sensor (Fig. 7). A magnet attached to the ground moves relative to an armature mounted on the inertial mass so that the relative motion of the armature and magnet shifts magnetic flux from one pair of gaps to the other. Coils wrapped on the armature are connected to a galvanometer to record the motions. Amplification of as much as 200,000 is possible at short periods with this system, sufficient to detect background microseismic unrest at almost every place on the Earth's surface. The inertial mass in the Benioff seismometer is not a part of a pendulum but is mounted on wire springs.

One other type of inertial seismometer is of interest, although it is more rarely used (Benioff, 1955). One plate of a capacitor is placed on the inertial mass and the other plate is fixed to the ground. Motion of the ground causes a variation in capacitance. Using this as part of a capacitance bridge, any relative motion of the plates can be used to drive a galvanometer. The variable capacitance may also be used to alter the frequency of an oscillator and a discriminator used to convert this frequency into a signal proportional to the ground motion.

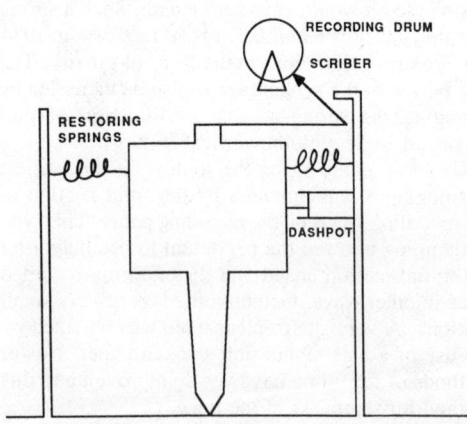

FIGURE 5. Principle of Wiechert's inverted-pendulum seismometer.

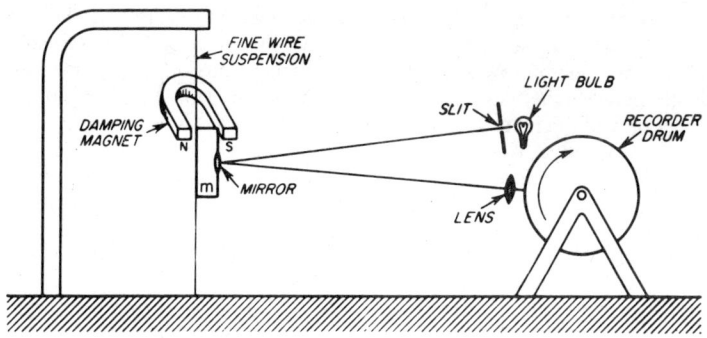

FIGURE 6. Principle of the Wood-Anderson torsion seismometer (Howell, 1959).

Modern Seismograph Systems

Today, direct recording seismographs have been replaced, except for strong-motion recorders, by devices that filter or otherwise modify the record of ground motion by the interposition of an electronic system between the seismometer and the recorder. This system may be simply a resistive or inductive coupling designed to damp both the seismometer and recording galvanometer to a desired degree or it may include a wide variety of amplifiers and filters.

The use of artificially generated waves in geophysical exploration placed new requirements on recording systems. Exploration systems need to be portable, so that instruments must be lightweight and rugged for easy transportation and quick set-up in the field. The problem was made more difficult by

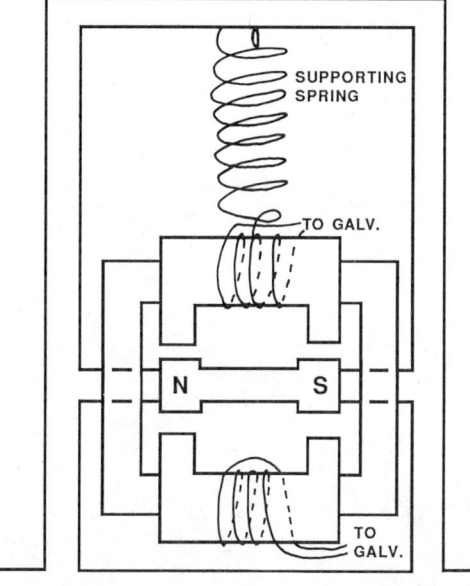

FIGURE 7. Principle of the Benioff variable reluctance seismometer.

the necessity of using vertical-component seismometers because this component contains the strongest useful pulses of energy. The moving-coil Galitzin seismometer was easily adapted to this need, although more sensitive Benioff variable-reluctance seismometers were tried for a while.

Early exploration seismic methods involved recording seismic waves from small explosions several miles from the recording instruments. The important seismic pulses were often very weak. Sufficient magnification with portable seismometers was most easily achieved by adding an electronic amplifier between the seismometer and the galvanometer, especially because the earliest galvanometers used were single strand instruments with a low sensitivity. The necessity of long wires between seismometer and recorder contributed to the need for amplifiers. To avoid picking up stray induced voltages, the impedance of the system had to be kept low, which meant that low voltages were generated.

Once the need for an amplifier was accepted, it opened up the possibility of conveniently introducing a variety of devices into the system. Filters could be used to select those frequencies where the signal-to-noise ratio was highest. These could be easily varied to meet local needs. The amplification could be varied to suit the requirements of individual recording locations. For example, less amplification was needed for recording sites nearest the source of the seismic waves.

When reflection seismic surveying was introduced in 1921, it became desirable to record the first arrivals from shallow reflecting horizons with less amplification than later arrivals from deep horizons. At first, variation in the amplification was accomplished by potentiometers, which were mechanically adjusted to increase the fraction of the generated seismic signal fed to the recording galvanometers while the signals were arriving. This crude method was soon replaced, at first by electronically programmed increases in amplification during the recording period and eventually by feedback systems that used the average level of the signal to adjust the sensitivity to keep the level of the recorded signal roughly the same throughout the recording period. Reflections were recognized on the seismograms not by sudden changes in amplitude but by correlations of in-phase amplitudes between the traces of several recorders. The number of recording locations (traces) on a single seismogram was gradually increased from fewer than ten to one hundred or more.

The inclusion of complex electronic equipment in the recording system fostered the mixing of signals from more than one seismometer. In reflection seismology, useful energy reaches the seismometer vertically from below. Energy from any other direction tends to be largely interfering noise. By combining the output of several seismometers physically separated from one another, energy arriving from below tends to be reinforced whereas energy travelling horizontally is more random and tends to interfere destructively.

Magnetic tape as a recording medium opened up wider possibilities in the use of arrays. The first advantage of magnetic tape was that seismograms could be played back using different filters until the one was found that gave the best signal-to-noise ratio. In addition, the replays of different traces could be displaced in time to align signals of interest, focusing on different directions of arrival or compensating for small differences in the time required for the signals to reach different seismometers from one reflecting surface below ground. The same principle is used in arrays of earthquake seismometers to focus on waves coming from a particular direction.

Analog tape recordings were eventually replaced by digital tape recordings, permitting even wider scope in signal processing. The history of digital signal processing is part of the story of seismic signal interpretation rather than of seismic instruments (see *Seismic Signal Processing*). The processing of seismic signals to bring out their features optimally has kept seismic applications near the forefront of the development of digital computer systems.

With electronic recording systems, it became possible to transmit the output of the seismometer over a long distance to the recorder site by radio (Gane et al., 1949).

With the use of tape recording, it also became possible to select for preservation only that portion of the seismic signal of special interest and to ignore the rest, thus reducing the cost of recording paper or tape. A continuous loop of tape is used to record the output of the seismometer, wiping out the old record just before it passes again through the recording head. Whenever the signal exceeds a set level, a high speed recorder is turned on that picks up the signal on the loop of tape and rerecords it just before it is wiped out. Rerecording continues for a fixed time after the signal drops below the trigger level. This intermittent recording reduces the cost of making a permanent record of ground motion at sites where earthquakes are rare and makes it practical to obtain more detailed records of the largest events than would be practical with slower-speed continuous recordings. Recorders can be and are built to start with the onset of strong motions, but these tend to lose the first few seconds of ground movements before the amplitude reaches the triggering level.

It is also possible to record on magnetic tape, then to preserve only those tapes on which interesting earthquakes are recorded, reusing the other tapes.

Inertial seismographs have been built with a wide variety of free periods. For seismic exploration, free periods of one tenth to one thirtieth of a second are usually used because millisecond timing of pulse arrivals is best accomplished by using the highest-frequency pulses that can be easily generated and transmitted through the ground and because noise often exceeds the useful signal at lower frequencies. To determine earthquake arrival times as a means of studying Earth structure, seismometers with a free period of around one second are preferred because the energy in seismic waves usually decreases rapidly at shorter periods and microseismic noise peaks between one and ten seconds. For recording the complete spectrum of ground motions, it is desirable to record faithfully as wide a spectrum as possible and very long period seismometers are needed. Horizontal pendulums with periods as great as 100 seconds are built. The Press-Ewing seismometer is a widely used example (Fig. 8).

The longer the period of a horizontal-component pendulum seismometer, the more sensitive it is to ground tilt, and very long period seismometers tend to be unstable due to thermally generated tilts. This weakness can be partially overcome by connecting a capacitor across the output of an electromagnetic seismometer as proposed by Pierre Bernard in 1955. Using a high-impedance winding in the seismometer coil, the effective period of a 30-second Press-Ewing seismometer can be increased to as great as 412 seconds with a 200 microfarad capacitor. Damping

FIGURE 8. Press-Ewing seismometer. (Courtesy Lamont-Doherty Geological Observatory)

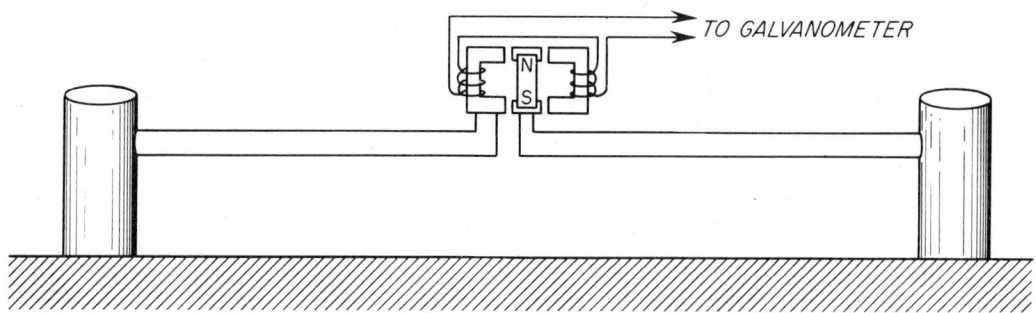

FIGURE 9. Principle of the Benioff strain seismometer (Howell, 1959).

is increased and sensitivity is reduced by the shunt; but the loss of sensitivity is easily overcome by the use of an electronic amplifier between the seismometer and recording system. This arrangement has the stability of a 30-second pendulum with the low-pass frequency limit expanded by a factor of more than ten. It is a common practice to use several different recorders sensitive to different spectra with a single seismometer separating the signals in the coupling electronic networks.

The greatest stability can be obtained by using feedback that maintains the seismometer pendulum in a fixed position. The ground acceleration is then measured by noting the amount of force required to hold the pendulum in its normal rest position. Such instruments were first built to study the free-oscillation modes of the earth, with periods in the tens of minutes. According to E. Weilandt (1983), nearly all wide-band seismometers built since 1973 make use of the force-balance principle. Force-feedback systems are also used in borehole seismometers where space for motion of the pendulum is limited.

Strain Seismographs

The first strain seismographs were developed at nearly the same time as inertial seismographs. John Milne measured the relative displacement between two posts by a mechanical-lever device with a magnification of six beginning in 1882.

In 1900 E. Oddone obtained a gain of 300 using a hydraulic lever wherein a large piston mounted on one pier presses into a fluid reservoir with a narrow relief opening on another pier. This principle can be used to detect earthquake pressure waves by means of water-level gauges in wells. The passage of earthquake waves is frequently recorded by such instruments.

Hugo Benioff began building strain seismographs in the 1930s. His seismographs differed from Milne's largely in size, in where they were installed, and in the transducers used to detect the relative motion of the two piers. Benioff's initial strain seismograph consisted of two pairs of piers at right angles 20 and 50 feet apart, with a steel tube, suspended by fine fibers, extending from one pier nearly to the other, and a variable reluctance sensor between the end of each rod and the second pier (Fig. 9). Only horizontal components are normally used because vertical strain is zero at the free surface of the Earth (neglecting the small resistance of the air). When operated with a velocity transducer and galvanometer recorder, the strain seismograph has a frequency-response characteristic that is identical to that of a displacement-transducer pendulum seismograph having a period and damping constant equal to the period and damping constant of the galvanometer. Thus, it is a good recorder of all periods shorter than the galvanometer free period. The variable-reluctance bridge makes high sensitivity possible. This system is sensitive to ambient temperature variations and was constructed in an underground tunnel and surrounded by an insulated housing. Temperature sensitivity was reduced by replacing the steel rod with fused-quartz glass in later models. The strain gauge has the advantage over a pendulum seismometer that it is insensitive to tilt of the ground. Absolute strains as small as .001 mm in a 24.08 meter length or 4.15×10^{-8} can be observed using a microscope.

A variety of recorders has been used in connection with Benioff strain seismometers. Variable-reluctance transducers have been linked to galvanometers with periods of 40 to 180 seconds. A double-bowstring mirror suspended between the fused-quartz rod and a pillar gave a 1-mm deflection for a strain of 4.15×10^{-9}, but was somewhat unstable. It was replaced by a capacitance-bridge transducer that provided a sensitivity of 1-mm for 5.18×10^{-10} strain (1 mm in 2000 km).

Simple electrical-resistance strain gauges cemented onto rock cores or the surface of scale models are also used to detect seismic waves. As the dimensions of the rock surface to which the strain gauge is attached change, the resistance of the strain-gauge changes. By making the resistance a part of a wheatstone bridge, high sensitivity can be obtained. This type of strain gauge is particularly useful in laboratory experiments because of its very small size. Such strain gauges are also used in field experiments where it is desired to place the seismometer in a narrow borehole with minimum disturbance of the ground. The strain gauge is cemented in place and not recovered after the experiment. Inertial seismometers are more commonly used in large boreholes such as oil wells.

B. F. HOWELL, JR.

References

Anderson, J. A., and H. O. Wood, 1925, Description and theory of the torsion seismometer, *Seismol Soc. America Bull.* **15**, 1-72.

Benioff, H., 1955, Earthquake seismographs and associated instruments, *Adv. in Geophysics* **2**, 219-275.

Davison, C., 1927, *The Founders of Seismology*. Cambridge: University Press, 240p.

Dewey, J., and P. Byerly, 1969, The early history of seismometry (to 1900), *Seismol. Soc. America Bull.* **59**, 183-227.

Gane, P. G., H. J. Logie, and J. H. Stephen, 1949, Triggered teleseismic recording equipment, *Seismol Soc. America Bull.* **39**, 117-143.

Howell, B. F., Jr., 1959, *Introduction to Geophysics*. New York: McGraw Hill, 399p.

Milne, J., 1899, *Earthquakes and Other Earth Movements*, 4th ed. New York: D. Appleton and Co., 376p.

Weilandt, E., 1983, Design principles of electronic inertial seismometers, in H. Kanamori, ed., *Earthquakes: Observations, Theory and Interpretation*. New York: North Holland, pp. 354-365.

Cross-references: Controlled Source Seismology; Earthquake Seismology; Exploration Seismology; Seismic Instrumentation; Seismograms: Interpretation; Strong Motion Seismology.

SEISMICITY: INTRAPLATE

Early in this century, seismologists noted that most earthquakes are located along zones of high seismicity, which we now identify as plate boundaries. These interplate earthquakes reflect slip between lithospheric plates or stresses set up in the immediate region of the plate boundary. Smaller numbers of intraplate earthquakes, which occur on the interiors of plates, provide constraints for a number of important tectonic questions, and thus command interest out of proportion to their limited number. Classical plate tectonics suggests the lithosphere is made up of a number of rigid, nondeforming plates. As intraplate seismicity demonstrates deformation within plates, the distribution and size of such earthquakes allows us to estimate the deviation from the rigid plate model and identify regions with anomalously high intraplate deformation rates. The distribution of intraplate earthquakes with depth provides constraints on the rheology and temperature structure of the lithosphere, as the maximum depth of seismicity is limited by temperature-dependent rock rheologies. Finally, the focal mechanisms of intraplate earthquakes give direct evidence of the state of stress within the lithospheric plates, which in turn is related to the viscosity structure and flow pattern within the mantle, the thermal structure of the lithosphere, and the magnitude of stresses applied at plate boundaries.

Intraplate earthquakes also generate considerable interest because of their possible danger to human life and property. Although much smaller and less frequent than subduction zone earthquakes, large intraplate events may cause tremendous damage. Intraplate earthquakes are often more damaging than plate margin earthquakes of the same size because intraplate regions generally have lower seismic attenuation and thus the destruction occurs over a larger area. In addition, because of the low seismicity levels and infrequency of major earthquakes, seismic hazards are extremely difficult to estimate for intraplate regions. Thus intraplate earthquakes may strike locations that are not well prepared for a major earthquake, possibly resulting in increased destruction.

Definition

Considerable confusion exists over the definition of the term *intraplate earthquake*. In a broad sense, the term can be applied to any earthquake not occurring along a fault bounding two lithospheric plates. This definition may be less than ideal; for example, it would classify most earthquakes comprising the Wadati-Benioff zone as intraplate, whereas the earthquakes are directly related to the nearby plate boundary and are better discussed in that context.

A much more restricted definition excludes all earthquakes that seem related to plate boundary processes. Such "mid-plate" events are often defined as being at least a certain distance, often 500 km, from the nearest plate boundary. However, there is some question about what constitutes a plate boundary, particularly for diffusely deforming zones that may show substantial seismicity but lack morphologic features generally associated with plate boundaries. As discussed later, some of these regions, often described as intraplate, may be better described as diffuse plate boundaries. Further difficulties arise because stress modeling suggests the intraplate stress field may be highly dependent on plate boundary processes, even thousands of kilometers from the boundary.

Thus, a completely satisfactory definition of an intraplate earthquake may not be possible. For the purposes of this discussion, we will adopt a rather loose definition, omitting earthquakes directly associated with the classical plate boundaries, such as events occurring in the Wadati-Benioff zone or outer rise of a subduction zone, while retaining events from regions traditionally described as intraplate, but which are perhaps better described as diffuse plate boundaries.

Magnitudes and Scaling Relations of Intraplate Earthquakes

The largest intraplate earthquakes since the development of instrumental seismology have occurred in parts of central Asia and the Indian Ocean thought to be deforming in response to the continental convergence between India and Asia occurring in the Himalayas. These earthquakes include the 1905 Mongolian earthquakes (M 8.25), the 1957 Gobi-Altai Mongolian earthquake (moment 1.8×10^{28} dyne·cm), and the 1920 Kansu, China event (moment 3×10^{28} dyne·cm). Large oceanic events have occurred in diffusely deforming regions of the Indian Ocean (the 1928 Ninetyeast Ridge and 1983 Chagos Bank events, both with moments of about 4×10^{27} dyne·cm).

The preceding events, which constitute the largest documented intraplate earthquakes, are located in broad, diffuse regions undergoing intense deformation. Large earthquakes that are truly midplate are more rare. The classic examples of midplate earthquakes are the 1811–1812 New Madrid, Missouri, events. Because the sequence occurred in a sparsely populated region before the invention of the seismograph, details of the New Madrid events are incomplete. However, estimates of the size of the earthquakes based on felt reports indicate a series of 3 large earthquakes; the moment of the largest may have been as large as 8×10^{27} dyne·cm. These earthquakes were felt over much of the eastern United States (Fig. 1) and would have been devastating had the epicentral region been heavily populated. The anomalously large felt area was due in part to low seismic attenuation in eastern North America relative to plate margins, where most other large earthquakes occur (Nuttli, 1973).

This discussion suggests that the moments of the largest intraplate earthquakes are probably in the range of $1-3 \times 10^{28}$ dyne·cm. As the largest subduction zone event, the 1960 Chilean earthquake had a seismic moment of approximately 2×10^{30} dyne·cm: The largest intraplate earthquakes are approximately 100 times smaller than the largest subduction zone events. However, the largest intra-

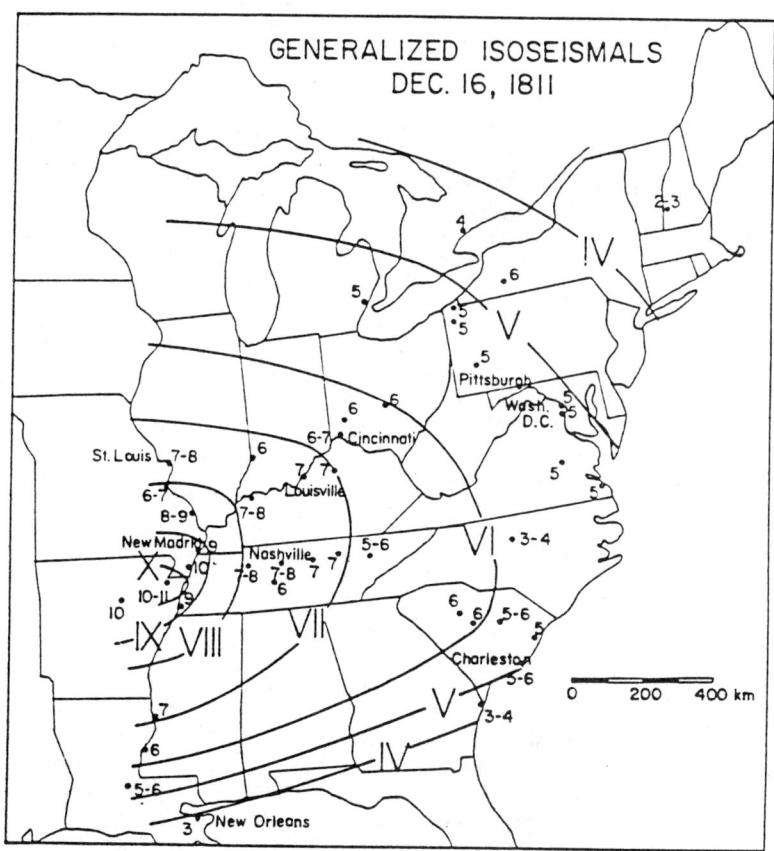

FIGURE 1. Isoseismal map showing the effects of the 16 December 1811, earthquake (Nuttli, 1973).

plate events are of the same order of magnitude as the largest strike-slip boundary events. This limit may reflect bounds on the physical dimensions of rupture along a steeply dipping fault imposed by the thickness of the high strength (and thus seismically active) portion of the lithosphere.

Several investigators have noted that scaling relations between seismic moment and various fault parameters determined for intraplate earthquakes differ from those for interplate events. H. Kanamori and D. L. Anderson found that large intraplate earthquakes are characterized by systematically higher stress drops (approximately 100 bars) than interplate earthquakes (approximately 30 bars). O. W. Nuttli has suggested the higher stress drops result because stress drop increases with earthquake size for intraplate earthquakes, rather than remaining constant, as generally observed for interplate events. Since the stress drop is proportional to the ratio of the displacement to the fault length, high stress drop implies that large intraplate earthquakes have greater displacement, and thus seismic moment, than interplate events with the same fault size.

Geographical Distribution of Intraplate Seismicity

Surveys of intraplate seismicity by several investigators give strong evidence that many intraplate earthquakes represent the reactivation of zones of weakness in the presence of a local or regional stress field. These zones of weakness may include topographic and bathymetric features, as well as preexisting faults, failed rifts, and suture zones. Careful examination of recorded seismicity shows that even many apparently isolated epicenters represent repeated shocks at the same location, sometimes separated in time by several decades, suggesting that intraplate deformation takes place through repeated slip along isolated active faults.

Focal mechanisms of intraplate earthquakes show a regional variability undoubtedly related to the sources of stress in the lithosphere. Some regions display a broad, regionally consistent stress pattern suggestive of plate-wide stresses. For example, over most of the eastern and central United States, stress data compiled by M. L. Zoback and M. Zoback suggest a largely E-W compressional stress field. This stress field may result from either the "ridge push" force, the horizontal force that results from the thickening and subsidence of the oceanic lithosphere as it cools, or from aesthenospheric resistance to the motion of the North American plate. Modeling studies suggest that a substantial component of ridge push stresses are needed to fit focal mechanism stress data in older oceanic lithosphere (Richardson et al., 1979). In other regions stresses may be more localized; S. Stein has suggested that large intraplate earthquakes along the passive margin of eastern Canada may result from stresses induced by deglaciation. Thermoelastic stresses may also play an important role, particularly in oceanic seismicity (see Intraplate Earthquakes and Lithospheric Thermal Structure).

Global compilations of intraplate seismicity also show several broad regions with extremely high rates of intraplate seismicity, suggesting that intraplate earthquakes may delineate major deformation zones within the larger lithospheric plates. Several continental regions, including the Great Basin and Central Asia, have long been known to display a diffuse pattern of seismicity. In Asia, extensive seismicity results from the deformation of the Asian continent as a result of the collision with the Indian plate. A series of left lateral strike-slip faults hundreds of kilometers north of the actual India-Asia boundary apparently accommodate some of the convergence. In North America, several magnitude seven earthquakes in the Great Basin and the Intermountain Seismic Belt, as well as considerable geologic evidence, indicate intense intraplate deformation.

Recent systematic studies of oceanic intraplate seismicity suggest similar zones of active intraplate seismicity and deformation within oceanic plates. These active regions include the proposed Caroline Plate, where bathymetry suggests a possible incipient subduction zone, and the region between the Lesser Antilles and the mid-Atlantic ridge, which probably marks a diffuse North America-South America plate boundary.

The most active source of oceanic intraplate seismicity is in the Indian Ocean between the Central Indian Ridge and the Sumatra Trench, where seven events of magnitude 7 or greater have occurred since 1910. The seismicity pattern (Fig. 2, top left) is diffuse and indicates a complex pattern of stresses. Intense deformation is also indicated by marine seismic profiling records and gravity data showing faulting and folding. Reexamination of plate motion data (spreading rates, transform fault orientations, and interplate earthquake slip vectors) indicates the deformation along this zone is resolvable as a separate plate boundary, with relative motion on the order of 1 cm•yr^{-1} (Fig. 2, top right) (Wiens et al., 1985). This boundary is more significant than the previously assumed India-Arabia boundary along the Owen Fracture Zone; given the new boundary the nearly aseismic Owen boundary is no longer required by the data, indicating that the diffuse zone separates the Australian Plate from an Indo-Arabian plate. The Australia-Arabia rotation pole is located just east of Chagos Ridge, as suggested by the N-S extension near the Central Indian Ridge and the N-S compression to the east.

Finite element stress modeling (Fig. 2, bottom) provides insight into the cause of the diffuse boundary. This modeling, which incorporates the variation of slab driving forces with the age of the subducting lithosphere, shows the particular combination of resistance in the Himalayan Convergent

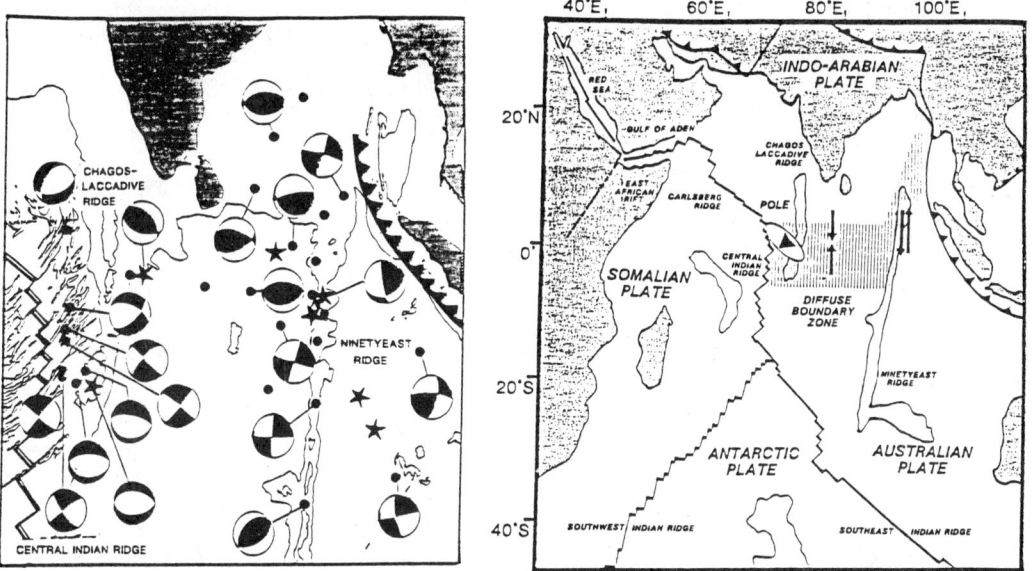

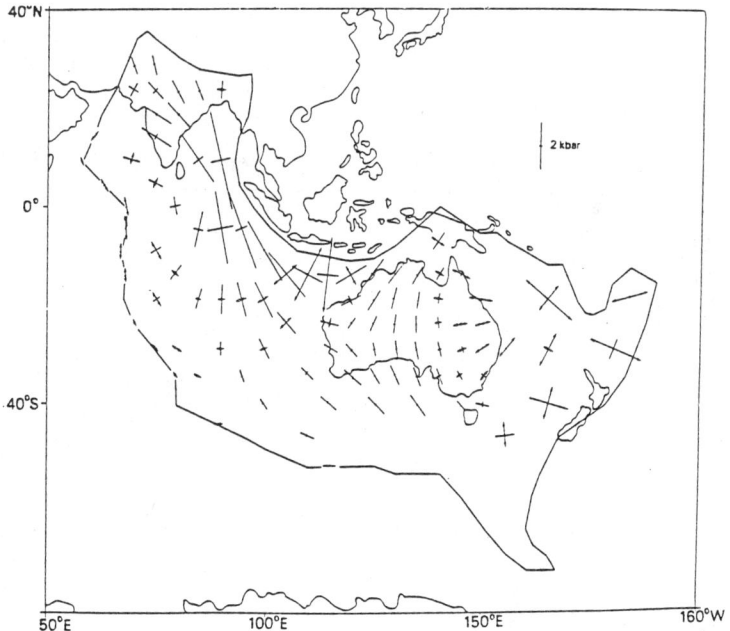

FIGURE 2. (Top left) Locations and focal mechanisms of earthquakes along the diffuse plate boundary in the Indian Ocean. (Top right) Schematic diagram of the Indian Ocean plate geometry determined by Wiens et al., (1985). (Bottom) Stress modeling for the Indian plate showing large compressive stresses in the diffuse boundary region (Cloetingh and Wortel, 1986).

Zone and subduction in the Java Trench focuses large compressional stresses in the region of the northern Ninetyeast Ridge (Cloetingh and Wortel, 1986). Although the future evolution of this boundary is uncertain, its development in response to the unique tectonic situation in the Indian Ocean may have important implications for the initiation of convergent plate boundaries in general.

Intraplate Earthquakes and Lithospheric Thermal Structure

Many overall features of intraplate seismicity, including seismicity level, focal depth, and focal mechanism, reflect the thermal and mechanical evolution of the lithosphere. In general, simpler thermal and structural histories make these relations easier to define for oceanic regions than for continental regions. The maximum depth of oceanic intraplate earthquakes increases with increasing lithospheric age (Fig. 3) and appears to be approximately limited by the 750°C isotherm according to plate cooling models of lithospheric temperature (Wiens and Stein, 1983). Thus the thickness of the lithosphere in which intraplate earthquakes occur roughly corresponds to the flexural elastic thickness but is much less than most estimates of lithospheric thickness from surface wave dispersion. The limiting temperature agrees with the results of olivine rheologies from rock mechanics experiments, which predict rapid weakening at these temperatures.

Similar relations are observed in continental regions. The maximum depth of small earthquakes is inversely correlated with heat flow and geothermal gradient, implying a connection between temperature and depth of faulting (Sibson, 1982). Estimates of the limiting temperature of continental seismicity are about 250–400°C, suggesting that the continental crust is significantly weaker at high temperatures than the oceanic upper mantle. Several diffusely deforming regions, mostly in Central Asia, show seismicity concentrations in the uppermost crust and in the upper mantle, separated by an aseismic region in the lower crust. Chen and Molnar (1983) suggest this effect indicates that the temperature near the Moho is above the limiting temperature for seismicity in the lower crust but below that for the stronger upper mantle.

The stresses indicated by the focal mechanisms of intraplate earthquakes also show a dependence on lithospheric age (Fig. 4). Older oceanic lithosphere shows thrust and strike slip faulting, indicating horizontal compression, whereas young oceanic lithosphere also shows some events indicating horizontal tension. The compression in older lithosphere is generally interpreted as reflecting the effect of "ridge push" stresses caused by the elevation of

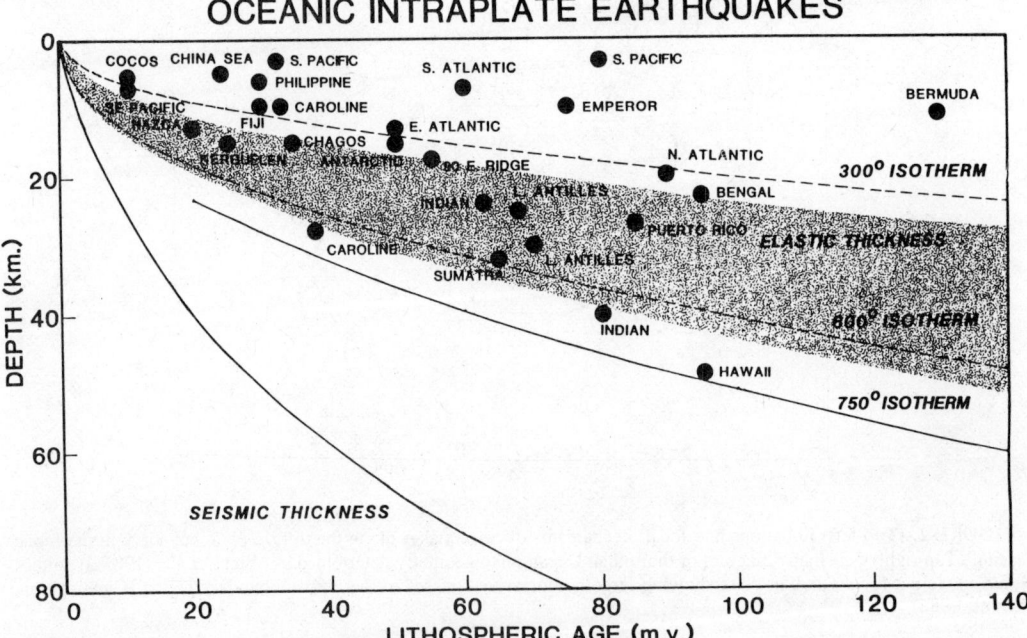

FIGURE 3. Depths of oceanic intraplate earthquakes as a function of lithospheric age, together with isotherms derived from a plate cooling thermal model. The depths increase with lithospheric age and are approximately limited by the 750°C isotherm. Oceanic intraplate earthquakes occur at depths comparable to the flexural elastic thickness but less than the seismic thickness of the lithosphere (Wiens and Stein, 1983).

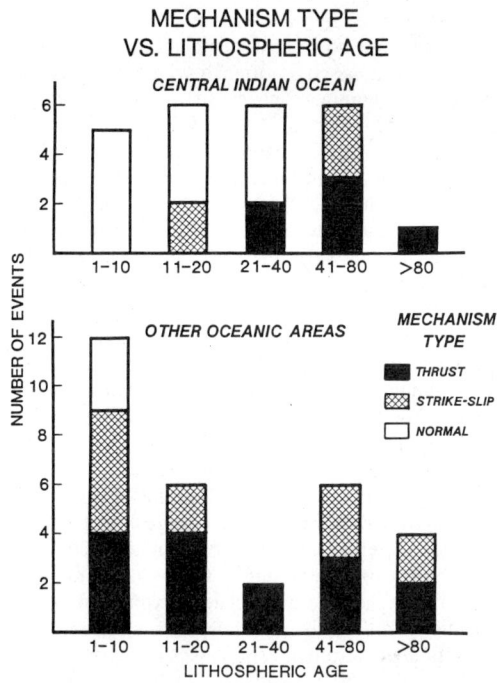

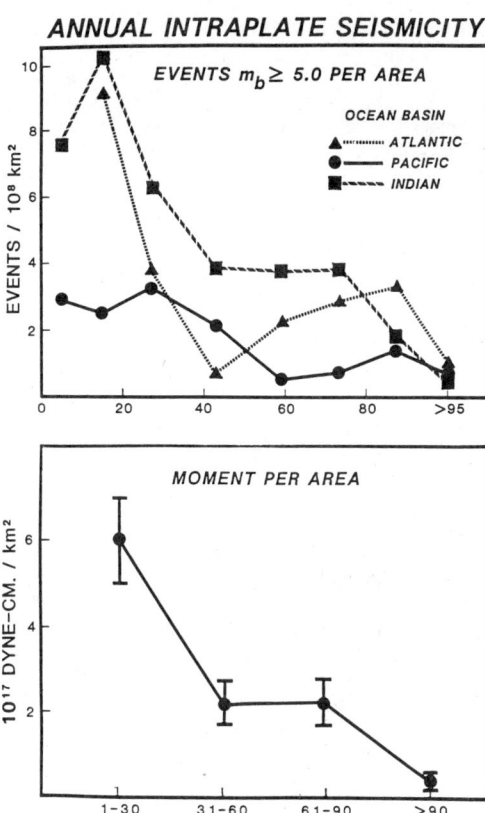

FIGURE 4. Faulting type as a function of lithospheric age for oceanic intraplate seismicity (Wiens and Stein, 1984). Older oceanic lithosphere shows thrust and strike-slip faulting, indicating horizontal compression dominates. Younger lithosphere shows a heterogeneity of faulting styles. The Indian Ocean data is shown separately because of its anomalous tectonic setting.

FIGURE 5. Oceanic intraplate seismicity per area as a function of lithospheric age (Wiens and Stein, 1984). Seismicity is highest in young lithosphere and decreases with lithospheric age, possibly suggesting that thermoelastic stresses produce significant seismicity.

the midocean ridges. The observed compression also places upper limits on the magnitude of viscous drag forces at the base of oceanic plates. Because the ridge push stress increases with the age of the plate, this stress is small near the midocean ridges, where thrust, strike-slip, and normal faulting events are observed.

Further evidence for a systematic difference in the stress field between young and older oceanic lithosphere comes from the seismicity rates. Figure 5 shows that the number of intraplate earthquakes per unit area of the oceanic lithosphere decreases with lithospheric age, a trend also observed for the total seismic moment (Fig. 5, bottom). Seismicity as a function of lithospheric volume shows an even more dramatic decrease with age. The high seismicity rate in young oceanic lithosphere may indicate that thermoelastic stress, caused by the contraction of the oceanic lithosphere as it cools, is an important component in producing intraplate earthquakes in young oceanic lithosphere. Additional evidence for the importance of thermoelastic stress in young oceanic lithosphere comes from the depths of earthquakes, which show normal faulting events located deeper than thrust faulting events. This stratification of compressional and extensional stresses is a notable result of stress models for the oceanic lithosphere, which include the effect of thermoelastic stress (Bratt et al., 1985). However, the earthquake evidence is somewhat equivocal, as most of the earthquakes making up this dataset are located in the Indian Ocean where large-scale deformation is occurring, suggesting the earthquakes may not be indicative of those found in typical intraplate regions.

DOUGLAS A. WIENS

References

Bratt, S. R., E. A. Bergman, and S. C. Solomon, 1985, Thermoelastic stress: how important as a cause of earthquakes in young lithosphere?, *Jour. Geophys. Research* **90,** 10249–10260.

Chen, W. P., and P. Molnar, 1983, The depth distribution of intercontinental and intraplate earthquakes and its implications for the thermal and mechanical properties of the lithosphere, *Jour. Geophys. Research* **88,** 4183–4214.

Cloetingh, S., and R. Wortel, 1986, Stress in the Indo-Australian plate, *Tectonophysics* **132**, 49-67.
Nuttli, O. W., 1973, The Mississippi Valley earthquakes of 1811 and 1812: Intensities, ground motion and magnitudes, *Seismol. Soc. America Bull.* **63**, 227-248.
Richardson, R. M., S. C. Solomon, and N. H. Sleep, 1979, Tectonic stress in the plates, *Rev. Geophysics Space Physics* **17**, 981-1008.
Sibson, R. H., 1982, Fault zone models, heat flow, and the depth distribution of earthquakes in the continental crust of the United States, *Seismol. Soc. America Bull.* **72**, 151-163.
Wiens, D. A., and S. Stein, 1983, Age dependence of oceanic intraplate seismicity and implications for lithospheric evolution, *Jour. Geophys. Research* **88**, 6455-6468.
Wiens, D. A., and S. Stein, 1984, Intraplate seismicity and stresses in young oceanic lithosphere, *Jour. Geophys. Research* **89**, 11442-11464.
Wiens, D. A., C. Demets, R. Gordon, S. Stein, D. Argus, J. Engeln, P. Lundgren, D. Quible, C. Stein, S. Weinstein, and D. Woods, 1985, A diffuse plate boundary model for Indian Ocean tectonics, *Geophys. Research Letters* **12**, 429-432.

Cross-references: Earthquake Mechanisms and Plate Tectonics; Earthquakes: Hazards and Predictions; Earthquakes: Magnitude, Energy, and Intensity; Earthquakes: Volcanogenic; Earthquakes and Crustal Deformation; Earthquake Seismology; Earthquakes and Seismicity; Seismicity and Plate Tectonics; Stress in the Earth's Lithosphere.

SEISMICITY: MID-OCEAN RIDGE

Because the oceanic lithosphere is formed at spreading centers, mid-ocean ridge seismicity directly reflects the evolution of the oceanic lithosphere. The mechanisms of these earthquakes show the basic kinematics of ridges and transforms. Moreover, this seismicity provides key evidence for the thermal-mechanical process that controls the evolution of the oceanic lithosphere.

Figure 1 contrasts the distribution of earthquakes on a portion of the Mid-Atlantic Ridge and the East Pacific Rise. On the Mid-Atlantic Ridge both the north-south trending ridge crest segments and the east-west trending transforms are seismically active. New lithosphere is formed at the ridges by seafloor spreading and then moves away. At the ridge crest, an axial valley forms parallel to the ridge trend. Normal faulting earthquakes are found to occur in this region with nodal planes parallel to the ridge trend. The transform faults are the sites of relative motion between lithosphere moving in opposite directions on either side. Thus, strike-slip faulting with one nodal plane parallel to the transform trend is seen. Earthquakes occur almost exclusively on the active segment of the transform fault between the two ridge segments, even though an inactive extension, known as a *fracture zone*, extends to either side. Although no relative plate motion occurs on the fracture zone, it is often marked by a distinct topographic feature. The sense of motion on the transform fault, right or left lateral, depends only on the direction of ridge segment offset.

The seismicity pattern is somewhat different along the East Pacific Rise. Here earthquakes occur on the transform faults, with the appropriate strike-slip mechanisms, but very few earthquakes occur on the ridge crest itself. This is probably related to the observation that the East Pacific Rise has an axial high rather than the axial valley that occurs along the Mid-Atlantic Ridge. The difference between the two is thought to be due to the difference in spreading rates. Slow spreading ridges, with half-spreading rates less than about 6 cm · yr^{-1}, usually have axial valleys, whereas fast spreading ridges (half rates greater than about 6 cm · yr^{-1}) almost always have axial highs and thus do not have ridge crest faulting.

Important effects are observed in the depths of ridge crest earthquakes and in their seismic moment on slow spreading ridges. The maximum depth of seismicity in the oceanic lithosphere is known to be temperature controlled; earthquakes occur only down to about the 750°C isotherm, in accord with the predictions of experimental rock mechanics. On mid-ocean ridges, the maximum depth of seismicity decreases with spreading rate. Moreover, the maximum seismic moments of ridge crest normal faulting earthquakes (Fig. 2) and transform fault earthquakes (Fig. 3) decrease with spreading rate. Moreover, a decrease in the seismic moment of ridge crest normal faulting earthquakes is observed (Fig. 2). Thus, for a given depth, increasing the spreading rate increases the temperature of the ridge crest and decreases the area available for faulting. These relationships are also seen on transform faults (Fig. 3). In general, centroid depths are limited to a shallow zone.

The difference between fast and slow spreading ridges, and the nature of the earthquakes associated with them, is a consequence of the thermal evolution of the oceanic lithosphere. This process can be described by using an elegant, very simple, but powerful model for the formation of the lithosphere by hot material at the ridge, which cools as the plate moves away.

In this model material at the ridge at mantle temperature T_m (about 1100°C) is brought to the ocean floor, which is at temperature T_s. The material is then transported away at velocity u while its upper surface remains at T_s (Fig. 4). If we neglect horizontal heat conduction, compared with vertical heat conduction, this problem is the same as the sudden cooling of a half-space—a standard problem in which the surface of a half-space, at temperature $T = T_m$ is suddenly cooled to T_s at time $t = 0$. If the surface is maintained at T_s, the half-space cools with

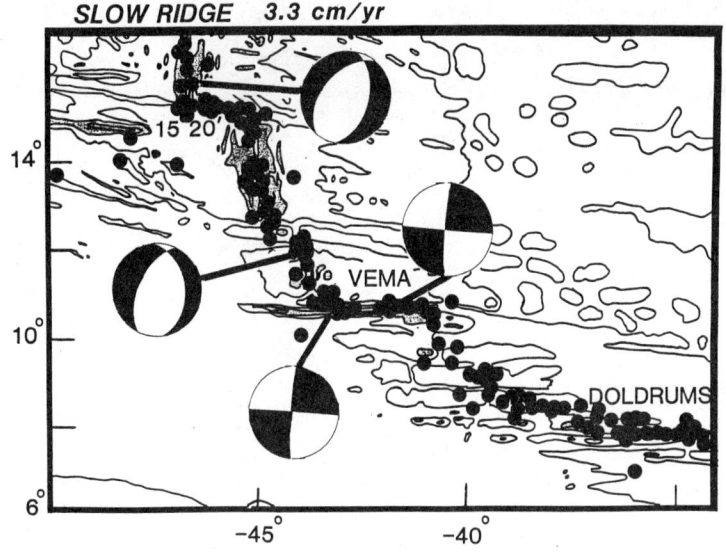

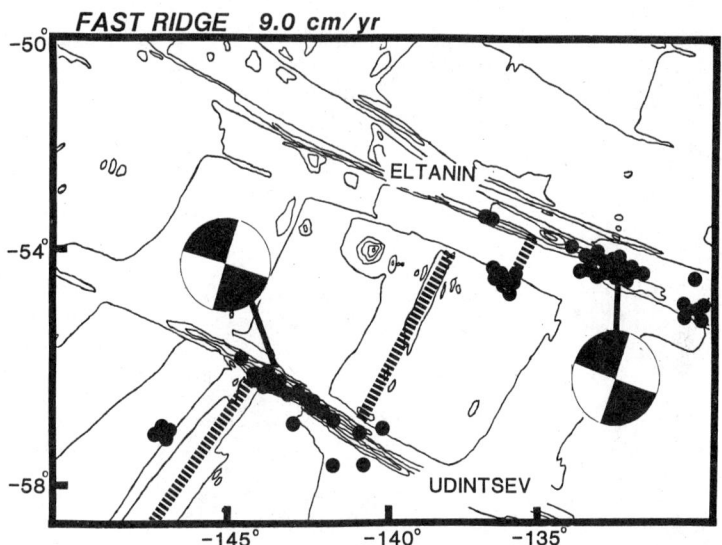

FIGURE 1. Maps contrasting faulting on slow and fast spreading centers: (top) The Mid-Atlantic ridge has earthquakes both on the active transform and ridge segments. Strike-slip faulting on a near vertical plane parallel to the transform azimuth is characteristic. On the ridge segments, normal faulting with nodal planes parallel to the ridge trend is seen. (bottom) The East Pacific Rise has strike-slip earthquakes on the transform segments, but no ridge axis normal fault events. Mechanisms from Engeln et al. (1986), Huang et al. (1986), and Stewart and Okal (1983).

time. The temperature as a function of depth and time is given by the one-dimensional heat flow equation

$$\frac{\partial T}{\partial t} = \frac{k}{\rho C_p} \frac{\partial^2 T}{\partial z^2} = \kappa \frac{\partial^2 T}{\partial z^2}$$

κ, the thermal diffusivity, measures the rate at which heat is conducted. It has units of distance squared divided by time and is defined as $\kappa = k/\rho C_p$, where k is the thermal conductivity, ρ is the density, and C_p is the specific heat at constant pressure.

The solution, assuming that any column of lithosphere cools this way and that $T_s = 0°C$, is

$$T(z, t) = T_m \, erf\left(\frac{z}{2\sqrt{\kappa t}}\right)$$

RIDGE CREST EARTHQUAKES

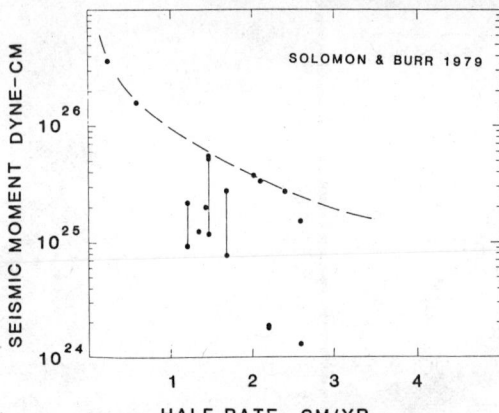

FIGURE 2. Decrease in maximum seismic moment of ridge crest normal fault earthquakes with spreading rate (Solomon and Burr, 1979).

TRANSFORM FAULT EARTHQUAKES

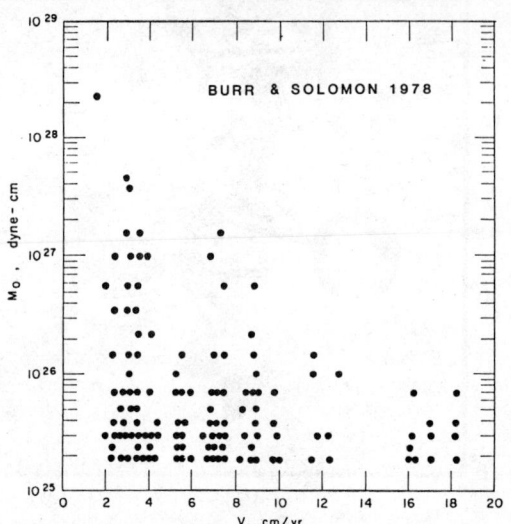

FIGURE 3. Seismic moment versus spreading rate for oceanic transforms (Solomon and Burr, 1979). The maximum moment decreases with spreading rate as expected from thermal considerations.

which gives the temperature at a depth z for material of age t. Since the age of the lithosphere is its distance from the ridge divided by the half-spreading rate v, the temperature as a function of position is just

$$T(z, x) = T_m \, erf\left(\frac{z}{2\sqrt{\kappa x/u}}\right)$$

Isotherms, lines of constant temperature in the plate, have the simple form

$$z_c = 2c \sqrt{\kappa t}$$

so the depth to a given temperature increases as the square root of lithospheric age (Fig. 5). The thickness of the lithosphere is an elusive concept, since this thickness is defined by mechanical properties. In many cases these properties coincide with an isotherm, typically about 1100°C, which can be considered the base of a thermal lithosphere whose thickness increases as the square root of plate age.

This simple cooling model predicts a variety of observed properties. In particular, it predicts the increase in ocean depth and decrease in heat flow, both of which vary as the square root of lithospheric age. Moreover, the depth to the low-velocity zone determined from the dispersion of surface waves, the maximum depth to which earthquakes occur within the oceanic lithosphere, and the thickness of elastic lithosphere inferred from the flexure due to loads such as seamounts all increase with age.

The thermal model also explains a great deal about earthquakes on ridges and transforms. The differ-

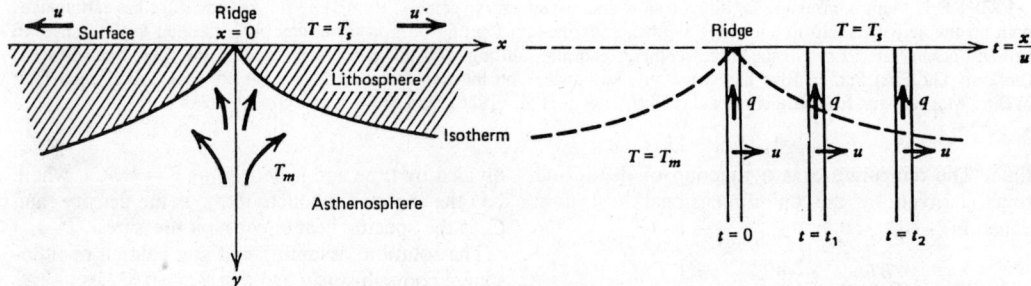

FIGURE 4. The cooling of the plate as it moves away from the ridge axis is the primary tectonic process in the oceanic lithosphere. Since a given column moves away from the ridge faster than heat is conducted in the horizontal direction, the cooling in the vertical direction can be treated as a one-dimensional problem (Turcotte and Schubert, 1982).

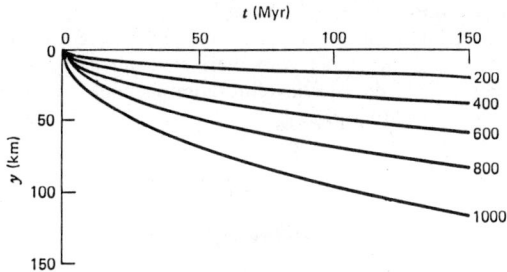

FIGURE 5. Predicted isotherms in the oceanic lithosphere, showing the square root of age cooling and thickening (Turcotte and Schubert, 1982).

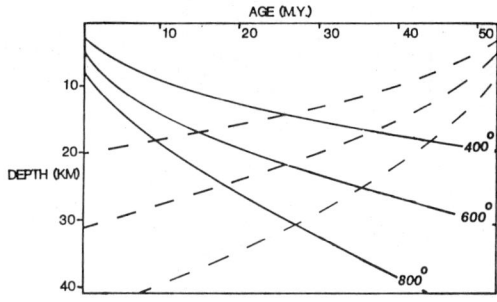

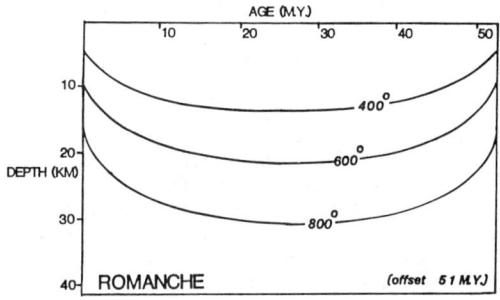

FIGURE 7. Thermal structure of the Romanche Transform. (top) temperatures on either flank predicted by the cooling half-space model; (bottom) average temperature distribution along the transform (Engeln et al., 1985).

ence between slow and fast spreading ridges is thought to be due to the different thermal structures. Figure 6 contrasts simple models for fast and slow ridges. At a given distance from the ridge, fast spreading produces younger and thinner lithosphere than does slow spreading. If the region below the 1185°C isotherm and above a depth of 5 km is considered to be a magma chamber through which material can rise, the area near the axis of a fast ridge has a larger magma chamber than does the slow ridge. As a result, the lithosphere moving away from a fast spreading ridge is more easily replaced than that for a slow ridge. Thus, in contrast to the axial valley and normal faulting earthquakes on a slow ridge, a fast ridge has an axial high and no earthquakes.

The relation between maximum seismic moment and spreading rate is another consequence, if we assume the area available for faulting is that above a given isotherm. Since isotherms shallow with increasing spreading rate, the depth of ridge crest earthquakes and their maximum moment should decrease. Similarly, since the thermal structure of a transform fault should be essentially the average of the expected temperature on the two sides (Fig. 7), the area available for faulting should decrease with spreading rate.

The depth of seismicity has interesting implications for the slip process on transforms. Given the depth of faulting, seismic slip rates can be inferred from total seismic moment release on a transform:

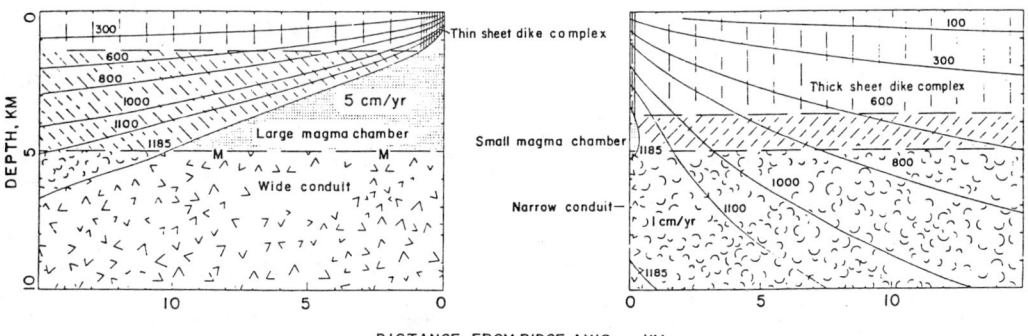

FIGURE 6. Petrological model for the difference between fast and slow spreading ridges (Sleep and Rosendahl, 1979). Fast ridges with large magma chambers replace lithosphere at the ridge quickly, so no volume deficit exists and no axial valley forms. Slow ridges have small chambers and cannot replace lithosphere as quickly as it spreads away, creating axial valleys.

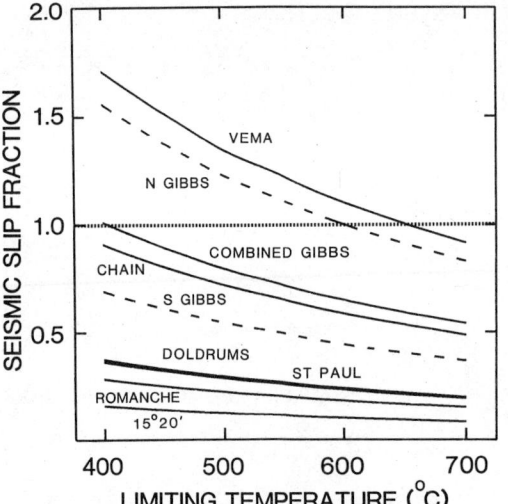

FIGURE 8. Fraction of seismic slip (seismic slip rate divided by predicted plate motion) versus limiting isotherm for major Atlantic transforms. All transforms except the Vema have a significant proportion of aseismic slip. The Vema may have a larger than estimated fault area due to tectonic complications. Because a short offset in the Gibbs Transform makes it unclear whether it should be regarded as one or two transforms, results are shown for both cases (Engeln et al., 1985).

Seismic slip rate

$$= \frac{total\ seismic\ moment}{(fault\ area)(rigidity)(time\ window)}$$

Fault areas corresponding to the region above different isotherms can be calculated numerically for each transform and compared with those predicted by plate motions. Assuming different possible limiting temperatures for seismic failure, seismic moment releases for six major Atlantic transforms covering the period 1920–1979 account for varying fractions of the slip predicted by the plate motion (Fig. 8). Some aseismic slip must be occurring if the time period sampled is long enough to be representative.

It is unclear what causes the variation in the seismic fraction of slip among transforms. The Vema Transform is a clear example; inferred seismic slip is substantially greater than predicted unless faulting extends to the 650°C isotherm. Transforms may differ substantially; the Vema may have an unusually long recurrence time or be mechanically anomalous due to its complex tectonic history.

A great deal remains to be learned about spreading centers, though the basic picture is well understood. One very powerful tool now being used is microearthquake surveys, which use arrays of ocean bottom seismometers to obtain precise locations. The larger earthquakes are also actively being studied through body wave and surface wave methods. Reflection and refraction seismology are yielding a better image of the velocity structure. These results are being integrated with marine geophysical and petrological data to develop more accurate models of the spreading process.

SETH STEIN
DALE F. WOODS

References

Engeln, J. F., D. A. Wiens, and S. Stein, 1986, Mechanisms and depths of Atlantic transform earthquakes, *Jour. Geophys. Research* **91**, 548–577.

Huang, P. Y., S. C. Solomon, E. A. Bergman, and J. L. Nabelek, 1986, Focal depths and mechanisms of Mid-Atlantic Ridge earthquakes from body waveform inversion, *Jour. Geophys. Research* **91**, 579–598.

Sleep, N. H., and B. R. Rosendahl, 1979, Topography and tectonics of mid-ocean ridge axes, *Jour. Geophys. Research* **84**, 6831–6839.

Solomon, S. C., and N. C. Burr, 1979, The relationship of source parameters of ridge-crest and transform earthquakes to the thermal structure of oceanic lithosphere, *Tectonophysics* **55**, 107–126.

Stewart, L. M., and E. A. Okal, 1983, Seismicity and aseismic slip along the Eltanin Fracture Zone, *Jour. Geophys. Research* **88**, 10495–10507.

Turcotte, D. L., and G. Schubert, 1982, *Geodynamics: Applications of Continuum Physics to Geological Problems*. New York: Wiley.

Cross-references: Earthquake Mechanisms and Plate Tectonics; Earthquake Seismology; Heat Flow in the Earth; Lithosphere, Oceanic: Formation and Evolution; Seismicity and Plate Tectonics; Seismic Source: Observations; Seismic Source: Theory.

SEISMICITY: SUBDUCTION ZONE

Earthquakes do not occur randomly within the Earth: Most are confined to narrow continuous belts that wrap around the Earth. Plate tectonics proposes that the entire surface of the Earth is covered with large, rigid mosaiclike plates that move relative to each other causing tectonic activity at their boundaries. Most earthquakes concentrate and form narrow, belt-shaped seismic zones along these plate boundaries.

New oceanic plates are produced at the midocean ridges by the upwelling of mantle material. The Earth's surface does not expand by the production of the new oceanic plates, because an equal amount of material returns downward, back into the mantle. Generally, this consumption of the plates is made at *subduction zones*, where oceanic plates plunge downward into the mantle. At subduction zone two plates collide with each other. One plate must ride over the other, forcing the heavier one back into the mantle. Thus the heavier oceanic plate subducts beneath the lighter continental plate. This subduction of the oceanic plate causes high seismic activity along the zone. Indeed, most of the world's large earthquakes occur in subduction zones.

Shallow Seismicity in Subduction Zones

Along subduction zones, seismic activity is generally high at shallow depths (shallower than about 60–70 km). Two principal types of focal mechanisms have been found for these shallow earthquakes (Isacks et al., 1968). The first is characterized by a predominance of low-angle thrust faulting (type A in Fig. 1). Extremely high seismicity with this focal mechanism type is located beneath the landward slope of the ocean trench and forms a slablike zone that dips beneath the island arc or arclike structure. Based on the hypocenter distribution and the focal mechanism type mentioned above, this shallow seismicity can be interpreted to originate at the boundary between the subducting oceanic and overriding continental plates from stresses generated by the relative motion of the two converging lithospheric plates. The low-angle thrust faulting mechanisms characteristic for shallow earthquakes at subduction zones thus reflect directly the relative movements of the two converging plates and the downward motion of the subducting oceanic plate.

The second type is a less active shallow seismicity located beneath the trench and is characterized by a predominance of normal faulting (type B in Fig. 1). As a relatively strong oceanic plate with a certain thickness passes beneath the ocean trench, it bends abruptly and thus extensional stresses normal to the trench axis are expected near the surface on the convex side of the bend. The shallow seismicity with normal faulting beneath the trench is a manifestation of the abrupt bending of the subducting oceanic plate. As a result of the bending beneath the trench axis, the principal stress in the deeper part of the oceanic plate is expected to be compressional. In fact, some of the earthquakes that occurred beneath the oceanic trench are reported to have this type of focal mechanism with focal depths of about 40–50 km, although the number of this type earthquake is quite small. In addition to the foregoing seismic activities beneath the trench and the landward slope of it, in some subduction zones, shallow earthquakes occur beneath the inland area (type C in Fig. 1). This shallow seismicity, confined mainly to the crust, is secondary to the major activity along the main thrust zone beneath the landward slope of the trench.

Deep Seismic Zone (Wadati-Benioff Zone) and Distribution of Stress within Subducting Plate

The shallow seismicity along the main thrust zone is continuous with the deep seismic zone (Wadati-Benioff zone), which dips beneath the island arc or arc-like structure. However the simple underthrusting typical of the shallow earthquakes along the main thrust zone does not persist at depths deeper than about 60–70 km. This deeper seismicity does not originate at the boundary between the two converging plates but within the subducted plate itself. Figure 2 shows stress distributions in the subducting lithospheric plates for many subduction zones (Isacks and Molnar, 1971). Also shown are approximate cross-sectional shapes of the deep seismic zones. There are remarkable systematic differences in focal mechanism type between intermediate-depth (60–300 km) and deep (>300 km) earthquakes. Deep earthquakes are almost invariably characterized by "down-dip compression," that is, the pressure axis of focal mechanism solution is parallel to the local dip of the deep seismic zone. If the tension axis is parallel to the local dip of the deep seismic zone, we call it "down-dip extension." Intermediate-depth earthquakes are more variable and include down-dip compression, down-dip extension, and other focal mechanism types. If examined

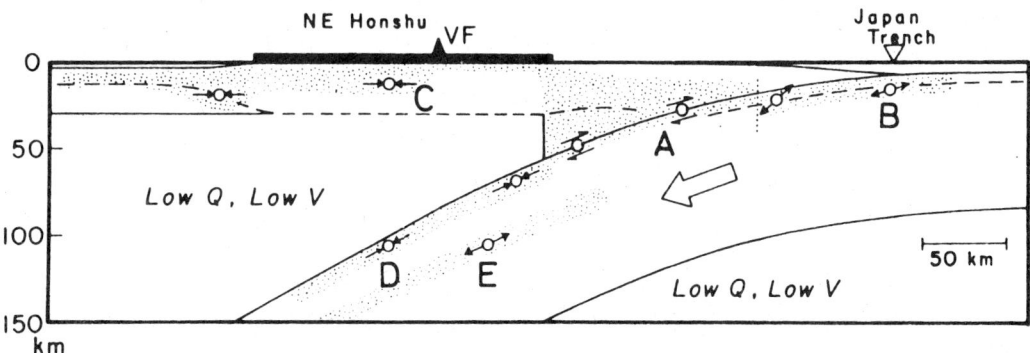

FIGURE 1. Schematic representation of vertical cross section of the crust and uppermost mantle beneath northeastern Honshu, Japan. Thick horizontal line and VF denote the land area and the volcanic front, respectively. The Pacific plate subducts beneath the land area at an angle of about 30° causing the seismic activity shown by the stippled areas. Focal mechanism types are also shown schematically. (After Yoshii, 1979)

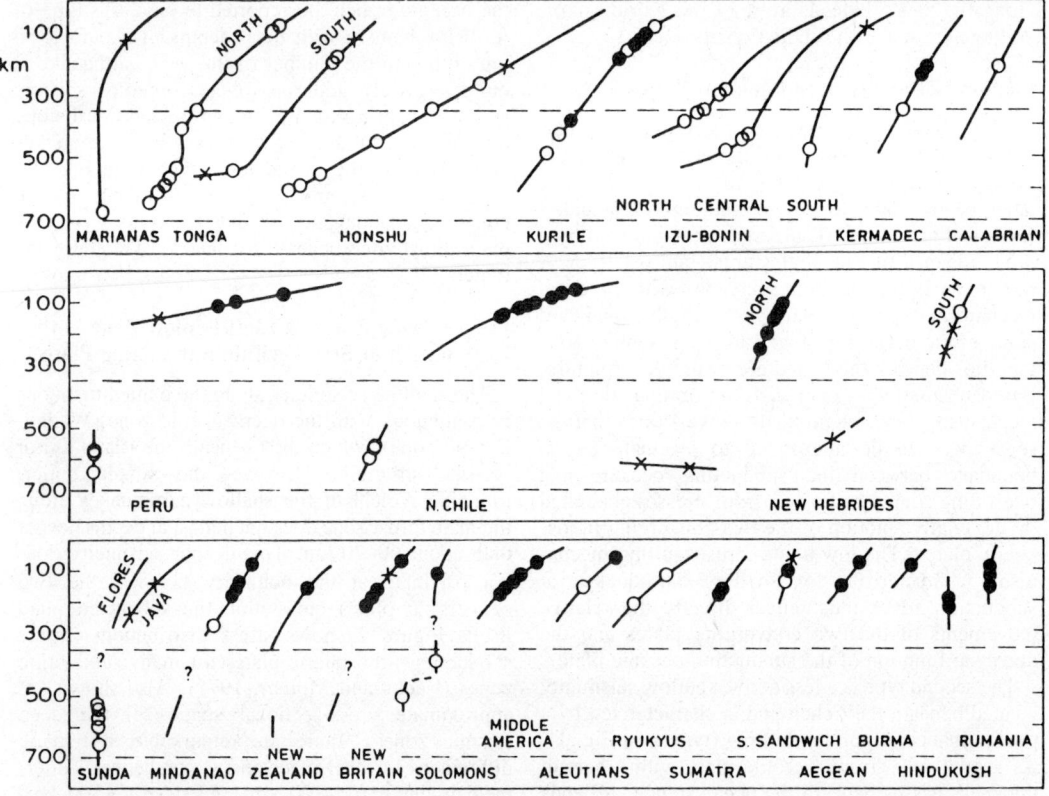

FIGURE 2. Global summary of the stress state within the deep seismic zone. Solid lines are approximate cross-sectional shapes of the deep seismic zones. Open circles indicate where earthquakes are compressional in the direction of the local dip of the deep seismic zone (down-dip compression). Solid circles are where earthquakes are extensional in the direction of the local dip of the zone (down-dip extension). Crosses indicate that neither the compression nor the tension axis is approximately parallel to the dip. (After Isacks and Molnar, 1971)

in detail, however, we can see some correlation between the focal mechanism type of intermediate-depth earthquakes and the distribution of seismic activity with depth. Down-dip extension is predominant in several subduction zones where either notable seismicity gaps are present between deep and intermediate-depth earthquakes (e.g., in Peru, northern Chile, and Vanuatu, New Hebrides; see also Fig. 4) or where deep earthquakes do not occur. On the contrary, down-dip compression is predominant where the deep seismic zone is continuous down to 600 km or so, though this correlation is not so clear.

Isacks and Molnar (1971) explained this remarkable feature of stress distributions within subducted plates (Fig. 3). The lithospheric oceanic plate, which is colder and hence heavier than the surrounding mantle, sinks into the asthenosphere under its own weight and is under extension (Fig. 3a). If the subducted oceanic plate penetrates the deeper part of the mantle, it encounters resistance to its downward motion and the stress state changes from down-dip extension to compression (Fig. 3b). The cause of this resistance might be either an increase in strength of the surrounding mantle with depth or a buoyant effect of the deeper portion of the subducted plate with lower density than the surrounding mantle.

The depth interval in which down-dip compression is predominant is bounded by the depths of about 350–400 km and 650–700 km. At these depths, phase changes of mantle material are expected to occur in the normal mantle. At about 400 km depth, seismic wave velocity increases sharply due to the olivine-spinel phase change (400-km discontinuity). Abrupt increase in seismic wave velocity appears again at about 650 km (650-km discontinuity). This velocity jump is presumed to be caused by the spinel-postspinel phase transition. The depths at which the two major phase changes actually occur within the subducted plate should be slightly different from those in the surrounding normal mantle. The depth of the shallower transition (olivine-spinel phase change) is thought to be elevated and that of the deeper transition (spinel-postspinel phase change) is presumed to be lowered within the subducted plate (Schubert et al., 1975). The additional load of dense

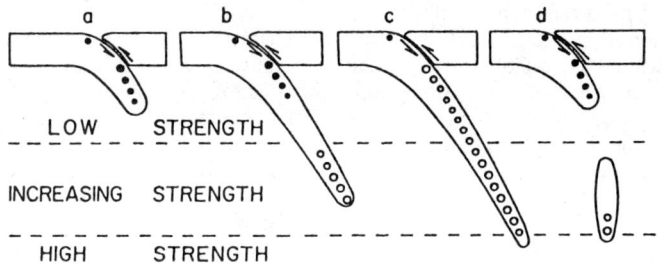

FIGURE 3. Distribution of stresses within the subducted plates where gravitational forces act on excess mass within the plates. Open and solid circles represent down-dip compression and down-dip extension, respectively. The relative amount of the seismic activity within the plate is qualitatively shown by the size of the circle. In (a) the plate sinks into the asthenosphere, and the load of excess mass is mainly supported by forces applied to the plate above the sinking portion. In (b) the plate penetrates the deeper portion of the upper mantle and encounters resistance to its downward motion. Part of the load is supported from below and part from above, the stress changing from extension to compression with depth. In (c) the entire load is supported from below, and the plate is under compression throughout. In (d) a piece has broken off. Horizontal dashed lines indicate possible phase changes in the upper mantle near 350–400 km and 600–700 km. (After Isacks and Molnar, 1971)

material, due to the elevation of the olivine-spinel phase boundary within the plate, further drives the lower portion of the plate (below about 350 km) downward. If the spinel-postspinel phase boundary is depressed within the subducted plate, an upward buoyant force will be generated, which is one possibility for the cause of the resistance to the sinking of the oceanic plate.

In several subduction zones, the deep seismic zones are not continuous but have remarkable seismicity gaps between intermediate-depth and deep earthquakes. (A clear example of the seismicity gap is seen in Fig. 4, which shows vertical cross section of the deep seismic zone beneath Vanuatu, New Hebrides). If the lower portion of the subducted plate becomes detached and is not connected with the upper portion, the predominant stresses within the upper portion of the plate will be down-dip exten-

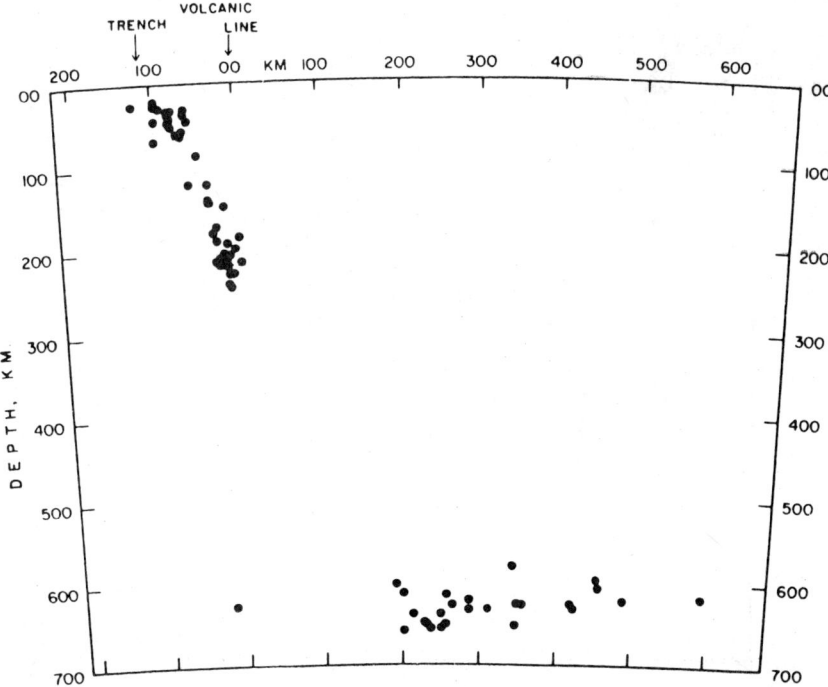

FIGURE 4. Vertical seismic section showing the deep seismic zone beneath Vanuatu, New Hebrides. A seismicity gap is clearly seen between about 300 km and 600 km. (After Pascal et al., 1978)

sion. On the contrary, the detached piece of the plate is under compression (Fig. 3d). Thus the deep seismic zones shown in the middle part and the lefthand side of the lower part of Fig. 2 are considered to correspond to b or d in Fig. 3. The sections in the upper part and the righthand side of the lower part of Fig. 2 correspond to c and a in Fig. 3, respectively.

Focal Depth Distribution and Depths of the Deepest Deep Earthquakes

Frequency distributions of seismicity for several subduction zones are shown in Fig. 5 as a function of focal depth (Isacks et al., 1968). In all subduction zones shown the seismicity is most active in the shallowest portion. This active seismicity decreases nearly exponentially as a function of depth in the upper 200 km. At greater depths the seismicity, when it is present, increases again and forms a fairly well-defined maximum in some depth range in the upper mantle. The variation of seismic activity with depth thus correlates well with the variation of focal mechanisms with depth (Fig. 2) and suggests that the cause of deep earthquakes is different from that of shallow or intermediate-depth earthquakes. Relatively high seismic energy release in the deeper portion of the deep seismic zone (between 350–400 km and 650–700 km) with down-dip compressional stress is explained by the elevation and depression of the two major phase boundaries (olivine-spinel phase change at about 400 km depth and spinel-postspinel phase change at about 650 km depth) within the subducted plate, which act to compress the plate between the two phase boundaries. The increasing strength of the surrounding mantle material with depth is another possibility for the cause of the high seismic energy release in this depth range.

Earthquakes do not occur in the lower mantle. The maximum depth of the deep seismicity in subduction zones is a fundamental observation that provides important information on mantle dynamics. As one can see from Fig. 2, the maximum depth of the deep

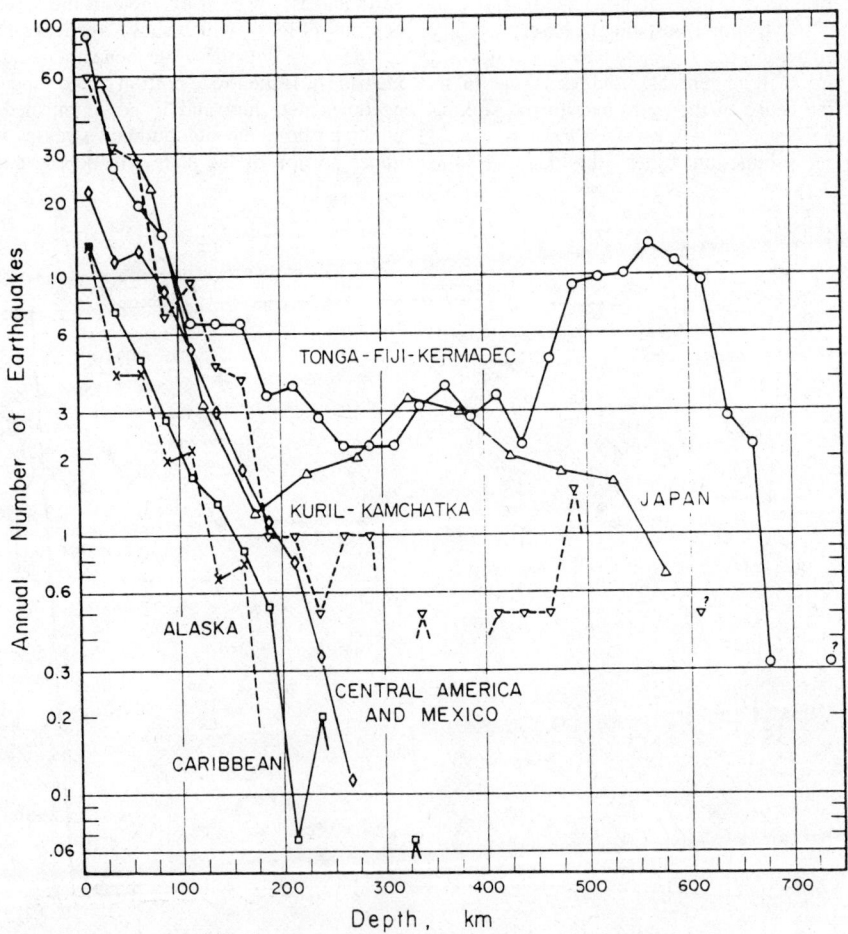

FIGURE 5. Focal depth distributions of earthquakes in several subduction zones. (After Isacks et al., 1968)

seismic zone is about 700 km. Recent reinvestigation on the depths of deep earthquakes reveals that the deepest reliable focal depths of deep earthquakes are at most 670–680 km and that the cessation of seismic activity at this depth is not gradual but rather abrupt (Stark and Frohlich, 1985). The following might explain this abrupt termination of the deep seismicity at about 670–680km in widely separated areas: (1) The lower mantle mechanically prevents the penetration by the subducting plate due to a large and sudden increase in strength of mantle material at the boundary between the upper and lower mantle. (2) As the subducting plate goes into the lower mantle, the spinel-postspinel phase change occurs and alters the mechanical properties of the subducted plate.

The first idea, that the lower mantle presents a barrier to the penetration of the subducting plate, can certainly explain the abrupt cessation of deep seismicity and also the predominance of down-dip compressional focal mechanisms for deep earthquakes. Seismic wave travel time analyses from deep earthquakes, however, provide evidence that the subducting plates, in several subduction zones, indeed penetrate the boundary between the upper and lower mantle and the penetration depth of the plates is at least several hundred kilometers below the deep seismicity cutoff (Creager and Jordan, 1986). This observation of aseismic extensions of the subducted plate seems to be inconsistent with the first explanation.

As discussed earlier in this entry, the elevation of the olivine-spinel phase boundary and the depression of the spinel-postspinel phase boundary within the subducted plate both act to compress the plate between the boundaries. Consequently the compressional stresses within the plate between these depths become relatively high, which might well explain the abrupt termination of the deep seismicity at about 670–680 km, the high seismic energy release between about 400 km and 650 km, and the down-dip compression predominating focal mechanisms for deep earthquakes. The deep seismicity cutoff could correspond to the depressed location of the spinel-postspinel phase boundary within the subducted plate, below which stresses are supposed to be too small to generate earthquakes. However, this hypothesis is not the only explanation for why the deep seismicity with down-dip compression predominating focal mechanisms is active between about 400 km and 650 km and stops abruptly at about 670–680 km. Unfortunately this important problem has not been completely solved and unique explanations are not possible, because the details of the kinetics of the phase changes that might occur at this depth range are not well known.

Geometry of Deep Seismic Zone

The approximate shapes of the deep seismic zones perpendicular to the trench axis for many subduction zones are summarized in Fig. 2, which indicates a marked variation in the geometry of the deep seismic zone. For example, the upper portion of the deep seismic zone beneath Peru dips at an extremely shallow angle. On the other hand, the dip of the upper portion of the deep seismic zone is much steeper beneath Vanuatu, New Hebrides. Such a variation in the geometry of the subducted oceanic plate (or the geometry of the deep seismic zone) is caused by several interdependent factors (Cross and Pilger, 1982). A rapid absolute motion of the overriding plate toward the trench and a rapid convergence rate both cause the shallow-angle subduction of the oceanic plate. When intraplate island-seamount chains, aseismic ridges, or oceanic plateaus begin to subduct beneath island arcs or arclike structures, they will also cause the low-angle subduction of the plate owing to their anomalously low densities compared with the normal oceanic plate. The age of the subducting oceanic plate and the accretion of sediment in the ocean trench are other factors affecting the subduction zone geometry. Young oceanic plate is relatively buoyant (because of its low density) and subducts at a relatively low angle. Accretion of sediment in the ocean trench, owing to the weight of the accretionary prism, depresses and flattens the shallowest portion of the subducting plate causing the trench axis to migrate seaward.

Along each subduction zone the configuration of the subducting plate also changes laterally. In some cases the lateral change in the dip of the subducted plate appears abruptly at the junction between the two well-defined discrete segments. The subducting Nazca plate beneath western South America shows a clear example of the abrupt change in the dip of the deep seismic zone (Fig. 6). The spatial distribution of hypocenters that occurred in this region clearly defines four segments of the deep seismic zones, in each of which the deep seismic zone has relatively uniform shape. The two segments beneath northern and central Peru (about $2°$–$15°$S) and beneath central Chile (about $27°$–$33°$) have anomalously small dips, whereas the other two segments beneath southern Peru and northern Chile (about $15°$–$27°$S), and beneath southern Chile (about $33°$–$45°$) have normal dip angles of about $30°$ (Isacks and Barazangi, 1977). Anomalously flat geometries beneath northern and central Peru and beneath central Chile might be related to the subduction of the Nazca ridge and the Juan Fernandez ridge, which have relatively low densities.

Local seismic network data reveal that, in the transition zone from the $30°$ dipping plate beneath southern Peru to the flat profile beneath central Peru, the subducting Nazca plate is contorted rather than torn over a 100-km-wide continuous lateral section at least in the upper 150 km depth range. The configuration of the contorted Nazca plate in the transition zone is schematically illustrated and is shown in Fig.

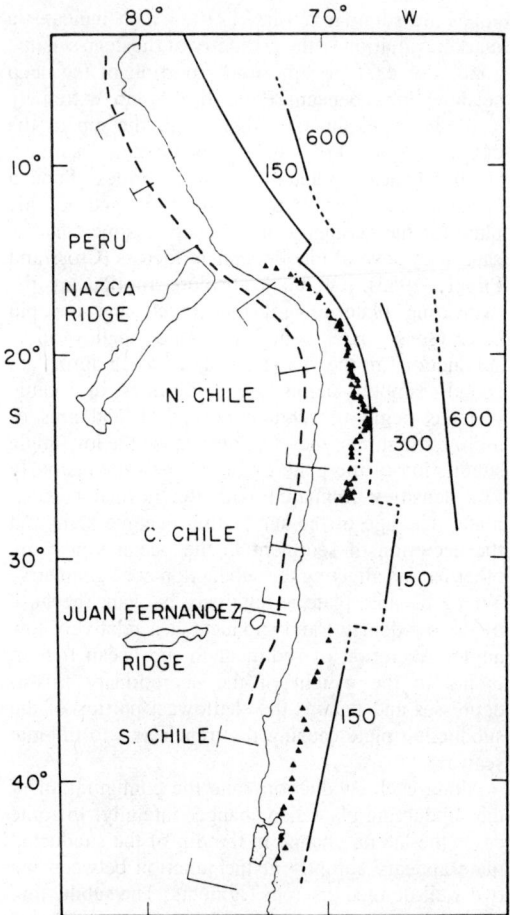

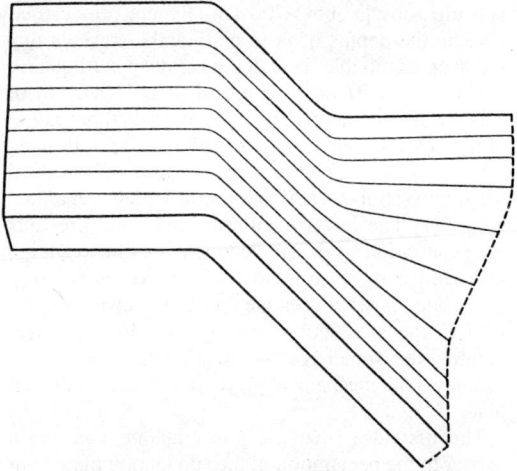

FIGURE 7. Schematic representation of the subducting Nazca plate in the transition zone from the nearly horizontal geometry beneath central Peru to the more steeply dipping one beneath southern Peru. The Nazca plate is not torn but contorted at least in the upper 150 km depth range. (After Hasegawa and Sacks, 1981).

FIGURE 6. Map showing contours of hypocentral depth to the top of the deep seismic zone beneath western South America. Dotted lines indicate that the contours are based on less data relative to the solid lines. Dashed line and solid triangles represent the trench axis and Quaternary volcanoes, respectively. The deep seismic zone is clearly divided into four segments in each of which the zone has relatively uniform shape. There is a remarkable correlation between the two flat segments of the deep seismic zone and the absence of Quaternary volcanoes on the overriding South American plate. (After Isacks and Barazangi, 1977)

7. In the other two transition zones beneath Chile we have no evidence for the contortion or the disruption of the subducting Nazca plate. Such a precise structure can be detected only from local seismic network data.

Double-Planed Deep Seismic Zone

Fine structure of the deep seismic zone beneath northeastern Honshu, Japan, is also obtained from local seismic network data (Hasegawa et al., 1978). Beneath this region the deep seismic zone is composed of two thin planes that are parallel to each other and are 30–40 km apart (Fig. 8). The shallower part of the upper seismic plane (shallower than 60 km) has low-angle thrust fault type (type A in Fig. 1), delineating the plate interface between the subducting oceanic plate and the overriding continental plate. The deeper part of the plane (deeper than 60km) is characterized by down-dip compression (type D in Fig. 1), while the lower seismic plane is characterized by down-dip extension (type E in Fig. 1). Subsequently, similar double seismic zones have been detected in a few other subduction zones where a sensitive local seismic network has been set up. These findings have partly updated the results of Isacks and Molnar (1971) on the distribution of stresses within the subducting plate.

Formation of such a double-planed deep seismic zone can be explained at least qualitatively by the unbending of the subducted oceanic plate (Engdahl and Scholz, 1977). As previously discussed, the subducting oceanic plate has undergone abrupt bending at shallow depths beneath the trench axis. The double-planed deep seismic zone appears at depths deeper than about 70 km. At these depths the oceanic plate is no longer bent and it subducts straightly with a constant dip angle. Thus the unbending of the plate should take place at these depths causing the compressional stresses near the surface and the extensional stresses deeper within the plate. The lower seismic plane will be formed along the lower edge of the elastic core, which occupies the upper portion of the plate.

Although this hypothesis is not the only explanation for the double seismic zone, the stress system producing the double seismic zone is probably quite

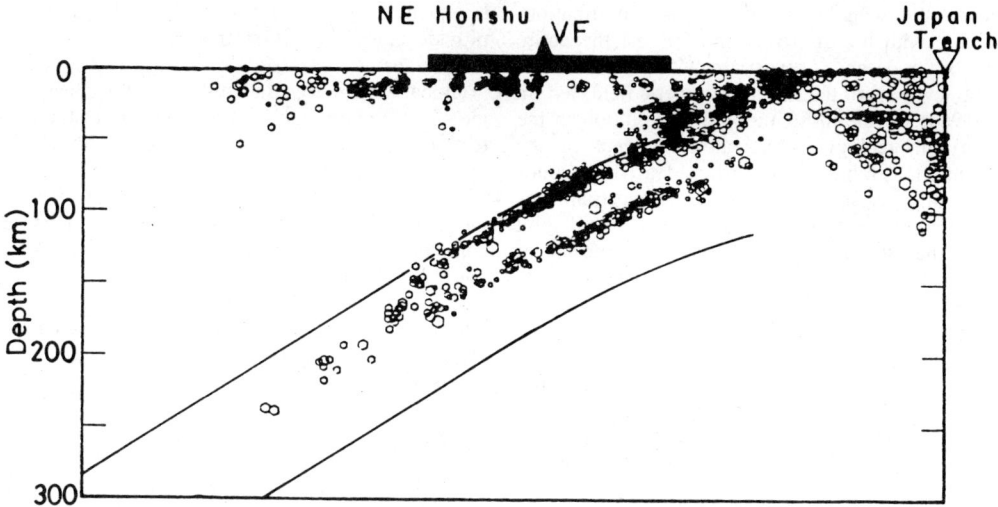

FIGURE 8. Vertical seismic section showing the double-planed deep seismic zone beneath northeastern Honshu, Japan. Thick horizontal line and VF denote the land area and the volcanic front, respectively. Focal mechanism types for these earthquakes are schematically shown in Fig. 1.

common in subduction zones, which may be the reason that the focal mechanisms for intermediate-depth earthquakes are of mixed type in many subduction zones (Fig. 2), suggesting the relative importance of such a stress system in the generation of earthquakes within the subducting plate. Precise observations to find such a fine structure as the double seismic zone will provide valuable information for further understanding the processes actually occurring at subduction zones.

AKIRA HASEGAWA

References

Creager, K. C., and T. H. Jordan, 1986, Slab penetration into the lower mantle beneath the Mariana and other island arcs of the northwest Pacific, *Jour. Geophys. Research* **91**, 3573-3589.

Cross, T. A., and R. H. Pilger, Jr., 1982, Controls of subduction geometry, location of magmatic arcs, and tectonics of arc and back-arc regions, *Geol. Soc. America Bull.* **93**, 545-562.

Engdahl, E. R., and C. H. Scholz, 1977, A double Benioff zone beneath the central Aleutians: An unbending of the lithosphere, *Geophys. Research Letters* **4**, 473-476.

Hasegawa, A., and I. S. Sacks, 1981, Subduction of the Nazca plate beneath Peru as determined from seismic observations, *Jour. Geophys. Research* **86**, 4971-4980.

Hasegawa, A., N. Umino, and A. Takagi, 1978, Double-planed deep seismic zone and upper-mantle structure in the northeastern Japan arc, *Royal Astron. Soc. Geophys. Jour.* **54**, 281-296. 1978.

Isacks, B. L., and M. Barazangi, 1977, Geometry of Benioff zones: Lateral segmentation and downward bending of the subducted lithosphere, in T. Talwani and W. Pitman, eds., *Island Arcs, Deep Sea Trenches, and Back-Arc Basins*, Maurice Ewing Ser., vol. 1. AGU Washington, D.C.: American Geophysical Union, 99-114.

Isacks, B. L., and P. Molnar, 1971, Distribution of stresses in the descending lithosphere from a global survey of focal-mechanism solutions of mantle earthquakes, *Rev. Geophysics Space Physics* **9**, 103-174.

Isacks, B. L., J. Oliver, and L. R. Sykes, 1968, Seismology and new global tectonics, *Jour. Geophys. Research*, **73**, 5855-5899.

Pascal, G., B. L. Isacks, M. Barazangi, and J. Dubois, 1978, Precise relocations of earthquakes and seismotectonics of the New Hebrides island arc, *Jour. Geophys. Research* **83**, 4957-4973.

Schubert, G., D. A. Yuen, and D. L. Turcotte, 1975, Role of phase transitions in a dynamic mantle, *Royal Astron. Geophys. Jour.* **42**, 705-735.

Stark, P. B., and C. Frohlich, 1985, The depths of the deepest deep earthquakes, *Jour. Geophys. Research* **90**, 1859-1869.

Yoshii, T., 1979, A detailed cross-section of the deep seismic zone beneath northeastern Honshu, Japan, *Tectonophysics* **55**, 349-360.

Cross-references: *Continental Collision Zones: Seismotectonics and Crustal Structure; Deep Earthquakes; Earthquake Mechanisms and Plate Tectonics; Earthquake Seismology; Earthquakes and Seismicity; Earth Structure: Global; Mantle, Upper: Structure; Mantle Discontinuities; Mineralogic Phase Transitions in the Earth; Ocean-Continent Transition: Structure; Seismicity and Plate Tectonics; Subduction Zones; Tsunamis.*

SEISMICITY AND PLATE TECTONICS

Much new information about Earth's seismicity and ocean bathymetry emerged during the two decades following World War II. This information, together with the discovery that almost the entire floor of the oceans is a vast areal "tape recording" of the reversals of the Earth's magnetic field, led, in the 1960s, to the most important revolution in the history of geology. The central concept of the revolution, stemming from earlier ideas about continental drift, isostasy, and mantle convection, was that the outermost 100 km thick shell of the solid Earth, the *lithosphere*, is divided into a mosaic of large and medium size "plates" that move about relative to one another as strong, nondeforming shells over a weak, viscously flowing substratum called the *asthenosphere*. The relative motions of the plates are manifested along their common boundaries. The concept of plate tectonics has had an enormous impact on geology by explaining major features of the Earth, such as volcanos, earthquakes, and the differentiation of the Earth's surface into ocean basins and continents, as the result of processes that operate at or near plate boundaries. The Earth's seismicity directly reveals the globe-encircling boundaries between the plates as continuous, narrow zones of concentrated seismic activity separated by vast areas of low seismicity within the plates. Studies of seismicity consequently played a key role in the development of plate tectonics and have continued to produce important new information about the processes of volcanism and crustal formation and modification that take place along plate boundaries.

The concept of a plate boundary as a very narrow zone of deformation—essentially a large through-going fault in the lithosphere—applies best to two types of boundaries: (1) oceanic transform faults that form the transverse fracture zones of the mid-ocean ridge system, and (2) the classic *subduction zone* that forms a plate boundary where the relative motions of the interacting plates are convergent and where one of the interacting plates is suboceanic. A third sharply defined plate boundary is the crustal zone of seafloor spreading where the plate motions are divergent and often approximately perpendicular to the ridge axis. The seismicity is not aligned along a single fault but is associated with multiple strands of normal faulting and magmatic processes within the narrow axial zone. The seismicity of these three types of plate boundaries together sharply outline Earth's major plates.

In contrast to these well-defined oceanic plate boundaries, the boundaries that go through continents or (continental fragments such as New Zealand) are broad zones of diffusely distributed deformation and seismicity. The record of instrumentally recorded seismicity is still too short to reveal completely the nature and extent of the deformations in the intracontinental zones, and in some areas even the location of the plate boundary remains enigmatic. The plate interiors, although relatively aseismic, contain many interesting isolated zones of seismicity in both oceanic and continental areas that provide clues about the stresses within the plates themselves.

Global seismicity can thus be grouped naturally into the following categories according to the plate tectonic context: (1) convergent plate boundaries where one of the plates is oceanic (subduction zones); (2) actively spreading ocean ridge systems, including spreading centers and transform faults; (3) continental zones of diffuse but substantial deformations that absorb the relative motions between the interacting plates; and (4) the seismicity of plate interiors. These divisions are considered separately after first discussing the types of information derivable from the study of earthquakes.

The Earthquake Source

Nearly all earthquakes represent a rapid release of shear strain within a small region; shallow earthquakes (0–70 km) are generally thought to represent rapid slippage of preexisting faults where the fault slip is parallel to the fault plane. The physical mechanism of intermediate depth and deep earthquakes (70–680 km) remains unknown, but the seismic waves radiated from earthquakes at all depths appear quite similar in most essential respects and are well accounted for by faultlike models based on the theory of elastic shear dislocations (see *Seismic Source: Theory* and *Seismic Source: Observations*).

The fault model for an earthquake is characterized by the orientation, size, and shape of the fault plane and by the direction, magnitude, and time-space development of fault slip across the plane. The quantities derivable from analyses of seismic waves included the following in order of decreasing reliability of the estimates: (1) *location of the hypocenter* (see *Earthquakes: Location Techniques*), the location of a point with coordinates in latitude, longitude and depth that is the best estimate of the initiation of the fault rupture; (2), the *magnitude* (see *Earthquakes: Magnitude, Energy, and Intensity*), a rough measure of earthquake size based on the logarithm of the amplitude of a specific seismic wave type (the amplitude is corrected for the distance the wave travels); (3) the *focal mechanism solution* (see *Earthquake Mechanisms*), the orientation in space of the strain released by the generation of seismic waves; (4) *seismic moment* (see *Seismic Source: Theory*), a measure of earthquake size derived from measurements of the long-period amplitudes of seismic waves; (5) *fault dimensions* (see *Seismic Source: Observations*), derived from analyses of the seismic wave forms, the spatial pattern of aftershock zones, and measurements of surface deformations and fault offsets near the source; and (6) *time-space history of rupturing or strain release* at the earthquake source (see *Seismic Source: Observations*), estimated from analyses of seismic wave forms. (1) and (3) define the location, orientation, and direction of slippage along major plate boundary faults, while (2), (4), and (5) yield estimates of the amounts and rates of seismic slippage along the boundaries and of the magnitude of stresses released by the earthquakes. (6), in combination with (4) and (5), has proven most useful in attacking the difficult and still unsolved problem of identifying the specific physical processes that produce an earthquake.

Location and Magnitude. Earthquake locations are determined by analyses of the arrival times at seismograph stations around the world of the first arriving compressional seismic waves (P waves) radiated from the source. In the early 1960s this process was computerized and centralized in two complementary efforts made to catalogue global seismicity: the more rapidly published *Preliminary Determination of Epicenters* (PDE), produced by a succession of U.S. government agencies (finally residing within the U.S. Geological Survey); and the more comprehensive publications of the International Seismological Centre (ISC), an internationally supported agency located in the United Kingdom. The ISC evolved from the International Seismological Service (ISS), the British organization that pioneered the cataloguing of global seismicity in the early part of the century. During the late 1950s and early 1960s a large expansion in the number and distribution of the world's seismograph stations occurred at about the same time that digital computers were coming into use in seismology. The confluence of these developments into the computerized locations and catalogues of the PDE and ISC produced an enormous increase in the rate of acquisition of information about global seismicity. For example, the ISC reported about 15,000 earthquakes per year during the late 1970s and early 1980s. The accuracy of locations reported in the bulletins varies according to many factors, but is probably not better than about 5–10 km unless the seismic zone is very well covered by a local network.

Magnitudes are determined by measurements of the amplitudes of certain seismic wave types; the particular magnitude scale depends upon the wave type chosen. The two most commonly used scales for global seismicity are based on the amplitudes of P waves (the body wave scale, m_b) and on the amplitudes of surface waves (the surface wave scale, M_s). Gutenberg and Richter's (1954) study of global seismicity remains the classic synthesis of global seismicity for the first half of the century, with uniform coverage of earthquakes larger than about $M_s = 7$. Since the early 1960s the computerized catalogues routinely include earthquakes as small as $m_b = 3.5$–4, although the catalogues are complete probably only for $m_b > 4.5$–5. The modern catalogues report magnitudes based on short-period P waves (the m_b scale) for nearly all earthquakes and magnitudes based on surface waves (the M_s scale) for large ($M_s > 5.5$) shallow earthquakes.

An outstanding characteristic of natural seismicity is the extremely large range of earthquake size. A continuum of sizes exists from the smallest earthquake detectable by sensitive instruments to the great earthquakes generated by fault slips of 10–20 m that occur over fault areas up to 200,000 km^2. The frequency of earthquake occurrence generally increases logarithmically with respect to decrease of magnitude. Thus, while the very great earthquakes ($M_s > 8.25$) occur on average only about once per decade, the small events ($m_b = 4$–5) routinely located by the global network number in the thousands per year. The methods of analysis and the interpretation of the results vary according to the size of the earthquakes studied. It is convenient to divide earthquake sizes into three major categories: small (M_s less than 5.25), medium ($M_s = 5.25$ to 7.5), and large ($M_s > 7.5$). These boundaries are not exact, but correspond very approximately to source dimensions less than 10 km, 10–100 km, and greater than 100 km, respectively.

Focal Mechanisms. One of the most significant breakthroughs in the study of the mechanisms of earthquakes was the deployment during the early 1960s of the World Wide Standardized Seismograph Network (WWSSN) by the U.S. government in cooperation with scientific organizations in numerous countries around the world. This network included over a hundred stations all with identical instrumentation and all routinely supplying recordings to a centralized distribution facility (the U.S. Department of Commerce's NOAA). Compared to most previous seismographs, the instrumental response of the WWSSN instrumentation was considerably improved in respect to low frequency or long period (3–300 seconds) seismic waves. The long period radiation integrates the detailed complexities of the seismic sources and is not so much affected by scattering by inhomogeneities in the Earth as are short period seismic waves. The high quality recordings provided abundant data to determine the overall orientations in space of the shear-strain released by the earthquake. Fault plane solutions for medium size earthquakes began to accumulate in substantial numbers in the mid- to late 1960s at about the same time that the most convincing evidence for seafloor spreading was emerging. This coincidence fueled the explosive development of plate tectonics during the late 1960s and early 1970s.

In the late 1970s, focal mechanism solutions (see *Earthquake Mechanisms*) began to be included on a routine basis in the PDE bulletins. Researchers at Harvard University and the U.S. Geological Survey developed computerized determinations of focal mechanism solutions that greatly increased the number of solutions obtained per year. The *centroid moment tensor solutions* are routinely determined for all earthquakes large enough for the analysis to be reliably performed. The solution is based on a more general model of the source mechanism than used in previous studies, and is obtained by analysis of the amplitudes and wave forms of many different seismic phases recorded by set of globally distributed, digitally recording seismograph stations. Most previous fault-plane solutions were determined only by analysis of the first motions of several of the larger seismic phases, primarily the first-arriving P phase. The moment tensor technique increased the number of solutions obtained per year by decreasing the smallest analyzable earthquake from about $M_s = 5.75$–6 for the older fault plane solutions to about $M_s = 5.25$ for the moment tensor solutions.

The high quality long period recordings of the WWSSN also were amenable to sophisticated analyses of the seismic wave forms. Instead of simply determining the spatial orientation of the strain release, seismologists could estimate the fault dimensions, amount of strain release or fault slip, and details of the rupturing process itself. The WWSSN data proved especially valuable for the study of the very largest earthquakes. Detailed study of the great Alaskan earthquake of 28 March 1964, for example, produced estimates of the dimensions of the fault area ruptured, the amount of fault offset, and some indications of the direction and time-space development of the rupturing. As the number of large earthquakes analyzed with modern data continues to accumulate, systematic relationships have emerged among the characteristics of the earthquake and the seismogenic plate boundaries. The earthquake characteristics include fault offset, fault area, recurrence times and stress release. These characteristics appear to be controlled partly by the geometry of the plate boundary, the rate and direction of the relative motion of the plates across the boundary, and major geological characteristics of the boundary.

The improvement in low frequency response led to a major improvement in the determination of gross earthquake size by using the amplitudes of the very low frequency waves radiated from the source. These very long wavelength waves are affected very much less than high frequency waves by the detailed complexities of seismic wave radiation and propagation, and effectively integrate the net result of the earthquake into a measure of size called the seismic moment (M_o). For a fault model of an earthquake M_o is equal to the product of fault area, the amount of fault slip, and the modulus of rigidity of the material around the fault. A combination of the older magnitude scales with seismic moment led to an extension of the classical surface wave scale (M_s) to earthquakes too large to be effectively measured by M_s. The new moment-based scale (M_w) reaches a value of 9.5 for the largest known earthquake, the great Chile earthquake of 22 May 1960.

Detailed analyses of the forms and amplitudes of seismic waves is time-consuming and difficult to do with the photographically recorded data of the WWSSN, but is greatly facilitated by digitally recorded data that can be fed directly into powerful computers (see *Seismic Instrumentation*). Global digital recording networks effectively began in the late 1970s with the upgrading of certain stations of the WWSSN and have since become the focus of major efforts in both the United States and France.

Local and Regional Seismic Networks. Concern about the hazards of large destructive earthquakes stimulated installation of local and regional seismograph networks designed to monitor very precisely the small earthquakes that are characteristically numerous in seismically active regions. These networks can resolve details of the seismicity at the kilometer and subkilometer scale inaccessible to the global network. By recording the very frequently occurring small earthquakes, the local and regional networks can attain a large sample of seismicity in a relatively short period of time. Networks in California, for example, include several hundred seismographs and routinely locate thousands of small earthquakes per year. The idea is that study of small earthquakes within a seismogenic zone will yield information about the fault geometry and processes of strain accumulation that are important for understanding the occurrence of the large destructive earthquakes. These networks have also provided important information about the relationships of earthquakes to plate boundaries and about earthquake activity in areas of low seismicity such as the interiors of plates.

Subduction Zones

A subduction zone is formed at a plate boundary where the two plates move toward one another (the motions are convergent), and where at least one of the plates is a suboceanic one created by seafloor spreading (see *Subduction Zones*). The suboceanic plate characteristically bends downward and descends into the interior beneath the overriding plate. A subduction zone is characterized by the parallel alignment of a deep oceanic trench and a zone of volcanos; in many subduction zones the volcanos are located on a submarine ridge and form an *island arc,* while in some areas, like western South America, the volcanic zone is built on the edge of a continent. The axis of the trench marks the surface expression of the giant fault zone—the zone of contact between the converging plates—which is inclined at angles of 10°–30° beneath the overriding plate. On one side of the trench the outer slope is formed by the down-bending suboceanic plate while on the opposite side the inner slope is the leading edge of the overriding island arc or continental margin. The zone of volcanos is located on the overriding plate about 50–200 km away from the axis of the trench. The material forming the volcanos appears to be produced by partial melting of the hot mantle rock located in the wedge between the inclined descending plate and the horizontal

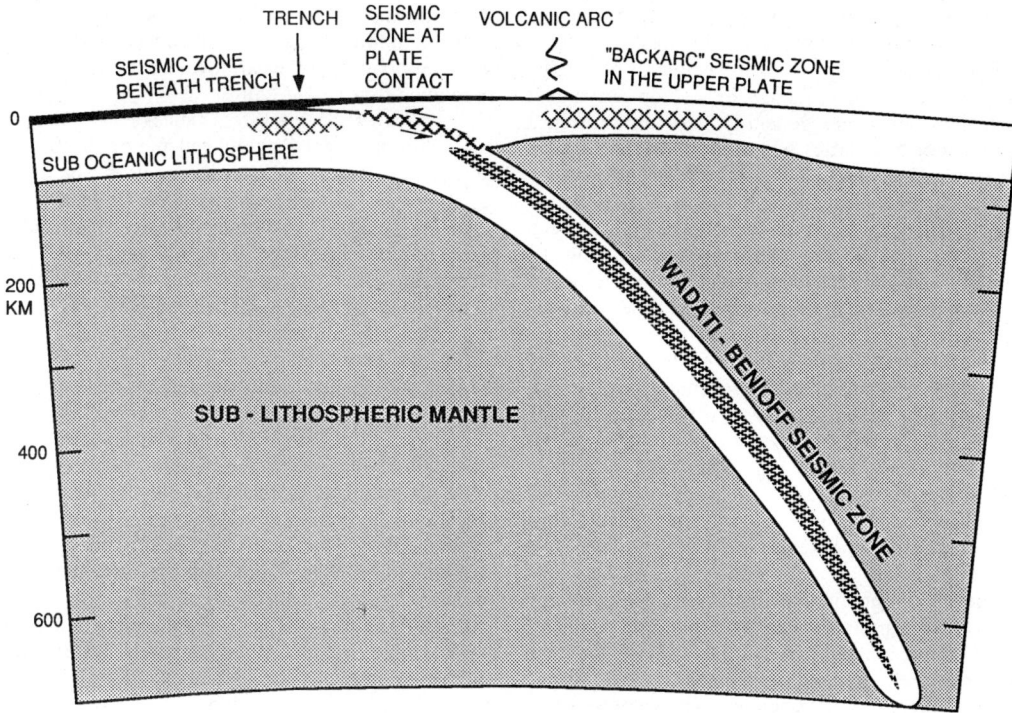

FIGURE 1. Scaled diagram showing vertical cross section through a subduction zone, a major type of convergent plate boundary where a suboceanic plate thrusts beneath the adjacent plate and descends into the mantle. The arrows show the motion of the subducted plate relative to the overriding plate, and the cross-hatched areas show the locations of the major seismic zones. The lithosphere in the upper or overriding plate is thinned above the descending suboceanic plate, an effect which tends to localize the deformation of the upper plate. In the island arcs of the western Pacific that upper plate zone of deformation is often extended to form rifted "back-arc" basins, while in western South America the continental upper plate is compressed and forms mountain belts (the Andean cordilleras and plateau) where the crust thickens.

overriding plate. The melting is probably stimulated by release of volatile material carried down with the descending suboceanic plate.

The Earth's largest and most frequently recurring earthquakes are located in the zone of contact between the converging plates of a subduction zone, while nearly all earthquakes located in the Earth deeper than about 70 km are located within the descending plates in subduction zones. In some subduction zones significant seismicity occurs also within the overriding plate. It is thus convenient to separate the seismicity into three groups according to location in (1) the interplate boundary, (2) the subducted plate, and (3) the overriding plate. These groups are illustrated in Fig. 1.

Plate Boundary. Subduction zones form as continuous, coherent segments over lengths of 500–1000 km; these segments are often linked together into longer systems, e.g., the system comprising southern Alaska through the Aleutians, the Kurile-Kamchatka arc, Japan and the Marianas island arc. The interplate boundary in a subduction zone thus has dimensions of the length of the subduction zone segment in which it is located, and has widths (measured down along the dip of the inclined fault zone) of between about 50 km and 200 km. The variation in width is determined partly by the geometry of the downbending subducted plate and to other less clearly understood properties including the effective thickness of the overriding plate and the effects of subducted plate morphology and composition on the "frictional" characteristics of the interplate boundary. Earthquakes occur at depths mainly between about 10 and 60 km along the inclined interplate fault zone, but most of the moderate sized events appear to have depths between about 15 km and 40 km.

An earthquake located along the plate boundary represents a rupture of a particular patch of the large fault zone. The size of a patch ruptured during a given earthquake increases with the magnitude of the event. Generally only the large earthquakes ($M_s > 7.5$) have dimensions comparable to the widths of the interplate boundary. The largest known rupture was that for the great Chile earthquake mentioned previously, which had a length of about 1000 km and a width of about 200 km (the entire width of the interplate fault zone beneath southern Chile).

The direction of fault slippage is determined by the direction of the *slip vectors* obtained from focal mechanism solutions. The slip vectors of both large and medium size earthquakes located on convergent plate boundaries are generally very uniform in a region. For boundaries where the motions between the converging plates can be estimated from analyses of seafloor spreading, the azimuths of the slip vectors are close to the directions of relative motion between the plates. Indeed, these directions are among the primary data that are used to determine a globally consistent system of plate motions. The vigorous seismicity of convergent plate boundaries produces good estimates of the directions of long-term global plate motions with samples of seismicity only several decades in duration, perhaps partly because of the generally high rate of relative motions of about 10 cm·yr^{-1} in many subduction zones. The agreement also argues for the validity of the approximation that the plate boundary is a single, simple fault zone accommodating the relative motions of the two converging plates.

The amount of slippage that occurs during a large earthquake can be inferred by combining estimates of seismic moment and of the size of the area ruptured. For example, a slippage of 20 m was estimated for the great Chile earthquake of 1960. In spite of the large numbers of small and medium size earthquakes, the infrequently occurring great earthquakes still account for nearly all of the seismic slippage along most plate boundaries. Thus estimates of rates of slippage depend on estimates of the recurrence time and sizes of the largest earthquakes. Studies of these factors have led to estimates of seismic slippage for many subduction zones that are in fair agreement with rates determined by analysis of the ocean floor "tape recorder." For the area of the Chilean earthquake, a recurrence time of 200 years would accommodate 10 cm·yr^{-1} convergence rate. This recurrence time is in accord with the historical record of great earthquakes in Chile.

Strain accumulates to some limiting value near the plate boundary and is then released by seismogenic fault slippage. The rate of plate convergence and the limiting value of strain accumulation thus determine the recurrence time of the largest earthquakes. The smaller earthquakes that occur along the plate boundary appear to be subsidiary to this cycle of strain accumulation and release; nevertheless, the patterns of occurrence of the small earthquakes in space and time may eventually provide keys to the prediction of the great earthquakes.

The utility of the concept of an earthquake cycle is demonstrated by the success of the concept of seismic *gaps* in forecasting large earthquakes. A gap is an area of a plate boundary that has not ruptured during some defined period of time during which adjacent areas have ruptured. The entire plate area is ruptured by a succession of events whose rupture areas are contiguous and together fill the total fault area. Information about the recurrence times for the particular unruptured area can then lead to estimates of the probability of the earthquake occurring within a specified time in the future.

The sizes of the largest earthquakes in a region depend on the strain accumulated and on the size of the potential rupture area. These factors vary from region to region. For example, the Chilean plate boundary appears to accommodate nearly all interplate motion by large earthquakes, while the Marianas island arc is quite anomalous in having no known great interplate earthquakes at all. This variability is partly accounted for by variations in the widths of plate boundaries, but the substantial deficiencies in seismic slippage that exist for regions such as the Marianas cannot be explained this way. A seismic slippage or *creep* may replace sudden, seismogenic slippage of the plate boundary. Creep could predominate over seismic slippage in a situation where the compressive stress component acting perpendicular to the fault plane is abnormally low, or where the physical properties of the interplate boundary favor creep instead of "stick-slip" behavior. The physical nature of the plate boundary may be greatly affected, for example, by water-filled sediments originally on the ocean floor that are carried into the plate boundary.

Subducted Plate. Earthquakes within the suboceanic plate occur at shallow depths near the trench axis, where the suboceanic plate is bending downward just prior to subduction, and within the subducted plate at nearly all depths from shallow depths just beneath the plate boundary to about 680 km. The maximum depth of 680 km, a cutoff depth below which no reliably located earthquake has so far been detected, may be related to the changes in the crystalline structure of mantle minerals that occur at the temperature and pressure at this depth in the Earth.

The earthquakes in the mantle form characteristically thin, slablike zones that are inclined beneath the island arc or continental margin at angles mostly between 30° and 60°, although angles from nearly vertical to nearly horizontal exist. The thicknesses of the zones appear in many areas to be about 30–40 km. Indeed, these *Wadati-Benioff zones* supplied the most direct and compelling evidence for the concept of subduction through studies of the propagation of seismic waves inside the zones and of the focal mechanisms of earthquakes forming the zones. Those studies revealed that a Wadati-Benioff zone is not a giant inclined fault plane but a zone where the material is stronger than the surrounding mantle material. The subducted plate is strong because it remains cooler than the surrounding mantle material through which it is descending. The rate of descent into the hot interior is rapid relative to the rate of heat conduction into the plate.

The focal mechanisms of earthquakes within the subducted plates reveal deformations that stretch, compress, or bend the plates. The bending stresses predominate in the region near the plate boundary where the plate bends from horizontal to an inclined alignment. Evidence has been found for both bending and elastic-plastic unbending effects. Beneath Honshu (Japan) the Wadati-Benioff zone at depths of about 70 km to 150 km consists of two thin parallel zones separated by about 30 km. The focal mechanisms in the two zones indicate compression in the upper zone and extension in the lower zone, as would be expected if the plate were unbending. Stretching or compression of the plate is dominant at greater depths where stresses in the plate appear to be determined by the combined effects of the negative buoyancy of the cold and dense material within the plate—gravity is pulling the plate downward—and the shearing tractions applied to the boundaries of the plate by its movement through the viscous material of the mantle. The prevalence of extensional stress at depths less than 400 km is taken to indicate the downward pull of the dense material sinking into the interior. The nearly universal dominance of compression aligned along the dip of the plate for earthquakes located below about 400-km depths is interpreted as evidence that the plates encounter increased resistance to downward motion near depths of 700 km. In some places focal mechanisms indicative of lateral stretching or compression are associated with sharp lateral changes in the dip or strike of the Wadati-Benioff zone.

Overriding Plate. The overriding plate is deforming substantially in some subduction zones. In these regions the overriding plate can be separated into two parts: a narrow and relatively undeformed *forearc zone* bounded on the oceanic side by the axis of the trench and on the opposite side by the zone of volcanos; and a *backarc zone* where the deformations of the overriding plate take place. The terms *forearc* and *backarc* have emerged from study of oceanic subduction zones, where the overriding plate includes an island arc, but the concepts apply equally well to subduction zones where the overriding plate is a continent, as, for example, is the case along western South America. The inclined subducted plate and the overriding plate define a wedge-shaped region of material located between the two plates. The forearc zone is approximately located above the plate boundary, while the backarc is located above the wedge. Convection of hot material in the wedge driven by the descent of the subducted plate appears able to heat the overriding plate and weaken it. The volcanism characteristic of subduction zones is a manifestation of the advection of heat into the upper plate.

Deformations of the heated and weakened backarc zone can be extensional, creating the back-arc basins found behind some of the western Pacific island arcs, or compressional, probably creating much of the Andean Cordilleras of western South America. In the eastern Indonesian subduction zone, the island of Sumatra (part of the overriding plate) has a prominent zone of strike slip deformation that follows the volcanic arc; farther north, this changes into a backarc extensional zone forming the Andaman Sea. In these areas of substantial backarc deformation the forearc forms a long narrow "plate" which moves relative to a stable interior part of the overriding plate. The map-view outline of the forearc strip may therefore change, or "bend," as has been suggested for the Marianas island arc of the western Pacific and the central Andes.

The forearc is generally a region of relatively low seismicity. The backarc seismicity varies greatly according to the setting and nature of the deformations, whether in an oceanic or continental plate and whether compressional, extensional, or strike slip. The compressional backarc zones of the Andes and of northern Honshu appear to generate larger and more frequent earthquakes than the extensional backarc basins behind the Tonga, Marianas and New Hebrides island arcs.

In neither the backarc areas of the overriding plates nor in the subducted plate are the earthquakes as large as the great plate boundary ruptures. This difference can be related to the geometry and dimensions of the seismogenic faults: the plate boundary fault dimensions can reach 200 km in width and 1000 km in length, while the multiple internal faults accommodating a broadly distributed deformation with a plate are each much smaller. One of the largest earthquakes within a subducted plate, the earthquake of 1933 located beneath the Japan trench offshore of Honshu and Kyushu islands, is estimated to have had dimensions near 70 km, which is large enough to have ruptured through a substantial fraction of the Pacific plate.

Collisions. The process described in the preceding section implies a kind of conveyor belt movement of oceanic plates into subduction zones. This process can be disrupted by the arrival of large oceanic plateaus or ridges, island arcs, or fragments of continents. The thick crusts of these terrains tend to resist subduction into the denser mantle material. The subduction zone can be "jammed" by this interaction with complex and often rapidly changing deformations developing in the overriding plate.

The most outstanding examples of such collisions are found in the the Solomons, New Britain-New Guinea, Indonesia, the Philippines and Taiwan areas of southeast Asia and the western Pacific. These areas exhibit subduction zones that are colliding with continental margins (Banda island arc, Taiwan), oceanic plateaus (Solomons, eastern Luzon), oppositely facing subduction zones (Moluccas), older island arc and continental fragments (western Luzon, New Guinea). Where fragments of the suboceanic plate are still being subducted a reasonably typical subduction zone can be identified. In areas of more advanced collision a remnant suboceanic plate can still be seen as a Wadati-Benioff zone. However, the generally diffuse deformational zones that develop after collision are often difficult to resolve with the available seismicity data alone. The seismicity and tectonics are transitional between simple subduction zones without collision and the complex and diffuse intracontinental zones of deformation. In addition to the compressional structures and related seismicity, strike-slip deformations are known to be a significant part of the deformations from geological studies of such striking features as the Philippine Fault. These zones serve as diffuse transform boundaries between the evolving convergent boundaries.

Ocean Ridges

The narrow linear belts of seismicity that delineate actively spreading oceanic ridges were in fact among the first indicators of the existence and continuity of the global submarine mountain belt. Improved resolution of the oceanic seismic zones revealed sharp right-angle bends that are associated with offset segments of the crest of the ridge. Seismicity occurs along the offset segments of ridge crests and along the fracture zones that connect the offset segments, but little seismicity is found along the extensions of the fracture zones beyond the offset segments. The ocean ridge seismicity is thus characterized by a relatively continuous narrow zone with right angle dog-leg bends connecting the various offset segments of ridge crest (see Fig. 1). The seismicity of ocean ridges is thus conveniently separated into two fairly distinct groups, the seismicity of fracture zones or transform faults and the seismicity of ridge crests.

Transform Faults. The focal mechanisms of earthquakes in fracture zones provide an elegant test of the concept of transform faulting as applied to the offsets of the ocean ridges (see *Earthquake Mechanisms and Plate Tectonics*). The key observations demonstrating the transform fault concept are (1) the strong concentration of seismicity along the part of the fracture zone between the offset ridge crest segments and (2) the sense of motion across the fracture zone indicated by the focal mechanism solutions. The model predicts that the sense of motion along the fracture zone is exactly the opposite to that expected for the alternative interpretation of the fracture zones as transcurrent faults producing the offset in the ridge crests. The sense of motion indicated by focal mechanism solutions is in excellent agreement with the concept of a transform fault connecting two offset segments of a spreading ridge, but is opposite that required if the faulting is considered to have produced the offset itself. The confinement of the seismicity to the portion of the fracture zone between the offset segments also discriminates in favor of the transform fault model.

Both the locations and the focal mechanism solutions indicate that the seismicity of fracture zones is characterized as slippage along a thin fault zone that forms a relatively simple plate boundary between two suboceanic plates. This characterization is similar to the interplate boundary of a subduction zone. In both cases, however, the characterization is limited by the resolving power of the observations, i.e., the 5–10 km precision in the earthquake locations that define the spatial extent of the fault zone.

Studies of transform fault earthquakes have progressed along two routes: detailed analyses of the seismic waveforms of moderate sized earthquakes and studies of small earthquakes recorded by networks of ocean bottom seismographs that are temporarily deployed in or very near the fracture zones. Accurately determined depths from both types of studies are generally less than 10 km beneath the ocean bottom. The shallow depths of earthquakes in oceanic transform faults is probably related to the thinness (vertical dimension) of the youthful lithospheric plates in areas of active seafloor spreading. The fact that the largest fracture zone earthquakes have magnitudes considerably smaller than the great interplate earthquakes in subduction zones is probably related to the small width of the seismogenic fault zone.

Ridge Crests. The focal mechanism solutions along ridge crests generally indicate normal faulting with horizontal extensional stress axes oriented perpendicular to the direction of spreading. Seismically active ridge crests are generally characterized by a central valley flanked by rugged ridges. The earthquakes are thought to occur along normal faults that are located on both sides of the central valley. The system of normal faults is related to the uplift and relative movements of blocks involved in the formation of the ridges flanking the central valley; the earthquakes reflect the pervasive extension in the region of the ridge crest, but are not associated with a single plate boundary fault; in fact, it is difficult to imagine a process of lithosphere and crustal creation along a ridge crest, which would involve a single simple fault boundary between the plates.

The seismic activity of ridge crests exhibits an inverse relationship to the rate of spreading. The least active ridge crest is the part of the East Pacific Rise located south of the equator where spreading rates are the highest known on the presently active ocean ridge system. Moreover, the P waves radiated from ridge crest earthquakes sometimes appear to have anomalously large fields of compressional first motions. This anomaly was first interpreted as evidence for a focal mechanism different from a shear fault, but later studies showed that the effect was actually due to the very shallow depths of the sources beneath the ocean floor, less than 5 km. Shallow source depths have also been found with ocean bottom seismograph networks and are interpreted to indicate only a very thin brittle layer at the top of the youthful (and hot) lithosphere just beginning to form along the ridge crest. Probably the thickness of the brittle layer and consequently the size and number of earthquakes increases with decrease of spreading rate. This relationship is also consistent with the fact that ridge crest seismicity is generally lower than fracture zone seismicity, since the lithosphere across fracture zones, although young, is older and thicker than that near ridge crests.

Another characteristic of ridge crest seismicity is the occurrence of numerous swarms of earthquakes. A swarm is distinguished from an aftershock sequence by the absence of a particularly large event in the cluster of earthquakes and by a slow buildup as well as slow decay in the rate of occurrence. Aftershock sequences typically commence sharply with a very high rate of occurrence following a large earthquake, with the rate of occurrence gradually tapering off in time (an approximately hyperbolic decrease with respect to time). Swarms appear characteristic of the seismicity of volcanic zones in many regions, and are thus probably associated with the pervasive magmatism that is characteristic of ridge crests.

Continental Plate Boundaries

Plate boundaries within continental regions appear significantly more complex and diffuse than those in which at least one of the interacting plates is oceanic. Even so, zones of large intracontinental deformation and high seismicity can be distinguished from vast areas of continental plates that are being very little deformed and are relatively stable. The intracontinental plate boundary zones can be classified into broad categories according to the predominant motions of the interacting plates that are accommodated by the deformations. Thus one can recognize predominantly transform, extensional, and convergent zones. The deformations within these zones are generally mixed, however, and many intracontinental zones include all three types of deformations. Major seismically active examples of intracontinental zones that absorb plate motions include the following:

Transform
  South Island, New Zealand
  Northern South America (Colombia, Venezuela)
  Western North America
  Gulf of Aden-Dead Sea-Turkey
Convergent
  Himalaya-Central Asia
  Zagros-E. Turkey-Caucasus
Extensional
  East African rift system

Some intracontinental zones are not clearly traceable from one major plate boundary to another, either by seismicity or by obvious evidence of surface deformations. Examples include the connection of the Arctic Ocean spreading ridge through Siberia and Asia to the subduction zones of the western Pacific, and the southern extension of the east African rift system. Motions accommodated across such zones may be relatively small, or possibly the zones diffuse gradually into very large areas with decreasing amounts of deformation per unit area, and are thus very difficult to map.

In the more seismically active intracontinental zones earthquake magnitudes are used to estimate the net deformation that is accommodated seismically. This estimation is derived from a summation of the moment tensors for the large events. The largest known events in intracontinental zones are associated with thrust faulting within the central Asian convergent zone. For example, the two great Mongolian earthquakes of 1905 each had magnitudes (M_w) of 8.4, and the Assam earthquake of 1950 had a magnitude of 8.6. These events are among the world's largest 20 events that have occurred during this century. For the Himalayan-Central Asian zone, summation of moments yields a net amount of deformation between the Indian and Eurasian plates that agrees with rates derived from analyses of the seafloor "tape recorder." The estimations are quite uncertain, however, because the instrumental record of seismicity is too short to sample the full range of deformations that are known from studies of Late Cenozoic structures.

A good example of the complexity of an intracontinental seismic zone is the transform system of the western United States. The San Andreas fault zone is only a part, although a major part, of a complex system of deformation that includes extension within the Basin and Range province, compression in the Transverse Ranges and in the coastal ranges of central California (the 1952 Kern County earthquake, the 1972 San Fernando earthquake, and the 1983 Coalinga earthquake all had thrust faulting focal mechanisms and manifested this compression), and multiple strands of strike-slip faulting oriented subparallel to the San Andreas. The Late Cenozoic tectonic evolution of Southern California is particularly complex, involving translation, rotation, and compression of small blocks and slivers caught within a broad zone of right-lateral shearing deformation. The San Andreas fault itself is a dominant component in this system, but appreciable deformations occur elsewhere in the zone. Increasing evidence indicates that the seismicity in Southern California and possibly in other transform zones within continents is largely confined to depths less than 15–20 km and is associated with an upper crustal layer that may be detached and deforming differently than deeper material. Depths in the range of 20–40 km have been found in several zones of compressional deformation (e.g., the Andes), which may indicate a thicker brittle layer in regions of crustal thickening.

The record of large earthquakes along the southern

San Andreas Fault zone has been extended back to about 10,000 B.P. by careful studies of stream deposits cut by the fault in several unique localities. These studies yield an average repeat time of 125 years for major earthquakes similar to the large Fort Tejon earthquake of 1857.

The intracontinental zones of central Asia, China, western North America, and areas around the Mediterranean have and continue to pose major hazards to human societies, particularly because the earthquakes can occur very close to, or even within, cities. In comparison, the interplate earthquakes of subduction zones are generally not so close to large cities, and the catastrophes caused by these earthquakes are not in proportion to their relatively great sizes and frequent occurrence. Historically, much of the damage and loss of life from great subduction zone earthquakes has been from the large tsunamis or giant sea waves generated by the earthquakes.

Intraplate Seismicity

The seismicity within plates, although very low, is quite detectable and of concern in respect to hazards to cities and critical facilities. The sporadic, rare occurrence of moderate size and large earthquakes makes intraplate seismicity very difficult to study. The factors controlling the location of the seismicity within plates remain largely unknown.

The most compelling evidence about the origin of intraplate earthquakes has to do with the directions of stress relief accommodated by the earthquakes. Focal mechanism solutions show uniformity in the orientation of stress over large regions, a result taken to indicate that the earthquakes reflect a global system of stresses within the lithosphere plates that are associated with the forces that move the plates. The focal mechanism solutions quite commonly have a horizontal axis of maximum compressive stress, a result indicative of a pervasive component of horizontal compressive stress within the plates. This result is corroborated in several continental regions by independent measurements of stress orientation in boreholes.

Direct association of intraplate earthquakes with mapped faults is rare, although this may be largely because of the subtle and often enigmatic evidence of intraplate fault activity. In one of the most active intraplate zones known, the seismic zone near New Madrid, Missouri, the seismicity is clearly associated with reactivation of a Precambrian fault zone in the basement beneath a cover of younger sediments. This discovery depended on a combination of several factors: the zone was active enough to produce a good sample of small earthquakes, a long record of seismicity was obtained by a local seismograph network, and extensive geophysical exploration was done of subsurface structure. The zone was the site of the moderately large earthquakes of 1811–1812, which were felt over a large part of the eastern United States.

Summary

Sharply defined, relatively simple fault zones form seismogenic plate boundaries where at least one of the plates is oceanic, while complex and diffuse zones of seismicity and associated crustal deformation characterize plate boundaries within continents or in collisional zones. This fundamental difference appears to be related to the very different mechanical behavior of continental and suboceanic lithosphere, a difference that is at least partly attributable to the large difference in thickness of the crust (about 5–6 km in the suboceanic lithosphere versus 30–70 km in the continental lithosphere). It is thought that crustal rock is inherently weaker than mantle rock, so that the continental lithosphere may be inherently weaker, but the more important factor may be the great heterogeneity of continental crust and lithosphere compared to suboceanic lithosphere. The heterogeneity causes complex redistribution of stress and deformation as a weakened area of a continental plate responds to stresses transmitted through the plates.

The seismicity of the simple plate boundaries has provided a powerful means to determine the relative motions between the plates and has revealed much of what we know about the subduction process. Unfortunately, the seismicity of intracontinental zones is too sporadic and still too poorly sampled to provide comparable information. The world's largest earthquakes are found in the interplate boundaries of subduction zones where the fault zone can have widths up to 100 to 200 km. Large earthquakes are also found in predominantly compressional zones of intracontinental deformation, where the upper brittle layer is perhaps relatively thick. The sizes of intracontinental rupture zones are limited, however, by the complexity and lack of continuity of the intraplate faults, so they do not reach the great size of interplate events in subduction zones.

BRYAN L. ISACKS

References

Gutenberg, B., and C. F. Richter, 1954, *The Seismicity of the Earth*, 2nd ed. Princeton, N.J.: Princeton University Press, 310p.

Isacks, B., J. Oliver, and L. R. Sykes, 1968, Seismology and the new global tectonics, *Jour. Geophys. Research* **73**, 5855–5899.

Isacks, B., and P. Molnar, 1971, Distribution of stresses in the descending lithosphere from a global survey of focal mechanism solutions of mantle earthquakes, *Rev. Geophysics Space Physics* **9**, 103–174.

Kanamori, H., 1977, The energy release in great earthquakes, *Jour. Geophys. Research* **82**, 2981–2987.

McCann, W. R., S. P. Nishenko, and L. R. Sykes, 1979, Seismic gaps and plate tectonics: seismic potential for major plate boundaries, *Pageoph.* **117**, 1082–1147.

Richter, C. F., 1958, *Elementary Seismology*. San Francisco: W. H. Freeman and Co., 768p.

Cross-references: *Continental Collision Zones: Seismotectonics and Crustal Structure; Deep Earthquakes; Earthquake Mechanisms; Earthquake Mechanisms and Plate Tectonics; Earthquakes: Hazards and Prediction; Earthquakes: Location Techniques; Earthquake Seismology; Earthquakes and Seismicity; Earth Structure: Global; Lithosphere, Oceanic: Formation and Evolution; Mantle, Upper: Structure; Mantle Convection and Plumes; Mantle Dynamics; Seismicity: Intraplate; Seismicity: Midocean Ridge; Seismicity: Subduction Zone; Seismic Source: Observations; Seismic Source: Theory; Seismology: History; Subduction Zones.*

SEISMIC MONITORING OF NUCLEAR EXPLOSIONS

Until 1957 all nuclear explosions were conducted near or above the surface of the Earth. But in that year, on September 19 in Nevada, the first underground nuclear explosion occurred. Code-named RAINIER, it was planned in part to see if nuclear testing could be done in an environment that was unhampered by the weather or by concerns over radioactive fallout but still fulfill all the needs of the test program. The resulting seismic data were studied intensively in both scientific and political arenas, setting a pattern that has continued ever since.

Over the last thirty years a special branch of seismology has developed associated with the monitoring of such explosions, most of which are weapons tests. As shown in Fig. 1, there have been hundreds of underground explosions, conducted (so far) by six countries.

As the first nuclear explosion in which the radioactive products were "contained" in the vicinity of the shot point, RAINIER presented a technical challenge to those who, for whatever reason, wished to obtain details of such testing activity. The importance of this challenge has had a major impact upon the development of geophysics. In particular, seismology grew rapidly as a science in the 1960s in response to the need to monitor explosions, in part because of the need to monitor any nuclear test-ban treaty that might be contemplated between the United States, the USSR, and other countries. For example, the first global network of calibrated seismographs as well as several regional networks (often called "arrays," if operated with centralized data handling) were set up to improve the capability of seismology as a whole, with funding derived from the need for improved explosion monitoring. With openly available data from this instrumentation, great improvements globally in earthquake location accuracy were achieved; studies of earthquake focal mechanisms

Lamont-Doherty Contribution Number 4388

became practical; and detailed features of deep Earth structure were discovered.

In general the problems of monitoring nuclear testing are more severe for tests conducted underground, than for tests in space, in the atmosphere, or underwater. In 1963 these problems appeared to many to be insuperable, a fact that contributed to the form of the Limited Test Ban Treaty, signed and ratified in that year, and banning signatories from nuclear testing in all environments except underground.

The detection of explosions, and their discrimination from earthquakes, is easier for events that are of a larger size. This fact influenced negotiation of the Threshold Test Ban Treaty in 1974, intended to take effect after March 1976, and prohibiting any underground nuclear weapons test having a yield exceeding 150 kilotons. (A kiloton, or kt, is an energy unit originally associated with the chemical energy of a thousand-ton TNT explosion; but it is now defined as 10^{12} calories. RAINIER, the first underground nuclear explosion, had a yield of 1.7 kt. The Hiroshima bomb was about 14 kt. Prior to negotiation of the 150 kt threshold, some underground tests had been well in excess of 1000 kt.) This Threshold Treaty introduced a third technical challenge to monitoring programs, beyond detection and discrimination, namely the question of yield estimation and whether this can be done accurately by seismic means.

In 1974 the USSR had ambitious plans to use nuclear explosions for redirecting major rivers from northerly to southerly flow, and for other civil engineering purposes. However, after 18 months of intensive negotiation it proved possible also to negotiate the Peaceful Nuclear Explosions Treaty of 1976, which extends the threshold of 150 kt on weapons test explosions to so-called Peaceful Nuclear Explosions (PNEs), defined as those nuclear explosions conducted off recognized weapons test sites. In addition this treaty permits PNE salvos with yield totals up to 1500 kt provided no individual explosion exceeds 150 kt. As of July 1989 these two treaties have not received the consent of the U.S. Senate, so they have not been ratified. Although both the United States and the USSR have declared their intention not to exceed the 150 kt threshold, there have been press reports that each country has raised questions as to whether the other side is indeed complying.

This brief history will indicate that technical capability to monitor explosions, or lack of such capability, plays a role in the development of policy options on weapons testing and/or arms control and the content of international treaties. The question of a Comprehensive Test Ban Treaty has been debated for thirty years, the fundamental issue being whether to assist or resist continued advances in nuclear weapons technology. It is also clear that related to

FIGURE 1. Annual numbers of nuclear explosions, detonated by six countries, from 16 July 1945, to 30 April 1986. Solid bars are used for atmospheric and underwater explosions; open bars for those underground.

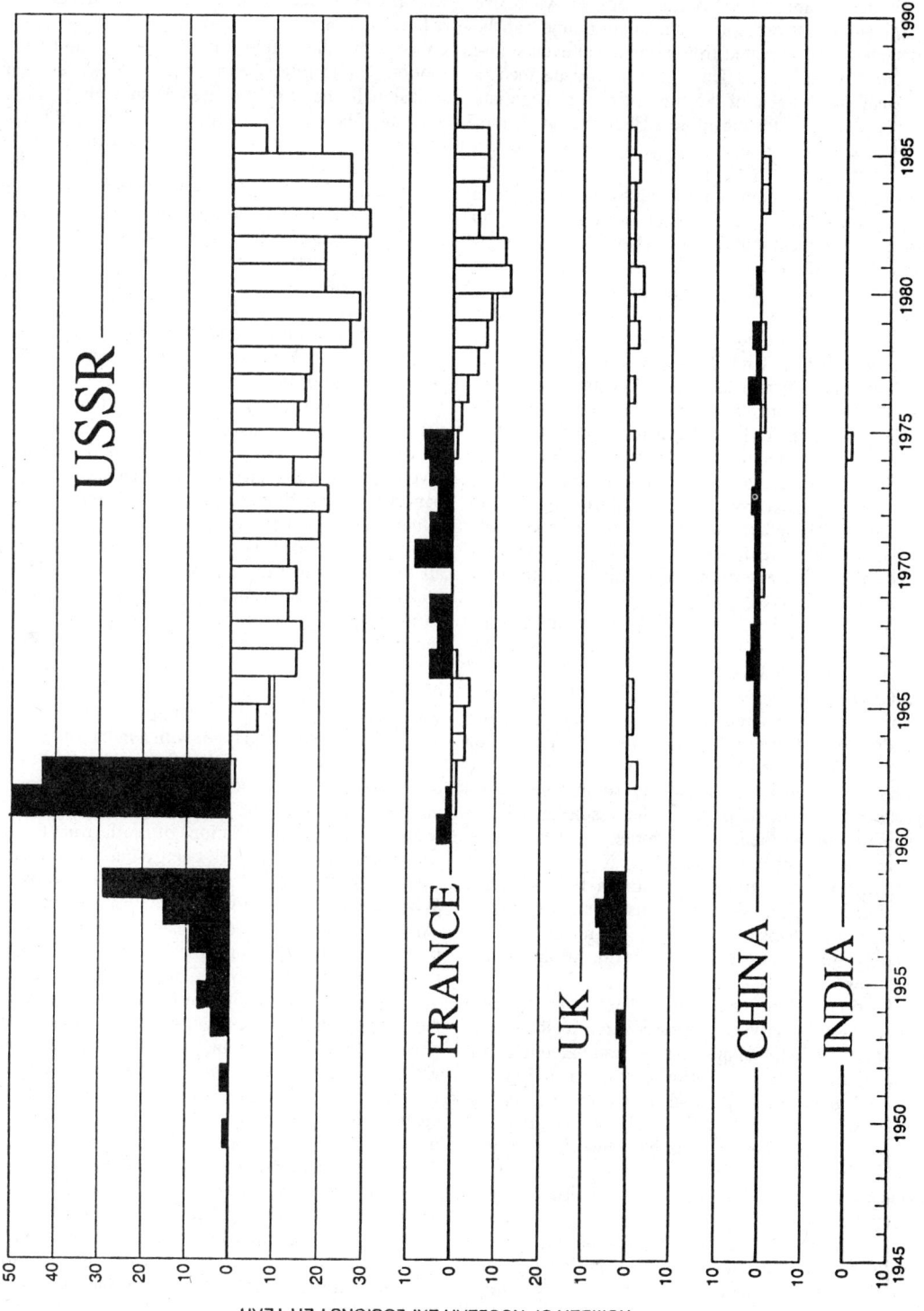

that debate are the technical issues of what size explosions that are conducted underground can be detected, discriminated from nonnuclear events, and accurately measured. Thus, if it is ever deemed in the national interest of the superpowers to negotiate a ban on future testing more stringent than treaties now in effect, a key technical question will be: down to what value of yield can monitoring be accomplished? In this connection, much improvement in monitoring capabilities has been made since the 150 kt threshold went into effect in 1976. Some seismologists claim now that there is no fundamental technical problem with monitoring explosions down to 1 kiloton, even if determined efforts at evasion must be considered. Some are more cautious, claiming that the United States (even with access to sophisticated seismic data gathered within the USSR) could monitor only down to about three times the size of RAINIER (i.e., down to about 5 kt). And some have asserted that it is possible to build cavities underground and use them to muffle and thus hide the seismic signal from a 30 kt explosion or even up to 100 kt. These latter assertions do not appear able to survive after review of the technical difficulties; but in fact, as assertions, one finds that they do continue to survive.

As Dr. Roger Hagengruber of the Sandia National Laboratory has noted: "The data provided by monitoring is viewed through a political lens" (interview in *The Financial Times*, Nov. 19, 1985). This fact was apparent at the very beginning of the underground test program, when preshot details for RAINIER were made available by the U.S. Atomic Energy Commission, ostensibly so that geophysicists could plan to pick up the resulting seismic signals and use them, for example, for the study of deep Earth structure. Whereas some scientists at an international scientific convention in 1957 were led by Professor Keith Bullen of Australia to organize a telegram of thanks to the AEC for releasing these details, the Soviet delegation walked out of the meeting.

Sections below make minimal reference to these political perspectives. Topics covered include basic properties of explosion signals as compared to earthquake signals; a description of the capabilities of certain global networks and of one array in particular; seismic estimation of explosion yield; and a brief review of methods that have been proposed as offering possibilities for clandestine weapons testing and thus of treaty evasion. Policy options for new limitations on nuclear testing and a presentation of much tutorial material on U.S. capability to monitor nuclear explosions are given in a report by the Office of Technology Assessment, U.S. Congress (1988).

Basic Properties of Explosion and Earthquake Signals

An underground explosion is a source principally of compressional waves having longitudinal motions, i.e., P waves, because the explosion products apply a fairly uniform pressure to the walls of the cavity created by the explosion. However, transverse motions indicating shearing, i.e., S waves, can occasionally be observed, due presumably to slight asymmetries of cavity loading, or to structural heterogeneities or the presence of initial shearing stresses in rock near the shot point. An earthquake, on the other hand, is generated typically as a result of massive rock failure in a shearing mode, in which it is common to think in terms of two blocks of material (within the crust, for shallow events) on opposite sides of a fault. The earthquake is the outcome of a spontaneous process of stress release, during which the two blocks move rapidly with respect to each other, sliding in frictional contact on the plane of the fault. Because of this shearing motion an earthquake radiates predominantly transverse motions, i.e., S waves. Though P waves from an earthquake are generated at a substantial fraction of the S-wave level, they are associated with a four-lobed radiation pattern of alternating compressions and rarefactions in the radiated first motions, rather than the relatively uniform pattern of P-wave compressions radiated in all directions out from an explosion source. These idealized radiation patterns, for P-waves from an earthquake and from an explosion, are shown schematically at the top of Fig. 2.

In addition to each type of body wave, each type of seismic source also generates seismic surface waves. For an explosion, the largest surface waves are usually of Rayleigh type, with a radiation pattern that usually varies little with azimuth. These features are presumed to be due to azimuthal symmetry of applied pressure on the cavity walls. Also, at most test sites, the lateral variations of Earth structure are insufficient to couple $P\text{-}SV$ energy into Love waves (the other principal category of surface waves). When significant Love waves are observed from an explosion, as does occasionally occur, this is presumed to result from some release (triggered by the explosion) of tectonic stresses present in the vicinity of the shot point. Earthquakes, on the other hand, being entirely a result of tectonic stress release, generate both Rayleigh and Love waves, each with a radiation pattern that varies azimuthally and that can be expressed trigonometrically in terms of the spatial orientation of the underlying fault and the direction of relative block motions across that fault. The picture that emerges is usually one of relatively simple surface waves from an explosion as compared to those from an earthquake; however, some explosion signals are complicated by the superposition of an earthquake-like component in the radiation pattern. A further characteristic of surface waves for the two types of source is that, for shallow earthquakes and explosions generating about the same level of body wave signals, earthquakes are far more efficient generators of surface waves. This feature can be seen in the lower panel of Figure 2, and is

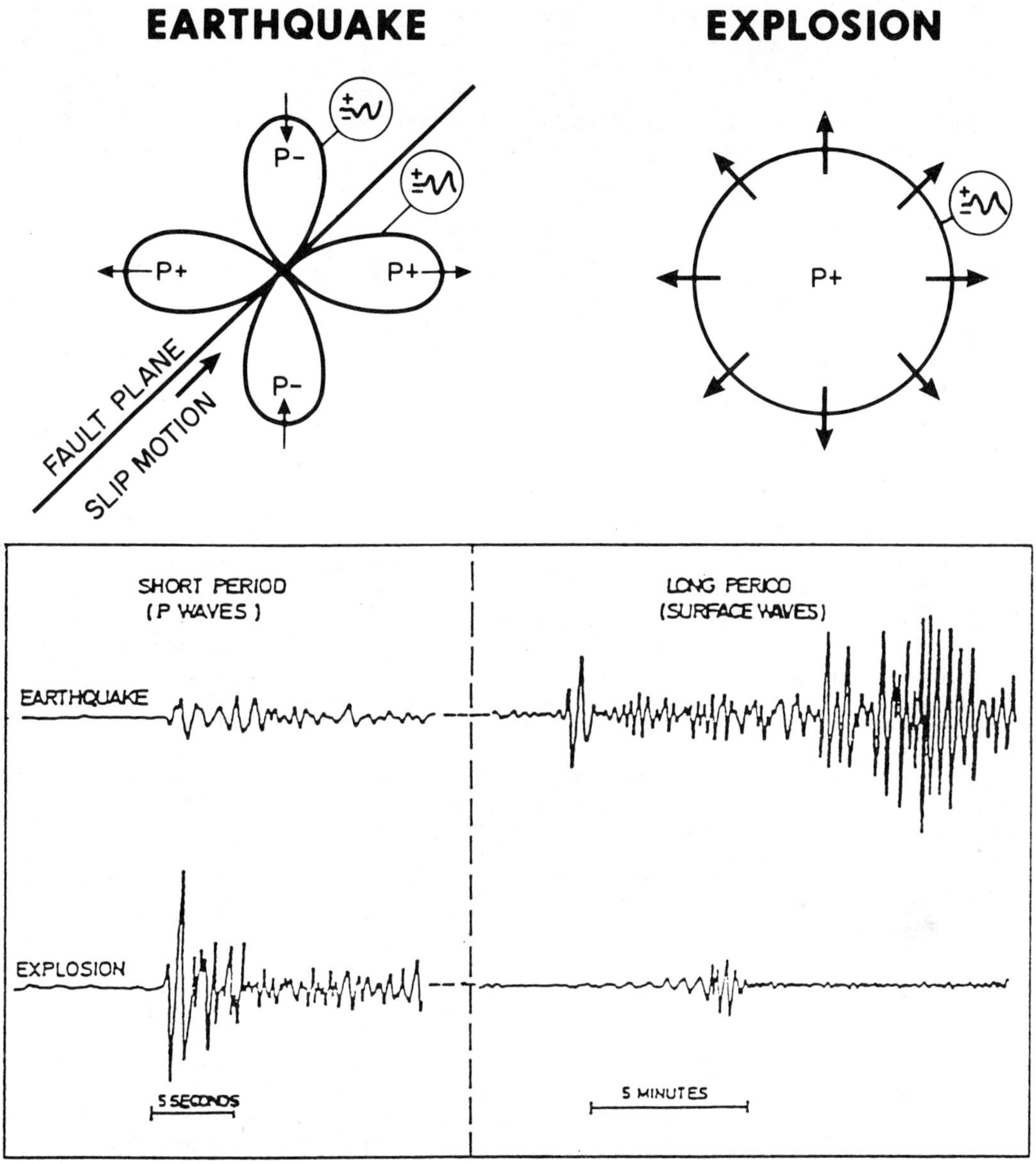

FIGURE 2. (*Upper*) The different radiation patterns for earthquakes and explosions. Earthquakes involve shear motion along a fault plane. Explosions are compressional sources of energy and radiate P waves equally in all directions. (*Lower*) Recorded signals (i.e., seismograms) of earthquakes and explosions have different characteristic features. Note the much stronger P-wave: surface-wave ratio, for the explosion as compared to the earthquake.

the basis of a routine method for discrimination between the two kinds of seismic event.

In practice, for different seismic events there is great variability in the number of points on the radiation pattern that are sampled. Thus, for an event of interest it is important to know from how many directions around the source and at what distances there is available seismic data. There are thousands of seismographs operated around the world by hundreds of different institutions in tens of different countries. For most seismic events, only a cursory glance at the data from a few stations is required for the trained analyst to conclude immediately that a particular event is an explosion or an earthquake. However, this does not address the real problems of monitoring, which often concern a small number of events that are relatively difficult to interpret. Thus, a good monitoring program must meet some kind of requirement that *all* types of earthquakes and explosions above a certain size are detected with high probability; are located accurately; are correctly identified as earthquake or explosion; and are measured adequately (i.e., the yield of explosions must be estimated to a certain standard of accuracy, that standard being set by considerations outside of seismology).

Seismologists characterize the size of seismic signals by means of logarithmic magnitude scales, with each scale based on a different type of seismic wave. A magnitude scale using teleseismic surface waves was first described about 50 years ago, based on the logarithm (to the base 10) of amplitude of maximum ground displacement due to surface waves with periods about 20 s. It is known as the M_s scale. (The word "teleseismic" refers to observations at a distance of more than about 2000 km from a seismic source.) Another widely used magnitude scale is that based on the amplitude of teleseismic P waves. Known as m_b, it entails measurement of ground motion at about 1 s period. As part of the assignment of M_s and m_b values, for a particular seismic event as recorded at a particular station, a standard correction is applied to account for the distance between the source and the receiver at which the data was obtained. Magnitudes range from about -3 for the smallest observable microearthquakes, up to above 8 for the largest earthquake. A one kiloton underground explosion has an m_b roughly about 4, and each year there are about 7500 earthquakes worldwide with $m_b \geq 4$ (Ringdal, 1985). Note that magnitude is only an empirical estimator of seismic event size. Although the use of seismic moment has superceded use of m_b and M_s in much of modern seismology, magnitude scales are still often used in discussion of seismic monitoring because this is a practical way to relate that discussion directly to properties of signal strength.

Having run broadly through these relevant features of seismic signals, let us next pay attention to specifics of explosion monitoring.

Detection

The Earth's surface continuously undergoes small motions, stimulated by ocean tides, winds, surf, and various activities of human society such as train and truck traffic, and pumping for oil. Seismic signals from earthquakes and explosions can be detected to the extent that they are stronger than this background noise, which varies greatly at different sites and in different frequency bands. Deployment of seismometers is more effective if sites can be found that are very quiet.

The signals of interest in seismic monitoring have a relatively impulsive nature. It follows that the detection of explosion signals can be done by comparing signal levels averaged over a short time (~ 2 seconds or less) with levels averaged over a longer period (~ 30 seconds), and continually updating the comparison. The analysis is traditionally done by eye, but it can be automated. Thus, a detector consists of one or more seismometers all in the same general location, plus some signal processor (human or a computer).

With many different detectors, the next technical problem is recognizing an association between signals collected at different geographic locations: that is, of relating them to some presumed common source such as an earthquake or an explosion (whose signals have been picked up by a network deployed over some region of the Earth or perhaps over the whole globe); and of locating that source. The basic method of location is the same as that by which we use our two ears to tell from which direction a sound is coming. Our brain analyzes the slight difference in the times at which a signal reaches our two ears. With a network of detectors of seismic signals, in effect each pair of detectors can be used to draw a line along which the source must lie. (The property of this line is that a source anywhere along it will generate the difference in signal arrival times actually seen at those two detectors.) With many such lines (one for each pair of detectors), the place where they all intersect must be the location of the source. For a good enough network and a strong enough source, the method has considerable redundancy. Signal detected at four or more stations, is usually enough to give a useful source location. A large underground explosion (150 kilotons) would be picked up at hundreds of seismometers around the world. A small explosion (1 kiloton) would be picked up on relatively few.

With many years of experience it has proven possible to assess the capabilities of a network of detectors, given the location and sensitivity of each seismometer and given also a characterization of noise levels at each site in the network. A useful way to describe these capabilities, both actual and predicted, is by contouring the globe in magnitude units, which give the threshold value of source strength above which a seismic source will be detected with high probability and at enough stations to be located by the network. Figure 3 shows the actual and the predicted value of these contours, for a particular network of 115 globally distributed stations, requiring 90% probability of detection (for events near the threshold level) at at least four stations, at least one of which must be at a teleseismic distance (Ringdal, 1985). This standard of detection capability is quite stringent. Though this network was "globally distributed," it suffered from relatively poor coverage in the Southern Hemisphere. Only a few useful stations could be found in Africa and South America. Thus, for this network, although the detection threshold has m_b in the range 3.9–4.3 in the Northern Hemisphere, the range is 4.2–4.8 in the Southern Hemisphere.

To improve detection capability for sources in some particular part of the world, one would need to add stations located at distances favorable for

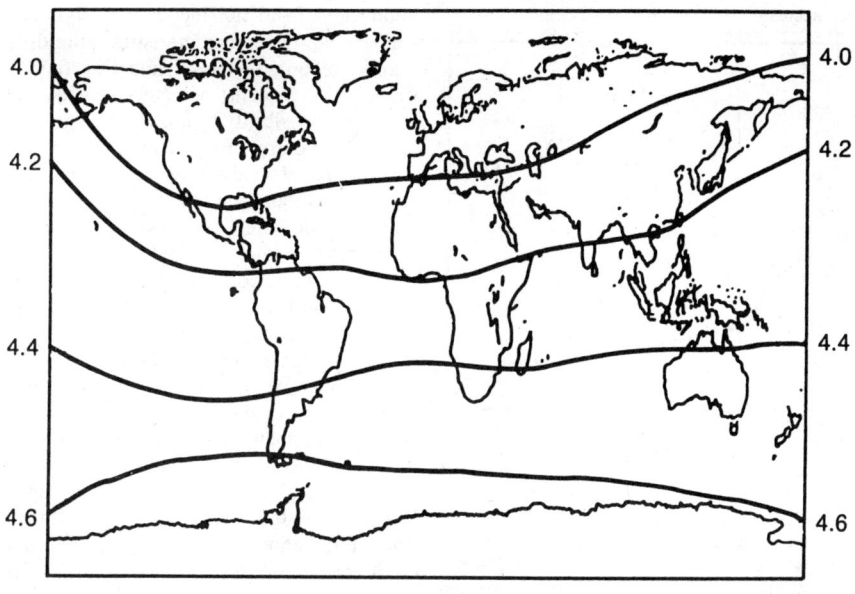

ESTIMATED

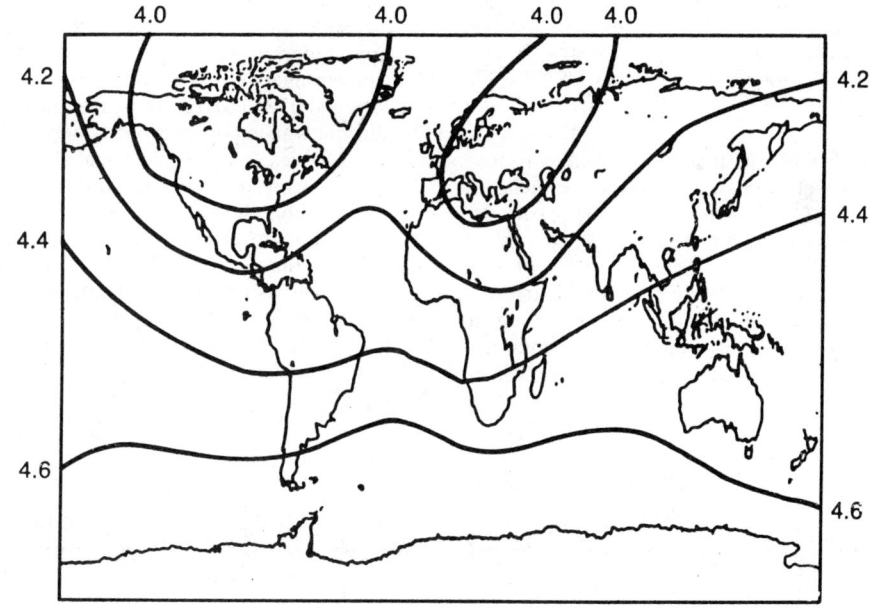

ACTUAL

FIGURE 3. Contours showing detection capability of a particular seismic network of existing stations. For planning purposes, it is important that the estimated capability is in good agreement with the actual capability. Here, the differences are only about 0.1 m_b unit. (From Ringdal, 1985.)

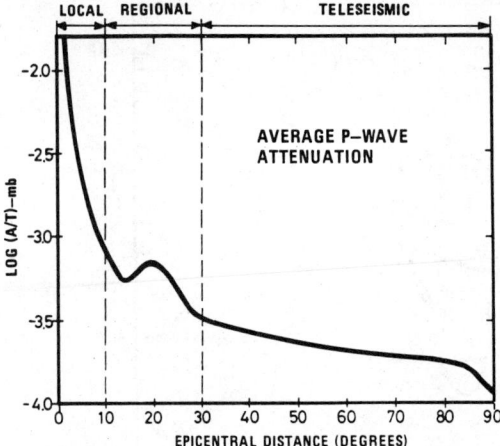

FIGURE 4. Attenuation of short-period signal amplitudes with distance. This figure is in m_b units, so it can be used to supply the distance correction needed to assign the m_b of a source as determined by observation at a particular distance. In forming log (A/T), A is the ground motion in microns, and T is the period of the largest part of the P-wave signal, in seconds. Distance here is measured in degrees for the angle subtended at the center of the Earth by the source and the receiver. One degree is thus about 110 km measured over the Earth's surface. The figure indicates that m_b is zero for an event with about 1 millimicron of ground motion, at 1 s period, about 1000 km from the source. Such a small event, however, could not in practice be detected at such a distance because Earth noise is too large, even at quiet sites.

picking up signals from that region. Figure 4 shows how signal level drops with distance and indicates the advantage of being able to use signals at so-called regional distances rather than having to rely on "teleseismic distances." The possibility of such improved detection capability bears on the question of whether in-country stations are required, on the territory of a nation being monitored for purposes of treaty compliance, or whether compliance can be monitored from outside of that country. Regional signals can have characteristics quite different from those recorded teleseismically (e.g., different frequency content; greater sensitivity to local geology; possibility of propagation guided purely by crustal structure). Empirically it is found that when explosions of a few tons are detonated at shallow depths, regional signals can be picked up several hundred kilometers away even for propagation paths that (from geological and geophysical considerations) would be considered unfavorable. An example is shown in Fig. 5.

The foregoing discussion has described how individual seismological stations may contribute to detection capability and some of the considerations that arise in evaluating a global network. Another practical element is introduced when one considers the capabilities of a group of seismometers in the same general region, all operated together as an "array," with central processing facilities that allow the multichannel signals to be filtered, delayed, and summed. Figure 6 shows the location of the Norwegian Seismic Array (NORSAR), and a map view of the 3800 km-long path to NORSAR from the main Soviet test site near the town of Semipalatinsk in East Kazakhstan, at about 50° N, 49° E. The lower part of the figure displays the first 20 s of the P-wave arrivals at individual NORSAR sensors from a Soviet nuclear explosion at this test site. Note that some sites appear more sensitive than others. For example, there are stronger signals at locations labelled 04C and 06C; and 05B and 07B are weaker than average. This illustrates an important property of seismic body wave amplitudes, namely their spatial variability, which here is presumably a result of focusing and defocusing within the crust and upper mantle.

Significant improvement in seismic signal-to-noise ratio can be achieved with a good array, as compared to this ratio for a single sensor. Thus, by adding the signals of Fig. 6 (after an appropriate time shift to bring all the P arrivals to a common reference time), the signal is enhanced if the P-wave first motion is

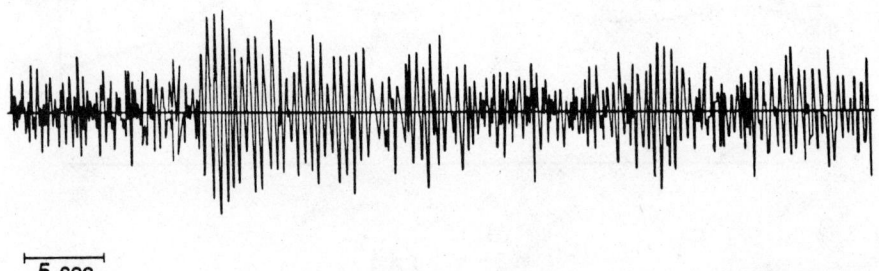

FIGURE 5. The seismogram at Lajitas, Texas, caused by a 5 ton chemical explosion about 725 km away. This signal has been filtered to pass frequencies above 2.5 Hz. The conclusion associated with this record was that "a single, borehole, vertical seismometer at Lajitas could detect a fully decoupled...nuclear explosion with a yield of about 0.2 kt at distances of the order of 750 to 1000 km. Processing...could be expected to lower this threshold." (Herrin, 1985, p. 525.)

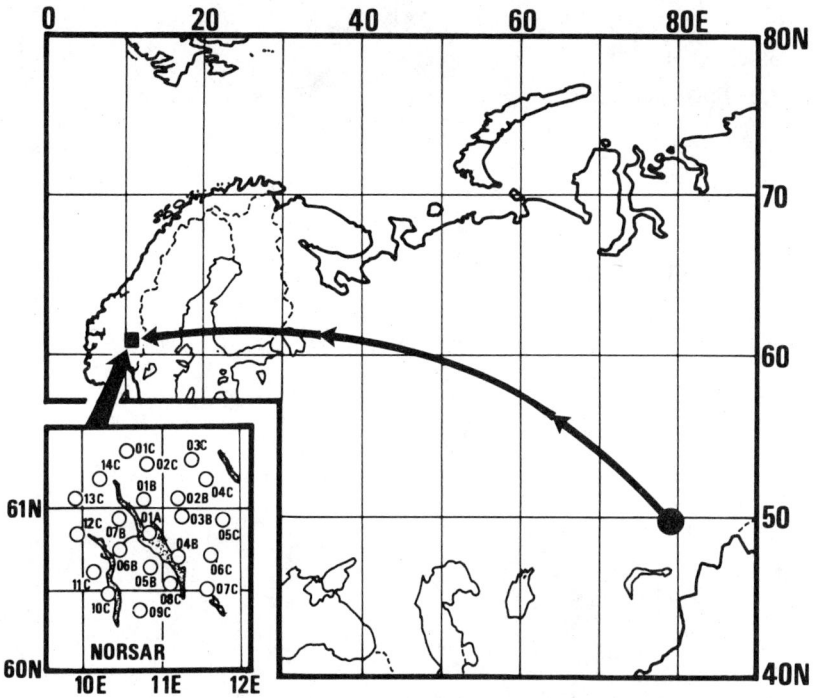

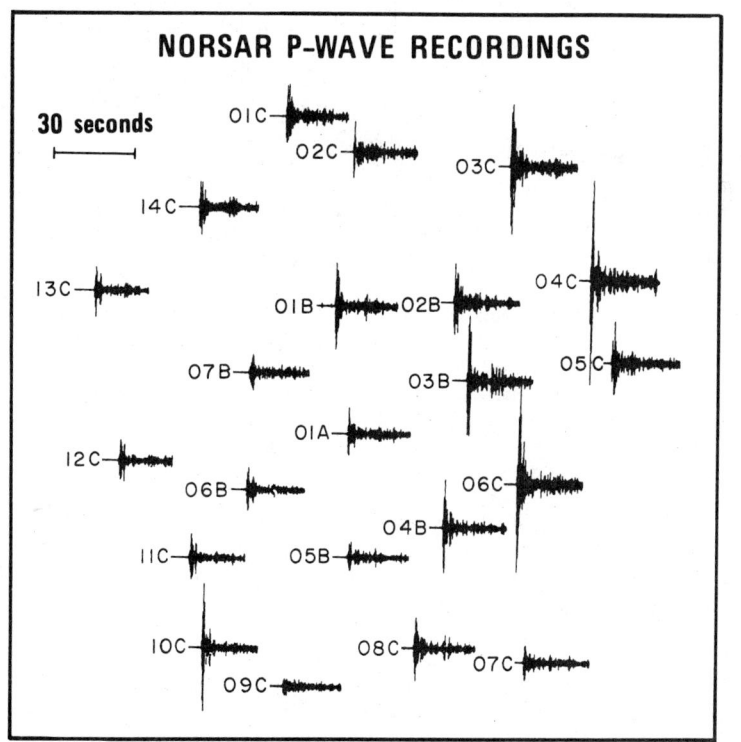

FIGURE 6. (*Upper*) Location of the seismic array known as NORSAR, and a propagation path from the main Soviet test site. NORSAR consists of seismometer installations located in a circular region about 200 km in diameter. (*Lower*) NORSAR seismograms for a nuclear explosion, arranged as for a map view of the individual seismometers.

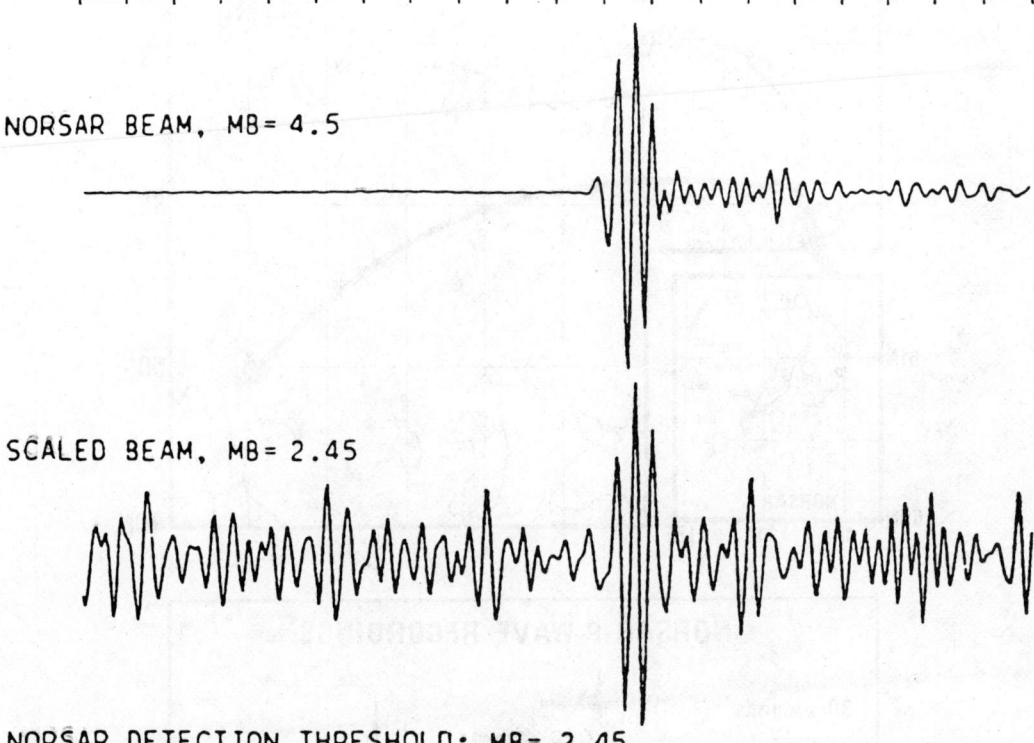

FIGURE 7. (*Upper*) The NORSAR beam for an East Kazakhstan explosion (m_b 4.5). (*Lower*) The beam for the signal has been scaled down (or equivalently, as shown here, the noise has been scaled up), to see how small an m_b could automatically be detected. The answer in this case is m_b 2.45. (NORSAR S.A.T.S.)

coherent across the array; and the relatively incoherent noise (which is apparent prior to P on each sensor) is suppressed. The outcome of such processing is often referred to as the "beam" formed by the array, which may be "steered" to "look" at different source regions by using the set of delays at each sensor appropriate for each source region, prior to signal summation. An example of the NORSAR beam is shown in the upper part of Fig. 7 for a Semipalatinsk explosion with m_b 4.5. Comparison of the signal strength with noise levels prior to the P arrival indicates here that events with much lower m_b would be detectable. It is natural to explore the detection threshold for NORSAR, applied to Semipalatinsk explosions, by experimenting to find how much this m_b 4.5 signal can be scaled down before it is lost in the preceding noise. Equivalently, one can scale *up* the noise, leaving the signal level unchanged as shown in the lower part of Fig. 7. It is found that the signal-to-noise amplitude ratio in the original seismogram is about 100 times greater than that required for automatic detection by the array. In this example, which gives a particularly favorable result, the detection threshold is estimated as only m_b 2.45. Typical thresholds achieved from such scaling experiments for this one array are reported in the following table:

Region	m_b Range
West of Ural Mountains	2.0–2.5 (possibly better)
Caspian Area	2.0–2.5
Semipalatinsk	2.5–3.0
Siberia	2.5–3.5.

The favorable sensitivity of site 06C at NORSAR has already been noted. In June 1985, after some years of careful preparation, a small array known as NORESS (Norwegian Regional Array System) was formally declared operational at this location. It consists of 25 seismometer sites over an area about 3 km across, linked by optical fiber to a central processing facility. At the top of Fig. 8 is shown a

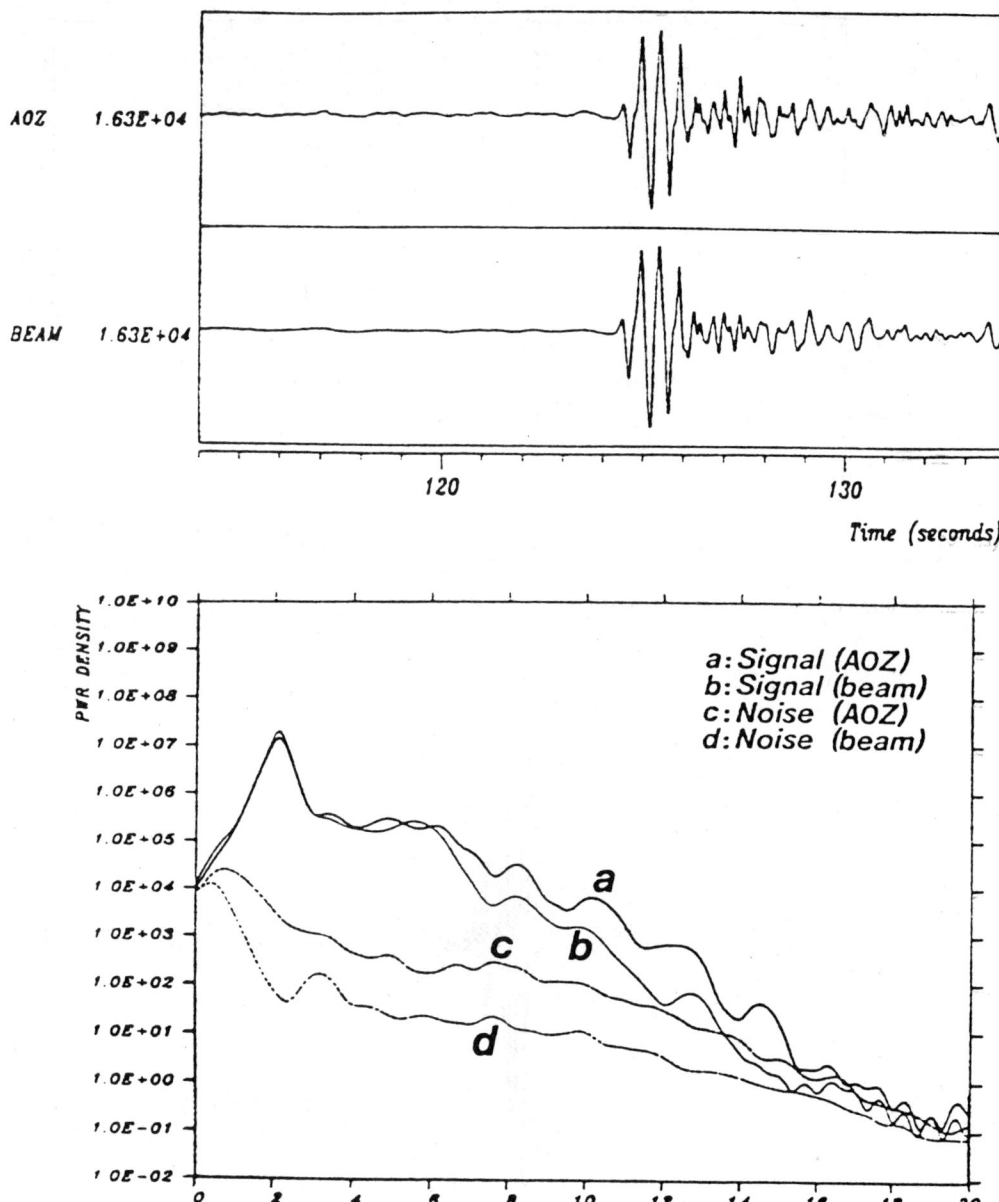

FIGURE 8. (*Upper*) NORESS seismograms, both single sensor and the beam, for an m_b 4.9 East Kazakhstan explosion on 25 July 1985. (*Lower*) Spectra of signal and noise. To see the benefit of an array, as compared to a single sensor, compare the signal-to-noise b–d with a–c. The improvement is at the lower end of the frequency range because of the need (for an effective array) to have a coherent signal. Coherency degenerates at high frequency. (NORSAR S.A.T.S.)

single sensor record, and the NORESS beam, for an East Kazakhstan explosion with m_b 4.9 on 25 July 1985. The spectra of these records are shown in the lower part of the figure, together with noise spectra. It is seen that even at this distance, nearly 4000 km from the source, there is significant signal above noise out to 15 Hz. The beam signal loss is negligible up to 6 Hz, and there is very good noise suppression in the beam around 2 Hz. (This latter property was a goal of the array design.) The remarkable detection capabilities of NORESS for teleseismic sources are further displayed in Fig. 9, which is for a small (m_b 3.5, presumably sub-kiloton) explosion near Semipalatinsk on 11 July 1985.

SEISMIC MONITORING OF NUCLEAR EXPLOSIONS

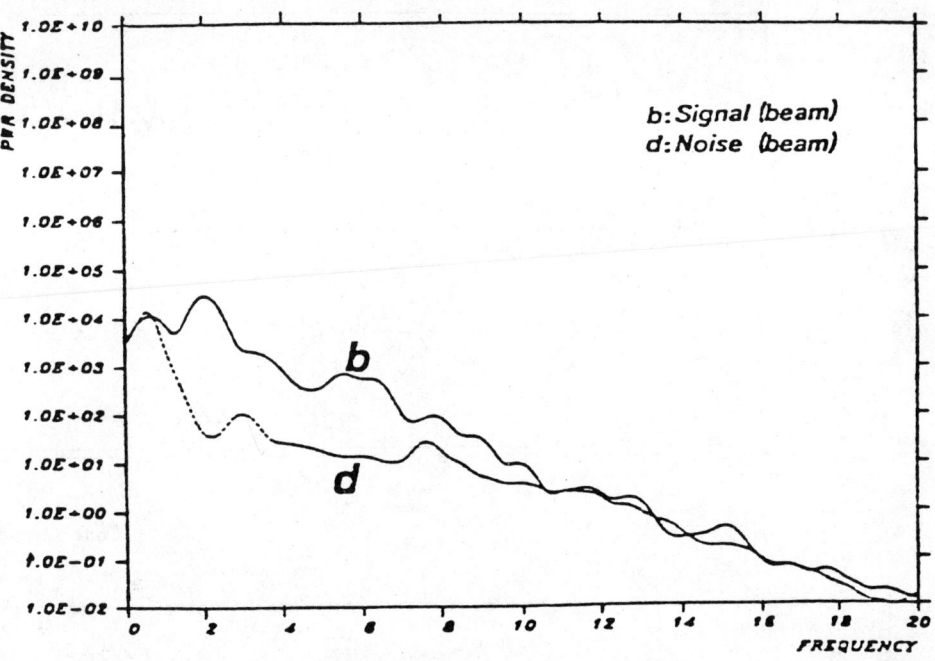

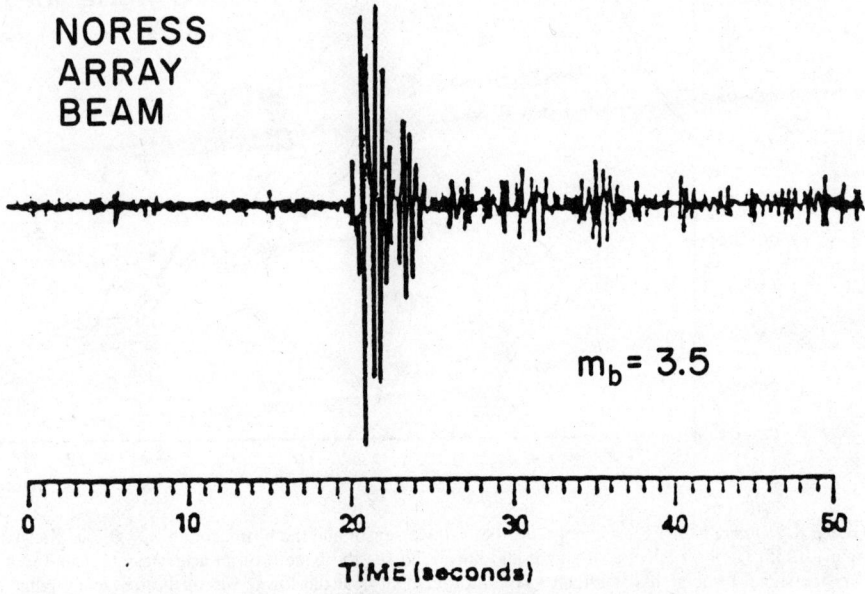

SIGNAL FROM SEMIPALATINSK, USSR, ON JULY 11, 1985.

FIGURE 9. Similar to Fig. 8, but for a much smaller (subkiloton) explosion. The spectrum shows favorable signal-to-noise from about 1.5 to 10 Hz. Filtering the beam to pass signal in this band, gives the seismogram shown here. (NORSAR S.A.T.S.)

NORESS was designed for signals from sources at "regional" distances, and the detection threshold for this array is down to about $m_b = 2.0–2.5$ at 1500 km epicentral distance. At regional distances, the seismic waves *Pg, Pn, Sn,* and *Lg* are commonly seen. (*Pg* propagates wholly within the crust; *Pn* and *Sn* travel in the top of the mantle, just below the Moho; and *Lg*, often the strongest signal, is guided largely by the crustal layer. See Fig. 10.) Figure 11 shows NORESS spectra out to 20 Hz, for a presumed quarry blast near Leningrad, a little over 1000 km away. *Lg* appears in a frequency band that differs from that of the P wave, which is observed out to 18 Hz.

It has recently been shown for a single sensor that NORESS signals have important information in even higher frequency bands. Some important results are shown in Fig. 12. They use data from an advanced recording system that permits analysis of frequencies even above 50 Hz. Remarkably, for a magnitude 3.0 event 500 km away, signal is about 20 dB above noise, all the way out to 50 Hz.

From this review of detection capability, it is apparent that the research frontier for seismic monitoring is represented by events roughly in the range $2 \leq m_b \leq 4$. Here, in contrast to large events, problems to be recognized include the much larger number of signals that must be processed, and the need to distinguish chemical explosions from nuclear explosions. Monitoring for a new test ban regime (for example, a low-yield threshold) would likely entail in-country networks of newly deployed seismometers in the United States, the USSR, and elsewhere.

Discrimination

Seismic methods for discriminating between explosions and earthquakes are based on interpretation of the event location; on the relative excitation of a variety of body waves and surface waves; and on properties of the signal spectrum associated with each of these two different types of source. Within these three broad categories, many different methods have been tried, with various degrees of success, over the years since the RAINIER explosion of 1957. As the capabilities of each method are probed, the question of interest is often: "Down to what size of seismic event does this method of discrimination work?" In some cases discrimination is unambiguous even at very small event size. (For example: however small an event, it may be presumed to be an earthquake if it is located at a depth greater than 15 km below the Earth's surface. Another example: even a small event will a priori be regarded with high probability as an explosion if it occurs in an

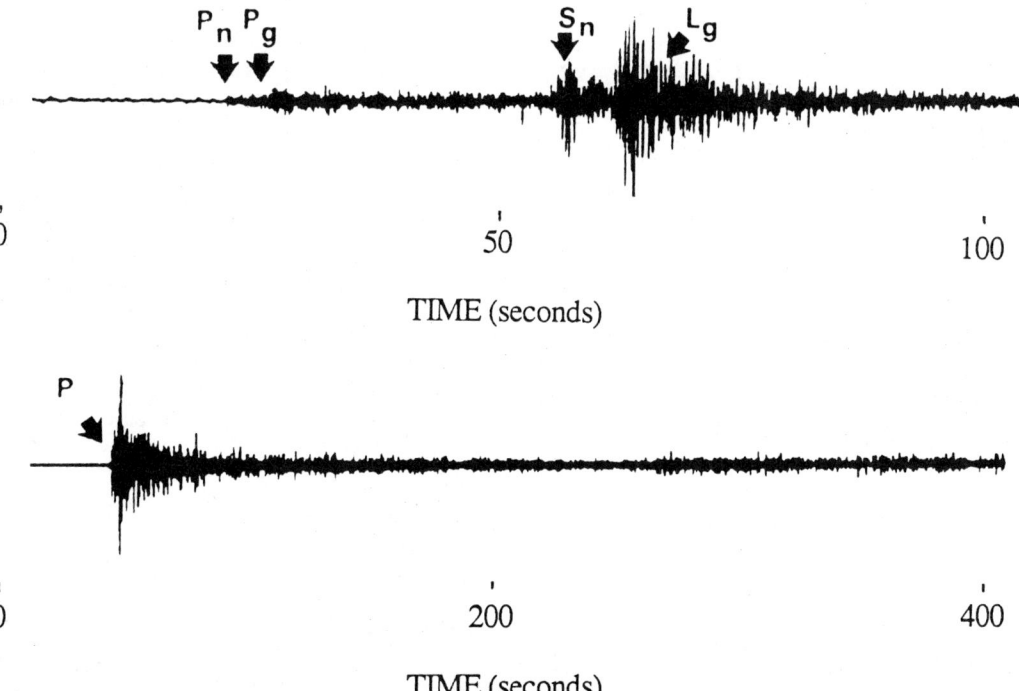

FIGURE 10. (*Upper*) Seismogram at a regional distance from the source. Signals here are associated with waves that propagate in the crust and uppermost mantle. (*Lower*) Seismogram at a teleseismic distance. This wave has propagated through the deep interior of the Earth.

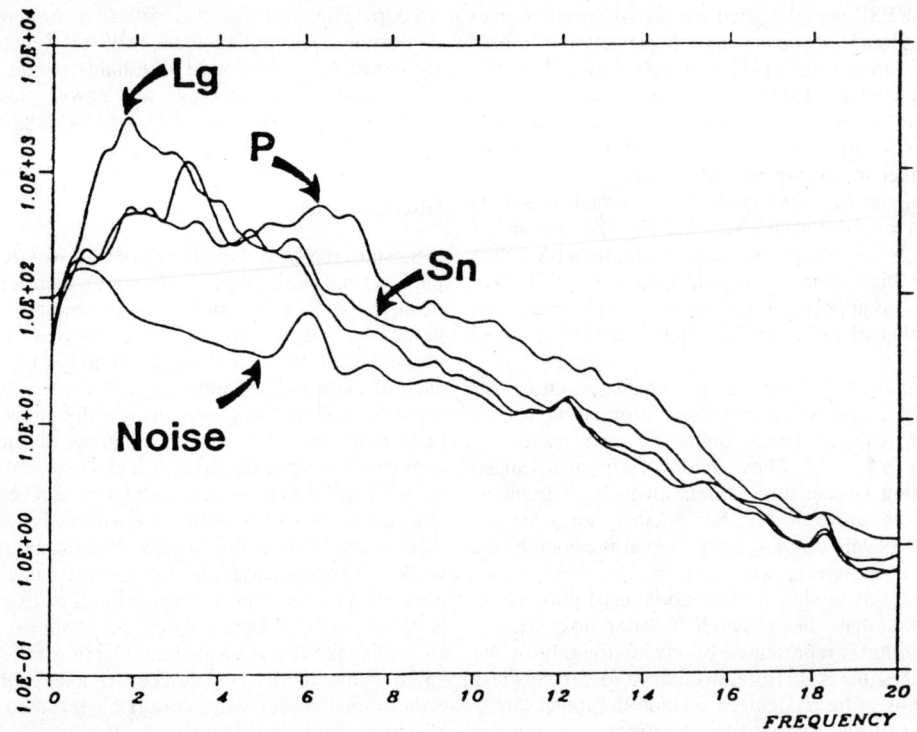

FIGURE 11. NORESS spectra of a presumed chemical explosion about 950 km away. Signal for Pn is above noise out to 18 Hz. (NORSAR S.A.T.S.)

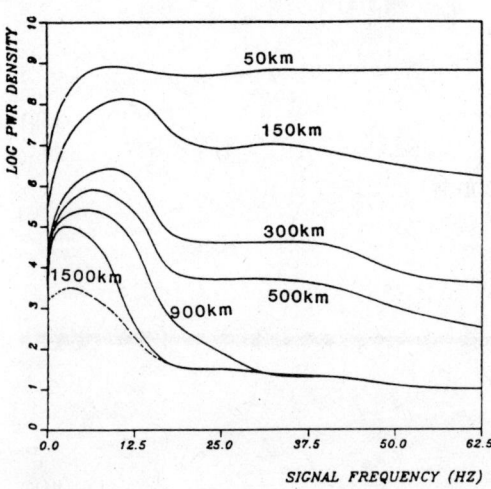

FIGURE 12. P-wave spectra obtained at a single NORESS sensor equipped with a special high-frequency recorder. The figure represents average features of about 100 regional events at various distances, all scaled to magnitude 3. A typical noise spectrum is shown as a dotted line. (NORSAR S.A.T.S.)

area that is geologically stable and that for decades has had no seismic activity.) In general the threshold for discrimination is higher than that for detection, perhaps by around 0.5 m_b units (Hannon, 1985).

The most useful methods for discrimination can be listed as follows:

1. Interpretation of the location. Is the event in a seismic or an aseismic area? Below the floor of an ocean? At depth below a continent?
2. Relative amplitude of body waves and surface waves (see Fig. 2). This relative amplitude can be studied by plotting the event of interest on an M_s:m_b diagram, such as the upper part of Fig. 13. The surface-wave amplitude is read typically from signals with period about 20 s, and the body-wave amplitude at about 1 s period. (Although effective for large enough events, an explosion with m_b much below 5 may not have large enough surface wave signals at teleseismic distances to apply this method dependably. A judgement must be made, whether or not to allow absence of signal to be used as evidence.)
3. Use of radiation patterns (see upper part of Fig. 2). Is the initial P-wave motion of the ground indicative of compression radiated to all directions from the source? Or, are dilatations

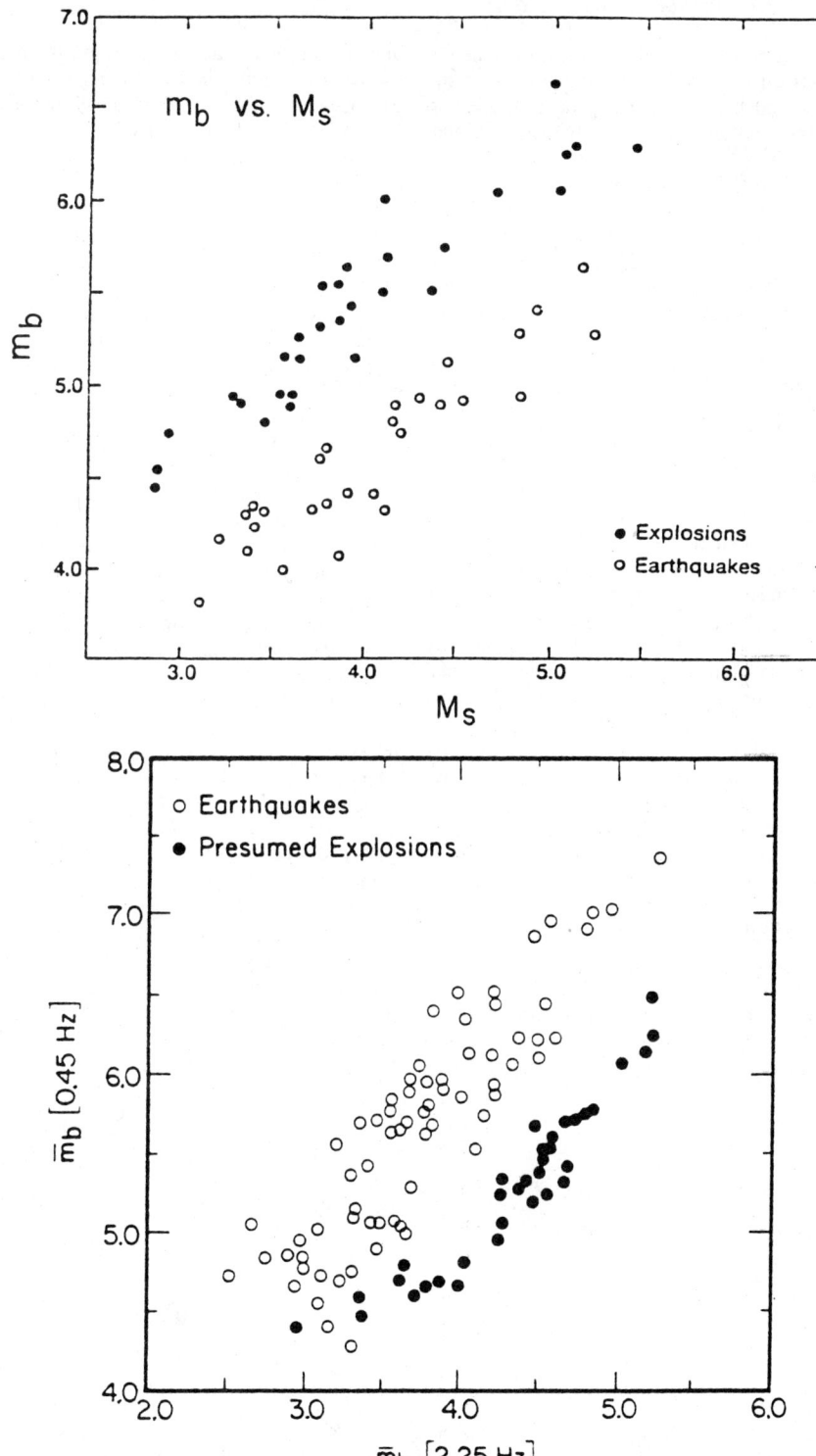

FIGURE 13. (*Upper*) M_s versus m_b for a suite of earthquakes and explosions (Marshall and Basham, 1972). (*Lower*) The body-wave magnitude is determined at two different frequencies (Stevens and Day, 1985). Earthquakes and explosions fall into separate populations on both these diagrams. Explosions are relatively efficient at generating higher frequencies.

recorded at some azimuths, indicating that the event has an earthquake-like radiation pattern? (This method has traditionally been applied to teleseisms and is under study for regional signals.)

4. Relative amplitude of body waves as recorded at high and at low frequencies. This method is being studied with frequencies in the range 0.5–50 Hz, to see if it can discriminate between very small earthquakes and explosions, even down to around m_b 2. The lower part of Fig. 13 gives an example.

Fortunately, the cumulative effect of applying several methods can result in unambiguous separation of earthquakes and explosions, in situations where no single discriminant is successful. Thus, in a suite of about 1000 seismic events with $m_b > 4.5$ recorded on only nine stations, for a 162-day period in 1972, more than 90% could be identified as earthquakes because their depth was 30 km or greater, and/or their epicenters were at sea by more than 25 km; the remainder were all identified as earthquakes by their observed $M_s:m_b$ ratio (Sykes et al., 1983). Incidentally, it appears that by the late 1960s–early 1970s, data available routinely to the international scientific community could be used for monitoring down to m_b 4.5 (Evernden, 1977). It is generally recognized that discrimination becomes significantly more difficult at magnitudes below m_b 4, because there are many more earthquakes to analyze (around 50,000 per year, between m_b 3 and m_b 4), with lower signal levels from each one (in comparison to an event with $m_b > 4$).

Yield Estimation

For a few tens of underground nuclear explosions, most of them at the Nevada Test Site, the yield has been announced by the country conducting the test. It has therefore been possible to calibrate observed seismic magnitudes for these tests against the announced yields, and an example is given in Fig. 14 using m_b values and yields reported for Nevada explosions in hard rock. Such a calibration curve can be applied to obtain a seismic yield estimate for Nevada explosions in hard rock with unannounced yield. However, several problems prevent use of a universal magnitude-yield relation. The first is that explosion energy couples differently into seismic signal, depending on the type of rock in the vicinity of the shot point. For example, coupling is much more efficient for hard water-saturated rock than it is for dry alluvium. The second problem is that seismic waves, once excited, tend to propagate more efficiently for sources in geologically stable regions than for sources in a region that is tectonically active. Thus, allowance must be made for physical and geological differences between test sites, and there can be variability even on a single test site. A third reason for the existence of many different magnitude-yield curves is that basically these curves are empirical, and can thus differ somewhat, depending on the details of the method used to assign m_b.

Therefore, some form of correction is usually required when applying a magnitude-yield curve developed for one test site to a new region. In this connection, it is of interest to note m_b and yield for the U.S. nuclear explosion LONGSHOT (conducted in 1965 in the volcanic breccias of an Aleutian island). The m_b value is about 5.9, corresponding to a yield of about 300 kt if the Nevada curve of Fig. 14 is applied directly. But the announced yield for LONGSHOT is 80 kt.

One way to obtain a calibration curve for the Aleutians is to add a correction to the Nevada values of m_b at a given yield, before the curve of Fig. 14 is used to supply a seismic yield estimate in this new location. The m_b correction, for a site differing from that where a calibration curve is directly available but presuming similar rock types in the vicinity of the two shot points, is called the *bias*. If the bias

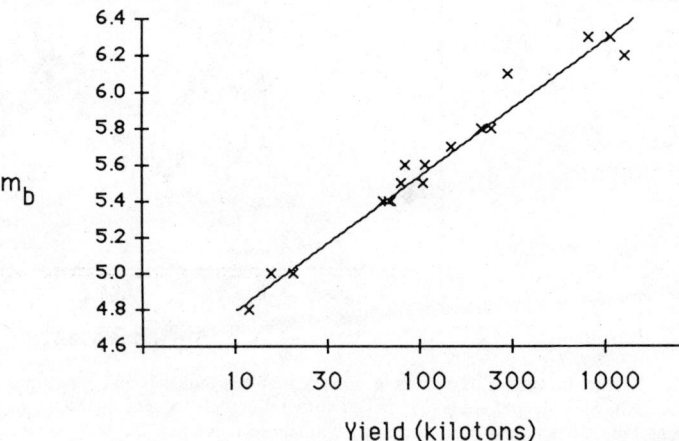

FIGURE 14. m_b versus announced yield, for 17 Nevada nuclear explosions in hard rock. The best-fitting straight line here is $m_b = 4.067 + 0.74 \log_{10}$ (Yield), from which the standard deviation of these points is 0.08 m_b units. (Data from Nuttli, 1986.)

correction is not applied, then, for a non-Nevada explosion, it can be presumed that a Nevada calibration curve will typically give too high a seismic yield estimate. This need for a bias correction arises because the Nevada Test Site is in a region of active tectonics, with significant episodes of volcanism in the last few million years, presumably associated with high temperatures within the upper mantle (and thus, anomalous attenuation of seismic waves propagating through the hot and partially molten upper layers of the Earth, 100 or 200 km in thickness beneath the western United States).

The existence of m_b bias has long been known in seismology in connection with what is called *station bias*. By this term is meant the systematic difference between mean m_b values (obtained for a particular seismic event by averaging reported m_b from many seismometers all over the globe), and m_b reported by just one station. For example, years of experience tell that the station BMO in Oregon (another region of active tectonism) has reported m_b values that for a given earthquake are typically about 0.3 units below the global average; and station KJN in Finland (in a stable shield region) reports values about 0.15 m_b units higher than the average. Their station bias values are thus -0.3 and $+0.15$ respectively. Station bias values commonly range over ± 0.4 m_b units, and it may be expected that source region bias (which is what must be applied when a standard m_b-yield curve is used for different source regions) will also range over about 0.8 m_b units.

At sites for which little direct calibration data are available, several methods have been proposed with which to estimate the m_b-yield relation. These methods include: the use of surface waves; the study of differences in body-wave spectra at high frequencies; and comparison with known m_b bias at sites with similar geology. A review by Sykes and Ruggi (1988) gives the magnitude-yield relation for the main Soviet test site at Semipalatinsk as

$$m_b = 4.262 + 0.973 \log_{10}(\text{Yield})$$

for yields up to 75 kt, and

$$m_b = 5.285 + 0.426 \log_{10}(\text{Yield})$$

for yields above 75 kt. In this connection, it is important to recognize that there is no such concept as an absolute, correct, seismic magnitude for a given event. Magnitude is an empirical measure, and it is important (for seismic yield estimation) that the event under study have its magnitude assigned by the same method used for the events that established the magnitude-yield relation.

In 1988, two special nuclear explosions were carried out by the United States and the USSR to assist in the evaluation of technologies to estimate yield. Because of their significance in light of the 1974 Threshold Test Ban Treaty, these explosions were called Joint Verification Experiments. One, in August, was conducted by the United States in Nevada and was monitored on-site by the USSR. Preliminary values of its magnitude and yield (measured on-site) were m_b around 5.4 ~ 5.5 (from stations reporting to the U.S. Geological Survey) and a yield "in the mid-140 kiloton range" (according to press reports, for example *The New York Times* of September 11 and October 30). The other explosion, in September, was a reciprocal experiment conducted by the USSR in East Kazakhstan and monitored on-site by the United States. Preliminary values of its magnitude and yield, from the same sources, were m_b around 6.0 ~ 6.1, and a yield (measured on-site) around 120 kilotons. If these preliminary values are confirmed, they give an excellent example of the bias factor. The two explosions were about the same yield, but the seismic signals from the East Kazakhstan explosion were about 0.6 m_b units higher than those from the explosion in Nevada, corresponding to a factor of four difference in seismic amplitudes. According to press reports also, the yields estimated from teleseismic signals proved quite accurate, after allowance for the bias, which had itself been estimated prior to these Joint Verification Experiments, and which can now be further refined.

Potential Methods for Evasion and Conclusions

Several methods have been proposed by which underground explosions could be concealed. One method is simply to make them small enough; but then there would be relatively little to learn from the point of view of a weapons designer. At significant yields, obviously the more important methods to explore are those that combine as many features as possible, designed to reduce seismic signal-to-noise ratios at all relevant monitoring stations. Proposed methods include emplacement of the nuclear device in material such as dry alluvium, to reduce the coupling of explosion energy into seismic signal; waiting until a sufficiently large natural earthquake occurs fairly near a test site, and then within a couple of minutes executing the weapons test so that its seismic signals would hopefully be swamped by the large and prolonged signals from the earthquake; and setting off a sequence of several explosions that are designed to simulate a natural earthquake signal. Careful study of each of these methods indicates that they are likely to be relatively ineffective in comparison with the method known as cavity decoupling, which we next discuss, and which appears to set the practical levels down to which seismic monitoring of nuclear explosions is possible.

When an underground explosive device is tightly packed into its hole ("tamped" or "fully coupled") and is detonated at sufficient depth to contain all radioactive products, a shock wave travels from the shot-point out into the surrounding rock some distance at speeds that exceed the normal P-wave speed. This nonlinear phenomenon reduces at sufficient distance from the shot-point, and thereafter the wave propagation can usually be regarded as linear. The so-called elastic radius for a tamped explosion, i.e. the radius beyond which wave propagation is linear, is roughly 100 meters times the cube root of the yield (in kilotons).

If, instead of being tamped, the explosion is set off inside a large underground cavity, then the shock wave set up in the rock can be weakened. The explosion is said to be decoupled, and theoretical work begun in 1958 has addressed the question of how much weaker the seismic signal might be made. If the explosion occurs in a sufficiently large cavity there is no nonlinear shock at all in the rock—that is, the walls of the cavity are never driven beyond their elastic limit. Such an explosion is called *fully decoupled*, and theoretical work plus experiment has indicated that signals can thereby be reduced by a factor around 100, compared to a tamped explosion. The cavity radius itself is the elastic radius for a fully-decoupled shot. For strong salt or granite, the cavity radius required for full decoupling has been estimated as about 20 m times the cube root of the yield. For weak salt, the required radius is about 50% greater.

At frequencies that have conventionally been used for seismic monitoring, the seismic signal strength is proportional (very roughly) to the volume within the elastic radius. From the results summarized earlier, we see that the elastic radius is about five times smaller for a fully decoupled shot than for one that is tamped. This reason is essentially why decoupling works and why it has been proposed as offering the technical possibility of a clandestine program of nuclear testing. However, the signal strength is not nearly so strongly reduced by decoupling at frequencies above that associated with resonances of the internal surface at the elastic radius. In practice, the frequency above which decoupling is likely to be substantially less effective is around 10 to 20 Hz, divided by the cube root of the yield (in kt).

Of course, a thorough discussion of decoupling as an evasion scenario should include several nonseismological considerations, including the military significance of being able to carry out nuclear tests up to various different yield levels (e.g., 0.1, 1, or 10 kt) and the political consequences if a clandestine test program were uncovered. Technical considerations include methods of (clandestine) cavity construction, and the capabilities of nonseismological surveillance techniques.

It would appear that at sufficiently low levels of explosion yield some kind of clandestine testing could be undertaken that would be undetectable by seismological means. However, the technical difficulties associated with cavity decoupling rise rapidly with increasing yield. They would be formidable at 1 kt, but consider the requirements for hiding a fully-decoupling 10 kt shot:

1. The volume of the cavity would need to be at least 500,000 m^3, which if spherical requires a cavity diameter of 100 m.
2. Though such a volume (albeit nonspherical and hence not so efficient for decoupling) could be created in certain geological formations by dissolving salt in water and pumping out brine, the cavity would have to be prevented from collapsing subsequent to a 10 kt blast. (Or, failing prevention, the evidence of any collapse would have to be concealed.)
3. Leakage of radioisotopes must be prevented.
4. Although decoupling by as much as a factor of 100 might occur in the frequency band conventionally used for monitoring, the decoupling factor drops by an order of magnitude at higher frequencies. From a fully-decoupled 10 kt shot, the seismic signal level around 10 Hz would likely be detectable at, for example, NORSAR/NORESS (and/or at any other advanced monitoring sites, including seismic stations at regional distances, that might be in place at the time of the explosion).

For these reasons, testing could be clandestine only if conducted at limited yields. Clearly, that limit is far below the current threshold of 150 kilotons, even with current monitoring capability. It is, however, difficult to reach precise conclusions on what future yield levels could be achieved in a hidden test program, because much would depend on the level of commitment to improvements in current monitoring. It would depend too on the level of effort postulated for the clandestine activity and on assessments of the political risk of detection, which could be perceived differently by the monitored and monitoring sides. While some assert that clandestine use of cavity decoupling would be so difficult to execute that it belongs to the realm of fantasy, others have been persuaded that the risk might indeed be manageable and that estimates of concealable yields under this evasion scenario must be made. Taking the latter more conservative approach, there appears to be a broad consensus among those seismologists who have studied the practical possibilities for evasion: given sophisticated monitoring that would include in-country arrays, it appears that explosions down to about the (decoupled) 5 kt level could be detected and recognized as explosions. This conclusion, however, does not presume that the promise of high frequency monitoring (above about 10 Hz) will be fulfilled. If recent results, described here in Figs. 11 and 12, are representative of what is possible with in-country stations, then, with sufficient effort, monitoring down to the (decoupled) 1 kt level appears achievable.

PAUL G. RICHARDS

References

Bache, T. C., 1982, Estimating the yield of underground nuclear explosions, *Seismol. Soc. America Bull.* **72**, S131–S168.
Evernden, J. E., 1977, Adequacy of routinely available data for identifying earthquakes of $m_b \geq 4.5$, *Seismol. Soc. America Bull.* **67**, 1099–1151.
Evernden, J. E., C. B. Archambeau, and E. Cranswick, 1986, An evaluation of seismic decoupling and underground nuclear test monitoring using high-frequency seismic data, *Rev. Geophysics* **24**, 143–215.
Hannon, W. J., 1985, Seismic verification of a comprehensive test ban, *Science* **227**, 251–257 (see also **228**, 792–794).
Herrin, E., 1985, Studies at the Lajitas Station, in A. U. Kerr, ed. *The VELA Program: A Twenty-Five Year Review of Basic Research.* Defense Advanced Research Projects Agency, 521–525.
Marshall, P. D., and P. W. Basham, 1972, Discrimination between earthquakes and underground explosions employing an improved M_s scale, *Royal Astron. Soc. Geophys. Jour.* **28**, 431–458.
NORSAR S.A.T.S., Semiannual technical summaries of the Royal Norwegian Council for Scientific and Industrial Research.
Nuttli, O. W., 1986, Yield estimates of Nevada test site explosions obtained from seismic Lg waves, *Jour. Geophys. Research* **91**, 2137–2151.
Ringdal, F., 1985, Study of magnitudes, seismicity and earthquake detectability using a global network, in A. U. Kerr, ed., *The VELA Program: A Twenty-Five Year Review of Basic Research.* Defense Advanced Research Projects Agency, 611–624.
Stevens, J. L., and S. M. Day, 1985, The physical basis of m_b: M_s and variable frequency magnitude methods for earthquake/explosion discrimination, *Jour. Geophys. Research* **90**, 3009–3020.
Sykes, L. R., J. F. Evernden, and I. S. Cifuentes, 1983, Seismic methods for verifying nuclear test bans, in D. W. Hafemeister and D. Schroeer, eds., *Physics, Technology and the Nuclear Arms Race,* American Institute of Physics Conference Proceedings #104. New York; American Institute of Physics.
Sykes, L. R., and S. Ruggi, 1988, Soviet underground nuclear testing: Inferences from seismic observations and historical perspective, in *Nuclear Weapons Data Book,* vol. IV, *Soviet Nuclear Weapons.* Cambridge, Mass.: Ballinger Publishing Company.
U.S. Congress, Office of Technology Assessment, 1988, *Seismic Verification of Nuclear Testing Treaties,* OTA-ISC-361. Washington, D.C.: U.S. Gov't. Printing Office.

Cross-References: *Earthquake Mechanisms; Earthquakes: Location Techniques; Earthquakes: Magnitude, Energy, and Intensity; Elasticity and Wave Propagation: Principles; Seismic Attenuation: Observation and Measurement; Seismic Instrumentation; Seismic Ray Theory; Seismic Source: Observations; Seismic Source: Theory.*

SEISMIC NOISE

The term *seismic noise* is used to denote the state of continuous unrest of the Earth's surface; it is measurable anywhere at any time. This continuous movement is a limiting factor in many scientific endeavors. Therefore, a basic knowledge of the nature and probable levels of the seismic background is necessary in order to assure the successful design and execution of many experiments. For instance, the measurement of the levels of the seismic background itself requires that the approximate levels of the noise be known beforehand in order to assure that the self noise of the instrumentation design will be sufficiently low to allow the resolution of the seismic noise in the ground. The ambient seismic background is the ultimate limiting factor in the seismologist's efforts to detect smaller and smaller events. Other examples of scientific fields in which earth noise is an important factor are electron microscopy and designing gravitational wave detectors. Earth noise is also becoming an increasingly important parameter in the defense industry, such as in inertial guidance system test and evaluation and advanced laser aiming systems, in which a small error in aiming direction caused by ground motion at the Earth's surface translates into a large error when aimed at a target in outer space.

The sources of Earth noise can be classified into two major subdivisions: those that result from mankind's activities (cultural) and those that occur naturally (microseisms). Some examples of cultural sources of seismic noise are the operation of mechanical machinery such as trains, electrical generating stations, vehicular traffic, etc. These activities create vibrations in the Earth's surface that can propagate for quite large distances to appear as seismic noise. The Earth's atmosphere acts as the primary source of microseismic energy by either interacting directly with the Earth's surface to generate seismic energy or secondarily through its driving of the oceans and other large water bodies. Moving water in rivers and streams also contributes to microseismic earth noise.

Cultural noise is primarily confined to the short period portion of the spectrum (periods less than one second) due to the inherent mechanical properties of most machines (finite physical size and relatively high rotational speed of most equipment). Noise of this type tends to attenuate fairly rapidly with distance. Therefore geographic isolation has historically been the most economical means of reducing the influence of cultural seismic noise on delicate endeavors. However, because of the ever increasing population density, geographic isolation is becoming harder and harder to achieve. In many cases, sophisticated mechanical isolation systems are now being employed to reduce the influence of cultural noise on delicate equipment.

Currently Known Lower Limits of the Microseismic Background

Throughout the history of seismology, the lowest known levels of microseismic noise have been steadily decreasing. Over the past 25 years, dramatic improvements in the inherent noise level of seismic equipment and improvements in instrument installation techniques have lowered known levels virtually across the entire seismic band (0.05 to 1000 s and beyond). In fact, 25 years ago very little was known about ground motion at periods greater than 20 s. This progress has been achieved as a result of advanced instrument research, which has been primarily concentrated at, but not limited to, three institutions. Research at very short period (0.05 to 1 s) performed at Southern Methodist University (Herrin, 1982) demonstrated that Earth background levels in this subband are significantly lower than had been previously believed. These results were chiefly the product of the development of very quiet instrumentation for this subband. Noise levels in the subband extending from 1 to 10 s have remained somewhat the same because the older instrumentation was capable of resolving the lower limits of Earth motion in this subband. However, improved installation techniques developed at the Lamont-Doherty Geological Observatory (Murphy and Savino, 1975), and at Teledyne Geotech (Fix, 1972) have dramatically lowered the known background levels between 10 s and 100 s. The new installation techniques involve methods of controlling and eliminating air motion around the sensors and burial of the sensors at significant depth below the Earth's surface to isolate the instruments from atmospherically generated nonpropagating surface noise. Using data from a superconducting gravimeter, Agnew and Berger (1978) have published spectra, which constitute the only measurements of seismic noise at very long periods (100 to 1,000,000 s).

A composite of the low-level microseismic background data by Herrin (1982) and Peterson (1980) and data from 100 to 1000 s by Agnew and Berger (1978) is presented in Fig. 1. These data are believed to be reliable estimates of the lowest levels of microseismic background likely to be found in the future anywhere on Earth, because modern instrument test and evaluation procedures have established that the data presented in this figure are true ground motion at the various sites at which they were measured. In addition, these data were measured at some of the quietest sites known. Small improvements over limited parts of the spectrum may be

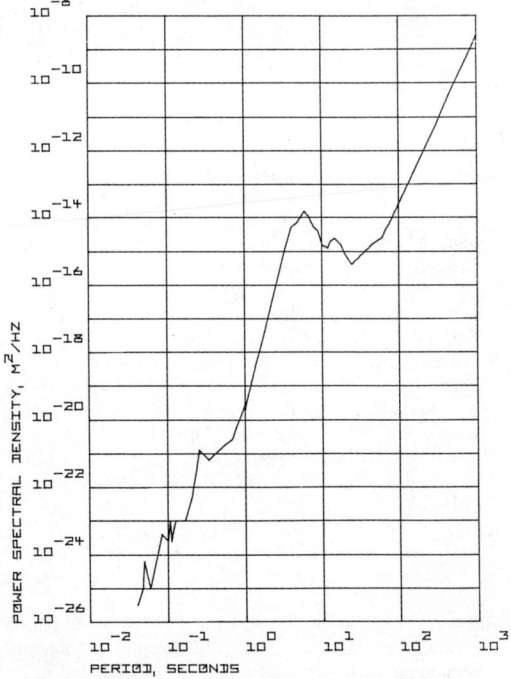

FIGURE 1. Power spectral density estimates of the lowest known levels of microseismic background.

found if super quiet sites are discovered in the future, but the changes should be minor. Therefore, the data presented in Fig. 1 can be assumed to be an estimate of the lowest possible limits of microseismic background over this spectral range.

Typical Range of the Microseismic Background

For many applications, estimates of the range over which the microseismic background might be expected to extend is more important than are the extreme lowest levels that it might reach. Typical levels are important because many systems must be designed to successfully operate over the entire range of microseismic background levels. Figure 2 presents composite estimates of the spectral data from fifteen stations located worldwide during a common time period (replotted in units of displacement from Holcomb, 1981). While these data are by no means comprehensive, they do present an idea of the typical limits between which the natural background may be expected to lie; the station distribution ranged from several stations located in the relatively quiet interior of large continents to a station located on the island of Guam, which had two typhoons raging in the relative vicinity. Notice that the microseismic background ranges over at least three decades of power for periods between 0.01 s and 10 s. By contrast, if we limit our data to that from four stations located in the interiors of four separate continents,

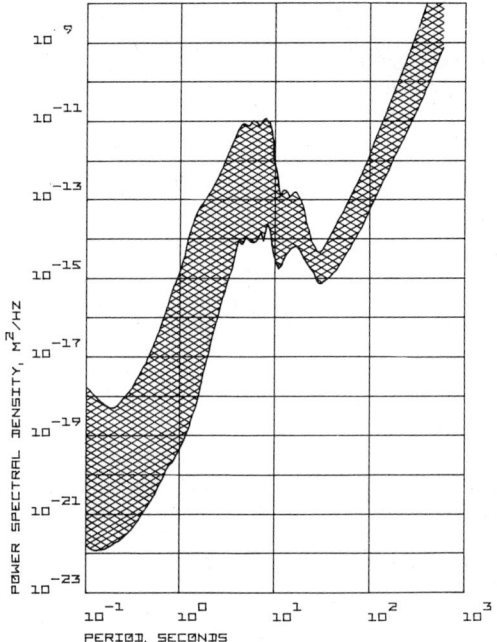

FIGURE 2. Power spectral density estimates of the range of the microseismic background from 15 stations located around the world.

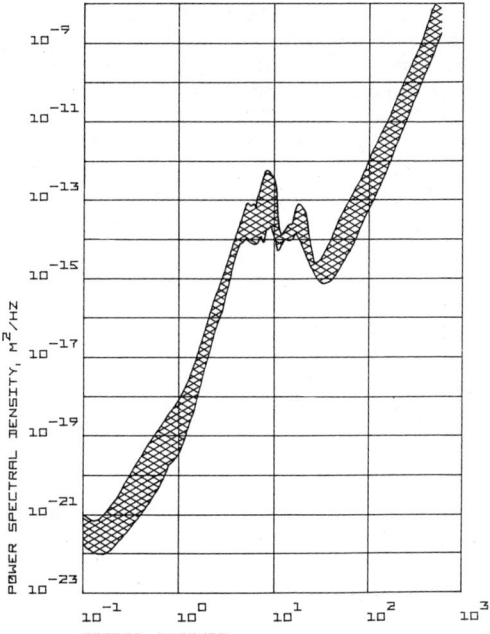

FIGURE 3. Power spectral density estimates of the range of the microseismic background from four stations located in the interiors of four separate continents.

the range of the seismic background is markedly reduced as in Fig. 3 (replotted in units of displacement from Holcomb, 1981). Here, the spectral densities from the four stations are approximately within a decade of one another across the entire seismic band. This figure is indicative of the typical levels of Earth background which should be found at any quiet, continentally-located, station installed at depth in competent rock.

Sources of the Microseismic Background

The question of the sources of the microseismic background has been the subject of heated controversy since the beginning of seismology. The solution to the problem has proven to be particularly elusive because of the complicated nature of the microseismic background. In general, microseismic background at any given site is comprised of energy generated from multiple simultaneous sources; the energy arrives at the site from several directions in the form of mixed wave types propagating in different modes. In addition, the energy is relatively continuous with time and in many cases very low level. The result is a phenomena that is very difficult to successfully model and quantitatively analyze. The temptation is to simplify the analysis by making assumptions that frequently lead to faulty conclusions. The history of the search for the origin of microseisms is extensive (see Iyer, 1964, and Hjortenberg, 1967 for comprehensive surveys of the older literature). Unfortunately, much of the literature has been rendered obsolete because modern instruments are much quieter over most of the seismic band and because digital data processing has permitted the quantitative analysis of realistic models of earth noise.

I do not mean to imply that all the answers are known by any means. The field is still active, and significant new results appear periodically in the literature. However, enough is known to formulate a fundamentally sound picture of the sources of the microseismic background between 0.01 and 1000 s.

The following explanation of the origin of the microseismic background is the author's interpretation of the existing literature. Dramatically opposing opinions can no doubt be found in the seismic community, particularly from a few of those who have worked in the field. However, widespread support will also be found for the ideas presented herein.

The combined influence of the atmosphere and wave action in bodies of water are the sources of the microseismic background from 0.01 to 1 s. Near coastlines, the ocean is the dominant source of seismic energy in this subband. However, this energy is attenuated exponentially with distance from the shoreline. At distances on the order of a few hundred kilometers from the coast, the ocean energy is attenuated to such an extent that the atmosphere is the major source of background. In these areas, the background in parts of this subband is generated by the direct action of the wind on obstructions such as trees and man-made structures, or directly on the ground itself. It may also be partially due to the action of the wind on large bodies of water such as lakes or reservoirs, if there are any in the vicinity, or to flowing water in rivers and streams. Regardless of the mechanism, the noise is primarily of quite local origin because of the rapid attenuation of seismic energy at these periods. There is frequently a diurnal character to noise in this subband because the atmosphere is usually calmer during the night. Because of changes in atmospheric activity, noise levels also tend to be higher, on the average, during the winter season than during the summer months. The details of the structure of the spectral background in this subband are frequently heavily site dependent due to the effects of the underlying site geology on wave propagation. The spectral structure of the seismic background in this subband has been used to model the geologic structure underlying the recording site. Near cities and industrial sites, cultural noise frequently overwhelms the background due to microseismic sources. Since much of the cultural noise has been edited out of the data presented in the figures herein, they do not present a true picture of the total noise background to be expected in this subband at such locations.

The microseismic noise in the subband extending from 1.0 to about 30 s is dominated by energy that is generated by the oceans of the Earth. Some atmospherically generated noise is present at the short-period end of this subband but the rest of the subband is dominated (at inland stations anyway) by two large spectral peaks both of which are now known to arise from processes in the oceans. The first peak, which is centered at approximately 6–9 seconds, is commonly referred to as the "secondary peak," and the second peak, which is centered at approximately 12–18 seconds, is commonly called the "primary peak." These peaks are delineated in Fig. 1. The argument over the probable sources of the energy in these two peaks has been heated and is the subject of numerous papers. The modern consensus is that the bulk of the energy in the primary peak is generated by the interaction of ocean waves on or near the coastlines of the world. In this process, ocean wave energy at a given period is converted directly into seismic energy with the same period. The spectral shape of the seismic energy is identical to the spectral shape of the ocean swell that creates it. The process, which generates the secondary peak, was first explained by Longuet-Higgins (1950). Energy in this peak is generated when ocean waves with the same period but traveling in opposite directions interact to produce seismic energy with a period equal to half that of the period of the ocean waves. This process can occur anywhere in the ocean; a coastline or shallow water is not required. However, secondary seismic energy is frequently generated along coastlines because ocean waves with nearly normal incidence to the coast are partially reflected back to sea, thereby creating the required "waves traveling in opposite directions with the same period" condition. Secondary microseisms are also generated in the wake of fast-moving storms. When these storms overtake and pass waves generated earlier, they generate waves traveling backward, head-on into the previously generated waves. Although it apparently has never been documented, there is no reason why waves propagating radially outward from two or more separate storms should not also interact in the open sea to generate secondary microseisms.

With the possible exception of sites that are located immediately on coastlines, the microseismic background beyond 30 s is dominated by the direct effects of the Earth's atmosphere. Noise levels in this part of the microseismic background are markedly higher during windy periods caused by the loading of the Earth's surface by air turbulence. Noise levels are also higher during the day than at night during no wind conditions because of the pressure variations generated by local convection cells generated by solar surface heating. The levels of noise in this subband can be significantly reduced by placing the sensors at depth either in old abandoned mines or in boreholes. Depths of the order of 100 m or more produce dramatic reductions in the noise level during windy conditions at periods up to 100 s and beyond.

The Future

The overall levels and general sources of the microseismic background are now well known. Future research efforts will be aimed at expanding the understanding of the details of the sources of the microseismic background and generation mechanisms both on land and in the sea. For example, the nature and sources of the fine structure in the spectra of the Earth's background at long period needs to be investigated. Finally, research and development of sophisticated systems for isolating equipment from the seismic background will become a much more active field as the state of the art expands in many scientific disciplines.

L. GARY HOLCOMB

References

Agnew, D. C., and J. Berger, 1978, Vertical seismic noise at very low, frequencies, *Jour. Geophys. Research* **83**, 5420–5424.

Fix, J. E., 1972, Ambient earth motion in the period range from 0.1 to 2560 sec., *Seismol. Soc. America Bull.* **62**, 1753–1760.

Herrin, E., 1982, The resolution of seismic instruments used in treaty verification research, *Seismol. Soc. America Bull.* **72**, part B, S61–S67.

Hjortenberg, E., 1967, *Bibliography of Microseisms, 1955-1964.* Copenhagen: Geodaetisk Instituts Skrifter 3.

Holcomb, L. G., 1981 The lower limits of seismic background noise levels, A Collection of Technical Papers, AIAA Guidance and Control Conference, August 19-21, 1981, Albuquerque, New Mexico. American Institute of Aeronautics and Astronautics.

Iyer, H. M., 1964, History and science of microseisms, *VESIAC State Art Rept. 4410-64-X.*

Longuet-Higgins, M. S., 1950, A theory of the origin of microseisms, *Royal Soc. London Philos. Trans. ser. A,* **243,** 1-35.

Murphy, A. J., and J. Savino, 1975, A comprehensive study of long-period (20 to 200 seconds) earth noise at the high-gain worldwide seismograph stations, *Seismol. Soc. America Bull.* **65,** 1827-1862.

Peterson, J., 1980, Preliminary observations of noise spectra at the SRO and ASRO stations, *U.S. Geol. Survey Open File Report 80-992.*

Cross-references: *Earthquake Seismology; Seismic Instrumentation; Seismic Instrumentation: History; Seismograms: Interpretation.*

SEISMIC PROPERTIES OF ROCKS

The past three decades have been marked by extensive geophysical exploration of the Earth's crust and upper mantle. During this period studies of lithospheric structure by reflection and refraction techniques may be counted among the major advances in solid earth geophysics. Furthermore, the most abundant, detailed, and reliable data on the Earth's interior come from these seismological investigations.

In many regions seismic velocities have been determined as a function of depth and propagation direction. Velocity usually increases with depth, however in some regions velocity reversals have been recognized. Laboratory measurements of the velocities of seismic waves in rocks are required for the interpretation of this field data. In this paper, recent advances in laboratory seismology are reviewed, the experimental methods are described and the current status is assessed—all from the point of view of the nonspecialist. Emphasis is placed on examining the information seismic data provides us on the mineralogy of the upper 100 km of the Earth.

Instrumentation

The technique for velocity measurements that is generally used in rock physics laboratories consists of the determination of compressional or shear wave travel time through cylindrical rock specimens. Transducers, which are excited by a pulse of 50 to 500 v and about 0.1 to 10 μs width, are placed on the ends of the rock core. The sending transducer converts the electrical pulse to a mechanical one, which is transmitted through the rock. The receiving transducer produces an electrical pulse upon reception of the incoming wave, which is amplified and displayed on a dual trace oscilloscope screen (Fig. 1). Once the system is calibrated for time delays, the travel time through the specimen is determined directly on the oscilloscope or with the use of a mercury delay line. Ceramic lead zirconate titanate (PZT) transducers of 1 MHz resonance frequencies are commonly used for compressional wave velocity measurements, while 1 MHz shear mode lead zirconate titanate or AC cut quartz transducers are used for shear-wave velocity measurements.

The mercury delay line consists of a fixed and a moveable transducer mounted in a column of mercury. The signal from the delay line is displayed

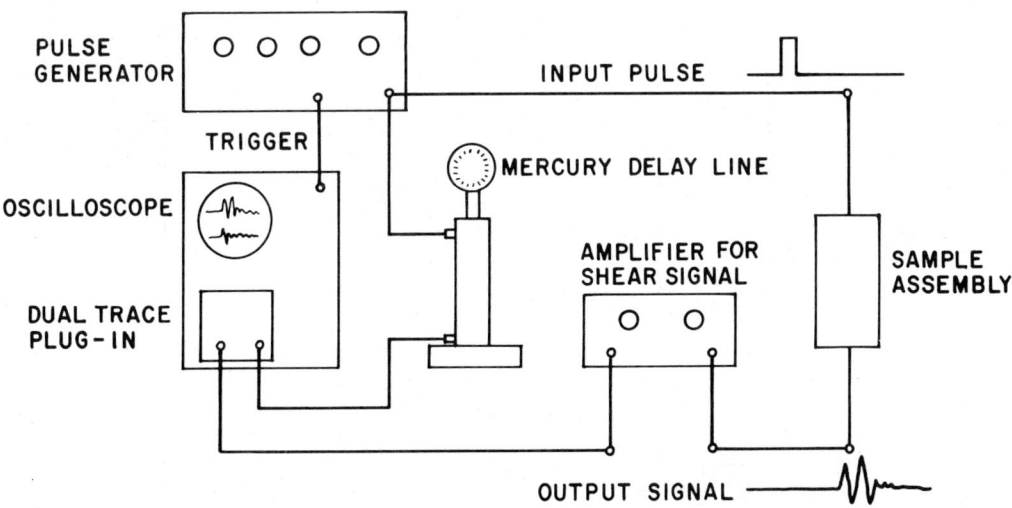

FIGURE 1. Schematic diagram of electronics for velocity measurements.

as a second trace on the oscilloscope. The distance between the two transducers is adjusted so the delay line signal is superimposed on the signal from the sample. With proper calibration the time of flight of the signal through the sample is then equivalent to that through the delay line and the rock velocity is simply calculated from the length of the mercury separating the transducers, the length of the sample, and the velocity of mercury. The major advantage of the delay line is that it increases the precision, especially for signals such as shear arrivals with slow rise times.

The rock specimens are typically cores 2.54 cm in diameter and 4 cm to 6 cm long. Bulk densities are calculated from core dimensions and weights. For measurements at elevated pressures, the cores are jacketed with copper foil and rubber tubing to prevent high pressure oil from entering microcracks and pores. To ground the transducers the core ends are either coated with silver conducting paint or a strip of brass is spot soldered to the copper jacket at each end. For measurements at high temperatures and pressures where gas is a pressure medium, the samples are usually encased in stainless steel. Velocity measurements as functions of confining and pore pressure require an additional pressure generating system and a more elaborate sample jacketing procedure, since the pore fluid must be isolated from the confining pressure fluid.

Analysis of Velocity Data

Comparisons of laboratory velocity data with seismically measured velocities are subject to many considerations. Influences of temperature and pressure must be taken into account before any attempt is made to infer mineralogy from seismic velocities. Laboratory velocities have been measured for a wide variety of rocks to hydrostatic pressures of 1000 MPa (equivalent to a depth in the Earth of approximately 35 km) and for several possible mantle rocks to 3000 MPa. Most crystalline rocks show rapid increases in velocity as pressure is increased to approximately 100 MPa; the rapid increases are followed by more gradual increases of velocity with increasing pressure (Fig. 2). The initial application of pressure affects the velocity by reducing microporosity, present as small cracks. Most of the microcracks in igneous and metamorphic rocks are closed at pressures between 100 and 1000 MPa.

Within the upper part of the Earth's crust, large-scale fracturing may have an analogous effect on seismic velocities. In deeper portions of the crust and in the upper mantle, however, it is likely that fractures and microcracks are no longer present because of the high confining pressures and metamorphic recrystallization. Thus seismic velocities in these deeper regions are more likely to present information on mineral composition.

In many sedimentary rocks, especially sandstones, porosity is much higher and consists of pore spaces between grains. This type of porosity, unlike microcracks, does not completely close at elevated pressures. Thus even at high pressure, velocities are much lower in quartz-rich sandstones than quartzites. In many sedimentary sections of the Earth's

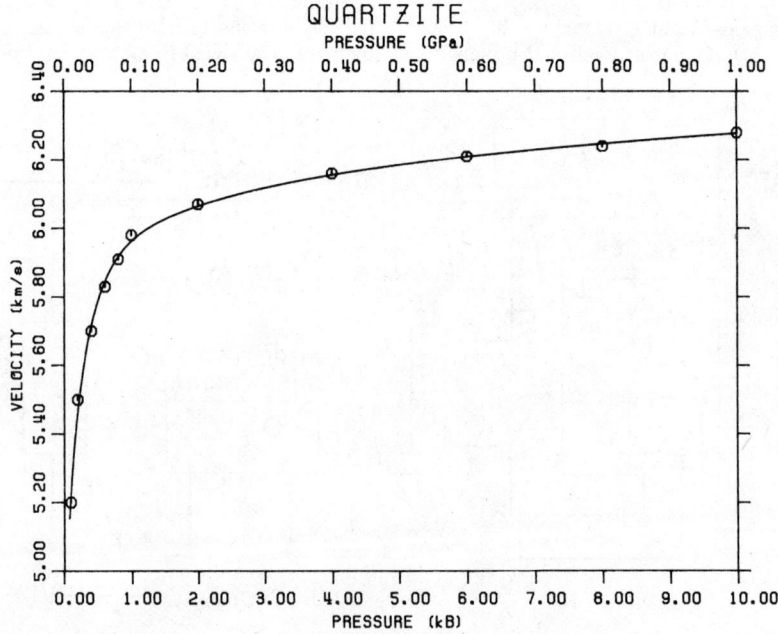

FIGURE 2. Compressional wave velocity versus pressure.

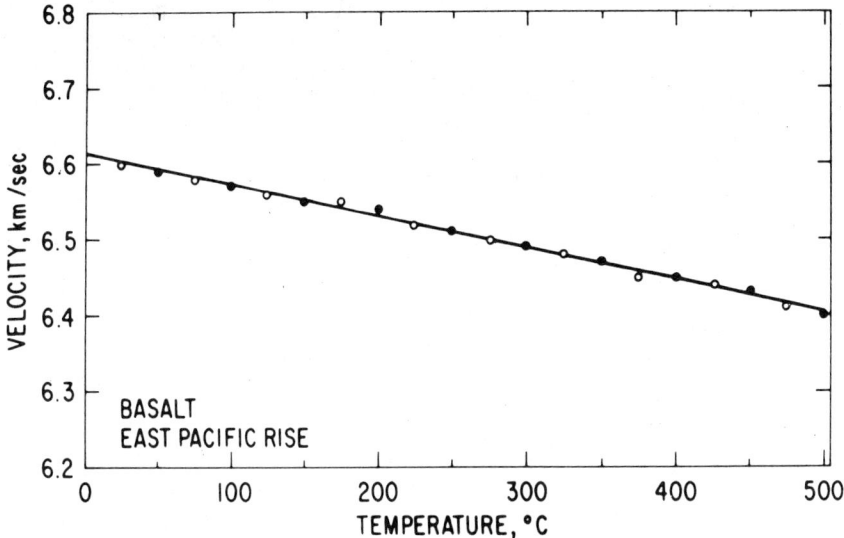

FIGURE 3. Compressional wave velocity versus temperature at 200 MPa confining pressure.

crust, porosity will have a dominant influence on seismic velocities. The velocities will, in addition, be influenced by the type of pore fluid present. For example, it is well known that water saturated rocks have higher velocities than rocks with dry pore spaces.

In addition to confining pressure produced by the weight of the overlying rock column, the pressure within pore spaces may have important effects on velocities. Laboratory velocity studies, mostly dealing with sedimentary rocks, have investigated the influence of fluid pressures applied to pore spaces as well as confining pressure on the external surface of the rocks. Usually velocity remains fairly constant if the internal and external pressures are changed by equivalent amounts.

For most common rock types, velocity decreases with increasing temperature. Experimental studies of the influence of temperature on velocity are complicated because at low pressures the rocks will often crack when heated, thereby giving anomalous velocities. Thus it is necessary to measure the temperature derivatives of velocities at relatively high pressures, usually on the order of a few hundred megapascals. An example of compressional wave velocity versus temperature for a common igneous rock is shown in Fig. 3.

At pressures high enough to eliminate the influ-

TABLE 1. Average Velocities of Rock Forming Minerals

Mineral	Density g/cm^3	Compressional Wave Velocity (km/s)	Shear Wave Velocity (km/s)
Perthite	2.57	5.74	3.13
Plagioclase (An$_9$)	2.61	6.07	3.40
Plagioclase (An$_{29}$)	2.64	6.30	3.44
Quartz	2.65	6.05	4.09
Plagioclase (An$_{53}$)	2.68	6.57	3.53
Calcite	2.71	6.53	3.36
Muscovite	2.79	5.78	3.33
Phlogopite	2.82	5.44	2.99
Biotite	3.05	5.26	2.87
Hornblende	3.12	6.81	3.72
Olivine (Fo$_{100}$)	3.22	8.57	5.02
Diopside	3.31	7.70	4.38
Olivine (Fo$_{93}$)	3.31	8.42	4.89
Augite	3.32	7.22	4.18
Bronzite	3.35	7.78	4.72
Garnet (grossularite)	3.60	8.72	5.07
Garnet (pyrope)	3.67	8.55	4.86
Garnet (almandite)	4.06	8.04	4.54

Source: Christensen, 1982

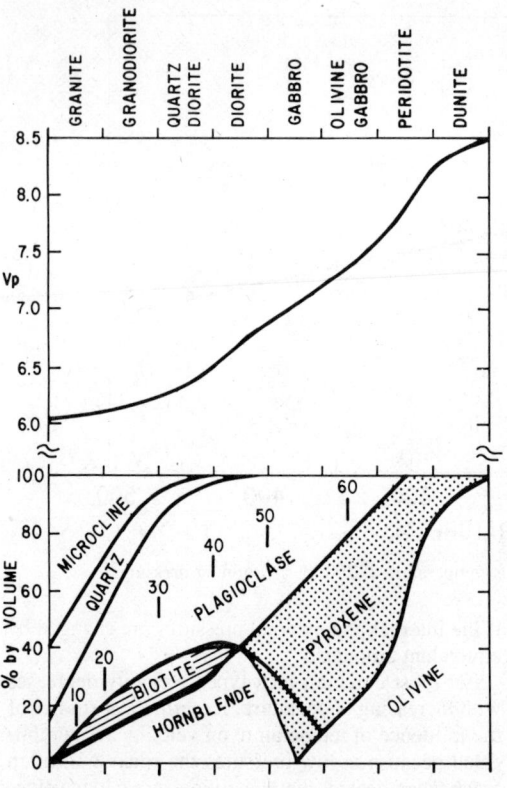

FIGURE 4. Compressional wave velocities at 200 MPa for common igneous rocks.

ence of cracks, velocities have been found to correlate well with mineralogy. Average velocities of randomly oriented aggregates of common rock forming minerals are given in Table 1. Note that the lower density minerals such as quartz and the alkali feldspars have appreciably lower velocities than the high density pyroxenes, olivine, and garnet. Systematic velocity changes associated with composition have also been observed within a given mineral group. For example, at 200 MPa compressional wave velocities in plagioclase feldspar increase from 6.0 km s^{-1} in ablite to 7.2 km s^{-1} for anorthite. In olivine velocities of 8.8 km s^{-1} for forsterite decrease to 6.8 km s^{-1} in fayalite.

It follows that velocities for specific rock types are frequently quite diagnostic; however, some variability for a specific rock type is observed because of variations of mineralogy common to rock classification schemes. This variability is illustrated in Fig. 4 for the common sequence of igneous rocks ranging from granite to dunite. Velocities in the range of granite to diorite generally correspond to reported upper continental crustal velocities, the velocities of gabbro are similar to lower crustal velocities, and peridotite and dunite velocities generally agree well with upper mantle compressional wave velocities.

In general, metamorphic rocks of mineral composition similar to igneous rocks have identical velocities, although this may be complicated by anisotropy as discussed later. Granitic gneisses and mafic granulites have average velocities similar to granites

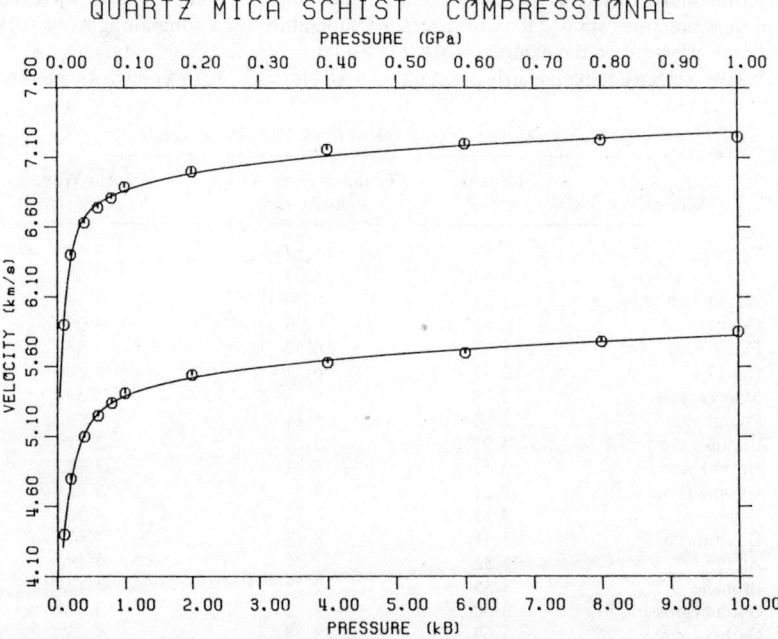

FIGURE 5. Compressional wave velocities measured perpendicular (lower velocities) and parallel (higher velocities) to schistosity in a quartz-mica schist.

and gabbros, respectively. Thus crustal velocities can be interpreted either in terms of igneous or metamorphic petrology.

Velocities in single crystals of the common rock forming minerals vary significantly with propagation direction. In general, for a given propagation direction in anisotropic media such as single crystals there are three waves, one compressional and two shear. Their vibration directions form an orthogonal set, which usually are not parallel or perpendicular to the propagation direction. The propagation of waves is related to the single crystal elastic constants through the Christoffel equation, which gives the three velocities for each direction as roots of a cubic equation. Details of wave propagation are related to the crystal symmetry. Most metamorphic rocks and some cumulate igneous rocks have preferred mineral orientations that are usually related to cleavage, foliation, or banding. It follows that many rocks are seismically anisotropic in a manner similar to single crystals. Compressional wave velocities vary with propagation direction (Fig. 5) and two shear waves travel in a given direction through the rock with different velocities. This latter property of anisotropic rocks, termed shear wave splitting (Fig. 6), has recently been observed in several crustal and upper mantle regions.

Discussion

During the past few years, the application of laboratory seismology to the interpretation of crustal and upper mantle seismic studies has provided much insight on the petrological nature of the Earth's crust and upper mantle. As was discussed in the previous section, compressional wave velocities can be correlated with mineralogical composition. Similar correlations exist for shear waves. For some rocks ratios of compressional to shear wave velocities provide diagnostic information on composition (e.g., see Christensen and Salisbury, 1975).

Early seismic refraction investigations usually presented velocity-depth models in terms of layered structures. Many recent studies have derived continuous velocity-depth relationships that appear more geologically reasonable. Once the effects of temperature and pressure are taken into account, these velocity gradients can be interpreted as changes in mineralogy or, at shallow crustal levels, changes in crack porosity. An excellent example is offered by combining recent oceanic crustal seismic studies with laboratory measurements of the velocity structure of ophiolites.

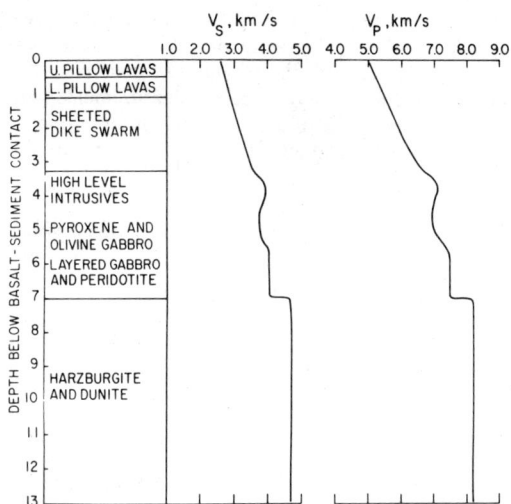

FIGURE 7. Shear-wave velocity (Vs) and compressional wave velocity (Vp) versus depth for the northern portion of the Samail ophiolite, Oman.

Ophiolites, believed to represent on land exposures of oceanic crust, usually contain a stratified sequence of rocks which from top to bottom consists of marine sediments, pillow basalts, dikes, gabbros, and ultramafic rocks. The laboratory velocity structure and petrology of a typical ophiolite sequence is shown in Fig. 7. The initial rapid increase of velocity extending to depths of 3 km, which was originally interpreted as an oceanic crustal seismic layer, originates from a decrease in porosity with depth. At depths between 4 and 7 km, velocities are fairly uniform. This region contains principally gabbro and metagabbroic rocks and corresponds to the main oceanic crustal layer (layer 3). The rapid increase in velocity encountered at the ultramafic section boundary is similar to observed velocity changes at the Mohorovičić discontinuity. Furthermore, the ultramafic sections of the ophiolites show 6% to 8% anisotropy originating from preferred orientation of olivine and pyroxene. This anisotropy correlates well with upper mantle seismic anisotropy observed by marine seismic refraction surveys.

New and exciting studies are currently underway on the seismic structure of the continental crust. In some continental regions, laboratory velocities have been used to correlate velocity discontinuities with changes of metamorphic grade and/or chemical

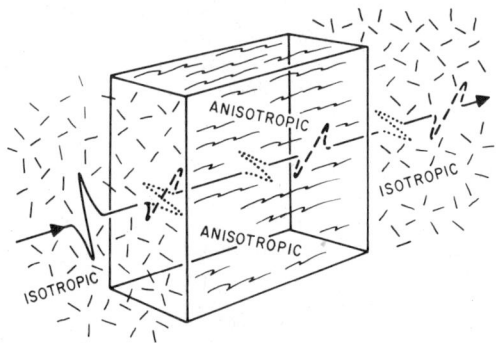

FIGURE 6. Shear-wave splitting.

composition (e.g., see Fountain, 1976). However, much remains to be learned about the nature of continental velocity discontinuities, gradients, and reversals. Laboratory velocity measurements will thus continue in the future to provide information on the mineralogy and chemical composition of the Earth.

NIKOLAS I. CHRISTENSEN

References

Birch, F., 1960, The velocity of compressional waves in rocks to 10 kilobars, 1, *Jour. Geophys. Research* **65**, 1083–1102.

Birch, F., 1961, The velocity of compressional waves in rocks to 10 kilobars, 2, *Jour. Geophys. Research* **66**, 2199–2224.

Christensen, N. I., 1979, Compressional wave velocities in rocks at high temperatures and pressures, critical thermal gradients, and crustal low-velocity zones, *Jour. Geophys. Research* **84**, 6849–6857.

Christensen, N. I., 1982, Seismic velocities, in R. S. Carmichael, ed., *Handbook of Physical Properties of Rocks*, Vol. II. Boca Raton, Fla.: CRC Press, 1–228.

Christensen, N. I., and M. H. Salisbury, 1975, Structure and constitution of the lower oceanic crust, *Rev. Geophysics Space Physics* **13**, 57–86.

Fountain, D. M., 1976, The Ivrea-Verbano and Strona Ceneri zones, northern Italy: a cross-section of the continental crust—new evidence from seismic velocities, *Tectonophysics* **33**, 145–165.

Kern, H., 1978, The effect of high temperature and high confining pressure on compressional wave velocities in quartz-bearing and quartz-free igneous and metamorphic rocks, *Tectonophysics* **44**, 185–203.

Salisbury, M. H., and N. I. Christensen, 1978, The seismic velocity structure of a traverse through the Bay of Islands ophiolite complex, Newfoundland, An exposure of oceanic crust and upper mantle, *Jour. Geophys. Research* **83**, 805–817.

Cross-references: *Continental Crustal Structure; Controlled Source Seismology; Deformation of Rocks and Minerals; Earth Structure: Global; Elasticity and Wave Propagation: Principles; Mantle Discontinuities; Mineralogic Phase Transitions in the Earth; Seismic Anisotropy in the Earth; Seismic Attenuation: Observations and Measurement; Seismology: History.*

SEISMIC RAY THEORY

High-frequency (HF) asymptotic methods have recently found broad applications in the numerical modeling and interpretation of observed seismic wave fields propagating in complex, two-dimensional (2-D) and three-dimensional (3-D) laterally varying layered structures. The simplest, yet very powerful, example of these HF methods is the *seismic ray method* (Babich, 1956; Karal and Keller, 1959). The great significance of the seismic ray method is its universality, effectiveness, conceptual clarity, and its ability to investigate various seismic body waves independently of other waves. Although its accuracy is limited, the seismic ray method is the only method, together with its extensions, able to give an approximate answer to many problems of HF seismic body waves propagating in structurally complex media.

Most of the equations presented in this article are applicable to an arbitrary HF multiply reflected seismic body wave propagating in a general 3-D laterally varying layered structure. The equations are quite concise and intelligible from a seismological point of view, due mainly to a broad use of matrix notation and transformation matrices.

Basic principles and applications of the seismic ray method for laterally varying 2-D and 3-D models are discussed in Aki and Richards (1980), Červený and Ravindra (1971), Červený et al. (1977), and Hanyga et al. (1984). For 1-D media, see Aki and Richards (1980) and Bullen and Bolt (1985). A more detailed derivation of most equations and results in this article and many useful references can be found in Červený (1985).

We use boldface letters to denote matrices; 3×3 matrices are denoted by a caret ($\wedge$) above the letter. If the same bold letter is used for 2×2 and 3×3 matrices with and without the caret (e.g., $\mathbf{M}$ and $\hat{\mathbf{M}}$), the matrix without the caret ($\mathbf{M}$) is the 2×2 left upper minor of the 3×3 matrix with a caret ($\hat{\mathbf{M}}$). The caret is also used for 3×1 column matrices (vectors in a 3-D space). Unless otherwise stated, the elements of a 3×1 column matrix represent Cartesian components of the corresponding vector. All other matrices (2×2, 4×4, etc.) do not have a caret. In component notation, uppercase indices indicate the numbers 1 and 2, lowercase indices the numbers 1, 2, and 3. Thus M_{IJ} denotes components of $\mathbf{M}$, M_{ij} the components of $\hat{\mathbf{M}}$. The Einstein summation convention is used throughout for indices written as subscripts.

The Ray Series

The *ray method*, also known as the *ray series method*, the *asymptotic ray theory* (ART) *method*, or the *geometric ray theory* (GRT) method, is based on the asymptotic HF solution of the elastodynamic equation for $\omega \to \infty$ (where ω is the prevailing circular frequency). Alternatively, the ray series may be also written for solutions that are nonanalytic along wavefronts.

The formal ray series solution (ansatz) is usually written in one of the equivalent forms:

$$u_k(x_j, t) = \mathrm{Re}\left\{ \sum_{n=0}^{\infty} U_k^{(n)}(x_j)\, F^{(n)}(t - T(x_j)) \right\} \quad (1)$$

$$u_k(x_j, t) = \text{Re}\left\{ F^{(0)}(t) * \sum_{n=0}^{\infty} U_k^{(n)}(x_j) h^{(n)}(t - T(x_j)) \right\} \quad (2)$$

Here $u_k(x_j, t)$ are the Cartesian components of the real-valued displacement vector $\hat{\mathbf{u}}(x_j, t)$, t is the time, x_j are Cartesian coordinates, the asterisk denotes convolution, and Re is the real part. The scalar real-valued function $T(x_j)$ of spatial coordinates x_j represents the *travel time* and is usually called *eikonal*, and the vectorial complex-valued functions $\hat{\mathbf{U}}^{(n)}(x_j)$ with Cartesian components $U_k^{(n)}(x_j)$ are called the *amplitude coefficients of the ray series*. It is assumed that $\hat{\mathbf{U}}^{(n)}$ and T depend on spatial coordinates only. The equation $T(x_j) = t$ represents a wavefront at time t. The vector perpendicular to the wavefront, $\hat{\mathbf{p}} = \nabla T$, is called the *slowness vector*. The complex-valued functions $F^{(n)}(\xi)$ are arbitrary, but they must satisfy the following conditions: (a) $F^{(n)}(\xi)$ are analytical signals; i.e., $F^{(n)}(\xi) = f^{(n)}(\xi) + ig^{(n)}(\xi)$, where i is the imaginary unit, and $f^{(n)}$ and $g^{(n)}$ are real-valued and form a Hilbert transform pair; (b) $F^{(n)}(\xi)$ are HF signals; i.e., their Fourier spectrum effectively vanishes for low frequencies; (c) $F^{(n)}(\xi)$ satisfy the relations $F^{(n)}(\xi) = \int_{-\infty}^{\xi} F^{(n-1)}(\theta) \, d\theta$, for $n = 1, 2, \ldots$. In Eq. 2, $h^{(0)}(\xi) = \delta(\xi)$, where $\delta(\xi)$ denotes the Dirac delta function. The functions $h^{(n)}$ satisfy condition (c); the function $F^{(0)}$ satisfies conditions (a) and (b). Condition (c) implies that the higher-order terms in the ray series are smoother than the lower-order terms. The ray series can be also written in the frequency domain; then $F^{(n)}(\xi) = \exp(-i\omega\xi)/(-i\omega)^n$.

Inserting the complex-valued form of the ray series, Eqs. 1 or 2, into the *elastodynamic equation*, we obtain a basic recurrence system of equations of the elastodynamic ray method (Červený and Ravindra, 1971). The basic system can be used to find successively the separated equations for $T(x_j)$, $\hat{\mathbf{U}}^{(0)}(x_j)$, $\hat{\mathbf{U}}^{(1)}(x_j)$, etc. Specifically, the *eikonal equations* are obtained for $T(x_j)$, and the *transport equations* are obtained for $\hat{\mathbf{U}}^{(n)}(x_j)$. These equations can be solved if certain initial conditions are specified, and they depend on the particular elastodynamic equation used. We shall consider isotropic perfectly elastic media without any prestress.

In seismological applications, usually only the first, *leading*, term of the ray series, also called the zero-order term, is considered. We do likewise, so we can omit the index n in Eqs. 1 and 2 and write the formal solution of the elastodynamic equation as

$$u_k(x_j, t) = \text{Re}\{ U_k(x_j) F(t - T(x_j)) \}$$

Notwithstanding, there are some applications in seismology in which the evaluation of higher-order terms of the ray series is useful (head waves, waves reflected at interfaces of higher order, etc.). The higher-order terms can be also used to assess the accuracy of the leading term of the ray series.

Eikonal Equation and Ray Tracing Systems

High-Frequency Separation of P and S Waves. Eikonal Equations From the first equation of the basic recurrence system of equations of the elastodynamic ray method, we obtain the *approximate HF separation* of the seismic wave field in a smoothly varying inhomogeneous medium into two waves, called *compressional* (P) and *shear* (S) waves.

The eikonal of the P wave satisfies the equation

$$\nabla T \cdot \nabla T = \frac{1}{\alpha^2(x_j)} \quad \text{with } \alpha = \left[\frac{\lambda + 2\mu}{\rho}\right]^{1/2} \quad (3)$$

where λ and μ are Lame's elastic parameters and ρ is the density. The P wave is polarized perpendicularly to the wavefront $T(x_j) = \text{const}$. Thus, in a HF approximation, the P wave is linearly polarized.

The eikonal of the S wave satisfies the equation

$$\nabla T \cdot \nabla T = \frac{1}{\beta^2(x_j)} \quad \text{with } \beta = \left[\frac{\mu}{\rho}\right]^{1/2} \quad (4)$$

The S wave is polarized in the plane tangent to the wavefront. The S wave is not generally linearly polarized, but it may be polarized elliptically in certain cases, even in the HF approximation.

Equations 3 and 4 are called the eikonal equations. The quantities $\alpha(x_j)$ and $\beta(x_j)$ represent local P and S velocities. They are assumed to be smooth functions of coordinates, and they do not depend on frequency, so the seismic body waves under consideration are *nondispersive* in the HF approximation.

Ray Tracing Systems. The eikonal equations may be solved by means of *characteristic trajectories* (characteristics for short). The characteristics of the eikonal equation can be identified as rays satisfying Fermat's principle and determining the direction of the flow of the HF part of the seismic wave field under consideration (see the section entitled "Alternative Approaches for Deriving Ray Tracing Systems"). Consider only one of the eikonal equations and denote the velocity by $V(x_j)$:

$$\nabla T \cdot \nabla T = \frac{1}{V^2(x_j)}; \quad \text{i.e.,}$$

$$\left(\frac{\partial T}{\partial x_1}\right)^2 + \left(\frac{\partial T}{\partial x_2}\right)^2 + \left(\frac{\partial T}{\partial x_3}\right)^2 = \frac{1}{V^2(x_j)} \quad (5)$$

V may be either α (for P waves) or β (for S waves). It follows from the eikonal equation that $|\hat{\mathbf{p}}| = 1/V$.

Assume that the ray is specified by the parametric equation $x_i = x_i(s)$, $i = 1, 2, 3$, where s is the arc length along the ray. Then the ray trajectory is a solution of the *ray tracing system*

$$\frac{dx_i}{ds} = Vp_i, \quad \frac{dp_i}{ds} = \frac{\partial}{\partial x_i}\left(\frac{1}{V}\right) \quad (6)$$

The travel time $T(s)$ is obtained from the relation $dT(s)/ds = 1/V$.

Instead of s, we can use some other monotonic variable along the ray, e.g., the travel time T, $V\,dT = ds$, or the variable σ, given by the relation

$$d\sigma = V\,ds = V^2\,dT \quad (7)$$

If we use σ, the ray tracing system (Eq. 6) and the equation for the travel time T read

$$\frac{dx_i}{d\sigma} = p_i, \quad \frac{dp_i}{d\sigma} = \frac{1}{2}\frac{\partial}{\partial x_i}\left(\frac{1}{V^2}\right), \quad \frac{dT}{d\sigma} = \frac{1}{V^2} \quad (8)$$

The initial conditions for the above ray tracing systems at a point O_0 of the ray, corresponding to $s = s_0$ or $\sigma = \sigma_0$, are

$$x_i = x_{i0}, \quad p_i = p_{i0}, \quad T = T_0 \quad (9)$$

where p_{i0} must satisfy Eq. 5 at O_0; i.e., $(p_{10})^2 + (p_{20})^2 + (p_{30})^2 = 1/V_0^2$, where $V_0 = V(x_{i0})$. It is simple to show that Eq. 5 is satisfied along the whole ray as soon as it is satisfied at one point of the ray. Note that p_{i0} can be expressed in terms of two takeoff angles δ_0 and ψ_0 at O_0: $p_{10} = V_0^{-1}\sin\delta_0\cos\psi_0$, $p_{20} = V_0^{-1}\sin\delta_0\sin\psi_0$, $p_{30} = V_0^{-1}\cos\delta_0$. Thus the initial conditions (Eq. 9) for the selected ray are expressed in terms of the initial direction (δ_0, ψ_0).

The ray tracing equations can be simply written in any curvilinear orthogonal coordinate system. In seismology, the spherical coordinate system is particularly important. For suitable ray tracing systems in curvilinear orthogonal coordinates, including spherical coordinates, see Červený (1987).

Rays across Interfaces. Assume that the ray is incident on a curved interface $\Sigma(x_i) = 0$ at a point Q. We denote by $\hat{N}$ the unit normal to Σ at Q. The slowness vector $\tilde{\mathbf{p}}(Q)$ of the ray of a generated reflected or transmitted wave (R/T wave) is given at Q by the following relation, which is a consequence of Snell's law (and called hereafter Snell's law):

$$\tilde{p}_i = p_i - \left\{(p_k N_k) \mp \epsilon\left[\left(\frac{1}{\tilde{V}}\right)^2 - \left(\frac{1}{V}\right)^2 + (p_k N_k)^2\right]^{1/2}\right\}N_i \quad (10)$$

where V and $\tilde{V}$ are velocities of the incident and R/T waves at Q, respectively, and $\epsilon = \text{sign}\,(p_i N_i)$ is the *orientation index*. The upper sign in Eq. 10 applies to the transmitted wave, the lower to the reflected wave.

Solution of the Ray Tracing System. The solution of a ray tracing system with specified initial conditions is usually called *initial value ray tracing*. Let us consider, e.g., the ray tracing system of Eq. 6 with initial conditions of Eq. 9. By the initial value ray tracing we obtain the Cartesian coordinates of points along the ray, $x_i = x_i(s)$, and the Cartesian components of the slowness vector, $p_i = p_i(s)$, at any point of the ray. We also determine the unit vector $\hat{\mathbf{t}}$ tangent to the ray, $t_i = Vp_i = dx_i/ds$, and dp_i/ds at any point on the ray.

Various numerical procedures, such as the Runge-Kutta or the Hamming predictor-corrector techniques, can be used to perform the initial value ray tracing. They give the solution with the required accuracy. The eikonal equation can then be used to test the computational accuracy of the slowness vector components along the ray

For certain simple distributions of velocity, the ray tracing system may be solved analytically. Analytical solutions are often used inside blocks or layers and in the cell approach, in which the layers or blocks are subdivided into individual cells, and the velocity within a cell is specified by a simple analytical law allowing analytical ray tracing. Most useful are velocity laws that do not generate first-order interfaces at cell boundaries.

For homogeneous media, the solution of the ray tracing system is fairly simple: The ray is a straight line, and the slowness vector is constant along the ray. Very popular is the velocity distribution with a constant gradient of velocity. In this case, the ray is a circle; the equations for the slowness vector and the travel time are, however, more complicated. The simplest analytical solutions are obtained for the quadratic slowness $(1/V)^2$ model by using Eq. 8. If the quadratic slowness is a linear function of Cartesian coordinates,

$$\frac{1}{V^2} = A + a_1 x_1 + a_2 x_2 + a_3 x_3$$

then we readily obtain the following simple, analytical, polynomial solution:

$$x_i(\sigma) = x_i(\sigma_0) + p_i(\sigma_0)(\sigma - \sigma_0) + \tfrac{1}{4}a_i(\sigma - \sigma_0)^2$$

$$p_i(\sigma) = p_i(\sigma_0) + \tfrac{1}{2}a_i(\sigma - \sigma_0)$$

$$T(\sigma) = T(\sigma_0) + (A + a_i x_i(\sigma_0))(\sigma - \sigma_0)$$

$$+ \tfrac{1}{2}a_i p_i(\sigma_0)(\sigma - \sigma_0)^2 + \tfrac{1}{12}a_i a_i(\sigma - \sigma_0)^3$$

This solution is especially suitable for triangular cells in 2-D models and for tetrahedron cells in 3-D models.

Similar polynomial approximations for rays and travel times can be found from Eq. 8 for more complex quadratic slowness models that are suitable for rectangular cells in 2-D models and for prismatic cells in 3-D models. See Červený (1987).

The ray may be specified not only by the initial point and initial direction at this point (as in the initial value ray tracing) but also by other conditions related to different points of the ray. Then we can speak of *boundary value ray tracing*. The most important case of boundary value ray tracing is *two-point ray tracing*, in which we seek the ray that connects two given points. Two-point ray tracing is considerably more complicated than initial value ray tracing, especially in 3-D structures. Several methods have been proposed to solve it. The *shooting method* exploits initial value ray tracing. An iterative loop in which the initial direction is changed is used to find the ray that passes through the second point. In the *bending method*, an initial ray path is guessed and then perturbed iteratively to satisfy the appropriate differential equations. In the *continuation method*, two-point ray tracing is first solved for some simpler structure and then the structure is gradually deformed. See also the section on dynamic ray tracing and propagator matrices for paraxial ray tracing.

Alternative Approaches for Deriving Ray Tracing Systems. The ray can be interpreted and derived in many ways. (a) The ray tracing system is equivalent to the Euler equations for the extremal curve of the Fermat functional. (b) The ray tracing system corresponds to Hamilton's equations for a suitably chosen Hamiltonian. (c) The rays are geodesics in a 3-D Riemannian space with the metric tensor g_{ij} given by $g_{11} = g_{22} = g_{33} = 1/V^2(x_j)$, $g_{ij} = 0$ for $i \neq j$. (d) The rays are trajectories along which the HF part of the energy of seismic body waves flows. In this way, the group velocity vector is tangent to the ray at any point on the ray. (e) In isotropic models the rays can be derived as orthogonal trajectories to wavefronts. (f) The ray tracing systems can also be derived from Snell's law (Eq. 10). This method is the simplest and seismologically most objective for deriving the ray tracing systems.

Ray Fields

We next consider a two-parameter system of rays and denote the ray parameters by γ_1 and γ_2. These parameters may be, e.g., the takeoff angles δ_0 and ψ_0 at a point source or the curvilinear coordinates along an initial surface. We define the *ray tube* as the family of rays with parameters within limits $[\gamma_1; \gamma_1 + d\gamma_1]$, $[\gamma_2; \gamma_2 + d\gamma_2]$. We introduce ray coordinates $\gamma_1, \gamma_2, \gamma_3$ in such a way that γ_1, γ_2 are the ray parameters and γ_3 is some monotonic parameter along the ray, e.g., the arc length s, travel time T, or σ (see Eq. 7). Here we use the arc length s as γ_3 (even though some expressions would be simpler in terms of σ). Note that the ray coordinate system is in general nonorthogonal. We denote the 3×3 transformation matrix from ray to Cartesian coordinates by $\hat{\mathbf{J}}$, with elements $J_{ij} = \partial x_i/\partial \gamma_j$. In the seismic ray method, especially in the computation of amplitudes, an important role is played by the Jacobian of the transformation $J = \det \hat{\mathbf{J}}$, called the *ray Jacobian*. The ray Jacobian measures the density of the ray field and the expansion and contraction of the ray tube.

The points at which the ray Jacobian vanishes are called the *caustic points*. At a first-order caustic point the ray tube shrinks to an elementary arc perpendicular to the direction of propagation. At this point,

$$J = \det \hat{\mathbf{J}} = 0, \quad \operatorname{tr} \hat{\mathbf{J}} \neq 1$$

At a second-order caustic point the ray tube shrinks to a point. At this point,

$$J = \det \hat{\mathbf{J}} = 0, \quad \operatorname{tr} \hat{\mathbf{J}} = 1$$

In space, caustic points form *caustic surfaces*, which are envelopes of rays that can take many possible forms. These surfaces, however, can be grouped into several basic types. The classification of these types is close to the theory of catastrophes, from which it also takes its terminology.

The elements of $\hat{\mathbf{J}}$ can be evaluated along the ray by solving an additional system of ordinary differential equations for $\partial x_i/\partial \gamma_j$ along the ray, also called the *dynamic ray tracing system*. Such a system can be easily derived from Eqs. 6 or 8, and we shall not write it here. A particularly simple dynamic ray tracing system is obtained if we use $\gamma_3 = \sigma$ as the variable along the ray (not s or T). Alternative dynamic ray tracing systems in ray-centered coordinates will be written in the section after the next.

Ray-Centered Coordinate System: Polarization Vectors

The ray-centered coordinate system q_1, q_2, q_3 is connected with a single ray Ω but not with the whole ray field. The coordinate q_3 corresponds to any monotonic variable along Ω (e.g., to s). Thus, Ω is identical to the coordinate axis q_3, along which $q_1 = q_2 = 0$. The coordinate axes q_1 and q_2 are straight lines perpendicular to Ω at q_3. They are selected to make the ray-centered coordinate system orthogonal and right-handed. This selection can be made as follows: We introduce the triplet of basis vectors of the ray-centered coordinate system connected with Ω by $\hat{\mathbf{e}}_1(s), \hat{\mathbf{e}}_2(s), \hat{\mathbf{t}}(s)$, where $\hat{\mathbf{t}}(s)$ is the unit vector tangent to Ω and $\hat{\mathbf{e}}_1(s), \hat{\mathbf{e}}_2(s)$ satisfy the following equations along the ray:

$$\frac{de_{1i}}{ds} = -V^2\left(e_{1k}\frac{dp_k}{ds}\right)p_i,$$

$$\frac{de_{2i}}{ds} = -V^2\left(e_{2k}\frac{dp_k}{ds}\right)p_i \quad (11)$$

In computations, only one of the vectors $\hat{\mathbf{e}}_1$, $\hat{\mathbf{e}}_2$ can be evaluated by using Eq. 11, since the second can be easily obtained from the relation $\hat{\mathbf{e}}_2 = \hat{\mathbf{t}} \times \hat{\mathbf{e}}_1$. The numerical solution of Eq. 11 is not difficult, since p_i and dp_i/ds are known from ray tracing. Equations 11 can be also reduced to one scalar equation.

The scale factors of the ray-centered coordinate system ($\gamma_3 = s$) are given by $h_1 = h_2 = 1$, $h_3 = 1 + V^{-1}[(\partial V/\partial q_1)q_1 + (\partial V/\partial q_2)q_2]$, where V and $\partial V/\partial q_I$ are taken at Ω.

The unit vectors $\hat{\mathbf{e}}_1$, $\hat{\mathbf{e}}_2$ determine the *polarization of S waves*. If the S wave has a direction of $\hat{\mathbf{e}}_1$ at any reference point of the ray, it points into the direction of $\hat{\mathbf{e}}_1(s)$ along the whole ray Ω (in a smoothly varying medium without interfaces). Therefore, these vectors are also called *polarization vectors*.

We now introduce three important *transformation matrices*. We denote the transformation matrix from ray-centered to Cartesian coordinates at Ω (for $q_I = 0$) by $\hat{\mathbf{H}}$. Its elements are given by the partial derivatives $H_{ij} = \partial x_i/\partial q_j$ for $q_I = 0$. Note that $\det \hat{\mathbf{H}} = 1$. Columns of $\hat{\mathbf{H}}$ correspond to the Cartesian components of the unit vectors $\hat{\mathbf{e}}_1$, $\hat{\mathbf{e}}_2$, $\mathbf{t}$.

Another important transformation matrix is $\hat{\mathbf{Q}}$, from ray to ray-centered coordinates at Ω. Its elements are given by the partial derivatives $Q_{ij} = \partial q_i/\partial \gamma_j$ for $q_I = 0$. It is obvious that $Q_{13} = Q_{23} = 0$ and $Q_{33} = 1$. It then immediately follows that $\det \hat{\mathbf{Q}} = \det \mathbf{Q}$. Since $\hat{\mathbf{J}} = \hat{\mathbf{H}}\hat{\mathbf{Q}}$, we can write

$$J = \det \hat{\mathbf{J}} = \det \hat{\mathbf{Q}} = \det \mathbf{Q}, \quad \operatorname{tr} \hat{\mathbf{J}} = \operatorname{tr} \mathbf{Q} + 1$$

Thus, the ray Jacobian is expressed in terms of the 2×2 matrix $\mathbf{Q}$. Note that $\mathbf{Q} = \mathbf{0}$ at the point source (central ray field).

Let us now denote the components of the slowness vector in ray-centered coordinates by $p_i^{(q)}$. On the central ray Ω, $p_1^{(q)} = p_2^{(q)} = 0$, $p_3^{(q)} = 1/V$. Outside Ω, however, $p_1^{(q)}$ and $p_2^{(q)}$ do not vanish. We introduce a matrix $\hat{\mathbf{P}}$ with elements $P_{ij} = \partial p_i^{(q)}/\partial \gamma_j$ for $q_I = 0$. Thus, $\hat{\mathbf{P}}$ is the transformation matrix from ray coordinates to phase space coordinates $p_i^{(q)}$. Note that $\mathbf{P} = \mathbf{0}$ at a point on Ω where $p_1^{(q)}$ and $p_2^{(q)}$ vanish in the vicinity of Ω, so the wavefront is locally plane (telescopic point).

The transformation matrices $\mathbf{Q}$ and $\mathbf{P}$ can be calculated by dynamic ray tracing along the ray Ω; see the next section.

Dynamic Ray Tracing

Dynamic Ray Tracing System. The dynamic ray tracing system can be derived and interpreted in several ways. One form, in general Cartesian coordinates, was discussed in a previous section. Another interpretation is based on the 2×2 transformation matrices $\mathbf{Q}$ and $\mathbf{P}$. It can be proved that $\mathbf{Q}$ and $\mathbf{P}$ satisfy the following system of matrix linear ordinary differential equations along Ω:

$$\frac{d\mathbf{Q}}{ds} = V\mathbf{P}, \quad \frac{d\mathbf{P}}{ds} = -V^{-2}\mathbf{V}\mathbf{Q} \quad (12)$$

The system 12 is usually called the *dynamic ray tracing system* or the *paraxial ray tracing system*. The 2×2 matrix $\mathbf{V}$ has elements $V_{IJ} = \partial^2 V/\partial q_I \partial q_J$ for $q_K = 0$. Note that $V_{IJ} = H_{kI}H_{lJ}(\partial^2 V/\partial x_k \partial x_l)$. The matrices $\mathbf{Q}$ and $\mathbf{P}$ are transformation matrices along the entire ray Ω, as soon as they have this meaning at one point of Ω.

We now introduce a 4×2 matrix

$$\mathbf{X} = \begin{pmatrix} \mathbf{Q} \\ \mathbf{P} \end{pmatrix}$$

Then Eq. 12 can be written in a more compact form, using a 4×4 matrix $\mathbf{S}$:

$$\frac{d\mathbf{X}}{ds} = \mathbf{S}\mathbf{X}, \text{ where } \mathbf{X} = \begin{pmatrix} \mathbf{Q} \\ \mathbf{P} \end{pmatrix},$$

$$\mathbf{S} = \begin{pmatrix} \mathbf{0} & V\mathbf{I} \\ -V^{-2}\mathbf{V} & \mathbf{0} \end{pmatrix}$$

The dynamic ray tracing system also represents the ray tracing system for paraxial rays (the rays close to Ω). We introduce a 4×1 matrix $\mathbf{W}(s) = (q_1, q_2, p_1^{(q)}, p_2^{(q)})^T$, where T denotes the transpose. Matrix $\mathbf{W}(s)$ specifies the ray-centered coordinates q_1, q_2 of the paraxial ray and the relevant components of the slowness vector. Then the paraxial ray tracing system reads

$$\frac{d\mathbf{W}}{ds} = \mathbf{S}\mathbf{W} \quad (13)$$

Ray Propagator Matrices. The system 13 has four linearly independent solutions. We introduce a 4×4 fundamental matrix $\mathbf{\Pi}(s, s_0)$ of linearly independent solutions of Eq. 13 specified at point $s = s_0$ on Ω by the initial condition

$$\mathbf{\Pi}(s_0, s_0) = \mathbf{I} \quad (14)$$

where $\mathbf{I}$ is a 4×4 identity matrix. We denote

$$\mathbf{\Pi}(s, s_0) = \begin{pmatrix} \mathbf{Q}_1(s, s_0) & \mathbf{Q}_2(s, s_0) \\ \mathbf{P}_1(s, s_0) & \mathbf{P}_2(s, s_0) \end{pmatrix} \quad (15)$$

where $\mathbf{Q}_1$, $\mathbf{P}_1$, $\mathbf{Q}_2$, and $\mathbf{P}_2$ are 2×2 matrices. $\mathbf{Q}_1$ and $\mathbf{P}_1$ are solutions of Eq. 12 for $\mathbf{Q}(s_0) = \mathbf{I}$, $\mathbf{P}(s_0) = \mathbf{0}$ (telescopic point solution). Similarly, $\mathbf{Q}_2$ and $\mathbf{P}_2$ are solutions of Eq. 12 for the initial conditions $\mathbf{Q}(s_0) = \mathbf{0}$, $\mathbf{P}(s_0) = \mathbf{I}$ (point source solution).

The matrices $\mathbf{Q}_1$, $\mathbf{Q}_2$, $\mathbf{P}_1$, $\mathbf{P}_2$ satisfy the following equations along the ray:

$$Q_1 Q_2^T - Q_2 Q_1^T = 0, \quad P_1 P_2^T - P_2 P_1^T = 0$$

$$Q_1 P_2^T - Q_2 P_1^T = I, \quad P_2 Q_1^T - P_1 Q_2^T = I$$

The fundamental matrix $\Pi(s, s_0)$ is also called the *ray propagator matrix* or the *ray propagator*. It has the following important properties:

$$\det \Pi(s, s_0) = \det \Pi(s_0, s_0) = 1$$

$$\Pi(s, s_0) = \Pi(s, s')\Pi(s', s_0) \quad (16)$$

$$\Pi(s_0, s) = \Pi^{-1}(s, s_0)$$

$$= \begin{pmatrix} P_2^T(s, s_0) & -Q_2^T(s, s_0) \\ -P_1^T(s, s_0) & Q_1^T(s, s_0) \end{pmatrix} \quad (17)$$

In Eq. 16, s' is an arbitrary point on Ω, not necessarily between s and s_0. Equation 16 can be used to connect the propagator matrices calculated independently along different segments of the ray.

Dynamic Ray Tracing across Curved Interfaces. Assume that Ω is incident at an interface Σ at the point Q. If the interface is first or second order, Π changes discontinuously across Σ. We introduce a right-handed local Cartesian coordinate system z_1, z_2, z_3 at Q with unit basis vectors $\hat{i}_1$, $\hat{i}_2$, $\hat{i}_3$. Usually, we introduce it in such a way that the z_3-axis coincides with the normal $\hat{N}$ to Σ at Q ($\hat{i}_3 = \hat{N}$), and the z_1 axis is tangent to Σ at Q and situated in the plane of incidence. Otherwise, the orientation of $\hat{i}_1$, $\hat{i}_2$, $\hat{i}_3$ may be arbitrary. We denote by $\hat{Z}$ the transformation matrix from the local Cartesian coordinate system z_i at Σ at Q to the general Cartesian coordinate system x_j, $Z_{ij} = \partial x_i/\partial z_j$ at Q. The columns of $\hat{Z}$ correspond to the Cartesian components of $\hat{i}_1$, $\hat{i}_2$, $\hat{i}_3$. Another transformation matrix we shall need is $\hat{G}$, from the ray-centered coordinate system q_i to the local Cartesian coordinate system z_j at Q, $\hat{G}_{ij} = \partial z_i/\partial q_j$ at Q. We can determine $\hat{G}$ simply as $\hat{G} = \hat{Z}^T \hat{H}$. We can also write $\hat{G} = \hat{G}^{\perp} \hat{G}^{\parallel}$ where $\hat{G}^{\perp}$ and $\hat{G}^{\parallel}$ are rotation matrices. $\hat{G}^{\perp}$ performs rotation about the ray in such a way to shift $\hat{e}_2$ into $\hat{i}_2$. The only nonvanishing components of $\hat{G}^{\perp}$ are $G_{11}^{\perp} = G_{22}^{\perp} = e_{2k} i_{2k}$, $G_{21}^{\perp} = -G_{12}^{\perp} = e_{1k} i_{2k}$, and $G_{33}^{\perp} = 1$. Similarly, $\hat{G}^{\parallel}$ performs the rotation about the local z_2-axis (unit vector $\hat{i}_2$), shifting into $\hat{i}_3$ the unit vector $\hat{t}$ tangent to the ray.

If we denote by $\mathbf{X}(Q)$ and $\tilde{\mathbf{X}}(Q)$ the 4×2 matrices

$$\begin{pmatrix} Q \\ P \end{pmatrix}$$

corresponding to the incident and selected R/T waves at Q, we obtain

$$\tilde{\mathbf{X}}(Q) = \mathbf{F}(Q)\mathbf{X}(Q),$$

$$\mathbf{F}(Q) = \begin{pmatrix} \tilde{G}^T & 0 \\ 0 & \tilde{G}^{-1} \end{pmatrix} \begin{pmatrix} I & 0 \\ uD + E - \tilde{E} & I \end{pmatrix}$$

$$\cdot \begin{pmatrix} (G^{-1})^T & 0 \\ 0 & G \end{pmatrix}$$

where $\mathbf{F}(Q)$ is a 4×4 matrix with all its elements determined at Q. The tilde quantities correspond to the R/T wave at Q, untilde quantities to the incident wave at Q; $\mathbf{D}$ is the 2×2 matrix of curvature of the interface Σ at Q, and

$$u = V^{-1} G_{33} - \tilde{V}^{-1} \tilde{G}_{33},$$

$$E_{IJ} = -V^{-2}\left(\frac{\partial V}{\partial q_K} G_{JK} G_{I3} + \frac{\partial V}{\partial q_3} G_{I3} G_{J3} \right.$$

$$\left. + \frac{\partial V}{\partial q_K} G_{IK} G_{J3} \right)$$

($\tilde{E}$ is similar, but all the quantities are written with a tilde.) Note that $\partial V/\partial q_i = H_{ki}(\partial V/\partial x_k)$ and, similarly, for $\partial \tilde{V}/\partial q_i$.

Let us now consider an arbitrary, multiply reflected, possibly converted wave in a 3-D laterally varying layered structure containing curved interfaces of first and second order $\Sigma_1, \Sigma_2, \ldots \Sigma_k$, including the Earth's surface. We again denote the selected ray by Ω, the initial point of the ray by O_0, the end point by O_s, and N points of R/T at various interfaces between O_0 and O_s by $Q_1, Q_2, \ldots Q_N$. Also, $Q_0 = O_0$. Individual segments of the ray may be either P or S. Then we obtain

$$\mathbf{X}(O_s) = \Pi(O_s, Q_N)$$

$$\cdot \prod_{i=N}^{1} [\mathbf{F}(Q_i)\Pi(Q_i, Q_{i-1})]\mathbf{X}(O_0)$$

where $\prod_{i=N}^{1} A_i$ denotes the matrix product $A_N \cdot A_{N-1} \cdots A_1$. In other words, the propagator matrix of an arbitrary, multiply R/T wave in a general 3-D layered structure is

$$\Pi(O_s, O_0) = \Pi(O_s, Q_N)$$

$$\cdot \prod_{i=N}^{1} [\mathbf{F}(Q_i)\Pi(Q_i, Q_{i-1})]$$

Applications of Dynamic Ray Tracing and Propagator Matrices

Important applications of dynamic ray tracing and of propagator matrices abound in seismology. One of the most important applications is computing the second derivatives of the travel time field along the ray and evaluating the paraxial travel times.

In ray-centered coordinates, the second derivatives of the travel time field in the direction perpendicular to the ray are fully described by the 2×2 symmetric matrix $\mathbf{M}(s)$, with elements $M_{IJ}(s) = \partial^2 T(q_1, q_2, s)/\partial q_I \partial q_J$, for $q_K = 0$, where $\mathbf{M} = \mathbf{PQ}^{-1}$. If $\mathbf{M}$ is specified at $s = s_0$ as $\mathbf{M}(s_0)$, we obtain at s,

$$\mathbf{M}(s) = [\mathbf{P}_1(s, s_0) + \mathbf{P}_2(s, s_0)\mathbf{M}(s_0)]$$
$$\cdot [\mathbf{Q}_1(s, s_0) + \mathbf{Q}_2(s, s_0)\mathbf{M}(s_0)]^{-1}$$

where $\mathbf{P}_1, \mathbf{P}_2, \mathbf{Q}_1, \mathbf{Q}_2$ are given by Eqs. 15 and 14. From $\mathbf{M}$, we obtain the 3×3 symmetric matrix

$$\hat{\mathbf{M}} = \begin{pmatrix} M_{11} & M_{12} & -V^{-2}(\partial V/\partial q_1) \\ M_{12} & M_{22} & -V^{-2}(\partial V/\partial q_2) \\ -V^{-2}(\partial V/\partial q_1) & -V^{-2}(\partial V/\partial q_2) & -V^{-2}(\partial V/\partial q_3) \end{pmatrix}$$

Recall that $\partial V/\partial q_i = H_{ki}(\partial V/\partial x_k)$. The 3×3 matrix of second derivatives of the travel time field with respect to Cartesian coordinates $\hat{\mathbf{N}}$, with elements $N_{ij} = \partial^2 T/\partial x_i \partial x_j$ at Ω, is

$$\hat{\mathbf{N}} = \hat{\mathbf{H}}\hat{\mathbf{M}}\hat{\mathbf{H}}^T$$

Note that $\mathbf{M}$ satisfies along the ray Ω the nonlinear matrix Riccati equation

$$\frac{d\mathbf{M}}{ds} + V\mathbf{M}^2 + V^{-2}\mathbf{V} = 0$$

Using the matrix of the second derivatives of the travel time field, we can write equations for the paraxial travel times. We wish to determine the travel time at a point S close to O_s (S is in general located outside Ω). We obtain

$$T(S) = T(O_s) + \hat{\mathbf{p}}^T(O_s)\hat{\mathbf{x}}(S, O_s)$$
$$+ \tfrac{1}{2}\hat{\mathbf{x}}^T(S, O_s)\hat{\mathbf{N}}(O_s)\hat{\mathbf{x}}(S, O_s) \quad (18)$$

where $\hat{\mathbf{x}}(S, O_s)$ is the column 3×1 matrix with elements $x_i(S) - x_i(O_s)$.

Dynamic ray tracing and the propagator ray matrix can be used in many other important seismological applications: for example, in initial value and boundary value ray tracing in the vicinity of the central ray, for evaluating geometrical spreading at O_s due to a point source at O_0, $|\det \mathbf{Q}_2(O_s, O_0)|^{1/2}$, for determining the curvature matrix of the wavefront, $\mathbf{K} = V\mathbf{M}$, for evaluating Fresnel volumes (physical rays), for evaluating paraxial approximation for the displacement vector (see below), and in the Gaussian beam and Maslov methods. They will surely find applications in the solution of various seismic inverse problems, in approaches based on linearization, and in seismic source studies. For a more detailed treatment see Červený (1985, 1987).

Vectorial Complex-Valued Amplitudes

The vectorial complex-valued amplitudes $\hat{\mathbf{U}}$ can be expressed either in Cartesian components $U_i^{(x)}$, or in ray-centered components $U_i^{(q)}$. To transform

$$\hat{\mathbf{U}}^{(q)} \equiv (U_1^{(q)}, U_2^{(q)}, U_3^{(q)})^T$$

into

$$\hat{\mathbf{U}}^{(x)} \equiv (U_1^{(x)}, U_2^{(x)}, U_3^{(x)})^T,$$

we can use the relation $\hat{\mathbf{U}}^{(x)} = \hat{\mathbf{H}}\hat{\mathbf{U}}^{(q)}$.

In the ray method, HF P and S waves are separated and propagate independently in smoothly varying media without interfaces. For P waves, $\hat{\mathbf{U}}^{(q)} = (0, 0, U_3^{(q)})^T$; for S waves, $\hat{\mathbf{U}}^{(q)} = (U_1^{(q)}, U_2^{(q)}, 0)^T$. We call $U_1^{(q)}$ the S_1-component and $U_2^{(q)}$ the S_2-component.

Transport Equation and Its Solution. In a smooth medium, $U_i^{(q)}$ can be evaluated along the ray by solving an ordinary first-order differential equation called the *transport equation*. The transport equation can be simply derived from the second equation of the basic recurrence system of equations of the elastodynamic ray method. It has the same form for all three components $U_i^{(q)}$:

$$\frac{dU_i^{(q)}}{ds} + \frac{1}{2}U_i^{(q)}\left(V\nabla^2 T + \frac{d}{ds}\ln \rho V^2\right) = 0$$

where $V = \alpha$ for P waves and $V = \beta$ for S waves. Components S_1 and S_2 are not coupled together in the ray-centered coordinate system (in smooth media without interfaces).

The transport equations can be easily solved analytically along the ray if we write $\nabla^2 T = J^{-1} d(J/V)/ds$. We again denote the initial point $s = s_0$ on the ray by O_0 and the end point by O_s. In a smooth medium without interfaces we obtain

$$\hat{U}^{(q)}(O_s) = \frac{\hat{\Psi}(\gamma_1, \gamma_2)}{[V(O_s)\rho(O_s)J(O_s)]^{1/2}}$$

$$= \left[\frac{V(O_0)\rho(O_0)J(O_0)}{V(O_s)\rho(O_s)J(O_s)}\right]^{1/2} \hat{U}^{(q)}(O_0) \quad (19)$$

Instead of $J(O_s)$ and $J(O_0)$, we can also write $\det \mathbf{Q}(O_s)$ and $\det \mathbf{Q}(O_0)$ in Eq. 19. The 3×1 column matrix $\hat{\Psi}$ is independent of s and is a function of γ_1, γ_2 only. It may be specified for various types of sources.

Paraxial Ray Approximation for the Displacement Vector. We can evaluate approximately the vectorial ray amplitudes at, and in the vicinity of, the point O_s on Ω if we introduce a correction to the direction of the displacement vector due to the curvature of the wavefront. Let us consider an arbitrary point S close to O_s, not necessarily situated in the plane perpendicular to Ω at O_s. If we also take into account the variation of amplitudes along Ω, we obtain

$$\hat{\mathbf{U}}^{(q)}(S) = \hat{\mathbf{B}}(S, O_s)\hat{\mathbf{U}}^{(q)}(O_s)$$

$$= \frac{1}{[V(O_s)\rho(O_s)J(O_s)]^{1/2}}\hat{\mathbf{B}}(S, O_s)\hat{\mathbf{\Psi}} \quad (20)$$

where

$$\hat{\mathbf{B}}(S, O_s)$$

$$= \begin{pmatrix} 1 - \tfrac{1}{2}B_d & 0 & \theta_1(S, O_s) \\ 0 & 1 - \tfrac{1}{2}B_d & \theta_2(S, O_s) \\ -\theta_1(S, O_s) & -\theta_2(S, O_s) & 1 - \tfrac{1}{2}B_d \end{pmatrix}$$

(21)

is the *paraxial approximation matrix* and

$$\hat{\boldsymbol{\theta}}(S, O_s) = V(O_s)\hat{\mathbf{M}}(O_s)\hat{\mathbf{H}}^{\mathrm{T}}(O_s)\hat{\mathbf{x}}(S, O_s),$$

$$\hat{\mathbf{x}}(S, O_s) = \hat{\mathbf{x}}(S) - \hat{\mathbf{x}}(O_s)$$

$$B_d(O_s) = \left\{ V(O_s)\,\mathrm{tr}\,\mathbf{M}(O_s) \right.$$

$$\left. + \left[\frac{d}{ds} \ln(V(s)\rho(s)) \right]_{O_s} \right\}$$

$$\cdot H_{i3}(O_s)x_i(S, O_s) \quad (22)$$

The elements θ_I express the variations of $\hat{\mathbf{U}}^{(q)}$ in the plane perpendicular to Ω at O_s, B_d along Ω, where B_d is a third-order correction.

Amplitudes across Interfaces. We consider only first-order interfaces, at which the velocities α, β and the density ρ are discontinuous, or at least one of them is discontinuous. If the ray impinges on the first-order interface, the leading term of the ray series becomes discontinuous. To determine the transformation of amplitudes across the interface, we must apply the boundary conditions at the interface.

Let us first consider that the unit vectors $\hat{\mathbf{e}}_2$ of the incident and R/T waves are perpendicular to the plane of incidence (i.e., they coincide with $\hat{\mathbf{i}}_2$, see "Dynamic Ray Tracing across Curved Interfaces"). Then the boundary conditions at the interface yield the following HF conclusions regarding the leading term of the ray series: (1) The P (or S_1) wave incident at an interface generates both P and S_1 reflected and transmitted waves, but not S_2 waves. (2) The S_2 wave (perpendicular to the plane of incidence) incident at an interface generates only S_2 reflected and transmitted waves, but not P and S_1 waves. (3) The ratio of any ray-centered component of the displacement vector of the R/T wave to the ray-centered component of the displacement vector of the incident wave at Q is

$$\tilde{U}_i^{(q)}(Q) = R_{ji}^D U_j^{(q)}(Q);$$

i.e., $\hat{\tilde{\mathbf{U}}}^{(q)}(Q) = (\hat{\mathbf{R}}^D)^{\mathrm{T}}\hat{\mathbf{U}}^{(q)}(Q)$

(23)

Here R_{ji}^D denotes the relevant displacement reflection/transmission coefficients of plane waves at a plane interface. They do not depend on the frequency, on the curvature of the wavefront of the incident wave, on the curvature of the interface at Q, or on the variations of velocities and densities in the vicinity of Q. They depend only on the local velocities and densities on both sides of Σ at Q and on the angle of incidence.

The 3×3 matrix $\hat{\mathbf{R}}^D$ of R/T coefficients has only one or two nonzero elements for a particular elementary wave under consideration:

Incident P – reflected/transmitted P: R_{33}^D

Incident P – reflected/transmitted S: R_{31}^D

Incident S – reflected/transmitted P: R_{13}^D

Incident S – reflected/transmitted S: R_{11}^D, R_{22}^D

Here R_{11}^D corresponds to the S_1S_1 coefficient, R_{22}^D to the S_2S_2 coefficient. Formulas for the R/T coefficients can be found in seismological literature (Červený and Ravindra, 1971). It is only necessary to take care of the proper signs of these coefficients, depending on the orientation of the unit vectors $\hat{\mathbf{e}}_1$, $\hat{\mathbf{e}}_2$ used for incident and R/T waves.

If the vectors $\hat{\mathbf{e}}_2$ of the incident and R/T waves are not perpendicular to the plane of incidence, we must rotate $\hat{\mathbf{e}}_1$, $\hat{\mathbf{e}}_2$ around Ω to shift $\hat{\mathbf{e}}_2$ into the proper position before applying Eq. 23. The rotation can be carried out with matrices $\hat{\mathbf{G}}^\perp$ (for the incident wave) and $\hat{\tilde{\mathbf{G}}}$ (for the R/T wave). The generally valid expressions are

$$\hat{\tilde{\mathbf{U}}}^{(q)}(Q) = (\hat{\tilde{\mathbf{G}}}^\perp(Q))^{\mathrm{T}}(\hat{\mathbf{R}}^D(Q))^{\mathrm{T}}\hat{\mathbf{G}}^\perp(Q)\hat{\mathbf{U}}^{(q)}(Q) \quad (24)$$

The generalization of Eqs. 19–24 for an arbitrary multiply reflected wave in a 3-D laterally varying layered structure is straightforward.

The ray method can even be used to study R/T waves generated at interfaces of higher order. Let us consider an interface of nth order at which the $(n-1)$st derivatives of velocities and density are discontinuous (the lower derivatives being continuous). Assume that the zero-order term of the incident wave $\hat{\mathbf{U}}^{(0)}$ is nonvanishing at Q; see Eq. 1. Then the monotypic transmitted wave is again of zero order, but other R/T waves are $(n-1)$st order. Thus the first nonvanishing terms in the ray series for these R/T waves contain $\hat{\mathbf{U}}^{(n-1)}$. The determination of the amplitudes of these R/T waves generated at an interface of higher order is more complicated than at the first-order interface.

Ray Theory Elastodynamic Green's Function

We give closed-form expressions for the ray theory elastodynamic Green's function, valid for an arbitrary, smooth, 3-D laterally varying layered structure.

We define the Green's function $\hat{\mathbf{G}}(\hat{\mathbf{x}}, t; \hat{\mathbf{x}}_0, t_0)$ with components $G_{mn}(\hat{\mathbf{x}}, t; \hat{\mathbf{x}}_0, t_0)$, $m, n = 1, 2, 3$, as follows: $G_{mn}(\hat{\mathbf{x}}, t; \hat{\mathbf{x}}_0, t_0)$ is the mth Cartesian component of the displacement vector at $\hat{\mathbf{x}}$ at time t, caused by the application of a single-force unit impulse in the direction of the nth Cartesian axis at the point $\hat{\mathbf{x}}_0$ at time t_0. We also denote the point $\hat{\mathbf{x}}_0$ by O_0 and the point $\hat{\mathbf{x}}$ by O_s. Then the Green's function is $\hat{\mathbf{G}}(O_s, t; O_0, t_0)$.

In ray theory, the complete wave field at O_s due to a point source at O_0 is composed of contributions that travel from O_0 to O_s along various ray trajectories Ω. These rays correspond, in general, to different elementary seismic body waves, such as reflected, refracted, multiply reflected, or converted. Certain contributions, however, can correspond even to multiple rays of the same elementary wave. Thus, we can write the ray expansion

$$\hat{\mathbf{G}}(O_s, t; O_0, t_0) = \sum_{(\Omega)} \hat{\mathbf{G}}^\Omega(O_s, t; O_0, t_0) \quad (25)$$

where $\hat{\mathbf{G}}^\Omega$ is the contribution corresponding to one selected ray Ω. In a laterally varying layered structure, the number of rays Ω connecting O_0 and O_s may be infinite, so only a partial ray expansion is possible, thus introducing additional errors to the ray computations.

For $\hat{\mathbf{G}}^\Omega(O_s, t; O_0, t_0)$, we can write an expression, valid for an arbitrary seismic body wave propagating in a 3-D layered structure (including multiply reflected converted waves) as

$$\hat{\mathbf{G}}^\Omega(O_s, t; O_0, t_0)$$
$$= \text{Re}\{\hat{\mathbf{A}}(O_s, O_0)\delta^A(t - t_0 - T(O_s, O_0))\} \quad (26)$$

where $\delta^A(\xi)$ is the analytical signal corresponding to the Dirac delta function, $\delta^A(\xi) = \delta(\xi) - i/\pi\xi$, $T(O_s, O_0)$ is the travel time of the given elementary wave along Ω from O_0 to O_s, and

$$\hat{\mathbf{A}}(O_s, O_0) = A\hat{\mathbf{C}}\hat{\mathbf{R}}\hat{\mathbf{E}}(\hat{\mathbf{C}}^0)^T \quad (27)$$

with

$$A = \frac{\exp(iT^c(O_s, O_0))}{4\pi[\rho(O_0)\rho(O_s)V(O_0)V(O_s)|\det \mathbf{Q}_2(O_s, O_0)|]^{1/2}} \quad (28)$$

In Eq. 28, $T^c(O_s, O_0)$ is the *phase shift due to caustics*, and $T^c(O_s, O_0) = -(\pi/2)k(O_s, O_0)$. The quantity $k(O_s, O_0)$ is called the *index of the ray trajectory* or the *KMAH index* (KMAH stands for Keller, Maslov, Arnold, and Hörmander; see Chapman, 1985). It is the number of caustic points along the ray trajectory Ω between O_0 and O_s, caustic points of the second order being considered twice. Introducing T^c, we can write $|\det \mathbf{Q}_2|$ instead of $\det \mathbf{Q}_2$ under the square root in the denominator. It is easy to see that A satisfies the principle of reciprocity; see Eq. 28.

The *complete receiver matrix* $\hat{\mathbf{C}}(O_s)$ is equal to $\hat{\mathbf{H}}(O_s)$ if the point O_s is located inside the medium, and to $\hat{\mathbf{Z}}\hat{\mathbf{R}}^c$ if O_s is located on the Earth's surface Σ. Here $\hat{\mathbf{Z}}$ is the transformation matrix from the local Cartesian coordinate system z_1, z_2, z_3 at Σ at the point of incidence Q to the general Cartesian coordinate system, introduced similarly as before. The *conversion matrix* $\hat{\mathbf{R}}^c$ contains the conversion coefficients and transforms the ray-centered component of the displacement vector of the incident P or S wave into the local Cartesian coordinate system at Σ at Q, taking also into account the generated reflected waves. Similarly, as in the case of R/T coefficients, care should be devoted to the correct signs of the conversion coefficients, which depend on the orientation of $\hat{\mathbf{e}}_1, \hat{\mathbf{e}}_2$ and on the orientation of the local Cartesian coordinate system basis vectors $\hat{\mathbf{i}}_1, \hat{\mathbf{i}}_2, \hat{\mathbf{i}}_3$. If O_s is situated at some inner interface Σ, $\hat{\mathbf{C}}$ is again given by $\hat{\mathbf{Z}}\hat{\mathbf{R}}^c$, but the relevant conversion coefficients are more complicated.

The *complete R/T matrix* $\hat{\mathbf{R}}$ includes the effects of reflections and transmissions at all R/T points between O_0 and O_s on the *spreading-free amplitudes*. Under the spreading-free amplitudes we understand $\hat{\mathbf{A}}(O_s, O_0)|\det \mathbf{Q}_2|^{1/2}$. Let us consider an arbitrary multiply reflected ray with N points of R/T between O_0 and O_s, denoted by $Q_1, Q_2, \ldots, Q_N$. Then

$$\hat{\mathbf{R}} = \prod_{i=N}^{1} \left\{ (\hat{\tilde{\mathbf{G}}}^\perp(Q_i))^T (\hat{\mathbf{R}}^D(Q_i))^T \hat{\mathbf{G}}^\perp(Q_i) \right.$$
$$\left. \cdot \left[\frac{\tilde{V}(Q_i)\tilde{\rho}(Q_i)\cos \tilde{\varphi}(Q_i)}{V(Q_i)\rho(Q_i)\cos \varphi(Q_i)}\right]^{1/2} \right\} \quad (29)$$

Here $\prod_{i=N}^{1} \hat{\mathbf{A}}_i$ denotes the matrix product $\hat{\mathbf{A}}_N\hat{\mathbf{A}}_{N-1} \cdots \hat{\mathbf{A}}_1$. The matrices in Eq. 29 have the same meaning as in Eq. 24; $\varphi(Q_i)$ and $\tilde{\varphi}(Q_i)$ denote the acute angles of incidence and of R/T, respectively. Note that $\hat{\mathbf{R}}$ satisfies the principle of reciprocity for any multiply reflected/transmitted ray between O_0 and O_s, including converted rays: $\hat{\mathbf{R}}(O_s, O_0) = \hat{\mathbf{R}}^T(O_0, O_s)$.

The *projection matrix* $\hat{\mathbf{E}}$ selects P and S waves at the source. For the wave leaving the source O_0 as a P wave, $E_{33} = 1$. For the wave leaving the source as an S wave, $E_{11} = E_{22} = 1$. All other elements of $\hat{\mathbf{E}}$ vanish. If we wish to separate the elementary waves that arrive at O_s as P or S waves, we can also introduce the projection matrix $\hat{\mathbf{E}}$ between $\hat{\mathbf{C}}$ and $\hat{\mathbf{R}}$ in Eq. 27.

The *complete source matrix* $\hat{\mathbf{C}}^0(O_0)$ has the same meaning as $\hat{\mathbf{C}}$, but it is applied to the source point O_0.

It is not difficult to see from Eq. 27 that the ray theory elastodynamic Green's function satisfies the principle of reciprocity in the following sense: $\hat{\mathbf{G}}(O_s, t; O_0, t_0) = \hat{\mathbf{G}}^T(O_0, t; O_s, t_0)$.

From Eqs. 25-29 we can also derive the paraxial ray approximation elastodynamic Green's function. Assume that the receiver is situated at S. To evaluate $\hat{\mathbf{G}}(S, t; O_0, t_0)$, we need not require that the rays of individual elementary waves, shot from O_0, pass exactly through S. They may end at O_s, located close to S. Then we obtain $\hat{\mathbf{G}}(S, t; O_0, t_0)$, using Eqs. 25-29, except $\hat{\mathbf{A}}(O_s, O_0)$ is multiplied on the left by the paraxial approximation matrix $\hat{\mathbf{B}}(S, O_s)$ (see Eq. 21), and $T(O_s, O_0)$ in Eq. 26 is replaced by $T(S, O_s)$, given by Eq. 18. Thus, the two-point ray tracing is not required in computing the paraxial ray approximation elastodynamic Green's function.

Moment-Tensor Point Sources

Using the Green's function, we can easily write compact expressions for point sources of more general types in 3-D laterally varying layered structures. Let us consider a *moment-tensor point source* at O_0 and denote the Cartesian components of the moment tensor by $M_{ij}(t)$. Then the Cartesian components of the displacement vector at O_s are determined by the expressions

$$u_n(O_s, t) = \dot{M}_{ij}(t) * \sum_{(\Omega)} G_{ni}^{\Omega}(O_s, t; O_0, 0) p_j(O_0) \tag{30}$$

Here $\dot{M}_{ij}$ denotes the time derivative of the moment tensor, the asterisk denotes convolution, and $p_j(O_0)$ is the component of the slowness vector at the source point O_0. This equation can be rewritten in many alternative forms, two of which are

$$u_n(O_s, t) = \text{Re} \left\{ \dot{M}_{ij}^A(t) * \sum_{(\Omega)} A_{ni}(O_s, O_0) p_j(O_0) \right.$$
$$\left. \cdot \delta(t - T(O_s, O_0)) \right\}$$
$$\hat{\mathbf{u}}(O_s, t) = \text{Re} \sum_{(\Omega)} [\hat{\mathbf{A}}(O_s, O_0)$$
$$\cdot \dot{\mathbf{M}}^A(t - T(O_s, O_0)) \hat{\mathbf{p}}(O_0)] \tag{31}$$

Here $\dot{M}_{ij}^A$ denotes the analytical signal corresponding to the time derivative of $M_{ij}(t)$.

Ray Synthetic Seismograms

Equations for the ray theory Green's function and for the moment-tensor point sources can be directly used to construct ray synthetic seismograms in 3-D laterally varying layered structures for various types of point sources and sources of finite extent.

Because ray theory can be applied to compute only the HF part of the seismic wave field, the resulting seismograms must be high-pass filtered to exclude the low-frequency contributions, if such contributions are included in the source time function. For example, only the HF part of $\dot{M}_{ij}^A$ should be used in Eq. 31, not its whole time history.

For reviews on generation and use of ray synthetic seismograms, see *J. Geophys.* **58**, Nos. 1-3, 1985.

Validity Conditions and Extensions of the Ray Method

The ray method is applicable only if the velocity distribution in the model and the interfaces in the vicinity of the ray under consideration are sufficiently smooth. Let us consider a prevailing wavelength λ of a considered wave. Then the two most important validity conditions are (a) $V/|\nabla V| \gg \lambda$ along the whole ray, (b) $R_1(Q) \gg \lambda$, $R_2(Q) \gg \lambda$ at all points Q of R/T, where R_1 and R_2 are the principal radii of the curvatures of the interfaces. More detailed validity conditions can be formulated in terms of Fresnel volumes.

The ray method has some other limitations, even in smooth media: It is of limited accuracy or even invalid in the singular regions of the ray field (caustic region, critical region, transition region between the shadow and illuminated zones, etc.). In its standard form, it cannot be used to investigate various diffracted waves, inhomogeneous waves, etc. See Chapman (1985).

Various modifications have been proposed to extend the applicability of the ray method; for example, local modifications of the ray method in the singular regions (Airy modification in the caustic region, Weber-Hermite modification in the critical region), the geometric theory of diffraction, the method of edge waves, etc. Recently, two extensions, which have broader possibilities, have been proposed: the Maslov method and the method of Gaussian beams.

The seismic ray method can be also used to study HF seismic wave propagation in anisotropic media, prestressed media, slightly dissipative media. One extension of the ray method, called the *space-time ray method*, can be used to investigate dispersive seismic waves and solve HF wave propagation problems with time-dependent boundary conditions.

V. ČERVENÝ

References

Aki, K., and P. G. Richards, 1980, *Quantitative Seismology*. San Francisco: Freeman.

Babich, V. M., 1956, Ray method of the computation of the intensity of wavefronts (in Russian), *Akad. Nauk SSSR Doklady* **110**, 355-357.

Bullen, K. E., and B. A. Bolt, 1985, *Theory of Seismology*. Cambridge: Cambridge University Press.

Červený, V., 1985, The application of ray tracing to the numerical modeling of seismic wavefields in complex structures, in G. Dohr, ed., *Seismic Shear Waves, Part A: Theory*. London: Geophysical Press, 1-124.

Červený, V., 1987, Ray tracing algorithms in three-dimensional laterally varying layered structures, in G. Nolet, ed., *Seismic Tomography*. Dordrecht: D. Reidel, 99-133.

Červený, V., I. A. Molotkov, and I. Pšenčík, 1977, *Ray Method in Seismology*. Praha: Universita Karlova.

Červený, V., and R. Ravindra, 1971, *Theory of Seismic Head Waves*. Toronto: University of Toronto Press.

Chapman, C. H., 1985, Ray theory and its extensions: WKBJ and Maslov seismograms, *Jour. Geophysics* **58**, 27-43.

Hanyga, A., E. Lenartowicz, and J. Pajchel, 1984, *Seismic Wave Propagation in the Earth*. Warszaw-Amsterdam: PWN-Elsevier.

Karal, F. C., and J. B. Keller, 1959, Elastic wave propagation in inhomogeneous media, *Acoust. Soc. America Jour.* **31**, 694-705.

Cross-references: *Elasticity and Wave Propagation: Principles; Elastic Waves in Homogeneous and Inhomogeneous Media; Seismic Diffraction; Seismic Wave Scattering; Seismograms: Synthetic.*

SEISMIC SIGNAL PROCESSING

The initial assumption upon which the following discussion is based is that our seismic data are safely stored on a digital magnetic tape and that a digital computer is humming quietly at our elbow. The data contain information about the source, data acquisition system, and various features of the subsurface of the Earth, and our task is to extract some portion of that information and display it with the greatest possible clarity. There will also be a component of the recorded data that is present even if the seismic source is never fired or the earthquake never occurs; for example, the seismic energy generated by the roots of trees on a windy day or the signal induced in the geophone coil by the magnetic field from a nearby power line. We can safely refer to this component of our data as "noise." However, the term "signal-generated noise" is frequently used, which implies that one portion of the energy radiated from the controlled source or earthquake contains useful information and the rest does not. This judgment is highly subjective, and the term "signal-to-noise ratio" requires some careful definition before it has any useful meaning. One man's noise is frequently another man's signal.

Much of what is to follow will be as familiar to a physicist, an electrical engineer, or an economist as it will be to a seismologist. However, it is possible to make two, admittedly rather sweeping, generalizations about seismic signal processing. First, coherent, repeatable signal-generated noise is a major problem. In a reflection seismic experiment, the dominant portion of the recorded energy travels directly from surface source to surface receiver, whereas the desired signals are the tiny echoes from buried reflecting interfaces. The signal-to-noise ratio will not be significantly improved by repeating the same experiment (constant source/receiver offset) several times and stacking because the signal-generated noise will be as repeatable as the desired signal. We need to perform the experiment many different ways (different source/receiver geometries) so that the signal-generated noise changes from one record to the next in a manner different from the change in the character of the desired reflected signals. We then have the opportunity to apply a selective stack of the data to enhance or suppress signals that vary with source/receiver offset in a particular way. The field procedures described in *Exploration Seismology* and *Controlled Source Seismology* are designed to facilitate this signal processing technique.

Second, the seismologist can only rarely assume that the input signal to the Earth is a known function. A seismic trace $x(t)$ can be represented as the convolution of a wavelet $w(t)$ with a reflectivity function $r(t)$ and additive noise $n(t)$:

$$x(t) = r(t) * w(t) + n(t) \qquad (1)$$

Normally, a signal processing scheme will be employed to extract either $r(t)$ or $w(t)$ from $x(t)$. If nothing is known about $r(t)$, $w(t)$, or $n(t)$, then this task might appear rather daunting. However, as in all ill-posed inverse problems, a solution is achieved by including additional a priori information or constraints. Then the ultimate success of the procedure depends on the validity of the assumptions made. In the following section, different deconvolution procedures are briefly described. Detailed descriptions of the methods can be found in the recommended texts. Two highly recommended articles that discuss the relative strengths and weaknesses of these and other methods are Jurkevics and Wiggins (1984) and White (1987).

Deconvolution

Wiener-Levinson Deconvolution. By far the most common method of computing a deconvolution operator is the Wiener-Levinson algorithm. If a signal $y(t)$ is input to a linear filter whose impulse response is $f(t)$, then the output $c(t)$ is the convolution of $y(t)$ with $f(t)$:

$$c(t) = y(t) * f(t)$$

The value of the filter $f(t)$ that transforms $y(t)$ into some desired output $d(t)$ can be determined by minimizing the sum of the squares of the differences between the digital samples of $c(t)$ and $d(t)$. In matrix notation,

$$\mathbf{d} = Y\mathbf{f} + \mathbf{e}$$

where the elements of matrix Y are the digital sample values of y arranged so that $Y\mathbf{f}$ represents the discrete convolution operation. By minimizing $|\mathbf{e}^T\mathbf{e}|$, we obtain the least squares normal equation

$$Y^T Y \mathbf{f} = Y^T \mathbf{d} \qquad (2)$$

where $Y^T Y$ is a square symmetric matrix whose rows or columns contain the discrete sample values of the autocorrelation function of the input signal. The main diagonal elements are the zero lag values. $Y^T \mathbf{d}$ represents the cross-correlation of the input with the desired output.

Return to Eq. 1; if a filter can be determined that transforms $w(t)$ into a spike, then, because convolution is a commutative process, the same filter will transform $r(t) * w(t)$ into $r(t)$ and $x(t)$ into our best estimate of $r(t)$. The beauty of Eq. 2 is that we do not need to know what $w(t)$ is, provided the following assumptions are true:

1. The Earth reflection coefficient series $r(t)$ is random and uncorrelated.
2. $w(t)$ is a minimum-phase wavelet.

For this discussion, it is sufficient to appreciate that of all possible wavelets with the same amplitude spectrum, the energy of a minimum-phase wavelet builds up fastest. Such a wavelet is sometimes described as *front-loaded*.

In addition, Eq. 1 is only valid if the layered earth is a linear system. Since the autocorrelation of a random and uncorrelated function is a spike at zero lag, $K(t)$, then the autocorrelation of $r(t) * w(t)$ is the autocorrelation of $w(t)$ convolved with $K(t)$; in other words, the autocorrelation of $w(t)$ times a scaling factor K. Hence, a good estimate of the autocorrelation matrix $Y^T Y$ for the wavelet $w(t)$ can be obtained by forming the autocorrelation of the known signal $x(t)$.

The right side of Eq. 2 still seems to present a problem, since it would appear that one needs to know $w(t)$ in order to form Y^T. However, since $\mathbf{d}$ is a spike at zero time, $Y^T \mathbf{d}$ is

$$\begin{bmatrix} w_0 w_1 w_2 \cdots w_n\ 0 \cdots\ 0 \cdots 0 \\ 0\ w_0 w_1 \cdots\ w_n\ 0 \cdots 0 \cdots 0 \\ \vdots \\ 0 \cdots\cdots\ 0\ w_0 \cdots\ w_{n-1} w_n \end{bmatrix} \begin{bmatrix} 1 \\ 0 \\ \vdots \\ 0 \end{bmatrix} = \begin{bmatrix} w_0 \\ 0 \\ \vdots \\ 0 \end{bmatrix}$$

Hence, the right side of Eq. 2 is simply a number in the first element of an otherwise zero vector. The resulting filter $\mathbf{f}$ will be causal. In other words, values of f_i have only been determined for times greater than zero.

The significance of the second assumption is that, if our wavelet is not minimum phase (front-loaded), it is impossible to turn the wavelet into a spike at zero time with a causal filter. However, by redefining time zero as, say t_3, the first three elements of $\mathbf{f}$ will become *anticipation* components (negative time components), in which case, Eq. 3 becomes

$$\begin{bmatrix} w_0 w_1 w_2 \cdots w_n\ 0\ 0 \cdots \\ \vdots \end{bmatrix} \begin{bmatrix} 0 \\ 0 \\ 0 \\ 1 \\ 0 \\ \vdots \end{bmatrix} = \begin{bmatrix} w_3 \\ w_2 \\ w_1 \\ w_0 \\ 0 \\ 0 \\ \vdots \end{bmatrix}$$

Under these circumstances, it is essential to know something about the wavelet shape, specifically the sample values from time zero up to the value equal to the maximum required anticipation component.

Predictive deconvolution is a means of removing reverberations from data. This type of signal-generated noise in seismic data is very common. Multiple reflections can result from a single layer within the Earth due to successive downgoing and upgoing reflections from the upper and lower interfaces. This effect is particularly pronounced within the water layer in a marine seismic experiment. If $m(t)$ is the impulse response of this reverberating component of the Earth, then our simple Earth model defined in Eq. 1 can be modified to

$$x(t) = r(t) * m(t) * w(t) + n(t)$$

Figure 1a shows a synthetic example of $m(t) * w(t)$. Since the later arrivals in this signal are entirely predictable from the first arrival, it is possible to use Eq. 2 to reduce $m(t) * w(t)$ to $w(t)$ or, more generally, to reduce $r(t) * m(t) * w(t)$ to $r(t) * w(t)$, provided that $r(t)$ is random and uncorrelated as before. In this case, it is necessary to assume that $m(t)$ is a minimum-phase sequence and $w(t)$ is a minimum-phase wavelet. Then, if α is the time delay between the reverberations in $m(t)$, the desired output $\mathbf{d}$ is the input shifted in time by α time units. Hence, $Y^T \mathbf{d}$ becomes the autocorrelation of the input signal starting at lag α:

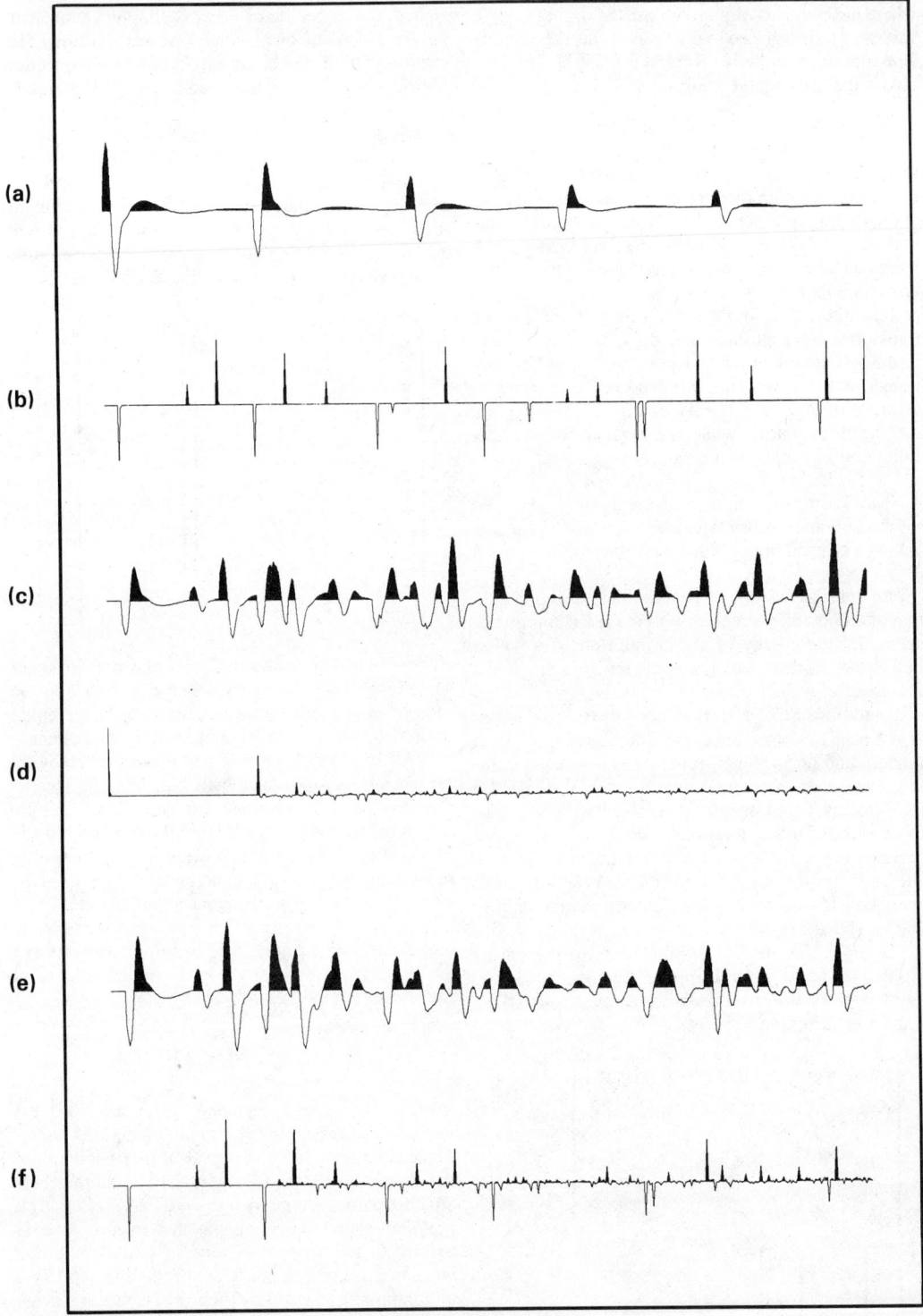

FIGURE 1. An example of predictive deconvolution: (a) a minimum phase wavelet convolved with the impulse response of the reverberating component of the Earth; (b) the reflection coefficient series $r(t)$; (c) the convolution of (a) with (b); (d) the prediction error filter; (e) the convolution of (d) with (c), the dereverberated trace; (f) (e) after wavelet deconvolution.

$$\begin{bmatrix} \phi_0 & \phi_1 & \cdots & \phi_n \\ \phi_1 & \phi_0 & \phi_1 & \cdots \\ \phi_2 & \phi_1 & \phi_0 & \phi_1 & \cdots \\ & & & & \end{bmatrix} \begin{bmatrix} f_0 \\ f_1 \\ \vdots \\ \end{bmatrix} = \begin{bmatrix} \phi_\alpha \\ \vdots \\ \phi_{n+\alpha} \end{bmatrix}$$

Finally, the *prediction error filter* is formed by negating the filter **f** and adding a 1 and $x - 1$ zeroes at the beginning:

$$\underbrace{1, 0, 0, 0, \ldots,}_{\alpha - 1} -f_0, -f_1, -f_2, \ldots$$

The process is illustrated in Fig. 1: (b) is the reflection coefficient series $r(t)$; (c) is the convolution of (a) with (b), $r(t) * m(t) * w(t)$; (d) is the prediction error filter, and (e) is the convolution of (d) with (c), the dereverberated trace. The result is more clearly presented by going a stage further and carrying out wavelet deconvolution on (e). The result (f) can then be compared with the reflection coefficient series (b).

It is instructive to observe what is going on in the autocorrelation domain. Figure 2a is the autocorrelation of $m(t)$, the impulse response of the reverberant component of the Earth. Figure 2b is the autocorrelation of the Earth reflection coefficient series, which ideally should be a spike at zero lag. However, this example is not completely random and uncorrelated, so we should not expect a perfect result. The predictive deconvolution has changed Fig. 2c, the autocorrelation of $r(t) * m(t) * w(t)$ into Fig. 2d.

The proportion of this article devoted to Wiener-Levinson deconvolution is a measure of its practical importance. Other important aspects of the procedure, not discussed, include the stabilization of the

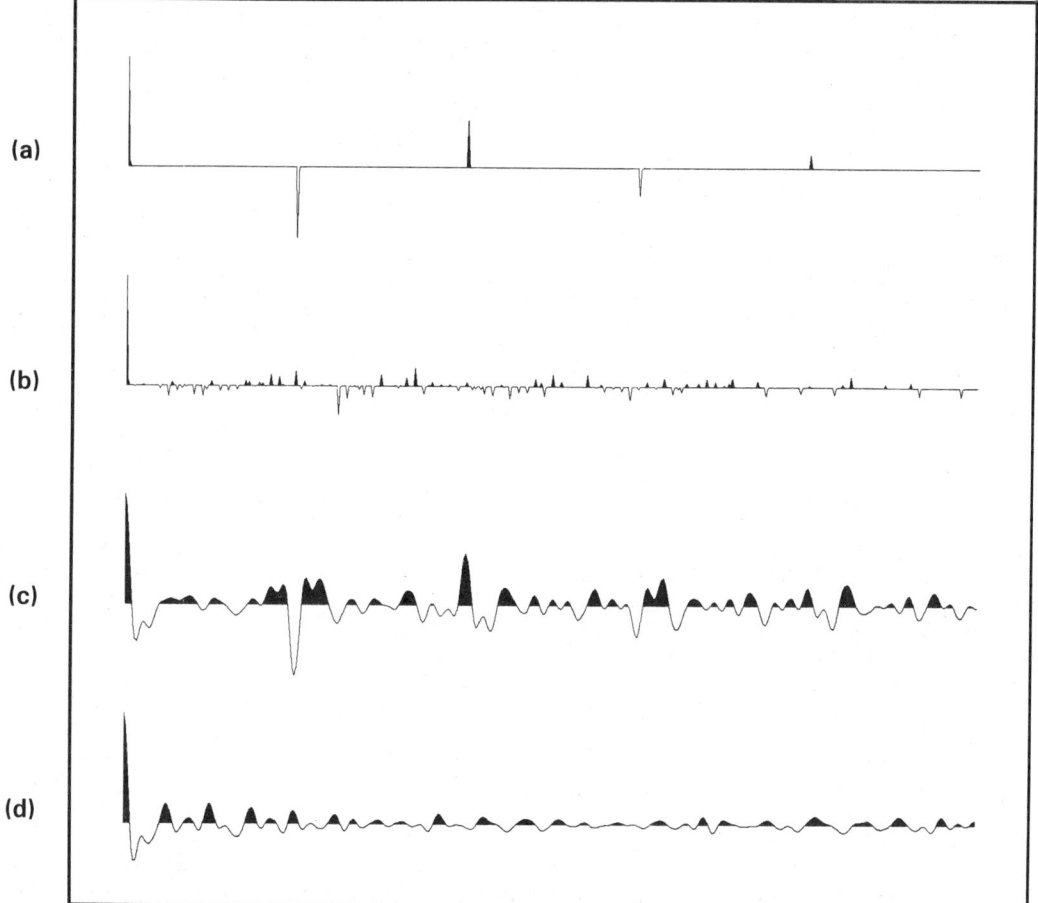

FIGURE 2. The autocorrelation domain: (a) the one-sided autocorrelation of the impulse response of the reverberating component of the Earth; (b) the one-sided autocorrelation of Fig. 1b; (c) the one-sided autocorrelation of Fig. 1c; (d) the one-sided autocorrelation of Fig. 1e.

procedure by the application of "white light," the relationship between wavelet and predictive deconvolution and the concept of "gapped" deconvolution, the effect of the finite bandwidth of the data and the effect of violating the assumptions upon which the method is based. See the references.

Homomorphic Deconvolution. Homomorphic deconvolution is a method of estimating the wavelet $w(t)$ that does not depend upon any assumptions about the wavelet phase. If we neglect the noise term in Eq. 1 for the time being, the Fourier transform of the signal $x(t)$ is the product of the Fourier transforms of $r(t)$ and $w(t)$.

$$X(w) = R(w)W(w) = |R|e^{i\phi_R} \cdot |W|e^{i\phi_W} \quad (4)$$

By taking the logarithm of the Fourier transform of $x(t)$, we see that the wavelet component and the reflectivity component of the signal becomes additive rather than convolved or multiplied together. Log spectra are called the *cepstral domain* values.

$$\text{Ln } X(w) = \text{Ln } |R| + i\phi_R + \text{Ln } |W| + i\phi_W \quad (5)$$

Unfortunately, the phantom term $e^{\pm i2\pi n}$ (where n is any integer) in Eq. 4 cannot be ignored. So the additional phase term $\pm 2i\pi n$ must be included on the right side of Eq. 5. The critical step in this algorithm is the "unwrapping" of the phase curve so that phase becomes a continuous function of frequency.

The assumption is now made that the wavelet is the same from trace to trace, but the reflectivity sequence is random. Hence, by stacking a number of traces in the cepstral domain, we should remove the reflectivity component in the averaging. Finally, the data are both inverse-cepstral-transformed and inverse-Fourier-transformed.

$$\exp[\text{Ln } \hat{X}(w)] \longrightarrow \hat{X}(w) \xrightarrow{\text{F.T.}} \hat{x}(t) \sim w(t)$$

(where ^ denotes the averaged value).

In addition to the dubious validity of the assumption about the reflectivity sequence, the main snag in the method seems to be the ambiguity in the phase values associated with very small amplitudes in the signal spectrum. However, in favorable conditions of signal to noise, the method has been shown to be effective.

Minimum Entropy Deconvolution. Studies on the statistical properties of reflectivity sequences deduced from borehole logs have shown that they have a tendency to be leptokurtic. *Kurtosis* is a measure of the "peakedness" of a distribution; a sequence with a narrow probability distribution (large kurtosis, leptokurtic) will have a limited number of large spikes, with most of the samples close to the zero mean. *Entropy* is a measure of the unpredictability of a time series. White noise is completely unpredictable. As a time series becomes more "spiky," its disorder or entropy is reduced. So as entropy decreases, kurtosis increases.

A multitrace wavelet deconvolution procedure has been formulated such that the entropy of the resultant signals is minimized (or kurtosis is maximized). If d_{ij} is the jth element of the ith trace of the desired output, then the deconvolution filter coefficients f_k must satisfy the equation

$$d_{ij} = \sum_k f_k x_{i,j-k}$$

subject to the constraint that the kurtosis K is maximized:

$$K = \sum_i K_i, \quad K_i = \sum_j d_{ij}^4 \left(\sum_j d_{ij}^2 \right)^{-2}$$

(Strictly speaking, K_i is called the *varimax norm*, and kurtosis is the normalized version of this function.)

Jurkevics and Wiggins (1984) show that, as one might expect, the method works well for sparse reflectivity series and less well when the series become denser. In other words, as in all inverse problems, the results are only as good as the model upon which the inversion is based.

Two-Dimensional Filtering

Methods described in this section involve the separation of two or more components of a seismic signal or signals in a two-dimensional space. The parameters of the two-dimensional space vary. In the first example, they are velocity and frequency, in the second, wavenumber and frequency and, in the third, time and time gradient (dt/dx). In all cases, the filtering is achieved by a simple multiplicative operation, blanking out the unwanted portions of the two-dimensional data. The key step is the transformation of the data into the appropriate domain and then inverse-transforming after the multiplicative filtering operation has been performed.

Moving Window Analysis. Moving window analysis is particularly useful for separating different modes of a dispersed wave type from one another. Figure 3 shows a synthetic example of a fundamental and first-higher-mode Rayleigh wave that are superimposed upon one another in the time domain. However, within any particular arrival time window, the fundamental mode has a frequency content different from the higher mode. This difference is the key to their successful separation.

The time domain signal is windowed with a narrow window centered at time t_i and Fourier-transformed. The resulting amplitude spectrum then forms the row of the time (or velocity) versus frequency (or period) matrix in Fig. 3 corresponding to time t_i. To achieve good frequency resolution, one requires a wide time window. Conversely, good temporal resolution (a narrow time window) results in poor frequency resolution. The optimal choice of time window, in order to achieve the best possible resolution in both the time and frequency domains, is the Gaussian function

$$w(t) = e^{-at} \xrightarrow{F.T.} \sqrt{\frac{\pi}{a}} e^{-\omega^2/4a^2}$$

Having determined the group velocity curves for the different modes from a plot like Fig. 3, we use the following effective means of isolating one of the modes:

1. Fourier-transform the original signal.
2. Each Fourier component, taken as a pure sinusoidal time function is windowed symmetrically by a cosine window with its midpoint at the group arrival time of that frequency (obtained from the determined group velocity curve).
3. The windowed sinusoidal components are added in the time domain to synthesize a time-varying filtered seismogram.

k-ω Filtering. Like the Wiener-Levinson algorithm, k-ω filtering is another workhorse in the seismic processing industry, and much has been written about it. March and Bailey (1983) is highly recommended.

The input data to the process are signals recorded by an array of receivers. These data in horizontal position–time space are Fourier-transformed to wave number–frequency space by Fourier-transforming each seismic trace (column of the two-dimensional matrix) to obtain a two-dimensional function in (x, ω) space and then Fourier-transforming each row to obtain a function in (k, ω) space. Since $k = \omega/c$, where c is velocity, a straight line in (x, t) space will transform to a straight line passing through the origin in (k, ω) space, with gradient c.

It follows that the interesting events with different dips (or apparent velocities) in Fig. 4a will map to distinct regions of k-ω space as shown in Fig. 4b. The method has been called *pie-slice* filtering because the $(k$-$\omega)$ window used to isolate energy falling within a particular range of values of apparent velocity will have the shape of a slice of pie with the vertex at the origin.

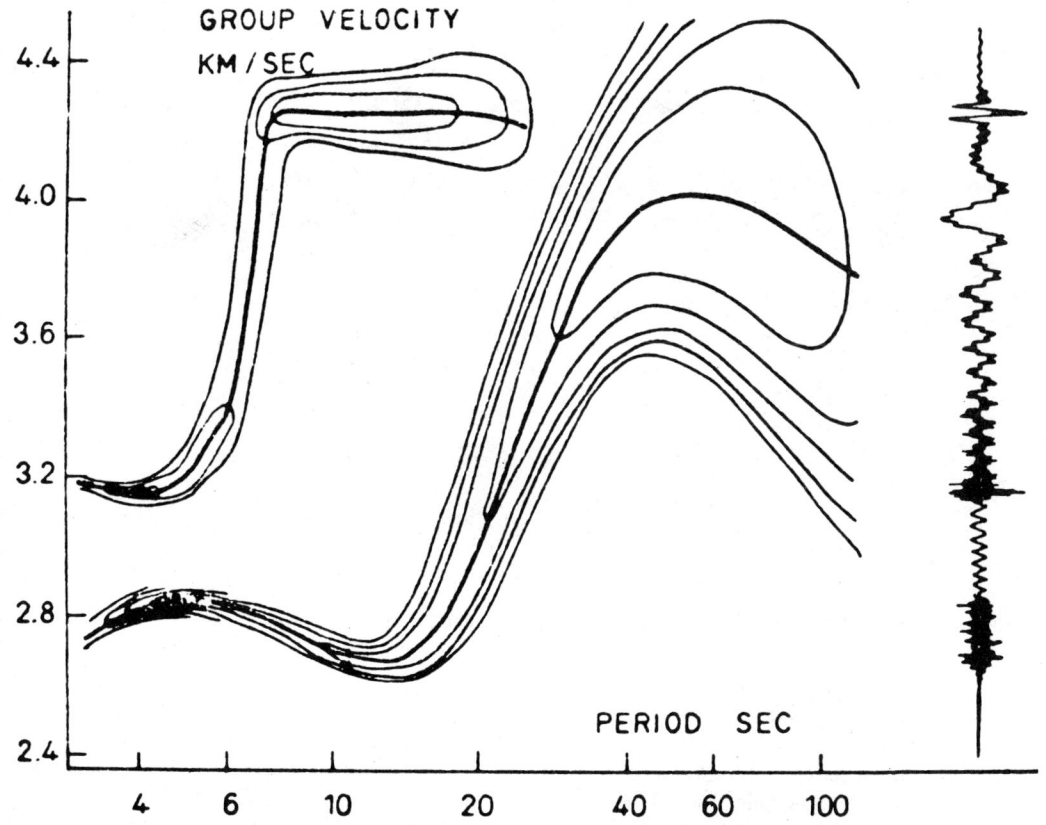

FIGURE 3. A synthetic example of a fundamental and first-higher-mode Rayleigh waves that are superimposed upon one another in the time domain and separated in the velocity/period domain. (After Dziewonski et al., 1969).

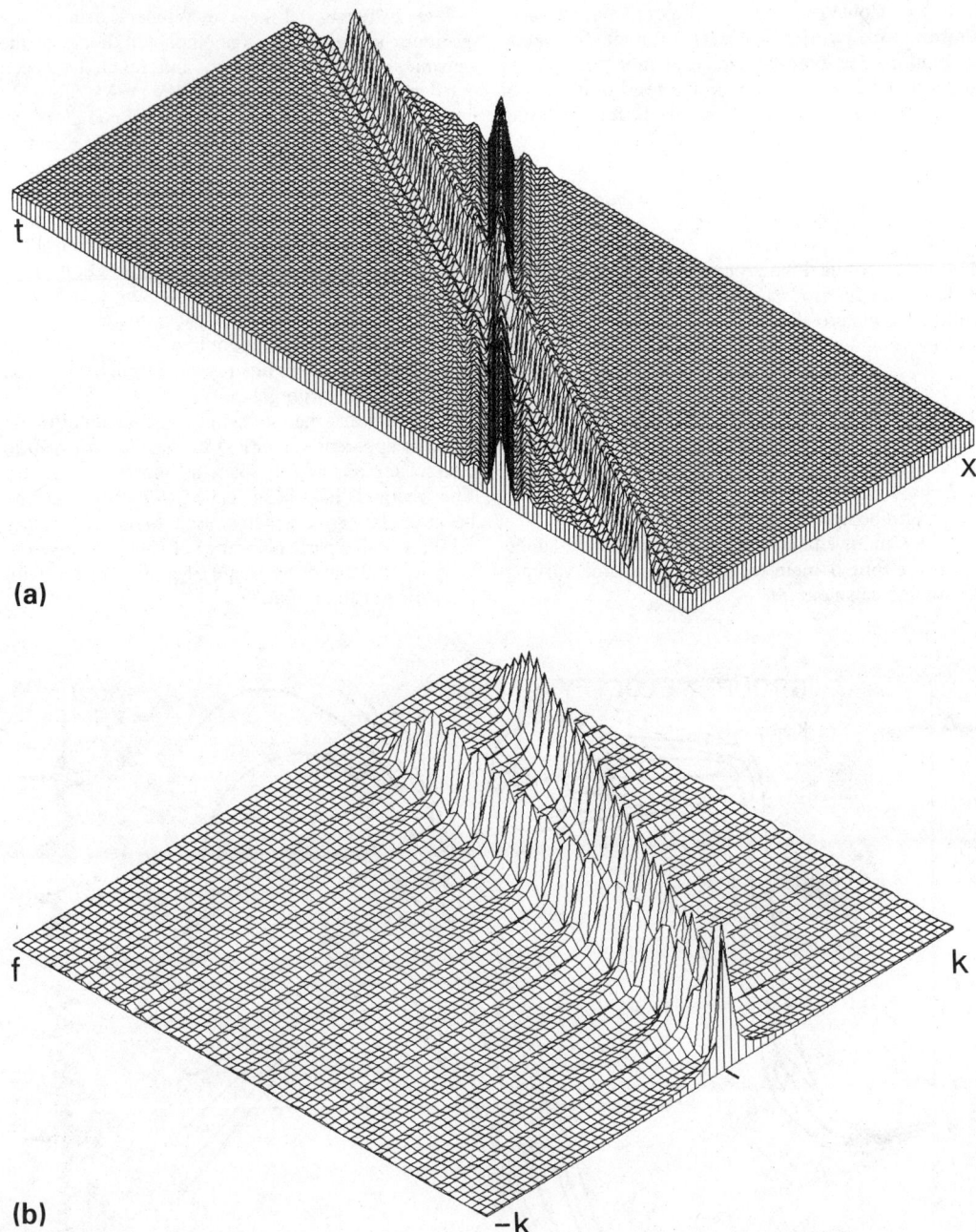

FIGURE 4. Two intersecting events in (x, t) space (a) transform to two separated events in (k, ω) space (b), where $\omega = 2\pi f$.

In practice, the separation of events with different apparent velocities may not be quite so straightforward. The Fourier transform of a box is a sine function. Consequently, if an event is abruptly truncated in the (x, t) domain, as in Fig. 5a, the event will exhibit a pattern of sidebands in the $(k\text{-}\omega)$ domain (Fig. 5b that, in Fig. 4, would overlap the energy from the second event).

The problem of "spatial aliasing" is illustrated in Fig. 6. Energy from a steeply dipping event in (x, t) space can wrap around $k\text{-}\omega$ space as in Fig. 6b, and separation of such an event using $k\text{-}\omega$ filtering becomes extremely difficult.

Slant Stack. *Slant stack* is a helpfully descriptive name for a procedure sometimes called $\tau\text{-}p$ or *Radon transformation*. Consider, for example, a

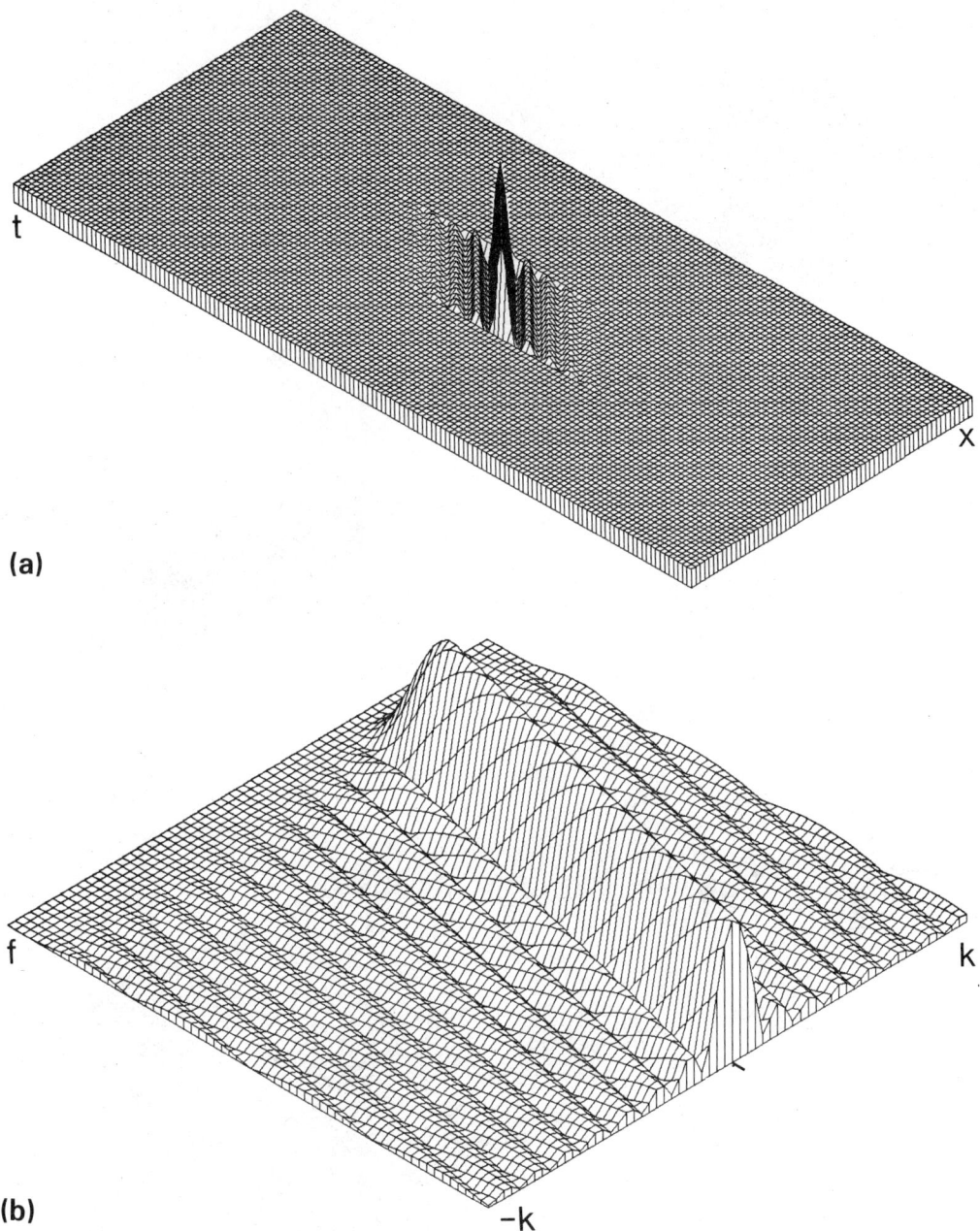

FIGURE 5. Spatial truncation of one of the events in Fig. 4a (a) results in broadening of the corresponding event in the k-ω domain (b).

common midpoint gather of seismic reflection data (shot-receiver offset, time space). A straight line through this space will be defined by $t = \tau + px$ where τ is the zero-offset intercept time and p is the gradient.

The idea is to stack energy that falls on such a line and to place the resultant at a point in (τ, p) space. So, in this transformation, a line maps to a point. In the example in Fig. 7, the direct wave A and head waves B and C map to positions A, B, and C in (τ, p) space while the reflected events map to the curves (shown as broken lines). The inverse transform from (τ, p) space to (t, x) space also involves a summation along a line since $\tau = t - px$ and $d\tau/dp = -x$.

The only difference is that, for reasons not immediately obvious, each column in (τ, p) space

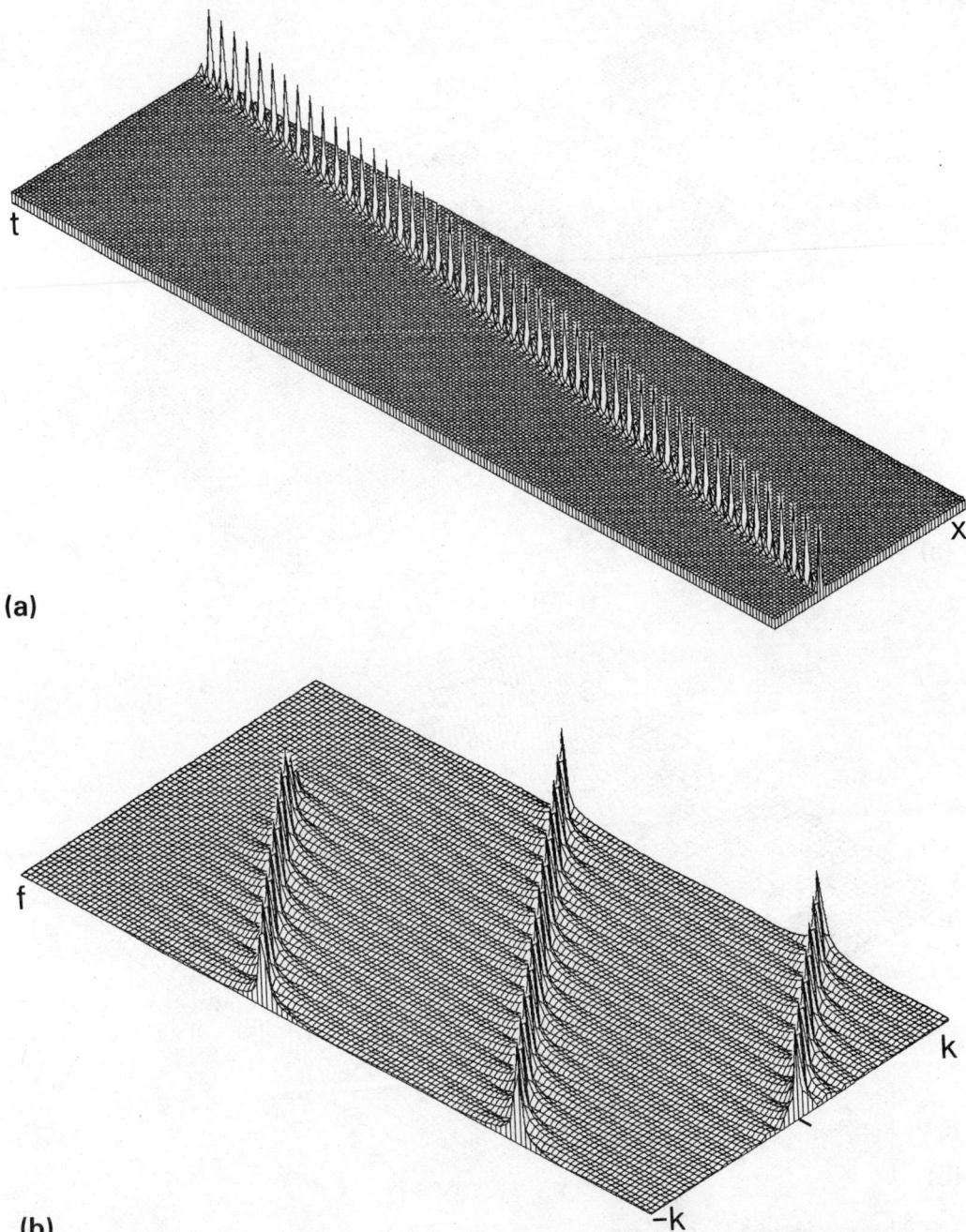

FIGURE 6. The steeply dipping event in (a) is spatially aliased. Wraparound occurs in k-ω space as in (b).

must first be convolved with the inverse Fourier transform of the modulus of ω (the angular frequency). For details, see Hatton et al. (1986).

One consequence of transforming data into (τ, p) space is that the energy in a single trace (τ, p_i) all arrives at the receivers with a constant time dip (dt/dx). It follows that, for a plane-layered Earth, all the energy in a single (τ, p_i) trace will have left the sources at a constant angle.

So, if a source radiation pattern (source wavelet) varies with azimuth, which is usually the case, one can legitimately attempt some deconvolution operation on a (τ, p_i) record involving the assumption of a constant wavelet $w(\tau)$, whereas the assumption of

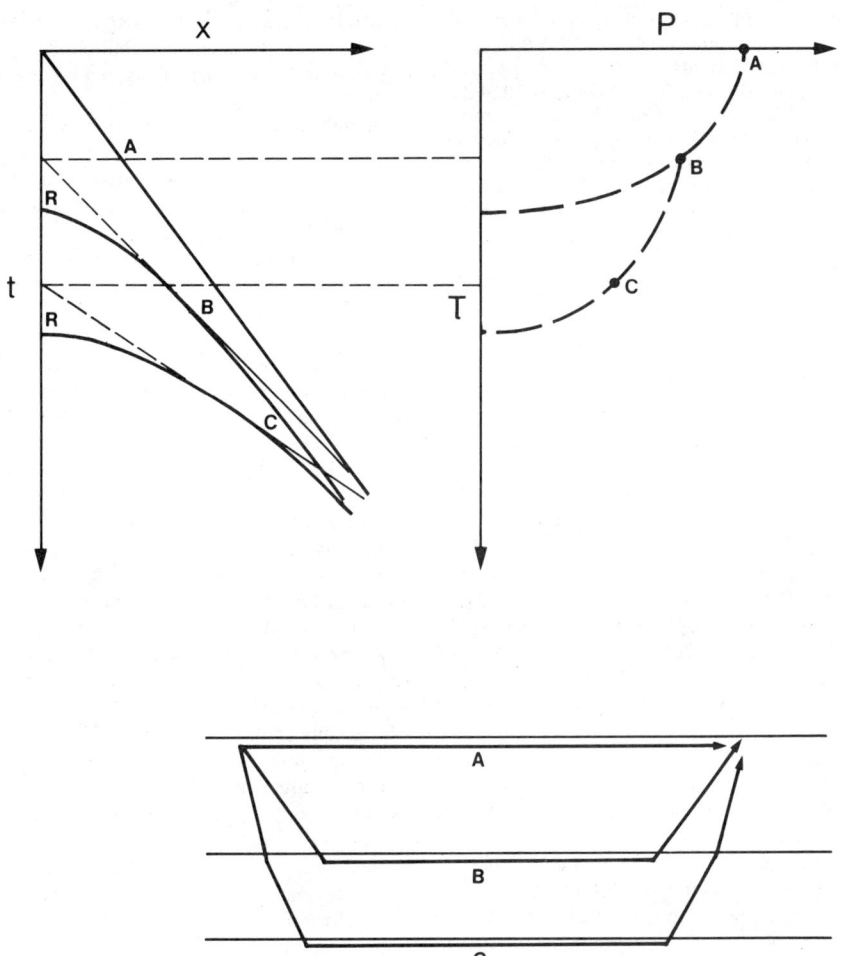

FIGURE 7. The τ-p transform. Events A, B, and C which have constant gradients in *x-t* space map to points in τ-p space. Reflected events (*R*) map to the curves (broken lines).

a constant $w(t)$ in a (t, x) record is not really valid. Hence there are potential advantages for both deconvolution and two-dimensional filtering of transforming into the slant stack domain.

Conclusion

Some arbitrary decisions have had to be made about topics to be excluded from this short article. Extensive coverage of seismic signal processing techniques is found in Kanasewich (1981). However, one vital aspect of signal processing is so deceptively straightforward that it rarely merits a mention in a standard text—the question of format of the final display: wiggle trace, wiggle trace with shading, variable intensity, color, and, if so, what colors. Hatton et al. (1986) discuss some of the more esoteric aspects of "cosmetic" adjustment to a seismic section so that the final product is "pleasing" to the eye. Subtle correlations in noisy data can sometimes be dramatically enhanced by a judicious choice of trace display type, scaling, and amplitude equalization. The final product of seismic signal processing is a picture, and any improvement in computer graphics software and hardware is of great significance to the seismologist.

M. H. WORTHINGTON

References

Dziewonski, A., S. Bloch, and M. Landisman, 1969, Techniques for analysis of transient seismic signals, *Seismol. Soc. America Bull.* **59** (1), 427–444.

Hatton, L., M. H. Worthington, and J. Makin, 1986, *Seismic Data Processing, Theory and Practice.* Osney Mead, Oxford: Blackwell Scientific Publications, 177p.

Jurkevics, A., and R. Wiggins, 1984, A critique of seismic deconvolution methods, *Geophysics* **49**, 2109–2116.

Kanasewich, E. R., 1981, *Time Sequence Analysis in Geophysics*, 3rd ed. Edmonton, Alta.: University of Alberta Press, 480p.

March, D. W., and A. D. Bailey, 1983, Two-dimensional transform and seismic processing, *First Break* **1,** 9–21.
Robinson, A. E., and S. Treitel, 1980, *Geophysical Signal Analysis.* Englewood Cliffs, N. J.: Prentice-Hall, 466p.
Sheriff, R. E., and L. P. Geldart, 1983, *Exploration Seismology,* vol. 2. *Data Processing and Interpretation.* Cambridge: Cambridge University Press, 221p.
White, R. E., 1987, Estimation problems in seismic deconvolution, in *Deconvolution and Inversion.* Oxford: Blackwell Scientific Publications.

Cross-references: *Controlled Source Seismology; Deep Seismic-Reflection Profiling; Exploration Seismology; Seismic Imaging; Seismic Instrumentation; Seismic Instrumentation: History; Seismic Noise; Seismic Wavefield Migration.*

SEISMIC SOURCE: OBSERVATIONS

Observational seismology dates back to antiquity. Large earthquakes that occur in populated regions are great natural catastrophes, and descriptions of earthquakes and their effects can be found in many ancient writings (see Richter, 1958, for several accounts). There are also ancient descriptions of other seismic phenomena: volcanic explosions, tsunamis, and landslides. Although these early observations of seismic phenomena are fascinating and scientifically important in some instances, this article concentrates on the developments in the modern era of seismology. A convenient demarkation of the modern era is the development and operation of seismographs in the decade from 1880 to 1890. The technological development of instruments that can record the mechanical vibrations of the Earth's surface gives seismology a purely quantitative observational basis with a direct connection to various aspects of the seismic source and the propagational aspects of the Earth. Most present-day seismological investigations of the seismic source are based on the analysis of seismic waves recorded by seismographs. The seismic waves are interpreted with the theory of elastodynamics. Progress in seismology results from a blend of new advances in instrumentation, additions to our theoretical framework, and the systematic analysis of earthquakes.

Despite the fundamental importance of seismograms in seismic source studies, many properties of seismic sources were first discovered through "macroseismic observations," a term that describes a variety of near-source observations:

    Geological mapping of fault traces and crustal deformation
    Direct measurement of shear dislocations across fault traces
    Geodetic measurements of the static deformation field associated with fault slip
    Relation between near-field shaking and fault slip.

Although these macroseismic observations can only be made for a few earthquakes, they yield insight into what we presume are general characteristics of earthquakes. Thus, this article first presents some features of the earthquake source based on macroseismic observations, followed by a discussion of how seismic waves are used to determine some of the important characteristics of earthquakes.

Although this article focuses on the earthquake source, it is worth mentioning the great range of seismic sources that are recorded by seismographs. Virtually any disturbance of the "resting" Earth with a time-scale less than an hour will produce seismic waves that could be recorded. Even a tree that falls in the forest generates seismic waves, though only a sensitive seismograph installed nearby would detect these waves.

The two main categories of seismic sources are transient and continuous. Transient sources have a total source duration of a few minutes or less; this duration cut-off corresponds to the duration of the largest earthquakes. Continuous sources may have a distinct spectral peak (such as microseisms) and hence appear as quasi-harmonic oscillations with durations of many hours or days. The main types of seismic sources that are routinely recorded are listed below:

Transient
    Earthquakes
    Explosions (caused by humans)
    Volcanic explosions
    Landslides
    Extraterrestrial impacts and explosions
Continuous
    Microseisms
    Atmospheric noise
    Volcanic harmonic tremor
    Underwater volcanic venting

Most seismologists are interested in the transient sources, consequently the continuous seismic sources are typically regarded as noise that limit our detection of the transient sources. Although the transient sources represent a wide variety of physical phenomena, modern seismic source theory provides a unifying theoretical framework that relates all of these diverse sources to the observed wavefield.

Earthquake sources are quite variable in their space-time occurrence—from isolated shocks to extensive aftershock sequences and earthquake swarms. Many reports associate earthquakes with a wide variety of phenomena; in some cases these observations are considered precursors to an earthquake. Often reported phenomena are the occurrence of earthquake "sounds" and "lights" (see Richter, 1958, for discussion). Other observations that have been associated with earthquakes or the earthquake process include variations in magnetic field, trace gases, gravity, groundwater, and animal behavior (see Richter, 1958; also references in Kasahara, 1981). The physical connections between these observations and the earthquake source vary between moderately understood to no obvious connection. Most seismologists study earthquakes within the physical theory of elastodynamics and this article follows the scientific progress in understanding the earthquake source within this context.

Earthquakes, Macroseismic Observations, and Elastic Rebound Hypothesis

Richter (1958) discusses several case histories of the association between earthquakes and faulting. These investigations led to the conclusion that faulting ground breakage and earthquake "shaking" were related. Aside from fault movement, other local macroseismic effects of an earthquake include destruction of cultural structures, secondary faulting and slumping, landslides and mudslides, and seiches. These important and sometimes impressive effects are triggered by the action of the primary faulting and the resultant near-field shaking. The main faulting associated with large earthquakes does not always extend up to the Earth's surface. For those earthquakes that do rupture to the surface, the surface trace of the fault and the magnitude and direction of the coseismic fault slip can be mapped. The results for many earthquakes can either be viewed as remarkably simple or complex. Simple in the sense that many fault traces are nearly linear over hundreds of kilometers (e.g., the 1906 great San Francisco and 1958 southeastern Alaska earthquakes) and the fault slip is in the same direction, albeit with spatially varying magnitude. Conversely, the results are complex because in some instances major faulting occurs on separate fault traces with highly nonuniform slip, and because even in detail along continuous simple traces, en echelon offsets and local deviations from the overall trend do occur. One consistent observation is that fault slip represents shear dislocation, i.e., the relative slip vector is in the plane of the fault surface. Gaping holes and crush zones are not large-scale features of observed faulting, although these effects can occur as secondary features.

Earthquakes occur on faults that often are major geological faults with dissimilar rock-types on either side of the fault that imply total accumulated displacements of many kilometers, several hundred kilometers in some cases. Earthquakes are part of the process associated with the large-scale motions of the lithosphere. This fact is confirmed by the geographical distribution of earthquakes (seismicity): large earthquakes occur as linear belts along major geological discontinuities. Thus in some cases earthquakes have important implications for geological time-scales, although the duration of earthquake fault slip has a time-scale of seconds to minutes.

Elastic Rebound Hypothesis. A careful study by Reid of the deformation associated with the great 1906 San Francisco earthquake resulted in the elastic rebound hypothesis. Observational evidence for the elastic rebound hypothesis comes from extending the measurements of coseismic displacement away from the fault trace (see examples in Kasahara, 1981). Relative to a reference frame far from the faulted region, the coseismic displacements of a vertical strike-slip fault decrease in magnitude perpendicular to the fault trace (Fig. 1). The strain change of the medium, which is given by the spatial derivative of the displacements, also decreases to zero away from the fault. (Detectable static strain offsets at great distances from an earthquake have been reported). The spatial decrease in coseismic displacement and strain shows that earthquake fault motion represents a localized strain change and the deformational properties of the surrounding earth play an important role. The observed shear strains are on the order of 10^{-4} to 10^{-3} adjacent to the fault zone. Thus, the strain changes associated with the earthquake cycle are infinitesimal and the equations for linear elasticity can be used everywhere except for along the fault zone itself. The characteristic distance of strain decline away from the fault trace is related to the depth of faulting; strike-slip faults typically show a characteristic depth of 10–20 km. As the average shear modulus of crustal rocks is $\approx 3\text{--}5 \times 10^{11}$ dyne/cm^2, the stress drop associated with a strain drop of 10^{-4} is between 10 and 100 bars (1 bar = 10^6 dyne/cm^2).

Measurements of coseismic strain change do not tell us the initial strain state. The elastic rebound hypothesis states that elastic strain slowly accumulates due to the slow relative motion of the lithosphere while some portion of the fault is "stuck." Eventually the fault zone ruptures and the dislocation reduces the elastic strain on both sides of the fault. Figure 1 displays the earthquake cycle as prescribed by the elastic rebound hypothesis. Although the elastic rebound hypothesis also applies to dip-slip earthquakes, observations of the surface deformation present a more complicated pattern. Geodetic strain measurements are an important component of earthquake seismology.

The elastic rebound hypothesis provides a physical model of earthquake occurrence that has been generally accepted: earthquakes are a strain/stress release process. As the shear strain increases due to lithospheric motions, shear stress will also increase throughout the source region. Shear stress at the fault zone eventually causes some localized failure, and a rupture front then sweeps across a certain portion of

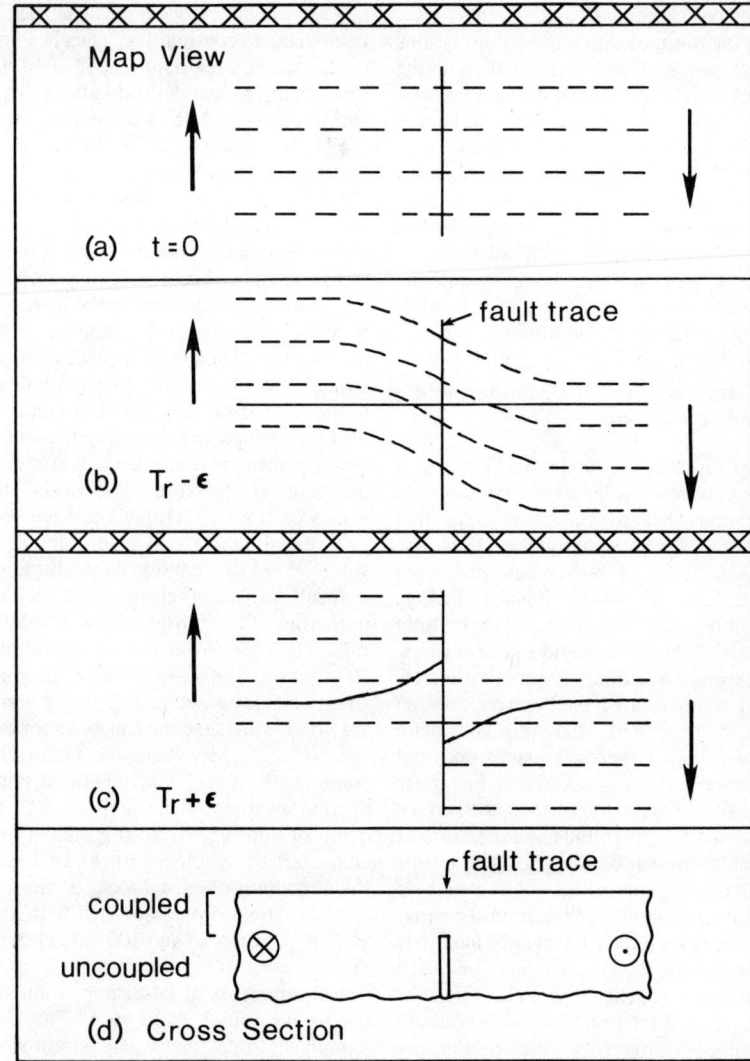

FIGURE 1. Elastic rebound hypothesis and the earthquake cycle. The strain cycle is illustrated with reference geodetic lines (dashed) across the fault trace in map view (a), (b), and (c). The top panel defines the reference zero-strain state just after a large earthquake (schematically represented by the cross-hatching). Time is arbitrarily set to zero in (a), T_r is the earthquake recurrence time, $-\epsilon/+\epsilon$ in (b) and (c) refer to just before/after the earthquake. Continued tectonic motion and aseismic creep at deeper depths slowly concentrate the strain at the fault zone. The fault zone then slips as an earthquake (cross-hatching between (b) and (c)) and the material on either side of the fault rebounds to a zero-strain state, though with a net displacement (compare (c) to (a)). If a new geodetic line is drawn across the fault just before the earthquake (solid line in (b)), then the coseismic deformation will reflect the strain rebound. The characteristic horizontal distance of the deformation is related to the depth of faulting (cross section in (d)).

the fault zone. During earthquake rupture, the fault zone is temporarily "unstuck" and slip is driven by the local level of shear stress. The slip reduces the applied stress level. It is not known whether the final stress level is near zero or remains at a high proportion of the initial stress level. If the strain/stress drop is complete, relative to the last earthquake cycle, then the coseismic displacement will equal the accumulated tectonic displacement since the previous earthquake. Earthquake displacement can thus be related to tectonic displacement. On the other hand, the strain/stress drop may be incomplete and without doubt varies spatially. It is difficult to ascertain the details of earthquake stress drop. The "stick-slip" behavior of earthquake faults is found in other physical systems; various laboratory-scale experiments have attempted to simulate this behavior (see Kasahara, 1981 for references).

Macroseismic observations are available for only a few of the shallow earthquakes, yet they are responsible for many of our ideas on the earthquake process. Earthquakes can occur at depths of ≈ 700 km in the deepest Wadati-Benioff zones associated with subduction. The controlling processes for deep seismicity may indeed be different than those for shallow tectonic seismicity.

From a macroseismic perspective, a larger earthquake consists of a larger faulted area and/or larger displacement. The intensity and duration of the near-field strong shaking generally increase as the fault area and displacement increase. Thus a kinematic measure of earthquake size that is clearly related to the total strain change within the source volume would be some combination of fault area and average displacement. A dynamic measure of earthquake size might be the integrated stress drop. Although earthquakes are a dynamic stress-release process, seismological observations are completely characterized by treating the earthquake as a kinematic process (i.e., the displacement history over the entire fault surface).

Earthquakes and Seismic Waves

Macroseismic observations of earthquake location, faulting geometry, fault length, and static displacements are available for only a trivial percentage of earthquakes. Since it is important to study earthquakes on a systematic global basis, the fundamental earthquake parameters must be determined for all earthquakes above a certain size. Seismic waves radiated by earthquakes provide the opportunity for studies of seismic source systematics.

Seismographic Instruments. Our ability to interpret seismic waves depends in part on the technological level of seismic instrumentation. The development of seismometry is outside the scope of this article (see *Seismic Instrumentation* and *Seismic Instrumentation: History*), but a few basic ideas are presented. Modern seismographic systems can record the waves from earthquakes halfway around the world that are too small to be locally felt. Indeed, one difficulty has been that seismographic systems are too sensitive with restricted dynamic range and consequently are saturated by the occasional large earthquake. This deficiency has been remedied with the use of force feedback systems.

Since there is a peak in the "noise" spectrum at a period between 5 to 10 s, seismologists eventually designed two instrument types that record nonoverlapping period bands on either side of the microseism band (see Fig. 2). These two instrument types define the period bands for long-period and short-period body waves. Seismologists have used instruments that are designed to measure ground acceleration, ground velocity, and strain, in addition to displacement. The earliest very-long period seismographs were a series of strain instruments that Benioff built and deployed. Benioff's seismographs recorded the longest periods observable for seismic waves: These waves are best characterized as the free oscillations of the Earth. Figure 2 shows the period bands that are used in seismological investigations.

Earthquake engineers are primarily interested in the near-field strong shaking that causes structural failure, hence they designed and installed special seismographic systems: accelerographs. These instruments are triggered by strong shaking and then record traces proportional to ground acceleration. These instruments have produced on-scale near-field recordings for several earthquakes, and ground accelerations of 1 g ($\approx 10^3$ cm/s^2) have been observed. As seismic source theory developed, seismologists became interested in these accelerograms for studies of the earthquake source (Aki and Richards, 1980).

The subsequent development of seismic wave studies follows the scheme in Fig. 3. The sequence approximately coincides with the historical development and is also a fairly logical development in that with each step we extract more information from the observed seismograms.

Travel Times and Earthquake Location. After an earthquake has been detected, the event must be

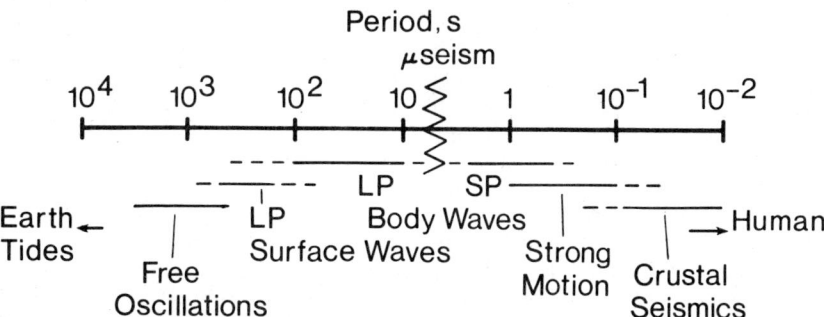

FIGURE 2. Seismic spectrum, the period range of mechanical vibrations relevant to seismological investigations. Frequency increases to the right, period is given in seconds. "Human" refers to the human perception band for acoustic waves, "μseism" shows the microseism noise band that separates body waves into short and long period bands.

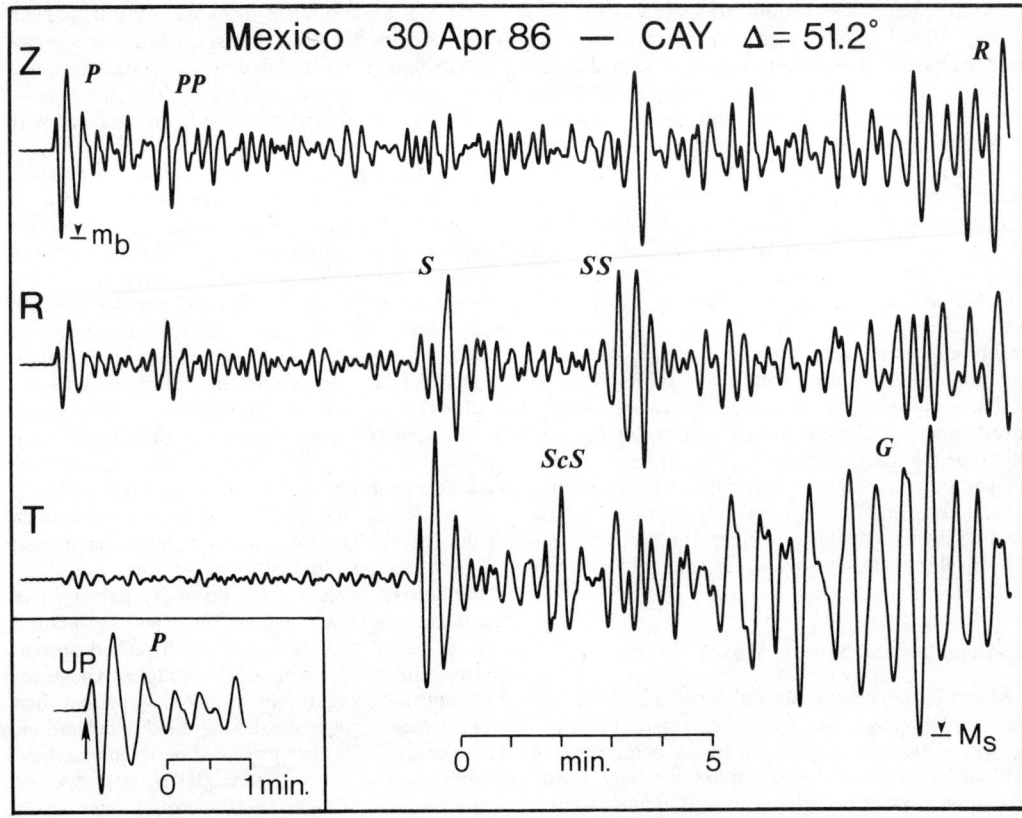

FIGURE 3. Seismograms and some parameters used in seismological analysis of sources. The three traces are the vertical (Z), radial (R), and transverse (T) components from the high-quality broad-band GEOSCOPE station at CAY recording the 30 April 1986 Mexico earthquake ($M_S = 7$). The inset trace is the vertical component P wave plotted at a different time scale (time scales are given in minutes). Various body wave phases (e.g. P and S) are easily identified, as are the Rayleigh (R) and Love (G) surface waves. Earthquake location depends on measurements of the P wave arrival time (arrow on inset trace). Magnitudes depend on amplitude measurements (i.e. m_b and M_S), while first-motion focal mechanisms depend on the initial polarity of the P wave. The compressional arrival is indicated by UP. Seismologists are now using the entire waveforms of selected phases to study earthquakes.

located in space and time, the first step in the analysis of seismic waves. The relative timing of the P wave arrivals provides information on the hypocenter location (point of rupture initiation), though the epicenter (projection of hypocenter onto Earth's surface) is typically better resolved. Global-scale earthquake location at a uniform magnitude threshold requires a global distribution of stations that are all using the same clock (timing errors should be less than 1 s). Hypocentral parameters have been determined routinely by various organizations since the early twentieth century. ISC and USGS seismicity catalogs are readily available. In addition, organizations that operate regional seismic networks are responsible for routine location of local earthquakes (see *Earthquakes: Location Techniques*).

Amplitudes and Earthquake Magnitude. The amplitudes of the radiated seismic waves are presumably related to the strength of the earthquake. Use of wave amplitudes requires a set of stations that are well-calibrated, i.e., the seismic trace is quantitatively and stably related to ground displacement or velocity. The Wood-Anderson instruments installed in Southern California in the 1920s provided a calibrated network in a seismically active area; Richter developed the local magnitude scale based on these recordings. This scale measures earthquake size with the log amplitude of the radiated waves. A critical aspect of the magnitude scale is the distance correction since geometric ray spreading and wave energy attenuation decrease the seismic wave amplitude with epicentral distance (distance between epicenter and station, typically measured in degrees of the angle between the vectors from the Earth's center to the epicenter and the station). Many stations are used to average over the azimuthal effects. The distance corrections vary between different regions even within North America.

Gutenberg and Richter (1949) then extended the concept of a local magnitude scale to a global scale using the amplitudes of both body waves (m_b) and surface waves (M_s). These magnitudes are still routinely calculated for earthquakes on a global scale, although the "correction" factors in the formula have been slightly modified through time.

Earthquake magnitude is clearly an arbitrary measure of earthquake size, but it serves as a widely accepted basis for comparing earthquake size. A significant contribution of Gutenberg and Richter (1949) was their effort to relate earthquake magnitude to a physical quantity: energy release. After a lengthy analysis, they proposed an empirical relationship between magnitude and earthquake energy. Although we now have better quality observations and a more direct theory for seismic energy release, the fundamental geophysical conclusions reached by Gutenberg and Richter are unchanged, e.g., the average annual rate of earthquake energy release is much less than the rate of heat flow through the Earth's surface (see *Earthquakes: Magnitude, Energy, and Intensity*).

Seismic Wave Radiation Pattern: Focal Mechanism. The seismic waves radiated by an earthquake source display systematic azimuthal changes related to the geometry of faulting. The simplest observation of the earthquake radiation pattern is the polarity of the initial P-wave arrival, although the orientation of initial S-wave particle motion is also used to determine "first-motion" focal mechanisms. The connection between shear dislocation geometry and P-wave polarity is schematically illustrated in Fig. 4. Imagine a spherical surface around the earthquake source, then the pattern of constant P-wave polarity divides the spherical surface into four equal-sized quadrants. This pattern can be easily characterized by the two orthogonal "nodal" planes that separate the compressional and dilatational quadrants. These nodal planes are directly related to faulting geometry: one nodal plane coincides with the fault plane while the slip vector coincides with the pole of the other nodal plane (sometimes referred to as the auxillary plane). Since the slip vector polarity is not known by plotting only the two nodal planes, the polarity of the quadrants must be specified. An alternative and compact representation of the focal mechanism is to specify the orientations of the orthogonal pressure and tension axes. The nodal planes and quadrant polarities are easily derived from these two axes. The null axis coincides with the line of intersection between the two nodal planes, and is orthogonal to the pressure and tension axes. Notice that all features of the focal mechanism are symmetric through the center of the focal sphere. Thus, all polarity observations can be plotted on one hemisphere and a stereographic projection can then be used to plot all information onto the equatorial plane.

The faulting geometry can be well-resolved if we

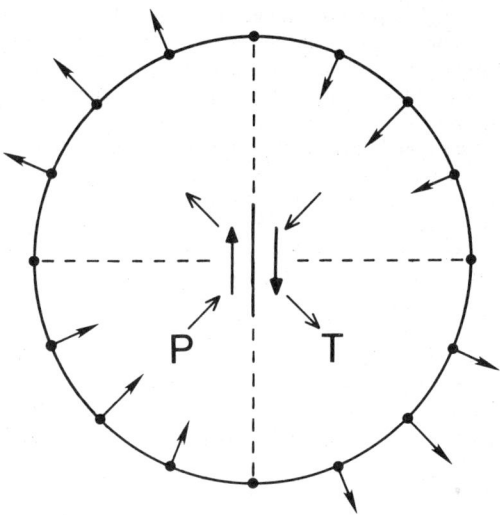

FIGURE 4. Schematic relationship between earthquake source and P wave first-motions. A shear dislocation, with sense of shear given by the bold arrows, generates a spherical P wavefront with a distinct radiation pattern. The initial motion of the P wave is indicated at several places around the wavefront by the inward (dilatational) and outward (compressional) directed arrows. The focal mechanism geometry can be specified either by the two nodal planes (dashed lines), or by the principal axes of compression (P) and tension (T).

observe first-motions from many P waves that completely sample the focal hemisphere. The first-motion data are plotted on the stereographic projection, and the orthogonal nodal planes that separate the compressional and dilatational observations are then found. When using first-motions from globally distributed stations, teleseismic P-wave raypaths all plot near the center of the stereoplot. Thus, nearly vertical nodal planes tend to be well-resolved but shallow-dipping nodal planes are poorly constrained (see *Earthquake Mechanisms*).

To conclude, first-motion mechanisms cannot always uniquely determine the fault geometry of an earthquake, nor can they detect changes in the focal mechanism during rupture duration. Also, polarity observations obviously do not indicate earthquake size or duration. To fully characterize the seismic source, the entire waveform must be used.

Earthquakes and Waveform Analysis

Seismic source theory provides a rigorous quantitative connection between the seismic source and the radiated waves. In order to generate synthetic seismograms that match the observed seismograms, good estimates of the Earth Green's functions must be available. The quantitative connection between source and seismograms allows the development of a formal inverse method to invert waveforms for various characteristics of the seismic source.

Although several techniques of waveform inversion are now highly advanced, the full range of possibilities has not yet been exploited. Waveform inversion will continue as one of the more fruitful research directions.

Physical Model of Earthquake Source. While the physical processes that occur within the fault zone are *not* linear elastic processes, the equations of elastodynamics describe the behavior of the material surrounding the fault zone. Thus to explain the radiated waves, it is only necessary to know the net effects of the fault zone processes as characterized by the tractions and displacement discontinuities applied to the internal surface (i.e., fault zone) of the elastic medium. In terms of the elastodynamic equations, these discontinuities can be replaced by a distribution of body forces within a purely elastic medium. A shear dislocation reduced to a point can be replaced by a double couple at that point. The strength of the double couple is the seismic moment, M_0. In terms of the kinematic fault parameters, the seismic moment is given by: $M_0 = \mu DA$, where μ is the shear modulus, D is the average displacement over fault area, A. The double couple is equivalent to compressional and dilatational force dipoles with equal strength. These dipoles correspond to the pressure and tension axes in Fig. 4. To examine the double couple point source in more detail, the cumulative seismic moment is a function of time: It is zero before the earthquake and achieves the final value when fault slip stops (see Fig. 5). The source time function that generates the teleseismic waves is the moment rate function, $\mathbf{f}(t)$. Note that a double couple is a special case of the possible combinations of force dipoles (see Ben-Menahem and Singh, 1981).

To delve deeper into seismic source theory, teleseismic waves are completely specified by the moment rate density tensor, $m_{jk}(t,x)$, which can vary with time (t) and space (x) throughout the source volume (see *Seismic Source: Theory* for development and discussion). The radiated seismic waves are always related to some integral of $m_{jk}(t,x)$ over the source region. It is consequently not possible to completely determine $m_{jk}(t,x)$ at all frequencies and spatial scales. The most reliable and important source description is the "point source" moment tensor, $\mathbf{M}_{jk}$, which is given by the integral of $m_{jk}(t,x)$ over source volume and time. The moment tensor specifies the overall source "strength" and average source geometry via the overall scaling and the relative size of the elements of $\mathbf{M}_{jk}$, respectively. If the point source moment tensor can be decomposed into just two non-zero orthogonal force dipoles that are equal but opposite in strength, then the overall source is equivalent to the above described point double couple or shear dislocation. For observations at sufficiently long periods, the "point source" moment tensor is the only aspect of the seismic

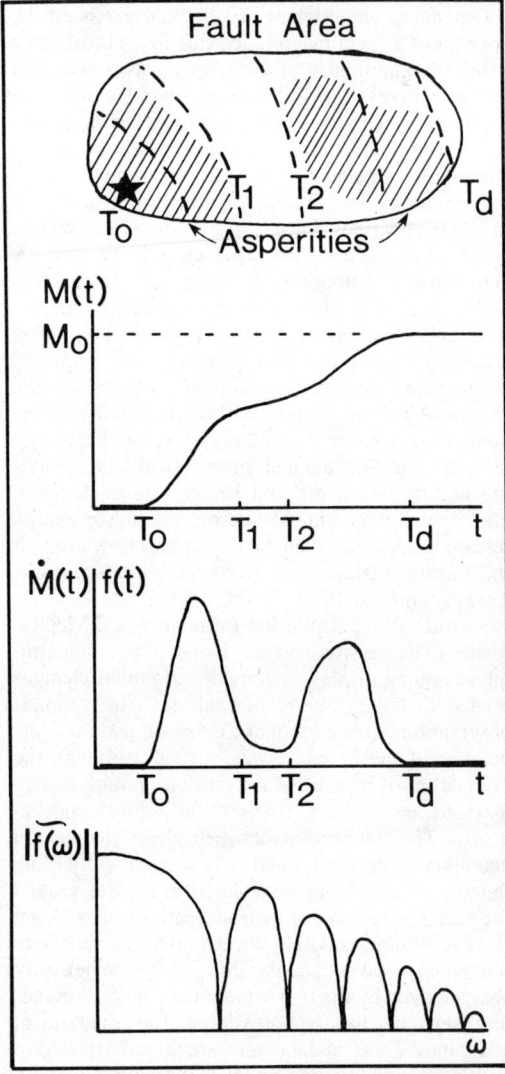

FIGURE 5. Seismic moment release and source time function. "Fault Area" shows a hypothetical rupture history where the rupture front starts at the hypocenter (star) and sweeps across the fault area. The dashed lines show the rupture front at various times. The fault area is heterogeneous, i.e. there are two regions of higher moment release: asperities. In the plot of $M(t)$, the seismic moment accumulates to reach the final value, M_0, at time T_d. Teleseismic waves are generated by the moment rate function, also referred to as the source time function: $\mathbf{f}(t) = \dot{M}(t)$. The integral of the source time function over source duration then returns the overall seismic moment. Note that the two asperities result in a "double-event" time function. Since $\mathbf{f}(t)$ is a one-sided function, the spectral amplitude of $\hat{\mathbf{f}}(\omega)$, the Fourier transform of $\mathbf{f}(t)$, achieves the maximum value at zero frequency. The first spectral zero is related to time function duration, and the high frequency decay asymptote is controlled by the details of moment release.

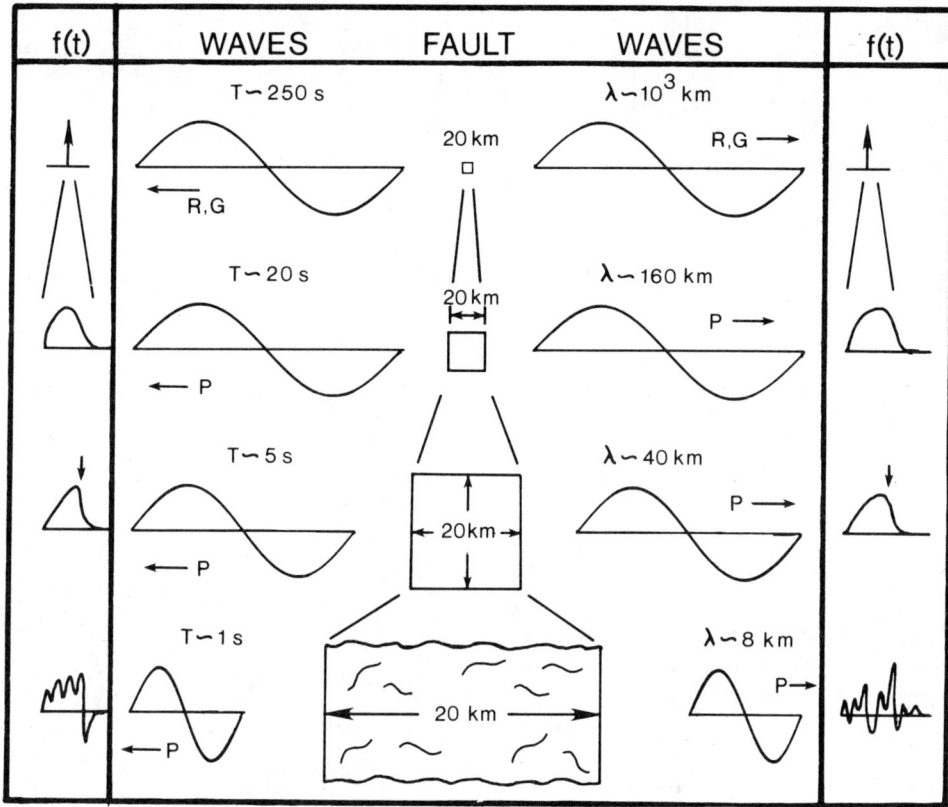

FIGURE 6. View of the earthquake source with different wave periods. Each level schematically shows the fault area and waves with period, T, and wavelength, λ. R, G, and P refer to Rayleigh, Love, and P waves, respectively. At each level, schematic source time functions are plotted for opposite azimuths to illustrate the increased resolution and complexity as the wave period decreases. The earthquake appears to be a point source with an impulse time function when viewed with very long period waves (top). The same earthquake appears to be a complicated finite source when viewed with short period waves (bottom).

source that can be resolved. One task for observational seismology is to use seismic waves to estimate the relative sizes of the force dipoles in $\mathbf{M}_{jk}$.

To understand a seismologist's view of the earthquake source, the best perspective is to "look" at a particular earthquake as a function of the seismic wave period (see Fig. 6). Akin to changing the magnification power of a telescope, we can "zoom in" on the earthquake by changing to shorter wave periods. First we prescribe the object to be examined: let the earthquake be a typical magnitude 7–7.5 event, with a fault dimension of ≈ 20 km and the faulting duration will be a few seconds. Since surface waves with a wave period of 250 s have a wavelength of ≈ 1000 km, the earthquake looks like a point source with respect to these very-long period waves. With a sufficient number of observations at 250 s, the seismic moment and average fault geometry of the earthquake can be reliably determined. Examples of these waves are illustrated in Fig. 7 by the multiple-orbit Rayleigh and Love waves from the 25 May 1981 Macquarie Ridge earthquake. We will now "zoom in" by looking at the earthquake with P waves in the long period band. In the period range of a few seconds to tens of seconds, we can resolve the primary feature of $\mathbf{f}(t)$, the duration of faulting (see second level in Fig. 6, and example in Fig. 8). Zooming down to periods of a few seconds, consistent azimuthal variations in the apparent duration of $\mathbf{f}(t)$ provide information on the overall length scale of faulting (third level in Fig. 6 and Fig. 8). The short period waves, periods of one second or less, with a wavelength equal to and less than the fault dimension "zoom in" on the space-time moment release variations within the fault area (see final level in Fig. 6 and example in Fig. 9). In addition to the source complications at shorter periods, the Green's functions are poorly known. Thus, seismograms recorded at different periods proceed from azimuthally coherent at the longest periods to quite complicated, maybe even incoherent, at the shortest periods.

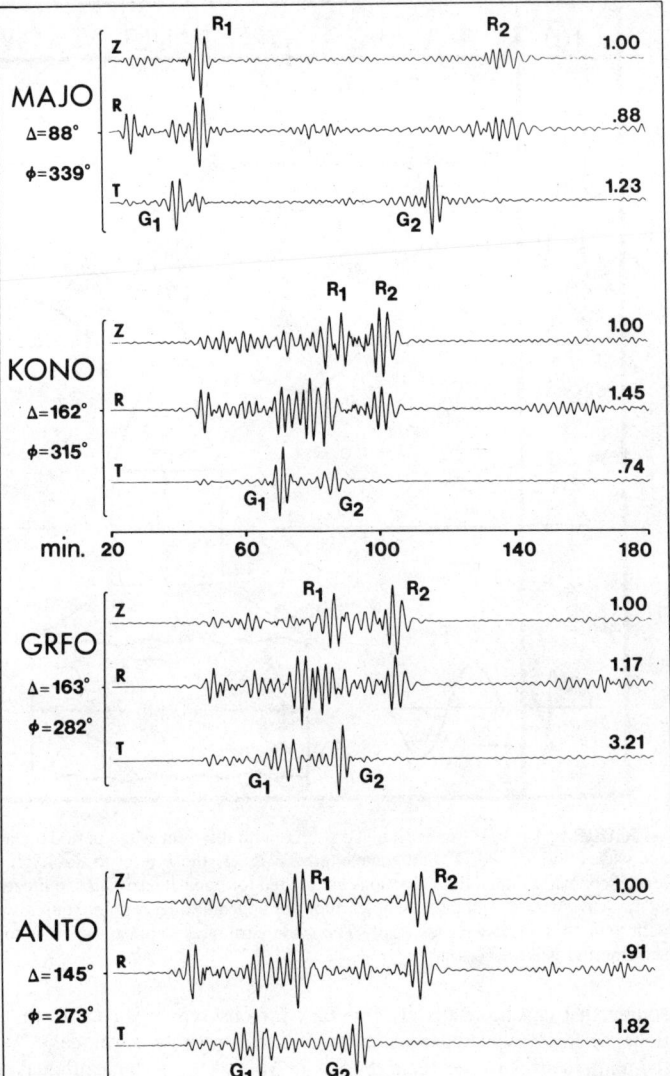

FIGURE 7. Long period surface waves of the 25 May 1981 Macquarie Ridge earthquake (M_S = 7.6). Three component long period GDSN seismograms are rotated into vertical (Z), radial (R), and transverse (T) components and filtered to enhance waves with a period of more than two minutes. Epicentral distance and azimuth are given for each station, the numbers at right refer to the relative amplitudes between the components. Time scale is minutes after origin time. Multiple-orbit Rayleigh (R1, R2) and Love (G1, G2) waves are well-recorded. The earthquake appears to be a point source to these waves. The relative amplitudes of the R and G phases constrain the faulting geometry to a right-lateral strike-slip mechanism with a fault strike of N30°E.

At the present, waveform analysis can reliably determine the seismic moment tensor and faulting duration, and in some cases it is possible to determine faulting dimension and long-wavelength variations in the spatial moment release. The variations in spatial moment release can be related to interpretative frameworks of earthquake occurrence such as the asperity model (see Kanamori, 1986).

Seismic Moment and Earthquake Scaling. Progress in seismic moment determination is one of the important recent advances in seismology. From the initial determination of seismic moment for the 1964 Niigata, Japan earthquake by K. Aki in 1966, the point source moment tensor and faulting duration are now systematically estimated for earthquakes with magnitude greater than 5, and the results are routinely published in the PDE and ISC catalogs.

The seismic moment tensor has joined hypocentral parameters and magnitudes as standard earthquake parameters (see Dziewonski and Woodhouse, 1983).

Seismic moment is the fundamental measure of earthquake size and it is particularly important to determine seismic moment for the largest earthquakes given the saturation of the M_s scale for large earthquakes. Comparative studies of global-scale moment release depend on reliable estimates of seismic moment. Kanamori and coworkers have estimated the seismic moment of nearly all great earthquakes in the twentieth century. Kanamori devised a new magnitude scale, M_w, that depends directly on seismic moment. Hence, the M_w scale does *not* saturate for the largest earthquakes. Kanamori defined the M_w scale to smoothly connect to the M_s scale at around magnitude 7.5 to 8.

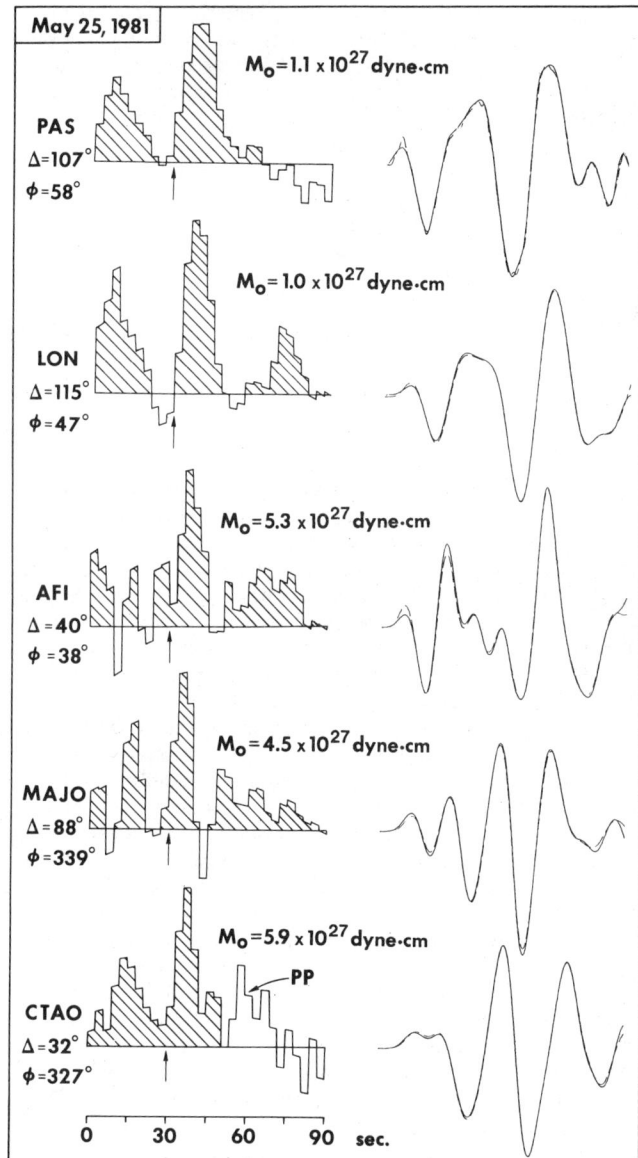

FIGURE 8. Long period P waves and source time functions of the 25 May 1981 Macquarie Ridge earthquake. Observed P wave seismograms are the solid traces at right, and station epicentral distance and azimuth are listed. Source time functions are deconvolved from the P waves and show a distinct "double-event" character. Second event initiation is indicated by arrows. Synthetic seismograms generated by the time functions are the dashed traces that overlay the observed seismograms. Some basic features of the earthquake moment release can be seen in these long period time functions.

Kanamori and coworkers also investigated the relationship between M_0, fault length or area, and the faulting duration for many earthquakes (see Kanamori, 1986, for references). While seismic moment and duration are readily determined from the seismic waves, most estimates of fault area are based on aftershock areas. One key conclusion, based on the strong positive correlation between $\log(M_0)$ and $\log(A)$, is that the static stress drop is nearly "constant," between 10 and 100 bars. (Note that this roughly agrees with the few macroseismic determinations of static stress drop.) There is of course considerable scatter about this order-of-magnitude "constant." There is also considerable debate as to whether there are systematic variations from the constant stress drop trend, i.e., the largest earthquakes have greater/lesser stress drops, or strike-slip events have greater/lesser stress drops. A fundamental limitation is that static stress drops are poorly determined. The static stress drop can be written in a generic form as:

$$\Delta\sigma = c\, M_0/A^{3/2}$$

where A is the characteristic fault area and c is a geometric shape factor of order one. The uncertainty in the estimates of M_0 based on long period waves is probably less than a factor of two; the uncertainty

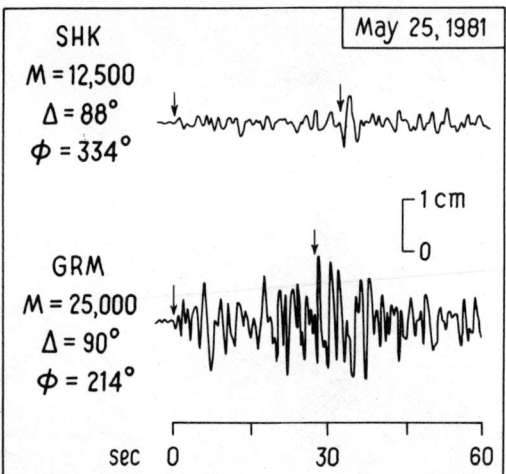

FIGURE 9. Short period P-wave seismograms of the 25 May 1981 Macquarie Ridge earthquake. The epicentral distance, azimuth, and magnification are listed for these WWSSN stations. Amplitude scale refers to original record amplitude for the listed magnifications. With this high resolution view of the earthquake, it is difficult to unambiguously correlate features of moment release; though the large amplitude second event initiation (arrows) appears as a coherent feature.

in A is difficult to ascertain, but can exceed a factor of two for small events. Estimates of $\Delta\sigma$ are thus uncertain to a factor of two even for large earthquakes, and this uncertainty can easily approach a factor of ten for smaller events.

Various estimates of the "dynamic" stress drop can be calculated by using the seismic moment in conjunction with some properties of the source time function, say duration or peak value (see *Seismic Source: Theory* for dissussion of the Brune stress drop). Dynamic stress drops can be determined solely from observations of seismic wave source time functions, though these stress drops do not possess the obvious interpretation as does the static stress drop. Overall, most calculated stress drops lie between ten bars and a few hundred bars, and intraplate events tend to display a higher stress drop than plate boundary events. The determination and interpretation of stress drops will continue as a primary activity in earthquake studies. Seismologists cannot yet reliably determine the actual variation in stress release across a ruptured fault.

Short Period Observations. Although short periods cannot be deterministically inverted for rupture process variations, short period radiation can be statistically characterized. Rather than look at the details of spectral amplitude and phase, the short period spectrum can be characterized by the slope of spectral decay. This high-frequency asymptote has a log slope variously estimated between -1 and -3. The latter slope would indicate a "smoother" rupture process. The asymptotic slope is related to the spatial heterogeneity spectrum of the rupture process, and this spatial heterogeneity is presumably related to the occurrence of smaller earthquakes on the same fault as the large earthquake. Thus there could be a correlation between background seismicity and the high frequency characteristics of a large earthquake. Some investigators are using the high frequency observations to infer the character of the earthquake rupture front. Another important reason to investigate scaling relations for the high-frequency asymptote is that the frequencies relevant to earthquake engineering lie in the high-frequency part of the spectrum for large earthquakes.

Summary

Seismic sources can be observed by a broad range of physical measurements. Despite the importance of macroseismic and geodetic observations, most earthquake research is based on recordings of seismic waves. Historically, seismologists first determined hypocentral parameters based on travel times, then magnitudes and focal mechanisms based on wave amplitudes and polarities, respectively. Progress in seismic source theory and instrumentation provide the potential for a complete kinematic description of seismic sources based on broad-band observations of the entire waveform. To date, the seismic moment and earthquake duration can be reliably and systematically determined. One active research trend is to "look inside" the earthquake source by using long to mid-period waves. Short period waves are statistically related to rupture and fault zone heterogeneity, and are extremely important to characterize for earthquake engineering purposes. A detailed view of the stress release mechanism must wait until reliable determinations of the spatial variations in stress drop are available. Future discoveries will undoubtedly continue to stem from the interaction between interpretative frameworks, seismic source theory, and new and improved observations.

Acknowledgments

M. Schell helped prepare several figures, and B. Romanowicz kindly provided GEOSCOPE data. The earthquake research program at the University of Michigan has been supported by grants from the National Science Foundation (EAR8351515 and EAR8720935).

LARRY J. RUFF

References

Aki, K., and P. Richards, 1980, *Quantitative Seismology, Theory and Methods*. San Francisco: W. H. Freeman & Company, 557p.
Ben-Menahem, A., and S. J. Singh, 1981, *Seismic Waves and Sources*. New York: Springer-Verlag, 1108p.

Dziewonski, A. M., and J. H. Woodhouse, 1983, Studies of the seismic source using normal-mode theory, in H. Kanamori and E. Boschi, eds., *Earthquakes: Observation, Theory, and Interpretation.* New York: Elsevier North-Holland, 45–137.

Gutenberg, B., and C. F. Richter, 1949, *Seismicity of the Earth.* Princeton, N.J.: Princeton University Press, 273p.

Kanamori, H., 1986, Rupture process of subduction zone earthquakes, *Ann. Rev. Earth and Planetary Sci.* **14**, 293–322.

Kasahara, K., 1981, *Earthquake Mechanics.* Cambridge: Cambridge University Press, 248p.

Richter, C. F., 1958 *Elementary Seismology.* San Francisco: W. H. Freeman & Company, 768p.

Cross-references: *Earthquake Mechanisms; Earthquake Mechanisms and Plate Tectonics; Earthquakes: Magnitude, Energy, and Intensity; Earthquakes and Crustal Deformation; Earthquake Seismology; Free Oscillations of the Earth; Seismic Instrumentation; Seismic Instrumentation: History; Seismic Monitoring of Nuclear Explosions; Seismic Source: Theory; Seismology: History; Surface Waves.*

SEISMIC SOURCE: THEORY

Geological, geodetical, and seismological evidence indicates that practically all shallow earthquakes are due to the fast propagation of a shear fracture on one or more well-defined fault surfaces. These faults have dimensions ranging from a few meters to hundreds of kilometers, depending on the size of the seismic event. Earthquakes occur in the colder sections of the lithosphere in response to slow but continuous accumulation of tectonic stresses due, in most places, to the relative motion of lithospheric plates. Earthquakes are usually localized at the boundaries between plates or on ancient zones of mechanical weakness. It has been determined from geodetic measurements that the deeper part of tectonic zones creeps continuously in response to plate motion. At the relatively cold temperatures that prevail at shallow depths, the lithosphere cannot deform sufficiently fast in response to plate motion, so stresses concentrate in the shallower layers of the lithosphere until they reach a certain threshold, after which continuous stress accumulation becomes unstable. Once the instability is reached, the earthquake occurs in the form of very rapidly spreading episode of slip on the fault surface. The main source of seismic radiation during an earthquake is the slip or dislocation of the fault, which for large earthquakes can reach several meters in a few seconds.

Seismic Radiation

Seismic radiation is calculated by using Green's functions, which are solutions of the elastodynamic wave equation for a point force $\mathbf{f}(t)$. Let $\mathbf{r}_0$ be the location of the point force. The radiation from the point force is then written in the form

$$u_i(\mathbf{r}, t) = \int_0^t G_{ij}(\mathbf{r}, t|\mathbf{r}_0, \tau) f_j(\tau)\, d\tau$$

where the tensor $\mathbf{G}$ is the Green's function. Computation of $\mathbf{G}$ is relatively simple for a homogeneous medium, but is extremely complicated for realistic heterogeneous models of the Earth. In the following we assume that it has been computed by one of the numerical or approximate methods described elsewhere in this encyclopedia.

In general, Green's functions contain two types of terms: near field and far field. The far-field terms are the usual P and S waves of classical seismic ray theory and surface waves channeled along the shallow Earth's waveguide. Far-field Green's functions present a characteristic dipolar symmetry in which the direction of the point force is the axis of the dipole. Near-field waves, on the other hand, are observed in a very small vicinity of the source at distances of about a wavelength. They are due to diffraction around the point source. The near-field waves decay faster with distance than do the classical body and surface waves, so their effect is important only in the study of long-period radiation in the vicinity of the source.

The Green's functions may be used to determine the radiation from a generalized distribution of body forces by means of the representation theorem:

$$u_i(\mathbf{r}, t) = \int_0^t \int_{V_0} G_{ij}(\mathbf{r}, t|\mathbf{r}_0, \tau) f_j(\mathbf{r}_0, \tau)\, dV_0\, d\tau \quad (1)$$

where V_0 denotes the region where body forces are distributed. The representation theorem, Eq. 1, may be extended to include forces applied on the boundary of the Earth or at internal discontinuities or faults. However, boundary forces are not usually involved in earthquake generation, and internal boundaries are more naturally introduced by defining appropriate moment-tensor distributions.

Seismic Moment Tensors

The most straightforward method to determine seismic radiation is to construct equivalent models of the seismic source. Instead of describing an earthquake by a detailed analysis of the slip on a fault, we *represent* the seismic event by a distribution of multipolar sources called *moment tensors*. Such an approach is objective since no assumption is made about the details of the rupture process, not even the existence of a fault plane.

In order to introduce moment tensors, we notice that seismic sources are of internal origin, so the externally applied point force used to generate Green's functions is not a mechanically acceptable representation of a seismic source. An acceptable source model may be constructed by requiring that the equivalent body force distribution satisfy the conditions

$$\int_V \mathbf{f}(\mathbf{r}) \, dV = 0, \quad \int_V \mathbf{r} \times \mathbf{f}(\mathbf{r}) \, dV = 0 \quad (2)$$

where V is the volume of the source. The first condition states that the resultant of the body force distribution is zero; the second states that the total torque of this force distribution is zero. An equivalent body force distribution that satisfies these two conditions is acceptable because it does not affect the motion of the Earth as a whole, which is the basic condition for the source to be of internal origin. A body force distribution that satisfies the conditions of Eq. 2 derives from the divergence of a symmetric tensor:

$$\mathbf{f}(\mathbf{r}, t) = -\nabla \cdot \mathbf{m}(\mathbf{r}, t)$$

where $\mathbf{m}$ is the seismic moment-tensor density. Moment tensors have the same symmetries as stress tensors. In fact, the moment tensor represents the inelastic stresses responsible for the generation of seismic waves; for this reason, some authors have proposed calling it the stress glut.

When an earthquake is observed in the far field and the wavelengths of the seismic waves are longer than the overall dimensions of the source, we can replace the seismic moment distribution by a concentrated source:

$$\mathbf{m}(\mathbf{r}, t) = \mathbf{M}(t)\delta(\mathbf{r} - \mathbf{r}_0)$$

where

$$\mathbf{M}(t) = \int_{V_0} \mathbf{m}(\mathbf{r}_0, t) \, dV_0$$

is the point moment tensor of the earthquake and $\mathbf{r}_0$ is the centroid of the moment-tensor distribution. The point moment tensor is a measure of the strength of the seismic source, which is independent of the details of the rupture process. Hence, it provides a very practical method for quantifying seismic sources. With the steadily increasing availability of three-component digital instruments, the point moment tensor has replaced magnitude as the most common means of measuring earthquakes.

Radiation of seismic waves from a point moment tensor is given by

$$u_i(\mathbf{r}, t) = \int_0^t \frac{\partial G_{ij}}{\partial r_k} (\mathbf{r}, t|\mathbf{r}_0, \tau) m_{jk}(\tau) \, d\tau \quad (3)$$

where summation over repeated indices is implicit. Far-field body waves radiated by a point moment tensor are characterized by a quadrupole geometry because, as shown by Eq. 3, radiation is proportional to the derivative of the Green's functions,

which themselves had a dipolar geometry. Since moment tensors are symmetrical (i.e., $m_{jk} = m_{kj}$), Eq. 3 may be conveniently rewritten in the form

$$u_i(\mathbf{r}, t) = \int_0^t H_{ijk}(\mathbf{r}, t|\mathbf{r}_0, \tau) m_{jk}(\tau) \, d\tau \quad (4)$$

where $H_{ijk} = \frac{1}{2}[G_{ij,k} + G_{ik,j}]$ is usually called the Green's function for a seismic point source.

Different elements in a point moment tensor may be interpreted in terms of force multipoles. Diagonal terms in $\mathbf{M}$ represent linear dipoles—two elementary coaxial forces oriented in opposite directions. They are equivalent to pure tension or compression in a stress tensor. The off-diagonal terms always appear in pairs because of the symmetry of moment tensors. They represent a plane quadrupole—i.e., two force couples that rotate in opposite directions. They correspond to pure shear in a stress tensor. This model applies for most earthquakes and is usually called the *double-couple source model*. In this model all elements in the moment tensor are zero except for a pair of off-diagonal terms. The size of these terms is called the *scalar seismic moment* of the earthquake. Radiation from a double-couple source model has a typical quadrupole form for both P and S waves. The quadrupole for P waves has two orthogonal nodal planes on which the amplitude of the P waves is zero. It has been verified with thousands of earthquakes that this pattern is the one that best matches P-wave radiation.

Radiation from extended sources may be calculated by summing the radiation from point sources. The representation theorem, Eq. 1, adapted to moment tensor sources is

$$u_i(\mathbf{r}, t) = \int_0^t \int_{V_0} H_{ijk}(\mathbf{r}, t|\mathbf{r}_0, \tau) m_{jk}(\mathbf{r}_0, \tau) \, dV_0 \, d\tau$$

where V_0 is the source volume.

Dislocation Model of Faulting

Most earthquakes are due to *faulting*, which may be described as time-dependent slip occurring across well-defined fault surfaces. Hence, a classical way of modeling an earthquake source is to define a dislocation or displacement discontinuity at an internal fault surface. Dislocation sources are easily described in terms of moment-tensor density distributions. A *dislocation* is simply a seismic moment distribution in the form of a very thin sheet:

$$\mathbf{m}(\mathbf{r}, t) = \mathbf{m}_s(\mathbf{r}, t)\delta[\mathbf{r} - \mathbf{r}(\Sigma)] \quad (5)$$

where $\delta(\mathbf{r})$ is Dirac's delta function and Σ is the fault surface. $\mathbf{m}_s$ is a surface density of moment release that measures the strength of seismic radiation from a point on the fault surface.

The seismic moment representation, Eq. 5, does

not contain slip explicitly; $\mathbf{m}_s$ may be related to slip on individual fault segments, considering the inelastic deformation that occurs in a very thin layer of width dH centered around the fault surface σ. For simplicity of notation we assume that the fault is plane and choose a coordinate system such that the z-axis is normal to the fault. Let us consider a cylindrical volume of thickness dH centered around a surface element $d\Sigma$. The inelastic strain ϵ inside this volume is related to the seismic moment density by

$$\mathbf{m}_s = \mu\epsilon\, dH \tag{6}$$

where μ is the elastic rigidity. The inelastic strain inside the elementary volume may be easily calculated in terms of the slip across the height dH. Assuming for simplicity that only the component Δu_x of slip is active, the only inelastic strain components that are different from zero are $\epsilon_{xz} = \epsilon_{zx} = \Delta u_x/dH$. Inserting this value of ϵ_s into Eq. 6, we get

$$(\mathbf{m}_s)_{xz} = (\mathbf{m}_s)_{zx} = \mu\,\Delta u_x \tag{7}$$

all other components of the surface moment tensor density being zero. Thus, the surface moment density of a fault is just the slip on the fault times the rigidity. This result was originally obtained by using Eq. 1, including internal discontinuities.

Radiation from dislocation models may be calculated by using Eq. 4. An explicit form of the radiation from a plane fault lying on the plane (xy) is

$$u_i(\mathbf{r}, t) = \int_0^t \int_\Sigma \mu(\mathbf{r}_0)\,\Delta u_x(\mathbf{r}_0, \tau) H_{ixz}(\mathbf{r}, t|\mathbf{r}_0, \tau)\, d\Sigma\, d\tau$$

where $\mathbf{r}_0$ is a vector on the fault plane. Radiation is controlled by the time-dependent distribution of slip on the fault.

The dislocation model in Eq. 7 is too general: Any arbitrary slip distribution on a fault may be described in this form. However, not every dislocation distribution is physically acceptable. For instance, Volterra dislocations—constant slip across the fault—are unacceptable because they produce interpenetration of matter at the edges of the fault. Additional restrictions on the distribution of slip on the fault may be obtained from the physics of rupture. The most obvious of these restrictions is that slip should be tapered near the edges of the fault surface, a condition naturally incorporated in the dynamic models discussed in the next section. A more compelling reason for considering simplified dislocation geometries is that observations cannot resolve all the details of rupture, so some a priori information is absolutely necessary to invert slip from observations.

As rupture develops, the equivalent total seismic moment of the fault,

$$M_0(t) = \mu \int_\Sigma \Delta u_x(\mathbf{r}_0, t)\, d\Sigma \tag{8}$$

increases and, once rupture comes to a stop, eventually reaches a static value usually designated M_0, the total seismic moment of the source. From Eq. 8 it can be seen that the total seismic moment for a dislocation source is simply the product of the elastic rigidity of the rocks surrounding the fault with the total slip $\Delta u(\mathbf{r})$ integrated on the fault plane. Seismic moment is a measure of the average slip on the fault times the total source area. Seismic moments for the largest earthquakes ever recorded are about 10^{23} N · m. Large events on island arcs have moments from 10^{20} N · m to 10^{22} N · m

Dynamic Source Models

The dislocation model as well as the seismic moment distributions are kinematic descriptions of the source processes. No information has been incorporated about the forces or stresses responsible for faulting. Dynamic models take into account the process of stress release that leads to fault slip. Faulting is due to a frictional instability in which the traction that holds fault surfaces together is rapidly reduced. Once the instability appears in a certain region of the fault, it will spread rapidly on the fault surface. The rupture, or shear crack, spreads with velocities that are comparable with those of the elastic waves in the medium. The frictional instability responsible for seismic faulting is sometimes called *stick-slip* by analogy with the intermittent slip observed in laboratory friction experiments. It is very likely that the same frictional instability responsibie for stick-slip in the laboratory occurs in an earthquake fault, but at much larger scale.

From a mechanical point of view we describe the earthquake as the reduction or drop in traction on the fault plane. Before the earthquake the fault plane is held together by spatially heterogeneous static friction. As the applied stress due to plate motion increases, the static friction at some point on the fault is eventually reached and slip starts to develop. In fact, slip probably starts at a very low rate before the maximum friction point is attained, and continues as accelerated creep once friction begins to decrease leading to the onset of instability. Once the instability appears, traction drops rapidly to kinematic friction and stays at this level until healing occurs on the fault. The difference between the applied stress before the earthquake and the kinematic friction is called the *stress drop*, which is the most important dynamic source parameter because it controls the total slip on the fault and seismic radiation in general.

The fundamental problem of crack or fault dynamics is to determine the slip function Δu corresponding to a given distribution of stress drop on the fault. Although a direct correlation between stresses and slip is expected, their detailed relationship is very sensitive to the geometry of faulting. To illustrate fault dynamics, we consider a very simple static circular shear crack model. Assuming a uniform stress drop $\Delta\sigma$ inside the fault, the total slip as a function of radius is

$$\Delta u(r) = \frac{24}{7\pi} \frac{\Delta\sigma}{\mu} \sqrt{a^2 - r^2}$$

for $r < a$, where a is the overall radius of the fault. This slip function has a characteristic elliptical shape, with an abrupt but continuous decrease of slip near the edge of the fault. This behavior of slip near the fault edge is a universal property of crack models with constant stress drop. The total seismic moment associated with this fault model is

$$M_0 = \tfrac{16}{7} \Delta\sigma \, a^3 \qquad (9)$$

Thus, the seismic moment is proportional to stress drop and to the cube of the source radius. This relationship, strictly valid for circular shear cracks, depends relatively weakly on fault geometry. It is possible to introduce correction factors for ellipticity, etc. These changes will affect only the numerical coefficient, not the linear dependence on slip and on the cube of the linear dimensions of the fault.

The circular shear-crack model has been applied to the interpretation of many earthquakes. Usually, the seismic moment is retrieved from body or surface wave studies, and the source radius is determined from body wave modeling, aftershock studies, or surface faulting. From Eq. 9 stress drop is estimated. Results from these studies are that M_0 and a^3 seem to be linearly related over a very wide range of seismic moments and that the regression coefficient depends very weakly on the region of the Earth where the relationship is being established. These results lead to the conclusion that to an accuracy of a factor of 2 or 3 stress drops are almost independent of earthquake size. Stress drops determined in this form appear to be very low, about 1–10 MPa, which is to be compared with shear strengths of 100 MPa for rocks in the laboratory. Two possible explanations of this result are that fault zones in the Earth are either very weak or that stress drop is a very small fraction of the ambient shear stress in the fault zone.

High-Frequency Seismic Radiation

The static model considered so far provide a very satisfactory description of low-frequency seismic radiation. At waves about the length of the overall source dimensions or shorter, the details of the rupture process control radiation. Determining the time history of slip is a difficult mathematical problem. Even for the simple model of a circular rupture growing self-similarly from the nucleation point to a final radius a, there is no analytical solution, and the problem has to be solved numerically. Several numerical methods have been used to generate solutions for dynamic crack models, of which finite differences and boundary integral equations are the most common. It appears from these simulations that the total duration $\langle T \rangle$ of a seismic event is approximately given by $\langle T \rangle = L/v_R$, where L is the characteristic linear dimension of the source and v_R is the average rupture velocity, which for most earthquakes is very close to the shear wave velocity. Duration controls the spectral width of seismic radiation. Radiation is approximately frequency independent at frequencies lower than a characteristic value ω_0, called the *corner frequency*, which is inversely proportional to duration and, consequently, to the linear dimension of the fault. The empirical relation

$$\omega_0 = 2.34 \frac{\beta}{L}$$

proposed by Brune, is used by most seismologists to interpret spectral observations in terms of source dimensions. At frequencies higher than ω_0, radiation is controlled by the details of the propagation of the rupture front. In simple computer simulations the strongest high-frequency waves are produced by the arrest of rupture. Rupture arrest may occur in two principal ways. First, rupture enters into a previously ruptured zone, and stress drop reduces to zero or may even become negative, (i.e., stress increases as the fault grows). Rupture in this case would stop smoothly, and it would produce weak stopping phases. The other possible stopping mechanism is that rupture runs into a physical or geometrical barrier like a fault kink or a material boundary between different rock species. In this case, fault arrest will be very abrupt, and strong stopping phases will be emitted by the stopping crack. Both mechanisms of rupture arrest probably occur in the Earth, although the second is probably very common since most observed spectra present evidence of abrupt rupture stop.

At high frequencies, seismic waves become increasingly incoherent, indicating that fault slip and stress drop are very heterogeneous in space. The study of the small-scale complexity of seismic sources is one of the most active fields of research in earthquake seismology. Understanding of this complexity is essential for predicting the acceleration waves that produce structural damage during earthquakes.

RAUL MADARIAGA

References

Aki, K., and P. G. Richards, 1980, *Quantitative Seismology*, vol. 2. San Francisco, Cal.: Freeman.
Boatwright, J., S. Das, and C. Scholtz, eds., 1986, *Proceeding of the Fifth Ewing Symposium*. Washington, D.C.: American Geophysical Union.

Brune, J. N., 1970, Tectonic stress and the spectra of seismic shear waves from earthquakes. *Jour. Geophys. Research* **75**, 4997–5009.

Kasahara, K., 1981, *Earthquake Mechanics*. Cambridge, U.K.: Cambridge University Press.

Madariaga, R. I., 1983, Earthquake source theory: A review, in *Earthquakes: Theory and Interpretation*. Rendiconti S. I. F., Course LXXXV, Bologna, Italy.

Rice, J. R., 1980, The mechanics of earthquake rupture, in A. M. Dziewonski and E. Boschi, eds., *Physics of the Earth's Interior*, Proceedings International School of Physics "Enrico Fermi." Amsterdam: North Holland.

Cross-references: *Earthquake Mechanisms; Earthquakes: Magnitude, Energy, and Intensity: Earthquake Seismology; Elasticity and Wave Propagation; Seismic Source: Observations; Seismology: History.*

SEISMIC TOMOGRAPHY

Seismic tomography is the technique of imaging zones of complex, heterogeneous velocity and/or attenuation structure of the Earth by using seismic waves. Seismic waves from natural sources (earthquakes) and artificial sources (explosions and Vibroseis) can be used in seismic tomography. The word "tomography" was coined by medical researchers and is derived from "tomograph" (meaning "slice picture"), a high-resolution, two-dimensional x-ray image of the part of a patient's body under examination. Several adjacent tomographs are stacked together to get a three-dimensional image of the body to identify tumors and other abnormalities. Seismic and medical tomography bear strong basic similarities. In both techniques, many crisscrossing rays from several sources illuminate the object from different directions and are recorded by a bank of receivers. In seismic imaging, in the present state-of-the-art, the variation in seismic velocities yields an image of the object, whereas in medical imaging (also called *computer-aided tomography* or *CAT scan*) variation in x-ray attenuation provides the image. Imaging of seismic attenuation anomalies, though theoretically feasible, is more complicated in practice than imaging velocity anomalies, and only a very few well-constrained studies on attenuation tomography (not discussed here) are available in the literature in contrast with the numerous published studies on velocity tomography. Because of the controlled locations of source and receiver and simplicity of wave propagation outside the anomalous body, medical tomography is far simpler than seismic tomography. In the seismic case, the source locations (if earthquakes) are not accurately known, often there is little control on locating receivers for optimal ray coverage, and the raypaths outside the volume being imaged are complex and sometimes nonlinearly dependent on the structure. On the positive side, however, the availability of a multiplicity of wave types in seismic tomography enables imaging of structures of diverse dimensions, from a few hundred meters to a few hundred kilometers in size.

Seismological use of the term "tomography" refers to a back-projection scheme in which a ray is traced back to its source and the observed "signal," be it travel time or amplitude deviation, is distributed along the ray path by a mathematical inverse procedure. The superposition of several such distributions reconstructs the image. In general, any technique that yields an image of an object by inversion of data generated by an experiment can be termed *tomography*. By this definition, the first x-ray picture or the first Earth model ever produced are strictly tomographic images. In this article, however, I will confine the discussion to the major new techniques used to delineate small-scale and large-scale two- and three-dimensional velocity anomalies. I have not discussed seismic reflection and refraction techniques because they do not generally use inverse procedures to construct the image. (For a discussion of these techniques see *Exploration Seismology* and *Controlled Source Seismology*.)

Historical Notes

The two major types of data used in modern seismic tomography, namely, body wave travel times and surface wave velocities, have been in use for several decades to construct regional and global Earth models. In the past, however, these studies basically yielded one-dimensional models in which P or S velocities were specified as a function of depth. Comparison or combination of several one-dimensional models provided information on the extent of the lateral heterogeneities. Such studies have provided a wealth of information on gross differences in Earth structure between continents and oceans and between diverse tectonic environments. In the related field of crustal structure studies, particularly as applied to oil exploration, data from multiple reflection and refraction profiles are the primary source of information to synthesize three-dimensional models of the heterogeneous structure of the Earth's crust.

The study of the three-dimensional velocity structure of the Earth received a big impetus around the middle of this century when it was recognized that seismic body waves from distant earthquakes seldom reached recording stations exactly at times predicted by available Earth models. The difference between the observed and predicted travel times, designated the *travel time residual*, could reach values of several seconds and was found to be a function of the geographic location of the seismic stations, epicentral distance of the earthquakes, and azimuth of the seismic waves. It was soon realized that the travel time residuals could be caused by lateral and vertical variations in Earth structure anywhere along the raypath: in the earthquake source regions, beneath the recording stations, and the raypaths in between. In the 1960s and 1970s, interpretation of travel time residuals consisted simply of assembling large data sets of travel time residuals from regional, continental, or global seismic networks, correcting for source and path effects by using a range of simple to sophisticated averaging techniques, and interpreting the resulting azimuth-independent and azimuth-dependent components of the "station terms" in terms of heterogeneous crust and mantle structure beneath the seismic network. Such early studies have resulted in the detection of strong lateral heterogeneities within the Earth. For example, in the United States analysis of teleseismic residual data showed that the tectonically active West has a relatively slow upper mantle, implying that it is hot and mobile compared with the fast upper mantle of the shield and stable platform of the East, interpreted to be relatively cold and solid. Detailed studies also showed evidence for upper-mantle seismic anisotropy related to mantle convection, which is one of the driving forces of plate tectonics (see *Seismic Anisotropy in the Earth*). These types of investigations all over the world during the 1960s and 1970s mark the beginnings of body wave tomography. The real breakthrough in body wave tomography occurred during 1974–1977 when Keiti Aki, a professor at the Massachusetts Institute of Technology, along with Anders Christoffersson of the University of Uppsala, Sweden, and Eystein Husebye of the Norwegian Seismic Array (NORSAR), Norway, developed an inversion technique for three-dimensional seismic modeling of the Earth, using teleseismic residuals. The publication of this technique, now popularly known as the ACH technique, almost coincided with publication by Keti Aki and William Lee of the U. S. Geological Survey of a parallel scheme for tomographic modeling using local earthquake data.

Surface wave studies from 1960 onward have revealed the existence of strong lateral-velocity heterogeneities. The inferences were based on distribution of phase and group velocities and computed regional velocity-depth functions. Development of formal tomographic imaging techniques using surface waves began in the early part of this decade by Don Anderson, Ichiro Nakanishi, Henri-Claude Nataf, and Toshiro Tanimoto of the California Institute of Technology, and Adam Dziewonski and John Woodhouse of Harvard University.

During the past decade, seismic tomography has been developing at an accelerated pace and can be expected to be an active area of seismological research for several decades to come. Over 100 large-scale and small-scale heterogeneous Earth models based on tomographic studies are now available. They provide insight into global tectonic processes, differences in deep structure between continents and oceans, and the presence of complex structures in volcanic zones, earthquake faults, and other diverse geological environments. For reviews see Aki (1982), Anderson and Dziewonski (1984), Dziewonski and Anderson (1984), Iyer (1984, 1988), and Thurber and Aki (1987). In the future, both experimental and theoretical developments in seismic tomography can be expected to undergo a high degree of sophistication and refinement.

Tomographic Methods

As mentioned earlier, setting aside the fast-developing field of image processing in controlled source seismology (seismic refraction and reflection), the two basic tomographic methods in use to image the Earth's heterogeneous velocity structure are those that use body waves and surface waves, respectively. Body wave tomography can further be subdivided into two groups, depending on whether the seismic sources are located outside or inside the modeled volume. The first group deals with the imaging of the crust and upper-mantle structure immediately beneath a seismic network by using waves from distant earthquakes (teleseisms), and with high-resolution imaging of the upper crust by using explosion-generated waves (as in NeHT experiments described later). The second group includes computation of local and regional velocity models using near earthquakes and global models using worldwide earthquakes. Even though the underlying mathematical principles are the same for all these techniques, there are significant differences in the formulation of the problems and in the computational procedures.

Regional Tomography Using Teleseismic Body Waves. The largest number of experimental case histories in seismic tomography involves the use of teleseismic travel time residuals to image the crust and mantle structure immediately beneath a seismic network. Seismic rays from *teleseisms*, which are earthquakes that occur at distances at least 20° from the recording station, reach the seismographs at angles of about 30° to the vertical for near teleseisms, to almost vertical for distant teleseisms (Fig. 1a). For a multiplicity of sources and recording stations, the ensemble of rays samples a volume that approximates a truncated cone (Fig. 1b). The first step in seismic tomography is to compute the travel time residual R_{ij} at the ith station for the jth event with respect to a known one-dimensional Earth model. We then perturb the model until we can absorb the travel time residuals in a new three-dimensional model. But first it is necessary to correct the travel time residuals for errors in origin time and hypocenter mislocations and for the velocity heterogeneities outside the modeled volume. This correction is made by simply subtracting from each R_{ij} the average event Residual $\bar{R}_j$ computed over the whole network. The resultant quantity, called the *relative residual* RR_{ij}, is therefore

$$RR_{ij} = R_{ij} - \bar{R}_j$$

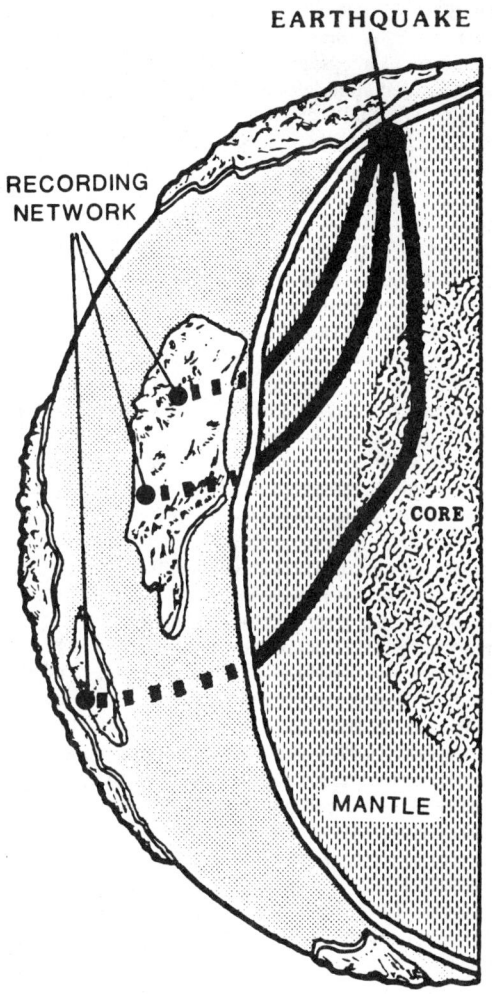
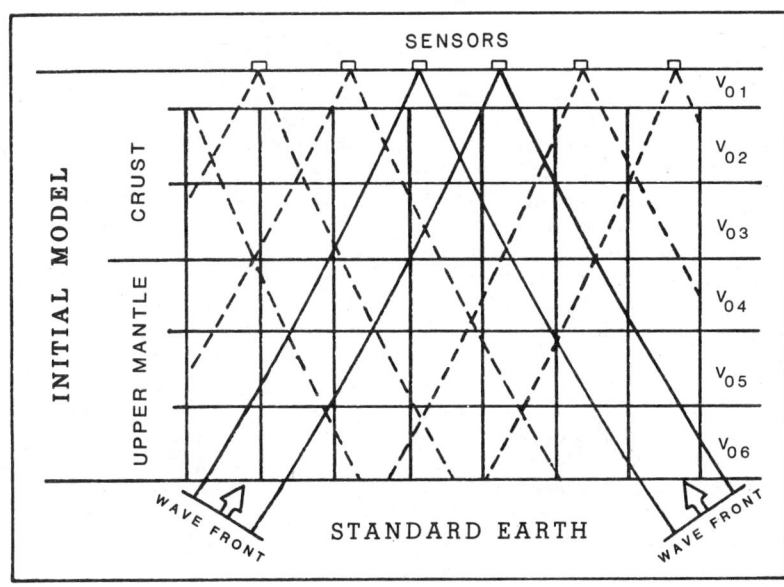

FIGURE 1. (a) Path of rays from a teleseismic source to recording seismic arrays through the Earth's mantle and core. (b) Division of modeling region into blocks for tomography using teleseismic waves. The initial model has constant layer velocities as indicated on the right-hand side.

where

$$\bar{R}_j = \frac{1}{N_j} \sum_{i=1}^{N_j} R_{ij}$$

N_j being the number of readings available for the jth event. RR_{ij} contains the required information on the three-dimensional velocity structure beneath the seismic network to a depth approximately equal to the average size of the network. The magnitudes of the residuals are a function of the amplitude of the velocity heterogeneities, and the shift in their spatial pattern as a function of event azimuth indicates the depth of the heterogeneities. Negative residuals indicate the existence of higher velocities than in the reference model used in computing the residuals, and vice versa. A rough rule of thumb on the amplitude of the heterogeneities is that a 10-km-long raypath in a zone of 10% velocity decrease causes about 0.2-s residual for P waves and about twice as much for S waves.

To image the three-dimensional structure from the teleseismic residual data, we divide the volume under the seismic network into several layers and each layer into blocks (Fig. 1b). Each block is assigned an unknown parameter, namely, the fractional deviation of slowness (reciprocal of velocity) from the mean slowness of the layer to which the block belongs. The mean slowness is specified a priori in the starting model. As the seismic ray travels toward the surface, it accumulates time perturbations, depending on the slowness perturbations in the blocks through which it travels. The observed relative residual at the surface is the sum of all these accumulated time perturbations. Thus, as many linear equations are generated as the number of observed residuals. They can be represented by the matrix equation

$$d = Gs \qquad (1)$$

where d is a vector containing RR_{ij}, G is a matrix containing the unperturbed travel times of ray segments in the blocks, and s is a vector containing the unknown fractional slowness perturbations. Equation 1 is of the standard type pertaining to linear systems frequently encountered in science and engineering. Many approaches are available to solve it. This equation represents an overdetermined system (that is, there are more data than unknowns) and is therefore most appropriately solved by a least squares method. Since this technique relies on a block formulation of a continuous medium, the solution is only an approximation of the real Earth, with some parts of the model less constrained than others, and still others possibly fully unconstrained. Backus and Gilbert (1967) introduced the useful concepts of resolution and error estimates to relate a particular solution to the real Earth. (For a detailed discussion of linear systems as applied to seismic tomography, see Aki and Richards, 1980). For solving Eq. 1, the two commonly used approaches in seismic tomography are the three-dimensional inversion technique developed by Aki et al. (1977), which we shall call the ACH 3-D inversion technique, or the ACH technique for short, and the medical analog of the back-projection technique developed by Humphreys et al. (1984) and Humphreys and Clayton (1988) at the California Institute of Technology, which we shall designate as the Caltech back-projection technique, or the Caltech technique for short.

ACH Technique. The conventional least squares solution of Eq. 1 to estimate s is nonunique because it has at least as many zero eigenvalues as the number of layers in the initial model. In other words, absolute velocities are unconstrained. Aki et al. (1977) used the generalized inverse and the damped least squares inverse to solve the problem. In the former, since small nonzero eigenvectors (i.e., poorly constrained parameters) are retained, the final model is rough or uneven due to the effects of the random errors. In the latter, small eigenvectors are smoothed out; hence, the resultant model, though smooth, suffers from poor resolution. Both methods, however, give essentially the same image, and the differences between them can be explained by the differences in the smoothing kernel specified by the resolution matrix $\mathcal{R}$, an important by-product of the inversions that needs to be understood to assess the reliability of the computed model. The matrix $\mathcal{R}$ is defined by the relationship

$$\hat{m} = \mathcal{R} m$$

which relates the true model m to the computed model $\hat{m}$, and is a symmetric singular matrix; its elements have values between $+1$ and -1, with the diagonals always positive. Each row of this matrix can be considered to be an averaging kernel relating all of m to one block of $\hat{m}$. Thus, the resolution matrix can be thought to be a window through which one sees a "fuzzy" picture of the real Earth.

The following points relevant to the ACH technique are, in general, true for other tomographic techniques as well. (1) The inversions provide only relative velocity perturbations and not absolute velocities. (2) For tomography using teleseismic residuals the final model is not critically dependent on the starting model, even though it will be appropriate to use the best available model. The starting model becomes particularly important to convert the modeled velocity perturbations into absolute velocities for interpretation in terms of physical properties of the Earth. (3) The better the cross fire between the rays, the better that block is modeled. (4) For teleseismic residual tomography, as was already mentioned, the depth of modeling is approximately equal to the aperture (horizontal dimensions) of the seismic network. (5) The size of the smallest resolvable anomaly is approximately equal to the wavelength of the seismic waves used or the station spacing, whichever is larger. For short-period (1-Hz) teleseismic P waves this limit is about 6 km. (6) Corrections for differences in elevation and geologic structure within the seismic network can be incorporated as part of the computations.

Caltech Technique. In the Caltech technique, as in the ACH technique, the model is discretized, but the simplification of the inversion algorithm enables the use of many small blocks, resulting in a final model that is less "blocky" looking than the model in the ACH technique. (There is no rule that only large blocks should be used in the ACH technique. The need for large block size is dictated by computational needs of the ACH technique involving manipulation of full matrices. In practice, using a fast computer, small blocks, and smoothing algorithms, ACH models often are as smooth as those derived by the Caltech technique.)

Following Humphreys and Clayton (1988), we can outline the mathematical scheme of the Caltech tomographic technique as follows.

$$t_r = \sum_b l_{rb} s_b$$

where t_r is the travel time residual associated with the rth ray, s_b is the slowness perturbation of the bth block, and l_{rb} is the length of the ray segment in the bth block. This equation can be written in matrix form as

$$t = Ls \qquad (2)$$

The classical least squares solution of Eq. 2 is

$$L^T L s = L^T t$$

where T denotes the transpose of the matrix. Up to this point there is no difference between ACH and Caltech techniques. Since for a large data set the computation of the leftside of the equation exceeds CPU memory resources of most computers, an approximation is made by initially using only the diagonal of $L^T L$. The solution can now be shown to be

$$s_b = \sum_r t_r l_{rb} \Big/ \sum_r l_{rb}^2$$

The computation of s_b from this equation is efficient and simple. As each ray is back-projected, contributions to the two sums are accumulated in separate computer memory spaces. Once this operation is completed, the slowness perturbation for each block is estimated. The computational efficiency of this back-projection scheme, however, is offset by the fact that the resulting image tends to be extremely blurred, especially along paths of high ray density.

Humphreys and Clayton (1988) use two classes of deblurring techniques, both of which are borrowed from medical tomography. The first is a deconvolutional technique, and the second is the simultaneous iterative reconstruction technique (SIRT). SIRT is an iterative scheme in which the difference between the observed residual and the residual computed from the latest inverse is back-projected, and the result is used to update the inverse.

One problem with this type of tomography is that the resolution matrix is not available as a by-product to assess the quality of the inversion. Humphreys and Clayton (1988) instead computes weights to the other blocks by unit velocity perturbation in a specific block. The resultant weights are used in a similar fashion to the resolution matrix of a conventional inversion.

Tomography Using Waves from Local and Regional Earthquakes. The main difference between the local earthquake case and the teleseismic case is that in the former the seismic sources are located in the modeling volume (Fig. 2). Since the source locations are not precisely known, perturbations to the origin times and hypocentral parameters must also be calculated in addition to the velocity perturbations. Aki and Lee (1976), using a simplified formulation of the local earthquake problem, were the first to show that this is possible for a three-dimensional case.

Let the initial estimates of the hypocenter for the jth earthquake be X_j^0, Y_j^0, Z_j^0, and let the origin time be T_j^0. Then the observed P travel time is

$$T_{ij}^{obs} = T_{ij}^{cal} + \left(\frac{\partial T}{\partial X}\right)_{ij} \Delta X_j + \left(\frac{\partial T}{\partial Y}\right)_{ij} \Delta Y_j$$
$$+ \left(\frac{\partial T}{\partial Z}\right)_{ij} \Delta Z_j + \Delta T_j + \sum_k T_{ij}^{(k)} F_k + E_{ij}$$

(3)

where T_{ij}^{cal} is the calculated P arrival time at the ith station based on a homogeneous initial model, $(\partial T/\partial X)_{ij}$, $(\partial T/\partial Y)_{ij}$, and $(\partial T/\partial Z)_{ij}$ are derivatives of the travel time with respect to the X-, Y-, and Z-coordinates calculated using the initial model, and ΔX_j, ΔY_j, ΔZ_j, and ΔT_j are the source perturbation terms. The sixth term incorporates the effects of the slowness perturbation on the travel time, $T_{ij}^{(k)}$ is the time the ray spends in the kth block, and F_k is the slowness perturbation in the block with respect to the initial Earth model. The last term, E_{ij}, is included to account for high-order terms and observational errors. The whole set of observed minus calculated arrival times can be arranged as a column vector τ, and the source and medium perturbations as a column vector χ. Then Eq. 3 can simply be written in matrix notation as

$$\tau = G\chi + \epsilon \qquad (4)$$

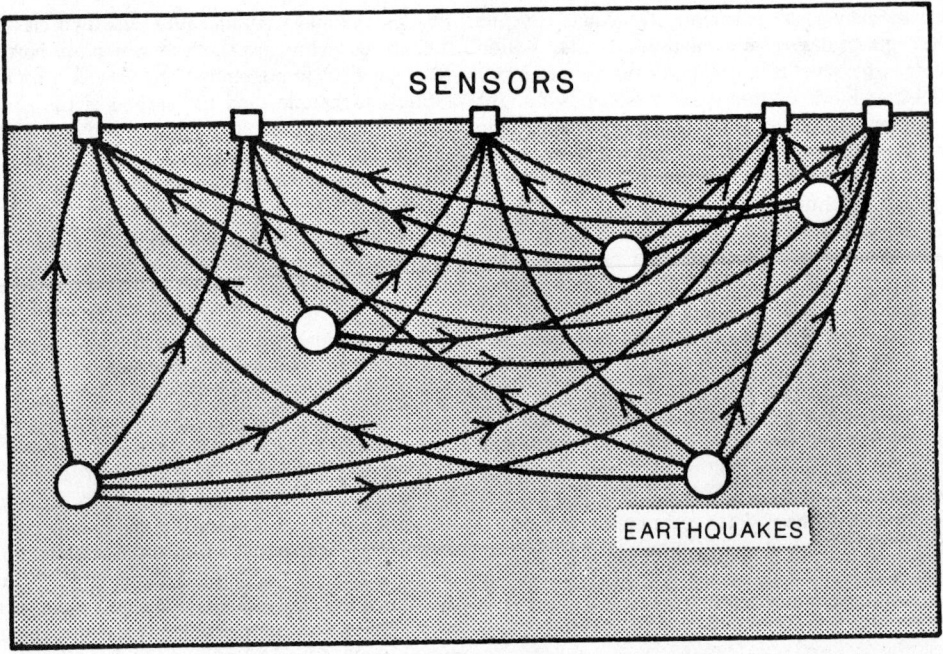

FIGURE 2. Schematic illustration of sources, receivers, and rays used in local earthquake tomography.

where G is a matrix containing the nonperturbation terms in the equation, partial derivatives and the T_{ij} term and ϵ is an error vector with components E_{ij}. Aki and Lee (1976) solved Eq. 4 by again using the damped least squares method.

The simultaneous solution for source and path parameters becomes computationally impractical for the large data sets commonly available for local earthquakes. Two approaches have been used to address this problem. The first method, parameter separation, was introduced by Pavlis and Booker (1980) and it uses orthogonal transformations for each event to annul the hypocenter components while retaining the velocity perturbations. The second method is similar to the Caltech tomographic technique. Thurber (1983) adopted parameter separation in conjunction with the ability to specify the initial velocity in terms of values at grid points rather than in layers. Using an approximate ray-tracing method, he solved Eq. 4 by iterating until the solution converged. The Caltech tomographic technique for local earthquake data is quite similar to that used for teleseismic waves. To handle the source problem, Hearn and Clayton (1986a, 1986b) solved separately for source residuals and receiver residuals, using the back-projection technique, and iteratively improved the model.

Global Tomographic Techniques. *Using Body Wave Travel Times.* The expansion of the world-wide seismic data base during the last two decades in conjunction with the availability of fast computers and the improvements in computational methods has made it possible to construct tomographic images of the whole Earth. One major source of such data is the International Seismological Centre in Edinburgh, Scotland, where almost a billion P travel time readings are compiled every year. Several researchers have taken advantage of this vast data base to compute whole Earth P-velocity models with progressive improvements in accuracy and resolution. Recent models derived from this data base are results of research carried out at the California Institute of Technology and Harvard University (see *Earth Structure, Global*).

R. P. Comer and Robert Clayton of the California Institute of Technology modeled the Earth's upper mantle by analyzing 1.7 million P readings from the world data catalog, using the Caltech technique. Dziewonski (1984) of Harvard University used about a half million P travel time residuals from 5000 teleseisms to model the Earth's lower mantle. Instead of using block parameterization, he solved for coefficients in a spherical harmonic expansion of velocity perturbation. The mathematical scheme involved in this computation is too complex to be discussed in detail here. Briefly, the approach is to represent the velocity perturbation $\delta V(r, \theta, \phi)$ at any point within the modeled volume of the Earth in terms of a triple summation in k, l, and m, respectively, of an expression involving spherical harmonic coefficients $_kA_l^m$ and $_kB_l^m$. The coefficients themselves are calculated by using the observed travel time residuals for numerous raypaths. The maximum angular order number L and radial order number K

are determined by the distribution of sources and receivers. In Dziewonski's inversion, these numbers are $K = 4$, providing a horizontal resolution of about 3000 km, and $L = 6$, providing a radial resolution of roughly 500 km. A multiple-iteration approach was used in the analysis, and velocity models, source locations, and station corrections were solved in each iteration. Dziewonski's P-velocity model is discussed in the next section. A variant of global-scale tomography of the Earth's mantle using body waves is the exciting new frontier of imaging the core-mantle boundary by the inversion of travel time residuals of waves reflected from and transmitted through the boundary (Morelli and Dziewonski, 1987).

Using Surface Wave Velocities. In recent years, global-scale tomography using seismic surface waves has emerged as a powerful tool to probe the three-dimensional S-velocity structure of the Earth's upper mantle. Surface waves are less sensitive to lateral heterogeneities than body waves are. Also they probe vertical structure in a different way than body waves do because the depth of penetration of surface waves depends upon their wavelengths. Surface wave modeling at present is limited to the upper mantle due to a paucity of ultra-long-period data and therefore complements the global lower mantle studies using body wave residuals. The power of surface wave imaging comes from the joint analysis of the two types of surface waves, namely Rayleigh waves and Love waves. Rayleigh waves are sensitive to vertically polarized shear waves, whereas Love waves are sensitive to horizontally polarized shear waves. Thus, comparison between the Rayleigh wave and Love wave velocity models provides quantitative information on the anisotropic structure of the upper mantle that is related to mantle flow.

The theoretical aspects of surface wave imaging are quite complex and will be discussed briefly. For details, see Nakanishi and Anderson (1983, 1984) and Nataf et al. (1986). In surface wave tomography, in addition to the usual linear system parameterization that is required to compute the spatial distribution of phase velocity perturbations as in conventional tomography, the dispersion (phase velocity as a function of frequency) that relates to the velocity and density distribution as a function of depth must also be modeled by the inversion. Conventional one-dimensional modeling using surface waves consists of computing the velocity-depth function by forward modeling or inversion of the dispersion relationship. In surface wave tomography, however, we must model the three-dimensional velocity structure of the Earth by using several thousand observed phase velocity deviations from phase velocities computed for a standard Earth model. This model is achieved by either expressing the phase velocities as a sum of velocity perturbations over area elements or by specifying the velocities as a spherical harmonic expansion. The data for these computations are phase velocity values measured over numerous intersecting paths (Fig. 3), and the final product is three-dimensional shear velocity perturbations.

Measurement of phase velocity of surface waves is a well-developed technique in seismology. New high-quality instrumentation (both analog and digital) that is presently used in global seismic networks has considerably enhanced the quality and quantity of these measurements and has made surface wave tomography possible. Phase velocity can be measured by the great circle, single-station, or two-station methods. In the great circle method, differential dispersion between a train of surface waves first recorded at a station and of the same train after it has traversed round the entire globe once is used; hence the technique is applicable only to global studies. In the single-station method, the surface wave dispersion is directly computed for the path from the source to the recording station; therefore, the technique is applicable for global and regional studies. A prerequisite, however, is an accurate knowledge of the earthquake source parameters so that the "starting phase" of the wave train can be calculated. The two-station method involves direct computation of the phase difference as a function of frequency for similar wave trains recorded at two stations. The few cases of surface wave tomography available at present use the great circle and single-station methods.

Woodhouse and Dziewonski (1984) have introduced a fundamentally different approach to surface wave tomography. Instead of inverting phase velocity dispersion data, they synthesize long-period seismograms directly by using parameters of a spherical harmonic representation of the Earth's global density and anisotropic shear velocity structure (see *Earth Structure, Global* and *Seismic Imaging*).

Tomography Results

The theoretical aspects of seismic tomography are fairly well understood. However, unlike in medical tomography, the lack of uniform source-receiver geometry and funding limitations makes it difficult to design optimal sesimic imaging experiments. Fortunately, there exists a global network of about 100 identical seismograph stations, each with matched sets of three-component short- and long-period instruments (the World Wide Standard Seismic Network or WWSSN). This network is supplemented by national seismic networks, some with dense arrays of instruments. The WWSSN and national networks together provide about a billion body wave arrival-time readings, which are compiled and published by the International Seismological Centre. This vast data base has been used for global-, continental-, and regional-scale tomography studies. Data from dense seismic networks in the

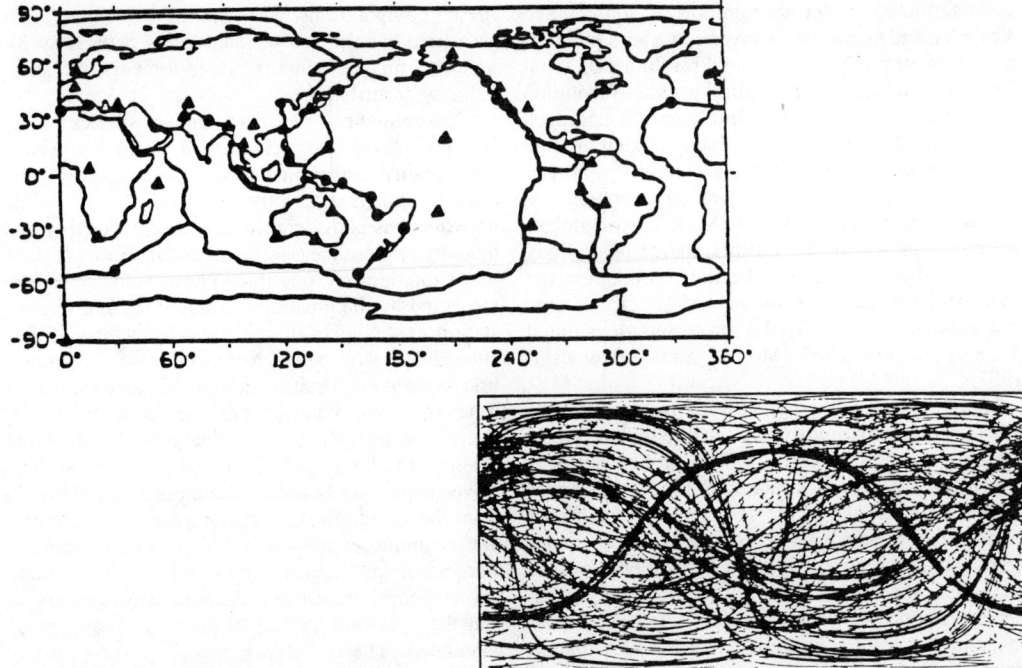

FIGURE 3. Typical configuration of sources, receivers, and raypaths used in surface wave tomography: (*top*) earthquake epicenters (circles) and seismograph stations (triangles); (*bottom*) Rayleigh wave paths connecting earthquakes and stations used for great circle phase velocity measurement. (Figure adapted from Nakanishi and Anderson, 1983, courtesy of the authors. Copyright American Geophysical Union)

United States, Japan, and Europe have also been used for seismic tomography. The data requirements for surface wave tomography are more stringent than for body wave tomography. About 50 digital worldwide stations provide data for these studies.

Altogether, a few case histories of tomographic studies using surface waves, about 50 studies using teleseismic P residuals and very few using S residuals, and about 20 using local earthquake data are available. Though these studies are by no means perfect in terms of density of instruments or quantity of data collected, together they emphasize the extreme heterogeneity of the Earth, whether we examine a small region, a large region, a whole continent, or the whole globe. These seismic imaging results are not new *discoveries*, but they reinforce and quantify the existence of such heterogeneous structure found from earlier seismological and geophysical studies and suspected from geologic and tectonic inferences. A representative sample of recent tomographic results is presented in the following sections.

Global Studies. Important new results have emerged from global tomographic studies using surface waves and body waves. Imaging using surface wave phase velocities and waveforms has given a three-dimensional picture of velocity variations in the upper mantle of the Earth (see *Earth Structure, Global*). For completeness, we summarize these results. The tomographic results 150-km deep confirm the inferences from global tectonics, namely, the existence of low velocities beneath mid-ocean ridges and fast velocities beneath continental shields and ocean basins. However, at a depth of 350 km the imaging finds high velocities beneath the shield that are weaker than those at 150 km and sometimes shifted. It also finds low velocities beneath the Red Sea rift and some of the hot spots. At 550 km, velocity anomalies from -2% to $+2\%$ still persist. Body wave studies in the lower mantle complement surface wave studies of the upper mantle. Velocity anomalies seem to be continuous over a large range of depths to the core-mantle boundary. An exciting new result is the large velocity perturbations revealed by seismic tomography near the core-mantle boundary, indicating mountainous "topography."

Probably the most tantalizing results that have come out of surface wave tomography relate to the identification of anisotropy in S-wave velocities, which have been interpreted in terms of mantle flow in the upper mantle (see *Seismic Anisotropy in the Earth*). Surface wave studies have enabled mapping of anisotropies in the horizontal and vertical directions, and provide a first-hand picture of convective flow in the mantle: horizontal flow in the direction of plate motion, upward flow around mid-ocean ridges associated with upwelling of hot material, and downward flow beneath old ocean basins due to sinking of cold material.

Continental-Scale Studies. P-wave tomographic studies for the United States and Europe using the worldwide data base have revealed the existence of a finer scale of heterogeneities than those indicated by global tomography. The main findings are that relatively high velocities are present beneath shields and old platforms to a depth of about 300 km, and relatively low velocities are present beneath tectonically active regions. For example, upper-mantle velocities are higher than normal in the eastern United States and the Scandinavian (Baltic) shield and lower than normal in the Basin and Range province in the western United States and the Mediterranean region in Europe. Other interesting findings include the discovery by Barbara Romanowicz of the Massachusetts Institute of Technology of high-velocity anomalies deeper than 250 km in the upper mantle in the western United States, inferred to be remnants of the Farallon Plate, which ceased subducting beneath California 10–20 Ma. In tomographic modeling for North America using S waves, Stephen Grand of the California Institute of Technology found a high-velocity root extending to a depth of about 400 km beneath the Canadian shield and intermediate velocities from 100 km to 350 km deep in the upper mantle of the eastern United States. The lowest velocities are found in the same depth range in the western United States. Grand's studies have also delineated an almost vertical, slablike, high-velocity anomaly extending from 700 km to 1700 km deep beneath the Caribbean. He speculates that this anomaly could also be the Farallon Plate.

Regional Studies. *A Transform Fault.* The San Andreas Fault in California is instrumented with over 1000 telemetered seismic stations covering its entire length from Mendocino in the north to the Mexican border in the south. The large quantity of teleseismic and local earthquake data collected by this network has been used for several small-scale and large-scale imaging studies of this transform boundary between the Pacific and American Plates. These studies have revealed a complex velocity structure in the crust and upper mantle of the San Andreas Fault. Of special interest are the upper-mantle heterogeneities in the depth range of 100–225 km (Figs. 4, 5), characterized by highs and lows with velocity perturbations in the ±3% range. One of the most intriguing of

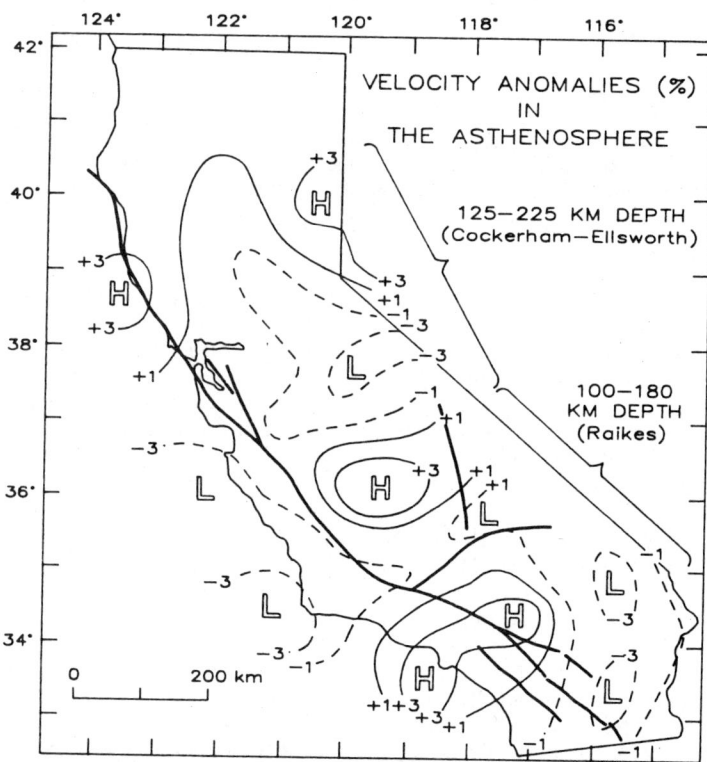

FIGURE 4. P-velocity anomalies in California 125–225 km deep from teleseismic residual tomography. Figure is a synthesis by Aki (1982) of available results. H and L represent high- and low-velocity anomalies, respectively. Contour interval is 2%. Dark lines represent the San Andreas Fault and associated faults. (Figure taken from Aki, 1982, courtesy of the author. Copyright American Geophysical Union)

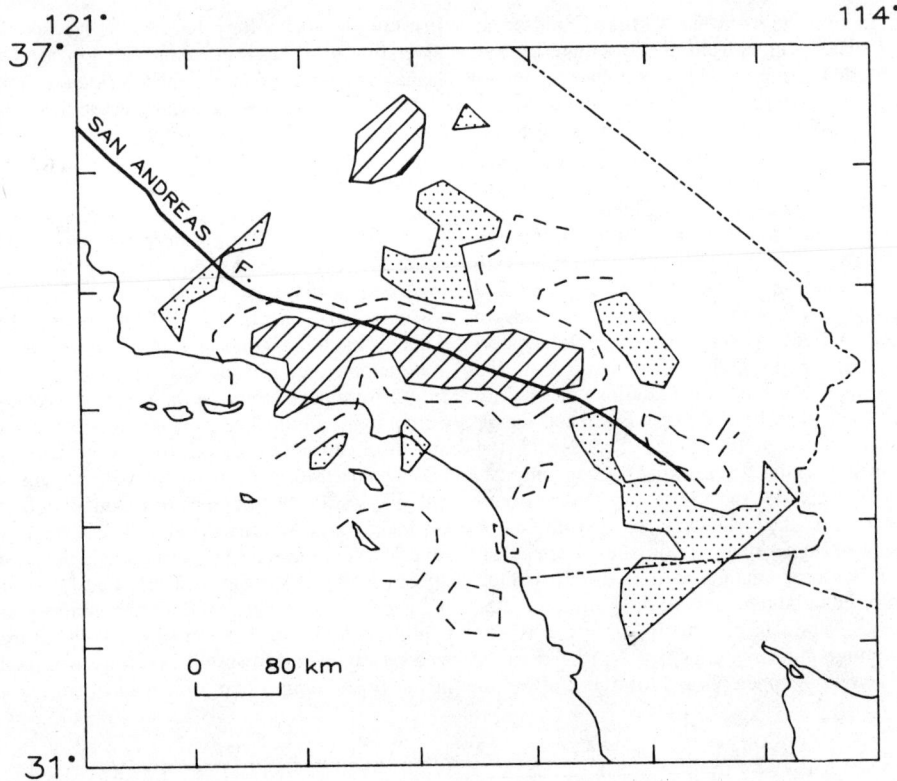

FIGURE 5. Tomography using teleseismic P residuals in Southern California: velocity anomalies at 100 km depth. Dotted areas are zones in which velocity perturbations are $< -1.5\%$, and hatched areas are zones in which velocity perturbations are $> +1.5\%$. Note that the two major anomalies are the high-velocity anomaly associated with the transverse ranges and the low-velocity anomaly associated with the Salton Trough spreading center. (Figure adapted from Humphreys et al., 1984, courtesy of the authors. Copyright American Geophysical Union)

these heterogeneities is the high-velocity anomaly that straddles the fault beneath the transverse ranges in Southern California. The anomaly is continuous across the fault, indicating that the upper mantle beneath the fault is mobile or ductile and periodically readjusts to the motion of the fault, or that the plate boundary at depth does not coincide with the surface expression of the fault.

Sections of the San Andreas Fault have also been imaged by using body waves from local and regional earthquakes. In Southern California, Hearn and Clayton (1986a,b), using P_g and P_n residuals, respectively, have imaged the upper crust and the whole crust and uppermost mantle in a region imaged from teleseismic travel times. The upper crust shows a clear velocity boundary corresponding to the surface trace of the San Andreas Fault (Fig. 6). On the other hand, the image of the whole crust and upper mantle (using P_n) dramatically differs from the upper crust. The fault boundaries are not seen, implying decoupling between the brittle, seismically active upper crust and the aseismic lower crust and upper mantle in this region. The Southern California studies, together with other tomographic studies in California, made possible by the high density of seismic stations and abundance of local earthquakes, are leading to an improved understanding of the structure and dynamics of earthquake faults.

A Continental-Margin Spreading Center. In probably the first deep three-dimensional modeling of a spreading center, Humphreys et al. (1984) have delineated a major zone of low velocity extending to a depth of 125 km beneath the Salton Trough in Imperial Valley, California (Fig. 5). They interpret this anomaly as hot upwelling material associated with the rifting.

A Continental Rift Zone. Continental rifts are enigmatic features on the surface of the Earth. Two of these rifts, the Rio Grande Rift in the United States and the East African Rift system have been the subjects of intensive probing by seismic tomography. In a joint study of the Rio Grande Rift by the U.S. Geological Survey and the University of California, Los Angeles, in 1982–1983, a 1000-km

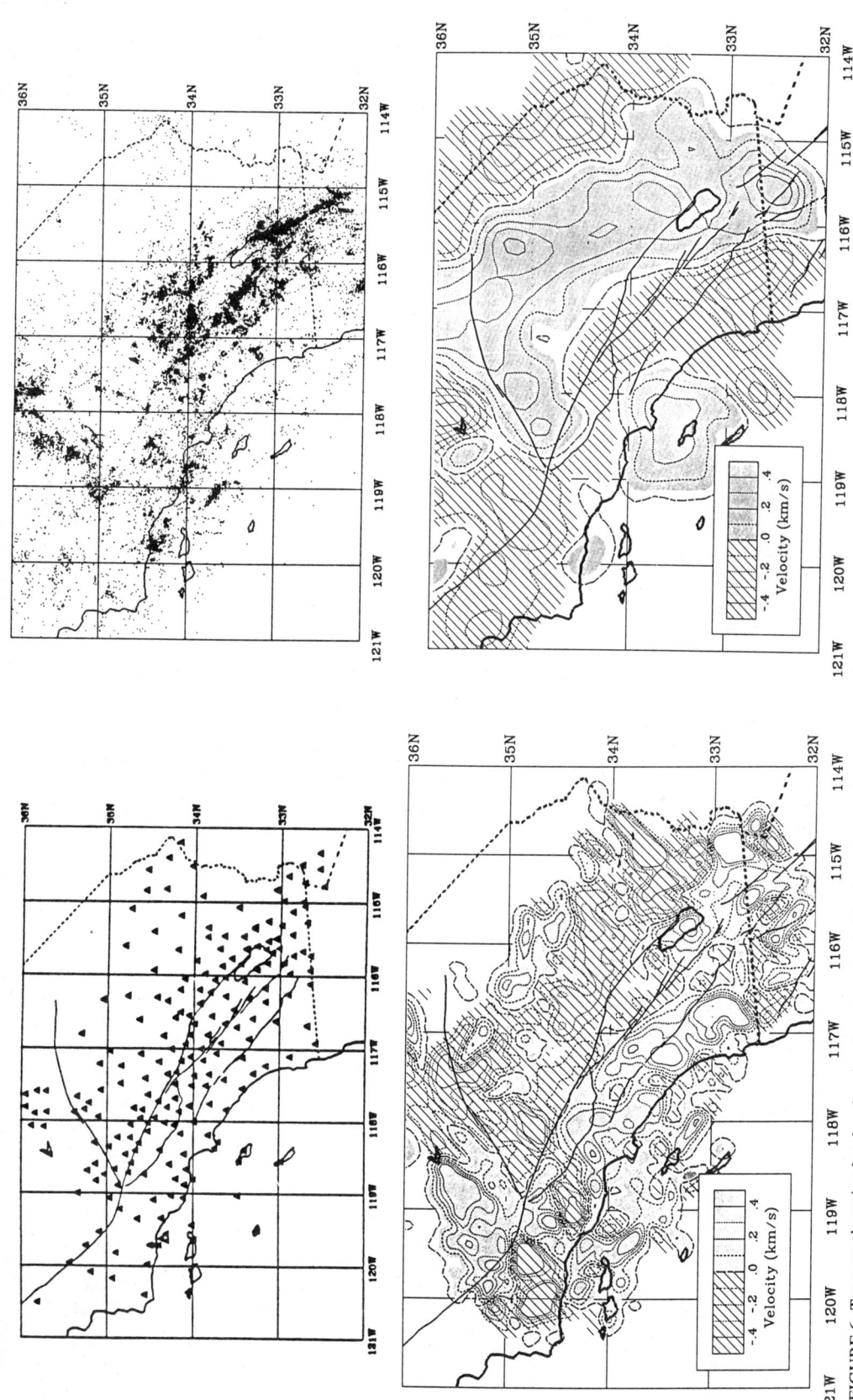

FIGURE 6. Tomography using local and regional earthquakes in Southern California: (*top left*) seismic network; (*top right*) earthquake epicenters; (*bottom left*) contour plot of P_g slowness for the upper crust; (*bottom right*) contour plot of P_n slowness for the crust and uppermost mantle. The *hatched* region indicates relatively low velocities, and the *shaded* regions are relatively fast. (Figures are taken from Hearn and Clayton, 1986a,b, courtesy of the authors. Copyright Seismological Society of America)

long, NW-SE profile was operated across the rift (Fig. 7). The tomographic model shows a low-velocity body in the upper mantle beneath the rift extending to 300 km on either side of the surface expression of the rift. This feature is interpreted as an asthenospheric upwarp, the cause or effect of the rift. Further studies by P. Davis and co-workers at the University of California, Los Angeles, in another section of the Rio Grande Rift and in the East African Rift system also reveal the existence of similar asthenospheric upwarps there. The implications of these findings are not fully understood in terms of the analogy between oceanic and continental rifts, nor can we determine from available data whether these rifts will eventually break up the respective continents. But these studies provide a beginning of seismological investigations into the fundamental global tectonic processes.

Subduction Zones. Subduction zones have long been recognized as key elements in global tectonics. The subducting slabs play a major role in driving lithospheric plates. In addition, they contain clues to the origin of the explosive andesitic volcanoes and the occurrence of destructive earthquakes along the plate boundaries. The most extensive tomographic imaging studies of subduction zones are those carried

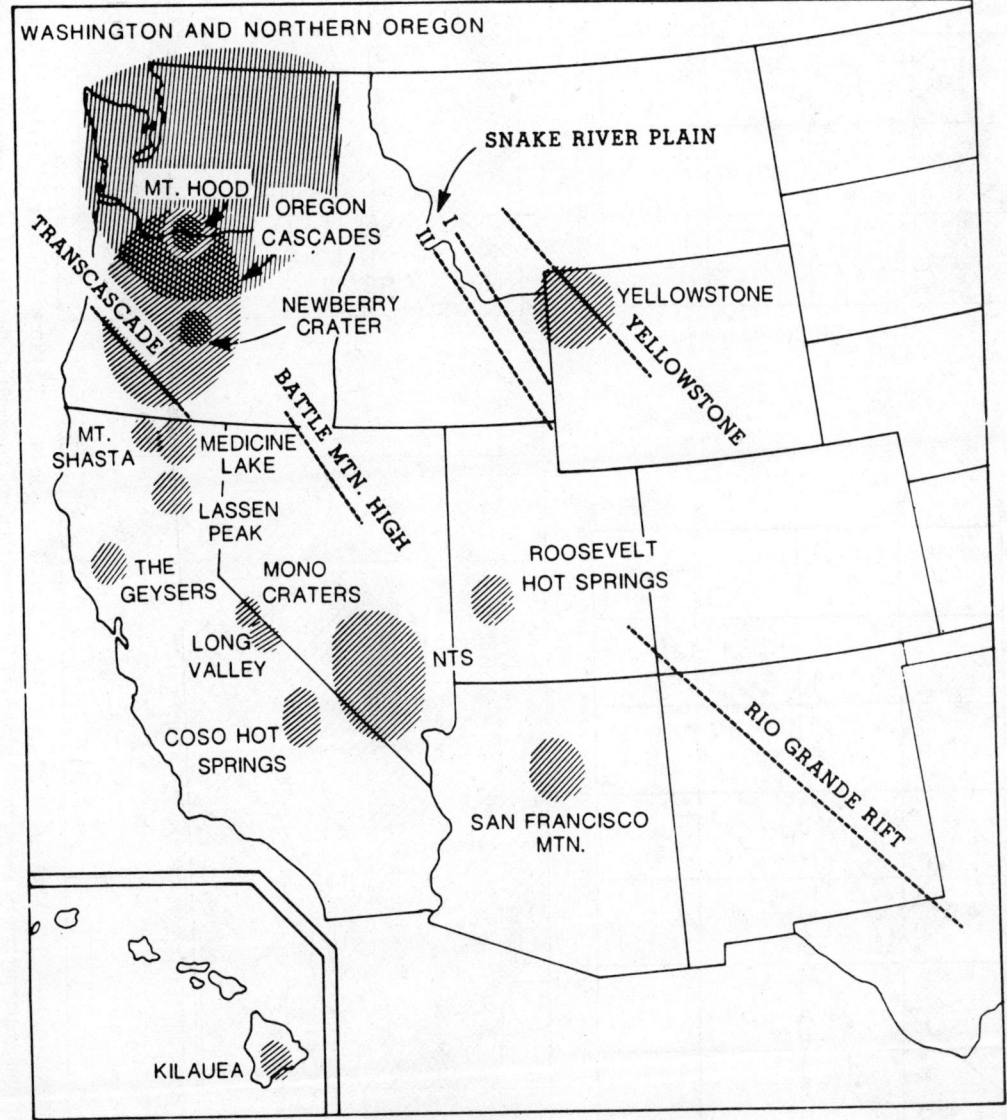

FIGURE 7. Locations of tomography experiments in volcanic and geothermal areas of western United States. Shaded zones indicate deployment of two-dimensional seismic arrays that yield three-dimensional velocity models, and lines indicate profiles that yield two-dimensional velocity models.

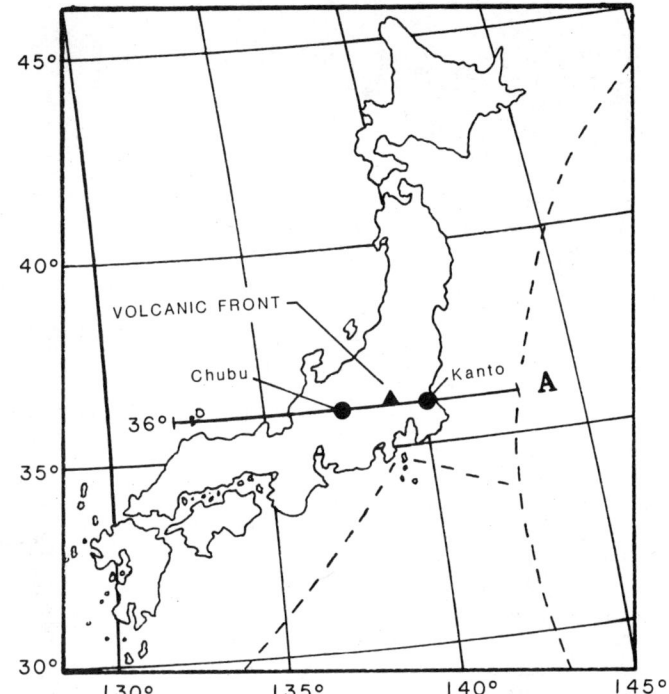
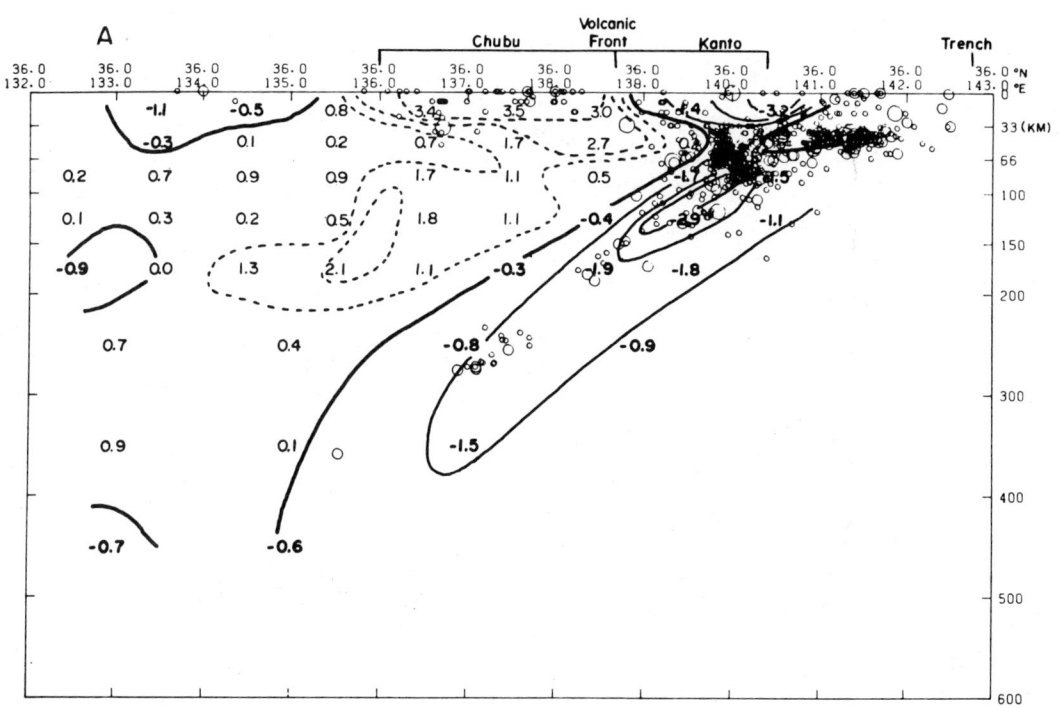

FIGURE 8. Tomography of the Japan subduction zone. (*top*) location map; trench axes are indicated by broken lines; (*bottom*) P-velocity perturbations along profile A together with earthquake hypocenters in the subduction zone. Contour interval is 1%. (Figure adapted from Hirahara, 1981, courtesy of the author. Copyright Elsevier Publishing Co., Amsterdam)

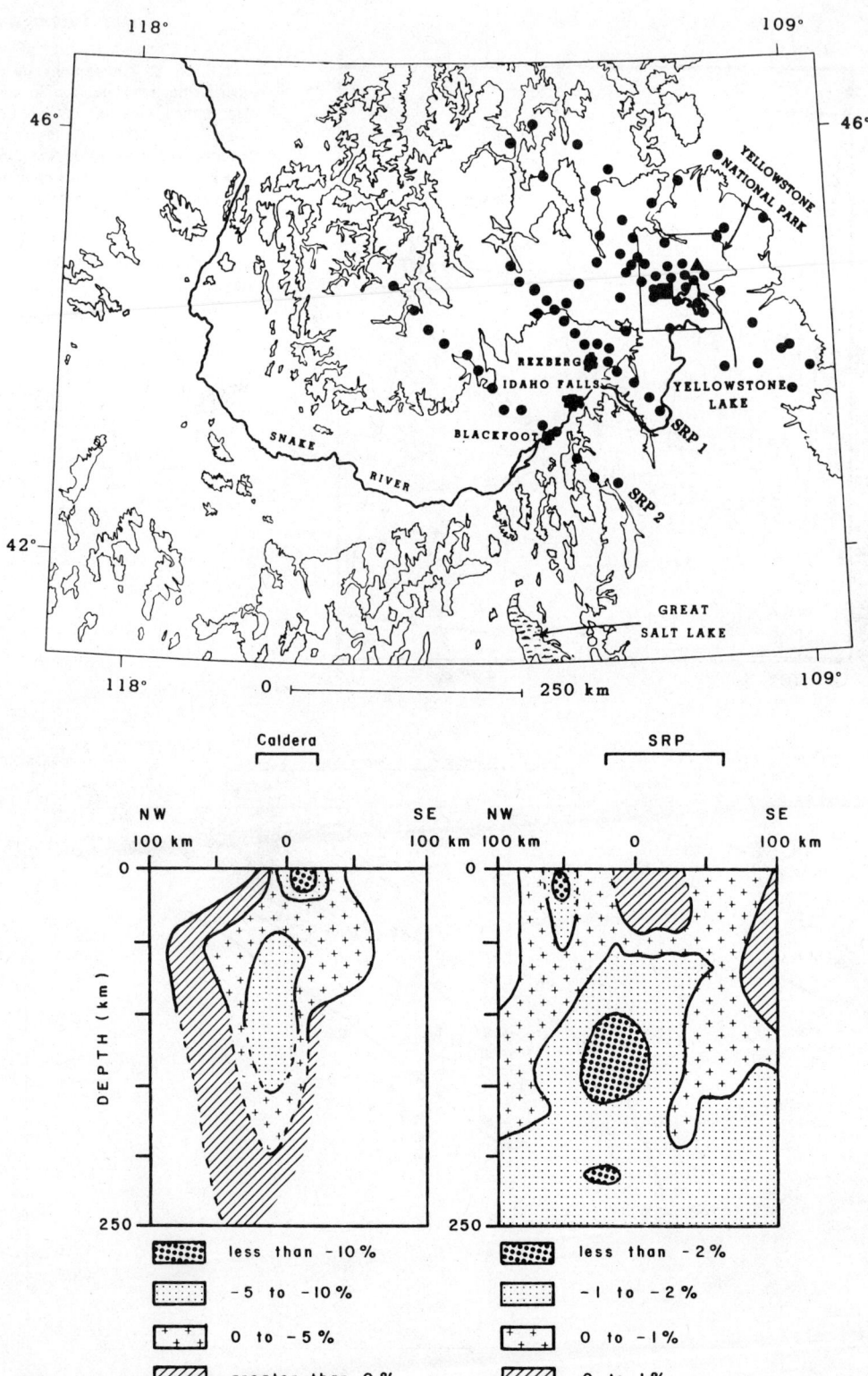

FIGURE 9. Seismic tomography of the Yellowstone hot-spot region: (*top*) locations of seismic stations; (*bottom*) velocity perturbation cross sections; (*left*) Yellowstone caldera, (*right*) Eastern Snake River Plain along profile SRP1. (Figures adapted from Evans, 1982, and Iyer, 1984)

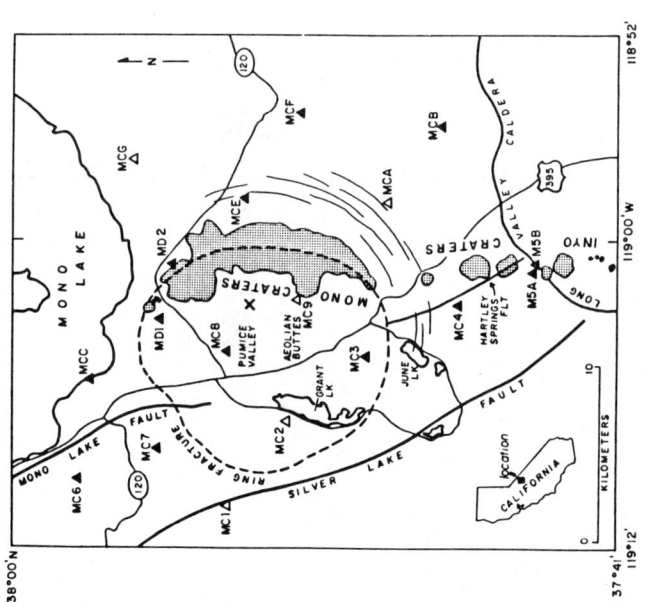

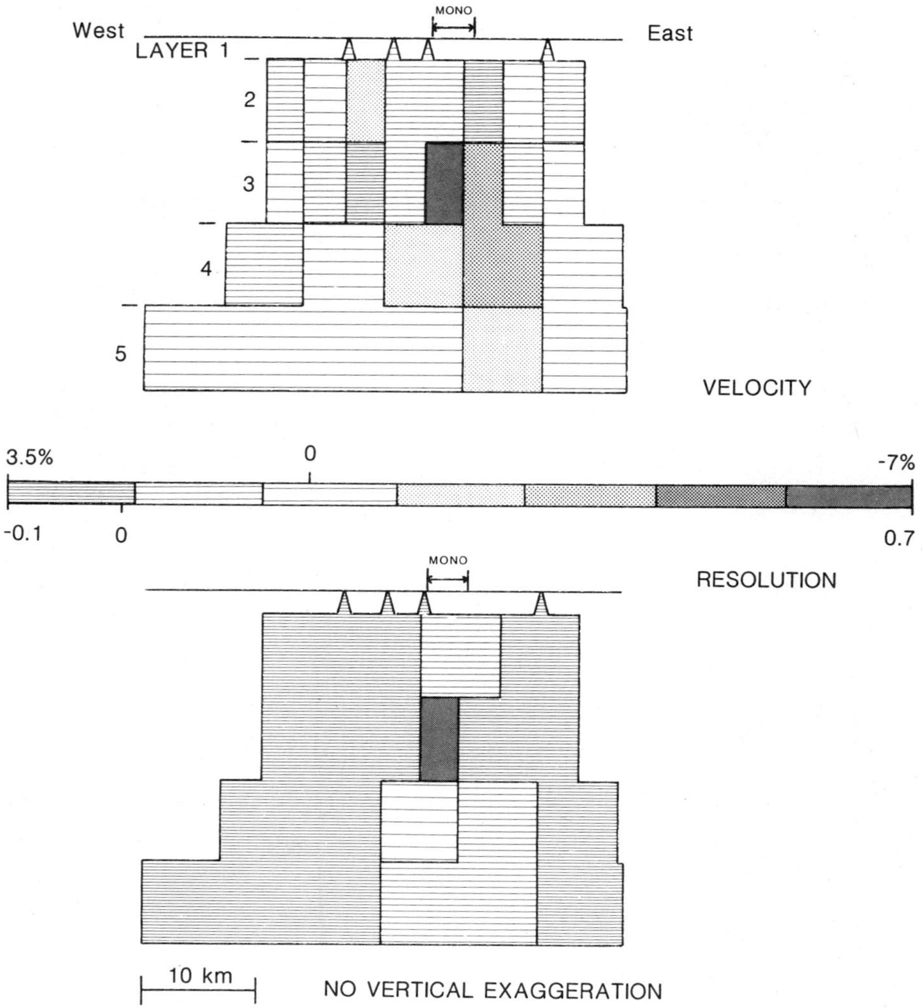

FIGURE 10. Seismic tomography of the Mono craters volcanic center in California. (*left*) Geographic location of the experiment, volcanic and tectonic features, and seismograph locations (triangles). (*right top*) Block model smoothed: blocks having same velocities are combined along an east-west line through Mono Craters. (*right bottom*) Row of resolution matrix corresponding to the most anomalous block in the velocity perturbation model. (Figure adapted from Achauer et al., 1986; copyright © 1986 by the American Geophysical Union.)

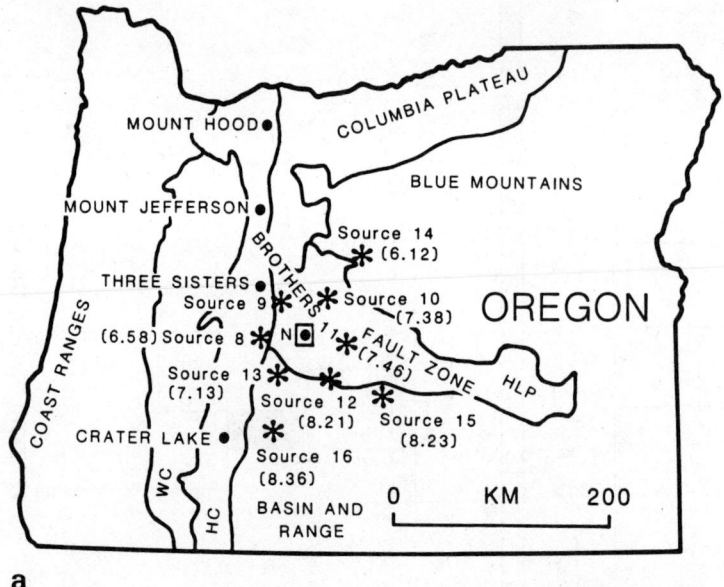

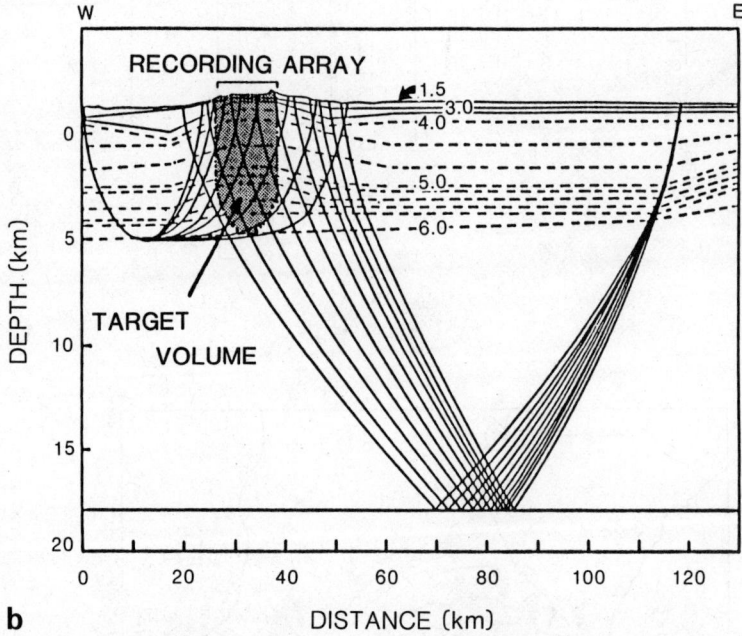

FIGURE 11. High-resolution seismic imaging experiment conducted by the U. S. Geological Survey, with collaboration from the U. S. Department of Energy, at Newberry volcano, Oregon. (a) Location of the experiment. Stars represent the chemical explosions set off, and square with dot is the location of the seismic array. (b) Conceptual drawings of typical raypaths used in the tomography. (c) (top) Velocity perturbation along an east-west cross section through the center of Newberry caldera. (bottom) Row of resolution matrix for the most anomalous velocity block near the center of the model. Note that the maximum diagonal value of resolution is 0.5 and the overflow from the scale is indicated by "0". (Figures are adapted from Achauer et al., 1988, courtesy of the authors; copyright © 1988 by the American Geophysical Union.)

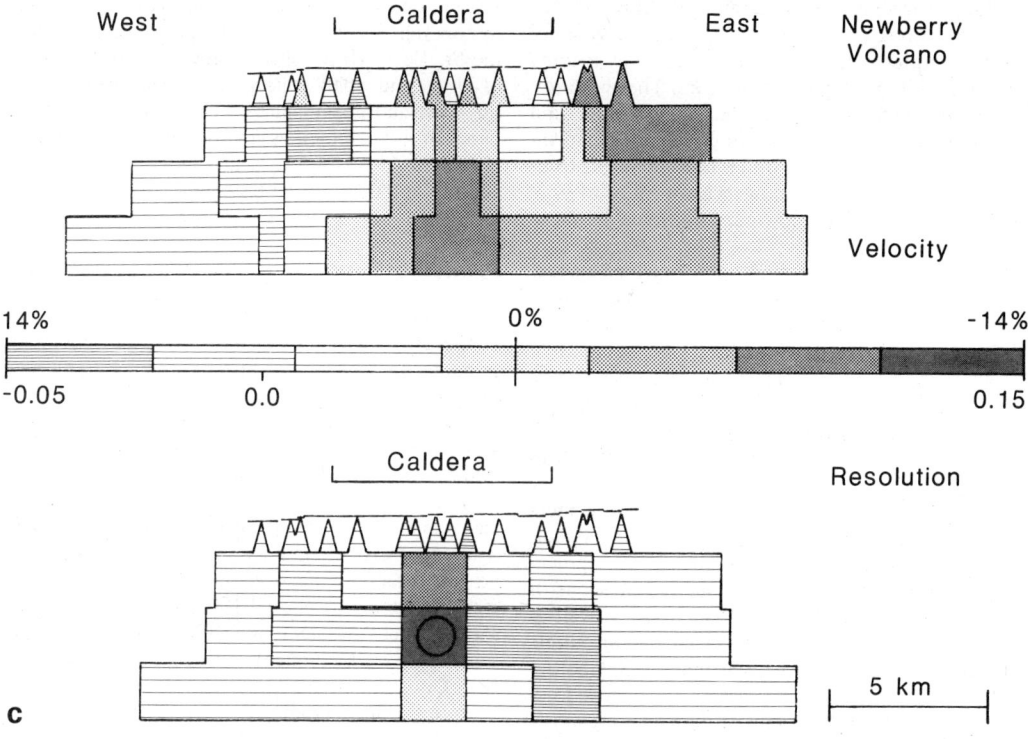

FIGURE 11. (*Continued*)

out in Japan with teleseismic and local earthquake travel time data collected from the country's dense seismic network. Research by Kazuro Hirahara has revealed the location and approximate thickness of the descending Pacific Plate beneath Japan and its complex structure at its margin with the Philippine Sea Plate (Fig. 8). In addition to the presence of high-velocity anomalies in the upper mantle associated with the plates, these studies have revealed the presence of deep low-velocity bodies beneath some of the young volcanoes and the geothermal region in Japan.

Seismic imaging of subduction zones in the United States is still at a preliminary stage. However, a few studies of the Pacific Northwest by the U. S. Geological Survey, the University of Washington, and the University of Oregon have revealed the orientation, thickness, and velocity structure of the subducting Juan de Fuca Plate, which, unlike many other plates, subducts without generating deep earthquakes beneath the Washington Cascades or any earthquakes beneath the Oregon Cascades.

Hot Spots. Several zones of young intraplate volcanism, which have associated with them characteristic geophysical anomalies, are found on the surface of the Earth. These zones, called *hot spots*, are believed to have deep thermal roots in the mantle, and leave visible scars on the Earth's surface as the lithospheric plates slide over them. Seismic tomography has been employed by scientists of the U. S. Geological Survey to image two of these hot spots, the Yellowstone Plateau, a continental hot spot, and Hawaii, an oceanic hot spot (see Fig. 7 for the locations). In addition, other investigators have imaged the upper crust of these regions in detail by using local earthquake waves. Results show the Yellowstone crust to be hot and probably partially molten with distinct magma reservoirs. A large magma body (zone of partial melt) also extends from the crust to at least 150 km beneath the 80-km by 40-km Yellowstone caldera (Fig. 9). Along the Eastern Snake River Plain, presumed to be the volcanic track left by the Yellowstone hot spot, the crust seems to be "healed," whereas magma still seems to be present in the upper mantle, supporting the theory that the Yellowstone–Eastern Snake River Plain volcanic system is associated with a migrating volcanic center with roots in the asthenosphere (Fig. 9). In the upper mantle beneath the island of Hawaii, an elongated zone of low velocities has been found, which coincides with the direction of the island chain (and hence the hot-spot track) and is interpreted to be partially molten rock. However, the crust, though showing velocity anomalies associated with repeated episodes of injection and cooling of magma, does not reveal the presence of substantial magma chambers. Tomographic imaging of the Kilauea volcano in Hawaii using local earthquakes, by

Clifford Thurber of the State University of New York at Stony Brook has revealed a small low-velocity zone, interpreted as the volcano's summit magma reservoir, within a high-velocity core. The similarity in deep structure between the Yellowstone and Hawaii hot spots suggests a similar deep mechanism for hot spots, whereas the difference in crustal structures shows that the transport mechanism of magma from some depth to the surface is different in these two regions.

Volcanic Centers. Several silicic volcanic centers in the western United States and several more mafic volcanoes of the Cascades have been studied by the U.S. Geological Survey under its geothermal program, using teleseismic travel time tomography (Fig. 7). In all of the six silicic centers investigated, crustal low-velocity zones of 100–1000 km^3 volume were found, which, when taken together with geologic and other geophysical data, imply that they are magma chambers (Fig. 10). Two of these centers are the Geysers–Clear Lake volcanic system associated with the largest geothermal power development in the world, and Long Valley caldera, the scene of an ongoing volcanic alert. In these areas, as well as in another geothermal area associated with silicic volcanism, the Coso geothermal area, U. S. Geological Survey and other investigators have carried out local earthquake tomography of the upper crust.

Unlike at these silicic centers, the volcanoes of the Cascade Range do not have identifiable low-velocity anomalies, indicating that large magma chambers are not present beneath them. These results show that the mechanisms of storage and transport of magma beneath intraplate silicic volcanic centers and in the volcanoes in subduction zones are different. In order to model small magmatic intrusions that may be present in the upper crust beneath the Cascade Range volcanoes, the U. S. Geological Survey carried out a high-resolution analog of teleseismic wave tomography at Newberry volcano in Oregon and at Medicine Lake volcano in California, using seismic waves generated by chemical explosions. In these experiments, based on an experimental procedure called the NeHT method after the French scientists Nercessian, Hirn, and Tarantola, who originally developed it to image the Mont Dore volcano in France (Nercessian et al., 1984), seismic rays refracted in the upper crust or reflected by a deep crustal layer are recorded by a dense array of seismographs deployed over the volcanic edifice. The rays from a ring of explosions illuminate the volume under study (Fig. 11). Results from the Newberry volcano indicate the presence of a 1–2-km-wide low-velocity anomaly 3–6 km beneath the summit caldera.

An Earthquake Source Region. The region in the vicinity of New Madrid, Missouri, where some of the largest earthquakes in coterminous United States occurred in 1811–1812, is also seismically the most active zone in the eastern United States. Imaging of this region by teleseismic and local earthquake wave tomography by Brian Mitchell and his colleagues of the St. Louis University has revealed the existence of large- and small-scale velocity heterogeneities in the crust and upper mantle. The major feature is a band of low velocities in the uppermost mantle correlating in position with the zone of seismic activity and the inferred basement rift system in the region.

Future

New developments in tomography are likely to be in the areas of improved theoretical and computational techniques, use of faster computers, and new instrumentation. Better modeling capabilities to reduce "smearing" in the final models, use of seismic wave packets, including multiple phases rather than first arrivals alone, capability to compute attenuation models in addition to velocity models, and simultaneous inversion of several geophysical data sets (seismic, gravity, magnetic, etc.) can all be expected to receive special attention from researchers. The proposed 100-station global digital seismic network and the 1000-station portable seismic network planned in the United States by the IRIS Consortium and similar developments in other countries are encouraging developments and promise to put in the hands of seismic tomographers facilities comparable to those of their medical counterparts.

Acknowledgments

I thank David Oppenheimer and John Evans for critically reviewing this manuscript and suggesting useful changes and Tim Hitchcock for preparing the figures.

H. M. IYER

References

Achauer, U., J. R. Evans, and D. A. Stauber, 1988, High-resolution tomography of compressional wave velocity structure at Newberry Volcano, Oregon Cascade Range, *Jour. Geophys. Research* **93**, 10135–10147.

Achauer, U., L. Greene, J. R. Evans, and H. M. Iyer, 1986, Nature of magma chamber underlying the Mono Craters area, eastern California, as determined from teleseismic travel time residuals, *Jour. Geophys. Research* **91**, 13873–13891.

Aki, K., 1982, Three dimensional seismic inhomogeneities in the lithosphere and asthenosphere, *Rev. Geophysics Space Physics* **20**, 161–170.

Aki, K., A. Christoffersson, and E. S. Husebye, 1977, Determination of the three dimensional seismic structure of the lithosphere, *Jour. Geophys. Research* **82**, 277–296.

Aki, K., and W. H. K. Lee, 1976, Determination of three dimensional velocity anomalies under a seismic array using first P arrival times from local earthquakes. Part I. A homogeneous initial earth model, *Jour. Geophys. Research* **81**, 4381–4399.

Aki, K., and P. Richards, 1980, *Quantitative Seismology: Theory and Methods*, vol. 2. San Francisco: W. H. Freeman and Co., 641-719.

Anderson, D. L., and A. M. Dziewonski, 1984, Seismic tomography, *Sci. Am.* **251**, 60-68.

Backus, G. E., and J. F. Gilbert, 1967, Numerical application of formalism for geophysical inverse problems, *Royal Astron Soc. Geophys. Jour.* **13**, 247-276.

Dziewonski, A. M., 1984, Mapping the lower mantle: determination of lateral heterogeneity in P velocity up to degree and order 6, *Jour. Geophys. Research* **89**, 5929-5952.

Dziewonski, A. M., and D. L. Anderson, 1984, Seismic tomography of the earth's interior, *Am. Scientist* **72**, 483-494.

Evans, J. R., 1982, Compressional wave velocity structure of the upper 350 km under the eastern Snake River Plain near Rexburg, Idaho, *Jour. Geophys. Research* **87**, 2654-2670.

Hearn, T. M., and R. W. Clayton, 1986a, Lateral velocity variations in Southern California. I. Results for the upper crust from P_g waves, *Seismol. Soc. America Bull.* **76**, 495-509.

Hearn, T. M., and R. W. Clayton, 1986b, Lateral velocity variations in Southern California. II. Results for the lower crust from P_n waves, *Seismol. Soc. America Bull.* **76**, 511-520.

Hirahara, K., 1981, Three dimensional seismic structure beneath southwest Japan, the subducting Philippine Sea Plate, *Tectonophysics* **79**, 1-44.

Humphreys, E., and R. W. Clayton, 1988, Adaptation of back projection tomography to seismic travel time problems, *Jour. Geophys. Research* **93**, 1073-1085.

Humphreys, E. R., R. W. Clayton, and B. H. Hager, 1984, A tomographic image of mantle structure beneath Southern California, *Geophys. Research Letters* **11**, 625-627.

Iyer, H. M., 1984, Geophysical evidence for the locations, shapes and sizes, and internal structures of magma chambers beneath regions of Quaternary volcanism, in S. Moorbath, R. N. Thompson, and E. R. Oxburgh, eds., Relative Contributions of Mantle, Oceanic Crust, and Continental Crust to Magma Genesis. *Philos. Trans. Royal Soc. London* **A310**, 473-510.

Iyer, H. M., 1988, Seismological detection and delineation of magma chambers beneath intraplate volcanic centers in Western U.S.A., in C. King and R. Scarpa, eds., *Modeling Volcanic Processes*. Braunschweig/Wiesbaden: Friedr. Vieweg & Sohn, 1-56.

Morelli, A., and A. M Dziewonski, 1987, Topography of the core-mantle boundary and lateral homogeneity of the liquid core, *Nature* **325**, 678-683.

Nakanishi, I., and D. L. Anderson, 1983, Measurements of mantle wave velocities and inversion for lateral heterogeneity and anisotropy. 1. Analysis of great circle phase velocities, *Jour. Geophys. Research* **88**, 10267-10283.

Nakanishi, I., and D. L. Anderson, 1984, Measurements of mantle wave velocities and inversion for lateral heterogeneity and anisotropy. 2. Analysis by single-station method, *Royal Astron. Soc. Geophys. Jour.* **78**, 573-618.

Nataf, H., I. Nakanishi, and D. L. Anderson, 1986, Measurements of mantle wave velocities and inversion for lateral heterogeneities and anisotropy. 3. Inversion, *Jour. Geophys. Research* **91**, 7261-7307.

Nercessian, A., A. Hirn, and A. Tarantola, 1984, Three-dimensional seismic transmission prospecting of the Mont Dore volcano, France,. *Geophys. Jour. Royal Astron. Soc.* **76**, 307-315.

Pavlis, G. L., and J. R. Booker, 1980, The mixed discrete continuous inverse problem: application to simultaneous determination of earthquake hypocenters and velocity structure, *Jour. Geophys. Research* **85**, 4801-4810.

Thurber, C. H., 1983, Earthquake locations and three-dimensional crustal structure in the Coyote Lake area, central California, *Jour. Geophys. Research* **88**, 8226-8236.

Thurber, C. H., and K. Aki, 1987, Three-dimensional seismic imaging, *Ann. Rev. Earth and Planetary Sci.* **15**, 115-139.

Woodhouse, J. H., and A. M. Dziewonski, 1984, Mapping the upper mantle: three dimensional modeling of earth structure by inversion of seismic wave forms, *Jour. Geophys. Research* **89**, 5953-5986.

Cross-references: *Earth Structure, Global; Mantle, Lower: Structure; Mantle, Upper: Structure; Mantle Convection and Plumes; Mantle Discontinuities; Mantle Dynamics; Mantle Viscosity; Seismic Imaging; Seismic Ray Theory; Seismic Wavefield Migration; Surface Waves.*

SEISMIC WAVEFIELD MIGRATION

Conventional time sections suffer from distortions and artifacts. *Migration* aims at correcting two of them: the misplacement of reflections in the section and the interference of diffractions with large parts of the display. It will be seen that, roughly speaking, the same cause, diffraction, can be considered as the cause for both these phenomena and therefore can be assigned the role of the villain of the piece. Consequently, the touchstone for an efficient action of a migration processing will be whether it is able to do a good job of focusing the hyperbolic-looking figures due to point diffractors.

Distortion of Seismic Images

Let us for simplicity consider seismic recordings performed with "zero offset," that is, with coincident sources and receivers. In such a situation, the waves strike all reflecting surfaces at normal incidence. It follows that, should there be even slight dips, the corresponding reflection points are not going to be located on the vertical of the shot point. For example, assume for simplicity that sound propagation velocity V is constant between surface and reflector (Fig. 1); then reflection point P is offset updip from the vertical of S by a distance $SP' = SP \sin \alpha = V(t/2)\sin \alpha$, where t is the two-way travel time from S to P and back, and α is the reflector's dip angle. To get an order of magnitude of the horizontal offset, take $V = 2500$ *m/s*, $t = 2$ s, $\alpha = 15°$. Then $SP' = 647$ m!

On the seismic section, the corresponding trace lies at the abscissa of its own shot point. All signals on

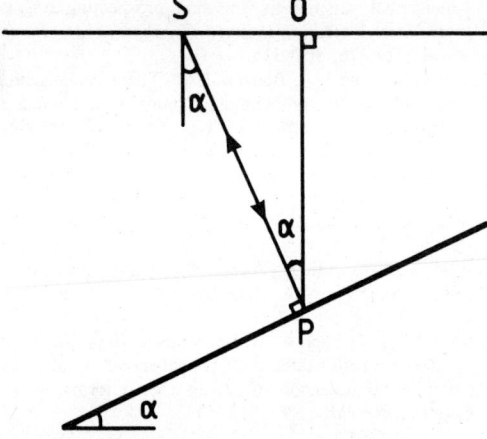

FIGURE 1. S is common location of both shot and receiver. Incident ray SP is reflected back on itself as PS. Velocity is assumed to be constant.

each trace therefore appear to have been generated in the Earth vertically below that shot point (Fig. 2), and the reflected energy is not displayed at the correct location. Let C_1 be the x scale factor and C_2 the time scale factor. Then $S_1'S_2'/S_1S_2 = C_1$ and $S_1'Q_1/(2\,S_1P_1/V) = C_2 = S_2'Q_2/(2S_2P_2/V)$. The lateral offset of reflection points P_1 and P_2 does not appear on the section: A rectilinear reflector segment P_1P_2 is portrayed as a rectilinear segment Q_1Q_2. The real dip angle $\alpha = \sin^{-1}[(S_1P_1 - S_2P_2)/S_1S_2]$ has been transformed into an apparent dip angle $\beta = \tan^{-1}[(S_1'Q_1 - S_2'Q_2)/S_1'S_2']$ with angles α and β related by the equation

$$\sin \alpha = V(C_1/C_2)\tan \beta \qquad (1)$$

with $C = V(C_1/C_2)$. This formula is the basis for several ancient migration methods.

Figure 3 shows the variation of α as a function of β when $C = 1$, for instance, if $V = 2500$ m/s, $C_1 = 4 \times 10^{-5}$, $C_2 = 10^{-1}$ m/s. For small dip angles, $\alpha \approx \beta$. For higher dips, $\beta < \alpha$ and β reaches a limit equal to $45°$ when the reflector is vertical ($\alpha = 90°$). Anticlines appear wider than in nature. Synclines look narrower than they actually are, and reflections from their flanks may even be represented as intersecting events on the section. An example can be seen in Fig. 4, where seismic alignments cross each other in an undeniably unnatural way.

Should velocity V between surface and reflector not be constant, as is most often the case, the same problems will arise, but the mathematics will only be slightly more complicated. Similar situations will be encountered for recordings with nonzero offsets. Then the reflection point for a shot-receiver pair will not lie on the vertical of the middle point of the pair. It usually suffers an updip offset.

Diffraction Figures

Besides the above-mentioned misrepresentation of reflecting surfaces, a problem arises when diffracting points or lines (most often related to faults) are present. Again assume a zero-offset recording arrangement. A given diffractor P (Fig. 5, page 1156) sends back energy toward the source, S, irrespective of length and orientation of SP. The diffracted event will be drawn on the section as a hyperbola if velocity V is constant:

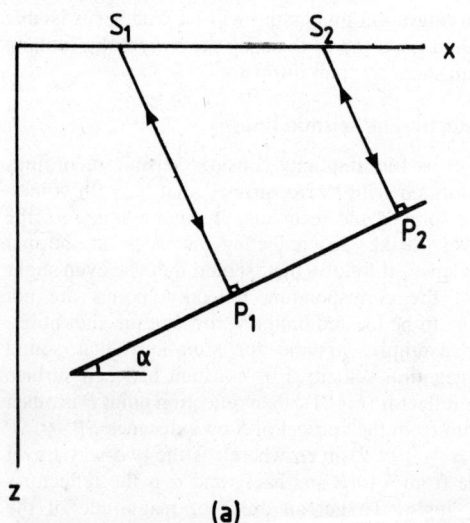
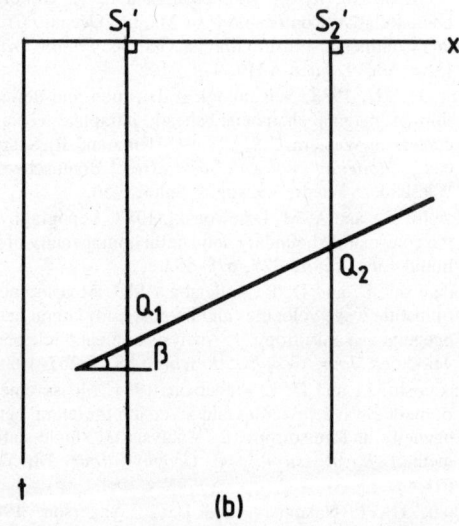

FIGURE 2. (a) In real Earth ($x =$ horizontal distance, $z =$ depth), waves emitted and recorded at surface points S_1 and S_2 are reflected at P_1 and P_2 on a reflector of dip angle α. (b) As displayed on the seismic section ($x =$ distance, $t =$ two-way time), corresponding seismic arrivals mimic a surface with dip angle β.

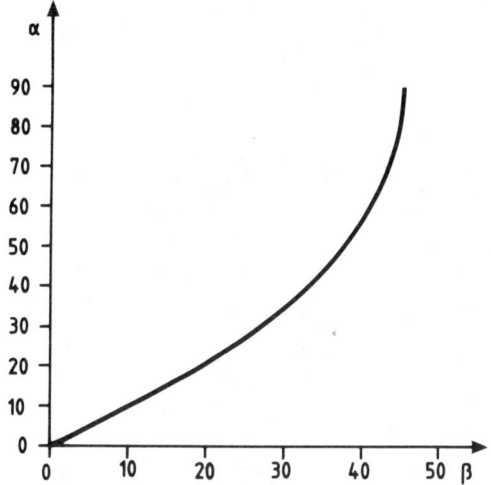

FIGURE 3. α as a function of β. Velocity V is assumed constant and $C = V(C_1/C_2) = 1$. Note that $\beta \leq \alpha$ and $\beta \leq 45°$.

$$t^2 - t_0^2 = 4X^2/V^2 \quad (2)$$

or as a hyperbolic-looking curve if velocity varies within the medium. A diffracting point—by definition a small physical discontinuity—is thus represented on the seismic section as a curve with infinite branches. Such figures will often interfere with each other or with concomitant reflections.

Note that migration of reflections and compression of diffraction curves are very closely related. As a matter of fact, if one is concerned only about propagation times, a reflector may be considered as a succession of diffractors. Collapsing their diffraction figures amounts to migrating all of the reflection points.

Necessity for Correcting These Deficiencies

Since seismic sections do not exhibit true dips and since the simplest geological structures are not represented faithfully, special steps have to be taken, especially since oil and gas exploration often requires an accurate determination of target x- and z-coordinates, particularly when slant or horizontal wells are drilled to reach the objectives.

On the other hand, the artifacts due to diffraction can cause a lot of confusion. It is therefore advantageous to collapse diffraction figures in order to improve the quality of the section. Fortunately, all modern processes that migrate reflectors also collapse the diffractions. Both problems are solved at once.

It is furthermore required that correct amplitude and character relationships be preserved, since the final goal is not only to produce exact isodepth and isopach maps but also to prepare data for a detailed quantitative interpretation in terms of lithology, fluid content, etc.

Migration by Geometrical Methods

Suppose a migration is performed. Then one can superimpose the unmigrated and migrated sections, both with x,t-coordinates (the t-coordinate of the migrated section corresponds to the vertical two-way travel time; i.e., $t = 2z/V$). Figure 6 shows the relationship between the migrated reflector element R and its counterpart R' on the unmigrated seismic section.

Figure 6 provides the basis for the most ancient migration method, which was in use even before the advent of digital computers. It was taking picked records as its raw material. The small, supposedly plane, reflector elements were characterized by their center reflection time and their apparent dip. The migration procedure was known as the *raypath and wavefront* method. Referring again to Fig. 6, we can envisage the principle of the migration of R' into R as the transport of Q toward P along the circle that has S as its center point. The correct location P is that which corresponds to the ray SP with the correct moveout or time gradient.

A chart was used, on which were displayed different wavefronts, labeled according to their propagation times, and different rays, labeled according to their time gradients. Reflector elements were moved to their migrated positions with, as coordinates on the chart, their two-way reflection time and the corresponding moveout. The use of the raypath-wavefront method thus necessitated the drawing of rays and wavefront families. In general, these curves were derived from a velocity-versus-depth distribution, which was derived from well velocity surveys. The migrated section was drawn by hand, each picked reflection element being migrated in turn. Another method, which is only a variant, was publicized by Hagedoorn as early as 1954 and is known by his name. The charts needed for this method are those of wavefronts and of diffraction curves.

Note that if the z-scale of the geological section is transformed into a time scale as described earlier, then the diffraction curve has its apex at the location of the migrated reflection point. Transforming the seismic section into a migrated time section will collapse the diffraction curve into its apex.

Migration of a Picked Zero-Offset Section

The migration of a zero-offset section relies on the *exploding reflector* model (Loewenthal et al., 1976), which is a simplified ray-tracing technique to obtain the arrival times on the zero-offset section. The idea is the following (Fig. 7): The reflectors are replaced by sources that will be set off at $t = 0$ and the velocity is halved. It is easy to generalize this idea when the velocity is not constant: We just get a more

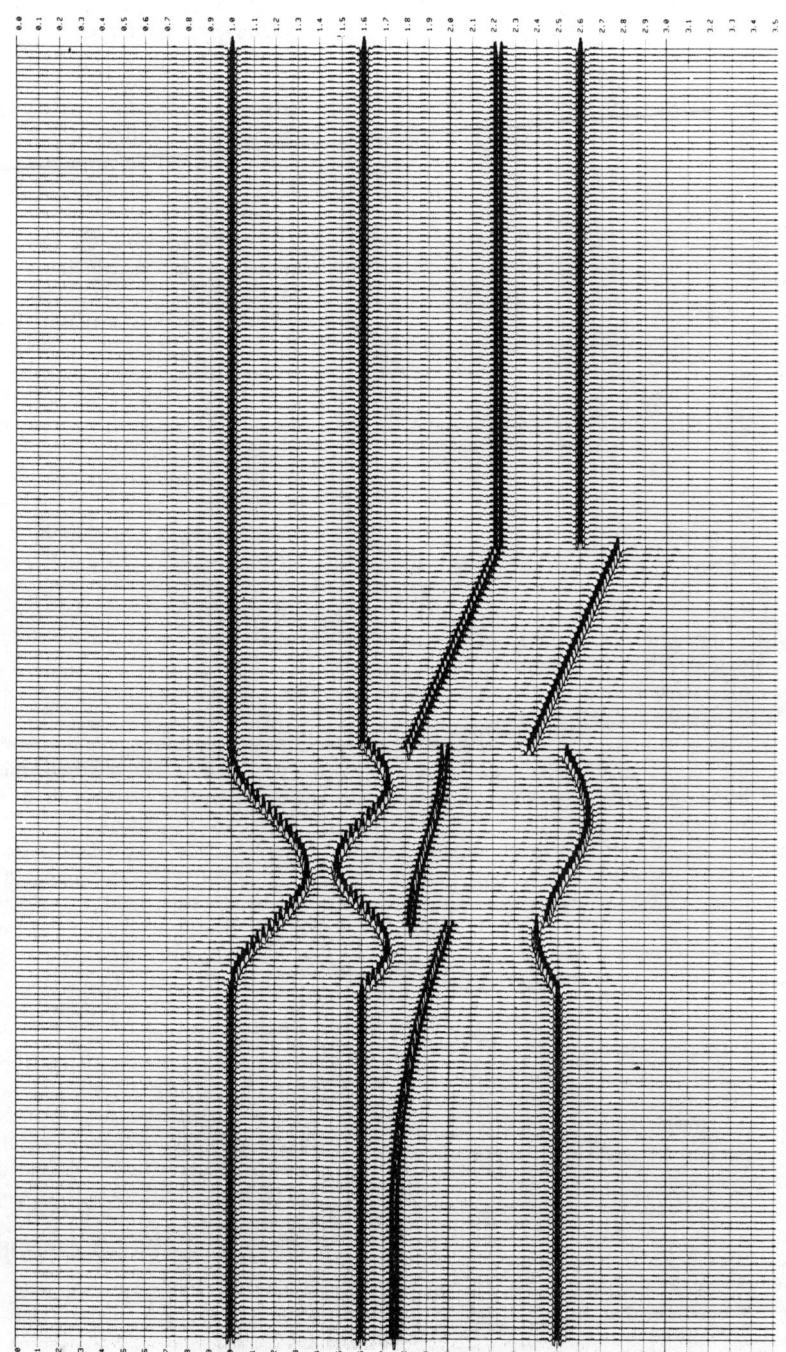

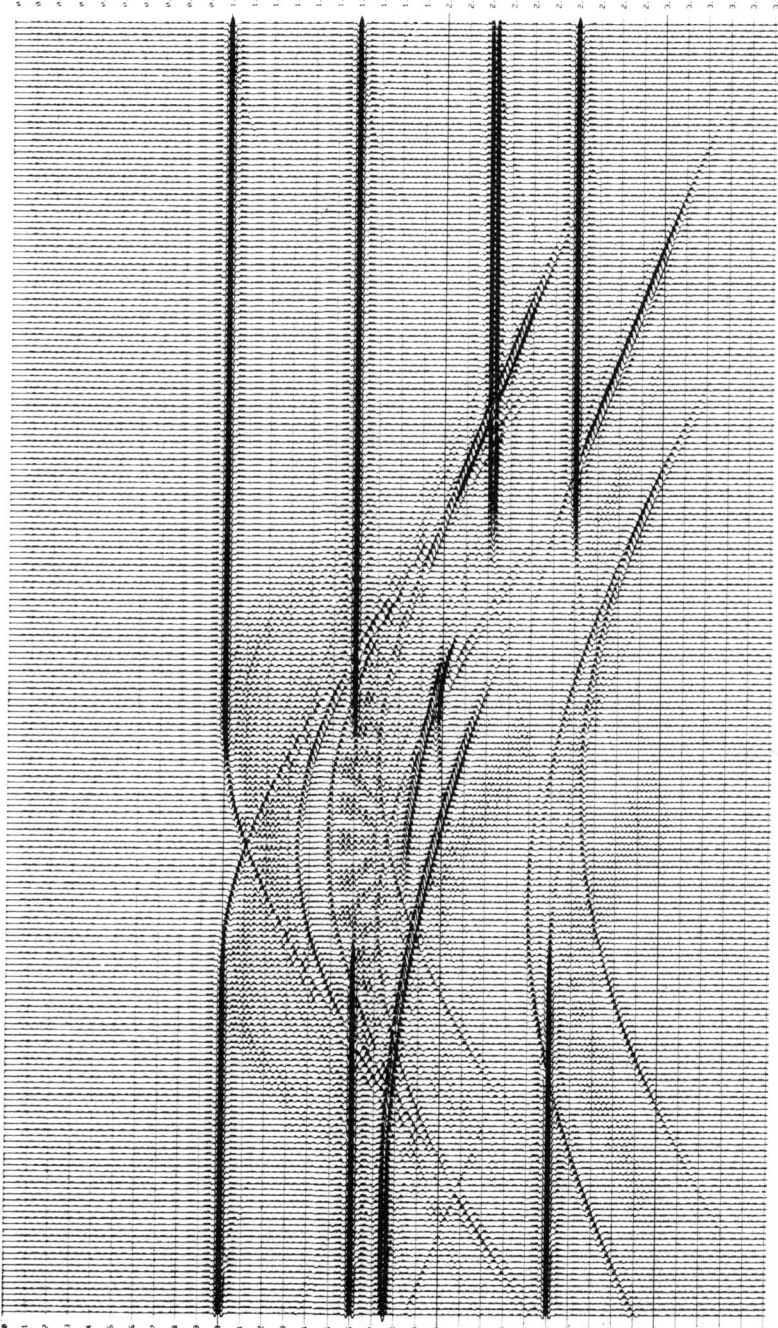

FIGURE 4. (top) A geological model; (bottom) the associated synthetic zero-offset seismic section (Courtesy of Seismograph Service Corporation, P.O. Box 1590, Tulsa, OK)

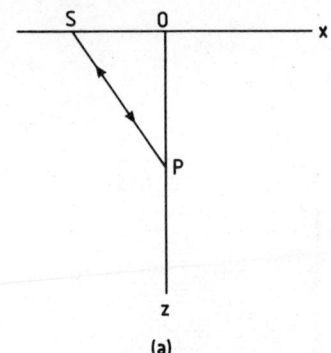
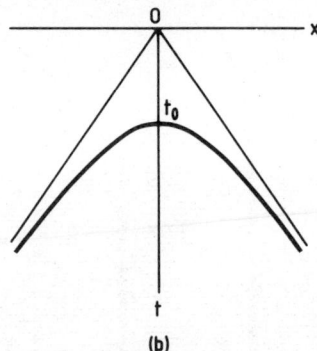

FIGURE 5. (a) Point P diffracts energy even when distance SO is large. (b) Time-distance curve on seismic section is a hyperbola when V is constant. Two-way time to apex, t_0, is $2\,PO/V$. Asymptotes are lines of equation $t = \pm 2x/V$ (with O as origin).

complicated ray tracing. This simplification is not always justified because it relies on the assumption that the downgoing raypath is the same as the upgoing one, and this assumption may not always be true.

The other important idea is that, if we consider a small piece of wavefront (sufficiently small so that it can be considered as plane) emerging at the surface with angle α, the arrival time representation of this piece of wavefront will be plane with an angle β given by Eq. 1 if we assume that velocity V is constant in the small area considered.

Therefore it is possible to derive from the arrival time curves the direction of the emerging ray for each receiver location and then, by classical ray tracing, to cause the upgoing wavefront to propagate backward. In other words, we can reconstruct the history of the picked wavefronts if the velocity distribution is known. Because we deal with the exploding-reflector model, the velocity to be used is half the real velocity. We are going to use the following *imaging principle*. Let us stop this backward wave propagation at time $t = 0$; we then obtain the image of the reflector, and we have achieved the migration of the zero-offset section.

This procedure is known as *migration of picked arrival times*. It is just a generalization of the older methods, which were used in order to perform this procedure by geometrical constructions, assuming that V depends only on depth z. This method is cheap and can be used to perform 3-D migrations (*contour migration*).

Suppose isochrons (lines of equal zero-offset reflection times) have been drawn for a given reflector. It can be shown that the emerging ray is located in the vertical plane of vector $\nabla \mathbf{t}$ and that the emergence angle i_0 (Fig. 8) is $\sin i_0 = (V_0/2)\,\nabla t$, where V_0 is the surface velocity and $\nabla t = |\nabla \mathbf{t}|$. Again, a classical backward ray tracing gives the location of the reflector at the position of the wavefront at $t = 0$. Dip and strike of the reflector can be deduced from the orientation of the unit vector along the ray at the reflector.

One drawback is that the required picking may be difficult if the quality of the data is poor. Modern codes allow an interactive picking. But the most important shortcoming is that only a very small part of the seismic information is used. Very few events are picked, and amplitude information is not used. This drawback motivated Claerbout and his group at Stanford University to derive a numerical procedure in order to perform the complete migration of zero-offset or prestack data. Surprisingly, the principle of these methods is more easily understood with prestack data with a simple enough velocity distribution.

FIGURE 6. Assume V constant, scales such that $V(C_1/C_2) = 1$, and data recorded with a zero-offset field arrangement. Unmigrated reflection point Q and migrated point P are located on diffraction hyperbola (H) of point P and on a circle that has S as its center point. R is a reflecting plane surface with dip α and R' its image with apparent dip β on the unmigrated section. R is tangent to the circle in P. R' is tangent to the hyperbola in Q.

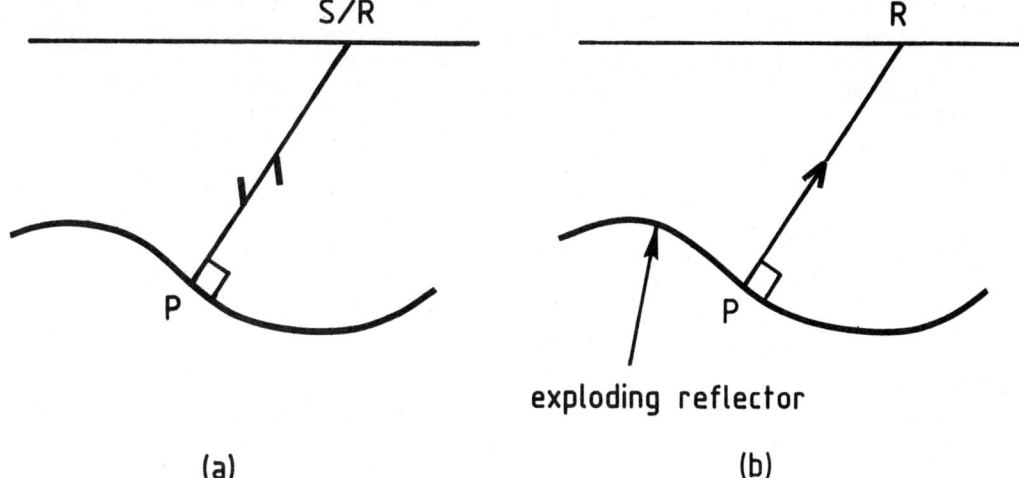

FIGURE 7. The exploding reflector model. Source S and receiver R at the same location. Propagation times are identical in both cases: (a) The wave leaves S at time zero, propagates with velocity V, is reflected at normal incidence at the reflection point P, and reaches R at time $t = 2SP/V$. (b) The wave leaves point P at time zero, propagates with velocity $V/2$, and eventually reaches receiver R at the same time as in (a).

Numerical Migration of Prestack Data

Let us assume that the wave propagation can be modeled by the classical scalar wave equation (acoustic approximation) and, for simplicity, that the density of the medium is constant. We introduce the following notations:

- $\mathbf{r}$ = vector of R^3 characterizing the location of a point
- t = time
- $V_a(\mathbf{r})$ = actual velocity distribution
- S = point defining location of seismic source
- $p_s(\mathbf{r}, t)$ = pressure field generated by seismic source at S
- $f_s(\mathbf{r}, t)$ = modelization of seismic source at S

With these assumptions, the pressure field p_s is the solution of the scalar wave equation

$$V_a^{-2} \frac{\partial^2 p_s}{\partial t^2} - \nabla^2 p_s = f_s \qquad (3)$$

Let the unknown velocity distribution $V_a(\mathbf{r})$ be split into two parts: $V_a(\mathbf{r}) = V(\mathbf{r}) + \delta V(\mathbf{r})$, where $V(\mathbf{r})$ is a smooth known velocity distribution and the $\delta V(\mathbf{r})$ are high-frequency heterogeneities that we want to identify.

We model the wave propagation using scattering theory. The total wavefield p_s is the superposition of a primary wave field p_s^0 and a secondary wave field δp_s resulting from the scattering of p_s^0 on the heterogeneities $\delta V(r)$:

$$p_s(\mathbf{r}, t) = p_s^0(\mathbf{r}, t) + \delta p_s(\mathbf{r}, t)$$

where p_s^0 and δp_s are, respectively, solutions of the equations

$$V^{-2} \frac{\partial^2 p_s^0}{\partial t^2} - \nabla^2 p_s^0 = f_s \qquad (4)$$

$$V^{-2} \frac{\partial^2 (\delta p_s)}{\partial t^2} - \nabla^2 (\delta p_s) = 2 V^{-3} \frac{\partial^2 p_s^0}{\partial t^2} \delta V \qquad (5)$$

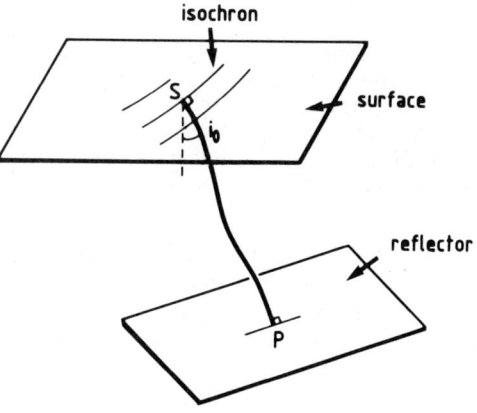

FIGURE 8. Emerging ray is perpendicular at S to the corresponding isochron; it therefore lies in the vertical plane of time-gradient vector ∇t. Angle i_0 is equal to $\sin^{-1}[(V_0/2)|\nabla t|]$. Ray tracing makes it possible to find the location of reflection point P and dip and strike of the reflector, knowing the velocity distribution.

The right side of Eq. 5 models the secondary sources in scattering terminology. Note that this approach for modeling is a linearization of the map $V_a \to p_s$, according to Eq. 3, about the reference medium defined by the distribution V; δp_s depends linearly on δV. We need to have V sufficiently close to V_s for this linearization to be justified. Because the reference medium is supposed to be smooth, we can easily understand that the scattered wave field consists of an upgoing part and a downgoing part that do not interfere with each other, the upgoing part being the interesting one because it is recorded.

Let us consider a perturbation δV with a support localized about a point D. Such a point will be called a *diffractor* or *scatterer*. Note that, because there is a linear dependency of δp_s on δV, we can study the effect of any perturbation $\delta V(\mathbf{r})$ by looking separately at the elementary perturbations associated with each diffractor and superposing all resulting effects.

The migration of prestack data relies on the following characterization property of the diffractors (imaging principle; Claerbout, 1971): The diffractor D is the point at which the primary wave field and the upgoing part of the scattered wave field are time-coincident (Fig. 9). Therefore if we have an estimate $\widetilde{\delta p_s}$ of the upgoing part of the scattered wavefield (see the section "Numerical Downward Continuation of the Data"), we can obtain an indicator of the existence of a diffractor at the point defined by $\mathbf{r}_0$ by computing the value at $t = 0$ of the cross-correlation between the estimated scattered field $\widetilde{\delta p_s}(\mathbf{r}_0, t)$ and the primary wave field $p_s^0(\mathbf{r}_0, t)$:

$$\widetilde{\delta V_s}(\mathbf{r}_0) = \int p_s^0(\mathbf{r}_0, \tau) \widetilde{\delta p_s}(\mathbf{r}_0, t + \tau) \, d\tau \quad \text{at } t = 0 \quad (6)$$

Note that there is a linear dependency between $\widetilde{\delta V_s}$ and the estimated upgoing scattered wavefield $\widetilde{\delta p_s}$. This estimate takes the amplitude of the seismic data into account. However, it cannot give a genuine approximation of the solution of the linearized inverse problem. The solution of Eq. 5 with

$$\delta V_s = \widetilde{\delta V_s}$$

is not the observed scattered wavefield at the measurement points. Greater detail about the relationship between migration and linearized inversion is found in Lailly (1984).

Moreover, even if the seismic excitation is short (a preliminary deconvolution of the data can help in this respect), the signal is still bandlimited and the estimate $\widetilde{\delta V_s}$ suffers from a lack of resolution, especially in the direction of the primary wavefront. Although the lack of resolution unavoidably results from the physics of the problem with bandlimited data, it is another difference between migration and linearized inversion, the latter yielding a better resolution. Equation 6 achieves a partial, bandlimited deconvolution of the scattered wave field by the primary wave field: Migration improves the resolution. For greater detail about migration viewed as a spatial deconvolution, see Berkhout (1982–1984). Note that, if we have at our disposal several shots that illuminate the scattering object under different directions, an improvement of the resolution will be obtained by adding (stacking) the estimates $\widetilde{\delta V_s}$ obtained for the different shots. The signal-to-noise ratio will be enhanced because of the redundancy available in the data. For a general discussion about spatial resolution, see Berkhout (1984).

As a consequence of the underlying linearization, it is very important to use a very good approximation of the actual velocity distribution as a reference medium; otherwise the estimates $\widetilde{\delta V_s}$ will give improperly located images of the diffractors, and the information will be completely destroyed when the uncorrelated estimates are being stacked.

Numerical Migration of Zero-Offset Data

From what has been described, we see that a purely geometrical description of migration is grossly inadequate. Bringing reflectors at their exact location is not enough. Diffracted events must be collapsed, and amplitude and spectral ratios have to be preserved. This preservation is especially important if the migrated data are expected to serve as the raw material for further "lithological" studies. The older migration methods did not take these effects into consideration. Modern migration processing, on the contrary, aims at achieving a more exact transformation concerning amplitudes. Hence, the widespread recourse to wave propagation theory, albeit under simplified forms.

The numerical migration of zero-offset data relies on a mathematical formulation of the exploding reflector model. The wavefield recorded at the surface is supposed to have been generated by sources located on the reflectors, which have an amplitude equal to the reflectivity and are detonated synchronously at time zero. As demonstrated by Cohen and Bleistein (1979), if the reference medium is unbounded (i.e., if the effect of the free surface has been removed) and has a constant velocity V, and if the seismic source is such that $f_s(\mathbf{r}, t) = \delta(\mathbf{r} - S) F(t)$, then the scattered wave field at each shot location S of the zero-offset line is related to the perturbation δV by

$$\delta p_s(\mathbf{r} = S; t) = \frac{1}{4\pi V^2} \int_0^t d\tau \, \frac{\theta(\mathbf{r} = S; \tau)}{\tau} \ddot{F}(t - \tau) \quad (7)$$

where $\ddot{F}$ is the second derivative of wavelet F and θ is the solution of

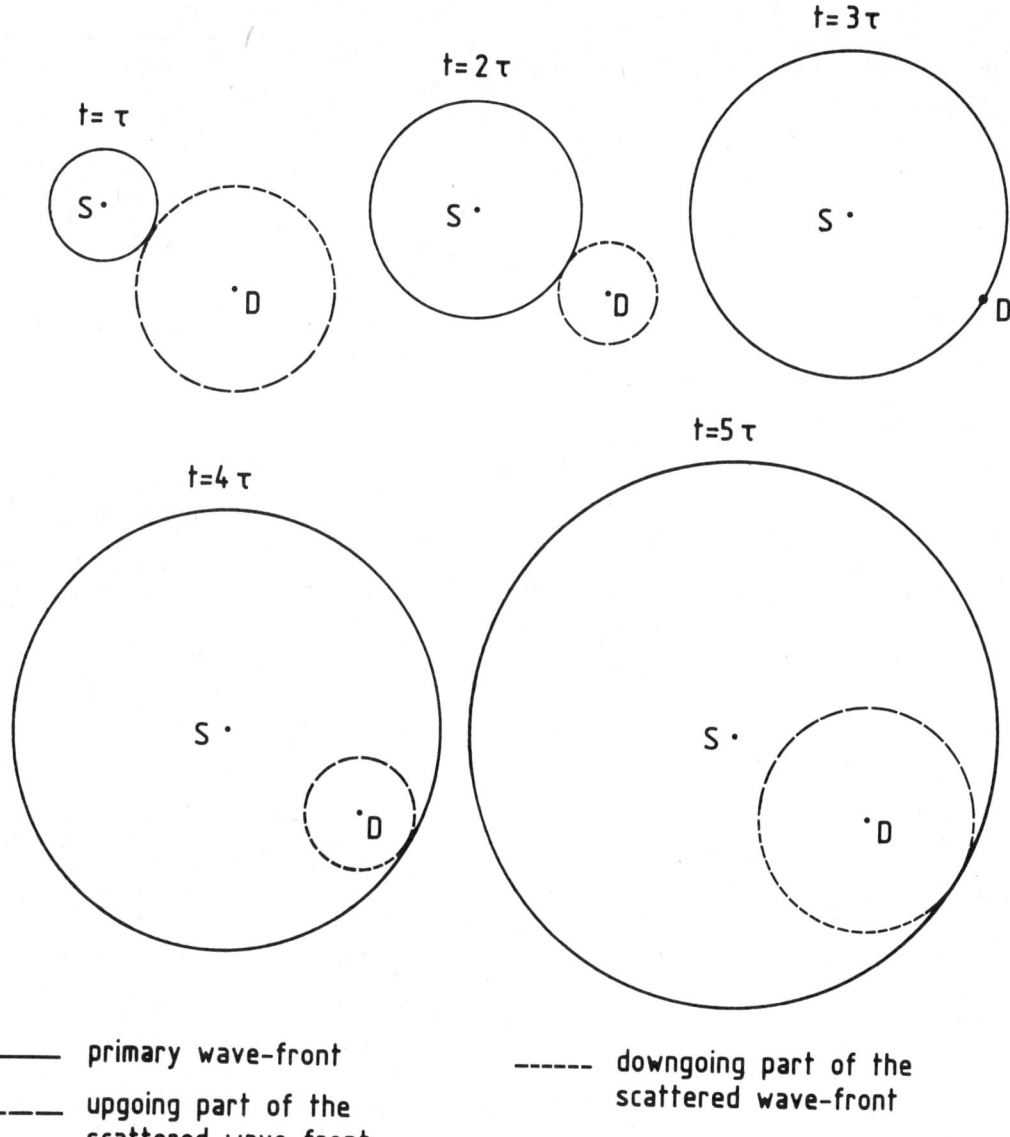

FIGURE 9. Imaging principle for a point diffractor D located at r. Wavefronts of primary and scattered wavefields have been displayed at times $t = \tau, 2\tau, 3\tau, 4\tau, 5\tau$. The primary wave reaches D at $t = 3\tau$. For $t < 3\tau$, the scattered wavefield is an imploding wave, whereas it is an exploding wave for $t > 3\tau$. It is completely focused at $t = 3\tau$. Imaging principle states that the primary wavefront is time-coincident with the upgoing part of the scattered wavefront.

$$\left(\frac{2}{V}\right)^2 \frac{\partial^2 \theta}{\partial t^2} - \nabla^2 \theta = 0, \quad \theta(t=0) = 0,$$

$$\frac{\partial \theta}{\partial t}(t=0) = \delta V \quad (8)$$

Note that the wave field θ is generated by the exploding reflector model: The medium is excited at time $t = 0$, and the propagation velocity is $V/2$. It is rather surprising, and very interesting from a practical standpoint, that the sequence of seismic experiments that are carried out along the line can be simulated by one experiment according to Eq. 8. Thus the cost of modeling a zero-offset section is no longer proportional to the number of shot points. This simplification of the forward problem is the reason why digital migration of zero-offset sections is so cheap compared with that of unstacked data. Unfortunately, the generalization of this result to the most interesting case, practically speaking, of a heterogeneous reference medium is rather troublesome and requires geometrical optics approximations.

The migration of zero-offset data endeavors to estimate δV, given δp_s, by the following procedure, which is similar to the migration of a picked section:

1. Estimate θ at the different shot locations. A deconvolution of the data by the wavelet $F(t)$ is required.
2. Estimate the wavefield θ everywhere in the substratum by a downward continuation.
3. Image the heterogeneities δV looking at $\partial \theta / \partial$ ($t = 0$) by use of the imaging principle.

This algorithm does not achieve the inversion of the linear map defined by Eqs. 7 and 8 (for greater detail see Lailly, 1984). As for the migration of unstacked data, the solution suffers amplitude distortions and lacks resolution.

Numerical Downward Continuation of the Data

In order to migrate prestack as well as zero-offset data, we have to reconstruct the history of the upgoing part of the scattered wavefield that was observed at the receiver locations. So we would like the wavefield q to satisfy

$$V^{-2} \frac{\partial^2 q}{\partial t^2} - \nabla^2 q = 0$$
$$q(z = 0) = g(x, t),$$
q made up of upgoing waves

where $g(x, t)$ denotes the seismic data recorded at $z = 0$. Note that the term "upgoing waves" is difficult to define accurately for a heterogeneous reference medium, even when it is smooth. Therefore the different methods commonly used to compute wave field q introduce approximations.

These methods are as follows:

Fourier (or f-k) methods may be applied to the case of a reference medium that depends on depth only. They use a two-dimensional (time and horizontal distance) Fourier transform, the sign of the vertical wavenumber being chosen such that harmonic waves propagate upward. The Wentzel-Kramers-Brillouin-Jeffreys (WKBJ) approximation may be used to take the velocity variations into account.

Kirchhoff methods compute the solution of the wave equation by using the classical Rayleigh integral recursively at planes with increasing depths, assuming that the velocity is locally constant (Berkhout, 1982). The upgoing nature of the wave fields results from the assumption that the sources are located under the horizontal plane where the wavefield is being computed.

Time-Reversal methods solve the wave equation backward in time, starting from zero final conditions. These conditions express the fact that the wave field, which is traveling upward, must have left the half-space $z > 0$ after a sufficiently large time t if we assume that the perturbations δV are not located too deeply.

One-Way Wave Equation methods consist in approximating the dispersion relation of the wave equation (Fig. 10), $Vk_z/\omega = [1 - V^2 k_x^2/\omega^2]^{1/2}$, by

$$\frac{Vk_z}{\omega} = 1 - \frac{V^2 k_x^2}{2\omega^2} \qquad (9)$$

an approximation valid for an upgoing harmonic plane wave $\exp i(\omega t - k_x x + k_z z)$ traveling in a nearly vertical direction, namely for $|Vk_x/\omega| \ll 1$.

For constant velocity V, the partial differential equation associated with Eq. 9 is

$$V^{-2} \frac{\partial^2 u}{\partial t^2} - V^{-1} \frac{\partial^2 u}{\partial z \partial t} - \frac{1}{2} \frac{\partial^2 u}{\partial x^2} = 0 \qquad (10)$$

Equation 10 is known as the parabolic or 15° paraxial approximation of the wave equation. As shown in Fig. 10, the approximation is correct for

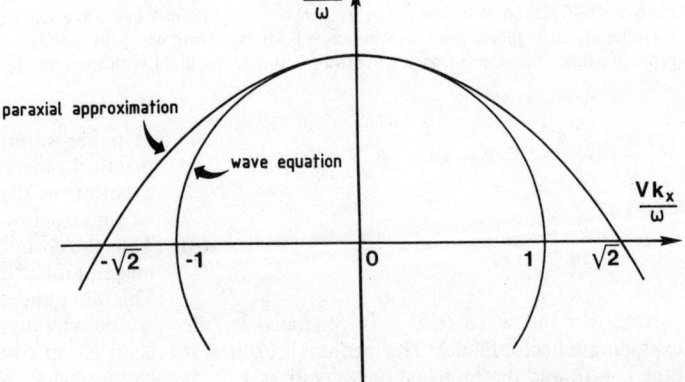

FIGURE 10. Slowness diagrams for the wave equation and the parabolic approximation. The dispersion relation of the wave equation is properly approximated for upgoing waves propagating in directions not farther than 15° from the vertical.

waves propagating in directions not farther than 15° from the vertical (for a distance of propagation less than 40 wavelengths).

Other ways of deriving this and more accurate paraxial approximations (yielding the so-called 45°, 60° equations) can be found in Tappert (1977) and Claerbout (1985). Compared to the complete wave equation, the paraxial approximation costs less when finite-difference techniques are used.

An important difficulty arises if one wants to generalize Eq. 9 to the case of laterally varying velocities. An introduction to this problem may be found in Hatton et al. (1981).

A commonly used, but rather unjustified, approximation is the time-migration procedure, as opposed to the more accurate depth migration. It consists, when the velocity varies laterally, in dropping some terms in the paraxial equation in order to simplify the computations. These terms are supposed to be small when the velocity distribution $V(x, z)$ can be approximated, for any x, by the same function $V(z)$, i.e., when the lateral variations of the velocity are small. The result is displayed with the time $\int_0^z dz/\bar{V}(z)$ as depth coordinate. For greater technical detail, see Claerbout (1985).

Figure 11 shows a zero-offset section and the velocity model that was used for the migration. The comparison of the time-migrated section with the depth-migrated section (Fig. 12) shows the artifacts introduced by the time migration: Reflectors on the depth-migrated section at the same depth for two different abcissae are imaged at different depths on the time-migrated section.

Whether a time or depth migration is performed, the finite-difference or finite-element techniques are very attractive tools for computing the solution of the paraxial approximation. The relative cheapness of both methods and their flexibility for handling complex reference media (Fig. 11) make them very popular. Such codes have also shown their efficiency for performing three-dimensional (3-D) migrations. The main drawback of these methods is the numerical dispersion (Fig. 13) that can create serious difficulties because it decreases the resolution and may lead to interpretation mistakes. A complete study of these problems as well as the description of the numerical schemes and algorithms used to solve the paraxial approximations are found in Brysk (1983), Claerbout (1985), and Dubrulle (1983).

Migration Strategy

The great variety of migration methods, not all of which have been described here, makes it necessary to look for and adopt an adequate strategy (for a detailed study, see Hosken and Deregowski, 1985). To give accurate results, a migration process must start with and lean on a correct knowledge of the velocity structure in the Earth. The velocity distribution first needed can be coarse. Only the main features need be known, because migration will provide the finer details. There is a kind of synergy between interpretation and migration. In difficult cases, good results can be obtained only if the strategy provides a close enough dialogue between interpreter and data processor. For instance, hard work and a sophisticated analysis were necessary to obtain a correct velocity model for the migration of the zero-offset data displayed in Fig. 11. New techniques, such as travel time inversion, may yield a genuine solution to this difficult problem.

As can be expected, the procedures ought to be adapted to the situation. A simple processing, such as a post-pick or contour map migration, may first be carried out in order to provide a rough approximation of the velocity model. Then advantage can be taken of the newly derived gross velocity distribution in order to obtain a better geometry and more accurate amplitudes. Finally, if a fair velocity model is available, a satisfactory depth migration can be expected. The estimation of the velocity model is clearly the crux of pre- or poststack depth migration.

Another field for selecting a strategy lies in choosing the migration method. Should one be in the presence of a quasi-tabular geology, then the use of a plain 2-D poststack time migration with a 15° finite-difference approximation or even, more simply, of a Fourier method will be in order. When dips are large, another approximation will have to be used. Poststack migration may still be carried out provided that a dip-moveout correction is first applied to the data (Bolondi and Rocca, 1985). When there are important lateral velocity variations in the overburden, it might be necessary to migrate unstacked data, even though the cost of the operation may seem high at first glance. In the most difficult situations, it may be relevant to adopt an iterative procedure.

Finally, one should not forget that 3-D data and a 3-D migration algorithm are, in general, necessary for the identification of a 3-D geological structure. Since seismic data acquisition and processing are usually 2-D, the interpreter is left with the task of correcting 2-D artifacts in a 3-D interpretation problem: 3-D ray tracing modeling might help in this very difficult situation.

Historical Development

During a long period of time, migration was performed by geometrical means. Only picked reflector elements could be redrawn at their correct location. The digital computer made it possible to manipulate entire sections, including all their events, reflections, diffractions, and noise, with their natural amplitude relationships.

Migration has become an integral part of the processing chain. Not only does it play its role in displaying reflectors with their correct forms and in getting rid of diffractions, but it also serves as an

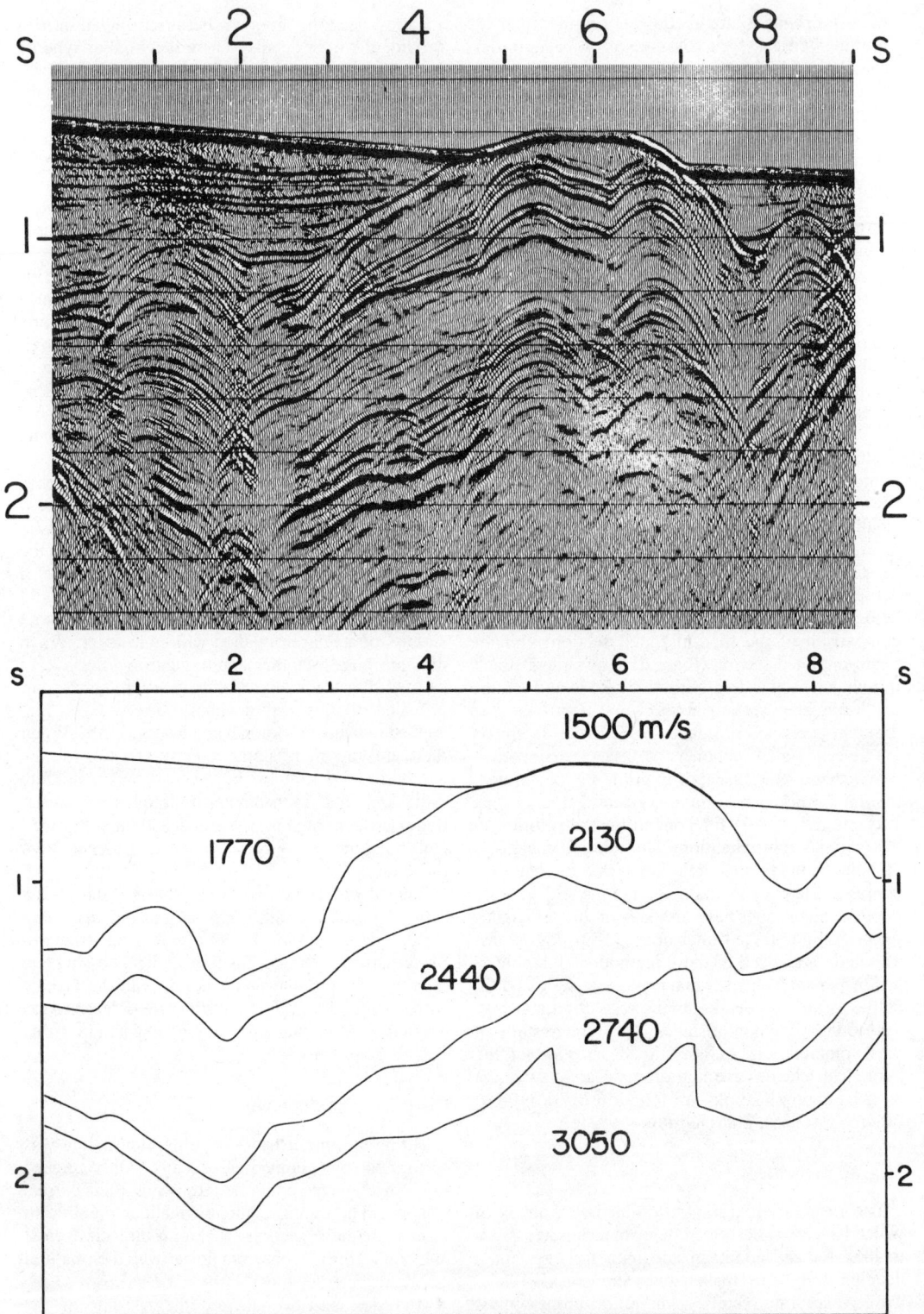

FIGURE 11. (top) A zero-offset section obtained from data recorded in the Santa Barbara Channel; (bottom) the velocity model used to migrate these data (Hatton, Larner and Gibson, 1981).

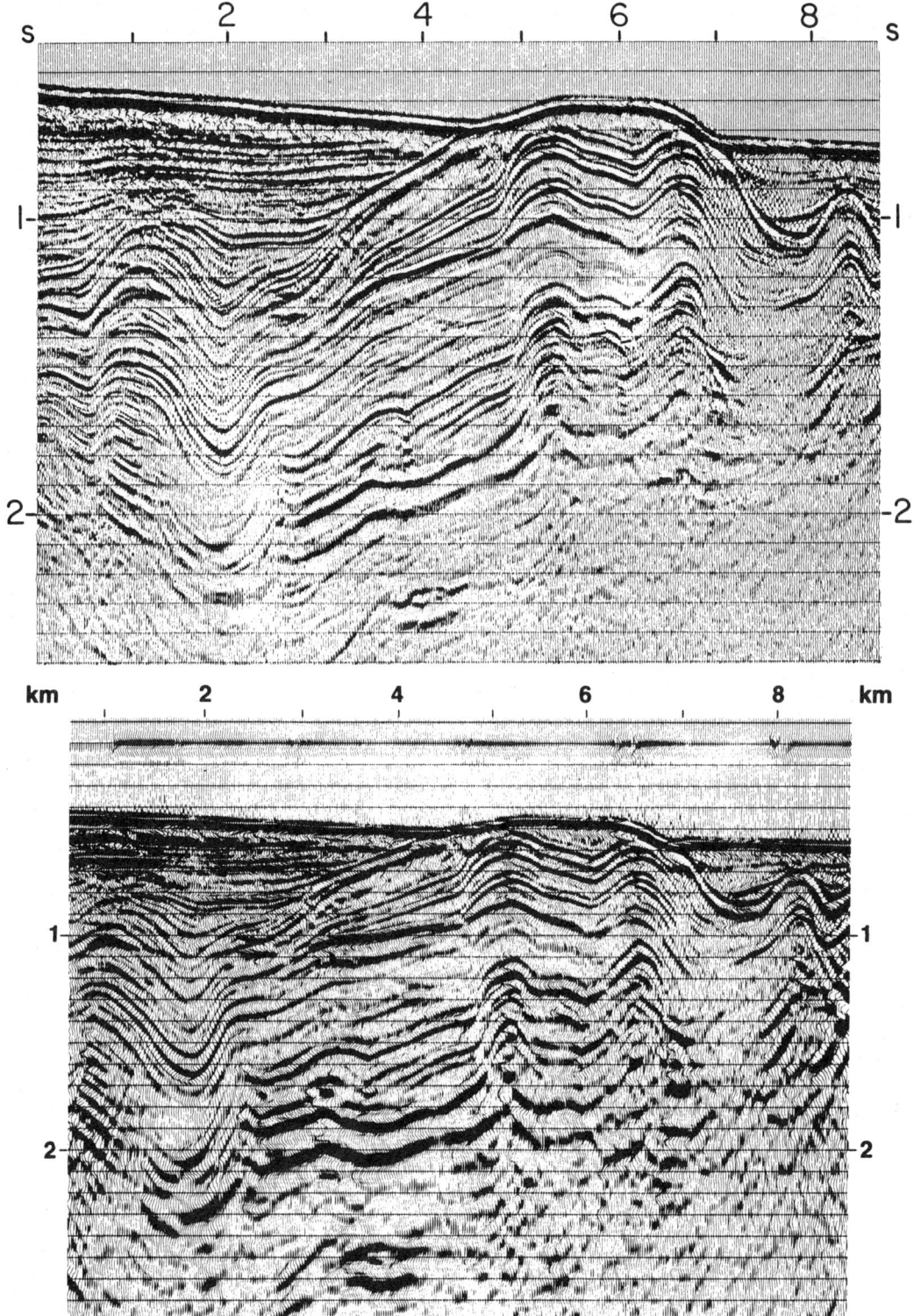

FIGURE 12. (top) Time-migrated section from data of Fig. 11; (bottom) corresponding depth-migrated section. Both migrations have yielded good images of synclines and anticlines. Reflectors are improperly located on the time-migrated section. On the depth-migrated section a decrease of resolution with depth can be observed due to the increase of wavelength in the high-velocity layers.

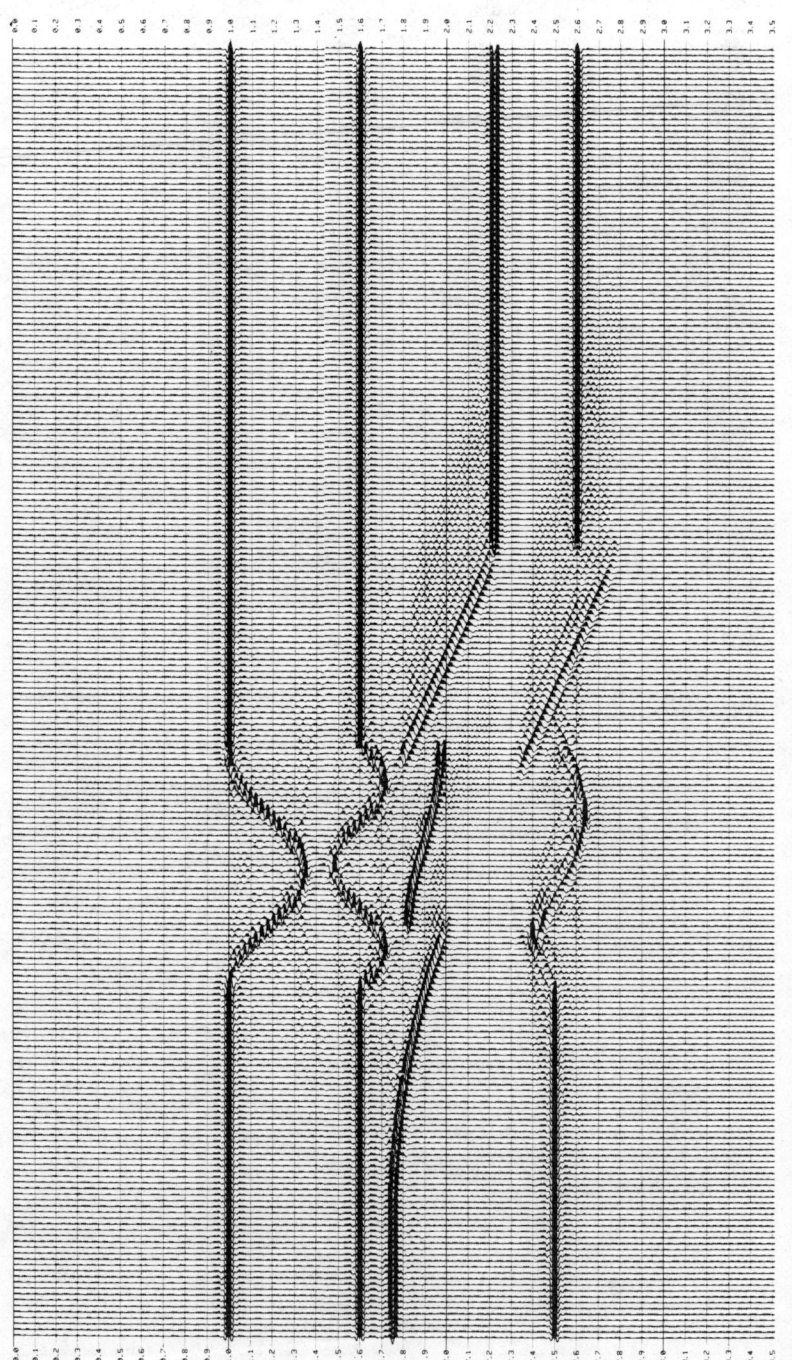

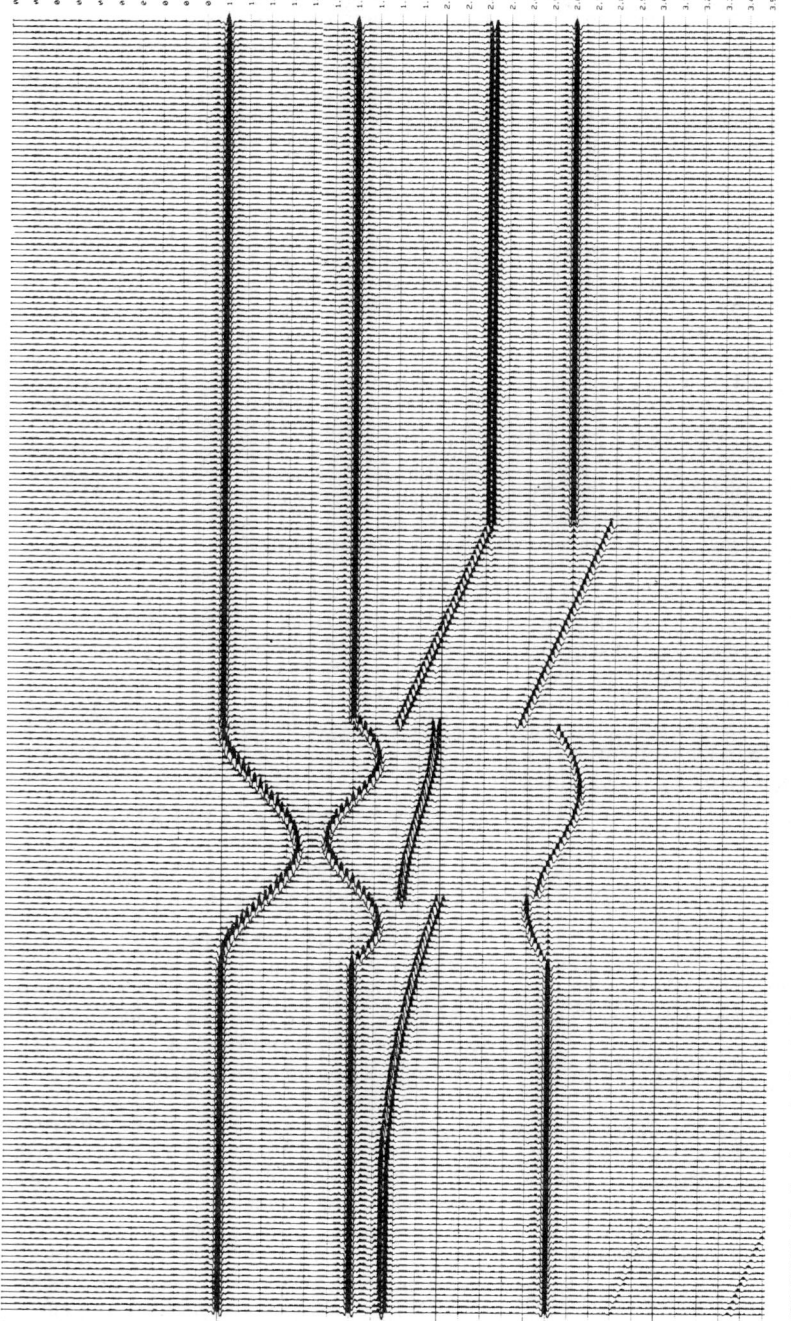

FIGURE 13. Sections obtained after migration of the time section displayed in Fig. 4: (top) 45° finite difference migration; (bottom) f-k migration. (Courtesy of Seismograph Service Corporation, P.O. Box 1590, Tulsa, OK)

efficient kind of stacking that improves signal-to-noise ratio and shrinks Fresnel zones. Ideally, the strength of the reflectors is preserved, a necessity for further quantitative studies.

Very early migration methods used special kinds of rules and protractors and rays and wavefront charts. Around the end of the 1960s digital methods began to be used (for instance, impulse holographic stacking for the migration of unstacked data). In 1970, Claerbout began publishing a long and epoch-making series of papers about wavefield continuation and migration (see Claerbout, 1971). He devised approximations of the scalar wave equation solutions by various ingenious difference equation schemes that made the calculations cheap enough for industrial use. Transform methods have also been proposed by Stolt, Gazdag, and others, which have been implemented in most processing software because of their simplicity. The many methods proposed, their variants, and their respective fields of application are described Hood (1981), Berkhout (1982, 1984), Claerbout (1985), and Stolt and Benson (1986). These references also provide different physical ways of understanding the basis for the migration of seismic data. A mathematical justification of this processing may be found in Miller et al. (1987).

The migration procedures now available are still rather imperfect. In today's practice, the medium is assumed to be isotropic, and head waves and converted waves are usually not taken into consideration. Hopefully, more sophisticated methods, tending toward complete data inversion and taking into account the 3-D nature of the problem, will be developed. Future generations of computers will be useful to provide such processing at an acceptable cost.

G. GRAU
P. LAILLY

References

Berkhout, A. J., 1982, *Seismic Migration*, vol. A. Amsterdam: Elsevier, 339p.

Berkhout, A. J., 1984, *Seismic Migration*, vol. B. Amsterdam: Elsevier, 274p.

Berkhout, A. J., 1984, *Seismic Resolution*. London-Amsterdam: Geophysical Press, 228p.

Bolondi, G., and F. Rocca, 1985, Normal moveout correction, offset continuation and prestack partial migration compared as prestack processes, in A. A. Fitch, ed., *Developments in Geophysical Exploration Methods—6*. London-New York: Elsevier, 27-70.

Brysk, H., 1983, Numerical analysis of the 45 degree finite difference equation for migration, *Geophysics* **48**, 532-542.

Claerbout, J. F., 1971, Toward a unified theory of reflector mapping, *Geophysics* **36**, 467-481.

Claerbout, J. F., 1985, *Imaging the Earth's Interior*. Oxford: Blackwell, 398p.

Cohen, J. K., and N. Bleistein, 1979, Velocity inversion procedure for acoustic waves, *Geophysics* **44**, 1077-1086.

Dubrulle, A. A., 1983, On numerical methods for migration in layered media, *Geophysical Prospecting* **31**, 237-264.

Hatton, L., K. L. Larner, and B. S. Gibson, 1981, Migration of seismic data from inhomogeneous media, *Geophysics* **46**, 751-767.

Hood, P., 1981, Migration, in A. A. Fitch, ed., *Developments in Geophysical Exploration Methods—2*. London: Applied Science Publishers, 151-230.

Hosken, J. W. J., and S. M. Deregowski, 1985, Migration strategy, *Geophysical Prospecting* **33**, 1-33.

Lailly, P., 1984, Migration methods: partial, but efficient solutions to the seismic inverse problem, in F. Santosa, Y. H. Pao, W. W. Symes, and C. Holland, eds., *Inverse Problems of Acoustic and Elastic Waves*. Philadelphia: SIAM.

Loewenthal, D., L. Lu, R. Roberson, and J. W. C. Sherwood, 1976, The wave equation applied to migration, *Geophysical Prospecting* **24**, 380-399.

Miller, D., M. Oristaglio, and G. Beylkin, 1987, A new slant on seismic imaging: migration and integral geometry, *Geophysics* **52**, 943-964.

Stolt, R. H., and A. K. Benson, 1986, *Seismic Migration*. London-Amsterdam: Geophysical Press, 382p.

Tappert, F. D., 1977, The parabolic approximation method, in J. B. Keller and J. S. Papadakis, eds., *Wave Propagation and Underwater Acoustics, Lecture Notes in Physics*, vol. 20. Berlin-Heidelberg-New York: Springer-Verlag, 224-287.

Cross-references: *Exploration Seismology; Finite-Difference Forward Modeling in Seismology; Seismic Diffraction; Seismic Imaging; Seismic Ray Theory; Seismic Signal Processing; Seismic Tomography; Seismic Wave Scattering.*

SEISMIC WAVE SCATTERING

The Spectrum of Lateral Heterogeneities of the Earth and the Scattering Regimes

The classical spherically symmetric (or layered) Earth model is undergoing a revolution. The Earth has been revealed to be laterally heterogeneous everywhere from the crust and mantle to the core, with scales from the size of rocks to the size of continents. These different scale heterogeneities are closely related to the most interesting phenomena in the Earth. For exploration geophysics, heterogeneities offer greater chances of finding oil and minerals. For structural and tectonic geophysics, heterogeneities reflect more geological or tectonic activities, such as earthquake faulting, plate subduction, continent collision, ridge activity, boundary interaction, plume or inclusion formations, and mantle convection. Seismic waves are among the most effective tools to explore and examine these heterogeneities. Figure 1 shows the strength-scale distribution of heterogeneities in the crust and mantle of the Earth, where $\bar{v}$ is the perturbation index of seismic (P or S) wave speed, defined as the rms value of relative

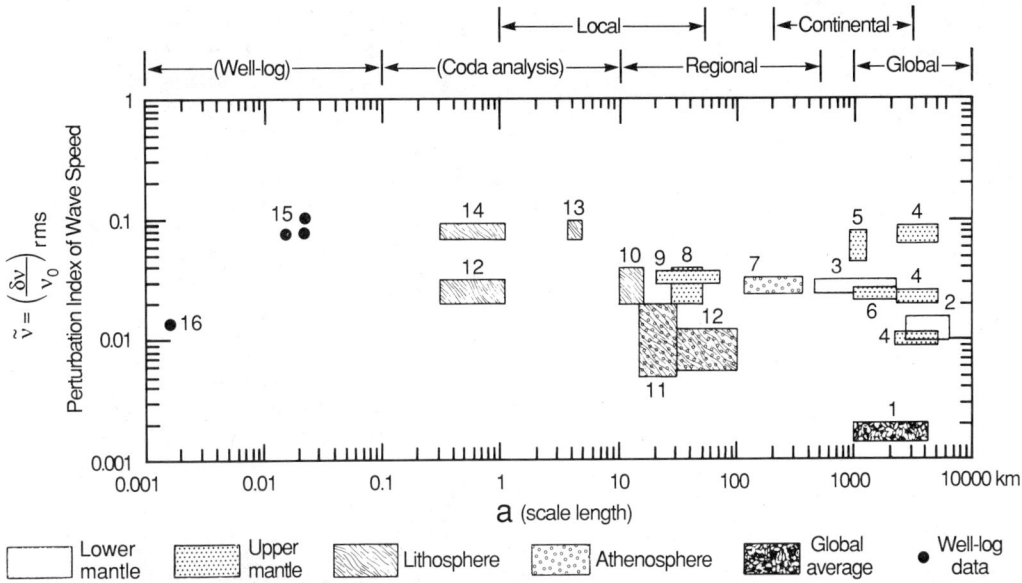

FIGURE 1. Strength-scale distribution of heterogeneities in the Earth. a is the scale length, and $\tilde{v} = (\delta v/v_0)_{rms}$ is the velocity perturbation index of the heterogeneities. (1) Global average, from the analysis of mode splitting of free oscillation (2) Lower mantle, from body wave inversion (3) Lower mantle, from body wave tomography (4) Upper mantle, from surface-wave waveform inversion (5) Upper mantle (Pacific plate), from surface wave full-wave inversion (6) Upper mantle (USA), from travel time inversion (7) Asthenosphere (Central USA, 125–225 km deep), from travel time inversion (8) Upper mantle (Southern California), from body wave tomography (9) Upper mantle, a summary, from travel time inversion (10) Lithosphere, from transmission fluctuation at LASA (11) Lithosphere, from transmission fluctuation at NORSAR (12) Lithosphere, from transmission fluctuation at NORSAR (13) Lithosphere, from coda wave analysis (14) Lithosphere, from coda wave analysis (15) crust, from acoustic well-log (16) Crust, from acoustic well-log. (From Wu and Aki 1988)

wave-speed variations over the measured region. The scales of heterogeneities revealed by seismic waves, not including the laboratory measurements of rock samples, span eight orders. These different scale heterogeneities have different effects on seismic waves. Global heterogeneities split the normal modes of the Earth's free oscillations excited by large earthquakes (see *Free Oscillations of the Earth*). Regional and local heterogeneities cause changes in waveform, travel time, and amplitude fluctuation. Heterogeneities near the core-mantle boundary (CMB) result in scattered waves in the shadow zone of the core. Small-scale local heterogeneities generate coda waves and cause apparent attenuation (scattering attenuation); fine-scale heterogeneities, such as cracks and thin layering, may cause effective anisotropy.

The effects of heterogeneities of different scales on seismic waves can be discussed in terms of different propagation regimes. For heterogeneities of scale a and strength $\tilde{v}$, the wave propagation regime can be characterized by three dimensionless variables: ka ($=2\pi a/\lambda$), L/a, and $\tilde{v}$, where k is the wave number, λ is the wavelength in the medium, and L is the propagation length or the extent of the heterogeneous region. Figure 2 shows the different propagation regimes and the validity regions of approximate analytical methods, which are as follows.

Quasi-Homogeneous: When $ka < 0.01$, the heterogeneities are too small to be seen by the waves. The medium can be treated as homogeneous with some effective parameters. The randomly distributed cracks may result in slower effective wave speed, and aligned cracks can cause effective anisotropy in wave speed.

Rayleigh Scattering: When $ka \ll 1$ (e.g., $ka < 0.1$), the scattered power is proportional to k^4. Scattering in this range can cause apparent attenuation.

Large-Angle Scattering: When $ka \approx 1$ (e.g., $0.1 < ka < 10$), the sizes of the heterogeneities are comparable to the wavelength, and scattering effects are highly significant. The incident power is scattered to different directions at large angles to the incident direction. This regime is also called *resonance scattering* or *Mie scattering*. Scattering in this range is most responsible for coda wave generation and scattering attenuation.

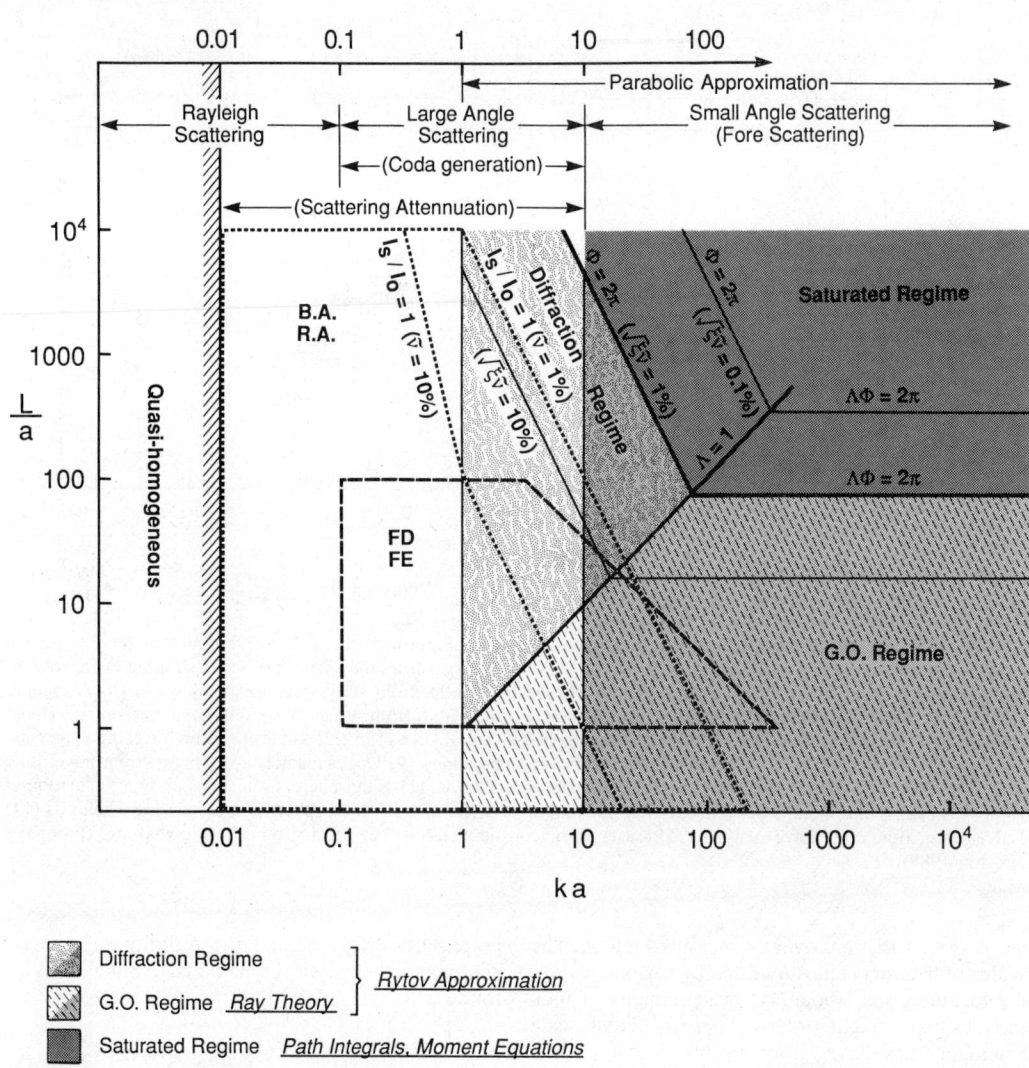

FIGURE 2. Scattering regimes and the region of validity of different approximation methods: a = scale length of the heterogeneities; L = propagation length of the wave; k = the wave number.

Small-Angle Scattering or *Forescattering:* When $ka \gg 1$, most scattered power is concentrated near the forward direction. In this case the parabolic approximation for the full-wave equation can be used. Parabolic approximations (PA) describe a one-way (in forward direction) wave propagation, since backscattered waves are very weak and therefore can be neglected. The use of PA is often extended to the region $ka > 1$ (as in Fig. 2). The lighter gray areas in Fig. 2 are the areas in which the PA may have large errors. Small-angle scattering becomes a focusing, diffraction, and interference problem. It produces travel time and amplitude fluctuations, which can then be used to estimate the medium-velocity distribution. Small-angle scattering can be further divided into three regimes depending on two parameters:

$$\Lambda = \frac{L_F^2}{a^2} = \frac{L}{ka^2} = \frac{L/a}{ka}$$

$$\Phi = \sqrt{\bar{v}^2 k^2 L a_L} = \bar{v}\xi(ka)\sqrt{\frac{L}{a}} \quad (1)$$

where Λ is the diffraction parameter, a dimensionless square Fresnel radius; $L_F = \sqrt{L/k}$ is the average

Fresnel radius along the propagation path L; Φ is the scattering strength parameter, which is approximately the rms phase fluctuations of waves passing through a distance L; a_L is the scale length along the propagation directions; and $\xi = a_L/a$ is the anisotropic ratio of the heterogeneities. The three subregimes are (Fig. 2):

Geometric Optics (GO) Regime: This regime, in which ray theory can be used, is limited by the lines $ka > 1$, $\Lambda < 1$, and $\Lambda\Phi < 2\pi$. Since $\Lambda < 1$, the average Fresnel radius along the path is smaller than the transverse scale of the heterogeneities; therefore diffractions may be neglected. However, if we use ray theory, $\Lambda\Phi < 2\pi$ must also be satisfied. From Eq. 1, $\Lambda\Phi = \sqrt{\xi}\,\tilde{v}(L/a)^{3/2}$, so the validity of ray theory is limited by the propagation length and the perturbation strength $\tilde{v}$. When $\Lambda\Phi > 2\pi$, the rays will split into interfering microrays, and the problem is more complicated.

Diffraction Regime: This regime is bounded by the lines $ka > 1$, $\Lambda > 1$, and $\Phi < 2\pi$, and, since $\Lambda > 1$, diffraction must be taken into account. The condition $\Phi < 2\pi$ limits the regime to weak (unsaturated) amplitude fluctuations. In the diffraction and GO regimes, the Rytov approximations (or the smooth perturbation approximations) can be used.

Saturated Regime: This regime is limited by the lines $\Phi > 2\pi$ and $\Lambda\Phi > 2\pi$. Amplitude fluctuations are saturated, and rays are split into numerous interfering microrays. There is no valid deterministic analytical method available. Path integrals and moment equations can be used for deriving statistical properties of waves.

For large-angle and Rayleigh scattering, the Born approximation can be used for weak scattering (when the scattered field is much weaker than the incident field). Figure 2 shows the confining lines of the Born approximation for $\tilde{v} = 1\%$ and $\tilde{v} = 10\%$. On those lines, the ratio of the scattered power to the incident power is $I_s/I_0 \approx 1$. When scattering is weak, the Rytov approximation reduces to the Born approximation. The Rytov approximation, however, can extend its validity region into the strong scattering regime for unsplit rays.

Methods of Study

Theoretical Analysis. In a source-free linear heterogeneous elastic medium, the equation of motion for displacement **u** is (Aki and Richards, 1980)

$$\rho(\mathbf{x})\ddot{u}_i = (\lambda(\mathbf{x})u_{j,j})_{,i} + [\mu(\mathbf{x})(u_{i,j} + u_{j,i})]_{,j} \quad (2)$$

where ρ is density, u_i are the components of the displacement, "$\cdot\cdot$" stands for $\partial/\partial t$, "$_{,j}$" stands for $\partial/\partial x_j$, a repeated index implies summations with respect to the index over the spatial dimensions, and λ and μ are the Lamé constants of the medium. Each of the numbers ρ, λ, μ may change with position **x**. For the sake of simplicity, we treat only isotropic media.

Boundary Matching Approach for Discontinuous Heterogeneities. Inside and outside the heterogeneities (inclusions), the media are homogeneous, and sharp discontinuities occur on boundaries. The scattering problem for a single inclusion can be formulated as a boundary value problem (either a system of partial differential equations having boundary conditions or a system of boundary integral equations). The problem for a complex medium with multiple inclusions can be attacked by multiple scattering theory based on the solution of the single-inclusion scattering problem.

Perturbation Method for Weak Scattering. Using the perturbation method, a heterogeneous medium is decomposed into a reference medium (the background medium) and perturbations. Suppose in Eq. 2 the material parameters can be expressed as

$$\rho(\mathbf{x}) = \rho_0 + \delta\rho(\mathbf{x}),$$
$$\lambda(\mathbf{x}) = \lambda_0 + \delta\lambda(\mathbf{x}),$$
$$\mu(\mathbf{x}) = \mu_0 + \delta\mu(\mathbf{x}) \quad (3)$$

where ρ_0 or λ_0 and μ_0 represent the reference medium, which is either homogeneous or slowly varying and continuous, and $\delta\rho$, $\delta\lambda$, and $\delta\mu$ are the deviations from their reference values. We decompose the total field as

$$\mathbf{u}(\mathbf{x}, t) = \mathbf{u}^0(\mathbf{x}, t) + \mathbf{U}(\mathbf{x}, t) \quad (4)$$

where the primary field u^0 is the field with no perturbation in the reference medium, and U is the scattered field. Substituting Eqs. 3 and 4 into 2 yields

$$\rho_0 \frac{\partial^2 U_i}{\partial t^2} - (\lambda_0 + \mu_0)(\nabla \cdot \mathbf{U})_{,i} - \mu_0 \nabla^2 U_i = Q_i$$
$$(5)$$

$$Q_i = -\delta\rho \frac{\partial^2 u_i^0}{\partial t^2} + (\delta\lambda + \delta\mu)(\nabla \cdot \mathbf{u}^0)_{,i} + \delta\mu \nabla^2 u_i^0$$
$$+ (\delta\lambda)_{,i}(\nabla \cdot \mathbf{u}^0) + (\delta\mu)_{,i}[u_{i,j}^0 + u_{j,i}^0] \quad (6)$$

Equation 5 is the wave equation in the reference medium (ρ_0, λ_0, μ_0), and $\mathbf{Q}(\mathbf{x}, t)$ represents the equivalent body forces due to the interaction of heterogeneities with the wave field. By introducing the Green's function of the reference medium $G_{ij}^0(\mathbf{x}, t|x', t')$ and applying the representation and divergence theorems, we can express the scattered field as

$$U_i = \int_V \{-\delta\rho \ddot{u}_j G^0_{ij} - [\delta_{jk}\delta\lambda u_{q,q} + \delta\mu(u_{j,k} + u_{k,j})] * G^0_{ij,k}\} d^3\mathbf{x}' \quad (7)$$

When the scattered field U is weak compared to the primary field u^0, we can approximate the total field u in Eq. 7 by the primary field u^0. This approximation is called the *Born approximation*. Equation 7 thus becomes an explicit formula for the scattered field.

High-Frequency Approximate Methods: Smooth Medium Approximation. When the scale of the heterogeneities is much larger than the wavelength ($ka \gg 1$), backscattering and large-angle scattering can be neglected. If the propagation distance is small such that the average Fresnel radius does not exceed the scale length a, then the propagation is in the GO regime (see Fig. 2), which considers only the focusing effect. Usually, the GO regime is excluded from scattering. However, when the propagation distance is long enough to have a Fresnel radius larger than a (in the diffraction regime), we must consider diffraction and interference effects, and thus must resort to the *Rytov approximation*, which is valid in both the diffraction and GO regimes.

High-Frequency Approximate Methods: Sharp Boundary Scattering Problems. When the heterogeneity is in sharp contrast to its host medium properties and is much larger than a wavelength but with sharp boundaries (edge, wedge, block, cavity, crack, etc.), several h-f approximation methods can be used. The geometric theory of diffraction (GTD) approximates the effect of sharp boundaries by adding diffraction rays from those boundaries to the regular ray solution. Diffraction coefficients of these rays are obtained from canonical solutions for different boundaries. As such, the range of application is limited by the availability of canonical solutions. GTD has been extended and applied to crack scattering problems.

When the heterogeneity is thin or composed of thin structures and its boundary surface is smooth compared with the wavelength, the physical optics approximation (POA) can be used for scattering problems. The POA (or Kirchhoff approximation) approximates field values on the boundary surface by a "black-white" distribution, (product of incident wave and reflection or transmission coefficient in lit region and null in unlit region). The scattered field is then calculated by the representation integral, which may be evaluated asymptotically in some cases.

Stochastic Approach and Statistical Properties of Heterogeneities. For very complex media, the deterministic approach is either intractable or impractical, and the stochastic approach then plays an important role. If the range-scale ratio $L/a \gg 1$ (see Fig. 2), this case is more suitable for the stochastic approach. Suppose the medium perturbations $\delta c(\mathbf{x})$ or $\delta\rho(\mathbf{x})$, $\delta\lambda(\mathbf{x})$, and $\delta\mu(\mathbf{x})$ are zero-mean Gaussian random variables. Then the heterogeneous medium becomes a random medium. A random medium is a family of innumerable heterogeneous media. Each member of the family, with a certain probability of existence, differs from the others in detailed structure but has some common statistical properties of the family, such as average size, strength, and power spectrum of the heterogeneities.

A random medium can be specified by the power spectra of its random parameters. Each random parameter, such as $\delta\rho(\mathbf{x})$, $\delta\lambda(\mathbf{x})$, $\delta\mu(\mathbf{x})$, or $\delta v(\mathbf{x})$, can be represented mathematically by a random field (or random function). The power spectrum $W(\mathbf{K})$ of a random field, where $\mathbf{K}$ is the wave number (spatial frequency) vector, is defined as the ensemble average of the power spectra of all the realizations of the random field (in the case of uniform random fields, their Fourier transforms are not defined and Fourier stieltjes integrals have to be introduced. See Appendix A of Ishimaru 1978). We can normalize $W(\mathbf{K})$ with the variance of the random parameter σ^2: $P(\mathbf{K}) = W(\mathbf{K})/\sigma^2$, where $P(\mathbf{K})$ is the *shape spectrum* of the random field. When the statistical properties of the random field are independent of space coordinates, the random field is called *uniform* and a correlation function $N(\mathbf{r})$, which is the Fourier transform of $P(\mathbf{K})$, can be defined. If the heterogeneities are distributed isotropically, then $N(r)$ and $P(K)$ are functions of scalars r and K, respectively. For example, for the exponential correlation function,

$$N(r) = \exp\left(-\frac{|r|}{a}\right), \quad P(K) = 8\pi \frac{a^3}{(1 + K^2 a^2)^2}$$

Numerical Simulation and Physical Model Experiment. Despite remarkable progress of the last two decades in the scattering theory of seismic waves, many complex problems do not even have approximate solutions. Numerical simulation of the propagation and scattering of seismic waves is now widely used. Commonly used methods include finite difference (FD) and finite element (FE) (especially the transient FE) (see *Finite-Difference Forward Modeling in Seismology*).

Physical model experimentation is another tool for studying seismic wave scattering (see *Seismology: Physical Model Studies*). Most experiments on elastic wave scattering use two-dimensional (2-D) models. The media are made from plates of aluminum or Fiberglass. Ultrasonic waves range from several tens of kilohertz to 1 MHz, with P wavelengths of 5–100 mm, S wavelengths of 3–60 mm. Discrete scatterers can be formed with holes of different shapes. Experiments performed include isotropic and nonisotropic scattering, scattering attenuation and the related phase delay, rough-surface scattering problems, such as transmission and reflection of Rayleigh waves passing across a wedge, and conversion between body waves and Rayleigh waves. Another category of model experiments is forescattering. Variation of the density and wave speed of the medium are obtained by changing the porosity of the plate (by varying the number density of regularly distributed small holes). After passing through the heterogeneous region of the plate, transmission fluctuations of phase and amplitude (along a transverse profile) are measured simultaneously with the apparent attenuation. In exploration seismology, model experiments have been widely used to study seismic wave scattering by salt domes, oil traps, or faulted strata.

Field Observation. See the last section, "Manifestations and Applications of Seismic Wave Scattering."

Basic Characteristics of Elastic Wave Scattering and the Scalar Wave Approximations

Elastic Wave Rayleigh Scattering. When the size of scatter is much smaller than wavelength, the whole heterogeneous body can be considered as a point scatterer. If we use spherical coordinates with polar axis in the incident direction x_1 (Fig. 3a), then for P-wave incidence we have the scattered P wave $^PU_r^P$ and S wave $^PU_{mer}^S$:

$$^PU_r^P = \frac{V}{4\pi} \frac{\omega^2}{\alpha_0^2} \left\{ \frac{\overline{\delta\rho}}{\rho_0} \cos\theta \right.$$
$$\left. - \frac{\overline{\delta\lambda}}{\lambda_0 + 2\mu_0} - \frac{2\overline{\delta\mu}}{\lambda_0 + 2\mu_0} \cos^2\theta \right\} \frac{1}{r} e^{-i\omega(t - r/\alpha_0)}$$

(8)

$$^PU_{mer}^S = \frac{V}{4\pi} \frac{\omega^2}{\alpha_0^2} \left(\frac{\alpha_0^2}{\beta_0^2}\right) \left\{ \frac{\overline{\delta\rho}}{\rho_0} \sin\theta \right.$$
$$\left. + \left(\frac{\beta_0}{\alpha_0}\right) \frac{\overline{\delta\mu}}{\mu_0} \sin 2\theta \right\} \frac{e^{-i\omega(t - r/\beta_0)}}{r}$$

(9)

where ω is the circular frequency, α_0 and β_0 are the P- and S-wave speeds, respectively, θ is the scattering angle (the angle between the incident and scattering directions), and r is the distance between the scatterer and the observer. The subscript r stands for the r-component, and mer stands for the meridian component. Because the problem is symmetric with respect to the polar axis, there is no latitudinal component of the S wave; i.e., $^PU_{lat}^S = 0$.

Figure 3a shows the equivalent forces and the scattered P and S waves produced by $\overline{\delta\rho}$, $\overline{\delta\lambda}$, and $\overline{\delta\mu}$ for plane P-wave incidence. We see that $\overline{\delta\rho}V$ generates a point single force, $\overline{\delta\lambda}V$ behaves like a point explosion (or contraction), and $\overline{\delta\mu}V$ behaves like a point normal crack (open or close).

Similarly, for a plane S wave incident in the x_1-direction and having its particle motion in the x_2-direction, the scattered P wave $^SU_r^P$ and the scattered S wave $^SU_{mer}^S$ and $^SU_{lat}^S$ can be written as

$$^SU_r^P = \frac{V}{4\pi} \frac{\omega^2}{\alpha_0^2} \left\{ \frac{\overline{\delta\rho}}{\rho_0} \cos\theta \right.$$
$$\left. - \left(\frac{\beta_0}{\alpha_0}\right) \frac{\overline{\delta\mu}}{\mu_0} \sin 2\theta \sin\phi \right\} \frac{1}{r} e^{-i\omega(t - r/\alpha_0)}$$

(10)

$$^SU_{mer}^S = -\frac{V}{4\pi} \frac{\omega^2}{\alpha_0^2} \left(\frac{\alpha_0}{\beta_0}\right)^2 \left\{ \frac{\overline{\delta\rho}}{\rho_0} \sin\theta \right.$$
$$\left. + \frac{\overline{\delta\mu}}{\mu_0} \cos 2\theta \sin\phi \right\} \frac{1}{r} e^{-i\omega(t - r/\beta_0)}$$

(11)

and

$$^SU_{lat}^S = \frac{V}{4\pi} \frac{\omega^2}{\alpha_0^2} \left(\frac{\alpha_0}{\beta_0}\right)^2 \frac{\overline{\delta\mu}}{\mu_0} \cos\theta \cos\phi \frac{1}{r} e^{-\omega(t - r/\beta_0)}$$

(12)

where the polar axis is taken in the direction of particle motion (y-axis, as in Fig. 3b). The equivalent forces and the scattered field produced by $\overline{\delta\rho}$ and $\overline{\delta\mu}$ are also shown in Fig. 3b. Note that $\overline{\delta\lambda}$ does not affect S-wave scattering, as expected. The density perturbation $\overline{\delta\rho}V$ still acts like a single force in the direction of particle motion. The scattered waves due to $\overline{\delta\mu}V$ are equivalent to the radiation fields of a point double couple, which is equivalent to a point dislocation (shear motion along a crack). In the figure, we decompose the scattered S wave due to $\overline{\delta\mu}$ into two parts, each of which corresponds to a pattern due to a single couple.

From Fig. 3 and the related formulas we can list several important features of elastic wave scattering.

1. Amplitudes of scattered waves have a frequency dependence of ω^2, so scattered power is proportional to ω^4, which is characteristic of Rayleigh scattering.
2. When $\mu = 0$, Eq. 8 is then the equation for acoustic wave scattering. In general, scattering patterns for elastic waves are much more complicated than those for acoustic waves.
3. From Fig. 3 we see that cross-coupled scattered waves (P-S coupling, $S_{mer} - S_{lat}$ coupling) are always away from the incident direction. Their maxima are perpendicular to the incident direction. Therefore, for pure forward scattering problems, the cross-coupled waves can be neglected.

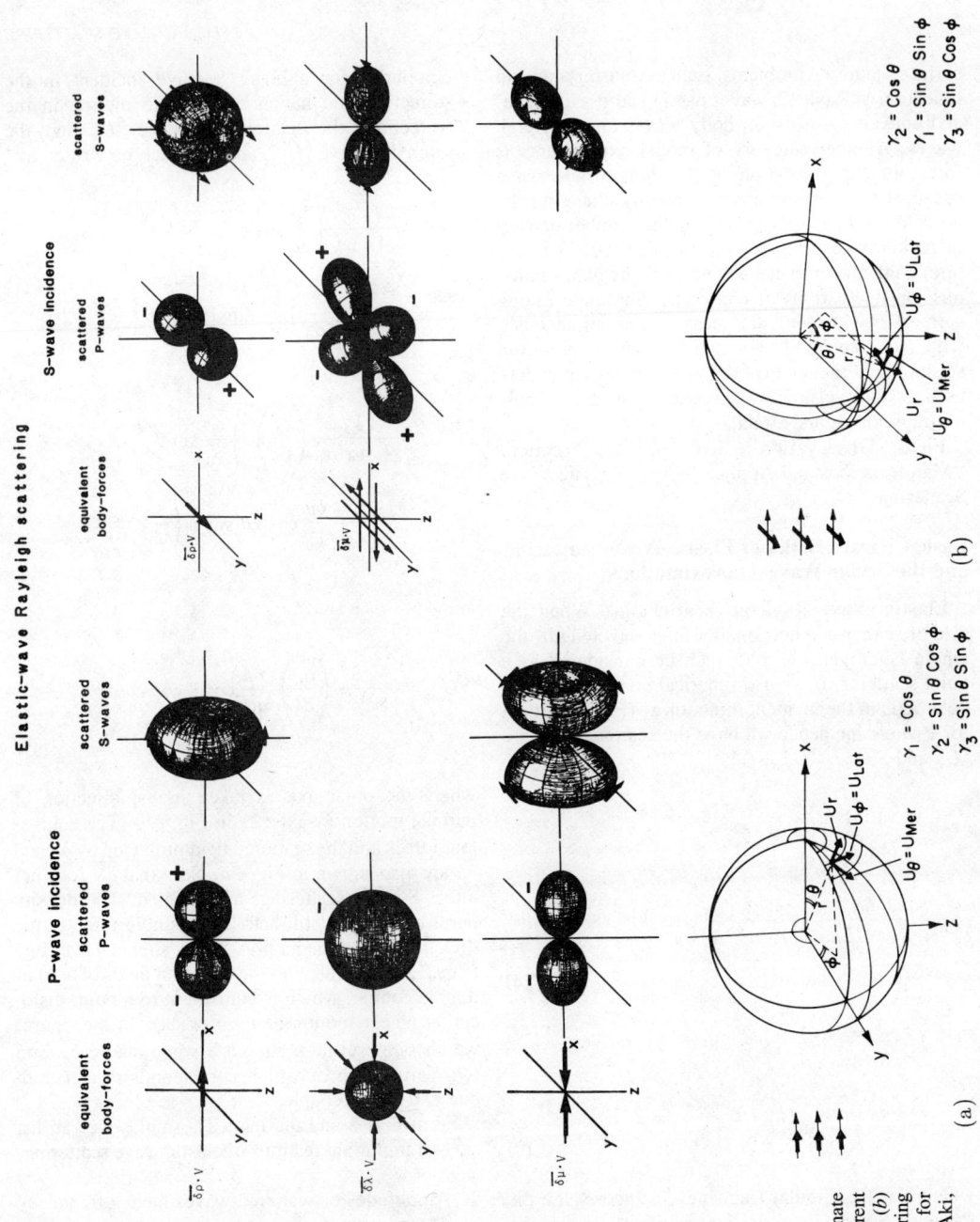

FIGURE 3. (a) Spherical coordinate system and scattering patterns for different equivalent forces for P-wave incidence; (b) spherical coordinate system and scattering patterns for different equivalent forces for S-wave incidence. (From Wu and Aki, 1985a, 584, 586)

4. From Eq. 8 we see that in the forward direction, $\cos\theta = 1$ and the parameter perturbations inside the braces become

$$\frac{\delta\rho}{\rho_0} - \frac{\delta\lambda + 2\delta\mu}{\lambda_0 + 2\mu_0} = -\frac{2\delta\alpha}{\alpha_0}$$

Therefore, scattering strength is proportional only to velocity perturbations. On the other hand, in the backward direction, $\cos\theta = -1$ and the parameter perturbations become

$$-\frac{\delta\rho}{\rho_0} - \frac{\delta\lambda + 2\delta\mu}{\lambda_0 + 2\mu_0} = -\frac{2\delta Z_p}{Z_{p0}}$$

where Z_p is P-wave impedance. Therefore, the strength of backscattered waves depends only on impedance perturbations. We see that forescattering and backscattering reveal quite different characteristics of a medium. The same conclusion can be drawn for common-mode S-S scattering (Eq. 11). In the forward direction ($\theta = \phi = 90°$) the strength depends only on S-wave velocity perturbations, whereas in the backward direction ($\theta = 90°$, $\phi = -90°$), the strength depends only on S-wave impedance perturbations.

5. Comparing the strengths of scattered P (Eqs. 8, 10) and S waves (Eqs. 9, 11, 12), we see that scattered S waves are α_0/β_0 times stronger than scattered P waves for similar perturbations. The reason is the S-wave impedance of a medium is always smaller than the P-wave impedance. After propagating and scattering for a long distance, scattered waves will be dominated by S waves. This is in agreement with the observations on coda waves.

Elastic Wave Rayleigh-Gans Scattering. When the wavelength is comparable to or smaller than the size of the inclusion, the equivalent source of scattering can no longer be regarded as a point source. Nevertheless, if the scattered field is still weaker than the incident field, the Born approximation can still be applied to the problem. This situation is known as *Rayleigh-Gans scattering*. From Eq. 7 we see that the scattered field is a superposition of waves scattered by all the volume elements of the heterogeneity, each of which is of Rayleigh scattering type.

If perturbations are formed by a nonuniform distribution of heterogeneities with the same composition, we can introduce a parameter distribution function $D(\mathbf{x})$ such that

$$\delta\rho(\mathbf{x}) = \delta\rho_0 D(\mathbf{x}), \quad \delta\lambda(\mathbf{x}) = \delta\lambda_0 D(\mathbf{x}),$$
$$\delta\mu(\mathbf{x}) = \delta\mu_0 D(\mathbf{x})$$

where $\delta\rho_0$, $\delta\lambda_0$, and $\delta\mu_0$ are the parameter perturbations at the center of the heterogeneity. The scattered field then can be written as

$$\mathbf{U} = \mathbf{C}(\delta\rho_0, \delta\lambda_0, \delta\mu_0) D(\tilde{\mathbf{k}}) \quad (13)$$

We call $\mathbf{C}$ the *composition factor*, which is the scattered field of elastic wave Rayleigh scattering by a unit volume of inclusion with perturbations $\delta\rho_0$, $\delta\lambda_0$, and $\delta\mu_0$, and D is called the *distribution factor* (or the *volume factor*, *form factor*, or *shape factor*), where

$$D(\tilde{\mathbf{k}}) = \int_V D(\mathbf{x}) e^{i\tilde{\mathbf{k}}\cdot\mathbf{x}} d^3x$$

is the 3-D complex spectrum of $D(\mathbf{x})$ and $\tilde{\mathbf{k}} = \mathbf{k}^{in} - \mathbf{k}^{sc}$, for $\mathbf{k}^{in} = k^{in}\hat{\mathbf{i}}$, $\mathbf{k}^{sc} = k^{sc}\hat{\mathbf{o}}$, where k^{in} and k^{sc} are the wave numbers of the incident and scattered waves, and $\hat{\mathbf{i}}$ and $\hat{\mathbf{o}}$ are the unit vectors in the incident and scattering directions. The resultant scattering pattern for a Rayleigh-Gans scattering is a product of a Rayleigh scattering pattern and a distribution factor pattern.

Elastic Wave Scattering by a Random Medium.
Mean Square Amplitudes of Scattered Waves. If $\delta\rho$, $\delta\lambda$, and $\delta\mu$ are from random heterogeneities with the same composition (the perturbations are totally correlated), we can use Eq. 13 to calculate the scattered fields for each realization and then take the ensemble average. In this case,

$$\langle|U|^2\rangle = \langle\mathbf{U}^*\cdot\mathbf{U}\rangle = \langle(\mathbf{C}^*\cdot\mathbf{C})(D^*D)\rangle$$
$$= \langle|\mathbf{C}|^2\rangle\langle D^*D\rangle = V\langle|\mathbf{C}|^2\rangle P(\tilde{\mathbf{k}}) \quad (14)$$

where $P(\tilde{\mathbf{k}})$ is the 3-D power spectrum of the parameter distribution function (shape spectrum):

$$P(\tilde{\mathbf{k}}) = \frac{1}{V}\langle D^*(\tilde{\mathbf{k}})D(\tilde{\mathbf{k}})\rangle$$

In deriving Eq. 14, we used the independence of C and D. Again we factored $\langle|\mathbf{U}|^2\rangle$ into a composition factor $\langle|\mathbf{C}|^2\rangle$ and a distribution factor $P(\tilde{\mathbf{k}})$. The composition factor is the mean square of the corresponding Rayleigh scattering amplitude. If the size of the random medium is much larger than the scale of the heterogeneities and if the heterogeneities have a statistically uniform distribution, we can consider approximately the random medium as a stationary (or uniform) random field, so a correlation function may be defined for the medium. Then the shape spectrum $P(\tilde{\mathbf{k}})$ is the Fourier transform of the correlation function $N(r)$.

Directional and Total Scattering Coefficients.
We define the directional scattering coefficient for P-P scattering as 4π times the mean scattered power in the $\hat{\mathbf{o}}$ direction per unit solid angle by a unit

volume of the random medium for a unit incident field (unit power flux density) in the $\hat{i}$ direction:

$$g^{\text{PP}}(\hat{o}, \hat{i}) = \frac{4\pi r^2}{A^2 V} \langle |^P U^P|^2 \rangle$$

In the same manner we can define the other scattering coefficients g^{PP}, g^{SP}, and g^{SS}. For P-wave incidence, if

$$\frac{\delta\rho}{\rho_0} = \epsilon, \quad \frac{\delta\lambda}{\lambda_0 + 2\mu_0} = \frac{m}{3}\epsilon, \quad \frac{\delta\mu}{\lambda_0 + 2\mu_0}$$

$$= \frac{n}{3}\epsilon, \quad \frac{\delta\mu}{\mu_0} = l\epsilon$$

we obtain

$$g^{\text{PP}}(\theta) = \frac{1}{4\pi} \epsilon^2 \left(\cos\theta - \frac{m}{3} - \frac{2n}{3}\cos^2\theta \right)^2 k_\alpha^4 P(\tilde{k})$$

$$g^{\text{PS}}(\theta) = \frac{1}{4\pi} \epsilon^2 \left(\sin\theta - l\frac{\beta_0}{\alpha_0}\sin 2\theta \right)^2 k_\beta^4 P(\tilde{k}) \quad (15)$$

The total scattering coefficient (or, put simply, scattering coefficient) g^P or put g^s is the mean total scattered power by a unit volume of a random medium for a unit incident P or S wave. For P-wave incidence

$$g^P = g^{\text{PP}} + g^{\text{PS}} = \frac{1}{4\pi} \int_{4\pi} g^{\text{PP}}(\theta) + g^{\text{PS}}(\theta) \, d\Omega$$

Wu and Aki (1985b) gave the results for exponential correlation functions. They showed that for low frequencies the scattering coefficients for elastic waves are more complicated and depend on all the perturbations of density and Lamé constants. On the other hand, the high-frequency asymptotics for both the elastic and scalar cases are equal, which implies that for weak perturbation of parameters, the travel time fluctuation in the forward direction will dominate the scattered field when the wavelength is very short compared to the sizes of heterogeneities.

Forescattering, Backscattering, and the Scalar Wave Approximation. In the nearly forward direction ($\theta \approx 0$), the directional scattering coefficients for P-wave incidence, Eq. 15 becomes

$$g^{\text{PP}}(\theta \approx 0) \approx \frac{1}{\pi} \left(\frac{\delta\alpha}{\alpha_0} \right)^2 k_\alpha^4 P(0),$$

$$g^{\text{PS}}(\theta \approx 0) \approx 0 \quad (16)$$

Thus, for forescattering, scattered waves are decoupled and the mean scattered power is proportional to the mean square of the P-velocity perturbation only. In this case the scalar wave approximation can be applied. In the backward direction ($\theta \approx \pi$),

$$g^{\text{PP}}(\theta \approx \pi) \approx \frac{1}{\pi} \left(\frac{\delta Z_p}{Z_{p0}} \right)^2 k_\alpha^4 P(2k_\alpha),$$

$$g^{\text{PS}}(\theta \approx \pi) \approx 0 \quad (17)$$

where $Z_p = \rho\alpha$ is the P-wave impedance. We see that backscattered waves are also decoupled and that the mean scattered power is proportional to the mean square perturbation of the P-wave impedance. The scalar wave approximation can be used in this case too. The same conclusion can be drawn for S-S scattering.

To see the differences between scalar and elastic wave scattering in Fig. 4, we compare the scalar case and the P-P scattering of the elastic case. Figures 4a, b, and c are for $\omega a/\alpha_0 = 0.1$, 1, and 10, respectively. For Rayleigh scattering (Fig. 4a) the scattering pattern of the scalar wave is isotropic (see the upper half-plane), whereas the pattern for elastic waves is quite unsymmetric, depending on the parameter combinations (the lower half-plane). When the wavelength becomes shorter than the scale length a, the scattered energy becomes more concentrated in the forward lobe, except for "impedance-type" scattering, which always has a big back lobe.

From Eqs. 16 and 17 we see that backscattering and forescattering sense not only different elastic properties of a medium but also different spectral components of the medium power spectrum. Forescattering feels only the dc component of the random medium $P(0)$ and is most sensitive to large-scale heterogeneities. This sensitivity can be understood physically: When waves pass through a random medium, the accumulated phase change is mainly determined by the large-scale variations—the effects from small variations tend to cancel each other. On the other hand, backscattering only senses $P(2k)$, the spectral component of the medium at $K = 2k$, which means that only the Fourier component of the medium with spatial period equal to half the wavelength of the incident field can be detected by backscattering. Therefore, it is most sensitive to heterogeneities with scale length comparable to the wavelength.

Manifestations and Applications of Seismic Wave Scattering

Transmission Fluctuations. It is a common observation that the travel times and amplitudes for an event fluctuate across a seismic array, such as the Norwegian seismic array (NORSAR), the large aperture seismic array (LASA) in Montana, USA, and other local or regional arrays. Fluctuation patterns may change drastically even for closely located events. For example, Fig. 5a shows the P-wave amplitude fluctuation patterns of NORSAR for two events separated by an incident angle near 4°. The two patterns are totally decorrelated. Figure

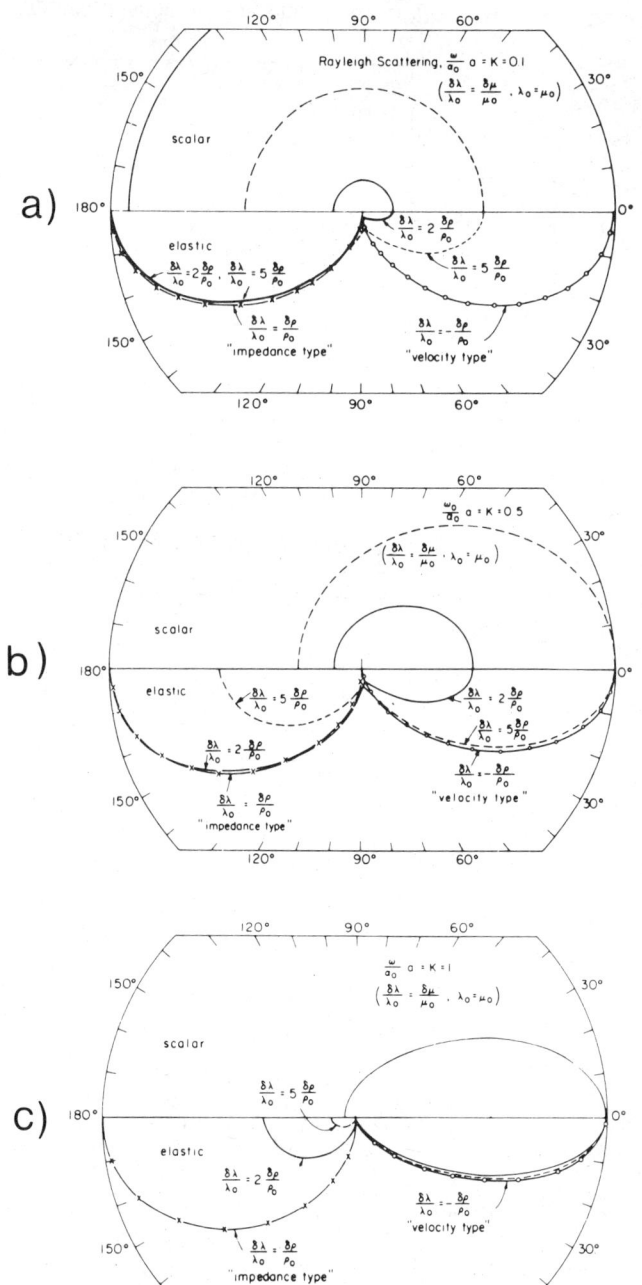

FIGURE 4. (a) Comparison of scalar wave scattering (upper half-plane) and elastic wave scattering (lower half-plane) for P-wave incidence when $K = \omega/\alpha_0\, a = 0.1$; (b) same as in (a), with $K = 1$; (c) same as in (a), with $K = 10$. (From Wu and Aki, 1985b, 10265)

5b shows the correlation coefficients of phase and log-amplitude fluctuation patterns of P waves as a function of separation of incident angles between events for NORSAR (from Flatté and Wu, 1988). The correlations are the averages over about 10,000 beam pairs (from 185 beams for amplitudes and 105 beams for phases). This sharp decorrelation with angle separations cannot be explained by the differences in time delay and absorption of different geometric paths. It must therefore be from the focusing, diffraction, and interference effects caused by lithospheric heterogeneities under the array.

For teleseismic events, the scattered waves measured from phase and amplitude fluctuations within an array are composed mostly of forward and small-angle scattered waves. From the last section we know that, for forescattering, the scattered energy is proportional to the scale of heterogeneities.

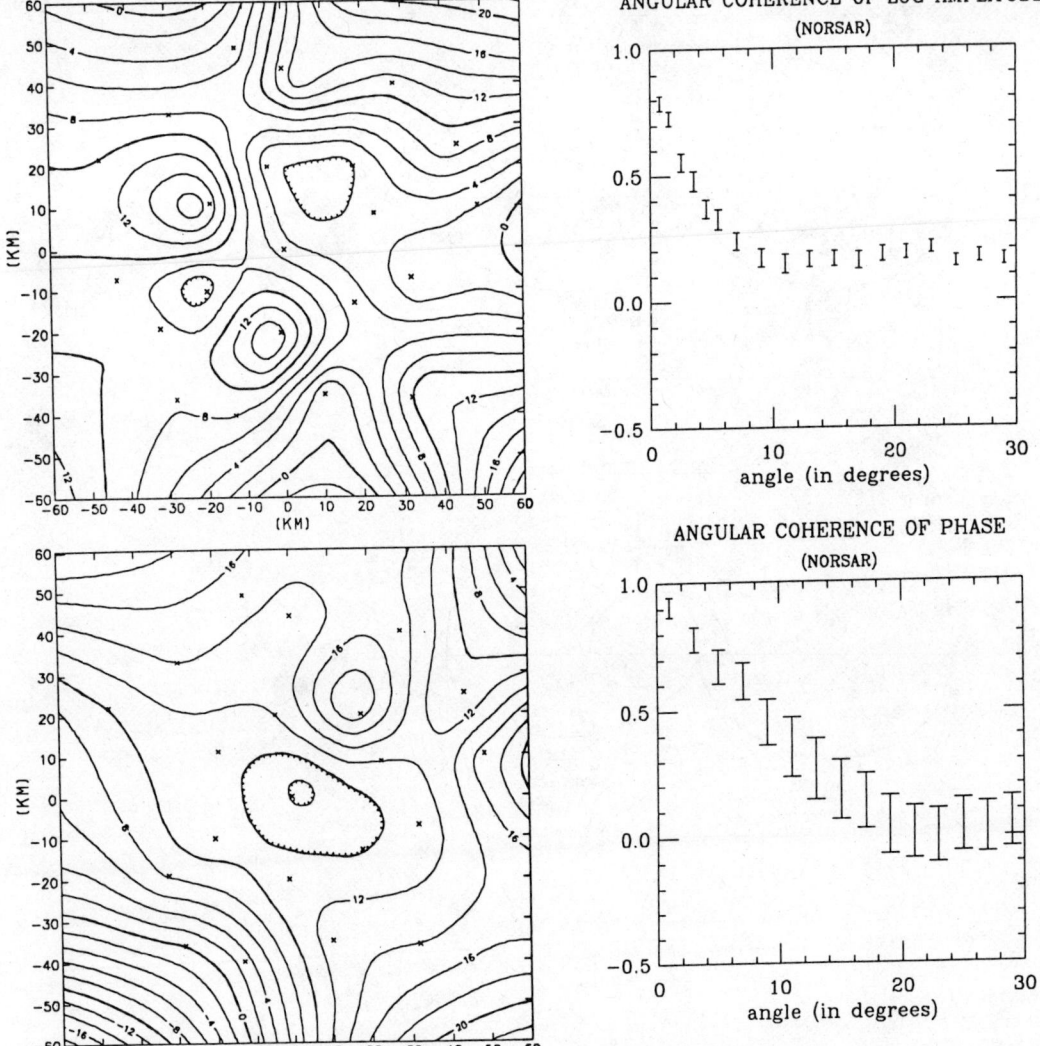

FIGURE 5. (a) P-wave amplitude fluctuation patterns of NORSAR for two neighboring beams. Beam 227: $\Delta = 35.1°$, azimuth = 117.0°; beam 236: $\Delta = 35.1°$, azimuth = 123.7°. Two patterns are totally decorrelated. (b) The observed log amplitude and phase coherences between two beams as functions of angular separation. (From Flatté and Wu, 1988)

Therefore, the contribution of scattering from small-scale heterogeneities is negligible. Only heterogeneities with scales satisfying $ka \gg 1$ need to be taken into consideration, and the PA can be used for this problem.

When the medium under the array has multiscale and complex heterogeneities, the stochastic approach is often desirable to obtain statistical properties of the heterogeneities (especially for the small-scale ones), such as the average scale, rms velocity perturbation, or the spectrum of the heterogeneities. Nikolaev (1972) used data of deep seismic sounding profiles to infer the velocity perturbation of the sampled region. Aki (1973) carried out the first study using data from a LASA. In the early studies, information utilized was rms phase and log A fluctuations of direct P arrivals, transverse correlation lengths of phase, and log A fluctuations. Limited by the amount of utilized information, the stochastic model of the heterogeneous media was set to be a uniform isotropic random medium with a Gaussian correlation function characterized by its rms relative velocities perturbation $\tilde{v}$ and average scale length a. For LASA the inferred random medium had $\tilde{v} = 4.0$–1.9%, $a \approx 10$–12 km, and a layer thickness of 60–130 km. For NORSAR $\tilde{v} = 0.5$–2.0%, $a = 30$ km, with a thickness of 100–250 km.

Recently Flatté and Wu (1988) have substantially expanded the utilized information in fluctuation observations. The entire transverse coherence function (TCF) of phase, log A fluctuations, and their cross are utilized. In addition, far more important is the utilization of the angular coherence functions (ACF) of phase, log A fluctuations, and their cross. ACF measures coherency between fluctuation patterns of an array for incident waves with different incident angles (from different events). The ACF measurement has several advantages over the TCF. For TCF, the transverse resolving power is limited by the station interval, but for ACF the limitation comes from the angular separation of the available events. Usually we have a much denser angular distribution of events compared with the sparse spatial distribution of stations. Furthermore, the existence of many events makes the estimations of ACF more robust than those of the TCF. Because of the great increase in the amount of utilized information, Flatté and Wu (1988) were able to take advantage of the recent progress in the theory of wave propagation in random media and to use more general models of random media. For the NORSAR data ($f = 2$ Hz) they used a two-layer (overlapped) random medium model. Each layer had a different perturbation strength and different power-law heterogeneity spectrum. The best model had the top layer bottomed at 200 km with a flat spectrum and the second layer located between 15 km and 250 km with a K^{-4} power-law spectrum, where K is spatial frequency (or wave number). The relative strengths of the two spectra are such that they cross at $K = 0.31$ km^{-1}. The spectrum obtained is bandlimited. The lower end is limited by the array aperture, $K_{min} = 2\pi/D$ with NORSAR diameter $D = 110$ km. The higher end is limited by the beam-forming process (subarray summation) and the wavelength. For the NORSAR problem, $K_{max} = 2\pi/5.5$ km. From the rms travel time fluctuation (0.135 s) and the rms log amplitude fluctuation (0.41 neper = 3.6 dB) of NORSAR data, the rms wave-speed perturbations for the first and second layers are $\tilde{v} \approx 0.9$–2.2% and $\tilde{v} \approx 0.5$–1.3%. The flat spectrum in the first layer means that the scale of heterogeneities represented by the spectrum is smaller than the 1-km limit of the passband. These are the small-scale heterogeneities of the lithosphere. The K^{-4} spectrum of the second layer means the related scale is larger than the 20-km upper limit. This spectrum is that of the large-scale heterogeneities in the lithosphere. In the upper crust (0–15 km) only small-scale heterogeneities exist. Below 200 km, down to about 250 km, only large-scale heterogeneities exist. This region could be the asthenosphere. In the middle (15–200 km) both large- and small-scale heterogeneities exist.

Upper-mantle heterogeneities are closely related to the mantle convection and plate dynamics. Understanding the nature and distribution of heterogeneities will help locate convection-active regions and discriminate between the chemical and thermal boundaries of the upper and lower mantles. Transmission fluctuation is a useful tool for exploring upper-mantle heterogeneities.

Coda Generation and Its Envelope Decay. The most striking manifestation of seismic wave scattering is the generation of coda waves. Coda waves are irregular arrivals following the major wave types (such as P, S, and surface waves) that show up on a seismogram as the tail. For local events (within 200–300 km) the tail of a seismogram after the S wave is usually called the *S coda*. For teleseismic events, the irregular arrivals after the P wave (the "neck") are called *P coda* waves. The origin and nature of the P coda is less well understood than those of the S coda. Here we discuss only the S coda.

Since K. Aki proposed in 1969 that the S coda of a local earthquake is composed of backscattered waves from numerous randomly distributed heterogeneities in the Earth, there have been coda studies from around the world (for a review see Herraiz and Espinosa, 1987) supporting this model. Today the scattering model of coda waves is well accepted. Questions remaining are: How deep are the heterogeneities distributed? How extensive is the contribution from surface topography scattering (by surface to body wave scattering)? How extensive is the contribution from multiples of thin layering structures? What is the best mathematical model of the envelope decay? Several important characteristics of coda waves that support the scattering model are as follows (Aki and Chouet, 1975; Aki, 1980).

1. The spectra of coda waves of different stations for an earthquake show very little difference, whereas the spectral difference of the direct S waves for different stations shows clearly because of the difference in travel distance and propagation path.
2. For a given local earthquake at an epicentral distance shorter than 100 km, the total duration of a seismogram is nearly independent of the epicentral distance or azimuth.
3. Coda envelope decays of bandpassed seismograms for different local earthquakes are similar and independent of the distance and the nature of the path between the epicenter and the station. For a local earthquake the coda level is quite stable and independent of the epicenter distance. Figure 6a shows the codas of the bandpassed (3.6–4.8 Hz) seismograms recorded at NORSAR for a local earthquake. The epicentral distance varies from few to more than 100 km, but the coda levels and envelope decays are similar.
4. The shape of coda envelope decay is also independent of magnitude, at least for earthquakes with $M < 6$.
5. Wave-number spectrum analysis shows the coda waves are coming from all directions (see Fig. 6b—note that the direct S is concentrated in the corresponding azimuth and travel with velocity 3.5 km/s).

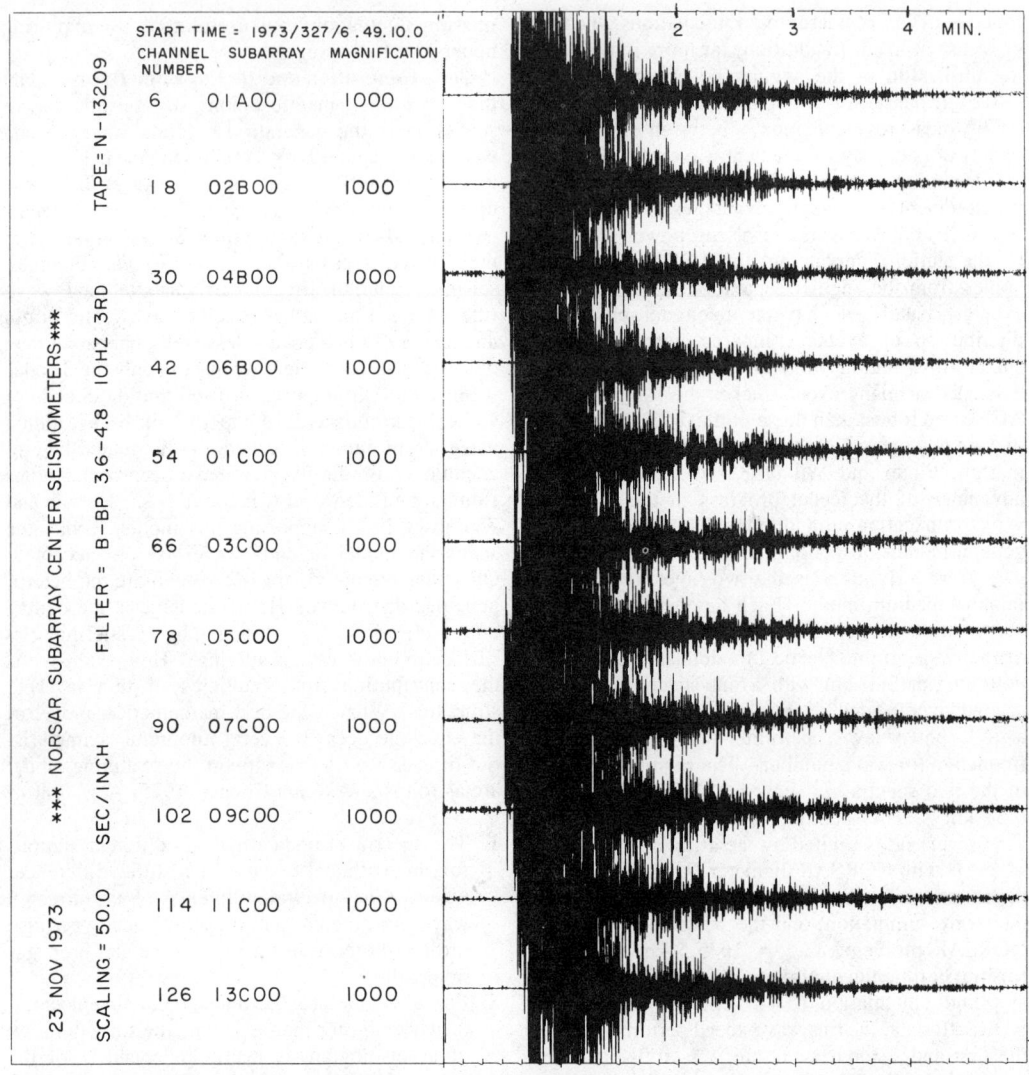

FIGURE 6(a). Short-period (bandpass from 3.6 to 4.8 Hz) record of a local earthquake at NORSAR near subarray 7C. The epicentral distance is a few kilometers to the closest subarray and more than 100 km to the farthest. The decay of coda power shows no dependence on the details of the different wave paths between the earthquake source and each subarray. (From Aki and Chouet, 1975, 3323)

6. The station response (site effect) of coda is very similar to that of an S wave but not a P wave. Also, measured coda Q behavior is nearly the same as S-wave Q. Therefore, coda is assumed to be composed mostly of S waves.

Coda Envelope Decay and Mathematical Models of Coda. Let g be the scattering coefficient of the medium, η_a the absorption coefficient, and Q_g, Q_a the corresponding Q's. Then we can define the *seismic albedo of the medium*:

$$B_0 = \frac{g}{g + \eta_a} = \frac{Q_g^{-1}}{Q_g^{-1} + Q_a^{-1}}$$

The parameter B_0 is crucial to estimating the importance of multiple scattering. When $B_0 < 0.5$, absorption is dominant and multiple scattering becomes less important. When $B_0 > 0.5$, scattering is dominant and multiple scattering may be important in the envelope calculation. If the lapse time t_c is less than τ_s (the scattering time or mean free time: $\tau_s = 1/gv_0$, where v_0 is wave speed), multiple scattering may still be neglected. When $t_c \gg \tau_s$, multiple scattering must be considered.

The most eye-catching examples of strong multiple scattering are seismograms from the moon. Figure 7 shows the record of one lunar event (from Dainty et al., 1974). The seismogram has spindle shape and

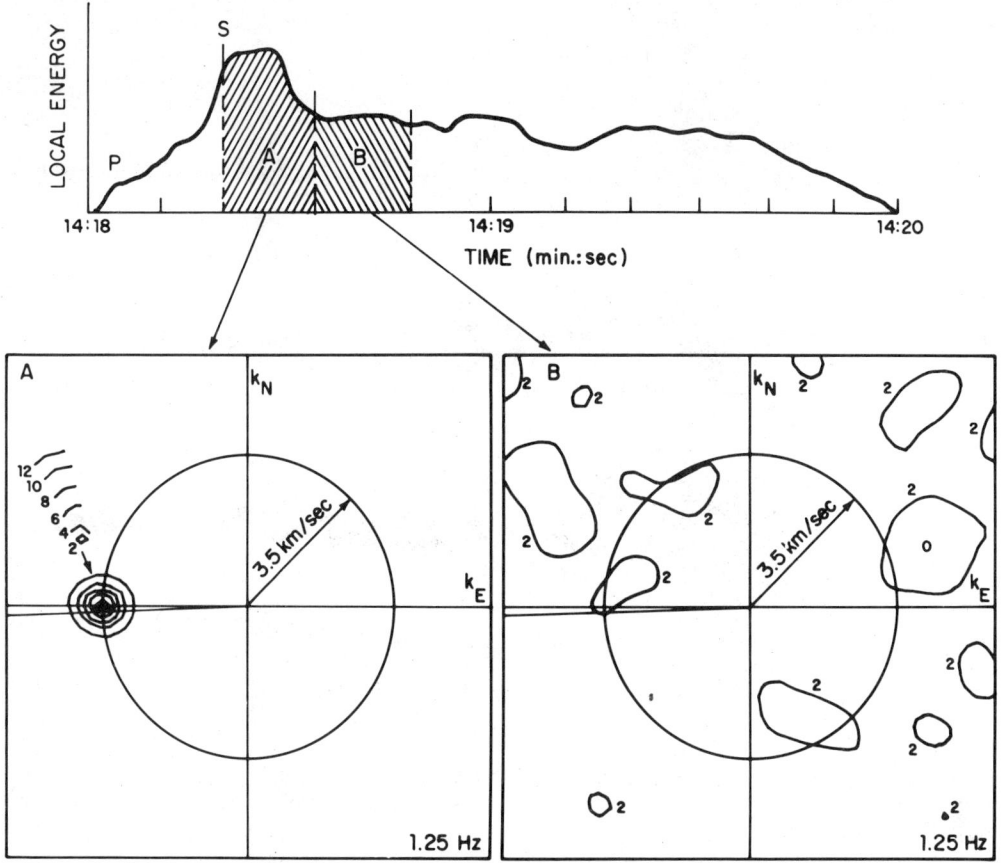

FIGURE 6(b). Wave-number spectra for S arrival and S coda in the frequency band 1–2 Hz for a strip-mining blast near LASA. (From Scheimer and Landers, 1974, reproduced from Aki and Chouet, 1975, 3324).

lasts for more than 1 h, because the crust of the Moon is very heterogeneous and has very little absorption. In the Earth the situation is different and varies from region to region.

The mathematical model for multiple scattering is quite complicated, but for two extreme cases simple models do exist: *single scattering theory* for small B_0 or small lapse time, and *diffusion theory* for strong scattering (large B_0 and long lapse time). The most widely used coda envelope model is the single backscattering model (Aki and Chouet, 1975; Aki, 1980), which is valid when the lapse time is greater than twice the S travel time. To remove the limitations on the lapse time and extend the model to the early coda, Sato (1977) proposed a single isotropic scattering model. Kopnichev (1977) considered the contribution of multiple isotropic scattering up to order 3. Gao et al. (1983) calculated multiple scattering up to order 7 and, with numerical fitting, obtained the approximate multiple isotropic scattering model. Another coda envelope model is the energy-flux model proposed by Frankel and Wennerberg (1987), based on the assumption of uniform distribution of coda energy behind the direct wavefront, which is observed in numerical simulations by finite difference. The single-scattering model predicts an additive apparent attenuation; that is, the apparent attenuation of the coda envelope is the sum of the absorption and scattering coefficients. In contrast, the energy flux and diffusion models predict the dominant role of absorption in coda envelope decay. Further study in theoretical modeling and experimentation are needed to solve the problem.

Another approach is to formulate the coda intensity-distance relation in the frequency-space domain by considering the multiple scattering effect. Wu (1985) derived the formulas using radiative transfer theory and found that the intensity-distance curve depends on the albedo B_0 of the medium, especially for $B > 0.5$. This dependence can be used to infer B_0 of the medium and therefore separate scattering attenuation from intrinsic attenuation.

Coda Q and S-Wave Q Measurements. Based on the scattering model, it is easy to measure Q_c from the coda envelope. After correcting geometric spreading, take the logarithm of the envelope. The slope of the curve is ωQ_c^{-1}. The starting time usually is from twice the t_s (S arrival time). For S-wave Q

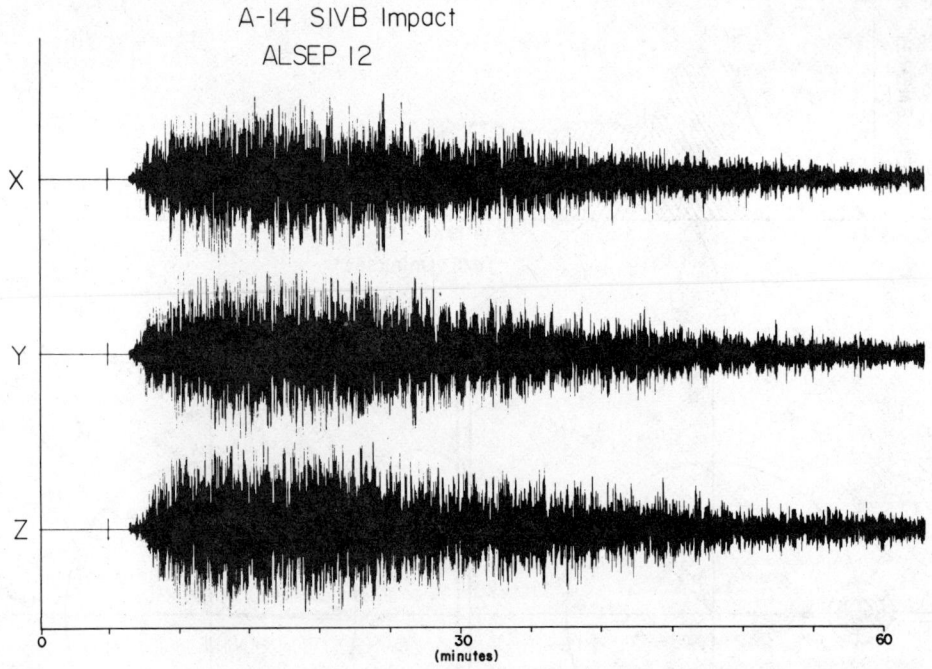

FIGURE 7. Seismograms of a lunar event from the impact of the S4B Saturn booster of *Appollo 14*. (From Dainty et al., 1974, 12)

measurements by the single-station method, a coda level at a fixed lapse time was used as a reference level of the source strength. Q_β^{-1} can be estimated from the measured S-to-coda amplitude ratio. Azimuthal variation is removed by averaging over a group of events in the same distance range from all directions.

All of these methods have been widely used to estimate Q_c and Q_β for different regions. Figures 8a and 8b show summarized results for Q_c^{-1} and Q_β^{-1} (together with some estimates by the L_g method). We see that Q_c^{-1} and Q_β^{-1} have similar frequency dependence. As expected, the Q_β^{-1} results show more fluctuating than do the Q_c^{-1} results. Coda excitation (the strength) can also be obtained. However, since $|S(w)|$ cannot be accurately determined, the estimation is less reliable than that of Q_c.

Coda wave analysis has many applications. Coda duration is widely used for magnitude determination. Coda amplitude can be used to estimate the seismic moment. Q_c measurement using coda waves can separate the path effect from the source effect and therefore gives the possibility of doing path correction to the source spectrum. Coda excitation measurement can provide information on lithospheric heterogeneities. One interesting phenomena, however, is the temporal change of coda decay. Many observations show changes of Q_c one to two years before a major earthquake. Since the measurement of coda Q is comparatively easy and coda Q change may be related to stress and rock property changes of the sampled volume by coda waves, coda decay monitoring seems to be a promising tool for earthquake prediction.

Scattering Attenuation. Scattering can decrease the amplitude of the received signal. This apparent attenuation is called *scattering attenuation*. In the literature, scattering Q is often used and is defined as $Q_{sc}^{-1} = 2(\eta_s/k)$, where Q_{sc} is scattering Q, η_s is the amplitude scattering attenuation coefficient, and k is the wave number. However, the scattering attenuation coefficient or scattering Q is not a well-defined quantity and is measurement-dependent. Unlike anelastic attenuation (intrinsic attenuation), which is the energy loss of the seismic wave by friction, scattering attenuation is only an energy redistribution in space and time. The wave energy is scattered by heterogeneities from the propagation direction to other directions, and the scattered energy may be rescattered back to the propagation direction and arrive at the receiver later. Therefore, depending on the size of the observation window (in space and time) and the measuring procedure, scattering attenuation may behave quite differently. The confusion and controversy in early applications of scattering attenuation are mainly caused by inappropriate applications of theory to measurements. Major progress in this field has been understanding the measurement dependency of scattering attenuations and the development of different approximate

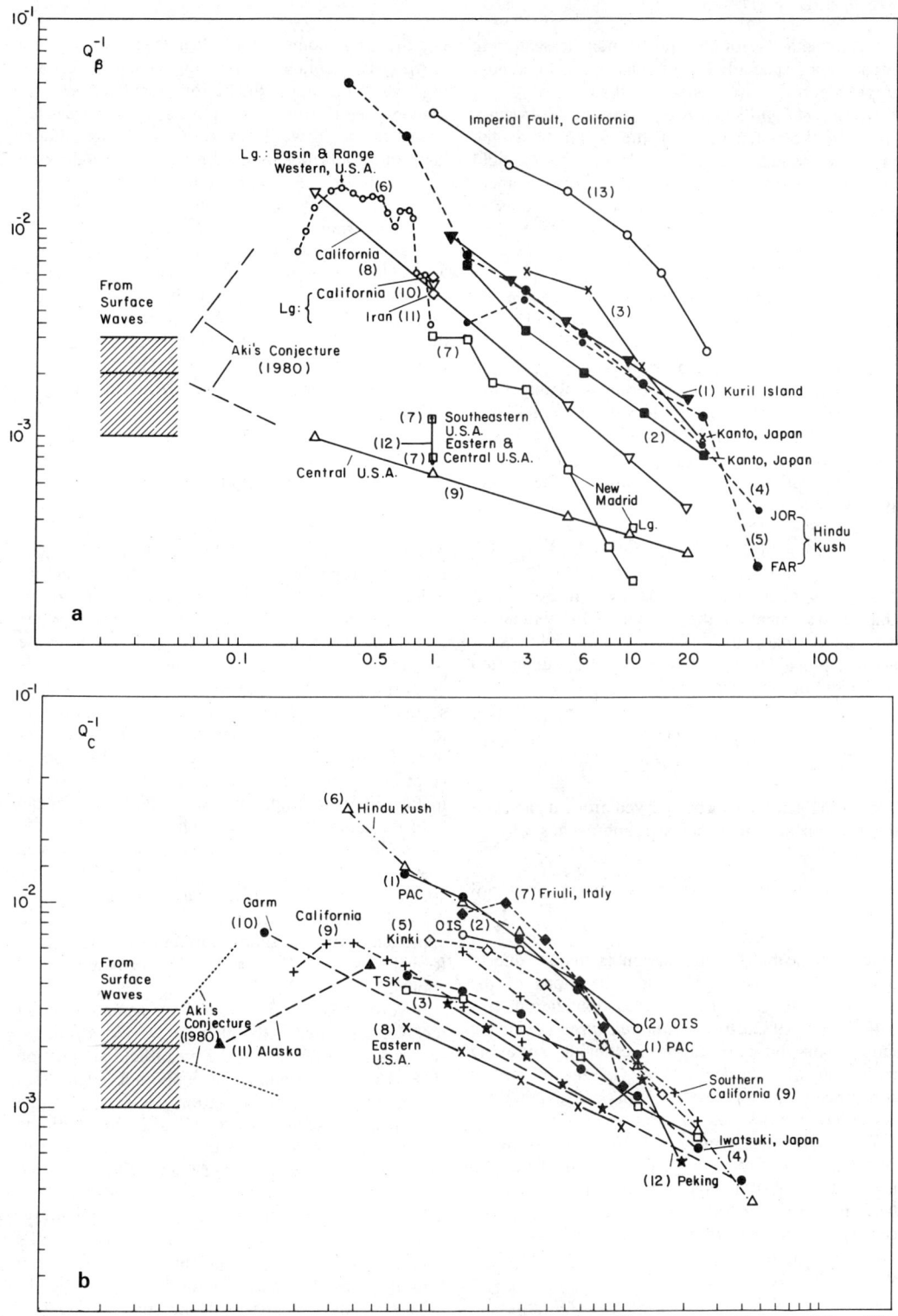

FIGURE 8. (*a*) A summary of measured Q_β^{-1} using the single station S-coda ratio method of Aki, together with some results from Lg analysis: (*b*) A summary of measured Q_c^{-1} from local coda analysis.

analytic methods for different kinds of scattering attenuation (especially for attenuation of coherency, amplitude, direct arrivals, and codas).

Mean Field Attenuation and Amplitude Attenuation. In the calculation of the mean field, the ensemble average is taken for the complex field (including phase and amplitude). The displacement of a random seismic wave u can be expressed as

$$u = Ae^{i\phi} = Ae^{i(\phi_0 + \delta\phi)}$$

where A is amplitude, ϕ is phase, ϕ_0 is the average phase delay after passing the random medium, and $\delta\phi$ is the random phase fluctuation caused by random heterogeneities. If the medium is purely elastic (no intrinsic attenuation) and scattering is mainly forescattering (neglecting energy loss due to large-angle scattering), then A can be considered as a constant and the only random part of the wave is $\delta\phi$. If $\delta\phi$ has a Gaussian probability distribution, the ensemble average of u is

$$\langle u \rangle = A \langle e^{i(\phi_0 + \delta\phi)} \rangle = Ae^{i\phi_0} e^{-\langle \delta\phi^2 \rangle / 2} \quad (18)$$

It can be seen from Eq. 18 that even if there is no amplitude attenuation, the mean field $\langle u \rangle$ will attenuate due to phase randomization of the field. For a known propagation distance R, the attenuation coefficient of the mean field is

$$\nu = \frac{\langle \delta\phi^2 \rangle}{2R} \quad (19)$$

The formula for ν can be derived from mean field theory. For high frequencies, it approaches

$$\nu = \frac{\tilde{v}^2 k^2 a}{2} \quad (20)$$

Since the mean field attenuation is mainly due to interference between different realizations of the random wave caused by phase randomization, ν is called (Wu, 1982a,b) the *randomization coefficient* rather than the attenuation coefficient, to avoid confusion. Since a decrease in $\langle u \rangle$ means loss of coherence of the wave field, ν can be also called the *decoherency coefficient* or, simply, *decoherence*.

To derive the average amplitude attenuation (AAA), we must remove the phase influence before taking the ensemble average, by multiplying the wave field by its complex conjugate. This operation corresponds to performing a phase correction for every realization of the ensemble before averaging. We get

$$\langle uu^* \rangle = \langle Ae^{i\phi} Ae^{-i\phi} \rangle = \langle A^2 \rangle$$

Therefore, to derive formulas for AAA, we have to solve the equation for the second moment $\langle uu^* \rangle$, not the first moment $\langle u \rangle$. Wu (1982a) and Sato (1982) have applied different approaches in deriving high-frequency approximate formulas for AAA (for a review see Herraiz and Espinosa, 1987). Scattered waves in the forward direction will be rescattered back into the propagation direction. Depending on the width of the observation time window, the forward scattered energy should be partly excluded from scattering attenuation. Wu used the multiple-forward-scattering approximation (or the single-backscattering approximation), which excluded all scattered waves in the forward half-space from the scattering loss. Later the approximation was modified to include only the forescattered waves within a certain critical angle θ_c as the recycled energy. For an exponential correlation function and $ka \gg 1$,

$$\eta_s \approx \cot \frac{\theta_c}{2} \frac{\tilde{v}^2}{4a} \quad (21)$$

where $\tilde{v} = (\delta v / v_0)_{\text{rms}}$. In contrast to Eq. 20, η_s is inversely proportional to the scale of heterogeneities a, and ν is proportional to a. These proportionalities are consistent with the physics of scattering. When a is smaller, there will be more scatterers along the propagation path. Since the number density is proportional to $1/a^3$ and the cross section of each scatter is a^2, the scattering loss will be proportional to $1/a$. On the other hand, larger a will cause greater phase fluctuations. Therefore, coherency loss should be proportional to a and increase with frequency, as in Eq. 20. Sato's approach is to modify the mean field formalism such that the phase (or travel time) fluctuation is excluded from scattering attenuation (Sato, 1982). He cut off the part of the heterogeneity spectrum below a critical wave number K_c and derived scattering attenuation by using the mean field approach (travel time corrected mean wave formalism). With $K_c = \pi/a$, the corresponding θ_c is about 29°.

In seismology, a single station is actually a point receiver compared with either the wavelength or the scale of heterogeneities. The mean field attenuation (MFA) cannot be applied to amplitude attenuation for signals from a single station. However, a large seismic array, such as NORSAR or LASA, with the aperture much larger than the scale of heterogeneities under concern, can offer a good opportunity to test the MFA. The beam-forming process (stacking the received signals for all the stations of the array) is similar to an ensemble-averaging process. The coherency attenuation coefficient ν can be estimated from the phase fluctuations over the array by Eq. 19 (assuming the thickness of the random layer is known). Figure 9a compares the measured ν for LASA and the theoretical prediction (from Wu, 1982b). We can see that ν increases with f^2 as predicted by theory. Figure 9b, on the other hand,

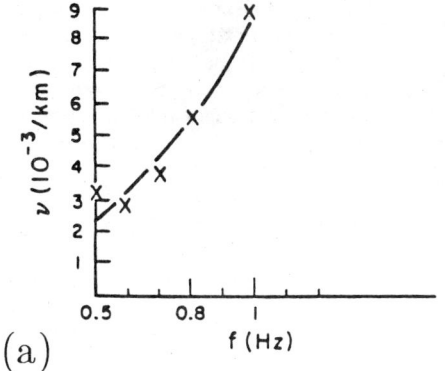

(a)

× ν_ϕ, the randomization coefficient calculated from the measured phase fluctuations

— ν, calculated from the medium parameters

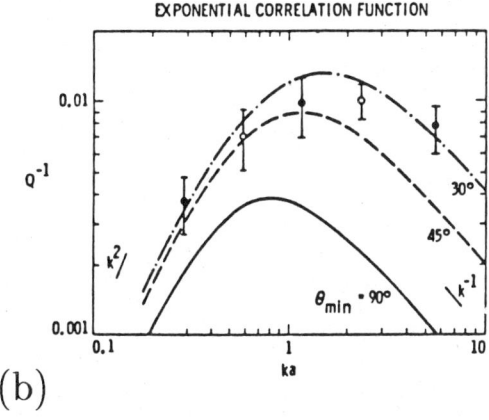

(b)

FIGURE 9. (a) Comparison of randomization coefficient ν_ϕ (the attenuation coefficient of mean field) from measured phase fluctuations by Aki at LASA (×) and calculated by the mean field formulation from the medium parameters (from Wu 1982b). (b) Comparison of the result of Q_s^{-1} (for the amplitude scattering attenuation) from the finite-difference simulations (circles) with the theoretical predictions by Wu and Sato. The random medium has an exponential correlation function. θ_{min} is the critical angle θ_c in Eq. 21. Note that the comparison is done for the 2-D case. (From Frankel and Clayton, 1986)

shows Q_{sc}^{-1} of the amplitude scattering attenuation from a numerical simulation for a random medium with an exponential correlation function (from Frankel and Clayton, 1986). The behavior of measured AAA is drastically different from MFA but agrees fairly well with Wu and Sato's theory.

The other problem is scattering attenuation of direct waves and of coda waves. The peak attenuation of first arrival is a problem of pulse propagation, which depends on both energy loss and the temporal coherency loss due to scattering. Meanwhile, coda wave attenuation is mainly an energy scattering loss problem, since coda waves have lost their coherency. However, measurement of coda attenuation (or coda Q) is model-dependent, as discussed in the previous section.

Scattering near the Core-Mantle Boundary. As we pointed out, heterogeneities in the lithosphere, especially of scales comparable to the wavelength, can produce significant coda waves, which can in turn reveal the properties of the heterogeneities. However, since scattered waves are one order smaller than major phases of direct waves, and the geometric spreading is also one order more severe than direct waves, coda waves scattered from heterogeneities in the deep mantle and near the CMB will be buried in the coda waves generated by near-surface heterogeneities and thus will be unobservable. Fortunately, the Earth is structured in such a way, mainly due to the sharp decrease of seismic velocity at the CMB, that for some range of distance (97° < Δ < 142°) the direct P wave has a geometric shadow. Therefore, in the shadow zone the scattered waves from the deep mantle or CMB will be observable in some cases and can be used to investigate the properties of the heterogeneities. In fact, the two major impacts to the traditional, spherically symmetric model of the Earth are triggered by evidence from seismic wave scattering. One is the observation of coda waves interpreted as backscattered waves from lithospheric heterogeneities (Aki, 1969); the other is the observation of the precursors to PKIKP in the shadow zone interpreted as scattered PKP waves from heterogeneities near the CMB (Haddon, 1972).

Precursors to PKIKP. PKIKP is the arrival passing from the source through the mantle, outer core, inner core, outer core, mantle, to the receiver (Fig. 10). Observations of short-period wave trains preceding PKIKP (can be earlier by as much as about 20 s) at epicentral distances between about 125° and 143° were well known. The widely accepted interpretation, up to 1972, was due to reflections or refractions from the postulated transition layer in the lower part of the outer core. However, the postulated transition layer had to have many layers and an increasingly complicated structure to account for all the observed arrivals. The decisive evidence for abandoning the transition layer hypothesis came from high-resolution wave-number spectral analysis of high-quality digital array data (from NORSAR and LASA), which allowed determination of precursor arrival directions. It was found that the directions of precursors are scattered around the direction of PKP (the wave passing through the mantle, outer core, mantle), indicating that precursors originate from a region close to the caustic (near the CMB) and not from transition layers in the core. Haddon (1972)

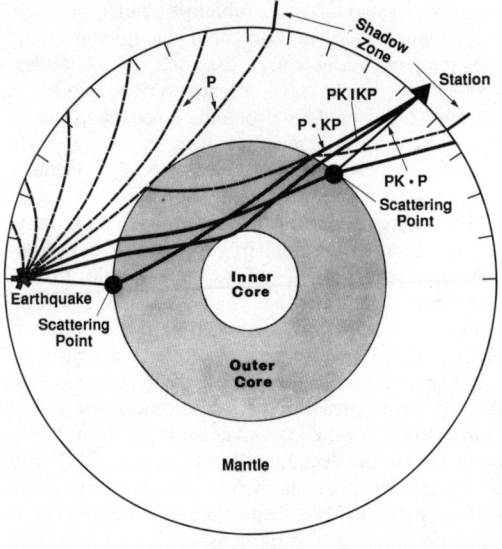

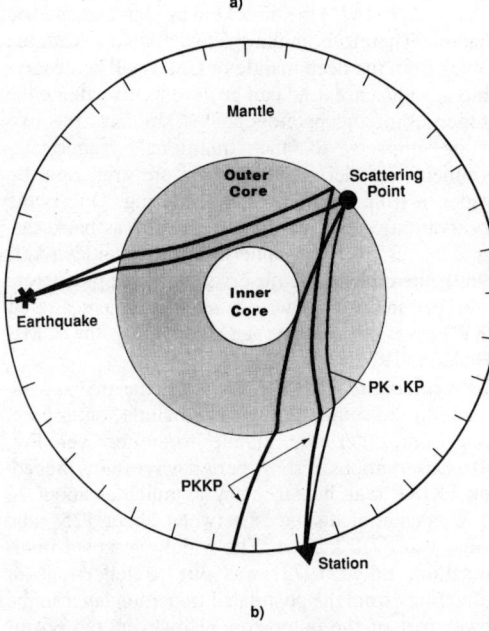

FIGURE 10. Diagram of wave paths for mantle-core waves and their scattered waves.

and Cleary and Haddon (1972) suggested that irregularities near the CMB could scatter the PKP waves into the shadow. They showed that the travel times and the slownesses (i.e., the arriving directions) calculated from scattering theory are all consistent with the observations. Figure 11a (from Cleary and Haddon, 1972) shows the comparison in travel time between the observations and the scattering theory. The observed precursors are assumed to be the scattered waves by heterogeneities near the CMB at both entrance (P·KP) and exit (PK· P) of the core. Here the dot represents the place in the path where scattering occurs. In the figure the lower curve is the minimum time curve corresponding to the scattering arrivals from the caustic at the CMB, and the upper curve is the minimum time curve for the caustic at 200 km above the CMB in the D″ layer. We see that the data are well confined by the two curves, implying that the precursors can be explained as scattered waves by the random heterogeneities in a layer about 200 km thick above the CMB. Figure 11b (from King et al., 1974) shows the comparison between the theoretical predictions and the observations of the slownesses (arriving directions) versus arriving times of the precursors for $\Delta = 130°$. The upper lobe (large slowness arrivals) has the arrivals scattered at the entrance of the core; in the lower lobe, they are scattered at the exit. The line in the middle (R) is the prediction of the transition layer hypothesis. It is clearly seen that the observations favor the scattering interpretation.

The amplitudes of scattered waves can be calculated by perturbation theory. Calculations showed that P-velocity variation of a few percent with scales of 10-30 km in a layer up to 200 km thick above the CMB could produce the observed amplitudes. Later it was shown that a slightly irregular topography (a few hundred meters high) could produce the same effect. Therefore, from the observations of forescattering of PKP only, the effects of volume heterogeneities near the CMB and of topography of the CMB are not distinguishable.

Precursors to PKKP. The PKKP wave interacts with the CMB at the entrance and exit of the core (similar to PKP) and at the underside reflection from the CMB (Fig. 10). However, backscattered waves from the underside reflection of the CMB would always arrive as precursors to the main phase itself because of the maximum time character of PKKP among PK·KP. Indeed Chang and Cleary (1978) and Doornbos (1980) found precursor wave trains up to 65 s earlier than PKKP. The precursors from backscattering of PK·KP cannot be explained satisfactorily by the scattering of volume heterogeneities above the CMB. This evidence is therefore in favor of the rough boundary hypothesis.

In summary, scattering from rough topography of CMB can explain the precursor phenomena to both PKIKP and PK·KP. However, the evidence did not exclude the existence of scattering from volume heterogeneities in the D″ layer. Other observations, such as PKPPKP, PcP, P_{diff}, and grazing angle diffraction may offer further opportunities to clarify the nature of heterogeneities near CMB.

Scattering by Surface Topography, Near-Surface Structures, and Deep Structures. Irregular topography can focus or defocus passing waves, scatter and couple different wave types (P and S, body, and surface waves), and therefore generate

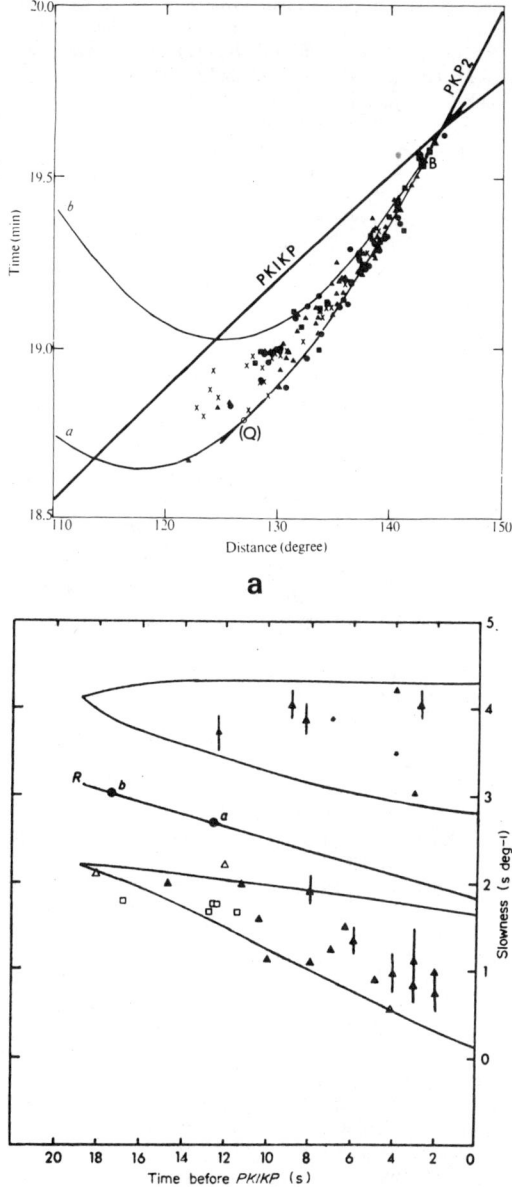

FIGURE 11. (a) Comparison in travel time between the observations and the scattering theory. The lower curve is the minimum time curve corresponding to the scattering arrivals from the caustic at the CMB; the upper curve is the minimum time curve for the caustic at 200 km above the CMB in the D″ layer. The heavy lines are the Jeffreys-Bullen PKP curves, with the surface caustic at point B. (From Cleary and Haddon, 1972) (b) Comparison between the measured slowness values and the predictions by the scattering theory for Δ = 130°. The upper and lower lobes are the areas of predicted slowness for the arrivals scattered at the entrance and exit of the core, respectively. The triangles and squares are observations at different regions (From King et al., 1974)

coda waves and drastically alter the waveform and amplitude of the direct arrivals. Low-velocity surface structures, such as filled basin-type structures, have the same effect as irregular topography. In addition, these structures can induce resonant reverberation on incoming (at receiver site) or outgoing waves (at source site). Finite-difference simulation shows that at the edge of basins large surface waves can be generated by incident body waves, even when direct surface waves may have been substantially decayed. In addition, Rayleigh waves incident on the edge of the basin in the source side can be converted into body waves and therefore produce secondary arrivals shortly after direct P-wave arrival.

In recent years, many studies, mainly by numerical simulation, have addressed the scattering problems of deep structures, such as ocean-continental transition zones, subducted slabs of the lithosphere, magma lenses, and other objects.

Crack Scattering and Effective Anisotropy. When a scattering medium is composed of randomly distributed, numerous scatterers, each of which is much smaller than the wavelength, the effect of an individual scatterer to wave propagation cannot be tracked and/or is not important; rather the collective behavior of the scatterers is of most interest. In this case a stochastic approach is preferable. Since the wavelength is much greater than the size of the scatterer, the wave field cannot discern each individual and only interacts simultaneously with many scatterers of different properties. In a sense, the wave field is taking an "ensemble average" while propagating through the medium. Therefore, the statistical characteristics of scattering are very stable for a single experiment, whereas in common seismological scattering problems, statistical quantities may have large fluctuations for different experiments. For this reason, we can call the scattering medium a "quasi-homogeneous" medium (see Fig. 2) with some effective medium parameters [e.g., effective attenuation or phase delay (dispersion) due to scattering, and albedo of a medium]. When individual scatterers (such as cracks) are anisotropic and oriented or aligned unsymmetrically, effective anisotropy may be formed. Shear wave splitting due to effective anisotropy by aligned cracks has been observed in many regions over the world and has many potential applications. Stochastic scattering (such as the mean field theory) is an important method for calculating the effective parameters.

Scattering and Nonlinearity. Usually propagation and scattering of seismic waves are treated as linear phenomena, and scattering characteristics are considered independent of wave amplitude (amplitude of strain), except very near to the source. This treatment is based on laboratory measurements that showed negligible nonlinearity. However, recent field measurements using controllable sources, such as Vibroseis, revealed surprising nonlinear effects of wave interaction. The cause of the discrepancy between field and laboratory measurements is attributed to the multiscale hierarchical heterogeneities in the lithosphere. Large-scale heterogeneities distort stress distribution and concentrate stresses in some volumes. In these high-stressed volumes, heterogeneities in smaller scales further concentrate the stresses into smaller volumes. Therefore each heterogeneity scale acts as a stress amplifier. In laboratory measurements rock samples are not subject to this stress amplification.

Scattering, absorption, and nonlinearity interact with each other and generate some complicated but interesting phenomena. Nonlinearity can pump energy from low frequency to high frequency and therefore produce apparent attenuation for the dominant frequency, which is not energy absorption but a redistribution to other frequencies. Scattering and absorption have opposite effects on nonlinearity: Absorption tends to reduce nonlinearity by decreasing the amplitude, but scattering enhances the effect of nonlinearity by increasing the length of the travel path and by magnifying the amplitudes in some places due to focusing. On the other hand, nonlinearity will enhance absorption by pumping energy to high-frequency waves. Therefore, nonlinearity introduces a new degree of freedom in medium characterization and gives us more tools to examine the medium, but it also leads to more severe nonuniqueness and ambiguity in interpreting the observations. For example, the frequency dependence of observed Q for short-period seismic waves in the lithosphere now has three candidates for the mechanism: intrinsic attenuation by some relaxation mechanism, scattering attenuation, and apparent nonlinear attenuation.

RU-SHAN WU

References

Aki, K., 1973, Scattering of P waves under the Montana Lasa, *Jour. Geophys. Research* **78**, 1334–1346.

Aki, K., 1980, Scattering and attenuation of shear waves in the lithosphere, *Jour. Geophys. Research* **85**, 6496–6504.

Aki, K., and A. B. Chouet, 1975, Origin of coda waves: source, attenuation and scattering effects, *Jour. Geophys. Research* **80**, 3322–3342.

Aki, K., and P. Richards, 1980, *Quantitative Seismology*, vol. 2. San Francisco: W. H. Freeman.

Chang, A. C., and J. R. Cleary, 1978, Precursors to PKKP, *Seis. Soc. Am. Bull.* **68**, 1059–1079.

Cleary, J. R., and R. A. W. Hadden, 1972, Seismic wave scattering near the core-mantle boundary: a new interpretation of precursors to PKP, *Nature* **240**, 549–511.

Dainty, A. M., M. N. Toksoz, K. R. Anderson, P. J. Pines, Y. Nakamura, and G. Latham, 1974, Seismic scattering and shallow structure of the moon in oceanus procellarum, *Moon* **9**, 11–29.

Doornbos, D. J., and J. C. Mondt, 1980, The interaction of elastic waves with a solid-liquid interface, with applications to the core-mantle boundary, *Pure and Applied Geophys.* **118**, 1293–1309.

Flatté, S. M., and R. S. Wu, 1988, Small-scale structure in the lithosphere and asthenosphere deduced from arrival-time and amplitude fluctuations at NORSAR, *Jour. Geophys. Research* **93**, 6601–6614.

Frankel, A., and R. W. Clayton, 1986, Finite difference simulations of seismic scattering: Implications for propagation of short-period seismic waves in the crust and models of crustal heterogeneity, *J. Geophys. Res.* **91**, 6465–6489.

Frankel, A., and L. Wennerberg, 1987, Energy-flux model of seismic coda: Separation of scattering and intrinsic attenuation, *Seismol. Soc. America, Bull.* **77**, 1223–1251.

Gao, L. S., N. N. Biswas, L. C. Lee, and K. Aki, 1983, Effects of multiple-scattering on coda waves in a three-dimensional medium, *Pure and Applied Geophys.* **121**, 3–15.

Haddon, R. A. W., 1972, Corrugations on the mantle-core boundary or transition layers between inner and outer cores? (abstract) *Trans. Am. Geophys. Union* **53**, 600.

Haddon, R. A. W., and J. R. Cleary, 1974, Evidence for scattering of seismic PKP waves near the mantle-core boundary, *Phys. Earth and Planetary Interiors* **8**, 211–234.

Herraiz, M., and A. F. Espinosa, 1987, Coda waves: a review, *Pure and Applied Geophys.* **125**, 500–577.

Hudson, J. A., 1977, Scattered waves in the coda of P, *Jour. Geophys.* **43**, 359–374.

Ishimaru, A., 1978, *Wave Propagation and Scattering in Random Media*, V. 1 and 2. New York: Academic Press.

King, D. W., R. A. W. Haddon, and J. R. Cleary, 1974, Array analysis of precursors to PKIKP in the distance range 128 deg to 142 deg, *Royal Astron. Soc. Geophys. Jour.* **37**, 157–173.

Knopoff, L., and J. A. Hudson, 1964, Scattering of elastic waves by small inhomogeneities, *Jour. Acoust. Soc. America* **36**, 338–343.

Kopnichev, T. F., 1977, The role of multiple scattering in the formation of a seismogram's tail, *Izv. Acad. Sci., USSR, Phys. Solid Earth* **13**, 394–398.

Nikolaev, A. V., 1972, *The Seismics of Heterogeneous and Turbid Media* (translated from Russian). Nauka, Moskow.

Sato, H., 1977, Energy propagation including scattering effects, single isotropic scattering approximation, *Jour. Phys. Earth* **25**, 27–41.

Sato, H., 1982, Attenuation of S waves in the lithosphere due to scattering by its random velocity structure, *Jour. Geophys. Research* **87**, 7779–7785.

Sato, H., 1984, Attenuation and envelope formation for three-component seismograms of small local, earthquakes in randomly inhomogeneous lithosphere, *Jour. Geophys. Research* **89**, 1221–1241.

Wu, R. S., 1982a, Attenuation of short period seismic waves due to scattering, *Geophys. Research Letters* **9**, 9–12.

Wu, R. S., 1982b, Mean field attenuation and amplitude attenuation due to wave scattering, *Wave Motion* **4**, 305–316.

Wu, R. S., 1985, Multiple scattering and energy transfer of seismic waves-separation of scattering effect from intrinsic attenuation—I. Theoretical modelling, *Royal Astron. Soc. Geophys. Jour.* **82**, 57–80.

Wu, R. S., and K. Aki, 1985a, Scattering characteristics of elastic waves by an elastic heterogeneity, *Geophysics* **50**, 582–595.

Wu, R. S., and K. Aki, 1985b, Elastic wave scattering by a random medium and the small scale inhomogeneities in the lithosphere, *Jour. Geophys. Res.* **90**(B12), 10261–10273.

Wu, R. S., and K. Aki, ed. "Scattering and Attenuation of Seismic Waves," (special issues of *Pure and Applied Geophys.*), Birkhäuser Verlag, vol. I, 1988; vol. II and III, 1989.

Cross-references: *Deep Seismic-Reflection Profiling; Earth Structure, Global; Elasticity and Wave Propagation: Principles; Elastic Waves in Homogeneous Media; Exploration Seismology; Finite-Difference Forward Modeling in Seismology; Inverse Theory and Methods: Seismology; Lunar Seismology; Seismic Attenuation: Observation and Measurement; Seismic Diffraction; Seismic Imaging; Seismic Monitoring of Nuclear Explosions; Seismic Noise; Seismic Ray Theory; Seismic Tomography; Seismic Wavefield Migration; Seismology: Physical Model Studies; Surface Waves.*

SEISMOGRAMS: INTERPRETATION

The Recording

A seismogram is the recording of earth displacement as a function of advancing time. The recording may be made on photosensitive paper by a moving beam of light that requires darkroom seismic vaults, or on wax-coated paper by a heated stylus, or on ordinary paper by pen and ink. Magnetic tapes are widely used for the recording of earthquakes at large-scale array installations. These data are then fed directly into digital computers for automatic analysis.

The seismogram can be recorded at any magnification, on any time scale, depending on the speed of motor that rotates the drum on which the recording paper or tape is mounted. Magnification of actual ground motion by means of mechanical or optical magnification, with or without other electromagnetic or electronic amplification is necessary for readable records.

Seismograms of the World-Wide Standard Seismograph Network (WWSSN) operated by the United States Geological Survey, Office of Earthquake Studies, are written photographically on three-drum recorders that permit the three components of ground motion to be recorded simultaneously. These components are vertical (Z), north-south (N), and east-west (E). The three-drum recorder also writes hour and minute marks, with radio signals for accurate time-keeping, as well as calibration pulses for amplification checks. Time is read in Universal Coordinated Time (UTC).

The Effect of Instrumentation on the Recording

The character of the seismogram is largely determined by the instrument response to ground motion. (See *Seismic Instrumentation*.) WWSSN instruments are of two types:

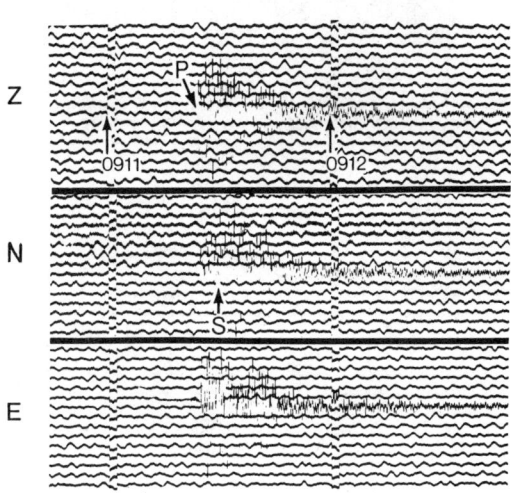

FIGURE 1. Short-period seismograms from GOL (Bergen Park, Colorado). Earthquake in NE Denver, distance 40 km from station, estimated depth 10 km. Magnitude (ML) 2.4 (GOL). (From Simon, 1981)

1. Short-period Benioff systems developed in the 1930s, where the free-period of the seismometer is 1.0 s, and the galvanometer is 0.75 s. Figure 1 is of a record made on the short-period system at the Cecil H. Green Geophysical Observatory (GOL) of the Colorado School of Mines, Bergen Park, Colorado.

2. Long-period systems developed by Press and Ewing in the 1950s, where the free-period of the seismometer is 30 s and that of the galvanometer is 100 s. Figure 2 is of a record made on the long-period system at the Lamont-Doherty Geological Observatory (PAL) of Columbia University, Palisades, New York. The seismometer period is 15 s.; the galvanometer period is 75 s for these instruments at this time. In the 1980s, instruments have been designed for broad-band reception of seismic waves, so that one set will serve the purpose of many sets of recording systems.

Interpretation of Seismograms

Analysis of earthquake recordings is necessary to determine the location of an *epicenter,* that point on the surface of the Earth where the shock occurred. The point under the surface at depth of occurrence, is called the *hypocenter* or focus. The distance or delta (Δ) of the earthquake from the recording station is obtained from the seismograms with the aid of a fundamental tool, the *seismic travel-time tables.* Curves made from the table for surface focus shocks devised empirically by Jeffreys and Bullen are seen in Fig. 3. Other tables for deep focus earthquakes developed by Gutenberg and Richter were published by Richter in 1958.

Phases on the seismogram are chosen by the inter-

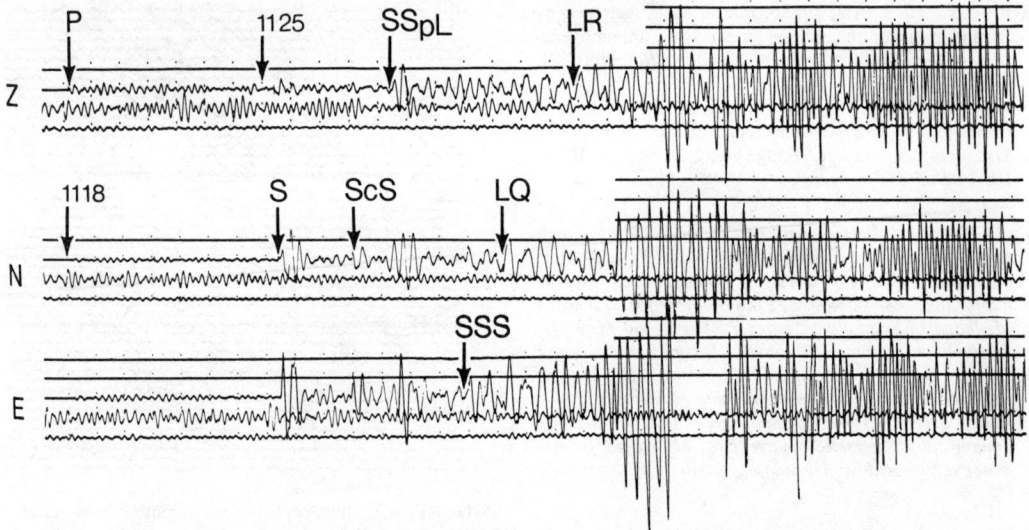

FIGURE 2. Long-period seismograms from PAL (Palisades, New York). Earthquake in Kodiak Island, Alaska, distance 5,700 km from station, depth 36 km, origin time: 11 09 15.3 UTC. Mag. (MS) 6 1/2–6 3/4 (PAL). (From Simon, 1981)

preter at the time of every measureable change in frequency and/or amplitude seen in the train of seismic activity. These phases are then compared with the arrival times on the travel-time curves. Increasing time interval between the P phase and the S phase denotes increasing distance from earthquake to station. Figure 1 is a "near" event, Fig. 2 is a "distant" earthquake. P and S phases are indicated as well as the time-scale on each record. Usually "near" events are of high frequency waves and recorded on short-period instruments, while more distant shocks have lower frequencies and are recorded on long-period instruments.

P waves, compressional-rarefactional waves best seen on vertical component seismograms, are of several types.

1. Guided waves that travel in channels in the upper layers of the earth's crust, designated as P_n, P^*, P_g, and P_L
2. Diffracted P beyond a distance of 105°
3. Multiple surface reflections of P waves, designated by PP, PPP, PPPP
4. P waves that graze the core of the earth or are reflected from it, as PcP, PcS, PcPPcP
5. P waves that pass through the core of the earth as PKP, PKKP, PKPPKP, etc.

S phases are of similarly corresponding types: S_n, S^*, S_g, SS, SSS, ScS, SKS, SKKS, etc. These are best seen on horizontal component seismograms. P and S waves occur also in various combinations with each other as PS, SP, PPS, SPP, etc.

Deep Earthquakes. Deep earthquakes are difficult to interpret. Turner published the first paper on deep-focus earthquakes in 1922 when he noted early-arriving phases from many earthquakes that could not otherwise be explained. (See references in Richter, 1958.) Figure 4 illustrates a deep earthquake in Argentina recorded on both short- and long-period seismograms at GOL. The time interval between the first P wave and its surface reflection, pP, indicates depth of focus. Most teleseismic, or distant, earthquakes with normal depth of focus (33 km) will have this reflection 9–11 s after the first arrival, P. The Jeffreys-Bullen Seismological Tables list pP-P times in seconds after P at different depths. The shock illustrated in Fig. 4 has a pP-P time of 123 s, indicating a depth of 590 km. The longer the time interval between the first two arrivals, the deeper the focus. Other phases also have deep reflections as sS, sSP, sSS illustrated. Confusion between pP and PP often occurs, resulting in misinterpretation of the deep earthquake. Usually surface waves in deep shocks are absent or sharply attenuated, as should be noted in Fig. 4.

Surface Waves. Most prominent on all long-period seismograms are surface waves (see *Surface Waves*). These vibrate in different modes and have been given various names by many investigators. However, there are two main types—Love (LQ) waves and Rayleigh (LR) waves, both illustrated in Fig. 2. Love waves are shear waves, traveling with a shearing motion, and are *transverse* waves (like S waves) recorded only by horizontal component seismograms. Rayleigh waves travel with retrograde elliptical motion along the path of propagation and are *longitudinal* waves (like P waves) best received by vertical seismometers but also registered on horizontal components.

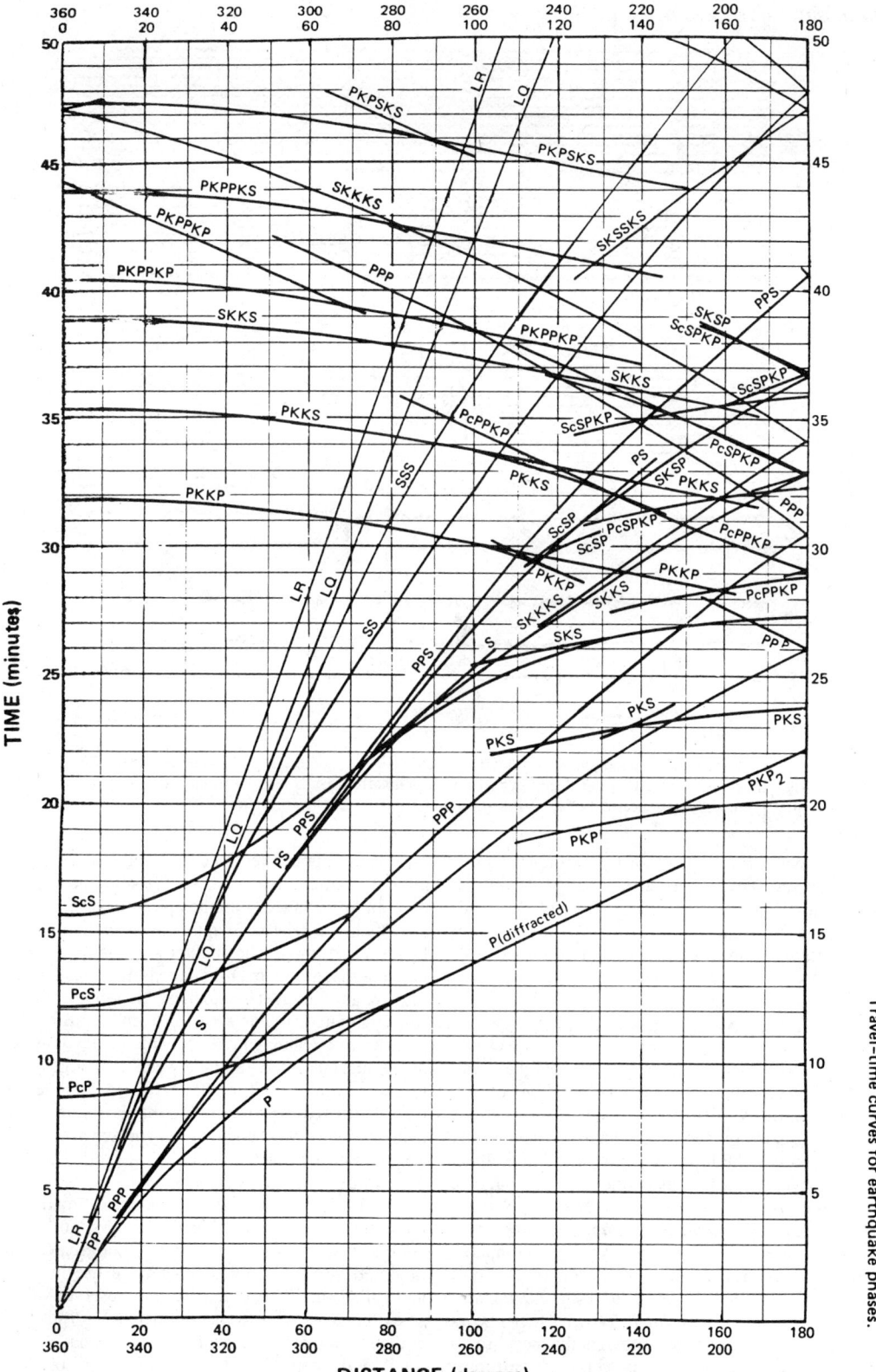

FIGURE 3. Seismic travel-time curves based on Jeffreys-Bullen Tables, modified from Jeffreys (1976). (From Simon, 1981)

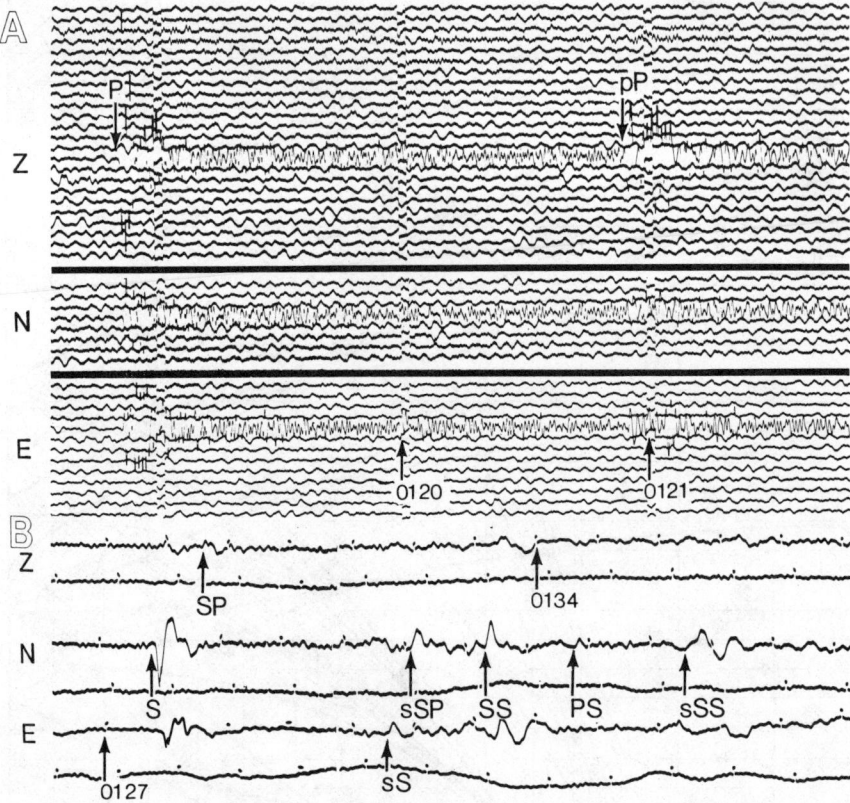

FIGURE 4. *A*. Short-period seismograms from GOL. Earthquake in Argentina, 8,675 km from station at a depth of 590 km. Origin time: 01 07 54.3 UTC. *B*. Long-period seismograms from GOL. Same earthquake as *A*. Note large body phases (P, pP, S, etc.) and absence of surface waves prominent in Fig. 2. (From Simon, 1981)

The direction, or *azimuth* of the earthquake from the recording station can be obtained from both P wave and surface wave particle motions. *Microseisms*, the constant "background" activity on seismograms, from 2 to 30 s period is not yet well understood.

Other Kinds of Work with Seismograms

The *mechanism* at the source of an earthquake can be determined from a study of the directions of first motions on many seismograms from distant stations at differing azimuths surrounding the epicenter of one large earthquake. Byerly, Stauder, Nuttli and others have evolved the *fault-plane* study for this purpose (see *Seismic Source: Theory; Seismic Source: Observations; Earthquake Mechanisms*).

The *magnitudes* of earthquakes are computed directly from measurements made on the seismogram using Richter's method (see *Earthquake Magnitude, Energy, and Intensity.*). Richter's work on California earthquakes showed the need for a uniform scale of magnitude as well as for a procedure for determining magnitude. Small shocks in populated areas cause more concern than large shocks in unpopulated areas. Calculation of actual ground motion recorded on seismograms as a function of distance from the epicenter gives a scale on which the relative magnitude of earthquakes is easily determined. The Richter scale is logarithmic in that a magnitude 6 earthquake is 10 times the size of a magnitude 5. The smallest identifiable shocks have magnitudes near zero, or even negative. The largest are over magnitude 8. This means that at a given distance, the ratio between the largest and smallest events is not less than 10^8. Magnitudes are now computed by different formulae for measurements made of both body waves and surface waves. For a more complete discussion see *Earthquakes: Magnitude, Energy, and Intensity.*

The *Modified Mercalli Scale* is an estimation of the *intensity* of an earthquake and is not to be confused with the Richter magnitude scale. The modified version adopted for general use in 1931, and called the MM scale, lists the observed *effects* of earthquakes with Roman numerals: I, "not felt" at the lower end of the scale to XII, "damage nearly total" at the upper end. Intensity maps, or isoseismal maps are made from "felt reports" based on the Modified Mercalli scale. This scale has no connec-

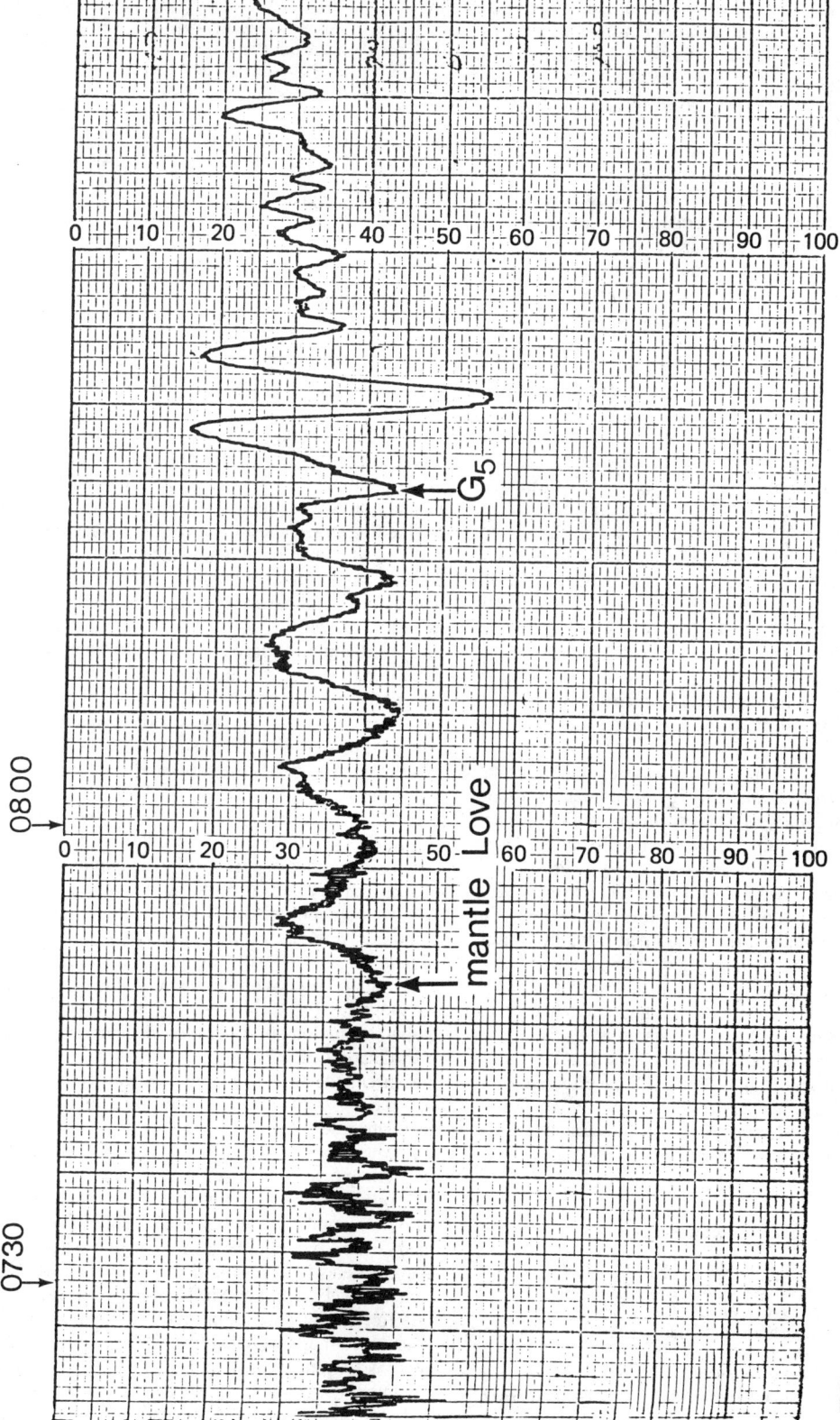

FIGURE 5. N-S unfiltered strainmeter record (GOL). Earthquake in southern Alaska, depth 33 km, magnitude (MS) 8.5 (GOL). Origin time: 03 36 14.2 UTC. Waves with periods of 10 to 6 minutes and velocities near 5 km s^{-1} are considered free oscillations of the Earth. The mantle Love wave (G5) has travelled 77,850 km around the Earth five times from the epicenter. (From Simon, 1981)

tion with work on seismograms but is included here to explain the frequently confused difference between magnitude and intensity. Japan and the USSR use different intensity scales.

The paths of seismic waves through and around the Earth produce characteristic wave patterns for oceanic and continental surface waves, or combinations of each. Seismic waves are changed by Earth structure, evidenced by changing velocities in the mantle and layers of the crust of the Earth. *Velocities* (km s^{-1}), both mantle and crustal, can be derived from the seismogram. Such studies began in the 1950s with the development of the long-period seismograph system and are now quite detailed (see, e.g., *Global Earth Structure*).

Mathematical equations of wave motion in various modes are used to build complex computer models of the Earth's layered crust. *Synthetic seismograms* produced by the computer in this way bear striking resemblance to those from real earthquakes. This work has become a sophisticated branch of theoretical seismology (see *Seismograms: Synthetic*).

The longest period instrument so far developed, the Earth *strainmeter*, records the longest earth waves, the *free oscillations* of the Earth with periods up to 100 hours. Figure 5 illustrates a strainmeter record from the Great Alaska earthquake of 28 March 1964, origin time: 03 36 UTC. The section of record shown is that of mantle Love waves the fifth time around the Earth, which have periods of from ten to six minutes and are considered free oscillations of the Earth.

Accumulation of vast quantities of accurately acquired and automatically analyzed data is already a fact of seismology. Statistical utilization of the data has been underway for several years now. These, along with other means of working with seismograms, may possibly lead to the prediction of earthquakes. Continuous study of the seismogram itself until every unknown, unexplained "wiggle" on the record is understood, will give more detailed knowledge of the Earth and its constant activity. Further development of instrumentation for other uses can be expected. The ocean-bottom seismograph operates on the floors of the oceans; similar systems have been sent to the Moon. All give seismologists more specialized data to better understand the Earth's origin, physical properties, and movements.

RUTH B. SIMON

References

Bullen, K. E., 1953, *An Introduction to the Theory of Seismology*. London: Cambridge University Press, 296p.
Jeffreys, H., 1976, *The Earth—Its Origin, History and Physical Constitution*. London: Cambridge University Press, 574p.
Oliver, J., 1962, A summary of observed seismic surface wave dispersion, *Seismol. Soc. America Bull.* **52,** 81-86.
Richter, C. F., 1958, *Elementary Seismology*. San Francisco: W. H. Freeman, 768p.
Simon, R. B., 1981, *Earthquake Interpretations, A Manual for Reading Seismograms*. Los Altos, Calif.: William Kaufmann, Inc., 150p.

Cross-references: *Earthquake Mechanisms; Earthquakes: Magnitude, Energy, and Intensity; Earthquake Seismology; Earth Structure, Global; Elasticity and Wave Propagation: Principles; Free Oscillations of the Earth; Seismic Instrumentation; Seismic Instrumentation: History; Seismic Noise; Seismic Source: Observations; Seismic Source: Theory; Seismology: History; Surface Waves.*

SEISMOGRAMS: SYNTHETIC

Synthetic seismograms have long been used in both the investigation of Earth structure and in the determination of earthquake source parameters. Until recently, however, there were relatively few methods for the routine generation of synthetic seismograms. The rapid development of more powerful and cheaper computers, coupled with significant theoretical advances in wave propagation, has resulted in a bewildering variety of different approaches. This article attempts to clarify the subject somewhat by explaining some of the theoretical background behind the more important methods, and by stressing the connections between the various techniques. In addition, this article attempts to answer the question of how to choose the most appropriate method for a particular problem.

In 1904 Horace Lamb constructed an analytical solution to the mathematical problem of finding the seismic wave field generated by an impulse applied to the surface of a uniform elastic half-space. Although seismologists investigated and solved more elaborate versions of this problem over the next several decades, very little progress was made in developing any method to calculate synthetic seismograms for a more realistic Earth model. The global seismologist was mostly concerned with the identification of a few primary seismic phases, and measurements of their travel times, for which simple ray theory with a one-dimensional model gave adequate results.

In the last 20 years, however, the nature of seismology has changed considerably. Several methods are now available for the routine computation of complete synthetic seismograms for a stratified Earth model. As a result, the spherically averaged elastic structure of the Earth as a whole is comparatively well known. Seismologists are therefore free to concentrate on more subtle lateral variations and other local features. In addition, the advent of extremely fast and relatively inexpensive computers, together with the increased availability of broadband, digital seismic data, has led to the development of sophisticated migration and inversion schemes, which produce high-resolution images of Earth structure from seismic waveforms. Such schemes require the calculation of accurate synthetic seismograms for a succession of increasingly elaborate models, and this requirement has in turn led to renewed interest in many theoretical aspects of wave propagation.

Although several methods exist for computing synthetic seismograms for a finite source region or propagating fracture, we discuss only a point source fixed at some point in the Earth. In the next section we discuss popular schemes for calculating seismograms for a one-dimensional Earth model, and explore some of the connections between these approaches. Whenever possible we avoid equations: Much more detail and plenty of equations can be found in the references quoted, as well as in the standard seismology texts such as Aki and Richards (1980) and Kennett (1983). We start with the "full synthetics" methods, which are used to calculate the complete response to a given source, and progress to the faster and more approximate asymptotic methods, which are used typically to investigate a specific seismic arrival. In the subsequent section we generalize to a laterally heterogeneous model and investigate the available synthetic seismogram techniques for it. In the last section we address the most important practical issue—the choice of method for a particular problem.

Synthetic Seismograms for Layered Media

Introduction. By far the most important class of methods for generating synthetic seismograms is that based on a one-dimensional (stratified) elastic model. Despite widely held beliefs to the contrary, this is not because seismologists think that the Earth really is perfectly vertically stratified. To a first approximation, however, such an assumption is extremely useful, both in global seismology and in exploration geophysics. In the first case, the obviously complex geologic features observed at the Earth's surface, and presumably persisting to significant depths, represent a relatively small perturbation to the basically layered model seen by long-wavelength seismic waves. In the second case, the reason is different, since even in a sedimentary environment the departures from a simple, layered structure may be very significant. The techniques of seismic processing, however, which include stacking traces from many receivers, are designed to minimize the effect of such lateral heterogeneity and to simulate the normal incidence reflection response from a stratified medium, even though this may be a gross distortion of the actual structure.

The second reason for making extensive use of one-dimensional methods is economic. The amount of computing time and storage required to generate synthetic seismograms for an arbitrary three-dimensionally varying model is immense. Thus, even where a layered model is known to be inadequate, it may be too expensive to compute seismograms for a more realistic heterogeneous structure.

Third, it is possible to retain much more physical insight into the process of wave propagation in a stratified medium, insight which may then be applied to the more general problem.

All the methods discussed here involve some mathematical transformation, typically one or more Fourier or Hankel transforms, of the original elastic wave equation. Such transformations between the time-distance domain and the frequency-slowness domain are crucial in understanding the physics of wave propagation in a layered medium. In the first section we show how features of the seismic response in the complex slowness plane translate into physical properties of the observed wave field. After this, we discuss normal mode summation, followed by reflectivity. Both techniques are used to generate complete synthetic seismograms with arbitrary precision. Next we discuss the generalized ray method, which is useful for computing accurate waveforms for specific seismic phases. After a brief reference to full-wave theory, we discuss the simplest technique, geometric ray theory, and finally the very versatile Wentzel-Kramers-Brillouin-Jeffreys (WKBJ) method. We provide a reference for each method, although no single reference fully describes both the theoretical and practical aspects of any given technique.

The Frequency-Slowness Domain. Before addressing specific methods for generating synthetic seismograms, we describe the connection between the physical time-distance domain and the more abstract frequency-slowness domain. A much more complete discussion of this topic can be found in Aki and Richards (1980, chap. 6).

We consider a horizontally layered elastic medium, having a point source at some depth, and define a cylindrical polar coordinate system (r, θ, z), where z is measured from the surface downward. To simplify the discussion, we assume that the source is cylindrically symmetric so that all quantities are independent of θ. The surface displacement $u(t, r, z = 0)$ is thus a function of time t and horizontal range r. We may write this displacement as the double transform

$$u(t, r, z) = \int_{-\infty}^{\infty} d\omega\, e^{i\omega t} \int_{-0}^{\infty} dp\, J_0(\omega p r) U(\omega, p, z),$$

where U, the transformed response, is a function of frequency ω, slowness p, and depth z, and J_0 is the usual Bessel function. The plane wave component U then satisfies an ordinary differential equation in z, with traction-free boundary conditions at the surface and an outgoing radiation condition at $z = \infty$. An important feature of the wave equation for a stratified model is that the original coupled differential equations for the three components of motion separate into one second-order equation for the SH (toroidal) component plus two coupled equations for the P-SV (spheroidal) system.

The various methods for calculating synthetic seismograms for a layered medium all start, in some sense, from this superposition of plane waves. The techniques vary in two ways: in how they solve the differential equation for U, and by what method (as well as in which order) the two transforms are evaluated. We discuss below the specific choice for each method and will limit the discussion here to general features of the response in the slowness plane for some arbitrary frequency.

We have seen that the seismic response in the frequency domain may be expressed as an integral over plane wave components U. The properties of this function in the complex slowness plane are of fundamental importance in calculating the time domain response, since they determine to what extent the path of integration may be deformed, as well as what approximation may be employed for the integrand. For a general layered model, incorporating smooth velocity variation as well as discontinuities in wave speed, the behavior of U is extremely complicated, with discontinuities generating branch cuts and resulting in multiple Riemann sheets, while particular geometric arrivals are associated with isolated saddle points. The simplest case is a finite, spherically layered model where the entire displacement field may be expressed as a sum of residue contributions from poles. In general, however, the seismogram is built up from a combination of geometric arrivals, head waves generated by branch line contributions, and trapped waves corresponding to isolated poles. Some knowledge of the behavior of the response in the frequency-slowness domain is thus crucial to understanding both the range of applicability and the expected accuracy of a particular method.

Normal Mode Summation. Two methods are commonly used to compute full synthetic seismograms for a stratified Earth model. When the wave field to be synthesized consists of relatively low frequency waves propagating over distances comparable with or greater than the Earth's circumference, the most appropriate technique is modal summation.

Modal summation (Takeuchi and Saito, 1972) exploits the fact that the Earth as a whole approximates a spherically stratified body, for which the response to any source may be written as a sum of normal modes of varying angular and radial order, each with its own characteristic frequency. The major task involved in calculating seismograms is thus to find these eigenfrequencies and related eigenfunctions, described in detail for both spherical and flat layered models by Takeuchi and Saito (1972). Although modal summation is very useful for synthesizing long-period seismograms, the number of modes required increases very rapidly as a function of frequency, and the computing cost generally becomes prohibitive for periods shorter than a few seconds.

Reflectivity. The most common alternative technique for generating full synthetics is the reflectivity method (Kennett and Kerry, 1979). In this approach, the Earth is modeled as a stack of uniform, horizontal layers. It might seem that this approach would introduce significant errors for the Earth as a whole, where the layering is spherical. Fortunately, however, there are mathematical mappings from spherical to flat geometry (the so-called Earth-flattening transformations) that preserve the wave kinematics and dynamics exactly for the SH system and relatively accurately, if not exactly, for the P-SV wave field. The response U in the frequency-slowness domain is calculated by decomposing the wave field within each layer into up- and downgoing components, choosing the correct combination at each interface to ensure continuity of stress and displacement. This procedure is repeated for a large suite of frequency and slowness values, and the double transform is then computed explicitly by using a numerical Hankel transform and a fast Fourier transform.

The original theoretical development was limited to a surface source, and the slowness integral was restricted to values corresponding to real angles of incidence on the structure. In the more general case of a buried source, the full wave field may be expressed in terms of reflection and transmission properties of those portions of the model above and below the source depth. In addition, by extending the slowness integral to a value slightly larger than the fundamental Rayleigh wave slowness, a full synthetic seismogram is computed, including higher-mode and surface wave effects.

Unlike modal summation, reflectivity can be used to compute synthetic seismograms for relatively short periods. As the frequency increases, however, more and more slowness values are needed to maintain accuracy. In addition, the computing time increases linearly with the number of layers in the model, which often needs to be large to approximate the real Earth closely. There is one mitigating factor, however; although the cost of computing the first seismogram can be extremely high, seismograms at other receivers for the same source are obtained by simply modifying the final Hankel transform, which forms a relatively trivial part of the calculation.

Generalized Ray Theory. The full frequency-slowness domain response U may be thought of as a sum of contributions arising from a variety of raypaths, corresponding to different routes of energy propagation within the Earth. Each such contribution is defined by the wave type (P or S) propagating within the various layers comprising the stratified model, together with appropriate reflection and transmission coefficients at the interfaces. In many portions of the resulting seismogram, the response is a superposition of many overlapping arrivals, as well as nongeometric contributions such as diffracted waves and modal phases. In other regions, however, clearly defined isolated arrivals are seen, which are associated with specific propagation paths. The generalized ray method (Helmberger, 1968) is an important technique for computing an accurate wave form for such an arrival.

The generalized ray method makes use of a theoretical technique, developed by Cagniard and modified by de Hoop, that involves a Laplace, rather than Fourier, transform over frequency. In this approach, the slowness p is replaced by a new variable $\theta(p)$, with the infinite integral along the real slowness axis being replaced by a finite, complex contour known as the *Cagniard path*. As a result, the two transformations over frequency and slowness are made to cancel each other out, leaving the time domain response as a single, finite integral. In fact, for a line source the solution is even simpler, since an analytic expression is obtained for the response. The most time-consuming part of a generalized ray calculation is the computation of the Cagniard path, which is defined by an implicit equation and varies for each individual ray as well as for each receiver location.

In typical applications, a handful of generalized rays will be computed and summed to synthesize a portion of a seismogram. Burdick and Orcutt (1979) have compared the results of generalized ray calculations with reflectivity seismograms and have found that the agreement is often extremely good.

Full-Wave Theory. In some circumstances, even though we are concerned with a particular seismic arrival, nongeometric effects must be taken into account in computing the wave field. Such is the case with a phase that has a turning point in the vicinity of a velocity discontinuity for some distance range. An example of this is the core phase SKS, where an infinite family of multiply reflected rays SK^2S, SK^3S, ... with very slowly decaying relative amplitudes is produced, many of which contribute to the observed seismogram.

Full-wave theory (Choy, 1977) uses a uniform asymptotic expression for that portion of the response describing such a family of arrivals, within the original spherical geometry of the Earth. In terms of the transforms already introduced, it may be characterized as a method where the real slowness integral is deformed to a path in the complex p plane while the frequency integration is left unchanged.

Although full-wave theory does not require an Earth-flattening transformation, some is still approximation involved because we assume the P and S waves are not strongly coupled above or below the discontinuity. Choy et al. (1980) have investigated the relative accuracy of full-wave theory and reflectivity and conclude that the seismograms produced are essentially identical. Full-wave theory thus represents a powerful technique for studying particular types of arrivals, when a full reflectivity calculation with an appropriate number of uniform layers would be unduly expensive.

Geometric Ray Theory. In many situations, the low-frequency energy generated by a seismic source is negligible or filtered out by the receiver, and a high-frequency or asymptotic approximation to the response produces a perfectly good representation of the observed signal. In such circumstances, or when accuracy of the waveform is not so important, the simplest method to employ is geometric ray theory (Aki and Richards, 1980).

The equations of geometric ray theory are usually derived by taking the frequency domain wave equation and solving it as an asymptotic series in $1/\omega$. An alternative theoretical approach is via the frequency-slowness domain. By expanding the lowest-order asymptotic expression for the response U about the appropriate saddle point, and evaluating the slowness integral via the steepest descent path, we obtain the same result. The resulting eikonal equation for the phase is readily solved in a layered model by numerical integration, and the ray parameter is adjusted until the ray in question passes through the receiver. The associated amplitude is given by an explicit expression, involving the derivative $d\Delta/dp$, where Δ is the range, and is simply evaluated numerically using closely spaced rays.

The WKBJ method. Although geometric ray theory is extremely simple to use and generally produces adequate results, under some circumstances it gives no answer (in a shadow zone) or the wrong answer (near a caustic). The WKBJ method developed by Chapman (1978) may be thought of as a more generally valid form of geometric ray theory.

There are three important features of the WKBJ method. The name itself derives from the fact that the response U in the frequency-slowness domain is given by the WKBJ asymptotic approximation. The most important theoretical aspect, however, is that the order of the transforms over slowness and frequency is reversed, with the frequency integral being performed first. Unlike geometric ray theory, the slowness integral is evaluated exactly, with no further approximation. As with generalized ray theory, the two integrals are forced in some sense to cancel each other out, and the resulting seismogram is obtained directly in the time domain. For the WKBJ seismogram, however, the path of integration in the slowness domain is along the real axis. The third feature is a practical consideration: The WKBJ method allows for implicit smoothing of the final seismogram, hence removing spurious caustics that might be introduced by whatever interpolation scheme is adopted for the velocity model.

Although, like geometric ray theory, the WKBJ method involves an asymptotic approximation, with the frequency-slowness domain response corresponding to the lowest-order term in $1/\omega$, the fact that no further error is introduced in the slowness integration leads to a more accurate, and more generally valid, result for the waveform.

Synthetic Seismograms for Laterally Varying Media

Introduction. For many applications of seismology, both in exploration geophysics and in global studies, it is simply inadequate to attempt to model the Earth as a layered medium. Indeed, the departure from stratification is often the most significant structural feature, whether it be an anticline forming a hydrocarbon trap, or a subducting plate marking the boundary of a global convection cell. In such cases, it is clearly necessary to be able to generate synthetic seismograms for a more general laterally heterogeneous model.

In contrast to the one-dimensional case, there is no exact method available for a more general model. Although extensions of the reflectivity and modal approaches have been developed for a laterally varying medium, the computer time associated with such schemes is prohibitive for most applications. There are, however, two types of method suitable for calculating synthetic seismograms in a heterogeneous model. The simplest and most widely used, particularly in exploration seismology, is geometric ray theory, which is conceptually simple and relatively inexpensive. In many applications, however, an accurate waveform is required, and the asymptotic approximations required are inadequate. In addition, there are many situations, such as caustics, in which geometric ray theory breaks down. Until recently, the only alternative methods were "brute-force" computational approaches, such as finite-difference and finite-element schemes, as well as more esoteric variants, such as pseudospectral methods. In recent years, however, theoretical developments in wave propagation have led to asymptotic methods of more general validity and greater precision than geometric ray theory, of which Gaussian beams and the "Maslov" method appear to be the most generally useful. We briefly describe these techniques and indicate the type of problem for which each might be appropriate.

Finite Difference/Finite Element. For some purposes, such as the inversion of seismic data to determine three-dimensional velocity structure, we need to compute accurate synthetic seimograms for a fully heterogeneous elastic model. In such cases, there is no alternative but to resort to some numerical scheme for solving the elastic equations of motion directly.

Many numerical methods have been devised, driven by the need for more efficient and more accurate computation, particularly within the realm of exploration seismology (see *Finite-Difference Forward Modeling in Seismology*). The two most popular approaches are finite-difference schemes (Boore, 1972) and finite elements (Lysmer and Drake, 1972), although other techniques, such as pseudospectral methods, have also been tried.

The limiting factor in such computational methods is computer storage, since an immense grid is needed to define a realistic three-dimensional model extending over many wavelengths. In addition, the important subject of error estimation for such algorithms is still an active area of research.

Geometric Ray Theory. When speed rather than extreme accuracy is the main consideration, an asymptotic method becomes more attractive than the computational schemes described above. The simplest, and hence most popular, technique is geometric ray theory (Cerveny et al., 1977) (see *Seismic Ray Theory*).

We have already looked at geometric ray theory for a layered model, and the basic philosophy is the same for a generally heterogeneous model. But there are additional complications in the three-dimensional case. In particular, it becomes much more difficult to identify the specific ray that links source with receiver, since each ray is defined by two independent parameters, corresponding to takeoff angles in the horizontal and vertical planes. A variety of schemes have been developed to solve this two-point boundary value problem, including the "bending" method, which improves iteratively on an initial ray that joins the two points but only satisfies the eikonal equation approximately, and "shooting" methods, which gradually change the takeoff angles to home in on the required target. A further complication is that a laterally heterogeneous velocity structure will typically give rise to more caustics than will a simple, layered model. Despite these disadvantages, the simplicity and speed of geometric ray theory make it an extremely attractive method for generating synthetic seismograms for a laterally heterogeneous model.

TABLE I. Synthetics Methods for a Stratified Model

Method	Frequency Range	Distance	Model Layering	CPU Time
Normal Modes	Low	Large	Spherical	Hours
Reflectivity	Moderate	Moderate	Flat Uniform	Hours
Generalized Ray	Arbitrary	Arbitrary	Uniform	1 Hour
Full Wave	Moderate	Fixed by Phase	Spherical	1 Hour
GRT	High	Arbitrary	Arbitrary	Minutes
WKBJ	Moderate–High	Arbitrary	Arbitrary	Minutes

TABLE 2. Synthetics Methods for a 2D or 3D Model

Method	Frequency Range	Distance	Model Type	CPU Time
FD/FE	Arbitrary	Few Wavelengths	Arbitrary	Many Hours
GRT	High	Arbitrary	Smooth	Minutes
Gaussian Beams	Moderate–High	Arbitrary	Smooth	Minutes
Maslov	Moderate–High	Arbitrary	Smooth	Minutes

Gaussian Beams. Although geometric ray theory is simple to use and produces relatively reliable seismograms in most cases, there are situations such as caustics where it fails altogether. In recent years, a number of more flexible and accurate asymptotic methods have been developed.

The Gaussian beam method (Cerveny et al., 1982) (also see *Seismic Ray Theory*) is based on the observation that it is possible to construct asymptotic solutions of the wave equation that are concentrated in the vicinity of a geometric ray. By decomposing the wave field at the source into a sum of such contributions, and propagating these contributions through the medium, we no longer need to solve the two-point boundary value problem encountered in geometric ray theory. The other major advantage of this technique is that the solution is valid in regions where geometric ray theory breaks down and it includes frequency-dependent effects missing from the simpler ray theoretical approach.

Maslov Theory. A related approach is the method of calculating synthetic seismograms based on the Maslov asymptotic theory (Chapman and Drummond, 1982). Unlike the Gaussian beam approach, which is somewhat ad hoc, the Maslov method sets up a rigorous theoretical framework based on an asymptotic Fourier transform. It is a generalization to a laterally heterogeneous medium of the WKBJ method. From a physical viewpoint, the source may be thought of as generating a superposition of constant-slowness wavefronts, which are propagated through the structure and recombined to form the response at the receiver.

Maslov theory, like the Gaussian beam method, is valid in regions where geometric ray theory breaks down. The caustics arising in geometric ray theory are not fundamental to the wave propagation process, but may be thought of as arising from the use of an inappropriate local coordinate system. By working in a hybrid distance/slowness domain, in which such singularities cannot occur, Maslov theory leads to results of much more general validity than for geometric ray theory.

Practical Considerations—How to Choose a Method

We have described several methods for calculating synthetic seismograms, making reference to the theoretical background and to the practical aspects of each technique. When choosing a method of computing seismograms, we must consider the source depth and receiver distance, the range of frequencies for which the response is required, the degree of accuracy required, and the amount of computing power available.

We can present only a superficial comparison of the various methods. Tables 1 and 2 list relative advantages and disadvantages. The descriptions are necessarily vague, however, and should not be viewed in any absolute sense. The category CPU time is intended only to give an order of magnitude idea of the computing time involved on a "small computer."

In conclusion, we should stress that none of these methods should be viewed merely as a "black box." The successful use of any such technique relies on some degree of insight into the underlying theory and, hence, on some appreciation of its limitations. Although a strong case can be made for standardization of computational schemes, this standardization should never be at the expense of such insight.

TIMOTHY J. CLARKE

References

Aki, K., and P. G. Richards, 1980, *Quantitative Seismology. Theory and Methods.* San Francisco: W. H. Freeman.

Boore, D. M., 1972, Finite difference methods in seismology, in B. A. Bolt, ed., *Methods in Computational Physics*, vol. 11. New York: Academic Press.

Burdick, L. J., and J. A. Orcutt, 1979, A comparison of the generalized ray and reflectivity methods of waveform synthesis, *Royal Astron. Soc. Geophys. Jour.* **58,** 261–278.

Cerveny, V., I. A. Molotkov, and I. Psencik, 1977, *Ray Method in Seismology.* Prague: Charles University Press.

Cerveny, V., M. M. Popov, and I. Psencik, 1982, Computation of wave fields in inhomogeneous media—Gaussian beam approach, *Royal Astron. Soc. Geophys. Jour.* **70,** 109–128.

Chapman, C. H., 1978, A new method for computing synthetic seismograms, *Royal Astron. Soc. Geophys. Jour.* **54,** 481–518.

Chapman, C. H., and R. Drummond, 1982, Body-wave seismograms in inhomogeneous media using Maslov asymptotic theory, *Seismol. Soc. America Bull.* **72,** S277–S317.

Choy, G. L., 1977, Theoretical seismograms of core phases calculated by frequency-dependent full wave theory, and their interpretation, *Royal Astron. Soc. Geophys. Jour.* **51,** 275–312.

Choy, G. L., V. F. Cormier, R. Kind, G. Müller, and P. G. Richards, 1980, A comparison of synthetic seismograms of core phases generated by the full wave theory and by the reflectivity method, *Royal Astron. Soc. Geophys. Jour.* **61**, 21-39.

Helmberger, D. V., 1968, The crust-mantle transition in the Bering Sea, *Seismol. Soc. America Bull.* **58**, 179-214.

Kennett, B. L. N., 1983, *Seismic Wave Propagation in Stratified Media.* Cambridge, England: Cambridge University Press.

Kennett, B. L. N., and N. J. Kerry, 1979, Seismic waves in a stratified half space, *Royal Astron. Soc. Geophys. Jour.* **57**, 557-583.

Lysmer, J., and L. A. Drake, 1972, A finite element method for seismology, in B. A. Bolt, ed., *Methods in Computational Physics*, vol. 11. New York: Academic Press.

Takeuchi, H., and M. Saito, 1972, Seismic surface waves, in B. A. Bolt, ed., *Methods in Computational Physics*, vol. 11. New York: Academic Press.

Cross-references: *Controlled Source Seismology; Earthquake Seismology; Earth Structure, Global; Elasticity and Wave Propagation: Principles; Elastic Waves in Homogeneous and Inhomogeneous Media; Exploration Seismology; Finite-Difference Forward Modeling in Seismology; Inverse Theory and Methods: Seismology; Seismic Imaging; Seismic Ray Theory; Seismic Wavefield Migration; Seismic Wave Scattering; Seismograms: Interpretation; Surface Waves.*

SEISMOLOGY: HISTORY

Although the scholarly study of the causes and effects of earthquakes can be traced into antiquity (Needham, 1959; Adams, 1938), seismology did not emerge as a separate science until the late nineteenth century; many of the most heavily studied branches of the subject began even more recently, in some cases only since the 1960s. This article reviews the growth of these different branches, and tries to explain both how and why they came to be. No attempt has been made to discuss all important results or to identify the earliest statement of any particular ideas; Stoneley (1967) gives a parallel view.

What seismologists do has been heavily influenced by the data and the interpretive methods that were available. Comparing nineteenth-century with recent reports on great earthquakes shows that the earlier reports mostly describe the effects of shaking at different places (seismic intensity), which in the modern reports is relegated to a few pages and an isoseismal map. This change reflects both the availability of other and more informative data and the use of more powerful methods to interpret them.

That more data are available is in large part because of developments in seismic recording. These are discussed in more detail in *Seismic Instrumentation: History* (see also Farrell, 1985) and only the highlights need be given here. The first step, from 1890 through about 1910, was the development of seismographs capable of faithfully recording (and precisely timing) waves from large distant earthquakes, the first demonstration of this being the recording in Germany of the Japanese earthquake of 18 April 1889, by E. von Rebeur-Paschwitz. Good recordings of smaller local earthquakes had to await the development of the Wood-Anderson seismometer in the 1920s. Accurate records of strong shaking, though a goal of the late nineteenth century, were not made until the development of strong-motion recorders in the 1930s (Hudson, 1983). Of course, for many studies a network is needed; instruments must not only exist but be available in sufficient numbers. The first global network was established by Milne in the 1890s, unfortunately with seismographs that soon became obsolete. Though international collaboration in seismology began in the 1900s (Rothé, 1981), the subsequent development of global seismic recording was largely a hodge-podge, with regional exceptions such as the Jesuit Seismological Association in North America (Macelwane, 1950). The most important advance in seismic instrumentation after the invention of the seismometer was the construction of the World-Wide Standard Seismograph Network (WWSSN) in the early 1960s, which created a global network of instruments with easily obtainable records and accurate calibration. Before the WWSSN, even the sense of motion on a seismogram could often be unreliable.

The routine production of well-calibrated seismograms was especially important because it coincided with another major change: In the 1960s new methods were introduced for analyzing seismic records and relating them to the structure of the Earth and the sources of seismic waves. Before this time, analysis of a seismogram usually meant timing the arrivals of bursts of energy and sometimes measuring their amplitude and period, all other information being discarded—not because it was meaningless, but because it was uninterpretable. The theory of elastic wave propagation had been applied to seismology by Hopkins in 1847; the difficulty in relating it to the data lay in the complexity of the problem, with complicated sources radiating through a heterogeneous medium. Only the simplest cases could be studied, and these largely reduced the seismogram to a series of discrete "arrivals." This situation changed with the advent of the digital computer, which made it possible to calculate wave propagation and earthquake rupture in more realistic conditions. At the same time digital processing of data made it possible for seismologists to use new concepts (that were largely imported from electrical engineering) to look at seismograms. The most powerful idea has been the study of seismic phenomena in the frequency domain, a procedure eased by the publication of the Fast Fourier Transform in 1965.

Neither the WWSSN nor the extensive use of computers could have happened without funding at much higher levels than were common before the 1960s. Such funds became available primarily because of two "applied" sides of seismology: *Seismic Monitoring of Nuclear Explosions* and *Earthquakes: Hazards and Predictions.* The first nuclear explosion (1945) was detected seismically (Bolt, 1976) and a U.S. monitoring program was initiated in the late 1940s. However, this had relatively little impact on seismological research until test-ban discussions in the late 1950s showed how little was known about many aspects of detecting and identifying explosions. The 1959 report by Berkner and others, which suggested substantially increased funding for seismology, was soon followed by large-scale support of the science, both within the United States and abroad, by the U.S. Defense Department Advanced Research Projects Agency, under Project VELA-UNIFORM, throughout the 1960s (Bates et al., 1982). As the plot of published papers (Fig. 1) shows, seismic research was greatly stimulated by the new tools these funds made available. Funding from this source shrank during the 1970s, being replaced to some extent by funds from earthquake prediction programs. Though predicting earthquakes was a hope of the pioneers of the science, by the 1950s it had come to be regarded by many as a will-o'-the-wisp (Richter, 1958). Partly because of social need, prediction programs were begun in Japan in 1965 and China in 1966. Early results from the Soviet Union, suggesting a systematic precursory change in seismic wave velocities, and the successful prediction of the 1975 Haicheng earthquake in China, have led to greatly increased support for this part of seismology.

As the name of the science suggests, the first seismologists were natural historians who specialized in the study of earthquakes. Their work included describing the effects of individual earthquakes, compiling lists of earthquakes in some region, and speculating on the causes of earthquakes. Descriptions of damage from shaking, all that was available to early workers, were systematized through the development of seismic intensity scales by de Rossi, Forel, and Mercalli, and by the use of isoseismal maps (Davison, 1927). These procedures became standard by 1900 and have remained largely unchanged, though discussions of damage have become less and less the province of the seismologist and more and more that of the earthquake engineer (for whom the behavior of the structure, rather than the motions beneath it, are of primary interest).

The listing of earthquakes in an area, or in modern terms the creation of a seismicity catalog, dates from the seventeenth century in Europe: Global compilations, to determine seismic and nonseismic regions of the world, had to await the nineteenth century. Most catalog-makers seek to make their lists as complete as possible, and here the influence of instrumental recording was profound by reducing the need for felt reports. On a global scale, this has meant seismicity maps unbiased by population distribution, the most notable early effort being that of Gutenberg and Richter; later work, while more

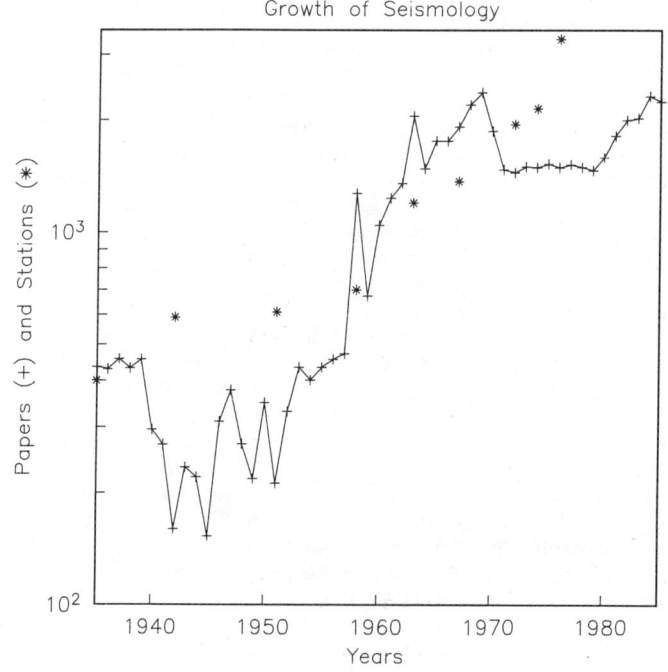

FIGURE 1. The annual number of papers entered in the *Bibliography of Seismology* (crosses connected by line), and (from Ambrayses and Melville, 1982, figure 4.1) an estimate of the total number of seismic stations in the world (stars).

accurate and complete, has not altered the basic picture of a few seismic belts. Especially for the deep ocean, where these belts could be seen to coincide with the mid-ocean rift, this observation was an important input to the ideas of plate tectonics. On a local scale, only instrumental records could reveal the patterns of the many unfelt earthquakes; a pioneering example was the southern California network established by H. O. Wood in the 1920s (Goodstein, 1984). The development of instrumental earthquake catalogs, with their improved coverage, also created both the need to quantify earthquake size and the data with which to do it. Richter's 1935 invention of a "magnitude" scale for local shocks, later extended by Gutenberg and Richter to teleseisms, soon became and largely remains the most used measure of earthquake size.

Early writers on earthquakes adduced many possible causes for tremors, including subterranean winds, steam explosions, chemical reactions, and the collapse of caverns (Adams, 1938). Many of these ideas were fostered by observations of the only other violent geological phenomena, volcanic eruptions, a connection fostered by the rough coincidence of seismic and volcanic belts. During the nineteenth century many of the older concepts became implausible and were gradually replaced by the idea that earthquakes were related to faulting (Howell, 1986). That an earthquake is nothing else but the release of accumulated elastic energy through fault motion was suggested by Gilbert in 1884. Geodetic data analyzed by Reid for the California earthquake of 1906 gave some of the first good evidence for this elastic-rebound theory, although the obscurity of his work abroad (Byerly, 1970) may have delayed widespread acceptance of this hypothesis. A full consensus on the importance of elastic rebound for earthquakes does not seem to have been reached for many years after Reid's work; for deep earthquakes especially, questions about the role of faulting have remained to the present.

By the first decade of the twentieth century many of the tools and ideas of modern earthquake studies were in place. At this time the development of sensitive seismographs created a whole new part of seismology: the study of elastic waves from earthquakes, to learn, not about the earthquake, but about the seismic wave velocities, and hence structure, within the Earth—a field previously notable for having more hypotheses than data (Brush, 1979). This side of seismology soon grew to equal if not overshadow the field of earthquake studies, since with the tools available before 1960 much could be learned about Earth structure but only a little more about earthquakes. Measurements from globally-distributed stations quickly led to the construction of travel-time tables for different phases, and these observations led to the discovery of the Earth's core by Oldham in 1906; in 1912 Gutenberg determined its size at close to the modern value. Seismic-wave evidence on the rigidity of the mantle, together with earth-tide data, then enabled Jeffreys, in 1926, to show that the core was fluid (Brush, 1980). Lehmann's 1936 proposal of an inner core established the last major structure to be discovered. The work of Jeffreys and Bullen throughout the 1930s resulted in travel-time tables accurate enough to still be a standard. The great achievement of seismology between 1900 and 1940 was thus to change our knowledge of the Earth's deep interior from near-total ignorance to something not far from our present knowledge. This rapid advance was in part possible because the great distances involved meant that timing requirements were relatively relaxed, and also because of the spherical symmetry of the Earth, which meant that an uneven global network gave fairly detailed coverage when mapped into epicentral distance. The tabulation of arrival times in the International Seismological Summary (Stoneley, 1970) was also a major aid.

Seismic-wave studies of shallow structure began at the same time as those of deep structure, the most notable early result being Mohorovičic's discovery that velocities increased abruptly at a depth of tens of kilometers, thus giving a new meaning to the term "the crust of the earth." Subsequent work on shallow structure proceeded more slowly than that on deeper structure, since dense enough arrays of instruments were available only in a few places; by 1940 enough data had been collected to establish the widespread nature of Mohorovičic's discontinuity. (Seismic measurements to depths of a few km for oil exploration began about 1920 and progressed rapidly, but this development was largely independent of crustal studies [Sheriff and Geldart, 1982; Bates et al., 1982]). After World War II many more studies of crustal structure were made using arrays of portable instruments; an important advance was the extension of these techniques into the deep oceans, where they showed a much thinner crust than beneath the continents.

Not only did refraction measurements for shallow structure flourish during the 1950s, but they were complemented by an increased use of seismic surface wave measurements, especially by Ewing's group at Columbia, who installed long-period seismographs worldwide. This advance in data collection was paralleled by Haskell's development of an efficient method to compute surface-wave dispersion for a layered structure. After the WWSSN made dispersion measurements possible over many paths, this method became common for establishing Earth structure in the crust and upper mantle.

The surface-wave dispersion studies involved measurements on the shape of the seismogram, not just first arrivals, and this was even more true of the next field of earth-structure study: the free oscillations of the Earth. Benioff claimed to have seen such oscillations after the 1952 Kamchatka earthquake, and this claim encouraged Alterman, Pekeris, and

Jarosch to compute the frequencies of such oscillations in a realistic Earth model. These theoretical results were ready when the Chilean earthquake of May 1960 provided a signal against which they could be compared. The subsequent identification (nearly simultaneously by several groups [Bullen, 1975]) of free-oscillation frequencies in their data could well be taken as the start of the modern era in seismology: Neither the data analysis nor the theoretical modeling would have been done without electronic computers. Free-oscillation studies at once became, and have remained, a significant field of seismology.

Computer techniques first impinged on seismology in the early 1960s by relating free oscillations and surface waves to structure. Only later (starting around 1970) did they begin to affect the more traditional studies of body waves and near-surface refractions, through the development of *Seismograms: Synthetic*. The drive to compute these accurately and cheaply in ever more realistic structures has made this a burgeoning theoretical area, with many applications.

Earthquake studies, long somewhat eclipsed by Earth-structure work, began to flourish in new directions in the 1960s. One of these directions owed little to computers but much to better data: the determination of earthquake fault planes from seismograms. This study had first been pursued in the 1920s by Nakano for local earthquakes in Japan and later by Byerly and Hodgson using teleseismic data, but the results had often been inconclusive and were hard to consider of more than local interest. This situation changed in the 1960s with the coming of the WWSSN data, which made such determinations more reliable, and with the development of plate tectonics, which showed that the instantaneous deformation seen in earthquakes agreed with the long-term results from magnetic anomalies. Seismic data thus came to be seen as tectonically meaningful, giving a considerable stimulus to methods of determining the nature of earthquakes from seismograms—another area where the increase in ability to solve wave propagation problems has had a major impact by making it possible to see which features of a seismogram are due to the source and which to the intervening medium. Many of the most recent observational studies also owe much to the establishment of dense networks of seismometers for local earthquake recording, and the greatly increased spread, since 1970, of arrays of strong-motion recorders (in the United States partly a byproduct of building code requirements).

Along with these observational studies, and sometimes in advance of them, was an enormous growth in theoretical models of the earthquake source, perhaps the area in seismology that has grown the most in the last few decades (Minster, 1985, gives a thorough review). Though elastic dislocation theory was developed by Volterra around the turn of the century, and though in retrospect its parallels with faulting seem obvious, early representations of the earthquake source instead concentrated on distributions of forces near the hypocenter. This concern with force distributions may have reflected scepticism about the relevance of faulting to earthquakes; in any case, the first studies in which earthquakes were modeled by dislocations in an elastic medium were not made until the 1950s, initially by Vvedenskaya.

Subsequent work in source theory has been extremely diverse. One trend has been the study of more complicated (and therefore presumably realistic) source models to examine how observations would depend on such features as rupture propagation and cessation. Another has been in attempts to include physically-based parameters, such as stress drop, in models. Still another research effort has been the development of the most general possible source that would be consistent with continuum mechanics, which has led to the moment-tensor representation now in common use. Perhaps more than any other part of seismology, the study of the seismic source shows how far the science has to go. A century of effort has established considerable knowledge about the Earth but we remain ignorant of many of the fundamental aspects of earthquake rupture. Like routine earthquake prediction, a full understanding of the rupture process is part of the future, rather than the past, of seismology.

DUNCAN CARR AGNEW

References

Adams, F. D., 1938, *The Birth and Development of the Geological Sciences*. Baltimore: Williams and Wilkins.

Ambrayses, N. N., and C. P Melville, 1982, *A History of Persian Earthquakes*. Cambridge: Cambridge University Press.

Bates, C. C., T. F. Gaskell, and R. B. Rice, 1982, *Geophysics in the Affairs of Man*. Oxford: Pergamon Press.

Bolt, B. A., 1976, *Nuclear Explosions and Earthquakes: the Parted Veil*. San Francisco: W. H. Freeman.

Brush, S. G., 1979, Nineteenth-century debates about the inside of the Earth: solid, liquid, or gas?, *Ann. Sci.* **36**, 225-254.

Brush, S. G., 1980, Discovery of the Earth's core, *Amer. Jour. Physics* **48**, 705-724.

Bullen, K. E., 1975, *The Earth's Density*. New York: Wiley.

Byerly, P., 1970, Book review, *Bull. Seismol. Soc. America* **60**, 2089.

Davison, C., 1927, *The Founders of Seismology*. Cambridge: Cambridge University Press.

Farrell, W. E., 1985, Sensors, systems, and arrays: seismic instrumentation under VELA-Uniform, in A. U. Kerr, ed., *The VELA Program: A Twenty-Five Year Review of Basic Research*. Washington, D.C.: Executive Graphic Services and Defense Advanced Research Projects Agency.

Goodstein, J. S., 1984, Waves in the earth: Seismology comes to southern California, *Hist. Stud. Phys. Sci.* **14**, 201-230.

Housner, G. W., 1983, Earthquake engineering—some early history, in D. E. Hudson, ed., *Golden Anniversary Workshop on Strong Motion Seismometry Proceedings.* Los Angeles: University of Southern California Department of Civil Engineering, 7-16.

Howell, B. F., 1986, History of ideas on the cause of earthquakes, *EOS (Am. Geophys. Union Trans)* **67**, 1323-1326.

Hudson, D. E., 1983, History of accelerograph development, in D. E. Hudson, ed., *Golden Anniversary Workshop on Strong Motion Seismometry Proceedings.* Los Angeles: University of Southern California Department of Civil Engineering, 29-56.

Lee, W. H. K., and S. W. Stewart, 1981, Principles and applications of microearthquake networks, *Adv. Geophys. Suppl. Ser.* **2**, 1-293.

Macelwane, J. B., 1950, *Jesuit Seismological Association: 1925-1950.* St. Louis: St. Louis University.

Minster, B., 1985, Twenty-five years of source theory, in A. U. Kerr, ed., *The VELA Program: A Twenty-Five Year Review of Basic Research.* Washington, D.C.: Executive Graphic Services and Defense Advanced Research Projects Agency, 67-116.

Needham, J., 1959, *Science and Civilization in China, Vol. III: The Sciences of the Heavens and the Earth.* Cambridge: Cambridge University Press.

Richter, C., 1958, *Elementary Seismology.* San Francisco: W. H. Freeman.

Rothé, J.-P., 1981, Fifty years of history of the International Association of Seismology (1901-1951), *Seismol. Soc. America Bull.* **71**, 905-923.

Sheriff, R. P., and L. P. Geldart, 1982, *Exploration Seismology: History, Theory, and Data Acquisition.* Cambridge: Cambridge University Press.

Stoneley, R., 1967, History of modern seismology, in S. K. Runcorn, ed., *International Dictionary of Geophysics.* Oxford: Pergamon, 724-729.

Stoneley, R., 1970, The history of the International Seismological Summary, *Royal Astron. Soc. Geophys. Jour.* **20**, 343-349.

Cross-references: *Controlled Source Seismology; Earthquake Seismology; Earth Structure, Global; Elasticity and Wave Propagation: Principles; Exploration Seismology; Seismic Instrumentation; Seismic Instrumentation: History; Seismograms: Interpretation.*

SEISMOLOGY: PHYSICAL MODEL STUDIES

Physical models are scale models, typically a few centimeters to meters in length, used as analogs of the Earth in wave propagation experiments. Solids, such as metals, ceramics, and plastics, and fluids, such as water, oils, and mercury, are used as analogs of rocks and magmas. Ultrasonic waves, typically in the megahertz band, are used as analogs of seismic waves.

Physical models can be divided into three categories, distinguished by the materials composing the model: (1) fluid models, in which only compressional waves can propagate: (2) three-dimensional solid models, which support both compressional and shear waves; and (3) thin-plate models, which act approximately as two-dimensional sheets (Oliver et al., 1954). This article will not discuss fluid models, since their applicability to seismological problems is limited. Three-dimensional solid models most closely approximate the real Earth. Plate models are used mainly because they are easier to fabricate than three-dimensional models.

Physical model studies of wave propagation are an alternative to numerical (that is, computer-based) methods and are used mainly in instances where numerical approaches are impractical. Since numerical methods are well developed for laterally homogeneous models (that is, models consisting of stacks of plane layers or spherical shells), physical models are used mainly to study laterally heterogeneous structures. One numerical technique that can also be used for laterally heterogeneous structures is the finite-difference method. This method works by integrating the differential equations of seismic wave motion on a discrete grid. Since the grid must be fine enough to adequately represent the smallest geologic structures and seismic wavelengths involved, and since any computer has a maximum number of grid points it can handle, finite-difference models are generally limited to small, smooth structures and short propagation paths. Physical models do not have these limitations (though they have others) and are useful in studying long propagation paths through very heterogeneous, rough structures.

The physics of wave propagation in perfectly elastic nongravitating bodies is such that time and distance can be exactly scaled. Thus, if the earth were a perfectly elastic nongravitating body, and if a scale model of the Earth could be prepared with exactly the same compressional and shear wave velocities and density structure as the real Earth, but at only one millionth of the size, then wave propagation in this body could be studied by using waves with frequencies a million times higher than those of typical seismic waves. Seismograms for the real and model Earths would agree exactly, apart from the scaling in time. It is not even necessary to model the actual numerical values of velocity and density, but only to model the ratios of velocities and densities and to include the proportionality factor in the time scaling.

On the other hand, the Earth is gravitating and slightly anelastic. In physical models studies there is no practical way to model self-gravitation, but since gravitation affects only the gravest free oscillations of the Earth, which are best studied by other methods, this limitation is usually not a problem. To properly model anelasticity, one must work with materials that have the same fractional absorption per wavelength (that is, quality factor) as do geologic materials. Both the Earth (at seismic frequencies) and many metals and plastics (at ultrasonic frequencies) have quality factors in the 50-1000 range, so this scaling is usually possible.

The most severe limitation in physical model studies is the preparation of models that are adequate analogs of the Earth. Some features, such as sharp boundaries between geologic units, rough surfaces, and small compact heterogeneities can be modeled very easily. Other features, such as slowly varying density gradients, are much more difficult to model. Such considerations usually lead to compromises in the realism of the model. For instance, a study of the effect of topography on the core-mantle boundary on the amplitudes of reflected seismic waves may model the topography of the boundary very realistically, but crudely model the Earth's mantle as a homogeneous solid without the increase in velocity with depth found in the real Earth.

Laboratory Apparatus

Typical equipment needed to perform ultrasonic experiments includes (see Fig. 1):

Ultrasonic Transducers for Generating and Detecting Ultrasound. Many broadband ultrasonic transducers are commercially available, including ones that preferentially excite compressional, shear, or surface waves (Fig. 2) (Krautkramer and Krautkramer, 1977). There are two general approaches for coupling the transducers to the model. In one approach, both the model and transducers are immersed in a tank of water, with the model and transducers separated by several centimeters. The advantage of this method is that the coupling is very reproducible. The geometry is similar to some real seismic experiments, especially those conducted in the ocean. The disadvantage is that the water does not support shear wave propagation and has a significant effect on the wave propagation. In the second method the transducers are attached directly to the model, usually with a spring clip, and coupled with a thin layer of fluid (for example, a light machine oil for compressional waves or honey for shear waves). Careful application of transducers leads to an amplitude repeatability of 10%. Better control of amplitudes (1–5%) can be achieved by referencing amplitudes of one seismic wave to another in the model; the latter wave then becomes an internal standard (Fig. 3).

Since the purpose of many model studies is to assess the effect of modifying structures in a limited part of the model, an appropriate internal standard can often be found. Most transducers have a radiation pattern that varies significantly with takeoff angle, since typical ultrasonic wavelengths are significantly smaller than transducer diameters (millimeters compared to centimeters), leading to wave interference (Fig. 4).

Pulse Generator for Driving the Source Transducer. Typical transducers are driven with pulses several hundred volts in amplitude with a duration of several hundred nanoseconds. The pulse generator should pulse at regular time intervals of about 1 s and should supply a synchronization pulse that can be used when recording the signals. Some pulse

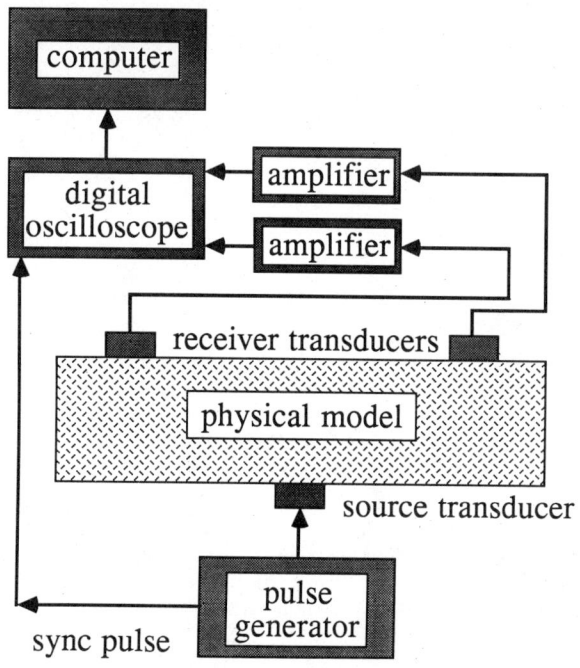

FIGURE 1. Schematic diagram of physical modeling equipment.

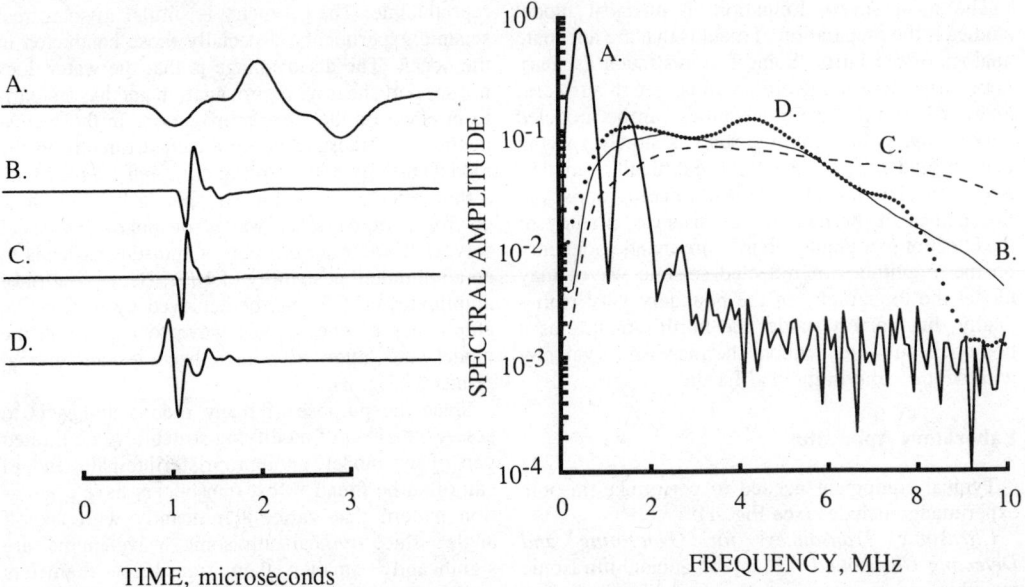

FIGURE 2. (Left) Responses for several commercially available broadband ultrasonic transducers. Seismograms are measured with two identical transducers at normal incidence across 10 cm of aluminum. To improve the signal-to-noise ratio, 128 pulses have been stacked: (A) Panametrics V101 compressional wave transducer, 0.5-MHz center frequency, 2.54-cm diameter; (B) Panametrics M110 compressional wave transducer, 5.0-MHz center frequency, 0.63-cm diameter; (C) Panametrics V112 compressional wave transducer, 10.0-MHz center frequency, 0.63-cm diameter. (D) Panametrics V156 shear wave transducer, 5.0-MHz center frequency, 0.63-cm diameter. (Right) Fourier spectral amplitude of transducers responses: (A), bold; (B), solid; (C), dashed; (D), dotted.

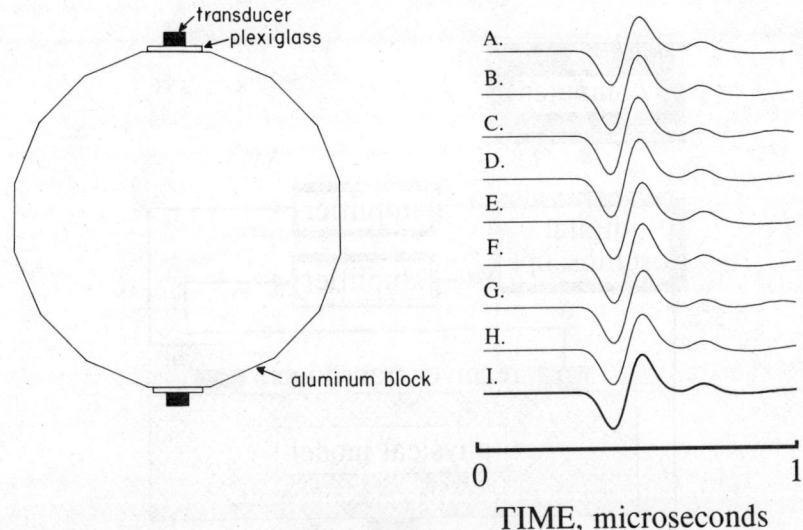

FIGURE 3. (Left) A test model consisting of a 16-sided, 15-cm-diameter polygon of 3.8-cm-thick aluminum. In order to provide for an internal standard, sheets of Plexiglass are bonded to the aluminum with epoxy resin. Transducers are attached to opposite faces of this model with spring clips and coupled with oil. The compressional wave transmitted from one transducer to another is observed, as are reflections from each transducer to the Plexiglass-aluminum boundary and back to itself. The reflected phases provide an internal standard, which permit 1–2% accuracy in determining the amplitude of the transmitted wave. (Right, A–H) The eight seismograms of transmitted waves, with amplitudes normalized by the amplitude of the reflected phases (not shown); (I) the average of seismograms A–H. Note that all the seismograms are very similar, which emphasizes the reproducibility of the method.

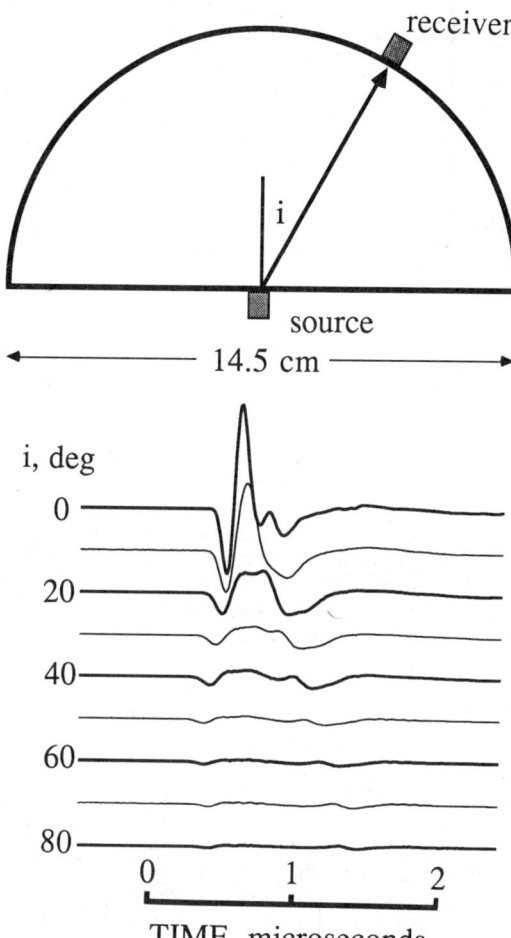

FIGURE 4. (Top) A test model consisting of a 5-cm-thick block of aluminum with a hemispherical upper surface is used to measure the radiation pattern of a 5-MHz compressional wave transducer (Panametrics M110). (Bottom) Seismograms observed for various takeoff angles i. Note that amplitude and pulse shape are strong functions of angle. This effect is due to the 6-mm diameter of the transducers, which are considerably larger than the 1-mm wavelength of 5-MHz elastic waves. At large takeoff angles there is considerable interference between waves arriving at opposite edges of the transducers.

generators have the ability to switch the transducer into a receiver shortly after it pulses, which can be useful for observing normal incidence reflections.

Instrumentation Voltage Amplifiers for Amplifying the Output of the Receiver Transducers. These amplifiers should have gains from 40 dB to 60 dB.

Digital Oscilloscope for Recording the Output of the Receiver Transducers. The oscilloscope should have a sampling rate of 100 MHz, a sample resolution of at least 8 bits. Signal dynamic range can be improved beyond 8 bits by digitally stacking several hundred repeated pulses (Fig. 5). Some digital oscilloscopes incorporate hardware that facilitates this kind of stacking.

Three-Dimensional Models

Model Preparation. Three-dimensional models best represent the Earth because they can support compressional waves, both polarizations of shear waves, and surface waves. But they are usually difficult to construct, which limits their use. The materials, usually metals, resins, glass, and plastics, are generally shaped into polyhedra chosen to represent features of the Earth. Since ultrasonic wavelengths can be as small as 0.1 mm, much effort must be expended to ensure that bonds between parts of the model are very smooth and adhesive layers are minimized. Since coupling transducers to the model is very important, both to ensure reproducibility and to maximize the energy input into the model, surfaces to which the transducers are attached must be very smooth. Polishing large surfaces of the model is a time-consuming task.

Uses of Three-Dimensional Models. Three-dimensional models have been used to investigate seismic processes of the Earth, including

1. Studying the reduction in amplitude of propagating compressional and shear waves due to scattering of the wave from three-dimensional heterogeneities (Fig. 6)
2. Studying the effects of scattering from rough surfaces and topography on propagating compressional and shear waves
3. Studying the effects of dipping layers, mountain ranges, and other laterally varying surface features of the Earth on surface waves.
4. Studying the reflection of body waves from salt domes, oil traps, faulted strata, and other buried features in the Earth (McDonald et al., 1983).

The first two uses study effects of stochastic structures (that is, random structures such as fracturing) on wave propagation; the last two study effects of deterministic structures (that is, systematic structures such as mountain ranges and volcanic intrusions) on wave propagation.

Two-Dimensional Plate Models

Theory. The motivation behind using plates is that, at frequencies corresponding to wavelengths many times the plate thickness, the plate behaves as if it were a two-dimensional medium of propagation. This two-dimensionality is only an approximation: What appear to be compressional and shear waves in the plate are actually "plate modes" that are somewhat dispersive. This dispersion can be characterized by the way the phase velocity $c(\omega)$ varies with angular frequency ω.

Let the plate surface be parallel to the xy plane, with the wave propagation in the x-direction and the

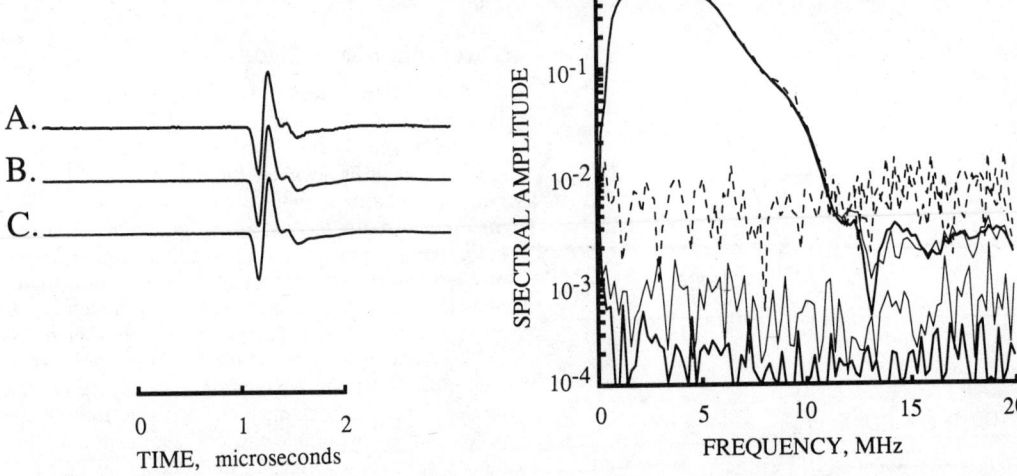

FIGURE 5. (Left) Seismograms for a compressional wave transmitted across 10 cm of aluminum. Trace A is a single pulse, recorded by an 8-bit digitizing oscilloscope. Traces B and C stack 128 and 1024 successive pulses, respectively, into 16-bit samples in order to improve the signal-to-noise ratio. (Right) Upper traces are the spectra of the seismograms at left (A, dashed; B, solid; C, bold). Lower traces are spectra of a portion of the signal immediately prior to the arrival of the wave and are an estimate of the noise level. Note that the stacking results in signal-to-noise levels better than 10:1 at the observed frequencies.

plane surfaces at $z = 0$ and $z = h$. Then the independent components of displacement **u** and stress τ can be arranged into a displacement-stress vector

$$\mathbf{f} = [u_x, u_y, u_z, \tau_{xz}, \tau_{yz}, \tau_{zz}]^T$$

The propagator matrix solution (Aki and Richards, 1980, sec. 7.2.2) to the wave equation gives

$$\mathbf{f}(z = 0) = \mathbf{P}\mathbf{f}(z = h) \qquad (1)$$

where **P** is a 6×6 propagator matrix. Since the stresses are zero on the plate surfaces, Eq. 1 can be interpreted as a dispersion relation for the plate modes. Since the shear motion parallel to the plate surface is decoupled from the compressional motion and shear motion normal to the plate surface, Eq. 1 simplifies to two dispersion relations:

$$P_{52} = 0 \qquad (2a)$$

for the decoupled shear motion and

$$P_{41}P_{63} - P_{43}P_{61} = 0 \qquad (2b)$$

for the coupled compressional-shear motion.

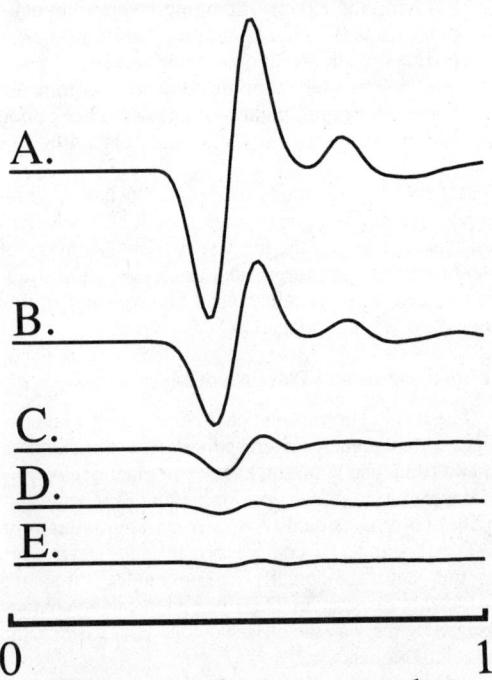

FIGURE 6. Example of a three-dimensional model experiment, the purpose of which is to study the effect of small heterogeneities on the amplitude of transmitted compressional waves. The model is the same as the one shown in Fig. 3, except that successively more heterogeneities are introduced into the model by drilling 1.3-mm-diameter holes normal to the surface of the polygon. As the concentration of the scatterers is increased from 0% to 3.75% volume fraction (seismograms A–E), the amplitude of the transmitted compressional wave decreases. These seismograms are actually averages of waves that traverse the eight directions across the polygon.

After substituting in detailed expressions for the propagator, Eq. 2a becomes

$$P_{52} = \nu(\omega)\mu \sinh \nu(\omega)h = 0$$

where $\nu(\omega) = \omega(c^{-2} - \beta^{-2})^{1/2}$, μ is the rigidity, and β is the shear velocity. One solution occurs when $\nu(\omega) = 0$, which gives the nondispersive fundamental mode with $c(\omega) = \beta$. This mode is used as a two-dimensional shear wave in plate experiments. Other solutions to Eq. 2a occur when $\sinh \nu(\omega)h = 0$, which gives the dispersive higher modes with

$$c(\omega) = \omega\left[\left(\frac{\omega}{\beta}\right)^2 - \left(\frac{n\pi}{h}\right)^2\right]^{-1},$$

$$n = 1, 2, 3, \ldots$$

These higher modes have low-frequency cutoffs at $\omega = n\pi\beta/h$. Plate experiments are confined to frequencies well below the lowest of these cutoffs.

The dispersion relation for the coupled compressional and shear waves is more complicated and can be solved numerically (Fig. 7). Its fundamental mode is approximately nondispersive at low frequencies

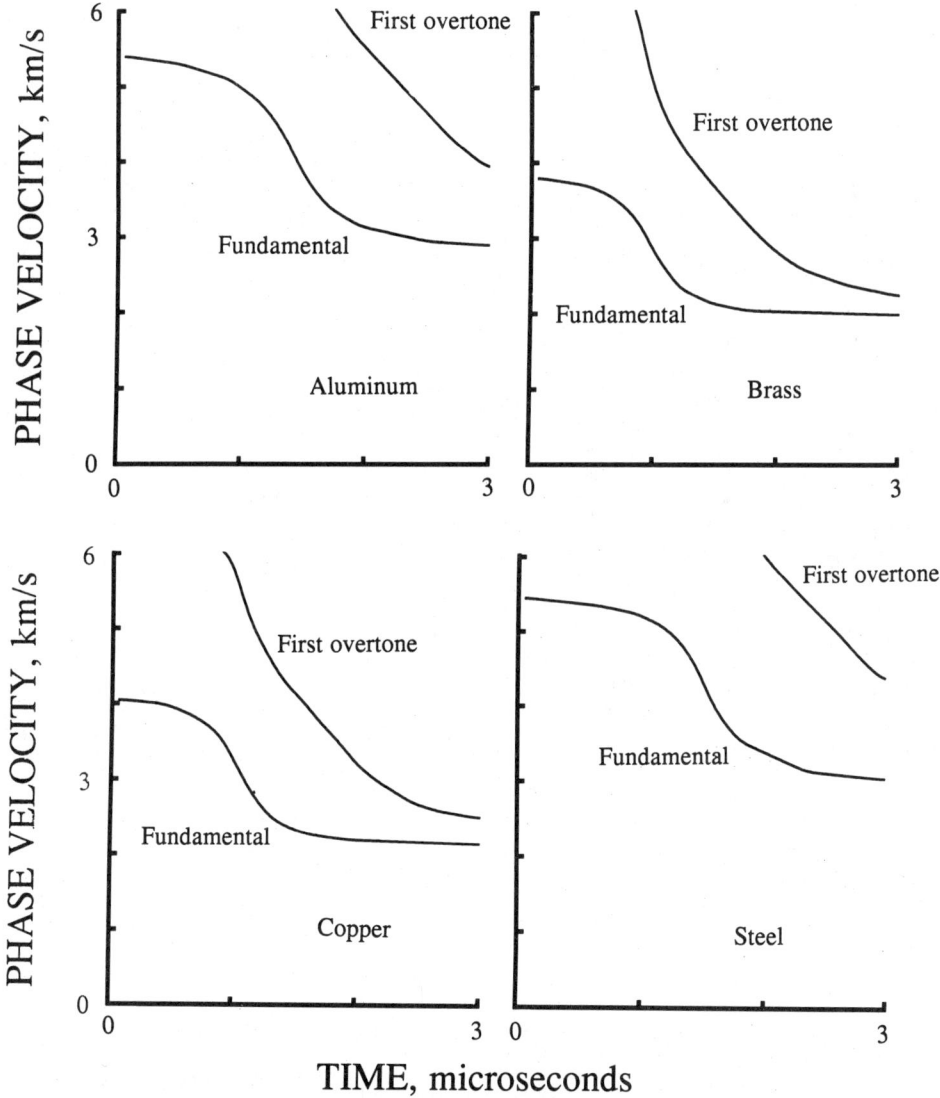

FIGURE 7. Phase velocity dispersion curves for the fundamental mode and first overtone of 1.59-mm- ($\frac{1}{16}$ in) thick aluminum, brass, copper, and steel plate. At sufficiently low frequencies the almost nondispersive fundamental mode and a nondispersive shear mode (not shown) are the only modes present. Plate experiments are limited to these low frequencies, which are less than about 1 MHz for 1.59-mm plates.

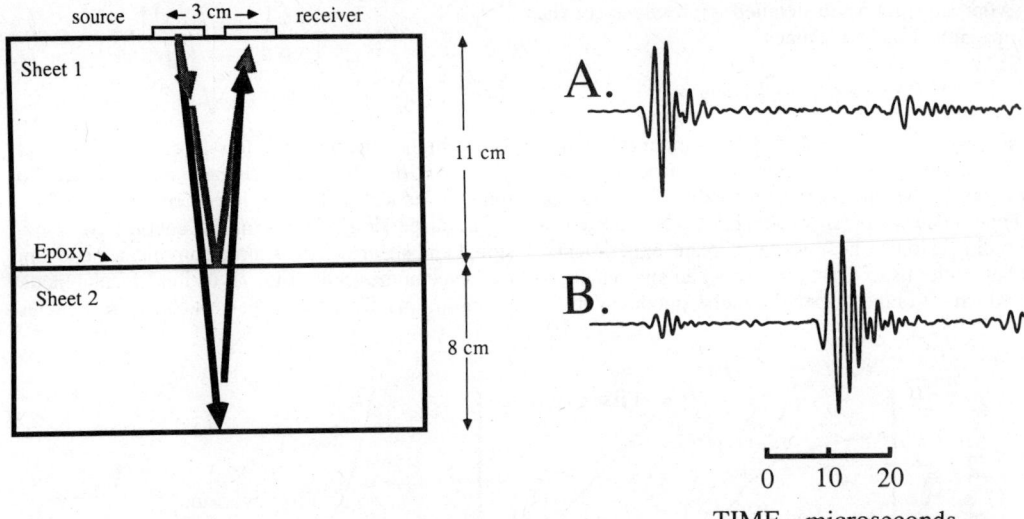

FIGURE 8. An experiment to examine the effect of bonding thin plates with epoxy resin. (Left) The edges of two 1.59-mm aluminum plates were polished and then bonded with epoxy resin: (Right, A) The seismogram of a reflected compressional wave before the lower plate was attached. (Right, B) The seismogram after the lower plate was attached. Note that a small reflection from the bonded interface is still present, even though there is no impedance contrast between the plates. However, about 95% of the wave energy is transmitted through the bond.

and is used as a two-dimensional compressional wave in plate experiments. The cutoff frequencies of the higher modes define an upper limit of frequencies that can be used in the model experiments. Plate models with a thickness of about 1 mm appear two-dimensional for frequencies less than about 1 MHz.

If a plate model were to be constructed to model a vertical plane through the Earth, the two fundamental modes would be analogous to P-SV waves in the Earth (that is, compressional and vertically polarized shear waves). The first overtone for the coupled compressional-shear waves has a particle motion similar to SH waves (that is, horizontally polarized shear waves), but is not present at low frequencies. Thus, SH waves cannot be modeled with plate experiments.

Model Preparation. Two-dimensional models only approximately represent a true two-dimensional medium of propagation. However, since these models are much easier to make than three-dimensional models, they are often useful. Cutting, shaping, and smoothing metal or other materials that are 1 to 2 mm thick (compared with three-dimensional models that are 4 cm or more thick) is relatively simple. In addition, it is relatively easy to bond plates together edge-to-edge and achieve good coupling (Fig. 8), whereas with three-dimensional models bonding sections is very difficult. Bonding is easier because the typical wavelengths used in plate experiments are a factor of 10 larger than those in three-dimensional models. However, even meter-size plates, with propagation paths hundreds of wavelengths long, are practical.

Uses of Two-Dimensional Models. Figure 9 shows the record sections of two circular plates with holes in their centers, a very small hole in Fig. 9a and a much larger hole in Fig. 9b. These record sections are a simple example of the use of plates to model the Earth (or other bodies). In both plates the metal represents the mantle and the hole represents the core, a very small core in Fig. 9a and a core scaled to the size of the Earth's core in Fig. 9b. Note the diffraction of the direct waves and the prominence of core reflections in Fig. 9b (relative to Fig. 9a). More scientifically relevant uses of two-dimensional models include

1. Studying the effects of very fractured lunar crust on the propagation of seismic waves.
2. Studying body wave interactions with scatterers

FIGURE 9. (A, Inset) A simple plate model consisting of a 1.27-mm-thick aluminum plate, with the shape of a circular disk with a small hole at its center. This model is a crude two-dimensional analog to a planet with a very small core: (A, main figure) A seismic record section for this model, with prominent phases labeled according to standard seismological practice. Transducers with a peak frequency of 0.5 MHz are used. The dispersive propagation in plates leads to somewhat ringy waveforms. (B, inset) A simple plate model of a planet with a large core. (B, main figure) A seismic record section for this model. Note that the reflections off the core-mantle boundary are very prominent compared to the model in A.

SEISMOLOGY: PHYSICAL MODEL STUDIES

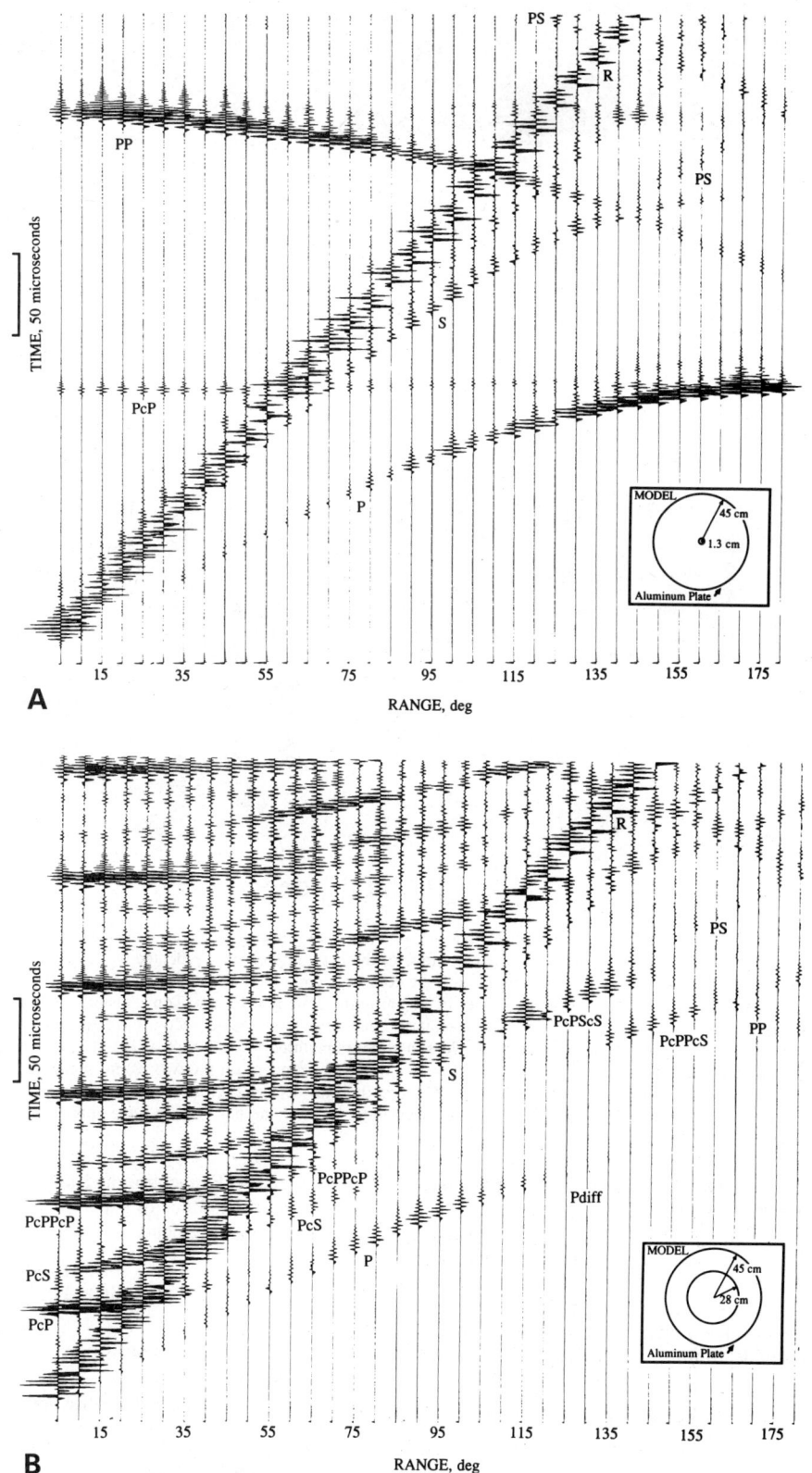

1209

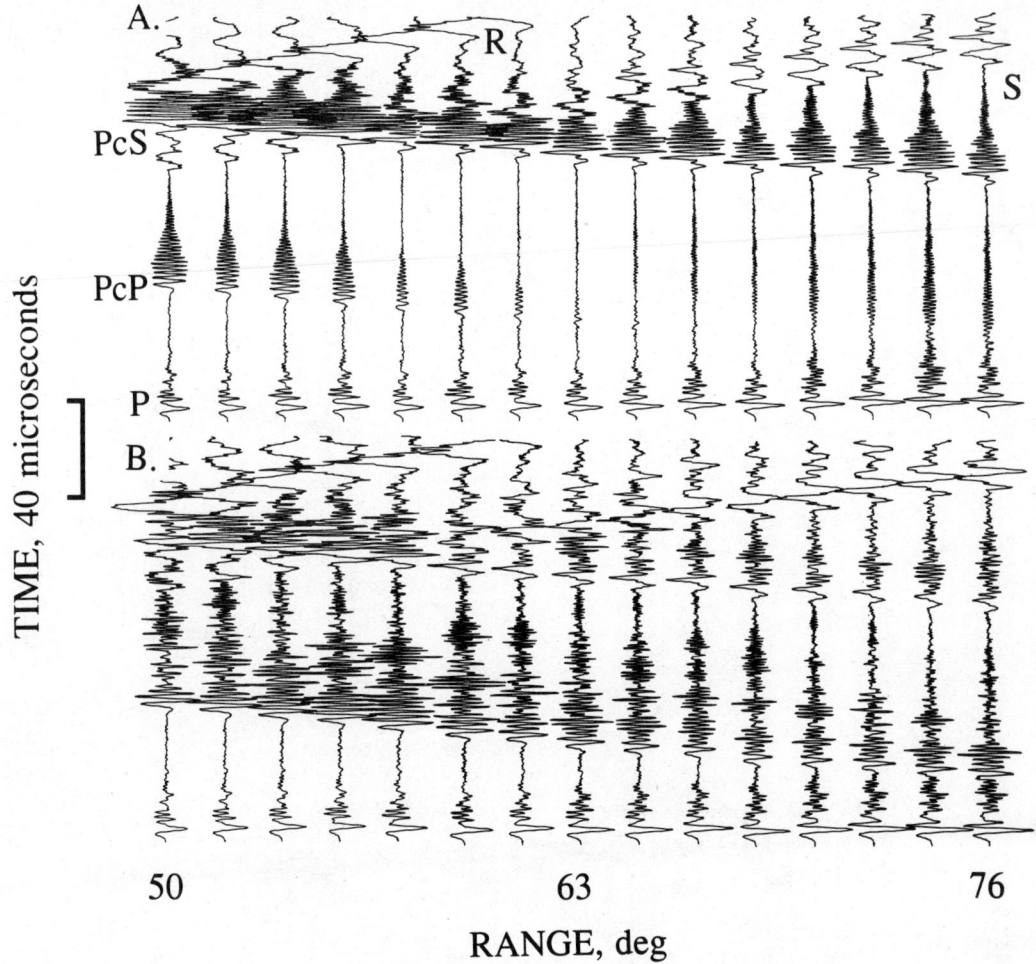

FIGURE 10. Example of a two-dimensional model experiment, the purpose of which is to study the effect of small irregularities on the core-mantle boundary on the amplitude of the PcP and PcS core-reflected waves. The Earth's mantle is modeled by a homogeneous sheet of aluminum, and the core by a void. Topography on the core consists of a random distribution of triangular hills, each of which scales to 14 km high in the real Earth: (Top) Seismic record section for a smooth core-mantle boundary; (bottom) rough core-mantle boundary. Both record sections are shifted in time so that their compressional wave first arrivals are aligned. Note that the transmitted phases, such as P and S, are not affected by the roughness on the core-mantle boundary and have very similar appearances on the two record sections. On the other hand, the PcP and PcS phases have been significantly scattered. The surface Rayleigh wave R was used as an internal amplitude standard, since it is unaffected by structures deep in the mantle and has an amplitude approximately independent of range.

at the core-mantle boundary (Menke, 1986) (Fig. 10) and in the upper mantle.
3. Studying the effects of structures such as sediment-filled valleys on propagating waves.

The first two uses study the effects of stochastic structures on wave propagation; the last use studies the effects of deterministic structures on wave propagation.

Acknowledgments

We gratefully acknowledge the support we have had over the years from the seismology division of the National Science Foundation and the ocean acoustics division of the Office of Naval Research for our physical model experiments.

WILLIAM MENKE
BRUCE DUBERDORFF

References

Aki, K., and P. G. Richards, 1980, *Quantitative Seismology, Theory and Methods.* San Francisco: W. H. Freeman and Co.

Krautkramer, J., and H. Krautkramer, 1977, *Ultrasonic Testing of Materials,* 2nd ed. Berlin: Springer-Verlag.

McDonald, J. A., G. H. F. Gardner, and F. J. Hilterman, eds., 1983, *Seismic Studies in Physical Modeling.* Boston: International Human Resources Development Corp.

Menke, W., 1986, Few 2-50 km corrugations on the core-mantle boundary, *Geophys. Research Letters* **13**, 1501-1504.

Oliver, J. A., F. Press, and M. Ewing, 1954, Two-dimensional model seismology, *Geophysics* **14**, 202-219.

Cross-references: *Elasticity and Wave Propagation: Principles; Seismic Attenuation: Observation and Measurement; Seismic Diffraction; Seismic Imaging; Seismic Ray Theory; Seismic Wave Scattering.*

SOLAR SYSTEM: ORIGINS

Because the origin of the solar system is not a reproducible experiment, hypotheses about solar system formation will never achieve the certainty accorded results in most other branches of science. Cosmogonical theories thus must be viewed with caution, but this should not detract from serious efforts to derive a temporal sequence of events, starting from plausible initial conditions, that may have led to the formation of our solar system. The theory of solar system origin outlined here is intended to approximate the current consensus of the cosmogonical community; the theory will undoubtedly evolve as our understanding of the basic processes improves.

Nebular Hypothesis

The central hypothesis of solar system origin, originally formulated by Immanuel Kant (1755) and independently by Pierre Simon Laplace (1796), is that the planets formed out of the same rotating, gaseous nebula that formed the Sun, the *presolar nebula*. Other possibilities can be shown to be impossible or highly improbable. For example, forming the planets from the outer layers of a preexisting Sun, following tidal deformation during a close encounter with another star, is both impossible because the high temperature of the Sun's atmosphere would result in dispersal rather than accumulation into planets, and highly improbable because of the great separation of stars compared to their diameters. Similarly, forming the planets elsewhere and then capturing them about the Sun suffers from the improbability of capturing the planets into orbits that all revolve about the Sun in the same direction and in nearly the same plane, as well as from the need to form the planets somewhere. Planet formation in the nebular hypothesis is inextricably linked to star formation.

Stars form from the gravitational contraction of interstellar clouds (nebulas), which are themselves composed in part of debris from earlier generations of stars. The interstellar cloud that formed our Sun contained elements produced inside progenitor stars by stellar nucleosynthesis and then injected into the interstellar medium through stellar winds and supernova explosions. The elements that constitute essentially all of the matter in the terrestrial planets resided primarily in dust grains in the presolar nebula. Dust grains constitute about 2% of the mass of interstellar clouds, with the bulk of the mass residing in hydrogen (about 77%) and helium (about 21%). In spite of their low mass fraction, dust grains regulate the nebula temperature by absorbing and scattering incoming optical radiation, and by radiating energy out of the nebula at infrared wavelengths. The densest clouds (number density about 10^4 to 10^6 cm^{-3}) can become quite cold (about 10 K) by radiating away their internal energy.

When an interstellar cloud is dense enough and cold enough, thermal pressure is unable to resist self-gravity, and the nebula contracts to a smaller size and higher density. If the interstellar nebula is threaded by substantial magnetic fields, then contraction is slowed by the inability of ionized components of the gas to move across the magnetic field lines, because collisions between the ionized and neutral components retard the contraction of the entire nebula. As the density increases, the nebula becomes opaque to ionizing radiation from external stars, decreasing the fractional ionization and hence the frictional drag between ions and neutrals, allowing the neutral portions of the nebula to contract even faster to higher density. This process of ambipolar diffusion produces a partially ionized, diffuse envelope containing significant magnetic fields and a neutral core where the magnetic field is increasingly dynamically unimportant.

Protostellar Collapse

Once the residual magnetic field becomes negligible, the nebula enters a dynamic collapse phase, where rapid contraction occurs at approximately supersonic velocities. Nebulas that are intermediate in evolution between dense interstellar clouds and stars are termed *protostars*. Protostellar collapse continues until portions of the nebula become so dense that the heat produced by the compression associated with the collapse can no longer be radiated away; thereafter the temperature begins to monotonically increase from the low temperatures associated with dense interstellar clouds to the high temperatures found in the solar interior ($\sim 10^7$ K). The number density undergoes an even more dramatic increase of a factor of about 10^{20}. The collapse forms a quasiequilibrium core surrounded by an infalling envelope of matter; the first core collapses further when it becomes hot enough to dissociate molecular hydrogen. A second, final core is formed, and the envelope matter either accretes onto this central protosun, or, because of conservation of angular momentum, falls onto an extended disk in the

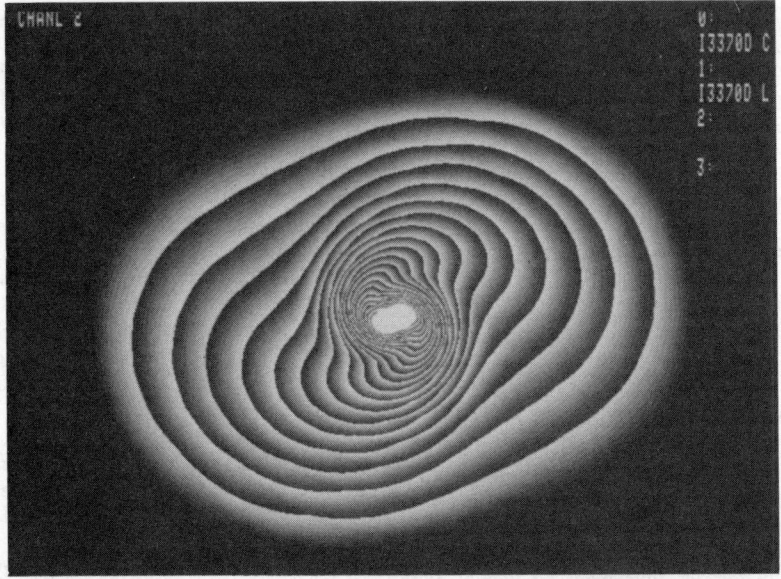

FIGURE 1. The solar system originated from the collapse of an interstellar cloud of gas and dust, which formed a hot, dense central protosun surrounded by a cooler, more diffuse solar nebula in which the planets formed. (Computer simulation by A. P. Boss, Carnegie Institution of Washington.)

equatorial plane of the protosun, to form the *solar nebula*.

The planets formed out of the solar nebula. The existence of the solar system is owed to the angular momentum content of the presolar nebula. Without any angular momentum, the presolar nebula may have collapsed to form a single star devoid of a preplanetary nebula that could evolve into our solar system. Too much angular momentum may produce a binary star system, however, which would be inconsistent with the solitary nature of our Sun. Because of this, the interstellar cloud that collapsed to form the solar system must have had relatively little angular momentum, though the precise value is uncertain.

Solar Nebula

The Sun contains about 99.9% of the mass of the solar system, but only about 2% of the total angular momentum, giving it a ratio of angular momentum to mass about 50,000 times smaller than that of the planets. The nebular hypothesis assumes that the matter that formed the Sun and the planets started off with roughly the same ratio of angular momentum to mass. Some of this angular momentum imbalance can be attributed to the loss of angular momentum to magnetic braking by the solar wind over the last 4.5 billion years (the approximate age of the Sun and the planets); the angular momentum is lost to the interstellar medium through the diffuse stream of matter ejected in the Sun's wind. A lower bound on the amount of angular momentum loss by the solar wind can be found by measuring the rotation rates of newly-formed stars of solar type, which have not had 4.5 billion years to spin down. Astronomical observations of such stars yield rotation rates about 10 times higher than that of the Sun, implying that soon after it was formed, the angular momentum to mass ratio of the Sun was at least 1/5000 that of the planets. Some of the imbalance can also be attributed to the fact that the low angular momentum gas in the presolar cloud would have preferentially collapsed toward the center of the nebula. However, the remaining imbalance requires that some physical process(es) operated in the solar nebula, transporting mass inward to the Sun and angular momentum outward to the solar nebula.

Though poorly understood at this time, several promising processes exist, namely viscous shear and gravitational torques, which may provide sufficient angular momentum transport to form a central Sun and a preplanetary disk out of an originally more homogeneous nebula. Viscous shear requires the presence of appreciable viscosity. While molecular viscosity is too feeble, the effective viscosity produced by turbulent motions (perhaps driven by convective instability in the solar nebula) may be sufficiently strong. Likewise, gravitational torques between asymmetric mass distributions in the solar nebula (e.g., a prolate protosun and a barlike or spiral solar nebula) can be highly effective in transporting angular momentum. Magnetic fields could also be important for angular momentum transport during this phase, but considering the presumed unimportance of the primordial magnetic field, significant magnetic fields require the existence of a well-defined protosun (capable of sustaining a magnetic dynamo) for their generation, as occurs in the later phase of angular momentum loss by the solar wind.

Loss of angular momentum from centrifugally supported inner regions will result in inward motion of mass, while gain of angular momentum in outer regions may move matter further outward. The early evolution of the solar nebula thus involved concentration of mass into the central protosun, and possibly expansion of the outer regions. At the same time, further accretion of gas and dust from the envelope of the presolar nebula will be adding matter to the solar nebula. Furthermore, the earliest phases of planetary formation could also be occurring: growth of the dust grains.

Dust Grain Coagulation

Astronomical observations of the extinction of star light by dust grains imply a mean dust grain size in the interstellar medium of about 0.1 micron. Interstellar clouds are so cold that the elements and molecules found in terrestrial rocks are all frozen onto the dust grains. Dust grains of this minuscule size (about 10^9 atoms) are the most fundamental building blocks for the terrestrial planets (Mercury, Venus, Earth and Moon, Mars) and the inferred rocky cores of the Jovian planets (Jupiter, Saturn, Uranus, Neptune).

For the terrestrial planets at least, the earliest phases of growth involved random (thermal) collisions between dust grains that resulted in grains sticking together because of intermolecular (van der Waals) forces, i.e., Brownian coagulation. During the protostellar collapse phase, collisions between gas molecules and dust grains are so frequent that the dust grains are forced to follow the motions of the gravitationally dominant gases. However, collisions between dust grains are still so rare that little growth by Brownian coagulation occurs. Because the grains effectively move along with the gas, collisions resulting from differential grain motion caused by the dependence of gas drag on grain size also occur infrequently. Coagulation rates can only become significant once the gas and dust grain density becomes high enough (more than 10^{10} molecules per cm^3), as happens later in the protostellar collapse phase. However, collapse velocities are so large that dust grains rapidly enter the solar nebula without undergoing much growth over their interstellar sizes.

Dust grain growth primarily occurs once the grains enter the solar nebula. Because the solar nebula is roughly in equilibrium, the gas is supported against collapse onto the protosun by both gas pressure gradients and rotational motion, while the dust grains are unaffected by the gas pressure gradients. Because the gas thus effectively feels a lower gravitational attraction toward the center than the dust grains, the gas rotates about the protosun slower than the grains, resulting in differential motion and gas drag on the grains. Gas drag cannot be too strong or last too long, or else the grains will spiral inward far enough to enter the protosun. Dust grains of different sizes will move through the nebula at different speeds, with the larger grains overtaking the smaller grains, because of the former's smaller deceleration (the gas drag force per unit mass is inversely proportional to radius). Relative motions between dust grains will produce collisions, and provided that the collisions occur at low enough speeds (velocities of about 1 km s^{-1} suffice to shatter dust grains), coagulation will occur. Turbulent motions of the gas, if present, will be imparted to the smaller dust grains, possibly resulting in a further enhancement of grain coagulation.

There are two basic ways in which planets could have formed in the solar nebula. The first begins with a gravitational instability of the dust grains in the nebula and the second with a gravitational instability of the gas. In either case, instability can occur when the self-gravitation of a disk overwhelms the disruptive effects of thermal pressure and rotational shear, allowing the disk to break up into a collection of self-gravitating bodies. The dust grain instability is generally favored as a starting point for forming the terrestrial planets, while the gas instability is a possible means for forming the Jovian planets. This reasoning may seem obvious, considering the largely rocky or gaseous compositions of the terrestrial and Jovian planets, respectively; more subtle reasons will become apparent in the discussion of each possibility.

Planetary Formation by Dust Grain Instability

Gas pressure gradients ensure that the gaseous component of the solar nebula maintains an equilibrium structure perpendicular to the nebula midplane, with a scale height perhaps one tenth (or more) of the heliocentric radius. As long as the gas density remains below a critical value (determined by the temperature and rotational structure of the disk), the gas disk will be stable. However, sedimentation of the dust grains into an even thinner disk in the midplane of the solar nebula occurs because of the negligibility of gas pressure support for the grains. The dust disk will become gravitationally unstable when it sediments to a sufficiently high dust density, i.e., when it becomes thin enough.

If the gaseous component of the nebula is turbulent, however, the smaller dust grains will be carried along with the turbulent gas, so that sedimentation may only occur during quiescent phases or once turbulence has stopped altogether. The degree, extent, and timing of turbulence in the solar nebula is quite uncertain. If turbulence is driven by thermal convection, then turbulence may occur so long as the dust grains have not grown in size by more than a factor of about 100; once that happens, the opacity of the solar nebula (caused by the dust grains) will no longer permit thermal convection (radiative losses will dominate). Evidently turbulence must cease once appreciable grain coagulation has occurred.

Once the solar nebula becomes quiescent, the grain sedimentation times are quite short (several thousand years at Earth to Jupiter orbital distances). The grains may accumulate to sizes on the order of cm to m during the sedimentation process. When portions of the dust grain disk become dense enough, gravitational instability can occur. The first gravitational instability produces clusters of dense *planetesimals* (3 g cm^{-3}, and about 0.1 km in size), supported in large part by the rotational spin produced by instability in a differentially rotating (shearing) nebula. The clusters are gravitationally unstable themselves, and collapse further to form planetesimals on the order of a few km in size, provided that gas drag and mutual collisions can remove their rotational spin. These gravitational instabilities occur rapidly, over times on the order of the orbital period. The initial planetesimals in the terrestrial planet region constituted a swarm of perhaps 10^{12} bodies on approximately circular orbits about the sun.

Further growth toward planetary size occurs through the accumulation of planetesimals following collisions. There are two distinct phases in the subsequent accumulation process. The first involves accumulation of the planetesimals just within a very narrow ring about the Sun (closely packed phase), where the collision probabilities in each ring can be accurately calculated ignoring the fact that each planetesimal is in orbit around the Sun. The second phase involves collisions between planetesimals located in different rings (loosely packed phase), where significant orbital evolution in radius and inclination occurs.

In the closely packed phase, the masses of the planetesimals are still small enough that mutual gravitational perturbations do not greatly perturb the nearly circular orbits. Collisions tend to result in accumulation of the debris, because kilometer-sized bodies are massive enough to gravitationally trap the debris resulting from collisions at the low relative velocities that occur on nearly circular orbits.

There are two possible extremes for the outcome of any accumulation process. Runaway accretion may occur, where one body rapidly grows by accreting the others. Runaway accretion in the closely packed phase seems to require the presence of an initial planetesimal mass distribution containing a seed mass, that is, one body whose mass is perhaps 10 ten times larger than the next largest body. Such a seed mass may experience a runaway because its large mass produces a gravitationally enhanced cross section for collisions with the smaller bodies.

Alternatively, growth in the closely packed phase may have been much more uniform, with most of the mass residing in a large number of planetesimals with nearly equal mass, until only a few bodies are left in each ring. In either case, however, it seems likely that the closely packed phase terminates with the formation of no more than a few bodies in each ring. Accumulation in the closely packed phase may be completed within about 10^4 years in the terrestrial planet zone, yielding planetesimals as large as about 10^{24} g (500 km). About 10^4 planetesimals of this size are needed to produce just the terrestrial planets. Because of the need for mutual gravitational forces to slowly perturb the initially circular orbits of the closely packed phase into the eccentric orbits of the loosely packed phase (thereby allowing further collisions), the next phase of evolution requires much more than 10^4 years.

Gas drag caused by the gaseous component of the solar nebula is not thought to have a major effect on the outcome of the closely packed phase. At some point the gaseous component must have been removed from the solar nebula or else completely accreted onto the planets or Sun. The early solar wind may have been strong enough to sweep residual gas and dust out of the planetesimal zone. Astronomical observations show that solar type stars undergo a phase of vigorous mass loss (T Tauri phase) roughly 10^6 years after their formation. The protosun may have experienced a sustained T Tauri phase, as well as more intense, episodic mass loss (e.g., FU Orionis outbursts or a bipolar flow phase), capable of removing the gas from the inner regions of the solar nebula. The scarcity of noble gases (e.g., Ne, Xe) in the Earth's atmosphere appears to be inconsistent with Earth formation in a gaseous nebula, where the primordial Earth would be expected to have a substantial atmosphere. If such a strong stellar wind was the cause of nebular gas dispersal, the inferred ages of T Tauri stars thus imply that the residual gas was removed after the closely packed phase, at the beginning of the loosely packed phase. These arguments suggest that further accumulation in the terrestrial zone occurred in the absence of dynamically significant gas.

Because of the increasing importance of self-gravitational forces for large planetesimals, collisions in the loosely packed phase should tend to trap the debris and increase the mass of the surviving body. Processes that could stall the accumulation process, such as fragmentation caused by tidal forces during near misses, or rotational fission of rapidly spinning bodies, are not thought to be important. Collisional fragmentation may modify the accumulation process, however. The details of collisions at high relative velocities (e.g., 10 km s^{-1}) between 1000 km (or larger) planetesimals are just now being investigated. On the other hand, if the planetesimals evolve toward a configuration of many widely separated bodies on roughly circular orbits (i.e., small relative velocities) further collisions will not occur and growth will stop. It is thought that the dynamics of accumulation in the loosely packed phase regulates the relative velocities of colliding planetesimals to intermediate values that allow accumulation to proceed.

Accumulation in the loosely packed phase involves significant migration of planetesimals in orbital radius because of the stochastic effects of multiple gravitational perturbations, especially during close encounters, resulting in mixing of parent bodies through the terrestrial planet zone. Because of the stochastic nature of this process, one can only demonstrate the probability of obtaining a terrestrial planet system similar to (but not exactly equivalent to) our own. The initial conditions defined by the end of the closely packed phase can indeed lead to a set of terrestrial planets with masses and orbital characteristics quite similar to our inner solar system. The stochastic nature of the final impacts may also account for the variation in rotational spin rates and directions among the planets. This final phase of accumulation requires on the order of 10^7 to 10^8 years in the terrestrial planet zone.

So far this subsection has concentrated on forming the terrestrial planets (see also *Earth and Moon: Origins*). Making the Jovian planets in this fashion may also be possible but encounters a severe time scale problem. If a dust layer instability and subsequent accumulation produces a massive planetesimal in the outer solar nebula, the planetesimal can directly acquire a gaseous envelope through rapid gravitational accretion of gas once the core reaches about 10 Earth masses. This scenario is an attractive means for explaining layering in the interiors of the Jovian planets, which are thought to have rock and ice cores of this mass surrounded by a much more massive hydrogen and helium envelope. Formation of the Jovian planets in this manner thus requires formation prior to removal of the solar nebula gases, and hence prior to the formation of the terrestrial planets. It is also advantageous to form Jupiter before the terrestrial planets, so that gravitational perturbations by Jupiter can eject planetesimals from the asteroid region, preventing the formation of another terrestrial planet there. However, most estimates of the time scale for planetesimal accumulation show a strong increase with distance from the sun (e.g., 10^8 to 10^9 years at Jupiter's orbit), in part because of the decreasing density and increasing orbital period. The time needed for accumulation in this fashion may even exceed the age of the solar system!

Rapid growth of planetesimal cores for the Jovian planets could occur if the density in the outer regions of the solar nebula was considerably higher than is usually assumed. If the solar nebula becomes too massive, however, the assumption of stability of the gaseous portion of the solar nebula may not be met. Thus the density must be enhanced only in just the right regions (i.e., around Jupiter and Saturn). The density structure of the solar nebula is too poorly understood at present to know if locally enhanced densities are reasonable.

Alternatively, rapid gas accretion onto planetesimal cores may become important for rocky cores with much less than 10 Earth masses, depending on the thermodynamics of the planetesimal atmosphere. Planetesimals of this reduced size can be accumulated much more rapidly, and once the planetesimals acquire massive gaseous envelopes, further accumulation through collisions may proceed quickly, before the gas is lost from the solar nebula. However, it may not be possible to subsequently assimilate enough rocky material to account for the inferred cores of the Jovian planets. It is also unknown whether this scenario would lead to a small number of Jovian planets with orbits similar to the Jovian planets in our solar system.

Planetary Formation by Gaseous Instability

Gravitational instability of the solar nebula gas depends not only on the density and temperature of the nebula, but also on the amount of mass in the central protosun. Large central masses (compared to the mass in the nebula) tend to stabilize the disk, through tidal forces on nascent instabilities and through rotational shear. The maximum amount of rotational shear occurs in a low mass, Keplerian nebula, where the gravity is primarily due to the protosun, resulting in the gas moving with angular velocity similar to the Keplerian orbits of planets. For reasonable temperatures (e.g., 100–300 K), it is thought that gaseous disks surrounding a 1 solar mass protosun must contain at least $\frac{1}{3}$ to 1 solar mass of additional matter in order to be gravitationally unstable. Disks that are not stabilized by a massive protosun can be unstable at much lower masses. Considering that gaseous nebula instability could very well be occurring at the same time that the protosun itself is small but increasing in mass, forming the Jovian planets through gaseous instability remains a possibility. Most details about gaseous instability are unknown at present, including what mass protoplanets would result from this type of instability.

Because the gaseous nebula instability is thought to occur preferentially in the outer regions of the solar nebula (where tidal forces of the protosun are minimized), the giant gaseous protoplanets initially may all be Jupiter-sized and located in the Jovian planet region. This explanation is clearly convenient for the formation of the Jovian planets, but not for the terrestrial planets. Forming the terrestrial planets in this manner requires forming a rocky core within the gaseous protoplanet, removing the gaseous envelope to reveal the core, and moving the core inward to the terrestrial planet zone.

Any dust grains or rocky planetesimals that had grown prior to the gas instability would be gravitationally included in the giant protoplanets, initially forming a roughly homogeneous mixture. If the rocky material is immiscible, it will sediment to the bottom of the gaseous envelope and form the necessary dense core. However, it may also be that the dust grains are miscible with H and He in giant protoplanet interiors, which would prevent precipitation of a rocky core, implying that the terrestrial planets, and even the Jovian planets (with their inferred rock and ice cores), could not have formed in this manner. Recent models of the interior of Jupiter imply that the rocky and icy matter may be more evenly distributed than was previously thought, with no more than a few Earth masses in the dense core and ten times as much dense matter (i.e., elements other than H and He) in the envelope, in which case this objection is reduced in importance for the Jovian planets. Core formation must still occur to some extent, though, if the terrestrial planets are to be formed by this means.

Tidal forces from the protosun and evaporation in the thermal bath of the inner solar nebula (heated by the protosun and ongoing accretion) have been proposed as mechanisms for removing the gaseous envelopes of protoplanets destined to become terrestrial planets. The latter process may require delicate timing: sedimentation to form the cores must occur prior to envelope stripping, and if left unchecked, thermal evaporation would dissipate all of the giant protoplanets, even proto-Jupiter. More critically, both of these stripping processes should result in increasing ratios of envelope to core mass with distance from the sun, contrary to what is inferred for the Jovian planets, where all four planets are thought to have roughly equal mass cores, yet Uranus and Neptune are much less massive than Jupiter and Saturn.

A mechanism for moving some of the protoplanets inward to become the terrestrial planets must also be specified. Alternatively, giant protoplanets may have formed right in the terrestrial planet zone. Gaseous instability in the terrestrial planet region requires such high gas densities that the total amount of gas in the nebula is significantly greater than what is necessary to account for the masses of the planets, making it that much harder to explain the subsequent removal of the excess gas. However, considering the observation of extremely energetic mass loss in young stars (with mass loss rates much higher than in T Tauri stars), removal of this excess gas may not be critical.

Minor Bodies

Besides the Sun, terrestrial planets, and Jovian planets, the solar system contains a much larger number of smaller bodies—satellites, asteroids, comets, and Pluto. The origin of the Moon is discussed in *Earth and Moon: Origins*.

There are two means of explaining the two types of satellites found around the Jovian planets. The regular satellites—which orbit in the same sense and in the equatorial plane of their planets, and which have a gradient in composition (rocky to icy moving outward)—formed as secondary bodies in the rotationally flattened, gaseous envelopes of the Jovian protoplanets, as the protoplanets contracted toward their present dimensions. The details are thought to be analogous to the processes by which the planets formed about the contracting protosun. The smaller, irregular satellites, on inclined orbits, often in a retrograde sense, with irregular compositions, probably formed as planetesimals orbiting the Sun and were gravitationally perturbed into Jovian-planet crossing orbits, there to be captured (by gas drag or collisions with other bodies) in the gaseous envelope of the Jovian protoplanets.

The asteroids are thought to be rocky planetesimals that did not manage to completely accumulate into protoplanets, because orbital perturbations by nearby Jupiter (presumably already formed) ejected many of the planetesimals from the region and excited the relative velocities of the survivors, encouraging collisional fragmentation rather than accumulation. Similarly, comets are largely icy planetesimals formed in the outer solar system, kicked farther out by orbital encounters with the growing Jovian planets. Considering its small size, anomalous composition, and high inclination orbit, which passes at times within that of Neptune, Pluto is likely to be an escaped satellite of Neptune, rather than a true planet.

Summary

The nebular hypothesis provides a self-consistent framework for understanding the origin of the solar system, and, like plate tectonics in modern geophysics, it is unlikely to be supplanted by a radically different concept. While we certainly expect many of the details of solar system cosmogony to change as we learn more, the origin of the solar system in its simplest terms is now reasonably well understood: the solar system formed 4.5 billion years ago from the collapse of a rotating interstellar cloud of gas and dust.

An outline exists for understanding the formation of the terrestrial planets by accumulation of planetesimals following the gravitational instability of a dust grain disk, and perhaps the formation of the Jovian planets directly through the gravitational instability of a gaseous disk. However, difficulties arise in trying to form the terrestrial planets through the Jovian planet mechanism, and vice versa. Perhaps it is not possible to explain the formation of the terrestrial and Jovian planets by a single mechanism. The temperature and density structure of the gaseous portions of the solar nebula may have favored gas stability in the inner regions and gas instability in the outer regions. Then both mechanisms may have operated: a dust layer instability in the inner regions, gaseous instability in the outer regions leading to rapid formation of the Jovian planets, followed by solar nebula removal, and final accumulation of planetesimals in the now gas-free inner region into the terrestrial planets.

ALAN P. BOSS

Bibliography

Black, D. C., and M. S. Matthews, eds., 1985, *Protostars and Planets II*. Tucson, Ariz.: University of Arizona Press, 1293 p.

Boss, A. P., 1987, Theory of collapse and protostar formation, in D. Hollenbach and H. Thronson, eds., *Summer School on Interstellar Processes*. Dordrecht: D. Reidel, 321 p.

Safronov, V. S., 1969, *Evolution of the Protoplanetary Nebula and Formation of the Earth and the Planets*. Moscow: Nauka, 206 p. (translation NASA TTF-677, 1972).

Cross-references: *Absolute Age Determination: Radiometric; Earth and Moon: Origins; Gravity Fields: Implications for Planetary Interiors; Meteorites: Nature and Origin; Planetary Geodesy; Planetary Magnetic Fields.*

SPHERICAL HARMONIC ANALYSIS

Spherical harmonic analysis is the procedure of representing a potential function by a sum of spherical harmonic functions. A potential function V is a solution to Laplace's equation, given here in spherical coordinates:

$$0 = \nabla^2 V = \frac{1}{r^2} \frac{\partial}{\partial r}\left(r^2 \frac{\partial V}{\partial r}\right)$$
$$+ \frac{1}{r^2 \sin\theta} \frac{\partial}{\partial \theta}\left(\sin\theta \frac{\partial V}{\partial \theta}\right)$$
$$+ \frac{1}{r^2 \sin^2\theta} \frac{\partial^2 V}{\partial \phi^2}$$

Using separation of variables and the principle of superposition, we can write the solution to Laplace's equation in the form

$$V = V_{\text{internal}} + V_{\text{external}} = V_i + V_e$$

where

$$V_i = a \sum_{n=0}^{\infty} \sum_{m=0}^{n} \left(\frac{a}{r}\right)^{n+1} (g_n^m \cos m\phi + h_n^m \sin m\phi) P_n^m(\cos\theta)$$

and V_e has the same form but with $(r/a)^n$ replacing $(a/r)^{n+1}$. Also, a is a reference radius (usually an Earth radius); r, θ, and ϕ are spherical coordinates with r as radial distance, θ as colatitude, and ϕ as longitude; and P_n^m is the associated Legendre function, or spherical function, of degree n and order m with the Schmidt normalization, also called *Schmidt's function*. Some authors write $P_n^m(\theta)$ in place of $P_n^m(\cos\theta)$. In applications, V_i is interpreted as arising from sources internal to the spherical shell under study, and V_e refers to external sources. This model assumes that the shell is source-free; i.e., in the case of geomagnetism, there are no currents or other magnetic sources within the volume being so modeled. Gravity studies omit the V_e term, and geomagnetic studies usually use only the first three external terms.

The coefficients g_n^m and h_n^m are the internal spherical harmonic coefficients, and those in the corresponding expression for V_e are the external spherical harmonic coefficients. In any reference to spherical harmonic coefficients, the normalization of the corresponding functions must be considered. Here the g_n^m and h_n^m are called *Gauss coefficients* with the Schmidt normalization.

Properties of Spherical Harmonic Functions

Separation of variables assumes a solution to Laplace's equation of the form $V(r, \theta, \phi) = R(r)S(\theta)T(\phi)$. Thus, we obtain the solutions

$$R(r) = \left(\frac{a}{r}\right)^{n+1} \text{ or } \left(\frac{r}{a}\right)^n$$

and $T(\phi) = \cos m\phi$ or $\sin m\phi$, and the differential equation

$$\frac{d}{d\mu}\left[(1 - \mu^2)\frac{dS}{d\mu}\right] + S\left[n(n+1) - \frac{m^2}{1 - \mu^2}\right] = 0$$

where $\mu = \cos\theta$. If $m = 0$, this is Legendre's equation with solution (here given with Neumann normalization; conversion to other normalizations is given later)

$$P_{n,0}(\mu) = P_n(\mu)$$
$$= \sum_{k=0}^{[n/2]} (-1)^k \frac{(2n - 2k)!}{2^n k!(n-k)!(n-2k)!} \mu^{n-2k}$$

where $[n/2]$ is the greatest integer less than or equal to $n/2$. Note that $P_0(\mu) = 1$. The Legendre polynomial $P_n(\mu)$ is an odd function if n is odd and is even if n is even.

If $m \neq 0$, differentiating Legendre's equation m times and using an appropriate integrating factor gives the solution for S in terms of the Legendre polynomial:

SPHERICAL HARMONIC ANALYSIS

$$P_{n,m}(\mu) = (1 - \mu)^{m/2} \frac{d^m P_n(\mu)}{d\mu^m},$$

for $m = 1, 2, \ldots, n$

Note that $P_{n,m} = 0$ if $m > n$. The foregoing formulas are not used in practice; computation is more efficient with the following recursion relations.

The functions $P_n^m(\cos \theta)\cos m\phi$ and $P_n^m(\cos \theta) \cdot \sin m\phi$ are called *spherical surface harmonics of degree n and order m* and are mutually orthogonal over the surface of a sphere; if any two different spherical surface harmonics are integrated over a spherical surface, the result is zero. For example,

$$\frac{1}{4\pi} \int_0^\pi \int_0^{2\pi} P_n^m(\cos \theta) \frac{\cos}{\sin} m\phi \, P_k^j$$

$$\cdot (\cos \theta) \frac{\cos}{\sin} j\phi \sin \theta \, d\phi \, d\theta = 0$$

unless $n = k$ and $m = j$ and the functions of ϕ are the same (both cosine or both sine).

If $m = 0$, $P_n^0(\cos \theta)$ are called *zonal surface harmonics* or *Legendre functions*, and their graphs divide the sphere into zones of positive or negative values separated by circles of latitude. If $m = n$, the spherical surface harmonics are called *sectorial surface harmonics*, with divisions of the sphere into sectors of constant sign bounded by meridians; if $n > m > 0$, they are called *tesseral surface harmonics*.

The functions $(a/r)^{n+1} P_n^m(\cos \theta)\cos m\phi$ and $(a/r)^{n+1} P_n^m(\cos \theta)\sin m\phi$ are called the *fundamental solutions* to Laplace's equation within a sphere. The corresponding functions with $(r/a)^n$ solve Laplace's equation external to a sphere.

Several normalizations are in common use. Schmidt functions are normalized so that the average square value over a spherical surface is $1/(2n + 1)$:

$$\frac{1}{4\pi} \int_0^\pi \int_0^{2\pi} \left[P_n^m(\cos \theta) \frac{\cos}{\sin} m\phi \right]^2$$

$$\cdot \sin \theta \, d\phi \, d\theta = \frac{1}{2n + 1}$$

Functions with Neumann normalization, denoted P_{nm}, and Gauss normalization, denoted P^{nm}, result in average square values of

$$\frac{1}{2(2n + 1)} \frac{(n + m)!}{(n - m)!}$$

and

$$\frac{1}{2(2n + 1)} \frac{(n + m)!(n - m)!}{[(2n - 1)!!]^2}$$

respectively, where $n! = n(n - 1)(n - 2)$ and $(2n - 1)!! = \text{product } (2n - 1)(2n - 3)(2n - 5)\cdots(7)(5)(3)(1)$. The functions $R_n^m = (2n + 1)^{1/2} P_n^m$ are fully normalized (the average square value is unity), and the corresponding surface harmonics are hence orthonormal.

The Gauss and Schmidt normalizations are most often used in geomagnetism. The Gauss, or Gauss-Laplace, functions are often used computationally by the relations

$$P^{0,0} = 1,$$

$$\frac{\partial P^{0,0}}{\partial \theta} = 0$$

$$P^{n,n} = (\sin \theta) P^{n-1, n-1},$$

$$\frac{\partial P^{n,n}}{\partial \theta} = (\sin \theta) \frac{\partial P^{n-1, n-1}}{\partial \theta} + (\cos \theta) P^{n-1, n-1}$$

where $n \geq 1$, and, for $m \neq n$,

$$P^{n,m} = (\cos \theta) P^{n-1, m} - K^{n,m} P^{n-2, m}$$

$$\frac{\partial P^{n,m}}{\partial \theta} = (\cos \theta) \frac{\partial P^{n-1, m}}{\partial \theta} - (\sin \theta) P^{n-1, m}$$

$$- K^{n,m} \frac{\partial P^{n-2, m}}{\partial \theta}$$

where

$$K^{n,m} = \frac{(n - 1)^2 - m^2}{(2n - 1)(2n - 3)}$$

Other relations may be found in Chapman and Bartels (1940, chap. 17).

To convert the Gauss functions to Schmidt normalization, multiply by $S^{n,m}$: $P_n^m = S^{n,m} P^{n,m}$, where

$$S^{0,0} = 1, \quad S^{n,0} = \frac{2n - 1}{n} S^{n-1, 0}$$

$$S^{n,m} = S^{n,m-1} \sqrt{\frac{(n - m + 1)J}{n + m}}$$

where $J = 2$ for $m = 1$ and $J = 1$ for $m > 1$. The Gauss normalization is hence useful for calculations, due to the simple recursion relations. However, the Schmidt normalization is the standard for publication in geomagnetism; the full normalization is used in gravitation.

Conversion to other normalizations is given by the formulas

$$P^{n,m} = \frac{(n-m)!}{(2n-1)!!} P_{n,m} \quad \text{Neumann to Gauss}$$

$$P_n^0 = P_{n,0} \quad \text{Neumann to Schmidt}$$

$$P_n^m = P_{n,m} \sqrt{2 \frac{(n-m)!}{(n+m)!}}$$

Conversion between normalizations is more efficiently calculated by means of recursion relations rather than from these definitions. The Gauss calculation and conversion to Schmidt normalization given above is one such example.

Nonlinear Weighted Least Squares

The method of nonlinear weighted least squares was applied by Jensen and Cain (1962) and Cain et al. (1967) to data of different types and of nonuniform location and distribution in position and time, taking into account data measurement errors. This method of obtaining spherical harmonic coefficients has dominated global geomagnetic modeling.

Least squares is a method of approximating data by a function, dependent on a set of parameters, so that the error, which is the sum of squares of differences between data and function, is minimized. The problem then is one of calculating the optimal set of parameters, and is equivalent to solving a system of linear equations. Weighted least squares refers to a least squares method in which the data are assigned weights, the larger weights indicating a tighter fit to the corresponding data.

Nonlinear refers to the use of a function that is nonlinearly dependent on the parameters. The nonlinear least squares method then requires an iterative, or repeated, solution of systems of linear equations, beginning with an initial set of parameters, and refining the parameters at each iteration. The modeling of geomagnetic intensity or angular data requires such a function.

Let $f(p_1, \ldots, p_K; r, \theta, \phi, t)$ be a function, also called a functional form, that depends on position and time variables r, θ, ϕ, and t, and on parameters $p_1, \ldots, p_K$. The function f is assumed to be nonlinear in the parameters, and is expanded in a power series about an initial set of parameters $q_1, \ldots, q_K$:

$$f(p_1, \ldots, p_K)$$
$$= f(q_1, \ldots, q_K)$$
$$+ \sum_{k=1}^{K} \frac{\partial f}{\partial p_k} (q_1, \ldots, q_K) \Delta p_k$$
$$+ \frac{1}{2} \sum_{k=1}^{K} \sum_{j=1}^{K} \frac{\partial f^2}{\partial p_k \partial p_j} (q_1, \ldots, q_K) \Delta p_k \Delta p_j$$
$$+ \text{higher-order terms}$$

where $\Delta p_k = p_k - q_k$ and where the position and time variables have been omitted. In the nonlinear least squares method, the Taylor series is truncated to first order, which automatically occurs whenever the function is linear in its parameters.

Let f_{pi} denote $f(p_1, \ldots, p_K; r_i, \theta_i, \phi_i, t_i)$, the value of f at the ith data location and at the parameters $p_1, \ldots, p_K$. Let f_{qi} denote the same for initial parameters $q_1, \ldots, q_K$. Let f_{oi} denote the dependent variable, or observation, at the ith data location. Let w_i be a weight for the ith observation. The error to be minimized is then

$$\sum_{i=1}^{I} w_i [f_{oi} - f_{pi}]^2 = \sum_{i=1}^{I} w_i [f_{oi} - f_{qi} - \sum_{k=1}^{K} \frac{\partial f_i}{\partial p_k}$$
$$\cdot (q_1, \ldots, q_K) \Delta p_k]^2$$
$$= \sum_{i=1}^{I} w_i [\Delta f_i - \sum_{k=1}^{K} \frac{\partial f_i}{\partial p_k}$$
$$\cdot (q_1, \ldots, q_K) \Delta p_k]^2$$

where the truncated power series for f_{pi} has been inserted and Δf_i denotes the residual $f_{oi} - f_{qi}$.

To minimize this error expression, differentiate it with respect to each Δp_k (or, equivalently, each p_k) and set to zero, thus obtaining K equations in K unknowns, $\Delta p_1, \ldots, \Delta p_K$. The new parameter values p_k are obtained via the relation $p_k = q_k + \Delta p_k$. The system of equations is

$$\sum_{i=1}^{I} w_i \sum_{k=1}^{K} \frac{\partial f_i}{\partial p_k} \frac{\partial f_i}{\partial p_j} \Delta p_k$$
$$= \sum_{i=1}^{I} w_i \Delta f_i \frac{\partial f_i}{\partial p_j}, \quad \text{for } j = 1, \ldots, K$$

where each derivative is evaluated at the initial parameter values of $q_1, \ldots, q_K$ and at the appropriate ith data point location.

The system can be conveniently written in matrix form. Let

$$Y = \begin{bmatrix} \Delta f_1 \\ \vdots \\ \Delta f_I \end{bmatrix} \quad \text{and} \quad \beta = \begin{bmatrix} \Delta p_1 \\ \vdots \\ \Delta p_K \end{bmatrix}$$

be column vectors of lengths I and K representing function residuals, and let

$$X = \begin{bmatrix} \frac{\partial f_1}{\partial p_1} & \cdots & \frac{\partial f_1}{\partial p_K} \\ \frac{\partial f_I}{\partial p_1} & \cdots & \frac{\partial f_I}{\partial p_K} \end{bmatrix} \quad \text{and} \quad W = \begin{bmatrix} w_1 & & 0 \\ & \ddots & \\ 0 & & w_I \end{bmatrix}$$

be the $I \times K$ matrix of derivatives $\partial f_i/\partial p_k$ calculated at the observation points $(r_i, \theta_i, \phi_i, t_i)$ and at the initial parameter values $q_1, \ldots, q_K$, and the $I \times I$ diagonal matrix of weights, respectively. More generally, $\mathbf{W}$ is the inverse of the variance-covariance matrix of the observations. However, in many applications, the observations are assumed to be independent, giving a diagonal matrix with entries equal to the reciprocals of the variances.

With this notation, the error becomes

$$(\mathbf{Y} - \mathbf{X}\boldsymbol{\beta})^T \mathbf{W}(\mathbf{Y} - \mathbf{X}\boldsymbol{\beta})$$

and, by differentiating with respect to each kth component of $\boldsymbol{\beta}$, the system of equations becomes

$$\mathbf{X}^T\mathbf{W}\mathbf{X}\boldsymbol{\beta} = \mathbf{X}^T\mathbf{W}\mathbf{Y}$$

Orthogonality; Integration

The orthogonality properties of the spherical surface harmonics may be used to recover the spherical harmonic coefficients if data are given on a spherical surface. The assumption is often made that the sources are entirely internal. For example, given the vertical derivative of the internal potential function (appropriate for gravity or geomagnetism; here the latter is used for illustration),

$$B_r = -\frac{\partial V}{\partial r}$$

$$= \sum_{n=0}^{\infty} (n+1) \sum_{m=0}^{\infty} \left(\frac{a}{r}\right)^{n+2}$$

$$\cdot (g_n^m \cos m\phi + h_n^m \sin m\phi) P_n^m(\cos \theta)$$

multiply, for example, by $p_k^j(\cos \theta)\cos j\phi$ and integrate over the surface of a sphere of fixed radius r, to get

$$\frac{1}{4\pi} \int_0^\pi \int_0^{2\pi} B_r(r, \theta, \phi) p_k^j(\cos \theta)\cos j\phi \sin \theta \, d\phi \, d\theta$$

$$= \frac{n+1}{2n+1} \left(\frac{a}{r}\right)^{n+2} g_n^m$$

The integral is approximated by a summation of the form

$$\sum_{s=1}^{S} \left[\sum_{t=1}^{T} B_r(r, \theta_s, \phi_t)\cos j\phi_t \right] p_k^j(\cos \theta_s) w_s$$

The inner summation is a finite Fourier transform and uses equal spacing in the ϕ variable. The outer summation is best calculated by Gauss-Legendre quadrature, which requires an unequal spacing in θ (based on the roots of a fixed Legendre polynomial) and the use of appropriate weights. The values of θ and the corresponding weights are found in many tables or may be calculated. This method was described by Franz Neumann (1838, 1879), who attributes it to Gauss. Workers have used simpler quadrature techniques, with uniform spacing in θ, but they assumed the residuals had a flat spatial power spectrum. The Gauss-Legendre quadrature is exact (given perfect data), whereas quadrature using equal spacing in θ is a convenient approximation, albeit not theoretically correct.

The integration method requires data on a global spherical grid. At present, values at grid points must be constructed via interpolation or averaging. However, in spite of not using the original data, this method is usable to a far higher degree and order than the least squares method.

Applications

In the least squares method developed by Jensen and Cain (1962), all components of the geomagnetic field are used simultaneously. (Earlier work involved the calculation of separate spherical harmonic models from each component, with an average model derived from the separate results.) Later developments allowed for expansion of the coefficients in terms of time, usually in quadratic or cubic polynomials. Secular variation data has been used as well (Schmitz and Cain, 1983). If external coefficients are used, they are usually severely limited, often to only three terms.

The least squares method is flexible, but the matrix for the system of equations is of order n^4, where n^* is the maximum degree and order retained in the truncated spherical harmonic expansion. Some savings may be made since the matrix is symmetric, and coefficients found to be statistically insignificant may be omitted; however, no geomagnetic model has yet been calculated beyond $n^* = 31$. Given the same data set, the least squares procedure gives different results when n^* is changed. The differences occur mainly in the coefficients of higher degree and order; hence a sufficiently large n^* should be chosen if the coefficients are of interest.

The integration method may be used to much higher degree and order, but does not have the generality of least squares. It is most easily applied to the radial component of the field, but has also been applied to the north-south and east-west components. The requirement of data on a fixed spherical surface is a problem for geomagnetic studies, even with satellite data, but not as severe a problem for gravity studies.

Other Methods

Gauss's method used integration in ϕ but replaced the integration in θ by least squares. This method overcomes the stringent requirements of the integration method as to location of the data on specified circles of latitude.

Indirect methods are those which calculate other

types of coefficients that may be converted into spherical harmonic coefficients. Examples are Fourier analysis in latitude and longitude and the harmonic spline method developed by Shure (1982), Shure et al. (1982), and Parker and Shure (1982). The latter derives field models that are optimally smoothed according to some criterion. An important application is to studies of the magnetic field at the core-mantle boundary.

Other methods for geomagnetism are in Barraclough (1978). Colombo (1981) describes several methods used in gravity, including least squares adjustment and least squares collocation.

An interesting variant of spherical harmonic analysis is in the application of fractional spherical harmonics to regional data. This method is spherical cap harmonic analysis, developed by Haines [1985a]. The region may be any reasonably caplike region of the globe, such as Canada (Haines, 1982b).

Other Concerns

The coordinate system need not be centered at the center of the Earth nor aligned with the rotation axis. The eccentric dipole has been used to give center and orientation of a coordinate system. The spatial power spectrum, analogous to the power spectrum of a time series in one variable derived from the squared amplitude of each Fourier component as a function of frequency, is

$$R_n = (n + 1) \sum_{m=0}^{n} [(g_n^m)^2 + (h_n^m)^2]$$

and has been used for the interpretation of the field. Since the individual coefficients at high degree are small, and since the rotation theorem (Courant and Hilbert, 1953, pp. 535ff) for surface harmonics states that coefficients of common degree are transformed under rotation into coefficients of the same degree, the spatial power spectrum gives a natural, rotation-invariant measure of the spherical harmonic coefficients. Based on its use in geomagnetism, the low-degree coefficients (below about 14) are considered to mainly represent the core field, higher whereas order coefficients represent the crustal field.

Backus (1970) showed that use of only scalar data, F, on a spherical surface is insufficient to determine the field. Such a Backus ambiguity (or perpendicular error effect) has been observed in modeling Magsat data (Stern et al., 1980).

D. SCHMITZ

References

Backus, G. E., 1970, Non-uniqueness of the external geomagnetic field determined by surface intensity measurements, *Jour. Geophys. Research* **75**, 6337–6341.

Barraclough, D. R., 1978, *Spherical Harmonic Models of the Geomagnetic Field*, Geomagnetic Bulletin 8, London: Institute of Geological Sciences.

Cain, J. C., S. J. Hendricks, R. A. Langel, and W. V. Hudson, 1967, A proposed model for the International Geomagnetic Reference Field—1965, *Jour. Geomagnetism and Geoelectricity* **19**, 335-355.

Chapman, S., and J. Bartels, 1940, *Geomagnetism*, vol. 2. London: Oxford University Press.

Colombo, O. L., 1981, Numerical Methods for Harmonic Analysis on the Sphere, *Ohio State Univ. Dept. Geod. Sci. Rept. No. 310*.

Courant, R., and D. Hilbert, 1953, *Methods of Mathematical Physics*, vol. 1. New York: Interscience Publishers.

Haines, G. V., 1985a, Spherical cap harmonic analysis, *Jour. Geophys. Research* **90**, 2583-2591.

Haines, G. V., 1985b, Magsat vertical field anomalies above 40°N from spherical cap harmonic analysis, *Jour. Geophys. Research* **90**, 2593-2598.

Jensen, D. C., and J. C. Cain, 1962, An interim geomagnetic field (abstract), *Jour. Geophys. Research* **67**, 3568-3569.

Neumann, F., 1838, Über eine neue Eigenschaft der Laplaceschen $Y^{(n)}$ und ihre Anwendung zur analytischen Darstellung derjenigen Phänomene, welche Funktionen der geographischen Länge und Breite sind (On a new property of the Laplacian $Y^{(n)}$ and its application to the analytical representation of those phenomena which are functions of geographical longitude and latitude, *Astron. Nachr.* **15**, 313-323 (in German). Also *Math. Ann.* **14**, 567-576, 1879.

Parker, R. L., and L. Shure, 1982, Efficient modeling of the Earth's magnetic field with harmonic splines, *Geophys. Research Letters* **9**, 812-815.

Schmitz, D. R., and J. C. Cain, 1983, Geomagnetic spherical harmonic analyses, 1. Techniques, *Jour. Geophys. Research* **88** (B2), 1222-1228.

Shure, L., 1982, Modern mathematical methods in geomagnetism, Ph.D. dissertation, Institute of Geophysics and Planetary Physics, University of California, San Diego.

Shure, L., R. L. Parker, and G. E. Backus, 1982, Harmonic splines for geomagnetic modelling, *Physics Earth and Planetary Interiors* **28**, 215-229.

Stern, D. P., R. A. Langel, and G. D. Mead, 1980, Backus effect observed by Magsat, *Geophys. Research Letters* **7**, 941-944.

Cross-references: *Earth's External Gravity Field; Earthquake Structure: Global; Geomagnetic Field: Asymmetries; Geomagnetic Field: Westward Drift; Geomagnetic Field, Main: Theory; Geomagnetic Field Analysis; Geomagnetic Secular Variation: Direction and Intensity; Gravity Fields: Implications for Planetary Interiors; Seismic Tomography; Surface Waves.*

STRESS IN THE EARTH'S LITHOSPHERE

The occurrence of earthquakes and volcanic eruptions are dramatic, often catastrophic, demonstrations of stresses within the lithosphere at work. Over geologic time these stresses give rise to plate motions and cause the uplift of mountains and subsidence of ocean basins. Understanding the distribution of forces in the lithosphere, and the evolution of those forces with time, is clearly fundamental to many problems in the varied disciplines of the Earth sciences.

Through refinement of plate tectonics theory over the past two decades, there has been a greatly improved understanding of the geometric and temporal movements between the major plates. To make the next quantum leap in understanding geologic phenomena, a greatly improved knowledge of *how* plate tectonics works will be required. This will require better knowledge of the forces that act in the crust and the strength and deformational properties of crustal materials. This knowledge is required to better understand the relative magnitudes of the stresses that drive (and inhibit) plate motion, to understand the dynamics of faulting along major plate boundaries and intraplate areas, and the overall mechanical, thermal, and rheological constraints on crustal deformation.

Definitions

Describing the magnitude and orientation of principal stresses in the Earth's lithosphere requires several fundamental definitions that are illustrated in Fig. 1. The stresses acting on a unit cube are a second-order tensor defined in terms of the direction of applied force and the orientation of the plane that force acts upon. In other words, S_{11} is the normal stress (force divided by area) acting in the x_1 direction on the plane perpendicular to the x_1 axis, S_{12} is

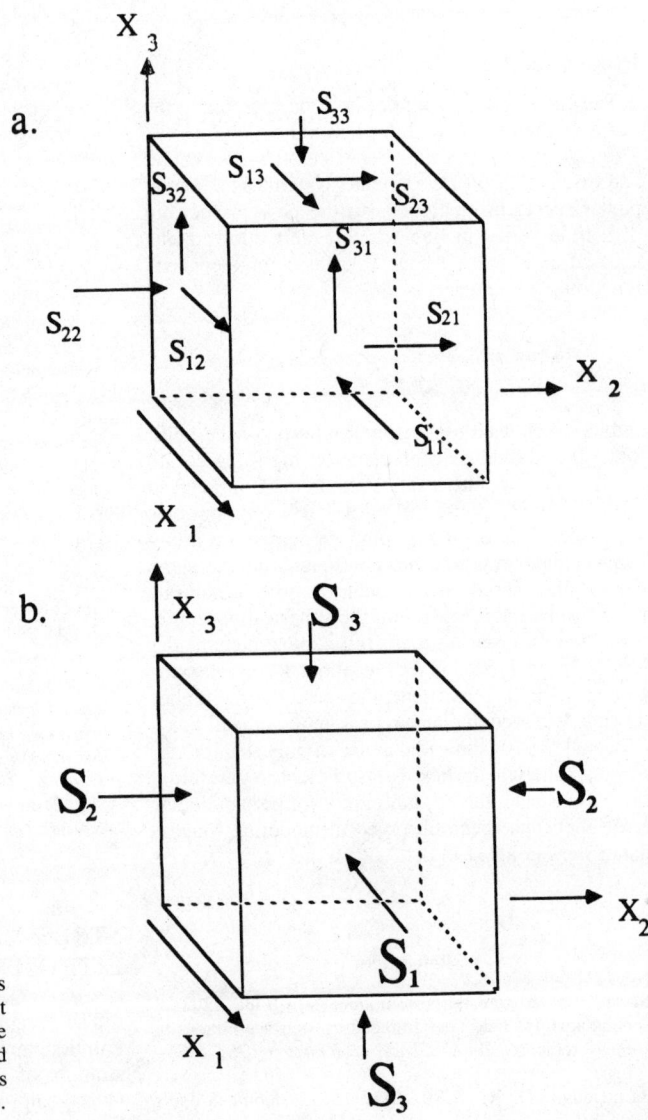

FIGURE 1. (*a*) Independent components of the stress tensor are defined with respect to a unit cube. (*b*) If X_1, X_2 and X_3 define principal planes, shear stresses vanish and the only nonzero components of the stress field are the principal stresses S_1, S_2, and S_3.

the component of shear stress acting in the x_1 direction on the plane perpendicular to the x_2 axis, and so on. We consider compressive stresses to be positive. Thus, the in situ stress field is described by a tensor with nine components

$$\begin{matrix} S_{11} & S_{12} & S_{13} \\ S_{21} & S_{22} & S_{23} \\ S_{31} & S_{32} & S_{33} \end{matrix}$$

Only six stress components are independent as equilibrium conditions require $S_{12} = S_{21}$, $S_{13} = S_{31}$ and $S_{23} = S_{32}$. Thus, the stress tensor is somewhat simplified but it is still obviously cumbersome to refer to a stress field described by six independent stress magnitudes and unconstrained orientations. Fortunately, in discussing the state of stress in the lithosphere, this situation can be considerably simplified. First, to describe the stress field completely we need only to refer to the orientation and magnitudes of the principal stresses. The principal stresses are the normal stresses that act on planes oriented such that shear stresses vanish. As shown in Fig. 1b, the X_1, X_2 and X_3 planes are principal planes and the only nonzero stress components are the principal stresses S_{11}, S_{22} and S_{33} which we refer to as S_1, S_2 and S_3, for simplicity. The second simplification in the description of the in situ stress field is the observation that within the lithosphere the principal planes appear to be oriented approximately horizontally and vertically (obvious exceptions are such areas as subducting slabs). In general, however, by simply referring to the orientation of either the maximum principal horizontal stress, S_{Hmax}, or the minimum principal horizontal stress, S_{hmin}, the orientation of the stress tensor is completely specified. Further, as one principal stress is essentially vertical, S_v, it can be described by the stress induced by the weight of the overlying rock

$$S_v \sim \int \rho(z) \, g \, dz$$

where $\rho(z)$ is the density of the overlying rock, g is the gravitational constant, and z is depth. For an average crustal density of 2.7 g/cm^3, the change in the vertical stress with depth is approximately 27 Mpa/km.

Basic Relationships

From simple faulting theory first elaborated by E. M. Anderson, it can be straightforwardly shown that in areas of extension, $S_v \geq S_{Hmax} \geq S_{hmin}$; in areas of strike slip faulting $S_{Hmax} \geq S_v \geq S_{hmin}$; and in areas of reverse faulting situ stress measurements $S_{Hmax} \gtrsim S_{hmin} \gtrsim S_v$. Thus, the relative magnitudes of the three principal stresses defines the tectonic style and, as S_v at depth can be estimated from average rock density, the vertical stress provides a convenient reference and the magnitudes of the horizontal principal stresses can be defined with respect to S_v.

As shown later, stresses at depth within the Earth's crust are always compressional, even in areas in extension. Thus, all of the values for stress magnitude in the previous paragraph are positive, compressional stresses. However, before we discuss stress magnitudes at depth it is necessary to discuss the effects of fluid pressures acting in the cracks and pores of rock. Hydrostatic fluid pressure, that is, that which results simply from an unconfined column of fluid extending from depth to the surface, increases at a rate of about 10 MPa/km. Numerous rock properties such as seismic velocity and compressibility, as well as the failure and frictional strength of rock, depend on the effective stress, σ, which is equal to the difference between the total stress S and the pore pressure, P. Thus, pore pressure reduces the effects of total stress at depth.

Using the concept of effective stress at depth, we can extend Anderson's faulting theory to predict stress magnitudes at depth through utilization of simplified two-dimensional Mohr-Coulomb failure theory. Two-dimensional faulting theory assumes that failure is only a function of the difference between the least and greatest principal stresses σ_1 and σ_3. In the case of normal faulting, faulting is controlled by σ_v, σ_{hmin} which correspond to σ_1 and σ_3, respectively, and σ_{Hmax} corresponds to σ_2 which is intermediate in value between σ_v and σ_{hmin} and does not influence faulting. Simplified Mohr-Coulomb failure theory indicates that frictional sliding occurs when the ratio of shear stress to effective normal stress on preexisting fault planes is equal to the coefficient of friction. As the coefficient of friction is relatively well-defined for most rocks, it can be shown that frictional sliding will occur when $\sigma_1/\sigma_3 \sim 3$. Based on this, in cases of hydrostatic pore pressure, the relationships defined above can be used to show that in extensional areas $S_{hmin} \sim 0.6\, S_v$, in reverse faulting areas $S_{Hmax} \sim 2.3\, S_v$ and in strike-slip faulting areas, when $S_v \sim \frac{1}{2}\,(S_{Hmax} + S_{hmin})$, $S_{Hmax} \sim 2.2\, S_{hmin}$. These simple relationships have been confirmed by in situ stress measurements to depths of about 2 km at a number of sites in intraplate areas. Whether or not these relationships apply to major plate boundaries such as the San Andreas fault and major subduction zones is the subject of ongoing research.

Measurement of Stress

Information on the state of stress in the lithosphere comes from a variety of sources—earthquake focal plane mechanisms, young geologic data on fault slip and volcanic alignments, in situ stress measurements, and stress-induced wellbore breakouts. Earthquake focal mechanisms are the most ubiquitous indicator of stress in the lithosphere. The pattern of seismic radiation from the focus of an earthquake permits construction of earthquake focal mechanisms (Fig. 2). Focal mechanisms contain information on the tectonic style (i.e., normal, strike-slip, or reverse faulting), and on the orientation of the fault plane and auxiliary plane and the orientation of the P (compressional), B (intermediate), and T ("tensional") axes, which are often assumed to be approximately equal to the orientations of σ_1, σ_2 and σ_3. Additional information is required from either local geology or the aftershock distribution to distinguish the fault plane from the auxiliary plane. As earthquakes occur on preexisting fault planes, the seismic radiation pattern is largely controlled by the orientation of the fault plane and not the in situ stress field. In other words, regardless of the ambient stress field, an earthquake focal plane mechanism always has the P and T axes at 45° to the fault plane and the B axes in the plane of the fault. The most useful way in which to use earthquake focal mechanisms to determine stress orientations is to consider average P, B, and T axes for earthquakes occurring on different faults within a limited region. The amplitude and frequency content of seismic waves radiating from an earthquake only give information about the magnitude of stress released in an earthquake (stress drop) and not the absolute stress levels. In general, stress drops are on the order of 1–10 MPa, a very small fraction of the shear stress that actually causes the earthquake if the Mohr-Coulomb theory outlined earlier is correct and laboratory-derived coefficients of friction are applicable to faults in situ.

In a manner similar to earthquake focal mechanisms, indications of recent fault slip can be used to

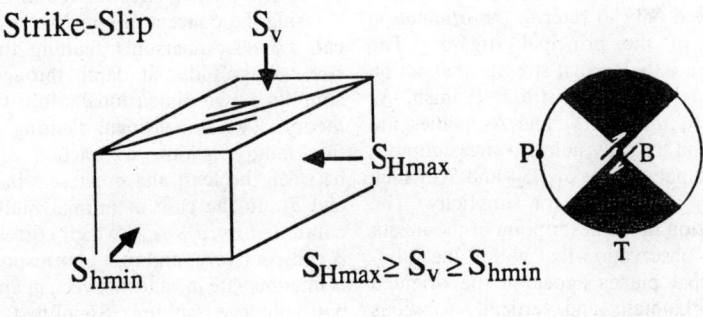

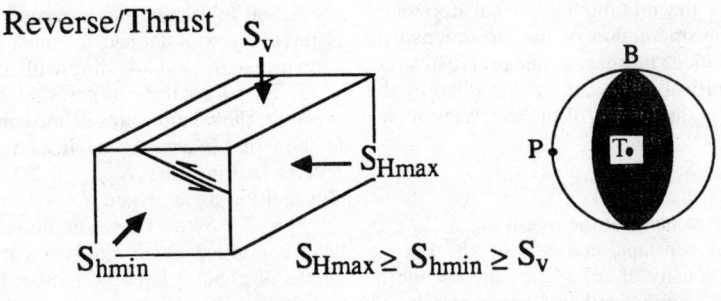

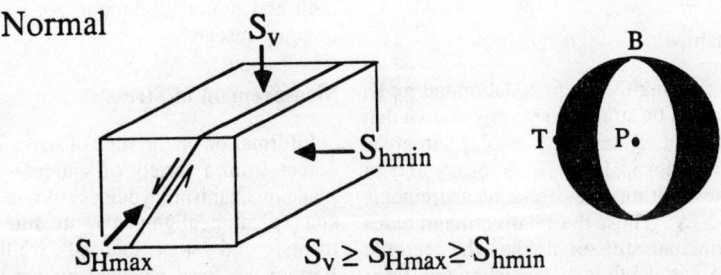

FIGURE 2. Simplified illustration of the various types of faulting, the relative magnitudes of the principal stresses and lower-hemisphere projection earthquake focal plane mechanisms.

determine stress orientation. Again, it is necessary to consider the directions of slip on a variety of planes to determine stress orientation uniquely from fault slip data.

When a borehole is drilled in rock, the stresses that were previously supported by the material that was exhumed is transferred to the surrounding material. In intact elastic rock, the resultant stress concentration around the wellbore is well understood from elastic theory. There are basically two ways in which the stress concentration around boreholes can be exploited to help measure in situ stresses. The first method is known as hydraulic fracturing and involves inducing a tensile fracture at the point around the hole where the concentrated hoop stress is least compressive. The second method occurs naturally in many boreholes because the compressive hoop stress around a hole can be large enough to exceed the strength of the rock. When this occurs, the rock around a portion of the wellbore fails and stress-induced wellbore breakouts form. Hydraulic fracturing and wellbore breakouts are the most important stress measurement methods because they are the only established methods for making measurements at depths more than a few tens of meters from the Earth's surface or a tunnel or shaft. We will limit ourselves to discussing these methods because, in general, only in situ stress measurements made at depths greater than 100 m seem to correlate reliably with earthquake focal plane mechanisms indicative of the tectonic stress field at midcrustal depths.

As illustrated on the left side of Fig. 3, the principle behind hydraulic fracturing stress measurements is based on the fact that a pressurized borehole will develop a tensile fracture at point around the hole where the concentrated hoop stress is least compressive. Thus, the fracture will develop at the azimuth of the maximum principal stress, S_{Hmax}, by determining the azimuth of the fracture at the wellbore the orientation of stress field can be determined. The hydraulic fracturing technique also has the capability to determine the magnitude of S_{Hmax} and S_{hmin}. In strike-slip and normal faulting areas, the fracture will propagate perpendicular to the minimum horizontal principal stress because it takes minimum energy to propagate an extension fracture perpendicular to the least force. It is thus straightforward to determine S_{hmin} from the pumping pressure and what is called the shut-in pressure, the pressure in the borehole immediately after pumping is stopped. From knowledge of S_{hmin}, the pressure it took to create the hydraulic fracture (the breakdown pressure) and an estimate of the tensile strength of the rock, it is also possible to estimate the magnitude of S_{Hmax} at depth. Young volcanic alignments and feeder dikes are useful as stress orientation indicators because magma is intruded into the crust much like a hydrofrac—the magma propagates in a plane perpendicular to the least principal compressive stress. Thus, the plane of a dike, or the inferred plane associated with an alignment of cinder cones, is measure of the principal stress directions at the time of intrusion.

Over the past decade, the widespread use of four-arm oriented calipers in the petroleum industry has resulted in appreciable information being available on stress orientation from stress-induced wellbore breakouts. As illustrated on the left side of Fig. 3, breakouts occur where the concentrated hoop stress around a hole is most compressive (at the azimuth of S_{hmin}) when the hoop stress around a hole exceeds the strength of the rock, causing the rock to fail and the hole to elongate in the direction of S_{hmin}. The right side of Fig. 3 shows a three-dimensional view of a hydraulically fractured section of a wellbore made by special digital processing of an acoustic scanning device similar to sonar. The image shows that as predicted by theory, the hydraulic fracture azimuth is orthogonal to that of the naturally-occurring breakouts in the holes. Two important advantages of wellbore breakouts is that in a given well many observations of stress orientation can be made and that data are widely available from literally thousands of wells around the world. In the next section, a great many of the data available worldwide come from wellbore breakouts in wells and boreholes.

Distribution of Crustal Stresses

In the past decade compilations of tectonic stress data have been made in North America, Europe and Scandinavia, Australia, Japan, and China and an initial global synthesis of these and data from intraplate oceanic areas is currently underway. The most complete data set is for the United States, which is used below for illustration.

Figure 4 shows a compilation of quality-ranked maximum horizontal compressive stress orientation data in the conterminous United States. The data come from wellbore breakouts, earthquake focal mechanisms, in situ stress measurements greater than 100 m depth, and young (generally < 2 my old) geologic indicators. Where different types of stress orientation data are available the correlation between the different data sources is quite good. Two conclusions can be drawn. First, the criterion used are approximately correct and data badly contaminated by nontectonic sources of stress have been effectively eliminated. By nontectonic sources of stress we mean stresses associated with near-surface effects such as weathering and topography or residual stresses reflecting past tectonic environments or a rock's thermal or diagenetic history. Second, no major changes in the orientation of the crustal stress field occur between the upper 4 km, where essentially all of the wellbore breakout and stress measurement data come from, and 4–15 km where the majority of focal mechanism data come from.

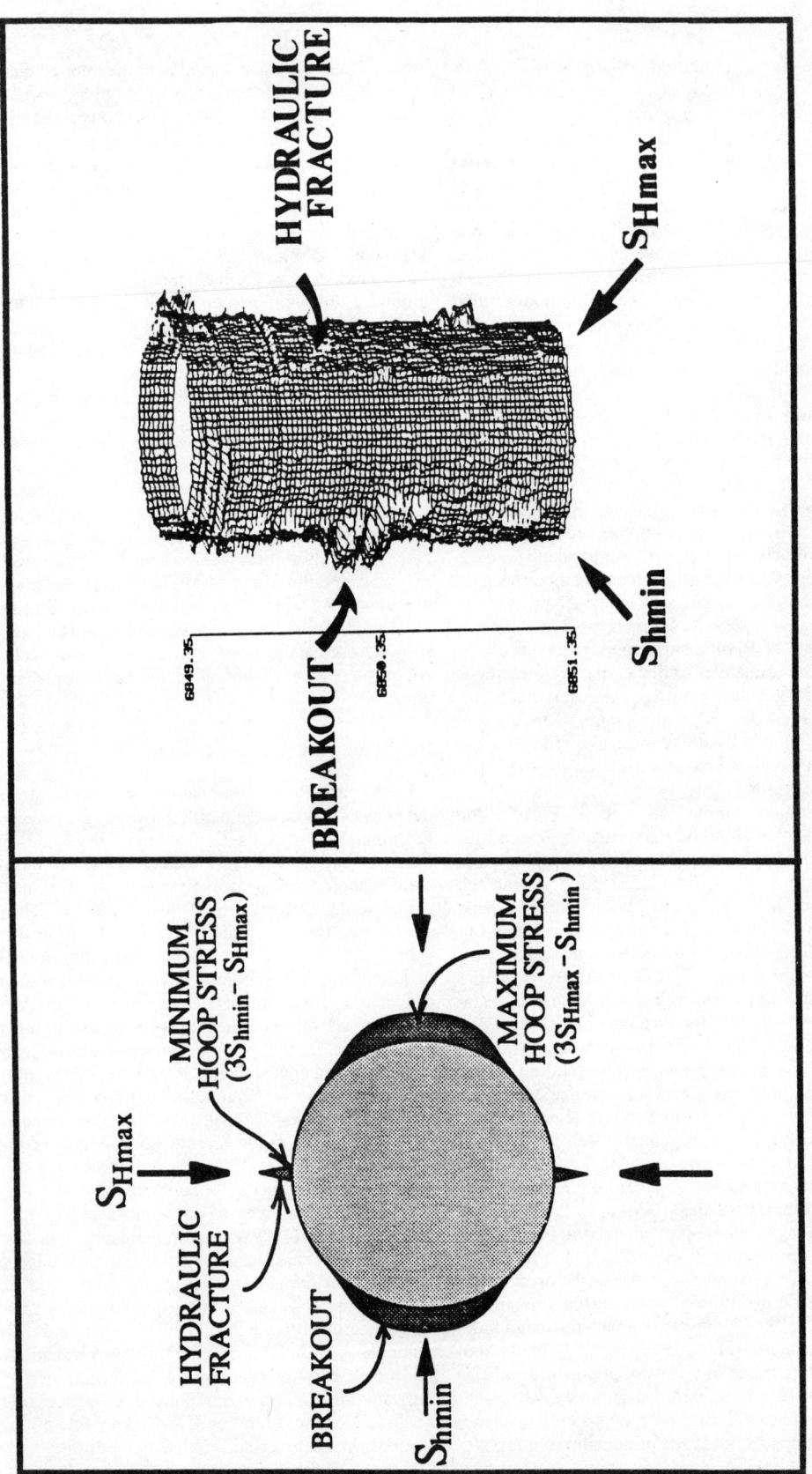

FIGURE 3. (*Left*) Schematic illustration that wellbore breakouts form at the azimuth of S_{hmin} where the concentration of circumferential "hoop" stress around the well is most compressive and hydraulic fractures form at the azimuth of S_{Hmax}, where the circumferential stress is least compressive. (*Right*) A three-dimensional view of a 60 cm-long interval of a well that contains both a man-made hydraulic fracture and naturally occurring breakouts. They are at right angles as predicted.

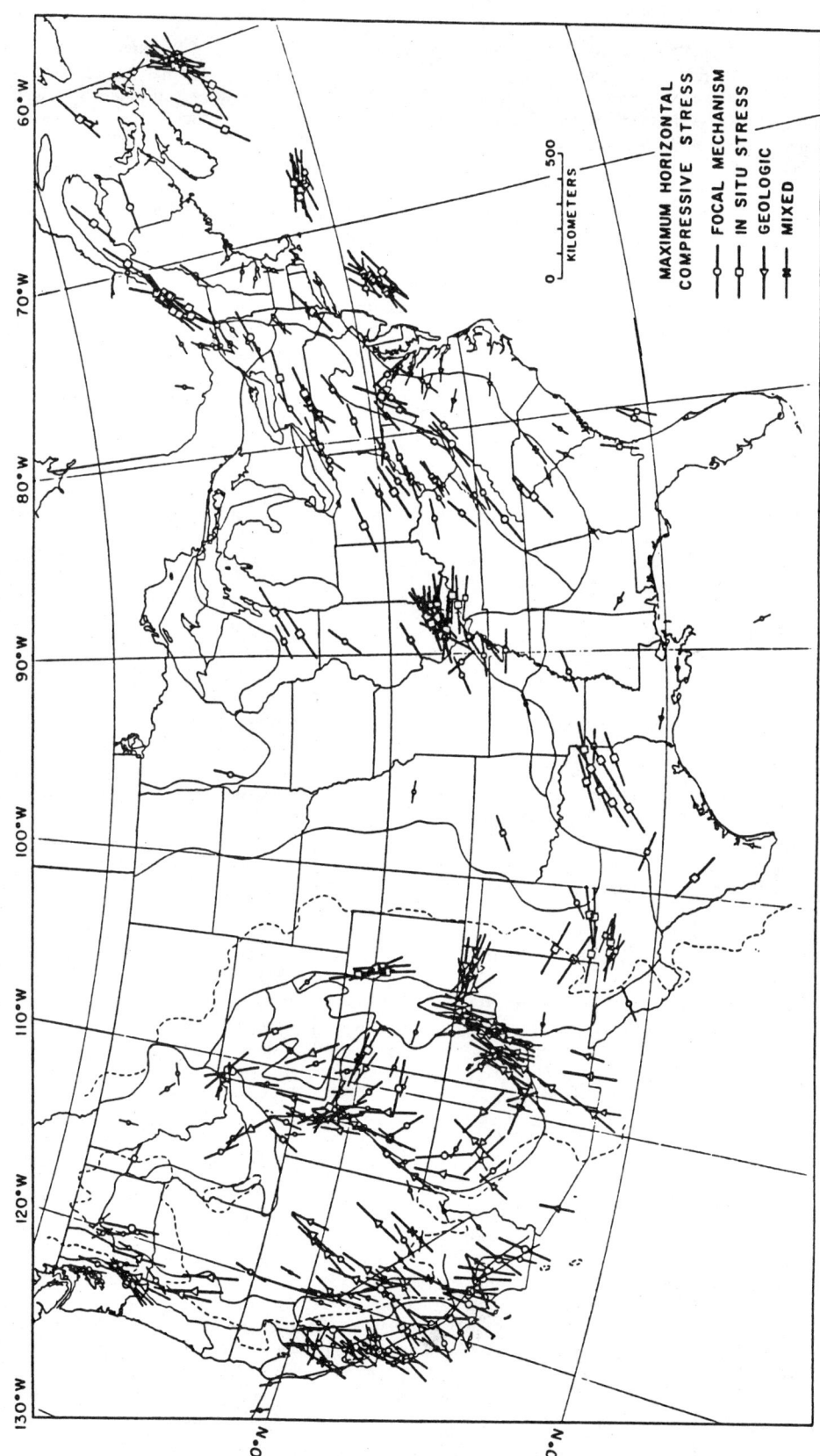

FIGURE 4. Orientation of maximum horizontal stress in the conterminous United States as determined from earthquake focal mechanisms, in situ stress measurements, and wellbore breakouts and geologic indicators. The length of the small bars indicate whether the data are of A, B, or C quality as scale derived to reflect both the reliability of the data and the degree to which it is thought to represent the tectonic stress field.

The relative uniformity of stress orientation and relative magnitudes is striking and permits mapping of regional stress fields. Figure 5 presents a generalized version of the U.S. stress map, which delineates provinces of relatively constant stress orientation and relative magnitude. Most of the central and eastern United States is part of an extremely broad "mid-plate" province characterized by NE to ENE horizontal compression. This province encompasses most of central and eastern Canada and probably extends close to the mid-Atlantic ridge. This NE to ENE orientation is approximately the direction of "ridge push" from the mid-Atlantic ridge and is also the direction of absolute plate motion suggesting that drag at the base of the lithosphere may be responsible for the observed stress field. These, and other, sources of tectonic stress are discussed later. Unfortunately, there is no way currently known to distinguish between these mechanisms. In the Gulf Coastal province the entire upper crust is dominated by the gulfward extension associated with growth faulting caused by heavy sedimentation. The state of stress in the mid- and lower-crust are unknown. In the western Cordillera, a number of distinct stress provinces are observed that correlate suprisingly well with many of the physiographic provinces. Distinct stress provinces are seen in (1) the Basin and Range and Rio Grande Rift provinces (combined as Cordillera Extension province), which is characterized by ~E-W extension, (2) the Colorado Plateau and Southern Great Plains provinces, both of which are characterized by NNE-SSW extension, (3) the Pacific Northwest province where N-S compression of unknown origin is observed, (4) the Cascadia Convergence province, which shows NE-SW compression due to subduction of the Juan de Fuca plate beneath North America, and (5) the San Andreas province characterized by NE-SW to N-S compression that may strongly be influenced by the San Andreas fault. Whereas the boundaries between distinct stress provinces are generally poorly defined, in some places, such as the boundary between the Rio Grande Rift and Southern Great Plains province, the change in the stress orientation is relatively rapid, apparently demonstrating the importance of shallow sources of stress in the crust in areas of active extension and volcanism.

Sources of Crustal Stress

Stresses in the lithosphere have both tectonic and nontectonic, or local, origin. The regional uniformity of the stress fields observed in Figs. 4 and 5 argue for tectonic origins. For many years, numerous workers suggested that residual stresses from past tectonic events may play an important role in defining the tectonic stress field. We have found no evidence for residual stresses at depth. If such stresses exist, they are only important in the upper few meters or tens of meters of the crust. Similarly, no evidence has been found that indicates that horizontal principal stresses result only from the weight of the overlying rock. This fallacious argument is based on the supposition that as a unit cube cannot expand horizontally due to imposition of a vertical stress (because a neighboring unit cube would be attempting to expand in the opposite direction) horizontal stresses at depth will be approximately $1/3$ of the vertical stress. Not only has such a stress state never been found in the Earth, the assumptions leading to the result are unjustified. The analysis basically assumes that the crust exists in the absence of gravity (or any other forces) before gravity is instantaneously "switched on." It is also physically unreasonable to have such a large difference between principal stresses. Intact rock will break when principal stress ratios approach 3 and recall that normal faulting will occur on pre-existing fault planes when the least horizontal stress is only about 60% of the value of the vertical stress. Thus, assuming that horizontal stresses are only about $1/3$ of the vertical stress is mechanically unrealistic. It is regrettable that this erroneous and misleading concept has made its way into the geological literature over the past two decades.

In this section only the primary sources of tectonic stress are discussed. Although it is possible to derive theoretically the significance of any particular source of stress, demonstrating its unique role in generating the stress field observed in any area is a difficult problem that will eventually be solved only by utilizing careful modeling with well-constrained observations.

Plate Driving Stresses. The sources of stress within the crust that come immediately to mind are those related to movement of the plates as well as the resistance to plate motion. Figure 6 illustrates these basic stresses as summarized by D. Forsyth and S. Uyeda. First, there are the ridge push force (F_{RP}) associated with the excess elevation of the mid-ocean ridges and the slab pull force (F_{SP}) resulting from the negative buoyancy of down-going slabs. It is clear that both of these sources contribute to plate motion and tend to act in the direction of plate motion. If there is flow in the upper asthenosphere, a positive drag force (F_{DF}) could be exerted on the lithosphere that would tend to drive plate motion, whereas if a plate is moving over a stable asthenosphere (such as may be the case beneath North America) the drag force (F_{CD}) would act to inhibit plate motion. In either case the drag force would result in stresses being transferred up into the lithosphere from its base. There are collision resistance forces (F_{CR}) resulting from the frictional resistance of a plate to subduction with the overlying plate or the collision between continental plates. As plates subduct into the viscous lower mantle additional slab resistive forces (F_{SR}) add to the collision resistance forces acting at shallow depth. Another force

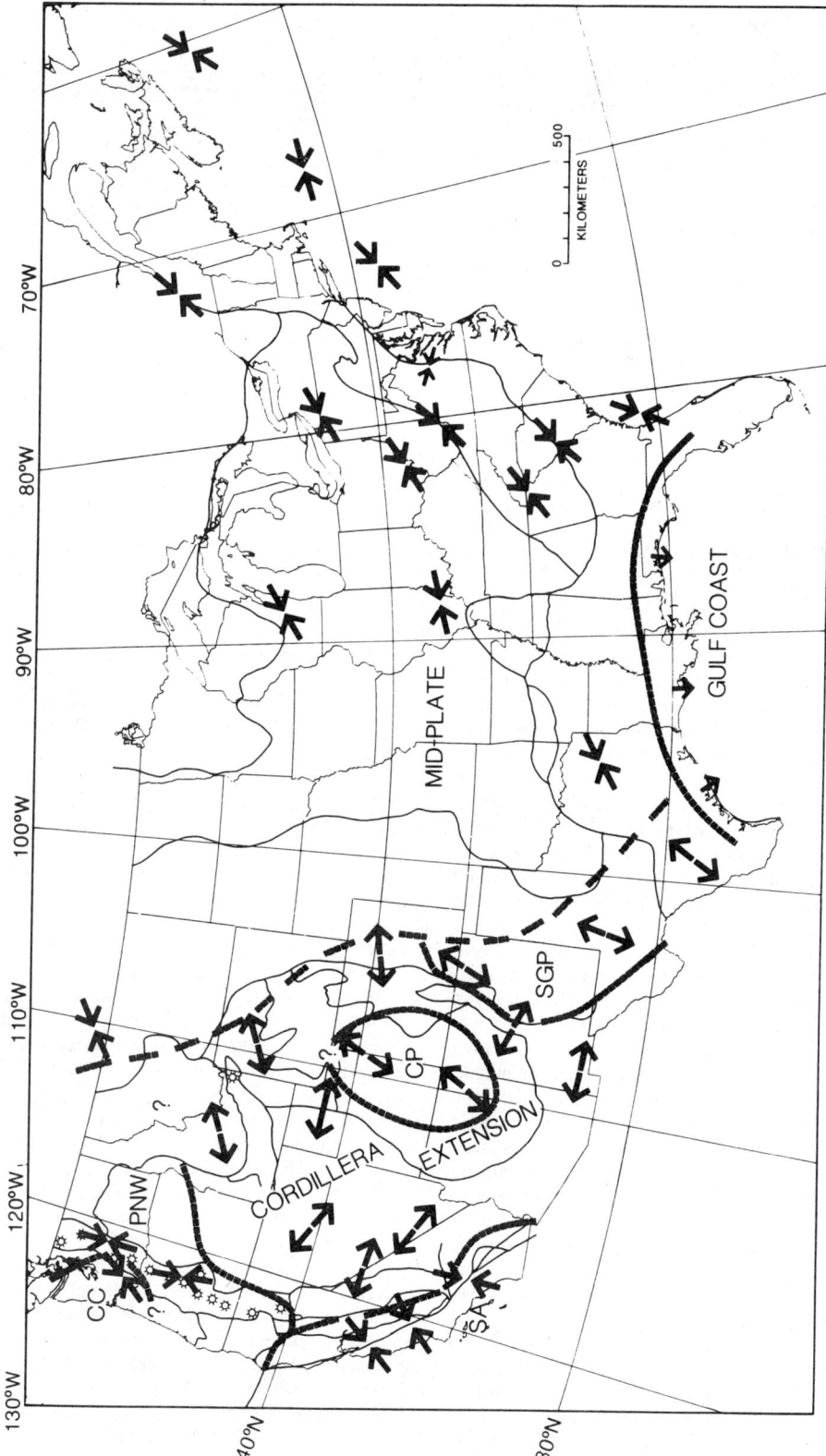

FIGURE 5. Interpretative stress map of the conterminous United States showing provinces of relatively uniform stress magnitude and orientation. Inward pointed arrows indicate a compressive stress regime (reverse and strike-slip faulting) and outward pointing arrow indicate an extensional (normal and strike-slip faulting).

1229

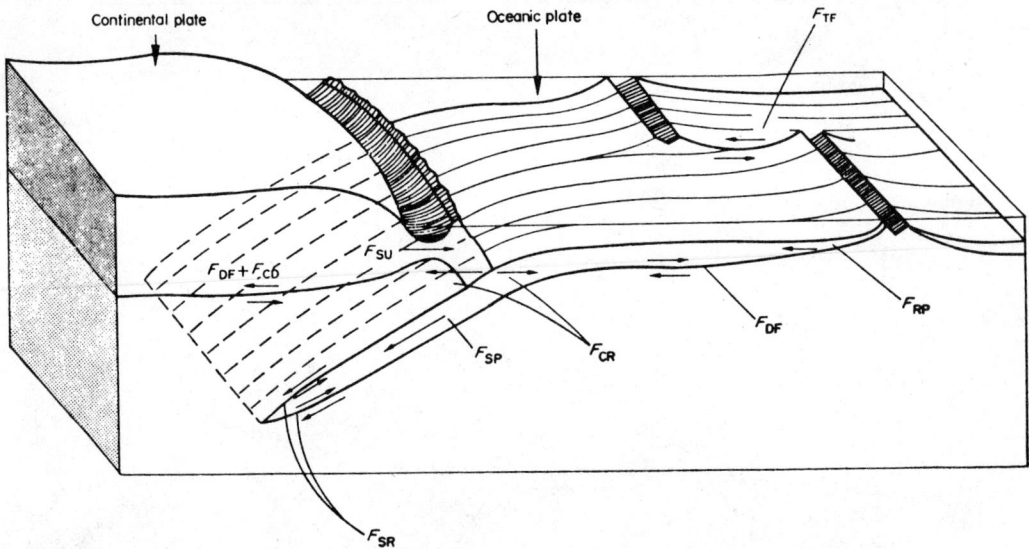

FIGURE 6. Schematic view of the various forces acting on an oceanic and continental plate. (From Forsyth and Uyeda, 1975)

resisting plate motion is that due to transform faults (F_{TF}), although the amount of transform resistance may be negligible. Finally, it has been proposed that a suction force (F_{SU}) acts on continental lithosphere near a subduction zone that tends to suck the continent toward the trench and result in back-arc spreading.

While it is possible to specify these various stresses associated with plate movement, both their relative and absolute importance in plate movement is not understood. While many people believe that either the ridge push or slab pull force is most important in causing plate motion, the Philippine Sea plate is moving without a ridge at its boundary and the North American plate is moving at a rate of about 2 cm/yr without being pulled down a trench. Thus, it is not clear that these forces are easily separable or that plate motion can be ascribed to a single dominating force. For example, general estimates of the magnitude of the ridge push force are about 10–20 MPa while the force due to slab pull has been estimated to be over 100 times that magnitude. Does this mean that slab pull is the most important force in causing plate motion? Even this question is not so easy to answer. First of all, the tensional strength of crustal materials is essentially zero and it is not possible to transmit large tensional stresses resulting from gravitational sinking of the slab that would tend to pull the plate down. This fact would suggest that there is a rough balance between the tensional slab pull force and the compressive forces resulting from the F_{CR} and F_{SR} resisting subduction. In other words, if resistance to plate motion balances the majority of the trench pull force, it is not clear whether the residual slab pull force is more or less significant than the ridge push force in causing plate motion. Furthermore, the role of asthenospheric motion in generating driving (or resisting) basal tractions is completely unknown, yet because of the enormous areas over which these forces act, the resultant stresses could potentially dominate all other sources of lithospheric stress. Thus, the fundamentally important question of what drives the plates remains a crucial question being addressed by considerable research. Modeling of the stress fields acting within the plates is an extremely important component of such research as it is possible to consider the interaction of the various source of stress as well as complex geometries and rheological properties. It is critically important to constrain such models, however, with the types of data presented in Fig. 4.

Thermal Stresses. As the lithosphere created at ridges cools, the induced tensional stresses cause a volumetric contraction of material. Such stresses would seem to be relatively unimportant as they are isotropic, that is, equal in all directions and shear stresses capable of causing crustal deformation are not induced. However, because the cooling of the lithosphere occurs more rapidly near the surface than at appreciable depths, when such cooling does occur at depth, the slow contraction of the lithosphere causes large horizontal compressional stresses to be induced at shallow depth. Although this stress mechanism is not renewable and cannot result in appreciable cumulative deformation, a number of significant mid-plate earthquakes have been apparently caused by this mechanism.

Lithospheric Flexure. Stresses are imparted by flexure of the lithosphere due to bending that occurs as a result of localized loads such as the Hawaiian/

Emperor seamount chain or resistance to subduction. The analyses of these processes have shown good correlation between topographic profiles and predictions of models of the lithosphere based on elastic analysis of beams. As in bending a beam, these models predict tensional stresses in the upper lithosphere and compressional stresses in the lower lithosphere. As mentioned earlier, the lithosphere cannot support tensional stresses and a variety of hypotheses have been put forth to overcome this aspect of the predicted stress field from flexural models. For example, some workers suggest that large horizontal compressional stresses exist that offset flexure-induced tension at shallow depth. To overcome the very large stresses predicted by such models in areas of greatest curvature, other workers have suggested that viscous behavior of the lower lithosphere attenuates stress magnitudes. Thus, understanding lithospheric rheology is crucial for understanding the exact manner in which flexure induces lithospheric stresses.

Topography. Perhaps the most obvious source of stress in the crust is that associated with topography. Using the formula previously presented for computing the magnitude of the vertical principal stress, it is straightforward to see that significant topography such as that associated with appreciable mountain chains induces significant vertical stress. For example, if we use an average elevation of 5 km for the high Himalayas, the vertical effective stress at sea level (assuming hydrostatic pore pressure at depth) is on the order of 100 MPa. Because of the limited strength of rock, appreciable horizontal stresses are required to support such a load. In the case of the Himalayas the horizontal stresses are those associated with collision of the Indian and Eurasian plates. In general, large stresses in the lithosphere are required to support such features as mountain belts and major volcanic piles such as the island of Hawaii. But while large crustal stresses clearly exist in areas of appreciable topography, topography alone cannot cause crustal deformation, because an equilibrium condition must exist, as in the example of the Himalayas between the topographic load and the horizontal stresses that support it. In other words one can think of topography as being associated with horizontal stresses but not causing them. In the case of the stresses induced by volcanic piles the case is somewhat more complicated. First, the rapidly induced vertical load can cause appreciable downslope movement, such as occurred with the 1978 Kanapala earthquake. The second result of a localized volcanic load is to cause lithospheric flexure as already discussed.

Topography also results in crustal deformation through the removal of topographic loads such as ice sheets. If an ice sheet has persisted for an extended period, the horizontal stresses are in equilibrium with the vertical load. Rapid unloading of the sheet results in horizontal stresses that are too large for the vertical stress thus enabling thrust faulting to occur. This mechanism is thought to be responsible for several large earthquakes in the Baffin Island area of northeastern Canada.

Summary

The state of stress in the brittle upper lithosphere is dominated by tectonic processes and limited by the strength of rock. Relatively uniform tectonic stresses appear to exist, and are possibly transmitted, over distances as large as thousands of kilometers. Yet the transitions between stress provinces can sometimes be quite rapid providing clear evidence of the importance of relatively shallow sources of stress in areas of active crustal extension volcanism. In general, principal tectonic stresses in the lithosphere are defined in vertical and horizontal planes. In intraplate areas, in situ stress measurements and inferences based on topography and flexure all suggest that shear stresses in the upper lithosphere are fairly large and seem controlled by the frictional strength of rock. Whether this principle also applies to major plate boundaries like the San Andreas fault is unknown, and various types of evidence suggest that, unlike the interiors of plates, plate boundaries might be quite weak. At sufficient depth below the brittle upper lithosphere, stresses are probably controlled by the ductile flow properties of the constituent rocks and minerals. The depth at which the transition from brittle to ductile behavior occurs is quite variable as temperature, fluid content, rock composition, and strain rate all have major influences on the deformational style and magnitudes of the stresses.

Inferring the state of stress in the lithosphere from modeling is crucially important. Yet modeling is critically dependent on poorly resolved issues involving lithospheric rheology and only weakly constrained by limited amounts of available data. Our ability to use data on the state of stress in the lithosphere to constrain the physics of lithospheric deformation and evolution is truly in its infancy.

MARK D. ZOBACK
MARY LOU ZOBACK

Bibliography

Anderson, E. M., 1951, *The Dynamics of Faulting and Dyke Formation with Applications to Britain,* 2nd ed. Edinburgh: Oliver and Boyd.

Brace, W. F., and D. L. Kohlstedt, 1980, Limits on lithospheric stress imposed by laboratory experiments, *Jour. Geophys. Research* **85,** 6248-6252.

Forsythe, D., and S. Uyeda, 1975, On the relative importance of the driving forces of plate motion, *Royal Astron. Soc. Geophys. Jour.* **43,** 163-199.

Jaeger, J. C., and N. G. W. Cook, 1969, *Fundamentals of Rock Mechanics,* 2nd ed. London: Methuen and Co.

Hanks, T. C., and C. B. Raleigh, 1980, The conference on magnitude of deviatoric stresses in the Earth's crust and uppermost mantle, *Jour. Geophys. Research* **85,** 6083-6085.

Kanamori, H., 1980, The state of stress in the Earth's lithosphere, *Physics of the Earth's Interior*, vol. 78. Bologna: Soc. Italiana di Fisica, 531–554.

Kirby, S. H., 1980, Tectonic stresses in the lithosphere: Constraints provided by the experimental deformation of rocks, *Jour. Geophys. Research* **85**, 6353–6363.

Meissner, R., 1986, *The Continental Crust*. Orlando, Fla.: Academic Press.

Richardson, R. M., S. C. Solomon, and N. H. Sleep, 1979, Tectonic stress in the plates, *Rev. Geophysics Space Physics* **17**, 981–1019.

Sibson, R. H., 1983, Continental fault structure and the shallow earthquake source, *Geol. Soc. London Jour.* **140**, 741–767.

Zoback, M. D., and J. H. Healy, 1984, Friction, faulting, and in-situ stress, *Annales Geophysicae* **2**, 689–698.

Zoback, M. L., and M. D. Zoback, 1987, Tectonic stress field of the continental U.S., in L. Pakiser and W. Mooney, eds., Geophysical Framework of the Continental United States, *Geol. Soc. America Mem.*, in press.

Cross-references: *Continental Collision Zones: Seismotectonics and Crustal Structural; Continental Drilling: Ultradeep; Continental Lithosphere; Continental Rifting; Crustal Movements and Tectonic Deformation; Deformation of Rocks and Minerals; Earthquake Mechanisms; Earthquake Mechanisms and Plate Tectonics; Earthquakes and Crustal Deformation; Lithosphere: Mechanical Properties; Mantle Viscosity; Seismicity: Intraplate; Seismicity and Plate Tectonics; Subduction Zones; Thin-Skin Tectonics.*

STRONG MOTION SEISMOLOGY

Strong motion seismology is the division of seismology concerned with strong ground shaking of earthquakes close to the epicenter, primarily with amplitudes capable of causing damage. Its scope includes the observation and empirical description of strong shaking, the study of the physical processes that control the characteristics of strong shaking, and the simulation of ground motion. There are two goals. The first in common with other branches of earthquake seismology is the scientific goal of understanding the phenomena. The second goal is to prevent death, injury, and economic loss from earthquakes by providing an accurate description of seismic hazards.

The strongest earthquake motions that have been recorded to date have peak accelerations between 1 g and 3 g, where 1 g ($\approx$ 980 cm $\cdot$ s^{-2}) is the acceleration of the Earth's gravity field. It is less clear what threshold of ground motion needs to be exceeded to be considered "strong motion." Many of the instruments currently in use are not able to resolve ground accelerations with amplitudes less than 10 cm $\cdot$ s^{-2}. People at rest are able to feel somewhat smaller motions, although lower levels do not cause damage except for very exceptional cases. However, strong motion seismologists do not confine their interests to levels above 10 cm $\cdot$ s^{-2} because seismic wave propagation through the earth is independent of amplitude, at least as a first approximation. Since weaker ground motions are much more common than motion above 10 cm $\cdot$ s^{-2}, an effective way to study the effects of the Earth on the ground motion is to gather and interpret data at these lower levels. In addition, one way to look at large earthquakes is as the sum of a large number of smaller ones. Thus ground motions with any amplitude can be employed to help understand strong ground motion.

Figure 1 suggests the size range of earthquakes that are the primary interest for strong motion seismology. It shows several statistics about killer earthquakes (i.e., earthquakes that kill one or more persons) during two decades, 1960–1979. The number of people killed is a general indicator of the extent of damage because the typical cause of death in an earthquake is from failure of a structure. Estimates of economic loss or property damage are much more ambiguous.

The figure shows that earthquakes with magnitude less than 5 are of minor concern; they are not known to damage structures of modern construction. Only a tiny fraction of such events has caused deaths, and the number of deaths associated with any one of these events is small. As the magnitude grows, both the destructive capability and average number of deaths per event also grows. Events with magnitudes between 6 and 7.5 are the most commonly responsible for deaths and damage, because events with magnitude over 8, with their immense destructive potential, are fortunately relatively uncommon.

Observation of strong motion is more difficult than the observations in other fields of seismology. An earthquake with magnitude greater than 6 can be recorded worldwide. Thus a seismologist who studies teleseisms can record on the order of 100 earthquakes per year, for interpretations of earth structure or tectonics. Similarly, local networks are generally set to the most sensitive level possible to detect and locate the smallest earthquakes, which are much more abundant than strong shaking. The sensitive instruments used for these two branches of earthquake studies are driven off scale by strong shaking, and thus their records cannot be used. A specialized instrument, the strong motion accelerograph, has been developed to record strong motion. A network of these specialized instruments must furthermore have the good fortune to be located close to the earthquake, and must be maintained, often for decades, in a state of readiness to record the rare strong shaking.

The physical processes that control the ground motion are generally separated into three parts: the excitation of seismic waves at the earthquake source; the effects of the Earth on the propagation of these waves including geometrical spreading, dispersion, scattering and attenuation; and the local site effects where scattering, topography, surficial geology, and

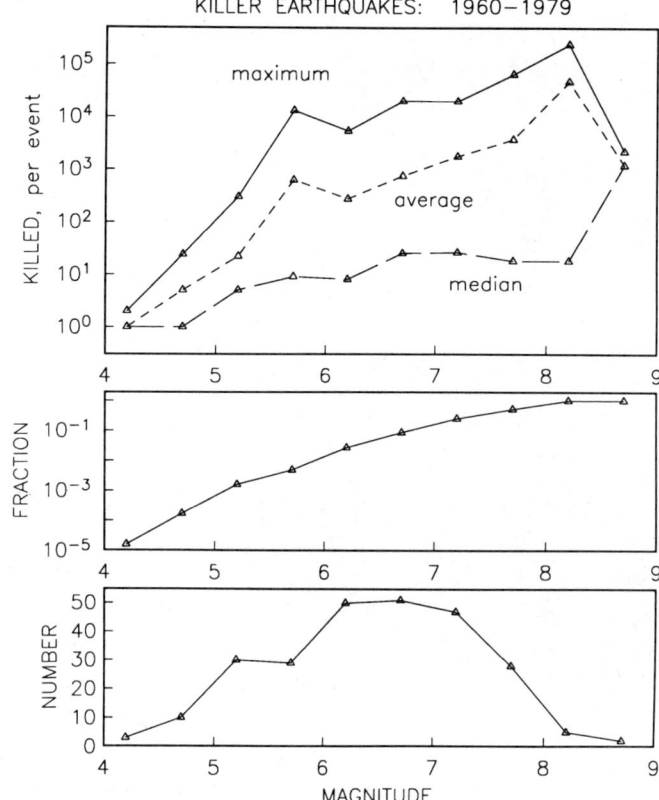

FIGURE 1. (*Top*) The mean, median, and maximum number of people killed per killer earthquake (an earthquake that kills one or more persons) as a function of earthquake magnitude. Statistics apply to the years 1960–1979, and are based on the catalog by Ganse and Nelson (1981). Events have been grouped into magnitude intervals of width 0.5 magnitude unit (i.e. 4.0–4.4; 4.5–4.9, . . ., 8.0–8.4, 8.5–8.9). (*Center*) Estimated fraction of the worldwide total number of earthquakes in each of the above magnitude groups that become killer earthquakes. (*Bottom*) Actual number of killer earthquakes 1960–1979, in each of the magnitude groups.

manmade structures further modify the incoming waves.

History

The founding father of the strong motion instrumentation program in the United States is John R. Freeman. Although he is better known for his contributions to hydraulics and the science of fire insurance, after the Tokyo, Japan earthquake of 1923, and the Santa Barbara, California and Montreal, Quebec earthquakes of 1925, he stimulated important early interactions between U.S. and Japanese institutions on earthquake engineering and wrote the first significant book in the English language on earthquake engineering, *Earthquake Damage and Earthquake Insurance*. He particularly recognized the urgent need for an instrument to record the strong shaking during earthquakes, and the result of his lobbying efforts was that the Coast and Geodetic Survey was authorized to develop and install such instruments in 1932. Nine months after the first instruments were installed, the first significant strong motion records were obtained from the 10 March 1933 Long Beach, California earthquake.

By the end of 1935, 51 accelerographs had been installed by the Coast and Geodetic Survey, and this number increased gradually to 71 stations in 1964. An explosive growth in the number of instruments was made possible by the development of commercially produced accelerographs recording on photographic film (the AR-240 in 1963, the RFT 250 in 1966, and the SMA-1 in 1969), the enactment of building ordinances in 1965 in Los Angeles and Beverly Hills that required accelerographs in all new buildings with heights of seven or more stories, and the increased interest following the March 1964, Alaska earthquake. The February 1971 San Fernando, California earthquake resulted in a major increase in the number of recordings (a total of 241 records were obtained in that earthquake). In the mid-1970s the first digital recording accelerographs were produced. In 1981, a total of over 2700 accelerographs were in operation in the United States. About 7000 had been manufactured by that time with the remainder in operation outside the United States, no longer in service, or operated by unidentified groups (Iwan, 1981).

Instrumentation

Strong motion seismology requires a specialized instrument called the strong motion accelerograph. The instrument is a self-contained unit in as compact a container as possible (Fig. 2). The sensor (accelerometer) is typically a damped spring-mass system.

FIGURE 2. Photograph of a Kinemetrics SMA-1 strong motion accelerograph, the most widely distributed seismic instrument in the world, with over 6600 produced through 1988. (Photo compliments of Kinemetrics, Inc.)

The spring is stiff, giving the system a high natural frequency (typically 25–50 Hz), with the parameters selected so that accelerations of 1–2 g (typically) will cause a deflection corresponding to full scale of the recording system. It is called an accelerometer because when the ground vibrates with frequencies less than the natural frequency of the sensor (and these frequencies usually predominate), the deflection of the sensor is proportional to the acceleration. Strong motion accelerographs invariably contain three accelerometers to measure the vertical and two perpendicular horizontal components of acceleration.

In accelerographs designed in the United States through the late 1970s, the recording medium was usually film. The enclosure was thus made into a light-proof box. A Japanese design used a pen to mechanically scratch a waxed paper. Newer designs convert an electrical signal to a digital format, which is recorded on a digital grade cassette or, beginning about 1984, stored in a digital memory within the unit. However, most of the accelerographs currently in operation still use the film recording, and film recorders are still a very common choice as a new instrument because of their initial lower cost and simplicity to operate.

To record frequencies of ground motion above 20 Hz, the film must move through the camera at a fairly high rate, 1 cm · s^{-1} being typical. The digital recorders sample at 100 to 200 samples per second. Thus to conserve recording medium, the accelerographs are all triggered, i.e., when an instrument detects strong motion above some threshold, it activates the recorder and turns the recorder off again when the strong shaking has stopped. Trigger mechanisms range from purely mechanical sensors or electromechanical sensors installed separately from the accelerometers to recent digital systems that make decisions based on signal amplitude coming from the sensors themselves. Usually, internal batteries operate the system, so that power failure will not cause a loss of data, but the batteries are kept fully charged by a trickle charging system to maintain readiness. Some units also contain timing systems, which have become more important as the analysis has become more sophisticated.

Archiving and Processing of Accelerograms

Significant accelerograms recorded on film formats are eventually digitized and distributed. A great deal of effort has been devoted to this process because of the importance for earthquake resistant design of accurate information. Although the instruments record acceleration, estimates of the ground velocity and displacement are also important for both engineering and geophysical research. Thus numerical integration of the records, and the accompanying issue of exactly where the true zero level of acceleration is located on the film traces, is of the utmost importance. From these film recordings, ground motions can generally be recovered for frequencies greater than about 0.2 Hz; this lower limit depends on the size of the earthquake and the type of instrument. At high frequencies, analog recordings are limited by the natural frequency of the accelerometer. On older styles of accelerographs, data is reliable to 10 or 15 Hz, while the more recent designs are reliable to about 25 Hz. Digital accelerographs are reliable for frequencies between about 0.05 Hz and 50 Hz.

Empirical Descriptions of Strong Ground Motion

Figure 3 shows as an example one of the most important strong motion accelerograms, obtained in the February 1971 San Fernando, California, earthquake. Figure 4 shows the Fourier amplitude spectrum for one of the components of this accelerogram. Figure 5 shows an idealized spectrum that has been annotated with the predominant physical influences in various frequency bands. Table 1 lists several of the most significant accelerograms.

Empirical descriptions of strong motion invariably incorporate a dependence on the earthquake size and distance to the active faulting. Although the seismic moment (see below) is a preferable measure of the overall size of an earthquake, the magnitude (see *Earthquakes: Magnitude, Energy, Intensity*) plays a major role in the empirical estimation of ground motion. Historically, magnitude was available, before the seismic moment, for correlation with ground motion characteristics. In addition, the magnitude is known for a much larger fraction of the historical earthquakes, which provide the bases for modern estimates of seismic hazard. Finally, seismic moment is determined from the asymptotic low-frequency limit of the spectrum while magnitude is found from essentially the same frequency band for all earthquakes. For larger earthquakes, because this asymptote is only reached for very low-frequency waves (e.g., less than 0.1 Hz), some scientists consider that a magnitude derived from a short period instrument, such as M_L or m_b (see *Earthquakes: Magnitude, Energy, Intensity*) provides a better basis to correlate with ground motion parameters of most importance for earthquake engineering since most structures are also more sensitive to these short periods than to very long periods.

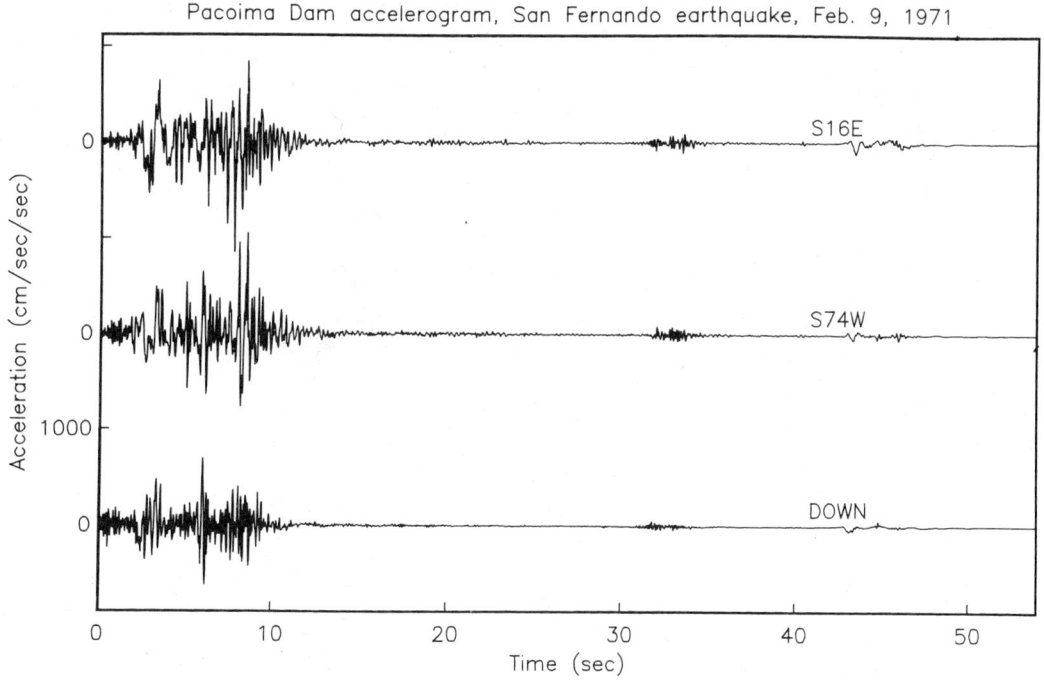

FIGURE 3. Strong-motion accelerogram recorded at Pacoima Dam during the San Fernando, California earthquake of 9 February 1971 (magnitude 6.5).

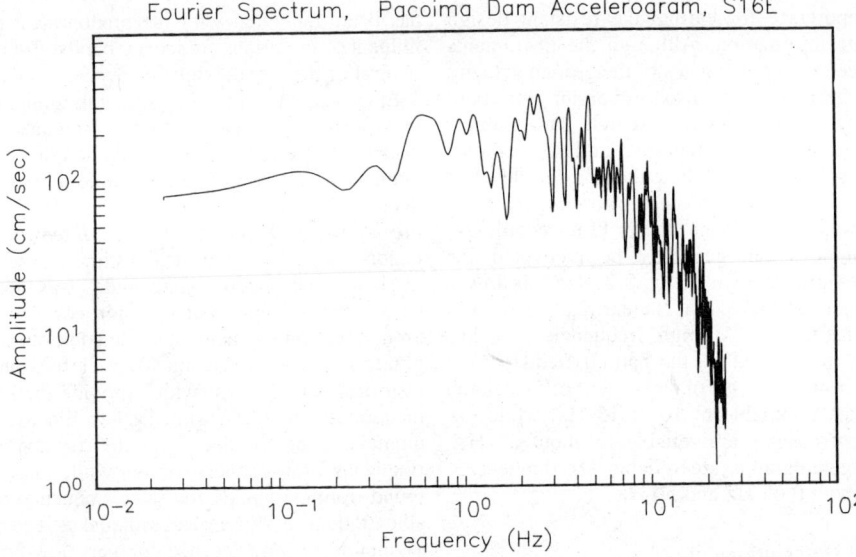

FIGURE 4. Fourier amplitude spectrum of acceleration corresponding to the S16°E component of Fig. 3.

Magnitude Dependence of the Character of Earthquake Motions. The character of earthquake motions changes as the size of the earthquake increases. Figure 6 shows an example of how accelerograms from several earthquakes, of different magnitudes, appear when plotted on a common scale. As the magnitude of the earthquake increases, the amplitudes of ground motion generally increase, and the durations increase dramatically. Figure 7 shows Fourier amplitude spectra corresponding to each of the accelerograms in Fig. 6. This figure shows that as the magnitude increases, the amplitudes of the low frequency waves increase dramatically, but the amplitudes of the high frequencies increase only slowly. These trends are well understood to follow from the increasing size of the active part of the fault, correlated with increased magnitude.

Parameters to Describe Strong Ground Motion. Because of the importance of strong motion recordings for earthquake engineering, a number of different parameters have come into use to represent various characteristics of strong motion recordings. The simplest of these is peak acceleration, which is easily obtained from analog records of the ground motions. After records are digitized, it is usual to obtain several additional parameters. Time domain parameters often include peak velocity, peak displacement, root-mean-square (rms) acceleration, and duration (for which there are several definitions).

In the frequency domain, the Fourier amplitude spectrum is generally determined. It is also very common to see members of a class of spectra known as response spectra, which are often more useful for engineering designs. Four types of response spectra are in common use: relative displacement (S_d), relative velocity (S_v), absolute acceleration (S_a), and pseudo-relative velocity (PSV). These response spectra are subsequently used as analogs for the response of a structure with the natural period of the oscillator on the abscissa.

All of these response spectra describe a peak response of a one-degree of freedom oscillator to the seismic excitation. Consider a damped oscillator, which operates on the same principle as an inertial seismograph (see *Seismic Instrumentation*). Let the undamped natural period of this oscillator be T_0 and let the fraction of critical damping be ζ. When the base of this oscillator is subjected to an acceleration, there is a relative displacement between the seismic mass and the base; the maximum value of this relative displacement is the value of the relative displacement spectrum (S_d) at the period and damping of the oscillator. Thus to calculate a relative displacement spectrum, it is necessary to calculate the response of a suite of oscillators to the accelerogram; typically this is done for about 100 different natural periods between 10 and 1/25 s. In a similar manner, the relative velocity response (S_v) has the value of the peak relative velocity between the seismic mass and the base. The absolute acceleration response (S_a) is the maximum acceleration of the seismic mass, which may, of course, differ from the maximum acceleration of the ground except when the natural period of the oscillator becomes very short. The maximum force exerted on the seismic mass is proportional to S_a. The pseudo-relative velocity, PSV, is obtained from S_d by $\text{PSV} = (2\pi/T_0)S_d$. PSV is proportional to the maximum kinetic energy associated with the oscillation of the seismic mass. The spectra are usually computed for a range

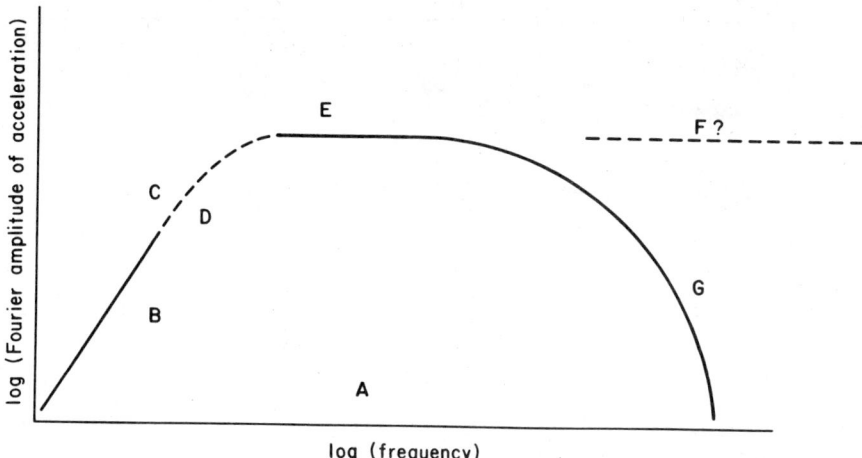

FIGURE 5. Qualitative illustration of phenomena controlling the seismic spectrum. A. Overall level is controlled by distance, R, from the seismic source. Geometrical spreading causes amplitudes to decrease at a rate between $R^{-1/2}$ and R^{-1}. B. Level below the corner frequency is controlled by seismic moment. C. Corner frequency is controlled by fault dimension. D. Level and shape at frequencies immediately above the corner frequency are controlled by sizes of largest asperities. E. Level of flat portion controlled by the level of stress and number of inhomogeneities on the fault. F. At the source, the spectrum might be nearly constant. G. Rolloff controlled by attenuation along the path, including significant attenuation in the last 5 km to 100 m of the path.

of damping, from 0% (undamped) to 20% of critical. The low damping is used because most manmade structures are similarly lightly damped.

To illustrate the range of values and spectral shapes, Fig. 8 gives PSV for a 5% damped oscillator for each of the accelerograms shown in Fig. 6. The tripartite axes used here is the usual presentation of PSV, allowing S_d and pseudo acceleration to be read also from the graph as a convenience for engineering applications.

Attenuation Relations. Based on past recordings of strong earthquake motion, consisting of data such as those in Figs. 6 to 8 the peak acceleration and other peak parameters can be described by an "attenuation relation" as a function of the earthquake magnitude, distance from the fault to the site, and sometimes other parameters. Figure 9 shows an example of one such relationship, developed by Joyner and Boore (1981) for California. The equation for this relationship is:

$$\log A = 1.97 + 0.249 M_w - \log r - 0.00255 r$$

where

$$r = (d^2 + 7.3^2)^{1/2} \quad 5.0 \leq M_w \leq 7.7$$

In this equation, the peak acceleration A is given in units of $cm \cdot s^{-2}$, M_w is the moment magnitude, d, in km is the horizontal distance to the map-view projection of the fault rupture.

TABLE 1. Significant Accelerograms (partial, selected list)

					Peak Accel. ($cm \cdot s^{-2}$)	
Date	Earthquake	M	Station	Distance (km)	horiz.	vert.
3/10/33	Long Beach	6.5	Long Beach, Calif.	27	193	280
5/1940	Imperial Valley	6.7	El Centro, Calif.	9	342	206
7/1952	Kern County	7.7	Taft, Calif.	43	176	103
6/1966	Parkfield	5.6	Parkfield Array #2	1	480	202
2/9/71	San Fernando	6.4	Pacoima Dam, Calif.	9	1148	696
5/17/76	Gazli, U.S.S.R.	7.0	Karakyr Point	20	738	1300
9/1978	Tabas, Iran	7.4	Tabas	3	911	732
10/15/79	Imperial Valley	6.3	Bonds Corner	6	974	460
			IV Array #6	1	1705	706
6/9/80	Mexicali Valley	6.2	Victoria	3	960	1078
9/19/85	Michoacan, Mexico	8.1	Caleta de Campos	20	141	89
			SCT, Mexico City	390	168	38

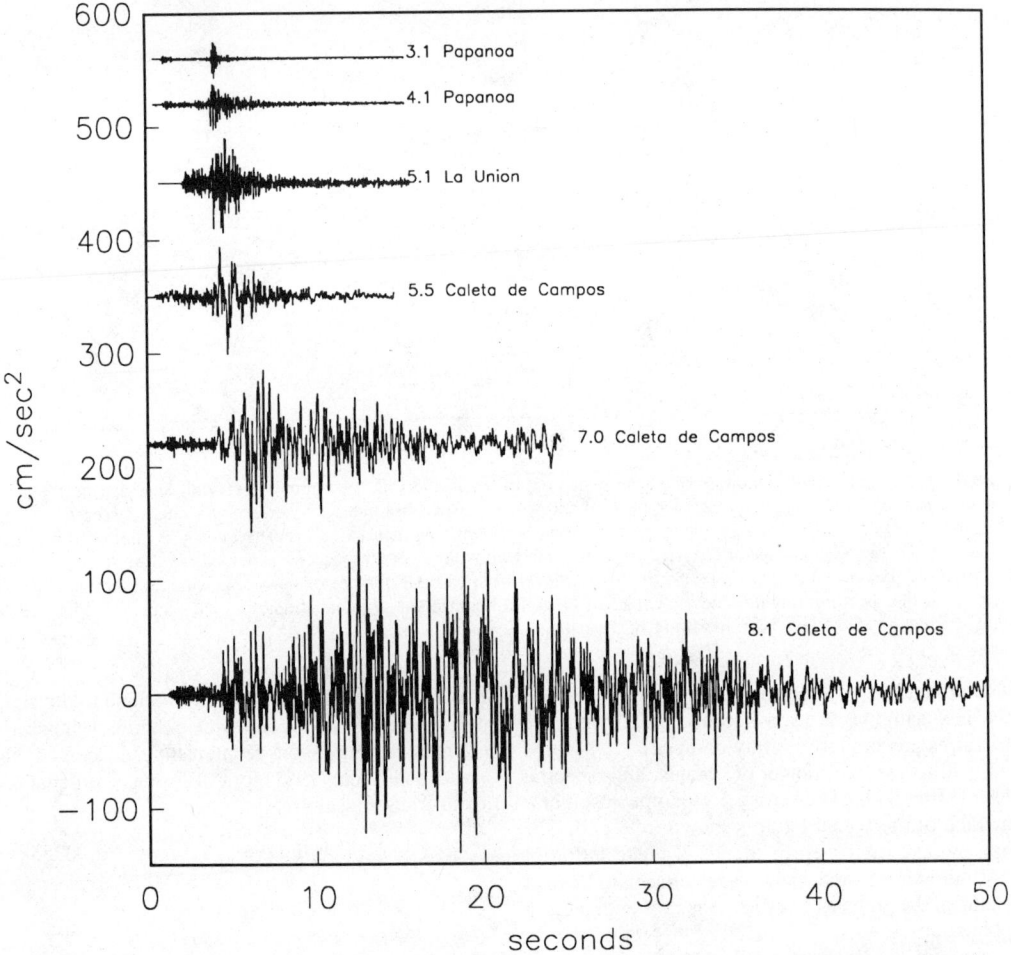

FIGURE 6. An example of accelerograms recorded in 1985 and 1986 on the Guerrero accelerograph array. All events have epicenters about 25 km from the station, and all stations are on hard rock. (From Anderson and Quaas, 1988)

The moment magnitude, M_w, is essentially a change of variable from seismic moment:

$$M_w = \tfrac{2}{3} (\log M_0 - 16.0)$$

where the units of M_0 are dyne-cm.

Naturally the attenuation relationships do not exactly describe peaks of past earthquake ground motions. The deviation of a datum from the predicted mean peak acceleration (the "residual") is treated as a random variable, and it is consistent with a lognormal distribution function out to two standard deviations at least. The standard deviation for the attenuation relation given above is 0.26, corresponding to a multiplicative factor of 1.8 (times the mean value) to obtain the value, which exceeds 84% of the data.

Peak acceleration is the most commonly employed parameter for these regressions, but peak velocity, peak displacement, spectral amplitudes, and duration have also been modeled in this manner.

Earthquake Source Effects on Strong Ground Motion

Earthquakes are caused when the rocks on opposite sides of a fault slip suddenly. The character of strong ground motions that result from this instability is strongly affected by the geometrical and dynamic characteristics of the active faulting. The geometrical characteristics include the size, shape, depth, and orientation of the fault area that slipped during the earthquake and the amount and direction of slip (see *Seismic Source: Theory* and *Seismic Source: Observation*). The orientation of the fault and the amount and direction of offset may be a variable over the fault surface. The important dynamic parameters include where on the fault the rupture initiated (hypocenter), how rapidly it spread over the fault

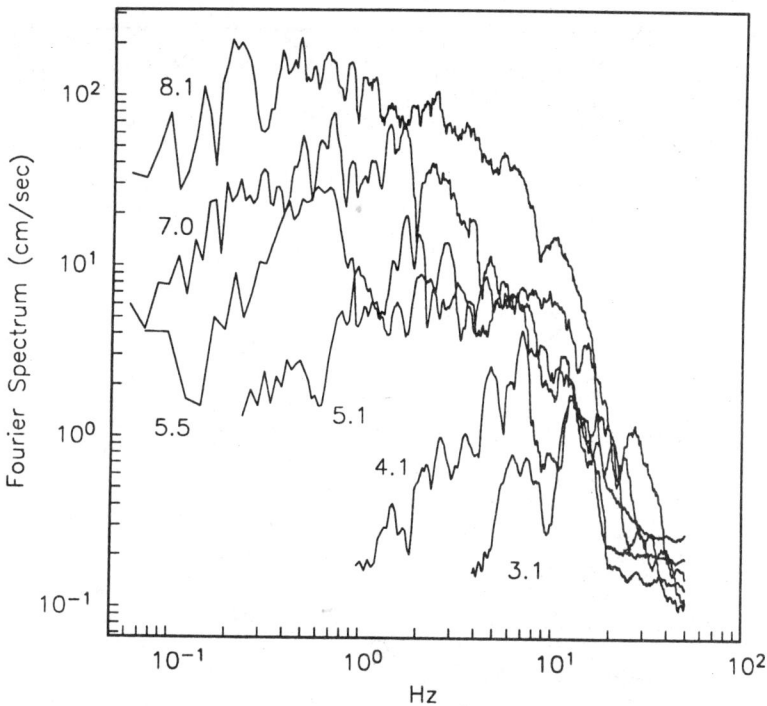

FIGURE 7. Smoothed Fourier amplitude spectra (whole record) corresponding to the accelerograms in Fig. 6. (From Anderson and Quaas, 1988)

(rupture velocity), how rapidly (rise time) and smoothly the slip took place at one point, and how coherently adjacent points on the fault moved. These dynamic parameters are functions of the stresses acting on the fault, the physical properties of the rock surrounding the fault, and the strength of the fault itself, and are variables over the dimension of a large fault during major earthquakes.

Seismic Moment. The seismic moment (see *Seismic Source: Theory* and *Earthquake Mechanisms*) is the best available single number to describe the size of an earthquake. It is given by the product of the fault area (e.g., length × width), average slip, and a characteristic of the rocks adjoining the fault called shear modulus. The moment has a fundamental effect on strong ground motion in determining the spectral amplitudes of low frequency ground motions (B, Fig. 5); conversely, it is often estimated from these ground motion observations. Table 2 gives examples of fault dimensions and seismic moments for a sample of United States earthquakes.

Because of the relationship of the moment to the size of the fault that caused the earthquake (length, width, dip and the slip distribution), and because the fault size may in some cases be estimated in advance from geological studies, these geometrical characteristics may be used to estimate the seismic moment of an anticipated earthquake.

Dynamic Characteristics of the Earthquake Rupture. Description of the spatial distribution of slip is currently only feasible for a very small subset of earthquakes that have occurred in exceptionally well-instrumented locations. Figure 10 shows the distribution of slip parameters inferred by Archuleta (1984), for the 15 October 1979 earthquake in Imperial Valley, California. During that earthquake, the dynamic rupture parameters and the slip distribution were apparently highly variable over both the length and depth of the fault plane. This variability was an important factor in determining the characteristics of the strong shaking during the earthquake.

Variability of the amount of offset on the fault may exist over a smaller scale than what is seen in Fig. 10. However, to resolve a significantly finer scale variability in slip distribution for an earthquake this size requires an impractical number of instruments. Also, the effect in radiated waves of an irregular slip function may be unresolvable from the effect of an irregular rupture velocity (Spudich and Frazier, 1984). Ultimately, one approach is to describe the more detailed spatial variability of slip and its effect on radiated waves using statistical techniques. For this approach, it is more convenient to consider the spectrum of radiated waves at the source (source spectrum) than it is to consider the details of the fault motions that caused the radiation. As an example, the high frequency spectrum (see E, Fig. 5) can be

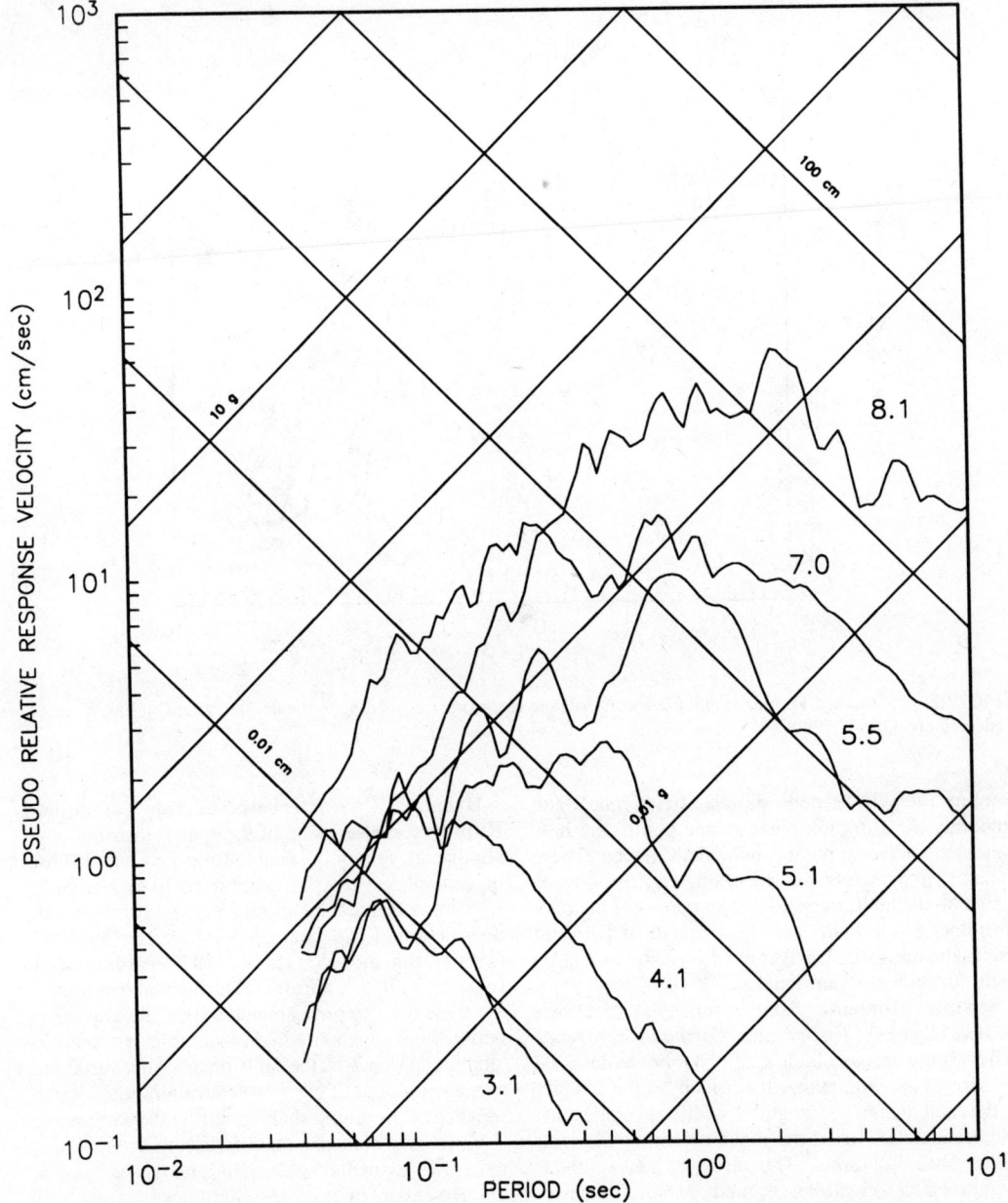

FIGURE 8. Response spectra (5% damped, Pseudorelative velocity) corresponding to the acceleration traces in Figure 6. No filtering of the digital accelerograms was necessary prior to preparation of this figure, but the response spectra have been truncated at long periods where noise becomes dominant. (From Anderson and Quaas, 1988)

modeled as the result of incoherent rupture of many asperities.

At even higher frequencies (G, Fig. 5) the falloff in the spectrum is apparently caused by attenuation between the source and receiver. However, the attenuation is not distributed uniformly between the source and the receiver. A good asymptotic approximation to the spectral shape of the high-frequency limit is exponential decay ($\sim e^{-\pi \kappa f}$) where the decay parameter, κ, is a function of distance and the site conditions of the station. Figure 11 shows, as an example, determination of κ at two stations, one on slightly weathered granite (Pinyon Flat in the southern California batholith) and the other on deep alluvium (Imperial Valley in the Colorado delta of southern California). Both stations show a system-

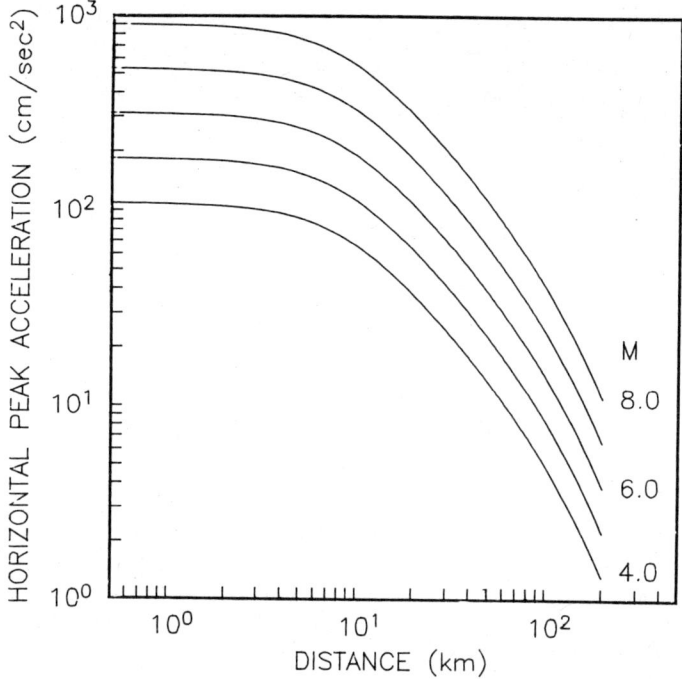

FIGURE 9. Average relationship of peak acceleration to the magnitude of an earthquake and the distance from earthquake source to site. (From Joyner and Boore, 1981)

atic increase in κ with distance, but the station on alluvium shows values offset to higher κ (more rapid decay) than the station on granite. This type of behavior can only be explained as an effect of attenuation. The intercept of κ (i.e., at zero distance) is controlled by the effect of attenuation in the sediments or weathered layer near the surface, while the increase of κ with distance appears to be the effect of lateral propagation.

Effects of Wave Propagation on Strong Ground Motion

Wave propagation studies play a significant role in strong motion seismology, both to understand accelerograms that have been recorded and to allow predictions of the likely characteristics of ground motions that might, in the future, affect a significant site. Complexities that are introduced into the strong ground motions by effects of wave propagation through the complex crust of the Earth are at least equal in importance to the complexities that are introduced at the source. It is commonly assumed that the effects of wave propagation on strong and on weak ground motion are the same. Although nonlinear effects resulting in slower velocities and more rapid attenuation of large amplitude waves are predicted for soft materials, field observational evidence for these are, so far, lacking and the problem remains unresolved. This review identifies

TABLE 2. Fault Dimensions and Seismic Moment for some Larger U. S. Earthquakes

Date	Location	M	Length[a] (km)	Width[a] (km)	Slip[a] (cm)	Moment[a] (dyne-cm)[b]
03-28-1964	Alaska	9.2	750	180	1215	8×10^{29}
04-18-1906	San Francisco	8	450	7	500	5×10^{27}
01-09-1857	So. California	8	380	12	400	6×10^{27}
05-02-1983	Coalinga, Calif.	6.7	25	15	20	2×10^{25}
02-09-1971	San Fernando, Calif.	6.4	16	16	120	1×10^{26}
10-15-1979	Imperial Valley, Calif.	6.3	30	10	30	3×10^{25}
06-28-1966	Parkfield, Calif.	5.6	38	7	30	1×10^{25}

[a]Sykes and Quittmeyer, 1981; Uhrhammer et al., 1983; Heaton, 1982).
[b]Moment is also often given in units of newton-meters. 1 nm = 10^7 dyne-cm.

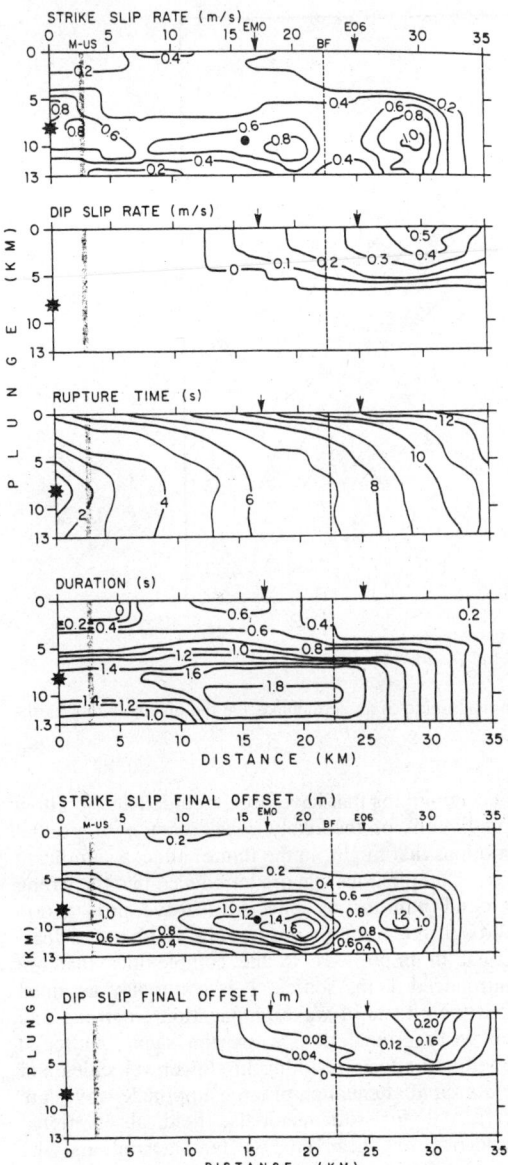

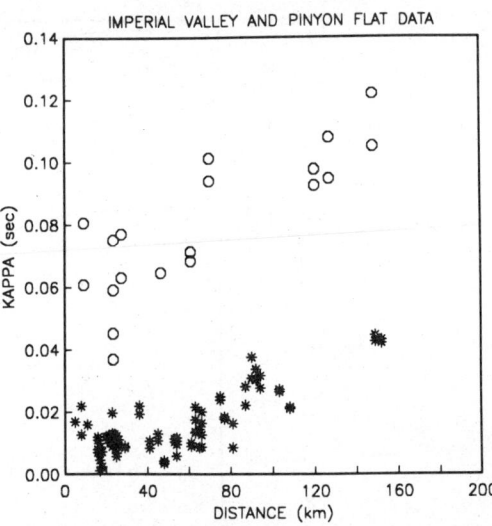

FIGURE 11. $\kappa(r)$ for Imperial Valley (open circles) and Anza (asterisks). The intercept is considered an effect of attenuation below the site, and the increase with distance an effect of attenuation along the path. (From Anderson, 1986)

FIGURE 10. Contours of the slip rate parameters on the Imperial fault plane during the Oct. 15, 1979 earthquake. Distances are measured along strike starting at the epicenter. The hypocenter is shown by the star. The down dip (plunge) coordinate starts at the Earth's surface. The Imperial fault plane dips 80°NE. For reference, the locations of the Mexico-U.S. border (M-US), Meloland (EMO), EO6, and the intersection of the Brawley fault (BF), which also ruptured during this earthquake, are shown. The solid circle indicates the hypocenter of the 2319:35 UT aftershock. (From Archuleta, 1984)

the major wave propagation effects that need to be considered for the precise computation of ground motion. These can be divided into four parts. The first is propagation of seismic waves through a flat layered Earth; the second is the effect of major systematic deviations from a flat layered model. The third is the effect of random variations in the velocity of the Earth on all scales, and the fourth is the effect of attenuation. The success in incorporating these effects depends on how thoroughly the seismic properties of the Earth between the earthquake source and the site are known. Depending on the complexity of the Earth, success may also be limited by computer capabilities to compute the response of the Earth.

The approximation of a flat (or radially) layered Earth is widely used in all aspects of seismology, and is usually quite successful for tasks such as locating earthquakes and describing arrival times of the various types of seismic waves on a global and regional scale (see *Elastic Waves in Homogeneous and Inhomogeneous Media*). There are several methods to calculate the response of such a medium to seismic excitation applied at a point (i.e., the Green's functions). At present the computational effort even for these seismograms is still large. These synthetic Green's functions are often used to determine motion on the causative fault, which is as consistent as possible with all observed ground motions. Observed seismic ground motions at low frequencies (i.e., less than 2 Hz can now be reproduced with this level of sophistication (Fig. 12). An active field of research at present is development of methods to handle wave propagation in an Earth with large-scale deviations from a flat-lying structure.

At high frequencies, it is not possible to describe the inhomogeneities of the Earth in sufficient detail

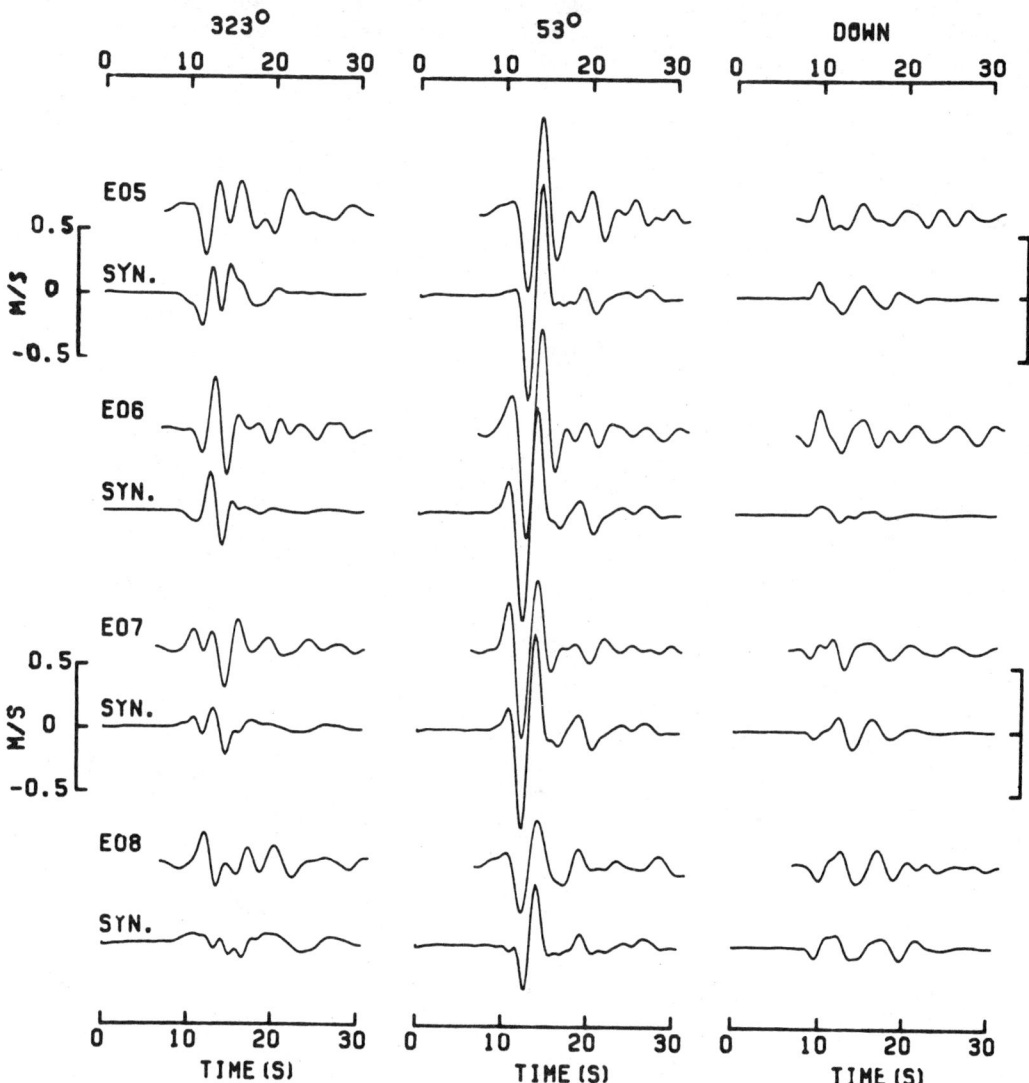

FIGURE 12. Comparison of observed and computed ground velocity at four sites in the October 1979 Imperial Valley earthquake. (From Archuleta, 1984)

to calculate the exact response explicitly, and according to current thought, this is not necessary either. Rather, most studies attempt to describe the effects of small random perturbations of the velocity on the passage of seismic waves.

A final aspect of wave propagation that is crucial to describing strong motion is attenuation (see *Seismic Attenuation: Observation and Measurement*), caused by the gradual absorption of energy from a wave as it travels through the Earth. Random scattering also causes one type of attenuation, and its effects still need to be sorted out from the anelastic absorption of energy by the Earth. A model for one of the effects of attenuation was discussed previously in connection with Fig. 11. The total effect of attenuation, however, has not yet been fully resolved, partly because of uncertainty in the source function and partly because other effects of wave propagation, especially site effects, can be large.

Site Effects. The term *site effects* is generally used to refer to wave propagation in the immediate vicinity of the site, as opposed to the propagation effects, which refer to the complete path from the source to receiver. The boundary between a site effect and a propagation effect is not always clear-cut, but it is useful to discuss them separately. Site effects can include modification of seismic waves by the local sedimentary cover, particularly where this local cover is not representative of the total path from the epicenter, the effect of alluvial valleys or basins, effect of local topography, and effects of the water table. Soil-structure interaction also might be considered a site effect, but it is not considered here because it represents modification of the ground motion by the artificial structures rather than by the Earth in the absence of such structures.

A classic type of site effects is illustrated in Fig. 13, which shows accelerograms from four locations in Mexico City recorded in the 19 September 1985 Mexico earthquake. The stations UNAM and VIV are least affected, while CDA and SCT have been strongly amplified at a period of two to three seconds to cause the large accelerations as shown, and cause the terrible toll of death and destruction from that event (Anderson et al., 1986). Numerous additional studies illustrate similar effects. Also, downhole measurements have illustrated the importance of the near-surface weathered layer. Propagation of body waves through a layered structure to the surface may be handled theoretically by the method of Haskell (1960).

Effects of local topography, including valleys, ridges, and small alluvial-filled valleys, may be grouped together. The results are that in general, topographic highs (e.g., ridges) tend to amplify the ground motions relative to a flat surface of the same material, and topographic lows (e.g., valleys) will deamplify the motion relative to a flat surface. Topographic lows, however, often are sites where sediments accumulate, with dual effects of amplifying the ground motion due to low rigidity, as discussed above, and trapping energy more effectively than a flat-layered structure. Ridges often are composed of relatively hard geological materials which, on the other hand, tend to be associated with lower amplitudes.

Synthetic Seismograms

For any assumed earthquake source, the combination of the above disciplines allows, in principle, computation of synthetic seismograms for any site. There are two approaches in doing this: one to solve the problem entirely by theoretical means, and the other to use small earthquakes with the idea that the seismograms obtained from these are a solution to

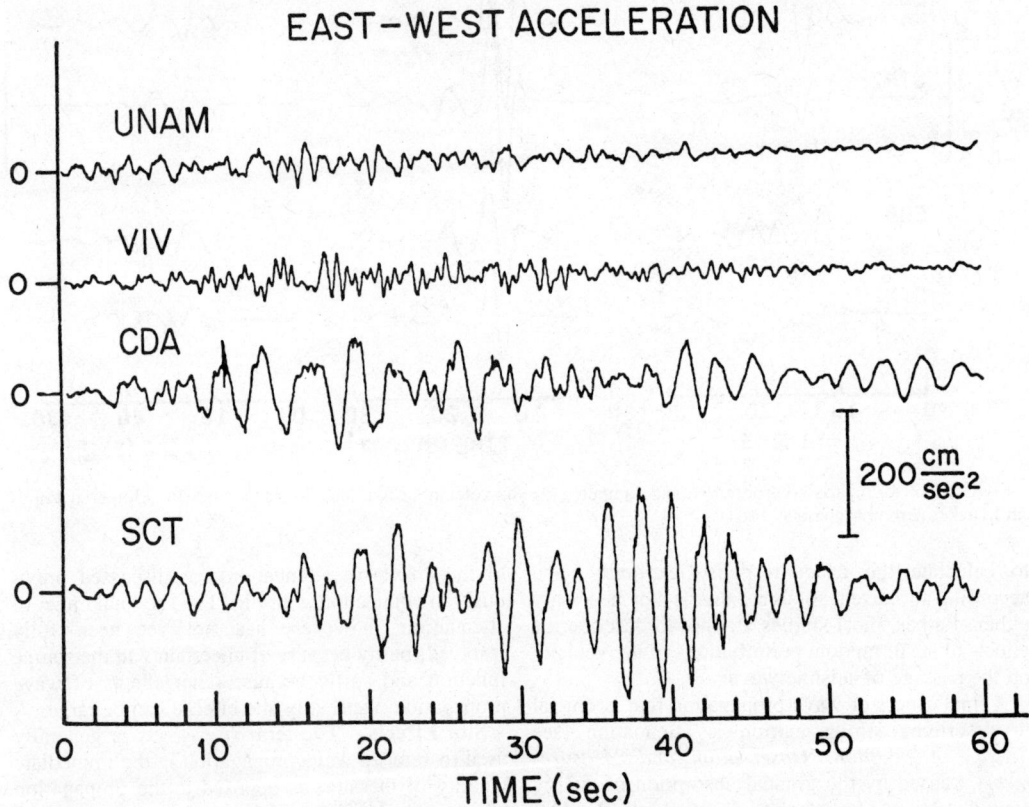

FIGURE 13. Most significant 1-minute segments of the E-W acceleration recorded on the free-field accelerographs in Mexico City. Complete accelerograms are longer; 100 seconds of motion preceded the segment of the SCT record shown here. No time correlation exists among these traces. (From Anderson et al., 1986)

the complete wave propagation problem. For both approaches, one needs to find a reasonable way to describe the earthquake source so that the resultant seismograms are realistic. Figure 14 shows that many characteristics of the seismograms from a past earthquake can be duplicated by these types of simulations. Special techniques are also available to simulate isolated features of ground motion close to the fault. An inexpensive final method which, like the regressions, does not use any geological information except the magnitude and distance is to generate a random time signal with an appropriate spectrum and duration. The ability to generate these synthetic seismograms is still being developed. At low frequencies, the amplitudes are reasonably well constrained (i.e., within a factor of two) and the duration can be predicted reasonably well. The amplitudes of high frequency shaking are still more uncertain.

Seismic Hazard Analysis

Seismic hazard analysis is concerned with the second goal of strong motion seismology, namely transmitting information to allow for informed decisions on earthquake resistant designs, governmental response to the hazards, and other societal impacts of earthquakes. A complete seismic hazard analysis requires the combination of two fundamentally different types of earthquake related information. The first is a description of the seismicity, by which is meant the locations, size distributions, and occurrence rates of earthquakes (see *Seismicity and Plate Tectonics; Seismicity: Subduction Zone; Seismicity: Intraplate*). The second is the estimation of the ground motions that result from each of the plausible earthquakes in the region, using techniques as described earlier in this article.

In its simplest form, a seismic hazard analysis might identify an extreme earthquake that might affect a region, such as a magnitude 8 earthquake on

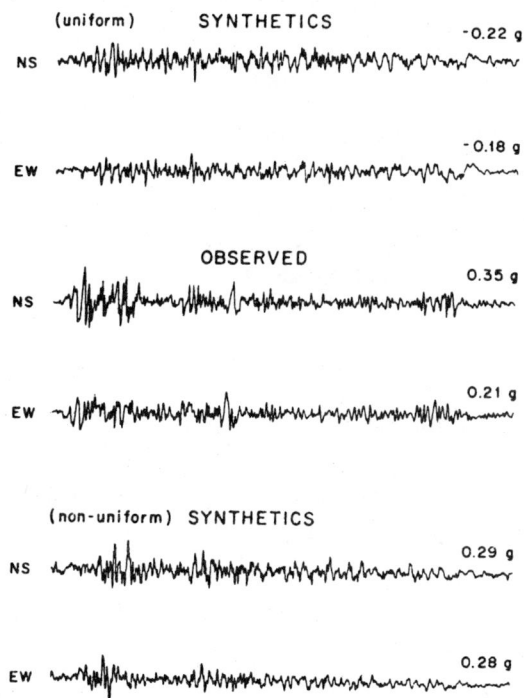

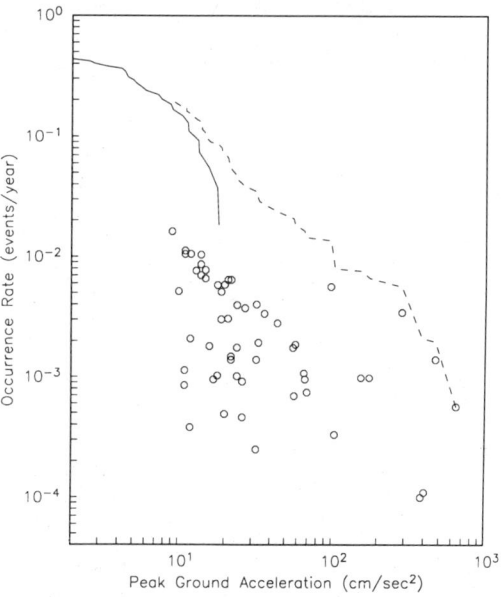

FIGURE 14. Observed and simulated acceleration records at El Centro station. The upper set of synthetics was obtained from a combination of the observed spectral amplitude for the El Centro earthquake and the phase information of records simulated for a $M_L \sim 6$ earthquake. The bottom set was obtained in the same way, but a nonuniform distribution of energy on the fault was considered. (From Munguia and Brune, 1984)

FIGURE 15. Example of a single-model probabilistic seismic hazard analysis. The solid line shows the pseudo-historical occurrence rate of peak acceleration as a function of mean annual occurrence rate, 1 January 1932 to 24 September 1986. This curve is based on actual earthquakes, but because many of these earthquakes were not recorded by strong motion accelerographs the peak accelerations for all of the historical earthquakes were estimated from a regression. The numbered data points represent specific hypothesized events; the size and occurrence rate of each hypothesized event is obtained from geological studies and these events taken together should account for the total seismic deformation rate. The peak acceleration is obtained from the regression. The dashed line is the cumulative expected rate of exceedance of peak acceleration, obtained by summing over the hypothesized events (Anderson et al., 1989).

the San Andreas fault for many of the sites in California, and then estimate the ground motions that might result from that earthquake. In a more complete form, called a probabilistic seismic hazard analysis (PSHA), a complete model of the seismicity is prepared, the amplitudes of ground motions that would result from each of the earthquakes in the model is calculated, and from these a cumulative distribution curve for the expected rate at which a ground motion level of interest would be exceeded is obtained (Fig. 15). Generally, there is uncertainty in the seismicity and the ground motion models. This uncertainty can be handled by a multiple-model PSHA in which an additional cumulative distribution curve is obtained for every model, or combination of models, under consideration. The resulting set of curves shows the impact of the uncertainties on the overall hazard estimates.

Summary

The research in strong motion is proceeding toward a capability to take any hypothetical earthquake and to generate seismograms and corresponding spectra that incorporate an understanding of all the physical phenomena as they apply in the region of interest. Research emphasis has been in developing and testing the quantitative tools necessary for this endeavor. There is reason for confidence that this physical approach is the best way to minimize the uncertainties in anticipating ground motion for future earthquakes.

JOHN G. ANDERSON

References

Anderson, J. G., 1986, Implication of attenuation for studies of the earthquake source, in *Earthquake Source Mechanics*, Am. Geophys. Union Geophys. Mon. 37, 311-318.

Anderson, J. G., and R. Quaas, 1988, Effect of magnitude on the character of strong ground motion: an example from the Guerrero, Mexico, strong motion network, *Earthquake Spectra* **4**, 635-646.

Anderson, J. G., P. Bodin, J. Brune, J. Prince, and S. Singh, 1986, Strong ground motion and source mechanism of the Mexico earthquake of September 19, 1985, *Science* **233**, 1043-1049.

Anderson, J. G., T. Rockwell, and D. Agnew, 1989, Study of seismic hazard in San Diego, *Earthquake Spectra* **5**, in press.

Archuleta, R. J., 1984, A faulting model for the 1979 Imperial Valley earthquake, *Jour. Geophys. Research* **89**, 4559-4585.

Freeman, J. R., 1932, *Earthquake Damage and Earthquake Insurance*. New York: McGraw-Hill.

Ganse, R. A., and J. B. Nelson, 1981, Catalog of Significant Earthquakes 2000 B.C.-1979. Report SE-27, World Data Center A for Solid Earth Geophysics. Boulder, Colo.: U.S. Department of Commerce.

Haskell, N. A., 1960, Crustal reflection of plane SH waves, *Jour. Geophys. Research* **65**, 4147-4150.

Heaton, T. H., 1982, The 1971 San Fernando earthquake: a double event?, *Seismol. Soc. America Bull.* **72**, 2037-2062.

Joyner, W. B., and D. M. Boore, 1981, Peak horizontal acceleration and velocity from strong-motion records including records from the 1979 Imperial Valley, California earthquake, *Seismol. Soc. America Bull.* **71**, 2011-2038.

Munguia, L., and J. N. Brune, 1984, Simulation of strong ground motion for earthquakes in Mexicali-Imperial Valley, *Geophys. Jour. Royal Astron Soc.* **79**, 747-771.

Spudich, P., and L. N. Frazer, 1984, Use of ray theory to calculate high-frequency radiation from earthquake sources having spatially variable rupture velocity and stress drop, *Seismol. Soc. America Bull.* **74**, 2061-2082.

Sykes, L. R., and R. C. Quittmeyer, 1981, Repeat times of great earthquakes along simple plate boundaries, in D. W. Simpson and P. G. Richards, eds., *Earthquake Prediction: An International Review*, M. Ewing Series 4. Washington, D.C.: American Geophysical Union, 217-247.

Uhrhammer, R. A., R. B. Darragh, and B. A. Bolt, 1983, The 1983 Coalinga earthquake sequence: May 2 through August 1, in J. H. Bennett and R. W. Sherburne, eds., *The 1983 Coalinga, California Earthquakes*, Spec. Publ. 66. Sacramento, Calif.: Calif. Dept. of Conservation, Div. of Mines and Geology, 221-240.

Bibliography

Aki, K., and P. G. Richards, 1980, *Quantitative Seismology Theory and Methods*, San Francisco: W. H. Freeman and Co.

Newmark, N. M., and W. J. Hall, 1982, *Earthquake Spectra and Design*. Berkeley, Calif.: Earthquake Engineering Research Institute Monograph, 103 pp.

Lomnitz, C., and E. Rosenblueth, eds., 1976, *Seismic Risk and Engineering Decisions*, Developments in Geotechnical Engineering, vol. 15. Amsterdam: Elsevier Scientific Publishing Co., 425 p.

Wiegel, R. L., ed., 1970, *Earthquake Engineering*. Englewood Cliffs, N.J.: Prentice-Hall, Inc., 517 p.

Cross-references: Earthquake Mechanisms; Earthquakes: Hazards and Prediction; Earthquakes and Crustal Deformation; Earthquake Seismology; Seismic Instrumentation; Seismic Instrumentation: History; Seismic Source: Observations; Seismic Source: Theory.

SUBDUCTION ZONES

The generation of new lithosphere at ocean spreading ridges (divergent plate margins) is balanced by the destruction of lithosphere at subduction zones (convergent plate margins). Ocean trenches, volcanic arcs and deep earthquakes that define inclined Benioff seismic zones occur at most sites where lithosphere capped by oceanic crust subducts into the interior of the earth at relative speeds from 10 to as much as 110 km Ma^{-1} (1 to 11 cm yr^{-1}). Geological and geophysical investigations have shown that at some sites of plate convergence, such as the Lesser Antilles (Fig. 1), the sediment carried on the incoming plate is largely accreted onto

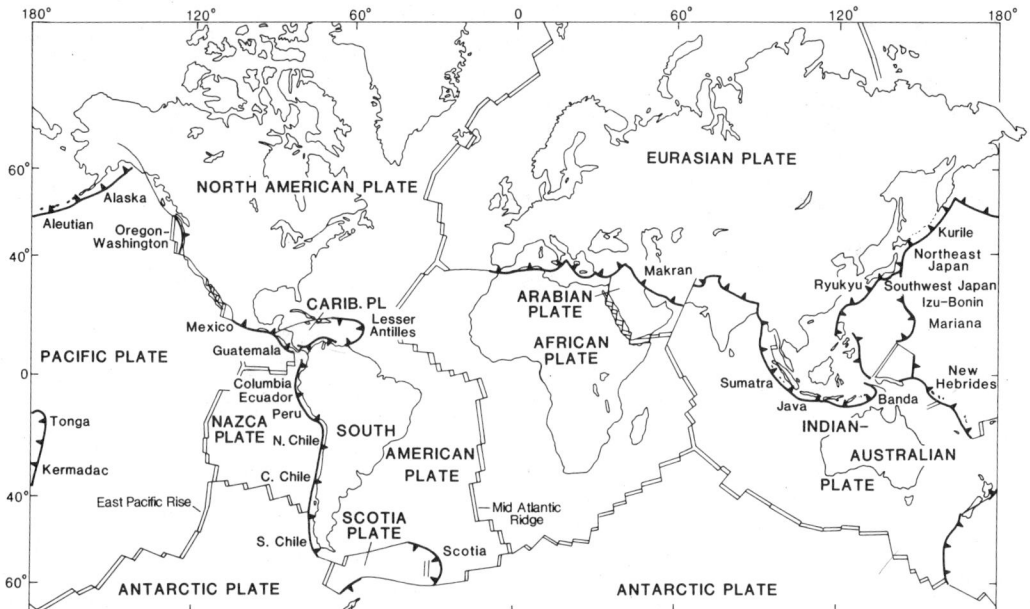

FIGURE 1. Map showing the location of subduction zones where plate convergence is faster than about 10 km Ma^{-1} (1 cm yr^{-1}). At most of these sites, an oceanic trench, volcanic arc and inclined Benioff seismic zone are present.

the leading edge of the overriding plate at depths of only a few kilometers, whereas at others, such as Mariana, not only is the incoming sediment completely subducted to unseen depths, but also some of the overriding plate is detached and deeply subducted as well. This article focuses on the geologic and geophysical aspects of subduction zones that indicate their diversity and bear on the geodynamics of the forearc region, that is the region between the volcanic arc, the trench axis, and the top of the descending plate down to depths of magmagenesis, approximately 125 km beneath the volcanic arc (Fig. 2a).

Tectonic Processes at Subduction Zones

Sediments and seamounts on the incoming plate are initially deformed in the zone of compression where bulldozerlike action causes the incoming sediment pile to thicken largely by km-scale thrusting and folding. An accretionary prism typically composed of variously deformed and metamorphosed sedimentary rocks and mafic seamounts can accumulate at the leading edge of the overriding plate. Subduction accretion is the general process of prism growth that occurs by offscraping and underplating (Fig. 2b). Offscraping leads to the trenchward growth or widening of the prism by the addition of material to its toe. Underplating is the process of addition of material to the bottom of the prism and overlying crystalline plate, which thickens and uplifts it.

Sediment subduction is the process by which unconsolidated sediment is transported beneath the overriding accretionary prism or crystalline block. The subducted sediment eventually is underplated, or returned to the inlet and offscraped, or transported to the depths of arc magmagenesis. Subduction erosion is the process in which rock is rasped or stoped from the base of the overriding block and dragged to greater depth. It can occur at depth at the same time subduction accretion is occurring at shallower levels, or vice versa.

Geologic and Geophysical Characteristics of Subduction Zones

General Relationships from Geophysical Studies. The topography, structure, and composition of the overriding block varies considerably from one margin to another (Dickinson and Seely, 1979). At some sites, such as Peru–Chile, it consists dominantly of continental crust; whereas at others, such as the Aleutians, it consists of oceanic crust; and at a few, such as Ecuador, continental crust gives way to oceanic crust both along and perpendicular to strike of the margin. On the arcward trench slope, a slope cover of seismically well-layered bedded sediments generally blankets the accretionary prism (or, where no prism is present, the overriding crystalline block). Upslope more than 20 km or so from the trench, slope strata are typically flat-lying and continuous, indicating that the underlying prism was not appreciably shortened by subduction-driven shortening since their deposition. Where large accretionary prisms are present, the slope cover is

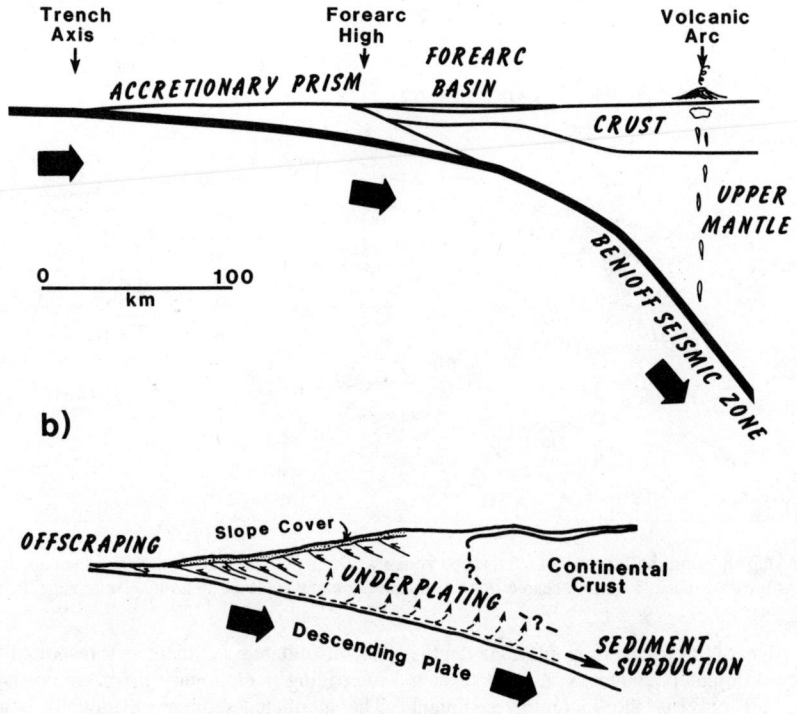

FIGURE 2. (*a*) Cross section illustrating the principle features of an idealized subduction zone. (*b*) Cross section of the upper 20 km of the subduction zone off Mexico illustrating the tectonic processes of prism accretion by offscraping and underplating and of deep sediment subduction. Offscraped sediments beneath the slope cover are thought to be imbricated along thrusts. Mass balance relations derived from seismic reflection and refraction studies and drill cores from Leg 66 of the Deep Sea Drilling Project indicate that since the mid-Tertiary about one-third of the incoming sediment has been offscraped, one-third has been underplanted, and one-third has been deeply subducted. (After Moore et al., 1982)

generally thin near the trench axis and thickens upslope, locally to as much as a kilometer or so—a relationship that indicates that the prism grew seaward with time. It laps onto the forearc high, or in some cases merges with seismically layered strata of the forearc basin that extend trenchward of the high. Horizons that are seismic reflectors in the slope cover generally are fairly continuous, although in places they may terminate abruptly against the arcward sides of areally restricted local sequences of reflectors as much as 1 to 2 km thick that are thought to be slope–basin deposits. In some cases these basin deposits broaden and thicken upslope; another relationship that indicates a prism grew seaward with time. The largest modern accretionary prisms are found off Makran, southwest Japan, and Lesser Antilles and are over 200 km wide.

The nature of large-scale structures within an accretionary prism or the overriding crystalline block are not readily determined. Off Mariana, parts of Peru–Chile, and Guatemala, seismic reflection profiles and deep sea drilling indicate that normal faults with offsets of hundreds of meters cut the overriding block. At these margins, subhorizontal extension of at least the upper portion of the overriding block approximately perpendicular to the trench has occurred concurrently with plate convergence.

Oceanic crust caps the descending lithospheric plate at nearly all active subduction zones. An exception occurs east of Java where recent subduction of the northern edge of the Australian continent causes the isostatic uplift of the accretionary prism and the emergence of the island of Timor. An ancient example is where the Indian continent entered the subduction zone along the southern margin of the Eurasian plate about 40 million years ago. The resultant continental collision and continued slow plate convergence has created the Himalayas. At normal subduction zones with accretionary prisms, the best multichannel seismic reflection profiles show the upper surface of the subducted oceanic crust can be traced as much as 50 km arcward of the trench. Seaward of the trench, particularly at margins lacking significant accretionary prisms, the oceanic crust is commonly broken by normal faults with scarps as much as a thousand meters high spaced kilometers apart. These normal faults can either be predominantly downthrown towards the trench, as off Japan, or form horsts and grabens, as off Peru–Chile.

At margins with accretionary prisms, the descending plate is isostatically depressed and its dip typically increases from a few degrees near the trench axis to approximately 10° beneath the forearc high. Earthquakes defining a Benioff seismic zone (Fig. 2a) are characteristic features of subduction zone plate boundaries (Isacks et al., 1968) and occur to maximum depths of 700 km at Tonga. The shape of the Benioff zone indicates that in most cases the plate dip steepens past the forearc high, typically attaining about 45° beneath the volcanic arc. Major earthquakes with thrust-type focal mechanisms occur where convergence speed is greater than about 20 km Ma^{-1} (2 cm yr^{-1}) and at depths as great as 60 km. They seem to occur very near the interface between the descending and overlying plates. However, only during a few great earthquakes, such as the one off Alaska in 1964, has ground breakage been found that indicates thrust-type displacements extended to the surface. Recent detailed studies show that seismicity actually varies greatly in type, magnitude, and distribution at different sites of plate convergence. Besides the thrusting motions near the plate interface, they show that normal-type earthquakes, probably due to plate bending, are also common in the top of the descending plate at some margins. The stress state and large-scale movement patterns within broad, thick accretionary prisms cannot be determined from studies of earthquakes because few appear to occur within them.

Seamounts up to several thousand meters high widely dot the ocean floors and locally form chains of islands. Similarly, scarps with relief up to about a thousand meters typically are present where oceanic transform faults offset lithosphere less than a few tens of millions of years old. Both are important in the history of subduction zones because they periodically impinge upon the overriding block. When this occurs, they can substantially modify trench sedimentation and accretion patterns by damming the axial flow of turbidity currents, can cause distortions of the overriding prism and crystalline plate, and may profoundly affect seismicity by acting as rough asperities along the plate interface.

Near-Trench Deformation Patterns from Seismic-Reflection Profiles. Most of the deformation of the incoming sediments and the accretionary prism, as measured by tilting of bedding of the uppermost strata, takes place within 10 to 30 km of the base of the trench slope (Karig, 1983). At most margins it is due to arcward-dipping, thrusting, and imbrication (Fig. 3a), although large-scale folding is an important process at several localities, such as Oregon–Washington, Makran, and southern Chile. At margins where the incoming sediments are thicker than about a kilometer, some of the flat-lying topmost incoming layers merge into those upslope. Examples of this thickening and uplift are found off Oregon–Washington, Alaska, the Aleutians, the Lesser Antilles, southwest Japan, Makran, Sumatra, and southern Chile. The only concrete evidence of possible detachment of a fragment of the top of the descending basement at an actively convergent margin is upthrust ridges of basaltic oceanic crust detected in sediment-poor regions of the Chile trench.

Where the incoming sediment is especially thick, no well-defined trench axis exists. Instead, the slope base marks the beginning of disruption, and in such cases is commonly termed a deformation front. At these margins, the basal few hundred meters of sediment can be traced arcward of the front with relatively little discernible disruption for tens of kilometers. Examples are the Lesser Antilles, Makran, and southwest Japan. A gently dipping fault or decollement must separate the upper more-deformed layers from the lower less-deformed ones.

Where the incoming sedimentary succession is thinner than about a kilometer, the trench axis is typically a well-defined trough as much as 20 km wide (Fig. 3b). At such sites the interpretation of seismic reflection profiles is much less clear. Examples are Java, Mexico, Guatemala, northeast Japan, parts of the Aleutians, most of Peru–Chile, and Mariana. Typically, all that can be detected with reflection methods is tilting or offset of sedimentary strata within a few kilometers of the base of the trench slope. Recent highly processed reflection profiles show that the top of the incoming trench fill off Peru and the central Aleutians is complexly imbricated and thickened by thrusting into slices 500–1000 m thick over a distance of 20–30 km arcward of the trench axis. The basal few hundred meters of the incoming sediment pile, which consist principally of pelagic deposits, appear to be carried far past the base of the trench slope.

Deformation Patterns and Rates from Deep Sea Drilling Project Cores. Deep sea drilling on trench slopes has reached maximum subbottom depths of about a kilometer (von Huene, 1984). Whether it has actually penetrated the base of the slope cover at any site is unclear, inasmuch as core recovery, particularly of sand, is generally poor and differentiation between deformed slope cover and offscraped trench deposits is difficult. Cores obtained off Oregon and Alaska (DSDP Leg 18) and southwest Japan (DSDP Leg 31) showed that sediment on the lower trench slope was identical to flat-lying sediment in the trench and were important early evidence for uplift of trench deposits at subduction zones. Other cores from off Japan (DSDP Legs 56

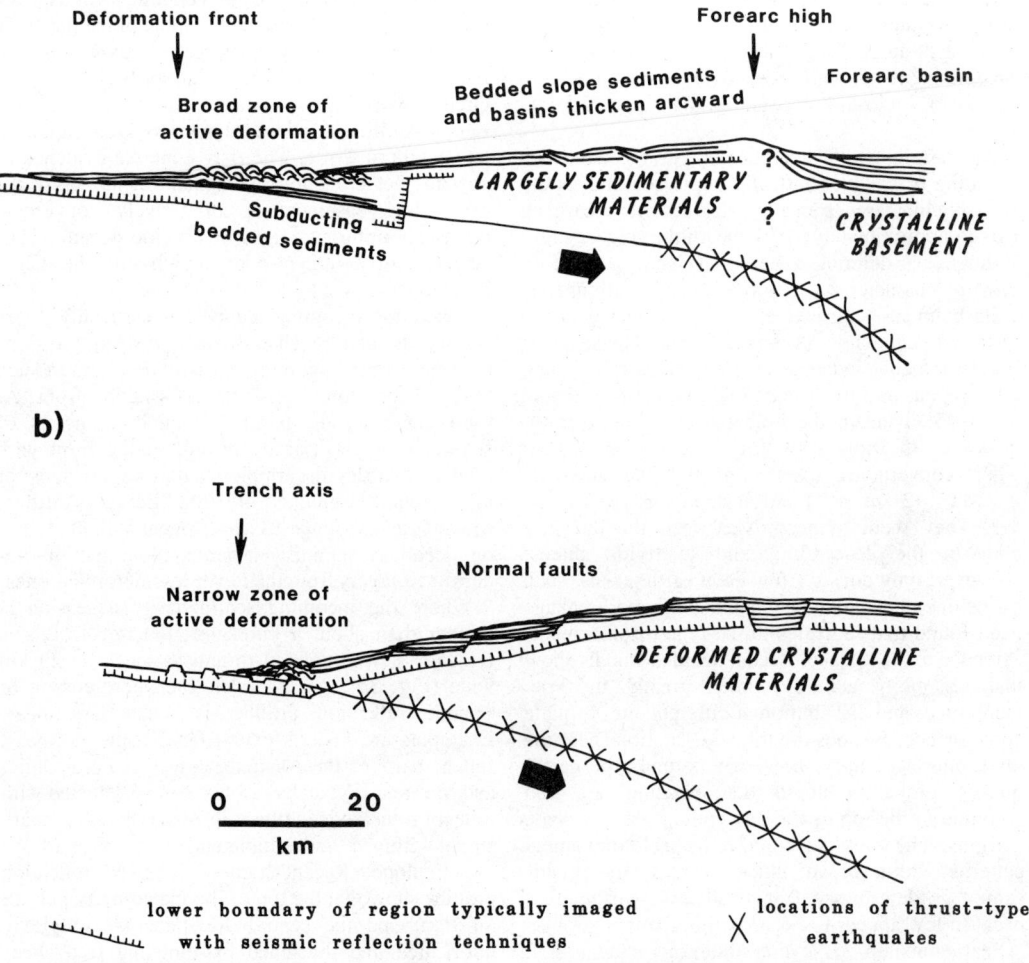

FIGURE 3. Composite schematic diagrams illustrating the geologic features discussed in the text that are commonly found in the forearc regions of subduction zones. (*a*) Margins with extensive subduction accretion. Examples of margins with most of these characteristics are Lesser Antilles and southwest Japan. (*b*) Margins with extensive subduction erosion. Examples are Mariana and Guatemala.

and 57), Mariana (DSDP Leg 60), Mexico (DSDP Leg 56), and Guatemala (DSDP Leg 67), Peru (DSDP Leg 112) showed increasing disruption with depth, with nearly vertical extension fractures being among the most common nonsedimentary structures. Most importantly, these deep sea drill cores provide critical data on the minimum age of the forearc block. They indicate either a long term lack of accretion and even subduction erosion off Mariana, northeast Japan, Guatemala, and Peru.

Short-term vertical movements of meters have occurred in forearc areas during seismic events, such as in and offshore Alaska during the great earthquake of 1964. Long-term rates of uplift, subsidence, or both, which are reflected in depth dependent benthic organisms, the stratigraphic position of the calcium carbonate compensation depth (CCD) and sedimentary facies patterns found in deep sea drill cores, are typically estimated to be around a few hundred meters per million years. The forearc regions off Japan, Mariana, Mexico, and Guatemala have undergone lengthy intervals of subsidence. Inasmuch as Japan and Mexico are presently undergoing uplift, it appears that both subsidence and uplift can occur during a period of continuous plate convergence.

Notably, the only thrust faults indicated by stratigraphic inversion have so far been penetrated by deep sea drilling at one actively convergent margin. They are near the base of the trench slope off the Lesser Antilles (DSDP Legs 78A and 110).

Deformation Patterns from Field Studies of Accretionary Complexes. One of the most problematic and distinctive features of many ancient accretionary complexes and the older portions of some active ones is the presence of rock bodies of chaotically mixed, mud-matrix melange. Examples include the huge Central Belt of the Franciscan Complex of California, the Uyak melange on Kodiak Island off Alaska, the Oyo melange on Nias Island off Sumatra, portions of the Shimanto Belt of southwest Japan, portions of the Basal Complex on Barbados Island in the Lesser Antilles. Many melanges in subduction complexes have a distinctive scaly cleavage that is clearly of tectonic origin. Many include exotic bodies of greenstone (locally pillowed, brecciated, or layered in flows), serpentinite, chert, or limestone, that typically comprise less than 10% of the volume of rock. These bodies are exotic in the sense that they were not simply deposited in situ with the voluminous terrigenous sediments that comprise the bulk of the prism.

Although the origin of specific bodies of melange, particularly ones bearing exotic blocks, is usually controversial, their common generation at subduction zones is widely recognized (Cowan, 1985). Some are the results of surficial submarine slumping or sliding (olistostromes). Others were formed by tectonic deformation within fault or shear zones. Even though these two melange-forming processes are distinctly different, many geologists have found it difficult to distinguish which was primarily responsible for forming specific melanges within subduction complexes, probably because similar appearing rock bodies can be produced by either process when they operate on poorly lithified sediment at low-temperatures and high pore-fluid pressures that keep effective stresses low.

Regardless of their origin, melanges will be difficult to image by seismic-reflection techniques, even near the surface, because of both lack of bedding continuity and scattering from included blocks. While it seems unlikely that block-bearing melange comprises large portions of the shallow levels of accretionary complexes where seismic imaging is good, as off Makran, southwest Japan, Oregon-Washington, and Lesser Antilles, the volumes of melange is unknown at the deeper levels of these margins that are not imaged and across most other active margins where seismic imaging is poor even at shallow levels.

Fault-bounded packets of well-bedded sediments are found in all accretionary complexes and in many places they are much more extensive than bodies or belts of chaotically mixed melange. Some of them are remnants of the slope cover or of slope basins. Others are fragments of offscraped trench fills. Although recently offscraped trench sediments can readily be distinguished from modern slope-basin deposits in seismic-reflection profiles across active margins, the distinction is much more difficult in ancient complexes and even in the older portions of active margins. Sedimentary sequences interpreted as offscraped trench-fill are usually faulted and folded, and bedding commonly dips steeply. Examples include portions of the Shimanto Belt of southwest Japan, the Ghost Rocks Formation of Kodiak Island off Alaska, the Chugach Formation and the Orca Group of southeast Alaska, and parts of the Coastal Belt of the Franciscan Complex of northern California. Portions of all of these terranes contain localized zones of intense deformation that can appropriately be categorized as melange. Although, as already noted, broad, open folds are commonly recognized in seismic profiles of sediments being offscraped near the trench axis, the folds seen in outcrop in ancient accretionary complexes are typically much smaller and tighter.

Sediment Subduction and Subduction Erosion

Most early studies of modern and ancient convergent plate margins focused on accretion. However, it has long been recognized that the volumes of rock derived from pelagic siliceous oozes and metalliferous sediments are much smaller in accretionary complexes than expected, were the incoming plate continually scraped clean of sediment during convergence. Apparently, these materials, which form the basal 200 m or so of the incoming sediment pile, are rarely accreted at levels that eventually become exposed by uplift and erosion; hence, they must be subducted to depths of at least 20–30 km.

Continental crust 1000–2000 million years old is found within a few tens of kilometers of the trench axis along parts of the Peru–Chile margin, indicating little if any cumulative subduction accretion in this region during the last 100 million year period of continuous plate convergence. Indeed, if present rates of convergence and thicknesses of trench fill have existed since the beginning of the Tertiary, considerable volumes of trench sediment are unaccounted for at many margins around the Pacific. Extensive sediment subduction is indicated. The presence of young arc-type volcanic or plutonic rocks in the forearc regions of Mariana and much of Peru-Chile indicate a progressive migration of the trench axis and volcanic arc, which can only be readily explained by complete sediment subduction along with some form of subduction erosion of the base of the overriding plate.

Recent trace element and isotopic studies of Pb, Sr, and Nd in volcanic deposits at several arcs seem to indicate the presence of recycled continentally derived material in the primary magma. Perhaps more definitively, the short-lived isotope ^{10}Be has been detected in arc volcanic rocks from parts of Central America and the Aleutians (Tera et al.,

1986). Because this isotope is abundant only in sediments less than a few million years old, partial derivation of some arc magmas from young, deeply subducted sediments seems required.

Sediment subduction to considerable depths and locally even subduction erosion of the overriding plate are now recognized as tectonic processes comparable in importance to accretion. The mechanism of these processes is particularly obscure however, because they destroy, rather than produce a rock record.

Blueschists and Subduction Zone Thermal Structure

Extensive tracts of metamorphic rocks known as blueschists are found on the arcward side of late Mesozoic and younger subduction complexes in New Caledonia, Japan, Alaska, California, and the Alps (Ernst, 1975). Blueschists are unusual rocks containing minerals such as lawsonite, aragonite, sodic amphibole, and sodic pyroxene, which indicate that metamorphic recrystallization occurred at extremely high-pressure and low-temperature conditions. Such conditions develop in subduction zones because plate convergence at speeds of tens of kilometers per million years (centimeters per year) convects cold lithosphere downward faster than the Earth's interior heat is conducted upward through it. As a result, after a few tens of millions of years of subduction, the front of the overriding plate cools and the local geothermal gradients become greatly depressed ($< 10°C/km$) as compared to typical gradients in the continents and ocean basins (25 to 35°C/km). Temperatures less than 200°C at depths of 30 km can be attained in subduction zones where plate convergence is fast (Fig. 4).

Once subduction ceases, the reestablishment of near-normal geothermal gradients will occur over a few tens of million years. Because the speed of conduction of heat through rocks is comparable to the speed of uplift of deep-seated rocks by erosion at the surface, the minerals in deep-seated blueschists become replaced by metamorphic minerals, which indicate more normal geothermal conditions. Wherever blueschists are preserved with little alteration, some form of synsubduction uplift to depths shallower than about 10 km is indicated.

Although the generation of blueschists at subduction zones seems so well understood that their presence is now one of the primary criteria used to recognize ancient sites of subduction, the relationship of their uplift and exposure to the growth of an accretionary prism is particularly obscure. Part of the problem is that no blueschist bedrock or fragment has yet been found by drilling or dredging on the trench slope of any actively convergent margin. Thus it appears that the emergence of blueschist bedrock over a substantial area at the surface typically occurs after subduction ceases. In any case, their preservation seems to directly indicate that subduction zones can be two-way streets, along which sediment is subducted down to, and uplifted back from, depths of 20–30 km or more.

Temperature is one of the primary factors controlling the geodynamic evolution of a subduction zone because of its strong effect on the mechanical properties of rock. Thermal modeling indicates that quasi-steady state low-temperature thermal conditions are attained after a few tens of millions of years of fast plate convergence. Once the base of the overriding plate is cooled, the mechanical response of subducted sediment can become relatively uniform down to depths of 30 km or more. Before cooling is extensive however, the large horizontal and vertical variations in temperature will strongly control the mechanical behavior of subducted rocks in this depth range.

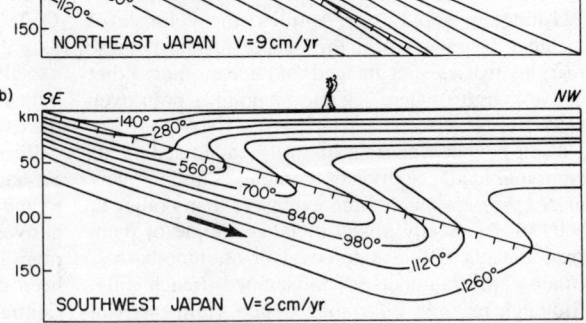

FIGURE 4. Cross sections illustrating the downbowed isotherms at a subduction zone that lead to the generation of the blueschists; a metamorphic rock containing minerals that indicate recrystallization occurred at unusually high-pressure and low-temperature conditions deep within the Earth (modified from Honda and Uyeda, 1983). (a) Northeast Japan with fast subduction. (b) Southwest Japan with slow subduction.

Tectonic Models for Subduction Zones

Subduction of oceanic-crust capped lithosphere into the interior of the Earth is the result of a thermally generated gravitative instability. Oceanic plates formed at spreading ridges become progressively denser as they cool, contract, and thicken. The primary manifestation of plate cooling is that the ocean floor subsides more than 2 km by the time it is 80 million years old. Oceanic lithosphere more than a few tens of millions of years old appears to be dense enough to be readily subductable. Another, perhaps more important effect is that the bulk density of the descending oceanic lithosphere increases greatly when the increase in pressure and temperature causes the basalts and gabbros within it to metamorphose into eclogite (composed largely of dense garnet and sodic pyroxene). This mineralogic change certainly enhances and may be the primary driving force of plate subduction.

The initiation and cessation of subduction occurs in several ways. One way, which is a normal consequence of steady state plate motion, is the transition of a transform plate margin into a subduction zone or vice versa by passage of a plate triple junction. This passage occurs progressively, perhaps over many tens of millions of years, as a triple junction "zippers" its way along the edge of a plate. It also occurs during times of major plate reorganization following the closing of an ocean basin by continental collision. At such times, new subduction zones may be initiated along extensive zones of major lithospheric weakness, such as young transform faults or ocean spreading ridges. Yet another way to initiate subduction is where mafic oceanic crust deeply buried by sediment at an old passive plate margin is metamorphosed into enough dense eclogite that the lithosphere in the region is gravitationally unstable. When this occurs, only a small change in regional tectonic stress conditions may cause a subduction zone to form.

Imbricate-Thrust Model for Subduction Accretion. Most models for the quasi-steady state tectonic behavior of subduction zones have focused on the process of accretion. In the early 1970s the imbricate-thrust model was developed through synthesis of geologic and geophysical data then available (Seely et al., 1974). The model envisions that thrust-bounded imbricate slices (or packets) of trench axis sedimentary deposits, oceanic pelagic deposits, and sometimes oceanic crustal fragments from the top of the descending plate form an accretionary prism that grows progressively seaward and also thickens because of continued subduction-driven shortening of the prism. Quantitative modeling of offscraping by imbrication of thrust sheets has derived from analogous models for continental thrust belts. More recent reflection and deep sea drilling data indicates that subduction-driven deformation is largely concentrated within a few tens of kilometers of the base of the inner trench slope; hence, many workers now envision that the thickening of a prism occurs by underplating of thrust-bounded packets of subducted sediment (duplexes) to its base. Davis et al. (1983), quantitatively modeled prism deformation by thrust faulting using an empirical Coulomb-failure criterion that was modified to account for the weakening effects of pore-fluid pressure.

The imbricate-thrust model for prism accretion has been widely applied to explain many field and geophysical observations at modern subduction zones that have particularly broad accretionary prisms, such as off Lesser Antilles, Oregon-Washington, and southwest Japan and some ancient accretionary complexes. It does not, however, address the processes of deep sediment subduction and subduction erosion indicated by field and isotopic studies and confirmed by deep sea drilling as tectonic processes just as important as accretion.

Subduction Channel Model for Subduction Accretion, Sediment Subduction, and Subduction Erosion. The subduction-channel model (Shreve and Cloos, 1986) is a comprehensive, quantitative theory for convergent margins where poorly consolidated sediment is present, and the convergence speed exceeds about 20 km Ma^{-1} (2 cm yr^{-1}) for a long enough time to cause depression of the isotherms in the forearc region. It accounts for the entire spectrum of tectonic response, from Mariana, where all incoming sediment is subducted and subduction erosion is removing the front of the overriding crystalline plate, to the Lesser Antilles, where all incoming sediment is nearly offscraped and a broad accretionary prism composed of thrust faulted and folded sediments is growing.

The model envisions that subducting sediment deforms approximately as a viscous fluid once it is dragged beneath the much more slowly deforming overriding crystalline plate or accretionary prism by the descending one. The deforming sediment between the two plates comprises a kind of shear zone, called the subduction channel (Fig. 5a). Deep subduction of some relatively low-density sediment occurs at all margins, because in the part of the channel near the top of the descending plate, the downward-directed shearing of the highly viscous sediment overcomes the upward-directed forces due to buoyancy and the adverse pressure gradient (Fig. 5b). The exact pattern of flow, the thickness of the channel, and the shearing stresses on the walls are governed by the sediment supply and density, the subduction speed, and the pressure gradient along the channel. The mechanical response of the foot and hanging wall blocks to the shearing stresses imposed on the channel walls determines the numbers and magnitudes of subduction zone earthquakes. Subduction erosion is likely where subduction speed is fast and the channel is narrow, because in such situations the shearing stresses on the walls may be sufficiently large to cause parts of the hanging wall

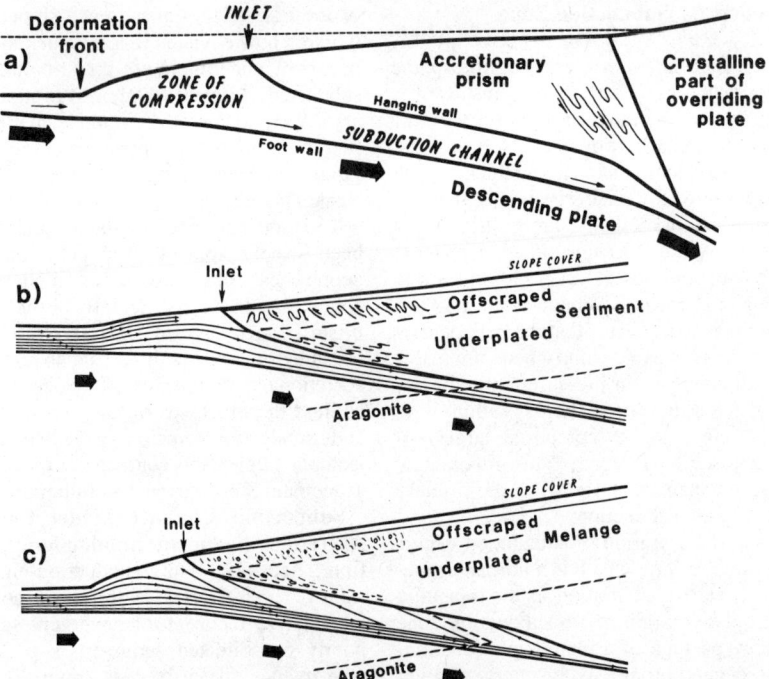

FIGURE 5. Subduction-channel model for prism accretion by offscraping and underplating and sediment subduction at convergent plate margins (after Shreve and Cloos, 1986). (a) Incoming sediment is deformed and thickened in the zone of compression. Some sediment is always dragged into the channel and probably carried to the depths of arc magmagenesis because the downward-directed shearing always overcomes the buoyancy and the adverse pressure gradient and prevents the channel from narrowing to zero thickness. The volume of incoming material in excess of that which can be dragged past the inlet is offscraped and builds the prism trenchward. (b) Streamlines within the incoming sediments and the subduction channel show movement patterns typical of margins where incoming sediment is offscraped and underplated. Subducted sediment can be underplated to the base of the overriding block because dewatering at the top of the channel causes it to compact and lithify and hence to decouple from underlying subduction-driven shearing movements. (c) Streamlines at a margin with upflow of subducted sediment along the top of the channel. This upflow generates thick bodies of intensely deformed and mixed subduction melange because of the large and complex shearing strains entailed by the reversal of direction.

block to become coupled to the downgoing material in the channel.

At the shallow depths imaged by seismic reflection methods, variations in lithology, bedding anisotropy, and pore fluid pressure control the spacing of faults and folds. At greater depths, the shearing becomes more uniformly distributed as the total deformation absorbed within the subducting sediment layer becomes larger. As the shearing becomes more penetrative, the mechanical response of the deforming sediment approaches that of a highly viscous fluid. Under appropriate conditions a thick zone of reverse flow, or upflow, can form along the roof of the channel (Fig. 5c). The upwelling material is very intensely deformed and becomes chaotically mixed, because of the large and complex shear strains entailed by the reversal of direction and is a subduction-generated melange. It can entrain and uplift exotic blocks such as pieces of seamounts dismembered during subduction and fragments of coarsely crystalline blueschists that were metamorphosed at intermediate temperatures during the early stages of subduction when the overriding plate was still hot.

Calculations indicate that the thickness of the subduction channel shear zone is highly variable. Typically it is less than about 500 m where the sediment supply is small and all incoming sediment is subducted, 500 m to a kilometer or so thick at margins with extensive offscraping of incoming sediment, and up to several kilometers thick where upflow occurs to form subduction melange.

Structural relations within bulldozed sediments in the zone of compression and thickening are envisioned to be similar to those in the generalized imbricate thrust model for accretion. However, the thrusted and folded sediments are not necessarily offscraped and accreted to the toe of the prism, because some deep sediment subduction always occurs. The critical factor is how the incoming sediment supply to the system (trench axis thickness times the subduction speed corrected for compaction and slope deposition) compares to the capacity of the system to drag sediment into the subduction channel shear zone. The opening, or inlet, to the channel is located at the leading edge of the overriding crystalline plate or accretionary prism. Typically it is 10-20 km upslope from the base of the trench slope, but can be farther upslope or downslope where sediment supply is much greater or smaller than the capacity at the inlet. Where incoming sediment supply is greatly in excess of inlet capacity because subduction speed is slow, sedimentation rate is high, or both, as off Lesser Antilles, southwest Japan, and Makran, the zone of compression can be many tens of kilometers wide and the transition from it to material that is permanently accreted to the front of the prism is gradational.

Underplating of material onto the roof of the channel thickens the overriding block and passively uplifts it without significant concurrent shortening. Underplating is envisioned to occur because dewatering at the roof of the channel causes compaction and lithification of the downgoing sediment (Fig. 5b) or upwelling subduction melange (Fig. 5c). The resultant drastic viscosity increase in turn causes decoupling from the less compacted and less lithified deforming material beneath. Decoupling may occur either by deactivation of discrete fault slip surfaces bounding packets of sediment centimeters to tens of meters thick when downgoing sediments are underplated or in more of a grain-by-grain manner when upwelling mud-matrix melange is involved. The rate of dewatering and underplating depends upon the permeability of the hanging wall block to whose base the sediments are accreted. Deep-seated underplating can drive the synsubduction uplift of extensive tracts of blueschists towards the surface.

MARK CLOOS

References

Cowan, D. S., 1985, Structural styles in Mesozoic and Cenozoic melanges in the Western Cordillera of North America, *Geol. Soc. America Bull.* **96**, 451-462.

Davis, D., J. Suppe, and F. A. Dahlen, 1983, Mechanics of fold-and-thrust belts and accretionary wedges, *Jour. Geophys. Research* **88**, 1153-1172.

Dickinson, W. R., and D. R. Seely, 1979, Structure and stratigraphy of forearc regions, *Am. Assoc. Petroleum Geologists Bull.* **62**, 2-31.

Ernst, W. G., 1975, Systematics of large-scale tectonics and age progressions in Alpine and circum-Pacific blueschist belts, *Tectonophysics* **26**, 229-246.

Honda, S., and S. Uyeda, 1983, Thermal process in subduction zone—a review and preliminary approach on the origin of arc volcanism, in D. Shimozuru and I. Yokoyama, eds., *Arc Volcanism: Physics and Tectonics*, Tokyo: Terra Scientific Publishing Company (TERRAPUB), 117-140.

Isacks, B., J. Oliver, and L. R. Sykes, 1968, Seismology and the new global tectonics, *Jour. Geophys. Research* **73**, 5855-5899.

Karig, D. E., 1983, Deformation in the forearc: implications for mountain belts, in K. J. Hsu, ed., *Mountain Building Processes*. London: Academic Press, 59-72.

Moore, J. C., J. S. Watkins, T. H. Shipley, K. J. McMillen, S. B. Bachman, and N. Lundberg, 1982, Geology and tectonic evolution of a juvenile accretionary terrane along a truncated convergent margin: synthesis of results from Leg 66 of the Deep Sea Drilling Project, southern Mexico, *Geol. Soc. America Bull.* **93**, 847-861.

Seely, D. R., P. R. Vail, and G. G. Walton, 1974, Trench slope model, in C. A. Burk and C. L. Drake, eds., *Geology of the Continental Margins*. New York: Springer-Verlag, 249-260.

Shreve, R. L., and M. Cloos, 1986, Dynamics of sediment subduction, melange formation, and prism accretion, *Jour. Geophys. Research* **91**, 10229-10245.

Tera, F., L. Brown, J. Morris, I. S. Sacks, J. Klein, and R. Middleton, 1986, Sediment incorporation in island-arc magmas: inferences from ^{10}Be, *Geochim. Cosmochim. Acta* **50**, 535-550.

von Huene, R., 1984, Tectonic processes along the front of modern convergent margins—research of the past decade, *Ann. Rev. Earth and Planetary Sci.* **12**, 359-381.

Cross-references: *Accretionary Tectonics: Examples from the North American Cordillera; Continental Collision Zones: Seismotectonics and Crustal Structure; Crustal Movements and Tectonic Deformation; Earthquake Mechanisms and Plate Tectonics; Earthquake Seismology; Far-Traveled Terranes; Mantle, Upper: Structure; Seismicity: Subduction Zone; Thin-Skin Tectonics.*

SURFACE WAVES

Surface waves propagate along boundaries parallel to the surface of the Earth, with wave motion confined largely to the vicinity of the free surface. Surface waves are the most prominent arrivals on most long period seismic recordings of shallow earthquakes (see Fig. 2 in *Seismograms: Interpretation* in this volume). For homogeneous layered media, there will be two principle types of surface waves, *Rayleigh waves* and *Love waves*, named respectively after Lord Rayleigh and A. E. H. Love who were first to treat the respective kinds of motion associated with each wave. The full mathematical treatment of surface waves, especially under "real earth" conditions is enormously complex and far beyond the scope of an article of this length. The reader will find more than ample mathematics in any of the standard texts on seismic wave propagation (e.g., Ewing et al., 1957, or Aki and Richards, 1980). The discussion here will focus on physical understanding of surface waves, the nature of their propagation in layered media, the ways in which they can be used to study the Earth, and some principal results. While a variety of other kinds of surface waves can exist in the Earth (e.g., Stoneley waves and leaky modes), the following discussion is confined to Rayleigh and Love waves. Any of the texts mentioned above will provide information on the minor surface waves.

Both Rayleigh waves and Love waves are *guided waves*. They can be distinguished from one another by the velocity with which they propagate and, more importantly, by their particle motion, shown schematically in Fig. 1. Rayleigh waves propagate along the Earth's surface with *retrograde* elliptical motion in the vertical plane of propagation. The amplitude of the particle motion tends to die out at depth. In a homogeneous medium with a Poisson's ratio of about 0.25 (commonly taken as the nominal value for rocks), Rayleigh waves will travel at a velocity of about 0.92 that of the shear velocity. Love waves propagate with a horizontally polarized shear (SH), or transverse, motion as shown in Fig. 1. Love waves, unlike Rayleigh waves, can exist only in a layered or otherwise heterogeneous medium. Their velocity for a given frequency is typically greater than the velocity of the Rayleigh wave. Both Rayleigh and Love waves are *spreading* or *dispersed* waves that in the Earth travel with group and phase velocities that tend to be higher for low frequencies than for high frequencies (see discussion of group velocity, phase velocity, and dispersion, later).

The propagation of surface waves is usually treated as an eigenvalue problem for which there are many possible solutions (modes) at any given frequency. The lowest order mode is termed the *fundamental mode*. The other modes are simply designated first order, second order, etc. *higher modes*. Surface wave modes are conceptually identical to organ pipe modes or other vibrational modes where fundamental tones accompanied by various overtones exist. Surface waves can also be treated as high order free oscillations, or normal modes, of the Earth (see *Free Oscillations of the Earth*). Except at extremely long periods ($T > 300$ s), however, surface waves are usually treated as traveling waves in a layered medium rather than as free oscillations of the Earth.

Surface Wave Propagation

Surface waves observed in the Earth are sinusoidal in character and *dispersive*. Wave dispersion is the spreading of the wavetrain in time due to the fact that different frequencies travel with different velocities. Because the velocities of different frequencies are related to the velocity-depth function of the medium, the dispersion of surface waves can be used to study the properties of the Earth's interior. In the early 1950s, Ewing and Press showed that the long wavetrains characteristic of Rayleigh waves across both land and sea could be explained as due entirely to dispersion in a layered earth. They proceeded to measure and analyze the dispersion curves (plots of group or phase velocity vs. period or frequency) for propagation paths across different tectonic regions of the Earth. From these analyses, they obtained approximate regional velocity structures as a function of depth in the Earth beneath continental and oceanic regions. Since those early studies, surface wave dispersion has been widely used for both regional and global studies of the outer 1000 km of the Earth.

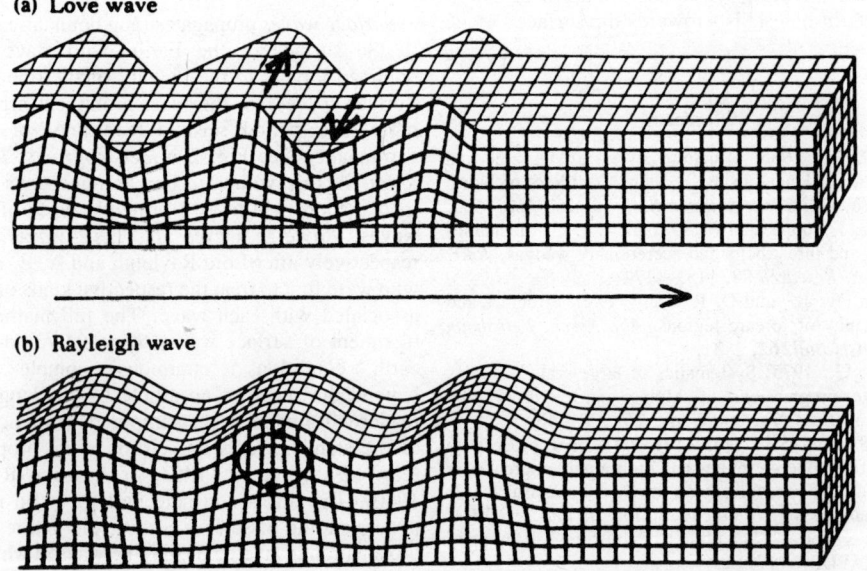

FIGURE 1. Particle motion for Rayleigh waves and Love waves. Rayleigh waves have retrograde particle motion confined to the vertical plane of propagation. Love have purely transverse motion in the horizontal plane. (After B. A. Bolt, *Nuclear Explosions and Earthquakes*, W. H. Freeman and Company; copyright © 1976.)

Group and Phase Velocity

Velocity of propagation of dispersive wavetrains can be described in terms of either *group velocity* or *phase velocity*. Phase velocity is the velocity of the wave at a *specific frequency*. This velocity, c, is defined as:

$$c = \omega/k \qquad (1)$$

where ω is angular frequency and k is wavenumber. Frequency is related to period of the wave, T, by $\omega = 2\pi/T$. Wavenumber equals $1/\lambda$, where λ is wavelength.

For body waves or other nondispersive waves, the phase velocity is the velocity of the wave in the medium. For surface (dispersive) waves, phase velocity is approximately the velocity of the peaks, troughs, or zero crossings of the wavetrain. As average velocity in the Earth typically increases as a function of depth, longer wavelengths (which sample to greater depths) will tend to have higher phase velocities.

Group velocity is the velocity with which surface waves of given frequency travel from the earthquake source to the seismograph. As the group velocity is associated with a *wavepacket*, it will be calculated in terms of a continuous rather than a discrete line spectrum. Group velocity, U, can be expressed in terms of frequency, ω, phase velocity, c, and wavenumber, k

$$U = \frac{d\omega}{dk} = c + k\frac{dc}{dk} \qquad (2)$$

At any given period, the group velocity of surface waves in the Earth will in general be less than the phase velocity. Some typical group velocity and phase velocity curves for simple layered media are shown in Fig. 2. The minimum in the group velocity curve is termed *Airy phase*, a major feature of many surface wave records. It is a stationary phase and therefore has substantially greater relative amplitude than the rest of the surface wavetrain. Note that waves with periods shorter than the period of the Airy phase will exhibit *reverse* group velocity dispersion, with the shorter periods traveling faster.

Measurement of Dispersion

Virtually all surface wave dispersion for both phase and group velocities is now determined by computer techniques. Group velocities between source and receiver can be measured by a variety of spectral filtering techniques. While frequency domain filtering techniques have equivalents in the time domain, most analysis is carried out in the frequency domain for reasons of computational efficiency. Perhaps the most commonly used method for group velocity determination is the *multiple filtering* technique introduced by Dziewonski et al., 1969. By that technique, the digitized record is Fourier transformed to the frequency domain and an appropriate instrument correction is applied for the

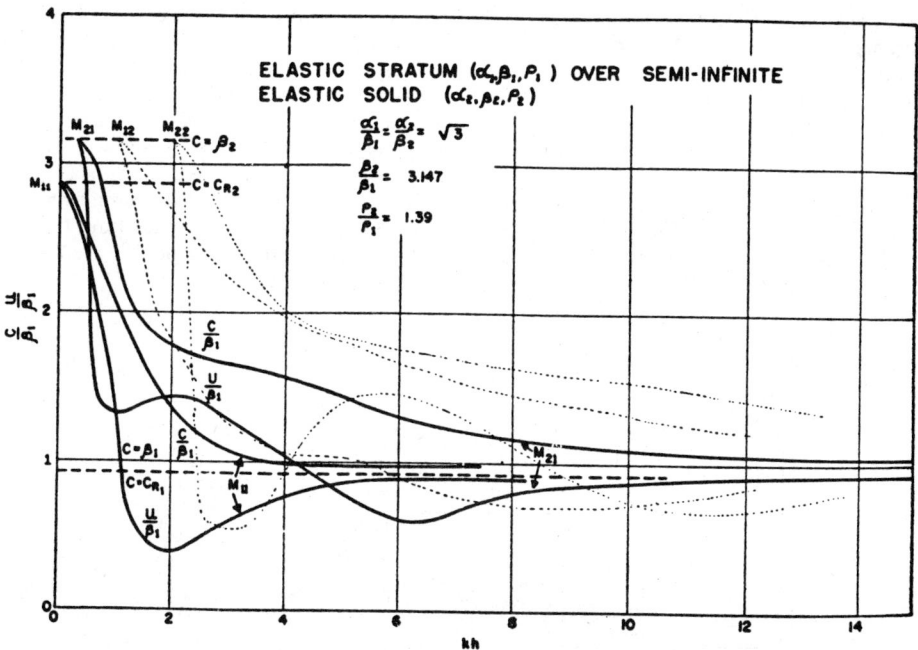

FIGURE 2. Example of Rayleigh wave group and phase velocity dispersion curves for the fundamental and first two higher modes in a medium consisting of a layer over a half-space. (From Tolstoy and Usdin, 1953)

seismograph response. The transformed seismogram is windowed over group arrival time (to minimize the effects of multipath interference) and then filtered over an array of specified center frequencies. The filter is a near-optimum zero phase shift Gaussian function, the width of which may be varied: A narrower bandpass for the filter will result in increased frequency resolution but poorer time resolution, and vice versa. The optimum bandwidth will depend upon the shape of the dispersion curve and the amplitude spectrum of the wavetrain. The filtered signal is inverse transformed back to the time domain. Instanteous amplitudes and phase angles are normalized so that their maxima are some value appropriate for printing on computer paper, usually 99, and the normalized numerical values are printed out on grids of time (or group velocity) versus period. The printed amplitudes are contoured and the group velocity curve for the wavetrain determined by tracing the maximum amplitude "ridge" on the period vs. group velocity diagram.

Phase velocity for regional studies is generally determined for pairs of matched stations aligned along the great circle path to the earthquake source, although single station phase velocity determinations and methods involving tripartite arrays are not uncommon. In the paired station methods, station spacings are typically small, and some effort is usually made to assure that the region between the stations is relatively homogeneous (i.e., does not contain significant tectonic boundaries). Thus the phase velocities can be referred to a single layered model.

The earliest phase velocity determinations were made by measuring the time delay between station pairs of peaks, troughs, or zero crossings for phases of different period in the surface wavetrain. Modern phase velocity measurements are based on spectral techniques. A much used method was developed by Block and Hales in 1968. By that technique, a group velocity dispersion curve is obtained for each station and an array of group velocity and period tabulated for that seismogram. Each seismogram is then windowed with a $\cos^2$ window of length five times the period of interest, centered on the group velocity corresponding to the specified period. The windowed seismograms are filtered using a very narrow bandpass filter and then time shifted with respect to one another and cross-multiplied over a specified range of phase velocities. Phase velocities are then corrected for differences in instrumental phase response (even for "matched" instruments) and normalized values of the product seismogram are printed onto a period-phase velocity grid (similar to that used for group velocity). Maximum positive values of the product seismogram will occur when the two records are in phase. The locus of maximum values on the phase velocity-period grid defines the phase velocity dispersion curve, although the choice of phase offset will be ambiguous by an integer wavelength. This ambiguity is seldom a problem, however, because the phase shift as a function of period must vary continuously and the choice of geophysically suitable values for phase velocity at very long periods is nearly always obvious. Figure 3 shows a typical processing flow diagram for obtaining phase velocity between a pair of stations by the method just described.

Because surface waves can be refracted out of plane of the great circle propagation path, two station phase velocity determinations can be in error. This problem can be overcome to some extent by use of a tripartite array for which one of the legs is parallel to the direction of propagation. It is not clear, however, that even a tripartite array can overcome the problems associated with significant out-of-plane refractions where different frequencies are affected differently and interference is common. L. Knopoff (1972) showed that results from the two-station method are superior to those obtained with an array in the presence of lateral inhomogeneity. As a consequence, most regional surface wave phase velocity studies are based on the two-station method.

Phase velocity can also be determined from data of a single station if the source-time function and focal mechanism of the earthquake are known. The

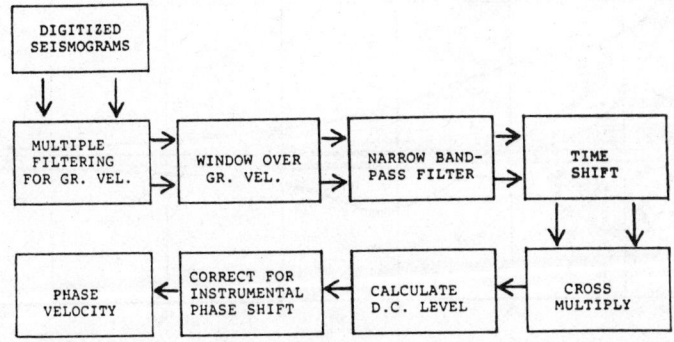

FIGURE 3. Schematic flow diagram for obtaining phase velocity by the two-station method introduced by S. Block and A. Hales in 1968.

single station method has proven to be a very powerful tool for very long period studies of global (mantle) structure (see later). For regional studies, however, source parameters are only rarely known sufficiently well to yield accurate phase velocity curves from the data, although where regional networks exist that can provide accurate locations and focal mechanisms for relatively small earthquakes, single station methods have been exploited effectively (e.g., Taylor and Patton, 1986).

Measurement of Attenuation

Surface waves have been used with varying degrees of success to measure attenuation (Q) in the Earth. The methods and results are discussed in some detail in *Seismic Attenuation: Observation and Measurement* (also, see Kovach, 1978). Surface wave attenuation measurements have been most useful for establishing global distribution of attenuation, as they depend for the most part on multiple passes of surface wavetrains as they circle the globe and are measured at a single station.

Inversion of Surface Wave Data

Standard inversion of surface wave data involves the matching of the observed dispersion curves to theoretical curves calculated for layered earth models. *All* models so obtained are non-unique to some extent as the problem is always underdetermined. Some major advances, however, have been made in the field of full waveform inversion where synthetic waveforms are matched to observed waveforms (see later discussion).

Inversion of phase and group velocity dispersion curves is based on standard inverse theory (see *Inverse Theory and Methods: Seismology*). A variety of methods starting with Thompson and Haskell in the early 1950s have been developed for calculating phase and group velocities of surface waves in layered media, corrected where necessary for sphericity of the Earth. Thus partial derivatives can be calculated for any Earth model and least squares iterative procedures applied to obtain a solution. The pioneering papers on linear inverse theory by G. Backus and F. Gilbert around 1970 made it possible to calculate the resolving kernels and errors associated with the various least squares solutions.

In practice, certain parameters are commonly assumed to be known (such as crustal thickness) or to be simple functions of other parameters (such as the ratio between compressional and shear velocity). Surface wave dispersion is most sensitive to shear velocities although for the higher modes in particular density can be important. Thus the number of the parameters can be reduced to allow modeling to focus on those parameters and regions of the earth of greatest interest. This reduction in number of parameters is partly the basis of the so-called hedgehog procedure developed by Keilis-Borok and Knopoff. The hedgehog program searches in a multidimensional parameter space for those seismic cross sections whose theoretical dispersion curves fall within a given uncertainty of the observations (Knopoff, 1972).

The basic problem of inversion of surface wave data is in the resolution of the method and the uniqueness of the derived earth models. One of the more comprehensive early efforts to determine quantitatively the degree of resolution obtainable from typical regional data was made by Der et al. (1970), based primarily on inaccuracies of measurement of group velocities of the fundamental and first higher modes of both Rayleigh and Love waves. Among points noted by Der was that over similar period ranges the Rayleigh waves provide considerably better resolution than do Love waves. When all four modes are combined, shear velocity resolution is about 0.05 km s^{-1} in the upper crust and about 0.08 in the lower crust. In the upper 50 km of the mantle, shear velocity resolution is about 0.1 km s^{-1}, and fundamental Love wave observations contribute little to resolving these upper mantle velocities. At depths of 150 km it is possible to resolve a layer 100 km with only 0.1 km s^{-1} accuracy if all four modes are used. For the same observational error, phase velocity measurements give a somewhat poorer resolution than do group velocity measurements. Knopoff (1972) showed for Rayleigh waves that the S-wave velocity at a depth of $0.4\lambda_R$ has the greatest effect on the shape of the phase velocity dispersion curve, where λ_R is the wavelength of the Rayleigh wave at the period of interest. Thus, for Rayleigh waves of period 20 s a phase velocity of 4 km s^{-1}, the depth of maximum influence is about 32 km, which means that for all periods greater than about 20 s, Rayleigh waves will be strongly influenced by the mantle in regions of "normal" continental crustal structure.

Apart from the poor resolving power of the fundamental Love mode, determination of the dispersion curve itself is highly uncertain. In particular, the group velocities of the fundamental and higher modes, particularly for oceanic paths, are nearly identical, making mode separation virtually impossible. The result is that mode interference produces spurious phase velocities, particularly at longer periods.

Results of Regional Studies

Despite the relatively small number of measurements of surface wave dispersion compared to body wave studies, much of what we know about the upper mantle, particularly shear-wave velocities, low velocity zones, and density stratification, comes from the analysis of surface waves, chiefly Rayleigh waves.

Knopoff (1972) divided the Earth into five distinct geologic provinces into which most surface wave

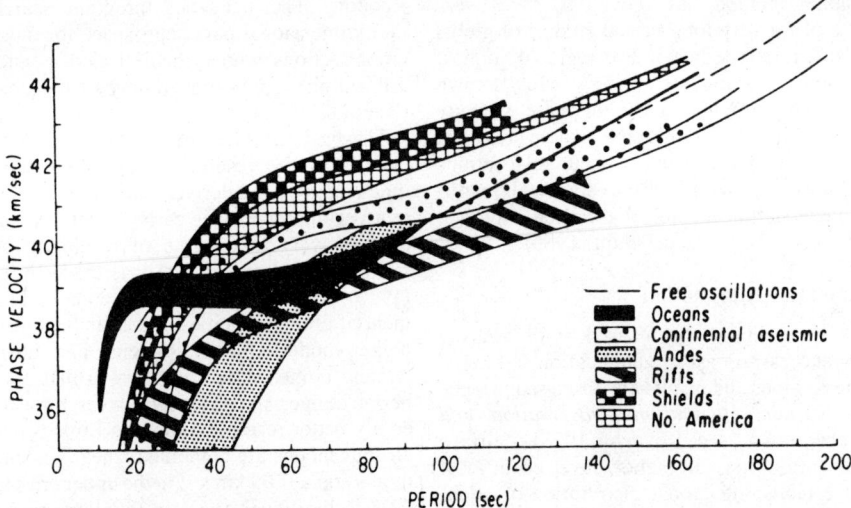

FIGURE 4. Representative phase velocity fields for the five principal tectonic provinces identified by Knopoff. (From Knopoff, 1972, with permission)

results can be categorized: (1) ancient Precambrian shields; (2) aseismic platforms; (3) rift zones; (4) mountainous regions; (5) oceanic regions. The summary figure of phase velocity curves published with Knopoff's paper still represents quite well the range of values found worldwide and is reproduced in Fig. 4. Knopoff examined the class of solutions applicable to each set of data. For shield structures, where phase velocities are higher than anywhere else in the world, the low-velocity channel for S waves, if present at all, must be very poorly developed. For aseismic platforms, the data require a well defined low velocity channel beginning at a depth of about 100 km. For rift structures, the data do not require a low velocity channel, but models without a channel all have low velocity material at the top of the mantle (low lid velocity). Mountainous structures are somewhat variable. For the Andes, the phase velocities are extremely low, reflecting the very thick crustal section. While the surface wave data certainly do not require a low velocity channel beneath the Andes, the full class of models consistent with the data have not been explored. The data for oceanic paths show very clearly that a low-velocity channel is required in the upper mantle below the lithosphere. As the lithosphere varies in thickness from just a few km near the ridge to about 100 km far from the ridge, the top of the asthenospheric low-velocity zone varies accordingly. In no instance do the surface wave data analyzed by Knopoff resolve the question of the base of the low-velocity channel.

Since the early 1970s, many additional regional surface wave studies have been done, particularly in Europe and the United States and in the ocean basins. While the results certainly add much information about the regions under study, the basic categorization of surface wave dispersion by geologic province has not changed significantly except for considerable refinement of oceanic models.

Global Surface Wave Results

Most measurements of "pure-path" regional structures of the kind described above are limited to periods of less than about 120 s. Thus, velocity and density resolution at depths greater than about 150 km is poor. To study structures deeper in the mantle requires measurements of phase and group velocities of surface waves in the period range 120–300 s. Such very long period waves may travel around the Earth many times on their great circle orbits. In standard nomenclature R_1 is the Rayleigh wave along the minor arc (shortest path) between source and receiver, R_2 travels along the major arc (the long path) between source and receiver, R_3 corresponds to travel along the minor arc plus one full global orbit, R_4 is for travel along the major arc plus one full global orbit, and so forth. The nomenclature is exactly equivalent for Love waves except that paths are designated by G_n.

In 1958 Y. Sato showed that the measurement of phase velocity for complete great circle orbits can be made without knowledge of the earthquake source or even instrumental response (since the same instrument is used to measure the phase). Thus phase velocities can be measured very accurately at a single station to periods of at least 250 or 300 s. These data, however, cannot be used to obtain local structure along the great circle paths without further constraints.

The fact that surface wave velocities are reasonably well correlated with geological province led naturally to an attempt to partition the globe into several *pure path* regions for the purpose of studying very long wavelength surface waves whose great circle orbits may cover several diverse tectonic provinces. This partitioning, termed *regionalization*, is premised on the assumption that large-scale heterogeneity in the upper mantle is sufficiently correlated with tectonic region that it can be considered to extend to great depth (several hundred km) beneath the surface geologic province with which it is associated.

The first attempts to regionalize very long period surface waves (100–350 s) were made by N. Toksöz and D. L. Anderson in 1966. All studies to about 1971 relied on simple divisions into "shield," "tectonic," and "oceanic." These divisions subsequently proved to be inadequate and many refinements in the regionalization of great circle Rayleigh wave phase velocities followed. These modified global regionalization schemes typically involved five or six categories of tectonic provinces. The global discretization is typically made on a 5° × 5° grid or larger. Regionalization of the oceans is a rather straightforward matter of categorizing regions according to age of the lithosphere. The regionalization of the continents proved to be a far more complex and controversial undertaking. There is also no general agreement as how to handle subduction zones. A comparison of various regionalization models proposed in the literature and a discussion of each in terms of its ability to fit observed Rayleigh wave data is contained in Souriau and Souriau (1983).

The advent of new global digital networks in the early 1980s brought a wealth of high quality digital data to seismology. This breakthrough meant that data could be analyzed on high-speed computers without going through the labor intensive and imprecise step of digitizing film records by hand. Not surprisingly, therefore, the 1980s also marked some of the most important advances in the history of surface wave analysis on a global scale. By the mid-1980s the blocklike regionalization methods for global studies had largely given way to low order spherical harmonic expansion representations of lateral heterogeneity. The spherical harmonic method is considered by many to be superior to the regionalization approach particularly for structures below about 300 km, where correlations with surface features begin to break down. On the other hand, regionalization schemes have the advantage that very accurate measurements can be performed (Nataf et al., 1986). For great circle observations alone G. Backus showed in the 1960s that only the even spherical harmonic components of the real heterogeneities can be retrieved from the data. This basic weakness can be circumvented by regionalization. In practice, therefore, some workers prefer to combine regionalization schemes with spherical harmonic representations. In many instances very long wavelength (mantle wave) studies of lateral heterogeneity are supplemented by body wave data to obtain full-scale inversions for mantle structure (e.g., see Dziewonski and Woodhouse, 1987, and *Earth Structure, Global* in this volume).

The most complete and most accurate method of surface wave inversion in a heterogeneous Earth is now recognized to be waveform fitting (see discussion in the section on waveform inversion). Waveform fitting is typically done using a relatively small number of large seismic events. I. Nakanishi and D. L. Anderson showed in 1982 that if the source parameters of the earthquake are known, then seismic waveforms of the mantle wave may be used to constrain local Earth structure independent of any regionalization. Woodhouse and Dziewonski (1984) obtained the earliest three-dimensional spherical harmonic models of upper mantle structure based on waveform inversion of mantle waves. Because lateral heterogeneity affects the determination of source parameters, Woodhouse and Dziewonski (1984) solved simultaneously for both source parameters and velocity structure of a heterogeneous Earth. Results from this kind of inversion and other inversions involving combined seismological data are described in more detail in *Earth Structure, Global*.

Higher Modes

Higher modes are simply the overtones of the surface wavetrain. It is useful for understanding higher modes to review some of the simplest mathematical bases for surface waves. For this purpose, we follow Aki and Richards (1980). Consider the one-dimensional problem of a Love wave propagating in a horizontal direction, x, with frequency ω, in a medium consisting of a homogeneous layer over a half-space.

$$\mathbf{u}(x, y, z, t) = \mathbf{Z} e^{i(kx - \omega t)} \qquad (3)$$

where $\mathbf{u}$ is displacement, k is wavenumber, t is time, and $\mathbf{Z}$ is a function of the depth, z. Coordinates are Cartesian, z positive downward, x positive in the direction of propagation.

To evaluate Eq. 3, we require that the waves satisfy the equations of motion and that the boundaries conditions be such that traction vanishes at the free surface, $z = 0$, and wave amplitudes vanish at $z = \infty$. Under these constraints, the problem becomes an eigenvalue problem such that nontrivial solutions of Eq. 3 exist only when for a given value of ω the corresponding wavenumbers are uniquely determined. This means that surface waves of a given frequency ω have discrete wave numbers (eigenvalues) $k_0(\omega)$, $k_1(\omega)$, etc. A major consequence of the eigenvalue relationship between ω and k is that the phase velocities of surface waves are fixed for a given frequency according the relationship

$$c_n = \omega / k_n$$

where subscript n denotes the nth mode. The *fundamental* surface wave mode corresponds to $n = 0$.

For any given frequency, ω, discrete modes exist. The higher modes have low end *cut-off frequencies* defined by

$$\omega_{cn} = \frac{n\pi\beta_1}{H} \bigg/ \left(1 - \frac{\beta_1^2}{\beta_2^2}\right) \qquad (4)$$

where β_1 and β_2 are shear-wave velocities of the layer and the half space respectively, and H is the layer thickness. Cutoff frequencies for higher mode dispersion curves based on a simple model of a layer over a half space are shown in Fig. 2.

Results

Regional Studies. While seismologists have long recognized that higher modes contain far more information about deep structures and density than do the fundamental modes, the quantitative study of higher modes without digitally recorded seismograms was a very formidable task. With the advent of high quality digital data from global stations and from broad-band arrays (such as the NARS array in Europe) it became possible by the early 1980s to exploit much more fully the information contained in the higher modes. The availability of high quality data recorded at global digital stations and on broadband stations on arrays such as the long linear NARS array in Europe has spurred the development of important new methods of analysis based on higher modes.

Higher modes figure prominently in regional studies of continental structures in that they sample to much greater depths than do fundamental modes and they provide substantially more information on density distribution with depth (Fig. 5). The increased depth penetration is particularly significant for regional studies as results from the fundamental mode alone are never sufficient to give unambiguous information on the existence or shape of mantle low velocity zones. When higher mode data are added, resolution as a function of depth improves markedly.

A major difficulty one encounters with higher modes is the fact that it is usually a multimode wavetrain that is excited. Because the modes are commonly not significantly separated in time, the problem becomes one of mode separation. Apart from waveform fitting, all analytical techniques are aimed at isolating the overtones in group velocity-phase velocity (UC) space. Two techniques in use today for multimode phase velocity data in the period range 10–100 s illustrate the basic concepts. The first method, applicable to data recorded along a linear array of stations, involves stacking (summing) of data to obtain a UC diagram on which energy is contoured for a given period (e.g., Nolet and Panza, 1976). If the length of the array is sufficiently long, there will be good mode separation so that each mode will stand out as a distinct "knot" on the UC diagram. In practice, diagrams are produced over the entire period range of interest and higher mode group

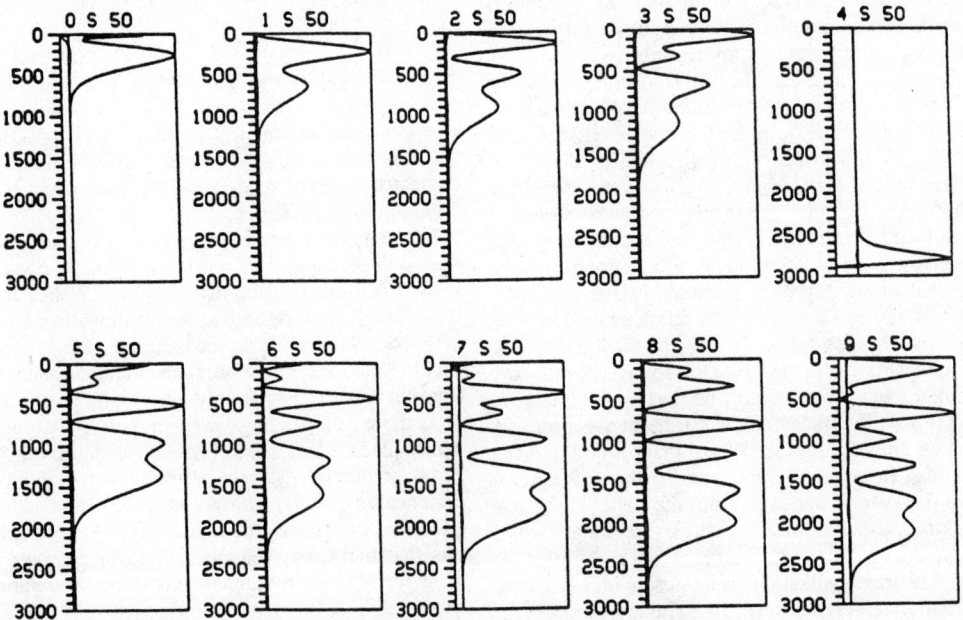

FIGURE 5. Frechet derivatives of the eigenfrequencies with respect to SV for the fundamental and first 9 overtones of the spheroidal modes with angular order 50. (From Tanimoto, 1987, with permission)

and phase velocity curves obtained directly. The second method, very similar to the first, but for application to single station data, was developed by Cara (1978). The method involves a single-station spatial filtering technique whereby a given overtone can be extracted from a set of multimode records and processed as a single-mode signal to get epicenter-station phase velocities.

Multimode phase velocity measurements along linear arrays require proper array geometry. The problems are of mode separation and identification: if the station spacing is too sparse, spatial aliasing can result in mode misidentification; for close station spacing where the total length of the array is short, however, interference may occur between modes so that they cannot be separated in UC space. Nolet and Panza (1976) have formalized the mode separation problem in terms of an "array response function." The width of the main lobe of the array response function, governed largely by the total length of the array, determines how well two or more modes that are close in wavenumber can be separated. For higher modes, array lengths of thousands of km may be required.

Among the major objectives of surface waves research is to improve the spatial resolution of the analysis. Most present methods of surface wave interpretation are based on ray approximations. The wavelength of surface waves, however, is sufficiently large to suggest that ray theory will produce inaccurate results wherever lateral structure varies over scales that are smaller than a wavelength (about 70 km at 20 s, 400 km at 100 s). This problem can be solved in principle by applying full wave theory to describe the surface wave wavefield, although there are serious computational limitations. Preliminary results from synthetic studies, however, suggest that the phase velocities of surface waves depend not only on the structure along the propagation path, but also on the structure of the surrounding region and on the geometry of the incoming wavefront.

Global Studies. The use of overtones in the analysis of three-dimensional velocity structure in the Earth came into play around the mid-1980s. Mantle structures obtained from very long period (250 s) fundamental mode surface waves cannot be resolved at depths greater than about 400 or 500 km. The depth to which structure can be reliably determined is substantially increased (to nearly 1000 km) by incorporating data from surface wave overtones in the inversion (Tanimoto, 1987). This can be seen in Fig. 5 where Frechet derivatives (a measure of the level of influence the material properties of the rock at a given depth have on the surface wave phase velocities) are plotted as a function of depth for the

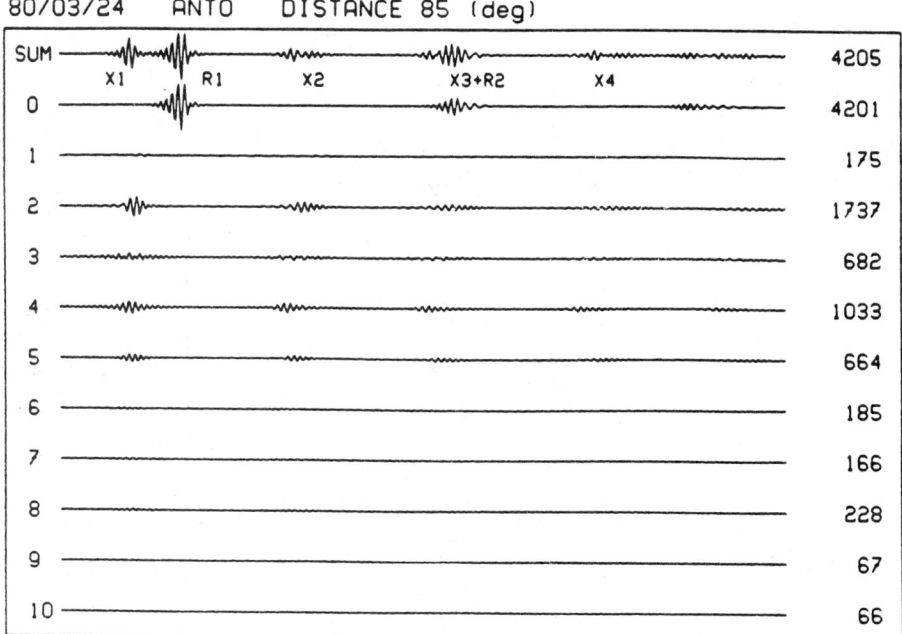

FIGURE 6. Synthetic radial component example showing higher mode (overtone) wavepackets for very long period surface waves. R_n denotes fundamental mode and X_n denotes overtone packets (see text for explanation). The top trace is the sum of fundamental and higher modes, the traces numbered 0–10 are contributions for the individual modes. Note that overtones higher than the fifth overtone contribute very little to the composite seismogram. Also note interference between X_3 and R_2. (From Tanimoto, 1987 with permission)

fundamental and various higher modes of angular order 50 (the corresponding surface wave period varies from 178 s for the fundamental to 73 s for the ninth overtone). It is quite apparent that higher modes are influenced to a far greater extent by very deep structure than is the fundamental mode.

Overtone studies have been limited primarily to the first four higher overtones (overtones 2, 3, 4, and 5), which are found to make up most of the energy in the overtone wave packets (Tanimoto, 1987). These particular overtones have similar group velocity, which means that they circle the globe as packets (see Fig. 6). By analogy with fundamental mode nomenclature, overtone wave packets are termed X_1, X_2, X_3, etc., corresponding to minor arc travel, major arc travel, minor arc travel plus one great circle orbit, etc. The higher mode packets travel much faster than does the fundamental mode, causing interference (such as R_2 with X_3 over some distance ranges). Because the overtone amplitudes on radial component seismograms tend to be large relative to amplitudes of the fundamental, inversion is typically done with radial seismograms in order to enhance the signal of the overtones.

Waveform Inversion

Waveform inversion for regional studies of surface waves is a relatively recent development. The phase perturbation approach to waveform inversion, initially developed by Lerner-Lam and Jordan (1983), is applicable to either regional or global surface wave analysis. Waveform inversion is achieved by constructing synthetic seismograms from an appropriate starting model, comparing (differencing) the synthetics directly with observed seismograms (in the time domain), and iteratively modifying the structure to minimize the residuals between synthetic and observed seismograms. The method allows for calculation of resolving kernels and error estimates of the solution.

Waveform inversion is a powerful analytical technique because it eliminates the need to obtain dispersion curves or even to identify or to separate

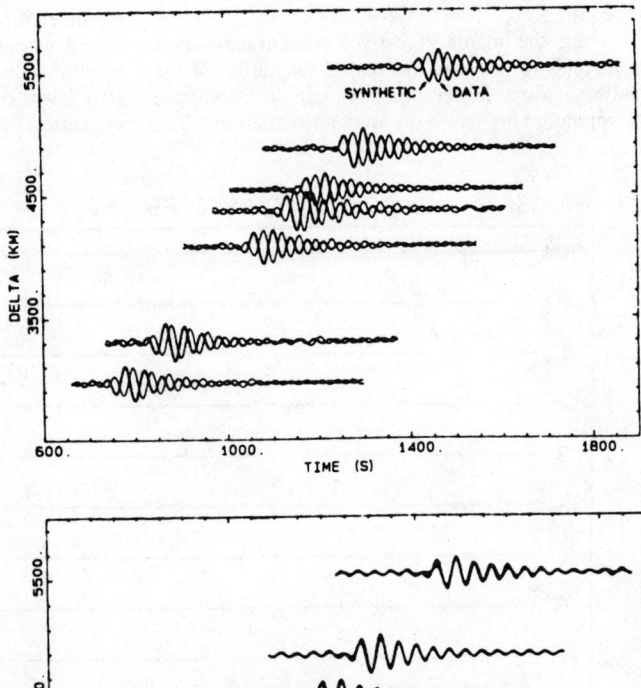

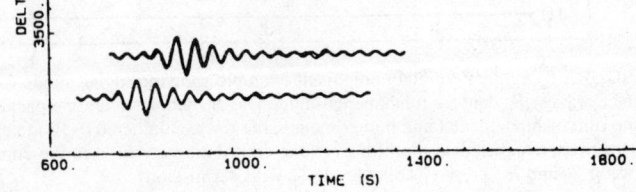

FIGURE 7. Waveform fitting of Rayleigh wave fundamental mode using nonlinear inversion for an event in NW Africa recorded across the NARS array. Figure on left shows waveform fit of the starting model, figure on the right shows waveform fit of the final model. (From Nolet et al., 1986 with permission)

modes. The successful inversion of the complete wavetrain means that all possible modes and periods have been considered. The method does have some important limitations, however, including the requirement of a starting model that is reasonably close to the actual structure. A major practical difficulty is to generate computationally fast and accurate synthetic seismograms for a heterogeneous Earth. Synthetics can be computed in various ways. For surface waves to periods of about 300 s the traveling wave formulation (e.g., Lerner-Lam and Jordan, 1983) is most computationally efficient. Complex normal mode computations to high order will produce good results but at great computational cost (see *Seismograms: Synthetic* and *Free Oscillations of the Earth*). Various approximations and simplifying assumptions must be made for all techniques and their effect on the final model evaluated. Lerner-Lam and Jordan discuss the approximations and sources of error of their method in some detail. Tanimoto (1987) has also considered the effect of various weighting schemes for global surface wave data (including overtones) and shows that they can lead to models with quite different patterns of velocity anomalies.

The waveform inversion method of Lerner-Lam and Jordan is a linearized technique and therefore imposes a limit on the frequency that can be handled.

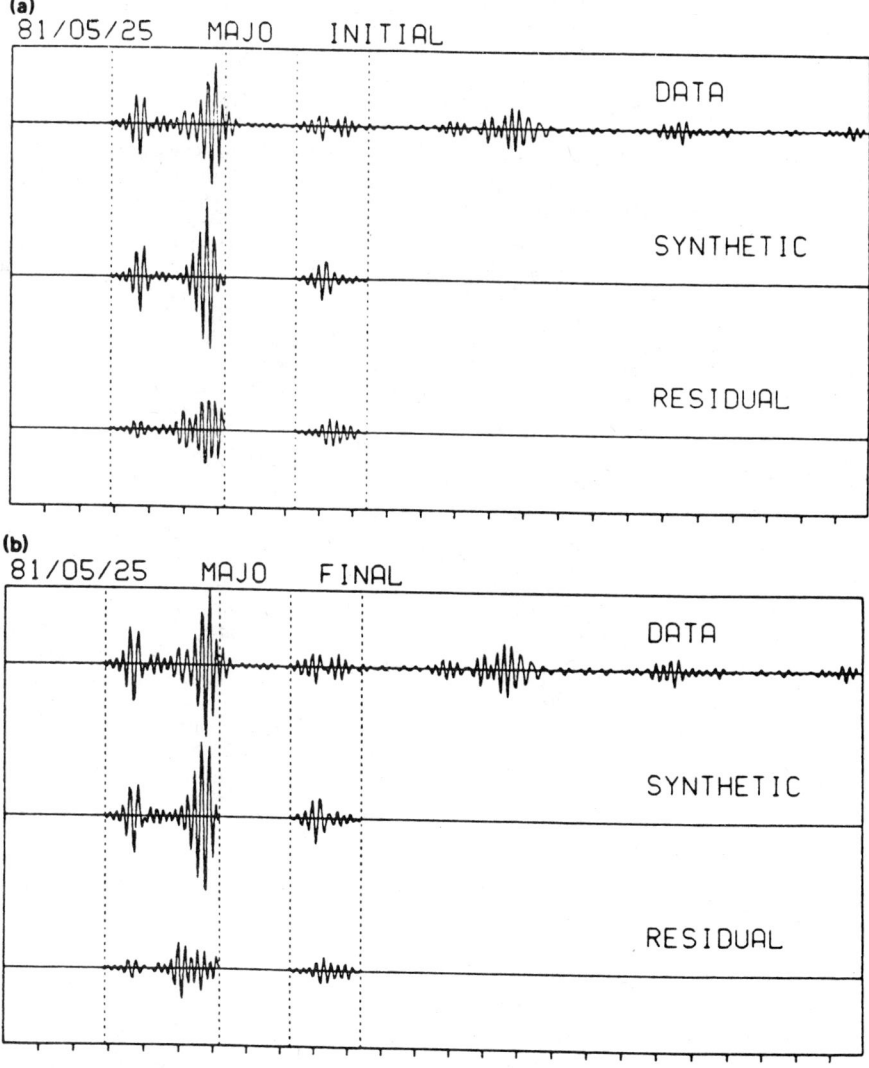

FIGURE 8. Example of waveform fitting of global long period overtone data showing observed seismograms, synthetic seismograms, and residual seismograms for minor and major arcs paths from an event in the South Pacific recorded in Japan. Top figure shows results for the starting model and the bottom figure is after inversion of the data. Tick marks are 10 minute intervals. (From Tanimoto, 1987 with permission)

This limitation is particularly serious for regional studies. To overcome the difficulties inherent in the linearized method, it was extended by Nolet et al. (1986) to include the conjugate gradient method for nonlinear waveform fitting. The nonlinear technique is robust in that it works well even for large errors in the starting model. An example of waveform fitting to a model of regionalized structure is shown in Fig. 7 for fundamental mode Rayleigh waves from an event in north Africa recorded across the NARS array.

Waveform inversion for global data is performed in a rather similar way, although commonly the inversion is made to include a focal mechanism as well as Earth structure (e.g., Woodhouse and Dziewonski, 1984). It has been shown that the accuracy of waveform modeling of Earth structure will depend quite critically on reliable determination of source parameters. In addition, uncertainty in the duration of the earthquake can be as much as 5 s, meaning that phase velocity cannot be determined accurately for minor arcs of distance less than about 50°. Tanimoto (1987) has applied full waveform inversion to fundamental and higher mode Rayleigh waves recorded on radial component seismograms. Synthetic seismograms were generated for an elliptical Earth model for all modes from the fundamental to the tenth overtone. An example showing synthetic and observed seismograms for the minor and major arc data only (delimited by dashed vertical lines) is shown in Fig. 8.

DAVID E. JAMES

References

Aki, K., and P. G. Richards, 1980, *Quantitative Seismology. Volume 1: Theory and Methods*. San Francisco: W. H. Freeman and Co., 557p.

Cara, M., 1978, Regional variations of higher Rayleigh-mode phase velocities: a spatial-filtering method, *Royal Astron. Soc. Geophys. Jour.* **54**, 439-460.

Der, Z., R. Masse, and M. Landisman, 1970, Effects of observational errors on the resolution of surface waves at intermediate distances, *Jour. Geophys. Research* **75**, 3399-3409.

Dziewonski, A., S. Block, and M. Landisman, 1969, A technique for the analysis of transient seismic signals, *Seismol. Soc. America Bull.* **59**, 427-444.

Dziewonski, A. M., and J. H. Woodhouse, 1987, Global images of the Earth's interior, *Science* **236**, 37-48.

Ewing, W. M., W. S. Jardetzky, and F. Press, 1957, *Elastic Waves in Layered Media*. New York: McGraw-Hill, 380p.

Kovach, R. L., 1978, Seismic surface waves and crustal and upper mantle structure, *Rev. Geophysics and Space Physics* **16**, 1-13.

Knopoff, L., 1972, Observation and inversion of surface-wave dispersion, *Tectonophysics* **13**, 497-519.

Lerner-Lam, A. L., and T. H. Jordan, 1983, Earth structure from fundamental and higher-mode waveform analysis, *Royal Astron. Soc. Geophys. Jour.* **75**, 759-797.

Nataf, H. C., I. Nakanishi, and D. L. Anderson, 1986, Measurements of mantle wave velocities and inversion for lateral heterogeneities and anisotropy 3. Inversion, *Jour. Geophys. Research* **91**, 7261-7307.

Nolet, G., and G. F. Panza, 1976, Array analysis of seismic surface waves: Limits and possibilities, *Pageoph.* **114**, 775-789.

Nolet, G., J. van Trier, and R. Huisman, 1986, A formalism for nonlinear inversion of seismic surface waves, *Geophys. Research Letters* **13**, 26-29.

Souriau, A., and M. Souriau, 1983, Test of tectonic models by great circle Rayleigh waves, *Royal Astron Soc. Geophys. Jour.* **73**, 533-551.

Tanimoto, T., 1987. The three-dimensional shear wave structure in the mantle by overtone waveform inversion—I. Radial seismogram inversion, *Royal Astron. Soc. Geophys. Jour.* **89**, 713-740.

Taylor, S. R., and H. J. Patton, 1986, Shear-wave structure from regionalized surface-wave dispersion in the Basin and Range, *Geophys. Research Letters* **13**, 30-33.

Tolstoy, I., and E. Usdin, 1953, Dispersive properties of stratified elastic and liquid media: a ray theory, *Geophysics*, **18**, 844-870.

Cross-references: *Continental Crustal Structure; Continental Lithosphere; Earthquake Seismology; Earth Structure, Global; Elasticity and Wave Propagation: Principles; Free Oscillations of the Earth; Inverse Theory and Methods: Seismology; Mantle, Lower: Structure; Mantle, Upper: Structure; Mantle Discontinuities; Seismic Anisotropy in the Earth; Seismic Attenuation: Observations and Measurement; Seismic Instrumentation; Seismic Instrumentation: History; Seismic Tomography; Seismograms: Interpretation; Seismograms: Synthetic; Seismology: History; Spherical Harmonic Analysis.*

THERMOREMANENCE

Thermoremanence (TRM) is the process in which a magnetic material becomes magnetized by virtue of cooling in the presence of an external applied magnetic field. That much of paleomagnetism works at all is due in large part to the peculiar properties of this process.

Consider, for instance, that the Earth's magnetic field strength is about 0.05 mT, about two orders of magnitude less than the field generated between the poles of a toy horseshoe magnet. Because many magnetic minerals are formed in the cooling of igneous rocks, and because the TRM acquired in this process can be strong and stable, the record of ancient magnetic fields recorded in igneous rocks may often be preserved to the present time (see *Natural Remanent Magnetization*).

The usefulness of some sedimentary rocks in paleomagnetic studies may also be due, indirectly, to the TRM process in that the strong magnetization of detrital grains may well have derived from a TRM in the parent igneous rock, even though the bulk magnetization of the sedimentary rock is reset at the time of deposition or shortly thereafter (see *Detrital Remanent Magnetization*).

The term "thermoremanence" used by itself generally implies a cooling from above the mineral's Curie temperature to room temperature in the presence of an external field. The external field may then be removed. This total TRM may be distinguished from a partial TRM, in which the cooling in the presence of an external field takes place through a more limited range of temperatures below the Curie point. In partial TRM experiments, cooling outside the specified range of temperatures is done in a zero external field.

The properties of thermoremanence may be summarized as follows. The direction of TRM acquired in an isotropic material is parallel to that of the external field. The strength of TRM acquired in an external field is greater than that of an isothermal remanence acquired in the same field, and is also far more stable in the presence of such destructive effects as partial alternating-field (af) demagnetization, partial thermal demagnetization, or the isothermal application of fields of opposite polarity to that of the remanence. These destructive effects form the basis for several of the techniques of magnetic cleaning used routinely in paleomagnetic studies (see *Rock Magnetism: Measuring Techniques and Apparatus*).

TRM varies linearly with the strength of the external field at low fields (generally less than 0.1 mT) and gradually saturates at higher fields. Both hyperbolic tangent and power-law relations have been used to describe the behavior of TRM (H_{ex}) between the linear and saturated regions.

Partial TRMs appear to be independent of one another. That is, the remanence acquired in a given range of cooling does not appear to affect or be affected by that acquired in any other cooling range. One result of this property is the empirical law of additivity of partial TRMs. This rule states that the sum of partial TRMs acquired in a set of nonoverlapping independent partial TRM acquisition experiments over the range from the Curie point to room temperature is identical to the total TRM that would be acquired in a single cooling process over the total range of temperatures.

Studies of partial TRM have shown that the remanence is acquired over a range of temperatures, but that many materials show a peak in partial TRM acquisition below the Curie point. The presence of this peak has given rise to the concept of a blocking temperature range, in which the magnetic configuration is said to be "blocked" or stabilized at elevated temperatures. As the material is cooled below the blocking range, the magnetization continues to increase as a result of the temperature dependence of saturation magnetization, an effect that accounts for much of the strength of thermoremanence.

TRM varies strongly with grain size of the magnetic mineral, with a stronger TRM being found in smaller grains. Experimental results show an approximate $d^{-0.7}$ dependence for magnetite, where d is grain diameter.

Variation of TRM properties is also due to the mineralogy and morphology of the grain. These variations are due to differences in Curie temperatures, saturation magnetizations, shape anisotropies, and defect structures of individual grains (see *Magnetic Properties of Minerals*).

The first comprehensive theoretical models of TRM were put forth between 1949 and 1955 by Louis Néel. He advanced two models: one for the single-domain size range, in which all atomic magnetizations within the grain are considered to be parallel; and another for the multidomain range, in which the magnetization is divided into a set of regions or domains, each with its own magnetization direction, by domain walls (see *Magnetic Domains*).

In Néel's single-domain model, the magnetization directions are constrained to a single preferred axis by either crystalline anisotropy or by a magnetic anisotropy induced by the shape of the grain. Though crystalline anisotropy in most magnetic minerals is not uniaxial, shape anisotropy often is and is generally felt to be the more important. Transitions of the domain magnetization from one direction along the preferred axis to the other (domain reversals) are governed by the combined effects of thermal excitation energy and the external magnetic field. At low temperatures and low external fields, the aniostropy energy dominates, producing a stable magnetization, but at elevated temperatures or fields the probability of a domain flipping to the opposite polarity becomes substantial.

Néel expressed domain flipping in terms of relaxation times for each of the two states: one with the magnetization pointing along the preferred axis ($\theta = 0$), and the other with it pointing in the opposite direction ($\theta = \pi$):

$$\frac{1}{\tau(0 \to \pi)} = C\left(1 + \frac{H}{H_C}\right)\left(1 - \frac{H}{H_C}\right)^{1/2}$$
$$\cdot \exp\left(-\frac{\nu M_s(H_C + H)^2}{2H_C kT}\right)$$

$$\frac{1}{\tau(\pi \to 0)} = C\left(1 - \frac{H}{H_C}\right)\left(1 - \frac{H}{H_C}\right)^{1/2}$$
$$\cdot \exp\left(-\frac{\nu M_s(H_C - H)^2}{2H_C kT}\right)$$

$$\frac{1}{\tau} = \frac{1}{\tau(0 \to \pi)} + \frac{1}{\tau(\pi \to 0)}$$

where τ is the relaxation time, H is the external field, H_C is the microscopic coercive force of the grain, ν is the grain volume, M_s is the saturation magnetization, k is the Boltzmann constant, and T is the Kelvin temperature. The constant C is a frequency factor related to the frequency of the attempts to surmount the energy barrier separating the two states.

Néel introduced the concept of a blocking temperature into his model by setting τ equal to the cooling time. Equations 1 may then be made to yield a temperature $T = T_B$, where T_B is the blocking temperature. For temperatures $T > T_B$, the system is "unblocked"; that is, the domain magnetization flips coherently and rapidly between the two opposite directions. This state has been called *superparamagnetic*, since stable remanence is not possible even though the temperature is below the Curie point. For $T < T_B$, however, the domain magnetization direction becomes frozen (blocked) with probabilities for each state determined by Eq. 1. Further cooling does nothing more than increase M_S.

Following this model, Néel derived an expression for a total TRM of the form

$$M_{\text{TRM}}(T) = M_S(T)\tanh\left(\frac{\nu M_B H}{kT_B}\right)$$

where T is the final temperature to which the material is cooled, and M_B and T_B are the saturation magnetization at the blocking temperature and the blocking temperature itself. Note that T_B is a function of H so the expression is more complex than it appears.

Néel's multidomain TRM theory was stated as an empirical model based on observed temperature dependencies of saturation magnetization and the coercive force of small magnetic grains. It has since been shown that his model is equivalent to one based upon the temperature dependencies of three energy terms: one each for the magnetostatic interaction between a single wall dividing a grain into two domains and the external field, the self-demagnetizing field of the grain, and the energy of interaction between the wall and a regular array of defects within the grain.

By itself, this model predicts a power-law dependence of TRM upon external field strength. Néel's introduction of an additional element in the form of a thermal fluctuation field produces a linear dependence of TRM upon external field strength at low fields. In essence, the thermal fluctuation field represents the effects of thermal excitations that prevent the wall from becoming trapped in extremely shallow local energy wells that are produced by the array of defects.

The Néel models predict the qualitative appearance of TRM acquisition data and that the blocking temperature decreases with increasing field strength. As a consequence of the latter effect, a TRM acquired in a strong field should be equivalent theoretically to an isothermal remanence acquired in that same field. This behavior is, in fact, observed.

As simple and fundamental as they are, the Néel models do not produce acceptable quantitative agreement with experiment. For instance, there is no apparent change in behavior at the single-domain-multidomain transition in grain size, and, indeed, most of the grains that contribute to stable remanence in rocks are above this threshold. In addition, Néel's single-domain theory predicts an approach to saturation at far weaker external fields than is observed.

The concept of pseudo-single-domain TRM was introduced in recognition of these discrepancies. On one level, the concept simply acknowledges that for grains of intermediate size (roughly 0.05–15 μm in magnetite), behavior is similar to that in the single-domain range (< 0.1 μm in magnetite), even though domain calculations predict the existence of domain walls, and they are observed to exist, in this range.

On another level, the concept calls for small irreducible moments to exist within the domain structure that behave for all intents and purposes as independent single-domain grains. These moments have been variously attributed to the effects of Barkhausen discreteness of wall positions, to small independently magnetized regions within the grain, and to the moments of domain walls themselves.

In addition to the acknowledgement of pseudo-single-domain behavior, attempts have been made to introduce additional elements into the Néel models to produce better agreement with observations. These attempts include allowing grain magnetizations to assume random angles with respect to the external field, the introduction of interactions between grains, and the calculation of more exact demagnetizing energies for multidomain grains, rather than simply assuming a constant demagnetizing factor for each grain. The latter consideration has led to the recognition that even in the absence of defects that might pin domain walls, minimum-energy states exist that have nonzero magnetizations when there are two or more domains in a grain. These states constitute essentially irreducible moments within a grain that can contribute to TRM acquisition, a process called *transdomain TRM*.

Recent observations have also shown that the nucleation of domain walls at the grain surface may in some cases be inhibited. At elevated temperatures, the number of domains is decreased, and a grain in the multidomain size range may become a single magnetic domain. As the temperature is lowered, the energy necessary to nucleate a domain wall may not be available, with the result that the grain reaches room temperature in a metastable single-domain state. Because of its volume, such a grain may contribute a disproportionate share of the total remanence.

In spite of these advances in understanding of the TRM process, it is fair to say that at present there is no satisfactory comprehensive theory of TRM acquisition for the grain-size ranges of interest to paleomagnetists. Although the fundamental physical mechanisms of magnetization described by Néel are unlikely to be displaced, the microscopic magnetic configurations and interactions that are involved in real grains probably have not yet been described adequately.

This state of affairs has come about partly due to the difficulty of preparing synthetic materials whose exact microscopic properties are fully characterized. Recent advances in the production of synthetic magnetic minerals offers the hope that this deficiency may be remedied in the not too distant future.

V. A. SCHMIDT

References

Dunlop, D. J., ed., 1977, Origin of thermoremanent magnetization. *Adv. Earth Planetary Sci.* **1**, 1–207.

Halgedahl, S., and M. Fuller, 1983, The dependence of magnetic domain structure upon magnetization state with emphasis upon nucleation as a mechanism for pseudo-single-domain behavior, *Jour. Geophys. Research* **88**, 6505–6522.

Moon, T., and R. T. Merrill, 1984, The magnetic moments of non-uniformly magnetized grains, *Physics Earth and Planetary Interiors* **34**, 186–194.

Moon, T., and R. T. Merrill, 1985, Nucleation theory and domain states in multidomain magnetic material, *Physics Earth and Planetary Interiors* **37**, 214–222.

Nagata, T., 1961, *Rock Magnetism*. Tokyo: Maruzen.

Néel, L., 1949, Theorie du trainage magnétique des ferromagnétiques en grain fins avec applications aux terres cuites, *Ann. Geophys.* **5**, 99–136.

Néel, L., 1955, Some theoretical aspects of rock magnetism, *Adv. Phys.* **4**, 191–242.

O'Reilly, W., 1984, *Rock and Mineral Magnetism*. Glasgow: Blackie & Son, Ltd. (distr. in USA by Chapman & Hall, NY).

Schmidt, V. A., 1973, A multidomain model of thermoremanence, *Earth and Planetary Sci. Lett.* **20**, 440–446.

Stacey, F. D., and S. K. Banerjee, 1974, *The Physical Principles of Rock Magnetism*. Amsterdam: Elsevier.

Wörm, H.-U., and H. Markert, 1987, The preparation of dispersed titanomagnetite particles by the glass-ceramic method, *Physics Earth and Planetary Interiors* **46**, 263–269.

Cross-references: Curie Temperature; Demagnetization; Geomagnetic Field, Main: Theory; Magnetic Domains; Magnetic Properties of Minerals; Magnetic Self-Reversal; Natural Remanent Magnetization (NRM); Rock Magnetism; Rock Magnetism: Measuring Techniques and Apparatus.

THIN-SKIN TECTONICS

Deformation of the solid Earth occurs in a great variety of styles, reflecting the wide gamut of possible physical conditions, material properties, strain rates, and strength anisotropies. It is generally assumed that styles of deformation are uniquely related to large scale (tectonic) and small scale (structural) fabrics and other recognizable characteristics. In fact, geologists have been quite successful in reconstructing deformation histories from snapshots of their cumulative effects by characterizing structure at a wide range of scales in ever increasing detail and by uniquely correlating characteristics of deformation and structural style.

Structural style is often found to be relatively homogeneous within domains or tectonic units that are typically bound by decoupling surfaces or major faults. In this context, thin-skin tectonics refers to a family of deformation systems and fault arrays in the upper lithosphere with distinct characteristics in terms of structural style and domain morphology. Thin-skin tectonics is largely controlled by subhorizontal decoupling layers or detachments. Structural domains decoupled from underlying tectonic units by these detachments tend to be thin vertically compared to their horizontal dimensions. These horizontal decoupling surfaces or layers generally coincide with a marked contrast in the style and amount of deformation. One prominent component of this deformation is usually shear strain or displacement on the detachment and the lateral transport of the hanging-wall block above this boundary with respect to the foot wall block below.

The concept of a detachment surface is fundamental to thin-skin tectonics. In some of the recent literature the term *detachment* is applied exclusively to cases where the overall pattern of deformation is horizontal extension. In this discussion, the term *detachment* is used with a more general meaning: a shallow-dipping fault or decoupling layer characterized by a displacement that generally decreases updip. This spatial variation of displacement on the detachment is accommodated by internal deformation on one or both sides of the detachment. In some cases the asymmetry in deformation across the detachment corresponds to a sharp contrast in rheology.

Typically, deformation is limited to the hanging-wall block and is concentrated in the thinner part of this block above the shallow portion of the detachment, while the footwall block behaves rigidly. Internal deformation in the hanging-wall block may completely absorb the displacement on the detachment in the updip direction, sometimes developing a blind detachment. In the down-dip direction deformation within the hanging-wall block tends to decrease so that the detachment degenerates into a regular fault separating media with similar mechanical properties and the tectonics ceases to be thin-skin.

Tectonic deformation is often coupled with other processes that affect the solid earth at or near the surface. Most important are the processes linked to the circulation of fluids and gasses, chiefly erosion and deposition. On one hand, the removal or addition of material at the surface affect the load and alter subsurface stresses that drive the deformation. On the other hand, the rates of erosion and deposition depend on relief and other factors that are often the result of tectonic deformation. Thus, tectonic deformation and material transport may form a feedback loop. Since erosion tends to remove weight from uplifting blocks and deposition tends to add weight to subsiding blocks, surface material transport and tectonic deformation are often in a positive feedback relationship.

Deformation is another concept that needs an introduction in the context of thin-skin tectonics. The geologist analyzes the long term cumulative effect of deformation. The geodesist measures the instantaneous effect of deformation and tends to consider deformation rate. Finally, from the seismological point of view, deformation is made up of two components, permanent deformation and elastic or recoverable deformation. Through the cycle of elastic strain accumulation and strain release by seismic fault-slip, the pattern of deformation and the contribution of the two components of strain tend to vary drastically. The integrated effect of the stick-slip cycle can then be related to the rate at which deformation is accumulated.

Since the locus of deformation can migrate through the material as the structure evolves, cumulative and instantaneous deformation generally differ. In fact, relatively high rates of deformation may occur where accumulated deformation is particularly low. This relationship tends to characterize zones where high tectonism coincides with zones of deposition. A good example is provided by accretionary wedges. These wedges occur along the edge of many overriding blocks at the plate boundary of convergent margins. The highest deformation rates occur in the newly accreted, relatively undeformed sediments closest to the toe of the overthrusting plate. Thin-skin deformation in extensional allochthons is also concentrated near their thinner ends, which are often characterized by a depositional environment.

The ubiquity and fundamental role of thin-skin deformation over a wide range of scales and tectonic regimes can be related to the tendency for mechanical properties in the lithosphere to be horizontally stratified. Different factors tend to contribute to this stratification at different scales. Generally, pressure, temperature and composition are strongly dependent on depth and vary relatively less laterally. These factors tend to control mechanical properties at the scale of the lithosphere. Lithology becomes an important factor at a smaller scale. Lithology is often horizontally stratified reflecting layering in undeformed sediments (stratigraphy), or reflecting stacks of structural units, such as thrust sheets, emplaced in previous orogenies (preexisting structure).

Elevated pore-fluid pressure (superhydrostatic) lowers the stress levels required for deformation. High pore pressure results from a reduction of pore space (consolidation) caused by deeper burial of sediments or by increased lateral stress. Superhydrostatic pore pressure tends to be dissipated by fluid flowing out of the system. High pore pressure transients, however, may persist through geologic time in the presence of aquicludes. Since permeability is coupled with lithology, pore pressure effects can be expected to enhance the horizontal layering of mechanical properties derived from stratigraphy and structure.

The fundamental characteristics of detachments and thin-skin tectonics are applicable to the three most common types of tectonic regimes: compressional, characterized by horizontal shortening and by a vertical least compressive stress axis; extentional, characterized by horizontal extension and by a vertical maximum compressive stress axis; transcurrent, characterized by horizontal shear and by vertical intermediate stress axis. The important role of thin-skin tectonics in compressional and extensional regimes has been documented in many cases covering a variety of settings and scales. Its role in regimes dominated by horizontal shear is now beginning to be recognized. Examples from the first two of these deformation regimes will be discussed.

Historical Perspective

In thin-skin tectonics, horizontal transport and shear on subhorizontal surfaces predominates over vertical movements. Thus, another name for thin-skin tectonics could be horizontal tectonics. The conceptual evolution of geology is characterized by a steady shift of emphasis from vertical to horizontal tectonics. This shift is still occurring at present.

The strong emphasis on vertical deformation that characterized the early stages in the evolution of geological science in the nineteenth century can be understood in terms of observational bias (e.g., vertical faults that intersect the surface and displace the horizontal layering are more detectable than horizontal faults parallel to stratification) and in terms of resistance against accepting large deformation in the crust (e.g., observed superpositions of deep water sediments on continental clastics required less deformation if accomplished by vertical rather than horizontal tectonics). The concept of isostasy became widely accepted after mountains were found to have roots. By this concept the Earth's outer shell is subdivided into blocks weakly coupled by subvertical faults. Tectonic forces act only at the base of the blocks causing differential vertical movements. This perturbation results in erosion, deposition, and gravity spreading, which redistribute mass at or near the surface and require continuing vertical movements to keep the blocks in isostatic equilibrium. The role of vertical tectonics as the fundamental mechanism for geologic deformation has been decreased by the weight of observational evidence for large scale horizontal deformation.

The wide acceptance of plate tectonics in the late 1960s can be considered a dramatic step in the transition from vertical to horizontal tectonics. The evolution of plate tectonics is symptomatic of the contrasting roles often played by evidence and conceptualization in geology, particularly regarding large-scale horizontal movements.

Plate tectonics predicts large horizontal displacements of lithospheric plates and provides the mechanism for tangential stresses, but it considers deformation only along narrow plate boundaries, not within the plates. Generally, the zones of high strain that define the plate boundaries are indeed relatively narrow, but in some cases, particularly where continental lithosphere is involved, these zones can be quite broad and diffuse (Molnar, 1988). The Basin and Range zone of extension and the Alpine-Himalaya convergence zone, for example, occupy large portions of the North American and the Asian plates, respectively.

Most deformation belts appear to include important thin-skin components. New data from the subsurface have tended to increase the significance of detachments between undeformed autochthonous (in place) "basement" and overriding blocks. Thus, belts of deformation may be wider at the surface than in the subsurface. Is it more appropriate to consider them zones of intraplate deformation or interplate boundaries with broad zones of thin-skin reactivation? One of the present challenges is to understand how wide or how narrow decoupling zones between plates are at various depths through the lithosphere, and how plate motion is best described in terms of surface deformation and thick versus thin-skin tectonics.

Thin-Skin Tectonics in Zones of Compression

Strain profiles across deformation boundaries where relative motion includes a component normal to the boundary tend to be asymmetrical, particularly where continental lithosphere is involved. This asymmetry is well developed across most convergence zones, is also documented in zones of extension and may be relevant along boundaries of transcurrence. The asymmetry derives from the propensity of the lithosphere for shear failure rather than bulk strain. Thus, a shallowly inclined zone of concentrated shear often accounts for most of the motion normal to the boundary. Basically this component of motion is accommodated by relative movement of a hanging-wall block or overriding slab, over a footwall block. As discussed earlier, the thinner part of the hanging-wall block does not behave as a rigid slab, but tends to deform internally in a pattern that can be described as thin-skin tectonics detached from the underlying footwall or basement.

Convergence zones are named oceanic or continental when the underriding lithosphere is oceanic or continental, respectively. Schematic illustrations of these regions are shown in Fig. 1. These two families of deformation belts are similar in many ways: the general comments about convergence zones in the introduction, for example, apply to both. In particular, thin-skin deformation characterizes the forward wedges of the overriding blocks in both types of convergence zones and are often called fold and thrust belts in continental zones and accretionary wedges in oceanic zones. Some illustrative cross sections across the Himalayan and Appalachian belts (Fig. 2) illustrate thin-skin tectonics associated with continental convergence. Figure 3 contains a line drawing from a COCORP line in the southern Appalachians and an interpretive block diagram from Cook and Oliver (1981). The similarity between the structures observed in these cross sections and those across accretionary wedges

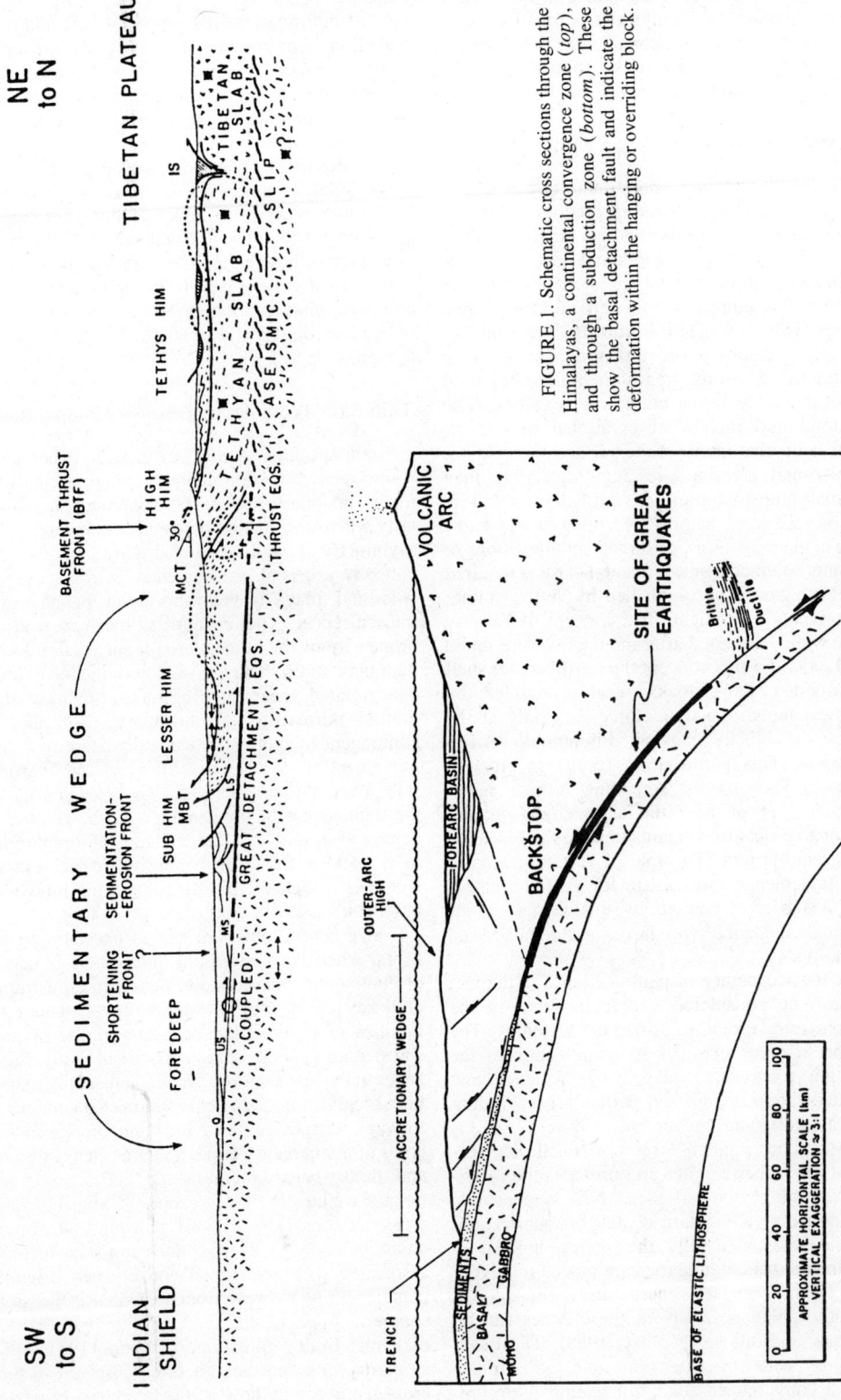

FIGURE 1. Schematic cross sections through the Himalayas, a continental convergence zone (*top*), and through a subduction zone (*bottom*). These show the basal detachment fault and indicate the deformation within the hanging or overriding block.

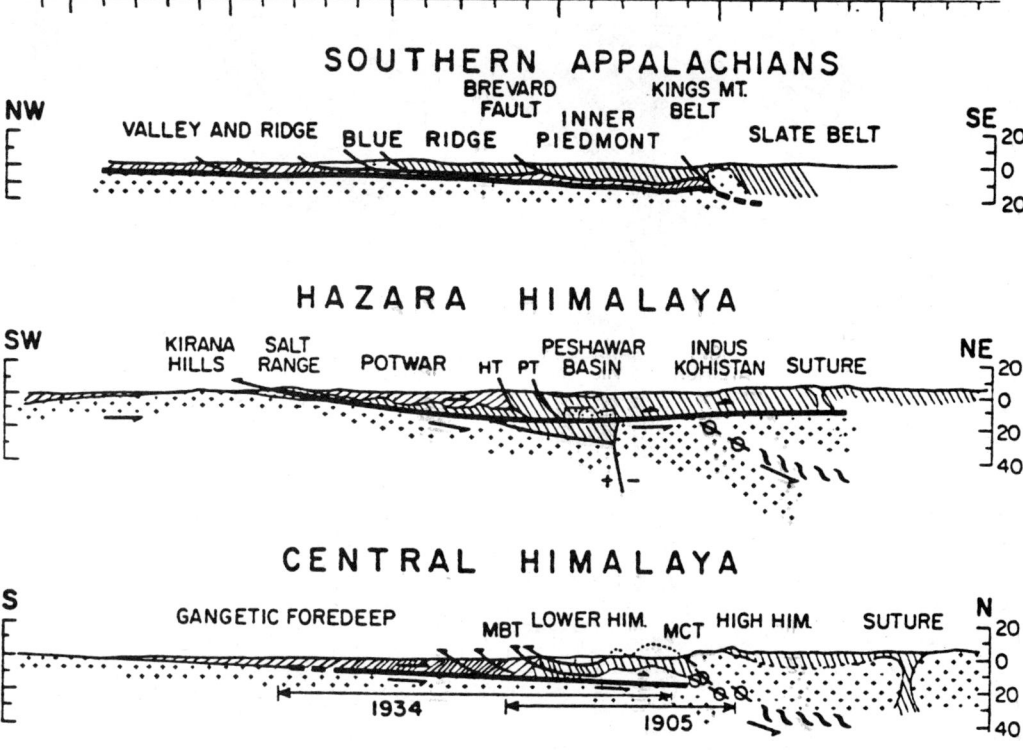

FIGURE 2. Three cross sections through continental convergence zones. (From Seeber and Armbruster, 1981)

can be seen by comparing these figures with Fig. 4, which shows line drawings from migrated seismic sections across accretionary wedges.

Models for the Development of Hanging Block Wedges. In collisional tectonic settings, whether ocean-ocean, continent-ocean, or continent-continent, a major detachment surface occurs above which the toe of the overriding or hanging block develops into a wedge-shaped tectonic unit with a characteristic pattern of internal deformation. The great similarities between these hanging block features suggests that the overall mechanical process behind their development is also similar. In all cases a décollement or detachment develops between the colliding blocks or plates along which large amounts of subhorizontal motion occurs. The hanging block deforms through horizontal shortening, both distributed as well as through folding and faulting. Motion along the thrust faults results in the accumulation of imbricate slices one on top of the next, producing an overall wedge shape with the narrow end in the direction of motion of the hanging block. The foot block, however, often remains relatively undeformed.

In the last century large thrust sheets were recognized by geologists working in the Alps, the Juras, and in Scotland. They found that displacements on subhorizontal faults of 100 km or more were demanded. Soon after, the mechanics through which relatively thin, broad sheets of rock could slide over a detachment became a source of controversy. Tectonic stress could not drive thin, relatively weak sheets of rock over distances many times their thickness without drastically altering their shape. The force that needs to be applied to the back of a thrust sheet to overcome the frictional resistance along its base is larger than the crushing strength of the rock being pushed. Hence, the rock of the thrust sheet would crush rather than slide. The mechanics regulating the motion of thrust sheets persisted as a perplexing enigma and is still not fully resolved.

A number of models have now been put forward to explain the mechanics of thrust faulting. For many years gravity sliding was accepted as the mechanism for thrust sheet movement. This model assumed that thrust sheets move entirely under the force of gravity, that is, the sheet essentially slides downhill. An important variation of this theory, called gravity spreading, suggested that a region of higher elevation would spread outward under the influence of gravity, allowing thrust sheets to slip updip as long as the overall mass increment was from higher toward lower elevations along the Earth's surface. Chapple (1978) pointed out some of the shortcomings of these models: most thrust sheets have been thrust uphill, not all regions of high elevation exhibit gravity sliding or spreading, and the cumulative amount of shortening is greatest in the interior of

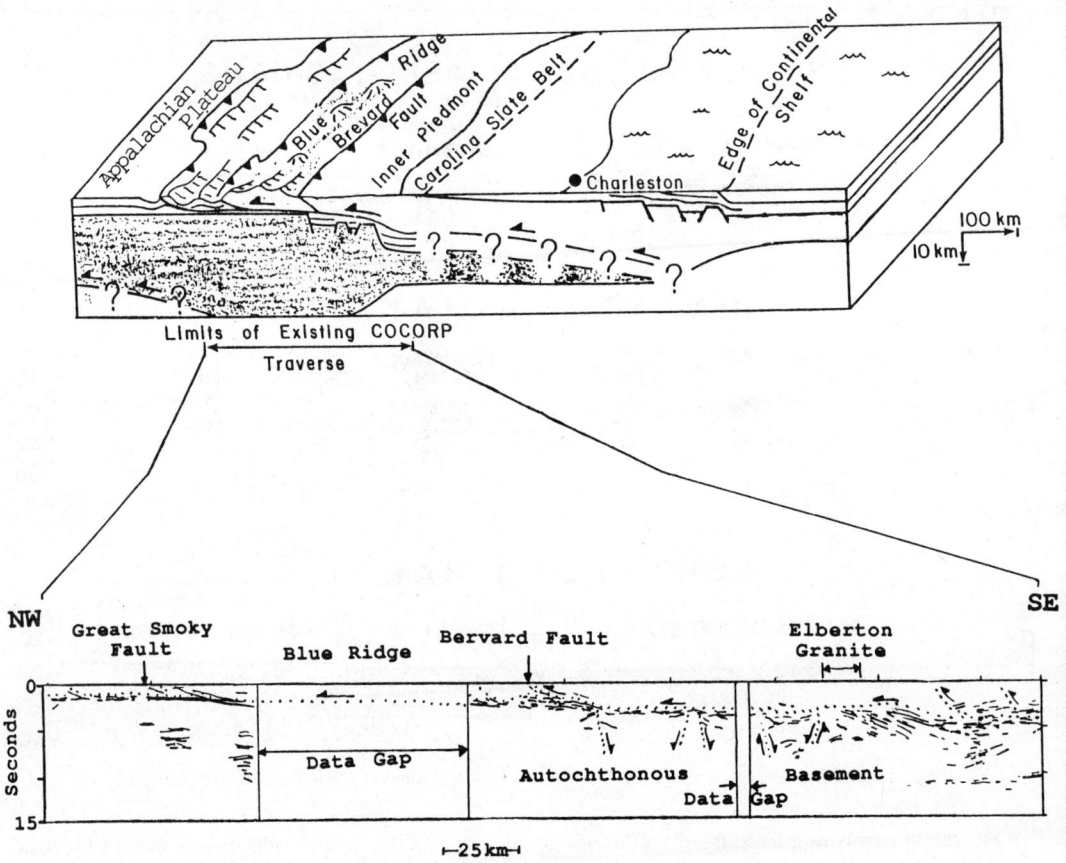

FIGURE 3. Interpretive block diagram (modified from Cook and Oliver, 1981) and line drawing from southern Appalachian COCORP seismic line (from Cook et al., 1979). Note the long detachment fault above which allochthonous material has been moved to the left.

most hanging blocks rather than at the toe as these models predict. Because of such problems, it is generally accepted today that gravity alone is insufficient to explain the motion of hanging block wedges.

A large advance in theory came from Hubbert and Rubey (1959), who showed that with sufficiently high fluid pressure maintained along a basal detachment large rectangular thrust sheets could be pushed from behind and remain intact. Shear stress on a fault is proportional not to just the normal stress, but to normal stress minus the fluid pressure, a quantity called the effective normal stress. Hubbert and Rubey determined the maximum length of rectangular thrust sheets that can be pushed horizontally without deformation as a function of the fluid pressure along the base. They also showed that large sheets are able to slide down gently inclined planes given sufficient fluid pressure. Though important, high fluid pressure alone does not solve the enigma of large thin sheets sliding distances many times their thickness.

Thrust belts are not composed of single rectangular sheets, but are composed of many imbricated and folded sheets. The overall shape of the deformed domain is that of a wedge when viewed in cross-section perpendicular to the belt. By 1890 a laboratory model had been used to show the development of a hanging block wedge from the overthrusting of imbricate slices (Cadell, 1890), and this model was used to explain the structures observed in the northwestern Highlands of Scotland. Another large advance in understanding thrust mechanics came with the work of Chapple (1978), who considered the wedge shape of thrust belts. He developed a model that included a wedge-shaped hanging block of relatively weak material undergoing pervasive defomation, decoupled from the underlying block by a weak basal layer. This model explained the observed basic characteristics of thrust sheets, including their sliding without deformation. Hanging block wedges are analogous to the wedge of snow or soil that develops in front of a bulldozer blade. The body deforms locally, progressively thickening at its back where it is pushed and thus forms a wedge shape. The wedge grows in taper through internal deformation until a critical taper is reached at which

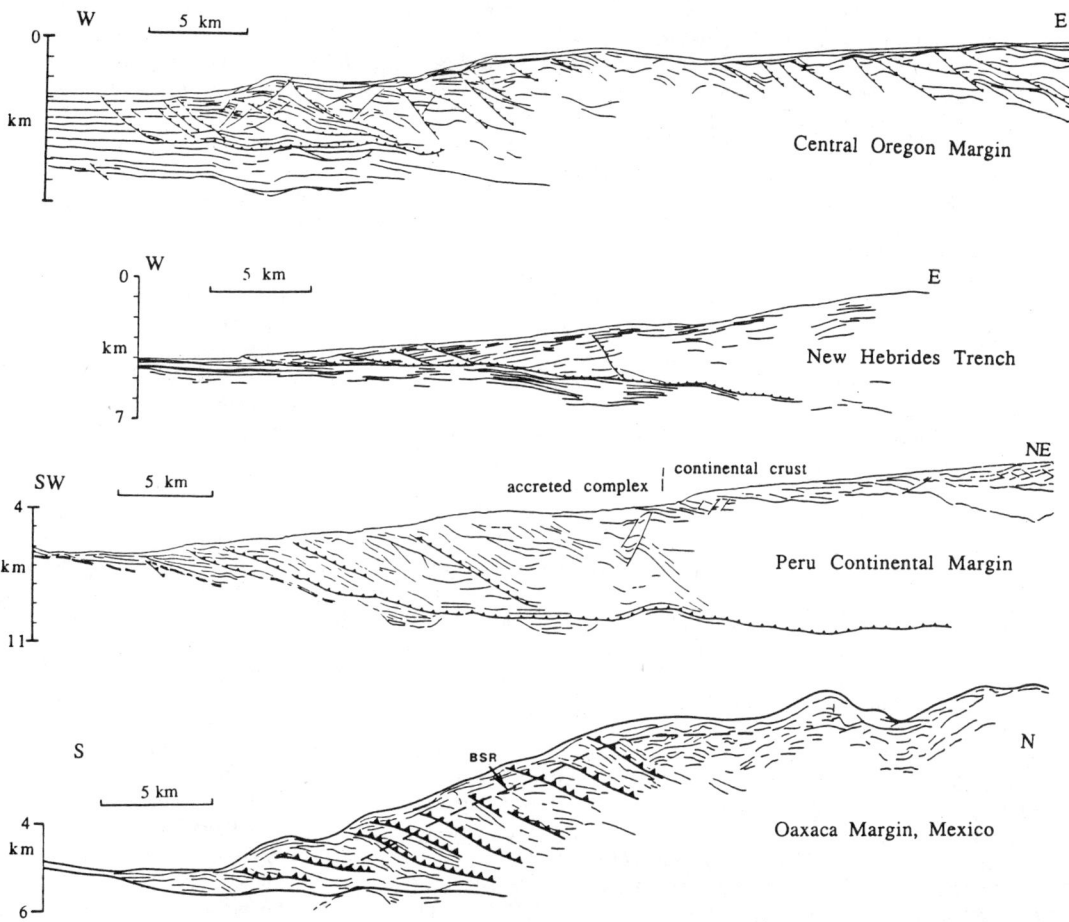

FIGURE 4. Line drawings from several migrated depth sections across accretionary wedges illustrating typical thin-skin tectonics associated with oceanic convergence. (From von Huene, 1987)

point the material is able to slide stably along its base. Similarly, in the case of thrust sheets the rock mass is tapered toward its toe because the total frictional resistance along the base decreases in that direction. Chapple assumed a plastic material was appropriate in his model, but the same critical taper theory applies to models with other rheologies. Other models have employed a Coulomb (brittle) material (Davis et al., 1983) and a viscous material (Emerman and Turcotte, 1983). The brittle theory is often applied to fold-and-thrust belts and accretionary wedges because these occur at shallow (10–15 km) depths where most rocks are brittle. Thus, the Coulomb critical-taper model of Davis et al. (1983) has gained wide acceptance today.

Examples of Collisional Thin-Skin Tectonics.

Continental Convergence. Uplift resulting in subaereal exposure and dissection has allowed geological exploration of continental convergence zones such as the Himalayas and the Appalachians. Recently, subsurface data have become available to constrain the three-dimensional structure of these belts, primarily seismic reflection data in the Appalachians, and earthquake data in the Himalayas. The Appalachians are a Paleozoic convergence zone subjected to many kilometers of denudation since the end of the convergence phase and the Himalayas are a convergence zone active since the Cenozoic. Both of these belts are considered to be the result of continental collision and both have absorbed shortening measured in the hundreds of kilometers.

In the following discussion H refers to the Himalayas and A refers to the Appalachians. The footwall or foreland, cratonic India (H) and cratonic North America (A), is flexed downward by the load of the overriding slab (hinterland) and forms a foreland basin (Indo-Gangetic Basin–H) filled with synorogenic sediments. Deformation in this basin may be small, in some cases consisting primarily of layer-parallel shortening above a blind detachment (Appalachian plateau). A sharp deformational front may mark the outer (toward the foreland) boundary of the fold and thrust belt proper (Allegheny Front–A). This belt typically includes two distinct structural domains, one involving synorogenic sediments of the foreland basin (Sub-Himalayas), the other preorogenic rocks (Lesser Himalayas). Pronounced differences in structural styles may reflect differences in lithology and rheology. The boundary between these two domains is marked by a major thrust emplacing older over younger rocks (Main Boundary Thrust–H; Blue Ridge Thrust–A). A major deep-rooted thrust (the Main Central Thrust–H; possibly the Brevard Fault and other deep rooted thrusts in the Piedmont at different stages of the evolution of the Appalachians) marks the boundary between the part of the overriding block or wedge that is shortening in a thin-skin mode and the thicker and deeper part of this wedge that is behaving rigidly (the High Himalayas and the Tibetan Plateau; 10+ kms of denudation in the Piedmont since the Paleozoic accounts for the current low topography in parts of the Appalachians structurally equivalent to the High Himalayas).

The concept of a master detachment below the Appalachian and the Himalayan fold belts had been gradually developing from interpretations of surface data. During the last decade these structural features were imaged with subsurface data: seismic reflection data in the Appalachians (COCORP) and earthquake data in the Himalayas (Tarbela network). The full extent of these structural features and of the related thin-skin deformation turned out to be beyond the most daring hypotheses. The large amounts of shortening, measured in hundreds of kilometers, which were resolved from surface structural data across the belts must be matched by forward transport of the allochthons. These large displacements on the detachments raise the possibility of foreland sediments buried deep below allochthonous crystalline rocks and fuel speculations on possible hydrocarbon reservoirs.

The large dimensions and high slip rates associated with master detachments of active convergence zones are pertinent to the issue of earthquake hazard. Studies of historic seismicity suggest that many active detachments accumulate much of their slip during major earthquakes and that most of the known great earthquakes were generated by detachments at convergence zones, both oceanic and continental. In fact, studies of large earthquakes have greatly contributed to the understanding of master detachments, which tend to be blind structures and not directly visible at the surface. A typical scenario would have a portion of a detachment, measured in several tens to a few hundred kilometers, locked for several tens to a few hundred years while the elastic strain accumulates in response to the continuing convergence. When the stress reaches a critical value, a rupture starts and propagates on the detachment releasing much of the accumulated elastic strain.

The repeat time for major slip events on the same portion of a detachment may be longer than the period covered by reliable reporting of earthquakes. Moreover, detailed earthquake data suggest that master faults tend to be aseismic at the small magnitude level during the interseismic period between major events. Thus, portions of active detachments that are currently accumulating elastic strain may be blind and invisible from direct surface observations, may not have generated large earthquakes during the historic period, and may not be readily recognized as active faults in three-dimensional representations of background seismicity. These potential sources of large destructive earthquakes, however, can now be placed in a well understood structural context, may be identified from a tectonic analysis of the orogen, and should be considered in assessments of earthquake hazard.

Oceanic Convergence: Subduction Zones. In subduction zones oceanic crust descends beneath continental crust or island arc crust along a major plate boundary thrust. In a simplistic view the overriding plate is analogous to a bulldozer blade and by scraping material from the top of the subducting plate a growing wedge of accreted material develops at the front of the overriding block. In reality, wide variation is observed among and along convergent margins in the nature and size of the structures. Some margins develop extremely large accretionary wedges, others have very small wedges, and some margins experience tectonic erosion of the overriding plate rather than accretion.

The shallowest part of the plate boundary is a thin-skin detachment surface, usually referred to as a décollement in accretionary wedge settings. Above this part of the plate boundary large deformation occurs in the accreted sediments of the hanging block, but sediment beneath the décollement remains relatively undeformed. At greater depths neither plate experiences large deformations, and the tectonics there are not characteristic of thin-skin deformation. Décollements have been widely observed (e.g., Fig. 4) on single and multi-channel seismic profiles since the mid-1970s (e.g., Seely et al., 1974; von Huene, 1987). These detachment surfaces have been traced under the upper plate for large distances, for example, up to 122 km in the Lesser Antilles where an extremely large accretionary wedge occurs (Westbrook et al., 1988). Strong evidence from drilling, hydrology, and mechanical considerations of the very narrow wedge taper commonly observed indicate that the décollement is the site of high pore-fluid pressures, probably greater than 90% of lithostatic pressure in some margins such as the Lesser Antilles.

Large variation is observed in the style of accretion and deformation among accretionary wedges.

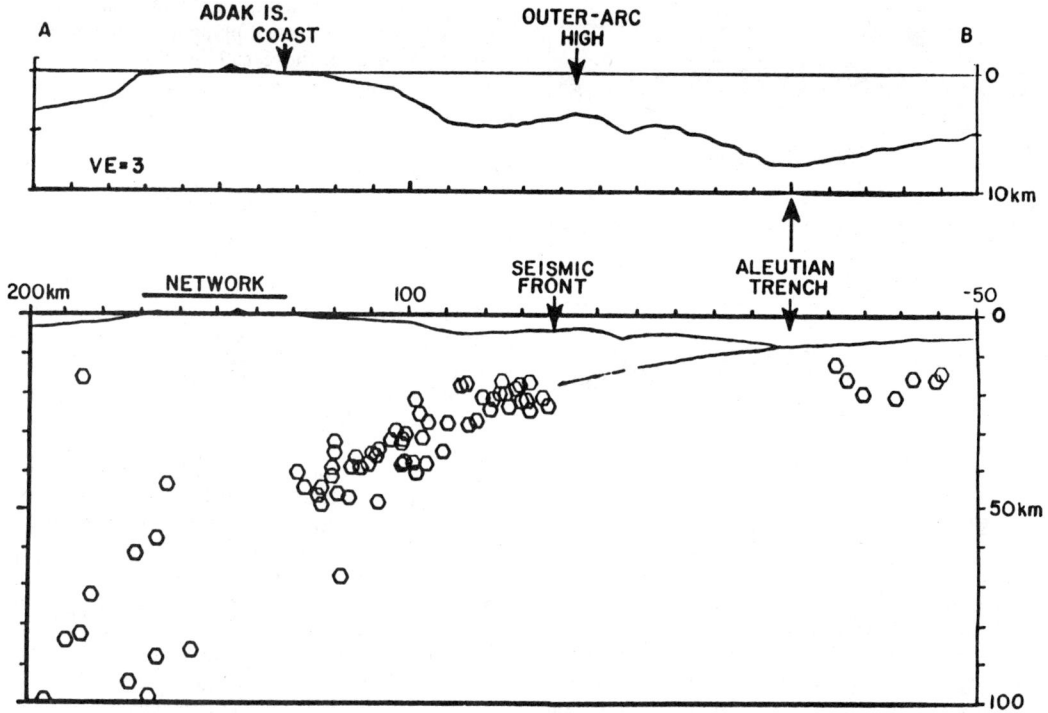

FIGURE 5. Cross section of topography and earthquake hypocenters through the central Aleutian Arc (from Byrne et al., 1988). Note that the shallowest part of the plate boundary, the décollement, is aseismic while many earthquakes are located along the deeper part of the plate boundary.

The variation is caused by differences in parameters such as the volume and type of sediment available for incorporation into the wedge, and in the convergence velocity and direction.

A typical cross section across a subduction zone (here the central Aleutian Arc) showing earthquake distribution and topography is shown in Fig. 5. The largest recorded earthquakes occur in subduction zones as thrust events along the plate boundary. These earthquakes commonly nucleate at depths of 30–40 km and rupture both updip and along strike. The shallowest part of the plate boundary is largely aseismic, however, as is the accretionary wedge overlying that part of the plate boundary. High pore-fluid pressure and the resultant low effective stress combined with the stable frictional slip properties of unconsolidated material leads to the observed aseismicity. In regions with large amounts of sediment a significant portion of the sediment column can be subducted past the trench beneath the décollement. Some of this sediment is then underplated to the base of the overriding plate and some is subducted to large depths. This excessive sediment may cause a larger portion of the shallow plate boundary to move aseismically resulting in a subduction zone that generates smaller earthquakes than it would if no sediment subduction were occurring.

Thin-Skin Tectonics in Zones of Extension

Major detachment faults are also very important in the extension of regions of continental lithosphere. These regions have thinned crust and normal fault geometries that indicate large amounts of extension. As indicated earlier, the hanging block overlying the detachment fault exhibits a deformation style characteristic of thin-skin tectonics. Initial crustal extension models assumed symmetric lithospheric stretching over large regions of the crust (e.g., McKenzie, 1978). These pure-shear models were produced to explain the crustal thinning, high heat flow, and subsidence observed in large sedimentary basins.

Most regions of large extension, however, have structural asymmetry on a wide range of scales, which has become the basis for a number of more recent extension models known broadly as simple-shear models. These models employ a large shallow dipping detachment fault or shear zone (e.g., Wernike, 1981; Lister et al., 1986; Buck et al., 1987), in which a brittle upper crust truncated below by a large low-angle detachment fault along which large displacement occurs (Fig. 6). Listric normal faults that sole out in the detachment cut the upper block into tilt blocks that move against each other in dominolike rotations (Fig. 7). The resulting shallow

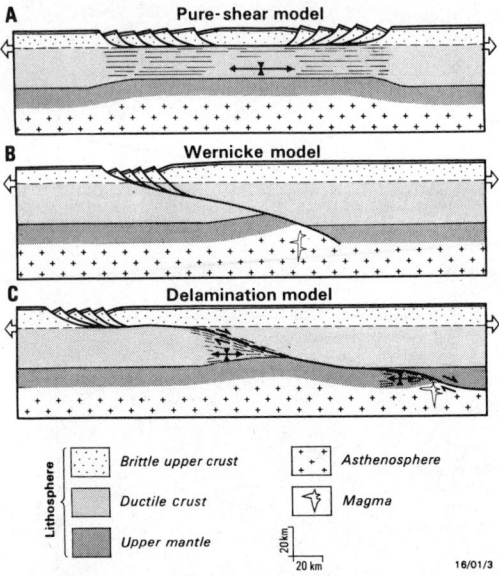

FIGURE 6. Schematic diagrams of three extension models. (From Lister et al., 1986)

that most of the earthquakes in extensional regimes are along relatively steeply dipping normal faults that seem to be beneath the detachment and the listric faults commonly observed in seismic lines and by geologic mapping.

Future Outlook

Recent new constraints on subsurface structure in the continental crust from a variety of techniques, particularly reflection profiling, has tended to emphasize the role played by large scale horizontal faulting and thin-skin tectonics in diverse tectonic environments. Most of the work, as reviewed here, pertains to zones of convergence and extension. In addition, recent work on the San Andreas fault system suggests that large scale detachment tectonics may be important there as well.

It is now clear that detachment faults extend over large distances and appear to have survived more than one orogenic event without offsets or discontinuities. Thus decoupling horizons in the upper crust may be inherited from previous orogenic events and may be reactivated in diverse tectonic regimes. In addition to crustal scale detachments that have primarily been addressed here, these features are also observed on the scale of rock outcrops and it appears that similar mechanics act on these smaller scales as well. "Basement" rock involved in near-surface deformation may have been transported far from its root. Thus, the width of lithospheric deformation may be narrower than the belts of surface deformation, or even than the inner part of these belts where crystalline and "basement" rocks are present. So far, subsurface data have tended to decrease the upper limit for the width of lithospheric boundary zones. Stripped of the thin-skin component, plate boundaries may appear much narrower than they do at the surface. A better understanding of lithospheric interactions, particularly at continental plate boundaries, depends largely on a better understanding of the degree of coupling between the tectonic regime in the upper crust and the rest of the lithosphere.

LEONARDO SEEBER
DANIEL BYRNE

structure is very similar to that observed in collisional settings. The lower plate is dragged up along the detachment fault in extensional settings rather than being thrust under as it is in collisional examples. Some models (e.g., Wernicke, 1981) favor a large normal fault that cuts the entire lithosphere, others (e.g., Lister et al., 1986), known as delamination models, use a detachment zone that runs horizontally beneath the brittle-ductile transition, steepens, then again runs horizontally at the base of the crust. At depth, these faults may become wide shear zones.

Recent modeling of the mechanics of extensional faulting using both simple and pure-shear models (Buck et al., 1987) has shown that each may be valid in different regions of extension. A form of pure-shear model appears to best fit the geophysical data from the northern Red Sea, while the simple shear models are more appropriate for other regions such as the Basin and Range. It is also becoming clear

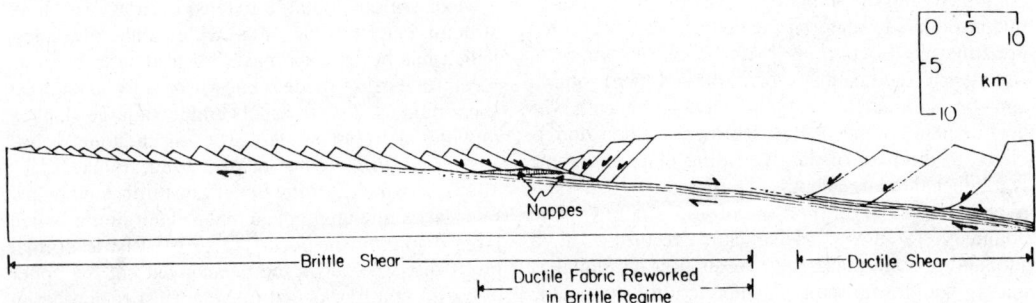

FIGURE 7. Schematic diagram showing the thin-skin structure in an extensional region. (From Wernike, 1981)

References

Buck, W. R., F. Martinez, M. S. Steckler, and J. R. Cochran, 1988, Thermal consequences of lithospheric extension: pure and simple, *Tectonics* **7**, 213–234.

Byrne, D. E., D. M. Davis, and L. R. Sykes, 1988, Loci and maximum size of thrust earthquakes and the mechanics of the shallow region of subduction zones, *Tectonics* **7**, 833–857.

Cadell, H. M., 1890 Experimental researches in mountain building, *Royal Soc. Edinburgh Trans.* **35**, 337–357.

Chapple, W. M., 1978, Mechanics of thin-skinned fold-and-thrust belts, *Geol. Soc. America Bull.* **89**, 1189–1198.

Cook, F. A., and J. E. Oliver, 1981, The late Precambrian-Early Paleozoic continental edge in the Appalachian orogen, *Am. Jour. Sci.* **281**, 993–1008.

Cook, F. A., D. S. Albaugh, L. D. Brown, S. Kaufman, J. E. Oliver, and R. D. Hatcher, Jr., 1979, Thin-skinned tectonics in the crystalline southern Appalachians; COCORP seismic-reflection profiling of the Blue Ridge and Piedmont, *Geology* **7**, 563–567.

Davis, D., J. Suppe, and F. A. Dahlen, 1983, Mechanics of fold-and-thrust belts and accretionary wedges, *Jour. Geophys. Research* **88**, 1153–1172.

Emerman, S., and D. Turcotte, 1983, A fluid model for the shape of accretionary wedges, *Earth Planetary Sci. Letters* **63**, 379–384.

Hubbert, M. K., and W. W. Rubey, 1959, Role of fluid pressure in mechanics of overthrust faulting, *Geol. Soc. America Bull.* **70**, 115–166.

Lister, G. S., M. A. Etheridge, and P. A. Symonds, 1986, Detachment faulting and the evolution of passive continental margins, *Geology* **14**, 246–250.

McKenzie, D., 1978, Some remarks on the development of basins, *Earth Planetary Sci. Letters* **40**, 25–32.

Molnar, P., 1988, Continental tectonics in the aftermath of plate tectonics, *Nature* **335**, 131–137.

Seeber, L., and J. G. Armbruster, 1981, Great detachment earthquakes along the Himalayan arc and long-term forecasting, in *Earthquake Prediction—An International Review*, Maurice Ewing Series 4. Washington, D.C.: American Geophysical Union, 259–277.

Seely, D. R., P. R. Vail, and G. G. Walton, 1974, Trench slope model, in C. A. Burke and C. L. Drake, eds., *The Geology of Continental Margins*. New York: Springer-Verlag, 249–260.

von Huene, R., 1987, Seismic images of modern convergent margin tectonic structure, AAPG studies in geology #26.

Wernike, B., 1981, Low-angle normal faults in the Basin and Range Province: nappe tectonics in an extending orogen, *Nature* **291**, 645–648.

Westbrook, G. K., J. W. Ladd, P. Buhl, N. Bangs, and G. Tiley, 1988, Cross section of an accretionary wedge: Barbados Ridge Complex, *Geology* **16**, 631–635.

Cross-references: *Accretionary Tectonics: Examples from the North American Cordillera; Brittle Phenomena; Continental Collision Zones: Seismotectonics and Crustal Structure; Continental Crustal Structure; Continental Drilling: Ultradeep; Continental Rifting: Types and Rates; Continental Rifting: Structural Traits; Crustal Movements and Tectonic Deformation; Deformation of Rocks and Minerals; Earthquakes and Crustal Deformation; Lithosphere: Mechanical Properties; Ocean-Continent Transition: Structure; Stress in the Earth's Lithosphere; Subduction Zones.*

TSUNAMIS

Tsunamis are a special type of oceanic surface gravity waves. Formally, they belong in the same fundamental classification as ordinary sea waves that can be observed daily at the beach. Tsunamis, however, are distinct in their mode of generation, in their characteristic period, and in their effect upon the shore where they impinge. Unlike ordinary waves generated by surface winds, tsunamis are produced by a shift in the position of the seafloor. Although ocean floor shifts can originate from undersea landslides and volcanic eruptions, the most common cause is submarine earthquakes. Reflecting this, tsunamis are sometimes called *seismic sea waves*. Compared with ordinary wind-driven waves, seismic sea waves have a much longer period and wavelength and, as a result, have a profoundly different effect on coastlines. Shoreline defenses such as riprap and breakwaters are designed to withstand storm waves with periods of 6–10 s and wavelengths of 100 m or so. With periods of 200–2000 s and wavelengths of tens of kilometers, tsunamis easily overwhelm most defenses.

To many, tsunamis conjure up thoughts of killer waves. Indeed, the amplitude of sea waves associated with the greatest earthquakes can be impressive. Open ocean heights of 5 m to 10 m are possible. Upon reaching shore, the waves shoal and are amplified by a factor of 2 or 3. Rightly so, the largest, damaging tsunamis come to mind when the subject is discussed; however, it is important to keep some perspective. Tsunamis over a meter or two in height are actually rare, needing about a magnitude 8 earthquake for their production. On a global average, less than one magnitude 8 earthquake occurs per year; and of these, perhaps one in ten is located under the ocean and capable of sea wave excitation. Just as in earthquake studies, the largest damaging events receive most of the publicity, but it is the smaller events that are often more useful scientifically, simply because they are much more frequent. A tsunami of just a few centimeters height is easily observable by modern pressure sensors even in the open sea. Waves of this size can be generated by earthquakes of magnitude 6.5, and occur several times per year.

The life of tsunami waves covers three phases: generation, propagation, and disposition at the receiver. This article concentrates on the generation of tsunamis, and emphasizes the seismic origin of the waves.

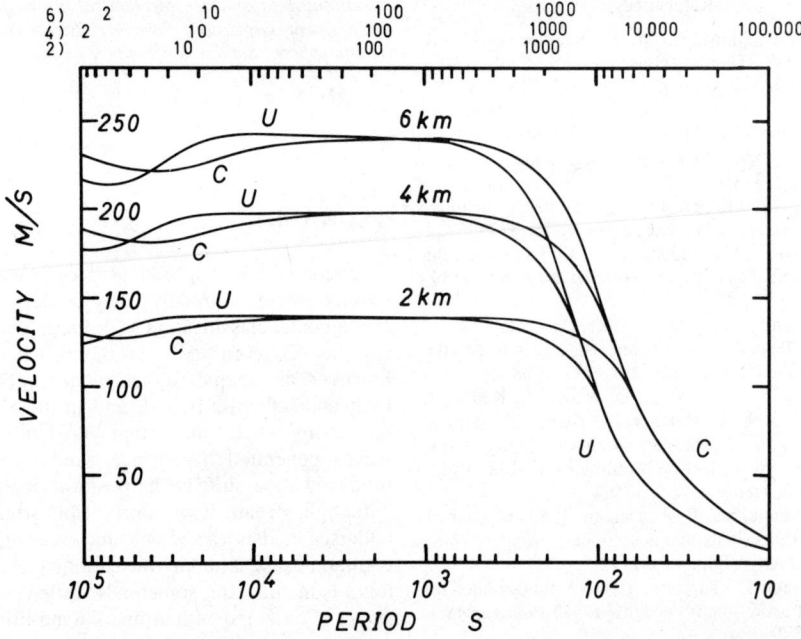

FIGURE 1. Tsunami dispersion curves for oceans 2, 4, and 6 km deep on a spherical Earth.

Characteristics of Tsunamis

Tsunamis are distinguished from other types of waves by their phase and group velocity, by their mode of energy partition, and by their distribution of particle displacement versus depth (eigenfunction shape). Figure 1 plots the phase (c) and group (u) velocity of tsunami waves on an elastic, spherically symmetric, self-gravitating Earth covered by uniform oceans 2, 4, and 6 km deep. Tsunamis travel quite rapidly, reaching speeds of 200–250 m · s^{-1} (720–900 km · hr^{-1}) in the open ocean. This speed is about 15 times the speed of ordinary waves at 10-s period and helps account for tsunamis' destructive power.

Largely, but not totally, gravity acts as the restoring force for tsunami motion. Some energy is stored elastically by compressing ocean water. In addition, because tsunami motions do penetrate the Earth, a small fraction is also stored by compressing and shearing the crustal layers below the seabed. Table 1 lists the fraction of potential energy stored gravitationally and elastically at four representative periods. At least 95% of the energy stored in tsunamis is of gravitational origin. With decreasing wave period, less and less energy is stored elasti-

TABLE 1. Energy Partition in Tsunami Modes

Period (s)		Gravitational P.E.	Compressional Elastic P.E.	Shear Elastic P.E.
14,300	Earth	0.001	0.014	0.022
	Ocean	0.957	0.006	0.000
	Total	0.958	0.020	0.022
1503	Earth	<0.001	0.002	0.004
	Ocean	0.987	0.006	0.000
	Total	0.988	0.008	0.004
150	Earth	≪0.001	0.001	0.001
	Ocean	0.996	0.002	0.000
	Total	0.996	0.003	0.001
50	Earth	≪0.001	≪0.001	≪0.001
	Ocean	1.000	≪0.001	0.000
	Total	1.000	≪0.001	≪0.001

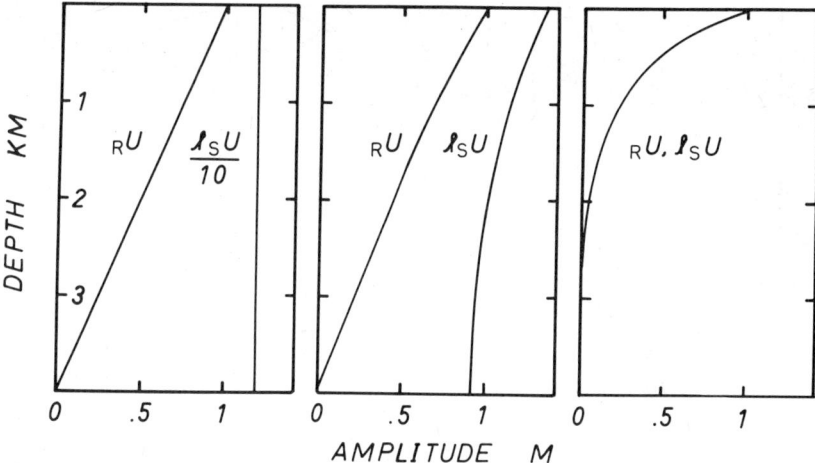

FIGURE 2. Vertical $_RU$ and horizontal l_SU tsunami eigenfunction components versus depth in a 4-km-deep ocean. Left to right are the tsunami periods 1503, 150, and 50 s. Vertical displacement at the ocean surface has been normalized to 1 m in each case.

cally. At 50 s period, the waves are virtually pure surface gravity modes of the ocean. The 50–100 s period range represents the high-frequency cutoff of tsunami excitation. Waves of these periods are so closely confined to the topmost kilometer or two of ocean that they cannot be excited by earthquake motions on the seafloor.

Two simplifying approximations are common in discussions of water waves of length λ in oceans of depth h: long-wave ($\lambda \gg h$) theory and short-wave ($\lambda \ll h$) theory. Long-wave theory predicts nondispersive propagation with $c = u = \sqrt{gh}$ and a linear variation of the eigenfunctions with depth in the water. Short-wave theory predicts dispersive propagation with

$$c = 2u = \sqrt{\lambda g/2\pi}$$

and exponentially decaying eigenfunctions with depth. It turns out that tsunamis have an intermediate character, behaving like shallow-water waves at their longest periods and like deep-water waves at their shortest periods. At intermediate frequencies, neither the long-wave nor short-wave simplification is adequate. A rigorous treatment in an elastic Earth requires a numerical approach that works directly with the equations of motion and not the simplified equations obtained through long- or short-wave theory.

Many properties of tsunami waves can be best understood by directly examining their eigenfunctions. Figure 2 plots vertical $_RU$ and horizontal l_SU amplitudes of displacement in an ocean 4 km deep. Left to right in the figure are the eigenfunctions at long (1503 s), intermediate (150 s), and short (50 s) tsunami periods. At 1503 s the tsunami has a wavelength of $\lambda = 297$ km and acts like a long wave.

The vertical displacement is maximum at the surface and small, but nonzero, at the seafloor. The horizontal displacement is constant and exceeds the vertical component by a factor of 10 or more. As the wave period is reduced to 150 s, λ decreases to 26 km—a length comparable to the ocean depth. Long-wave characteristics begin to break down, displacements are no longer linear functions of depth, and horizontal and vertical eigenfunctions more closely agree in amplitude. At 50 s the transition to deep-water waves is complete. Horizontal and vertical motions are equal and exponentially decreasing with depth.

Because tsunamis are produced by earthquake faults at the seafloor and below, it is most important to consider how the eigenfunctions behave, not only in the ocean but also in the elastic Earth. Figure 3 plots vertical and horizontal tsunami displacements versus depth below the seafloor for the same periods as in Fig. 2. Now, the eigenfunctions have similar shapes at each period. The two most revealing features of the eigenfunctions are their small amplitude and characteristic penetration depth. Low-frequency eigenfunctions have larger amplitudes and penetrate deeper into the Earth than do those at higher frequencies. The overall rapid decay of eigenfunction amplitude with increasing depth and frequency is the single most important feature governing the tsunamigenic effectiveness of earthquakes. Deep seismic sources (> 60 km) excite only the gravest tsunami periods. Even then, because of the small amplitudes of the eigenfunctions within the Earth (10^{-3} to 10^{-7} of surface displacements), earthquakes will have to be extremely large even in the best circumstances to excite sea waves to damaging levels. The need for very large earthquakes ($M > 8$) to generate a disastrous tsunami has been confirmed by historical observation.

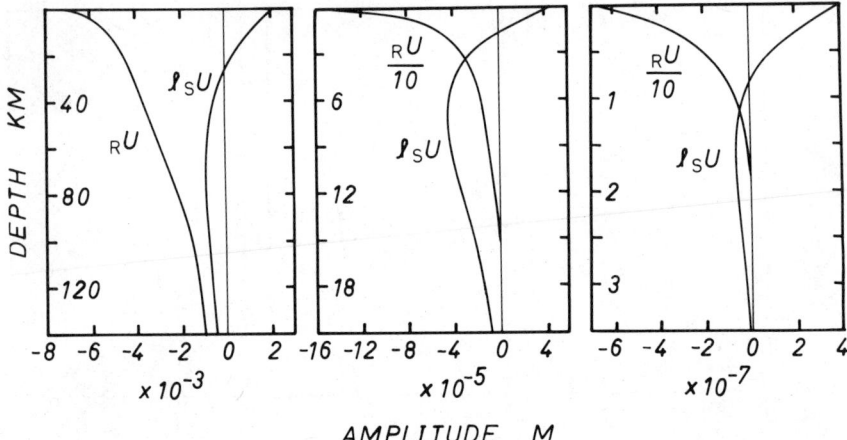

FIGURE 3. Tsunami eigenfunction components as a function of depth in the solid Earth. The modes and normalization are the same as in Fig. 2. Note the changing vertical scale.

Seismic Excitation of Tsunamis

Earthquakes produce most tsunamis. Not surprisingly, the parameters of earthquakes determine many of the characteristics of the generated sea waves. Earthquakes result from slip-on faults; although there can be dozens of parameters that describe the process, three are most important: earthquake moment, mechanism, and depth. (See *Seismic Source: Observations*; *Seismic Source: Theory*; *Earthquakes: Magnitude, Energy, and Intensity*.) The moment M_0 of an earthquake is a scalar measure of its intrinsic strength. Moment is the product of rigidity μ, fault area A, and average slip Δu. Moment and the more popular measure, magnitude, are related through various empirical formulas. One relation in common use defines a moment magnitude as $M_w = \frac{2}{3}(\log M_0 - 9.05)$. For magnitudes less than 7.5, M_w and surface wave magnitude M_s are more or less equivalent. For larger earthquakes, M_w and M_s can differ substantially: M_s, defined by the amplitude of 20-s surface waves, saturates near $M_s = 8.3$ and no longer increases with increasing earthquake moment. On the other hand, M_w, defined in terms of M_0, continues to grow with increasing moment, reaching values of 9.5 for the largest earthquakes known. Earthquake strength varies over a factor of 10^5 within the magnitude range $6.5 \leq M_w \leq 9.5$ (see Table 2). Even without detailed models of tsunami generation, it is safe to suppose that the larger the intrinsic size of the earthquake, the larger the tsunami, all else fixed.

Earthquake *mechanism* (see *Earthquake Mechanisms*) specifies the orientation of the fault surface, commonly assumed to be a segment of a plane. Earthquake fault mechanism can be summarized in three parameters: the strike and dip angles of the fault, and the direction angle of slip. The role of fault mechanism on tsunami production is not as obvious as the influence of moment; however, one might suspect that faults that generate large vertical displacements of the seafloor would be more effective than faults that generate large horizontal displacements.

Earthquake *depth* is self-explanatory. Because tsunami generation depends on the deformation of the seafloor, the separation of the earthquake fault from the bottom of the ocean is an important parameter. One could speculate that a deep earthquake would be less effective in tsunami production than a similar earthquake close to the ocean floor.

Synthetic Models of Tsunami Motions

To quantify our impressions concerning the role of earthquake parameters on tsunami generation, we must compute numerical, or synthetic, tsunami waveforms. One compact way to do this is to view tsunamis as one of the Earth's normal modes of vibration. Normal mode theory formulates the ith component of displacement at a position $\mathbf{r}$ resulting from a point source at position $\mathbf{r}_s$ as

TABLE 2. Variation of Earthquake Moment versus Magnitude

Magnitude M_w	Moment M_0 (Nm)	Fault Length (km)	Process Time (s)
6.5	6.3×10^{18}	32	11
7.0	3.5×10^{19}	56	19
7.5	2.0×10^{20}	100	33
8.0	1.1×10^{21}	180	59
8.5	6.3×10^{21}	320	110
9.0	3.5×10^{22}	560	190
9.5	2.0×10^{23}	1000	330

Note: Columns 3 and 4 give typical values for earthquake fault length and total duration of faulting.

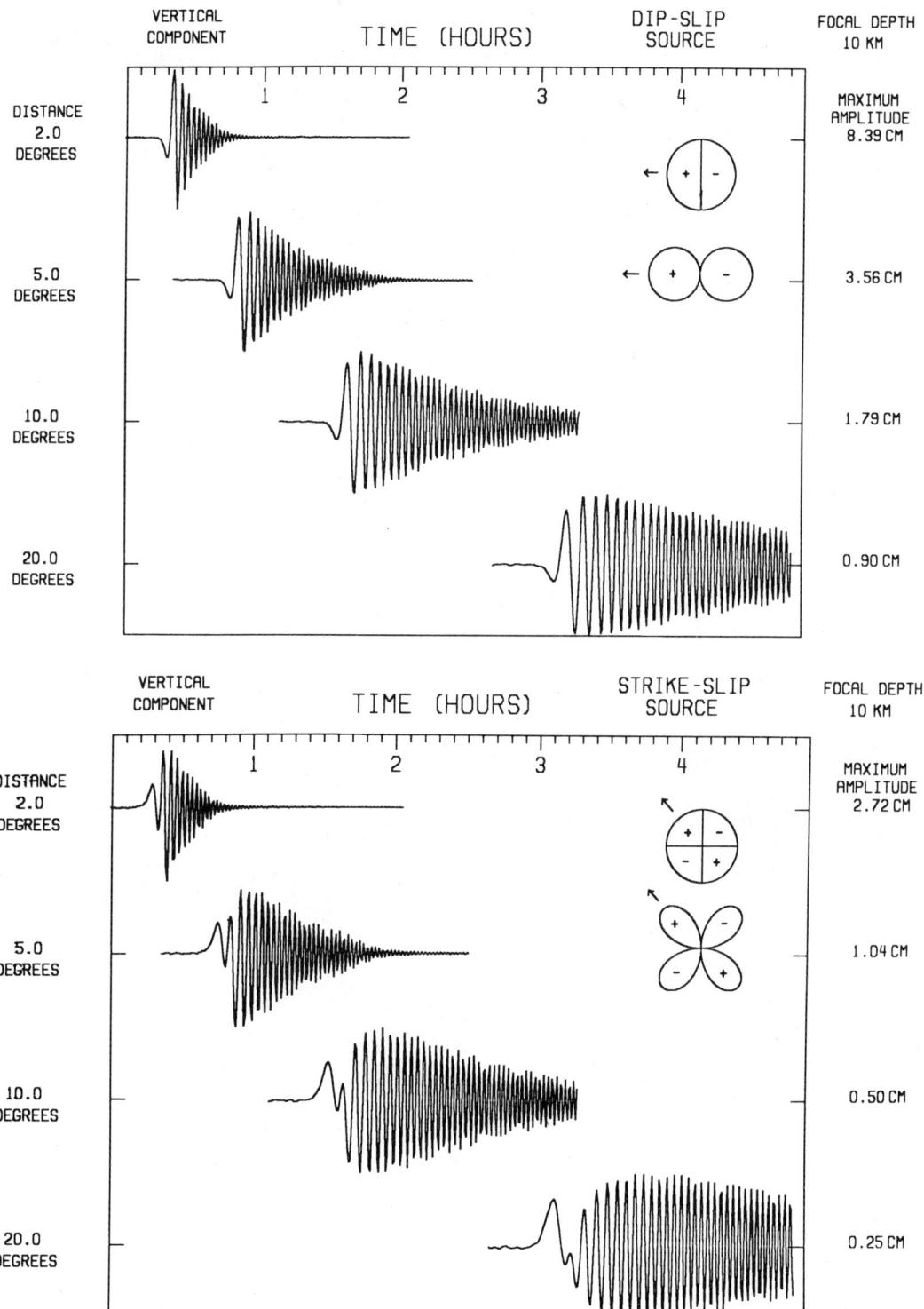

FIGURE 4. Synthetic record section of vertical tsunami motions at distances of 2°, 5°, 10°, and 20° from point dip-slip (*top*) and strike-slip (*bottom*) earthquakes of moment $M_0 = 10^{20}$ N·m and depth 10 km. Time runs for 5 hr, and the maximum amplitude of each trace is given in centimeters at the right. The lower half of the focal sphere and tsunami radiation pattern are shown in the upper right. Azimuth of observation ϕ is shown by the arrows. For other directions the waveforms should be scaled by $\sin \phi$ and $\sin 2\phi$, respectively.

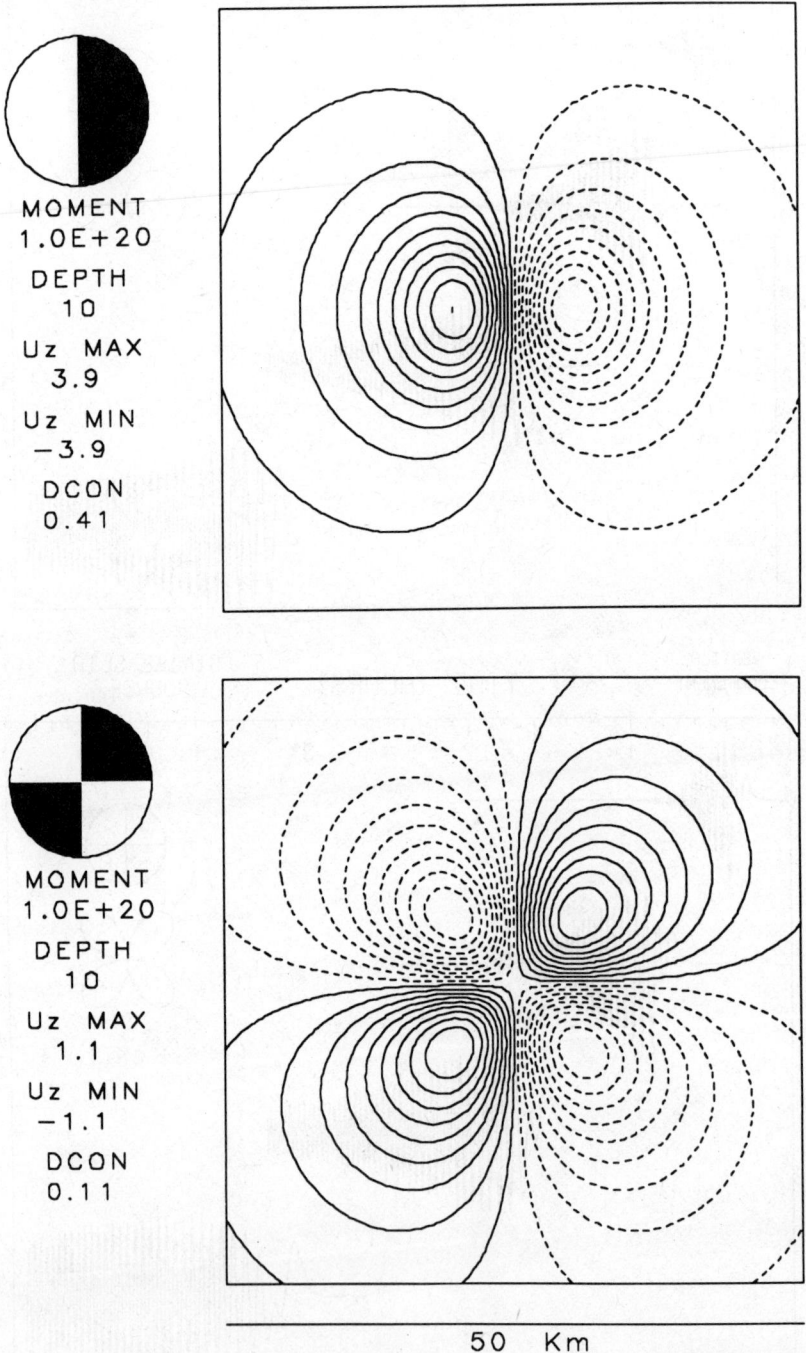

FIGURE 5. Contour map of static vertical displacements of the seafloor for the dip-slip (top) and strike-slip (bottom) sources of Fig. 4. The maps are 50 km square. The contour intervals are 41 cm (top) and 11 cm (bottom). Maximum excursions of the surface are 3.9 and 1.1 m, respectively.

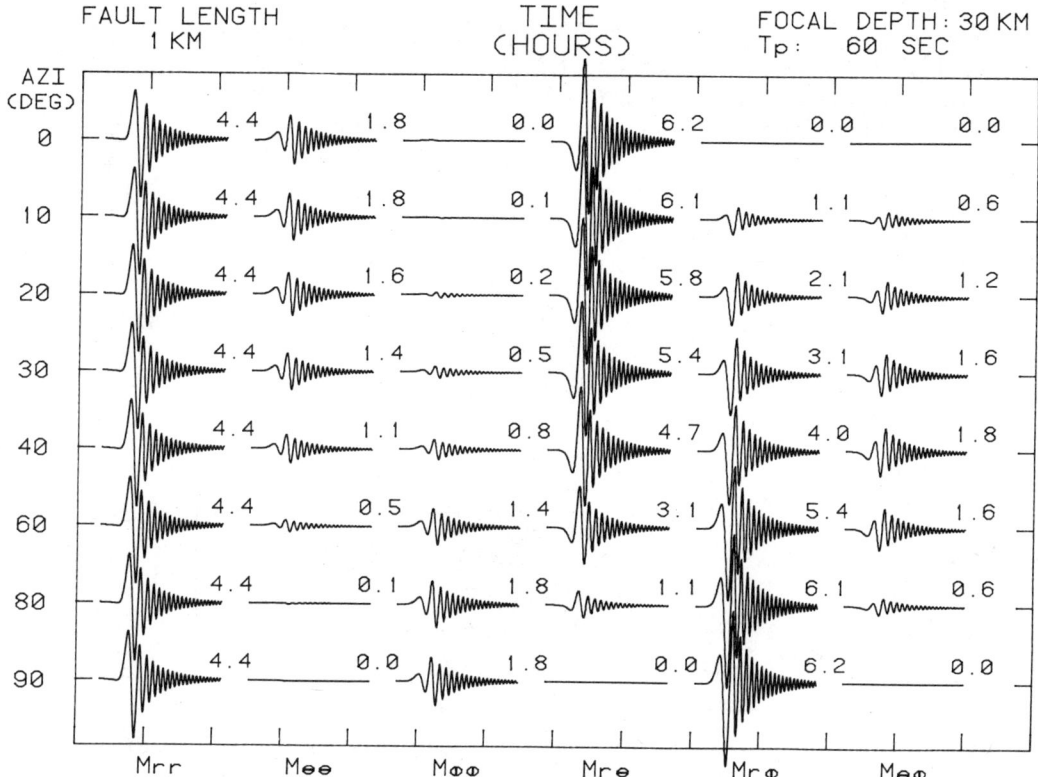

FIGURE 6. Theoretical vertical sea surface displacement versus azimuth at 3000 km distance from a point source buried at a depth of 30 km. The columns are the waves produced by the six moment tensor sources M_{rr}, $M_{\theta\theta}$, $M_{\phi\phi}$, $M_{r\theta} = M_{\theta r}$, $M_{r\phi} = M_{\phi r}$, and $M_{\theta\phi} = M_{\phi\theta}$. Each moment tensor component has a strength of 10^{21} N·m. The number to the right of each trace is the maximum pulse height in centimeters. Time marks are 1 hr apart, and all records start at $t = 3$ hr and 50 m.

$$u_i(\mathbf{r}, t) = \sum_{l=0}^{\infty} \sum_{m=-2}^{2} \mathbf{\hat{i}} \cdot \mathbf{U}_l^m(\mathbf{r}) \epsilon_{ijk}^m(r_s) \int_{-\infty}^{t} \dot{M}_{jk}(\hat{t})$$
$$[1 - \cos \omega_l(t - \hat{t})] d\hat{t} \quad (1)$$

Here l and m are the angular and azimuthal orders of a tsunami mode

$$\mathbf{U}_l^m(\mathbf{r}) = \mathbf{\hat{r}}_R U_l(r) Y_l^m(\theta, \phi) + {}_sU_l(r) \nabla_1 Y_l^m(\theta, \phi)$$

of frequency ω_l; $\mathbf{r}$, θ, and ϕ are the radial, epicentral, and azimuthal coordinates in a spherical frame with polar axis through $\mathbf{r}_s$; $Y_l^m(\theta, \phi)$ is a spherical harmonic, and ϵ_{jk} is the strain tensor for the mode evaluated at the source radius. The symmetric tensor $\dot{M}_{jk}$ is called the moment rate tensor; its six parameters give a mathematical representation of a point dislocation. In terms of the vector fault normal $\hat{n}$ and vector slip direction $\hat{a}$,

$$\dot{M}_{jk}(t) = \dot{M}_0(t)(\hat{a}_j \hat{n}_k + \hat{n}_j \hat{a}_k)$$

$\dot{M}_0(t)$ is the time history of moment release, and $M_0 = \int_{-\infty}^{\infty} \dot{M}_0(\hat{t}) d\hat{t}$.

All of the information relating earthquake parameters and tsunami features is contained in Eq. 1. The tsunami radiation pattern, for instance, is found in $\mathbf{U}_l^m \epsilon_l^m$, which includes terms such as $\sin m\phi$ and $\cos m\phi$. Point moment tensors only produce azimuthal orders $|m| \leq 2$. Thus, tsunamis from a point source will have radiation patterns with azimuthal dependence no more intricate than $\sin 2\phi$ or $\cos 2\phi$. Equation 1 also indicates that tsunami amplitudes from a point source are directly proportional to M_0.

Figure 4 shows 5 h of tsunami waveforms calculated at epicentral distances of $\theta = 2°$, $5°$, $10°$, and $20°$ (222, 555, 1100, 2200 km) from dip and strike-slip point sources of moment 10^{20} N·m ($M_w = 7.3$) buried 10 km deep. Sea waves from dip-slip and strike-slip sources have radiation patterns of $\sin \phi$ and $\sin 2\phi$, respectively. The waveforms for Fig. 4 are computed at the azimuth of maximum strength, $\phi = 270°$ and $\phi = 315°$. Frequency dispersion, with the long periods arriving first, is the most conspicuous feature of the waveforms. The waves have a

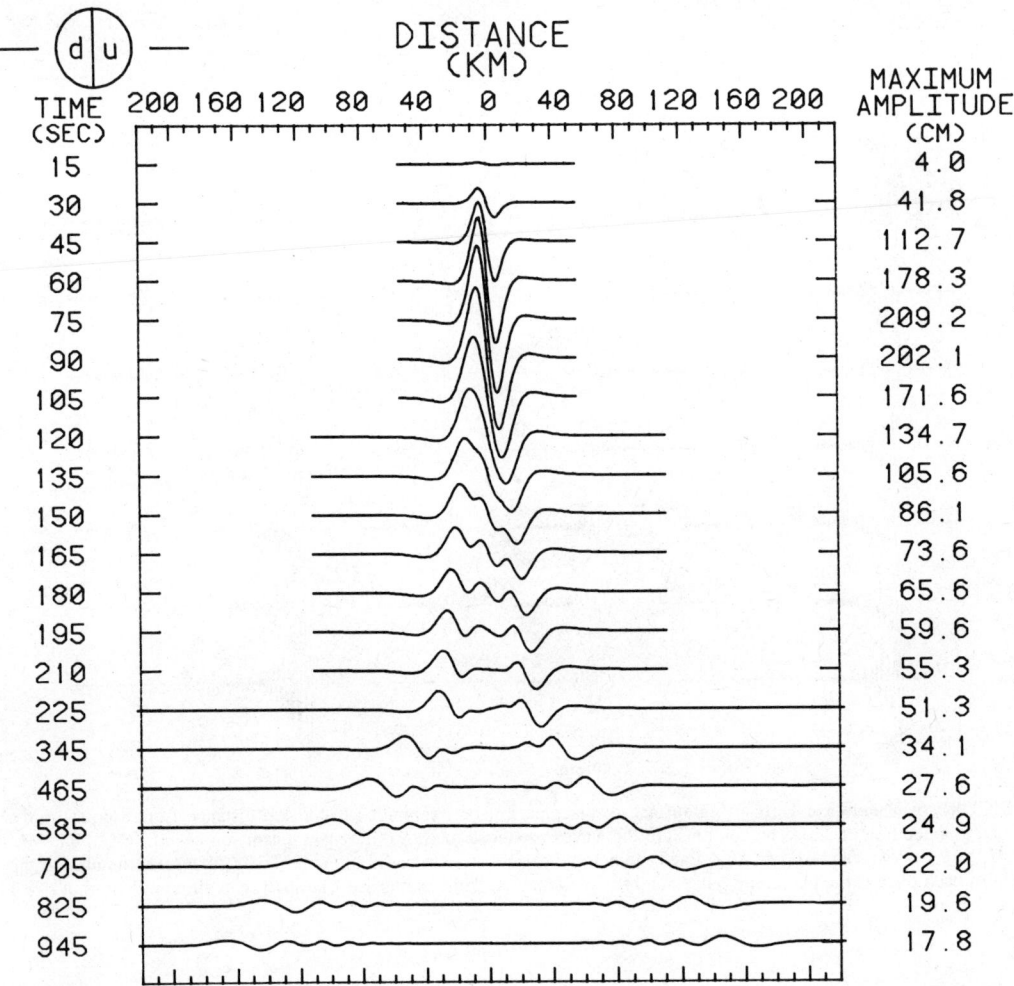

FIGURE 7. Cross sections showing vertical sea surface motions versus distance from a dip-slip event ($M_0 = 10^{20}$ N·m) buried 10 km below the seafloor. Cross sections are taken in a direction perpendicular to the fault plane, as indicated by the focal mechanism shown at the top. Snapshots are taken at 15-s intervals from 0 s to 225 s and at 120-s intervals from 225 s to 945 s. Elapsed time and maximum amplitudes in cm are given at the left and right sides.

rapid onset, reaching maximum height in the first cycle or two. Afterward, a slow decay, lasting an hour or more, is evident.

Note the contrast in tsunami amplitudes between the two earthquakes of Fig. 4. For point sources, dip-slips produce wave heights about four times larger than do strike-slips of equal moment. Perhaps the easiest way to understand the effect of earthquake mechanism is to consider the seafloor deformations of each fault. Figure 5 contours the static vertical displacement from point-dip and strike-slip earthquakes of moment 10^{20} N · m. The expected sin ϕ and sin 2ϕ dependence of the uplifts is obvious over a region 50 km wide. The most significant difference in the fields is the maximum vertical displacement; 1.1 m for the strike-slip versus 3.9 m for the dip-slip. It is no coincidence that the ratio of maximum vertical uplift for these two faults is the same as the ratio of tsunami heights in Fig. 4. After all, it is the vertical deformation that has the strongest influence on tsunami production.

Figure 6 shows seismic sea waves at 3000 km in a 4-km-deep ocean from each of the six independent moment tensor elements. The tsunami field from *any* fault plane orientation can be constructed from a linear combination of these columns. The moment tensors M_{rr}, $M_{\theta\theta}$, and $M_{\phi\phi}$ represent a vertical and two perpendicular horizontal vector dipole forces, $M_{r\theta}$ and $M_{r\phi}$ represent dip-slip dislocations separated by a 90° rotation in azimuth, and $M_{\theta\phi}$ represents a strike-slip dislocation. The vertical dipole force and the two dip-slip dislocations dominate tsunami production because these sources generate the greatest vertical displacement at the seafloor. From

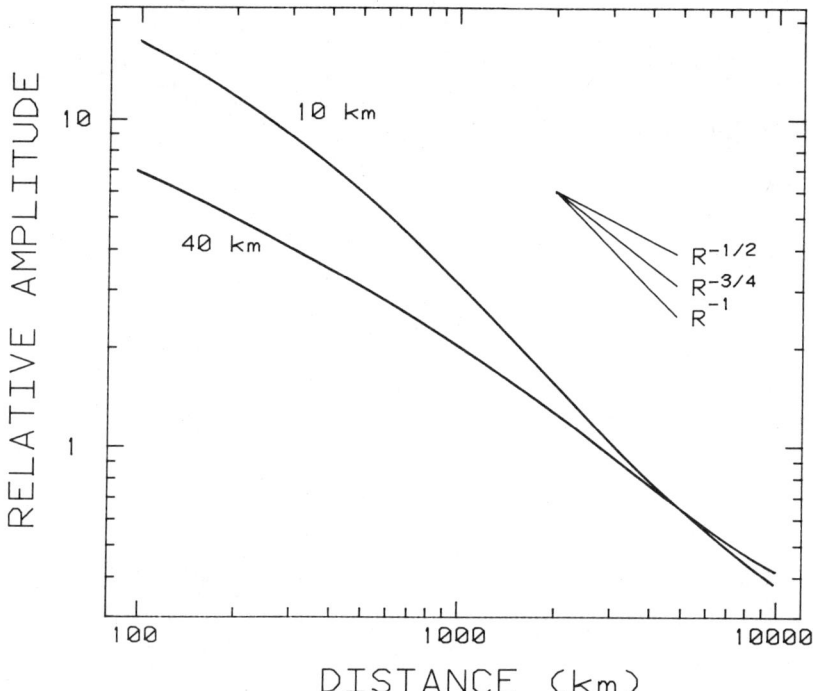

FIGURE 8. Computed variation of maximum tsunami height versus distance from 100 km to 10,000 km. Top and bottom curves show relative amplitudes of tsunamis produced from equivalent dip-slip faulting on vertical planes at 10 km and 40 km deep. Within a few hundred kilometers of the source, tsunamis generated from shallow faults are several times larger than those generated from the same fault at depth. Tsunami amplitudes equalize during propagation however, and beyond 2000 km source depth has only a weak influence on tsunami amplitudes.

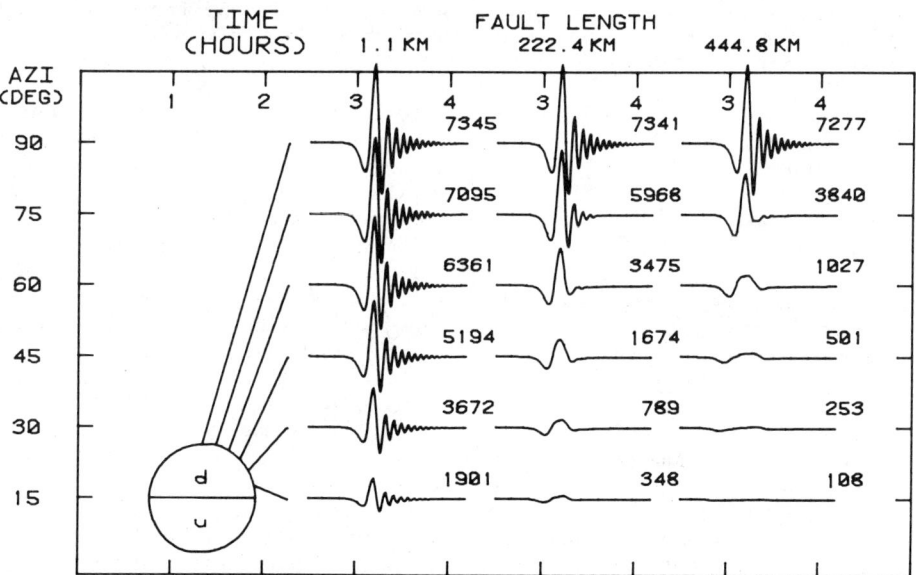

FIGURE 9. Theoretical vertical sea surface displacement 2223 km from three 40-km-deep dip-slip sources. Each source has a moment of 10^{20} N·m and a process time of 25 s. Columns record motions versus azimuth from the strike direction, assuming fault lengths of 1.1, 222, and 444 km. A map view of the focal mechanism with the azimuthal locations of the receivers appears in the lower left. The d and u denote hemispheres of downward and upward motion. The time scale is in hours, and the right two columns have each been shifted 120 min. True maximum amplitudes of the traces are given in microns. Note that at azimuths away from the beam direction ($\phi = 0°$), tsunami pulses are broader, smaller, and depleted of high frequencies relative to tsunamis from the 1.1-km source.

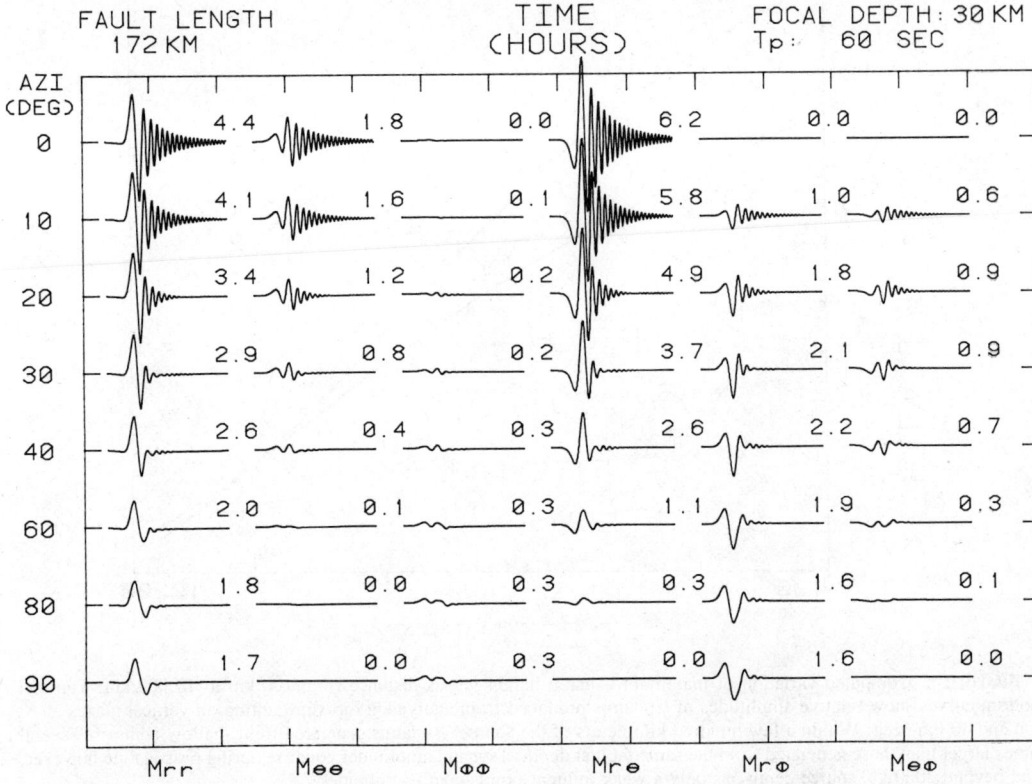

FIGURE 10. Theoretical vertical sea surface displacements for the six moment tensor sources now 172 km in length. All other features are the same as in Fig. 6. Note how waves propagating near the strike direction $\phi = 90°$ have been drastically reduced. For long faults, the only two moment tensor sources of significance are the vertical linear couple M_{rr} and the dip-slip double couple $M_{r\theta} = M_{\theta r}$.

this knowledge, it is useful to define the tsunami moment M_T as that fraction of the total earthquake moment M_0 that contributes significantly to tsunami production:

$$M_T = (M_{rr}^2 + M_{r\theta}^2 + M_{r\phi}^2)^{1/2}$$

Figure 7 illustrates a computed cross section of the sea surface versus time subsequent to a dip-slip earthquake of 25-s duration at a depth of 20 km. In essence, the figure is a "motion picture" of the birth of a sea wave. For the first 2 min after the beginning of the earthquake, a 50-km-wide surface disturbance grows in amplitude but remains stationary. Upon reaching full height, the leading edges of the wave begin to propagate outward. Characteristic tsunami dispersion only begins to be seen about 10 min after the event at a distance of 150 km. For shorelines close to tsunamigenic regions, such as Japan from earthquakes in the Japan trench, seismic sea waves arrive more or less as a single pulse. For areas more distant from the tsunami source area, such as Hawaii and the western United States from Alaskan earthquakes, tsunamis arrive with many oscillations, dispersion having spread out the pulse. Note that the rightward-traveling waves are opposite in sign to the leftward-traveling waves. Whether the first onset at a particular shoreline is a withdrawal or an inundation depends strictly on the relative position of the shore and the fault.

The effect of earthquake depth on tsunami production is most pronounced within a few hundred kilometers of the source. Because short-period tsunami eigenfunctions penetrate the solid Earth to only a few kilometers (see Fig. 3), shallow earthquakes excite higher-frequency tsunami modes than do deep earthquakes. Higher-frequency waves, however, travel more slowly than do longer-period waves. As a result, high-frequency waves from shallow earthquakes contribute to maximum tsunami height only for the length of time that the tsunami propagates as a single pulse. After a few hundred kilometers of travel, high-frequency waves drift to the back of the wave train (see Fig. 4) and no longer add to the tsunami maximum that occurs early in the first few cycles. The effect of source depth is summarized in Fig. 8, where maximum tsunami height from earthquakes of equivalent moment and mechanism, but different depths, are compared. At 100 km distance the 10-km-deep earthquake gener-

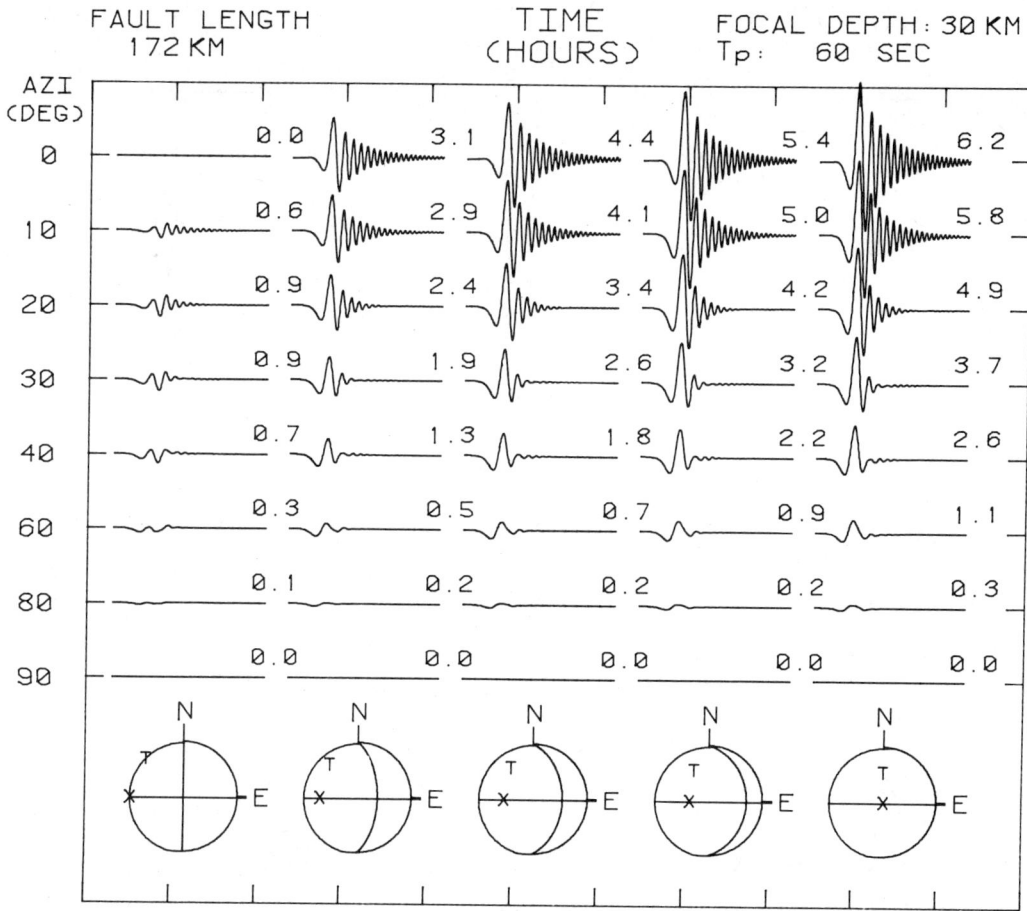

FIGURE 11. Theoretical vertical sea surface displacement at 3000 km for five different faults with moment 10^{21} N·m. X and T in the focal sphere diagrams mark the location of the slip vector and tension axis. The 172-km fault strikes west to east. Azimuth $\phi = 0°$ coincides with north. All other features are the same as in Fig. 10. Note the large increase in amplitude as the dislocation migrates from pure strike-slip to pure dip-slip on a vertical plane.

ates a wave about three times as large as the 40-km-deep event. Beyond 2000 km, the added high frequencies present in the shallow event have fallen behind and wave heights have equalized. Any earthquake depth less than 60 km appears to be equally efficient in tsunamigenesis at this distance.

Effect of Earthquake Physical Size

The foregoing discussions have dealt with point source models of earthquakes. Real earthquakes have finite physical size and duration of faulting, and conclusions about tsunami generation must take them into account. The third and fourth columns of Table 2 list earthquake fault lengths and process times for typical events of magnitude $6.5 \leq M_w \leq 9.5$. The largest earthquakes have fault lengths of several hundred kilometers and require several hundred seconds to rupture completely. Generally, these large earthquakes have lengths three or four times greater than their widths. For long, thin earthquake faults, tsunamis are preferentially beamed perpendicular to the fault length, regardless of the focal mechanism. Tsunamis radiated parallel to the fault tend to be eliminated as the fault length increases. The reduction in amplitude of tsunamis radiated along strike is very clear in Fig. 9, which shows the evolution in waveforms from a lengthening dip-slip source. Tsunamis from earthquakes over magnitude 8 (length ≈ 200 km) are emitted in a tight beam of about 60° width. The effect of fault length on the six moment tensor tsunami fields is illustrated in Fig. 10. In contrast to the point source of Fig. 6, only two moment tensor elements are significant for long faults: M_{rr}, the vertical linear couple, and $M_{r\theta}$, the dip-slip dislocation that strikes parallel to the length of the fault. One important conclusion from the study of finite sources is that, although long faults tend to focus waves at right angles to their length, the maximum amplitude of the generated waves is no larger than the maximum height expected from a

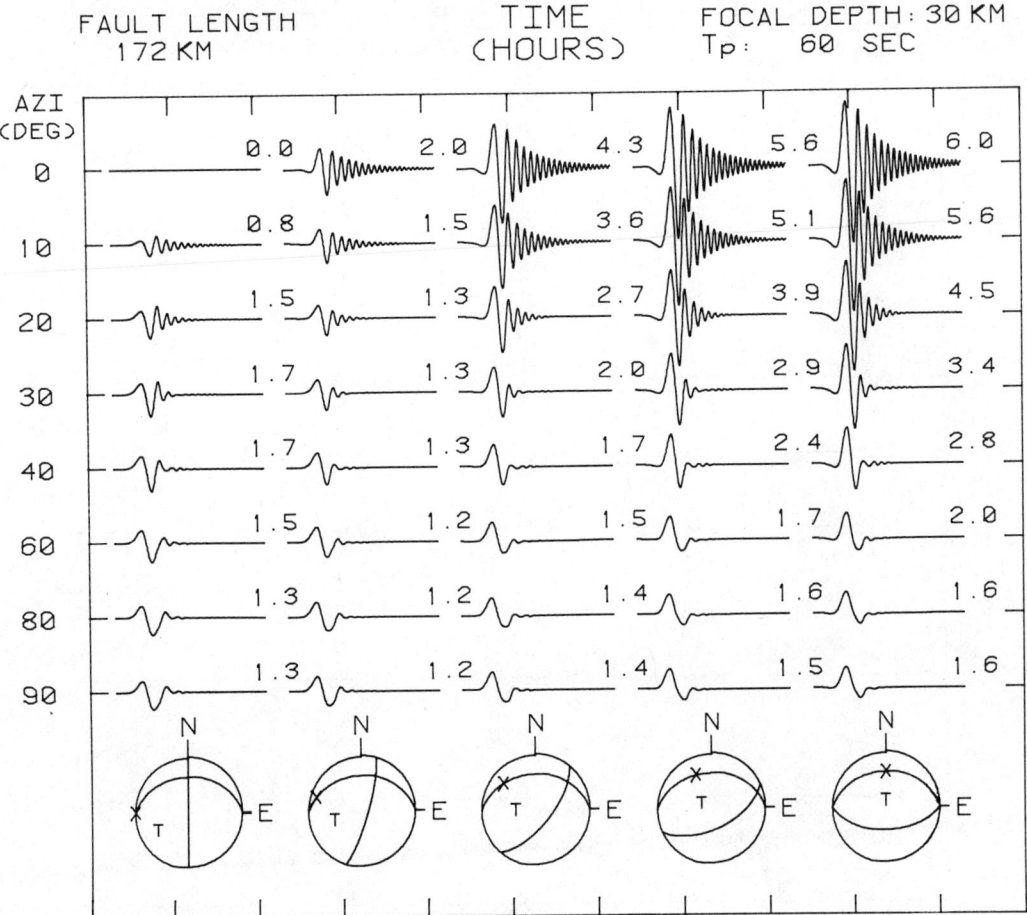

FIGURE 12. Tsunamis produced from earthquake dislocations that migrate from pure strike slip (left) to pure dip-slip (right) on a plane plunging 35° to the north. All other features are the same as in Fig. 11. Note how the sea wave amplitudes produced from these earthquakes are very similar to those in the respective columns of Fig. 11 despite the very different fault mechanisms.

point source of the same depth, moment, and mechanism. This result simplifies the task of tsunami forecasting.

Figures 11, 12, and 13 catalog three suites of earthquake focal mechanisms and their associated tsunamis. The set constitutes a ready reference in visual form of the expected tsunami field from a wide variety of earthquake faults. In the first two figures, the focal mechanism migrates from pure strike-slip to pure dip-slip. The fault surface dips 90° in Fig. 11 and 35° in Fig. 12. In both instances, tsunami amplitudes increase significantly, with increasing downdip component of the slip vector. For long faults, the ratio of maximum tsunami heights from strike-slip and dip-slip sources of equal moment can be as much as a factor of 10. Figure 13 shows the tsunamis generated by pure dip-slip on planes plunging from 0° to 90°. For long thin faults, the dip of the fault plane is not particularly significant.

Maximum Tsunami Amplitude

Of all the questions concerning earthquake parameters and tsunami generation, the one most commonly asked is: What is the largest-amplitude tsunami expected at distance X from an earthquake of magnitude Y? Figure 14 attempts to address this question. It plots maximum open-ocean tsunami height versus moment magnitude M_w for distances of 1000, 3000, 5000, and 7000 km. Maximum tsunami height is computed for the most efficient focal mechanism, at the most efficient depth, in the direction of maximum radiated strength. The earthquakes are assumed to have typical dimensions and time scales. Anomalous earthquakes with great slip over small faults, or ones that possess extraordinarily long process times, may generate tsunami amplitudes that differ somewhat from these results.

In the open ocean, maximum tsunami heights vary

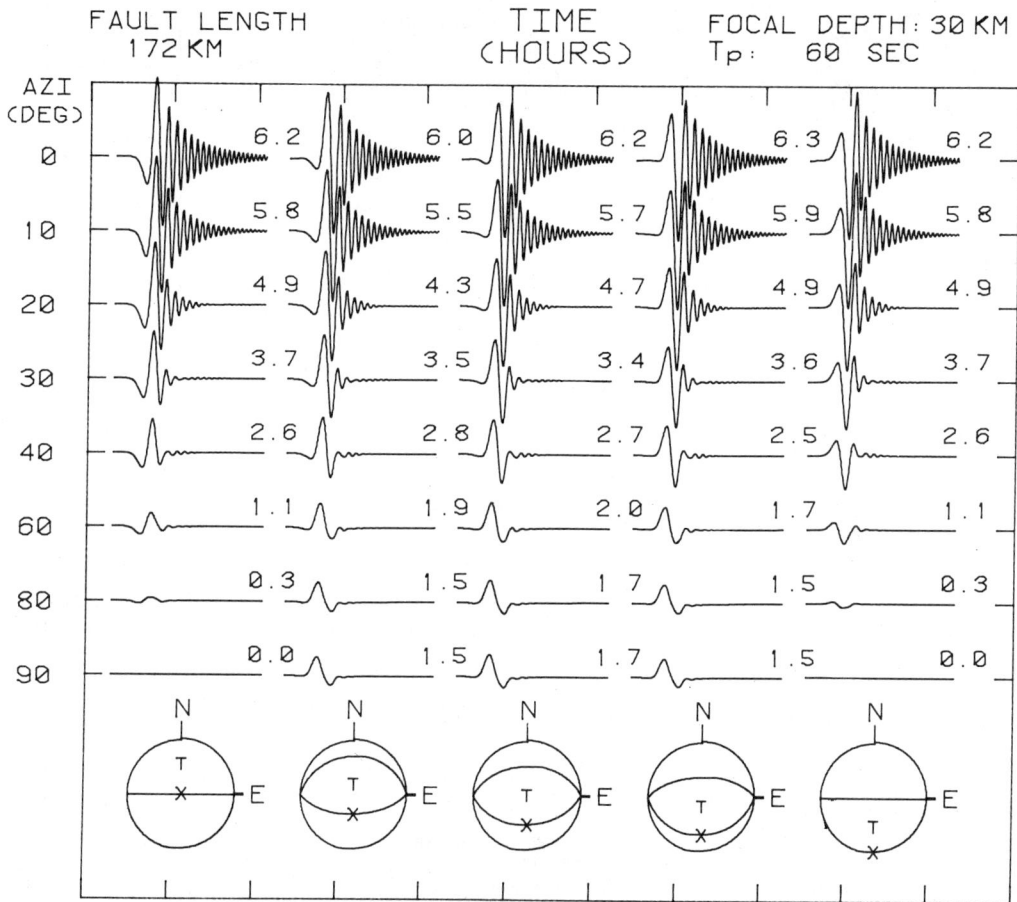

FIGURE 13. Tsunamis from five pure dip-slip earthquakes on fault planes plunging from 0° to 90°. Note how the amplitude of wave excitation is insensitive to the plunge of the fault plane.

from 10 cm to 10 m as M_w varies from 7.5 to 9.5. For lesser earthquakes ($M_w < 8.5$), maximum tsunami heights are proportional to $10^{1.5M_w}$. One unit increase in magnitude translates into about a factor of 30 in tsunami height. For truly great earthquakes ($M_w > 8.5$), the dependence on magnitude weakens slightly. These largest events are actually becoming too large in dimension and duration to efficiently generate tsunamis.

Tsunami Forecasting

With the availability of credible models of tsunami generation, one logical next step is tsunami forecasting. Forecast contrasts prediction in that the goal is not to state that a tsunami will occur at a certain place at a certain time weeks or months in the future. A forecast is more immediate. Its limited goal is to estimate the strength of a tsunami, assuming that a potentially dangerous earthquake has already occurred. Seismic waves contain information about the parameters of earthquake faulting, and

they travel 20 or 30 times faster than sea waves. At distant locations, there may be several hours between the arrival of the seismic waves and the arrival of the tsunami. This time could be spent analyzing the seismic waves, estimating earthquake parameters (namely moment, mechanism, and depth), and forecasting the expected heights of the oncoming wave with the aid of models of tsunami generation. Earthquake moment is by far the most important parameter in the forecast because it relates the strength of the earthquake at tsunami periods. Surface wave magnitude M_s is the most commonly quoted measure of earthquake size: M_s gauges the strength of earthquakes at 20 s period. Unfortunately, because the strength of large earthquakes at this period can differ greatly from their strength at tsunami periods (≈ 1000 s), M_s is a very poor indicator of tsunami potential.

A key element in tsunami forecasting is the reliable and rapid determination of seismic parameters. Although parameter estimation from single stations is possible, reliability is better assured if seismic data

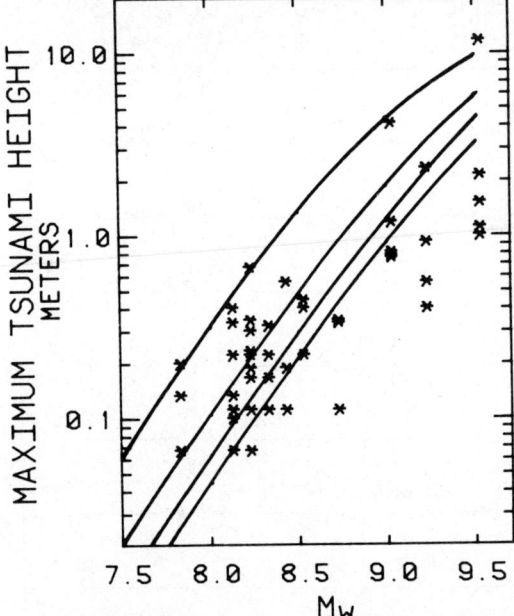

FIGURE 14. Plot of predicted maximum open ocean tsunami height versus earthquake magnitude M_w. The four lines quantify wave heights in a direction perpendicular to the strike of a vertical dip-slip fault at distances of 1000, 3000, 5000, and 7000 km. The source depth is 5 km. The stars are measurements of tsunami amplitude at distances greater than 1000 km from 14 different Pacific earthquakes. All of these should fall at or below the upper line, which was calculated by assuming optimal conditions for tsunami production and subsequent observation. When comparing these open ocean heights with observed tsunami run-ups, keep in mind the factor of 2 or 3 amplification that results from shoaling.

from widely distributed stations are available. Likewise, forecasts are useful only if they can be issued prior to the arrival of the sea wave. Thus, the best scenario requires that seismograms be assembled at a central laboratory from farflung stations, and that the collection be done quickly. It is well within current capabilities to collect the needed data within the required time restrictions by using satellite technology. Satellites could transmit data from anywhere in the Pacific basin to a central location for analysis within an hour or two after a major earthquake. Presently, most tsunami warnings are based on very limited knowledge of the parameters of seaquakes, often only the epicenter and an estimate of M_s. Not surprisingly, being based on such limited information, tsunami warnings have not been completely dependable. The cost of inaccurate forecasts is high, not only monetarily, in the expenses of activating emergency preparedness programs, but also in the erosion of public faith in forecasts of any type.

Proposals for establishing a global network of modern seismic stations linked in real time by satellites are well developed and have already been submitted to national funding agencies. Perhaps by the time of the next large seaquake, safety officials can definitely tell the public whether to leave the beach.

STEVEN N. WARD

Bibliography

Comer, R. P., 1984, The tsunami mode of a flat earth and its excitation by earthquake sources, *Royal Astron. Soc. Geophys. Jour.* **77**, 1–27.

Comer, R. P., 1984, Tsunami generation: a comparison of traditional and normal mode approaches, *Royal Astron. Soc. Geophys. Jour.* **77**, 29–41.

Okal, E. A., 1982, Mode-wave equivalence and other asymptotic problems in tsunami theory, *Physics Earth and Planetary Interiors* **30**, 1–11.

Okal, E. A., 1988, Seismic parameters controlling far-field tsunami amplitudes: A review, *Natural Hazards* **1**, 67–96.

Okal, E. A., and J. Talandier, 1986, T-wave duration, magnitudes and seismic moment of an earthquake; application to tsunami warning, *Jour. Phys. Earth* **34**, 19–42.

Ward, S. N., 1980, Relationships of tsunami generation and an earthquake source, *Jour. Phys. Earth* **28**, 441–474.

Ward, S. N., 1981, On tsunami nucleation: I. A point source, *Jour. Geophys. Research* **86**, 7895–7900.

Ward, S. N., 1982, On tsunami nucleation: II. An instantaneous modulated line source, *Physics Earth and Planetary Interiors* **27**, 273–285.

Ward, S. N., 1982, Earthquake mechanisms and tsunami generation—The Kurile Islands event of October 13, 1963, *Seismol. Soc. America Bull.* **72**, 759–777.

Cross-references: *Earthquake Mechanisms; Earthquakes: Hazards and Prediction; Earthquakes: Magnitude, Energy, and Intensity; Earthquake Seismology; Seismicity and Plate Tectonics; Seismic Source: Observations; Seismic Source: Theory; Seismology: History.*

V

VERY-LONG-BASELINE INTERFEROMETRY (VLBI)

Very-long-baseline interferometry (VLBI) is a powerful new space-based geodetic technique that has revolutionized our ability to determine the locations of points on the Earth's surface. A single-day VLBI observing session is sufficient to determine the vector separation between observatories on different continents with an accuracy of a few centimeters and to monitor the rotation, flexure, and tectonic deformation of the Earth with similar accuracy. Since the VLBI measurements are inherently tied to the locations of extragalactic radio sources, which form the most nearly inertial coordinate frame known, they have a level of accuracy and stability unmatched by any other measurement technique.

A VLBI observing session typically employs several stations that simultaneously record signals from extragalactic radio sources. The recordings are processed to determine the differences in the arrival times of the signal wavefronts at the various stations. These differential arrival times are then analyzed to determine the vector separations ("baselines") between the stations with an accuracy of a few centimeters or less. By monitoring the time variations in these baseline vector components it is also possible to determine the small irregularities in the rotation of the Earth and the motions caused by tectonic and elastic deformations of the Earth's crust. The observed differential arrival times can also be used to estimate the coordinates of the celestial radio sources with submilliarcsecond accuracy. The stability of this celestial coordinate frame is essential for long-term monitoring of the motions of the Earth with VLBI techniques.

Baseline Determinations

Routine VLBI determinations of baseline vectors have been conducted for several years on baselines spanning the continental United States and the Atlantic and Pacific basins. The baseline parameter most often studied is the length of the baseline since length is invariant under rotation and is unaffected by changes in Earth orientation. The repeatability of the baseline length determinations is one measure of the observation error levels. Repeatability at the level of about 2 cm is routinely achieved. Because the level of crustal strain due to plate motions and other tectonic processes is in the range of a few centimeters per year, these determinations are sufficiently accurate to observe such tectonic effects on a time scale of a few years. Several groups have reported observation of the strain resulting from the separation of the North American and Eurasian plates (Herring et al., 1986b; Carter et al., 1985), using observations taken between 1980 and 1984. The velocity observed was 19 mm/year with a conservative error estimate of ±10 mm/year, which compares very well with the expected plate motion of about 17 mm/year. We do not yet regard this evidence as conclusive for the detection of plate motion because of the relatively large uncertainty. However we do expect the uncertainty to decrease as the square root of increasing time (assuming that the measurement series continues uniformly), and at some point in the fairly near future we expect to be able to make a definitive statement about the rates of contemporary plate motion. Observations using baselines in the Pacific basin, where the expected strain rates are as large as 94 mm/year, were begun in 1984. Preliminary analyses of the first one or two years of data exhibit strain rates that are roughly consistent with the expected rates, and continued monitoring of these baselines should establish the values of contemporary strain rates on these baselines in just a few more years.

Simple error analysis of the VLBI delay observations indicates that the errors ought to be much smaller than the 2-cm scatter seen in the baseline length determinations. The observed scatter is believed to be caused in large part by unmodeled atmospheric refraction effects rather than by observation uncertainties. To test this hypothesis a series of measurements was made on a very short baseline (1.2 km) between the Westford and Haystack antennas in Massachusetts. On a baseline as short as this most of the atmospheric refraction variations are common to both antennas and thus largely cancel from the observations. The baseline determinations that resulted from these observing sessions exhibited a scatter of about 5 mm, quite close to the limiting accuracy expected from the VLBI observations (Rogers et al., 1978).

To further test the hypothesis that the measurement errors are dominated by atmospheric refraction effects and to probe the limits of the accuracy possible in modeling atmospheric effects, a group at Harvard's Center for Astrophysics has conducted a series of observing sessions using antennas in Westford, Massachusetts, and Mojave, California. The observing schedules have been carefully crafted to improve atmosphere sensitivity by including observations at very low elevation angles. Using the best available atmosphere modeling techniques including Kalman filter estimates of the stochastic behavior of the atmosphere, these researchers have been able to demonstrate subcentimeter repeatability for this baseline as well (Davis, 1986). These results suggest that atmospheric refraction effects can be modeled well enough that subcentimeter baseline accuracy may soon become routinely available.

Polar Motion

The Earth's spin axis moves relative to its solid crust with characteristic periods of one year (driven by seasonal atmospheric fluctuations) and a natural resonance period of 14 months (the so-called Chandler Wobble). VLBI determinations of this motion are compared with corresponding determinations from laser ranging to artificial satellites (SLR) in *Earth Orientation*. The agreement between the two sets of determinations is excellent; the RMS difference between the two sets of measurements is less than 2 ms of arc, or about 6 cm of displacement at the Earth's surface. This level of agreement between measurements made with radically different observing techniques places a strong bound on the total errors present in both techniques.

Continued monitoring of the pole position with this level of accuracy will provide data essential to determining the underlying mechanisms that excite the polar motion. (In the absence of excitation the motion would be expected to damp to zero.) In recent years earthquake displacements have received a great deal of attention as a possible excitation mechanism. The VLBI data are already sufficient to largely rule out this mechanism (Robertson et al., 1985a). Current research on polar motion excitation mechanisms is focused on atmospheric effects, that is, changes in the angular momentum and moment of

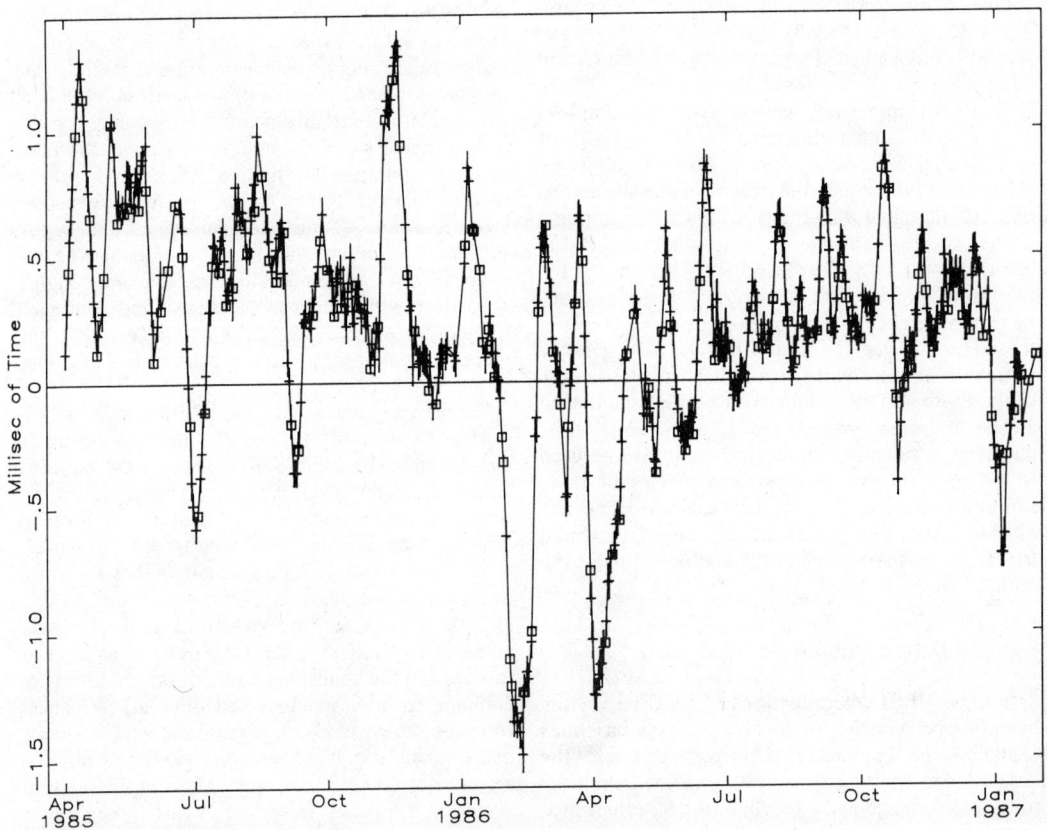

FIGURE 1. UT1 determinations from VLBI observations, shown as differences from the standard UT1 values tabulated by the Bureau International de l'Heure in Paris. 24-hour, four-baseline observing sessions are shown with squares, and 1-hour, single-baseline observing sessions are shown with crosses.

inertia of the Earth caused by phenomena such as changing wind patterns and snow-melt.

UT1

Irregularities in the Earth's rotational phase angle (sometimes called universal time or UT1) can be monitored by VLBI techniques with a level of accuracy and stability similar to the pole position determinations. Unfortunately the accuracy of these determinations cannot be directly verified by intercomparison with other observing techniques because there is no independent observing technique that can match the accuracy and long-term stability of the VLBI UT1 determinations. Therefore to determine the accuracy of the VLBI UT1 measurements by intercomparisons it is necessary to compare different sets of VLBI measurements. Such intercomparisons are useful only to the degree that the different VLBI determinations do not have common sources of error. It is possible to minimize the common error sources by intercomparing sets of VLBI measurements made with radically different baselines and observing schedules. Such an intercomparison between 4- station, 24-hour observing sessions and 2-station, 1-hour observing sessions has established the accuracy of the VLBI UT1 determinations at no worse than 0.1 ms of time (Robertson et al., 1985*b*). Some of the UT1 determinations from these observing sessions are shown in Fig. 1.

Nutation

VLBI measurements are also sensitive to the location of the Earth's spin axis in space (nutations). Researchers at Harvard's Center for Astrophysics discovered a discrepancy between the VLBI determinations of nutation and the standard IAU 1984 theoretical model. The VLBI nutation residuals, which have an annual signature with an amplitude of about 2 ms of arc, are shown in Fig. 2. Since the theoretical calculation was based on the most sophisticated model available for the internal structure of the Earth, these nutation residuals implied the existence of an error in that model. After detailed theoretical examination of the possible model changes that might influence the annual nutation, it was concluded that the most likely model error lay

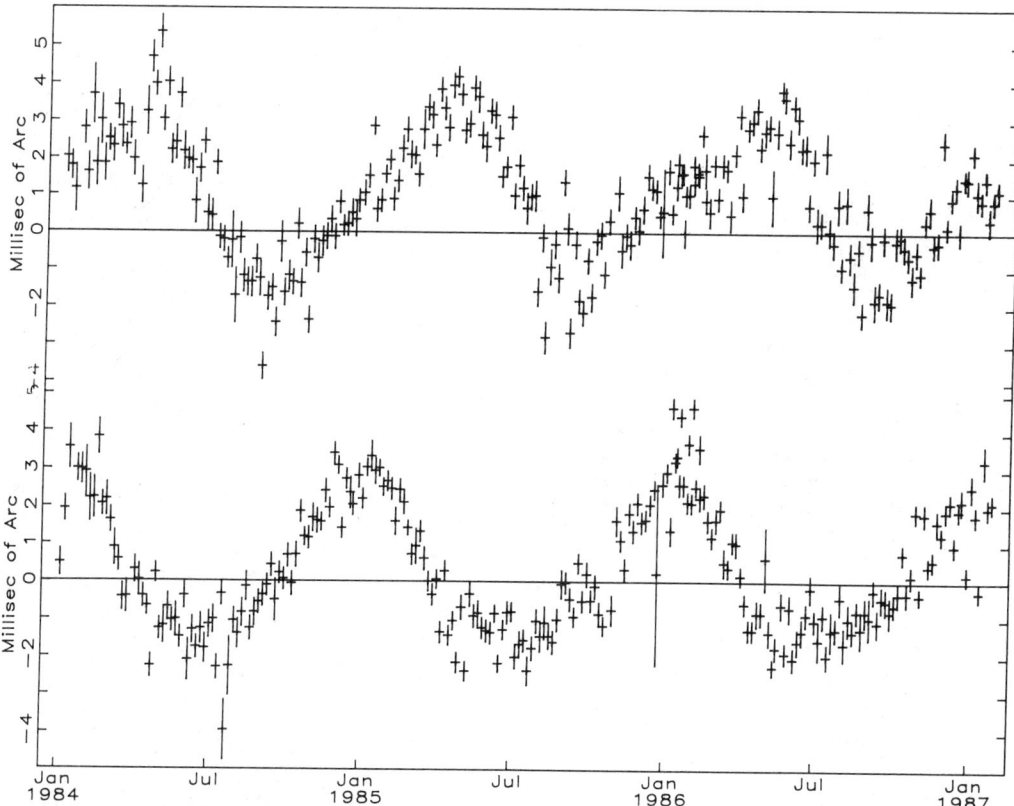

FIGURE 2. VLBI determinations of nutation in longitude (top) and obliquity (bottom) shown as differences from the standard IAU 1980 nutation model. The 2 ms annual signature discussed in the text shows clearly. The longitude values have been scaled by the sine of the obliquity.

in the magnitude of the rotational coupling between the Earth's fluid core and its solid mantle, and that an increase in the flattening of the core-mantle interface by about 0.5 km would produce the observed nutation residuals (Herring et al., 1986a; Gwinn et al., 1986). The problem of the shape of the core-mantle interface has received a lot of attention from seismologists recently (cf. Morelli and Dziewonski, 1987; Hager et al., 1985), and the additional constraint on this shape provided by VLBI observations is expected to prove useful in these studies.

Earth Tides

It is widely known that the gravitational fields of the Moon and Sun raise tides in the Earth's oceans. Less appreciated is the fact that the solid surface of the Earth also moves in response to these gravity fields with a characteristic amplitude of about 30 cm, not much less than that of the oceans. This flexure of the solid Earth affects the positions of the VLBI observing stations and can therefore be observed in the differential arrival time measurements. The tidal displacement measurements are commonly expressed in terms of a Love number, a dimensionless constant that defines the amplitude of the tide. VLBI Love number determinations have been shown to have a precision of better than a centimeter in terms of equivalent displacement (Herring et al., 1983, Carter et al., 1985).

Astrometry

The long-term stability of geodetic measurements made with VLBI is critically dependent on the stability of the celestial reference coordinate system, which is based on the coordinates of extragalactic radio sources. Those radio sources are believed to be at distances in excess of 10^9 light years and thus have negligible proper motions. Detailed studies of close pairs of radio sources have not revealed any motions of the sources at the level of micro-arc-seconds per year (cf. Bartel et al., 1986). Studies of the repeatability of VLBI source coordinate determinations indicate that the precision of those determinations is substantially better than 0.001 arc second (Robertson et al., 1986). These radio sources are therefore appropriate to define the most nearly inertial reference coordinate system known.

Conclusions

VLBI observations are contributing significantly to our understanding of fluctuations in the Earth's rotation and the rotational interaction of the Earth's mantle, fluid core, and atmosphere. They are beginning to attain sufficient time history to resolve the expected tectonic motions of the Earth's crustal plates and to observed strains within those plates. Further progress is expected as methods for modeling atmospheric refraction improve. Other improvements are expected as advances in electronics reduce the cost and improve the reliability of the equipment needed to make the observations and reduce the data. In the next few years we should continue to see important new results from the application of VLBI observations to geodesy and geophysics.

D. S. ROBERTSON

References

Bartel, N., T. A. Herring, M. I. Ratner, I. I. Shapiro, and B. E. Corey, 1986, VLBI Limits on the Proper Motion of the 'Core' of the Superluminal Quasar 3C345, *Nature* **319**, 733.

Carter, W. E., D. S. Robertson, and J. R. MacKay, 1985, Geodetic Radio Interferometric Surveying: Applications and Results, *Jour. Geophys. Research* **90**, 4577–4587.

Davis, J., 1986, Atmospheric Propagation Effects on Radio Interferometry, Ph.D. Diss., Massachusetts Institute of Technology.

Gwinn, C. R., T. A. Herring, and I. I. Shapiro, 1986, Geodesy by Radio Interferometry: Studies of the Forced Nutations of the Earth, Part II: Interpretation, *Jour. Geophys. Research* **91**, 4755–4765.

Hager, B. H., R. W. Clayton, M. A. Richards, R. P. Comer, and A. M. Dziewonski, 1985, Lower Mantle Heterogeneity, Dynamic Topography and the Geoid, *Nature* **313**, 541–545.

Herring, T. A., B. E. Corey, C. C. Counselman III, I. I. Shapiro, A. E. E. Rogers, A. R. Whitney, T. A. Clark, C. A. Knight, C. Ma, J. W. Ryan, B. R. Schupler, N. R. Vandenberg, B. O. Ronnang, G. Lundqvist, E. Elgered, J. A. Campbell, and P. Richards, 1983, Determination of Tidal Parameters from VLBI Observations, in J. Kuo, ed., *Proceedings of the Ninth International Symposium on Earth Tides*. Stuttgart: E. Schweizerbart'sche Verlagsbuchhandlung, 205–214.

Herring, T. A., C. R. Gwinn, and I. I. Shapiro, 1986a, Geodesy by Radio Interferometry: Studies of the Forced Nutations of the Earth, Part I: Data Analysis, *Jour. Geophys. Research* **91**, 4745–4754.

Herring, T. A., I. I. Shapiro, T. A. Clark, C. Ma, J. W. Ryan, B. R. Schupler, C. A. Knight, G. Lundqvist, D. B. Shaffer, N. R. Vandenberg, B. E. Corey, H. F. Hinteregger, A. E. E. Rogers, J. C. Webber, A. R. Whitney, G. Elgered, B. O. Ronnang, and J. L. Davis, 1986b, Geodesy by Radio Interferometry: Evidence for Contemporary Plate Motion, *Jour. Geophys. Research* **91**, 8341–8347.

Morelli, A., and A. Dziewonski, 1987, Topography of the Core-Mantle Boundary and Lateral Homogeneity of the Liquid Core, *Nature* **325**, 678–683.

Robertson, D. S., W. E. Carter, B. D. Tapley, B. E. Schutz, and R. J. Eanes, 1985a, Polar Motion Measurements: Sub-Decimeter Accuracy Verified by Intercomparison, *Science* **229**, 1259–1261.

Robertson, D. S., W. E. Carter, J. A. Campbell, and H. Schuh, 1985b, Daily UT1 Determinations from IRIS Very Long Baseline Interferometry, *Nature* **316**, 424–427.

Robertson, D. S., F. W. Fallon, and W. E. Carter, 1986, Celestial Reference Coordinate Systems: Sub-millisecond of Arc Repeatability Demonstrated with VLBI Observations, *Astron. Jour.* **91**, 1456–1462.

Rogers, A. E. E., C. A. Knight, H. F. Hinteregger, A. R. Whitney, C. C. Counselman, I. I. Shapiro, S. Gourevitch, and T. A. Clark, 1978, Geodesy by Radio Interferometry: Determination of a 1.24 km Baseline with Approximately 5 mm Repeatability, *Jour. Geophys. Research* **83**, 325, 334.

Cross-references: *Doppler Positioning: Satellite; Earth Orientation; Geodesy: Satellite; Geodetic Reference Systems; Global Positioning System (GPS); Mantle Convection and Plumes; Mantle Dynamics; Satellite Altimetry; Satellite Laser Positioning.*

VISCOUS REMANENT MAGNETIZATION (VRM) AND VISCOUS REMAGNETIZATION

Magnetic viscosity is the gradual change of magnetization with time in an applied magnetic field **H**. A ferromagnetic substance exposed briefly to a field acquires *isothermal remanent magnetization* (IRM). A longer exposure produces additional remanence, called *viscous remanent magnetization* (VRM): the longer the exposure time t, the stronger the VRM. *Viscous remagnetization* is the time-dependent change, in response to a change in the direction or intensity of **H**, of VRM or other remanences like thermoremanent magnetization (TRM), depositional remanent magnetization (DRM), or chemical remanent magnetization (CRM). (See *Thermoremanence; Detrital Remanent Magnetization; Chemical Remanent Magnetization*.) *Viscous decay* or *demagnetization* is the response to zero-field conditions (frequent in laboratory experiments but never found in nature).

The natural remanent magnetization (NRM) of virtually all rocks includes a low-temperature VRM produced by the Earth's magnetic field during the 0.7-Ma Brunhes normal epoch (see *Natural Remanent Magnetization*). This VRM obscures older primary NRMs; its removal is the main purpose of paleomagnetic cleaning methods like alternating field (af) and thermal demagnetization (see *Demagnetization*.) Brunhes-epoch VRM is easily recognized because it roughly parallels the local geomagnetic field. It is usually weaker than TRM, CRM, or DRM produced in similar fields (≤ 0.2 mT throughout geological history), and it is also usually "softer" or more easily cleaned than these older NRMs. However, VRM in magnetically hard minerals like hematite, pyrrhotite, titanomagnetite, and iron is not easily af cleaned.

Slowly cooled plutons and rocks reheated orogenically or by burial in sedimentary or volcanic piles acquire viscous partial thermoremanent magnetization (VPTRM). VPTRM blurs the distinction between purely thermal and purely viscous magnetization or remagnetization processes. Cooling times of millions or tens of millions of years encompass several field polarity reversals. (See *Geomagnetic Polarity Reversals: Theory and Models; Geomagnetic Polarity Reversals: Observations; Geomagnetic Reversal Sequence: Statistical Structure*.) VPTRM is thus potentially a bipolar remanence of reduced net intensity. Very slowly cooled rocks may even track apparent polar wander due to continental drift in different blocking temperature fractions of VPTRM. Unless acquired since late Mesozoic, VPTRM does not parallel the present field, nor is it easily af or thermally cleaned. On the plus side, it constitutes a paleomagnetic signal rather than noise, although dating the time of remagnetization is a difficult problem, usually requiring ^{40}Ar/^{39}Ar mineral ages.

The fundamental mechanism of VRM, low-temperature viscous remagnetization, and VPTRM is the gradual approach of net magnetization **M** to equilibrium in field **H** at temperature T, aided by thermal fluctuations of magnetization in individual ferromagnetic crystals. These phenomena are therefore often called *thermal fluctuation aftereffects*. If the approach to equilibrium were not extremely sluggish in most crystals at ordinary temperatures, there would be no NRM and no paleomagnetic method.

Domain walls in multidomain crystals are stabilized by lattice defects and will move if destabilized by the slow diffusion of these defects. This *diffusion aftereffect* may be significant over geologically short times, particularly in titanomagnetites.

Thermal Fluctuation Theory

Néel's (1949, 1955) theory of magnetic viscosity considers an ensemble of N aligned uniaxial single-domain particles, each with volume V, anisotropy constant K, and spontaneous magnetization M_s. At time t, n^+ particles have moments $m^+ = +VM_s$ aligned with a field **H** applied at $t = 0$, and n^- particles have moments $m^- = -VM_s$ opposed to **H**. The reduced average magnetization $M(t)/M_s$ is given by the distribution function $(n^+ - n^-)/N = n(t)/N$, with

$$\frac{dn}{dt} = -\frac{n(t) - n_{eq}}{\tau} \quad (1)$$

Equation 1 is called the *relaxation equation*, and τ is the *relaxation time*. The equilibrium Boltzmann distribution n_{eq}/N attained when $t \to \infty$ is tanh (VM_sH/kT) in this case. When $H = 0$, $n_{eq} = 0$, and Eq. 1 resembles the radioactive decay equation, with the important difference that τ is not a constant. Since transitions from m^- to m^+, and vice versa, are thermally activated,

$$\frac{1}{\tau} = C \exp\left(\frac{-VK}{kT}\right) \quad (2)$$

C being about 10^{10} s^{-1}.

At constant temperature, the solution to the relaxation equation is

$$n(t) = n_0 \exp\left(\frac{-t}{\tau}\right) + n_{eq}\left[1 - \exp\left(\frac{-t}{\tau}\right)\right] \quad (3)$$

n_0 being the $t = 0$ value of n. The first term describes viscous decay of the ensemble distribution function (and magnetization) from its initial state, and the second term describes viscous acquisition or approach to a new equilibrium. It is often convenient to treat the acquisition and decay terms in viscous remagnetization separately, although they involve the same grains.

Some insight can be gained by replacing Eq. 3 by the step function

$$n = \begin{cases} n_0, & t < \tau \\ n_{eq}, & t > \tau \end{cases}$$

This function embodies Néel's blocking approximation. The initial frozen out-of-equilibrium state is imagined to suddenly unblock (i.e., relax) to the equilibrium state when $t \to \tau$ or $\tau \to t$. This picture works well in modeling TRM or CRM, because τ depends exponentially on T and V. Blocking temperatures and blocking volumes are therefore quite sharp. A blocking picture is less precise, but nevertheless helpful, in modeling viscous changes.

A real rock, even an idealized one containing only one significant magnetic phase exclusively of single-domain size, contains many ensembles. The grain distribution $f(V, K) \, dV \, dK$ is usually poorly known. The formal equation for viscous acquisition of total sample moment is

$$m(t) = \iint VM_s n_{eq}$$
$$\cdot \left[1 - \exp\left(\frac{-t}{\tau}\right)\right] f(V, K) \, dV \, dK \quad (4)$$

A similar equation describes viscous decay. If $f(V, K)$ is known, $m(t)$ can be found by numerical integration. Analytic solutions are possible only for special forms of $f(V, K)$.

Some Fundamental Questions

For most ferromagnetic materials on a laboratory time scale, viscous changes are proportional to log t. Viscous acquisition and decay coefficients, S_a and S_d are the values of $dM/d \log t$ for growth of M for time t_a in a weak field **H**, starting from a demagnetized state, followed by decay of $M(t_a)$ in zero field for a time $t_d = t_a$. The fundamental questions to be addressed are to what extent viscous changes are really logarithmic with time and whether $S_d = -S_a$, as is usually assumed.

The superposition of exponential relaxations of many ensembles of varying τ results in an overall log t relaxation, provided log τ is uniformly distributed. According to Eq. 2, for T constant, a uniform distribution of log τ corresponds to a uniform distribution $f(V, K)$, which is reasonable for narrow ranges of V and K, corresponding to only a few decades of τ. Replacing $f(V, K) \, dV \, dK$ in Eq. 4 by $f(\tau) \, d\tau = $ const. $d \log \tau$ leads to a log t decay of m (see Dunlop, 1973 for details).

Another approach (Walton, 1980) is to calculate $S_a = t(dM/dt)$ and S_d directly. If we replace K by a suitable average value and approximate n_{eq}/N by VM_sH/kT, since H is small, Eq. 4 leads to

$$S_a = \iint \frac{M_s^2 H}{kT} \frac{t}{\tau}\left[1 - \exp\left(\frac{-t}{\tau}\right)\right] V^2 Nf(V) \, dV$$

The integrand is of the form $x(1 - e^{-x})$, with x itself exponential. It peaks sharply at $x = 1$ or $t = \tau$, which is the blocking condition. All nonexponential factors can be assigned their blocking values [e.g., $V \to V_B = (kT/K) \log Ct$] and taken outside the integral, giving

$$S_a = Ik^2 N\left(\frac{M_s^2 T^2}{K^3}\right) H (\log Ct)^2 f\left(\frac{kT}{K} \log Ct\right) \quad (5)$$

the integral I being approximately 1.

Assuming shape anisotropy, $K(T) \propto M_s^2(T)$. If the grain distribution is uniform over the range of ensembles affected in time t_a, we have

$$S_a = C'H\left(\frac{T}{M_s^2}\right)^2 (\log Ct)^2 \quad (6)$$

whereas if $f(V) \propto 1/V$ (Walton, 1980), as for the larger grains in a log-normal distribution,

$$S_a = C''H\frac{T}{M_s^2} \log Ct \quad (7)$$

C', C'' being constants. Either equation predicts a log t dependence for VRM intensity for times such that log $t \ll \log C \approx 23$, i.e., for a few decades of t, as observed. Deviations from log t dependence will become obvious even at short times for very viscous samples [those with a large $f(V_B)$] and at longer times for all samples. Deviations will appear earlier if Eq. 6, rather than Eq. 7, applies.

The decay process is precisely reciprocal to the acquisition process. However, the initial state is more complicated. Ensembles with $\tau \ll t_a$ will be completely relaxed; their initial state when $H \to 0$ will be $n(t_a) = n_{eq}$. But ensembles with $\tau \approx t_a$ will be only partially relaxed; their initial state for decay will be $n(t_a) = pn_{eq}$, where $p = 1 - \exp(-t_a/\tau)$ from Eq. 3. A fraction p of this initial n will decay in time $t_d = t_a$, so $S_d = -pS_a$ for such ensembles.

Overall we expect S_d to be slightly less than S_a, the exact difference depending on $f(V)$.

VRM of Multidomain and Interacting Grains

Viscous magnetization due to thermal activation of domain walls has been treated theoretically in the weak-field limit by Néel, in fields comparable to the coercive force by Street and Woolley and by Gaunt, and in the presence of internal demagnetizing fields by Stacey (see Dunlop, 1973 for a review). All theories predict a log t dependence of VRM for geologically short times. The coefficients S_a and S_d vary with temperature, either as T/M_s or as $(T/M_s)^{1/2}$. The theories resemble single-domain formulations, with V in Eq. 2 replaced by the volume V_{act} activated in a single Barkhausen jump. The frequency constant C is about 10^{10} s^{-1}, as for single-domain particles. Magnetic viscosity due to activation of entire walls should be negligible, since V_{act} is much greater than V for single-domain particles. It must be that segments of walls are activated past single pinning sites. This phenomenon is indeed observed in large crystals of titanomagnetite (see *Magnetic Domains*).

Magnetite, titanomagnetite, and pyrrhotite crystals often fail to renucleate equilibrium domain structures following saturation. A possible mechanism for magnetic viscosity in multidomain grains would be the thermally activated nucleation of domain walls in parts of the crystal where the internal field is just below the critical nucleation field. This process has been termed *transdomain VRM*.

Interacting single-domain particles have been treated theoretically in two different ways. One method is to use a distribution of $V_{act} > V$ in Eq. 2, since particles act collectively. A single ensemble of identical particles, if strongly interacting, should have log t behavior. Another approach is to include the interactions via a random field that affects each particle. Walton and Dunlop (1985) predicted significant deviation from log t behavior if interactions are strong. The theory gave a good fit to acquisition and decay data for interacting single-domain magnetite crystals with a known $f(V)$.

Experimental Results

Room temperature viscosity coefficients S_a and S_d have their highest values in single-domain grains just above superparamagnetic size ($V \gtrsim V_B$). In magnetite, the only mineral studied in sufficient detail to discern broad grain-size trends, the peak in VRM is around 30–40 nm (Fig. 1). Coefficients S_a and S_d fall sharply (by about a factor of 5) in only slightly larger single-domain grains (≤ 100 nm) and remain at these levels throughout the multidomain range, with perhaps a gradual increase in the 4–15-μm range. Since samples greater than 1 μm in size are known to have adhering fine particles, it is not certain that the intrinsic size dependence of viscous effects has been isolated. VRM may well be due to a fraction of particles much smaller than the norm, even in the single-domain samples.

Typical viscous acquisition and decay data for magnetite grains just above single-domain size

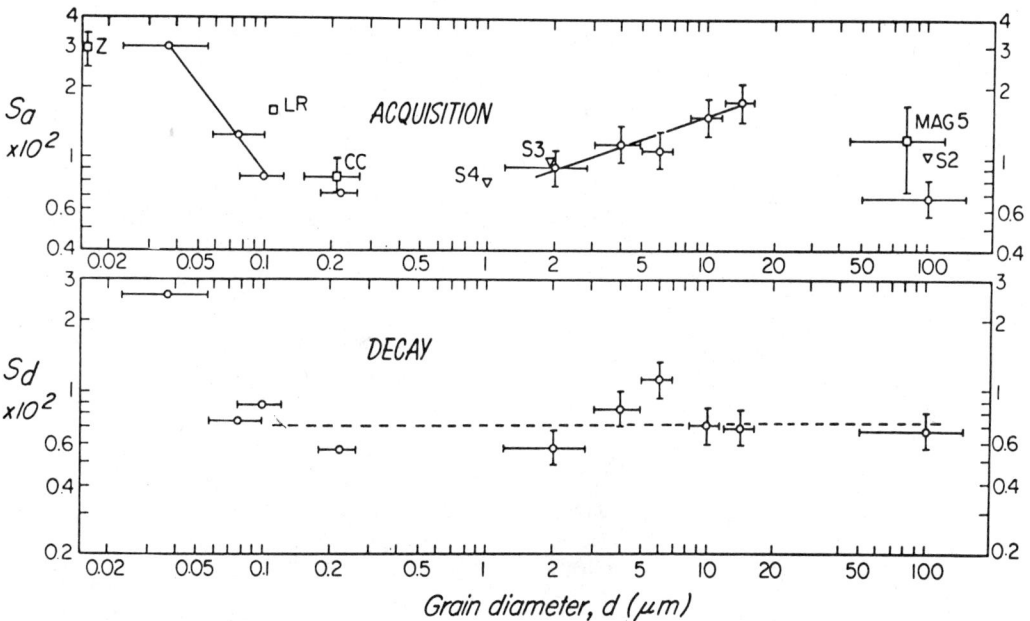

FIGURE 1. Room-temperature viscosity coefficients S_a and S_d in units of emu · cm^{-3} or kA · m^{-1} for magnetites of various grain sizes. (After Dunlop, 1983)

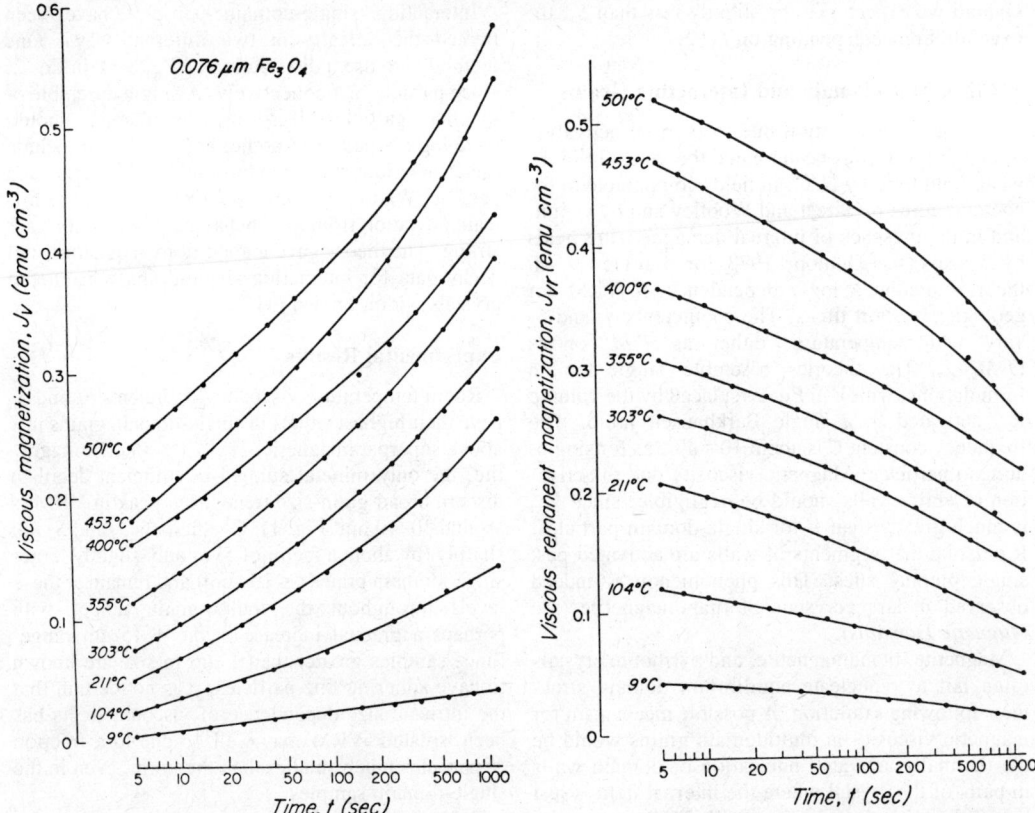

FIGURE 2. Viscous acquisition and decay curves at various temperatures for magnetite with average grain size 76 nm exposed to a field $H = 10$ Oe or 1 mT. (After Dunlop, 1983)

appear in Fig. 2. For experimental convenience, viscous induced magnetization (VIM) rather than VRM was monitored in the growth curves. For each decay run, $t = 0$ marks the instant H (here 1 mT) was switched off at the end of the acquisition run.

Over short time spans (2.4 decades of t in Fig. 2), room temperature acquisition and decay are proportional to log t. Over longer times and at higher temperatures, M is a quadratic function of log t. The general pattern in single-domain and small multidomain material is that non-log t behavior becomes apparent whenever a sufficiently large fraction of the grain distribution is revealed by t or T. Shimizu's (1960) viscosity data for multidomain magnetites greater than 1 μm in size went as log t at all temperatures. It is not clear whether this is fundamental multidomain behavior or results from weak viscosity.

The inherent temperature dependence of viscosity coefficients, in Eqs. 6 and 7, for example, tends to be obscured by the variability of the grain distribution, which is scanned with a narrow thermal "window," as expressed by the $f[(kT/K)\log Ct]$ factor in Eq. 5. Very different temperature dependences of S_a and S_d have been reported for magnetite by different authors. A good theoretical match to single-domain data like those of Fig. 2 was obtained (Walton, 1983) by using the measured $f(V)$ of the sample to evaluate Eq. 5.

Viscosity studies have been extended to very short times by measuring the dependence of initial susceptibility χ on the frequency f of an applied alternating field $\tilde{H}$. For soils containing single-domain magnetite and maghemite, $d\chi/d \log f$ = const. No frequency dependence was detectable for multidomain magnetite.

Experiment substantiates the theoretical prediction that VRM intensity is proportional to field strength H for weak fields (e.g., Eqs. 6 and 7). The most detailed studies are by Creer (1957) for single-domain hematite and Le Borgne (1960) for soils containing single-domain magnetite and maghemite. Linearity with H was observed over the range 0.05–1 mT.

Strong fields, such as those used in af cleaning, have an additional effect on viscosity. They lower the energy barrier KV to rotation or the corresponding barrier to domain wall motion, thereby directly affecting τ. In the single-domain case, Eq. 2 becomes

$$\frac{1}{\tau_H} = C_H \exp\left[-\frac{VK}{kT}\left(1 - \frac{HM_s}{2K}\right)^2\right] \quad (8)$$

where $C_H \approx C$. One consequence is "hard" VRM in minerals with high K/M_s, like hematite, pyrrhotite, elongated single-domain iron and many compositions of titanomagnetite. VRM is acquired under weak-field conditions where Eq. 2 applies but is af demagnetized by a strong field $\tilde{H}_c$ to which Eq. 8 applies. Using the blocking approximations $\tau = t_a$ for the last ensemble to acquire VRM and $\tau_H = 1/f$ when $H = \tilde{H}_c$ for af cleaning of this same ensemble, we have, by combining Eqs. 2 and 8,

$$\tilde{H}_c = \frac{2K}{M_s}\left[1 - \left(\frac{\log(C_H/f)}{\log Ct_a}\right)^{1/2}\right] \quad (9)$$

The resistance of VRM to af cleaning is thus proportional to K/M_s, the other factor being independent of mineral properties. This prediction is borne out experimentally. Room temperature VRM in magnetite is easily af cleaned, usually by fields less than 10 mT, but in hematite, cleaning fields may exceed 100 mT, the limit of many commercial demagnetizers.

Experimentally, the rate at which VRM or VIM is acquired is comparable to, but usually slightly greater than, the rate at which it subsequently decays. The difference between S_a and S_d is much increased if VRM is acquired in a field H that significantly lowers energy barriers (Eq. 8). The VRM will then contain ensembles with $\tau_H < t_a$ but $\tau > t_d$ in zero-field decay.

Viscosity coefficients depend on the initial state of magnetization. If the initial state is an af or thermally cleaned one, the coefficients also decrease markedly with "wait time" after demagnetization. The former observation is consistent with thermal fluctuation models; the latter seems to require another mechanism, such as diffusion aftereffect. For a detailed review of experimental results, see Dunlop (1973) and Moskowitz (1985).

Applications

VRM as a Dating Method. Viscous remagnetization can be used as a dating method. Heller and Markert (1973) sampled dolerite blocks that had been reoriented in Roman times in building Hadrian's Wall in northern England. The VRM formed in the Earth's field since that time was resolved from the surviving stable primary NRM by af demagnetization, and a multidomain analog of Eq. 9 was used to determine log Ct_a. Two of the three blocks tested gave reasonable estimates of elapsed time: $t_a = 1.6$–1.8 ka.

The resolution of this method is determined by $dM/dt = (dM/d \log t) \, t^{-1} = S_a/t_a$. There are two problems. First, S_a must be constant for time periods of geological length, contrary to usual laboratory observations. The success of Heller and Markert's experiment suggests that multidomain grains may succeed in this respect where single-domain grains fail. Second, the resolution is inversely proportional to the length of time to be measured, so only geologically recent events are datable.

Granulometry by Magnetic Viscosity. Magnetic viscosity is a sensitive probe of narrow (V, K) bands of the grain distribution. If K has a narrow spread compared with V, inversion of Eq. 4 or its decay analog by Laplace transforms yields $f(\tau)$, i.e., $f(V)$. A cruder estimate, which is adequate in most applications, is to use a blocking approximation as in Eq. 5 to obtain $f(V_B) = f(kT \log Ct/K)$. Small V_B can be accessed by using the frequency dependence of susceptibility. Large V_B can be activated above room temperature.

Simulation of Brunhes Epoch VRM. In very viscous rocks, particularly unoriented or partially oriented seafloor samples, one would like to know if a large part of the NRM is VRM produced during the past 0.7 Ma. One hypothesized origin of the Cretaceous quiet zones in linear anomaly patterns over the Atlantic, for example, is viscous remagnetization. Since relaxation times of approximately 1 Ma are inaccessible in the laboratory, two methods have been used. The first is simple extrapolation, assuming S_a constant. This assumption is certainly unjustified with massive doleritic flows, the most viscous shallow oceanic rocks (Lowrie and Kent, 1978).

The second method is to activate long relaxation times in the laboratory in mild heatings that do not exceed the $\approx 200°C$ titanomagnetite Curie temperatures or cause mineralogical alteration. The principle is simple. The last ensemble to be activated at T_0 in time t_a will also be the last ensemble activated in a laboratory time t_L at T. Equation 2 applies to both situations. Assuming $K(T) \propto M_s^2(T)$ as with single-domain shape anisotropy,

$$\frac{(\log Ct_L)T}{M_s^2(T)} = \frac{(\log Ct_a)T_0}{M_s^2(T_0)} \quad (10)$$

Although this relation works well when hematite is the NRM carrier, it does not seem to hold for magnetite. Figure 3 compares (t_L, T) and (t_a, T_0) for limestone boulders reoriented and deposited in glacial drumlins about 10 ka ago. The points do not obey Eq. 10 (using M_s for magnetite), which predicts reactivation in about 1 hr at 200°C. Instead $t_L = 1$ hr for $T = 275°C$, in accord with a relation proposed by Walton (1980):

$$\frac{(\log Ct_L)^2 T}{M_s^2(T)} = \frac{(\log Ct_a)^2 T_0}{M_s^2(T_0)} \quad (11)$$

This relation is obtained by integrating Eq. 7. It answers the question: What exposure time t_L to field

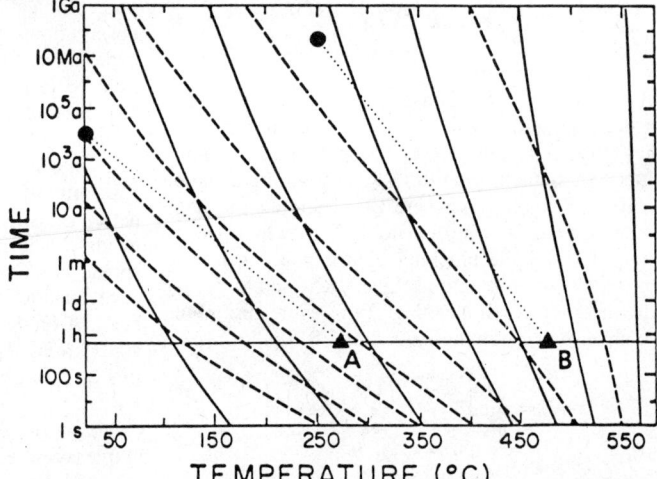

FIGURE 3. Matching (t, T) conditions for remagnetization in nature (circles) and thermal demagnetization in the laboratory (triangles), determined for low- and high-temperature thermoviscous NRMs of limestone samples. Theoretical t-T relations follow Eqs. 10 or 11 of the text (solid and dashed curves, respectively). (After Kent, 1985)

H at temperature T will produce the *same VRM intensity* as exposure t_a to the same H at T_0? This question is different than the one posed earlier: What exposure t_L at T will reactivate *the same ensembles* as exposure t_a at T_0? Since S_a increases with rising temperature even if $f(V_B)$ is constant (Eq. 5), Eqs. 10 and 11 will always give different answers for t_L. It is puzzling that Eq. 11 appears to give the truer answer.

Thermoviscous Remagnetization. A closely allied problem is the temperature T_B required for short-term thermal demagnetization of VPTRM acquired during slow cooling in nature. Thermoviscous magnetization is difficult to treat theoretically, since T is not constant in Eqs. 1 and 2 and the cooling history is often poorly known. If a simple blocking approximation is used (Pullaiah et al., 1975), the result is similar to Eq. 10 with T replaced by T_B and (t_a, T_0) replaced by (t_R, T_R), the estimated cooling time and average temperature.

Since T_R may be quite high in burial orogens, thermoviscous NRM "overprints" require much higher temperature cleaning than do Brunhes epoch overprints. For VPTRMs in which cooling rate and maximum T_R can be estimated (T_R from the conodont alteration index, for example), (t_L, T_B) in thermal cleaning follow a relation akin to Eq. 11 rather than Eq. 10. Figure 3 gives an example.

Cooling-Rate Dependence of TRM. The theoretical approaches of the previous section also give estimates of the intensity of slowly cooled TRM (i.e., VPTRM). In single-domain grains, about 5% increase in intensity is predicted for the slowest versus the fastest practical laboratory cooling rates. These predictions are borne out quite well by experiment (Fox and Aitken, 1980).

VRM in Paleointensity Determination. The Thellier-type paleointensity determinations (see *Archaeomagnetism* and *Paleomagnetic Field: Intensity*) in which the thermal decay of NRM is compared to the complementary growth of TRM, must be corrected for the difference between natural and laboratory cooling rates. Walton (1984) points out a more direct connection between time and paleofield intensity. Since S_a, and similarly S_d, are proportional to H, whatever the details of the grain distribution (Eqs. 6, 7), a comparison of the viscous decay of NRM and the viscous growth of VRM in a laboratory field yields an estimate of the ancient field. The drawback of the method is that VRM utilizes a very small fraction of the grain distribution, whereas TRM utilizes it all. The advantage is that replicate viscous decay experiments at a series of temperatures are sensitive indicators of mineral alteration that might go undetected in a Thellier experiment.

How to Clean "Hard" VRM. Red marls studied by Ouliac (1976) possessed "hard" VRMs parallel to the present Earth's field that were not completely removed by af demagnetization to about 100 mT. The VRMs were completely erased by thermal demagnetization to $\leq 340°C$, as predicted (for hematite) by Eq. 10.

VRM in Rocks

Lunar rocks owe their magnetic properties to metallic iron. Lunar soils and low-grade breccias contain ultrafine (≤ 20 nm) single-domain iron particles and are extremely viscous. High-grade breccias, basalts, and anorthosites have much weaker viscosity originating in multidomain iron. For a full discussion, see Dunlop (1973).

Terrestrial submarine basalts are rather weakly viscous. However, coarse-grained massive flows can be extremely viscous (Lowrie and Kent, 1978). The VRM is likely due to large homogeneous titanomagnetite grains with soft, easily moved domain walls. VRM of oxidized titanomagnetites is very dependent on the degree of oxidation (Özdemir and Banerjee, 1981).

For Tertiary and early Quaternary subaerial basalts, Brunhes epoch VRM averages about one fourth of total NRM intensity (Prevot, 1981). Soft VRM is due to homogeneous multidomain titanomagnetite grains. Relatively hard VRM is carried by single-domain size magnetite-ilmenite intergrowths formed by oxyexsolution. Thermal cleaning to no more than 300°C erases all Brunhes epoch VRM.

Red sediments of all types have relatively strong and hard VRMs. Creer's (1957) study remains the classic. Among the magnetite-bearing sediments, limestones have been the most carefully studied.

Plutonic rocks, particularly intermediate and felsic ones, tend to be quite viscous. The VRM is due to multidomain magnetite, often in grains of about 100 μm in size and is soft to af cleaning ($\tilde{H}_c < 10$ mT). VRM and stable NRM may reside in different magnetite fractions.

Soils and baked clays (e.g., pottery and bricks) owe their viscosity to fine single-domain magnetite and maghemite. The classic study is Le Borgne (1960).

DAVID J. DUNLOP

References

Creer, K. M., 1957, The remanent magnetization of unstable Keuper marls, *Royal. Soc. London Philos. Trans.* **A250,** 130-143.

Dunlop, D. J., 1973, Theory of the magnetic viscosity of lunar and terrestrial rocks, *Rev. Geophysics Space Physics* **11,** 855-901.

Dunlop, D. J., 1983, Viscous magnetization of 0.04-100 μm magnetites, *Royal Astron. Soc. Geophys. Jour.* **74,** 667-687.

Fox, J. M. W., and M. J. Aitken, 1980, Cooling-rate dependence of thermoremanent magnetisation, *Nature* **283,** 462-463.

Heller, F., and H. Markert, 1973, The age of viscous remanent magnetization of Hadrian's Wall (Northern England), *Royal Astron. Soc. Geophys. Jour.* **31,** 395-406.

Kent, D. V., 1985, Thermoviscous remagnetization in some Appalachian limestones, *Geophys. Research Letters* **12,** 805-808.

Le Borgne, E., 1960, Etude expérimentale du traînage magnétique dans le cas d'un ensemble de grains magnétiques très fins dispersés dans une substance non magnétique, *Ann. Géophys.* **16,** 445-494.

Lowrie, W., and D. V. Kent, 1978, Characteristics of VRM in oceanic basalts, *Jour. Geophys.* **44,** 297-315.

Moskowitz, B. M., 1985, Magnetic viscosity, diffusion after-effect, and disaccommodation in natural and synthetic samples, *Royal Astron. Soc. Geophys. Jour.* **82,** 143-161.

Néel, L., 1949, Théorie du traînage magnétique des ferromagnétiques en grains fins avec applications aux terres cuites, *Ann. Géophys.* **5,** 99-136.

Néel, L., 1955, Some theoretical aspects of rock magnetism, *Advan. Phys.* **4,** 191-243.

Ouliac, M., 1976, Removal of secondary magnetization from natural remanent magnetization of sedimentary rocks: alternating field or thermal demagnetization technique? *Earth Planetary Sci. Letters* **29,** 65-70.

Özdemir, Ö., and S. K. Banerjee, 1981, An experimental study of magnetic viscosity in synthetic monodomain titanomaghemites: implications for the magnetization of the ocean crust, *Jour. Geophys. Research* **86,** 11864-11868.

Prévot, M., 1981, Some aspects of magnetic viscosity in subaerial and submarine volcanic rocks, *Royal Astron. Soc. Geophys. Jour.* **66,** 169-192.

Pullaiah, G., E. Irving, K. L. Buchan, and D. J. Dunlop, 1975, Magnetization changes caused by burial and uplift, *Earth Planetary Sci. Letters* **28,** 133-143.

Shimizu, Y., 1960, Magnetic viscosity of magnetite, *Jour. Geomagnetism and Geoelectricity* **11,** 125-138.

Stacey, F. D., 1963, The physical theory of rock magnetism, *Adv. Physics* **12,** 45-133.

Stacey, F. D., and S. K. Banerjee, 1974, *The Physical Principles of Rock Magnetism.* Amsterdam: Elsevier, 195 p.

Walton, D., 1980, Time-temperature relations in the magnetization of assemblies of single domain grains, *Nature* **286,** 245-247.

Walton, D., 1983, Viscous magnetization, *Nature* **305,** 616-619.

Walton, D., 1984, Re-evaluation of Greek archaeomagnitudes, *Nature* **310,** 740-743.

Walton, D., and D. J. Dunlop, 1985, The magnetization of a random assembly of interacting moments, *Solid State Communications* **53,** 359-362.

Cross-references: *Chemical Remanent Magnetization; Curie Temperature; Demagnetization; Detrital Remanent Magnetization (DRM); Geomagnetic Field, Main: Theory; Geomagnetic Polarity Reversals: Observations; Geomagnetic Polarity Reversals: Theory and Models; Geomagnetic Reversal Sequence: Statistical Structure; Magnetic Domains; Magnetic Properties of Minerals; Magnetic Self-Reversal; Natural Remanent Magnetization (NRM); Rock Magnetism; Rock Magnetization: Measuring Techniques and Apparatus; Thermoremanence.*

APPENDIX

The International System of Units

The international system of units, abbreviated SI for *Système International* (d'Unités), was defined and given official status by the 11th General Conference on Weights and Measures, 1960. The complete listing of SI units and various other supplementary units may be found in *The International System of Units (SI)* (National Bureau of Standards Special Publication 330, 1981 edition) and in *The International System of Units: Physical Constants and Conversion Factors*, second revision, 1973, by E. A. Mechtly (National Aeronautics and Space Administration Publication SP-7012). Both publications are published by and available through the U.S. Government Printing Office (Washington, D.C. 20402, U.S.A.).

The SI system is comprised of seven *base* units as well as a number of *derived* and *supplementary* units. The following compilation includes tables of the SI units (Tables 1-7) as well as tables of certain important units that are outside the SI system (Tables 8-11) but are still in common use. The tabular data are from NBS Special Publication 330, 1981 edition.

TABLE 1. SI Base Units

Quantity	SI Unit Name	SI Unit Symbol	Quantity	SI Unit Name	SI Unit Symbol
Length	meter[a]	m	Thermodynamic temperature	kelvin[e]	K
Mass	kilogram[b]	kg	Amount of substance	mole[f]	mol
Time	second[c]	s	Luminous intensity	candela[g]	cd
Electric current	ampere[d]	A			

[a] A meter is the length equal to 1 650 763.73 wavelengths in vacuum of the radiation corresponding to the transition between the levels $2p_{10}$ and $5d_5$ of the krypton-86 atom.

[b] A kilogram is equal to the mass of the international prototype of the kilogram.

[c] A second is the duration of 9 192 631 770 periods of the radiation corresponding to the transition between the two hyperfine levels of the ground state of the cesium-133 atom.

[d] An ampere is that constant current which, if maintained in two straight parallel conductors of infinite length, of negligible circular cross section, and placed 1 meter apart in vacuum, would produce between these conductors a force equal to 2×10^{-7} newton per meter of length.

[e] A kelvin is the fraction 1/273.16 of the thermodynamic temperature of the triple point of water.

[f] A mole is the amount of substance of a system that contains as many elementary entities as there are atoms in 0.012 kilogram of carbon 12.

[g] A candela is the luminous intensity, in a given direction, of a source that emits monochromatic radiation of frequency 540×10^{12} hertz and that has a radiant intensity in that direction of (1/683) watt per steradian.

TABLE 2. Examples of SI Derived Units Expressed in Terms of Base Units

Quantity	SI Unit	
	Name	Symbol
Area	square meter	m^2
Volume	cubic meter	m^3
Speed, velocity	meter per second	m/s
Acceleration	meter per second squared	m/s^2
Wave number	1 per meter	m^{-1}
Density, mass density	kilogram per cubic meter	kg/m^3
Specific volume	cubic meter per kilogram	m^3/kg
Current density	ampere per square meter	A/m^2
Magnetic field strength	ampere per meter	A/m
Concentration (of amount of substance)	mole per cubic meter	mol/m^3
Luminance	candela per square meter	cd/m^2

TABLE 3. SI Derived Units with Special Names

Quantity	SI Unit			
	Name	Symbol	Expression in terms of other units	Expression in terms of SI base units
Frequency	hertz	Hz		s^{-1}
Force	newton	N		$m \cdot kg \cdot s^{-2}$
Pressure, stress	pascal	Pa	N/m^2	$m^{-1} \cdot kg \cdot s^{-2}$
Energy, work, quantity of heat	joule	J	$N \cdot m$	$m^2 \cdot kg \cdot s^{-2}$
Power, radiant flux	watt	W	J/s	$m^2 \cdot kg \cdot s^{-3}$
Electric charge, quantity of electricity	coulomb	C		$s \cdot A$
Electric potential, potental difference, electromotive force	volt	V	W/A	$m^2 \cdot kg \cdot s^{-3} \cdot A^{-1}$
Capacitance	farad	F	C/V	$m^{-2} \cdot kg^{-1} \cdot s^4 \cdot A^2$
Electric resistance	ohm	Ω	V/A	$m^2 \cdot kg \cdot s^{-3} \cdot A^{-2}$
Electric conductance	siemens	S	A/V	$m^{-2} \cdot kg^{-1} \cdot s^3 \cdot A^2$
Magnetic flux	weber	Wb	$V \cdot s$	$m^2 \cdot kg \cdot s^{-2} \cdot A^{-1}$
Magnetic flux density	tesla	T	Wb/m^2	$kg \cdot s^{-2} \cdot A^{-1}$
Inductance	henry	H	Wb/A	$m^2 \cdot kg \cdot s^{-2} \cdot A^{-2}$
Celsius temperature	degree Celsius	°C		K
Luminous flux	lumen	lm		$cd \cdot sr$
Illuminance	lux	lx	lm/m^2	$m^{-2} \cdot cd \cdot sr$

TABLE 4. Examples of SI Derived Units Expressed by Means of Special Names

Quantity	SI Unit Name	SI Unit Symbol	Expression in terms of SI base units
Dynamic viscosity	pascal second	Pa·s	$m^{-1} \cdot kg \cdot s^{-1}$
Moment of force	newton meter	N·m	$m^2 \cdot kg \cdot s^{-2}$
Surface tension	newton per meter	N/m	$kg \cdot s^{-2}$
Heat flux density, irradiance	watt per square meter	W/m²	$kg \cdot s^{-3}$
Heat capacity, entropy	joule per kelvin	J/K	$m^2 \cdot kg \cdot s^{-2} \cdot K^{-1}$
Specific heat capacity, specific entropy	joule per kilogram kelvin	J/(kg·K)	$m^2 \cdot s^{-2} \cdot K^{-1}$
Specific energy	joule per kilogram	J/kg	$m^2 \cdot s^{-2}$
Thermal conductivity	watt per meter kelvin	W/(m·K)	$m \cdot kg \cdot s^{-3} \cdot K^{-1}$
Energy density	joule per cubic meter	J/m³	$m^{-1} \cdot kg \cdot s^{-2}$
Electric field strength	volt per meter	V/m	$m \cdot kg \cdot s^{-3} \cdot A^{-1}$
Electric charge density	coulomb per cubic meter	C/m³	$m^{-3} \cdot s \cdot A$
Electric flux density	coulomb per square meter	C/m²	$m^{-2} \cdot s \cdot A$
Permittivity	farad per meter	F/m	$m^{-3} \cdot kg^{-1} \cdot s^4 \cdot A^2$
Permeability	henry per meter	H/m	$m \cdot kg \cdot s^{-2} \cdot A^{-2}$
Molar energy	joule per mole	J/mol	$m^2 \cdot kg \cdot s^{-2} \cdot mol^{-1}$
Molar entropy, molar heat capacity	joule per mole kelvin	J/(mol·K)	$m^2 \cdot kg \cdot s^{-2} \cdot K^{-1} \cdot mol^{-1}$
Exposure (x and γ rays)	coulomb per kilogram	C/kg	$kg^{-1} \cdot s \cdot A$
Absorbed dose rate	gray per second	Gy/s	$m^2 \cdot s^{-3}$

TABLE 5. SI Supplementary Units

Quantity	SI Unit Name	SI Unit Symbol
Plane angle	radian	rad
Solid angle	steradian	sr

TABLE 6. Examples of SI Derived Units Formed by Using Supplementary Units

Quantity	SI Unit Name	SI Unit Symbol
Angular velocity	radian per second	rad/s
Angular acceleration	radian per second squared	rad/s²
Radiant intensity	watt per steradian	W/sr
Radiance	watt per square meter steradian	$W \cdot m^{-2} \cdot sr^{-1}$

TABLE 7. SI Prefixes

Factor	Prefix	Symbol	Factor	Prefix	Symbol
10^{18}	exa	E	10^{-1}	deci	d
10^{15}	peta	P	10^{-2}	centi	c
10^{12}	tera	T	10^{-3}	milli	m
10^{9}	giga	G	10^{-6}	micro	μ
10^{6}	mega	M	10^{-9}	nano	n
10^{3}	kilo	k	10^{-12}	pico	p
10^{2}	hecto	h	10^{-15}	femto	f
10^{1}	deka	da	10^{-18}	atto	a

Note: NBS Special Publication 330 (p. 12) states "Among the base units of the International System, the unit of mass is the only one whose name, for historical reasons, contains a prefix. Names of decimal multiples and sub-multiples of the unit of mass are formed by attaching prefixes to the word 'gram'." Thus, *for example*, 10^{-6} kg = 1 milligram, (*not* 1 microkilogram).

TABLE 8. Units in Use with the International System

Name	Symbol	Value in SI unit	
Minute	min	1 min	= 60 s
Hour	h	1 h	= 60 min = 3 600 s
Day	d	1 d	= 24 h = 86 400 s
Degree	°	1°	= $(\pi/180)$ rad
Minute	'	1'	= $(1/60)° = (\pi/10\,800)$ rad
Second	"	1"	= $(1/60)' = (\pi/648\,000)$ rad
Liter	l, L	1 L	= 1 dm^3 = 10^{-3} m^3
Metric ton	t	1 t	= 10^3 kg

TABLE 9. Units in Use Temporarily with the International System

Name	Symbol	Value in SI unit
Nautical mile		1 nautical mile = 1 852 m
Knot		1 nautical mile per hour = (1852/3600) m/s
Ångström	Å	1 Å = 0.1 nm = 10^{-10} m
Are	a	1 a = 1 dam^2 = 10^2 m^2
Hectare	ha	1 ha = 1 hm^2 = 10^4 m^2
Barn	b	1 b = 100 fm^2 = 10^{-28} m^2
Bar	bar	1 bar = 0.1 MPa = 10^5 Pa
Gal	Gal	1 Gal = 1 cm/s^2 = 10^{-2} m/s^2
Curie	Ci	1 Ci = 3.7×10^{10} Bq
Roentgen	R	1 R = 2.58×10^{-4} C/kg
Rad	rad	1 rad = 1 cGy = 10^{-2} Gy
Rem	rem	1 rem = 1 cSv = 10^{-2} Sv

TABLE 10. CGS Units with Special Names

Name	Symbol	Value in SI unit
Erg	erg	1 erg = 10^{-7} J
Dyne	dyn	1 dyn = 10^{-5} N
Poise	P	1 P = 1 dyn·s/cm^2 = 0.1 Pa·s
Stokes	St	1 St = 1 cm^2/s = 10^{-4} m^2/s
Gauss	Gs, G	1 Gs corresponds to 10^{-4} T
Oersted	Oe	1 Oe corresponds to $\dfrac{1000}{4\pi}$ A/m
Maxwell	Mx	1 Mx corresponds to 10^{-8} Wb
Stilb	sb	1 sb = 1 cd/cm^2 = 10^4 cd/m^2
Phot	ph	1 ph = 10^4 lx

TABLE 11. Other Units Generally Deprecated

Name	Value in SI unit
Fermi	1 fermi = 1 fm = 10^{-15} m
Metric carat	1 metric carat = 200 mg = 2×10^{-4} kg
Torr	1 torr = $\dfrac{101\,325}{760}$ Pa
Standard atmosphere (atm)	1 atm = 101 325 Pa
Kilogram-force (kgf)	1 kgf = 9.806 65 N
Calorie (cal)	
Micron (μ)	1 μ = 1 μm = 10^{-6} m
X unit	
Stere (st)	1 st = 1 m^3
Gamma (γ)	1 γ = 1 nT = 10^{-9} T
γ	1 γ = 1 μg = 10^{-9} kg
λ	1 λ = 1 μL = 10^{-6} L = 10^{-9} m^3

AUTHOR CITATION INDEX

Abalakin, V. K., 938
Abbott, J. G., 20
Abdallah, A., 166
Abe, K., 260, 314
Abramowitz, M., 430
Achauer, U., 1150
Achenbach, J. D., 1023
Acheson, D. J., 584, 739
Acuna, M. H., 522
Adams, F. D., 1201
Adams, O. S., 469
Ade-Hall, J. M., 173
Adler, R. K., 844
Aggarwal, Y. P., 267
Agnew, D., 1246
Agnew, D. C., 1037, 1092
Ahrens, T. J., 854, 969
Aitken, M. J., 37, 1303
Akaogi, M., 854
Akasofu, S.-I., 498
Aki, K., 89, 245, 267, 288, 294, 307, 314, 367, 372, 430, 654, 1023, 1107, 1128, 1132, 1150, 1151, 1186, 1187, 1197, 1210, 1266
Aki, N., 1246
Akimoto, S., 733, 854, 909
Alasia, F., 9
Albaugh, D. S., 1279
Albright, L., 897
Alford, R. M., 430, 431
Alldredge, L. R., 498, 577
Allegre, C. J., 819
Allen, C., 314
Almond, M., 909
Altamini, A., 482
Alterman, Z., 372, 430, 431
Alvarez, W., 495
Ambrayses, N. N., 74, 1201
Anderle, R., 482
Anderle, R. J., 222
Anders, E., 847
Anderson, D. L., 358, 367, 674, 698, 775, 787, 1005, 1017, 1018, 1151, 1266
Anderson, E. M., 1231
Anderson, J. A., 1044
Anderson, J. D., 938
Anderson, J. G., 1246
Anderson, K., 675
Anderson, K. R., 1186
Anderson, L. M., 1018
Anderson, R. N., 646
Anderson, W. L., 139
Ando, M., 1005
Angus-Leppan, P. V., 489
ANS, 288

Anufriev, A. P., 159
Aoki, H., 1005
Appel, E., 720
Archambeau, C. B., 1089
Archuleta, R. J., 1246
Argus, D., 1050
Argus, D. F., 923
Arkani-Hamed, J., 215
Armbruster, J., 75
Armbruster, J. G., 1279
Armijo, R., 75, 901
Arnautov, G. P., 9
Arnold, K., 215
As, J. A., 961
Asada, T., 1005
Asano, S., 1005
Ashby, M. F., 201
Asudeh, I., 74
Atkinson, B. K., 47
Aubin, J.-P., 26
Avetisjan, R. A., 806
Aydin, A., 126

Babich, V. M., 1107
Bache, T. C., 1089
Bachman, S. B., 1255
Bachtadse, V., 900
Backus, G. E., 159, 160, 372, 522, 534, 584, 654, 1151, 1221
Bagin, V. I., 57
Bailey, A. D., 1118
Baker, T. F., 363
Bakun, W. H., 267
Balazs, E. I., 474
Baldwin, A. L., 299
Ballard, R. D., 668
Balling, N., 166
Bally, A., 74
Bally, A. W., 126, 166, 190
Balmer, H., 592
Bame, D. A., 294
Bamford, D., 1005
Banerjee, S. K., 173, 706, 720, 721, 733, 888, 950, 994, 1269, 1303
Bangs, N., 1279
Barazangi, M., 74, 75, 89, 148, 166, 190, 260, 871, 1061
Barker, F. S., 577
Barker, J. S., 245, 294
Barnes, R. T. H., 160
Barraclough, D. R., 160, 577, 1221
Barrick, D. E., 968
Bartel, N., 1296
Bartelme, N., 458

Bartels, J., 495, 498, 522, 534, 592, 770, 1221
Barton, C. E., 739, 888
Barton, P. J., 871
Basaltic Volcanism Study Project, 698
Basham, P. W., 1089
Bass, J. D., 787
Bates, C. C., 1201
Bath, M., 288
Bauer, L. A., 510, 577
Beavan, R. J., 294
Beck, M. E., 900
Beck, M. E., Jr., 888
Becker, J. J., 57, 720
Becquerel, H., 5
Beetz, W. von, 57
Behle, A., 431
Behrensmeyer, A. K., 746
Ben-Avraham, Z., 407
Bendat, J. S., 622, 770
Bender, P. L., 166
Benioff, H., 245, 1044
Ben-Israel, A., 654
Ben-Menahem, A., 1128
Bennett, G., 267
Benson, A. K., 1166
Benthien, R. H., 57
Benton, E. R., 584
Berberian, M., 74
Berbert, J., 988
Berger, J., 1092
Berggren, W. A., 495, 560, 746
Bergman, E. A., 260, 1049, 1054
Berkhout, A. J., 1166
Bernstein, R. L., 968
Berry, B. S., 201, 1018
Berry, R. M., 474
Beske-Diehl, S., 994
Bessel, F. W., 453
Beylkin, G., 1166
Bierman, G. J., 465
Bills, B. G., 938
Bindschadler, R. A., 969
Birch, F., 854, 1098
Bird, P., 97
Bishop, T. N., 393
Biswas, N. N., 1186
Bjornsson, A., 166
Blackett, P. M. S., 57, 909, 961
Blakely, R. J., 617, 746, 993
Blakemore, R., 890
Bleil, U., 173
Bleistein, N., 1166
Bloch, S., 1117
Block, S., 1266
Bloom, A. L., 507

1309

AUTHOR INDEX

Bloxham, J., 577
Boatwright, J., 1132
Bock, G., 806
Bodin, P., 1246
Bodine, J. H., 260, 633, 660, 1005
Bogue, S. W., 546
Bois, C., 148
Bolondi, G., 1166
Bolt, B. A., 271, 288, 307, 358, 1107, 1201, 1246
Bomford, G., 447, 469, 482, 489
Bonatti, E., 104
Booker, J. R., 272, 510, 770, 1151
Boore, D. M., 430, 1197, 1246
Borevsky, L. V., 92
Born, G. H., 968, 969
Boss, A. P., 230
Bott, M. H. P., 126, 646, 871
Boucher, C., 222, 482
Boulanger, Yu. D., 9
Bourles, D., 218
Bowen, A. N., 871
Bowen, N. L., 698
Boyd, D., 33
Boyd, F. R., 97
Brace, W. F., 47, 201, 1231
Braginsky, S. I., 159, 160, 517, 584, 739
Bratt, S. R., 1049
Brecher, A., 897
Breiner, S., 507
Brenner, A. C., 969
Bretagnon, P., 938
Briden, J. C., 211
Brillinger, D. R., 314
Brocher, T., 667
Brock, A., 888
Brocker, T. M., 148
Brosche, P., 160
Brown, E. T., 48
Brown, G. S., 969, 1023
Brown, J. M., 854
Brown, L., 190, 1255
Brown, L. D., 89, 148, 1279
Brown, O. B., 968
Brown, W. F., 720
Brune, J. N., 1133, 1246
Brunner, F. K., 489
Brush, S. G., 1201
Brysk, H., 1166
Bube, K. P., 393
Buchan, K. L., 1303
Buck, W. R., 1279
Buddington, A. F., 173
Buhl, P., 667, 1279
Bukowinski, M. S. T., 787
Buland, R., 271
Bullard, E. C., 160, 385, 510, 577, 584, 909
Bullen, K. E., 307, 358, 359, 1107, 1192, 1201
Bunterbarth, G., 646
Burchfiel, B. C., 20, 126, 871
Burckle, L. H., 746
Burdick, L. J., 1197
Burford, R. O., 267, 268
Burke, B. F., 945

Burke, K., 104
Burnett, M., 668
Burnham, R. J. P., 37
Burr, N. C., 1054
Bursa, M., 938
Buselli, G., 139
Busse, F. H., 517, 739, 802
Butler, R. F., 673, 897
Byerlee, J. D., 48
Byerly, P., 1044, 1201
Byrne, D. E., 1279

Cadell, H. M., 1279
Cagniard, L., 372, 770
Cain, J. C., 33, 522, 988, 1221
Cain, J. E., 988
Cameron, R. L., 969
Campbell, J. A., 1296
Campbell, W. H., 534
Cande, S. C., 495, 746, 875
Cannizzo, L., 9
Cara, M., 1005, 1266
Carlheim-Gyllenskold, V., 510
Carlson, R. W., 802
Carmichael, C. M., 875
Carmichael, I. S. E., 173
Carsey, F. D., 969
Carter, W. E., 239, 478, 1296
Casadevall, T. J., 295
Castle, R. O., 474
Cathles, L. M., 1023
Cazenave, A., 160
Cazes, M., 148
Cerutti, G., 9
Cerveny, V., 367, 372, 1108, 1197
Chakrabarty, S. K., 1037
Chambers, R., 1023
Champion, D. E., 888
Chandrasekhar, S., 739
Chang, A. C., 1023, 1186
Channell, J., 890
Channell, J. E. T., 901
Chapman, C. H., 271, 1023, 1108, 1197
Chapman, D. S., 97
Chapman, S., 495, 498, 522, 534, 592, 770, 1221
Chapple, W. M., 260, 1279
Chase, C. G., 260, 407
Chave, A. D., 139, 770
Cheesman, S. J., 139
Chelton, D. B., 968
Chen, W.-P., 74, 201, 660, 1049
Chenet, P. Y., 126
Cheney, R. E., 968, 969
Cheng, C. H., 675
Chengfa, C., 74
Chernyshev, N. M., 148
Chesnokov, E. M., 787
Chevallier, R., 173
Chikazumi, S., 720
Chinnery, M. A., 299
Chou, T. A., 245
Choudhury, S. N. R., 1037
Chouet, A. B., 1186
Chouet, B., 294
Chovitz, B., 385

Chovitz, B. H., 410
Choy, G. L., 322, 1017, 1197, 1198
Christensen, N. I., 667, 1005, 1098
Christensen, U., 802
Christensen, U. R., 378
Christodoulidis, D. C., 166
Christoffersson, A., 1150
Christoskow, L. V., 288
Christy, J. W., 938
Cifuentes, I. S., 1089
Cisowski, S. M., 672, 673, 897
Cita, M. B., 746
Claerbout, J. C., 430
Claerbout, J. F., 393, 1166
Clark, D., 469
Clark, T. A., 1296, 1297
Clarke, A. R., 469
Clarke, J., 770
Clayton, R., 430
Clayton, R. W., 160, 431, 626, 819, 1151, 1186, 1296
Cleary, J. R., 775, 1023, 1186
Cleary, M. P., 48
Clegg, J. A., 909
Clemence, G. M., 482
Clement, B. M., 546
Clerc, G., 507
Cloetingh, S., 1050
Cloos, M., 1255
Cochran, J. R., 104, 1279
Coe, R. S., 546, 555, 876, 888
Cohen, J. K., 1166
Colin, L., 945
Collinson, D. W., 57, 211, 672, 706, 888, 961
Colombo, O. L., 457, 1221
Comer, R. P., 626, 819, 1292, 1296
Coney, P. J., 20, 407, 900
Connery, J. E. P., 522
Cook, A. H., 9
Cook, F. A., 1279
Cook, N. G. W., 48, 1231
Cooper, A. K., 407
Corey, B. E., 1296
Cormier, V. F., 322, 1017, 1198
Couch, R., 495
Counselman, C. C., 1297
Counselman, C. C., III, 1296
Courant, R., 1221
Courtillot, V., 160, 166, 577
Cowan, D. S., 1255
Cox, A., 57, 260, 307, 407, 546, 555, 909, 923, 993
Cox, A. V., 495, 560, 746, 888
Craig, H., 267
Crampin, S., 787, 1005
Crandall, C. L., 469
Cranswick, E., 1089
Crase, E., 1032
Creager, K., 1061
Creager, K. C., 358, 775, 787
Creer, K. M., 318, 739, 909, 961, 1303
Cross, T. A., 1061

Crosson, R. S., 294
Crough, S. T., 819
Cruz, J. Y., 457
Cudlip, W., 969
Cuong, P. G., 517
Cutler, R. T., 393

Dablain, M. A., 430
Dagley, P., 876
Dahlen, F. A., 440, 441, 1005, 1255, 1279
Dainty, A., 675
Dainty, A. M., 1186
Dalrymple, G. B., 746
Dalrymple, G. R., 546
Damotte, B., 148
Darot, M., 201
Darragh, R. B., 1246
Das, S., 1132
Davies, G. F., 378, 775, 802, 819
Davies, M. E., 938
Davis, D., 1255, 1279
Davis, D. M., 1279
Davis, G. A., 20
Davis, J., 1296
Davis, J. L., 1296
Davis, J. P., 440
Davis, P. J., 26
Davis, T. N., 498
Davis, W. M., 988
Davison, C., 1044, 1201
Dawson, J. B., 97
Day, R., 211
Day, S. M., 1089
de Charpal, O., 75
Decker, L. B., 222
Decker, R. W., 294
Dehlinger, P., 616
Delambre, J. B. J., 453
Demarest, H. H., 900
DeMets, C., 923, 1050
Denham, C. R., 888
Der, Z., 1266
Der, Z. A., 1017
Deregowski, S. M., 1166
deSitter, W., 385
Dessler, A. J., 945
Detrick, R. S., 667
De Vaucouleurs, G., 938
Dewey, J., 1044
Dewey, J. W., 271
Diament, M., 633
Dickey, J. O., 834
Dickey, L. O., 239
Dickinson, W. R., 1255
Dickson, D. C., 746
Dickson, G. O., 495, 993
Dieterich, J. H., 48
Dimri, V., 622
DiPippo, R., 598
Dix, C. H., 148
Dixon, A. E., 775
Dobrin, M. B., 393, 617
Dodd, R. T., 847
Dodson, R., 546
Doell, R. R., 546, 746, 909
Dohr, G., 393

Dohr, G. P., 148
Donahue, T. M., 945
Donovan, T. J., 57
Doornbos, D. J., 775, 1023, 1186
Dorman, J., 166, 675, 871
Dorman, L., 668
Douglas, A., 1037
Douglas, B. C., 968, 969
Douglas, B. J., 1005
Downey, W. S., 37
Dowty, E., 733
Drake, L. A., 1198
Drake, M., 626
Dreyfus Rawson, M., 746
Drummond, R., 1197
Dubois, J., 1061
Dubois, R. L., 37
Dubrulle, A. A., 1166
Ducarme, B., 322
Duchesneau, T. D., 478
Ducruix, J., 160, 577
Duda, S. J., 288
Duennebier, F., 675
Dunlop, D. J., 58, 211, 1269, 1303
Dunn, J. R., 546
Dunn, P. J., 166
Durgin, C. M., 469
Dziewonski, A., 294, 1117, 1266, 1296
Dziewonski, A. M., 215, 245, 358, 359, 440, 441, 457, 458, 626, 775, 787, 788, 819, 834, 1005, 1129, 1151, 1296
Dzurisin, D., 294, 295

Eanes, B. J., 834
Eanes, R. J., 239, 1296
Eaton, G. P., 126
Ebihara, M., 847
Edwards, R. N., 139
Egorkin, A. V., 148
Eissler, H. K., 294
Eittreim, S. L., 871
Ekstrom, G., 294
Elachi, C., 968
Eldredge, S., 900
Elgered, G., 1296
Eliason, E., 969
Elmore, R. D., 57
Eltayeb, I. A., 739
Emerman, S., 1279
Emerman, S. H., 431
Endo, E. T., 294, 295
Engdahl, E. R., 806, 1061
Engebretson, D. C., 407, 888
Engeln, J., 1050
Engeln, J. F., 260, 923, 1054
Engquist, B., 430
Enz, U., 720
Epstein, E. F., 478
Eremeev, V. F., 331
Ernst, W. G., 1255
Espinosa, A. F., 1186
Esteva, L., 260
Estey, L. H., 1005
Etheridge, M. A., 871, 1279

Eubanks, T. M., 239
Evans, J. R., 1150, 1151
Evans, M. E., 211
Evernden, J. E., 1089
Evernden, J. F., 1089
Ewing, J., 668
Ewing, M., 633, 675, 1211
Ewing, W. M., 372, 1266
Eysteinsson, H., 770

Faber, S., 806
Faller, J. E., 9
Fallon, F. W., 1296
Farrell, W. E., 1201
Faure, G., 5
Fearn, D. R., 739
Federal Geodetic Control Committee, 474, 478
Fedotov, S. A., 267
Fehler, M., 294
Feissel, M., 222, 482
Feng, Y. Y., 9
Ferrari, A. J., 938
Ferrazzini, V., 294
Filson, J., 1037
Fischer, I., 453
Fisher, R. A., 888
Fisher, R. M., 673
Fiske, R. S., 295
Fix, J. E., 1092
Flatte, S. M., 1186
Fleming, J. A., 534
Fleming, R. H., 453
Fletcher, E. J., 720
Flinn, E. A., 806
Florescu, V., 720
Fluh, E., 89
Flynn, J. J., 495, 560, 746
Foner, S., 507
Ford, P. G., 969
Forgey, R. L., 57
Forsberg, R., 457, 622
Forsyth, D. W., 260, 660, 667, 1005
Forsythe, D., 1231
Forte, A. M., 834
Foster, J. H., 746, 961
Fountain, D. M., 1098
Fournier, R. O., 598
Fowler, S. R., 871
Fox, J. M. W., 1303
Fox, P. J., 173
Francheteau, J., 668
Frankel, A., 1186
Franklin, K. L., 945
Frazer, L. N., 1246
Freedman, C., 160, 510, 577, 584
Freeman, J. R., 1246
Freeman, R., 57, 890, 901
Friedlander, G., 5
Frohlich, C., 1061
Froidevaux, C., 1005
Frost, H. J., 201
Fuchs, H., 458
Fuchs, K., 806, 1005
Fuhrman, M., 897
Fujiwara, Y., 961

AUTHOR INDEX

Fullagar, P. K., 139
Fuller, M., 507, 546, 555, 672, 720, 721, 1269
Fuller, M. D., 706, 730, 961

Gabrielse, H., 20
Galdeano, A., 148
Galson, D., 646
Gamble, T. D., 770
Gane, P. G., 1044
Gangi, A. F., 367
Ganse, R. A., 1246
Gao, J. L., 9
Gao, L. S., 1186
Gardner, G. H. F., 1211
Gaskell, T. F., 1201
Gass, I. G., 97
Gauss, C. F., 453
Gautesen, A. K., 1023
Gehrels, G. E., 20
Gehrels, T., 945
Geiger, L., 271
Geissman, J. W., 900
Geldart, L., 89
Geldart, L. P., 190, 1118, 1202
Gellman, H., 160, 510, 577, 584
Gemperle, R., 720
Geodynamics Program Office, National Aeronautics and Space Administration, 166
Geophysics Study Committee, National Research Council, 166
Gerald, C. F., 431
Gerver, M., 654
Giardini, D., 322, 358, 359, 440
Gibbs, A. D., 126
Gibson, B. S., 431, 1166
Giese, P., 89
Gilbert, F., 322, 359, 372, 440, 654, 834, 1037
Gilbert, G. K., 267
Gilbert, J. F., 1151
Gire, C., 160, 577
Given, J. W., 294, 1017
Glatzmaier, G. A., 517
Glen, W., 909, 923
Gnoinski, A., 507
Goins, N. R., 675
Goldstein, C. M., 1037
Golub, G. H., 431
Gomez-Trevino, E., 139
Goodstein, J. S., 1201
Gordey, S. P., 20
Gordon, R., 1050
Gordon, R. G., 923
Goree, W. S., 507, 961
Gorter, E. W., 733
Gose, W. A., 673, 897
Gossett, F. R., 469
Goto, K., 720
Goubau, W. M., 770
Gough, D. I., 507
Gourevitch, S., 1297
Gracie, G., 26, 469
Graham, J. W., 706, 909
Grant, F. S., 33

Greene, L., 1150
Greenspan, H. P., 160
Greville, T. N. E., 654
Griffiths, D. F., 431
Griffiths, D. H., 217
Gromme, C. S., 546, 555, 876
Groten, E., 447
Grove, T. L., 698
Grow, J. A., 871
Gubbins, D., 517, 555, 577
Gudmundsson, J. S., 598
Gueguen, Y., 201
Guennoc, P., 75
Gunst, R. H., 498
Guo, Y. G., 9
Gurnis, M., 787, 819
Gutenberg, B., 288, 307, 314, 1017, 1070, 1129
Guzkowska, M. A. J., 969
Gwinn, C. R., 626, 1296

Ha, J., 806
Haasbroek, N. D., 453
Haddon, R. A. W., 1023, 1186
Haenel, R., 646
Hage, H., 322
Hagee, V. L., 888
Hager, B., 1005
Hager, B. H., 160, 215, 260, 626, 627, 698, 819, 1151, 1296
Hahn, A., 522
Haigh, G., 57
Haines, E., 260
Haines, G. V., 1221
Hales, A. L., 457
Halgedahl, S., 1269
Halgedahl, S. L., 720, 721
Hall, W. J., 1246
Halley, E., 510
Halls, C. H., 875
Halls, H. C., 211, 495
Hamano, Y., 961
Hamilton, N., 706
Hammond, J. A., 9
Hammond, S. R., 746
Hanada, H., 9
Hanafy, M., 458
Hanks, T., 267
Hanks, T. C., 288, 1231
Hannah, J., 474
Hannon, W. J., 1089
Hanson, R. J., 654
Hanyga, A., 1108
Harding, T. P., 126
Hargraves, R. B., 173, 646, 698, 733
Harjes, H.-P., 288
Harland, W. B., 495
Harradon, H. D., 592
Harrington, R. S., 938
Harrison, C. G. A., 993
Harrison, J. C., 363
Hart, R. B., 260, 307, 746, 923
Hartmann, W. K., 230
Hasegawa, A., 181, 1061
Hasegawa, T., 1005
Haskell, N. A., 1246

Hatcher, R. D., Jr., 1279
Hatton, L., 1117, 1166
Hayford, J. F., 299
Hayne, G. S., 969
Hays, J. F., 698
Hazard, D. L., 534
Head, J. W., 626
Healy, J. H., 1232
Hearn, T. M., 1151
Heath, G. R., 57
Heaton, T. H., 1246
Heirtzler, J. R., 495, 746, 993
Heiskanen, W., 410, 622
Heiskanen, W. A., 331, 457, 616
Hejda, P., 961
Heller, F., 733, 890, 891, 1303
Heller, W., 622
Hellevege, K.-H., 730
Helmberger, D. V., 431, 775, 1198
Helmert, F. R., 453
Hendricks, S. J., 522, 1221
Henry, S. G., 211
Henshaw, P. C., 890
Henson, I. H., 440
Henyey, T. L., 888
Herbert-Gustar, L. K., 307
Hermance, J. F., 770
Herndon, J. M., 897
Herraiz, M., 1186
Herrero-Bervera, E., 546
Herrin, E., 1089, 1092
Herring, T. A., 626, 1296
Herrmann, R. B., 245
Herron, E. M., 495, 746, 993
Hess, H., 1005
Hide, R., 160, 510, 739
Higgins, G., 318
Higgins, G. H., 584
Hilbert, D., 1221
Hilde, T. W. C., 407
Hildreth, E. W., 698
Hileman, J., 314
Hillhouse, J. W., 407, 546
Hilterman, F. J., 1211
Hinteregger, H. F., 1296, 1297
Hipkin, R. G., 787
Hirahara, K., 1005, 1151
Hirata, T., 314
Hirn, A., 148, 1151
Hirvonen, R. A., 469
Hjortenberg, E., 1093
Hodder, B. M., 577
Hodgson, C. V., 469
Hoek, E., 48
Hoffman, K. A., 211, 546, 555
Hofmann, A. W., 775, 787, 819
Hohmann, G. W., 139
Holbrook, W. S., 148
Holcomb, L. G., 1093
Holcomb, R., 888
Holcomb, R. T., 295
Holdahl, S. R., 474
Holmes, A., 5
Holroyd, M. T., 33
Holt, B., 969
Honda, H., 245
Honda, S., 1255

1312

AUTHOR INDEX

Honnorez, B. M., 173
Hood, L. L., 673, 897
Hood, P., 507, 1166
Hood, P. J., 33
Hopfield, H. S., 222, 489
Hormander, L., 331, 457
Hornafius, J. S., 57
Hosken, J. W. J., 1166
Hoskins, J. K., 9
Hosmer, G. L., 469
Hothem, L. D., 222
Hotine, M., 469
Housner, G. W., 1202
Houtz, R. E., 668
Howard, K. L., 900
Howe, S. L., 1037
Howell, B. F., 1202
Howell, B. F., Jr., 1044
Hoye, G. S., 673
Hradilek, L., 489
Hrouda, F., 706
Huang, C. Q., 9
Huang, D. L., 9
Huang, P. Y., 260, 1054
Hubbard, T. P., 173
Hubbard, W. B., 938
Hubbert, H., 688
Hubbert, M. K., 1279
Hudson, D. E., 1202
Hudson, J. A., 372, 1186
Hudson, W. V., 522, 1221
Hufen, J.-H., 522
Hughes, T. J., 969
Huisman, R., 1266
Humphreys, E., 1151
Hunten, D. M., 945
Hurtig, E., 288
Hurukawa, N., 1005
Hurwitz, L., 592
Husebye, E. S., 1150

IAGA Division I, 498
IAGA Division I Working Group 1, 577
Iida, K., 314
Iijima, T., 988
Ikami, A., 1005
Ilan, A., 431
Iliff, R. L., 9
Imamura, A., 267
Inatani, H., 1005
International Association of Geodesy, 410
Irifune, T., 854
Irons, H. R., 507
Irving, E., 217, 495, 909, 1303
Isacks, B., 74, 314, 787, 1070, 1255
Isacks, B. L., 181, 1061
Ishibashi, K., 267
Ishida, M., 314*
Ishikawa, Y., 733, 1005
Ishimaru, A., 1186
Ishimoto, M., 314, 367
Israeli, M., 431
Ito, E., 854

Ito, H., 546
Iyer, H. M., 1093, 1150, 1151

Jachens, R. C., 617
Jackson, D. D., 1017
Jackson, I., 806
Jackson, I. N. S., 201
Jackson, J., 74
Jacobs, J., 950
Jacobs, J. A., 318, 322, 546, 555
Jaeger, J. C., 48, 1231
James, D. E., 148
James, I. N., 160
Jardetsky, W. S., 372, 1266
Jarosch, H., 372
Jarvis, G. T., 787, 802
Jault, D., 160
Jaupart, C., 646
Jeanloz, R., 359, 775, 787, 806
Jeffreys, H., 271, 359, 385, 1192
Jelenska, M., 961
Jelinek, V., 961
Jenkins, R. E., 222
Jensen, D. C., 1221
Jentsch, M., 1005
Johnsen, G., 166
Johnson, E. A., 909
Johnson, G. L., 668
Johnson, H. P., 57, 994
Johnson, M. W., 453
Johnston, D. H., 1018
Jones, D. L., 20, 407, 900
Joppen, M., 871
Jordan, S., 622
Jordan, T. H., 97, 260, 271, 358, 440, 775, 787, 977, 1005, 1061, 1266
Joyner, W. B., 1246
Julian, B. R., 245, 294
Jupp, D. L. B., 139
Jurkevics, A., 1117

Kadinsky-Cade, K., 74
Kagan, Y. Y., 314
Kahmen, H., 489
Kalish, E. N., 9
Kalish, Ye. N., 9
Kamas, D., 1037
Kanamori, H., 294, 314, 315, 367, 909, 923, 1018, 1070, 1129, 1232
Kanasewich, E. R., 1117
Kanninen, M. F., 48
Kappelmeyer, O., 646
Kappus, M., 9
Karal, F. C., 430, 1108
Karig, D. E., 1255
Karlin, R., 57, 890
Karner, G. D., 633, 660
Karnik, V., 288
Kasahara, K., 307, 1005, 1129, 1133
Kasser, M., 166
Kato, T., 854
Katsura, I., 891
Kaufman, A. A., 139
Kaufman, S., 1279

Kaula, W. M., 626
Kawai, N., 961
Kawasaki, I., 1005
Keilis-Borok, V., 288
Kelleher, J., 314
Kelleher, J. A., 267
Keller, G. V., 139
Keller, J. B., 1023, 1108
Kellogg, O. D., 331
Kelly, K. R., 430, 431
Kemp, D. V., 688
Kennedy, G. C., 318, 584
Kennedy, J. W., 5
Kennett, B., 668
Kennett, B. L. N., 307, 372, 1023, 1198
Kent, D. V., 173, 218, 495, 546, 560, 746, 890, 1303
Kent, J. T., 211
Kern, H., 1098
Kerridge, D. J., 160
Kerridge, J. F., 847
Kerry, N. J., 1198
Kershner, R. B., 222
Keys, D., 89
Khan, M. A., 385
Kidd, W., 75
Kind, R., 1198
King, D. W., 1186
King, J., 888
King, R., 482
King, R. F., 217, 961
Kinoshita, H., 961
Kinzler, R. J., 698
Kirby, S. H., 48, 201, 660, 1232
Kirschvink, J. L., 211, 891
Kirwan, A. D., 969
Kisslinger, C., 245, 267
Kittel, C., 720
Klein, F. W., 294
Klein, J., 1255
Kligfield, R., 706
Klosko, S. M., 166
Knapp, D. G., 592
Knight, C. A., 215, 1296, 1297
Knight, J. H., 139
Knopoff, L., 314, 367, 1186, 1266
Kobayashi, K., 57
Koblinsky, C. J., 969
Koch, D. A., 770
Koenigsberger, J. G., 57
Kogan, M. G., 633
Kohlstedt, D. L., 47, 201, 1231
Kojima, H., 720
Kolenkiewicz, R., 166
Kondorskaya, N. V., 288
Kono, M., 875, 961
Kooy, C., 720
Kopnichev, T. F., 1186
Koronkevitch, V. P., 9
Kosloff, D., 1032
Kouba, J., 222, 223
Kovach, R. L., 674, 675, 1266
Kovacheva, M., 888
Koyanagi, R. Y., 294, 295
Kozlovsky, Ye. A., 92
Krakiwsky, E., 410, 617

1313

AUTHOR INDEX

Krakiwsky, E. J., 26, 447
Krarup, T., 331
Krause, F., 517, 584
Krause, J., 267
Krautkramer, H., 1210
Krautkramer, J., 1210
Kreyszig, E., 26
Kroenke, L., 407
Krohn, C. E., 1037
Kronenberg, A. K., 660
Kropacek, V., 961
Kubotera, A., 294
Kukkamaki, T. J., 474
Kulikov, T. B., 92
Kummer, B., 431
Kuo, B. Y., 660

LaBrecque, J. L., 495, 746, 994
Labreque, J. L., 875
Ladd, J. W., 1279
Lai, C., 555
Lailly, P., 1166
Laj, C., 547
Lajoie, J. J., 139
Lajoie, K., 888
Lamb, H., 372
Lambeck, K., 239, 627
Lammlein, D., 675
Lamontagne, Y., 139
Lanczos, C., 654, 770
Landisman, M., 1117, 1266
Langan, R. T., 393
Langel, R. A., 160, 522, 988, 989, 1221
Langseth, M. G., 746, 994
Langston, C. A., 245
Lapwood, E. R., 359, 457
Larner, K., 1023
Larner, K. L., 1166
Larson, E. E., 57, 897
Larson, R. L., 407
LASE Study Group, 871
Latham, G., 675, 1186
Latham, G. V., 267
Laursen, V., 507
Lawson, C. A., 733
Lawson, C. H., 654
Lay, T., 294, 318, 322, 775
Le Borgne, E., 1303
Le Dain, A., 166
Le Dain, A. Y., 901
Lederle, T., 938
Lee, L. C., 1186
Lee, S. H., 888
Lee, T., 139
Lee, W. H. K., 271, 294, 314, 646, 654, 1037, 1150, 1202
Lees, A. C., 787
Le Grand, H. E., 909
Leland, O. M., 469
LeMouel, J.-L., 160, 577
Lenartowicz, E., 1108
Lenzen, V. F., 9
LePichon, X., 495, 746, 993
Lepine, J., 166
Lerner-Lam, A. L., 1266
Leroy, C. F., 222

Levander, A. R., 431
Leven, J. H., 806
Leveque, J.-J., 1005
Levi, S., 495, 890, 994
Levy, E., 555
Levy, E. H., 897
Levy, G. M., 139
Lewis, R., 139
Li, D. X., 9
Li, X.-D., 322, 358, 359, 440
Liddicoat, J. C., 888
Lieske, J. H., 938
Likhite, S. D., 961
Lilley, B. A., 720
Lindh, A. G., 260, 267
Lindsley, D. H., 173
Lipa, B. J., 968
Lipman, P. W., 295
Lister, C. R. B., 668, 994
Lister, G. S., 871, 1279
Liu, H.-P., 1018
Liu, L., 854
Llewellyn, P. G., 495
Lockwood, J. P., 295
Loewenthal, D., 431, 1032, 1166
Logie, H. J., 1044
Lomax, R., 211
Lomnitz, C., 1246
Longuet-Higgins, M. S., 1093
Longwell, C. R., 909
Loper, D. E., 318, 739, 775, 802
Loriot, G. B., 969
Love, A. E. H., 367, 372
Love, P. L., 393
Lowenstam, H. A., 891
Lowes, F. J., 534
Lowrie, W., 495, 546, 560, 706, 890, 891, 1303
Lu, L., 1166
Lufeng, M., 239
Lund, S. P., 888
Lundberg, N., 1255
Lundgren, P., 1050
Lundqvist, G., 1296
Lyle, M., 57
Lynn, W., 1023
Lyon-Caen, H., 74
Lysmer, J., 1198

Ma, C., 1296
MacArthur, J. L., 968
McAdoo, D. C., 968
McBirney, A. R., 698
McCabe, C., 57, 901
McCann, W. R., 267, 314, 1070
McCarthy, D., 482
McClain, J. S., 668
McClure, T. F., 222
McDonald, J. A., 1211
Macdonald, K. C., 668
McDonald, K. L., 498
McDougall, J., 37
McElfresh, T. W., 1017
McElhinny, M. W., 211, 495, 534, 546, 555, 577, 584, 592, 857, 875, 876, 888, 901, 945, 950
Macelwane, J. B., 1202

McFadden, P. L., 211, 495, 555, 560, 577, 875, 876, 888
McGeary, S. E., 407
McGill, G. E., 969
McGlechin, T. R., 166
McGoogan, J. T., 969
McGrath, P. H., 33
MacGregor, I. D., 97
Macias, E. S., 5
McIlhinny, M., 522
McIlwain, C. E., 498
Mackay, J. R., 478, 1296
McKenzie, D., 89, 104, 1279
McKenzie, D. P., 74, 378, 617, 819, 871
McLaughlin, K. L., 1018
McMaken, H., 1023
McMillen, K. J., 1255
Macnae, J. C., 139
McNally, K. C., 314
McNeill, J. D., 139
McNutt, M. K., 633, 660
McNutt, S. R., 294
McWilliams, M., 888
Madariaga, R., 431
Madariaga, R. I., 1133
Madden, T., 577
Madden, T. R., 160
Madrid, R., 20
Madsen, J., 667
Mahrer, K. D., 431
Major, A., 217
Makin, J., 1117
Malin, S. R. C., 160, 577
Maling, D. H., 844
Malkus, W. V. R., 739
Mallett, R., 307
Malone, S. D., 294, 295
Malone, S. M., 294
Malumyan, L. M., 57
Mangnani, M. H., 201
Mankinen, E. A., 546, 555, 876
Manton, W. I., 97
March, D. W., 1118
Mardia, K. V., 211
Marfurt, K. J., 431
Marianiuk, J., 507
Markert, H., 1269, 1303
Markushevich, V., 654
Marlow, M. S., 407
Marsh, B. D., 688
Marsh, J. G., 969
Marshall, M., 57
Marshall, P. D., 1089
Marson, I., 9
Martin, T. V., 969
Martinez, F., 1279
Marussi, A., 447
Mascle, A., 148
Mason, B., 673
Mason, I. M., 969
Masse, R., 1266
Masters, G., 322, 440
Masters, T. G., 517
Masuda, K., 48
Masuda, T., 1005
Masursky, H., 969

AUTHOR INDEX

Mathieu, S., 173
Matsushita, S., 534
Matte, P., 148
Matthews, D. H., 89, 994
Matthews, M. S., 847, 945
Mattis, D. G., 950
Matumoto, T., 267
Mauersberger, P., 522
Maurain, Ch., 57
Maxey, M. R., 688
Mayhew, M. A., 988
Mead, G. D., 1221
Mehringer, P. J., Jr., 888
Meiklejohn, W. H., 57
Meissl, P., 26
Meissner, R., 89, 148, 1232
Melbourne, W., 482
Melchior, P., 318, 322, 363
Melson, W. G., 673
Melton, B., 1037
Melville, C., 74
Melville, C. P., 1201
Menard, H. W., 668, 923, 994
Menke, W., 359, 393, 654, 770, 1018, 1211
Mercanton, P., 909
Merriam, M., 469
Merrill, R. T., 57, 495, 522, 534, 546, 555, 560, 577, 584, 592, 721, 888, 890, 945, 950, 1269
Metcalf, M., 721
Meyer, H. O. A., 97
Meyer, J., 522
Middleton, R., 1255
Mikhail, E. M., 26, 469
Mikhailova, N. G., 806
Miller, D., 1166
Miller, E. L., 20
Miller, J. M., 5
Miller, L., 969
Miller, L. S., 969
Miller, M. M., 20
Milne, J., 1044
Minakami, T., 295
Minster, B., 1202
Minster, J. B., 260, 977, 1005, 1018
Miracky, R. F., 770
Mitchell, A. C., 592
Mitchell, A. R., 431
Mitchell, B. J., 1005
Mitchell, H. C., 469
Mithen, D. P., 126
Miyashiro, A., 89
Mizutani, H., 48
Moffatt, H. K., 517
Mogi, K., 267, 314
Mohr, P. A., 104
Molnar, P., 74, 75, 166, 181, 201, 260, 267, 660, 787, 1049, 1061, 1070, 1279
Molodensky, M. S., 331
Molotkov, I. A., 1108
Molyneux, L., 961
Mondt, J. C., 775, 1023, 1186
Monger, J. W. H., 20, 21, 407, 900

Montadert, L., 75, 126
Moon, T., 721, 1269
Mooney, W. D., 148
Moore, J. C., 1255
Moore, J. G., 295
Moores, E. M., 668
Mora, P., 1032
Moran, J. M., 239
Morelli, A., 359, 1151, 1296
Morgan, J., 871
Morgan, P., 97, 646
Morgan, W. J., 75
Moritz, H., 239, 331, 410, 457, 616, 622
Moriya, T., 1005
Morley, L. W., 994
Morner, N., 166
Moroz, V. I., 945
Morris, D. E., 555
Morris, J., 1255
Morrison, L. V., 160
Morton, K. W., 431
Moskowitz, B., 994
Moskowitz, B. M., 721, 1303
Mott, P. A., 307
Mueller, I. I., 239, 447, 482
Muffler, L. J. P., 598
Muirhead, K., 806
Muirhead, K. J., 775
Muller, G., 806, 1198
Muller, R. A., 555
Mullins, C. E., 37
Multhauf, R. P., 9
Munguia, L., 1246
Munyan, A. C., 909
Muramatsu, K., 1005
Murphy, A. J., 1093
Murphy, J., 988
Murphy, T., 909
Mussett, A. E., 876
Mussetter, W., 469
Muth, L. A., 584
Mutter, J., 667
Mutter, J. C., 871

Nabelek, J. L., 260, 1054
Nabighian, M. N., 139
Nagata, T., 673, 733, 857, 875, 897, 909, 950, 961, 1269
Nakamura, Y., 675, 1186
Nakanishi, I., 1005, 1151, 1266
Nakata, J. S., 294
Nataf, H., 1151
Nataf, H. C., 1266
National Academy of Sciences, 92
National Oceanic and Atmospheric Administration, 478
Navrotsky, A., 854
Needham, J., 453, 592, 1202
Neel, L., 57, 58, 733, 1269, 1303
Neel, L. E. F., 909
Nelson, H. R., Jr., 393
Nelson, J. B., 1246
Nelson, J. H., 592
Nercessian, A., 1151
Nersesov, I. L., 267
Ness, G., 495

Ness, N. F., 522
Nettleton, L. L., 617
Neumann, F., 1221
Newhall, C. G., 295
Newman, G. A., 139
Newmark, N. M., 1246
Newton, R. G., 654
Newton, R. R., 222
Ni, J., 74, 75, 260
Nicholls, J., 173
Nicolas, A., 201
Niebauer, T. M., 9
Nigi, J., 622
Nikolaev, A. V., 1186
Nishenko, S. P., 267, 314, 1070
Nishimura, C. E., 1005
Nishitani, T., 961
Nixon, J., 160, 510, 577, 584
Nixon, P. H., 97
Noble, M., 1032
Nolet, G., 148, 359, 787, 1266
Nord, G. L., 733
Nordquist, J., 314
NORSAR, 1089
Nortmann, R., 288
Noson, L. L., 294
Nowick, A. S., 201, 1018
Noyes, R. M., 698
Nur, A., 126, 267, 407
Nuttli, O. W., 1050, 1089

O'Connell, R., 1005
O'Connell, R. J., 215
O'Donnell, A., 1017
Office of Management and Budget, 478
Ogata, Y., 314
O'Hare, S., 923
Ohtake, M., 267
Okada, H., 97, 1005
Okada, Y., 299
Okal, E. A., 260, 1054, 1292
Okubo, P. G., 48
Oldenberg, D. W., 139
Oliver, J., 74, 267, 314, 901, 1061, 1070, 1192, 1255
Oliver, J. A., 1211
Oliver, J. E., 1279
Olsen, J., 507
Olson, P., 555, 802, 888
Omori, F., 314
Opdyke, N. D., 173, 546, 746
Orcutt, J. A., 359, 667, 668, 1005, 1023, 1197
O'Reilly, W., 173, 706, 730, 950, 1269
Oristaglio, M., 1166
Orowan, E., 201
Orszag, S. A., 431
Ouliac, M., 1303
Ousley, G., 988
Owens, W. H., 706
Oxburgh, E. R., 802
Ozdemir, O., 58, 1303

Page, R., 315
Paik, H. J., 457

AUTHOR INDEX

Pajchel, J., 1108
Palmer, H. C., 173
Panel on Crustal Movement Measurements, National Research Council, 166
Panza, G. F., 1266
Park, J., 440
Park, J. K., 211
Parker, E. N., 517, 555
Parker, R. L., 440, 633, 654, 1221
Parkinson, W. D., 507, 534, 577, 592
Parmentier, E. M., 660
Parry, J. H., 961
Parsons, B., 626, 646, 668
Parsons, B. E., 617
Pascal, G., 1061
Paterson, M. S., 48, 201
Paterson, N. R., 33, 617
Patton, H. J., 1266
Paulding, B. W., 47
Pavlenkova, N. A., 148
Pavlin, G. B., 245
Pavlis, G. L., 148, 272, 654, 1151
Payne, M. A., 217
Peacor, D. R., 57, 901
Pearce, G. W., 673
Peddie, N. W., 522, 534
Pedersen, J., 770
Pekeris, C. L., 372
Peltier, W. R., 787, 802, 834
Peltzer, G., 75, 901
Pesonen, L. J., 495, 875
Petersen, N., 173, 721, 733, 891
Peterson, J., 1093
Pettey, J. E., 239
Pettingill, G. H., 969
Pfiffner, O. A., 706
Phillips, J. D., 495
Phillips, R. G., 627
Phillips, R. J., 230
Phinney, R. A., 1023
Pica, A., 1032
Pickton, C. A. G., 495
Piersol, A. G., 622, 770
Pilger, R. H., Jr., 1061
Pines, P. J., 1186
Pitman, W. C., III, 495, 746, 993
Pitt, A. M., 598
Plesinger, A., 1037
Podrouzkova, Z., 961
Poirier, J-P., 201
Pollack, H. N., 97, 646
Pollard, D. D., 48
Popelar, C. H., 48
Popelar, J., 223
Popov, M. M., 1197
Potemra, T. A., 988, 989
Press, F., 372, 675, 1211, 1266
Prestcott, C., 871
Prevot, M., 546, 555, 876, 1303
Price, R. A., 21, 407
Primdahl, F., 507
Prince, J., 1246
Proctor, M. R. E., 739
Prohdehl, C., 89, 1005
Prothero, W., 668, 1037

Psencik, I., 367, 1108, 1197
Pullaiah, G., 495, 1303
Purcaru, G., 267

Qiu, X. M., 9
Quaas, R., 1246
Quible, D., 1050
Quittmeyer, R., 75
Quittmeyer, R. C., 315, 1246

Raasch, G. O., 909
Radler, F.-H., 517
Raefsky, A., 260
Rahdhakrishnamurty, C., 961
Raiche, A. P., 139
Rainsford, H. F., 469
Raisbeck, G. M., 218
Raitt, R. W., 668
Raj, R., 201
Raleigh, B., 267
Raleigh, C. B., 92, 1231
Ramdohr, P., 897
Ramey, J. W., 295
Ranganayaki, R. P., 897
Raoult, J. F., 148
Rapley, C. G., 969
Rapp, R. H., 457, 622, 938
Ratner, M. I., 1296
Ravindra, R., 372
Rayleigh, Lord, 372
Raymond, C. A., 994
Rees, A. I., 217, 706, 961
Reeves, C. V., 33, 617
Reford, M. S., 33
Regan, J., 1005
Regan, R. D., 33, 988
Reid, H. F., 267, 299, 307, 315
Reid, M. J., 239
Reilly, W. I., 844
Reitzel, J. S., 507
Resch, G. M., 489
Reshef, M., 1032
Resnick, J. R., 393
Reynolds, W. F., 469
Rhodes, P., 721
Ribe, N. M., 633
Rice, J. R., 48, 1133
Rice, R. B., 1201
Richards, M. A., 626, 627, 819, 1296
Richards, P., 1128, 1151, 1186, 1296
Richards, P. G., 245, 267, 288, 307, 367, 372, 430, 1018, 1023, 1107, 1132, 1197, 1198, 1210, 1246, 1266
Richardson, R. M., 1050, 1232
Richardus, P., 469, 844
Richmond, A. D., 510
Richter, C., 1202
Richter, C. F., 267, 288, 307, 314, 315, 1070, 1071, 1129, 1192
Richter, F., 819
Richter, F. M., 378, 775
Richtmyer, R. D., 431
Rigden, S. M., 854
Rikitake, T., 267, 307, 517

Ringdal, F., 1089
Ringwood, A., 787
Ringwood, A. E., 230, 854, 1005
Ringwood, E. A., 806
Ritzwoller, M., 322, 440
Riuscetti, M., 295
Riznichenko, Yu. V., 288
Roberson, R., 1166
Roberts, A. A., 57
Roberts, D., 75
Roberts, N., 495, 555, 876
Roberts, P. H., 584, 739
Robertson, D. S., 239, 478, 1296
Robineau, B., 166
Robinson, A. E., 1118
Robinson, E. A., 393
Robinson, S., 888
Rocca, F., 1166
Rockwell, T., 1246
Rodgers, P. W., 1037
Rodriguez, M., 260
Rogers, A. E. E., 1296, 1297
Rolph, T. C., 876
Ronnang, B. O., 1296
Rosen, R. D., 239
Rosenberg, M., 720
Rosenblueth, E., 1246
Rosendahl, B. R., 126, 1054
Ross, C. A., 21
Ross, J. R. P., 21
Rothe, J.-P., 1202
Roufosse, M., 617
Rowe, M. W., 897
Rowlands, G., 721
Rowlett, H., 314
Rowley, J. C., 92
Roy, J. L., 211
Rubey, W. W., 1279
Rubincam, D. P., 834
Ruegg, J., 166
Ruff, L., 315
Ruggi, S., 1089
Rumble III, D., 730
Runcorn, S. K., 318, 510, 672, 909, 961
Rundle, J. B., 300
Russell, C. T., 945
Rutherford, E., 5
Rutter, H., 139
Ryan, J. W., 1296
Ryan, M. P., 295
Ryan, W. B. F., 746
Rybach, L., 646

Sacks, I. S., 97, 806, 1061, 1255
Sadovsky, M. A., 267
Saffer, B., 57
Sahasrabudhe, P. W., 961
Saito, M., 1198
Saito, T., 746
Sakuma, A., 9
Saleeby, J. B., 20, 21
Salisbury, M. H., 667, 1098
Salstein, D. A., 239
Sands, R. W., 9
Sanso, F., 331, 457
Sarda, P., 819

Sarkar, D., 288
Sasagawa, G., 9
Sasaki, Y., 1005
Sass, J. H., 646
Sassen, R., 57
Sato, H., 1186
Savage, J., 267
Savage, J. C., 166
Savarensky, E. F., 288
Savino, J., 314, 1093
Sawamoto, H., 854
Sawaoka, A., 961
Schick, R., 295
Schmidt, P. W., 211
Schmidt, V. A., 546
Schmidt, W., 431
Schmitz, D. R., 522, 1221
Schneider, W. A., 393
Schnetzler, C. C., 989
Schoeps, D., 215
Scholl, D. W., 407
Scholte, J. G. J., 1023
Scholtz, C., 267, 1132
Scholz, C., 47
Scholz, C. H., 201, 315, 1061
Scholz, H., 48
Schubert, G., 378, 517, 646, 802, 1005, 1054, 1061
Schuh, F., 92
Schuh, H., 1296
Schuh, W.-D., 458
Schulkes, J. A., 733
Schultz, B. E., 834
Schupler, B. R., 1296
Schutz, B. E., 239, 1296
Schwarz, K. P., 622
Schwee, L. J., 507
Schwerer, F. C., 673
Sclater, J. G., 626, 646, 668
Scotese, C. R., 901
Scotese, R., 57
Scott, S., 584
Scrase, F. J., 315
Seeber, L., 75, 1279
Seely, D. R., 1255, 1279
Segall, P., 48
Seidelman, P. K., 938
Seidl, D., 288, 295
Senanayake, W., 577
Senanayake, W. E., 876
Sengor, A. M., 89
Sengor, A. M. C., 75, 104
Sepplin, T. O., 223
Serson, P. H., 507
Settle, M., 988
Shaffer, D. B., 1296
Shankland, T. J., 854
Shapiro, I. I., 626, 1296, 1297
Shaw, H. R., 688
Shaw, J., 495, 546, 876
Shcheglov, S. N., 9
Shearer, P., 668
Shearer, P. M., 359, 1005
Shebalin, N. V., 288
Sheffels, B., 633
Sheriff, R., 89
Sheriff, R. E., 190, 1118

Sheriff, R. P., 1202
Sherwood, J. W. C., 1166
Shimamura, H., 1005
Shimizaki, K., 314
Shimizu, Y., 1303
Shipley, T. H., 1255
Shreve, R. L., 1255
Shtolts, E. V., 720
Shudofsky, G. N., 260
Shuey, R. T., 393
Shure, L., 1221
Sibson, R. H., 1050, 1232
Sibuet, J.-C., 75
Siebert, M., 522
Sieh, K. E., 315
Sigurdsson, S., 166
Silberling, N. J., 407
Silgado, F. E., 267
Silver, P. G., 440, 802
Simon, R. B., 1192
Simpson, D. W., 267
Simpson, R. W., 617
Sinclair, A. T., 938
Singer, H., 802
Singh, S., 1246
Singh, S. J., 1128
Singh, S. K., 260
Sipkin, S. A., 97, 245
Sjogren, W. L., 938
Skiles, D. D., 510, 888
Sleep, N. H., 260, 668, 1050, 1054, 1232
Smith, A. G., 495
Smith, B. A., 938
Smith, C., 89
Smith, D., 482
Smith, D. E., 166
Smith, G. L., 871
Smith, G. M., 994
Smith, M. L., 1005
Smith, R. B., 260
Snoke, J. A., 97, 806
Snyder, D. B., 75
Snyder, J. P., 844
Soffel, H. C., 720, 721
Solomon, S. C., 260, 1049, 1050, 1054, 1232
Solov'yov, S. L., 288
Sonett, C. P., 673, 897
Souriau, A., 1266
Souriau, M., 1266
Soward, A. M., 739
Spakman, W., 787
Speith, M. A., 160
Spence, G. D., 871
Spera, F. J., 688
Spilker, J. J., Jr., 601
Spindler, D. A., 393
Spohn, T., 378, 517
Sprague, D., 646
Spudich, P., 1246
Spudich, P. K. P., 668
Srnka, L. J., 673
Stacey, F., 950
Stacey, F. D., 318, 646, 706, 733, 775, 1269, 1303
Stanley, H. R., 969

Stansell, T. A., 223
Stark, P. B., 1061
Stauber, D. A., 1150
Staudacher, T., 819
Stauder, S. J. W., 245
Stechschulte, V. C., 315
Steckler, M. S., 260, 633, 660, 1005, 1279
Stegun, I. A., 430
Stein, A., 89
Stein, C., 1050
Stein, S., 260, 660, 923, 1050, 1054
Steinhart, J. S., 148
Steketee, J. A., 300
Stephen, J. H., 1044
Stephens, R. A., 431
Stephenson, A., 672
Steppe, J. A., 239
Sterba, O., 961
Stern, D. P., 1221
Sternberg, R. S., 888
Stevens, C. F., 20
Stevens, C. H., 21
Stevens, J. L., 1089
Stevenson, D., 322
Stevenson, D. J., 378, 495, 517, 938, 945
Stewart, L. M., 260, 1054
Stewart, R. R., 1018
Stewart, S. W., 271, 294, 654, 1037, 1202
Stewartson, K., 739
Stiller, H., 288
Stoffa, P., 871
Stolper, E. M., 698, 854
Stolt, R. H., 1166
Stone, E. C., 498
Stoneley, R., 1202
Strangway, D. W., 673, 897
Stuart, W. D., 267
Stuart, W. F., 507
Stubbs, P. H. S., 909
Stupavsky, M., 211
Stus, Yu. F., 9
Suetsugu, D., 1005
Sugiura, M., 498
Sugiura, N., 897
Sundermann, J., 160
Sunkel, H., 457, 458
Suppe, J., 1255, 1279
Sutcliffe, W. D., 469
Sutton, G., 675
Suyehiro, K., 1005
Suzuki, Z., 181
Sverdrup, H. A., 453
Sverdrup, K. A., 271
Swanson, D. A., 295
Swithinbank, C. W. M., 969
Sykes, L. R., 267, 314, 315, 1061, 1070, 1089, 1246, 1255, 1279
Symonds, P. A., 871, 1279
Symons, D. T. A., 211
Synnott, S. P., 938
Syono, Y., 721, 733
Szymanski, A., 507
Szymczak, R., 721

AUTHOR INDEX

Tachinaka, H., 577
Takagi, A., 181, 1061
Takeuchi, H., 909, 923, 1198
Talandier, J., 1292
Talwani, M., 633, 746, 871, 994
Tanigawa, W. R., 294
Tanimoto, T., 1005, 1266
Tapley, B., 482
Tapley, B. D., 239, 834, 1296
Tappert, F. D., 1166
Tapponnier, P., 75, 166, 901
Tarantola, A., 166, 1032, 1151
Tarasyuk, V. G., 9
Tarling, D. H., 37, 888
Tate, D. R., 9
Tauxe, L., 746
Taylor, G. J., 230
Taylor, L. A., 673
Taylor, P. T., 989
Taylor, S. R., 673, 1266
Teifel, V. G., 938
Telford, W., 89
Tempelman-Kluit, D. J., 20, 21
Tera, F., 1255
Thatcher, W., 166, 300
Thellier, E., 58, 876
Thellier, O., 58
Theyer, F., 546, 746
Thomas, R. H., 969
Thompson, A. B., 806
Thompson, G. A., 407
Thompson, R., 37, 510, 577, 888
Thomson, J. J., 5
Thorbergsson, G., 166
Thurber, C. H., 654, 1151
Tiley, G., 1279
Tilling, R. I., 295
Timur, A., 1018
Tite, M. S., 37
Tivey, M. A., 994
Tjuflin, Y. S., 938
Tocher, D., 315
Todd, A., 746
Todhunter, I., 453
Toksoz, M. N., 215, 1018, 1186
Toksoz, N., 675
Torge, W., 410
Torreilles, G., 148
Torrence, M. H., 166
Torreson, O. W., 909
Tosha, T., 961
Tough, J. G., 584
Toy, K. M., 359
Trefethen, L. N., 431
Treitel, S., 393, 431, 1118
Trigg, D. H., 507
Trorey, A. W., 1023
Tryggvason, E., 166
Tscherning, C. C., 457, 458, 622
Tsuboi, C., 616
Tsubokawa, T., 9
Tucholka, P., 547, 739
Tullis, J., 201
Tullis, T. E., 48, 201
Turcotte, D., 1279
Turcotte, D. L., 378, 646, 802, 819, 1054, 1061

Turner, G. M., 888
Turner, H. H., 315
Turner, R., 267

Uhrhammer, R. A., 1246
Ulrich, W., 378
Umino, N., 181, 1061
Ungar, A., 431
U.S. Congress, 1089
Usami, T., 359
Utsu, T., 181, 315
Uyeda, S., 260, 407, 733, 871, 909, 923, 1231, 1255

Vail, P. R., 1255, 1279
Valdiya, K. S., 75
Valet, J-P., 547, 555
Vali, H., 891
Van Bemmelen, W., 511
van Couvering, J. A., 495, 560, 746
Vandenberg, N. R., 1296
Van der Voo, R., 57, 378, 900, 901
Vanek, J., 288
Vanicek, P., 26, 410, 447, 474, 617
Van Loan, C. F., 431
van Trier, J., 1266
Vartanyan, G. S., 92
Vassiliou, A. A., 622
Vassiliou, M. S., 260
Vening Meinesz, F. A., 616
Vera, E., 667
Verall, R. A., 201
Verhoogen, J., 318
Verosub, K. L., 217, 218, 888
Vestine, E. H., 511, 577
Vicente, R., 482
Vidale, J. E., 431
Vincent, E. A., 173
Vine, F. J., 994
Vinnik, L. P., 806
Virieux, J., 431
Vlaar, N. J., 787
Vogt, P. R., 668
Von Dobeneck, T., 891
von Huene, R., 1255, 1279
Voorhies, C. V., 160
Vozoff, K., 139, 770

Wadati, K., 181, 315
Wai, W., 1023
Wait, J. R., 139, 770
Walcott, R. I., 166
Walker, B. M., 673
Walker, D., 698
Walker, T. R., 57
Walsh, E. J., 968
Walters, R., 495
Walton, D., 1303
Walton, G. G., 1255, 1279
Walton, M., 92
Ward, J. C., 730

Ward, R. W., 431
Ward, S. N., 1292
Ware, J. A., 394
Warren, S. E., 37
Wasilewski, P. J., 897
Wasson, J. T., 847, 897
Watanabe, H., 584
Waters, K. H., 394
Waterstraat, P., 888
Watkins, J. S., 675, 1255
Watkins, N. D., 173, 218
Watson, D. E., 897
Watson, E. B., 698
Watts, A. B., 260, 617, 633, 660, 1005
Weaver, C. S., 294, 295
Webb, E. K., 489
Webber, J. C., 1296
Webster, W. J., Jr., 989
Weidner, D. J., 787, 854
Weilandt, E., 1044
Weiner, A. T., 201
Weinstein, S., 1050
Wells, D., 465
Wenk, H.-R., 359
Wennerberg, L., 1186
Wenzel, H.-G., 458
Werner, J., 923
Wernicke, B., 126, 871, 1279
Wesnousky, S. G., 315
West, G. F., 139
Westbrook, G. K., 871, 1279
Wetherill, G. W., 230
Wever, Th., 89
White, A. A., 160
White, D. E., 598
White, R. E., 1118
White, R. S., 871
White, W. M., 775, 787
Whitehead, J. A., 802
Whitham, 511
Whitney, A. R., 1296, 1297
Wiegel, R. L., 308, 1246
Wiens, D. A., 260, 660, 1050, 1054
Wieser, M., 458
Wiggins, R., 1117
Wilks, S. S., 26
Williams, D. C., 489
Williams, I., 546
Williams, I. S., 555
Williams, J. G., 834
Williams, T. M., 969
Willmore, P. L., 1037
Wilson, C. A., 160
Wilson, R. L., 173, 211, 888
Windisch, C. C., 746, 994
Woirgard, J., 201
Wong, T.-F., 48
Wood, H. O., 1044
Wood, J. A., 847
Wood, R. M., 909
Woodhouse, J. H., 245, 322, 358, 359, 440, 441, 458, 788, 1005, 1129, 1151, 1266
Woods, D., 1050
Woolard, E. W., 482

AUTHOR INDEX

Working Group on California Earthquake Probabilities, 268
Worm, H.-U., 1269
Wortel, M. J. R., 787
Wortel, R., 788, 1050
Worthington, M. H., 1117
Wright, A. E., 217
Wright, C., 775
Wright, J. E., 20
Wright, T. W., 469
Wu, F., 267
Wu, R. S., 1186, 1187
Wuenschel, P. C., 394
Wunsch, C., 968
Wu Yu, M., 546
Wyld, H. W., 393
Wyss, M., 267, 268, 288

Yagi, T., 854
Yamada, H., 854
Yamada, I., 48
Yamada, T., 1005
Yamazaki, T., 891
Yamerone, C. A., 968
Yang, M., 1023
Yiou, F., 218
Yoder, C. F., 834
Yoder, H. S., 698
Yoshida, M., 961
Yoshii, T., 1061
Young, G. M., 474
Yu, G.-K., 1005
Yuen, D. A., 688, 802, 1005, 1061
Yukutake, T., 522, 577, 584
Yurkin, M. I., 331

Zagniy, G., 888
Zatopek, A., 288
Zeleny, M., 720
Zelinka, T., 961
Zener, C., 1018
Zhang, G. Y., 9
Zhao, W.-L., 75
Zhou, J. H., 9
Zijderveld, J. D. A., 211
Zilkoski, D. B., 474
Zmuda, A. J., 534, 989
Zoback, M. D., 1232
Zoback, M. L., 1232
Zuganov, S. K., 148
Zumberge, M. A., 9
Zwalley, H. J., 969

SUBJECT INDEX

Boldface is used to indicate the main entries and the pages on which they begin. *Italic* page numbers designate the more important passages.

ABSOLUTE AGE DETERMINATION, 1
ABSOLUTE GRAVITY MEASUREMENTS, 5
free fall, 5
pendulum, 5
Absolute gravity meters, 5
absolute accuracy of, 9
Absolute plate motions, 921
Accreted terranes, 10. See also North America Cordillera
Accretionary prism, 10, 1247
ACCRETIONARY TECTONICS, 9
Achondrites, 845, 896
ACORP, 181
Active margins, 860
Acton, G., 923
Adiabatic fusion, 691
ADJUSTMENT METHODS, 21
AEROMAGNETIC SURVEYING, 26
Aeromagnetic survey operations, 28
calibration, 29
magnetic compensation, 28
survey specifications, 29
Aeromagnetic survey systems, 27
Afar rift, 162
Aftershocks, earthquake, 301
Agnew, Duncan Carr, 1037, 1201
Airborne magnetometers, 27
fluxgate, 27
optical absorption, 27
proton precession, 27
Airborne Positioning Systems, 28
Airy caustic, 1021
Airy phase, 1257
Aki, K., 367
Albers equal-area conic projection, 843
Alkali-vapor magnetometer, 981
Alldredge, L. R., 498
Allende meteorite, 893
Allochthonous terranes, 12
Amalgamation, 10
Anderle, R. J., 601
Anderson, John G., 1246
Anelastic deformation, 191
Angus-Leppan, P. V., 489
Apparent polar wander paths, 897, 910
Archaean paradox, 377
Archaeomagnetic dating, 34

ARCHAEOMAGNETISM, 33
Arnold, K., 215
Aseismic continuation of subducted slabs, 784
Aseismic deformation, 297
Asperities, 1124
Astatic magnetometer, 951
Asteroids, 1216
Asthenosphere, 1062
Astrometry, 1296
Astronomical reference frames, 480
Asymmetric spreading, 916
Atmospheric angular momentum, 237
Attenuation and anisotropy, seismic, 336
attenuation operator, 1011
Autocorrelation, 1111
Auxiliary fault plane, 241
Azimuthal anisotropy, 995
inversions for upper mantle structure, 1002
mantle flow, 1001
Azimuths and bearings, 465

Backarc zone, 1067
Ballistic magnetometer, 956
Banerjee, Subir K., 993
Barazangi, Muawia, 74
Barraclough, D. R., 592
Barton, C. E., 577
Basin and Range Province, 101
Be^{10}, 1251
Becquerel, Henri, 1
Ben-Avraham, Z., 407
Benioff strain seismometer, 1043
Benioff variable reluctance seismometer, 1041
Benioff zone. See Wadati–Benioff zone
Bills, B. G., 938
Biomagnetism, 950
Biot's theory, 41
BIRPS, 181
Blackett, P. M. S., 901
Blocking temperature, 50, 166, 202, 1268
Blueschists, 1252
Bolt, B. A., 307
Border fault systems, 104
Boss, A. P., 230, 1217
Bossler, J. D., 478

Bouguer gravity anomaly, 606
Braginsky, S. I., 739
Brittle failure, 38
effects of pore fluid, 41
BRITTLE PHENOMENA, 38
Broadcast ephemerides, 218
Brunhes–Matuyama reversal, 539, 550
Buland, R., 271
Bullard, E. C., 901
Busse, F. H., 517
Byerlee's law, 46
Byrne, Daniel, 1278

Cache Creek terrane, 15
Cain, J. C., 522
CALCRUST, 181
Carbonaceous chondrites, 893
Carlson, R., 5
Carnegie Institution of Washington, 528, 591
Carter, W. E., 239
Cauchy, 364
Celestial reference coordinate system, 1296
Červený, V., 1107
Chandler wobble, 233, 1294
Channell, J. E. T., 890
Characteristic remanent magnetization, 203
CHEMICAL REMANENT MAGNETIZATION (CRM), 49, 856
Chondrites, 844
Chondritic Earth model, 372
Chovitz, B. H., 410
Christensen, N. I., 1098
Christensen, U., 378
Cisowski, S. M., 672, 897
Claerbout, J., 1166
Clairaut, A. C., 324
Cloos, M., 1255
CMP gathers, 385
COCORP, 81, 144, 181
Coda waves, 1177
diffusion theory, 1179
Q, 1179
single scattering theory, 1179
Common mid-point (CMP) method, 144, 385
Compensated linear vector dipole, 996

Compensation depth, 627
Comprehensive Test Ban Treaty, *1071*
Conductive heat flow, 634
Conglomerate test, 203
Conrad discontinuity, 75
Consistency test, 203
Continental collision, *10*, 1067
CONTINENTAL COLLISION ZONES, 58
  evolutionary model, *73*
CONTINENTAL CRUSTAL STRUCTURE, 75
  continental cratons and platforms, 79
  Phanerozoic belts, 83
  seismic measurements, *77*
  upper crustal structure, 78
Continental drift, *904*
CONTINENTAL DRILLING, 89
CONTINENTAL LITHOSPHERE, 92, *183*
  chemical boundary layer, 94
  elastic thickness, 92
  mantle composition, 94
  mechanical properties, *657*
  rheology, *199*
Continental plate boundaries, *1069*
CONTINENTAL RIFTING, 97, 1142
  flexure, 111
  infrastructure, 109
  rates, 103
CONTINENTAL RIFTS, 104
Continuous inverse theory, 652
CONTROLLED SOURCE ELECTROMAGNETIC MAPPING OF THE CRUST, 126
CONTROLLED SOURCE SEISMOLOGY, 139
Convective heat flow, 634
Convective instability, *788*
Converted seismic waves, 95
Core-mantle boundary, 834
  scattering near, 1183
CORE-MANTLE COUPLING, 148
Core-mantle coupling mechanisms, 150
  magnetic coupling, *152*
  topographic coupling, *156*
  viscous coupling, *151*
Cormier, Vernon F., *1017*
Corner frequency, *1132*
Coseismic deformation, *295*
Coulomb criterion, 40
Cox, A., *907*, *990*
Critical Rayleigh number, 789
Critical reflections, *140*
Crosson, R. S., *294*
Crustal evolution, 84
Crustal isostasy, 624
CRUSTAL MOVEMENTS AND TECTONIC DEFORMATION, 160
Crustal seismic reflectivity, 87

Cryogenic magnetometers, *505*
Crystal fractionation, *694*
CURIE TEMPERATURE, 166, 947
  primary and secondary, *166*
Current channeling, 757
Cylindrical equal-area projection, 841

Dalrymple, G. B., 990
Daly, Reginald Aldworth, 676
Davies, G. F., 819
Davis, P., 440
Deconvolution, 387, *1108*
DEEP EARTHQUAKES, 174
  causes, *180*
  earthquake mechanism, *177*
  physical process, *179*
  source process, *177*
Deep Sea Drilling Program (DSDP), *989*
DEEP SEISMIC-REFLECTION PROFILING, 181
Deep seismic sounding method (DSS), *140*
Deep seismic zone, *174*, *1055*
  geometry of, *1059*
  stress distributions, 1056
DEFORMATION OF ROCKS AND MINERALS, 191
DEKORP, 181
De Magnete, 511
DEMAGNETIZATION, *201*
  alternating field, *204*
  chemical, *205*
  thermal, *204*
Demultiplexing, 387
DENSITY DISTRIBUTION IN THE EARTH, 211
Detachment, 1270
DETRITAL REMANENT MAGNETISM (DRM), 215, 856
De Voogd, B., *190*
Diamagnetism, 946
Diamond anvil cell, 848
Diapir chain, 801
Dietz, R. S., *912*, *989*
Diffraction
  around an objective, 1019
  at a caustic, 1021
  at a plane boundary, 1020
  by topographic relief, 1021
Diffuse plate boundaries, 920
Diffusional creep, 196
Digital recorder, 1036
Discrete inverse theory, 649
Dislocation creep, 195
Dislocation glide, 195
Dislocation model of faulting, *1130*
Dislocation surface, 295
Displaced terrains, 899
Displacement, *363*
Doell, R. R., *907*, *990*
Domain structure, magnetic, 716
Domain wall, magnetic, 709

Doornbos, Durk J., 1023
Doppler navigation systems, 28
Doppler pole positions, 222
DOPPLER POSITIONING, 218
DOSECC, 91
Double couple, 240, *1124*
Double-couple source model, *1130*
Double-planed deep seismic zone, *1060*
Double seismic zone, *176*
Doubly diffusive convection, *696*
D″ region, *770*
Duberdorff, Bruce, 1210
Ductile behavior, 38
Duda, S. J., *288*
Dunlop, David J., *1303*
Dynamic corrections, *389*
Dynamic source models, seismic, *1131*
Dynamo hypothesis, *511*
Dynamos, α-ω, *513*
Dziewonski, A. M., 358

Earth
  geodesy, *934*
  gravity field, 384, 408, 454, 602
  thermal history, *374*
EARTH AND MOON: ORIGINS, 224
  age, *4*
  oldest rocks, 4
EARTH ORIENTATION, 231
Earthquake cycle, 1120
Earthquake energy, 277
Earthquake engineering, 306
Earthquake location, 1121
Earthquake magnitude, 1122
EARTHQUAKE MECHANISMS, 239
EARTHQUAKE MECHANISMS AND PLATE TECTONICS, 245
  mid-ocean ridge, 247
Earthquake parameters, *303*
Earthquake prediction, *261*, *266*, 304
Earthquakes. *See also* Deep earthquakes
  focal mechanisms, 1063
  ground accelerations, 1232
  interplate, *262*
  intraplate, *1044*
  location and magnitude, 1063
  low-frequency, *289*
  magnitude, *274*
  magnitude calibrating functions, *273*
  magnitude distributions, *312*
  spatial distributions, *308*
  tectonic, *289*
  temporal distributions, *308*
  world's largest (table), *311*
EARTHQUAKES: HAZARDS AND PREDICTIONS, 260
EARTHQUAKES: LOCATION TECHNIQUES, 268

SUBJECT INDEX

EARTHQUAKES: MAGNITUDE, ENERGY, AND INTENSITY, 272
EARTHQUAKES: VOLCANOGENIC, 288
EARTHQUAKES AND CRUSTAL DEFORMATION, 295
EARTHQUAKE SEISMOLOGY, 300
Earthquake source, 1062, *1118*
  physical model of, *1124*
EARTHQUAKES AND SEISMICITY, 308
Earthquake swarms, *301*
EARTH STRUCTURE, GLOBAL, 331
EARTH TIDES, 359, 1296
EARTH'S CORE, 315
  composition, *316*
  dynamics, *317*
  energetics, *317*
  seismic diffraction, *1018*
  structure, *316*
EARTH'S CORE: STRUCTURE, 318
Earth's crust, 27
EARTH'S EXTERNAL GRAVITY FIELD, 322
Earth's gravity field, *623*
East African rift, *97*, 254
East Coast Magnetic Anomaly (ECMA), 869
East Pacific Rise, 1050
ECORS, 181
Edwards, R. N., 139
Eikonal equation, 1099
Einstein, A., 678
ELASTICITY AND WAVE PROPAGATION, 363
Elastic rebound hypothesis, *1119*
ELASTIC WAVES IN HOMOGENEOUS AND INHOMOGENEOUS MEDIA, 367
Electrical conductivity, in the Earth, *127*
Electromagnetic (CSEM) mapping, *126*
  layered Earth problem, 127
  three-dimensional effects, 134
Electronic distance measurement, *487*
Ellipsoidal Earth, 836
Elsasser, W. M., 901
ENERGY BUDGET OF THE EARTH, 372
EQUILIBRIUM FIGURE OF THE EARTH, 378
  shape, 384
  size, 384
Eruption prediction. *See* Volcanic prediction
Euler pole, *246*
Euler vector, *246*
Ewing, J. Alfred, 1039
Ewing, Maurice, 661
Ewing's seismometer, 1039

Expanding Spread Profiling (ESPs), 147, 865
EXPLORATION SEISMOLOGY, 385
  data acquisition, *387*
  data processing, *387*
External gravitational potential, 378

Far-field displacement, 241
FAR-TRAVELLED TERRANES, 395
Fault parameters, seismic, 241
Faults, types, 1224
Faye anomalies, 612
Ferromagnetism, 34
Fifth Fundamental Catalog, 481
FIGURE OF THE EARTH, 407
  time variations, *410*
Filtering
  k-ω, 1113
  moving window analysis, 1112
  two-dimensional, 1112
FINITE-DIFFERENCE FORWARD MODELING IN SEISMOLOGY, 410
Fischer, I. K., 453
Flattening, of the Earth, *408*
Fluids, deep crustal, *189*
Fluxgate magnetometers, *952*, *980*
Focal mechanism solutions, *239*
  continental earthquakes, *254*
  subduction zone, *248*
Focal sphere, *241*
Fold test, 203
Forbes, A. J., 507
Forearc zone, *1067*
Forensic seismology, 304
Foreshocks, *265*
Forsyth, D. W., 660
400 km discontinuity, *777*
Fracture mechanics, *43*
  crack propagation, 43
Fracture strength, 38
Fracture zone, *248*
Franciscan Complex, 1251
Free-air gravity anomaly, *604*
FREE OSCILLATIONS OF THE EARTH, *319*, 336, **431**
  forward problem, 432
  splitting, 435
Fresnel shadow, 1018
Frozen-flux reversals, 552
Frozen flux theory, *153*
Fuller, M., 672

Gal, 5
Galileo, 5
Galitzin, B., 1040
Galvanometer, *1035*
Gauss, Carl Friedrich, 589
Gauss coefficients, 518
Geiger, L., 268
Geiger's method, 268
GEODESY: GEOMETRIC, 442
GEODESY: HISTORICAL INTRODUCTION, 447
  ellipsoidal Earth, 449
  geoid, 451

GEODESY; PHYSICAL, 453
GEODESY: SATELLITE, 458
Geodetic boundary value problem, *325*
Geodetic datums, *444*
GEODETIC GROUND POSITIONING, 465
GEODETIC LEVELING, 470
GEODETIC NETWORKS AND CONTROL SURVEYS, 474
  datums, standards, and specifications, 475
  geodetic network, 475
  survey techniques, 477
  types of, 476
Geodetic reference frames, *480*
Geodetic reference models, *408*
GEODETIC REFERENCE SYSTEMS, 478
  astronomical, 479
  geodetic, 479
GEODETIC REFRACTION, 482
  atmospheric boundary layer and surface layer, 483
  leveling refraction, 486
  satellite Doppler observations, 488
  trigonometric heighting, 485
  types of solution, 483
  water vapor radiometer, 483
Geodynamo, *511*
  energetics, 513
  kinematic dynamo theory, 511
  magnetohydrodynamic theory, 514
  time dependence, 516
Geoid, *407*
GEOMAGNETIC FIELD: ASYMMETRIES, 489
GEOMAGNETIC FIELD: ELEMENTS, 495
GEOMAGNETIC FIELD: MEASUREMENT, 498
GEOMAGNETIC FIELD: WESTWARD DRIFT, 507
Geomagnetic field, coordinate systems, 497
Geomagnetic field, main, 523
GEOMAGNETIC FIELD, MAIN: THEORY, 511
GEOMAGNETIC FIELD ANALYSIS, 517
Geomagnetic instruments, 524
Geomagnetic "jerks," 561
GEOMAGNETIC MEASUREMENT TECHNIQUES AND SURVEYS, 522
  satellite, 532
Geomagnetic polarity reversals, *490*, 910
  flooding models, 549
  geometry of transitional field, 539
  intensity of transitional field, 541
  periodicities, 559
  polarity asymmetries, *490*
  processes, *553*

standing field model, *549*
standing and flooding field models, *536*
time scale, *740, 913*
transitional field, *535*
GEOMAGNETIC POLARITY REVERSALS: OBSERVATIONS, 535
GEOMAGNETIC POLARITY REVERSALS: THEORY AND MODELS, 547
GEOMAGNETIC REVERSAL SEQUENCE: STATISTICAL STRUCTURE, 556
Geomagnetic secular variation, *521*
  acceleration impulses (jerks), *574*
  of the dipole field, *569*
  dipole and nondipole contributions, *567*
  discovery of, *586*
  motion of the magnetic poles, *574*
  standing and drifting components, *567*
  westward drift, *566*
GEOMAGNETIC SECULAR VARIATION: DIRECTION AND INTENSITY, 560
GEOMAGNETIC SECULAR VARIATION: THEORY, 578
Geomagnetism
  external and induced fields, *519*
  theories of, *902*
GEOMAGNETISM: HISTORICAL INTRODUCTION, 584
Geomancy, *585*
Geometric satellite geodesy, *458*
Geometric spreading, *1006*
GEOS-3, *965*
GEOSAT, *966*
Geostrophic flow, *155*
Geotherm, *635*
GEOTHERMAL ENERGY, 592
  geochemical and geophysical exploration techniques, *597*
  utilization, *595*
Geothermal gradient, *635*
Gilbert, William, *586*
Glacial isostatic adjustment, *820*
Global laser network, *971*
GLOBAL POSITIONING SYSTEM (GPS), 460, 473, 477, 598
  error sources, *463*
  geodetic positioning, *461*
  receivers, *601*
  system description, *599*
Gobi-Altai Mongolian earthquake of 1957, *1045*
Golconda allochthon, *12*
Gordon, R. G., *923*
Graham, J. W., *901*
Grau, G., *1166*
Gravitational acceleration, g, *5*
Gravity anomalies, computation formulas, *614*

GRAVITY ANOMALIES: INTERPRETATION, 601
GRAVITY ANOMALIES: STATISTICAL ANALYSIS, 617
Gravity field data, *454*
  data processing, *456*
GRAVITY FIELDS: IMPLICATIONS FOR PLANETARY INTERIORS, 622
Gravity fields, planetary
  Mars, *625*
  Moon, *624*
  Venus, *624*
Gravity and gravitation, definitions, *453*
GRAVITY AND ISOSTASY, 627
  admittance technique, *630*
  forward modeling, *629*
  inverse methods, *630*
Gravity and topography, *658*
Greenwich mean sidereal time, *479*
Green's Function, *327, 364, 1129*
  plane wave decomposition, *366*
  realistic Earth models, *366*
Group velocity, *1257*
Grout, Frank Fitch, *676*
Guffanti, M., *598*
Guided waves. *See* Surface waves
Gutenberg, B., *1200*
Gwanmesia, G., *854*

Hales, A., *804*
Half-graben, *104*
Halgedahl, S. L., *720*
Halley, Edmond, *507, 587*
Hanging block wedges, *1273*
Haniford, V., *854*
Hasegawa, Akira, *1061*
Hatch, R. R., *464*
Hatcher, R. D., Jr., *92*
Hawaiian plume, *639*
Heat capacity, of the Earth, *373*
HEAT FLOW IN THE EARTH, 634
  continental, *640*
  oceanic, *636*
Heat sources and sinks, *372*
Heat transport, in the Earth, *373*
Hematite-Illmenite, *725*
  Curie point temperature, *726*
  magnetocrystalline anisotropy, *728*
  magnetostriction, *728*
Heng, Chang, *1037*
Henriksen, S., *931*
Hermance, J. F., *770*
Hess, H. H., *912, 989*
High pressure technology, *847*
Himalayan Collision zone, *62, 255*
Hoffman, K. A., *555*
Holcomb, L. Gary, *1092*
Holmes, Arthur, *1, 788, 807*
Homogeneous medium, seismic wave propagation, *368*
Homomorphic deconvolution, *1112*
Hood, P., *33*

Horizontal positioning, *443*
Hot spots, *921, 1149*
Howell, B. F., Jr., *1044*
Humboldt, *588*
Huygens, *5, 324*
Hydraulic fracturing, *1225*
Hydrothermal system, *593*
Hypocenter, *268*

Induction magnetometers, *502*
Indus-Tsangpo suture, *58*
Inelastic deformation, *38*
Inner core boundary, *320*
Insular (Wrangellian) Super-Terrane, *18*
  Alexander terrane, *18*
  Peninsular terrane, *18*
  Wrangellia, *18*
Intermontane (Stikinian) Super-Terrane, *14*
International Geomagnetic Reference Field (IGRF), *523, 563*
International Geophysical Year (IGY), *591*
International Seismological Centre (ISC), *278*
Interplate earthquakes, *262*
Intracontinental seismic zone, *1069*
Intraplate deformation, *164, 919*
Intraplate earthquake, *1044*
  lithospheric thermal structure, *1048*
  magnitudes and scaling relations, *1045*
  oceanic, *1048*
Intraplate seismicity, *1070*
Intrinsic attenuation, *1006*
INVERSE THEORY AND METHODS: SEISMOLOGY, 647
Ionospheric refraction, *219*
Iranian Plateau, seismotectonics, *70*
Isacks, Bryan L., *1070*
Island arcs, seismicity, *174*
Isochrons, *2*
  mineral, *2*
  whole-rock, *2*
Isostasy
  continental, *632*
  oceanic, *631*
Isostatic compensation models, *627*
Isostatic gravity anomalies, *609*
Isothermal remanent magnetization (IRM), *855*
Isotope geology, *2*
Isotopes, radioactive, *1*
Iyer, H. M., *1150*

James, D. E., *97, 148, 393, 1266*
Jeffreys, H., *268, 1200*
Jekeli, C., *331*
Joint hypocentral determination, *271*
Jovian planets, *1215*
Jupiter
  geodesy, *936*
  magnetic field, *942*

Kant, Immanuel, 1211
Kaula, W. M., 626
Kawasaki, Ichiro, 1005
Keen, C., 190
Kennett, B. L. N., 372
Khan, M. A., 385
Koblinsky, C. J., 968
Koenigsberger ratio, Q_n, 990
Kootenay terrane, 12
Kouba, J., 222
Kovach, R. L., 674
Kropacek, V., 961

Lageos, 969
Lailly, P., 1166
Laj, C., 546
Lake Tanganyika Rift Zone, 113
Lamb, Horace, 368, *1192*
Lambert conformal conic projection, 840
Langel, Robert A., 988
Langereis, C. G., 211
Laplace, Pierre Simon, 1211
Laplace's equation, 1217
LASE project, 865
Lawson, C. A., 733
Lay, T., 775
Least squares collocation, 329
Lee, T., 847
Legendre functions, 1218
Le Grand, H. E., 909
Lehmann discontinuity, 777
Length of the day (lod), *148*
  decade variations, 149
Levander, A. R., 430
Leveling, geodetic, *470*
Levi, S., 57
Linssen, J. H., 211
LITHOPROBE, 181
LITHOSPHERE: MECHANICAL PROPERTIES, 655
LITHOSPHERE, OCEANIC: FORMATION AND EVOLUTION, 660
  anisotropy, 662
  thickness, 1004
Lithosphere, thermal, 93
Lithospheric doubling, 781
Lithospheric plate, definition, *1062*
Lithospheric strength, 624
LONGSHOT, 1086
Loper, D. E., 318
Lord Rayleigh, *368*, 1255
Love, A. E. H., 368, 1255
Love waves, *302*, *1255*
Lower crustal reflections, *187*
Lower mantle
  dynamics, 774
  lateral heterogeneity, *773*
Low-frequency earthquakes, *289*
Lowrie, W., 706
Low velocity zone, seismic, 333, 776
  oceanic, 1004
  crustal, *78*
Low viscosity zone, 808
Lunar formation, *227*

LUNAR PALEOMAGNETISM, 668
  breccias, 669
  mare basalts, 668
  origin, 671
  paleointensity, 670
LUNAR SEISMOLOGY, 673
Lunar tidal force, *359*
Lund, S. P., 888

m_b, 275
M_s, 274
McCarthy, D. D., 482
McClain, J. S., 667
McDougall, I., 990
McElhinny, M. W., 495
McFadden, P. L., 495, 560
Mackwell, S. J., 201
Macroseismic observations, 1118
McNally, K. C., 314
McNutt, M., 633
MAC waves, 735
Madariaga, R., 1132
Magma
  assimilation, *695*
  genesis, 689
  heat transfer, *688*
  immiscibility, *695*
  mixing, 695
  tectonic setting, 689
Magma boundary layers, 696
MAGMA PHYSICS, 676
  convective style, 684
  heat transfer, 683
MAGMATIC PROCESSES, 688
  plumbing, 693
Magmatic stoping, 676
Magmatic viscosity, 679
MAGNETIC ANALYSIS OF ROCK FABRIC, 698
Magnetic anisotropy, of minerals, 699
Magnetic anomaly maps, satellite, 985
Magnetic balances, *958*
Magnetic charts, earliest, 587
Magnetic coercivity, 51
MAGNETIC DOMAINS, 706, 949
  dependence upon grain size, 710
  patterns, 713
  wall pinning, 717
Magnetic fabric analysis
  igneous rocks, 704
  metamorphic and deformed rocks, 705
  methods, 699
  sedimentary rocks, 702
Magnetic field, crustal, 27
Magnetic inclination, discovery of, 586
Magnetic indices, *498*
Magnetic instruments, *501*
Magnetic observatories, 526, 588
MAGNETIC PROPERTIES OF MINERALS, 721
Magnetic Reynolds number, 159

MAGNETIC SELF-REVERSAL, 730
Magnetic shape anisotropy, 948
Magnetic source-layer of the ocean crust, 992
Magnetic storm, *748*
  discovery of, 587
Magnetic stripes, on ocean floor, *989*
Magnetic viscosity, *1297*
Magnetocrystalline anisotropy, 708, *947*
Magnetohydrodynamics (MHD), *149*
MAGNETOHYDRODYNAMIC WAVES WITHIN THE EARTH, 733
  torsional oscillations, 737
Magnetometer arrays, 532
Magnetometers
  airborne, *27*
  alkali-vapor, *981*
  astatic, *951*
  ballistic, *956*
  cryogenic, *505*
  fluxgate, *952*, *980*
  induction, *502*
  proton precession, *981*
  resonance, *502*
  saturable core, *502*
  spinner, *956*
  torsion, *501*, *958*
  vibrating sample, *956*
Magnetosphere, 978
Magnetostatic energy, 708
MAGNETOSTRATIGRAPHY, 740
Magnetostrictive anisotropy, *948*
MAGNETOTELLURIC METHOD: FUNDAMENTAL CONCEPTS, 746
  field system, *761*
Magnitude, earthquake, *274*
Magnitude calibrating functions, earthquakes, *273*
MAGSAT, 532, 983
Mantle
  chemical heterogeneities, 813
  conductivity, 575
  mineral phases, *849*
  plumes, *798*, *811*
MANTLE, LOWER: STRUCTURE, 770
MANTLE, UPPER: STRUCTURE, 775
MANTLE CONVECTION AND PLUMES, 788
  boundary layer theory, 791
  role of the transition zone, 795
  small-scale modes, 811
  structure of, *793*
MANTLE DISCONTINUITIES, 802
MANTLE DYNAMICS, 806
MANTLE VISCOSITY, 819
  anomalous gravitational field, 824

SUBJECT INDEX

MAP PROJECTIONS IN GEODESY, 834
Marginal seas, *399*
Marine geodesy, history, *452*
Marine magnetic anomalies, *990*
Mariner 10, 939
Mariner 2, 940
Mariner 4, 941
Marquardt's algorithm, 270
Mars
　geodesy, *935*
　magnetic field, *941*
Marsh, B. D., 688
Marsh, J. G., 968
Masters, G., 322
Mean Earth ellipsoid, 613
Meissner, R., 89
Melanges, 1251
Menke, William, 1210
Mercalli intensity scale, *287*, 1190
Mercator projection, 837
Mercury
　geodesy, *933*
　magnetic field, *939*
Merrill, R. T., *495*, 950
METEORITES: NATURE AND ORIGIN, 844
　ages, 3
　magnetic classification, 891
　paleomagnetic intensity, 893
Microplates, 899
Microseismic noise, *1090*
　sources of, 1091
Mid-Atlantic Ridge, 1050
Mid-ocean ridge, slow and fast spreading, 1053
Migration. *See* Seismic wavefield migration
Miller, M. M., 20
Milne, John, 1038
MINERALOGIC PHASE TRANSFORMATIONS IN THE EARTH, 847
　seismic signature, 853
Minimum entropy deconvolution, 1112
Mohorovičić, A., 332
Mohorovičić discontinuity, 75, *190*
Mohr criterion, 41
Moment of inertia, core-mantle system, 149
Moon
　geodesy, *935*
　magnetic field, *944*
　velocity structure and composition, 674
Moonquake, 300, *673*
Mora, Peter, 1032
Morgan, P., 646
Mullender, T. A. T., 211
Multiple filtering, 1257
Muting, 389

NAD 1983, 836
Nagata, T., 857
National Earthquake Information Service (NEIS), U.S.A., 277

NATURAL REMANENT MAGNETIZATION (NRM), 34, 855
NAVSTAR, 458. *See also* Global positioning system
Navy navigation satellite system, 218, 458
Néel, L., 49, 166, *903*, *1267*
　multidomain TRM theory, 1268
　single-domain model, 1268
Néel temperature. *See* Curie temperature
Neptune, geodesy, *937*
New Madrid earthquakes, 1045
Newton, Isaac, 323
Nishenko, S. P., 267
Nodal planes, 1123
Nodal surface, 242
Nolet, G., 787
Nontidal acceleration, of planetary rotation, 824
NORESS, 1081
Normal faults, 109
Normal gravity formula, 602
NORSAR, *1078*, 1174
North American Cordillera, *10*
　accreted terranes, 10
　Cache Creek terrane, 15
　Golconda allochthon, 12
　Intermontane (Stikinian) Super-Terrane, 14
　Kootenay terrane, 12
　Roberts Mountains allochthon, 12
　Slide Mountain terrane, 14
　Stikine terrane, 15
North American Datum (NAD) 1927, 836
Nuclear explosions, *1071*
　detection, *1076*
　methods for concealing, 1087
　seismic discrimination, *1083*
　yield estimation, *1086*
Nuclear Test Ban Treaty, *304*
Null axis, 241
Nur, A. 407
Nutation, Earth, 1295

Oblique Mercator projection, 838
Oblique spreading, 916
OCEAN–CONTINENT TRANSITION: STRUCTURE, 858
Ocean Drilling Program (ODP), 989
Oceanic crust, *184*, *661*
　formation, 663
　hydrothermal circulation in, 665
Oceanic lithosphere
　cooling model, *1052*
　mechanical properties, *656*
　rheology, *199*
Oceanic plateaus, *395*
　migration and consumption, 395
Ocean ridges, 1068
Okhotsk Sea, *403*
Oldham, R. D., 332, *1200*
Olivine, seismic velocities, 996
Ophiolites, 83, 1097

Orcutt, J. A., 667
Ordinary chondrites, 895
O'Reilly, W., 730

P_n velocity anisotropy, 998
Pacific nondipole low, 573
Pacifica Continent, *405*
Paleo-heat flow, 643
Paleointensities, geomagnetic, 34, 541
Paleointensity determinations, control of, 55
　in igneous rocks, 54
　in sedimentary rocks, 52
　marine magnetic anomalies, 54
Paleomagnetic euler poles, 921
PALEOMAGNETIC FIELD: INTENSITY, 872
PALEOMAGNETIC SECULAR VARIATION, 876
　models, 885
　time-series analysis, 880
PALEOMAGNETISM: DEEP-SEA SEDIMENTS, 889
PALEOMAGNETISM: METEORITES, 891
PALEOMAGNETISM: OROGENIC BELTS, 897
PALEOMAGNETISM AND CONTINENTAL DRIFT: HISTORICAL INTRODUCTION, 901
PALEOMAGNETISM AND PLATE TECTONICS, 909
Paleorifts, 102
Pangaea, 910
Paramagnetism, 946
Parkfield earthquake, 267
Partial melting, *691*
　geochemical consequences, 692
Passive margins, 864
　seismic measurements, 865
Pavlis, G. L., 654
Peddie, N. W., 534
Peltier, W. R., 834
Pendulum, 5
Pendulum seismograph, 1039
Perovskite, *851*
Petersen, N., 173
Phase adjustment, 26
Phase velocity, seismic, *1257*
PHOTOGRAMMETRY, 923
Planetary dynamo, 938
Planetary formation, *1213*
　dust grain instability, 1213
　by gaseous instability, 1215
PLANETARY GEODESY, 931
PLANETARY MAGNETIC FIELDS, 938
Planetesimals, *224*, 1214
Plastic deformation, *192*
Plate boundary, 245, *1062*
Plate motions, mantle dynamics, 809
Plate tectonics, *910*
　crustal movements, *160*
Pluto, geodesy, *937*

1325

SUBJECT INDEX

Point gravity anomaly, 617
Polar motion, *232, 1294*
Polar Orbiting Geophysical Observatories (POGOs), 591
Polar wander, 826
Polyconic projection, 843
Positioning and reference systems, *443*
Postglacial isostatic rebound, 164
Postglacial sea levels, 821
Power-law creep equation, 195
Prandtl number, 682, *789*
Precession and nutation, *232*
Precursors (earthquakes), *265*
Predictive deconvolution, *1109*
Preliminary reference Earth model (PREM), *332*
Presolar nebula, 1211
Press-Ewing seismometer, 1042
Pressure axis, *241*
Primary magnetization, *947*
Principal component analysis, 210
Projection theorem, 23
Project MAGNET, 591
Propagating rifts, 916
Proton precession magnetometer, *981*
Protostars, 1211
Protostellar collapse, *1211*
Proudman-Taylor theorem, 158
P wave, *302*

Q, *1007*
  in mantle, *176*
Quasi-compressional waves, *995*
Quasi-shear waves, *995*

Radiative heat transfer, 634
Radioactive decay, *1, 372*
  half-life, 3
Radioactive systems, *3*
  geochronology, 3
Radiometric dating, *1*
  isotopic "resetting," 4
Radon transformation, *1114*
RAINIER, 1071
Raitt, Russell, 661
Rapp, R. H., 622
Ray fields, 1101
Rayleigh, Lord, 1255
Rayleigh-Benard convection, *789*
Rayleigh-Gans scattering, *1173*
Rayleigh number, 373, 682, *789*
Rayleigh scattering, 1167, *1171*
Rayleigh waves, *302*, 1255
  azimuthal anisotropy, 998
Ray parameter, seismic, 332
Ray propagator matrices, 1102
Ray series method. *See* Seismic ray theory
Ray synthetic seismograms, 1107
Ray tracing, dynamic, *1102*
  across curved interfaces, 1103
Recurrence intervals, seismic, 263
Reference coordinate system, geodesy, 445

Reference ellipsoid, *444*
Reflection seismology, *142*
Refraction seismology, *140*
Regional magnetic anomalies, *32*
Reilinger, R. E., 166
Relative gravity meters, 9
Relaxation spectrum, 1007
Remanent magnetization, *946*
Remsberg, A., 854
Resonance magnetometers, *502*
Reversal test, 203
Reversed magnetism, 903
Rheology
  of lithosphere, *199, 655*
  of magma, 678
  mantle, *200*
Richards, Paul G., 1089
Ridge crests, 1068
Ridge offsets, 404
Rifted continental margin, *185*
Rifts
  active, 98
  Dead Sea Rift, 102
  Gulf of California, 102
  linking, *111*
  magmatism, 98
  morphology, *101*
  and orogenic belts, 84
  passive, 98
  Red Sea, 103
  Rhinegraben, 102
  Rio Grande, 102
Roberts, P., 159
Roberts Mountains allochthon, 12
Robertson, D. S., 1296
Rock friction, 45
ROCK MAGNETISM, 946
ROCK MAGNETISM: MEASURING TECHNIQUES AND APPARATUS, 950
Rosendahl, B. R., 104, 126
Ross, James Clark, 589
Rotation, of the Earth, *231*
Ruff, Larry J., 1128
Runcorn, S. K., 788, 903
Russell, C. T., 944
Rutherford, Ernest, 1

Sabine, Edward, 590
Saleeby, J. B., 20
San Andreas fault, 163, 295, 1069, 1141
San Francisco earthquake, 1906, *300*
Sasatani, T., 181
SATELLITE ALTIMETRY, 962
  altimeter systems, 962
  applications, 965
SATELLITE LASER POSITIONING, 969
  results, 973
Satellite magnetic instrumentation, *980*
Sputnik, 982
SATELLITE MAGNETIC MEASUREMENTS, 977

Satellite orbits, sun-synchronous, 979
Saturable core magnetometers, *502*
Saturn
  geodesy, *936*
  magnetic field, *942*
Scattering attenuation, *1008, 1180*
Schmidt, V. A., 1269
Schmitz, D., 1221
ScS polarization anisotropy, 999
Seafloor spreading, *912*
SEAFLOOR SPREADING: MAGNETIC EVIDENCE, 989
Sea level datum, 470
SEASAT, 966
Secondary magnetization, *947*
Seeber, Leonardo, 1278
SEISMIC ANISOTROPY IN THE EARTH, 994, *1097*
  definition, 994
  stress-strain relation, 994
  wave propagation, 994
Seismic arrays, 1036
SEISMIC ATTENUATION: OBSERVATION AND MEASUREMENT, 1005
  Born approximation, 1023
  free oscillations, *1013*
  frequency dependence, *1014*
  scattering, 1018
  strain dependence, 1016
Seismic crustal studies, *139*
SEISMIC DIFFRACTION, 1018
Seismic gaps, 262, 1066
Seismic group recorders, 392
Seismic hazard analysis, 1245
SEISMIC IMAGING, 1024
  the forward problem, 1024
  the inverse problem, 1025
SEISMIC INSTRUMENTATION, 1033
SEISMIC INSTRUMENTATION: HISTORY, 1037
Seismic intensity, 275, *287*
SEISMICITY: INTRAPLATE, 1044
SEISMICITY: MID-OCEAN RIDGE, 1050
SEISMICITY: SUBDUCTION ZONE, 1054
SEISMICITY AND PLATE TECTONICS, 1061
Seismic migration, 391, *1151*
  downward continuation of the data, *1160*
  Fourier (or f-k) methods, *1160*
  geometrical methods, *1153*
  Kirchhoff methods, *1160*
  picked zero-offset section, *1153*
  of prestack data, 1157
  time-reversal methods, *1160*
Seismic models, physical, *1202*
  laboratory apparatus, 1203
  three-dimensional, 1205
  two-dimensional plate models, 1205

Seismic moment, *240*, *1126*
  in anisotropic medium, 996
  tensors, *1129*
**SEISMIC MONITORING OF
  NUCLEAR EXPLOSIONS,
  1071**
Seismic networks, local and
  regional, 1064
SEISMIC NOISE, 1089
**SEISMIC PROPERTIES OF
  ROCKS, 1093**
Seismic radiation, 1129
SEISMIC RAY THEORY, 1098
  elastodynamic Green's function,
    1106
Seismic recording, 1035
Seismic scattering, mean free paths,
  1009
Seismic sea waves. *See* Tsunamis
**SEISMIC SIGNAL
  PROCESSING, 1108**
**SEISMIC SOURCE: OBSERVA-
  TIONS, 1118**
**SEISMIC SOURCE: THEORY,
  1129**
SEISMIC TOMOGRAPHY, 147,
  346, **1133**
  ACH technique, *1136*
  Caltech back-projection
    technique, *1136*
  methods, 1134
  results, 1139
  volcanic centers, 1150
Seismic travel-time tables, *1187*
Seismic velocities
  laboratory measurements, *1093*
  versus pressure, *1094*
**SEISMIC WAVEFIELD
  MIGRATION, 1151**
Seismic wave radiation pattern,
  1123
**SEISMIC WAVE SCAT-
  TERING, 1166**
  by a random medium, 1173
**SEISMOGRAMS: INTERPRE-
  TATION, 1187**
SEISMOGRAMS: SYNTHETIC,
  303, **1192**, *1193*
  finite difference/finite element,
    1196
  full-wave theory, *1195*
  Gaussian beams, *1197*
  generalized ray theory, *1194*
  geometric ray theory, *1196*
  Maslov theory, *1197*
  normal mode summation, *1194*
  reflectivity, *1194*
  the WKBJ method, *1195*
Seismograph, *1036*
Seismographic instruments, 1121
Seismographs, *1037*
  inertial, 1038
Seismograph systems, 281
Seismological Society of America,
  261
Seismological Society of Japan,
  261

SEISMOLOGY: HISTORY, 1198
**SEISMOLOGY: PHYSICAL
  MODEL STUDIES, 1202**
Seismometer, *1033*
Seismoscopes, 1037
Sequential adjustment, 26
Shaw, J., 875
Shear-wave splitting, 1097
Shen Kua, 585
SH-SV polarization anisotropy, of
  surface waves, 999
Sidereal time, 479
Sign bit recording, 145
Simon, Ruth B., 1192
Single-couple model, *240*
Site effects, 1243
650 km discontinuity, *779*
Slab pull, 253
Slant stack, *1114*
Slide Mountain terrane, 14
Slip-weakening, 46
Smith, D. E., 977
Snay, R. A., 299
Snoke, J. A., 245
Snyder, J. P., 844
Solar nebula, 224, *1212*
  dust grain coagulation, 1213
SOLAR SYSTEM: ORIGINS,
  224, **1211**
  nebular hypothesis, 1211
Solar tidal force, *359*
Soret effect, *697*
Spatial aliasing, 148
Spectral magnitude, seismic, 278,
  286
**SPHERICAL HARMONIC
  ANALYSIS,** 326, **1217**
Spherically symmetric Earth model
  (SSEM), *331*
Spinel α-β transition, *851*
Spinner magnetometers, *956*
Sputnik I, 458
SQUID magnetometers, *952*
Starlette, 969
Static fatigue, 42
Statics corrections, seismic, *389*
Station corrections, seismic, *271*
Stein, S., 260, 1054
Stereographic projection, 841
Stick-slip faulting, 1120, 1131
Stikine terrane, 15
Stochastic dispersion, 1008
Stoughton, H. W., 469
Strain seismographs, 1043
Stress, crustal, *1225*
  sources of, 1228
  topography, 1231
Stress, tectonic, *1228*
  lithospheric flexure, 1230
  plate motion, 1228
  thermal, 1230
Stress drops, *1131*, *1224*
**STRESS IN THE EARTH'S
  LITHOSPHERE, 1221**
  measurement of, 1223
Stress and strain, 364
Stress-strain hysteresis, 1006

Stress tensor, *1222*
**STRONG MOTION SEIS-
  MOLOGY, 1232**
  accelerograph, *1233*
  earthquake source effects, 1238
  effects of wave propagation, 1241
  ground motion, *1236*
  recorders, 1036
  synthetic seismograms, 1244
Subduction accretion, *1247*
  models, 1253
Subduction underplating, 1247
SUBDUCTION ZONES, 862,
  1064, **1246**
  focal depth distribution and
    deepest earthquakes, 1058
  geologic and geophysical charac-
    teristics, 1247
  olivine-spinel phase boundary,
    1059
  plate boundary, 1065
  sediment subduction and subduc-
    tion erosion, 1251
  seismicity, 784
  seismicity gap, *1057*
  shallow seismicity, *1055*
  spinel-postspinel phase boundary,
    1059
  tectonic models for, *1253*
  tectonic processes, *1247*
  thermal structure, *642*, *1252*
Sünkel, H., 457
Super-terrane, 10
SURFACE WAVES, 95, *371*,
  995, 1188, **1255**
  attenuation, 1259
  dispersion curves, *1256*
  fundamental mode, *1256*
  global, 1260
  higher modes, *1256*
  phase velocity curves, 1260
  regional
  tripartite arrays, 1258
Suspect terranes, 9
SV_n velocity anisotropy, 998
S wave, 302
Synthetic magnitude calibrating
  functions, seismic, 280

Talwani, M., 871
Tarantola, Albert, 1032
Tarling, D. H., 37, 990
Tauer, J., 961
Tauxe, L., 746
Tectonic earthquakes, *289*
Tectonic erosion, 10
Tectonothermal age, 641
Tectosphere model, *93*, *780*
Temperature-depletion compensa-
  tion hypothesis, *93*
Tension axis, *241*
Terrain correction, 607
Terrane attrition, 10
Terrestrial heat flow, *635*
Terrestrial planet, accumulation,
  224
Thermal isostasy, 624

SUBJECT INDEX

Thermal plumes, *791*
THERMOREMANENCE, 34, *856*, **1267**
Thermoviscous remagnetization, *1302*
THIN-SKIN TECTONICS, 1269
  convergence zones, 1271
  in zones of extension, 1277
Thompson, R., 510
Thomson, J. J., 1
Thomson, William (Lord Kelvin), 1
Three-dimensional Earth structure, 346
Threshold Test Ban Treaty, *1071*
Thrust belts, *1274*
Tibetan Plateau, 59
Tidal waves. *See* Tsunamis)
Titanomaghemites, *724*
  Curie point temperature, 724
Titanomagnetites, *721*
  Curie point temperature, 722
  magnetocrystalline anisotropy, 723
  magnetostriction-strain anisotropy, 723
Tomography. *See* Seismic tomography
TOPEX altimeter, 964
Torsion magnetometer, *501*, *958*
Transdomain TRM, 1269
Transfer fault, *109*
Transform fault, *247*, 913, 1053, 1068
Transform plate boundary, *163*
TRANSIT, 218
Transit system, 458
  error sources, 459
  geodetic positioning, 459
Transport equation, 1104
Transverse Mercator projection, 838
Transverse seismic anisotropy, *994*
Travel-time curves, *1189*
Travel time residuals, 1133
Triangulation, geodetic, *467*
Tropospheric refraction, 219
True polar wander, 921
TSUNAMIS, 1279
  forecasting, 1291
  maximum amplitude, 1290
  seismic excitation, 1282

Ultradeep drill holes, 89
Ultramafic nodules, continental, 94
United Services for Earthquake Research (ESSN), USSR, 277
Universal Time, 479
Upper mantle
  anelasticity, 779
  constitution, 779
  lateral heterogeneity, 781
  seismic anisotropy, 787
  transition zone, 778
Uranus
  geodesy, *936*
  magnetic field, *944*
Urey ratio, 373
UT1 length-of-day (LOD), *236*, 1295

Van der Voo, R., 900
Vaniček, P., 26
VELA-UNIFORM, 1199
Velocities, of rock forming minerals, *1095*
Velocity dispersion, 1008
Velocity filtering, seismic, 140
Vening-Meinesz, F. A., 860
Venus
  geodesy, *933*
  magnetic field, *940*
Verhoogen, J., 788
Verosub, K. L., 217
Vertical positioning, *444*
Vertical seismic profiling (VSP), *393*, *1031*
VERY-LONG-BASELINE INTERFEROMETRY (VLBI), *477*, 598, **1293**
Vibrating-sample magnetometer, *956*
Vibroseis, *387*
Vine, Fred, 912
Vine-Matthews-Morley hypothesis, 989
Vinnik, L. P., 806
Virtual geomagnetic pole, 535
Viscosity, mantle, 812
Viscosity, temperature-dependent, *794*
Viscous demagnetization, *1297*
Viscous magnetization, 856

VISCOUS REMANENT MAGNETIZATION (VRM) AND VISCOUS REMAGNETIZATION, 1297
Volcanic gaps, *398*
Volcanic prediction, *293*
Volcanic tremor, *289*

Wadati, 174
Wadati-Benioff zone, 253, *1055*, *1066*, 1249
Wahr, J. M., 363
Walker, D., 698
Ward, Steven N., 1292
Waveform inversion, seismic, 1264
Wavepacket, 1257
Wegener, Alfred, 910
Weidner, D. J., 854
West Philippine Basin, 404
Westward drift, 507
  theory of, 578
Wever, Th., 89
Wiechert, Emil, 1039
Wiener-Levinson deconvolution, *1108*
Wiens, Douglas A., 1049
Wilcox, L. E., 617
Wong, T-F., 47
Wood, H. O., 1200
Wood-Anderson torsion seismometer, 1041
Woods, D. F., 260, 1054
World Data Centers, 534
World Wide Standardized Seismograph network (WWSSN), 1063, *1198*
Worthington, M. H., 1117
Wrangellia, 899
Wu, Ru-Shan, 1186

Young, G., 474
Yukutake, T., 584

Zagros Collision Zone, 67, 255
Zelinka, T., 961
Zero-offset trace, 144
Zhu, S.-Y., 447
Zijderveld, J. D. A., 211
Zoback, Mark D., 1231
Zoback, Mary Lou, 1231
Zumberge, M. A., 9